INTERNATIONAL ENERGY AGENCY
AGENCE INTERNATIONALE DE L'ENERGIE

ENERGY STATISTICS OF NON-OECD COUNTRIES

1996 - 1997

STATISTIQUES DE L'ENERGIE DES PAYS NON-MEMBRES

1999 Edition

INTERNATIONAL ENERGY AGENCY

9, RUE DE LA FÉDÉRATION, 75739 PARIS CEDEX 15, FRANCE

The International Energy Agency (IEA) is an autonomous body which was established in November 1974 within the framework of the Organisation for Economic Co-operation and Development (OECD) to implement an international energy programme.

It carries out a comprehensive programme of energy co-operation among twenty four* of the OECD's twenty nine Member countries. The basic aims of the IEA are:

- To maintain and improve systems for coping with oil supply disruptions;
- To promote rational energy policies in a global context through co-operative relations with non-member countries, industry and international organisations;
- To operate a permanent information system on the international oil market;
- To improve the world's energy supply and demand structure by developing alternative energy sources and increasing the efficiency of energy use;
- To assist in the integration of environmental and energy policies.

**IEA Member countries: Australia, Austria, Belgium, Canada, Denmark, Finland, France, Germany, Greece, Hungary, Ireland, Italy, Japan, Luxembourg, the Netherlands, New Zealand, Norway, Portugal, Spain, Sweden, Switzerland, Turkey, the United Kingdom, the United States. The European Commission also takes part in the work of the IEA.*

ORGANISATION FOR ECONOMIC CO-OPERATION AND DEVELOPMENT

Pursuant to Article 1 of the Convention signed in Paris on 14th December 1960, and which came into force on 30th September 1961, the Organisation for Economic Co-operation and Development (OECD) shall promote policies designed:

- To achieve the highest sustainable economic growth and employment and a rising standard of living in Member countries, while maintaining financial stability, and thus to contribute to the development of the world economy;
- To contribute to sound economic expansion in Member as well as non-member countries in the process of economic development; and
- To contribute to the expansion of world trade on a multilateral, non-discriminatory basis in accordance with international obligations.

The original Member countries of the OECD are Austria, Belgium, Canada, Denmark, France, Germany, Greece, Iceland, Ireland, Italy, Luxembourg, the Netherlands, Norway, Portugal, Spain, Sweden, Switzerland, Turkey, the United Kingdom and the United States. The following countries became Members subsequently through accession at the dates indicated hereafter: Japan (28th April 1964), Finland (28th January 1969), Australia (7th June 1971), New Zealand (29th May 1973), Mexico (18th May 1994), the Czech Republic (21st December 1995), Hungary (7th May 1996), Poland (22nd November 1996) and the Republic of Korea (12th December 1996). The Commission of the European Communities takes part in the work of the OECD (Article 13 of the OECD Convention).

AGENCE INTERNATIONALE DE L'ÉNERGIE

9, RUE DE LA FEDERATION, 75739 PARIS CEDEX 15, FRANCE

L'agence internationale de l'énergie (AIE) est un organe autonome institué en novembre 1974 dans le cadre de l'Organisation de coopération et de développement économiques (OCDE) afin de mettre en œuvre un programme international de l'énergie.

Elle applique un programme général de coopération dans le domaine de l'énergie entre vingt-quatre* des vingt-neuf pays Membres de l'OCDE. Les objectifs fondamentaux de l'AIE sont les suivants :

- tenir à jour et améliorer des systèmes permettant de faire face à des perturbations des approvisionnements pétroliers ;
- œuvrer en faveur de politiques énergétiques rationnelles dans un contexte mondial grâce à des relations de coopération avec les pays non membres, l'industrie et les organisations internationales ;
- gérer un système d'information continue sur le marché international du pétrole ;
- améliorer la structure de l'offre et de la demande mondiales d'énergie en favorisant la mise en valeur de sources d'énergie de substitution et une utilisation plus rationnelle de l'énergie ;
- contribuer à l'intégration des politiques d'énergie et d'environnement.

** Pays Membres de l'AIE : Allemagne, Australie, Autriche, Belgique, Canada, Danemark, Espagne, États-Unis, Finlande, France, Grèce, Hongrie, Irlande, Italie, Japon, Luxembourg, Norvège, Nouvelle-Zélande, Pays-Bas, Portugal, Royaume-Uni, Suède, Suisse et Turquie. La Commission des Communautés européennes participe également aux travaux de l'AIE.*

ORGANISATION DE COOPÉRATION ET DE DÉVELOPPEMENT ÉCONOMIQUES

En vertu de l'article 1er de la Convention signée le 14 décembre 1960, à Paris, et entrée en vigueur le 30 septembre 1961, l'Organisation de Coopération et de Développement Économiques (OCDE) a pour objectif de promouvoir des politiques visant :

- à réaliser la plus forte expansion de l'économie et de l'emploi et une progression du niveau de vie dans les pays Membres, tout en maintenant la stabilité financière, et à contribuer ainsi au développement de l'économie mondiale ;
- à contribuer à une saine expansion économique dans les pays Membres, ainsi que les pays non membres, en voie de développement économique ;
- à contribuer à l'expansion du commerce mondial sur une base multilatérale et non discriminatoire conformément aux obligations internationales.

Les pays Membres originaires de l'OCDE sont : l'Allemagne, l'Autriche, la Belgique, le Canada, le Danemark, l'Espagne, les États-Unis, la France, la Grèce, l'Irlande, l'Islande, l'Italie, le Luxembourg, la Norvège, les Pays-Bas, le Portugal, le Royaume-Uni, la Suède, la Suisse et la Turquie. Les pays suivants sont ultérieurement devenus Membres par adhésion aux dates indiquées ci-après : le Japon (28 avril 1964), la Finlande (28 janvier 1969), l'Australie (7 juin 1971), la Nouvelle-Zélande (29 mai 1973), le Mexique (18 mai 1994), la République tchèque (21 décembre 1995), la Hongrie (7 mai 1996), la Pologne (22 novembre 1996) et la République de Corée (12 décembre 1996). La Commission des Communautés européennes participe aux travaux de l'OCDE (article 13 de la Convention de l'OCDE).

TABLE OF CONTENTS

SUMMARY TABLES

PETROLEUM PRODUCTS

TABLE DES MATIERES

TABLEAUX RECAPITULATIFS

PRODUITS PETROLIERS

ABBREVIATIONS

Btu:	British thermal unit
GWh:	gigawatt hour
kcal:	kilocalorie
kg:	kilogramme
kJ:	kilojoule
Mt:	million tonnes
m^3:	cubic metre
t:	metric ton = tonne = 1000 kg
TJ:	terajoule
toe:	tonne of oil equivalent = 10^7 kcal
CHP:	combined heat and power
GCV:	gross calorific value
HHV:	higher heating value = GCV
LHV:	lower heating value = NCV
NCV:	net calorific value
PPP:	purchasing power parity
EU:	European Union
IEA:	International Energy Agency
OECD:	Organisation for Economic Co-Operation and Development
OLADE:	Organización Latino Americana de Energía
UN:	United Nations
IPCC:	Intergovernmental Panel on Climate Change
ISIC:	International Standard Industrial Classification
UNIPEDE:	International Union of Producers and Distributers of Electrical Energy
-	not applicable, nil or not available

Note	**See multilingual pullout, including Chinese, at the end of the publication.**
Attention	**Voir le dépliant en plusieurs langues à la fin du présent recueil.**
Achtung	**Aufklappbarer Text auf der letzten Umschlagseite.**
Attenzione	**Riferirsi al glossario poliglotta alla fine del libro.**
注意	巻末の日本語の折り込みページを参照
Nota	**Véase el glosario plurilingüe al final del libro.**
Примеч.	Смотрите многоязычный словарь в конце книги.

ABREVIATIONS

Btu :	British thermal unit
GWh :	gigawatt-heure
kcal :	kilocalorie
kg :	kilogramme
kJ :	kilojoule
Mt :	million de tonnes
m^3 :	mètre cube
t :	tonne métrique = 1000 kg
tep :	tonne d' équivalent pétrole = 10^7 kcal
TJ :	terajoule
PCI :	pouvoir calorifique inférieur
PCS :	pouvoir calorifique supérieur
PPA :	parité de pouvoir d'achat
AIE :	Agence internationale de l' énergie
OCDE :	Organisation de coopération et de développement économiques
OLADE :	Organización Latino Americana de Energía
ONU :	Organisation des nations unies
UE :	Union européenne
CITI :	Classification internationale type par industrie
GIEC :	Groupe d'experts intergouvernemental sur l' évolution du climat
UNIPEDE :	Union Internationale des Producteurs et Distributeurs d'Energie Electrique
-	sans objet, néant ou non disponible

Attention	**Voir le dépliant en plusieurs langues, y compris en chinois, à la fin du présent recueil.**
Note	**See multilingual pullout at the end of the publication.**
Achtung	**Aufklappbarer Text auf der letzten Umschlagseite.**
Attenzione	**Riferirsi al glossario poliglotta alla fine del libro.**
注意	巻末の日本語の折り込みページを参照
Nota	**Véase el glosario plurilingüe al final del libro.**
Примеч.	Смотрите многоязычный словарь в конце книги.

INTRODUCTION

This new publication offers the same in-depth statistical coverage as the homonymous publication for OECD countries. It provides detailed statistics on production, trade and consumption for each source of energy for more than 100 non-OECD countries, and main regions, including developing countries, Central and Eastern European countries and the former USSR. The consistency and complementarity of OECD and non-OECD countries' statistics ensure an accurate picture of the global energy situation.

The data shown for the Member countries of the Economic Commission for Europe of the United Nations (ECE-UN) are based on information provided in four annual questionnaires common to the OECD, the ECE-UN and the European Union: "Oil", "Natural Gas", "Solid Fuels, Wastes and Manufactured Gases", and "Electricity and Heat" completed by the national administrations. The commodity balances for the other countries are based on national energy data of heterogeneous nature which have been converted and adjusted to fit the IEA's format. This volume has been prepared in close collaboration with other international organisations including the Organizacíon Latino Americana De Energía (OLADE), the Asia Pacific Energy Research Centre, the Statistical Office of the United Nations and the Forestry Department of the Food and Agriculture Organisation of the United Nations. It draws upon and complements the extensive work of the United Nations in the field of world energy statistics.

While every effort is made to ensure the accuracy of the data, quality is not homogeneous throughout the publication. Special methodological issues arise in a number of countries. In some countries data are based on secondary sources, and where incomplete or unavailable, on estimates. In general, data are likely to be more accurate for production, trade and total consumption than for individual sectors in transformation or final consumption. Commodity balances are presented in two formats reflecting the degree of detail available. Moreover, the breakdown by fuel of electricity and heat production in the transformation sector provided in the *Energy Statistics of OECD Countries* is not shown in this publication. General issues of data quality, country notes and individual country data should be consulted when using regional aggregates.

A companion volume – *Energy Balances of Non-OECD Countries* – presents corresponding data in comprehensive balances expressed in a common unit, million tonnes of oil equivalent (Mtoe), with 1 toe = 10^7 kcal = 41.868 gigajoules.

The IEA published *World Energy Statistics and Balances* in 1989 and 1990 and *Energy Statistics and Balances of Non-OECD Countries* from 1991 to 1998. This book, for the first time published as a separate volume, revises and updates those publications.

Energy data on OECD and non-OECD countries are collected by the team in the Energy Statistics Division (ESD) of the IEA Secretariat, headed by Mr Jean-Yves Garnier. Non-OECD countries statistics are currently the responsibility of Mr Coleman Nee, Mr Bruno Castellano, and Mr Yannis Yaxas. Ms Nina Kousnetzoff has overall editorial responsibility. Secretarial support was supplied by Ms Sharon Michel and Ms Susan Stolarow.

Complete supply and consumption data from 1971 to 1997 are available on diskettes suitable for use on IBM-compatible personal computers. An order form has been provided in the back of this publication.

Enquiries about data or methodology should be addressed to the head of the non-OECD Countries Section, Energy Statistics Division, at:

Telephone: (+33-1) 40-57-66-34
Fax: (+33-1) 40-57-66-49
E-mail: wed@iea.org

INTRODUCTION

Cette nouvelle publication offre le même niveau de détail statistique que la publication homonyme pour les pays de l'OCDE. Elle fournit, pour chaque source d'énergie, des statistiques détaillées sur la production, les échanges et la consommation dans plus de 100 pays ne faisant pas partie de l'OCDE ainsi que pour plusieurs régions qui comprennent des pays en développement, des pays d'Europe centrale et orientale et l'ex-URSS. La compatibilité et la complémentarité des statistiques des pays membres de l'OCDE et des pays ne faisant pas partie de l'OCDE garantissent la qualité de l'image de la situation énergétique mondiale.

Les données publiées sur les pays membres de la Commission Economique pour l'Europe des Nations Unies (CEE-ONU) sont basées sur les informations recueillies dans quatre questionnaires annuels communs de l'OCDE, de la CEE-ONU et de l'Union Européenne : « Pétrole », « Gaz naturel », « Combustibles solides, déchets et gaz manufacturés » et « Electricité et chaleur », remplis par les administrations nationales. Les bilans par produit des autres pays sont basés sur des statistiques énergétiques nationales de nature hétérogène qui ont été converties et harmonisées afin d'être conformes au format de l'AIE. Ce document a été préparé en étroite collaboration avec d'autres organisations internationales dont l'Organización Latino Americana De Energía (OLADE), le Asia Pacific Energy Research Centre, le Bureau de Statistiques des Nations Unies, et le Département des forêts de l'Organisation des Nations Unies pour l'Alimentation et l'Agriculture. Il utilise et complète le travail considérable déjà accompli par les Nations Unies dans le domaine des statistiques énergétiques.

Bien que tout ait été mis en œuvre pour assurer l'exactitude de ces données, la qualité des chiffres de cette publication n'est pas toujours homogène. Des problèmes méthodologiques particuliers se posent dans un certain nombre de pays. Par ailleurs pour certains pays les données proviennent de sources secondaires et, lorsque les données manquent, sont basées sur des estimations. D'une façon générale, les chiffres sont sans doute plus exacts en ce qui concerne la production, les échanges et la consommation totale que pour les secteurs désagrégés de la transformation ou de la consommation finale. Les bilans par produit sont présentés en deux formats différents selon les détails disponibles. De plus, la ventilation par combustible de la production d'électricité et de chaleur du secteur transformation fournie dans les *Statistiques de l'énergie des pays de l'OCDE* n'est pas présentée dans la présente publication. Il convient, lors de l'utilisation des agrégats régionaux, de consulter également la note générale sur la qualité des données, les notes relatives aux différents pays et les données par pays.

Un recueil complémentaire - *Bilans énergétiques des pays non-membres 1996-1997* - contient des données équivalentes et exprimées sous la forme de bilans globaux dans une unité commune, à savoir en millions de tonnes d'équivalent pétrole (Mtep), sur la base de 1 tep = 10^7 kcal = 41,868 gigajoules.

L'AIE a publié *Statistiques et bilans énergétiques mondiaux* en 1989 et 1990 et *Statistiques et bilans énergétiques des pays non-membres* de 1991 à 1998. Le présent recueil, publié pour la première fois en volume séparé, met à jour et révise les précédentes éditions.

Les données énergétiques sur les pays membres et non-membres de l'OCDE sont collectées par la Division des statistiques énergétiques (ESD) du Secrétariat de l'AIE, dirigée par M. Jean-Yves Garnier. M. Coleman Nee, M. Bruno Castellano et M. Yannis Yaxas sont actuellement responsables des statistiques des pays non-membres de l'OCDE. Mme Nina Kousnetzoff est responsable de la publication. Mme Sharon Michel et Mme Susan Stolarow ont assuré le secrétariat d'édition et la saisie des textes.

Des données complètes sur l'offre et la demande sont disponibles, pour les années 1971 à 1997, sur disquettes exploitables par des ordinateurs personnels compatibles IBM. Un formulaire de commande est fourni à la fin de cet ouvrage.

Les demandes de renseignements sur les données ou la méthodologie doivent être adressées au chef de la section des pays non-membres, division des statistiques de l'énergie :

Téléphone : (+33-1) 40-57-66-34
Fax : (+33-1) 40-57-66-49
E-mail : wed@iea.org

PART I: METHODOLOGY

PARTIE I : METHODOLOGIE

1. ISSUES OF DATA QUALITY

A. Methodology

Considerable effort has been made to ensure that data presented in this publication adhere to the IEA definitions contained in the General Notes (Part I.2 and 3). These definitions are used by most of the international organisations that collect energy statistics. Nevertheless, the national energy statistics which are reported to international organisations are often collected using criteria and definitions which differ, sometimes considerably, from those employed by the international organisations. The extent to which the IEA Secretariat has identified these differences and, where possible, adjusted the data to meet international definitions, is outlined below. Recognized anomalies occurring in specific countries are presented in Part I.6, Country Notes and Sources. The Country Notes simply identify some of the more important and obvious deviations from IEA methodology in certain countries and are by no means a comprehensive list of anomalies by country.

B. Estimation

In addition to any adjustments undertaken to compensate for differences in definitions, estimations are sometimes required to complete major aggregates from which key statistics are missing. Examples may be found in the more detailed accounts given below. Except as concerns combustible renewables and waste (see E.5 below), it has been the Secretariat's aim to provide all of the elements of commodity balances down to the level of Final Consumption for all countries and years. This entails providing the elements of supply as well as inputs of primary fuels and outputs of secondary fuels to and from the main transformation activities such as oil refining and electricity generation. This has often required estimations prepared after consultation with national statistical offices, oil companies, electricity utilities and national energy experts.

C. Time Series and Political Changes

Commodity balances for the republics of the former USSR have been constructed since 1992. These balances have been constructed from official data and, where necessary, estimates have been made based on information obtained from industry sources and other international organisations. Summary tables and commodity balances for the country 'former USSR' continue to be published. While it remains difficult to collect data for all of the former USSR states on a consistent and common basis the regional aggregate forms a link with the past showing the overall effects of adaptation to the new conditions.

Commodity balances for the individual countries of former Yugoslavia also appear in this publication. These have been compiled from official data. Again, in the interest of maintaining time series, figures for the country 'former Yugoslavia' continue to be published.

Commodity balances for the Slovak Republic have been constructed since 1971. For the years 1989 to 1997, the balances for the Slovak Republic have been constructed from official data. For the years 1971 to 1988, estimates have been made where necessary (see Country Notes and Sources).

Energy statistics for some countries undergo countinuous changes in their coverage or methodology so that, for example, it is possible to present more detailed energy accounts for China beginning in 1980. Consequently, "breaks in series" are considered to be unavoidable.

The IEA Secretariat reviews its databases each year. In the light of new assessments, important revisions are made to the time series of individual countries during the course of this review. Therefore, data in this publication have been substantially revised with respect to previous editions.

D. Classification of Fuel Uses

National statistics sources often lack adequate information on the consumption of fuels in different categories of end use. Many countries do not conduct annual surveys of the fuel consumption in the main sectors of economic activity and consequently published data are based on out-of-date surveys. Sectoral disaggregation of consumption for individual countries should therefore be interpreted with caution.

Previous to the reforms undertaken in the 1990's the sectoral classification of fuel consumption in the transition economies (eastern Europe and the countries of the former USSR) and China differed greatly from that practiced in market economies. Sectoral consumption was defined according to the economic branch to which the user of the fuel belonged rather than according to the purpose or use of the fuel. Consumption of gasoline in the vehicle fleet of an enterprise attached to the economic branch 'iron and steel' was classified as industrial consumption of gasoline in the iron and steel industry. Where possible, the data have been adjusted to fit international classifications, for example, all gasoline is assumed to be consumed in the 'transport' sector and is reported in this publication as such. However, it has not been possible to reclassify products other than gasoline and jet fuel with the same facility, and few adjustments have been made to the remaining products.

E. Specific Issues by Fuel

1. Oil

The IEA Secretariat collects comprehensive statistics for oil supply and use, including refineries' own use of oil, oil delivered to ships' bunkers and oil for use as petrochemical feedstock. National statistics often do not report these amounts. Reported production of refined products frequently refers to net rather than gross refinery output and consumption of oil products may be limited to sales to domestic markets and may not include deliveries to international shipping or aircraft. Oil consumed as petrochemical feedstock in integrated refinery/petrochemical complexes is often not included in available official statistics.

Where possible, the Secretariat, in consultation with the oil industry, makes estimates of these unreported data. In the absence of any other indication of refinery fuel use the consumption has been estimated to be about 5 per cent of refinery throughput and has been divided equally between refinery gas and heavy fuel oil.

For a description of the nature of the adjustments made to the sectoral consumption of oil products, see above section 'Classification of Fuel Uses'.

2. Natural Gas

The IEA defines natural gas production as marketable production, i.e. that which is net of field losses, flaring, venting and reinjection. Furthermore, natural gas should be comprised mainly of methane and other gases such as ethane and higher hydrocarbons should be reported under the heading of 'oil'.

Readily available data for natural gas, however, often do not identify the separate elements of field losses, flaring, venting and reinjection. Moreover, reported data are frequently unaccompanied by adequate definitions so that it is difficult or impossible to identify gas at the different stages of its separation into dry gas (methane) and the heavier fractions in gas separation facilities.

Natural gas supply and demand statistics are normally reported in volumetric units and it is difficult to obtain accurate data on the calorific value of the gas. Where the heat value is unknown, a general gross calorific value of 38 TJ/million m3 has been applied.

It should also be noted that reliable consumption data for natural gas at a disaggregated level are often difficult to locate. This is especially true of some of the larger natural gas consuming countries in the Middle East. Industrial use of natural gas for these countries is therefore frequently missing from data appearing in this publication.

3. Electricity

As defined in the General Notes (Part I.2), an autoproducer of electricity is an establishment which, in addition to its main activities, generates electricity wholly or partly for its own use. Data on the basis of this definition are frequently unknown in non-OECD countries. In such cases the fuels inputs for autoproduced electricity are reported in the appropriate end use sector.

When statistics of the production of electricity from inputs of Combustible renewables and waste are available, they are included in total electricity production. These data are not comprehensive; for example, much of the electricity generated from waste biomass in sugar refining remains unreported.

Inputs of fuels for electricity generation are estimated (when unreported) using information on electricity output, fuel efficiency and the type of generation capacity.

4. Heat

The transition economies (eastern Europe and countries of the former USSR) and China employed a methodology with respect to heat that set them apart from common practice in market economies. The approach taken was to allocate the transformation of primary fuels (coal, oil and gas) by industry into heat *for consumption on site* to the transformation activity *Heat Production*, not to industrial consumption as in the IEA methodology[1]. The transformation output of *Heat* was then allocated to the various end use sectors. The losses occurring in the transformation of fuels into heat in industry were not included in final consumption of industry. Although a number of countries have recently switched to the practice of the international organisations, this important distinction reduces the validity of cross-country comparisons of sectoral end use consumption between transition economies and market economies.

5. Combustible Renewables and Waste

The IEA publishes production and domestic supply of combustible renewables and waste for all non-OECD countries and all regions for the years 1974 to 1997 (1971 to 1997 on diskettes).

Data shown are often from secondary sources, inconsistent and may be of questionable quality, which makes comparisons between countries difficult. The historical data of many countries derive from surveys which were often irregular, irreconcilable and conducted at the local rather than national level making them incomparable between regions and time. Where historical series are incomplete or unavailable, they have been estimated using the methodology consistent with the projection framework of the IEA's 1998 edition of *World Energy Outlook* (September 1998). First, nationwide biomass and wastes domestic supply per capita was compiled or estimated for 1995. Secondly, per capita supply for the years 1971 to 1994 was estimated, using a log/log equation with either GDP per capita or the percentage of urban population as the exogenous variable, depending on the region. Thirdly, total biomass and waste supply for the years 1996 and 1997 was estimated assuming a growth rate either constant, or equal to the population growth, or following the 1971-1994 estimation. However, these estimated time-series should be treated very cautiously.

The chart below* attempts to provide a broad indication of the data quality and estimation methodology by region.

Region	Main source of data	Data Quality	Exogenous Variables
Africa	FAO database and AfDB	Low	% urban population
Latin America	National and OLADE	High	None
Asia	Surveys	High to Low	GDP per capita
Non-OECD Europe	Questionnaires and FAO	High to Medium	None
Former USSR	National and questionnaires and FAO	High to Medium	None
Middle East	FAO	Medium to Low	None

AfDB: African Development Bank

1 The international methodology restricts the inclusion of heat in transformation sector to that sold to third parties. See definition in Part I.2.

Although the methods used for estimations are consistent with those used for biomass in the *World Energy Outlook 1998*, minor numerical discrepancies occur for a number of reasons. The reasons include later data revisions and the level of country disaggregations employed.

For the years 1994 to 1997, balances down to final consumption by end-use for individual products or product categories have been compiled for all the countries. The data for the years 1996 and 1997 are shown in the Annual Tables. Charcoal production is shown in the Summary Tables. The figures confirm the importance of vegetal fuels in the energy sector of many developing countries.

The IEA hopes that the inclusion of these data will encourage national administrations and other agencies active in the field to enhance the level and quality of data collection and coverage. More details on the methodology used by country may be provided on request, and comments are welcome.

2. GENERAL NOTES

The tables include all "commercial" sources of energy, both primary (hard coal, brown coal/lignite, peat, natural gas, crude oil, NGL, hydro, geothermal/solar, wind, tide, etc. and nuclear power) and secondary (coal products, manufactured gases, petroleum products, electricity and heat). Data also include various sources of combustible renewables and waste, such as solid biomass and animal products, gas/liquids from biomass, municipal waste and industrial waste.

Each table is divided into three main parts: the first showing *supply* elements, the second showing the *transformation* and *energy* sectors, and the third showing *final consumption* broken down into the various end-use sectors.

The following description refers to the layout of the full commodity balance type presentation. In the case of the regional and abbreviated tables, the definition of the products is the same. However, sectoral breakdown has been restricted to main totals (Energy Sector, etc.) and to selected main categories (Iron and Steel, etc.).

A. Supply

The first part of the basic energy balance shows the following elements of supply:

Production
\+ *Inputs from other sources*
\+ *Imports*
\- *Exports*
\- *International marine bunkers*
± *Stock changes*
= *Domestic supply*

1. Production

Production refers to the quantities of fuels extracted or produced, calculated after any operation for removal of inert matter or impurities (e.g. sulphur from natural gas).

2. Inputs from Other Sources

All inputs of origin other than primary energy sources explicitly recognised in the tables are listed under inputs *from other sources*, e.g. under crude oil: inputs of origin other than crude oil and NGL such as hydrogen, synthetic crude oil (including mineral oil extracted from bituminous minerals such as shales, bituminous sand, etc.); under additives: benzol, alcohol and methanol produced from natural gas; under refinery feedstocks: backflows from the petrochemical industry used as refinery feedstocks; under hard coal: recovered slurries, middlings, recuperated coal dust and other low-grade coal products that cannot be classified according to type of coal from which they are obtained; under gas works gas: natural gas, refinery gas, and LPG, that are treated or mixed in gas works (i.e. gas works gas produced from sources other than coal).

3. Imports and Exports

Imports and exports comprise amounts having crossed the national territorial boundaries of the country whether or not customs clearance has taken place.

a) Coal

Imports and exports comprise the amount of fuels obtained from or supplied to other countries, whether or not there is an economic or customs union between the relevant countries. Coal in transit should not be included.

b) Oil and Gas

Quantities of crude oil and oil products imported or exported under processing agreements (i.e. refining on account) are included. Quantities of oil in transit are excluded. Crude oil, NGL and natural gas are reported as coming from the country of origin; refinery feedstocks and oil products are reported as coming from the country of last consignment.

Re-exports of oil imported for processing within bonded areas are shown as an export of product from the processing country to the final destination.

c) Electricity

Amounts are considered as imported or exported when they have crossed the national territorial boundaries of the country.

4. International Marine Bunkers

International marine bunkers cover those quantities delivered to sea-going ships of all flags, including warships. Consumption by ships engaged in transport in inland and coastal waters and by fishing vessels in all waters is not included. See definitions of transport (Section 2.C.2) and agriculture (Section 2.C.3).

5. Stock Changes

Stock changes reflect the difference between opening stock levels on the first day of the year and closing levels on the last day of the year of stocks on national territory held by producers, importers, energy transformation industries and large consumers. Oil and gas stock changes in pipelines are not taken into account. With the exception of large users mentioned above, changes in final users' stocks are not taken into account. A stock build is shown as a negative number, and a stock draw as a positive number.

6. Domestic Supply

Domestic supply is defined as production + inputs from other sources + imports - exports - international marine bunkers ± stock changes.

7. Transfers

Transfers comprise *interproduct transfers*, *products transferred* and *recycled products*.

Interproduct transfers result from reclassification of products either because their specification has changed or because they are blended into another product, e.g. kerosene may be reclassified as gasoil after blending with the latter in order to meet its winter diesel specification. The net balance of *interproduct transfers* is zero.

Products transferred is intended for petroleum products imported for further processing in refineries. For example, fuel oil imported for upgrading in a refinery is transferred to the feedstocks category.

Recycled products are finished products which pass a second time through the marketing network, **after** having been once delivered to final consumers (e.g. used lubricants which are reprocessed).

8. Statistical Differences

Statistical difference is defined as deliveries to final consumption + use for transformation and consumption within the energy sector + distribution losses – domestic supply – transfers. Statistical differences arise because the data for the individual components of supply are often derived from different data sources by the national administration. Furthermore, the inclusion of changes in some large consumers' stocks in the supply part of the balance introduces distortions which also contribute to the statistical differences.

B. Transformation and Energy Sectors

The *transformation sector* comprises the conversion of primary forms of energy to secondary and further transformation (e.g. coking coal to coke, crude oil to petroleum products, heavy fuel oil to electricity).

The *energy sector* comprises the amount of fuels used by the energy producing industries (e.g. for heating, lighting and operation of all equipment used in the extraction process, for traction and for distribution).

The following categories are distinguished in the *transformation* and *energy* sectors:

1. Transformation Sector

- *Electricity plants* (refers to plants which are designed to produce electricity only). If one or more units of the plant is a CHP unit (and the inputs and outputs can not be distinguished on a unit basis) then the whole plant is designated as a CHP plant. Both public[1] and autoproducer[2] plants are included here.
- *Combined heat and power plants* (refers to plants which are designed to produce both heat and electricity). UNIPEDE refers to these as co-generation power stations. If possible, fuel inputs and electricity/heat outputs are on a unit basis rather than on a plant basis. However, if data are not available on a unit basis, the convention for defining a CHP plant noted above should be adopted. Both public and autoproducer plants are included here. *Note that for autoproducer's CHP plants, all fuel inputs to electricity production are taken into account, while only the part of fuel inputs to heat* ***sold*** *is shown. Fuel inputs for the production of heat consumed within the autoproducer's establishment are* ***not*** *included here but are included with figures for the final consumption of fuels in the appropriate consuming sector.*
- *Heat plants* (refers to plants [including heat pumps and electric boilers] designed to produce heat only and who sell heat to a third party [e.g. residential, commercial or industrial consumers] under the provisions of a contract). Both public and autoproducer plants are included here.
- *Blast furnaces/Gas works* (covers the quantities of fuels used for the production of town gas, blast furnace gas and oxygen steel furnace gas). The production of pig-iron from iron ore in blast furnaces uses fuels for supporting the blast furnace charge and providing heat and carbon for the reduction of the iron ore. Accounting for the calorific content of the fuels entering the process is a complex matter as transformation (into blast furnace gas) and consumption (heat of combustion) occur simultaneously. Some carbon is also retained in the pig-iron; almost all of this reappears later in the oxygen steel furnace gas (or converter gas) when the pig-iron is converted to steel. In principle, the quantities of all fuels (e.g. pulverised coal injection (PCI) coal, coke oven coke, natural gas and oil) entering blast furnaces and the quantity of blast furnace gas and oxygen steel furnace gas produced are collected. However, except for coke oven coke inputs, the data are often uncertain or incomplete. The Secretariat then needs to split these inputs into the transformation and consumption components. The transformation component is shown in the row *blast furnaces/gas works* in the column appropriate for the fuel, and the consumption component is shown in the row *iron and steel* in final consumption in the column appropriate for the fuel. Starting with the present edition, the Secretariat decided to assume an 80/20 split of coke oven coke between the transformation and consumption components. This results in more of the energy inputs appearing in the *blast furnaces/gas works* row and less appearing in the *iron and steel* row than in the previous editions. This split has been estimated based on the results of the model developed for OECD countries.[3] It provides a more consistent balancing of the carbon input into and output from the blast furnaces when the IEA data are used to calculate CO_2 emissions from fuel combustion using the Intergovernmental Panel on Climate Change (IPCC) methodology as published in the *Revised 1996 IPCC Guidelines for National Greenhouse Gas Inventories.*[4]
- *Coke/Patent fuel/BKB plants* (covers the use of fuels for the manufacture of coke, coke oven gas, patent fuels and BKB).

---

1 Public supply undertakings generate electricity and/or heat for sale to third parties, *as their primary activity*. They may be privately or publicly owned. Note that the sale need not take place through the public grid.

2 Autoproducer undertakings generate electricity and/or heat, wholly or partly for their own use as an activity which supports their primary activity. They may be privately or publicly owned.

3 Please refer to *Energy Statistics of OECD Countries* for more information on the blast furnace model used for the OECD countries.

4 The *Revised 1996 IPCC Guidelines for National Greenhouse Gas Inventories* are available from the OECD (see inside back cover for sales outlets) or on the Internet (http://www.iea.org/ipcc.htm).

- *Petroleum refineries* (covers the use of hydrocarbons for the manufacture of finished petroleum products).
- *Petrochemical industry* (covers backflows returned from the petrochemical sector). Note, backflows from oil products that are used for non-energy purposes (i.e. white spirit and lubricants) are not included here, but in non-energy use.
- *Liquefaction* (includes diverse liquefaction processes, such as coal and natural gas liquefaction in South Africa).
- *Other transformation sector* (covers charcoal burners losses and other non-specified transformation).

2. Energy Sector

Energy producing industries' own use includes energy consumed by transformation industries for heating, pumping, traction and lighting purposes [ISIC[1] Divisions 10, 11, 12, 23 and 40]:

- *Coal mines* (hard coal and lignite);
- *Oil and gas extraction* (flared gas is not included; includes use of gas by Liquified Natural Gas plants);
- *Petroleum refineries*;
- *Electricity, CHP and heat plants*;
- *Pumped storage* (electricity consumed in hydro-electric plants);
- *Other energy sector* (including own consumption in patent fuel plants, coke ovens, gas works, BKB and lignite coke plants as well as the non-specified energy sector's use).

3. Distribution losses

Distribution losses include losses in gas distribution, electricity transmission, and coal transport. It may also include unaccounted for use of crude oil and petroleum products.

1 International Standard Industrial Classification of All Economic Activities, Series M, No. 4/Rev. 3, United Nations, New York, 1990.

C. Final Consumption

The term *final consumption* (equal to the sum of end-use sectors' consumption) implies that energy used for transformation and for own use of the energy producing industries is excluded. Final consumption reflects for the most part deliveries to consumers (see note on stock changes in Section 2.A.5).

In final consumption, petrochemical feedstocks are covered under **industry** as an *of which* item under chemical industry for those oil products that are principally used for energy purposes. Separated from these are the other oil products that are mainly used for non-energy purposes (see non-energy use in Section 2.C.4), which are shown in the rows for non-energy uses and included only in **total final consumption**. Backflows from the petrochemical industry are not included in final consumption (see Sections 2.A.2 and 2.B.1).

1. Industry Sector

Consumption of the *industry sector* is specified in the following sub-sectors (energy used for transport by industry is not included here but is reported under transport):

- *Iron and steel industry* [ISIC Group 271 and Class 2731];
- *Chemical industry* [ISIC Division 24];
  of which: petrochemical feedstocks. The petrochemical industry includes cracking and reforming processes for the purpose of producing ethylene, propylene, butylene, synthesis gas, aromatics, butadene and other hydrocarbon-based raw materials in processes such as steam cracking, aromatics plants and steam reforming. [Part of ISIC Group 241]; See feedstocks under Section 2.C.4 (Non-energy use).
- *Non-ferrous metals* basic industries [ISIC Group 272 and Class 2732];
- *Non-metallic mineral* products such as glass, ceramic, cement, etc. [ISIC Division 26];
- *Transport equipment* [ISIC Divisions 34 and 35];
- *Machinery*. Fabricated metal products, machinery and equipment other than transport equipment [ISIC Divisions 28, 29, 30, 31 and 32];

- *Mining (excluding fuels) and quarrying* [ISIC Divisions 13 and 14];
- *Food and tobacco* [ISIC Divisions 15 and 16];
- *Paper, pulp and print* [ISIC Divisions 21 and 22];
- *Wood and wood products* (other than pulp and paper) [ISIC Division 20];
- *Construction* [ISIC Division 45];
- *Textile and leather* [ISIC Divisions 17, 18 and 19];
- *Non-specified* (any manufacturing industry not included above) [ISIC Divisions 25, 33, 36 and 37];

Note: Most countries have difficulties supplying an industrial breakdown for all fuels. In these cases, the *non-specified* industry row has been used. *Regional aggregates of industrial consumption should therefore be used with caution.*

2. Transport Sector

Consumption in the *Transport sector* covers all transport activity regardless of the economic sector to which it is contributing [ISIC Divisions 60, 61 and 62], and is divided into the following sub-sectors:

- *Air*: Deliveries of aviation fuels to international civil aviation and to all domestic air transport, commercial, private, agricultural, military, etc. It also includes use for purposes other than flying, e.g. bench testing of engines, but not use of fuels for road transport by airline companies;
- *Road*: All fuels used in road vehicles (including military) as well as agricultural and industrial highway use. Excludes motor gasoline used in stationary engines, and diesel oil for use in tractors that are not for highway use;
- *Rail*: All quantities used in rail traffic, including industrial railways;
- *Pipeline transport*: Energy used for transport of materials by pipeline;
- *Internal navigation* (including small craft and coastal vessels not purchasing their bunker requirements under international marine bunker contracts). Fuel used for ocean, coastal and inland fishing should be included in agriculture;
- *Non-specified.*

3. Other Sectors

- *Agriculture*: Defined as all deliveries to users classified as agriculture, hunting and forestry by the ISIC, and therefore includes energy consumed by such users whether for traction (excluding agricultural highway use), power or heating (agricultural and domestic). Also includes fuels used for ocean, coastal and inland fishing. ISIC Divisions 01, 02 and 05;
- *Commercial and public services*: All activities coming into ISIC Divisions 41, 50, 51, 52, 55, 63, 64, 65, 66, 67, 70, 71, 72, 73, 74, 75, 80, 85, 90, 91, 92, 93 and 99;
- *Residential*: All consumption by households, excluding fuels used for transport. Includes households with employed persons (ISIC Division 95) which is a small part of total residential consumption;
- *Non-specified*: Includes all fuel use not elsewhere specified (e.g. military fuel consumption with the exception of transport fuels in the domestic air and road sectors and consumption in the above-designated categories for which separate figures have not been provided).

4. Non-energy use

Non-energy use covers use of *other* petroleum products such as white spirit, paraffin waxes, lubricants, bitumen and other products (see Sections 3.C.10 and 3.C.11). It also includes the non-energy use of coal (excluding peat). They are shown separately in final consumption under the heading *non-energy use*. It is assumed that the use of these products is exclusively non-energy use. It should be noted that petroleum coke is shown as *non-energy use* only when there is evidence of such use, otherwise it is shown under energy use in industry or in other sectors.

Feedstocks for petrochemical industry are accounted for in industry under chemical industry (row 31) and shown separately under: *of which: feedstocks* (row 32). This covers all oil and gas, including naphtha except the *other petroleum products* listed in Section 3.C.11.

3. NOTES ON ENERGY SOURCES

A. Coal

The heading *coal mines* refers only to coal which is used directly within the coal industry. It excludes coal burned in pithead power stations (included under *transformation* - electricity plants) and free allocations to miners and their families (considered as part of household consumption and therefore included under *other sectors* - residential).

1. Coking Coal

Coking coal refers to coal with a quality that allows the production of a coke suitable to support a blast furnace charge. Its gross calorific value is greater than 23 865 kJ/kg (5 700 kcal/kg) on an ash-free but moist basis.

2. Other Bituminous Coal and Anthracite

Other bituminous coal is used for steam raising and space heating purposes and includes all anthracite coals and bituminous coals not included under coking coal. Its gross calorific value is greater than 23 865 kJ/kg (5 700 kcal/kg), but usually lower than that of coking coal.

3. Sub-Bituminous Coal

Non-agglomerating coals with a gross calorific value between 17 435 kJ/kg (4 165 kcal/kg) and 23 865 kJ/kg (5 700 kcal/kg) containing more than 31 per cent volatile matter on a dry mineral matter free basis.

4. Lignite

Lignite is a non-agglomerating coal with a gross calorific value of less than 17 435 kJ/kg (4 165 kcal/kg), and greater than 31 per cent volatile matter on a dry mineral matter free basis. Oil shale is also included in this category.

5. Peat

Combustible soft, porous or compressed, fossil sedimentary deposit of plant origin with high water content (up to 90 per cent in the raw state), easily cut, of light to dark brown colour. Peat used for non-energy purposes is not included.

6. Coke Oven Coke and Gas Coke

Coke oven coke is the solid product obtained from the carbonisation of coal, principally coking coal, at high temperature. It is low in moisture content and volatile matter. Also included are semi-coke, a solid product obtained from the carbonisation of coal at a low temperature, lignite coke and semi-coke made from lignite. The heading *other energy sector* represents consumption at the coking plants themselves. Consumption in the iron and steel industry does not include coke converted into blast furnace gas. To obtain the total consumption of coke oven coke in the iron and steel industry, the quantities converted into blast furnace gas have to be added (thesc are shown under *blast furnaces/gas works*).

Gas coke is a by-product of hard coal used for the production of town gas in gas works. Gas coke is used for heating purposes. *Other energy sector* data represent consumption of gas coke at gas works.

7. Patent Fuel and Brown Coal / Peat Briquettes (BKB)

Patent fuel is a composition fuel manufactured from coal fines with the addition of a binding agent (pitch). The amount of patent fuel produced is, therefore, slightly higher than the actual amount of coal consumed in the transformation process. Consumption of patent fuels during the patent fuel manufacturing process is shown under *other energy sector.*

BKB are composition fuels manufactured from brown coal, produced by briquetting under high pressure. These figures include peat briquettes, dried lignite fines and dust, and brown coal breeze. The heading *other energy sector* includes consumption by briquetting plants.

B. Crude Oil, NGL, Refinery Feedstocks

Petroleum refineries under *transformation* shows inputs of crude oil, NGL, refinery feedstocks, additives and other hydrocarbons into the refining process.

1. Crude Oil

Crude oil is a mineral oil consisting of a mixture of hydrocarbons of natural origin, being yellow to black in colour, of variable density and viscosity. It also includes lease condensate (separator liquids) which are recovered from gaseous hydrocarbons in lease separation facilities.

Other hydrocarbons, including synthetic crude oil, mineral oils extracted from bituminous minerals such as shales, bituminous sand, etc., and oils from coal liquefaction are included in the row *from other sources*. See Section 2.A.2.

Emulsified oils (e.g. orimulsion) are included here.

2. Natural Gas Liquids (NGL)

NGLs are the liquid or liquefied hydrocarbons produced in the manufacture, purification and stabilisation of natural gas. These are those portions of natural gas which are recovered as liquids in separators, field facilities, or gas processing plants. NGLs include but are not limited to ethane, propane, butane, pentane, natural gasoline and condensate. They may also include small quantities of non-hydrocarbons.

3. Refinery Feedstocks

A refinery feedstock is a product or a combination of products derived from crude oil and destined for further processing other than blending in the refining industry. It is transformed into one or more components and/or finished products. This definition covers those finished products imported for refinery intake and those returned from the petrochemical industry to the refining industry.

4. Additives

Additives are non-hydrocarbon substances added to or blended with a product to modify its properties, for example, to improve its combustion characteristics. Alcohols and ethers (MTBE, methyl tertiary-butyl ether) and chemical alloys such as tetraethyl lead are included here. Ethanol is not included here, but under *Gas/Liquids from Biomass.*

C. Petroleum Products

Petroleum products are any oil-based products which can be obtained by distillation and are normally used outside the refining industry. The exceptions to this are those finished products which are classified as refinery feedstocks above.

Production of petroleum products shows gross refinery output for each product.

Refinery fuel (row *petroleum refineries*, under *energy sector*) represents consumption of petroleum products, both intermediate and finished, within refineries, e.g. for heating, lighting, traction, etc.

1. Refinery Gas (not liquefied)

Refinery gas is defined as non-condensable gas obtained during distillation of crude oil or treatment of oil products (e.g. cracking) in refineries. It consists mainly of hydrogen, methane, ethane and olefins. It also includes gases which are returned from the petrochemical industry. Refinery gas production refers to gross production. Own consumption is shown separately under *petroleum refineries* in the *energy* sector.

2. Liquefied Petroleum Gases (LPG) and Ethane

These are the light hydrocarbons fraction of the paraffin series, derived from refinery processes, crude oil stabilisation plants and natural gas processing plants comprising propane (C_3H_8) and butane (C_4H_{10}) or a combination of the two. They are normally liquefied under pressure for transportation and storage.

Ethane is a naturally gaseous straight-chain hydrocarbon (C_2H_6). It is a colourless paraffinic gas which is extracted from natural gas and refinery gas streams.

3. Motor Gasoline

This is light hydrocarbon oil for use in internal combustion engines such as motor vehicles, excluding aircraft. Motor gasoline is distilled between 35°C and 215°C and is used as a fuel for land based spark ignition engines. Motor gasoline may include additives, oxygenates and octane enhancers, including lead compounds such as TEL (Tetraethyl lead) and TML (tetramethyl lead).

4. Aviation Gasoline

Aviation gasoline is motor spirit prepared especially for aviation piston engines, with an octane number suited to the engine, a freezing point of -60°C, and a distillation range usually within the limits of 30°C and 180°C.

5. Jet Fuel

This category comprises both gasoline and kerosene type jet fuels meeting specifications for use in aviation turbine power units.

a) Gasoline type jet fuel

This includes all light hydrocarbon oils for use in aviation turbine power units. They distil between 100°C and 250°C. It is obtained by blending kerosenes and gasoline or naphthas in such a way that the aromatic content does not exceed 25 per cent in volume, and the vapour pressure is between 13.7 kPa and 20.6 kPa. Additives can be included to improve fuel stability and combustibility.

b) Kerosene type jet fuel

This is medium distillate used for aviation turbine power units. It has the same distillation characteristics and flash point as kerosene (between 150°C and 300°C but not generally above 250°C). In addition, it has particular specifications (such as freezing point) which are established by the International Air Transport Association (IATA).

6. Kerosene

Kerosene comprises refined petroleum distillate intermediate in volatility between gasoline and gas/diesel oil. It is a medium oil distilling between 150°C and 300°C.

7. Gas/Diesel Oil (Distillate Fuel Oil)

Gas/diesel oil includes heavy gas oils. Gas oils are obtained from the lowest fraction from atmospheric distillation of crude oil, while heavy gas oils are obtained by vacuum redistillation of the residual from atmospheric distillation. Gas/diesel oil distils between 180°C and 380°C. Several grades are available depending on uses: diesel oil for diesel compression ignition (cars, trucks, marine, etc.), light heating oil for industrial and commercial uses, and other gas oil including heavy gas oils which distil between 380°C and 540°C and which are used as petrochemical feedstocks.

8. Heavy Fuel Oil (Residual)

This heading defines oils that make up the distillation residue. It comprises all residual fuel oils, including those obtained by blending. Its kinematic viscosity is above 10 cSt at 80°C. The flash point is always above 50°C and the density is always more than 0.90 kg/l.

9. Naphtha

Naphtha is a feedstock destined either for the petrochemical industry (e.g. ethylene manufacture or aromatics production) or for gasoline production by reforming or isomerisation within the refinery. Naphtha comprises material in the 30°C and 210°C distillation range or part of this range. Naphtha imported for blending is shown as an import of naphtha, then shown in the transfers row as a negative entry for naphtha and a positive entry for the corresponding finished product (e.g. gasoline).

10. Petroleum Coke

Petroleum coke is defined as a black solid residue, obtained mainly by cracking and carbonising of residue feedstocks, tar and pitches in processes such as delayed coking or fluid coking. It consists mainly of carbon (90 to 95 per cent) and has a low ash content. It is used as a feedstock in coke ovens for the steel industry, for heating purposes, for electrode manufacture and for production of chemicals. The two most important qualities are "green coke" and "calcinated coke". This category also includes "catalyst coke" deposited on the catalyst during refining processes: this coke is not recoverable and is usually burned as refinery fuel.

11. Other Petroleum Products

The category *other petroleum products* groups together white spirit and SBP, lubricants, bitumen, paraffin waxes and others.

a) White Spirit and SBP:

White spirit and SBP are refined distillate intermediates with a distillation in the naphtha/kerosene range.

They are sub-divided as:

i) Industrial Spirit (SBP): Light oils distilling between 30°C and 200°C, with a temperature difference between 5 per cent volume and 90 per cent volume distillation points, including losses, of not more than 60°C. In other words, SBP is a light oil of narrower cut than motor spirit. There are 7 or 8 grades of industrial spirit, depending on the position of the cut in the distillation range defined above.

ii) White Spirit: Industrial spirit with a flash point above 30°C. The distillation range of white spirit is 135°C to 200°C.

b) Lubricants:

Lubricants are hydrocarbons produced from distillate or residue; they are mainly used to reduce friction between bearing surfaces. This category includes all finished grades of lubricating oil, from spindle oil to cylinder oil, and those used in greases, including motor oils and all grades of lubricating oil base stocks.

c) Bitumen:

Solid, semi-solid or viscous hydrocarbon with a colloidal structure, being brown to black in colour, obtained as a residue in the distillation of crude oil, vacuum distillation of oil residues from atmospheric distillation. Bitumen is often referred to as asphalt and is primarily used for surfacing of roads and for roofing material. This category includes fluidized and cut back bitumen.

d) Paraffin Waxes:

Saturated aliphatic hydrocarbons (with the general formula C_nH_{2n+2}). These waxes are residues extracted when dewaxing lubricant oils, and they have a crystalline structure with carbon number greater than 12. Their main characteristics are that they are colourless, odourless and translucent, with a melting point above 45°C.

e) Others:

Includes the petroleum products not classified above, for example: tar, sulphur, and grease. This category also includes aromatics (e.g. BTX or benzene, toluene and xylene) and olefins (e.g. propylene) produced within refineries.

D. Gases

The figures for these four categories of gas are all expressed in terajoules based on **gross calorific values**.

1. Natural Gas

Natural gas comprises gases, occurring in underground deposits, whether liquefied or gaseous, consisting mainly of methane. It includes both "non-associated" gas originating from fields producing only hydrocarbons in gaseous form, and "associated" gas produced in association with crude oil as well as methane recovered from coal mines (colliery gas).

Production is measured after purification and extraction of NGL and sulphur, and excludes re-injected gas, quantities vented or flared. It includes gas consumed by gas processing plants and gas transported by pipeline.

2. Gas Works Gas

Gas works gas covers all types of gas produced in public utility or private plants, whose main purpose is the manufacture, transport and distribution of gas. It includes gas produced by carbonisation (including gas produced by coke ovens and transferred to gas works), by total gasification (with or without enrichment with oil products), by cracking of natural gas, and by reforming and simple mixing of gases and/or air. This heading also includes substitute natural gas, which is a high calorific value gas manufactured by chemical conversion of a hydrocarbon fossil fuel.

3. Coke Oven Gas

Coke oven gas is obtained as a by-product of the manufacture of coke oven coke for the production of iron and steel.

4. Blast Furnace Gas

Blast furnace gas is produced during the combustion of coke in blast furnaces in the iron and steel industry. It is recovered and used as a fuel partly within the plant and partly in other steel industry processes or in power stations equipped to burn it. Also included here is oxygen steel furnace gas which is obtained as a by-product of the production of steel in an oxygen furnace and is recovered on leaving the furnace. The gas is also known as converter gas or LD gas.

E. Combustible Renewables and Waste

The figures for these four categories of fuels are all expressed in terajoules based on **net calorific values**.

1. Solid Biomass and Animal Products

Biomass is defined as any plant matter used directly as fuel or converted into other forms before combustion. Included are wood, vegetal waste (including wood waste and crops used for energy production), animal materials/wastes, sulphite lyes, also known as "black liquor" (an alkaline spent liquor from the digesters in the production of sulphate or soda pulp during the manufacture of paper where the energy content derives from the lignin removed from the wood pulp) and other solid biomass.

Charcoal produced from solid biomass is also included here. Since charcoal is a secondary product, its treatment is slightly different than that of the other primary biomass. Production of charcoal (an output in the transformation process) is offset by the inputs of primary biomass into the charcoal production process. The losses from this process are included in the row *other transformation sector*. Other supply (e.g. trade and stock changes) as well as consumption are aggregated directly with the primary biomass. However, in some countries, only the primary biomass is reported. Reported production of charcoal is available separately in the Summary Tables at the end of the publication.

2. Gas/Liquids from Biomass

Biomass gases are derived principally from the anaerobic fermentation of biomass and solid wastes and combusted to produce heat and/or power. Included in this category are landfill gas and sludge gas (sewage gas and gas from animal slurries). Bio-additives such as ethanol are also included in this category.

3. Municipal Waste

Municipal waste consists of products that are combusted directly to produce heat and/or power and comprises wastes produced by the residential, commercial and public services sectors that are collected by local authorities for disposal in a central location. Hospital waste is included in this category.

4. Industrial Waste

Industrial waste consists of solid and liquid products (e.g. tyres) combusted directly, usually in specialised plants, to produce heat and/or power and that are not reported in the category *solid biomass and animal products*.

F. Electricity and Heat

1. Electricity

Gross electricity production is measured at the terminals of all alternator sets in a station; it therefore includes the energy taken by station auxiliaries and losses in transformers that are considered integral parts of the station.

The difference between gross and net production is generally calculated as 7 per cent for conventional thermal stations, 1 per cent for hydro stations, and

6 per cent for nuclear, geothermal and solar stations. Hydro stations' production includes production from pumped storage plants.

2. Heat

In recent years, the production of heat for sale has been increasing in importance. To reflect this, heat production represents all heat production from public CHP and heat plants as well as heat sold by autoproducer CHP and heat plants to third parties.

Corresponding fuels to produce quantities of heat for sale are being recorded in the transformation sector under the rows *CHP plants* and *Heat plants*. The use of fuels for heat which is not sold is recorded under the sectors in which the fuel use occurs.

4. NOTES ON SUMMARY TABLES

Four summary tables provide data in a format which differs from the Annual Tables:

Production of Charcoal

Charcoal is a secondary biomass product and refers to the solid residue, consisting mainly of carbon, derived from the distillation of wood or other biomass products in the absence of air. Please note that in the Annual Tables, charcoal is included under Solid biomass (see Notes on Energy Sources, Part .3).

Oil Consumption by Product, in thousand tonnes and in million barrels per day

Rows 1 to 8 show direct inland consumption by product, excluding refinery throughput and refinery energy use, but including distribution losses. *Motor gasoline* includes liquids from biomass. *Other products* cover crude oil, other hydrocarbons, refinery gas, petroleum coke, white spirit and SBP, lubricants, bitumen, paraffin waxes and others such as tar, sulphur, grease, as well as aromatics (e.g. BTX or benzene, toluene and xylene) and olefins (e.g. propylene) produced within refineries.

Row 9, *Refinery fuel*, shows petroleum refineries own use of petroleum fuels, which include mainly refinery gas, gas/diesel oil and heavy fuel oil.

Row 11, *Bunkers*, shows international marine bunkers consumption of liquid fuels, which include mainly gas/diesel oil and heavy fuel oil.

The following conversion factors are used, unless otherwise indicated, for all countries and all years.

	Barrels per tonne
Refinery gas	8.00
Ethane	16.85
LPG	11.60
Naphtha	8.50
	(8.90 OECD Europe)
Aviation Gasoline	8.90
Motor Gasoline	8.53
	(8.45 OECD Europe)
Jet Gasoline	7.93
	(8.25 OECD)
Jet Kerosene	7.93
	(7.88 OECD Europe)
Other Kerosene	7.74
	(7.88 OECD Europe)
Gas/Diesel Oil	7.46
Heavy Fuel Oil	6.66
	(6.45 OECD Europe)
White Spirit	7.00
	(8.46 OECD)
Lubricants	7.09
Bitumen	6.08
Paraffin Waxes	7.00
	(7.85 OECD)
Petroleum Coke	5.50
Non Specified Products	7.00
	(8.00 OECD)

Consumption of Electricity

Covers electricity use in transformation, energy sector and final consumption, but excludes distribution losses.

5. GEOGRAPHICAL COVERAGE

- **Africa** includes Algeria, Angola, Benin, Cameroon, Congo, Democratic Republic of Congo, Egypt, Ethiopia, Gabon, Ghana, Ivory Coast, Kenya, Libya, Morocco, Mozambique, Nigeria, Senegal, South Africa, Sudan, Tanzania, Tunisia, Zambia, Zimbabwe and **Other Africa**.
- **Other Africa** includes Botswana, Burkina-Faso, Burundi, Cape Verde, Central African Republic, Chad, Djibouti, Equatorial Guinea, Gambia, Guinea, Guinea-Bissau, Lesotho, Liberia, Madagascar, Malawi, Mali, Mauritania, Mauritius, Niger, Réunion, Rwanda, Sao Tome-Principe, Seychelles, Sierra Leone, Somalia, Swaziland, Togo and Uganda.
- **Latin America** includes Argentina, Bolivia, Brazil, Chile, Colombia, Costa Rica, Cuba, Dominican Republic, Ecuador, El Salvador, Guatemala, Haiti, Honduras, Jamaica, Netherlands Antilles, Nicaragua, Panama, Paraguay, Peru, Trinidad and Tobago, Uruguay, Venezuela and **Other Latin America**.
- **Other Latin America** includes Antigua and Barbuda, Bahamas, Barbados, Belize, Bermuda, Dominica, French Guiana, Grenada, Guadeloupe, Guyana, Martinique, St. Kitts-Nevis-Anguilla, Saint Lucia, St. Vincent-Grenadines and Surinam.
- **Asia** includes Bangladesh, Brunei, Chinese Taipei, India, Indonesia, DPR of Korea, Malaysia, Myanmar, Nepal, Pakistan, Philippines, Singapore, Sri Lanka, Thailand, Vietnam and **Other Asia**.
- **Other Asia and Other Oceania** include Afghanistan, Bhutan, Fiji, French Polynesia, Kiribati, Maldives, New Caledonia, Papua New Guinea, Samoa, Solomon Islands and Vanuatu.
- **China** includes the People's Republic of China and Hong Kong (China).
- **Non-OECD Europe** includes Albania, Bosnia-Herzegovina, Bulgaria, Croatia, Cyprus, the Former Yugoslav Republic of Macedonia (FYROM), Gibraltar, Malta, Romania, Slovak Republic, Slovenia and the Federal Republic of Yugoslavia.
- **Former USSR** includes Armenia, Azerbaijan, Belarus, Estonia, Georgia, Kazakhstan, Kyrgyzstan, Latvia, Lithuania, Moldova, Russia, Tajikistan, Turkmenistan, Ukraine and Uzbekistan.
- **Middle East** includes Bahrain, Iran, Iraq, Israel, Jordan, Kuwait, Lebanon, Oman, Qatar, Saudi Arabia, Syria, the United Arab Emirates and Yemen.
- The **Organisation for Economic Co-Operation and Development (OECD)** includes Australia, Austria, Belgium, Canada, the Czech Republic, Denmark, Finland, France, Germany, Greece, Hungary, Iceland, Ireland, Italy, Japan, Korea, Luxembourg, Mexico, the Netherlands, New Zealand, Norway, Poland, Portugal, Spain, Sweden, Switzerland, Turkey, the United Kingdom and the United States.

  Within OECD:

  Denmark excludes Greenland and the Danish Faroes.

France includes Monaco, and excludes overseas departments (Martinique, Guadeloupe, French Polynesia and Réunion).

Germany includes the new federal states of Germany.

Italy includes San Marino and the Vatican.

Japan includes Okinawa.

The **Netherlands** excludes Surinam and the Netherlands Antilles.

Portugal includes the Azores and Madeira.

Spain includes the Canary Islands.

Switzerland includes Liechtenstein.

United States includes Puerto Rico, Guam, the Virgin Islands and the Hawaiian Free Trade Zone.

- **The Organisation of the Petroleum Exporting Countries (OPEC)** includes Algeria, Indonesia, Iran, Iraq, Kuwait, Libya, Nigeria, Qatar, Saudi Arabia, the United Arab Emirates and Venezuela.

Please note that the following countries have not been considered due to lack of data:

Africa: Comoros, Namibia, Saint Helena and Western Sahara.

America: Aruba, British Virgin Islands, Caymen Islands, Falkland Islands, Montserrat, Saint Pierre-Miquelon and Turks and Caicos Islands;

Asia and Oceania: American Samoa, Cambodia, Christmas Island, Cook Islands, Laos, Macau, Mongolia, Nauru, Niue, Pacific Islands (US Trust), Tonga and Wake Island.

6. COUNTRY NOTES AND SOURCES

NOTES ET SOURCES PAR PAYS

General References – Références Générales

Annual Bulletin of Coal Statistics for Europe, Economic Commission for Europe (ECE), New York, 1994.

Annual Bulletin of Electric Energy Statistics for Europe, Economic Commission for Europe (ECE), New York, 1994.

Annual Bulletin of Gas Statistics for Europe, Economic Commission for Europe (ECE), New York, 1994.

Annual Bulletin of General Energy Statistics for Europe, Economic Commission for Europe (ECE), New York, 1994.

Annual Report July 1991- June 1992, South African Development Community (SADC), Gaborone, 1993.

Annual Statistical Bulletin 1997, Organization of Petroleum Exporting Countries (OPEC), Vienna, 1998.

APEC Energy Database, Mosaic/ World Wide Web site: http://www.ieej.or.jp/apec/database

Arab Oil and Gas Directory 1998, Arab Petroleum Research Centre, Paris, 1999.

ASEAN Energy Review 1995 Edition, ASEAN-EC Energy Management Training and Research Centre (AEEMTRC), Jakarta, 1996.

Base CHELEM-PIB, Centre d'Etudes Prospectives et d'Informations Internationles (CEPII) Paris, 1999.

Eastern Bloc Energy, Tadcaster, various issues to May 1999.

Energy Indicators of Developing Member Countries, Asian Development Bank (ADB), Manila, 1994.

Energy Information Administration (EIA) Mosaic/World Wide Web site: http://www.eia.doe.gov

Energy-Economic Information System (SIEE), Latin American Energy Organization (OLADE), Ecuador, 1999.

Energy Statistics Yearbook 1990, South African Development Community (SADC), Luanda, 1992.

Energy Statistics Yearbook 1996, United Nations, New York, 1998.

Food and Agriculture Organisation of the United Nations, 1999 Forestry Data (via Internet), Rome, 1999.

Forests and Biomass Sub-sector in Africa, African Energy Programme of the African Development Bank, Abidjan, 1996.

International Coal Report, various issues to May 1999.

International Energy Annual 1990, 1991, 1992, 1993, Energy Information Administration (EIA), Washington, D.C., 1991-1994.

International Energy Data Report 1992, World Energy Council, London, 1993.

Les Centrales Nucléaires dans le Monde Commissariat à l'Énergie Atomique, Paris, 1998.

Middle East Economic Survey (MEES), Nicosia, various issues to June 1999.

Natural Gas in the World, 1998 Survey, Cedigaz, Paris, 1998.

Notes d'Information et Statistiques, Banque Centrale des Etats de l'Afrique de l'Ouest, Dakar, 1995.

Pétrole 1994, Comité Professionnel du Pétrole (CPDP), Paris, 1995.

PIW's Global Oil Stocks & Balances, New York, various issues to June 1995.

PlanEcon Energy Outlook for Eastern Europe and the Former Soviet Republics, Washington, October 1998.

Prospects of Arab Petroleum Refining Industry, Organization of Arab Petroleum Exporting Countries (OAPEC), Kuwait, 1990.

Review of Wood Energy Data in RWEDP Member Countries, Regional Wood Energy Development Programme in Asia, Food and Agriculture Organisation of the United Nations, Bangkok, 1997.

Statistical Handbook 1993 States of the Former USSR, The World Bank, Washington, 1993.

Statistical Yearbook of the Member States of the CMEA, Council of Mutual Economic Assistance (CMEA), Moscow, 1985 and 1990.

The United Nations Energy Statistics Database 1996, United Nations Statistical Office (UNSO), New York, 1998.

World Development Indicators 1999 on CD-ROM, The World Bank, Washington, 1999.

Note:

- The OLADE database was used for most of the Latin American countries.

- For the period 1971 to 1996, the UN database was the only source of information for the eleven individual countries which are not listed below, and for the regions Other Africa, Other Latin America and Other Asia and Oceania. It was also used in a number of other countries as a complementary source.

Albania

Large quantities of oil widely reported to have moved through Albania into former Yugoslavia are not included in oil trade for 1993. Although estimated to represent up to 100 pour cent of underlying domestic consumption, no reliable figures for this trade were available. Series have been revised since 1990.

Sources 1971-1997:

Combustible Renewables and Waste:

The UN Energy Statistics Database, 1995, UN ECE Energy Questionnaires and Secretariat estimates.

Sources up to 1995:

Aide Memoire of World Bank Mission to Albania May/June 1991.

The UN Energy Statistics Database, 1994.

UN ECE Energy Questionnaires 1994 and 1995.

Algeria

Sources 1971-1997:

Combustible Renewables and Waste:

The UN Energy Statistics Database, 1996.

Sources 1992-1997:

Direct communication to the Secretariat from the Ministry of Industry and Energy, Information Systems Management Department, Algiers.

Sources up to 1991:

Bilan Energétique National, Gouvernement Algérien, Algiers, 1984.

Algérie Energie, No 6, Ministère de l'Energie et des Industries Chimiques et Pétrochimiques, Algiers, 1979 to 1983.

Annuaire Statistique de l'Algérie 1980-1984, Office National des Statistiques, Algiers, 1985.

Angola

Sources 1971-1997:

Combustible Renewables and Waste:

Secretariat estimates based on 1991 data from African Energy Programme of the African Development Bank, *Forests and Biomass Sub-sector in Africa,* Abidjan, 1996.

Sources 1992-1997:

Direct communications to the Secretariat from oil industry sources.

Eskom Annual Statistical Yearbook 1993, 1994, 1995 (Johannesburg, 1994, 1995, 1996) citing Empresa Nacional de Electricidade, Luanda as source.

The UN Energy Statistics Database, 1995.

Sources up to 1991:

Le Pétrole et l'Industrie Pétrolière en Angola en 1985, Poste d'Expansion Economique de Luanda, Luanda, 1985.

Argentina

Sources 1971-1997:

Combustible Renewables and Waste:

SIEE, (OLADE).

Sources 1992-1997:

Secretaría de Energía Wide Web site: http://www.mecon.ar/energia/

Direct communication to the Secretariat from the Ministry of Economy and Public Services, Secretariat of Energy, Buenos Aires.

Annuario Estadístico de La Republica Argentina, Instituto Nacional de Estadistica y Censos, Buenos Aires, September 1997.

Sources up to 1991:

Anuario de Combustibles, Ministerio de Obras y Servicios Públicos, Secretaria de Energía, Buenos Aires, 1980 to 1984, 1986, 1988 to 1992, 1995, 1997.

Combustibles Boletin Mensual, Ministerio de Obras y Servicios Públicos, Secretaria de Energía, Buenos Aires, various editions.

Natural Gas Projection up to 2000, Gas del Estado Argentina, Buenos Aires, 1970, 1984 to 1986.

Anuario Estadístico de la Republica Argentina 1970-1981, Instituto Nacional de Estadístico y Censos, Secretaria de Planificación, Buenos Aires, 1982.

Anuario Energía Eléctrica, Ministerio de Obras y Servicios Públicos, Secretaria de Energía, Buenos Aires, 1987 to 1990.

Balance Energetico Nacional 1970-1985, Ministerio de Obras y Servicios Públicos, Secretaria de Energía, Buenos Aires, 1986.

Plan Energetico Nacional 1986-2000, Ministerio de Obras y Servicios Públicos, Secretaria de Energía, Subsecretaria de Planificación Energetica, Buenos Aires, 1985.

Anuario Estadístico, Yacimientos Petrolíferos Fiscales, Buenos Aires, 1984 to 1987.

Memoria Y Balance General, Yacimientos Petrolíferos Fiscales, Buenos Aires, 1984 to 1986.

Bahrain

Sources 1992-1997:

Direct communication to the Secretariat from oil industry sources.

Statistical Abstract, 1994, Council of Ministers, Control Statistics Organisation, Bahrain, 1995.

The UN Energy Statistics Database, 1995.

Sources up to 1991:

1986 Annual Report, Bahrain Monetary Agency, Bahrain, 1987.

B.S.C. Annual Report, Bahrain Petroleum Company, Bahrain, 1982, 1983 and 1984.

Foreign Trade Statistics, Council of Ministers, Central Statistics Organisation, Bahrain, 1985.

Bahrain in Figures, Council of Ministers, Central Statistics Organisation, Bahrain, 1983, 1984 and 1985.

Statistical Abstract 1990, Council of Ministers, Central Statistics Organisation, Bahrain, 1991.

Bangladesh

Energy statistics are reported for a fiscal year.

Sources 1971-1997:

Combustible Renewables and Waste:

Secretariat estimates based on a per capita average consumption from various surveys and studies.

Sources 1996-1997:

Statistical Yearbook of Bangladesh 1996,1997, 7th Edition, Ministry of Planning, Bangladesh Bureau of Statistics, Dhaka, 1997,1998.

Direct communication to the Secretariat from oil and gas industry sources and electricity utility.

Sources 1992-1995:

Statistical Pocket Book of Bangladesh, Ministry of Planning, Bangladesh Bureau of Statistics, Dhaka, 1986 to 1996.

The UN Energy Statistics Database, 1995.

Sources up to 1991:

Bangladesh Energy Balances 1976-1981, Government of Bangladesh, Dhaka, 1982.

Statistical Yearbook of Bangladesh 1991, Government of Bangladesh, Dhaka, 1976 to 1991.

Monthly Statistical Bulletin of Bangladesh, Ministry of Planning, Bangladesh Bureau of Statistics, Statistics Division, Dhaka, June 1986 and October 1989.

Benin

Sources 1971-1997:

Combustible Renewables and Waste:

Secretariat estimates based on 1991 data from *Forests and Biomass Sub-sector in Africa,* African Energy Programme of the African Development Bank, Abidjan, 1996.

Sources up to 1997:

Direct communication to the Secretariat from the Electricity utility 1998, 1999.

Direct communication to the Secretariat from the Direction de l' Energie, Cotonou, 1999.

The UN Energy Statistics Database, 1995.

Rapport sur l'Etat de l'Economie Nationale, Ministère de l' Economie, Cotonou, septembre 1993.

Bolivia

Sources 1971-1997:

Combustible Renewables and Waste:

SIEE, (OLADE).

Sources 1992-1997:

Informe Estadístico 1992, 1993, 1994, 1995, 1996 and 1997, Yacimientos Petrolíferos Fiscales Bolivianos, La Paz, 1993, 1994, 1995, 1996, 1997, 1998.

Memoria Anual 1992, Yacimientos Petrolíferos Fiscales Bolivianos, La Paz, 1993.

Sources up to 1991:

Boletin Estadístico 1973-1985, Banco Central de Bolivia, Division de Estudios Económicos, La Paz, 1986.

Diez Anos de Estadística Petrolera en Bolivia 1976-1986, Dirección de Planeamiento, Division de Estadística, La Paz, 1987.

Empresa Nacional de Electricidad S.A. 1986 Ende Memoria, Empresa Nacional de Electricidad, La Paz, 1987.

Brazil

Sources 1971-1997:

Combustible Renewables and Waste:

Ministério de Minas e Energia.

Sources 1992-1997:

Balanço Energético Nacional, Ministério de Minas e Energia, Brasilia, 1993, 1994, 1995, 1996, 1997, 1998.

Direct communication to the Secretariat from Petrobrás, Commercial Department, Marine Fuels Division.

Sources up to 1991:

Balanço Energético Nacional, Ministério de Minas e Energia, Brasilia, 1983 to 1992.

Anuario Estatistico, Conselho Nacional do Petroleo, Diretoria de Planejamento, Coordenadoria de Estatistica, Brasilia, 1982, 1987, 1988.

Brunei

Historical series have been considerably revised using direct communication from the Office of the Prime Minister, Petroleum Unit.

Sources 1971-1997:

Combustible Renewables and Waste:

The UN Energy Statistics Database, 1995.

Sources 1990-1997:

Direct communication to the secretariat from the Office of the Prime Minister, Petroleum Unit 1999.

Direct communication to the secretariat from the Ministry of Development, Electrical Services Department 1999.

Brunei Statistical Yearbook, 1992 to 1994, Ministry of Finance, Statistics Section, Brunei, 1993, 1995.

Direct communication to the Secretariat from the UN Energy Statistics Unit.

Sources up to 1991:

Fifth National Development Plan 1986-1990, Ministry of Finance, Economic Planning Unit, Bandar Seri Bagawan, 1985.

Bulgaria

See Part I.1, Issues of Data Quality above.

Sources 1971-1997:

Combustible Renewables and Waste:

The UN Energy Statistics Database, 1995 and UN ECE Energy Questionnaires.

Sources 1992-1997:

UN ECE Energy Questionnaires.

Energy Balances, National Statistical Institute, Sofia, 1995.

Sources up to 1991:

Energy Development of Bulgaria, Government of Bulgaria, Sofia, 1980 and 1984.

Energy in Bulgaria, Government of Bulgaria, Sofia, 1980 to 1983.

General Statistics in the Republic of Bulgaria 1989/1990, Government of Bulgaria, Sofia, 1991.

Cameroon

Sources 1971-1997:

Combustible Renewables and Waste:

Secretariat estimates based on 1991 data from *Forests and Biomass Sub-sector in Africa,* African Energy Programme of the African Development Bank, Abidjan, 1996.

Sources up to 1997:

Direct communication to the Secretariat from oil industry sources and the electricity utility.

The UN Energy Statistics Database, 1995.

Chile

Sources 1971-1997:

Combustible Renewables and Waste:

Comisión Nacional de Energía.

Sources 1992-1997:

Balance Nacional de Energía 1995, 1997 Comisión Nacional de Energía, Santiago, 1996, 1998.

Balance Nacional de Energía 1977-1996, Comisión Nacional de Energía, Santiago, September 1997.

Balance de Energía Preliminar 1993, Comisión Nacional de Energía, Santiago, 1994.

Balance de Energía 1973 - 1992, Comisión Nacional de Energía, Santiago, 1993, *1975-1994*, Santiago, 1995.

Sources up to 1991

Compendio Estadístico Chile 1985, Ministerio de Economía, Fomento Y Reconstrucción, Instituto Nacional de Estadísticas, Santiago, 1986.

China

See Part I.1, Issues of Data Quality above.

Coal production statistics refer to unwashed and unscreened coal. IEA coal statistics normally refer to coal after washing and screening for the removal of inorganic matter.

It is known that much of the agricultural use of diesel in China is for transport purposes but the national data have not been adjusted by the IEA Secretariat to take account of this.

See Part I.1, Issues of Data Quality for further information.

Sources 1971-1997:

Combustible Renewables and Waste:

Secretariat estimates based on a per capita average consumption from various surveys and studies.

Sources 1992-1997:

Energy Balances of China 1997, obtained from the APEC Energy Database.

Energy Balances of China, provided to the Secretariat by the State Statistical Bureau for 1993, 1994, 1995 and 1996.

China Statistical Yearbook 1995, State Statistical Bureau of the People's Republic of China, Beijing, 1995.

China Energy Databook, Lawrence Berkeley National Laboratory, Berkeley, 1996.

1995 Energy Report of China, State Planning Commission of the People's Republic of China, Beijing, 1995.

China Petroleum Newsletter Monthly, China Petroleum Information Institute, Beijing, various issues to June 1995.

China OGP, China Oil Gas and Petrochemicals/Xinhua News Agency, Beijing, various issues to June 1995.

Petroleum Data Monthly, China Oil Gas and Petrochemicals/Xinhua News Agency, Beijing, various issues to June 1995.

Energy of China, China International Book Trading Co, Beijing, various issues to June 1995.

Energy Commodity Account of China, Asian Development Bank, Manila, 1994.

China's Customs Statistics, State Statistical Bureau, Economic Information & Agency, Beijing, various editions to 1995.

China's Customs Statistics, General Administration of Customs, PRC, Economic Information and Agency, Hong Kong, various editions from 1991 to 1995.

Statistical Yearbook of China, State Statistical Bureau of the People's Republic of China, Economic Information & Agency, Hong Kong, various editions from 1981 to 1994.

Statistical Communique, State Statistical Bureau of the People's Republic of China, Hong Kong, March 1995.

China's Downstream Oil Industry in Transition, Fesharaki Associates Consulting and Technical Services, Inc., Honolulu, 1993.

Sources up to 1991:

Outline of Rational Utilization and Conservation of Energy in China, Bureau of Energy Conservation State Planning Commission, Beijing, June 1987.

China Coal Industry Yearbook, Ministry of Coal Industry, People's Republic of China, Beijing, 1983, 1984 and 1985.

Energy in China 1989, Ministry of Energy, People's Republic of China, Beijing, 1990.

Electric Industry in China in 1987, Ministry of Water Resources and Electric Power, Department of Planning, Beijing, 1988.

Statistical Yearbook of China 1991, State Statistical Bureau of the People's Republic of China, Beijing, 1992.

China: A Statistics Survey 1975-1984, State Statistical Bureau, Beijing, 1985.

China Petro-Chemical Corporation (SINOPEC) Annual Report, SINOPEC, Beijing, 1987.

Almanac of China's Foreign Economic Relations and Trade, The Editorial Board of the Almanac, Beijing, 1986.

Chinese Taipei

Autoproducer electricity includes only the inputs and outputs of the iron and steel industry and waste disposal plants. Energy used by the other autoproducers (mostly industrial cogeneration) was counted as final consumption.

Sources 1971-1997:

Combustible Renewables and Waste:

The UN Energy Statistics Database, 1995.

Sources 1992-1997:

Energy Balances in Taiwan, Ministry of Economic Affairs, Taipei, 1992 to 1997.

Yearbook of Energy Statistics, Ministry of Trade Industry and Energy, Taipei, 1996.

Sources up to 1991:

Industry of Free China 1975-1985, Council for Economic Planning and Development, Taipei, 1986.

Taiwan Statistical Data Book 1954-1985, Council for Economic Planning and Development, Taipei, 1986.

Energy Policy for the Taiwan Area, Ministry of Economic Affairs, Energy Committee, Taipei, 1984.

The Energy Situation in Taiwan, Ministry of Economic Affairs, Energy Committee, Taipei, 1986, 1987, 1988 and 1992.

Energy Balances in Taiwan, Ministry of Economic Affairs, Taipei, 1984 to 1991.

Energy Indicators Quarterly, Taiwan Area, Ministry of Economic Affairs, Taipei, 1986.

Taipower 1987 Annual Report, Taipower, Taipei, 1988.

Energy Data Report, 1986, WEC National Committee, Taipei, 1986.

Colombia

Sources 1971-1997:

Combustible Renewables and Waste:

Ministry of Mines and Energy, Energy Information Department.

Sources 1992-1997:

Direct communication to the Secretariat from the Ministry of Mines and Energy, Energy Information Department, Bogotá.

Sources up to 1991:

Estadísticas Basicas del Sector Carbón, Carbocol, Oficina de Planeación, Bogotá, various editions from 1980 to 1988.

Colombia Estadística 1985, DANE, Bogotá, 1970 to 1983 and 1987.

Empresa Colombiana de Petróleos, Informe Anual, Empresa Colombiana de Petróleos, Bogotá, 1979, 1980, 1981 and 1985.

Estadísticas de la Industria Petrolera Colombiana Bogota 1979-1984, Empresa Colombiana de Petróleos, Bogotá, 1985.

Informe Estadístico Sector Eléctrico Colombiano, Government of Colombia, Bogotá, 1987 and 1988.

La Electrificacion en Colombia 1984-1985, Instituto Colombiano de Energía Electrica, Bogotá, 1986.

Balances Energéticos 1975-1986, Ministerio de Minas y Energía, Bogota, 1987.

Energía y Minas Para el Progreso Social 1982-1986, Ministerio de Minas y Energía, Bogota, 1987.

Estadísticas Minero-Energéticas 1940-1990, Ministerio de Minas y Energía, Bogota, 1990.

Boletin Minero-Energético, Ministerio de Minas y Energía, Bogota, December 1991.

Congo

Sources 1971-1997:

Combustible Renewables and Waste:

Secretariat estimates based on 1991 data from *Forests and Biomass Sub-sector in Africa,* African Energy Programme of the African Development Bank, Abidjan, 1996.

Sources up to 1997:

Annual Statistical Yearbook 1993, 1994, 1995, Eskom, Johannesburg, 1994, 1995, 1996, citing Empresa Nacional de Electricidade, Luanda as source.

L'Energie en Afrique, IEPE/ENDA, Paris, 1995, in turn sourced from the Direction des Etudes et de la Planification, Ministère des Mines et de l'Energie, and the Société Congolaise de Raffinage, Brazzaville.

Direct communication to the Secretariat from the UN Energy Statistics Unit.

Democratic Republic of Congo

Sources 1971-1997:

Combustible Renewables and Waste:

Secretariat estimates based on 1991 data from *Forests and Biomass Sub-sector in Africa,* African Energy Programme of the African Development Bank, Abidjan, 1996.

Sources up to 1997:

L'Energie en Afrique, IEPE/ENDA, Paris, 1995, in turn sourced from the *Annuaire Statistique Energétique 1990,* Communauté Economique des Pays des Grands Lacs, Bujumbura, 1990.

The UN Energy Statistics Database, 1996.

Cuba

Historical series have been considerably revised using *Anuario Estadístico de Cuba 1996,* from Oficina Nacional de Estadísticas.

Sources 1971-1997:

Combustible Renewables and Waste:

SIEE, (OLADE).

Anuario Estadístico de Cuba 1996, Oficina Nacional de Estadísticas, Havana, 1998.

Sources up to 1991:

Compendio estadístico de energía de Cuba 1989, Comite Estatal de Estadísticas, Havana, 1989.

Anuario Estadístico de Cuba, Comite Estatal de Estadísticas, Havana, various editions from 1978 to 1987.

Cyprus

Sources 1971-1997:

Combustible Renewables and Waste:

UN ECE Energy Questionnaires and Secretariat estimates.

Sources 1992-1997:

UN ECE Energy Questionnaires.

Electricity Authority of Cyprus Annual Report 1988, 1992, 1996, Electricity Authority of Cyprus, Nicosia, 1989, 1993, 1997.

Industrial Statistics 1988, Ministry of Finance, Department of Statistics, Nicosia, 1989.

Ecuador

Sources 1971-1997:

Combustible Renewables and Waste:

Ministerio de Energia y Minas and *SIEE*, OLADE.

Sources 1996:

Sector Energético Ecuatoriano, Edición No 7, Direccion de Planificacion, Ministerio de Energia y Minas, Quito, November 1997.

Sources 1992-1995:

Balance Energético Nacional 1995, Ministerio de Energia y Minas, Quito, December 1996.

Balances Energéticos 1988-1994, Ministerio de Energia y Minas, Quito, 1996.

Sources up to 1991: Ministerio de Energia y Minas,

Cuentas Nacionales, Banco Central del Ecuador, Quito, various editions from 1982 to 1987.

Memoria 1980-1984, Banco Central del Ecuador, Quito, 1985.

Ecuadorian Energy Balances 1974-1986, Instituto Nacional de Energía, Quito, 1987.

Informacion Estadística Mensual, No. 1610, Instituto Nacional de Energía, Quito, 1988.

Plan Maestro de Electrificación de Ecuador, Ministerio de Energía y Minas, Quito,1989.

Egypt

Sources 1971-1996:

Combustible Renewables and Waste:

The UN Energy Statistics Database, 1995.

Sources 1992-1997:

1995, 1997 Annual Report, Ministry of Petroleum, Egyptian General Petroleum Corporation, Cairo, 1996,1998.

Annual Report of Electricity Statistics 1996/1997, Ministry of Electricity and Energy, Egyptian Electricity Authority, Cairo, 1998.

Arab Oil and Gas, The Arab Petroleum Research Center, Paris, October 1997.

Middle East Economic Survey, Middle East Petroleum and Economic Publications, Nicosia, February 1994, June 1996, March 1998.

Direct submisson to the Secretariat from the Ministry of Petroleum.

A Survey of the Egyptian Oil Industry 1993, Embassy of the United States of America in Cairo, Cairo, 1994.

Sources up to 1991:

Annual Report of Electricity Statistics 1990/1991, Ministry of Electricity and Energy, Egyptian Electricity Authority, Cairo, 1992.

Statistical Yearbook of the Arab Republic of Egypt, Central Agency for Public Mobilisation and Statistics, Cairo, 1977 to 1986.

L'Electricité, l'Energie, et le Pétrole, République Arabe d'Egypte, Organisme Général de l'Information, Cairo, 1990.

Annual Report, The Egyptian General Petroleum Corporation, Cairo, 1985.

Ethiopia

Ethiopia is including Eritrea.

Sources 1971-1997:

Combustible Renewables and Waste:

Secretariat estimates based on 1992 data from Eshetu, L. and Bogale, W., *Power Restructuring in Ethiopia*, AFREPREN, Nairobi, 1996.

Sources 1992-1997

Direct communication to the Secretariat from the Ministry of Economic Development and Co-Operation, Addis Ababa, 1998, 1999.

Direct communication to the Secretariat from the UN Energy Statistics Unit.

Sources up to 1991:

Ten Years of Petroleum Imports, Refinery Products, and Exports, Ministry of Mines & Energy, Addis Ababa, 1989.

Energy Balance for the Year 1984, Ministry of Mines & Energy, Addis Ababa, 1985.

1983 Annual Report, National Bank of Ethiopia, Addis Ababa, 1984.

Quarterly Bulletin, National Bank of Ethiopia, Addis Ababa, various editions from 1980 to 1985.

Gabon

Sources 1971-1997:

Combustible Renewables and Waste:

Secretariat estimates based on 1991 data from *Forests and Biomass Sub-sector in Africa,* African Energy Programme of the African Development Bank, Abidjan, 1996.

Direct communication from the oil industry.

Sources 1992-1997:

Tableau de Bord de l' Economie, Situation 1997, Perspectives 1998-1999, Direction Générale de l'Economie, Ministère des Finance, de l' Economie, du Budget et des participations, chargé de la privatisation, Mai 1998.

Direct communication to the Secretariat from the Société Gabonaise de Raffinage, Port Gentil, 1997.

Rapport d'Activité Banque Gabonaise de Développement, Libreville, 1985, 1990, 1992 and 1993.

The UN Energy Statistics Database, 1995.

Sources up to 1991:

Tableau de Bord de l'Economie, Situation 1983 Perspective 1984-85, Ministère de l'Economie et des Finances, Direction Générale de l'Economie, Libreville, 1984.

Ghana

Sources 1971-1997:

Combustible Renewables and Waste:

Ministry of Mines and Energy and the UN Energy Statistics Database, 1995.

Sources 1992-1996

Direct communication to the Secretariat from the UN Energy Statistics Unit.

National Energy Statistics, Ministry of Energy and Mines, Accra, 1997.

Quarterly Digest of Statistics, Government of Ghana, Statistical Services, Accra, March 1990, March 1991, March 1992, March 1995.

Sources up to 1991:

Energy Balances, Volta River Authority, Accra, various editions from 1970 to 1985.

Hong Kong, China

Sources 1971-1997:

Combustible Renewables and Waste:

The UN Energy Statistics Database, 1995 and Secretariat estimates.

Sources 1992-1996:

Hong Kong Energy Statistics - Annual Report, Census and Statistics Department, Hong Kong, various editions to 1997.

Hong Kong Energy Statistics - Quarterly Supplement, Census and Statistics Department, Hong Kong, various editions to 1997.

Hong Kong Monthly Digest of Statistics, Census and Statistics Department, Hong Kong, various editions to 1994.

India

Natural gas supply and demand data are reported by fiscal year. Oil supply data are reported by calendar year, and the oil consumption data have been pro-rated using available fiscal year end use detail. Autoproducer's electricity has been assumed to be consumed in the industrial sector.

Sources 1971-1997:

Combustible Renewables and Waste:

Secretariat estimates based on a per capita average consumption from various surveys and studies.

Sources 1992-1997:

Indian Oil and Gas, Ministry of Petroleum and Natural Gas, Economics & Statistics Division, New Delhi, January 1985 to March 1998.

Direct communication to the Secretariat from the Ministry of Petroleum and Natural Gas.

Coal Directory of India, 1992-1993, 1993-1994, 1995-1996, 1996-1997, 1997-1998, Ministry of Coal, Coal Controller's Organization, Calcutta, 1994, 1995, 1996, 1997, 1998.

Annual Review of Coal Statistics, 1993-1994, 1995-1996, 1996-1997, 1997-1998, Ministry of Coal, Coal Controller's Organization, Calcutta, 1995, 1997, 1998, 1999.

Annual Report 1994-1995, 1995-1996, 1998-1999, Ministry of Coal, New Delhi, 1995, 1996, 1999.

Indian Petroleum and Natural Gas Statistics, Ministry of Petroleum and Natural Gas, Economics & Statistics Division, New Delhi, 1985 to 1997.

Energy Data Directory, Yearbook "TEDDY", and *Annual Report,* Tata Energy Research Institute "TERI", New Delhi, 1986, 1987, 1988, 1990, 1994/1995 and 1995/1996.

General Review, Public Electricity Supply, India Statistics, Central Electricity Authority, New Delhi, 1982 to 1985, 1995, 1996.

Monthly Abstract of Statistics, Ministry of Planning, Central Statistics Organisation, Department of Statistics, New Delhi, various editions from 1984 to March 1998.

Annual Report 1993-1994, 1998-1999 Ministry of Petroleum and Natural Gas, New Delhi, 1995, 1999.

India Monthly Coal Statistics, Ministry of Energy, Department of Coal, Coal Controllers Organisation, New Delhi, various editions to June 1996.

Annual Report 1994-1996, 1998-1999 Ministry of Energy, Department of Non-Conventional Energy, New Delhi, 1996, 1999.

India's Energy Sector, July 1995, , Center for Monitoring Indian Economy PVT Ltd., Bombay, 1995.

Monthly Review of the Indian Economy, Center for Monitoring Indian Economy PVT Ltd., New Delhi, various issues from 1994 to June 1999.

Sources up to 1991:

Indian Oil Corporation Limited 1987-88 Annual Report, Indian Oil Corporation Limited, New Delhi, 1989-1992.

Report 1986-87, Ministry of Energy, Department of Coal, New Delhi, 1981 to 1987.

Annual Report 1986-1987, Ministry of Energy, Department of Non-Conventional Energy, New Delhi, 1987.

Economic Survey, Ministry of Finance, New Delhi, various editions from 1975 to 1986.

Statistical Outline of India, Ministry of Finance, New Delhi, 1983, 1984, 1986, and 1987.

Monthly Coal Bulletin, vol xxxvi no.2., Ministry of Labour, Directorate General of Mines Safety, New Delhi, February 1986.

Indonesia

Although it has been estimated that electricity output by autoproducers is approximately equal to that generated by the national electricity utility, since this amount is not known with any certainty, autoproducer output refers only to that amount sold to the public electricity grid.

Sources 1971-1997:

Combustible Renewables and Waste:

The UN Energy Statistics Database, 1996 and Secretariat estimates.

Sources 1992-1997:

Direct communication to the Secretariat from the Ministry of Mines and Energy, Directorate General of Electricity and Energy Development.

The Petroleum Report Indonesia, U.S. Embassy in Jakarta, Jakarta, 1986 to 1996.

Oil Statistics of Indonesia, Direktorat Jenderal Minyak Dan Gas Bumi, Jakarta, 1981 to December 1998.

Statistik Dan Informasi Ketenagalistrikan Dan Energi, Direktorat Jenderal Listrik Dan Pengembangan Energi, Jakarta, December 1998.

Mining and Energy in Indonesia, 1995, Ministry of Mines and Energy, Jakarta, 1995.

Sources up to 1991:

Indonesian Financial Statistics, Bank of Indonesia, Jakarta, 1982.

Indikator Ekonomi 1980-1985, Biro Pusat Statistik, Jakarta, 1986.

Statistical Yearbook of Indonesia, Biro Pusat Statistik, Jakarta, 1978 to 1984, 1992.

Statistik Pertambangan Umum, 1973 - 1985, Biro Pusat Statistik, Jakarta, 1986.

Energy Planning for Development in Indonesia, Directorate General for Power, Ministry of Mines and Energy, Jakarta, 1981.

Commercial Information, Electric Power Corporation, Perusahaan Umum Listrik Negara, Jakarta, 1984 and 1985.

Iran

Energy statistics are reported on a fiscal year basis.

Sources 1971-1997:

Combustible Renewables and Waste:

The UN Energy Statistics Database, 1996 and FAO, Forestry Statistics, 1999 (via Internet).

Sources 1992-1997:

Direct communication to the Secretariat from the Ministry of Petroleum, 1999.

Direct communication to the Secretariat from the Ministry of Energy, Office of Deputy Minister for Energy, Teheran 1998.

Electric Power in Iran, Ministry of Energy, Power Planning Bureau, Statistics Section, Teheran, 1992.

Sources up to 1991:

Electric Power in Iran, Ministry of Energy, Power Planning Bureau, Statistics Section, Teheran, 1967 to 1977, 1988, 1990, 1991.

Israel

For reasons of confidentiality the annual "*Energy in Israel*" only reports aggregated data for kerosene and jet kerosene.

Sources 1971-1996:

Combustible Renewables and Waste:

FAO, Forestry Statistics, 1999.

Sources 1992-1997:

Direct communication to the Secretariat from the Ministry of Energy and Infrastructure, Jerusalem.

Energy in Israel, Ministry of Energy and Infrastructure, Central Bureau of Statistics, Jerusalem, 1992, 1993, 1994, 1995, 1996, 1997.

Statistical Report 1993, 1994, 1995, The Israel Electric Corporation, Haifa, April 1994, May 1995, April 1996.

Statistical Results 1992, The Israel Electric Corporation, Haifa, June 1993.

Central Bureau of Statistics (CBS) World Wide Web site: http://www.cbs.gov.il

Sources up to 1991:

Energy in Israel, Ministry of Energy and Infrastructure, Central Bureau of Statistics, Jerusalem, 1975 to 1991.

Statistical Abstract of Israel, Ministry of Energy and Infrastructure, Central Bureau of Statistics, Jerusalem, 1985.

Supplement to Monthly Bulletin of Statistics, Ministry of Energy and Infrastructure, Central Bureau of Statistics, Jerusalem, various editions from 1984 to 1986.

Ivory Coast

Sources 1971-1996:

Combustible Renewables and Waste:

Secretariat estimates based on 1991 data from *Forests and Biomass Sub-sector in Africa,* African Energy Programme of the African Development Bank, Abidjan, 1996.

Sources 1996-1997:

La Côte d'Ivoire en chiffres, Ministère de l'Economie et des Finances, edition 1996-97.

Direct communication to the Secretariat from oil industry and the Ministry of Energy, Abidjan, July 1998.

Sources 1992-1995:

Direct communication to the Secretariat from the Bureau des Economies d'Energie.

L'Energie en Afrique, IEPE/ENDA, Paris, 1995, in turn sourced from the Ministère des Mines et de L'Energie, Abidjan.

The UN Energy Statistics database, 1995.

Sources up to 1991:

Etudes & Conjoncture 1982 - 1986, Ministère de l'Economie et des Finances, Direction de la Planification et de la Prévision, Abidjan, 1987.

Jamaica

Sources 1971-1997:

Combustible Renewables and Waste:

SIEE, (OLADE).

Sources 1992-1994:

Economic and Social Survey Jamaica 1992 to 1995, Planning Institute of Jamaica, Kingston, 1993, 1994, April 1995, April 1996.

Sources up to 1991:

National Energy Outlook 1985-1989, Petroleum Corporation of Jamaica, Economics and Planning Division, Kingston, 1985.

Energy and Economic Review, Petroleum Corporation of Jamaica, Energy Economics Department, Kingston, September 1986, December 1986 and March 1987.

Production Statistics 1988, Planning Institute of Jamaica, Kingston,1989.

Statistical Digest, Research and Development Division, Bank of Jamaica, Kingston, 1984, 1985, 1986, 1989, 1990 and 1991.

Jordan

Sources 1971-1997:

Combustible Renewables and Waste:

FAO, Forestry Statistics, 1999 (via Internet).

Sources 1992-1997:

Annual Report 1992, 1993, 1995, 1996, Jordan Electricity Authority, Amman, 1993, 1994, 1996, 1997.

Energy and Electricity in Jordan 1992, 1993, 1994, 1995, Jordan Electricity Authority, Amman, 1993, 1994, 1995, 1996.

Statistical Yearbook, 1994, Department of Statistics, Amman, 1995.

Sources up to 1991:

Monthly Statistical Bulletin, Central Bank of Jordan, Department of Research Studies, Amman, various issues.

Statistical Yearbook, Department of Statistics, Amman, 1985, 1986 and 1988.

1986 Annual Report, Ministry of Energy and Mineral Resources, Amman, 1987.

1989 Annual Report, Ministry of Energy and Mineral Resources, Amman, 1990.

Kenya

Sources 1971-1997:

Combustible Renewables and Waste:

Secretariat estimates based on 1991 data from *Forests and Biomass Sub-sector in Africa,* African Energy Programme of the African Development Bank, Abidjan, 1996.

Sources 1992-1997:

Economic Survey, 1995, 1996, 1997, 1998, Central Bureau of Statistics, Nairobi.

The UN Energy Statistics Database, 1996.

Sources up to 1991:

Economic Survey, Government of Kenya, Nairobi, 1989.

Economic Survey 1991, Ministry of Planning and National Development, Central Bureau of Statistics, Nairobi, 1992.

Kenya Statistical Digest, Ministry of Planning and National Development, Central Bureau of Statistics, Nairobi, 1988.

Kuwait

Data include 50 per cent of Neutral Zone output.

Sources 1971-1997:

Combustible Renewables and Waste:

FAO, Forestry Statistics, 1999 (via Internet).

Sources 1992-1997:

Direct communication to the Secretariat from the Ministry of Oil, Safat.

Monthly Digest of Statistics, Ministry of Planning, Central Statistical Office, Kuwait, 1998.

A Survey of the Kuwait Oil Industry, Embassy of the United States of America in Kuwait City, Kuwait,1993.

Twelfth Annual Report 1991-1992, Kuwait Petroleum Corporation, Kuwait, 1993.

Sources up to 1991:

Quarterly Statistical Bulletin, Central Bank of Kuwait, Kuwait, various editions from 1986 and 1987.

The Kuwaiti Economy, Central Bank of Kuwait, Kuwait, various editions from 1980 to 1985.

Annual Statistical Abstract, Ministry of Planning, Central Statistical Office, Kuwait, 1986 and 1989.

Monthly Digest of Statistics, Ministry of Planning, Central Statistical Office, Kuwait, various editions from 1986 to 1990.

Economic and Financial Bulletin Monthly, Central Bank of Kuwait, Kuwait, various editions from 1983 to 1986.

Kuwait in Figures, The National Bank of Kuwait, Kuwait, 1986 and 1987.

Lebanon

There was no refinery production in 1993, 1994, 1995, 1996 and 1997.

Sources 1971-1997:

Combustible Renewables and Waste:

FAO, Forestry Statistics, 1999 (via Internet) and Secretariat estimates.

Sources 1992-1997:

L'Energie au Liban, Les Bilans Energétiques en 1997, Association Libanaise pour la Maîtrise de l'Energie, Beirut,1998.

L'Energie au Liban, le Défi, Association Libanaise pour la Maîtrise de l'Energie, Beirut, décembre 1996.

Les Bilans Energétiques au Liban, Association Libanaise pour la Maîtrise de l'Energie, Beirut, décembre 1994.

Direct communication to the Secretariat from Association Libanaise pour la Maîtrise de l'Energie.

Libya

Sources 1971-1997:

Combustible Renewables and Waste:

The UN Energy Statistics Database, 1996.

Sources up to 1991:

Statistical Abstract of Libya, 19th vol, Government of Libya, Tripoli, 1983.

Malaysia

Sources 1971-1997:

Combustible Renewables and Waste:

The UN Energy Statistics Database, 1996 and FAO, Forestry Statistics, 1999 (via Internet).

Sources 1992-1997:

Direct communication to the Secretariat from the Ministry of Energy, Telecommunications and Posts.

Information Malaysia, Ministry Telekom Dan Pos Malaysia, Kuala Lumpur, 1995, 1996.

National Energy Balance, Ministry of Energy, Telecommunications and Posts, Kuala Lumpur, 1980-1995.

Sources up to 1991:

National Energy Balances Malaysia, Ministry Telekom Dan Pos Malaysia, Kuala Lumpur, 1978 to 1991.

Malta

UN ECE Questionnaire on Oil 1995 to 1997.

UN ECE Questionnaire on Coal 1994 and 1995.

UN ECE Questionnaire on Electricity and Heat, 1994 to 1997.

Morocco

Sources 1971-1997:

Combustible Renewables and Waste:

The UN Energy Statistics Database, 1996.

Sources 1992-1997:

Annuaire Statistique du Maroc, Ministère du Plan, Direction de la Statistique, Rabat, 1980, 1984, 1986 to 1998.

Electricity consumption by economic sector by direct communication from the Ministère du Plan, Rabat.

Sources up to 1991:

Rapport d'Activité du Secteur Pétrolier 1983, Ministère de l'Energie et des Mines, Direction de l'Energie, Rabat, 1984.

Rapport sur les Données Energétiques Nationales 1979-1981, Ministère de l'Energie et des Mines, Rabat, 1982.

Le Maroc en Chiffres 1986, Ministère du Plan, Direction de la Statistique, Rabat, 1987.

Rapport d'Activité 1992, Office National de l'Electricité, Casablanca, 1993.

Rapport Annuel, Office National de Recherches et d'Exploitations Pétrolieres, Maroc, 1984.

Mozambique

Sources 1971-1996:

Combustible Renewables and Waste:

Secretariat estimates based on 1991 data from *Forests and Biomass Sub-sector in Africa,* African Energy Programme of the African Development Bank, Abidjan, 1996.

Sources 1992-1997:

Direct communication to the Secretariat from the Ministry of Energy and Mineral Resources, (Maputo June 1998, April 1999) and the electricity utility.

Annual Statistical Yearbook 1993, 1994, 1995, Eskom, Johannesburg, 1994, 1995, 1996, citing Electricidade de Mozambique, Maputo, as source.

The UN Energy Statistics Database, 1995.

Myanmar

Sources 1971-1997:

Combustible Renewables and Waste:

Secretariat estimates based on 1990 data from *UNDP Sixth Country Programme Union of Myanmar,* World Bank, Programme Sectoral Review of Energy, by Sousing John, et. al., Washington, D.C., 1991.

Sources 1992-1997:

Direct communications to the Secretariat from the Ministry of Energy, Planning Department, Rangoon, 1996, 1997 and 1998, 1999.

Review of the Financial Economic and Social Conditions, Ministry of National Planning and Economic Development, Central Statistical Organization, Rangoon, 1995, 1996.

Statistical Yearbook, Ministry of National Planning and Economic Development, Central Statistical Organization, Rangoon, 1995, 1996.

The UN Energy Statistics Database, 1995.

Sources up to 1991:

Sectoral Energy Demand in Myanmar, UNDP Economic and Social Commission for Asia and The Pacific, Bangkok, 1992.

Selected Monthly Economic Indicators, paper no. 3, Ministry of Planning and Finance, Central Statistical Organization, Rangoon, 1989.

Nepal

Energy statistics are reported for a fiscal year.

Sources 1971-1997:

Combustible Renewables and Waste:

Water and Energy Commission Secretariat (WECS), Ministry of Water Resources.

Sources 1992-1997:

Asian Energy News, Asian Institute of Technology, Pathumthani, November 1997.

Nepal and the World, a Statistical Profile, Federation of Nepalese Chambers of Commerce and Industrie, Kathmandou, 1997.

Energy Balance Sheet of Nepal 1981-1992, Ministry of Water Resources, Water and Energy Commission, Katmandou, 1993.

Energy Synopsis Report 1994/95, Ministry of Water Resources, Water and Energy Commission, Katmandou, 1996.

Netherlands Antilles

Sources 1992-1994:

Direct communication to the Secretariat from the Central Bureau of Statistics, Fort Amsterdam, Curaçao.

Nicaragua

Historical series have been considerably revised using direct communication from Instituto Nicaraguense de Energía, Dirección General de Hidrocarburos.

Sources 1971-1997:

Combustible Renewables and Waste:

SIEE, (OLADE).

Direct submission to the Secretariat from the Instituto Nicaraguense de Energía, Dirección General de Hidrocarburos, Managua, 1999.

Informe Annual 1996: Datos Estadisticos del Sector Electrico, INE, Managua, 1999.

Direct submission to the Secretariat from the Electricity Utility.

Nigeria

Direct submission to the Secretariat from the Energy Commission of Nigeria.

Sources 1971-1997:

Combustible Renewables and Waste:

Secretariat estimates based on 1991 data from *Forests and Biomass Sub-sector in Africa,* African Energy Programme of the African Development Bank, Abidjan, 1996.

Sources 1992-1997:

Annual Report and statement of Accounts 1995, Central Bank of Nigeria, Lagos, 1996.

Direct communication from the oil industry. Statistical difference probably includes oil products smuggled into neighbouring countries for consumption.

Nigerian Petroleum News, Energy Publications, monthly reports, various issues up to May 1998.

Sources up to 1991:

Annual Report and Statement of Accounts, Central Bank of Nigeria, Lagos, various editions from 1981 to 1987.

Basic Energy Statistics for Nigeria, Nigerian National Petroleum Corporation, Lagos, 1984.

NNPC Annual Statistical Bulletin, Nigerian National Petroleum Corporation, Lagos, 1983 to 1987.

The Economic and Financial Review, Central Bank of Nigeria, Lagos, various editions.

Oman

Sources 1992-1997:

Direct communication to the Secretariat from the Ministry of Petroleum and Minerals, Muscat, October 1997, October 1998.

Direct communication to the Secretariat from the Ministry of Electricity & Water, Office of the Under Secretary, Ruwi, September 1998.

Quarterly Bulletin December 1994, Central Bank of Oman, Muscat, February 1995.

Annual Report 1992, Central Bank of Oman, Muscat, 1993.

Statistical Yearbook, 1994, 1995, 1996, 1997 Ministry of Development, Muscat, August 1995, August 1996, August 1997 and August 1998.

Sources up to 1991:

Quarterly Bulletin, Central Bank of Oman, Muscat, 1986, 1987, 1989 and 1995.

Annual Report to His Majesty the Sultan of Oman, Department of Information and Public Affairs, Petroleum Development, Muscat, 1981, 1982, and 1984.

Oman Facts and Figures 1986, Directorate General of National Statistics, Development Council, Technical Secretariat, Muscat, 1987.

Quarterly Bulletin on Main Economic Indicators, Directorate General of National Statistics, Muscat, March 1989.

Statistical Yearbook, Directorate General of National Statistics, Development Council, Muscat, 1985, 1986, 1988 and 1992.

Pakistan

Energy statistics are reported on a fiscal year basis.

Sources 1971-1997:

Combustible Renewables and Waste:

Secretariat estimates based on 1991 data from "Household Energy Strategy Study (HESS)" of 1991.

Sources 1992-1997:

Energy Year Book, Ministry of Petroleum and Natural Resources, Directorate General of New and Renewable Energy Resources, Islamabad, various editions from 1979 to 1998.

Pakistan Economic Survey 1994-1995, 1996, 1997, Government of Pakistan, Finance Division, Islamabad, 1995, 1997, 1998.

Statistical Supplement 1993/1994, Finance Division, Economic Adviser's Wing, Government of Pakistan, Islamabad, 1995.

Sources up to 1991:

As above.

Monthly Statistical Bulletin, no. 12, Federal Bureau of Statistics, Islamabad, December 1989.

1986 Bulletin, The State Bank of Pakistan, Islamabad, 1987.

Panama

Sources 1971-1997:

Combustible Renewables and Waste:

SIEE, (OLADE) and "Balance Energetico Nacional."

Sources 1992-1994:

Direct communication to the Secretariat from the Electricity and Hydro Resources Institute, Panama.

Sources up to 1991:

Balance Energetico Nacional, Serie Historica 1970-80, CONADE Peica Programa Energético del Istmo Centro Americano, Comisión Nacional de Energía, Panama, 1981.

Paraguay

The Itaipu hydroelectric plant, operating since 1984 and located on the Paraná river (which forms the border of Brazil and Paraguay) was formed as a joint venture between Eletrobrás and the Paraguayan government. Production is equally shared between Brazil and Paraguay. Consumption in Paraguay accounts for less than 5 per cent of the total power produced, the remaining 45 per cent is exported to Brazil, and has been accounted as such. See Brazil country notes for supply historical data from Itaipu.

Sources 1971-1996:

Combustible Renewables and Waste:

SIEE, (OLADE) and Secretariat estimates.

Sources 1997:

Direct communication to the Secretariat from the Electricity utility.

Sources up to 1991:

Boletin Estadístico no. 316, Banco Central del Paraguay, Departamento de Estudios Económicos, Asuncion, 1984.

Importaciones Por Productos Principales 1964-1986, Banco Central del Paraguay, Asuncion, 1987.

Peru

Sources 1971-1996:

Combustible Renewables and Waste:

SIEE, (OLADE).

Sources 1992-1997:

Direct communication to the Secretariat from the Ministry of Energy and Mines, Lima.

Balance Nacional de Energía 1997, Ministerio de Energía y Minas, Oficina Técnica de Energía, República del Perú.

Sources up to 1991:

Resena Económica, Banco Central de Reserva del Peru, Lima, 1984 and 1985.

Balance Nacional de Energía, Ministerio de Energía y Minas, Oficina Sectorial de Planificación, Lima, various editions from 1978 to 1987.

Annual Report, Petróleos del Peru, Lima, 1983 and 1984.

Estadísticas de las Operaciones Exploración/Producción 1985, Petróleos del Peru, Lima, 1986.

Philippines

Sources 1971-1997:

Combustible Renewables and Waste:

Department of Energy, National Economic and Development Authority and the UN Energy Statistics Database, 1995.

Sources 1996-1997:

Philippine Energy Bulletin 1997, 1998, Department of Energy, Metro Manila, 1998, 1999.

Sources 1992-1995:

Direct communication to the Secretariat from the Office of Energy Affairs, Metro Manila.

Philippine Statistical Yearbook 1977-1983, and *1993,* National Economic and Development Authority, Manila, 1978-1984, and 1994.

The APEC Energy Statistics 1994, Tokyo, October 1996.

The UN Energy Statistics Database, 1995.

Sources up to 1991:

1990 Power Developmemt Program (1990 - 2005), National Power Corporation, Manila, 1990.

Philippine Medium-term Energy Plan 1988 - 1992, Office of Energy Affairs, Manila, 1989.

1985 and *1989 Annual Report,* National Power Corporation, Manila, 1986, 1990.

Philippine Economic Indicators, National Economic and Development Authority, Manila, various editions of 1985.

Accomplishment Report: Energy Self-Reliance 1973-1983, Ministry of Energy, Manila, 1984.

Industrial Energy Profiles 1972-1979, vol. 1-4, Ministry of Energy, Manila, 1980.

National Energy Program, Ministry of Energy, Manila, 1982-1987 and 1986-1990.

Philippine Statistics 1974-1981, Ministry of Energy, Manila, 1982.

Energy Statistics, National Economic and Development Authority, Manila, 1983.

Quarterly Review, Office of Energy Affairs, Manila, various editions.

Qatar

Sources 1971-1997:

Combustible Renewables and Waste:

FAO, Forestry Statistics, 1998 (via Internet).

Sources 1992-1997:

Annual Statistical Abstract, Presidency of the Council of Ministers, Central Statistical Office, Doha, July 1994, July 1995, July 1996, July 1997, July 1998.

The UN Energy Statistics Database, 1995.

Sources up to 1991:

Qatar General Petroleum Corporation 1981-1985, General Petroleum Corporation, Doha, 1986.

Economic Survey of Qatar 1990, Ministry of Economy and Commerce, Department of Economic Affairs, Doha, 1991.

Statistical Report 1987 Electricity & Water, Ministry of Electricity, Doha, 1988.

State of Qatar Seventh Annual Report 1983, Qatar Monetary Agency, Department of Research and Statistics, Doha, 1984.

Romania

See Part I.1, Issues of Data Quality, above.

Sources 1971-1997:

Combustible Renewables and Waste:

UN ECE Energy Questionnaires and Secretariat estimates.

Sources 1992-1997:

UN ECE Energy Questionnaires.

Buletin Statistic de Informare Publica, Comisia Nationala Pentru Statistica, Bucharest, various issues to June 1995.

Renel Information Bulletin, Romanian Electricity Authority, Bucharest, 1990, 1991, 1992, 1993, 1994.

Sources up to 1991:

Anuarul Statistic al Republicii Socialiste Romania, Comisia Nationala Pentru Statistica, Bucharest, 1984, 1985, 1986, 1990 and 1991.

Saudi Arabia

The data include 50 per cent of Neutral Zone output.

Sources 1971-1997:

Combustible Renewables and Waste:

FAO, Forestry Statistics, 1999 (via Internet).

Sources 1992-1997:

Direct submissions from oil industry sources.

A Survey of the Saudi Arabian Oil Industry 1993, Embassy of the United States of America in Riyadh, Riyadh, January 1994.

Electricity Growth and Development in the Kingdom of Saudi Arabia up to the year of 1416H. (1996G.), Ministry of Industry and Electricity, Riyadh, 1997.

Sources up to 1991:

Annual Reports, ARAMCO, various issues.

Petroleum Statistical Bulletin 1983, Ministry of Petroleum and Mineral Resources, Riyadh, 1984.

Achievement of the Development Plans 1970-1984, Ministry of Planning, Riyadh, 1985.

The 1st, 2nd, 3rd and 4th Development Plans, Ministry of Planning, Riyadh, 1970, 1975, 1980 and 1985.

Annual Report, Saudi Arabian Monetary Agency, Research and Statistics Department, Riyadh, 1984, 1985, 1986, 1988 and 1989.

Statistical Summary, Saudi Arabian Monetary Agency, Research and Statistics Department, Riyadh, 1986.

Senegal

Sources 1971-1997:

Combustible Renewables and Waste:

Secretariat estimates based on 1994 data from *Forests and Biomass Sub-sector in Africa,* African Energy Programme of the African Development Bank, Abidjan, 1996, and from direct communication with ENDA, Senegal.

Sources 1992-1997:

Direct communication to the Secretariat from the Ministère de l'Energie, des Mines et de l'Industrie, Direction de l'Energie, Dakar, 1998, 1998.

Report of Senegal on the Inventory of Greenhouse Gases Sources, Ministère de l'Environnement et de la Protection de la Nature, Dakar, 1994.

Direct communication to the Secretariat from ENDA - Energy Program, Dakar, 1997.

Direct communicatons from oil industry sources.

The UN Energy Statistics Database, 1995.

Sources up to 1991:

Situation Economique 1985, Ministère de l'Economie et des Finances, Direction de la Statistique, Senegal, 1986.

Singapore

Official Singapore trade statistics do not show oil trade between Indonesia and Singapore. The quantity of this trade for crude oil has been estimated.

Sources 1971-1997:

Combustible Renewables and Waste:

The UN Energy Statistics Database, 1996.

Sources 1992-1997:

Direct communication to the Secretariat from the Public Utilities Board.

Direct submissions from oil industry sources.

Yearbook of Statistics Singapore 1993, 1995, 1996, Department of Statistics, Singapore, 1994, 1996, 1997.

The Strategist Oil Report, Singapore, various issues up to March 1999.

Singapore Trade Statistics, Department of Statistics, Singapore, various editions from 1985 to 1995.

Petroleum in Singapore 1993/1994, Petroleum Intelligence Weekly, Singapore, 1994.

AEEMTRC, 1996.

Sources up to 1991:

Monthly Digest of Statistics, Department of Statistics, Singapore, various editions from 1987 and 1989.

Yearbook of Statistics Singapore 1975/1985, Department of Statistics, Singapore, 1986.

Asean Oil Movements and Factors Affecting Intra-Asean Oil Trade, Institute of Southeast Asian Studies, Singapore, 1988.

The Changing Structure of the Oil Market and Its Implications for Singapore's Oil Industry, Institute of Southeast Asian Studies, Singapore, 1988.

Public Utilities Board Annual Report (1986 and 1989), Public Utilities Board, Singapore, 1987 and 1990.

Slovak Republic

See Part I.1, Issues of Data Quality, above.

Sources 1971-1997:

Combustible Renewables and Waste:

UN ECE Energy Questionnaires, the UN Energy Statistics Database, 1995 and from the Slovak government as a result of IEA in-depth energy survey.

Sources 1980-1997:

Direct submission from the Power Research Institute (EGU), Bratislava. Primary source: Statistical Office of the Slovak republic:

1989-1995: complete balances.

1980-1988: electricity and heat balances; primary balances for coal, oil, gas and electricity.

UN ECE Energy Questionnaires 1995 to 1997.

Source 1971-1988:

Statistical Yearbook of the Czechoslovak Socialist Republic, Federal Czech and Slovak Statistical Office, Prague, 1981 to 1991.

South Africa

In the second half of 1995, the IEA undertook an in-depth energy survey of South Africa and the results were published in *Energy Policies of South Africa,* IEA/OECD, Paris, 1996. As a result of the survey, much of the data have been revised and improved. Natural gas production began in 1993; all gas is liquefied and refined into petroleum products. Input of coal to coal liquefaction is shown in the coal column of the balance, and the crude produced appears as an output in the crude column; this crude is included in refinery input.

Sources 1971-1997:

Combustible Renewables and Waste:

South African Energy Statistics 1950-1989, No. 1, National Energy Council, Pretoria, 1989 and Secretariat estimates.

Sources 1993-1997:

Direct submission from the Institute for Energy Studies, Rand Afrikaans University, Pretoria, 1998, 1999.

Digest of South African Energy Statistics 1998, Department of Minerals and Energy, Pretoria, 1999

Direct submissions from the Department of Mineral and Energy Affairs, Pretoria.

Direct submissions from the Energy Research Institute, University of Cape Town.

Eskom Annual Report, Electricity Supply Commission (ESKOM), South Africa, 1989 to 1994.

Statistical Yearbook, Electricity Supply Commission (ESKOM), South Africa, 1983 to 1994.

South Africa's Mineral Industry, Department of Mineral and Energy Affairs, Braamfontein, 1995.

South African Energy Statistics, 1950-1993, Department of Mineral and Energy Affairs, Pretoria, 1995.

Wholesale Trade Sales of Petroleum Products, Central Statistical Service, Pretoria, 1995.

South African Coal Statistics 1994, South African Coal Report, Randburg, 1995.

Energy Balances in South Africa 1970-1993, Energy Research Institute, Plumstead, 1995.

Sources up to 1991:

Statistical News Release 1981-1985, Central Statistical Service, South Africa, various editions from 1986 to 1989.

Annual Report Energy Affairs 1985, Department of Mineral and Energy Affairs, Pretoria, 1986.

Energy Projections for South Africa (1985 Balance), Institute for Energy Studies, Rand Afrikaans University, South Africa, 1986.

Sri Lanka

Sources 1971-1997:

Combustible Renewables and Waste:

Energy Conservation Fund and Ceylon Electricity Board.

Sources 1992-1997:

Sri Lanka Energy Balance 1997, Energy Conservation Fund, Colombo, November 1997.

Annual Report 1993, Central Bank of Sri Lanka, Colombo, July 1994.

Direct communication to the Secretariat from the Ceylon Electricity Board, *Sri Lanka Energy Balances, 1994.*

Sources up to 1991:

Energy Balance Sheet 1991, 1992, Energy Unit, Ceylon Electricity Board, Colombo, 1992, 1993.

Bulletin 1989, Central Bank of Sri Lanka, Colombo, July 1989.

Bulletin (monthly), Central Bank of Sri Lanka, Colombo, May 1992.

Sectoral Energy Demand in Sri Lanka, UNDP Economic and Social Commission for Asia and The Pacific, Bangkok, 1992.

External Trade Statistics 1992, Government of Sri Lanka, Colombo, 1993.

Sudan

Sources 1971-1997:

Combustible Renewables and Waste:

Secretariat estimates based on 1990 data from Bhagavan, M.R., Editor, *Energy Utilities and Institutions in Africa*, AFREPREN, Nairobi, 1996.

Sources 1992-1997:

Direct communication to the Secretariat from the Ministry of Energy and Mines, Khartoum, June 1998, April 1999.

Sources up to 1991:

Foreign Trade Statistical Digest 1990, Government of Sudan, Khartoum, 1991.

Syria

Sources 1971-1997:

Combustible Renewables and Waste:

FAO, Forestry Statistics, 1998 (via Internet) and Secretariat estimates.

Sources 1992-1997:

Statistical Abstract 1992-1997, Office of the Prime Minister, Central Bureau of Statistics, Damascus, 1993, 1996, 1997, 1998.

The UN Energy Statistics Database, 1995.

Sources up to 1991:

Quarterly Bulletin, Central Bank of Syria, Research Department, Damascus, 1984.

Tanzania

Sources 1971-1997:

Combustible Renewables and Waste:

Secretariat estimates based on 1990 data from *Energy Statistics Yearbook 1990,* SADC, Luanda, 1992.

Sources up to 1997:

Direct communication to the Secretariat from the electricity utility.

Tanzanian Economic Trends, Economic Research Bureau, University of Dar-es-Salaam, 1991.

Thailand

Sources 1971-1997:

Combustible Renewables and Waste:

Thailand Energy Situation, Ministry of Science, Technology and Energy, National Energy Administration.

Sources 1992-1997:

Electric Power in Thailand, Ministry of Science, Technology and Energy, National Energy Administration, Bangkok, 1985, 1986, 1988 to 1998.

Oil and Thailand, Ministry of Science, Technology and Energy, National Energy Administration, Bangkok, 1979 to 1998.

Thailand Energy Situation, Ministry of Science, Technology and Energy, National Energy Administration, Bangkok, 1978 to 1998.

Trinidad-and-Tobago

Sources 1971-1997:

Combustible Renewables and Waste:

SIEE, (OLADE).

Sources 1992-1997:

Direct communication to the Secretariat from the Ministry of Energy and Natural Resources, Port of Spain.

Annual Economic Survey 1994, 1995, Central Bank of Trinidad and Tobago, Port of Spain 1995, 1996.

Petroleum Industry Monthly Bulletin, Ministry of Energy and Natural Resources, Port of Spain, various issues to 1998.

Sources up to 1991:

Annual Statistical Digest, Central Statistical Office, Port of Spain, 1983 and 1984.

History And Forecast, Electricity Commission, Port of Spain, 1987.

Annual Report, Ministry of Energy and Natural Resources, Port of Spain, 1985 and 1986.

The National Energy Balances 1979-1983, Ministry of Energy and Natural Resources, Port of Spain, 1984.

Trinidad and Tobago Electricity Commission Annual Report, Trinidad and Tobago Electricity Commission, Port of Spain, 1984 and 1985.

Tunisia

Sources 1971-1997:

Combustible Renewables and Waste:

Secretariat estimates based on 1991 data from *Analyse du Bilan de Bois d'Energie et Identification d'un Plan d'Action,* Ministry of Agriculture, Tunis, 1998.

Sources 1992-1997:

Energy balance and electricity consumption and production data provided by direct submission from the Observatoire National de l'Energie, Agence pour la Maîtrise de l'Energie, Tunis.

Sources up to 1991:

Bilan Energétique de l'Année 1991, Banque Centrale de Tunisie, Tunis, September 1992.

Rapport d'Activité 1990, Observatoire National de l'Energie, Agence pour la Maîtrise de l'Energie, Tunis, 1991.

Rapport Annuel 1990, Banque Centrale de Tunisie, Tunis, 1991.

Activités du Secteur Pétrolier en Tunisie, Banque Centrale de Tunisie, Tunis, 1987.

Statistiques Financières, Banque Centrale de Tunisie, Tunis, 1986.

Entreprise Tunisienne d'Activités Pétrolières (ETAP), Tunis, 1987.

Annuaire Statistique de la Tunisie, Institut National de la Statistique, Ministère du Plan, Tunis, 1985 and 1986.

L'Economie de la Tunisie en Chiffres, Institut National de la Statistique, Tunis, 1984 and 1985.

Activités et Comptes de Gestion, Société Tunisienne de l'Electricité et du Gaz, Tunis, 1987.

United Arab Emirates

Estimates of annual sales for marine bunkers in the facilities offshore of Fujairah in the UAE have been made in consultation with the oil industry.

Sources 1992-1997:

Combustible Renewables and Waste:

FAO, Forestry Statistics, 1999 (via Internet).

Sources 1993-1997:

Statistical Yearbook 1995, Department of Planning, Abu Dhabi, November 1996.

Sources up to 1992:

Abu Dhabi National Oil Company, 1985 Annual Report, Abu Dhabi National Oil Company, Abu Dhabi, 1986.

United Arab Emirates Statistical Review 1981, Ministry of Petroleum and Mineral Resources, Abu Dhabi, 1982.

Annual Statistical Abstract, Ministry of Planning, Central Statistical Department, Abu Dhabi, various editions from 1980 to 1993.

Uruguay

The power produced from the Salto Grande hydroelectric plant, operating since 1980 and located on the Uruguay river (natural border of Argentina and Uruguay), is equally shared between the two countries. Electricity exports include power produced in Salto Grande and exported to Argentina.

Sources 1971-1997:

Combustible Renewables and Waste:

Direccion Nacional de Energia and *SIEE* (OLADE).

Sources 1992-1997:

Balance Energetico Nacional, Ministerio de Industria, Energía y Mineria, Dirección Nacional de Energía, Montevideo, 1992, 1993, 1994, 1995 1996 and 1997.

Sources up to 1991:

UTE Memoria Anual, Administración Nacional de Usinas y Trasmiones Eléctricas, Montevideo, 1978.

Boletín Estadístico 1975-1985, Banco Central del Uruguay, Departemento de Investigaciones Económicas, Montevideo, 1986.

Informaciones y Estadísticas Nacionales e Internationales no. 48, Centro de Estadísticas Nacionales y Comercio Internacional, Montevideo, 1989.

Boletín Mensual Energético, Ministerio de Industria y Energía, Dirección Nacional de Energía, Montevideo, various editions from 1986 to 1987.

Balance Energetico Nacional, Ministerio de Industria, Energía y Mineria, Dirección Nacional de Energía, Montevideo, 1981, 1985, 1986, 1990, 1991.

Former USSR

Coal production statistics refer to unwashed and unscreened coal. IEA coal statistics normally refer to coal after washing and screening for the removal of inorganic matter. Also see notes under 'Classification of Fuel Uses' and 'Heat', above.

The energy balances presented for the former USSR include Secretariat estimates of fuel consumption in the main categories of transformation. These estimates are based on secondary sources and on isolated references in FSU literature.

Sources 1991 to 1997 for NIS:

External trade of the region Former USSR: *PlanEcon Energy Outlook for the Former Soviet Republics,* Washington, June 1995 and 1996 and *PlanEcon Energy Outlook for Eastern Europe and the Former Soviet Republics,* Washington, September 1997 and October 1998.

Foreign Scouting Service, Commonwealth of Independent States, IHS Energy Group – IEDS Petroconsultants, Geneva, April 1999.

Statistical Bulletin, various editions, The State Committee of Statistics of the CIS, Moscow, 1993 and 1994.

External Trade of the Independent Republics and the Baltic States in 1991, The State Committee of Statistics of the CIS, Moscow, 1992.

Statistical Yearbook, 1993, 1994 and 1995, The State Committee of Statistics of the CIS, Moscow, 1994, 1995 and 1996.

Sources up to 1990 for the Former USSR:

Statistical Yearbook, The State Committee for Statistics of the USSR, Moscow, various editions from 1980 to 1989.

External Trade of the Independent Republics and the Baltic States, 1990 and 1991, The State Committee of Statistics of the CIS, Moscow, 1992.

External Trade of the USSR, annual and quarterly, various editions, The State Committee of Statistics of the USSR, Moscow, 1986 to 1990.

External Trade of the USSR, annual and quarterly, various editions, The State Committee of Statistics of the USSR, Moscow, 1986 to 1990.

CIR Staff Paper no. 14, 28, 29, 30, 32 and 36, Center for International Research, U.S. Bureau of the Census, Washington, 1986, 1987 and 1988.

Yearbook on Foreign Trade, The Ministry of Foreign Trade, Moscow, 1986.

Armenia

Data for Combustible renewables and waste: FAO, Forestry Statistics, 1998 (via Internet) and Secretariat estimates.

Direct communications to the Secretariat from the Ministry of Energy and Fuels, 1992 and 1996.

UN ECE Questionnaire on Electricity, 1992 to 1997.

UN ECE Questionnaire on Coal, 1995 and 1996.

Azerbaijan

Data for Combustible renewables and waste: FAO, Forestry Statistics, 1998 (via Internet) and Secretariat estimates.

Direct communications to the Secretariat from the Ministry of Economics, June 1999.

Direct communications to the Secretariat from the State Committee of Statistics and Analysis, May 1993, June 1994, May 1996 and June 1998 .

UN ECE Energy Questionnaires, 1992 to 1994.

UN ECE Questionnaires on Oil, Gas and Electricity 1995 and 1996.

Belarus

Data for Combustible renewables and waste: UN ECE Energy Questionnaires and Secretariat estimates.

Direct communication to the Secretariat from the Ministry of Statistics and Analysis, May 1996.

UN ECE Energy Questionnaires, 1990 to 1997.

Estonia

Data for Combustible renewables and waste: UN ECE Energy Questionnaires and Secretariat estimates.

Up to 1994, process heat from final consumption is included in the transformation sector.

UN ECE Questionnaires (Summary), 1970 to 1990.

UN ECE Questionnaire on Coal, 1990.

Statistical Office of Estonia, *Energy Balances, 1994.*

UN ECE Energy Questionnaires, 1991 to 1997.

Georgia

Data for Combustible renewables and waste. The UN Energy Statistics Database, 1995.

Official energy balance of Georgia 1990-1997, Ministry of Economy and Ministry of Energy, Tbilissi, November 1998.

Energy outlook of Georgia 1998-2010, draft, Georgian Energy Research Institute, Tbilissi.

Energy outlook of Georgia 1993-2015, Georgian Public Institute Techinform, Tbilissi.

Direct communication to the Secretariat from the Socio-Economical Information Committee, 1992.

Kazakhstan

Data for Combustible renewables and waste: FAO, Forestry Statistics, 1998 (via Internet).

Direct communication to the Secretariat from the State Committee of Statistics and Analysis, August 1994.

UN ECE Energy Questionnaires, 1993 to 1995.

Kyrgyzstan

Data for Combustible renewables and waste: The UN Energy Statistics Database, 1995.

Direct communication to the Secretariat, 1992.

UN ECE Energy Questionnaires, 1993 to 1996.

Latvia

Data for Combustible renewables and waste: UN ECE Energy Questionnaires and Secretariat estimates.

Balance of Latvian Energy Sources 1991 and 1992, EC PHARE Project Implementation Unit, Riga, 1994.

UN ECE Energy Questionnaire on Coal, 1993 to 1997.

UN ECE Questionnaire on Natural Gas, 1992 to 1997.

UN ECE Questionnaire on Electricity and Heat, 1993 to 1997.

UN ECE Questionnaire on Oil 1995 to 1997.

Lithuania

Data for Combustible renewables and waste: UN ECE Energy Questionnaires and Secretariat estimates.

Direct communications to the Secretariat of the Energy Agency, February 1994 and May 1996.

UN ECE Energy Questionnaires, 1992 to 1997.

Balances of electricity, heat, fuel and energy in Lithuania 1991 – 1993, Lithuanian Ministry of Energy Agency, Vilnius.

Lithuania Power Demand and Supply Options, The World Bank, Washington, 1993.

Energy Conservation Potentials in Lithuania and Latvia, Riso National Laboratory, Roskilde, 1992.

Energy in Lithuania (Power, Heat and Fuel Balances 1980 - 1992), Lithuanian Energy Institute, Vilnius, 1993.

Moldova

Data for Combustible renewables and waste: The UN Energy Statistics Database, 1995.

Direct communication to the Secretariat from the Ministry of Industry and Energy, July 1992.

UN ECE Questionnaire on Electricity and Heat, 1991 to 1997.

UN ECE Questionnaire on Coal, 1992 to 1997.

UN ECE Questionnaire on Oil, 1993 to 1997.

UN ECE Questionnaire on Natural Gas, 1991 to 1997.

Russia

Data for Combustible renewables and waste: The State Committee of Statistics of Russian Federation and Secretariat estimates.

Process heat from final consumption is included in the transformation sector.

Oil and Natural Gas: Direct communication to the Secretariat from the State Committee of Statistics of Russia, May 1995.

Energy trade: Direct communication to the Secretariat from the State Committee of Statistics of Russia, July 1994.

UN ECE Questionnaire on Coal, 1992 to 1996.

UN ECE Questionnaire on Natural Gas, 1991 to 1996.

UN ECE Questionnaire on Electricity and Heat, 1991 to 1996.

UN ECE Questionnaires on Oil, 1991 to 1996.

Statistical Yearbook of Russia 1994, 1997 and 1998, The State Committee of Statistics, Moscow, 1994, 1997 and 1998.

The Russian Federation in 1992, Statistical Yearbook, The State Committee of Statistics of Russia, Moscow, 1993.

Russian Federation External Trade, annual and quarterly various editions, The State Committee of Statistics of Russia, Moscow.

Statistical Bulletin, various editions, The State Committee of Statistics of the CIS, Moscow, 1993 and 1994.

Statistical Bulletin n° 3, The State Committee of Statistics of Russia, Moscow, 1992.

Fuel and Energy Balance of Russia 1990, The State Committee of Statistics of Russia, Moscow, 1991.

Direct communication to the Secretariat, June 1998.

Energetika, Energo-Atomisdat, Moscow, 1981 and 1987.

Tajikistan

UN ECE Questionnaire on Coal, 1994 and 1995.

UN ECE Questionnaire on Natural Gas, 1994 and 1995.

UN ECE Questionnaire on Electricity and Heat, 1994 and 1995.

UN ECE Questionnaire on Oil, 1994.

Ukraine

Coal production statistics refer to unwashed and unscreened coal.

Data for Combustible renewables and waste: Statistical Office in Kiev, World Bank and Secretariat estimates.

Direct communication to the Secretariat from the Ministry of Statistics, the Coal Ministry, the National Dispatching Company, November 1995.

Coal: Direct communications to the Secretariat from the State Mining University of Ukraine, May 1995 and 1996.

Natural Gas: Direct communication to the Secretariat from Ukrgazprom, February 1995.

Direct communication to the Secretariat from the Ministry of Statistics of the Ukraine, July 1994.

UN ECE Questionnaire on Coal, 1991, 1992 and 1995 to 1997.

UN ECE Questionnaire on Oil, 1992 and 1995 to 1997.

UN ECE Questionnaire on Natural Gas, 1992 and 1995 to 1997.

UN ECE Questionnaire on Electricity and Heat, 1991, 1992, and 1994 to 1997.

Ukraine in 1992, Statistical Handbook, Ministry of Statistics of the Ukraine, Kiev, 1993.

Ukraine Power Demand and Supply Options, The World Bank, Washington, 1993.

Power Industry in Ukraine, Ministry of Power and Electrification, Kiev, 1994.

Energy Issues Paper, Ministry of Economy, March 1995.

Ukraine Energy Sector Statistical Review 1993, 1994, 1995, 1996 and 1997, World Bank Regional Office, Kiev, 1994, 1995, 1996, 1997 and 1998.

Global Energy Saving Strategy for Ukraine, Commission of the European Communities, TACIS, Madrid, July 1995.

Uzbekistan

Up to 1994, crude oil includes NGL.

Direct communications to the Secretariat from the Institute of Power Engineering and Automation, March 1994 and June 1996.

UN ECE Questionnaires 1995 to 1997.

Venezuela

Sources 1971-1997:

Combustible Renewables and Waste:

The UN Energy Statistics Database, 1996.

Sources 1992-1997:

Direct communication to the Secretariat from the Ministry of Energy and Mines.

Petróleo y Otros Datos Estadísticos, Dirección General Sectorial de Hidrocarburos, Dirección de Economía de Hidrocarburos, Caracas, 1993, 1994, 1995 and 1997.

Sources up to 1991:

Petróleo y Otros Datos Estadísticos, Dirección General Sectorial de Hidrocarburos, Dirección de Planificación y Economía de Hidrocarburos, Caracas, 1983 to 1991.

Balance Energetico Consolidado de Venezuela 1970-1984, Ministerio de Energía y Minas, Dirección General Sectorial de Energía, División de Programación Energetica, Caracas, 1986.

Compendio Estadístico del Sector Eléctrico, Ministerio de Energía y Minas, Dirección de Electricidad, Carbón y Otras Energías, Caracas, 1984, 1989, 1990 and 1991.

Memoria Y Cuenta, Ministerio de Energía y Minas, Caracas, 1991.

Petróleos de Venezuela S.A. 1985 Annual Report, Petróleos de Venezuela, Caracas, 1991.

Vietnam

Sources 1971-1997:

Combustible Renewables and Waste:

Secretariat estimates based on 1992 data from *Vietnam Rural and Household Energy Issues and Options: Report No. 161/94,* World Bank, ESMAP, Washington, D.C., 1994.

Direct communications to the Secretariat from the Center for Energy-Environment Research and Development, Pathumthami, 1997, 1998 and 1999.

Data were supplied by RWEDP in Asia, Bangkok, Thailand.

Direct communication from the Oil Industry.

Sectoral Energy Demand in Vietnam, UNDP Economic and Social Commission for Asia and The Pacific, Bangkok, 1992.

Energy Commodity Account of Vietnam 1992, Asian Development Bank, Manila, 1994.

World Economic Problems No 2 (20), National Centre for Social Sciences of the S.R. Vietnam, Institute of World Economy, Hanoi, 1993.

Vietnam Energy Review, Institute of Energy, Hanoï, 1995, 1997, 1998.

Yemen

Sources 1971-1996:

Combustible Renewables and Waste:

The UN Energy Statistics Database, 1995 and Secretariat estimates.

Sources 1992-1997:

Statistical Yearbook 1993, 1994,1995, 1996, 1997, Ministry of Planning and Development, Central Statistical Organization, Republic of Yemen, Yemen, 1994, 1995, 1996 and 1997.

Statistical Indicators in the Electricity Sector, Ministry of Planning and Development, Central Statistical Organization, Republic of Yemen, Yemen, 1993.

Sources up to 1991:

Statistical Yearbook, Government of Yemen Arab Republic, Yemen, 1988.

Former Yugoslavia

From 1992, the energy balance for former Yugoslavia has been prepared by adding together the basic statistics of it's component countries: Croatia, Slovenia, Former Yugoslav Republic of Macedonia, Bosnia-Herzegovina, and the Federal Republic of Yugoslavia. In the case of the latter two, energy data beyond basic production flows are extremely poor. Trade between the component countries has been discounted. Oil imports into Former Yugoslavia breaking the UN embargo of Serbia & Montenegro are theoretically included, although the absolute levels of this trade are not known with any certainty. As such the balance for former Yugoslavia should be seen as an estimation that is necessary for completing regional aggregates.

Sources up to 1991:

Statisticki Godisnjak Yugoslavije, Socijalisticka Federativna Rebublika Jugoslavija, Savezni Zavod Za Statistiku, Beograd, 1985 to 1991.

Indeks, Socijalisticka Federativna Rebublika Jugoslavija, Beograd, 1990, 1991, 1992.

Bosnia-Herzegovina

Combustible Renewables and Waste: The UN Energy Statistics Database, 1995.

UN ECE Energy Questionnaires 1993, 1994.

Croatia

UN ECE Energy Questionnaires 1993 to 1997.

Former Yugoslav Republic of Macedonia (FYROM)

Combustible Renewables and Waste: The UN Energy Statistics Database, 1995 and FAO, Forestry Statistics, 1998 (via Internet).

UN ECE Energy Questionnaires, 1993 to 1995.

Slovenia

UN ECE Energy Questionnaires 1993 to 1997.

Zambia

Sources 1971-1997:

Combustible Renewables and Waste:

Secretariat estimates based on 1991 data from *Forests and Biomass Sub-sector in Africa,* African Energy Programme of the African Development Bank, Abidjan, 1996.

Sources 1992-1996:

Direction Communication to the Secretariat from oil sources.

Annual Statistical Yearbook 1993,1994 and 1995 (*Consumption in Zambia 1978-1983),* Eskom, Lusaka, 1984.

Zimbabwe

Sources 1971-1997:

Combustible Renewables and Waste:

Secretariat estimates based on 1991 data from *Forests and Biomass Sub-sector in Africa,* African Energy Programme of the African Development Bank, Abidjan, 1996.

Sources 1996-1997:

Direct communication to the Secretariat from the Ministry of Environment and Tourism, Harare, 1999.

Direct communication to the Secretariat from the electricity utility.

Electricity Statistics Information, Central Statistical Office, Causeway, February 1998.

Sources 1992-1995:

Eskom Annual Statistical Yearbook 1993, 1995 and 1995, Johannesburg, 1994, 1995, 1996, citing Zimbabwe Electricity Supply Authority, Harare as source.

The UN Energy Statistics Database, 1995.

Sources up to 1991:

Zimbabwe Statistical Yearbook 1986, Central Statistical Office, Harare, 1990.

Quarterly Digest of Statistics, Central Statistical Office, Harare, 1990.

Zimbabwe Electricity Supply Authority Annual Report, Zimbabwe Electricity Supply Authority, Harare, 1986 to 1991.

Other Africa, Other Latin America, and Other Asia and Oceania

The series for these 'sum' countries are made up by simple addition of their component countries. Intra-component country trade is therefore included as part of total trade, although to truly represent trade to and from the grouping, intra-component country movements should be discounted. Trade is therefore likely to be overstated.

1. DE LA QUALITE DES DONNEES

A. Méthodologie

Des efforts considérables ont été déployés pour veiller à ce que les données présentées dans cette publication correspondent aux définitions de l'AIE figurant dans les Notes générales (parties I.2 et 3). Ces définitions sont utilisées par la plupart des organisations internationales qui recueillent des statistiques sur l'énergie. Toutefois, les statistiques nationales sur l'énergie communiquées à ces organisations sont souvent fondées sur des critères et définitions qui diffèrent, parfois sensiblement, de ceux qu'elles emploient. On trouvera ci-après des informations sur les différences repérées par le Secrétariat de l'AIE et sur les ajustements qu'il a pu, le cas échéant, opérer pour faire correspondre les données aux définitions internationales. Les différences notoires relevées dans certains pays sont signalées dans la partie I.6, Notes et sources par pays. Les notes par pays indiquent seulement quelques-unes des déviations les plus importantes et les plus évidentes constatées dans certains pays par rapport à la méthodologie de l'AIE, et ne constituent en aucun cas une liste exhaustive des différences et anomalies par pays.

B. Estimation

Outre les ajustements opérés pour compenser les différences de définition, des estimations se révèlent parfois nécessaires pour compléter certains agrégats importants pour lesquels il manque des statistiques essentielles. On en trouvera des exemples dans les notes plus détaillées proposées ci-dessous. Sauf en ce qui concerne les énergies combustibles renouvelables et les déchets (voir plus loin E.5), le Secrétariat s'est attaché à fournir tous les éléments des bilans par produit jusqu'à la Consommation finale pour tous les pays et toutes les années. Il a fallu pour cela indiquer les éléments de l'offre ainsi que les entrées de combustibles primaires et les sorties de combustibles secondaires correspondant aux principales activités de transformation telles que le raffinage du pétrole et la production d'électricité. Des estimations ont souvent dû être établies, après consultation d'offices statistiques nationaux, de compagnies pétrolières, de compagnies d'électricité et d'experts nationaux en matière d'énergie.

C. Séries chronologiques et changements politiques

Les bilans par produit des républiques de l'ex-URSS ont été établis à partir de 1992. Ces bilans ont été établis à l'aide de données officielles et, lorsque cela a été nécessaire, des estimations ont été effectuées à partir de données obtenues auprès de diverses industries et d'autres organisations internationales. Il est à noter que les Tableaux récapitulatifs et les bilans par produit de la région "ex-URSS" continuent à être publiés. En effet tant qu'il demeurera difficile de collecter de façon cohérente et homogène des données pour tous les pays de l'ex-URSS, l'agrégat régional servira de point de comparaison avec la situation antérieure et d'indicateur d'adaptation au nouvel environnement.

Les bilans par produit des pays de l'ex-République yougoslave apparaissent également dans cette publication. Ils ont été établis à partir de données officielles. Ici encore, pour ne pas interrompre les séries temporelles, les chiffres de "l'ex-Yougoslavie" continuent à être publiés.

Les bilans par produit de la République slovaque ont été établis à partir de 1971. Pour les années 1989 à 1997, les bilans de la République slovaque ont été établis à partir de données officielles. Pour les années 1971 à 1988, des estimations ont été faites lorsque c'était nécessaire (voir Notes et sources par pays).

Les statistiques énergétiques de certains pays ont subi diverses modifications soit au niveau des données prises en compte soit de la méthodologie utilisée, de sorte que des données énergétiques plus détaillées ont pu être présentées, par exemple, pour la Chine à partir de 1980. En conséquence, certaines "ruptures de séries" apparaissent inévitables.

Le Secrétariat de l'AIE réexamine ses bases de données chaque année. A la lumière des nouvelles évaluations, d'importantes révisions sont apportées aux séries chronologiques des pays au cours de cet examen. En conséquence, les données de la présente publication ont été sensiblement révisées par rapport aux précédentes éditions.

D. Classification des utilisations des combustibles

Les sources statistiques nationales manquent souvent d'informations adéquates concernant la consommation de combustibles suivant les différentes catégories d'utilisations finales. Beaucoup de pays ne font pas d'enquêtes annuelles de la consommation de combustibles dans les principaux secteurs de l'économie, c'est pourquoi les données publiées sont fondées sur des enquêtes anciennes. La ventilation par secteurs de la consommation dans les différents pays doit donc être interprétée avec prudence.

Avant les réformes entreprises dans les années 1990, la classification sectorielle de la consommation de combustibles dans les économies en transition (d'Europe de l'Est et de l'ex-URSS) et en Chine était très différente de celle pratiquée dans les économies de marché. La consommation sectorielle était définie d'après la branche de l'économie à laquelle appartenait l'utilisateur du combustible et non d'après la vocation ou l'usage du combustible. Par exemple, la consommation d'essence des véhicules d'une entreprise appartenant au secteur de la sidérurgie faisait partie de la consommation industrielle d'essence du secteur de la sidérurgie. Les données ont été ajustées dans la mesure du possible pour correspondre aux classifications internationales: par exemple, la totalité de la consommation d'essence est supposée relever du secteur des transports et est reportée comme telle dans la présente publication. Cependant, il n'a pas été possible de reclasser tous les produits aussi facilement que l'essence et le carburéacteur, et peu d'ajustements ont été opérés sur les autres produits.

E. Problèmes particuliers par combustible

1. Pétrole

Le Secrétariat de l'AIE recueille des statistiques très complètes concernant l'offre et la consommation de pétrole, y compris l'autoconsommation des raffineries, les soutages maritimes et le pétrole utilisé en pétrochimie. Les statistiques nationales font rarement état de ces chiffres. La production indiquée de produits raffinés correspond souvent à la production nette, et non brute, des raffineries et la consommation de produits pétroliers se limite parfois aux ventes sur les marchés intérieurs et ne comprend pas les livraisons aux transports internationaux maritimes ou aériens. Le pétrole utilisé comme produit de base en pétrochimie dans des complexes intégrés de raffinage/pétrochimie n'est souvent pas comptabilisé dans les statistiques officielles disponibles.

Lorsque cela est possible, le Secrétariat procède, en consultation avec l'industrie pétrolière, à une estimation des données non communiquées. En l'absence de toute autre indication sur l'utilisation de combustibles par les raffineries, la consommation des raffineries a été estimée à environ 5 pour cent des entrées en raffinerie et divisée à parts égales entre le gaz de raffinerie et le fioul lourd.

Pour une description des ajustements apportés à la consommation sectorielle de produits pétroliers, se reporter au paragraphe ci-devant intitulé "Classification des utilisations des combustibles".

2. Gaz naturel

L'AIE définit la production de gaz naturel comme la production commercialisable, c'est-à-dire nette des pertes à l'extraction et des quantités brûlées à la torchère, rejetées et réinjectées. De plus, le gaz naturel doit être constitué essentiellement de méthane, et les autres gaz tels que l'éthane et les hydrocarbures à chaîne plus longue doivent être comptabilisés dans la rubrique "pétrole".

Toutefois, les données directement disponibles sur le gaz naturel n'indiquent souvent pas le détail des pertes à l'extraction ainsi que des quantités brûlées à la torchère, rejetées et réinjectées. Les données communiquées sont rarement accompagnées de définitions adéquates, c'est pourquoi il est difficile, voire impossible, d'identifier le gaz aux différents stades de sa séparation en gaz sec (méthane) et en fractions plus lourdes dans les installations de séparation.

Les statistiques concernant l'offre et la demande de gaz naturel sont généralement exprimées en unités de volume et il est difficile d'obtenir des données précises sur le pouvoir calorifique du gaz. Lorsque ce chiffre n'est pas connu, un pouvoir calorifique supérieur de 38 TJ/million de m3 a été appliqué.

Signalons par ailleurs qu'il est souvent difficile d'obtenir des données fiables sur la ventilation de la consommation de gaz naturel. Cette constatation s'applique tout particulièrement à certains des plus grands pays consommateurs de gaz naturel du Moyen-Orient. Pour cette raison, la consommation industrielle de gaz naturel de ces pays n'est souvent pas indiquée dans les données figurant dans la présente publication.

3. Electricité

Comme l'indiquent les Notes générales (partie I.2), on appelle autoproducteur d'électricité une entreprise qui, en plus de ses activités principales, produit de l'énergie électrique destinée en totalité ou en partie à couvrir ses besoins propres. Les pays non membres de l'OCDE disposent rarement de données répondant à cette définition. Pour ces pays, les entrées de combustibles correspondant à l'autoproduction d'électricité sont indiquées dans la rubrique utilisation finale industrielle.

Lorsqu'il existe des statistiques concernant la production d'électricité à partir d'énergies renouvelables combustibles et de déchets, elles sont inclues dans la production totale d'électricité. Ces données ne sont pas complètes : par exemple, l'électricité produite à partir des déchets végétaux des sucreries n'est généralement pas indiquée.

Les entrées de combustibles utilisés pour produire de l'électricité sont estimées (lorsqu'elles ne sont pas communiquées) à l'aide de données sur la production d'électricité, le rendement énergétique et le type d'installation de production.

4. Chaleur

Les économies en transition (d'Europe de l'Est et de l'ex-URSS) et la Chine utilisaient, s'agissant de la chaleur, une méthodologie qui les démarquait des pratiques communément adoptées dans les économies de marché. Cette méthodologie consistait à comptabiliser la transformation par l'industrie de combustibles primaires (charbon, pétrole et gaz) en chaleur destinée à la *consommation sur place* dans l'activité de transformation Production de chaleur et non dans la consommation industrielle, comme le prévoit la méthodologie de l'AIE[1] ; la production de Chaleur était ensuite répartie entre les différents secteurs d'utilisation finale. Les pertes survenant pendant la transformation des combustibles en chaleur par l'industrie n'étaient pas comptabilisées dans la consommation industrielle finale. Bien que plusieurs pays aient récemment adopté la pratique des organisations internationales, cette différence non négligeable limite la validité des comparaisons entre les pays en transition et les économies de marché pour ce qui est de la consommation finale par secteur.

1 La méthodologie internationale ne comptabilise la chaleur dans le secteur de transformation que dans la mesure où elle est vendue à des tiers. Voir la définition dans la partie I.2.

5. Energies renouvelables combustibles et déchets

L'AIE publie les données de production et d'approvisionnement intérieur en énergies renouvelables combustibles et en déchets de tous les pays non-membres de l'OCDE et toutes les régions pour les années 1974 à 1997 (1971 à 1997 sur disquettes).

Ces données proviennent souvent de sources secondaires, elles ne sont pas harmonisées et leur qualité est douteuse, ce qui rend difficile les comparaisons entre les pays. Les données historiques de nombreux pays proviennent d'enquêtes souvent irrégulières, non harmonisées et menées à un niveau local plutôt que national. Elles ne sont donc comparables ni entre régions ni dans le temps. Lorsque des séries chronologiques étaient incomplètes ou non disponibles, le Secrétariat a procédé à l'estimation des données selon une méthodologie compatible avec le cadre prévisionnel de l'édition 1998 de l'ouvrage de l'AIE intitulé *World Energy Outlook* (septembre 1998). Premièrement, l'approvisionnement intérieur par habitant de biomasse et de déchets a été calculé ou estimé pour 1995. Deuxièmement, l'approvisionnement par habitant a été estimé pour les années 1971 à 1994 en utilisant une équation log/log avec comme variable exogène soit le PIB par habitant, soit le taux d'urbanisation, selon la région. Troisièmement, l'approvisionnement total de biomasse et de déchets pour les années 1996 et 1997 a été estimé en supposant un taux de croissance soit constant, soit égal à la croissance de la population, soit conforme à l'estimation sur la période 1971-1994. Cependant, ces séries historiques estimées doivent être traitées avec prudence.

Le tableau ci-dessous fournit des indications sommaires sur la qualité des données et la méthode d'estimation par région.

Région	**Principales sources de données**	**Qualité des données**	**Variable exogène**
Afrique	base de données FAO et BAfD	Faible	taux d'urbanisation
Amérique latine	nationales et OLADE	Élevée	aucune
Asie	enquêtes	Élevée à faible	PIB per habitant
Europe non-OCDE	questionnaires et FAO	Élevée à moyenne	aucune
ex-USSR	nationales et questionnaires et FAO	Élevée à moyenne	aucune
Moyen-Orient	FAO	Moyenne à faible	aucune

BAfD: Banque Africaine de Développement.

Bien qu'un soin tout particulier ait été apporté à la compatibilité des estimations avec celles réalisées dans la partie biomasse du *World Energy Outlook 1998*, des différences mineures subsistent pour certaines raisons. Parmi ces raisons, il convient de souligner des révisions plus tardives des données utilisées ainsi que des niveaux différents de désagrégations par pays.

Pour les années 1994 à 1997, tous les éléments des bilans jusqu'à la consommation par usage ont été compilés par produit ou catégorie de produits pour tous les pays. Les données pour les années 1996 et 1997 sont fournies dans les Tableaux annuels. La production de charbon de bois est fournie dans les Tableaux récapitulatifs. Les valeurs confirment l'importance des énergies d'origine végétale dans le secteur énergétique de nombreux pays en voie de développement.

L'AIE espère que l'inclusion de ces données encouragera les administrations et les autres agences spécialisées à améliorer la qualité et étendre la couverture de la collecte des données. Plus de détails sur la méthode suivie par pays peuvent être obtenus sur demande, et les commentaires sont bienvenus.

2. NOTES GENERALES

Les tableaux couvrent l'ensemble des sources "commerciales" d'énergie, c'est-à-dire à la fois les sources primaires (houille, lignite, tourbe, gaz naturel, pétrole brut, liquides de gaz naturel (LGN) et énergies hydraulique, géothermique/solaire, éolienne, marémotrice, etc. ainsi que l'énergie nucléaire) et les sources secondaires (dérivés du charbon, gaz manufacturés, produits pétroliers, électricité et chaleur). Des données sont également indiquées pour diverses sources d'énergies renouvelables combustibles et de déchets, telles que la biomasse solide et les produits d'origine animale, les gaz/liquides tirés de la biomasse, les déchets urbains et les déchets industriels.

Chaque tableau est divisé en trois parties principales: la première récapitulant les éléments de *l'approvisionnement*; la deuxième faisant état des secteurs *transformation* et *énergie*; et la troisième indiquant la *consommation finale* ventilée entre les divers secteurs d'utilisation finale.

La description ci-dessus concerne les bilans par produit complets. Dans le cas des bilans régionaux et des bilans agrégés les définitions des produits sont les mêmes que pour les bilans complets. Cependant, la ventilation par secteur est réduite aux principaux totaux (secteur énergie, etc.) et à des catégories de consommation choisies (sidérurgie, etc.).

A. Approvisionnement

La première partie du bilan énergétique de base fournit les éléments suivants de l'approvisionnement:

Production
\+ *Apports d'autres sources*
\+ *Importations*
\- *Exportations*
\- *Soutages maritimes internationaux*
± *Variations des stocks*
= *Approvisionnement intérieur*

1. Production

La *production* comprend les quantités de combustibles extraites ou produites, après extraction des matières inertes ou des impuretés (par exemple, après extraction du soufre contenu dans le gaz naturel).

2. Apports d'autres sources

La rubrique *apports d'autres sources* couvre tous les apports de produits dont l'origine ne correspond pas explicitement aux définitions des sources d'énergie primaire figurant dans les tableaux ; par exemple, pour la catégorie pétrole brut : les produits provenant d'autres sources que le pétrole brut ou les LGN, comme l'hydrogène, le pétrole brut de synthèse (y compris les huiles minérales extraites de minéraux bitumineux tels que les schistes, les sables asphaltiques, etc.) ; pour les additifs : le benzol, l'alcool et le méthanol produits à partir du gaz naturel ; pour les produits d'alimentation des raffineries : les quantités renvoyées par l'industrie pétrochimique et utilisées comme produits d'alimentation des raffineries ; pour la houille : les schlamms et les mixtes, la poussière récupérée et d'autres produits charbonniers de basse qualité, qui ne sont pas classables d'après le type de charbon d'origine ; pour le gaz d'usine à gaz : le gaz naturel, le gaz de raffinerie et le GPL, traités ou mélangés dans les usines à gaz (c'est-à-dire, gaz d'usine à gaz produit à partir d'autres sources que le charbon).

3. Importations et exportations

La rubrique *importations* et *exportations* désigne les quantités de produits ayant franchi les frontières du territoire national, que le dédouanement ait été effectué ou non.

a) Charbon

Les *importations* et *exportations* comprennent les quantités de combustibles obtenues d'autres pays ou fournies à d'autres pays, qu'il existe ou non une union économique ou douanière entre les pays en question. Le charbon en transit n'est pas pris en compte.

b) Pétrole et gaz

Cette rubrique comprend les quantités de pétrole brut et de produits pétroliers importées ou exportées au titre d'accords de traitement (à savoir, raffinage à façon). Les quantités de pétrole en transit ne sont pas prises en compte. Le pétrole brut, les LGN et le gaz naturel sont répertoriés en fonction de leur pays d'origine. Pour les produits d'alimentation des raffineries et les produits pétroliers, en revanche, c'est le dernier pays de provenance qui est pris en compte.

Les réexportations de pétrole importé pour raffinage en zone franche sont comptabilisées dans les exportations de produits pétroliers par le pays de raffinage vers le pays de destination finale.

c) Electricité

Les quantités sont considérées comme importées ou exportées lorsqu'elles ont franchi les limites territoriales du pays.

4. Soutages maritimes internationaux

Les *soutages maritimes internationaux* correspondent aux quantités fournies aux navires de haute mer, y compris les navires de guerre, quel que soit leur pavillon. La consommation des navires assurant le transport par cabotage ou navigation intérieure et des navires de pêche n'est pas comprise. Voir ci-dessous les définitions des secteurs des transports (section 2.C.2) et de l'agriculture (section 2.C.3).

5. Variations des stocks

Les *variations des stocks* expriment la différence enregistrée entre le premier jour et le dernier jour de l'année dans le niveau des stocks détenus sur le territoire national par les producteurs, les importateurs, les entreprises de transformation de l'énergie et les gros consommateurs. Les variations des quantités de pétrole et de gaz stockées dans les oléoducs et les gazoducs ne sont pas prises en compte. Sauf chez les gros consommateurs susmentionnés, les variations des stocks des utilisateurs finals ne sont pas comptabilisées. Une augmentation des stocks est indiquée par un chiffre négatif, tandis qu'une diminution apparaît sous la forme d'un chiffre positif.

6. Approvisionnement intérieur

L'*approvisionnement intérieur* est ainsi défini : production + apports d'autres sources + importations - exportations - soutages maritimes internationaux ± variations des stocks.

7. Transferts

La rubrique *transferts* comprend les lignes *transferts entre produits*, *produits transférés* et *produits recyclés*.

Les *transferts entre produits* visent les produits dont le classement a changé soit parce que leurs spécifications ont été modifiées soit parce qu'ils ont été mélangés pour former un autre produit. Ainsi, le kérosène peut être reclassé comme gazole après mélange avec ce dernier produit pour obtenir un gazole conforme aux spécifications hivernales. Le solde net des *transferts entre produits* est nul.

Les *produits transférés* sont des produits pétroliers importés pour subir un traitement complémentaire dans des raffineries. Par exemple : le fioul importé pour conversion dans une raffinerie est transféré dans la catégorie des produits d'alimentation.

Les *produits recyclés* sont des produits finis qui sont remis dans le circuit commercial, **après** avoir été livrés une première fois au consommateur final (par exemple les lubrifiants usés qui sont régénérés).

8. Ecarts statistiques

L'écart statistique est défini comme les livraisons destinées à la consommation finale + les quantités utilisées pour la transformation et la consommation dans le secteur de l'énergie + les pertes de distribution – l'approvisionnement intérieur – les transferts. En effet, les écarts statistiques proviennent des données relatives aux différentes composantes de l'approvisionnement qui sont souvent tirées par l'administration nationale de sources différentes. En outre, la prise en compte des variations des stocks de certains gros consommateurs dans la partie approvisionnement du bilan crée des distorsions qui contribuent aux écarts statistiques.

B. Secteurs transformation et énergie

Le *secteur transformation* englobe les activités de transformation des formes d'énergie primaire en énergie secondaire, et de transformation ultérieure (par exemple, celle du charbon à coke en coke, du pétrole brut en produits pétroliers, du fioul lourd en électricité).

Le *secteur énergie* englobe les quantités de combustibles utilisées par les industries productrices d'énergie (par exemple, pour le chauffage, l'éclairage et le fonctionnement de tous les équipements intervenant dans le processus d'extraction, ou encore pour la traction et la distribution).

Dans les secteurs *transformation* et *énergie* on distingue les catégories suivantes :

1. Secteur Transformation

- *Centrales électriques* (désigne les centrales conçues pour produire uniquement de l'électricité). Si une unité ou plus de la centrale est une installation de cogénération (et que l'on ne peut pas comptabiliser séparément, sur une base unitaire, les combustibles utilisés et la production), elle est considérée comme une centrale de cogénération. Tant les centrales publiques[1] que les installations des autoproducteurs[2] entrent dans cette rubrique.
- *Centrales de cogénération chaleur/électricité* (désigne les centrales conçues pour produire de la chaleur et de l'électricité). L'UNIPEDE les appelle "installations de production combinée d'énergie électrique et de chaleur". Dans la mesure du possible, les consommations de combustibles et les productions de chaleur/électricité doivent être exprimées sur la base des unités plutôt que des centrales. Cependant, à défaut de données disponibles exprimées sur une base unitaire, il convient d'adopter la convention indiquée ci-dessus pour la définition d'une centrale de cogénération. *On notera que, dans le cas des installations de cogénération chaleur/électricité des autoproducteurs, sont comptabilisés tous les combustibles utilisés pour la production d'électricité, tandis que seule la partie des combustibles utilisés pour la production de chaleur* ***vendue*** *est indiquée. Les combustibles utilisés pour la production de la chaleur destinée à la consommation interne des autoproducteurs* ***ne sont pas*** *comptabilisés dans cette rubrique mais dans les données concernant la consommation finale de combustibles du secteur de consommation approprié.*
- *Centrales calogènes* (désigne les installations [pompes à chaleur et chaudières électriques comprises] conçues pour produire uniquement de la chaleur et qui en vendent à des tiers [par exemple, consommateurs des secteurs résidentiel, commercial ou industriel] selon les termes d'un contrat). Cette rubrique comprend aussi bien les centrales publiques que les installations des autoproducteurs.
- *Hauts-fourneaux et Usines à gaz* (couvrant les quantités de combustibles utilisées pour la production de gaz de ville, de gaz de haut fourneau et de gaz de convertisseur à oxygène). Dans la production de fonte brute à partir de minerai de fer dans les hauts-fourneaux, les combustibles sont utilisés pour la charge des hauts-fourneaux et pour l'apport de chaleur et de carbone nécessaires à la réduction du minerai. Comptabiliser le pouvoir calorifique des combustibles qui entrent dans ce procédé est une tâche complexe, car la transformation (en gaz de haut fourneau) et la consommation (de chaleur de combustion) interviennent simultanément. Il se produit aussi une rétention de carbone dans la fonte brute; la quasi-totalité de ce carbone réapparaît ultérieurement dans le

---

1 La production publique désigne les installations dont la *principale activité* est la production d'électricité et/ou de chaleur pour la vente à des tiers. Elles peuvent appartenir au secteur privé ou public. Il convient de noter que les ventes ne se font pas nécessairement par l'intermédiaire du réseau public.

2 L'autoproduction désigne les installations qui produisent de l'électricité et/ou de la chaleur, en totalité ou en partie pour leur consommation propre, en tant qu'activité qui contribue à leur activité principale. Elles peuvent appartenir au secteur privé ou public.

gaz de convertisseur à oxygène (appelé aussi gaz de convertisseur ou gaz LD) lorsque la fonte brute est transformée en acier. En principe, les quantités de tous les combustibles (par exemple, charbon utilisé pour l'injection de charbon pulvérisé, coke de four à coke, gaz naturel et fioul) entrant dans les hauts-fourneaux et la quantité de gaz de haut fourneau et gaz de convertisseur à oxygène produite sont recueillies. Cependant, à l'exception des entrées de coke de four, les données sont souvent douteuses ou incomplètes. Le Secrétariat doit répartir ces apports entre les secteurs de transformation et de consommation. L'élément correspondant à la transformation est indiqué dans la ligne *hauts-fourneaux et usines à gaz* à la colonne appropriée pour le combustible dont il s'agit, et l'élément correspondant à la consommation est indiqué dans la ligne *sidérurgie,* à la consommation finale, à la colonne appropriée pour le combustible dont il s'agit. Avec la présente publication, le Secrétariat a décidé de supposer une ventilation des entrées de coke de four de 80% et 20% respectivement entre les secteurs de transformation et de consommation. Le résultat de ce changement, par rapport aux éditions précédentes, est une augmentation de la quantité d'énergie entrant à la ligne *hauts-fourneaux et usines à gaz* et une diminution à la ligne *sidérurgie*. Cette ventilation a été estimée à partir des résultats du modèle appliqué aux pays de l'OCDE.[1] Elle fournit un équilibrage plus cohérent entre les entrées de carbone dans les hauts fourneaux et les sorties correspondantes, lorsque les données de l'AIE sont utilisées pour calculer les émissions de CO_2 dues à la combustion, en utilisant la méthodologie du Groupe d'experts intergouvernemental sur l'évolution du climat (GIEC) telle qu'elle est publiée dans *Lignes directrices révisées en 1996 du GIEC pour les inventaires nationaux de gaz à effet de serre*[2].

1 Voir *Statistiques de l'énergie des pays de l'OCDE* pour plus d'informations sur le modèle d'estimation des hauts fourneaux appliqué aux pays de l'OCDE.

2 La publication *Lignes directrices révisées en 1996 du GIEC pour les inventaires nationaux de gaz à effet de serre* est disponible auprès de l'OCDE (voir la page de couverture intérieure pour les points de vente) ou sur Internet (http://www.iea.org/ipcc.htm).

- *Cokeries/Fabriques d'agglomérés/Fabriques de briquettes de lignite* (couvrant les combustibles utilisés pour la production de coke, de gaz de cokerie, d'agglomérés et de briquettes de lignite [BKB]).
- *Raffineries de pétrole* (couvrant les hydrocarbures utilisés pour la production de produits pétroliers finis).
- *Industrie pétrochimique* (couvrant les produits renvoyés aux raffineries par l'industrie pétrochimique). Il convient de noter que les retours en raffinerie des produits pétroliers utilisés à des fins non énergétiques (i.e. white spirit et lubrifiants) ne sont pas inclus sous cette rubrique, mais sous utilisations non énergétiques.
- La *liquéfaction* (comprend divers procédés de liquéfaction, notamment la liquéfaction de charbon et de gaz naturel en Afrique du Sud).
- *Secteur transformation - autres* (couvrant les pertes des charbonnières et les autres transformations non spécifiées).

2. Secteur énergie

La consommation propre du secteur de la production d'énergie comprend l'énergie consommée par les industries de transformation pour la chauffe, le pompage, la traction et l'éclairage [Divisions 10, 11, 12, 23 et 40 de la CITI[3]] :

- *mines de charbon* (houille et lignite) ;
- *extraction de pétrole et de gaz* (le gaz brûlé à la torche n'est pas compris; comprend l'utilisation de gaz par les installations de gaz naturel liquefié) ;
- *raffineries de pétrole* ;
- *centrales électriques, centrales de cogénération et centrales calogènes* ;
- *énergie absorbée par le pompage* (électricité consommée dans les centrales hydrauliques) ;
- *secteur énergie-autres* (comprend la consommation propre des fabriques d'agglomérés, des cokeries, des usines à gaz, des

3 Classification internationale type par industrie de toutes les branches d'activité économique, Série M, No. 4/Rév. 3, Nations Unies, New York, 1990.

fabriques de briquettes et de coke de lignite, ainsi que les utilisations non spécifiées du secteur énergie).

3. Pertes de distribution

Les *pertes de distribution* incluent les pertes enregistrées lors de la distribution du gaz, du transport de l'électricité et du transport du charbon. Peut également inclure des usages non comptabilisés de pétrole brut et de produits pétroliers.

C. Consommation finale

Le terme *consommation finale* (qui correspond à la somme des consommations des secteurs d'utilisation finale) signifie que l'énergie utilisée pour la transformation et pour la consommation propre des industries productrices d'énergie est exclue. La consommation finale recouvre la majeure partie des livraisons aux consommateurs (voir la note sur les variations des stocks à la section 2.A.5).

Dans la consommation finale, les produits d'alimentation de l'industrie pétrochimique sont inclus dans le secteur **industrie** dans une sous-catégorie de la rubrique *industrie chimique* pour les produits pétroliers qui sont utilisés essentiellement à des fins énergétiques. En revanche, les produits pétroliers qui sont principalement utilisés à des fins non énergétiques (voir utilisations non énergétiques à la section 2.C.4), sont indiqués sous les rubriques *utilisations non énergétiques* et inclus uniquement dans la **consommation finale totale**. Les retours de produits de l'industrie pétrochimique ne sont pas pris en compte dans la consommation finale (voir les sections 2.A.2 et 2.B.1).

1. Secteur industrie

La consommation du *secteur industrie* est répartie entre les sous-secteurs suivants (l'énergie utilisée par l'industrie pour le transport n'est pas prise en compte ici mais figure dans la rubrique transports) :

- *Sidérurgie* [Groupe 271 et Classe 2731 de la CITI] ;
- *Industrie chimique* [Division 24 de la CITI] ; *dont* : produits d'alimentation de l'industrie pétrochimique. L'industrie pétrochimique comprend les opérations de craquage et de reformage destinées à la production de l'éthylène, du propylène, du butylène, du gaz de synthèse, des aromatiques, du butadiène et d'autres matières premières à base d'hydrocarbures [partie du Groupe 241 de la CITI] ; voir produits d'alimentation, dans la section 2.C.4 (Utilisations non énergétiques) ;
- Industries de base des *métaux non ferreux* [Groupe 272 et Classe 2732 de la CITI] ;
- *Produits minéraux non métalliques* tels que verre, céramiques, ciment, etc. [Division 26 de la CITI] ;
- *Matériel de transport* [Divisions 34 et 35 de la CITI] ;
- *Construction mécanique*. Fabrication d'ouvrages en métaux, de machines et de matériels à l'exclusion du matériel de transport [Divisions 28, 29, 30, 31 et 32 de la CITI] ;
- *Industries extractives* (*à l'exception des combustibles*) [Divisions 13 et 14 de la CITI] ;
- *Industrie alimentaire et tabacs* [Divisions 15 et 16 de la CITI] ;
- *Papier, pâte à papier et imprimerie* [Divisions 21 et 22 de la CITI] ;
- *Bois et produits dérivés* (à exclusion de la pâte à papier et du papier) [Division 20 de la CITI] ;
- *Construction* [Division 45 de la CITI] ;
- *Textiles et cuir* [Divisions 17, 18 et 19 de la CITI] ;
- *Non spécifiés* (tout autre secteur industriel non spécifié précédemment) [Divisions 25, 33, 36 et 37 de la CITI].

Note : La plupart des pays éprouvent des difficultés pour fournir une ventilation par branche d'activité pour tous les combustibles. Dans ces cas, la rubrique *non spécifiés* a été utilisée. *Les agrégats régionaux de la consommation industrielle doivent donc être employés avec précaution.*

2. Secteur transports

La consommation dans le *secteur transports* couvre toutes les activités de transport quel que soit le secteur économique concerné [Divisions 60, 61 et 62 de la CITI], et elle est ventilée entre les différents sous-secteurs suivants :

- *Transport aérien* : Livraisons de carburants aviation à l'aviation civile internationale et pour toutes les activités de transport aérien intérieur, à savoir commerciales, privées, agricoles, militaires, etc. Comprend également les quantités utilisées à des fins autres que le vol proprement dit, par exemple, l'essai de moteurs au banc, mais non le carburant utilisé par les compagnies aériennes pour le transport routier ;

- *Transport routier* : La totalité des carburants utilisés dans les véhicules routiers (militaires compris) ainsi que le carburant consommé par les transports agricoles et industriels sur route. Ne tient pas compte de l'essence moteur employée dans les moteurs fixes, ni du gazole utilisé par les tracteurs ailleurs que sur route ;

- *Transport ferroviaire* : Toutes les quantités utilisées par le trafic ferroviaire, y compris par les chemins de fer industriels ;

- *Transport par conduites* : L'énergie utilisée pour le transport de substances par conduites ;

- *Navigation intérieure* (y compris la consommation des petites embarcations et des bateaux de cabotage n'achetant pas leur soutage aux termes de contrats de soutages maritimes internationaux). Le carburant utilisé pour la pêche en haute mer, le long du littoral et dans les eaux intérieures doit être comptabilisé dans le secteur agriculture ;

- *Non spécifiés.*

3. Autres secteurs

- *Agriculture* : Cette rubrique couvre, par définition, toutes les livraisons aux usagers classés dans les rubriques agriculture, chasse et sylviculture de la CITI, et comprend donc les produits énergétiques consommés par ces usagers que ce soit pour la traction automobile (à l'exception des carburants utilisés par les engins agricoles sur route), pour la production d'énergie ou le chauffage (dans les secteurs agricole ou résidentiel). Elle comprend aussi les carburants utilisés pour la pêche en haute mer, le long du littoral et dans les eaux intérieures. Divisions 01, 02 et 05 de la CITI ;

- *Services marchands et publics* : Cette rubrique recouvre toutes les activités qui relèvent des Divisions 41, 50, 51, 52, 55, 63, 64, 65, 66, 67, 70, 71, 72, 73, 74, 75, 80, 85, 90, 91, 92, 93 et 99 ;

- *Résidentiel* : Cette rubrique couvre toutes les quantités consommées par les ménages, à l'exception des combustibles utilisés dans les transports. Elle comprend les ménages employant du personnel domestique (Division 95 de la CITI), ce qui représente une faible part de la consommation résidentielle totale ;

- *Non spécifiés* : Cette rubrique couvre toutes les quantités de combustibles consommées qui n'ont pas été précisées ailleurs (par exemple, la consommation de combustibles pour les activités militaires, à l'exclusion des carburants dans les secteurs du transport routier et du transport aérien intérieur, et la consommation dans les catégories précitées pour lesquelles des données ventilées n'ont pas été fournies).

4. Utilisations non énergétiques

Les utilisations non énergétiques comprennent la consommation des *autres* produits pétroliers, notamment white spirit, paraffines, lubrifiants, bitume et produits divers (voir les sections 3.C.10 et 3.C.11). Ils incluent également les utilisations non énergétiques du charbon (excepté pour la tourbe). Ces produits se trouvent ventilés à part, dans la consommation finale, sous la rubrique *utilisations non énergétiques*. Il est présumé que l'usage de ces produits est strictement non énergétique. Il convient de noter que le coke de pétrole ne figure sous la rubrique *utilisations non énergétiques* que si cette utilisation est prouvée ; dans le cas contraire ce produit est comptabilisé avec les utilisations énergétiques dans l'industrie ou dans les autres secteurs.

Les chiffres concernant les produits d'alimentation de l'industrie pétrochimique sont comptabilisés dans le secteur de l'industrie, au titre de l'industrie chimique (ligne 31) et figurent séparément à la rubrique *dont : produits d'alimentation* (ligne 32). Sont compris dans cette rubrique tous les produits pétroliers, y compris le naphta, à l'exception des *autres produits pétroliers* énumérés à la section 3.C.11.

3. NOTES CONCERNANT LES SOURCES D'ENERGIE

A. Charbon

La rubrique *mines de charbon* ne comprend que le charbon utilisé directement par les charbonnages. Elle exclut le charbon consommé par les centrales électriques minières (compris dans la rubrique *transformation* - centrales électriques) et les quantités de charbon allouées gratuitement aux mineurs et à leurs familles (considérées comme consommation des ménages et classées de ce fait dans la rubrique *autres secteurs* - résidentiel).

1. Charbon à coke

On appelle charbon à coke un charbon d'une qualité permettant la production d'un coke susceptible d'être utilisé dans les hauts-fourneaux. Son pouvoir calorifique supérieur dépasse 23 865 kJ/kg (5 700 kcal/kg), valeur mesurée pour un combustible exempt de cendres, mais humide.

2. Autres charbons bitumineux et anthracite

Les autres charbons bitumineux sont utilisés pour la production de vapeur et pour le chauffage des locaux. Cette catégorie comprend tous les charbons anthraciteux et bitumineux autres que les charbons à coke. Son pouvoir calorifique supérieur dépasse 23 865 kJ/kg (5 700 kcal/kg), mais est généralement inférieur à celui du charbon à coke.

3. Charbons sous-bitumineux

On appelle charbons sous-bitumineux les charbons non agglutinants d'un pouvoir calorifique supérieur compris entre 17 435 kJ/kg (4 165 kcal/kg) et 23 865 kJ/kg (5 700 kcal/kg), contenant plus de 31 pour cent de matières volatiles sur produit sec exempt de matières minérales.

4. Lignite

Le lignite est un charbon non agglutinant dont le pouvoir calorifique supérieur n'atteint pas 17 435 kJ/kg (4 165 kcal/kg), et qui contient plus de 31 pour cent de matières volatiles sur produit sec exempt de matières minérales. Les schistes bitumineux sont également inclus dans cette catégorie.

5. Tourbe

Sédiment fossile d'origine végétale poreux ou comprimé, combustible à haute teneur en eau (jusqu'à 90 pour cent sur brut), facilement rayé, de couleur brun clair à brun foncé. La tourbe utilisée à des fins non énergétiques n'est pas prise en compte.

6. Coke de four à coke (coke de cokerie) et coke d'usine à gaz

Le coke de cokerie est un produit solide obtenu par carbonisation à haute température du charbon, et surtout du charbon à coke ; la teneur en eau et en matières volatiles est faible. Le semi-coke, produit solide obtenu par carbonisation à basse température, le coke et le semi-coke de lignite sont également inclus dans cette rubrique. La rubrique *secteur énergie-autres* représente la consommation interne des cokeries. La consommation de l'industrie sidérurgique ne comprend pas le coke transformé en gaz de haut fourneau. Pour obtenir la consommation totale de coke de cokerie de l'industrie sidérurgique, il faut ajouter les quantités de coke transformées en gaz de haut fourneau (elles apparaissent sous la rubrique *hauts-fourneaux et usines à gaz*).

Le coke d'usine à gaz est un sous-produit de la houille utilisée pour la production de gaz de ville dans les usines à gaz. Il est principalement utilisé pour le chauffage. La rubrique *secteur énergie-autres* couvre la consommation de coke de gaz dans les usines à gaz.

7. Agglomérés et briquettes de lignite (et de tourbe) (BKB)

Les agglomérés sont des combustibles composites fabriqués à partir de fines de charbon par moulage avec adjonction d'un liant tel que le brai. La quantité d'agglomérés produite est donc légèrement supérieure au tonnage de houille effectivement utilisé à cet effet. La consommation d'agglomérés durant le processus de fabrication des agglomérés apparaît sous la rubrique *secteur énergie-autres*.

Les briquettes de lignite (BKB) sont des combustibles composites fabriqués à partir du lignite, et agglomérés sous haute pression. Ces données couvrent les briquettes de tourbe, le lignite séché, la poussière de lignite et les fines de lignite. La consommation des usines de briquettes est comprise dans la rubrique *secteur énergie-autres*.

B. Pétrole brut, liquides de gaz naturel et produits d'alimentation des raffineries

Sous le titre *transformation*, la rubrique *raffineries de pétrole* indique les quantités de pétrole brut, de LGN, de produits d'alimentation des raffineries, d'additifs et d'autres hydrocarbures qui sont utilisées dans le processus de raffinage.

1. Pétrole brut

C'est une huile minérale, constituée d'un mélange d'hydrocarbures d'origine naturelle. Sa couleur va du jaune au noir, sa densité et sa viscosité sont variables. Cette catégorie comprend aussi les condensats (provenant des séparateurs) directement récupérés sur les périmètres d'exploitation des hydrocarbures gazeux dans les installations de séparation des phases liquide et gazeuse.

Les autres hydrocarbures, notamment le pétrole brut synthétique, les huiles minérales extraites des roches bitumineuses telles que schistes, sables asphaltiques, etc. ainsi que les huiles issues de la liquéfaction du charbon figurent à la ligne *autres sources*. Voir la section 2.A.2.

Les huiles émulsionnées (par exemple, l'orimulsion) sont prises en compte dans cette catégorie.

2. Liquides de gaz naturel (LGN)

Les LGN sont des hydrocarbures liquides ou liquéfiés obtenus pendant le traitement, la purification et la stabilisation du gaz naturel. Il s'agit des fractions de gaz naturel qui sont récupérées sous forme liquide dans les installations de séparation, dans les installations sur les gisements ou dans les usines de traitement du gaz. Les LGN comprennent l'éthane, le propane, le butane, le pentane, l'essence naturelle et les condensats, sans que la liste soit limitative. Ils peuvent aussi inclure certaines quantités de substances autres que des hydrocarbures.

3. Produits d'alimentation des raffineries

C'est un produit ou une combinaison de produits dérivés du pétrole brut et destinés à subir un traitement ultérieur autre qu'un mélange dans l'industrie du raffinage. Il est transformé en un ou plusieurs constituants et/ou produits finis. Cette définition recouvre les produits finis qui sont importés pour la consommation des raffineries et ceux qui sont renvoyés par l'industrie pétrochimique aux raffineries.

4. Additifs

Les additifs sont des substances autres que des hydrocarbures qui sont ajoutées ou mélangées à un produit afin de modifier ses propriétés, pour améliorer par exemple ses caractéristiques lors de la combustion. Les alcools et les éthers (MTBE ou méthyl tertio-butyl éther), ou des substances telles que le plomb tétraéthyle, sont comptabilisés dans cette rubrique. L'éthanol n'y figure pas, mais apparaît à la rubrique *gaz/liquides tirés de la biomasse*.

C. Produits pétroliers

Ce sont tous les produits dérivés du pétrole qui peuvent être obtenus par distillation et qui sont, en général, utilisés en dehors de l'industrie du raffinage.

Les produits finis classés comme produits d'alimentation des raffineries (voir ci-dessus) n'entrent pas dans cette catégorie.

Les données sur la *production* de produits pétroliers font apparaître, pour chaque produit, la production brute des raffineries.

La consommation de combustibles des raffineries (figurant dans le *secteur énergie*, ligne *raffineries de pétrole*) représente leur consommation de produits pétroliers, qu'il s'agisse de produits intermédiaires ou de produits finis, utilisés par exemple pour le chauffage, l'éclairage, la traction, etc.

1. Gaz de raffinerie (non liquéfiés)

Cette catégorie couvre, par définition, les gaz non condensables obtenus dans les raffineries lors de la distillation du pétrole brut ou du traitement des produits pétroliers (par craquage, par exemple). Il s'agit principalement d'hydrogène, de méthane, d'éthane et d'oléfines. Sont compris également les gaz retournés aux raffineries par l'industrie pétrochimique. La production de gaz de raffinerie correspond à la production brute. La consommation propre des raffineries est comptabilisée séparément à la ligne *raffineries de pétrole* dans la rubrique *secteur énergie*.

2. Gaz de pétrole liquéfiés (GPL) et éthane

Il s'agit des fractions légères d'hydrocarbures paraffiniques qui s'obtiennent lors du raffinage ainsi que dans les installations de stabilisation du pétrole brut et de traitement du gaz naturel. Ce sont le propane (C_3H_8) et le butane (C_4H_{10}) ou un mélange de ces deux hydrocarbures. Ils sont généralement liquéfiés sous pression pour le transport et le stockage.

L'éthane (C_2H_6) est un hydrocarbure à chaîne droite, gazeux à l'état naturel. C'est un gaz paraffinique incolore que l'on extrait du gaz naturel et des gaz de raffinerie.

3. Essence moteur

C'est un hydrocarbure léger utilisé dans les moteurs à combustion interne, tels que ceux des véhicules à moteur, à l'exception des aéronefs. L'essence moteur est distillée entre 35°C et 215°C et utilisée comme carburant pour les moteurs terrestres à allumage commandé. L'essence moteur peut contenir des additifs, des composés oxygénés et des additifs améliorant l'indice d'octane, notamment des composés plombés comme le PTE (plomb tétraéthyle) et le PTM (plomb tétraméthyle).

4. Essence aviation

Il s'agit d'une essence spécialement préparée pour les moteurs à pistons des avions, avec un indice d'octane adapté au moteur, un point de congélation de -60°C et un intervalle de distillation habituellement compris entre 30°C et 180°C.

5. Carburéacteurs

Cette catégorie comprend les carburéacteurs type essence et les carburéacteurs type kérosène, qui répondent aux spécifications d'utilisation pour les turbomoteurs pour avion.

a) Carburéacteur type essence

Cette catégorie comprend tous les hydrocarbures légers utilisés dans les turbomoteurs pour avion. Ils distillent entre 100°C et 250°C. Ils sont obtenus par mélange de kérosène et d'essence ou de naphtas, de manière à ce que la teneur en aromatiques soit égale ou inférieure à 25 pour cent en volume, et que la pression de vapeur se situe entre 13,7 kPa et 20,6 kPa. Des additifs peuvent être ajoutés afin d'accroître la stabilité et la combustibilité du carburant.

b) Carburéacteur type kérosène

C'est un distillat moyen utilisé dans les turbomoteurs pour avion, qui répond aux mêmes caractéristiques de distillation et présente le même point d'éclair que le kérosène (entre 150°C et 300°C, mais ne dépassant pas 250°C en général). De plus, il est conforme à des spécifications particulières (concernant notamment le point de congélation), définies par l'Association du transport aérien international (IATA).

6. Kérosène

Le kérosène comprend les distillats de pétrole raffiné dont la volatilité est comprise entre celle de l'essence et celle du gazole/carburant diesel. C'est une huile moyenne qui distille entre 150°C et 300°C.

7. Gazole/carburant diesel (Distillat de coupe intermédiaire)

Les gazoles/carburants diesel sont des huiles lourdes. Les gazoles sont extraits de la dernière fraction issue de la distillation atmosphérique du pétrole brut, tandis que les gazoles lourds sont obtenus par redistillation sous vide du résidu de la distillation atmosphérique. Le gazole/carburant diesel distille entre 180°C et 380°C. Plusieurs qualités sont disponibles, selon l'utilisation : gazole pour moteur diesel à allumage par compression (automobiles, poids lourds, bateaux, etc.), fioul léger pour le chauffage des locaux industriels et commerciaux, et autres gazoles, y compris les huiles lourdes distillant entre 380°C et 540°C utilisées comme produit d'alimentation dans l'industrie pétrochimique.

8. Fioul lourd (résiduel)

Ce sont les huiles lourdes constituant le résidu de distillation. La définition englobe tous les fiouls résiduels (y compris ceux obtenus par mélange). La viscosité cinétique est supérieure à 10 cSt à 80°C. Le point d'éclair est toujours supérieur à 50°C, et la densité est toujours supérieure à 0,9 kg/l.

9. Naphtas

Les naphtas sont un produit d'alimentation des raffineries destiné soit à l'industrie pétrochimique (par exemple, fabrication d'éthylène ou production de composés aromatiques) soit à la production d'essence par reformage ou isomérisation dans la raffinerie. Les naphtas correspondent aux fractions distillant entre 30°C et 210°C ou sur une partie de cette plage de température. Les naphtas importés pour mélange doivent être indiqués dans les importations, puis repris à la ligne *transferts*, affectés d'un signe négatif pour les naphtas, et d'un signe positif pour les produits finis correspondants (par exemple, essence).

10. Coke de pétrole

Le coke de pétrole est un résidu solide noir brillant, obtenu principalement par craquage et carbonisation de résidus de produits d'alimentation, de goudrons et de poix, dans des procédés tels que la cokéfaction différée ou la cokéfaction fluide. Il se compose essentiellement de carbone (90 à 95 pour cent) et brûle en laissant peu de cendres. Il est employé comme produit d'alimentation dans les cokeries des usines sidérurgiques, pour la chauffe, pour la fabrication d'électrodes et pour la production de substances chimiques. Les deux qualités les plus importantes de coke sont le coke de pétrole et le coke de pétrole calciné. Cette catégorie comprend également le coke de catalyse, qui se dépose sur le catalyseur pendant les opérations de raffinage ; ce coke n'est pas récupérable, et il est en général brûlé comme combustible dans les raffineries.

11. Autres produits pétroliers

La catégorie *autres produits pétroliers* regroupe les white spirit et SBP, les lubrifiants, le bitume, les paraffines et d'autres produits.

a) White spirit et essences spéciales (SBP)

Ce sont des distillats intermédiaires raffinés, dont l'intervalle de distillation se situe entre celui des naphtas et celui du kérosène.

Ils se subdivisent en :

i) ***Essences spéciales (SBP) :*** Huiles légères distillant entre 30°C et 200°C et dont l'écart de température entre les points de distillation de 5 pour cent et 90 pour cent en volume, y compris les pertes, est inférieur ou égal à 60°C. En d'autres termes, il s'agit d'une huile légère, de coupe plus étroite que celle des essences moteur. On distingue 7 ou 8 qualités d'essences spéciales, selon la position de la coupe dans l'intervalle de distillation défini plus haut.

ii) ***White spirit :*** Essence industrielle dont le point d'éclair est supérieur à 30°C. L'intervalle de distillation du white spirit est compris entre 135°C et 200°C.

b) Lubrifiants

Les lubrifiants sont des hydrocarbures obtenus à partir de distillats ou de résidus ; ils sont principalement utilisés pour réduire les frottements entre surfaces d'appui. Cette catégorie comprend tous les grades d'huiles lubrifiantes, depuis les spindles jusqu'aux huiles à cylindres, et les huiles entrant dans les graisses, y compris les huiles moteur et tous les grades d'huiles de base pour lubrifiants.

c) Bitume

Hydrocarbure solide, semi-solide ou visqueux, à structure colloïdale, de couleur brune à noire ; c'est un résidu de la distillation du pétrole brut obtenu par distillation sous vide des huiles résiduelles de distillation atmosphérique. Le bitume est aussi souvent appelé asphalte, et il est principalement employé pour le revêtement des routes et pour les matériaux de toiture. Cette catégorie couvre le bitume fluidisé et le cutback.

d) Paraffines

Hydrocarbures aliphatiques saturés (dont la formule générale est C_nH_{2n+2}). Les paraffines sont des résidus du déparaffinage des huiles lubrifiantes ; elles présentent une structure cristalline avec C > 12 et, pour principales caractéristiques, d'être incolores, inodores et translucides ainsi que d'avoir un point de fusion supérieur à 45°C.

e) Autres

Tous les produits pétroliers qui ne sont pas classés ci-dessus, par exemple, le goudron, le soufre et la graisse. Cette catégorie comprend également les composés aromatiques (par exemple, BTX ou benzène, toluène et xylènes) et les oléfines (par exemple, propylène) produits dans les raffineries.

D. Gaz

Les valeurs relatives aux quatre catégories de gaz sont toutes exprimées en térajoules, sur la base du **pouvoir calorifique supérieur**.

1. Gaz naturel

Le gaz naturel est constitué de gaz, méthane essentiellement, sous forme liquide ou gazeuse, extraits de gisements naturels souterrains. Il peut s'agir aussi bien de gaz "non associé" provenant de gisements qui produisent uniquement des hydrocarbures sous forme gazeuse, que de gaz "associé" obtenu en même temps que le pétrole brut, ou de méthane récupéré dans les mines de charbon (grisou).

La production est mesurée après élimination des impuretés et extraction des LGN et du soufre. Elle exclut les quantités de gaz réinjectées et les quantités rejetées ou brûlées à la torchère. Elle comprend les quantités de gaz utilisées dans l'industrie gazière et le gaz transporté par gazoduc.

2. Gaz d'usine à gaz

Cette catégorie couvre tous les types de gaz produits dans les usines des entreprises publiques ou privées ayant pour principal objet la production, le transport et la distribution de gaz. Cette catégorie comprend le gaz produit par carbonisation (y compris le gaz produit dans les fours à coke et transféré aux usines à gaz), par gazéification totale avec ou sans enrichissement au moyen de produits pétroliers, par craquage du gaz naturel ou par reformage et simple mélange avec d'autres gaz et/ou de l'air. Cette rubrique recouvre également le gaz naturel de substitution dont le pouvoir calorifique est élevé, et qui est produit par conversion chimique d'hydrocarbures.

3. Gaz de cokerie

Le gaz de cokerie est un sous-produit de la fabrication du coke de cokerie utilisé en sidérurgie.

4. Gaz de haut fourneau

Le gaz de haut fourneau est obtenu lors de la combustion du coke dans les hauts-fourneaux de l'industrie sidérurgique. Il est récupéré et utilisé comme combustible, en partie dans l'usine même, et en partie pour d'autres procédés de l'industrie sidérurgique, ou encore dans des centrales électriques dotées d'équipements adaptés pour en brûler. Cette rubrique comprend également le gaz obtenu comme sous-produit lors de l'élaboration de l'acier dans les fours à oxygène ou convertisseurs basiques avec soufflage d'oxygène, qui est récupéré à la sortie du gueulard. Ce gaz est également appelé gaz de convertisseur ou gaz LD.

E. Energies renouvelables combustibles et déchets

Les données concernant ces quatre catégories de combustibles sont toutes exprimées en térajoules, sur la base du **pouvoir calorifique inférieur**.

1. Biomasse solide et produits d'origine animale

La biomasse est, par définition, toute matière végétale utilisée directement comme combustible ou transformée avant de la brûler sous une autre forme. Elle comprend le bois, les déchets végétaux (y compris les déchets de bois et les cultures destinées à la production d'énergie), les matières/déchets d'origine animale, les lessives sulfitiques (résidus de fabrication de la pâte à papier), également appelées "liqueur noire" (liqueur alcaline de rejet des digesteurs lors de la production de pâte au sulfate ou à la soude dans le procédé d'élaboration du papier, dont le contenu énergétique provient de la lignine extraite de la pâte chimique) et autre biomasse solide.

Le charbon de bois produit à partir de la biomasse solide est également inclus ici. Etant donné que le charbon de bois est un produit secondaire, son traitement est légèrement différent de celui de la biomasse primaire. La production de charbon de bois (un produit du processus de transformation) est compensée par les consommations correspondantes de biomasse primaire dans le processus de production. Les pertes lors de ce processus sont reportées dans la ligne *Secteur transformation – autres*. Les autres approvisionnements (par exemple le commerce et les variations de stocks) ainsi que la consommation de charbon de bois sont agrégés directement avec la biomasse primaire. Cependant, dans quelques pays, seule la biomasse primaire est reportée. La production de charbon de bois reportée est disponible séparement dans les Tableaux sommaires à la fin de la publication.

2. Gaz/liquides tirés de la biomasse

Ce sont les gaz qui sont produits principalement par fermentation anaérobie de biomasse et de déchets solides et brûlés pour produire de la chaleur et/ou de l'énergie électrique. Cette catégorie comprend les gaz de décharge et les gaz de digestion des boues (gaz issus des eaux usées et des lisiers). Les additifs tirés de la biomasse (ou bio-additifs) comme l'éthanol entrent également dans cette catégorie.

3. Déchets urbains et assimilés

Les déchets urbains correspondent aux produits brûlés directement pour produire de la chaleur et/ou de l'énergie électrique, dont notamment les déchets des secteurs résidentiel et commercial ainsi que du secteur des services publics, qui sont recueillis par les autorités municipales pour leur élimination dans des installations centralisées. Les déchets hospitaliers entrent dans cette catégorie.

4. Déchets industriels

Il s'agit de produits liquides et solides brûlés directement, généralement dans des installations spécialisées, pour produire de la chaleur et/ou de l'énergie électrique et qui ne sont pas notifiés dans la catégorie *biomasse solide et produits d'origine animale*.

F. Electricité et chaleur

1. Electricité

La production brute d'électricité est mesurée aux bornes de tous les groupes d'alternateurs d'une centrale. Elle comprend donc l'énergie absorbée par les équipements auxiliaires et les pertes dans les transformateurs qui sont considérés comme faisant partie intégrante de la centrale.

La différence entre production nette et brute est généralement évaluée à 7 pour cent dans les centrales thermiques classiques, à 1 pour cent dans les centrales hydroélectriques et à 6 pour cent dans les centrales nucléaires, géothermiques ou solaires. La production hydraulique comprend la production des centrales à accumulation par pompage (également appelées centrales de pompage).

2. Chaleur

La production de chaleur destinée à la vente acquiert une importance grandissante depuis quelques années. Pour tenir compte de cette évolution, la production de chaleur représente toute la production publique de chaleur provenant des centrales de cogénération chaleur/électricité et calogènes ainsi que la chaleur vendue à des tiers par les centrales de cogénération chaleur/électricité et calogènes des autoproducteurs. Les quantités correspondantes de combustibles utilisées pour produire la chaleur destinée à la vente sont indiquées, dans le secteur transformation, aux lignes *installations de cogénération* et *centrales calogènes*. Les quantités de combustibles utilisées pour produire la chaleur qui n'est pas vendue sont rapportées dans les secteurs où cette consommation a lieu.

4. NOTES CONCERNANT LES TABLEAUX RECAPITULATIFS

Quatre tableaux récapitulatifs fournissent des données dont le format diffère de celui des Tableaux annuels :

Production de charbon de bois

Le charbon de bois est un produit secondaire de la biomasse, qui se compose principalement de carbone, issu de la combustion lente et incomplète du bois ou d'autres produits végétaux en l'absence d'air. On notera que dans les tableaux annuels, le charbon de bois est inclus dans la biomasse solide (voir les Notes concernant les sources d'énergie, partie I.3).

Consommation de pétrole par produit, en milliers de tonnes et millions de barils-jour

Les lignes 1 à 8 fournissent la consommation intérieure, non compris les entrées en raffinerie et la consommation d'énergie par les raffineries, mais y compris les pertes de distribution. *L'essence moteur* comprend les liquides tirés de la biomasse. Les *autres produits* comprennent le pétrole brut, les autres hydrocarbures, le gaz de raffinerie, le coke de pétrole, le white spirit et les essences spéciales, les lubrifiants, le bitume, les paraffines et les autres produits tels que le goudron, le soufre, la graisse ainsi que les composés aromatiques (par exemple, BTX ou benzène, toluène et xylène) et les oléfines (par exemple, propylène) produits dans les raffineries.

A la ligne 9, *combustibles de raffinerie* indique la consommation propre de combustibles pétroliers par les raffineries, qui comprennent principalement des gaz de raffinerie, du gazole-carburant diesel et du fioul lourd.

La ligne 11 indique les *soutages* maritimes internationaux de combustibles liquides, qui comprennent Principalement du gazole-carburant diesel et du fioul lourd.

Les coefficients de conversion suivants sont valables pour tous les pays et pour toutes les années.

	Barils/tonne
Gaz de raffinerie	8.00
Ethane	16.85
GPL	11.60
Naphta	8.50
	(8.90 OCDE Europe)
Essence aviation	8.90
Essence moteur	8.53
	(8.45 OCDE Europe)
Carburéacteur type essence	7.93
	(8.25 OCDE)
Carburéacteur type kérosène	7.93
	(7.88 OCDE Europe)
Autre kérosène	7.74
	(7.88 OCDE Europe)
Gazole/carburant diesel	7.46
Fioul lourd	6.66
	(6.45 OCDE Europe)
White Spirit	7.00
	(8.46 OCDE)
Lubrifiants	7.09
Bitume	6.08
Paraffines	7.00
	(7.85 OCDE)
Coke de pétrole	5.50
Autres produits non spécifiés	7.00
	(8.00 OCDE)

Consommation d'électricité

Comprend l'utilisation de l'électricité dans les secteurs de transformation, d'énergie et de consommation finale, mais ne comprend pas les pertes de distribution.

5. COUVERTURE GEOGRAPHIQUE

- **L'Afrique** comprend l'Afrique du Sud, l'Algérie, l'Angola, le Bénin, le Cameroun, le Congo, la République démocratique du Congo, l'Egypte, l'Ethiopie, le Gabon, le Ghana, la Côte d'Ivoire, le Kenya, la Libye, le Maroc, le Mozambique, le Nigéria, le Sénégal, le Soudan, la Tanzanie, la Tunisie, la Zambie, le Zimbabwe et les **autres pays d'Afrique.**

- **Les autres pays d'Afrique** comprennent le Botswana, le Burkina Faso, le Burundi, le Cap-Vert, la République centrafricaine, Djibouti, la Gambie, la Guinée, la Guinée-Bissau, la Guinée équatoriale, le Lesotho, le Libéria, Madagascar, le Malawi, le Mali, la Mauritanie, Maurice, le Niger, l'Ouganda, la Réunion, le Rwanda, Sao Tomé-et-Principe, les Seychelles, la Sierra Leone, la Somalie, le Swaziland, le Tchad et le Togo.

- **L'Amérique latine** comprend les Antilles néerlandaises, l'Argentine, la Bolivie, le Brésil, le Chili, la Colombie, le Costa Rica, Cuba, la République dominicaine, El Salvador, l'Equateur, le Guatemala, Haïti, le Honduras, la Jamaïque, le Nicaragua, Panama, le Paraguay, le Pérou, Trinité-et-Tobago, l'Uruguay, le Venezuela et les **autres pays d'Amérique latine.**

- **Les autres pays d'Amérique latine** comprennent Antigua-et-Barbuda, les Bahamas, la Barbade, le Belize, les Bermudes, la Dominique, la Grenade, la Guadeloupe, le Guyana, la Guyane française, la Martinique, Saint-Kitts-et-Nevis et Anguilla, Sainte-Lucie, Saint-Vincent-et-les-Grenadines et le Surinam.

- **L'Asie** comprend le Bangladesh, Brunei, la République populaire démocratique de Corée, l'Inde, l'Indonésie, la Malaisie, Myanmar, le Népal, le Pakistan, les Philippines, Singapour, le Sri Lanka, le Taipei chinois, la Thaïlande, le Viêt-Nam et les **autres pays d'Asie**.

- **Les autres pays d'Asie et d'Océanie** comprennent l'Afghanistan, le Bhoutan, les Fidji, Kiribati, les Maldives, la Nouvelle-Calédonie, la Papouasie-Nouvelle-Guinée, la Polynésie française, le Samoa, les Iles Salomon et Vanuatu.

- **La Chine** comprend la République populaire de Chine et Hong Kong (Chine).

- **Le Moyen-Orient** comprend l'Arabie saoudite, Bahreïn, les Emirats arabes unis, l'Iran, l'Iraq, Israël, la Jordanie, le Koweït, le Liban, Oman, le Qatar, la Syrie et le Yémen.

- **La région Europe hors OCDE** comprend l'Albanie, la Bosnie-Herzégovine, la Bulgarie, Chypre, la Croatie, Gibraltar, l'ex-République yougoslave de Macédoine (FYROM), Malte, la Roumanie, la Slovaquie, la Slovénie et la République fédérative de Yougoslavie.

- **L'ex-URSS** comprend l'Arménie, l'Azerbaïdjan, le Bélarus, l'Estonie, la Géorgie, le Kazakhstan, le Kirghizistan, la Lettonie, la Lituanie, la Moldavie, l'Ouzbékistan, la Fédération de Russie, le Tadjikistan, le Turkménistan et l'Ukraine.

- L'**Organisation de coopération et de développement économiques (OCDE)** comprend l'Allemagne, l'Australie, l'Autriche, la Belgique, le Canada, la Corée, le Danemark, l'Espagne, les Etats-Unis, la Finlande, la France, la Grèce, la Hongrie, l'Irlande, l'Islande, l'Italie, le Japon, le Luxembourg, le Mexique, la Norvège, la Nouvelle-Zélande, les Pays-Bas, la Pologne, le Portugal, la République tchèque, le Royaume-Uni, la Suède, la Suisse et la Turquie.

- Dans la zone de l'OCDE :

  L'**Allemagne** tient compte des nouveaux Länder à partir de 1970.

  Le Groenland et les Iles Féroé danoises ne sont pas pris en compte dans les données relatives au **Danemark**.

  L'**Espagne** englobe les Iles Canaries.

  Les **Etats-Unis** englobent Porto-Rico, Guam et les Iles Vierges ainsi que la zone franche d'Hawaï.

  Dans les données relatives à la **France**, Monaco est pris en compte, mais non les départements d'outre-mer (Martinique, Guadeloupe, Polynésie française et Ile de la Réunion).

  L'**Italie** englobe Saint-Marin et le Vatican.

  Le **Japon** englobe Okinawa.

  Ni le Surinam ni les Antilles néerlandaises ne sont pris en compte dans les données relatives aux **Pays-Bas**.

  Le **Portugal** englobe les Açores et l'Ile de Madère.

  La **Suisse** englobe le Liechtenstein.

- **L'Organisation des pays exportateurs de pétrole (OPEP)** comprend l'Algérie, l'Indonésie, l'Iran, l'Irak, le Koweit, la Libye, le Nigéria, Qatar, l'Arabie saoudite, les Emirats arabes unis et le Vénézuela.

  On notera que les pays suivants n'ont pas été pris en compte par suite d'un manque de données :

  Afrique : Comores, Namibie, Sainte-Hélène et Sahara Occidental.

  Amérique : Aruba, Iles Vierges Britanniques, Iles Caïmanes, Iles Falkland, Montserrat, Saint-Pierre et Miquelon et les Iles Turks et Caïcos.

  Asie et Océanie : Samoa américaines, Cambodge, Ile Christmas, Iles Cook, Laos, Macao, Mongolie, Nauru, Nioué, Iles du Pacifique (total. amér.), Tonga et Ile de Wake.

PART II:
STATISTICAL DATA

PARTIE II
DONNEES STATISTIQUES

ANNUAL TABLES

TABLEAUX ANNUELS

1996-1997

World / Monde : 1996

SUPPLY AND CONSUMPTION *APPROVISIONNEMENT ET DEMANDE*	Coal / *Charbon* (1000 tonnes)							Oil / *Pétrole* (1000 tonnes)			
	Coking Coal *Charbon à coke*	Other Bit. Coal *Autres charb. bit.*	Sub-Bit. Coal *Charbon sous-bit.*	Lignite *Lignite*	Peat *Tourbe*	Oven and Gas Coke *Coke de four/gaz*	Pat. Fuel and BKB *Agg./briq. de lignite*	Crude Oil *Pétrole brut*	NGL *LGN*	Feed-stocks *Produits d'aliment.*	Additives *Additifs*
Production	554079	2871189	440269	815223	24625	367003	28890	3140517	209248	-	9344
From Other Sources	51	3777	560	-	-	72	-	823	-	20264	233
Imports	190085	308636	1042	9612	-	18261	1507	1747354	16372	51978	2668
Exports	-179782	-331529	-9020	-2424	-192	-20609	-1725	-1683561	-62263	-7229	-491
Stock Changes	-1926	42950	1022	2150	620	-1292	211	-10239	-38	-3015	-125
DOMESTIC SUPPLY	**562507**	**2895023**	**433873**	**824561**	**25053**	**363435**	**28883**	**3194894**	**163319**	**61998**	**11629**
Intl. Marine Bunkers	-	-	-	-	-	-	-	-	-	-	-
Transfers	-	-	-	-	-	-	-	-275	-109293	45272	-177
Statistical Differences	-15089	91076	4832	7709	-1241	-22730	-304	16475	-1981	770	438
TRANSFORMATION	**532988**	**1934004**	**418883**	**768970**	**18190**	**230755**	**1331**	**3190386**	**43252**	**108040**	**11890**
Electricity Plants	17326	1514231	371605	543552	5613	79	751	26991	182	-	-
CHP Plants	35	312080	43345	176245	7075	-	262	1913	-	-	-
Heat Plants	10	37672	1169	23243	1833	607	151	115	-	-	-
Blast Furnaces/Gas Works	9142	23421	1082	-	-	229842	167	-	307	-	-
Coke/Pat. Fuel/BKB Plants	506475	18074	1682	25930	3669	227	-	-	-	-	-
Petroleum Refineries	-	-	-	-	-	-	-	3169600	42763	108040	11890
Petrochemical Industry	-	-	-	-	-	-	-	-	-	-	-
Liquefaction	-	28526	-	-	-	-	-	-8233	-	-	-
Other Transform. Sector	-	-	-	-	-	-	-	-	-	-	-
ENERGY SECTOR	**111**	**69476**	**28**	**864**	**223**	**2461**	**255**	**6935**	**526**	**-**	**-**
Coal Mines	108	34899	28	376	109	554	40	14	-	-	-
Oil and Gas Extraction	-	2449	-	-	-	17	-	3429	514	-	-
Petroleum Refineries	-	189	-	-	-	-	-	3355	-	-	-
Electr., CHP+Heat Plants	-	23531	-	59	4	32	28	137	-	-	-
Pumped Storage (Elec.)	-	-	-	-	-	-	-	-	-	-	-
Other Energy Sector	3	8408	-	429	110	1858	187	-	12	-	-
Distribution Losses	9	30109	35	4480	151	49	1	6096	98	-	-
FINAL CONSUMPTION	**14310**	**952510**	**19759**	**57956**	**5248**	**107440**	**26992**	**7677**	**8169**	**-**	**-**
INDUSTRY SECTOR	**13635**	**660215**	**16897**	**36457**	**1336**	**85261**	**3001**	**4727**	**8169**	**-**	**-**
Iron and Steel	-	-	-	-	-	-	-	-	-	-	-
Chemical and Petrochem.	-	-	-	-	-	-	-	-	-	-	-
of which: Feedstocks	-	-	-	-	-	-	-	-	-	-	-
Non-Ferrous Metals	-	-	-	-	-	-	-	-	-	-	-
Non-Metallic Minerals	-	-	-	-	-	-	-	-	-	-	-
Transport Equipment	-	-	-	-	-	-	-	-	-	-	-
Machinery	-	-	-	-	-	-	-	-	-	-	-
Mining and Quarrying	-	-	-	-	-	-	-	-	-	-	-
Food and Tobacco	-	-	-	-	-	-	-	-	-	-	-
Paper, Pulp and Print	-	-	-	-	-	-	-	-	-	-	-
Wood and Wood Products	-	-	-	-	-	-	-	-	-	-	-
Construction	-	-	-	-	-	-	-	-	-	-	-
Textile and Leather	-	-	-	-	-	-	-	-	-	-	-
Non-specified	-	-	-	-	-	-	-	-	-	-	-
TRANSPORT SECTOR	**12**	**11086**	**227**	**502**	**1**	**23**	**40**	**9**	**-**	**-**	**-**
Air	-	-	-	-	-	-	-	-	-	-	-
Road	-	-	-	-	-	-	-	-	-	-	-
Rail	-	-	-	-	-	-	-	-	-	-	-
Pipeline Transport	-	-	-	-	-	-	-	-	-	-	-
Internal Navigation	-	-	-	-	-	-	-	-	-	-	-
Non-specified	-	-	-	-	-	-	-	-	-	-	-
OTHER SECTORS	**-**	**-**	**-**	**-**	**-**	**-**	**-**	**-**	**-**	**-**	**-**
Agriculture	-	-	-	-	-	-	-	-	-	-	-
Comm. and Publ. Services	-	-	-	-	-	-	-	-	-	-	-
Residential	-	-	-	-	-	-	-	-	-	-	-
Non-specified	-	-	-	-	-	-	-	-	-	-	-
NON-ENERGY USE	**232**	**24745**	**-**	**478**	**-**	**13613**	**187**	**1532**	**-**	**-**	**-**
in Industry/Trans./Energy	-	-	-	-	-	-	-	-	-	-	-
in Transport	-	-	-	-	-	-	-	-	-	-	-
in Other Sectors	-	-	-	-	-	-	-	-	-	-	-

World / Monde : 1996

SUPPLY AND CONSUMPTION *APPROVISIONNEMENT ET DEMANDE*	Oil cont. / *Pétrole cont.* (1000 tonnes)										
	Refinery Gas *Gaz de raffinerie*	LPG + Ethane *GPL + éthane*	Motor Gasoline *Essence moteur*	Aviation Gasoline *Essence aviation*	Jet Fuel *Carbu-réacteurs*	Kerosene *Kérosène*	Gas/ Diesel *Gazole*	Heavy Fuel Oil *Fioul lourd*	Naphtha *Naphta*	Petrol. Coke *Coke de pétrole*	Other Prod. *Autres prod.*
Production	95300	81723	801421	1773	205526	87814	931175	643412	143291	59576	200574
From Other Sources	-	-	-	-	-	-	-	-	-	-	-
Imports	13	51012	82309	612	28854	20892	175565	167717	62312	16232	31726
Exports	-	-29759	-98298	-314	-46163	-12317	-196924	-191937	-48834	-20970	-41059
Stock Changes	2	783	1863	18	-79	-1235	756	236	33	-243	1214
DOMESTIC SUPPLY	**95315**	**103759**	**787295**	**2089**	**188138**	**95154**	**910572**	**619428**	**156802**	**54595**	**192455**
Intl. Marine Bunkers	-	-	-280	-	-21	-	-25501	-106034	-	-	-386
Transfers	-4	88976	2525	165	-629	-2956	558	-6714	-7858	-187	-8243
Statistical Differences	-113	9	-1747	-6	-3057	-3417	2756	-4203	-1757	-118	4610
TRANSFORMATION	**4009**	**5332**	**628**	**-**	**-**	**525**	**37533**	**259178**	**16891**	**3163**	**3314**
Electricity Plants	707	642	37	-	-	72	34133	187699	1086	1670	466
CHP Plants	2300	613	-	-	-	9	798	21183	16	204	2283
Heat Plants	653	64	-	-	-	14	1235	45849	41	-	133
Blast Furnaces/Gas Works	167	2834	-	-	-	-	122	2749	1143	304	-
Coke/Pat. Fuel/BKB Plants	-	-	-	-	-	-	-	20	-	945	-
Petroleum Refineries	-	12	68	-	-	-	-	-	-	-	-
Petrochemical Industry	182	1166	523	-	-	430	1245	1677	14605	40	432
Liquefaction	-	-	-	-	-	-	-	-	-	-	-
Other Transform. Sector	-	1	-	-	-	-	-	1	-	-	-
ENERGY SECTOR	**87609**	**2940**	**1706**	**-**	**5**	**566**	**8699**	**50269**	**361**	**24300**	**6057**
Coal Mines	-	3	4	-	-	61	1928	283	-	-	6
Oil and Gas Extraction	65	99	1081	-	-	9	3074	1282	-	1305	11
Petroleum Refineries	87483	2777	609	-	2	455	2423	45771	224	22884	5525
Electr., CHP+Heat Plants	2	4	11	-	-	12	823	1433	-	-	12
Pumped Storage (Elec.)	-	-	-	-	-	-	-	-	-	-	-
Other Energy Sector	59	57	1	-	3	29	451	1500	137	111	503
Distribution Losses	-	170	299	-	20	18	406	134	248	-	177
FINAL CONSUMPTION	**3580**	**184302**	**785160**	**2248**	**184406**	**87672**	**841747**	**192896**	**129687**	**26827**	**178888**
INDUSTRY SECTOR	**3553**	**89925**	**7265**	**9**	**286**	**9126**	**95490**	**147216**	**129686**	**10902**	**6998**
Iron and Steel	-	-	-	-	-	-	-	-	-	-	-
Chemical and Petrochem.	-	-	-	-	-	-	-	-	-	-	-
of which: Feedstocks	-	-	-	-	-	-	-	-	-	-	-
Non-Ferrous Metals	-	-	-	-	-	-	-	-	-	-	-
Non-Metallic Minerals	-	-	-	-	-	-	-	-	-	-	-
Transport Equipment	-	-	-	-	-	-	-	-	-	-	-
Machinery	-	-	-	-	-	-	-	-	-	-	-
Mining and Quarrying	-	-	-	-	-	-	-	-	-	-	-
Food and Tobacco	-	-	-	-	-	-	-	-	-	-	-
Paper, Pulp and Print	-	-	-	-	-	-	-	-	-	-	-
Wood and Wood Products	-	-	-	-	-	-	-	-	-	-	-
Construction	-	-	-	-	-	-	-	-	-	-	-
Textile and Leather	-	-	-	-	-	-	-	-	-	-	-
Non-specified	-	-	-	-	-	-	-	-	-	-	-
TRANSPORT SECTOR	**15**	**9264**	**764252**	**2178**	**183411**	**853**	**476951**	**15309**	**1**	**-**	**585**
Air	-	-	28	2175	183406	717	3	-	-	-	-
Road	2	9081	760404	3	-	58	426226	278	1	-	156
Rail	-	-	-	-	-	-	-	-	-	-	-
Pipeline Transport	-	-	-	-	-	-	-	-	-	-	-
Internal Navigation	-	-	-	-	-	-	-	-	-	-	-
Non-specified	-	-	-	-	-	-	-	-	-	-	-
OTHER SECTORS	**-**	**-**	**-**	**-**	**-**	**-**	**-**	**-**	**-**	**-**	**-**
Agriculture	-	-	-	-	-	-	-	-	-	-	-
Comm. and Publ. Services	-	-	-	-	-	-	-	-	-	-	-
Residential	-	-	-	-	-	-	-	-	-	-	-
Non-specified	-	-	-	-	-	-	-	-	-	-	-
NON-ENERGY USE	**-**	**1935**	**269**	**3**	**1**	**44**	**6**	**12**	**-**	**15845**	**169670**
in Industry/Transf./Energy	-	-	-	-	-	-	-	-	-	-	-
in Transport	-	-	-	-	-	-	-	-	-	-	-
in Other Sectors	-	-	-	-	-	-	-	-	-	-	-

World / Monde : 1996

SUPPLY AND CONSUMPTION *APPROVISIONNEMENT ET DEMANDE*	Gas / *Gaz* (TJ)				Comb. Renew. & Waste / *En. Re. Comb. & Déchets* (TJ)				(GWh)	(TJ)
	Natural Gas *Gaz naturel*	Gas Works *Usines à gaz*	Coke Ovens *Cokeries*	Blast Furnaces *Hauts fourneaux*	Solid Biomass *Biomasse solide*	Gas/Liquids from Biomass *Gaz/Liquides tirés de biomasse*	Municipal Waste *Déchets urbains*	Industrial Waste *Déchets industriels*	Electricity *Electricité*	Heat *Chaleur*
Production	88834104	1173722	1746785	2408397	42624550	512363	530173	377331	13727549	12017245
From Other Sources	-	37217	-	-	3341	34448	-	16038	-	-
Imports	20720654	-	-	-	16679	31512	-	209	423456	122
Exports	-20704720	-	-	-	-28717	-4523	-	-6799	-424456	-122
Stock Changes	-731789	2	-	-	-6483	-3184	-	382	-	-
DOMESTIC SUPPLY	**88118249**	**1210941**	**1746785**	**2408397**	**42609369**	**570616**	**530173**	**387161**	**13726549**	**12017245**
Intl. Marine Bunkers	-	-	-	-	-	-	-	-	-	-
Transfers	-	-	-	-	-	-6533	-	-	-	-
Statistical Differences	-3789	-2441	-13970	15362	456	-10727	-12913	-7116	7425	27067
TRANSFORMATION	**29319988**	**17006**	**351443**	**690956**	**4112497**	**117519**	**512396**	**258880**	**6102**	**-**
Electricity Plants	13151043	3022	167513	442253	477853	59297	303827	57733	-	-
CHP Plants	11845962	1321	103625	216280	1539382	16228	158836	127822	-	-
Heat Plants	3310496	12663	58444	32224	251039	1519	49733	73318	-	-
Blast Furnaces/Gas Works	860109	-	18861	-	1959	-	-	-	-	-
Coke/Pat. Fuel/BKB Plants	4177	-	3000	199	4567	-	-	7	-	-
Petroleum Refineries	3200	-	-	-	105	40475	-	-	-	-
Petrochemical Industry	-	-	-	-	-	-	-	-	-	-
Liquefaction	127624	-	-	-	-	-	-	-	-	-
Other Transform. Sector	17377	-	-	-	1837592	-	-	-	6102	-
ENERGY SECTOR	**9919876**	**17179**	**340288**	**156231**	**297**	**1911**	**98**	**1961**	**1331733**	**973559**
Coal Mines	1136402	28	2691	-	78	-	-	31	107849	85256
Oil and Gas Extraction	6717777	-	-	-	-	-	-	-	128961	167664
Petroleum Refineries	1539825	3048	1960	-	-	209	-	141	156072	486908
Electr., CHP+Heat Plants	13508	459	2713	5230	219	408	98	1	773189	88331
Pumped Storage (Elec.)	-	-	-	-	-	-	-	-	90965	-
Other Energy Sector	512364	13644	332924	151001	-	1294	-	1788	74697	145400
Distribution Losses	993185	7594	8405	32307	7494	-	59	29	1156049	705796
FINAL CONSUMPTION	**47881411**	**1166721**	**1032679**	**1544265**	**38489537**	**433926**	**4707**	**119175**	**11240090**	**10364957**
INDUSTRY SECTOR	**21280278**	**437656**	**963301**	**1541667**	**5360887**	**6415**	**409**	**104211**	**4803898**	**4276492**
Iron and Steel	-	-	-	-	-	-	-	-	-	-
Chemical and Petrochem.	-	-	-	-	-	-	-	-	-	-
of which: Feedstocks	-	-	-	-	-	-	-	-	-	-
Non-Ferrous Metals	-	-	-	-	-	-	-	-	-	-
Non-Metallic Minerals	-	-	-	-	-	-	-	-	-	-
Transport Equipment	-	-	-	-	-	-	-	-	-	-
Machinery	-	-	-	-	-	-	-	-	-	-
Mining and Quarrying	-	-	-	-	-	-	-	-	-	-
Food and Tobacco	-	-	-	-	-	-	-	-	-	-
Paper, Pulp and Print	-	-	-	-	-	-	-	-	-	-
Wood and Wood Products	-	-	-	-	-	-	-	-	-	-
Construction	-	-	-	-	-	-	-	-	-	-
Textile and Leather	-	-	-	-	-	-	-	-	-	-
Non-specified	-	-	-	-	-	-	-	-	-	-
TRANSPORT SECTOR	**1873990**	**32**	**-**	**-**	**1087**	**337879**	**-**	**37**	**218746**	**-**
Air	-	14	-	-	-	-	-	-	178	-
Road	86810	-	-	-	-	337879	-	-	169	-
Rail	-	-	-	-	-	-	-	-	-	-
Pipeline Transport	-	-	-	-	-	-	-	-	-	-
Internal Navigation	-	-	-	-	-	-	-	-	-	-
Non-specified	-	-	-	-	-	-	-	-	-	-
OTHER SECTORS	**-**	**-**	**-**	**-**	**-**	**-**	**-**	**-**	**-**	**-**
Agriculture	-	-	-	-	-	-	-	-	-	-
Comm. and Publ. Services	-	-	-	-	-	-	-	-	-	-
Residential	-	-	-	-	-	-	-	-	-	-
Non-specified	-	-	-	-	-	-	-	-	-	-
NON-ENERGY USE	**-**	**-**	**3807**	**667**	**-**	**-**	**-**	**-**	**-**	**-**
in Industry/Transf./Energy	-	-	-	-	-	-	-	-	-	-
in Transport	-	-	-	-	-	-	-	-	-	-
in Other Sectors	-	-	-	-	-	-	-	-	-	-

World / Monde : 1997

SUPPLY AND CONSUMPTION / *APPROVISIONNEMENT ET DEMANDE*	Coal / *Charbon* (1000 tonnes)							Oil / *Pétrole* (1000 tonnes)			
	Coking Coal / *Charbon à coke*	Other Bit. Coal / *Autres charb. bit.*	Sub-Bit. Coal / *Charbon sous-bit.*	Lignite / *Lignite*	Peat / *Tourbe*	Oven and Gas Coke / *Coke de four/gaz*	Pat. Fuel and BKB / *Agg./briq. de lignite*	Crude Oil / *Pétrole brut*	NGL / *LGN*	Feed-stocks / *Produits d'aliment.*	Additives / *Additifs*
Production	552691	2901563	442900	798227	22399	366006	26013	3235790	214217	-	9879
From Other Sources	150	2982	427	-	-	80	-	963	-	24779	184
Imports	190135	321860	687	7671	9	19044	1193	1850814	16742	49372	3110
Exports	-179234	-342332	-8416	-1994	-158	-24464	-1551	-1766218	-63715	-7426	-457
Stock Changes	-1276	-31499	5087	-125	-753	2125	-23	-16890	65	-534	120
DOMESTIC SUPPLY	**562466**	**2852574**	**440685**	**803779**	**21497**	**362791**	**25632**	**3304459**	**167309**	**66191**	**12836**
Intl. Marine Bunkers	-	-	-	-	-	-	-	-	-	-	-
Transfers	-	-	-	-	-	-	-	-293	-108759	44343	-267
Statistical Differences	-15747	71260	12660	13126	-	-7359	50	495	-4928	1430	416
TRANSFORMATION	**531222**	**1938504**	**434683**	**757704**	**16831**	**241407**	**1079**	**3287365**	**43779**	**111964**	**12985**
Electricity Plants	16479	1541124	393097	543202	5571	3	597	24461	41	-	-
CHP Plants	42	292494	37736	170377	5929	-	200	2063	-	-	-
Heat Plants	45	33081	960	22199	1764	563	144	113	-	-	-
Blast Furnaces/Gas Works	8824	25658	1214	-	-	240627	138	-	219	-	-
Coke/Pat. Fuel/BKB Plants	505832	18303	1676	21926	3567	214	-	-	-	-	-
Petroleum Refineries	-	-	-	-	-	-	-	3268790	43519	111964	12985
Petrochemical Industry	-	-	-	-	-	-	-	-	-	-	-
Liquefaction	-	27766	-	-	-	-	-	-8062	-	-	-
Other Transform. Sector	-	78	-	-	-	-	-	-	-	-	-
ENERGY SECTOR	**213**	**77009**	**24**	**1417**	**175**	**2407**	**109**	**6951**	**457**	**-**	**-**
Coal Mines	80	38125	24	320	45	549	36	15	-	-	-
Oil and Gas Extraction	-	2774	-	-	-	17	-	3591	420	-	-
Petroleum Refineries	-	175	-	187	-	-	-	3191	-	-	-
Electr., CHP+Heat Plants	-	26618	-	42	-	29	24	151	-	-	-
Pumped Storage (Elec.)	-	-	-	-	-	-	-	-	-	-	-
Other Energy Sector	133	9317	-	868	130	1812	49	3	37	-	-
Distribution Losses	10	28133	29	4195	218	46	-	2535	28	-	-
FINAL CONSUMPTION	**15274**	**880188**	**18609**	**53589**	**4273**	**111572**	**24494**	**7810**	**9358**	**-**	**-**
INDUSTRY SECTOR	**14860**	**619847**	**15568**	**34415**	**1220**	**88741**	**3202**	**4815**	**9358**	**-**	**-**
Iron and Steel	-	-	-	-	-	-	-	-	-	-	-
Chemical and Petrochem.	-	-	-	-	-	-	-	-	-	-	-
of which: Feedstocks	-	-	-	-	-	-	-	-	-	-	-
Non-Ferrous Metals	-	-	-	-	-	-	-	-	-	-	-
Non-Metallic Minerals	-	-	-	-	-	-	-	-	-	-	-
Transport Equipment	-	-	-	-	-	-	-	-	-	-	-
Machinery	-	-	-	-	-	-	-	-	-	-	-
Mining and Quarrying	-	-	-	-	-	-	-	-	-	-	-
Food and Tobacco	-	-	-	-	-	-	-	-	-	-	-
Paper, Pulp and Print	-	-	-	-	-	-	-	-	-	-	-
Wood and Wood Products	-	-	-	-	-	-	-	-	-	-	-
Construction	-	-	-	-	-	-	-	-	-	-	-
Textile and Leather	-	-	-	-	-	-	-	-	-	-	-
Non-specified	-	-	-	-	-	-	-	-	-	-	-
TRANSPORT SECTOR	**8**	**13087**	**286**	**492**	**1**	**10**	**27**	**9**	**-**	**-**	**-**
Air	-	-	-	-	-	-	-	-	-	-	-
Road	-	-	-	-	-	-	-	-	-	-	-
Rail	-	-	-	-	-	-	-	-	-	-	-
Pipeline Transport	-	-	-	-	-	-	-	-	-	-	-
Internal Navigation	-	-	-	-	-	-	-	-	-	-	-
Non-specified	-	-	-	-	-	-	-	-	-	-	-
OTHER SECTORS	**-**	**-**	**-**	**-**	**-**	**-**	**-**	**-**	**-**	**-**	**-**
Agriculture	-	-	-	-	-	-	-	-	-	-	-
Comm. and Publ. Services	-	-	-	-	-	-	-	-	-	-	-
Residential	-	-	-	-	-	-	-	-	-	-	-
Non-specified	-	-	-	-	-	-	-	-	-	-	-
NON-ENERGY USE	**240**	**23012**	**-**	**385**	**-**	**13931**	**199**	**1537**	**-**	**-**	**-**
in Industry/Trans./Energy	-	-	-	-	-	-	-	-	-	-	-
in Transport	-	-	-	-	-	-	-	-	-	-	-
in Other Sectors	-	-	-	-	-	-	-	-	-	-	-

World / Monde : 1997

SUPPLY AND CONSUMPTION	Oil cont. / *Pétrole cont.* (1000 tonnes)										
	Refinery Gas	LPG + Ethane	Motor Gasoline	Aviation Gasoline	Jet Fuel	Kerosene	Gas/ Diesel	Heavy Fuel Oil	Naphtha	Petrol. Coke	Other Prod.
APPROVISIONNEMENT ET DEMANDE	*Gaz de raffinerie*	*GPL + éthane*	*Essence moteur*	*Essence aviation*	*Carbu-réacteurs*	*Kérosène*	*Gazole*	*Fioul lourd*	*Naphta*	*Coke de pétrole*	*Autres prod.*
Production	98186	84362	820432	1681	209329	88389	962367	649301	155625	61797	215524
From Other Sources	-	-	-	-	-	-	-	-	-	-	-
Imports	11	53152	84053	644	30924	18834	180928	164372	66248	17914	37727
Exports	-	-31356	-104603	-207	-44563	-12060	-201911	-199248	-50752	-22231	-43791
Stock Changes	5	-1951	-1765	72	-1367	-528	-5612	2596	-628	-822	-330
DOMESTIC SUPPLY	**98202**	**104207**	**798117**	**2190**	**194323**	**94635**	**935772**	**617021**	**170493**	**56658**	**209130**
Intl. Marine Bunkers	-	-	-274	-	-14	-	-27017	-107701	-	-	-344
Transfers	191	89666	4319	194	-869	-2335	-550	-8656	-8272	-6	-6489
Statistical Differences	532	1105	-4333	-2	-3455	-3708	1897	-457	-101	51	4217
TRANSFORMATION	**4313**	**4875**	**682**	**-**	**-**	**598**	**42193**	**254838**	**19945**	**3826**	**4008**
Electricity Plants	707	529	38	-	-	76	38911	189995	1135	2226	450
CHP Plants	2349	607	-	-	-	9	616	18774	31	248	2574
Heat Plants	759	51	-	-	-	17	1303	41717	41	-	200
Blast Furnaces/Gas Works	262	2467	-	-	-	-	134	2903	990	403	-
Coke/Pat. Fuel/BKB Plants	-	-	-	-	-	-	-	20	-	882	-
Petroleum Refineries	-	12	69	-	-	-	-	-	-	-	-
Petrochemical Industry	236	1208	575	-	-	496	1229	1428	17748	67	784
Liquefaction	-	-	-	-	-	-	-	-	-	-	-
Other Transform. Sector	-	1	-	-	-	-	-	1	-	-	-
ENERGY SECTOR	**90154**	**2905**	**1299**	**-**	**6**	**497**	**8670**	**49983**	**207**	**24348**	**7105**
Coal Mines	-	2	2	-	-	100	1948	293	-	-	2
Oil and Gas Extraction	94	112	1020	-	-	18	2984	1641	3	1235	25
Petroleum Refineries	89883	2706	248	-	3	300	2496	44942	200	23001	6522
Electr., CHP+Heat Plants	3	5	12	-	-	25	756	1749	-	-	-
Pumped Storage (Elec.)	-	-	-	-	-	-	-	-	-	-	-
Other Energy Sector	174	80	17	-	3	54	486	1358	4	112	556
Distribution Losses	2	265	261	-	17	7	154	117	231	-	276
FINAL CONSUMPTION	**4456**	**186933**	**795587**	**2382**	**189962**	**87490**	**859085**	**195269**	**141737**	**28529**	**195125**
INDUSTRY SECTOR	**4427**	**91213**	**7019**	**7**	**282**	**9225**	**96131**	**146557**	**141737**	**12607**	**7851**
Iron and Steel	-	-	-	-	-	-	-	-	-	-	-
Chemical and Petrochem.	-	-	-	-	-	-	-	-	-	-	-
of which: Feedstocks	-	-	-	-	-	-	-	-	-	-	-
Non-Ferrous Metals	-	-	-	-	-	-	-	-	-	-	-
Non-Metallic Minerals	-	-	-	-	-	-	-	-	-	-	-
Transport Equipment	-	-	-	-	-	-	-	-	-	-	-
Machinery	-	-	-	-	-	-	-	-	-	-	-
Mining and Quarrying	-	-	-	-	-	-	-	-	-	-	-
Food and Tobacco	-	-	-	-	-	-	-	-	-	-	-
Paper, Pulp and Print	-	-	-	-	-	-	-	-	-	-	-
Wood and Wood Products	-	-	-	-	-	-	-	-	-	-	-
Construction	-	-	-	-	-	-	-	-	-	-	-
Textile and Leather	-	-	-	-	-	-	-	-	-	-	-
Non-specified	-	-	-	-	-	-	-	-	-	-	-
TRANSPORT SECTOR	**16**	**10068**	**775391**	**2312**	**189031**	**953**	**493387**	**19728**	**-**	**-**	**649**
Air	-	-	29	2309	189031	828	2	-	-	-	1
Road	2	9879	771165	3	-	53	442777	366	-	-	192
Rail	-	-	-	-	-	-	-	-	-	-	-
Pipeline Transport	-	-	-	-	-	-	-	-	-	-	-
Internal Navigation	-	-	-	-	-	-	-	-	-	-	-
Non-specified	-	-	-	-	-	-	-	-	-	-	-
OTHER SECTORS	**-**	**-**	**-**	**-**	**-**	**-**	**-**	**-**	**-**	**-**	**-**
Agriculture	-	-	-	-	-	-	-	-	-	-	-
Comm. and Publ. Services	-	-	-	-	-	-	-	-	-	-	-
Residential	-	-	-	-	-	-	-	-	-	-	-
Non-specified	-	-	-	-	-	-	-	-	-	-	-
NON-ENERGY USE	**-**	**1509**	**273**	**2**	**1**	**25**	**6**	**11**	**-**	**15807**	**184714**
in Industry/Transf./Energy	-	-	-	-	-	-	-	-	-	-	-
in Transport	-	-	-	-	-	-	-	-	-	-	-
in Other Sectors	-	-	-	-	-	-	-	-	-	-	-

World / Monde : 1997

SUPPLY AND CONSUMPTION / *APPROVISIONNEMENT ET DEMANDE*	Gas / *Gaz* (TJ)				Comb. Renew. & Waste / *En. Re. Comb. & Déchets* (TJ)				(GWh)	(TJ)
	Natural Gas / *Gaz naturel*	Gas Works / *Usines à gaz*	Coke Ovens / *Cokeries*	Blast Furnaces / *Hauts fourneaux*	Solid Biomass / *Biomasse solide*	Gas/Liquids from Biomass / *Gaz/Liquides tirés de biomasse*	Municipal Waste / *Déchets urbains*	Industrial Waste / *Déchets industriels*	Electricity / *Electricité*	Heat / *Chaleur*
Production	89193031	1192399	1786278	2508657	42919596	622268	541185	408761	14021340	11283682
From Other Sources	-	22712	-	-	3341	35489	-	12146	-	-
Imports	20508592	1	-	-	20042	21826	-	-	429230	145
Exports	-20506426	-	-	-	-33343	-28685	-	-6799	-425543	-145
Stock Changes	-262158	-	-	-	-2837	-50103	-	57	-	-
DOMESTIC SUPPLY	**88933039**	**1215112**	**1786278**	**2508657**	**42906798**	**600795**	**541185**	**414165**	**14025027**	**11283682**
Intl. Marine Bunkers	-	-	-	-	-	-	-	-	-	-
Transfers	-	-	-	-	-	-18700	-	-	-	-
Statistical Differences	-55360	-4220	-44360	15516	-8089	-17811	-16430	-7593	-4340	19585
TRANSFORMATION	**30039909**	**19435**	**346482**	**770341**	**4201477**	**117343**	**520945**	**290471**	**6967**	**-**
Electricity Plants	14079802	5249	167898	470615	497051	58108	305724	62188	-	-
CHP Plants	11786985	1478	95051	265674	1542821	19273	162637	146231	-	-
Heat Plants	3185595	12708	63502	33905	269800	1719	52584	82037	-	-
Blast Furnaces/Gas Works	851402	-	17031	-	1932	-	-	-	-	-
Coke/Pat. Fuel/BKB Plants	3813	-	3000	147	5242	-	-	15	-	-
Petroleum Refineries	3434	-	-	-	105	38243	-	-	-	-
Petrochemical Industry	-	-	-	-	-	-	-	-	-	-
Liquefaction	111206	-	-	-	-	-	-	-	-	-
Other Transform. Sector	17672	-	-	-	1884526	-	-	-	6967	-
ENERGY SECTOR	**10473291**	**21636**	**344054**	**165744**	**229**	**2897**	**100**	**1968**	**1366160**	**919056**
Coal Mines	1133797	-	2608	-	174	-	-	24	114843	79200
Oil and Gas Extraction	7165394	-	-	-	-	-	-	-	135655	150199
Petroleum Refineries	1653548	3289	1862	149	-	1180	-	18	160812	461731
Electr., CHP+Heat Plants	12654	1114	2883	5141	55	423	100	-	784232	91957
Pumped Storage (Elec.)	-	-	-	-	-	-	-	-	92823	-
Other Energy Sector	507898	17233	336701	160454	-	1294	-	1926	77795	135969
Distribution Losses	940086	8792	5593	27319	7521	-	60	29	1174567	660153
FINAL CONSUMPTION	**47424393**	**1161029**	**1045789**	**1560769**	**38690852**	**444044**	**3650**	**114104**	**11472993**	**9724058**
INDUSTRY SECTOR	**21396495**	**460148**	**969724**	**1558194**	**5516956**	**7655**	**420**	**100025**	**4913428**	**4096512**
Iron and Steel	-	-	-	-	-	-	-	-	-	-
Chemical and Petrochem.	-	-	-	-	-	-	-	-	-	-
of which: Feedstocks	-	-	-	-	-	-	-	-	-	-
Non-Ferrous Metals	-	-	-	-	-	-	-	-	-	-
Non-Metallic Minerals	-	-	-	-	-	-	-	-	-	-
Transport Equipment	-	-	-	-	-	-	-	-	-	-
Machinery	-	-	-	-	-	-	-	-	-	-
Mining and Quarrying	-	-	-	-	-	-	-	-	-	-
Food and Tobacco	-	-	-	-	-	-	-	-	-	-
Paper, Pulp and Print	-	-	-	-	-	-	-	-	-	-
Wood and Wood Products	-	-	-	-	-	-	-	-	-	-
Construction	-	-	-	-	-	-	-	-	-	-
Textile and Leather	-	-	-	-	-	-	-	-	-	-
Non-specified	-	-	-	-	-	-	-	-	-	-
TRANSPORT SECTOR	**1918568**	**29**	**-**	**-**	**1304**	**346554**	**-**	**37**	**222298**	**-**
Air	-	12	-	-	-	-	-	-	212	-
Road	85768	-	-	-	-	346554	-	-	931	-
Rail	-	-	-	-	-	-	-	-	-	-
Pipeline Transport	-	-	-	-	-	-	-	-	-	-
Internal Navigation	-	-	-	-	-	-	-	-	-	-
Non-specified	-	-	-	-	-	-	-	-	-	-
OTHER SECTORS	**-**	**-**	**-**	**-**	**-**	**-**	**-**	**-**	**-**	**-**
Agriculture	-	-	-	-	-	-	-	-	-	-
Comm. and Publ. Services	-	-	-	-	-	-	-	-	-	-
Residential	-	-	-	-	-	-	-	-	-	-
Non-specified	-	-	-	-	-	-	-	-	-	-
NON-ENERGY USE	**-**	**-**	**4322**	**677**	**-**	**-**	**-**	**-**	**-**	**-**
in Industry/Transf./Energy	-	-	-	-	-	-	-	-	-	-
in Transport	-	-	-	-	-	-	-	-	-	-
in Other Sectors	-	-	-	-	-	-	-	-	-	-

OECD Total / Total OCDE : 1996

	Coal / *Charbon* (1000 tonnes)							Oil / *Pétrole* (1000 tonnes)			
SUPPLY AND CONSUMPTION	Coking Coal	Other Bit. Coal	Sub-Bit. Coal	Lignite	Peat	Oven and Gas Coke	Pat. Fuel and BKB	Crude Oil	NGL	Feed-stocks	Additives
APPROVISIONNEMENT ET DEMANDE	*Charbon à coke*	*Autres charb. bit.*	*Charbon sous-bit.*	*Lignite*	*Tourbe*	*Coke de four/gaz*	*Agg./briq. de lignite*	*Pétrole brut*	*LGN*	*Produits d'aliment.*	*Additifs*
Production	258351	824507	428590	536274	15470	144715	12106	887939	106690	-	9169
From Other Sources	51	3777	560	-	-	72	-	98	-	20264	233
Imports	139618	213553	657	2939	-	12654	1152	1353805	16176	50664	2558
Exports	-169413	-125847	-9020	-94	-49	-10808	-1624	-366582	-20548	-7229	-491
Intl. Marine Bunkers	-	-	-	-	-	-	-	-	-	-	-
Stock Changes	-2716	15318	1280	1659	-286	-104	153	2571	-6	-3016	-123
DOMESTIC SUPPLY	**225891**	**931308**	**422067**	**540778**	**15135**	**146529**	**11787**	**1877831**	**102312**	**60683**	**11346**
Transfers	-	-	-	-	-	-	-	-199	-65635	43243	-177
Statistical Differences	-2578	11660	4727	10369	-952	-852	31	9828	-1975	770	438
TRANSFORMATION	**219278**	**810667**	**415417**	**526548**	**11488**	**110050**	**1015**	**1886538**	**31939**	**104696**	**11607**
Electricity Plants	17326	662532	369626	426760	5613	-	741	15701	182	-	-
CHP Plants	35	116791	41906	77702	4417	-	103	1317	-	-	-
Heat Plants	5	14174	1121	334	752	245	4	-	-	-	-
Blast Furnaces/Gas Works	9142	12281	1082	-	-	109613	167	-	-	-	-
Coke/Pat. Fuel/BKB Plants	192770	4889	1682	21752	706	192	-	-	-	-	-
Petroleum Refineries	-	-	-	-	-	-	-	1870502	31757	104696	11607
Petrochemical Industry	-	-	-	-	-	-	-	-	-	-	-
Liquefaction	-	-	-	-	-	-	-	-982	-	-	-
Other Transform. Sector	-	-	-	-	-	-	-	-	-	-	-
ENERGY SECTOR	**95**	**3399**	**8**	**466**	**-**	**336**	**189**	**894**	**512**	**-**	**-**
Coal Mines	92	2610	8	37	-	3	2	-	-	-	-
Oil and Gas Extraction	-	-	-	-	-	-	-	894	512	-	-
Petroleum Refineries	-	189	-	-	-	-	-	-	-	-	-
Electr., CHP+Heat Plants	-	268	-	-	-	4	-	-	-	-	-
Pumped Storage (Elec.)	-	-	-	-	-	-	-	-	-	-	-
Other Energy Sector	3	332	-	429	-	329	187	-	-	-	-
Distribution Losses	9	1	35	1	-	5	-	-	-	-	-
FINAL CONSUMPTION	**3931**	**128901**	**11334**	**24132**	**2695**	**35286**	**10614**	**28**	**2251**	**-**	**-**
INDUSTRY SECTOR	**3491**	**97719**	**9046**	**15273**	**1261**	**29344**	**2865**	**28**	**2251**	**-**	**-**
Iron and Steel	2979	8979	732	3	-	23471	13	-	-	-	-
Chemical and Petrochem.	13	12139	2277	7295	33	584	361	28	2251	-	-
of which: Feedstocks	-	-	-	-	-	-	-	-	*2251*	-	-
Non-Ferrous Metals	8	1729	2118	443	-	1061	41	-	-	-	-
Non-Metallic Minerals	13	40782	832	486	-	1256	1810	-	-	-	-
Transport Equipment	-	651	50	27	-	73	57	-	-	-	-
Machinery	12	2155	358	112	-	613	11	-	-	-	-
Mining and Quarrying	-	576	410	10	-	467	18	-	-	-	-
Food and Tobacco	157	6934	1067	1429	-	398	335	-	-	-	-
Paper, Pulp and Print	-	7943	253	307	8	38	132	-	-	-	-
Wood and Wood Products	-	480	43	50	-	3	11	-	-	-	-
Construction	173	1466	60	1507	-	102	-	-	-	-	-
Textile and Leather	-	1502	323	304	-	23	30	-	-	-	-
Non-specified	136	12383	523	3300	1220	1255	46	-	-	-	-
TRANSPORT SECTOR	**11**	**20**	**222**	**4**	**-**	**19**	**9**	**-**	**-**	**-**	**-**
Air	-	-	-	-	-	-	-	-	-	-	-
Road	-	-	-	-	-	-	-	-	-	-	-
Rail	11	19	30	4	-	19	9	-	-	-	-
Pipeline Transport	-	-	-	-	-	-	-	-	-	-	-
Internal Navigation	-	1	192	-	-	-	-	-	-	-	-
Non-specified	-	-	-	-	-	-	-	-	-	-	-
OTHER SECTORS	**197**	**30838**	**2066**	**8629**	**1434**	**5267**	**7556**	**-**	**-**	**-**	**-**
Agriculture	-	2782	180	283	60	195	39	-	-	-	-
Comm. and Publ. Services	1	2481	364	926	30	2626	1606	-	-	-	-
Residential	196	25342	1522	7383	1344	2444	5888	-	-	-	-
Non-specified	-	233	-	37	-	2	23	-	-	-	-
NON-ENERGY USE	**232**	**324**	**-**	**226**	**-**	**656**	**184**	**-**	**-**	**-**	**-**
in Industry/Trans./Energy	2	322	-	226	-	656	174	-	-	-	-
in Transport	-	-	-	-	-	-	-	-	-	-	-
in Other Sectors	230	2	-	-	-	-	10	-	-	-	-

OECD Total / Total OCDE : 1996

SUPPLY AND CONSUMPTION / *APPROVISIONNEMENT ET DEMANDE*	Oil cont. / *Pétrole cont.* (1000 tonnes)										
	Refinery Gas / *Gaz de raffinerie*	LPG + Ethane / *GPL + éthane*	Motor Gasoline / *Essence moteur*	Aviation Gasoline / *Essence aviation*	Jet Fuel / *Carbu-réacteurs*	Kerosene / *Kérosène*	Gas/ Diesel / *Gazole*	Heavy Fuel Oil / *Fioul lourd*	Naphtha / *Naphta*	Petrol. Coke / *Coke de pétrole*	Other Prod. / *Autres prod.*
Production	71719	50944	593470	1193	135885	41780	560959	280032	81557	56605	124664
From Other Sources	-	-	-	-	-	-	-	-	-	-	-
Imports	-	35653	54082	122	17690	8890	88115	71181	56827	15947	23032
Exports	-	-7789	-55856	-192	-16999	-1720	-87225	-60234	-20538	-20646	-22347
Intl. Marine Bunkers	-	-	-	-	-21	-	-18219	-60510	-	-	-358
Stock Changes	-	915	1624	2	99	-782	-1992	67	197	-280	1463
DOMESTIC SUPPLY	**71719**	**79723**	**593320**	**1125**	**136654**	**48168**	**541638**	**230536**	**118043**	**51626**	**126454**
Transfers	-102	58518	-6096	166	-567	-2888	473	-7104	-7353	-187	-9531
Statistical Differences	-47	-1	-478	-	-685	-876	2801	-1220	289	-158	1386
TRANSFORMATION	**1494**	**4586**	**523**	**-**	**-**	**443**	**9475**	**113767**	**16113**	**3163**	**782**
Electricity Plants	437	602	-	-	-	3	7137	100965	1066	1670	-
CHP Plants	708	61	-	-	-	-	794	7315	16	204	350
Heat Plants	-	29	-	-	-	10	177	1766	-	-	-
Blast Furnaces/Gas Works	167	2728	-	-	-	-	122	2024	426	304	-
Coke/Pat. Fuel/BKB Plants	-	-	-	-	-	-	-	20	-	945	-
Petroleum Refineries	-	-	-	-	-	-	-	-	-	-	-
Petrochemical Industry	182	1166	523	-	-	430	1245	1677	14605	40	432
Liquefaction	-	-	-	-	-	-	-	-	-	-	-
Other Transform. Sector	-	-	-	-	-	-	-	-	-	-	-
ENERGY SECTOR	**68386**	**1695**	**595**	**-**	**1**	**428**	**2448**	**23736**	**177**	**24232**	**3588**
Coal Mines	-	3	1	-	-	-	537	28	-	-	2
Oil and Gas Extraction	-	40	-	-	-	1	181	164	-	1305	-
Petroleum Refineries	68386	1649	591	-	1	427	1416	23181	177	22832	3486
Electr., CHP+Heat Plants	-	-	2	-	-	-	54	259	-	-	-
Pumped Storage (Elec.)	-	-	-	-	-	-	-	-	-	-	-
Other Energy Sector	-	3	1	-	-	-	260	104	-	95	100
Distribution Losses	-	19	92	-	16	5	51	78	-	-	-
FINAL CONSUMPTION	**1690**	**131940**	**585536**	**1291**	**135385**	**43528**	**532938**	**84631**	**94689**	**23886**	**113939**
INDUSTRY SECTOR	**1690**	**80764**	**2934**	**3**	**67**	**7437**	**56885**	**62911**	**94689**	**10363**	**888**
Iron and Steel	-	1677	-	-	-	271	2233	5407	-	64	-
Chemical and Petrochem.	1660	57542	897	-	-	1228	6881	14406	94460	809	865
of which: Feedstocks	*706*	*56969*	*892*	-	-	*887*	*3883*	*2523*	*93020*	-	-
Non-Ferrous Metals	-	748	-	-	-	179	985	2400	27	28	-
Non-Metallic Minerals	-	1714	1	-	-	100	4767	8453	3	5959	21
Transport Equipment	-	754	28	3	52	5	1242	719	-	-	-
Machinery	-	2560	37	-	15	382	3358	1591	-	221	-
Mining and Quarrying	-	368	2	-	-	53	3798	1038	-	-	2
Food and Tobacco	2	630	11	-	-	9	5532	7344	-	-	-
Paper, Pulp and Print	28	1059	2	-	-	107	1941	11014	-	140	-
Wood and Wood Products	-	156	1	-	-	3	2910	455	-	-	-
Construction	-	140	784	-	-	1136	9114	515	-	-	-
Textile and Leather	-	782	1	-	-	20	2013	4416	-	21	-
Non-specified	-	12634	1170	-	-	3944	12111	5153	199	3121	-
TRANSPORT SECTOR	**-**	**8674**	**578986**	**1288**	**134674**	**78**	**299300**	**8271**	**-**	**-**	**6**
Air	-	-	28	1288	134674	-	1	-	-	-	-
Road	-	8613	575786	-	-	54	269808	99	-	-	6
Rail	-	-	-	-	-	14	17529	45	-	-	-
Pipeline Transport	-	-	-	-	-	-	49	1	-	-	-
Internal Navigation	-	1	3172	-	-	3	9419	7818	-	-	-
Non-specified	-	60	-	-	-	7	2494	308	-	-	-
OTHER SECTORS	**-**	**42502**	**3616**	**-**	**644**	**36013**	**176753**	**13449**	**-**	**80**	**13**
Agriculture	-	2430	2786	-	-	3672	44415	1727	-	9	6
Comm. and Publ. Services	-	8919	720	-	1	12995	51504	7657	-	10	7
Residential	-	29976	110	-	-	19237	78693	3558	-	61	-
Non-specified	-	1177	-	-	643	109	2141	507	-	-	-
NON-ENERGY USE	**-**	**-**	**-**	**-**	**-**	**-**	**-**	**-**	**-**	**13443**	**113032**
in Industry/Transf./Energy	-	-	-	-	-	-	-	-	-	13443	102472
in Transport	-	-	-	-	-	-	-	-	-	-	7486
in Other Sectors	-	-	-	-	-	-	-	-	-	-	3074

OECD Total / Total OCDE : 1996

SUPPLY AND CONSUMPTION / *APPROVISIONNEMENT ET DEMANDE*	Gas / *Gaz* (TJ)				Comb. Renew. & Waste / *En. Re. Comb. & Déchets* (TJ)				(GWh)	(TJ)
	Natural Gas / *Gaz naturel*	Gas Works / *Usines à gaz*	Coke Ovens / *Cokeries*	Blast Furnaces / *Hauts fourneaux*	Solid Biomass / *Biomasse solide*	Gas/Liquids from Biomass / *Gaz/Liquides tirés de biomasse*	Municipal Waste / *Déchets urbains*	Industrial Waste / *Déchets industriels*	Electricity / *Electricité*	Heat / *Chaleur*
Production	40385164	951874	969259	1551403	5613023	185766	530173	304357	8782843	2311185
From Other Sources	-	25637	-	-	-	34448	-	16038	-	-
Imports	15137878	-	-	-	12035	4083	-	-	277231	122
Exports	-7136545	-	-	-	-255	-	-	-	-271336	-122
Intl. Marine Bunkers	-	-	-	-	-	-	-	-	-	-
Stock Changes	-179214	2	-	-	-318	-462	-	407	-	-
DOMESTIC SUPPLY	**48207283**	**977513**	**969259**	**1551403**	**5624485**	**223835**	**530173**	**320802**	**8788738**	**2311185**
Transfers	-	-	-	-	-	-6533	-	-	-	-
Statistical Differences	-32925	-1911	-12966	-25476	-539	-10727	-12913	-3511	-	-
TRANSFORMATION	**12365294**	**12516**	**255934**	**529755**	**1832989**	**117463**	**512396**	**240797**	**4809**	**-**
Electricity Plants	7297794	-	135933	361080	313047	59297	303827	57733	-	-
CHP Plants	3929295	3	98694	167616	1447479	16228	158836	117508	-	-
Heat Plants	240175	12513	2564	860	65696	1463	49733	65549	-	-
Blast Furnaces/Gas Works	822031	-	15743	-	-	-	-	-	-	-
Coke/Pat. Fuel/BKB Plants	4177	-	3000	199	4501	-	-	7	-	-
Petroleum Refineries	-	-	-	-	-	40475	-	-	-	-
Petrochemical Industry	-	-	-	-	-	-	-	-	-	-
Liquefaction	55810	-	-	-	-	-	-	-	-	-
Other Transform. Sector	16012	-	-	-	2266	-	-	-	4809	-
ENERGY SECTOR	**3876702**	**10430**	**276162**	**146803**	**3**	**617**	**98**	**173**	**797782**	**134672**
Coal Mines	12478	28	2437	-	1	-	-	31	41355	15507
Oil and Gas Extraction	2805361	-	-	-	-	-	-	-	55241	1
Petroleum Refineries	968810	2995	1960	-	-	209	-	141	92791	54747
Electr., CHP+Heat Plants	752	-	-	-	2	408	98	1	480202	22657
Pumped Storage (Elec.)	-	-	-	-	-	-	-	-	81202	-
Other Energy Sector	89301	7407	271765	146803	-	-	-	-	46991	41760
Distribution Losses	183722	6006	4344	21874	-	-	59	-	563816	176648
FINAL CONSUMPTION	**31748640**	**946650**	**419853**	**827495**	**3790954**	**88495**	**4707**	**76321**	**7422331**	**1999865**
INDUSTRY SECTOR	**12834331**	**317414**	**418060**	**827395**	**1518692**	**6415**	**409**	**67221**	**2955244**	**613552**
Iron and Steel	1229758	57811	384151	822308	21	-	-	498	334503	16282
Chemical and Petrochem.	5035289	55895	8495	1871	19813	2365	-	12088	560219	202104
of which: Feedstocks	*965821*	-	-	-	-	-	-	-	-	-
Non-Ferrous Metals	542003	15693	221	54	2527	2426	-	132	260909	4459
Non-Metallic Minerals	1166980	16410	12012	1873	20598	15	-	3823	154868	5441
Transport Equipment	277067	83	148	1	60	-	-	54	121009	25272
Machinery	766241	58593	7451	-	362	55	-	111	282130	16143
Mining and Quarrying	182878	6	2246	-	44	-	-	222	102890	4761
Food and Tobacco	1157215	41040	984	-	167479	588	-	255	200029	34864
Paper, Pulp and Print	1212168	26937	369	-	783847	90	-	1816	365377	52377
Wood and Wood Products	79130	20	-	-	425449	-	-	-	57738	6261
Construction	18822	7	432	-	277	1	-	20	15152	1638
Textile and Leather	328527	7698	147	-	5641	-	-	2	109563	18202
Non-specified	838253	37221	1404	1288	92574	875	409	48200	390857	225748
TRANSPORT SECTOR	**1051166**	**1**	**-**	**-**	**60**	**42707**	**-**	**8**	**103705**	**-**
Air	-	-	-	-	-	-	-	-	-	-
Road	25470	-	-	-	-	42707	-	-	-	-
Rail	-	1	-	-	5	-	-	8	88392	-
Pipeline Transport	1023916	-	-	-	-	-	-	-	3866	-
Internal Navigation	-	-	-	-	55	-	-	-	-	-
Non-specified	1780	-	-	-	-	-	-	-	11447	-
OTHER SECTORS	**17863143**	**629235**	**1269**	**-**	**2272202**	**39373**	**4298**	**9092**	**4363382**	**1386313**
Agriculture	237631	51	-	-	61015	-	12	104	76077	11726
Comm. and Publ. Services	5118435	212002	38	-	61776	36896	2908	601	1972858	305592
Residential	11887328	417140	1231	-	2091739	-	1378	-	2297774	941471
Non-specified	619749	42	-	-	57672	2477	-	8387	16673	127524
NON-ENERGY USE	**-**	**-**	**524**	**100**	**-**	**-**	**-**	**-**	**-**	**-**
in Industry/Transf./Energy	-	-	524	100	-	-	-	-	-	-
in Transport	-	-	-	-	-	-	-	-	-	-
in Other Sectors	-	-	-	-	-	-	-	-	-	-

OECD Total / Total OCDE : 1997

SUPPLY AND CONSUMPTION / *APPROVISIONNEMENT ET DEMANDE*	Coal / *Charbon* (1000 tonnes)							Oil / *Pétrole* (1000 tonnes)			
	Coking Coal / *Charbon à coke*	Other Bit. Coal / *Autres charb. bit.*	Sub-Bit. Coal / *Charbon sous-bit.*	Lignite / *Lignite*	Peat / *Tourbe*	Oven and Gas Coke / *Coke de four/gaz*	Pat. Fuel and BKB / *Agg./briq. de lignite*	Crude Oil / *Pétrole brut*	NGL / *LGN*	Feed-stocks / *Produits d'aliment.*	Additives / *Additifs*
Production	257485	852821	431866	531134	14584	143882	10131	900104	105899	-	9701
From Other Sources	150	2982	427	-	-	80	-	318	-	24779	184
Imports	140117	229737	387	3178	-	13765	958	1436696	16727	47303	2999
Exports	-169595	-130080	-8415	-128	-56	-11466	-1422	-386392	-21918	-7426	-457
Intl. Marine Bunkers	-	-	-	-	-	-	-	-	-	-	-
Stock Changes	-418	-17566	4944	939	-763	1883	-15	-4180	46	-456	121
DOMESTIC SUPPLY	**227739**	**937894**	**429209**	**535123**	**13765**	**148144**	**9652**	**1946546**	**100754**	**64200**	**12548**
Transfers	-	-	-	-	-	-	-	-212	-65328	41858	-267
Statistical Differences	-3954	2754	12541	12983	-	-234	44	-5718	-1581	1430	416
TRANSFORMATION	**217710**	**813436**	**431063**	**523218**	**11313**	**111347**	**772**	**1939668**	**31157**	**107488**	**12697**
Electricity Plants	16479	669905	390917	428020	5571	-	579	12540	41	-	-
CHP Plants	42	112113	36345	76788	4419	-	52	1465	-	-	-
Heat Plants	45	12855	911	360	700	188	3	-	-	-	-
Blast Furnaces/Gas Works	8824	12709	1214	-	-	110977	138	-	-	-	-
Coke/Pat. Fuel/BKB Plants	192320	5854	1676	18050	623	182	-	-	-	-	-
Petroleum Refineries	-	-	-	-	-	-	-	1926474	31116	107488	12697
Petrochemical Industry	-	-	-	-	-	-	-	-	-	-	-
Liquefaction	-	-	-	-	-	-	-	-811	-	-	-
Other Transform. Sector	-	-	-	-	-	-	-	-	-	-	-
ENERGY SECTOR	**171**	**3108**	**7**	**1094**	**-**	**317**	**50**	**922**	**419**	**-**	**-**
Coal Mines	67	2461	7	38	-	1	2	-	-	-	-
Oil and Gas Extraction	-	-	-	-	-	-	-	922	419	-	-
Petroleum Refineries	-	175	-	187	-	-	-	-	-	-	-
Electr., CHP+Heat Plants	-	302	-	1	-	2	-	-	-	-	-
Pumped Storage (Elec.)	-	-	-	-	-	-	-	-	-	-	-
Other Energy Sector	104	170	-	868	-	314	48	-	-	-	-
Distribution Losses	10	-	29	-	-	5	-	-	-	-	-
FINAL CONSUMPTION	**5894**	**124104**	**10651**	**23794**	**2452**	**36241**	**8874**	**26**	**2269**	**-**	**-**
INDUSTRY SECTOR	**5483**	**95134**	**7978**	**14835**	**1165**	**30344**	**3041**	**26**	**2269**	**-**	**-**
Iron and Steel	4206	10553	616	1	-	24861	12	-	-	-	-
Chemical and Petrochem.	122	12262	1904	6424	26	505	386	26	2269	-	-
of which: Feedstocks	-	-	-	-	-	-	-	-	*2269*	-	-
Non-Ferrous Metals	8	1689	2038	359	-	1132	50	-	-	-	-
Non-Metallic Minerals	18	39566	887	419	-	1095	1927	-	-	-	-
Transport Equipment	-	515	76	-	-	61	12	-	-	-	-
Machinery	7	1910	374	84	-	494	6	-	-	-	-
Mining and Quarrying	-	571	426	3	-	554	32	-	-	-	-
Food and Tobacco	183	7083	497	1442	-	388	322	-	-	-	-
Paper, Pulp and Print	-	8386	345	235	-	36	205	-	-	-	-
Wood and Wood Products	-	470	27	48	-	3	11	-	-	-	-
Construction	425	1771	41	1366	-	72	-	-	-	-	-
Textile and Leather	1	1117	324	400	-	23	27	-	-	-	-
Non-specified	513	9241	423	4054	1139	1120	51	-	-	-	-
TRANSPORT SECTOR	**7**	**27**	**282**	**2**	**-**	**6**	**1**	**-**	**-**	**-**	**-**
Air	-	-	-	-	-	-	-	-	-	-	-
Road	-	-	-	-	-	-	-	-	-	-	-
Rail	7	27	74	2	-	6	1	-	-	-	-
Pipeline Transport	-	-	-	-	-	-	-	-	-	-	-
Internal Navigation	-	-	208	-	-	-	-	-	-	-	-
Non-specified	-	-	-	-	-	-	-	-	-	-	-
OTHER SECTORS	**164**	**28610**	**2391**	**8806**	**1287**	**5273**	**5636**	**-**	**-**	**-**	**-**
Agriculture	12	2458	127	280	40	220	34	-	-	-	-
Comm. and Publ. Services	4	2496	249	958	15	2532	1094	-	-	-	-
Residential	148	23416	2015	7542	1232	2518	4495	-	-	-	-
Non-specified	-	240	-	26	-	3	13	-	-	-	-
NON-ENERGY USE	**240**	**333**	**-**	**151**	**-**	**618**	**196**	**-**	**-**	**-**	**-**
in Industry/Trans./Energy	3	333	-	151	-	618	186	-	-	-	-
in Transport	-	-	-	-	-	-	-	-	-	-	-
in Other Sectors	237	-	-	-	-	-	10	-	-	-	-

OECD Total / Total OCDE : 1997

	Oil cont. / *Pétrole cont.* (1000 tonnes)										
SUPPLY AND CONSUMPTION	Refinery Gas	LPG + Ethane	Motor Gasoline	Aviation Gasoline	Jet Fuel	Kerosene	Gas/ Diesel	Heavy Fuel Oil	Naphtha	Petrol. Coke	Other Prod.
APPROVISIONNEMENT ET DEMANDE	*Gaz de raffinerie*	*GPL + éthane*	*Essence moteur*	*Essence aviation*	*Carbu- réacteurs*	*Kérosène*	*Gazole*	*Fioul lourd*	*Naphta*	*Coke de pétrole*	*Autres prod.*
Production	72997	51907	603401	1160	139533	43916	577982	284394	92753	58875	134881
From Other Sources	-	-	-	-	-	-	-	-	-	-	-
Imports	-	36897	58021	127	17781	5534	83906	64465	58289	17245	27033
Exports	-	-7187	-61399	-180	-15268	-2535	-93055	-65180	-22890	-21942	-24578
Intl. Marine Bunkers	-	-	-	-	-14	-	-18877	-61106	-	-	-340
Stock Changes	2	-1703	-1711	90	-1584	-57	-4478	2176	-495	-749	-231
DOMESTIC SUPPLY	**72999**	**79914**	**598312**	**1197**	**140448**	**46858**	**545478**	**224749**	**127657**	**53429**	**136765**
Transfers	100	57283	-2997	195	-804	-2242	-688	-8985	-6735	-6	-8058
Statistical Differences	-51	1763	-2165	-2	-855	-827	1250	1771	1016	-58	1655
TRANSFORMATION	**1524**	**4231**	**575**	**-**	**-**	**509**	**8338**	**114985**	**19245**	**3826**	**1399**
Electricity Plants	387	522	-	-	-	4	6202	103381	1115	2226	-
CHP Plants	639	13	-	-	-	-	613	6570	31	248	615
Heat Plants	-	40	-	-	-	9	160	1467	-	-	-
Blast Furnaces/Gas Works	262	2448	-	-	-	-	134	2146	351	403	-
Coke/Pat. Fuel/BKB Plants	-	-	-	-	-	-	-	20	-	882	-
Petroleum Refineries	-	-	-	-	-	-	-	-	-	-	-
Petrochemical Industry	236	1208	575	-	-	496	1229	1401	17748	67	784
Liquefaction	-	-	-	-	-	-	-	-	-	-	-
Other Transform. Sector	-	-	-	-	-	-	-	-	-	-	-
ENERGY SECTOR	**69291**	**1370**	**226**	**-**	**2**	**260**	**2483**	**21962**	**162**	**24228**	**4321**
Coal Mines	-	2	-	-	-	-	564	24	-	-	2
Oil and Gas Extraction	-	37	-	-	-	3	194	231	3	1235	25
Petroleum Refineries	69291	1313	222	-	2	257	1397	21359	159	22986	4234
Electr., CHP+Heat Plants	-	-	3	-	-	-	53	282	-	-	-
Pumped Storage (Elec.)	-	-	-	-	-	-	-	-	-	-	-
Other Energy Sector	-	18	1	-	-	-	275	66	-	7	60
Distribution Losses	-	18	88	-	15	1	45	65	-	-	-
FINAL CONSUMPTION	**2233**	**133341**	**592261**	**1390**	**138772**	**43019**	**535174**	**80523**	**102531**	**25311**	**124642**
INDUSTRY SECTOR	**2233**	**82025**	**2988**	**2**	**75**	**7566**	**56379**	**59709**	**102531**	**12106**	**1194**
Iron and Steel	-	1656	-	-	-	266	2284	6169	-	259	-
Chemical and Petrochem.	2233	58879	899	-	-	1383	6884	12580	102342	1067	1133
of which: Feedstocks	*1050*	*58351*	*896*	-	-	*1037*	*4067*	*1779*	*101318*	-	-
Non-Ferrous Metals	-	712	1	-	-	181	928	2222	25	24	-
Non-Metallic Minerals	-	1703	4	-	-	96	4784	8160	3	7062	38
Transport Equipment	-	734	30	2	62	4	1200	626	-	-	-
Machinery	-	2533	40	-	13	382	3027	1532	-	334	18
Mining and Quarrying	-	425	1	-	-	90	3787	970	-	-	2
Food and Tobacco	-	642	9	-	-	10	5456	6707	-	-	-
Paper, Pulp and Print	-	1048	3	-	-	117	1987	10205	-	131	-
Wood and Wood Products	-	151	1	-	-	2	2958	427	-	-	-
Construction	-	150	806	-	-	956	8402	547	-	-	2
Textile and Leather	-	766	1	-	-	24	1928	4235	-	29	-
Non-specified	-	12626	1193	-	-	4055	12754	5329	161	3200	1
TRANSPORT SECTOR	**-**	**9439**	**585644**	**1388**	**138121**	**67**	**307865**	**8823**	**-**	**-**	**-**
Air	-	-	27	1388	138121	-	-	-	-	-	-
Road	-	9378	582409	-	-	47	279536	84	-	-	-
Rail	-	-	-	-	-	13	16750	43	-	-	-
Pipeline Transport	-	-	-	-	-	-	11	-	-	-	-
Internal Navigation	-	1	3208	-	-	4	9044	8399	-	-	-
Non-specified	-	60	-	-	-	3	2524	297	-	-	-
OTHER SECTORS	**-**	**41877**	**3629**	**-**	**576**	**35386**	**170930**	**11991**	**-**	**115**	**19**
Agriculture	-	2460	2788	-	-	3463	44724	1596	-	10	-
Comm. and Publ. Services	-	9097	731	-	-	13402	47317	7108	-	13	12
Residential	-	29458	110	-	-	18344	76802	2995	-	92	-
Non-specified	-	862	-	-	576	177	2087	292	-	-	7
NON-ENERGY USE	**-**	**-**	**-**	**-**	**-**	**-**	**-**	**-**	**-**	**13090**	**123429**
in Industry/Transf./Energy	-	-	-	-	-	-	-	-	-	13090	112394
in Transport	-	-	-	-	-	-	-	-	-	-	7803
in Other Sectors	-	-	-	-	-	-	-	-	-	-	3232

OECD Total / Total OCDE : 1997

	Gas / *Gaz* (TJ)				Comb. Renew. & Waste / *En. Re. Comb. & Déchets* (TJ)				(GWh)	(TJ)
SUPPLY AND CONSUMPTION	Natural Gas	Gas Works	Coke Ovens	Blast Furnaces	Solid Biomass	Gas/Liquids from Biomass	Municipal Waste	Industrial Waste	Electricity	Heat
APPROVISIONNEMENT ET DEMANDE	*Gaz naturel*	*Usines à gaz*	*Cokeries*	*Hauts fourneaux*	*Biomasse solide*	*Gaz/Liquides tirés de biomasse*	*Déchets urbains*	*Déchets industriels*	*Electricité*	*Chaleur*
Production	40496052	955937	951739	1572186	5537883	228497	541185	336422	8905535	2287147
From Other Sources	-	14452	-	-	-	35489	-	12146	-	-
Imports	15469360	-	-	-	15592	3484	-	-	279574	145
Exports	-7245894	-	-	-	-242	-	-	-	-269510	-145
Intl. Marine Bunkers	-	-	-	-	-	-	-	-	-	-
Stock Changes	-258255	-	-	-	2368	-3076	-	57	-	-
DOMESTIC SUPPLY	**48461263**	**970389**	**951739**	**1572186**	**5555601**	**264394**	**541185**	**348625**	**8915599**	**2287147**
Transfers	-	-	-	-	-	-18700	-	-	-	-
Statistical Differences	-174971	-4406	-16960	-31863	-4535	-17811	-16430	-3988	-	-
TRANSFORMATION	**12966750**	**13454**	**237483**	**596644**	**1858108**	**117285**	**520945**	**273020**	**5065**	**-**
Electricity Plants	7904516	895	128553	376628	315875	58108	305724	62188	-	-
CHP Plants	3966909	23	89574	218986	1451473	19273	162637	135793	-	-
Heat Plants	223944	12536	2647	883	83170	1661	52584	75024	-	-
Blast Furnaces/Gas Works	811828	-	13709	-	-	-	-	-	-	-
Coke/Pat. Fuel/BKB Plants	3813	-	3000	147	5204	-	-	15	-	-
Petroleum Refineries	-	-	-	-	-	38243	-	-	-	-
Petrochemical Industry	-	-	-	-	-	-	-	-	-	-
Liquefaction	39392	-	-	-	-	-	-	-	-	-
Other Transform. Sector	16348	-	-	-	2386	-	-	-	5065	-
ENERGY SECTOR	**3984176**	**4939**	**275077**	**152633**	**3**	**1603**	**100**	**180**	**793273**	**135623**
Coal Mines	12691	-	2338	-	1	-	-	24	40703	16822
Oil and Gas Extraction	2868721	-	-	-	-	-	-	-	55559	15
Petroleum Refineries	1031570	3282	1862	149	-	1180	-	18	93135	55553
Electr., CHP+Heat Plants	927	-	-	-	2	423	100	-	473644	23475
Pumped Storage (Elec.)	-	-	-	-	-	-	-	-	82682	-
Other Energy Sector	70267	1657	270877	152484	-	-	-	138	47550	39758
Distribution Losses	158899	7500	2213	21611	-	-	60	-	553288	179795
FINAL CONSUMPTION	**31176467**	**940090**	**420006**	**769435**	**3692955**	**108995**	**3650**	**71437**	**7563973**	**1971729**
INDUSTRY SECTOR	**12895947**	**337504**	**417980**	**769315**	**1574620**	**7655**	**420**	**63215**	**3046035**	**625345**
Iron and Steel	1240724	59648	387100	764307	20	-	-	2792	347193	15886
Chemical and Petrochem.	5042949	59380	6476	2163	22042	3001	-	15703	572102	216279
of which: Feedstocks	*975518*	-	-	-	-	-	-	-	-	-
Non-Ferrous Metals	548686	15935	209	59	2424	2499	-	33	264847	3202
Non-Metallic Minerals	1181378	16417	12183	2134	21420	17	-	6659	155604	5523
Transport Equipment	276849	49	158	1	59	-	-	3	124547	25601
Machinery	771639	66658	6659	1	398	37	-	167	291798	15002
Mining and Quarrying	177901	1	2210	-	126	-	-	221	102672	4934
Food and Tobacco	1162948	43561	931	-	182973	914	-	518	204975	37179
Paper, Pulp and Print	1227832	28357	349	-	817830	83	-	2270	375492	51963
Wood and Wood Products	80738	173	-	-	428942	-	-	985	59383	7114
Construction	17928	196	473	-	267	1	-	9	13548	2014
Textile and Leather	331957	8162	140	-	5008	-	-	8	114638	18175
Non-specified	834418	38967	1092	650	93111	1103	420	33847	419236	222473
TRANSPORT SECTOR	**1096803**	**-**	**-**	**-**	**60**	**62033**	**-**	**8**	**104549**	**-**
Air	-	-	-	-	-	-	-	-	-	-
Road	24530	-	-	-	-	62033	-	-	-	-
Rail	-	-	-	-	5	-	-	8	89183	-
Pipeline Transport	1070639	-	-	-	-	-	-	-	4045	-
Internal Navigation	-	-	-	-	55	-	-	-	-	-
Non-specified	1634	-	-	-	-	-	-	-	11321	-
OTHER SECTORS	**17183717**	**602586**	**1072**	**-**	**2118275**	**39307**	**3230**	**8214**	**4413389**	**1346384**
Agriculture	238146	23	-	-	59703	61	20	-	75942	12284
Comm. and Publ. Services	5137443	213264	141	-	60098	36677	2452	604	2034941	298391
Residential	11264253	389247	931	-	1944806	10	758	-	2287235	901125
Non-specified	543875	52	-	-	53668	2559	-	7610	15271	134584
NON-ENERGY USE	**-**	**-**	**954**	**120**	**-**	**-**	**-**	**-**	**-**	**-**
in Industry/Transf./Energy	-	-	954	120	-	-	-	-	-	-
in Transport	-	-	-	-	-	-	-	-	-	-
in Other Sectors	-	-	-	-	-	-	-	-	-	-

Non-OECD Total / Total non-OCDE : 1996

SUPPLY AND CONSUMPTION / *APPROVISIONNEMENT ET DEMANDE*	Coal / *Charbon* (1000 tonnes)							Oil / *Pétrole* (1000 tonnes)			
	Coking Coal / *Charbon à coke*	Other Bit. Coal / *Autres charb. bit.*	Sub-Bit. Coal / *Charbon sous-bit.*	Lignite / *Lignite*	Peat / *Tourbe*	Oven and Gas Coke / *Coke de four/gaz*	Pat. Fuel and BKB / *Agg./briq. de lignite*	Crude Oil / *Pétrole brut*	NGL / *LGN*	Feed-stocks / *Produits d'aliment.*	Additives / *Additifs*
Production	295728	2046682	11679	278949	9155	222288	16784	2252578	102558	-	175
From Other Sources	-	-	-	-	-	-	-	725	-	-	-
Imports	50467	95083	385	6673	-	5607	355	393549	196	1314	110
Exports	-10369	-205682	-	-2330	-143	-9801	-101	-1316979	-41715	-	-
Intl. Marine Bunkers	-	-	-	-	-	-	-	-	-	-	-
Stock Changes	790	27632	-258	491	906	-1188	58	-12810	-32	1	-2
DOMESTIC SUPPLY	**336616**	**1963715**	**11806**	**283783**	**9918**	**216906**	**17096**	**1317063**	**61007**	**1315**	**283**
Transfers	-	-	-	-	-	-	-	-76	-43658	2029	-
Statistical Differences	-12511	79416	105	-2660	-289	-21878	-335	6647	-6	-	-
TRANSFORMATION	**313710**	**1123337**	**3466**	**242422**	**6702**	**120705**	**316**	**1303848**	**11313**	**3344**	**283**
Electricity Plants	-	851699	1979	116792	-	79	10	11290	-	-	-
CHP Plants	-	195289	1439	98543	2658	-	159	596	-	-	-
Heat Plants	5	23498	48	22909	1081	362	147	115	-	-	-
Blast Furnaces/Gas Works	-	11140	-	-	-	120229	-	-	307	-	-
Coke/Pat. Fuel/BKB Plants	313705	13185	-	4178	2963	35	-	-	-	-	-
Petroleum Refineries	-	-	-	-	-	-	-	1299098	11006	3344	283
Petrochemical Industry	-	-	-	-	-	-	-	-	-	-	-
Liquefaction	-	28526	-	-	-	-	-	-7251	-	-	-
Other Transform. Sector	-	-	-	-	-	-	-	-	-	-	-
ENERGY SECTOR	**16**	**66077**	**20**	**398**	**223**	**2125**	**66**	**6041**	**14**	**-**	**-**
Coal Mines	16	32289	20	339	109	551	38	14	-	-	-
Oil and Gas Extraction	-	2449	-	-	-	17	-	2535	2	-	-
Petroleum Refineries	-	-	-	-	-	-	-	3355	-	-	-
Electr., CHP+Heat Plants	-	23263	-	59	4	28	28	137	-	-	-
Pumped Storage (Elec.)	-	-	-	-	-	-	-	-	-	-	-
Other Energy Sector	-	8076	-	-	110	1529	-	-	12	-	-
Distribution Losses	-	30108	-	4479	151	44	1	6096	98	-	-
FINAL CONSUMPTION	**10379**	**823609**	**8425**	**33824**	**2553**	**72154**	**16378**	**7649**	**5918**	**-**	**-**
INDUSTRY SECTOR	**10144**	**562496**	**7851**	**21184**	**75**	**55917**	**136**	**4699**	**5918**	**-**	**-**
Iron and Steel	-	-	-	-	-	-	-	-	-	-	-
Chemical and Petrochem.	-	-	-	-	-	-	-	-	-	-	-
of which: Feedstocks	-	-	-	-	-	-	-	-	-	-	-
Non-Ferrous Metals	-	-	-	-	-	-	-	-	-	-	-
Non-Metallic Minerals	-	-	-	-	-	-	-	-	-	-	-
Transport Equipment	-	-	-	-	-	-	-	-	-	-	-
Machinery	-	-	-	-	-	-	-	-	-	-	-
Mining and Quarrying	-	-	-	-	-	-	-	-	-	-	-
Food and Tobacco	-	-	-	-	-	-	-	-	-	-	-
Paper, Pulp and Print	-	-	-	-	-	-	-	-	-	-	-
Wood and Wood Products	-	-	-	-	-	-	-	-	-	-	-
Construction	-	-	-	-	-	-	-	-	-	-	-
Textile and Leather	-	-	-	-	-	-	-	-	-	-	-
Non-specified	-	-	-	-	-	-	-	-	-	-	-
TRANSPORT SECTOR	**1**	**11066**	**5**	**498**	**1**	**4**	**31**	**9**	**-**	**-**	**-**
Air	-	-	-	-	-	-	-	-	-	-	-
Road	-	-	-	-	-	-	-	-	-	-	-
Rail	-	-	-	-	-	-	-	-	-	-	-
Pipeline Transport	-	-	-	-	-	-	-	-	-	-	-
Internal Navigation	-	-	-	-	-	-	-	-	-	-	-
Non-specified	-	-	-	-	-	-	-	-	-	-	-
OTHER SECTORS	**-**	**-**	**-**	**-**	**-**	**-**	**-**	**-**	**-**	**-**	**-**
Agriculture	-	-	-	-	-	-	-	-	-	-	-
Comm. and Publ. Services	-	-	-	-	-	-	-	-	-	-	-
Residential	-	-	-	-	-	-	-	-	-	-	-
Non-specified	-	-	-	-	-	-	-	-	-	-	-
NON-ENERGY USE	**-**	**24421**	**-**	**252**	**-**	**12957**	**3**	**1532**	**-**	**-**	**-**
in Industry/Trans./Energy	-	-	-	-	-	-	-	-	-	-	-
in Transport	-	-	-	-	-	-	-	-	-	-	-
in Other Sectors	-	-	-	-	-	-	-	-	-	-	-

Non-OECD Total / Total non-OCDE : 1996

SUPPLY AND CONSUMPTION / *APPROVISIONNEMENT ET DEMANDE*	Oil cont. / *Pétrole cont.* (1000 tonnes)										
	Refinery Gas / *Gaz de raffinerie*	LPG + Ethane / *GPL + éthane*	Motor Gasoline / *Essence moteur*	Aviation Gasoline / *Essence aviation*	Jet Fuel / *Carbu-réacteurs*	Kerosene / *Kérosène*	Gas/ Diesel / *Gazole*	Heavy Fuel Oil / *Fioul lourd*	Naphtha / *Naphta*	Petrol. Coke / *Coke de pétrole*	Other Prod. / *Autres prod.*
Production	23581	30779	207951	580	69641	46034	370216	363380	61734	2971	75910
From Other Sources	-	-	-	-	-	-	-	-	-	-	-
Imports	13	15359	28227	490	11164	12002	87450	96536	5485	285	8694
Exports	-	-21970	-42442	-122	-29164	-10597	-109699	-131703	-28296	-324	-18712
Intl. Marine Bunkers	-	-	-280	-	-	-	-7282	-45524	-	-	-28
Stock Changes	2	-132	239	16	-178	-453	2748	169	-164	37	-249
DOMESTIC SUPPLY	**23596**	**24036**	**193695**	**964**	**51463**	**46986**	**343433**	**282858**	**38759**	**2969**	**65615**
Transfers	98	30458	8621	-1	-62	-68	85	390	-505	-	1288
Statistical Differences	-66	10	-1269	-6	-2372	-2541	-45	-2983	-2046	40	3224
TRANSFORMATION	**2515**	**746**	**105**	**-**	**-**	**82**	**28058**	**145411**	**778**	**-**	**2532**
Electricity Plants	270	40	37	-	-	69	26996	86734	20	-	466
CHP Plants	1592	552	-	-	-	9	4	13868	-	-	1933
Heat Plants	653	35	-	-	-	4	1058	44083	41	-	133
Blast Furnaces/Gas Works	-	106	-	-	-	-	-	725	717	-	-
Coke/Pat. Fuel/BKB Plants	-	-	-	-	-	-	-	-	-	-	-
Petroleum Refineries	-	12	68	-	-	-	-	-	-	-	-
Petrochemical Industry	-	-	-	-	-	-	-	-	-	-	-
Liquefaction	-	-	-	-	-	-	-	-	-	-	-
Other Transform. Sector	-	1	-	-	-	-	-	1	-	-	-
ENERGY SECTOR	**19223**	**1245**	**1111**	**-**	**4**	**138**	**6251**	**26533**	**184**	**68**	**2469**
Coal Mines	-	-	3	-	-	61	1391	255	-	-	4
Oil and Gas Extraction	65	59	1081	-	-	8	2893	1118	-	-	11
Petroleum Refineries	19097	1128	18	-	1	28	1007	22590	47	52	2039
Electr., CHP+Heat Plants	2	4	9	-	-	12	769	1174	-	-	12
Pumped Storage (Elec.)	-	-	-	-	-	-	-	-	-	-	-
Other Energy Sector	59	54	-	-	3	29	191	1396	137	16	403
Distribution Losses	-	151	207	-	4	13	355	56	248	-	177
FINAL CONSUMPTION	**1890**	**52362**	**199624**	**957**	**49021**	**44144**	**308809**	**108265**	**34998**	**2941**	**64949**
INDUSTRY SECTOR	**1863**	**9161**	**4331**	**6**	**219**	**1689**	**38605**	**84305**	**34997**	**539**	**6110**
Iron and Steel	-	-	-	-	-	-	-	-	-	-	-
Chemical and Petrochem.	-	-	-	-	-	-	-	-	-	-	-
of which: Feedstocks	-	-	-	-	-	-	-	-	-	-	-
Non-Ferrous Metals	-	-	-	-	-	-	-	-	-	-	-
Non-Metallic Minerals	-	-	-	-	-	-	-	-	-	-	-
Transport Equipment	-	-	-	-	-	-	-	-	-	-	-
Machinery	-	-	-	-	-	-	-	-	-	-	-
Mining and Quarrying	-	-	-	-	-	-	-	-	-	-	-
Food and Tobacco	-	-	-	-	-	-	-	-	-	-	-
Paper, Pulp and Print	-	-	-	-	-	-	-	-	-	-	-
Wood and Wood Products	-	-	-	-	-	-	-	-	-	-	-
Construction	-	-	-	-	-	-	-	-	-	-	-
Textile and Leather	-	-	-	-	-	-	-	-	-	-	-
Non-specified	-	-	-	-	-	-	-	-	-	-	-
TRANSPORT SECTOR	**15**	**590**	**185266**	**890**	**48737**	**775**	**177651**	**7038**	**1**	**-**	**579**
Air	-	-	-	887	48732	717	2	-	-	-	-
Road	2	468	184618	3	-	4	156418	179	1	-	150
Rail	-	-	-	-	-	-	-	-	-	-	-
Pipeline Transport	-	-	-	-	-	-	-	-	-	-	-
Internal Navigation	-	-	-	-	-	-	-	-	-	-	-
Non-specified	-	-	-	-	-	-	-	-	-	-	-
OTHER SECTORS	**-**	**-**	**-**	**-**	**-**	**-**	**-**	**-**	**-**	**-**	**-**
Agriculture	-	-	-	-	-	-	-	-	-	-	-
Comm. and Publ. Services	-	-	-	-	-	-	-	-	-	-	-
Residential	-	-	-	-	-	-	-	-	-	-	-
Non-specified	-	-	-	-	-	-	-	-	-	-	-
NON-ENERGY USE	**-**	**1935**	**269**	**3**	**1**	**44**	**6**	**12**	**-**	**2402**	**56638**
in Industry/Transf./Energy	-	-	-	-	-	-	-	-	-	-	-
in Transport	-	-	-	-	-	-	-	-	-	-	-
in Other Sectors	-	-	-	-	-	-	-	-	-	-	-

Non-OECD Total / Total non-OCDE : 1996

	Gas / *Gaz* (TJ)				Comb. Renew. & Waste / *En. Re. Comb. & Déchets* (TJ)				(GWh)	(TJ)
SUPPLY AND CONSUMPTION	Natural Gas	Gas Works	Coke Ovens	Blast Furnaces	Solid Biomass	Gas/Liquids from Biomass	Municipal Waste	Industrial Waste	Electricity	Heat
APPROVISIONNEMENT ET DEMANDE	*Gaz naturel*	*Usines à gaz*	*Cokeries*	*Hauts fourneaux*	*Biomasse solide*	*Gaz/Liquides tirés de biomasse*	*Déchets urbains*	*Déchets industriels*	*Electricité*	*Chaleur*
Production	48448940	221848	777526	856994	37011527	326597	-	72974	4944706	9706060
From Other Sources	-	11580	-	-	3341	-	-	-	-	-
Imports	5582776	-	-	-	4644	27429	-	209	146225	-
Exports	-13568175	-	-	-	-28462	-4523	-	-6799	-153120	-
Intl. Marine Bunkers	-	-	-	-	-	-	-	-	-	-
Stock Changes	-552575	-	-	-	-6165	-2722	-	-25	-	-
DOMESTIC SUPPLY	**39910966**	**233428**	**777526**	**856994**	**36984884**	**346781**	**-**	**66359**	**4937811**	**9706060**
Transfers	-	-	-	-	-	-	-	-	-	-
Statistical Differences	29136	-530	-1004	40838	995	-	-	-3605	7425	27067
TRANSFORMATION	**16954694**	**4490**	**95509**	**161201**	**2279508**	**56**	**-**	**18083**	**1293**	**-**
Electricity Plants	5853249	3022	31580	81173	164806	-	-	-	-	-
CHP Plants	7916667	1318	4931	48664	91903	-	-	10314	-	-
Heat Plants	3070321	150	55880	31364	185343	56	-	7769	-	-
Blast Furnaces/Gas Works	38078	-	3118	-	1959	-	-	-	-	-
Coke/Pat. Fuel/BKB Plants	-	-	-	-	66	-	-	-	-	-
Petroleum Refineries	3200	-	-	-	105	-	-	-	-	-
Petrochemical Industry	-	-	-	-	-	-	-	-	-	-
Liquefaction	71814	-	-	-	-	-	-	-	-	-
Other Transform. Sector	1365	-	-	-	1835326	-	-	-	1293	-
ENERGY SECTOR	**6043174**	**6749**	**64126**	**9428**	**294**	**1294**	**-**	**1788**	**533951**	**838887**
Coal Mines	1123924	-	254	-	77	-	-	-	66494	69749
Oil and Gas Extraction	3912416	-	-	-	-	-	-	-	73720	167663
Petroleum Refineries	571015	53	-	-	-	-	-	-	63281	432161
Electr., CHP+Heat Plants	12756	459	2713	5230	217	-	-	-	292987	65674
Pumped Storage (Elec.)	-	-	-	-	-	-	-	-	9763	-
Other Energy Sector	423063	6237	61159	4198	-	1294	-	1788	27706	103640
Distribution Losses	809463	1588	4061	10433	7494	-	-	29	592233	529148
FINAL CONSUMPTION	**16132771**	**220071**	**612826**	**716770**	**34698583**	**345431**	**-**	**42854**	**3817759**	**8365092**
INDUSTRY SECTOR	**8445947**	**120242**	**545241**	**714272**	**3842195**	**-**	**-**	**36990**	**1848654**	**3662940**
Iron and Steel	-	-	-	-	-	-	-	-	-	-
Chemical and Petrochem.	-	-	-	-	-	-	-	-	-	-
of which: Feedstocks	-	-	-	-	-	-	-	-	-	-
Non-Ferrous Metals	-	-	-	-	-	-	-	-	-	-
Non-Metallic Minerals	-	-	-	-	-	-	-	-	-	-
Transport Equipment	-	-	-	-	-	-	-	-	-	-
Machinery	-	-	-	-	-	-	-	-	-	-
Mining and Quarrying	-	-	-	-	-	-	-	-	-	-
Food and Tobacco	-	-	-	-	-	-	-	-	-	-
Paper, Pulp and Print	-	-	-	-	-	-	-	-	-	-
Wood and Wood Products	-	-	-	-	-	-	-	-	-	-
Construction	-	-	-	-	-	-	-	-	-	-
Textile and Leather	-	-	-	-	-	-	-	-	-	-
Non-specified	-	-	-	-	-	-	-	-	-	-
TRANSPORT SECTOR	**822824**	**31**	**-**	**-**	**1027**	**295172**	**-**	**29**	**115041**	**-**
Air	-	14	-	-	-	-	-	-	178	-
Road	61340	-	-	-	-	295172	-	-	169	-
Rail	-	-	-	-	-	-	-	-	-	-
Pipeline Transport	-	-	-	-	-	-	-	-	-	-
Internal Navigation	-	-	-	-	-	-	-	-	-	-
Non-specified	-	-	-	-	-	-	-	-	-	-
OTHER SECTORS	**-**	**-**	**-**	**-**	**-**	**-**	**-**	**-**	**-**	**-**
Agriculture	-	-	-	-	-	-	-	-	-	-
Comm. and Publ. Services	-	-	-	-	-	-	-	-	-	-
Residential	-	-	-	-	-	-	-	-	-	-
Non-specified	-	-	-	-	-	-	-	-	-	-
NON-ENERGY USE	**-**	**-**	**3283**	**567**	**-**	**-**	**-**	**-**	**-**	**-**
in Industry/Transf./Energy	-	-	-	-	-	-	-	-	-	-
in Transport	-	-	-	-	-	-	-	-	-	-
in Other Sectors	-	-	-	-	-	-	-	-	-	-

Non-OECD Total / Total non-OCDE : 1997

	Coal / *Charbon* (1000 tonnes)							Oil / *Pétrole* (1000 tonnes)			
SUPPLY AND CONSUMPTION ***APPROVISIONNEMENT ET DEMANDE***	Coking Coal *Charbon à coke*	Other Bit. Coal *Autres charb. bit.*	Sub-Bit. Coal *Charbon sous-bit.*	Lignite *Lignite*	Peat *Tourbe*	Oven and Gas Coke *Coke de four/gaz*	Pat. Fuel and BKB *Agg./briq. de lignite*	Crude Oil *Pétrole brut*	NGL *LGN*	Feed-stocks *Produits d'aliment.*	Additives *Additifs*
Production	295206	2048742	11034	267093	7815	222124	15882	2335686	108318	-	178
From Other Sources	-	-	-	-	-	-	-	645	-	-	-
Imports	50018	92123	300	4493	9	5279	235	414118	15	2069	111
Exports	-9639	-212252	-1	-1866	-102	-12998	-129	-1379826	-41797	-	-
Intl. Marine Bunkers	-	-	-	-	-	-	-	-	-	-	-
Stock Changes	-858	-13933	143	-1064	10	242	-8	-12710	19	-78	-1
DOMESTIC SUPPLY	**334727**	**1914680**	**11476**	**268656**	**7732**	**214647**	**15980**	**1357913**	**66555**	**1991**	**288**
Transfers	-	-	-	-	-	-	-	-81	-43431	2485	-
Statistical Differences	-11793	68506	119	143	-	-7125	6	6213	-3347	-	-
TRANSFORMATION	**313512**	**1125068**	**3620**	**234486**	**5518**	**130060**	**307**	**1347697**	**12622**	**4476**	**288**
Electricity Plants	-	871219	2180	115182	-	3	18	11921	-	-	-
CHP Plants	-	180381	1391	93589	1510	-	148	598	-	-	-
Heat Plants	-	20226	49	21839	1064	375	141	113	-	-	-
Blast Furnaces/Gas Works	-	12949	-	-	-	129650	-	-	219	-	-
Coke/Pat. Fuel/BKB Plants	313512	12449	-	3876	2944	32	-	-	-	-	-
Petroleum Refineries	-	-	-	-	-	-	-	1342316	12403	4476	288
Petrochemical Industry	-	-	-	-	-	-	-	-	-	-	-
Liquefaction	-	27766	-	-	-	-	-	-7251	-	-	-
Other Transform. Sector	-	78	-	-	-	-	-	-	-	-	-
ENERGY SECTOR	**42**	**73901**	**17**	**323**	**175**	**2090**	**59**	**6029**	**38**	**-**	**-**
Coal Mines	13	35664	17	282	45	548	34	15	-	-	-
Oil and Gas Extraction	-	2774	-	-	-	17	-	2669	1	-	-
Petroleum Refineries	-	-	-	-	-	-	-	3191	-	-	-
Electr., CHP+Heat Plants	-	26316	-	41	-	27	24	151	-	-	-
Pumped Storage (Elec.)	-	-	-	-	-	-	-	-	-	-	-
Other Energy Sector	29	9147	-	-	130	1498	1	3	37	-	-
Distribution Losses	-	28133	-	4195	218	41	-	2535	28	-	-
FINAL CONSUMPTION	**9380**	**756084**	**7958**	**29795**	**1821**	**75331**	**15620**	**7784**	**7089**	**-**	**-**
INDUSTRY SECTOR	**9377**	**524713**	**7590**	**19580**	**55**	**58397**	**161**	**4789**	**7089**	**-**	**-**
Iron and Steel	-	-	-	-	-	-	-	-	-	-	-
Chemical and Petrochem.	-	-	-	-	-	-	-	-	-	-	-
of which: Feedstocks	-	-	-	-	-	-	-	-	-	-	-
Non-Ferrous Metals	-	-	-	-	-	-	-	-	-	-	-
Non-Metallic Minerals	-	-	-	-	-	-	-	-	-	-	-
Transport Equipment	-	-	-	-	-	-	-	-	-	-	-
Machinery	-	-	-	-	-	-	-	-	-	-	-
Mining and Quarrying	-	-	-	-	-	-	-	-	-	-	-
Food and Tobacco	-	-	-	-	-	-	-	-	-	-	-
Paper, Pulp and Print	-	-	-	-	-	-	-	-	-	-	-
Wood and Wood Products	-	-	-	-	-	-	-	-	-	-	-
Construction	-	-	-	-	-	-	-	-	-	-	-
Textile and Leather	-	-	-	-	-	-	-	-	-	-	-
Non-specified	-	-	-	-	-	-	-	-	-	-	-
TRANSPORT SECTOR	**1**	**13060**	**4**	**490**	**1**	**4**	**26**	**9**	**-**	**-**	**-**
Air	-	-	-	-	-	-	-	-	-	-	-
Road	-	-	-	-	-	-	-	-	-	-	-
Rail	-	-	-	-	-	-	-	-	-	-	-
Pipeline Transport	-	-	-	-	-	-	-	-	-	-	-
Internal Navigation	-	-	-	-	-	-	-	-	-	-	-
Non-specified	-	-	-	-	-	-	-	-	-	-	-
OTHER SECTORS	**-**	**-**	**-**	**-**	**-**	**-**	**-**	**-**	**-**	**-**	**-**
Agriculture	-	-	-	-	-	-	-	-	-	-	-
Comm. and Publ. Services	-	-	-	-	-	-	-	-	-	-	-
Residential	-	-	-	-	-	-	-	-	-	-	-
Non-specified	-	-	-	-	-	-	-	-	-	-	-
NON-ENERGY USE	**-**	**22679**	**-**	**234**	**-**	**13313**	**3**	**1537**	**-**	**-**	**-**
in Industry/Trans./Energy	-	-	-	-	-	-	-	-	-	-	-
in Transport	-	-	-	-	-	-	-	-	-	-	-
in Other Sectors	-	-	-	-	-	-	-	-	-	-	-

Non-OECD Total / Total non-OCDE : 1997

SUPPLY AND CONSUMPTION / *APPROVISIONNEMENT ET DEMANDE*	Oil cont. / *Pétrole cont.* (1000 tonnes)										
	Refinery Gas / *Gaz de raffinerie*	LPG + Ethane / *GPL + éthane*	Motor Gasoline / *Essence moteur*	Aviation Gasoline / *Essence aviation*	Jet Fuel / *Carbu- réacteurs*	Kerosene / *Kérosène*	Gas/ Diesel / *Gazole*	Heavy Fuel Oil / *Fioul lourd*	Naphtha / *Naphta*	Petrol. Coke / *Coke de pétrole*	Other Prod. / *Autres prod.*
Production	25189	32455	217031	521	69796	44473	384385	364907	62872	2922	80643
From Other Sources	-	-	-	-	-	-	-	-	-	-	-
Imports	11	16255	26032	517	13143	13300	97022	99907	7959	669	10694
Exports	-	-24169	-43204	-27	-29295	-9525	-108856	-134068	-27862	-289	-19213
Intl. Marine Bunkers	-	-	-274	-	-	-	-8140	-46595	-	-	-4
Stock Changes	3	-248	-54	-18	217	-471	-1134	420	-133	-73	-99
DOMESTIC SUPPLY	**25203**	**24293**	**199531**	**993**	**53861**	**47777**	**363277**	**284571**	**42836**	**3229**	**72021**
Transfers	91	32383	7316	-1	-65	-93	138	329	-1537	-	1569
Statistical Differences	583	-658	-2168	-	-2600	-2881	647	-2228	-1117	109	2562
TRANSFORMATION	**2789**	**644**	**107**	**-**	**-**	**89**	**33855**	**139853**	**700**	**-**	**2609**
Electricity Plants	320	7	38	-	-	72	32709	86614	20	-	450
CHP Plants	1710	594	-	-	-	9	3	12204	-	-	1959
Heat Plants	759	11	-	-	-	8	1143	40250	41	-	200
Blast Furnaces/Gas Works	-	19	-	-	-	-	-	757	639	-	-
Coke/Pat. Fuel/BKB Plants	-	-	-	-	-	-	-	-	-	-	-
Petroleum Refineries	-	12	69	-	-	-	-	-	-	-	-
Petrochemical Industry	-	-	-	-	-	-	-	27	-	-	-
Liquefaction	-	-	-	-	-	-	-	-	-	-	-
Other Transform. Sector	-	1	-	-	-	-	-	1	-	-	-
ENERGY SECTOR	**20863**	**1535**	**1073**	**-**	**4**	**237**	**6187**	**28021**	**45**	**120**	**2784**
Coal Mines	-	-	2	-	-	100	1384	269	-	-	-
Oil and Gas Extraction	94	75	1020	-	-	15	2790	1410	-	-	-
Petroleum Refineries	20592	1393	26	-	1	43	1099	23583	41	15	2288
Electr., CHP+Heat Plants	3	5	9	-	-	25	703	1467	-	-	-
Pumped Storage (Elec.)	-	-	-	-	-	-	-	-	-	-	-
Other Energy Sector	174	62	16	-	3	54	211	1292	4	105	496
Distribution Losses	2	247	173	-	2	6	109	52	231	-	276
FINAL CONSUMPTION	**2223**	**53592**	**203326**	**992**	**51190**	**44471**	**323911**	**114746**	**39206**	**3218**	**70483**
INDUSTRY SECTOR	**2194**	**9188**	**4031**	**5**	**207**	**1659**	**39752**	**86848**	**39206**	**501**	**6657**
Iron and Steel	-	-	-	-	-	-	-	-	-	-	-
Chemical and Petrochem.	-	-	-	-	-	-	-	-	-	-	-
of which: Feedstocks	-	-	-	-	-	-	-	-	-	-	-
Non-Ferrous Metals	-	-	-	-	-	-	-	-	-	-	-
Non-Metallic Minerals	-	-	-	-	-	-	-	-	-	-	-
Transport Equipment	-	-	-	-	-	-	-	-	-	-	-
Machinery	-	-	-	-	-	-	-	-	-	-	-
Mining and Quarrying	-	-	-	-	-	-	-	-	-	-	-
Food and Tobacco	-	-	-	-	-	-	-	-	-	-	-
Paper, Pulp and Print	-	-	-	-	-	-	-	-	-	-	-
Wood and Wood Products	-	-	-	-	-	-	-	-	-	-	-
Construction	-	-	-	-	-	-	-	-	-	-	-
Textile and Leather	-	-	-	-	-	-	-	-	-	-	-
Non-specified	-	-	-	-	-	-	-	-	-	-	-
TRANSPORT SECTOR	**16**	**629**	**189747**	**924**	**50910**	**886**	**185522**	**10905**	**-**	**-**	**649**
Air	-	-	2	921	50910	828	2	-	-	-	1
Road	2	501	188756	3	-	6	163241	282	-	-	192
Rail	-	-	-	-	-	-	-	-	-	-	-
Pipeline Transport	-	-	-	-	-	-	-	-	-	-	-
Internal Navigation	-	-	-	-	-	-	-	-	-	-	-
Non-specified	-	-	-	-	-	-	-	-	-	-	-
OTHER SECTORS	**-**	**-**	**-**	**-**	**-**	**-**	**-**	**-**	**-**	**-**	**-**
Agriculture	-	-	-	-	-	-	-	-	-	-	-
Comm. and Publ. Services	-	-	-	-	-	-	-	-	-	-	-
Residential	-	-	-	-	-	-	-	-	-	-	-
Non-specified	-	-	-	-	-	-	-	-	-	-	-
NON-ENERGY USE	**-**	**1509**	**273**	**2**	**1**	**25**	**6**	**11**	**-**	**2717**	**61285**
in Industry/Transf./Energy	-	-	-	-	-	-	-	-	-	-	-
in Transport	-	-	-	-	-	-	-	-	-	-	-
in Other Sectors	-	-	-	-	-	-	-	-	-	-	-

Non-OECD Total / Total non-OCDE : 1997

	Gas / *Gaz* (TJ)				Comb. Renew. & Waste / *En. Re. Comb. & Déchets* (TJ)				(GWh)	(TJ)
SUPPLY AND CONSUMPTION	Natural Gas	Gas Works	Coke Ovens	Blast Furnaces	Solid Biomass	Gas/Liquids from Biomass	Municipal Waste	Industrial Waste	Electricity	Heat
APPROVISIONNEMENT ET DEMANDE	*Gaz naturel*	*Usines à gaz*	*Cokeries*	*Hauts fourneaux*	*Biomasse solide*	*Gaz/Liquides tirés de biomasse*	*Déchets urbains*	*Déchets industriels*	*Electricité*	*Chaleur*
Production	48696979	236462	834539	936471	37381713	393771	-	72339	5115805	8996535
From Other Sources	-	8260	-	-	3341	-	-	-	-	-
Imports	5039232	1	-	-	4450	18342	-	-	149656	-
Exports	-13260532	-	-	-	-33101	-28685	-	-6799	-156033	-
Intl. Marine Bunkers	-	-	-	-	-	-	-	-	-	-
Stock Changes	-3903	-	-	-	-5205	-47027	-	-	-	-
DOMESTIC SUPPLY	**40471776**	**244723**	**834539**	**936471**	**37351197**	**336401**	**-**	**65540**	**5109428**	**8996535**
Transfers	-	-	-	-	-	-	-	-	-	-
Statistical Differences	119611	186	-27400	47379	-3554	-	-	-3605	-4340	19585
TRANSFORMATION	**17073159**	**5981**	**108999**	**173697**	**2343369**	**58**	**-**	**17451**	**1902**	**-**
Electricity Plants	6175286	4354	39345	93987	181176	-	-	-	-	-
CHP Plants	7820076	1455	5477	46688	91348	-	-	10438	-	-
Heat Plants	2961651	172	60855	33022	186630	58	-	7013	-	-
Blast Furnaces/Gas Works	39574	-	3322	-	1932	-	-	-	-	-
Coke/Pat. Fuel/BKB Plants	-	-	-	-	38	-	-	-	-	-
Petroleum Refineries	3434	-	-	-	105	-	-	-	-	-
Petrochemical Industry	-	-	-	-	-	-	-	-	-	-
Liquefaction	71814	-	-	-	-	-	-	-	-	-
Other Transform. Sector	1324	-	-	-	1882140	-	-	-	1902	-
ENERGY SECTOR	**6489115**	**16697**	**68977**	**13111**	**226**	**1294**	**-**	**1788**	**572887**	**783433**
Coal Mines	1121106	-	270	-	173	-	-	-	74140	62378
Oil and Gas Extraction	4296673	-	-	-	-	-	-	-	80096	150184
Petroleum Refineries	621978	7	-	-	-	-	-	-	67677	406178
Electr., CHP+Heat Plants	11727	1114	2883	5141	53	-	-	-	310588	68482
Pumped Storage (Elec.)	-	-	-	-	-	-	-	-	10141	-
Other Energy Sector	437631	15576	65824	7970	-	1294	-	1788	30245	96211
Distribution Losses	781187	1292	3380	5708	7521	-	-	29	621279	480358
FINAL CONSUMPTION	**16247926**	**220939**	**625783**	**791334**	**34997897**	**335049**	**-**	**42667**	**3909020**	**7752329**
INDUSTRY SECTOR	**8500548**	**122644**	**551744**	**788879**	**3942336**	**-**	**-**	**36810**	**1867393**	**3471167**
Iron and Steel	-	-	-	-	-	-	-	-	-	-
Chemical and Petrochem.	-	-	-	-	-	-	-	-	-	-
of which: Feedstocks	-	-	-	-	-	-	-	-	-	-
Non-Ferrous Metals	-	-	-	-	-	-	-	-	-	-
Non-Metallic Minerals	-	-	-	-	-	-	-	-	-	-
Transport Equipment	-	-	-	-	-	-	-	-	-	-
Machinery	-	-	-	-	-	-	-	-	-	-
Mining and Quarrying	-	-	-	-	-	-	-	-	-	-
Food and Tobacco	-	-	-	-	-	-	-	-	-	-
Paper, Pulp and Print	-	-	-	-	-	-	-	-	-	-
Wood and Wood Products	-	-	-	-	-	-	-	-	-	-
Construction	-	-	-	-	-	-	-	-	-	-
Textile and Leather	-	-	-	-	-	-	-	-	-	-
Non-specified	-	-	-	-	-	-	-	-	-	-
TRANSPORT SECTOR	**821765**	**29**	**-**	**-**	**1244**	**284521**	**-**	**29**	**117749**	**-**
Air	-	12	-	-	-	-	-	-	212	-
Road	61238	-	-	-	-	284521	-	-	931	-
Rail	-	-	-	-	-	-	-	-	-	-
Pipeline Transport	-	-	-	-	-	-	-	-	-	-
Internal Navigation	-	-	-	-	-	-	-	-	-	-
Non-specified	-	-	-	-	-	-	-	-	-	-
OTHER SECTORS	**-**	**-**	**-**	**-**	**-**	**-**	**-**	**-**	**-**	**-**
Agriculture	-	-	-	-	-	-	-	-	-	-
Comm. and Publ. Services	-	-	-	-	-	-	-	-	-	-
Residential	-	-	-	-	-	-	-	-	-	-
Non-specified	-	-	-	-	-	-	-	-	-	-
NON-ENERGY USE	**-**	**-**	**3368**	**557**	**-**	**-**	**-**	**-**	**-**	**-**
in Industry/Transf./Energy	-	-	-	-	-	-	-	-	-	-
in Transport	-	-	-	-	-	-	-	-	-	-
in Other Sectors	-	-	-	-	-	-	-	-	-	-

Africa / Afrique

SUPPLY AND CONSUMPTION 1996	Coal (1000 tonnes)							Oil (1000 tonnes)			
	Coking Coal	Other Bit. Coal	Sub-Bit. Coal	Lignite	Peat	Oven and Gas Coke	Pat. Fuel and BKB	Crude Oil	NGL	Feed-stocks	Additives
Production	4129	208897	-	-	12	5826	-	329471	30468	-	-
Imports	2726	2987	-	-	-	285	85	29928	-	-	-
Exports	-	-60382	-	-	-	-503	-	-243101	-16303	-	-
Intl. Marine Bunkers	-	-	-	-	-	-	-	-	-	-	-
Stock Changes	1	2837	-	-	-	82	-	-8936	-26	-	-
DOMESTIC SUPPLY	**6856**	**154339**	**-**	**-**	**12**	**5690**	**85**	**107362**	**14139**	**-**	**-**
Transfers and Stat. Diff.	-45	139	-	-	-1	-323	-	8482	-8678	-	-
TRANSFORMATION	**6811**	**130065**	**-**	**-**	**-**	**4152**	**-**	**113195**	**5399**	**-**	**-**
Electricity and CHP Plants	-	96533	-	-	-	-	-	-	-	-	-
Petroleum Refineries	-	-	-	-	-	-	-	120446	5399	-	-
Other Transform. Sector	6811	33532	-	-	-	4152	-	-7251	-	-	-
ENERGY SECTOR	**-**	**-**	**-**	**-**	**-**	**41**	**-**	**2227**	**-**	**-**	**-**
DISTRIBUTION LOSSES	**-**	**-**	**-**	**-**	**-**	**-**	**-**	**394**	**62**	**-**	**-**
FINAL CONSUMPTION	**-**	**24413**	**-**	**-**	**11**	**1174**	**85**	**28**	**-**	**-**	**-**
INDUSTRY SECTOR	**-**	**12074**	**-**	**-**	**-**	**1172**	**-**	**28**	**-**	**-**	**-**
Iron and Steel	-	-	-	-	-	-	-	-	-	-	-
Chemical and Petrochem.	-	-	-	-	-	-	-	-	-	-	-
Non-Metallic Minerals	-	-	-	-	-	-	-	-	-	-	-
Non-specified	-	-	-	-	-	-	-	-	-	-	-
TRANSPORT SECTOR	**-**	**53**	**-**	**-**	**-**	**1**	**-**	**-**	**-**	**-**	**-**
Air	-	-	-	-	-	-	-	-	-	-	-
Road	-	-	-	-	-	-	-	-	-	-	-
Non-specified	-	-	-	-	-	-	-	-	-	-	-
OTHER SECTORS	**-**	**-**	**-**	**-**	**-**	**-**	**-**	**-**	**-**	**-**	**-**
Agriculture	-	-	-	-	-	-	-	-	-	-	-
Comm. and Publ. Services	-	-	-	-	-	-	-	-	-	-	-
Residential	-	-	-	-	-	-	-	-	-	-	-
Non-specified	-	-	-	-	-	-	-	-	-	-	-
NON-ENERGY USE	**-**	**7625**	**-**	**-**	**-**	**-**	**-**	**-**	**-**	**-**	**-**

APPROVISIONNEMENT ET DEMANDE 1997	*Charbon* (1000 tonnes)							*Pétrole* (1000 tonnes)			
	Charbon à coke	*Autres charb. bit.*	*Charbon sous-bit.*	*Lignite*	*Tourbe*	*Coke de four/gaz*	*Agg./briq. de lignite*	*Pétrole brut*	*LGN*	*Produits d'aliment.*	*Additifs*
Production	4243	222099	-	-	12	5625	-	339462	31779	-	-
Imports	2517	3017	-	-	-	344	85	29059	-	-	-
Exports	-	-64319	-	-	-	-496	-	-251070	-16029	-	-
Intl. Marine Bunkers	-	-	-	-	-	-	-	-	-	-	-
Stock Changes	-108	-2445	-	-	-	-60	-	-6963	-	-	-
DOMESTIC SUPPLY	**6652**	**158352**	**-**	**-**	**12**	**5413**	**85**	**110488**	**15750**	**-**	**-**
Transfers and Stat. Diff.	-9	1861	-	-	-	-326	-	6592	-9934	-	-
TRANSFORMATION	**6643**	**134313**	**-**	**-**	**-**	**3934**	**-**	**114796**	**5816**	**-**	**-**
Electricity and CHP Plants	-	101160	-	-	-	-	-	-	-	-	-
Petroleum Refineries	-	-	-	-	-	-	-	122047	5816	-	-
Other Transform. Sector	6643	33153	-	-	-	3934	-	-7251	-	-	-
ENERGY SECTOR	**-**	**-**	**-**	**-**	**-**	**41**	**-**	**1939**	**-**	**-**	**-**
DISTRIBUTION LOSSES	**-**	**-**	**-**	**-**	**-**	**-**	**-**	**317**	**-**	**-**	**-**
FINAL CONSUMPTION	**-**	**25900**	**-**	**-**	**12**	**1112**	**85**	**28**	**-**	**-**	**-**
INDUSTRY SECTOR	**-**	**13540**	**-**	**-**	**-**	**1110**	**-**	**28**	**-**	**-**	**-**
Iron and Steel	-	-	-	-	-	-	-	-	-	-	-
Chemical and Petrochem.	-	-	-	-	-	-	-	-	-	-	-
Non-Metallic Minerals	-	-	-	-	-	-	-	-	-	-	-
Non-specified	-	-	-	-	-	-	-	-	-	-	-
TRANSPORT SECTOR	**-**	**33**	**-**	**-**	**-**	**1**	**-**	**-**	**-**	**-**	**-**
Air	-	-	-	-	-	-	-	-	-	-	-
Road	-	-	-	-	-	-	-	-	-	-	-
Non-specified	-	-	-	-	-	-	-	-	-	-	-
OTHER SECTORS	**-**	**-**	**-**	**-**	**-**	**-**	**-**	**-**	**-**	**-**	**-**
Agriculture	-	-	-	-	-	-	-	-	-	-	-
Comm. and Publ. Services	-	-	-	-	-	-	-	-	-	-	-
Residential	-	-	-	-	-	-	-	-	-	-	-
Non-specified	-	-	-	-	-	-	-	-	-	-	-
NON-ENERGY USE	**-**	**7617**	**-**	**-**	**-**	**-**	**-**	**-**	**-**	**-**	**-**

Africa / Afrique

SUPPLY AND CONSUMPTION 1996	Oil cont. (1000 tonnes)										
	Refinery Gas	LPG + Ethane	Motor Gasoline	Aviation Gasoline	Jet Fuel	Kerosene	Gas/ Diesel	Heavy Fuel Oil	Naphtha	Petrol. Coke	Other Prod.
Production	2072	2230	23168	165	6491	6297	31648	36644	7689	156	3844
Imports	-	1319	2877	282	1610	776	7168	2885	24	2	1027
Exports	-	-4596	-1865	-99	-1975	-412	-6711	-11291	-8154	-	-330
Intl. Marine Bunkers	-	-	-	-	-	-	-1547	-6987	-	-	-20
Stock Changes	-	4	-367	-	80	-19	-234	-69	23	-	42
DOMESTIC SUPPLY	**2072**	**-1043**	**23813**	**348**	**6206**	**6642**	**30324**	**21182**	**-418**	**158**	**4563**
Transfers and Stat. Diff.	-	6617	-2265	-3	-269	-1818	-1780	-2312	1400	-	1292
TRANSFORMATION	**-**	**-**	**-**	**-**	**-**	**66**	**3380**	**10279**	**-**	**-**	**-**
Electricity and CHP Plants	-	-	-	-	-	66	3380	10279	-	-	-
Petroleum Refineries	-	-	-	-	-	-	-	-	-	-	-
Other Transform. Sector	-	-	-	-	-	-	-	-	-	-	-
ENERGY SECTOR	**2001**	**-**	**-**	**-**	**-**	**-**	**210**	**1181**	**-**	**-**	**-**
DISTRIBUTION LOSSES	**-**	**48**	**-**	**-**	**1**	**-**	**1**	**-**	**-**	**-**	**-**
FINAL CONSUMPTION	**71**	**5526**	**21548**	**345**	**5936**	**4758**	**24953**	**7410**	**982**	**158**	**5855**
INDUSTRY SECTOR	**71**	**393**	**38**	**-**	**-**	**283**	**4719**	**6295**	**982**	**2**	**30**
Iron and Steel	-	-	-	-	-	-	-	-	-	-	-
Chemical and Petrochem.	-	-	-	-	-	-	-	-	-	-	-
Non-Metallic Minerals	-	-	-	-	-	-	-	-	-	-	-
Non-specified	-	-	-	-	-	-	-	-	-	-	-
TRANSPORT SECTOR	**-**	**72**	**21288**	**345**	**5936**	**13**	**11962**	**25**	**-**	**-**	**-**
Air	-	-	-	342	5936	-	-	-	-	-	-
Road	-	72	21246	3	-	-	11505	-	-	-	-
Non-specified	-	-	-	-	-	-	-	-	-	-	-
OTHER SECTORS	**-**	**-**	**-**	**-**	**-**	**-**	**-**	**-**	**-**	**-**	**-**
Agriculture	-	-	-	-	-	-	-	-	-	-	-
Comm. and Publ. Services	-	-	-	-	-	-	-	-	-	-	-
Residential	-	-	-	-	-	-	-	-	-	-	-
Non-specified	-	-	-	-	-	-	-	-	-	-	-
NON-ENERGY USE	**-**	**189**	**-**	**-**	**-**	**-**	**-**	**-**	**-**	**156**	**5761**

APPROVISIONNEMENT ET DEMANDE 1997	*Pétrole cont.* (1000 tonnes)										
	Gaz de raffinerie	*GPL + éthane*	*Essence moteur*	*Essence aviation*	*Carbu- réacteurs*	*Kérosène*	*Gazole*	*Fioul lourd*	*Naphta*	*Coke de pétrole*	*Autres prod.*
Production	2052	2419	22406	80	6754	6138	31433	36611	9043	122	3778
Imports	-	1636	2768	277	1662	804	6954	3026	24	2	911
Exports	-	-5720	-1152	-	-1920	-283	-3856	-11094	-9130	-	-143
Intl. Marine Bunkers	-	-	-	-	-	-	-1541	-6226	-	-	-
Stock Changes	-	6	-86	-	79	13	-260	55	-23	-	55
DOMESTIC SUPPLY	**2052**	**-1659**	**23936**	**357**	**6575**	**6672**	**32730**	**22372**	**-86**	**124**	**4601**
Transfers and Stat. Diff.	-	7653	-2239	-3	-256	-1813	-1843	-2276	1102	-	1309
TRANSFORMATION	**-**	**-**	**-**	**-**	**-**	**69**	**3608**	**11010**	**-**	**-**	**-**
Electricity and CHP Plants	-	-	-	-	-	69	3608	11010	-	-	-
Petroleum Refineries	-	-	-	-	-	-	-	-	-	-	-
Other Transform. Sector	-	-	-	-	-	-	-	-	-	-	-
ENERGY SECTOR	**1988**	**-**	**-**	**-**	**-**	**-**	**223**	**1222**	**-**	**-**	**-**
DISTRIBUTION LOSSES	**-**	**103**	**-**	**-**	**-**	**-**	**-**	**1**	**-**	**-**	**-**
FINAL CONSUMPTION	**64**	**5891**	**21697**	**354**	**6319**	**4790**	**27056**	**7863**	**1016**	**124**	**5910**
INDUSTRY SECTOR	**64**	**426**	**39**	**-**	**-**	**310**	**4953**	**6821**	**1016**	**2**	**31**
Iron and Steel	-	-	-	-	-	-	-	-	-	-	-
Chemical and Petrochem.	-	-	-	-	-	-	-	-	-	-	-
Non-Metallic Minerals	-	-	-	-	-	-	-	-	-	-	-
Non-specified	-	-	-	-	-	-	-	-	-	-	-
TRANSPORT SECTOR	**-**	**77**	**21450**	**354**	**6319**	**11**	**12240**	**17**	**-**	**-**	**-**
Air	-	-	-	351	6319	-	-	-	-	-	-
Road	-	77	21022	3	-	-	11782	-	-	-	-
Non-specified	-	-	-	-	-	-	-	-	-	-	-
OTHER SECTORS	**-**	**-**	**-**	**-**	**-**	**-**	**-**	**-**	**-**	**-**	**-**
Agriculture	-	-	-	-	-	-	-	-	-	-	-
Comm. and Publ. Services	-	-	-	-	-	-	-	-	-	-	-
Residential	-	-	-	-	-	-	-	-	-	-	-
Non-specified	-	-	-	-	-	-	-	-	-	-	-
NON-ENERGY USE	**-**	**189**	**-**	**-**	**-**	**-**	**-**	**-**	**-**	**122**	**5812**

Africa / Afrique

SUPPLY AND	Gas (TJ)				Comb. Renew. & Waste (TJ)				(GWh)	(TJ)
CONSUMPTION 1996	Natural Gas	Gas Works	Coke Ovens	Blast Furnaces	Solid Biomass	Gas/Liquids from Biomass	Municipal Waste	Industrial Waste	Electricity	Heat
Production	3663504	30325	26461	41923	9465965	-	-	-	384973	-
Imports	56858	-	-	-	30	-	-	-	6225	-
Exports	-1881548	-	-	-	-9920	-	-	-	-8276	-
Intl. Marine Bunkers	-	-	-	-	-	-	-	-	-	-
Stock Changes	-	-	-	-	-	-	-	-	-	-
DOMESTIC SUPPLY	**1838814**	**30325**	**26461**	**41923**	**9456074**	**-**	**-**	**-**	**382922**	**-**
Transfers and Stat. Diff.	-7052	-1	42	-7319	-4	-	-	-	1559	-
TRANSFORMATION	**796010**	**-**	**-**	**-**	**1029048**	**-**	**-**	**-**	**-**	**-**
Electricity and CHP Plants	724196	-	-	-	-	-	-	-	-	-
Petroleum Refineries	-	-	-	-	-	-	-	-	-	-
Other Transform. Sector	71814	-	-	-	1029048	-	-	-	-	-
ENERGY SECTOR	**480438**	**53**	**-**	**-**	**-**	**-**	**-**	**-**	**41221**	**-**
DISTRIBUTION LOSSES	**28925**	**-**	**512**	**7789**	**-**	**-**	**-**	**-**	**37437**	**-**
FINAL CONSUMPTION	**526389**	**30271**	**25991**	**26815**	**8427022**	**-**	**-**	**-**	**305823**	**-**
INDUSTRY SECTOR	**361942**	**28588**	**25991**	**26815**	**856063**	**-**	**-**	**-**	**153600**	**-**
Iron and Steel	-	-	-	-	-	-	-	-	-	-
Chemical and Petrochem.	-	-	-	-	-	-	-	-	-	-
Non-Metallic Minerals	-	-	-	-	-	-	-	-	-	-
Non-specified	-	-	-	-	-	-	-	-	-	-
TRANSPORT SECTOR	**26382**	**14**	**-**	**-**	**-**	**-**	**-**	**-**	**4965**	**-**
Air	-	14	-	-	-	-	-	-	13	-
Road	-	-	-	-	-	-	-	-	8	-
Non-specified	-	-	-	-	-	-	-	-	-	-
OTHER SECTORS	**-**	**-**	**-**	**-**	**-**	**-**	**-**	**-**	**-**	**-**
Agriculture	-	-	-	-	-	-	-	-	-	-
Comm. and Publ. Services	-	-	-	-	-	-	-	-	-	-
Residential	-	-	-	-	-	-	-	-	-	-
Non-specified	-	-	-	-	-	-	-	-	-	-
NON-ENERGY USE	**-**	**-**	**-**	**-**	**-**	**-**	**-**	**-**	**-**	**-**

APPROVISIONNEMENT	*Gaz* (TJ)				*En. Re. Comb. & Déchets* (TJ)				(GWh)	(TJ)
ET DEMANDE 1997	*Gaz naturel*	*Usines à gaz*	*Cokeries*	*Hauts fourneaux*	*Biomasse solide*	*Gaz/Liquides tirés de biomasse*	*Déchets urbains*	*Déchets industriels*	*Electricité*	*Chaleur*
Production	4105138	30262	25620	32839	9698314	-	-	-	401808	-
Imports	28894	-	-	-	31	-	-	-	7179	-
Exports	-2218278	-	-	-	-10081	-	-	-	-9694	-
Intl. Marine Bunkers	-	-	-	-	-	-	-	-	-	-
Stock Changes	-	-	-	-	-	-	-	-	-	-
DOMESTIC SUPPLY	**1915754**	**30262**	**25620**	**32839**	**9688263**	**-**	**-**	**-**	**399293**	**-**
Transfers and Stat. Diff.	-6129	-	18	-7321	-8	-	-	-	-88	-
TRANSFORMATION	**820360**	**-**	**-**	**-**	**1050148**	**-**	**-**	**-**	**-**	**-**
Electricity and CHP Plants	748546	-	-	-	-	-	-	-	-	-
Petroleum Refineries	-	-	-	-	-	-	-	-	-	-
Other Transform. Sector	71814	-	-	-	1050148	-	-	-	-	-
ENERGY SECTOR	**514123**	**7**	**-**	**-**	**-**	**-**	**-**	**-**	**42240**	**-**
DISTRIBUTION LOSSES	**28360**	**-**	**184**	**2582**	**-**	**-**	**-**	**-**	**38601**	**-**
FINAL CONSUMPTION	**546782**	**30255**	**25454**	**22936**	**8638108**	**-**	**-**	**-**	**318364**	**-**
INDUSTRY SECTOR	**381023**	**28469**	**25454**	**22936**	**878107**	**-**	**-**	**-**	**158754**	**-**
Iron and Steel	-	-	-	-	-	-	-	-	-	-
Chemical and Petrochem.	-	-	-	-	-	-	-	-	-	-
Non-Metallic Minerals	-	-	-	-	-	-	-	-	-	-
Non-specified	-	-	-	-	-	-	-	-	-	-
TRANSPORT SECTOR	**31361**	**12**	**-**	**-**	**-**	**-**	**-**	**-**	**5278**	**-**
Air	-	12	-	-	-	-	-	-	14	-
Road	-	-	-	-	-	-	-	-	8	-
Non-specified	-	-	-	-	-	-	-	-	-	-
OTHER SECTORS	**-**	**-**	**-**	**-**	**-**	**-**	**-**	**-**	**-**	**-**
Agriculture	-	-	-	-	-	-	-	-	-	-
Comm. and Publ. Services	-	-	-	-	-	-	-	-	-	-
Residential	-	-	-	-	-	-	-	-	-	-
Non-specified	-	-	-	-	-	-	-	-	-	-
NON-ENERGY USE	**-**	**-**	**-**	**-**	**-**	**-**	**-**	**-**	**-**	**-**

Latin America / Amérique latine

SUPPLY AND CONSUMPTION 1996	Coal (1000 tonnes)							Oil (1000 tonnes)			
	Coking Coal	Other Bit. Coal	Sub-Bit. Coal	Lignite	Peat	Oven and Gas Coke	Pat. Fuel and BKB	Crude Oil	NGL	Feed-stocks	Additives
Production	1909	37819	-	-	-	10606	-	305131	9118	-	-
Imports	14660	3637	-	-	-	2062	2	68348	-	-	-
Exports	-800	-27433	-	-	-	-13	-	-151261	-414	-	-
Intl. Marine Bunkers	-	-	-	-	-	-	-	-	-	-	-
Stock Changes	-486	-36	-	-	-	-360	-	-3287	-	-	-
DOMESTIC SUPPLY	**15283**	**13987**	**-**	**-**	**-**	**12295**	**2**	**218931**	**8704**	**-**	**-**
Transfers and Stat. Diff.	-	627	-	-	-	21	-	-736	-7632	277	-
TRANSFORMATION	**13687**	**8801**	**-**	**-**	**-**	**9363**	**-**	**212299**	**719**	**277**	**-**
Electricity and CHP Plants	-	8463	-	-	-	76	-	92	-	-	-
Petroleum Refineries	-	-	-	-	-	-	-	212207	719	277	-
Other Transform. Sector	13687	338	-	-	-	9287	-	-	-	-	-
ENERGY SECTOR	**-**	**8**	**-**	**-**	**-**	**2**	**-**	**152**	**-**	**-**	**-**
DISTRIBUTION LOSSES	**-**	**114**	**-**	**-**	**-**	**26**	**-**	**3612**	**-**	**-**	**-**
FINAL CONSUMPTION	**1596**	**5691**	**-**	**-**	**-**	**2925**	**2**	**2132**	**353**	**-**	**-**
INDUSTRY SECTOR	**1596**	**5338**	**-**	**-**	**-**	**2730**	**-**	**2022**	**353**	**-**	**-**
Iron and Steel	-	-	-	-	-	-	-	-	-	-	-
Chemical and Petrochem.	-	-	-	-	-	-	-	-	-	-	-
Non-Metallic Minerals	-	-	-	-	-	-	-	-	-	-	-
Non-specified	-	-	-	-	-	-	-	-	-	-	-
TRANSPORT SECTOR	**-**	**1**	**-**	**-**	**-**	**-**	**-**	**-**	**-**	**-**	**-**
Air	-	-	-	-	-	-	-	-	-	-	-
Road	-	-	-	-	-	-	-	-	-	-	-
Non-specified	-	-	-	-	-	-	-	-	-	-	-
OTHER SECTORS	**-**	**-**	**-**	**-**	**-**	**-**	**-**	**-**	**-**	**-**	**-**
Agriculture	-	-	-	-	-	-	-	-	-	-	-
Comm. and Publ. Services	-	-	-	-	-	-	-	-	-	-	-
Residential	-	-	-	-	-	-	-	-	-	-	-
Non-specified	-	-	-	-	-	-	-	-	-	-	-
NON-ENERGY USE	**-**	**-**	**-**	**-**	**-**	**195**	**-**	**-**	**-**	**-**	**-**

APPROVISIONNEMENT ET DEMANDE 1997	*Charbon* (1000 tonnes)							*Pétrole* (1000 tonnes)			
	Charbon à coke	*Autres charb. bit.*	*Charbon sous-bit.*	*Lignite*	*Tourbe*	*Coke de four/gaz*	*Agg./briq. de lignite*	*Pétrole brut*	*LGN*	*Produits d'aliment.*	*Additifs*
Production	1761	41441	-	-	-	10393	-	318407	8970	-	-
Imports	13873	4716	-	-	-	2041	2	71730	-	-	-
Exports	-800	-31147	-	-	-	-18	-	-162186	-284	-	-
Intl. Marine Bunkers	-	-	-	-	-	-	-	-	-	-	-
Stock Changes	255	223	-	-	-	-132	-	1111	-	-	-
DOMESTIC SUPPLY	**15089**	**15233**	**-**	**-**	**-**	**12284**	**2**	**229062**	**8686**	**-**	**-**
Transfers and Stat. Diff.	-	1169	-	-	-	-28	-	-1991	-7531	266	-
TRANSFORMATION	**13161**	**9989**	**-**	**-**	**-**	**9329**	**-**	**224671**	**782**	**266**	**-**
Electricity and CHP Plants	-	9729	-	-	-	-	-	100	-	-	-
Petroleum Refineries	-	-	-	-	-	-	-	224571	782	266	-
Other Transform. Sector	13161	260	-	-	-	9329	-	-	-	-	-
ENERGY SECTOR	**-**	**-**	**-**	**-**	**-**	**-**	**-**	**161**	**-**	**-**	**-**
DISTRIBUTION LOSSES	**-**	**122**	**-**	**-**	**-**	**22**	**-**	**74**	**-**	**-**	**-**
FINAL CONSUMPTION	**1928**	**6291**	**-**	**-**	**-**	**2905**	**2**	**2165**	**373**	**-**	**-**
INDUSTRY SECTOR	**1928**	**5917**	**-**	**-**	**-**	**2646**	**-**	**2054**	**373**	**-**	**-**
Iron and Steel	-	-	-	-	-	-	-	-	-	-	-
Chemical and Petrochem.	-	-	-	-	-	-	-	-	-	-	-
Non-Metallic Minerals	-	-	-	-	-	-	-	-	-	-	-
Non-specified	-	-	-	-	-	-	-	-	-	-	-
TRANSPORT SECTOR	**-**	**1**	**-**	**-**	**-**	**-**	**-**	**-**	**-**	**-**	**-**
Air	-	-	-	-	-	-	-	-	-	-	-
Road	-	-	-	-	-	-	-	-	-	-	-
Non-specified	-	-	-	-	-	-	-	-	-	-	-
OTHER SECTORS	**-**	**-**	**-**	**-**	**-**	**-**	**-**	**-**	**-**	**-**	**-**
Agriculture	-	-	-	-	-	-	-	-	-	-	-
Comm. and Publ. Services	-	-	-	-	-	-	-	-	-	-	-
Residential	-	-	-	-	-	-	-	-	-	-	-
Non-specified	-	-	-	-	-	-	-	-	-	-	-
NON-ENERGY USE	**-**	**-**	**-**	**-**	**-**	**259**	**-**	**-**	**-**	**-**	**-**

Latin America / Amérique latine

SUPPLY AND CONSUMPTION 1996	Oil cont. (1000 tonnes)										
	Refinery Gas	LPG + Ethane	Motor Gasoline	Aviation Gasoline	Jet Fuel	Kerosene	Gas/ Diesel	Heavy Fuel Oil	Naphtha	Petrol. Coke	Other Prod.
Production	2360	5920	42898	127	11990	2988	60246	51238	6306	1149	14790
Imports	-	4316	5377	85	1063	1042	13927	12086	2237	-	975
Exports	-	-2584	-10216	-20	-5255	-778	-14020	-25990	-610	-24	-5629
Intl. Marine Bunkers	-	-	-	-	-	-	-1425	-5607	-	-	-1
Stock Changes	-	-107	-641	1	-82	-190	-62	-95	-186	-	9
DOMESTIC SUPPLY	**2360**	**7545**	**37418**	**193**	**7716**	**3062**	**58666**	**31632**	**7747**	**1125**	**10144**
Transfers and Stat. Diff.	-1	5452	1478	1	-492	34	615	-1086	-2450	-2	1918
TRANSFORMATION	**24**	**15**	**97**	**-**	**-**	**-**	**4047**	**11687**	**200**	**-**	**206**
Electricity and CHP Plants	24	-	29	-	-	-	4047	11551	-	-	206
Petroleum Refineries	-	12	68	-	-	-	-	-	-	-	-
Other Transform. Sector	-	3	-	-	-	-	-	136	200	-	-
ENERGY SECTOR	**2335**	**35**	**15**	**-**	**3**	**32**	**738**	**2606**	**9**	**-**	**2028**
DISTRIBUTION LOSSES	**-**	**28**	**55**	**-**	**-**	**12**	**260**	**11**	**25**	**-**	**51**
FINAL CONSUMPTION	**-**	**12919**	**38729**	**194**	**7221**	**3052**	**54236**	**16242**	**5063**	**1123**	**9777**
INDUSTRY SECTOR	**-**	**2398**	**407**	**-**	**-**	**398**	**4905**	**12751**	**5063**	**15**	**1624**
Iron and Steel	-	-	-	-	-	-	-	-	-	-	-
Chemical and Petrochem.	-	-	-	-	-	-	-	-	-	-	-
Non-Metallic Minerals	-	-	-	-	-	-	-	-	-	-	-
Non-specified	-	-	-	-	-	-	-	-	-	-	-
TRANSPORT SECTOR	**-**	**123**	**37605**	**136**	**7193**	**-**	**38338**	**1342**	**-**	**-**	**-**
Air	-	-	-	136	7193	-	-	-	-	-	-
Road	-	41	37523	-	-	-	36149	-	-	-	-
Non-specified	-	-	-	-	-	-	-	-	-	-	-
OTHER SECTORS	**-**	**-**	**-**	**-**	**-**	**-**	**-**	**-**	**-**	**-**	**-**
Agriculture	-	-	-	-	-	-	-	-	-	-	-
Comm. and Publ. Services	-	-	-	-	-	-	-	-	-	-	-
Residential	-	-	-	-	-	-	-	-	-	-	-
Non-specified	-	-	-	-	-	-	-	-	-	-	-
NON-ENERGY USE	**-**	**-**	**87**	**-**	**-**	**32**	**-**	**-**	**-**	**1108**	**8110**

APPROVISIONNEMENT ET DEMANDE 1997	*Pétrole cont.* (1000 tonnes)										
	Gaz de raffinerie	*GPL + éthane*	*Essence moteur*	*Essence aviation*	*Carbu-réacteurs*	*Kérosène*	*Gazole*	*Fioul lourd*	*Naphta*	*Coke de pétrole*	*Autres prod.*
Production	2336	6070	46677	130	12250	2558	64086	54591	5722	1133	15433
Imports	-	4621	4990	118	1217	1187	15125	11346	3158	-	784
Exports	-	-1972	-11844	-21	-5321	-839	-15293	-27394	-728	-24	-5420
Intl. Marine Bunkers	-	-	-	-	-	-	-1523	-5888	-	-	-
Stock Changes	-	-76	-243	-20	-13	32	327	508	-35	-	64
DOMESTIC SUPPLY	**2336**	**8643**	**39580**	**207**	**8133**	**2938**	**62722**	**33163**	**8117**	**1109**	**10861**
Transfers and Stat. Diff.	-88	5207	1378	3	-345	-6	-32	-1595	-1711	-	2088
TRANSFORMATION	**30**	**19**	**99**	**-**	**-**	**-**	**4494**	**11629**	**186**	**-**	**228**
Electricity and CHP Plants	30	-	30	-	-	-	4494	11497	-	-	228
Petroleum Refineries	-	12	69	-	-	-	-	-	-	-	-
Other Transform. Sector	-	7	-	-	-	-	-	132	186	-	-
ENERGY SECTOR	**2218**	**39**	**23**	**-**	**3**	**27**	**867**	**2974**	**4**	**-**	**2270**
DISTRIBUTION LOSSES	**-**	**31**	**59**	**-**	**-**	**-**	**31**	**12**	**8**	**-**	**136**
FINAL CONSUMPTION	**-**	**13761**	**40777**	**210**	**7785**	**2905**	**57298**	**16953**	**6208**	**1109**	**10315**
INDUSTRY SECTOR	**-**	**2857**	**424**	**-**	**-**	**380**	**5342**	**13282**	**6208**	**16**	**1794**
Iron and Steel	-	-	-	-	-	-	-	-	-	-	-
Chemical and Petrochem.	-	-	-	-	-	-	-	-	-	-	-
Non-Metallic Minerals	-	-	-	-	-	-	-	-	-	-	-
Non-specified	-	-	-	-	-	-	-	-	-	-	-
TRANSPORT SECTOR	**-**	**152**	**39629**	**149**	**7744**	**-**	**40596**	**1618**	**-**	**-**	**-**
Air	-	-	-	149	7744	-	-	-	-	-	-
Road	-	51	39544	-	-	-	38301	-	-	-	-
Non-specified	-	-	-	-	-	-	-	-	-	-	-
OTHER SECTORS	**-**	**-**	**-**	**-**	**-**	**-**	**-**	**-**	**-**	**-**	**-**
Agriculture	-	-	-	-	-	-	-	-	-	-	-
Comm. and Publ. Services	-	-	-	-	-	-	-	-	-	-	-
Residential	-	-	-	-	-	-	-	-	-	-	-
Non-specified	-	-	-	-	-	-	-	-	-	-	-
NON-ENERGY USE	**-**	**-**	**88**	**-**	**-**	**14**	**-**	**-**	**-**	**1093**	**8493**

Latin America / Amérique latine

SUPPLY AND CONSUMPTION 1996	Gas (TJ) Natural Gas	Gas Works	Coke Ovens	Blast Furnaces	Comb. Renew. & Waste (TJ) Solid Biomass	Gas/Liquids from Biomass	Municipal Waste	Industrial Waste	(GWh) Electricity	(TJ) Heat
Production	3357132	21975	85772	19847	3093171	276282	-	-	657212	-
Imports	81891	-	-	-	298	27429	-	-	41082	-
Exports	-82293	-	-	-	-87	-4523	-	-	-44784	-
Intl. Marine Bunkers	-	-	-	-	-	-	-	-	-	-
Stock Changes	-	-	-	-	-6533	-2722	-	-	-	-
DOMESTIC SUPPLY	**3356730**	**21975**	**85772**	**19847**	**3086849**	**296466**	**-**	**-**	**653510**	**-**
Transfers and Stat. Diff.	-22733	-1	-1	42	-1638	-	-	-	-2916	-
TRANSFORMATION	**860294**	**3022**	**7508**	**7995**	**401913**	**-**	**-**	**-**	**-**	**-**
Electricity and CHP Plants	822216	3022	6181	7995	155885	-	-	-	-	-
Petroleum Refineries	-	-	-	-	105	-	-	-	-	-
Other Transform. Sector	38078	-	1327	-	245923	-	-	-	-	-
ENERGY SECTOR	**798127**	**721**	**18826**	**373**	**-**	**1294**	**-**	**-**	**19473**	**-**
DISTRIBUTION LOSSES	**128051**	**1220**	**3236**	**966**	**1453**	**-**	**-**	**-**	**107231**	**-**
FINAL CONSUMPTION	**1547525**	**17011**	**56201**	**10555**	**2681845**	**295172**	**-**	**-**	**523890**	**-**
INDUSTRY SECTOR	**1155640**	**1778**	**56201**	**10555**	**1333594**	**-**	**-**	**-**	**237977**	**-**
Iron and Steel	-	-	-	-	-	-	-	-	-	-
Chemical and Petrochem.	-	-	-	-	-	-	-	-	-	-
Non-Metallic Minerals	-	-	-	-	-	-	-	-	-	-
Non-specified	-	-	-	-	-	-	-	-	-	-
TRANSPORT SECTOR	**46073**	**17**	**-**	**-**	**347**	**295172**	**-**	**-**	**1997**	**-**
Air	-	-	-	-	-	-	-	-	-	-
Road	43882	-	-	-	-	295172	-	-	90	-
Non-specified	-	-	-	-	-	-	-	-	-	-
OTHER SECTORS	**-**	**-**	**-**	**-**	**-**	**-**	**-**	**-**	**-**	**-**
Agriculture	-	-	-	-	-	-	-	-	-	-
Comm. and Publ. Services	-	-	-	-	-	-	-	-	-	-
Residential	-	-	-	-	-	-	-	-	-	-
Non-specified	-	-	-	-	-	-	-	-	-	-
NON-ENERGY USE	**-**	**-**	**-**	**-**	**-**	**-**	**-**	**-**	**-**	**-**

APPROVISIONNEMENT ET DEMANDE 1997	*Gaz (TJ) Gaz naturel*	*Usines à gaz*	*Cokeries*	*Hauts fourneaux*	*En. Re. Comb. & Déchets (TJ) Biomasse solide*	*Gaz/Liquides tirés de biomasse*	*Déchets urbains*	*Déchets industriels*	(GWh) *Electricité*	(TJ) *Chaleur*
Production	3718400	29571	83642	21162	3081162	343185	-	-	688297	-
Imports	90999	1	-	-	382	18342	-	-	46862	-
Exports	-117769	-	-	-	-213	-28685	-	-	-46738	-
Intl. Marine Bunkers	-	-	-	-	-	-	-	-	-	-
Stock Changes	-	-	-	-	-6616	-47027	-	-	-	-
DOMESTIC SUPPLY	**3691630**	**29572**	**83642**	**21162**	**3074715**	**285815**	**-**	**-**	**688421**	**-**
Transfers and Stat. Diff.	-41229	-2	-1	-1	-2449	-	-	-	-2303	-
TRANSFORMATION	**845698**	**4354**	**7987**	**7778**	**407964**	**-**	**-**	**-**	**-**	**-**
Electricity and CHP Plants	806124	4354	6502	7778	165012	-	-	-	-	-
Petroleum Refineries	-	-	-	-	105	-	-	-	-	-
Other Transform. Sector	39574	-	1485	-	242847	-	-	-	-	-
ENERGY SECTOR	**976802**	**2184**	**17702**	**373**	**-**	**1294**	**-**	**-**	**21103**	**-**
DISTRIBUTION LOSSES	**128587**	**619**	**2864**	**957**	**1453**	**-**	**-**	**-**	**111868**	**-**
FINAL CONSUMPTION	**1699314**	**22413**	**55088**	**12053**	**2662849**	**284521**	**-**	**-**	**553147**	**-**
INDUSTRY SECTOR	**1268414**	**5945**	**55088**	**12053**	**1381318**	**-**	**-**	**-**	**251765**	**-**
Iron and Steel	-	-	-	-	-	-	-	-	-	-
Chemical and Petrochem.	-	-	-	-	-	-	-	-	-	-
Non-Metallic Minerals	-	-	-	-	-	-	-	-	-	-
Non-specified	-	-	-	-	-	-	-	-	-	-
TRANSPORT SECTOR	**53290**	**17**	**-**	**-**	**347**	**284521**	**-**	**-**	**2908**	**-**
Air	-	-	-	-	-	-	-	-	-	-
Road	50984	-	-	-	-	284521	-	-	848	-
Non-specified	-	-	-	-	-	-	-	-	-	-
OTHER SECTORS	**-**	**-**	**-**	**-**	**-**	**-**	**-**	**-**	**-**	**-**
Agriculture	-	-	-	-	-	-	-	-	-	-
Comm. and Publ. Services	-	-	-	-	-	-	-	-	-	-
Residential	-	-	-	-	-	-	-	-	-	-
Non-specified	-	-	-	-	-	-	-	-	-	-
NON-ENERGY USE	**-**	**-**	**-**	**-**	**-**	**-**	**-**	**-**	**-**	**-**

Asia (excluding China) / Asie (Chine non incluse)

SUPPLY AND CONSUMPTION 1996	Coal (1000 tonnes)							Oil (1000 tonnes)			
	Coking Coal	Other Bit. Coal	Sub-Bit. Coal	Lignite	Peat	Oven and Gas Coke	Pat. Fuel and BKB	Crude Oil	NGL	Feed-stocks	Additives
Production	9205	366872	7769	44160	-	17873	-	163973	11897	-	62
Imports	16767	37517	-	1	-	786	80	188324	-	860	101
Exports	-	-41496	-	-	-	-14	-	-76348	-1069	-	-
Intl. Marine Bunkers	-	-	-	-	-	-	-	-	-	-	-
Stock Changes	-126	2888	-	-491	-	-216	-	-697	-	-	-5
DOMESTIC SUPPLY	**25846**	**365781**	**7769**	**43670**	**-**	**18429**	**80**	**275252**	**10828**	**860**	**158**
Transfers and Stat. Diff.	514	2242	-	-1	-	1	-	1090	-9932	1752	-
TRANSFORMATION	**26360**	**240112**	**-**	**34012**	**-**	**14688**	**-**	**276009**	**94**	**2612**	**158**
Electricity and CHP Plants	-	240112	-	34012	-	-	-	-	-	-	-
Petroleum Refineries	-	-	-	-	-	-	-	276009	94	2612	158
Other Transform. Sector	26360	-	-	-	-	14688	-	-	-	-	-
ENERGY SECTOR	**-**	**4069**	**-**	**-**	**-**	**-**	**-**	**-**	**-**	**-**	**-**
DISTRIBUTION LOSSES	**-**	**-**	**-**	**-**	**-**	**-**	**-**	**100**	**2**	**-**	**-**
FINAL CONSUMPTION	**-**	**123842**	**7769**	**9657**	**-**	**3742**	**80**	**233**	**800**	**-**	**-**
INDUSTRY SECTOR	**-**	**122344**	**7769**	**9650**	**-**	**3672**	**80**	**233**	**800**	**-**	**-**
Iron and Steel	-	-	-	-	-	-	-	-	-	-	-
Chemical and Petrochem.	-	-	-	-	-	-	-	-	-	-	-
Non-Metallic Minerals	-	-	-	-	-	-	-	-	-	-	-
Non-specified	-	-	-	-	-	-	-	-	-	-	-
TRANSPORT SECTOR	**-**	**156**	**-**	**-**	**-**	**-**	**-**	**-**	**-**	**-**	**-**
Air	-	-	-	-	-	-	-	-	-	-	-
Road	-	-	-	-	-	-	-	-	-	-	-
Non-specified	-	-	-	-	-	-	-	-	-	-	-
OTHER SECTORS	**-**	**-**	**-**	**-**	**-**	**-**	**-**	**-**	**-**	**-**	**-**
Agriculture	-	-	-	-	-	-	-	-	-	-	-
Comm. and Publ. Services	-	-	-	-	-	-	-	-	-	-	-
Residential	-	-	-	-	-	-	-	-	-	-	-
Non-specified	-	-	-	-	-	-	-	-	-	-	-
NON-ENERGY USE	**-**	**-**	**-**	**7**	**-**	**-**	**-**	**-**	**-**	**-**	**-**

APPROVISIONNEMENT ET DEMANDE 1997	*Charbon* (1000 tonnes)							*Pétrole* (1000 tonnes)			
	Charbon à coke	*Autres charb. bit.*	*Charbon sous-bit.*	*Lignite*	*Tourbe*	*Coke de four/gaz*	*Agg./briq. de lignite*	*Pétrole brut*	*LGN*	*Produits d'aliment.*	*Additifs*
Production	8523	384691	7536	46489	-	18632	-	166169	14062	-	62
Imports	18719	40430	-	-	-	679	126	191721	-	1291	101
Exports	-	-46275	-	-	-	-9	-	-77768	-1378	-	-
Intl. Marine Bunkers	-	-	-	-	-	-	-	-	-	-	-
Stock Changes	-612	77	-	-779	-	90	-	33	-	-	-
DOMESTIC SUPPLY	**26630**	**378923**	**7536**	**45710**	**-**	**19392**	**126**	**280155**	**12684**	**1291**	**163**
Transfers and Stat. Diff.	500	2504	-	-1	-	-	-	5128	-11911	2219	-
TRANSFORMATION	**27130**	**252044**	**-**	**36042**	**-**	**15457**	**13**	**284939**	**14**	**3510**	**163**
Electricity and CHP Plants	-	252044	-	36042	-	-	13	-	-	-	-
Petroleum Refineries	-	-	-	-	-	-	-	284939	14	3510	163
Other Transform. Sector	27130	-	-	-	-	15457	-	-	-	-	-
ENERGY SECTOR	**-**	**3925**	**-**	**-**	**-**	**-**	**-**	**-**	**-**	**-**	**-**
DISTRIBUTION LOSSES	**-**	**-**	**-**	**-**	**-**	**-**	**-**	**67**	**14**	**-**	**-**
FINAL CONSUMPTION	**-**	**125458**	**7536**	**9667**	**-**	**3935**	**113**	**277**	**745**	**-**	**-**
INDUSTRY SECTOR	**-**	**123933**	**7536**	**9665**	**-**	**3865**	**113**	**277**	**745**	**-**	**-**
Iron and Steel	-	-	-	-	-	-	-	-	-	-	-
Chemical and Petrochem.	-	-	-	-	-	-	-	-	-	-	-
Non-Metallic Minerals	-	-	-	-	-	-	-	-	-	-	-
Non-specified	-	-	-	-	-	-	-	-	-	-	-
TRANSPORT SECTOR	**-**	**75**	**-**	**-**	**-**	**-**	**-**	**-**	**-**	**-**	**-**
Air	-	-	-	-	-	-	-	-	-	-	-
Road	-	-	-	-	-	-	-	-	-	-	-
Non-specified	-	-	-	-	-	-	-	-	-	-	-
OTHER SECTORS	**-**	**-**	**-**	**-**	**-**	**-**	**-**	**-**	**-**	**-**	**-**
Agriculture	-	-	-	-	-	-	-	-	-	-	-
Comm. and Publ. Services	-	-	-	-	-	-	-	-	-	-	-
Residential	-	-	-	-	-	-	-	-	-	-	-
Non-specified	-	-	-	-	-	-	-	-	-	-	-
NON-ENERGY USE	**-**	**-**	**-**	**2**	**-**	**-**	**-**	**-**	**-**	**-**	**-**

Asia (excluding China) / Asie (Chine non incluse)

SUPPLY AND CONSUMPTION 1996	Oil cont. (1000 tonnes)										
	Refinery Gas	LPG + Ethane	Motor Gasoline	Aviation Gasoline	Jet Fuel	Kerosene	Gas/ Diesel	Heavy Fuel Oil	Naphtha	Petrol. Coke	Other Prod.
Production	4825	5522	33683	4	19459	16117	84714	71054	16992	299	13399
Imports	-	3071	8786	53	4078	7757	38853	38915	2205	10	3114
Exports	-	-4374	-8418	-1	-7588	-2070	-20827	-22131	-8223	-26	-6080
Intl. Marine Bunkers	-	-	-	-	-	-	-2783	-16225	-	-	-6
Stock Changes	-	3	371	4	-116	-275	64	-822	-5	21	-190
DOMESTIC SUPPLY	**4825**	**4222**	**34422**	**60**	**15833**	**21529**	**100021**	**70791**	**10969**	**304**	**10237**
Transfers and Stat. Diff.	74	8512	882	-2	-1549	-1054	2339	-5401	-1367	84	1188
TRANSFORMATION	**8**	**94**	**-**	**-**	**-**	**-**	**7829**	**27852**	**-**	**-**	**-**
Electricity and CHP Plants	8	-	-	-	-	-	7829	27843	-	-	-
Petroleum Refineries	-	-	-	-	-	-	-	-	-	-	-
Other Transform. Sector	-	94	-	-	-	-	-	9	-	-	-
ENERGY SECTOR	**4891**	**11**	**2**	**-**	**1**	**27**	**1077**	**7541**	**47**	**15**	**11**
DISTRIBUTION LOSSES	**-**	**55**	**5**	**-**	**2**	**-**	**1**	**1**	**-**	**-**	**-**
FINAL CONSUMPTION	**-**	**12574**	**35297**	**58**	**14281**	**20448**	**93453**	**29996**	**9555**	**373**	**11414**
INDUSTRY SECTOR	**-**	**3511**	**204**	**-**	**-**	**262**	**14586**	**26433**	**9555**	**19**	**928**
Iron and Steel	-	-	-	-	-	-	-	-	-	-	-
Chemical and Petrochem.	-	-	-	-	-	-	-	-	-	-	-
Non-Metallic Minerals	-	-	-	-	-	-	-	-	-	-	-
Non-specified	-	-	-	-	-	-	-	-	-	-	-
TRANSPORT SECTOR	**-**	**142**	**35007**	**58**	**14281**	**8**	**68221**	**1047**	**-**	**-**	**-**
Air	-	-	-	58	14281	-	2	-	-	-	-
Road	-	142	34987	-	-	-	63603	16	-	-	-
Non-specified	-	-	-	-	-	-	-	-	-	-	-
OTHER SECTORS	**-**	**-**	**-**	**-**	**-**	**-**	**-**	**-**	**-**	**-**	**-**
Agriculture	-	-	-	-	-	-	-	-	-	-	-
Comm. and Publ. Services	-	-	-	-	-	-	-	-	-	-	-
Residential	-	-	-	-	-	-	-	-	-	-	-
Non-specified	-	-	-	-	-	-	-	-	-	-	-
NON-ENERGY USE	**-**	**1249**	**-**	**-**	**-**	**-**	**-**	**-**	**-**	**354**	**10486**

APPROVISIONNEMENT ET DEMANDE 1997	*Pétrole cont.* (1000 tonnes)										
	Gaz de raffinerie	*GPL + éthane*	*Essence moteur*	*Essence aviation*	*Carbu- réacteurs*	*Kérosène*	*Gazole*	*Fioul lourd*	*Naphta*	*Coke de pétrole*	*Autres prod.*
Production	5009	5916	36068	7	18725	15937	88126	72826	16017	308	13926
Imports	-	3164	7917	53	5277	7134	42408	39179	3791	10	2892
Exports	-	-4182	-7877	-1	-7230	-751	-20623	-21142	-6687	-9	-6042
Intl. Marine Bunkers	-	-	-	-	-	-	-3572	-17729	-	-	-3
Stock Changes	-	-210	541	1	95	-366	214	1058	-67	9	-252
DOMESTIC SUPPLY	**5009**	**4688**	**36649**	**60**	**16867**	**21954**	**106553**	**74192**	**13054**	**318**	**10521**
Transfers and Stat. Diff.	91	7881	-130	-1	-1602	-894	1286	-3255	-1722	7	-284
TRANSFORMATION	**9**	**-**	**-**	**-**	**-**	**-**	**8375**	**30138**	**-**	**-**	**-**
Electricity and CHP Plants	9	-	-	-	-	-	8375	30129	-	-	-
Petroleum Refineries	-	-	-	-	-	-	-	-	-	-	-
Other Transform. Sector	-	-	-	-	-	-	-	9	-	-	-
ENERGY SECTOR	**5091**	**11**	**2**	**-**	**1**	**29**	**1128**	**7556**	**41**	**15**	**12**
DISTRIBUTION LOSSES	**-**	**85**	**5**	**-**	**1**	**1**	**-**	**-**	**-**	**-**	**-**
FINAL CONSUMPTION	**-**	**12473**	**36512**	**59**	**15263**	**21030**	**98336**	**33243**	**11291**	**310**	**10225**
INDUSTRY SECTOR	**-**	**3141**	**107**	**-**	**-**	**271**	**15452**	**29408**	**11291**	**28**	**978**
Iron and Steel	-	-	-	-	-	-	-	-	-	-	-
Chemical and Petrochem.	-	-	-	-	-	-	-	-	-	-	-
Non-Metallic Minerals	-	-	-	-	-	-	-	-	-	-	-
Non-specified	-	-	-	-	-	-	-	-	-	-	-
TRANSPORT SECTOR	**-**	**129**	**36334**	**59**	**15263**	**8**	**72087**	**1257**	**-**	**-**	**-**
Air	-	-	-	59	15263	-	2	-	-	-	-
Road	-	129	36329	-	-	-	67229	6	-	-	-
Non-specified	-	-	-	-	-	-	-	-	-	-	-
OTHER SECTORS	**-**	**-**	**-**	**-**	**-**	**-**	**-**	**-**	**-**	**-**	**-**
Agriculture	-	-	-	-	-	-	-	-	-	-	-
Comm. and Publ. Services	-	-	-	-	-	-	-	-	-	-	-
Residential	-	-	-	-	-	-	-	-	-	-	-
Non-specified	-	-	-	-	-	-	-	-	-	-	-
NON-ENERGY USE	**-**	**1317**	**-**	**-**	**-**	**-**	**-**	**-**	**-**	**282**	**9247**

Asia (excluding China) / Asie (Chine non incluse)

SUPPLY AND CONSUMPTION 1996	Gas (TJ)				Comb. Renew. & Waste (TJ)				(GWh)	(TJ)
	Natural Gas	Gas Works	Coke Ovens	Blast Furnaces	Solid Biomass	Gas/Liquids from Biomass	Municipal Waste	Industrial Waste	Electricity	Heat
Production	7169945	26456	30731	29104	14758921	-	-	-	976040	-
Imports	202611	-	-	-	905	-	-	-	2794	-
Exports	-2500946	-	-	-	-5411	-	-	-	-1929	-
Intl. Marine Bunkers		-	-	-	-	-	-	-	-	-
Stock Changes	-11320	-	-	-	-1	-	-	-	-	-
DOMESTIC SUPPLY	**4860290**	**26456**	**30731**	**29104**	**14754414**	**-**	**-**	**-**	**976905**	**-**
Transfers and Stat. Diff.	26572	12	1	-1	368	-	-	-	7913	-
TRANSFORMATION	**1797529**	**-**	**4516**	**9612**	**562790**	**-**	**-**	**-**	**-**	**-**
Electricity and CHP Plants	1797220	-	4516	9612	8921	-	-	-	-	-
Petroleum Refineries	-	-	-	-	-	-	-	-	-	-
Other Transform. Sector	309	-	-	-	553869	-	-	-	-	-
ENERGY SECTOR	**1251517**	**-**	**-**	**-**	**-**	**-**	**-**	**-**	**57886**	**-**
DISTRIBUTION LOSSES	**74183**	**233**	**-**	**139**	**-**	**-**	**-**	**-**	**164623**	**-**
FINAL CONSUMPTION	**1763633**	**26235**	**26216**	**19352**	**14191992**	**-**	**-**	**-**	**762309**	**-**
INDUSTRY SECTOR	**1460194**	**2098**	**26216**	**19352**	**1599340**	**-**	**-**	**-**	**349100**	**-**
Iron and Steel	-	-	-	-	-	-	-	-	-	-
Chemical and Petrochem.	-	-	-	-	-	-	-	-	-	-
Non-Metallic Minerals	-	-	-	-	-	-	-	-	-	-
Non-specified	-	-	-	-	-	-	-	-	-	-
TRANSPORT SECTOR	**782**	**-**	**-**	**-**	**-**	**-**	**-**	**-**	**7459**	**-**
Air	-	-	-	-	-	-	-	-	165	-
Road	215	-	-	-	-	-	-	-	17	-
Non-specified	-	-	-	-	-	-	-	-	-	-
OTHER SECTORS	**-**	**-**	**-**	**-**	**-**	**-**	**-**	**-**	**-**	**-**
Agriculture	-	-	-	-	-	-	-	-	-	-
Comm. and Publ. Services	-	-	-	-	-	-	-	-	-	-
Residential	-	-	-	-	-	-	-	-	-	-
Non-specified	-	-	-	-	-	-	-	-	-	-
NON-ENERGY USE	**-**	**-**	**-**	**-**	**-**	**-**	**-**	**-**	**-**	**-**

APPROVISIONNEMENT ET DEMANDE 1997	*Gaz* (TJ)				*En. Re. Comb. & Déchets* (TJ)				(GWh)	(TJ)
	Gaz naturel	*Usines à gaz*	*Cokeries*	*Hauts fourneaux*	*Biomasse solide*	*Gaz/Liquides tirés de biomasse*	*Déchets urbains*	*Déchets industriels*	*Electricité*	*Chaleur*
Production	7449141	26596	43386	40718	14884323	-	-	-	1053430	-
Imports	233693	-	-	-	618	-	-	-	2689	-
Exports	-2451169	-	-	-	-5296	-	-	-	-1946	-
Intl. Marine Bunkers	-	-	-	-	-	-	-	-	-	-
Stock Changes	-14487	-	-	-	1	-	-	-	-	-
DOMESTIC SUPPLY	**5217178**	**26596**	**43386**	**40718**	**14879646**	**-**	**-**	**-**	**1054173**	**-**
Transfers and Stat. Diff.	-46082	189	-3	1	372	-	-	-	-1333	-
TRANSFORMATION	**2017398**	**-**	**6084**	**12367**	**598078**	**-**	**-**	**-**	**-**	**-**
Electricity and CHP Plants	2017083	-	6084	12367	16039	-	-	-	-	-
Petroleum Refineries	-	-	-	-	-	-	-	-	-	-
Other Transform. Sector	315	-	-	-	582039	-	-	-	-	-
ENERGY SECTOR	**1267318**	**-**	**-**	**-**	**-**	**-**	**-**	**-**	**61172**	**-**
DISTRIBUTION LOSSES	**68612**	**488**	**-**	**397**	**-**	**-**	**-**	**-**	**170218**	**-**
FINAL CONSUMPTION	**1817768**	**26297**	**37299**	**27955**	**14283310**	**-**	**-**	**-**	**821450**	**-**
INDUSTRY SECTOR	**1495121**	**2098**	**37299**	**27955**	**1625986**	**-**	**-**	**-**	**375204**	**-**
Iron and Steel	-	-	-	-	-	-	-	-	-	-
Chemical and Petrochem.	-	-	-	-	-	-	-	-	-	-
Non-Metallic Minerals	-	-	-	-	-	-	-	-	-	-
Non-specified	-	-	-	-	-	-	-	-	-	-
TRANSPORT SECTOR	**758**	**-**	**-**	**-**	**-**	**-**	**-**	**-**	**7987**	**-**
Air	-	-	-	-	-	-	-	-	198	-
Road	232	-	-	-	-	-	-	-	18	-
Non-specified	-	-	-	-	-	-	-	-	-	-
OTHER SECTORS	**-**	**-**	**-**	**-**	**-**	**-**	**-**	**-**	**-**	**-**
Agriculture	-	-	-	-	-	-	-	-	-	-
Comm. and Publ. Services	-	-	-	-	-	-	-	-	-	-
Residential	-	-	-	-	-	-	-	-	-	-
Non-specified	-	-	-	-	-	-	-	-	-	-
NON-ENERGY USE	**-**	**-**	**-**	**-**	**-**	**-**	**-**	**-**	**-**	**-**

China / Chine : 1996

SUPPLY AND CONSUMPTION / *APPROVISIONNEMENT ET DEMANDE*	Coal / *Charbon* (1000 tonnes)							Oil / *Pétrole* (1000 tonnes)			
	Coking Coal / *Charbon à coke*	Other Bit. Coal / *Autres charb. bit.*	Sub-Bit. Coal / *Charbon sous-bit.*	Lignite / *Lignite*	Peat / *Tourbe*	Oven and Gas Coke / *Coke de four/gaz*	Pat. Fuel and BKB / *Agg./briq. de lignite*	Crude Oil / *Pétrole brut*	NGL / *LGN*	Feed-stocks / *Produits d'aliment.*	Additives / *Additifs*
Production	184558	1212141	-	-	-	139371	10260	157334	-	-	-
From Other Sources	-	-	-	-	-	-	-	-	-	-	-
Imports	-	9986	-	-	-	252	-	22617	-	-	-
Exports	-	-36484	-	-	-	-7994	-	-20402	-	-	-
Intl. Marine Bunkers	-	-	-	-	-	-	-	-	-	-	-
Stock Changes	-	8687	-	-	-	-1003	-	-341	-	-	-
DOMESTIC SUPPLY	**184558**	**1194330**	**-**	**-**	**-**	**130626**	**10260**	**159208**	**-**	**-**	**-**
Transfers	-	-	-	-	-	-	-	-	-	-	-
Statistical Differences	-	76873	-	-	-	-19288	-355	-557	-	-	-
TRANSFORMATION	**184558**	**574595**	**-**	**-**	**-**	**59993**	**-**	**152129**	**-**	**-**	**-**
Electricity Plants	-	496146	-	-	-	-	-	682	-	-	-
CHP Plants	-	63657	-	-	-	-	-	-	-	-	-
Heat Plants	-	-	-	-	-	-	-	108	-	-	-
Blast Furnaces/Gas Works	-	5820	-	-	-	59993	-	-	-	-	-
Coke/Pat. Fuel/BKB Plants	184558	8972	-	-	-	-	-	-	-	-	-
Petroleum Refineries	-	-	-	-	-	-	-	151339	-	-	-
Petrochemical Industry	-	-	-	-	-	-	-	-	-	-	-
Liquefaction	-	-	-	-	-	-	-	-	-	-	-
Other Transform. Sector	-	-	-	-	-	-	-	-	-	-	-
ENERGY SECTOR	**-**	**61175**	**-**	**-**	**-**	**1954**	**-**	**2932**	**-**	**-**	**-**
Coal Mines	-	27505	-	-	-	423	-	14	-	-	-
Oil and Gas Extraction	-	2449	-	-	-	17	-	1691	-	-	-
Petroleum Refineries	-	-	-	-	-	-	-	1090	-	-	-
Electr., CHP+Heat Plants	-	23145	-	-	-	28	-	137	-	-	-
Pumped Storage (Elec.)	-	-	-	-	-	-	-	-	-	-	-
Other Energy Sector	-	8076	-	-	-	1486	-	-	-	-	-
Distribution Losses	-	21976	-	-	-	-	-	1599	-	-	-
FINAL CONSUMPTION	**-**	**613457**	**-**	**-**	**-**	**49391**	**9905**	**1991**	**-**	**-**	**-**
INDUSTRY SECTOR	**-**	**402992**	**-**	**-**	**-**	**36796**	**-**	**1623**	**-**	**-**	**-**
Iron and Steel	-	42033	-	-	-	16632	-	49	-	-	-
Chemical and Petrochem.	-	86721	-	-	-	6266	-	1461	-	-	-
of which: Feedstocks	-	-	-	-	-	-	-	*274*	-	-	-
Non-Ferrous Metals	-	9395	-	-	-	2435	-	5	-	-	-
Non-Metallic Minerals	-	135275	-	-	-	2951	-	60	-	-	-
Transport Equipment	-	7073	-	-	-	372	-	1	-	-	-
Machinery	-	24189	-	-	-	5324	-	8	-	-	-
Mining and Quarrying	-	5798	-	-	-	1792	-	2	-	-	-
Food and Tobacco	-	34120	-	-	-	412	-	5	-	-	-
Paper, Pulp and Print	-	17714	-	-	-	25	-	1	-	-	-
Wood and Wood Products	-	5471	-	-	-	21	-	-	-	-	-
Construction	-	4464	-	-	-	139	-	26	-	-	-
Textile and Leather	-	22730	-	-	-	90	-	3	-	-	-
Non-specified	-	8009	-	-	-	337	-	2	-	-	-
TRANSPORT SECTOR	**-**	**10164**	**-**	**-**	**-**	**-**	**-**	**-**	**-**	**-**	**-**
Air	-	-	-	-	-	-	-	-	-	-	-
Road	-	-	-	-	-	-	-	-	-	-	-
Rail	-	10091	-	-	-	-	-	-	-	-	-
Pipeline Transport	-	-	-	-	-	-	-	-	-	-	-
Internal Navigation	-	73	-	-	-	-	-	-	-	-	-
Non-specified	-	-	-	-	-	-	-	-	-	-	-
OTHER SECTORS	**-**	**184214**	**-**	**-**	**-**	**2930**	**9905**	**368**	**-**	**-**	**-**
Agriculture	-	19173	-	-	-	1198	-	110	-	-	-
Comm. and Publ. Services	-	30953	-	-	-	470	-	258	-	-	-
Residential	-	134088	-	-	-	1262	9905	-	-	-	-
Non-specified	-	-	-	-	-	-	-	-	-	-	-
NON-ENERGY USE	**-**	**16087**	**-**	**-**	**-**	**9665**	**-**	**-**	**-**	**-**	**-**
in Industry/Trans./Energy	-	16087	-	-	-	9665	-	-	-	-	-
in Transport	-	-	-	-	-	-	-	-	-	-	-
in Other Sectors	-	-	-	-	-	-	-	-	-	-	-

China / Chine : 1996

SUPPLY AND CONSUMPTION / *APPROVISIONNEMENT ET DEMANDE*	Oil cont. / *Pétrole cont.* (1000 tonnes)										
	Refinery Gas / *Gaz de raffinerie*	LPG + Ethane / *GPL + éthane*	Motor Gasoline / *Essence moteur*	Aviation Gasoline / *Essence aviation*	Jet Fuel / *Carbu-réacteurs*	Kerosene / *Kérosène*	Gas/ Diesel / *Gazole*	Heavy Fuel Oil / *Fioul lourd*	Naphtha / *Naphta*	Petrol. Coke / *Coke de pétrole*	Other Prod. / *Autres prod.*
Production	4501	6059	32636	110	2746	2637	44190	25045	15892	-	14812
From Other Sources	-	-	-	-	-	-	-	-	-	-	-
Imports	-	3737	531	-	3250	716	13660	13440	517	-	1542
Exports	-	-357	-1425	-	-120	-744	-7331	-2562	-	-	-1465
Intl. Marine Bunkers	-	-	-280	-	-	-	-939	-2130	-	-	-
Stock Changes	-	-46	287	-	-107	48	740	129	-	-	-
DOMESTIC SUPPLY	**4501**	**9393**	**31749**	**110**	**5769**	**2657**	**50320**	**33922**	**16409**	**-**	**14889**
Transfers	-	-	-	-	-	-	-	-	-	-	-
Statistical Differences	20	79	1	-	-	487	-795	4725	-	-	-
TRANSFORMATION	**876**	**71**	**8**	**-**	**-**	**-**	**2842**	**15330**	**517**	**-**	**-**
Electricity Plants	236	40	8	-	-	-	2543	11471	-	-	-
CHP Plants	-	-	-	-	-	-	-	-	-	-	-
Heat Plants	640	31	-	-	-	-	299	3278	-	-	-
Blast Furnaces/Gas Works	-	-	-	-	-	-	-	581	517	-	-
Coke/Pat. Fuel/BKB Plants	-	-	-	-	-	-	-	-	-	-	-
Petroleum Refineries	-	-	-	-	-	-	-	-	-	-	-
Petrochemical Industry	-	-	-	-	-	-	-	-	-	-	-
Liquefaction	-	-	-	-	-	-	-	-	-	-	-
Other Transform. Sector	-	-	-	-	-	-	-	-	-	-	-
ENERGY SECTOR	**2746**	**1084**	**-**	**-**	**-**	**74**	**3291**	**5366**	**-**	**-**	**-**
Coal Mines	-	-	-	-	-	34	310	16	-	-	-
Oil and Gas Extraction	65	57	-	-	-	7	1910	1089	-	-	-
Petroleum Refineries	2679	994	-	-	-	20	327	3033	-	-	-
Electr., CHP+Heat Plants	2	4	-	-	-	12	719	1121	-	-	-
Pumped Storage (Elec.)	-	-	-	-	-	-	-	-	-	-	-
Other Energy Sector	-	29	-	-	-	1	25	107	-	-	-
Distribution Losses	-	14	-	-	-	-	-	-	-	-	-
FINAL CONSUMPTION	**899**	**8303**	**31742**	**110**	**5769**	**3070**	**43392**	**17951**	**15892**	**-**	**14889**
INDUSTRY SECTOR	**899**	**824**	**-**	**-**	**-**	**406**	**6297**	**14429**	**15892**	**-**	**-**
Iron and Steel	1	7	-	-	-	5	428	3487	-	-	-
Chemical and Petrochem.	800	475	-	-	-	76	1294	5433	15892	-	-
of which: Feedstocks	*243*	*73*	-	-	-	*12*	*19*	*961*	*15892*	-	-
Non-Ferrous Metals	-	4	-	-	-	7	176	626	-	-	-
Non-Metallic Minerals	7	88	-	-	-	25	1150	3010	-	-	-
Transport Equipment	3	7	-	-	-	40	222	143	-	-	-
Machinery	5	48	-	-	-	79	754	633	-	-	-
Mining and Quarrying	-	-	-	-	-	7	296	26	-	-	-
Food and Tobacco	-	18	-	-	-	31	329	283	-	-	-
Paper, Pulp and Print	79	1	-	-	-	43	137	95	-	-	-
Wood and Wood Products	-	-	-	-	-	13	118	23	-	-	-
Construction	-	4	-	-	-	50	935	140	-	-	-
Textile and Leather	-	10	-	-	-	22	300	388	-	-	-
Non-specified	4	162	-	-	-	8	158	142	-	-	-
TRANSPORT SECTOR	**-**	**1**	**31601**	**110**	**5769**	**-**	**19679**	**2217**	**-**	**-**	**-**
Air	-	-	-	110	5769	-	-	-	-	-	-
Road	-	1	31601	-	-	-	12452	76	-	-	-
Rail	-	-	-	-	-	-	3452	10	-	-	-
Pipeline Transport	-	-	-	-	-	-	9	58	-	-	-
Internal Navigation	-	-	-	-	-	-	3718	2073	-	-	-
Non-specified	-	-	-	-	-	-	48	-	-	-	-
OTHER SECTORS	**-**	**7478**	**-**	**-**	**-**	**2664**	**17416**	**1305**	**-**	**-**	**-**
Agriculture	-	-	-	-	-	17	10284	25	-	-	-
Comm. and Publ. Services	-	443	-	-	-	1945	6911	1280	-	-	-
Residential	-	7035	-	-	-	702	221	-	-	-	-
Non-specified	-	-	-	-	-	-	-	-	-	-	-
NON-ENERGY USE	**-**	**-**	**141**	**-**	**-**	**-**	**-**	**-**	**-**	**-**	**14889**
in Industry/Transf./Energy	-	-	141	-	-	-	-	-	-	-	14889
in Transport	-	-	-	-	-	-	-	-	-	-	-
in Other Sectors	-	-	-	-	-	-	-	-	-	-	-

China / Chine : 1996

SUPPLY AND CONSUMPTION / *APPROVISIONNEMENT ET DEMANDE*	Gas / *Gaz* (TJ)				Comb. Renew. & Waste / *En. Re. Comb. & Déchets* (TJ)				(GWh)	(TJ)
	Natural Gas / *Gaz naturel*	Gas Works / *Usines à gaz*	Coke Ovens / *Cokeries*	Blast Furnaces / *Hauts fourneaux*	Solid Biomass / *Biomasse solide*	Gas/Liquids from Biomass / *Gaz/Liquides tirés de biomasse*	Municipal Waste / *Déchets urbains*	Industrial Waste / *Déchets industriels*	Electricity / *Electricité*	Heat / *Chaleur*
Production	871307	137522	434287	481306	8626511	50259	-	-	1108459	1148301
From Other Sources	-	-	-	-	-	-	-	-	-	-
Imports	68823	-	-	-	433	-	-	-	7902	-
Exports	-68823	-	-	-	-58	-	-	-	-4243	-
Intl. Marine Bunkers	-	-	-	-	-	-	-	-	-	-
Stock Changes	-	-	-	-	-	-	-	-	-	-
DOMESTIC SUPPLY	**871307**	**137522**	**434287**	**481306**	**8626886**	**50259**	**-**	**-**	**1112118**	**1148301**
Transfers	-	-	-	-	-	-	-	-	-	-
Statistical Differences	-1614	-540	-818	47758	-	-	-	-	1	-2212
TRANSFORMATION	**33528**	**-**	**34681**	**77336**	**-**	**-**	**-**	**-**	**-**	**-**
Electricity Plants	32272	-	20883	62459	-	-	-	-	-	-
CHP Plants	-	-	-	-	-	-	-	-	-	-
Heat Plants	1256	-	13798	14877	-	-	-	-	-	-
Blast Furnaces/Gas Works	-	-	-	-	-	-	-	-	-	-
Coke/Pat. Fuel/BKB Plants	-	-	-	-	-	-	-	-	-	-
Petroleum Refineries	-	-	-	-	-	-	-	-	-	-
Petrochemical Industry	-	-	-	-	-	-	-	-	-	-
Liquefaction	-	-	-	-	-	-	-	-	-	-
Other Transform. Sector	-	-	-	-	-	-	-	-	-	-
ENERGY SECTOR	**165042**	**5975**	**32984**	**-**	**-**	**-**	**-**	**-**	**160912**	**240205**
Coal Mines	-	-	254	-	-	-	-	-	39135	1867
Oil and Gas Extraction	129002	-	-	-	-	-	-	-	25817	9951
Petroleum Refineries	34741	-	-	-	-	-	-	-	15565	195260
Electr., CHP+Heat Plants	173	459	2713	-	-	-	-	-	79315	31192
Pumped Storage (Elec.)	-	-	-	-	-	-	-	-	-	-
Other Energy Sector	1126	5516	30017	-	-	-	-	-	1080	1935
Distribution Losses	24562	-	-	-	-	-	-	-	78311	14869
FINAL CONSUMPTION	**646561**	**131007**	**365804**	**451728**	**8626886**	**50259**	**-**	**-**	**872896**	**891015**
INDUSTRY SECTOR	**545196**	**84088**	**301535**	**451728**	**-**	**-**	**-**	**-**	**589919**	**695292**
Iron and Steel	17934	36528	268765	451728	-	-	-	-	98965	111727
Chemical and Petrochem.	371326	2913	20278	-	-	-	-	-	147946	279228
of which: Feedstocks	*208188*	-	-	-	-	-	-	-	-	-
Non-Ferrous Metals	1733	9487	137	-	-	-	-	-	46523	31737
Non-Metallic Minerals	13125	8195	6948	-	-	-	-	-	65543	7600
Transport Equipment	1559	2079	293	-	-	-	-	-	16903	23426
Machinery	24821	15987	4567	-	-	-	-	-	51179	27758
Mining and Quarrying	2686	-	-	-	-	-	-	-	18957	18679
Food and Tobacco	5415	112	137	-	-	-	-	-	35299	53078
Paper, Pulp and Print	520	19	-	-	-	-	-	-	21889	40550
Wood and Wood Products	-	-	-	-	-	-	-	-	5690	4021
Construction	6541	-	117	-	-	-	-	-	17448	1577
Textile and Leather	26381	262	254	-	-	-	-	-	45836	84598
Non-specified	73155	8506	39	-	-	-	-	-	17741	11313
TRANSPORT SECTOR	**4938**	**-**	**-**	**-**	**-**	**-**	**-**	**-**	**11846**	**-**
Air	-	-	-	-	-	-	-	-	-	-
Road	4938	-	-	-	-	-	-	-	-	-
Rail	-	-	-	-	-	-	-	-	10806	-
Pipeline Transport	-	-	-	-	-	-	-	-	1040	-
Internal Navigation	-	-	-	-	-	-	-	-	-	-
Non-specified	-	-	-	-	-	-	-	-	-	-
OTHER SECTORS	**96427**	**46919**	**64269**	**-**	**8626886**	**50259**	**-**	**-**	**271131**	**195723**
Agriculture	996	-	-	-	-	-	-	-	61829	217
Comm. and Publ. Services	10007	12970	17077	-	-	-	-	-	88023	25620
Residential	85424	33949	47192	-	8626511	50259	-	-	121279	169886
Non-specified	-	-	-	-	375	-	-	-	-	-
NON-ENERGY USE	**-**	**-**	**-**	**-**	**-**	**-**	**-**	**-**	**-**	**-**
in Industry/Transf./Energy	-	-	-	-	-	-	-	-	-	-
in Transport	-	-	-	-	-	-	-	-	-	-
in Other Sectors	-	-	-	-	-	-	-	-	-	-

China / Chine : 1997

SUPPLY AND CONSUMPTION / *APPROVISIONNEMENT ET DEMANDE*	Coal / *Charbon* (1000 tonnes)							Oil / *Pétrole* (1000 tonnes)			
	Coking Coal / *Charbon à coke*	Other Bit. Coal / *Autres charb. bit.*	Sub-Bit. Coal / *Charbon sous-bit.*	Lignite / *Lignite*	Peat / *Tourbe*	Oven and Gas Coke / *Coke de four/gaz*	Pat. Fuel and BKB / *Agg./briq. de lignite*	Crude Oil / *Pétrole brut*	NGL / *LGN*	Feed-stocks / *Produits d'aliment.*	Additives / *Additifs*
Production	185189	1187631	-	-	-	139957	9610	160741	-	-	-
From Other Sources	-	-	-	-	-	-	-	-	-	-	-
Imports	-	7724	-	-	-	252	-	35470	-	-	-
Exports	-	-30730	-	-	-	-10581	-	-19829	-	-	-
Intl. Marine Bunkers	-	-	-	-	-	-	-	-	-	-	-
Stock Changes	-	-12511	-	-	-	361	-	-1386	-	-	-
DOMESTIC SUPPLY	**185189**	**1152114**	**-**	**-**	**-**	**129989**	**9610**	**174996**	**-**	**-**	**-**
Transfers	-	-	-	-	-	-	-	-	-	-	-
Statistical Differences	-	62696	-	-	-	-6843	-	-1058	-	-	-
TRANSFORMATION	**185189**	**576240**	**-**	**-**	**-**	**68977**	**-**	**166865**	**-**	**-**	**-**
Electricity Plants	-	496796	-	-	-	-	-	684	-	-	-
CHP Plants	-	63714	-	-	-	-	-	-	-	-	-
Heat Plants	-	-	-	-	-	-	-	108	-	-	-
Blast Furnaces/Gas Works	-	7326	-	-	-	68977	-	-	-	-	-
Coke/Pat. Fuel/BKB Plants	185189	8404	-	-	-	-	-	-	-	-	-
Petroleum Refineries	-	-	-	-	-	-	-	166073	-	-	-
Petrochemical Industry	-	-	-	-	-	-	-	-	-	-	-
Liquefaction	-	-	-	-	-	-	-	-	-	-	-
Other Transform. Sector	-	-	-	-	-	-	-	-	-	-	-
ENERGY SECTOR	**-**	**69289**	**-**	**-**	**-**	**1916**	**-**	**3223**	**-**	**-**	**-**
Coal Mines	-	31153	-	-	-	415	-	15	-	-	-
Oil and Gas Extraction	-	2774	-	-	-	17	-	1859	-	-	-
Petroleum Refineries	-	-	-	-	-	-	-	1198	-	-	-
Electr., CHP+Heat Plants	-	26215	-	-	-	27	-	151	-	-	-
Pumped Storage (Elec.)	-	-	-	-	-	-	-	-	-	-	-
Other Energy Sector	-	9147	-	-	-	1457	-	-	-	-	-
Distribution Losses	-	21214	-	-	-	-	-	1758	-	-	-
FINAL CONSUMPTION	**-**	**548067**	**-**	**-**	**-**	**52253**	**9610**	**2092**	**-**	**-**	**-**
INDUSTRY SECTOR	**-**	**362229**	**-**	**-**	**-**	**39286**	**-**	**1687**	**-**	**-**	**-**
Iron and Steel	-	37781	-	-	-	18993	-	51	-	-	-
Chemical and Petrochem.	-	77949	-	-	-	6306	-	1520	-	-	-
of which: Feedstocks	-	-	-	-	-	-	-	*285*	-	-	-
Non-Ferrous Metals	-	8445	-	-	-	2451	-	5	-	-	-
Non-Metallic Minerals	-	121591	-	-	-	2970	-	62	-	-	-
Transport Equipment	-	6358	-	-	-	374	-	1	-	-	-
Machinery	-	21742	-	-	-	5358	-	8	-	-	-
Mining and Quarrying	-	5212	-	-	-	1803	-	2	-	-	-
Food and Tobacco	-	30669	-	-	-	415	-	5	-	-	-
Paper, Pulp and Print	-	15922	-	-	-	25	-	1	-	-	-
Wood and Wood Products	-	4918	-	-	-	21	-	-	-	-	-
Construction	-	4012	-	-	-	140	-	27	-	-	-
Textile and Leather	-	20431	-	-	-	91	-	3	-	-	-
Non-specified	-	7199	-	-	-	339	-	2	-	-	-
TRANSPORT SECTOR	**-**	**12370**	**-**	**-**	**-**	**-**	**-**	**-**	**-**	**-**	**-**
Air	-	-	-	-	-	-	-	-	-	-	-
Road	-	-	-	-	-	-	-	-	-	-	-
Rail	-	12281	-	-	-	-	-	-	-	-	-
Pipeline Transport	-	-	-	-	-	-	-	-	-	-	-
Internal Navigation	-	89	-	-	-	-	-	-	-	-	-
Non-specified	-	-	-	-	-	-	-	-	-	-	-
OTHER SECTORS	**-**	**159008**	**-**	**-**	**-**	**3240**	**9610**	**405**	**-**	**-**	**-**
Agriculture	-	19267	-	-	-	1447	-	121	-	-	-
Comm. and Publ. Services	-	26208	-	-	-	487	-	284	-	-	-
Residential	-	113533	-	-	-	1306	9610	-	-	-	-
Non-specified	-	-	-	-	-	-	-	-	-	-	-
NON-ENERGY USE	**-**	**14460**	**-**	**-**	**-**	**9727**	**-**	**-**	**-**	**-**	**-**
in Industry/Trans./Energy	-	14460	-	-	-	9727	-	-	-	-	-
in Transport	-	-	-	-	-	-	-	-	-	-	-
in Other Sectors	-	-	-	-	-	-	-	-	-	-	-

China / Chine : 1997

SUPPLY AND CONSUMPTION / *APPROVISIONNEMENT ET DEMANDE*	Oil cont. / *Pétrole cont.* (1000 tonnes)										
	Refinery Gas / *Gaz de raffinerie*	LPG + Ethane / *GPL + éthane*	Motor Gasoline / *Essence moteur*	Aviation Gasoline / *Essence aviation*	Jet Fuel / *Carbu- réacteurs*	Kerosene / *Kérosène*	Gas/ Diesel / *Gazole*	Heavy Fuel Oil / *Fioul lourd*	Naphtha / *Naphta*	Petrol. Coke / *Coke de pétrole*	Other Prod. / *Autres prod.*
Production	5341	6679	35060	132	3303	2826	49245	23112	17073	-	17814
From Other Sources	-	-	-	-	-	-	-	-	-	-	-
Imports	-	3748	554	-	3438	1450	17982	17785	453	-	4019
Exports	-	-410	-1905	-	-265	-740	-9663	-2838	-	-	-1680
Intl. Marine Bunkers	-	-	-274	-	-	-	-852	-1975	-	-	-
Stock Changes	-	49	-568	-	28	-145	-1869	-85	-	-	-
DOMESTIC SUPPLY	**5341**	**10066**	**32867**	**132**	**6504**	**3391**	**54843**	**35999**	**17526**	**-**	**20153**
Transfers	-	-	-	-	-	-	-	-	-	-	-
Statistical Differences	728	162	407	-	1	-270	2052	3943	-	-	-
TRANSFORMATION	**1039**	**12**	**8**	**-**	**-**	**-**	**6981**	**11896**	**453**	**-**	**-**
Electricity Plants	280	7	8	-	-	-	6654	8777	-	-	-
CHP Plants	-	-	-	-	-	-	-	-	-	-	-
Heat Plants	759	5	-	-	-	-	327	2502	-	-	-
Blast Furnaces/Gas Works	-	-	-	-	-	-	-	617	453	-	-
Coke/Pat. Fuel/BKB Plants	-	-	-	-	-	-	-	-	-	-	-
Petroleum Refineries	-	-	-	-	-	-	-	-	-	-	-
Petrochemical Industry	-	-	-	-	-	-	-	-	-	-	-
Liquefaction	-	-	-	-	-	-	-	-	-	-	-
Other Transform. Sector	-	-	-	-	-	-	-	-	-	-	-
ENERGY SECTOR	**3963**	**1395**	**-**	**-**	**-**	**154**	**3029**	**6790**	**-**	**-**	**-**
Coal Mines	-	-	-	-	-	71	285	20	-	-	-
Oil and Gas Extraction	94	73	-	-	-	15	1758	1378	-	-	-
Petroleum Refineries	3866	1280	-	-	-	41	301	3838	-	-	-
Electr., CHP+Heat Plants	3	5	-	-	-	25	662	1419	-	-	-
Pumped Storage (Elec.)	-	-	-	-	-	-	-	-	-	-	-
Other Energy Sector	-	37	-	-	-	2	23	135	-	-	-
Distribution Losses	-	18	-	-	-	-	-	-	-	-	-
FINAL CONSUMPTION	**1067**	**8803**	**33266**	**132**	**6505**	**2967**	**46885**	**21256**	**17073**	**-**	**20153**
INDUSTRY SECTOR	**1067**	**825**	**-**	**-**	**-**	**355**	**5867**	**13613**	**17073**	**-**	**-**
Iron and Steel	1	7	-	-	-	4	399	3295	-	-	-
Chemical and Petrochem.	949	487	-	-	-	67	1205	5134	17073	-	-
of which: Feedstocks	*288*	*75*	-	-	-	-	*20*	*908*	*17073*	-	-
Non-Ferrous Metals	-	4	-	-	-	6	164	592	-	-	-
Non-Metallic Minerals	8	90	-	-	-	22	1072	2844	-	-	-
Transport Equipment	4	7	-	-	-	35	207	135	-	-	-
Machinery	6	49	-	-	-	69	702	598	-	-	-
Mining and Quarrying	-	-	-	-	-	6	276	25	-	-	-
Food and Tobacco	-	18	-	-	-	27	307	267	-	-	-
Paper, Pulp and Print	94	1	-	-	-	38	128	90	-	-	-
Wood and Wood Products	-	-	-	-	-	11	110	22	-	-	-
Construction	-	4	-	-	-	44	870	132	-	-	-
Textile and Leather	-	10	-	-	-	19	280	367	-	-	-
Non-specified	5	148	-	-	-	7	147	112	-	-	-
TRANSPORT SECTOR	**-**	**16**	**33120**	**132**	**6505**	**-**	**21168**	**5752**	**-**	**-**	**-**
Air	-	-	-	132	6505	-	-	-	-	-	-
Road	-	16	33120	-	-	-	13262	197	-	-	-
Rail	-	-	-	-	-	-	3776	26	-	-	-
Pipeline Transport	-	-	-	-	-	-	10	151	-	-	-
Internal Navigation	-	-	-	-	-	-	4067	5378	-	-	-
Non-specified	-	-	-	-	-	-	53	-	-	-	-
OTHER SECTORS	**-**	**7962**	**-**	**-**	**-**	**2612**	**19850**	**1891**	**-**	**-**	**-**
Agriculture	-	-	-	-	-	14	10757	29	-	-	-
Comm. and Publ. Services	-	472	-	-	-	1910	8815	1862	-	-	-
Residential	-	7490	-	-	-	688	278	-	-	-	-
Non-specified	-	-	-	-	-	-	-	-	-	-	-
NON-ENERGY USE	**-**	**-**	**146**	**-**	**-**	**-**	**-**	**-**	**-**	**-**	**20153**
in Industry/Transf./Energy	-	-	146	-	-	-	-	-	-	-	20153
in Transport	-	-	-	-	-	-	-	-	-	-	-
in Other Sectors	-	-	-	-	-	-	-	-	-	-	-

China / Chine : 1997

	Gas / *Gaz* (TJ)				Comb. Renew. & Waste / *En. Re. Comb. & Déchets* (TJ)				(GWh)	(TJ)
SUPPLY AND CONSUMPTION	Natural Gas	Gas Works	Coke Ovens	Blast Furnaces	Solid Biomass	Gas/Liquids from Biomass	Municipal Waste	Industrial Waste	Electricity	Heat
APPROVISIONNEMENT ET DEMANDE	*Gaz naturel*	*Usines à gaz*	*Cokeries*	*Hauts fourneaux*	*Biomasse solide*	*Gaz/Liquides tirés de biomasse*	*Déchets urbains*	*Déchets industriels*	*Electricité*	*Chaleur*
Production	983449	144256	488338	553382	8672689	50528	-	-	1163416	1175695
From Other Sources	-	-	-	-	-	-	-	-	-	-
Imports	107462	-	-	-	433	-	-	-	7965	-
Exports	-107462	-	-	-	-58	-	-	-	-7763	-
Intl. Marine Bunkers	-	-	-	-	-	-	-	-	-	-
Stock Changes	-	-	-	-	-	-	-	-	-	-
DOMESTIC SUPPLY	**983449**	**144256**	**488338**	**553382**	**8673064**	**50528**	**-**	**-**	**1163618**	**1175695**
Transfers	-	-	-	-	-	-	-	-	-	-
Statistical Differences	5381	-1	-28084	54910	-	-	-	-	2	12201
TRANSFORMATION	**92181**	**-**	**44439**	**88917**	**-**	**-**	**-**	**-**	**-**	**-**
Electricity Plants	88727	-	26759	71812	-	-	-	-	-	-
CHP Plants	-	-	-	-	-	-	-	-	-	-
Heat Plants	3454	-	17680	17105	-	-	-	-	-	-
Blast Furnaces/Gas Works	-	-	-	-	-	-	-	-	-	-
Coke/Pat. Fuel/BKB Plants	-	-	-	-	-	-	-	-	-	-
Petroleum Refineries	-	-	-	-	-	-	-	-	-	-
Petrochemical Industry	-	-	-	-	-	-	-	-	-	-
Liquefaction	-	-	-	-	-	-	-	-	-	-
Other Transform. Sector	-	-	-	-	-	-	-	-	-	-
ENERGY SECTOR	**247085**	**14506**	**35053**	**-**	**-**	**-**	**-**	**-**	**194284**	**239323**
Coal Mines	-	-	270	-	-	-	-	-	47251	1860
Oil and Gas Extraction	193129	-	-	-	-	-	-	-	31171	9914
Petroleum Refineries	52011	-	-	-	-	-	-	-	18793	194544
Electr., CHP+Heat Plants	259	1114	2883	-	-	-	-	-	95765	31077
Pumped Storage (Elec.)	-	-	-	-	-	-	-	-	-	-
Other Energy Sector	1686	13392	31900	-	-	-	-	-	1304	1928
Distribution Losses	36771	-	-	-	-	-	-	-	93674	14814
FINAL CONSUMPTION	**612793**	**129749**	**380762**	**519375**	**8673064**	**50528**	**-**	**-**	**875662**	**933759**
INDUSTRY SECTOR	**508945**	**82407**	**310125**	**519375**	**-**	**-**	**-**	**-**	**564288**	**727461**
Iron and Steel	15114	35793	276422	519375	-	-	-	-	94670	116897
Chemical and Petrochem.	312950	2854	20856	-	-	-	-	-	141525	292147
of which: Feedstocks	*175459*	-	-	-	-	-	-	-	-	-
Non-Ferrous Metals	1460	9296	141	-	-	-	-	-	44504	33205
Non-Metallic Minerals	11062	8030	7146	-	-	-	-	-	62699	7952
Transport Equipment	1314	2037	301	-	-	-	-	-	16169	24510
Machinery	20919	15665	4697	-	-	-	-	-	48958	29042
Mining and Quarrying	2264	-	-	-	-	-	-	-	18134	19543
Food and Tobacco	4564	110	141	-	-	-	-	-	33767	55534
Paper, Pulp and Print	438	19	-	-	-	-	-	-	20939	42427
Wood and Wood Products	-	-	-	-	-	-	-	-	5443	4207
Construction	5513	-	120	-	-	-	-	-	16691	1649
Textile and Leather	22234	257	261	-	-	-	-	-	43847	88512
Non-specified	111113	8346	40	-	-	-	-	-	16942	11836
TRANSPORT SECTOR	**850**	**-**	**-**	**-**	**-**	**-**	**-**	**-**	**15318**	**-**
Air	-	-	-	-	-	-	-	-	-	-
Road	850	-	-	-	-	-	-	-	-	-
Rail	-	-	-	-	-	-	-	-	13973	-
Pipeline Transport	-	-	-	-	-	-	-	-	1345	-
Internal Navigation	-	-	-	-	-	-	-	-	-	-
Non-specified	-	-	-	-	-	-	-	-	-	-
OTHER SECTORS	**102998**	**47342**	**70637**	**-**	**8673064**	**50528**	**-**	**-**	**296056**	**206298**
Agriculture	459	-	-	-	-	-	-	-	63977	533
Comm. and Publ. Services	10752	13356	18769	-	-	-	-	-	97344	32763
Residential	91787	33986	51868	-	8672689	50528	-	-	134735	173002
Non-specified	-	-	-	-	375	-	-	-	-	-
NON-ENERGY USE	**-**	**-**	**-**	**-**	**-**	**-**	**-**	**-**	**-**	**-**
in Industry/Transf./Energy	-	-	-	-	-	-	-	-	-	-
in Transport	-	-	-	-	-	-	-	-	-	-
in Other Sectors	-	-	-	-	-	-	-	-	-	-

Non-OECD Europe / Europe non-OCDE

SUPPLY AND CONSUMPTION 1996	Coal (1000 tonnes)							Oil (1000 tonnes)			
	Coking Coal	Other Bit. Coal	Sub-Bit. Coal	Lignite	Peat	Oven and Gas Coke	Pat. Fuel and BKB	Crude Oil	NGL	Feed-stocks	Additives
Production	312	1291	3890	123671	2	5948	1201	9717	455	-	113
Imports	8333	5145	385	4308	-	460	5	26358	-	69	1
Exports	-	-10	-	-29	-	-339	-	-111	-	-	-
Intl. Marine Bunkers	-	-	-	-	-	-	-	-	-	-	-
Stock Changes	414	-908	-258	-1840	2	191	-2	-411	-	1	3
DOMESTIC SUPPLY	**9059**	**5518**	**4017**	**126110**	**4**	**6260**	**1204**	**35553**	**455**	**70**	**117**
Transfers and Stat. Diff.	-45	15	105	46	-1	-58	20	4	-293	-	-
TRANSFORMATION	**8291**	**4260**	**3466**	**117188**	**-**	**3944**	**155**	**35198**	**133**	**70**	**117**
Electricity and CHP Plants	-	4171	3418	114033	-	-	149	-	-	-	-
Petroleum Refineries	-	-	-	-	-	-	-	35197	133	70	117
Other Transform. Sector	8291	89	48	3155	-	3944	6	1	-	-	-
ENERGY SECTOR	**16**	**69**	**20**	**169**	**-**	**-**	**4**	**176**	**12**	**-**	**-**
DISTRIBUTION LOSSES	**-**	**4**	**-**	**301**	**-**	**-**	**-**	**183**	**17**	**-**	**-**
FINAL CONSUMPTION	**707**	**1200**	**636**	**8498**	**3**	**2258**	**1065**	**-**	**-**	**-**	**-**
INDUSTRY SECTOR	**472**	**912**	**82**	**4367**	**-**	**1614**	**9**	**-**	**-**	**-**	**-**
Iron and Steel	-	-	-	-	-	-	-	-	-	-	-
Chemical and Petrochem.	-	-	-	-	-	-	-	-	-	-	-
Non-Metallic Minerals	-	-	-	-	-	-	-	-	-	-	-
Non-specified	-	-	-	-	-	-	-	-	-	-	-
TRANSPORT SECTOR	**1**	**2**	**5**	**293**	**-**	**-**	**2**	**-**	**-**	**-**	**-**
Air	-	-	-	-	-	-	-	-	-	-	-
Road	-	-	-	-	-	-	-	-	-	-	-
Non-specified	-	-	-	-	-	-	-	-	-	-	-
OTHER SECTORS	**-**	**-**	**-**	**-**	**-**	**-**	**-**	**-**	**-**	**-**	**-**
Agriculture	-	-	-	-	-	-	-	-	-	-	-
Comm. and Publ. Services	-	-	-	-	-	-	-	-	-	-	-
Residential	-	-	-	-	-	-	-	-	-	-	-
Non-specified	-	-	-	-	-	-	-	-	-	-	-
NON-ENERGY USE	**-**	**15**	**-**	**-**	**-**	**568**	**-**	**-**	**-**	**-**	**-**

APPROVISIONNEMENT ET DEMANDE 1997	*Charbon* (1000 tonnes)							*Pétrole* (1000 tonnes)			
	Charbon à coke	*Autres charb. bit.*	*Charbon sous-bit.*	*Lignite*	*Tourbe*	*Coke de four/gaz*	*Agg./briq. de lignite*	*Pétrole brut*	*LGN*	*Produits d'aliment.*	*Additifs*
Production	324	1672	3488	116014	-	6211	1082	9445	488	-	104
Imports	9314	5205	300	2336	8	593	5	25384	-	77	1
Exports	-	-3	-1	-9	-	-172	-	-40	-	-	-
Intl. Marine Bunkers	-	-	-	-	-	-	-	-	-	-	-
Stock Changes	-393	62	143	-170	-	-16	5	9	-5	-4	-1
DOMESTIC SUPPLY	**9245**	**6936**	**3930**	**118171**	**8**	**6616**	**1092**	**34798**	**483**	**73**	**104**
Transfers and Stat. Diff.	-97	-248	119	144	-	72	6	-17	-318	-	-
TRANSFORMATION	**8760**	**5334**	**3620**	**110516**	**-**	**4322**	**139**	**34522**	**128**	**73**	**104**
Electricity and CHP Plants	-	5241	3571	107763	-	-	135	-	-	-	-
Petroleum Refineries	-	-	-	-	-	-	-	34521	128	73	104
Other Transform. Sector	8760	93	49	2753	-	4322	4	1	-	-	-
ENERGY SECTOR	**42**	**23**	**17**	**133**	**-**	**-**	**2**	**148**	**37**	**-**	**-**
DISTRIBUTION LOSSES	**-**	**5**	**-**	**300**	**-**	**-**	**-**	**109**	**-**	**-**	**-**
FINAL CONSUMPTION	**346**	**1326**	**412**	**7366**	**8**	**2366**	**957**	**2**	**-**	**-**	**-**
INDUSTRY SECTOR	**343**	**953**	**54**	**4072**	**8**	**1568**	**6**	**2**	**-**	**-**	**-**
Iron and Steel	-	-	-	-	-	-	-	-	-	-	-
Chemical and Petrochem.	-	-	-	-	-	-	-	-	-	-	-
Non-Metallic Minerals	-	-	-	-	-	-	-	-	-	-	-
Non-specified	-	-	-	-	-	-	-	-	-	-	-
TRANSPORT SECTOR	**1**	**5**	**4**	**294**	**-**	**-**	**1**	**-**	**-**	**-**	**-**
Air	-	-	-	-	-	-	-	-	-	-	-
Road	-	-	-	-	-	-	-	-	-	-	-
Non-specified	-	-	-	-	-	-	-	-	-	-	-
OTHER SECTORS	**-**	**-**	**-**	**-**	**-**	**-**	**-**	**-**	**-**	**-**	**-**
Agriculture	-	-	-	-	-	-	-	-	-	-	-
Comm. and Publ. Services	-	-	-	-	-	-	-	-	-	-	-
Residential	-	-	-	-	-	-	-	-	-	-	-
Non-specified	-	-	-	-	-	-	-	-	-	-	-
NON-ENERGY USE	**-**	**14**	**-**	**-**	**-**	**698**	**-**	**-**	**-**	**-**	**-**

Non-OECD Europe / Europe non-OCDE

SUPPLY AND CONSUMPTION 1996	Oil cont. (1000 tonnes)										
	Refinery Gas	LPG + Ethane	Motor Gasoline	Aviation Gasoline	Jet Fuel	Kerosene	Gas/ Diesel	Heavy Fuel Oil	Naphtha	Petrol. Coke	Other Prod.
Production	1259	636	6855	5	487	234	10849	7962	2461	658	2827
Imports	-	123	1652	2	570	19	2311	5936	352	153	285
Exports	-	-200	-3007	-	-153	-40	-3808	-901	-261	-232	-268
Intl. Marine Bunkers	-	-	-	-	-	-	-216	-1075	-	-	-1
Stock Changes	2	2	137	-	5	2	87	-64	4	16	-2
DOMESTIC SUPPLY	**1261**	**561**	**5637**	**7**	**909**	**215**	**9223**	**11858**	**2556**	**595**	**2841**
Transfers and Stat. Diff.	-61	59	-68	-3	4	-29	421	58	-298	-42	-145
TRANSFORMATION	**79**	**11**	**-**	**-**	**-**	**-**	**91**	**8125**	**61**	**-**	**352**
Electricity and CHP Plants	66	-	-	-	-	-	71	6659	20	-	227
Petroleum Refineries	-	-	-	-	-	-	-	-	-	-	-
Other Transform. Sector	13	11	-	-	-	-	20	1466	41	-	125
ENERGY SECTOR	**906**	**30**	**14**	**-**	**-**	**2**	**127**	**947**	**128**	**53**	**246**
DISTRIBUTION LOSSES	**-**	**-**	**121**	**-**	**-**	**-**	**83**	**37**	**223**	**-**	**123**
FINAL CONSUMPTION	**215**	**579**	**5434**	**4**	**913**	**184**	**9343**	**2807**	**1846**	**500**	**1975**
INDUSTRY SECTOR	**215**	**173**	**138**	**-**	**-**	**6**	**1271**	**2427**	**1845**	**500**	**313**
Iron and Steel	-	-	-	-	-	-	-	-	-	-	-
Chemical and Petrochem.	-	-	-	-	-	-	-	-	-	-	-
Non-Metallic Minerals	-	-	-	-	-	-	-	-	-	-	-
Non-specified	-	-	-	-	-	-	-	-	-	-	-
TRANSPORT SECTOR	**-**	**12**	**5221**	**4**	**881**	**6**	**5633**	**109**	**1**	**-**	**3**
Air	-	-	-	4	877	3	-	-	-	-	-
Road	-	12	5216	-	-	2	5139	8	1	-	3
Non-specified	-	-	-	-	-	-	-	-	-	-	-
OTHER SECTORS	**-**	**-**	**-**	**-**	**-**	**-**	**-**	**-**	**-**	**-**	**-**
Agriculture	-	-	-	-	-	-	-	-	-	-	-
Comm. and Publ. Services	-	-	-	-	-	-	-	-	-	-	-
Residential	-	-	-	-	-	-	-	-	-	-	-
Non-specified	-	-	-	-	-	-	-	-	-	-	-
NON-ENERGY USE	**-**	**-**	**-**	**-**	**-**	**-**	**-**	**-**	**-**	**-**	**1576**

APPROVISIONNEMENT ET DEMANDE 1997	*Pétrole cont.* (1000 tonnes)										
	Gaz de raffinerie	*GPL + éthane*	*Essence moteur*	*Essence aviation*	*Carbu- réacteurs*	*Kérosène*	*Gazole*	*Fioul lourd*	*Naphta*	*Coke de pétrole*	*Autres prod.*
Production	1330	619	7047	3	497	228	10632	7802	2359	499	2547
Imports	-	126	1952	1	650	57	2608	6255	358	157	262
Exports	-	-180	-2885	-	-97	-23	-3587	-771	-301	-246	-253
Intl. Marine Bunkers	-	-	-	-	-	-	-240	-1156	-	-	-1
Stock Changes	3	4	-341	1	54	-12	-257	-536	-8	-36	26
DOMESTIC SUPPLY	**1333**	**569**	**5773**	**5**	**1104**	**250**	**9156**	**11594**	**2408**	**374**	**2581**
Transfers and Stat. Diff.	-57	202	-1	-	-101	9	207	138	-329	102	677
TRANSFORMATION	**69**	**14**	**-**	**-**	**-**	**-**	**90**	**7830**	**61**	**-**	**333**
Electricity and CHP Plants	69	-	-	-	-	-	52	6620	20	-	149
Petroleum Refineries	-	-	-	-	-	-	-	-	-	-	-
Other Transform. Sector	-	14	-	-	-	-	38	1210	41	-	184
ENERGY SECTOR	**873**	**-**	**31**	**-**	**-**	**24**	**101**	**813**	**-**	**21**	**402**
DISTRIBUTION LOSSES	**-**	**2**	**87**	**-**	**-**	**4**	**63**	**36**	**223**	**-**	**140**
FINAL CONSUMPTION	**334**	**755**	**5654**	**5**	**1003**	**231**	**9109**	**3053**	**1795**	**455**	**2383**
INDUSTRY SECTOR	**334**	**214**	**115**	**-**	**-**	**9**	**1073**	**2428**	**1795**	**455**	**419**
Iron and Steel	-	-	-	-	-	-	-	-	-	-	-
Chemical and Petrochem.	-	-	-	-	-	-	-	-	-	-	-
Non-Metallic Minerals	-	-	-	-	-	-	-	-	-	-	-
Non-specified	-	-	-	-	-	-	-	-	-	-	-
TRANSPORT SECTOR	**-**	**13**	**5411**	**5**	**977**	**8**	**5409**	**257**	**-**	**-**	**17**
Air	-	-	2	5	977	3	-	-	-	-	1
Road	-	13	5403	-	-	4	4842	6	-	-	15
Non-specified	-	-	-	-	-	-	-	-	-	-	-
OTHER SECTORS	**-**	**-**	**-**	**-**	**-**	**-**	**-**	**-**	**-**	**-**	**-**
Agriculture	-	-	-	-	-	-	-	-	-	-	-
Comm. and Publ. Services	-	-	-	-	-	-	-	-	-	-	-
Residential	-	-	-	-	-	-	-	-	-	-	-
Non-specified	-	-	-	-	-	-	-	-	-	-	-
NON-ENERGY USE	**-**	**-**	**-**	**-**	**-**	**-**	**-**	**-**	**-**	**-**	**1626**

Non-OECD Europe / Europe non-OCDE

SUPPLY AND CONSUMPTION 1996	Gas (TJ)				Comb. Renew. & Waste (TJ)				(GWh)	(TJ)
	Natural Gas	Gas Works	Coke Ovens	Blast Furnaces	Solid Biomass	Gas/Liquids from Biomass	Municipal Waste	Industrial Waste	Electricity	Heat
Production	747706	436	41205	53204	267695	-	-	1835	209728	528384
Imports	873411	-	-	-	1461	-	-	-	15392	-
Exports	-	-	-	-	-653	-	-	-	-10524	-
Intl. Marine Bunkers	-	-	-	-	-	-	-	-	-	-
Stock Changes	4722	-	-	-	-329	-	-	-	-	-
DOMESTIC SUPPLY	**1625839**	**436**	**41205**	**53204**	**268174**	**-**	**-**	**1835**	**214596**	**528384**
Transfers and Stat. Diff.	2568	-	-228	358	2416	-	-	-	-308	29279
TRANSFORMATION	**568735**	**-**	**4931**	**13165**	**5695**	**-**	**-**	**1835**	**1000**	**-**
Electricity and CHP Plants	427431	-	4931	12872	754	-	-	613	-	-
Petroleum Refineries	-	-	-	-	-	-	-	-	-	-
Other Transform. Sector	141304	-	-	293	4941	-	-	1222	1000	-
ENERGY SECTOR	**64696**	**-**	**7826**	**3825**	**294**	**-**	**-**	**-**	**29720**	**64342**
DISTRIBUTION LOSSES	**37311**	**11**	**313**	**1539**	**81**	**-**	**-**	**-**	**25738**	**57447**
FINAL CONSUMPTION	**957665**	**425**	**27907**	**35033**	**264520**	**-**	**-**	**-**	**157830**	**435874**
INDUSTRY SECTOR	**673615**	**148**	**27907**	**35033**	**12305**	**-**	**-**	**-**	**64652**	**145104**
Iron and Steel	-	-	-	-	-	-	-	-	-	-
Chemical and Petrochem.	-	-	-	-	-	-	-	-	-	-
Non-Metallic Minerals	-	-	-	-	-	-	-	-	-	-
Non-specified	-	-	-	-	-	-	-	-	-	-
TRANSPORT SECTOR	**4102**	**-**	**-**	**-**	**252**	**-**	**-**	**-**	**4844**	**-**
Air	-	-	-	-	-	-	-	-	-	-
Road	3939	-	-	-	-	-	-	-	54	-
Non-specified	-	-	-	-	-	-	-	-	-	-
OTHER SECTORS	**-**	**-**	**-**	**-**	**-**	**-**	**-**	**-**	**-**	**-**
Agriculture	-	-	-	-	-	-	-	-	-	-
Comm. and Publ. Services	-	-	-	-	-	-	-	-	-	-
Residential	-	-	-	-	-	-	-	-	-	-
Non-specified	-	-	-	-	-	-	-	-	-	-
NON-ENERGY USE	**-**	**-**	**-**	**-**	**-**	**-**	**-**	**-**	**-**	**-**

APPROVISIONNEMENT ET DEMANDE 1997	*Gaz* (TJ)				*En. Re. Comb. & Déchets* (TJ)				(GWh)	(TJ)
	Gaz naturel	*Usines à gaz*	*Cokeries*	*Hauts fourneaux*	*Biomasse solide*	*Gaz/Liquides tirés de biomasse*	*Déchets urbains*	*Déchets industriels*	*Electricité*	*Chaleur*
Production	658039	464	43674	60400	204162	-	-	1198	206673	494551
Imports	769511	-	-	-	1505	-	-	-	14667	-
Exports	-	-	-	-	-777	-	-	-	-11257	-
Intl. Marine Bunkers	-	-	-	-	-	-	-	-	-	-
Stock Changes	288	-	-	-	-255	-	-	-	-	-
DOMESTIC SUPPLY	**1427838**	**464**	**43674**	**60400**	**204635**	**-**	**-**	**1198**	**210083**	**494551**
Transfers and Stat. Diff.	40701	-	670	-210	-1322	-	-	-	-527	7384
TRANSFORMATION	**454437**	**-**	**5477**	**12449**	**4504**	**-**	**-**	**1198**	**1688**	**-**
Electricity and CHP Plants	317200	-	5477	12449	282	-	-	737	-	-
Petroleum Refineries	-	-	-	-	-	-	-	-	-	-
Other Transform. Sector	137237	-	-	-	4222	-	-	461	1688	-
ENERGY SECTOR	**98688**	**-**	**11615**	**7597**	**191**	**-**	**-**	**-**	**26440**	**61765**
DISTRIBUTION LOSSES	**15510**	**49**	**332**	**1772**	**85**	**-**	**-**	**-**	**25616**	**52171**
FINAL CONSUMPTION	**899904**	**415**	**26920**	**38372**	**198533**	**-**	**-**	**-**	**155812**	**387999**
INDUSTRY SECTOR	**601383**	**175**	**26920**	**38372**	**15221**	**-**	**-**	**-**	**64960**	**125336**
Iron and Steel	-	-	-	-	-	-	-	-	-	-
Chemical and Petrochem.	-	-	-	-	-	-	-	-	-	-
Non-Metallic Minerals	-	-	-	-	-	-	-	-	-	-
Non-specified	-	-	-	-	-	-	-	-	-	-
TRANSPORT SECTOR	**1400**	**-**	**-**	**-**	**426**	**-**	**-**	**-**	**4631**	**-**
Air	-	-	-	-	-	-	-	-	-	-
Road	1298	-	-	-	-	-	-	-	57	-
Non-specified	-	-	-	-	-	-	-	-	-	-
OTHER SECTORS	**-**	**-**	**-**	**-**	**-**	**-**	**-**	**-**	**-**	**-**
Agriculture	-	-	-	-	-	-	-	-	-	-
Comm. and Publ. Services	-	-	-	-	-	-	-	-	-	-
Residential	-	-	-	-	-	-	-	-	-	-
Non-specified	-	-	-	-	-	-	-	-	-	-
NON-ENERGY USE	**-**	**-**	**-**	**-**	**-**	**-**	**-**	**-**	**-**	**-**

Former USSR / Ex-URSS : 1996

SUPPLY AND CONSUMPTION / *APPROVISIONNEMENT ET DEMANDE*	Coal / *Charbon* (1000 tonnes)							Oil / *Pétrole* (1000 tonnes)			
	Coking Coal / *Charbon à coke*	Other Bit. Coal / *Autres charb. bit.*	Sub-Bit. Coal / *Charbon sous-bit.*	Lignite / *Lignite*	Peat / *Tourbe*	Oven and Gas Coke / *Coke de four/gaz*	Pat. Fuel and BKB / *Agg./briq. de lignite*	Crude Oil / *Pétrole brut*	NGL / *LGN*	Feed-stocks / *Produits d'aliment.*	Additives / *Additifs*
Production	95213	219142	20	110697	9141	41774	5323	350368	-	-	-
From Other Sources	-	-	-	-	-	-	-	376	-	-	-
Imports	500	700	-	-	-	-	-	-	-	-	-
Exports	-2725	-16030	-	-	-	-	-	-111478	-	-	-
Intl. Marine Bunkers	-	-	-	-	-	-	-	-	-	-	-
Stock Changes	1026	17403	-	2885	761	940	142	4873	-	-	-
DOMESTIC SUPPLY	**94014**	**221215**	**20**	**113582**	**9902**	**42714**	**5465**	**244139**	**-**	**-**	**-**
Transfers	-	-	-	-	-	-	-	-76	-	-	-
Statistical Differences	-12935	-721	-	-2705	-287	-2231	-	-1569	-	-	-
TRANSFORMATION	**73003**	**157694**	**-**	**90801**	**6702**	**27851**	**161**	**238448**	**-**	**-**	**-**
Electricity Plants	-	740	-	2166	-	3	-	-	-	-	-
CHP Plants	-	129356	-	64703	2658	-	20	596	-	-	-
Heat Plants	-	23409	-	22451	1081	355	141	6	-	-	-
Blast Furnaces/Gas Works	-	-	-	-	-	27458	-	307	-	-	-
Coke/Pat. Fuel/BKB Plants	73003	4189	-	1481	2963	35	-	-	-	-	-
Petroleum Refineries	-	-	-	-	-	-	-	237539	-	-	-
Petrochemical Industry	-	-	-	-	-	-	-	-	-	-	-
Liquefaction	-	-	-	-	-	-	-	-	-	-	-
Other Transform. Sector	-	-	-	-	-	-	-	-	-	-	-
ENERGY SECTOR	**-**	**756**	**-**	**229**	**223**	**128**	**62**	**556**	**-**	**-**	**-**
Coal Mines	-	638	-	182	109	128	35	-	-	-	-
Oil and Gas Extraction	-	-	-	-	-	-	-	556	-	-	-
Petroleum Refineries	-	-	-	-	-	-	-	-	-	-	-
Electr., CHP+Heat Plants	-	118	-	47	4	-	27	-	-	-	-
Pumped Storage (Elec.)	-	-	-	-	-	-	-	-	-	-	-
Other Energy Sector	-	-	-	-	110	-	-	-	-	-	-
Distribution Losses	-	8014	-	4178	151	18	1	225	-	-	-
FINAL CONSUMPTION	**8076**	**54030**	**20**	**15669**	**2539**	**12486**	**5241**	**3265**	**-**	**-**	**-**
INDUSTRY SECTOR	**8076**	**17860**	**-**	**7167**	**75**	**9755**	**47**	**793**	**-**	**-**	**-**
Iron and Steel	-	-	-	-	-	-	-	-	-	-	-
Chemical and Petrochem.	-	-	-	-	-	-	-	-	-	-	-
of which: Feedstocks	-	-	-	-	-	-	-	-	-	-	-
Non-Ferrous Metals	-	-	-	-	-	-	-	-	-	-	-
Non-Metallic Minerals	-	-	-	-	-	-	-	-	-	-	-
Transport Equipment	-	-	-	-	-	-	-	-	-	-	-
Machinery	-	-	-	-	-	-	-	-	-	-	-
Mining and Quarrying	-	-	-	-	-	-	-	-	-	-	-
Food and Tobacco	-	-	-	-	-	-	-	-	-	-	-
Paper, Pulp and Print	-	-	-	-	-	-	-	-	-	-	-
Wood and Wood Products	-	-	-	-	-	-	-	-	-	-	-
Construction	-	-	-	-	-	-	-	-	-	-	-
Textile and Leather	-	-	-	-	-	-	-	-	-	-	-
Non-specified	-	-	-	-	-	-	-	-	-	-	-
TRANSPORT SECTOR	**-**	**690**	**-**	**205**	**1**	**3**	**29**	**9**	**-**	**-**	**-**
Air	-	-	-	-	-	-	-	-	-	-	-
Road	-	-	-	-	-	-	-	-	-	-	-
Rail	-	-	-	-	-	-	-	-	-	-	-
Pipeline Transport	-	-	-	-	-	-	-	-	-	-	-
Internal Navigation	-	-	-	-	-	-	-	-	-	-	-
Non-specified	-	-	-	-	-	-	-	-	-	-	-
OTHER SECTORS	**-**	**-**	**-**	**-**	**-**	**-**	**-**	**-**	**-**	**-**	**-**
Agriculture	-	-	-	-	-	-	-	-	-	-	-
Comm. and Publ. Services	-	-	-	-	-	-	-	-	-	-	-
Residential	-	-	-	-	-	-	-	-	-	-	-
Non-specified	-	-	-	-	-	-	-	-	-	-	-
NON-ENERGY USE	**-**	**694**	**-**	**245**	**-**	**2529**	**3**	**1532**	**-**	**-**	**-**
in Industry/Trans./Energy	-	-	-	-	-	-	-	-	-	-	-
in Transport	-	-	-	-	-	-	-	-	-	-	-
in Other Sectors	-	-	-	-	-	-	-	-	-	-	-

Former USSR / Ex-URSS : 1996

SUPPLY AND CONSUMPTION / *APPROVISIONNEMENT ET DEMANDE*	Oil cont. / *Pétrole cont.* (1000 tonnes)										
	Refinery Gas / *Gaz de raffinerie*	LPG + Ethane / *GPL + éthane*	Motor Gasoline / *Essence moteur*	Aviation Gasoline / *Essence aviation*	Jet Fuel / *Carbu-réacteurs*	Kerosene / *Kérosène*	Gas/ Diesel / *Gazole*	Heavy Fuel Oil / *Fioul lourd*	Naphtha / *Naphta*	Petrol. Coke / *Coke de pétrole*	Other Prod. / *Autres prod.*
Production	5401	5553	36937	44	10682	795	63411	88183	21	709	19223
From Other Sources	-	-	-	-	-	-	-	-	-	-	-
Imports	-	-	1753	-	-	-	806	2439	-	-	-
Exports	-	-	-2003	-	-	-43	-29768	-26329	-	-	-2500
Intl. Marine Bunkers	-	-	-	-	-	-	-35	-58	-	-	-
Stock Changes	13	1733	798	72	-1060	22	4668	2412	-21	78	196
DOMESTIC SUPPLY	**5414**	**7286**	**37485**	**116**	**9622**	**774**	**39082**	**66647**	**-**	**787**	**16919**
Transfers	-	27	-	-	-	-	-	1	-	-	75
Statistical Differences	-	-	-	-	-	-	-	-	-	-	-
TRANSFORMATION	**1528**	**555**	**-**	**-**	**-**	**13**	**825**	**52008**	**-**	**-**	**1974**
Electricity Plants	-	-	-	-	-	-	86	3160	-	-	260
CHP Plants	1528	552	-	-	-	9	-	9509	-	-	1706
Heat Plants	-	3	-	-	-	4	739	39339	-	-	8
Blast Furnaces/Gas Works	-	-	-	-	-	-	-	-	-	-	-
Coke/Pat. Fuel/BKB Plants	-	-	-	-	-	-	-	-	-	-	-
Petroleum Refineries	-	-	-	-	-	-	-	-	-	-	-
Petrochemical Industry	-	-	-	-	-	-	-	-	-	-	-
Liquefaction	-	-	-	-	-	-	-	-	-	-	-
Other Transform. Sector	-	-	-	-	-	-	-	-	-	-	-
ENERGY SECTOR	**3181**	**19**	**1080**	**-**	**-**	**3**	**87**	**1375**	**-**	**-**	**184**
Coal Mines	-	-	1	-	-	-	6	-	-	-	-
Oil and Gas Extraction	-	-	1079	-	-	-	-	-	-	-	-
Petroleum Refineries	3181	-	-	-	-	-	-	574	-	-	-
Electr., CHP+Heat Plants	-	-	-	-	-	-	3	30	-	-	-
Pumped Storage (Elec.)	-	-	-	-	-	-	-	-	-	-	-
Other Energy Sector	-	19	-	-	-	3	78	771	-	-	184
Distribution Losses	-	6	26	-	1	1	10	7	-	-	3
FINAL CONSUMPTION	**705**	**6733**	**36379**	**116**	**9621**	**757**	**38160**	**13258**	**-**	**787**	**14833**
INDUSTRY SECTOR	**678**	**1824**	**3544**	**6**	**219**	**71**	**3894**	**5749**	**-**	**3**	**3215**
Iron and Steel	-	-	-	-	-	-	-	-	-	-	-
Chemical and Petrochem.	-	-	-	-	-	-	-	-	-	-	-
of which: Feedstocks	-	-	-	-	-	-	-	-	-	-	-
Non-Ferrous Metals	-	-	-	-	-	-	-	-	-	-	-
Non-Metallic Minerals	-	-	-	-	-	-	-	-	-	-	-
Transport Equipment	-	-	-	-	-	-	-	-	-	-	-
Machinery	-	-	-	-	-	-	-	-	-	-	-
Mining and Quarrying	-	-	-	-	-	-	-	-	-	-	-
Food and Tobacco	-	-	-	-	-	-	-	-	-	-	-
Paper, Pulp and Print	-	-	-	-	-	-	-	-	-	-	-
Wood and Wood Products	-	-	-	-	-	-	-	-	-	-	-
Construction	-	-	-	-	-	-	-	-	-	-	-
Textile and Leather	-	-	-	-	-	-	-	-	-	-	-
Non-specified	-	-	-	-	-	-	-	-	-	-	-
TRANSPORT SECTOR	**15**	**186**	**24049**	**107**	**9397**	**34**	**14131**	**2298**	**-**	**-**	**576**
Air	-	-	-	107	9396	-	-	-	-	-	-
Road	2	146	23550	-	-	2	7883	79	-	-	147
Rail	-	-	-	-	-	-	-	-	-	-	-
Pipeline Transport	-	-	-	-	-	-	-	-	-	-	-
Internal Navigation	-	-	-	-	-	-	-	-	-	-	-
Non-specified	-	-	-	-	-	-	-	-	-	-	-
OTHER SECTORS	**-**	**-**	**-**	**-**	**-**	**-**	**-**	**-**	**-**	**-**	**-**
Agriculture	-	-	-	-	-	-	-	-	-	-	-
Comm. and Publ. Services	-	-	-	-	-	-	-	-	-	-	-
Residential	-	-	-	-	-	-	-	-	-	-	-
Non-specified	-	-	-	-	-	-	-	-	-	-	-
NON-ENERGY USE	**-**	**497**	**41**	**3**	**1**	**12**	**6**	**12**	**-**	**784**	**9610**
in Industry/Transf./Energy	-	-	-	-	-	-	-	-	-	-	-
in Transport	-	-	-	-	-	-	-	-	-	-	-
in Other Sectors	-	-	-	-	-	-	-	-	-	-	-

Former USSR / Ex-URSS : 1996

	Gas / *Gaz* (TJ)				Comb. Renew. & Waste / *En. Re. Comb. & Déchets* (TJ)				(GWh)	(TJ)
SUPPLY AND CONSUMPTION ***APPROVISIONNEMENT ET DEMANDE***	Natural Gas *Gaz naturel*	Gas Works *Usines à gaz*	Coke Ovens *Cokeries*	Blast Furnaces *Hauts fourneaux*	Solid Biomass *Biomasse solide*	Gas/Liquids from Biomass *Gaz/Liquides tirés de biomasse*	Municipal Waste *Déchets urbains*	Industrial Waste *Déchets industriels*	Electricity *Electricité*	Heat *Chaleur*
Production	26663548	5134	159070	231610	756545	56	-	71139	1262525	8029375
From Other Sources	-	11580	-	-	3341	-	-	-	-	-
Imports	-	-	-	-	-	-	-	-	-	-
Exports	-4655950	-	-	-	-12156	-	-	-6590	-10236	-
Intl. Marine Bunkers	-	-	-	-	-	-	-	-	-	-
Stock Changes	-350671	-	-	-	699	-	-	-25	-	-
DOMESTIC SUPPLY	**21656927**	**16714**	**159070**	**231610**	**748429**	**56**	**-**	**64524**	**1252289**	**8029375**
Transfers	-	-	-	-	-	-	-	-	-	-
Statistical Differences	-117111	-	-	-	-147	-	-	-3605	985	-
TRANSFORMATION	**11393757**	**1468**	**43873**	**53093**	**274381**	**56**	**-**	**16248**	**293**	**-**
Electricity Plants	905609	-	-	-	-	-	-	-	-	-
CHP Plants	7556131	1318	-	36899	91149	-	-	9701	-	-
Heat Plants	2928817	150	42082	16194	182557	56	-	6547	-	-
Blast Furnaces/Gas Works	-	-	1791	-	-	-	-	-	-	-
Coke/Pat. Fuel/BKB Plants	-	-	-	-	-	-	-	-	-	-
Petroleum Refineries	3200	-	-	-	-	-	-	-	-	-
Petrochemical Industry	-	-	-	-	-	-	-	-	-	-
Liquefaction	-	-	-	-	-	-	-	-	-	-
Other Transform. Sector	-	-	-	-	675	-	-	-	293	-
ENERGY SECTOR	**1365809**	**-**	**4490**	**5230**	**-**	**-**	**-**	**1788**	**196062**	**534340**
Coal Mines	1123333	-	-	-	-	-	-	-	21009	66200
Oil and Gas Extraction	178584	-	-	-	-	-	-	-	43957	157300
Petroleum Refineries	-	-	-	-	-	-	-	-	13878	219894
Electr., CHP+Heat Plants	7341	-	-	5230	-	-	-	-	93209	6060
Pumped Storage (Elec.)	-	-	-	-	-	-	-	-	764	-
Other Energy Sector	56551	-	4490	-	-	-	-	1788	23245	84886
Distribution Losses	473174	124	-	-	5960	-	-	29	147234	456832
FINAL CONSUMPTION	**8307076**	**15122**	**110707**	**173287**	**467941**	**-**	**-**	**42854**	**909685**	**7038203**
INDUSTRY SECTOR	**2897620**	**3542**	**107391**	**170789**	**33104**	**-**	**-**	**36990**	**402647**	**2822544**
Iron and Steel	-	-	-	-	-	-	-	-	-	-
Chemical and Petrochem.	-	-	-	-	-	-	-	-	-	-
of which: Feedstocks	-	-	-	-	-	-	-	-	-	-
Non-Ferrous Metals	-	-	-	-	-	-	-	-	-	-
Non-Metallic Minerals	-	-	-	-	-	-	-	-	-	-
Transport Equipment	-	-	-	-	-	-	-	-	-	-
Machinery	-	-	-	-	-	-	-	-	-	-
Mining and Quarrying	-	-	-	-	-	-	-	-	-	-
Food and Tobacco	-	-	-	-	-	-	-	-	-	-
Paper, Pulp and Print	-	-	-	-	-	-	-	-	-	-
Wood and Wood Products	-	-	-	-	-	-	-	-	-	-
Construction	-	-	-	-	-	-	-	-	-	-
Textile and Leather	-	-	-	-	-	-	-	-	-	-
Non-specified	-	-	-	-	-	-	-	-	-	-
TRANSPORT SECTOR	**740547**	**-**	**-**	**-**	**428**	**-**	**-**	**29**	**83057**	**-**
Air	-	-	-	-	-	-	-	-	-	-
Road	8366	-	-	-	-	-	-	-	-	-
Rail	-	-	-	-	-	-	-	-	-	-
Pipeline Transport	-	-	-	-	-	-	-	-	-	-
Internal Navigation	-	-	-	-	-	-	-	-	-	-
Non-specified	-	-	-	-	-	-	-	-	-	-
OTHER SECTORS	**-**	**-**	**-**	**-**	**-**	**-**	**-**	**-**	**-**	**-**
Agriculture	-	-	-	-	-	-	-	-	-	-
Comm. and Publ. Services	-	-	-	-	-	-	-	-	-	-
Residential	-	-	-	-	-	-	-	-	-	-
Non-specified	-	-	-	-	-	-	-	-	-	-
NON-ENERGY USE	**-**	**-**	**3283**	**567**	**-**	**-**	**-**	**-**	**-**	**-**
in Industry/Transf./Energy	-	-	-	-	-	-	-	-	-	-
in Transport	-	-	-	-	-	-	-	-	-	-
in Other Sectors	-	-	-	-	-	-	-	-	-	-

Former USSR / Ex-URSS : 1997

	Coal / *Charbon* (1000 tonnes)							Oil / *Pétrole* (1000 tonnes)			
SUPPLY AND CONSUMPTION	Coking Coal	Other Bit. Coal	Sub-Bit. Coal	Lignite	Peat	Oven and Gas Coke	Pat. Fuel and BKB	Crude Oil	NGL	Feed-stocks	Additives
APPROVISIONNEMENT ET DEMANDE	*Charbon à coke*	*Autres charb. bit.*	*Charbon sous-bit.*	*Lignite*	*Tourbe*	*Coke de four/gaz*	*Agg./briq. de lignite*	*Pétrole brut*	*LGN*	*Produits d'aliment.*	*Additifs*
Production	94764	210688	10	104169	7803	40416	5190	358153	-	-	-
From Other Sources	-	-	-	-	-	-	-	396	-	-	-
Imports	700	2500	-	-	-	-	-	-	-	-	-
Exports	-3100	-17000	-	-	-	-	-	-115130	-	-	-
Intl. Marine Bunkers	-	-	-	-	-	-	-	-	-	-	-
Stock Changes	-1442	-2454	-	185	-91	-355	-125	2235	-	-	-
DOMESTIC SUPPLY	**90922**	**193734**	**10**	**104354**	**7712**	**40061**	**5065**	**245654**	**-**	**-**	**-**
Transfers	-	-	-	-	-	-	-	-81	-	-	-
Statistical Differences	-12187	281	-	-	-	-	-	-2451	-	-	-
TRANSFORMATION	**71629**	**138495**	**-**	**87507**	**5518**	**27327**	**155**	**239119**	**-**	**-**	**-**
Electricity Plants	-	281	-	2106	-	3	-	-	-	-	-
CHP Plants	-	113982	-	62439	1510	-	18	598	-	-	-
Heat Plants	-	20133	-	21535	1064	369	137	4	-	-	-
Blast Furnaces/Gas Works	-	-	-	-	-	26923	-	219	-	-	-
Coke/Pat. Fuel/BKB Plants	71629	4021	-	1427	2944	32	-	-	-	-	-
Petroleum Refineries	-	-	-	-	-	-	-	238298	-	-	-
Petrochemical Industry	-	-	-	-	-	-	-	-	-	-	-
Liquefaction	-	-	-	-	-	-	-	-	-	-	-
Other Transform. Sector	-	78	-	-	-	-	-	-	-	-	-
ENERGY SECTOR	**-**	**664**	**-**	**190**	**175**	**133**	**57**	**559**	**-**	**-**	**-**
Coal Mines	-	564	-	160	45	133	32	-	-	-	-
Oil and Gas Extraction	-	-	-	-	-	-	-	556	-	-	-
Petroleum Refineries	-	-	-	-	-	-	-	-	-	-	-
Electr., CHP+Heat Plants	-	100	-	30	-	-	24	-	-	-	-
Pumped Storage (Elec.)	-	-	-	-	-	-	-	-	-	-	-
Other Energy Sector	-	-	-	-	130	-	1	3	-	-	-
Distribution Losses	-	6792	-	3895	218	19	-	224	-	-	-
FINAL CONSUMPTION	**7106**	**48064**	**10**	**12762**	**1801**	**12582**	**4853**	**3220**	**-**	**-**	**-**
INDUSTRY SECTOR	**7106**	**17163**	**-**	**5843**	**47**	**9744**	**42**	**741**	**-**	**-**	**-**
Iron and Steel	-	-	-	-	-	-	-	-	-	-	-
Chemical and Petrochem.	-	-	-	-	-	-	-	-	-	-	-
of which: Feedstocks	-	-	-	-	-	-	-	-	-	-	-
Non-Ferrous Metals	-	-	-	-	-	-	-	-	-	-	-
Non-Metallic Minerals	-	-	-	-	-	-	-	-	-	-	-
Transport Equipment	-	-	-	-	-	-	-	-	-	-	-
Machinery	-	-	-	-	-	-	-	-	-	-	-
Mining and Quarrying	-	-	-	-	-	-	-	-	-	-	-
Food and Tobacco	-	-	-	-	-	-	-	-	-	-	-
Paper, Pulp and Print	-	-	-	-	-	-	-	-	-	-	-
Wood and Wood Products	-	-	-	-	-	-	-	-	-	-	-
Construction	-	-	-	-	-	-	-	-	-	-	-
Textile and Leather	-	-	-	-	-	-	-	-	-	-	-
Non-specified	-	-	-	-	-	-	-	-	-	-	-
TRANSPORT SECTOR	**-**	**576**	**-**	**196**	**1**	**3**	**25**	**9**	**-**	**-**	**-**
Air	-	-	-	-	-	-	-	-	-	-	-
Road	-	-	-	-	-	-	-	-	-	-	-
Rail	-	-	-	-	-	-	-	-	-	-	-
Pipeline Transport	-	-	-	-	-	-	-	-	-	-	-
Internal Navigation	-	-	-	-	-	-	-	-	-	-	-
Non-specified	-	-	-	-	-	-	-	-	-	-	-
OTHER SECTORS	**-**	**-**	**-**	**-**	**-**	**-**	**-**	**-**	**-**	**-**	**-**
Agriculture	-	-	-	-	-	-	-	-	-	-	-
Comm. and Publ. Services	-	-	-	-	-	-	-	-	-	-	-
Residential	-	-	-	-	-	-	-	-	-	-	-
Non-specified	-	-	-	-	-	-	-	-	-	-	-
NON-ENERGY USE	**-**	**588**	**-**	**232**	**-**	**2629**	**3**	**1537**	**-**	**-**	**-**
in Industry/Trans./Energy	-	-	-	-	-	-	-	-	-	-	-
in Transport	-	-	-	-	-	-	-	-	-	-	-
in Other Sectors	-	-	-	-	-	-	-	-	-	-	-

Former USSR / Ex-URSS : 1997

SUPPLY AND CONSUMPTION *APPROVISIONNEMENT ET DEMANDE*	Oil cont. / *Pétrole cont.* (1000 tonnes)										
	Refinery Gas *Gaz de raffinerie*	LPG + Ethane *GPL + éthane*	Motor Gasoline *Essence moteur*	Aviation Gasoline *Essence aviation*	Jet Fuel *Carbu-réacteurs*	Kerosene *Kérosène*	Gas/ Diesel *Gazole*	Heavy Fuel Oil *Fioul lourd*	Naphtha *Naphta*	Petrol. Coke *Coke de pétrole*	Other Prod. *Autres prod.*
Production	5924	5864	38323	44	9667	809	64026	85880	26	860	19988
From Other Sources	-	-	-	-	-	-	-	-	-	-	-
Imports	-	-	1680	-	-	-	770	2340	-	-	-
Exports	-	-	-4200	-	-	-	-29600	-28000	-	-	-2500
Intl. Marine Bunkers	-	-	-	-	-	-	-38	-126	-	-	-
Stock Changes	11	681	-643	58	-773	-12	4199	915	-26	444	-50
DOMESTIC SUPPLY	**5935**	**6545**	**35160**	**102**	**8894**	**797**	**39357**	**61009**	**-**	**1304**	**17438**
Transfers	-	32	-	-	-	-	-	1	-	-	49
Statistical Differences	-	-	-	-	-	-	-	-	-	-	-
TRANSFORMATION	**1642**	**599**	**-**	**-**	**-**	**17**	**858**	**47341**	**-**	**-**	**2048**
Electricity Plants	-	-	-	-	-	-	80	2872	-	-	222
CHP Plants	1642	594	-	-	-	9	-	7904	-	-	1810
Heat Plants	-	5	-	-	-	8	778	36565	-	-	16
Blast Furnaces/Gas Works	-	-	-	-	-	-	-	-	-	-	-
Coke/Pat. Fuel/BKB Plants	-	-	-	-	-	-	-	-	-	-	-
Petroleum Refineries	-	-	-	-	-	-	-	-	-	-	-
Petrochemical Industry	-	-	-	-	-	-	-	-	-	-	-
Liquefaction	-	-	-	-	-	-	-	-	-	-	-
Other Transform. Sector	-	-	-	-	-	-	-	-	-	-	-
ENERGY SECTOR	**3533**	**22**	**1017**	**-**	**-**	**3**	**94**	**1262**	**-**	**84**	**100**
Coal Mines	-	-	-	-	-	-	6	-	-	-	-
Oil and Gas Extraction	-	-	1017	-	-	-	-	-	-	-	-
Petroleum Refineries	3359	-	-	-	-	-	-	473	-	-	-
Electr., CHP+Heat Plants	-	-	-	-	-	-	5	29	-	-	-
Pumped Storage (Elec.)	-	-	-	-	-	-	-	-	-	-	-
Other Energy Sector	174	22	-	-	-	3	83	760	-	84	100
Distribution Losses	2	8	22	-	1	1	15	3	-	-	-
FINAL CONSUMPTION	**758**	**5948**	**34121**	**102**	**8893**	**776**	**38390**	**12404**	**-**	**1220**	**15339**
INDUSTRY SECTOR	**729**	**1684**	**3346**	**5**	**207**	**78**	**3885**	**5639**	**-**	**-**	**3435**
Iron and Steel	-	-	-	-	-	-	-	-	-	-	-
Chemical and Petrochem.	-	-	-	-	-	-	-	-	-	-	-
of which: Feedstocks	-	-	-	-	-	-	-	-	-	-	-
Non-Ferrous Metals	-	-	-	-	-	-	-	-	-	-	-
Non-Metallic Minerals	-	-	-	-	-	-	-	-	-	-	-
Transport Equipment	-	-	-	-	-	-	-	-	-	-	-
Machinery	-	-	-	-	-	-	-	-	-	-	-
Mining and Quarrying	-	-	-	-	-	-	-	-	-	-	-
Food and Tobacco	-	-	-	-	-	-	-	-	-	-	-
Paper, Pulp and Print	-	-	-	-	-	-	-	-	-	-	-
Wood and Wood Products	-	-	-	-	-	-	-	-	-	-	-
Construction	-	-	-	-	-	-	-	-	-	-	-
Textile and Leather	-	-	-	-	-	-	-	-	-	-	-
Non-specified	-	-	-	-	-	-	-	-	-	-	-
TRANSPORT SECTOR	**16**	**187**	**22504**	**95**	**8680**	**34**	**14304**	**2004**	**-**	**-**	**632**
Air	-	-	-	95	8680	-	-	-	-	-	-
Road	2	160	22039	-	-	2	8107	73	-	-	177
Rail	-	-	-	-	-	-	-	-	-	-	-
Pipeline Transport	-	-	-	-	-	-	-	-	-	-	-
Internal Navigation	-	-	-	-	-	-	-	-	-	-	-
Non-specified	-	-	-	-	-	-	-	-	-	-	-
OTHER SECTORS	**-**	**-**	**-**	**-**	**-**	**-**	**-**	**-**	**-**	**-**	**-**
Agriculture	-	-	-	-	-	-	-	-	-	-	-
Comm. and Publ. Services	-	-	-	-	-	-	-	-	-	-	-
Residential	-	-	-	-	-	-	-	-	-	-	-
Non-specified	-	-	-	-	-	-	-	-	-	-	-
NON-ENERGY USE	**-**	**3**	**39**	**2**	**1**	**11**	**6**	**11**	**-**	**1220**	**9796**
in Industry/Transf./Energy	-	-	-	-	-	-	-	-	-	-	-
in Transport	-	-	-	-	-	-	-	-	-	-	-
in Other Sectors	-	-	-	-	-	-	-	-	-	-	-

Former USSR / Ex-URSS : 1997

SUPPLY AND CONSUMPTION *APPROVISIONNEMENT ET DEMANDE*	Gas / *Gaz* (TJ) Natural Gas *Gaz naturel*	Gas Works *Usines à gaz*	Coke Ovens *Cokeries*	Blast Furnaces *Hauts fourneaux*	Comb. Renew. & Waste / *En. Re. Comb. & Déchets* (TJ) Solid Biomass *Biomasse solide*	Gas/Liquids from Biomass *Gaz/Liquides tirés de biomasse*	Municipal Waste *Déchets urbains*	Industrial Waste *Déchets industriels*	(GWh) Electricity *Electricité*	(TJ) Heat *Chaleur*
Production	25209024	5313	149879	227970	798239	58	-	71141	1236059	7326289
From Other Sources	-	8260	-	-	3341	-	-	-	-	-
Imports	-	-	-	-	-	-	-	-	-	-
Exports	-4399590	-	-	-	-	-	-	-	-7888	-
Intl. Marine Bunkers	-	-	-	-	-	-	-	-	-	-
Stock Changes	301653	-	-	-	-14858	-	-	-6799	-	-
DOMESTIC SUPPLY	**21111087**	**13573**	**149879**	**227970**	**786722**	**58**	**-**	**64342**	**1228171**	**7326289**
Transfers	-	-	-	-	-	-	-	-	-	-
Statistical Differences	-7163	-	-	-	-147	-	-	-3605	-176	-
TRANSFORMATION	**11246473**	**1627**	**45012**	**52186**	**276893**	**58**	**-**	**16253**	**214**	**-**
Electricity Plants	874763	-	-	-	-	-	-	-	-	-
CHP Plants	7546307	1455	-	36269	91191	-	-	9701	-	-
Heat Plants	2821969	172	43175	15917	185027	58	-	6552	-	-
Blast Furnaces/Gas Works	-	-	1837	-	-	-	-	-	-	-
Coke/Pat. Fuel/BKB Plants	-	-	-	-	-	-	-	-	-	-
Petroleum Refineries	3434	-	-	-	-	-	-	-	-	-
Petrochemical Industry	-	-	-	-	-	-	-	-	-	-
Liquefaction	-	-	-	-	-	-	-	-	-	-
Other Transform. Sector	-	-	-	-	675	-	-	-	214	-
ENERGY SECTOR	**1350808**	**-**	**4607**	**5141**	**35**	**-**	**-**	**1788**	**197413**	**482345**
Coal Mines	1120853	-	-	-	9	-	-	-	20530	59030
Oil and Gas Extraction	169312	-	-	-	-	-	-	-	45007	140270
Petroleum Refineries	-	-	-	-	-	-	-	-	14424	195627
Electr., CHP+Heat Plants	9120	-	-	5141	26	-	-	-	93243	6197
Pumped Storage (Elec.)	-	-	-	-	-	-	-	-	663	-
Other Energy Sector	51523	-	4607	-	-	-	-	1788	23546	81221
Distribution Losses	460090	136	-	-	5983	-	-	29	144784	413373
FINAL CONSUMPTION	**8046553**	**11810**	**100260**	**170643**	**503664**	**-**	**-**	**42667**	**885584**	**6430571**
INDUSTRY SECTOR	**2749781**	**3550**	**96858**	**168188**	**33915**	**-**	**-**	**36810**	**399234**	**2618370**
Iron and Steel	-	-	-	-	-	-	-	-	-	-
Chemical and Petrochem.	-	-	-	-	-	-	-	-	-	-
of which: Feedstocks	-	-	-	-	-	-	-	-	-	-
Non-Ferrous Metals	-	-	-	-	-	-	-	-	-	-
Non-Metallic Minerals	-	-	-	-	-	-	-	-	-	-
Transport Equipment	-	-	-	-	-	-	-	-	-	-
Machinery	-	-	-	-	-	-	-	-	-	-
Mining and Quarrying	-	-	-	-	-	-	-	-	-	-
Food and Tobacco	-	-	-	-	-	-	-	-	-	-
Paper, Pulp and Print	-	-	-	-	-	-	-	-	-	-
Wood and Wood Products	-	-	-	-	-	-	-	-	-	-
Construction	-	-	-	-	-	-	-	-	-	-
Textile and Leather	-	-	-	-	-	-	-	-	-	-
Non-specified	-	-	-	-	-	-	-	-	-	-
TRANSPORT SECTOR	**734106**	**-**	**-**	**-**	**471**	**-**	**-**	**29**	**80734**	**-**
Air	-	-	-	-	-	-	-	-	-	-
Road	7874	-	-	-	-	-	-	-	-	-
Rail	-	-	-	-	-	-	-	-	-	-
Pipeline Transport	-	-	-	-	-	-	-	-	-	-
Internal Navigation	-	-	-	-	-	-	-	-	-	-
Non-specified	-	-	-	-	-	-	-	-	-	-
OTHER SECTORS	**-**	**-**	**-**	**-**	**-**	**-**	**-**	**-**	**-**	**-**
Agriculture	-	-	-	-	-	-	-	-	-	-
Comm. and Publ. Services	-	-	-	-	-	-	-	-	-	-
Residential	-	-	-	-	-	-	-	-	-	-
Non-specified	-	-	-	-	-	-	-	-	-	-
NON-ENERGY USE	**-**	**-**	**3368**	**557**	**-**	**-**	**-**	**-**	**-**	**-**
in Industry/Transf./Energy	-	-	-	-	-	-	-	-	-	-
in Transport	-	-	-	-	-	-	-	-	-	-
in Other Sectors	-	-	-	-	-	-	-	-	-	-

Middle East / Moyen-Orient

SUPPLY AND CONSUMPTION 1996	Coal (1000 tonnes)							Oil (1000 tonnes)			
	Coking Coal	Other Bit. Coal	Sub-Bit. Coal	Lignite	Peat	Oven and Gas Coke	Pat. Fuel and BKB	Crude Oil	NGL	Feed-stocks	Additives
Production	402	520	-	421	-	890	-	942279	45274	-	-
Imports	598	7356	-	-	-	2	-	24412	-	-	-
Exports	-	-13	-	-	-	-	-	-685575	-21724	-	-
Intl. Marine Bunkers	-	-	-	-	-	-	-	-	-	-	-
Stock Changes	-	682	-	-	-	-	-	-774	-	-	-
DOMESTIC SUPPLY	**1000**	**8545**	**-**	**421**	**-**	**892**	**-**	**280342**	**23550**	**-**	**-**
Transfers and Stat. Diff.	-	241	-	-	-	-	-	-71	-17125	-	-
TRANSFORMATION	**1000**	**7810**	**-**	**421**	**-**	**714**	**-**	**280271**	**1660**	**-**	**-**
Electricity and CHP Plants	-	7810	-	421	-	-	-	10516	-	-	-
Petroleum Refineries	-	-	-	-	-	-	-	269755	1660	-	-
Other Transform. Sector	1000	-	-	-	-	714	-	-	-	-	-
ENERGY SECTOR	-	-	-	-	-	-	-	-	-	-	-
DISTRIBUTION LOSSES	-	-	-	-	-	-	-	-	-	-	-
FINAL CONSUMPTION	**-**	**976**	**-**	**-**	**-**	**178**	**-**	**-**	**4765**	**-**	**-**
INDUSTRY SECTOR	**-**	**976**	**-**	**-**	**-**	**178**	**-**	**-**	**4765**	**-**	**-**
Iron and Steel	-	-	-	-	-	-	-	-	-	-	-
Chemical and Petrochem.	-	-	-	-	-	-	-	-	-	-	-
Non-Metallic Minerals	-	-	-	-	-	-	-	-	-	-	-
Non-specified	-	-	-	-	-	-	-	-	-	-	-
TRANSPORT SECTOR	-	-	-	-	-	-	-	-	-	-	-
Air	-	-	-	-	-	-	-	-	-	-	-
Road	-	-	-	-	-	-	-	-	-	-	-
Non-specified	-	-	-	-	-	-	-	-	-	-	-
OTHER SECTORS	-	-	-	-	-	-	-	-	-	-	-
Agriculture	-	-	-	-	-	-	-	-	-	-	-
Comm. and Publ. Services	-	-	-	-	-	-	-	-	-	-	-
Residential	-	-	-	-	-	-	-	-	-	-	-
Non-specified	-	-	-	-	-	-	-	-	-	-	-
NON-ENERGY USE	-	-	-	-	-	-	-	-	-	-	-

APPROVISIONNEMENT ET DEMANDE 1997	*Charbon* (1000 tonnes)							*Pétrole* (1000 tonnes)			
	Charbon à coke	*Autres charb. bit.*	*Charbon sous-bit.*	*Lignite*	*Tourbe*	*Coke de four/gaz*	*Agg./briq. de lignite*	*Pétrole brut*	*LGN*	*Produits d'aliment.*	*Additifs*
Production	402	520	-	421	-	890	-	989137	47452	-	-
Imports	598	8285	-	-	-	2	-	27102	-	-	-
Exports	-	-13	-	-	-	-	-	-726440	-22017	-	-
Intl. Marine Bunkers	-	-	-	-	-	-	-	-	-	-	-
Stock Changes	-	596	-	-	-	-	-	-2874	-	-	-
DOMESTIC SUPPLY	**1000**	**9388**	**-**	**421**	**-**	**892**	**-**	**286925**	**25435**	**-**	**-**
Transfers and Stat. Diff.	-	243	-	-	-	-	-	-	-17074	-	-
TRANSFORMATION	**1000**	**8653**	**-**	**421**	**-**	**714**	**-**	**286925**	**2390**	**-**	**-**
Electricity and CHP Plants	-	8653	-	421	-	-	-	11137	-	-	-
Petroleum Refineries	-	-	-	-	-	-	-	275788	2390	-	-
Other Transform. Sector	1000	-	-	-	-	714	-	-	-	-	-
ENERGY SECTOR	-	-	-	-	-	-	-	-	-	-	-
DISTRIBUTION LOSSES	-	-	-	-	-	-	-	-	-	-	-
FINAL CONSUMPTION	**-**	**978**	**-**	**-**	**-**	**178**	**-**	**-**	**5971**	**-**	**-**
INDUSTRY SECTOR	**-**	**978**	**-**	**-**	**-**	**178**	**-**	**-**	**5971**	**-**	**-**
Iron and Steel	-	-	-	-	-	-	-	-	-	-	-
Chemical and Petrochem.	-	-	-	-	-	-	-	-	-	-	-
Non-Metallic Minerals	-	-	-	-	-	-	-	-	-	-	-
Non-specified	-	-	-	-	-	-	-	-	-	-	-
TRANSPORT SECTOR	-	-	-	-	-	-	-	-	-	-	-
Air	-	-	-	-	-	-	-	-	-	-	-
Road	-	-	-	-	-	-	-	-	-	-	-
Non-specified	-	-	-	-	-	-	-	-	-	-	-
OTHER SECTORS	-	-	-	-	-	-	-	-	-	-	-
Agriculture	-	-	-	-	-	-	-	-	-	-	-
Comm. and Publ. Services	-	-	-	-	-	-	-	-	-	-	-
Residential	-	-	-	-	-	-	-	-	-	-	-
Non-specified	-	-	-	-	-	-	-	-	-	-	-
NON-ENERGY USE	-	-	-	-	-	-	-	-	-	-	-

Middle East / Moyen-Orient

SUPPLY AND CONSUMPTION 1996	Oil cont. (1000 tonnes)										
	Refinery Gas	LPG + Ethane	Motor Gasoline	Aviation Gasoline	Jet Fuel	Kerosene	Gas/ Diesel	Heavy Fuel Oil	Naphtha	Petrol. Coke	Other Prod.
Production	3163	4859	31774	125	17786	16966	75158	83254	12373	-	7015
Imports	-	863	2921	5	138	1640	6873	16440	150	-	694
Exports	-	-9637	-11516	-	-12509	-6489	-26680	-39426	-11027	-	-1646
Intl. Marine Bunkers	-	-	-	-	-	-	-337	-13442	-	-	-
Stock Changes	-	-13	-8	-	-7	-10	783	-	-	-	-41
DOMESTIC SUPPLY	**3163**	**-3928**	**23171**	**130**	**5408**	**12107**	**55797**	**46826**	**1496**	**-**	**6022**
Transfers and Stat. Diff.	-	9722	7324	-	-128	-229	-760	1422	164	-	184
TRANSFORMATION	**-**	**-**	**-**	**-**	**-**	**3**	**9044**	**20130**	**-**	**-**	**-**
Electricity and CHP Plants	-	-	-	-	-	3	9044	20130	-	-	-
Petroleum Refineries	-	-	-	-	-	-	-	-	-	-	-
Other Transform. Sector	-	-	-	-	-	-	-	-	-	-	-
ENERGY SECTOR	**3163**	**66**	**-**	**-**	**-**	**-**	**721**	**7517**	**-**	**-**	**-**
DISTRIBUTION LOSSES	**-**	**-**	**-**	**-**	**-**	**-**	**-**	**-**	**-**	**-**	**-**
FINAL CONSUMPTION	**-**	**5728**	**30495**	**130**	**5280**	**11875**	**45272**	**20601**	**1660**	**-**	**6206**
INDUSTRY SECTOR	**-**	**38**	**-**	**-**	**-**	**263**	**2933**	**16221**	**1660**	**-**	**-**
Iron and Steel	-	-	-	-	-	-	-	-	-	-	-
Chemical and Petrochem.	-	-	-	-	-	-	-	-	-	-	-
Non-Metallic Minerals	-	-	-	-	-	-	-	-	-	-	-
Non-specified	-	-	-	-	-	-	-	-	-	-	-
TRANSPORT SECTOR	**-**	**54**	**30495**	**130**	**5280**	**714**	**19687**	**-**	**-**	**-**	**-**
Air	-	-	-	130	5280	714	-	-	-	-	-
Road	-	54	30495	-	-	-	19687	-	-	-	-
Non-specified	-	-	-	-	-	-	-	-	-	-	-
OTHER SECTORS	**-**	**-**	**-**	**-**	**-**	**-**	**-**	**-**	**-**	**-**	**-**
Agriculture	-	-	-	-	-	-	-	-	-	-	-
Comm. and Publ. Services	-	-	-	-	-	-	-	-	-	-	-
Residential	-	-	-	-	-	-	-	-	-	-	-
Non-specified	-	-	-	-	-	-	-	-	-	-	-
NON-ENERGY USE	**-**	**-**	**-**	**-**	**-**	**-**	**-**	**-**	**-**	**-**	**6206**

APPROVISIONNEMENT ET DEMANDE 1997	*Pétrole cont.* (1000 tonnes)										
	Gaz de raffinerie	*GPL + éthane*	*Essence moteur*	*Essence aviation*	*Carbu- réacteurs*	*Kérosène*	*Gazole*	*Fioul lourd*	*Naphta*	*Coke de pétrole*	*Autres prod.*
Production	3197	4888	31450	125	18600	15977	76837	84085	12632	-	7157
Imports	-	1058	3285	5	180	2619	8446	16791	175	-	528
Exports	-	-10486	-9198	-	-12995	-6821	-27205	-41139	-10990	-	-1819
Intl. Marine Bunkers	-	-	-	-	-	-	-374	-13495	-	-	-
Stock Changes	-	-19	29	-	-1	-	212	-	-	-	-
DOMESTIC SUPPLY	**3197**	**-4559**	**25566**	**130**	**5784**	**11775**	**57916**	**46242**	**1817**	**-**	**5866**
Transfers and Stat. Diff.	-	10588	5733	-	-362	-	-885	1145	6	-	292
TRANSFORMATION	**-**	**-**	**-**	**-**	**-**	**3**	**9449**	**20009**	**-**	**-**	**-**
Electricity and CHP Plants	-	-	-	-	-	3	9449	20009	-	-	-
Petroleum Refineries	-	-	-	-	-	-	-	-	-	-	-
Other Transform. Sector	-	-	-	-	-	-	-	-	-	-	-
ENERGY SECTOR	**3197**	**68**	**-**	**-**	**-**	**-**	**745**	**7404**	**-**	**-**	**-**
DISTRIBUTION LOSSES	**-**	**-**	**-**	**-**	**-**	**-**	**-**	**-**	**-**	**-**	**-**
FINAL CONSUMPTION	**-**	**5961**	**31299**	**130**	**5422**	**11772**	**46837**	**19974**	**1823**	**-**	**6158**
INDUSTRY SECTOR	**-**	**41**	**-**	**-**	**-**	**256**	**3180**	**15657**	**1823**	**-**	**-**
Iron and Steel	-	-	-	-	-	-	-	-	-	-	-
Chemical and Petrochem.	-	-	-	-	-	-	-	-	-	-	-
Non-Metallic Minerals	-	-	-	-	-	-	-	-	-	-	-
Non-specified	-	-	-	-	-	-	-	-	-	-	-
TRANSPORT SECTOR	**-**	**55**	**31299**	**130**	**5422**	**825**	**19718**	**-**	**-**	**-**	**-**
Air	-	-	-	130	5422	825	-	-	-	-	-
Road	-	55	31299	-	-	-	19718	-	-	-	-
Non-specified	-	-	-	-	-	-	-	-	-	-	-
OTHER SECTORS	**-**	**-**	**-**	**-**	**-**	**-**	**-**	**-**	**-**	**-**	**-**
Agriculture	-	-	-	-	-	-	-	-	-	-	-
Comm. and Publ. Services	-	-	-	-	-	-	-	-	-	-	-
Residential	-	-	-	-	-	-	-	-	-	-	-
Non-specified	-	-	-	-	-	-	-	-	-	-	-
NON-ENERGY USE	**-**	**-**	**-**	**-**	**-**	**-**	**-**	**-**	**-**	**-**	**6158**

Middle East / Moyen-Orient

SUPPLY AND CONSUMPTION 1996	Gas (TJ)				Comb. Renew. & Waste (TJ)				(GWh)	(TJ)
	Natural Gas	Gas Works	Coke Ovens	Blast Furnaces	Solid Biomass	Gas/Liquids from Biomass	Municipal Waste	Industrial Waste	Electricity	Heat
Production	5975798	-	-	-	42719	-	-	-	345769	-
Imports	9310	-	-	-	1341	-	-	-	683	-
Exports	-284049	-	-	-	-2	-	-	-	-981	-
Intl. Marine Bunkers	-	-	-	-	-	-	-	-	-	-
Stock Changes	-	-	-	-	-	-	-	-	-	-
DOMESTIC SUPPLY	**5701059**	-	-	-	**44058**	-	-	-	**345471**	-
Transfers and Stat. Diff.	148506	-	-	-	-	-	-	-	191	-
TRANSFORMATION	**1504841**	-	-	-	**5681**	-	-	-	-	-
Electricity and CHP Plants	1504841	-	-	-	-	-	-	-	-	-
Petroleum Refineries	-	-	-	-	-	-	-	-	-	-
Other Transform. Sector	-	-	-	-	5681	-	-	-	-	-
ENERGY SECTOR	**1917545**	-	-	-	-	-	-	-	**28677**	-
DISTRIBUTION LOSSES	**43257**	-	-	-	-	-	-	-	**31659**	-
FINAL CONSUMPTION	**2383922**	-	-	-	**38377**	-	-	-	**285326**	-
INDUSTRY SECTOR	**1351740**	-	-	-	**7789**	-	-	-	**50759**	-
Iron and Steel	-	-	-	-	-	-	-	-	-	-
Chemical and Petrochem.	-	-	-	-	-	-	-	-	-	-
Non-Metallic Minerals	-	-	-	-	-	-	-	-	-	-
Non-specified	-	-	-	-	-	-	-	-	-	-
TRANSPORT SECTOR	-	-	-	-	-	-	-	-	**873**	-
Air	-	-	-	-	-	-	-	-	-	-
Road	-	-	-	-	-	-	-	-	-	-
Non-specified	-	-	-	-	-	-	-	-	-	-
OTHER SECTORS	-	-	-	-	-	-	-	-	-	-
Agriculture	-	-	-	-	-	-	-	-	-	-
Comm. and Publ. Services	-	-	-	-	-	-	-	-	-	-
Residential	-	-	-	-	-	-	-	-	-	-
Non-specified	-	-	-	-	-	-	-	-	-	-
NON-ENERGY USE	-	-	-	-	-	-	-	-	-	-

APPROVISIONNEMENT ET DEMANDE 1997	*Gaz* (TJ)				*En. Re. Comb. & Déchets* (TJ)				(GWh)	(TJ)
	Gaz naturel	*Usines à gaz*	*Cokeries*	*Hauts fourneaux*	*Biomasse solide*	*Gaz/Liquides tirés de biomasse*	*Déchets urbains*	*Déchets industriels*	*Electricité*	*Chaleur*
Production	6573788	-	-	-	42824	-	-	-	366122	-
Imports	9310	-	-	-	1330	-	-	-	608	-
Exports	-458258	-	-	-	-2	-	-	-	-1061	-
Intl. Marine Bunkers	-	-	-	-	-	-	-	-	-	-
Stock Changes	-	-	-	-	-	-	-	-	-	-
DOMESTIC SUPPLY	**6124840**	-	-	-	**44152**	-	-	-	**365669**	-
Transfers and Stat. Diff.	174132	-	-	-	-	-	-	-	85	-
TRANSFORMATION	**1596612**	-	-	-	**5782**	-	-	-	-	-
Electricity and CHP Plants	1596612	-	-	-	-	-	-	-	-	-
Petroleum Refineries	-	-	-	-	-	-	-	-	-	-
Other Transform. Sector	-	-	-	-	5782	-	-	-	-	-
ENERGY SECTOR	**2034291**	-	-	-	-	-	-	-	**30235**	-
DISTRIBUTION LOSSES	**43257**	-	-	-	-	-	-	-	**36518**	-
FINAL CONSUMPTION	**2624812**	-	-	-	**38369**	-	-	-	**299001**	-
INDUSTRY SECTOR	**1495881**	-	-	-	**7789**	-	-	-	**53188**	-
Iron and Steel	-	-	-	-	-	-	-	-	-	-
Chemical and Petrochem.	-	-	-	-	-	-	-	-	-	-
Non-Metallic Minerals	-	-	-	-	-	-	-	-	-	-
Non-specified	-	-	-	-	-	-	-	-	-	-
TRANSPORT SECTOR	-	-	-	-	-	-	-	-	**893**	-
Air	-	-	-	-	-	-	-	-	-	-
Road	-	-	-	-	-	-	-	-	-	-
Non-specified	-	-	-	-	-	-	-	-	-	-
OTHER SECTORS	-	-	-	-	-	-	-	-	-	-
Agriculture	-	-	-	-	-	-	-	-	-	-
Comm. and Publ. Services	-	-	-	-	-	-	-	-	-	-
Residential	-	-	-	-	-	-	-	-	-	-
Non-specified	-	-	-	-	-	-	-	-	-	-
NON-ENERGY USE	-	-	-	-	-	-	-	-	-	-

Albania / Albanie

SUPPLY AND CONSUMPTION 1996	Coal (1000 tonnes)							Oil (1000 tonnes)			
	Coking Coal	Other Bit. Coal	Sub-Bit. Coal	Lignite	Peat	Oven and Gas Coke	Pat. Fuel and BKB	Crude Oil	NGL	Feed-stocks	Additives
Production	-	-	-	101	-	-	-	488	1	-	-
Imports	-	-	-	-	-	-	-	5	-	-	-
Exports	-	-	-	-	-	-	-	-	-	-	-
Intl. Marine Bunkers	-	-	-	-	-	-	-	-	-	-	-
Stock Changes	-	-	-	-	-	-	-	-	-	-	-
DOMESTIC SUPPLY	**-**	**-**	**-**	**101**	**-**	**-**	**-**	**493**	**1**	**-**	**-**
Transfers and Stat. Diff.	-	-	-	-	-	-	-	-	-	-	-
TRANSFORMATION	**-**	**-**	**-**	**-**	**-**	**-**	**-**	**493**	**1**	**-**	**-**
Electricity and CHP Plants	-	-	-	-	-	-	-	-	-	-	-
Petroleum Refineries	-	-	-	-	-	-	-	493	1	-	-
Other Transform. Sector	-	-	-	-	-	-	-	-	-	-	-
ENERGY SECTOR	**-**	**-**	**-**	**-**	**-**	**-**	**-**	**-**	**-**	**-**	**-**
DISTRIBUTION LOSSES	**-**	**-**	**-**	**-**	**-**	**-**	**-**	**-**	**-**	**-**	**-**
FINAL CONSUMPTION	**-**	**-**	**-**	**101**	**-**	**-**	**-**	**-**	**-**	**-**	**-**
INDUSTRY SECTOR	**-**	**-**	**-**	**-**	**-**	**-**	**-**	**-**	**-**	**-**	**-**
Iron and Steel	-	-	-	-	-	-	-	-	-	-	-
Chemical and Petrochem.	-	-	-	-	-	-	-	-	-	-	-
Non-Metallic Minerals	-	-	-	-	-	-	-	-	-	-	-
Non-specified	-	-	-	-	-	-	-	-	-	-	-
TRANSPORT SECTOR	**-**	**-**	**-**	**-**	**-**	**-**	**-**	**-**	**-**	**-**	**-**
Air	-	-	-	-	-	-	-	-	-	-	-
Road	-	-	-	-	-	-	-	-	-	-	-
Non-specified	-	-	-	-	-	-	-	-	-	-	-
OTHER SECTORS	**-**	**-**	**-**	**101**	**-**	**-**	**-**	**-**	**-**	**-**	**-**
Agriculture	-	-	-	-	-	-	-	-	-	-	-
Comm. and Publ. Services	-	-	-	101	-	-	-	-	-	-	-
Residential	-	-	-	-	-	-	-	-	-	-	-
Non-specified	-	-	-	-	-	-	-	-	-	-	-
NON-ENERGY USE	**-**	**-**	**-**	**-**	**-**	**-**	**-**	**-**	**-**	**-**	**-**

APPROVISIONNEMENT ET DEMANDE 1997	*Charbon* (1000 tonnes)							*Pétrole* (1000 tonnes)			
	Charbon à coke	*Autres charb. bit.*	*Charbon sous-bit.*	*Lignite*	*Tourbe*	*Coke de four/gaz*	*Agg./briq. de lignite*	*Pétrole brut*	*LGN*	*Produits d'aliment.*	*Additifs*
Production	-	-	-	70	-	-	-	360	1	-	-
Imports	-	-	-	-	-	-	-	6	-	-	-
Exports	-	-	-	-	-	-	-	-	-	-	-
Intl. Marine Bunkers	-	-	-	-	-	-	-	-	-	-	-
Stock Changes	-	-	-	-	-	-	-	-	-	-	-
DOMESTIC SUPPLY	**-**	**-**	**-**	**70**	**-**	**-**	**-**	**366**	**1**	**-**	**-**
Transfers and Stat. Diff.	-	-	-	-	-	-	-	-	-	-	-
TRANSFORMATION	**-**	**-**	**-**	**-**	**-**	**-**	**-**	**366**	**1**	**-**	**-**
Electricity and CHP Plants	-	-	-	-	-	-	-	-	-	-	-
Petroleum Refineries	-	-	-	-	-	-	-	366	1	-	-
Other Transform. Sector	-	-	-	-	-	-	-	-	-	-	-
ENERGY SECTOR	**-**	**-**	**-**	**-**	**-**	**-**	**-**	**-**	**-**	**-**	**-**
DISTRIBUTION LOSSES	**-**	**-**	**-**	**-**	**-**	**-**	**-**	**-**	**-**	**-**	**-**
FINAL CONSUMPTION	**-**	**-**	**-**	**70**	**-**	**-**	**-**	**-**	**-**	**-**	**-**
INDUSTRY SECTOR	**-**	**-**	**-**	**-**	**-**	**-**	**-**	**-**	**-**	**-**	**-**
Iron and Steel	-	-	-	-	-	-	-	-	-	-	-
Chemical and Petrochem.	-	-	-	-	-	-	-	-	-	-	-
Non-Metallic Minerals	-	-	-	-	-	-	-	-	-	-	-
Non-specified	-	-	-	-	-	-	-	-	-	-	-
TRANSPORT SECTOR	**-**	**-**	**-**	**-**	**-**	**-**	**-**	**-**	**-**	**-**	**-**
Air	-	-	-	-	-	-	-	-	-	-	-
Road	-	-	-	-	-	-	-	-	-	-	-
Non-specified	-	-	-	-	-	-	-	-	-	-	-
OTHER SECTORS	**-**	**-**	**-**	**70**	**-**	**-**	**-**	**-**	**-**	**-**	**-**
Agriculture	-	-	-	-	-	-	-	-	-	-	-
Comm. and Publ. Services	-	-	-	70	-	-	-	-	-	-	-
Residential	-	-	-	-	-	-	-	-	-	-	-
Non-specified	-	-	-	-	-	-	-	-	-	-	-
NON-ENERGY USE	**-**	**-**	**-**	**-**	**-**	**-**	**-**	**-**	**-**	**-**	**-**

Albania / Albanie

SUPPLY AND CONSUMPTION 1996	Oil cont. (1000 tonnes)										
	Refinery Gas	LPG + Ethane	Motor Gasoline	Aviation Gasoline	Jet Fuel	Kerosene	Gas/ Diesel	Heavy Fuel Oil	Naphtha	Petrol. Coke	Other Prod.
Production	40	-	74	-	-	69	108	46	-	61	36
Imports	-	-	99	-	-	-	-	-	-	-	-
Exports	-	-	-	-	-	-	-	-	-	-	-
Intl. Marine Bunkers	-	-	-	-	-	-	-	-	-	-	-
Stock Changes	-	-	-	-	-	-	-	-	-	-	-
DOMESTIC SUPPLY	**40**	**-**	**173**	**-**	**-**	**69**	**108**	**46**	**-**	**61**	**36**
Transfers and Stat. Diff.	-	-	-	-	-	-	-	-	-	-	-
TRANSFORMATION	**-**	**-**	**-**	**-**	**-**	**-**	**-**	**46**	**-**	**-**	**-**
Electricity and CHP Plants	-	-	-	-	-	-	-	46	-	-	-
Petroleum Refineries	-	-	-	-	-	-	-	-	-	-	-
Other Transform. Sector	-	-	-	-	-	-	-	-	-	-	-
ENERGY SECTOR	**40**	**-**	**-**	**-**	**-**	**-**	**-**	**-**	**-**	**-**	**-**
DISTRIBUTION LOSSES	**-**	**-**	**-**	**-**	**-**	**-**	**-**	**-**	**-**	**-**	**-**
FINAL CONSUMPTION	**-**	**-**	**173**	**-**	**-**	**69**	**108**	**-**	**-**	**61**	**36**
INDUSTRY SECTOR	**-**	**-**	**-**	**-**	**-**	**-**	**-**	**-**	**-**	**61**	**-**
Iron and Steel	-	-	-	-	-	-	-	-	-	-	-
Chemical and Petrochem.	-	-	-	-	-	-	-	-	-	-	-
Non-Metallic Minerals	-	-	-	-	-	-	-	-	-	-	-
Non-specified	-	-	-	-	-	-	-	-	-	61	-
TRANSPORT SECTOR	**-**	**-**	**173**	**-**	**-**	**-**	**108**	**-**	**-**	**-**	**-**
Air	-	-	-	-	-	-	-	-	-	-	-
Road	-	-	173	-	-	-	108	-	-	-	-
Non-specified	-	-	-	-	-	-	-	-	-	-	-
OTHER SECTORS	**-**	**-**	**-**	**-**	**-**	**69**	**-**	**-**	**-**	**-**	**-**
Agriculture	-	-	-	-	-	-	-	-	-	-	-
Comm. and Publ. Services	-	-	-	-	-	-	-	-	-	-	-
Residential	-	-	-	-	-	69	-	-	-	-	-
Non-specified	-	-	-	-	-	-	-	-	-	-	-
NON-ENERGY USE	**-**	**-**	**-**	**-**	**-**	**-**	**-**	**-**	**-**	**-**	**36**

APPROVISIONNEMENT ET DEMANDE 1997	*Pétrole cont.* (1000 tonnes)										
	Gaz de raffinerie	*GPL + éthane*	*Essence moteur*	*Essence aviation*	*Carbu- réacteurs*	*Kérosène*	*Gazole*	*Fioul lourd*	*Naphta*	*Coke de pétrole*	*Autres prod.*
Production	30	-	74	-	-	57	107	39	-	29	14
Imports	-	-	105	-	-	-	-	-	-	-	-
Exports	-	-	-	-	-	-	-	-	-	-	-
Intl. Marine Bunkers	-	-	-	-	-	-	-	-	-	-	-
Stock Changes	-	-	-	-	-	-	-	-	-	-	-
DOMESTIC SUPPLY	**30**	**-**	**179**	**-**	**-**	**57**	**107**	**39**	**-**	**29**	**14**
Transfers and Stat. Diff.	-	-	-	-	-	-	-	-	-	-	-
TRANSFORMATION	**-**	**-**	**-**	**-**	**-**	**-**	**-**	**39**	**-**	**-**	**-**
Electricity and CHP Plants	-	-	-	-	-	-	-	39	-	-	-
Petroleum Refineries	-	-	-	-	-	-	-	-	-	-	-
Other Transform. Sector	-	-	-	-	-	-	-	-	-	-	-
ENERGY SECTOR	**30**	**-**	**-**	**-**	**-**	**-**	**-**	**-**	**-**	**-**	**-**
DISTRIBUTION LOSSES	**-**	**-**	**-**	**-**	**-**	**-**	**-**	**-**	**-**	**-**	**-**
FINAL CONSUMPTION	**-**	**-**	**179**	**-**	**-**	**57**	**107**	**-**	**-**	**29**	**14**
INDUSTRY SECTOR	**-**	**-**	**-**	**-**	**-**	**-**	**-**	**-**	**-**	**29**	**-**
Iron and Steel	-	-	-	-	-	-	-	-	-	-	-
Chemical and Petrochem.	-	-	-	-	-	-	-	-	-	-	-
Non-Metallic Minerals	-	-	-	-	-	-	-	-	-	-	-
Non-specified	-	-	-	-	-	-	-	-	-	29	-
TRANSPORT SECTOR	**-**	**-**	**179**	**-**	**-**	**-**	**107**	**-**	**-**	**-**	**-**
Air	-	-	-	-	-	-	-	-	-	-	-
Road	-	-	179	-	-	-	107	-	-	-	-
Non-specified	-	-	-	-	-	-	-	-	-	-	-
OTHER SECTORS	**-**	**-**	**-**	**-**	**-**	**57**	**-**	**-**	**-**	**-**	**-**
Agriculture	-	-	-	-	-	-	-	-	-	-	-
Comm. and Publ. Services	-	-	-	-	-	-	-	-	-	-	-
Residential	-	-	-	-	-	57	-	-	-	-	-
Non-specified	-	-	-	-	-	-	-	-	-	-	-
NON-ENERGY USE	**-**	**-**	**-**	**-**	**-**	**-**	**-**	**-**	**-**	**-**	**14**

Albania / Albanie

SUPPLY AND CONSUMPTION 1996	Gas (TJ)				Comb. Renew. & Waste (TJ)				(GWh)	(TJ)
	Natural Gas	Gas Works	Coke Ovens	Blast Furnaces	Solid Biomass	Gas/Liquids from Biomass	Municipal Waste	Industrial Waste	Electricity	Heat
Production	894	-	-	-	2500	-	-	-	5926	1189
Imports	-	-	-	-	-	-	-	-	200	-
Exports	-	-	-	-	-	-	-	-	-	-
Intl. Marine Bunkers	-	-	-	-	-	-	-	-	-	-
Stock Changes	-	-	-	-	-	-	-	-	-	-
DOMESTIC SUPPLY	**894**	**-**	**-**	**-**	**2500**	**-**	**-**	**-**	**6126**	**1189**
Transfers and Stat. Diff.	-	-	-	-	-	-	-	-	-	-
TRANSFORMATION	**-**	**-**	**-**	**-**	**-**	**-**	**-**	**-**	**-**	**-**
Electricity and CHP Plants	-	-	-	-	-	-	-	-	-	-
Petroleum Refineries	-	-	-	-	-	-	-	-	-	-
Other Transform. Sector	-	-	-	-	-	-	-	-	-	-
ENERGY SECTOR	**-**	**-**	**-**	**-**	**-**	**-**	**-**	**-**	**50**	**1048**
DISTRIBUTION LOSSES	**-**	**-**	**-**	**-**	**-**	**-**	**-**	**-**	**3105**	**-**
FINAL CONSUMPTION	**894**	**-**	**-**	**-**	**2500**	**-**	**-**	**-**	**2971**	**141**
INDUSTRY SECTOR	**735**	**-**	**-**	**-**	**-**	**-**	**-**	**-**	**797**	**-**
Iron and Steel	-	-	-	-	-	-	-	-	-	-
Chemical and Petrochem.	319	-	-	-	-	-	-	-	-	-
Non-Metallic Minerals	32	-	-	-	-	-	-	-	-	-
Non-specified	384	-	-	-	-	-	-	-	797	-
TRANSPORT SECTOR	**-**	**-**	**-**	**-**	**-**	**-**	**-**	**-**	**-**	**-**
Air	-	-	-	-	-	-	-	-	-	-
Road	-	-	-	-	-	-	-	-	-	-
Non-specified	-	-	-	-	-	-	-	-	-	-
OTHER SECTORS	**159**	**-**	**-**	**-**	**2500**	**-**	**-**	**-**	**2174**	**141**
Agriculture	-	-	-	-	-	-	-	-	54	-
Comm. and Publ. Services	-	-	-	-	-	-	-	-	47	-
Residential	127	-	-	-	2500	-	-	-	1278	141
Non-specified	32	-	-	-	-	-	-	-	795	-
NON-ENERGY USE	**-**	**-**	**-**	**-**	**-**	**-**	**-**	**-**	**-**	**-**

APPROVISIONNEMENT ET DEMANDE 1997	*Gaz* (TJ)				*En. Re. Comb. & Déchets* (TJ)				(GWh)	(TJ)
	Gaz naturel	*Usines à gaz*	*Cokeries*	*Hauts fourneaux*	*Biomasse solide*	*Gaz/Liquides tirés de biomasse*	*Déchets urbains*	*Déchets industriels*	*Electricité*	*Chaleur*
Production	700	-	-	-	2500	-	-	-	5600	1189
Imports	-	-	-	-	-	-	-	-	200	-
Exports	-	-	-	-	-	-	-	-	-	-
Intl. Marine Bunkers	-	-	-	-	-	-	-	-	-	-
Stock Changes	-	-	-	-	-	-	-	-	-	-
DOMESTIC SUPPLY	**700**	**-**	**-**	**-**	**2500**	**-**	**-**	**-**	**5800**	**1189**
Transfers and Stat. Diff.	-	-	-	-	-	-	-	-	-	-
TRANSFORMATION	**-**	**-**	**-**	**-**	**-**	**-**	**-**	**-**	**-**	**-**
Electricity and CHP Plants	-	-	-	-	-	-	-	-	-	-
Petroleum Refineries	-	-	-	-	-	-	-	-	-	-
Other Transform. Sector	-	-	-	-	-	-	-	-	-	-
ENERGY SECTOR	**-**	**-**	**-**	**-**	**-**	**-**	**-**	**-**	**47**	**1048**
DISTRIBUTION LOSSES	**-**	**-**	**-**	**-**	**-**	**-**	**-**	**-**	**2940**	**-**
FINAL CONSUMPTION	**700**	**-**	**-**	**-**	**2500**	**-**	**-**	**-**	**2813**	**141**
INDUSTRY SECTOR	**583**	**-**	**-**	**-**	**-**	**-**	**-**	**-**	**755**	**-**
Iron and Steel	-	-	-	-	-	-	-	-	-	-
Chemical and Petrochem.	233	-	-	-	-	-	-	-	-	-
Non-Metallic Minerals	39	-	-	-	-	-	-	-	-	-
Non-specified	311	-	-	-	-	-	-	-	755	-
TRANSPORT SECTOR	**-**	**-**	**-**	**-**	**-**	**-**	**-**	**-**	**-**	**-**
Air	-	-	-	-	-	-	-	-	-	-
Road	-	-	-	-	-	-	-	-	-	-
Non-specified	-	-	-	-	-	-	-	-	-	-
OTHER SECTORS	**117**	**-**	**-**	**-**	**2500**	**-**	**-**	**-**	**2058**	**141**
Agriculture	-	-	-	-	-	-	-	-	51	-
Comm. and Publ. Services	-	-	-	-	-	-	-	-	44	-
Residential	78	-	-	-	2500	-	-	-	1210	141
Non-specified	39	-	-	-	-	-	-	-	753	-
NON-ENERGY USE	**-**	**-**	**-**	**-**	**-**	**-**	**-**	**-**	**-**	**-**

Algeria / Algérie : 1996

SUPPLY AND CONSUMPTION / *APPROVISIONNEMENT ET DEMANDE*	Coal / *Charbon* (1000 tonnes)							Oil / *Pétrole* (1000 tonnes)			
	Coking Coal / *Charbon à coke*	Other Bit. Coal / *Autres charb. bit.*	Sub-Bit. Coal / *Charbon sous-bit.*	Lignite / *Lignite*	Peat / *Tourbe*	Oven and Gas Coke / *Coke de four/gaz*	Pat. Fuel and BKB / *Agg./briq. de lignite*	Crude Oil / *Pétrole brut*	NGL / *LGN*	Feed-stocks / *Produits d'aliment.*	Additives / *Additifs*
Production	-	-	-	-	-	555	-	37470	21252	-	-
From Other Sources	-	-	-	-	-	-	-	-	-	-	-
Imports	761	-	-	-	-	-	-	352	-	-	-
Exports	-	-	-	-	-	-7	-	-18342	-16303	-	-
Intl. Marine Bunkers	-	-	-	-	-	-	-	-	-	-	-
Stock Changes	1	-	-	-	-	82	-	55	-26	-	-
DOMESTIC SUPPLY	**762**	**-**	**-**	**-**	**-**	**630**	**-**	**19535**	**4923**	**-**	**-**
Transfers	-	-	-	-	-	-	-	-	-4923	-	-
Statistical Differences	-45	-	-	-	-	-1	-	222	62	-	-
TRANSFORMATION	**717**	**-**	**-**	**-**	**-**	**497**	**-**	**18888**	**-**	**-**	**-**
Electricity Plants	-	-	-	-	-	-	-	-	-	-	-
CHP Plants	-	-	-	-	-	-	-	-	-	-	-
Heat Plants	-	-	-	-	-	-	-	-	-	-	-
Blast Furnaces/Gas Works	-	-	-	-	-	497	-	-	-	-	-
Coke/Pat. Fuel/BKB Plants	717	-	-	-	-	-	-	-	-	-	-
Petroleum Refineries	-	-	-	-	-	-	-	18888	-	-	-
Petrochemical Industry	-	-	-	-	-	-	-	-	-	-	-
Liquefaction	-	-	-	-	-	-	-	-	-	-	-
Other Transform. Sector	-	-	-	-	-	-	-	-	-	-	-
ENERGY SECTOR	**-**	**-**	**-**	**-**	**-**	**-**	**-**	**447**	**-**	**-**	**-**
Coal Mines	-	-	-	-	-	-	-	-	-	-	-
Oil and Gas Extraction	-	-	-	-	-	-	-	31	-	-	-
Petroleum Refineries	-	-	-	-	-	-	-	416	-	-	-
Electr., CHP+Heat Plants	-	-	-	-	-	-	-	-	-	-	-
Pumped Storage (Elec.)	-	-	-	-	-	-	-	-	-	-	-
Other Energy Sector	-	-	-	-	-	-	-	-	-	-	-
Distribution Losses	-	-	-	-	-	-	-	394	62	-	-
FINAL CONSUMPTION	**-**	**-**	**-**	**-**	**-**	**132**	**-**	**28**	**-**	**-**	**-**
INDUSTRY SECTOR	**-**	**-**	**-**	**-**	**-**	**132**	**-**	**28**	**-**	**-**	**-**
Iron and Steel	-	-	-	-	-	124	-	-	-	-	-
Chemical and Petrochem.	-	-	-	-	-	-	-	-	-	-	-
of which: Feedstocks	-	-	-	-	-	-	-	-	-	-	-
Non-Ferrous Metals	-	-	-	-	-	-	-	-	-	-	-
Non-Metallic Minerals	-	-	-	-	-	-	-	-	-	-	-
Transport Equipment	-	-	-	-	-	-	-	-	-	-	-
Machinery	-	-	-	-	-	-	-	-	-	-	-
Mining and Quarrying	-	-	-	-	-	-	-	-	-	-	-
Food and Tobacco	-	-	-	-	-	-	-	-	-	-	-
Paper, Pulp and Print	-	-	-	-	-	-	-	-	-	-	-
Wood and Wood Products	-	-	-	-	-	-	-	-	-	-	-
Construction	-	-	-	-	-	-	-	-	-	-	-
Textile and Leather	-	-	-	-	-	-	-	-	-	-	-
Non-specified	-	-	-	-	-	8	-	28	-	-	-
TRANSPORT SECTOR	**-**	**-**	**-**	**-**	**-**	**-**	**-**	**-**	**-**	**-**	**-**
Air	-	-	-	-	-	-	-	-	-	-	-
Road	-	-	-	-	-	-	-	-	-	-	-
Rail	-	-	-	-	-	-	-	-	-	-	-
Pipeline Transport	-	-	-	-	-	-	-	-	-	-	-
Internal Navigation	-	-	-	-	-	-	-	-	-	-	-
Non-specified	-	-	-	-	-	-	-	-	-	-	-
OTHER SECTORS	**-**	**-**	**-**	**-**	**-**	**-**	**-**	**-**	**-**	**-**	**-**
Agriculture	-	-	-	-	-	-	-	-	-	-	-
Comm. and Publ. Services	-	-	-	-	-	-	-	-	-	-	-
Residential	-	-	-	-	-	-	-	-	-	-	-
Non-specified	-	-	-	-	-	-	-	-	-	-	-
NON-ENERGY USE	**-**	**-**	**-**	**-**	**-**	**-**	**-**	**-**	**-**	**-**	**-**
in Industry/Trans./Energy	-	-	-	-	-	-	-	-	-	-	-
in Transport	-	-	-	-	-	-	-	-	-	-	-
in Other Sectors	-	-	-	-	-	-	-	-	-	-	-

Algeria / Algérie : 1996

SUPPLY AND CONSUMPTION / *APPROVISIONNEMENT ET DEMANDE*	Oil cont. / *Pétrole cont.* (1000 tonnes)										
	Refinery Gas / *Gaz de raffinerie*	LPG + Ethane / *GPL + éthane*	Motor Gasoline / *Essence moteur*	Aviation Gasoline / *Essence aviation*	Jet Fuel / *Carbu-réacteurs*	Kerosene / *Kérosène*	Gas/ Diesel / *Gazole*	Heavy Fuel Oil / *Fioul lourd*	Naphtha / *Naphta*	Petrol. Coke / *Coke de pétrole*	Other Prod. / *Autres prod.*
Production	242	645	2336	-	1104	10	6143	4983	3464	-	324
From Other Sources	-	-	-	-	-	-	-	-	-	-	-
Imports	-	-	-	-	-	-	-	120	-	-	18
Exports	-	-3935	-324	-	-745	-	-2969	-4822	-3785	-	-
Intl. Marine Bunkers	-	-	-	-	-	-	-105	-229	-	-	-
Stock Changes	-	-	11	-	-	-	-132	-29	-	-	-5
DOMESTIC SUPPLY	**242**	**-3290**	**2023**	**-**	**359**	**10**	**2937**	**23**	**-321**	**-**	**337**
Transfers	-	4923	-	-	-	-	-	-	-	-	-
Statistical Differences	-	-53	-	-	-57	-	-	-	321	-	-
TRANSFORMATION	**-**	**-**	**-**	**-**	**-**	**-**	**206**	**23**	**-**	**-**	**-**
Electricity Plants	-	-	-	-	-	-	206	23	-	-	-
CHP Plants	-	-	-	-	-	-	-	-	-	-	-
Heat Plants	-	-	-	-	-	-	-	-	-	-	-
Blast Furnaces/Gas Works	-	-	-	-	-	-	-	-	-	-	-
Coke/Pat. Fuel/BKB Plants	-	-	-	-	-	-	-	-	-	-	-
Petroleum Refineries	-	-	-	-	-	-	-	-	-	-	-
Petrochemical Industry	-	-	-	-	-	-	-	-	-	-	-
Liquefaction	-	-	-	-	-	-	-	-	-	-	-
Other Transform. Sector	-	-	-	-	-	-	-	-	-	-	-
ENERGY SECTOR	**242**	**-**	**-**	**-**	**-**	**-**	**-**	**-**	**-**	**-**	**-**
Coal Mines	-	-	-	-	-	-	-	-	-	-	-
Oil and Gas Extraction	-	-	-	-	-	-	-	-	-	-	-
Petroleum Refineries	242	-	-	-	-	-	-	-	-	-	-
Electr., CHP+Heat Plants	-	-	-	-	-	-	-	-	-	-	-
Pumped Storage (Elec.)	-	-	-	-	-	-	-	-	-	-	-
Other Energy Sector	-	-	-	-	-	-	-	-	-	-	-
Distribution Losses	-	-	-	-	-	-	-	-	-	-	-
FINAL CONSUMPTION	**-**	**1580**	**2023**	**-**	**302**	**10**	**2731**	**-**	**-**	**-**	**337**
INDUSTRY SECTOR	**-**	**14**	**-**	**-**	**-**	**-**	**-**	**-**	**-**	**-**	**-**
Iron and Steel	-	-	-	-	-	-	-	-	-	-	-
Chemical and Petrochem.	-	-	-	-	-	-	-	-	-	-	-
of which: Feedstocks	-	-	-	-	-	-	-	-	-	-	-
Non-Ferrous Metals	-	-	-	-	-	-	-	-	-	-	-
Non-Metallic Minerals	-	-	-	-	-	-	-	-	-	-	-
Transport Equipment	-	-	-	-	-	-	-	-	-	-	-
Machinery	-	-	-	-	-	-	-	-	-	-	-
Mining and Quarrying	-	-	-	-	-	-	-	-	-	-	-
Food and Tobacco	-	-	-	-	-	-	-	-	-	-	-
Paper, Pulp and Print	-	-	-	-	-	-	-	-	-	-	-
Wood and Wood Products	-	-	-	-	-	-	-	-	-	-	-
Construction	-	-	-	-	-	-	-	-	-	-	-
Textile and Leather	-	-	-	-	-	-	-	-	-	-	-
Non-specified	-	14	-	-	-	-	-	-	-	-	-
TRANSPORT SECTOR	**-**	**70**	**2023**	**-**	**302**	**-**	**-**	**-**	**-**	**-**	**-**
Air	-	-	-	-	302	-	-	-	-	-	-
Road	-	70	2023	-	-	-	-	-	-	-	-
Rail	-	-	-	-	-	-	-	-	-	-	-
Pipeline Transport	-	-	-	-	-	-	-	-	-	-	-
Internal Navigation	-	-	-	-	-	-	-	-	-	-	-
Non-specified	-	-	-	-	-	-	-	-	-	-	-
OTHER SECTORS	**-**	**1307**	**-**	**-**	**-**	**10**	**2731**	**-**	**-**	**-**	**-**
Agriculture	-	-	-	-	-	-	-	-	-	-	-
Comm. and Publ. Services	-	-	-	-	-	-	-	-	-	-	-
Residential	-	1307	-	-	-	10	-	-	-	-	-
Non-specified	-	-	-	-	-	-	2731	-	-	-	-
NON-ENERGY USE	**-**	**189**	**-**	**-**	**-**	**-**	**-**	**-**	**-**	**-**	**337**
in Industry/Transf./Energy	-	189	-	-	-	-	-	-	-	-	337
in Transport	-	-	-	-	-	-	-	-	-	-	-
in Other Sectors	-	-	-	-	-	-	-	-	-	-	-

Algeria / Algérie : 1996

SUPPLY AND CONSUMPTION / *APPROVISIONNEMENT ET DEMANDE*	Gas / *Gaz* (TJ)				Comb. Renew. & Waste / *En. Re. Comb. & Déchets* (TJ)				(GWh)	(TJ)
	Natural Gas / *Gaz naturel*	Gas Works / *Usines à gaz*	Coke Ovens / *Cokeries*	Blast Furnaces / *Hauts fourneaux*	Solid Biomass / *Biomasse solide*	Gas/Liquids from Biomass / *Gaz/Liquides tirés de biomasse*	Municipal Waste / *Déchets urbains*	Industrial Waste / *Déchets industriels*	Electricity / *Electricité*	Heat / *Chaleur*
Production	2558440	-	1024	9715	21264	-	-	-	20650	-
From Other Sources	-	-	-	-	-	-	-	-	-	-
Imports	-	-	-	-	-	-	-	-	280	-
Exports	-1825284	-	-	-	-	-	-	-	-422	-
Intl. Marine Bunkers	-	-	-	-	-	-	-	-	-	-
Stock Changes	-	-	-	-	-	-	-	-	-	-
DOMESTIC SUPPLY	**733156**	**-**	**1024**	**9715**	**21264**	**-**	**-**	**-**	**20508**	**-**
Transfers	-	-	-	-	-	-	-	-	-	-
Statistical Differences	2049	-	46	42	-	-	-	-	3	-
TRANSFORMATION	**260422**	**-**	**-**	**-**	**-**	**-**	**-**	**-**	**-**	**-**
Electricity Plants	260422	-	-	-	-	-	-	-	-	-
CHP Plants	-	-	-	-	-	-	-	-	-	-
Heat Plants	-	-	-	-	-	-	-	-	-	-
Blast Furnaces/Gas Works	-	-	-	-	-	-	-	-	-	-
Coke/Pat. Fuel/BKB Plants	-	-	-	-	-	-	-	-	-	-
Petroleum Refineries	-	-	-	-	-	-	-	-	-	-
Petrochemical Industry	-	-	-	-	-	-	-	-	-	-
Liquefaction	-	-	-	-	-	-	-	-	-	-
Other Transform. Sector	-	-	-	-	-	-	-	-	-	-
ENERGY SECTOR	**241439**	**-**	**-**	**-**	**-**	**-**	**-**	**-**	**1662**	**-**
Coal Mines	-	-	-	-	-	-	-	-	-	-
Oil and Gas Extraction	210637	-	-	-	-	-	-	-	257	-
Petroleum Refineries	23590	-	-	-	-	-	-	-	152	-
Electr., CHP+Heat Plants	-	-	-	-	-	-	-	-	1253	-
Pumped Storage (Elec.)	-	-	-	-	-	-	-	-	-	-
Other Energy Sector	7212	-	-	-	-	-	-	-	-	-
Distribution Losses	7445	-	512	7789	-	-	-	-	3814	-
FINAL CONSUMPTION	**225899**	**-**	**558**	**1968**	**21264**	**-**	**-**	**-**	**15035**	**-**
INDUSTRY SECTOR	**106971**	**-**	**558**	**1968**	**-**	**-**	**-**	**-**	**6259**	**-**
Iron and Steel	11400	-	558	1968	-	-	-	-	625	-
Chemical and Petrochem.	39271	-	-	-	-	-	-	-	720	-
of which: Feedstocks	*37875*	-	-	-	-	-	-	-	-	-
Non-Ferrous Metals	-	-	-	-	-	-	-	-	-	-
Non-Metallic Minerals	-	-	-	-	-	-	-	-	-	-
Transport Equipment	-	-	-	-	-	-	-	-	-	-
Machinery	-	-	-	-	-	-	-	-	-	-
Mining and Quarrying	-	-	-	-	-	-	-	-	-	-
Food and Tobacco	-	-	-	-	-	-	-	-	-	-
Paper, Pulp and Print	-	-	-	-	-	-	-	-	-	-
Wood and Wood Products	-	-	-	-	-	-	-	-	-	-
Construction	46901	-	-	-	-	-	-	-	1432	-
Textile and Leather	-	-	-	-	-	-	-	-	-	-
Non-specified	9399	-	-	-	-	-	-	-	3482	-
TRANSPORT SECTOR	**26382**	**-**	**-**	**-**	**-**	**-**	**-**	**-**	**338**	**-**
Air	-	-	-	-	-	-	-	-	-	-
Road	-	-	-	-	-	-	-	-	-	-
Rail	-	-	-	-	-	-	-	-	230	-
Pipeline Transport	26382	-	-	-	-	-	-	-	108	-
Internal Navigation	-	-	-	-	-	-	-	-	-	-
Non-specified	-	-	-	-	-	-	-	-	-	-
OTHER SECTORS	**92546**	**-**	**-**	**-**	**21264**	**-**	**-**	**-**	**8438**	**-**
Agriculture	-	-	-	-	-	-	-	-	-	-
Comm. and Publ. Services	-	-	-	-	-	-	-	-	-	-
Residential	92546	-	-	-	21264	-	-	-	8438	-
Non-specified	-	-	-	-	-	-	-	-	-	-
NON-ENERGY USE	**-**	**-**	**-**	**-**	**-**	**-**	**-**	**-**	**-**	**-**
in Industry/Transf./Energy	-	-	-	-	-	-	-	-	-	-
in Transport	-	-	-	-	-	-	-	-	-	-
in Other Sectors	-	-	-	-	-	-	-	-	-	-

Algeria / Algérie : 1997

SUPPLY AND CONSUMPTION *APPROVISIONNEMENT ET DEMANDE*	Coal / *Charbon* (1000 tonnes)							Oil / *Pétrole* (1000 tonnes)			
	Coking Coal *Charbon à coke*	Other Bit. Coal *Autres charb. bit.*	Sub-Bit. Coal *Charbon sous-bit.*	Lignite *Lignite*	Peat *Tourbe*	Oven and Gas Coke *Coke de four/gaz*	Pat. Fuel and BKB *Agg./briq. de lignite*	Crude Oil *Pétrole brut*	NGL *LGN*	Feed-stocks *Produits d'aliment.*	Additives *Additifs*
Production	-	-	-	-	-	354	-	37654	21988	-	-
From Other Sources	-	-	-	-	-	-	-	-	-	-	-
Imports	552	-	-	-	-	59	-	253	-	-	-
Exports	-	-	-	-	-	-	-	-17354	-16029	-	-
Intl. Marine Bunkers	-	-	-	-	-	-	-	-	-	-	-
Stock Changes	-108	-	-	-	-	-60	-	-	-	-	-
DOMESTIC SUPPLY	**444**	**-**	**-**	**-**	**-**	**353**	**-**	**20553**	**5959**	**-**	**-**
Transfers	-	-	-	-	-	-	-	-	-5959	-	-
Statistical Differences	-9	-	-	-	-	-4	-	730	-	-	-
TRANSFORMATION	**435**	**-**	**-**	**-**	**-**	**279**	**-**	**20582**	**-**	**-**	**-**
Electricity Plants	-	-	-	-	-	-	-	-	-	-	-
CHP Plants	-	-	-	-	-	-	-	-	-	-	-
Heat Plants	-	-	-	-	-	-	-	-	-	-	-
Blast Furnaces/Gas Works	-	-	-	-	-	279	-	-	-	-	-
Coke/Pat. Fuel/BKB Plants	435	-	-	-	-	-	-	-	-	-	-
Petroleum Refineries	-	-	-	-	-	-	-	20582	-	-	-
Petrochemical Industry	-	-	-	-	-	-	-	-	-	-	-
Liquefaction	-	-	-	-	-	-	-	-	-	-	-
Other Transform. Sector	-	-	-	-	-	-	-	-	-	-	-
ENERGY SECTOR	**-**	**-**	**-**	**-**	**-**	**-**	**-**	**356**	**-**	**-**	**-**
Coal Mines	-	-	-	-	-	-	-	-	-	-	-
Oil and Gas Extraction	-	-	-	-	-	-	-	25	-	-	-
Petroleum Refineries	-	-	-	-	-	-	-	331	-	-	-
Electr., CHP+Heat Plants	-	-	-	-	-	-	-	-	-	-	-
Pumped Storage (Elec.)	-	-	-	-	-	-	-	-	-	-	-
Other Energy Sector	-	-	-	-	-	-	-	-	-	-	-
Distribution Losses	-	-	-	-	-	-	-	317	-	-	-
FINAL CONSUMPTION	**-**	**-**	**-**	**-**	**-**	**70**	**-**	**28**	**-**	**-**	**-**
INDUSTRY SECTOR	**-**	**-**	**-**	**-**	**-**	**70**	**-**	**28**	**-**	**-**	**-**
Iron and Steel	-	-	-	-	-	70	-	-	-	-	-
Chemical and Petrochem.	-	-	-	-	-	-	-	-	-	-	-
of which: Feedstocks	-	-	-	-	-	-	-	-	-	-	-
Non-Ferrous Metals	-	-	-	-	-	-	-	-	-	-	-
Non-Metallic Minerals	-	-	-	-	-	-	-	-	-	-	-
Transport Equipment	-	-	-	-	-	-	-	-	-	-	-
Machinery	-	-	-	-	-	-	-	-	-	-	-
Mining and Quarrying	-	-	-	-	-	-	-	-	-	-	-
Food and Tobacco	-	-	-	-	-	-	-	-	-	-	-
Paper, Pulp and Print	-	-	-	-	-	-	-	-	-	-	-
Wood and Wood Products	-	-	-	-	-	-	-	-	-	-	-
Construction	-	-	-	-	-	-	-	-	-	-	-
Textile and Leather	-	-	-	-	-	-	-	-	-	-	-
Non-specified	-	-	-	-	-	-	-	28	-	-	-
TRANSPORT SECTOR	**-**	**-**	**-**	**-**	**-**	**-**	**-**	**-**	**-**	**-**	**-**
Air	-	-	-	-	-	-	-	-	-	-	-
Road	-	-	-	-	-	-	-	-	-	-	-
Rail	-	-	-	-	-	-	-	-	-	-	-
Pipeline Transport	-	-	-	-	-	-	-	-	-	-	-
Internal Navigation	-	-	-	-	-	-	-	-	-	-	-
Non-specified	-	-	-	-	-	-	-	-	-	-	-
OTHER SECTORS	**-**	**-**	**-**	**-**	**-**	**-**	**-**	**-**	**-**	**-**	**-**
Agriculture	-	-	-	-	-	-	-	-	-	-	-
Comm. and Publ. Services	-	-	-	-	-	-	-	-	-	-	-
Residential	-	-	-	-	-	-	-	-	-	-	-
Non-specified	-	-	-	-	-	-	-	-	-	-	-
NON-ENERGY USE	**-**	**-**	**-**	**-**	**-**	**-**	**-**	**-**	**-**	**-**	**-**
in Industry/Trans./Energy	-	-	-	-	-	-	-	-	-	-	-
in Transport	-	-	-	-	-	-	-	-	-	-	-
in Other Sectors	-	-	-	-	-	-	-	-	-	-	-

Algeria / Algérie : 1997

SUPPLY AND CONSUMPTION *APPROVISIONNEMENT ET DEMANDE*	Oil cont. / *Pétrole cont.* (1000 tonnes)										
	Refinery Gas *Gaz de raffinerie*	LPG + Ethane *GPL + éthane*	Motor Gasoline *Essence moteur*	Aviation Gasoline *Essence aviation*	Jet Fuel *Carbu- réacteurs*	Kerosene *Kérosène*	Gas/ Diesel *Gazole*	Heavy Fuel Oil *Fioul lourd*	Naphtha *Naphta*	Petrol. Coke *Coke de pétrole*	Other Prod. *Autres prod.*
Production	242	741	1989	-	1197	10	6523	5556	4043	-	281
From Other Sources	-	-	-	-	-	-	-	-	-	-	-
Imports	-	-	-	-	-	-	-	83	-	-	18
Exports	-	-5045	-242	-	-797	-	-1693	-5325	-4043	-	-
Intl. Marine Bunkers	-	-	-	-	-	-	-105	-125	-	-	-
Stock Changes	-	-	-	-	-	-	-125	-105	-	-	-
DOMESTIC SUPPLY	**242**	**-4304**	**1747**	**-**	**400**	**10**	**4600**	**84**	**-**	**-**	**299**
Transfers	-	5959	-	-	-	-	-	-	-	-	-
Statistical Differences	-	-54	-	-	-70	-	-	-51	-	-	-
TRANSFORMATION	**-**	**-**	**-**	**-**	**-**	**-**	**206**	**33**	**-**	**-**	**-**
Electricity Plants	-	-	-	-	-	-	206	33	-	-	-
CHP Plants	-	-	-	-	-	-	-	-	-	-	-
Heat Plants	-	-	-	-	-	-	-	-	-	-	-
Blast Furnaces/Gas Works	-	-	-	-	-	-	-	-	-	-	-
Coke/Pat. Fuel/BKB Plants	-	-	-	-	-	-	-	-	-	-	-
Petroleum Refineries	-	-	-	-	-	-	-	-	-	-	-
Petrochemical Industry	-	-	-	-	-	-	-	-	-	-	-
Liquefaction	-	-	-	-	-	-	-	-	-	-	-
Other Transform. Sector	-	-	-	-	-	-	-	-	-	-	-
ENERGY SECTOR	**242**	**-**	**-**	**-**	**-**	**-**	**-**	**-**	**-**	**-**	**-**
Coal Mines	-	-	-	-	-	-	-	-	-	-	-
Oil and Gas Extraction	-	-	-	-	-	-	-	-	-	-	-
Petroleum Refineries	242	-	-	-	-	-	-	-	-	-	-
Electr., CHP+Heat Plants	-	-	-	-	-	-	-	-	-	-	-
Pumped Storage (Elec.)	-	-	-	-	-	-	-	-	-	-	-
Other Energy Sector	-	-	-	-	-	-	-	-	-	-	-
Distribution Losses	-	-	-	-	-	-	-	-	-	-	-
FINAL CONSUMPTION	**-**	**1601**	**1747**	**-**	**330**	**10**	**4394**	**-**	**-**	**-**	**299**
INDUSTRY SECTOR	**-**	**14**	**-**	**-**	**-**	**-**	**-**	**-**	**-**	**-**	**-**
Iron and Steel	-	-	-	-	-	-	-	-	-	-	-
Chemical and Petrochem.	-	-	-	-	-	-	-	-	-	-	-
of which: Feedstocks	-	-	-	-	-	-	-	-	-	-	-
Non-Ferrous Metals	-	-	-	-	-	-	-	-	-	-	-
Non-Metallic Minerals	-	-	-	-	-	-	-	-	-	-	-
Transport Equipment	-	-	-	-	-	-	-	-	-	-	-
Machinery	-	-	-	-	-	-	-	-	-	-	-
Mining and Quarrying	-	-	-	-	-	-	-	-	-	-	-
Food and Tobacco	-	-	-	-	-	-	-	-	-	-	-
Paper, Pulp and Print	-	-	-	-	-	-	-	-	-	-	-
Wood and Wood Products	-	-	-	-	-	-	-	-	-	-	-
Construction	-	-	-	-	-	-	-	-	-	-	-
Textile and Leather	-	-	-	-	-	-	-	-	-	-	-
Non-specified	-	14	-	-	-	-	-	-	-	-	-
TRANSPORT SECTOR	**-**	**71**	**1747**	**-**	**330**	**-**	**-**	**-**	**-**	**-**	**-**
Air	-	-	-	-	330	-	-	-	-	-	-
Road	-	71	1747	-	-	-	-	-	-	-	-
Rail	-	-	-	-	-	-	-	-	-	-	-
Pipeline Transport	-	-	-	-	-	-	-	-	-	-	-
Internal Navigation	-	-	-	-	-	-	-	-	-	-	-
Non-specified	-	-	-	-	-	-	-	-	-	-	-
OTHER SECTORS	**-**	**1327**	**-**	**-**	**-**	**10**	**4394**	**-**	**-**	**-**	**-**
Agriculture	-	-	-	-	-	-	-	-	-	-	-
Comm. and Publ. Services	-	-	-	-	-	-	-	-	-	-	-
Residential	-	1327	-	-	-	10	-	-	-	-	-
Non-specified	-	-	-	-	-	-	4394	-	-	-	-
NON-ENERGY USE	**-**	**189**	**-**	**-**	**-**	**-**	**-**	**-**	**-**	**-**	**299**
in Industry/Transf./Energy	-	189	-	-	-	-	-	-	-	-	299
in Transport	-	-	-	-	-	-	-	-	-	-	-
in Other Sectors	-	-	-	-	-	-	-	-	-	-	-

Algeria / Algérie : 1997

SUPPLY AND CONSUMPTION / *APPROVISIONNEMENT ET DEMANDE*	Gas / *Gaz* (TJ)				Comb. Renew. & Waste / *En. Re. Comb. & Déchets* (TJ)				(GWh)	(TJ)
	Natural Gas / *Gaz naturel*	Gas Works / *Usines à gaz*	Coke Ovens / *Cokeries*	Blast Furnaces / *Hauts fourneaux*	Solid Biomass / *Biomasse solide*	Gas/Liquids from Biomass / *Gaz/Liquides tirés de biomasse*	Municipal Waste / *Déchets urbains*	Industrial Waste / *Déchets industriels*	Electricity / *Electricité*	Heat / *Chaleur*
Production	2949330	-	544	4510	21732	-	-	-	21685	-
From Other Sources	-	-	-	-	-	-	-	-	-	-
Imports	-	-	-	-	-	-	-	-	312	-
Exports	-2162014	-	-	-	-	-	-	-	-313	-
Intl. Marine Bunkers	-	-	-	-	-	-	-	-	-	-
Stock Changes	-	-	-	-	-	-	-	-	-	-
DOMESTIC SUPPLY	**787316**	**-**	**544**	**4510**	**21732**	**-**	**-**	**-**	**21684**	**-**
Transfers	-	-	-	-	-	-	-	-	-	-
Statistical Differences	-3070	-	22	40	-	-	-	-	-	-
TRANSFORMATION	**274474**	**-**	**-**	**-**	**-**	**-**	**-**	**-**	**-**	**-**
Electricity Plants	274474	-	-	-	-	-	-	-	-	-
CHP Plants	-	-	-	-	-	-	-	-	-	-
Heat Plants	-	-	-	-	-	-	-	-	-	-
Blast Furnaces/Gas Works	-	-	-	-	-	-	-	-	-	-
Coke/Pat. Fuel/BKB Plants	-	-	-	-	-	-	-	-	-	-
Petroleum Refineries	-	-	-	-	-	-	-	-	-	-
Petrochemical Industry	-	-	-	-	-	-	-	-	-	-
Liquefaction	-	-	-	-	-	-	-	-	-	-
Other Transform. Sector	-	-	-	-	-	-	-	-	-	-
ENERGY SECTOR	**267449**	**-**	**-**	**-**	**-**	**-**	**-**	**-**	**1926**	**-**
Coal Mines	-	-	-	-	-	-	-	-	-	-
Oil and Gas Extraction	236600	-	-	-	-	-	-	-	297	-
Petroleum Refineries	23823	-	-	-	-	-	-	-	191	-
Electr., CHP+Heat Plants	-	-	-	-	-	-	-	-	1438	-
Pumped Storage (Elec.)	-	-	-	-	-	-	-	-	-	-
Other Energy Sector	7026	-	-	-	-	-	-	-	-	-
Distribution Losses	8143	-	184	2582	-	-	-	-	3178	-
FINAL CONSUMPTION	**234180**	**-**	**382**	**1968**	**21732**	**-**	**-**	**-**	**16580**	**-**
INDUSTRY SECTOR	**116415**	**-**	**382**	**1968**	**-**	**-**	**-**	**-**	**7685**	**-**
Iron and Steel	10655	-	382	1968	-	-	-	-	529	-
Chemical and Petrochem.	53043	-	-	-	-	-	-	-	771	-
of which: Feedstocks	*51740*	-	-	-	-	-	-	-	-	-
Non-Ferrous Metals	-	-	-	-	-	-	-	-	-	-
Non-Metallic Minerals	-	-	-	-	-	-	-	-	-	-
Transport Equipment	-	-	-	-	-	-	-	-	-	-
Machinery	-	-	-	-	-	-	-	-	-	-
Mining and Quarrying	-	-	-	-	-	-	-	-	-	-
Food and Tobacco	-	-	-	-	-	-	-	-	-	-
Paper, Pulp and Print	-	-	-	-	-	-	-	-	-	-
Wood and Wood Products	-	-	-	-	-	-	-	-	-	-
Construction	43877	-	-	-	-	-	-	-	1383	-
Textile and Leather	-	-	-	-	-	-	-	-	-	-
Non-specified	8840	-	-	-	-	-	-	-	5002	-
TRANSPORT SECTOR	**31361**	**-**	**-**	**-**	**-**	**-**	**-**	**-**	**354**	**-**
Air	-	-	-	-	-	-	-	-	-	-
Road	-	-	-	-	-	-	-	-	-	-
Rail	-	-	-	-	-	-	-	-	267	-
Pipeline Transport	31361	-	-	-	-	-	-	-	87	-
Internal Navigation	-	-	-	-	-	-	-	-	-	-
Non-specified	-	-	-	-	-	-	-	-	-	-
OTHER SECTORS	**86404**	**-**	**-**	**-**	**21732**	**-**	**-**	**-**	**8541**	**-**
Agriculture	-	-	-	-	-	-	-	-	-	-
Comm. and Publ. Services	-	-	-	-	-	-	-	-	-	-
Residential	86404	-	-	-	21732	-	-	-	8541	-
Non-specified	-	-	-	-	-	-	-	-	-	-
NON-ENERGY USE	**-**	**-**	**-**	**-**	**-**	**-**	**-**	**-**	**-**	**-**
in Industry/Transf./Energy	-	-	-	-	-	-	-	-	-	-
in Transport	-	-	-	-	-	-	-	-	-	-
in Other Sectors	-	-	-	-	-	-	-	-	-	-

Angola

SUPPLY AND CONSUMPTION 1996	Coal (1000 tonnes)							Oil (1000 tonnes)			
	Coking Coal	Other Bit. Coal	Sub-Bit. Coal	Lignite	Peat	Oven and Gas Coke	Pat. Fuel and BKB	Crude Oil	NGL	Feed-stocks	Additives
Production	-	-	-	-	-	-	-	34689	-	-	-
Imports	-	-	-	-	-	-	-	-	-	-	-
Exports	-	-	-	-	-	-	-	-33024	-	-	-
Intl. Marine Bunkers	-	-	-	-	-	-	-	-	-	-	-
Stock Changes	-	-	-	-	-	-	-	-	-	-	-
DOMESTIC SUPPLY	-	-	-	-	-	-	-	**1665**	-	-	-
Transfers and Stat. Diff.	-	-	-	-	-	-	-	-1	-	-	-
TRANSFORMATION	-	-	-	-	-	-	-	**1600**	-	-	-
Electricity and CHP Plants	-	-	-	-	-	-	-	-	-	-	-
Petroleum Refineries	-	-	-	-	-	-	-	1600	-	-	-
Other Transform. Sector	-	-	-	-	-	-	-	-	-	-	-
ENERGY SECTOR	-	-	-	-	-	-	-	**64**	-	-	-
DISTRIBUTION LOSSES	-	-	-	-	-	-	-	-	-	-	-
FINAL CONSUMPTION	-	-	-	-	-	-	-	-	-	-	-
INDUSTRY SECTOR	-	-	-	-	-	-	-	-	-	-	-
Iron and Steel	-	-	-	-	-	-	-	-	-	-	-
Chemical and Petrochem.	-	-	-	-	-	-	-	-	-	-	-
Non-Metallic Minerals	-	-	-	-	-	-	-	-	-	-	-
Non-specified	-	-	-	-	-	-	-	-	-	-	-
TRANSPORT SECTOR	-	-	-	-	-	-	-	-	-	-	-
Air	-	-	-	-	-	-	-	-	-	-	-
Road	-	-	-	-	-	-	-	-	-	-	-
Non-specified	-	-	-	-	-	-	-	-	-	-	-
OTHER SECTORS	-	-	-	-	-	-	-	-	-	-	-
Agriculture	-	-	-	-	-	-	-	-	-	-	-
Comm. and Publ. Services	-	-	-	-	-	-	-	-	-	-	-
Residential	-	-	-	-	-	-	-	-	-	-	-
Non-specified	-	-	-	-	-	-	-	-	-	-	-
NON-ENERGY USE	-	-	-	-	-	-	-	-	-	-	-

APPROVISIONNEMENT ET DEMANDE 1997	*Charbon* (1000 tonnes)							*Pétrole* (1000 tonnes)			
	Charbon à coke	*Autres charb. bit.*	*Charbon sous-bit.*	*Lignite*	*Tourbe*	*Coke de four/gaz*	*Agg./briq. de lignite*	*Pétrole brut*	*LGN*	*Produits d'aliment.*	*Additifs*
Production	-	-	-	-	-	-	-	34966	-	-	-
Imports	-	-	-	-	-	-	-	-	-	-	-
Exports	-	-	-	-	-	-	-	-32983	-	-	-
Intl. Marine Bunkers	-	-	-	-	-	-	-	-	-	-	-
Stock Changes	-	-	-	-	-	-	-	-	-	-	-
DOMESTIC SUPPLY	-	-	-	-	-	-	-	**1983**	-	-	-
Transfers and Stat. Diff.	-	-	-	-	-	-	-	1	-	-	-
TRANSFORMATION	-	-	-	-	-	-	-	**1918**	-	-	-
Electricity and CHP Plants	-	-	-	-	-	-	-	-	-	-	-
Petroleum Refineries	-	-	-	-	-	-	-	1918	-	-	-
Other Transform. Sector	-	-	-	-	-	-	-	-	-	-	-
ENERGY SECTOR	-	-	-	-	-	-	-	**66**	-	-	-
DISTRIBUTION LOSSES	-	-	-	-	-	-	-	-	-	-	-
FINAL CONSUMPTION	-	-	-	-	-	-	-	-	-	-	-
INDUSTRY SECTOR	-	-	-	-	-	-	-	-	-	-	-
Iron and Steel	-	-	-	-	-	-	-	-	-	-	-
Chemical and Petrochem.	-	-	-	-	-	-	-	-	-	-	-
Non-Metallic Minerals	-	-	-	-	-	-	-	-	-	-	-
Non-specified	-	-	-	-	-	-	-	-	-	-	-
TRANSPORT SECTOR	-	-	-	-	-	-	-	-	-	-	-
Air	-	-	-	-	-	-	-	-	-	-	-
Road	-	-	-	-	-	-	-	-	-	-	-
Non-specified	-	-	-	-	-	-	-	-	-	-	-
OTHER SECTORS	-	-	-	-	-	-	-	-	-	-	-
Agriculture	-	-	-	-	-	-	-	-	-	-	-
Comm. and Publ. Services	-	-	-	-	-	-	-	-	-	-	-
Residential	-	-	-	-	-	-	-	-	-	-	-
Non-specified	-	-	-	-	-	-	-	-	-	-	-
NON-ENERGY USE	-	-	-	-	-	-	-	-	-	-	-

Angola

SUPPLY AND CONSUMPTION 1996	Oil cont. (1000 tonnes)										
	Refinery Gas	LPG + Ethane	Motor Gasoline	Aviation Gasoline	Jet Fuel	Kerosene	Gas/ Diesel	Heavy Fuel Oil	Naphtha	Petrol. Coke	Other Prod.
Production	43	20	122	-	178	56	355	711	7	-	32
Imports	-	7	7	-	-	-	7	7	-	-	-
Exports	-	-	-40	-	-	-	-5	-85	-	-	-
Intl. Marine Bunkers	-	-	-	-	-	-	-230	-430	-	-	-
Stock Changes	-	-	-	-	-	-	-	-	-	-	-
DOMESTIC SUPPLY	**43**	**27**	**89**	**-**	**178**	**56**	**127**	**203**	**7**	**-**	**32**
Transfers and Stat. Diff.	-	-	-	-	-	-	-	-	-	-	-
TRANSFORMATION	**-**	**-**	**-**	**-**	**-**	**-**	**28**	**203**	**-**	**-**	**-**
Electricity and CHP Plants	-	-	-	-	-	-	28	203	-	-	-
Petroleum Refineries	-	-	-	-	-	-	-	-	-	-	-
Other Transform. Sector	-	-	-	-	-	-	-	-	-	-	-
ENERGY SECTOR	**43**	**-**	**-**	**-**	**-**	**-**	**-**	**-**	**-**	**-**	**-**
DISTRIBUTION LOSSES	**-**	**-**	**-**	**-**	**-**	**-**	**-**	**-**	**-**	**-**	**-**
FINAL CONSUMPTION	**-**	**27**	**89**	**-**	**178**	**56**	**99**	**-**	**7**	**-**	**32**
INDUSTRY SECTOR	**-**	**-**	**-**	**-**	**-**	**-**	**-**	**-**	**7**	**-**	**-**
Iron and Steel	-	-	-	-	-	-	-	-	-	-	-
Chemical and Petrochem.	-	-	-	-	-	-	-	-	7	-	-
Non-Metallic Minerals	-	-	-	-	-	-	-	-	-	-	-
Non-specified	-	-	-	-	-	-	-	-	-	-	-
TRANSPORT SECTOR	**-**	**-**	**89**	**-**	**178**	**-**	**99**	**-**	**-**	**-**	**-**
Air	-	-	-	-	178	-	-	-	-	-	-
Road	-	-	89	-	-	-	99	-	-	-	-
Non-specified	-	-	-	-	-	-	-	-	-	-	-
OTHER SECTORS	**-**	**27**	**-**	**-**	**-**	**56**	**-**	**-**	**-**	**-**	**-**
Agriculture	-	-	-	-	-	-	-	-	-	-	-
Comm. and Publ. Services	-	-	-	-	-	-	-	-	-	-	-
Residential	-	27	-	-	-	56	-	-	-	-	-
Non-specified	-	-	-	-	-	-	-	-	-	-	-
NON-ENERGY USE	**-**	**-**	**-**	**-**	**-**	**-**	**-**	**-**	**-**	**-**	**32**

APPROVISIONNEMENT ET DEMANDE 1997	*Pétrole cont.* (1000 tonnes)										
	Gaz de raffinerie	*GPL + éthane*	*Essence moteur*	*Essence aviation*	*Carbu- réacteurs*	*Kérosène*	*Gazole*	*Fioul lourd*	*Naphta*	*Coke de pétrole*	*Autres prod.*
Production	44	35	97	-	306	24	486	699	170	-	19
Imports	-	-	7	-	-	-	56	-	-	-	-
Exports	-	-	-5	-	-15	-	-50	-85	-155	-	-
Intl. Marine Bunkers	-	-	-	-	-	-	-230	-430	-	-	-
Stock Changes	-	-	-	-	-	-	-	-	-	-	-
DOMESTIC SUPPLY	**44**	**35**	**99**	**-**	**291**	**24**	**262**	**184**	**15**	**-**	**19**
Transfers and Stat. Diff.	-	-	-	-	-	32	-	33	-	-	13
TRANSFORMATION	**-**	**-**	**-**	**-**	**-**	**-**	**58**	**217**	**-**	**-**	**-**
Electricity and CHP Plants	-	-	-	-	-	-	58	217	-	-	-
Petroleum Refineries	-	-	-	-	-	-	-	-	-	-	-
Other Transform. Sector	-	-	-	-	-	-	-	-	-	-	-
ENERGY SECTOR	**44**	**-**	**-**	**-**	**-**	**-**	**-**	**-**	**-**	**-**	**-**
DISTRIBUTION LOSSES	**-**	**-**	**-**	**-**	**-**	**-**	**-**	**-**	**-**	**-**	**-**
FINAL CONSUMPTION	**-**	**35**	**99**	**-**	**291**	**56**	**204**	**-**	**15**	**-**	**32**
INDUSTRY SECTOR	**-**	**-**	**-**	**-**	**-**	**-**	**-**	**-**	**15**	**-**	**-**
Iron and Steel	-	-	-	-	-	-	-	-	-	-	-
Chemical and Petrochem.	-	-	-	-	-	-	-	-	15	-	-
Non-Metallic Minerals	-	-	-	-	-	-	-	-	-	-	-
Non-specified	-	-	-	-	-	-	-	-	-	-	-
TRANSPORT SECTOR	**-**	**-**	**99**	**-**	**291**	**-**	**204**	**-**	**-**	**-**	**-**
Air	-	-	-	-	291	-	-	-	-	-	-
Road	-	-	99	-	-	-	204	-	-	-	-
Non-specified	-	-	-	-	-	-	-	-	-	-	-
OTHER SECTORS	**-**	**35**	**-**	**-**	**-**	**56**	**-**	**-**	**-**	**-**	**-**
Agriculture	-	-	-	-	-	-	-	-	-	-	-
Comm. and Publ. Services	-	-	-	-	-	-	-	-	-	-	-
Residential	-	35	-	-	-	56	-	-	-	-	-
Non-specified	-	-	-	-	-	-	-	-	-	-	-
NON-ENERGY USE	**-**	**-**	**-**	**-**	**-**	**-**	**-**	**-**	**-**	**-**	**32**

Angola

SUPPLY AND	Gas (TJ)				Comb. Renew. & Waste (TJ)				(GWh)	(TJ)
CONSUMPTION 1996	Natural Gas	Gas Works	Coke Ovens	Blast Furnaces	Solid Biomass	Gas/Liquids from Biomass	Municipal Waste	Industrial Waste	Electricity	Heat
Production	21280	-	-	-	210714	-	-	-	1027	-
Imports	-	-	-	-	-	-	-	-	-	-
Exports	-	-	-	-	-	-	-	-	-	-
Intl. Marine Bunkers	-	-	-	-	-	-	-	-	-	-
Stock Changes	-	-	-	-	-	-	-	-	-	-
DOMESTIC SUPPLY	**21280**	**-**	**-**	**-**	**210714**	**-**	**-**	**-**	**1027**	**-**
Transfers and Stat. Diff.	-	-	-	-	-	-	-	-	-	-
TRANSFORMATION	**-**	**-**	**-**	**-**	**55533**	**-**	**-**	**-**	**-**	**-**
Electricity and CHP Plants	-	-	-	-	-	-	-	-	-	-
Petroleum Refineries	-	-	-	-	-	-	-	-	-	-
Other Transform. Sector	-	-	-	-	55533	-	-	-	-	-
ENERGY SECTOR	**-**	**-**	**-**	**-**	**-**	**-**	**-**	**-**	**45**	**-**
DISTRIBUTION LOSSES	**-**	**-**	**-**	**-**	**-**	**-**	**-**	**-**	**292**	**-**
FINAL CONSUMPTION	**21280**	**-**	**-**	**-**	**155181**	**-**	**-**	**-**	**690**	**-**
INDUSTRY SECTOR	**21280**	**-**	**-**	**-**	**3633**	**-**	**-**	**-**	**216**	**-**
Iron and Steel	-	-	-	-	-	-	-	-	-	-
Chemical and Petrochem.	-	-	-	-	-	-	-	-	-	-
Non-Metallic Minerals	-	-	-	-	-	-	-	-	-	-
Non-specified	21280	-	-	-	3633	-	-	-	216	-
TRANSPORT SECTOR	**-**	**-**	**-**	**-**	**-**	**-**	**-**	**-**	**-**	**-**
Air	-	-	-	-	-	-	-	-	-	-
Road	-	-	-	-	-	-	-	-	-	-
Non-specified	-	-	-	-	-	-	-	-	-	-
OTHER SECTORS	**-**	**-**	**-**	**-**	**151548**	**-**	**-**	**-**	**474**	**-**
Agriculture	-	-	-	-	-	-	-	-	-	-
Comm. and Publ. Services	-	-	-	-	-	-	-	-	-	-
Residential	-	-	-	-	151548	-	-	-	474	-
Non-specified	-	-	-	-	-	-	-	-	-	-
NON-ENERGY USE	**-**	**-**	**-**	**-**	**-**	**-**	**-**	**-**	**-**	**-**

APPROVISIONNEMENT	*Gaz* (TJ)				*En. Re. Comb. & Déchets* (TJ)				(GWh)	(TJ)
ET DEMANDE 1997	*Gaz naturel*	*Usines à gaz*	*Cokeries*	*Hauts fourneaux*	*Biomasse solide*	*Gaz/Liquides tirés de biomasse*	*Déchets urbains*	*Déchets industriels*	*Electricité*	*Chaleur*
Production	21660	-	-	-	216825	-	-	-	1109	-
Imports	-	-	-	-	-	-	-	-	-	-
Exports	-	-	-	-	-	-	-	-	-	-
Intl. Marine Bunkers	-	-	-	-	-	-	-	-	-	-
Stock Changes	-	-	-	-	-	-	-	-	-	-
DOMESTIC SUPPLY	**21660**	**-**	**-**	**-**	**216825**	**-**	**-**	**-**	**1109**	**-**
Transfers and Stat. Diff.	-	-	-	-	-	-	-	-	-	-
TRANSFORMATION	**-**	**-**	**-**	**-**	**57144**	**-**	**-**	**-**	**-**	**-**
Electricity and CHP Plants	-	-	-	-	-	-	-	-	-	-
Petroleum Refineries	-	-	-	-	-	-	-	-	-	-
Other Transform. Sector	-	-	-	-	57144	-	-	-	-	-
ENERGY SECTOR	**-**	**-**	**-**	**-**	**-**	**-**	**-**	**-**	**49**	**-**
DISTRIBUTION LOSSES	**-**	**-**	**-**	**-**	**-**	**-**	**-**	**-**	**315**	**-**
FINAL CONSUMPTION	**21660**	**-**	**-**	**-**	**159681**	**-**	**-**	**-**	**745**	**-**
INDUSTRY SECTOR	**21660**	**-**	**-**	**-**	**3738**	**-**	**-**	**-**	**233**	**-**
Iron and Steel	-	-	-	-	-	-	-	-	-	-
Chemical and Petrochem.	-	-	-	-	-	-	-	-	-	-
Non-Metallic Minerals	-	-	-	-	-	-	-	-	-	-
Non-specified	21660	-	-	-	3738	-	-	-	233	-
TRANSPORT SECTOR	**-**	**-**	**-**	**-**	**-**	**-**	**-**	**-**	**-**	**-**
Air	-	-	-	-	-	-	-	-	-	-
Road	-	-	-	-	-	-	-	-	-	-
Non-specified	-	-	-	-	-	-	-	-	-	-
OTHER SECTORS	**-**	**-**	**-**	**-**	**155943**	**-**	**-**	**-**	**512**	**-**
Agriculture	-	-	-	-	-	-	-	-	-	-
Comm. and Publ. Services	-	-	-	-	-	-	-	-	-	-
Residential	-	-	-	-	155943	-	-	-	512	-
Non-specified	-	-	-	-	-	-	-	-	-	-
NON-ENERGY USE	**-**	**-**	**-**	**-**	**-**	**-**	**-**	**-**	**-**	**-**

Argentina / Argentine : 1996

SUPPLY AND CONSUMPTION / *APPROVISIONNEMENT ET DEMANDE*	Coal / *Charbon* (1000 tonnes)							Oil / *Pétrole* (1000 tonnes)			
	Coking Coal / *Charbon à coke*	Other Bit. Coal / *Autres charb. bit.*	Sub-Bit. Coal / *Charbon sous-bit.*	Lignite / *Lignite*	Peat / *Tourbe*	Oven and Gas Coke / *Coke de four/gaz*	Pat. Fuel and BKB / *Agg./briq. de lignite*	Crude Oil / *Pétrole brut*	NGL / *LGN*	Feed-stocks / *Produits d'aliment.*	Additives / *Additifs*
Production	-	310	-	-	-	741	-	40311	1640	-	-
From Other Sources	-	-	-	-	-	-	-	-	-	-	-
Imports	1049	-	-	-	-	-	-	751	-	-	-
Exports	-	-	-	-	-	-	-	-16691	-	-	-
Intl. Marine Bunkers	-	-	-	-	-	-	-	-	-	-	-
Stock Changes	-78	-	-	-	-	-6	-	-74	-	-	-
DOMESTIC SUPPLY	**971**	**310**	**-**	**-**	**-**	**735**	**-**	**24297**	**1640**	**-**	**-**
Transfers	-	-	-	-	-	-	-	-	-1287	-	-
Statistical Differences	-	305	-	-	-	-4	-	213	-	-	-
TRANSFORMATION	**971**	**607**	**-**	**-**	**-**	**585**	**-**	**24482**	**-**	**-**	**-**
Electricity Plants	-	607	-	-	-	-	-	-	-	-	-
CHP Plants	-	-	-	-	-	-	-	-	-	-	-
Heat Plants	-	-	-	-	-	-	-	-	-	-	-
Blast Furnaces/Gas Works	-	-	-	-	-	585	-	-	-	-	-
Coke/Pat. Fuel/BKB Plants	971	-	-	-	-	-	-	-	-	-	-
Petroleum Refineries	-	-	-	-	-	-	-	24482	-	-	-
Petrochemical Industry	-	-	-	-	-	-	-	-	-	-	-
Liquefaction	-	-	-	-	-	-	-	-	-	-	-
Other Transform. Sector	-	-	-	-	-	-	-	-	-	-	-
ENERGY SECTOR	**-**	**8**	**-**	**-**	**-**	**-**	**-**	**28**	**-**	**-**	**-**
Coal Mines	-	8	-	-	-	-	-	-	-	-	-
Oil and Gas Extraction	-	-	-	-	-	-	-	28	-	-	-
Petroleum Refineries	-	-	-	-	-	-	-	-	-	-	-
Electr., CHP+Heat Plants	-	-	-	-	-	-	-	-	-	-	-
Pumped Storage (Elec.)	-	-	-	-	-	-	-	-	-	-	-
Other Energy Sector	-	-	-	-	-	-	-	-	-	-	-
Distribution Losses	-	-	-	-	-	-	-	-	-	-	-
FINAL CONSUMPTION	**-**	**-**	**-**	**-**	**-**	**146**	**-**	**-**	**353**	**-**	**-**
INDUSTRY SECTOR	**-**	**-**	**-**	**-**	**-**	**146**	**-**	**-**	**353**	**-**	**-**
Iron and Steel	-	-	-	-	-	146	-	-	-	-	-
Chemical and Petrochem.	-	-	-	-	-	-	-	-	353	-	-
of which: Feedstocks	-	-	-	-	-	-	-	-	*353*	-	-
Non-Ferrous Metals	-	-	-	-	-	-	-	-	-	-	-
Non-Metallic Minerals	-	-	-	-	-	-	-	-	-	-	-
Transport Equipment	-	-	-	-	-	-	-	-	-	-	-
Machinery	-	-	-	-	-	-	-	-	-	-	-
Mining and Quarrying	-	-	-	-	-	-	-	-	-	-	-
Food and Tobacco	-	-	-	-	-	-	-	-	-	-	-
Paper, Pulp and Print	-	-	-	-	-	-	-	-	-	-	-
Wood and Wood Products	-	-	-	-	-	-	-	-	-	-	-
Construction	-	-	-	-	-	-	-	-	-	-	-
Textile and Leather	-	-	-	-	-	-	-	-	-	-	-
Non-specified	-	-	-	-	-	-	-	-	-	-	-
TRANSPORT SECTOR	**-**	**-**	**-**	**-**	**-**	**-**	**-**	**-**	**-**	**-**	**-**
Air	-	-	-	-	-	-	-	-	-	-	-
Road	-	-	-	-	-	-	-	-	-	-	-
Rail	-	-	-	-	-	-	-	-	-	-	-
Pipeline Transport	-	-	-	-	-	-	-	-	-	-	-
Internal Navigation	-	-	-	-	-	-	-	-	-	-	-
Non-specified	-	-	-	-	-	-	-	-	-	-	-
OTHER SECTORS	**-**	**-**	**-**	**-**	**-**	**-**	**-**	**-**	**-**	**-**	**-**
Agriculture	-	-	-	-	-	-	-	-	-	-	-
Comm. and Publ. Services	-	-	-	-	-	-	-	-	-	-	-
Residential	-	-	-	-	-	-	-	-	-	-	-
Non-specified	-	-	-	-	-	-	-	-	-	-	-
NON-ENERGY USE	**-**	**-**	**-**	**-**	**-**	**-**	**-**	**-**	**-**	**-**	**-**
in Industry/Trans./Energy	-	-	-	-	-	-	-	-	-	-	-
in Transport	-	-	-	-	-	-	-	-	-	-	-
in Other Sectors	-	-	-	-	-	-	-	-	-	-	-

Argentina / Argentine : 1996

	Oil cont. / *Pétrole cont.* (1000 tonnes)										
SUPPLY AND CONSUMPTION	Refinery Gas	LPG + Ethane	Motor Gasoline	Aviation Gasoline	Jet Fuel	Kerosene	Gas/ Diesel	Heavy Fuel Oil	Naphtha	Petrol. Coke	Other Prod.
APPROVISIONNEMENT ET DEMANDE	*Gaz de raffinerie*	*GPL + éthane*	*Essence moteur*	*Essence aviation*	*Carbu- réacteurs*	*Kérosène*	*Gazole*	*Fioul lourd*	*Naphta*	*Coke de pétrole*	*Autres prod.*
Production	509	892	5177	5	1244	201	9575	1924	1805	1134	808
From Other Sources	-	-	-	-	-	-	-	-	-	-	-
Imports	-	-	270	4	96	78	828	294	28	-	67
Exports	-	-595	-1066	-	-28	-	-711	-372	-610	-24	-55
Intl. Marine Bunkers	-	-	-	-	-	-	-204	-374	-	-	-
Stock Changes	-	4	-	-	-	-7	-25	32	-	-	-
DOMESTIC SUPPLY	**509**	**301**	**4381**	**9**	**1312**	**272**	**9463**	**1504**	**1223**	**1110**	**820**
Transfers	-	1053	234	-	-	-	-	-	-	-	-
Statistical Differences	-1	-39	15	2	-245	67	6	383	-1087	-2	-140
TRANSFORMATION	**24**	-	-	-	-	-	**212**	**760**	-	-	-
Electricity Plants	24	-	-	-	-	-	212	760	-	-	-
CHP Plants	-	-	-	-	-	-	-	-	-	-	-
Heat Plants	-	-	-	-	-	-	-	-	-	-	-
Blast Furnaces/Gas Works	-	-	-	-	-	-	-	-	-	-	-
Coke/Pat. Fuel/BKB Plants	-	-	-	-	-	-	-	-	-	-	-
Petroleum Refineries	-	-	-	-	-	-	-	-	-	-	-
Petrochemical Industry	-	-	-	-	-	-	-	-	-	-	-
Liquefaction	-	-	-	-	-	-	-	-	-	-	-
Other Transform. Sector	-	-	-	-	-	-	-	-	-	-	-
ENERGY SECTOR	**484**	-	-	-	-	-	**21**	**501**	-	-	-
Coal Mines	-	-	-	-	-	-	-	-	-	-	-
Oil and Gas Extraction	-	-	-	-	-	-	-	-	-	-	-
Petroleum Refineries	484	-	-	-	-	-	21	501	-	-	-
Electr., CHP+Heat Plants	-	-	-	-	-	-	-	-	-	-	-
Pumped Storage (Elec.)	-	-	-	-	-	-	-	-	-	-	-
Other Energy Sector	-	-	-	-	-	-	-	-	-	-	-
Distribution Losses	-	-	-	-	-	-	232	-	-	-	-
FINAL CONSUMPTION	-	**1315**	**4630**	**11**	**1067**	**339**	**9004**	**626**	**136**	**1108**	**680**
INDUSTRY SECTOR	-	**500**	-	-	-	-	**91**	**442**	**136**	-	-
Iron and Steel	-	-	-	-	-	-	-	-	-	-	-
Chemical and Petrochem.	-	452	-	-	-	-	-	-	136	-	-
of which: Feedstocks	-	*452*	-	-	-	-	-	-	*136*	-	-
Non-Ferrous Metals	-	-	-	-	-	-	-	-	-	-	-
Non-Metallic Minerals	-	-	-	-	-	-	-	-	-	-	-
Transport Equipment	-	-	-	-	-	-	-	-	-	-	-
Machinery	-	-	-	-	-	-	-	-	-	-	-
Mining and Quarrying	-	-	-	-	-	-	-	-	-	-	-
Food and Tobacco	-	-	-	-	-	-	-	-	-	-	-
Paper, Pulp and Print	-	-	-	-	-	-	-	-	-	-	-
Wood and Wood Products	-	-	-	-	-	-	-	-	-	-	-
Construction	-	-	-	-	-	-	-	-	-	-	-
Textile and Leather	-	-	-	-	-	-	-	-	-	-	-
Non-specified	-	48	-	-	-	-	91	442	-	-	-
TRANSPORT SECTOR	-	-	**4630**	**11**	**1067**	-	**6226**	**95**	-	-	-
Air	-	-	-	11	1067	-	-	-	-	-	-
Road	-	-	4630	-	-	-	6226	-	-	-	-
Rail	-	-	-	-	-	-	-	-	-	-	-
Pipeline Transport	-	-	-	-	-	-	-	-	-	-	-
Internal Navigation	-	-	-	-	-	-	-	95	-	-	-
Non-specified	-	-	-	-	-	-	-	-	-	-	-
OTHER SECTORS	-	**815**	-	-	-	**339**	**2687**	**89**	-	-	-
Agriculture	-	-	-	-	-	-	2602	-	-	-	-
Comm. and Publ. Services	-	11	-	-	-	-	85	89	-	-	-
Residential	-	804	-	-	-	339	-	-	-	-	-
Non-specified	-	-	-	-	-	-	-	-	-	-	-
NON-ENERGY USE	-	-	-	-	-	-	-	-	-	**1108**	**680**
in Industry/Transf./Energy	-	-	-	-	-	-	-	-	-	1108	680
in Transport	-	-	-	-	-	-	-	-	-	-	-
in Other Sectors	-	-	-	-	-	-	-	-	-	-	-

Argentina / Argentine : 1996

	Gas / *Gaz* (TJ)				Comb. Renew. & Waste / *En. Re. Comb. & Déchets* (TJ)				(GWh)	(TJ)
SUPPLY AND CONSUMPTION	Natural Gas	Gas Works	Coke Ovens	Blast Furnaces	Solid Biomass	Gas/Liquids from Biomass	Municipal Waste	Industrial Waste	Electricity	Heat
APPROVISIONNEMENT ET DEMANDE	*Gaz naturel*	*Usines à gaz*	*Cokeries*	*Hauts fourneaux*	*Biomasse solide*	*Gaz/Liquides tirés de biomasse*	*Déchets urbains*	*Déchets industriels*	*Electricité*	*Chaleur*
Production	1195328	-	7631	11055	112664	-	-	-	69759	-
From Other Sources	-	-	-	-	-	-	-	-	-	-
Imports	81891	-	-	-	-	-	-	-	3663	-
Exports	-	-	-	-	-	-	-	-	-302	-
Intl. Marine Bunkers	-	-	-	-	-	-	-	-	-	-
Stock Changes	-	-	-	-	-	-	-	-	-	-
DOMESTIC SUPPLY	**1277219**	**-**	**7631**	**11055**	**112664**	**-**	**-**	**-**	**73120**	**-**
Transfers	-	-	-	-	-	-	-	-	-	-
Statistical Differences	-	-	-	42	-	-	-	-	-3842	-
TRANSFORMATION	**388936**	**-**	**1815**	**6030**	**18807**	**-**	**-**	**-**	**-**	**-**
Electricity Plants	388936	-	1815	6030	5655	-	-	-	-	-
CHP Plants	-	-	-	-	-	-	-	-	-	-
Heat Plants	-	-	-	-	-	-	-	-	-	-
Blast Furnaces/Gas Works	-	-	-	-	1959	-	-	-	-	-
Coke/Pat. Fuel/BKB Plants	-	-	-	-	-	-	-	-	-	-
Petroleum Refineries	-	-	-	-	-	-	-	-	-	-
Petrochemical Industry	-	-	-	-	-	-	-	-	-	-
Liquefaction	-	-	-	-	-	-	-	-	-	-
Other Transform. Sector	-	-	-	-	11193	-	-	-	-	-
ENERGY SECTOR	**173274**	**-**	**-**	**-**	**-**	**-**	**-**	**-**	**2239**	**-**
Coal Mines	-	-	-	-	-	-	-	-	-	-
Oil and Gas Extraction	173274	-	-	-	-	-	-	-	-	-
Petroleum Refineries	-	-	-	-	-	-	-	-	-	-
Electr., CHP+Heat Plants	-	-	-	-	-	-	-	-	2239	-
Pumped Storage (Elec.)	-	-	-	-	-	-	-	-	-	-
Other Energy Sector	-	-	-	-	-	-	-	-	-	-
Distribution Losses	117903	-	-	335	-	-	-	-	12759	-
FINAL CONSUMPTION	**597106**	**-**	**5816**	**4732**	**93857**	**-**	**-**	**-**	**54280**	**-**
INDUSTRY SECTOR	**274428**	**-**	**5816**	**4732**	**75149**	**-**	**-**	**-**	**22276**	**-**
Iron and Steel	-	-	5816	-	-	-	-	-	-	-
Chemical and Petrochem.	9213	-	-	-	-	-	-	-	-	-
of which: Feedstocks	*9213*	-	-	-	-	-	-	-	-	-
Non-Ferrous Metals	-	-	-	-	-	-	-	-	-	-
Non-Metallic Minerals	-	-	-	-	-	-	-	-	-	-
Transport Equipment	-	-	-	-	-	-	-	-	-	-
Machinery	-	-	-	-	-	-	-	-	-	-
Mining and Quarrying	-	-	-	-	-	-	-	-	-	-
Food and Tobacco	-	-	-	-	-	-	-	-	-	-
Paper, Pulp and Print	-	-	-	-	-	-	-	-	-	-
Wood and Wood Products	-	-	-	-	-	-	-	-	-	-
Construction	-	-	-	-	-	-	-	-	-	-
Textile and Leather	-	-	-	-	-	-	-	-	-	-
Non-specified	265215	-	-	4732	75149	-	-	-	22276	-
TRANSPORT SECTOR	**42155**	**-**	**-**	**-**	**-**	**-**	**-**	**-**	**420**	**-**
Air	-	-	-	-	-	-	-	-	-	-
Road	42155	-	-	-	-	-	-	-	-	-
Rail	-	-	-	-	-	-	-	-	420	-
Pipeline Transport	-	-	-	-	-	-	-	-	-	-
Internal Navigation	-	-	-	-	-	-	-	-	-	-
Non-specified	-	-	-	-	-	-	-	-	-	-
OTHER SECTORS	**280523**	**-**	**-**	**-**	**18708**	**-**	**-**	**-**	**31584**	**-**
Agriculture	-	-	-	-	-	-	-	-	469	-
Comm. and Publ. Services	51089	-	-	-	-	-	-	-	12857	-
Residential	229434	-	-	-	16151	-	-	-	17629	-
Non-specified	-	-	-	-	2557	-	-	-	629	-
NON-ENERGY USE	**-**	**-**	**-**	**-**	**-**	**-**	**-**	**-**	**-**	**-**
in Industry/Transf./Energy	-	-	-	-	-	-	-	-	-	-
in Transport	-	-	-	-	-	-	-	-	-	-
in Other Sectors	-	-	-	-	-	-	-	-	-	-

Argentina / Argentine : 1997

SUPPLY AND CONSUMPTION / *APPROVISIONNEMENT ET DEMANDE*	Coal / *Charbon* (1000 tonnes)							Oil / *Pétrole* (1000 tonnes)			
	Coking Coal / *Charbon à coke*	Other Bit. Coal / *Autres charb. bit.*	Sub-Bit. Coal / *Charbon sous-bit.*	Lignite / *Lignite*	Peat / *Tourbe*	Oven and Gas Coke / *Coke de four/gaz*	Pat. Fuel and BKB / *Agg./briq. de lignite*	Crude Oil / *Pétrole brut*	NGL / *LGN*	Feed-stocks / *Produits d'aliment.*	Additives / *Additifs*
Production	-	251	-	-	-	757	-	42837	1513	-	-
From Other Sources	-	-	-	-	-	-	-	-	-	-	-
Imports	881	-	-	-	-	-	-	935	-	-	-
Exports	-	-	-	-	-	-	-	-17104	-	-	-
Intl. Marine Bunkers	-	-	-	-	-	-	-	-	-	-	-
Stock Changes	111	-	-	-	-	12	-	22	-	-	-
DOMESTIC SUPPLY	**992**	**251**	**-**	**-**	**-**	**769**	**-**	**26690**	**1513**	**-**	**-**
Transfers	-	-	-	-	-	-	-	-	-1140	-	-
Statistical Differences	-	53	-	-	-	-23	-	136	-	-	-
TRANSFORMATION	**992**	**304**	**-**	**-**	**-**	**597**	**-**	**26826**	**-**	**-**	**-**
Electricity Plants	-	304	-	-	-	-	-	-	-	-	-
CHP Plants	-	-	-	-	-	-	-	-	-	-	-
Heat Plants	-	-	-	-	-	-	-	-	-	-	-
Blast Furnaces/Gas Works	-	-	-	-	-	597	-	-	-	-	-
Coke/Pat. Fuel/BKB Plants	992	-	-	-	-	-	-	-	-	-	-
Petroleum Refineries	-	-	-	-	-	-	-	26826	-	-	-
Petrochemical Industry	-	-	-	-	-	-	-	-	-	-	-
Liquefaction	-	-	-	-	-	-	-	-	-	-	-
Other Transform. Sector	-	-	-	-	-	-	-	-	-	-	-
ENERGY SECTOR	**-**	**-**	**-**	**-**	**-**	**-**	**-**	**-**	**-**	**-**	**-**
Coal Mines	-	-	-	-	-	-	-	-	-	-	-
Oil and Gas Extraction	-	-	-	-	-	-	-	-	-	-	-
Petroleum Refineries	-	-	-	-	-	-	-	-	-	-	-
Electr., CHP+Heat Plants	-	-	-	-	-	-	-	-	-	-	-
Pumped Storage (Elec.)	-	-	-	-	-	-	-	-	-	-	-
Other Energy Sector	-	-	-	-	-	-	-	-	-	-	-
Distribution Losses	-	-	-	-	-	-	-	-	-	-	-
FINAL CONSUMPTION	**-**	**-**	**-**	**-**	**-**	**149**	**-**	**-**	**373**	**-**	**-**
INDUSTRY SECTOR	**-**	**-**	**-**	**-**	**-**	**149**	**-**	**-**	**373**	**-**	**-**
Iron and Steel	-	-	-	-	-	149	-	-	-	-	-
Chemical and Petrochem.	-	-	-	-	-	-	-	-	373	-	-
of which: Feedstocks	-	-	-	-	-	-	-	-	*373*	-	-
Non-Ferrous Metals	-	-	-	-	-	-	-	-	-	-	-
Non-Metallic Minerals	-	-	-	-	-	-	-	-	-	-	-
Transport Equipment	-	-	-	-	-	-	-	-	-	-	-
Machinery	-	-	-	-	-	-	-	-	-	-	-
Mining and Quarrying	-	-	-	-	-	-	-	-	-	-	-
Food and Tobacco	-	-	-	-	-	-	-	-	-	-	-
Paper, Pulp and Print	-	-	-	-	-	-	-	-	-	-	-
Wood and Wood Products	-	-	-	-	-	-	-	-	-	-	-
Construction	-	-	-	-	-	-	-	-	-	-	-
Textile and Leather	-	-	-	-	-	-	-	-	-	-	-
Non-specified	-	-	-	-	-	-	-	-	-	-	-
TRANSPORT SECTOR	**-**	**-**	**-**	**-**	**-**	**-**	**-**	**-**	**-**	**-**	**-**
Air	-	-	-	-	-	-	-	-	-	-	-
Road	-	-	-	-	-	-	-	-	-	-	-
Rail	-	-	-	-	-	-	-	-	-	-	-
Pipeline Transport	-	-	-	-	-	-	-	-	-	-	-
Internal Navigation	-	-	-	-	-	-	-	-	-	-	-
Non-specified	-	-	-	-	-	-	-	-	-	-	-
OTHER SECTORS	**-**	**-**	**-**	**-**	**-**	**-**	**-**	**-**	**-**	**-**	**-**
Agriculture	-	-	-	-	-	-	-	-	-	-	-
Comm. and Publ. Services	-	-	-	-	-	-	-	-	-	-	-
Residential	-	-	-	-	-	-	-	-	-	-	-
Non-specified	-	-	-	-	-	-	-	-	-	-	-
NON-ENERGY USE	**-**	**-**	**-**	**-**	**-**	**-**	**-**	**-**	**-**	**-**	**-**
in Industry/Trans./Energy	-	-	-	-	-	-	-	-	-	-	-
in Transport	-	-	-	-	-	-	-	-	-	-	-
in Other Sectors	-	-	-	-	-	-	-	-	-	-	-

Argentina / Argentine : 1997

SUPPLY AND CONSUMPTION / *APPROVISIONNEMENT ET DEMANDE*	Oil cont. / *Pétrole cont.* (1000 tonnes)										
	Refinery Gas / *Gaz de raffinerie*	LPG + Ethane / *GPL + éthane*	Motor Gasoline / *Essence moteur*	Aviation Gasoline / *Essence aviation*	Jet Fuel / *Carbu-réacteurs*	Kerosene / *Kérosène*	Gas/ Diesel / *Gazole*	Heavy Fuel Oil / *Fioul lourd*	Naphtha / *Naphta*	Petrol. Coke / *Coke de pétrole*	Other Prod. / *Autres prod.*
Production	429	804	5747	5	1138	182	10491	2649	833	1117	797
From Other Sources	-	-	-	-	-	-	-	-	-	-	-
Imports	-	24	150	5	113	78	703	122	-	-	70
Exports	-	-610	-1820	-	-44	-	-1063	-443	-726	-24	-33
Intl. Marine Bunkers	-	-	-	-	-	-	-204	-374	-	-	-
Stock Changes	-	-	-	-	-	-	102	135	-	-	-
DOMESTIC SUPPLY	**429**	**218**	**4077**	**10**	**1207**	**260**	**10029**	**2089**	**107**	**1093**	**834**
Transfers	-	1134	228	-	-	-	-	-	-	-	-
Statistical Differences	-88	-32	-	-	-	-	-349	-393	-	-	-146
TRANSFORMATION	**30**	**-**	**-**	**-**	**-**	**-**	**127**	**520**	**-**	**-**	**-**
Electricity Plants	30	-	-	-	-	-	127	520	-	-	-
CHP Plants	-	-	-	-	-	-	-	-	-	-	-
Heat Plants	-	-	-	-	-	-	-	-	-	-	-
Blast Furnaces/Gas Works	-	-	-	-	-	-	-	-	-	-	-
Coke/Pat. Fuel/BKB Plants	-	-	-	-	-	-	-	-	-	-	-
Petroleum Refineries	-	-	-	-	-	-	-	-	-	-	-
Petrochemical Industry	-	-	-	-	-	-	-	-	-	-	-
Liquefaction	-	-	-	-	-	-	-	-	-	-	-
Other Transform. Sector	-	-	-	-	-	-	-	-	-	-	-
ENERGY SECTOR	**311**	**-**	**-**	**-**	**-**	**-**	**84**	**699**	**-**	**-**	**-**
Coal Mines	-	-	-	-	-	-	-	-	-	-	-
Oil and Gas Extraction	-	-	-	-	-	-	-	-	-	-	-
Petroleum Refineries	311	-	-	-	-	-	84	699	-	-	-
Electr., CHP+Heat Plants	-	-	-	-	-	-	-	-	-	-	-
Pumped Storage (Elec.)	-	-	-	-	-	-	-	-	-	-	-
Other Energy Sector	-	-	-	-	-	-	-	-	-	-	-
Distribution Losses	-	-	-	-	-	-	-	-	-	-	-
FINAL CONSUMPTION	**-**	**1320**	**4305**	**10**	**1207**	**260**	**9469**	**477**	**107**	**1093**	**688**
INDUSTRY SECTOR	**-**	**500**	**-**	**-**	**-**	**-**	**60**	**345**	**107**	**-**	**-**
Iron and Steel	-	-	-	-	-	-	-	-	-	-	-
Chemical and Petrochem.	-	478	-	-	-	-	-	-	107	-	-
of which: Feedstocks	-	*478*	-	-	-	-	-	-	*107*	-	-
Non-Ferrous Metals	-	-	-	-	-	-	-	-	-	-	-
Non-Metallic Minerals	-	-	-	-	-	-	-	-	-	-	-
Transport Equipment	-	-	-	-	-	-	-	-	-	-	-
Machinery	-	-	-	-	-	-	-	-	-	-	-
Mining and Quarrying	-	-	-	-	-	-	-	-	-	-	-
Food and Tobacco	-	-	-	-	-	-	-	-	-	-	-
Paper, Pulp and Print	-	-	-	-	-	-	-	-	-	-	-
Wood and Wood Products	-	-	-	-	-	-	-	-	-	-	-
Construction	-	-	-	-	-	-	-	-	-	-	-
Textile and Leather	-	-	-	-	-	-	-	-	-	-	-
Non-specified	-	22	-	-	-	-	60	345	-	-	-
TRANSPORT SECTOR	**-**	**-**	**4305**	**10**	**1207**	**-**	**6696**	**75**	**-**	**-**	**-**
Air	-	-	-	10	1207	-	-	-	-	-	-
Road	-	-	4305	-	-	-	6696	-	-	-	-
Rail	-	-	-	-	-	-	-	-	-	-	-
Pipeline Transport	-	-	-	-	-	-	-	-	-	-	-
Internal Navigation	-	-	-	-	-	-	-	75	-	-	-
Non-specified	-	-	-	-	-	-	-	-	-	-	-
OTHER SECTORS	**-**	**820**	**-**	**-**	**-**	**260**	**2713**	**57**	**-**	**-**	**-**
Agriculture	-	-	-	-	-	-	2644	-	-	-	-
Comm. and Publ. Services	-	14	-	-	-	-	69	57	-	-	-
Residential	-	806	-	-	-	260	-	-	-	-	-
Non-specified	-	-	-	-	-	-	-	-	-	-	-
NON-ENERGY USE	**-**	**-**	**-**	**-**	**-**	**-**	**-**	**-**	**-**	**1093**	**688**
in Industry/Transf./Energy	-	-	-	-	-	-	-	-	-	1093	688
in Transport	-	-	-	-	-	-	-	-	-	-	-
in Other Sectors	-	-	-	-	-	-	-	-	-	-	-

Argentina / Argentine : 1997

	Gas / *Gaz* (TJ)				Comb. Renew. & Waste / *En. Re. Comb. & Déchets* (TJ)				(GWh)	(TJ)
SUPPLY AND CONSUMPTION	Natural Gas	Gas Works	Coke Ovens	Blast Furnaces	Solid Biomass	Gas/Liquids from Biomass	Municipal Waste	Industrial Waste	Electricity	Heat
APPROVISIONNEMENT ET DEMANDE	*Gaz naturel*	*Usines à gaz*	*Cokeries*	*Hauts fourneaux*	*Biomasse solide*	*Gaz/Liquides tirés de biomasse*	*Déchets urbains*	*Déchets industriels*	*Electricité*	*Chaleur*
Production	1305680	-	7678	11935	110851	-	-	-	71947	-
From Other Sources	-	-	-	-	-	-	-	-	-	-
Imports	65653	-	-	-	-	-	-	-	5465	-
Exports	-28529	-	-	-	-	-	-	-	-279	-
Intl. Marine Bunkers	-	-	-	-	-	-	-	-	-	-
Stock Changes	-	-	-	-	-	-	-	-	-	-
DOMESTIC SUPPLY	**1342804**	**-**	**7678**	**11935**	**110851**	**-**	**-**	**-**	**77133**	**-**
Transfers	-	-	-	-	-	-	-	-	-	-
Statistical Differences	-	-	-1	-1	7	-	-	-	-3644	-
TRANSFORMATION	**373244**	**-**	**1826**	**5946**	**12026**	**-**	**-**	**-**	**-**	**-**
Electricity Plants	373244	-	1826	5946	4190	-	-	-	-	-
CHP Plants	-	-	-	-	-	-	-	-	-	-
Heat Plants	-	-	-	-	-	-	-	-	-	-
Blast Furnaces/Gas Works	-	-	-	-	1932	-	-	-	-	-
Coke/Pat. Fuel/BKB Plants	-	-	-	-	-	-	-	-	-	-
Petroleum Refineries	-	-	-	-	-	-	-	-	-	-
Petrochemical Industry	-	-	-	-	-	-	-	-	-	-
Liquefaction	-	-	-	-	-	-	-	-	-	-
Other Transform. Sector	-	-	-	-	5904	-	-	-	-	-
ENERGY SECTOR	**201560**	**-**	**-**	**-**	**-**	**-**	**-**	**-**	**2644**	**-**
Coal Mines	-	-	-	-	-	-	-	-	-	-
Oil and Gas Extraction	201560	-	-	-	-	-	-	-	-	-
Petroleum Refineries	-	-	-	-	-	-	-	-	-	-
Electr., CHP+Heat Plants	-	-	-	-	-	-	-	-	2644	-
Pumped Storage (Elec.)	-	-	-	-	-	-	-	-	-	-
Other Energy Sector	-	-	-	-	-	-	-	-	-	-
Distribution Losses	125685	-	-	377	-	-	-	-	12565	-
FINAL CONSUMPTION	**642315**	**-**	**5851**	**5611**	**98832**	**-**	**-**	**-**	**58280**	**-**
INDUSTRY SECTOR	**290166**	**-**	**5851**	**5611**	**82348**	**-**	**-**	**-**	**24105**	**-**
Iron and Steel	-	-	5851	-	-	-	-	-	-	-
Chemical and Petrochem.	9741	-	-	-	-	-	-	-	-	-
of which: Feedstocks	*9741*	-	-	-	-	-	-	-	-	-
Non-Ferrous Metals	-	-	-	-	-	-	-	-	-	-
Non-Metallic Minerals	-	-	-	-	-	-	-	-	-	-
Transport Equipment	-	-	-	-	-	-	-	-	-	-
Machinery	-	-	-	-	-	-	-	-	-	-
Mining and Quarrying	-	-	-	-	-	-	-	-	-	-
Food and Tobacco	-	-	-	-	-	-	-	-	-	-
Paper, Pulp and Print	-	-	-	-	-	-	-	-	-	-
Wood and Wood Products	-	-	-	-	-	-	-	-	-	-
Construction	-	-	-	-	-	-	-	-	-	-
Textile and Leather	-	-	-	-	-	-	-	-	-	-
Non-specified	280425	-	-	5611	82348	-	-	-	24105	-
TRANSPORT SECTOR	**48880**	**-**	**-**	**-**	**-**	**-**	**-**	**-**	**478**	**-**
Air	-	-	-	-	-	-	-	-	-	-
Road	48880	-	-	-	-	-	-	-	-	-
Rail	-	-	-	-	-	-	-	-	478	-
Pipeline Transport	-	-	-	-	-	-	-	-	-	-
Internal Navigation	-	-	-	-	-	-	-	-	-	-
Non-specified	-	-	-	-	-	-	-	-	-	-
OTHER SECTORS	**303269**	**-**	**-**	**-**	**16484**	**-**	**-**	**-**	**33697**	**-**
Agriculture	-	-	-	-	-	-	-	-	481	-
Comm. and Publ. Services	66768	-	-	-	-	-	-	-	13636	-
Residential	236501	-	-	-	13927	-	-	-	18862	-
Non-specified	-	-	-	-	2557	-	-	-	718	-
NON-ENERGY USE	**-**	**-**	**-**	**-**	**-**	**-**	**-**	**-**	**-**	**-**
in Industry/Transf./Energy	-	-	-	-	-	-	-	-	-	-
in Transport	-	-	-	-	-	-	-	-	-	-
in Other Sectors	-	-	-	-	-	-	-	-	-	-

Armenia / Arménie

SUPPLY AND CONSUMPTION 1996	Coal (1000 tonnes)							Oil (1000 tonnes)			
	Coking Coal	Other Bit. Coal	Sub-Bit. Coal	Lignite	Peat	Oven and Gas Coke	Pat. Fuel and BKB	Crude Oil	NGL	Feed-stocks	Additives
Production	-	-	-	-	-	-	-	-	-	-	-
Imports	-	5	-	-	-	-	-	-	-	-	-
Exports	-	-	-	-	-	-	-	-	-	-	-
Intl. Marine Bunkers	-	-	-	-	-	-	-	-	-	-	-
Stock Changes	-	-	-	-	-	-	-	-	-	-	-
DOMESTIC SUPPLY	**-**	**5**	**-**	**-**	**-**	**-**	**-**	**-**	**-**	**-**	**-**
Transfers and Stat. Diff.	-	-	-	-	-	-	-	-	-	-	-
TRANSFORMATION	**-**	**-**	**-**	**-**	**-**	**-**	**-**	**-**	**-**	**-**	**-**
Electricity and CHP Plants	-	-	-	-	-	-	-	-	-	-	-
Petroleum Refineries	-	-	-	-	-	-	-	-	-	-	-
Other Transform. Sector	-	-	-	-	-	-	-	-	-	-	-
ENERGY SECTOR	**-**	**-**	**-**	**-**	**-**	**-**	**-**	**-**	**-**	**-**	**-**
DISTRIBUTION LOSSES	**-**	**-**	**-**	**-**	**-**	**-**	**-**	**-**	**-**	**-**	**-**
FINAL CONSUMPTION	**-**	**5**	**-**	**-**	**-**	**-**	**-**	**-**	**-**	**-**	**-**
INDUSTRY SECTOR	**-**	**-**	**-**	**-**	**-**	**-**	**-**	**-**	**-**	**-**	**-**
Iron and Steel	-	-	-	-	-	-	-	-	-	-	-
Chemical and Petrochem.	-	-	-	-	-	-	-	-	-	-	-
Non-Metallic Minerals	-	-	-	-	-	-	-	-	-	-	-
Non-specified	-	-	-	-	-	-	-	-	-	-	-
TRANSPORT SECTOR	**-**	**-**	**-**	**-**	**-**	**-**	**-**	**-**	**-**	**-**	**-**
Air	-	-	-	-	-	-	-	-	-	-	-
Road	-	-	-	-	-	-	-	-	-	-	-
Non-specified	-	-	-	-	-	-	-	-	-	-	-
OTHER SECTORS	**-**	**5**	**-**	**-**	**-**	**-**	**-**	**-**	**-**	**-**	**-**
Agriculture	-	-	-	-	-	-	-	-	-	-	-
Comm. and Publ. Services	-	-	-	-	-	-	-	-	-	-	-
Residential	-	5	-	-	-	-	-	-	-	-	-
Non-specified	-	-	-	-	-	-	-	-	-	-	-
NON-ENERGY USE	**-**	**-**	**-**	**-**	**-**	**-**	**-**	**-**	**-**	**-**	**-**

APPROVISIONNEMENT ET DEMANDE 1997	*Charbon* (1000 tonnes)							*Pétrole* (1000 tonnes)			
	Charbon à coke	*Autres charb. bit.*	*Charbon sous-bit.*	*Lignite*	*Tourbe*	*Coke de four/gaz*	*Agg./briq. de lignite*	*Pétrole brut*	*LGN*	*Produits d'aliment.*	*Additifs*
Production	-	-	-	-	-	-	-	-	-	-	-
Imports	-	5	-	-	-	-	-	-	-	-	-
Exports	-	-	-	-	-	-	-	-	-	-	-
Intl. Marine Bunkers	-	-	-	-	-	-	-	-	-	-	-
Stock Changes	-	-	-	-	-	-	-	-	-	-	-
DOMESTIC SUPPLY	**-**	**5**	**-**	**-**	**-**	**-**	**-**	**-**	**-**	**-**	**-**
Transfers and Stat. Diff.	-	-	-	-	-	-	-	-	-	-	-
TRANSFORMATION	**-**	**-**	**-**	**-**	**-**	**-**	**-**	**-**	**-**	**-**	**-**
Electricity and CHP Plants	-	-	-	-	-	-	-	-	-	-	-
Petroleum Refineries	-	-	-	-	-	-	-	-	-	-	-
Other Transform. Sector	-	-	-	-	-	-	-	-	-	-	-
ENERGY SECTOR	**-**	**-**	**-**	**-**	**-**	**-**	**-**	**-**	**-**	**-**	**-**
DISTRIBUTION LOSSES	**-**	**-**	**-**	**-**	**-**	**-**	**-**	**-**	**-**	**-**	**-**
FINAL CONSUMPTION	**-**	**5**	**-**	**-**	**-**	**-**	**-**	**-**	**-**	**-**	**-**
INDUSTRY SECTOR	**-**	**-**	**-**	**-**	**-**	**-**	**-**	**-**	**-**	**-**	**-**
Iron and Steel	-	-	-	-	-	-	-	-	-	-	-
Chemical and Petrochem.	-	-	-	-	-	-	-	-	-	-	-
Non-Metallic Minerals	-	-	-	-	-	-	-	-	-	-	-
Non-specified	-	-	-	-	-	-	-	-	-	-	-
TRANSPORT SECTOR	**-**	**-**	**-**	**-**	**-**	**-**	**-**	**-**	**-**	**-**	**-**
Air	-	-	-	-	-	-	-	-	-	-	-
Road	-	-	-	-	-	-	-	-	-	-	-
Non-specified	-	-	-	-	-	-	-	-	-	-	-
OTHER SECTORS	**-**	**5**	**-**	**-**	**-**	**-**	**-**	**-**	**-**	**-**	**-**
Agriculture	-	-	-	-	-	-	-	-	-	-	-
Comm. and Publ. Services	-	-	-	-	-	-	-	-	-	-	-
Residential	-	5	-	-	-	-	-	-	-	-	-
Non-specified	-	-	-	-	-	-	-	-	-	-	-
NON-ENERGY USE	**-**	**-**	**-**	**-**	**-**	**-**	**-**	**-**	**-**	**-**	**-**

Armenia / Arménie

SUPPLY AND CONSUMPTION 1996	Oil cont. (1000 tonnes)										
	Refinery Gas	LPG + Ethane	Motor Gasoline	Aviation Gasoline	Jet Fuel	Kerosene	Gas/ Diesel	Heavy Fuel Oil	Naphtha	Petrol. Coke	Other Prod.
Production	-	-	-	-	-	-	-	-	-	-	-
Imports	-	1	20	-	19	2	34	63	-	-	15
Exports	-	-	-	-	-	-	-	-	-	-	-
Intl. Marine Bunkers	-	-	-	-	-	-	-	-	-	-	-
Stock Changes	-	-	-	-	-	-	-	-	-	-	-
DOMESTIC SUPPLY	**-**	**1**	**20**	**-**	**19**	**2**	**34**	**63**	**-**	**-**	**15**
Transfers and Stat. Diff.	-	-	-	-	-	-	-	-	-	-	-
TRANSFORMATION	**-**	**-**	**-**	**-**	**-**	**-**	**-**	**63**	**-**	**-**	**-**
Electricity and CHP Plants	-	-	-	-	-	-	-	63	-	-	-
Petroleum Refineries	-	-	-	-	-	-	-	-	-	-	-
Other Transform. Sector	-	-	-	-	-	-	-	-	-	-	-
ENERGY SECTOR	**-**	**-**	**-**	**-**	**-**	**-**	**-**	**-**	**-**	**-**	**-**
DISTRIBUTION LOSSES	**-**	**-**	**-**	**-**	**-**	**-**	**-**	**-**	**-**	**-**	**-**
FINAL CONSUMPTION	**-**	**1**	**20**	**-**	**19**	**2**	**34**	**-**	**-**	**-**	**15**
INDUSTRY SECTOR	**-**	**-**	**-**	**-**	**-**	**-**	**-**	**-**	**-**	**-**	**-**
Iron and Steel	-	-	-	-	-	-	-	-	-	-	-
Chemical and Petrochem.	-	-	-	-	-	-	-	-	-	-	-
Non-Metallic Minerals	-	-	-	-	-	-	-	-	-	-	-
Non-specified	-	-	-	-	-	-	-	-	-	-	-
TRANSPORT SECTOR	**-**	**-**	**20**	**-**	**19**	**-**	**-**	**-**	**-**	**-**	**-**
Air	-	-	-	-	19	-	-	-	-	-	-
Road	-	-	20	-	-	-	-	-	-	-	-
Non-specified	-	-	-	-	-	-	-	-	-	-	-
OTHER SECTORS	**-**	**1**	**-**	**-**	**-**	**2**	**34**	**-**	**-**	**-**	**-**
Agriculture	-	-	-	-	-	-	-	-	-	-	-
Comm. and Publ. Services	-	-	-	-	-	-	-	-	-	-	-
Residential	-	-	-	-	-	2	-	-	-	-	-
Non-specified	-	1	-	-	-	-	34	-	-	-	-
NON-ENERGY USE	**-**	**-**	**-**	**-**	**-**	**-**	**-**	**-**	**-**	**-**	**15**

APPROVISIONNEMENT ET DEMANDE 1997	*Pétrole cont.* (1000 tonnes)										
	Gaz de raffinerie	*GPL + éthane*	*Essence moteur*	*Essence aviation*	*Carbu- réacteurs*	*Kérosène*	*Gazole*	*Fioul lourd*	*Naphta*	*Coke de pétrole*	*Autres prod.*
Production	-	-	-	-	-	-	-	-	-	-	-
Imports	-	1	20	-	19	2	34	63	-	-	15
Exports	-	-	-	-	-	-	-	-	-	-	-
Intl. Marine Bunkers	-	-	-	-	-	-	-	-	-	-	-
Stock Changes	-	-	-	-	-	-	-	-	-	-	-
DOMESTIC SUPPLY	**-**	**1**	**20**	**-**	**19**	**2**	**34**	**63**	**-**	**-**	**15**
Transfers and Stat. Diff.	-	-	-	-	-	-	-	-	-	-	-
TRANSFORMATION	**-**	**-**	**-**	**-**	**-**	**-**	**-**	**63**	**-**	**-**	**-**
Electricity and CHP Plants	-	-	-	-	-	-	-	63	-	-	-
Petroleum Refineries	-	-	-	-	-	-	-	-	-	-	-
Other Transform. Sector	-	-	-	-	-	-	-	-	-	-	-
ENERGY SECTOR	**-**	**-**	**-**	**-**	**-**	**-**	**-**	**-**	**-**	**-**	**-**
DISTRIBUTION LOSSES	**-**	**-**	**-**	**-**	**-**	**-**	**-**	**-**	**-**	**-**	**-**
FINAL CONSUMPTION	**-**	**1**	**20**	**-**	**19**	**2**	**34**	**-**	**-**	**-**	**15**
INDUSTRY SECTOR	**-**	**-**	**-**	**-**	**-**	**-**	**-**	**-**	**-**	**-**	**-**
Iron and Steel	-	-	-	-	-	-	-	-	-	-	-
Chemical and Petrochem.	-	-	-	-	-	-	-	-	-	-	-
Non-Metallic Minerals	-	-	-	-	-	-	-	-	-	-	-
Non-specified	-	-	-	-	-	-	-	-	-	-	-
TRANSPORT SECTOR	**-**	**-**	**20**	**-**	**19**	**-**	**-**	**-**	**-**	**-**	**-**
Air	-	-	-	-	19	-	-	-	-	-	-
Road	-	-	20	-	-	-	-	-	-	-	-
Non-specified	-	-	-	-	-	-	-	-	-	-	-
OTHER SECTORS	**-**	**1**	**-**	**-**	**-**	**2**	**34**	**-**	**-**	**-**	**-**
Agriculture	-	-	-	-	-	-	-	-	-	-	-
Comm. and Publ. Services	-	-	-	-	-	-	-	-	-	-	-
Residential	-	-	-	-	-	2	-	-	-	-	-
Non-specified	-	1	-	-	-	-	34	-	-	-	-
NON-ENERGY USE	**-**	**-**	**-**	**-**	**-**	**-**	**-**	**-**	**-**	**-**	**15**

Armenia / Arménie

SUPPLY AND CONSUMPTION 1996	Gas (TJ)				Comb. Renew. & Waste (TJ)				(GWh)	(TJ)
	Natural Gas	Gas Works	Coke Ovens	Blast Furnaces	Solid Biomass	Gas/Liquids from Biomass	Municipal Waste	Industrial Waste	Electricity	Heat
Production	-	-	-	-	39	-	-	-	6214	3762
Imports	41470	-	-	-	-	-	-	-	-	-
Exports	-	-	-	-	-	-	-	-	-	-
Intl. Marine Bunkers	-	-	-	-	-	-	-	-	-	-
Stock Changes	-	-	-	-	-	-	-	-	-	-
DOMESTIC SUPPLY	**41470**	**-**	**-**	**-**	**39**	**-**	**-**	**-**	**6214**	**3762**
Transfers and Stat. Diff.	-	-	-	-	-	-	-	-	-	-
TRANSFORMATION	**31876**	**-**	**-**	**-**	**-**	**-**	**-**	**-**	**-**	**-**
Electricity and CHP Plants	31876	-	-	-	-	-	-	-	-	-
Petroleum Refineries	-	-	-	-	-	-	-	-	-	-
Other Transform. Sector	-	-	-	-	-	-	-	-	-	-
ENERGY SECTOR	**-**	**-**	**-**	**-**	**-**	**-**	**-**	**-**	**452**	**-**
DISTRIBUTION LOSSES	**-**	**-**	**-**	**-**	**-**	**-**	**-**	**-**	**2349**	**752**
FINAL CONSUMPTION	**9594**	**-**	**-**	**-**	**39**	**-**	**-**	**-**	**3413**	**3010**
INDUSTRY SECTOR	**4955**	**-**	**-**	**-**	**-**	**-**	**-**	**-**	**815**	**1254**
Iron and Steel	-	-	-	-	-	-	-	-	2	-
Chemical and Petrochem.	-	-	-	-	-	-	-	-	168	-
Non-Metallic Minerals	-	-	-	-	-	-	-	-	81	-
Non-specified	4955	-	-	-	-	-	-	-	564	1254
TRANSPORT SECTOR	**-**	**-**	**-**	**-**	**-**	**-**	**-**	**-**	**173**	**-**
Air	-	-	-	-	-	-	-	-	-	-
Road	-	-	-	-	-	-	-	-	-	-
Non-specified	-	-	-	-	-	-	-	-	173	-
OTHER SECTORS	**4639**	**-**	**-**	**-**	**39**	**-**	**-**	**-**	**2425**	**1756**
Agriculture	633	-	-	-	-	-	-	-	303	-
Comm. and Publ. Services	-	-	-	-	-	-	-	-	1023	-
Residential	-	-	-	-	-	-	-	-	1084	1756
Non-specified	4006	-	-	-	39	-	-	-	15	-
NON-ENERGY USE	**-**	**-**	**-**	**-**	**-**	**-**	**-**	**-**	**-**	**-**

APPROVISIONNEMENT ET DEMANDE 1997	*Gaz* (TJ)				*En. Re. Comb. & Déchets* (TJ)				(GWh)	(TJ)
	Gaz naturel	*Usines à gaz*	*Cokeries*	*Hauts fourneaux*	*Biomasse solide*	*Gaz/Liquides tirés de biomasse*	*Déchets urbains*	*Déchets industriels*	*Electricité*	*Chaleur*
Production	-	-	-	-	39	-	-	-	6021	3762
Imports	51649	-	-	-	-	-	-	-	-	-
Exports	-	-	-	-	-	-	-	-	-	-
Intl. Marine Bunkers	-	-	-	-	-	-	-	-	-	-
Stock Changes	-	-	-	-	-	-	-	-	-	-
DOMESTIC SUPPLY	**51649**	**-**	**-**	**-**	**39**	**-**	**-**	**-**	**6021**	**3762**
Transfers and Stat. Diff.	-7163	-	-	-	-	-	-	-	-	-
TRANSFORMATION	**34194**	**-**	**-**	**-**	**-**	**-**	**-**	**-**	**-**	**-**
Electricity and CHP Plants	34194	-	-	-	-	-	-	-	-	-
Petroleum Refineries	-	-	-	-	-	-	-	-	-	-
Other Transform. Sector	-	-	-	-	-	-	-	-	-	-
ENERGY SECTOR	**-**	**-**	**-**	**-**	**-**	**-**	**-**	**-**	**451**	**-**
DISTRIBUTION LOSSES	**-**	**-**	**-**	**-**	**-**	**-**	**-**	**-**	**1251**	**752**
FINAL CONSUMPTION	**10292**	**-**	**-**	**-**	**39**	**-**	**-**	**-**	**4319**	**3010**
INDUSTRY SECTOR	**5316**	**-**	**-**	**-**	**-**	**-**	**-**	**-**	**724**	**1254**
Iron and Steel	-	-	-	-	-	-	-	-	3	-
Chemical and Petrochem.	-	-	-	-	-	-	-	-	200	-
Non-Metallic Minerals	-	-	-	-	-	-	-	-	96	-
Non-specified	5316	-	-	-	-	-	-	-	425	1254
TRANSPORT SECTOR	**-**	**-**	**-**	**-**	**-**	**-**	**-**	**-**	**152**	**-**
Air	-	-	-	-	-	-	-	-	-	-
Road	-	-	-	-	-	-	-	-	-	-
Non-specified	-	-	-	-	-	-	-	-	152	-
OTHER SECTORS	**4976**	**-**	**-**	**-**	**39**	**-**	**-**	**-**	**3443**	**1756**
Agriculture	678	-	-	-	-	-	-	-	245	-
Comm. and Publ. Services	-	-	-	-	-	-	-	-	330	-
Residential	-	-	-	-	-	-	-	-	2557	1756
Non-specified	4298	-	-	-	39	-	-	-	311	-
NON-ENERGY USE	**-**	**-**	**-**	**-**	**-**	**-**	**-**	**-**	**-**	**-**

Azerbaijan / Azerbaïdjan

SUPPLY AND CONSUMPTION 1996	Coal (1000 tonnes)							Oil (1000 tonnes)			
	Coking Coal	Other Bit. Coal	Sub-Bit. Coal	Lignite	Peat	Oven and Gas Coke	Pat. Fuel and BKB	Crude Oil	NGL	Feed-stocks	Additives
Production	-	-	-	-	-	-	-	8890	210	-	-
Imports	-	6	-	-	-	-	-	-	-	-	-
Exports	-	-	-	-	-	-	-	-	-	-	-
Intl. Marine Bunkers	-	-	-	-	-	-	-	-	-	-	-
Stock Changes	-	-	-	-	-	-	-	-	-	-	-
DOMESTIC SUPPLY	**-**	**6**	**-**	**-**	**-**	**-**	**-**	**8890**	**210**	**-**	**-**
Transfers and Stat. Diff.	-	-	-	-	-	-	-	-	-	-	-
TRANSFORMATION	**-**	**-**	**-**	**-**	**-**	**-**	**-**	**8890**	**210**	**-**	**-**
Electricity and CHP Plants	-	-	-	-	-	-	-	-	-	-	-
Petroleum Refineries	-	-	-	-	-	-	-	8890	210	-	-
Other Transform. Sector	-	-	-	-	-	-	-	-	-	-	-
ENERGY SECTOR	**-**	**-**	**-**	**-**	**-**	**-**	**-**	**-**	**-**	**-**	**-**
DISTRIBUTION LOSSES	**-**	**-**	**-**	**-**	**-**	**-**	**-**	**-**	**-**	**-**	**-**
FINAL CONSUMPTION	**-**	**6**	**-**	**-**	**-**	**-**	**-**	**-**	**-**	**-**	**-**
INDUSTRY SECTOR	**-**	**-**	**-**	**-**	**-**	**-**	**-**	**-**	**-**	**-**	**-**
Iron and Steel	-	-	-	-	-	-	-	-	-	-	-
Chemical and Petrochem.	-	-	-	-	-	-	-	-	-	-	-
Non-Metallic Minerals	-	-	-	-	-	-	-	-	-	-	-
Non-specified	-	-	-	-	-	-	-	-	-	-	-
TRANSPORT SECTOR	**-**	**6**	**-**	**-**	**-**	**-**	**-**	**-**	**-**	**-**	**-**
Air	-	-	-	-	-	-	-	-	-	-	-
Road	-	-	-	-	-	-	-	-	-	-	-
Non-specified	-	6	-	-	-	-	-	-	-	-	-
OTHER SECTORS	**-**	**-**	**-**	**-**	**-**	**-**	**-**	**-**	**-**	**-**	**-**
Agriculture	-	-	-	-	-	-	-	-	-	-	-
Comm. and Publ. Services	-	-	-	-	-	-	-	-	-	-	-
Residential	-	-	-	-	-	-	-	-	-	-	-
Non-specified	-	-	-	-	-	-	-	-	-	-	-
NON-ENERGY USE	**-**	**-**	**-**	**-**	**-**	**-**	**-**	**-**	**-**	**-**	**-**

APPROVISIONNEMENT ET DEMANDE 1997	*Charbon* (1000 tonnes)							*Pétrole* (1000 tonnes)			
	Charbon à coke	*Autres charb. bit.*	*Charbon sous-bit.*	*Lignite*	*Tourbe*	*Coke de four/gaz*	*Agg./briq. de lignite*	*Pétrole brut*	*LGN*	*Produits d'aliment.*	*Additifs*
Production	-	-	-	-	-	-	-	8822	200	-	-
Imports	-	6	-	-	-	-	-	260	-	-	-
Exports	-	-	-	-	-	-	-	-300	-	-	-
Intl. Marine Bunkers	-	-	-	-	-	-	-	-	-	-	-
Stock Changes	-	-	-	-	-	-	-	-	-	-	-
DOMESTIC SUPPLY	**-**	**6**	**-**	**-**	**-**	**-**	**-**	**8782**	**200**	**-**	**-**
Transfers and Stat. Diff.	-	-	-	-	-	-	-	-	-	-	-
TRANSFORMATION	**-**	**-**	**-**	**-**	**-**	**-**	**-**	**8782**	**200**	**-**	**-**
Electricity and CHP Plants	-	-	-	-	-	-	-	-	-	-	-
Petroleum Refineries	-	-	-	-	-	-	-	8782	200	-	-
Other Transform. Sector	-	-	-	-	-	-	-	-	-	-	-
ENERGY SECTOR	**-**	**-**	**-**	**-**	**-**	**-**	**-**	**-**	**-**	**-**	**-**
DISTRIBUTION LOSSES	**-**	**-**	**-**	**-**	**-**	**-**	**-**	**-**	**-**	**-**	**-**
FINAL CONSUMPTION	**-**	**6**	**-**	**-**	**-**	**-**	**-**	**-**	**-**	**-**	**-**
INDUSTRY SECTOR	**-**	**-**	**-**	**-**	**-**	**-**	**-**	**-**	**-**	**-**	**-**
Iron and Steel	-	-	-	-	-	-	-	-	-	-	-
Chemical and Petrochem.	-	-	-	-	-	-	-	-	-	-	-
Non-Metallic Minerals	-	-	-	-	-	-	-	-	-	-	-
Non-specified	-	-	-	-	-	-	-	-	-	-	-
TRANSPORT SECTOR	**-**	**6**	**-**	**-**	**-**	**-**	**-**	**-**	**-**	**-**	**-**
Air	-	-	-	-	-	-	-	-	-	-	-
Road	-	-	-	-	-	-	-	-	-	-	-
Non-specified	-	6	-	-	-	-	-	-	-	-	-
OTHER SECTORS	**-**	**-**	**-**	**-**	**-**	**-**	**-**	**-**	**-**	**-**	**-**
Agriculture	-	-	-	-	-	-	-	-	-	-	-
Comm. and Publ. Services	-	-	-	-	-	-	-	-	-	-	-
Residential	-	-	-	-	-	-	-	-	-	-	-
Non-specified	-	-	-	-	-	-	-	-	-	-	-
NON-ENERGY USE	**-**	**-**	**-**	**-**	**-**	**-**	**-**	**-**	**-**	**-**	**-**

Azerbaijan / Azerbaïdjan

SUPPLY AND CONSUMPTION 1996	Oil cont. (1000 tonnes)										
	Refinery Gas	LPG + Ethane	Motor Gasoline	Aviation Gasoline	Jet Fuel	Kerosene	Gas/ Diesel	Heavy Fuel Oil	Naphtha	Petrol. Coke	Other Prod.
Production	100	4	737	1	527	150	2111	3961	-	59	306
Imports	-	-	-	-	-	-	12	66	-	-	6
Exports	-	-	-	-	-	-34	-1821	-100	-	-9	-279
Intl. Marine Bunkers	-	-	-	-	-	-	-	-	-	-	-
Stock Changes	-	-	-	-	-	-	-	-	-	-	-
DOMESTIC SUPPLY	**100**	**4**	**737**	**1**	**527**	**116**	**302**	**3927**	**-**	**50**	**33**
Transfers and Stat. Diff.	-	-	-	-	-	-	-	-	-	-	-
TRANSFORMATION	**-**	**-**	**-**	**-**	**-**	**-**	**-**	**3697**	**-**	**-**	**-**
Electricity and CHP Plants	-	-	-	-	-	-	-	3697	-	-	-
Petroleum Refineries	-	-	-	-	-	-	-	-	-	-	-
Other Transform. Sector	-	-	-	-	-	-	-	-	-	-	-
ENERGY SECTOR	**100**	**-**	**-**	**-**	**-**	**-**	**-**	**-**	**-**	**-**	**-**
DISTRIBUTION LOSSES	**-**	**-**	**-**	**-**	**-**	**-**	**-**	**-**	**-**	**-**	**-**
FINAL CONSUMPTION	**-**	**4**	**737**	**1**	**527**	**116**	**302**	**230**	**-**	**50**	**33**
INDUSTRY SECTOR	**-**	**-**	**-**	**-**	**-**	**-**	**-**	**157**	**-**	**-**	**-**
Iron and Steel	-	-	-	-	-	-	-	-	-	-	-
Chemical and Petrochem.	-	-	-	-	-	-	-	-	-	-	-
Non-Metallic Minerals	-	-	-	-	-	-	-	-	-	-	-
Non-specified	-	-	-	-	-	-	-	157	-	-	-
TRANSPORT SECTOR	**-**	**-**	**737**	**1**	**527**	**-**	**-**	**9**	**-**	**-**	**-**
Air	-	-	-	1	527	-	-	-	-	-	-
Road	-	-	737	-	-	-	-	-	-	-	-
Non-specified	-	-	-	-	-	-	-	9	-	-	-
OTHER SECTORS	**-**	**4**	**-**	**-**	**-**	**116**	**302**	**64**	**-**	**-**	**-**
Agriculture	-	-	-	-	-	-	-	40	-	-	-
Comm. and Publ. Services	-	-	-	-	-	-	-	13	-	-	-
Residential	-	-	-	-	-	-	-	11	-	-	-
Non-specified	-	4	-	-	-	116	302	-	-	-	-
NON-ENERGY USE	**-**	**-**	**-**	**-**	**-**	**-**	**-**	**-**	**-**	**50**	**33**

APPROVISIONNEMENT ET DEMANDE 1997	*Pétrole cont.* (1000 tonnes)										
	Gaz de raffinerie	*GPL + éthane*	*Essence moteur*	*Essence aviation*	*Carbu- réacteurs*	*Kérosène*	*Gazole*	*Fioul lourd*	*Naphta*	*Coke de pétrole*	*Autres prod.*
Production	100	3	800	1	450	150	1900	3900	-	50	168
Imports	-	-	-	-	-	-	68	-	-	-	4
Exports	-	-	-308	-	-	-	-1629	-	-	-	-139
Intl. Marine Bunkers	-	-	-	-	-	-	-	-	-	-	-
Stock Changes	-	-	-	-	-	-	-	-	-	-	-
DOMESTIC SUPPLY	**100**	**3**	**492**	**1**	**450**	**150**	**339**	**3900**	**-**	**50**	**33**
Transfers and Stat. Diff.	-	-	-	-	-	-	-	-	-	-	-
TRANSFORMATION	**-**	**-**	**-**	**-**	**-**	**-**	**-**	**3683**	**-**	**-**	**-**
Electricity and CHP Plants	-	-	-	-	-	-	-	3683	-	-	-
Petroleum Refineries	-	-	-	-	-	-	-	-	-	-	-
Other Transform. Sector	-	-	-	-	-	-	-	-	-	-	-
ENERGY SECTOR	**100**	**-**	**-**	**-**	**-**	**-**	**-**	**-**	**-**	**-**	**-**
DISTRIBUTION LOSSES	**-**	**-**	**-**	**-**	**-**	**-**	**-**	**-**	**-**	**-**	**-**
FINAL CONSUMPTION	**-**	**3**	**492**	**1**	**450**	**150**	**339**	**217**	**-**	**50**	**33**
INDUSTRY SECTOR	**-**	**-**	**-**	**-**	**-**	**-**	**-**	**150**	**-**	**-**	**-**
Iron and Steel	-	-	-	-	-	-	-	-	-	-	-
Chemical and Petrochem.	-	-	-	-	-	-	-	-	-	-	-
Non-Metallic Minerals	-	-	-	-	-	-	-	-	-	-	-
Non-specified	-	-	-	-	-	-	-	150	-	-	-
TRANSPORT SECTOR	**-**	**-**	**492**	**1**	**450**	**-**	**-**	**10**	**-**	**-**	**-**
Air	-	-	-	1	450	-	-	-	-	-	-
Road	-	-	492	-	-	-	-	-	-	-	-
Non-specified	-	-	-	-	-	-	-	10	-	-	-
OTHER SECTORS	**-**	**3**	**-**	**-**	**-**	**150**	**339**	**57**	**-**	**-**	**-**
Agriculture	-	-	-	-	-	-	-	35	-	-	-
Comm. and Publ. Services	-	-	-	-	-	-	-	12	-	-	-
Residential	-	-	-	-	-	-	-	10	-	-	-
Non-specified	-	3	-	-	-	150	339	-	-	-	-
NON-ENERGY USE	**-**	**-**	**-**	**-**	**-**	**-**	**-**	**-**	**-**	**50**	**33**

Azerbaijan / Azerbaïdjan

SUPPLY AND CONSUMPTION 1996	Gas (TJ)				Comb. Renew. & Waste (TJ)				(GWh)	(TJ)
	Natural Gas	Gas Works	Coke Ovens	Blast Furnaces	Solid Biomass	Gas/Liquids from Biomass	Municipal Waste	Industrial Waste	Electricity	Heat
Production	237698	-	-	-	68	-	-	-	17088	31067
Imports	905	-	-	-	117	-	-	-	1022	-
Exports	-	-	-	-	-	-	-	-	-580	-
Intl. Marine Bunkers	-	-	-	-	-	-	-	-	-	-
Stock Changes	-	-	-	-	-	-	-	-	-	-
DOMESTIC SUPPLY	**238603**	**-**	**-**	**-**	**185**	**-**	**-**	**-**	**17530**	**31067**
Transfers and Stat. Diff.	-	-	-	-	-	-	-	-	1064	-
TRANSFORMATION	**71592**	**-**	**-**	**-**	**-**	**-**	**-**	**-**	**-**	**-**
Electricity and CHP Plants	71592	-	-	-	-	-	-	-	-	-
Petroleum Refineries	-	-	-	-	-	-	-	-	-	-
Other Transform. Sector	-	-	-	-	-	-	-	-	-	-
ENERGY SECTOR	**-**	**-**	**-**	**-**	**-**	**-**	**-**	**-**	**1044**	**-**
DISTRIBUTION LOSSES	**-**	**-**	**-**	**-**	**-**	**-**	**-**	**-**	**3785**	**-**
FINAL CONSUMPTION	**167011**	**-**	**-**	**-**	**185**	**-**	**-**	**-**	**13765**	**31067**
INDUSTRY SECTOR	**69952**	**-**	**-**	**-**	**-**	**-**	**-**	**-**	**4796**	**31067**
Iron and Steel	-	-	-	-	-	-	-	-	36	-
Chemical and Petrochem.	-	-	-	-	-	-	-	-	511	-
Non-Metallic Minerals	-	-	-	-	-	-	-	-	77	-
Non-specified	69952	-	-	-	-	-	-	-	4172	31067
TRANSPORT SECTOR	**1659**	**-**	**-**	**-**	**-**	**-**	**-**	**-**	**455**	**-**
Air	-	-	-	-	-	-	-	-	-	-
Road	-	-	-	-	-	-	-	-	-	-
Non-specified	1659	-	-	-	-	-	-	-	455	-
OTHER SECTORS	**95400**	**-**	**-**	**-**	**185**	**-**	**-**	**-**	**8514**	**-**
Agriculture	22300	-	-	-	-	-	-	-	4267	-
Comm. and Publ. Services	-	-	-	-	-	-	-	-	866	-
Residential	73100	-	-	-	-	-	-	-	3251	-
Non-specified	-	-	-	-	185	-	-	-	130	-
NON-ENERGY USE	**-**	**-**	**-**	**-**	**-**	**-**	**-**	**-**	**-**	**-**

APPROVISIONNEMENT ET DEMANDE 1997	*Gaz (TJ)*				*En. Re. Comb. & Déchets (TJ)*				(GWh)	(TJ)
	Gaz naturel	*Usines à gaz*	*Cokeries*	*Hauts fourneaux*	*Biomasse solide*	*Gaz/Liquides tirés de biomasse*	*Déchets urbains*	*Déchets industriels*	*Electricité*	*Chaleur*
Production	224692	-	-	-	68	-	-	-	16800	31067
Imports	-	-	-	-	117	-	-	-	1400	-
Exports	-	-	-	-	-	-	-	-	-600	-
Intl. Marine Bunkers	-	-	-	-	-	-	-	-	-	-
Stock Changes	-	-	-	-	-	-	-	-	-	-
DOMESTIC SUPPLY	**224692**	**-**	**-**	**-**	**185**	**-**	**-**	**-**	**17600**	**31067**
Transfers and Stat. Diff.	-	-	-	-	-	-	-	-	-	-
TRANSFORMATION	**62959**	**-**	**-**	**-**	**-**	**-**	**-**	**-**	**-**	**-**
Electricity and CHP Plants	62959	-	-	-	-	-	-	-	-	-
Petroleum Refineries	-	-	-	-	-	-	-	-	-	-
Other Transform. Sector	-	-	-	-	-	-	-	-	-	-
ENERGY SECTOR	**-**	**-**	**-**	**-**	**-**	**-**	**-**	**-**	**1020**	**-**
DISTRIBUTION LOSSES	**-**	**-**	**-**	**-**	**-**	**-**	**-**	**-**	**3800**	**-**
FINAL CONSUMPTION	**161733**	**-**	**-**	**-**	**185**	**-**	**-**	**-**	**12780**	**31067**
INDUSTRY SECTOR	**70250**	**-**	**-**	**-**	**-**	**-**	**-**	**-**	**4453**	**31067**
Iron and Steel	-	-	-	-	-	-	-	-	33	-
Chemical and Petrochem.	-	-	-	-	-	-	-	-	474	-
Non-Metallic Minerals	-	-	-	-	-	-	-	-	72	-
Non-specified	70250	-	-	-	-	-	-	-	3874	31067
TRANSPORT SECTOR	**1583**	**-**	**-**	**-**	**-**	**-**	**-**	**-**	**422**	**-**
Air	-	-	-	-	-	-	-	-	-	-
Road	-	-	-	-	-	-	-	-	-	-
Non-specified	1583	-	-	-	-	-	-	-	422	-
OTHER SECTORS	**89900**	**-**	**-**	**-**	**185**	**-**	**-**	**-**	**7905**	**-**
Agriculture	20800	-	-	-	-	-	-	-	3962	-
Comm. and Publ. Services	-	-	-	-	-	-	-	-	804	-
Residential	69100	-	-	-	-	-	-	-	3018	-
Non-specified	-	-	-	-	185	-	-	-	121	-
NON-ENERGY USE	**-**	**-**	**-**	**-**	**-**	**-**	**-**	**-**	**-**	**-**

Bahrain / Bahrein

SUPPLY AND CONSUMPTION 1996	Coal (1000 tonnes)							Oil (1000 tonnes)			
	Coking Coal	Other Bit. Coal	Sub-Bit. Coal	Lignite	Peat	Oven and Gas Coke	Pat. Fuel and BKB	Crude Oil	NGL	Feed-stocks	Additives
Production	-	-	-	-	-	-	-	1947	382	-	-
Imports	-	-	-	-	-	-	-	11050	-	-	-
Exports	-	-	-	-	-	-	-	-	-106	-	-
Intl. Marine Bunkers	-	-	-	-	-	-	-	-	-	-	-
Stock Changes	-	-	-	-	-	-	-	-	-	-	-
DOMESTIC SUPPLY	-	-	-	-	-	-	-	**12997**	**276**	-	-
Transfers and Stat. Diff.	-	-	-	-	-	-	-	85	-276	-	-
TRANSFORMATION	-	-	-	-	-	-	-	**13082**	-	-	-
Electricity and CHP Plants	-	-	-	-	-	-	-	-	-	-	-
Petroleum Refineries	-	-	-	-	-	-	-	13082	-	-	-
Other Transform. Sector	-	-	-	-	-	-	-	-	-	-	-
ENERGY SECTOR	-	-	-	-	-	-	-	-	-	-	-
DISTRIBUTION LOSSES	-	-	-	-	-	-	-	-	-	-	-
FINAL CONSUMPTION	-	-	-	-	-	-	-	-	-	-	-
INDUSTRY SECTOR	-	-	-	-	-	-	-	-	-	-	-
Iron and Steel	-	-	-	-	-	-	-	-	-	-	-
Chemical and Petrochem.	-	-	-	-	-	-	-	-	-	-	-
Non-Metallic Minerals	-	-	-	-	-	-	-	-	-	-	-
Non-specified	-	-	-	-	-	-	-	-	-	-	-
TRANSPORT SECTOR	-	-	-	-	-	-	-	-	-	-	-
Air	-	-	-	-	-	-	-	-	-	-	-
Road	-	-	-	-	-	-	-	-	-	-	-
Non-specified	-	-	-	-	-	-	-	-	-	-	-
OTHER SECTORS	-	-	-	-	-	-	-	-	-	-	-
Agriculture	-	-	-	-	-	-	-	-	-	-	-
Comm. and Publ. Services	-	-	-	-	-	-	-	-	-	-	-
Residential	-	-	-	-	-	-	-	-	-	-	-
Non-specified	-	-	-	-	-	-	-	-	-	-	-
NON-ENERGY USE	-	-	-	-	-	-	-	-	-	-	-

APPROVISIONNEMENT ET DEMANDE 1997	*Charbon* (1000 tonnes)							*Pétrole* (1000 tonnes)			
	Charbon à coke	*Autres charb. bit.*	*Charbon sous-bit.*	*Lignite*	*Tourbe*	*Coke de four/gaz*	*Agg./briq. de lignite*	*Pétrole brut*	*LGN*	*Produits d'aliment.*	*Additifs*
Production	-	-	-	-	-	-	-	1984	344	-	-
Imports	-	-	-	-	-	-	-	12070	-	-	-
Exports	-	-	-	-	-	-	-	-	-76	-	-
Intl. Marine Bunkers	-	-	-	-	-	-	-	-	-	-	-
Stock Changes	-	-	-	-	-	-	-	-	-	-	-
DOMESTIC SUPPLY	-	-	-	-	-	-	-	**14054**	**268**	-	-
Transfers and Stat. Diff.	-	-	-	-	-	-	-	-	-268	-	-
TRANSFORMATION	-	-	-	-	-	-	-	**14054**	-	-	-
Electricity and CHP Plants	-	-	-	-	-	-	-	-	-	-	-
Petroleum Refineries	-	-	-	-	-	-	-	14054	-	-	-
Other Transform. Sector	-	-	-	-	-	-	-	-	-	-	-
ENERGY SECTOR	-	-	-	-	-	-	-	-	-	-	-
DISTRIBUTION LOSSES	-	-	-	-	-	-	-	-	-	-	-
FINAL CONSUMPTION	-	-	-	-	-	-	-	-	-	-	-
INDUSTRY SECTOR	-	-	-	-	-	-	-	-	-	-	-
Iron and Steel	-	-	-	-	-	-	-	-	-	-	-
Chemical and Petrochem.	-	-	-	-	-	-	-	-	-	-	-
Non-Metallic Minerals	-	-	-	-	-	-	-	-	-	-	-
Non-specified	-	-	-	-	-	-	-	-	-	-	-
TRANSPORT SECTOR	-	-	-	-	-	-	-	-	-	-	-
Air	-	-	-	-	-	-	-	-	-	-	-
Road	-	-	-	-	-	-	-	-	-	-	-
Non-specified	-	-	-	-	-	-	-	-	-	-	-
OTHER SECTORS	-	-	-	-	-	-	-	-	-	-	-
Agriculture	-	-	-	-	-	-	-	-	-	-	-
Comm. and Publ. Services	-	-	-	-	-	-	-	-	-	-	-
Residential	-	-	-	-	-	-	-	-	-	-	-
Non-specified	-	-	-	-	-	-	-	-	-	-	-
NON-ENERGY USE	-	-	-	-	-	-	-	-	-	-	-

Bahrain / Bahrein

SUPPLY AND CONSUMPTION 1996	Oil cont. (1000 tonnes)										
	Refinery Gas	LPG + Ethane	Motor Gasoline	Aviation Gasoline	Jet Fuel	Kerosene	Gas/ Diesel	Heavy Fuel Oil	Naphtha	Petrol. Coke	Other Prod.
Production	49	19	948	-	751	1396	3423	2470	1627	-	456
Imports	-	5	-	-	-	-	-	-	-	-	-
Exports	-	-	-603	-	-481	-1373	-4092	-2397	-2139	-	-427
Intl. Marine Bunkers	-	-	-	-	-	-	-37	-73	-	-	-
Stock Changes	-	-	-	-	-	-	803	-	-	-	-
DOMESTIC SUPPLY	**49**	**24**	**345**	**-**	**270**	**23**	**97**	**-**	**-512**	**-**	**29**
Transfers and Stat. Diff.	-	-	-1	-	1	-	-	-	512	-	-
TRANSFORMATION	**-**	**-**	**-**	**-**	**-**	**-**	**-**	**-**	**-**	**-**	**-**
Electricity and CHP Plants	-	-	-	-	-	-	-	-	-	-	-
Petroleum Refineries	-	-	-	-	-	-	-	-	-	-	-
Other Transform. Sector	-	-	-	-	-	-	-	-	-	-	-
ENERGY SECTOR	**49**	**-**	**-**	**-**	**-**	**-**	**-**	**-**	**-**	**-**	**-**
DISTRIBUTION LOSSES	**-**	**-**	**-**	**-**	**-**	**-**	**-**	**-**	**-**	**-**	**-**
FINAL CONSUMPTION	**-**	**24**	**344**	**-**	**271**	**23**	**97**	**-**	**-**	**-**	**29**
INDUSTRY SECTOR	**-**	**-**	**-**	**-**	**-**	**-**	**-**	**-**	**-**	**-**	**-**
Iron and Steel	-	-	-	-	-	-	-	-	-	-	-
Chemical and Petrochem.	-	-	-	-	-	-	-	-	-	-	-
Non-Metallic Minerals	-	-	-	-	-	-	-	-	-	-	-
Non-specified	-	-	-	-	-	-	-	-	-	-	-
TRANSPORT SECTOR	**-**	**-**	**344**	**-**	**271**	**-**	**97**	**-**	**-**	**-**	**-**
Air	-	-	-	-	271	-	-	-	-	-	-
Road	-	-	344	-	-	-	97	-	-	-	-
Non-specified	-	-	-	-	-	-	-	-	-	-	-
OTHER SECTORS	**-**	**24**	**-**	**-**	**-**	**23**	**-**	**-**	**-**	**-**	**-**
Agriculture	-	-	-	-	-	-	-	-	-	-	-
Comm. and Publ. Services	-	-	-	-	-	-	-	-	-	-	-
Residential	-	24	-	-	-	23	-	-	-	-	-
Non-specified	-	-	-	-	-	-	-	-	-	-	-
NON-ENERGY USE	**-**	**-**	**-**	**-**	**-**	**-**	**-**	**-**	**-**	**-**	**29**

APPROVISIONNEMENT ET DEMANDE 1997	*Pétrole cont.* (1000 tonnes)										
	Gaz de raffinerie	*GPL + éthane*	*Essence moteur*	*Essence aviation*	*Carbu- réacteurs*	*Kérosène*	*Gazole*	*Fioul lourd*	*Naphta*	*Coke de pétrole*	*Autres prod.*
Production	53	26	1061	-	818	1520	3747	2295	1715	-	425
Imports	-	-	-	-	-	-	-	-	-	-	-
Exports	-	-2	-540	-	-566	-1496	-3793	-2222	-1982	-	-396
Intl. Marine Bunkers	-	-	-	-	-	-	-37	-73	-	-	-
Stock Changes	-	-	-	-	-	-	184	-	-	-	-
DOMESTIC SUPPLY	**53**	**24**	**521**	**-**	**252**	**24**	**101**	**-**	**-267**	**-**	**29**
Transfers and Stat. Diff.	-	-	-	-	-	-1	-	-	267	-	-
TRANSFORMATION	**-**	**-**	**-**	**-**	**-**	**-**	**-**	**-**	**-**	**-**	**-**
Electricity and CHP Plants	-	-	-	-	-	-	-	-	-	-	-
Petroleum Refineries	-	-	-	-	-	-	-	-	-	-	-
Other Transform. Sector	-	-	-	-	-	-	-	-	-	-	-
ENERGY SECTOR	**53**	**-**	**-**	**-**	**-**	**-**	**-**	**-**	**-**	**-**	**-**
DISTRIBUTION LOSSES	**-**	**-**	**-**	**-**	**-**	**-**	**-**	**-**	**-**	**-**	**-**
FINAL CONSUMPTION	**-**	**24**	**521**	**-**	**252**	**23**	**101**	**-**	**-**	**-**	**29**
INDUSTRY SECTOR	**-**	**-**	**-**	**-**	**-**	**-**	**-**	**-**	**-**	**-**	**-**
Iron and Steel	-	-	-	-	-	-	-	-	-	-	-
Chemical and Petrochem.	-	-	-	-	-	-	-	-	-	-	-
Non-Metallic Minerals	-	-	-	-	-	-	-	-	-	-	-
Non-specified	-	-	-	-	-	-	-	-	-	-	-
TRANSPORT SECTOR	**-**	**-**	**521**	**-**	**252**	**-**	**101**	**-**	**-**	**-**	**-**
Air	-	-	-	-	252	-	-	-	-	-	-
Road	-	-	521	-	-	-	101	-	-	-	-
Non-specified	-	-	-	-	-	-	-	-	-	-	-
OTHER SECTORS	**-**	**24**	**-**	**-**	**-**	**23**	**-**	**-**	**-**	**-**	**-**
Agriculture	-	-	-	-	-	-	-	-	-	-	-
Comm. and Publ. Services	-	-	-	-	-	-	-	-	-	-	-
Residential	-	24	-	-	-	23	-	-	-	-	-
Non-specified	-	-	-	-	-	-	-	-	-	-	-
NON-ENERGY USE	**-**	**-**	**-**	**-**	**-**	**-**	**-**	**-**	**-**	**-**	**29**

Bahrain / Bahrein

SUPPLY AND CONSUMPTION 1996	Gas (TJ)				Comb. Renew. & Waste (TJ)				(GWh)	(TJ)
	Natural Gas	Gas Works	Coke Ovens	Blast Furnaces	Solid Biomass	Gas/Liquids from Biomass	Municipal Waste	Industrial Waste	Electricity	Heat
Production	245157	-	-	-	-	-	-	-	4771	-
Imports	-	-	-	-	-	-	-	-	-	-
Exports	-	-	-	-	-	-	-	-	-	-
Intl. Marine Bunkers	-	-	-	-	-	-	-	-	-	-
Stock Changes	-	-	-	-	-	-	-	-	-	-
DOMESTIC SUPPLY	**245157**	-	-	-	-	-	-	-	**4771**	-
Transfers and Stat. Diff.	1	-	-	-	-	-	-	-	-	-
TRANSFORMATION	**73449**	-	-	-	-	-	-	-	-	-
Electricity and CHP Plants	73449	-	-	-	-	-	-	-	-	-
Petroleum Refineries	-	-	-	-	-	-	-	-	-	-
Other Transform. Sector	-	-	-	-	-	-	-	-	-	-
ENERGY SECTOR	**43613**	-	-	-	-	-	-	-	**311**	-
DISTRIBUTION LOSSES	-	-	-	-	-	-	-	-	**172**	-
FINAL CONSUMPTION	**128096**	-	-	-	-	-	-	-	**4288**	-
INDUSTRY SECTOR	**128096**	-	-	-	-	-	-	-	**701**	-
Iron and Steel	5826	-	-	-	-	-	-	-	-	-
Chemical and Petrochem.	54218	-	-	-	-	-	-	-	-	-
Non-Metallic Minerals	-	-	-	-	-	-	-	-	-	-
Non-specified	68052	-	-	-	-	-	-	-	701	-
TRANSPORT SECTOR	-	-	-	-	-	-	-	-	-	-
Air	-	-	-	-	-	-	-	-	-	-
Road	-	-	-	-	-	-	-	-	-	-
Non-specified	-	-	-	-	-	-	-	-	-	-
OTHER SECTORS	-	-	-	-	-	-	-	-	**3587**	-
Agriculture	-	-	-	-	-	-	-	-	28	-
Comm. and Publ. Services	-	-	-	-	-	-	-	-	1117	-
Residential	-	-	-	-	-	-	-	-	2442	-
Non-specified	-	-	-	-	-	-	-	-	-	-
NON-ENERGY USE	-	-	-	-	-	-	-	-	-	-

APPROVISIONNEMENT ET DEMANDE 1997	*Gaz* (TJ)				*En. Re. Comb. & Déchets* (TJ)				(GWh)	(TJ)
	Gaz naturel	*Usines à gaz*	*Cokeries*	*Hauts fourneaux*	*Biomasse solide*	*Gaz/Liquides tirés de biomasse*	*Déchets urbains*	*Déchets industriels*	*Electricité*	*Chaleur*
Production	237915	-	-	-	-	-	-	-	4924	-
Imports	-	-	-	-	-	-	-	-	-	-
Exports	-	-	-	-	-	-	-	-	-	-
Intl. Marine Bunkers	-	-	-	-	-	-	-	-	-	-
Stock Changes	-	-	-	-	-	-	-	-	-	-
DOMESTIC SUPPLY	**237915**	-	-	-	-	-	-	-	**4924**	-
Transfers and Stat. Diff.	-1	-	-	-	-	-	-	-	-1	-
TRANSFORMATION	**73303**	-	-	-	-	-	-	-	-	-
Electricity and CHP Plants	73303	-	-	-	-	-	-	-	-	-
Petroleum Refineries	-	-	-	-	-	-	-	-	-	-
Other Transform. Sector	-	-	-	-	-	-	-	-	-	-
ENERGY SECTOR	**41810**	-	-	-	-	-	-	-	**321**	-
DISTRIBUTION LOSSES	-	-	-	-	-	-	-	-	**177**	-
FINAL CONSUMPTION	**122801**	-	-	-	-	-	-	-	**4425**	-
INDUSTRY SECTOR	**122801**	-	-	-	-	-	-	-	**723**	-
Iron and Steel	5585	-	-	-	-	-	-	-	-	-
Chemical and Petrochem.	51977	-	-	-	-	-	-	-	-	-
Non-Metallic Minerals	-	-	-	-	-	-	-	-	-	-
Non-specified	65239	-	-	-	-	-	-	-	723	-
TRANSPORT SECTOR	-	-	-	-	-	-	-	-	-	-
Air	-	-	-	-	-	-	-	-	-	-
Road	-	-	-	-	-	-	-	-	-	-
Non-specified	-	-	-	-	-	-	-	-	-	-
OTHER SECTORS	-	-	-	-	-	-	-	-	**3702**	-
Agriculture	-	-	-	-	-	-	-	-	29	-
Comm. and Publ. Services	-	-	-	-	-	-	-	-	1153	-
Residential	-	-	-	-	-	-	-	-	2520	-
Non-specified	-	-	-	-	-	-	-	-	-	-
NON-ENERGY USE	-	-	-	-	-	-	-	-	-	-

Bangladesh

SUPPLY AND CONSUMPTION 1996	Coal (1000 tonnes)							Oil (1000 tonnes)			
	Coking Coal	Other Bit. Coal	Sub-Bit. Coal	Lignite	Peat	Oven and Gas Coke	Pat. Fuel and BKB	Crude Oil	NGL	Feed-stocks	Additives
Production	-	-	-	-	-	-	-	10	-	-	-
Imports	-	-	-	-	-	-	-	752	-	-	-
Exports	-	-	-	-	-	-	-	-	-	-	-
Intl. Marine Bunkers	-	-	-	-	-	-	-	-	-	-	-
Stock Changes	-	-	-	-	-	-	-	-	-	-	-
DOMESTIC SUPPLY	**-**	**-**	**-**	**-**	**-**	**-**	**-**	**762**	**-**	**-**	**-**
Transfers and Stat. Diff.	-	-	-	-	-	-	-	-	-	359	-
TRANSFORMATION	**-**	**-**	**-**	**-**	**-**	**-**	**-**	**762**	**-**	**359**	**-**
Electricity and CHP Plants	-	-	-	-	-	-	-	-	-	-	-
Petroleum Refineries	-	-	-	-	-	-	-	762	-	359	-
Other Transform. Sector	-	-	-	-	-	-	-	-	-	-	-
ENERGY SECTOR	**-**	**-**	**-**	**-**	**-**	**-**	**-**	**-**	**-**	**-**	**-**
DISTRIBUTION LOSSES	**-**	**-**	**-**	**-**	**-**	**-**	**-**	**-**	**-**	**-**	**-**
FINAL CONSUMPTION	**-**	**-**	**-**	**-**	**-**	**-**	**-**	**-**	**-**	**-**	**-**
INDUSTRY SECTOR	**-**	**-**	**-**	**-**	**-**	**-**	**-**	**-**	**-**	**-**	**-**
Iron and Steel	-	-	-	-	-	-	-	-	-	-	-
Chemical and Petrochem.	-	-	-	-	-	-	-	-	-	-	-
Non-Metallic Minerals	-	-	-	-	-	-	-	-	-	-	-
Non-specified	-	-	-	-	-	-	-	-	-	-	-
TRANSPORT SECTOR	**-**	**-**	**-**	**-**	**-**	**-**	**-**	**-**	**-**	**-**	**-**
Air	-	-	-	-	-	-	-	-	-	-	-
Road	-	-	-	-	-	-	-	-	-	-	-
Non-specified	-	-	-	-	-	-	-	-	-	-	-
OTHER SECTORS	**-**	**-**	**-**	**-**	**-**	**-**	**-**	**-**	**-**	**-**	**-**
Agriculture	-	-	-	-	-	-	-	-	-	-	-
Comm. and Publ. Services	-	-	-	-	-	-	-	-	-	-	-
Residential	-	-	-	-	-	-	-	-	-	-	-
Non-specified	-	-	-	-	-	-	-	-	-	-	-
NON-ENERGY USE	**-**	**-**	**-**	**-**	**-**	**-**	**-**	**-**	**-**	**-**	**-**

APPROVISIONNEMENT ET DEMANDE 1997	*Charbon* (1000 tonnes)							*Pétrole* (1000 tonnes)			
	Charbon à coke	*Autres charb. bit.*	*Charbon sous-bit.*	*Lignite*	*Tourbe*	*Coke de four/gaz*	*Agg./briq. de lignite*	*Pétrole brut*	*LGN*	*Produits d'aliment.*	*Additifs*
Production	-	-	-	-	-	-	-	7	-	-	-
Imports	-	-	-	-	-	-	-	763	-	-	-
Exports	-	-	-	-	-	-	-	-	-	-	-
Intl. Marine Bunkers	-	-	-	-	-	-	-	-	-	-	-
Stock Changes	-	-	-	-	-	-	-	-	-	-	-
DOMESTIC SUPPLY	**-**	**-**	**-**	**-**	**-**	**-**	**-**	**770**	**-**	**-**	**-**
Transfers and Stat. Diff.	-	-	-	-	-	-	-	-1	-	362	-
TRANSFORMATION	**-**	**-**	**-**	**-**	**-**	**-**	**-**	**769**	**-**	**362**	**-**
Electricity and CHP Plants	-	-	-	-	-	-	-	-	-	-	-
Petroleum Refineries	-	-	-	-	-	-	-	769	-	362	-
Other Transform. Sector	-	-	-	-	-	-	-	-	-	-	-
ENERGY SECTOR	**-**	**-**	**-**	**-**	**-**	**-**	**-**	**-**	**-**	**-**	**-**
DISTRIBUTION LOSSES	**-**	**-**	**-**	**-**	**-**	**-**	**-**	**-**	**-**	**-**	**-**
FINAL CONSUMPTION	**-**	**-**	**-**	**-**	**-**	**-**	**-**	**-**	**-**	**-**	**-**
INDUSTRY SECTOR	**-**	**-**	**-**	**-**	**-**	**-**	**-**	**-**	**-**	**-**	**-**
Iron and Steel	-	-	-	-	-	-	-	-	-	-	-
Chemical and Petrochem.	-	-	-	-	-	-	-	-	-	-	-
Non-Metallic Minerals	-	-	-	-	-	-	-	-	-	-	-
Non-specified	-	-	-	-	-	-	-	-	-	-	-
TRANSPORT SECTOR	**-**	**-**	**-**	**-**	**-**	**-**	**-**	**-**	**-**	**-**	**-**
Air	-	-	-	-	-	-	-	-	-	-	-
Road	-	-	-	-	-	-	-	-	-	-	-
Non-specified	-	-	-	-	-	-	-	-	-	-	-
OTHER SECTORS	**-**	**-**	**-**	**-**	**-**	**-**	**-**	**-**	**-**	**-**	**-**
Agriculture	-	-	-	-	-	-	-	-	-	-	-
Comm. and Publ. Services	-	-	-	-	-	-	-	-	-	-	-
Residential	-	-	-	-	-	-	-	-	-	-	-
Non-specified	-	-	-	-	-	-	-	-	-	-	-
NON-ENERGY USE	**-**	**-**	**-**	**-**	**-**	**-**	**-**	**-**	**-**	**-**	**-**

Bangladesh

SUPPLY AND CONSUMPTION 1996	Oil cont. (1000 tonnes)										
	Refinery Gas	LPG + Ethane	Motor Gasoline	Aviation Gasoline	Jet Fuel	Kerosene	Gas/ Diesel	Heavy Fuel Oil	Naphtha	Petrol. Coke	Other Prod.
Production	26	15	119	-	1	274	240	76	32	-	222
Imports	-	-	76	-	98	179	975	228	40	-	34
Exports	-	-	-	-	-	-	-	-	-	-	-
Intl. Marine Bunkers	-	-	-	-	-	-	-	-7	-	-	-
Stock Changes	-	-	-	-	-	-	-	-	-	-	-
DOMESTIC SUPPLY	**26**	**15**	**195**	**-**	**99**	**453**	**1215**	**297**	**72**	**-**	**256**
Transfers and Stat. Diff.	-	-	-	-	-	-	87	3	-	-	-
TRANSFORMATION	**-**	**-**	**-**	**-**	**-**	**-**	**52**	**164**	**-**	**-**	**-**
Electricity and CHP Plants	-	-	-	-	-	-	52	164	-	-	-
Petroleum Refineries	-	-	-	-	-	-	-	-	-	-	-
Other Transform. Sector	-	-	-	-	-	-	-	-	-	-	-
ENERGY SECTOR	**26**	**-**	**-**	**-**	**-**	**-**	**-**	**36**	**-**	**-**	**-**
DISTRIBUTION LOSSES	**-**	**-**	**-**	**-**	**-**	**-**	**-**	**-**	**-**	**-**	**-**
FINAL CONSUMPTION	**-**	**15**	**195**	**-**	**99**	**453**	**1250**	**100**	**72**	**-**	**256**
INDUSTRY SECTOR	**-**	**-**	**-**	**-**	**-**	**-**	**42**	**100**	**72**	**-**	**-**
Iron and Steel	-	-	-	-	-	-	-	-	-	-	-
Chemical and Petrochem.	-	-	-	-	-	-	-	-	-	-	-
Non-Metallic Minerals	-	-	-	-	-	-	-	-	-	-	-
Non-specified	-	-	-	-	-	-	42	100	72	-	-
TRANSPORT SECTOR	**-**	**-**	**195**	**-**	**99**	**-**	**779**	**-**	**-**	**-**	**-**
Air	-	-	-	-	99	-	-	-	-	-	-
Road	-	-	195	-	-	-	499	-	-	-	-
Non-specified	-	-	-	-	-	-	280	-	-	-	-
OTHER SECTORS	**-**	**15**	**-**	**-**	**-**	**453**	**429**	**-**	**-**	**-**	**-**
Agriculture	-	-	-	-	-	-	429	-	-	-	-
Comm. and Publ. Services	-	-	-	-	-	-	-	-	-	-	-
Residential	-	15	-	-	-	453	-	-	-	-	-
Non-specified	-	-	-	-	-	-	-	-	-	-	-
NON-ENERGY USE	**-**	**-**	**-**	**-**	**-**	**-**	**-**	**-**	**-**	**-**	**256**

APPROVISIONNEMENT ET DEMANDE 1997	*Pétrole cont.* (1000 tonnes)										
	Gaz de raffinerie	*GPL + éthane*	*Essence moteur*	*Essence aviation*	*Carbu-réacteurs*	*Kérosène*	*Gazole*	*Fioul lourd*	*Naphta*	*Coke de pétrole*	*Autres prod.*
Production	26	15	121	-	1	274	248	76	32	-	222
Imports	-	1	77	-	101	193	974	202	41	-	34
Exports	-	-	-	-	-	-	-	-	-	-	-
Intl. Marine Bunkers	-	-	-	-	-	-	-	-7	-	-	-
Stock Changes	-	-	-	-	-	-	-	-	-	-	-
DOMESTIC SUPPLY	**26**	**16**	**198**	**-**	**102**	**467**	**1222**	**271**	**73**	**-**	**256**
Transfers and Stat. Diff.	-	2	-	-	-	1	87	2	-	-	-
TRANSFORMATION	**-**	**-**	**-**	**-**	**-**	**-**	**52**	**149**	**-**	**-**	**-**
Electricity and CHP Plants	-	-	-	-	-	-	52	149	-	-	-
Petroleum Refineries	-	-	-	-	-	-	-	-	-	-	-
Other Transform. Sector	-	-	-	-	-	-	-	-	-	-	-
ENERGY SECTOR	**26**	**-**	**-**	**-**	**-**	**-**	**-**	**33**	**-**	**-**	**-**
DISTRIBUTION LOSSES	**-**	**-**	**-**	**-**	**-**	**-**	**-**	**-**	**-**	**-**	**-**
FINAL CONSUMPTION	**-**	**18**	**198**	**-**	**102**	**468**	**1257**	**91**	**73**	**-**	**256**
INDUSTRY SECTOR	**-**	**-**	**-**	**-**	**-**	**-**	**42**	**91**	**73**	**-**	**-**
Iron and Steel	-	-	-	-	-	-	-	-	-	-	-
Chemical and Petrochem.	-	-	-	-	-	-	-	-	-	-	-
Non-Metallic Minerals	-	-	-	-	-	-	-	-	-	-	-
Non-specified	-	-	-	-	-	-	42	91	73	-	-
TRANSPORT SECTOR	**-**	**-**	**198**	**-**	**102**	**-**	**784**	**-**	**-**	**-**	**-**
Air	-	-	-	-	102	-	-	-	-	-	-
Road	-	-	198	-	-	-	502	-	-	-	-
Non-specified	-	-	-	-	-	-	282	-	-	-	-
OTHER SECTORS	**-**	**18**	**-**	**-**	**-**	**468**	**431**	**-**	**-**	**-**	**-**
Agriculture	-	-	-	-	-	-	431	-	-	-	-
Comm. and Publ. Services	-	-	-	-	-	-	-	-	-	-	-
Residential	-	18	-	-	-	468	-	-	-	-	-
Non-specified	-	-	-	-	-	-	-	-	-	-	-
NON-ENERGY USE	**-**	**-**	**-**	**-**	**-**	**-**	**-**	**-**	**-**	**-**	**256**

Bangladesh

SUPPLY AND	Gas (TJ)				Comb. Renew. & Waste (TJ)				(GWh)	(TJ)
CONSUMPTION 1996	Natural Gas	Gas Works	Coke Ovens	Blast Furnaces	Solid Biomass	Gas/Liquids from Biomass	Municipal Waste	Industrial Waste	Electricity	Heat
Production	274361	-	-	-	665910	-	-	-	11474	-
Imports	-	-	-	-	-	-	-	-	-	-
Exports	-	-	-	-	-	-	-	-	-	-
Intl. Marine Bunkers	-	-	-	-	-	-	-	-	-	-
Stock Changes	-	-	-	-	-	-	-	-	-	-
DOMESTIC SUPPLY	**274361**	-	-	-	**665910**	-	-	-	**11474**	-
Transfers and Stat. Diff.	-2	-	-	-	-	-	-	-	3	-
TRANSFORMATION	**113211**	-	-	-	-	-	-	-	-	-
Electricity and CHP Plants	113211	-	-	-	-	-	-	-	-	-
Petroleum Refineries	-	-	-	-	-	-	-	-	-	-
Other Transform. Sector	-	-	-	-	-	-	-	-	-	-
ENERGY SECTOR	-	-	-	-	-	-	-	-	**644**	-
DISTRIBUTION LOSSES	**18053**	-	-	-	-	-	-	-	**1837**	-
FINAL CONSUMPTION	**143095**	-	-	-	**665910**	-	-	-	**8996**	-
INDUSTRY SECTOR	**117765**	-	-	-	**201791**	-	-	-	**6982**	-
Iron and Steel	33	-	-	-	-	-	-	-	-	-
Chemical and Petrochem.	89011	-	-	-	-	-	-	-	-	-
Non-Metallic Minerals	-	-	-	-	-	-	-	-	-	-
Non-specified	28721	-	-	-	201791	-	-	-	6982	-
TRANSPORT SECTOR	-	-	-	-	-	-	-	-	-	-
Air	-	-	-	-	-	-	-	-	-	-
Road	-	-	-	-	-	-	-	-	-	-
Non-specified	-	-	-	-	-	-	-	-	-	-
OTHER SECTORS	**25330**	-	-	-	**464119**	-	-	-	**2014**	-
Agriculture	-	-	-	-	-	-	-	-	172	-
Comm. and Publ. Services	3182	-	-	-	-	-	-	-	421	-
Residential	22148	-	-	-	-	-	-	-	1341	-
Non-specified	-	-	-	-	464119	-	-	-	80	-
NON-ENERGY USE	-	-	-	-	-	-	-	-	-	-

APPROVISIONNEMENT	*Gaz* (TJ)				*En. Re. Comb. & Déchets* (TJ)				(GWh)	(TJ)
ET DEMANDE 1997	*Gaz naturel*	*Usines à gaz*	*Cokeries*	*Hauts fourneaux*	*Biomasse solide*	*Gaz/Liquides tirés de biomasse*	*Déchets urbains*	*Déchets industriels*	*Electricité*	*Chaleur*
Production	269618	-	-	-	671327	-	-	-	11858	-
Imports	-	-	-	-	-	-	-	-	-	-
Exports	-	-	-	-	-	-	-	-	-	-
Intl. Marine Bunkers	-	-	-	-	-	-	-	-	-	-
Stock Changes	-	-	-	-	-	-	-	-	-	-
DOMESTIC SUPPLY	**269618**	-	-	-	**671327**	-	-	-	**11858**	-
Transfers and Stat. Diff.	-	-	-	-	1	-	-	-	1	-
TRANSFORMATION	**114451**	-	-	-	-	-	-	-	-	-
Electricity and CHP Plants	114451	-	-	-	-	-	-	-	-	-
Petroleum Refineries	-	-	-	-	-	-	-	-	-	-
Other Transform. Sector	-	-	-	-	-	-	-	-	-	-
ENERGY SECTOR	-	-	-	-	-	-	-	-	**615**	-
DISTRIBUTION LOSSES	**17038**	-	-	-	-	-	-	-	**1796**	-
FINAL CONSUMPTION	**138129**	-	-	-	**671328**	-	-	-	**9448**	-
INDUSTRY SECTOR	**111167**	-	-	-	**203433**	-	-	-	**7486**	-
Iron and Steel	36	-	-	-	-	-	-	-	-	-
Chemical and Petrochem.	80448	-	-	-	-	-	-	-	-	-
Non-Metallic Minerals	-	-	-	-	-	-	-	-	-	-
Non-specified	30683	-	-	-	203433	-	-	-	7486	-
TRANSPORT SECTOR	-	-	-	-	-	-	-	-	-	-
Air	-	-	-	-	-	-	-	-	-	-
Road	-	-	-	-	-	-	-	-	-	-
Non-specified	-	-	-	-	-	-	-	-	-	-
OTHER SECTORS	**26962**	-	-	-	**467895**	-	-	-	**1962**	-
Agriculture	-	-	-	-	-	-	-	-	154	-
Comm. and Publ. Services	3393	-	-	-	-	-	-	-	413	-
Residential	23569	-	-	-	-	-	-	-	1319	-
Non-specified	-	-	-	-	467895	-	-	-	76	-
NON-ENERGY USE	-	-	-	-	-	-	-	-	-	-

Belarus / Bélarus : 1996

SUPPLY AND CONSUMPTION / *APPROVISIONNEMENT ET DEMANDE*	Coal / *Charbon* (1000 tonnes)							Oil / *Pétrole* (1000 tonnes)			
	Coking Coal / *Charbon à coke*	Other Bit. Coal / *Autres charb. bit.*	Sub-Bit. Coal / *Charbon sous-bit.*	Lignite / *Lignite*	Peat / *Tourbe*	Oven and Gas Coke / *Coke de four/gaz*	Pat. Fuel and BKB / *Agg./briq. de lignite*	Crude Oil / *Pétrole brut*	NGL / *LGN*	Feed-stocks / *Produits d'aliment.*	Additives / *Additifs*
Production	-	-	-	-	2846	-	1515	1860	-	-	-
From Other Sources	-	-	-	-	-	-	-	-	-	-	-
Imports	-	997	-	-	-	-	-	10645	-	-	-
Exports	-	-	-	-	-101	-	-60	-300	-	-	-
Intl. Marine Bunkers	-	-	-	-	-	-	-	-	-	-	-
Stock Changes	-	119	-	-	13	-	24	250	-	-	-
DOMESTIC SUPPLY	**-**	**1116**	**-**	**-**	**2758**	**-**	**1479**	**12455**	**-**	**-**	**-**
Transfers	-	-	-	-	-	-	-	-	-	-	-
Statistical Differences	-	-	-	-	-	-	-	-	-	-	-
TRANSFORMATION	**-**	**337**	**-**	**-**	**2589**	**-**	**94**	**11760**	**-**	**-**	**-**
Electricity Plants	-	-	-	-	-	-	-	-	-	-	-
CHP Plants	-	-	-	-	-	-	-	-	-	-	-
Heat Plants	-	337	-	-	341	-	94	-	-	-	-
Blast Furnaces/Gas Works	-	-	-	-	-	-	-	-	-	-	-
Coke/Pat. Fuel/BKB Plants	-	-	-	-	2248	-	-	-	-	-	-
Petroleum Refineries	-	-	-	-	-	-	-	11760	-	-	-
Petrochemical Industry	-	-	-	-	-	-	-	-	-	-	-
Liquefaction	-	-	-	-	-	-	-	-	-	-	-
Other Transform. Sector	-	-	-	-	-	-	-	-	-	-	-
ENERGY SECTOR	**-**	**1**	**-**	**-**	**26**	**-**	**-**	**-**	**-**	**-**	**-**
Coal Mines	-	-	-	-	-	-	-	-	-	-	-
Oil and Gas Extraction	-	-	-	-	-	-	-	-	-	-	-
Petroleum Refineries	-	-	-	-	-	-	-	-	-	-	-
Electr., CHP+Heat Plants	-	1	-	-	-	-	-	-	-	-	-
Pumped Storage (Elec.)	-	-	-	-	-	-	-	-	-	-	-
Other Energy Sector	-	-	-	-	26	-	-	-	-	-	-
Distribution Losses	-	-	-	-	-	-	-	143	-	-	-
FINAL CONSUMPTION	**-**	**778**	**-**	**-**	**143**	**-**	**1385**	**552**	**-**	**-**	**-**
INDUSTRY SECTOR	**-**	**21**	**-**	**-**	**22**	**-**	**9**	**552**	**-**	**-**	**-**
Iron and Steel	-	-	-	-	-	-	-	-	-	-	-
Chemical and Petrochem.	-	-	-	-	-	-	-	552	-	-	-
of which: Feedstocks	-	-	-	-	-	-	-	*552*	-	-	-
Non-Ferrous Metals	-	-	-	-	-	-	-	-	-	-	-
Non-Metallic Minerals	-	-	-	-	-	-	-	-	-	-	-
Transport Equipment	-	-	-	-	-	-	-	-	-	-	-
Machinery	-	8	-	-	1	-	1	-	-	-	-
Mining and Quarrying	-	-	-	-	-	-	-	-	-	-	-
Food and Tobacco	-	-	-	-	-	-	-	-	-	-	-
Paper, Pulp and Print	-	-	-	-	-	-	1	-	-	-	-
Wood and Wood Products	-	-	-	-	-	-	-	-	-	-	-
Construction	-	12	-	-	21	-	6	-	-	-	-
Textile and Leather	-	1	-	-	-	-	1	-	-	-	-
Non-specified	-	-	-	-	-	-	-	-	-	-	-
TRANSPORT SECTOR	**-**	**26**	**-**	**-**	**-**	**-**	**-**	**-**	**-**	**-**	**-**
Air	-	-	-	-	-	-	-	-	-	-	-
Road	-	-	-	-	-	-	-	-	-	-	-
Rail	-	25	-	-	-	-	-	-	-	-	-
Pipeline Transport	-	-	-	-	-	-	-	-	-	-	-
Internal Navigation	-	-	-	-	-	-	-	-	-	-	-
Non-specified	-	1	-	-	-	-	-	-	-	-	-
OTHER SECTORS	**-**	**731**	**-**	**-**	**121**	**-**	**1376**	**-**	**-**	**-**	**-**
Agriculture	-	10	-	-	2	-	5	-	-	-	-
Comm. and Publ. Services	-	-	-	-	-	-	1	-	-	-	-
Residential	-	89	-	-	8	-	1182	-	-	-	-
Non-specified	-	632	-	-	111	-	188	-	-	-	-
NON-ENERGY USE	**-**	**-**	**-**	**-**	**-**	**-**	**-**	**-**	**-**	**-**	**-**
in Industry/Trans./Energy	-	-	-	-	-	-	-	-	-	-	-
in Transport	-	-	-	-	-	-	-	-	-	-	-
in Other Sectors	-	-	-	-	-	-	-	-	-	-	-

Belarus / Bélarus : 1996

SUPPLY AND CONSUMPTION *APPROVISIONNEMENT ET DEMANDE*	Oil cont. / *Pétrole cont.* (1000 tonnes)										
	Refinery Gas *Gaz de raffinerie*	LPG + Ethane *GPL + éthane*	Motor Gasoline *Essence moteur*	Aviation Gasoline *Essence aviation*	Jet Fuel *Carbu-réacteurs*	Kerosene *Kérosène*	Gas/ Diesel *Gazole*	Heavy Fuel Oil *Fioul lourd*	Naphtha *Naphta*	Petrol. Coke *Coke de pétrole*	Other Prod. *Autres prod.*
Production	500	200	1816	-	-	22	3170	4812	-	-	680
From Other Sources	-	-	-	-	-	-	-	-	-	-	-
Imports	-	115	72	-	-	-	18	16	-	3	20
Exports	-	-5	-683	-	-	-6	-1470	-901	-	-	-209
Intl. Marine Bunkers	-	-	-	-	-	-	-	-	-	-	-
Stock Changes	-	6	100	-	-	-	182	784	-	-	-
DOMESTIC SUPPLY	**500**	**316**	**1305**	**-**	**-**	**16**	**1900**	**4711**	**-**	**3**	**491**
Transfers	-	-	-	-	-	-	-	-	-	-	-
Statistical Differences	-	-	-	-	-	-	-	-	-	-	-
TRANSFORMATION	**-**	**2**	**-**	**-**	**-**	**-**	**14**	**3930**	**-**	**-**	**-**
Electricity Plants	-	-	-	-	-	-	-	511	-	-	-
CHP Plants	-	-	-	-	-	-	-	1922	-	-	-
Heat Plants	-	2	-	-	-	-	14	1497	-	-	-
Blast Furnaces/Gas Works	-	-	-	-	-	-	-	-	-	-	-
Coke/Pat. Fuel/BKB Plants	-	-	-	-	-	-	-	-	-	-	-
Petroleum Refineries	-	-	-	-	-	-	-	-	-	-	-
Petrochemical Industry	-	-	-	-	-	-	-	-	-	-	-
Liquefaction	-	-	-	-	-	-	-	-	-	-	-
Other Transform. Sector	-	-	-	-	-	-	-	-	-	-	-
ENERGY SECTOR	**500**	**-**	**-**	**-**	**-**	**-**	**-**	**-**	**-**	**-**	**-**
Coal Mines	-	-	-	-	-	-	-	-	-	-	-
Oil and Gas Extraction	-	-	-	-	-	-	-	-	-	-	-
Petroleum Refineries	500	-	-	-	-	-	-	-	-	-	-
Electr., CHP+Heat Plants	-	-	-	-	-	-	-	-	-	-	-
Pumped Storage (Elec.)	-	-	-	-	-	-	-	-	-	-	-
Other Energy Sector	-	-	-	-	-	-	-	-	-	-	-
Distribution Losses	-	2	17	-	-	-	7	4	-	-	-
FINAL CONSUMPTION	**-**	**312**	**1288**	**-**	**-**	**16**	**1879**	**777**	**-**	**3**	**491**
INDUSTRY SECTOR	**-**	**7**	**16**	**-**	**-**	**16**	**227**	**623**	**-**	**3**	**-**
Iron and Steel	-	-	-	-	-	-	-	1	-	-	-
Chemical and Petrochem.	-	1	1	-	-	16	9	238	-	3	-
of which: Feedstocks	-	-	-	-	-	-	*2*	-	-	-	-
Non-Ferrous Metals	-	-	-	-	-	-	-	-	-	-	-
Non-Metallic Minerals	-	-	-	-	-	-	-	-	-	-	-
Transport Equipment	-	-	-	-	-	-	-	-	-	-	-
Machinery	-	1	2	-	-	-	9	4	-	-	-
Mining and Quarrying	-	-	-	-	-	-	-	-	-	-	-
Food and Tobacco	-	-	-	-	-	-	11	-	-	-	-
Paper, Pulp and Print	-	-	2	-	-	-	7	1	-	-	-
Wood and Wood Products	-	-	-	-	-	-	-	-	-	-	-
Construction	-	1	9	-	-	-	82	241	-	-	-
Textile and Leather	-	-	-	-	-	-	2	-	-	-	-
Non-specified	-	4	2	-	-	-	107	138	-	-	-
TRANSPORT SECTOR	**-**	**17**	**801**	**-**	**-**	**-**	**922**	**1**	**-**	**-**	**-**
Air	-	-	-	-	-	-	-	-	-	-	-
Road	-	17	792	-	-	-	567	-	-	-	-
Rail	-	-	-	-	-	-	320	1	-	-	-
Pipeline Transport	-	-	-	-	-	-	1	-	-	-	-
Internal Navigation	-	-	-	-	-	-	18	-	-	-	-
Non-specified	-	-	9	-	-	-	16	-	-	-	-
OTHER SECTORS	**-**	**288**	**471**	**-**	**-**	**-**	**730**	**153**	**-**	**-**	**-**
Agriculture	-	2	18	-	-	-	537	10	-	-	-
Comm. and Publ. Services	-	-	-	-	-	-	-	-	-	-	-
Residential	-	284	418	-	-	-	182	-	-	-	-
Non-specified	-	2	35	-	-	-	11	143	-	-	-
NON-ENERGY USE	**-**	**-**	**-**	**-**	**-**	**-**	**-**	**-**	**-**	**-**	**491**
in Industry/Transf./Energy	-	-	-	-	-	-	-	-	-	-	491
in Transport	-	-	-	-	-	-	-	-	-	-	-
in Other Sectors	-	-	-	-	-	-	-	-	-	-	-

Belarus / Bélarus : 1996

SUPPLY AND CONSUMPTION / *APPROVISIONNEMENT ET DEMANDE*	Gas / *Gaz* (TJ)				Comb. Renew. & Waste / *En. Re. Comb. & Déchets* (TJ)				(GWh)	(TJ)
	Natural Gas / *Gaz naturel*	Gas Works / *Usines à gaz*	Coke Ovens / *Cokeries*	Blast Furnaces / *Hauts fourneaux*	Solid Biomass / *Biomasse solide*	Gas/Liquids from Biomass / *Gaz/Liquides tirés de biomasse*	Municipal Waste / *Déchets urbains*	Industrial Waste / *Déchets industriels*	Electricity / *Electricité*	Heat / *Chaleur*
Production	9617	-	-	-	21596	-	-	-	23728	326570
From Other Sources	-	-	-	-	-	-	-	-	-	-
Imports	554032	-	-	-	-	-	-	-	11144	-
Exports	-	-	-	-	-	-	-	-	-2601	-
Intl. Marine Bunkers	-	-	-	-	-	-	-	-	-	-
Stock Changes	-270	-	-	-	-	-	-	-	-	-
DOMESTIC SUPPLY	**563379**	**-**	**-**	**-**	**21596**	**-**	**-**	**-**	**32271**	**326570**
Transfers	-	-	-	-	-	-	-	-	-	-
Statistical Differences	-	-	-	-	-	-	-	-	-	-
TRANSFORMATION	**415109**	**-**	**-**	**-**	**6124**	**-**	**-**	**-**	**-**	**-**
Electricity Plants	90994	-	-	-	-	-	-	-	-	-
CHP Plants	176811	-	-	-	-	-	-	-	-	-
Heat Plants	147304	-	-	-	6124	-	-	-	-	-
Blast Furnaces/Gas Works	-	-	-	-	-	-	-	-	-	-
Coke/Pat. Fuel/BKB Plants	-	-	-	-	-	-	-	-	-	-
Petroleum Refineries	-	-	-	-	-	-	-	-	-	-
Petrochemical Industry	-	-	-	-	-	-	-	-	-	-
Liquefaction	-	-	-	-	-	-	-	-	-	-
Other Transform. Sector	-	-	-	-	-	-	-	-	-	-
ENERGY SECTOR	**-**	**-**	**-**	**-**	**-**	**-**	**-**	**-**	**3015**	**-**
Coal Mines	-	-	-	-	-	-	-	-	-	-
Oil and Gas Extraction	-	-	-	-	-	-	-	-	161	-
Petroleum Refineries	-	-	-	-	-	-	-	-	574	-
Electr., CHP+Heat Plants	-	-	-	-	-	-	-	-	2170	-
Pumped Storage (Elec.)	-	-	-	-	-	-	-	-	-	-
Other Energy Sector	-	-	-	-	-	-	-	-	110	-
Distribution Losses	5987	-	-	-	-	-	-	-	3757	20934
FINAL CONSUMPTION	**142283**	**-**	**-**	**-**	**15472**	**-**	**-**	**-**	**25499**	**305636**
INDUSTRY SECTOR	**76008**	**-**	**-**	**-**	**352**	**-**	**-**	**-**	**10785**	**125604**
Iron and Steel	2279	-	-	-	-	-	-	-	915	-
Chemical and Petrochem.	53839	-	-	-	-	-	-	-	3869	41868
of which: Feedstocks	*48162*	-	-	-	-	-	-	-	-	-
Non-Ferrous Metals	-	-	-	-	-	-	-	-	7	-
Non-Metallic Minerals	11471	-	-	-	-	-	-	-	936	8374
Transport Equipment	-	-	-	-	-	-	-	-	23	-
Machinery	4132	-	-	-	-	-	-	-	2306	16747
Mining and Quarrying	-	-	-	-	-	-	-	-	-	-
Food and Tobacco	1236	-	-	-	88	-	-	-	841	12560
Paper, Pulp and Print	-	-	-	-	-	-	-	-	207	8374
Wood and Wood Products	348	-	-	-	235	-	-	-	301	8373
Construction	-	-	-	-	29	-	-	-	344	4187
Textile and Leather	347	-	-	-	-	-	-	-	510	8374
Non-specified	2356	-	-	-	-	-	-	-	526	16747
TRANSPORT SECTOR	**3244**	**-**	**-**	**-**	**88**	**-**	**-**	**-**	**1824**	**-**
Air	-	-	-	-	-	-	-	-	-	-
Road	1468	-	-	-	-	-	-	-	-	-
Rail	-	-	-	-	-	-	-	-	701	-
Pipeline Transport	1776	-	-	-	-	-	-	-	785	-
Internal Navigation	-	-	-	-	-	-	-	-	-	-
Non-specified	-	-	-	-	88	-	-	-	338	-
OTHER SECTORS	**63031**	**-**	**-**	**-**	**15032**	**-**	**-**	**-**	**12890**	**180032**
Agriculture	1893	-	-	-	352	-	-	-	2585	12560
Comm. and Publ. Services	-	-	-	-	-	-	-	-	3608	-
Residential	42368	-	-	-	11750	-	-	-	5077	154912
Non-specified	18770	-	-	-	2930	-	-	-	1620	12560
NON-ENERGY USE	**-**	**-**	**-**	**-**	**-**	**-**	**-**	**-**	**-**	**-**
in Industry/Transf./Energy	-	-	-	-	-	-	-	-	-	-
in Transport	-	-	-	-	-	-	-	-	-	-
in Other Sectors	-	-	-	-	-	-	-	-	-	-

Belarus / Bélarus : 1997

SUPPLY AND CONSUMPTION / *APPROVISIONNEMENT ET DEMANDE*	Coal / *Charbon* (1000 tonnes)							Oil / *Pétrole* (1000 tonnes)			
	Coking Coal / *Charbon à coke*	Other Bit. Coal / *Autres charb. bit.*	Sub-Bit. Coal / *Charbon sous-bit.*	Lignite / *Lignite*	Peat / *Tourbe*	Oven and Gas Coke / *Coke de four/gaz*	Pat. Fuel and BKB / *Agg./briq. de lignite*	Crude Oil / *Pétrole brut*	NGL / *LGN*	Feed-stocks / *Produits d'aliment.*	Additives / *Additifs*
Production	-	-	-	-	2763	-	1515	1822	-	-	-
From Other Sources	-	-	-	-	-	-	-	-	-	-	-
Imports	-	786	-	-	-	-	-	10461	-	-	-
Exports	-	-	-	-	-96	-	-79	-400	-	-	-
Intl. Marine Bunkers	-	-	-	-	-	-	-	-	-	-	-
Stock Changes	-	24	-	-	-1	-	-2	-95	-	-	-
DOMESTIC SUPPLY	**-**	**810**	**-**	**-**	**2666**	**-**	**1434**	**11788**	**-**	**-**	**-**
Transfers	-	-	-	-	-	-	-	-	-	-	-
Statistical Differences	-	-	-	-	-	-	-	72	-	-	-
TRANSFORMATION	**-**	**252**	**-**	**-**	**2508**	**-**	**95**	**11199**	**-**	**-**	**-**
Electricity Plants	-	-	-	-	-	-	-	-	-	-	-
CHP Plants	-	-	-	-	-	-	-	-	-	-	-
Heat Plants	-	252	-	-	260	-	95	-	-	-	-
Blast Furnaces/Gas Works	-	-	-	-	-	-	-	-	-	-	-
Coke/Pat. Fuel/BKB Plants	-	-	-	-	2248	-	-	-	-	-	-
Petroleum Refineries	-	-	-	-	-	-	-	11199	-	-	-
Petrochemical Industry	-	-	-	-	-	-	-	-	-	-	-
Liquefaction	-	-	-	-	-	-	-	-	-	-	-
Other Transform. Sector	-	-	-	-	-	-	-	-	-	-	-
ENERGY SECTOR	**-**	**1**	**-**	**-**	**26**	**-**	**-**	**-**	**-**	**-**	**-**
Coal Mines	-	-	-	-	-	-	-	-	-	-	-
Oil and Gas Extraction	-	-	-	-	-	-	-	-	-	-	-
Petroleum Refineries	-	-	-	-	-	-	-	-	-	-	-
Electr., CHP+Heat Plants	-	1	-	-	-	-	-	-	-	-	-
Pumped Storage (Elec.)	-	-	-	-	-	-	-	-	-	-	-
Other Energy Sector	-	-	-	-	26	-	-	-	-	-	-
Distribution Losses	-	-	-	-	-	-	-	135	-	-	-
FINAL CONSUMPTION	**-**	**557**	**-**	**-**	**132**	**-**	**1339**	**526**	**-**	**-**	**-**
INDUSTRY SECTOR	**-**	**18**	**-**	**-**	**19**	**-**	**9**	**526**	**-**	**-**	**-**
Iron and Steel	-	-	-	-	-	-	-	-	-	-	-
Chemical and Petrochem.	-	1	-	-	-	-	-	526	-	-	-
of which: Feedstocks	-	-	-	-	-	-	-	*526*	-	-	-
Non-Ferrous Metals	-	-	-	-	-	-	-	-	-	-	-
Non-Metallic Minerals	-	-	-	-	-	-	-	-	-	-	-
Transport Equipment	-	-	-	-	-	-	-	-	-	-	-
Machinery	-	-	-	-	2	-	1	-	-	-	-
Mining and Quarrying	-	-	-	-	-	-	-	-	-	-	-
Food and Tobacco	-	-	-	-	-	-	-	-	-	-	-
Paper, Pulp and Print	-	2	-	-	1	-	1	-	-	-	-
Wood and Wood Products	-	-	-	-	-	-	-	-	-	-	-
Construction	-	13	-	-	16	-	6	-	-	-	-
Textile and Leather	-	1	-	-	-	-	1	-	-	-	-
Non-specified	-	1	-	-	-	-	-	-	-	-	-
TRANSPORT SECTOR	**-**	**12**	**-**	**-**	**-**	**-**	**-**	**-**	**-**	**-**	**-**
Air	-	-	-	-	-	-	-	-	-	-	-
Road	-	-	-	-	-	-	-	-	-	-	-
Rail	-	10	-	-	-	-	-	-	-	-	-
Pipeline Transport	-	-	-	-	-	-	-	-	-	-	-
Internal Navigation	-	-	-	-	-	-	-	-	-	-	-
Non-specified	-	2	-	-	-	-	-	-	-	-	-
OTHER SECTORS	**-**	**526**	**-**	**-**	**113**	**-**	**1330**	**-**	**-**	**-**	**-**
Agriculture	-	6	-	-	3	-	5	-	-	-	-
Comm. and Publ. Services	-	1	-	-	-	-	1	-	-	-	-
Residential	-	66	-	-	3	-	1177	-	-	-	-
Non-specified	-	453	-	-	107	-	147	-	-	-	-
NON-ENERGY USE	**-**	**1**	**-**	**-**	**-**	**-**	**-**	**-**	**-**	**-**	**-**
in Industry/Trans./Energy	-	1	-	-	-	-	-	-	-	-	-
in Transport	-	-	-	-	-	-	-	-	-	-	-
in Other Sectors	-	-	-	-	-	-	-	-	-	-	-

Belarus / Bélarus : 1997

SUPPLY AND CONSUMPTION / *APPROVISIONNEMENT ET DEMANDE*	Oil cont. / *Pétrole cont.* (1000 tonnes)										
	Refinery Gas / *Gaz de raffinerie*	LPG + Ethane / *GPL + éthane*	Motor Gasoline / *Essence moteur*	Aviation Gasoline / *Essence aviation*	Jet Fuel / *Carbu-réacteurs*	Kerosene / *Kérosène*	Gas/ Diesel / *Gazole*	Heavy Fuel Oil / *Fioul lourd*	Naphtha / *Naphta*	Petrol. Coke / *Coke de pétrole*	Other Prod. / *Autres prod.*
Production	500	138	1954	-	-	86	3115	4524	-	-	755
From Other Sources	-	-	-	-	-	-	-	-	-	-	-
Imports	-	173	56	-	-	1	33	27	-	-	31
Exports	-	-	-666	-	-	-63	-990	-713	-	-	-224
Intl. Marine Bunkers	-	-	-	-	-	-	-	-	-	-	-
Stock Changes	-	4	-97	-	-	-	-242	-133	-	-	-
DOMESTIC SUPPLY	**500**	**315**	**1247**	**-**	**-**	**24**	**1916**	**3705**	**-**	**-**	**562**
Transfers	-	-	-	-	-	-	-	-	-	-	-
Statistical Differences	-	-	-	-	-	-	-	-	-	-	-
TRANSFORMATION	**-**	**4**	**-**	**-**	**-**	**-**	**42**	**2759**	**-**	**-**	**-**
Electricity Plants	-	-	-	-	-	-	-	314	-	-	-
CHP Plants	-	-	-	-	-	-	-	1050	-	-	-
Heat Plants	-	4	-	-	-	-	42	1395	-	-	-
Blast Furnaces/Gas Works	-	-	-	-	-	-	-	-	-	-	-
Coke/Pat. Fuel/BKB Plants	-	-	-	-	-	-	-	-	-	-	-
Petroleum Refineries	-	-	-	-	-	-	-	-	-	-	-
Petrochemical Industry	-	-	-	-	-	-	-	-	-	-	-
Liquefaction	-	-	-	-	-	-	-	-	-	-	-
Other Transform. Sector	-	-	-	-	-	-	-	-	-	-	-
ENERGY SECTOR	**500**	**-**	**-**	**-**	**-**	**-**	**-**	**-**	**-**	**-**	**-**
Coal Mines	-	-	-	-	-	-	-	-	-	-	-
Oil and Gas Extraction	-	-	-	-	-	-	-	-	-	-	-
Petroleum Refineries	500	-	-	-	-	-	-	-	-	-	-
Electr., CHP+Heat Plants	-	-	-	-	-	-	-	-	-	-	-
Pumped Storage (Elec.)	-	-	-	-	-	-	-	-	-	-	-
Other Energy Sector	-	-	-	-	-	-	-	-	-	-	-
Distribution Losses	-	2	13	-	-	-	7	3	-	-	-
FINAL CONSUMPTION	**-**	**309**	**1234**	**-**	**-**	**24**	**1867**	**943**	**-**	**-**	**562**
INDUSTRY SECTOR	**-**	**4**	**11**	**-**	**-**	**24**	**207**	**768**	**-**	**-**	**-**
Iron and Steel	-	-	-	-	-	-	2	1	-	-	-
Chemical and Petrochem.	-	-	1	-	-	24	4	215	-	-	-
of which: Feedstocks	-	-	-	-	-	-	*2*	-	-	-	-
Non-Ferrous Metals	-	-	-	-	-	-	-	-	-	-	-
Non-Metallic Minerals	-	-	-	-	-	-	-	-	-	-	-
Transport Equipment	-	-	-	-	-	-	-	-	-	-	-
Machinery	-	1	1	-	-	-	7	14	-	-	-
Mining and Quarrying	-	-	-	-	-	-	-	-	-	-	-
Food and Tobacco	-	-	-	-	-	-	2	-	-	-	-
Paper, Pulp and Print	-	-	2	-	-	-	7	1	-	-	-
Wood and Wood Products	-	-	-	-	-	-	-	-	-	-	-
Construction	-	1	7	-	-	-	91	264	-	-	-
Textile and Leather	-	-	-	-	-	-	1	-	-	-	-
Non-specified	-	2	-	-	-	-	93	273	-	-	-
TRANSPORT SECTOR	**-**	**20**	**811**	**-**	**-**	**-**	**964**	**1**	**-**	**-**	**-**
Air	-	-	-	-	-	-	-	-	-	-	-
Road	-	20	808	-	-	-	593	-	-	-	-
Rail	-	-	-	-	-	-	334	1	-	-	-
Pipeline Transport	-	-	-	-	-	-	1	-	-	-	-
Internal Navigation	-	-	-	-	-	-	19	-	-	-	-
Non-specified	-	-	3	-	-	-	17	-	-	-	-
OTHER SECTORS	**-**	**285**	**412**	**-**	**-**	**-**	**696**	**174**	**-**	**-**	**-**
Agriculture	-	1	14	-	-	-	526	4	-	-	-
Comm. and Publ. Services	-	-	-	-	-	-	-	-	-	-	-
Residential	-	284	398	-	-	-	161	-	-	-	-
Non-specified	-	-	-	-	-	-	9	170	-	-	-
NON-ENERGY USE	**-**	**-**	**-**	**-**	**-**	**-**	**-**	**-**	**-**	**-**	**562**
in Industry/Transf./Energy	-	-	-	-	-	-	-	-	-	-	562
in Transport	-	-	-	-	-	-	-	-	-	-	-
in Other Sectors	-	-	-	-	-	-	-	-	-	-	-

Belarus / Bélarus : 1997

SUPPLY AND CONSUMPTION / *APPROVISIONNEMENT ET DEMANDE*	Gas / *Gaz* (TJ)				Comb. Renew. & Waste / *En. Re. Comb. & Déchets* (TJ)				(GWh)	(TJ)
	Natural Gas / *Gaz naturel*	Gas Works / *Usines à gaz*	Coke Ovens / *Cokeries*	Blast Furnaces / *Hauts fourneaux*	Solid Biomass / *Biomasse solide*	Gas/Liquids from Biomass / *Gaz/Liquides tirés de biomasse*	Municipal Waste / *Déchets urbains*	Industrial Waste / *Déchets industriels*	Electricity / *Electricité*	Heat / *Chaleur*
Production	9501	-	-	-	28687	-	-	-	26057	334944
From Other Sources	-	-	-	-	-	-	-	-	-	-
Imports	627260	-	-	-	-	-	-	-	10308	-
Exports	-	-	-	-	-	-	-	-	-2688	-
Intl. Marine Bunkers	-	-	-	-	-	-	-	-	-	-
Stock Changes	4248	-	-	-	-	-	-	-	-	-
DOMESTIC SUPPLY	**641009**	**-**	**-**	**-**	**28687**	**-**	**-**	**-**	**33677**	**334944**
Transfers	-	-	-	-	-	-	-	-	-	-
Statistical Differences	-	-	-	-	-	-	-	-	-	-
TRANSFORMATION	**494632**	**-**	**-**	**-**	**5831**	**-**	**-**	**-**	**-**	**-**
Electricity Plants	119998	-	-	-	-	-	-	-	-	-
CHP Plants	214661	-	-	-	-	-	-	-	-	-
Heat Plants	159973	-	-	-	5831	-	-	-	-	-
Blast Furnaces/Gas Works	-	-	-	-	-	-	-	-	-	-
Coke/Pat. Fuel/BKB Plants	-	-	-	-	-	-	-	-	-	-
Petroleum Refineries	-	-	-	-	-	-	-	-	-	-
Petrochemical Industry	-	-	-	-	-	-	-	-	-	-
Liquefaction	-	-	-	-	-	-	-	-	-	-
Other Transform. Sector	-	-	-	-	-	-	-	-	-	-
ENERGY SECTOR	**-**	**-**	**-**	**-**	**-**	**-**	**-**	**-**	**3112**	**4187**
Coal Mines	-	-	-	-	-	-	-	-	-	-
Oil and Gas Extraction	-	-	-	-	-	-	-	-	158	-
Petroleum Refineries	-	-	-	-	-	-	-	-	643	-
Electr., CHP+Heat Plants	-	-	-	-	-	-	-	-	2198	-
Pumped Storage (Elec.)	-	-	-	-	-	-	-	-	-	-
Other Energy Sector	-	-	-	-	-	-	-	-	113	4187
Distribution Losses	5214	-	-	-	-	-	-	-	3801	20934
FINAL CONSUMPTION	**141163**	**-**	**-**	**-**	**22856**	**-**	**-**	**-**	**26764**	**309823**
INDUSTRY SECTOR	**73150**	**-**	**-**	**-**	**469**	**-**	**-**	**-**	**12569**	**125604**
Iron and Steel	3167	-	-	-	-	-	-	-	1100	-
Chemical and Petrochem.	46694	-	-	-	-	-	-	-	4635	46054
of which: Feedstocks	*39394*	-	-	-	-	-	-	-	-	-
Non-Ferrous Metals	-	-	-	-	-	-	-	-	8	-
Non-Metallic Minerals	13054	-	-	-	-	-	-	-	984	8374
Transport Equipment	-	-	-	-	-	-	-	-	31	-
Machinery	4364	-	-	-	-	-	-	-	2502	20933
Mining and Quarrying	-	-	-	-	-	-	-	-	-	-
Food and Tobacco	850	-	-	-	-	-	-	-	898	12560
Paper, Pulp and Print	-	-	-	-	29	-	-	-	239	8374
Wood and Wood Products	1468	-	-	-	-	-	-	-	355	8374
Construction	386	-	-	-	264	-	-	-	360	4187
Textile and Leather	425	-	-	-	117	-	-	-	580	8374
Non-specified	2742	-	-	-	59	-	-	-	877	8374
TRANSPORT SECTOR	**3128**	**-**	**-**	**-**	**147**	**-**	**-**	**-**	**1715**	**-**
Air	-	-	-	-	-	-	-	-	-	-
Road	1390	-	-	-	-	-	-	-	-	-
Rail	-	-	-	-	29	-	-	-	726	-
Pipeline Transport	1738	-	-	-	-	-	-	-	643	-
Internal Navigation	-	-	-	-	-	-	-	-	-	-
Non-specified	-	-	-	-	118	-	-	-	346	-
OTHER SECTORS	**64885**	**-**	**-**	**-**	**22240**	**-**	**-**	**-**	**12480**	**184219**
Agriculture	1777	-	-	-	176	-	-	-	2471	12560
Comm. and Publ. Services	-	-	-	-	-	-	-	-	2979	-
Residential	42909	-	-	-	18460	-	-	-	5347	163285
Non-specified	20199	-	-	-	3604	-	-	-	1683	8374
NON-ENERGY USE	**-**	**-**	**-**	**-**	**-**	**-**	**-**	**-**	**-**	**-**
in Industry/Transf./Energy	-	-	-	-	-	-	-	-	-	-
in Transport	-	-	-	-	-	-	-	-	-	-
in Other Sectors	-	-	-	-	-	-	-	-	-	-

Benin / Bénin

SUPPLY AND CONSUMPTION 1996	Coal (1000 tonnes)							Oil (1000 tonnes)			
	Coking Coal	Other Bit. Coal	Sub-Bit. Coal	Lignite	Peat	Oven and Gas Coke	Pat. Fuel and BKB	Crude Oil	NGL	Feed-stocks	Additives
Production	-	-	-	-	-	-	-	72	-	-	-
Imports	-	-	-	-	-	-	-	-	-	-	-
Exports	-	-	-	-	-	-	-	-75	-	-	-
Intl. Marine Bunkers	-	-	-	-	-	-	-	-	-	-	-
Stock Changes	-	-	-	-	-	-	-	3	-	-	-
DOMESTIC SUPPLY	-	-	-	-	-	-	-	-	-	-	-
Transfers and Stat. Diff.	-	-	-	-	-	-	-	-	-	-	-
TRANSFORMATION	-	-	-	-	-	-	-	-	-	-	-
Electricity and CHP Plants	-	-	-	-	-	-	-	-	-	-	-
Petroleum Refineries	-	-	-	-	-	-	-	-	-	-	-
Other Transform. Sector	-	-	-	-	-	-	-	-	-	-	-
ENERGY SECTOR	-	-	-	-	-	-	-	-	-	-	-
DISTRIBUTION LOSSES	-	-	-	-	-	-	-	-	-	-	-
FINAL CONSUMPTION	-	-	-	-	-	-	-	-	-	-	-
INDUSTRY SECTOR	-	-	-	-	-	-	-	-	-	-	-
Iron and Steel	-	-	-	-	-	-	-	-	-	-	-
Chemical and Petrochem.	-	-	-	-	-	-	-	-	-	-	-
Non-Metallic Minerals	-	-	-	-	-	-	-	-	-	-	-
Non-specified	-	-	-	-	-	-	-	-	-	-	-
TRANSPORT SECTOR	-	-	-	-	-	-	-	-	-	-	-
Air	-	-	-	-	-	-	-	-	-	-	-
Road	-	-	-	-	-	-	-	-	-	-	-
Non-specified	-	-	-	-	-	-	-	-	-	-	-
OTHER SECTORS	-	-	-	-	-	-	-	-	-	-	-
Agriculture	-	-	-	-	-	-	-	-	-	-	-
Comm. and Publ. Services	-	-	-	-	-	-	-	-	-	-	-
Residential	-	-	-	-	-	-	-	-	-	-	-
Non-specified	-	-	-	-	-	-	-	-	-	-	-
NON-ENERGY USE	-	-	-	-	-	-	-	-	-	-	-

APPROVISIONNEMENT ET DEMANDE 1997	*Charbon* (1000 tonnes)							*Pétrole* (1000 tonnes)			
	Charbon à coke	*Autres charb. bit.*	*Charbon sous-bit.*	*Lignite*	*Tourbe*	*Coke de four/gaz*	*Agg./briq. de lignite*	*Pétrole brut*	*LGN*	*Produits d'aliment.*	*Additifs*
Production	-	-	-	-	-	-	-	66	-	-	-
Imports	-	-	-	-	-	-	-	-	-	-	-
Exports	-	-	-	-	-	-	-	-64	-	-	-
Intl. Marine Bunkers	-	-	-	-	-	-	-	-	-	-	-
Stock Changes	-	-	-	-	-	-	-	-2	-	-	-
DOMESTIC SUPPLY	-	-	-	-	-	-	-	-	-	-	-
Transfers and Stat. Diff.	-	-	-	-	-	-	-	-	-	-	-
TRANSFORMATION	-	-	-	-	-	-	-	-	-	-	-
Electricity and CHP Plants	-	-	-	-	-	-	-	-	-	-	-
Petroleum Refineries	-	-	-	-	-	-	-	-	-	-	-
Other Transform. Sector	-	-	-	-	-	-	-	-	-	-	-
ENERGY SECTOR	-	-	-	-	-	-	-	-	-	-	-
DISTRIBUTION LOSSES	-	-	-	-	-	-	-	-	-	-	-
FINAL CONSUMPTION	-	-	-	-	-	-	-	-	-	-	-
INDUSTRY SECTOR	-	-	-	-	-	-	-	-	-	-	-
Iron and Steel	-	-	-	-	-	-	-	-	-	-	-
Chemical and Petrochem.	-	-	-	-	-	-	-	-	-	-	-
Non-Metallic Minerals	-	-	-	-	-	-	-	-	-	-	-
Non-specified	-	-	-	-	-	-	-	-	-	-	-
TRANSPORT SECTOR	-	-	-	-	-	-	-	-	-	-	-
Air	-	-	-	-	-	-	-	-	-	-	-
Road	-	-	-	-	-	-	-	-	-	-	-
Non-specified	-	-	-	-	-	-	-	-	-	-	-
OTHER SECTORS	-	-	-	-	-	-	-	-	-	-	-
Agriculture	-	-	-	-	-	-	-	-	-	-	-
Comm. and Publ. Services	-	-	-	-	-	-	-	-	-	-	-
Residential	-	-	-	-	-	-	-	-	-	-	-
Non-specified	-	-	-	-	-	-	-	-	-	-	-
NON-ENERGY USE	-	-	-	-	-	-	-	-	-	-	-

Benin / Bénin

SUPPLY AND CONSUMPTION 1996	Oil cont. (1000 tonnes)										
	Refinery Gas	LPG + Ethane	Motor Gasoline	Aviation Gasoline	Jet Fuel	Kerosene	Gas/ Diesel	Heavy Fuel Oil	Naphtha	Petrol. Coke	Other Prod.
Production	-	-	-	-	-	-	-	-	-	-	-
Imports	-	1	97	-	56	39	90	14	-	-	-
Exports	-	-	-	-	-	-	-1	-	-	-	-
Intl. Marine Bunkers	-	-	-	-	-	-	-5	-	-	-	-
Stock Changes	-	-	7	-	-	-	18	4	-	-	-
DOMESTIC SUPPLY	**-**	**1**	**104**	**-**	**56**	**39**	**102**	**18**	**-**	**-**	**-**
Transfers and Stat. Diff.	-	-	1	-	-	-	-	1	-	-	-
TRANSFORMATION	**-**	**-**	**-**	**-**	**-**	**-**	**6**	**-**	**-**	**-**	**-**
Electricity and CHP Plants	-	-	-	-	-	-	6	-	-	-	-
Petroleum Refineries	-	-	-	-	-	-	-	-	-	-	-
Other Transform. Sector	-	-	-	-	-	-	-	-	-	-	-
ENERGY SECTOR	**-**	**-**	**-**	**-**	**-**	**-**	**-**	**-**	**-**	**-**	**-**
DISTRIBUTION LOSSES	**-**	**-**	**-**	**-**	**-**	**-**	**-**	**-**	**-**	**-**	**-**
FINAL CONSUMPTION	**-**	**1**	**105**	**-**	**56**	**39**	**96**	**19**	**-**	**-**	**-**
INDUSTRY SECTOR	**-**	**-**	**-**	**-**	**-**	**-**	**-**	**19**	**-**	**-**	**-**
Iron and Steel	-	-	-	-	-	-	-	-	-	-	-
Chemical and Petrochem.	-	-	-	-	-	-	-	-	-	-	-
Non-Metallic Minerals	-	-	-	-	-	-	-	-	-	-	-
Non-specified	-	-	-	-	-	-	-	19	-	-	-
TRANSPORT SECTOR	**-**	**-**	**105**	**-**	**56**	**-**	**96**	**-**	**-**	**-**	**-**
Air	-	-	-	-	56	-	-	-	-	-	-
Road	-	-	105	-	-	-	96	-	-	-	-
Non-specified	-	-	-	-	-	-	-	-	-	-	-
OTHER SECTORS	**-**	**1**	**-**	**-**	**-**	**39**	**-**	**-**	**-**	**-**	**-**
Agriculture	-	-	-	-	-	-	-	-	-	-	-
Comm. and Publ. Services	-	-	-	-	-	-	-	-	-	-	-
Residential	-	1	-	-	-	39	-	-	-	-	-
Non-specified	-	-	-	-	-	-	-	-	-	-	-
NON-ENERGY USE	**-**	**-**	**-**	**-**	**-**	**-**	**-**	**-**	**-**	**-**	**-**

APPROVISIONNEMENT ET DEMANDE 1997	*Pétrole cont.* (1000 tonnes)										
	Gaz de raffinerie	*GPL + éthane*	*Essence moteur*	*Essence aviation*	*Carbu- réacteurs*	*Kérosène*	*Gazole*	*Fioul lourd*	*Naphta*	*Coke de pétrole*	*Autres prod.*
Production	-	-	-	-	-	-	-	-	-	-	-
Imports	-	1	131	-	49	34	71	18	-	-	-
Exports	-	-	-	-	-	-	-4	-1	-	-	-
Intl. Marine Bunkers	-	-	-	-	-	-	-5	-	-	-	-
Stock Changes	-	-	-	-	-	-	23	-2	-	-	-
DOMESTIC SUPPLY	**-**	**1**	**131**	**-**	**49**	**34**	**85**	**15**	**-**	**-**	**-**
Transfers and Stat. Diff.	-	-	-5	-	-	-	-	4	-	-	-
TRANSFORMATION	**-**	**-**	**-**	**-**	**-**	**-**	**6**	**-**	**-**	**-**	**-**
Electricity and CHP Plants	-	-	-	-	-	-	6	-	-	-	-
Petroleum Refineries	-	-	-	-	-	-	-	-	-	-	-
Other Transform. Sector	-	-	-	-	-	-	-	-	-	-	-
ENERGY SECTOR	**-**	**-**	**-**	**-**	**-**	**-**	**-**	**-**	**-**	**-**	**-**
DISTRIBUTION LOSSES	**-**	**-**	**-**	**-**	**-**	**-**	**-**	**-**	**-**	**-**	**-**
FINAL CONSUMPTION	**-**	**1**	**126**	**-**	**49**	**34**	**79**	**19**	**-**	**-**	**-**
INDUSTRY SECTOR	**-**	**-**	**-**	**-**	**-**	**-**	**-**	**19**	**-**	**-**	**-**
Iron and Steel	-	-	-	-	-	-	-	-	-	-	-
Chemical and Petrochem.	-	-	-	-	-	-	-	-	-	-	-
Non-Metallic Minerals	-	-	-	-	-	-	-	-	-	-	-
Non-specified	-	-	-	-	-	-	-	19	-	-	-
TRANSPORT SECTOR	**-**	**-**	**126**	**-**	**49**	**-**	**79**	**-**	**-**	**-**	**-**
Air	-	-	-	-	49	-	-	-	-	-	-
Road	-	-	126	-	-	-	79	-	-	-	-
Non-specified	-	-	-	-	-	-	-	-	-	-	-
OTHER SECTORS	**-**	**1**	**-**	**-**	**-**	**34**	**-**	**-**	**-**	**-**	**-**
Agriculture	-	-	-	-	-	-	-	-	-	-	-
Comm. and Publ. Services	-	-	-	-	-	-	-	-	-	-	-
Residential	-	1	-	-	-	34	-	-	-	-	-
Non-specified	-	-	-	-	-	-	-	-	-	-	-
NON-ENERGY USE	**-**	**-**	**-**	**-**	**-**	**-**	**-**	**-**	**-**	**-**	**-**

Benin / Bénin

SUPPLY AND CONSUMPTION 1996	Gas (TJ)				Comb. Renew. & Waste (TJ)				(GWh)	(TJ)
	Natural Gas	Gas Works	Coke Ovens	Blast Furnaces	Solid Biomass	Gas/Liquids from Biomass	Municipal Waste	Industrial Waste	Electricity	Heat
Production	-	-	-	-	75044	-	-	-	47	-
Imports	-	-	-	-	-	-	-	-	264	-
Exports	-	-	-	-	-	-	-	-	-	-
Intl. Marine Bunkers	-	-	-	-	-	-	-	-	-	-
Stock Changes	-	-	-	-	-	-	-	-	-	-
DOMESTIC SUPPLY	-	-	-	-	**75044**	-	-	-	**311**	-
Transfers and Stat. Diff.	-	-	-	-	-1	-	-	-	-	-
TRANSFORMATION	-	-	-	-	**1532**	-	-	-	-	-
Electricity and CHP Plants	-	-	-	-	-	-	-	-	-	-
Petroleum Refineries	-	-	-	-	-	-	-	-	-	-
Other Transform. Sector	-	-	-	-	1532	-	-	-	-	-
ENERGY SECTOR	-	-	-	-	-	-	-	-	-	-
DISTRIBUTION LOSSES	-	-	-	-	-	-	-	-	**41**	-
FINAL CONSUMPTION	-	-	-	-	**73511**	-	-	-	**270**	-
INDUSTRY SECTOR	-	-	-	-	**13783**	-	-	-	**130**	-
Iron and Steel	-	-	-	-	-	-	-	-	-	-
Chemical and Petrochem.	-	-	-	-	-	-	-	-	-	-
Non-Metallic Minerals	-	-	-	-	-	-	-	-	-	-
Non-specified	-	-	-	-	13783	-	-	-	130	-
TRANSPORT SECTOR	-	-	-	-	-	-	-	-	-	-
Air	-	-	-	-	-	-	-	-	-	-
Road	-	-	-	-	-	-	-	-	-	-
Non-specified	-	-	-	-	-	-	-	-	-	-
OTHER SECTORS	-	-	-	-	**59728**	-	-	-	**140**	-
Agriculture	-	-	-	-	-	-	-	-	-	-
Comm. and Publ. Services	-	-	-	-	-	-	-	-	8	-
Residential	-	-	-	-	59728	-	-	-	132	-
Non-specified	-	-	-	-	-	-	-	-	-	-
NON-ENERGY USE	-	-	-	-	-	-	-	-	-	-

APPROVISIONNEMENT ET DEMANDE 1997	*Gaz* (TJ)				*En. Re. Comb. & Déchets* (TJ)				(GWh)	(TJ)
	Gaz naturel	*Usines à gaz*	*Cokeries*	*Hauts fourneaux*	*Biomasse solide*	*Gaz/Liquides tirés de biomasse*	*Déchets urbains*	*Déchets industriels*	*Electricité*	*Chaleur*
Production	-	-	-	-	76620	-	-	-	50	-
Imports	-	-	-	-	-	-	-	-	238	-
Exports	-	-	-	-	-	-	-	-	-	-
Intl. Marine Bunkers	-	-	-	-	-	-	-	-	-	-
Stock Changes	-	-	-	-	-	-	-	-	-	-
DOMESTIC SUPPLY	-	-	-	-	**76620**	-	-	-	**288**	-
Transfers and Stat. Diff.	-	-	-	-	-1	-	-	-	-	-
TRANSFORMATION	-	-	-	-	**1564**	-	-	-	-	-
Electricity and CHP Plants	-	-	-	-	-	-	-	-	-	-
Petroleum Refineries	-	-	-	-	-	-	-	-	-	-
Other Transform. Sector	-	-	-	-	1564	-	-	-	-	-
ENERGY SECTOR	-	-	-	-	-	-	-	-	-	-
DISTRIBUTION LOSSES	-	-	-	-	-	-	-	-	**38**	-
FINAL CONSUMPTION	-	-	-	-	**75055**	-	-	-	**250**	-
INDUSTRY SECTOR	-	-	-	-	**14072**	-	-	-	**121**	-
Iron and Steel	-	-	-	-	-	-	-	-	-	-
Chemical and Petrochem.	-	-	-	-	-	-	-	-	-	-
Non-Metallic Minerals	-	-	-	-	-	-	-	-	-	-
Non-specified	-	-	-	-	14072	-	-	-	121	-
TRANSPORT SECTOR	-	-	-	-	-	-	-	-	-	-
Air	-	-	-	-	-	-	-	-	-	-
Road	-	-	-	-	-	-	-	-	-	-
Non-specified	-	-	-	-	-	-	-	-	-	-
OTHER SECTORS	-	-	-	-	**60983**	-	-	-	**129**	-
Agriculture	-	-	-	-	-	-	-	-	-	-
Comm. and Publ. Services	-	-	-	-	-	-	-	-	11	-
Residential	-	-	-	-	60983	-	-	-	118	-
Non-specified	-	-	-	-	-	-	-	-	-	-
NON-ENERGY USE	-	-	-	-	-	-	-	-	-	-

Bolivia / Bolivie

SUPPLY AND CONSUMPTION 1996	Coal (1000 tonnes)							Oil (1000 tonnes)			
	Coking Coal	Other Bit. Coal	Sub-Bit. Coal	Lignite	Peat	Oven and Gas Coke	Pat. Fuel and BKB	Crude Oil	NGL	Feed-stocks	Additives
Production	-	-	-	-	-	-	-	1424	312	-	-
Imports	-	-	-	-	-	-	-	-	-	-	-
Exports	-	-	-	-	-	-	-	-	-	-	-
Intl. Marine Bunkers	-	-	-	-	-	-	-	-	-	-	-
Stock Changes	-	-	-	-	-	-	-	40	-	-	-
DOMESTIC SUPPLY	**-**	**-**	**-**	**-**	**-**	**-**	**-**	**1464**	**312**	**-**	**-**
Transfers and Stat. Diff.	-	-	-	-	-	-	-	136	-210	-	-
TRANSFORMATION	**-**	**-**	**-**	**-**	**-**	**-**	**-**	**1600**	**102**	**-**	**-**
Electricity and CHP Plants	-	-	-	-	-	-	-	-	-	-	-
Petroleum Refineries	-	-	-	-	-	-	-	1600	102	-	-
Other Transform. Sector	-	-	-	-	-	-	-	-	-	-	-
ENERGY SECTOR	**-**	**-**	**-**	**-**	**-**	**-**	**-**	**-**	**-**	**-**	**-**
DISTRIBUTION LOSSES	**-**	**-**	**-**	**-**	**-**	**-**	**-**	**-**	**-**	**-**	**-**
FINAL CONSUMPTION	**-**	**-**	**-**	**-**	**-**	**-**	**-**	**-**	**-**	**-**	**-**
INDUSTRY SECTOR	**-**	**-**	**-**	**-**	**-**	**-**	**-**	**-**	**-**	**-**	**-**
Iron and Steel	-	-	-	-	-	-	-	-	-	-	-
Chemical and Petrochem.	-	-	-	-	-	-	-	-	-	-	-
Non-Metallic Minerals	-	-	-	-	-	-	-	-	-	-	-
Non-specified	-	-	-	-	-	-	-	-	-	-	-
TRANSPORT SECTOR	**-**	**-**	**-**	**-**	**-**	**-**	**-**	**-**	**-**	**-**	**-**
Air	-	-	-	-	-	-	-	-	-	-	-
Road	-	-	-	-	-	-	-	-	-	-	-
Non-specified	-	-	-	-	-	-	-	-	-	-	-
OTHER SECTORS	**-**	**-**	**-**	**-**	**-**	**-**	**-**	**-**	**-**	**-**	**-**
Agriculture	-	-	-	-	-	-	-	-	-	-	-
Comm. and Publ. Services	-	-	-	-	-	-	-	-	-	-	-
Residential	-	-	-	-	-	-	-	-	-	-	-
Non-specified	-	-	-	-	-	-	-	-	-	-	-
NON-ENERGY USE	**-**	**-**	**-**	**-**	**-**	**-**	**-**	**-**	**-**	**-**	**-**

APPROVISIONNEMENT ET DEMANDE 1997	*Charbon* (1000 tonnes)							*Pétrole* (1000 tonnes)			
	Charbon à coke	*Autres charb. bit.*	*Charbon sous-bit.*	*Lignite*	*Tourbe*	*Coke de four/gaz*	*Agg./briq. de lignite*	*Pétrole brut*	*LGN*	*Produits d'aliment.*	*Additifs*
Production	-	-	-	-	-	-	-	1460	353	-	-
Imports	-	-	-	-	-	-	-	-	-	-	-
Exports	-	-	-	-	-	-	-	-	-	-	-
Intl. Marine Bunkers	-	-	-	-	-	-	-	-	-	-	-
Stock Changes	-	-	-	-	-	-	-	-	-	-	-
DOMESTIC SUPPLY	**-**	**-**	**-**	**-**	**-**	**-**	**-**	**1460**	**353**	**-**	**-**
Transfers and Stat. Diff.	-	-	-	-	-	-	-	209	-251	-	-
TRANSFORMATION	**-**	**-**	**-**	**-**	**-**	**-**	**-**	**1669**	**102**	**-**	**-**
Electricity and CHP Plants	-	-	-	-	-	-	-	-	-	-	-
Petroleum Refineries	-	-	-	-	-	-	-	1669	102	-	-
Other Transform. Sector	-	-	-	-	-	-	-	-	-	-	-
ENERGY SECTOR	**-**	**-**	**-**	**-**	**-**	**-**	**-**	**-**	**-**	**-**	**-**
DISTRIBUTION LOSSES	**-**	**-**	**-**	**-**	**-**	**-**	**-**	**-**	**-**	**-**	**-**
FINAL CONSUMPTION	**-**	**-**	**-**	**-**	**-**	**-**	**-**	**-**	**-**	**-**	**-**
INDUSTRY SECTOR	**-**	**-**	**-**	**-**	**-**	**-**	**-**	**-**	**-**	**-**	**-**
Iron and Steel	-	-	-	-	-	-	-	-	-	-	-
Chemical and Petrochem.	-	-	-	-	-	-	-	-	-	-	-
Non-Metallic Minerals	-	-	-	-	-	-	-	-	-	-	-
Non-specified	-	-	-	-	-	-	-	-	-	-	-
TRANSPORT SECTOR	**-**	**-**	**-**	**-**	**-**	**-**	**-**	**-**	**-**	**-**	**-**
Air	-	-	-	-	-	-	-	-	-	-	-
Road	-	-	-	-	-	-	-	-	-	-	-
Non-specified	-	-	-	-	-	-	-	-	-	-	-
OTHER SECTORS	**-**	**-**	**-**	**-**	**-**	**-**	**-**	**-**	**-**	**-**	**-**
Agriculture	-	-	-	-	-	-	-	-	-	-	-
Comm. and Publ. Services	-	-	-	-	-	-	-	-	-	-	-
Residential	-	-	-	-	-	-	-	-	-	-	-
Non-specified	-	-	-	-	-	-	-	-	-	-	-
NON-ENERGY USE	**-**	**-**	**-**	**-**	**-**	**-**	**-**	**-**	**-**	**-**	**-**

Bolivia / Bolivie

SUPPLY AND CONSUMPTION 1996	Oil cont. (1000 tonnes)										
	Refinery Gas	LPG + Ethane	Motor Gasoline	Aviation Gasoline	Jet Fuel	Kerosene	Gas/ Diesel	Heavy Fuel Oil	Naphtha	Petrol. Coke	Other Prod.
Production	84	50	431	4	122	31	400	1	-	-	428
Imports	-	-	-	-	-	-	133	-	-	-	-
Exports	-	-	-	-	-	-	-	-	-	-	-
Intl. Marine Bunkers	-	-	-	-	-	-	-	-	-	-	-
Stock Changes	-	-11	-	-	2	-6	-	-	-	-	-
DOMESTIC SUPPLY	**84**	**39**	**431**	**4**	**124**	**25**	**533**	**1**	**-**	**-**	**428**
Transfers and Stat. Diff.	-	210	-	-	-	-	55	-	-	-	-417
TRANSFORMATION	**-**	**-**	**-**	**-**	**-**	**-**	**56**	**-**	**-**	**-**	**-**
Electricity and CHP Plants	-	-	-	-	-	-	56	-	-	-	-
Petroleum Refineries	-	-	-	-	-	-	-	-	-	-	-
Other Transform. Sector	-	-	-	-	-	-	-	-	-	-	-
ENERGY SECTOR	**84**	**-**	**3**	**-**	**-**	**-**	**18**	**-**	**-**	**-**	**-**
DISTRIBUTION LOSSES	**-**	**-**	**-**	**-**	**-**	**-**	**-**	**-**	**-**	**-**	**-**
FINAL CONSUMPTION	**-**	**249**	**428**	**4**	**124**	**25**	**514**	**1**	**-**	**-**	**11**
INDUSTRY SECTOR	**-**	**7**	**-**	**-**	**-**	**7**	**67**	**1**	**-**	**-**	**-**
Iron and Steel	-	-	-	-	-	-	-	-	-	-	-
Chemical and Petrochem.	-	-	-	-	-	-	-	-	-	-	-
Non-Metallic Minerals	-	-	-	-	-	-	-	-	-	-	-
Non-specified	-	7	-	-	-	7	67	1	-	-	-
TRANSPORT SECTOR	**-**	**-**	**428**	**4**	**124**	**-**	**398**	**-**	**-**	**-**	**-**
Air	-	-	-	4	124	-	-	-	-	-	-
Road	-	-	428	-	-	-	359	-	-	-	-
Non-specified	-	-	-	-	-	-	39	-	-	-	-
OTHER SECTORS	**-**	**242**	**-**	**-**	**-**	**18**	**49**	**-**	**-**	**-**	**-**
Agriculture	-	-	-	-	-	-	49	-	-	-	-
Comm. and Publ. Services	-	-	-	-	-	-	-	-	-	-	-
Residential	-	242	-	-	-	18	-	-	-	-	-
Non-specified	-	-	-	-	-	-	-	-	-	-	-
NON-ENERGY USE	**-**	**-**	**-**	**-**	**-**	**-**	**-**	**-**	**-**	**-**	**11**

APPROVISIONNEMENT ET DEMANDE 1997	*Pétrole cont.* (1000 tonnes)										
	Gaz de raffinerie	*GPL + éthane*	*Essence moteur*	*Essence aviation*	*Carbu- réacteurs*	*Kérosène*	*Gazole*	*Fioul lourd*	*Naphta*	*Coke de pétrole*	*Autres prod.*
Production	84	57	467	4	138	25	322	1	-	-	510
Imports	-	-	-	-	-	-	235	-	-	-	-
Exports	-	-	-	-	-	-	-	-	-	-	-
Intl. Marine Bunkers	-	-	-	-	-	-	-	-	-	-	-
Stock Changes	-	-16	-	-	-6	-	-	-	-	-	-
DOMESTIC SUPPLY	**84**	**41**	**467**	**4**	**132**	**25**	**557**	**1**	**-**	**-**	**510**
Transfers and Stat. Diff.	-	250	-	-	-	-2	-	-	-	-	-497
TRANSFORMATION	**-**	**-**	**-**	**-**	**-**	**-**	**56**	**-**	**-**	**-**	**-**
Electricity and CHP Plants	-	-	-	-	-	-	56	-	-	-	-
Petroleum Refineries	-	-	-	-	-	-	-	-	-	-	-
Other Transform. Sector	-	-	-	-	-	-	-	-	-	-	-
ENERGY SECTOR	**84**	**-**	**11**	**-**	**-**	**-**	**23**	**-**	**-**	**-**	**-**
DISTRIBUTION LOSSES	**-**	**-**	**-**	**-**	**-**	**-**	**-**	**-**	**-**	**-**	**-**
FINAL CONSUMPTION	**-**	**291**	**456**	**4**	**132**	**23**	**478**	**1**	**-**	**-**	**13**
INDUSTRY SECTOR	**-**	**8**	**-**	**-**	**-**	**6**	**62**	**1**	**-**	**-**	**-**
Iron and Steel	-	-	-	-	-	-	-	-	-	-	-
Chemical and Petrochem.	-	-	-	-	-	-	-	-	-	-	-
Non-Metallic Minerals	-	-	-	-	-	-	-	-	-	-	-
Non-specified	-	8	-	-	-	6	62	1	-	-	-
TRANSPORT SECTOR	**-**	**-**	**456**	**4**	**132**	**-**	**370**	**-**	**-**	**-**	**-**
Air	-	-	-	4	132	-	-	-	-	-	-
Road	-	-	456	-	-	-	334	-	-	-	-
Non-specified	-	-	-	-	-	-	36	-	-	-	-
OTHER SECTORS	**-**	**283**	**-**	**-**	**-**	**17**	**46**	**-**	**-**	**-**	**-**
Agriculture	-	-	-	-	-	-	46	-	-	-	-
Comm. and Publ. Services	-	-	-	-	-	-	-	-	-	-	-
Residential	-	283	-	-	-	17	-	-	-	-	-
Non-specified	-	-	-	-	-	-	-	-	-	-	-
NON-ENERGY USE	**-**	**-**	**-**	**-**	**-**	**-**	**-**	**-**	**-**	**-**	**13**

Bolivia / Bolivie

SUPPLY AND CONSUMPTION 1996	Gas (TJ)				Comb. Renew. & Waste (TJ)				(GWh)	(TJ)
	Natural Gas	Gas Works	Coke Ovens	Blast Furnaces	Solid Biomass	Gas/Liquids from Biomass	Municipal Waste	Industrial Waste	Electricity	Heat
Production	116023	-	-	-	32413	-	-	-	3232	-
Imports	-	-	-	-	-	-	-	-	-	-
Exports	-82293	-	-	-	-	-	-	-	-2	-
Intl. Marine Bunkers	-	-	-	-	-	-	-	-	-	-
Stock Changes	-	-	-	-	-	-	-	-	-	-
DOMESTIC SUPPLY	**33730**	-	-	-	**32413**	-	-	-	**3230**	-
Transfers and Stat. Diff.	11079	-	-	-	-	-	-	-	-	-
TRANSFORMATION	**15409**	-	-	-	**2128**	-	-	-	-	-
Electricity and CHP Plants	15409	-	-	-	994	-	-	-	-	-
Petroleum Refineries	-	-	-	-	-	-	-	-	-	-
Other Transform. Sector	-	-	-	-	1134	-	-	-	-	-
ENERGY SECTOR	**9727**	-	-	-	-	-	-	-	**38**	-
DISTRIBUTION LOSSES	**7775**	-	-	-	-	-	-	-	**379**	-
FINAL CONSUMPTION	**11898**	-	-	-	**30285**	-	-	-	**2813**	-
INDUSTRY SECTOR	**11820**	-	-	-	**7112**	-	-	-	**1143**	-
Iron and Steel	-	-	-	-	-	-	-	-	-	-
Chemical and Petrochem.	-	-	-	-	-	-	-	-	-	-
Non-Metallic Minerals	-	-	-	-	-	-	-	-	-	-
Non-specified	11820	-	-	-	7112	-	-	-	1143	-
TRANSPORT SECTOR	-	-	-	-	-	-	-	-	-	-
Air	-	-	-	-	-	-	-	-	-	-
Road	-	-	-	-	-	-	-	-	-	-
Non-specified	-	-	-	-	-	-	-	-	-	-
OTHER SECTORS	**78**	-	-	-	**23173**	-	-	-	**1670**	-
Agriculture	-	-	-	-	-	-	-	-	-	-
Comm. and Publ. Services	-	-	-	-	-	-	-	-	492	-
Residential	78	-	-	-	23173	-	-	-	1178	-
Non-specified	-	-	-	-	-	-	-	-	-	-
NON-ENERGY USE	-	-	-	-	-	-	-	-	-	-

APPROVISIONNEMENT ET DEMANDE 1997	*Gaz* (TJ)				*En. Re. Comb. & Déchets* (TJ)				(GWh)	(TJ)
	Gaz naturel	*Usines à gaz*	*Cokeries*	*Hauts fourneaux*	*Biomasse solide*	*Gaz/Liquides tirés de biomasse*	*Déchets urbains*	*Déchets industriels*	*Electricité*	*Chaleur*
Production	139720	-	-	-	36672	-	-	-	3433	-
Imports	-	-	-	-	-	-	-	-	8	-
Exports	-89240	-	-	-	-	-	-	-	-4	-
Intl. Marine Bunkers	-	-	-	-	-	-	-	-	-	-
Stock Changes	-	-	-	-	-	-	-	-	-	-
DOMESTIC SUPPLY	**50480**	-	-	-	**36672**	-	-	-	**3437**	-
Transfers and Stat. Diff.	-9851	-	-	-	-	-	-	-	1	-
TRANSFORMATION	**14248**	-	-	-	**2797**	-	-	-	-	-
Electricity and CHP Plants	14248	-	-	-	1133	-	-	-	-	-
Petroleum Refineries	-	-	-	-	-	-	-	-	-	-
Other Transform. Sector	-	-	-	-	1664	-	-	-	-	-
ENERGY SECTOR	**8279**	-	-	-	-	-	-	-	**28**	-
DISTRIBUTION LOSSES	**2018**	-	-	-	-	-	-	-	**370**	-
FINAL CONSUMPTION	**16084**	-	-	-	**33875**	-	-	-	**3040**	-
INDUSTRY SECTOR	**16006**	-	-	-	**12766**	-	-	-	**1235**	-
Iron and Steel	-	-	-	-	-	-	-	-	-	-
Chemical and Petrochem.	-	-	-	-	-	-	-	-	-	-
Non-Metallic Minerals	-	-	-	-	-	-	-	-	-	-
Non-specified	16006	-	-	-	12766	-	-	-	1235	-
TRANSPORT SECTOR	-	-	-	-	-	-	-	-	-	-
Air	-	-	-	-	-	-	-	-	-	-
Road	-	-	-	-	-	-	-	-	-	-
Non-specified	-	-	-	-	-	-	-	-	-	-
OTHER SECTORS	**78**	-	-	-	**21109**	-	-	-	**1805**	-
Agriculture	-	-	-	-	-	-	-	-	-	-
Comm. and Publ. Services	-	-	-	-	-	-	-	-	532	-
Residential	78	-	-	-	21109	-	-	-	1273	-
Non-specified	-	-	-	-	-	-	-	-	-	-
NON-ENERGY USE	-	-	-	-	-	-	-	-	-	-

Bosnia-Herzegovina / Bosnie-Herzégovine

SUPPLY AND CONSUMPTION 1996	Coal (1000 tonnes)							Oil (1000 tonnes)			
	Coking Coal	Other Bit. Coal	Sub-Bit. Coal	Lignite	Peat	Oven and Gas Coke	Pat. Fuel and BKB	Crude Oil	NGL	Feed-stocks	Additives
Production	-	-	-	1640	-	-	-	-	-	-	-
Imports	-	-	-	-	-	-	-	-	-	-	-
Exports	-	-	-	-	-	-	-	-	-	-	-
Intl. Marine Bunkers	-	-	-	-	-	-	-	-	-	-	-
Stock Changes	-	-	-	-	-	-	-	-	-	-	-
DOMESTIC SUPPLY	**-**	**-**	**-**	**1640**	**-**	**-**	**-**	**-**	**-**	**-**	**-**
Transfers and Stat. Diff.	-	-	-	-	-	-	-	-	-	-	-
TRANSFORMATION	**-**	**-**	**-**	**1350**	**-**	**-**	**-**	**-**	**-**	**-**	**-**
Electricity and CHP Plants	-	-	-	1350	-	-	-	-	-	-	-
Petroleum Refineries	-	-	-	-	-	-	-	-	-	-	-
Other Transform. Sector	-	-	-	-	-	-	-	-	-	-	-
ENERGY SECTOR	**-**	**-**	**-**	**-**	**-**	**-**	**-**	**-**	**-**	**-**	**-**
DISTRIBUTION LOSSES	**-**	**-**	**-**	**-**	**-**	**-**	**-**	**-**	**-**	**-**	**-**
FINAL CONSUMPTION	**-**	**-**	**-**	**290**	**-**	**-**	**-**	**-**	**-**	**-**	**-**
INDUSTRY SECTOR	**-**	**-**	**-**	**-**	**-**	**-**	**-**	**-**	**-**	**-**	**-**
Iron and Steel	-	-	-	-	-	-	-	-	-	-	-
Chemical and Petrochem.	-	-	-	-	-	-	-	-	-	-	-
Non-Metallic Minerals	-	-	-	-	-	-	-	-	-	-	-
Non-specified	-	-	-	-	-	-	-	-	-	-	-
TRANSPORT SECTOR	**-**	**-**	**-**	**290**	**-**	**-**	**-**	**-**	**-**	**-**	**-**
Air	-	-	-	-	-	-	-	-	-	-	-
Road	-	-	-	-	-	-	-	-	-	-	-
Non-specified	-	-	-	290	-	-	-	-	-	-	-
OTHER SECTORS	**-**	**-**	**-**	**-**	**-**	**-**	**-**	**-**	**-**	**-**	**-**
Agriculture	-	-	-	-	-	-	-	-	-	-	-
Comm. and Publ. Services	-	-	-	-	-	-	-	-	-	-	-
Residential	-	-	-	-	-	-	-	-	-	-	-
Non-specified	-	-	-	-	-	-	-	-	-	-	-
NON-ENERGY USE	**-**	**-**	**-**	**-**	**-**	**-**	**-**	**-**	**-**	**-**	**-**

APPROVISIONNEMENT ET DEMANDE 1997	*Charbon* (1000 tonnes)							*Pétrole* (1000 tonnes)			
	Charbon à coke	*Autres charb. bit.*	*Charbon sous-bit.*	*Lignite*	*Tourbe*	*Coke de four/gaz*	*Agg./briq. de lignite*	*Pétrole brut*	*LGN*	*Produits d'aliment.*	*Additifs*
Production	-	-	-	1640	-	-	-	-	-	-	-
Imports	-	-	-	-	-	-	-	-	-	-	-
Exports	-	-	-	-	-	-	-	-	-	-	-
Intl. Marine Bunkers	-	-	-	-	-	-	-	-	-	-	-
Stock Changes	-	-	-	-	-	-	-	-	-	-	-
DOMESTIC SUPPLY	**-**	**-**	**-**	**1640**	**-**	**-**	**-**	**-**	**-**	**-**	**-**
Transfers and Stat. Diff.	-	-	-	-	-	-	-	-	-	-	-
TRANSFORMATION	**-**	**-**	**-**	**1350**	**-**	**-**	**-**	**-**	**-**	**-**	**-**
Electricity and CHP Plants	-	-	-	1350	-	-	-	-	-	-	-
Petroleum Refineries	-	-	-	-	-	-	-	-	-	-	-
Other Transform. Sector	-	-	-	-	-	-	-	-	-	-	-
ENERGY SECTOR	**-**	**-**	**-**	**-**	**-**	**-**	**-**	**-**	**-**	**-**	**-**
DISTRIBUTION LOSSES	**-**	**-**	**-**	**-**	**-**	**-**	**-**	**-**	**-**	**-**	**-**
FINAL CONSUMPTION	**-**	**-**	**-**	**290**	**-**	**-**	**-**	**-**	**-**	**-**	**-**
INDUSTRY SECTOR	**-**	**-**	**-**	**-**	**-**	**-**	**-**	**-**	**-**	**-**	**-**
Iron and Steel	-	-	-	-	-	-	-	-	-	-	-
Chemical and Petrochem.	-	-	-	-	-	-	-	-	-	-	-
Non-Metallic Minerals	-	-	-	-	-	-	-	-	-	-	-
Non-specified	-	-	-	-	-	-	-	-	-	-	-
TRANSPORT SECTOR	**-**	**-**	**-**	**290**	**-**	**-**	**-**	**-**	**-**	**-**	**-**
Air	-	-	-	-	-	-	-	-	-	-	-
Road	-	-	-	-	-	-	-	-	-	-	-
Non-specified	-	-	-	290	-	-	-	-	-	-	-
OTHER SECTORS	**-**	**-**	**-**	**-**	**-**	**-**	**-**	**-**	**-**	**-**	**-**
Agriculture	-	-	-	-	-	-	-	-	-	-	-
Comm. and Publ. Services	-	-	-	-	-	-	-	-	-	-	-
Residential	-	-	-	-	-	-	-	-	-	-	-
Non-specified	-	-	-	-	-	-	-	-	-	-	-
NON-ENERGY USE	**-**	**-**	**-**	**-**	**-**	**-**	**-**	**-**	**-**	**-**	**-**

Bosnia-Herzegovina / Bosnie-Herzégovine

SUPPLY AND CONSUMPTION 1996	Oil cont. (1000 tonnes)										
	Refinery Gas	LPG + Ethane	Motor Gasoline	Aviation Gasoline	Jet Fuel	Kerosene	Gas/ Diesel	Heavy Fuel Oil	Naphtha	Petrol. Coke	Other Prod.
Production	-	-	-	-	-	-	-	-	-	-	-
Imports	-	-	115	-	36	-	269	96	298	-	45
Exports	-	-	-	-	-	-	-	-	-	-	-
Intl. Marine Bunkers	-	-	-	-	-	-	-	-	-	-	-
Stock Changes	-	-	-	-	-	-	-	-	-	-	-
DOMESTIC SUPPLY	**-**	**-**	**115**	**-**	**36**	**-**	**269**	**96**	**298**	**-**	**45**
Transfers and Stat. Diff.	-	-	-	-	-	-	-	-	-	-	-
TRANSFORMATION	**-**	**-**	**-**	**-**	**-**	**-**	**12**	**60**	**61**	**-**	**-**
Electricity and CHP Plants	-	-	-	-	-	-	12	-	20	-	-
Petroleum Refineries	-	-	-	-	-	-	-	-	-	-	-
Other Transform. Sector	-	-	-	-	-	-	-	60	41	-	-
ENERGY SECTOR	**-**	**-**	**-**	**-**	**-**	**-**	**-**	**-**	**-**	**-**	**-**
DISTRIBUTION LOSSES	**-**	**-**	**87**	**-**	**-**	**-**	**62**	**36**	**223**	**-**	**-**
FINAL CONSUMPTION	**-**	**-**	**28**	**-**	**36**	**-**	**195**	**-**	**14**	**-**	**45**
INDUSTRY SECTOR	**-**	**-**	**-**	**-**	**-**	**-**	**-**	**-**	**14**	**-**	**45**
Iron and Steel	-	-	-	-	-	-	-	-	7	-	16
Chemical and Petrochem.	-	-	-	-	-	-	-	-	-	-	-
Non-Metallic Minerals	-	-	-	-	-	-	-	-	-	-	-
Non-specified	-	-	-	-	-	-	-	-	7	-	29
TRANSPORT SECTOR	**-**	**-**	**28**	**-**	**36**	**-**	**195**	**-**	**-**	**-**	**-**
Air	-	-	-	-	36	-	-	-	-	-	-
Road	-	-	28	-	-	-	191	-	-	-	-
Non-specified	-	-	-	-	-	-	4	-	-	-	-
OTHER SECTORS	**-**	**-**	**-**	**-**	**-**	**-**	**-**	**-**	**-**	**-**	**-**
Agriculture	-	-	-	-	-	-	-	-	-	-	-
Comm. and Publ. Services	-	-	-	-	-	-	-	-	-	-	-
Residential	-	-	-	-	-	-	-	-	-	-	-
Non-specified	-	-	-	-	-	-	-	-	-	-	-
NON-ENERGY USE	**-**	**-**	**-**	**-**	**-**	**-**	**-**	**-**	**-**	**-**	**-**

APPROVISIONNEMENT ET DEMANDE 1997	*Pétrole cont.* (1000 tonnes)										
	Gaz de raffinerie	*GPL + éthane*	*Essence moteur*	*Essence aviation*	*Carbu- réacteurs*	*Kérosène*	*Gazole*	*Fioul lourd*	*Naphta*	*Coke de pétrole*	*Autres prod.*
Production	-	-	-	-	-	-	-	-	-	-	-
Imports	-	-	115	-	36	-	269	96	298	-	45
Exports	-	-	-	-	-	-	-	-	-	-	-
Intl. Marine Bunkers	-	-	-	-	-	-	-	-	-	-	-
Stock Changes	-	-	-	-	-	-	-	-	-	-	-
DOMESTIC SUPPLY	**-**	**-**	**115**	**-**	**36**	**-**	**269**	**96**	**298**	**-**	**45**
Transfers and Stat. Diff.	-	-	-	-	-	-	-	-	-	-	-
TRANSFORMATION	**-**	**-**	**-**	**-**	**-**	**-**	**12**	**60**	**61**	**-**	**-**
Electricity and CHP Plants	-	-	-	-	-	-	12	-	20	-	-
Petroleum Refineries	-	-	-	-	-	-	-	-	-	-	-
Other Transform. Sector	-	-	-	-	-	-	-	60	41	-	-
ENERGY SECTOR	**-**	**-**	**-**	**-**	**-**	**-**	**-**	**-**	**-**	**-**	**-**
DISTRIBUTION LOSSES	**-**	**-**	**87**	**-**	**-**	**-**	**62**	**36**	**223**	**-**	**-**
FINAL CONSUMPTION	**-**	**-**	**28**	**-**	**36**	**-**	**195**	**-**	**14**	**-**	**45**
INDUSTRY SECTOR	**-**	**-**	**-**	**-**	**-**	**-**	**-**	**-**	**14**	**-**	**45**
Iron and Steel	-	-	-	-	-	-	-	-	7	-	16
Chemical and Petrochem.	-	-	-	-	-	-	-	-	-	-	-
Non-Metallic Minerals	-	-	-	-	-	-	-	-	-	-	-
Non-specified	-	-	-	-	-	-	-	-	7	-	29
TRANSPORT SECTOR	**-**	**-**	**28**	**-**	**36**	**-**	**195**	**-**	**-**	**-**	**-**
Air	-	-	-	-	36	-	-	-	-	-	-
Road	-	-	28	-	-	-	191	-	-	-	-
Non-specified	-	-	-	-	-	-	4	-	-	-	-
OTHER SECTORS	**-**	**-**	**-**	**-**	**-**	**-**	**-**	**-**	**-**	**-**	**-**
Agriculture	-	-	-	-	-	-	-	-	-	-	-
Comm. and Publ. Services	-	-	-	-	-	-	-	-	-	-	-
Residential	-	-	-	-	-	-	-	-	-	-	-
Non-specified	-	-	-	-	-	-	-	-	-	-	-
NON-ENERGY USE	**-**	**-**	**-**	**-**	**-**	**-**	**-**	**-**	**-**	**-**	**-**

Bosnia-Herzegovina / Bosnie-Herzégovine

SUPPLY AND CONSUMPTION 1996	Gas (TJ)				Comb. Renew. & Waste (TJ)				(GWh)	(TJ)
	Natural Gas	Gas Works	Coke Ovens	Blast Furnaces	Solid Biomass	Gas/Liquids from Biomass	Municipal Waste	Industrial Waste	Electricity	Heat
Production	-	-	-	-	6500	-	-	-	2203	966
Imports	9834	-	-	-	-	-	-	-	387	-
Exports	-	-	-	-	-	-	-	-	-182	-
Intl. Marine Bunkers	-	-	-	-	-	-	-	-	-	-
Stock Changes	-	-	-	-	-	-	-	-	-	-
DOMESTIC SUPPLY	**9834**	**-**	**-**	**-**	**6500**	**-**	**-**	**-**	**2408**	**966**
Transfers and Stat. Diff.	6584	-	-	-	-	-	-	-	-	-
TRANSFORMATION	**15918**	**-**	**-**	**-**	**-**	**-**	**-**	**-**	**-**	**-**
Electricity and CHP Plants	-	-	-	-	-	-	-	-	-	-
Petroleum Refineries	-	-	-	-	-	-	-	-	-	-
Other Transform. Sector	15918	-	-	-	-	-	-	-	-	-
ENERGY SECTOR	**-**	**-**	**-**	**-**	**-**	**-**	**-**	**-**	**157**	**-**
DISTRIBUTION LOSSES	**500**	**-**	**-**	**-**	**-**	**-**	**-**	**-**	**512**	**-**
FINAL CONSUMPTION	**-**	**-**	**-**	**-**	**6500**	**-**	**-**	**-**	**1739**	**966**
INDUSTRY SECTOR	**-**	**-**	**-**	**-**	**-**	**-**	**-**	**-**	**-**	**-**
Iron and Steel	-	-	-	-	-	-	-	-	-	-
Chemical and Petrochem.	-	-	-	-	-	-	-	-	-	-
Non-Metallic Minerals	-	-	-	-	-	-	-	-	-	-
Non-specified	-	-	-	-	-	-	-	-	-	-
TRANSPORT SECTOR	**-**	**-**	**-**	**-**	**-**	**-**	**-**	**-**	**-**	**-**
Air	-	-	-	-	-	-	-	-	-	-
Road	-	-	-	-	-	-	-	-	-	-
Non-specified	-	-	-	-	-	-	-	-	-	-
OTHER SECTORS	**-**	**-**	**-**	**-**	**6500**	**-**	**-**	**-**	**1739**	**966**
Agriculture	-	-	-	-	-	-	-	-	-	-
Comm. and Publ. Services	-	-	-	-	-	-	-	-	-	-
Residential	-	-	-	-	6500	-	-	-	-	-
Non-specified	-	-	-	-	-	-	-	-	1739	966
NON-ENERGY USE	**-**	**-**	**-**	**-**	**-**	**-**	**-**	**-**	**-**	**-**

APPROVISIONNEMENT ET DEMANDE 1997	*Gaz* (TJ)				*En. Re. Comb. & Déchets* (TJ)				(GWh)	(TJ)
	Gaz naturel	*Usines à gaz*	*Cokeries*	*Hauts fourneaux*	*Biomasse solide*	*Gaz/Liquides tirés de biomasse*	*Déchets urbains*	*Déchets industriels*	*Electricité*	*Chaleur*
Production	-	-	-	-	6500	-	-	-	2203	966
Imports	9834	-	-	-	-	-	-	-	387	-
Exports	-	-	-	-	-	-	-	-	-182	-
Intl. Marine Bunkers	-	-	-	-	-	-	-	-	-	-
Stock Changes	-	-	-	-	-	-	-	-	-	-
DOMESTIC SUPPLY	**9834**	**-**	**-**	**-**	**6500**	**-**	**-**	**-**	**2408**	**966**
Transfers and Stat. Diff.	6584	-	-	-	-	-	-	-	-	-
TRANSFORMATION	**15918**	**-**	**-**	**-**	**-**	**-**	**-**	**-**	**-**	**-**
Electricity and CHP Plants	-	-	-	-	-	-	-	-	-	-
Petroleum Refineries	-	-	-	-	-	-	-	-	-	-
Other Transform. Sector	15918	-	-	-	-	-	-	-	-	-
ENERGY SECTOR	**-**	**-**	**-**	**-**	**-**	**-**	**-**	**-**	**157**	**-**
DISTRIBUTION LOSSES	**500**	**-**	**-**	**-**	**-**	**-**	**-**	**-**	**512**	**-**
FINAL CONSUMPTION	**-**	**-**	**-**	**-**	**6500**	**-**	**-**	**-**	**1739**	**966**
INDUSTRY SECTOR	**-**	**-**	**-**	**-**	**-**	**-**	**-**	**-**	**-**	**-**
Iron and Steel	-	-	-	-	-	-	-	-	-	-
Chemical and Petrochem.	-	-	-	-	-	-	-	-	-	-
Non-Metallic Minerals	-	-	-	-	-	-	-	-	-	-
Non-specified	-	-	-	-	-	-	-	-	-	-
TRANSPORT SECTOR	**-**	**-**	**-**	**-**	**-**	**-**	**-**	**-**	**-**	**-**
Air	-	-	-	-	-	-	-	-	-	-
Road	-	-	-	-	-	-	-	-	-	-
Non-specified	-	-	-	-	-	-	-	-	-	-
OTHER SECTORS	**-**	**-**	**-**	**-**	**6500**	**-**	**-**	**-**	**1739**	**966**
Agriculture	-	-	-	-	-	-	-	-	-	-
Comm. and Publ. Services	-	-	-	-	-	-	-	-	-	-
Residential	-	-	-	-	6500	-	-	-	-	-
Non-specified	-	-	-	-	-	-	-	-	1739	966
NON-ENERGY USE	**-**	**-**	**-**	**-**	**-**	**-**	**-**	**-**	**-**	**-**

Brazil / Brésil : 1996

SUPPLY AND CONSUMPTION / *APPROVISIONNEMENT ET DEMANDE*	Coal / *Charbon* (1000 tonnes)							Oil / *Pétrole* (1000 tonnes)			
	Coking Coal / *Charbon à coke*	Other Bit. Coal / *Autres charb. bit.*	Sub-Bit. Coal / *Charbon sous-bit.*	Lignite / *Lignite*	Peat / *Tourbe*	Oven and Gas Coke / *Coke de four/gaz*	Pat. Fuel and BKB / *Agg./briq. de lignite*	Crude Oil / *Pétrole brut*	NGL / *LGN*	Feed-stocks / *Produits d'aliment.*	Additives / *Additifs*
Production	133	4672	-	-	-	8667	-	39401	977	-	-
From Other Sources	-	-	-	-	-	-	-	-	-	-	-
Imports	12847	-	-	-	-	1716	-	28594	-	-	-
Exports	-	-	-	-	-	-	-	-104	-	-	-
Intl. Marine Bunkers	-	-	-	-	-	-	-	-	-	-	-
Stock Changes	-408	248	-	-	-	-204	-	-1885	-	-	-
DOMESTIC SUPPLY	**12572**	**4920**	**-**	**-**	**-**	**10179**	**-**	**66006**	**977**	**-**	**-**
Transfers	-	-	-	-	-	-	-	-	-447	-	-
Statistical Differences	-	-	-	-	-	-	-	-	-	-	-
TRANSFORMATION	**10976**	**3643**	**-**	**-**	**-**	**7763**	**-**	**66006**	**530**	**-**	**-**
Electricity Plants	-	3643	-	-	-	28	-	-	-	-	-
CHP Plants	-	-	-	-	-	-	-	-	-	-	-
Heat Plants	-	-	-	-	-	-	-	-	-	-	-
Blast Furnaces/Gas Works	-	-	-	-	-	7735	-	-	-	-	-
Coke/Pat. Fuel/BKB Plants	10976	-	-	-	-	-	-	-	-	-	-
Petroleum Refineries	-	-	-	-	-	-	-	66006	530	-	-
Petrochemical Industry	-	-	-	-	-	-	-	-	-	-	-
Liquefaction	-	-	-	-	-	-	-	-	-	-	-
Other Transform. Sector	-	-	-	-	-	-	-	-	-	-	-
ENERGY SECTOR	**-**	**-**	**-**	**-**	**-**	**2**	**-**	**-**	**-**	**-**	**-**
Coal Mines	-	-	-	-	-	-	-	-	-	-	-
Oil and Gas Extraction	-	-	-	-	-	-	-	-	-	-	-
Petroleum Refineries	-	-	-	-	-	-	-	-	-	-	-
Electr., CHP+Heat Plants	-	-	-	-	-	-	-	-	-	-	-
Pumped Storage (Elec.)	-	-	-	-	-	-	-	-	-	-	-
Other Energy Sector	-	-	-	-	-	2	-	-	-	-	-
Distribution Losses	-	-	-	-	-	-	-	-	-	-	-
FINAL CONSUMPTION	**1596**	**1277**	**-**	**-**	**-**	**2414**	**-**	**-**	**-**	**-**	**-**
INDUSTRY SECTOR	**1596**	**1277**	**-**	**-**	**-**	**2219**	**-**	**-**	**-**	**-**	**-**
Iron and Steel	1596	17	-	-	-	1934	-	-	-	-	-
Chemical and Petrochem.	-	328	-	-	-	-	-	-	-	-	-
of which: Feedstocks	-	-	-	-	-	-	-	-	-	-	-
Non-Ferrous Metals	-	-	-	-	-	240	-	-	-	-	-
Non-Metallic Minerals	-	526	-	-	-	30	-	-	-	-	-
Transport Equipment	-	-	-	-	-	-	-	-	-	-	-
Machinery	-	-	-	-	-	-	-	-	-	-	-
Mining and Quarrying	-	-	-	-	-	15	-	-	-	-	-
Food and Tobacco	-	190	-	-	-	-	-	-	-	-	-
Paper, Pulp and Print	-	195	-	-	-	-	-	-	-	-	-
Wood and Wood Products	-	-	-	-	-	-	-	-	-	-	-
Construction	-	-	-	-	-	-	-	-	-	-	-
Textile and Leather	-	5	-	-	-	-	-	-	-	-	-
Non-specified	-	16	-	-	-	-	-	-	-	-	-
TRANSPORT SECTOR	**-**	**-**	**-**	**-**	**-**	**-**	**-**	**-**	**-**	**-**	**-**
Air	-	-	-	-	-	-	-	-	-	-	-
Road	-	-	-	-	-	-	-	-	-	-	-
Rail	-	-	-	-	-	-	-	-	-	-	-
Pipeline Transport	-	-	-	-	-	-	-	-	-	-	-
Internal Navigation	-	-	-	-	-	-	-	-	-	-	-
Non-specified	-	-	-	-	-	-	-	-	-	-	-
OTHER SECTORS	**-**	**-**	**-**	**-**	**-**	**-**	**-**	**-**	**-**	**-**	**-**
Agriculture	-	-	-	-	-	-	-	-	-	-	-
Comm. and Publ. Services	-	-	-	-	-	-	-	-	-	-	-
Residential	-	-	-	-	-	-	-	-	-	-	-
Non-specified	-	-	-	-	-	-	-	-	-	-	-
NON-ENERGY USE	**-**	**-**	**-**	**-**	**-**	**195**	**-**	**-**	**-**	**-**	**-**
in Industry/Trans./Energy	-	-	-	-	-	195	-	-	-	-	-
in Transport	-	-	-	-	-	-	-	-	-	-	-
in Other Sectors	-	-	-	-	-	-	-	-	-	-	-

Brazil / Brésil : 1996

SUPPLY AND CONSUMPTION / *APPROVISIONNEMENT ET DEMANDE*	Oil cont. / *Pétrole cont.* (1000 tonnes)										
	Refinery Gas / *Gaz de raffinerie*	LPG + Ethane / *GPL + éthane*	Motor Gasoline / *Essence moteur*	Aviation Gasoline / *Essence aviation*	Jet Fuel / *Carbu- réacteurs*	Kerosene / *Kérosène*	Gas/ Diesel / *Gazole*	Heavy Fuel Oil / *Fioul lourd*	Naphtha / *Naphta*	Petrol. Coke / *Coke de pétrole*	Other Prod. / *Autres prod.*
Production	-	3110	10061	42	2237	393	21086	12306	4257	-	5874
From Other Sources	-	-	-	-	-	-	-	-	-	-	-
Imports	-	2277	605	-	-	508	3871	1496	2209	-	366
Exports	-	-	-384	-	-	-715	-44	-858	-	-	-250
Intl. Marine Bunkers	-	-	-	-	-	-	-279	-1073	-	-	-
Stock Changes	-	4	-57	-	-	-66	-31	-138	-117	-	-30
DOMESTIC SUPPLY	**-**	**5391**	**10225**	**42**	**2237**	**120**	**24603**	**11733**	**6349**	**-**	**5960**
Transfers	-	346	334	-	-	-	-	-	-1432	-	1102
Statistical Differences	-	-31	-35	-	-	-5	-28	159	81	-	42
TRANSFORMATION	**-**	**-**	**-**	**-**	**-**	**-**	**807**	**1127**	**43**	**-**	**206**
Electricity Plants	-	-	-	-	-	-	807	1127	-	-	206
CHP Plants	-	-	-	-	-	-	-	-	-	-	-
Heat Plants	-	-	-	-	-	-	-	-	-	-	-
Blast Furnaces/Gas Works	-	-	-	-	-	-	-	-	43	-	-
Coke/Pat. Fuel/BKB Plants	-	-	-	-	-	-	-	-	-	-	-
Petroleum Refineries	-	-	-	-	-	-	-	-	-	-	-
Petrochemical Industry	-	-	-	-	-	-	-	-	-	-	-
Liquefaction	-	-	-	-	-	-	-	-	-	-	-
Other Transform. Sector	-	-	-	-	-	-	-	-	-	-	-
ENERGY SECTOR	**-**	**9**	**-**	**-**	**-**	**2**	**154**	**1340**	**9**	**-**	**2027**
Coal Mines	-	-	-	-	-	-	-	-	-	-	-
Oil and Gas Extraction	-	-	-	-	-	-	-	-	-	-	-
Petroleum Refineries	-	9	-	-	-	2	154	1340	-	-	2027
Electr., CHP+Heat Plants	-	-	-	-	-	-	-	-	-	-	-
Pumped Storage (Elec.)	-	-	-	-	-	-	-	-	-	-	-
Other Energy Sector	-	-	-	-	-	-	-	-	9	-	-
Distribution Losses	-	28	52	-	-	11	23	10	25	-	51
FINAL CONSUMPTION	**-**	**5669**	**10472**	**42**	**2237**	**102**	**23591**	**9415**	**4921**	**-**	**4820**
INDUSTRY SECTOR	**-**	**406**	**-**	**-**	**-**	**22**	**413**	**7924**	**4921**	**-**	**1624**
Iron and Steel	-	49	-	-	-	7	11	345	-	-	5
Chemical and Petrochem.	-	12	-	-	-	-	89	1557	4921	-	1067
of which: Feedstocks	-	-	-	-	-	-	-	-	*4921*	-	-
Non-Ferrous Metals	-	32	-	-	-	-	27	628	-	-	452
Non-Metallic Minerals	-	183	-	-	-	2	20	1817	-	-	42
Transport Equipment	-	-	-	-	-	-	-	-	-	-	-
Machinery	-	-	-	-	-	-	-	-	-	-	-
Mining and Quarrying	-	1	-	-	-	2	109	550	-	-	9
Food and Tobacco	-	29	-	-	-	4	32	950	-	-	-
Paper, Pulp and Print	-	10	-	-	-	2	23	844	-	-	35
Wood and Wood Products	-	-	-	-	-	-	-	-	-	-	-
Construction	-	-	-	-	-	-	-	-	-	-	-
Textile and Leather	-	2	-	-	-	-	3	275	-	-	-
Non-specified	-	88	-	-	-	5	99	958	-	-	14
TRANSPORT SECTOR	**-**	**-**	**10472**	**42**	**2237**	**-**	**18981**	**895**	**-**	**-**	**-**
Air	-	-	-	42	2237	-	-	-	-	-	-
Road	-	-	10472	-	-	-	18269	-	-	-	-
Rail	-	-	-	-	-	-	367	-	-	-	-
Pipeline Transport	-	-	-	-	-	-	-	-	-	-	-
Internal Navigation	-	-	-	-	-	-	345	895	-	-	-
Non-specified	-	-	-	-	-	-	-	-	-	-	-
OTHER SECTORS	**-**	**5263**	**-**	**-**	**-**	**48**	**4197**	**596**	**-**	**-**	**15**
Agriculture	-	1	-	-	-	-	4059	73	-	-	-
Comm. and Publ. Services	-	165	-	-	-	1	138	523	-	-	15
Residential	-	5097	-	-	-	47	-	-	-	-	-
Non-specified	-	-	-	-	-	-	-	-	-	-	-
NON-ENERGY USE	**-**	**-**	**-**	**-**	**-**	**32**	**-**	**-**	**-**	**-**	**3181**
in Industry/Transf./Energy	-	-	-	-	-	32	-	-	-	-	3181
in Transport	-	-	-	-	-	-	-	-	-	-	-
in Other Sectors	-	-	-	-	-	-	-	-	-	-	-

Brazil / Brésil : 1996

SUPPLY AND CONSUMPTION / *APPROVISIONNEMENT ET DEMANDE*	Gas / *Gaz* (TJ)				Comb. Renew. & Waste / *En. Re. Comb. & Déchets* (TJ)				(GWh)	(TJ)
	Natural Gas / *Gaz naturel*	Gas Works / *Usines à gaz*	Coke Ovens / *Cokeries*	Blast Furnaces / *Hauts fourneaux*	Solid Biomass / *Biomasse solide*	Gas/Liquids from Biomass / *Gaz/Liquides tirés de biomasse*	Municipal Waste / *Déchets urbains*	Industrial Waste / *Déchets industriels*	Electricity / *Electricité*	Heat / *Chaleur*
Production	222269	6558	69830	-	1374439	271315	-	-	291204	-
From Other Sources	-	-	-	-	-	-	-	-	-	-
Imports	-	-	-	-	209	27429	-	-	36566	-
Exports	-	-	-	-	-	-4523	-	-	-8	-
Intl. Marine Bunkers	-	-	-	-	-	-	-	-	-	-
Stock Changes	-	-	-	-	-6533	-2722	-	-	-	-
DOMESTIC SUPPLY	**222269**	**6558**	**69830**	**-**	**1368115**	**291499**	**-**	**-**	**327762**	**-**
Transfers	-	-	-	-	-	-	-	-	-	-
Statistical Differences	511	-1	-1	-	-3	-	-	-	40	-
TRANSFORMATION	**14656**	**-**	**4366**	**-**	**247553**	**-**	**-**	**-**	**-**	**-**
Electricity Plants	10608	-	4366	-	69999	-	-	-	-	-
CHP Plants	-	-	-	-	-	-	-	-	-	-
Heat Plants	-	-	-	-	-	-	-	-	-	-
Blast Furnaces/Gas Works	4048	-	-	-	-	-	-	-	-	-
Coke/Pat. Fuel/BKB Plants	-	-	-	-	-	-	-	-	-	-
Petroleum Refineries	-	-	-	-	-	-	-	-	-	-
Petrochemical Industry	-	-	-	-	-	-	-	-	-	-
Liquefaction	-	-	-	-	-	-	-	-	-	-
Other Transform. Sector	-	-	-	-	177554	-	-	-	-	-
ENERGY SECTOR	**50484**	**18**	**16841**	**-**	**-**	**-**	**-**	**-**	**9022**	**-**
Coal Mines	-	-	-	-	-	-	-	-	-	-
Oil and Gas Extraction	50484	-	-	-	-	-	-	-	-	-
Petroleum Refineries	-	-	-	-	-	-	-	-	-	-
Electr., CHP+Heat Plants	-	-	-	-	-	-	-	-	9022	-
Pumped Storage (Elec.)	-	-	-	-	-	-	-	-	-	-
Other Energy Sector	-	18	16841	-	-	-	-	-	-	-
Distribution Losses	2373	1220	1804	-	-	-	-	-	50117	-
FINAL CONSUMPTION	**155267**	**5319**	**46818**	**-**	**1120559**	**291499**	**-**	**-**	**268663**	**-**
INDUSTRY SECTOR	**148986**	**146**	**46818**	**-**	**772863**	**-**	**-**	**-**	**129755**	**-**
Iron and Steel	33547	55	46818	-	126214	-	-	-	21167	-
Chemical and Petrochem.	62349	-	-	-	5360	-	-	-	15121	-
of which: Feedstocks	*34990*	-	-	-	-	-	-	-	-	-
Non-Ferrous Metals	1163	-	-	-	25753	-	-	-	28685	-
Non-Metallic Minerals	5304	-	-	-	82789	-	-	-	5828	-
Transport Equipment	-	-	-	-	-	-	-	-	-	-
Machinery	-	-	-	-	-	-	-	-	-	-
Mining and Quarrying	4979	-	-	-	-	-	-	-	5849	-
Food and Tobacco	11493	18	-	-	374121	-	-	-	13630	-
Paper, Pulp and Print	6468	-	-	-	129146	-	-	-	9954	-
Wood and Wood Products	-	-	-	-	-	-	-	-	-	-
Construction	-	-	-	-	-	-	-	-	-	-
Textile and Leather	8422	-	-	-	4397	-	-	-	5980	-
Non-specified	15261	73	-	-	25083	-	-	-	23541	-
TRANSPORT SECTOR	**1442**	**-**	**-**	**-**	**-**	**291499**	**-**	**-**	**1150**	**-**
Air	-	-	-	-	-	-	-	-	-	-
Road	1442	-	-	-	-	291499	-	-	-	-
Rail	-	-	-	-	-	-	-	-	1150	-
Pipeline Transport	-	-	-	-	-	-	-	-	-	-
Internal Navigation	-	-	-	-	-	-	-	-	-	-
Non-specified	-	-	-	-	-	-	-	-	-	-
OTHER SECTORS	**4839**	**5173**	**-**	**-**	**347696**	**-**	**-**	**-**	**137758**	**-**
Agriculture	-	-	-	-	77638	-	-	-	9852	-
Comm. and Publ. Services	1954	1457	-	-	6365	-	-	-	58850	-
Residential	2885	3716	-	-	263693	-	-	-	69056	-
Non-specified	-	-	-	-	-	-	-	-	-	-
NON-ENERGY USE	**-**	**-**	**-**	**-**	**-**	**-**	**-**	**-**	**-**	**-**
in Industry/Transf./Energy	-	-	-	-	-	-	-	-	-	-
in Transport	-	-	-	-	-	-	-	-	-	-
in Other Sectors	-	-	-	-	-	-	-	-	-	-

Brazil / Brésil : 1997

SUPPLY AND CONSUMPTION *APPROVISIONNEMENT ET DEMANDE*	Coal / *Charbon* (1000 tonnes)							Oil / *Pétrole* (1000 tonnes)			
	Coking Coal *Charbon à coke*	Other Bit. Coal *Autres charb. bit.*	Sub-Bit. Coal *Charbon sous-bit.*	Lignite *Lignite*	Peat *Tourbe*	Oven and Gas Coke *Coke de four/gaz*	Pat. Fuel and BKB *Agg./briq. de lignite*	Crude Oil *Pétrole brut*	NGL *LGN*	Feed-stocks *Produits d'aliment.*	Additives *Additifs*
Production	90	5557	-	-	-	8553	-	42191	1041	-	-
From Other Sources	-	-	-	-	-	-	-	-	-	-	-
Imports	12256	-	-	-	-	1708	-	29529	-	-	-
Exports	-	-	-	-	-	-	-	-128	-	-	-
Intl. Marine Bunkers	-	-	-	-	-	-	-	-	-	-	-
Stock Changes	144	-282	-	-	-	-105	-	-1442	-	-	-
DOMESTIC SUPPLY	**12490**	**5275**	**-**	**-**	**-**	**10156**	**-**	**70150**	**1041**	**-**	**-**
Transfers	-	-	-	-	-	-	-	-	-437	-	-
Statistical Differences	-	-	-	-	-	-	-	-	-	-	-
TRANSFORMATION	**10562**	**4266**	**-**	**-**	**-**	**7725**	**-**	**70150**	**604**	**-**	**-**
Electricity Plants	-	4266	-	-	-	-	-	-	-	-	-
CHP Plants	-	-	-	-	-	-	-	-	-	-	-
Heat Plants	-	-	-	-	-	-	-	-	-	-	-
Blast Furnaces/Gas Works	-	-	-	-	-	7725	-	-	-	-	-
Coke/Pat. Fuel/BKB Plants	10562	-	-	-	-	-	-	-	-	-	-
Petroleum Refineries	-	-	-	-	-	-	-	70150	604	-	-
Petrochemical Industry	-	-	-	-	-	-	-	-	-	-	-
Liquefaction	-	-	-	-	-	-	-	-	-	-	-
Other Transform. Sector	-	-	-	-	-	-	-	-	-	-	-
ENERGY SECTOR	**-**	**-**	**-**	**-**	**-**	**-**	**-**	**-**	**-**	**-**	**-**
Coal Mines	-	-	-	-	-	-	-	-	-	-	-
Oil and Gas Extraction	-	-	-	-	-	-	-	-	-	-	-
Petroleum Refineries	-	-	-	-	-	-	-	-	-	-	-
Electr., CHP+Heat Plants	-	-	-	-	-	-	-	-	-	-	-
Pumped Storage (Elec.)	-	-	-	-	-	-	-	-	-	-	-
Other Energy Sector	-	-	-	-	-	-	-	-	-	-	-
Distribution Losses	-	-	-	-	-	-	-	-	-	-	-
FINAL CONSUMPTION	**1928**	**1009**	**-**	**-**	**-**	**2431**	**-**	**-**	**-**	**-**	**-**
INDUSTRY SECTOR	**1928**	**1009**	**-**	**-**	**-**	**2172**	**-**	**-**	**-**	**-**	**-**
Iron and Steel	1928	17	-	-	-	1931	-	-	-	-	-
Chemical and Petrochem.	-	315	-	-	-	-	-	-	-	-	-
of which: Feedstocks	-	-	-	-	-	-	-	-	-	-	-
Non-Ferrous Metals	-	24	-	-	-	201	-	-	-	-	-
Non-Metallic Minerals	-	308	-	-	-	40	-	-	-	-	-
Transport Equipment	-	-	-	-	-	-	-	-	-	-	-
Machinery	-	-	-	-	-	-	-	-	-	-	-
Mining and Quarrying	-	-	-	-	-	-	-	-	-	-	-
Food and Tobacco	-	154	-	-	-	-	-	-	-	-	-
Paper, Pulp and Print	-	182	-	-	-	-	-	-	-	-	-
Wood and Wood Products	-	-	-	-	-	-	-	-	-	-	-
Construction	-	-	-	-	-	-	-	-	-	-	-
Textile and Leather	-	4	-	-	-	-	-	-	-	-	-
Non-specified	-	5	-	-	-	-	-	-	-	-	-
TRANSPORT SECTOR	**-**	**-**	**-**	**-**	**-**	**-**	**-**	**-**	**-**	**-**	**-**
Air	-	-	-	-	-	-	-	-	-	-	-
Road	-	-	-	-	-	-	-	-	-	-	-
Rail	-	-	-	-	-	-	-	-	-	-	-
Pipeline Transport	-	-	-	-	-	-	-	-	-	-	-
Internal Navigation	-	-	-	-	-	-	-	-	-	-	-
Non-specified	-	-	-	-	-	-	-	-	-	-	-
OTHER SECTORS	**-**	**-**	**-**	**-**	**-**	**-**	**-**	**-**	**-**	**-**	**-**
Agriculture	-	-	-	-	-	-	-	-	-	-	-
Comm. and Publ. Services	-	-	-	-	-	-	-	-	-	-	-
Residential	-	-	-	-	-	-	-	-	-	-	-
Non-specified	-	-	-	-	-	-	-	-	-	-	-
NON-ENERGY USE	**-**	**-**	**-**	**-**	**-**	**259**	**-**	**-**	**-**	**-**	**-**
in Industry/Trans./Energy	-	-	-	-	-	259	-	-	-	-	-
in Transport	-	-	-	-	-	-	-	-	-	-	-
in Other Sectors	-	-	-	-	-	-	-	-	-	-	-

Brazil / Brésil : 1997

SUPPLY AND CONSUMPTION *APPROVISIONNEMENT ET DEMANDE*	Oil cont. / *Pétrole cont.* (1000 tonnes)										
	Refinery Gas *Gaz de raffinerie*	LPG + Ethane *GPL + éthane*	Motor Gasoline *Essence moteur*	Aviation Gasoline *Essence aviation*	Jet Fuel *Carbu-réacteurs*	Kerosene *Kérosène*	Gas/ Diesel *Gazole*	Heavy Fuel Oil *Fioul lourd*	Naphtha *Naphta*	Petrol. Coke *Coke de pétrole*	Other Prod. *Autres prod.*
Production	-	3300	11406	48	2517	60	21999	13695	4583	-	6378
From Other Sources	-	-	-	-	-	-	-	-	-	-	-
Imports	-	2386	249	-	-	696	4633	424	3158	-	163
Exports	-	-3	-420	-	-	-781	-186	-775	-	-	-63
Intl. Marine Bunkers	-	-	-	-	-	-	-419	-1289	-	-	-
Stock Changes	-	-52	27	-	-	55	59	161	107	-	73
DOMESTIC SUPPLY	**-**	**5631**	**11262**	**48**	**2517**	**30**	**26086**	**12216**	**7848**	**-**	**6551**
Transfers	-	440	219	-	-	-	31	-	-1620	-	1253
Statistical Differences	-	-146	-24	-	-	40	-30	244	-87	-	-77
TRANSFORMATION	**-**	**-**	**-**	**-**	**-**	**-**	**1105**	**1016**	**43**	**-**	**228**
Electricity Plants	-	-	-	-	-	-	1105	1016	-	-	228
CHP Plants	-	-	-	-	-	-	-	-	-	-	-
Heat Plants	-	-	-	-	-	-	-	-	-	-	-
Blast Furnaces/Gas Works	-	-	-	-	-	-	-	-	43	-	-
Coke/Pat. Fuel/BKB Plants	-	-	-	-	-	-	-	-	-	-	-
Petroleum Refineries	-	-	-	-	-	-	-	-	-	-	-
Petrochemical Industry	-	-	-	-	-	-	-	-	-	-	-
Liquefaction	-	-	-	-	-	-	-	-	-	-	-
Other Transform. Sector	-	-	-	-	-	-	-	-	-	-	-
ENERGY SECTOR	**-**	**15**	**-**	**-**	**-**	**1**	**167**	**1580**	**4**	**-**	**2253**
Coal Mines	-	-	-	-	-	-	-	-	-	-	-
Oil and Gas Extraction	-	-	-	-	-	-	-	-	-	-	-
Petroleum Refineries	-	15	-	-	-	1	167	1580	-	-	2253
Electr., CHP+Heat Plants	-	-	-	-	-	-	-	-	-	-	-
Pumped Storage (Elec.)	-	-	-	-	-	-	-	-	-	-	-
Other Energy Sector	-	-	-	-	-	-	-	-	4	-	-
Distribution Losses	-	30	57	-	-	-	25	11	8	-	136
FINAL CONSUMPTION	**-**	**5880**	**11400**	**48**	**2517**	**69**	**24790**	**9853**	**6086**	**-**	**5110**
INDUSTRY SECTOR	**-**	**540**	**-**	**-**	**-**	**28**	**434**	**8179**	**6086**	**-**	**1794**
Iron and Steel	-	58	-	-	-	5	10	290	-	-	78
Chemical and Petrochem.	-	14	-	-	-	10	88	1719	6086	-	1163
of which: Feedstocks	-	-	-	-	-	-	-	-	*6086*	-	-
Non-Ferrous Metals	-	39	-	-	-	-	28	636	-	-	453
Non-Metallic Minerals	-	248	-	-	-	2	24	2005	-	-	53
Transport Equipment	-	-	-	-	-	-	-	-	-	-	-
Machinery	-	-	-	-	-	-	-	-	-	-	-
Mining and Quarrying	-	3	-	-	-	2	112	535	-	-	31
Food and Tobacco	-	31	-	-	-	2	33	918	-	-	-
Paper, Pulp and Print	-	13	-	-	-	2	20	828	-	-	-
Wood and Wood Products	-	-	-	-	-	-	-	-	-	-	-
Construction	-	-	-	-	-	-	-	-	-	-	-
Textile and Leather	-	3	-	-	-	-	3	277	-	-	-
Non-specified	-	131	-	-	-	5	116	971	-	-	16
TRANSPORT SECTOR	**-**	**-**	**11400**	**48**	**2517**	**-**	**19977**	**1119**	**-**	**-**	**-**
Air	-	-	-	48	2517	-	-	-	-	-	-
Road	-	-	11400	-	-	-	19206	-	-	-	-
Rail	-	-	-	-	-	-	342	-	-	-	-
Pipeline Transport	-	-	-	-	-	-	-	-	-	-	-
Internal Navigation	-	-	-	-	-	-	429	1119	-	-	-
Non-specified	-	-	-	-	-	-	-	-	-	-	-
OTHER SECTORS	**-**	**5340**	**-**	**-**	**-**	**27**	**4379**	**555**	**-**	**-**	**-**
Agriculture	-	3	-	-	-	-	4180	72	-	-	-
Comm. and Publ. Services	-	216	-	-	-	-	199	483	-	-	-
Residential	-	5121	-	-	-	27	-	-	-	-	-
Non-specified	-	-	-	-	-	-	-	-	-	-	-
NON-ENERGY USE	**-**	**-**	**-**	**-**	**-**	**14**	**-**	**-**	**-**	**-**	**3316**
in Industry/Transf./Energy	-	-	-	-	-	14	-	-	-	-	3316
in Transport	-	-	-	-	-	-	-	-	-	-	-
in Other Sectors	-	-	-	-	-	-	-	-	-	-	-

Brazil / Brésil : 1997

SUPPLY AND CONSUMPTION / *APPROVISIONNEMENT ET DEMANDE*	Gas / *Gaz* (TJ)				Comb. Renew. & Waste / *En. Re. Comb. & Déchets* (TJ)				(GWh)	(TJ)
	Natural Gas / *Gaz naturel*	Gas Works / *Usines à gaz*	Coke Ovens / *Cokeries*	Blast Furnaces / *Hauts fourneaux*	Solid Biomass / *Biomasse solide*	Gas/Liquids from Biomass / *Gaz/Liquides tirés de biomasse*	Municipal Waste / *Déchets urbains*	Industrial Waste / *Déchets industriels*	Electricity / *Electricité*	Heat / *Chaleur*
Production	246045	5793	67036	-	1419445	338191	-	-	307302	-
From Other Sources	-	-	-	-	-	-	-	-	-	-
Imports	-	-	-	-	293	18342	-	-	40478	-
Exports	-	-	-	-	-126	-28685	-	-	-8	-
Intl. Marine Bunkers	-	-	-	-	-	-	-	-	-	-
Stock Changes	-	-	-	-	-6616	-47027	-	-	-	-
DOMESTIC SUPPLY	**246045**	**5793**	**67036**	-	**1412996**	**280821**	-	-	**347772**	-
Transfers	-	-	-	-	-	-	-	-	-	-
Statistical Differences	-1862	-2	-1	-	-	-	-	-	-	-
TRANSFORMATION	**15587**	-	**4676**	-	**255684**	-	-	-	-	-
Electricity Plants	11772	-	4676	-	74821	-	-	-	-	-
CHP Plants	-	-	-	-	-	-	-	-	-	-
Heat Plants	-	-	-	-	-	-	-	-	-	-
Blast Furnaces/Gas Works	3815	-	-	-	-	-	-	-	-	-
Coke/Pat. Fuel/BKB Plants	-	-	-	-	-	-	-	-	-	-
Petroleum Refineries	-	-	-	-	-	-	-	-	-	-
Petrochemical Industry	-	-	-	-	-	-	-	-	-	-
Liquefaction	-	-	-	-	-	-	-	-	-	-
Other Transform. Sector	-	-	-	-	180863	-	-	-	-	-
ENERGY SECTOR	**53974**	**528**	**15677**	-	-	-	-	-	**10171**	-
Coal Mines	-	-	-	-	-	-	-	-	-	-
Oil and Gas Extraction	53974	-	-	-	-	-	-	-	-	-
Petroleum Refineries	-	-	-	-	-	-	-	-	-	-
Electr., CHP+Heat Plants	-	-	-	-	-	-	-	-	10171	-
Pumped Storage (Elec.)	-	-	-	-	-	-	-	-	-	-
Other Energy Sector	-	528	15677	-	-	-	-	-	-	-
Distribution Losses	884	619	1261	-	-	-	-	-	52248	-
FINAL CONSUMPTION	**173738**	**4644**	**45421**	-	**1157312**	**280821**	-	-	**285353**	-
INDUSTRY SECTOR	**164991**	**90**	**45421**	-	**807396**	-	-	-	**135702**	-
Iron and Steel	32105	36	45421	-	132203	-	-	-	21540	-
Chemical and Petrochem.	69095	18	-	-	4564	-	-	-	16479	-
of which: Feedstocks	*33175*	-	-	-	-	-	-	-	-	-
Non-Ferrous Metals	1629	-	-	-	24288	-	-	-	28809	-
Non-Metallic Minerals	9259	-	-	-	82705	-	-	-	6410	-
Transport Equipment	-	-	-	-	-	-	-	-	-	-
Machinery	-	-	-	-	-	-	-	-	-	-
Mining and Quarrying	7026	-	-	-	-	-	-	-	6308	-
Food and Tobacco	12702	-	-	-	403015	-	-	-	14248	-
Paper, Pulp and Print	6654	-	-	-	131517	-	-	-	10281	-
Wood and Wood Products	-	-	-	-	-	-	-	-	-	-
Construction	-	-	-	-	-	-	-	-	-	-
Textile and Leather	8747	-	-	-	4188	-	-	-	5847	-
Non-specified	17774	36	-	-	24916	-	-	-	25780	-
TRANSPORT SECTOR	**1861**	-	-	-	-	**280821**	-	-	**1160**	-
Air	-	-	-	-	-	-	-	-	-	-
Road	1861	-	-	-	-	280821	-	-	-	-
Rail	-	-	-	-	-	-	-	-	1160	-
Pipeline Transport	-	-	-	-	-	-	-	-	-	-
Internal Navigation	-	-	-	-	-	-	-	-	-	-
Non-specified	-	-	-	-	-	-	-	-	-	-
OTHER SECTORS	**6886**	**4554**	-	-	**349916**	-	-	-	**148491**	-
Agriculture	-	-	-	-	76800	-	-	-	10635	-
Comm. and Publ. Services	3676	1129	-	-	6282	-	-	-	63760	-
Residential	3210	3425	-	-	266834	-	-	-	74096	-
Non-specified	-	-	-	-	-	-	-	-	-	-
NON-ENERGY USE	-	-	-	-	-	-	-	-	-	-
in Industry/Transf./Energy	-	-	-	-	-	-	-	-	-	-
in Transport	-	-	-	-	-	-	-	-	-	-
in Other Sectors	-	-	-	-	-	-	-	-	-	-

Brunei

SUPPLY AND CONSUMPTION 1996	Coal (1000 tonnes)							Oil (1000 tonnes)			
	Coking Coal	Other Bit. Coal	Sub-Bit. Coal	Lignite	Peat	Oven and Gas Coke	Pat. Fuel and BKB	Crude Oil	NGL	Feed-stocks	Additives
Production	-	-	-	-	-	-	-	7938	650	-	-
Imports	-	-	-	-	-	-	-	-	-	-	-
Exports	-	-	-	-	-	-	-	-7732	-642	-	-
Intl. Marine Bunkers	-	-	-	-	-	-	-	-	-	-	-
Stock Changes	-	-	-	-	-	-	-	85	-	-	-
DOMESTIC SUPPLY	-	-	-	-	-	-	-	**291**	**8**	-	-
Transfers and Stat. Diff.	-	-	-	-	-	-	-	255	-8	-	-
TRANSFORMATION	-	-	-	-	-	-	-	**546**	-	-	-
Electricity and CHP Plants	-	-	-	-	-	-	-	-	-	-	-
Petroleum Refineries	-	-	-	-	-	-	-	546	-	-	-
Other Transform. Sector	-	-	-	-	-	-	-	-	-	-	-
ENERGY SECTOR	-	-	-	-	-	-	-	-	-	-	-
DISTRIBUTION LOSSES	-	-	-	-	-	-	-	-	-	-	-
FINAL CONSUMPTION	-	-	-	-	-	-	-	-	-	-	-
INDUSTRY SECTOR	-	-	-	-	-	-	-	-	-	-	-
Iron and Steel	-	-	-	-	-	-	-	-	-	-	-
Chemical and Petrochem.	-	-	-	-	-	-	-	-	-	-	-
Non-Metallic Minerals	-	-	-	-	-	-	-	-	-	-	-
Non-specified	-	-	-	-	-	-	-	-	-	-	-
TRANSPORT SECTOR	-	-	-	-	-	-	-	-	-	-	-
Air	-	-	-	-	-	-	-	-	-	-	-
Road	-	-	-	-	-	-	-	-	-	-	-
Non-specified	-	-	-	-	-	-	-	-	-	-	-
OTHER SECTORS	-	-	-	-	-	-	-	-	-	-	-
Agriculture	-	-	-	-	-	-	-	-	-	-	-
Comm. and Publ. Services	-	-	-	-	-	-	-	-	-	-	-
Residential	-	-	-	-	-	-	-	-	-	-	-
Non-specified	-	-	-	-	-	-	-	-	-	-	-
NON-ENERGY USE	-	-	-	-	-	-	-	-	-	-	-

APPROVISIONNEMENT ET DEMANDE 1997	*Charbon* (1000 tonnes)							*Pétrole* (1000 tonnes)			
	Charbon à coke	*Autres charb. bit.*	*Charbon sous-bit.*	*Lignite*	*Tourbe*	*Coke de four/gaz*	*Agg./briq. de lignite*	*Pétrole brut*	*LGN*	*Produits d'aliment.*	*Additifs*
Production	-	-	-	-	-	-	-	7825	715	-	-
Imports	-	-	-	-	-	-	-	-	-	-	-
Exports	-	-	-	-	-	-	-	-7300	-699	-	-
Intl. Marine Bunkers	-	-	-	-	-	-	-	-	-	-	-
Stock Changes	-	-	-	-	-	-	-	-174	-	-	-
DOMESTIC SUPPLY	-	-	-	-	-	-	-	**351**	**16**	-	-
Transfers and Stat. Diff.	-	-	-	-	-	-	-	252	-16	-	-
TRANSFORMATION	-	-	-	-	-	-	-	**603**	-	-	-
Electricity and CHP Plants	-	-	-	-	-	-	-	-	-	-	-
Petroleum Refineries	-	-	-	-	-	-	-	603	-	-	-
Other Transform. Sector	-	-	-	-	-	-	-	-	-	-	-
ENERGY SECTOR	-	-	-	-	-	-	-	-	-	-	-
DISTRIBUTION LOSSES	-	-	-	-	-	-	-	-	-	-	-
FINAL CONSUMPTION	-	-	-	-	-	-	-	-	-	-	-
INDUSTRY SECTOR	-	-	-	-	-	-	-	-	-	-	-
Iron and Steel	-	-	-	-	-	-	-	-	-	-	-
Chemical and Petrochem.	-	-	-	-	-	-	-	-	-	-	-
Non-Metallic Minerals	-	-	-	-	-	-	-	-	-	-	-
Non-specified	-	-	-	-	-	-	-	-	-	-	-
TRANSPORT SECTOR	-	-	-	-	-	-	-	-	-	-	-
Air	-	-	-	-	-	-	-	-	-	-	-
Road	-	-	-	-	-	-	-	-	-	-	-
Non-specified	-	-	-	-	-	-	-	-	-	-	-
OTHER SECTORS	-	-	-	-	-	-	-	-	-	-	-
Agriculture	-	-	-	-	-	-	-	-	-	-	-
Comm. and Publ. Services	-	-	-	-	-	-	-	-	-	-	-
Residential	-	-	-	-	-	-	-	-	-	-	-
Non-specified	-	-	-	-	-	-	-	-	-	-	-
NON-ENERGY USE	-	-	-	-	-	-	-	-	-	-	-

Brunei

SUPPLY AND CONSUMPTION 1996	Oil cont. (1000 tonnes)										
	Refinery Gas	LPG + Ethane	Motor Gasoline	Aviation Gasoline	Jet Fuel	Kerosene	Gas/ Diesel	Heavy Fuel Oil	Naphtha	Petrol. Coke	Other Prod.
Production	55	9	189	-	58	5	127	1	6	-	-
Imports	-	-	6	-	17	-	38	-	-	-	28
Exports	-	-	-	-	-	-	-	-	-	-	-1
Intl. Marine Bunkers	-	-	-	-	-	-	-	-	-	-	-
Stock Changes	-	-	-	-	-4	-	-	-	-	-	-
DOMESTIC SUPPLY	**55**	**9**	**195**	**-**	**71**	**5**	**165**	**1**	**6**	**-**	**27**
Transfers and Stat. Diff.	-	8	-	-	1	-	-15	-	-	-	-
TRANSFORMATION	**-**	**-**	**-**	**-**	**-**	**-**	**4**	**-**	**-**	**-**	**-**
Electricity and CHP Plants	-	-	-	-	-	-	4	-	-	-	-
Petroleum Refineries	-	-	-	-	-	-	-	-	-	-	-
Other Transform. Sector	-	-	-	-	-	-	-	-	-	-	-
ENERGY SECTOR	**55**	**-**	**-**	**-**	**-**	**-**	**-**	**-**	**-**	**-**	**-**
DISTRIBUTION LOSSES	**-**	**-**	**4**	**-**	**1**	**-**	**-**	**-**	**-**	**-**	**-**
FINAL CONSUMPTION	**-**	**17**	**191**	**-**	**71**	**5**	**146**	**1**	**6**	**-**	**27**
INDUSTRY SECTOR	**-**	**-**	**-**	**-**	**-**	**-**	**69**	**1**	**6**	**-**	**-**
Iron and Steel	-	-	-	-	-	-	-	-	-	-	-
Chemical and Petrochem.	-	-	-	-	-	-	-	-	6	-	-
Non-Metallic Minerals	-	-	-	-	-	-	-	-	-	-	-
Non-specified	-	-	-	-	-	-	69	1	-	-	-
TRANSPORT SECTOR	**-**	**-**	**191**	**-**	**71**	**3**	**76**	**-**	**-**	**-**	**-**
Air	-	-	-	-	71	-	-	-	-	-	-
Road	-	-	191	-	-	-	76	-	-	-	-
Non-specified	-	-	-	-	-	3	-	-	-	-	-
OTHER SECTORS	**-**	**17**	**-**	**-**	**-**	**2**	**1**	**-**	**-**	**-**	**-**
Agriculture	-	-	-	-	-	-	-	-	-	-	-
Comm. and Publ. Services	-	-	-	-	-	-	-	-	-	-	-
Residential	-	17	-	-	-	2	1	-	-	-	-
Non-specified	-	-	-	-	-	-	-	-	-	-	-
NON-ENERGY USE	**-**	**-**	**-**	**-**	**-**	**-**	**-**	**-**	**-**	**-**	**27**

APPROVISIONNEMENT ET DEMANDE 1997	*Pétrole cont.* (1000 tonnes)										
	Gaz de raffinerie	*GPL + éthane*	*Essence moteur*	*Essence aviation*	*Carbu- réacteurs*	*Kérosène*	*Gazole*	*Fioul lourd*	*Naphta*	*Coke de pétrole*	*Autres prod.*
Production	59	2	194	-	68	4	142	1	6	-	-
Imports	-	-	7	-	28	-	43	-	-	-	25
Exports	-	-	-	-	-	-	-	-	-	-	-1
Intl. Marine Bunkers	-	-	-	-	-	-	-	-	-	-	-
Stock Changes	-	-	-	-	-12	-	-2	-	-	-	-
DOMESTIC SUPPLY	**59**	**2**	**201**	**-**	**84**	**4**	**183**	**1**	**6**	**-**	**24**
Transfers and Stat. Diff.	-	16	-1	-	1	-1	-10	-	-	-	-
TRANSFORMATION	**-**	**-**	**-**	**-**	**-**	**-**	**5**	**-**	**-**	**-**	**-**
Electricity and CHP Plants	-	-	-	-	-	-	5	-	-	-	-
Petroleum Refineries	-	-	-	-	-	-	-	-	-	-	-
Other Transform. Sector	-	-	-	-	-	-	-	-	-	-	-
ENERGY SECTOR	**59**	**-**	**-**	**-**	**-**	**-**	**-**	**-**	**-**	**-**	**-**
DISTRIBUTION LOSSES	**-**	**-**	**5**	**-**	**1**	**-**	**-**	**-**	**-**	**-**	**-**
FINAL CONSUMPTION	**-**	**18**	**195**	**-**	**84**	**3**	**168**	**1**	**6**	**-**	**24**
INDUSTRY SECTOR	**-**	**-**	**-**	**-**	**-**	**-**	**81**	**1**	**6**	**-**	**-**
Iron and Steel	-	-	-	-	-	-	-	-	-	-	-
Chemical and Petrochem.	-	-	-	-	-	-	-	-	6	-	-
Non-Metallic Minerals	-	-	-	-	-	-	-	-	-	-	-
Non-specified	-	-	-	-	-	-	81	1	-	-	-
TRANSPORT SECTOR	**-**	**-**	**195**	**-**	**84**	**2**	**84**	**-**	**-**	**-**	**-**
Air	-	-	-	-	84	-	-	-	-	-	-
Road	-	-	195	-	-	-	84	-	-	-	-
Non-specified	-	-	-	-	-	2	-	-	-	-	-
OTHER SECTORS	**-**	**18**	**-**	**-**	**-**	**1**	**3**	**-**	**-**	**-**	**-**
Agriculture	-	-	-	-	-	-	1	-	-	-	-
Comm. and Publ. Services	-	-	-	-	-	-	-	-	-	-	-
Residential	-	18	-	-	-	1	2	-	-	-	-
Non-specified	-	-	-	-	-	-	-	-	-	-	-
NON-ENERGY USE	**-**	**-**	**-**	**-**	**-**	**-**	**-**	**-**	**-**	**-**	**24**

Brunei

SUPPLY AND CONSUMPTION 1996	Gas (TJ)				Comb. Renew. & Waste (TJ)				(GWh)	(TJ)
	Natural Gas	Gas Works	Coke Ovens	Blast Furnaces	Solid Biomass	Gas/Liquids from Biomass	Municipal Waste	Industrial Waste	Electricity	Heat
Production	412233	-	-	-	772	-	-	-	2123	-
Imports	-	-	-	-	-	-	-	-	-	-
Exports	-338687	-	-	-	-	-	-	-	-	-
Intl. Marine Bunkers	-	-	-	-	-	-	-	-	-	-
Stock Changes	-	-	-	-	-	-	-	-	-	-
DOMESTIC SUPPLY	**73546**	-	-	-	**772**	-	-	-	**2123**	-
Transfers and Stat. Diff.	15228	-	-	-	-	-	-	-	-1	-
TRANSFORMATION	**34246**	-	-	-	-	-	-	-	-	-
Electricity and CHP Plants	34246	-	-	-	-	-	-	-	-	-
Petroleum Refineries	-	-	-	-	-	-	-	-	-	-
Other Transform. Sector	-	-	-	-	-	-	-	-	-	-
ENERGY SECTOR	**32250**	-	-	-	-	-	-	-	**86**	-
DISTRIBUTION LOSSES	**22278**	-	-	-	-	-	-	-	**43**	-
FINAL CONSUMPTION	-	-	-	-	**772**	-	-	-	**1993**	-
INDUSTRY SECTOR	-	-	-	-	-	-	-	-	**281**	-
Iron and Steel	-	-	-	-	-	-	-	-	-	-
Chemical and Petrochem.	-	-	-	-	-	-	-	-	-	-
Non-Metallic Minerals	-	-	-	-	-	-	-	-	-	-
Non-specified	-	-	-	-	-	-	-	-	281	-
TRANSPORT SECTOR	-	-	-	-	-	-	-	-	-	-
Air	-	-	-	-	-	-	-	-	-	-
Road	-	-	-	-	-	-	-	-	-	-
Non-specified	-	-	-	-	-	-	-	-	-	-
OTHER SECTORS	-	-	-	-	**772**	-	-	-	**1712**	-
Agriculture	-	-	-	-	-	-	-	-	-	-
Comm. and Publ. Services	-	-	-	-	-	-	-	-	1227	-
Residential	-	-	-	-	-	-	-	-	485	-
Non-specified	-	-	-	-	772	-	-	-	-	-
NON-ENERGY USE	-	-	-	-	-	-	-	-	-	-

APPROVISIONNEMENT ET DEMANDE 1997	*Gaz* (TJ)				*En. Re. Comb. & Déchets* (TJ)				(GWh)	(TJ)
	Gaz naturel	*Usines à gaz*	*Cokeries*	*Hauts fourneaux*	*Biomasse solide*	*Gaz/Liquides tirés de biomasse*	*Déchets urbains*	*Déchets industriels*	*Electricité*	*Chaleur*
Production	412149	-	-	-	772	-	-	-	2407	-
Imports	-	-	-	-	-	-	-	-	-	-
Exports	-336577	-	-	-	-	-	-	-	-	-
Intl. Marine Bunkers	-	-	-	-	-	-	-	-	-	-
Stock Changes	-	-	-	-	-	-	-	-	-	-
DOMESTIC SUPPLY	**75572**	-	-	-	**772**	-	-	-	**2407**	-
Transfers and Stat. Diff.	14996	-	-	-	-	-	-	-	1	-
TRANSFORMATION	**38848**	-	-	-	-	-	-	-	-	-
Electricity and CHP Plants	38848	-	-	-	-	-	-	-	-	-
Petroleum Refineries	-	-	-	-	-	-	-	-	-	-
Other Transform. Sector	-	-	-	-	-	-	-	-	-	-
ENERGY SECTOR	**32711**	-	-	-	-	-	-	-	**58**	-
DISTRIBUTION LOSSES	**19009**	-	-	-	-	-	-	-	**29**	-
FINAL CONSUMPTION	-	-	-	-	**772**	-	-	-	**2321**	-
INDUSTRY SECTOR	-	-	-	-	-	-	-	-	**263**	-
Iron and Steel	-	-	-	-	-	-	-	-	-	-
Chemical and Petrochem.	-	-	-	-	-	-	-	-	-	-
Non-Metallic Minerals	-	-	-	-	-	-	-	-	-	-
Non-specified	-	-	-	-	-	-	-	-	263	-
TRANSPORT SECTOR	-	-	-	-	-	-	-	-	-	-
Air	-	-	-	-	-	-	-	-	-	-
Road	-	-	-	-	-	-	-	-	-	-
Non-specified	-	-	-	-	-	-	-	-	-	-
OTHER SECTORS	-	-	-	-	**772**	-	-	-	**2058**	-
Agriculture	-	-	-	-	-	-	-	-	-	-
Comm. and Publ. Services	-	-	-	-	-	-	-	-	1522	-
Residential	-	-	-	-	-	-	-	-	536	-
Non-specified	-	-	-	-	772	-	-	-	-	-
NON-ENERGY USE	-	-	-	-	-	-	-	-	-	-

Bulgaria / Bulgarie : 1996

SUPPLY AND CONSUMPTION / *APPROVISIONNEMENT ET DEMANDE*	Coal / *Charbon* (1000 tonnes)							Oil / *Pétrole* (1000 tonnes)			
	Coking Coal / *Charbon à coke*	Other Bit. Coal / *Autres charb. bit.*	Sub-Bit. Coal / *Charbon sous-bit.*	Lignite / *Lignite*	Peat / *Tourbe*	Oven and Gas Coke / *Coke de four/gaz*	Pat. Fuel and BKB / *Agg./briq. de lignite*	Crude Oil / *Pétrole brut*	NGL / *LGN*	Feed-stocks / *Produits d'aliment.*	Additives / *Additifs*
Production	-	138	3060	28104	-	1087	1201	32	-	-	-
From Other Sources	-	-	-	-	-	-	-	-	-	-	-
Imports	1438	2275	-	-	-	141	-	6996	-	-	-
Exports	-	-	-	-	-	-6	-	-	-	-	-
Intl. Marine Bunkers	-	-	-	-	-	-	-	-	-	-	-
Stock Changes	53	-396	-171	-282	-	20	-2	-74	-	-	-
DOMESTIC SUPPLY	**1491**	**2017**	**2889**	**27822**	**-**	**1242**	**1199**	**6954**	**-**	**-**	**-**
Transfers	-	-	-	-	-	-	-	-	-	-	-
Statistical Differences	-	19	105	81	-	-23	15	1	-	-	-
TRANSFORMATION	**1491**	**1911**	**2391**	**27705**	**-**	**833**	**154**	**6955**	**-**	**-**	**-**
Electricity Plants	-	311	1450	21663	-	-	10	-	-	-	-
CHP Plants	-	1582	894	3333	-	-	139	-	-	-	-
Heat Plants	-	18	47	12	-	-	5	1	-	-	-
Blast Furnaces/Gas Works	-	-	-	-	-	833	-	-	-	-	-
Coke/Pat. Fuel/BKB Plants	1491	-	-	2697	-	-	-	-	-	-	-
Petroleum Refineries	-	-	-	-	-	-	-	6954	-	-	-
Petrochemical Industry	-	-	-	-	-	-	-	-	-	-	-
Liquefaction	-	-	-	-	-	-	-	-	-	-	-
Other Transform. Sector	-	-	-	-	-	-	-	-	-	-	-
ENERGY SECTOR	**-**	**-**	**15**	**14**	**-**	**-**	**3**	**-**	**-**	**-**	**-**
Coal Mines	-	-	15	14	-	-	3	-	-	-	-
Oil and Gas Extraction	-	-	-	-	-	-	-	-	-	-	-
Petroleum Refineries	-	-	-	-	-	-	-	-	-	-	-
Electr., CHP+Heat Plants	-	-	-	-	-	-	-	-	-	-	-
Pumped Storage (Elec.)	-	-	-	-	-	-	-	-	-	-	-
Other Energy Sector	-	-	-	-	-	-	-	-	-	-	-
Distribution Losses	-	-	-	-	-	-	-	-	-	-	-
FINAL CONSUMPTION	**-**	**125**	**588**	**184**	**-**	**386**	**1057**	**-**	**-**	**-**	**-**
INDUSTRY SECTOR	**-**	**120**	**72**	**5**	**-**	**386**	**6**	**-**	**-**	**-**	**-**
Iron and Steel	-	19	-	-	-	208	-	-	-	-	-
Chemical and Petrochem.	-	14	-	-	-	96	-	-	-	-	-
of which: Feedstocks	-	-	-	-	-	-	-	-	-	-	-
Non-Ferrous Metals	-	-	-	-	-	72	-	-	-	-	-
Non-Metallic Minerals	-	66	57	4	-	-	3	-	-	-	-
Transport Equipment	-	-	-	-	-	1	-	-	-	-	-
Machinery	-	2	1	-	-	8	1	-	-	-	-
Mining and Quarrying	-	15	1	-	-	-	-	-	-	-	-
Food and Tobacco	-	1	3	1	-	1	-	-	-	-	-
Paper, Pulp and Print	-	-	-	-	-	-	-	-	-	-	-
Wood and Wood Products	-	-	-	-	-	-	-	-	-	-	-
Construction	-	2	7	-	-	-	1	-	-	-	-
Textile and Leather	-	1	2	-	-	-	1	-	-	-	-
Non-specified	-	-	1	-	-	-	-	-	-	-	-
TRANSPORT SECTOR	**-**	**1**	**4**	**1**	**-**	**-**	**2**	**-**	**-**	**-**	**-**
Air	-	-	-	-	-	-	-	-	-	-	-
Road	-	-	-	-	-	-	-	-	-	-	-
Rail	-	1	4	1	-	-	2	-	-	-	-
Pipeline Transport	-	-	-	-	-	-	-	-	-	-	-
Internal Navigation	-	-	-	-	-	-	-	-	-	-	-
Non-specified	-	-	-	-	-	-	-	-	-	-	-
OTHER SECTORS	**-**	**4**	**512**	**178**	**-**	**-**	**1049**	**-**	**-**	**-**	**-**
Agriculture	-	1	2	-	-	-	3	-	-	-	-
Comm. and Publ. Services	-	3	2	3	-	-	3	-	-	-	-
Residential	-	-	450	162	-	-	1042	-	-	-	-
Non-specified	-	-	58	13	-	-	1	-	-	-	-
NON-ENERGY USE	**-**	**-**	**-**	**-**	**-**	**-**	**-**	**-**	**-**	**-**	**-**
in Industry/Trans./Energy	-	-	-	-	-	-	-	-	-	-	-
in Transport	-	-	-	-	-	-	-	-	-	-	-
in Other Sectors	-	-	-	-	-	-	-	-	-	-	-

Bulgaria / Bulgarie : 1996

SUPPLY AND CONSUMPTION / *APPROVISIONNEMENT ET DEMANDE*	Oil cont. / *Pétrole cont.* (1000 tonnes)										
	Refinery Gas / *Gaz de raffinerie*	LPG + Ethane / *GPL + éthane*	Motor Gasoline / *Essence moteur*	Aviation Gasoline / *Essence aviation*	Jet Fuel / *Carbu-réacteurs*	Kerosene / *Kérosène*	Gas/ Diesel / *Gazole*	Heavy Fuel Oil / *Fioul lourd*	Naphtha / *Naphta*	Petrol. Coke / *Coke de pétrole*	Other Prod. / *Autres prod.*
Production	137	93	1266	-	184	-	2351	1735	642	-	314
From Other Sources	-	-	-	-	-	-	-	-	-	-	-
Imports	-	-	1	1	81	-	85	215	54	-	28
Exports	-	-26	-298	-	-41	-	-1190	-29	-18	-	-6
Intl. Marine Bunkers	-	-	-	-	-	-	-55	-184	-	-	-
Stock Changes	-	-	-10	-	-7	-	29	-48	-	-	-
DOMESTIC SUPPLY	**137**	**67**	**959**	**1**	**217**	**-**	**1220**	**1689**	**678**	**-**	**336**
Transfers	-	-	-	-	-	-	-	-1	-	-	1
Statistical Differences	-	-	-27	-	2	-	-51	75	-	-	-
TRANSFORMATION	**15**	**-**	**-**	**-**	**-**	**-**	**9**	**1360**	**-**	**-**	**189**
Electricity Plants	-	-	-	-	-	-	-	29	-	-	-
CHP Plants	15	-	-	-	-	-	2	541	-	-	189
Heat Plants	-	-	-	-	-	-	7	790	-	-	-
Blast Furnaces/Gas Works	-	-	-	-	-	-	-	-	-	-	-
Coke/Pat. Fuel/BKB Plants	-	-	-	-	-	-	-	-	-	-	-
Petroleum Refineries	-	-	-	-	-	-	-	-	-	-	-
Petrochemical Industry	-	-	-	-	-	-	-	-	-	-	-
Liquefaction	-	-	-	-	-	-	-	-	-	-	-
Other Transform. Sector	-	-	-	-	-	-	-	-	-	-	-
ENERGY SECTOR	**101**	**-**	**9**	**-**	**-**	**-**	**28**	**150**	**-**	**-**	**-**
Coal Mines	-	-	2	-	-	-	21	-	-	-	-
Oil and Gas Extraction	-	-	1	-	-	-	2	-	-	-	-
Petroleum Refineries	101	-	-	-	-	-	-	149	-	-	-
Electr., CHP+Heat Plants	-	-	6	-	-	-	5	1	-	-	-
Pumped Storage (Elec.)	-	-	-	-	-	-	-	-	-	-	-
Other Energy Sector	-	-	-	-	-	-	-	-	-	-	-
Distribution Losses	-	-	3	-	-	-	1	1	-	-	-
FINAL CONSUMPTION	**21**	**67**	**920**	**1**	**219**	**-**	**1131**	**252**	**678**	**-**	**148**
INDUSTRY SECTOR	**21**	**10**	**78**	**-**	**-**	**-**	**217**	**227**	**678**	**-**	**137**
Iron and Steel	-	-	2	-	-	-	11	13	-	-	-
Chemical and Petrochem.	21	-	5	-	-	-	13	36	678	-	93
of which: Feedstocks	-	-	-	-	-	-	-	-	*678*	-	-
Non-Ferrous Metals	-	1	-	-	-	-	3	35	-	-	-
Non-Metallic Minerals	-	5	3	-	-	-	18	101	-	-	-
Transport Equipment	-	-	1	-	-	-	4	1	-	-	-
Machinery	-	3	12	-	-	-	12	6	-	-	-
Mining and Quarrying	-	-	2	-	-	-	38	6	-	-	-
Food and Tobacco	-	-	13	-	-	-	23	12	-	-	-
Paper, Pulp and Print	-	-	1	-	-	-	2	3	-	-	-
Wood and Wood Products	-	1	1	-	-	-	2	3	-	-	-
Construction	-	-	26	-	-	-	81	8	-	-	-
Textile and Leather	-	-	5	-	-	-	5	2	-	-	-
Non-specified	-	-	7	-	-	-	5	1	-	-	44
TRANSPORT SECTOR	**-**	**-**	**822**	**1**	**187**	**-**	**399**	**9**	**-**	**-**	**-**
Air	-	-	-	1	183	-	-	-	-	-	-
Road	-	-	818	-	-	-	352	6	-	-	-
Rail	-	-	-	-	-	-	13	2	-	-	-
Pipeline Transport	-	-	-	-	-	-	-	-	-	-	-
Internal Navigation	-	-	-	-	-	-	6	-	-	-	-
Non-specified	-	-	4	-	4	-	28	1	-	-	-
OTHER SECTORS	**-**	**57**	**20**	**-**	**32**	**-**	**515**	**16**	**-**	**-**	**-**
Agriculture	-	-	20	-	-	-	270	8	-	-	-
Comm. and Publ. Services	-	-	-	-	-	-	39	4	-	-	-
Residential	-	57	-	-	-	-	172	-	-	-	-
Non-specified	-	-	-	-	32	-	34	4	-	-	-
NON-ENERGY USE	**-**	**-**	**-**	**-**	**-**	**-**	**-**	**-**	**-**	**-**	**11**
in Industry/Transf./Energy	-	-	-	-	-	-	-	-	-	-	11
in Transport	-	-	-	-	-	-	-	-	-	-	-
in Other Sectors	-	-	-	-	-	-	-	-	-	-	-

Bulgaria / Bulgarie : 1996

	Gas / *Gaz* (TJ)				Comb. Renew. & Waste / *En. Re. Comb. & Déchets* (TJ)				(GWh)	(TJ)
SUPPLY AND CONSUMPTION *APPROVISIONNEMENT ET DEMANDE*	Natural Gas *Gaz naturel*	Gas Works *Usines à gaz*	Coke Ovens *Cokeries*	Blast Furnaces *Hauts fourneaux*	Solid Biomass *Biomasse solide*	Gas/Liquids from Biomass *Gaz/Liquides tirés de biomasse*	Municipal Waste *Déchets urbains*	Industrial Waste *Déchets industriels*	Electricity *Electricité*	Heat *Chaleur*
Production	1534	-	8256	10946	10471	-	-	-	42716	140120
From Other Sources	-	-	-	-	-	-	-	-	-	-
Imports	220040	-	-	-	-	-	-	-	1803	-
Exports	-	-	-	-	-306	-	-	-	-2252	-
Intl. Marine Bunkers	-	-	-	-	-	-	-	-	-	-
Stock Changes	-3981	-	-	-	-117	-	-	-	-	-
DOMESTIC SUPPLY	**217593**	**-**	**8256**	**10946**	**10048**	**-**	**-**	**-**	**42267**	**140120**
Transfers	-	-	-	-	-	-	-	-	-	-
Statistical Differences	-1256	-	-	-	-250	-	-	-	-308	7925
TRANSFORMATION	**109736**	**-**	**2450**	**4127**	**1551**	**-**	**-**	**-**	**-**	**-**
Electricity Plants	-	-	-	-	-	-	-	-	-	-
CHP Plants	75388	-	2450	4127	-	-	-	-	-	-
Heat Plants	33292	-	-	-	961	-	-	-	-	-
Blast Furnaces/Gas Works	-	-	-	-	-	-	-	-	-	-
Coke/Pat. Fuel/BKB Plants	-	-	-	-	-	-	-	-	-	-
Petroleum Refineries	-	-	-	-	-	-	-	-	-	-
Petrochemical Industry	-	-	-	-	-	-	-	-	-	-
Liquefaction	-	-	-	-	-	-	-	-	-	-
Other Transform. Sector	1056	-	-	-	590	-	-	-	-	-
ENERGY SECTOR	**3712**	**-**	**4268**	**-**	**1**	**-**	**-**	**-**	**6492**	**16025**
Coal Mines	-	-	-	-	1	-	-	-	757	832
Oil and Gas Extraction	113	-	-	-	-	-	-	-	8	23
Petroleum Refineries	3595	-	-	-	-	-	-	-	347	3923
Electr., CHP+Heat Plants	4	-	-	-	-	-	-	-	4282	1833
Pumped Storage (Elec.)	-	-	-	-	-	-	-	-	-	-
Other Energy Sector	-	-	4268	-	-	-	-	-	1098	9414
Distribution Losses	4350	-	-	-	22	-	-	-	5577	9540
FINAL CONSUMPTION	**98539**	**-**	**1538**	**6819**	**8224**	**-**	**-**	**-**	**29890**	**122480**
INDUSTRY SECTOR	**96901**	**-**	**1538**	**6819**	**196**	**-**	**-**	**-**	**12258**	**82810**
Iron and Steel	12941	-	1538	6819	2	-	-	-	1969	5218
Chemical and Petrochem.	56813	-	-	-	7	-	-	-	3583	41657
of which: Feedstocks	*29387*	-	-	-	-	-	-	-	-	-
Non-Ferrous Metals	172	-	-	-	3	-	-	-	956	1668
Non-Metallic Minerals	25618	-	-	-	8	-	-	-	948	4534
Transport Equipment	-	-	-	-	7	-	-	-	166	375
Machinery	1128	-	-	-	38	-	-	-	1393	4372
Mining and Quarrying	-	-	-	-	5	-	-	-	868	560
Food and Tobacco	152	-	-	-	27	-	-	-	1103	11531
Paper, Pulp and Print	64	-	-	-	1	-	-	-	342	4879
Wood and Wood Products	-	-	-	-	47	-	-	-	128	2049
Construction	6	-	-	-	30	-	-	-	250	130
Textile and Leather	-	-	-	-	13	-	-	-	471	5026
Non-specified	7	-	-	-	8	-	-	-	81	811
TRANSPORT SECTOR	**-**	**-**	**-**	**-**	**10**	**-**	**-**	**-**	**811**	**-**
Air	-	-	-	-	-	-	-	-	-	-
Road	-	-	-	-	-	-	-	-	-	-
Rail	-	-	-	-	10	-	-	-	425	-
Pipeline Transport	-	-	-	-	-	-	-	-	-	-
Internal Navigation	-	-	-	-	-	-	-	-	-	-
Non-specified	-	-	-	-	-	-	-	-	386	-
OTHER SECTORS	**1638**	**-**	**-**	**-**	**8018**	**-**	**-**	**-**	**16821**	**39670**
Agriculture	-	-	-	-	92	-	-	-	600	3236
Comm. and Publ. Services	1466	-	-	-	112	-	-	-	1637	886
Residential	-	-	-	-	7181	-	-	-	11486	27827
Non-specified	172	-	-	-	633	-	-	-	3098	7721
NON-ENERGY USE	**-**	**-**	**-**	**-**	**-**	**-**	**-**	**-**	**-**	**-**
in Industry/Transf./Energy	-	-	-	-	-	-	-	-	-	-
in Transport	-	-	-	-	-	-	-	-	-	-
in Other Sectors	-	-	-	-	-	-	-	-	-	-

Bulgaria / Bulgarie : 1997

SUPPLY AND CONSUMPTION / *APPROVISIONNEMENT ET DEMANDE*	Coal / *Charbon* (1000 tonnes)							Oil / *Pétrole* (1000 tonnes)			
	Coking Coal / *Charbon à coke*	Other Bit. Coal / *Autres charb. bit.*	Sub-Bit. Coal / *Charbon sous-bit.*	Lignite / *Lignite*	Peat / *Tourbe*	Oven and Gas Coke / *Coke de four/gaz*	Pat. Fuel and BKB / *Agg./briq. de lignite*	Crude Oil / *Pétrole brut*	NGL / *LGN*	Feed-stocks / *Produits d'aliment.*	Additives / *Additifs*
Production	-	102	2677	26929	-	1165	1082	28	-	-	-
From Other Sources	-	-	-	-	-	-	-	-	-	-	-
Imports	1683	2029	-	-	-	153	-	5888	-	-	-
Exports	-	-	-	-	-	-5	-	-	-	-	-
Intl. Marine Bunkers	-	-	-	-	-	-	-	-	-	-	-
Stock Changes	-27	298	85	237	-	-12	2	49	-	-	-
DOMESTIC SUPPLY	**1656**	**2429**	**2762**	**27166**	**-**	**1301**	**1084**	**5965**	**-**	**-**	**-**
Transfers	-	-	-	-	-	-	-	-	-	-	-
Statistical Differences	-	33	119	206	-	9	6	1	-	-	-
TRANSFORMATION	**1656**	**2370**	**2483**	**27230**	**-**	**918**	**139**	**5966**	**-**	**-**	**-**
Electricity Plants	-	147	1555	21057	-	-	5	-	-	-	-
CHP Plants	-	2209	879	3713	-	-	130	-	-	-	-
Heat Plants	-	14	49	11	-	-	4	1	-	-	-
Blast Furnaces/Gas Works	-	-	-	-	-	918	-	-	-	-	-
Coke/Pat. Fuel/BKB Plants	1656	-	-	2449	-	-	-	-	-	-	-
Petroleum Refineries	-	-	-	-	-	-	-	5965	-	-	-
Petrochemical Industry	-	-	-	-	-	-	-	-	-	-	-
Liquefaction	-	-	-	-	-	-	-	-	-	-	-
Other Transform. Sector	-	-	-	-	-	-	-	-	-	-	-
ENERGY SECTOR	**-**	**1**	**14**	**14**	**-**	**-**	**2**	**-**	**-**	**-**	**-**
Coal Mines	-	1	14	14	-	-	2	-	-	-	-
Oil and Gas Extraction	-	-	-	-	-	-	-	-	-	-	-
Petroleum Refineries	-	-	-	-	-	-	-	-	-	-	-
Electr., CHP+Heat Plants	-	-	-	-	-	-	-	-	-	-	-
Pumped Storage (Elec.)	-	-	-	-	-	-	-	-	-	-	-
Other Energy Sector	-	-	-	-	-	-	-	-	-	-	-
Distribution Losses	-	-	-	-	-	-	-	-	-	-	-
FINAL CONSUMPTION	**-**	**91**	**384**	**128**	**-**	**392**	**949**	**-**	**-**	**-**	**-**
INDUSTRY SECTOR	**-**	**86**	**47**	**3**	**-**	**392**	**5**	**-**	**-**	**-**	**-**
Iron and Steel	-	13	-	-	-	229	-	-	-	-	-
Chemical and Petrochem.	-	15	1	-	-	85	-	-	-	-	-
of which: Feedstocks	-	-	-	-	-	-	-	-	-	-	-
Non-Ferrous Metals	-	-	-	-	-	70	-	-	-	-	-
Non-Metallic Minerals	-	46	35	2	-	-	1	-	-	-	-
Transport Equipment	-	-	-	-	-	-	-	-	-	-	-
Machinery	-	3	2	-	-	7	1	-	-	-	-
Mining and Quarrying	-	6	-	-	-	-	-	-	-	-	-
Food and Tobacco	-	1	1	1	-	1	1	-	-	-	-
Paper, Pulp and Print	-	-	-	-	-	-	-	-	-	-	-
Wood and Wood Products	-	-	1	-	-	-	-	-	-	-	-
Construction	-	-	6	-	-	-	1	-	-	-	-
Textile and Leather	-	1	1	-	-	-	1	-	-	-	-
Non-specified	-	1	-	-	-	-	-	-	-	-	-
TRANSPORT SECTOR	**-**	**2**	**3**	**-**	**-**	**-**	**1**	**-**	**-**	**-**	**-**
Air	-	-	-	-	-	-	-	-	-	-	-
Road	-	-	-	-	-	-	-	-	-	-	-
Rail	-	2	3	-	-	-	1	-	-	-	-
Pipeline Transport	-	-	-	-	-	-	-	-	-	-	-
Internal Navigation	-	-	-	-	-	-	-	-	-	-	-
Non-specified	-	-	-	-	-	-	-	-	-	-	-
OTHER SECTORS	**-**	**3**	**334**	**125**	**-**	**-**	**943**	**-**	**-**	**-**	**-**
Agriculture	-	1	-	-	-	-	2	-	-	-	-
Comm. and Publ. Services	-	2	1	3	-	-	1	-	-	-	-
Residential	-	-	306	122	-	-	939	-	-	-	-
Non-specified	-	-	27	-	-	-	1	-	-	-	-
NON-ENERGY USE	**-**	**-**	**-**	**-**	**-**	**-**	**-**	**-**	**-**	**-**	**-**
in Industry/Trans./Energy	-	-	-	-	-	-	-	-	-	-	-
in Transport	-	-	-	-	-	-	-	-	-	-	-
in Other Sectors	-	-	-	-	-	-	-	-	-	-	-

Bulgaria / Bulgarie : 1997

SUPPLY AND CONSUMPTION / *APPROVISIONNEMENT ET DEMANDE*	Oil cont. / *Pétrole cont.* (1000 tonnes)										
	Refinery Gas / *Gaz de raffinerie*	LPG + Ethane / *GPL + éthane*	Motor Gasoline / *Essence moteur*	Aviation Gasoline / *Essence aviation*	Jet Fuel / *Carbu-réacteurs*	Kerosene / *Kérosène*	Gas/ Diesel / *Gazole*	Heavy Fuel Oil / *Fioul lourd*	Naphtha / *Naphta*	Petrol. Coke / *Coke de pétrole*	Other Prod. / *Autres prod.*
Production	123	84	1112	-	152	-	2040	1477	615	-	227
From Other Sources	-	-	-	-	-	-	-	-	-	-	-
Imports	-	1	2	-	91	-	133	322	58	-	27
Exports	-	-	-453	-	-33	-	-1001	-30	-	-	-1
Intl. Marine Bunkers	-	-	-	-	-	-	-82	-258	-	-	-
Stock Changes	-	2	-20	1	18	-	-77	27	-	-	-
DOMESTIC SUPPLY	**123**	**87**	**641**	**1**	**228**	**-**	**1013**	**1538**	**673**	**-**	**253**
Transfers	-	-	-	-	-	-	-	-	-	-	-
Statistical Differences	-	-	-32	-	-2	-	3	53	-	-	-
TRANSFORMATION	**6**	**-**	**-**	**-**	**-**	**-**	**27**	**1195**	**-**	**-**	**149**
Electricity Plants	-	-	-	-	-	-	-	17	-	-	-
CHP Plants	6	-	-	-	-	-	2	562	-	-	149
Heat Plants	-	-	-	-	-	-	25	589	-	-	-
Blast Furnaces/Gas Works	-	-	-	-	-	-	-	-	-	-	-
Coke/Pat. Fuel/BKB Plants	-	-	-	-	-	-	-	-	-	-	-
Petroleum Refineries	-	-	-	-	-	-	-	-	-	-	-
Petrochemical Industry	-	-	-	-	-	-	-	27	-	-	-
Liquefaction	-	-	-	-	-	-	-	-	-	-	-
Other Transform. Sector	-	-	-	-	-	-	-	-	-	-	-
ENERGY SECTOR	**94**	**-**	**9**	**-**	**-**	**-**	**26**	**84**	**-**	**-**	**6**
Coal Mines	-	-	2	-	-	-	19	-	-	-	-
Oil and Gas Extraction	-	-	1	-	-	-	1	-	-	-	-
Petroleum Refineries	94	-	-	-	-	-	-	84	-	-	6
Electr., CHP+Heat Plants	-	-	6	-	-	-	6	-	-	-	-
Pumped Storage (Elec.)	-	-	-	-	-	-	-	-	-	-	-
Other Energy Sector	-	-	-	-	-	-	-	-	-	-	-
Distribution Losses	-	2	-	-	-	-	1	-	-	-	-
FINAL CONSUMPTION	**23**	**85**	**600**	**1**	**226**	**-**	**962**	**312**	**673**	**-**	**98**
INDUSTRY SECTOR	**23**	**12**	**68**	**-**	**-**	**-**	**195**	**296**	**673**	**-**	**74**
Iron and Steel	-	-	1	-	-	-	12	10	-	-	-
Chemical and Petrochem.	23	-	5	-	-	-	11	36	673	-	74
of which: Feedstocks	-	-	-	-	-	-	-	-	*673*	-	-
Non-Ferrous Metals	-	1	1	-	-	-	4	34	-	-	-
Non-Metallic Minerals	-	6	2	-	-	-	12	161	-	-	-
Transport Equipment	-	1	1	-	-	-	3	1	-	-	-
Machinery	-	3	8	-	-	-	10	6	-	-	-
Mining and Quarrying	-	-	2	-	-	-	38	7	-	-	-
Food and Tobacco	-	1	12	-	-	-	21	16	-	-	-
Paper, Pulp and Print	-	-	1	-	-	-	2	2	-	-	-
Wood and Wood Products	-	-	1	-	-	-	4	3	-	-	-
Construction	-	-	25	-	-	-	69	7	-	-	-
Textile and Leather	-	-	3	-	-	-	4	3	-	-	-
Non-specified	-	-	6	-	-	-	5	10	-	-	-
TRANSPORT SECTOR	**-**	**1**	**513**	**1**	**200**	**-**	**302**	**7**	**-**	**-**	**4**
Air	-	-	-	1	200	-	-	-	-	-	1
Road	-	1	509	-	-	-	260	6	-	-	2
Rail	-	-	-	-	-	-	1	-	-	-	1
Pipeline Transport	-	-	-	-	-	-	-	-	-	-	-
Internal Navigation	-	-	-	-	-	-	2	-	-	-	-
Non-specified	-	-	4	-	-	-	39	1	-	-	-
OTHER SECTORS	**-**	**72**	**19**	**-**	**26**	**-**	**465**	**9**	**-**	**-**	**11**
Agriculture	-	-	19	-	-	-	253	7	-	-	1
Comm. and Publ. Services	-	-	-	-	-	-	25	2	-	-	1
Residential	-	72	-	-	-	-	153	-	-	-	8
Non-specified	-	-	-	-	26	-	34	-	-	-	1
NON-ENERGY USE	**-**	**-**	**-**	**-**	**-**	**-**	**-**	**-**	**-**	**-**	**9**
in Industry/Transf./Energy	-	-	-	-	-	-	-	-	-	-	9
in Transport	-	-	-	-	-	-	-	-	-	-	-
in Other Sectors	-	-	-	-	-	-	-	-	-	-	-

Bulgaria / Bulgarie : 1997

	Gas / *Gaz* (TJ)				Comb. Renew. & Waste / *En. Re. Comb. & Déchets* (TJ)				(GWh)	(TJ)
SUPPLY AND CONSUMPTION *APPROVISIONNEMENT ET DEMANDE*	Natural Gas *Gaz naturel*	Gas Works *Usines à gaz*	Coke Ovens *Cokeries*	Blast Furnaces *Hauts fourneaux*	Solid Biomass *Biomasse solide*	Gas/Liquids from Biomass *Gaz/Liquides tirés de biomasse*	Municipal Waste *Déchets urbains*	Industrial Waste *Déchets industriels*	Electricity *Electricité*	Heat *Chaleur*
Production	1307	-	8663	13962	10492	-	-	-	42803	123844
From Other Sources	-	-	-	-	-	-	-	-	-	-
Imports	179192	-	-	-	-	-	-	-	785	-
Exports	-	-	-	-	-430	-	-	-	-4335	-
Intl. Marine Bunkers	-	-	-	-	-	-	-	-	-	-
Stock Changes	-8387	-	-	-	-25	-	-	-	-	-
DOMESTIC SUPPLY	**172112**	**-**	**8663**	**13962**	**10037**	**-**	**-**	**-**	**39253**	**123844**
Transfers	-	-	-	-	-	-	-	-	-	-
Statistical Differences	857	-	-	-	-192	-	-	-	-242	9077
TRANSFORMATION	**86201**	**-**	**2507**	**5448**	**1566**	**-**	**-**	**-**	**-**	**-**
Electricity Plants	-	-	-	-	-	-	-	-	-	-
CHP Plants	54617	-	2507	5448	-	-	-	-	-	-
Heat Plants	30575	-	-	-	937	-	-	-	-	-
Blast Furnaces/Gas Works	-	-	-	-	-	-	-	-	-	-
Coke/Pat. Fuel/BKB Plants	-	-	-	-	-	-	-	-	-	-
Petroleum Refineries	-	-	-	-	-	-	-	-	-	-
Petrochemical Industry	-	-	-	-	-	-	-	-	-	-
Liquefaction	-	-	-	-	-	-	-	-	-	-
Other Transform. Sector	1009	-	-	-	629	-	-	-	-	-
ENERGY SECTOR	**2990**	**-**	**4737**	**-**	**1**	**-**	**-**	**-**	**6374**	**13643**
Coal Mines	-	-	-	-	-	-	-	-	800	873
Oil and Gas Extraction	12	-	-	-	-	-	-	-	12	-
Petroleum Refineries	2973	-	-	-	-	-	-	-	309	3334
Electr., CHP+Heat Plants	5	-	-	-	1	-	-	-	4390	2468
Pumped Storage (Elec.)	-	-	-	-	-	-	-	-	-	-
Other Energy Sector	-	-	4737	-	-	-	-	-	863	6968
Distribution Losses	2000	-	-	-	85	-	-	-	6010	8808
FINAL CONSUMPTION	**81778**	**-**	**1419**	**8514**	**8193**	**-**	**-**	**-**	**26627**	**110470**
INDUSTRY SECTOR	**81202**	**-**	**1419**	**8514**	**173**	**-**	**-**	**-**	**11737**	**75514**
Iron and Steel	13676	-	1419	8514	2	-	-	-	1928	5594
Chemical and Petrochem.	46105	-	-	-	3	-	-	-	3413	39232
of which: Feedstocks	*25189*	-	-	-	-	-	-	-	-	-
Non-Ferrous Metals	167	-	-	-	4	-	-	-	787	1834
Non-Metallic Minerals	19879	-	-	-	9	-	-	-	856	3746
Transport Equipment	-	-	-	-	7	-	-	-	159	202
Machinery	1134	-	-	-	39	-	-	-	1146	3486
Mining and Quarrying	-	-	-	-	5	-	-	-	1008	1004
Food and Tobacco	171	-	-	-	29	-	-	-	731	9637
Paper, Pulp and Print	59	-	-	-	1	-	-	-	285	4038
Wood and Wood Products	-	-	-	-	14	-	-	-	123	1465
Construction	8	-	-	-	27	-	-	-	303	126
Textile and Leather	1	-	-	-	15	-	-	-	462	4687
Non-specified	2	-	-	-	18	-	-	-	536	463
TRANSPORT SECTOR	**-**	**-**	**-**	**-**	**14**	**-**	**-**	**-**	**657**	**-**
Air	-	-	-	-	-	-	-	-	-	-
Road	-	-	-	-	-	-	-	-	-	-
Rail	-	-	-	-	14	-	-	-	18	-
Pipeline Transport	-	-	-	-	-	-	-	-	-	-
Internal Navigation	-	-	-	-	-	-	-	-	-	-
Non-specified	-	-	-	-	-	-	-	-	639	-
OTHER SECTORS	**576**	**-**	**-**	**-**	**8006**	**-**	**-**	**-**	**14233**	**34956**
Agriculture	-	-	-	-	218	-	-	-	355	1564
Comm. and Publ. Services	525	-	-	-	87	-	-	-	1359	594
Residential	-	-	-	-	7521	-	-	-	9885	25989
Non-specified	51	-	-	-	180	-	-	-	2634	6809
NON-ENERGY USE	**-**	**-**	**-**	**-**	**-**	**-**	**-**	**-**	**-**	**-**
in Industry/Transf./Energy	-	-	-	-	-	-	-	-	-	-
in Transport	-	-	-	-	-	-	-	-	-	-
in Other Sectors	-	-	-	-	-	-	-	-	-	-

Cameroon / Cameroun

SUPPLY AND CONSUMPTION 1996	Coal (1000 tonnes)							Oil (1000 tonnes)			
	Coking Coal	Other Bit. Coal	Sub-Bit. Coal	Lignite	Peat	Oven and Gas Coke	Pat. Fuel and BKB	Crude Oil	NGL	Feed-stocks	Additives
Production	-	-	-	-	-	-	-	5984	-	-	-
Imports	-	-	-	-	-	-	-	-	-	-	-
Exports	-	-	-	-	-	-	-	-5104	-	-	-
Intl. Marine Bunkers	-	-	-	-	-	-	-	-	-	-	-
Stock Changes	-	-	-	-	-	-	-	-	-	-	-
DOMESTIC SUPPLY	-	-	-	-	-	-	-	**880**	-	-	-
Transfers and Stat. Diff.	-	-	-	-	-	-	-	-	-	-	-
TRANSFORMATION	-	-	-	-	-	-	-	**880**	-	-	-
Electricity and CHP Plants	-	-	-	-	-	-	-	-	-	-	-
Petroleum Refineries	-	-	-	-	-	-	-	880	-	-	-
Other Transform. Sector	-	-	-	-	-	-	-	-	-	-	-
ENERGY SECTOR	-	-	-	-	-	-	-	-	-	-	-
DISTRIBUTION LOSSES	-	-	-	-	-	-	-	-	-	-	-
FINAL CONSUMPTION	-	-	-	-	-	-	-	-	-	-	-
INDUSTRY SECTOR	-	-	-	-	-	-	-	-	-	-	-
Iron and Steel	-	-	-	-	-	-	-	-	-	-	-
Chemical and Petrochem.	-	-	-	-	-	-	-	-	-	-	-
Non-Metallic Minerals	-	-	-	-	-	-	-	-	-	-	-
Non-specified	-	-	-	-	-	-	-	-	-	-	-
TRANSPORT SECTOR	-	-	-	-	-	-	-	-	-	-	-
Air	-	-	-	-	-	-	-	-	-	-	-
Road	-	-	-	-	-	-	-	-	-	-	-
Non-specified	-	-	-	-	-	-	-	-	-	-	-
OTHER SECTORS	-	-	-	-	-	-	-	-	-	-	-
Agriculture	-	-	-	-	-	-	-	-	-	-	-
Comm. and Publ. Services	-	-	-	-	-	-	-	-	-	-	-
Residential	-	-	-	-	-	-	-	-	-	-	-
Non-specified	-	-	-	-	-	-	-	-	-	-	-
NON-ENERGY USE	-	-	-	-	-	-	-	-	-	-	-

APPROVISIONNEMENT ET DEMANDE 1997	*Charbon* (1000 tonnes)							*Pétrole* (1000 tonnes)			
	Charbon à coke	*Autres charb. bit.*	*Charbon sous-bit.*	*Lignite*	*Tourbe*	*Coke de four/gaz*	*Agg./briq. de lignite*	*Pétrole brut*	*LGN*	*Produits d'aliment.*	*Additifs*
Production	-	-	-	-	-	-	-	6269	-	-	-
Imports	-	-	-	-	-	-	-	-	-	-	-
Exports	-	-	-	-	-	-	-	-5361	-	-	-
Intl. Marine Bunkers	-	-	-	-	-	-	-	-	-	-	-
Stock Changes	-	-	-	-	-	-	-	-	-	-	-
DOMESTIC SUPPLY	-	-	-	-	-	-	-	**908**	-	-	-
Transfers and Stat. Diff.	-	-	-	-	-	-	-	-	-	-	-
TRANSFORMATION	-	-	-	-	-	-	-	**908**	-	-	-
Electricity and CHP Plants	-	-	-	-	-	-	-	-	-	-	-
Petroleum Refineries	-	-	-	-	-	-	-	908	-	-	-
Other Transform. Sector	-	-	-	-	-	-	-	-	-	-	-
ENERGY SECTOR	-	-	-	-	-	-	-	-	-	-	-
DISTRIBUTION LOSSES	-	-	-	-	-	-	-	-	-	-	-
FINAL CONSUMPTION	-	-	-	-	-	-	-	-	-	-	-
INDUSTRY SECTOR	-	-	-	-	-	-	-	-	-	-	-
Iron and Steel	-	-	-	-	-	-	-	-	-	-	-
Chemical and Petrochem.	-	-	-	-	-	-	-	-	-	-	-
Non-Metallic Minerals	-	-	-	-	-	-	-	-	-	-	-
Non-specified	-	-	-	-	-	-	-	-	-	-	-
TRANSPORT SECTOR	-	-	-	-	-	-	-	-	-	-	-
Air	-	-	-	-	-	-	-	-	-	-	-
Road	-	-	-	-	-	-	-	-	-	-	-
Non-specified	-	-	-	-	-	-	-	-	-	-	-
OTHER SECTORS	-	-	-	-	-	-	-	-	-	-	-
Agriculture	-	-	-	-	-	-	-	-	-	-	-
Comm. and Publ. Services	-	-	-	-	-	-	-	-	-	-	-
Residential	-	-	-	-	-	-	-	-	-	-	-
Non-specified	-	-	-	-	-	-	-	-	-	-	-
NON-ENERGY USE	-	-	-	-	-	-	-	-	-	-	-

Cameroon / Cameroun

SUPPLY AND CONSUMPTION 1996	Oil cont. (1000 tonnes)										
	Refinery Gas	LPG + Ethane	Motor Gasoline	Aviation Gasoline	Jet Fuel	Kerosene	Gas/ Diesel	Heavy Fuel Oil	Naphtha	Petrol. Coke	Other Prod.
Production	-	18	234	1	55	98	360	51	-	-	31
Imports	-	-	-	-	-	4	5	-	-	-	-
Exports	-	-	-	-	-	-1	-4	-	-	-	-
Intl. Marine Bunkers	-	-	-	-	-	-	-44	-	-	-	-
Stock Changes	-	-	-	-	-	-	-	-	-	-	-
DOMESTIC SUPPLY	**-**	**18**	**234**	**1**	**55**	**101**	**317**	**51**	**-**	**-**	**31**
Transfers and Stat. Diff.	-	-	-	-	-	-1	1	-	-	-	-
TRANSFORMATION	**-**	**-**	**-**	**-**	**-**	**-**	**9**	**-**	**-**	**-**	**-**
Electricity and CHP Plants	-	-	-	-	-	-	9	-	-	-	-
Petroleum Refineries	-	-	-	-	-	-	-	-	-	-	-
Other Transform. Sector	-	-	-	-	-	-	-	-	-	-	-
ENERGY SECTOR	**-**	**-**	**-**	**-**	**-**	**-**	**-**	**-**	**-**	**-**	**-**
DISTRIBUTION LOSSES	**-**	**-**	**-**	**-**	**-**	**-**	**-**	**-**	**-**	**-**	**-**
FINAL CONSUMPTION	**-**	**18**	**234**	**1**	**55**	**100**	**309**	**51**	**-**	**-**	**31**
INDUSTRY SECTOR	**-**	**-**	**-**	**-**	**-**	**-**	**-**	**51**	**-**	**-**	**-**
Iron and Steel	-	-	-	-	-	-	-	-	-	-	-
Chemical and Petrochem.	-	-	-	-	-	-	-	-	-	-	-
Non-Metallic Minerals	-	-	-	-	-	-	-	-	-	-	-
Non-specified	-	-	-	-	-	-	-	51	-	-	-
TRANSPORT SECTOR	**-**	**-**	**234**	**1**	**55**	**-**	**309**	**-**	**-**	**-**	**-**
Air	-	-	-	1	55	-	-	-	-	-	-
Road	-	-	234	-	-	-	309	-	-	-	-
Non-specified	-	-	-	-	-	-	-	-	-	-	-
OTHER SECTORS	**-**	**18**	**-**	**-**	**-**	**100**	**-**	**-**	**-**	**-**	**-**
Agriculture	-	-	-	-	-	-	-	-	-	-	-
Comm. and Publ. Services	-	-	-	-	-	-	-	-	-	-	-
Residential	-	18	-	-	-	100	-	-	-	-	-
Non-specified	-	-	-	-	-	-	-	-	-	-	-
NON-ENERGY USE	**-**	**-**	**-**	**-**	**-**	**-**	**-**	**-**	**-**	**-**	**31**

APPROVISIONNEMENT ET DEMANDE 1997	*Pétrole cont.* (1000 tonnes)										
	Gaz de raffinerie	*GPL + éthane*	*Essence moteur*	*Essence aviation*	*Carbu- réacteurs*	*Kérosène*	*Gazole*	*Fioul lourd*	*Naphta*	*Coke de pétrole*	*Autres prod.*
Production	-	24	234	-	51	99	385	49	-	-	31
Imports	-	-	2	-	-	5	7	-	-	-	-
Exports	-	-	-3	-	-	-7	-7	-	-	-	-
Intl. Marine Bunkers	-	-	-	-	-	-	-53	-	-	-	-
Stock Changes	-	-	-	-	-	-	-	-	-	-	-
DOMESTIC SUPPLY	**-**	**24**	**233**	**-**	**51**	**97**	**332**	**49**	**-**	**-**	**31**
Transfers and Stat. Diff.	-	-	-	-	-	-	-1	-	-	-	-
TRANSFORMATION	**-**	**-**	**-**	**-**	**-**	**-**	**9**	**-**	**-**	**-**	**-**
Electricity and CHP Plants	-	-	-	-	-	-	9	-	-	-	-
Petroleum Refineries	-	-	-	-	-	-	-	-	-	-	-
Other Transform. Sector	-	-	-	-	-	-	-	-	-	-	-
ENERGY SECTOR	**-**	**-**	**-**	**-**	**-**	**-**	**-**	**-**	**-**	**-**	**-**
DISTRIBUTION LOSSES	**-**	**-**	**-**	**-**	**-**	**-**	**-**	**-**	**-**	**-**	**-**
FINAL CONSUMPTION	**-**	**24**	**233**	**-**	**51**	**97**	**322**	**49**	**-**	**-**	**31**
INDUSTRY SECTOR	**-**	**-**	**-**	**-**	**-**	**-**	**-**	**49**	**-**	**-**	**-**
Iron and Steel	-	-	-	-	-	-	-	-	-	-	-
Chemical and Petrochem.	-	-	-	-	-	-	-	-	-	-	-
Non-Metallic Minerals	-	-	-	-	-	-	-	-	-	-	-
Non-specified	-	-	-	-	-	-	-	49	-	-	-
TRANSPORT SECTOR	**-**	**-**	**233**	**-**	**51**	**-**	**322**	**-**	**-**	**-**	**-**
Air	-	-	-	-	51	-	-	-	-	-	-
Road	-	-	233	-	-	-	322	-	-	-	-
Non-specified	-	-	-	-	-	-	-	-	-	-	-
OTHER SECTORS	**-**	**24**	**-**	**-**	**-**	**97**	**-**	**-**	**-**	**-**	**-**
Agriculture	-	-	-	-	-	-	-	-	-	-	-
Comm. and Publ. Services	-	-	-	-	-	-	-	-	-	-	-
Residential	-	24	-	-	-	97	-	-	-	-	-
Non-specified	-	-	-	-	-	-	-	-	-	-	-
NON-ENERGY USE	**-**	**-**	**-**	**-**	**-**	**-**	**-**	**-**	**-**	**-**	**31**

Cameroon / Cameroun

SUPPLY AND CONSUMPTION 1996	Gas (TJ)				Comb. Renew. & Waste (TJ)				(GWh)	(TJ)
	Natural Gas	Gas Works	Coke Ovens	Blast Furnaces	Solid Biomass	Gas/Liquids from Biomass	Municipal Waste	Industrial Waste	Electricity	Heat
Production	-	-	-	-	188307	-	-	-	2902	-
Imports	-	-	-	-	-	-	-	-	-	-
Exports	-	-	-	-	-	-	-	-	-	-
Intl. Marine Bunkers	-	-	-	-	-	-	-	-	-	-
Stock Changes	-	-	-	-	-	-	-	-	-	-
DOMESTIC SUPPLY	-	-	-	-	**188307**	-	-	-	**2902**	-
Transfers and Stat. Diff.	-	-	-	-	-	-	-	-	-	-
TRANSFORMATION	-	-	-	-	**8541**	-	-	-	-	-
Electricity and CHP Plants	-	-	-	-	-	-	-	-	-	-
Petroleum Refineries	-	-	-	-	-	-	-	-	-	-
Other Transform. Sector	-	-	-	-	8541	-	-	-	-	-
ENERGY SECTOR	-	-	-	-	-	-	-	-	-	-
DISTRIBUTION LOSSES	-	-	-	-	-	-	-	-	**587**	-
FINAL CONSUMPTION	-	-	-	-	**179766**	-	-	-	**2315**	-
INDUSTRY SECTOR	-	-	-	-	**30870**	-	-	-	**1315**	-
Iron and Steel	-	-	-	-	-	-	-	-	-	-
Chemical and Petrochem.	-	-	-	-	-	-	-	-	-	-
Non-Metallic Minerals	-	-	-	-	-	-	-	-	-	-
Non-specified	-	-	-	-	30870	-	-	-	1315	-
TRANSPORT SECTOR	-	-	-	-	-	-	-	-	-	-
Air	-	-	-	-	-	-	-	-	-	-
Road	-	-	-	-	-	-	-	-	-	-
Non-specified	-	-	-	-	-	-	-	-	-	-
OTHER SECTORS	-	-	-	-	**148896**	-	-	-	**1000**	-
Agriculture	-	-	-	-	-	-	-	-	-	-
Comm. and Publ. Services	-	-	-	-	-	-	-	-	274	-
Residential	-	-	-	-	148896	-	-	-	326	-
Non-specified	-	-	-	-	-	-	-	-	400	-
NON-ENERGY USE	-	-	-	-	-	-	-	-	-	-

APPROVISIONNEMENT ET DEMANDE 1997	*Gaz* (TJ)				*En. Re. Comb. & Déchets* (TJ)				(GWh)	(TJ)
	Gaz naturel	*Usines à gaz*	*Cokeries*	*Hauts fourneaux*	*Biomasse solide*	*Gaz/Liquides tirés de biomasse*	*Déchets urbains*	*Déchets industriels*	*Electricité*	*Chaleur*
Production	-	-	-	-	193768	-	-	-	3128	-
Imports	-	-	-	-	-	-	-	-	-	-
Exports	-	-	-	-	-	-	-	-	-	-
Intl. Marine Bunkers	-	-	-	-	-	-	-	-	-	-
Stock Changes	-	-	-	-	-	-	-	-	-	-
DOMESTIC SUPPLY	-	-	-	-	**193768**	-	-	-	**3128**	-
Transfers and Stat. Diff.	-	-	-	-	-1	-	-	-	-1	-
TRANSFORMATION	-	-	-	-	**8788**	-	-	-	-	-
Electricity and CHP Plants	-	-	-	-	-	-	-	-	-	-
Petroleum Refineries	-	-	-	-	-	-	-	-	-	-
Other Transform. Sector	-	-	-	-	8788	-	-	-	-	-
ENERGY SECTOR	-	-	-	-	-	-	-	-	-	-
DISTRIBUTION LOSSES	-	-	-	-	-	-	-	-	**611**	-
FINAL CONSUMPTION	-	-	-	-	**184979**	-	-	-	**2516**	-
INDUSTRY SECTOR	-	-	-	-	**31765**	-	-	-	**1428**	-
Iron and Steel	-	-	-	-	-	-	-	-	-	-
Chemical and Petrochem.	-	-	-	-	-	-	-	-	-	-
Non-Metallic Minerals	-	-	-	-	-	-	-	-	-	-
Non-specified	-	-	-	-	31765	-	-	-	1428	-
TRANSPORT SECTOR	-	-	-	-	-	-	-	-	-	-
Air	-	-	-	-	-	-	-	-	-	-
Road	-	-	-	-	-	-	-	-	-	-
Non-specified	-	-	-	-	-	-	-	-	-	-
OTHER SECTORS	-	-	-	-	**153214**	-	-	-	**1088**	-
Agriculture	-	-	-	-	-	-	-	-	-	-
Comm. and Publ. Services	-	-	-	-	-	-	-	-	298	-
Residential	-	-	-	-	153214	-	-	-	355	-
Non-specified	-	-	-	-	-	-	-	-	435	-
NON-ENERGY USE	-	-	-	-	-	-	-	-	-	-

Chile / Chili : 1996

SUPPLY AND CONSUMPTION / *APPROVISIONNEMENT ET DEMANDE*	Coal / *Charbon* (1000 tonnes)							Oil / *Pétrole* (1000 tonnes)			
	Coking Coal / *Charbon à coke*	Other Bit. Coal / *Autres charb. bit.*	Sub-Bit. Coal / *Charbon sous-bit.*	Lignite / *Lignite*	Peat / *Tourbe*	Oven and Gas Coke / *Coke de four/gaz*	Pat. Fuel and BKB / *Agg./briq. de lignite*	Crude Oil / *Pétrole brut*	NGL / *LGN*	Feed-stocks / *Produits d'aliment.*	Additives / *Additifs*
Production	-	1004	-	-	-	515	-	361	214	-	-
From Other Sources	-	-	-	-	-	-	-	-	-	-	-
Imports	713	2882	-	-	-	85	-	7597	-	-	-
Exports	-	-	-	-	-	-13	-	-	-	-	-
Intl. Marine Bunkers	-	-	-	-	-	-	-	-	-	-	-
Stock Changes	-	100	-	-	-	71	-	59	-	-	-
DOMESTIC SUPPLY	**713**	**3986**	**-**	**-**	**-**	**658**	**-**	**8017**	**214**	**-**	**-**
Transfers	-	-	-	-	-	-	-	-	-143	-	-
Statistical Differences	-	-3	-	-	-	-	-	-	-	-	-
TRANSFORMATION	**713**	**3220**	**-**	**-**	**-**	**446**	**-**	**8017**	**71**	**-**	**-**
Electricity Plants	-	3210	-	-	-	48	-	-	-	-	-
CHP Plants	-	-	-	-	-	-	-	-	-	-	-
Heat Plants	-	-	-	-	-	-	-	-	-	-	-
Blast Furnaces/Gas Works	-	10	-	-	-	398	-	-	-	-	-
Coke/Pat. Fuel/BKB Plants	713	-	-	-	-	-	-	-	-	-	-
Petroleum Refineries	-	-	-	-	-	-	-	8017	71	-	-
Petrochemical Industry	-	-	-	-	-	-	-	-	-	-	-
Liquefaction	-	-	-	-	-	-	-	-	-	-	-
Other Transform. Sector	-	-	-	-	-	-	-	-	-	-	-
ENERGY SECTOR	**-**	**-**	**-**	**-**	**-**	**-**	**-**	**-**	**-**	**-**	**-**
Coal Mines	-	-	-	-	-	-	-	-	-	-	-
Oil and Gas Extraction	-	-	-	-	-	-	-	-	-	-	-
Petroleum Refineries	-	-	-	-	-	-	-	-	-	-	-
Electr., CHP+Heat Plants	-	-	-	-	-	-	-	-	-	-	-
Pumped Storage (Elec.)	-	-	-	-	-	-	-	-	-	-	-
Other Energy Sector	-	-	-	-	-	-	-	-	-	-	-
Distribution Losses	-	-	-	-	-	-	-	-	-	-	-
FINAL CONSUMPTION	**-**	**763**	**-**	**-**	**-**	**212**	**-**	**-**	**-**	**-**	**-**
INDUSTRY SECTOR	**-**	**635**	**-**	**-**	**-**	**212**	**-**	**-**	**-**	**-**	**-**
Iron and Steel	-	78	-	-	-	100	-	-	-	-	-
Chemical and Petrochem.	-	-	-	-	-	-	-	-	-	-	-
of which: Feedstocks	-	-	-	-	-	-	-	-	-	-	-
Non-Ferrous Metals	-	-	-	-	-	-	-	-	-	-	-
Non-Metallic Minerals	-	187	-	-	-	92	-	-	-	-	-
Transport Equipment	-	-	-	-	-	-	-	-	-	-	-
Machinery	-	-	-	-	-	-	-	-	-	-	-
Mining and Quarrying	-	26	-	-	-	1	-	-	-	-	-
Food and Tobacco	-	155	-	-	-	9	-	-	-	-	-
Paper, Pulp and Print	-	2	-	-	-	-	-	-	-	-	-
Wood and Wood Products	-	-	-	-	-	-	-	-	-	-	-
Construction	-	-	-	-	-	-	-	-	-	-	-
Textile and Leather	-	-	-	-	-	-	-	-	-	-	-
Non-specified	-	187	-	-	-	10	-	-	-	-	-
TRANSPORT SECTOR	**-**	**-**	**-**	**-**	**-**	**-**	**-**	**-**	**-**	**-**	**-**
Air	-	-	-	-	-	-	-	-	-	-	-
Road	-	-	-	-	-	-	-	-	-	-	-
Rail	-	-	-	-	-	-	-	-	-	-	-
Pipeline Transport	-	-	-	-	-	-	-	-	-	-	-
Internal Navigation	-	-	-	-	-	-	-	-	-	-	-
Non-specified	-	-	-	-	-	-	-	-	-	-	-
OTHER SECTORS	**-**	**128**	**-**	**-**	**-**	**-**	**-**	**-**	**-**	**-**	**-**
Agriculture	-	116	-	-	-	-	-	-	-	-	-
Comm. and Publ. Services	-	-	-	-	-	-	-	-	-	-	-
Residential	-	12	-	-	-	-	-	-	-	-	-
Non-specified	-	-	-	-	-	-	-	-	-	-	-
NON-ENERGY USE	**-**	**-**	**-**	**-**	**-**	**-**	**-**	**-**	**-**	**-**	**-**
in Industry/Trans./Energy	-	-	-	-	-	-	-	-	-	-	-
in Transport	-	-	-	-	-	-	-	-	-	-	-
in Other Sectors	-	-	-	-	-	-	-	-	-	-	-

Chile / Chili : 1996

SUPPLY AND CONSUMPTION / *APPROVISIONNEMENT ET DEMANDE*	Oil cont. / *Pétrole cont.* (1000 tonnes)										
	Refinery Gas / *Gaz de raffinerie*	LPG + Ethane / *GPL + éthane*	Motor Gasoline / *Essence moteur*	Aviation Gasoline / *Essence aviation*	Jet Fuel / *Carbu-réacteurs*	Kerosene / *Kérosène*	Gas/ Diesel / *Gazole*	Heavy Fuel Oil / *Fioul lourd*	Naphtha / *Naphta*	Petrol. Coke / *Coke de pétrole*	Other Prod. / *Autres prod.*
Production	509	323	1847	9	373	315	2823	1655	133	-	-
From Other Sources	-	-	-	-	-	-	-	-	-	-	-
Imports	-	518	315	-	113	83	858	481	-	-	-
Exports	-	-	-12	-2	-	-	-30	-	-	-	-
Intl. Marine Bunkers	-	-	-	-	-	-	-	-254	-	-	-
Stock Changes	-	-87	1	-2	-51	-79	-140	-49	-71	-	-
DOMESTIC SUPPLY	**509**	**754**	**2151**	**5**	**435**	**319**	**3511**	**1833**	**62**	**-**	**-**
Transfers	-	143	-	-	-	-	-	-	-	-	-
Statistical Differences	-	1	-	1	-	1	-2	9	-	-	-
TRANSFORMATION	**-**	**2**	**-**	**-**	**-**	**-**	**78**	**547**	**58**	**-**	**-**
Electricity Plants	-	-	-	-	-	-	78	547	-	-	-
CHP Plants	-	-	-	-	-	-	-	-	-	-	-
Heat Plants	-	-	-	-	-	-	-	-	-	-	-
Blast Furnaces/Gas Works	-	2	-	-	-	-	-	-	58	-	-
Coke/Pat. Fuel/BKB Plants	-	-	-	-	-	-	-	-	-	-	-
Petroleum Refineries	-	-	-	-	-	-	-	-	-	-	-
Petrochemical Industry	-	-	-	-	-	-	-	-	-	-	-
Liquefaction	-	-	-	-	-	-	-	-	-	-	-
Other Transform. Sector	-	-	-	-	-	-	-	-	-	-	-
ENERGY SECTOR	**509**	**2**	**-**	**-**	**-**	**-**	**15**	**148**	**-**	**-**	**-**
Coal Mines	-	-	-	-	-	-	-	-	-	-	-
Oil and Gas Extraction	-	2	-	-	-	-	-	-	-	-	-
Petroleum Refineries	509	-	-	-	-	-	15	148	-	-	-
Electr., CHP+Heat Plants	-	-	-	-	-	-	-	-	-	-	-
Pumped Storage (Elec.)	-	-	-	-	-	-	-	-	-	-	-
Other Energy Sector	-	-	-	-	-	-	-	-	-	-	-
Distribution Losses	-	-	-	-	-	-	-	-	-	-	-
FINAL CONSUMPTION	**-**	**894**	**2151**	**6**	**435**	**320**	**3416**	**1147**	**4**	**-**	**-**
INDUSTRY SECTOR	**-**	**104**	**-**	**-**	**-**	**148**	**999**	**987**	**4**	**-**	**-**
Iron and Steel	-	-	-	-	-	-	28	40	-	-	-
Chemical and Petrochem.	-	-	-	-	-	-	-	4	-	-	-
of which: Feedstocks	-	-	-	-	-	-	-	-	-	-	-
Non-Ferrous Metals	-	-	-	-	-	-	240	-	-	-	-
Non-Metallic Minerals	-	-	-	-	-	-	4	79	-	-	-
Transport Equipment	-	-	-	-	-	-	-	-	-	-	-
Machinery	-	-	-	-	-	-	-	-	-	-	-
Mining and Quarrying	-	1	-	-	-	11	19	287	-	-	-
Food and Tobacco	-	-	-	-	-	-	-	11	-	-	-
Paper, Pulp and Print	-	2	-	-	-	-	3	154	-	-	-
Wood and Wood Products	-	-	-	-	-	-	-	-	-	-	-
Construction	-	-	-	-	-	-	-	-	-	-	-
Textile and Leather	-	-	-	-	-	-	-	-	-	-	-
Non-specified	-	101	-	-	-	137	705	412	4	-	-
TRANSPORT SECTOR	**-**	**-**	**2151**	**6**	**435**	**-**	**2320**	**-**	**-**	**-**	**-**
Air	-	-	-	6	435	-	-	-	-	-	-
Road	-	-	2151	-	-	-	1893	-	-	-	-
Rail	-	-	-	-	-	-	14	-	-	-	-
Pipeline Transport	-	-	-	-	-	-	-	-	-	-	-
Internal Navigation	-	-	-	-	-	-	413	-	-	-	-
Non-specified	-	-	-	-	-	-	-	-	-	-	-
OTHER SECTORS	**-**	**790**	**-**	**-**	**-**	**172**	**97**	**160**	**-**	**-**	**-**
Agriculture	-	1	-	-	-	-	19	136	-	-	-
Comm. and Publ. Services	-	-	-	-	-	-	-	-	-	-	-
Residential	-	789	-	-	-	172	-	-	-	-	-
Non-specified	-	-	-	-	-	-	78	24	-	-	-
NON-ENERGY USE	**-**	**-**	**-**	**-**	**-**	**-**	**-**	**-**	**-**	**-**	**-**
in Industry/Transf./Energy	-	-	-	-	-	-	-	-	-	-	-
in Transport	-	-	-	-	-	-	-	-	-	-	-
in Other Sectors	-	-	-	-	-	-	-	-	-	-	-

Chile / Chili : 1996

	Gas / *Gaz* (TJ)				Comb. Renew. & Waste / *En. Re. Comb. & Déchets* (TJ)				(GWh)	(TJ)
SUPPLY AND CONSUMPTION / ***APPROVISIONNEMENT ET DEMANDE***	Natural Gas / *Gaz naturel*	Gas Works / *Usines à gaz*	Coke Ovens / *Cokeries*	Blast Furnaces / *Hauts fourneaux*	Solid Biomass / *Biomasse solide*	Gas/Liquids from Biomass / *Gaz/Liquides tirés de biomasse*	Municipal Waste / *Déchets urbains*	Industrial Waste / *Déchets industriels*	Electricity / *Electricité*	Heat / *Chaleur*
Production	64720	2989	6092	5497	153967	1294	-	-	30790	-
From Other Sources	-	-	-	-	-	-	-	-	-	-
Imports	-	-	-	-	-	-	-	-	-	-
Exports	-	-	-	-	-	-	-	-	-	-
Intl. Marine Bunkers	-	-	-	-	-	-	-	-	-	-
Stock Changes	-	-	-	-	-	-	-	-	-	-
DOMESTIC SUPPLY	**64720**	**2989**	**6092**	**5497**	**153967**	**1294**	**-**	**-**	**30790**	**-**
Transfers	-	-	-	-	-	-	-	-	-	-
Statistical Differences	1	-	-	-	-	-	-	-	-	-
TRANSFORMATION	**3513**	**-**	**1327**	**-**	**13350**	**-**	**-**	**-**	**-**	**-**
Electricity Plants	3513	-	-	-	13350	-	-	-	-	-
CHP Plants	-	-	-	-	-	-	-	-	-	-
Heat Plants	-	-	-	-	-	-	-	-	-	-
Blast Furnaces/Gas Works	-	-	1327	-	-	-	-	-	-	-
Coke/Pat. Fuel/BKB Plants	-	-	-	-	-	-	-	-	-	-
Petroleum Refineries	-	-	-	-	-	-	-	-	-	-
Petrochemical Industry	-	-	-	-	-	-	-	-	-	-
Liquefaction	-	-	-	-	-	-	-	-	-	-
Other Transform. Sector	-	-	-	-	-	-	-	-	-	-
ENERGY SECTOR	**17543**	**-**	**-**	**373**	**-**	**1294**	**-**	**-**	**1225**	**-**
Coal Mines	-	-	-	-	-	-	-	-	51	-
Oil and Gas Extraction	17543	-	-	-	-	-	-	-	-	-
Petroleum Refineries	-	-	-	-	-	-	-	-	245	-
Electr., CHP+Heat Plants	-	-	-	-	-	-	-	-	860	-
Pumped Storage (Elec.)	-	-	-	-	-	-	-	-	-	-
Other Energy Sector	-	-	-	373	-	1294	-	-	69	-
Distribution Losses	-	-	1432	343	-	-	-	-	2688	-
FINAL CONSUMPTION	**43665**	**2989**	**3333**	**4781**	**140617**	**-**	**-**	**-**	**26877**	**-**
INDUSTRY SECTOR	**34973**	**-**	**3333**	**4781**	**32487**	**-**	**-**	**-**	**18349**	**-**
Iron and Steel	-	-	3333	4145	-	-	-	-	861	-
Chemical and Petrochem.	34466	-	-	-	-	-	-	-	397	-
of which: Feedstocks	-	-	-	-	-	-	-	-	-	-
Non-Ferrous Metals	-	-	-	-	-	-	-	-	7431	-
Non-Metallic Minerals	-	-	-	-	-	-	-	-	435	-
Transport Equipment	-	-	-	-	-	-	-	-	-	-
Machinery	-	-	-	-	-	-	-	-	-	-
Mining and Quarrying	-	-	-	-	-	-	-	-	211	-
Food and Tobacco	-	-	-	-	-	-	-	-	100	-
Paper, Pulp and Print	-	-	-	-	18288	-	-	-	2502	-
Wood and Wood Products	-	-	-	-	-	-	-	-	-	-
Construction	-	-	-	-	-	-	-	-	-	-
Textile and Leather	-	-	-	-	-	-	-	-	-	-
Non-specified	507	-	-	636	14199	-	-	-	6412	-
TRANSPORT SECTOR	**285**	**-**	**-**	**-**	**-**	**-**	**-**	**-**	**201**	**-**
Air	-	-	-	-	-	-	-	-	-	-
Road	285	-	-	-	-	-	-	-	90	-
Rail	-	-	-	-	-	-	-	-	111	-
Pipeline Transport	-	-	-	-	-	-	-	-	-	-
Internal Navigation	-	-	-	-	-	-	-	-	-	-
Non-specified	-	-	-	-	-	-	-	-	-	-
OTHER SECTORS	**8407**	**2989**	**-**	**-**	**108130**	**-**	**-**	**-**	**8327**	**-**
Agriculture	-	-	-	-	-	-	-	-	143	-
Comm. and Publ. Services	-	-	-	-	-	-	-	-	-	-
Residential	8407	2989	-	-	108130	-	-	-	8184	-
Non-specified	-	-	-	-	-	-	-	-	-	-
NON-ENERGY USE	**-**	**-**	**-**	**-**	**-**	**-**	**-**	**-**	**-**	**-**
in Industry/Transf./Energy	-	-	-	-	-	-	-	-	-	-
in Transport	-	-	-	-	-	-	-	-	-	-
in Other Sectors	-	-	-	-	-	-	-	-	-	-

Chile / Chili : 1997

SUPPLY AND CONSUMPTION / *APPROVISIONNEMENT ET DEMANDE*	Coal / *Charbon* (1000 tonnes)							Oil / *Pétrole* (1000 tonnes)			
	Coking Coal / *Charbon à coke*	Other Bit. Coal / *Autres charb. bit.*	Sub-Bit. Coal / *Charbon sous-bit.*	Lignite / *Lignite*	Peat / *Tourbe*	Oven and Gas Coke / *Coke de four/gaz*	Pat. Fuel and BKB / *Agg./briq. de lignite*	Crude Oil / *Pétrole brut*	NGL / *LGN*	Feed-stocks / *Produits d'aliment.*	Additives / *Additifs*
Production	-	1044	-	-	-	473	-	270	221	-	-
From Other Sources	-	-	-	-	-	-	-	-	-	-	-
Imports	685	3940	-	-	-	64	-	8224	-	-	-
Exports	-	-	-	-	-	-18	-	-	-	-	-
Intl. Marine Bunkers	-	-	-	-	-	-	-	-	-	-	-
Stock Changes	-	448	-	-	-	81	-	-77	-	-	-
DOMESTIC SUPPLY	**685**	**5432**	**-**	**-**	**-**	**600**	**-**	**8417**	**221**	**-**	**-**
Transfers	-	-	-	-	-	-	-	-	-145	-	-
Statistical Differences	-	-	-	-	-	-	-	-	-	-	-
TRANSFORMATION	**685**	**3949**	**-**	**-**	**-**	**417**	**-**	**8417**	**76**	**-**	**-**
Electricity Plants	-	3944	-	-	-	-	-	-	-	-	-
CHP Plants	-	-	-	-	-	-	-	-	-	-	-
Heat Plants	-	-	-	-	-	-	-	-	-	-	-
Blast Furnaces/Gas Works	-	5	-	-	-	417	-	-	-	-	-
Coke/Pat. Fuel/BKB Plants	685	-	-	-	-	-	-	-	-	-	-
Petroleum Refineries	-	-	-	-	-	-	-	8417	76	-	-
Petrochemical Industry	-	-	-	-	-	-	-	-	-	-	-
Liquefaction	-	-	-	-	-	-	-	-	-	-	-
Other Transform. Sector	-	-	-	-	-	-	-	-	-	-	-
ENERGY SECTOR	**-**	**-**	**-**	**-**	**-**	**-**	**-**	**-**	**-**	**-**	**-**
Coal Mines	-	-	-	-	-	-	-	-	-	-	-
Oil and Gas Extraction	-	-	-	-	-	-	-	-	-	-	-
Petroleum Refineries	-	-	-	-	-	-	-	-	-	-	-
Electr., CHP+Heat Plants	-	-	-	-	-	-	-	-	-	-	-
Pumped Storage (Elec.)	-	-	-	-	-	-	-	-	-	-	-
Other Energy Sector	-	-	-	-	-	-	-	-	-	-	-
Distribution Losses	-	-	-	-	-	-	-	-	-	-	-
FINAL CONSUMPTION	**-**	**1483**	**-**	**-**	**-**	**183**	**-**	**-**	**-**	**-**	**-**
INDUSTRY SECTOR	**-**	**1326**	**-**	**-**	**-**	**183**	**-**	**-**	**-**	**-**	**-**
Iron and Steel	-	77	-	-	-	104	-	-	-	-	-
Chemical and Petrochem.	-	-	-	-	-	-	-	-	-	-	-
of which: Feedstocks	-	-	-	-	-	-	-	-	-	-	-
Non-Ferrous Metals	-	-	-	-	-	-	-	-	-	-	-
Non-Metallic Minerals	-	231	-	-	-	62	-	-	-	-	-
Transport Equipment	-	-	-	-	-	-	-	-	-	-	-
Machinery	-	-	-	-	-	-	-	-	-	-	-
Mining and Quarrying	-	31	-	-	-	-	-	-	-	-	-
Food and Tobacco	-	83	-	-	-	8	-	-	-	-	-
Paper, Pulp and Print	-	1	-	-	-	-	-	-	-	-	-
Wood and Wood Products	-	-	-	-	-	-	-	-	-	-	-
Construction	-	-	-	-	-	-	-	-	-	-	-
Textile and Leather	-	-	-	-	-	-	-	-	-	-	-
Non-specified	-	903	-	-	-	9	-	-	-	-	-
TRANSPORT SECTOR	**-**	**-**	**-**	**-**	**-**	**-**	**-**	**-**	**-**	**-**	**-**
Air	-	-	-	-	-	-	-	-	-	-	-
Road	-	-	-	-	-	-	-	-	-	-	-
Rail	-	-	-	-	-	-	-	-	-	-	-
Pipeline Transport	-	-	-	-	-	-	-	-	-	-	-
Internal Navigation	-	-	-	-	-	-	-	-	-	-	-
Non-specified	-	-	-	-	-	-	-	-	-	-	-
OTHER SECTORS	**-**	**157**	**-**	**-**	**-**	**-**	**-**	**-**	**-**	**-**	**-**
Agriculture	-	138	-	-	-	-	-	-	-	-	-
Comm. and Publ. Services	-	-	-	-	-	-	-	-	-	-	-
Residential	-	19	-	-	-	-	-	-	-	-	-
Non-specified	-	-	-	-	-	-	-	-	-	-	-
NON-ENERGY USE	**-**	**-**	**-**	**-**	**-**	**-**	**-**	**-**	**-**	**-**	**-**
in Industry/Trans./Energy	-	-	-	-	-	-	-	-	-	-	-
in Transport	-	-	-	-	-	-	-	-	-	-	-
in Other Sectors	-	-	-	-	-	-	-	-	-	-	-

Chile / Chili : 1997

SUPPLY AND CONSUMPTION / *APPROVISIONNEMENT ET DEMANDE*	Oil cont. / *Pétrole cont.* (1000 tonnes)										
	Refinery Gas / *Gaz de raffinerie*	LPG + Ethane / *GPL + éthane*	Motor Gasoline / *Essence moteur*	Aviation Gasoline / *Essence aviation*	Jet Fuel / *Carbu-réacteurs*	Kerosene / *Kérosène*	Gas/ Diesel / *Gazole*	Heavy Fuel Oil / *Fioul lourd*	Naphtha / *Naphta*	Petrol. Coke / *Coke de pétrole*	Other Prod. / *Autres prod.*
Production	557	324	1964	8	543	271	2937	1663	197	-	-
From Other Sources	-	-	-	-	-	-	-	-	-	-	-
Imports	-	510	296	-	97	-	945	404	-	-	-
Exports	-	-25	-55	-3	-5	-	-41	-	-	-	-
Intl. Marine Bunkers	-	-	-	-	-	-	-	-308	-	-	-
Stock Changes	-	-17	32	1	-85	47	139	20	-143	-	-
DOMESTIC SUPPLY	**557**	**792**	**2237**	**6**	**550**	**318**	**3980**	**1779**	**54**	**-**	**-**
Transfers	-	145	-	-	-	-	-	-	-	-	-
Statistical Differences	-	-	-	1	-	1	-	50	-	-	-
TRANSFORMATION	**-**	**2**	**-**	**-**	**-**	**-**	**193**	**414**	**41**	**-**	**-**
Electricity Plants	-	-	-	-	-	-	193	414	-	-	-
CHP Plants	-	-	-	-	-	-	-	-	-	-	-
Heat Plants	-	-	-	-	-	-	-	-	-	-	-
Blast Furnaces/Gas Works	-	2	-	-	-	-	-	-	41	-	-
Coke/Pat. Fuel/BKB Plants	-	-	-	-	-	-	-	-	-	-	-
Petroleum Refineries	-	-	-	-	-	-	-	-	-	-	-
Petrochemical Industry	-	-	-	-	-	-	-	-	-	-	-
Liquefaction	-	-	-	-	-	-	-	-	-	-	-
Other Transform. Sector	-	-	-	-	-	-	-	-	-	-	-
ENERGY SECTOR	**557**	**2**	**-**	**-**	**-**	**-**	**12**	**77**	**-**	**-**	**-**
Coal Mines	-	-	-	-	-	-	-	-	-	-	-
Oil and Gas Extraction	-	2	-	-	-	-	-	-	-	-	-
Petroleum Refineries	557	-	-	-	-	-	12	77	-	-	-
Electr., CHP+Heat Plants	-	-	-	-	-	-	-	-	-	-	-
Pumped Storage (Elec.)	-	-	-	-	-	-	-	-	-	-	-
Other Energy Sector	-	-	-	-	-	-	-	-	-	-	-
Distribution Losses	-	-	-	-	-	-	-	-	-	-	-
FINAL CONSUMPTION	**-**	**933**	**2237**	**7**	**550**	**319**	**3775**	**1338**	**13**	**-**	**-**
INDUSTRY SECTOR	**-**	**124**	**-**	**-**	**-**	**137**	**1222**	**1196**	**13**	**-**	**-**
Iron and Steel	-	-	-	-	-	-	29	40	-	-	-
Chemical and Petrochem.	-	-	-	-	-	-	-	4	-	-	-
of which: Feedstocks	-	-	-	-	-	-	-	-	-	-	-
Non-Ferrous Metals	-	-	-	-	-	-	292	-	-	-	-
Non-Metallic Minerals	-	-	-	-	-	-	4	55	-	-	-
Transport Equipment	-	-	-	-	-	-	-	-	-	-	-
Machinery	-	-	-	-	-	-	-	-	-	-	-
Mining and Quarrying	-	3	-	-	-	11	22	313	-	-	-
Food and Tobacco	-	-	-	-	-	-	-	12	-	-	-
Paper, Pulp and Print	-	1	-	-	-	-	3	148	-	-	-
Wood and Wood Products	-	-	-	-	-	-	-	-	-	-	-
Construction	-	-	-	-	-	-	-	-	-	-	-
Textile and Leather	-	-	-	-	-	-	-	-	-	-	-
Non-specified	-	120	-	-	-	126	872	624	13	-	-
TRANSPORT SECTOR	**-**	**-**	**2237**	**7**	**550**	**-**	**2435**	**-**	**-**	**-**	**-**
Air	-	-	-	7	550	-	-	-	-	-	-
Road	-	-	2237	-	-	-	1999	-	-	-	-
Rail	-	-	-	-	-	-	13	-	-	-	-
Pipeline Transport	-	-	-	-	-	-	-	-	-	-	-
Internal Navigation	-	-	-	-	-	-	423	-	-	-	-
Non-specified	-	-	-	-	-	-	-	-	-	-	-
OTHER SECTORS	**-**	**809**	**-**	**-**	**-**	**182**	**118**	**142**	**-**	**-**	**-**
Agriculture	-	-	-	-	-	-	23	121	-	-	-
Comm. and Publ. Services	-	106	-	-	-	-	-	-	-	-	-
Residential	-	703	-	-	-	182	-	-	-	-	-
Non-specified	-	-	-	-	-	-	95	21	-	-	-
NON-ENERGY USE	**-**	**-**	**-**	**-**	**-**	**-**	**-**	**-**	**-**	**-**	**-**
in Industry/Transf./Energy	-	-	-	-	-	-	-	-	-	-	-
in Transport	-	-	-	-	-	-	-	-	-	-	-
in Other Sectors	-	-	-	-	-	-	-	-	-	-	-

Chile / Chili : 1997

	Gas / *Gaz* (TJ)				Comb. Renew. & Waste / *En. Re. Comb. & Déchets* (TJ)				(GWh)	(TJ)
SUPPLY AND CONSUMPTION *APPROVISIONNEMENT ET DEMANDE*	Natural Gas *Gaz naturel*	Gas Works *Usines à gaz*	Coke Ovens *Cokeries*	Blast Furnaces *Hauts fourneaux*	Solid Biomass *Biomasse solide*	Gas/Liquids from Biomass *Gaz/Liquides tirés de biomasse*	Municipal Waste *Déchets urbains*	Industrial Waste *Déchets industriels*	Electricity *Electricité*	Heat *Chaleur*
Production	75334	2892	6819	5602	153275	1294	-	-	33994	-
From Other Sources	-	-	-	-	-	-	-	-	-	-
Imports	25346	-	-	-	-	-	-	-	-	-
Exports	-	-	-	-	-	-	-	-	-	-
Intl. Marine Bunkers	-	-	-	-	-	-	-	-	-	-
Stock Changes	-	-	-	-	-	-	-	-	-	-
DOMESTIC SUPPLY	**100680**	**2892**	**6819**	**5602**	**153275**	**1294**	**-**	**-**	**33994**	**-**
Transfers	-	-	-	-	-	-	-	-	-	-
Statistical Differences	3	-	-	-	-6	-	-	-	-	-
TRANSFORMATION	**7683**	**-**	**1485**	**23**	**13720**	**-**	**-**	**-**	**-**	**-**
Electricity Plants	7683	-	-	23	13720	-	-	-	-	-
CHP Plants	-	-	-	-	-	-	-	-	-	-
Heat Plants	-	-	-	-	-	-	-	-	-	-
Blast Furnaces/Gas Works	-	-	1485	-	-	-	-	-	-	-
Coke/Pat. Fuel/BKB Plants	-	-	-	-	-	-	-	-	-	-
Petroleum Refineries	-	-	-	-	-	-	-	-	-	-
Petrochemical Industry	-	-	-	-	-	-	-	-	-	-
Liquefaction	-	-	-	-	-	-	-	-	-	-
Other Transform. Sector	-	-	-	-	-	-	-	-	-	-
ENERGY SECTOR	**17653**	**-**	**-**	**373**	**-**	**1294**	**-**	**-**	**1362**	**-**
Coal Mines	-	-	-	-	-	-	-	-	43	-
Oil and Gas Extraction	17653	-	-	-	-	-	-	-	-	-
Petroleum Refineries	-	-	-	-	-	-	-	-	263	-
Electr., CHP+Heat Plants	-	-	-	-	-	-	-	-	946	-
Pumped Storage (Elec.)	-	-	-	-	-	-	-	-	-	-
Other Energy Sector	-	-	-	373	-	1294	-	-	110	-
Distribution Losses	-	-	1603	329	-	-	-	-	3222	-
FINAL CONSUMPTION	**75347**	**2892**	**3731**	**4877**	**139549**	**-**	**-**	**-**	**29410**	**-**
INDUSTRY SECTOR	**66028**	**-**	**3731**	**4877**	**26014**	**-**	**-**	**-**	**20276**	**-**
Iron and Steel	-	-	3731	3905	-	-	-	-	317	-
Chemical and Petrochem.	53915	-	-	-	-	-	-	-	571	-
of which: Feedstocks	-	-	-	-	-	-	-	-	-	-
Non-Ferrous Metals	-	-	-	-	-	-	-	-	8826	-
Non-Metallic Minerals	-	-	-	-	-	-	-	-	474	-
Transport Equipment	-	-	-	-	-	-	-	-	-	-
Machinery	-	-	-	-	-	-	-	-	-	-
Mining and Quarrying	-	-	-	-	-	-	-	-	236	-
Food and Tobacco	-	-	-	-	-	-	-	-	100	-
Paper, Pulp and Print	-	-	-	-	12422	-	-	-	2289	-
Wood and Wood Products	-	-	-	-	-	-	-	-	-	-
Construction	-	-	-	-	-	-	-	-	-	-
Textile and Leather	-	-	-	-	-	-	-	-	-	-
Non-specified	12113	-	-	972	13592	-	-	-	7463	-
TRANSPORT SECTOR	**243**	**-**	**-**	**-**	**-**	**-**	**-**	**-**	**211**	**-**
Air	-	-	-	-	-	-	-	-	-	-
Road	243	-	-	-	-	-	-	-	96	-
Rail	-	-	-	-	-	-	-	-	115	-
Pipeline Transport	-	-	-	-	-	-	-	-	-	-
Internal Navigation	-	-	-	-	-	-	-	-	-	-
Non-specified	-	-	-	-	-	-	-	-	-	-
OTHER SECTORS	**9076**	**2892**	**-**	**-**	**113535**	**-**	**-**	**-**	**8923**	**-**
Agriculture	-	-	-	-	-	-	-	-	173	-
Comm. and Publ. Services	1654	-	-	-	-	-	-	-	-	-
Residential	7422	2892	-	-	113535	-	-	-	8750	-
Non-specified	-	-	-	-	-	-	-	-	-	-
NON-ENERGY USE	**-**	**-**	**-**	**-**	**-**	**-**	**-**	**-**	**-**	**-**
in Industry/Transf./Energy	-	-	-	-	-	-	-	-	-	-
in Transport	-	-	-	-	-	-	-	-	-	-
in Other Sectors	-	-	-	-	-	-	-	-	-	-

People's Republic of China / République populaire de Chine : 1996

SUPPLY AND CONSUMPTION / *APPROVISIONNEMENT ET DEMANDE*	Coal / *Charbon* (1000 tonnes)							Oil / *Pétrole* (1000 tonnes)			
	Coking Coal / *Charbon à coke*	Other Bit. Coal / *Autres charb. bit.*	Sub-Bit. Coal / *Charbon sous-bit.*	Lignite / *Lignite*	Peat / *Tourbe*	Oven and Gas Coke / *Coke de four/gaz*	Pat. Fuel and BKB / *Agg./briq. de lignite*	Crude Oil / *Pétrole brut*	NGL / *LGN*	Feed-stocks / *Produits d'aliment.*	Additives / *Additifs*
Production	184558	1212141	-	-	-	139371	10260	157334	-	-	-
From Other Sources	-	-	-	-	-	-	-	-	-	-	-
Imports	-	3217	-	-	-	252	-	22617	-	-	-
Exports	-	-36484	-	-	-	-7994	-	-20402	-	-	-
Intl. Marine Bunkers	-	-	-	-	-	-	-	-	-	-	-
Stock Changes	-	8687	-	-	-	-1003	-	-341	-	-	-
DOMESTIC SUPPLY	**184558**	**1187561**	**-**	**-**	**-**	**130626**	**10260**	**159208**	**-**	**-**	**-**
Transfers	-	-	-	-	-	-	-	-	-	-	-
Statistical Differences	-	75582	-	-	-	-19288	-355	-557	-	-	-
TRANSFORMATION	**184558**	**566535**	**-**	**-**	**-**	**59993**	**-**	**152129**	**-**	**-**	**-**
Electricity Plants	-	488086	-	-	-	-	-	682	-	-	-
CHP Plants	-	63657	-	-	-	-	-	-	-	-	-
Heat Plants	-	-	-	-	-	-	-	108	-	-	-
Blast Furnaces/Gas Works	-	5820	-	-	-	59993	-	-	-	-	-
Coke/Pat. Fuel/BKB Plants	184558	8972	-	-	-	-	-	-	-	-	-
Petroleum Refineries	-	-	-	-	-	-	-	151339	-	-	-
Petrochemical Industry	-	-	-	-	-	-	-	-	-	-	-
Liquefaction	-	-	-	-	-	-	-	-	-	-	-
Other Transform. Sector	-	-	-	-	-	-	-	-	-	-	-
ENERGY SECTOR	**-**	**61175**	**-**	**-**	**-**	**1954**	**-**	**2932**	**-**	**-**	**-**
Coal Mines	-	27505	-	-	-	423	-	14	-	-	-
Oil and Gas Extraction	-	2449	-	-	-	17	-	1691	-	-	-
Petroleum Refineries	-	-	-	-	-	-	-	1090	-	-	-
Electr., CHP+Heat Plants	-	23145	-	-	-	28	-	137	-	-	-
Pumped Storage (Elec.)	-	-	-	-	-	-	-	-	-	-	-
Other Energy Sector	-	8076	-	-	-	1486	-	-	-	-	-
Distribution Losses	-	21976	-	-	-	-	-	1599	-	-	-
FINAL CONSUMPTION	**-**	**613457**	**-**	**-**	**-**	**49391**	**9905**	**1991**	**-**	**-**	**-**
INDUSTRY SECTOR	**-**	**402992**	**-**	**-**	**-**	**36796**	**-**	**1623**	**-**	**-**	**-**
Iron and Steel	-	42033	-	-	-	16632	-	49	-	-	-
Chemical and Petrochem.	-	86721	-	-	-	6266	-	1461	-	-	-
of which: Feedstocks	-	-	-	-	-	-	-	*274*	-	-	-
Non-Ferrous Metals	-	9395	-	-	-	2435	-	5	-	-	-
Non-Metallic Minerals	-	135275	-	-	-	2951	-	60	-	-	-
Transport Equipment	-	7073	-	-	-	372	-	1	-	-	-
Machinery	-	24189	-	-	-	5324	-	8	-	-	-
Mining and Quarrying	-	5798	-	-	-	1792	-	2	-	-	-
Food and Tobacco	-	34120	-	-	-	412	-	5	-	-	-
Paper, Pulp and Print	-	17714	-	-	-	25	-	1	-	-	-
Wood and Wood Products	-	5471	-	-	-	21	-	-	-	-	-
Construction	-	4464	-	-	-	139	-	26	-	-	-
Textile and Leather	-	22730	-	-	-	90	-	3	-	-	-
Non-specified	-	8009	-	-	-	337	-	2	-	-	-
TRANSPORT SECTOR	**-**	**10164**	**-**	**-**	**-**	**-**	**-**	**-**	**-**	**-**	**-**
Air	-	-	-	-	-	-	-	-	-	-	-
Road	-	-	-	-	-	-	-	-	-	-	-
Rail	-	10091	-	-	-	-	-	-	-	-	-
Pipeline Transport	-	-	-	-	-	-	-	-	-	-	-
Internal Navigation	-	73	-	-	-	-	-	-	-	-	-
Non-specified	-	-	-	-	-	-	-	-	-	-	-
OTHER SECTORS	**-**	**184214**	**-**	**-**	**-**	**2930**	**9905**	**368**	**-**	**-**	**-**
Agriculture	-	19173	-	-	-	1198	-	110	-	-	-
Comm. and Publ. Services	-	30953	-	-	-	470	-	258	-	-	-
Residential	-	134088	-	-	-	1262	9905	-	-	-	-
Non-specified	-	-	-	-	-	-	-	-	-	-	-
NON-ENERGY USE	**-**	**16087**	**-**	**-**	**-**	**9665**	**-**	**-**	**-**	**-**	**-**
in Industry/Trans./Energy	-	16087	-	-	-	9665	-	-	-	-	-
in Transport	-	-	-	-	-	-	-	-	-	-	-
in Other Sectors	-	-	-	-	-	-	-	-	-	-	-

People's Republic of China / République populaire de Chine : 1996

SUPPLY AND CONSUMPTION / *APPROVISIONNEMENT ET DEMANDE*	Oil cont. / *Pétrole cont.* (1000 tonnes)										
	Refinery Gas / *Gaz de raffinerie*	LPG + Ethane / *GPL + éthane*	Motor Gasoline / *Essence moteur*	Aviation Gasoline / *Essence aviation*	Jet Fuel / *Carbu-réacteurs*	Kerosene / *Kérosène*	Gas/ Diesel / *Gazole*	Heavy Fuel Oil / *Fioul lourd*	Naphtha / *Naphta*	Petrol. Coke / *Coke de pétrole*	Other Prod. / *Autres prod.*
Production	4501	6059	32636	110	2746	2637	44190	25045	15892	-	14812
From Other Sources	-	-	-	-	-	-	-	-	-	-	-
Imports	-	3550	79	-	-	659	4651	9426	-	-	1065
Exports	-	-333	-1314	-	-	-744	-1574	-366	-	-	-1173
Intl. Marine Bunkers	-	-	-280	-	-	-	-350	-342	-	-	-
Stock Changes	-	-45	293	-	-	50	800	60	-	-	-
DOMESTIC SUPPLY	**4501**	**9231**	**31414**	**110**	**2746**	**2602**	**47717**	**33823**	**15892**	**-**	**14704**
Transfers	-	-	-	-	-	-	-	-	-	-	-
Statistical Differences	20	80	1	-	-	487	-796	4724	-	-	-
TRANSFORMATION	**876**	**71**	**8**	**-**	**-**	**-**	**2832**	**15271**	**-**	**-**	**-**
Electricity Plants	236	40	8	-	-	-	2533	11412	-	-	-
CHP Plants	-	-	-	-	-	-	-	-	-	-	-
Heat Plants	640	31	-	-	-	-	299	3278	-	-	-
Blast Furnaces/Gas Works	-	-	-	-	-	-	-	581	-	-	-
Coke/Pat. Fuel/BKB Plants	-	-	-	-	-	-	-	-	-	-	-
Petroleum Refineries	-	-	-	-	-	-	-	-	-	-	-
Petrochemical Industry	-	-	-	-	-	-	-	-	-	-	-
Liquefaction	-	-	-	-	-	-	-	-	-	-	-
Other Transform. Sector	-	-	-	-	-	-	-	-	-	-	-
ENERGY SECTOR	**2746**	**1084**	**-**	**-**	**-**	**74**	**3291**	**5366**	**-**	**-**	**-**
Coal Mines	-	-	-	-	-	34	310	16	-	-	-
Oil and Gas Extraction	65	57	-	-	-	7	1910	1089	-	-	-
Petroleum Refineries	2679	994	-	-	-	20	327	3033	-	-	-
Electr., CHP+Heat Plants	2	4	-	-	-	12	719	1121	-	-	-
Pumped Storage (Elec.)	-	-	-	-	-	-	-	-	-	-	-
Other Energy Sector	-	29	-	-	-	1	25	107	-	-	-
Distribution Losses	-	14	-	-	-	-	-	-	-	-	-
FINAL CONSUMPTION	**899**	**8142**	**31407**	**110**	**2746**	**3015**	**40798**	**17910**	**15892**	**-**	**14704**
INDUSTRY SECTOR	**899**	**664**	**-**	**-**	**-**	**406**	**6297**	**14388**	**15892**	**-**	**-**
Iron and Steel	1	7	-	-	-	5	428	3487	-	-	-
Chemical and Petrochem.	800	475	-	-	-	76	1294	5433	15892	-	-
of which: Feedstocks	*243*	*73*	-	-	-	*12*	*19*	*961*	*15892*	-	-
Non-Ferrous Metals	-	4	-	-	-	7	176	626	-	-	-
Non-Metallic Minerals	7	88	-	-	-	25	1150	3010	-	-	-
Transport Equipment	3	7	-	-	-	40	222	143	-	-	-
Machinery	5	48	-	-	-	79	754	633	-	-	-
Mining and Quarrying	-	-	-	-	-	7	296	26	-	-	-
Food and Tobacco	-	18	-	-	-	31	329	283	-	-	-
Paper, Pulp and Print	79	1	-	-	-	43	137	95	-	-	-
Wood and Wood Products	-	-	-	-	-	13	118	23	-	-	-
Construction	-	4	-	-	-	50	935	140	-	-	-
Textile and Leather	-	10	-	-	-	22	300	388	-	-	-
Non-specified	4	2	-	-	-	8	158	101	-	-	-
TRANSPORT SECTOR	**-**	**1**	**31266**	**110**	**2746**	**-**	**17085**	**2217**	**-**	**-**	**-**
Air	-	-	-	110	2746	-	-	-	-	-	-
Road	-	1	31266	-	-	-	9858	76	-	-	-
Rail	-	-	-	-	-	-	3452	10	-	-	-
Pipeline Transport	-	-	-	-	-	-	9	58	-	-	-
Internal Navigation	-	-	-	-	-	-	3718	2073	-	-	-
Non-specified	-	-	-	-	-	-	48	-	-	-	-
OTHER SECTORS	**-**	**7477**	**-**	**-**	**-**	**2609**	**17416**	**1305**	**-**	**-**	**-**
Agriculture	-	-	-	-	-	17	10284	25	-	-	-
Comm. and Publ. Services	-	442	-	-	-	1945	6911	1280	-	-	-
Residential	-	7035	-	-	-	647	221	-	-	-	-
Non-specified	-	-	-	-	-	-	-	-	-	-	-
NON-ENERGY USE	**-**	**-**	**141**	**-**	**-**	**-**	**-**	**-**	**-**	**-**	**14704**
in Industry/Transf./Energy	-	-	141	-	-	-	-	-	-	-	14704
in Transport	-	-	-	-	-	-	-	-	-	-	-
in Other Sectors	-	-	-	-	-	-	-	-	-	-	-

People's Republic of China / République populaire de Chine : 1996

SUPPLY AND CONSUMPTION / *APPROVISIONNEMENT ET DEMANDE*	Gas / *Gaz* (TJ)				Comb. Renew. & Waste / *En. Re. Comb. & Déchets* (TJ)				(GWh)	(TJ)
	Natural Gas / *Gaz naturel*	Gas Works / *Usines à gaz*	Coke Ovens / *Cokeries*	Blast Furnaces / *Hauts fourneaux*	Solid Biomass / *Biomasse solide*	Gas/Liquids from Biomass / *Gaz/Liquides tirés de biomasse*	Municipal Waste / *Déchets urbains*	Industrial Waste / *Déchets industriels*	Electricity / *Electricité*	Heat / *Chaleur*
Production	871307	114533	434287	481306	8624499	50259	-	-	1080017	1148301
From Other Sources	-	-	-	-	-	-	-	-	-	-
Imports	-	-	-	-	-	-	-	-	124	-
Exports	-68823	-	-	-	-	-	-	-	-3712	-
Intl. Marine Bunkers	-	-	-	-	-	-	-	-	-	-
Stock Changes	-	-	-	-	-	-	-	-	-	-
DOMESTIC SUPPLY	**802484**	**114533**	**434287**	**481306**	**8624499**	**50259**	**-**	**-**	**1076429**	**1148301**
Transfers	-	-	-	-	-	-	-	-	-	-
Statistical Differences	-1614	-540	-818	47758	-	-	-	-	-	-2212
TRANSFORMATION	**33528**	**-**	**34681**	**77336**	**-**	**-**	**-**	**-**	**-**	**-**
Electricity Plants	32272	-	20883	62459	-	-	-	-	-	-
CHP Plants	-	-	-	-	-	-	-	-	-	-
Heat Plants	1256	-	13798	14877	-	-	-	-	-	-
Blast Furnaces/Gas Works	-	-	-	-	-	-	-	-	-	-
Coke/Pat. Fuel/BKB Plants	-	-	-	-	-	-	-	-	-	-
Petroleum Refineries	-	-	-	-	-	-	-	-	-	-
Petrochemical Industry	-	-	-	-	-	-	-	-	-	-
Liquefaction	-	-	-	-	-	-	-	-	-	-
Other Transform. Sector	-	-	-	-	-	-	-	-	-	-
ENERGY SECTOR	**165042**	**5975**	**32984**	**-**	**-**	**-**	**-**	**-**	**160912**	**240205**
Coal Mines	-	-	254	-	-	-	-	-	39135	1867
Oil and Gas Extraction	129002	-	-	-	-	-	-	-	25817	9951
Petroleum Refineries	34741	-	-	-	-	-	-	-	15565	195260
Electr., CHP+Heat Plants	173	459	2713	-	-	-	-	-	79315	31192
Pumped Storage (Elec.)	-	-	-	-	-	-	-	-	-	-
Other Energy Sector	1126	5516	30017	-	-	-	-	-	1080	1935
Distribution Losses	24562	-	-	-	-	-	-	-	74257	14869
FINAL CONSUMPTION	**577738**	**108018**	**365804**	**451728**	**8624499**	**50259**	**-**	**-**	**841260**	**891015**
INDUSTRY SECTOR	**476373**	**83170**	**301535**	**451728**	**-**	**-**	**-**	**-**	**584381**	**695292**
Iron and Steel	17934	36528	268765	451728	-	-	-	-	98965	111727
Chemical and Petrochem.	371326	2913	20278	-	-	-	-	-	147946	279228
of which: Feedstocks	*208188*	-	-	-	-	-	-	-	-	-
Non-Ferrous Metals	1733	9487	137	-	-	-	-	-	46523	31737
Non-Metallic Minerals	13125	8195	6948	-	-	-	-	-	65543	7600
Transport Equipment	1559	2079	293	-	-	-	-	-	16903	23426
Machinery	24821	15987	4567	-	-	-	-	-	51179	27758
Mining and Quarrying	2686	-	-	-	-	-	-	-	18957	18679
Food and Tobacco	5415	112	137	-	-	-	-	-	35299	53078
Paper, Pulp and Print	520	19	-	-	-	-	-	-	21889	40550
Wood and Wood Products	-	-	-	-	-	-	-	-	5690	4021
Construction	6541	-	117	-	-	-	-	-	17448	1577
Textile and Leather	26381	262	254	-	-	-	-	-	45836	84598
Non-specified	4332	7588	39	-	-	-	-	-	12203	11313
TRANSPORT SECTOR	**4938**	**-**	**-**	**-**	**-**	**-**	**-**	**-**	**11846**	**-**
Air	-	-	-	-	-	-	-	-	-	-
Road	4938	-	-	-	-	-	-	-	-	-
Rail	-	-	-	-	-	-	-	-	10806	-
Pipeline Transport	-	-	-	-	-	-	-	-	1040	-
Internal Navigation	-	-	-	-	-	-	-	-	-	-
Non-specified	-	-	-	-	-	-	-	-	-	-
OTHER SECTORS	**96427**	**24848**	**64269**	**-**	**8624499**	**50259**	**-**	**-**	**245033**	**195723**
Agriculture	996	-	-	-	-	-	-	-	61829	217
Comm. and Publ. Services	10007	2885	17077	-	-	-	-	-	70035	25620
Residential	85424	21963	47192	-	8624499	50259	-	-	113169	169886
Non-specified	-	-	-	-	-	-	-	-	-	-
NON-ENERGY USE	**-**	**-**	**-**	**-**	**-**	**-**	**-**	**-**	**-**	**-**
in Industry/Transf./Energy	-	-	-	-	-	-	-	-	-	-
in Transport	-	-	-	-	-	-	-	-	-	-
in Other Sectors	-	-	-	-	-	-	-	-	-	-

People's Republic of China / République populaire de Chine : 1997

SUPPLY AND CONSUMPTION / *APPROVISIONNEMENT ET DEMANDE*	Coal / *Charbon* (1000 tonnes)							Oil / *Pétrole* (1000 tonnes)			
	Coking Coal / *Charbon à coke*	Other Bit. Coal / *Autres charb. bit.*	Sub-Bit. Coal / *Charbon sous-bit.*	Lignite / *Lignite*	Peat / *Tourbe*	Oven and Gas Coke / *Coke de four/gaz*	Pat. Fuel and BKB / *Agg./briq. de lignite*	Crude Oil / *Pétrole brut*	NGL / *LGN*	Feed-stocks / *Produits d'aliment.*	Additives / *Additifs*
Production	185189	1187631	-	-	-	139957	9610	160741	-	-	-
From Other Sources	-	-	-	-	-	-	-	-	-	-	-
Imports	-	2013	-	-	-	252	-	35470	-	-	-
Exports	-	-30730	-	-	-	-10581	-	-19829	-	-	-
Intl. Marine Bunkers	-	-	-	-	-	-	-	-	-	-	-
Stock Changes	-	-12511	-	-	-	361	-	-1386	-	-	-
DOMESTIC SUPPLY	**185189**	**1146403**	**-**	**-**	**-**	**129989**	**9610**	**174996**	**-**	**-**	**-**
Transfers	-	-	-	-	-	-	-	-	-	-	-
Statistical Differences	-	60135	-	-	-	-6843	-	-1058	-	-	-
TRANSFORMATION	**185189**	**567968**	**-**	**-**	**-**	**68977**	**-**	**166865**	**-**	**-**	**-**
Electricity Plants	-	488524	-	-	-	-	-	684	-	-	-
CHP Plants	-	63714	-	-	-	-	-	-	-	-	-
Heat Plants	-	-	-	-	-	-	-	108	-	-	-
Blast Furnaces/Gas Works	-	7326	-	-	-	68977	-	-	-	-	-
Coke/Pat. Fuel/BKB Plants	185189	8404	-	-	-	-	-	-	-	-	-
Petroleum Refineries	-	-	-	-	-	-	-	166073	-	-	-
Petrochemical Industry	-	-	-	-	-	-	-	-	-	-	-
Liquefaction	-	-	-	-	-	-	-	-	-	-	-
Other Transform. Sector	-	-	-	-	-	-	-	-	-	-	-
ENERGY SECTOR	**-**	**69289**	**-**	**-**	**-**	**1916**	**-**	**3223**	**-**	**-**	**-**
Coal Mines	-	31153	-	-	-	415	-	15	-	-	-
Oil and Gas Extraction	-	2774	-	-	-	17	-	1859	-	-	-
Petroleum Refineries	-	-	-	-	-	-	-	1198	-	-	-
Electr., CHP+Heat Plants	-	26215	-	-	-	27	-	151	-	-	-
Pumped Storage (Elec.)	-	-	-	-	-	-	-	-	-	-	-
Other Energy Sector	-	9147	-	-	-	1457	-	-	-	-	-
Distribution Losses	-	21214	-	-	-	-	-	1758	-	-	-
FINAL CONSUMPTION	**-**	**548067**	**-**	**-**	**-**	**52253**	**9610**	**2092**	**-**	**-**	**-**
INDUSTRY SECTOR	**-**	**362229**	**-**	**-**	**-**	**39286**	**-**	**1687**	**-**	**-**	**-**
Iron and Steel	-	37781	-	-	-	18993	-	51	-	-	-
Chemical and Petrochem.	-	77949	-	-	-	6306	-	1520	-	-	-
of which: Feedstocks	-	-	-	-	-	-	-	*285*	-	-	-
Non-Ferrous Metals	-	8445	-	-	-	2451	-	5	-	-	-
Non-Metallic Minerals	-	121591	-	-	-	2970	-	62	-	-	-
Transport Equipment	-	6358	-	-	-	374	-	1	-	-	-
Machinery	-	21742	-	-	-	5358	-	8	-	-	-
Mining and Quarrying	-	5212	-	-	-	1803	-	2	-	-	-
Food and Tobacco	-	30669	-	-	-	415	-	5	-	-	-
Paper, Pulp and Print	-	15922	-	-	-	25	-	1	-	-	-
Wood and Wood Products	-	4918	-	-	-	21	-	-	-	-	-
Construction	-	4012	-	-	-	140	-	27	-	-	-
Textile and Leather	-	20431	-	-	-	91	-	3	-	-	-
Non-specified	-	7199	-	-	-	339	-	2	-	-	-
TRANSPORT SECTOR	**-**	**12370**	**-**	**-**	**-**	**-**	**-**	**-**	**-**	**-**	**-**
Air	-	-	-	-	-	-	-	-	-	-	-
Road	-	-	-	-	-	-	-	-	-	-	-
Rail	-	12281	-	-	-	-	-	-	-	-	-
Pipeline Transport	-	-	-	-	-	-	-	-	-	-	-
Internal Navigation	-	89	-	-	-	-	-	-	-	-	-
Non-specified	-	-	-	-	-	-	-	-	-	-	-
OTHER SECTORS	**-**	**159008**	**-**	**-**	**-**	**3240**	**9610**	**405**	**-**	**-**	**-**
Agriculture	-	19267	-	-	-	1447	-	121	-	-	-
Comm. and Publ. Services	-	26208	-	-	-	487	-	284	-	-	-
Residential	-	113533	-	-	-	1306	9610	-	-	-	-
Non-specified	-	-	-	-	-	-	-	-	-	-	-
NON-ENERGY USE	**-**	**14460**	**-**	**-**	**-**	**9727**	**-**	**-**	**-**	**-**	**-**
in Industry/Trans./Energy	-	14460	-	-	-	9727	-	-	-	-	-
in Transport	-	-	-	-	-	-	-	-	-	-	-
in Other Sectors	-	-	-	-	-	-	-	-	-	-	-

People's Republic of China / République populaire de Chine : 1997

	Oil cont. / *Pétrole cont.* (1000 tonnes)										
SUPPLY AND CONSUMPTION ***APPROVISIONNEMENT ET DEMANDE***	Refinery Gas *Gaz de raffinerie*	LPG + Ethane *GPL + éthane*	Motor Gasoline *Essence moteur*	Aviation Gasoline *Essence aviation*	Jet Fuel *Carbu- réacteurs*	Kerosene *Kérosène*	Gas/ Diesel *Gazole*	Heavy Fuel Oil *Fioul lourd*	Naphtha *Naphta*	Petrol. Coke *Coke de pétrole*	Other Prod. *Autres prod.*
Production	5341	6679	35060	132	3303	2826	49245	23112	17073	-	17814
From Other Sources	-	-	-	-	-	-	-	-	-	-	-
Imports	-	3582	84	-	-	1381	7428	13711	-	-	3582
Exports	-	-392	-1782	-	-	-723	-2321	-517	-	-	-1557
Intl. Marine Bunkers	-	-	-274	-	-	-	-343	-335	-	-	-
Stock Changes	-	51	-564	-	-	-146	-1820	-57	-	-	-
DOMESTIC SUPPLY	**5341**	**9920**	**32524**	**132**	**3303**	**3338**	**52189**	**35914**	**17073**	**-**	**19839**
Transfers	-	-	-	-	-	-	-	-	-	-	-
Statistical Differences	728	161	406	-	-	-270	2052	3943	-	-	1
TRANSFORMATION	**1039**	**12**	**8**	**-**	**-**	**-**	**6965**	**11828**	**-**	**-**	**-**
Electricity Plants	280	7	8	-	-	-	6638	8709	-	-	-
CHP Plants	-	-	-	-	-	-	-	-	-	-	-
Heat Plants	759	5	-	-	-	-	327	2502	-	-	-
Blast Furnaces/Gas Works	-	-	-	-	-	-	-	617	-	-	-
Coke/Pat. Fuel/BKB Plants	-	-	-	-	-	-	-	-	-	-	-
Petroleum Refineries	-	-	-	-	-	-	-	-	-	-	-
Petrochemical Industry	-	-	-	-	-	-	-	-	-	-	-
Liquefaction	-	-	-	-	-	-	-	-	-	-	-
Other Transform. Sector	-	-	-	-	-	-	-	-	-	-	-
ENERGY SECTOR	**3963**	**1395**	**-**	**-**	**-**	**154**	**3029**	**6790**	**-**	**-**	**-**
Coal Mines	-	-	-	-	-	71	285	20	-	-	-
Oil and Gas Extraction	94	73	-	-	-	15	1758	1378	-	-	-
Petroleum Refineries	3866	1280	-	-	-	41	301	3838	-	-	-
Electr., CHP+Heat Plants	3	5	-	-	-	25	662	1419	-	-	-
Pumped Storage (Elec.)	-	-	-	-	-	-	-	-	-	-	-
Other Energy Sector	-	37	-	-	-	2	23	135	-	-	-
Distribution Losses	-	18	-	-	-	-	-	-	-	-	-
FINAL CONSUMPTION	**1067**	**8656**	**32922**	**132**	**3303**	**2914**	**44247**	**21239**	**17073**	**-**	**19840**
INDUSTRY SECTOR	**1067**	**679**	**-**	**-**	**-**	**355**	**5867**	**13596**	**17073**	**-**	**-**
Iron and Steel	1	7	-	-	-	4	399	3295	-	-	-
Chemical and Petrochem.	949	487	-	-	-	67	1205	5134	17073	-	-
of which: Feedstocks	*288*	*75*	-	-	-	-	*20*	*908*	*17073*	-	-
Non-Ferrous Metals	-	4	-	-	-	6	164	592	-	-	-
Non-Metallic Minerals	8	90	-	-	-	22	1072	2844	-	-	-
Transport Equipment	4	7	-	-	-	35	207	135	-	-	-
Machinery	6	49	-	-	-	69	702	598	-	-	-
Mining and Quarrying	-	-	-	-	-	6	276	25	-	-	-
Food and Tobacco	-	18	-	-	-	27	307	267	-	-	-
Paper, Pulp and Print	94	1	-	-	-	38	128	90	-	-	-
Wood and Wood Products	-	-	-	-	-	11	110	22	-	-	-
Construction	-	4	-	-	-	44	870	132	-	-	-
Textile and Leather	-	10	-	-	-	19	280	367	-	-	-
Non-specified	5	2	-	-	-	7	147	95	-	-	-
TRANSPORT SECTOR	**-**	**16**	**32776**	**132**	**3303**	**-**	**18530**	**5752**	**-**	**-**	**-**
Air	-	-	-	132	3303	-	-	-	-	-	-
Road	-	16	32776	-	-	-	10624	197	-	-	-
Rail	-	-	-	-	-	-	3776	26	-	-	-
Pipeline Transport	-	-	-	-	-	-	10	151	-	-	-
Internal Navigation	-	-	-	-	-	-	4067	5378	-	-	-
Non-specified	-	-	-	-	-	-	53	-	-	-	-
OTHER SECTORS	**-**	**7961**	**-**	**-**	**-**	**2559**	**19850**	**1891**	**-**	**-**	**-**
Agriculture	-	-	-	-	-	14	10757	29	-	-	-
Comm. and Publ. Services	-	471	-	-	-	1910	8815	1862	-	-	-
Residential	-	7490	-	-	-	635	278	-	-	-	-
Non-specified	-	-	-	-	-	-	-	-	-	-	-
NON-ENERGY USE	**-**	**-**	**146**	**-**	**-**	**-**	**-**	**-**	**-**	**-**	**19840**
in Industry/Transf./Energy	-	-	146	-	-	-	-	-	-	-	19840
in Transport	-	-	-	-	-	-	-	-	-	-	-
in Other Sectors	-	-	-	-	-	-	-	-	-	-	-

People's Republic of China / République populaire de Chine : 1997

SUPPLY AND CONSUMPTION	Gas / *Gaz* (TJ)				Comb. Renew. & Waste / *En. Re. Comb. & Déchets* (TJ)				(GWh)	(TJ)
SUPPLY AND CONSUMPTION / ***APPROVISIONNEMENT ET DEMANDE***	Natural Gas / *Gaz naturel*	Gas Works / *Usines à gaz*	Coke Ovens / *Cokeries*	Blast Furnaces / *Hauts fourneaux*	Solid Biomass / *Biomasse solide*	Gas/Liquids from Biomass / *Gaz/Liquides tirés de biomasse*	Municipal Waste / *Déchets urbains*	Industrial Waste / *Déchets industriels*	Electricity / *Electricité*	Heat / *Chaleur*
Production	983449	120351	488338	553382	8670677	50528	-	-	1134471	1175695
From Other Sources	-	-	-	-	-	-	-	-	-	-
Imports	-	-	-	-	-	-	-	-	89	-
Exports	-107462	-	-	-	-	-	-	-	-7204	-
Intl. Marine Bunkers	-	-	-	-	-	-	-	-	-	-
Stock Changes	-	-	-	-	-	-	-	-	-	-
DOMESTIC SUPPLY	**875987**	**120351**	**488338**	**553382**	**8670677**	**50528**	**-**	**-**	**1127356**	**1175695**
Transfers	-	-	-	-	-	-	-	-	-	-
Statistical Differences	5381	-1	-28084	54910	-	-	-	-	2	12201
TRANSFORMATION	**92181**	**-**	**44439**	**88917**	**-**	**-**	**-**	**-**	**-**	**-**
Electricity Plants	88727	-	26759	71812	-	-	-	-	-	-
CHP Plants	-	-	-	-	-	-	-	-	-	-
Heat Plants	3454	-	17680	17105	-	-	-	-	-	-
Blast Furnaces/Gas Works	-	-	-	-	-	-	-	-	-	-
Coke/Pat. Fuel/BKB Plants	-	-	-	-	-	-	-	-	-	-
Petroleum Refineries	-	-	-	-	-	-	-	-	-	-
Petrochemical Industry	-	-	-	-	-	-	-	-	-	-
Liquefaction	-	-	-	-	-	-	-	-	-	-
Other Transform. Sector	-	-	-	-	-	-	-	-	-	-
ENERGY SECTOR	**247085**	**14506**	**35053**	**-**	**-**	**-**	**-**	**-**	**194284**	**239323**
Coal Mines	-	-	270	-	-	-	-	-	47251	1860
Oil and Gas Extraction	193129	-	-	-	-	-	-	-	31171	9914
Petroleum Refineries	52011	-	-	-	-	-	-	-	18793	194544
Electr., CHP+Heat Plants	259	1114	2883	-	-	-	-	-	95765	31077
Pumped Storage (Elec.)	-	-	-	-	-	-	-	-	-	-
Other Energy Sector	1686	13392	31900	-	-	-	-	-	1304	1928
Distribution Losses	36771	-	-	-	-	-	-	-	89657	14814
FINAL CONSUMPTION	**505331**	**105844**	**380762**	**519375**	**8670677**	**50528**	**-**	**-**	**843417**	**933759**
INDUSTRY SECTOR	**401483**	**81496**	**310125**	**519375**	**-**	**-**	**-**	**-**	**559020**	**727461**
Iron and Steel	15114	35793	276422	519375	-	-	-	-	94670	116897
Chemical and Petrochem.	312950	2854	20856	-	-	-	-	-	141525	292147
of which: Feedstocks	*175459*	-	-	-	-	-	-	-	-	-
Non-Ferrous Metals	1460	9296	141	-	-	-	-	-	44504	33205
Non-Metallic Minerals	11062	8030	7146	-	-	-	-	-	62699	7952
Transport Equipment	1314	2037	301	-	-	-	-	-	16169	24510
Machinery	20919	15665	4697	-	-	-	-	-	48958	29042
Mining and Quarrying	2264	-	-	-	-	-	-	-	18134	19543
Food and Tobacco	4564	110	141	-	-	-	-	-	33767	55534
Paper, Pulp and Print	438	19	-	-	-	-	-	-	20939	42427
Wood and Wood Products	-	-	-	-	-	-	-	-	5443	4207
Construction	5513	-	120	-	-	-	-	-	16691	1649
Textile and Leather	22234	257	261	-	-	-	-	-	43847	88512
Non-specified	3651	7435	40	-	-	-	-	-	11674	11836
TRANSPORT SECTOR	**850**	**-**	**-**	**-**	**-**	**-**	**-**	**-**	**15318**	**-**
Air	-	-	-	-	-	-	-	-	-	-
Road	850	-	-	-	-	-	-	-	-	-
Rail	-	-	-	-	-	-	-	-	13973	-
Pipeline Transport	-	-	-	-	-	-	-	-	1345	-
Internal Navigation	-	-	-	-	-	-	-	-	-	-
Non-specified	-	-	-	-	-	-	-	-	-	-
OTHER SECTORS	**102998**	**24348**	**70637**	**-**	**8670677**	**50528**	**-**	**-**	**269079**	**206298**
Agriculture	459	-	-	-	-	-	-	-	63977	533
Comm. and Publ. Services	10752	2827	18769	-	-	-	-	-	78406	32763
Residential	91787	21521	51868	-	8670677	50528	-	-	126696	173002
Non-specified	-	-	-	-	-	-	-	-	-	-
NON-ENERGY USE	**-**	**-**	**-**	**-**	**-**	**-**	**-**	**-**	**-**	**-**
in Industry/Transf./Energy	-	-	-	-	-	-	-	-	-	-
in Transport	-	-	-	-	-	-	-	-	-	-
in Other Sectors	-	-	-	-	-	-	-	-	-	-

Chinese Taipei / Taipei chinois : 1996

SUPPLY AND CONSUMPTION / *APPROVISIONNEMENT ET DEMANDE*	Coal / *Charbon* (1000 tonnes)							Oil / *Pétrole* (1000 tonnes)			
	Coking Coal / *Charbon à coke*	Other Bit. Coal / *Autres charb. bit.*	Sub-Bit. Coal / *Charbon sous-bit.*	Lignite / *Lignite*	Peat / *Tourbe*	Oven and Gas Coke / *Coke de four/gaz*	Pat. Fuel and BKB / *Agg./briq. de lignite*	Crude Oil / *Pétrole brut*	NGL / *LGN*	Feed-stocks / *Produits d'aliment.*	Additives / *Additifs*
Production	-	147	-	-	-	3063	-	55	-	-	-
From Other Sources	-	-	-	-	-	-	-	-	-	-	-
Imports	4112	26976	-	-	-	53	-	32522	-	-	-
Exports	-	-	-	-	-	-	-	-	-	-	-
Intl. Marine Bunkers	-	-	-	-	-	-	-	-	-	-	-
Stock Changes	-126	-1114	-	-	-	-216	-	-548	-	-	-
DOMESTIC SUPPLY	**3986**	**26009**	**-**	**-**	**-**	**2900**	**-**	**32029**	**-**	**-**	**-**
Transfers	-	-	-	-	-	-	-	-	-	1393	-
Statistical Differences	-	-2	-	-	-	-1	-	-1	-	-	-
TRANSFORMATION	**3986**	**19862**	**-**	**-**	**-**	**2319**	**-**	**32028**	**-**	**1393**	**-**
Electricity Plants	-	19862	-	-	-	-	-	-	-	-	-
CHP Plants	-	-	-	-	-	-	-	-	-	-	-
Heat Plants	-	-	-	-	-	-	-	-	-	-	-
Blast Furnaces/Gas Works	-	-	-	-	-	2319	-	-	-	-	-
Coke/Pat. Fuel/BKB Plants	3986	-	-	-	-	-	-	-	-	-	-
Petroleum Refineries	-	-	-	-	-	-	-	32028	-	1393	-
Petrochemical Industry	-	-	-	-	-	-	-	-	-	-	-
Liquefaction	-	-	-	-	-	-	-	-	-	-	-
Other Transform. Sector	-	-	-	-	-	-	-	-	-	-	-
ENERGY SECTOR	**-**	**-**	**-**	**-**	**-**	**-**	**-**	**-**	**-**	**-**	**-**
Coal Mines	-	-	-	-	-	-	-	-	-	-	-
Oil and Gas Extraction	-	-	-	-	-	-	-	-	-	-	-
Petroleum Refineries	-	-	-	-	-	-	-	-	-	-	-
Electr., CHP+Heat Plants	-	-	-	-	-	-	-	-	-	-	-
Pumped Storage (Elec.)	-	-	-	-	-	-	-	-	-	-	-
Other Energy Sector	-	-	-	-	-	-	-	-	-	-	-
Distribution Losses	-	-	-	-	-	-	-	-	-	-	-
FINAL CONSUMPTION	**-**	**6145**	**-**	**-**	**-**	**580**	**-**	**-**	**-**	**-**	**-**
INDUSTRY SECTOR	**-**	**6145**	**-**	**-**	**-**	**580**	**-**	**-**	**-**	**-**	**-**
Iron and Steel	-	839	-	-	-	580	-	-	-	-	-
Chemical and Petrochem.	-	1647	-	-	-	-	-	-	-	-	-
of which: Feedstocks	-	-	-	-	-	-	-	-	-	-	-
Non-Ferrous Metals	-	-	-	-	-	-	-	-	-	-	-
Non-Metallic Minerals	-	3122	-	-	-	-	-	-	-	-	-
Transport Equipment	-	-	-	-	-	-	-	-	-	-	-
Machinery	-	-	-	-	-	-	-	-	-	-	-
Mining and Quarrying	-	-	-	-	-	-	-	-	-	-	-
Food and Tobacco	-	-	-	-	-	-	-	-	-	-	-
Paper, Pulp and Print	-	421	-	-	-	-	-	-	-	-	-
Wood and Wood Products	-	-	-	-	-	-	-	-	-	-	-
Construction	-	-	-	-	-	-	-	-	-	-	-
Textile and Leather	-	116	-	-	-	-	-	-	-	-	-
Non-specified	-	-	-	-	-	-	-	-	-	-	-
TRANSPORT SECTOR	**-**	**-**	**-**	**-**	**-**	**-**	**-**	**-**	**-**	**-**	**-**
Air	-	-	-	-	-	-	-	-	-	-	-
Road	-	-	-	-	-	-	-	-	-	-	-
Rail	-	-	-	-	-	-	-	-	-	-	-
Pipeline Transport	-	-	-	-	-	-	-	-	-	-	-
Internal Navigation	-	-	-	-	-	-	-	-	-	-	-
Non-specified	-	-	-	-	-	-	-	-	-	-	-
OTHER SECTORS	**-**	**-**	**-**	**-**	**-**	**-**	**-**	**-**	**-**	**-**	**-**
Agriculture	-	-	-	-	-	-	-	-	-	-	-
Comm. and Publ. Services	-	-	-	-	-	-	-	-	-	-	-
Residential	-	-	-	-	-	-	-	-	-	-	-
Non-specified	-	-	-	-	-	-	-	-	-	-	-
NON-ENERGY USE	**-**	**-**	**-**	**-**	**-**	**-**	**-**	**-**	**-**	**-**	**-**
in Industry/Trans./Energy	-	-	-	-	-	-	-	-	-	-	-
in Transport	-	-	-	-	-	-	-	-	-	-	-
in Other Sectors	-	-	-	-	-	-	-	-	-	-	-

Chinese Taipei / Taipei chinois : 1996

SUPPLY AND CONSUMPTION / *APPROVISIONNEMENT ET DEMANDE*	Oil cont. / *Pétrole cont.* (1000 tonnes)										
	Refinery Gas / *Gaz de raffinerie*	LPG + Ethane / *GPL + éthane*	Motor Gasoline / *Essence moteur*	Aviation Gasoline / *Essence aviation*	Jet Fuel / *Carbu-réacteurs*	Kerosene / *Kérosène*	Gas/ Diesel / *Gazole*	Heavy Fuel Oil / *Fioul lourd*	Naphtha / *Naphta*	Petrol. Coke / *Coke de pétrole*	Other Prod. / *Autres prod.*
Production	1030	765	4465	2	1352	408	5580	13685	3656	-	2110
From Other Sources	-	-	-	-	-	-	-	-	-	-	-
Imports	-	812	1548	-	553	-	-	2286	864	-	803
Exports	-	-1	-	-	-	-	-879	-892	-112	-	-238
Intl. Marine Bunkers	-	-	-	-	-	-	-273	-2090	-	-	-1
Stock Changes	-	-57	186	-	-4	-276	229	-358	32	-	-193
DOMESTIC SUPPLY	**1030**	**1519**	**6199**	**2**	**1901**	**132**	**4657**	**12631**	**4440**	**-**	**2481**
Transfers	-	-	-	-1	-62	-93	-8	-91	-872	-	-265
Statistical Differences	-1	1	-	-1	-	1	-2	-1	-	-	-2
TRANSFORMATION	**8**	**-**	**-**	**-**	**-**	**-**	**436**	**5718**	**-**	**-**	**-**
Electricity Plants	8	-	-	-	-	-	436	5718	-	-	-
CHP Plants	-	-	-	-	-	-	-	-	-	-	-
Heat Plants	-	-	-	-	-	-	-	-	-	-	-
Blast Furnaces/Gas Works	-	-	-	-	-	-	-	-	-	-	-
Coke/Pat. Fuel/BKB Plants	-	-	-	-	-	-	-	-	-	-	-
Petroleum Refineries	-	-	-	-	-	-	-	-	-	-	-
Petrochemical Industry	-	-	-	-	-	-	-	-	-	-	-
Liquefaction	-	-	-	-	-	-	-	-	-	-	-
Other Transform. Sector	-	-	-	-	-	-	-	-	-	-	-
ENERGY SECTOR	**1021**	**11**	**2**	**-**	**1**	**-**	**30**	**484**	**-**	**-**	**-**
Coal Mines	-	-	-	-	-	-	-	-	-	-	-
Oil and Gas Extraction	-	-	-	-	-	-	-	-	-	-	-
Petroleum Refineries	1021	11	2	-	1	-	30	476	-	-	-
Electr., CHP+Heat Plants	-	-	-	-	-	-	-	-	-	-	-
Pumped Storage (Elec.)	-	-	-	-	-	-	-	-	-	-	-
Other Energy Sector	-	-	-	-	-	-	-	8	-	-	-
Distribution Losses	-	-	-	-	-	-	-	-	-	-	-
FINAL CONSUMPTION	**-**	**1509**	**6197**	**-**	**1838**	**40**	**4181**	**6337**	**3568**	**-**	**2214**
INDUSTRY SECTOR	**-**	**443**	**3**	**-**	**-**	**10**	**162**	**5712**	**3568**	**-**	**928**
Iron and Steel	-	34	-	-	-	4	8	593	-	-	-
Chemical and Petrochem.	-	37	-	-	-	1	26	2111	3568	-	928
of which: Feedstocks	-	-	-	-	-	-	-	-	*3568*	-	*928*
Non-Ferrous Metals	-	17	-	-	-	1	2	90	-	-	-
Non-Metallic Minerals	-	99	-	-	-	-	32	448	-	-	-
Transport Equipment	-	16	-	-	-	-	8	-	-	-	-
Machinery	-	14	-	-	-	-	9	17	-	-	-
Mining and Quarrying	-	-	-	-	-	-	15	15	-	-	-
Food and Tobacco	-	10	-	-	-	1	13	410	-	-	-
Paper, Pulp and Print	-	13	-	-	-	-	2	419	-	-	-
Wood and Wood Products	-	-	-	-	-	-	-	-	-	-	-
Construction	-	-	-	-	-	-	39	132	-	-	-
Textile and Leather	-	7	-	-	-	-	3	1157	-	-	-
Non-specified	-	196	3	-	-	3	5	320	-	-	-
TRANSPORT SECTOR	**-**	**6**	**6163**	**-**	**1838**	**-**	**2839**	**221**	**-**	**-**	**-**
Air	-	-	-	-	1838	-	2	-	-	-	-
Road	-	6	6163	-	-	-	2729	11	-	-	-
Rail	-	-	-	-	-	-	6	4	-	-	-
Pipeline Transport	-	-	-	-	-	-	-	-	-	-	-
Internal Navigation	-	-	-	-	-	-	102	206	-	-	-
Non-specified	-	-	-	-	-	-	-	-	-	-	-
OTHER SECTORS	**-**	**1060**	**31**	**-**	**-**	**30**	**1180**	**404**	**-**	**-**	**-**
Agriculture	-	-	-	-	-	-	777	92	-	-	-
Comm. and Publ. Services	-	-	30	-	-	28	320	285	-	-	-
Residential	-	1060	-	-	-	-	-	-	-	-	-
Non-specified	-	-	1	-	-	2	83	27	-	-	-
NON-ENERGY USE	**-**	**-**	**-**	**-**	**-**	**-**	**-**	**-**	**-**	**-**	**1286**
in Industry/Transf./Energy	-	-	-	-	-	-	-	-	-	-	1286
in Transport	-	-	-	-	-	-	-	-	-	-	-
in Other Sectors	-	-	-	-	-	-	-	-	-	-	-

Chinese Taipei / Taipei chinois : 1996

SUPPLY AND CONSUMPTION / *APPROVISIONNEMENT ET DEMANDE*	Gas / *Gaz* (TJ)				Comb. Renew. & Waste / *En. Re. Comb. & Déchets* (TJ)				(GWh)	(TJ)
	Natural Gas / *Gaz naturel*	Gas Works / *Usines à gaz*	Coke Ovens / *Cokeries*	Blast Furnaces / *Hauts fourneaux*	Solid Biomass / *Biomasse solide*	Gas/Liquids from Biomass / *Gaz/Liquides tirés de biomasse*	Municipal Waste / *Déchets urbains*	Industrial Waste / *Déchets industriels*	Electricity / *Electricité*	Heat / *Chaleur*
Production	33574	-	30731	29104	537	-	-	-	135277	-
From Other Sources	-	-	-	-	-	-	-	-	-	-
Imports	142400	-	-	-	-	-	-	-	-	-
Exports	-	-	-	-	-	-	-	-	-	-
Intl. Marine Bunkers	-	-	-	-	-	-	-	-	-	-
Stock Changes	-3998	-	-	-	-	-	-	-	-	-
DOMESTIC SUPPLY	**171976**	**-**	**30731**	**29104**	**537**	**-**	**-**	**-**	**135277**	**-**
Transfers	-	-	-	-	-	-	-	-	-	-
Statistical Differences	-1	-	1	-1	-	-	-	-	6684	-
TRANSFORMATION	**63441**	**-**	**4516**	**9612**	**-**	**-**	**-**	**-**	**-**	**-**
Electricity Plants	63132	-	4516	9612	-	-	-	-	-	-
CHP Plants	-	-	-	-	-	-	-	-	-	-
Heat Plants	-	-	-	-	-	-	-	-	-	-
Blast Furnaces/Gas Works	-	-	-	-	-	-	-	-	-	-
Coke/Pat. Fuel/BKB Plants	-	-	-	-	-	-	-	-	-	-
Petroleum Refineries	-	-	-	-	-	-	-	-	-	-
Petrochemical Industry	-	-	-	-	-	-	-	-	-	-
Liquefaction	-	-	-	-	-	-	-	-	-	-
Other Transform. Sector	309	-	-	-	-	-	-	-	-	-
ENERGY SECTOR	**31261**	**-**	**-**	**-**	**-**	**-**	**-**	**-**	**14258**	**-**
Coal Mines	-	-	-	-	-	-	-	-	8	-
Oil and Gas Extraction	-	-	-	-	-	-	-	-	6	-
Petroleum Refineries	572	-	-	-	-	-	-	-	2446	-
Electr., CHP+Heat Plants	-	-	-	-	-	-	-	-	5610	-
Pumped Storage (Elec.)	-	-	-	-	-	-	-	-	5549	-
Other Energy Sector	30689	-	-	-	-	-	-	-	639	-
Distribution Losses	2450	-	-	139	-	-	-	-	7656	-
FINAL CONSUMPTION	**74823**	**-**	**26216**	**19352**	**537**	**-**	**-**	**-**	**120047**	**-**
INDUSTRY SECTOR	**39959**	**-**	**26216**	**19352**	**-**	**-**	**-**	**-**	**62002**	**-**
Iron and Steel	4612	-	26216	19352	-	-	-	-	7868	-
Chemical and Petrochem.	19730	-	-	-	-	-	-	-	17894	-
of which: Feedstocks	*18299*	-	-	-	-	-	-	-	-	-
Non-Ferrous Metals	-	-	-	-	-	-	-	-	344	-
Non-Metallic Minerals	12053	-	-	-	-	-	-	-	4563	-
Transport Equipment	-	-	-	-	-	-	-	-	1335	-
Machinery	2347	-	-	-	-	-	-	-	11893	-
Mining and Quarrying	-	-	-	-	-	-	-	-	170	-
Food and Tobacco	416	-	-	-	-	-	-	-	3080	-
Paper, Pulp and Print	-	-	-	-	-	-	-	-	3904	-
Wood and Wood Products	-	-	-	-	-	-	-	-	570	-
Construction	-	-	-	-	-	-	-	-	494	-
Textile and Leather	793	-	-	-	-	-	-	-	8011	-
Non-specified	8	-	-	-	-	-	-	-	1876	-
TRANSPORT SECTOR	**-**	**-**	**-**	**-**	**-**	**-**	**-**	**-**	**513**	**-**
Air	-	-	-	-	-	-	-	-	165	-
Road	-	-	-	-	-	-	-	-	17	-
Rail	-	-	-	-	-	-	-	-	329	-
Pipeline Transport	-	-	-	-	-	-	-	-	-	-
Internal Navigation	-	-	-	-	-	-	-	-	2	-
Non-specified	-	-	-	-	-	-	-	-	-	-
OTHER SECTORS	**34864**	**-**	**-**	**-**	**537**	**-**	**-**	**-**	**57532**	**-**
Agriculture	-	-	-	-	-	-	-	-	2223	-
Comm. and Publ. Services	6393	-	-	-	-	-	-	-	16786	-
Residential	27264	-	-	-	-	-	-	-	27580	-
Non-specified	1207	-	-	-	537	-	-	-	10943	-
NON-ENERGY USE	**-**	**-**	**-**	**-**	**-**	**-**	**-**	**-**	**-**	**-**
in Industry/Transf./Energy	-	-	-	-	-	-	-	-	-	-
in Transport	-	-	-	-	-	-	-	-	-	-
in Other Sectors	-	-	-	-	-	-	-	-	-	-

Chinese Taipei / Taipei chinois : 1997

SUPPLY AND CONSUMPTION / *APPROVISIONNEMENT ET DEMANDE*	Coal / *Charbon* (1000 tonnes)							Oil / *Pétrole* (1000 tonnes)			
	Coking Coal / *Charbon à coke*	Other Bit. Coal / *Autres charb. bit.*	Sub-Bit. Coal / *Charbon sous-bit.*	Lignite / *Lignite*	Peat / *Tourbe*	Oven and Gas Coke / *Coke de four/gaz*	Pat. Fuel and BKB / *Agg./briq. de lignite*	Crude Oil / *Pétrole brut*	NGL / *LGN*	Feed-stocks / *Produits d'aliment.*	Additives / *Additifs*
Production	-	99	-	-	-	3993	-	48	-	-	-
From Other Sources	-	-	-	-	-	-	-	-	-	-	-
Imports	6300	29952	-	-	-	-	-	32366	-	-	-
Exports	-	-	-	-	-	-	-	-	-	-	-
Intl. Marine Bunkers	-	-	-	-	-	-	-	-	-	-	-
Stock Changes	-612	-1954	-	-	-	90	-	224	-	-	-
DOMESTIC SUPPLY	**5688**	**28097**	**-**	**-**	**-**	**4083**	**-**	**32638**	**-**	**-**	**-**
Transfers	-	-	-	-	-	-	-	-	-	1857	-
Statistical Differences	1	-	-	-	-	-	-	-	-	-	-
TRANSFORMATION	**5689**	**22201**	**-**	**-**	**-**	**3266**	**-**	**32638**	**-**	**1857**	**-**
Electricity Plants	-	22201	-	-	-	-	-	-	-	-	-
CHP Plants	-	-	-	-	-	-	-	-	-	-	-
Heat Plants	-	-	-	-	-	-	-	-	-	-	-
Blast Furnaces/Gas Works	-	-	-	-	-	3266	-	-	-	-	-
Coke/Pat. Fuel/BKB Plants	5689	-	-	-	-	-	-	-	-	-	-
Petroleum Refineries	-	-	-	-	-	-	-	32638	-	1857	-
Petrochemical Industry	-	-	-	-	-	-	-	-	-	-	-
Liquefaction	-	-	-	-	-	-	-	-	-	-	-
Other Transform. Sector	-	-	-	-	-	-	-	-	-	-	-
ENERGY SECTOR	**-**	**-**	**-**	**-**	**-**	**-**	**-**	**-**	**-**	**-**	**-**
Coal Mines	-	-	-	-	-	-	-	-	-	-	-
Oil and Gas Extraction	-	-	-	-	-	-	-	-	-	-	-
Petroleum Refineries	-	-	-	-	-	-	-	-	-	-	-
Electr., CHP+Heat Plants	-	-	-	-	-	-	-	-	-	-	-
Pumped Storage (Elec.)	-	-	-	-	-	-	-	-	-	-	-
Other Energy Sector	-	-	-	-	-	-	-	-	-	-	-
Distribution Losses	-	-	-	-	-	-	-	-	-	-	-
FINAL CONSUMPTION	**-**	**5896**	**-**	**-**	**-**	**817**	**-**	**-**	**-**	**-**	**-**
INDUSTRY SECTOR	**-**	**5896**	**-**	**-**	**-**	**817**	**-**	**-**	**-**	**-**	**-**
Iron and Steel	-	888	-	-	-	817	-	-	-	-	-
Chemical and Petrochem.	-	1592	-	-	-	-	-	-	-	-	-
of which: Feedstocks	-	-	-	-	-	-	-	-	-	-	-
Non-Ferrous Metals	-	-	-	-	-	-	-	-	-	-	-
Non-Metallic Minerals	-	2849	-	-	-	-	-	-	-	-	-
Transport Equipment	-	-	-	-	-	-	-	-	-	-	-
Machinery	-	-	-	-	-	-	-	-	-	-	-
Mining and Quarrying	-	-	-	-	-	-	-	-	-	-	-
Food and Tobacco	-	-	-	-	-	-	-	-	-	-	-
Paper, Pulp and Print	-	490	-	-	-	-	-	-	-	-	-
Wood and Wood Products	-	-	-	-	-	-	-	-	-	-	-
Construction	-	-	-	-	-	-	-	-	-	-	-
Textile and Leather	-	77	-	-	-	-	-	-	-	-	-
Non-specified	-	-	-	-	-	-	-	-	-	-	-
TRANSPORT SECTOR	**-**	**-**	**-**	**-**	**-**	**-**	**-**	**-**	**-**	**-**	**-**
Air	-	-	-	-	-	-	-	-	-	-	-
Road	-	-	-	-	-	-	-	-	-	-	-
Rail	-	-	-	-	-	-	-	-	-	-	-
Pipeline Transport	-	-	-	-	-	-	-	-	-	-	-
Internal Navigation	-	-	-	-	-	-	-	-	-	-	-
Non-specified	-	-	-	-	-	-	-	-	-	-	-
OTHER SECTORS	**-**	**-**	**-**	**-**	**-**	**-**	**-**	**-**	**-**	**-**	**-**
Agriculture	-	-	-	-	-	-	-	-	-	-	-
Comm. and Publ. Services	-	-	-	-	-	-	-	-	-	-	-
Residential	-	-	-	-	-	-	-	-	-	-	-
Non-specified	-	-	-	-	-	-	-	-	-	-	-
NON-ENERGY USE	**-**	**-**	**-**	**-**	**-**	**-**	**-**	**-**	**-**	**-**	**-**
in Industry/Trans./Energy	-	-	-	-	-	-	-	-	-	-	-
in Transport	-	-	-	-	-	-	-	-	-	-	-
in Other Sectors	-	-	-	-	-	-	-	-	-	-	-

Chinese Taipei / Taipei chinois : 1997

SUPPLY AND CONSUMPTION / *APPROVISIONNEMENT ET DEMANDE*	Oil cont. / *Pétrole cont.* (1000 tonnes)										
	Refinery Gas / *Gaz de raffinerie*	LPG + Ethane / *GPL + éthane*	Motor Gasoline / *Essence moteur*	Aviation Gasoline / *Essence aviation*	Jet Fuel / *Carbu-réacteurs*	Kerosene / *Kérosène*	Gas/ Diesel / *Gazole*	Heavy Fuel Oil / *Fioul lourd*	Naphtha / *Naphta*	Petrol. Coke / *Coke de pétrole*	Other Prod. / *Autres prod.*
Production	1043	808	4655	-	1501	465	5777	13698	3899	-	2321
From Other Sources	-	-	-	-	-	-	-	-	-	-	-
Imports	-	788	1151	-	416	-	-	3237	1381	-	757
Exports	-	-1	-	-	-	-	-1188	-1369	-76	-	-267
Intl. Marine Bunkers	-	-	-	-	-	-	-207	-2607	-	-	-2
Stock Changes	-	-58	529	1	23	-316	-106	475	-57	-	-236
DOMESTIC SUPPLY	**1043**	**1537**	**6335**	**1**	**1940**	**149**	**4276**	**13434**	**5147**	**-**	**2573**
Transfers	-	-	-19	-1	-65	-123	-1	-78	-1296	-	-292
Statistical Differences	-	-	-	-	-1	3	-	-2	-	-	2
TRANSFORMATION	**9**	**-**	**-**	**-**	**-**	**-**	**136**	**6306**	**-**	**-**	**-**
Electricity Plants	9	-	-	-	-	-	136	6306	-	-	-
CHP Plants	-	-	-	-	-	-	-	-	-	-	-
Heat Plants	-	-	-	-	-	-	-	-	-	-	-
Blast Furnaces/Gas Works	-	-	-	-	-	-	-	-	-	-	-
Coke/Pat. Fuel/BKB Plants	-	-	-	-	-	-	-	-	-	-	-
Petroleum Refineries	-	-	-	-	-	-	-	-	-	-	-
Petrochemical Industry	-	-	-	-	-	-	-	-	-	-	-
Liquefaction	-	-	-	-	-	-	-	-	-	-	-
Other Transform. Sector	-	-	-	-	-	-	-	-	-	-	-
ENERGY SECTOR	**1034**	**11**	**2**	**-**	**1**	**-**	**33**	**513**	**-**	**-**	**-**
Coal Mines	-	-	-	-	-	-	-	-	-	-	-
Oil and Gas Extraction	-	-	-	-	-	-	-	-	-	-	-
Petroleum Refineries	1034	11	2	-	1	-	32	506	-	-	-
Electr., CHP+Heat Plants	-	-	-	-	-	-	-	-	-	-	-
Pumped Storage (Elec.)	-	-	-	-	-	-	-	-	-	-	-
Other Energy Sector	-	-	-	-	-	-	1	7	-	-	-
Distribution Losses	-	-	-	-	-	-	-	-	-	-	-
FINAL CONSUMPTION	**-**	**1526**	**6314**	**-**	**1873**	**29**	**4106**	**6535**	**3851**	**-**	**2283**
INDUSTRY SECTOR	**-**	**423**	**4**	**-**	**-**	**11**	**172**	**6012**	**3851**	**-**	**978**
Iron and Steel	-	37	-	-	-	4	8	600	-	-	-
Chemical and Petrochem.	-	33	-	-	-	1	27	2291	3851	-	978
of which: Feedstocks	-	-	-	-	-	-	-	-	*3851*	-	*978*
Non-Ferrous Metals	-	19	-	-	-	2	3	104	-	-	-
Non-Metallic Minerals	-	81	-	-	-	-	33	458	-	-	-
Transport Equipment	-	16	-	-	-	-	11	-	-	-	-
Machinery	-	16	-	-	-	-	9	16	-	-	-
Mining and Quarrying	-	-	-	-	-	-	16	19	-	-	-
Food and Tobacco	-	9	-	-	-	1	12	410	-	-	-
Paper, Pulp and Print	-	13	-	-	-	-	2	397	-	-	-
Wood and Wood Products	-	-	-	-	-	-	-	-	-	-	-
Construction	-	-	-	-	-	-	42	120	-	-	-
Textile and Leather	-	6	-	-	-	-	3	1252	-	-	-
Non-specified	-	193	4	-	-	3	6	345	-	-	-
TRANSPORT SECTOR	**-**	**15**	**6293**	**-**	**1873**	**-**	**2954**	**188**	**-**	**-**	**-**
Air	-	-	-	-	1873	-	2	-	-	-	-
Road	-	15	6293	-	-	-	2812	1	-	-	-
Rail	-	-	-	-	-	-	-	-	-	-	-
Pipeline Transport	-	-	-	-	-	-	-	-	-	-	-
Internal Navigation	-	-	-	-	-	-	140	187	-	-	-
Non-specified	-	-	-	-	-	-	-	-	-	-	-
OTHER SECTORS	**-**	**1088**	**17**	**-**	**-**	**18**	**980**	**335**	**-**	**-**	**-**
Agriculture	-	-	-	-	-	-	676	92	-	-	-
Comm. and Publ. Services	-	-	17	-	-	16	215	216	-	-	-
Residential	-	1088	-	-	-	-	-	-	-	-	-
Non-specified	-	-	-	-	-	2	89	27	-	-	-
NON-ENERGY USE	**-**	**-**	**-**	**-**	**-**	**-**	**-**	**-**	**-**	**-**	**1305**
in Industry/Transf./Energy	-	-	-	-	-	-	-	-	-	-	1305
in Transport	-	-	-	-	-	-	-	-	-	-	-
in Other Sectors	-	-	-	-	-	-	-	-	-	-	-

Chinese Taipei / Taipei chinois : 1997

	Gas / *Gaz* (TJ)				Comb. Renew. & Waste / *En. Re. Comb. & Déchets* (TJ)				(GWh)	(TJ)
SUPPLY AND CONSUMPTION *APPROVISIONNEMENT ET DEMANDE*	Natural Gas *Gaz naturel*	Gas Works *Usines à gaz*	Coke Ovens *Cokeries*	Blast Furnaces *Hauts fourneaux*	Solid Biomass *Biomasse solide*	Gas/Liquids from Biomass *Gaz/Liquides tirés de biomasse*	Municipal Waste *Déchets urbains*	Industrial Waste *Déchets industriels*	Electricity *Electricité*	Heat *Chaleur*
Production	32036	-	43386	40718	537	-	-	-	153294	-
From Other Sources	-	-	-	-	-	-	-	-	-	-
Imports	174343	-	-	-	-	-	-	-	-	-
Exports	-	-	-	-	-	-	-	-	-	-
Intl. Marine Bunkers	-	-	-	-	-	-	-	-	-	-
Stock Changes	-7634	-	-	-	-	-	-	-	-	-
DOMESTIC SUPPLY	**198745**	**-**	**43386**	**40718**	**537**	**-**	**-**	**-**	**153294**	**-**
Transfers	-	-	-	-	-	-	-	-	-	-
Statistical Differences	1	-	-3	1	-	-	-	-	-3107	-
TRANSFORMATION	**89104**	**-**	**6084**	**12367**	**-**	**-**	**-**	**-**	**-**	**-**
Electricity Plants	88789	-	6084	12367	-	-	-	-	-	-
CHP Plants	-	-	-	-	-	-	-	-	-	-
Heat Plants	-	-	-	-	-	-	-	-	-	-
Blast Furnaces/Gas Works	-	-	-	-	-	-	-	-	-	-
Coke/Pat. Fuel/BKB Plants	-	-	-	-	-	-	-	-	-	-
Petroleum Refineries	-	-	-	-	-	-	-	-	-	-
Petrochemical Industry	-	-	-	-	-	-	-	-	-	-
Liquefaction	-	-	-	-	-	-	-	-	-	-
Other Transform. Sector	315	-	-	-	-	-	-	-	-	-
ENERGY SECTOR	**28953**	**-**	**-**	**-**	**-**	**-**	**-**	**-**	**14668**	**-**
Coal Mines	-	-	-	-	-	-	-	-	8	-
Oil and Gas Extraction	-	-	-	-	-	-	-	-	6	-
Petroleum Refineries	342	-	-	-	-	-	-	-	2462	-
Electr., CHP+Heat Plants	-	-	-	-	-	-	-	-	6017	-
Pumped Storage (Elec.)	-	-	-	-	-	-	-	-	5507	-
Other Energy Sector	28611	-	-	-	-	-	-	-	668	-
Distribution Losses	5041	-	-	397	-	-	-	-	6896	-
FINAL CONSUMPTION	**75648**	**-**	**37299**	**27955**	**537**	**-**	**-**	**-**	**128623**	**-**
INDUSTRY SECTOR	**40423**	**-**	**37299**	**27955**	**-**	**-**	**-**	**-**	**67914**	**-**
Iron and Steel	5397	-	37299	27955	-	-	-	-	9533	-
Chemical and Petrochem.	19538	-	-	-	-	-	-	-	19361	-
of which: Feedstocks	*17673*	-	-	-	-	-	-	-	-	-
Non-Ferrous Metals	-	-	-	-	-	-	-	-	387	-
Non-Metallic Minerals	11544	-	-	-	-	-	-	-	4552	-
Transport Equipment	-	-	-	-	-	-	-	-	1399	-
Machinery	2497	-	-	-	-	-	-	-	13932	-
Mining and Quarrying	-	-	-	-	-	-	-	-	178	-
Food and Tobacco	467	-	-	-	-	-	-	-	3082	-
Paper, Pulp and Print	-	-	-	-	-	-	-	-	3959	-
Wood and Wood Products	-	-	-	-	-	-	-	-	580	-
Construction	-	-	-	-	-	-	-	-	491	-
Textile and Leather	971	-	-	-	-	-	-	-	8435	-
Non-specified	9	-	-	-	-	-	-	-	2025	-
TRANSPORT SECTOR	**-**	**-**	**-**	**-**	**-**	**-**	**-**	**-**	**589**	**-**
Air	-	-	-	-	-	-	-	-	198	-
Road	-	-	-	-	-	-	-	-	18	-
Rail	-	-	-	-	-	-	-	-	370	-
Pipeline Transport	-	-	-	-	-	-	-	-	-	-
Internal Navigation	-	-	-	-	-	-	-	-	3	-
Non-specified	-	-	-	-	-	-	-	-	-	-
OTHER SECTORS	**35225**	**-**	**-**	**-**	**537**	**-**	**-**	**-**	**60120**	**-**
Agriculture	-	-	-	-	-	-	-	-	2264	-
Comm. and Publ. Services	6586	-	-	-	-	-	-	-	18038	-
Residential	27463	-	-	-	-	-	-	-	27841	-
Non-specified	1176	-	-	-	537	-	-	-	11977	-
NON-ENERGY USE	**-**	**-**	**-**	**-**	**-**	**-**	**-**	**-**	**-**	**-**
in Industry/Transf./Energy	-	-	-	-	-	-	-	-	-	-
in Transport	-	-	-	-	-	-	-	-	-	-
in Other Sectors	-	-	-	-	-	-	-	-	-	-

Colombia / Colombie : 1996

SUPPLY AND CONSUMPTION / *APPROVISIONNEMENT ET DEMANDE*	Coal / *Charbon* (1000 tonnes)							Oil / *Pétrole* (1000 tonnes)			
	Coking Coal / *Charbon à coke*	Other Bit. Coal / *Autres charb. bit.*	Sub-Bit. Coal / *Charbon sous-bit.*	Lignite / *Lignite*	Peat / *Tourbe*	Oven and Gas Coke / *Coke de four/gaz*	Pat. Fuel and BKB / *Agg./briq. de lignite*	Crude Oil / *Pétrole brut*	NGL / *LGN*	Feed-stocks / *Produits d'aliment.*	Additives / *Additifs*
Production	1776	28289	-	-	-	683	-	32312	329	-	-
From Other Sources	-	-	-	-	-	-	-	-	-	-	-
Imports	-	-	-	-	-	-	-	-	-	-	-
Exports	-800	-23981	-	-	-	-	-	-16607	-	-	-
Intl. Marine Bunkers	-	-	-	-	-	-	-	-	-	-	-
Stock Changes	-	-365	-	-	-	-136	-	-296	-	-	-
DOMESTIC SUPPLY	**976**	**3943**	**-**	**-**	**-**	**547**	**-**	**15409**	**329**	**-**	**-**
Transfers	-	-	-	-	-	-	-	-	-329	-	-
Statistical Differences	-	319	-	-	-	-1	-	2	-	-	-
TRANSFORMATION	**976**	**1203**	**-**	**-**	**-**	**414**	**-**	**14490**	**-**	**-**	**-**
Electricity Plants	-	875	-	-	-	-	-	14	-	-	-
CHP Plants	-	-	-	-	-	-	-	-	-	-	-
Heat Plants	-	-	-	-	-	-	-	-	-	-	-
Blast Furnaces/Gas Works	-	328	-	-	-	414	-	-	-	-	-
Coke/Pat. Fuel/BKB Plants	976	-	-	-	-	-	-	-	-	-	-
Petroleum Refineries	-	-	-	-	-	-	-	14476	-	-	-
Petrochemical Industry	-	-	-	-	-	-	-	-	-	-	-
Liquefaction	-	-	-	-	-	-	-	-	-	-	-
Other Transform. Sector	-	-	-	-	-	-	-	-	-	-	-
ENERGY SECTOR	**-**	**-**	**-**	**-**	**-**	**-**	**-**	**62**	**-**	**-**	**-**
Coal Mines	-	-	-	-	-	-	-	-	-	-	-
Oil and Gas Extraction	-	-	-	-	-	-	-	-	-	-	-
Petroleum Refineries	-	-	-	-	-	-	-	62	-	-	-
Electr., CHP+Heat Plants	-	-	-	-	-	-	-	-	-	-	-
Pumped Storage (Elec.)	-	-	-	-	-	-	-	-	-	-	-
Other Energy Sector	-	-	-	-	-	-	-	-	-	-	-
Distribution Losses	-	114	-	-	-	26	-	-	-	-	-
FINAL CONSUMPTION	**-**	**2945**	**-**	**-**	**-**	**106**	**-**	**859**	**-**	**-**	**-**
INDUSTRY SECTOR	**-**	**2728**	**-**	**-**	**-**	**106**	**-**	**847**	**-**	**-**	**-**
Iron and Steel	-	-	-	-	-	103	-	52	-	-	-
Chemical and Petrochem.	-	154	-	-	-	2	-	82	-	-	-
of which: Feedstocks	-	-	-	-	-	-	-	-	-	-	-
Non-Ferrous Metals	-	495	-	-	-	-	-	-	-	-	-
Non-Metallic Minerals	-	1075	-	-	-	-	-	162	-	-	-
Transport Equipment	-	-	-	-	-	1	-	-	-	-	-
Machinery	-	-	-	-	-	-	-	72	-	-	-
Mining and Quarrying	-	-	-	-	-	-	-	-	-	-	-
Food and Tobacco	-	205	-	-	-	-	-	214	-	-	-
Paper, Pulp and Print	-	461	-	-	-	-	-	1	-	-	-
Wood and Wood Products	-	1	-	-	-	-	-	106	-	-	-
Construction	-	-	-	-	-	-	-	13	-	-	-
Textile and Leather	-	337	-	-	-	-	-	93	-	-	-
Non-specified	-	-	-	-	-	-	-	52	-	-	-
TRANSPORT SECTOR	**-**	**1**	**-**	**-**	**-**	**-**	**-**	**-**	**-**	**-**	**-**
Air	-	-	-	-	-	-	-	-	-	-	-
Road	-	-	-	-	-	-	-	-	-	-	-
Rail	-	1	-	-	-	-	-	-	-	-	-
Pipeline Transport	-	-	-	-	-	-	-	-	-	-	-
Internal Navigation	-	-	-	-	-	-	-	-	-	-	-
Non-specified	-	-	-	-	-	-	-	-	-	-	-
OTHER SECTORS	**-**	**216**	**-**	**-**	**-**	**-**	**-**	**12**	**-**	**-**	**-**
Agriculture	-	-	-	-	-	-	-	5	-	-	-
Comm. and Publ. Services	-	-	-	-	-	-	-	7	-	-	-
Residential	-	216	-	-	-	-	-	-	-	-	-
Non-specified	-	-	-	-	-	-	-	-	-	-	-
NON-ENERGY USE	**-**	**-**	**-**	**-**	**-**	**-**	**-**	**-**	**-**	**-**	**-**
in Industry/Trans./Energy	-	-	-	-	-	-	-	-	-	-	-
in Transport	-	-	-	-	-	-	-	-	-	-	-
in Other Sectors	-	-	-	-	-	-	-	-	-	-	-

Colombia / Colombie : 1996

SUPPLY AND CONSUMPTION / *APPROVISIONNEMENT ET DEMANDE*	Oil cont. / *Pétrole cont.* (1000 tonnes)										
	Refinery Gas / *Gaz de raffinerie*	LPG + Ethane / *GPL + éthane*	Motor Gasoline / *Essence moteur*	Aviation Gasoline / *Essence aviation*	Jet Fuel / *Carbu-réacteurs*	Kerosene / *Kérosène*	Gas/ Diesel / *Gazole*	Heavy Fuel Oil / *Fioul lourd*	Naphtha / *Naphta*	Petrol. Coke / *Coke de pétrole*	Other Prod. / *Autres prod.*
Production	565	507	4562	21	819	160	3093	2445	-	-	1827
From Other Sources	-	-	-	-	-	-	-	-	-	-	-
Imports	-	2	1042	-	6	-	2	80	-	-	-
Exports	-	-29	-	-	-83	-	-587	-2316	-	-	-4
Intl. Marine Bunkers	-	-	-	-	-	-	-172	-17	-	-	-
Stock Changes	-	-	-122	-	-46	-	223	-41	-	-	43
DOMESTIC SUPPLY	**565**	**480**	**5482**	**21**	**696**	**160**	**2559**	**151**	**-**	**-**	**1866**
Transfers	-	179	150	-	-	-	-	-	-	-	-
Statistical Differences	-	-	5	-	-23	-1	-2	1	-	-	-39
TRANSFORMATION	**-**	**12**	**68**	**-**	**-**	**-**	**32**	**25**	**-**	**-**	**-**
Electricity Plants	-	-	-	-	-	-	32	25	-	-	-
CHP Plants	-	-	-	-	-	-	-	-	-	-	-
Heat Plants	-	-	-	-	-	-	-	-	-	-	-
Blast Furnaces/Gas Works	-	-	-	-	-	-	-	-	-	-	-
Coke/Pat. Fuel/BKB Plants	-	-	-	-	-	-	-	-	-	-	-
Petroleum Refineries	-	12	68	-	-	-	-	-	-	-	-
Petrochemical Industry	-	-	-	-	-	-	-	-	-	-	-
Liquefaction	-	-	-	-	-	-	-	-	-	-	-
Other Transform. Sector	-	-	-	-	-	-	-	-	-	-	-
ENERGY SECTOR	**565**	**5**	**7**	**-**	**-**	**1**	**15**	**44**	**-**	**-**	**-**
Coal Mines	-	-	-	-	-	-	-	-	-	-	-
Oil and Gas Extraction	-	-	-	-	-	-	-	-	-	-	-
Petroleum Refineries	565	5	7	-	-	1	15	44	-	-	-
Electr., CHP+Heat Plants	-	-	-	-	-	-	-	-	-	-	-
Pumped Storage (Elec.)	-	-	-	-	-	-	-	-	-	-	-
Other Energy Sector	-	-	-	-	-	-	-	-	-	-	-
Distribution Losses	-	-	-	-	-	-	-	-	-	-	-
FINAL CONSUMPTION	**-**	**642**	**5562**	**21**	**673**	**158**	**2510**	**83**	**-**	**-**	**1827**
INDUSTRY SECTOR	**-**	**60**	**58**	**-**	**-**	**91**	**429**	**81**	**-**	**-**	**-**
Iron and Steel	-	11	-	-	-	10	23	-	-	-	-
Chemical and Petrochem.	-	6	-	-	-	13	44	2	-	-	-
of which: Feedstocks	-	-	-	-	-	-	-	-	-	-	-
Non-Ferrous Metals	-	-	-	-	-	-	-	-	-	-	-
Non-Metallic Minerals	-	17	-	-	-	53	55	1	-	-	-
Transport Equipment	-	-	-	-	-	-	-	-	-	-	-
Machinery	-	2	-	-	-	-	16	57	-	-	-
Mining and Quarrying	-	-	-	-	-	-	-	-	-	-	-
Food and Tobacco	-	17	-	-	-	5	119	8	-	-	-
Paper, Pulp and Print	-	3	-	-	-	2	6	10	-	-	-
Wood and Wood Products	-	-	-	-	-	2	2	-	-	-	-
Construction	-	-	58	-	-	-	17	-	-	-	-
Textile and Leather	-	4	-	-	-	6	27	3	-	-	-
Non-specified	-	-	-	-	-	-	120	-	-	-	-
TRANSPORT SECTOR	**-**	**-**	**5423**	**21**	**673**	**-**	**1353**	**-**	**-**	**-**	**-**
Air	-	-	-	21	673	-	-	-	-	-	-
Road	-	-	5342	-	-	-	1280	-	-	-	-
Rail	-	-	-	-	-	-	-	-	-	-	-
Pipeline Transport	-	-	-	-	-	-	-	-	-	-	-
Internal Navigation	-	-	81	-	-	-	73	-	-	-	-
Non-specified	-	-	-	-	-	-	-	-	-	-	-
OTHER SECTORS	**-**	**582**	**81**	**-**	**-**	**67**	**728**	**2**	**-**	**-**	**-**
Agriculture	-	-	8	-	-	1	427	1	-	-	-
Comm. and Publ. Services	-	50	-	-	-	-	301	1	-	-	-
Residential	-	532	73	-	-	66	-	-	-	-	-
Non-specified	-	-	-	-	-	-	-	-	-	-	-
NON-ENERGY USE	**-**	**-**	**-**	**-**	**-**	**-**	**-**	**-**	**-**	**-**	**1827**
in Industry/Transf./Energy	-	-	-	-	-	-	-	-	-	-	1827
in Transport	-	-	-	-	-	-	-	-	-	-	-
in Other Sectors	-	-	-	-	-	-	-	-	-	-	-

Colombia / Colombie : 1996

SUPPLY AND CONSUMPTION / *APPROVISIONNEMENT ET DEMANDE*	Gas / *Gaz* (TJ)				Comb. Renew. & Waste / *En. Re. Comb. & Déchets* (TJ)				(GWh)	(TJ)
	Natural Gas / *Gaz naturel*	Gas Works / *Usines à gaz*	Coke Ovens / *Cokeries*	Blast Furnaces / *Hauts fourneaux*	Solid Biomass / *Biomasse solide*	Gas/Liquids from Biomass / *Gaz/Liquides tirés de biomasse*	Municipal Waste / *Déchets urbains*	Industrial Waste / *Déchets industriels*	Electricity / *Electricité*	Heat / *Chaleur*
Production	202081	-	2219	2609	288427	-	-	-	44605	-
From Other Sources	-	-	-	-	-	-	-	-	-	-
Imports	-	-	-	-	-	-	-	-	164	-
Exports	-	-	-	-	-	-	-	-	-3	-
Intl. Marine Bunkers	-	-	-	-	-	-	-	-	-	-
Stock Changes	-	-	-	-	-	-	-	-	-	-
DOMESTIC SUPPLY	**202081**	**-**	**2219**	**2609**	**288427**	**-**	**-**	**-**	**44766**	**-**
Transfers	-	-	-	-	-	-	-	-	-	-
Statistical Differences	-1458	-	-	-	45	-	-	-	1886	-
TRANSFORMATION	**70515**	**-**	**-**	**1965**	**9865**	**-**	**-**	**-**	**-**	**-**
Electricity Plants	70515	-	-	1965	5357	-	-	-	-	-
CHP Plants	-	-	-	-	-	-	-	-	-	-
Heat Plants	-	-	-	-	-	-	-	-	-	-
Blast Furnaces/Gas Works	-	-	-	-	-	-	-	-	-	-
Coke/Pat. Fuel/BKB Plants	-	-	-	-	-	-	-	-	-	-
Petroleum Refineries	-	-	-	-	-	-	-	-	-	-
Petrochemical Industry	-	-	-	-	-	-	-	-	-	-
Liquefaction	-	-	-	-	-	-	-	-	-	-
Other Transform. Sector	-	-	-	-	4508	-	-	-	-	-
ENERGY SECTOR	**64041**	**-**	**1985**	**-**	**-**	**-**	**-**	**-**	**703**	**-**
Coal Mines	-	-	-	-	-	-	-	-	-	-
Oil and Gas Extraction	21297	-	-	-	-	-	-	-	-	-
Petroleum Refineries	42744	-	-	-	-	-	-	-	-	-
Electr., CHP+Heat Plants	-	-	-	-	-	-	-	-	703	-
Pumped Storage (Elec.)	-	-	-	-	-	-	-	-	-	-
Other Energy Sector	-	-	1985	-	-	-	-	-	-	-
Distribution Losses	-	-	-	288	-	-	-	-	9743	-
FINAL CONSUMPTION	**66067**	**-**	**234**	**356**	**278607**	**-**	**-**	**-**	**36206**	**-**
INDUSTRY SECTOR	**48677**	**-**	**234**	**356**	**73502**	**-**	**-**	**-**	**11332**	**-**
Iron and Steel	1354	-	234	356	-	-	-	-	1907	-
Chemical and Petrochem.	20524	-	-	-	1139	-	-	-	1645	-
of which: Feedstocks	-	-	-	-	-	-	-	-	-	-
Non-Ferrous Metals	-	-	-	-	7	-	-	-	-	-
Non-Metallic Minerals	18638	-	-	-	737	-	-	-	1415	-
Transport Equipment	-	-	-	-	-	-	-	-	-	-
Machinery	1228	-	-	-	7	-	-	-	464	-
Mining and Quarrying	-	-	-	-	-	-	-	-	914	-
Food and Tobacco	3075	-	-	-	62033	-	-	-	1933	-
Paper, Pulp and Print	2323	-	-	-	9522	-	-	-	874	-
Wood and Wood Products	818	-	-	-	-	-	-	-	150	-
Construction	-	-	-	-	-	-	-	-	56	-
Textile and Leather	507	-	-	-	57	-	-	-	1583	-
Non-specified	210	-	-	-	-	-	-	-	391	-
TRANSPORT SECTOR	**2191**	**-**	**-**	**-**	**-**	**-**	**-**	**-**	**-**	**-**
Air	-	-	-	-	-	-	-	-	-	-
Road	-	-	-	-	-	-	-	-	-	-
Rail	-	-	-	-	-	-	-	-	-	-
Pipeline Transport	-	-	-	-	-	-	-	-	-	-
Internal Navigation	-	-	-	-	-	-	-	-	-	-
Non-specified	2191	-	-	-	-	-	-	-	-	-
OTHER SECTORS	**15199**	**-**	**-**	**-**	**205105**	**-**	**-**	**-**	**24874**	**-**
Agriculture	-	-	-	-	54816	-	-	-	376	-
Comm. and Publ. Services	2280	-	-	-	-	-	-	-	6600	-
Residential	12919	-	-	-	149547	-	-	-	14746	-
Non-specified	-	-	-	-	742	-	-	-	3152	-
NON-ENERGY USE	**-**	**-**	**-**	**-**	**-**	**-**	**-**	**-**	**-**	**-**
in Industry/Transf./Energy	-	-	-	-	-	-	-	-	-	-
in Transport	-	-	-	-	-	-	-	-	-	-
in Other Sectors	-	-	-	-	-	-	-	-	-	-

Colombia / Colombie : 1997

SUPPLY AND CONSUMPTION / *APPROVISIONNEMENT ET DEMANDE*	Coal / *Charbon* (1000 tonnes)							Oil / *Pétrole* (1000 tonnes)			
	Coking Coal / *Charbon à coke*	Other Bit. Coal / *Autres charb. bit.*	Sub-Bit. Coal / *Charbon sous-bit.*	Lignite / *Lignite*	Peat / *Tourbe*	Oven and Gas Coke / *Coke de four/gaz*	Pat. Fuel and BKB / *Agg./briq. de lignite*	Crude Oil / *Pétrole brut*	NGL / *LGN*	Feed-stocks / *Produits d'aliment.*	Additives / *Additifs*
Production	1671	28996	-	-	-	610	-	33636	225	-	-
From Other Sources	-	-	-	-	-	-	-	-	-	-	-
Imports	-	-	-	-	-	-	-	-	-	-	-
Exports	-800	-25696	-	-	-	-	-	-17867	-	-	-
Intl. Marine Bunkers	-	-	-	-	-	-	-	-	-	-	-
Stock Changes	-	57	-	-	-	-92	-	-158	-	-	-
DOMESTIC SUPPLY	**871**	**3357**	**-**	**-**	**-**	**518**	**-**	**15611**	**225**	**-**	**-**
Transfers	-	-	-	-	-	-	-	-	-225	-	-
Statistical Differences	-	1116	-	-	-	-	-	-299	-	-	-
TRANSFORMATION	**871**	**1331**	**-**	**-**	**-**	**394**	**-**	**14307**	**-**	**-**	**-**
Electricity Plants	-	1076	-	-	-	-	-	18	-	-	-
CHP Plants	-	-	-	-	-	-	-	-	-	-	-
Heat Plants	-	-	-	-	-	-	-	-	-	-	-
Blast Furnaces/Gas Works	-	255	-	-	-	394	-	-	-	-	-
Coke/Pat. Fuel/BKB Plants	871	-	-	-	-	-	-	-	-	-	-
Petroleum Refineries	-	-	-	-	-	-	-	14289	-	-	-
Petrochemical Industry	-	-	-	-	-	-	-	-	-	-	-
Liquefaction	-	-	-	-	-	-	-	-	-	-	-
Other Transform. Sector	-	-	-	-	-	-	-	-	-	-	-
ENERGY SECTOR	**-**	**-**	**-**	**-**	**-**	**-**	**-**	**71**	**-**	**-**	**-**
Coal Mines	-	-	-	-	-	-	-	-	-	-	-
Oil and Gas Extraction	-	-	-	-	-	-	-	-	-	-	-
Petroleum Refineries	-	-	-	-	-	-	-	71	-	-	-
Electr., CHP+Heat Plants	-	-	-	-	-	-	-	-	-	-	-
Pumped Storage (Elec.)	-	-	-	-	-	-	-	-	-	-	-
Other Energy Sector	-	-	-	-	-	-	-	-	-	-	-
Distribution Losses	-	122	-	-	-	22	-	-	-	-	-
FINAL CONSUMPTION	**-**	**3020**	**-**	**-**	**-**	**102**	**-**	**934**	**-**	**-**	**-**
INDUSTRY SECTOR	**-**	**2809**	**-**	**-**	**-**	**102**	**-**	**921**	**-**	**-**	**-**
Iron and Steel	-	-	-	-	-	99	-	89	-	-	-
Chemical and Petrochem.	-	159	-	-	-	2	-	56	-	-	-
of which: Feedstocks	-	-	-	-	-	-	-	-	-	-	-
Non-Ferrous Metals	-	510	-	-	-	-	-	-	-	-	-
Non-Metallic Minerals	-	1107	-	-	-	-	-	178	-	-	-
Transport Equipment	-	-	-	-	-	1	-	-	-	-	-
Machinery	-	-	-	-	-	-	-	79	-	-	-
Mining and Quarrying	-	-	-	-	-	-	-	-	-	-	-
Food and Tobacco	-	211	-	-	-	-	-	233	-	-	-
Paper, Pulp and Print	-	475	-	-	-	-	-	1	-	-	-
Wood and Wood Products	-	-	-	-	-	-	-	116	-	-	-
Construction	-	-	-	-	-	-	-	12	-	-	-
Textile and Leather	-	347	-	-	-	-	-	101	-	-	-
Non-specified	-	-	-	-	-	-	-	56	-	-	-
TRANSPORT SECTOR	**-**	**1**	**-**	**-**	**-**	**-**	**-**	**-**	**-**	**-**	**-**
Air	-	-	-	-	-	-	-	-	-	-	-
Road	-	-	-	-	-	-	-	-	-	-	-
Rail	-	1	-	-	-	-	-	-	-	-	-
Pipeline Transport	-	-	-	-	-	-	-	-	-	-	-
Internal Navigation	-	-	-	-	-	-	-	-	-	-	-
Non-specified	-	-	-	-	-	-	-	-	-	-	-
OTHER SECTORS	**-**	**210**	**-**	**-**	**-**	**-**	**-**	**13**	**-**	**-**	**-**
Agriculture	-	-	-	-	-	-	-	5	-	-	-
Comm. and Publ. Services	-	-	-	-	-	-	-	8	-	-	-
Residential	-	210	-	-	-	-	-	-	-	-	-
Non-specified	-	-	-	-	-	-	-	-	-	-	-
NON-ENERGY USE	**-**	**-**	**-**	**-**	**-**	**-**	**-**	**-**	**-**	**-**	**-**
in Industry/Trans./Energy	-	-	-	-	-	-	-	-	-	-	-
in Transport	-	-	-	-	-	-	-	-	-	-	-
in Other Sectors	-	-	-	-	-	-	-	-	-	-	-

Colombia / Colombie : 1997

SUPPLY AND CONSUMPTION / *APPROVISIONNEMENT ET DEMANDE*	Oil cont. / *Pétrole cont.* (1000 tonnes)										
	Refinery Gas / *Gaz de raffinerie*	LPG + Ethane / *GPL + éthane*	Motor Gasoline / *Essence moteur*	Aviation Gasoline / *Essence aviation*	Jet Fuel / *Carbu- réacteurs*	Kerosene / *Kérosène*	Gas/ Diesel / *Gazole*	Heavy Fuel Oil / *Fioul lourd*	Naphtha / *Naphta*	Petrol. Coke / *Coke de pétrole*	Other Prod. / *Autres prod.*
Production	568	509	4413	20	617	134	3105	2489	-	-	1686
From Other Sources	-	-	-	-	-	-	-	-	-	-	-
Imports	-	-	1306	-	6	-	1	5	-	-	-
Exports	-	-3	-	-	-41	-	-426	-2356	-	-	-
Intl. Marine Bunkers	-	-	-	-	-	-	-164	-6	-	-	-
Stock Changes	-	-	-113	-	27	-	-	-38	-	-	-
DOMESTIC SUPPLY	**568**	**506**	**5606**	**20**	**609**	**134**	**2516**	**94**	**-**	**-**	**1686**
Transfers	-	-	134	-	-	-	-	-	-	-	-
Statistical Differences	-	2	68	-	-2	-	90	134	-	-	239
TRANSFORMATION	**-**	**12**	**69**	**-**	**-**	**-**	**32**	**22**	**-**	**-**	**-**
Electricity Plants	-	-	-	-	-	-	32	22	-	-	-
CHP Plants	-	-	-	-	-	-	-	-	-	-	-
Heat Plants	-	-	-	-	-	-	-	-	-	-	-
Blast Furnaces/Gas Works	-	-	-	-	-	-	-	-	-	-	-
Coke/Pat. Fuel/BKB Plants	-	-	-	-	-	-	-	-	-	-	-
Petroleum Refineries	-	12	69	-	-	-	-	-	-	-	-
Petrochemical Industry	-	-	-	-	-	-	-	-	-	-	-
Liquefaction	-	-	-	-	-	-	-	-	-	-	-
Other Transform. Sector	-	-	-	-	-	-	-	-	-	-	-
ENERGY SECTOR	**568**	**4**	**7**	**-**	**-**	**1**	**13**	**40**	**-**	**-**	**-**
Coal Mines	-	-	-	-	-	-	-	-	-	-	-
Oil and Gas Extraction	-	-	-	-	-	-	-	-	-	-	-
Petroleum Refineries	568	4	7	-	-	1	13	40	-	-	-
Electr., CHP+Heat Plants	-	-	-	-	-	-	-	-	-	-	-
Pumped Storage (Elec.)	-	-	-	-	-	-	-	-	-	-	-
Other Energy Sector	-	-	-	-	-	-	-	-	-	-	-
Distribution Losses	-	-	-	-	-	-	-	-	-	-	-
FINAL CONSUMPTION	**-**	**492**	**5732**	**20**	**607**	**133**	**2561**	**166**	**-**	**-**	**1925**
INDUSTRY SECTOR	**-**	**46**	**59**	**-**	**-**	**77**	**419**	**150**	**-**	**-**	**-**
Iron and Steel	-	8	-	-	-	8	22	-	-	-	-
Chemical and Petrochem.	-	5	-	-	-	11	43	4	-	-	-
of which: Feedstocks	-	-	-	-	-	-	-	-	-	-	-
Non-Ferrous Metals	-	-	-	-	-	-	-	-	-	-	-
Non-Metallic Minerals	-	13	-	-	-	45	54	2	-	-	-
Transport Equipment	-	-	-	-	-	-	-	-	-	-	-
Machinery	-	2	-	-	-	-	16	105	-	-	-
Mining and Quarrying	-	-	-	-	-	-	-	-	-	-	-
Food and Tobacco	-	13	-	-	-	4	116	15	-	-	-
Paper, Pulp and Print	-	2	-	-	-	2	6	18	-	-	-
Wood and Wood Products	-	-	-	-	-	2	2	-	-	-	-
Construction	-	-	59	-	-	-	17	-	-	-	-
Textile and Leather	-	3	-	-	-	5	26	6	-	-	-
Non-specified	-	-	-	-	-	-	117	-	-	-	-
TRANSPORT SECTOR	**-**	**-**	**5601**	**20**	**607**	**-**	**1387**	**-**	**-**	**-**	**-**
Air	-	-	-	20	607	-	-	-	-	-	-
Road	-	-	5517	-	-	-	1312	-	-	-	-
Rail	-	-	-	-	-	-	-	-	-	-	-
Pipeline Transport	-	-	-	-	-	-	-	-	-	-	-
Internal Navigation	-	-	84	-	-	-	75	-	-	-	-
Non-specified	-	-	-	-	-	-	-	-	-	-	-
OTHER SECTORS	**-**	**446**	**72**	**-**	**-**	**56**	**755**	**16**	**-**	**-**	**-**
Agriculture	-	-	-	-	-	1	430	2	-	-	-
Comm. and Publ. Services	-	38	-	-	-	-	305	14	-	-	-
Residential	-	408	63	-	-	55	-	-	-	-	-
Non-specified	-	-	9	-	-	-	20	-	-	-	-
NON-ENERGY USE	**-**	**-**	**-**	**-**	**-**	**-**	**-**	**-**	**-**	**-**	**1925**
in Industry/Transf./Energy	-	-	-	-	-	-	-	-	-	-	1925
in Transport	-	-	-	-	-	-	-	-	-	-	-
in Other Sectors	-	-	-	-	-	-	-	-	-	-	-

Colombia / Colombie : 1997

SUPPLY AND CONSUMPTION / *APPROVISIONNEMENT ET DEMANDE*	Gas / *Gaz* (TJ)				Comb. Renew. & Waste / *En. Re. Comb. & Déchets* (TJ)				(GWh)	(TJ)
	Natural Gas / *Gaz naturel*	Gas Works / *Usines à gaz*	Coke Ovens / *Cokeries*	Blast Furnaces / *Hauts fourneaux*	Solid Biomass / *Biomasse solide*	Gas/Liquids from Biomass / *Gaz/Liquides tirés de biomasse*	Municipal Waste / *Déchets urbains*	Industrial Waste / *Déchets industriels*	Electricity / *Electricité*	Heat / *Chaleur*
Production	248458	-	2109	2544	220566	-	-	-	46115	-
From Other Sources	-	-	-	-	-	-	-	-	-	-
Imports	-	-	-	-	-	-	-	-	199	-
Exports	-	-	-	-	-	-	-	-	-	-
Intl. Marine Bunkers	-	-	-	-	-	-	-	-	-	-
Stock Changes	-	-	-	-	-	-	-	-	-	-
DOMESTIC SUPPLY	**248458**	**-**	**2109**	**2544**	**220566**	**-**	**-**	**-**	**46314**	**-**
Transfers	-	-	-	-	-	-	-	-	-	-
Statistical Differences	-1017	-	1	-	-546	-	-	-	69	-
TRANSFORMATION	**112562**	**-**	**-**	**1809**	**7722**	**-**	**-**	**-**	**-**	**-**
Electricity Plants	112562	-	-	1809	5419	-	-	-	-	-
CHP Plants	-	-	-	-	-	-	-	-	-	-
Heat Plants	-	-	-	-	-	-	-	-	-	-
Blast Furnaces/Gas Works	-	-	-	-	-	-	-	-	-	-
Coke/Pat. Fuel/BKB Plants	-	-	-	-	-	-	-	-	-	-
Petroleum Refineries	-	-	-	-	-	-	-	-	-	-
Petrochemical Industry	-	-	-	-	-	-	-	-	-	-
Liquefaction	-	-	-	-	-	-	-	-	-	-
Other Transform. Sector	-	-	-	-	2303	-	-	-	-	-
ENERGY SECTOR	**66709**	**-**	**2025**	**-**	**-**	**-**	**-**	**-**	**768**	**-**
Coal Mines	-	-	-	-	-	-	-	-	-	-
Oil and Gas Extraction	20962	-	-	-	-	-	-	-	-	-
Petroleum Refineries	45747	-	-	-	-	-	-	-	-	-
Electr., CHP+Heat Plants	-	-	-	-	-	-	-	-	768	-
Pumped Storage (Elec.)	-	-	-	-	-	-	-	-	-	-
Other Energy Sector	-	-	2025	-	-	-	-	-	-	-
Distribution Losses	-	-	-	251	-	-	-	-	10183	-
FINAL CONSUMPTION	**68170**	**-**	**85**	**484**	**212298**	**-**	**-**	**-**	**35432**	**-**
INDUSTRY SECTOR	**46999**	**-**	**85**	**484**	**72494**	**-**	**-**	**-**	**11283**	**-**
Iron and Steel	1464	-	85	484	-	-	-	-	1899	-
Chemical and Petrochem.	20273	-	-	-	1484	-	-	-	1638	-
of which: Feedstocks	-	-	-	-	-	-	-	-	-	-
Non-Ferrous Metals	-	-	-	-	6	-	-	-	-	-
Non-Metallic Minerals	18453	-	-	-	651	-	-	-	1409	-
Transport Equipment	-	-	-	-	-	-	-	-	-	-
Machinery	-	-	-	-	6	-	-	-	462	-
Mining and Quarrying	-	-	-	-	-	-	-	-	910	-
Food and Tobacco	2884	-	-	-	60748	-	-	-	1925	-
Paper, Pulp and Print	2396	-	-	-	9526	-	-	-	870	-
Wood and Wood Products	943	-	-	-	-	-	-	-	149	-
Construction	-	-	-	-	-	-	-	-	56	-
Textile and Leather	477	-	-	-	58	-	-	-	1576	-
Non-specified	109	-	-	-	15	-	-	-	389	-
TRANSPORT SECTOR	**2306**	**-**	**-**	**-**	**-**	**-**	**-**	**-**	**43**	**-**
Air	-	-	-	-	-	-	-	-	-	-
Road	-	-	-	-	-	-	-	-	-	-
Rail	-	-	-	-	-	-	-	-	43	-
Pipeline Transport	-	-	-	-	-	-	-	-	-	-
Internal Navigation	-	-	-	-	-	-	-	-	-	-
Non-specified	2306	-	-	-	-	-	-	-	-	-
OTHER SECTORS	**18865**	**-**	**-**	**-**	**139804**	**-**	**-**	**-**	**24106**	**-**
Agriculture	-	-	-	-	49326	-	-	-	308	-
Comm. and Publ. Services	2830	-	-	-	-	-	-	-	6795	-
Residential	16035	-	-	-	89636	-	-	-	14728	-
Non-specified	-	-	-	-	842	-	-	-	2275	-
NON-ENERGY USE	**-**	**-**	**-**	**-**	**-**	**-**	**-**	**-**	**-**	**-**
in Industry/Transf./Energy	-	-	-	-	-	-	-	-	-	-
in Transport	-	-	-	-	-	-	-	-	-	-
in Other Sectors	-	-	-	-	-	-	-	-	-	-

Congo

SUPPLY AND CONSUMPTION 1996	Coal (1000 tonnes)							Oil (1000 tonnes)			
	Coking Coal	Other Bit. Coal	Sub-Bit. Coal	Lignite	Peat	Oven and Gas Coke	Pat. Fuel and BKB	Crude Oil	NGL	Feed-stocks	Additives
Production	-	-	-	-	-	-	-	10370	-	-	-
Imports	-	-	-	-	-	-	-	-	-	-	-
Exports	-	-	-	-	-	-	-	-8996	-	-	-
Intl. Marine Bunkers	-	-	-	-	-	-	-	-	-	-	-
Stock Changes	-	-	-	-	-	-	-	-764	-	-	-
DOMESTIC SUPPLY	-	-	-	-	-	-	-	**610**	-	-	-
Transfers and Stat. Diff.	-	-	-	-	-	-	-	-	-	-	-
TRANSFORMATION	-	-	-	-	-	-	-	**610**	-	-	-
Electricity and CHP Plants	-	-	-	-	-	-	-	-	-	-	-
Petroleum Refineries	-	-	-	-	-	-	-	610	-	-	-
Other Transform. Sector	-	-	-	-	-	-	-	-	-	-	-
ENERGY SECTOR	-	-	-	-	-	-	-	-	-	-	-
DISTRIBUTION LOSSES	-	-	-	-	-	-	-	-	-	-	-
FINAL CONSUMPTION	-	-	-	-	-	-	-	-	-	-	-
INDUSTRY SECTOR	-	-	-	-	-	-	-	-	-	-	-
Iron and Steel	-	-	-	-	-	-	-	-	-	-	-
Chemical and Petrochem.	-	-	-	-	-	-	-	-	-	-	-
Non-Metallic Minerals	-	-	-	-	-	-	-	-	-	-	-
Non-specified	-	-	-	-	-	-	-	-	-	-	-
TRANSPORT SECTOR	-	-	-	-	-	-	-	-	-	-	-
Air	-	-	-	-	-	-	-	-	-	-	-
Road	-	-	-	-	-	-	-	-	-	-	-
Non-specified	-	-	-	-	-	-	-	-	-	-	-
OTHER SECTORS	-	-	-	-	-	-	-	-	-	-	-
Agriculture	-	-	-	-	-	-	-	-	-	-	-
Comm. and Publ. Services	-	-	-	-	-	-	-	-	-	-	-
Residential	-	-	-	-	-	-	-	-	-	-	-
Non-specified	-	-	-	-	-	-	-	-	-	-	-
NON-ENERGY USE	-	-	-	-	-	-	-	-	-	-	-

APPROVISIONNEMENT ET DEMANDE 1997	*Charbon* (1000 tonnes)							*Pétrole* (1000 tonnes)			
	Charbon à coke	*Autres charb. bit.*	*Charbon sous-bit.*	*Lignite*	*Tourbe*	*Coke de four/gaz*	*Agg./briq. de lignite*	*Pétrole brut*	*LGN*	*Produits d'aliment.*	*Additifs*
Production	-	-	-	-	-	-	-	12330	165	-	-
Imports	-	-	-	-	-	-	-	-	-	-	-
Exports	-	-	-	-	-	-	-	-11720	-	-	-
Intl. Marine Bunkers	-	-	-	-	-	-	-	-	-	-	-
Stock Changes	-	-	-	-	-	-	-	-	-	-	-
DOMESTIC SUPPLY	-	-	-	-	-	-	-	**610**	**165**	-	-
Transfers and Stat. Diff.	-	-	-	-	-	-	-	-	-165	-	-
TRANSFORMATION	-	-	-	-	-	-	-	**610**	-	-	-
Electricity and CHP Plants	-	-	-	-	-	-	-	-	-	-	-
Petroleum Refineries	-	-	-	-	-	-	-	610	-	-	-
Other Transform. Sector	-	-	-	-	-	-	-	-	-	-	-
ENERGY SECTOR	-	-	-	-	-	-	-	-	-	-	-
DISTRIBUTION LOSSES	-	-	-	-	-	-	-	-	-	-	-
FINAL CONSUMPTION	-	-	-	-	-	-	-	-	-	-	-
INDUSTRY SECTOR	-	-	-	-	-	-	-	-	-	-	-
Iron and Steel	-	-	-	-	-	-	-	-	-	-	-
Chemical and Petrochem.	-	-	-	-	-	-	-	-	-	-	-
Non-Metallic Minerals	-	-	-	-	-	-	-	-	-	-	-
Non-specified	-	-	-	-	-	-	-	-	-	-	-
TRANSPORT SECTOR	-	-	-	-	-	-	-	-	-	-	-
Air	-	-	-	-	-	-	-	-	-	-	-
Road	-	-	-	-	-	-	-	-	-	-	-
Non-specified	-	-	-	-	-	-	-	-	-	-	-
OTHER SECTORS	-	-	-	-	-	-	-	-	-	-	-
Agriculture	-	-	-	-	-	-	-	-	-	-	-
Comm. and Publ. Services	-	-	-	-	-	-	-	-	-	-	-
Residential	-	-	-	-	-	-	-	-	-	-	-
Non-specified	-	-	-	-	-	-	-	-	-	-	-
NON-ENERGY USE	-	-	-	-	-	-	-	-	-	-	-

Congo

SUPPLY AND CONSUMPTION 1996	Oil cont. (1000 tonnes)										
	Refinery Gas	LPG + Ethane	Motor Gasoline	Aviation Gasoline	Jet Fuel	Kerosene	Gas/ Diesel	Heavy Fuel Oil	Naphtha	Petrol. Coke	Other Prod.
Production	-	4	55	57	16	50	92	340	-	-	11
Imports	-	-	-	-	-	-	-	5	-	-	10
Exports	-	-	-3	-	-4	-	-10	-293	-	-	-
Intl. Marine Bunkers	-	-	-	-	-	-	-	-10	-	-	-
Stock Changes	-	-	-	-	-	-	-	-	-	-	-
DOMESTIC SUPPLY	**-**	**4**	**52**	**57**	**12**	**50**	**82**	**42**	**-**	**-**	**21**
Transfers and Stat. Diff.	-	-1	1	-	-	-2	-	-	-	-	-1
TRANSFORMATION	**-**	**-**	**-**	**-**	**-**	**-**	**1**	**-**	**-**	**-**	**-**
Electricity and CHP Plants	-	-	-	-	-	-	1	-	-	-	-
Petroleum Refineries	-	-	-	-	-	-	-	-	-	-	-
Other Transform. Sector	-	-	-	-	-	-	-	-	-	-	-
ENERGY SECTOR	**-**	**-**	**-**	**-**	**-**	**-**	**-**	**38**	**-**	**-**	**-**
DISTRIBUTION LOSSES	**-**	**-**	**-**	**-**	**-**	**-**	**-**	**-**	**-**	**-**	**-**
FINAL CONSUMPTION	**-**	**3**	**53**	**57**	**12**	**48**	**81**	**4**	**-**	**-**	**20**
INDUSTRY SECTOR	**-**	**-**	**-**	**-**	**-**	**-**	**-**	**4**	**-**	**-**	**-**
Iron and Steel	-	-	-	-	-	-	-	-	-	-	-
Chemical and Petrochem.	-	-	-	-	-	-	-	-	-	-	-
Non-Metallic Minerals	-	-	-	-	-	-	-	-	-	-	-
Non-specified	-	-	-	-	-	-	-	4	-	-	-
TRANSPORT SECTOR	**-**	**-**	**53**	**57**	**12**	**-**	**81**	**-**	**-**	**-**	**-**
Air	-	-	-	57	12	-	-	-	-	-	-
Road	-	-	53	-	-	-	81	-	-	-	-
Non-specified	-	-	-	-	-	-	-	-	-	-	-
OTHER SECTORS	**-**	**3**	**-**	**-**	**-**	**48**	**-**	**-**	**-**	**-**	**-**
Agriculture	-	-	-	-	-	-	-	-	-	-	-
Comm. and Publ. Services	-	-	-	-	-	-	-	-	-	-	-
Residential	-	3	-	-	-	48	-	-	-	-	-
Non-specified	-	-	-	-	-	-	-	-	-	-	-
NON-ENERGY USE	**-**	**-**	**-**	**-**	**-**	**-**	**-**	**-**	**-**	**-**	**20**

APPROVISIONNEMENT ET DEMANDE 1997	*Pétrole cont.* (1000 tonnes)										
	Gaz de raffinerie	*GPL + éthane*	*Essence moteur*	*Essence aviation*	*Carbu- réacteurs*	*Kérosène*	*Gazole*	*Fioul lourd*	*Naphta*	*Coke de pétrole*	*Autres prod.*
Production	-	4	55	57	16	50	92	340	-	-	11
Imports	-	-	-	-	-	-	-	5	-	-	10
Exports	-	-	-3	-	-4	-	-10	-293	-	-	-
Intl. Marine Bunkers	-	-	-	-	-	-	-	-10	-	-	-
Stock Changes	-	-	-	-	-	-	-	-	-	-	-
DOMESTIC SUPPLY	**-**	**4**	**52**	**57**	**12**	**50**	**82**	**42**	**-**	**-**	**21**
Transfers and Stat. Diff.	-	-1	1	-	-	-2	-	-	-	-	-1
TRANSFORMATION	**-**	**-**	**-**	**-**	**-**	**-**	**1**	**-**	**-**	**-**	**-**
Electricity and CHP Plants	-	-	-	-	-	-	1	-	-	-	-
Petroleum Refineries	-	-	-	-	-	-	-	-	-	-	-
Other Transform. Sector	-	-	-	-	-	-	-	-	-	-	-
ENERGY SECTOR	**-**	**-**	**-**	**-**	**-**	**-**	**-**	**38**	**-**	**-**	**-**
DISTRIBUTION LOSSES	**-**	**-**	**-**	**-**	**-**	**-**	**-**	**-**	**-**	**-**	**-**
FINAL CONSUMPTION	**-**	**3**	**53**	**57**	**12**	**48**	**81**	**4**	**-**	**-**	**20**
INDUSTRY SECTOR	**-**	**-**	**-**	**-**	**-**	**-**	**-**	**4**	**-**	**-**	**-**
Iron and Steel	-	-	-	-	-	-	-	-	-	-	-
Chemical and Petrochem.	-	-	-	-	-	-	-	-	-	-	-
Non-Metallic Minerals	-	-	-	-	-	-	-	-	-	-	-
Non-specified	-	-	-	-	-	-	-	4	-	-	-
TRANSPORT SECTOR	**-**	**-**	**53**	**57**	**12**	**-**	**81**	**-**	**-**	**-**	**-**
Air	-	-	-	57	12	-	-	-	-	-	-
Road	-	-	53	-	-	-	81	-	-	-	-
Non-specified	-	-	-	-	-	-	-	-	-	-	-
OTHER SECTORS	**-**	**3**	**-**	**-**	**-**	**48**	**-**	**-**	**-**	**-**	**-**
Agriculture	-	-	-	-	-	-	-	-	-	-	-
Comm. and Publ. Services	-	-	-	-	-	-	-	-	-	-	-
Residential	-	3	-	-	-	48	-	-	-	-	-
Non-specified	-	-	-	-	-	-	-	-	-	-	-
NON-ENERGY USE	**-**	**-**	**-**	**-**	**-**	**-**	**-**	**-**	**-**	**-**	**20**

Congo

SUPPLY AND CONSUMPTION 1996	Gas (TJ)				Comb. Renew. & Waste (TJ)				(GWh)	(TJ)
	Natural Gas	Gas Works	Coke Ovens	Blast Furnaces	Solid Biomass	Gas/Liquids from Biomass	Municipal Waste	Industrial Waste	Electricity	Heat
Production	131	-	-	-	35294	-	-	-	435	-
Imports	-	-	-	-	-	-	-	-	115	-
Exports	-	-	-	-	-	-	-	-	-	-
Intl. Marine Bunkers	-	-	-	-	-	-	-	-	-	-
Stock Changes	-	-	-	-	-	-	-	-	-	-
DOMESTIC SUPPLY	**131**	**-**	**-**	**-**	**35294**	**-**	**-**	**-**	**550**	**-**
Transfers and Stat. Diff.	-	-	-	-	-1	-	-	-	-	-
TRANSFORMATION	**131**	**-**	**-**	**-**	**1381**	**-**	**-**	**-**	**-**	**-**
Electricity and CHP Plants	131	-	-	-	-	-	-	-	-	-
Petroleum Refineries	-	-	-	-	-	-	-	-	-	-
Other Transform. Sector	-	-	-	-	1381	-	-	-	-	-
ENERGY SECTOR	**-**	**-**	**-**	**-**	**-**	**-**	**-**	**-**	**7**	**-**
DISTRIBUTION LOSSES	**-**	**-**	**-**	**-**	**-**	**-**	**-**	**-**	**4**	**-**
FINAL CONSUMPTION	**-**	**-**	**-**	**-**	**33912**	**-**	**-**	**-**	**539**	**-**
INDUSTRY SECTOR	**-**	**-**	**-**	**-**	**5626**	**-**	**-**	**-**	**288**	**-**
Iron and Steel	-	-	-	-	-	-	-	-	-	-
Chemical and Petrochem.	-	-	-	-	-	-	-	-	-	-
Non-Metallic Minerals	-	-	-	-	-	-	-	-	-	-
Non-specified	-	-	-	-	5626	-	-	-	288	-
TRANSPORT SECTOR	**-**	**-**	**-**	**-**	**-**	**-**	**-**	**-**	**-**	**-**
Air	-	-	-	-	-	-	-	-	-	-
Road	-	-	-	-	-	-	-	-	-	-
Non-specified	-	-	-	-	-	-	-	-	-	-
OTHER SECTORS	**-**	**-**	**-**	**-**	**28286**	**-**	**-**	**-**	**251**	**-**
Agriculture	-	-	-	-	-	-	-	-	-	-
Comm. and Publ. Services	-	-	-	-	-	-	-	-	-	-
Residential	-	-	-	-	28286	-	-	-	251	-
Non-specified	-	-	-	-	-	-	-	-	-	-
NON-ENERGY USE	**-**	**-**	**-**	**-**	**-**	**-**	**-**	**-**	**-**	**-**

APPROVISIONNEMENT ET DEMANDE 1997	*Gaz* (TJ)				*En. Re. Comb. & Déchets* (TJ)				(GWh)	(TJ)
	Gaz naturel	*Usines à gaz*	*Cokeries*	*Hauts fourneaux*	*Biomasse solide*	*Gaz/Liquides tirés de biomasse*	*Déchets urbains*	*Déchets industriels*	*Electricité*	*Chaleur*
Production	131	-	-	-	36106	-	-	-	431	-
Imports	-	-	-	-	-	-	-	-	114	-
Exports	-	-	-	-	-	-	-	-	-	-
Intl. Marine Bunkers	-	-	-	-	-	-	-	-	-	-
Stock Changes	-	-	-	-	-	-	-	-	-	-
DOMESTIC SUPPLY	**131**	**-**	**-**	**-**	**36106**	**-**	**-**	**-**	**545**	**-**
Transfers and Stat. Diff.	-	-	-	-	-2	-	-	-	-	-
TRANSFORMATION	**131**	**-**	**-**	**-**	**1412**	**-**	**-**	**-**	**-**	**-**
Electricity and CHP Plants	131	-	-	-	-	-	-	-	-	-
Petroleum Refineries	-	-	-	-	-	-	-	-	-	-
Other Transform. Sector	-	-	-	-	1412	-	-	-	-	-
ENERGY SECTOR	**-**	**-**	**-**	**-**	**-**	**-**	**-**	**-**	**7**	**-**
DISTRIBUTION LOSSES	**-**	**-**	**-**	**-**	**-**	**-**	**-**	**-**	**4**	**-**
FINAL CONSUMPTION	**-**	**-**	**-**	**-**	**34692**	**-**	**-**	**-**	**534**	**-**
INDUSTRY SECTOR	**-**	**-**	**-**	**-**	**5755**	**-**	**-**	**-**	**285**	**-**
Iron and Steel	-	-	-	-	-	-	-	-	-	-
Chemical and Petrochem.	-	-	-	-	-	-	-	-	-	-
Non-Metallic Minerals	-	-	-	-	-	-	-	-	-	-
Non-specified	-	-	-	-	5755	-	-	-	285	-
TRANSPORT SECTOR	**-**	**-**	**-**	**-**	**-**	**-**	**-**	**-**	**-**	**-**
Air	-	-	-	-	-	-	-	-	-	-
Road	-	-	-	-	-	-	-	-	-	-
Non-specified	-	-	-	-	-	-	-	-	-	-
OTHER SECTORS	**-**	**-**	**-**	**-**	**28937**	**-**	**-**	**-**	**249**	**-**
Agriculture	-	-	-	-	-	-	-	-	-	-
Comm. and Publ. Services	-	-	-	-	-	-	-	-	-	-
Residential	-	-	-	-	28937	-	-	-	249	-
Non-specified	-	-	-	-	-	-	-	-	-	-
NON-ENERGY USE	**-**	**-**	**-**	**-**	**-**	**-**	**-**	**-**	**-**	**-**

Democratic Republic of Congo / République démocratique du Congo

SUPPLY AND CONSUMPTION 1996	Coal (1000 tonnes)							Oil (1000 tonnes)			
	Coking Coal	Other Bit. Coal	Sub-Bit. Coal	Lignite	Peat	Oven and Gas Coke	Pat. Fuel and BKB	Crude Oil	NGL	Feed-stocks	Additives
Production	-	93	-	-	-	-	-	1371	-	-	-
Imports	-	43	-	-	-	153	85	119	-	-	-
Exports	-	-	-	-	-	-	-	-1079	-	-	-
Intl. Marine Bunkers	-	-	-	-	-	-	-	-	-	-	-
Stock Changes	-	-	-	-	-	-	-	-	-	-	-
DOMESTIC SUPPLY	**-**	**136**	**-**	**-**	**-**	**153**	**85**	**411**	**-**	**-**	**-**
Transfers and Stat. Diff.	-	-	-	-	-	-	-	-47	-	-	-
TRANSFORMATION	**-**	**-**	**-**	**-**	**-**	**122**	**-**	**364**	**-**	**-**	**-**
Electricity and CHP Plants	-	-	-	-	-	-	-	-	-	-	-
Petroleum Refineries	-	-	-	-	-	-	-	364	-	-	-
Other Transform. Sector	-	-	-	-	-	122	-	-	-	-	-
ENERGY SECTOR	**-**	**-**	**-**	**-**	**-**	**-**	**-**	**-**	**-**	**-**	**-**
DISTRIBUTION LOSSES	**-**	**-**	**-**	**-**	**-**	**-**	**-**	**-**	**-**	**-**	**-**
FINAL CONSUMPTION	**-**	**136**	**-**	**-**	**-**	**31**	**85**	**-**	**-**	**-**	**-**
INDUSTRY SECTOR	**-**	**136**	**-**	**-**	**-**	**31**	**-**	**-**	**-**	**-**	**-**
Iron and Steel	-	-	-	-	-	31	-	-	-	-	-
Chemical and Petrochem.	-	-	-	-	-	-	-	-	-	-	-
Non-Metallic Minerals	-	-	-	-	-	-	-	-	-	-	-
Non-specified	-	136	-	-	-	-	-	-	-	-	-
TRANSPORT SECTOR	**-**	**-**	**-**	**-**	**-**	**-**	**-**	**-**	**-**	**-**	**-**
Air	-	-	-	-	-	-	-	-	-	-	-
Road	-	-	-	-	-	-	-	-	-	-	-
Non-specified	-	-	-	-	-	-	-	-	-	-	-
OTHER SECTORS	**-**	**-**	**-**	**-**	**-**	**-**	**85**	**-**	**-**	**-**	**-**
Agriculture	-	-	-	-	-	-	-	-	-	-	-
Comm. and Publ. Services	-	-	-	-	-	-	-	-	-	-	-
Residential	-	-	-	-	-	-	85	-	-	-	-
Non-specified	-	-	-	-	-	-	-	-	-	-	-
NON-ENERGY USE	**-**	**-**	**-**	**-**	**-**	**-**	**-**	**-**	**-**	**-**	**-**

APPROVISIONNEMENT ET DEMANDE 1997	*Charbon* (1000 tonnes)							*Pétrole* (1000 tonnes)			
	Charbon à coke	*Autres charb. bit.*	*Charbon sous-bit.*	*Lignite*	*Tourbe*	*Coke de four/gaz*	*Agg./briq. de lignite*	*Pétrole brut*	*LGN*	*Produits d'aliment.*	*Additifs*
Production	-	93	-	-	-	-	-	1306	-	-	-
Imports	-	43	-	-	-	153	85	119	-	-	-
Exports	-	-	-	-	-	-	-	-1014	-	-	-
Intl. Marine Bunkers	-	-	-	-	-	-	-	-	-	-	-
Stock Changes	-	-	-	-	-	-	-	-	-	-	-
DOMESTIC SUPPLY	**-**	**136**	**-**	**-**	**-**	**153**	**85**	**411**	**-**	**-**	**-**
Transfers and Stat. Diff.	-	-	-	-	-	-	-	-47	-	-	-
TRANSFORMATION	**-**	**-**	**-**	**-**	**-**	**122**	**-**	**364**	**-**	**-**	**-**
Electricity and CHP Plants	-	-	-	-	-	-	-	-	-	-	-
Petroleum Refineries	-	-	-	-	-	-	-	364	-	-	-
Other Transform. Sector	-	-	-	-	-	122	-	-	-	-	-
ENERGY SECTOR	**-**	**-**	**-**	**-**	**-**	**-**	**-**	**-**	**-**	**-**	**-**
DISTRIBUTION LOSSES	**-**	**-**	**-**	**-**	**-**	**-**	**-**	**-**	**-**	**-**	**-**
FINAL CONSUMPTION	**-**	**136**	**-**	**-**	**-**	**31**	**85**	**-**	**-**	**-**	**-**
INDUSTRY SECTOR	**-**	**136**	**-**	**-**	**-**	**31**	**-**	**-**	**-**	**-**	**-**
Iron and Steel	-	-	-	-	-	31	-	-	-	-	-
Chemical and Petrochem.	-	-	-	-	-	-	-	-	-	-	-
Non-Metallic Minerals	-	-	-	-	-	-	-	-	-	-	-
Non-specified	-	136	-	-	-	-	-	-	-	-	-
TRANSPORT SECTOR	**-**	**-**	**-**	**-**	**-**	**-**	**-**	**-**	**-**	**-**	**-**
Air	-	-	-	-	-	-	-	-	-	-	-
Road	-	-	-	-	-	-	-	-	-	-	-
Non-specified	-	-	-	-	-	-	-	-	-	-	-
OTHER SECTORS	**-**	**-**	**-**	**-**	**-**	**-**	**85**	**-**	**-**	**-**	**-**
Agriculture	-	-	-	-	-	-	-	-	-	-	-
Comm. and Publ. Services	-	-	-	-	-	-	-	-	-	-	-
Residential	-	-	-	-	-	-	85	-	-	-	-
Non-specified	-	-	-	-	-	-	-	-	-	-	-
NON-ENERGY USE	**-**	**-**	**-**	**-**	**-**	**-**	**-**	**-**	**-**	**-**	**-**

Democratic Republic of Congo / République démocratique du Congo

SUPPLY AND CONSUMPTION 1996	Oil cont. (1000 tonnes)										
	Refinery Gas	LPG + Ethane	Motor Gasoline	Aviation Gasoline	Jet Fuel	Kerosene	Gas/ Diesel	Heavy Fuel Oil	Naphtha	Petrol. Coke	Other Prod.
Production	7	-	54	-	23	47	65	134	-	-	12
Imports	-	12	181	21	124	66	272	212	-	-	47
Exports	-	-	-	-	-	-	-	-26	-	-	-
Intl. Marine Bunkers	-	-	-	-	-	-	-	-35	-	-	-
Stock Changes	-	1	-	-	-	-	-	-	-	-	-
DOMESTIC SUPPLY	**7**	**13**	**235**	**21**	**147**	**113**	**337**	**285**	**-**	**-**	**59**
Transfers and Stat. Diff.	-	-	-	-3	-6	-	4	4	-	-	-
TRANSFORMATION	**-**	**-**	**-**	**-**	**-**	**-**	**30**	**16**	**-**	**-**	**-**
Electricity and CHP Plants	-	-	-	-	-	-	30	16	-	-	-
Petroleum Refineries	-	-	-	-	-	-	-	-	-	-	-
Other Transform. Sector	-	-	-	-	-	-	-	-	-	-	-
ENERGY SECTOR	**7**	**-**	**-**	**-**	**-**	**-**	**-**	**9**	**-**	**-**	**-**
DISTRIBUTION LOSSES	**-**	**-**	**-**	**-**	**-**	**-**	**-**	**-**	**-**	**-**	**-**
FINAL CONSUMPTION	**-**	**13**	**235**	**18**	**141**	**113**	**311**	**264**	**-**	**-**	**59**
INDUSTRY SECTOR	**-**	**2**	**-**	**-**	**-**	**-**	**-**	**56**	**-**	**-**	**-**
Iron and Steel	-	-	-	-	-	-	-	-	-	-	-
Chemical and Petrochem.	-	-	-	-	-	-	-	-	-	-	-
Non-Metallic Minerals	-	-	-	-	-	-	-	-	-	-	-
Non-specified	-	2	-	-	-	-	-	56	-	-	-
TRANSPORT SECTOR	**-**	**-**	**235**	**18**	**141**	**-**	**311**	**-**	**-**	**-**	**-**
Air	-	-	-	18	141	-	-	-	-	-	-
Road	-	-	235	-	-	-	311	-	-	-	-
Non-specified	-	-	-	-	-	-	-	-	-	-	-
OTHER SECTORS	**-**	**11**	**-**	**-**	**-**	**113**	**-**	**208**	**-**	**-**	**-**
Agriculture	-	-	-	-	-	-	-	-	-	-	-
Comm. and Publ. Services	-	-	-	-	-	-	-	-	-	-	-
Residential	-	11	-	-	-	113	-	-	-	-	-
Non-specified	-	-	-	-	-	-	-	208	-	-	-
NON-ENERGY USE	**-**	**-**	**-**	**-**	**-**	**-**	**-**	**-**	**-**	**-**	**59**

APPROVISIONNEMENT ET DEMANDE 1997	*Pétrole cont.* (1000 tonnes)										
	Gaz de raffinerie	*GPL + éthane*	*Essence moteur*	*Essence aviation*	*Carbu- réacteurs*	*Kérosène*	*Gazole*	*Fioul lourd*	*Naphta*	*Coke de pétrole*	*Autres prod.*
Production	7	-	54	-	23	47	65	134	-	-	12
Imports	-	12	181	21	124	66	272	212	-	-	47
Exports	-	-	-	-	-	-	-	-26	-	-	-
Intl. Marine Bunkers	-	-	-	-	-	-	-	-35	-	-	-
Stock Changes	-	1	-	-	-	-	-	-	-	-	-
DOMESTIC SUPPLY	**7**	**13**	**235**	**21**	**147**	**113**	**337**	**285**	**-**	**-**	**59**
Transfers and Stat. Diff.	-	-	-	-3	-6	-	4	4	-	-	-
TRANSFORMATION	**-**	**-**	**-**	**-**	**-**	**-**	**30**	**16**	**-**	**-**	**-**
Electricity and CHP Plants	-	-	-	-	-	-	30	16	-	-	-
Petroleum Refineries	-	-	-	-	-	-	-	-	-	-	-
Other Transform. Sector	-	-	-	-	-	-	-	-	-	-	-
ENERGY SECTOR	**7**	**-**	**-**	**-**	**-**	**-**	**-**	**9**	**-**	**-**	**-**
DISTRIBUTION LOSSES	**-**	**-**	**-**	**-**	**-**	**-**	**-**	**-**	**-**	**-**	**-**
FINAL CONSUMPTION	**-**	**13**	**235**	**18**	**141**	**113**	**311**	**264**	**-**	**-**	**59**
INDUSTRY SECTOR	**-**	**2**	**-**	**-**	**-**	**-**	**-**	**56**	**-**	**-**	**-**
Iron and Steel	-	-	-	-	-	-	-	-	-	-	-
Chemical and Petrochem.	-	-	-	-	-	-	-	-	-	-	-
Non-Metallic Minerals	-	-	-	-	-	-	-	-	-	-	-
Non-specified	-	2	-	-	-	-	-	56	-	-	-
TRANSPORT SECTOR	**-**	**-**	**235**	**18**	**141**	**-**	**311**	**-**	**-**	**-**	**-**
Air	-	-	-	18	141	-	-	-	-	-	-
Road	-	-	235	-	-	-	311	-	-	-	-
Non-specified	-	-	-	-	-	-	-	-	-	-	-
OTHER SECTORS	**-**	**11**	**-**	**-**	**-**	**113**	**-**	**208**	**-**	**-**	**-**
Agriculture	-	-	-	-	-	-	-	-	-	-	-
Comm. and Publ. Services	-	-	-	-	-	-	-	-	-	-	-
Residential	-	11	-	-	-	113	-	-	-	-	-
Non-specified	-	-	-	-	-	-	-	208	-	-	-
NON-ENERGY USE	**-**	**-**	**-**	**-**	**-**	**-**	**-**	**-**	**-**	**-**	**59**

Democratic Republic of Congo / République démocratique du Congo

SUPPLY AND CONSUMPTION 1996	Gas (TJ)				Comb. Renew. & Waste (TJ)				(GWh)	(TJ)
	Natural Gas	Gas Works	Coke Ovens	Blast Furnaces	Solid Biomass	Gas/Liquids from Biomass	Municipal Waste	Industrial Waste	Electricity	Heat
Production	-	-	-	-	506712	-	-	-	6261	-
Imports	-	-	-	-	-	-	-	-	59	-
Exports	-	-	-	-	-	-	-	-	-199	-
Intl. Marine Bunkers	-	-	-	-	-	-	-	-	-	-
Stock Changes	-	-	-	-	-	-	-	-	-	-
DOMESTIC SUPPLY	**-**	**-**	**-**	**-**	**506712**	**-**	**-**	**-**	**6121**	**-**
Transfers and Stat. Diff.	-	-	-	-	-	-	-	-	-26	-
TRANSFORMATION	**-**	**-**	**-**	**-**	**23220**	**-**	**-**	**-**	**-**	**-**
Electricity and CHP Plants	-	-	-	-	-	-	-	-	-	-
Petroleum Refineries	-	-	-	-	-	-	-	-	-	-
Other Transform. Sector	-	-	-	-	23220	-	-	-	-	-
ENERGY SECTOR	**-**	**-**	**-**	**-**	**-**	**-**	**-**	**-**	**25**	**-**
DISTRIBUTION LOSSES	**-**	**-**	**-**	**-**	**-**	**-**	**-**	**-**	**205**	**-**
FINAL CONSUMPTION	**-**	**-**	**-**	**-**	**483492**	**-**	**-**	**-**	**5865**	**-**
INDUSTRY SECTOR	**-**	**-**	**-**	**-**	**102168**	**-**	**-**	**-**	**3806**	**-**
Iron and Steel	-	-	-	-	-	-	-	-	-	-
Chemical and Petrochem.	-	-	-	-	-	-	-	-	-	-
Non-Metallic Minerals	-	-	-	-	-	-	-	-	-	-
Non-specified	-	-	-	-	102168	-	-	-	3806	-
TRANSPORT SECTOR	**-**	**-**	**-**	**-**	**-**	**-**	**-**	**-**	**-**	**-**
Air	-	-	-	-	-	-	-	-	-	-
Road	-	-	-	-	-	-	-	-	-	-
Non-specified	-	-	-	-	-	-	-	-	-	-
OTHER SECTORS	**-**	**-**	**-**	**-**	**381324**	**-**	**-**	**-**	**2059**	**-**
Agriculture	-	-	-	-	-	-	-	-	-	-
Comm. and Publ. Services	-	-	-	-	-	-	-	-	-	-
Residential	-	-	-	-	381324	-	-	-	1953	-
Non-specified	-	-	-	-	-	-	-	-	106	-
NON-ENERGY USE	**-**	**-**	**-**	**-**	**-**	**-**	**-**	**-**	**-**	**-**

APPROVISIONNEMENT ET DEMANDE 1997	*Gaz* (TJ)				*En. Re. Comb. & Déchets* (TJ)				(GWh)	(TJ)
	Gaz naturel	*Usines à gaz*	*Cokeries*	*Hauts fourneaux*	*Biomasse solide*	*Gaz/Liquides tirés de biomasse*	*Déchets urbains*	*Déchets industriels*	*Electricité*	*Chaleur*
Production	-	-	-	-	522927	-	-	-	6010	-
Imports	-	-	-	-	-	-	-	-	59	-
Exports	-	-	-	-	-	-	-	-	-199	-
Intl. Marine Bunkers	-	-	-	-	-	-	-	-	-	-
Stock Changes	-	-	-	-	-	-	-	-	-	-
DOMESTIC SUPPLY	**-**	**-**	**-**	**-**	**522927**	**-**	**-**	**-**	**5870**	**-**
Transfers and Stat. Diff.	-	-	-	-	-	-	-	-	-24	-
TRANSFORMATION	**-**	**-**	**-**	**-**	**23963**	**-**	**-**	**-**	**-**	**-**
Electricity and CHP Plants	-	-	-	-	-	-	-	-	-	-
Petroleum Refineries	-	-	-	-	-	-	-	-	-	-
Other Transform. Sector	-	-	-	-	23963	-	-	-	-	-
ENERGY SECTOR	**-**	**-**	**-**	**-**	**-**	**-**	**-**	**-**	**24**	**-**
DISTRIBUTION LOSSES	**-**	**-**	**-**	**-**	**-**	**-**	**-**	**-**	**197**	**-**
FINAL CONSUMPTION	**-**	**-**	**-**	**-**	**498964**	**-**	**-**	**-**	**5625**	**-**
INDUSTRY SECTOR	**-**	**-**	**-**	**-**	**105437**	**-**	**-**	**-**	**3650**	**-**
Iron and Steel	-	-	-	-	-	-	-	-	-	-
Chemical and Petrochem.	-	-	-	-	-	-	-	-	-	-
Non-Metallic Minerals	-	-	-	-	-	-	-	-	-	-
Non-specified	-	-	-	-	105437	-	-	-	3650	-
TRANSPORT SECTOR	**-**	**-**	**-**	**-**	**-**	**-**	**-**	**-**	**-**	**-**
Air	-	-	-	-	-	-	-	-	-	-
Road	-	-	-	-	-	-	-	-	-	-
Non-specified	-	-	-	-	-	-	-	-	-	-
OTHER SECTORS	**-**	**-**	**-**	**-**	**393527**	**-**	**-**	**-**	**1975**	**-**
Agriculture	-	-	-	-	-	-	-	-	-	-
Comm. and Publ. Services	-	-	-	-	-	-	-	-	-	-
Residential	-	-	-	-	393527	-	-	-	1873	-
Non-specified	-	-	-	-	-	-	-	-	102	-
NON-ENERGY USE	**-**	**-**	**-**	**-**	**-**	**-**	**-**	**-**	**-**	**-**

Costa Rica

SUPPLY AND CONSUMPTION 1996	Coal (1000 tonnes)							Oil (1000 tonnes)			
	Coking Coal	Other Bit. Coal	Sub-Bit. Coal	Lignite	Peat	Oven and Gas Coke	Pat. Fuel and BKB	Crude Oil	NGL	Feed-stocks	Additives
Production	-	-	-	-	-	-	-	-	-	-	-
Imports	-	-	-	-	-	12	-	622	-	-	-
Exports	-	-	-	-	-	-	-	-	-	-	-
Intl. Marine Bunkers	-	-	-	-	-	-	-	-	-	-	-
Stock Changes	-	-	-	-	-	-	-	-5	-	-	-
DOMESTIC SUPPLY	**-**	**-**	**-**	**-**	**-**	**12**	**-**	**617**	**-**	**-**	**-**
Transfers and Stat. Diff.	-	-	-	-	-	-	-	17	-	-	-
TRANSFORMATION	**-**	**-**	**-**	**-**	**-**	**10**	**-**	**627**	**-**	**-**	**-**
Electricity and CHP Plants	-	-	-	-	-	-	-	-	-	-	-
Petroleum Refineries	-	-	-	-	-	-	-	627	-	-	-
Other Transform. Sector	-	-	-	-	-	10	-	-	-	-	-
ENERGY SECTOR	**-**	**-**	**-**	**-**	**-**	**-**	**-**	**-**	**-**	**-**	**-**
DISTRIBUTION LOSSES	**-**	**-**	**-**	**-**	**-**	**-**	**-**	**7**	**-**	**-**	**-**
FINAL CONSUMPTION	**-**	**-**	**-**	**-**	**-**	**2**	**-**	**-**	**-**	**-**	**-**
INDUSTRY SECTOR	**-**	**-**	**-**	**-**	**-**	**2**	**-**	**-**	**-**	**-**	**-**
Iron and Steel	-	-	-	-	-	2	-	-	-	-	-
Chemical and Petrochem.	-	-	-	-	-	-	-	-	-	-	-
Non-Metallic Minerals	-	-	-	-	-	-	-	-	-	-	-
Non-specified	-	-	-	-	-	-	-	-	-	-	-
TRANSPORT SECTOR	**-**	**-**	**-**	**-**	**-**	**-**	**-**	**-**	**-**	**-**	**-**
Air	-	-	-	-	-	-	-	-	-	-	-
Road	-	-	-	-	-	-	-	-	-	-	-
Non-specified	-	-	-	-	-	-	-	-	-	-	-
OTHER SECTORS	**-**	**-**	**-**	**-**	**-**	**-**	**-**	**-**	**-**	**-**	**-**
Agriculture	-	-	-	-	-	-	-	-	-	-	-
Comm. and Publ. Services	-	-	-	-	-	-	-	-	-	-	-
Residential	-	-	-	-	-	-	-	-	-	-	-
Non-specified	-	-	-	-	-	-	-	-	-	-	-
NON-ENERGY USE	**-**	**-**	**-**	**-**	**-**	**-**	**-**	**-**	**-**	**-**	**-**

APPROVISIONNEMENT ET DEMANDE 1997	*Charbon* (1000 tonnes)							*Pétrole* (1000 tonnes)			
	Charbon à coke	*Autres charb. bit.*	*Charbon sous-bit.*	*Lignite*	*Tourbe*	*Coke de four/gaz*	*Agg./briq. de lignite*	*Pétrole brut*	*LGN*	*Produits d'aliment.*	*Additifs*
Production	-	-	-	-	-	-	-	-	-	-	-
Imports	-	-	-	-	-	-	-	626	-	-	-
Exports	-	-	-	-	-	-	-	-	-	-	-
Intl. Marine Bunkers	-	-	-	-	-	-	-	-	-	-	-
Stock Changes	-	-	-	-	-	-	-	9	-	-	-
DOMESTIC SUPPLY	**-**	**-**	**-**	**-**	**-**	**-**	**-**	**635**	**-**	**-**	**-**
Transfers and Stat. Diff.	-	-	-	-	-	-	-	-12	-	-	-
TRANSFORMATION	**-**	**-**	**-**	**-**	**-**	**-**	**-**	**615**	**-**	**-**	**-**
Electricity and CHP Plants	-	-	-	-	-	-	-	-	-	-	-
Petroleum Refineries	-	-	-	-	-	-	-	615	-	-	-
Other Transform. Sector	-	-	-	-	-	-	-	-	-	-	-
ENERGY SECTOR	**-**	**-**	**-**	**-**	**-**	**-**	**-**	**-**	**-**	**-**	**-**
DISTRIBUTION LOSSES	**-**	**-**	**-**	**-**	**-**	**-**	**-**	**8**	**-**	**-**	**-**
FINAL CONSUMPTION	**-**	**-**	**-**	**-**	**-**	**-**	**-**	**-**	**-**	**-**	**-**
INDUSTRY SECTOR	**-**	**-**	**-**	**-**	**-**	**-**	**-**	**-**	**-**	**-**	**-**
Iron and Steel	-	-	-	-	-	-	-	-	-	-	-
Chemical and Petrochem.	-	-	-	-	-	-	-	-	-	-	-
Non-Metallic Minerals	-	-	-	-	-	-	-	-	-	-	-
Non-specified	-	-	-	-	-	-	-	-	-	-	-
TRANSPORT SECTOR	**-**	**-**	**-**	**-**	**-**	**-**	**-**	**-**	**-**	**-**	**-**
Air	-	-	-	-	-	-	-	-	-	-	-
Road	-	-	-	-	-	-	-	-	-	-	-
Non-specified	-	-	-	-	-	-	-	-	-	-	-
OTHER SECTORS	**-**	**-**	**-**	**-**	**-**	**-**	**-**	**-**	**-**	**-**	**-**
Agriculture	-	-	-	-	-	-	-	-	-	-	-
Comm. and Publ. Services	-	-	-	-	-	-	-	-	-	-	-
Residential	-	-	-	-	-	-	-	-	-	-	-
Non-specified	-	-	-	-	-	-	-	-	-	-	-
NON-ENERGY USE	**-**	**-**	**-**	**-**	**-**	**-**	**-**	**-**	**-**	**-**	**-**

Costa Rica

SUPPLY AND CONSUMPTION 1996	Oil cont. (1000 tonnes)										
	Refinery Gas	LPG + Ethane	Motor Gasoline	Aviation Gasoline	Jet Fuel	Kerosene	Gas/ Diesel	Heavy Fuel Oil	Naphtha	Petrol. Coke	Other Prod.
Production	-	2	76	-	30	3	177	302	12	-	22
Imports	-	44	336	5	68	-	448	39	-	-	-
Exports	-	-	-	-	-16	-	-14	-129	-	-	-
Intl. Marine Bunkers	-	-	-	-	-	-	-	-	-	-	-
Stock Changes	-	-2	-7	1	-1	-2	-	-4	2	-	-
DOMESTIC SUPPLY	**-**	**44**	**405**	**6**	**81**	**1**	**611**	**208**	**14**	**-**	**22**
Transfers and Stat. Diff.	-	2	32	-2	9	6	17	-	-12	-	-3
TRANSFORMATION	**-**	**-**	**-**	**-**	**-**	**-**	**112**	**21**	**-**	**-**	**-**
Electricity and CHP Plants	-	-	-	-	-	-	112	21	-	-	-
Petroleum Refineries	-	-	-	-	-	-	-	-	-	-	-
Other Transform. Sector	-	-	-	-	-	-	-	-	-	-	-
ENERGY SECTOR	**-**	**-**	**-**	**-**	**-**	**-**	**1**	**29**	**-**	**-**	**-**
DISTRIBUTION LOSSES	**-**	**-**	**-**	**-**	**-**	**-**	**-**	**-**	**-**	**-**	**-**
FINAL CONSUMPTION	**-**	**46**	**437**	**4**	**90**	**7**	**515**	**158**	**2**	**-**	**19**
INDUSTRY SECTOR	**-**	**12**	**-**	**-**	**-**	**2**	**27**	**155**	**2**	**-**	**-**
Iron and Steel	-	-	-	-	-	-	-	-	-	-	-
Chemical and Petrochem.	-	-	-	-	-	1	9	13	2	-	-
Non-Metallic Minerals	-	-	-	-	-	-	-	-	-	-	-
Non-specified	-	12	-	-	-	1	18	142	-	-	-
TRANSPORT SECTOR	**-**	**-**	**429**	**4**	**90**	**-**	**411**	**-**	**-**	**-**	**-**
Air	-	-	-	4	90	-	-	-	-	-	-
Road	-	-	428	-	-	-	390	-	-	-	-
Non-specified	-	-	1	-	-	-	21	-	-	-	-
OTHER SECTORS	**-**	**34**	**8**	**-**	**-**	**5**	**77**	**3**	**-**	**-**	**-**
Agriculture	-	-	8	-	-	-	60	2	-	-	-
Comm. and Publ. Services	-	6	-	-	-	3	17	1	-	-	-
Residential	-	28	-	-	-	2	-	-	-	-	-
Non-specified	-	-	-	-	-	-	-	-	-	-	-
NON-ENERGY USE	**-**	**-**	**-**	**-**	**-**	**-**	**-**	**-**	**-**	**-**	**19**

APPROVISIONNEMENT ET DEMANDE 1997	*Pétrole cont.* (1000 tonnes)										
	Gaz de raffinerie	*GPL + éthane*	*Essence moteur*	*Essence aviation*	*Carbu- réacteurs*	*Kérosène*	*Gazole*	*Fioul lourd*	*Naphta*	*Coke de pétrole*	*Autres prod.*
Production	-	2	80	-	29	6	171	292	7	-	26
Imports	-	52	398	3	72	-	383	32	-	-	-
Exports	-	-	-	-	-21	-	-10	-83	-	-	-
Intl. Marine Bunkers	-	-	-	-	-	-	-	-	-	-	-
Stock Changes	-	2	6	-1	-3	1	-17	13	-	-	-2
DOMESTIC SUPPLY	**-**	**56**	**484**	**2**	**77**	**7**	**527**	**254**	**7**	**-**	**24**
Transfers and Stat. Diff.	-	-6	-37	2	23	1	57	-33	-5	-	-1
TRANSFORMATION	**-**	**-**	**-**	**-**	**-**	**-**	**45**	**15**	**-**	**-**	**-**
Electricity and CHP Plants	-	-	-	-	-	-	45	15	-	-	-
Petroleum Refineries	-	-	-	-	-	-	-	-	-	-	-
Other Transform. Sector	-	-	-	-	-	-	-	-	-	-	-
ENERGY SECTOR	**-**	**-**	**-**	**-**	**-**	**-**	**1**	**29**	**-**	**-**	**-**
DISTRIBUTION LOSSES	**-**	**-**	**-**	**-**	**-**	**-**	**-**	**-**	**-**	**-**	**-**
FINAL CONSUMPTION	**-**	**50**	**447**	**4**	**100**	**8**	**538**	**177**	**2**	**-**	**23**
INDUSTRY SECTOR	**-**	**13**	**-**	**-**	**-**	**2**	**33**	**174**	**2**	**-**	**-**
Iron and Steel	-	-	-	-	-	-	-	-	-	-	-
Chemical and Petrochem.	-	-	-	-	-	1	11	15	2	-	-
Non-Metallic Minerals	-	-	-	-	-	-	-	-	-	-	-
Non-specified	-	13	-	-	-	1	22	159	-	-	-
TRANSPORT SECTOR	**-**	**-**	**439**	**4**	**99**	**-**	**422**	**-**	**-**	**-**	**-**
Air	-	-	-	4	99	-	-	-	-	-	-
Road	-	-	438	-	-	-	397	-	-	-	-
Non-specified	-	-	1	-	-	-	25	-	-	-	-
OTHER SECTORS	**-**	**37**	**8**	**-**	**1**	**6**	**83**	**3**	**-**	**-**	**-**
Agriculture	-	-	8	-	1	-	65	2	-	-	-
Comm. and Publ. Services	-	7	-	-	-	4	18	1	-	-	-
Residential	-	30	-	-	-	2	-	-	-	-	-
Non-specified	-	-	-	-	-	-	-	-	-	-	-
NON-ENERGY USE	**-**	**-**	**-**	**-**	**-**	**-**	**-**	**-**	**-**	**-**	**23**

Costa Rica

SUPPLY AND CONSUMPTION 1996	Gas (TJ)				Comb. Renew. & Waste (TJ)				(GWh)	(TJ)
	Natural Gas	Gas Works	Coke Ovens	Blast Furnaces	Solid Biomass	Gas/Liquids from Biomass	Municipal Waste	Industrial Waste	Electricity	Heat
Production	-	-	-	-	15660	-	-	-	4791	-
Imports	-	-	-	-	-	-	-	-	233	-
Exports	-	-	-	-	-	-	-	-	-109	-
Intl. Marine Bunkers	-	-	-	-	-	-	-	-	-	-
Stock Changes	-	-	-	-	-	-	-	-	-	-
DOMESTIC SUPPLY	-	-	-	-	**15660**	-	-	-	**4915**	-
Transfers and Stat. Diff.	-	-	-	-	-	-	-	-	-79	-
TRANSFORMATION	-	-	-	-	**278**	-	-	-	-	-
Electricity and CHP Plants	-	-	-	-	-	-	-	-	-	-
Petroleum Refineries	-	-	-	-	-	-	-	-	-	-
Other Transform. Sector	-	-	-	-	278	-	-	-	-	-
ENERGY SECTOR	-	-	-	-	-	-	-	-	**15**	-
DISTRIBUTION LOSSES	-	-	-	-	-	-	-	-	**386**	-
FINAL CONSUMPTION	-	-	-	-	**15382**	-	-	-	**4435**	-
INDUSTRY SECTOR	-	-	-	-	**7307**	-	-	-	**1021**	-
Iron and Steel	-	-	-	-	-	-	-	-	-	-
Chemical and Petrochem.	-	-	-	-	-	-	-	-	204	-
Non-Metallic Minerals	-	-	-	-	-	-	-	-	-	-
Non-specified	-	-	-	-	7307	-	-	-	817	-
TRANSPORT SECTOR	-	-	-	-	-	-	-	-	-	-
Air	-	-	-	-	-	-	-	-	-	-
Road	-	-	-	-	-	-	-	-	-	-
Non-specified	-	-	-	-	-	-	-	-	-	-
OTHER SECTORS	-	-	-	-	**8075**	-	-	-	**3414**	-
Agriculture	-	-	-	-	-	-	-	-	6	-
Comm. and Publ. Services	-	-	-	-	27	-	-	-	1365	-
Residential	-	-	-	-	8048	-	-	-	2043	-
Non-specified	-	-	-	-	-	-	-	-	-	-
NON-ENERGY USE	-	-	-	-	-	-	-	-	-	-

APPROVISIONNEMENT ET DEMANDE 1997	*Gaz* (TJ)				*En. Re. Comb. & Déchets* (TJ)				(GWh)	(TJ)
	Gaz naturel	*Usines à gaz*	*Cokeries*	*Hauts fourneaux*	*Biomasse solide*	*Gaz/Liquides tirés de biomasse*	*Déchets urbains*	*Déchets industriels*	*Electricité*	*Chaleur*
Production	-	-	-	-	11274	-	-	-	5599	-
Imports	-	-	-	-	-	-	-	-	107	-
Exports	-	-	-	-	-	-	-	-	-241	-
Intl. Marine Bunkers	-	-	-	-	-	-	-	-	-	-
Stock Changes	-	-	-	-	-	-	-	-	-	-
DOMESTIC SUPPLY	-	-	-	-	**11274**	-	-	-	**5465**	-
Transfers and Stat. Diff.	-	-	-	-	-	-	-	-	-341	-
TRANSFORMATION	-	-	-	-	**278**	-	-	-	-	-
Electricity and CHP Plants	-	-	-	-	-	-	-	-	-	-
Petroleum Refineries	-	-	-	-	-	-	-	-	-	-
Other Transform. Sector	-	-	-	-	278	-	-	-	-	-
ENERGY SECTOR	-	-	-	-	-	-	-	-	**18**	-
DISTRIBUTION LOSSES	-	-	-	-	-	-	-	-	**420**	-
FINAL CONSUMPTION	-	-	-	-	**10996**	-	-	-	**4686**	-
INDUSTRY SECTOR	-	-	-	-	**7674**	-	-	-	**1096**	-
Iron and Steel	-	-	-	-	-	-	-	-	-	-
Chemical and Petrochem.	-	-	-	-	-	-	-	-	219	-
Non-Metallic Minerals	-	-	-	-	-	-	-	-	-	-
Non-specified	-	-	-	-	7674	-	-	-	877	-
TRANSPORT SECTOR	-	-	-	-	-	-	-	-	-	-
Air	-	-	-	-	-	-	-	-	-	-
Road	-	-	-	-	-	-	-	-	-	-
Non-specified	-	-	-	-	-	-	-	-	-	-
OTHER SECTORS	-	-	-	-	**3322**	-	-	-	**3590**	-
Agriculture	-	-	-	-	-	-	-	-	7	-
Comm. and Publ. Services	-	-	-	-	-	-	-	-	1464	-
Residential	-	-	-	-	3322	-	-	-	2119	-
Non-specified	-	-	-	-	-	-	-	-	-	-
NON-ENERGY USE	-	-	-	-	-	-	-	-	-	-

Croatia / Croatie : 1996

	Coal / *Charbon* (1000 tonnes)							Oil / *Pétrole* (1000 tonnes)			
SUPPLY AND CONSUMPTION ***APPROVISIONNEMENT ET DEMANDE***	Coking Coal *Charbon à coke*	Other Bit. Coal *Autres charb. bit.*	Sub-Bit. Coal *Charbon sous-bit.*	Lignite *Lignite*	Peat *Tourbe*	Oven and Gas Coke *Coke de four/gaz*	Pat. Fuel and BKB *Agg./briq. de lignite*	Crude Oil *Pétrole brut*	NGL *LGN*	Feed-stocks *Produits d'aliment.*	Additives *Additifs*
Production	-	64	-	2	-	-	-	1469	227	-	-
From Other Sources	-	-	-	-	-	-	-	-	-	-	-
Imports	-	51	-	147	-	21	-	3624	-	-	-
Exports	-	-	-	-	-	-	-	-111	-	-	-
Intl. Marine Bunkers	-	-	-	-	-	-	-	-	-	-	-
Stock Changes	-	2	-	-	-	5	-	-53	-	-	-
DOMESTIC SUPPLY	**-**	**117**	**-**	**149**	**-**	**26**	**-**	**4929**	**227**	**-**	**-**
Transfers	-	-	-	-	-	-	-	-	-227	-	-
Statistical Differences	-	-	-	-	-	-	-	-	-	-	-
TRANSFORMATION	**-**	**55**	**-**	**6**	**-**	**-**	**-**	**4842**	**-**	**-**	**-**
Electricity Plants	-	55	-	-	-	-	-	-	-	-	-
CHP Plants	-	-	-	6	-	-	-	-	-	-	-
Heat Plants	-	-	-	-	-	-	-	-	-	-	-
Blast Furnaces/Gas Works	-	-	-	-	-	-	-	-	-	-	-
Coke/Pat. Fuel/BKB Plants	-	-	-	-	-	-	-	-	-	-	-
Petroleum Refineries	-	-	-	-	-	-	-	4842	-	-	-
Petrochemical Industry	-	-	-	-	-	-	-	-	-	-	-
Liquefaction	-	-	-	-	-	-	-	-	-	-	-
Other Transform. Sector	-	-	-	-	-	-	-	-	-	-	-
ENERGY SECTOR	**-**	**1**	**-**	**-**	**-**	**-**	**-**	**87**	**-**	**-**	**-**
Coal Mines	-	1	-	-	-	-	-	-	-	-	-
Oil and Gas Extraction	-	-	-	-	-	-	-	87	-	-	-
Petroleum Refineries	-	-	-	-	-	-	-	-	-	-	-
Electr., CHP+Heat Plants	-	-	-	-	-	-	-	-	-	-	-
Pumped Storage (Elec.)	-	-	-	-	-	-	-	-	-	-	-
Other Energy Sector	-	-	-	-	-	-	-	-	-	-	-
Distribution Losses	-	-	-	-	-	-	-	-	-	-	-
FINAL CONSUMPTION	**-**	**61**	**-**	**143**	**-**	**26**	**-**	**-**	**-**	**-**	**-**
INDUSTRY SECTOR	**-**	**61**	**-**	**106**	**-**	**26**	**-**	**-**	**-**	**-**	**-**
Iron and Steel	-	-	-	-	-	9	-	-	-	-	-
Chemical and Petrochem.	-	-	-	13	-	-	-	-	-	-	-
of which: Feedstocks	-	-	-	-	-	-	-	-	-	-	-
Non-Ferrous Metals	-	-	-	-	-	-	-	-	-	-	-
Non-Metallic Minerals	-	59	-	5	-	13	-	-	-	-	-
Transport Equipment	-	-	-	-	-	-	-	-	-	-	-
Machinery	-	-	-	-	-	1	-	-	-	-	-
Mining and Quarrying	-	-	-	-	-	-	-	-	-	-	-
Food and Tobacco	-	2	-	86	-	-	-	-	-	-	-
Paper, Pulp and Print	-	-	-	-	-	-	-	-	-	-	-
Wood and Wood Products	-	-	-	-	-	3	-	-	-	-	-
Construction	-	-	-	-	-	-	-	-	-	-	-
Textile and Leather	-	-	-	2	-	-	-	-	-	-	-
Non-specified	-	-	-	-	-	-	-	-	-	-	-
TRANSPORT SECTOR	**-**	**-**	**-**	**-**	**-**	**-**	**-**	**-**	**-**	**-**	**-**
Air	-	-	-	-	-	-	-	-	-	-	-
Road	-	-	-	-	-	-	-	-	-	-	-
Rail	-	-	-	-	-	-	-	-	-	-	-
Pipeline Transport	-	-	-	-	-	-	-	-	-	-	-
Internal Navigation	-	-	-	-	-	-	-	-	-	-	-
Non-specified	-	-	-	-	-	-	-	-	-	-	-
OTHER SECTORS	**-**	**-**	**-**	**37**	**-**	**-**	**-**	**-**	**-**	**-**	**-**
Agriculture	-	-	-	-	-	-	-	-	-	-	-
Comm. and Publ. Services	-	-	-	14	-	-	-	-	-	-	-
Residential	-	-	-	23	-	-	-	-	-	-	-
Non-specified	-	-	-	-	-	-	-	-	-	-	-
NON-ENERGY USE	**-**	**-**	**-**	**-**	**-**	**-**	**-**	**-**	**-**	**-**	**-**
in Industry/Trans./Energy	-	-	-	-	-	-	-	-	-	-	-
in Transport	-	-	-	-	-	-	-	-	-	-	-
in Other Sectors	-	-	-	-	-	-	-	-	-	-	-

Croatia / Croatie : 1996

	Oil cont. / *Pétrole cont.* (1000 tonnes)										
SUPPLY AND CONSUMPTION ***APPROVISIONNEMENT ET DEMANDE***	Refinery Gas *Gaz de raffinerie*	LPG + Ethane *GPL + éthane*	Motor Gasoline *Essence moteur*	Aviation Gasoline *Essence aviation*	Jet Fuel *Carbu-réacteurs*	Kerosene *Kérosène*	Gas/ Diesel *Gazole*	Heavy Fuel Oil *Fioul lourd*	Naphtha *Naphta*	Petrol. Coke *Coke de pétrole*	Other Prod. *Autres prod.*
Production	195	166	1024	-	83	9	1434	1340	257	46	252
From Other Sources	-	-	-	-	-	-	-	-	-	-	-
Imports	-	3	6	-	8	-	74	115	-	-	45
Exports	-	-137	-420	-	-17	-2	-451	-154	-182	-9	-180
Intl. Marine Bunkers	-	-	-	-	-	-	-12	-17	-	-	-
Stock Changes	-	-	-10	-	4	-	-22	-58	8	-	-28
DOMESTIC SUPPLY	**195**	**32**	**600**	**-**	**78**	**7**	**1023**	**1226**	**83**	**37**	**89**
Transfers	-	177	-	-	-	-	-	-	50	-	-
Statistical Differences	-	-	-	-	-	-	-	-	-	-	-
TRANSFORMATION	**5**	**11**	**-**	**-**	**-**	**-**	**27**	**775**	**-**	**-**	**-**
Electricity Plants	-	-	-	-	-	-	21	385	-	-	-
CHP Plants	5	-	-	-	-	-	1	359	-	-	-
Heat Plants	-	1	-	-	-	-	5	31	-	-	-
Blast Furnaces/Gas Works	-	10	-	-	-	-	-	-	-	-	-
Coke/Pat. Fuel/BKB Plants	-	-	-	-	-	-	-	-	-	-	-
Petroleum Refineries	-	-	-	-	-	-	-	-	-	-	-
Petrochemical Industry	-	-	-	-	-	-	-	-	-	-	-
Liquefaction	-	-	-	-	-	-	-	-	-	-	-
Other Transform. Sector	-	-	-	-	-	-	-	-	-	-	-
ENERGY SECTOR	**190**	**30**	**-**	**-**	**-**	**-**	**1**	**228**	**-**	**37**	**-**
Coal Mines	-	-	-	-	-	-	-	-	-	-	-
Oil and Gas Extraction	-	-	-	-	-	-	1	-	-	-	-
Petroleum Refineries	190	30	-	-	-	-	-	228	-	37	-
Electr., CHP+Heat Plants	-	-	-	-	-	-	-	-	-	-	-
Pumped Storage (Elec.)	-	-	-	-	-	-	-	-	-	-	-
Other Energy Sector	-	-	-	-	-	-	-	-	-	-	-
Distribution Losses	-	-	-	-	-	-	-	-	-	-	-
FINAL CONSUMPTION	**-**	**168**	**600**	**-**	**78**	**7**	**995**	**223**	**133**	**-**	**89**
INDUSTRY SECTOR	**-**	**102**	**8**	**-**	**-**	**-**	**89**	**203**	**133**	**-**	**-**
Iron and Steel	-	1	-	-	-	-	3	5	-	-	-
Chemical and Petrochem.	-	94	-	-	-	-	2	52	133	-	-
of which: Feedstocks	-	*86*	-	-	-	-	-	-	*133*	-	-
Non-Ferrous Metals	-	1	-	-	-	-	2	1	-	-	-
Non-Metallic Minerals	-	3	-	-	-	-	3	66	-	-	-
Transport Equipment	-	1	-	-	-	-	-	2	-	-	-
Machinery	-	1	-	-	-	-	2	1	-	-	-
Mining and Quarrying	-	-	-	-	-	-	8	3	-	-	-
Food and Tobacco	-	1	-	-	-	-	16	44	-	-	-
Paper, Pulp and Print	-	-	-	-	-	-	2	7	-	-	-
Wood and Wood Products	-	-	-	-	-	-	1	2	-	-	-
Construction	-	-	8	-	-	-	47	-	-	-	-
Textile and Leather	-	-	-	-	-	-	3	19	-	-	-
Non-specified	-	-	-	-	-	-	-	1	-	-	-
TRANSPORT SECTOR	**-**	**12**	**585**	**-**	**78**	**-**	**506**	**8**	**-**	**-**	**-**
Air	-	-	-	-	78	-	-	-	-	-	-
Road	-	12	584	-	-	-	433	2	-	-	-
Rail	-	-	-	-	-	-	28	1	-	-	-
Pipeline Transport	-	-	-	-	-	-	-	-	-	-	-
Internal Navigation	-	-	1	-	-	-	22	5	-	-	-
Non-specified	-	-	-	-	-	-	23	-	-	-	-
OTHER SECTORS	**-**	**54**	**7**	**-**	**-**	**7**	**400**	**12**	**-**	**-**	**-**
Agriculture	-	3	7	-	-	-	138	5	-	-	-
Comm. and Publ. Services	-	2	-	-	-	-	89	1	-	-	-
Residential	-	49	-	-	-	7	173	6	-	-	-
Non-specified	-	-	-	-	-	-	-	-	-	-	-
NON-ENERGY USE	**-**	**-**	**-**	**-**	**-**	**-**	**-**	**-**	**-**	**-**	**89**
in Industry/Transf./Energy	-	-	-	-	-	-	-	-	-	-	68
in Transport	-	-	-	-	-	-	-	-	-	-	18
in Other Sectors	-	-	-	-	-	-	-	-	-	-	3

Croatia / Croatie : 1996

SUPPLY AND CONSUMPTION / *APPROVISIONNEMENT ET DEMANDE*	Gas / *Gaz* (TJ)				Comb. Renew. & Waste / *En. Re. Comb. & Déchets* (TJ)				(GWh)	(TJ)
	Natural Gas / *Gaz naturel*	Gas Works / *Usines à gaz*	Coke Ovens / *Cokeries*	Blast Furnaces / *Hauts fourneaux*	Solid Biomass / *Biomasse solide*	Gas/Liquids from Biomass / *Gaz/Liquides tirés de biomasse*	Municipal Waste / *Déchets urbains*	Industrial Waste / *Déchets industriels*	Electricity / *Electricité*	Heat / *Chaleur*
Production	67853	436	-	-	13680	-	-	100	10548	13737
From Other Sources	-	-	-	-	-	-	-	-	-	-
Imports	33402	-	-	-	-	-	-	-	3960	-
Exports	-	-	-	-	-	-	-	-	-1630	-
Intl. Marine Bunkers	-	-	-	-	-	-	-	-	-	-
Stock Changes	-426	-	-	-	-	-	-	-	-	-
DOMESTIC SUPPLY	**100829**	**436**	**-**	**-**	**13680**	**-**	**-**	**100**	**12878**	**13737**
Transfers	-	-	-	-	-	-	-	-	-	-
Statistical Differences	-	-	-	-	-	-	-	-	-	-
TRANSFORMATION	**19969**	**-**	**-**	**-**	**-**	**-**	**-**	**100**	**-**	**-**
Electricity Plants	8643	-	-	-	-	-	-	-	-	-
CHP Plants	9772	-	-	-	-	-	-	100	-	-
Heat Plants	1554	-	-	-	-	-	-	-	-	-
Blast Furnaces/Gas Works	-	-	-	-	-	-	-	-	-	-
Coke/Pat. Fuel/BKB Plants	-	-	-	-	-	-	-	-	-	-
Petroleum Refineries	-	-	-	-	-	-	-	-	-	-
Petrochemical Industry	-	-	-	-	-	-	-	-	-	-
Liquefaction	-	-	-	-	-	-	-	-	-	-
Other Transform. Sector	-	-	-	-	-	-	-	-	-	-
ENERGY SECTOR	**10838**	**-**	**-**	**-**	**-**	**-**	**-**	**-**	**686**	**906**
Coal Mines	-	-	-	-	-	-	-	-	11	-
Oil and Gas Extraction	10530	-	-	-	-	-	-	-	115	-
Petroleum Refineries	247	-	-	-	-	-	-	-	272	-
Electr., CHP+Heat Plants	61	-	-	-	-	-	-	-	288	906
Pumped Storage (Elec.)	-	-	-	-	-	-	-	-	-	-
Other Energy Sector	-	-	-	-	-	-	-	-	-	-
Distribution Losses	3830	11	-	-	-	-	-	-	1908	1918
FINAL CONSUMPTION	**66192**	**425**	**-**	**-**	**13680**	**-**	**-**	**-**	**10284**	**10913**
INDUSTRY SECTOR	**42871**	**148**	**-**	**-**	**-**	**-**	**-**	**-**	**2651**	**3703**
Iron and Steel	1871	26	-	-	-	-	-	-	236	-
Chemical and Petrochem.	25090	-	-	-	-	-	-	-	542	721
of which: Feedstocks	*20265*	-	-	-	-	-	-	-	-	-
Non-Ferrous Metals	-	-	-	-	-	-	-	-	69	-
Non-Metallic Minerals	7291	44	-	-	-	-	-	-	426	110
Transport Equipment	21	46	-	-	-	-	-	-	116	21
Machinery	292	11	-	-	-	-	-	-	122	270
Mining and Quarrying	450	-	-	-	-	-	-	-	42	30
Food and Tobacco	4200	-	-	-	-	-	-	-	366	1096
Paper, Pulp and Print	1808	-	-	-	-	-	-	-	211	271
Wood and Wood Products	472	-	-	-	-	-	-	-	169	396
Construction	-	-	-	-	-	-	-	-	108	-
Textile and Leather	1261	-	-	-	-	-	-	-	185	609
Non-specified	115	21	-	-	-	-	-	-	59	179
TRANSPORT SECTOR	**-**	**-**	**-**	**-**	**-**	**-**	**-**	**-**	**242**	**-**
Air	-	-	-	-	-	-	-	-	-	-
Road	-	-	-	-	-	-	-	-	-	-
Rail	-	-	-	-	-	-	-	-	138	-
Pipeline Transport	-	-	-	-	-	-	-	-	-	-
Internal Navigation	-	-	-	-	-	-	-	-	-	-
Non-specified	-	-	-	-	-	-	-	-	104	-
OTHER SECTORS	**23321**	**277**	**-**	**-**	**13680**	**-**	**-**	**-**	**7391**	**7210**
Agriculture	958	-	-	-	-	-	-	-	69	27
Comm. and Publ. Services	4655	70	-	-	-	-	-	-	2424	825
Residential	17708	207	-	-	13680	-	-	-	4898	6358
Non-specified	-	-	-	-	-	-	-	-	-	-
NON-ENERGY USE	**-**	**-**	**-**	**-**	**-**	**-**	**-**	**-**	**-**	**-**
in Industry/Transf./Energy	-	-	-	-	-	-	-	-	-	-
in Transport	-	-	-	-	-	-	-	-	-	-
in Other Sectors	-	-	-	-	-	-	-	-	-	-

Croatia / Croatie : 1997

SUPPLY AND CONSUMPTION / *APPROVISIONNEMENT ET DEMANDE*	Coal / *Charbon* (1000 tonnes)							Oil / *Pétrole* (1000 tonnes)			
	Coking Coal / *Charbon à coke*	Other Bit. Coal / *Autres charb. bit.*	Sub-Bit. Coal / *Charbon sous-bit.*	Lignite / *Lignite*	Peat / *Tourbe*	Oven and Gas Coke / *Coke de four/gaz*	Pat. Fuel and BKB / *Agg./briq. de lignite*	Crude Oil / *Pétrole brut*	NGL / *LGN*	Feed-stocks / *Produits d'aliment.*	Additives / *Additifs*
Production	-	49	-	-	-	-	-	1496	252	-	-
From Other Sources	-	-	-	-	-	-	-	-	-	-	-
Imports	-	125	-	133	-	33	-	3699	-	-	-
Exports	-	-3	-	-	-	-	-	-40	-	-	-
Intl. Marine Bunkers	-	-	-	-	-	-	-	-	-	-	-
Stock Changes	-	115	-	-	-	-3	-	-43	-	-	-
DOMESTIC SUPPLY	**-**	**286**	**-**	**133**	**-**	**30**	**-**	**5112**	**252**	**-**	**-**
Transfers	-	-	-	-	-	-	-	-	-252	-	-
Statistical Differences	-	-	-	-	-	-	-	-	-	-	-
TRANSFORMATION	**-**	**230**	**-**	**17**	**-**	**-**	**-**	**5112**	**-**	**-**	**-**
Electricity Plants	-	230	-	-	-	-	-	-	-	-	-
CHP Plants	-	-	-	17	-	-	-	-	-	-	-
Heat Plants	-	-	-	-	-	-	-	-	-	-	-
Blast Furnaces/Gas Works	-	-	-	-	-	-	-	-	-	-	-
Coke/Pat. Fuel/BKB Plants	-	-	-	-	-	-	-	-	-	-	-
Petroleum Refineries	-	-	-	-	-	-	-	5112	-	-	-
Petrochemical Industry	-	-	-	-	-	-	-	-	-	-	-
Liquefaction	-	-	-	-	-	-	-	-	-	-	-
Other Transform. Sector	-	-	-	-	-	-	-	-	-	-	-
ENERGY SECTOR	**-**	**-**	**-**	**-**	**-**	**-**	**-**	**-**	**-**	**-**	**-**
Coal Mines	-	-	-	-	-	-	-	-	-	-	-
Oil and Gas Extraction	-	-	-	-	-	-	-	-	-	-	-
Petroleum Refineries	-	-	-	-	-	-	-	-	-	-	-
Electr., CHP+Heat Plants	-	-	-	-	-	-	-	-	-	-	-
Pumped Storage (Elec.)	-	-	-	-	-	-	-	-	-	-	-
Other Energy Sector	-	-	-	-	-	-	-	-	-	-	-
Distribution Losses	-	-	-	-	-	-	-	-	-	-	-
FINAL CONSUMPTION	**-**	**56**	**-**	**116**	**-**	**30**	**-**	**-**	**-**	**-**	**-**
INDUSTRY SECTOR	**-**	**56**	**-**	**84**	**-**	**30**	**-**	**-**	**-**	**-**	**-**
Iron and Steel	-	-	-	-	-	17	-	-	-	-	-
Chemical and Petrochem.	-	-	-	6	-	-	-	-	-	-	-
of which: Feedstocks	-	-	-	-	-	-	-	-	-	-	-
Non-Ferrous Metals	-	-	-	-	-	-	-	-	-	-	-
Non-Metallic Minerals	-	53	-	4	-	6	-	-	-	-	-
Transport Equipment	-	-	-	-	-	-	-	-	-	-	-
Machinery	-	-	-	-	-	-	-	-	-	-	-
Mining and Quarrying	-	-	-	-	-	-	-	-	-	-	-
Food and Tobacco	-	3	-	73	-	7	-	-	-	-	-
Paper, Pulp and Print	-	-	-	-	-	-	-	-	-	-	-
Wood and Wood Products	-	-	-	-	-	-	-	-	-	-	-
Construction	-	-	-	-	-	-	-	-	-	-	-
Textile and Leather	-	-	-	-	-	-	-	-	-	-	-
Non-specified	-	-	-	1	-	-	-	-	-	-	-
TRANSPORT SECTOR	**-**	**-**	**-**	**-**	**-**	**-**	**-**	**-**	**-**	**-**	**-**
Air	-	-	-	-	-	-	-	-	-	-	-
Road	-	-	-	-	-	-	-	-	-	-	-
Rail	-	-	-	-	-	-	-	-	-	-	-
Pipeline Transport	-	-	-	-	-	-	-	-	-	-	-
Internal Navigation	-	-	-	-	-	-	-	-	-	-	-
Non-specified	-	-	-	-	-	-	-	-	-	-	-
OTHER SECTORS	**-**	**-**	**-**	**32**	**-**	**-**	**-**	**-**	**-**	**-**	**-**
Agriculture	-	-	-	-	-	-	-	-	-	-	-
Comm. and Publ. Services	-	-	-	12	-	-	-	-	-	-	-
Residential	-	-	-	20	-	-	-	-	-	-	-
Non-specified	-	-	-	-	-	-	-	-	-	-	-
NON-ENERGY USE	**-**	**-**	**-**	**-**	**-**	**-**	**-**	**-**	**-**	**-**	**-**
in Industry/Trans./Energy	-	-	-	-	-	-	-	-	-	-	-
in Transport	-	-	-	-	-	-	-	-	-	-	-
in Other Sectors	-	-	-	-	-	-	-	-	-	-	-

Croatia / Croatie : 1997

	Oil cont. / *Pétrole cont.* (1000 tonnes)										
SUPPLY AND CONSUMPTION	Refinery Gas	LPG + Ethane	Motor Gasoline	Aviation Gasoline	Jet Fuel	Kerosene	Gas/ Diesel	Heavy Fuel Oil	Naphtha	Petrol. Coke	Other Prod.
APPROVISIONNEMENT ET DEMANDE	*Gaz de raffinerie*	*GPL + éthane*	*Essence moteur*	*Essence aviation*	*Carbu- réacteurs*	*Kérosène*	*Gazole*	*Fioul lourd*	*Naphta*	*Coke de pétrole*	*Autres prod.*
Production	202	158	1089	-	95	5	1492	1484	182	28	323
From Other Sources	-	-	-	-	-	-	-	-	-	-	-
Imports	-	2	67	-	25	2	224	54	-	15	42
Exports	-	-146	-455	-	-34	-1	-503	-127	-238	-29	-185
Intl. Marine Bunkers	-	-	-	-	-	-	-7	-17	-	-	-
Stock Changes	-	1	-23	-	-4	-	-16	-74	2	1	21
DOMESTIC SUPPLY	**202**	**15**	**678**	**-**	**82**	**6**	**1190**	**1320**	**-54**	**15**	**201**
Transfers	-	198	-	-	-	-	-	-	54	-	-
Statistical Differences	-	-	-	-	-	-	-	-	-	-	-
TRANSFORMATION	**4**	**14**	**-**	**-**	**-**	**-**	**10**	**749**	**-**	**-**	**-**
Electricity Plants	-	-	-	-	-	-	4	487	-	-	-
CHP Plants	4	-	-	-	-	-	-	225	-	-	-
Heat Plants	-	1	-	-	-	-	6	37	-	-	-
Blast Furnaces/Gas Works	-	13	-	-	-	-	-	-	-	-	-
Coke/Pat. Fuel/BKB Plants	-	-	-	-	-	-	-	-	-	-	-
Petroleum Refineries	-	-	-	-	-	-	-	-	-	-	-
Petrochemical Industry	-	-	-	-	-	-	-	-	-	-	-
Liquefaction	-	-	-	-	-	-	-	-	-	-	-
Other Transform. Sector	-	-	-	-	-	-	-	-	-	-	-
ENERGY SECTOR	**198**	**-**	**-**	**-**	**-**	**-**	**4**	**281**	**-**	**-**	**-**
Coal Mines	-	-	-	-	-	-	-	-	-	-	-
Oil and Gas Extraction	-	-	-	-	-	-	3	-	-	-	-
Petroleum Refineries	198	-	-	-	-	-	-	281	-	-	-
Electr., CHP+Heat Plants	-	-	-	-	-	-	1	-	-	-	-
Pumped Storage (Elec.)	-	-	-	-	-	-	-	-	-	-	-
Other Energy Sector	-	-	-	-	-	-	-	-	-	-	-
Distribution Losses	-	-	-	-	-	-	-	-	-	-	-
FINAL CONSUMPTION	**-**	**199**	**678**	**-**	**82**	**6**	**1176**	**290**	**-**	**15**	**201**
INDUSTRY SECTOR	**-**	**121**	**11**	**-**	**-**	**-**	**122**	**251**	**-**	**15**	**-**
Iron and Steel	-	1	-	-	-	-	3	5	-	-	-
Chemical and Petrochem.	-	102	-	-	-	-	2	72	-	-	-
of which: Feedstocks	-	*94*	-	-	-	-	-	-	-	-	-
Non-Ferrous Metals	-	-	-	-	-	-	3	1	-	-	-
Non-Metallic Minerals	-	3	-	-	-	-	6	87	-	15	-
Transport Equipment	-	2	-	-	-	-	-	2	-	-	-
Machinery	-	1	-	-	-	-	2	6	-	-	-
Mining and Quarrying	-	1	-	-	-	-	10	2	-	-	-
Food and Tobacco	-	4	-	-	-	-	14	50	-	-	-
Paper, Pulp and Print	-	-	-	-	-	-	1	3	-	-	-
Wood and Wood Products	-	2	-	-	-	-	1	1	-	-	-
Construction	-	1	11	-	-	-	78	-	-	-	-
Textile and Leather	-	1	-	-	-	-	1	18	-	-	-
Non-specified	-	3	-	-	-	-	1	4	-	-	-
TRANSPORT SECTOR	**-**	**12**	**661**	**-**	**82**	**-**	**570**	**11**	**-**	**-**	**-**
Air	-	-	-	-	82	-	-	-	-	-	-
Road	-	12	661	-	-	-	487	-	-	-	-
Rail	-	-	-	-	-	-	31	-	-	-	-
Pipeline Transport	-	-	-	-	-	-	-	-	-	-	-
Internal Navigation	-	-	-	-	-	-	26	11	-	-	-
Non-specified	-	-	-	-	-	-	26	-	-	-	-
OTHER SECTORS	**-**	**66**	**6**	**-**	**-**	**6**	**484**	**28**	**-**	**-**	**-**
Agriculture	-	2	6	-	-	-	166	2	-	-	-
Comm. and Publ. Services	-	3	-	-	-	1	109	7	-	-	-
Residential	-	61	-	-	-	5	209	19	-	-	-
Non-specified	-	-	-	-	-	-	-	-	-	-	-
NON-ENERGY USE	**-**	**-**	**-**	**-**	**-**	**-**	**-**	**-**	**-**	**-**	**201**
in Industry/Transf./Energy	-	-	-	-	-	-	-	-	-	-	171
in Transport	-	-	-	-	-	-	-	-	-	-	26
in Other Sectors	-	-	-	-	-	-	-	-	-	-	4

Croatia / Croatie : 1997

SUPPLY AND CONSUMPTION / *APPROVISIONNEMENT ET DEMANDE*	Gas / *Gaz* (TJ)				Comb. Renew. & Waste / *En. Re. Comb. & Déchets* (TJ)				(GWh)	(TJ)
	Natural Gas / *Gaz naturel*	Gas Works / *Usines à gaz*	Coke Ovens / *Cokeries*	Blast Furnaces / *Hauts fourneaux*	Solid Biomass / *Biomasse solide*	Gas/Liquids from Biomass / *Gaz/Liquides tirés de biomasse*	Municipal Waste / *Déchets urbains*	Industrial Waste / *Déchets industriels*	Electricity / *Electricité*	Heat / *Chaleur*
Production	65254	464	-	-	13545	-	-	10	9685	13327
From Other Sources	-	-	-	-	-	-	-	-	-	-
Imports	39729	-	-	-	-	-	-	-	4608	-
Exports	-	-	-	-	-	-	-	-	-660	-
Intl. Marine Bunkers	-	-	-	-	-	-	-	-	-	-
Stock Changes	-464	-	-	-	-	-	-	-	-	-
DOMESTIC SUPPLY	**104519**	**464**	-	-	**13545**	-	-	**10**	**13633**	**13327**
Transfers	-	-	-	-	-	-	-	-	-	-
Statistical Differences	-	-	-	-	-	-	-	-	-	-
TRANSFORMATION	**20767**	-	-	-	-	-	-	**10**	-	-
Electricity Plants	4856	-	-	-	-	-	-	-	-	-
CHP Plants	14433	-	-	-	-	-	-	10	-	-
Heat Plants	1478	-	-	-	-	-	-	-	-	-
Blast Furnaces/Gas Works	-	-	-	-	-	-	-	-	-	-
Coke/Pat. Fuel/BKB Plants	-	-	-	-	-	-	-	-	-	-
Petroleum Refineries	-	-	-	-	-	-	-	-	-	-
Petrochemical Industry	-	-	-	-	-	-	-	-	-	-
Liquefaction	-	-	-	-	-	-	-	-	-	-
Other Transform. Sector	-	-	-	-	-	-	-	-	-	-
ENERGY SECTOR	**10431**	-	-	-	-	-	-	-	**791**	**1259**
Coal Mines	-	-	-	-	-	-	-	-	12	-
Oil and Gas Extraction	10264	-	-	-	-	-	-	-	119	-
Petroleum Refineries	133	-	-	-	-	-	-	-	257	-
Electr., CHP+Heat Plants	34	-	-	-	-	-	-	-	386	1259
Pumped Storage (Elec.)	-	-	-	-	-	-	-	-	-	-
Other Energy Sector	-	-	-	-	-	-	-	-	17	-
Distribution Losses	2725	49	-	-	-	-	-	-	1822	1838
FINAL CONSUMPTION	**70596**	**415**	-	-	**13545**	-	-	-	**11020**	**10230**
INDUSTRY SECTOR	**46930**	**175**	-	-	-	-	-	-	**3012**	**3188**
Iron and Steel	1809	19	-	-	-	-	-	-	276	-
Chemical and Petrochem.	27413	-	-	-	-	-	-	-	536	958
of which: Feedstocks	*22268*	-	-	-	-	-	-	-	-	-
Non-Ferrous Metals	201	-	-	-	-	-	-	-	55	-
Non-Metallic Minerals	7615	32	-	-	-	-	-	-	448	125
Transport Equipment	19	-	-	-	-	-	-	-	116	28
Machinery	612	-	-	-	-	-	-	-	179	379
Mining and Quarrying	395	-	-	-	-	-	-	-	41	30
Food and Tobacco	4914	-	-	-	-	-	-	-	402	310
Paper, Pulp and Print	2212	-	-	-	-	-	-	-	236	665
Wood and Wood Products	315	-	-	-	-	-	-	-	136	-
Construction	-	-	-	-	-	-	-	-	235	-
Textile and Leather	1193	-	-	-	-	-	-	-	187	693
Non-specified	232	124	-	-	-	-	-	-	165	-
TRANSPORT SECTOR	-	-	-	-	-	-	-	-	**248**	-
Air	-	-	-	-	-	-	-	-	-	-
Road	-	-	-	-	-	-	-	-	-	-
Rail	-	-	-	-	-	-	-	-	140	-
Pipeline Transport	-	-	-	-	-	-	-	-	10	-
Internal Navigation	-	-	-	-	-	-	-	-	-	-
Non-specified	-	-	-	-	-	-	-	-	98	-
OTHER SECTORS	**23666**	**240**	-	-	**13545**	-	-	-	**7760**	**7042**
Agriculture	855	-	-	-	-	-	-	-	60	-
Comm. and Publ. Services	4313	30	-	-	-	-	-	-	2510	898
Residential	18498	210	-	-	13545	-	-	-	5190	6144
Non-specified	-	-	-	-	-	-	-	-	-	-
NON-ENERGY USE	-	-	-	-	-	-	-	-	-	-
in Industry/Transf./Energy	-	-	-	-	-	-	-	-	-	-
in Transport	-	-	-	-	-	-	-	-	-	-
in Other Sectors	-	-	-	-	-	-	-	-	-	-

Cuba

SUPPLY AND CONSUMPTION 1996	Coal (1000 tonnes)							Oil (1000 tonnes)			
	Coking Coal	Other Bit. Coal	Sub-Bit. Coal	Lignite	Peat	Oven and Gas Coke	Pat. Fuel and BKB	Crude Oil	NGL	Feed-stocks	Additives
Production	-	-	-	-	-	-	-	1476	-	-	-
Imports	-	1	-	-	-	16	-	1636	-	-	-
Exports	-	-	-	-	-	-	-	-	-	-	-
Intl. Marine Bunkers	-	-	-	-	-	-	-	-	-	-	-
Stock Changes	-	-	-	-	-	-	-	-	-	-	-
DOMESTIC SUPPLY	**-**	**1**	**-**	**-**	**-**	**16**	**-**	**3112**	**-**	**-**	**-**
Transfers and Stat. Diff.	-	11	-	-	-	26	-	364	-	-	-
TRANSFORMATION	**-**	**-**	**-**	**-**	**-**	**37**	**-**	**2301**	**-**	**-**	**-**
Electricity and CHP Plants	-	-	-	-	-	-	-	69	-	-	-
Petroleum Refineries	-	-	-	-	-	-	-	2232	-	-	-
Other Transform. Sector	-	-	-	-	-	37	-	-	-	-	-
ENERGY SECTOR	**-**	**-**	**-**	**-**	**-**	**-**	**-**	**-**	**-**	**-**	**-**
DISTRIBUTION LOSSES	**-**	**-**	**-**	**-**	**-**	**-**	**-**	**-**	**-**	**-**	**-**
FINAL CONSUMPTION	**-**	**12**	**-**	**-**	**-**	**5**	**-**	**1175**	**-**	**-**	**-**
INDUSTRY SECTOR	**-**	**12**	**-**	**-**	**-**	**5**	**-**	**1175**	**-**	**-**	**-**
Iron and Steel	-	12	-	-	-	5	-	-	-	-	-
Chemical and Petrochem.	-	-	-	-	-	-	-	-	-	-	-
Non-Metallic Minerals	-	-	-	-	-	-	-	-	-	-	-
Non-specified	-	-	-	-	-	-	-	1175	-	-	-
TRANSPORT SECTOR	**-**	**-**	**-**	**-**	**-**	**-**	**-**	**-**	**-**	**-**	**-**
Air	-	-	-	-	-	-	-	-	-	-	-
Road	-	-	-	-	-	-	-	-	-	-	-
Non-specified	-	-	-	-	-	-	-	-	-	-	-
OTHER SECTORS	**-**	**-**	**-**	**-**	**-**	**-**	**-**	**-**	**-**	**-**	**-**
Agriculture	-	-	-	-	-	-	-	-	-	-	-
Comm. and Publ. Services	-	-	-	-	-	-	-	-	-	-	-
Residential	-	-	-	-	-	-	-	-	-	-	-
Non-specified	-	-	-	-	-	-	-	-	-	-	-
NON-ENERGY USE	**-**	**-**	**-**	**-**	**-**	**-**	**-**	**-**	**-**	**-**	**-**

APPROVISIONNEMENT ET DEMANDE 1997	*Charbon* (1000 tonnes)							*Pétrole* (1000 tonnes)			
	Charbon à coke	*Autres charb. bit.*	*Charbon sous-bit.*	*Lignite*	*Tourbe*	*Coke de four/gaz*	*Agg./briq. de lignite*	*Pétrole brut*	*LGN*	*Produits d'aliment.*	*Additifs*
Production	-	-	-	-	-	-	-	1714	-	-	-
Imports	-	42	-	-	-	32	-	1940	-	-	-
Exports	-	-	-	-	-	-	-	-	-	-	-
Intl. Marine Bunkers	-	-	-	-	-	-	-	-	-	-	-
Stock Changes	-	-	-	-	-	-	-	-	-	-	-
DOMESTIC SUPPLY	**-**	**42**	**-**	**-**	**-**	**32**	**-**	**3654**	**-**	**-**	**-**
Transfers and Stat. Diff.	-	-	-	-	-	-	-	-1	-	-	-
TRANSFORMATION	**-**	**-**	**-**	**-**	**-**	**27**	**-**	**2520**	**-**	**-**	**-**
Electricity and CHP Plants	-	-	-	-	-	-	-	73	-	-	-
Petroleum Refineries	-	-	-	-	-	-	-	2447	-	-	-
Other Transform. Sector	-	-	-	-	-	27	-	-	-	-	-
ENERGY SECTOR	**-**	**-**	**-**	**-**	**-**	**-**	**-**	**-**	**-**	**-**	**-**
DISTRIBUTION LOSSES	**-**	**-**	**-**	**-**	**-**	**-**	**-**	**-**	**-**	**-**	**-**
FINAL CONSUMPTION	**-**	**42**	**-**	**-**	**-**	**5**	**-**	**1133**	**-**	**-**	**-**
INDUSTRY SECTOR	**-**	**42**	**-**	**-**	**-**	**5**	**-**	**1133**	**-**	**-**	**-**
Iron and Steel	-	42	-	-	-	5	-	-	-	-	-
Chemical and Petrochem.	-	-	-	-	-	-	-	-	-	-	-
Non-Metallic Minerals	-	-	-	-	-	-	-	-	-	-	-
Non-specified	-	-	-	-	-	-	-	1133	-	-	-
TRANSPORT SECTOR	**-**	**-**	**-**	**-**	**-**	**-**	**-**	**-**	**-**	**-**	**-**
Air	-	-	-	-	-	-	-	-	-	-	-
Road	-	-	-	-	-	-	-	-	-	-	-
Non-specified	-	-	-	-	-	-	-	-	-	-	-
OTHER SECTORS	**-**	**-**	**-**	**-**	**-**	**-**	**-**	**-**	**-**	**-**	**-**
Agriculture	-	-	-	-	-	-	-	-	-	-	-
Comm. and Publ. Services	-	-	-	-	-	-	-	-	-	-	-
Residential	-	-	-	-	-	-	-	-	-	-	-
Non-specified	-	-	-	-	-	-	-	-	-	-	-
NON-ENERGY USE	**-**	**-**	**-**	**-**	**-**	**-**	**-**	**-**	**-**	**-**	**-**

Cuba

SUPPLY AND CONSUMPTION 1996	Oil cont. (1000 tonnes)										
	Refinery Gas	LPG + Ethane	Motor Gasoline	Aviation Gasoline	Jet Fuel	Kerosene	Gas/ Diesel	Heavy Fuel Oil	Naphtha	Petrol. Coke	Other Prod.
Production	22	61	359	-	-	222	212	862	83	15	286
Imports	-	32	58	-	37	-	1530	2912	-	-	133
Exports	-	-	-	-	-	-	-	-	-	-	-
Intl. Marine Bunkers	-	-	-	-	-	-	-15	-89	-	-	-1
Stock Changes	-	-7	-34	-	223	1	-72	24	-	-	-
DOMESTIC SUPPLY	**22**	**86**	**383**	**-**	**260**	**223**	**1655**	**3709**	**83**	**15**	**418**
Transfers and Stat. Diff.	-	1	-	-	1	3	-	1	-	-	-93
TRANSFORMATION	**-**	**-**	**-**	**-**	**-**	**-**	**33**	**3519**	**83**	**-**	**-**
Electricity and CHP Plants	-	-	-	-	-	-	33	3383	-	-	-
Petroleum Refineries	-	-	-	-	-	-	-	-	-	-	-
Other Transform. Sector	-	-	-	-	-	-	-	136	83	-	-
ENERGY SECTOR	**22**	**-**	**-**	**-**	**-**	**3**	**15**	**-**	**-**	**-**	**-**
DISTRIBUTION LOSSES	**-**	**-**	**-**	**-**	**-**	**-**	**5**	**-**	**-**	**-**	**-**
FINAL CONSUMPTION	**-**	**87**	**383**	**-**	**261**	**223**	**1602**	**191**	**-**	**15**	**325**
INDUSTRY SECTOR	**-**	**9**	**-**	**-**	**-**	**3**	**836**	**187**	**-**	**15**	**-**
Iron and Steel	-	-	-	-	-	-	-	-	-	-	-
Chemical and Petrochem.	-	-	-	-	-	-	-	-	-	-	-
Non-Metallic Minerals	-	-	-	-	-	-	-	-	-	-	-
Non-specified	-	9	-	-	-	3	836	187	-	15	-
TRANSPORT SECTOR	**-**	**-**	**383**	**-**	**261**	**-**	**236**	**-**	**-**	**-**	**-**
Air	-	-	-	-	261	-	-	-	-	-	-
Road	-	-	383	-	-	-	236	-	-	-	-
Non-specified	-	-	-	-	-	-	-	-	-	-	-
OTHER SECTORS	**-**	**78**	**-**	**-**	**-**	**220**	**530**	**4**	**-**	**-**	**-**
Agriculture	-	-	-	-	-	-	250	4	-	-	-
Comm. and Publ. Services	-	8	-	-	-	3	-	-	-	-	-
Residential	-	66	-	-	-	217	-	-	-	-	-
Non-specified	-	4	-	-	-	-	280	-	-	-	-
NON-ENERGY USE	**-**	**-**	**-**	**-**	**-**	**-**	**-**	**-**	**-**	**-**	**325**

APPROVISIONNEMENT ET DEMANDE 1997	*Pétrole cont.* (1000 tonnes)										
	Gaz de raffinerie	*GPL + éthane*	*Essence moteur*	*Essence aviation*	*Carbu- réacteurs*	*Kérosène*	*Gazole*	*Fioul lourd*	*Naphta*	*Coke de pétrole*	*Autres prod.*
Production	21	56	386	-	-	240	213	866	91	16	313
Imports	-	33	62	-	66	-	1539	3247	-	-	145
Exports	-	-	-	-	-	-	-	-	-	-	-
Intl. Marine Bunkers	-	-	-	-	-	-	-15	-98	-	-	-
Stock Changes	-	-1	-37	-	213	-24	-47	40	-	-	-
DOMESTIC SUPPLY	**21**	**88**	**411**	**-**	**279**	**216**	**1690**	**4055**	**91**	**16**	**458**
Transfers and Stat. Diff.	-	-	-	-	-	-	1	-	-	-	-120
TRANSFORMATION	**-**	**-**	**-**	**-**	**-**	**-**	**35**	**3846**	**91**	**-**	**-**
Electricity and CHP Plants	-	-	-	-	-	-	35	3714	-	-	-
Petroleum Refineries	-	-	-	-	-	-	-	-	-	-	-
Other Transform. Sector	-	-	-	-	-	-	-	132	91	-	-
ENERGY SECTOR	**21**	**-**	**-**	**-**	**-**	**-**	**15**	**-**	**-**	**-**	**-**
DISTRIBUTION LOSSES	**-**	**-**	**-**	**-**	**-**	**-**	**5**	**-**	**-**	**-**	**-**
FINAL CONSUMPTION	**-**	**88**	**411**	**-**	**279**	**216**	**1636**	**209**	**-**	**16**	**338**
INDUSTRY SECTOR	**-**	**10**	**-**	**-**	**-**	**2**	**858**	**204**	**-**	**16**	**-**
Iron and Steel	-	-	-	-	-	-	-	-	-	-	-
Chemical and Petrochem.	-	-	-	-	-	-	-	-	-	-	-
Non-Metallic Minerals	-	-	-	-	-	-	-	-	-	-	-
Non-specified	-	10	-	-	-	2	858	204	-	16	-
TRANSPORT SECTOR	**-**	**-**	**411**	**-**	**279**	**-**	**248**	**-**	**-**	**-**	**-**
Air	-	-	-	-	279	-	-	-	-	-	-
Road	-	-	411	-	-	-	248	-	-	-	-
Non-specified	-	-	-	-	-	-	-	-	-	-	-
OTHER SECTORS	**-**	**78**	**-**	**-**	**-**	**214**	**530**	**5**	**-**	**-**	**-**
Agriculture	-	-	-	-	-	-	250	5	-	-	-
Comm. and Publ. Services	-	8	-	-	-	3	-	-	-	-	-
Residential	-	68	-	-	-	211	-	-	-	-	-
Non-specified	-	2	-	-	-	-	280	-	-	-	-
NON-ENERGY USE	**-**	**-**	**-**	**-**	**-**	**-**	**-**	**-**	**-**	**-**	**338**

Cuba

SUPPLY AND CONSUMPTION 1996	Gas (TJ)				Comb. Renew. & Waste (TJ)				(GWh)	(TJ)
	Natural Gas	Gas Works	Coke Ovens	Blast Furnaces	Solid Biomass	Gas/Liquids from Biomass	Municipal Waste	Industrial Waste	Electricity	Heat
Production	733	6190	-	-	225915	3223	-	-	13236	-
Imports	-	-	-	-	-	-	-	-	-	-
Exports	-	-	-	-	-	-	-	-	-	-
Intl. Marine Bunkers	-	-	-	-	-	-	-	-	-	-
Stock Changes	-	-	-	-	-	-	-	-	-	-
DOMESTIC SUPPLY	**733**	**6190**	**-**	**-**	**225915**	**3223**	**-**	**-**	**13236**	**-**
Transfers and Stat. Diff.	-11	-	-	-	-	-	-	-	1	-
TRANSFORMATION	**114**	**-**	**-**	**-**	**32045**	**-**	**-**	**-**	**-**	**-**
Electricity and CHP Plants	114	-	-	-	28724	-	-	-	-	-
Petroleum Refineries	-	-	-	-	-	-	-	-	-	-
Other Transform. Sector	-	-	-	-	3321	-	-	-	-	-
ENERGY SECTOR	**-**	**-**	**-**	**-**	**-**	**-**	**-**	**-**	**867**	**-**
DISTRIBUTION LOSSES	**-**	**-**	**-**	**-**	**1453**	**-**	**-**	**-**	**2576**	**-**
FINAL CONSUMPTION	**608**	**6190**	**-**	**-**	**192417**	**3223**	**-**	**-**	**9794**	**-**
INDUSTRY SECTOR	**547**	**558**	**-**	**-**	**189491**	**-**	**-**	**-**	**3976**	**-**
Iron and Steel	-	-	-	-	-	-	-	-	103	-
Chemical and Petrochem.	-	-	-	-	-	-	-	-	315	-
Non-Metallic Minerals	-	-	-	-	-	-	-	-	-	-
Non-specified	547	558	-	-	189491	-	-	-	3558	-
TRANSPORT SECTOR	**-**	**17**	**-**	**-**	**-**	**3223**	**-**	**-**	**69**	**-**
Air	-	-	-	-	-	-	-	-	-	-
Road	-	-	-	-	-	3223	-	-	-	-
Non-specified	-	17	-	-	-	-	-	-	69	-
OTHER SECTORS	**61**	**5615**	**-**	**-**	**2926**	**-**	**-**	**-**	**5749**	**-**
Agriculture	61	3	-	-	1256	-	-	-	191	-
Comm. and Publ. Services	-	-	-	-	424	-	-	-	2188	-
Residential	-	5612	-	-	1246	-	-	-	3370	-
Non-specified	-	-	-	-	-	-	-	-	-	-
NON-ENERGY USE	**-**	**-**	**-**	**-**	**-**	**-**	**-**	**-**	**-**	**-**

APPROVISIONNEMENT ET DEMANDE 1997	*Gaz* (TJ)				*En. Re. Comb. & Déchets* (TJ)				(GWh)	(TJ)
	Gaz naturel	*Usines à gaz*	*Cokeries*	*Hauts fourneaux*	*Biomasse solide*	*Gaz/Liquides tirés de biomasse*	*Déchets urbains*	*Déchets industriels*	*Electricité*	*Chaleur*
Production	850	6190	-	-	228921	3223	-	-	14087	-
Imports	-	-	-	-	-	-	-	-	-	-
Exports	-	-	-	-	-	-	-	-	-	-
Intl. Marine Bunkers	-	-	-	-	-	-	-	-	-	-
Stock Changes	-	-	-	-	-	-	-	-	-	-
DOMESTIC SUPPLY	**850**	**6190**	**-**	**-**	**228921**	**3223**	**-**	**-**	**14087**	**-**
Transfers and Stat. Diff.	-2	-	-	-	-154	-	-	-	714	-
TRANSFORMATION	**114**	**-**	**-**	**-**	**32839**	**-**	**-**	**-**	**-**	**-**
Electricity and CHP Plants	114	-	-	-	29402	-	-	-	-	-
Petroleum Refineries	-	-	-	-	-	-	-	-	-	-
Other Transform. Sector	-	-	-	-	3437	-	-	-	-	-
ENERGY SECTOR	**-**	**-**	**-**	**-**	**-**	**-**	**-**	**-**	**855**	**-**
DISTRIBUTION LOSSES	**-**	**-**	**-**	**-**	**1453**	**-**	**-**	**-**	**2398**	**-**
FINAL CONSUMPTION	**734**	**6190**	**-**	**-**	**194475**	**3223**	**-**	**-**	**11548**	**-**
INDUSTRY SECTOR	**673**	**558**	**-**	**-**	**191370**	**-**	**-**	**-**	**5258**	**-**
Iron and Steel	-	-	-	-	-	-	-	-	137	-
Chemical and Petrochem.	-	-	-	-	-	-	-	-	416	-
Non-Metallic Minerals	-	-	-	-	-	-	-	-	-	-
Non-specified	673	558	-	-	191370	-	-	-	4705	-
TRANSPORT SECTOR	**-**	**17**	**-**	**-**	**-**	**3223**	**-**	**-**	**91**	**-**
Air	-	-	-	-	-	-	-	-	-	-
Road	-	-	-	-	-	3223	-	-	-	-
Non-specified	-	17	-	-	-	-	-	-	91	-
OTHER SECTORS	**61**	**5615**	**-**	**-**	**3105**	**-**	**-**	**-**	**6199**	**-**
Agriculture	61	3	-	-	1357	-	-	-	210	-
Comm. and Publ. Services	-	-	-	-	424	-	-	-	2220	-
Residential	-	5612	-	-	1324	-	-	-	3769	-
Non-specified	-	-	-	-	-	-	-	-	-	-
NON-ENERGY USE	**-**	**-**	**-**	**-**	**-**	**-**	**-**	**-**	**-**	**-**

Cyprus / Chypre

SUPPLY AND CONSUMPTION 1996	Coal (1000 tonnes)							Oil (1000 tonnes)			
	Coking Coal	Other Bit. Coal	Sub-Bit. Coal	Lignite	Peat	Oven and Gas Coke	Pat. Fuel and BKB	Crude Oil	NGL	Feed-stocks	Additives
Production	-	-	-	-	-	-	-	-	-	-	-
Imports	-	17	-	-	-	-	-	804	-	-	-
Exports	-	-	-	-	-	-	-	-	-	-	-
Intl. Marine Bunkers	-	-	-	-	-	-	-	-	-	-	-
Stock Changes	-	-	-	-	-	-	-	-44	-	-	-
DOMESTIC SUPPLY	**-**	**17**	**-**	**-**	**-**	**-**	**-**	**760**	**-**	**-**	**-**
Transfers and Stat. Diff.	-	-	-	-	-	-	-	-	-	-	-
TRANSFORMATION	**-**	**-**	**-**	**-**	**-**	**-**	**-**	**760**	**-**	**-**	**-**
Electricity and CHP Plants	-	-	-	-	-	-	-	-	-	-	-
Petroleum Refineries	-	-	-	-	-	-	-	760	-	-	-
Other Transform. Sector	-	-	-	-	-	-	-	-	-	-	-
ENERGY SECTOR	**-**	**-**	**-**	**-**	**-**	**-**	**-**	**-**	**-**	**-**	**-**
DISTRIBUTION LOSSES	**-**	**-**	**-**	**-**	**-**	**-**	**-**	**-**	**-**	**-**	**-**
FINAL CONSUMPTION	**-**	**17**	**-**	**-**	**-**	**-**	**-**	**-**	**-**	**-**	**-**
INDUSTRY SECTOR	**-**	**17**	**-**	**-**	**-**	**-**	**-**	**-**	**-**	**-**	**-**
Iron and Steel	-	-	-	-	-	-	-	-	-	-	-
Chemical and Petrochem.	-	-	-	-	-	-	-	-	-	-	-
Non-Metallic Minerals	-	17	-	-	-	-	-	-	-	-	-
Non-specified	-	-	-	-	-	-	-	-	-	-	-
TRANSPORT SECTOR	**-**	**-**	**-**	**-**	**-**	**-**	**-**	**-**	**-**	**-**	**-**
Air	-	-	-	-	-	-	-	-	-	-	-
Road	-	-	-	-	-	-	-	-	-	-	-
Non-specified	-	-	-	-	-	-	-	-	-	-	-
OTHER SECTORS	**-**	**-**	**-**	**-**	**-**	**-**	**-**	**-**	**-**	**-**	**-**
Agriculture	-	-	-	-	-	-	-	-	-	-	-
Comm. and Publ. Services	-	-	-	-	-	-	-	-	-	-	-
Residential	-	-	-	-	-	-	-	-	-	-	-
Non-specified	-	-	-	-	-	-	-	-	-	-	-
NON-ENERGY USE	**-**	**-**	**-**	**-**	**-**	**-**	**-**	**-**	**-**	**-**	**-**

APPROVISIONNEMENT ET DEMANDE 1997	*Charbon* (1000 tonnes)							*Pétrole* (1000 tonnes)			
	Charbon à coke	*Autres charb. bit.*	*Charbon sous-bit.*	*Lignite*	*Tourbe*	*Coke de four/gaz*	*Agg./briq. de lignite*	*Pétrole brut*	*LGN*	*Produits d'aliment.*	*Additifs*
Production	-	-	-	-	-	-	-	-	-	-	-
Imports	-	26	-	-	8	-	-	1039	-	-	-
Exports	-	-	-	-	-	-	-	-	-	-	-
Intl. Marine Bunkers	-	-	-	-	-	-	-	-	-	-	-
Stock Changes	-	-7	-	-	-	-	-	4	-	-	-
DOMESTIC SUPPLY	**-**	**19**	**-**	**-**	**8**	**-**	**-**	**1043**	**-**	**-**	**-**
Transfers and Stat. Diff.	-	-	-	-	-	-	-	-	-	-	-
TRANSFORMATION	**-**	**-**	**-**	**-**	**-**	**-**	**-**	**1043**	**-**	**-**	**-**
Electricity and CHP Plants	-	-	-	-	-	-	-	-	-	-	-
Petroleum Refineries	-	-	-	-	-	-	-	1043	-	-	-
Other Transform. Sector	-	-	-	-	-	-	-	-	-	-	-
ENERGY SECTOR	**-**	**-**	**-**	**-**	**-**	**-**	**-**	**-**	**-**	**-**	**-**
DISTRIBUTION LOSSES	**-**	**-**	**-**	**-**	**-**	**-**	**-**	**-**	**-**	**-**	**-**
FINAL CONSUMPTION	**-**	**19**	**-**	**-**	**8**	**-**	**-**	**-**	**-**	**-**	**-**
INDUSTRY SECTOR	**-**	**19**	**-**	**-**	**8**	**-**	**-**	**-**	**-**	**-**	**-**
Iron and Steel	-	-	-	-	-	-	-	-	-	-	-
Chemical and Petrochem.	-	-	-	-	-	-	-	-	-	-	-
Non-Metallic Minerals	-	19	-	-	-	-	-	-	-	-	-
Non-specified	-	-	-	-	8	-	-	-	-	-	-
TRANSPORT SECTOR	**-**	**-**	**-**	**-**	**-**	**-**	**-**	**-**	**-**	**-**	**-**
Air	-	-	-	-	-	-	-	-	-	-	-
Road	-	-	-	-	-	-	-	-	-	-	-
Non-specified	-	-	-	-	-	-	-	-	-	-	-
OTHER SECTORS	**-**	**-**	**-**	**-**	**-**	**-**	**-**	**-**	**-**	**-**	**-**
Agriculture	-	-	-	-	-	-	-	-	-	-	-
Comm. and Publ. Services	-	-	-	-	-	-	-	-	-	-	-
Residential	-	-	-	-	-	-	-	-	-	-	-
Non-specified	-	-	-	-	-	-	-	-	-	-	-
NON-ENERGY USE	**-**	**-**	**-**	**-**	**-**	**-**	**-**	**-**	**-**	**-**	**-**

Cyprus / Chypre

SUPPLY AND CONSUMPTION 1996	Oil cont. (1000 tonnes)										
	Refinery Gas	LPG + Ethane	Motor Gasoline	Aviation Gasoline	Jet Fuel	Kerosene	Gas/ Diesel	Heavy Fuel Oil	Naphtha	Petrol. Coke	Other Prod.
Production	12	26	95	-	3	18	258	316	-	-	30
Imports	-	24	99	-	286	-	225	599	-	153	39
Exports	-	-	-	-	-	-	-	-22	-	-	-
Intl. Marine Bunkers	-	-	-	-	-	-	-25	-65	-	-	-1
Stock Changes	-	-1	-5	-	3	-	-3	52	-	-6	-
DOMESTIC SUPPLY	**12**	**49**	**189**	**-**	**292**	**18**	**455**	**880**	**-**	**147**	**68**
Transfers and Stat. Diff.	-	2	-3	-	-43	-	10	-50	-	-	2
TRANSFORMATION	**-**	**-**	**-**	**-**	**-**	**-**	**6**	**703**	**-**	**-**	**-**
Electricity and CHP Plants	-	-	-	-	-	-	6	703	-	-	-
Petroleum Refineries	-	-	-	-	-	-	-	-	-	-	-
Other Transform. Sector	-	-	-	-	-	-	-	-	-	-	-
ENERGY SECTOR	**12**	**-**	**-**	**-**	**-**	**-**	**-**	**16**	**-**	**-**	**-**
DISTRIBUTION LOSSES	**-**	**-**	**-**	**-**	**-**	**-**	**-**	**-**	**-**	**-**	**-**
FINAL CONSUMPTION	**-**	**51**	**186**	**-**	**249**	**18**	**459**	**111**	**-**	**147**	**70**
INDUSTRY SECTOR	**-**	**-**	**-**	**-**	**-**	**-**	**161**	**111**	**-**	**147**	**-**
Iron and Steel	-	-	-	-	-	-	-	-	-	-	-
Chemical and Petrochem.	-	-	-	-	-	-	-	-	-	-	-
Non-Metallic Minerals	-	-	-	-	-	-	-	111	-	147	-
Non-specified	-	-	-	-	-	-	161	-	-	-	-
TRANSPORT SECTOR	**-**	**-**	**186**	**-**	**249**	**-**	**298**	**-**	**-**	**-**	**-**
Air	-	-	-	-	249	-	-	-	-	-	-
Road	-	-	186	-	-	-	298	-	-	-	-
Non-specified	-	-	-	-	-	-	-	-	-	-	-
OTHER SECTORS	**-**	**51**	**-**	**-**	**-**	**18**	**-**	**-**	**-**	**-**	**-**
Agriculture	-	-	-	-	-	-	-	-	-	-	-
Comm. and Publ. Services	-	-	-	-	-	-	-	-	-	-	-
Residential	-	51	-	-	-	18	-	-	-	-	-
Non-specified	-	-	-	-	-	-	-	-	-	-	-
NON-ENERGY USE	**-**	**-**	**-**	**-**	**-**	**-**	**-**	**-**	**-**	**-**	**70**

APPROVISIONNEMENT ET DEMANDE 1997	*Pétrole cont.* (1000 tonnes)										
	Gaz de raffinerie	*GPL + éthane*	*Essence moteur*	*Essence aviation*	*Carbu-réacteurs*	*Kérosène*	*Gazole*	*Fioul lourd*	*Naphta*	*Coke de pétrole*	*Autres prod.*
Production	16	32	141	-	5	20	365	421	-	-	38
Imports	-	17	50	-	252	-	146	476	-	142	34
Exports	-	-	-	-	-	-	-	-	-	-	-
Intl. Marine Bunkers	-	-	-	-	-	-	-27	-71	-	-	-1
Stock Changes	-	-	3	-	-5	-	-8	-3	-	10	2
DOMESTIC SUPPLY	**16**	**49**	**194**	**-**	**252**	**20**	**476**	**823**	**-**	**152**	**73**
Transfers and Stat. Diff.	-	3	-3	-	-7	-	13	4	-	-	2
TRANSFORMATION	**-**	**-**	**-**	**-**	**-**	**-**	**6**	**743**	**-**	**-**	**-**
Electricity and CHP Plants	-	-	-	-	-	-	6	743	-	-	-
Petroleum Refineries	-	-	-	-	-	-	-	-	-	-	-
Other Transform. Sector	-	-	-	-	-	-	-	-	-	-	-
ENERGY SECTOR	**16**	**-**	**-**	**-**	**-**	**-**	**-**	**14**	**-**	**-**	**-**
DISTRIBUTION LOSSES	**-**	**-**	**-**	**-**	**-**	**-**	**-**	**-**	**-**	**-**	**-**
FINAL CONSUMPTION	**-**	**52**	**191**	**-**	**245**	**20**	**483**	**70**	**-**	**152**	**75**
INDUSTRY SECTOR	**-**	**-**	**-**	**-**	**-**	**-**	**169**	**70**	**-**	**152**	**-**
Iron and Steel	-	-	-	-	-	-	-	-	-	-	-
Chemical and Petrochem.	-	-	-	-	-	-	-	-	-	-	-
Non-Metallic Minerals	-	-	-	-	-	-	-	70	-	152	-
Non-specified	-	-	-	-	-	-	169	-	-	-	-
TRANSPORT SECTOR	**-**	**-**	**191**	**-**	**245**	**-**	**314**	**-**	**-**	**-**	**-**
Air	-	-	-	-	245	-	-	-	-	-	-
Road	-	-	191	-	-	-	314	-	-	-	-
Non-specified	-	-	-	-	-	-	-	-	-	-	-
OTHER SECTORS	**-**	**52**	**-**	**-**	**-**	**20**	**-**	**-**	**-**	**-**	**-**
Agriculture	-	-	-	-	-	-	-	-	-	-	-
Comm. and Publ. Services	-	-	-	-	-	-	-	-	-	-	-
Residential	-	52	-	-	-	20	-	-	-	-	-
Non-specified	-	-	-	-	-	-	-	-	-	-	-
NON-ENERGY USE	**-**	**-**	**-**	**-**	**-**	**-**	**-**	**-**	**-**	**-**	**75**

Cyprus / Chypre

SUPPLY AND CONSUMPTION 1996	Gas (TJ) Natural Gas	Gas Works	Coke Ovens	Blast Furnaces	Comb. Renew. & Waste (TJ) Solid Biomass	Gas/Liquids from Biomass	Municipal Waste	Industrial Waste	(GWh) Electricity	(TJ) Heat
Production	-	-	-	-	596	-	-	-	2592	-
Imports	-	-	-	-	75	-	-	-	-	-
Exports	-	-	-	-	-	-	-	-	-	-
Intl. Marine Bunkers	-	-	-	-	-	-	-	-	-	-
Stock Changes	-	-	-	-	-	-	-	-	-	-
DOMESTIC SUPPLY	**-**	**-**	**-**	**-**	**671**	**-**	**-**	**-**	**2592**	**-**
Transfers and Stat. Diff.	-	-	-	-	-	-	-	-	-	-
TRANSFORMATION	**-**	**-**	**-**	**-**	**345**	**-**	**-**	**-**	**-**	**-**
Electricity and CHP Plants	-	-	-	-	-	-	-	-	-	-
Petroleum Refineries	-	-	-	-	-	-	-	-	-	-
Other Transform. Sector	-	-	-	-	345	-	-	-	-	-
ENERGY SECTOR	**-**	**-**	**-**	**-**	**-**	**-**	**-**	**-**	**144**	**-**
DISTRIBUTION LOSSES	**-**	**-**	**-**	**-**	**-**	**-**	**-**	**-**	**149**	**-**
FINAL CONSUMPTION	**-**	**-**	**-**	**-**	**326**	**-**	**-**	**-**	**2299**	**-**
INDUSTRY SECTOR	**-**	**-**	**-**	**-**	**-**	**-**	**-**	**-**	**403**	**-**
Iron and Steel	-	-	-	-	-	-	-	-	-	-
Chemical and Petrochem.	-	-	-	-	-	-	-	-	43	-
Non-Metallic Minerals	-	-	-	-	-	-	-	-	168	-
Non-specified	-	-	-	-	-	-	-	-	192	-
TRANSPORT SECTOR	**-**	**-**	**-**	**-**	**-**	**-**	**-**	**-**	**37**	**-**
Air	-	-	-	-	-	-	-	-	-	-
Road	-	-	-	-	-	-	-	-	-	-
Non-specified	-	-	-	-	-	-	-	-	37	-
OTHER SECTORS	**-**	**-**	**-**	**-**	**326**	**-**	**-**	**-**	**1859**	**-**
Agriculture	-	-	-	-	-	-	-	-	79	-
Comm. and Publ. Services	-	-	-	-	-	-	-	-	814	-
Residential	-	-	-	-	261	-	-	-	824	-
Non-specified	-	-	-	-	65	-	-	-	142	-
NON-ENERGY USE	**-**	**-**	**-**	**-**	**-**	**-**	**-**	**-**	**-**	**-**

APPROVISIONNEMENT ET DEMANDE 1997	*Gaz (TJ) Gaz naturel*	*Usines à gaz*	*Cokeries*	*Hauts fourneaux*	*En. Re. Comb. & Déchets (TJ) Biomasse solide*	*Gaz/Liquides tirés de biomasse*	*Déchets urbains*	*Déchets industriels*	*(GWh) Electricité*	*(TJ) Chaleur*
Production	-	-	-	-	602	-	-	-	2711	-
Imports	-	-	-	-	119	-	-	-	-	-
Exports	-	-	-	-	-	-	-	-	-	-
Intl. Marine Bunkers	-	-	-	-	-	-	-	-	-	-
Stock Changes	-	-	-	-	-	-	-	-	-	-
DOMESTIC SUPPLY	**-**	**-**	**-**	**-**	**721**	**-**	**-**	**-**	**2711**	**-**
Transfers and Stat. Diff.	-	-	-	-	-	-	-	-	-	-
TRANSFORMATION	**-**	**-**	**-**	**-**	**431**	**-**	**-**	**-**	**-**	**-**
Electricity and CHP Plants	-	-	-	-	-	-	-	-	-	-
Petroleum Refineries	-	-	-	-	-	-	-	-	-	-
Other Transform. Sector	-	-	-	-	431	-	-	-	-	-
ENERGY SECTOR	**-**	**-**	**-**	**-**	**-**	**-**	**-**	**-**	**156**	**-**
DISTRIBUTION LOSSES	**-**	**-**	**-**	**-**	**-**	**-**	**-**	**-**	**173**	**-**
FINAL CONSUMPTION	**-**	**-**	**-**	**-**	**290**	**-**	**-**	**-**	**2382**	**-**
INDUSTRY SECTOR	**-**	**-**	**-**	**-**	**-**	**-**	**-**	**-**	**395**	**-**
Iron and Steel	-	-	-	-	-	-	-	-	-	-
Chemical and Petrochem.	-	-	-	-	-	-	-	-	11	-
Non-Metallic Minerals	-	-	-	-	-	-	-	-	165	-
Non-specified	-	-	-	-	-	-	-	-	219	-
TRANSPORT SECTOR	**-**	**-**	**-**	**-**	**-**	**-**	**-**	**-**	**21**	**-**
Air	-	-	-	-	-	-	-	-	-	-
Road	-	-	-	-	-	-	-	-	-	-
Non-specified	-	-	-	-	-	-	-	-	21	-
OTHER SECTORS	**-**	**-**	**-**	**-**	**290**	**-**	**-**	**-**	**1966**	**-**
Agriculture	-	-	-	-	-	-	-	-	76	-
Comm. and Publ. Services	-	-	-	-	-	-	-	-	1000	-
Residential	-	-	-	-	230	-	-	-	834	-
Non-specified	-	-	-	-	60	-	-	-	56	-
NON-ENERGY USE	**-**	**-**	**-**	**-**	**-**	**-**	**-**	**-**	**-**	**-**

Dominican Republic / République dominicaine

SUPPLY AND CONSUMPTION 1996	Coal (1000 tonnes)							Oil (1000 tonnes)			
	Coking Coal	Other Bit. Coal	Sub-Bit. Coal	Lignite	Peat	Oven and Gas Coke	Pat. Fuel and BKB	Crude Oil	NGL	Feed-stocks	Additives
Production	-	-	-	-	-	-	-	-	-	-	-
Imports	-	128	-	-	-	-	-	2110	-	-	-
Exports	-	-	-	-	-	-	-	-	-	-	-
Intl. Marine Bunkers	-	-	-	-	-	-	-	-	-	-	-
Stock Changes	-	-	-	-	-	-	-	-	-	-	-
DOMESTIC SUPPLY	**-**	**128**	**-**	**-**	**-**	**-**	**-**	**2110**	**-**	**-**	**-**
Transfers and Stat. Diff.	-	-	-	-	-	-	-	-	-	-	-
TRANSFORMATION	**-**	**128**	**-**	**-**	**-**	**-**	**-**	**2110**	**-**	**-**	**-**
Electricity and CHP Plants	-	128	-	-	-	-	-	-	-	-	-
Petroleum Refineries	-	-	-	-	-	-	-	2110	-	-	-
Other Transform. Sector	-	-	-	-	-	-	-	-	-	-	-
ENERGY SECTOR	**-**	**-**	**-**	**-**	**-**	**-**	**-**	**-**	**-**	**-**	**-**
DISTRIBUTION LOSSES	**-**	**-**	**-**	**-**	**-**	**-**	**-**	**-**	**-**	**-**	**-**
FINAL CONSUMPTION	**-**	**-**	**-**	**-**	**-**	**-**	**-**	**-**	**-**	**-**	**-**
INDUSTRY SECTOR	**-**	**-**	**-**	**-**	**-**	**-**	**-**	**-**	**-**	**-**	**-**
Iron and Steel	-	-	-	-	-	-	-	-	-	-	-
Chemical and Petrochem.	-	-	-	-	-	-	-	-	-	-	-
Non-Metallic Minerals	-	-	-	-	-	-	-	-	-	-	-
Non-specified	-	-	-	-	-	-	-	-	-	-	-
TRANSPORT SECTOR	**-**	**-**	**-**	**-**	**-**	**-**	**-**	**-**	**-**	**-**	**-**
Air	-	-	-	-	-	-	-	-	-	-	-
Road	-	-	-	-	-	-	-	-	-	-	-
Non-specified	-	-	-	-	-	-	-	-	-	-	-
OTHER SECTORS	**-**	**-**	**-**	**-**	**-**	**-**	**-**	**-**	**-**	**-**	**-**
Agriculture	-	-	-	-	-	-	-	-	-	-	-
Comm. and Publ. Services	-	-	-	-	-	-	-	-	-	-	-
Residential	-	-	-	-	-	-	-	-	-	-	-
Non-specified	-	-	-	-	-	-	-	-	-	-	-
NON-ENERGY USE	**-**	**-**	**-**	**-**	**-**	**-**	**-**	**-**	**-**	**-**	**-**

APPROVISIONNEMENT ET DEMANDE 1997	*Charbon* (1000 tonnes)							*Pétrole* (1000 tonnes)			
	Charbon à coke	*Autres charb. bit.*	*Charbon sous-bit.*	*Lignite*	*Tourbe*	*Coke de four/gaz*	*Agg./briq. de lignite*	*Pétrole brut*	*LGN*	*Produits d'aliment.*	*Additifs*
Production	-	-	-	-	-	-	-	-	-	-	-
Imports	-	139	-	-	-	-	-	2214	-	-	-
Exports	-	-	-	-	-	-	-	-	-	-	-
Intl. Marine Bunkers	-	-	-	-	-	-	-	-	-	-	-
Stock Changes	-	-	-	-	-	-	-	-1	-	-	-
DOMESTIC SUPPLY	**-**	**139**	**-**	**-**	**-**	**-**	**-**	**2213**	**-**	**-**	**-**
Transfers and Stat. Diff.	-	-	-	-	-	-	-	-	-	-	-
TRANSFORMATION	**-**	**139**	**-**	**-**	**-**	**-**	**-**	**2213**	**-**	**-**	**-**
Electricity and CHP Plants	-	139	-	-	-	-	-	-	-	-	-
Petroleum Refineries	-	-	-	-	-	-	-	2213	-	-	-
Other Transform. Sector	-	-	-	-	-	-	-	-	-	-	-
ENERGY SECTOR	**-**	**-**	**-**	**-**	**-**	**-**	**-**	**-**	**-**	**-**	**-**
DISTRIBUTION LOSSES	**-**	**-**	**-**	**-**	**-**	**-**	**-**	**-**	**-**	**-**	**-**
FINAL CONSUMPTION	**-**	**-**	**-**	**-**	**-**	**-**	**-**	**-**	**-**	**-**	**-**
INDUSTRY SECTOR	**-**	**-**	**-**	**-**	**-**	**-**	**-**	**-**	**-**	**-**	**-**
Iron and Steel	-	-	-	-	-	-	-	-	-	-	-
Chemical and Petrochem.	-	-	-	-	-	-	-	-	-	-	-
Non-Metallic Minerals	-	-	-	-	-	-	-	-	-	-	-
Non-specified	-	-	-	-	-	-	-	-	-	-	-
TRANSPORT SECTOR	**-**	**-**	**-**	**-**	**-**	**-**	**-**	**-**	**-**	**-**	**-**
Air	-	-	-	-	-	-	-	-	-	-	-
Road	-	-	-	-	-	-	-	-	-	-	-
Non-specified	-	-	-	-	-	-	-	-	-	-	-
OTHER SECTORS	**-**	**-**	**-**	**-**	**-**	**-**	**-**	**-**	**-**	**-**	**-**
Agriculture	-	-	-	-	-	-	-	-	-	-	-
Comm. and Publ. Services	-	-	-	-	-	-	-	-	-	-	-
Residential	-	-	-	-	-	-	-	-	-	-	-
Non-specified	-	-	-	-	-	-	-	-	-	-	-
NON-ENERGY USE	**-**	**-**	**-**	**-**	**-**	**-**	**-**	**-**	**-**	**-**	**-**

Dominican Republic / République dominicaine

SUPPLY AND CONSUMPTION 1996	Oil cont. (1000 tonnes)										
	Refinery Gas	LPG + Ethane	Motor Gasoline	Aviation Gasoline	Jet Fuel	Kerosene	Gas/ Diesel	Heavy Fuel Oil	Naphtha	Petrol. Coke	Other Prod.
Production	-	39	362	-	45	188	434	1038	-	-	-
Imports	-	301	247	-	14	59	521	352	-	-	-
Exports	-	-	-	-	-	-	-	-	-	-	-
Intl. Marine Bunkers	-	-	-	-	-	-	-	-	-	-	-
Stock Changes	-	-	-	-	-	-	7	3	-	-	-
DOMESTIC SUPPLY	**-**	**340**	**609**	**-**	**59**	**247**	**962**	**1393**	**-**	**-**	**-**
Transfers and Stat. Diff.	-	-	-	-	-	-	-2	-1	-	-	-
TRANSFORMATION	**-**	**-**	**29**	**-**	**-**	**-**	**326**	**1220**	**-**	**-**	**-**
Electricity and CHP Plants	-	-	29	-	-	-	326	1220	-	-	-
Petroleum Refineries	-	-	-	-	-	-	-	-	-	-	-
Other Transform. Sector	-	-	-	-	-	-	-	-	-	-	-
ENERGY SECTOR	**-**	**-**	**-**	**-**	**-**	**-**	**-**	**-**	**-**	**-**	**-**
DISTRIBUTION LOSSES	**-**	**-**	**-**	**-**	**-**	**-**	**-**	**-**	**-**	**-**	**-**
FINAL CONSUMPTION	**-**	**340**	**580**	**-**	**59**	**247**	**634**	**172**	**-**	**-**	**-**
INDUSTRY SECTOR	**-**	**31**	**-**	**-**	**-**	**-**	**153**	**165**	**-**	**-**	**-**
Iron and Steel	-	-	-	-	-	-	-	-	-	-	-
Chemical and Petrochem.	-	-	-	-	-	-	-	-	-	-	-
Non-Metallic Minerals	-	-	-	-	-	-	-	-	-	-	-
Non-specified	-	31	-	-	-	-	153	165	-	-	-
TRANSPORT SECTOR	**-**	**72**	**493**	**-**	**59**	**-**	**404**	**7**	**-**	**-**	**-**
Air	-	-	-	-	59	-	-	-	-	-	-
Road	-	-	493	-	-	-	-	-	-	-	-
Non-specified	-	72	-	-	-	-	404	7	-	-	-
OTHER SECTORS	**-**	**237**	**-**	**-**	**-**	**247**	**77**	**-**	**-**	**-**	**-**
Agriculture	-	-	-	-	-	-	35	-	-	-	-
Comm. and Publ. Services	-	-	-	-	-	-	-	-	-	-	-
Residential	-	237	-	-	-	247	42	-	-	-	-
Non-specified	-	-	-	-	-	-	-	-	-	-	-
NON-ENERGY USE	**-**	**-**	**87**	**-**	**-**	**-**	**-**	**-**	**-**	**-**	**-**

APPROVISIONNEMENT ET DEMANDE 1997	*Pétrole cont.* (1000 tonnes)										
	Gaz de raffinerie	*GPL + éthane*	*Essence moteur*	*Essence aviation*	*Carbu- réacteurs*	*Kérosène*	*Gazole*	*Fioul lourd*	*Naphta*	*Coke de pétrole*	*Autres prod.*
Production	-	36	369	-	47	200	449	1100	-	-	-
Imports	-	389	252	-	15	63	538	373	-	-	-
Exports	-	-	-	-	-	-	-	-	-	-	-
Intl. Marine Bunkers	-	-	-	-	-	-	-	-	-	-	-
Stock Changes	-	-	-	-	-	-	7	3	-	-	-
DOMESTIC SUPPLY	**-**	**425**	**621**	**-**	**62**	**263**	**994**	**1476**	**-**	**-**	**-**
Transfers and Stat. Diff.	-	-	-	-	-	-	1	1	-	-	-
TRANSFORMATION	**-**	**-**	**30**	**-**	**-**	**-**	**338**	**1293**	**-**	**-**	**-**
Electricity and CHP Plants	-	-	30	-	-	-	338	1293	-	-	-
Petroleum Refineries	-	-	-	-	-	-	-	-	-	-	-
Other Transform. Sector	-	-	-	-	-	-	-	-	-	-	-
ENERGY SECTOR	**-**	**-**	**-**	**-**	**-**	**-**	**-**	**-**	**-**	**-**	**-**
DISTRIBUTION LOSSES	**-**	**-**	**-**	**-**	**-**	**-**	**-**	**-**	**-**	**-**	**-**
FINAL CONSUMPTION	**-**	**425**	**591**	**-**	**62**	**263**	**657**	**184**	**-**	**-**	**-**
INDUSTRY SECTOR	**-**	**38**	**-**	**-**	**-**	**-**	**159**	**176**	**-**	**-**	**-**
Iron and Steel	-	-	-	-	-	-	-	-	-	-	-
Chemical and Petrochem.	-	-	-	-	-	-	-	-	-	-	-
Non-Metallic Minerals	-	-	-	-	-	-	-	-	-	-	-
Non-specified	-	38	-	-	-	-	159	176	-	-	-
TRANSPORT SECTOR	**-**	**90**	**503**	**-**	**62**	**-**	**418**	**8**	**-**	**-**	**-**
Air	-	-	-	-	62	-	-	-	-	-	-
Road	-	-	503	-	-	-	-	-	-	-	-
Non-specified	-	90	-	-	-	-	418	8	-	-	-
OTHER SECTORS	**-**	**297**	**-**	**-**	**-**	**263**	**80**	**-**	**-**	**-**	**-**
Agriculture	-	-	-	-	-	-	36	-	-	-	-
Comm. and Publ. Services	-	-	-	-	-	-	-	-	-	-	-
Residential	-	297	-	-	-	263	44	-	-	-	-
Non-specified	-	-	-	-	-	-	-	-	-	-	-
NON-ENERGY USE	**-**	**-**	**88**	**-**	**-**	**-**	**-**	**-**	**-**	**-**	**-**

Dominican Republic / République dominicaine

SUPPLY AND CONSUMPTION 1996	Gas (TJ)				Comb. Renew. & Waste (TJ)				(GWh)	(TJ)
	Natural Gas	Gas Works	Coke Ovens	Blast Furnaces	Solid Biomass	Gas/Liquids from Biomass	Municipal Waste	Industrial Waste	Electricity	Heat
Production	-	-	-	-	54544	-	-	-	6847	-
Imports	-	-	-	-	-	-	-	-	-	-
Exports	-	-	-	-	-	-	-	-	-	-
Intl. Marine Bunkers	-	-	-	-	-	-	-	-	-	-
Stock Changes	-	-	-	-	-	-	-	-	-	-
DOMESTIC SUPPLY	**-**	**-**	**-**	**-**	**54544**	**-**	**-**	**-**	**6847**	**-**
Transfers and Stat. Diff.	-	-	-	-	-	-	-	-	-	-
TRANSFORMATION	**-**	**-**	**-**	**-**	**9286**	**-**	**-**	**-**	**-**	**-**
Electricity and CHP Plants	-	-	-	-	843	-	-	-	-	-
Petroleum Refineries	-	-	-	-	-	-	-	-	-	-
Other Transform. Sector	-	-	-	-	8443	-	-	-	-	-
ENERGY SECTOR	**-**	**-**	**-**	**-**	**-**	**-**	**-**	**-**	**262**	**-**
DISTRIBUTION LOSSES	**-**	**-**	**-**	**-**	**-**	**-**	**-**	**-**	**1741**	**-**
FINAL CONSUMPTION	**-**	**-**	**-**	**-**	**45258**	**-**	**-**	**-**	**4844**	**-**
INDUSTRY SECTOR	**-**	**-**	**-**	**-**	**12052**	**-**	**-**	**-**	**1243**	**-**
Iron and Steel	-	-	-	-	-	-	-	-	-	-
Chemical and Petrochem.	-	-	-	-	-	-	-	-	-	-
Non-Metallic Minerals	-	-	-	-	-	-	-	-	-	-
Non-specified	-	-	-	-	12052	-	-	-	1243	-
TRANSPORT SECTOR	**-**	**-**	**-**	**-**	**-**	**-**	**-**	**-**	**-**	**-**
Air	-	-	-	-	-	-	-	-	-	-
Road	-	-	-	-	-	-	-	-	-	-
Non-specified	-	-	-	-	-	-	-	-	-	-
OTHER SECTORS	**-**	**-**	**-**	**-**	**33206**	**-**	**-**	**-**	**3601**	**-**
Agriculture	-	-	-	-	-	-	-	-	-	-
Comm. and Publ. Services	-	-	-	-	-	-	-	-	-	-
Residential	-	-	-	-	33206	-	-	-	3601	-
Non-specified	-	-	-	-	-	-	-	-	-	-
NON-ENERGY USE	**-**	**-**	**-**	**-**	**-**	**-**	**-**	**-**	**-**	**-**

APPROVISIONNEMENT ET DEMANDE 1997	*Gaz* (TJ)				*En. Re. Comb. & Déchets* (TJ)				(GWh)	(TJ)
	Gaz naturel	*Usines à gaz*	*Cokeries*	*Hauts fourneaux*	*Biomasse solide*	*Gaz/Liquides tirés de biomasse*	*Déchets urbains*	*Déchets industriels*	*Electricité*	*Chaleur*
Production	-	-	-	-	54785	-	-	-	7335	-
Imports	-	-	-	-	-	-	-	-	-	-
Exports	-	-	-	-	-	-	-	-	-	-
Intl. Marine Bunkers	-	-	-	-	-	-	-	-	-	-
Stock Changes	-	-	-	-	-	-	-	-	-	-
DOMESTIC SUPPLY	**-**	**-**	**-**	**-**	**54785**	**-**	**-**	**-**	**7335**	**-**
Transfers and Stat. Diff.	-	-	-	-	-	-	-	-	-	-
TRANSFORMATION	**-**	**-**	**-**	**-**	**8809**	**-**	**-**	**-**	**-**	**-**
Electricity and CHP Plants	-	-	-	-	854	-	-	-	-	-
Petroleum Refineries	-	-	-	-	-	-	-	-	-	-
Other Transform. Sector	-	-	-	-	7955	-	-	-	-	-
ENERGY SECTOR	**-**	**-**	**-**	**-**	**-**	**-**	**-**	**-**	**276**	**-**
DISTRIBUTION LOSSES	**-**	**-**	**-**	**-**	**-**	**-**	**-**	**-**	**2036**	**-**
FINAL CONSUMPTION	**-**	**-**	**-**	**-**	**45976**	**-**	**-**	**-**	**5023**	**-**
INDUSTRY SECTOR	**-**	**-**	**-**	**-**	**12116**	**-**	**-**	**-**	**1392**	**-**
Iron and Steel	-	-	-	-	-	-	-	-	-	-
Chemical and Petrochem.	-	-	-	-	-	-	-	-	-	-
Non-Metallic Minerals	-	-	-	-	-	-	-	-	-	-
Non-specified	-	-	-	-	12116	-	-	-	1392	-
TRANSPORT SECTOR	**-**	**-**	**-**	**-**	**-**	**-**	**-**	**-**	**-**	**-**
Air	-	-	-	-	-	-	-	-	-	-
Road	-	-	-	-	-	-	-	-	-	-
Non-specified	-	-	-	-	-	-	-	-	-	-
OTHER SECTORS	**-**	**-**	**-**	**-**	**33860**	**-**	**-**	**-**	**3631**	**-**
Agriculture	-	-	-	-	-	-	-	-	-	-
Comm. and Publ. Services	-	-	-	-	-	-	-	-	-	-
Residential	-	-	-	-	33860	-	-	-	3631	-
Non-specified	-	-	-	-	-	-	-	-	-	-
NON-ENERGY USE	**-**	**-**	**-**	**-**	**-**	**-**	**-**	**-**	**-**	**-**

Ecuador / Equateur

SUPPLY AND CONSUMPTION 1996	Coal (1000 tonnes)							Oil (1000 tonnes)			
	Coking Coal	Other Bit. Coal	Sub-Bit. Coal	Lignite	Peat	Oven and Gas Coke	Pat. Fuel and BKB	Crude Oil	NGL	Feed-stocks	Additives
Production	-	-	-	-	-	-	-	19791	305	-	-
Imports	-	-	-	-	-	-	-	-	-	-	-
Exports	-	-	-	-	-	-	-	-12054	-	-	-
Intl. Marine Bunkers	-	-	-	-	-	-	-	-	-	-	-
Stock Changes	-	-	-	-	-	-	-	-205	-	-	-
DOMESTIC SUPPLY	**-**	**-**	**-**	**-**	**-**	**-**	**-**	**7532**	**305**	**-**	**-**
Transfers and Stat. Diff.	-	-	-	-	-	-	-	-37	-305	277	-
TRANSFORMATION	**-**	**-**	**-**	**-**	**-**	**-**	**-**	**7373**	**-**	**277**	**-**
Electricity and CHP Plants	-	-	-	-	-	-	-	-	-	-	-
Petroleum Refineries	-	-	-	-	-	-	-	7373	-	277	-
Other Transform. Sector	-	-	-	-	-	-	-	-	-	-	-
ENERGY SECTOR	**-**	**-**	**-**	**-**	**-**	**-**	**-**	**55**	**-**	**-**	**-**
DISTRIBUTION LOSSES	**-**	**-**	**-**	**-**	**-**	**-**	**-**	**67**	**-**	**-**	**-**
FINAL CONSUMPTION	**-**	**-**	**-**	**-**	**-**	**-**	**-**	**-**	**-**	**-**	**-**
INDUSTRY SECTOR	**-**	**-**	**-**	**-**	**-**	**-**	**-**	**-**	**-**	**-**	**-**
Iron and Steel	-	-	-	-	-	-	-	-	-	-	-
Chemical and Petrochem.	-	-	-	-	-	-	-	-	-	-	-
Non-Metallic Minerals	-	-	-	-	-	-	-	-	-	-	-
Non-specified	-	-	-	-	-	-	-	-	-	-	-
TRANSPORT SECTOR	**-**	**-**	**-**	**-**	**-**	**-**	**-**	**-**	**-**	**-**	**-**
Air	-	-	-	-	-	-	-	-	-	-	-
Road	-	-	-	-	-	-	-	-	-	-	-
Non-specified	-	-	-	-	-	-	-	-	-	-	-
OTHER SECTORS	**-**	**-**	**-**	**-**	**-**	**-**	**-**	**-**	**-**	**-**	**-**
Agriculture	-	-	-	-	-	-	-	-	-	-	-
Comm. and Publ. Services	-	-	-	-	-	-	-	-	-	-	-
Residential	-	-	-	-	-	-	-	-	-	-	-
Non-specified	-	-	-	-	-	-	-	-	-	-	-
NON-ENERGY USE	**-**	**-**	**-**	**-**	**-**	**-**	**-**	**-**	**-**	**-**	**-**

APPROVISIONNEMENT ET DEMANDE 1997	*Charbon* (1000 tonnes)							*Pétrole* (1000 tonnes)			
	Charbon à coke	*Autres charb. bit.*	*Charbon sous-bit.*	*Lignite*	*Tourbe*	*Coke de four/gaz*	*Agg./briq. de lignite*	*Pétrole brut*	*LGN*	*Produits d'aliment.*	*Additifs*
Production	-	-	-	-	-	-	-	20763	305	-	-
Imports	-	-	-	-	-	-	-	-	-	-	-
Exports	-	-	-	-	-	-	-	-13376	-	-	-
Intl. Marine Bunkers	-	-	-	-	-	-	-	-	-	-	-
Stock Changes	-	-	-	-	-	-	-	-25	-	-	-
DOMESTIC SUPPLY	**-**	**-**	**-**	**-**	**-**	**-**	**-**	**7362**	**305**	**-**	**-**
Transfers and Stat. Diff.	-	-	-	-	-	-	-	267	-305	266	-
TRANSFORMATION	**-**	**-**	**-**	**-**	**-**	**-**	**-**	**7481**	**-**	**266**	**-**
Electricity and CHP Plants	-	-	-	-	-	-	-	-	-	-	-
Petroleum Refineries	-	-	-	-	-	-	-	7481	-	266	-
Other Transform. Sector	-	-	-	-	-	-	-	-	-	-	-
ENERGY SECTOR	**-**	**-**	**-**	**-**	**-**	**-**	**-**	**82**	**-**	**-**	**-**
DISTRIBUTION LOSSES	**-**	**-**	**-**	**-**	**-**	**-**	**-**	**66**	**-**	**-**	**-**
FINAL CONSUMPTION	**-**	**-**	**-**	**-**	**-**	**-**	**-**	**-**	**-**	**-**	**-**
INDUSTRY SECTOR	**-**	**-**	**-**	**-**	**-**	**-**	**-**	**-**	**-**	**-**	**-**
Iron and Steel	-	-	-	-	-	-	-	-	-	-	-
Chemical and Petrochem.	-	-	-	-	-	-	-	-	-	-	-
Non-Metallic Minerals	-	-	-	-	-	-	-	-	-	-	-
Non-specified	-	-	-	-	-	-	-	-	-	-	-
TRANSPORT SECTOR	**-**	**-**	**-**	**-**	**-**	**-**	**-**	**-**	**-**	**-**	**-**
Air	-	-	-	-	-	-	-	-	-	-	-
Road	-	-	-	-	-	-	-	-	-	-	-
Non-specified	-	-	-	-	-	-	-	-	-	-	-
OTHER SECTORS	**-**	**-**	**-**	**-**	**-**	**-**	**-**	**-**	**-**	**-**	**-**
Agriculture	-	-	-	-	-	-	-	-	-	-	-
Comm. and Publ. Services	-	-	-	-	-	-	-	-	-	-	-
Residential	-	-	-	-	-	-	-	-	-	-	-
Non-specified	-	-	-	-	-	-	-	-	-	-	-
NON-ENERGY USE	**-**	**-**	**-**	**-**	**-**	**-**	**-**	**-**	**-**	**-**	**-**

Ecuador / Equateur

SUPPLY AND CONSUMPTION 1996	Oil cont. (1000 tonnes)										
	Refinery Gas	LPG + Ethane	Motor Gasoline	Aviation Gasoline	Jet Fuel	Kerosene	Gas/ Diesel	Heavy Fuel Oil	Naphtha	Petrol. Coke	Other Prod.
Production	-	138	1576	7	205	50	1538	3430	-	-	132
Imports	-	338	-	-	3	-	349	-	-	-	-
Exports	-	-	-	-	-	-	-	-1411	-	-	-
Intl. Marine Bunkers	-	-	-	-	-	-	-	-287	-	-	-
Stock Changes	-	-1	-177	-	-1	-3	47	1	-	-	-
DOMESTIC SUPPLY	**-**	**475**	**1399**	**7**	**207**	**47**	**1934**	**1733**	**-**	**-**	**132**
Transfers and Stat. Diff.	-	101	-65	-	1	-17	173	-664	-	-	-
TRANSFORMATION	**-**	**-**	**-**	**-**	**-**	**-**	**260**	**521**	**-**	**-**	**-**
Electricity and CHP Plants	-	-	-	-	-	-	260	521	-	-	-
Petroleum Refineries	-	-	-	-	-	-	-	-	-	-	-
Other Transform. Sector	-	-	-	-	-	-	-	-	-	-	-
ENERGY SECTOR	**-**	**-**	**4**	**-**	**-**	**-**	**1**	**26**	**-**	**-**	**-**
DISTRIBUTION LOSSES	**-**	**-**	**-**	**-**	**-**	**-**	**-**	**-**	**-**	**-**	**-**
FINAL CONSUMPTION	**-**	**576**	**1330**	**7**	**208**	**30**	**1846**	**522**	**-**	**-**	**132**
INDUSTRY SECTOR	**-**	**28**	**186**	**-**	**-**	**7**	**206**	**240**	**-**	**-**	**-**
Iron and Steel	-	-	-	-	-	-	-	-	-	-	-
Chemical and Petrochem.	-	-	-	-	-	-	-	-	-	-	-
Non-Metallic Minerals	-	-	-	-	-	-	-	-	-	-	-
Non-specified	-	28	186	-	-	7	206	240	-	-	-
TRANSPORT SECTOR	**-**	**-**	**1030**	**7**	**208**	**-**	**1170**	**274**	**-**	**-**	**-**
Air	-	-	-	7	208	-	-	-	-	-	-
Road	-	-	1030	-	-	-	944	-	-	-	-
Non-specified	-	-	-	-	-	-	226	274	-	-	-
OTHER SECTORS	**-**	**548**	**114**	**-**	**-**	**23**	**470**	**8**	**-**	**-**	**-**
Agriculture	-	17	30	-	-	3	284	-	-	-	-
Comm. and Publ. Services	-	-	45	-	-	1	186	8	-	-	-
Residential	-	508	39	-	-	10	-	-	-	-	-
Non-specified	-	23	-	-	-	9	-	-	-	-	-
NON-ENERGY USE	**-**	**-**	**-**	**-**	**-**	**-**	**-**	**-**	**-**	**-**	**132**

APPROVISIONNEMENT ET DEMANDE 1997	*Pétrole cont.* (1000 tonnes)										
	Gaz de raffinerie	*GPL + éthane*	*Essence moteur*	*Essence aviation*	*Carbu- réacteurs*	*Kérosène*	*Gazole*	*Fioul lourd*	*Naphta*	*Coke de pétrole*	*Autres prod.*
Production	-	112	1371	7	247	50	1740	3546	-	-	123
Imports	-	410	-	-	5	-	440	-	-	-	-
Exports	-	-	-	-	-	-	-21	-1554	-	-	-
Intl. Marine Bunkers	-	-	-	-	-	-	-	-316	-	-	-
Stock Changes	-	-	-	-	-	-3	31	-14	-	-	-
DOMESTIC SUPPLY	**-**	**522**	**1371**	**7**	**252**	**47**	**2190**	**1662**	**-**	**-**	**123**
Transfers and Stat. Diff.	-	109	-64	-	-40	-17	20	-478	-	-	-
TRANSFORMATION	**-**	**-**	**-**	**-**	**-**	**-**	**198**	**488**	**-**	**-**	**-**
Electricity and CHP Plants	-	-	-	-	-	-	198	488	-	-	-
Petroleum Refineries	-	-	-	-	-	-	-	-	-	-	-
Other Transform. Sector	-	-	-	-	-	-	-	-	-	-	-
ENERGY SECTOR	**-**	**-**	**4**	**-**	**-**	**-**	**1**	**26**	**-**	**-**	**-**
DISTRIBUTION LOSSES	**-**	**-**	**-**	**-**	**-**	**-**	**-**	**-**	**-**	**-**	**-**
FINAL CONSUMPTION	**-**	**631**	**1303**	**7**	**212**	**30**	**2011**	**670**	**-**	**-**	**123**
INDUSTRY SECTOR	**-**	**31**	**182**	**-**	**-**	**7**	**229**	**309**	**-**	**-**	**-**
Iron and Steel	-	-	-	-	-	-	-	-	-	-	-
Chemical and Petrochem.	-	-	-	-	-	-	-	-	-	-	-
Non-Metallic Minerals	-	-	-	-	-	-	-	-	-	-	-
Non-specified	-	31	182	-	-	7	229	309	-	-	-
TRANSPORT SECTOR	**-**	**-**	**1009**	**7**	**212**	**-**	**1267**	**348**	**-**	**-**	**-**
Air	-	-	-	7	212	-	-	-	-	-	-
Road	-	-	1009	-	-	-	1022	-	-	-	-
Non-specified	-	-	-	-	-	-	245	348	-	-	-
OTHER SECTORS	**-**	**600**	**112**	**-**	**-**	**23**	**515**	**13**	**-**	**-**	**-**
Agriculture	-	18	30	-	-	3	310	-	-	-	-
Comm. and Publ. Services	-	-	44	-	-	1	205	13	-	-	-
Residential	-	557	38	-	-	10	-	-	-	-	-
Non-specified	-	25	-	-	-	9	-	-	-	-	-
NON-ENERGY USE	**-**	**-**	**-**	**-**	**-**	**-**	**-**	**-**	**-**	**-**	**123**

Ecuador / Equateur

SUPPLY AND CONSUMPTION 1996	Gas (TJ)				Comb. Renew. & Waste (TJ)				(GWh)	(TJ)
	Natural Gas	Gas Works	Coke Ovens	Blast Furnaces	Solid Biomass	Gas/Liquids from Biomass	Municipal Waste	Industrial Waste	Electricity	Heat
Production	-	-	-	-	50913	-	-	-	9260	-
Imports	-	-	-	-	-	-	-	-	-	-
Exports	-	-	-	-	-	-	-	-	-	-
Intl. Marine Bunkers	-	-	-	-	-	-	-	-	-	-
Stock Changes	-	-	-	-	-	-	-	-	-	-
DOMESTIC SUPPLY	-	-	-	-	**50913**	-	-	-	**9260**	-
Transfers and Stat. Diff.	-	-	-	-	-	-	-	-	-	-
TRANSFORMATION	-	-	-	-	-	-	-	-	-	-
Electricity and CHP Plants	-	-	-	-	-	-	-	-	-	-
Petroleum Refineries	-	-	-	-	-	-	-	-	-	-
Other Transform. Sector	-	-	-	-	-	-	-	-	-	-
ENERGY SECTOR	-	-	-	-	-	-	-	-	**318**	-
DISTRIBUTION LOSSES	-	-	-	-	-	-	-	-	**1915**	-
FINAL CONSUMPTION	-	-	-	-	**50913**	-	-	-	**7027**	-
INDUSTRY SECTOR	-	-	-	-	**12679**	-	-	-	**2262**	-
Iron and Steel	-	-	-	-	-	-	-	-	-	-
Chemical and Petrochem.	-	-	-	-	-	-	-	-	-	-
Non-Metallic Minerals	-	-	-	-	-	-	-	-	-	-
Non-specified	-	-	-	-	12679	-	-	-	2262	-
TRANSPORT SECTOR	-	-	-	-	-	-	-	-	-	-
Air	-	-	-	-	-	-	-	-	-	-
Road	-	-	-	-	-	-	-	-	-	-
Non-specified	-	-	-	-	-	-	-	-	-	-
OTHER SECTORS	-	-	-	-	**38234**	-	-	-	**4765**	-
Agriculture	-	-	-	-	-	-	-	-	-	-
Comm. and Publ. Services	-	-	-	-	-	-	-	-	2059	-
Residential	-	-	-	-	38234	-	-	-	2656	-
Non-specified	-	-	-	-	-	-	-	-	50	-
NON-ENERGY USE	-	-	-	-	-	-	-	-	-	-

APPROVISIONNEMENT ET DEMANDE 1997	*Gaz* (TJ)				*En. Re. Comb. & Déchets* (TJ)				(GWh)	(TJ)
	Gaz naturel	*Usines à gaz*	*Cokeries*	*Hauts fourneaux*	*Biomasse solide*	*Gaz/Liquides tirés de biomasse*	*Déchets urbains*	*Déchets industriels*	*Electricité*	*Chaleur*
Production	-	-	-	-	47608	-	-	-	9595	-
Imports	-	-	-	-	-	-	-	-	-	-
Exports	-	-	-	-	-	-	-	-	-	-
Intl. Marine Bunkers	-	-	-	-	-	-	-	-	-	-
Stock Changes	-	-	-	-	-	-	-	-	-	-
DOMESTIC SUPPLY	-	-	-	-	**47608**	-	-	-	**9595**	-
Transfers and Stat. Diff.	-	-	-	-	-	-	-	-	-12	-
TRANSFORMATION	-	-	-	-	-	-	-	-	-	-
Electricity and CHP Plants	-	-	-	-	-	-	-	-	-	-
Petroleum Refineries	-	-	-	-	-	-	-	-	-	-
Other Transform. Sector	-	-	-	-	-	-	-	-	-	-
ENERGY SECTOR	-	-	-	-	-	-	-	-	**128**	-
DISTRIBUTION LOSSES	-	-	-	-	-	-	-	-	**2163**	-
FINAL CONSUMPTION	-	-	-	-	**47608**	-	-	-	**7292**	-
INDUSTRY SECTOR	-	-	-	-	**13343**	-	-	-	**2361**	-
Iron and Steel	-	-	-	-	-	-	-	-	-	-
Chemical and Petrochem.	-	-	-	-	-	-	-	-	-	-
Non-Metallic Minerals	-	-	-	-	-	-	-	-	-	-
Non-specified	-	-	-	-	13343	-	-	-	2361	-
TRANSPORT SECTOR	-	-	-	-	-	-	-	-	-	-
Air	-	-	-	-	-	-	-	-	-	-
Road	-	-	-	-	-	-	-	-	-	-
Non-specified	-	-	-	-	-	-	-	-	-	-
OTHER SECTORS	-	-	-	-	**34265**	-	-	-	**4931**	-
Agriculture	-	-	-	-	-	-	-	-	-	-
Comm. and Publ. Services	-	-	-	-	-	-	-	-	2151	-
Residential	-	-	-	-	34265	-	-	-	2733	-
Non-specified	-	-	-	-	-	-	-	-	47	-
NON-ENERGY USE	-	-	-	-	-	-	-	-	-	-

Egypt / Egypte

SUPPLY AND CONSUMPTION 1996	Coal (1000 tonnes)							Oil (1000 tonnes)			
	Coking Coal	Other Bit. Coal	Sub-Bit. Coal	Lignite	Peat	Oven and Gas Coke	Pat. Fuel and BKB	Crude Oil	NGL	Feed-stocks	Additives
Production	-	-	-	-	-	1860	-	42692	2388	-	-
Imports	1540	-	-	-	-	1	-	-	-	-	-
Exports	-	-	-	-	-	-466	-	-6728	-	-	-
Intl. Marine Bunkers	-	-	-	-	-	-	-	-	-	-	-
Stock Changes	-	-	-	-	-	-	-	-7653	-	-	-
DOMESTIC SUPPLY	**1540**	-	-	-	-	**1395**	-	**28311**	**2388**	-	-
Transfers and Stat. Diff.	-	-	-	-	-	-322	-	-	-975	-	-
TRANSFORMATION	**1540**	-	-	-	-	**803**	-	**28311**	**1413**	-	-
Electricity and CHP Plants	-	-	-	-	-	-	-	-	-	-	-
Petroleum Refineries	-	-	-	-	-	-	-	28311	1413	-	-
Other Transform. Sector	1540	-	-	-	-	803	-	-	-	-	-
ENERGY SECTOR	-	-	-	-	-	-	-	-	-	-	-
DISTRIBUTION LOSSES	-	-	-	-	-	-	-	-	-	-	-
FINAL CONSUMPTION	-	-	-	-	-	**270**	-	-	-	-	-
INDUSTRY SECTOR	-	-	-	-	-	**270**	-	-	-	-	-
Iron and Steel	-	-	-	-	-	270	-	-	-	-	-
Chemical and Petrochem.	-	-	-	-	-	-	-	-	-	-	-
Non-Metallic Minerals	-	-	-	-	-	-	-	-	-	-	-
Non-specified	-	-	-	-	-	-	-	-	-	-	-
TRANSPORT SECTOR	-	-	-	-	-	-	-	-	-	-	-
Air	-	-	-	-	-	-	-	-	-	-	-
Road	-	-	-	-	-	-	-	-	-	-	-
Non-specified	-	-	-	-	-	-	-	-	-	-	-
OTHER SECTORS	-	-	-	-	-	-	-	-	-	-	-
Agriculture	-	-	-	-	-	-	-	-	-	-	-
Comm. and Publ. Services	-	-	-	-	-	-	-	-	-	-	-
Residential	-	-	-	-	-	-	-	-	-	-	-
Non-specified	-	-	-	-	-	-	-	-	-	-	-
NON-ENERGY USE	-	-	-	-	-	-	-	-	-	-	-

APPROVISIONNEMENT ET DEMANDE 1997	*Charbon* (1000 tonnes)							*Pétrole* (1000 tonnes)			
	Charbon à coke	*Autres charb. bit.*	*Charbon sous-bit.*	*Lignite*	*Tourbe*	*Coke de four/gaz*	*Agg./briq. de lignite*	*Pétrole brut*	*LGN*	*Produits d'aliment.*	*Additifs*
Production	-	-	-	-	-	1860	-	41312	2625	-	-
Imports	1540	-	-	-	-	1	-	-	-	-	-
Exports	-	-	-	-	-	-466	-	-6143	-	-	-
Intl. Marine Bunkers	-	-	-	-	-	-	-	-	-	-	-
Stock Changes	-	-	-	-	-	-	-	-6520	-	-	-
DOMESTIC SUPPLY	**1540**	-	-	-	-	**1395**	-	**28649**	**2625**	-	-
Transfers and Stat. Diff.	-	-	-	-	-	-322	-	-	-982	-	-
TRANSFORMATION	**1540**	-	-	-	-	**803**	-	**28649**	**1643**	-	-
Electricity and CHP Plants	-	-	-	-	-	-	-	-	-	-	-
Petroleum Refineries	-	-	-	-	-	-	-	28649	1643	-	-
Other Transform. Sector	1540	-	-	-	-	803	-	-	-	-	-
ENERGY SECTOR	-	-	-	-	-	-	-	-	-	-	-
DISTRIBUTION LOSSES	-	-	-	-	-	-	-	-	-	-	-
FINAL CONSUMPTION	-	-	-	-	-	**270**	-	-	-	-	-
INDUSTRY SECTOR	-	-	-	-	-	**270**	-	-	-	-	-
Iron and Steel	-	-	-	-	-	270	-	-	-	-	-
Chemical and Petrochem.	-	-	-	-	-	-	-	-	-	-	-
Non-Metallic Minerals	-	-	-	-	-	-	-	-	-	-	-
Non-specified	-	-	-	-	-	-	-	-	-	-	-
TRANSPORT SECTOR	-	-	-	-	-	-	-	-	-	-	-
Air	-	-	-	-	-	-	-	-	-	-	-
Road	-	-	-	-	-	-	-	-	-	-	-
Non-specified	-	-	-	-	-	-	-	-	-	-	-
OTHER SECTORS	-	-	-	-	-	-	-	-	-	-	-
Agriculture	-	-	-	-	-	-	-	-	-	-	-
Comm. and Publ. Services	-	-	-	-	-	-	-	-	-	-	-
Residential	-	-	-	-	-	-	-	-	-	-	-
Non-specified	-	-	-	-	-	-	-	-	-	-	-
NON-ENERGY USE	-	-	-	-	-	-	-	-	-	-	-

Egypt / Egypte

SUPPLY AND CONSUMPTION 1996	Oil cont. (1000 tonnes)										
	Refinery Gas	LPG + Ethane	Motor Gasoline	Aviation Gasoline	Jet Fuel	Kerosene	Gas/ Diesel	Heavy Fuel Oil	Naphtha	Petrol. Coke	Other Prod.
Production	439	443	2153	-	814	1266	5652	12886	2665	156	1141
Imports	-	189	75	-	-	-	830	-	-	-	35
Exports	-	-	-	-	-	-	-	-2620	-2665	-	-
Intl. Marine Bunkers	-	-	-	-	-	-	-293	-2783	-	-	-
Stock Changes	-	-	-215	-	-1	-	-138	-	-	-	30
DOMESTIC SUPPLY	**439**	**632**	**2013**	**-**	**813**	**1266**	**6051**	**7483**	**-**	**156**	**1206**
Transfers and Stat. Diff.	-	978	-	-	-	-	-2	282	-	-	-22
TRANSFORMATION	**-**	**-**	**-**	**-**	**-**	**-**	**798**	**4353**	**-**	**-**	**-**
Electricity and CHP Plants	-	-	-	-	-	-	798	4353	-	-	-
Petroleum Refineries	-	-	-	-	-	-	-	-	-	-	-
Other Transform. Sector	-	-	-	-	-	-	-	-	-	-	-
ENERGY SECTOR	**439**	**-**	**-**	**-**	**-**	**-**	**170**	**435**	**-**	**-**	**-**
DISTRIBUTION LOSSES	**-**	**48**	**-**	**-**	**-**	**-**	**-**	**-**	**-**	**-**	**-**
FINAL CONSUMPTION	**-**	**1562**	**2013**	**-**	**813**	**1266**	**5081**	**2977**	**-**	**156**	**1184**
INDUSTRY SECTOR	**-**	**115**	**-**	**-**	**-**	**122**	**3210**	**2977**	**-**	**-**	**-**
Iron and Steel	-	-	-	-	-	-	-	-	-	-	-
Chemical and Petrochem.	-	-	-	-	-	-	-	-	-	-	-
Non-Metallic Minerals	-	-	-	-	-	-	-	-	-	-	-
Non-specified	-	115	-	-	-	122	3210	2977	-	-	-
TRANSPORT SECTOR	**-**	**-**	**2013**	**-**	**813**	**-**	**1871**	**-**	**-**	**-**	**-**
Air	-	-	-	-	813	-	-	-	-	-	-
Road	-	-	2013	-	-	-	1871	-	-	-	-
Non-specified	-	-	-	-	-	-	-	-	-	-	-
OTHER SECTORS	**-**	**1447**	**-**	**-**	**-**	**1144**	**-**	**-**	**-**	**-**	**-**
Agriculture	-	-	-	-	-	-	-	-	-	-	-
Comm. and Publ. Services	-	-	-	-	-	-	-	-	-	-	-
Residential	-	1447	-	-	-	1144	-	-	-	-	-
Non-specified	-	-	-	-	-	-	-	-	-	-	-
NON-ENERGY USE	**-**	**-**	**-**	**-**	**-**	**-**	**-**	**-**	**-**	**156**	**1184**

APPROVISIONNEMENT ET DEMANDE 1997	*Pétrole cont.* (1000 tonnes)										
	Gaz de raffinerie	*GPL + éthane*	*Essence moteur*	*Essence aviation*	*Carbu- réacteurs*	*Kérosène*	*Gazole*	*Fioul lourd*	*Naphta*	*Coke de pétrole*	*Autres prod.*
Production	423	459	1747	-	883	1197	5867	12914	3071	122	1238
Imports	-	420	158	-	-	-	855	180	-	-	29
Exports	-	-	-	-	-50	-	-	-2001	-3071	-	-
Intl. Marine Bunkers	-	-	-	-	-	-	-276	-2753	-	-	-
Stock Changes	-	-	175	-	31	-	39	-	-	-	39
DOMESTIC SUPPLY	**423**	**879**	**2080**	**-**	**864**	**1197**	**6485**	**8340**	**-**	**122**	**1306**
Transfers and Stat. Diff.	-	986	-	-	-	-	-2	314	-	-	-26
TRANSFORMATION	**-**	**-**	**-**	**-**	**-**	**-**	**855**	**4851**	**-**	**-**	**-**
Electricity and CHP Plants	-	-	-	-	-	-	855	4851	-	-	-
Petroleum Refineries	-	-	-	-	-	-	-	-	-	-	-
Other Transform. Sector	-	-	-	-	-	-	-	-	-	-	-
ENERGY SECTOR	**423**	**-**	**-**	**-**	**-**	**-**	**182**	**485**	**-**	**-**	**-**
DISTRIBUTION LOSSES	**-**	**103**	**-**	**-**	**-**	**-**	**-**	**-**	**-**	**-**	**-**
FINAL CONSUMPTION	**-**	**1762**	**2080**	**-**	**864**	**1197**	**5446**	**3318**	**-**	**122**	**1280**
INDUSTRY SECTOR	**-**	**130**	**-**	**-**	**-**	**115**	**3441**	**3318**	**-**	**-**	**-**
Iron and Steel	-	-	-	-	-	-	-	-	-	-	-
Chemical and Petrochem.	-	-	-	-	-	-	-	-	-	-	-
Non-Metallic Minerals	-	-	-	-	-	-	-	-	-	-	-
Non-specified	-	130	-	-	-	115	3441	3318	-	-	-
TRANSPORT SECTOR	**-**	**-**	**2080**	**-**	**864**	**-**	**2005**	**-**	**-**	**-**	**-**
Air	-	-	-	-	864	-	-	-	-	-	-
Road	-	-	2080	-	-	-	2005	-	-	-	-
Non-specified	-	-	-	-	-	-	-	-	-	-	-
OTHER SECTORS	**-**	**1632**	**-**	**-**	**-**	**1082**	**-**	**-**	**-**	**-**	**-**
Agriculture	-	-	-	-	-	-	-	-	-	-	-
Comm. and Publ. Services	-	-	-	-	-	-	-	-	-	-	-
Residential	-	1632	-	-	-	1082	-	-	-	-	-
Non-specified	-	-	-	-	-	-	-	-	-	-	-
NON-ENERGY USE	**-**	**-**	**-**	**-**	**-**	**-**	**-**	**-**	**-**	**122**	**1280**

Egypt / Egypte

SUPPLY AND CONSUMPTION 1996	Gas (TJ)				Comb. Renew. & Waste (TJ)				(GWh)	(TJ)
	Natural Gas	Gas Works	Coke Ovens	Blast Furnaces	Solid Biomass	Gas/Liquids from Biomass	Municipal Waste	Industrial Waste	Electricity	Heat
Production	503495	360	-	-	51597	-	-	-	54444	-
Imports	-	-	-	-	30	-	-	-	-	-
Exports	-	-	-	-	-767	-	-	-	-	-
Intl. Marine Bunkers	-	-	-	-	-	-	-	-	-	-
Stock Changes	-	-	-	-	-	-	-	-	-	-
DOMESTIC SUPPLY	**503495**	**360**	**-**	**-**	**50859**	**-**	**-**	**-**	**54444**	**-**
Transfers and Stat. Diff.	-8469	-	-	-	-	-	-	-	-250	-
TRANSFORMATION	**315970**	**-**	**-**	**-**	**-**	**-**	**-**	**-**	**-**	**-**
Electricity and CHP Plants	315970	-	-	-	-	-	-	-	-	-
Petroleum Refineries	-	-	-	-	-	-	-	-	-	-
Other Transform. Sector	-	-	-	-	-	-	-	-	-	-
ENERGY SECTOR	**32604**	**-**	**-**	**-**	**-**	**-**	**-**	**-**	**2129**	**-**
DISTRIBUTION LOSSES	**-**	**-**	**-**	**-**	**-**	**-**	**-**	**-**	**6285**	**-**
FINAL CONSUMPTION	**146452**	**360**	**-**	**-**	**50859**	**-**	**-**	**-**	**45780**	**-**
INDUSTRY SECTOR	**106058**	**-**	**-**	**-**	**25961**	**-**	**-**	**-**	**21578**	**-**
Iron and Steel	-	-	-	-	-	-	-	-	-	-
Chemical and Petrochem.	67678	-	-	-	-	-	-	-	-	-
Non-Metallic Minerals	-	-	-	-	-	-	-	-	-	-
Non-specified	38380	-	-	-	25961	-	-	-	21578	-
TRANSPORT SECTOR	**-**	**-**	**-**	**-**	**-**	**-**	**-**	**-**	**-**	**-**
Air	-	-	-	-	-	-	-	-	-	-
Road	-	-	-	-	-	-	-	-	-	-
Non-specified	-	-	-	-	-	-	-	-	-	-
OTHER SECTORS	**40394**	**360**	**-**	**-**	**24898**	**-**	**-**	**-**	**24202**	**-**
Agriculture	-	-	-	-	-	-	-	-	1638	-
Comm. and Publ. Services	-	-	-	-	-	-	-	-	-	-
Residential	31616	360	-	-	24898	-	-	-	16710	-
Non-specified	8778	-	-	-	-	-	-	-	5854	-
NON-ENERGY USE	**-**	**-**	**-**	**-**	**-**	**-**	**-**	**-**	**-**	**-**

APPROVISIONNEMENT ET DEMANDE 1997	*Gaz* (TJ)				*En. Re. Comb. & Déchets* (TJ)				(GWh)	(TJ)
	Gaz naturel	*Usines à gaz*	*Cokeries*	*Hauts fourneaux*	*Biomasse solide*	*Gaz/Liquides tirés de biomasse*	*Déchets urbains*	*Déchets industriels*	*Electricité*	*Chaleur*
Production	515137	360	-	-	52577	-	-	-	57656	-
Imports	-	-	-	-	31	-	-	-	-	-
Exports	-	-	-	-	-782	-	-	-	-	-
Intl. Marine Bunkers	-	-	-	-	-	-	-	-	-	-
Stock Changes	-	-	-	-	-	-	-	-	-	-
DOMESTIC SUPPLY	**515137**	**360**	**-**	**-**	**51825**	**-**	**-**	**-**	**57656**	**-**
Transfers and Stat. Diff.	-7723	-	-	-	-	-	-	-	-263	-
TRANSFORMATION	**321974**	**-**	**-**	**-**	**-**	**-**	**-**	**-**	**-**	**-**
Electricity and CHP Plants	321974	-	-	-	-	-	-	-	-	-
Petroleum Refineries	-	-	-	-	-	-	-	-	-	-
Other Transform. Sector	-	-	-	-	-	-	-	-	-	-
ENERGY SECTOR	**31312**	**-**	**-**	**-**	**-**	**-**	**-**	**-**	**2254**	**-**
DISTRIBUTION LOSSES	**-**	**-**	**-**	**-**	**-**	**-**	**-**	**-**	**6656**	**-**
FINAL CONSUMPTION	**154128**	**360**	**-**	**-**	**51825**	**-**	**-**	**-**	**48483**	**-**
INDUSTRY SECTOR	**111492**	**-**	**-**	**-**	**26454**	**-**	**-**	**-**	**22851**	**-**
Iron and Steel	-	-	-	-	-	-	-	-	-	-
Chemical and Petrochem.	63536	-	-	-	-	-	-	-	-	-
Non-Metallic Minerals	-	-	-	-	-	-	-	-	-	-
Non-specified	47956	-	-	-	26454	-	-	-	22851	-
TRANSPORT SECTOR	**-**	**-**	**-**	**-**	**-**	**-**	**-**	**-**	**-**	**-**
Air	-	-	-	-	-	-	-	-	-	-
Road	-	-	-	-	-	-	-	-	-	-
Non-specified	-	-	-	-	-	-	-	-	-	-
OTHER SECTORS	**42636**	**360**	**-**	**-**	**25371**	**-**	**-**	**-**	**25632**	**-**
Agriculture	-	-	-	-	-	-	-	-	1735	-
Comm. and Publ. Services	-	-	-	-	-	-	-	-	2	-
Residential	32376	360	-	-	25371	-	-	-	17695	-
Non-specified	10260	-	-	-	-	-	-	-	6200	-
NON-ENERGY USE	**-**	**-**	**-**	**-**	**-**	**-**	**-**	**-**	**-**	**-**

El Salvador

SUPPLY AND CONSUMPTION 1996	Coal (1000 tonnes)							Oil (1000 tonnes)			
	Coking Coal	Other Bit. Coal	Sub-Bit. Coal	Lignite	Peat	Oven and Gas Coke	Pat. Fuel and BKB	Crude Oil	NGL	Feed-stocks	Additives
Production	-	-	-	-	-	-	-	-	-	-	-
Imports	-	-	-	-	-	1	-	785	-	-	-
Exports	-	-	-	-	-	-	-	-	-	-	-
Intl. Marine Bunkers	-	-	-	-	-	-	-	-	-	-	-
Stock Changes	-	-	-	-	-	-	-	-	-	-	-
DOMESTIC SUPPLY	-	-	-	-	-	**1**	-	**785**	-	-	-
Transfers and Stat. Diff.	-	-	-	-	-	-	-	8	-	-	-
TRANSFORMATION	-	-	-	-	-	**1**	-	**793**	-	-	-
Electricity and CHP Plants	-	-	-	-	-	-	-	-	-	-	-
Petroleum Refineries	-	-	-	-	-	-	-	793	-	-	-
Other Transform. Sector	-	-	-	-	-	1	-	-	-	-	-
ENERGY SECTOR	-	-	-	-	-	-	-	-	-	-	-
DISTRIBUTION LOSSES	-	-	-	-	-	-	-	-	-	-	-
FINAL CONSUMPTION	-	-	-	-	-	-	-	-	-	-	-
INDUSTRY SECTOR	-	-	-	-	-	-	-	-	-	-	-
Iron and Steel	-	-	-	-	-	-	-	-	-	-	-
Chemical and Petrochem.	-	-	-	-	-	-	-	-	-	-	-
Non-Metallic Minerals	-	-	-	-	-	-	-	-	-	-	-
Non-specified	-	-	-	-	-	-	-	-	-	-	-
TRANSPORT SECTOR	-	-	-	-	-	-	-	-	-	-	-
Air	-	-	-	-	-	-	-	-	-	-	-
Road	-	-	-	-	-	-	-	-	-	-	-
Non-specified	-	-	-	-	-	-	-	-	-	-	-
OTHER SECTORS	-	-	-	-	-	-	-	-	-	-	-
Agriculture	-	-	-	-	-	-	-	-	-	-	-
Comm. and Publ. Services	-	-	-	-	-	-	-	-	-	-	-
Residential	-	-	-	-	-	-	-	-	-	-	-
Non-specified	-	-	-	-	-	-	-	-	-	-	-
NON-ENERGY USE	-	-	-	-	-	-	-	-	-	-	-

APPROVISIONNEMENT ET DEMANDE 1997	*Charbon* (1000 tonnes)							*Pétrole* (1000 tonnes)			
	Charbon à coke	*Autres charb. bit.*	*Charbon sous-bit.*	*Lignite*	*Tourbe*	*Coke de four/gaz*	*Agg./briq. de lignite*	*Pétrole brut*	*LGN*	*Produits d'aliment.*	*Additifs*
Production	-	-	-	-	-	-	-	-	-	-	-
Imports	-	-	-	-	-	1	-	804	-	-	-
Exports	-	-	-	-	-	-	-	-	-	-	-
Intl. Marine Bunkers	-	-	-	-	-	-	-	-	-	-	-
Stock Changes	-	-	-	-	-	-	-	13	-	-	-
DOMESTIC SUPPLY	-	-	-	-	-	**1**	-	**817**	-	-	-
Transfers and Stat. Diff.	-	-	-	-	-	-	-	112	-	-	-
TRANSFORMATION	-	-	-	-	-	**1**	-	**929**	-	-	-
Electricity and CHP Plants	-	-	-	-	-	-	-	-	-	-	-
Petroleum Refineries	-	-	-	-	-	-	-	929	-	-	-
Other Transform. Sector	-	-	-	-	-	1	-	-	-	-	-
ENERGY SECTOR	-	-	-	-	-	-	-	-	-	-	-
DISTRIBUTION LOSSES	-	-	-	-	-	-	-	-	-	-	-
FINAL CONSUMPTION	-	-	-	-	-	-	-	-	-	-	-
INDUSTRY SECTOR	-	-	-	-	-	-	-	-	-	-	-
Iron and Steel	-	-	-	-	-	-	-	-	-	-	-
Chemical and Petrochem.	-	-	-	-	-	-	-	-	-	-	-
Non-Metallic Minerals	-	-	-	-	-	-	-	-	-	-	-
Non-specified	-	-	-	-	-	-	-	-	-	-	-
TRANSPORT SECTOR	-	-	-	-	-	-	-	-	-	-	-
Air	-	-	-	-	-	-	-	-	-	-	-
Road	-	-	-	-	-	-	-	-	-	-	-
Non-specified	-	-	-	-	-	-	-	-	-	-	-
OTHER SECTORS	-	-	-	-	-	-	-	-	-	-	-
Agriculture	-	-	-	-	-	-	-	-	-	-	-
Comm. and Publ. Services	-	-	-	-	-	-	-	-	-	-	-
Residential	-	-	-	-	-	-	-	-	-	-	-
Non-specified	-	-	-	-	-	-	-	-	-	-	-
NON-ENERGY USE	-	-	-	-	-	-	-	-	-	-	-

El Salvador

SUPPLY AND CONSUMPTION 1996	Oil cont. (1000 tonnes)										
	Refinery Gas	LPG + Ethane	Motor Gasoline	Aviation Gasoline	Jet Fuel	Kerosene	Gas/ Diesel	Heavy Fuel Oil	Naphtha	Petrol. Coke	Other Prod.
Production	18	14	199	-	47	18	103	307	-	-	45
Imports	-	81	107	-	3	1	425	151	-	-	6
Exports	-	-	-4	-	-4	-2	-	-112	-	-	-14
Intl. Marine Bunkers	-	-	-	-	-	-	-	-	-	-	-
Stock Changes	-	-	-17	-	1	-	-9	36	-	-	-2
DOMESTIC SUPPLY	**18**	**95**	**285**	**-**	**47**	**17**	**519**	**382**	**-**	**-**	**35**
Transfers and Stat. Diff.	-	1	-1	-	1	1	-6	-	-	-	10
TRANSFORMATION	**-**	**-**	**-**	**-**	**-**	**-**	**89**	**183**	**-**	**-**	**-**
Electricity and CHP Plants	-	-	-	-	-	-	89	183	-	-	-
Petroleum Refineries	-	-	-	-	-	-	-	-	-	-	-
Other Transform. Sector	-	-	-	-	-	-	-	-	-	-	-
ENERGY SECTOR	**18**	**-**	**-**	**-**	**-**	**-**	**-**	**-**	**-**	**-**	**-**
DISTRIBUTION LOSSES	**-**	**-**	**-**	**-**	**-**	**-**	**-**	**-**	**-**	**-**	**-**
FINAL CONSUMPTION	**-**	**96**	**284**	**-**	**48**	**18**	**424**	**199**	**-**	**-**	**45**
INDUSTRY SECTOR	**-**	**26**	**-**	**-**	**-**	**3**	**53**	**196**	**-**	**-**	**-**
Iron and Steel	-	-	-	-	-	-	-	-	-	-	-
Chemical and Petrochem.	-	-	-	-	-	-	-	-	-	-	-
Non-Metallic Minerals	-	-	-	-	-	-	-	-	-	-	-
Non-specified	-	26	-	-	-	3	53	196	-	-	-
TRANSPORT SECTOR	**-**	**-**	**284**	**-**	**48**	**-**	**366**	**-**	**-**	**-**	**-**
Air	-	-	-	-	48	-	-	-	-	-	-
Road	-	-	284	-	-	-	366	-	-	-	-
Non-specified	-	-	-	-	-	-	-	-	-	-	-
OTHER SECTORS	**-**	**70**	**-**	**-**	**-**	**15**	**5**	**3**	**-**	**-**	**-**
Agriculture	-	-	-	-	-	-	-	-	-	-	-
Comm. and Publ. Services	-	-	-	-	-	-	-	-	-	-	-
Residential	-	70	-	-	-	15	5	3	-	-	-
Non-specified	-	-	-	-	-	-	-	-	-	-	-
NON-ENERGY USE	**-**	**-**	**-**	**-**	**-**	**-**	**-**	**-**	**-**	**-**	**45**

APPROVISIONNEMENT ET DEMANDE 1997	*Pétrole cont.* (1000 tonnes)										
	Gaz de raffinerie	*GPL + éthane*	*Essence moteur*	*Essence aviation*	*Carbu- réacteurs*	*Kérosène*	*Gazole*	*Fioul lourd*	*Naphta*	*Coke de pétrole*	*Autres prod.*
Production	19	17	216	-	48	19	248	291	-	-	32
Imports	-	87	113	-	3	1	297	97	-	-	8
Exports	-	-	-5	-	-3	-1	-	-48	-	-	-10
Intl. Marine Bunkers	-	-	-	-	-	-	-	-	-	-	-
Stock Changes	-	2	-16	-	1	-	10	47	-	-	7
DOMESTIC SUPPLY	**19**	**106**	**308**	**-**	**49**	**19**	**555**	**387**	**-**	**-**	**37**
Transfers and Stat. Diff.	-	-1	-	-	-1	-1	-6	-	-	-	-5
TRANSFORMATION	**-**	**-**	**-**	**-**	**-**	**-**	**96**	**185**	**-**	**-**	**-**
Electricity and CHP Plants	-	-	-	-	-	-	96	185	-	-	-
Petroleum Refineries	-	-	-	-	-	-	-	-	-	-	-
Other Transform. Sector	-	-	-	-	-	-	-	-	-	-	-
ENERGY SECTOR	**19**	**-**	**-**	**-**	**-**	**-**	**-**	**-**	**-**	**-**	**-**
DISTRIBUTION LOSSES	**-**	**-**	**-**	**-**	**-**	**-**	**-**	**-**	**-**	**-**	**-**
FINAL CONSUMPTION	**-**	**105**	**308**	**-**	**48**	**18**	**453**	**202**	**-**	**-**	**32**
INDUSTRY SECTOR	**-**	**29**	**-**	**-**	**-**	**3**	**57**	**199**	**-**	**-**	**-**
Iron and Steel	-	-	-	-	-	-	-	-	-	-	-
Chemical and Petrochem.	-	-	-	-	-	-	-	-	-	-	-
Non-Metallic Minerals	-	-	-	-	-	-	-	-	-	-	-
Non-specified	-	29	-	-	-	3	57	199	-	-	-
TRANSPORT SECTOR	**-**	**-**	**308**	**-**	**48**	**-**	**390**	**-**	**-**	**-**	**-**
Air	-	-	-	-	48	-	-	-	-	-	-
Road	-	-	308	-	-	-	390	-	-	-	-
Non-specified	-	-	-	-	-	-	-	-	-	-	-
OTHER SECTORS	**-**	**76**	**-**	**-**	**-**	**15**	**6**	**3**	**-**	**-**	**-**
Agriculture	-	-	-	-	-	-	-	-	-	-	-
Comm. and Publ. Services	-	-	-	-	-	-	-	-	-	-	-
Residential	-	76	-	-	-	15	6	3	-	-	-
Non-specified	-	-	-	-	-	-	-	-	-	-	-
NON-ENERGY USE	**-**	**-**	**-**	**-**	**-**	**-**	**-**	**-**	**-**	**-**	**32**

El Salvador

SUPPLY AND CONSUMPTION 1996	Gas (TJ)				Comb. Renew. & Waste (TJ)				(GWh)	(TJ)
	Natural Gas	Gas Works	Coke Ovens	Blast Furnaces	Solid Biomass	Gas/Liquids from Biomass	Municipal Waste	Industrial Waste	Electricity	Heat
Production	-	-	-	-	77965	-	-	-	3418	-
Imports	-	-	-	-	-	-	-	-	21	-
Exports	-	-	-	-	-	-	-	-	-	-
Intl. Marine Bunkers	-	-	-	-	-	-	-	-	-	-
Stock Changes	-	-	-	-	-	-	-	-	-	-
DOMESTIC SUPPLY	**-**	**-**	**-**	**-**	**77965**	**-**	**-**	**-**	**3439**	**-**
Transfers and Stat. Diff.	-	-	-	-	-1679	-	-	-	-	-
TRANSFORMATION	**-**	**-**	**-**	**-**	**3829**	**-**	**-**	**-**	**-**	**-**
Electricity and CHP Plants	-	-	-	-	3306	-	-	-	-	-
Petroleum Refineries	-	-	-	-	-	-	-	-	-	-
Other Transform. Sector	-	-	-	-	523	-	-	-	-	-
ENERGY SECTOR	**-**	**-**	**-**	**-**	**-**	**-**	**-**	**-**	**51**	**-**
DISTRIBUTION LOSSES	**-**	**-**	**-**	**-**	**-**	**-**	**-**	**-**	**462**	**-**
FINAL CONSUMPTION	**-**	**-**	**-**	**-**	**72457**	**-**	**-**	**-**	**2926**	**-**
INDUSTRY SECTOR	**-**	**-**	**-**	**-**	**12359**	**-**	**-**	**-**	**842**	**-**
Iron and Steel	-	-	-	-	-	-	-	-	-	-
Chemical and Petrochem.	-	-	-	-	-	-	-	-	-	-
Non-Metallic Minerals	-	-	-	-	-	-	-	-	-	-
Non-specified	-	-	-	-	12359	-	-	-	842	-
TRANSPORT SECTOR	**-**	**-**	**-**	**-**	**-**	**-**	**-**	**-**	**-**	**-**
Air	-	-	-	-	-	-	-	-	-	-
Road	-	-	-	-	-	-	-	-	-	-
Non-specified	-	-	-	-	-	-	-	-	-	-
OTHER SECTORS	**-**	**-**	**-**	**-**	**60098**	**-**	**-**	**-**	**2084**	**-**
Agriculture	-	-	-	-	-	-	-	-	-	-
Comm. and Publ. Services	-	-	-	-	-	-	-	-	1026	-
Residential	-	-	-	-	60098	-	-	-	1058	-
Non-specified	-	-	-	-	-	-	-	-	-	-
NON-ENERGY USE	**-**	**-**	**-**	**-**	**-**	**-**	**-**	**-**	**-**	**-**

APPROVISIONNEMENT ET DEMANDE 1997	*Gaz* (TJ)				*En. Re. Comb. & Déchets* (TJ)				(GWh)	(TJ)
	Gaz naturel	*Usines à gaz*	*Cokeries*	*Hauts fourneaux*	*Biomasse solide*	*Gaz/Liquides tirés de biomasse*	*Déchets urbains*	*Déchets industriels*	*Electricité*	*Chaleur*
Production	-	-	-	-	80660	-	-	-	3643	-
Imports	-	-	-	-	-	-	-	-	88	-
Exports	-	-	-	-	-	-	-	-	-	-
Intl. Marine Bunkers	-	-	-	-	-	-	-	-	-	-
Stock Changes	-	-	-	-	-	-	-	-	-	-
DOMESTIC SUPPLY	**-**	**-**	**-**	**-**	**80660**	**-**	**-**	**-**	**3731**	**-**
Transfers and Stat. Diff.	-	-	-	-	-1790	-	-	-	-1	-
TRANSFORMATION	**-**	**-**	**-**	**-**	**4089**	**-**	**-**	**-**	**-**	**-**
Electricity and CHP Plants	-	-	-	-	3521	-	-	-	-	-
Petroleum Refineries	-	-	-	-	-	-	-	-	-	-
Other Transform. Sector	-	-	-	-	568	-	-	-	-	-
ENERGY SECTOR	**-**	**-**	**-**	**-**	**-**	**-**	**-**	**-**	**66**	**-**
DISTRIBUTION LOSSES	**-**	**-**	**-**	**-**	**-**	**-**	**-**	**-**	**480**	**-**
FINAL CONSUMPTION	**-**	**-**	**-**	**-**	**74781**	**-**	**-**	**-**	**3184**	**-**
INDUSTRY SECTOR	**-**	**-**	**-**	**-**	**12890**	**-**	**-**	**-**	**905**	**-**
Iron and Steel	-	-	-	-	-	-	-	-	-	-
Chemical and Petrochem.	-	-	-	-	-	-	-	-	-	-
Non-Metallic Minerals	-	-	-	-	-	-	-	-	-	-
Non-specified	-	-	-	-	12890	-	-	-	905	-
TRANSPORT SECTOR	**-**	**-**	**-**	**-**	**-**	**-**	**-**	**-**	**-**	**-**
Air	-	-	-	-	-	-	-	-	-	-
Road	-	-	-	-	-	-	-	-	-	-
Non-specified	-	-	-	-	-	-	-	-	-	-
OTHER SECTORS	**-**	**-**	**-**	**-**	**61891**	**-**	**-**	**-**	**2279**	**-**
Agriculture	-	-	-	-	-	-	-	-	-	-
Comm. and Publ. Services	-	-	-	-	-	-	-	-	1130	-
Residential	-	-	-	-	61891	-	-	-	1149	-
Non-specified	-	-	-	-	-	-	-	-	-	-
NON-ENERGY USE	**-**	**-**	**-**	**-**	**-**	**-**	**-**	**-**	**-**	**-**

Estonia / Estonie : 1996

	Coal / *Charbon* (1000 tonnes)							Oil / *Pétrole* (1000 tonnes)			
SUPPLY AND CONSUMPTION	Coking Coal	Other Bit. Coal	Sub-Bit. Coal	Lignite	Peat	Oven and Gas Coke	Pat. Fuel and BKB	Crude Oil	NGL	Feed-stocks	Additives
APPROVISIONNEMENT ET DEMANDE	*Charbon à coke*	*Autres charb. bit.*	*Charbon sous-bit.*	*Lignite*	*Tourbe*	*Coke de four/gaz*	*Agg./briq. de lignite*	*Pétrole brut*	*LGN*	*Produits d'aliment.*	*Additifs*
Production	-	-	-	12297	700	44	162	-	-	-	-
From Other Sources	-	-	-	-	-	-	-	376	-	-	-
Imports	-	94	-	829	-	1	-	-	-	-	-
Exports	-	-	-	-3	-1	-45	-41	-	-	-	-
Intl. Marine Bunkers	-	-	-	-	-	-	-	-	-	-	-
Stock Changes	-	21	-	473	-64	1	2	-3	-	-	-
DOMESTIC SUPPLY	**-**	**115**	**-**	**13596**	**635**	**1**	**123**	**373**	**-**	**-**	**-**
Transfers	-	-	-	-	-	-	-	-	-	-	-
Statistical Differences	-	-	-	-	-	-	-	-	-	-	-
TRANSFORMATION	**-**	**26**	**-**	**12999**	**448**	**-**	**5**	**373**	**-**	**-**	**-**
Electricity Plants	-	-	-	-	-	-	-	-	-	-	-
CHP Plants	-	-	-	12816	25	-	-	-	-	-	-
Heat Plants	-	26	-	19	98	-	5	-	-	-	-
Blast Furnaces/Gas Works	-	-	-	-	-	-	-	-	-	-	-
Coke/Pat. Fuel/BKB Plants	-	-	-	164	325	-	-	-	-	-	-
Petroleum Refineries	-	-	-	-	-	-	-	373	-	-	-
Petrochemical Industry	-	-	-	-	-	-	-	-	-	-	-
Liquefaction	-	-	-	-	-	-	-	-	-	-	-
Other Transform. Sector	-	-	-	-	-	-	-	-	-	-	-
ENERGY SECTOR	**-**	**-**	**-**	**-**	**84**	**-**	**-**	**-**	**-**	**-**	**-**
Coal Mines	-	-	-	-	-	-	-	-	-	-	-
Oil and Gas Extraction	-	-	-	-	-	-	-	-	-	-	-
Petroleum Refineries	-	-	-	-	-	-	-	-	-	-	-
Electr., CHP+Heat Plants	-	-	-	-	-	-	-	-	-	-	-
Pumped Storage (Elec.)	-	-	-	-	-	-	-	-	-	-	-
Other Energy Sector	-	-	-	-	84	-	-	-	-	-	-
Distribution Losses	-	1	-	1	81	-	1	-	-	-	-
FINAL CONSUMPTION	**-**	**88**	**-**	**596**	**22**	**1**	**117**	**-**	**-**	**-**	**-**
INDUSTRY SECTOR	**-**	**37**	**-**	**498**	**14**	**1**	**4**	**-**	**-**	**-**	**-**
Iron and Steel	-	4	-	-	-	1	-	-	-	-	-
Chemical and Petrochem.	-	-	-	-	-	-	-	-	-	-	-
of which: Feedstocks	-	-	-	-	-	-	-	-	-	-	-
Non-Ferrous Metals	-	-	-	-	-	-	-	-	-	-	-
Non-Metallic Minerals	-	12	-	498	1	-	4	-	-	-	-
Transport Equipment	-	2	-	-	-	-	-	-	-	-	-
Machinery	-	3	-	-	-	-	-	-	-	-	-
Mining and Quarrying	-	-	-	-	-	-	-	-	-	-	-
Food and Tobacco	-	3	-	-	10	-	-	-	-	-	-
Paper, Pulp and Print	-	-	-	-	-	-	-	-	-	-	-
Wood and Wood Products	-	-	-	-	-	-	-	-	-	-	-
Construction	-	7	-	-	2	-	-	-	-	-	-
Textile and Leather	-	4	-	-	-	-	-	-	-	-	-
Non-specified	-	2	-	-	1	-	-	-	-	-	-
TRANSPORT SECTOR	**-**	**2**	**-**	**-**	**-**	**-**	**1**	**-**	**-**	**-**	**-**
Air	-	-	-	-	-	-	-	-	-	-	-
Road	-	-	-	-	-	-	-	-	-	-	-
Rail	-	2	-	-	-	-	1	-	-	-	-
Pipeline Transport	-	-	-	-	-	-	-	-	-	-	-
Internal Navigation	-	-	-	-	-	-	-	-	-	-	-
Non-specified	-	-	-	-	-	-	-	-	-	-	-
OTHER SECTORS	**-**	**49**	**-**	**-**	**8**	**-**	**112**	**-**	**-**	**-**	**-**
Agriculture	-	5	-	-	5	-	-	-	-	-	-
Comm. and Publ. Services	-	-	-	-	-	-	-	-	-	-	-
Residential	-	44	-	-	3	-	112	-	-	-	-
Non-specified	-	-	-	-	-	-	-	-	-	-	-
NON-ENERGY USE	**-**	**-**	**-**	**98**	**-**	**-**	**-**	**-**	**-**	**-**	**-**
in Industry/Trans./Energy	-	-	-	98	-	-	-	-	-	-	-
in Transport	-	-	-	-	-	-	-	-	-	-	-
in Other Sectors	-	-	-	-	-	-	-	-	-	-	-

Estonia / Estonie : 1996

SUPPLY AND CONSUMPTION / *APPROVISIONNEMENT ET DEMANDE*	Oil cont. / *Pétrole cont.* (1000 tonnes)										
	Refinery Gas / *Gaz de raffinerie*	LPG + Ethane / *GPL + éthane*	Motor Gasoline / *Essence moteur*	Aviation Gasoline / *Essence aviation*	Jet Fuel / *Carbu-réacteurs*	Kerosene / *Kérosène*	Gas/ Diesel / *Gazole*	Heavy Fuel Oil / *Fioul lourd*	Naphtha / *Naphta*	Petrol. Coke / *Coke de pétrole*	Other Prod. / *Autres prod.*
Production	-	-	1	-	-	-	-	343	-	-	7
From Other Sources	-	-	-	-	-	-	-	-	-	-	-
Imports	-	7	339	-	28	-	435	478	-	-	77
Exports	-	-	-37	-	-11	-	-44	-194	-	-	-41
Intl. Marine Bunkers	-	-	-	-	-	-	-35	-58	-	-	-
Stock Changes	-	-	-22	-	-1	-	18	-21	-	-	3
DOMESTIC SUPPLY	**-**	**7**	**281**	**-**	**16**	**-**	**374**	**548**	**-**	**-**	**46**
Transfers	-	-	-	-	-	-	-	1	-	-	-
Statistical Differences	-	-	-	-	-	-	-	-	-	-	-
TRANSFORMATION	**-**	**-**	**-**	**-**	**-**	**-**	**5**	**296**	**-**	**-**	**-**
Electricity Plants	-	-	-	-	-	-	-	-	-	-	-
CHP Plants	-	-	-	-	-	-	-	88	-	-	-
Heat Plants	-	-	-	-	-	-	5	208	-	-	-
Blast Furnaces/Gas Works	-	-	-	-	-	-	-	-	-	-	-
Coke/Pat. Fuel/BKB Plants	-	-	-	-	-	-	-	-	-	-	-
Petroleum Refineries	-	-	-	-	-	-	-	-	-	-	-
Petrochemical Industry	-	-	-	-	-	-	-	-	-	-	-
Liquefaction	-	-	-	-	-	-	-	-	-	-	-
Other Transform. Sector	-	-	-	-	-	-	-	-	-	-	-
ENERGY SECTOR	**-**	**-**	**1**	**-**	**-**	**-**	**10**	**-**	**-**	**-**	**-**
Coal Mines	-	-	1	-	-	-	6	-	-	-	-
Oil and Gas Extraction	-	-	-	-	-	-	-	-	-	-	-
Petroleum Refineries	-	-	-	-	-	-	-	-	-	-	-
Electr., CHP+Heat Plants	-	-	-	-	-	-	3	-	-	-	-
Pumped Storage (Elec.)	-	-	-	-	-	-	-	-	-	-	-
Other Energy Sector	-	-	-	-	-	-	1	-	-	-	-
Distribution Losses	-	-	1	-	-	-	1	2	-	-	-
FINAL CONSUMPTION	**-**	**7**	**279**	**-**	**16**	**-**	**358**	**251**	**-**	**-**	**46**
INDUSTRY SECTOR	**-**	**-**	**1**	**-**	**-**	**-**	**51**	**203**	**-**	**-**	**-**
Iron and Steel	-	-	-	-	-	-	-	-	-	-	-
Chemical and Petrochem.	-	-	-	-	-	-	-	53	-	-	-
of which: Feedstocks	-	-	-	-	-	-	-	-	-	-	-
Non-Ferrous Metals	-	-	-	-	-	-	-	-	-	-	-
Non-Metallic Minerals	-	-	-	-	-	-	2	8	-	-	-
Transport Equipment	-	-	-	-	-	-	-	3	-	-	-
Machinery	-	-	-	-	-	-	3	4	-	-	-
Mining and Quarrying	-	-	-	-	-	-	-	-	-	-	-
Food and Tobacco	-	-	-	-	-	-	34	71	-	-	-
Paper, Pulp and Print	-	-	-	-	-	-	-	9	-	-	-
Wood and Wood Products	-	-	-	-	-	-	1	25	-	-	-
Construction	-	-	1	-	-	-	10	6	-	-	-
Textile and Leather	-	-	-	-	-	-	-	9	-	-	-
Non-specified	-	-	-	-	-	-	1	15	-	-	-
TRANSPORT SECTOR	**-**	**1**	**276**	**-**	**16**	**-**	**209**	**8**	**-**	**-**	**-**
Air	-	-	-	-	16	-	-	-	-	-	-
Road	-	1	276	-	-	-	166	3	-	-	-
Rail	-	-	-	-	-	-	36	5	-	-	-
Pipeline Transport	-	-	-	-	-	-	-	-	-	-	-
Internal Navigation	-	-	-	-	-	-	7	-	-	-	-
Non-specified	-	-	-	-	-	-	-	-	-	-	-
OTHER SECTORS	**-**	**6**	**2**	**-**	**-**	**-**	**98**	**40**	**-**	**-**	**-**
Agriculture	-	-	2	-	-	-	40	19	-	-	-
Comm. and Publ. Services	-	-	-	-	-	-	14	21	-	-	-
Residential	-	6	-	-	-	-	44	-	-	-	-
Non-specified	-	-	-	-	-	-	-	-	-	-	-
NON-ENERGY USE	**-**	**-**	**-**	**-**	**-**	**-**	**-**	**-**	**-**	**-**	**46**
in Industry/Transf./Energy	-	-	-	-	-	-	-	-	-	-	39
in Transport	-	-	-	-	-	-	-	-	-	-	5
in Other Sectors	-	-	-	-	-	-	-	-	-	-	2

Estonia / Estonie : 1996

SUPPLY AND CONSUMPTION / *APPROVISIONNEMENT ET DEMANDE*	Gas / *Gaz* (TJ) Natural Gas / *Gaz naturel*	Gas Works / *Usines à gaz*	Coke Ovens / *Cokeries*	Blast Furnaces / *Hauts fourneaux*	Comb. Renew. & Waste / *En. Re. Comb. & Déchets* (TJ) Solid Biomass / *Biomasse solide*	Gas/Liquids from Biomass / *Gaz/Liquides tirés de biomasse*	Municipal Waste / *Déchets urbains*	Industrial Waste / *Déchets industriels*	(GWh) Electricity / *Electricité*	(TJ) Heat / *Chaleur*
Production	-	5134	-	-	24689	56	-	-	9103	33132
From Other Sources	-	-	-	-	-	-	-	-	-	-
Imports	29585	-	-	-	-	-	-	-	240	-
Exports	-	-	-	-	-378	-	-	-	-1100	-
Intl. Marine Bunkers	-	-	-	-	-	-	-	-	-	-
Stock Changes	-	-	-	-	57	-	-	-	-	-
DOMESTIC SUPPLY	**29585**	**5134**	**-**	**-**	**24368**	**56**	**-**	**-**	**8243**	**33132**
Transfers	-	-	-	-	-	-	-	-	-	-
Statistical Differences	-	-	-	-	-	-	-	-	-	-
TRANSFORMATION	**15209**	**1468**	**-**	**-**	**2610**	**56**	**-**	**-**	**134**	**-**
Electricity Plants	-	-	-	-	-	-	-	-	-	-
CHP Plants	3539	1318	-	-	32	-	-	-	-	-
Heat Plants	11670	150	-	-	2578	56	-	-	-	-
Blast Furnaces/Gas Works	-	-	-	-	-	-	-	-	-	-
Coke/Pat. Fuel/BKB Plants	-	-	-	-	-	-	-	-	-	-
Petroleum Refineries	-	-	-	-	-	-	-	-	-	-
Petrochemical Industry	-	-	-	-	-	-	-	-	-	-
Liquefaction	-	-	-	-	-	-	-	-	-	-
Other Transform. Sector	-	-	-	-	-	-	-	-	134	-
ENERGY SECTOR	**-**	**-**	**-**	**-**	**-**	**-**	**-**	**-**	**1571**	**-**
Coal Mines	-	-	-	-	-	-	-	-	273	-
Oil and Gas Extraction	-	-	-	-	-	-	-	-	-	-
Petroleum Refineries	-	-	-	-	-	-	-	-	-	-
Electr., CHP+Heat Plants	-	-	-	-	-	-	-	-	1116	-
Pumped Storage (Elec.)	-	-	-	-	-	-	-	-	-	-
Other Energy Sector	-	-	-	-	-	-	-	-	182	-
Distribution Losses	39	124	-	-	9	-	-	-	1710	6950
FINAL CONSUMPTION	**14337**	**3542**	**-**	**-**	**21749**	**-**	**-**	**-**	**4828**	**26182**
INDUSTRY SECTOR	**12167**	**3542**	**-**	**-**	**4035**	**-**	**-**	**-**	**1907**	**2151**
Iron and Steel	-	-	-	-	-	-	-	-	44	-
Chemical and Petrochem.	8471	3542	-	-	-	-	-	-	548	650
of which: Feedstocks	*7659*	-	-	-	-	-	-	-	-	-
Non-Ferrous Metals	-	-	-	-	-	-	-	-	-	-
Non-Metallic Minerals	937	-	-	-	40	-	-	-	181	46
Transport Equipment	23	-	-	-	12	-	-	-	78	22
Machinery	108	-	-	-	48	-	-	-	155	57
Mining and Quarrying	-	-	-	-	2	-	-	-	19	-
Food and Tobacco	1266	-	-	-	177	-	-	-	244	542
Paper, Pulp and Print	834	-	-	-	967	-	-	-	58	178
Wood and Wood Products	5	-	-	-	807	-	-	-	140	163
Construction	22	-	-	-	56	-	-	-	119	49
Textile and Leather	181	-	-	-	24	-	-	-	196	233
Non-specified	320	-	-	-	1902	-	-	-	125	211
TRANSPORT SECTOR	**33**	**-**	**-**	**-**	**8**	**-**	**-**	**-**	**105**	**-**
Air	-	-	-	-	-	-	-	-	-	-
Road	33	-	-	-	-	-	-	-	-	-
Rail	-	-	-	-	8	-	-	-	19	-
Pipeline Transport	-	-	-	-	-	-	-	-	-	-
Internal Navigation	-	-	-	-	-	-	-	-	-	-
Non-specified	-	-	-	-	-	-	-	-	86	-
OTHER SECTORS	**2137**	**-**	**-**	**-**	**17706**	**-**	**-**	**-**	**2816**	**24031**
Agriculture	145	-	-	-	70	-	-	-	341	-
Comm. and Publ. Services	297	-	-	-	93	-	-	-	1241	3114
Residential	1695	-	-	-	17543	-	-	-	1234	20917
Non-specified	-	-	-	-	-	-	-	-	-	-
NON-ENERGY USE	**-**	**-**	**-**	**-**	**-**	**-**	**-**	**-**	**-**	**-**
in Industry/Transf./Energy	-	-	-	-	-	-	-	-	-	-
in Transport	-	-	-	-	-	-	-	-	-	-
in Other Sectors	-	-	-	-	-	-	-	-	-	-

Estonia / Estonie : 1997

SUPPLY AND CONSUMPTION / *APPROVISIONNEMENT ET DEMANDE*	Coal / *Charbon* (1000 tonnes)							Oil / *Pétrole* (1000 tonnes)			
	Coking Coal / *Charbon à coke*	Other Bit. Coal / *Autres charb. bit.*	Sub-Bit. Coal / *Charbon sous-bit.*	Lignite / *Lignite*	Peat / *Tourbe*	Oven and Gas Coke / *Coke de four/gaz*	Pat. Fuel and BKB / *Agg./briq. de lignite*	Crude Oil / *Pétrole brut*	NGL / *LGN*	Feed-stocks / *Produits d'aliment.*	Additives / *Additifs*
Production	-	-	-	11895	571	42	129	-	-	-	-
From Other Sources	-	-	-	-	-	-	-	396	-	-	-
Imports	-	48	-	1534	-	1	-	-	-	-	-
Exports	-	-	-	-16	-4	-35	-50	-	-	-	-
Intl. Marine Bunkers	-	-	-	-	-	-	-	-	-	-	-
Stock Changes	-	50	-	50	67	-7	-12	5	-	-	-
DOMESTIC SUPPLY	**-**	**98**	**-**	**13463**	**634**	**1**	**67**	**401**	**-**	**-**	**-**
Transfers	-	-	-	-	-	-	-	-	-	-	-
Statistical Differences	-	-	-	-	-	-	-	-	-	-	-
TRANSFORMATION	**-**	**27**	**-**	**13029**	**381**	**-**	**3**	**401**	**-**	**-**	**-**
Electricity Plants	-	-	-	-	-	-	-	-	-	-	-
CHP Plants	-	-	-	12786	14	-	-	-	-	-	-
Heat Plants	-	27	-	76	118	-	3	-	-	-	-
Blast Furnaces/Gas Works	-	-	-	-	-	-	-	-	-	-	-
Coke/Pat. Fuel/BKB Plants	-	-	-	167	249	-	-	-	-	-	-
Petroleum Refineries	-	-	-	-	-	-	-	401	-	-	-
Petrochemical Industry	-	-	-	-	-	-	-	-	-	-	-
Liquefaction	-	-	-	-	-	-	-	-	-	-	-
Other Transform. Sector	-	-	-	-	-	-	-	-	-	-	-
ENERGY SECTOR	**-**	**-**	**-**	**-**	**72**	**-**	**1**	**-**	**-**	**-**	**-**
Coal Mines	-	-	-	-	-	-	-	-	-	-	-
Oil and Gas Extraction	-	-	-	-	-	-	-	-	-	-	-
Petroleum Refineries	-	-	-	-	-	-	-	-	-	-	-
Electr., CHP+Heat Plants	-	-	-	-	-	-	-	-	-	-	-
Pumped Storage (Elec.)	-	-	-	-	-	-	-	-	-	-	-
Other Energy Sector	-	-	-	-	72	-	1	-	-	-	-
Distribution Losses	-	-	-	-	163	-	-	-	-	-	-
FINAL CONSUMPTION	**-**	**71**	**-**	**434**	**18**	**1**	**63**	**-**	**-**	**-**	**-**
INDUSTRY SECTOR	**-**	**10**	**-**	**342**	**7**	**1**	**3**	**-**	**-**	**-**	**-**
Iron and Steel	-	-	-	-	-	-	-	-	-	-	-
Chemical and Petrochem.	-	-	-	10	-	-	-	-	-	-	-
of which: Feedstocks	-	-	-	-	-	-	-	-	-	-	-
Non-Ferrous Metals	-	-	-	-	-	-	-	-	-	-	-
Non-Metallic Minerals	-	6	-	326	-	-	3	-	-	-	-
Transport Equipment	-	-	-	-	-	-	-	-	-	-	-
Machinery	-	-	-	-	-	1	-	-	-	-	-
Mining and Quarrying	-	-	-	-	-	-	-	-	-	-	-
Food and Tobacco	-	2	-	-	5	-	-	-	-	-	-
Paper, Pulp and Print	-	-	-	-	-	-	-	-	-	-	-
Wood and Wood Products	-	-	-	5	-	-	-	-	-	-	-
Construction	-	1	-	-	1	-	-	-	-	-	-
Textile and Leather	-	1	-	1	-	-	-	-	-	-	-
Non-specified	-	-	-	-	1	-	-	-	-	-	-
TRANSPORT SECTOR	**-**	**2**	**-**	**-**	**-**	**-**	**-**	**-**	**-**	**-**	**-**
Air	-	-	-	-	-	-	-	-	-	-	-
Road	-	-	-	-	-	-	-	-	-	-	-
Rail	-	2	-	-	-	-	-	-	-	-	-
Pipeline Transport	-	-	-	-	-	-	-	-	-	-	-
Internal Navigation	-	-	-	-	-	-	-	-	-	-	-
Non-specified	-	-	-	-	-	-	-	-	-	-	-
OTHER SECTORS	**-**	**59**	**-**	**1**	**11**	**-**	**60**	**-**	**-**	**-**	**-**
Agriculture	-	2	-	1	6	-	-	-	-	-	-
Comm. and Publ. Services	-	11	-	-	2	-	-	-	-	-	-
Residential	-	46	-	-	3	-	60	-	-	-	-
Non-specified	-	-	-	-	-	-	-	-	-	-	-
NON-ENERGY USE	**-**	**-**	**-**	**91**	**-**	**-**	**-**	**-**	**-**	**-**	**-**
in Industry/Trans./Energy	-	-	-	91	-	-	-	-	-	-	-
in Transport	-	-	-	-	-	-	-	-	-	-	-
in Other Sectors	-	-	-	-	-	-	-	-	-	-	-

Estonia / Estonie : 1997

	Oil cont. / *Pétrole cont.* (1000 tonnes)										
SUPPLY AND CONSUMPTION / ***APPROVISIONNEMENT ET DEMANDE***	Refinery Gas / *Gaz de raffinerie*	LPG + Ethane / *GPL + éthane*	Motor Gasoline / *Essence moteur*	Aviation Gasoline / *Essence aviation*	Jet Fuel / *Carbu- réacteurs*	Kerosene / *Kérosène*	Gas/ Diesel / *Gazole*	Heavy Fuel Oil / *Fioul lourd*	Naphtha / *Naphta*	Petrol. Coke / *Coke de pétrole*	Other Prod. / *Autres prod.*
Production	-	-	6	-	-	-	-	367	-	-	7
From Other Sources	-	-	-	-	-	-	-	-	-	-	-
Imports	-	9	423	2	27	-	584	514	-	-	100
Exports	-	-	-136	-	-9	-	-176	-370	-	-	-61
Intl. Marine Bunkers	-	-	-	-	-	-	-31	-71	-	-	-
Stock Changes	-	-	12	-	2	-	1	43	-	-	-4
DOMESTIC SUPPLY	**-**	**9**	**305**	**2**	**20**	**-**	**378**	**483**	**-**	**-**	**42**
Transfers	-	-	-	-	-	-	-	1	-	-	-
Statistical Differences	-	-	-	-	-	-	-	-	-	-	-
TRANSFORMATION	**-**	**-**	**-**	**-**	**-**	**-**	**12**	**295**	**-**	**-**	**-**
Electricity Plants	-	-	-	-	-	-	-	-	-	-	-
CHP Plants	-	-	-	-	-	-	-	98	-	-	-
Heat Plants	-	-	-	-	-	-	12	197	-	-	-
Blast Furnaces/Gas Works	-	-	-	-	-	-	-	-	-	-	-
Coke/Pat. Fuel/BKB Plants	-	-	-	-	-	-	-	-	-	-	-
Petroleum Refineries	-	-	-	-	-	-	-	-	-	-	-
Petrochemical Industry	-	-	-	-	-	-	-	-	-	-	-
Liquefaction	-	-	-	-	-	-	-	-	-	-	-
Other Transform. Sector	-	-	-	-	-	-	-	-	-	-	-
ENERGY SECTOR	**-**	**-**	**-**	**-**	**-**	**-**	**12**	**2**	**-**	**-**	**-**
Coal Mines	-	-	-	-	-	-	6	-	-	-	-
Oil and Gas Extraction	-	-	-	-	-	-	-	-	-	-	-
Petroleum Refineries	-	-	-	-	-	-	-	-	-	-	-
Electr., CHP+Heat Plants	-	-	-	-	-	-	4	2	-	-	-
Pumped Storage (Elec.)	-	-	-	-	-	-	-	-	-	-	-
Other Energy Sector	-	-	-	-	-	-	2	-	-	-	-
Distribution Losses	-	1	-	-	-	-	-	-	-	-	-
FINAL CONSUMPTION	**-**	**8**	**305**	**2**	**20**	**-**	**354**	**187**	**-**	**-**	**42**
INDUSTRY SECTOR	**-**	**1**	**1**	**-**	**-**	**-**	**47**	**154**	**-**	**-**	**-**
Iron and Steel	-	-	-	-	-	-	-	-	-	-	-
Chemical and Petrochem.	-	-	-	-	-	-	-	42	-	-	-
of which: Feedstocks	-	-	-	-	-	-	-	-	-	-	-
Non-Ferrous Metals	-	-	-	-	-	-	-	-	-	-	-
Non-Metallic Minerals	-	-	-	-	-	-	2	2	-	-	-
Transport Equipment	-	-	-	-	-	-	-	1	-	-	-
Machinery	-	-	-	-	-	-	1	1	-	-	-
Mining and Quarrying	-	-	-	-	-	-	-	-	-	-	-
Food and Tobacco	-	-	-	-	-	-	31	61	-	-	-
Paper, Pulp and Print	-	-	-	-	-	-	-	6	-	-	-
Wood and Wood Products	-	-	-	-	-	-	1	17	-	-	-
Construction	-	-	1	-	-	-	11	4	-	-	-
Textile and Leather	-	-	-	-	-	-	-	6	-	-	-
Non-specified	-	1	-	-	-	-	1	14	-	-	-
TRANSPORT SECTOR	**-**	**1**	**302**	**2**	**20**	**-**	**201**	**5**	**-**	**-**	**-**
Air	-	-	-	2	20	-	-	-	-	-	-
Road	-	1	302	-	-	-	162	2	-	-	-
Rail	-	-	-	-	-	-	33	3	-	-	-
Pipeline Transport	-	-	-	-	-	-	-	-	-	-	-
Internal Navigation	-	-	-	-	-	-	6	-	-	-	-
Non-specified	-	-	-	-	-	-	-	-	-	-	-
OTHER SECTORS	**-**	**6**	**2**	**-**	**-**	**-**	**106**	**28**	**-**	**-**	**-**
Agriculture	-	-	2	-	-	-	37	9	-	-	-
Comm. and Publ. Services	-	-	-	-	-	-	12	19	-	-	-
Residential	-	6	-	-	-	-	57	-	-	-	-
Non-specified	-	-	-	-	-	-	-	-	-	-	-
NON-ENERGY USE	**-**	**-**	**-**	**-**	**-**	**-**	**-**	**-**	**-**	**-**	**42**
in Industry/Transf./Energy	-	-	-	-	-	-	-	-	-	-	37
in Transport	-	-	-	-	-	-	-	-	-	-	3
in Other Sectors	-	-	-	-	-	-	-	-	-	-	2

Estonia / Estonie : 1997

SUPPLY AND CONSUMPTION / *APPROVISIONNEMENT ET DEMANDE*	Gas / *Gaz* (TJ)				Comb. Renew. & Waste / *En. Re. Comb. & Déchets* (TJ)				(GWh)	(TJ)
	Natural Gas / *Gaz naturel*	Gas Works / *Usines à gaz*	Coke Ovens / *Cokeries*	Blast Furnaces / *Hauts fourneaux*	Solid Biomass / *Biomasse solide*	Gas/Liquids from Biomass / *Gaz/Liquides tirés de biomasse*	Municipal Waste / *Déchets urbains*	Industrial Waste / *Déchets industriels*	Electricity / *Electricité*	Heat / *Chaleur*
Production	-	5313	-	-	26118	58	-	-	9218	32593
From Other Sources	-	-	-	-	-	-	-	-	-	-
Imports	28747	-	-	-	-	-	-	-	210	-
Exports	-	-	-	-	-1540	-	-	-	-1184	-
Intl. Marine Bunkers	-	-	-	-	-	-	-	-	-	-
Stock Changes	-	-	-	-	157	-	-	-	-	-
DOMESTIC SUPPLY	**28747**	**5313**	**-**	**-**	**24735**	**58**	**-**	**-**	**8244**	**32593**
Transfers	-	-	-	-	-	-	-	-	-	-
Statistical Differences	-	-	-	-	-	-	-	-	-	-
TRANSFORMATION	**14627**	**1627**	**-**	**-**	**3213**	**58**	**-**	**-**	**105**	**-**
Electricity Plants	-	-	-	-	-	-	-	-	-	-
CHP Plants	3370	1455	-	-	74	-	-	-	-	-
Heat Plants	11257	172	-	-	3139	58	-	-	-	-
Blast Furnaces/Gas Works	-	-	-	-	-	-	-	-	-	-
Coke/Pat. Fuel/BKB Plants	-	-	-	-	-	-	-	-	-	-
Petroleum Refineries	-	-	-	-	-	-	-	-	-	-
Petrochemical Industry	-	-	-	-	-	-	-	-	-	-
Liquefaction	-	-	-	-	-	-	-	-	-	-
Other Transform. Sector	-	-	-	-	-	-	-	-	105	-
ENERGY SECTOR	**-**	**-**	**-**	**-**	**9**	**-**	**-**	**-**	**1576**	**-**
Coal Mines	-	-	-	-	9	-	-	-	275	-
Oil and Gas Extraction	-	-	-	-	-	-	-	-	-	-
Petroleum Refineries	-	-	-	-	-	-	-	-	-	-
Electr., CHP+Heat Plants	-	-	-	-	-	-	-	-	1153	-
Pumped Storage (Elec.)	-	-	-	-	-	-	-	-	-	-
Other Energy Sector	-	-	-	-	-	-	-	-	148	-
Distribution Losses	53	136	-	-	15	-	-	-	1510	5589
FINAL CONSUMPTION	**14067**	**3550**	**-**	**-**	**21498**	**-**	**-**	**-**	**5053**	**27004**
INDUSTRY SECTOR	**11942**	**3550**	**-**	**-**	**2540**	**-**	**-**	**-**	**2115**	**2662**
Iron and Steel	-	-	-	-	-	-	-	-	7	-
Chemical and Petrochem.	8330	3550	-	-	-	-	-	-	520	700
of which: Feedstocks	*7535*	-	-	-	-	-	-	-	-	-
Non-Ferrous Metals	-	-	-	-	-	-	-	-	-	-
Non-Metallic Minerals	917	-	-	-	23	-	-	-	259	35
Transport Equipment	20	-	-	-	7	-	-	-	63	24
Machinery	108	-	-	-	30	-	-	-	150	67
Mining and Quarrying	-	-	-	-	1	-	-	-	124	-
Food and Tobacco	1239	-	-	-	113	-	-	-	249	689
Paper, Pulp and Print	815	-	-	-	370	-	-	-	85	274
Wood and Wood Products	2	-	-	-	642	-	-	-	130	218
Construction	20	-	-	-	36	-	-	-	104	51
Textile and Leather	177	-	-	-	16	-	-	-	206	298
Non-specified	314	-	-	-	1302	-	-	-	218	306
TRANSPORT SECTOR	**34**	**-**	**-**	**-**	**1**	**-**	**-**	**-**	**108**	**-**
Air	-	-	-	-	-	-	-	-	-	-
Road	34	-	-	-	-	-	-	-	-	-
Rail	-	-	-	-	1	-	-	-	14	-
Pipeline Transport	-	-	-	-	-	-	-	-	-	-
Internal Navigation	-	-	-	-	-	-	-	-	-	-
Non-specified	-	-	-	-	-	-	-	-	94	-
OTHER SECTORS	**2091**	**-**	**-**	**-**	**18957**	**-**	**-**	**-**	**2830**	**24342**
Agriculture	143	-	-	-	234	-	-	-	229	-
Comm. and Publ. Services	266	-	-	-	410	-	-	-	1396	3203
Residential	1682	-	-	-	18313	-	-	-	1205	21139
Non-specified	-	-	-	-	-	-	-	-	-	-
NON-ENERGY USE	**-**	**-**	**-**	**-**	**-**	**-**	**-**	**-**	**-**	**-**
in Industry/Transf./Energy	-	-	-	-	-	-	-	-	-	-
in Transport	-	-	-	-	-	-	-	-	-	-
in Other Sectors	-	-	-	-	-	-	-	-	-	-

Ethiopia / Ethiopie

SUPPLY AND CONSUMPTION 1996	Coal (1000 tonnes)							Oil (1000 tonnes)			
	Coking Coal	Other Bit. Coal	Sub-Bit. Coal	Lignite	Peat	Oven and Gas Coke	Pat. Fuel and BKB	Crude Oil	NGL	Feed-stocks	Additives
Production	-	-	-	-	-	-	-	-	-	-	-
Imports	-	-	-	-	-	-	-	758	-	-	-
Exports	-	-	-	-	-	-	-	-	-	-	-
Intl. Marine Bunkers	-	-	-	-	-	-	-	-	-	-	-
Stock Changes	-	-	-	-	-	-	-	-1	-	-	-
DOMESTIC SUPPLY	**-**	**-**	**-**	**-**	**-**	**-**	**-**	**757**	**-**	**-**	**-**
Transfers and Stat. Diff.	-	-	-	-	-	-	-	91	-	-	-
TRANSFORMATION	**-**	**-**	**-**	**-**	**-**	**-**	**-**	**848**	**-**	**-**	**-**
Electricity and CHP Plants	-	-	-	-	-	-	-	-	-	-	-
Petroleum Refineries	-	-	-	-	-	-	-	848	-	-	-
Other Transform. Sector	-	-	-	-	-	-	-	-	-	-	-
ENERGY SECTOR	**-**	**-**	**-**	**-**	**-**	**-**	**-**	**-**	**-**	**-**	**-**
DISTRIBUTION LOSSES	**-**	**-**	**-**	**-**	**-**	**-**	**-**	**-**	**-**	**-**	**-**
FINAL CONSUMPTION	**-**	**-**	**-**	**-**	**-**	**-**	**-**	**-**	**-**	**-**	**-**
INDUSTRY SECTOR	**-**	**-**	**-**	**-**	**-**	**-**	**-**	**-**	**-**	**-**	**-**
Iron and Steel	-	-	-	-	-	-	-	-	-	-	-
Chemical and Petrochem.	-	-	-	-	-	-	-	-	-	-	-
Non-Metallic Minerals	-	-	-	-	-	-	-	-	-	-	-
Non-specified	-	-	-	-	-	-	-	-	-	-	-
TRANSPORT SECTOR	**-**	**-**	**-**	**-**	**-**	**-**	**-**	**-**	**-**	**-**	**-**
Air	-	-	-	-	-	-	-	-	-	-	-
Road	-	-	-	-	-	-	-	-	-	-	-
Non-specified	-	-	-	-	-	-	-	-	-	-	-
OTHER SECTORS	**-**	**-**	**-**	**-**	**-**	**-**	**-**	**-**	**-**	**-**	**-**
Agriculture	-	-	-	-	-	-	-	-	-	-	-
Comm. and Publ. Services	-	-	-	-	-	-	-	-	-	-	-
Residential	-	-	-	-	-	-	-	-	-	-	-
Non-specified	-	-	-	-	-	-	-	-	-	-	-
NON-ENERGY USE	**-**	**-**	**-**	**-**	**-**	**-**	**-**	**-**	**-**	**-**	**-**

APPROVISIONNEMENT ET DEMANDE 1997	*Charbon* (1000 tonnes)							*Pétrole* (1000 tonnes)			
	Charbon à coke	*Autres charb. bit.*	*Charbon sous-bit.*	*Lignite*	*Tourbe*	*Coke de four/gaz*	*Agg./briq. de lignite*	*Pétrole brut*	*LGN*	*Produits d'aliment.*	*Additifs*
Production	-	-	-	-	-	-	-	-	-	-	-
Imports	-	-	-	-	-	-	-	463	-	-	-
Exports	-	-	-	-	-	-	-	-	-	-	-
Intl. Marine Bunkers	-	-	-	-	-	-	-	-	-	-	-
Stock Changes	-	-	-	-	-	-	-	-	-	-	-
DOMESTIC SUPPLY	**-**	**-**	**-**	**-**	**-**	**-**	**-**	**463**	**-**	**-**	**-**
Transfers and Stat. Diff.	-	-	-	-	-	-	-	-	-	-	-
TRANSFORMATION	**-**	**-**	**-**	**-**	**-**	**-**	**-**	**463**	**-**	**-**	**-**
Electricity and CHP Plants	-	-	-	-	-	-	-	-	-	-	-
Petroleum Refineries	-	-	-	-	-	-	-	463	-	-	-
Other Transform. Sector	-	-	-	-	-	-	-	-	-	-	-
ENERGY SECTOR	**-**	**-**	**-**	**-**	**-**	**-**	**-**	**-**	**-**	**-**	**-**
DISTRIBUTION LOSSES	**-**	**-**	**-**	**-**	**-**	**-**	**-**	**-**	**-**	**-**	**-**
FINAL CONSUMPTION	**-**	**-**	**-**	**-**	**-**	**-**	**-**	**-**	**-**	**-**	**-**
INDUSTRY SECTOR	**-**	**-**	**-**	**-**	**-**	**-**	**-**	**-**	**-**	**-**	**-**
Iron and Steel	-	-	-	-	-	-	-	-	-	-	-
Chemical and Petrochem.	-	-	-	-	-	-	-	-	-	-	-
Non-Metallic Minerals	-	-	-	-	-	-	-	-	-	-	-
Non-specified	-	-	-	-	-	-	-	-	-	-	-
TRANSPORT SECTOR	**-**	**-**	**-**	**-**	**-**	**-**	**-**	**-**	**-**	**-**	**-**
Air	-	-	-	-	-	-	-	-	-	-	-
Road	-	-	-	-	-	-	-	-	-	-	-
Non-specified	-	-	-	-	-	-	-	-	-	-	-
OTHER SECTORS	**-**	**-**	**-**	**-**	**-**	**-**	**-**	**-**	**-**	**-**	**-**
Agriculture	-	-	-	-	-	-	-	-	-	-	-
Comm. and Publ. Services	-	-	-	-	-	-	-	-	-	-	-
Residential	-	-	-	-	-	-	-	-	-	-	-
Non-specified	-	-	-	-	-	-	-	-	-	-	-
NON-ENERGY USE	**-**	**-**	**-**	**-**	**-**	**-**	**-**	**-**	**-**	**-**	**-**

Ethiopia / Ethiopie

SUPPLY AND CONSUMPTION 1996	Oil cont. (1000 tonnes)										
	Refinery Gas	LPG + Ethane	Motor Gasoline	Aviation Gasoline	Jet Fuel	Kerosene	Gas/ Diesel	Heavy Fuel Oil	Naphtha	Petrol. Coke	Other Prod.
Production	4	7	125	-	60	12	211	334	-	-	12
Imports	-	-	59	6	200	-	375	-	-	-	40
Exports	-	-	-	-	-	-	-28	-40	-	-	-
Intl. Marine Bunkers	-	-	-	-	-	-	-4	-149	-	-	-
Stock Changes	-	-	-	-	-	-	-	-	-	-	-
DOMESTIC SUPPLY	**4**	**7**	**184**	**6**	**260**	**12**	**554**	**145**	**-**	**-**	**52**
Transfers and Stat. Diff.	-	-	-4	-	-	-	1	1	-	-	-
TRANSFORMATION	**-**	**-**	**-**	**-**	**-**	**-**	**43**	**33**	**-**	**-**	**-**
Electricity and CHP Plants	-	-	-	-	-	-	43	33	-	-	-
Petroleum Refineries	-	-	-	-	-	-	-	-	-	-	-
Other Transform. Sector	-	-	-	-	-	-	-	-	-	-	-
ENERGY SECTOR	**4**	**-**	**-**	**-**	**-**	**-**	**-**	**28**	**-**	**-**	**-**
DISTRIBUTION LOSSES	**-**	**-**	**-**	**-**	**-**	**-**	**-**	**-**	**-**	**-**	**-**
FINAL CONSUMPTION	**-**	**7**	**180**	**6**	**260**	**12**	**512**	**85**	**-**	**-**	**52**
INDUSTRY SECTOR	**-**	**-**	**-**	**-**	**-**	**-**	**121**	**85**	**-**	**-**	**-**
Iron and Steel	-	-	-	-	-	-	-	-	-	-	-
Chemical and Petrochem.	-	-	-	-	-	-	-	-	-	-	-
Non-Metallic Minerals	-	-	-	-	-	-	-	-	-	-	-
Non-specified	-	-	-	-	-	-	121	85	-	-	-
TRANSPORT SECTOR	**-**	**-**	**112**	**6**	**260**	**-**	**311**	**-**	**-**	**-**	**-**
Air	-	-	-	6	260	-	-	-	-	-	-
Road	-	-	112	-	-	-	311	-	-	-	-
Non-specified	-	-	-	-	-	-	-	-	-	-	-
OTHER SECTORS	**-**	**7**	**68**	**-**	**-**	**12**	**80**	**-**	**-**	**-**	**-**
Agriculture	-	-	-	-	-	-	33	-	-	-	-
Comm. and Publ. Services	-	-	-	-	-	-	14	-	-	-	-
Residential	-	7	19	-	-	12	-	-	-	-	-
Non-specified	-	-	49	-	-	-	33	-	-	-	-
NON-ENERGY USE	**-**	**-**	**-**	**-**	**-**	**-**	**-**	**-**	**-**	**-**	**52**

APPROVISIONNEMENT ET DEMANDE 1997	*Pétrole cont.* (1000 tonnes)										
	Gaz de raffinerie	*GPL + éthane*	*Essence moteur*	*Essence aviation*	*Carbu- réacteurs*	*Kérosène*	*Gazole*	*Fioul lourd*	*Naphta*	*Coke de pétrole*	*Autres prod.*
Production	2	3	71	-	36	8	116	178	-	-	12
Imports	-	-	79	6	229	-	49	69	-	-	40
Exports	-	-	-	-	-	-	-28	-52	-	-	-
Intl. Marine Bunkers	-	-	-	-	-	-	-	-70	-	-	-
Stock Changes	-	-	-	-	-	-	-	-	-	-	-
DOMESTIC SUPPLY	**2**	**3**	**150**	**6**	**265**	**8**	**137**	**125**	**-**	**-**	**52**
Transfers and Stat. Diff.	-	-	-3	-	1	-	-1	-	-	-	-
TRANSFORMATION	**-**	**-**	**-**	**-**	**-**	**-**	**10**	**28**	**-**	**-**	**-**
Electricity and CHP Plants	-	-	-	-	-	-	10	28	-	-	-
Petroleum Refineries	-	-	-	-	-	-	-	-	-	-	-
Other Transform. Sector	-	-	-	-	-	-	-	-	-	-	-
ENERGY SECTOR	**2**	**-**	**-**	**-**	**-**	**-**	**-**	**24**	**-**	**-**	**-**
DISTRIBUTION LOSSES	**-**	**-**	**-**	**-**	**-**	**-**	**-**	**-**	**-**	**-**	**-**
FINAL CONSUMPTION	**-**	**3**	**147**	**6**	**266**	**8**	**126**	**73**	**-**	**-**	**52**
INDUSTRY SECTOR	**-**	**-**	**-**	**-**	**-**	**-**	**30**	**73**	**-**	**-**	**-**
Iron and Steel	-	-	-	-	-	-	-	-	-	-	-
Chemical and Petrochem.	-	-	-	-	-	-	-	-	-	-	-
Non-Metallic Minerals	-	-	-	-	-	-	-	-	-	-	-
Non-specified	-	-	-	-	-	-	30	73	-	-	-
TRANSPORT SECTOR	**-**	**-**	**91**	**6**	**266**	**-**	**77**	**-**	**-**	**-**	**-**
Air	-	-	-	6	266	-	-	-	-	-	-
Road	-	-	91	-	-	-	77	-	-	-	-
Non-specified	-	-	-	-	-	-	-	-	-	-	-
OTHER SECTORS	**-**	**3**	**56**	**-**	**-**	**8**	**19**	**-**	**-**	**-**	**-**
Agriculture	-	-	-	-	-	-	8	-	-	-	-
Comm. and Publ. Services	-	-	-	-	-	-	3	-	-	-	-
Residential	-	3	16	-	-	8	-	-	-	-	-
Non-specified	-	-	40	-	-	-	8	-	-	-	-
NON-ENERGY USE	**-**	**-**	**-**	**-**	**-**	**-**	**-**	**-**	**-**	**-**	**52**

Ethiopia / Ethiopie

SUPPLY AND CONSUMPTION 1996	Gas (TJ)				Comb. Renew. & Waste (TJ)				(GWh)	(TJ)
	Natural Gas	Gas Works	Coke Ovens	Blast Furnaces	Solid Biomass	Gas/Liquids from Biomass	Municipal Waste	Industrial Waste	Electricity	Heat
Production	-	-	-	-	660100	-	-	-	1316	-
Imports	-	-	-	-	-	-	-	-	-	-
Exports	-	-	-	-	-	-	-	-	-	-
Intl. Marine Bunkers	-	-	-	-	-	-	-	-	-	-
Stock Changes	-	-	-	-	-	-	-	-	-	-
DOMESTIC SUPPLY	**-**	**-**	**-**	**-**	**660100**	**-**	**-**	**-**	**1316**	**-**
Transfers and Stat. Diff.	-	-	-	-	-	-	-	-	-35	-
TRANSFORMATION	**-**	**-**	**-**	**-**	**22140**	**-**	**-**	**-**	**-**	**-**
Electricity and CHP Plants	-	-	-	-	-	-	-	-	-	-
Petroleum Refineries	-	-	-	-	-	-	-	-	-	-
Other Transform. Sector	-	-	-	-	22140	-	-	-	-	-
ENERGY SECTOR	**-**	**-**	**-**	**-**	**-**	**-**	**-**	**-**	**12**	**-**
DISTRIBUTION LOSSES	**-**	**-**	**-**	**-**	**-**	**-**	**-**	**-**	**15**	**-**
FINAL CONSUMPTION	**-**	**-**	**-**	**-**	**637960**	**-**	**-**	**-**	**1254**	**-**
INDUSTRY SECTOR	**-**	**-**	**-**	**-**	**-**	**-**	**-**	**-**	**815**	**-**
Iron and Steel	-	-	-	-	-	-	-	-	-	-
Chemical and Petrochem.	-	-	-	-	-	-	-	-	-	-
Non-Metallic Minerals	-	-	-	-	-	-	-	-	-	-
Non-specified	-	-	-	-	-	-	-	-	815	-
TRANSPORT SECTOR	**-**	**-**	**-**	**-**	**-**	**-**	**-**	**-**	**-**	**-**
Air	-	-	-	-	-	-	-	-	-	-
Road	-	-	-	-	-	-	-	-	-	-
Non-specified	-	-	-	-	-	-	-	-	-	-
OTHER SECTORS	**-**	**-**	**-**	**-**	**637960**	**-**	**-**	**-**	**439**	**-**
Agriculture	-	-	-	-	-	-	-	-	-	-
Comm. and Publ. Services	-	-	-	-	-	-	-	-	12	-
Residential	-	-	-	-	-	-	-	-	427	-
Non-specified	-	-	-	-	637960	-	-	-	-	-
NON-ENERGY USE	**-**	**-**	**-**	**-**	**-**	**-**	**-**	**-**	**-**	**-**

APPROVISIONNEMENT ET DEMANDE 1997	*Gaz* (TJ)				*En. Re. Comb. & Déchets* (TJ)				(GWh)	(TJ)
	Gaz naturel	*Usines à gaz*	*Cokeries*	*Hauts fourneaux*	*Biomasse solide*	*Gaz/Liquides tirés de biomasse*	*Déchets urbains*	*Déchets industriels*	*Electricité*	*Chaleur*
Production	-	-	-	-	675282	-	-	-	1340	-
Imports	-	-	-	-	-	-	-	-	-	-
Exports	-	-	-	-	-	-	-	-	-	-
Intl. Marine Bunkers	-	-	-	-	-	-	-	-	-	-
Stock Changes	-	-	-	-	-	-	-	-	-	-
DOMESTIC SUPPLY	**-**	**-**	**-**	**-**	**675282**	**-**	**-**	**-**	**1340**	**-**
Transfers and Stat. Diff.	-	-	-	-	1	-	-	-	-36	-
TRANSFORMATION	**-**	**-**	**-**	**-**	**22650**	**-**	**-**	**-**	**-**	**-**
Electricity and CHP Plants	-	-	-	-	-	-	-	-	-	-
Petroleum Refineries	-	-	-	-	-	-	-	-	-	-
Other Transform. Sector	-	-	-	-	22650	-	-	-	-	-
ENERGY SECTOR	**-**	**-**	**-**	**-**	**-**	**-**	**-**	**-**	**12**	**-**
DISTRIBUTION LOSSES	**-**	**-**	**-**	**-**	**-**	**-**	**-**	**-**	**15**	**-**
FINAL CONSUMPTION	**-**	**-**	**-**	**-**	**652633**	**-**	**-**	**-**	**1277**	**-**
INDUSTRY SECTOR	**-**	**-**	**-**	**-**	**-**	**-**	**-**	**-**	**830**	**-**
Iron and Steel	-	-	-	-	-	-	-	-	-	-
Chemical and Petrochem.	-	-	-	-	-	-	-	-	-	-
Non-Metallic Minerals	-	-	-	-	-	-	-	-	-	-
Non-specified	-	-	-	-	-	-	-	-	830	-
TRANSPORT SECTOR	**-**	**-**	**-**	**-**	**-**	**-**	**-**	**-**	**-**	**-**
Air	-	-	-	-	-	-	-	-	-	-
Road	-	-	-	-	-	-	-	-	-	-
Non-specified	-	-	-	-	-	-	-	-	-	-
OTHER SECTORS	**-**	**-**	**-**	**-**	**652633**	**-**	**-**	**-**	**447**	**-**
Agriculture	-	-	-	-	-	-	-	-	-	-
Comm. and Publ. Services	-	-	-	-	-	-	-	-	12	-
Residential	-	-	-	-	-	-	-	-	435	-
Non-specified	-	-	-	-	652633	-	-	-	-	-
NON-ENERGY USE	**-**	**-**	**-**	**-**	**-**	**-**	**-**	**-**	**-**	**-**

Gabon

SUPPLY AND CONSUMPTION 1996	Coal (1000 tonnes)							Oil (1000 tonnes)			
	Coking Coal	Other Bit. Coal	Sub-Bit. Coal	Lignite	Peat	Oven and Gas Coke	Pat. Fuel and BKB	Crude Oil	NGL	Feed-stocks	Additives
Production	-	-	-	-	-	-	-	18277	-	-	-
Imports	-	-	-	-	-	-	-	-	-	-	-
Exports	-	-	-	-	-	-	-	-17363	-	-	-
Intl. Marine Bunkers	-	-	-	-	-	-	-	-	-	-	-
Stock Changes	-	-	-	-	-	-	-	-188	-	-	-
DOMESTIC SUPPLY	-	-	-	-	-	-	-	**726**	-	-	-
Transfers and Stat. Diff.	-	-	-	-	-	-	-	-	-	-	-
TRANSFORMATION	-	-	-	-	-	-	-	**726**	-	-	-
Electricity and CHP Plants	-	-	-	-	-	-	-	-	-	-	-
Petroleum Refineries	-	-	-	-	-	-	-	726	-	-	-
Other Transform. Sector	-	-	-	-	-	-	-	-	-	-	-
ENERGY SECTOR	-	-	-	-	-	-	-	-	-	-	-
DISTRIBUTION LOSSES	-	-	-	-	-	-	-	-	-	-	-
FINAL CONSUMPTION	-	-	-	-	-	-	-	-	-	-	-
INDUSTRY SECTOR	-	-	-	-	-	-	-	-	-	-	-
Iron and Steel	-	-	-	-	-	-	-	-	-	-	-
Chemical and Petrochem.	-	-	-	-	-	-	-	-	-	-	-
Non-Metallic Minerals	-	-	-	-	-	-	-	-	-	-	-
Non-specified	-	-	-	-	-	-	-	-	-	-	-
TRANSPORT SECTOR	-	-	-	-	-	-	-	-	-	-	-
Air	-	-	-	-	-	-	-	-	-	-	-
Road	-	-	-	-	-	-	-	-	-	-	-
Non-specified	-	-	-	-	-	-	-	-	-	-	-
OTHER SECTORS	-	-	-	-	-	-	-	-	-	-	-
Agriculture	-	-	-	-	-	-	-	-	-	-	-
Comm. and Publ. Services	-	-	-	-	-	-	-	-	-	-	-
Residential	-	-	-	-	-	-	-	-	-	-	-
Non-specified	-	-	-	-	-	-	-	-	-	-	-
NON-ENERGY USE	-	-	-	-	-	-	-	-	-	-	-

APPROVISIONNEMENT ET DEMANDE 1997	*Charbon* (1000 tonnes)							*Pétrole* (1000 tonnes)			
	Charbon à coke	*Autres charb. bit.*	*Charbon sous-bit.*	*Lignite*	*Tourbe*	*Coke de four/gaz*	*Agg./briq. de lignite*	*Pétrole brut*	*LGN*	*Produits d'aliment.*	*Additifs*
Production	-	-	-	-	-	-	-	18462	-	-	-
Imports	-	-	-	-	-	-	-	-	-	-	-
Exports	-	-	-	-	-	-	-	-17539	-	-	-
Intl. Marine Bunkers	-	-	-	-	-	-	-	-	-	-	-
Stock Changes	-	-	-	-	-	-	-	-198	-	-	-
DOMESTIC SUPPLY	-	-	-	-	-	-	-	**725**	-	-	-
Transfers and Stat. Diff.	-	-	-	-	-	-	-	-	-	-	-
TRANSFORMATION	-	-	-	-	-	-	-	**725**	-	-	-
Electricity and CHP Plants	-	-	-	-	-	-	-	-	-	-	-
Petroleum Refineries	-	-	-	-	-	-	-	725	-	-	-
Other Transform. Sector	-	-	-	-	-	-	-	-	-	-	-
ENERGY SECTOR	-	-	-	-	-	-	-	-	-	-	-
DISTRIBUTION LOSSES	-	-	-	-	-	-	-	-	-	-	-
FINAL CONSUMPTION	-	-	-	-	-	-	-	-	-	-	-
INDUSTRY SECTOR	-	-	-	-	-	-	-	-	-	-	-
Iron and Steel	-	-	-	-	-	-	-	-	-	-	-
Chemical and Petrochem.	-	-	-	-	-	-	-	-	-	-	-
Non-Metallic Minerals	-	-	-	-	-	-	-	-	-	-	-
Non-specified	-	-	-	-	-	-	-	-	-	-	-
TRANSPORT SECTOR	-	-	-	-	-	-	-	-	-	-	-
Air	-	-	-	-	-	-	-	-	-	-	-
Road	-	-	-	-	-	-	-	-	-	-	-
Non-specified	-	-	-	-	-	-	-	-	-	-	-
OTHER SECTORS	-	-	-	-	-	-	-	-	-	-	-
Agriculture	-	-	-	-	-	-	-	-	-	-	-
Comm. and Publ. Services	-	-	-	-	-	-	-	-	-	-	-
Residential	-	-	-	-	-	-	-	-	-	-	-
Non-specified	-	-	-	-	-	-	-	-	-	-	-
NON-ENERGY USE	-	-	-	-	-	-	-	-	-	-	-

Gabon

SUPPLY AND CONSUMPTION 1996	Oil cont. (1000 tonnes)										
	Refinery Gas	LPG + Ethane	Motor Gasoline	Aviation Gasoline	Jet Fuel	Kerosene	Gas/ Diesel	Heavy Fuel Oil	Naphtha	Petrol. Coke	Other Prod.
Production	30	8	52	-	44	24	165	277	72	-	49
Imports	-	6	-	13	9	5	69	-	-	-	15
Exports	-	-	-18	-	-	-	-	-170	-62	-	-
Intl. Marine Bunkers	-	-	-	-	-	-	-	-70	-	-	-
Stock Changes	-	-	-	-	1	-	-2	-	-	-	-
DOMESTIC SUPPLY	**30**	**14**	**34**	**13**	**54**	**29**	**232**	**37**	**10**	**-**	**64**
Transfers and Stat. Diff.	-	1	-	-	-	-	-	-	-	-	-
TRANSFORMATION	**-**	**-**	**-**	**-**	**-**	**-**	**44**	**-**	**-**	**-**	**-**
Electricity and CHP Plants	-	-	-	-	-	-	44	-	-	-	-
Petroleum Refineries	-	-	-	-	-	-	-	-	-	-	-
Other Transform. Sector	-	-	-	-	-	-	-	-	-	-	-
ENERGY SECTOR	**30**	**-**	**-**	**-**	**-**	**-**	**-**	**-**	**-**	**-**	**-**
DISTRIBUTION LOSSES	**-**	**-**	**-**	**-**	**-**	**-**	**-**	**-**	**-**	**-**	**-**
FINAL CONSUMPTION	**-**	**15**	**34**	**13**	**54**	**29**	**188**	**37**	**10**	**-**	**64**
INDUSTRY SECTOR	**-**	**-**	**-**	**-**	**-**	**-**	**67**	**37**	**10**	**-**	**-**
Iron and Steel	-	-	-	-	-	-	-	-	-	-	-
Chemical and Petrochem.	-	-	-	-	-	-	-	-	10	-	-
Non-Metallic Minerals	-	-	-	-	-	-	-	-	-	-	-
Non-specified	-	-	-	-	-	-	67	37	-	-	-
TRANSPORT SECTOR	**-**	**-**	**34**	**13**	**54**	**-**	**110**	**-**	**-**	**-**	**-**
Air	-	-	-	13	54	-	-	-	-	-	-
Road	-	-	34	-	-	-	89	-	-	-	-
Non-specified	-	-	-	-	-	-	21	-	-	-	-
OTHER SECTORS	**-**	**15**	**-**	**-**	**-**	**29**	**11**	**-**	**-**	**-**	**-**
Agriculture	-	-	-	-	-	-	-	-	-	-	-
Comm. and Publ. Services	-	-	-	-	-	-	-	-	-	-	-
Residential	-	15	-	-	-	29	-	-	-	-	-
Non-specified	-	-	-	-	-	-	11	-	-	-	-
NON-ENERGY USE	**-**	**-**	**-**	**-**	**-**	**-**	**-**	**-**	**-**	**-**	**64**

APPROVISIONNEMENT ET DEMANDE 1997	*Pétrole cont.* (1000 tonnes)										
	Gaz de raffinerie	*GPL + éthane*	*Essence moteur*	*Essence aviation*	*Carbu- réacteurs*	*Kérosène*	*Gazole*	*Fioul lourd*	*Naphta*	*Coke de pétrole*	*Autres prod.*
Production	29	17	56	-	43	24	162	272	71	-	43
Imports	-	1	-	13	37	6	137	-	-	-	16
Exports	-	-1	-15	-	-	-	-	-186	-61	-	-
Intl. Marine Bunkers	-	-	-	-	-	-	-	-54	-	-	-
Stock Changes	-	-	-	-	-	-	-	-	-	-	-
DOMESTIC SUPPLY	**29**	**17**	**41**	**13**	**80**	**30**	**299**	**32**	**10**	**-**	**59**
Transfers and Stat. Diff.	-	-	-	-	-	-	-1	-	-	-	-
TRANSFORMATION	**-**	**-**	**-**	**-**	**-**	**-**	**57**	**-**	**-**	**-**	**-**
Electricity and CHP Plants	-	-	-	-	-	-	57	-	-	-	-
Petroleum Refineries	-	-	-	-	-	-	-	-	-	-	-
Other Transform. Sector	-	-	-	-	-	-	-	-	-	-	-
ENERGY SECTOR	**29**	**-**	**-**	**-**	**-**	**-**	**-**	**-**	**-**	**-**	**-**
DISTRIBUTION LOSSES	**-**	**-**	**-**	**-**	**-**	**-**	**-**	**-**	**-**	**-**	**-**
FINAL CONSUMPTION	**-**	**17**	**41**	**13**	**80**	**30**	**241**	**32**	**10**	**-**	**59**
INDUSTRY SECTOR	**-**	**-**	**-**	**-**	**-**	**-**	**86**	**32**	**10**	**-**	**-**
Iron and Steel	-	-	-	-	-	-	-	-	-	-	-
Chemical and Petrochem.	-	-	-	-	-	-	-	-	10	-	-
Non-Metallic Minerals	-	-	-	-	-	-	-	-	-	-	-
Non-specified	-	-	-	-	-	-	86	32	-	-	-
TRANSPORT SECTOR	**-**	**-**	**41**	**13**	**80**	**-**	**141**	**-**	**-**	**-**	**-**
Air	-	-	-	13	80	-	-	-	-	-	-
Road	-	-	41	-	-	-	114	-	-	-	-
Non-specified	-	-	-	-	-	-	27	-	-	-	-
OTHER SECTORS	**-**	**17**	**-**	**-**	**-**	**30**	**14**	**-**	**-**	**-**	**-**
Agriculture	-	-	-	-	-	-	-	-	-	-	-
Comm. and Publ. Services	-	-	-	-	-	-	-	-	-	-	-
Residential	-	17	-	-	-	30	-	-	-	-	-
Non-specified	-	-	-	-	-	-	14	-	-	-	-
NON-ENERGY USE	**-**	**-**	**-**	**-**	**-**	**-**	**-**	**-**	**-**	**-**	**59**

Gabon

SUPPLY AND CONSUMPTION 1996	Gas (TJ)				Comb. Renew. & Waste (TJ)				(GWh)	(TJ)
	Natural Gas	Gas Works	Coke Ovens	Blast Furnaces	Solid Biomass	Gas/Liquids from Biomass	Municipal Waste	Industrial Waste	Electricity	Heat
Production	3317	-	-	-	35525	-	-	-	970	-
Imports	-	-	-	-	-	-	-	-	-	-
Exports	-	-	-	-		-	-	-	-	-
Intl. Marine Bunkers	-	-	-	-		-	-	-		-
Stock Changes	-	-	-	-	-	-	-	-	-	-
DOMESTIC SUPPLY	**3317**	**-**	**-**	**-**	**35525**	**-**	**-**	**-**	**970**	**-**
Transfers and Stat. Diff.	-	-	-	-	-	-	-	-	1	-
TRANSFORMATION	**1990**	**-**	**-**	**-**	**-**	**-**	**-**	**-**	**-**	**-**
Electricity and CHP Plants	1990	-	-	-	-	-	-	-	-	-
Petroleum Refineries	-	-	-	-	-	-	-	-	-	-
Other Transform. Sector	-	-	-	-	-	-	-	-	-	-
ENERGY SECTOR	**1298**	**-**	**-**	**-**	**-**	**-**	**-**	**-**	**39**	**-**
DISTRIBUTION LOSSES	**-**	**-**	**-**	**-**	**-**	**-**	**-**	**-**	**97**	**-**
FINAL CONSUMPTION	**29**	**-**	**-**	**-**	**35525**	**-**	**-**	**-**	**835**	**-**
INDUSTRY SECTOR	**29**	**-**	**-**	**-**	**7105**	**-**	**-**	**-**	**434**	**-**
Iron and Steel	-	-	-	-	-	-	-	-	-	-
Chemical and Petrochem.	-	-	-	-	-	-	-	-	-	-
Non-Metallic Minerals	-	-	-	-	-	-	-	-	-	-
Non-specified	29	-	-	-	7105	-	-	-	434	-
TRANSPORT SECTOR	**-**	**-**	**-**	**-**	**-**	**-**	**-**	**-**	**-**	**-**
Air	-	-	-	-	-	-	-	-	-	-
Road	-	-	-	-	-	-	-	-	-	-
Non-specified	-	-	-	-	-	-	-	-	-	-
OTHER SECTORS	**-**	**-**	**-**	**-**	**28420**	**-**	**-**	**-**	**401**	**-**
Agriculture	-	-	-	-	-	-	-	-	-	-
Comm. and Publ. Services	-	-	-	-	-	-	-	-	155	-
Residential	-	-	-	-	28420	-	-	-	246	-
Non-specified	-	-	-	-	-	-	-	-	-	-
NON-ENERGY USE	**-**	**-**	**-**	**-**	**-**	**-**	**-**	**-**	**-**	**-**

APPROVISIONNEMENT ET DEMANDE 1997	*Gaz (TJ)*				*En. Re. Comb. & Déchets (TJ)*				(GWh)	(TJ)
	Gaz naturel	*Usines à gaz*	*Cokeries*	*Hauts fourneaux*	*Biomasse solide*	*Gaz/Liquides tirés de biomasse*	*Déchets urbains*	*Déchets industriels*	*Electricité*	*Chaleur*
Production	3162	-	-	-	36022	-	-	-	1007	-
Imports	-	-	-	-	-	-	-	-	-	-
Exports	-	-	-	-	-	-	-	-	-	-
Intl. Marine Bunkers	-	-	-	-	-	-	-	-	-	-
Stock Changes	-	-	-	-	-	-	-	-	-	-
DOMESTIC SUPPLY	**3162**	**-**	**-**	**-**	**36022**	**-**	**-**	**-**	**1007**	**-**
Transfers and Stat. Diff.	-	-	-	-	-	-	-	-	-	-
TRANSFORMATION	**1897**	**-**	**-**	**-**	**-**	**-**	**-**	**-**	**-**	**-**
Electricity and CHP Plants	1897	-	-	-	-	-	-	-	-	-
Petroleum Refineries	-	-	-	-	-	-	-	-	-	-
Other Transform. Sector	-	-	-	-	-	-	-	-	-	-
ENERGY SECTOR	**1237**	**-**	**-**	**-**	**-**	**-**	**-**	**-**	**39**	**-**
DISTRIBUTION LOSSES	**-**	**-**	**-**	**-**	**-**	**-**	**-**	**-**	**101**	**-**
FINAL CONSUMPTION	**28**	**-**	**-**	**-**	**36022**	**-**	**-**	**-**	**867**	**-**
INDUSTRY SECTOR	**28**	**-**	**-**	**-**	**7204**	**-**	**-**	**-**	**451**	**-**
Iron and Steel	-	-	-	-	-	-	-	-	-	-
Chemical and Petrochem.	-	-	-	-	-	-	-	-	-	-
Non-Metallic Minerals	-	-	-	-	-	-	-	-	-	-
Non-specified	28	-	-	-	7204	-	-	-	451	-
TRANSPORT SECTOR	**-**	**-**	**-**	**-**	**-**	**-**	**-**	**-**	**-**	**-**
Air	-	-	-	-	-	-	-	-	-	-
Road	-	-	-	-	-	-	-	-	-	-
Non-specified	-	-	-	-	-	-	-	-	-	-
OTHER SECTORS	**-**	**-**	**-**	**-**	**28818**	**-**	**-**	**-**	**416**	**-**
Agriculture	-	-	-	-	-	-	-	-	-	-
Comm. and Publ. Services	-	-	-	-	-	-	-	-	161	-
Residential	-	-	-	-	28818	-	-	-	255	-
Non-specified	-	-	-	-	-	-	-	-	-	-
NON-ENERGY USE	**-**	**-**	**-**	**-**	**-**	**-**	**-**	**-**	**-**	**-**

Georgia / Géorgie

SUPPLY AND CONSUMPTION 1996	Coal (1000 tonnes)							Oil (1000 tonnes)			
	Coking Coal	Other Bit. Coal	Sub-Bit. Coal	Lignite	Peat	Oven and Gas Coke	Pat. Fuel and BKB	Crude Oil	NGL	Feed-stocks	Additives
Production	-	23	-	-	-	-	-	128	-	-	-
Imports	-	49	-	-	-	-	-	-	-	-	-
Exports	-	-	-	-	-	-	-	-113	-	-	-
Intl. Marine Bunkers	-	-	-	-	-	-	-	-	-	-	-
Stock Changes	-	-	-	-	-	-	-	-	-	-	-
DOMESTIC SUPPLY	**-**	**72**	**-**	**-**	**-**	**-**	**-**	**15**	**-**	**-**	**-**
Transfers and Stat. Diff.	-	-	-	-	-	-	-	-	-	-	-
TRANSFORMATION	**-**	**-**	**-**	**-**	**-**	**-**	**-**	**15**	**-**	**-**	**-**
Electricity and CHP Plants	-	-	-	-	-	-	-	-	-	-	-
Petroleum Refineries	-	-	-	-	-	-	-	15	-	-	-
Other Transform. Sector	-	-	-	-	-	-	-	-	-	-	-
ENERGY SECTOR	**-**	**-**	**-**	**-**	**-**	**-**	**-**	**-**	**-**	**-**	**-**
DISTRIBUTION LOSSES	**-**	**-**	**-**	**-**	**-**	**-**	**-**	**-**	**-**	**-**	**-**
FINAL CONSUMPTION	**-**	**72**	**-**	**-**	**-**	**-**	**-**	**-**	**-**	**-**	**-**
INDUSTRY SECTOR	**-**	**-**	**-**	**-**	**-**	**-**	**-**	**-**	**-**	**-**	**-**
Iron and Steel	-	-	-	-	-	-	-	-	-	-	-
Chemical and Petrochem.	-	-	-	-	-	-	-	-	-	-	-
Non-Metallic Minerals	-	-	-	-	-	-	-	-	-	-	-
Non-specified	-	-	-	-	-	-	-	-	-	-	-
TRANSPORT SECTOR	**-**	**-**	**-**	**-**	**-**	**-**	**-**	**-**	**-**	**-**	**-**
Air	-	-	-	-	-	-	-	-	-	-	-
Road	-	-	-	-	-	-	-	-	-	-	-
Non-specified	-	-	-	-	-	-	-	-	-	-	-
OTHER SECTORS	**-**	**72**	**-**	**-**	**-**	**-**	**-**	**-**	**-**	**-**	**-**
Agriculture	-	-	-	-	-	-	-	-	-	-	-
Comm. and Publ. Services	-	-	-	-	-	-	-	-	-	-	-
Residential	-	-	-	-	-	-	-	-	-	-	-
Non-specified	-	72	-	-	-	-	-	-	-	-	-
NON-ENERGY USE	**-**	**-**	**-**	**-**	**-**	**-**	**-**	**-**	**-**	**-**	**-**

APPROVISIONNEMENT ET DEMANDE 1997	*Charbon* (1000 tonnes)							*Pétrole* (1000 tonnes)			
	Charbon à coke	*Autres charb. bit.*	*Charbon sous-bit.*	*Lignite*	*Tourbe*	*Coke de four/gaz*	*Agg./briq. de lignite*	*Pétrole brut*	*LGN*	*Produits d'aliment.*	*Additifs*
Production	-	5	-	-	-	-	-	134	-	-	-
Imports	-	-	-	-	-	-	-	27	-	-	-
Exports	-	-	-	-	-	-	-	-131	-	-	-
Intl. Marine Bunkers	-	-	-	-	-	-	-	-	-	-	-
Stock Changes	-	-	-	-	-	-	-	-	-	-	-
DOMESTIC SUPPLY	**-**	**5**	**-**	**-**	**-**	**-**	**-**	**30**	**-**	**-**	**-**
Transfers and Stat. Diff.	-	-	-	-	-	-	-	-	-	-	-
TRANSFORMATION	**-**	**-**	**-**	**-**	**-**	**-**	**-**	**27**	**-**	**-**	**-**
Electricity and CHP Plants	-	-	-	-	-	-	-	-	-	-	-
Petroleum Refineries	-	-	-	-	-	-	-	27	-	-	-
Other Transform. Sector	-	-	-	-	-	-	-	-	-	-	-
ENERGY SECTOR	**-**	**-**	**-**	**-**	**-**	**-**	**-**	**3**	**-**	**-**	**-**
DISTRIBUTION LOSSES	**-**	**-**	**-**	**-**	**-**	**-**	**-**	**-**	**-**	**-**	**-**
FINAL CONSUMPTION	**-**	**5**	**-**	**-**	**-**	**-**	**-**	**-**	**-**	**-**	**-**
INDUSTRY SECTOR	**-**	**-**	**-**	**-**	**-**	**-**	**-**	**-**	**-**	**-**	**-**
Iron and Steel	-	-	-	-	-	-	-	-	-	-	-
Chemical and Petrochem.	-	-	-	-	-	-	-	-	-	-	-
Non-Metallic Minerals	-	-	-	-	-	-	-	-	-	-	-
Non-specified	-	-	-	-	-	-	-	-	-	-	-
TRANSPORT SECTOR	**-**	**-**	**-**	**-**	**-**	**-**	**-**	**-**	**-**	**-**	**-**
Air	-	-	-	-	-	-	-	-	-	-	-
Road	-	-	-	-	-	-	-	-	-	-	-
Non-specified	-	-	-	-	-	-	-	-	-	-	-
OTHER SECTORS	**-**	**5**	**-**	**-**	**-**	**-**	**-**	**-**	**-**	**-**	**-**
Agriculture	-	-	-	-	-	-	-	-	-	-	-
Comm. and Publ. Services	-	-	-	-	-	-	-	-	-	-	-
Residential	-	-	-	-	-	-	-	-	-	-	-
Non-specified	-	5	-	-	-	-	-	-	-	-	-
NON-ENERGY USE	**-**	**-**	**-**	**-**	**-**	**-**	**-**	**-**	**-**	**-**	**-**

Georgia / Géorgie

SUPPLY AND CONSUMPTION 1996	Oil cont. (1000 tonnes)										
	Refinery Gas	LPG + Ethane	Motor Gasoline	Aviation Gasoline	Jet Fuel	Kerosene	Gas/ Diesel	Heavy Fuel Oil	Naphtha	Petrol. Coke	Other Prod.
Production	1	-	-	-	-	-	4	7	-	-	1
Imports	13	15	447	-	52	35	180	124	-	-	17
Exports	-	-	-	-	-	-	-	-5	-	-	-
Intl. Marine Bunkers	-	-	-	-	-	-	-	-	-	-	-
Stock Changes	-	-	-9	-	-6	-5	-	-29	-	-	-
DOMESTIC SUPPLY	**14**	**15**	**438**	**-**	**46**	**30**	**184**	**97**	**-**	**-**	**18**
Transfers and Stat. Diff.	-	-	-	-	-	-	-	-	-	-	-
TRANSFORMATION	**-**	**-**	**-**	**-**	**-**	**-**	**-**	**89**	**-**	**-**	**-**
Electricity and CHP Plants	-	-	-	-	-	-	-	89	-	-	-
Petroleum Refineries	-	-	-	-	-	-	-	-	-	-	-
Other Transform. Sector	-	-	-	-	-	-	-	-	-	-	-
ENERGY SECTOR	**14**	**-**	**-**	**-**	**-**	**-**	**-**	**-**	**-**	**-**	**-**
DISTRIBUTION LOSSES	**-**	**-**	**-**	**-**	**-**	**-**	**-**	**-**	**-**	**-**	**-**
FINAL CONSUMPTION	**-**	**15**	**438**	**-**	**46**	**30**	**184**	**8**	**-**	**-**	**18**
INDUSTRY SECTOR	**-**	**-**	**21**	**-**	**-**	**-**	**36**	**1**	**-**	**-**	**1**
Iron and Steel	-	-	-	-	-	-	-	-	-	-	-
Chemical and Petrochem.	-	-	-	-	-	-	-	-	-	-	-
Non-Metallic Minerals	-	-	-	-	-	-	-	-	-	-	-
Non-specified	-	-	21	-	-	-	36	1	-	-	1
TRANSPORT SECTOR	**-**	**-**	**372**	**-**	**46**	**-**	**76**	**7**	**-**	**-**	**4**
Air	-	-	-	-	46	-	-	-	-	-	-
Road	-	-	372	-	-	-	76	-	-	-	-
Non-specified	-	-	-	-	-	-	-	7	-	-	4
OTHER SECTORS	**-**	**15**	**45**	**-**	**-**	**30**	**72**	**-**	**-**	**-**	**13**
Agriculture	-	-	11	-	-	-	65	-	-	-	3
Comm. and Publ. Services	-	-	14	-	-	-	2	-	-	-	-
Residential	-	15	-	-	-	30	2	-	-	-	4
Non-specified	-	-	20	-	-	-	3	-	-	-	6
NON-ENERGY USE	**-**	**-**	**-**	**-**	**-**	**-**	**-**	**-**	**-**	**-**	**-**

APPROVISIONNEMENT ET DEMANDE 1997	*Pétrole cont.* (1000 tonnes)										
	Gaz de raffinerie	*GPL + éthane*	*Essence moteur*	*Essence aviation*	*Carbu- réacteurs*	*Kérosène*	*Gazole*	*Fioul lourd*	*Naphta*	*Coke de pétrole*	*Autres prod.*
Production	1	-	5	-	-	-	7	10	-	-	1
Imports	11	42	456	-	48	28	236	75	-	-	50
Exports	-	-	-2	-	-	-	-	-3	-	-	-
Intl. Marine Bunkers	-	-	-	-	-	-	-	-	-	-	-
Stock Changes	-	-3	13	-	-	3	-77	3	-	-	-
DOMESTIC SUPPLY	**12**	**39**	**472**	**-**	**48**	**31**	**166**	**85**	**-**	**-**	**51**
Transfers and Stat. Diff.	-	-	-	-	-	-	-	-	-	-	-
TRANSFORMATION	**-**	**-**	**-**	**-**	**-**	**-**	**-**	**85**	**-**	**-**	**-**
Electricity and CHP Plants	-	-	-	-	-	-	-	85	-	-	-
Petroleum Refineries	-	-	-	-	-	-	-	-	-	-	-
Other Transform. Sector	-	-	-	-	-	-	-	-	-	-	-
ENERGY SECTOR	**12**	**-**	**-**	**-**	**-**	**-**	**-**	**-**	**-**	**-**	**-**
DISTRIBUTION LOSSES	**-**	**-**	**-**	**-**	**-**	**-**	**-**	**-**	**-**	**-**	**-**
FINAL CONSUMPTION	**-**	**39**	**472**	**-**	**48**	**31**	**166**	**-**	**-**	**-**	**51**
INDUSTRY SECTOR	**-**	**-**	**28**	**-**	**-**	**-**	**33**	**-**	**-**	**-**	**1**
Iron and Steel	-	-	-	-	-	-	-	-	-	-	-
Chemical and Petrochem.	-	-	-	-	-	-	-	-	-	-	-
Non-Metallic Minerals	-	-	-	-	-	-	-	-	-	-	-
Non-specified	-	-	28	-	-	-	33	-	-	-	1
TRANSPORT SECTOR	**-**	**-**	**363**	**-**	**48**	**-**	**62**	**-**	**-**	**-**	**3**
Air	-	-	-	-	48	-	-	-	-	-	-
Road	-	-	363	-	-	-	62	-	-	-	-
Non-specified	-	-	-	-	-	-	-	-	-	-	3
OTHER SECTORS	**-**	**39**	**81**	**-**	**-**	**31**	**71**	**-**	**-**	**-**	**47**
Agriculture	-	-	16	-	-	-	58	-	-	-	3
Comm. and Publ. Services	-	-	65	-	-	-	9	-	-	-	-
Residential	-	39	-	-	-	31	-	-	-	-	3
Non-specified	-	-	-	-	-	-	4	-	-	-	41
NON-ENERGY USE	**-**	**-**	**-**	**-**	**-**	**-**	**-**	**-**	**-**	**-**	**-**

Georgia / Géorgie

SUPPLY AND CONSUMPTION 1996	Gas (TJ)				Comb. Renew. & Waste (TJ)				(GWh)	(TJ)
	Natural Gas	Gas Works	Coke Ovens	Blast Furnaces	Solid Biomass	Gas/Liquids from Biomass	Municipal Waste	Industrial Waste	Electricity	Heat
Production	113	-	-	-	1465	-	-	-	7226	19950
Imports	29368	-	-	-	-	-	-	-	221	-
Exports	-	-	-	-	-	-	-	-	-135	-
Intl. Marine Bunkers	-	-	-	-	-	-	-	-	-	-
Stock Changes	-	-	-	-	-	-	-	-	-	-
DOMESTIC SUPPLY	**29481**	**-**	**-**	**-**	**1465**	**-**	**-**	**-**	**7312**	**19950**
Transfers and Stat. Diff.	-	-	-	-	-	-	-	-	-	-
TRANSFORMATION	**13308**	**-**	**-**	**-**	**-**	**-**	**-**	**-**	**-**	**-**
Electricity and CHP Plants	13308	-	-	-	-	-	-	-	-	-
Petroleum Refineries	-	-	-	-	-	-	-	-	-	-
Other Transform. Sector	-	-	-	-	-	-	-	-	-	-
ENERGY SECTOR	**-**	**-**	**-**	**-**	**-**	**-**	**-**	**-**	**-**	**-**
DISTRIBUTION LOSSES	**-**	**-**	**-**	**-**	**-**	**-**	**-**	**-**	**1074**	**-**
FINAL CONSUMPTION	**16173**	**-**	**-**	**-**	**1465**	**-**	**-**	**-**	**6238**	**19950**
INDUSTRY SECTOR	**11574**	**-**	**-**	**-**	**-**	**-**	**-**	**-**	**876**	**-**
Iron and Steel	3506	-	-	-	-	-	-	-	324	-
Chemical and Petrochem.	6673	-	-	-	-	-	-	-	-	-
Non-Metallic Minerals	-	-	-	-	-	-	-	-	-	-
Non-specified	1395	-	-	-	-	-	-	-	552	-
TRANSPORT SECTOR	**-**	**-**	**-**	**-**	**-**	**-**	**-**	**-**	**250**	**-**
Air	-	-	-	-	-	-	-	-	-	-
Road	-	-	-	-	-	-	-	-	-	-
Non-specified	-	-	-	-	-	-	-	-	250	-
OTHER SECTORS	**4599**	**-**	**-**	**-**	**1465**	**-**	**-**	**-**	**5112**	**19950**
Agriculture	-	-	-	-	-	-	-	-	26	-
Comm. and Publ. Services	-	-	-	-	-	-	-	-	2447	-
Residential	4599	-	-	-	-	-	-	-	2639	-
Non-specified	-	-	-	-	1465	-	-	-	-	19950
NON-ENERGY USE	**-**	**-**	**-**	**-**	**-**	**-**	**-**	**-**	**-**	**-**

APPROVISIONNEMENT ET DEMANDE 1997	*Gaz* (TJ)				*En. Re. Comb. & Déchets* (TJ)				(GWh)	(TJ)
	Gaz naturel	*Usines à gaz*	*Cokeries*	*Hauts fourneaux*	*Biomasse solide*	*Gaz/Liquides tirés de biomasse*	*Déchets urbains*	*Déchets industriels*	*Electricité*	*Chaleur*
Production	113	-	-	-	1465	-	-	-	7172	19950
Imports	35589	-	-	-	-	-	-	-	653	-
Exports	-	-	-	-	-	-	-	-	-462	-
Intl. Marine Bunkers	-	-	-	-	-	-	-	-	-	-
Stock Changes	-	-	-	-	-	-	-	-	-	-
DOMESTIC SUPPLY	**35702**	**-**	**-**	**-**	**1465**	**-**	**-**	**-**	**7363**	**19950**
Transfers and Stat. Diff.	-	-	-	-	-	-	-	-	-	-
TRANSFORMATION	**14100**	**-**	**-**	**-**	**-**	**-**	**-**	**-**	**-**	**-**
Electricity and CHP Plants	14100	-	-	-	-	-	-	-	-	-
Petroleum Refineries	-	-	-	-	-	-	-	-	-	-
Other Transform. Sector	-	-	-	-	-	-	-	-	-	-
ENERGY SECTOR	**-**	**-**	**-**	**-**	**-**	**-**	**-**	**-**	**-**	**-**
DISTRIBUTION LOSSES	**-**	**-**	**-**	**-**	**-**	**-**	**-**	**-**	**1163**	**-**
FINAL CONSUMPTION	**21602**	**-**	**-**	**-**	**1465**	**-**	**-**	**-**	**6200**	**19950**
INDUSTRY SECTOR	**16324**	**-**	**-**	**-**	**-**	**-**	**-**	**-**	**828**	**-**
Iron and Steel	2752	-	-	-	-	-	-	-	285	-
Chemical and Petrochem.	7766	-	-	-	-	-	-	-	-	-
Non-Metallic Minerals	-	-	-	-	-	-	-	-	-	-
Non-specified	5806	-	-	-	-	-	-	-	543	-
TRANSPORT SECTOR	**-**	**-**	**-**	**-**	**-**	**-**	**-**	**-**	**231**	**-**
Air	-	-	-	-	-	-	-	-	-	-
Road	-	-	-	-	-	-	-	-	-	-
Non-specified	-	-	-	-	-	-	-	-	231	-
OTHER SECTORS	**5278**	**-**	**-**	**-**	**1465**	**-**	**-**	**-**	**5141**	**19950**
Agriculture	-	-	-	-	-	-	-	-	14	-
Comm. and Publ. Services	-	-	-	-	-	-	-	-	2534	-
Residential	5278	-	-	-	-	-	-	-	2593	-
Non-specified	-	-	-	-	1465	-	-	-	-	19950
NON-ENERGY USE	**-**	**-**	**-**	**-**	**-**	**-**	**-**	**-**	**-**	**-**

Ghana

SUPPLY AND CONSUMPTION 1996	Coal (1000 tonnes)							Oil (1000 tonnes)			
	Coking Coal	Other Bit. Coal	Sub-Bit. Coal	Lignite	Peat	Oven and Gas Coke	Pat. Fuel and BKB	Crude Oil	NGL	Feed-stocks	Additives
Production	-	-	-	-	-	-	-	355	-	-	-
Imports	-	3	-	-	-	-	-	1098	-	-	-
Exports	-	-	-	-	-	-	-	-	-	-	-
Intl. Marine Bunkers	-	-	-	-	-	-	-	-	-	-	-
Stock Changes	-	-	-	-	-	-	-	-2	-	-	-
DOMESTIC SUPPLY	**-**	**3**	**-**	**-**	**-**	**-**	**-**	**1451**	**-**	**-**	**-**
Transfers and Stat. Diff.	-	-	-	-	-	-	-	-416	-	-	-
TRANSFORMATION	**-**	**-**	**-**	**-**	**-**	**-**	**-**	**1035**	**-**	**-**	**-**
Electricity and CHP Plants	-	-	-	-	-	-	-	-	-	-	-
Petroleum Refineries	-	-	-	-	-	-	-	1035	-	-	-
Other Transform. Sector	-	-	-	-	-	-	-	-	-	-	-
ENERGY SECTOR	**-**	**-**	**-**	**-**	**-**	**-**	**-**	**-**	**-**	**-**	**-**
DISTRIBUTION LOSSES	**-**	**-**	**-**	**-**	**-**	**-**	**-**	**-**	**-**	**-**	**-**
FINAL CONSUMPTION	**-**	**3**	**-**	**-**	**-**	**-**	**-**	**-**	**-**	**-**	**-**
INDUSTRY SECTOR	**-**	**-**	**-**	**-**	**-**	**-**	**-**	**-**	**-**	**-**	**-**
Iron and Steel	-	-	-	-	-	-	-	-	-	-	-
Chemical and Petrochem.	-	-	-	-	-	-	-	-	-	-	-
Non-Metallic Minerals	-	-	-	-	-	-	-	-	-	-	-
Non-specified	-	-	-	-	-	-	-	-	-	-	-
TRANSPORT SECTOR	**-**	**-**	**-**	**-**	**-**	**-**	**-**	**-**	**-**	**-**	**-**
Air	-	-	-	-	-	-	-	-	-	-	-
Road	-	-	-	-	-	-	-	-	-	-	-
Non-specified	-	-	-	-	-	-	-	-	-	-	-
OTHER SECTORS	**-**	**3**	**-**	**-**	**-**	**-**	**-**	**-**	**-**	**-**	**-**
Agriculture	-	-	-	-	-	-	-	-	-	-	-
Comm. and Publ. Services	-	-	-	-	-	-	-	-	-	-	-
Residential	-	3	-	-	-	-	-	-	-	-	-
Non-specified	-	-	-	-	-	-	-	-	-	-	-
NON-ENERGY USE	**-**	**-**	**-**	**-**	**-**	**-**	**-**	**-**	**-**	**-**	**-**

APPROVISIONNEMENT ET DEMANDE 1997	*Charbon* (1000 tonnes)							*Pétrole* (1000 tonnes)			
	Charbon à coke	*Autres charb. bit.*	*Charbon sous-bit.*	*Lignite*	*Tourbe*	*Coke de four/gaz*	*Agg./briq. de lignite*	*Pétrole brut*	*LGN*	*Produits d'aliment.*	*Additifs*
Production	-	-	-	-	-	-	-	355	-	-	-
Imports	-	3	-	-	-	-	-	1098	-	-	-
Exports	-	-	-	-	-	-	-	-	-	-	-
Intl. Marine Bunkers	-	-	-	-	-	-	-	-	-	-	-
Stock Changes	-	-	-	-	-	-	-	-2	-	-	-
DOMESTIC SUPPLY	**-**	**3**	**-**	**-**	**-**	**-**	**-**	**1451**	**-**	**-**	**-**
Transfers and Stat. Diff.	-	-	-	-	-	-	-	-416	-	-	-
TRANSFORMATION	**-**	**-**	**-**	**-**	**-**	**-**	**-**	**1035**	**-**	**-**	**-**
Electricity and CHP Plants	-	-	-	-	-	-	-	-	-	-	-
Petroleum Refineries	-	-	-	-	-	-	-	1035	-	-	-
Other Transform. Sector	-	-	-	-	-	-	-	-	-	-	-
ENERGY SECTOR	**-**	**-**	**-**	**-**	**-**	**-**	**-**	**-**	**-**	**-**	**-**
DISTRIBUTION LOSSES	**-**	**-**	**-**	**-**	**-**	**-**	**-**	**-**	**-**	**-**	**-**
FINAL CONSUMPTION	**-**	**3**	**-**	**-**	**-**	**-**	**-**	**-**	**-**	**-**	**-**
INDUSTRY SECTOR	**-**	**-**	**-**	**-**	**-**	**-**	**-**	**-**	**-**	**-**	**-**
Iron and Steel	-	-	-	-	-	-	-	-	-	-	-
Chemical and Petrochem.	-	-	-	-	-	-	-	-	-	-	-
Non-Metallic Minerals	-	-	-	-	-	-	-	-	-	-	-
Non-specified	-	-	-	-	-	-	-	-	-	-	-
TRANSPORT SECTOR	**-**	**-**	**-**	**-**	**-**	**-**	**-**	**-**	**-**	**-**	**-**
Air	-	-	-	-	-	-	-	-	-	-	-
Road	-	-	-	-	-	-	-	-	-	-	-
Non-specified	-	-	-	-	-	-	-	-	-	-	-
OTHER SECTORS	**-**	**3**	**-**	**-**	**-**	**-**	**-**	**-**	**-**	**-**	**-**
Agriculture	-	-	-	-	-	-	-	-	-	-	-
Comm. and Publ. Services	-	-	-	-	-	-	-	-	-	-	-
Residential	-	3	-	-	-	-	-	-	-	-	-
Non-specified	-	-	-	-	-	-	-	-	-	-	-
NON-ENERGY USE	**-**	**-**	**-**	**-**	**-**	**-**	**-**	**-**	**-**	**-**	**-**

Ghana

SUPPLY AND CONSUMPTION 1996	Oil cont. (1000 tonnes)										
	Refinery Gas	LPG + Ethane	Motor Gasoline	Aviation Gasoline	Jet Fuel	Kerosene	Gas/ Diesel	Heavy Fuel Oil	Naphtha	Petrol. Coke	Other Prod.
Production	55	12	287	-	52	132	315	254	-	-	54
Imports	-	-	71	5	-	-	101	-	-	-	6
Exports	-	-3	-	-	-	-	-	-201	-	-	-
Intl. Marine Bunkers	-	-	-	-	-	-	-25	-	-	-	-
Stock Changes	-	-	-	-	-	-	-	-	-	-	-
DOMESTIC SUPPLY	**55**	**9**	**358**	**5**	**52**	**132**	**391**	**53**	**-**	**-**	**60**
Transfers and Stat. Diff.	-	21	-1	-	-	-	-1	-	-	-	-
TRANSFORMATION	**-**	**-**	**-**	**-**	**-**	**-**	**20**	**-**	**-**	**-**	**-**
Electricity and CHP Plants	-	-	-	-	-	-	20	-	-	-	-
Petroleum Refineries	-	-	-	-	-	-	-	-	-	-	-
Other Transform. Sector	-	-	-	-	-	-	-	-	-	-	-
ENERGY SECTOR	**55**	**-**	**-**	**-**	**-**	**-**	**-**	**-**	**-**	**-**	**-**
DISTRIBUTION LOSSES	**-**	**-**	**-**	**-**	**-**	**-**	**-**	**-**	**-**	**-**	**-**
FINAL CONSUMPTION	**-**	**30**	**357**	**5**	**52**	**132**	**370**	**53**	**-**	**-**	**60**
INDUSTRY SECTOR	**-**	**12**	**-**	**-**	**-**	**-**	**38**	**53**	**-**	**-**	**-**
Iron and Steel	-	-	-	-	-	-	-	-	-	-	-
Chemical and Petrochem.	-	-	-	-	-	-	-	-	-	-	-
Non-Metallic Minerals	-	-	-	-	-	-	-	-	-	-	-
Non-specified	-	12	-	-	-	-	38	53	-	-	-
TRANSPORT SECTOR	**-**	**-**	**357**	**5**	**52**	**-**	**247**	**-**	**-**	**-**	**-**
Air	-	-	-	5	52	-	-	-	-	-	-
Road	-	-	357	-	-	-	212	-	-	-	-
Non-specified	-	-	-	-	-	-	35	-	-	-	-
OTHER SECTORS	**-**	**18**	**-**	**-**	**-**	**132**	**85**	**-**	**-**	**-**	**-**
Agriculture	-	-	-	-	-	-	49	-	-	-	-
Comm. and Publ. Services	-	-	-	-	-	-	36	-	-	-	-
Residential	-	18	-	-	-	132	-	-	-	-	-
Non-specified	-	-	-	-	-	-	-	-	-	-	-
NON-ENERGY USE	**-**	**-**	**-**	**-**	**-**	**-**	**-**	**-**	**-**	**-**	**60**

APPROVISIONNEMENT ET DEMANDE 1997	*Pétrole cont.* (1000 tonnes)										
	Gaz de raffinerie	*GPL + éthane*	*Essence moteur*	*Essence aviation*	*Carbu- réacteurs*	*Kérosène*	*Gazole*	*Fioul lourd*	*Naphta*	*Coke de pétrole*	*Autres prod.*
Production	55	12	287	-	52	132	315	254	-	-	54
Imports	-	-	71	5	-	-	101	-	-	-	6
Exports	-	-3	-	-	-	-	-	-201	-	-	-
Intl. Marine Bunkers	-	-	-	-	-	-	-25	-	-	-	-
Stock Changes	-	-	-	-	-	-	-	-	-	-	-
DOMESTIC SUPPLY	**55**	**9**	**358**	**5**	**52**	**132**	**391**	**53**	**-**	**-**	**60**
Transfers and Stat. Diff.	-	21	-1	-	-	-	-1	-	-	-	-
TRANSFORMATION	**-**	**-**	**-**	**-**	**-**	**-**	**20**	**-**	**-**	**-**	**-**
Electricity and CHP Plants	-	-	-	-	-	-	20	-	-	-	-
Petroleum Refineries	-	-	-	-	-	-	-	-	-	-	-
Other Transform. Sector	-	-	-	-	-	-	-	-	-	-	-
ENERGY SECTOR	**55**	**-**	**-**	**-**	**-**	**-**	**-**	**-**	**-**	**-**	**-**
DISTRIBUTION LOSSES	**-**	**-**	**-**	**-**	**-**	**-**	**-**	**-**	**-**	**-**	**-**
FINAL CONSUMPTION	**-**	**30**	**357**	**5**	**52**	**132**	**370**	**53**	**-**	**-**	**60**
INDUSTRY SECTOR	**-**	**12**	**-**	**-**	**-**	**-**	**38**	**53**	**-**	**-**	**-**
Iron and Steel	-	-	-	-	-	-	-	-	-	-	-
Chemical and Petrochem.	-	-	-	-	-	-	-	-	-	-	-
Non-Metallic Minerals	-	-	-	-	-	-	-	-	-	-	-
Non-specified	-	12	-	-	-	-	38	53	-	-	-
TRANSPORT SECTOR	**-**	**-**	**357**	**5**	**52**	**-**	**247**	**-**	**-**	**-**	**-**
Air	-	-	-	5	52	-	-	-	-	-	-
Road	-	-	357	-	-	-	212	-	-	-	-
Non-specified	-	-	-	-	-	-	35	-	-	-	-
OTHER SECTORS	**-**	**18**	**-**	**-**	**-**	**132**	**85**	**-**	**-**	**-**	**-**
Agriculture	-	-	-	-	-	-	49	-	-	-	-
Comm. and Publ. Services	-	-	-	-	-	-	36	-	-	-	-
Residential	-	18	-	-	-	132	-	-	-	-	-
Non-specified	-	-	-	-	-	-	-	-	-	-	-
NON-ENERGY USE	**-**	**-**	**-**	**-**	**-**	**-**	**-**	**-**	**-**	**-**	**60**

Ghana

SUPPLY AND CONSUMPTION 1996	Gas (TJ)				Comb. Renew. & Waste (TJ)				(GWh)	(TJ)
	Natural Gas	Gas Works	Coke Ovens	Blast Furnaces	Solid Biomass	Gas/Liquids from Biomass	Municipal Waste	Industrial Waste	Electricity	Heat
Production	-	-	-	-	202752	-	-	-	5987	-
Imports	-	-	-	-	-	-	-	-	61	-
Exports	-	-	-	-	-	-	-	-	-458	-
Intl. Marine Bunkers	-	-	-	-	-	-	-	-	-	-
Stock Changes	-	-	-	-	-	-	-	-	-	-
DOMESTIC SUPPLY	-	-	-	-	**202752**	-	-	-	**5590**	-
Transfers and Stat. Diff.	-	-	-	-	-	-	-	-	648	-
TRANSFORMATION	-	-	-	-	**49562**	-	-	-	-	-
Electricity and CHP Plants	-	-	-	-	-	-	-	-	-	-
Petroleum Refineries	-	-	-	-	-	-	-	-	-	-
Other Transform. Sector	-	-	-	-	49562	-	-	-	-	-
ENERGY SECTOR	-	-	-	-	-	-	-	-	**1408**	-
DISTRIBUTION LOSSES	-	-	-	-	-	-	-	-	**13**	-
FINAL CONSUMPTION	-	-	-	-	**153190**	-	-	-	**4817**	-
INDUSTRY SECTOR	-	-	-	-	**13926**	-	-	-	**3384**	-
Iron and Steel	-	-	-	-	-	-	-	-	-	-
Chemical and Petrochem.	-	-	-	-	-	-	-	-	-	-
Non-Metallic Minerals	-	-	-	-	-	-	-	-	-	-
Non-specified	-	-	-	-	13926	-	-	-	3384	-
TRANSPORT SECTOR	-	-	-	-	-	-	-	-	-	-
Air	-	-	-	-	-	-	-	-	-	-
Road	-	-	-	-	-	-	-	-	-	-
Non-specified	-	-	-	-	-	-	-	-	-	-
OTHER SECTORS	-	-	-	-	**139264**	-	-	-	**1433**	-
Agriculture	-	-	-	-	-	-	-	-	-	-
Comm. and Publ. Services	-	-	-	-	-	-	-	-	-	-
Residential	-	-	-	-	139264	-	-	-	1426	-
Non-specified	-	-	-	-	-	-	-	-	7	-
NON-ENERGY USE	-	-	-	-	-	-	-	-	-	-

APPROVISIONNEMENT ET DEMANDE 1997	*Gaz (TJ)*				*En. Re. Comb. & Déchets (TJ)*				(GWh)	(TJ)
	Gaz naturel	*Usines à gaz*	*Cokeries*	*Hauts fourneaux*	*Biomasse solide*	*Gaz/Liquides tirés de biomasse*	*Déchets urbains*	*Déchets industriels*	*Electricité*	*Chaleur*
Production	-	-	-	-	207415	-	-	-	6155	-
Imports	-	-	-	-	-	-	-	-	61	-
Exports	-	-	-	-	-	-	-	-	-458	-
Intl. Marine Bunkers	-	-	-	-	-	-	-	-	-	-
Stock Changes	-	-	-	-	-	-	-	-	-	-
DOMESTIC SUPPLY	-	-	-	-	**207415**	-	-	-	**5758**	-
Transfers and Stat. Diff.	-	-	-	-	-	-	-	-	667	-
TRANSFORMATION	-	-	-	-	**50702**	-	-	-	-	-
Electricity and CHP Plants	-	-	-	-	-	-	-	-	-	-
Petroleum Refineries	-	-	-	-	-	-	-	-	-	-
Other Transform. Sector	-	-	-	-	50702	-	-	-	-	-
ENERGY SECTOR	-	-	-	-	-	-	-	-	**1451**	-
DISTRIBUTION LOSSES	-	-	-	-	-	-	-	-	**13**	-
FINAL CONSUMPTION	-	-	-	-	**156713**	-	-	-	**4961**	-
INDUSTRY SECTOR	-	-	-	-	**14246**	-	-	-	**3486**	-
Iron and Steel	-	-	-	-	-	-	-	-	-	-
Chemical and Petrochem.	-	-	-	-	-	-	-	-	-	-
Non-Metallic Minerals	-	-	-	-	-	-	-	-	-	-
Non-specified	-	-	-	-	14246	-	-	-	3486	-
TRANSPORT SECTOR	-	-	-	-	-	-	-	-	-	-
Air	-	-	-	-	-	-	-	-	-	-
Road	-	-	-	-	-	-	-	-	-	-
Non-specified	-	-	-	-	-	-	-	-	-	-
OTHER SECTORS	-	-	-	-	**142467**	-	-	-	**1475**	-
Agriculture	-	-	-	-	-	-	-	-	-	-
Comm. and Publ. Services	-	-	-	-	-	-	-	-	-	-
Residential	-	-	-	-	142467	-	-	-	1475	-
Non-specified	-	-	-	-	-	-	-	-	-	-
NON-ENERGY USE	-	-	-	-	-	-	-	-	-	-

Gibraltar

SUPPLY AND CONSUMPTION 1996	Coal (1000 tonnes)							Oil (1000 tonnes)			
	Coking Coal	Other Bit. Coal	Sub-Bit. Coal	Lignite	Peat	Oven and Gas Coke	Pat. Fuel and BKB	Crude Oil	NGL	Feed-stocks	Additives
Production	-	-	-	-	-	-	-	-	-	-	-
Imports	-	-	-	-	-	-	-	-	-	-	-
Exports	-	-	-	-	-	-	-	-	-	-	-
Intl. Marine Bunkers	-	-	-	-	-	-	-	-	-	-	-
Stock Changes	-	-	-	-	-	-	-	-	-	-	-
DOMESTIC SUPPLY	-	-	-	-	-	-	-	-	-	-	-
Transfers and Stat. Diff.	-	-	-	-	-	-	-	-	-	-	-
TRANSFORMATION	-	-	-	-	-	-	-	-	-	-	-
Electricity and CHP Plants	-	-	-	-	-	-	-	-	-	-	-
Petroleum Refineries	-	-	-	-	-	-	-	-	-	-	-
Other Transform. Sector	-	-	-	-	-	-	-	-	-	-	-
ENERGY SECTOR	-	-	-	-	-	-	-	-	-	-	-
DISTRIBUTION LOSSES	-	-	-	-	-	-	-	-	-	-	-
FINAL CONSUMPTION	-	-	-	-	-	-	-	-	-	-	-
INDUSTRY SECTOR	-	-	-	-	-	-	-	-	-	-	-
Iron and Steel	-	-	-	-	-	-	-	-	-	-	-
Chemical and Petrochem.	-	-	-	-	-	-	-	-	-	-	-
Non-Metallic Minerals	-	-	-	-	-	-	-	-	-	-	-
Non-specified	-	-	-	-	-	-	-	-	-	-	-
TRANSPORT SECTOR	-	-	-	-	-	-	-	-	-	-	-
Air	-	-	-	-	-	-	-	-	-	-	-
Road	-	-	-	-	-	-	-	-	-	-	-
Non-specified	-	-	-	-	-	-	-	-	-	-	-
OTHER SECTORS	-	-	-	-	-	-	-	-	-	-	-
Agriculture	-	-	-	-	-	-	-	-	-	-	-
Comm. and Publ. Services	-	-	-	-	-	-	-	-	-	-	-
Residential	-	-	-	-	-	-	-	-	-	-	-
Non-specified	-	-	-	-	-	-	-	-	-	-	-
NON-ENERGY USE	-	-	-	-	-	-	-	-	-	-	-

APPROVISIONNEMENT ET DEMANDE 1997	*Charbon* (1000 tonnes)							*Pétrole* (1000 tonnes)			
	Charbon à coke	*Autres charb. bit.*	*Charbon sous-bit.*	*Lignite*	*Tourbe*	*Coke de four/gaz*	*Agg./briq. de lignite*	*Pétrole brut*	*LGN*	*Produits d'aliment.*	*Additifs*
Production	-	-	-	-	-	-	-	-	-	-	-
Imports	-	-	-	-	-	-	-	-	-	-	-
Exports	-	-	-	-	-	-	-	-	-	-	-
Intl. Marine Bunkers	-	-	-	-	-	-	-	-	-	-	-
Stock Changes	-	-	-	-	-	-	-	-	-	-	-
DOMESTIC SUPPLY	-	-	-	-	-	-	-	-	-	-	-
Transfers and Stat. Diff.	-	-	-	-	-	-	-	-	-	-	-
TRANSFORMATION	-	-	-	-	-	-	-	-	-	-	-
Electricity and CHP Plants	-	-	-	-	-	-	-	-	-	-	-
Petroleum Refineries	-	-	-	-	-	-	-	-	-	-	-
Other Transform. Sector	-	-	-	-	-	-	-	-	-	-	-
ENERGY SECTOR	-	-	-	-	-	-	-	-	-	-	-
DISTRIBUTION LOSSES	-	-	-	-	-	-	-	-	-	-	-
FINAL CONSUMPTION	-	-	-	-	-	-	-	-	-	-	-
INDUSTRY SECTOR	-	-	-	-	-	-	-	-	-	-	-
Iron and Steel	-	-	-	-	-	-	-	-	-	-	-
Chemical and Petrochem.	-	-	-	-	-	-	-	-	-	-	-
Non-Metallic Minerals	-	-	-	-	-	-	-	-	-	-	-
Non-specified	-	-	-	-	-	-	-	-	-	-	-
TRANSPORT SECTOR	-	-	-	-	-	-	-	-	-	-	-
Air	-	-	-	-	-	-	-	-	-	-	-
Road	-	-	-	-	-	-	-	-	-	-	-
Non-specified	-	-	-	-	-	-	-	-	-	-	-
OTHER SECTORS	-	-	-	-	-	-	-	-	-	-	-
Agriculture	-	-	-	-	-	-	-	-	-	-	-
Comm. and Publ. Services	-	-	-	-	-	-	-	-	-	-	-
Residential	-	-	-	-	-	-	-	-	-	-	-
Non-specified	-	-	-	-	-	-	-	-	-	-	-
NON-ENERGY USE	-	-	-	-	-	-	-	-	-	-	-

Gibraltar

SUPPLY AND CONSUMPTION 1996	Oil cont. (1000 tonnes)										
	Refinery Gas	LPG + Ethane	Motor Gasoline	Aviation Gasoline	Jet Fuel	Kerosene	Gas/ Diesel	Heavy Fuel Oil	Naphtha	Petrol. Coke	Other Prod.
Production	-	-	-	-	-	-	-	-	-	-	-
Imports	-	-	16	-	4	-	152	800	-	-	15
Exports	-	-	-	-	-	-	-	-	-	-	-
Intl. Marine Bunkers	-	-	-	-	-	-	-104	-750	-	-	-
Stock Changes	-	-	-	-	-	-	-	-	-	-	-
DOMESTIC SUPPLY	**-**	**-**	**16**	**-**	**4**	**-**	**48**	**50**	**-**	**-**	**15**
Transfers and Stat. Diff.	-	-	-	-	-	-	-	-	-	-	-
TRANSFORMATION	**-**	**-**	**-**	**-**	**-**	**-**	**-**	**50**	**-**	**-**	**-**
Electricity and CHP Plants	-	-	-	-	-	-	-	50	-	-	-
Petroleum Refineries	-	-	-	-	-	-	-	-	-	-	-
Other Transform. Sector	-	-	-	-	-	-	-	-	-	-	-
ENERGY SECTOR	**-**	**-**	**-**	**-**	**-**	**-**	**-**	**-**	**-**	**-**	**-**
DISTRIBUTION LOSSES	**-**	**-**	**-**	**-**	**-**	**-**	**-**	**-**	**-**	**-**	**-**
FINAL CONSUMPTION	**-**	**-**	**16**	**-**	**4**	**-**	**48**	**-**	**-**	**-**	**15**
INDUSTRY SECTOR	**-**	**-**	**-**	**-**	**-**	**-**	**-**	**-**	**-**	**-**	**-**
Iron and Steel	-	-	-	-	-	-	-	-	-	-	-
Chemical and Petrochem.	-	-	-	-	-	-	-	-	-	-	-
Non-Metallic Minerals	-	-	-	-	-	-	-	-	-	-	-
Non-specified	-	-	-	-	-	-	-	-	-	-	-
TRANSPORT SECTOR	**-**	**-**	**16**	**-**	**4**	**-**	**48**	**-**	**-**	**-**	**-**
Air	-	-	-	-	4	-	-	-	-	-	-
Road	-	-	16	-	-	-	48	-	-	-	-
Non-specified	-	-	-	-	-	-	-	-	-	-	-
OTHER SECTORS	**-**	**-**	**-**	**-**	**-**	**-**	**-**	**-**	**-**	**-**	**-**
Agriculture	-	-	-	-	-	-	-	-	-	-	-
Comm. and Publ. Services	-	-	-	-	-	-	-	-	-	-	-
Residential	-	-	-	-	-	-	-	-	-	-	-
Non-specified	-	-	-	-	-	-	-	-	-	-	-
NON-ENERGY USE	**-**	**-**	**-**	**-**	**-**	**-**	**-**	**-**	**-**	**-**	**15**

APPROVISIONNEMENT ET DEMANDE 1997	*Pétrole cont.* (1000 tonnes)										
	Gaz de raffinerie	*GPL + éthane*	*Essence moteur*	*Essence aviation*	*Carbu-réacteurs*	*Kérosène*	*Gazole*	*Fioul lourd*	*Naphta*	*Coke de pétrole*	*Autres prod.*
Production	-	-	-	-	-	-	-	-	-	-	-
Imports	-	-	16	-	4	-	152	800	-	-	15
Exports	-	-	-	-	-	-	-	-	-	-	-
Intl. Marine Bunkers	-	-	-	-	-	-	-104	-750	-	-	-
Stock Changes	-	-	-	-	-	-	-	-	-	-	-
DOMESTIC SUPPLY	**-**	**-**	**16**	**-**	**4**	**-**	**48**	**50**	**-**	**-**	**15**
Transfers and Stat. Diff.	-	-	-	-	-	-	-	-	-	-	-
TRANSFORMATION	**-**	**-**	**-**	**-**	**-**	**-**	**-**	**50**	**-**	**-**	**-**
Electricity and CHP Plants	-	-	-	-	-	-	-	50	-	-	-
Petroleum Refineries	-	-	-	-	-	-	-	-	-	-	-
Other Transform. Sector	-	-	-	-	-	-	-	-	-	-	-
ENERGY SECTOR	**-**	**-**	**-**	**-**	**-**	**-**	**-**	**-**	**-**	**-**	**-**
DISTRIBUTION LOSSES	**-**	**-**	**-**	**-**	**-**	**-**	**-**	**-**	**-**	**-**	**-**
FINAL CONSUMPTION	**-**	**-**	**16**	**-**	**4**	**-**	**48**	**-**	**-**	**-**	**15**
INDUSTRY SECTOR	**-**	**-**	**-**	**-**	**-**	**-**	**-**	**-**	**-**	**-**	**-**
Iron and Steel	-	-	-	-	-	-	-	-	-	-	-
Chemical and Petrochem.	-	-	-	-	-	-	-	-	-	-	-
Non-Metallic Minerals	-	-	-	-	-	-	-	-	-	-	-
Non-specified	-	-	-	-	-	-	-	-	-	-	-
TRANSPORT SECTOR	**-**	**-**	**16**	**-**	**4**	**-**	**48**	**-**	**-**	**-**	**-**
Air	-	-	-	-	4	-	-	-	-	-	-
Road	-	-	16	-	-	-	48	-	-	-	-
Non-specified	-	-	-	-	-	-	-	-	-	-	-
OTHER SECTORS	**-**	**-**	**-**	**-**	**-**	**-**	**-**	**-**	**-**	**-**	**-**
Agriculture	-	-	-	-	-	-	-	-	-	-	-
Comm. and Publ. Services	-	-	-	-	-	-	-	-	-	-	-
Residential	-	-	-	-	-	-	-	-	-	-	-
Non-specified	-	-	-	-	-	-	-	-	-	-	-
NON-ENERGY USE	**-**	**-**	**-**	**-**	**-**	**-**	**-**	**-**	**-**	**-**	**15**

Gibraltar

SUPPLY AND CONSUMPTION 1996	Gas (TJ)				Comb. Renew. & Waste (TJ)				(GWh)	(TJ)
	Natural Gas	Gas Works	Coke Ovens	Blast Furnaces	Solid Biomass	Gas/Liquids from Biomass	Municipal Waste	Industrial Waste	Electricity	Heat
Production	-	-	-	-	-	-	-	-	93	-
Imports	-	-	-	-	-	-	-	-	-	-
Exports	-	-	-	-	-	-	-	-	-	-
Intl. Marine Bunkers	-	-	-	-	-	-	-	-	-	-
Stock Changes	-	-	-	-	-	-	-	-	-	-
DOMESTIC SUPPLY	**-**	**-**	**-**	**-**	**-**	**-**	**-**	**-**	**93**	**-**
Transfers and Stat. Diff.	-	-	-	-	-	-	-	-	-	-
TRANSFORMATION	**-**	**-**	**-**	**-**	**-**	**-**	**-**	**-**	**-**	**-**
Electricity and CHP Plants	-	-	-	-	-	-	-	-	-	-
Petroleum Refineries	-	-	-	-	-	-	-	-	-	-
Other Transform. Sector	-	-	-	-	-	-	-	-	-	-
ENERGY SECTOR	**-**	**-**	**-**	**-**	**-**	**-**	**-**	**-**	**5**	**-**
DISTRIBUTION LOSSES	**-**	**-**	**-**	**-**	**-**	**-**	**-**	**-**	**-**	**-**
FINAL CONSUMPTION	**-**	**-**	**-**	**-**	**-**	**-**	**-**	**-**	**88**	**-**
INDUSTRY SECTOR	**-**	**-**	**-**	**-**	**-**	**-**	**-**	**-**	**-**	**-**
Iron and Steel	-	-	-	-	-	-	-	-	-	-
Chemical and Petrochem.	-	-	-	-	-	-	-	-	-	-
Non-Metallic Minerals	-	-	-	-	-	-	-	-	-	-
Non-specified	-	-	-	-	-	-	-	-	-	-
TRANSPORT SECTOR	**-**	**-**	**-**	**-**	**-**	**-**	**-**	**-**	**-**	**-**
Air	-	-	-	-	-	-	-	-	-	-
Road	-	-	-	-	-	-	-	-	-	-
Non-specified	-	-	-	-	-	-	-	-	-	-
OTHER SECTORS	**-**	**-**	**-**	**-**	**-**	**-**	**-**	**-**	**88**	**-**
Agriculture	-	-	-	-	-	-	-	-	-	-
Comm. and Publ. Services	-	-	-	-	-	-	-	-	-	-
Residential	-	-	-	-	-	-	-	-	-	-
Non-specified	-	-	-	-	-	-	-	-	88	-
NON-ENERGY USE	**-**	**-**	**-**	**-**	**-**	**-**	**-**	**-**	**-**	**-**

APPROVISIONNEMENT ET DEMANDE 1997	*Gaz* (TJ)				*En. Re. Comb. & Déchets* (TJ)				(GWh)	(TJ)
	Gaz naturel	*Usines à gaz*	*Cokeries*	*Hauts fourneaux*	*Biomasse solide*	*Gaz/Liquides tirés de biomasse*	*Déchets urbains*	*Déchets industriels*	*Electricité*	*Chaleur*
Production	-	-	-	-	-	-	-	-	93	-
Imports	-	-	-	-	-	-	-	-	-	-
Exports	-	-	-	-	-	-	-	-	-	-
Intl. Marine Bunkers	-	-	-	-	-	-	-	-	-	-
Stock Changes	-	-	-	-	-	-	-	-	-	-
DOMESTIC SUPPLY	**-**	**-**	**-**	**-**	**-**	**-**	**-**	**-**	**93**	**-**
Transfers and Stat. Diff.	-	-	-	-	-	-	-	-	-	-
TRANSFORMATION	**-**	**-**	**-**	**-**	**-**	**-**	**-**	**-**	**-**	**-**
Electricity and CHP Plants	-	-	-	-	-	-	-	-	-	-
Petroleum Refineries	-	-	-	-	-	-	-	-	-	-
Other Transform. Sector	-	-	-	-	-	-	-	-	-	-
ENERGY SECTOR	**-**	**-**	**-**	**-**	**-**	**-**	**-**	**-**	**5**	**-**
DISTRIBUTION LOSSES	**-**	**-**	**-**	**-**	**-**	**-**	**-**	**-**	**-**	**-**
FINAL CONSUMPTION	**-**	**-**	**-**	**-**	**-**	**-**	**-**	**-**	**88**	**-**
INDUSTRY SECTOR	**-**	**-**	**-**	**-**	**-**	**-**	**-**	**-**	**-**	**-**
Iron and Steel	-	-	-	-	-	-	-	-	-	-
Chemical and Petrochem.	-	-	-	-	-	-	-	-	-	-
Non-Metallic Minerals	-	-	-	-	-	-	-	-	-	-
Non-specified	-	-	-	-	-	-	-	-	-	-
TRANSPORT SECTOR	**-**	**-**	**-**	**-**	**-**	**-**	**-**	**-**	**-**	**-**
Air	-	-	-	-	-	-	-	-	-	-
Road	-	-	-	-	-	-	-	-	-	-
Non-specified	-	-	-	-	-	-	-	-	-	-
OTHER SECTORS	**-**	**-**	**-**	**-**	**-**	**-**	**-**	**-**	**88**	**-**
Agriculture	-	-	-	-	-	-	-	-	-	-
Comm. and Publ. Services	-	-	-	-	-	-	-	-	-	-
Residential	-	-	-	-	-	-	-	-	-	-
Non-specified	-	-	-	-	-	-	-	-	88	-
NON-ENERGY USE	**-**	**-**	**-**	**-**	**-**	**-**	**-**	**-**	**-**	**-**

Guatemala

SUPPLY AND CONSUMPTION 1996	Coal (1000 tonnes)							Oil (1000 tonnes)			
	Coking Coal	Other Bit. Coal	Sub-Bit. Coal	Lignite	Peat	Oven and Gas Coke	Pat. Fuel and BKB	Crude Oil	NGL	Feed-stocks	Additives
Production	-	-	-	-	-	-	-	795	-	-	-
Imports	-	-	-	-	-	-	-	794	-	-	-
Exports	-	-	-	-	-	-	-	-718	-	-	-
Intl. Marine Bunkers	-	-	-	-	-	-	-	-	-	-	-
Stock Changes	-	-	-	-	-	-	-	-	-	-	-
DOMESTIC SUPPLY	**-**	**-**	**-**	**-**	**-**	**-**	**-**	**871**	**-**	**-**	**-**
Transfers and Stat. Diff.	-	-	-	-	-	-	-	-80	-	-	-
TRANSFORMATION	**-**	**-**	**-**	**-**	**-**	**-**	**-**	**791**	**-**	**-**	**-**
Electricity and CHP Plants	-	-	-	-	-	-	-	-	-	-	-
Petroleum Refineries	-	-	-	-	-	-	-	791	-	-	-
Other Transform. Sector	-	-	-	-	-	-	-	-	-	-	-
ENERGY SECTOR	**-**	**-**	**-**	**-**	**-**	**-**	**-**	**-**	**-**	**-**	**-**
DISTRIBUTION LOSSES	**-**	**-**	**-**	**-**	**-**	**-**	**-**	**-**	**-**	**-**	**-**
FINAL CONSUMPTION	**-**	**-**	**-**	**-**	**-**	**-**	**-**	**-**	**-**	**-**	**-**
INDUSTRY SECTOR	**-**	**-**	**-**	**-**	**-**	**-**	**-**	**-**	**-**	**-**	**-**
Iron and Steel	-	-	-	-	-	-	-	-	-	-	-
Chemical and Petrochem.	-	-	-	-	-	-	-	-	-	-	-
Non-Metallic Minerals	-	-	-	-	-	-	-	-	-	-	-
Non-specified	-	-	-	-	-	-	-	-	-	-	-
TRANSPORT SECTOR	**-**	**-**	**-**	**-**	**-**	**-**	**-**	**-**	**-**	**-**	**-**
Air	-	-	-	-	-	-	-	-	-	-	-
Road	-	-	-	-	-	-	-	-	-	-	-
Non-specified	-	-	-	-	-	-	-	-	-	-	-
OTHER SECTORS	**-**	**-**	**-**	**-**	**-**	**-**	**-**	**-**	**-**	**-**	**-**
Agriculture	-	-	-	-	-	-	-	-	-	-	-
Comm. and Publ. Services	-	-	-	-	-	-	-	-	-	-	-
Residential	-	-	-	-	-	-	-	-	-	-	-
Non-specified	-	-	-	-	-	-	-	-	-	-	-
NON-ENERGY USE	**-**	**-**	**-**	**-**	**-**	**-**	**-**	**-**	**-**	**-**	**-**

APPROVISIONNEMENT ET DEMANDE 1997	*Charbon* (1000 tonnes)							*Pétrole* (1000 tonnes)			
	Charbon à coke	*Autres charb. bit.*	*Charbon sous-bit.*	*Lignite*	*Tourbe*	*Coke de four/gaz*	*Agg./briq. de lignite*	*Pétrole brut*	*LGN*	*Produits d'aliment.*	*Additifs*
Production	-	-	-	-	-	-	-	1065	-	-	-
Imports	-	-	-	-	-	-	-	839	-	-	-
Exports	-	-	-	-	-	-	-	-974	-	-	-
Intl. Marine Bunkers	-	-	-	-	-	-	-	-	-	-	-
Stock Changes	-	-	-	-	-	-	-	-	-	-	-
DOMESTIC SUPPLY	**-**	**-**	**-**	**-**	**-**	**-**	**-**	**930**	**-**	**-**	**-**
Transfers and Stat. Diff.	-	-	-	-	-	-	-	-83	-	-	-
TRANSFORMATION	**-**	**-**	**-**	**-**	**-**	**-**	**-**	**847**	**-**	**-**	**-**
Electricity and CHP Plants	-	-	-	-	-	-	-	-	-	-	-
Petroleum Refineries	-	-	-	-	-	-	-	847	-	-	-
Other Transform. Sector	-	-	-	-	-	-	-	-	-	-	-
ENERGY SECTOR	**-**	**-**	**-**	**-**	**-**	**-**	**-**	**-**	**-**	**-**	**-**
DISTRIBUTION LOSSES	**-**	**-**	**-**	**-**	**-**	**-**	**-**	**-**	**-**	**-**	**-**
FINAL CONSUMPTION	**-**	**-**	**-**	**-**	**-**	**-**	**-**	**-**	**-**	**-**	**-**
INDUSTRY SECTOR	**-**	**-**	**-**	**-**	**-**	**-**	**-**	**-**	**-**	**-**	**-**
Iron and Steel	-	-	-	-	-	-	-	-	-	-	-
Chemical and Petrochem.	-	-	-	-	-	-	-	-	-	-	-
Non-Metallic Minerals	-	-	-	-	-	-	-	-	-	-	-
Non-specified	-	-	-	-	-	-	-	-	-	-	-
TRANSPORT SECTOR	**-**	**-**	**-**	**-**	**-**	**-**	**-**	**-**	**-**	**-**	**-**
Air	-	-	-	-	-	-	-	-	-	-	-
Road	-	-	-	-	-	-	-	-	-	-	-
Non-specified	-	-	-	-	-	-	-	-	-	-	-
OTHER SECTORS	**-**	**-**	**-**	**-**	**-**	**-**	**-**	**-**	**-**	**-**	**-**
Agriculture	-	-	-	-	-	-	-	-	-	-	-
Comm. and Publ. Services	-	-	-	-	-	-	-	-	-	-	-
Residential	-	-	-	-	-	-	-	-	-	-	-
Non-specified	-	-	-	-	-	-	-	-	-	-	-
NON-ENERGY USE	**-**	**-**	**-**	**-**	**-**	**-**	**-**	**-**	**-**	**-**	**-**

Guatemala

SUPPLY AND CONSUMPTION 1996	Oil cont. (1000 tonnes)										
	Refinery Gas	LPG + Ethane	Motor Gasoline	Aviation Gasoline	Jet Fuel	Kerosene	Gas/ Diesel	Heavy Fuel Oil	Naphtha	Petrol. Coke	Other Prod.
Production	20	7	133	-	16	16	248	277	-	-	68
Imports	-	144	411	-	29	28	506	199	-	-	28
Exports	-	-	-	-	-	-	-	-	-	-	-27
Intl. Marine Bunkers	-	-	-	-	-	-	-120	-	-	-	-
Stock Changes	-	2	-18	-	3	3	-4	28	-	-	-
DOMESTIC SUPPLY	**20**	**153**	**526**	**-**	**48**	**47**	**630**	**504**	**-**	**-**	**69**
Transfers and Stat. Diff.	-	-	-	-	-1	-1	-	-1	-	-	-1
TRANSFORMATION	**-**	**-**	**-**	**-**	**-**	**-**	**90**	**216**	**-**	**-**	**-**
Electricity and CHP Plants	-	-	-	-	-	-	90	216	-	-	-
Petroleum Refineries	-	-	-	-	-	-	-	-	-	-	-
Other Transform. Sector	-	-	-	-	-	-	-	-	-	-	-
ENERGY SECTOR	**20**	**-**	**-**	**-**	**-**	**-**	**-**	**-**	**-**	**-**	**-**
DISTRIBUTION LOSSES	**-**	**-**	**-**	**-**	**-**	**-**	**-**	**-**	**-**	**-**	**-**
FINAL CONSUMPTION	**-**	**153**	**526**	**-**	**47**	**46**	**540**	**287**	**-**	**-**	**68**
INDUSTRY SECTOR	**-**	**20**	**-**	**-**	**-**	**5**	**80**	**284**	**-**	**-**	**-**
Iron and Steel	-	-	-	-	-	-	-	-	-	-	-
Chemical and Petrochem.	-	-	-	-	-	-	-	-	-	-	-
Non-Metallic Minerals	-	-	-	-	-	-	-	-	-	-	-
Non-specified	-	20	-	-	-	5	80	284	-	-	-
TRANSPORT SECTOR	**-**	**1**	**526**	**-**	**47**	**-**	**383**	**-**	**-**	**-**	**-**
Air	-	-	-	-	47	-	-	-	-	-	-
Road	-	-	526	-	-	-	383	-	-	-	-
Non-specified	-	1	-	-	-	-	-	-	-	-	-
OTHER SECTORS	**-**	**132**	**-**	**-**	**-**	**41**	**77**	**3**	**-**	**-**	**-**
Agriculture	-	2	-	-	-	-	32	-	-	-	-
Comm. and Publ. Services	-	22	-	-	-	1	45	3	-	-	-
Residential	-	108	-	-	-	40	-	-	-	-	-
Non-specified	-	-	-	-	-	-	-	-	-	-	-
NON-ENERGY USE	**-**	**-**	**-**	**-**	**-**	**-**	**-**	**-**	**-**	**-**	**68**

APPROVISIONNEMENT ET DEMANDE 1997	*Pétrole cont.* (1000 tonnes)										
	Gaz de raffinerie	*GPL + éthane*	*Essence moteur*	*Essence aviation*	*Carbu- réacteurs*	*Kérosène*	*Gazole*	*Fioul lourd*	*Naphta*	*Coke de pétrole*	*Autres prod.*
Production	21	7	139	-	17	16	266	283	-	-	73
Imports	-	148	430	-	31	31	542	203	-	-	30
Exports	-	-	-	-	-	-	-	-	-	-	-29
Intl. Marine Bunkers	-	-	-	-	-	-	-120	-	-	-	-
Stock Changes	-	2	-19	-	3	3	-5	29	-	-	-
DOMESTIC SUPPLY	**21**	**157**	**550**	**-**	**51**	**50**	**683**	**515**	**-**	**-**	**74**
Transfers and Stat. Diff.	-	2	-	-	-1	-1	1	-	-	-	-1
TRANSFORMATION	**-**	**-**	**-**	**-**	**-**	**-**	**97**	**221**	**-**	**-**	**-**
Electricity and CHP Plants	-	-	-	-	-	-	97	221	-	-	-
Petroleum Refineries	-	-	-	-	-	-	-	-	-	-	-
Other Transform. Sector	-	-	-	-	-	-	-	-	-	-	-
ENERGY SECTOR	**21**	**-**	**-**	**-**	**-**	**-**	**-**	**-**	**-**	**-**	**-**
DISTRIBUTION LOSSES	**-**	**-**	**-**	**-**	**-**	**-**	**-**	**-**	**-**	**-**	**-**
FINAL CONSUMPTION	**-**	**159**	**550**	**-**	**50**	**49**	**587**	**294**	**-**	**-**	**73**
INDUSTRY SECTOR	**-**	**21**	**-**	**-**	**-**	**5**	**86**	**290**	**-**	**-**	**-**
Iron and Steel	-	-	-	-	-	-	-	-	-	-	-
Chemical and Petrochem.	-	-	-	-	-	-	-	-	-	-	-
Non-Metallic Minerals	-	-	-	-	-	-	-	-	-	-	-
Non-specified	-	21	-	-	-	5	86	290	-	-	-
TRANSPORT SECTOR	**-**	**1**	**550**	**-**	**50**	**-**	**418**	**-**	**-**	**-**	**-**
Air	-	-	-	-	50	-	-	-	-	-	-
Road	-	-	550	-	-	-	418	-	-	-	-
Non-specified	-	1	-	-	-	-	-	-	-	-	-
OTHER SECTORS	**-**	**137**	**-**	**-**	**-**	**44**	**83**	**4**	**-**	**-**	**-**
Agriculture	-	2	-	-	-	-	34	-	-	-	-
Comm. and Publ. Services	-	23	-	-	-	1	49	4	-	-	-
Residential	-	112	-	-	-	43	-	-	-	-	-
Non-specified	-	-	-	-	-	-	-	-	-	-	-
NON-ENERGY USE	**-**	**-**	**-**	**-**	**-**	**-**	**-**	**-**	**-**	**-**	**73**

Guatemala

SUPPLY AND CONSUMPTION 1996	Gas (TJ)				Comb. Renew. & Waste (TJ)				(GWh)	(TJ)
	Natural Gas	Gas Works	Coke Ovens	Blast Furnaces	Solid Biomass	Gas/Liquids from Biomass	Municipal Waste	Industrial Waste	Electricity	Heat
Production	-	-	-	-	124866	-	-	-	4247	-
Imports	-	-	-	-	-	-	-	-	-	-
Exports	-	-	-	-	-	-	-	-	-	-
Intl. Marine Bunkers	-	-	-	-	-	-	-	-	-	-
Stock Changes	-	-	-	-	-	-	-	-	-	-
DOMESTIC SUPPLY	-	-	-	-	**124866**	-	-	-	**4247**	-
Transfers and Stat. Diff.	-	-	-	-	-	-	-	-	-1	-
TRANSFORMATION	-	-	-	-	**6316**	-	-	-	-	-
Electricity and CHP Plants	-	-	-	-	3962	-	-	-	-	-
Petroleum Refineries	-	-	-	-	105	-	-	-	-	-
Other Transform. Sector	-	-	-	-	2249	-	-	-	-	-
ENERGY SECTOR	-	-	-	-	-	-	-	-	**15**	-
DISTRIBUTION LOSSES	-	-	-	-	-	-	-	-	**549**	-
FINAL CONSUMPTION	-	-	-	-	**118550**	-	-	-	**3682**	-
INDUSTRY SECTOR	-	-	-	-	**7561**	-	-	-	**1247**	-
Iron and Steel	-	-	-	-	-	-	-	-	-	-
Chemical and Petrochem.	-	-	-	-	-	-	-	-	-	-
Non-Metallic Minerals	-	-	-	-	-	-	-	-	-	-
Non-specified	-	-	-	-	7561	-	-	-	1247	-
TRANSPORT SECTOR	-	-	-	-	-	-	-	-	-	-
Air	-	-	-	-	-	-	-	-	-	-
Road	-	-	-	-	-	-	-	-	-	-
Non-specified	-	-	-	-	-	-	-	-	-	-
OTHER SECTORS	-	-	-	-	**110989**	-	-	-	**2435**	-
Agriculture	-	-	-	-	-	-	-	-	-	-
Comm. and Publ. Services	-	-	-	-	-	-	-	-	1242	-
Residential	-	-	-	-	110989	-	-	-	1193	-
Non-specified	-	-	-	-	-	-	-	-	-	-
NON-ENERGY USE	-	-	-	-	-	-	-	-	-	-

APPROVISIONNEMENT ET DEMANDE 1997	*Gaz* (TJ)				*En. Re. Comb. & Déchets* (TJ)				(GWh)	(TJ)
	Gaz naturel	*Usines à gaz*	*Cokeries*	*Hauts fourneaux*	*Biomasse solide*	*Gaz/Liquides tirés de biomasse*	*Déchets urbains*	*Déchets industriels*	*Electricité*	*Chaleur*
Production	-	-	-	-	126879	-	-	-	4897	-
Imports	-	-	-	-	-	-	-	-	-	-
Exports	-	-	-	-	-	-	-	-	-	-
Intl. Marine Bunkers	-	-	-	-	-	-	-	-	-	-
Stock Changes	-	-	-	-	-	-	-	-	-	-
DOMESTIC SUPPLY	-	-	-	-	**126879**	-	-	-	**4897**	-
Transfers and Stat. Diff.	-	-	-	-	-	-	-	-	-	-
TRANSFORMATION	-	-	-	-	**6403**	-	-	-	-	-
Electricity and CHP Plants	-	-	-	-	4032	-	-	-	-	-
Petroleum Refineries	-	-	-	-	105	-	-	-	-	-
Other Transform. Sector	-	-	-	-	2266	-	-	-	-	-
ENERGY SECTOR	-	-	-	-	-	-	-	-	**17**	-
DISTRIBUTION LOSSES	-	-	-	-	-	-	-	-	**634**	-
FINAL CONSUMPTION	-	-	-	-	**120475**	-	-	-	**4246**	-
INDUSTRY SECTOR	-	-	-	-	**7679**	-	-	-	**1438**	-
Iron and Steel	-	-	-	-	-	-	-	-	-	-
Chemical and Petrochem.	-	-	-	-	-	-	-	-	-	-
Non-Metallic Minerals	-	-	-	-	-	-	-	-	-	-
Non-specified	-	-	-	-	7679	-	-	-	1438	-
TRANSPORT SECTOR	-	-	-	-	-	-	-	-	-	-
Air	-	-	-	-	-	-	-	-	-	-
Road	-	-	-	-	-	-	-	-	-	-
Non-specified	-	-	-	-	-	-	-	-	-	-
OTHER SECTORS	-	-	-	-	**112796**	-	-	-	**2808**	-
Agriculture	-	-	-	-	-	-	-	-	-	-
Comm. and Publ. Services	-	-	-	-	-	-	-	-	1432	-
Residential	-	-	-	-	112796	-	-	-	1376	-
Non-specified	-	-	-	-	-	-	-	-	-	-
NON-ENERGY USE	-	-	-	-	-	-	-	-	-	-

Haiti

SUPPLY AND CONSUMPTION 1996	Coal (1000 tonnes)							Oil (1000 tonnes)			
	Coking Coal	Other Bit. Coal	Sub-Bit. Coal	Lignite	Peat	Oven and Gas Coke	Pat. Fuel and BKB	Crude Oil	NGL	Feed-stocks	Additives
Production	-	-	-	-	-	-	-	-	-	-	-
Imports	-	-	-	-	-	-	-	-	-	-	-
Exports	-	-	-	-	-	-	-	-	-	-	-
Intl. Marine Bunkers	-	-	-	-	-	-	-	-	-	-	-
Stock Changes	-	-	-	-	-	-	-	-	-	-	-
DOMESTIC SUPPLY	-	-	-	-	-	-	-	-	-	-	-
Transfers and Stat. Diff.	-	-	-	-	-	-	-	-	-	-	-
TRANSFORMATION	-	-	-	-	-	-	-	-	-	-	-
Electricity and CHP Plants	-	-	-	-	-	-	-	-	-	-	-
Petroleum Refineries	-	-	-	-	-	-	-	-	-	-	-
Other Transform. Sector	-	-	-	-	-	-	-	-	-	-	-
ENERGY SECTOR	-	-	-	-	-	-	-	-	-	-	-
DISTRIBUTION LOSSES	-	-	-	-	-	-	-	-	-	-	-
FINAL CONSUMPTION	-	-	-	-	-	-	-	-	-	-	-
INDUSTRY SECTOR	-	-	-	-	-	-	-	-	-	-	-
Iron and Steel	-	-	-	-	-	-	-	-	-	-	-
Chemical and Petrochem.	-	-	-	-	-	-	-	-	-	-	-
Non-Metallic Minerals	-	-	-	-	-	-	-	-	-	-	-
Non-specified	-	-	-	-	-	-	-	-	-	-	-
TRANSPORT SECTOR	-	-	-	-	-	-	-	-	-	-	-
Air	-	-	-	-	-	-	-	-	-	-	-
Road	-	-	-	-	-	-	-	-	-	-	-
Non-specified	-	-	-	-	-	-	-	-	-	-	-
OTHER SECTORS	-	-	-	-	-	-	-	-	-	-	-
Agriculture	-	-	-	-	-	-	-	-	-	-	-
Comm. and Publ. Services	-	-	-	-	-	-	-	-	-	-	-
Residential	-	-	-	-	-	-	-	-	-	-	-
Non-specified	-	-	-	-	-	-	-	-	-	-	-
NON-ENERGY USE	-	-	-	-	-	-	-	-	-	-	-

APPROVISIONNEMENT ET DEMANDE 1997	*Charbon* (1000 tonnes)							*Pétrole* (1000 tonnes)			
	Charbon à coke	*Autres charb. bit.*	*Charbon sous-bit.*	*Lignite*	*Tourbe*	*Coke de four/gaz*	*Agg./briq. de lignite*	*Pétrole brut*	*LGN*	*Produits d'aliment.*	*Additifs*
Production	-	-	-	-	-	-	-	-	-	-	-
Imports	-	-	-	-	-	-	-	-	-	-	-
Exports	-	-	-	-	-	-	-	-	-	-	-
Intl. Marine Bunkers	-	-	-	-	-	-	-	-	-	-	-
Stock Changes	-	-	-	-	-	-	-	-	-	-	-
DOMESTIC SUPPLY	-	-	-	-	-	-	-	-	-	-	-
Transfers and Stat. Diff.	-	-	-	-	-	-	-	-	-	-	-
TRANSFORMATION	-	-	-	-	-	-	-	-	-	-	-
Electricity and CHP Plants	-	-	-	-	-	-	-	-	-	-	-
Petroleum Refineries	-	-	-	-	-	-	-	-	-	-	-
Other Transform. Sector	-	-	-	-	-	-	-	-	-	-	-
ENERGY SECTOR	-	-	-	-	-	-	-	-	-	-	-
DISTRIBUTION LOSSES	-	-	-	-	-	-	-	-	-	-	-
FINAL CONSUMPTION	-	-	-	-	-	-	-	-	-	-	-
INDUSTRY SECTOR	-	-	-	-	-	-	-	-	-	-	-
Iron and Steel	-	-	-	-	-	-	-	-	-	-	-
Chemical and Petrochem.	-	-	-	-	-	-	-	-	-	-	-
Non-Metallic Minerals	-	-	-	-	-	-	-	-	-	-	-
Non-specified	-	-	-	-	-	-	-	-	-	-	-
TRANSPORT SECTOR	-	-	-	-	-	-	-	-	-	-	-
Air	-	-	-	-	-	-	-	-	-	-	-
Road	-	-	-	-	-	-	-	-	-	-	-
Non-specified	-	-	-	-	-	-	-	-	-	-	-
OTHER SECTORS	-	-	-	-	-	-	-	-	-	-	-
Agriculture	-	-	-	-	-	-	-	-	-	-	-
Comm. and Publ. Services	-	-	-	-	-	-	-	-	-	-	-
Residential	-	-	-	-	-	-	-	-	-	-	-
Non-specified	-	-	-	-	-	-	-	-	-	-	-
NON-ENERGY USE	-	-	-	-	-	-	-	-	-	-	-

Haiti

SUPPLY AND CONSUMPTION 1996	Oil cont. (1000 tonnes)										
	Refinery Gas	LPG + Ethane	Motor Gasoline	Aviation Gasoline	Jet Fuel	Kerosene	Gas/ Diesel	Heavy Fuel Oil	Naphtha	Petrol. Coke	Other Prod.
Production	-	-	-	-	-	-	-	-	-	-	-
Imports	-	6	77	-	19	28	197	24	-	-	2
Exports	-	-	-	-	-	-	-	-	-	-	-
Intl. Marine Bunkers	-	-	-	-	-	-	-	-	-	-	-
Stock Changes	-	-	-	-	-	-	-	-	-	-	-
DOMESTIC SUPPLY	**-**	**6**	**77**	**-**	**19**	**28**	**197**	**24**	**-**	**-**	**2**
Transfers and Stat. Diff.	-	-	1	-	-	-	1	-	-	-	-
TRANSFORMATION	**-**	**-**	**-**	**-**	**-**	**-**	**68**	**8**	**-**	**-**	**-**
Electricity and CHP Plants	-	-	-	-	-	-	68	8	-	-	-
Petroleum Refineries	-	-	-	-	-	-	-	-	-	-	-
Other Transform. Sector	-	-	-	-	-	-	-	-	-	-	-
ENERGY SECTOR	**-**	**-**	**-**	**-**	**-**	**-**	**-**	**-**	**-**	**-**	**-**
DISTRIBUTION LOSSES	**-**	**-**	**-**	**-**	**-**	**-**	**-**	**-**	**-**	**-**	**-**
FINAL CONSUMPTION	**-**	**6**	**78**	**-**	**19**	**28**	**130**	**16**	**-**	**-**	**2**
INDUSTRY SECTOR	**-**	**-**	**-**	**-**	**-**	**3**	**30**	**16**	**-**	**-**	**-**
Iron and Steel	-	-	-	-	-	-	-	-	-	-	-
Chemical and Petrochem.	-	-	-	-	-	-	-	-	-	-	-
Non-Metallic Minerals	-	-	-	-	-	-	-	-	-	-	-
Non-specified	-	-	-	-	-	3	30	16	-	-	-
TRANSPORT SECTOR	**-**	**-**	**78**	**-**	**19**	**-**	**100**	**-**	**-**	**-**	**-**
Air	-	-	-	-	19	-	-	-	-	-	-
Road	-	-	78	-	-	-	-	-	-	-	-
Non-specified	-	-	-	-	-	-	100	-	-	-	-
OTHER SECTORS	**-**	**6**	**-**	**-**	**-**	**25**	**-**	**-**	**-**	**-**	**-**
Agriculture	-	-	-	-	-	-	-	-	-	-	-
Comm. and Publ. Services	-	-	-	-	-	-	-	-	-	-	-
Residential	-	6	-	-	-	25	-	-	-	-	-
Non-specified	-	-	-	-	-	-	-	-	-	-	-
NON-ENERGY USE	**-**	**-**	**-**	**-**	**-**	**-**	**-**	**-**	**-**	**-**	**2**

APPROVISIONNEMENT ET DEMANDE 1997	*Pétrole cont.* (1000 tonnes)										
	Gaz de raffinerie	*GPL + éthane*	*Essence moteur*	*Essence aviation*	*Carbu- réacteurs*	*Kérosène*	*Gazole*	*Fioul lourd*	*Naphta*	*Coke de pétrole*	*Autres prod.*
Production	-	-	-	-	-	-	-	-	-	-	-
Imports	-	6	85	-	19	34	209	106	-	-	9
Exports	-	-	-	-	-	-	-	-	-	-	-
Intl. Marine Bunkers	-	-	-	-	-	-	-	-	-	-	-
Stock Changes	-	-	-	-	-	-	-	-	-	-	-
DOMESTIC SUPPLY	**-**	**6**	**85**	**-**	**19**	**34**	**209**	**106**	**-**	**-**	**9**
Transfers and Stat. Diff.	-	-	-	-	-	1	-	-38	-	-	-
TRANSFORMATION	**-**	**-**	**-**	**-**	**-**	**-**	**50**	**60**	**-**	**-**	**-**
Electricity and CHP Plants	-	-	-	-	-	-	50	60	-	-	-
Petroleum Refineries	-	-	-	-	-	-	-	-	-	-	-
Other Transform. Sector	-	-	-	-	-	-	-	-	-	-	-
ENERGY SECTOR	**-**	**-**	**-**	**-**	**-**	**-**	**-**	**-**	**-**	**-**	**-**
DISTRIBUTION LOSSES	**-**	**-**	**-**	**-**	**-**	**-**	**-**	**-**	**-**	**-**	**-**
FINAL CONSUMPTION	**-**	**6**	**85**	**-**	**19**	**35**	**159**	**8**	**-**	**-**	**9**
INDUSTRY SECTOR	**-**	**-**	**-**	**-**	**-**	**8**	**61**	**8**	**-**	**-**	**-**
Iron and Steel	-	-	-	-	-	-	-	-	-	-	-
Chemical and Petrochem.	-	-	-	-	-	-	-	-	-	-	-
Non-Metallic Minerals	-	-	-	-	-	-	-	-	-	-	-
Non-specified	-	-	-	-	-	8	61	8	-	-	-
TRANSPORT SECTOR	**-**	**-**	**85**	**-**	**19**	**-**	**98**	**-**	**-**	**-**	**-**
Air	-	-	-	-	19	-	-	-	-	-	-
Road	-	-	85	-	-	-	-	-	-	-	-
Non-specified	-	-	-	-	-	-	98	-	-	-	-
OTHER SECTORS	**-**	**6**	**-**	**-**	**-**	**27**	**-**	**-**	**-**	**-**	**-**
Agriculture	-	-	-	-	-	-	-	-	-	-	-
Comm. and Publ. Services	-	-	-	-	-	-	-	-	-	-	-
Residential	-	6	-	-	-	27	-	-	-	-	-
Non-specified	-	-	-	-	-	-	-	-	-	-	-
NON-ENERGY USE	**-**	**-**	**-**	**-**	**-**	**-**	**-**	**-**	**-**	**-**	**9**

Haiti

SUPPLY AND CONSUMPTION 1996	Gas (TJ)				Comb. Renew. & Waste (TJ)				(GWh)	(TJ)
	Natural Gas	Gas Works	Coke Ovens	Blast Furnaces	Solid Biomass	Gas/Liquids from Biomass	Municipal Waste	Industrial Waste	Electricity	Heat
Production	-	-	-	-	65949	-	-	-	625	-
Imports	-	-	-	-	-	-	-	-	-	-
Exports	-	-	-	-	-	-	-	-	-	-
Intl. Marine Bunkers	-	-	-	-	-	-	-	-	-	-
Stock Changes	-	-	-	-	-	-	-	-	-	-
DOMESTIC SUPPLY	**-**	**-**	**-**	**-**	**65949**	**-**	**-**	**-**	**625**	**-**
Transfers and Stat. Diff.	-	-	-	-	-	-	-	-	-23	-
TRANSFORMATION	**-**	**-**	**-**	**-**	**12504**	**-**	**-**	**-**	**-**	**-**
Electricity and CHP Plants	-	-	-	-	400	-	-	-	-	-
Petroleum Refineries	-	-	-	-	-	-	-	-	-	-
Other Transform. Sector	-	-	-	-	12104	-	-	-	-	-
ENERGY SECTOR	**-**	**-**	**-**	**-**	**-**	**-**	**-**	**-**	**12**	**-**
DISTRIBUTION LOSSES	**-**	**-**	**-**	**-**	**-**	**-**	**-**	**-**	**338**	**-**
FINAL CONSUMPTION	**-**	**-**	**-**	**-**	**53445**	**-**	**-**	**-**	**252**	**-**
INDUSTRY SECTOR	**-**	**-**	**-**	**-**	**2954**	**-**	**-**	**-**	**113**	**-**
Iron and Steel	-	-	-	-	-	-	-	-	-	-
Chemical and Petrochem.	-	-	-	-	-	-	-	-	-	-
Non-Metallic Minerals	-	-	-	-	-	-	-	-	-	-
Non-specified	-	-	-	-	2954	-	-	-	113	-
TRANSPORT SECTOR	**-**	**-**	**-**	**-**	**-**	**-**	**-**	**-**	**-**	**-**
Air	-	-	-	-	-	-	-	-	-	-
Road	-	-	-	-	-	-	-	-	-	-
Non-specified	-	-	-	-	-	-	-	-	-	-
OTHER SECTORS	**-**	**-**	**-**	**-**	**50491**	**-**	**-**	**-**	**139**	**-**
Agriculture	-	-	-	-	-	-	-	-	-	-
Comm. and Publ. Services	-	-	-	-	2022	-	-	-	23	-
Residential	-	-	-	-	48469	-	-	-	116	-
Non-specified	-	-	-	-	-	-	-	-	-	-
NON-ENERGY USE	**-**	**-**	**-**	**-**	**-**	**-**	**-**	**-**	**-**	**-**

APPROVISIONNEMENT ET DEMANDE 1997	*Gaz* (TJ)				*En. Re. Comb. & Déchets* (TJ)				(GWh)	(TJ)
	Gaz naturel	*Usines à gaz*	*Cokeries*	*Hauts fourneaux*	*Biomasse solide*	*Gaz/Liquides tirés de biomasse*	*Déchets urbains*	*Déchets industriels*	*Electricité*	*Chaleur*
Production	-	-	-	-	53647	-	-	-	625	-
Imports	-	-	-	-	-	-	-	-	-	-
Exports	-	-	-	-	-	-	-	-	-	-
Intl. Marine Bunkers	-	-	-	-	-	-	-	-	-	-
Stock Changes	-	-	-	-	-	-	-	-	-	-
DOMESTIC SUPPLY	**-**	**-**	**-**	**-**	**53647**	**-**	**-**	**-**	**625**	**-**
Transfers and Stat. Diff.	-	-	-	-	1	-	-	-	-35	-
TRANSFORMATION	**-**	**-**	**-**	**-**	**12921**	**-**	**-**	**-**	**-**	**-**
Electricity and CHP Plants	-	-	-	-	400	-	-	-	-	-
Petroleum Refineries	-	-	-	-	-	-	-	-	-	-
Other Transform. Sector	-	-	-	-	12521	-	-	-	-	-
ENERGY SECTOR	**-**	**-**	**-**	**-**	**-**	**-**	**-**	**-**	**12**	**-**
DISTRIBUTION LOSSES	**-**	**-**	**-**	**-**	**-**	**-**	**-**	**-**	**267**	**-**
FINAL CONSUMPTION	**-**	**-**	**-**	**-**	**40727**	**-**	**-**	**-**	**311**	**-**
INDUSTRY SECTOR	**-**	**-**	**-**	**-**	**2767**	**-**	**-**	**-**	**125**	**-**
Iron and Steel	-	-	-	-	-	-	-	-	-	-
Chemical and Petrochem.	-	-	-	-	-	-	-	-	-	-
Non-Metallic Minerals	-	-	-	-	-	-	-	-	-	-
Non-specified	-	-	-	-	2767	-	-	-	125	-
TRANSPORT SECTOR	**-**	**-**	**-**	**-**	**-**	**-**	**-**	**-**	**-**	**-**
Air	-	-	-	-	-	-	-	-	-	-
Road	-	-	-	-	-	-	-	-	-	-
Non-specified	-	-	-	-	-	-	-	-	-	-
OTHER SECTORS	**-**	**-**	**-**	**-**	**37960**	**-**	**-**	**-**	**186**	**-**
Agriculture	-	-	-	-	-	-	-	-	-	-
Comm. and Publ. Services	-	-	-	-	2092	-	-	-	105	-
Residential	-	-	-	-	35868	-	-	-	81	-
Non-specified	-	-	-	-	-	-	-	-	-	-
NON-ENERGY USE	**-**	**-**	**-**	**-**	**-**	**-**	**-**	**-**	**-**	**-**

Honduras

SUPPLY AND CONSUMPTION 1996	Coal (1000 tonnes)							Oil (1000 tonnes)			
	Coking Coal	Other Bit. Coal	Sub-Bit. Coal	Lignite	Peat	Oven and Gas Coke	Pat. Fuel and BKB	Crude Oil	NGL	Feed-stocks	Additives
Production	-	-	-	-	-	-	-	-	-	-	-
Imports	-	-	-	-	-	1	-	-	-	-	-
Exports	-	-	-	-	-	-	-	-	-	-	-
Intl. Marine Bunkers	-	-	-	-	-	-	-	-	-	-	-
Stock Changes	-	-	-	-	-	-	-	-	-	-	-
DOMESTIC SUPPLY	-	-	-	-	-	1	-	-	-	-	-
Transfers and Stat. Diff.	-	-	-	-	-	-	-	-	-	-	-
TRANSFORMATION	-	-	-	-	-	1	-	-	-	-	-
Electricity and CHP Plants	-	-	-	-	-	-	-	-	-	-	-
Petroleum Refineries	-	-	-	-	-	-	-	-	-	-	-
Other Transform. Sector	-	-	-	-	-	1	-	-	-	-	-
ENERGY SECTOR	-	-	-	-	-	-	-	-	-	-	-
DISTRIBUTION LOSSES	-	-	-	-	-	-	-	-	-	-	-
FINAL CONSUMPTION	-	-	-	-	-	-	-	-	-	-	-
INDUSTRY SECTOR	-	-	-	-	-	-	-	-	-	-	-
Iron and Steel	-	-	-	-	-	-	-	-	-	-	-
Chemical and Petrochem.	-	-	-	-	-	-	-	-	-	-	-
Non-Metallic Minerals	-	-	-	-	-	-	-	-	-	-	-
Non-specified	-	-	-	-	-	-	-	-	-	-	-
TRANSPORT SECTOR	-	-	-	-	-	-	-	-	-	-	-
Air	-	-	-	-	-	-	-	-	-	-	-
Road	-	-	-	-	-	-	-	-	-	-	-
Non-specified	-	-	-	-	-	-	-	-	-	-	-
OTHER SECTORS	-	-	-	-	-	-	-	-	-	-	-
Agriculture	-	-	-	-	-	-	-	-	-	-	-
Comm. and Publ. Services	-	-	-	-	-	-	-	-	-	-	-
Residential	-	-	-	-	-	-	-	-	-	-	-
Non-specified	-	-	-	-	-	-	-	-	-	-	-
NON-ENERGY USE	-	-	-	-	-	-	-	-	-	-	-

APPROVISIONNEMENT ET DEMANDE 1997	*Charbon* (1000 tonnes)							*Pétrole* (1000 tonnes)			
	Charbon à coke	*Autres charb. bit.*	*Charbon sous-bit.*	*Lignite*	*Tourbe*	*Coke de four/gaz*	*Agg./briq. de lignite*	*Pétrole brut*	*LGN*	*Produits d'aliment.*	*Additifs*
Production	-	-	-	-	-	-	-	-	-	-	-
Imports	-	-	-	-	-	1	-	-	-	-	-
Exports	-	-	-	-	-	-	-	-	-	-	-
Intl. Marine Bunkers	-	-	-	-	-	-	-	-	-	-	-
Stock Changes	-	-	-	-	-	-	-	-	-	-	-
DOMESTIC SUPPLY	-	-	-	-	-	1	-	-	-	-	-
Transfers and Stat. Diff.	-	-	-	-	-	-	-	-	-	-	-
TRANSFORMATION	-	-	-	-	-	1	-	-	-	-	-
Electricity and CHP Plants	-	-	-	-	-	-	-	-	-	-	-
Petroleum Refineries	-	-	-	-	-	-	-	-	-	-	-
Other Transform. Sector	-	-	-	-	-	1	-	-	-	-	-
ENERGY SECTOR	-	-	-	-	-	-	-	-	-	-	-
DISTRIBUTION LOSSES	-	-	-	-	-	-	-	-	-	-	-
FINAL CONSUMPTION	-	-	-	-	-	-	-	-	-	-	-
INDUSTRY SECTOR	-	-	-	-	-	-	-	-	-	-	-
Iron and Steel	-	-	-	-	-	-	-	-	-	-	-
Chemical and Petrochem.	-	-	-	-	-	-	-	-	-	-	-
Non-Metallic Minerals	-	-	-	-	-	-	-	-	-	-	-
Non-specified	-	-	-	-	-	-	-	-	-	-	-
TRANSPORT SECTOR	-	-	-	-	-	-	-	-	-	-	-
Air	-	-	-	-	-	-	-	-	-	-	-
Road	-	-	-	-	-	-	-	-	-	-	-
Non-specified	-	-	-	-	-	-	-	-	-	-	-
OTHER SECTORS	-	-	-	-	-	-	-	-	-	-	-
Agriculture	-	-	-	-	-	-	-	-	-	-	-
Comm. and Publ. Services	-	-	-	-	-	-	-	-	-	-	-
Residential	-	-	-	-	-	-	-	-	-	-	-
Non-specified	-	-	-	-	-	-	-	-	-	-	-
NON-ENERGY USE	-	-	-	-	-	-	-	-	-	-	-

Honduras

SUPPLY AND CONSUMPTION 1996	Oil cont. (1000 tonnes)										
	Refinery Gas	LPG + Ethane	Motor Gasoline	Aviation Gasoline	Jet Fuel	Kerosene	Gas/ Diesel	Heavy Fuel Oil	Naphtha	Petrol. Coke	Other Prod.
Production	-	-	-	-	-	-	-	-	-	-	-
Imports	-	45	253	-	22	52	476	314	-	-	-
Exports	-	-	-	-	-	-	-	-	-	-	-
Intl. Marine Bunkers	-	-	-	-	-	-	-	-	-	-	-
Stock Changes	-	-7	-42	-	-3	-7	14	17	-	-	-
DOMESTIC SUPPLY	**-**	**38**	**211**	**-**	**19**	**45**	**490**	**331**	**-**	**-**	**-**
Transfers and Stat. Diff.	-	-	1	-	5	-	1	-2	-	-	-
TRANSFORMATION	**-**	**-**	**-**	**-**	**-**	**-**	**4**	**-**	**-**	**-**	**-**
Electricity and CHP Plants	-	-	-	-	-	-	4	-	-	-	-
Petroleum Refineries	-	-	-	-	-	-	-	-	-	-	-
Other Transform. Sector	-	-	-	-	-	-	-	-	-	-	-
ENERGY SECTOR	**-**	**-**	**-**	**-**	**-**	**-**	**-**	**-**	**-**	**-**	**-**
DISTRIBUTION LOSSES	**-**	**-**	**-**	**-**	**-**	**-**	**-**	**-**	**-**	**-**	**-**
FINAL CONSUMPTION	**-**	**38**	**212**	**-**	**24**	**45**	**487**	**329**	**-**	**-**	**-**
INDUSTRY SECTOR	**-**	**2**	**12**	**-**	**-**	**-**	**155**	**315**	**-**	**-**	**-**
Iron and Steel	-	-	-	-	-	-	-	-	-	-	-
Chemical and Petrochem.	-	-	-	-	-	-	-	-	-	-	-
Non-Metallic Minerals	-	-	-	-	-	-	-	-	-	-	-
Non-specified	-	2	12	-	-	-	155	315	-	-	-
TRANSPORT SECTOR	**-**	**-**	**188**	**-**	**24**	**-**	**297**	**-**	**-**	**-**	**-**
Air	-	-	-	-	24	-	-	-	-	-	-
Road	-	-	188	-	-	-	297	-	-	-	-
Non-specified	-	-	-	-	-	-	-	-	-	-	-
OTHER SECTORS	**-**	**36**	**12**	**-**	**-**	**45**	**35**	**14**	**-**	**-**	**-**
Agriculture	-	-	-	-	-	-	3	2	-	-	-
Comm. and Publ. Services	-	-	12	-	-	27	32	12	-	-	-
Residential	-	36	-	-	-	18	-	-	-	-	-
Non-specified	-	-	-	-	-	-	-	-	-	-	-
NON-ENERGY USE	**-**	**-**	**-**	**-**	**-**	**-**	**-**	**-**	**-**	**-**	**-**

APPROVISIONNEMENT ET DEMANDE 1997	*Pétrole cont.* (1000 tonnes)										
	Gaz de raffinerie	*GPL + éthane*	*Essence moteur*	*Essence aviation*	*Carbu- réacteurs*	*Kérosène*	*Gazole*	*Fioul lourd*	*Naphta*	*Coke de pétrole*	*Autres prod.*
Production	-	-	-	-	-	-	-	-	-	-	-
Imports	-	41	254	-	22	52	478	269	-	-	-
Exports	-	-	-	-	-	-	-	-	-	-	-
Intl. Marine Bunkers	-	-	-	-	-	-	-	-	-	-	-
Stock Changes	-	2	-40	-	-3	-6	17	64	-	-	-
DOMESTIC SUPPLY	**-**	**43**	**214**	**-**	**19**	**46**	**495**	**333**	**-**	**-**	**-**
Transfers and Stat. Diff.	-	-	1	-	5	-	-1	1	-	-	-
TRANSFORMATION	**-**	**-**	**-**	**-**	**-**	**-**	**4**	**-**	**-**	**-**	**-**
Electricity and CHP Plants	-	-	-	-	-	-	4	-	-	-	-
Petroleum Refineries	-	-	-	-	-	-	-	-	-	-	-
Other Transform. Sector	-	-	-	-	-	-	-	-	-	-	-
ENERGY SECTOR	**-**	**-**	**-**	**-**	**-**	**-**	**-**	**-**	**-**	**-**	**-**
DISTRIBUTION LOSSES	**-**	**-**	**-**	**-**	**-**	**-**	**-**	**-**	**-**	**-**	**-**
FINAL CONSUMPTION	**-**	**43**	**215**	**-**	**24**	**46**	**490**	**334**	**-**	**-**	**-**
INDUSTRY SECTOR	**-**	**3**	**13**	**-**	**-**	**-**	**156**	**319**	**-**	**-**	**-**
Iron and Steel	-	-	-	-	-	-	-	-	-	-	-
Chemical and Petrochem.	-	-	-	-	-	-	-	-	-	-	-
Non-Metallic Minerals	-	-	-	-	-	-	-	-	-	-	-
Non-specified	-	3	13	-	-	-	156	319	-	-	-
TRANSPORT SECTOR	**-**	**-**	**190**	**-**	**24**	**-**	**299**	**-**	**-**	**-**	**-**
Air	-	-	-	-	24	-	-	-	-	-	-
Road	-	-	190	-	-	-	299	-	-	-	-
Non-specified	-	-	-	-	-	-	-	-	-	-	-
OTHER SECTORS	**-**	**40**	**12**	**-**	**-**	**46**	**35**	**15**	**-**	**-**	**-**
Agriculture	-	-	-	-	-	-	3	2	-	-	-
Comm. and Publ. Services	-	-	12	-	-	27	32	13	-	-	-
Residential	-	40	-	-	-	19	-	-	-	-	-
Non-specified	-	-	-	-	-	-	-	-	-	-	-
NON-ENERGY USE	**-**	**-**	**-**	**-**	**-**	**-**	**-**	**-**	**-**	**-**	**-**

Honduras

SUPPLY AND CONSUMPTION 1996	Gas (TJ)				Comb. Renew. & Waste (TJ)				(GWh)	(TJ)
	Natural Gas	Gas Works	Coke Ovens	Blast Furnaces	Solid Biomass	Gas/Liquids from Biomass	Municipal Waste	Industrial Waste	Electricity	Heat
Production	-	-	-	-	63744	-	-	-	3059	-
Imports	-	-	-	-	-	-	-	-	-	-
Exports	-	-	-	-	-	-	-	-	-	-
Intl. Marine Bunkers	-	-	-	-	-	-	-	-	-	-
Stock Changes	-	-	-	-	-	-	-	-	-	-
DOMESTIC SUPPLY	-	-	-	-	**63744**	-	-	-	**3059**	-
Transfers and Stat. Diff.	-	-	-	-	-	-	-	-	-48	-
TRANSFORMATION	-	-	-	-	**515**	-	-	-	-	-
Electricity and CHP Plants	-	-	-	-	-	-	-	-	-	-
Petroleum Refineries	-	-	-	-	-	-	-	-	-	-
Other Transform. Sector	-	-	-	-	515	-	-	-	-	-
ENERGY SECTOR	-	-	-	-	-	-	-	-	**11**	-
DISTRIBUTION LOSSES	-	-	-	-	-	-	-	-	**673**	-
FINAL CONSUMPTION	-	-	-	-	**63229**	-	-	-	**2327**	-
INDUSTRY SECTOR	-	-	-	-	**11685**	-	-	-	**778**	-
Iron and Steel	-	-	-	-	-	-	-	-	-	-
Chemical and Petrochem.	-	-	-	-	-	-	-	-	-	-
Non-Metallic Minerals	-	-	-	-	-	-	-	-	-	-
Non-specified	-	-	-	-	11685	-	-	-	778	-
TRANSPORT SECTOR	-	-	-	-	-	-	-	-	-	-
Air	-	-	-	-	-	-	-	-	-	-
Road	-	-	-	-	-	-	-	-	-	-
Non-specified	-	-	-	-	-	-	-	-	-	-
OTHER SECTORS	-	-	-	-	**51544**	-	-	-	**1549**	-
Agriculture	-	-	-	-	-	-	-	-	-	-
Comm. and Publ. Services	-	-	-	-	-	-	-	-	681	-
Residential	-	-	-	-	51544	-	-	-	868	-
Non-specified	-	-	-	-	-	-	-	-	-	-
NON-ENERGY USE	-	-	-	-	-	-	-	-	-	-

APPROVISIONNEMENT ET DEMANDE 1997	*Gaz* (TJ)				*En. Re. Comb. & Déchets* (TJ)				(GWh)	(TJ)
	Gaz naturel	*Usines à gaz*	*Cokeries*	*Hauts fourneaux*	*Biomasse solide*	*Gaz/Liquides tirés de biomasse*	*Déchets urbains*	*Déchets industriels*	*Electricité*	*Chaleur*
Production	-	-	-	-	72204	-	-	-	3294	-
Imports	-	-	-	-	-	-	-	-	-	-
Exports	-	-	-	-	-	-	-	-	-	-
Intl. Marine Bunkers	-	-	-	-	-	-	-	-	-	-
Stock Changes	-	-	-	-	-	-	-	-	-	-
DOMESTIC SUPPLY	-	-	-	-	**72204**	-	-	-	**3294**	-
Transfers and Stat. Diff.	-	-	-	-	-	-	-	-	-21	-
TRANSFORMATION	-	-	-	-	**651**	-	-	-	-	-
Electricity and CHP Plants	-	-	-	-	-	-	-	-	-	-
Petroleum Refineries	-	-	-	-	-	-	-	-	-	-
Other Transform. Sector	-	-	-	-	651	-	-	-	-	-
ENERGY SECTOR	-	-	-	-	-	-	-	-	**12**	-
DISTRIBUTION LOSSES	-	-	-	-	-	-	-	-	**800**	-
FINAL CONSUMPTION	-	-	-	-	**71553**	-	-	-	**2461**	-
INDUSTRY SECTOR	-	-	-	-	**12533**	-	-	-	**732**	-
Iron and Steel	-	-	-	-	-	-	-	-	-	-
Chemical and Petrochem.	-	-	-	-	-	-	-	-	-	-
Non-Metallic Minerals	-	-	-	-	-	-	-	-	-	-
Non-specified	-	-	-	-	12533	-	-	-	732	-
TRANSPORT SECTOR	-	-	-	-	-	-	-	-	-	-
Air	-	-	-	-	-	-	-	-	-	-
Road	-	-	-	-	-	-	-	-	-	-
Non-specified	-	-	-	-	-	-	-	-	-	-
OTHER SECTORS	-	-	-	-	**59020**	-	-	-	**1729**	-
Agriculture	-	-	-	-	-	-	-	-	-	-
Comm. and Publ. Services	-	-	-	-	-	-	-	-	745	-
Residential	-	-	-	-	59020	-	-	-	984	-
Non-specified	-	-	-	-	-	-	-	-	-	-
NON-ENERGY USE	-	-	-	-	-	-	-	-	-	-

Hong Kong, China / Hong Kong, Chine : 1996

	Coal / *Charbon* (1000 tonnes)							Oil / *Pétrole* (1000 tonnes)			
SUPPLY AND CONSUMPTION *APPROVISIONNEMENT ET DEMANDE*	Coking Coal *Charbon à coke*	Other Bit. Coal *Autres charb. bit.*	Sub-Bit. Coal *Charbon sous-bit.*	Lignite *Lignite*	Peat *Tourbe*	Oven and Gas Coke *Coke de four/gaz*	Pat. Fuel and BKB *Agg./briq. de lignite*	Crude Oil *Pétrole brut*	NGL *LGN*	Feed-stocks *Produits d'aliment.*	Additives *Additifs*
Production	-	-	-	-	-	-	-	-	-	-	-
From Other Sources	-	-	-	-	-	-	-	-	-	-	-
Imports	-	6769	-	-	-	-	-	-	-	-	-
Exports	-	-	-	-	-	-	-	-	-	-	-
Intl. Marine Bunkers	-	-	-	-	-	-	-	-	-	-	-
Stock Changes	-	-	-	-	-	-	-	-	-	-	-
DOMESTIC SUPPLY	**-**	**6769**	**-**	**-**	**-**	**-**	**-**	**-**	**-**	**-**	**-**
Transfers	-	-	-	-	-	-	-	-	-	-	-
Statistical Differences	-	1291	-	-	-	-	-	-	-	-	-
TRANSFORMATION	**-**	**8060**	**-**	**-**	**-**	**-**	**-**	**-**	**-**	**-**	**-**
Electricity Plants	-	8060	-	-	-	-	-	-	-	-	-
CHP Plants	-	-	-	-	-	-	-	-	-	-	-
Heat Plants	-	-	-	-	-	-	-	-	-	-	-
Blast Furnaces/Gas Works	-	-	-	-	-	-	-	-	-	-	-
Coke/Pat. Fuel/BKB Plants	-	-	-	-	-	-	-	-	-	-	-
Petroleum Refineries	-	-	-	-	-	-	-	-	-	-	-
Petrochemical Industry	-	-	-	-	-	-	-	-	-	-	-
Liquefaction	-	-	-	-	-	-	-	-	-	-	-
Other Transform. Sector	-	-	-	-	-	-	-	-	-	-	-
ENERGY SECTOR	**-**	**-**	**-**	**-**	**-**	**-**	**-**	**-**	**-**	**-**	**-**
Coal Mines	-	-	-	-	-	-	-	-	-	-	-
Oil and Gas Extraction	-	-	-	-	-	-	-	-	-	-	-
Petroleum Refineries	-	-	-	-	-	-	-	-	-	-	-
Electr., CHP+Heat Plants	-	-	-	-	-	-	-	-	-	-	-
Pumped Storage (Elec.)	-	-	-	-	-	-	-	-	-	-	-
Other Energy Sector	-	-	-	-	-	-	-	-	-	-	-
Distribution Losses	-	-	-	-	-	-	-	-	-	-	-
FINAL CONSUMPTION	**-**	**-**	**-**	**-**	**-**	**-**	**-**	**-**	**-**	**-**	**-**
INDUSTRY SECTOR	**-**	**-**	**-**	**-**	**-**	**-**	**-**	**-**	**-**	**-**	**-**
Iron and Steel	-	-	-	-	-	-	-	-	-	-	-
Chemical and Petrochem.	-	-	-	-	-	-	-	-	-	-	-
of which: Feedstocks	-	-	-	-	-	-	-	-	-	-	-
Non-Ferrous Metals	-	-	-	-	-	-	-	-	-	-	-
Non-Metallic Minerals	-	-	-	-	-	-	-	-	-	-	-
Transport Equipment	-	-	-	-	-	-	-	-	-	-	-
Machinery	-	-	-	-	-	-	-	-	-	-	-
Mining and Quarrying	-	-	-	-	-	-	-	-	-	-	-
Food and Tobacco	-	-	-	-	-	-	-	-	-	-	-
Paper, Pulp and Print	-	-	-	-	-	-	-	-	-	-	-
Wood and Wood Products	-	-	-	-	-	-	-	-	-	-	-
Construction	-	-	-	-	-	-	-	-	-	-	-
Textile and Leather	-	-	-	-	-	-	-	-	-	-	-
Non-specified	-	-	-	-	-	-	-	-	-	-	-
TRANSPORT SECTOR	**-**	**-**	**-**	**-**	**-**	**-**	**-**	**-**	**-**	**-**	**-**
Air	-	-	-	-	-	-	-	-	-	-	-
Road	-	-	-	-	-	-	-	-	-	-	-
Rail	-	-	-	-	-	-	-	-	-	-	-
Pipeline Transport	-	-	-	-	-	-	-	-	-	-	-
Internal Navigation	-	-	-	-	-	-	-	-	-	-	-
Non-specified	-	-	-	-	-	-	-	-	-	-	-
OTHER SECTORS	**-**	**-**	**-**	**-**	**-**	**-**	**-**	**-**	**-**	**-**	**-**
Agriculture	-	-	-	-	-	-	-	-	-	-	-
Comm. and Publ. Services	-	-	-	-	-	-	-	-	-	-	-
Residential	-	-	-	-	-	-	-	-	-	-	-
Non-specified	-	-	-	-	-	-	-	-	-	-	-
NON-ENERGY USE	**-**	**-**	**-**	**-**	**-**	**-**	**-**	**-**	**-**	**-**	**-**
in Industry/Trans./Energy	-	-	-	-	-	-	-	-	-	-	-
in Transport	-	-	-	-	-	-	-	-	-	-	-
in Other Sectors	-	-	-	-	-	-	-	-	-	-	-

Hong Kong, China / Hong Kong, Chine : 1996

SUPPLY AND CONSUMPTION / *APPROVISIONNEMENT ET DEMANDE*	Oil cont. / *Pétrole cont.* (1000 tonnes)										
	Refinery Gas / *Gaz de raffinerie*	LPG + Ethane / *GPL + éthane*	Motor Gasoline / *Essence moteur*	Aviation Gasoline / *Essence aviation*	Jet Fuel / *Carbu-réacteurs*	Kerosene / *Kérosène*	Gas/ Diesel / *Gazole*	Heavy Fuel Oil / *Fioul lourd*	Naphtha / *Naphta*	Petrol. Coke / *Coke de pétrole*	Other Prod. / *Autres prod.*
Production	-	-	-	-	-	-	-	-	-	-	-
From Other Sources	-	-	-	-	-	-	-	-	-	-	-
Imports	-	187	452	-	3250	57	9009	4014	517	-	477
Exports	-	-24	-111	-	-120	-	-5757	-2196	-	-	-292
Intl. Marine Bunkers	-	-	-	-	-	-	-589	-1788	-	-	-
Stock Changes	-	-1	-6	-	-107	-2	-60	69	-	-	-
DOMESTIC SUPPLY	**-**	**162**	**335**	**-**	**3023**	**55**	**2603**	**99**	**517**	**-**	**185**
Transfers	-	-	-	-	-	-	-	-	-	-	-
Statistical Differences	-	-1	-	-	-	-	1	1	-	-	-
TRANSFORMATION	**-**	**-**	**-**	**-**	**-**	**-**	**10**	**59**	**517**	**-**	**-**
Electricity Plants	-	-	-	-	-	-	10	59	-	-	-
CHP Plants	-	-	-	-	-	-	-	-	-	-	-
Heat Plants	-	-	-	-	-	-	-	-	-	-	-
Blast Furnaces/Gas Works	-	-	-	-	-	-	-	-	517	-	-
Coke/Pat. Fuel/BKB Plants	-	-	-	-	-	-	-	-	-	-	-
Petroleum Refineries	-	-	-	-	-	-	-	-	-	-	-
Petrochemical Industry	-	-	-	-	-	-	-	-	-	-	-
Liquefaction	-	-	-	-	-	-	-	-	-	-	-
Other Transform. Sector	-	-	-	-	-	-	-	-	-	-	-
ENERGY SECTOR	**-**	**-**	**-**	**-**	**-**	**-**	**-**	**-**	**-**	**-**	**-**
Coal Mines	-	-	-	-	-	-	-	-	-	-	-
Oil and Gas Extraction	-	-	-	-	-	-	-	-	-	-	-
Petroleum Refineries	-	-	-	-	-	-	-	-	-	-	-
Electr., CHP+Heat Plants	-	-	-	-	-	-	-	-	-	-	-
Pumped Storage (Elec.)	-	-	-	-	-	-	-	-	-	-	-
Other Energy Sector	-	-	-	-	-	-	-	-	-	-	-
Distribution Losses	-	-	-	-	-	-	-	-	-	-	-
FINAL CONSUMPTION	**-**	**161**	**335**	**-**	**3023**	**55**	**2594**	**41**	**-**	**-**	**185**
INDUSTRY SECTOR	**-**	**160**	**-**	**-**	**-**	**-**	**-**	**41**	**-**	**-**	**-**
Iron and Steel	-	-	-	-	-	-	-	-	-	-	-
Chemical and Petrochem.	-	-	-	-	-	-	-	-	-	-	-
of which: Feedstocks	-	-	-	-	-	-	-	-	-	-	-
Non-Ferrous Metals	-	-	-	-	-	-	-	-	-	-	-
Non-Metallic Minerals	-	-	-	-	-	-	-	-	-	-	-
Transport Equipment	-	-	-	-	-	-	-	-	-	-	-
Machinery	-	-	-	-	-	-	-	-	-	-	-
Mining and Quarrying	-	-	-	-	-	-	-	-	-	-	-
Food and Tobacco	-	-	-	-	-	-	-	-	-	-	-
Paper, Pulp and Print	-	-	-	-	-	-	-	-	-	-	-
Wood and Wood Products	-	-	-	-	-	-	-	-	-	-	-
Construction	-	-	-	-	-	-	-	-	-	-	-
Textile and Leather	-	-	-	-	-	-	-	-	-	-	-
Non-specified	-	160	-	-	-	-	-	41	-	-	-
TRANSPORT SECTOR	**-**	**-**	**335**	**-**	**3023**	**-**	**2594**	**-**	**-**	**-**	**-**
Air	-	-	-	-	3023	-	-	-	-	-	-
Road	-	-	335	-	-	-	2594	-	-	-	-
Rail	-	-	-	-	-	-	-	-	-	-	-
Pipeline Transport	-	-	-	-	-	-	-	-	-	-	-
Internal Navigation	-	-	-	-	-	-	-	-	-	-	-
Non-specified	-	-	-	-	-	-	-	-	-	-	-
OTHER SECTORS	**-**	**1**	**-**	**-**	**-**	**55**	**-**	**-**	**-**	**-**	**-**
Agriculture	-	-	-	-	-	-	-	-	-	-	-
Comm. and Publ. Services	-	1	-	-	-	-	-	-	-	-	-
Residential	-	-	-	-	-	55	-	-	-	-	-
Non-specified	-	-	-	-	-	-	-	-	-	-	-
NON-ENERGY USE	**-**	**-**	**-**	**-**	**-**	**-**	**-**	**-**	**-**	**-**	**185**
in Industry/Transf./Energy	-	-	-	-	-	-	-	-	-	-	185
in Transport	-	-	-	-	-	-	-	-	-	-	-
in Other Sectors	-	-	-	-	-	-	-	-	-	-	-

Hong Kong, China / Hong Kong, Chine : 1996

SUPPLY AND CONSUMPTION / *APPROVISIONNEMENT ET DEMANDE*	Gas / *Gaz* (TJ) Natural Gas / *Gaz naturel*	Gas Works / *Usines à gaz*	Coke Ovens / *Cokeries*	Blast Furnaces / *Hauts fourneaux*	Comb. Renew. & Waste / *En. Re. Comb. & Déchets* (TJ) Solid Biomass / *Biomasse solide*	Gas/Liquids from Biomass / *Gaz/Liquides tirés de biomasse*	Municipal Waste / *Déchets urbains*	Industrial Waste / *Déchets industriels*	(GWh) Electricity / *Electricité*	(TJ) Heat / *Chaleur*
Production	-	22989	-	-	2012	-	-	-	28442	-
From Other Sources	-	-	-	-	-	-	-	-	-	-
Imports	68823	-	-	-	433	-	-	-	7778	-
Exports	-	-	-	-	-58	-	-	-	-531	-
Intl. Marine Bunkers	-	-	-	-	-	-	-	-	-	-
Stock Changes	-	-	-	-	-	-	-	-	-	-
DOMESTIC SUPPLY	**68823**	**22989**	**-**	**-**	**2387**	**-**	**-**	**-**	**35689**	**-**
Transfers	-	-	-	-	-	-	-	-	-	-
Statistical Differences	-	-	-	-	-	-	-	-	1	-
TRANSFORMATION	**-**	**-**	**-**	**-**	**-**	**-**	**-**	**-**	**-**	**-**
Electricity Plants	-	-	-	-	-	-	-	-	-	-
CHP Plants	-	-	-	-	-	-	-	-	-	-
Heat Plants	-	-	-	-	-	-	-	-	-	-
Blast Furnaces/Gas Works	-	-	-	-	-	-	-	-	-	-
Coke/Pat. Fuel/BKB Plants	-	-	-	-	-	-	-	-	-	-
Petroleum Refineries	-	-	-	-	-	-	-	-	-	-
Petrochemical Industry	-	-	-	-	-	-	-	-	-	-
Liquefaction	-	-	-	-	-	-	-	-	-	-
Other Transform. Sector	-	-	-	-	-	-	-	-	-	-
ENERGY SECTOR	**-**	**-**	**-**	**-**	**-**	**-**	**-**	**-**	**-**	**-**
Coal Mines	-	-	-	-	-	-	-	-	-	-
Oil and Gas Extraction	-	-	-	-	-	-	-	-	-	-
Petroleum Refineries	-	-	-	-	-	-	-	-	-	-
Electr., CHP+Heat Plants	-	-	-	-	-	-	-	-	-	-
Pumped Storage (Elec.)	-	-	-	-	-	-	-	-	-	-
Other Energy Sector	-	-	-	-	-	-	-	-	-	-
Distribution Losses	-	-	-	-	-	-	-	-	4054	-
FINAL CONSUMPTION	**68823**	**22989**	**-**	**-**	**2387**	**-**	**-**	**-**	**31636**	**-**
INDUSTRY SECTOR	**68823**	**918**	**-**	**-**	**-**	**-**	**-**	**-**	**5538**	**-**
Iron and Steel	-	-	-	-	-	-	-	-	-	-
Chemical and Petrochem.	-	-	-	-	-	-	-	-	-	-
of which: Feedstocks	-	-	-	-	-	-	-	-	-	-
Non-Ferrous Metals	-	-	-	-	-	-	-	-	-	-
Non-Metallic Minerals	-	-	-	-	-	-	-	-	-	-
Transport Equipment	-	-	-	-	-	-	-	-	-	-
Machinery	-	-	-	-	-	-	-	-	-	-
Mining and Quarrying	-	-	-	-	-	-	-	-	-	-
Food and Tobacco	-	-	-	-	-	-	-	-	-	-
Paper, Pulp and Print	-	-	-	-	-	-	-	-	-	-
Wood and Wood Products	-	-	-	-	-	-	-	-	-	-
Construction	-	-	-	-	-	-	-	-	-	-
Textile and Leather	-	-	-	-	-	-	-	-	-	-
Non-specified	68823	918	-	-	-	-	-	-	5538	-
TRANSPORT SECTOR	**-**	**-**	**-**	**-**	**-**	**-**	**-**	**-**	**-**	**-**
Air	-	-	-	-	-	-	-	-	-	-
Road	-	-	-	-	-	-	-	-	-	-
Rail	-	-	-	-	-	-	-	-	-	-
Pipeline Transport	-	-	-	-	-	-	-	-	-	-
Internal Navigation	-	-	-	-	-	-	-	-	-	-
Non-specified	-	-	-	-	-	-	-	-	-	-
OTHER SECTORS	**-**	**22071**	**-**	**-**	**2387**	**-**	**-**	**-**	**26098**	**-**
Agriculture	-	-	-	-	-	-	-	-	-	-
Comm. and Publ. Services	-	10085	-	-	-	-	-	-	17988	-
Residential	-	11986	-	-	2012	-	-	-	8110	-
Non-specified	-	-	-	-	375	-	-	-	-	-
NON-ENERGY USE	**-**	**-**	**-**	**-**	**-**	**-**	**-**	**-**	**-**	**-**
in Industry/Transf./Energy	-	-	-	-	-	-	-	-	-	-
in Transport	-	-	-	-	-	-	-	-	-	-
in Other Sectors	-	-	-	-	-	-	-	-	-	-

Hong Kong, China / Hong Kong, Chine : 1997

SUPPLY AND CONSUMPTION *APPROVISIONNEMENT ET DEMANDE*	Coal / *Charbon* (1000 tonnes)							Oil / *Pétrole* (1000 tonnes)			
	Coking Coal *Charbon à coke*	Other Bit. Coal *Autres charb. bit.*	Sub-Bit. Coal *Charbon sous-bit.*	Lignite *Lignite*	Peat *Tourbe*	Oven and Gas Coke *Coke de four/gaz*	Pat. Fuel and BKB *Agg./briq. de lignite*	Crude Oil *Pétrole brut*	NGL *LGN*	Feed-stocks *Produits d'aliment.*	Additives *Additifs*
Production	-	-	-	-	-	-	-	-	-	-	-
From Other Sources	-	-	-	-	-	-	-	-	-	-	-
Imports	-	5711	-	-	-	-	-	-	-	-	-
Exports	-	-	-	-	-	-	-	-	-	-	-
Intl. Marine Bunkers	-	-	-	-	-	-	-	-	-	-	-
Stock Changes	-	-	-	-	-	-	-	-	-	-	-
DOMESTIC SUPPLY	**-**	**5711**	**-**	**-**	**-**	**-**	**-**	**-**	**-**	**-**	**-**
Transfers	-	-	-	-	-	-	-	-	-	-	-
Statistical Differences	-	2561	-	-	-	-	-	-	-	-	-
TRANSFORMATION	**-**	**8272**	**-**	**-**	**-**	**-**	**-**	**-**	**-**	**-**	**-**
Electricity Plants	-	8272	-	-	-	-	-	-	-	-	-
CHP Plants	-	-	-	-	-	-	-	-	-	-	-
Heat Plants	-	-	-	-	-	-	-	-	-	-	-
Blast Furnaces/Gas Works	-	-	-	-	-	-	-	-	-	-	-
Coke/Pat. Fuel/BKB Plants	-	-	-	-	-	-	-	-	-	-	-
Petroleum Refineries	-	-	-	-	-	-	-	-	-	-	-
Petrochemical Industry	-	-	-	-	-	-	-	-	-	-	-
Liquefaction	-	-	-	-	-	-	-	-	-	-	-
Other Transform. Sector	-	-	-	-	-	-	-	-	-	-	-
ENERGY SECTOR	**-**	**-**	**-**	**-**	**-**	**-**	**-**	**-**	**-**	**-**	**-**
Coal Mines	-	-	-	-	-	-	-	-	-	-	-
Oil and Gas Extraction	-	-	-	-	-	-	-	-	-	-	-
Petroleum Refineries	-	-	-	-	-	-	-	-	-	-	-
Electr., CHP+Heat Plants	-	-	-	-	-	-	-	-	-	-	-
Pumped Storage (Elec.)	-	-	-	-	-	-	-	-	-	-	-
Other Energy Sector	-	-	-	-	-	-	-	-	-	-	-
Distribution Losses	-	-	-	-	-	-	-	-	-	-	-
FINAL CONSUMPTION	**-**	**-**	**-**	**-**	**-**	**-**	**-**	**-**	**-**	**-**	**-**
INDUSTRY SECTOR	**-**	**-**	**-**	**-**	**-**	**-**	**-**	**-**	**-**	**-**	**-**
Iron and Steel	-	-	-	-	-	-	-	-	-	-	-
Chemical and Petrochem.	-	-	-	-	-	-	-	-	-	-	-
of which: Feedstocks	-	-	-	-	-	-	-	-	-	-	-
Non-Ferrous Metals	-	-	-	-	-	-	-	-	-	-	-
Non-Metallic Minerals	-	-	-	-	-	-	-	-	-	-	-
Transport Equipment	-	-	-	-	-	-	-	-	-	-	-
Machinery	-	-	-	-	-	-	-	-	-	-	-
Mining and Quarrying	-	-	-	-	-	-	-	-	-	-	-
Food and Tobacco	-	-	-	-	-	-	-	-	-	-	-
Paper, Pulp and Print	-	-	-	-	-	-	-	-	-	-	-
Wood and Wood Products	-	-	-	-	-	-	-	-	-	-	-
Construction	-	-	-	-	-	-	-	-	-	-	-
Textile and Leather	-	-	-	-	-	-	-	-	-	-	-
Non-specified	-	-	-	-	-	-	-	-	-	-	-
TRANSPORT SECTOR	**-**	**-**	**-**	**-**	**-**	**-**	**-**	**-**	**-**	**-**	**-**
Air	-	-	-	-	-	-	-	-	-	-	-
Road	-	-	-	-	-	-	-	-	-	-	-
Rail	-	-	-	-	-	-	-	-	-	-	-
Pipeline Transport	-	-	-	-	-	-	-	-	-	-	-
Internal Navigation	-	-	-	-	-	-	-	-	-	-	-
Non-specified	-	-	-	-	-	-	-	-	-	-	-
OTHER SECTORS	**-**	**-**	**-**	**-**	**-**	**-**	**-**	**-**	**-**	**-**	**-**
Agriculture	-	-	-	-	-	-	-	-	-	-	-
Comm. and Publ. Services	-	-	-	-	-	-	-	-	-	-	-
Residential	-	-	-	-	-	-	-	-	-	-	-
Non-specified	-	-	-	-	-	-	-	-	-	-	-
NON-ENERGY USE	**-**	**-**	**-**	**-**	**-**	**-**	**-**	**-**	**-**	**-**	**-**
in Industry/Trans./Energy	-	-	-	-	-	-	-	-	-	-	-
in Transport	-	-	-	-	-	-	-	-	-	-	-
in Other Sectors	-	-	-	-	-	-	-	-	-	-	-

Hong Kong, China / Hong Kong, Chine : 1997

	Oil cont. / *Pétrole cont.* (1000 tonnes)										
SUPPLY AND CONSUMPTION ***APPROVISIONNEMENT ET DEMANDE***	Refinery Gas *Gaz de raffinerie*	LPG + Ethane *GPL + éthane*	Motor Gasoline *Essence moteur*	Aviation Gasoline *Essence aviation*	Jet Fuel *Carbu-réacteurs*	Kerosene *Kérosène*	Gas/ Diesel *Gazole*	Heavy Fuel Oil *Fioul lourd*	Naphtha *Naphta*	Petrol. Coke *Coke de pétrole*	Other Prod. *Autres prod.*
Production	-	-	-	-	-	-	-	-	-	-	-
From Other Sources	-	-	-	-	-	-	-	-	-	-	-
Imports	-	166	470	-	3438	69	10554	4074	453	-	437
Exports	-	-18	-123	-	-265	-17	-7342	-2321	-	-	-123
Intl. Marine Bunkers	-	-	-	-	-	-	-509	-1640	-	-	-
Stock Changes	-	-2	-4	-	28	1	-49	-28	-	-	-
DOMESTIC SUPPLY	**-**	**146**	**343**	**-**	**3201**	**53**	**2654**	**85**	**453**	**-**	**314**
Transfers	-	-	-	-	-	-	-	-	-	-	-
Statistical Differences	-	1	1	-	1	-	-	-	-	-	-1
TRANSFORMATION	**-**	**-**	**-**	**-**	**-**	**-**	**16**	**68**	**453**	**-**	**-**
Electricity Plants	-	-	-	-	-	-	16	68	-	-	-
CHP Plants	-	-	-	-	-	-	-	-	-	-	-
Heat Plants	-	-	-	-	-	-	-	-	-	-	-
Blast Furnaces/Gas Works	-	-	-	-	-	-	-	-	453	-	-
Coke/Pat. Fuel/BKB Plants	-	-	-	-	-	-	-	-	-	-	-
Petroleum Refineries	-	-	-	-	-	-	-	-	-	-	-
Petrochemical Industry	-	-	-	-	-	-	-	-	-	-	-
Liquefaction	-	-	-	-	-	-	-	-	-	-	-
Other Transform. Sector	-	-	-	-	-	-	-	-	-	-	-
ENERGY SECTOR	**-**	**-**	**-**	**-**	**-**	**-**	**-**	**-**	**-**	**-**	**-**
Coal Mines	-	-	-	-	-	-	-	-	-	-	-
Oil and Gas Extraction	-	-	-	-	-	-	-	-	-	-	-
Petroleum Refineries	-	-	-	-	-	-	-	-	-	-	-
Electr., CHP+Heat Plants	-	-	-	-	-	-	-	-	-	-	-
Pumped Storage (Elec.)	-	-	-	-	-	-	-	-	-	-	-
Other Energy Sector	-	-	-	-	-	-	-	-	-	-	-
Distribution Losses	-	-	-	-	-	-	-	-	-	-	-
FINAL CONSUMPTION	**-**	**147**	**344**	**-**	**3202**	**53**	**2638**	**17**	**-**	**-**	**313**
INDUSTRY SECTOR	**-**	**146**	**-**	**-**	**-**	**-**	**-**	**17**	**-**	**-**	**-**
Iron and Steel	-	-	-	-	-	-	-	-	-	-	-
Chemical and Petrochem.	-	-	-	-	-	-	-	-	-	-	-
of which: Feedstocks	-	-	-	-	-	-	-	-	-	-	-
Non-Ferrous Metals	-	-	-	-	-	-	-	-	-	-	-
Non-Metallic Minerals	-	-	-	-	-	-	-	-	-	-	-
Transport Equipment	-	-	-	-	-	-	-	-	-	-	-
Machinery	-	-	-	-	-	-	-	-	-	-	-
Mining and Quarrying	-	-	-	-	-	-	-	-	-	-	-
Food and Tobacco	-	-	-	-	-	-	-	-	-	-	-
Paper, Pulp and Print	-	-	-	-	-	-	-	-	-	-	-
Wood and Wood Products	-	-	-	-	-	-	-	-	-	-	-
Construction	-	-	-	-	-	-	-	-	-	-	-
Textile and Leather	-	-	-	-	-	-	-	-	-	-	-
Non-specified	-	146	-	-	-	-	-	17	-	-	-
TRANSPORT SECTOR	**-**	**-**	**344**	**-**	**3202**	**-**	**2638**	**-**	**-**	**-**	**-**
Air	-	-	-	-	3202	-	-	-	-	-	-
Road	-	-	344	-	-	-	2638	-	-	-	-
Rail	-	-	-	-	-	-	-	-	-	-	-
Pipeline Transport	-	-	-	-	-	-	-	-	-	-	-
Internal Navigation	-	-	-	-	-	-	-	-	-	-	-
Non-specified	-	-	-	-	-	-	-	-	-	-	-
OTHER SECTORS	**-**	**1**	**-**	**-**	**-**	**53**	**-**	**-**	**-**	**-**	**-**
Agriculture	-	-	-	-	-	-	-	-	-	-	-
Comm. and Publ. Services	-	1	-	-	-	-	-	-	-	-	-
Residential	-	-	-	-	-	53	-	-	-	-	-
Non-specified	-	-	-	-	-	-	-	-	-	-	-
NON-ENERGY USE	**-**	**-**	**-**	**-**	**-**	**-**	**-**	**-**	**-**	**-**	**313**
in Industry/Transf./Energy	-	-	-	-	-	-	-	-	-	-	313
in Transport	-	-	-	-	-	-	-	-	-	-	-
in Other Sectors	-	-	-	-	-	-	-	-	-	-	-

Hong Kong, China / Hong Kong, Chine : 1997

SUPPLY AND CONSUMPTION / *APPROVISIONNEMENT ET DEMANDE*	Gas / *Gaz* (TJ)				Comb. Renew. & Waste / *En. Re. Comb. & Déchets* (TJ)				(GWh)	(TJ)
	Natural Gas / *Gaz naturel*	Gas Works / *Usines à gaz*	Coke Ovens / *Cokeries*	Blast Furnaces / *Hauts fourneaux*	Solid Biomass / *Biomasse solide*	Gas/Liquids from Biomass / *Gaz/Liquides tirés de biomasse*	Municipal Waste / *Déchets urbains*	Industrial Waste / *Déchets industriels*	Electricity / *Electricité*	Heat / *Chaleur*
Production	-	23905	-	-	2012	-	-	-	28945	-
From Other Sources	-	-	-	-	-	-	-	-	-	-
Imports	107462	-	-	-	433	-	-	-	7876	-
Exports	-	-	-	-	-58	-	-	-	-559	-
Intl. Marine Bunkers	-	-	-	-	-	-	-	-	-	-
Stock Changes	-	-	-	-	-	-	-	-	-	-
DOMESTIC SUPPLY	**107462**	**23905**	**-**	**-**	**2387**	**-**	**-**	**-**	**36262**	**-**
Transfers	-	-	-	-	-	-	-	-	-	-
Statistical Differences	-	-	-	-	-	-	-	-	-	-
TRANSFORMATION	**-**	**-**	**-**	**-**	**-**	**-**	**-**	**-**	**-**	**-**
Electricity Plants	-	-	-	-	-	-	-	-	-	-
CHP Plants	-	-	-	-	-	-	-	-	-	-
Heat Plants	-	-	-	-	-	-	-	-	-	-
Blast Furnaces/Gas Works	-	-	-	-	-	-	-	-	-	-
Coke/Pat. Fuel/BKB Plants	-	-	-	-	-	-	-	-	-	-
Petroleum Refineries	-	-	-	-	-	-	-	-	-	-
Petrochemical Industry	-	-	-	-	-	-	-	-	-	-
Liquefaction	-	-	-	-	-	-	-	-	-	-
Other Transform. Sector	-	-	-	-	-	-	-	-	-	-
ENERGY SECTOR	**-**	**-**	**-**	**-**	**-**	**-**	**-**	**-**	**-**	**-**
Coal Mines	-	-	-	-	-	-	-	-	-	-
Oil and Gas Extraction	-	-	-	-	-	-	-	-	-	-
Petroleum Refineries	-	-	-	-	-	-	-	-	-	-
Electr., CHP+Heat Plants	-	-	-	-	-	-	-	-	-	-
Pumped Storage (Elec.)	-	-	-	-	-	-	-	-	-	-
Other Energy Sector	-	-	-	-	-	-	-	-	-	-
Distribution Losses	-	-	-	-	-	-	-	-	4017	-
FINAL CONSUMPTION	**107462**	**23905**	**-**	**-**	**2387**	**-**	**-**	**-**	**32245**	**-**
INDUSTRY SECTOR	**107462**	**911**	**-**	**-**	**-**	**-**	**-**	**-**	**5268**	**-**
Iron and Steel	-	-	-	-	-	-	-	-	-	-
Chemical and Petrochem.	-	-	-	-	-	-	-	-	-	-
of which: Feedstocks	-	-	-	-	-	-	-	-	-	-
Non-Ferrous Metals	-	-	-	-	-	-	-	-	-	-
Non-Metallic Minerals	-	-	-	-	-	-	-	-	-	-
Transport Equipment	-	-	-	-	-	-	-	-	-	-
Machinery	-	-	-	-	-	-	-	-	-	-
Mining and Quarrying	-	-	-	-	-	-	-	-	-	-
Food and Tobacco	-	-	-	-	-	-	-	-	-	-
Paper, Pulp and Print	-	-	-	-	-	-	-	-	-	-
Wood and Wood Products	-	-	-	-	-	-	-	-	-	-
Construction	-	-	-	-	-	-	-	-	-	-
Textile and Leather	-	-	-	-	-	-	-	-	-	-
Non-specified	107462	911	-	-	-	-	-	-	5268	-
TRANSPORT SECTOR	**-**	**-**	**-**	**-**	**-**	**-**	**-**	**-**	**-**	**-**
Air	-	-	-	-	-	-	-	-	-	-
Road	-	-	-	-	-	-	-	-	-	-
Rail	-	-	-	-	-	-	-	-	-	-
Pipeline Transport	-	-	-	-	-	-	-	-	-	-
Internal Navigation	-	-	-	-	-	-	-	-	-	-
Non-specified	-	-	-	-	-	-	-	-	-	-
OTHER SECTORS	**-**	**22994**	**-**	**-**	**2387**	**-**	**-**	**-**	**26977**	**-**
Agriculture	-	-	-	-	-	-	-	-	-	-
Comm. and Publ. Services	-	10529	-	-	-	-	-	-	18938	-
Residential	-	12465	-	-	2012	-	-	-	8039	-
Non-specified	-	-	-	-	375	-	-	-	-	-
NON-ENERGY USE	**-**	**-**	**-**	**-**	**-**	**-**	**-**	**-**	**-**	**-**
in Industry/Transf./Energy	-	-	-	-	-	-	-	-	-	-
in Transport	-	-	-	-	-	-	-	-	-	-
in Other Sectors	-	-	-	-	-	-	-	-	-	-

India / Inde : 1996

SUPPLY AND CONSUMPTION / *APPROVISIONNEMENT ET DEMANDE*	Coal / *Charbon* (1000 tonnes)							Oil / *Pétrole* (1000 tonnes)			
	Coking Coal / *Charbon à coke*	Other Bit. Coal / *Autres charb. bit.*	Sub-Bit. Coal / *Charbon sous-bit.*	Lignite / *Lignite*	Peat / *Tourbe*	Oven and Gas Coke / *Coke de four/gaz*	Pat. Fuel and BKB / *Agg./briq. de lignite*	Crude Oil / *Pétrole brut*	NGL / *LGN*	Feed-stocks / *Produits d'aliment.*	Additives / *Additifs*
Production	6436	279150	-	22640	-	11150	-	31480	2662	-	-
From Other Sources	-	-	-	-	-	-	-	-	-	-	-
Imports	9050	400	-	-	-	-	-	33904	-	-	-
Exports	-	-130	-	-	-	-	-	-	-405	-	-
Intl. Marine Bunkers	-	-	-	-	-	-	-	-	-	-	-
Stock Changes	-	3620	-	-	-	-	-	-	-	-	-
DOMESTIC SUPPLY	**15486**	**283040**	**-**	**22640**	**-**	**11150**	**-**	**65384**	**2257**	**-**	**-**
Transfers	-	-	-	-	-	-	-	-	-1489	-	-
Statistical Differences	-	1000	-	-	-	-	-	-2373	-	-	-
TRANSFORMATION	**15486**	**199620**	**-**	**17580**	**-**	**8864**	**-**	**63011**	**-**	**-**	**-**
Electricity Plants	-	199620	-	17580	-	-	-	-	-	-	-
CHP Plants	-	-	-	-	-	-	-	-	-	-	-
Heat Plants	-	-	-	-	-	-	-	-	-	-	-
Blast Furnaces/Gas Works	-	-	-	-	-	8864	-	-	-	-	-
Coke/Pat. Fuel/BKB Plants	15486	-	-	-	-	-	-	-	-	-	-
Petroleum Refineries	-	-	-	-	-	-	-	63011	-	-	-
Petrochemical Industry	-	-	-	-	-	-	-	-	-	-	-
Liquefaction	-	-	-	-	-	-	-	-	-	-	-
Other Transform. Sector	-	-	-	-	-	-	-	-	-	-	-
ENERGY SECTOR	**-**	**3430**	**-**	**-**	**-**	**-**	**-**	**-**	**-**	**-**	**-**
Coal Mines	-	3430	-	-	-	-	-	-	-	-	-
Oil and Gas Extraction	-	-	-	-	-	-	-	-	-	-	-
Petroleum Refineries	-	-	-	-	-	-	-	-	-	-	-
Electr., CHP+Heat Plants	-	-	-	-	-	-	-	-	-	-	-
Pumped Storage (Elec.)	-	-	-	-	-	-	-	-	-	-	-
Other Energy Sector	-	-	-	-	-	-	-	-	-	-	-
Distribution Losses	-	-	-	-	-	-	-	-	-	-	-
FINAL CONSUMPTION	**-**	**80990**	**-**	**5060**	**-**	**2286**	**-**	**-**	**768**	**-**	**-**
INDUSTRY SECTOR	**-**	**80850**	**-**	**5060**	**-**	**2216**	**-**	**-**	**768**	**-**	**-**
Iron and Steel	-	17530	-	-	-	2216	-	-	-	-	-
Chemical and Petrochem.	-	-	-	5060	-	-	-	-	768	-	-
of which: Feedstocks	-	-	-	-	-	-	-	-	*768*	-	-
Non-Ferrous Metals	-	-	-	-	-	-	-	-	-	-	-
Non-Metallic Minerals	-	10080	-	-	-	-	-	-	-	-	-
Transport Equipment	-	-	-	-	-	-	-	-	-	-	-
Machinery	-	-	-	-	-	-	-	-	-	-	-
Mining and Quarrying	-	-	-	-	-	-	-	-	-	-	-
Food and Tobacco	-	-	-	-	-	-	-	-	-	-	-
Paper, Pulp and Print	-	3510	-	-	-	-	-	-	-	-	-
Wood and Wood Products	-	-	-	-	-	-	-	-	-	-	-
Construction	-	1010	-	-	-	-	-	-	-	-	-
Textile and Leather	-	1310	-	-	-	-	-	-	-	-	-
Non-specified	-	47410	-	-	-	-	-	-	-	-	-
TRANSPORT SECTOR	**-**	**140**	**-**	**-**	**-**	**-**	**-**	**-**	**-**	**-**	**-**
Air	-	-	-	-	-	-	-	-	-	-	-
Road	-	-	-	-	-	-	-	-	-	-	-
Rail	-	140	-	-	-	-	-	-	-	-	-
Pipeline Transport	-	-	-	-	-	-	-	-	-	-	-
Internal Navigation	-	-	-	-	-	-	-	-	-	-	-
Non-specified	-	-	-	-	-	-	-	-	-	-	-
OTHER SECTORS	**-**	**-**	**-**	**-**	**-**	**70**	**-**	**-**	**-**	**-**	**-**
Agriculture	-	-	-	-	-	-	-	-	-	-	-
Comm. and Publ. Services	-	-	-	-	-	-	-	-	-	-	-
Residential	-	-	-	-	-	70	-	-	-	-	-
Non-specified	-	-	-	-	-	-	-	-	-	-	-
NON-ENERGY USE	**-**	**-**	**-**	**-**	**-**	**-**	**-**	**-**	**-**	**-**	**-**
in Industry/Trans./Energy	-	-	-	-	-	-	-	-	-	-	-
in Transport	-	-	-	-	-	-	-	-	-	-	-
in Other Sectors	-	-	-	-	-	-	-	-	-	-	-

India / Inde : 1996

SUPPLY AND CONSUMPTION *APPROVISIONNEMENT ET DEMANDE*	Oil cont. / *Pétrole cont.* (1000 tonnes)										
	Refinery Gas *Gaz de raffinerie*	LPG + Ethane *GPL + éthane*	Motor Gasoline *Essence moteur*	Aviation Gasoline *Essence aviation*	Jet Fuel *Carbu-réacteurs*	Kerosene *Kérosène*	Gas/ Diesel *Gazole*	Heavy Fuel Oil *Fioul lourd*	Naphtha *Naphta*	Petrol. Coke *Coke de pétrole*	Other Prod. *Autres prod.*
Production	1141	1568	4818	-	2119	6396	23648	12959	6379	274	3342
From Other Sources	-	-	-	-	-	-	-	-	-	-	-
Imports	-	1036	448	26	150	4281	13405	694	229	-	-
Exports	-	-	-	-	-	-	-	-48	-2492	-	-
Intl. Marine Bunkers	-	-	-	-	-	-	-204	-335	-	-	-
Stock Changes	-	-	-	-	-	-	-	-	-	-	-
DOMESTIC SUPPLY	**1141**	**2604**	**5266**	**26**	**2269**	**10677**	**36849**	**13270**	**4116**	**274**	**3342**
Transfers	-	1489	-	-	-	-	-	-	-	-	-
Statistical Differences	-	-	-301	-	-104	-1100	1	-1	-493	80	1485
TRANSFORMATION	-	-	-	-	-	-	**400**	**2200**	-	-	-
Electricity Plants	-	-	-	-	-	-	400	2200	-	-	-
CHP Plants	-	-	-	-	-	-	-	-	-	-	-
Heat Plants	-	-	-	-	-	-	-	-	-	-	-
Blast Furnaces/Gas Works	-	-	-	-	-	-	-	-	-	-	-
Coke/Pat. Fuel/BKB Plants	-	-	-	-	-	-	-	-	-	-	-
Petroleum Refineries	-	-	-	-	-	-	-	-	-	-	-
Petrochemical Industry	-	-	-	-	-	-	-	-	-	-	-
Liquefaction	-	-	-	-	-	-	-	-	-	-	-
Other Transform. Sector	-	-	-	-	-	-	-	-	-	-	-
ENERGY SECTOR	**1141**	-	-	-	-	-	**735**	**2315**	-	-	-
Coal Mines	-	-	-	-	-	-	735	-	-	-	-
Oil and Gas Extraction	-	-	-	-	-	-	-	-	-	-	-
Petroleum Refineries	1141	-	-	-	-	-	-	2315	-	-	-
Electr., CHP+Heat Plants	-	-	-	-	-	-	-	-	-	-	-
Pumped Storage (Elec.)	-	-	-	-	-	-	-	-	-	-	-
Other Energy Sector	-	-	-	-	-	-	-	-	-	-	-
Distribution Losses	-	-	-	-	-	-	-	-	-	-	-
FINAL CONSUMPTION	-	**4093**	**4965**	**26**	**2165**	**9577**	**35715**	**8754**	**3623**	**354**	**4827**
INDUSTRY SECTOR	-	**731**	-	-	-	-	**3764**	**7367**	**3623**	-	-
Iron and Steel	-	-	-	-	-	-	221	631	-	-	-
Chemical and Petrochem.	-	-	-	-	-	-	388	4393	3623	-	-
of which: Feedstocks	-	-	-	-	-	-	-	-	*3623*	-	-
Non-Ferrous Metals	-	-	-	-	-	-	12	167	-	-	-
Non-Metallic Minerals	-	-	-	-	-	-	361	360	-	-	-
Transport Equipment	-	-	-	-	-	-	-	-	-	-	-
Machinery	-	-	-	-	-	-	362	219	-	-	-
Mining and Quarrying	-	-	-	-	-	-	724	79	-	-	-
Food and Tobacco	-	-	-	-	-	-	802	265	-	-	-
Paper, Pulp and Print	-	-	-	-	-	-	-	-	-	-	-
Wood and Wood Products	-	-	-	-	-	-	-	-	-	-	-
Construction	-	-	-	-	-	-	-	-	-	-	-
Textile and Leather	-	-	-	-	-	-	324	804	-	-	-
Non-specified	-	731	-	-	-	-	570	449	-	-	-
TRANSPORT SECTOR	-	-	**4965**	**26**	**2165**	-	**30329**	**399**	-	-	-
Air	-	-	-	26	2165	-	-	-	-	-	-
Road	-	-	4965	-	-	-	28282	5	-	-	-
Rail	-	-	-	-	-	-	1710	62	-	-	-
Pipeline Transport	-	-	-	-	-	-	-	-	-	-	-
Internal Navigation	-	-	-	-	-	-	328	332	-	-	-
Non-specified	-	-	-	-	-	-	9	-	-	-	-
OTHER SECTORS	-	**3362**	-	-	-	**9577**	**1622**	**988**	-	-	-
Agriculture	-	-	-	-	-	-	-	345	-	-	-
Comm. and Publ. Services	-	-	-	-	-	-	-	-	-	-	-
Residential	-	3362	-	-	-	9577	1622	464	-	-	-
Non-specified	-	-	-	-	-	-	-	179	-	-	-
NON-ENERGY USE	-	-	-	-	-	-	-	-	-	**354**	**4827**
in Industry/Transf./Energy	-	-	-	-	-	-	-	-	-	354	4827
in Transport	-	-	-	-	-	-	-	-	-	-	-
in Other Sectors	-	-	-	-	-	-	-	-	-	-	-

India / Inde : 1996

	Gas / *Gaz* (TJ)				Comb. Renew. & Waste / *En. Re. Comb. & Déchets* (TJ)				(GWh)	(TJ)
SUPPLY AND CONSUMPTION *APPROVISIONNEMENT ET DEMANDE*	Natural Gas *Gaz naturel*	Gas Works *Usines à gaz*	Coke Ovens *Cokeries*	Blast Furnaces *Hauts fourneaux*	Solid Biomass *Biomasse solide*	Gas/Liquids from Biomass *Gaz/Liquides tirés de biomasse*	Municipal Waste *Déchets urbains*	Industrial Waste *Déchets industriels*	Electricity *Electricité*	Heat *Chaleur*
Production	805113	-	-	-	7976282	-	-	-	435075	-
From Other Sources	-	-	-	-	-	-	-	-	-	-
Imports	-	-	-	-	-	-	-	-	1675	-
Exports	-	-	-	-	-	-	-	-	-130	-
Intl. Marine Bunkers	-	-	-	-	-	-	-	-	-	-
Stock Changes	-	-	-	-	-	-	-	-	-	-
DOMESTIC SUPPLY	**805113**	**-**	**-**	**-**	**7976282**	**-**	**-**	**-**	**436620**	**-**
Transfers	-	-	-	-	-	-	-	-	-	-
Statistical Differences	1	-	-	-	-1	-	-	-	-3347	-
TRANSFORMATION	**278638**	**-**	**-**	**-**	**-**	**-**	**-**	**-**	**-**	**-**
Electricity Plants	278638	-	-	-	-	-	-	-	-	-
CHP Plants	-	-	-	-	-	-	-	-	-	-
Heat Plants	-	-	-	-	-	-	-	-	-	-
Blast Furnaces/Gas Works	-	-	-	-	-	-	-	-	-	-
Coke/Pat. Fuel/BKB Plants	-	-	-	-	-	-	-	-	-	-
Petroleum Refineries	-	-	-	-	-	-	-	-	-	-
Petrochemical Industry	-	-	-	-	-	-	-	-	-	-
Liquefaction	-	-	-	-	-	-	-	-	-	-
Other Transform. Sector	-	-	-	-	-	-	-	-	-	-
ENERGY SECTOR	**102684**	**-**	**-**	**-**	**-**	**-**	**-**	**-**	**27760**	**-**
Coal Mines	-	-	-	-	-	-	-	-	-	-
Oil and Gas Extraction	-	-	-	-	-	-	-	-	-	-
Petroleum Refineries	102684	-	-	-	-	-	-	-	-	-
Electr., CHP+Heat Plants	-	-	-	-	-	-	-	-	27760	-
Pumped Storage (Elec.)	-	-	-	-	-	-	-	-	-	-
Other Energy Sector	-	-	-	-	-	-	-	-	-	-
Distribution Losses	-	-	-	-	-	-	-	-	77727	-
FINAL CONSUMPTION	**423792**	**-**	**-**	**-**	**7976281**	**-**	**-**	**-**	**327786**	**-**
INDUSTRY SECTOR	**406527**	**-**	**-**	**-**	**908690**	**-**	**-**	**-**	**149608**	**-**
Iron and Steel	-	-	-	-	-	-	-	-	11840	-
Chemical and Petrochem.	397308	-	-	-	-	-	-	-	6863	-
of which: Feedstocks	*27709*	-	-	-	-	-	-	-	-	-
Non-Ferrous Metals	-	-	-	-	-	-	-	-	-	-
Non-Metallic Minerals	-	-	-	-	-	-	-	-	-	-
Transport Equipment	-	-	-	-	-	-	-	-	-	-
Machinery	-	-	-	-	-	-	-	-	-	-
Mining and Quarrying	-	-	-	-	-	-	-	-	-	-
Food and Tobacco	-	-	-	-	-	-	-	-	-	-
Paper, Pulp and Print	-	-	-	-	-	-	-	-	-	-
Wood and Wood Products	-	-	-	-	-	-	-	-	-	-
Construction	-	-	-	-	-	-	-	-	-	-
Textile and Leather	-	-	-	-	-	-	-	-	10375	-
Non-specified	9219	-	-	-	908690	-	-	-	120530	-
TRANSPORT SECTOR	**-**	**-**	**-**	**-**	**-**	**-**	**-**	**-**	**6575**	**-**
Air	-	-	-	-	-	-	-	-	-	-
Road	-	-	-	-	-	-	-	-	-	-
Rail	-	-	-	-	-	-	-	-	6575	-
Pipeline Transport	-	-	-	-	-	-	-	-	-	-
Internal Navigation	-	-	-	-	-	-	-	-	-	-
Non-specified	-	-	-	-	-	-	-	-	-	-
OTHER SECTORS	**17265**	**-**	**-**	**-**	**7067591**	**-**	**-**	**-**	**171603**	**-**
Agriculture	7140	-	-	-	-	-	-	-	88585	-
Comm. and Publ. Services	-	-	-	-	-	-	-	-	25784	-
Residential	10125	-	-	-	7067591	-	-	-	53525	-
Non-specified	-	-	-	-	-	-	-	-	3709	-
NON-ENERGY USE	**-**	**-**	**-**	**-**	**-**	**-**	**-**	**-**	**-**	**-**
in Industry/Transf./Energy	-	-	-	-	-	-	-	-	-	-
in Transport	-	-	-	-	-	-	-	-	-	-
in Other Sectors	-	-	-	-	-	-	-	-	-	-

India / Inde : 1997

SUPPLY AND CONSUMPTION / *APPROVISIONNEMENT ET DEMANDE*	Coal / *Charbon* (1000 tonnes)							Oil / *Pétrole* (1000 tonnes)			
	Coking Coal / *Charbon à coke*	Other Bit. Coal / *Autres charb. bit.*	Sub-Bit. Coal / *Charbon sous-bit.*	Lignite / *Lignite*	Peat / *Tourbe*	Oven and Gas Coke / *Coke de four/gaz*	Pat. Fuel and BKB / *Agg./briq. de lignite*	Crude Oil / *Pétrole brut*	NGL / *LGN*	Feed-stocks / *Produits d'aliment.*	Additives / *Additifs*
Production	5837	291333	-	23050	-	11150	-	33826	2937	-	-
From Other Sources	-	-	-	-	-	-	-	-	-	-	-
Imports	9010	440	-	-	-	-	-	32367	-	-	-
Exports	-	-60	-	-	-	-	-	-	-594	-	-
Intl. Marine Bunkers	-	-	-	-	-	-	-	-	-	-	-
Stock Changes	-	550	-	-	-	-	-	-	-	-	-
DOMESTIC SUPPLY	**14847**	**292263**	-	**23050**	-	**11150**	-	**66193**	**2343**	-	-
Transfers	-	-	-	-	-	-	-	-	-1686	-	-
Statistical Differences	-	-	-	-	-	-	-	-	1	-	-
TRANSFORMATION	**14847**	**205530**	-	**17898**	-	**8864**	-	**66193**	-	-	-
Electricity Plants	-	205530	-	17898	-	-	-	-	-	-	-
CHP Plants	-	-	-	-	-	-	-	-	-	-	-
Heat Plants	-	-	-	-	-	-	-	-	-	-	-
Blast Furnaces/Gas Works	-	-	-	-	-	8864	-	-	-	-	-
Coke/Pat. Fuel/BKB Plants	14847	-	-	-	-	-	-	-	-	-	-
Petroleum Refineries	-	-	-	-	-	-	-	66193	-	-	-
Petrochemical Industry	-	-	-	-	-	-	-	-	-	-	-
Liquefaction	-	-	-	-	-	-	-	-	-	-	-
Other Transform. Sector	-	-	-	-	-	-	-	-	-	-	-
ENERGY SECTOR	-	**3090**	-	-	-	-	-	-	-	-	-
Coal Mines	-	3090	-	-	-	-	-	-	-	-	-
Oil and Gas Extraction	-	-	-	-	-	-	-	-	-	-	-
Petroleum Refineries	-	-	-	-	-	-	-	-	-	-	-
Electr., CHP+Heat Plants	-	-	-	-	-	-	-	-	-	-	-
Pumped Storage (Elec.)	-	-	-	-	-	-	-	-	-	-	-
Other Energy Sector	-	-	-	-	-	-	-	-	-	-	-
Distribution Losses	-	-	-	-	-	-	-	-	-	-	-
FINAL CONSUMPTION	-	**83643**	-	**5152**	-	**2286**	-	-	**658**	-	-
INDUSTRY SECTOR	-	**83583**	-	**5152**	-	**2216**	-	-	**658**	-	-
Iron and Steel	-	17800	-	-	-	2216	-	-	-	-	-
Chemical and Petrochem.	-	-	-	5152	-	-	-	-	658	-	-
of which: Feedstocks	-	-	-	-	-	-	-	-	*658*	-	-
Non-Ferrous Metals	-	-	-	-	-	-	-	-	-	-	-
Non-Metallic Minerals	-	10080	-	-	-	-	-	-	-	-	-
Transport Equipment	-	-	-	-	-	-	-	-	-	-	-
Machinery	-	-	-	-	-	-	-	-	-	-	-
Mining and Quarrying	-	-	-	-	-	-	-	-	-	-	-
Food and Tobacco	-	-	-	-	-	-	-	-	-	-	-
Paper, Pulp and Print	-	3430	-	-	-	-	-	-	-	-	-
Wood and Wood Products	-	-	-	-	-	-	-	-	-	-	-
Construction	-	1600	-	-	-	-	-	-	-	-	-
Textile and Leather	-	1360	-	-	-	-	-	-	-	-	-
Non-specified	-	49313	-	-	-	-	-	-	-	-	-
TRANSPORT SECTOR	-	**60**	-	-	-	-	-	-	-	-	-
Air	-	-	-	-	-	-	-	-	-	-	-
Road	-	-	-	-	-	-	-	-	-	-	-
Rail	-	60	-	-	-	-	-	-	-	-	-
Pipeline Transport	-	-	-	-	-	-	-	-	-	-	-
Internal Navigation	-	-	-	-	-	-	-	-	-	-	-
Non-specified	-	-	-	-	-	-	-	-	-	-	-
OTHER SECTORS	-	-	-	-	-	**70**	-	-	-	-	-
Agriculture	-	-	-	-	-	-	-	-	-	-	-
Comm. and Publ. Services	-	-	-	-	-	-	-	-	-	-	-
Residential	-	-	-	-	-	70	-	-	-	-	-
Non-specified	-	-	-	-	-	-	-	-	-	-	-
NON-ENERGY USE	-	-	-	-	-	-	-	-	-	-	-
in Industry/Trans./Energy	-	-	-	-	-	-	-	-	-	-	-
in Transport	-	-	-	-	-	-	-	-	-	-	-
in Other Sectors	-	-	-	-	-	-	-	-	-	-	-

India / Inde : 1997

SUPPLY AND CONSUMPTION *APPROVISIONNEMENT ET DEMANDE*	Oil cont. / *Pétrole cont.* (1000 tonnes)										
	Refinery Gas *Gaz de raffinerie*	LPG + Ethane *GPL + éthane*	Motor Gasoline *Essence moteur*	Aviation Gasoline *Essence aviation*	Jet Fuel *Carbu-réacteurs*	Kerosene *Kérosène*	Gas/ Diesel *Gazole*	Heavy Fuel Oil *Fioul lourd*	Naphtha *Naphta*	Petrol. Coke *Coke de pétrole*	Other Prod. *Autres prod.*
Production	1299	1636	4967	-	2147	6701	24600	13943	6358	282	3877
From Other Sources	-	-	-	-	-	-	-	-	-	-	-
Imports	-	861	-	26	55	3812	14075	141	229	-	141
Exports	-	-22	-	-	-	-	-	-197	-1407	-	-
Intl. Marine Bunkers	-	-	-	-	-	-	-191	-335	-	-	-
Stock Changes	-	-	-	-	-	-	-	-	-	-	-
DOMESTIC SUPPLY	**1299**	**2475**	**4967**	**26**	**2202**	**10513**	**38484**	**13552**	**5180**	**282**	**4018**
Transfers	-	1686	-	-	-	-	-	-	-	-	-
Statistical Differences	-	-1	-	-	-	-644	1	1	-426	-	-14
TRANSFORMATION	**-**	**-**	**-**	**-**	**-**	**-**	**400**	**2200**	**-**	**-**	**-**
Electricity Plants	-	-	-	-	-	-	400	2200	-	-	-
CHP Plants	-	-	-	-	-	-	-	-	-	-	-
Heat Plants	-	-	-	-	-	-	-	-	-	-	-
Blast Furnaces/Gas Works	-	-	-	-	-	-	-	-	-	-	-
Coke/Pat. Fuel/BKB Plants	-	-	-	-	-	-	-	-	-	-	-
Petroleum Refineries	-	-	-	-	-	-	-	-	-	-	-
Petrochemical Industry	-	-	-	-	-	-	-	-	-	-	-
Liquefaction	-	-	-	-	-	-	-	-	-	-	-
Other Transform. Sector	-	-	-	-	-	-	-	-	-	-	-
ENERGY SECTOR	**1299**	**-**	**-**	**-**	**-**	**-**	**735**	**2315**	**-**	**-**	**-**
Coal Mines	-	-	-	-	-	-	735	-	-	-	-
Oil and Gas Extraction	-	-	-	-	-	-	-	-	-	-	-
Petroleum Refineries	1299	-	-	-	-	-	-	2315	-	-	-
Electr., CHP+Heat Plants	-	-	-	-	-	-	-	-	-	-	-
Pumped Storage (Elec.)	-	-	-	-	-	-	-	-	-	-	-
Other Energy Sector	-	-	-	-	-	-	-	-	-	-	-
Distribution Losses	-	-	-	-	-	-	-	-	-	-	-
FINAL CONSUMPTION	**-**	**4160**	**4967**	**26**	**2202**	**9869**	**37350**	**9038**	**4754**	**282**	**4004**
INDUSTRY SECTOR	**-**	**743**	**-**	**-**	**-**	**-**	**3938**	**7606**	**4754**	**-**	**-**
Iron and Steel	-	-	-	-	-	-	231	651	-	-	-
Chemical and Petrochem.	-	-	-	-	-	-	406	4535	4754	-	-
of which: Feedstocks	-	-	-	-	-	-	-	-	*4754*	-	-
Non-Ferrous Metals	-	-	-	-	-	-	13	172	-	-	-
Non-Metallic Minerals	-	-	-	-	-	-	378	372	-	-	-
Transport Equipment	-	-	-	-	-	-	-	-	-	-	-
Machinery	-	-	-	-	-	-	379	226	-	-	-
Mining and Quarrying	-	-	-	-	-	-	757	82	-	-	-
Food and Tobacco	-	-	-	-	-	-	839	274	-	-	-
Paper, Pulp and Print	-	-	-	-	-	-	-	-	-	-	-
Wood and Wood Products	-	-	-	-	-	-	-	-	-	-	-
Construction	-	-	-	-	-	-	-	-	-	-	-
Textile and Leather	-	-	-	-	-	-	339	830	-	-	-
Non-specified	-	743	-	-	-	-	596	464	-	-	-
TRANSPORT SECTOR	**-**	**-**	**4967**	**26**	**2202**	**-**	**31716**	**412**	**-**	**-**	**-**
Air	-	-	-	26	2202	-	-	-	-	-	-
Road	-	-	4967	-	-	-	29576	5	-	-	-
Rail	-	-	-	-	-	-	1788	64	-	-	-
Pipeline Transport	-	-	-	-	-	-	-	-	-	-	-
Internal Navigation	-	-	-	-	-	-	343	343	-	-	-
Non-specified	-	-	-	-	-	-	9	-	-	-	-
OTHER SECTORS	**-**	**3417**	**-**	**-**	**-**	**9869**	**1696**	**1020**	**-**	**-**	**-**
Agriculture	-	-	-	-	-	-	-	356	-	-	-
Comm. and Publ. Services	-	-	-	-	-	-	-	-	-	-	-
Residential	-	3417	-	-	-	9869	1696	479	-	-	-
Non-specified	-	-	-	-	-	-	-	185	-	-	-
NON-ENERGY USE	**-**	**-**	**-**	**-**	**-**	**-**	**-**	**-**	**-**	**282**	**4004**
in Industry/Transf./Energy	-	-	-	-	-	-	-	-	-	282	4004
in Transport	-	-	-	-	-	-	-	-	-	-	-
in Other Sectors	-	-	-	-	-	-	-	-	-	-	-

India / Inde : 1997

	Gas / *Gaz* (TJ)				Comb. Renew. & Waste / *En. Re. Comb. & Déchets* (TJ)				(GWh)	(TJ)
SUPPLY AND CONSUMPTION *APPROVISIONNEMENT ET DEMANDE*	Natural Gas *Gaz naturel*	Gas Works *Usines à gaz*	Coke Ovens *Cokeries*	Blast Furnaces *Hauts fourneaux*	Solid Biomass *Biomasse solide*	Gas/Liquids from Biomass *Gaz/Liquides tirés de biomasse*	Municipal Waste *Déchets urbains*	Industrial Waste *Déchets industriels*	Electricity *Electricité*	Heat *Chaleur*
Production	828972	-	-	-	8073745	-	-	-	463402	-
From Other Sources	-	-	-	-	-	-	-	-	-	-
Imports	-	-	-	-	-	-	-	-	1675	-
Exports	-	-	-	-	-	-	-	-	-130	-
Intl. Marine Bunkers	-	-	-	-	-	-	-	-	-	-
Stock Changes	-	-	-	-	-	-	-	-	-	-
DOMESTIC SUPPLY	**828972**	**-**	**-**	**-**	**8073745**	**-**	**-**	**-**	**464947**	**-**
Transfers	-	-	-	-	-	-	-	-	-	-
Statistical Differences	-1	-	-	-	-	-	-	-	-3565	-
TRANSFORMATION	**282673**	**-**	**-**	**-**	**-**	**-**	**-**	**-**	**-**	**-**
Electricity Plants	282673	-	-	-	-	-	-	-	-	-
CHP Plants	-	-	-	-	-	-	-	-	-	-
Heat Plants	-	-	-	-	-	-	-	-	-	-
Blast Furnaces/Gas Works	-	-	-	-	-	-	-	-	-	-
Coke/Pat. Fuel/BKB Plants	-	-	-	-	-	-	-	-	-	-
Petroleum Refineries	-	-	-	-	-	-	-	-	-	-
Petrochemical Industry	-	-	-	-	-	-	-	-	-	-
Liquefaction	-	-	-	-	-	-	-	-	-	-
Other Transform. Sector	-	-	-	-	-	-	-	-	-	-
ENERGY SECTOR	**107740**	**-**	**-**	**-**	**-**	**-**	**-**	**-**	**29561**	**-**
Coal Mines	-	-	-	-	-	-	-	-	-	-
Oil and Gas Extraction	-	-	-	-	-	-	-	-	-	-
Petroleum Refineries	107740	-	-	-	-	-	-	-	-	-
Electr., CHP+Heat Plants	-	-	-	-	-	-	-	-	29561	-
Pumped Storage (Elec.)	-	-	-	-	-	-	-	-	-	-
Other Energy Sector	-	-	-	-	-	-	-	-	-	-
Distribution Losses	-	-	-	-	-	-	-	-	82770	-
FINAL CONSUMPTION	**438558**	**-**	**-**	**-**	**8073745**	**-**	**-**	**-**	**349051**	**-**
INDUSTRY SECTOR	**419730**	**-**	**-**	**-**	**919794**	**-**	**-**	**-**	**159313**	**-**
Iron and Steel	-	-	-	-	-	-	-	-	12608	-
Chemical and Petrochem.	409189	-	-	-	-	-	-	-	7308	-
of which: Feedstocks	*29755*	-	-	-	-	-	-	-	-	-
Non-Ferrous Metals	-	-	-	-	-	-	-	-	-	-
Non-Metallic Minerals	-	-	-	-	-	-	-	-	-	-
Transport Equipment	-	-	-	-	-	-	-	-	-	-
Machinery	-	-	-	-	-	-	-	-	-	-
Mining and Quarrying	-	-	-	-	-	-	-	-	-	-
Food and Tobacco	-	-	-	-	-	-	-	-	-	-
Paper, Pulp and Print	-	-	-	-	-	-	-	-	-	-
Wood and Wood Products	-	-	-	-	-	-	-	-	-	-
Construction	-	-	-	-	-	-	-	-	-	-
Textile and Leather	-	-	-	-	-	-	-	-	11048	-
Non-specified	10541	-	-	-	919794	-	-	-	128349	-
TRANSPORT SECTOR	**-**	**-**	**-**	**-**	**-**	**-**	**-**	**-**	**7002**	**-**
Air	-	-	-	-	-	-	-	-	-	-
Road	-	-	-	-	-	-	-	-	-	-
Rail	-	-	-	-	-	-	-	-	7002	-
Pipeline Transport	-	-	-	-	-	-	-	-	-	-
Internal Navigation	-	-	-	-	-	-	-	-	-	-
Non-specified	-	-	-	-	-	-	-	-	-	-
OTHER SECTORS	**18828**	**-**	**-**	**-**	**7153951**	**-**	**-**	**-**	**182736**	**-**
Agriculture	8362	-	-	-	-	-	-	-	94332	-
Comm. and Publ. Services	-	-	-	-	-	-	-	-	27457	-
Residential	10466	-	-	-	7153951	-	-	-	56997	-
Non-specified	-	-	-	-	-	-	-	-	3950	-
NON-ENERGY USE	**-**	**-**	**-**	**-**	**-**	**-**	**-**	**-**	**-**	**-**
in Industry/Transf./Energy	-	-	-	-	-	-	-	-	-	-
in Transport	-	-	-	-	-	-	-	-	-	-
in Other Sectors	-	-	-	-	-	-	-	-	-	-

Indonesia / Indonésie : 1996

SUPPLY AND CONSUMPTION / *APPROVISIONNEMENT ET DEMANDE*	Coal / *Charbon* (1000 tonnes)							Oil / *Pétrole* (1000 tonnes)			
	Coking Coal / *Charbon à coke*	Other Bit. Coal / *Autres charb. bit.*	Sub-Bit. Coal / *Charbon sous-bit.*	Lignite / *Lignite*	Peat / *Tourbe*	Oven and Gas Coke / *Coke de four/gaz*	Pat. Fuel and BKB / *Agg./briq. de lignite*	Crude Oil / *Pétrole brut*	NGL / *LGN*	Feed-stocks / *Produits d'aliment.*	Additives / *Additifs*
Production	-	50157	-	-	-	-	-	74113	2930	-	-
From Other Sources	-	-	-	-	-	-	-	-	-	-	-
Imports	-	411	-	-	-	-	-	8954	-	-	-
Exports	-	-36443	-	-	-	-	-	-38061	-	-	-
Intl. Marine Bunkers	-	-	-	-	-	-	-	-	-	-	-
Stock Changes	-	413	-	-	-	-	-	-	-	-	-
DOMESTIC SUPPLY	**-**	**14538**	**-**	**-**	**-**	**-**	**-**	**45006**	**2930**	**-**	**-**
Transfers	-	-	-	-	-	-	-	-	-2930	-	-
Statistical Differences	-	-1960	-	-	-	-	-	319	-	-	-
TRANSFORMATION	**-**	**7456**	**-**	**-**	**-**	**-**	**-**	**45325**	**-**	**-**	**-**
Electricity Plants	-	7456	-	-	-	-	-	-	-	-	-
CHP Plants	-	-	-	-	-	-	-	-	-	-	-
Heat Plants	-	-	-	-	-	-	-	-	-	-	-
Blast Furnaces/Gas Works	-	-	-	-	-	-	-	-	-	-	-
Coke/Pat. Fuel/BKB Plants	-	-	-	-	-	-	-	-	-	-	-
Petroleum Refineries	-	-	-	-	-	-	-	45325	-	-	-
Petrochemical Industry	-	-	-	-	-	-	-	-	-	-	-
Liquefaction	-	-	-	-	-	-	-	-	-	-	-
Other Transform. Sector	-	-	-	-	-	-	-	-	-	-	-
ENERGY SECTOR	**-**	**-**	**-**	**-**	**-**	**-**	**-**	**-**	**-**	**-**	**-**
Coal Mines	-	-	-	-	-	-	-	-	-	-	-
Oil and Gas Extraction	-	-	-	-	-	-	-	-	-	-	-
Petroleum Refineries	-	-	-	-	-	-	-	-	-	-	-
Electr., CHP+Heat Plants	-	-	-	-	-	-	-	-	-	-	-
Pumped Storage (Elec.)	-	-	-	-	-	-	-	-	-	-	-
Other Energy Sector	-	-	-	-	-	-	-	-	-	-	-
Distribution Losses	-	-	-	-	-	-	-	-	-	-	-
FINAL CONSUMPTION	**-**	**5122**	**-**	**-**	**-**	**-**	**-**	**-**	**-**	**-**	**-**
INDUSTRY SECTOR	**-**	**5122**	**-**	**-**	**-**	**-**	**-**	**-**	**-**	**-**	**-**
Iron and Steel	-	-	-	-	-	-	-	-	-	-	-
Chemical and Petrochem.	-	-	-	-	-	-	-	-	-	-	-
of which: Feedstocks	-	-	-	-	-	-	-	-	-	-	-
Non-Ferrous Metals	-	-	-	-	-	-	-	-	-	-	-
Non-Metallic Minerals	-	5122	-	-	-	-	-	-	-	-	-
Transport Equipment	-	-	-	-	-	-	-	-	-	-	-
Machinery	-	-	-	-	-	-	-	-	-	-	-
Mining and Quarrying	-	-	-	-	-	-	-	-	-	-	-
Food and Tobacco	-	-	-	-	-	-	-	-	-	-	-
Paper, Pulp and Print	-	-	-	-	-	-	-	-	-	-	-
Wood and Wood Products	-	-	-	-	-	-	-	-	-	-	-
Construction	-	-	-	-	-	-	-	-	-	-	-
Textile and Leather	-	-	-	-	-	-	-	-	-	-	-
Non-specified	-	-	-	-	-	-	-	-	-	-	-
TRANSPORT SECTOR	**-**	**-**	**-**	**-**	**-**	**-**	**-**	**-**	**-**	**-**	**-**
Air	-	-	-	-	-	-	-	-	-	-	-
Road	-	-	-	-	-	-	-	-	-	-	-
Rail	-	-	-	-	-	-	-	-	-	-	-
Pipeline Transport	-	-	-	-	-	-	-	-	-	-	-
Internal Navigation	-	-	-	-	-	-	-	-	-	-	-
Non-specified	-	-	-	-	-	-	-	-	-	-	-
OTHER SECTORS	**-**	**-**	**-**	**-**	**-**	**-**	**-**	**-**	**-**	**-**	**-**
Agriculture	-	-	-	-	-	-	-	-	-	-	-
Comm. and Publ. Services	-	-	-	-	-	-	-	-	-	-	-
Residential	-	-	-	-	-	-	-	-	-	-	-
Non-specified	-	-	-	-	-	-	-	-	-	-	-
NON-ENERGY USE	**-**	**-**	**-**	**-**	**-**	**-**	**-**	**-**	**-**	**-**	**-**
in Industry/Trans./Energy	-	-	-	-	-	-	-	-	-	-	-
in Transport	-	-	-	-	-	-	-	-	-	-	-
in Other Sectors	-	-	-	-	-	-	-	-	-	-	-

Indonesia / Indonésie : 1996

SUPPLY AND CONSUMPTION / *APPROVISIONNEMENT ET DEMANDE*	Oil cont. / *Pétrole cont.* (1000 tonnes)										
	Refinery Gas / *Gaz de raffinerie*	LPG + Ethane / *GPL + éthane*	Motor Gasoline / *Essence moteur*	Aviation Gasoline / *Essence aviation*	Jet Fuel / *Carbu-réacteurs*	Kerosene / *Kérosène*	Gas/ Diesel / *Gazole*	Heavy Fuel Oil / *Fioul lourd*	Naphtha / *Naphta*	Petrol. Coke / *Coke de pétrole*	Other Prod. / *Autres prod.*
Production	591	511	6834	2	1143	6851	12492	12087	1181	-	1500
From Other Sources	-	-	-	-	-	-	-	-	-	-	-
Imports	-	-	877	-	1566	1836	3911	1843	-	-	-
Exports	-	-2720	-315	-	-297	-739	-422	-8241	-883	-	-153
Intl. Marine Bunkers	-	-	-	-	-	-	-24	-320	-	-	-
Stock Changes	-	-	-	4	-	-	-	-	-	-	-
DOMESTIC SUPPLY	**591**	**-2209**	**7396**	**6**	**2412**	**7948**	**15957**	**5369**	**298**	**-**	**1347**
Transfers	-	2930	-	-	-	-	-	-	-	-	-
Statistical Differences	-	-	-	-	-814	-	1441	-324	-	-	-217
TRANSFORMATION	**-**	**-**	**-**	**-**	**-**	**-**	**2240**	**1019**	**-**	**-**	**-**
Electricity Plants	-	-	-	-	-	-	2240	1019	-	-	-
CHP Plants	-	-	-	-	-	-	-	-	-	-	-
Heat Plants	-	-	-	-	-	-	-	-	-	-	-
Blast Furnaces/Gas Works	-	-	-	-	-	-	-	-	-	-	-
Coke/Pat. Fuel/BKB Plants	-	-	-	-	-	-	-	-	-	-	-
Petroleum Refineries	-	-	-	-	-	-	-	-	-	-	-
Petrochemical Industry	-	-	-	-	-	-	-	-	-	-	-
Liquefaction	-	-	-	-	-	-	-	-	-	-	-
Other Transform. Sector	-	-	-	-	-	-	-	-	-	-	-
ENERGY SECTOR	**591**	**-**	**-**	**-**	**-**	**-**	**45**	**1322**	**-**	**-**	**-**
Coal Mines	-	-	-	-	-	-	-	-	-	-	-
Oil and Gas Extraction	-	-	-	-	-	-	-	-	-	-	-
Petroleum Refineries	591	-	-	-	-	-	45	1322	-	-	-
Electr., CHP+Heat Plants	-	-	-	-	-	-	-	-	-	-	-
Pumped Storage (Elec.)	-	-	-	-	-	-	-	-	-	-	-
Other Energy Sector	-	-	-	-	-	-	-	-	-	-	-
Distribution Losses	-	-	-	-	-	-	-	-	-	-	-
FINAL CONSUMPTION	**-**	**721**	**7396**	**6**	**1598**	**7948**	**15113**	**2704**	**298**	**-**	**1130**
INDUSTRY SECTOR	**-**	**313**	**-**	**-**	**-**	**93**	**4526**	**2420**	**298**	**-**	**-**
Iron and Steel	-	-	-	-	-	-	201	668	-	-	-
Chemical and Petrochem.	-	-	-	-	-	-	419	413	298	-	-
of which: Feedstocks	-	-	-	-	-	-	-	-	-	-	-
Non-Ferrous Metals	-	-	-	-	-	-	-	-	-	-	-
Non-Metallic Minerals	-	-	-	-	-	-	611	517	-	-	-
Transport Equipment	-	-	-	-	-	-	-	-	-	-	-
Machinery	-	-	-	-	-	-	67	-	-	-	-
Mining and Quarrying	-	-	-	-	-	-	915	117	-	-	-
Food and Tobacco	-	-	-	-	-	-	526	162	-	-	-
Paper, Pulp and Print	-	-	-	-	-	-	-	-	-	-	-
Wood and Wood Products	-	-	-	-	-	-	-	-	-	-	-
Construction	-	-	-	-	-	-	326	-	-	-	-
Textile and Leather	-	-	-	-	-	-	1012	168	-	-	-
Non-specified	-	313	-	-	-	93	449	375	-	-	-
TRANSPORT SECTOR	**-**	**-**	**7396**	**6**	**1598**	**-**	**8774**	**179**	**-**	**-**	**-**
Air	-	-	-	6	1598	-	-	-	-	-	-
Road	-	-	7396	-	-	-	7467	-	-	-	-
Rail	-	-	-	-	-	-	-	-	-	-	-
Pipeline Transport	-	-	-	-	-	-	-	-	-	-	-
Internal Navigation	-	-	-	-	-	-	1307	179	-	-	-
Non-specified	-	-	-	-	-	-	-	-	-	-	-
OTHER SECTORS	**-**	**408**	**-**	**-**	**-**	**7855**	**1813**	**105**	**-**	**-**	**-**
Agriculture	-	-	-	-	-	-	1514	105	-	-	-
Comm. and Publ. Services	-	-	-	-	-	-	299	-	-	-	-
Residential	-	408	-	-	-	7855	-	-	-	-	-
Non-specified	-	-	-	-	-	-	-	-	-	-	-
NON-ENERGY USE	**-**	**-**	**-**	**-**	**-**	**-**	**-**	**-**	**-**	**-**	**1130**
in Industry/Transf./Energy	-	-	-	-	-	-	-	-	-	-	1130
in Transport	-	-	-	-	-	-	-	-	-	-	-
in Other Sectors	-	-	-	-	-	-	-	-	-	-	-

Indonesia / Indonésie : 1996

SUPPLY AND CONSUMPTION / *APPROVISIONNEMENT ET DEMANDE*	Gas / *Gaz* (TJ)				Comb. Renew. & Waste / *En. Re. Comb. & Déchets* (TJ)				(GWh)	(TJ)
	Natural Gas / *Gaz naturel*	Gas Works / *Usines à gaz*	Coke Ovens / *Cokeries*	Blast Furnaces / *Hauts fourneaux*	Solid Biomass / *Biomasse solide*	Gas/Liquids from Biomass / *Gaz/Liquides tirés de biomasse*	Municipal Waste / *Déchets urbains*	Industrial Waste / *Déchets industriels*	Electricity / *Electricité*	Heat / *Chaleur*
Production	2931762	22000	-	-	1867465	-	-	-	67062	-
From Other Sources	-	-	-	-	-	-	-	-	-	-
Imports	-	-	-	-	-	-	-	-	-	-
Exports	-1450383	-	-	-	-4256	-	-	-	-	-
Intl. Marine Bunkers	-	-	-	-	-	-	-	-	-	-
Stock Changes	-	-	-	-	-	-	-	-	-	-
DOMESTIC SUPPLY	**1481379**	**22000**	**-**	**-**	**1863209**	**-**	**-**	**-**	**67062**	**-**
Transfers	-	-	-	-	-	-	-	-	-	-
Statistical Differences	-1538	-	-	-	371	-	-	-	2426	-
TRANSFORMATION	**268634**	**-**	**-**	**-**	**5094**	**-**	**-**	**-**	**-**	**-**
Electricity Plants	268634	-	-	-	-	-	-	-	-	-
CHP Plants	-	-	-	-	-	-	-	-	-	-
Heat Plants	-	-	-	-	-	-	-	-	-	-
Blast Furnaces/Gas Works	-	-	-	-	-	-	-	-	-	-
Coke/Pat. Fuel/BKB Plants	-	-	-	-	-	-	-	-	-	-
Petroleum Refineries	-	-	-	-	-	-	-	-	-	-
Petrochemical Industry	-	-	-	-	-	-	-	-	-	-
Liquefaction	-	-	-	-	-	-	-	-	-	-
Other Transform. Sector	-	-	-	-	5094	-	-	-	-	-
ENERGY SECTOR	**799319**	**-**	**-**	**-**	**-**	**-**	**-**	**-**	**2805**	**-**
Coal Mines	-	-	-	-	-	-	-	-	-	-
Oil and Gas Extraction	753945	-	-	-	-	-	-	-	-	-
Petroleum Refineries	45374	-	-	-	-	-	-	-	-	-
Electr., CHP+Heat Plants	-	-	-	-	-	-	-	-	2805	-
Pumped Storage (Elec.)	-	-	-	-	-	-	-	-	-	-
Other Energy Sector	-	-	-	-	-	-	-	-	-	-
Distribution Losses	-	-	-	-	-	-	-	-	8280	-
FINAL CONSUMPTION	**411888**	**22000**	**-**	**-**	**1858486**	**-**	**-**	**-**	**58403**	**-**
INDUSTRY SECTOR	**352049**	**-**	**-**	**-**	**-**	**-**	**-**	**-**	**28410**	**-**
Iron and Steel	49403	-	-	-	-	-	-	-	-	-
Chemical and Petrochem.	269939	-	-	-	-	-	-	-	-	-
of which: Feedstocks	*269939*	-	-	-	-	-	-	-	-	-
Non-Ferrous Metals	-	-	-	-	-	-	-	-	-	-
Non-Metallic Minerals	6036	-	-	-	-	-	-	-	-	-
Transport Equipment	-	-	-	-	-	-	-	-	-	-
Machinery	-	-	-	-	-	-	-	-	-	-
Mining and Quarrying	-	-	-	-	-	-	-	-	-	-
Food and Tobacco	-	-	-	-	-	-	-	-	-	-
Paper, Pulp and Print	-	-	-	-	-	-	-	-	-	-
Wood and Wood Products	-	-	-	-	-	-	-	-	-	-
Construction	-	-	-	-	-	-	-	-	-	-
Textile and Leather	-	-	-	-	-	-	-	-	-	-
Non-specified	26671	-	-	-	-	-	-	-	28410	-
TRANSPORT SECTOR	**-**	**-**	**-**	**-**	**-**	**-**	**-**	**-**	**-**	**-**
Air	-	-	-	-	-	-	-	-	-	-
Road	-	-	-	-	-	-	-	-	-	-
Rail	-	-	-	-	-	-	-	-	-	-
Pipeline Transport	-	-	-	-	-	-	-	-	-	-
Internal Navigation	-	-	-	-	-	-	-	-	-	-
Non-specified	-	-	-	-	-	-	-	-	-	-
OTHER SECTORS	**59839**	**22000**	**-**	**-**	**1858486**	**-**	**-**	**-**	**29993**	**-**
Agriculture	-	-	-	-	-	-	-	-	-	-
Comm. and Publ. Services	-	11000	-	-	-	-	-	-	6155	-
Residential	59839	11000	-	-	1858486	-	-	-	20536	-
Non-specified	-	-	-	-	-	-	-	-	3302	-
NON-ENERGY USE	**-**	**-**	**-**	**-**	**-**	**-**	**-**	**-**	**-**	**-**
in Industry/Transf./Energy	-	-	-	-	-	-	-	-	-	-
in Transport	-	-	-	-	-	-	-	-	-	-
in Other Sectors	-	-	-	-	-	-	-	-	-	-

Indonesia / Indonésie : 1997

SUPPLY AND CONSUMPTION / *APPROVISIONNEMENT ET DEMANDE*	Coal / *Charbon* (1000 tonnes)							Oil / *Pétrole* (1000 tonnes)			
	Coking Coal / *Charbon à coke*	Other Bit. Coal / *Autres charb. bit.*	Sub-Bit. Coal / *Charbon sous-bit.*	Lignite / *Lignite*	Peat / *Tourbe*	Oven and Gas Coke / *Coke de four/gaz*	Pat. Fuel and BKB / *Agg./briq. de lignite*	Crude Oil / *Pétrole brut*	NGL / *LGN*	Feed-stocks / *Produits d'aliment.*	Additives / *Additifs*
Production	-	55102	-	-	-	-	-	72797	3045	-	-
From Other Sources	-	-	-	-	-	-	-	-	-	-	-
Imports	-	385	-	-	-	-	-	8544	-	-	-
Exports	-	-41474	-	-	-	-	-	-38649	-	-	-
Intl. Marine Bunkers	-	-	-	-	-	-	-	-	-	-	-
Stock Changes	-	1507	-	-	-	-	-	-	-	-	-
DOMESTIC SUPPLY	**-**	**15520**	**-**	**-**	**-**	**-**	**-**	**42692**	**3045**	**-**	**-**
Transfers	-	-	-	-	-	-	-	-	-2179	-	-
Statistical Differences	-	-1179	-	-	-	-	-	3318	-866	-	-
TRANSFORMATION	**-**	**9888**	**-**	**-**	**-**	**-**	**-**	**46010**	**-**	**-**	**-**
Electricity Plants	-	9888	-	-	-	-	-	-	-	-	-
CHP Plants	-	-	-	-	-	-	-	-	-	-	-
Heat Plants	-	-	-	-	-	-	-	-	-	-	-
Blast Furnaces/Gas Works	-	-	-	-	-	-	-	-	-	-	-
Coke/Pat. Fuel/BKB Plants	-	-	-	-	-	-	-	-	-	-	-
Petroleum Refineries	-	-	-	-	-	-	-	46010	-	-	-
Petrochemical Industry	-	-	-	-	-	-	-	-	-	-	-
Liquefaction	-	-	-	-	-	-	-	-	-	-	-
Other Transform. Sector	-	-	-	-	-	-	-	-	-	-	-
ENERGY SECTOR	**-**	**-**	**-**	**-**	**-**	**-**	**-**	**-**	**-**	**-**	**-**
Coal Mines	-	-	-	-	-	-	-	-	-	-	-
Oil and Gas Extraction	-	-	-	-	-	-	-	-	-	-	-
Petroleum Refineries	-	-	-	-	-	-	-	-	-	-	-
Electr., CHP+Heat Plants	-	-	-	-	-	-	-	-	-	-	-
Pumped Storage (Elec.)	-	-	-	-	-	-	-	-	-	-	-
Other Energy Sector	-	-	-	-	-	-	-	-	-	-	-
Distribution Losses	-	-	-	-	-	-	-	-	-	-	-
FINAL CONSUMPTION	**-**	**4453**	**-**	**-**	**-**	**-**	**-**	**-**	**-**	**-**	**-**
INDUSTRY SECTOR	**-**	**4453**	**-**	**-**	**-**	**-**	**-**	**-**	**-**	**-**	**-**
Iron and Steel	-	-	-	-	-	-	-	-	-	-	-
Chemical and Petrochem.	-	-	-	-	-	-	-	-	-	-	-
of which: Feedstocks	-	-	-	-	-	-	-	-	-	-	-
Non-Ferrous Metals	-	-	-	-	-	-	-	-	-	-	-
Non-Metallic Minerals	-	4453	-	-	-	-	-	-	-	-	-
Transport Equipment	-	-	-	-	-	-	-	-	-	-	-
Machinery	-	-	-	-	-	-	-	-	-	-	-
Mining and Quarrying	-	-	-	-	-	-	-	-	-	-	-
Food and Tobacco	-	-	-	-	-	-	-	-	-	-	-
Paper, Pulp and Print	-	-	-	-	-	-	-	-	-	-	-
Wood and Wood Products	-	-	-	-	-	-	-	-	-	-	-
Construction	-	-	-	-	-	-	-	-	-	-	-
Textile and Leather	-	-	-	-	-	-	-	-	-	-	-
Non-specified	-	-	-	-	-	-	-	-	-	-	-
TRANSPORT SECTOR	**-**	**-**	**-**	**-**	**-**	**-**	**-**	**-**	**-**	**-**	**-**
Air	-	-	-	-	-	-	-	-	-	-	-
Road	-	-	-	-	-	-	-	-	-	-	-
Rail	-	-	-	-	-	-	-	-	-	-	-
Pipeline Transport	-	-	-	-	-	-	-	-	-	-	-
Internal Navigation	-	-	-	-	-	-	-	-	-	-	-
Non-specified	-	-	-	-	-	-	-	-	-	-	-
OTHER SECTORS	**-**	**-**	**-**	**-**	**-**	**-**	**-**	**-**	**-**	**-**	**-**
Agriculture	-	-	-	-	-	-	-	-	-	-	-
Comm. and Publ. Services	-	-	-	-	-	-	-	-	-	-	-
Residential	-	-	-	-	-	-	-	-	-	-	-
Non-specified	-	-	-	-	-	-	-	-	-	-	-
NON-ENERGY USE	**-**	**-**	**-**	**-**	**-**	**-**	**-**	**-**	**-**	**-**	**-**
in Industry/Trans./Energy	-	-	-	-	-	-	-	-	-	-	-
in Transport	-	-	-	-	-	-	-	-	-	-	-
in Other Sectors	-	-	-	-	-	-	-	-	-	-	-

Indonesia / Indonésie : 1997

SUPPLY AND CONSUMPTION / *APPROVISIONNEMENT ET DEMANDE*	Oil cont. / *Pétrole cont.* (1000 tonnes)										
	Refinery Gas / *Gaz de raffinerie*	LPG + Ethane / *GPL + éthane*	Motor Gasoline / *Essence moteur*	Aviation Gasoline / *Essence aviation*	Jet Fuel / *Carbu-réacteurs*	Kerosene / *Kérosène*	Gas/ Diesel / *Gazole*	Heavy Fuel Oil / *Fioul lourd*	Naphtha / *Naphta*	Petrol. Coke / *Coke de pétrole*	Other Prod. / *Autres prod.*
Production	581	518	7950	7	969	6266	12258	12835	1082	-	1500
From Other Sources	-	-	-	-	-	-	-	-	-	-	-
Imports	-	-	-	-	2360	1826	7265	2498	-	-	-
Exports	-	-2133	-515	-	-60	-	-	-6236	-893	-	-153
Intl. Marine Bunkers	-	-	-	-	-	-	-12	-332	-	-	-
Stock Changes	-	-	-	-	-	-	-	-	-	-	-
DOMESTIC SUPPLY	**581**	**-1615**	**7435**	**7**	**3269**	**8092**	**19511**	**8765**	**189**	**-**	**1347**
Transfers	-	2179	-	-	-	-	-	-	-	-	-
Statistical Differences	-	1	-	-	-1598	1	-	1	-	-	-217
TRANSFORMATION	**-**	**-**	**-**	**-**	**-**	**-**	**3239**	**1573**	**-**	**-**	**-**
Electricity Plants	-	-	-	-	-	-	3239	1573	-	-	-
CHP Plants	-	-	-	-	-	-	-	-	-	-	-
Heat Plants	-	-	-	-	-	-	-	-	-	-	-
Blast Furnaces/Gas Works	-	-	-	-	-	-	-	-	-	-	-
Coke/Pat. Fuel/BKB Plants	-	-	-	-	-	-	-	-	-	-	-
Petroleum Refineries	-	-	-	-	-	-	-	-	-	-	-
Petrochemical Industry	-	-	-	-	-	-	-	-	-	-	-
Liquefaction	-	-	-	-	-	-	-	-	-	-	-
Other Transform. Sector	-	-	-	-	-	-	-	-	-	-	-
ENERGY SECTOR	**581**	**-**	**-**	**-**	**-**	**-**	**45**	**1322**	**-**	**-**	**-**
Coal Mines	-	-	-	-	-	-	-	-	-	-	-
Oil and Gas Extraction	-	-	-	-	-	-	-	-	-	-	-
Petroleum Refineries	581	-	-	-	-	-	45	1322	-	-	-
Electr., CHP+Heat Plants	-	-	-	-	-	-	-	-	-	-	-
Pumped Storage (Elec.)	-	-	-	-	-	-	-	-	-	-	-
Other Energy Sector	-	-	-	-	-	-	-	-	-	-	-
Distribution Losses	-	-	-	-	-	-	-	-	-	-	-
FINAL CONSUMPTION	**-**	**565**	**7435**	**7**	**1671**	**8093**	**16227**	**5871**	**189**	**-**	**1130**
INDUSTRY SECTOR	**-**	**245**	**-**	**-**	**-**	**95**	**4860**	**5254**	**189**	**-**	**-**
Iron and Steel	-	-	-	-	-	-	216	1450	-	-	-
Chemical and Petrochem.	-	-	-	-	-	-	450	897	189	-	-
of which: Feedstocks	-	-	-	-	-	-	-	-	-	-	-
Non-Ferrous Metals	-	-	-	-	-	-	-	-	-	-	-
Non-Metallic Minerals	-	-	-	-	-	-	656	1122	-	-	-
Transport Equipment	-	-	-	-	-	-	-	-	-	-	-
Machinery	-	-	-	-	-	-	72	-	-	-	-
Mining and Quarrying	-	-	-	-	-	-	982	254	-	-	-
Food and Tobacco	-	-	-	-	-	-	565	352	-	-	-
Paper, Pulp and Print	-	-	-	-	-	-	-	-	-	-	-
Wood and Wood Products	-	-	-	-	-	-	-	-	-	-	-
Construction	-	-	-	-	-	-	350	-	-	-	-
Textile and Leather	-	-	-	-	-	-	1087	365	-	-	-
Non-specified	-	245	-	-	-	95	482	814	-	-	-
TRANSPORT SECTOR	**-**	**-**	**7435**	**7**	**1671**	**-**	**9420**	**389**	**-**	**-**	**-**
Air	-	-	-	7	1671	-	-	-	-	-	-
Road	-	-	7435	-	-	-	8017	-	-	-	-
Rail	-	-	-	-	-	-	-	-	-	-	-
Pipeline Transport	-	-	-	-	-	-	-	-	-	-	-
Internal Navigation	-	-	-	-	-	-	1403	389	-	-	-
Non-specified	-	-	-	-	-	-	-	-	-	-	-
OTHER SECTORS	**-**	**320**	**-**	**-**	**-**	**7998**	**1947**	**228**	**-**	**-**	**-**
Agriculture	-	-	-	-	-	-	1626	228	-	-	-
Comm. and Publ. Services	-	-	-	-	-	-	321	-	-	-	-
Residential	-	320	-	-	-	7998	-	-	-	-	-
Non-specified	-	-	-	-	-	-	-	-	-	-	-
NON-ENERGY USE	**-**	**-**	**-**	**-**	**-**	**-**	**-**	**-**	**-**	**-**	**1130**
in Industry/Transf./Energy	-	-	-	-	-	-	-	-	-	-	1130
in Transport	-	-	-	-	-	-	-	-	-	-	-
in Other Sectors	-	-	-	-	-	-	-	-	-	-	-

Indonesia / Indonésie : 1997

SUPPLY AND CONSUMPTION / *APPROVISIONNEMENT ET DEMANDE*	Gas / *Gaz* (TJ)				Comb. Renew. & Waste / *En. Re. Comb. & Déchets* (TJ)				(GWh)	(TJ)
	Natural Gas / *Gaz naturel*	Gas Works / *Usines à gaz*	Coke Ovens / *Cokeries*	Blast Furnaces / *Hauts fourneaux*	Solid Biomass / *Biomasse solide*	Gas/Liquids from Biomass / *Gaz/Liquides tirés de biomasse*	Municipal Waste / *Déchets urbains*	Industrial Waste / *Déchets industriels*	Electricity / *Electricité*	Heat / *Chaleur*
Production	2933615	22000	-	-	1867465	-	-	-	74832	-
From Other Sources	-	-	-	-	-	-	-	-	-	-
Imports	-	-	-	-	-	-	-	-	-	-
Exports	-1467535	-	-	-	-4256	-	-	-	-	-
Intl. Marine Bunkers	-	-	-	-	-	-	-	-	-	-
Stock Changes	-	-	-	-	-	-	-	-	-	-
DOMESTIC SUPPLY	**1466080**	**22000**	**-**	**-**	**1863209**	**-**	**-**	**-**	**74832**	**-**
Transfers	-	-	-	-	-	-	-	-	-	-
Statistical Differences	3553	-	-	-	371	-	-	-	2681	-
TRANSFORMATION	**260729**	**-**	**-**	**-**	**5094**	**-**	**-**	**-**	**-**	**-**
Electricity Plants	260729	-	-	-	-	-	-	-	-	-
CHP Plants	-	-	-	-	-	-	-	-	-	-
Heat Plants	-	-	-	-	-	-	-	-	-	-
Blast Furnaces/Gas Works	-	-	-	-	-	-	-	-	-	-
Coke/Pat. Fuel/BKB Plants	-	-	-	-	-	-	-	-	-	-
Petroleum Refineries	-	-	-	-	-	-	-	-	-	-
Petrochemical Industry	-	-	-	-	-	-	-	-	-	-
Liquefaction	-	-	-	-	-	-	-	-	-	-
Other Transform. Sector	-	-	-	-	5094	-	-	-	-	-
ENERGY SECTOR	**771544**	**-**	**-**	**-**	**-**	**-**	**-**	**-**	**2919**	**-**
Coal Mines	-	-	-	-	-	-	-	-	-	-
Oil and Gas Extraction	723608	-	-	-	-	-	-	-	-	-
Petroleum Refineries	47936	-	-	-	-	-	-	-	-	-
Electr., CHP+Heat Plants	-	-	-	-	-	-	-	-	2919	-
Pumped Storage (Elec.)	-	-	-	-	-	-	-	-	-	-
Other Energy Sector	-	-	-	-	-	-	-	-	-	-
Distribution Losses	-	-	-	-	-	-	-	-	8616	-
FINAL CONSUMPTION	**437360**	**22000**	**-**	**-**	**1858486**	**-**	**-**	**-**	**65978**	**-**
INDUSTRY SECTOR	**366163**	**-**	**-**	**-**	**-**	**-**	**-**	**-**	**31270**	**-**
Iron and Steel	45821	-	-	-	-	-	-	-	-	-
Chemical and Petrochem.	278630	-	-	-	-	-	-	-	-	-
of which: Feedstocks	*278630*	-	-	-	-	-	-	-	-	-
Non-Ferrous Metals	-	-	-	-	-	-	-	-	-	-
Non-Metallic Minerals	5190	-	-	-	-	-	-	-	-	-
Transport Equipment	-	-	-	-	-	-	-	-	-	-
Machinery	-	-	-	-	-	-	-	-	-	-
Mining and Quarrying	-	-	-	-	-	-	-	-	-	-
Food and Tobacco	-	-	-	-	-	-	-	-	-	-
Paper, Pulp and Print	-	-	-	-	-	-	-	-	-	-
Wood and Wood Products	-	-	-	-	-	-	-	-	-	-
Construction	-	-	-	-	-	-	-	-	-	-
Textile and Leather	-	-	-	-	-	-	-	-	-	-
Non-specified	36522	-	-	-	-	-	-	-	31270	-
TRANSPORT SECTOR	**-**	**-**	**-**	**-**	**-**	**-**	**-**	**-**	**-**	**-**
Air	-	-	-	-	-	-	-	-	-	-
Road	-	-	-	-	-	-	-	-	-	-
Rail	-	-	-	-	-	-	-	-	-	-
Pipeline Transport	-	-	-	-	-	-	-	-	-	-
Internal Navigation	-	-	-	-	-	-	-	-	-	-
Non-specified	-	-	-	-	-	-	-	-	-	-
OTHER SECTORS	**71197**	**22000**	**-**	**-**	**1858486**	**-**	**-**	**-**	**34708**	**-**
Agriculture	-	-	-	-	-	-	-	-	-	-
Comm. and Publ. Services	-	11000	-	-	-	-	-	-	7160	-
Residential	71197	11000	-	-	1858486	-	-	-	23889	-
Non-specified	-	-	-	-	-	-	-	-	3659	-
NON-ENERGY USE	**-**	**-**	**-**	**-**	**-**	**-**	**-**	**-**	**-**	**-**
in Industry/Transf./Energy	-	-	-	-	-	-	-	-	-	-
in Transport	-	-	-	-	-	-	-	-	-	-
in Other Sectors	-	-	-	-	-	-	-	-	-	-

Iran

SUPPLY AND CONSUMPTION 1996	Coal (1000 tonnes)							Oil (1000 tonnes)			
	Coking Coal	Other Bit. Coal	Sub-Bit. Coal	Lignite	Peat	Oven and Gas Coke	Pat. Fuel and BKB	Crude Oil	NGL	Feed-stocks	Additives
Production	402	520	-	-	-	890	-	182152	1660	-	-
Imports	598	-	-	-	-	-	-	-	-	-	-
Exports	-	-13	-	-	-	-	-	-119631	-	-	-
Intl. Marine Bunkers	-	-	-	-	-	-	-	-	-	-	-
Stock Changes	-	-	-	-	-	-	-	-3625	-	-	-
DOMESTIC SUPPLY	**1000**	**507**	**-**	**-**	**-**	**890**	**-**	**58896**	**1660**	**-**	**-**
Transfers and Stat. Diff.	-	243	-	-	-	-	-	1	-	-	-
TRANSFORMATION	**1000**	**-**	**-**	**-**	**-**	**712**	**-**	**58897**	**1660**	**-**	**-**
Electricity and CHP Plants	-	-	-	-	-	-	-	-	-	-	-
Petroleum Refineries	-	-	-	-	-	-	-	58897	1660	-	-
Other Transform. Sector	1000	-	-	-	-	712	-	-	-	-	-
ENERGY SECTOR	**-**	**-**	**-**	**-**	**-**	**-**	**-**	**-**	**-**	**-**	**-**
DISTRIBUTION LOSSES	**-**	**-**	**-**	**-**	**-**	**-**	**-**	**-**	**-**	**-**	**-**
FINAL CONSUMPTION	**-**	**750**	**-**	**-**	**-**	**178**	**-**	**-**	**-**	**-**	**-**
INDUSTRY SECTOR	**-**	**750**	**-**	**-**	**-**	**178**	**-**	**-**	**-**	**-**	**-**
Iron and Steel	-	-	-	-	-	178	-	-	-	-	-
Chemical and Petrochem.	-	-	-	-	-	-	-	-	-	-	-
Non-Metallic Minerals	-	-	-	-	-	-	-	-	-	-	-
Non-specified	-	750	-	-	-	-	-	-	-	-	-
TRANSPORT SECTOR	**-**	**-**	**-**	**-**	**-**	**-**	**-**	**-**	**-**	**-**	**-**
Air	-	-	-	-	-	-	-	-	-	-	-
Road	-	-	-	-	-	-	-	-	-	-	-
Non-specified	-	-	-	-	-	-	-	-	-	-	-
OTHER SECTORS	**-**	**-**	**-**	**-**	**-**	**-**	**-**	**-**	**-**	**-**	**-**
Agriculture	-	-	-	-	-	-	-	-	-	-	-
Comm. and Publ. Services	-	-	-	-	-	-	-	-	-	-	-
Residential	-	-	-	-	-	-	-	-	-	-	-
Non-specified	-	-	-	-	-	-	-	-	-	-	-
NON-ENERGY USE	**-**	**-**	**-**	**-**	**-**	**-**	**-**	**-**	**-**	**-**	**-**

APPROVISIONNEMENT ET DEMANDE 1997	*Charbon* (1000 tonnes)							*Pétrole* (1000 tonnes)			
	Charbon à coke	*Autres charb. bit.*	*Charbon sous-bit.*	*Lignite*	*Tourbe*	*Coke de four/gaz*	*Agg./briq. de lignite*	*Pétrole brut*	*LGN*	*Produits d'aliment.*	*Additifs*
Production	402	520	-	-	-	890	-	178726	2390	-	-
Imports	598	-	-	-	-	-	-	-	-	-	-
Exports	-	-13	-	-	-	-	-	-119829	-	-	-
Intl. Marine Bunkers	-	-	-	-	-	-	-	-	-	-	-
Stock Changes	-	-	-	-	-	-	-	199	-	-	-
DOMESTIC SUPPLY	**1000**	**507**	**-**	**-**	**-**	**890**	**-**	**59096**	**2390**	**-**	**-**
Transfers and Stat. Diff.	-	243	-	-	-	-	-	-1	-	-	-
TRANSFORMATION	**1000**	**-**	**-**	**-**	**-**	**712**	**-**	**59095**	**2390**	**-**	**-**
Electricity and CHP Plants	-	-	-	-	-	-	-	-	-	-	-
Petroleum Refineries	-	-	-	-	-	-	-	59095	2390	-	-
Other Transform. Sector	1000	-	-	-	-	712	-	-	-	-	-
ENERGY SECTOR	**-**	**-**	**-**	**-**	**-**	**-**	**-**	**-**	**-**	**-**	**-**
DISTRIBUTION LOSSES	**-**	**-**	**-**	**-**	**-**	**-**	**-**	**-**	**-**	**-**	**-**
FINAL CONSUMPTION	**-**	**750**	**-**	**-**	**-**	**178**	**-**	**-**	**-**	**-**	**-**
INDUSTRY SECTOR	**-**	**750**	**-**	**-**	**-**	**178**	**-**	**-**	**-**	**-**	**-**
Iron and Steel	-	-	-	-	-	178	-	-	-	-	-
Chemical and Petrochem.	-	-	-	-	-	-	-	-	-	-	-
Non-Metallic Minerals	-	-	-	-	-	-	-	-	-	-	-
Non-specified	-	750	-	-	-	-	-	-	-	-	-
TRANSPORT SECTOR	**-**	**-**	**-**	**-**	**-**	**-**	**-**	**-**	**-**	**-**	**-**
Air	-	-	-	-	-	-	-	-	-	-	-
Road	-	-	-	-	-	-	-	-	-	-	-
Non-specified	-	-	-	-	-	-	-	-	-	-	-
OTHER SECTORS	**-**	**-**	**-**	**-**	**-**	**-**	**-**	**-**	**-**	**-**	**-**
Agriculture	-	-	-	-	-	-	-	-	-	-	-
Comm. and Publ. Services	-	-	-	-	-	-	-	-	-	-	-
Residential	-	-	-	-	-	-	-	-	-	-	-
Non-specified	-	-	-	-	-	-	-	-	-	-	-
NON-ENERGY USE	**-**	**-**	**-**	**-**	**-**	**-**	**-**	**-**	**-**	**-**	**-**

Iran

SUPPLY AND CONSUMPTION 1996	Oil cont. (1000 tonnes)										
	Refinery Gas	LPG + Ethane	Motor Gasoline	Aviation Gasoline	Jet Fuel	Kerosene	Gas/ Diesel	Heavy Fuel Oil	Naphtha	Petrol. Coke	Other Prod.
Production	-	1328	6515	117	568	7500	15520	20623	692	-	3325
Imports	-	553	1495	-	-	1558	5221	-	-	-	-
Exports	-	-	-	-	-	-	-	-4048	-344	-	-265
Intl. Marine Bunkers	-	-	-	-	-	-	-90	-465	-	-	-
Stock Changes	-	-	-	-	-	-	-	-	-	-	-
DOMESTIC SUPPLY	**-**	**1881**	**8010**	**117**	**568**	**9058**	**20651**	**16110**	**348**	**-**	**3060**
Transfers and Stat. Diff.	-	1	-	-	-	-	1	-1	-	-	-
TRANSFORMATION	**-**	**-**	**-**	**-**	**-**	**3**	**1663**	**7067**	**-**	**-**	**-**
Electricity and CHP Plants	-	-	-	-	-	3	1663	7067	-	-	-
Petroleum Refineries	-	-	-	-	-	-	-	-	-	-	-
Other Transform. Sector	-	-	-	-	-	-	-	-	-	-	-
ENERGY SECTOR	**-**	**66**	**-**	**-**	**-**	**-**	**361**	**1861**	**-**	**-**	**-**
DISTRIBUTION LOSSES	**-**	**-**	**-**	**-**	**-**	**-**	**-**	**-**	**-**	**-**	**-**
FINAL CONSUMPTION	**-**	**1816**	**8010**	**117**	**568**	**9055**	**18628**	**7181**	**348**	**-**	**3060**
INDUSTRY SECTOR	**-**	**12**	**-**	**-**	**-**	**259**	**2215**	**4899**	**348**	**-**	**-**
Iron and Steel	-	-	-	-	-	-	-	-	-	-	-
Chemical and Petrochem.	-	-	-	-	-	-	-	-	348	-	-
Non-Metallic Minerals	-	-	-	-	-	-	-	-	-	-	-
Non-specified	-	12	-	-	-	259	2215	4899	-	-	-
TRANSPORT SECTOR	**-**	**54**	**8010**	**117**	**568**	**-**	**9742**	**-**	**-**	**-**	**-**
Air	-	-	-	117	568	-	-	-	-	-	-
Road	-	54	8010	-	-	-	9742	-	-	-	-
Non-specified	-	-	-	-	-	-	-	-	-	-	-
OTHER SECTORS	**-**	**1750**	**-**	**-**	**-**	**8796**	**6671**	**2282**	**-**	**-**	**-**
Agriculture	-	-	-	-	-	268	3676	-	-	-	-
Comm. and Publ. Services	-	72	-	-	-	495	966	2282	-	-	-
Residential	-	1678	-	-	-	8033	2029	-	-	-	-
Non-specified	-	-	-	-	-	-	-	-	-	-	-
NON-ENERGY USE	**-**	**-**	**-**	**-**	**-**	**-**	**-**	**-**	**-**	**-**	**3060**

APPROVISIONNEMENT ET DEMANDE 1997	*Pétrole cont.* (1000 tonnes)										
	Gaz de raffinerie	*GPL + éthane*	*Essence moteur*	*Essence aviation*	*Carbu- réacteurs*	*Kérosène*	*Gazole*	*Fioul lourd*	*Naphta*	*Coke de pétrole*	*Autres prod.*
Production	-	1246	6500	117	570	6260	15388	21418	695	-	3332
Imports	-	687	1797	-	-	2537	6146	-	-	-	-
Exports	-	-	-	-	-	-	-	-5801	-344	-	-265
Intl. Marine Bunkers	-	-	-	-	-	-	-90	-465	-	-	-
Stock Changes	-	-	-	-	-	-	-	-	-	-	-
DOMESTIC SUPPLY	**-**	**1933**	**8297**	**117**	**570**	**8797**	**21444**	**15152**	**351**	**-**	**3067**
Transfers and Stat. Diff.	-	1	-	-	-	1	1	-1	-	-	-
TRANSFORMATION	**-**	**-**	**-**	**-**	**-**	**3**	**1727**	**6647**	**-**	**-**	**-**
Electricity and CHP Plants	-	-	-	-	-	3	1727	6647	-	-	-
Petroleum Refineries	-	-	-	-	-	-	-	-	-	-	-
Other Transform. Sector	-	-	-	-	-	-	-	-	-	-	-
ENERGY SECTOR	**-**	**68**	**-**	**-**	**-**	**-**	**375**	**1750**	**-**	**-**	**-**
DISTRIBUTION LOSSES	**-**	**-**	**-**	**-**	**-**	**-**	**-**	**-**	**-**	**-**	**-**
FINAL CONSUMPTION	**-**	**1866**	**8297**	**117**	**570**	**8795**	**19343**	**6754**	**351**	**-**	**3067**
INDUSTRY SECTOR	**-**	**12**	**-**	**-**	**-**	**252**	**2300**	**4608**	**351**	**-**	**-**
Iron and Steel	-	-	-	-	-	-	-	-	-	-	-
Chemical and Petrochem.	-	-	-	-	-	-	-	-	351	-	-
Non-Metallic Minerals	-	-	-	-	-	-	-	-	-	-	-
Non-specified	-	12	-	-	-	252	2300	4608	-	-	-
TRANSPORT SECTOR	**-**	**55**	**8297**	**117**	**570**	**-**	**10116**	**-**	**-**	**-**	**-**
Air	-	-	-	117	570	-	-	-	-	-	-
Road	-	55	8297	-	-	-	10116	-	-	-	-
Non-specified	-	-	-	-	-	-	-	-	-	-	-
OTHER SECTORS	**-**	**1799**	**-**	**-**	**-**	**8543**	**6927**	**2146**	**-**	**-**	**-**
Agriculture	-	-	-	-	-	260	3817	-	-	-	-
Comm. and Publ. Services	-	74	-	-	-	481	1003	2146	-	-	-
Residential	-	1725	-	-	-	7802	2107	-	-	-	-
Non-specified	-	-	-	-	-	-	-	-	-	-	-
NON-ENERGY USE	**-**	**-**	**-**	**-**	**-**	**-**	**-**	**-**	**-**	**-**	**3067**

Iran

SUPPLY AND CONSUMPTION 1996	Gas (TJ)				Comb. Renew. & Waste (TJ)				(GWh)	(TJ)
	Natural Gas	Gas Works	Coke Ovens	Blast Furnaces	Solid Biomass	Gas/Liquids from Biomass	Municipal Waste	Industrial Waste	Electricity	Heat
Production	1590376	-	-	-	32914	-	-	-	90851	-
Imports	-	-	-	-	-	-	-	-	-	-
Exports	-3800	-	-	-	-	-	-	-	-	-
Intl. Marine Bunkers	-	-	-	-	-	-	-	-	-	-
Stock Changes	-	-	-	-	-	-	-	-	-	-
DOMESTIC SUPPLY	**1586576**	**-**	**-**	**-**	**32914**	**-**	**-**	**-**	**90851**	**-**
Transfers and Stat. Diff.	187079	-	-	-	-	-	-	-	-	-
TRANSFORMATION	**490779**	**-**	**-**	**-**	**2852**	**-**	**-**	**-**	**-**	**-**
Electricity and CHP Plants	490779	-	-	-	-	-	-	-	-	-
Petroleum Refineries	-	-	-	-	-	-	-	-	-	-
Other Transform. Sector	-	-	-	-	2852	-	-	-	-	-
ENERGY SECTOR	**52240**	**-**	**-**	**-**	**-**	**-**	**-**	**-**	**4005**	**-**
DISTRIBUTION LOSSES	**43257**	**-**	**-**	**-**	**-**	**-**	**-**	**-**	**18379**	**-**
FINAL CONSUMPTION	**1187379**	**-**	**-**	**-**	**30062**	**-**	**-**	**-**	**68467**	**-**
INDUSTRY SECTOR	**706609**	**-**	**-**	**-**	**7789**	**-**	**-**	**-**	**21721**	**-**
Iron and Steel	-	-	-	-	-	-	-	-	-	-
Chemical and Petrochem.	188484	-	-	-	-	-	-	-	-	-
Non-Metallic Minerals	-	-	-	-	-	-	-	-	-	-
Non-specified	518125	-	-	-	7789	-	-	-	21721	-
TRANSPORT SECTOR	**-**	**-**	**-**	**-**	**-**	**-**	**-**	**-**	**-**	**-**
Air	-	-	-	-	-	-	-	-	-	-
Road	-	-	-	-	-	-	-	-	-	-
Non-specified	-	-	-	-	-	-	-	-	-	-
OTHER SECTORS	**480770**	**-**	**-**	**-**	**22273**	**-**	**-**	**-**	**46746**	**-**
Agriculture	-	-	-	-	-	-	-	-	5731	-
Comm. and Publ. Services	63396	-	-	-	-	-	-	-	15966	-
Residential	417374	-	-	-	-	-	-	-	22244	-
Non-specified	-	-	-	-	22273	-	-	-	2805	-
NON-ENERGY USE	**-**	**-**	**-**	**-**	**-**	**-**	**-**	**-**	**-**	**-**

APPROVISIONNEMENT ET DEMANDE 1997	*Gaz* (TJ)				*En. Re. Comb. & Déchets* (TJ)				(GWh)	(TJ)
	Gaz naturel	*Usines à gaz*	*Cokeries*	*Hauts fourneaux*	*Biomasse solide*	*Gaz/Liquides tirés de biomasse*	*Déchets urbains*	*Déchets industriels*	*Electricité*	*Chaleur*
Production	1786762	-	-	-	32914	-	-	-	95794	-
Imports	-	-	-	-	-	-	-	-	-	-
Exports	-3800	-	-	-	-	-	-	-	-	-
Intl. Marine Bunkers	-	-	-	-	-	-	-	-	-	-
Stock Changes	-	-	-	-	-	-	-	-	-	-
DOMESTIC SUPPLY	**1782962**	**-**	**-**	**-**	**32914**	**-**	**-**	**-**	**95794**	**-**
Transfers and Stat. Diff.	206994	-	-	-	-	-	-	-	1	-
TRANSFORMATION	**552126**	**-**	**-**	**-**	**2959**	**-**	**-**	**-**	**-**	**-**
Electricity and CHP Plants	552126	-	-	-	-	-	-	-	-	-
Petroleum Refineries	-	-	-	-	-	-	-	-	-	-
Other Transform. Sector	-	-	-	-	2959	-	-	-	-	-
ENERGY SECTOR	**58770**	**-**	**-**	**-**	**-**	**-**	**-**	**-**	**4282**	**-**
DISTRIBUTION LOSSES	**43257**	**-**	**-**	**-**	**-**	**-**	**-**	**-**	**20649**	**-**
FINAL CONSUMPTION	**1335803**	**-**	**-**	**-**	**29955**	**-**	**-**	**-**	**70864**	**-**
INDUSTRY SECTOR	**794936**	**-**	**-**	**-**	**7789**	**-**	**-**	**-**	**22481**	**-**
Iron and Steel	-	-	-	-	-	-	-	-	-	-
Chemical and Petrochem.	212045	-	-	-	-	-	-	-	-	-
Non-Metallic Minerals	-	-	-	-	-	-	-	-	-	-
Non-specified	582891	-	-	-	7789	-	-	-	22481	-
TRANSPORT SECTOR	**-**	**-**	**-**	**-**	**-**	**-**	**-**	**-**	**-**	**-**
Air	-	-	-	-	-	-	-	-	-	-
Road	-	-	-	-	-	-	-	-	-	-
Non-specified	-	-	-	-	-	-	-	-	-	-
OTHER SECTORS	**540867**	**-**	**-**	**-**	**22166**	**-**	**-**	**-**	**48383**	**-**
Agriculture	-	-	-	-	-	-	-	-	5932	-
Comm. and Publ. Services	71321	-	-	-	-	-	-	-	16525	-
Residential	469546	-	-	-	-	-	-	-	23023	-
Non-specified	-	-	-	-	22166	-	-	-	2903	-
NON-ENERGY USE	**-**	**-**	**-**	**-**	**-**	**-**	**-**	**-**	**-**	**-**

Iraq / Irak

SUPPLY AND CONSUMPTION 1996	Coal (1000 tonnes)							Oil (1000 tonnes)			
	Coking Coal	Other Bit. Coal	Sub-Bit. Coal	Lignite	Peat	Oven and Gas Coke	Pat. Fuel and BKB	Crude Oil	NGL	Feed-stocks	Additives
Production	-	-	-	-	-	-	-	28374	166	-	-
Imports	-	-	-	-	-	-	-	-	-	-	-
Exports	-	-	-	-	-	-	-	-4327	-	-	-
Intl. Marine Bunkers	-	-	-	-	-	-	-	-	-	-	-
Stock Changes	-	-	-	-	-	-	-	-	-	-	-
DOMESTIC SUPPLY	**-**	**-**	**-**	**-**	**-**	**-**	**-**	**24047**	**166**	**-**	**-**
Transfers and Stat. Diff.	-	-	-	-	-	-	-	270	-166	-	-
TRANSFORMATION	**-**	**-**	**-**	**-**	**-**	**-**	**-**	**24317**	**-**	**-**	**-**
Electricity and CHP Plants	-	-	-	-	-	-	-	1800	-	-	-
Petroleum Refineries	-	-	-	-	-	-	-	22517	-	-	-
Other Transform. Sector	-	-	-	-	-	-	-	-	-	-	-
ENERGY SECTOR	**-**	**-**	**-**	**-**	**-**	**-**	**-**	**-**	**-**	**-**	**-**
DISTRIBUTION LOSSES	**-**	**-**	**-**	**-**	**-**	**-**	**-**	**-**	**-**	**-**	**-**
FINAL CONSUMPTION	**-**	**-**	**-**	**-**	**-**	**-**	**-**	**-**	**-**	**-**	**-**
INDUSTRY SECTOR	**-**	**-**	**-**	**-**	**-**	**-**	**-**	**-**	**-**	**-**	**-**
Iron and Steel	-	-	-	-	-	-	-	-	-	-	-
Chemical and Petrochem.	-	-	-	-	-	-	-	-	-	-	-
Non-Metallic Minerals	-	-	-	-	-	-	-	-	-	-	-
Non-specified	-	-	-	-	-	-	-	-	-	-	-
TRANSPORT SECTOR	**-**	**-**	**-**	**-**	**-**	**-**	**-**	**-**	**-**	**-**	**-**
Air	-	-	-	-	-	-	-	-	-	-	-
Road	-	-	-	-	-	-	-	-	-	-	-
Non-specified	-	-	-	-	-	-	-	-	-	-	-
OTHER SECTORS	**-**	**-**	**-**	**-**	**-**	**-**	**-**	**-**	**-**	**-**	**-**
Agriculture	-	-	-	-	-	-	-	-	-	-	-
Comm. and Publ. Services	-	-	-	-	-	-	-	-	-	-	-
Residential	-	-	-	-	-	-	-	-	-	-	-
Non-specified	-	-	-	-	-	-	-	-	-	-	-
NON-ENERGY USE	**-**	**-**	**-**	**-**	**-**	**-**	**-**	**-**	**-**	**-**	**-**

APPROVISIONNEMENT ET DEMANDE 1997	*Charbon* (1000 tonnes)							*Pétrole* (1000 tonnes)			
	Charbon à coke	*Autres charb. bit.*	*Charbon sous-bit.*	*Lignite*	*Tourbe*	*Coke de four/gaz*	*Agg./briq. de lignite*	*Pétrole brut*	*LGN*	*Produits d'aliment.*	*Additifs*
Production	-	-	-	-	-	-	-	56502	498	-	-
Imports	-	-	-	-	-	-	-	-	-	-	-
Exports	-	-	-	-	-	-	-	-31593	-	-	-
Intl. Marine Bunkers	-	-	-	-	-	-	-	-	-	-	-
Stock Changes	-	-	-	-	-	-	-	-	-	-	-
DOMESTIC SUPPLY	**-**	**-**	**-**	**-**	**-**	**-**	**-**	**24909**	**498**	**-**	**-**
Transfers and Stat. Diff.	-	-	-	-	-	-	-	-	-498	-	-
TRANSFORMATION	**-**	**-**	**-**	**-**	**-**	**-**	**-**	**24909**	**-**	**-**	**-**
Electricity and CHP Plants	-	-	-	-	-	-	-	1844	-	-	-
Petroleum Refineries	-	-	-	-	-	-	-	23065	-	-	-
Other Transform. Sector	-	-	-	-	-	-	-	-	-	-	-
ENERGY SECTOR	**-**	**-**	**-**	**-**	**-**	**-**	**-**	**-**	**-**	**-**	**-**
DISTRIBUTION LOSSES	**-**	**-**	**-**	**-**	**-**	**-**	**-**	**-**	**-**	**-**	**-**
FINAL CONSUMPTION	**-**	**-**	**-**	**-**	**-**	**-**	**-**	**-**	**-**	**-**	**-**
INDUSTRY SECTOR	**-**	**-**	**-**	**-**	**-**	**-**	**-**	**-**	**-**	**-**	**-**
Iron and Steel	-	-	-	-	-	-	-	-	-	-	-
Chemical and Petrochem.	-	-	-	-	-	-	-	-	-	-	-
Non-Metallic Minerals	-	-	-	-	-	-	-	-	-	-	-
Non-specified	-	-	-	-	-	-	-	-	-	-	-
TRANSPORT SECTOR	**-**	**-**	**-**	**-**	**-**	**-**	**-**	**-**	**-**	**-**	**-**
Air	-	-	-	-	-	-	-	-	-	-	-
Road	-	-	-	-	-	-	-	-	-	-	-
Non-specified	-	-	-	-	-	-	-	-	-	-	-
OTHER SECTORS	**-**	**-**	**-**	**-**	**-**	**-**	**-**	**-**	**-**	**-**	**-**
Agriculture	-	-	-	-	-	-	-	-	-	-	-
Comm. and Publ. Services	-	-	-	-	-	-	-	-	-	-	-
Residential	-	-	-	-	-	-	-	-	-	-	-
Non-specified	-	-	-	-	-	-	-	-	-	-	-
NON-ENERGY USE	**-**	**-**	**-**	**-**	**-**	**-**	**-**	**-**	**-**	**-**	**-**

Iraq / Irak

SUPPLY AND CONSUMPTION 1996	Oil cont. (1000 tonnes)										
	Refinery Gas	LPG + Ethane	Motor Gasoline	Aviation Gasoline	Jet Fuel	Kerosene	Gas/ Diesel	Heavy Fuel Oil	Naphtha	Petrol. Coke	Other Prod.
Production	450	979	2927	-	541	977	6608	7495	482	-	757
Imports	-	-	-	-	-	-	-	-	-	-	-
Exports	-	-	-	-	-141	-22	-1569	-1389	-	-	-
Intl. Marine Bunkers	-	-	-	-	-	-	-	-	-	-	-
Stock Changes	-	-	-	-	-	-	-	-	-	-	-
DOMESTIC SUPPLY	**450**	**979**	**2927**	**-**	**400**	**955**	**5039**	**6106**	**482**	**-**	**757**
Transfers and Stat. Diff.	-	152	-69	-	12	-7	283	183	-5	-	-7
TRANSFORMATION	**-**	**-**	**-**	**-**	**-**	**-**	**-**	**3400**	**-**	**-**	**-**
Electricity and CHP Plants	-	-	-	-	-	-	-	3400	-	-	-
Petroleum Refineries	-	-	-	-	-	-	-	-	-	-	-
Other Transform. Sector	-	-	-	-	-	-	-	-	-	-	-
ENERGY SECTOR	**450**	**-**	**-**	**-**	**-**	**-**	**-**	**1149**	**-**	**-**	**-**
DISTRIBUTION LOSSES	**-**	**-**	**-**	**-**	**-**	**-**	**-**	**-**	**-**	**-**	**-**
FINAL CONSUMPTION	**-**	**1131**	**2858**	**-**	**412**	**948**	**5322**	**1740**	**477**	**-**	**750**
INDUSTRY SECTOR	**-**	**-**	**-**	**-**	**-**	**-**	**-**	**1740**	**477**	**-**	**-**
Iron and Steel	-	-	-	-	-	-	-	-	-	-	-
Chemical and Petrochem.	-	-	-	-	-	-	-	-	477	-	-
Non-Metallic Minerals	-	-	-	-	-	-	-	-	-	-	-
Non-specified	-	-	-	-	-	-	-	1740	-	-	-
TRANSPORT SECTOR	**-**	**-**	**2858**	**-**	**412**	**-**	**5322**	**-**	**-**	**-**	**-**
Air	-	-	-	-	412	-	-	-	-	-	-
Road	-	-	2858	-	-	-	5322	-	-	-	-
Non-specified	-	-	-	-	-	-	-	-	-	-	-
OTHER SECTORS	**-**	**1131**	**-**	**-**	**-**	**948**	**-**	**-**	**-**	**-**	**-**
Agriculture	-	-	-	-	-	-	-	-	-	-	-
Comm. and Publ. Services	-	-	-	-	-	-	-	-	-	-	-
Residential	-	1131	-	-	-	948	-	-	-	-	-
Non-specified	-	-	-	-	-	-	-	-	-	-	-
NON-ENERGY USE	**-**	**-**	**-**	**-**	**-**	**-**	**-**	**-**	**-**	**-**	**750**

APPROVISIONNEMENT ET DEMANDE 1997	*Pétrole cont.* (1000 tonnes)										
	Gaz de raffinerie	*GPL + éthane*	*Essence moteur*	*Essence aviation*	*Carbu-réacteurs*	*Kérosène*	*Gazole*	*Fioul lourd*	*Naphta*	*Coke de pétrole*	*Autres prod.*
Production	461	1003	2998	-	554	1001	6769	7677	494	-	775
Imports	-	-	-	-	-	-	-	-	-	-	-
Exports	-	-	-	-	-131	-28	-1302	-1209	-	-	-
Intl. Marine Bunkers	-	-	-	-	-	-	-	-	-	-	-
Stock Changes	-	-	-	-	-	-	-	-	-	-	-
DOMESTIC SUPPLY	**461**	**1003**	**2998**	**-**	**423**	**973**	**5467**	**6468**	**494**	**-**	**775**
Transfers and Stat. Diff.	-	156	-62	-	-	-	-	1	-5	-	-7
TRANSFORMATION	**-**	**-**	**-**	**-**	**-**	**-**	**-**	**3497**	**-**	**-**	**-**
Electricity and CHP Plants	-	-	-	-	-	-	-	3497	-	-	-
Petroleum Refineries	-	-	-	-	-	-	-	-	-	-	-
Other Transform. Sector	-	-	-	-	-	-	-	-	-	-	-
ENERGY SECTOR	**461**	**-**	**-**	**-**	**-**	**-**	**-**	**1182**	**-**	**-**	**-**
DISTRIBUTION LOSSES	**-**	**-**	**-**	**-**	**-**	**-**	**-**	**-**	**-**	**-**	**-**
FINAL CONSUMPTION	**-**	**1159**	**2936**	**-**	**423**	**973**	**5467**	**1790**	**489**	**-**	**768**
INDUSTRY SECTOR	**-**	**-**	**-**	**-**	**-**	**-**	**-**	**1790**	**489**	**-**	**-**
Iron and Steel	-	-	-	-	-	-	-	-	-	-	-
Chemical and Petrochem.	-	-	-	-	-	-	-	-	489	-	-
Non-Metallic Minerals	-	-	-	-	-	-	-	-	-	-	-
Non-specified	-	-	-	-	-	-	-	1790	-	-	-
TRANSPORT SECTOR	**-**	**-**	**2936**	**-**	**423**	**-**	**5467**	**-**	**-**	**-**	**-**
Air	-	-	-	-	423	-	-	-	-	-	-
Road	-	-	2936	-	-	-	5467	-	-	-	-
Non-specified	-	-	-	-	-	-	-	-	-	-	-
OTHER SECTORS	**-**	**1159**	**-**	**-**	**-**	**973**	**-**	**-**	**-**	**-**	**-**
Agriculture	-	-	-	-	-	-	-	-	-	-	-
Comm. and Publ. Services	-	-	-	-	-	-	-	-	-	-	-
Residential	-	1159	-	-	-	973	-	-	-	-	-
Non-specified	-	-	-	-	-	-	-	-	-	-	-
NON-ENERGY USE	**-**	**-**	**-**	**-**	**-**	**-**	**-**	**-**	**-**	**-**	**768**

Iraq / Irak

SUPPLY AND CONSUMPTION 1996	Gas (TJ)				Comb. Renew. & Waste (TJ)				(GWh)	(TJ)
	Natural Gas	Gas Works	Coke Ovens	Blast Furnaces	Solid Biomass	Gas/Liquids from Biomass	Municipal Waste	Industrial Waste	Electricity	Heat
Production	154126	-	-	-	1100	-	-	-	29000	-
Imports	-	-	-	-	-	-	-	-	-	-
Exports	-	-	-	-	-	-	-	-	-	-
Intl. Marine Bunkers	-	-	-	-	-	-	-	-	-	-
Stock Changes	-	-	-	-	-	-	-	-	-	-
DOMESTIC SUPPLY	**154126**	**-**	**-**	**-**	**1100**	**-**	**-**	**-**	**29000**	**-**
Transfers and Stat. Diff.	-	-	-	-	-	-	-	-	-	-
TRANSFORMATION	**-**	**-**	**-**	**-**	**476**	**-**	**-**	**-**	**-**	**-**
Electricity and CHP Plants	-	-	-	-	-	-	-	-	-	-
Petroleum Refineries	-	-	-	-	-	-	-	-	-	-
Other Transform. Sector	-	-	-	-	476	-	-	-	-	-
ENERGY SECTOR	**-**	**-**	**-**	**-**	**-**	**-**	**-**	**-**	**-**	**-**
DISTRIBUTION LOSSES	**-**	**-**	**-**	**-**	**-**	**-**	**-**	**-**	**-**	**-**
FINAL CONSUMPTION	**154126**	**-**	**-**	**-**	**624**	**-**	**-**	**-**	**29000**	**-**
INDUSTRY SECTOR	**154126**	**-**	**-**	**-**	**-**	**-**	**-**	**-**	**-**	**-**
Iron and Steel	-	-	-	-	-	-	-	-	-	-
Chemical and Petrochem.	-	-	-	-	-	-	-	-	-	-
Non-Metallic Minerals	-	-	-	-	-	-	-	-	-	-
Non-specified	154126	-	-	-	-	-	-	-	-	-
TRANSPORT SECTOR	**-**	**-**	**-**	**-**	**-**	**-**	**-**	**-**	**-**	**-**
Air	-	-	-	-	-	-	-	-	-	-
Road	-	-	-	-	-	-	-	-	-	-
Non-specified	-	-	-	-	-	-	-	-	-	-
OTHER SECTORS	**-**	**-**	**-**	**-**	**624**	**-**	**-**	**-**	**29000**	**-**
Agriculture	-	-	-	-	-	-	-	-	-	-
Comm. and Publ. Services	-	-	-	-	-	-	-	-	-	-
Residential	-	-	-	-	-	-	-	-	-	-
Non-specified	-	-	-	-	624	-	-	-	29000	-
NON-ENERGY USE	**-**	**-**	**-**	**-**	**-**	**-**	**-**	**-**	**-**	**-**

APPROVISIONNEMENT ET DEMANDE 1997	*Gaz (TJ)*				*En. Re. Comb. & Déchets (TJ)*				*(GWh)*	*(TJ)*
	Gaz naturel	*Usines à gaz*	*Cokeries*	*Hauts fourneaux*	*Biomasse solide*	*Gaz/Liquides tirés de biomasse*	*Déchets urbains*	*Déchets industriels*	*Electricité*	*Chaleur*
Production	172203	-	-	-	1100	-	-	-	29561	-
Imports	-	-	-	-	-	-	-	-	-	-
Exports	-	-	-	-	-	-	-	-	-	-
Intl. Marine Bunkers	-	-	-	-	-	-	-	-	-	-
Stock Changes	-	-	-	-	-	-	-	-	-	-
DOMESTIC SUPPLY	**172203**	**-**	**-**	**-**	**1100**	**-**	**-**	**-**	**29561**	**-**
Transfers and Stat. Diff.	-	-	-	-	-	-	-	-	-	-
TRANSFORMATION	**-**	**-**	**-**	**-**	**476**	**-**	**-**	**-**	**-**	**-**
Electricity and CHP Plants	-	-	-	-	-	-	-	-	-	-
Petroleum Refineries	-	-	-	-	-	-	-	-	-	-
Other Transform. Sector	-	-	-	-	476	-	-	-	-	-
ENERGY SECTOR	**-**	**-**	**-**	**-**	**-**	**-**	**-**	**-**	**-**	**-**
DISTRIBUTION LOSSES	**-**	**-**	**-**	**-**	**-**	**-**	**-**	**-**	**-**	**-**
FINAL CONSUMPTION	**172203**	**-**	**-**	**-**	**624**	**-**	**-**	**-**	**29561**	**-**
INDUSTRY SECTOR	**172203**	**-**	**-**	**-**	**-**	**-**	**-**	**-**	**-**	**-**
Iron and Steel	-	-	-	-	-	-	-	-	-	-
Chemical and Petrochem.	-	-	-	-	-	-	-	-	-	-
Non-Metallic Minerals	-	-	-	-	-	-	-	-	-	-
Non-specified	172203	-	-	-	-	-	-	-	-	-
TRANSPORT SECTOR	**-**	**-**	**-**	**-**	**-**	**-**	**-**	**-**	**-**	**-**
Air	-	-	-	-	-	-	-	-	-	-
Road	-	-	-	-	-	-	-	-	-	-
Non-specified	-	-	-	-	-	-	-	-	-	-
OTHER SECTORS	**-**	**-**	**-**	**-**	**624**	**-**	**-**	**-**	**29561**	**-**
Agriculture	-	-	-	-	-	-	-	-	-	-
Comm. and Publ. Services	-	-	-	-	-	-	-	-	-	-
Residential	-	-	-	-	-	-	-	-	-	-
Non-specified	-	-	-	-	624	-	-	-	29561	-
NON-ENERGY USE	**-**	**-**	**-**	**-**	**-**	**-**	**-**	**-**	**-**	**-**

Israel / Israël : 1996

SUPPLY AND CONSUMPTION / *APPROVISIONNEMENT ET DEMANDE*	Coal / *Charbon* (1000 tonnes)							Oil / *Pétrole* (1000 tonnes)			
	Coking Coal / *Charbon à coke*	Other Bit. Coal / *Autres charb. bit.*	Sub-Bit. Coal / *Charbon sous-bit.*	Lignite / *Lignite*	Peat / *Tourbe*	Oven and Gas Coke / *Coke de four/gaz*	Pat. Fuel and BKB / *Agg./briq. de lignite*	Crude Oil / *Pétrole brut*	NGL / *LGN*	Feed-stocks / *Produits d'aliment.*	Additives / *Additifs*
Production	-	-	-	421	-	-	-	4	21	-	-
From Other Sources	-	-	-	-	-	-	-	-	-	-	-
Imports	-	7156	-	-	-	-	-	10090	-	-	-
Exports	-	-	-	-	-	-	-	-	-	-	-
Intl. Marine Bunkers	-	-	-	-	-	-	-	-	-	-	-
Stock Changes	-	682	-	-	-	-	-	-	-	-	-
DOMESTIC SUPPLY	**-**	**7838**	**-**	**421**	**-**	**-**	**-**	**10094**	**21**	**-**	**-**
Transfers	-	-	-	-	-	-	-	-	-	-	-
Statistical Differences	-	-2	-	-	-	-	-	-	-	-	-
TRANSFORMATION	**-**	**7810**	**-**	**421**	**-**	**-**	**-**	**10094**	**-**	**-**	**-**
Electricity Plants	-	7810	-	421	-	-	-	-	-	-	-
CHP Plants	-	-	-	-	-	-	-	-	-	-	-
Heat Plants	-	-	-	-	-	-	-	-	-	-	-
Blast Furnaces/Gas Works	-	-	-	-	-	-	-	-	-	-	-
Coke/Pat. Fuel/BKB Plants	-	-	-	-	-	-	-	-	-	-	-
Petroleum Refineries	-	-	-	-	-	-	-	10094	-	-	-
Petrochemical Industry	-	-	-	-	-	-	-	-	-	-	-
Liquefaction	-	-	-	-	-	-	-	-	-	-	-
Other Transform. Sector	-	-	-	-	-	-	-	-	-	-	-
ENERGY SECTOR	**-**	**-**	**-**	**-**	**-**	**-**	**-**	**-**	**-**	**-**	**-**
Coal Mines	-	-	-	-	-	-	-	-	-	-	-
Oil and Gas Extraction	-	-	-	-	-	-	-	-	-	-	-
Petroleum Refineries	-	-	-	-	-	-	-	-	-	-	-
Electr., CHP+Heat Plants	-	-	-	-	-	-	-	-	-	-	-
Pumped Storage (Elec.)	-	-	-	-	-	-	-	-	-	-	-
Other Energy Sector	-	-	-	-	-	-	-	-	-	-	-
Distribution Losses	-	-	-	-	-	-	-	-	-	-	-
FINAL CONSUMPTION	**-**	**26**	**-**	**-**	**-**	**-**	**-**	**-**	**21**	**-**	**-**
INDUSTRY SECTOR	**-**	**26**	**-**	**-**	**-**	**-**	**-**	**-**	**21**	**-**	**-**
Iron and Steel	-	-	-	-	-	-	-	-	-	-	-
Chemical and Petrochem.	-	6	-	-	-	-	-	-	21	-	-
of which: Feedstocks	-	-	-	-	-	-	-	-	*21*	-	-
Non-Ferrous Metals	-	-	-	-	-	-	-	-	-	-	-
Non-Metallic Minerals	-	-	-	-	-	-	-	-	-	-	-
Transport Equipment	-	-	-	-	-	-	-	-	-	-	-
Machinery	-	-	-	-	-	-	-	-	-	-	-
Mining and Quarrying	-	-	-	-	-	-	-	-	-	-	-
Food and Tobacco	-	-	-	-	-	-	-	-	-	-	-
Paper, Pulp and Print	-	-	-	-	-	-	-	-	-	-	-
Wood and Wood Products	-	-	-	-	-	-	-	-	-	-	-
Construction	-	-	-	-	-	-	-	-	-	-	-
Textile and Leather	-	-	-	-	-	-	-	-	-	-	-
Non-specified	-	20	-	-	-	-	-	-	-	-	-
TRANSPORT SECTOR	**-**	**-**	**-**	**-**	**-**	**-**	**-**	**-**	**-**	**-**	**-**
Air	-	-	-	-	-	-	-	-	-	-	-
Road	-	-	-	-	-	-	-	-	-	-	-
Rail	-	-	-	-	-	-	-	-	-	-	-
Pipeline Transport	-	-	-	-	-	-	-	-	-	-	-
Internal Navigation	-	-	-	-	-	-	-	-	-	-	-
Non-specified	-	-	-	-	-	-	-	-	-	-	-
OTHER SECTORS	**-**	**-**	**-**	**-**	**-**	**-**	**-**	**-**	**-**	**-**	**-**
Agriculture	-	-	-	-	-	-	-	-	-	-	-
Comm. and Publ. Services	-	-	-	-	-	-	-	-	-	-	-
Residential	-	-	-	-	-	-	-	-	-	-	-
Non-specified	-	-	-	-	-	-	-	-	-	-	-
NON-ENERGY USE	**-**	**-**	**-**	**-**	**-**	**-**	**-**	**-**	**-**	**-**	**-**
in Industry/Trans./Energy	-	-	-	-	-	-	-	-	-	-	-
in Transport	-	-	-	-	-	-	-	-	-	-	-
in Other Sectors	-	-	-	-	-	-	-	-	-	-	-

Israel / Israël : 1996

	Oil cont. / *Pétrole cont.* (1000 tonnes)										
SUPPLY AND CONSUMPTION *APPROVISIONNEMENT ET DEMANDE*	Refinery Gas *Gaz de raffinerie*	LPG + Ethane *GPL + éthane*	Motor Gasoline *Essence moteur*	Aviation Gasoline *Essence aviation*	Jet Fuel *Carbu- réacteurs*	Kerosene *Kérosène*	Gas/ Diesel *Gazole*	Heavy Fuel Oil *Fioul lourd*	Naphtha *Naphta*	Petrol. Coke *Coke de pétrole*	Other Prod. *Autres prod.*
Production	-	298	1801	-	-	1236	2494	3168	636	-	-
From Other Sources	-	-	-	-	-	-	-	-	-	-	-
Imports	-	-	8	-	-	-	-	1406	150	-	569
Exports	-	-61	-166	-	-	-10	-761	-344	-22	-	-
Intl. Marine Bunkers	-	-	-	-	-	-	-40	-54	-	-	-
Stock Changes	-	-	-	-	-	-	-	-	-	-	-42
DOMESTIC SUPPLY	**-**	**237**	**1643**	**-**	**-**	**1226**	**1693**	**4176**	**764**	**-**	**527**
Transfers	-	-	-	-	-	-	-	-	-	-	-
Statistical Differences	-	57	386	-	-	-233	303	66	-34	-	-
TRANSFORMATION	**-**	**-**	**-**	**-**	**-**	**-**	**128**	**2150**	**-**	**-**	**-**
Electricity Plants	-	-	-	-	-	-	128	2150	-	-	-
CHP Plants	-	-	-	-	-	-	-	-	-	-	-
Heat Plants	-	-	-	-	-	-	-	-	-	-	-
Blast Furnaces/Gas Works	-	-	-	-	-	-	-	-	-	-	-
Coke/Pat. Fuel/BKB Plants	-	-	-	-	-	-	-	-	-	-	-
Petroleum Refineries	-	-	-	-	-	-	-	-	-	-	-
Petrochemical Industry	-	-	-	-	-	-	-	-	-	-	-
Liquefaction	-	-	-	-	-	-	-	-	-	-	-
Other Transform. Sector	-	-	-	-	-	-	-	-	-	-	-
ENERGY SECTOR	**-**	**-**	**-**	**-**	**-**	**-**	**-**	**500**	**-**	**-**	**-**
Coal Mines	-	-	-	-	-	-	-	-	-	-	-
Oil and Gas Extraction	-	-	-	-	-	-	-	-	-	-	-
Petroleum Refineries	-	-	-	-	-	-	-	500	-	-	-
Electr., CHP+Heat Plants	-	-	-	-	-	-	-	-	-	-	-
Pumped Storage (Elec.)	-	-	-	-	-	-	-	-	-	-	-
Other Energy Sector	-	-	-	-	-	-	-	-	-	-	-
Distribution Losses	-	-	-	-	-	-	-	-	-	-	-
FINAL CONSUMPTION	**-**	**294**	**2029**	**-**	**-**	**993**	**1868**	**1592**	**730**	**-**	**527**
INDUSTRY SECTOR	**-**	**-**	**-**	**-**	**-**	**-**	**126**	**1046**	**730**	**-**	**-**
Iron and Steel	-	-	-	-	-	-	-	-	-	-	-
Chemical and Petrochem.	-	-	-	-	-	-	-	-	730	-	-
of which: Feedstocks	-	-	-	-	-	-	-	-	*730*	-	-
Non-Ferrous Metals	-	-	-	-	-	-	-	-	-	-	-
Non-Metallic Minerals	-	-	-	-	-	-	-	-	-	-	-
Transport Equipment	-	-	-	-	-	-	-	-	-	-	-
Machinery	-	-	-	-	-	-	-	-	-	-	-
Mining and Quarrying	-	-	-	-	-	-	-	-	-	-	-
Food and Tobacco	-	-	-	-	-	-	-	-	-	-	-
Paper, Pulp and Print	-	-	-	-	-	-	-	-	-	-	-
Wood and Wood Products	-	-	-	-	-	-	-	-	-	-	-
Construction	-	-	-	-	-	-	-	-	-	-	-
Textile and Leather	-	-	-	-	-	-	-	-	-	-	-
Non-specified	-	-	-	-	-	-	126	1046	-	-	-
TRANSPORT SECTOR	**-**	**-**	**2029**	**-**	**-**	**714**	**837**	**-**	**-**	**-**	**-**
Air	-	-	-	-	-	714	-	-	-	-	-
Road	-	-	2029	-	-	-	837	-	-	-	-
Rail	-	-	-	-	-	-	-	-	-	-	-
Pipeline Transport	-	-	-	-	-	-	-	-	-	-	-
Internal Navigation	-	-	-	-	-	-	-	-	-	-	-
Non-specified	-	-	-	-	-	-	-	-	-	-	-
OTHER SECTORS	**-**	**294**	**-**	**-**	**-**	**279**	**905**	**546**	**-**	**-**	**-**
Agriculture	-	-	-	-	-	-	-	-	-	-	-
Comm. and Publ. Services	-	-	-	-	-	-	-	-	-	-	-
Residential	-	294	-	-	-	279	-	-	-	-	-
Non-specified	-	-	-	-	-	-	905	546	-	-	-
NON-ENERGY USE	**-**	**-**	**-**	**-**	**-**	**-**	**-**	**-**	**-**	**-**	**527**
in Industry/Transf./Energy	-	-	-	-	-	-	-	-	-	-	527
in Transport	-	-	-	-	-	-	-	-	-	-	-
in Other Sectors	-	-	-	-	-	-	-	-	-	-	-

Israel / Israël : 1996

	Gas / *Gaz* (TJ)				Comb. Renew. & Waste / *En. Re. Comb. & Déchets* (TJ)				(GWh)	(TJ)
SUPPLY AND CONSUMPTION ***APPROVISIONNEMENT ET DEMANDE***	Natural Gas *Gaz naturel*	Gas Works *Usines à gaz*	Coke Ovens *Cokeries*	Blast Furnaces *Hauts fourneaux*	Solid Biomass *Biomasse solide*	Gas/Liquids from Biomass *Gaz/Liquides tirés de biomasse*	Municipal Waste *Déchets urbains*	Industrial Waste *Déchets industriels*	Electricity *Electricité*	Heat *Chaleur*
Production	572	-	-	-	127	-	-	-	32497	-
From Other Sources	-	-	-	-	-	-	-	-	-	-
Imports	-	-	-	-	115	-	-	-	-	-
Exports	-	-	-	-	-	-	-	-	-979	-
Intl. Marine Bunkers	-	-	-	-	-	-	-	-	-	-
Stock Changes	-	-	-	-	-	-	-	-	-	-
DOMESTIC SUPPLY	**572**	**-**	**-**	**-**	**242**	**-**	**-**	**-**	**31518**	**-**
Transfers	-	-	-	-	-	-	-	-	-	-
Statistical Differences	5	-	-	-	-	-	-	-	112	-
TRANSFORMATION	**200**	**-**	**-**	**-**	**-**	**-**	**-**	**-**	**-**	**-**
Electricity Plants	200	-	-	-	-	-	-	-	-	-
CHP Plants	-	-	-	-	-	-	-	-	-	-
Heat Plants	-	-	-	-	-	-	-	-	-	-
Blast Furnaces/Gas Works	-	-	-	-	-	-	-	-	-	-
Coke/Pat. Fuel/BKB Plants	-	-	-	-	-	-	-	-	-	-
Petroleum Refineries	-	-	-	-	-	-	-	-	-	-
Petrochemical Industry	-	-	-	-	-	-	-	-	-	-
Liquefaction	-	-	-	-	-	-	-	-	-	-
Other Transform. Sector	-	-	-	-	-	-	-	-	-	-
ENERGY SECTOR	**-**	**-**	**-**	**-**	**-**	**-**	**-**	**-**	**1402**	**-**
Coal Mines	-	-	-	-	-	-	-	-	-	-
Oil and Gas Extraction	-	-	-	-	-	-	-	-	-	-
Petroleum Refineries	-	-	-	-	-	-	-	-	-	-
Electr., CHP+Heat Plants	-	-	-	-	-	-	-	-	1402	-
Pumped Storage (Elec.)	-	-	-	-	-	-	-	-	-	-
Other Energy Sector	-	-	-	-	-	-	-	-	-	-
Distribution Losses	-	-	-	-	-	-	-	-	1305	-
FINAL CONSUMPTION	**377**	**-**	**-**	**-**	**242**	**-**	**-**	**-**	**28923**	**-**
INDUSTRY SECTOR	**372**	**-**	**-**	**-**	**-**	**-**	**-**	**-**	**7826**	**-**
Iron and Steel	-	-	-	-	-	-	-	-	966	-
Chemical and Petrochem.	372	-	-	-	-	-	-	-	2224	-
of which: Feedstocks	*372*	-	-	-	-	-	-	-	-	-
Non-Ferrous Metals	-	-	-	-	-	-	-	-	-	-
Non-Metallic Minerals	-	-	-	-	-	-	-	-	716	-
Transport Equipment	-	-	-	-	-	-	-	-	279	-
Machinery	-	-	-	-	-	-	-	-	110	-
Mining and Quarrying	-	-	-	-	-	-	-	-	433	-
Food and Tobacco	-	-	-	-	-	-	-	-	1004	-
Paper, Pulp and Print	-	-	-	-	-	-	-	-	306	-
Wood and Wood Products	-	-	-	-	-	-	-	-	137	-
Construction	-	-	-	-	-	-	-	-	144	-
Textile and Leather	-	-	-	-	-	-	-	-	571	-
Non-specified	-	-	-	-	-	-	-	-	936	-
TRANSPORT SECTOR	**-**	**-**	**-**	**-**	**-**	**-**	**-**	**-**	**873**	**-**
Air	-	-	-	-	-	-	-	-	-	-
Road	-	-	-	-	-	-	-	-	-	-
Rail	-	-	-	-	-	-	-	-	-	-
Pipeline Transport	-	-	-	-	-	-	-	-	-	-
Internal Navigation	-	-	-	-	-	-	-	-	-	-
Non-specified	-	-	-	-	-	-	-	-	873	-
OTHER SECTORS	**5**	**-**	**-**	**-**	**242**	**-**	**-**	**-**	**20224**	**-**
Agriculture	-	-	-	-	-	-	-	-	1433	-
Comm. and Publ. Services	-	-	-	-	-	-	-	-	6141	-
Residential	5	-	-	-	127	-	-	-	8639	-
Non-specified	-	-	-	-	115	-	-	-	4011	-
NON-ENERGY USE	**-**	**-**	**-**	**-**	**-**	**-**	**-**	**-**	**-**	**-**
in Industry/Transf./Energy	-	-	-	-	-	-	-	-	-	-
in Transport	-	-	-	-	-	-	-	-	-	-
in Other Sectors	-	-	-	-	-	-	-	-	-	-

Israel / Israël : 1997

SUPPLY AND CONSUMPTION *APPROVISIONNEMENT ET DEMANDE*	Coal / *Charbon* (1000 tonnes)							Oil / *Pétrole* (1000 tonnes)			
	Coking Coal *Charbon à coke*	Other Bit. Coal *Autres charb. bit.*	Sub-Bit. Coal *Charbon sous-bit.*	Lignite *Lignite*	Peat *Tourbe*	Oven and Gas Coke *Coke de four/gaz*	Pat. Fuel and BKB *Agg./briq. de lignite*	Crude Oil *Pétrole brut*	NGL *LGN*	Feed-stocks *Produits d'aliment.*	Additives *Additifs*
Production	-	-	-	421	-	-	-	5	21	-	-
From Other Sources	-	-	-	-	-	-	-	-	-	-	-
Imports	-	8085	-	-	-	-	-	11543	-	-	-
Exports	-	-	-	-	-	-	-	-	-	-	-
Intl. Marine Bunkers	-	-	-	-	-	-	-	-	-	-	-
Stock Changes	-	596	-	-	-	-	-	-	-	-	-
DOMESTIC SUPPLY	**-**	**8681**	**-**	**421**	**-**	**-**	**-**	**11548**	**21**	**-**	**-**
Transfers	-	-	-	-	-	-	-	-	-	-	-
Statistical Differences	-	-	-	-	-	-	-	-	-	-	-
TRANSFORMATION	**-**	**8653**	**-**	**421**	**-**	**-**	**-**	**11548**	**-**	**-**	**-**
Electricity Plants	-	8653	-	421	-	-	-	-	-	-	-
CHP Plants	-	-	-	-	-	-	-	-	-	-	-
Heat Plants	-	-	-	-	-	-	-	-	-	-	-
Blast Furnaces/Gas Works	-	-	-	-	-	-	-	-	-	-	-
Coke/Pat. Fuel/BKB Plants	-	-	-	-	-	-	-	-	-	-	-
Petroleum Refineries	-	-	-	-	-	-	-	11548	-	-	-
Petrochemical Industry	-	-	-	-	-	-	-	-	-	-	-
Liquefaction	-	-	-	-	-	-	-	-	-	-	-
Other Transform. Sector	-	-	-	-	-	-	-	-	-	-	-
ENERGY SECTOR	**-**	**-**	**-**	**-**	**-**	**-**	**-**	**-**	**-**	**-**	**-**
Coal Mines	-	-	-	-	-	-	-	-	-	-	-
Oil and Gas Extraction	-	-	-	-	-	-	-	-	-	-	-
Petroleum Refineries	-	-	-	-	-	-	-	-	-	-	-
Electr., CHP+Heat Plants	-	-	-	-	-	-	-	-	-	-	-
Pumped Storage (Elec.)	-	-	-	-	-	-	-	-	-	-	-
Other Energy Sector	-	-	-	-	-	-	-	-	-	-	-
Distribution Losses	-	-	-	-	-	-	-	-	-	-	-
FINAL CONSUMPTION	**-**	**28**	**-**	**-**	**-**	**-**	**-**	**-**	**21**	**-**	**-**
INDUSTRY SECTOR	**-**	**28**	**-**	**-**	**-**	**-**	**-**	**-**	**21**	**-**	**-**
Iron and Steel	-	-	-	-	-	-	-	-	-	-	-
Chemical and Petrochem.	-	6	-	-	-	-	-	-	21	-	-
of which: Feedstocks	-	-	-	-	-	-	-	-	*21*	-	-
Non-Ferrous Metals	-	-	-	-	-	-	-	-	-	-	-
Non-Metallic Minerals	-	-	-	-	-	-	-	-	-	-	-
Transport Equipment	-	-	-	-	-	-	-	-	-	-	-
Machinery	-	-	-	-	-	-	-	-	-	-	-
Mining and Quarrying	-	-	-	-	-	-	-	-	-	-	-
Food and Tobacco	-	-	-	-	-	-	-	-	-	-	-
Paper, Pulp and Print	-	-	-	-	-	-	-	-	-	-	-
Wood and Wood Products	-	-	-	-	-	-	-	-	-	-	-
Construction	-	-	-	-	-	-	-	-	-	-	-
Textile and Leather	-	-	-	-	-	-	-	-	-	-	-
Non-specified	-	22	-	-	-	-	-	-	-	-	-
TRANSPORT SECTOR	**-**	**-**	**-**	**-**	**-**	**-**	**-**	**-**	**-**	**-**	**-**
Air	-	-	-	-	-	-	-	-	-	-	-
Road	-	-	-	-	-	-	-	-	-	-	-
Rail	-	-	-	-	-	-	-	-	-	-	-
Pipeline Transport	-	-	-	-	-	-	-	-	-	-	-
Internal Navigation	-	-	-	-	-	-	-	-	-	-	-
Non-specified	-	-	-	-	-	-	-	-	-	-	-
OTHER SECTORS	**-**	**-**	**-**	**-**	**-**	**-**	**-**	**-**	**-**	**-**	**-**
Agriculture	-	-	-	-	-	-	-	-	-	-	-
Comm. and Publ. Services	-	-	-	-	-	-	-	-	-	-	-
Residential	-	-	-	-	-	-	-	-	-	-	-
Non-specified	-	-	-	-	-	-	-	-	-	-	-
NON-ENERGY USE	**-**	**-**	**-**	**-**	**-**	**-**	**-**	**-**	**-**	**-**	**-**
in Industry/Trans./Energy	-	-	-	-	-	-	-	-	-	-	-
in Transport	-	-	-	-	-	-	-	-	-	-	-
in Other Sectors	-	-	-	-	-	-	-	-	-	-	-

Israel / Israël : 1997

SUPPLY AND CONSUMPTION / *APPROVISIONNEMENT ET DEMANDE*	Oil cont. / *Pétrole cont.* (1000 tonnes)										
	Refinery Gas / *Gaz de raffinerie*	LPG + Ethane / *GPL + éthane*	Motor Gasoline / *Essence moteur*	Aviation Gasoline / *Essence aviation*	Jet Fuel / *Carbu-réacteurs*	Kerosene / *Kérosène*	Gas/ Diesel / *Gazole*	Heavy Fuel Oil / *Fioul lourd*	Naphtha / *Naphta*	Petrol. Coke / *Coke de pétrole*	Other Prod. / *Autres prod.*
Production	-	341	2061	-	-	1414	2853	3624	728	-	-
From Other Sources	-	-	-	-	-	-	-	-	-	-	-
Imports	-	-	8	-	-	-	-	1609	175	-	424
Exports	-	-61	-166	-	-	-311	-989	-447	-28	-	-
Intl. Marine Bunkers	-	-	-	-	-	-	-77	-104	-	-	-
Stock Changes	-	-	-	-	-	-	-	-	-	-	-
DOMESTIC SUPPLY	**-**	**280**	**1903**	**-**	**-**	**1103**	**1787**	**4682**	**875**	**-**	**424**
Transfers	-	-	-	-	-	-	-	-	-	-	-
Statistical Differences	-	-	-	-	-	1	1	-	-	-	-
TRANSFORMATION	**-**	**-**	**-**	**-**	**-**	**-**	**127**	**2143**	**-**	**-**	**-**
Electricity Plants	-	-	-	-	-	-	127	2143	-	-	-
CHP Plants	-	-	-	-	-	-	-	-	-	-	-
Heat Plants	-	-	-	-	-	-	-	-	-	-	-
Blast Furnaces/Gas Works	-	-	-	-	-	-	-	-	-	-	-
Coke/Pat. Fuel/BKB Plants	-	-	-	-	-	-	-	-	-	-	-
Petroleum Refineries	-	-	-	-	-	-	-	-	-	-	-
Petrochemical Industry	-	-	-	-	-	-	-	-	-	-	-
Liquefaction	-	-	-	-	-	-	-	-	-	-	-
Other Transform. Sector	-	-	-	-	-	-	-	-	-	-	-
ENERGY SECTOR	**-**	**-**	**-**	**-**	**-**	**-**	**-**	**500**	**-**	**-**	**-**
Coal Mines	-	-	-	-	-	-	-	-	-	-	-
Oil and Gas Extraction	-	-	-	-	-	-	-	-	-	-	-
Petroleum Refineries	-	-	-	-	-	-	-	500	-	-	-
Electr., CHP+Heat Plants	-	-	-	-	-	-	-	-	-	-	-
Pumped Storage (Elec.)	-	-	-	-	-	-	-	-	-	-	-
Other Energy Sector	-	-	-	-	-	-	-	-	-	-	-
Distribution Losses	-	-	-	-	-	-	-	-	-	-	-
FINAL CONSUMPTION	**-**	**280**	**1903**	**-**	**-**	**1104**	**1661**	**2039**	**875**	**-**	**424**
INDUSTRY SECTOR	**-**	**-**	**-**	**-**	**-**	**-**	**112**	**1340**	**875**	**-**	**-**
Iron and Steel	-	-	-	-	-	-	-	-	-	-	-
Chemical and Petrochem.	-	-	-	-	-	-	-	-	875	-	-
of which: Feedstocks	-	-	-	-	-	-	-	-	*875*	-	-
Non-Ferrous Metals	-	-	-	-	-	-	-	-	-	-	-
Non-Metallic Minerals	-	-	-	-	-	-	-	-	-	-	-
Transport Equipment	-	-	-	-	-	-	-	-	-	-	-
Machinery	-	-	-	-	-	-	-	-	-	-	-
Mining and Quarrying	-	-	-	-	-	-	-	-	-	-	-
Food and Tobacco	-	-	-	-	-	-	-	-	-	-	-
Paper, Pulp and Print	-	-	-	-	-	-	-	-	-	-	-
Wood and Wood Products	-	-	-	-	-	-	-	-	-	-	-
Construction	-	-	-	-	-	-	-	-	-	-	-
Textile and Leather	-	-	-	-	-	-	-	-	-	-	-
Non-specified	-	-	-	-	-	-	112	1340	-	-	-
TRANSPORT SECTOR	**-**	**-**	**1903**	**-**	**-**	**825**	**744**	**-**	**-**	**-**	**-**
Air	-	-	-	-	-	825	-	-	-	-	-
Road	-	-	1903	-	-	-	744	-	-	-	-
Rail	-	-	-	-	-	-	-	-	-	-	-
Pipeline Transport	-	-	-	-	-	-	-	-	-	-	-
Internal Navigation	-	-	-	-	-	-	-	-	-	-	-
Non-specified	-	-	-	-	-	-	-	-	-	-	-
OTHER SECTORS	**-**	**280**	**-**	**-**	**-**	**279**	**805**	**699**	**-**	**-**	**-**
Agriculture	-	-	-	-	-	-	-	-	-	-	-
Comm. and Publ. Services	-	-	-	-	-	-	-	-	-	-	-
Residential	-	280	-	-	-	279	-	-	-	-	-
Non-specified	-	-	-	-	-	-	805	699	-	-	-
NON-ENERGY USE	**-**	**-**	**-**	**-**	**-**	**-**	**-**	**-**	**-**	**-**	**424**
in Industry/Transf./Energy	-	-	-	-	-	-	-	-	-	-	424
in Transport	-	-	-	-	-	-	-	-	-	-	-
in Other Sectors	-	-	-	-	-	-	-	-	-	-	-

Israel / Israël : 1997

SUPPLY AND CONSUMPTION / *APPROVISIONNEMENT ET DEMANDE*	Gas / *Gaz* (TJ)				Comb. Renew. & Waste / *En. Re. Comb. & Déchets* (TJ)				(GWh)	(TJ)
	Natural Gas / *Gaz naturel*	Gas Works / *Usines à gaz*	Coke Ovens / *Cokeries*	Blast Furnaces / *Hauts fourneaux*	Solid Biomass / *Biomasse solide*	Gas/Liquids from Biomass / *Gaz/Liquides tirés de biomasse*	Municipal Waste / *Déchets urbains*	Industrial Waste / *Déchets industriels*	Electricity / *Electricité*	Heat / *Chaleur*
Production	605	-	-	-	127	-	-	-	35098	-
From Other Sources	-	-	-	-	-	-	-	-	-	-
Imports	-	-	-	-	115	-	-	-	-	-
Exports	-	-	-	-	-	-	-	-	-1061	-
Intl. Marine Bunkers	-	-	-	-	-	-	-	-	-	-
Stock Changes	-	-	-	-	-	-	-	-	-	-
DOMESTIC SUPPLY	**605**	**-**	**-**	**-**	**242**	**-**	**-**	**-**	**34037**	**-**
Transfers	-	-	-	-	-	-	-	-	-	-
Statistical Differences	-1	-	-	-	-	-	-	-	-	-
TRANSFORMATION	**209**	**-**	**-**	**-**	**-**	**-**	**-**	**-**	**-**	**-**
Electricity Plants	209	-	-	-	-	-	-	-	-	-
CHP Plants	-	-	-	-	-	-	-	-	-	-
Heat Plants	-	-	-	-	-	-	-	-	-	-
Blast Furnaces/Gas Works	-	-	-	-	-	-	-	-	-	-
Coke/Pat. Fuel/BKB Plants	-	-	-	-	-	-	-	-	-	-
Petroleum Refineries	-	-	-	-	-	-	-	-	-	-
Petrochemical Industry	-	-	-	-	-	-	-	-	-	-
Liquefaction	-	-	-	-	-	-	-	-	-	-
Other Transform. Sector	-	-	-	-	-	-	-	-	-	-
ENERGY SECTOR	**-**	**-**	**-**	**-**	**-**	**-**	**-**	**-**	**1462**	**-**
Coal Mines	-	-	-	-	-	-	-	-	-	-
Oil and Gas Extraction	-	-	-	-	-	-	-	-	-	-
Petroleum Refineries	-	-	-	-	-	-	-	-	-	-
Electr., CHP+Heat Plants	-	-	-	-	-	-	-	-	1462	-
Pumped Storage (Elec.)	-	-	-	-	-	-	-	-	-	-
Other Energy Sector	-	-	-	-	-	-	-	-	-	-
Distribution Losses	-	-	-	-	-	-	-	-	2990	-
FINAL CONSUMPTION	**395**	**-**	**-**	**-**	**242**	**-**	**-**	**-**	**29585**	**-**
INDUSTRY SECTOR	**390**	**-**	**-**	**-**	**-**	**-**	**-**	**-**	**8004**	**-**
Iron and Steel	-	-	-	-	-	-	-	-	988	-
Chemical and Petrochem.	390	-	-	-	-	-	-	-	2275	-
of which: Feedstocks	*390*	-	-	-	-	-	-	-	-	-
Non-Ferrous Metals	-	-	-	-	-	-	-	-	-	-
Non-Metallic Minerals	-	-	-	-	-	-	-	-	732	-
Transport Equipment	-	-	-	-	-	-	-	-	285	-
Machinery	-	-	-	-	-	-	-	-	113	-
Mining and Quarrying	-	-	-	-	-	-	-	-	443	-
Food and Tobacco	-	-	-	-	-	-	-	-	1027	-
Paper, Pulp and Print	-	-	-	-	-	-	-	-	313	-
Wood and Wood Products	-	-	-	-	-	-	-	-	140	-
Construction	-	-	-	-	-	-	-	-	147	-
Textile and Leather	-	-	-	-	-	-	-	-	584	-
Non-specified	-	-	-	-	-	-	-	-	957	-
TRANSPORT SECTOR	**-**	**-**	**-**	**-**	**-**	**-**	**-**	**-**	**893**	**-**
Air	-	-	-	-	-	-	-	-	-	-
Road	-	-	-	-	-	-	-	-	-	-
Rail	-	-	-	-	-	-	-	-	-	-
Pipeline Transport	-	-	-	-	-	-	-	-	-	-
Internal Navigation	-	-	-	-	-	-	-	-	-	-
Non-specified	-	-	-	-	-	-	-	-	893	-
OTHER SECTORS	**5**	**-**	**-**	**-**	**242**	**-**	**-**	**-**	**20688**	**-**
Agriculture	-	-	-	-	-	-	-	-	1466	-
Comm. and Publ. Services	-	-	-	-	-	-	-	-	6282	-
Residential	5	-	-	-	127	-	-	-	8837	-
Non-specified	-	-	-	-	115	-	-	-	4103	-
NON-ENERGY USE	**-**	**-**	**-**	**-**	**-**	**-**	**-**	**-**	**-**	**-**
in Industry/Transf./Energy	-	-	-	-	-	-	-	-	-	-
in Transport	-	-	-	-	-	-	-	-	-	-
in Other Sectors	-	-	-	-	-	-	-	-	-	-

Ivory Coast / Côte d'Ivoire

SUPPLY AND CONSUMPTION 1996	Coal (1000 tonnes)							Oil (1000 tonnes)			
	Coking Coal	Other Bit. Coal	Sub-Bit. Coal	Lignite	Peat	Oven and Gas Coke	Pat. Fuel and BKB	Crude Oil	NGL	Feed-stocks	Additives
Production	-	-	-	-	-	-	-	884	-	-	-
Imports	-	-	-	-	-	-	-	2041	-	-	-
Exports	-	-	-	-	-	-	-	-	-	-	-
Intl. Marine Bunkers	-	-	-	-	-	-	-	-	-	-	-
Stock Changes	-	-	-	-	-	-	-	-608	-	-	-
DOMESTIC SUPPLY	-	-	-	-	-	-	-	**2317**	-	-	-
Transfers and Stat. Diff.	-	-	-	-	-	-	-	-	-	-	-
TRANSFORMATION	-	-	-	-	-	-	-	**2317**	-	-	-
Electricity and CHP Plants	-	-	-	-	-	-	-	-	-	-	-
Petroleum Refineries	-	-	-	-	-	-	-	2317	-	-	-
Other Transform. Sector	-	-	-	-	-	-	-	-	-	-	-
ENERGY SECTOR	-	-	-	-	-	-	-	-	-	-	-
DISTRIBUTION LOSSES	-	-	-	-	-	-	-	-	-	-	-
FINAL CONSUMPTION	-	-	-	-	-	-	-	-	-	-	-
INDUSTRY SECTOR	-	-	-	-	-	-	-	-	-	-	-
Iron and Steel	-	-	-	-	-	-	-	-	-	-	-
Chemical and Petrochem.	-	-	-	-	-	-	-	-	-	-	-
Non-Metallic Minerals	-	-	-	-	-	-	-	-	-	-	-
Non-specified	-	-	-	-	-	-	-	-	-	-	-
TRANSPORT SECTOR	-	-	-	-	-	-	-	-	-	-	-
Air	-	-	-	-	-	-	-	-	-	-	-
Road	-	-	-	-	-	-	-	-	-	-	-
Non-specified	-	-	-	-	-	-	-	-	-	-	-
OTHER SECTORS	-	-	-	-	-	-	-	-	-	-	-
Agriculture	-	-	-	-	-	-	-	-	-	-	-
Comm. and Publ. Services	-	-	-	-	-	-	-	-	-	-	-
Residential	-	-	-	-	-	-	-	-	-	-	-
Non-specified	-	-	-	-	-	-	-	-	-	-	-
NON-ENERGY USE	-	-	-	-	-	-	-	-	-	-	-

APPROVISIONNEMENT ET DEMANDE 1997	*Charbon* (1000 tonnes)							*Pétrole* (1000 tonnes)			
	Charbon à coke	*Autres charb. bit.*	*Charbon sous-bit.*	*Lignite*	*Tourbe*	*Coke de four/gaz*	*Agg./briq. de lignite*	*Pétrole brut*	*LGN*	*Produits d'aliment.*	*Additifs*
Production	-	-	-	-	-	-	-	789	-	-	-
Imports	-	-	-	-	-	-	-	1719	-	-	-
Exports	-	-	-	-	-	-	-	-	-	-	-
Intl. Marine Bunkers	-	-	-	-	-	-	-	-	-	-	-
Stock Changes	-	-	-	-	-	-	-	-	-	-	-
DOMESTIC SUPPLY	-	-	-	-	-	-	-	**2508**	-	-	-
Transfers and Stat. Diff.	-	-	-	-	-	-	-	-	-	-	-
TRANSFORMATION	-	-	-	-	-	-	-	**2508**	-	-	-
Electricity and CHP Plants	-	-	-	-	-	-	-	-	-	-	-
Petroleum Refineries	-	-	-	-	-	-	-	2508	-	-	-
Other Transform. Sector	-	-	-	-	-	-	-	-	-	-	-
ENERGY SECTOR	-	-	-	-	-	-	-	-	-	-	-
DISTRIBUTION LOSSES	-	-	-	-	-	-	-	-	-	-	-
FINAL CONSUMPTION	-	-	-	-	-	-	-	-	-	-	-
INDUSTRY SECTOR	-	-	-	-	-	-	-	-	-	-	-
Iron and Steel	-	-	-	-	-	-	-	-	-	-	-
Chemical and Petrochem.	-	-	-	-	-	-	-	-	-	-	-
Non-Metallic Minerals	-	-	-	-	-	-	-	-	-	-	-
Non-specified	-	-	-	-	-	-	-	-	-	-	-
TRANSPORT SECTOR	-	-	-	-	-	-	-	-	-	-	-
Air	-	-	-	-	-	-	-	-	-	-	-
Road	-	-	-	-	-	-	-	-	-	-	-
Non-specified	-	-	-	-	-	-	-	-	-	-	-
OTHER SECTORS	-	-	-	-	-	-	-	-	-	-	-
Agriculture	-	-	-	-	-	-	-	-	-	-	-
Comm. and Publ. Services	-	-	-	-	-	-	-	-	-	-	-
Residential	-	-	-	-	-	-	-	-	-	-	-
Non-specified	-	-	-	-	-	-	-	-	-	-	-
NON-ENERGY USE	-	-	-	-	-	-	-	-	-	-	-

Ivory Coast / Côte d'Ivoire

SUPPLY AND CONSUMPTION 1996	Oil cont. (1000 tonnes)										
	Refinery Gas	LPG + Ethane	Motor Gasoline	Aviation Gasoline	Jet Fuel	Kerosene	Gas/ Diesel	Heavy Fuel Oil	Naphtha	Petrol. Coke	Other Prod.
Production	54	23	514	-	117	229	447	652	-	-	99
Imports	-	23	153	-	3	24	114	-	-	-	-
Exports	-	-10	-457	-	-38	-149	-	-291	-	-	-44
Intl. Marine Bunkers	-	-	-	-	-	-	-10	-162	-	-	-
Stock Changes	-	-	-54	-	23	-40	-	-	-	-	-
DOMESTIC SUPPLY	**54**	**36**	**156**	**-**	**105**	**64**	**551**	**199**	**-**	**-**	**55**
Transfers and Stat. Diff.	-	-1	-	-	-5	1	1	1	-	-	-
TRANSFORMATION	**-**	**-**	**-**	**-**	**-**	**-**	**164**	**85**	**-**	**-**	**-**
Electricity and CHP Plants	-	-	-	-	-	-	164	85	-	-	-
Petroleum Refineries	-	-	-	-	-	-	-	-	-	-	-
Other Transform. Sector	-	-	-	-	-	-	-	-	-	-	-
ENERGY SECTOR	**54**	**-**	**-**	**-**	**-**	**-**	**-**	**52**	**-**	**-**	**-**
DISTRIBUTION LOSSES	**-**	**-**	**-**	**-**	**-**	**-**	**-**	**-**	**-**	**-**	**-**
FINAL CONSUMPTION	**-**	**35**	**156**	**-**	**100**	**65**	**388**	**63**	**-**	**-**	**55**
INDUSTRY SECTOR	**-**	**8**	**-**	**-**	**-**	**5**	**63**	**60**	**-**	**-**	**-**
Iron and Steel	-	-	-	-	-	-	-	-	-	-	-
Chemical and Petrochem.	-	-	-	-	-	-	-	-	-	-	-
Non-Metallic Minerals	-	-	-	-	-	-	-	-	-	-	-
Non-specified	-	8	-	-	-	5	63	60	-	-	-
TRANSPORT SECTOR	**-**	**-**	**156**	**-**	**100**	**-**	**283**	**3**	**-**	**-**	**-**
Air	-	-	-	-	100	-	-	-	-	-	-
Road	-	-	156	-	-	-	270	-	-	-	-
Non-specified	-	-	-	-	-	-	13	3	-	-	-
OTHER SECTORS	**-**	**27**	**-**	**-**	**-**	**60**	**42**	**-**	**-**	**-**	**-**
Agriculture	-	-	-	-	-	-	26	-	-	-	-
Comm. and Publ. Services	-	-	-	-	-	-	16	-	-	-	-
Residential	-	27	-	-	-	60	-	-	-	-	-
Non-specified	-	-	-	-	-	-	-	-	-	-	-
NON-ENERGY USE	**-**	**-**	**-**	**-**	**-**	**-**	**-**	**-**	**-**	**-**	**55**

APPROVISIONNEMENT ET DEMANDE 1997	*Pétrole cont.* (1000 tonnes)										
	Gaz de raffinerie	*GPL + éthane*	*Essence moteur*	*Essence aviation*	*Carbu- réacteurs*	*Kérosène*	*Gazole*	*Fioul lourd*	*Naphta*	*Coke de pétrole*	*Autres prod.*
Production	58	25	556	-	127	248	484	706	-	-	108
Imports	-	23	153	-	3	24	99	-	-	-	-
Exports	-	-10	-560	-	-26	-207	-	-335	-	-	-44
Intl. Marine Bunkers	-	-	-	-	-	-	-10	-162	-	-	-
Stock Changes	-	-	-	-	-	-	-	-	-	-	-
DOMESTIC SUPPLY	**58**	**38**	**149**	**-**	**104**	**65**	**573**	**209**	**-**	**-**	**64**
Transfers and Stat. Diff.	-	-1	1	-	-	-	-	-1	-	-	-
TRANSFORMATION	**-**	**-**	**-**	**-**	**-**	**-**	**171**	**88**	**-**	**-**	**-**
Electricity and CHP Plants	-	-	-	-	-	-	171	88	-	-	-
Petroleum Refineries	-	-	-	-	-	-	-	-	-	-	-
Other Transform. Sector	-	-	-	-	-	-	-	-	-	-	-
ENERGY SECTOR	**58**	**-**	**-**	**-**	**-**	**-**	**-**	**54**	**-**	**-**	**-**
DISTRIBUTION LOSSES	**-**	**-**	**-**	**-**	**-**	**-**	**-**	**-**	**-**	**-**	**-**
FINAL CONSUMPTION	**-**	**37**	**150**	**-**	**104**	**65**	**402**	**66**	**-**	**-**	**64**
INDUSTRY SECTOR	**-**	**8**	**-**	**-**	**-**	**5**	**66**	**62**	**-**	**-**	**-**
Iron and Steel	-	-	-	-	-	-	-	-	-	-	-
Chemical and Petrochem.	-	-	-	-	-	-	-	-	-	-	-
Non-Metallic Minerals	-	-	-	-	-	-	-	-	-	-	-
Non-specified	-	8	-	-	-	5	66	62	-	-	-
TRANSPORT SECTOR	**-**	**-**	**150**	**-**	**104**	**-**	**293**	**4**	**-**	**-**	**-**
Air	-	-	-	-	104	-	-	-	-	-	-
Road	-	-	150	-	-	-	280	-	-	-	-
Non-specified	-	-	-	-	-	-	13	4	-	-	-
OTHER SECTORS	**-**	**29**	**-**	**-**	**-**	**60**	**43**	**-**	**-**	**-**	**-**
Agriculture	-	-	-	-	-	-	27	-	-	-	-
Comm. and Publ. Services	-	-	-	-	-	-	16	-	-	-	-
Residential	-	29	-	-	-	60	-	-	-	-	-
Non-specified	-	-	-	-	-	-	-	-	-	-	-
NON-ENERGY USE	**-**	**-**	**-**	**-**	**-**	**-**	**-**	**-**	**-**	**-**	**64**

Ivory Coast / Côte d'Ivoire

SUPPLY AND CONSUMPTION 1996	Gas (TJ)				Comb. Renew. & Waste (TJ)				(GWh)	(TJ)
	Natural Gas	Gas Works	Coke Ovens	Blast Furnaces	Solid Biomass	Gas/Liquids from Biomass	Municipal Waste	Industrial Waste	Electricity	Heat
Production	-	-	-	-	159960	-	-	-	3015	-
Imports	-	-	-	-	-	-	-	-	34	-
Exports	-	-	-	-	-	-	-	-	-	-
Intl. Marine Bunkers	-	-	-	-	-	-	-	-	-	-
Stock Changes	-	-	-	-	-	-	-	-	-	-
DOMESTIC SUPPLY	-	-	-	-	**159960**	-	-	-	**3049**	-
Transfers and Stat. Diff.	-	-	-	-	-	-	-	-	-	-
TRANSFORMATION	-	-	-	-	**59598**	-	-	-	-	-
Electricity and CHP Plants	-	-	-	-	-	-	-	-	-	-
Petroleum Refineries	-	-	-	-	-	-	-	-	-	-
Other Transform. Sector	-	-	-	-	59598	-	-	-	-	-
ENERGY SECTOR	-	-	-	-	-	-	-	-	**140**	-
DISTRIBUTION LOSSES	-	-	-	-	-	-	-	-	**488**	-
FINAL CONSUMPTION	-	-	-	-	**100362**	-	-	-	**2421**	-
INDUSTRY SECTOR	-	-	-	-	-	-	-	-	**1204**	-
Iron and Steel	-	-	-	-	-	-	-	-	-	-
Chemical and Petrochem.	-	-	-	-	-	-	-	-	216	-
Non-Metallic Minerals	-	-	-	-	-	-	-	-	-	-
Non-specified	-	-	-	-	-	-	-	-	988	-
TRANSPORT SECTOR	-	-	-	-	-	-	-	-	-	-
Air	-	-	-	-	-	-	-	-	-	-
Road	-	-	-	-	-	-	-	-	-	-
Non-specified	-	-	-	-	-	-	-	-	-	-
OTHER SECTORS	-	-	-	-	**100362**	-	-	-	**1217**	-
Agriculture	-	-	-	-	-	-	-	-	180	-
Comm. and Publ. Services	-	-	-	-	13210	-	-	-	1037	-
Residential	-	-	-	-	87152	-	-	-	-	-
Non-specified	-	-	-	-	-	-	-	-	-	-
NON-ENERGY USE	-	-	-	-	-	-	-	-	-	-

APPROVISIONNEMENT ET DEMANDE 1997	*Gaz* (TJ)				*En. Re. Comb. & Déchets* (TJ)				(GWh)	(TJ)
	Gaz naturel	*Usines à gaz*	*Cokeries*	*Hauts fourneaux*	*Biomasse solide*	*Gaz/Liquides tirés de biomasse*	*Déchets urbains*	*Déchets industriels*	*Electricité*	*Chaleur*
Production	-	-	-	-	164599	-	-	-	3213	-
Imports	-	-	-	-	-	-	-	-	34	-
Exports	-	-	-	-	-	-	-	-	-	-
Intl. Marine Bunkers	-	-	-	-	-	-	-	-	-	-
Stock Changes	-	-	-	-	-	-	-	-	-	-
DOMESTIC SUPPLY	-	-	-	-	**164599**	-	-	-	**3247**	-
Transfers and Stat. Diff.	-	-	-	-	-1	-	-	-	-	-
TRANSFORMATION	-	-	-	-	**61326**	-	-	-	-	-
Electricity and CHP Plants	-	-	-	-	-	-	-	-	-	-
Petroleum Refineries	-	-	-	-	-	-	-	-	-	-
Other Transform. Sector	-	-	-	-	61326	-	-	-	-	-
ENERGY SECTOR	-	-	-	-	-	-	-	-	**150**	-
DISTRIBUTION LOSSES	-	-	-	-	-	-	-	-	**520**	-
FINAL CONSUMPTION	-	-	-	-	**103272**	-	-	-	**2577**	-
INDUSTRY SECTOR	-	-	-	-	-	-	-	-	**1281**	-
Iron and Steel	-	-	-	-	-	-	-	-	-	-
Chemical and Petrochem.	-	-	-	-	-	-	-	-	230	-
Non-Metallic Minerals	-	-	-	-	-	-	-	-	-	-
Non-specified	-	-	-	-	-	-	-	-	1051	-
TRANSPORT SECTOR	-	-	-	-	-	-	-	-	-	-
Air	-	-	-	-	-	-	-	-	-	-
Road	-	-	-	-	-	-	-	-	-	-
Non-specified	-	-	-	-	-	-	-	-	-	-
OTHER SECTORS	-	-	-	-	**103272**	-	-	-	**1296**	-
Agriculture	-	-	-	-	-	-	-	-	192	-
Comm. and Publ. Services	-	-	-	-	13593	-	-	-	1104	-
Residential	-	-	-	-	89679	-	-	-	-	-
Non-specified	-	-	-	-	-	-	-	-	-	-
NON-ENERGY USE	-	-	-	-	-	-	-	-	-	-

Jamaica / Jamaïque

SUPPLY AND CONSUMPTION 1996	Coal (1000 tonnes)							Oil (1000 tonnes)			
	Coking Coal	Other Bit. Coal	Sub-Bit. Coal	Lignite	Peat	Oven and Gas Coke	Pat. Fuel and BKB	Crude Oil	NGL	Feed-stocks	Additives
Production	-	-	-	-	-	-	-	-	-	-	-
Imports	-	64	-	-	-	-	-	1030	-	-	-
Exports	-	-	-	-	-	-	-	-	-	-	-
Intl. Marine Bunkers	-	-	-	-	-	-	-	-	-	-	-
Stock Changes	-	-	-	-	-	-	-	40	-	-	-
DOMESTIC SUPPLY	**-**	**64**	**-**	**-**	**-**	**-**	**-**	**1070**	**-**	**-**	**-**
Transfers and Stat. Diff.	-	-	-	-	-	-	-	-	-	-	-
TRANSFORMATION	**-**	**-**	**-**	**-**	**-**	**-**	**-**	**1070**	**-**	**-**	**-**
Electricity and CHP Plants	-	-	-	-	-	-	-	-	-	-	-
Petroleum Refineries	-	-	-	-	-	-	-	1070	-	-	-
Other Transform. Sector	-	-	-	-	-	-	-	-	-	-	-
ENERGY SECTOR	**-**	**-**	**-**	**-**	**-**	**-**	**-**	**-**	**-**	**-**	**-**
DISTRIBUTION LOSSES	**-**	**-**	**-**	**-**	**-**	**-**	**-**	**-**	**-**	**-**	**-**
FINAL CONSUMPTION	**-**	**64**	**-**	**-**	**-**	**-**	**-**	**-**	**-**	**-**	**-**
INDUSTRY SECTOR	**-**	**64**	**-**	**-**	**-**	**-**	**-**	**-**	**-**	**-**	**-**
Iron and Steel	-	-	-	-	-	-	-	-	-	-	-
Chemical and Petrochem.	-	-	-	-	-	-	-	-	-	-	-
Non-Metallic Minerals	-	64	-	-	-	-	-	-	-	-	-
Non-specified	-	-	-	-	-	-	-	-	-	-	-
TRANSPORT SECTOR	**-**	**-**	**-**	**-**	**-**	**-**	**-**	**-**	**-**	**-**	**-**
Air	-	-	-	-	-	-	-	-	-	-	-
Road	-	-	-	-	-	-	-	-	-	-	-
Non-specified	-	-	-	-	-	-	-	-	-	-	-
OTHER SECTORS	**-**	**-**	**-**	**-**	**-**	**-**	**-**	**-**	**-**	**-**	**-**
Agriculture	-	-	-	-	-	-	-	-	-	-	-
Comm. and Publ. Services	-	-	-	-	-	-	-	-	-	-	-
Residential	-	-	-	-	-	-	-	-	-	-	-
Non-specified	-	-	-	-	-	-	-	-	-	-	-
NON-ENERGY USE	**-**	**-**	**-**	**-**	**-**	**-**	**-**	**-**	**-**	**-**	**-**

APPROVISIONNEMENT ET DEMANDE 1997	*Charbon* (1000 tonnes)							*Pétrole* (1000 tonnes)			
	Charbon à coke	*Autres charb. bit.*	*Charbon sous-bit.*	*Lignite*	*Tourbe*	*Coke de four/gaz*	*Agg./briq. de lignite*	*Pétrole brut*	*LGN*	*Produits d'aliment.*	*Additifs*
Production	-	-	-	-	-	-	-	-	-	-	-
Imports	-	67	-	-	-	-	-	1125	-	-	-
Exports	-	-	-	-	-	-	-	-	-	-	-
Intl. Marine Bunkers	-	-	-	-	-	-	-	-	-	-	-
Stock Changes	-	-	-	-	-	-	-	44	-	-	-
DOMESTIC SUPPLY	**-**	**67**	**-**	**-**	**-**	**-**	**-**	**1169**	**-**	**-**	**-**
Transfers and Stat. Diff.	-	-	-	-	-	-	-	-	-	-	-
TRANSFORMATION	**-**	**-**	**-**	**-**	**-**	**-**	**-**	**1169**	**-**	**-**	**-**
Electricity and CHP Plants	-	-	-	-	-	-	-	-	-	-	-
Petroleum Refineries	-	-	-	-	-	-	-	1169	-	-	-
Other Transform. Sector	-	-	-	-	-	-	-	-	-	-	-
ENERGY SECTOR	**-**	**-**	**-**	**-**	**-**	**-**	**-**	**-**	**-**	**-**	**-**
DISTRIBUTION LOSSES	**-**	**-**	**-**	**-**	**-**	**-**	**-**	**-**	**-**	**-**	**-**
FINAL CONSUMPTION	**-**	**67**	**-**	**-**	**-**	**-**	**-**	**-**	**-**	**-**	**-**
INDUSTRY SECTOR	**-**	**67**	**-**	**-**	**-**	**-**	**-**	**-**	**-**	**-**	**-**
Iron and Steel	-	-	-	-	-	-	-	-	-	-	-
Chemical and Petrochem.	-	-	-	-	-	-	-	-	-	-	-
Non-Metallic Minerals	-	67	-	-	-	-	-	-	-	-	-
Non-specified	-	-	-	-	-	-	-	-	-	-	-
TRANSPORT SECTOR	**-**	**-**	**-**	**-**	**-**	**-**	**-**	**-**	**-**	**-**	**-**
Air	-	-	-	-	-	-	-	-	-	-	-
Road	-	-	-	-	-	-	-	-	-	-	-
Non-specified	-	-	-	-	-	-	-	-	-	-	-
OTHER SECTORS	**-**	**-**	**-**	**-**	**-**	**-**	**-**	**-**	**-**	**-**	**-**
Agriculture	-	-	-	-	-	-	-	-	-	-	-
Comm. and Publ. Services	-	-	-	-	-	-	-	-	-	-	-
Residential	-	-	-	-	-	-	-	-	-	-	-
Non-specified	-	-	-	-	-	-	-	-	-	-	-
NON-ENERGY USE	**-**	**-**	**-**	**-**	**-**	**-**	**-**	**-**	**-**	**-**	**-**

Jamaica / Jamaïque

SUPPLY AND CONSUMPTION 1996	Oil cont. (1000 tonnes)										
	Refinery Gas	LPG + Ethane	Motor Gasoline	Aviation Gasoline	Jet Fuel	Kerosene	Gas/ Diesel	Heavy Fuel Oil	Naphtha	Petrol. Coke	Other Prod.
Production	-	8	165	-	72	22	266	492	-	-	18
Imports	-	56	203	1	134	42	264	1393	-	-	4
Exports	-	-	-	-	-	-	-19	-37	-	-	-10
Intl. Marine Bunkers	-	-	-	-	-	-	-37	-	-	-	-
Stock Changes	-	-	7	-	2	1	24	41	-	-	-
DOMESTIC SUPPLY	**-**	**64**	**375**	**1**	**208**	**65**	**498**	**1889**	**-**	**-**	**12**
Transfers and Stat. Diff.	-	1	-	-	1	5	1	2	-	-	-
TRANSFORMATION	**-**	**-**	**-**	**-**	**-**	**-**	**240**	**1526**	**-**	**-**	**-**
Electricity and CHP Plants	-	-	-	-	-	-	240	1526	-	-	-
Petroleum Refineries	-	-	-	-	-	-	-	-	-	-	-
Other Transform. Sector	-	-	-	-	-	-	-	-	-	-	-
ENERGY SECTOR	**-**	**-**	**-**	**-**	**3**	**6**	**-**	**2**	**-**	**-**	**-**
DISTRIBUTION LOSSES	**-**	**-**	**-**	**-**	**-**	**-**	**-**	**-**	**-**	**-**	**-**
FINAL CONSUMPTION	**-**	**65**	**375**	**1**	**206**	**64**	**259**	**363**	**-**	**-**	**12**
INDUSTRY SECTOR	**-**	**-**	**-**	**-**	**-**	**-**	**92**	**98**	**-**	**-**	**-**
Iron and Steel	-	-	-	-	-	-	-	-	-	-	-
Chemical and Petrochem.	-	-	-	-	-	-	-	-	-	-	-
Non-Metallic Minerals	-	-	-	-	-	-	-	-	-	-	-
Non-specified	-	-	-	-	-	-	92	98	-	-	-
TRANSPORT SECTOR	**-**	**-**	**375**	**1**	**206**	**-**	**157**	**12**	**-**	**-**	**-**
Air	-	-	-	1	206	-	-	-	-	-	-
Road	-	-	375	-	-	-	-	-	-	-	-
Non-specified	-	-	-	-	-	-	157	12	-	-	-
OTHER SECTORS	**-**	**65**	**-**	**-**	**-**	**64**	**10**	**253**	**-**	**-**	**-**
Agriculture	-	1	-	-	-	-	6	245	-	-	-
Comm. and Publ. Services	-	8	-	-	-	9	4	8	-	-	-
Residential	-	56	-	-	-	55	-	-	-	-	-
Non-specified	-	-	-	-	-	-	-	-	-	-	-
NON-ENERGY USE	**-**	**-**	**-**	**-**	**-**	**-**	**-**	**-**	**-**	**-**	**12**

APPROVISIONNEMENT ET DEMANDE 1997	*Pétrole cont.* (1000 tonnes)										
	Gaz de raffinerie	*GPL + éthane*	*Essence moteur*	*Essence aviation*	*Carbu- réacteurs*	*Kérosène*	*Gazole*	*Fioul lourd*	*Naphta*	*Coke de pétrole*	*Autres prod.*
Production	-	9	179	-	82	26	273	506	-	-	18
Imports	-	62	220	1	153	48	271	1431	-	-	4
Exports	-	-	-	-	-	-	-20	-38	-	-	-10
Intl. Marine Bunkers	-	-	-	-	-	-	-37	-	-	-	-
Stock Changes	-	-	8	-	2	1	25	40	-	-	-
DOMESTIC SUPPLY	**-**	**71**	**407**	**1**	**237**	**75**	**512**	**1939**	**-**	**-**	**12**
Transfers and Stat. Diff.	-	1	-	-	-	4	-	-2	-	-	-
TRANSFORMATION	**-**	**-**	**-**	**-**	**-**	**-**	**245**	**1566**	**-**	**-**	**-**
Electricity and CHP Plants	-	-	-	-	-	-	245	1566	-	-	-
Petroleum Refineries	-	-	-	-	-	-	-	-	-	-	-
Other Transform. Sector	-	-	-	-	-	-	-	-	-	-	-
ENERGY SECTOR	**-**	**-**	**-**	**-**	**3**	**7**	**-**	**2**	**-**	**-**	**-**
DISTRIBUTION LOSSES	**-**	**-**	**-**	**-**	**-**	**-**	**-**	**-**	**-**	**-**	**-**
FINAL CONSUMPTION	**-**	**72**	**407**	**1**	**234**	**72**	**267**	**369**	**-**	**-**	**12**
INDUSTRY SECTOR	**-**	**-**	**-**	**-**	**-**	**-**	**95**	**100**	**-**	**-**	**-**
Iron and Steel	-	-	-	-	-	-	-	-	-	-	-
Chemical and Petrochem.	-	-	-	-	-	-	-	-	-	-	-
Non-Metallic Minerals	-	-	-	-	-	-	-	-	-	-	-
Non-specified	-	-	-	-	-	-	95	100	-	-	-
TRANSPORT SECTOR	**-**	**-**	**407**	**1**	**234**	**-**	**162**	**12**	**-**	**-**	**-**
Air	-	-	-	1	234	-	-	-	-	-	-
Road	-	-	407	-	-	-	-	-	-	-	-
Non-specified	-	-	-	-	-	-	162	12	-	-	-
OTHER SECTORS	**-**	**72**	**-**	**-**	**-**	**72**	**10**	**257**	**-**	**-**	**-**
Agriculture	-	1	-	-	-	-	6	249	-	-	-
Comm. and Publ. Services	-	9	-	-	-	10	4	8	-	-	-
Residential	-	62	-	-	-	62	-	-	-	-	-
Non-specified	-	-	-	-	-	-	-	-	-	-	-
NON-ENERGY USE	**-**	**-**	**-**	**-**	**-**	**-**	**-**	**-**	**-**	**-**	**12**

Jamaica / Jamaïque

SUPPLY AND CONSUMPTION 1996	Gas (TJ)				Comb. Renew. & Waste (TJ)				(GWh)	(TJ)
	Natural Gas	Gas Works	Coke Ovens	Blast Furnaces	Solid Biomass	Gas/Liquids from Biomass	Municipal Waste	Industrial Waste	Electricity	Heat
Production	-	-	-	-	22452	-	-	-	6038	-
Imports	-	-	-	-	-	-	-	-	-	-
Exports	-	-	-	-	-	-	-	-	-	-
Intl. Marine Bunkers	-	-	-	-	-	-	-	-	-	-
Stock Changes	-	-	-	-	-	-	-	-	-	-
DOMESTIC SUPPLY	-	-	-	-	**22452**	-	-	-	**6038**	-
Transfers and Stat. Diff.	-	-	-	-	-	-	-	-	-19	-
TRANSFORMATION	-	-	-	-	**19595**	-	-	-	-	-
Electricity and CHP Plants	-	-	-	-	11570	-	-	-	-	-
Petroleum Refineries	-	-	-	-	-	-	-	-	-	-
Other Transform. Sector	-	-	-	-	8025	-	-	-	-	-
ENERGY SECTOR	-	-	-	-	-	-	-	-	**15**	-
DISTRIBUTION LOSSES	-	-	-	-	-	-	-	-	**653**	-
FINAL CONSUMPTION	-	-	-	-	**2857**	-	-	-	**5351**	-
INDUSTRY SECTOR	-	-	-	-	-	-	-	-	**3425**	-
Iron and Steel	-	-	-	-	-	-	-	-	-	-
Chemical and Petrochem.	-	-	-	-	-	-	-	-	-	-
Non-Metallic Minerals	-	-	-	-	-	-	-	-	-	-
Non-specified	-	-	-	-	-	-	-	-	3425	-
TRANSPORT SECTOR	-	-	-	-	-	-	-	-	-	-
Air	-	-	-	-	-	-	-	-	-	-
Road	-	-	-	-	-	-	-	-	-	-
Non-specified	-	-	-	-	-	-	-	-	-	-
OTHER SECTORS	-	-	-	-	**2857**	-	-	-	**1926**	-
Agriculture	-	-	-	-	-	-	-	-	-	-
Comm. and Publ. Services	-	-	-	-	-	-	-	-	1134	-
Residential	-	-	-	-	2857	-	-	-	774	-
Non-specified	-	-	-	-	-	-	-	-	18	-
NON-ENERGY USE	-	-	-	-	-	-	-	-	-	-

APPROVISIONNEMENT ET DEMANDE 1997	*Gaz* (TJ)				*En. Re. Comb. & Déchets* (TJ)				(GWh)	(TJ)
	Gaz naturel	*Usines à gaz*	*Cokeries*	*Hauts fourneaux*	*Biomasse solide*	*Gaz/Liquides tirés de biomasse*	*Déchets urbains*	*Déchets industriels*	*Electricité*	*Chaleur*
Production	-	-	-	-	24501	-	-	-	6255	-
Imports	-	-	-	-	-	-	-	-	-	-
Exports	-	-	-	-	-	-	-	-	-	-
Intl. Marine Bunkers	-	-	-	-	-	-	-	-	-	-
Stock Changes	-	-	-	-	-	-	-	-	-	-
DOMESTIC SUPPLY	-	-	-	-	**24501**	-	-	-	**6255**	-
Transfers and Stat. Diff.	-	-	-	-	-	-	-	-	-20	-
TRANSFORMATION	-	-	-	-	**21556**	-	-	-	-	-
Electricity and CHP Plants	-	-	-	-	13272	-	-	-	-	-
Petroleum Refineries	-	-	-	-	-	-	-	-	-	-
Other Transform. Sector	-	-	-	-	8284	-	-	-	-	-
ENERGY SECTOR	-	-	-	-	-	-	-	-	**16**	-
DISTRIBUTION LOSSES	-	-	-	-	-	-	-	-	**676**	-
FINAL CONSUMPTION	-	-	-	-	**2945**	-	-	-	**5543**	-
INDUSTRY SECTOR	-	-	-	-	-	-	-	-	**3548**	-
Iron and Steel	-	-	-	-	-	-	-	-	-	-
Chemical and Petrochem.	-	-	-	-	-	-	-	-	-	-
Non-Metallic Minerals	-	-	-	-	-	-	-	-	-	-
Non-specified	-	-	-	-	-	-	-	-	3548	-
TRANSPORT SECTOR	-	-	-	-	-	-	-	-	-	-
Air	-	-	-	-	-	-	-	-	-	-
Road	-	-	-	-	-	-	-	-	-	-
Non-specified	-	-	-	-	-	-	-	-	-	-
OTHER SECTORS	-	-	-	-	**2945**	-	-	-	**1995**	-
Agriculture	-	-	-	-	-	-	-	-	-	-
Comm. and Publ. Services	-	-	-	-	-	-	-	-	1175	-
Residential	-	-	-	-	2945	-	-	-	802	-
Non-specified	-	-	-	-	-	-	-	-	18	-
NON-ENERGY USE	-	-	-	-	-	-	-	-	-	-

Jordan / Jordanie

SUPPLY AND CONSUMPTION 1996	Coal (1000 tonnes) Coking Coal	Other Bit. Coal	Sub-Bit. Coal	Lignite	Peat	Oven and Gas Coke	Pat. Fuel and BKB	Oil (1000 tonnes) Crude Oil	NGL	Feed-stocks	Additives
Production	-	-	-	-	-	-	-	2	-	-	-
Imports	-	-	-	-	-	-	-	3272	-	-	-
Exports	-	-	-	-	-	-	-	-	-	-	-
Intl. Marine Bunkers	-	-	-	-	-	-	-	-	-	-	-
Stock Changes	-	-	-	-	-	-	-	-	-	-	-
DOMESTIC SUPPLY	-	-	-	-	-	-	-	**3274**	-	-	-
Transfers and Stat. Diff.	-	-	-	-	-	-	-	-	-	-	-
TRANSFORMATION	-	-	-	-	-	-	-	**3274**	-	-	-
Electricity and CHP Plants	-	-	-	-	-	-	-	-	-	-	-
Petroleum Refineries	-	-	-	-	-	-	-	3274	-	-	-
Other Transform. Sector	-	-	-	-	-	-	-	-	-	-	-
ENERGY SECTOR	-	-	-	-	-	-	-	-	-	-	-
DISTRIBUTION LOSSES	-	-	-	-	-	-	-	-	-	-	-
FINAL CONSUMPTION	-	-	-	-	-	-	-	-	-	-	-
INDUSTRY SECTOR	-	-	-	-	-	-	-	-	-	-	-
Iron and Steel	-	-	-	-	-	-	-	-	-	-	-
Chemical and Petrochem.	-	-	-	-	-	-	-	-	-	-	-
Non-Metallic Minerals	-	-	-	-	-	-	-	-	-	-	-
Non-specified	-	-	-	-	-	-	-	-	-	-	-
TRANSPORT SECTOR	-	-	-	-	-	-	-	-	-	-	-
Air	-	-	-	-	-	-	-	-	-	-	-
Road	-	-	-	-	-	-	-	-	-	-	-
Non-specified	-	-	-	-	-	-	-	-	-	-	-
OTHER SECTORS	-	-	-	-	-	-	-	-	-	-	-
Agriculture	-	-	-	-	-	-	-	-	-	-	-
Comm. and Publ. Services	-	-	-	-	-	-	-	-	-	-	-
Residential	-	-	-	-	-	-	-	-	-	-	-
Non-specified	-	-	-	-	-	-	-	-	-	-	-
NON-ENERGY USE	-	-	-	-	-	-	-	-	-	-	-

APPROVISIONNEMENT ET DEMANDE 1997	*Charbon* (1000 tonnes) *Charbon à coke*	*Autres charb. bit.*	*Charbon sous-bit.*	*Lignite*	*Tourbe*	*Coke de four/gaz*	*Agg./briq. de lignite*	*Pétrole* (1000 tonnes) *Pétrole brut*	*LGN*	*Produits d'aliment.*	*Additifs*
Production	-	-	-	-	-	-	-	-	-	-	-
Imports	-	-	-	-	-	-	-	3489	-	-	-
Exports	-	-	-	-	-	-	-	-	-	-	-
Intl. Marine Bunkers	-	-	-	-	-	-	-	-	-	-	-
Stock Changes	-	-	-	-	-	-	-	-	-	-	-
DOMESTIC SUPPLY	-	-	-	-	-	-	-	**3489**	-	-	-
Transfers and Stat. Diff.	-	-	-	-	-	-	-	-	-	-	-
TRANSFORMATION	-	-	-	-	-	-	-	**3489**	-	-	-
Electricity and CHP Plants	-	-	-	-	-	-	-	-	-	-	-
Petroleum Refineries	-	-	-	-	-	-	-	3489	-	-	-
Other Transform. Sector	-	-	-	-	-	-	-	-	-	-	-
ENERGY SECTOR	-	-	-	-	-	-	-	-	-	-	-
DISTRIBUTION LOSSES	-	-	-	-	-	-	-	-	-	-	-
FINAL CONSUMPTION	-	-	-	-	-	-	-	-	-	-	-
INDUSTRY SECTOR	-	-	-	-	-	-	-	-	-	-	-
Iron and Steel	-	-	-	-	-	-	-	-	-	-	-
Chemical and Petrochem.	-	-	-	-	-	-	-	-	-	-	-
Non-Metallic Minerals	-	-	-	-	-	-	-	-	-	-	-
Non-specified	-	-	-	-	-	-	-	-	-	-	-
TRANSPORT SECTOR	-	-	-	-	-	-	-	-	-	-	-
Air	-	-	-	-	-	-	-	-	-	-	-
Road	-	-	-	-	-	-	-	-	-	-	-
Non-specified	-	-	-	-	-	-	-	-	-	-	-
OTHER SECTORS	-	-	-	-	-	-	-	-	-	-	-
Agriculture	-	-	-	-	-	-	-	-	-	-	-
Comm. and Publ. Services	-	-	-	-	-	-	-	-	-	-	-
Residential	-	-	-	-	-	-	-	-	-	-	-
Non-specified	-	-	-	-	-	-	-	-	-	-	-
NON-ENERGY USE	-	-	-	-	-	-	-	-	-	-	-

Jordan / Jordanie

SUPPLY AND CONSUMPTION 1996	Oil cont. (1000 tonnes)										
	Refinery Gas	LPG + Ethane	Motor Gasoline	Aviation Gasoline	Jet Fuel	Kerosene	Gas/ Diesel	Heavy Fuel Oil	Naphtha	Petrol. Coke	Other Prod.
Production	61	134	501	8	299	203	854	1021	-	-	171
Imports	-	87	-	3	-	-	252	646	-	-	-
Exports	-	-	-	-	-	-	-	-	-	-	-
Intl. Marine Bunkers	-	-	-	-	-	-	-	-	-	-	-
Stock Changes	-	-	9	-	-5	-10	-	-	-	-	1
DOMESTIC SUPPLY	**61**	**221**	**510**	**11**	**294**	**193**	**1106**	**1667**	**-**	**-**	**172**
Transfers and Stat. Diff.	-	-	-	-	-	-	-	224	-	-	-
TRANSFORMATION	**-**	**-**	**-**	**-**	**-**	**-**	**274**	**1246**	**-**	**-**	**-**
Electricity and CHP Plants	-	-	-	-	-	-	274	1246	-	-	-
Petroleum Refineries	-	-	-	-	-	-	-	-	-	-	-
Other Transform. Sector	-	-	-	-	-	-	-	-	-	-	-
ENERGY SECTOR	**61**	**-**	**-**	**-**	**-**	**-**	**2**	**140**	**-**	**-**	**-**
DISTRIBUTION LOSSES	**-**	**-**	**-**	**-**	**-**	**-**	**-**	**-**	**-**	**-**	**-**
FINAL CONSUMPTION	**-**	**221**	**510**	**11**	**294**	**193**	**830**	**505**	**-**	**-**	**172**
INDUSTRY SECTOR	**-**	**-**	**-**	**-**	**-**	**-**	**-**	**505**	**-**	**-**	**-**
Iron and Steel	-	-	-	-	-	-	-	-	-	-	-
Chemical and Petrochem.	-	-	-	-	-	-	-	-	-	-	-
Non-Metallic Minerals	-	-	-	-	-	-	-	-	-	-	-
Non-specified	-	-	-	-	-	-	-	505	-	-	-
TRANSPORT SECTOR	**-**	**-**	**510**	**11**	**294**	**-**	**467**	**-**	**-**	**-**	**-**
Air	-	-	-	11	294	-	-	-	-	-	-
Road	-	-	510	-	-	-	467	-	-	-	-
Non-specified	-	-	-	-	-	-	-	-	-	-	-
OTHER SECTORS	**-**	**221**	**-**	**-**	**-**	**193**	**363**	**-**	**-**	**-**	**-**
Agriculture	-	15	-	-	-	-	-	-	-	-	-
Comm. and Publ. Services	-	-	-	-	-	-	-	-	-	-	-
Residential	-	190	-	-	-	185	35	-	-	-	-
Non-specified	-	16	-	-	-	8	328	-	-	-	-
NON-ENERGY USE	**-**	**-**	**-**	**-**	**-**	**-**	**-**	**-**	**-**	**-**	**172**

APPROVISIONNEMENT ET DEMANDE 1997	*Pétrole cont.* (1000 tonnes)										
	Gaz de raffinerie	*GPL + éthane*	*Essence moteur*	*Essence aviation*	*Carbu- réacteurs*	*Kérosène*	*Gazole*	*Fioul lourd*	*Naphta*	*Coke de pétrole*	*Autres prod.*
Production	65	143	535	8	319	216	913	1084	-	-	182
Imports	-	100	-	3	-	-	269	686	-	-	-
Exports	-	-	-	-	-	-	-	-	-	-	-
Intl. Marine Bunkers	-	-	-	-	-	-	-	-	-	-	-
Stock Changes	-	-	-	-	-	-	-	-	-	-	-
DOMESTIC SUPPLY	**65**	**243**	**535**	**11**	**319**	**216**	**1182**	**1770**	**-**	**-**	**182**
Transfers and Stat. Diff.	-	-	-	-	-	-	-	238	-	-	-
TRANSFORMATION	**-**	**-**	**-**	**-**	**-**	**-**	**293**	**1323**	**-**	**-**	**-**
Electricity and CHP Plants	-	-	-	-	-	-	293	1323	-	-	-
Petroleum Refineries	-	-	-	-	-	-	-	-	-	-	-
Other Transform. Sector	-	-	-	-	-	-	-	-	-	-	-
ENERGY SECTOR	**65**	**-**	**-**	**-**	**-**	**-**	**2**	**149**	**-**	**-**	**-**
DISTRIBUTION LOSSES	**-**	**-**	**-**	**-**	**-**	**-**	**-**	**-**	**-**	**-**	**-**
FINAL CONSUMPTION	**-**	**243**	**535**	**11**	**319**	**216**	**887**	**536**	**-**	**-**	**182**
INDUSTRY SECTOR	**-**	**-**	**-**	**-**	**-**	**-**	**-**	**536**	**-**	**-**	**-**
Iron and Steel	-	-	-	-	-	-	-	-	-	-	-
Chemical and Petrochem.	-	-	-	-	-	-	-	-	-	-	-
Non-Metallic Minerals	-	-	-	-	-	-	-	-	-	-	-
Non-specified	-	-	-	-	-	-	-	536	-	-	-
TRANSPORT SECTOR	**-**	**-**	**535**	**11**	**319**	**-**	**499**	**-**	**-**	**-**	**-**
Air	-	-	-	11	319	-	-	-	-	-	-
Road	-	-	535	-	-	-	499	-	-	-	-
Non-specified	-	-	-	-	-	-	-	-	-	-	-
OTHER SECTORS	**-**	**243**	**-**	**-**	**-**	**216**	**388**	**-**	**-**	**-**	**-**
Agriculture	-	17	-	-	-	-	-	-	-	-	-
Comm. and Publ. Services	-	-	-	-	-	-	-	-	-	-	-
Residential	-	208	-	-	-	207	37	-	-	-	-
Non-specified	-	18	-	-	-	9	351	-	-	-	-
NON-ENERGY USE	**-**	**-**	**-**	**-**	**-**	**-**	**-**	**-**	**-**	**-**	**182**

Jordan / Jordanie

SUPPLY AND CONSUMPTION 1996	Gas (TJ)				Comb. Renew. & Waste (TJ)				(GWh)	(TJ)
	Natural Gas	Gas Works	Coke Ovens	Blast Furnaces	Solid Biomass	Gas/Liquids from Biomass	Municipal Waste	Industrial Waste	Electricity	Heat
Production	8171	-	-	-	78	-	-	-	6058	-
Imports	-	-	-	-	20	-	-	-	-	-
Exports	-	-	-	-	-	-	-	-	-2	-
Intl. Marine Bunkers	-	-	-	-	-	-	-	-	-	-
Stock Changes	-	-	-	-	-	-	-	-	-	-
DOMESTIC SUPPLY	**8171**	-	-	-	**98**	-	-	-	**6056**	-
Transfers and Stat. Diff.	-	-	-	-	-	-	-	-	80	-
TRANSFORMATION	**8171**	-	-	-	-	-	-	-	-	-
Electricity and CHP Plants	8171	-	-	-	-	-	-	-	-	-
Petroleum Refineries	-	-	-	-	-	-	-	-	-	-
Other Transform. Sector	-	-	-	-	-	-	-	-	-	-
ENERGY SECTOR	-	-	-	-	-	-	-	-	**419**	-
DISTRIBUTION LOSSES	-	-	-	-	-	-	-	-	**594**	-
FINAL CONSUMPTION	-	-	-	-	**98**	-	-	-	**5123**	-
INDUSTRY SECTOR	-	-	-	-	-	-	-	-	**1773**	-
Iron and Steel	-	-	-	-	-	-	-	-	14	-
Chemical and Petrochem.	-	-	-	-	-	-	-	-	186	-
Non-Metallic Minerals	-	-	-	-	-	-	-	-	379	-
Non-specified	-	-	-	-	-	-	-	-	1194	-
TRANSPORT SECTOR	-	-	-	-	-	-	-	-	-	-
Air	-	-	-	-	-	-	-	-	-	-
Road	-	-	-	-	-	-	-	-	-	-
Non-specified	-	-	-	-	-	-	-	-	-	-
OTHER SECTORS	-	-	-	-	**98**	-	-	-	**3350**	-
Agriculture	-	-	-	-	-	-	-	-	921	-
Comm. and Publ. Services	-	-	-	-	-	-	-	-	867	-
Residential	-	-	-	-	78	-	-	-	1562	-
Non-specified	-	-	-	-	20	-	-	-	-	-
NON-ENERGY USE	-	-	-	-	-	-	-	-	-	-

APPROVISIONNEMENT ET DEMANDE 1997	*Gaz* (TJ)				*En. Re. Comb. & Déchets* (TJ)				(GWh)	(TJ)
	Gaz naturel	*Usines à gaz*	*Cokeries*	*Hauts fourneaux*	*Biomasse solide*	*Gaz/Liquides tirés de biomasse*	*Déchets urbains*	*Déchets industriels*	*Electricité*	*Chaleur*
Production	8776	-	-	-	91	-	-	-	6273	-
Imports	-	-	-	-	20	-	-	-	-	-
Exports	-	-	-	-	-	-	-	-	-	-
Intl. Marine Bunkers	-	-	-	-	-	-	-	-	-	-
Stock Changes	-	-	-	-	-	-	-	-	-	-
DOMESTIC SUPPLY	**8776**	-	-	-	**111**	-	-	-	**6273**	-
Transfers and Stat. Diff.	-	-	-	-	-	-	-	-	83	-
TRANSFORMATION	**8776**	-	-	-	-	-	-	-	-	-
Electricity and CHP Plants	8776	-	-	-	-	-	-	-	-	-
Petroleum Refineries	-	-	-	-	-	-	-	-	-	-
Other Transform. Sector	-	-	-	-	-	-	-	-	-	-
ENERGY SECTOR	-	-	-	-	-	-	-	-	**434**	-
DISTRIBUTION LOSSES	-	-	-	-	-	-	-	-	**615**	-
FINAL CONSUMPTION	-	-	-	-	**111**	-	-	-	**5307**	-
INDUSTRY SECTOR	-	-	-	-	-	-	-	-	**1837**	-
Iron and Steel	-	-	-	-	-	-	-	-	15	-
Chemical and Petrochem.	-	-	-	-	-	-	-	-	193	-
Non-Metallic Minerals	-	-	-	-	-	-	-	-	393	-
Non-specified	-	-	-	-	-	-	-	-	1236	-
TRANSPORT SECTOR	-	-	-	-	-	-	-	-	-	-
Air	-	-	-	-	-	-	-	-	-	-
Road	-	-	-	-	-	-	-	-	-	-
Non-specified	-	-	-	-	-	-	-	-	-	-
OTHER SECTORS	-	-	-	-	**111**	-	-	-	**3470**	-
Agriculture	-	-	-	-	-	-	-	-	954	-
Comm. and Publ. Services	-	-	-	-	-	-	-	-	898	-
Residential	-	-	-	-	91	-	-	-	1618	-
Non-specified	-	-	-	-	20	-	-	-	-	-
NON-ENERGY USE	-	-	-	-	-	-	-	-	-	-

Kazakhstan

SUPPLY AND CONSUMPTION 1996	Coal (1000 tonnes)							Oil (1000 tonnes)			
	Coking Coal	Other Bit. Coal	Sub-Bit. Coal	Lignite	Peat	Oven and Gas Coke	Pat. Fuel and BKB	Crude Oil	NGL	Feed-stocks	Additives
Production	10986	62254	-	3591	-	-	-	21050	1925	-	-
Imports	90	1010	-	85	-	842	-	3041	118	-	-
Exports	-3000	-18700	-	-230	-	-14	-	-12684	-1711	-	-
Intl. Marine Bunkers	-	-	-	-	-	-	-	-	-	-	-
Stock Changes	-	-	-	-	-	-	-	-	-	-	-
DOMESTIC SUPPLY	**8076**	**44564**	**-**	**3446**	**-**	**828**	**-**	**11407**	**332**	**-**	**-**
Transfers and Stat. Diff.	-	-	-	-	-	-	-	-	-	-	-
TRANSFORMATION	**-**	**39480**	**-**	**-**	**-**	**662**	**-**	**11407**	**332**	**-**	**-**
Electricity and CHP Plants	-	39480	-	-	-	-	-	-	-	-	-
Petroleum Refineries	-	-	-	-	-	-	-	11407	332	-	-
Other Transform. Sector	-	-	-	-	-	662	-	-	-	-	-
ENERGY SECTOR	**-**	**-**	**-**	**-**	**-**	**-**	**-**	**-**	**-**	**-**	**-**
DISTRIBUTION LOSSES	**-**	**-**	**-**	**-**	**-**	**-**	**-**	**-**	**-**	**-**	**-**
FINAL CONSUMPTION	**8076**	**5084**	**-**	**3446**	**-**	**166**	**-**	**-**	**-**	**-**	**-**
INDUSTRY SECTOR	**8076**	**5084**	**-**	**3446**	**-**	**166**	**-**	**-**	**-**	**-**	**-**
Iron and Steel	-	-	-	-	-	166	-	-	-	-	-
Chemical and Petrochem.	-	-	-	-	-	-	-	-	-	-	-
Non-Metallic Minerals	-	-	-	-	-	-	-	-	-	-	-
Non-specified	8076	5084	-	3446	-	-	-	-	-	-	-
TRANSPORT SECTOR	**-**	**-**	**-**	**-**	**-**	**-**	**-**	**-**	**-**	**-**	**-**
Air	-	-	-	-	-	-	-	-	-	-	-
Road	-	-	-	-	-	-	-	-	-	-	-
Non-specified	-	-	-	-	-	-	-	-	-	-	-
OTHER SECTORS	**-**	**-**	**-**	**-**	**-**	**-**	**-**	**-**	**-**	**-**	**-**
Agriculture	-	-	-	-	-	-	-	-	-	-	-
Comm. and Publ. Services	-	-	-	-	-	-	-	-	-	-	-
Residential	-	-	-	-	-	-	-	-	-	-	-
Non-specified	-	-	-	-	-	-	-	-	-	-	-
NON-ENERGY USE	**-**	**-**	**-**	**-**	**-**	**-**	**-**	**-**	**-**	**-**	**-**

APPROVISIONNEMENT ET DEMANDE 1997	*Charbon* (1000 tonnes)							*Pétrole* (1000 tonnes)			
	Charbon à coke	*Autres charb. bit.*	*Charbon sous-bit.*	*Lignite*	*Tourbe*	*Coke de four/gaz*	*Agg./briq. de lignite*	*Pétrole brut*	*LGN*	*Produits d'aliment.*	*Additifs*
Production	10526	59648	-	2473	-	-	-	23409	2040	-	-
Imports	80	920	-	61	-	842	-	608	10	-	-
Exports	-3500	-21400	-	-91	-	-14	-	-14900	-1500	-	-
Intl. Marine Bunkers	-	-	-	-	-	-	-	-	-	-	-
Stock Changes	-	-	-	-	-	-	-	-	-	-	-
DOMESTIC SUPPLY	**7106**	**39168**	**-**	**2443**	**-**	**828**	**-**	**9117**	**550**	**-**	**-**
Transfers and Stat. Diff.	-	-	-	-	-	-	-	-	-	-	-
TRANSFORMATION	**-**	**34705**	**-**	**-**	**-**	**662**	**-**	**9117**	**550**	**-**	**-**
Electricity and CHP Plants	-	34705	-	-	-	-	-	-	-	-	-
Petroleum Refineries	-	-	-	-	-	-	-	9117	550	-	-
Other Transform. Sector	-	-	-	-	-	662	-	-	-	-	-
ENERGY SECTOR	**-**	**-**	**-**	**-**	**-**	**-**	**-**	**-**	**-**	**-**	**-**
DISTRIBUTION LOSSES	**-**	**-**	**-**	**-**	**-**	**-**	**-**	**-**	**-**	**-**	**-**
FINAL CONSUMPTION	**7106**	**4463**	**-**	**2443**	**-**	**166**	**-**	**-**	**-**	**-**	**-**
INDUSTRY SECTOR	**7106**	**4463**	**-**	**2443**	**-**	**166**	**-**	**-**	**-**	**-**	**-**
Iron and Steel	-	-	-	-	-	166	-	-	-	-	-
Chemical and Petrochem.	-	-	-	-	-	-	-	-	-	-	-
Non-Metallic Minerals	-	-	-	-	-	-	-	-	-	-	-
Non-specified	7106	4463	-	2443	-	-	-	-	-	-	-
TRANSPORT SECTOR	**-**	**-**	**-**	**-**	**-**	**-**	**-**	**-**	**-**	**-**	**-**
Air	-	-	-	-	-	-	-	-	-	-	-
Road	-	-	-	-	-	-	-	-	-	-	-
Non-specified	-	-	-	-	-	-	-	-	-	-	-
OTHER SECTORS	**-**	**-**	**-**	**-**	**-**	**-**	**-**	**-**	**-**	**-**	**-**
Agriculture	-	-	-	-	-	-	-	-	-	-	-
Comm. and Publ. Services	-	-	-	-	-	-	-	-	-	-	-
Residential	-	-	-	-	-	-	-	-	-	-	-
Non-specified	-	-	-	-	-	-	-	-	-	-	-
NON-ENERGY USE	**-**	**-**	**-**	**-**	**-**	**-**	**-**	**-**	**-**	**-**	**-**

Kazakhstan

SUPPLY AND CONSUMPTION 1996	Oil cont. (1000 tonnes)										
	Refinery Gas	LPG + Ethane	Motor Gasoline	Aviation Gasoline	Jet Fuel	Kerosene	Gas/ Diesel	Heavy Fuel Oil	Naphtha	Petrol. Coke	Other Prod.
Production	115	60	2291	-	314	432	3295	3900	-	-	539
Imports	-	89	172	2	68	2	176	190	-	-	165
Exports	-	-32	-249	-	-5	-24	-1310	-530	-	-	-2
Intl. Marine Bunkers	-	-	-	-	-	-	-	-	-	-	-
Stock Changes	-	-	-	-	-	-	-	-	-	-	-
DOMESTIC SUPPLY	**115**	**117**	**2214**	**2**	**377**	**410**	**2161**	**3560**	**-**	**-**	**702**
Transfers and Stat. Diff.	-	-	-	-	-	-	-	-	-	-	-
TRANSFORMATION	**-**	**-**	**-**	**-**	**-**	**-**	**-**	**1594**	**-**	**-**	**-**
Electricity and CHP Plants	-	-	-	-	-	-	-	1594	-	-	-
Petroleum Refineries	-	-	-	-	-	-	-	-	-	-	-
Other Transform. Sector	-	-	-	-	-	-	-	-	-	-	-
ENERGY SECTOR	**115**	**-**	**-**	**-**	**-**	**-**	**-**	**350**	**-**	**-**	**-**
DISTRIBUTION LOSSES	**-**	**-**	**-**	**-**	**-**	**-**	**-**	**-**	**-**	**-**	**-**
FINAL CONSUMPTION	**-**	**117**	**2214**	**2**	**377**	**410**	**2161**	**1616**	**-**	**-**	**702**
INDUSTRY SECTOR	**-**	**-**	**-**	**-**	**-**	**-**	**-**	**-**	**-**	**-**	**-**
Iron and Steel	-	-	-	-	-	-	-	-	-	-	-
Chemical and Petrochem.	-	-	-	-	-	-	-	-	-	-	-
Non-Metallic Minerals	-	-	-	-	-	-	-	-	-	-	-
Non-specified	-	-	-	-	-	-	-	-	-	-	-
TRANSPORT SECTOR	**-**	**-**	**2214**	**2**	**377**	**-**	**-**	**-**	**-**	**-**	**-**
Air	-	-	-	2	377	-	-	-	-	-	-
Road	-	-	2214	-	-	-	-	-	-	-	-
Non-specified	-	-	-	-	-	-	-	-	-	-	-
OTHER SECTORS	**-**	**117**	**-**	**-**	**-**	**410**	**2161**	**1616**	**-**	**-**	**596**
Agriculture	-	-	-	-	-	-	-	-	-	-	-
Comm. and Publ. Services	-	-	-	-	-	-	-	-	-	-	-
Residential	-	-	-	-	-	-	-	-	-	-	-
Non-specified	-	117	-	-	-	410	2161	1616	-	-	596
NON-ENERGY USE	**-**	**-**	**-**	**-**	**-**	**-**	**-**	**-**	**-**	**-**	**106**

APPROVISIONNEMENT ET DEMANDE 1997	*Pétrole cont.* (1000 tonnes)										
	Gaz de raffinerie	*GPL + éthane*	*Essence moteur*	*Essence aviation*	*Carbu- réacteurs*	*Kérosène*	*Gazole*	*Fioul lourd*	*Naphta*	*Coke de pétrole*	*Autres prod.*
Production	95	60	1781	-	200	370	2838	3000	-	-	668
Imports	-	85	194	2	103	-	77	150	-	-	87
Exports	-	-30	-77	-	-	-	-719	-350	-	-	-125
Intl. Marine Bunkers	-	-	-	-	-	-	-	-	-	-	-
Stock Changes	-	-	-	-	-	-	-	-	-	-	-
DOMESTIC SUPPLY	**95**	**115**	**1898**	**2**	**303**	**370**	**2196**	**2800**	**-**	**-**	**630**
Transfers and Stat. Diff.	-	-	-	-	-	-	-	-	-	-	-
TRANSFORMATION	**-**	**-**	**-**	**-**	**-**	**-**	**-**	**1256**	**-**	**-**	**-**
Electricity and CHP Plants	-	-	-	-	-	-	-	1256	-	-	-
Petroleum Refineries	-	-	-	-	-	-	-	-	-	-	-
Other Transform. Sector	-	-	-	-	-	-	-	-	-	-	-
ENERGY SECTOR	**95**	**-**	**-**	**-**	**-**	**-**	**-**	**270**	**-**	**-**	**-**
DISTRIBUTION LOSSES	**-**	**-**	**-**	**-**	**-**	**-**	**-**	**-**	**-**	**-**	**-**
FINAL CONSUMPTION	**-**	**115**	**1898**	**2**	**303**	**370**	**2196**	**1274**	**-**	**-**	**630**
INDUSTRY SECTOR	**-**	**-**	**-**	**-**	**-**	**-**	**-**	**-**	**-**	**-**	**-**
Iron and Steel	-	-	-	-	-	-	-	-	-	-	-
Chemical and Petrochem.	-	-	-	-	-	-	-	-	-	-	-
Non-Metallic Minerals	-	-	-	-	-	-	-	-	-	-	-
Non-specified	-	-	-	-	-	-	-	-	-	-	-
TRANSPORT SECTOR	**-**	**-**	**1898**	**2**	**303**	**-**	**-**	**-**	**-**	**-**	**-**
Air	-	-	-	2	303	-	-	-	-	-	-
Road	-	-	1898	-	-	-	-	-	-	-	-
Non-specified	-	-	-	-	-	-	-	-	-	-	-
OTHER SECTORS	**-**	**115**	**-**	**-**	**-**	**370**	**2196**	**1274**	**-**	**-**	**543**
Agriculture	-	-	-	-	-	-	-	-	-	-	-
Comm. and Publ. Services	-	-	-	-	-	-	-	-	-	-	-
Residential	-	-	-	-	-	-	-	-	-	-	-
Non-specified	-	115	-	-	-	370	2196	1274	-	-	543
NON-ENERGY USE	**-**	**-**	**-**	**-**	**-**	**-**	**-**	**-**	**-**	**-**	**87**

Kazakhstan

SUPPLY AND CONSUMPTION 1996	Gas (TJ) Natural Gas	Gas Works	Coke Ovens	Blast Furnaces	Comb. Renew. & Waste (TJ) Solid Biomass	Gas/Liquids from Biomass	Municipal Waste	Industrial Waste	(GWh) Electricity	(TJ) Heat
Production	245955	-	-	-	3100	-	-	-	58657	256
Imports	207161	-	-	-	-	-	-	-	6854	-
Exports	-88293	-	-	-	-	-	-	-	-	-
Intl. Marine Bunkers	-	-	-	-	-	-	-	-	-	-
Stock Changes	-	-	-	-	-	-	-	-	-	-
DOMESTIC SUPPLY	**364823**	**-**	**-**	**-**	**3100**	**-**	**-**	**-**	**65511**	**256**
Transfers and Stat. Diff.	-	-	-	-	-	-	-	-	-	-
TRANSFORMATION	**94137**	**-**	**-**	**-**	**-**	**-**	**-**	**-**	**-**	**-**
Electricity and CHP Plants	94137	-	-	-	-	-	-	-	-	-
Petroleum Refineries	-	-	-	-	-	-	-	-	-	-
Other Transform. Sector	-	-	-	-	-	-	-	-	-	-
ENERGY SECTOR	**58737**	**-**	**-**	**-**	**-**	**-**	**-**	**-**	**10919**	**-**
DISTRIBUTION LOSSES	**2149**	**-**	**-**	**-**	**-**	**-**	**-**	**-**	**8976**	**-**
FINAL CONSUMPTION	**209800**	**-**	**-**	**-**	**3100**	**-**	**-**	**-**	**45616**	**256**
INDUSTRY SECTOR	**-**	**-**	**-**	**-**	**-**	**-**	**-**	**-**	**20854**	**-**
Iron and Steel	-	-	-	-	-	-	-	-	8000	-
Chemical and Petrochem.	-	-	-	-	-	-	-	-	3142	-
Non-Metallic Minerals	-	-	-	-	-	-	-	-	-	-
Non-specified	-	-	-	-	-	-	-	-	9712	-
TRANSPORT SECTOR	**-**	**-**	**-**	**-**	**-**	**-**	**-**	**-**	**3432**	**-**
Air	-	-	-	-	-	-	-	-	-	-
Road	-	-	-	-	-	-	-	-	-	-
Non-specified	-	-	-	-	-	-	-	-	3432	-
OTHER SECTORS	**209800**	**-**	**-**	**-**	**3100**	**-**	**-**	**-**	**21330**	**256**
Agriculture	-	-	-	-	-	-	-	-	11049	-
Comm. and Publ. Services	-	-	-	-	-	-	-	-	-	-
Residential	-	-	-	-	-	-	-	-	6061	-
Non-specified	209800	-	-	-	3100	-	-	-	4220	256
NON-ENERGY USE	**-**	**-**	**-**	**-**	**-**	**-**	**-**	**-**	**-**	**-**

APPROVISIONNEMENT ET DEMANDE 1997	*Gaz* (TJ) *Gaz naturel*	*Usines à gaz*	*Cokeries*	*Hauts fourneaux*	*En. Re. Comb. & Déchets* (TJ) *Biomasse solide*	*Gaz/Liquides tirés de biomasse*	*Déchets urbains*	*Déchets industriels*	(GWh) *Electricité*	(TJ) *Chaleur*
Production	305898	-	-	-	3100	-	-	-	52000	256
Imports	82940	-	-	-	-	-	-	-	6700	-
Exports	-100282	-	-	-	-	-	-	-	-	-
Intl. Marine Bunkers	-	-	-	-	-	-	-	-	-	-
Stock Changes	-	-	-	-	-	-	-	-	-	-
DOMESTIC SUPPLY	**288556**	**-**	**-**	**-**	**3100**	**-**	**-**	**-**	**58700**	**256**
Transfers and Stat. Diff.	-	-	-	-	-	-	-	-	-	-
TRANSFORMATION	**74458**	**-**	**-**	**-**	**-**	**-**	**-**	**-**	**-**	**-**
Electricity and CHP Plants	74458	-	-	-	-	-	-	-	-	-
Petroleum Refineries	-	-	-	-	-	-	-	-	-	-
Other Transform. Sector	-	-	-	-	-	-	-	-	-	-
ENERGY SECTOR	**46446**	**-**	**-**	**-**	**-**	**-**	**-**	**-**	**9784**	**-**
DISTRIBUTION LOSSES	**1697**	**-**	**-**	**-**	**-**	**-**	**-**	**-**	**8043**	**-**
FINAL CONSUMPTION	**165955**	**-**	**-**	**-**	**3100**	**-**	**-**	**-**	**40873**	**256**
INDUSTRY SECTOR	**-**	**-**	**-**	**-**	**-**	**-**	**-**	**-**	**18686**	**-**
Iron and Steel	-	-	-	-	-	-	-	-	7168	-
Chemical and Petrochem.	-	-	-	-	-	-	-	-	2815	-
Non-Metallic Minerals	-	-	-	-	-	-	-	-	-	-
Non-specified	-	-	-	-	-	-	-	-	8703	-
TRANSPORT SECTOR	**-**	**-**	**-**	**-**	**-**	**-**	**-**	**-**	**3075**	**-**
Air	-	-	-	-	-	-	-	-	-	-
Road	-	-	-	-	-	-	-	-	-	-
Non-specified	-	-	-	-	-	-	-	-	3075	-
OTHER SECTORS	**165955**	**-**	**-**	**-**	**3100**	**-**	**-**	**-**	**19112**	**256**
Agriculture	-	-	-	-	-	-	-	-	9900	-
Comm. and Publ. Services	-	-	-	-	-	-	-	-	-	-
Residential	-	-	-	-	-	-	-	-	5431	-
Non-specified	165955	-	-	-	3100	-	-	-	3781	256
NON-ENERGY USE	**-**	**-**	**-**	**-**	**-**	**-**	**-**	**-**	**-**	**-**

Kenya

SUPPLY AND CONSUMPTION 1996	Coal (1000 tonnes)							Oil (1000 tonnes)			
	Coking Coal	Other Bit. Coal	Sub-Bit. Coal	Lignite	Peat	Oven and Gas Coke	Pat. Fuel and BKB	Crude Oil	NGL	Feed-stocks	Additives
Production	-	-	-	-	-	-	-	-	-	-	-
Imports	-	89	-	-	-	1	-	1413	-	-	-
Exports	-	-	-	-	-	-	-	-	-	-	-
Intl. Marine Bunkers	-	-	-	-	-	-	-	-	-	-	-
Stock Changes	-	-	-	-	-	-	-	-	-	-	-
DOMESTIC SUPPLY	**-**	**89**	**-**	**-**	**-**	**1**	**-**	**1413**	**-**	**-**	**-**
Transfers and Stat. Diff.	-	-	-	-	-	-	-	430	-	-	-
TRANSFORMATION	**-**	**-**	**-**	**-**	**-**	**-**	**-**	**1761**	**-**	**-**	**-**
Electricity and CHP Plants	-	-	-	-	-	-	-	-	-	-	-
Petroleum Refineries	-	-	-	-	-	-	-	1761	-	-	-
Other Transform. Sector	-	-	-	-	-	-	-	-	-	-	-
ENERGY SECTOR	**-**	**-**	**-**	**-**	**-**	**-**	**-**	**82**	**-**	**-**	**-**
DISTRIBUTION LOSSES	**-**	**-**	**-**	**-**	**-**	**-**	**-**	**-**	**-**	**-**	**-**
FINAL CONSUMPTION	**-**	**89**	**-**	**-**	**-**	**1**	**-**	**-**	**-**	**-**	**-**
INDUSTRY SECTOR	**-**	**89**	**-**	**-**	**-**	**-**	**-**	**-**	**-**	**-**	**-**
Iron and Steel	-	-	-	-	-	-	-	-	-	-	-
Chemical and Petrochem.	-	-	-	-	-	-	-	-	-	-	-
Non-Metallic Minerals	-	89	-	-	-	-	-	-	-	-	-
Non-specified	-	-	-	-	-	-	-	-	-	-	-
TRANSPORT SECTOR	**-**	**-**	**-**	**-**	**-**	**-**	**-**	**-**	**-**	**-**	**-**
Air	-	-	-	-	-	-	-	-	-	-	-
Road	-	-	-	-	-	-	-	-	-	-	-
Non-specified	-	-	-	-	-	-	-	-	-	-	-
OTHER SECTORS	**-**	**-**	**-**	**-**	**-**	**1**	**-**	**-**	**-**	**-**	**-**
Agriculture	-	-	-	-	-	-	-	-	-	-	-
Comm. and Publ. Services	-	-	-	-	-	-	-	-	-	-	-
Residential	-	-	-	-	-	1	-	-	-	-	-
Non-specified	-	-	-	-	-	-	-	-	-	-	-
NON-ENERGY USE	**-**	**-**	**-**	**-**	**-**	**-**	**-**	**-**	**-**	**-**	**-**

APPROVISIONNEMENT ET DEMANDE 1997	*Charbon* (1000 tonnes)							*Pétrole* (1000 tonnes)			
	Charbon à coke	*Autres charb. bit.*	*Charbon sous-bit.*	*Lignite*	*Tourbe*	*Coke de four/gaz*	*Agg./briq. de lignite*	*Pétrole brut*	*LGN*	*Produits d'aliment.*	*Additifs*
Production	-	-	-	-	-	-	-	-	-	-	-
Imports	-	92	-	-	-	1	-	1834	-	-	-
Exports	-	-	-	-	-	-	-	-	-	-	-
Intl. Marine Bunkers	-	-	-	-	-	-	-	-	-	-	-
Stock Changes	-	-	-	-	-	-	-	-	-	-	-
DOMESTIC SUPPLY	**-**	**92**	**-**	**-**	**-**	**1**	**-**	**1834**	**-**	**-**	**-**
Transfers and Stat. Diff.	-	-	-	-	-	-	-	-94	-	-	-
TRANSFORMATION	**-**	**-**	**-**	**-**	**-**	**-**	**-**	**1647**	**-**	**-**	**-**
Electricity and CHP Plants	-	-	-	-	-	-	-	-	-	-	-
Petroleum Refineries	-	-	-	-	-	-	-	1647	-	-	-
Other Transform. Sector	-	-	-	-	-	-	-	-	-	-	-
ENERGY SECTOR	**-**	**-**	**-**	**-**	**-**	**-**	**-**	**93**	**-**	**-**	**-**
DISTRIBUTION LOSSES	**-**	**-**	**-**	**-**	**-**	**-**	**-**	**-**	**-**	**-**	**-**
FINAL CONSUMPTION	**-**	**92**	**-**	**-**	**-**	**1**	**-**	**-**	**-**	**-**	**-**
INDUSTRY SECTOR	**-**	**92**	**-**	**-**	**-**	**-**	**-**	**-**	**-**	**-**	**-**
Iron and Steel	-	-	-	-	-	-	-	-	-	-	-
Chemical and Petrochem.	-	-	-	-	-	-	-	-	-	-	-
Non-Metallic Minerals	-	92	-	-	-	-	-	-	-	-	-
Non-specified	-	-	-	-	-	-	-	-	-	-	-
TRANSPORT SECTOR	**-**	**-**	**-**	**-**	**-**	**-**	**-**	**-**	**-**	**-**	**-**
Air	-	-	-	-	-	-	-	-	-	-	-
Road	-	-	-	-	-	-	-	-	-	-	-
Non-specified	-	-	-	-	-	-	-	-	-	-	-
OTHER SECTORS	**-**	**-**	**-**	**-**	**-**	**1**	**-**	**-**	**-**	**-**	**-**
Agriculture	-	-	-	-	-	-	-	-	-	-	-
Comm. and Publ. Services	-	-	-	-	-	-	-	-	-	-	-
Residential	-	-	-	-	-	1	-	-	-	-	-
Non-specified	-	-	-	-	-	-	-	-	-	-	-
NON-ENERGY USE	**-**	**-**	**-**	**-**	**-**	**-**	**-**	**-**	**-**	**-**	**-**

Kenya

SUPPLY AND CONSUMPTION 1996	Oil cont. (1000 tonnes)										
	Refinery Gas	LPG + Ethane	Motor Gasoline	Aviation Gasoline	Jet Fuel	Kerosene	Gas/ Diesel	Heavy Fuel Oil	Naphtha	Petrol. Coke	Other Prod.
Production	61	27	309	-	266	114	422	512	-	-	22
Imports	-	-	149	5	180	48	374	173	-	-	28
Exports	-	-	-58	-	-5	-	-94	-70	-	-	-1
Intl. Marine Bunkers	-	-	-	-	-	-	-29	-191	-	-	-
Stock Changes	-	4	-	-	-	-	-	-	-	-	-
DOMESTIC SUPPLY	**61**	**31**	**400**	**5**	**441**	**162**	**673**	**424**	**-**	**-**	**49**
Transfers and Stat. Diff.	-	-	-1	-	1	-	1	-1	-	-	-
TRANSFORMATION	**-**	**-**	**-**	**-**	**-**	**-**	**33**	**100**	**-**	**-**	**-**
Electricity and CHP Plants	-	-	-	-	-	-	33	100	-	-	-
Petroleum Refineries	-	-	-	-	-	-	-	-	-	-	-
Other Transform. Sector	-	-	-	-	-	-	-	-	-	-	-
ENERGY SECTOR	**61**	**-**	**-**	**-**	**-**	**-**	**-**	**30**	**-**	**-**	**-**
DISTRIBUTION LOSSES	**-**	**-**	**-**	**-**	**-**	**-**	**-**	**-**	**-**	**-**	**-**
FINAL CONSUMPTION	**-**	**31**	**399**	**5**	**442**	**162**	**641**	**293**	**-**	**-**	**49**
INDUSTRY SECTOR	**-**	**9**	**14**	**-**	**-**	**17**	**124**	**247**	**-**	**-**	**-**
Iron and Steel	-	-	-	-	-	-	-	-	-	-	-
Chemical and Petrochem.	-	-	-	-	-	-	-	-	-	-	-
Non-Metallic Minerals	-	-	-	-	-	-	-	-	-	-	-
Non-specified	-	9	14	-	-	17	124	247	-	-	-
TRANSPORT SECTOR	**-**	**-**	**367**	**5**	**442**	**-**	**433**	**4**	**-**	**-**	**-**
Air	-	-	-	5	442	-	-	-	-	-	-
Road	-	-	367	-	-	-	397	-	-	-	-
Non-specified	-	-	-	-	-	-	36	4	-	-	-
OTHER SECTORS	**-**	**22**	**18**	**-**	**-**	**145**	**84**	**42**	**-**	**-**	**-**
Agriculture	-	-	5	-	-	-	36	37	-	-	-
Comm. and Publ. Services	-	-	-	-	-	-	-	-	-	-	-
Residential	-	16	-	-	-	145	-	-	-	-	-
Non-specified	-	6	13	-	-	-	48	5	-	-	-
NON-ENERGY USE	**-**	**-**	**-**	**-**	**-**	**-**	**-**	**-**	**-**	**-**	**49**

APPROVISIONNEMENT ET DEMANDE 1997	*Pétrole cont.* (1000 tonnes)										
	Gaz de raffinerie	*GPL + éthane*	*Essence moteur*	*Essence aviation*	*Carbu- réacteurs*	*Kérosène*	*Gazole*	*Fioul lourd*	*Naphta*	*Coke de pétrole*	*Autres prod.*
Production	61	24	273	-	234	101	412	499	-	-	19
Imports	-	-	138	4	197	48	348	161	-	-	25
Exports	-	-	-20	-	-	-	-67	-82	-	-	-1
Intl. Marine Bunkers	-	-	-	-	-	-	-29	-191	-	-	-
Stock Changes	-	7	-	-	-	-	-	-	-	-	-
DOMESTIC SUPPLY	**61**	**31**	**391**	**4**	**431**	**149**	**664**	**387**	**-**	**-**	**43**
Transfers and Stat. Diff.	-	-	-	-	-	-	-2	-1	-	-	-
TRANSFORMATION	**-**	**-**	**-**	**-**	**-**	**-**	**33**	**100**	**-**	**-**	**-**
Electricity and CHP Plants	-	-	-	-	-	-	33	100	-	-	-
Petroleum Refineries	-	-	-	-	-	-	-	-	-	-	-
Other Transform. Sector	-	-	-	-	-	-	-	-	-	-	-
ENERGY SECTOR	**61**	**-**	**-**	**-**	**-**	**-**	**-**	**30**	**-**	**-**	**-**
DISTRIBUTION LOSSES	**-**	**-**	**-**	**-**	**-**	**-**	**-**	**-**	**-**	**-**	**-**
FINAL CONSUMPTION	**-**	**31**	**391**	**4**	**431**	**149**	**629**	**256**	**-**	**-**	**43**
INDUSTRY SECTOR	**-**	**9**	**14**	**-**	**-**	**16**	**122**	**216**	**-**	**-**	**-**
Iron and Steel	-	-	-	-	-	-	-	-	-	-	-
Chemical and Petrochem.	-	-	-	-	-	-	-	-	-	-	-
Non-Metallic Minerals	-	-	-	-	-	-	-	-	-	-	-
Non-specified	-	9	14	-	-	16	122	216	-	-	-
TRANSPORT SECTOR	**-**	**-**	**359**	**4**	**431**	**-**	**425**	**4**	**-**	**-**	**-**
Air	-	-	-	4	431	-	-	-	-	-	-
Road	-	-	359	-	-	-	390	-	-	-	-
Non-specified	-	-	-	-	-	-	35	4	-	-	-
OTHER SECTORS	**-**	**22**	**18**	**-**	**-**	**133**	**82**	**36**	**-**	**-**	**-**
Agriculture	-	-	5	-	-	-	35	32	-	-	-
Comm. and Publ. Services	-	-	-	-	-	-	-	-	-	-	-
Residential	-	16	-	-	-	133	-	-	-	-	-
Non-specified	-	6	13	-	-	-	47	4	-	-	-
NON-ENERGY USE	**-**	**-**	**-**	**-**	**-**	**-**	**-**	**-**	**-**	**-**	**43**

Kenya

SUPPLY AND CONSUMPTION 1996	Gas (TJ)				Comb. Renew. & Waste (TJ)				(GWh)	(TJ)
	Natural Gas	Gas Works	Coke Ovens	Blast Furnaces	Solid Biomass	Gas/Liquids from Biomass	Municipal Waste	Industrial Waste	Electricity	Heat
Production	-	-	-	-	452565	-	-	-	4040	-
Imports	-	-	-	-	-	-	-	-	149	-
Exports	-	-	-	-	-	-	-	-	-	-
Intl. Marine Bunkers	-	-	-	-	-	-	-	-	-	-
Stock Changes	-	-	-	-	-	-	-	-	-	-
DOMESTIC SUPPLY	-	-	-	-	**452565**	-	-	-	**4189**	-
Transfers and Stat. Diff.	-	-	-	-	-1	-	-	-	2	-
TRANSFORMATION	-	-	-	-	**128142**	-	-	-	-	-
Electricity and CHP Plants	-	-	-	-	-	-	-	-	-	-
Petroleum Refineries	-	-	-	-	-	-	-	-	-	-
Other Transform. Sector	-	-	-	-	128142	-	-	-	-	-
ENERGY SECTOR	-	-	-	-	-	-	-	-	**48**	-
DISTRIBUTION LOSSES	-	-	-	-	-	-	-	-	**639**	-
FINAL CONSUMPTION	-	-	-	-	**324422**	-	-	-	**3504**	-
INDUSTRY SECTOR	-	-	-	-	**22475**	-	-	-	**2137**	-
Iron and Steel	-	-	-	-	-	-	-	-	-	-
Chemical and Petrochem.	-	-	-	-	-	-	-	-	-	-
Non-Metallic Minerals	-	-	-	-	-	-	-	-	-	-
Non-specified	-	-	-	-	22475	-	-	-	2137	-
TRANSPORT SECTOR	-	-	-	-	-	-	-	-	-	-
Air	-	-	-	-	-	-	-	-	-	-
Road	-	-	-	-	-	-	-	-	-	-
Non-specified	-	-	-	-	-	-	-	-	-	-
OTHER SECTORS	-	-	-	-	**301947**	-	-	-	**1367**	-
Agriculture	-	-	-	-	24408	-	-	-	-	-
Comm. and Publ. Services	-	-	-	-	1932	-	-	-	298	-
Residential	-	-	-	-	275607	-	-	-	1069	-
Non-specified	-	-	-	-	-	-	-	-	-	-
NON-ENERGY USE	-	-	-	-	-	-	-	-	-	-

APPROVISIONNEMENT ET DEMANDE 1997	*Gaz* (TJ)				*En. Re. Comb. & Déchets* (TJ)				(GWh)	(TJ)
	Gaz naturel	*Usines à gaz*	*Cokeries*	*Hauts fourneaux*	*Biomasse solide*	*Gaz/Liquides tirés de biomasse*	*Déchets urbains*	*Déchets industriels*	*Electricité*	*Chaleur*
Production	-	-	-	-	461164	-	-	-	4238	-
Imports	-	-	-	-	-	-	-	-	144	-
Exports	-	-	-	-	-	-	-	-	-	-
Intl. Marine Bunkers	-	-	-	-	-	-	-	-	-	-
Stock Changes	-	-	-	-	-	-	-	-	-	-
DOMESTIC SUPPLY	-	-	-	-	**461164**	-	-	-	**4382**	-
Transfers and Stat. Diff.	-	-	-	-	-2	-	-	-	-	-
TRANSFORMATION	-	-	-	-	**130576**	-	-	-	-	-
Electricity and CHP Plants	-	-	-	-	-	-	-	-	-	-
Petroleum Refineries	-	-	-	-	-	-	-	-	-	-
Other Transform. Sector	-	-	-	-	130576	-	-	-	-	-
ENERGY SECTOR	-	-	-	-	-	-	-	-	**50**	-
DISTRIBUTION LOSSES	-	-	-	-	-	-	-	-	**707**	-
FINAL CONSUMPTION	-	-	-	-	**330586**	-	-	-	**3625**	-
INDUSTRY SECTOR	-	-	-	-	**22902**	-	-	-	**2263**	-
Iron and Steel	-	-	-	-	-	-	-	-	-	-
Chemical and Petrochem.	-	-	-	-	-	-	-	-	-	-
Non-Metallic Minerals	-	-	-	-	-	-	-	-	-	-
Non-specified	-	-	-	-	22902	-	-	-	2263	-
TRANSPORT SECTOR	-	-	-	-	-	-	-	-	-	-
Air	-	-	-	-	-	-	-	-	-	-
Road	-	-	-	-	-	-	-	-	-	-
Non-specified	-	-	-	-	-	-	-	-	-	-
OTHER SECTORS	-	-	-	-	**307684**	-	-	-	**1362**	-
Agriculture	-	-	-	-	24872	-	-	-	-	-
Comm. and Publ. Services	-	-	-	-	1969	-	-	-	246	-
Residential	-	-	-	-	280843	-	-	-	1116	-
Non-specified	-	-	-	-	-	-	-	-	-	-
NON-ENERGY USE	-	-	-	-	-	-	-	-	-	-

Dem. People's Rep. of Korea / Rép. pop. dém. de Corée

SUPPLY AND CONSUMPTION 1996	Coal (1000 tonnes)							Oil (1000 tonnes)			
	Coking Coal	Other Bit. Coal	Sub-Bit. Coal	Lignite	Peat	Oven and Gas Coke	Pat. Fuel and BKB	Crude Oil	NGL	Feed-stocks	Additives
Production	2769	21367	7769	-	-	2904	-	-	-	-	-
Imports	2525	-	-	-	-	303	-	1121	-	-	-
Exports	-	-505	-	-	-	-	-	-	-	-	-
Intl. Marine Bunkers	-	-	-	-	-	-	-	-	-	-	-
Stock Changes	-	-	-	-	-	-	-	-	-	-	-
DOMESTIC SUPPLY	**5294**	**20862**	**7769**	**-**	**-**	**3207**	**-**	**1121**	**-**	**-**	**-**
Transfers and Stat. Diff.	514	3631	-	-	-	-	-	-	-	-	-
TRANSFORMATION	**5808**	**8080**	**-**	**-**	**-**	**2566**	**-**	**1121**	**-**	**-**	**-**
Electricity and CHP Plants	-	8080	-	-	-	-	-	-	-	-	-
Petroleum Refineries	-	-	-	-	-	-	-	1121	-	-	-
Other Transform. Sector	5808	-	-	-	-	2566	-	-	-	-	-
ENERGY SECTOR	**-**	**-**	**-**	**-**	**-**	**-**	**-**	**-**	**-**	**-**	**-**
DISTRIBUTION LOSSES	**-**	**-**	**-**	**-**	**-**	**-**	**-**	**-**	**-**	**-**	**-**
FINAL CONSUMPTION	**-**	**16413**	**7769**	**-**	**-**	**641**	**-**	**-**	**-**	**-**	**-**
INDUSTRY SECTOR	**-**	**16413**	**7769**	**-**	**-**	**641**	**-**	**-**	**-**	**-**	**-**
Iron and Steel	-	-	-	-	-	641	-	-	-	-	-
Chemical and Petrochem.	-	-	-	-	-	-	-	-	-	-	-
Non-Metallic Minerals	-	-	-	-	-	-	-	-	-	-	-
Non-specified	-	16413	7769	-	-	-	-	-	-	-	-
TRANSPORT SECTOR	**-**	**-**	**-**	**-**	**-**	**-**	**-**	**-**	**-**	**-**	**-**
Air	-	-	-	-	-	-	-	-	-	-	-
Road	-	-	-	-	-	-	-	-	-	-	-
Non-specified	-	-	-	-	-	-	-	-	-	-	-
OTHER SECTORS	**-**	**-**	**-**	**-**	**-**	**-**	**-**	**-**	**-**	**-**	**-**
Agriculture	-	-	-	-	-	-	-	-	-	-	-
Comm. and Publ. Services	-	-	-	-	-	-	-	-	-	-	-
Residential	-	-	-	-	-	-	-	-	-	-	-
Non-specified	-	-	-	-	-	-	-	-	-	-	-
NON-ENERGY USE	**-**	**-**	**-**	**-**	**-**	**-**	**-**	**-**	**-**	**-**	**-**

APPROVISIONNEMENT ET DEMANDE 1997	*Charbon* (1000 tonnes)							*Pétrole* (1000 tonnes)			
	Charbon à coke	*Autres charb. bit.*	*Charbon sous-bit.*	*Lignite*	*Tourbe*	*Coke de four/gaz*	*Agg./briq. de lignite*	*Pétrole brut*	*LGN*	*Produits d'aliment.*	*Additifs*
Production	2686	20726	7536	-	-	2817	-	-	-	-	-
Imports	2449	-	-	-	-	294	-	1087	-	-	-
Exports	-	-490	-	-	-	-	-	-	-	-	-
Intl. Marine Bunkers	-	-	-	-	-	-	-	-	-	-	-
Stock Changes	-	-	-	-	-	-	-	-	-	-	-
DOMESTIC SUPPLY	**5135**	**20236**	**7536**	**-**	**-**	**3111**	**-**	**1087**	**-**	**-**	**-**
Transfers and Stat. Diff.	499	3523	-	-	-	-	-	-	-	-	-
TRANSFORMATION	**5634**	**7838**	**-**	**-**	**-**	**2489**	**-**	**1087**	**-**	**-**	**-**
Electricity and CHP Plants	-	7838	-	-	-	-	-	-	-	-	-
Petroleum Refineries	-	-	-	-	-	-	-	1087	-	-	-
Other Transform. Sector	5634	-	-	-	-	2489	-	-	-	-	-
ENERGY SECTOR	**-**	**-**	**-**	**-**	**-**	**-**	**-**	**-**	**-**	**-**	**-**
DISTRIBUTION LOSSES	**-**	**-**	**-**	**-**	**-**	**-**	**-**	**-**	**-**	**-**	**-**
FINAL CONSUMPTION	**-**	**15921**	**7536**	**-**	**-**	**622**	**-**	**-**	**-**	**-**	**-**
INDUSTRY SECTOR	**-**	**15921**	**7536**	**-**	**-**	**622**	**-**	**-**	**-**	**-**	**-**
Iron and Steel	-	-	-	-	-	622	-	-	-	-	-
Chemical and Petrochem.	-	-	-	-	-	-	-	-	-	-	-
Non-Metallic Minerals	-	-	-	-	-	-	-	-	-	-	-
Non-specified	-	15921	7536	-	-	-	-	-	-	-	-
TRANSPORT SECTOR	**-**	**-**	**-**	**-**	**-**	**-**	**-**	**-**	**-**	**-**	**-**
Air	-	-	-	-	-	-	-	-	-	-	-
Road	-	-	-	-	-	-	-	-	-	-	-
Non-specified	-	-	-	-	-	-	-	-	-	-	-
OTHER SECTORS	**-**	**-**	**-**	**-**	**-**	**-**	**-**	**-**	**-**	**-**	**-**
Agriculture	-	-	-	-	-	-	-	-	-	-	-
Comm. and Publ. Services	-	-	-	-	-	-	-	-	-	-	-
Residential	-	-	-	-	-	-	-	-	-	-	-
Non-specified	-	-	-	-	-	-	-	-	-	-	-
NON-ENERGY USE	**-**	**-**	**-**	**-**	**-**	**-**	**-**	**-**	**-**	**-**	**-**

Dem. People's Rep. of Korea / Rép. pop. dém. de Corée

SUPPLY AND CONSUMPTION 1996	Oil cont. (1000 tonnes)										
	Refinery Gas	LPG + Ethane	Motor Gasoline	Aviation Gasoline	Jet Fuel	Kerosene	Gas/ Diesel	Heavy Fuel Oil	Naphtha	Petrol. Coke	Other Prod.
Production	25	-	368	-	-	81	387	251	-	-	-
Imports	-	-	44	-	-	40	229	59	-	-	-
Exports	-	-	-	-	-	-	-	-	-	-	-
Intl. Marine Bunkers	-	-	-	-	-	-	-	-	-	-	-
Stock Changes	-	-	-	-	-	-	-	-	-	-	-
DOMESTIC SUPPLY	**25**	**-**	**412**	**-**	**-**	**121**	**616**	**310**	**-**	**-**	**-**
Transfers and Stat. Diff.	-	-	-31	-	-	-4	-25	-11	-	-	-
TRANSFORMATION	**-**	**-**	**-**	**-**	**-**	**-**	**-**	**-**	**-**	**-**	**-**
Electricity and CHP Plants	-	-	-	-	-	-	-	-	-	-	-
Petroleum Refineries	-	-	-	-	-	-	-	-	-	-	-
Other Transform. Sector	-	-	-	-	-	-	-	-	-	-	-
ENERGY SECTOR	**25**	**-**	**-**	**-**	**-**	**-**	**-**	**101**	**-**	**-**	**-**
DISTRIBUTION LOSSES	**-**	**-**	**-**	**-**	**-**	**-**	**-**	**-**	**-**	**-**	**-**
FINAL CONSUMPTION	**-**	**-**	**381**	**-**	**-**	**117**	**591**	**198**	**-**	**-**	**-**
INDUSTRY SECTOR	**-**	**-**	**-**	**-**	**-**	**-**	**-**	**198**	**-**	**-**	**-**
Iron and Steel	-	-	-	-	-	-	-	-	-	-	-
Chemical and Petrochem.	-	-	-	-	-	-	-	-	-	-	-
Non-Metallic Minerals	-	-	-	-	-	-	-	-	-	-	-
Non-specified	-	-	-	-	-	-	-	198	-	-	-
TRANSPORT SECTOR	**-**	**-**	**381**	**-**	**-**	**-**	**591**	**-**	**-**	**-**	**-**
Air	-	-	-	-	-	-	-	-	-	-	-
Road	-	-	381	-	-	-	591	-	-	-	-
Non-specified	-	-	-	-	-	-	-	-	-	-	-
OTHER SECTORS	**-**	**-**	**-**	**-**	**-**	**117**	**-**	**-**	**-**	**-**	**-**
Agriculture	-	-	-	-	-	-	-	-	-	-	-
Comm. and Publ. Services	-	-	-	-	-	-	-	-	-	-	-
Residential	-	-	-	-	-	117	-	-	-	-	-
Non-specified	-	-	-	-	-	-	-	-	-	-	-
NON-ENERGY USE	**-**	**-**	**-**	**-**	**-**	**-**	**-**	**-**	**-**	**-**	**-**

APPROVISIONNEMENT ET DEMANDE 1997	*Pétrole cont.* (1000 tonnes)										
	Gaz de raffinerie	*GPL + éthane*	*Essence moteur*	*Essence aviation*	*Carbu-réacteurs*	*Kérosène*	*Gazole*	*Fioul lourd*	*Naphta*	*Coke de pétrole*	*Autres prod.*
Production	24	-	357	-	-	79	375	243	-	-	-
Imports	-	-	43	-	-	39	222	57	-	-	-
Exports	-	-	-	-	-	-	-	-	-	-	-
Intl. Marine Bunkers	-	-	-	-	-	-	-	-	-	-	-
Stock Changes	-	-	-	-	-	-	-	-	-	-	-
DOMESTIC SUPPLY	**24**	**-**	**400**	**-**	**-**	**118**	**597**	**300**	**-**	**-**	**-**
Transfers and Stat. Diff.	-	-	-30	-	-	-5	-24	-10	-	-	-
TRANSFORMATION	**-**	**-**	**-**	**-**	**-**	**-**	**-**	**-**	**-**	**-**	**-**
Electricity and CHP Plants	-	-	-	-	-	-	-	-	-	-	-
Petroleum Refineries	-	-	-	-	-	-	-	-	-	-	-
Other Transform. Sector	-	-	-	-	-	-	-	-	-	-	-
ENERGY SECTOR	**24**	**-**	**-**	**-**	**-**	**-**	**-**	**98**	**-**	**-**	**-**
DISTRIBUTION LOSSES	**-**	**-**	**-**	**-**	**-**	**-**	**-**	**-**	**-**	**-**	**-**
FINAL CONSUMPTION	**-**	**-**	**370**	**-**	**-**	**113**	**573**	**192**	**-**	**-**	**-**
INDUSTRY SECTOR	**-**	**-**	**-**	**-**	**-**	**-**	**-**	**192**	**-**	**-**	**-**
Iron and Steel	-	-	-	-	-	-	-	-	-	-	-
Chemical and Petrochem.	-	-	-	-	-	-	-	-	-	-	-
Non-Metallic Minerals	-	-	-	-	-	-	-	-	-	-	-
Non-specified	-	-	-	-	-	-	-	192	-	-	-
TRANSPORT SECTOR	**-**	**-**	**370**	**-**	**-**	**-**	**573**	**-**	**-**	**-**	**-**
Air	-	-	-	-	-	-	-	-	-	-	-
Road	-	-	370	-	-	-	573	-	-	-	-
Non-specified	-	-	-	-	-	-	-	-	-	-	-
OTHER SECTORS	**-**	**-**	**-**	**-**	**-**	**113**	**-**	**-**	**-**	**-**	**-**
Agriculture	-	-	-	-	-	-	-	-	-	-	-
Comm. and Publ. Services	-	-	-	-	-	-	-	-	-	-	-
Residential	-	-	-	-	-	113	-	-	-	-	-
Non-specified	-	-	-	-	-	-	-	-	-	-	-
NON-ENERGY USE	**-**	**-**	**-**	**-**	**-**	**-**	**-**	**-**	**-**	**-**	**-**

Dem. People's Rep. of Korea / Rép. pop. dém. de Corée

SUPPLY AND CONSUMPTION 1996	Gas (TJ)				Comb. Renew. & Waste (TJ)				(GWh)	(TJ)
	Natural Gas	Gas Works	Coke Ovens	Blast Furnaces	Solid Biomass	Gas/Liquids from Biomass	Municipal Waste	Industrial Waste	Electricity	Heat
Production	-	-	-	-	42676	-	-	-	35036	-
Imports	-	-	-	-	-	-	-	-	-	-
Exports	-	-	-	-	-	-	-	-	-	-
Intl. Marine Bunkers	-	-	-	-	-	-	-	-	-	-
Stock Changes	-	-	-	-	-	-	-	-	-	-
DOMESTIC SUPPLY	**-**	**-**	**-**	**-**	**42676**	**-**	**-**	**-**	**35036**	**-**
Transfers and Stat. Diff.	-	-	-	-	-	-	-	-	-8	-
TRANSFORMATION	**-**	**-**	**-**	**-**	**-**	**-**	**-**	**-**	**-**	**-**
Electricity and CHP Plants	-	-	-	-	-	-	-	-	-	-
Petroleum Refineries	-	-	-	-	-	-	-	-	-	-
Other Transform. Sector	-	-	-	-	-	-	-	-	-	-
ENERGY SECTOR	**-**	**-**	**-**	**-**	**-**	**-**	**-**	**-**	**-**	**-**
DISTRIBUTION LOSSES	**-**	**-**	**-**	**-**	**-**	**-**	**-**	**-**	**29424**	**-**
FINAL CONSUMPTION	**-**	**-**	**-**	**-**	**42676**	**-**	**-**	**-**	**5604**	**-**
INDUSTRY SECTOR	**-**	**-**	**-**	**-**	**-**	**-**	**-**	**-**	**-**	**-**
Iron and Steel	-	-	-	-	-	-	-	-	-	-
Chemical and Petrochem.	-	-	-	-	-	-	-	-	-	-
Non-Metallic Minerals	-	-	-	-	-	-	-	-	-	-
Non-specified	-	-	-	-	-	-	-	-	-	-
TRANSPORT SECTOR	**-**	**-**	**-**	**-**	**-**	**-**	**-**	**-**	**-**	**-**
Air	-	-	-	-	-	-	-	-	-	-
Road	-	-	-	-	-	-	-	-	-	-
Non-specified	-	-	-	-	-	-	-	-	-	-
OTHER SECTORS	**-**	**-**	**-**	**-**	**42676**	**-**	**-**	**-**	**5604**	**-**
Agriculture	-	-	-	-	-	-	-	-	-	-
Comm. and Publ. Services	-	-	-	-	-	-	-	-	-	-
Residential	-	-	-	-	-	-	-	-	-	-
Non-specified	-	-	-	-	42676	-	-	-	5604	-
NON-ENERGY USE	**-**	**-**	**-**	**-**	**-**	**-**	**-**	**-**	**-**	**-**

APPROVISIONNEMENT ET DEMANDE 1997	*Gaz* (TJ)				*En. Re. Comb. & Déchets* (TJ)				(GWh)	(TJ)
	Gaz naturel	*Usines à gaz*	*Cokeries*	*Hauts fourneaux*	*Biomasse solide*	*Gaz/Liquides tirés de biomasse*	*Déchets urbains*	*Déchets industriels*	*Electricité*	*Chaleur*
Production	-	-	-	-	42676	-	-	-	33985	-
Imports	-	-	-	-	-	-	-	-	-	-
Exports	-	-	-	-	-	-	-	-	-	-
Intl. Marine Bunkers	-	-	-	-	-	-	-	-	-	-
Stock Changes	-	-	-	-	-	-	-	-	-	-
DOMESTIC SUPPLY	**-**	**-**	**-**	**-**	**42676**	**-**	**-**	**-**	**33985**	**-**
Transfers and Stat. Diff.	-	-	-	-	-	-	-	-	-8	-
TRANSFORMATION	**-**	**-**	**-**	**-**	**-**	**-**	**-**	**-**	**-**	**-**
Electricity and CHP Plants	-	-	-	-	-	-	-	-	-	-
Petroleum Refineries	-	-	-	-	-	-	-	-	-	-
Other Transform. Sector	-	-	-	-	-	-	-	-	-	-
ENERGY SECTOR	**-**	**-**	**-**	**-**	**-**	**-**	**-**	**-**	**-**	**-**
DISTRIBUTION LOSSES	**-**	**-**	**-**	**-**	**-**	**-**	**-**	**-**	**28541**	**-**
FINAL CONSUMPTION	**-**	**-**	**-**	**-**	**42676**	**-**	**-**	**-**	**5436**	**-**
INDUSTRY SECTOR	**-**	**-**	**-**	**-**	**-**	**-**	**-**	**-**	**-**	**-**
Iron and Steel	-	-	-	-	-	-	-	-	-	-
Chemical and Petrochem.	-	-	-	-	-	-	-	-	-	-
Non-Metallic Minerals	-	-	-	-	-	-	-	-	-	-
Non-specified	-	-	-	-	-	-	-	-	-	-
TRANSPORT SECTOR	**-**	**-**	**-**	**-**	**-**	**-**	**-**	**-**	**-**	**-**
Air	-	-	-	-	-	-	-	-	-	-
Road	-	-	-	-	-	-	-	-	-	-
Non-specified	-	-	-	-	-	-	-	-	-	-
OTHER SECTORS	**-**	**-**	**-**	**-**	**42676**	**-**	**-**	**-**	**5436**	**-**
Agriculture	-	-	-	-	-	-	-	-	-	-
Comm. and Publ. Services	-	-	-	-	-	-	-	-	-	-
Residential	-	-	-	-	-	-	-	-	-	-
Non-specified	-	-	-	-	42676	-	-	-	5436	-
NON-ENERGY USE	**-**	**-**	**-**	**-**	**-**	**-**	**-**	**-**	**-**	**-**

Kuwait / Koweit

SUPPLY AND CONSUMPTION 1996	Coal (1000 tonnes)							Oil (1000 tonnes)			
	Coking Coal	Other Bit. Coal	Sub-Bit. Coal	Lignite	Peat	Oven and Gas Coke	Pat. Fuel and BKB	Crude Oil	NGL	Feed-stocks	Additives
Production	-	-	-	-	-	-	-	103264	3494	-	-
Imports	-	-	-	-	-	-	-	-	-	-	-
Exports	-	-	-	-	-	-	-	-63641	-	-	-
Intl. Marine Bunkers	-	-	-	-	-	-	-	-	-	-	-
Stock Changes	-	-	-	-	-	-	-	-	-	-	-
DOMESTIC SUPPLY	-	-	-	-	-	-	-	**39623**	**3494**	-	-
Transfers and Stat. Diff.	-	-	-	-	-	-	-	804	-3494	-	-
TRANSFORMATION	-	-	-	-	-	-	-	**40427**	-	-	-
Electricity and CHP Plants	-	-	-	-	-	-	-	-	-	-	-
Petroleum Refineries	-	-	-	-	-	-	-	40427	-	-	-
Other Transform. Sector	-	-	-	-	-	-	-	-	-	-	-
ENERGY SECTOR	-	-	-	-	-	-	-	-	-	-	-
DISTRIBUTION LOSSES	-	-	-	-	-	-	-	-	-	-	-
FINAL CONSUMPTION	-	-	-	-	-	-	-	-	-	-	-
INDUSTRY SECTOR	-	-	-	-	-	-	-	-	-	-	-
Iron and Steel	-	-	-	-	-	-	-	-	-	-	-
Chemical and Petrochem.	-	-	-	-	-	-	-	-	-	-	-
Non-Metallic Minerals	-	-	-	-	-	-	-	-	-	-	-
Non-specified	-	-	-	-	-	-	-	-	-	-	-
TRANSPORT SECTOR	-	-	-	-	-	-	-	-	-	-	-
Air	-	-	-	-	-	-	-	-	-	-	-
Road	-	-	-	-	-	-	-	-	-	-	-
Non-specified	-	-	-	-	-	-	-	-	-	-	-
OTHER SECTORS	-	-	-	-	-	-	-	-	-	-	-
Agriculture	-	-	-	-	-	-	-	-	-	-	-
Comm. and Publ. Services	-	-	-	-	-	-	-	-	-	-	-
Residential	-	-	-	-	-	-	-	-	-	-	-
Non-specified	-	-	-	-	-	-	-	-	-	-	-
NON-ENERGY USE	-	-	-	-	-	-	-	-	-	-	-

APPROVISIONNEMENT ET DEMANDE 1997	*Charbon* (1000 tonnes)							*Pétrole* (1000 tonnes)			
	Charbon à coke	*Autres charb. bit.*	*Charbon sous-bit.*	*Lignite*	*Tourbe*	*Coke de four/gaz*	*Agg./briq. de lignite*	*Pétrole brut*	*LGN*	*Produits d'aliment.*	*Additifs*
Production	-	-	-	-	-	-	-	106040	3827	-	-
Imports	-	-	-	-	-	-	-	-	-	-	-
Exports	-	-	-	-	-	-	-	-60896	-	-	-
Intl. Marine Bunkers	-	-	-	-	-	-	-	-	-	-	-
Stock Changes	-	-	-	-	-	-	-	-	-	-	-
DOMESTIC SUPPLY	-	-	-	-	-	-	-	**45144**	**3827**	-	-
Transfers and Stat. Diff.	-	-	-	-	-	-	-	-	-3827	-	-
TRANSFORMATION	-	-	-	-	-	-	-	**45144**	-	-	-
Electricity and CHP Plants	-	-	-	-	-	-	-	-	-	-	-
Petroleum Refineries	-	-	-	-	-	-	-	45144	-	-	-
Other Transform. Sector	-	-	-	-	-	-	-	-	-	-	-
ENERGY SECTOR	-	-	-	-	-	-	-	-	-	-	-
DISTRIBUTION LOSSES	-	-	-	-	-	-	-	-	-	-	-
FINAL CONSUMPTION	-	-	-	-	-	-	-	-	-	-	-
INDUSTRY SECTOR	-	-	-	-	-	-	-	-	-	-	-
Iron and Steel	-	-	-	-	-	-	-	-	-	-	-
Chemical and Petrochem.	-	-	-	-	-	-	-	-	-	-	-
Non-Metallic Minerals	-	-	-	-	-	-	-	-	-	-	-
Non-specified	-	-	-	-	-	-	-	-	-	-	-
TRANSPORT SECTOR	-	-	-	-	-	-	-	-	-	-	-
Air	-	-	-	-	-	-	-	-	-	-	-
Road	-	-	-	-	-	-	-	-	-	-	-
Non-specified	-	-	-	-	-	-	-	-	-	-	-
OTHER SECTORS	-	-	-	-	-	-	-	-	-	-	-
Agriculture	-	-	-	-	-	-	-	-	-	-	-
Comm. and Publ. Services	-	-	-	-	-	-	-	-	-	-	-
Residential	-	-	-	-	-	-	-	-	-	-	-
Non-specified	-	-	-	-	-	-	-	-	-	-	-
NON-ENERGY USE	-	-	-	-	-	-	-	-	-	-	-

Kuwait / Koweit

SUPPLY AND CONSUMPTION 1996	Oil cont. (1000 tonnes)										
	Refinery Gas	LPG + Ethane	Motor Gasoline	Aviation Gasoline	Jet Fuel	Kerosene	Gas/ Diesel	Heavy Fuel Oil	Naphtha	Petrol. Coke	Other Prod.
Production	334	-	1957	-	7727	1417	11328	9616	2303	-	1073
Imports	-	-	-	1	-	-	-	-	-	-	-
Exports	-	-3408	-399	-	-7339	-1392	-10674	-6206	-2303	-	-908
Intl. Marine Bunkers	-	-	-	-	-	-	-40	-150	-	-	-
Stock Changes	-	-	-	-	-	-	-	-	-	-	-
DOMESTIC SUPPLY	**334**	**-3408**	**1558**	**1**	**388**	**25**	**614**	**3260**	**-**	**-**	**165**
Transfers and Stat. Diff.	-	3494	-	-	-	-	-	-	-	-	-80
TRANSFORMATION	**-**	**-**	**-**	**-**	**-**	**-**	**200**	**900**	**-**	**-**	**-**
Electricity and CHP Plants	-	-	-	-	-	-	200	900	-	-	-
Petroleum Refineries	-	-	-	-	-	-	-	-	-	-	-
Other Transform. Sector	-	-	-	-	-	-	-	-	-	-	-
ENERGY SECTOR	**334**	**-**	**-**	**-**	**-**	**-**	**-**	**1108**	**-**	**-**	**-**
DISTRIBUTION LOSSES	**-**	**-**	**-**	**-**	**-**	**-**	**-**	**-**	**-**	**-**	**-**
FINAL CONSUMPTION	**-**	**86**	**1558**	**1**	**388**	**25**	**414**	**1252**	**-**	**-**	**85**
INDUSTRY SECTOR	**-**	**-**	**-**	**-**	**-**	**-**	**246**	**1252**	**-**	**-**	**-**
Iron and Steel	-	-	-	-	-	-	-	-	-	-	-
Chemical and Petrochem.	-	-	-	-	-	-	-	-	-	-	-
Non-Metallic Minerals	-	-	-	-	-	-	-	-	-	-	-
Non-specified	-	-	-	-	-	-	246	1252	-	-	-
TRANSPORT SECTOR	**-**	**-**	**1558**	**1**	**388**	**-**	**168**	**-**	**-**	**-**	**-**
Air	-	-	-	1	388	-	-	-	-	-	-
Road	-	-	1558	-	-	-	168	-	-	-	-
Non-specified	-	-	-	-	-	-	-	-	-	-	-
OTHER SECTORS	**-**	**86**	**-**	**-**	**-**	**25**	**-**	**-**	**-**	**-**	**-**
Agriculture	-	-	-	-	-	-	-	-	-	-	-
Comm. and Publ. Services	-	-	-	-	-	-	-	-	-	-	-
Residential	-	86	-	-	-	25	-	-	-	-	-
Non-specified	-	-	-	-	-	-	-	-	-	-	-
NON-ENERGY USE	**-**	**-**	**-**	**-**	**-**	**-**	**-**	**-**	**-**	**-**	**85**

APPROVISIONNEMENT ET DEMANDE 1997	*Pétrole cont.* (1000 tonnes)										
	Gaz de raffinerie	*GPL + éthane*	*Essence moteur*	*Essence aviation*	*Carbu- réacteurs*	*Kérosène*	*Gazole*	*Fioul lourd*	*Naphta*	*Coke de pétrole*	*Autres prod.*
Production	373	-	1832	-	8156	1496	13116	11623	2512	-	1153
Imports	-	-	-	1	-	-	-	-	-	-	-
Exports	-	-3728	-133	-	-7732	-1468	-12379	-7521	-2512	-	-1058
Intl. Marine Bunkers	-	-	-	-	-	-	-40	-150	-	-	-
Stock Changes	-	-	-	-	-	-	-	-	-	-	-
DOMESTIC SUPPLY	**373**	**-3728**	**1699**	**1**	**424**	**28**	**697**	**3952**	**-**	**-**	**95**
Transfers and Stat. Diff.	-	3827	1	-	-	-1	-	-	-	-	-
TRANSFORMATION	**-**	**-**	**-**	**-**	**-**	**-**	**227**	**1091**	**-**	**-**	**-**
Electricity and CHP Plants	-	-	-	-	-	-	227	1091	-	-	-
Petroleum Refineries	-	-	-	-	-	-	-	-	-	-	-
Other Transform. Sector	-	-	-	-	-	-	-	-	-	-	-
ENERGY SECTOR	**373**	**-**	**-**	**-**	**-**	**-**	**-**	**1343**	**-**	**-**	**-**
DISTRIBUTION LOSSES	**-**	**-**	**-**	**-**	**-**	**-**	**-**	**-**	**-**	**-**	**-**
FINAL CONSUMPTION	**-**	**99**	**1700**	**1**	**424**	**27**	**470**	**1518**	**-**	**-**	**95**
INDUSTRY SECTOR	**-**	**-**	**-**	**-**	**-**	**-**	**279**	**1518**	**-**	**-**	**-**
Iron and Steel	-	-	-	-	-	-	-	-	-	-	-
Chemical and Petrochem.	-	-	-	-	-	-	-	-	-	-	-
Non-Metallic Minerals	-	-	-	-	-	-	-	-	-	-	-
Non-specified	-	-	-	-	-	-	279	1518	-	-	-
TRANSPORT SECTOR	**-**	**-**	**1700**	**1**	**424**	**-**	**191**	**-**	**-**	**-**	**-**
Air	-	-	-	1	424	-	-	-	-	-	-
Road	-	-	1700	-	-	-	191	-	-	-	-
Non-specified	-	-	-	-	-	-	-	-	-	-	-
OTHER SECTORS	**-**	**99**	**-**	**-**	**-**	**27**	**-**	**-**	**-**	**-**	**-**
Agriculture	-	-	-	-	-	-	-	-	-	-	-
Comm. and Publ. Services	-	-	-	-	-	-	-	-	-	-	-
Residential	-	99	-	-	-	27	-	-	-	-	-
Non-specified	-	-	-	-	-	-	-	-	-	-	-
NON-ENERGY USE	**-**	**-**	**-**	**-**	**-**	**-**	**-**	**-**	**-**	**-**	**95**

Kuwait / Koweit

SUPPLY AND CONSUMPTION 1996	Gas (TJ)				Comb. Renew. & Waste (TJ)				(GWh)	(TJ)
	Natural Gas	Gas Works	Coke Ovens	Blast Furnaces	Solid Biomass	Gas/Liquids from Biomass	Municipal Waste	Industrial Waste	Electricity	Heat
Production	206194	-	-	-	-	-	-	-	25475	-
Imports	-	-	-	-	165	-	-	-	-	-
Exports	-	-	-	-	-	-	-	-	-	-
Intl. Marine Bunkers	-	-	-	-	-	-	-	-	-	-
Stock Changes	-	-	-	-	-	-	-	-	-	-
DOMESTIC SUPPLY	**206194**	**-**	**-**	**-**	**165**	**-**	**-**	**-**	**25475**	**-**
Transfers and Stat. Diff.	-	-	-	-	-	-	-	-	-	-
TRANSFORMATION	**206194**	**-**	**-**	**-**	**-**	**-**	**-**	**-**	**-**	**-**
Electricity and CHP Plants	206194	-	-	-	-	-	-	-	-	-
Petroleum Refineries	-	-	-	-	-	-	-	-	-	-
Other Transform. Sector	-	-	-	-	-	-	-	-	-	-
ENERGY SECTOR	**-**	**-**	**-**	**-**	**-**	**-**	**-**	**-**	**3740**	**-**
DISTRIBUTION LOSSES	**-**	**-**	**-**	**-**	**-**	**-**	**-**	**-**	**-**	**-**
FINAL CONSUMPTION	**-**	**-**	**-**	**-**	**165**	**-**	**-**	**-**	**21735**	**-**
INDUSTRY SECTOR	**-**	**-**	**-**	**-**	**-**	**-**	**-**	**-**	**-**	**-**
Iron and Steel	-	-	-	-	-	-	-	-	-	-
Chemical and Petrochem.	-	-	-	-	-	-	-	-	-	-
Non-Metallic Minerals	-	-	-	-	-	-	-	-	-	-
Non-specified	-	-	-	-	-	-	-	-	-	-
TRANSPORT SECTOR	**-**	**-**	**-**	**-**	**-**	**-**	**-**	**-**	**-**	**-**
Air	-	-	-	-	-	-	-	-	-	-
Road	-	-	-	-	-	-	-	-	-	-
Non-specified	-	-	-	-	-	-	-	-	-	-
OTHER SECTORS	**-**	**-**	**-**	**-**	**165**	**-**	**-**	**-**	**21735**	**-**
Agriculture	-	-	-	-	-	-	-	-	-	-
Comm. and Publ. Services	-	-	-	-	-	-	-	-	-	-
Residential	-	-	-	-	165	-	-	-	21735	-
Non-specified	-	-	-	-	-	-	-	-	-	-
NON-ENERGY USE	**-**	**-**	**-**	**-**	**-**	**-**	**-**	**-**	**-**	**-**

APPROVISIONNEMENT ET DEMANDE 1997	*Gaz* (TJ)				*En. Re. Comb. & Déchets* (TJ)				(GWh)	(TJ)
	Gaz naturel	*Usines à gaz*	*Cokeries*	*Hauts fourneaux*	*Biomasse solide*	*Gaz/Liquides tirés de biomasse*	*Déchets urbains*	*Déchets industriels*	*Electricité*	*Chaleur*
Production	207299	-	-	-	-	-	-	-	27091	-
Imports	-	-	-	-	165	-	-	-	-	-
Exports	-	-	-	-	-	-	-	-	-	-
Intl. Marine Bunkers	-	-	-	-	-	-	-	-	-	-
Stock Changes	-	-	-	-	-	-	-	-	-	-
DOMESTIC SUPPLY	**207299**	**-**	**-**	**-**	**165**	**-**	**-**	**-**	**27091**	**-**
Transfers and Stat. Diff.	-	-	-	-	-	-	-	-	-	-
TRANSFORMATION	**207299**	**-**	**-**	**-**	**-**	**-**	**-**	**-**	**-**	**-**
Electricity and CHP Plants	207299	-	-	-	-	-	-	-	-	-
Petroleum Refineries	-	-	-	-	-	-	-	-	-	-
Other Transform. Sector	-	-	-	-	-	-	-	-	-	-
ENERGY SECTOR	**-**	**-**	**-**	**-**	**-**	**-**	**-**	**-**	**3781**	**-**
DISTRIBUTION LOSSES	**-**	**-**	**-**	**-**	**-**	**-**	**-**	**-**	**-**	**-**
FINAL CONSUMPTION	**-**	**-**	**-**	**-**	**165**	**-**	**-**	**-**	**23310**	**-**
INDUSTRY SECTOR	**-**	**-**	**-**	**-**	**-**	**-**	**-**	**-**	**-**	**-**
Iron and Steel	-	-	-	-	-	-	-	-	-	-
Chemical and Petrochem.	-	-	-	-	-	-	-	-	-	-
Non-Metallic Minerals	-	-	-	-	-	-	-	-	-	-
Non-specified	-	-	-	-	-	-	-	-	-	-
TRANSPORT SECTOR	**-**	**-**	**-**	**-**	**-**	**-**	**-**	**-**	**-**	**-**
Air	-	-	-	-	-	-	-	-	-	-
Road	-	-	-	-	-	-	-	-	-	-
Non-specified	-	-	-	-	-	-	-	-	-	-
OTHER SECTORS	**-**	**-**	**-**	**-**	**165**	**-**	**-**	**-**	**23310**	**-**
Agriculture	-	-	-	-	-	-	-	-	-	-
Comm. and Publ. Services	-	-	-	-	-	-	-	-	-	-
Residential	-	-	-	-	165	-	-	-	23310	-
Non-specified	-	-	-	-	-	-	-	-	-	-
NON-ENERGY USE	**-**	**-**	**-**	**-**	**-**	**-**	**-**	**-**	**-**	**-**

Kyrgyzstan / Kirghizistan

SUPPLY AND CONSUMPTION 1996	Coal (1000 tonnes)							Oil (1000 tonnes)			
	Coking Coal	Other Bit. Coal	Sub-Bit. Coal	Lignite	Peat	Oven and Gas Coke	Pat. Fuel and BKB	Crude Oil	NGL	Feed-stocks	Additives
Production	-	370	-	280	-	-	-	100	-	-	-
Imports	-	793	-	52	-	-	-	12	-	-	-
Exports	-	-92	-	-	-	-	-	-21	-	-	-
Intl. Marine Bunkers	-	-	-	-	-	-	-	-	-	-	-
Stock Changes	-	-	-	-	-	-	-	-	-	-	-
DOMESTIC SUPPLY	**-**	**1071**	**-**	**332**	**-**	**-**	**-**	**91**	**-**	**-**	**-**
Transfers and Stat. Diff.	-	-	-	-	-	-	-	-	-	-	-
TRANSFORMATION	**-**	**490**	**-**	**-**	**-**	**-**	**-**	**91**	**-**	**-**	**-**
Electricity and CHP Plants	-	490	-	-	-	-	-	-	-	-	-
Petroleum Refineries	-	-	-	-	-	-	-	91	-	-	-
Other Transform. Sector	-	-	-	-	-	-	-	-	-	-	-
ENERGY SECTOR	**-**	**-**	**-**	**-**	**-**	**-**	**-**	**-**	**-**	**-**	**-**
DISTRIBUTION LOSSES	**-**	**-**	**-**	**-**	**-**	**-**	**-**	**-**	**-**	**-**	**-**
FINAL CONSUMPTION	**-**	**581**	**-**	**332**	**-**	**-**	**-**	**-**	**-**	**-**	**-**
INDUSTRY SECTOR	**-**	**581**	**-**	**332**	**-**	**-**	**-**	**-**	**-**	**-**	**-**
Iron and Steel	-	-	-	-	-	-	-	-	-	-	-
Chemical and Petrochem.	-	-	-	-	-	-	-	-	-	-	-
Non-Metallic Minerals	-	-	-	-	-	-	-	-	-	-	-
Non-specified	-	581	-	332	-	-	-	-	-	-	-
TRANSPORT SECTOR	**-**	**-**	**-**	**-**	**-**	**-**	**-**	**-**	**-**	**-**	**-**
Air	-	-	-	-	-	-	-	-	-	-	-
Road	-	-	-	-	-	-	-	-	-	-	-
Non-specified	-	-	-	-	-	-	-	-	-	-	-
OTHER SECTORS	**-**	**-**	**-**	**-**	**-**	**-**	**-**	**-**	**-**	**-**	**-**
Agriculture	-	-	-	-	-	-	-	-	-	-	-
Comm. and Publ. Services	-	-	-	-	-	-	-	-	-	-	-
Residential	-	-	-	-	-	-	-	-	-	-	-
Non-specified	-	-	-	-	-	-	-	-	-	-	-
NON-ENERGY USE	**-**	**-**	**-**	**-**	**-**	**-**	**-**	**-**	**-**	**-**	**-**

APPROVISIONNEMENT ET DEMANDE 1997	*Charbon* (1000 tonnes)							*Pétrole* (1000 tonnes)			
	Charbon à coke	*Autres charb. bit.*	*Charbon sous-bit.*	*Lignite*	*Tourbe*	*Coke de four/gaz*	*Agg./briq. de lignite*	*Pétrole brut*	*LGN*	*Produits d'aliment.*	*Additifs*
Production	-	500	-	280	-	-	-	85	-	-	-
Imports	-	1500	-	50	-	-	-	65	-	-	-
Exports	-	-100	-	-	-	-	-	-40	-	-	-
Intl. Marine Bunkers	-	-	-	-	-	-	-	-	-	-	-
Stock Changes	-	-	-	-	-	-	-	-	-	-	-
DOMESTIC SUPPLY	**-**	**1900**	**-**	**330**	**-**	**-**	**-**	**110**	**-**	**-**	**-**
Transfers and Stat. Diff.	-	-	-	-	-	-	-	-	-	-	-
TRANSFORMATION	**-**	**510**	**-**	**-**	**-**	**-**	**-**	**110**	**-**	**-**	**-**
Electricity and CHP Plants	-	510	-	-	-	-	-	-	-	-	-
Petroleum Refineries	-	-	-	-	-	-	-	110	-	-	-
Other Transform. Sector	-	-	-	-	-	-	-	-	-	-	-
ENERGY SECTOR	**-**	**-**	**-**	**-**	**-**	**-**	**-**	**-**	**-**	**-**	**-**
DISTRIBUTION LOSSES	**-**	**-**	**-**	**-**	**-**	**-**	**-**	**-**	**-**	**-**	**-**
FINAL CONSUMPTION	**-**	**1390**	**-**	**330**	**-**	**-**	**-**	**-**	**-**	**-**	**-**
INDUSTRY SECTOR	**-**	**1390**	**-**	**330**	**-**	**-**	**-**	**-**	**-**	**-**	**-**
Iron and Steel	-	-	-	-	-	-	-	-	-	-	-
Chemical and Petrochem.	-	-	-	-	-	-	-	-	-	-	-
Non-Metallic Minerals	-	-	-	-	-	-	-	-	-	-	-
Non-specified	-	1390	-	330	-	-	-	-	-	-	-
TRANSPORT SECTOR	**-**	**-**	**-**	**-**	**-**	**-**	**-**	**-**	**-**	**-**	**-**
Air	-	-	-	-	-	-	-	-	-	-	-
Road	-	-	-	-	-	-	-	-	-	-	-
Non-specified	-	-	-	-	-	-	-	-	-	-	-
OTHER SECTORS	**-**	**-**	**-**	**-**	**-**	**-**	**-**	**-**	**-**	**-**	**-**
Agriculture	-	-	-	-	-	-	-	-	-	-	-
Comm. and Publ. Services	-	-	-	-	-	-	-	-	-	-	-
Residential	-	-	-	-	-	-	-	-	-	-	-
Non-specified	-	-	-	-	-	-	-	-	-	-	-
NON-ENERGY USE	**-**	**-**	**-**	**-**	**-**	**-**	**-**	**-**	**-**	**-**	**-**

Kyrgyzstan / Kirghizistan

SUPPLY AND CONSUMPTION 1996	Oil cont. (1000 tonnes)										
	Refinery Gas	LPG + Ethane	Motor Gasoline	Aviation Gasoline	Jet Fuel	Kerosene	Gas/ Diesel	Heavy Fuel Oil	Naphtha	Petrol. Coke	Other Prod.
Production	-	-	-	-	-	-	-	-	-	-	91
Imports	-	-	230	21	68	-	145	100	-	-	13
Exports	-	-	-	-	-	-	-1	-	-	-	-
Intl. Marine Bunkers	-	-	-	-	-	-	-	-	-	-	-
Stock Changes	-	-	-	-	-	-	-	-	-	-	-
DOMESTIC SUPPLY	**-**	**-**	**230**	**21**	**68**	**-**	**144**	**100**	**-**	**-**	**104**
Transfers and Stat. Diff.	-	-	-	-	-	-	-	-	-	-	-
TRANSFORMATION	**-**	**-**	**-**	**-**	**-**	**-**	**-**	**-**	**-**	**-**	**-**
Electricity and CHP Plants	-	-	-	-	-	-	-	-	-	-	-
Petroleum Refineries	-	-	-	-	-	-	-	-	-	-	-
Other Transform. Sector	-	-	-	-	-	-	-	-	-	-	-
ENERGY SECTOR	**-**	**-**	**-**	**-**	**-**	**-**	**-**	**-**	**-**	**-**	**-**
DISTRIBUTION LOSSES	**-**	**-**	**-**	**-**	**-**	**-**	**-**	**-**	**-**	**-**	**-**
FINAL CONSUMPTION	**-**	**-**	**230**	**21**	**68**	**-**	**144**	**100**	**-**	**-**	**104**
INDUSTRY SECTOR	**-**	**-**	**-**	**-**	**-**	**-**	**-**	**-**	**-**	**-**	**-**
Iron and Steel	-	-	-	-	-	-	-	-	-	-	-
Chemical and Petrochem.	-	-	-	-	-	-	-	-	-	-	-
Non-Metallic Minerals	-	-	-	-	-	-	-	-	-	-	-
Non-specified	-	-	-	-	-	-	-	-	-	-	-
TRANSPORT SECTOR	**-**	**-**	**230**	**21**	**68**	**-**	**-**	**-**	**-**	**-**	**-**
Air	-	-	-	21	68	-	-	-	-	-	-
Road	-	-	230	-	-	-	-	-	-	-	-
Non-specified	-	-	-	-	-	-	-	-	-	-	-
OTHER SECTORS	**-**	**-**	**-**	**-**	**-**	**-**	**144**	**100**	**-**	**-**	**91**
Agriculture	-	-	-	-	-	-	-	-	-	-	-
Comm. and Publ. Services	-	-	-	-	-	-	-	-	-	-	-
Residential	-	-	-	-	-	-	-	-	-	-	-
Non-specified	-	-	-	-	-	-	144	100	-	-	91
NON-ENERGY USE	**-**	**-**	**-**	**-**	**-**	**-**	**-**	**-**	**-**	**-**	**13**

APPROVISIONNEMENT ET DEMANDE 1997	*Pétrole cont.* (1000 tonnes)										
	Gaz de raffinerie	*GPL + éthane*	*Essence moteur*	*Essence aviation*	*Carbu- réacteurs*	*Kérosène*	*Gazole*	*Fioul lourd*	*Naphta*	*Coke de pétrole*	*Autres prod.*
Production	-	-	-	-	-	-	-	-	-	-	110
Imports	-	-	115	18	45	-	100	88	-	-	12
Exports	-	-	-	-	-	-	-	-	-	-	-
Intl. Marine Bunkers	-	-	-	-	-	-	-	-	-	-	-
Stock Changes	-	-	-	-	-	-	-	-	-	-	-
DOMESTIC SUPPLY	**-**	**-**	**115**	**18**	**45**	**-**	**100**	**88**	**-**	**-**	**122**
Transfers and Stat. Diff.	-	-	-	-	-	-	-	-	-	-	-
TRANSFORMATION	**-**	**-**	**-**	**-**	**-**	**-**	**-**	**-**	**-**	**-**	**-**
Electricity and CHP Plants	-	-	-	-	-	-	-	-	-	-	-
Petroleum Refineries	-	-	-	-	-	-	-	-	-	-	-
Other Transform. Sector	-	-	-	-	-	-	-	-	-	-	-
ENERGY SECTOR	**-**	**-**	**-**	**-**	**-**	**-**	**-**	**-**	**-**	**-**	**-**
DISTRIBUTION LOSSES	**-**	**-**	**-**	**-**	**-**	**-**	**-**	**-**	**-**	**-**	**-**
FINAL CONSUMPTION	**-**	**-**	**115**	**18**	**45**	**-**	**100**	**88**	**-**	**-**	**122**
INDUSTRY SECTOR	**-**	**-**	**-**	**-**	**-**	**-**	**-**	**-**	**-**	**-**	**-**
Iron and Steel	-	-	-	-	-	-	-	-	-	-	-
Chemical and Petrochem.	-	-	-	-	-	-	-	-	-	-	-
Non-Metallic Minerals	-	-	-	-	-	-	-	-	-	-	-
Non-specified	-	-	-	-	-	-	-	-	-	-	-
TRANSPORT SECTOR	**-**	**-**	**115**	**18**	**45**	**-**	**-**	**-**	**-**	**-**	**-**
Air	-	-	-	18	45	-	-	-	-	-	-
Road	-	-	115	-	-	-	-	-	-	-	-
Non-specified	-	-	-	-	-	-	-	-	-	-	-
OTHER SECTORS	**-**	**-**	**-**	**-**	**-**	**-**	**100**	**88**	**-**	**-**	**110**
Agriculture	-	-	-	-	-	-	-	-	-	-	-
Comm. and Publ. Services	-	-	-	-	-	-	-	-	-	-	-
Residential	-	-	-	-	-	-	-	-	-	-	-
Non-specified	-	-	-	-	-	-	100	88	-	-	110
NON-ENERGY USE	**-**	**-**	**-**	**-**	**-**	**-**	**-**	**-**	**-**	**-**	**12**

Kyrgyzstan / Kirghizistan

SUPPLY AND CONSUMPTION 1996	Gas (TJ)				Comb. Renew. & Waste (TJ)				(GWh)	(TJ)
	Natural Gas	Gas Works	Coke Ovens	Blast Furnaces	Solid Biomass	Gas/Liquids from Biomass	Municipal Waste	Industrial Waste	Electricity	Heat
Production	1015	-	-	-	150	-	-	-	13758	19162
Imports	40074	-	-	-	-	-	-	-	7116	-
Exports	-	-	-	-	-	-	-	-	-9196	-
Intl. Marine Bunkers	-	-	-	-	-	-	-	-	-	-
Stock Changes	-	-	-	-	-	-	-	-	-	-
DOMESTIC SUPPLY	**41089**	**-**	**-**	**-**	**150**	**-**	**-**	**-**	**11678**	**19162**
Transfers and Stat. Diff.	-	-	-	-	-	-	-	-	-79	-
TRANSFORMATION	**23608**	**-**	**-**	**-**	**-**	**-**	**-**	**-**	**-**	**-**
Electricity and CHP Plants	23608	-	-	-	-	-	-	-	-	-
Petroleum Refineries	-	-	-	-	-	-	-	-	-	-
Other Transform. Sector	-	-	-	-	-	-	-	-	-	-
ENERGY SECTOR	**-**	**-**	**-**	**-**	**-**	**-**	**-**	**-**	**279**	**-**
DISTRIBUTION LOSSES	**-**	**-**	**-**	**-**	**-**	**-**	**-**	**-**	**4551**	**-**
FINAL CONSUMPTION	**17481**	**-**	**-**	**-**	**150**	**-**	**-**	**-**	**6769**	**19162**
INDUSTRY SECTOR	**-**	**-**	**-**	**-**	**-**	**-**	**-**	**-**	**1589**	**-**
Iron and Steel	-	-	-	-	-	-	-	-	-	-
Chemical and Petrochem.	-	-	-	-	-	-	-	-	5	-
Non-Metallic Minerals	-	-	-	-	-	-	-	-	-	-
Non-specified	-	-	-	-	-	-	-	-	1584	-
TRANSPORT SECTOR	**-**	**-**	**-**	**-**	**-**	**-**	**-**	**-**	**127**	**-**
Air	-	-	-	-	-	-	-	-	-	-
Road	-	-	-	-	-	-	-	-	-	-
Non-specified	-	-	-	-	-	-	-	-	127	-
OTHER SECTORS	**17481**	**-**	**-**	**-**	**150**	**-**	**-**	**-**	**5053**	**19162**
Agriculture	-	-	-	-	-	-	-	-	2631	-
Comm. and Publ. Services	-	-	-	-	-	-	-	-	-	-
Residential	-	-	-	-	-	-	-	-	1701	-
Non-specified	17481	-	-	-	150	-	-	-	721	19162
NON-ENERGY USE	**-**	**-**	**-**	**-**	**-**	**-**	**-**	**-**	**-**	**-**

APPROVISIONNEMENT ET DEMANDE 1997	*Gaz (TJ)*				*En. Re. Comb. & Déchets (TJ)*				(GWh)	(TJ)
	Gaz naturel	*Usines à gaz*	*Cokeries*	*Hauts fourneaux*	*Biomasse solide*	*Gaz/Liquides tirés de biomasse*	*Déchets urbains*	*Déchets industriels*	*Electricité*	*Chaleur*
Production	1561	-	-	-	150	-	-	-	12600	19162
Imports	22241	-	-	-	-	-	-	-	7000	-
Exports	-	-	-	-	-	-	-	-	-8700	-
Intl. Marine Bunkers	-	-	-	-	-	-	-	-	-	-
Stock Changes	-	-	-	-	-	-	-	-	-	-
DOMESTIC SUPPLY	**23802**	**-**	**-**	**-**	**150**	**-**	**-**	**-**	**10900**	**19162**
Transfers and Stat. Diff.	-	-	-	-	-	-	-	-	-	-
TRANSFORMATION	**14437**	**-**	**-**	**-**	**-**	**-**	**-**	**-**	**-**	**-**
Electricity and CHP Plants	14437	-	-	-	-	-	-	-	-	-
Petroleum Refineries	-	-	-	-	-	-	-	-	-	-
Other Transform. Sector	-	-	-	-	-	-	-	-	-	-
ENERGY SECTOR	**-**	**-**	**-**	**-**	**-**	**-**	**-**	**-**	**262**	**-**
DISTRIBUTION LOSSES	**-**	**-**	**-**	**-**	**-**	**-**	**-**	**-**	**4277**	**-**
FINAL CONSUMPTION	**9365**	**-**	**-**	**-**	**150**	**-**	**-**	**-**	**6361**	**19162**
INDUSTRY SECTOR	**-**	**-**	**-**	**-**	**-**	**-**	**-**	**-**	**1493**	**-**
Iron and Steel	-	-	-	-	-	-	-	-	-	-
Chemical and Petrochem.	-	-	-	-	-	-	-	-	5	-
Non-Metallic Minerals	-	-	-	-	-	-	-	-	-	-
Non-specified	-	-	-	-	-	-	-	-	1488	-
TRANSPORT SECTOR	**-**	**-**	**-**	**-**	**-**	**-**	**-**	**-**	**119**	**-**
Air	-	-	-	-	-	-	-	-	-	-
Road	-	-	-	-	-	-	-	-	-	-
Non-specified	-	-	-	-	-	-	-	-	119	-
OTHER SECTORS	**9365**	**-**	**-**	**-**	**150**	**-**	**-**	**-**	**4749**	**19162**
Agriculture	-	-	-	-	-	-	-	-	2472	-
Comm. and Publ. Services	-	-	-	-	-	-	-	-	-	-
Residential	-	-	-	-	-	-	-	-	1599	-
Non-specified	9365	-	-	-	150	-	-	-	678	19162
NON-ENERGY USE	**-**	**-**	**-**	**-**	**-**	**-**	**-**	**-**	**-**	**-**

Latvia / Lettonie : 1996

SUPPLY AND CONSUMPTION / *APPROVISIONNEMENT ET DEMANDE*	Coal / *Charbon* (1000 tonnes)							Oil / *Pétrole* (1000 tonnes)			
	Coking Coal / *Charbon à coke*	Other Bit. Coal / *Autres charb. bit.*	Sub-Bit. Coal / *Charbon sous-bit.*	Lignite / *Lignite*	Peat / *Tourbe*	Oven and Gas Coke / *Coke de four/gaz*	Pat. Fuel and BKB / *Agg./briq. de lignite*	Crude Oil / *Pétrole brut*	NGL / *LGN*	Feed-stocks / *Produits d'aliment.*	Additives / *Additifs*
Production	-	-	-	-	389	-	28	-	-	-	-
From Other Sources	-	-	-	-	-	-	-	-	-	-	-
Imports	-	213	-	-	-	8	-	-	-	-	-
Exports	-	-5	-	-	-	-	-	-	-	-	-
Intl. Marine Bunkers	-	-	-	-	-	-	-	-	-	-	-
Stock Changes	-	31	-	-	-41	-	-	-	-	-	-
DOMESTIC SUPPLY	**-**	**239**	**-**	**-**	**348**	**8**	**28**	**-**	**-**	**-**	**-**
Transfers	-	-	-	-	-	-	-	-	-	-	-
Statistical Differences	-	-	-	-	-	-	-	-	-	-	-
TRANSFORMATION	**-**	**76**	**-**	**-**	**340**	**-**	**7**	**-**	**-**	**-**	**-**
Electricity Plants	-	-	-	-	-	-	-	-	-	-	-
CHP Plants	-	-	-	-	234	-	-	-	-	-	-
Heat Plants	-	76	-	-	63	-	7	-	-	-	-
Blast Furnaces/Gas Works	-	-	-	-	-	-	-	-	-	-	-
Coke/Pat. Fuel/BKB Plants	-	-	-	-	43	-	-	-	-	-	-
Petroleum Refineries	-	-	-	-	-	-	-	-	-	-	-
Petrochemical Industry	-	-	-	-	-	-	-	-	-	-	-
Liquefaction	-	-	-	-	-	-	-	-	-	-	-
Other Transform. Sector	-	-	-	-	-	-	-	-	-	-	-
ENERGY SECTOR	**-**	**-**	**-**	**-**	**-**	**-**	**-**	**-**	**-**	**-**	**-**
Coal Mines	-	-	-	-	-	-	-	-	-	-	-
Oil and Gas Extraction	-	-	-	-	-	-	-	-	-	-	-
Petroleum Refineries	-	-	-	-	-	-	-	-	-	-	-
Electr., CHP+Heat Plants	-	-	-	-	-	-	-	-	-	-	-
Pumped Storage (Elec.)	-	-	-	-	-	-	-	-	-	-	-
Other Energy Sector	-	-	-	-	-	-	-	-	-	-	-
Distribution Losses	-	-	-	-	-	-	-	-	-	-	-
FINAL CONSUMPTION	**-**	**163**	**-**	**-**	**8**	**8**	**21**	**-**	**-**	**-**	**-**
INDUSTRY SECTOR	**-**	**11**	**-**	**-**	**2**	**-**	**1**	**-**	**-**	**-**	**-**
Iron and Steel	-	-	-	-	-	-	-	-	-	-	-
Chemical and Petrochem.	-	-	-	-	-	-	-	-	-	-	-
of which: Feedstocks	-	-	-	-	-	-	-	-	-	-	-
Non-Ferrous Metals	-	-	-	-	-	-	-	-	-	-	-
Non-Metallic Minerals	-	2	-	-	-	-	-	-	-	-	-
Transport Equipment	-	-	-	-	-	-	-	-	-	-	-
Machinery	-	-	-	-	-	-	-	-	-	-	-
Mining and Quarrying	-	-	-	-	2	-	1	-	-	-	-
Food and Tobacco	-	5	-	-	-	-	-	-	-	-	-
Paper, Pulp and Print	-	-	-	-	-	-	-	-	-	-	-
Wood and Wood Products	-	-	-	-	-	-	-	-	-	-	-
Construction	-	1	-	-	-	-	-	-	-	-	-
Textile and Leather	-	1	-	-	-	-	-	-	-	-	-
Non-specified	-	2	-	-	-	-	-	-	-	-	-
TRANSPORT SECTOR	**-**	**-**	**-**	**-**	**-**	**-**	**-**	**-**	**-**	**-**	**-**
Air	-	-	-	-	-	-	-	-	-	-	-
Road	-	-	-	-	-	-	-	-	-	-	-
Rail	-	-	-	-	-	-	-	-	-	-	-
Pipeline Transport	-	-	-	-	-	-	-	-	-	-	-
Internal Navigation	-	-	-	-	-	-	-	-	-	-	-
Non-specified	-	-	-	-	-	-	-	-	-	-	-
OTHER SECTORS	**-**	**152**	**-**	**-**	**6**	**-**	**20**	**-**	**-**	**-**	**-**
Agriculture	-	6	-	-	-	-	1	-	-	-	-
Comm. and Publ. Services	-	87	-	-	4	-	6	-	-	-	-
Residential	-	59	-	-	2	-	13	-	-	-	-
Non-specified	-	-	-	-	-	-	-	-	-	-	-
NON-ENERGY USE	**-**	**-**	**-**	**-**	**-**	**8**	**-**	**-**	**-**	**-**	**-**
in Industry/Trans./Energy	-	-	-	-	-	8	-	-	-	-	-
in Transport	-	-	-	-	-	-	-	-	-	-	-
in Other Sectors	-	-	-	-	-	-	-	-	-	-	-

Latvia / Lettonie : 1996

SUPPLY AND CONSUMPTION / *APPROVISIONNEMENT ET DEMANDE*	Oil cont. / *Pétrole cont.* (1000 tonnes)										
	Refinery Gas / *Gaz de raffinerie*	LPG + Ethane / *GPL + éthane*	Motor Gasoline / *Essence moteur*	Aviation Gasoline / *Essence aviation*	Jet Fuel / *Carbu-réacteurs*	Kerosene / *Kérosène*	Gas/ Diesel / *Gazole*	Heavy Fuel Oil / *Fioul lourd*	Naphtha / *Naphta*	Petrol. Coke / *Coke de pétrole*	Other Prod. / *Autres prod.*
Production	-	-	-	-	-	-	-	-	-	-	-
From Other Sources	-	-	-	-	-	-	-	-	-	-	-
Imports	-	31	448	-	44	4	640	1176	-	-	3
Exports	-	-	-11	-	-	-	-	-	-	-	-26
Intl. Marine Bunkers	-	-	-	-	-	-	-	-	-	-	-
Stock Changes	-	4	-3	-	-10	-	-209	-35	-	-	-
DOMESTIC SUPPLY	**-**	**35**	**434**	**-**	**34**	**4**	**431**	**1141**	**-**	**-**	**-23**
Transfers	-	-	-	-	-	-	-	-	-	-	26
Statistical Differences	-	-	-	-	-	-	-	-	-	-	-
TRANSFORMATION	**-**	**-**	**-**	**-**	**-**	**-**	**2**	**1021**	**-**	**-**	**2**
Electricity Plants	-	-	-	-	-	-	-	-	-	-	-
CHP Plants	-	-	-	-	-	-	-	303	-	-	2
Heat Plants	-	-	-	-	-	-	2	718	-	-	-
Blast Furnaces/Gas Works	-	-	-	-	-	-	-	-	-	-	-
Coke/Pat. Fuel/BKB Plants	-	-	-	-	-	-	-	-	-	-	-
Petroleum Refineries	-	-	-	-	-	-	-	-	-	-	-
Petrochemical Industry	-	-	-	-	-	-	-	-	-	-	-
Liquefaction	-	-	-	-	-	-	-	-	-	-	-
Other Transform. Sector	-	-	-	-	-	-	-	-	-	-	-
ENERGY SECTOR	**-**	**-**	**-**	**-**	**-**	**-**	**-**	**-**	**-**	**-**	**-**
Coal Mines	-	-	-	-	-	-	-	-	-	-	-
Oil and Gas Extraction	-	-	-	-	-	-	-	-	-	-	-
Petroleum Refineries	-	-	-	-	-	-	-	-	-	-	-
Electr., CHP+Heat Plants	-	-	-	-	-	-	-	-	-	-	-
Pumped Storage (Elec.)	-	-	-	-	-	-	-	-	-	-	-
Other Energy Sector	-	-	-	-	-	-	-	-	-	-	-
Distribution Losses	-	-	-	-	-	-	-	-	-	-	-
FINAL CONSUMPTION	**-**	**35**	**434**	**-**	**34**	**4**	**429**	**120**	**-**	**-**	**1**
INDUSTRY SECTOR	**-**	**3**	**3**	**-**	**1**	**2**	**31**	**23**	**-**	**-**	**1**
Iron and Steel	-	-	-	-	-	-	-	6	-	-	-
Chemical and Petrochem.	-	-	-	-	-	-	-	-	-	-	-
of which: Feedstocks	-	-	-	-	-	-	-	-	-	-	-
Non-Ferrous Metals	-	-	-	-	-	-	-	-	-	-	-
Non-Metallic Minerals	-	-	-	-	-	-	1	8	-	-	-
Transport Equipment	-	-	-	-	-	-	7	3	-	-	-
Machinery	-	-	-	-	-	-	-	-	-	-	-
Mining and Quarrying	-	-	-	-	-	-	-	1	-	-	-
Food and Tobacco	-	-	-	-	-	1	12	-	-	-	1
Paper, Pulp and Print	-	-	-	-	-	-	-	2	-	-	-
Wood and Wood Products	-	-	-	-	-	-	-	-	-	-	-
Construction	-	-	2	-	-	-	5	2	-	-	-
Textile and Leather	-	-	-	-	-	-	1	-	-	-	-
Non-specified	-	3	1	-	1	1	5	1	-	-	-
TRANSPORT SECTOR	**-**	**2**	**429**	**-**	**32**	**-**	**246**	**77**	**-**	**-**	**-**
Air	-	-	-	-	32	-	-	-	-	-	-
Road	-	2	429	-	-	-	148	-	-	-	-
Rail	-	-	-	-	-	-	76	-	-	-	-
Pipeline Transport	-	-	-	-	-	-	-	-	-	-	-
Internal Navigation	-	-	-	-	-	-	22	77	-	-	-
Non-specified	-	-	-	-	-	-	-	-	-	-	-
OTHER SECTORS	**-**	**30**	**2**	**-**	**1**	**2**	**152**	**20**	**-**	**-**	**-**
Agriculture	-	-	2	-	-	1	62	16	-	-	-
Comm. and Publ. Services	-	3	-	-	1	-	22	4	-	-	-
Residential	-	27	-	-	-	-	68	-	-	-	-
Non-specified	-	-	-	-	-	1	-	-	-	-	-
NON-ENERGY USE	**-**	**-**	**-**	**-**	**-**	**-**	**-**	**-**	**-**	**-**	**-**
in Industry/Transf./Energy	-	-	-	-	-	-	-	-	-	-	-
in Transport	-	-	-	-	-	-	-	-	-	-	-
in Other Sectors	-	-	-	-	-	-	-	-	-	-	-

Latvia / Lettonie : 1996

SUPPLY AND CONSUMPTION / *APPROVISIONNEMENT ET DEMANDE*	Gas / *Gaz* (TJ)				Comb. Renew. & Waste / *En. Re. Comb. & Déchets* (TJ)				(GWh)	(TJ)
	Natural Gas / *Gaz naturel*	Gas Works / *Usines à gaz*	Coke Ovens / *Cokeries*	Blast Furnaces / *Hauts fourneaux*	Solid Biomass / *Biomasse solide*	Gas/Liquids from Biomass / *Gaz/Liquides tirés de biomasse*	Municipal Waste / *Déchets urbains*	Industrial Waste / *Déchets industriels*	Electricity / *Electricité*	Heat / *Chaleur*
Production	-	-	-	-	31814	-	-	-	3124	54925
From Other Sources	-	-	-	-	-	-	-	-	-	-
Imports	40692	-	-	-	-	-	-	-	3438	-
Exports	-	-	-	-	-6718	-	-	-	-211	-
Intl. Marine Bunkers	-	-	-	-	-	-	-	-	-	-
Stock Changes	-38	-	-	-	-485	-	-	-	-	-
DOMESTIC SUPPLY	**40654**	**-**	**-**	**-**	**24611**	**-**	**-**	**-**	**6351**	**54925**
Transfers	-	-	-	-	-	-	-	-	-	-
Statistical Differences	-	-	-	-	-	-	-	-	-	-
TRANSFORMATION	**16740**	**-**	**-**	**-**	**7762**	**-**	**-**	**-**	**4**	**-**
Electricity Plants	-	-	-	-	-	-	-	-	-	-
CHP Plants	11172	-	-	-	-	-	-	-	-	-
Heat Plants	5568	-	-	-	7762	-	-	-	-	-
Blast Furnaces/Gas Works	-	-	-	-	-	-	-	-	-	-
Coke/Pat. Fuel/BKB Plants	-	-	-	-	-	-	-	-	-	-
Petroleum Refineries	-	-	-	-	-	-	-	-	-	-
Petrochemical Industry	-	-	-	-	-	-	-	-	-	-
Liquefaction	-	-	-	-	-	-	-	-	-	-
Other Transform. Sector	-	-	-	-	-	-	-	-	4	-
ENERGY SECTOR	**1158**	**-**	**-**	**-**	**-**	**-**	**-**	**-**	**449**	**1946**
Coal Mines	-	-	-	-	-	-	-	-	-	-
Oil and Gas Extraction	-	-	-	-	-	-	-	-	-	-
Petroleum Refineries	-	-	-	-	-	-	-	-	-	-
Electr., CHP+Heat Plants	-	-	-	-	-	-	-	-	160	1060
Pumped Storage (Elec.)	-	-	-	-	-	-	-	-	-	-
Other Energy Sector	1158	-	-	-	-	-	-	-	289	886
Distribution Losses	1121	-	-	-	-	-	-	-	1456	6890
FINAL CONSUMPTION	**21635**	**-**	**-**	**-**	**16849**	**-**	**-**	**-**	**4442**	**46089**
INDUSTRY SECTOR	**10089**	**-**	**-**	**-**	**2055**	**-**	**-**	**-**	**1697**	**13851**
Iron and Steel	2797	-	-	-	-	-	-	-	79	882
Chemical and Petrochem.	254	-	-	-	12	-	-	-	210	1353
of which: Feedstocks	-	-	-	-	-	-	-	-	-	-
Non-Ferrous Metals	-	-	-	-	-	-	-	-	-	-
Non-Metallic Minerals	1358	-	-	-	22	-	-	-	101	590
Transport Equipment	384	-	-	-	-	-	-	-	77	478
Machinery	88	-	-	-	157	-	-	-	159	585
Mining and Quarrying	4	-	-	-	-	-	-	-	8	12
Food and Tobacco	2075	-	-	-	488	-	-	-	278	5199
Paper, Pulp and Print	132	-	-	-	21	-	-	-	37	257
Wood and Wood Products	536	-	-	-	1133	-	-	-	204	1646
Construction	19	-	-	-	42	-	-	-	-	1212
Textile and Leather	330	-	-	-	52	-	-	-	194	1200
Non-specified	2112	-	-	-	128	-	-	-	350	437
TRANSPORT SECTOR	**35**	**-**	**-**	**-**	**-**	**-**	**-**	**-**	**178**	**-**
Air	-	-	-	-	-	-	-	-	-	-
Road	35	-	-	-	-	-	-	-	-	-
Rail	-	-	-	-	-	-	-	-	52	-
Pipeline Transport	-	-	-	-	-	-	-	-	35	-
Internal Navigation	-	-	-	-	-	-	-	-	-	-
Non-specified	-	-	-	-	-	-	-	-	91	-
OTHER SECTORS	**11511**	**-**	**-**	**-**	**14794**	**-**	**-**	**-**	**2567**	**32238**
Agriculture	1049	-	-	-	787	-	-	-	176	386
Comm. and Publ. Services	6240	-	-	-	4093	-	-	-	1298	3546
Residential	4222	-	-	-	9914	-	-	-	1093	28306
Non-specified	-	-	-	-	-	-	-	-	-	-
NON-ENERGY USE	**-**	**-**	**-**	**-**	**-**	**-**	**-**	**-**	**-**	**-**
in Industry/Transf./Energy	-	-	-	-	-	-	-	-	-	-
in Transport	-	-	-	-	-	-	-	-	-	-
in Other Sectors	-	-	-	-	-	-	-	-	-	-

Latvia / Lettonie : 1997

	Coal / *Charbon* (1000 tonnes)							Oil / *Pétrole* (1000 tonnes)			
SUPPLY AND CONSUMPTION	Coking Coal	Other Bit. Coal	Sub-Bit. Coal	Lignite	Peat	Oven and Gas Coke	Pat. Fuel and BKB	Crude Oil	NGL	Feed-stocks	Additives
APPROVISIONNEMENT ET DEMANDE	*Charbon à coke*	*Autres charb. bit.*	*Charbon sous-bit.*	*Lignite*	*Tourbe*	*Coke de four/gaz*	*Agg./briq. de lignite*	*Pétrole brut*	*LGN*	*Produits d'aliment.*	*Additifs*
Production	-	-	-	-	391	-	22	-	-	-	-
From Other Sources	-	-	-	-	-	-	-	-	-	-	-
Imports	-	218	-	-	-	12	-	-	-	-	-
Exports	-	-2	-	-	-	-	-	-	-	-	-
Intl. Marine Bunkers	-	-	-	-	-	-	-	-	-	-	-
Stock Changes	-	-20	-	-	-46	-	-	-	-	-	-
DOMESTIC SUPPLY	**-**	**196**	**-**	**-**	**345**	**12**	**22**	**-**	**-**	**-**	**-**
Transfers	-	-	-	-	-	-	-	-	-	-	-
Statistical Differences	-	-	-	-	-	-	-	-	-	-	-
TRANSFORMATION	**-**	**40**	**-**	**-**	**307**	**-**	**7**	**-**	**-**	**-**	**-**
Electricity Plants	-	-	-	-	-	-	-	-	-	-	-
CHP Plants	-	-	-	-	228	-	-	-	-	-	-
Heat Plants	-	40	-	-	43	-	7	-	-	-	-
Blast Furnaces/Gas Works	-	-	-	-	-	-	-	-	-	-	-
Coke/Pat. Fuel/BKB Plants	-	-	-	-	36	-	-	-	-	-	-
Petroleum Refineries	-	-	-	-	-	-	-	-	-	-	-
Petrochemical Industry	-	-	-	-	-	-	-	-	-	-	-
Liquefaction	-	-	-	-	-	-	-	-	-	-	-
Other Transform. Sector	-	-	-	-	-	-	-	-	-	-	-
ENERGY SECTOR	**-**	**-**	**-**	**-**	**32**	**-**	**-**	**-**	**-**	**-**	**-**
Coal Mines	-	-	-	-	-	-	-	-	-	-	-
Oil and Gas Extraction	-	-	-	-	-	-	-	-	-	-	-
Petroleum Refineries	-	-	-	-	-	-	-	-	-	-	-
Electr., CHP+Heat Plants	-	-	-	-	-	-	-	-	-	-	-
Pumped Storage (Elec.)	-	-	-	-	-	-	-	-	-	-	-
Other Energy Sector	-	-	-	-	32	-	-	-	-	-	-
Distribution Losses	-	-	-	-	-	-	-	-	-	-	-
FINAL CONSUMPTION	**-**	**156**	**-**	**-**	**6**	**12**	**15**	**-**	**-**	**-**	**-**
INDUSTRY SECTOR	**-**	**12**	**-**	**-**	**1**	**2**	**1**	**-**	**-**	**-**	**-**
Iron and Steel	-	-	-	-	-	-	-	-	-	-	-
Chemical and Petrochem.	-	-	-	-	-	-	-	-	-	-	-
of which: Feedstocks	-	-	-	-	-	-	-	-	-	-	-
Non-Ferrous Metals	-	-	-	-	-	-	-	-	-	-	-
Non-Metallic Minerals	-	3	-	-	1	-	-	-	-	-	-
Transport Equipment	-	-	-	-	-	-	-	-	-	-	-
Machinery	-	-	-	-	-	-	-	-	-	-	-
Mining and Quarrying	-	-	-	-	-	-	-	-	-	-	-
Food and Tobacco	-	5	-	-	-	2	1	-	-	-	-
Paper, Pulp and Print	-	2	-	-	-	-	-	-	-	-	-
Wood and Wood Products	-	-	-	-	-	-	-	-	-	-	-
Construction	-	1	-	-	-	-	-	-	-	-	-
Textile and Leather	-	1	-	-	-	-	-	-	-	-	-
Non-specified	-	-	-	-	-	-	-	-	-	-	-
TRANSPORT SECTOR	**-**	**-**	**-**	**-**	**-**	**-**	**-**	**-**	**-**	**-**	**-**
Air	-	-	-	-	-	-	-	-	-	-	-
Road	-	-	-	-	-	-	-	-	-	-	-
Rail	-	-	-	-	-	-	-	-	-	-	-
Pipeline Transport	-	-	-	-	-	-	-	-	-	-	-
Internal Navigation	-	-	-	-	-	-	-	-	-	-	-
Non-specified	-	-	-	-	-	-	-	-	-	-	-
OTHER SECTORS	**-**	**144**	**-**	**-**	**5**	**-**	**14**	**-**	**-**	**-**	**-**
Agriculture	-	5	-	-	-	-	1	-	-	-	-
Comm. and Publ. Services	-	79	-	-	1	-	4	-	-	-	-
Residential	-	60	-	-	4	-	9	-	-	-	-
Non-specified	-	-	-	-	-	-	-	-	-	-	-
NON-ENERGY USE	**-**	**-**	**-**	**-**	**-**	**10**	**-**	**-**	**-**	**-**	**-**
in Industry/Trans./Energy	-	-	-	-	-	10	-	-	-	-	-
in Transport	-	-	-	-	-	-	-	-	-	-	-
in Other Sectors	-	-	-	-	-	-	-	-	-	-	-

Latvia / Lettonie : 1997

SUPPLY AND CONSUMPTION / *APPROVISIONNEMENT ET DEMANDE*	Oil cont. / *Pétrole cont.* (1000 tonnes)										
	Refinery Gas / *Gaz de raffinerie*	LPG + Ethane / *GPL + éthane*	Motor Gasoline / *Essence moteur*	Aviation Gasoline / *Essence aviation*	Jet Fuel / *Carbu-réacteurs*	Kerosene / *Kérosène*	Gas/ Diesel / *Gazole*	Heavy Fuel Oil / *Fioul lourd*	Naphtha / *Naphta*	Petrol. Coke / *Coke de pétrole*	Other Prod. / *Autres prod.*
Production	-	-	-	-	-	-	-	-	-	-	-
From Other Sources	-	-	-	-	-	-	-	-	-	-	-
Imports	-	39	365	-	41	4	190	878	-	-	44
Exports	-	-	-5	-	-	-	-	-	-	-	-
Intl. Marine Bunkers	-	-	-	-	-	-	-	-	-	-	-
Stock Changes	-	-1	14	-	-3	-	214	-46	-	-	-3
DOMESTIC SUPPLY	**-**	**38**	**374**	**-**	**38**	**4**	**404**	**832**	**-**	**-**	**41**
Transfers	-	-	-	-	-	-	-	-	-	-	-
Statistical Differences	-	-	-	-	-	-	-	-	-	-	-
TRANSFORMATION	**-**	**-**	**-**	**-**	**-**	**1**	**7**	**534**	**-**	**-**	**1**
Electricity Plants	-	-	-	-	-	-	-	-	-	-	-
CHP Plants	-	-	-	-	-	-	-	152	-	-	-
Heat Plants	-	-	-	-	-	1	7	382	-	-	1
Blast Furnaces/Gas Works	-	-	-	-	-	-	-	-	-	-	-
Coke/Pat. Fuel/BKB Plants	-	-	-	-	-	-	-	-	-	-	-
Petroleum Refineries	-	-	-	-	-	-	-	-	-	-	-
Petrochemical Industry	-	-	-	-	-	-	-	-	-	-	-
Liquefaction	-	-	-	-	-	-	-	-	-	-	-
Other Transform. Sector	-	-	-	-	-	-	-	-	-	-	-
ENERGY SECTOR	**-**	**1**	**-**	**-**	**-**	**-**	**3**	**4**	**-**	**-**	**-**
Coal Mines	-	-	-	-	-	-	-	-	-	-	-
Oil and Gas Extraction	-	-	-	-	-	-	-	-	-	-	-
Petroleum Refineries	-	-	-	-	-	-	-	-	-	-	-
Electr., CHP+Heat Plants	-	-	-	-	-	-	1	-	-	-	-
Pumped Storage (Elec.)	-	-	-	-	-	-	-	-	-	-	-
Other Energy Sector	-	1	-	-	-	-	2	4	-	-	-
Distribution Losses	-	1	-	-	-	-	-	-	-	-	-
FINAL CONSUMPTION	**-**	**36**	**374**	**-**	**38**	**3**	**394**	**294**	**-**	**-**	**40**
INDUSTRY SECTOR	**-**	**1**	**2**	**-**	**2**	**2**	**41**	**216**	**-**	**-**	**19**
Iron and Steel	-	-	-	-	-	-	-	8	-	-	19
Chemical and Petrochem.	-	-	-	-	-	-	-	47	-	-	-
of which: Feedstocks	-	-	-	-	-	-	-	-	-	-	-
Non-Ferrous Metals	-	-	-	-	-	-	-	-	-	-	-
Non-Metallic Minerals	-	-	-	-	-	1	1	54	-	-	-
Transport Equipment	-	-	-	-	-	-	2	3	-	-	-
Machinery	-	-	-	-	-	-	1	5	-	-	-
Mining and Quarrying	-	-	-	-	-	-	-	-	-	-	-
Food and Tobacco	-	1	-	-	2	1	19	61	-	-	-
Paper, Pulp and Print	-	-	-	-	-	-	-	-	-	-	-
Wood and Wood Products	-	-	1	-	-	-	3	21	-	-	-
Construction	-	-	1	-	-	-	11	3	-	-	-
Textile and Leather	-	-	-	-	-	-	4	13	-	-	-
Non-specified	-	-	-	-	-	-	-	1	-	-	-
TRANSPORT SECTOR	**-**	**2**	**368**	**-**	**32**	**-**	**285**	**52**	**-**	**-**	**21**
Air	-	-	-	-	32	-	-	-	-	-	-
Road	-	2	368	-	-	-	185	-	-	-	21
Rail	-	-	-	-	-	-	80	-	-	-	-
Pipeline Transport	-	-	-	-	-	-	-	-	-	-	-
Internal Navigation	-	-	-	-	-	-	20	52	-	-	-
Non-specified	-	-	-	-	-	-	-	-	-	-	-
OTHER SECTORS	**-**	**33**	**4**	**-**	**4**	**1**	**68**	**26**	**-**	**-**	**-**
Agriculture	-	-	2	-	1	-	55	6	-	-	-
Comm. and Publ. Services	-	4	2	-	1	1	13	20	-	-	-
Residential	-	29	-	-	2	-	-	-	-	-	-
Non-specified	-	-	-	-	-	-	-	-	-	-	-
NON-ENERGY USE	**-**	**-**	**-**	**-**	**-**	**-**	**-**	**-**	**-**	**-**	**-**
in Industry/Transf./Energy	-	-	-	-	-	-	-	-	-	-	-
in Transport	-	-	-	-	-	-	-	-	-	-	-
in Other Sectors	-	-	-	-	-	-	-	-	-	-	-

Latvia / Lettonie : 1997

	Gas / *Gaz* (TJ)				Comb. Renew. & Waste / *En. Re. Comb. & Déchets* (TJ)				(GWh)	(TJ)
SUPPLY AND CONSUMPTION	Natural Gas	Gas Works	Coke Ovens	Blast Furnaces	Solid Biomass	Gas/Liquids from Biomass	Municipal Waste	Industrial Waste	Electricity	Heat
APPROVISIONNEMENT ET DEMANDE	*Gaz naturel*	*Usines à gaz*	*Cokeries*	*Hauts fourneaux*	*Biomasse solide*	*Gaz/Liquides tirés de biomasse*	*Déchets urbains*	*Déchets industriels*	*Electricité*	*Chaleur*
Production	-	-	-	-	54585	-	-	-	4501	46540
From Other Sources	-	-	-	-	-	-	-	-	-	-
Imports	49249	-	-	-	-	-	-	-	1824	-
Exports	-	-	-	-	-9869	-	-	-	-1	-
Intl. Marine Bunkers	-	-	-	-	-	-	-	-	-	-
Stock Changes	299	-	-	-	556	-	-	-	-	-
DOMESTIC SUPPLY	**49548**	**-**	**-**	**-**	**45272**	**-**	**-**	**-**	**6324**	**46540**
Transfers	-	-	-	-	-	-	-	-	-	-
Statistical Differences	-	-	-	-	-	-	-	-	-176	-
TRANSFORMATION	**31949**	**-**	**-**	**-**	**9936**	**-**	**-**	**-**	**6**	**-**
Electricity Plants	-	-	-	-	-	-	-	-	-	-
CHP Plants	22932	-	-	-	-	-	-	-	-	-
Heat Plants	9017	-	-	-	9936	-	-	-	-	-
Blast Furnaces/Gas Works	-	-	-	-	-	-	-	-	-	-
Coke/Pat. Fuel/BKB Plants	-	-	-	-	-	-	-	-	-	-
Petroleum Refineries	-	-	-	-	-	-	-	-	-	-
Petrochemical Industry	-	-	-	-	-	-	-	-	-	-
Liquefaction	-	-	-	-	-	-	-	-	-	-
Other Transform. Sector	-	-	-	-	-	-	-	-	6	-
ENERGY SECTOR	**638**	**-**	**-**	**-**	**26**	**-**	**-**	**-**	**478**	**1565**
Coal Mines	-	-	-	-	-	-	-	-	-	-
Oil and Gas Extraction	-	-	-	-	-	-	-	-	-	-
Petroleum Refineries	-	-	-	-	-	-	-	-	-	-
Electr., CHP+Heat Plants	-	-	-	-	26	-	-	-	174	1197
Pumped Storage (Elec.)	-	-	-	-	-	-	-	-	-	-
Other Energy Sector	638	-	-	-	-	-	-	-	304	368
Distribution Losses	1160	-	-	-	-	-	-	-	1324	6752
FINAL CONSUMPTION	**15801**	**-**	**-**	**-**	**35310**	**-**	**-**	**-**	**4340**	**38223**
INDUSTRY SECTOR	**9962**	**-**	**-**	**-**	**4167**	**-**	**-**	**-**	**1484**	**11384**
Iron and Steel	4396	-	-	-	-	-	-	-	102	725
Chemical and Petrochem.	212	-	-	-	110	-	-	-	202	1111
of which: Feedstocks	-	-	-	-	-	-	-	-	-	-
Non-Ferrous Metals	-	-	-	-	-	-	-	-	-	-
Non-Metallic Minerals	711	-	-	-	20	-	-	-	98	485
Transport Equipment	124	-	-	-	2	-	-	-	74	393
Machinery	82	-	-	-	197	-	-	-	173	481
Mining and Quarrying	3	-	-	-	5	-	-	-	6	10
Food and Tobacco	3127	-	-	-	610	-	-	-	325	4273
Paper, Pulp and Print	117	-	-	-	22	-	-	-	24	211
Wood and Wood Products	569	-	-	-	2492	-	-	-	179	1353
Construction	22	-	-	-	355	-	-	-	21	996
Textile and Leather	519	-	-	-	34	-	-	-	215	986
Non-specified	80	-	-	-	320	-	-	-	65	360
TRANSPORT SECTOR	**32**	**-**	**-**	**-**	**-**	**-**	**-**	**-**	**176**	**-**
Air	-	-	-	-	-	-	-	-	-	-
Road	32	-	-	-	-	-	-	-	-	-
Rail	-	-	-	-	-	-	-	-	54	-
Pipeline Transport	-	-	-	-	-	-	-	-	37	-
Internal Navigation	-	-	-	-	-	-	-	-	-	-
Non-specified	-	-	-	-	-	-	-	-	85	-
OTHER SECTORS	**5807**	**-**	**-**	**-**	**31143**	**-**	**-**	**-**	**2680**	**26839**
Agriculture	647	-	-	-	529	-	-	-	187	15
Comm. and Publ. Services	1722	-	-	-	3700	-	-	-	1411	2263
Residential	3438	-	-	-	26914	-	-	-	1082	24561
Non-specified	-	-	-	-	-	-	-	-	-	-
NON-ENERGY USE	**-**	**-**	**-**	**-**	**-**	**-**	**-**	**-**	**-**	**-**
in Industry/Transf./Energy	-	-	-	-	-	-	-	-	-	-
in Transport	-	-	-	-	-	-	-	-	-	-
in Other Sectors	-	-	-	-	-	-	-	-	-	-

Lebanon / Liban

SUPPLY AND CONSUMPTION 1996	Coal (1000 tonnes)							Oil (1000 tonnes)			
	Coking Coal	Other Bit. Coal	Sub-Bit. Coal	Lignite	Peat	Oven and Gas Coke	Pat. Fuel and BKB	Crude Oil	NGL	Feed-stocks	Additives
Production	-	-	-	-	-	-	-	-	-	-	-
Imports	-	200	-	-	-	-	-	-	-	-	-
Exports	-	-	-	-	-	-	-	-	-	-	-
Intl. Marine Bunkers	-	-	-	-	-	-	-	-	-	-	-
Stock Changes	-	-	-	-	-	-	-	-	-	-	-
DOMESTIC SUPPLY	**-**	**200**	**-**	**-**	**-**	**-**	**-**	**-**	**-**	**-**	**-**
Transfers and Stat. Diff.	-	-	-	-	-	-	-	-	-	-	-
TRANSFORMATION	**-**	**-**	**-**	**-**	**-**	**-**	**-**	**-**	**-**	**-**	**-**
Electricity and CHP Plants	-	-	-	-	-	-	-	-	-	-	-
Petroleum Refineries	-	-	-	-	-	-	-	-	-	-	-
Other Transform. Sector	-	-	-	-	-	-	-	-	-	-	-
ENERGY SECTOR	**-**	**-**	**-**	**-**	**-**	**-**	**-**	**-**	**-**	**-**	**-**
DISTRIBUTION LOSSES	**-**	**-**	**-**	**-**	**-**	**-**	**-**	**-**	**-**	**-**	**-**
FINAL CONSUMPTION	**-**	**200**	**-**	**-**	**-**	**-**	**-**	**-**	**-**	**-**	**-**
INDUSTRY SECTOR	**-**	**200**	**-**	**-**	**-**	**-**	**-**	**-**	**-**	**-**	**-**
Iron and Steel	-	-	-	-	-	-	-	-	-	-	-
Chemical and Petrochem.	-	-	-	-	-	-	-	-	-	-	-
Non-Metallic Minerals	-	200	-	-	-	-	-	-	-	-	-
Non-specified	-	-	-	-	-	-	-	-	-	-	-
TRANSPORT SECTOR	**-**	**-**	**-**	**-**	**-**	**-**	**-**	**-**	**-**	**-**	**-**
Air	-	-	-	-	-	-	-	-	-	-	-
Road	-	-	-	-	-	-	-	-	-	-	-
Non-specified	-	-	-	-	-	-	-	-	-	-	-
OTHER SECTORS	**-**	**-**	**-**	**-**	**-**	**-**	**-**	**-**	**-**	**-**	**-**
Agriculture	-	-	-	-	-	-	-	-	-	-	-
Comm. and Publ. Services	-	-	-	-	-	-	-	-	-	-	-
Residential	-	-	-	-	-	-	-	-	-	-	-
Non-specified	-	-	-	-	-	-	-	-	-	-	-
NON-ENERGY USE	**-**	**-**	**-**	**-**	**-**	**-**	**-**	**-**	**-**	**-**	**-**

APPROVISIONNEMENT ET DEMANDE 1997	*Charbon* (1000 tonnes)							*Pétrole* (1000 tonnes)			
	Charbon à coke	*Autres charb. bit.*	*Charbon sous-bit.*	*Lignite*	*Tourbe*	*Coke de four/gaz*	*Agg./briq. de lignite*	*Pétrole brut*	*LGN*	*Produits d'aliment.*	*Additifs*
Production	-	-	-	-	-	-	-	-	-	-	-
Imports	-	200	-	-	-	-	-	-	-	-	-
Exports	-	-	-	-	-	-	-	-	-	-	-
Intl. Marine Bunkers	-	-	-	-	-	-	-	-	-	-	-
Stock Changes	-	-	-	-	-	-	-	-	-	-	-
DOMESTIC SUPPLY	**-**	**200**	**-**	**-**	**-**	**-**	**-**	**-**	**-**	**-**	**-**
Transfers and Stat. Diff.	-	-	-	-	-	-	-	-	-	-	-
TRANSFORMATION	**-**	**-**	**-**	**-**	**-**	**-**	**-**	**-**	**-**	**-**	**-**
Electricity and CHP Plants	-	-	-	-	-	-	-	-	-	-	-
Petroleum Refineries	-	-	-	-	-	-	-	-	-	-	-
Other Transform. Sector	-	-	-	-	-	-	-	-	-	-	-
ENERGY SECTOR	**-**	**-**	**-**	**-**	**-**	**-**	**-**	**-**	**-**	**-**	**-**
DISTRIBUTION LOSSES	**-**	**-**	**-**	**-**	**-**	**-**	**-**	**-**	**-**	**-**	**-**
FINAL CONSUMPTION	**-**	**200**	**-**	**-**	**-**	**-**	**-**	**-**	**-**	**-**	**-**
INDUSTRY SECTOR	**-**	**200**	**-**	**-**	**-**	**-**	**-**	**-**	**-**	**-**	**-**
Iron and Steel	-	-	-	-	-	-	-	-	-	-	-
Chemical and Petrochem.	-	-	-	-	-	-	-	-	-	-	-
Non-Metallic Minerals	-	200	-	-	-	-	-	-	-	-	-
Non-specified	-	-	-	-	-	-	-	-	-	-	-
TRANSPORT SECTOR	**-**	**-**	**-**	**-**	**-**	**-**	**-**	**-**	**-**	**-**	**-**
Air	-	-	-	-	-	-	-	-	-	-	-
Road	-	-	-	-	-	-	-	-	-	-	-
Non-specified	-	-	-	-	-	-	-	-	-	-	-
OTHER SECTORS	**-**	**-**	**-**	**-**	**-**	**-**	**-**	**-**	**-**	**-**	**-**
Agriculture	-	-	-	-	-	-	-	-	-	-	-
Comm. and Publ. Services	-	-	-	-	-	-	-	-	-	-	-
Residential	-	-	-	-	-	-	-	-	-	-	-
Non-specified	-	-	-	-	-	-	-	-	-	-	-
NON-ENERGY USE	**-**	**-**	**-**	**-**	**-**	**-**	**-**	**-**	**-**	**-**	**-**

Lebanon / Liban

SUPPLY AND CONSUMPTION 1996	Oil cont. (1000 tonnes)										
	Refinery Gas	LPG + Ethane	Motor Gasoline	Aviation Gasoline	Jet Fuel	Kerosene	Gas/ Diesel	Heavy Fuel Oil	Naphtha	Petrol. Coke	Other Prod.
Production	-	-	-	-	-	-	-	-	-	-	-
Imports	-	124	1379	-	107	4	930	1623	-	-	109
Exports	-	-	-	-	-	-	-	-	-	-	-
Intl. Marine Bunkers	-	-	-	-	-	-	-	-	-	-	-
Stock Changes	-	-	-	-	-	-	-	-	-	-	-
DOMESTIC SUPPLY	**-**	**124**	**1379**	**-**	**107**	**4**	**930**	**1623**	**-**	**-**	**109**
Transfers and Stat. Diff.	-	-	-	-	-	-	-	-	-	-	-
TRANSFORMATION	**-**	**-**	**-**	**-**	**-**	**-**	**195**	**1342**	**-**	**-**	**-**
Electricity and CHP Plants	-	-	-	-	-	-	195	1342	-	-	-
Petroleum Refineries	-	-	-	-	-	-	-	-	-	-	-
Other Transform. Sector	-	-	-	-	-	-	-	-	-	-	-
ENERGY SECTOR	**-**	**-**	**-**	**-**	**-**	**-**	**-**	**-**	**-**	**-**	**-**
DISTRIBUTION LOSSES	**-**	**-**	**-**	**-**	**-**	**-**	**-**	**-**	**-**	**-**	**-**
FINAL CONSUMPTION	**-**	**124**	**1379**	**-**	**107**	**4**	**735**	**281**	**-**	**-**	**109**
INDUSTRY SECTOR	**-**	**-**	**-**	**-**	**-**	**-**	**346**	**281**	**-**	**-**	**-**
Iron and Steel	-	-	-	-	-	-	-	-	-	-	-
Chemical and Petrochem.	-	-	-	-	-	-	-	-	-	-	-
Non-Metallic Minerals	-	-	-	-	-	-	-	-	-	-	-
Non-specified	-	-	-	-	-	-	346	281	-	-	-
TRANSPORT SECTOR	**-**	**-**	**1379**	**-**	**107**	**-**	**15**	**-**	**-**	**-**	**-**
Air	-	-	-	-	107	-	-	-	-	-	-
Road	-	-	1379	-	-	-	15	-	-	-	-
Non-specified	-	-	-	-	-	-	-	-	-	-	-
OTHER SECTORS	**-**	**124**	**-**	**-**	**-**	**4**	**374**	**-**	**-**	**-**	**-**
Agriculture	-	-	-	-	-	-	-	-	-	-	-
Comm. and Publ. Services	-	-	-	-	-	-	-	-	-	-	-
Residential	-	124	-	-	-	4	374	-	-	-	-
Non-specified	-	-	-	-	-	-	-	-	-	-	-
NON-ENERGY USE	**-**	**-**	**-**	**-**	**-**	**-**	**-**	**-**	**-**	**-**	**109**

APPROVISIONNEMENT ET DEMANDE 1997	*Pétrole cont.* (1000 tonnes)										
	Gaz de raffinerie	*GPL + éthane*	*Essence moteur*	*Essence aviation*	*Carbu- réacteurs*	*Kérosène*	*Gazole*	*Fioul lourd*	*Naphta*	*Coke de pétrole*	*Autres prod.*
Production	-	-	-	-	-	-	-	-	-	-	-
Imports	-	141	1310	-	109	4	1376	1731	-	-	88
Exports	-	-	-	-	-	-	-	-	-	-	-
Intl. Marine Bunkers	-	-	-	-	-	-	-	-	-	-	-
Stock Changes	-	-	-	-	-	-	-	-	-	-	-
DOMESTIC SUPPLY	**-**	**141**	**1310**	**-**	**109**	**4**	**1376**	**1731**	**-**	**-**	**88**
Transfers and Stat. Diff.	-	-	-	-	-	-	-	-	-	-	-
TRANSFORMATION	**-**	**-**	**-**	**-**	**-**	**-**	**338**	**1393**	**-**	**-**	**-**
Electricity and CHP Plants	-	-	-	-	-	-	338	1393	-	-	-
Petroleum Refineries	-	-	-	-	-	-	-	-	-	-	-
Other Transform. Sector	-	-	-	-	-	-	-	-	-	-	-
ENERGY SECTOR	**-**	**-**	**-**	**-**	**-**	**-**	**-**	**-**	**-**	**-**	**-**
DISTRIBUTION LOSSES	**-**	**-**	**-**	**-**	**-**	**-**	**-**	**-**	**-**	**-**	**-**
FINAL CONSUMPTION	**-**	**141**	**1310**	**-**	**109**	**4**	**1038**	**338**	**-**	**-**	**88**
INDUSTRY SECTOR	**-**	**-**	**-**	**-**	**-**	**-**	**489**	**338**	**-**	**-**	**-**
Iron and Steel	-	-	-	-	-	-	-	-	-	-	-
Chemical and Petrochem.	-	-	-	-	-	-	-	-	-	-	-
Non-Metallic Minerals	-	-	-	-	-	-	-	-	-	-	-
Non-specified	-	-	-	-	-	-	489	338	-	-	-
TRANSPORT SECTOR	**-**	**-**	**1310**	**-**	**109**	**-**	**21**	**-**	**-**	**-**	**-**
Air	-	-	-	-	109	-	-	-	-	-	-
Road	-	-	1310	-	-	-	21	-	-	-	-
Non-specified	-	-	-	-	-	-	-	-	-	-	-
OTHER SECTORS	**-**	**141**	**-**	**-**	**-**	**4**	**528**	**-**	**-**	**-**	**-**
Agriculture	-	-	-	-	-	-	-	-	-	-	-
Comm. and Publ. Services	-	-	-	-	-	-	-	-	-	-	-
Residential	-	141	-	-	-	4	528	-	-	-	-
Non-specified	-	-	-	-	-	-	-	-	-	-	-
NON-ENERGY USE	**-**	**-**	**-**	**-**	**-**	**-**	**-**	**-**	**-**	**-**	**88**

Lebanon / Liban

SUPPLY AND	Gas (TJ)				Comb. Renew. & Waste (TJ)				(GWh)	(TJ)
CONSUMPTION 1996	Natural Gas	Gas Works	Coke Ovens	Blast Furnaces	Solid Biomass	Gas/Liquids from Biomass	Municipal Waste	Industrial Waste	Electricity	Heat
Production	-	-	-	-	5060	-	-	-	6965	-
Imports	-	-	-	-	61	-	-	-	683	-
Exports	-	-	-	-	-	-	-	-	-	-
Intl. Marine Bunkers	-	-	-	-	-	-	-	-	-	-
Stock Changes	-	-	-	-	-	-	-	-	-	-
DOMESTIC SUPPLY	**-**	**-**	**-**	**-**	**5121**	**-**	**-**	**-**	**7648**	**-**
Transfers and Stat. Diff.	-	-	-	-	-	-	-	-	-	-
TRANSFORMATION	**-**	**-**	**-**	**-**	**653**	**-**	**-**	**-**	**-**	**-**
Electricity and CHP Plants	-	-	-	-	-	-	-	-	-	-
Petroleum Refineries	-	-	-	-	-	-	-	-	-	-
Other Transform. Sector	-	-	-	-	653	-	-	-	-	-
ENERGY SECTOR	**-**	**-**	**-**	**-**	**-**	**-**	**-**	**-**	**-**	**-**
DISTRIBUTION LOSSES	**-**	**-**	**-**	**-**	**-**	**-**	**-**	**-**	**918**	**-**
FINAL CONSUMPTION	**-**	**-**	**-**	**-**	**4468**	**-**	**-**	**-**	**6730**	**-**
INDUSTRY SECTOR	**-**	**-**	**-**	**-**	**-**	**-**	**-**	**-**	**1759**	**-**
Iron and Steel	-	-	-	-	-	-	-	-	-	-
Chemical and Petrochem.	-	-	-	-	-	-	-	-	-	-
Non-Metallic Minerals	-	-	-	-	-	-	-	-	-	-
Non-specified	-	-	-	-	-	-	-	-	1759	-
TRANSPORT SECTOR	**-**	**-**	**-**	**-**	**-**	**-**	**-**	**-**	**-**	**-**
Air	-	-	-	-	-	-	-	-	-	-
Road	-	-	-	-	-	-	-	-	-	-
Non-specified	-	-	-	-	-	-	-	-	-	-
OTHER SECTORS	**-**	**-**	**-**	**-**	**4468**	**-**	**-**	**-**	**4971**	**-**
Agriculture	-	-	-	-	-	-	-	-	-	-
Comm. and Publ. Services	-	-	-	-	-	-	-	-	918	-
Residential	-	-	-	-	-	-	-	-	3671	-
Non-specified	-	-	-	-	4468	-	-	-	382	-
NON-ENERGY USE	**-**	**-**	**-**	**-**	**-**	**-**	**-**	**-**	**-**	**-**

APPROVISIONNEMENT	*Gaz* (TJ)				*En. Re. Comb. & Déchets* (TJ)				(GWh)	(TJ)
ET DEMANDE 1997	*Gaz naturel*	*Usines à gaz*	*Cokeries*	*Hauts fourneaux*	*Biomasse solide*	*Gaz/Liquides tirés de biomasse*	*Déchets urbains*	*Déchets industriels*	*Electricité*	*Chaleur*
Production	-	-	-	-	5152	-	-	-	8515	-
Imports	-	-	-	-	64	-	-	-	608	-
Exports	-	-	-	-	-	-	-	-	-	-
Intl. Marine Bunkers	-	-	-	-	-	-	-	-	-	-
Stock Changes	-	-	-	-	-	-	-	-	-	-
DOMESTIC SUPPLY	**-**	**-**	**-**	**-**	**5216**	**-**	**-**	**-**	**9123**	**-**
Transfers and Stat. Diff.	-	-	-	-	-	-	-	-	-	-
TRANSFORMATION	**-**	**-**	**-**	**-**	**647**	**-**	**-**	**-**	**-**	**-**
Electricity and CHP Plants	-	-	-	-	-	-	-	-	-	-
Petroleum Refineries	-	-	-	-	-	-	-	-	-	-
Other Transform. Sector	-	-	-	-	647	-	-	-	-	-
ENERGY SECTOR	**-**	**-**	**-**	**-**	**-**	**-**	**-**	**-**	**-**	**-**
DISTRIBUTION LOSSES	**-**	**-**	**-**	**-**	**-**	**-**	**-**	**-**	**1123**	**-**
FINAL CONSUMPTION	**-**	**-**	**-**	**-**	**4568**	**-**	**-**	**-**	**8000**	**-**
INDUSTRY SECTOR	**-**	**-**	**-**	**-**	**-**	**-**	**-**	**-**	**2100**	**-**
Iron and Steel	-	-	-	-	-	-	-	-	-	-
Chemical and Petrochem.	-	-	-	-	-	-	-	-	-	-
Non-Metallic Minerals	-	-	-	-	-	-	-	-	-	-
Non-specified	-	-	-	-	-	-	-	-	2100	-
TRANSPORT SECTOR	**-**	**-**	**-**	**-**	**-**	**-**	**-**	**-**	**-**	**-**
Air	-	-	-	-	-	-	-	-	-	-
Road	-	-	-	-	-	-	-	-	-	-
Non-specified	-	-	-	-	-	-	-	-	-	-
OTHER SECTORS	**-**	**-**	**-**	**-**	**4568**	**-**	**-**	**-**	**5900**	**-**
Agriculture	-	-	-	-	-	-	-	-	-	-
Comm. and Publ. Services	-	-	-	-	-	-	-	-	1340	-
Residential	-	-	-	-	-	-	-	-	3050	-
Non-specified	-	-	-	-	4568	-	-	-	1510	-
NON-ENERGY USE	**-**	**-**	**-**	**-**	**-**	**-**	**-**	**-**	**-**	**-**

Libya / Libye

SUPPLY AND CONSUMPTION 1996	Coal (1000 tonnes)							Oil (1000 tonnes)			
	Coking Coal	Other Bit. Coal	Sub-Bit. Coal	Lignite	Peat	Oven and Gas Coke	Pat. Fuel and BKB	Crude Oil	NGL	Feed-stocks	Additives
Production	-	-	-	-	-	-	-	67384	2721	-	-
Imports	-	-	-	-	-	-	-	-	-	-	-
Exports	-	-	-	-	-	-	-	-53909	-	-	-
Intl. Marine Bunkers	-	-	-	-	-	-	-	-	-	-	-
Stock Changes	-	-	-	-	-	-	-	-	-	-	-
DOMESTIC SUPPLY	-	-	-	-	-	-	-	**13475**	**2721**	-	-
Transfers and Stat. Diff.	-	-	-	-	-	-	-	3365	-2721	-	-
TRANSFORMATION	-	-	-	-	-	-	-	**16840**	-	-	-
Electricity and CHP Plants	-	-	-	-	-	-	-	-	-	-	-
Petroleum Refineries	-	-	-	-	-	-	-	16840	-	-	-
Other Transform. Sector	-	-	-	-	-	-	-	-	-	-	-
ENERGY SECTOR	-	-	-	-	-	-	-	-	-	-	-
DISTRIBUTION LOSSES	-	-	-	-	-	-	-	-	-	-	-
FINAL CONSUMPTION	-	-	-	-	-	-	-	-	-	-	-
INDUSTRY SECTOR	-	-	-	-	-	-	-	-	-	-	-
Iron and Steel	-	-	-	-	-	-	-	-	-	-	-
Chemical and Petrochem.	-	-	-	-	-	-	-	-	-	-	-
Non-Metallic Minerals	-	-	-	-	-	-	-	-	-	-	-
Non-specified	-	-	-	-	-	-	-	-	-	-	-
TRANSPORT SECTOR	-	-	-	-	-	-	-	-	-	-	-
Air	-	-	-	-	-	-	-	-	-	-	-
Road	-	-	-	-	-	-	-	-	-	-	-
Non-specified	-	-	-	-	-	-	-	-	-	-	-
OTHER SECTORS	-	-	-	-	-	-	-	-	-	-	-
Agriculture	-	-	-	-	-	-	-	-	-	-	-
Comm. and Publ. Services	-	-	-	-	-	-	-	-	-	-	-
Residential	-	-	-	-	-	-	-	-	-	-	-
Non-specified	-	-	-	-	-	-	-	-	-	-	-
NON-ENERGY USE	-	-	-	-	-	-	-	-	-	-	-

APPROVISIONNEMENT ET DEMANDE 1997	*Charbon* (1000 tonnes)							*Pétrole* (1000 tonnes)			
	Charbon à coke	*Autres charb. bit.*	*Charbon sous-bit.*	*Lignite*	*Tourbe*	*Coke de four/gaz*	*Agg./briq. de lignite*	*Pétrole brut*	*LGN*	*Produits d'aliment.*	*Additifs*
Production	-	-	-	-	-	-	-	68425	2721	-	-
Imports	-	-	-	-	-	-	-	-	-	-	-
Exports	-	-	-	-	-	-	-	-54691	-	-	-
Intl. Marine Bunkers	-	-	-	-	-	-	-	-	-	-	-
Stock Changes	-	-	-	-	-	-	-	-	-	-	-
DOMESTIC SUPPLY	-	-	-	-	-	-	-	**13734**	**2721**	-	-
Transfers and Stat. Diff.	-	-	-	-	-	-	-	3398	-2721	-	-
TRANSFORMATION	-	-	-	-	-	-	-	**17132**	-	-	-
Electricity and CHP Plants	-	-	-	-	-	-	-	-	-	-	-
Petroleum Refineries	-	-	-	-	-	-	-	17132	-	-	-
Other Transform. Sector	-	-	-	-	-	-	-	-	-	-	-
ENERGY SECTOR	-	-	-	-	-	-	-	-	-	-	-
DISTRIBUTION LOSSES	-	-	-	-	-	-	-	-	-	-	-
FINAL CONSUMPTION	-	-	-	-	-	-	-	-	-	-	-
INDUSTRY SECTOR	-	-	-	-	-	-	-	-	-	-	-
Iron and Steel	-	-	-	-	-	-	-	-	-	-	-
Chemical and Petrochem.	-	-	-	-	-	-	-	-	-	-	-
Non-Metallic Minerals	-	-	-	-	-	-	-	-	-	-	-
Non-specified	-	-	-	-	-	-	-	-	-	-	-
TRANSPORT SECTOR	-	-	-	-	-	-	-	-	-	-	-
Air	-	-	-	-	-	-	-	-	-	-	-
Road	-	-	-	-	-	-	-	-	-	-	-
Non-specified	-	-	-	-	-	-	-	-	-	-	-
OTHER SECTORS	-	-	-	-	-	-	-	-	-	-	-
Agriculture	-	-	-	-	-	-	-	-	-	-	-
Comm. and Publ. Services	-	-	-	-	-	-	-	-	-	-	-
Residential	-	-	-	-	-	-	-	-	-	-	-
Non-specified	-	-	-	-	-	-	-	-	-	-	-
NON-ENERGY USE	-	-	-	-	-	-	-	-	-	-	-

Libya / Libye

SUPPLY AND CONSUMPTION 1996	Oil cont. (1000 tonnes)										
	Refinery Gas	LPG + Ethane	Motor Gasoline	Aviation Gasoline	Jet Fuel	Kerosene	Gas/ Diesel	Heavy Fuel Oil	Naphtha	Petrol. Coke	Other Prod.
Production	553	238	1957	3	1423	259	4244	4799	1139	-	115
Imports	-	-	-	-	-	-	-	-	-	-	30
Exports	-	-449	-250	-	-980	-56	-1722	-1207	-1285	-	-
Intl. Marine Bunkers	-	-	-	-	-	-	-	-90	-	-	-
Stock Changes	-	-	-	-	-	-	-	-	-	-	-
DOMESTIC SUPPLY	**553**	**-211**	**1707**	**3**	**443**	**203**	**2522**	**3502**	**-146**	**-**	**145**
Transfers and Stat. Diff.	-	622	-20	-	-128	53	-190	-356	1079	-	29
TRANSFORMATION	**-**	**-**	**-**	**-**	**-**	**-**	**1000**	**2663**	**-**	**-**	**-**
Electricity and CHP Plants	-	-	-	-	-	-	1000	2663	-	-	-
Petroleum Refineries	-	-	-	-	-	-	-	-	-	-	-
Other Transform. Sector	-	-	-	-	-	-	-	-	-	-	-
ENERGY SECTOR	**553**	**-**	**-**	**-**	**-**	**-**	**-**	**173**	**-**	**-**	**-**
DISTRIBUTION LOSSES	**-**	**-**	**-**	**-**	**-**	**-**	**-**	**-**	**-**	**-**	**-**
FINAL CONSUMPTION	**-**	**411**	**1687**	**3**	**315**	**256**	**1332**	**310**	**933**	**-**	**174**
INDUSTRY SECTOR	**-**	**-**	**-**	**-**	**-**	**-**	**-**	**310**	**933**	**-**	**-**
Iron and Steel	-	-	-	-	-	-	-	-	-	-	-
Chemical and Petrochem.	-	-	-	-	-	-	-	-	933	-	-
Non-Metallic Minerals	-	-	-	-	-	-	-	-	-	-	-
Non-specified	-	-	-	-	-	-	-	310	-	-	-
TRANSPORT SECTOR	**-**	**-**	**1687**	**3**	**315**	**-**	**1332**	**-**	**-**	**-**	**-**
Air	-	-	-	-	315	-	-	-	-	-	-
Road	-	-	1687	3	-	-	1332	-	-	-	-
Non-specified	-	-	-	-	-	-	-	-	-	-	-
OTHER SECTORS	**-**	**411**	**-**	**-**	**-**	**256**	**-**	**-**	**-**	**-**	**-**
Agriculture	-	-	-	-	-	-	-	-	-	-	-
Comm. and Publ. Services	-	-	-	-	-	-	-	-	-	-	-
Residential	-	411	-	-	-	256	-	-	-	-	-
Non-specified	-	-	-	-	-	-	-	-	-	-	-
NON-ENERGY USE	**-**	**-**	**-**	**-**	**-**	**-**	**-**	**-**	**-**	**-**	**174**

APPROVISIONNEMENT ET DEMANDE 1997	*Pétrole cont.* (1000 tonnes)										
	Gaz de raffinerie	*GPL + éthane*	*Essence moteur*	*Essence aviation*	*Carbu- réacteurs*	*Kérosène*	*Gazole*	*Fioul lourd*	*Naphta*	*Coke de pétrole*	*Autres prod.*
Production	553	263	1970	3	1430	274	4272	4834	1143	-	119
Imports	-	-	-	-	-	-	-	-	-	-	30
Exports	-	-464	-258	-	-1017	-56	-1778	-1253	-1333	-	-
Intl. Marine Bunkers	-	-	-	-	-	-	-	-90	-	-	-
Stock Changes	-	-	-	-	-	-	-	-	-	-	-
DOMESTIC SUPPLY	**553**	**-201**	**1712**	**3**	**413**	**218**	**2494**	**3491**	**-190**	**-**	**149**
Transfers and Stat. Diff.	-	623	63	-	-117	53	-217	-210	1149	-	30
TRANSFORMATION	**-**	**-**	**-**	**-**	**-**	**-**	**1000**	**2663**	**-**	**-**	**-**
Electricity and CHP Plants	-	-	-	-	-	-	1000	2663	-	-	-
Petroleum Refineries	-	-	-	-	-	-	-	-	-	-	-
Other Transform. Sector	-	-	-	-	-	-	-	-	-	-	-
ENERGY SECTOR	**553**	**-**	**-**	**-**	**-**	**-**	**-**	**173**	**-**	**-**	**-**
DISTRIBUTION LOSSES	**-**	**-**	**-**	**-**	**-**	**-**	**-**	**-**	**-**	**-**	**-**
FINAL CONSUMPTION	**-**	**422**	**1775**	**3**	**296**	**271**	**1277**	**445**	**959**	**-**	**179**
INDUSTRY SECTOR	**-**	**-**	**-**	**-**	**-**	**-**	**-**	**445**	**959**	**-**	**-**
Iron and Steel	-	-	-	-	-	-	-	-	-	-	-
Chemical and Petrochem.	-	-	-	-	-	-	-	-	959	-	-
Non-Metallic Minerals	-	-	-	-	-	-	-	-	-	-	-
Non-specified	-	-	-	-	-	-	-	445	-	-	-
TRANSPORT SECTOR	**-**	**-**	**1775**	**3**	**296**	**-**	**1277**	**-**	**-**	**-**	**-**
Air	-	-	-	-	296	-	-	-	-	-	-
Road	-	-	1775	3	-	-	1277	-	-	-	-
Non-specified	-	-	-	-	-	-	-	-	-	-	-
OTHER SECTORS	**-**	**422**	**-**	**-**	**-**	**271**	**-**	**-**	**-**	**-**	**-**
Agriculture	-	-	-	-	-	-	-	-	-	-	-
Comm. and Publ. Services	-	-	-	-	-	-	-	-	-	-	-
Residential	-	422	-	-	-	271	-	-	-	-	-
Non-specified	-	-	-	-	-	-	-	-	-	-	-
NON-ENERGY USE	**-**	**-**	**-**	**-**	**-**	**-**	**-**	**-**	**-**	**-**	**179**

Libya / Libye

SUPPLY AND CONSUMPTION 1996	Gas (TJ)				Comb. Renew. & Waste (TJ)				(GWh)	(TJ)
	Natural Gas	Gas Works	Coke Ovens	Blast Furnaces	Solid Biomass	Gas/Liquids from Biomass	Municipal Waste	Industrial Waste	Electricity	Heat
Production	261471	-	-	-	5236	-	-	-	18180	-
Imports	-	-	-	-	-	-	-	-	-	-
Exports	-56264	-	-	-	-	-	-	-	-	-
Intl. Marine Bunkers	-	-	-	-	-	-	-	-	-	-
Stock Changes	-	-	-	-	-	-	-	-	-	-
DOMESTIC SUPPLY	**205207**	**-**	**-**	**-**	**5236**	**-**	**-**	**-**	**18180**	**-**
Transfers and Stat. Diff.	-3285	-	-	-	-	-	-	-	-	-
TRANSFORMATION	**-**	**-**	**-**	**-**	**-**	**-**	**-**	**-**	**-**	**-**
Electricity and CHP Plants	-	-	-	-	-	-	-	-	-	-
Petroleum Refineries	-	-	-	-	-	-	-	-	-	-
Other Transform. Sector	-	-	-	-	-	-	-	-	-	-
ENERGY SECTOR	**128396**	**-**	**-**	**-**	**-**	**-**	**-**	**-**	**-**	**-**
DISTRIBUTION LOSSES	**-**	**-**	**-**	**-**	**-**	**-**	**-**	**-**	**-**	**-**
FINAL CONSUMPTION	**73526**	**-**	**-**	**-**	**5236**	**-**	**-**	**-**	**18180**	**-**
INDUSTRY SECTOR	**73526**	**-**	**-**	**-**	**-**	**-**	**-**	**-**	**-**	**-**
Iron and Steel	-	-	-	-	-	-	-	-	-	-
Chemical and Petrochem.	73526	-	-	-	-	-	-	-	-	-
Non-Metallic Minerals	-	-	-	-	-	-	-	-	-	-
Non-specified	-	-	-	-	-	-	-	-	-	-
TRANSPORT SECTOR	**-**	**-**	**-**	**-**	**-**	**-**	**-**	**-**	**-**	**-**
Air	-	-	-	-	-	-	-	-	-	-
Road	-	-	-	-	-	-	-	-	-	-
Non-specified	-	-	-	-	-	-	-	-	-	-
OTHER SECTORS	**-**	**-**	**-**	**-**	**5236**	**-**	**-**	**-**	**18180**	**-**
Agriculture	-	-	-	-	-	-	-	-	-	-
Comm. and Publ. Services	-	-	-	-	-	-	-	-	-	-
Residential	-	-	-	-	5236	-	-	-	-	-
Non-specified	-	-	-	-	-	-	-	-	18180	-
NON-ENERGY USE	**-**	**-**	**-**	**-**	**-**	**-**	**-**	**-**	**-**	**-**

APPROVISIONNEMENT ET DEMANDE 1997	*Gaz (TJ)*				*En. Re. Comb. & Déchets (TJ)*				(GWh)	(TJ)
	Gaz naturel	*Usines à gaz*	*Cokeries*	*Hauts fourneaux*	*Biomasse solide*	*Gaz/Liquides tirés de biomasse*	*Déchets urbains*	*Déchets industriels*	*Electricité*	*Chaleur*
Production	267580	-	-	-	5236	-	-	-	18180	-
Imports	-	-	-	-	-	-	-	-	-	-
Exports	-56264	-	-	-	-	-	-	-	-	-
Intl. Marine Bunkers	-	-	-	-	-	-	-	-	-	-
Stock Changes	-	-	-	-	-	-	-	-	-	-
DOMESTIC SUPPLY	**211316**	**-**	**-**	**-**	**5236**	**-**	**-**	**-**	**18180**	**-**
Transfers and Stat. Diff.	4804	-	-	-	-	-	-	-	-	-
TRANSFORMATION	**-**	**-**	**-**	**-**	**-**	**-**	**-**	**-**	**-**	**-**
Electricity and CHP Plants	-	-	-	-	-	-	-	-	-	-
Petroleum Refineries	-	-	-	-	-	-	-	-	-	-
Other Transform. Sector	-	-	-	-	-	-	-	-	-	-
ENERGY SECTOR	**137424**	**-**	**-**	**-**	**-**	**-**	**-**	**-**	**-**	**-**
DISTRIBUTION LOSSES	**-**	**-**	**-**	**-**	**-**	**-**	**-**	**-**	**-**	**-**
FINAL CONSUMPTION	**78696**	**-**	**-**	**-**	**5236**	**-**	**-**	**-**	**18180**	**-**
INDUSTRY SECTOR	**78696**	**-**	**-**	**-**	**-**	**-**	**-**	**-**	**-**	**-**
Iron and Steel	-	-	-	-	-	-	-	-	-	-
Chemical and Petrochem.	78696	-	-	-	-	-	-	-	-	-
Non-Metallic Minerals	-	-	-	-	-	-	-	-	-	-
Non-specified	-	-	-	-	-	-	-	-	-	-
TRANSPORT SECTOR	**-**	**-**	**-**	**-**	**-**	**-**	**-**	**-**	**-**	**-**
Air	-	-	-	-	-	-	-	-	-	-
Road	-	-	-	-	-	-	-	-	-	-
Non-specified	-	-	-	-	-	-	-	-	-	-
OTHER SECTORS	**-**	**-**	**-**	**-**	**5236**	**-**	**-**	**-**	**18180**	**-**
Agriculture	-	-	-	-	-	-	-	-	-	-
Comm. and Publ. Services	-	-	-	-	-	-	-	-	-	-
Residential	-	-	-	-	5236	-	-	-	-	-
Non-specified	-	-	-	-	-	-	-	-	18180	-
NON-ENERGY USE	**-**	**-**	**-**	**-**	**-**	**-**	**-**	**-**	**-**	**-**

Lithuania / Lituanie : 1996

SUPPLY AND CONSUMPTION / *APPROVISIONNEMENT ET DEMANDE*	Coal / *Charbon* (1000 tonnes)							Oil / *Pétrole* (1000 tonnes)			
	Coking Coal / *Charbon à coke*	Other Bit. Coal / *Autres charb. bit.*	Sub-Bit. Coal / *Charbon sous-bit.*	Lignite / *Lignite*	Peat / *Tourbe*	Oven and Gas Coke / *Coke de four/gaz*	Pat. Fuel and BKB / *Agg./briq. de lignite*	Crude Oil / *Pétrole brut*	NGL / *LGN*	Feed-stocks / *Produits d'aliment.*	Additives / *Additifs*
Production	-	-	-	-	77	-	21	155	-	-	-
From Other Sources	-	-	-	-	-	-	-	-	-	-	-
Imports	-	414	-	-	-	-	3	3786	-	385	-
Exports	-	-6	-	-	-	-	-	-123	-	-	-
Intl. Marine Bunkers	-	-	-	-	-	-	-	-	-	-	-
Stock Changes	-	-70	-	-	-1	-	-	-6	-	-	-
DOMESTIC SUPPLY	**-**	**338**	**-**	**-**	**76**	**-**	**24**	**3812**	**-**	**385**	**-**
Transfers	-	-	-	-	-	-	-	-49	-	-	-
Statistical Differences	-	-	-	-	-	-	-	-	-	-	-
TRANSFORMATION	**-**	**32**	**-**	**-**	**37**	**-**	**-**	**3757**	**-**	**385**	**-**
Electricity Plants	-	-	-	-	-	-	-	-	-	-	-
CHP Plants	-	-	-	-	-	-	-	-	-	-	-
Heat Plants	-	32	-	-	15	-	-	6	-	-	-
Blast Furnaces/Gas Works	-	-	-	-	-	-	-	-	-	-	-
Coke/Pat. Fuel/BKB Plants	-	-	-	-	22	-	-	-	-	-	-
Petroleum Refineries	-	-	-	-	-	-	-	3751	-	385	-
Petrochemical Industry	-	-	-	-	-	-	-	-	-	-	-
Liquefaction	-	-	-	-	-	-	-	-	-	-	-
Other Transform. Sector	-	-	-	-	-	-	-	-	-	-	-
ENERGY SECTOR	**-**	**-**	**-**	**-**	**4**	**-**	**-**	**-**	**-**	**-**	**-**
Coal Mines	-	-	-	-	-	-	-	-	-	-	-
Oil and Gas Extraction	-	-	-	-	-	-	-	-	-	-	-
Petroleum Refineries	-	-	-	-	-	-	-	-	-	-	-
Electr., CHP+Heat Plants	-	-	-	-	4	-	-	-	-	-	-
Pumped Storage (Elec.)	-	-	-	-	-	-	-	-	-	-	-
Other Energy Sector	-	-	-	-	-	-	-	-	-	-	-
Distribution Losses	-	3	-	-	1	-	-	-	-	-	-
FINAL CONSUMPTION	**-**	**303**	**-**	**-**	**34**	**-**	**24**	**6**	**-**	**-**	**-**
INDUSTRY SECTOR	**-**	**16**	**-**	**-**	**15**	**-**	**1**	**3**	**-**	**-**	**-**
Iron and Steel	-	-	-	-	-	-	-	-	-	-	-
Chemical and Petrochem.	-	-	-	-	-	-	-	-	-	-	-
of which: Feedstocks	-	-	-	-	-	-	-	-	-	-	-
Non-Ferrous Metals	-	-	-	-	-	-	-	-	-	-	-
Non-Metallic Minerals	-	10	-	-	15	-	1	1	-	-	-
Transport Equipment	-	-	-	-	-	-	-	-	-	-	-
Machinery	-	-	-	-	-	-	-	-	-	-	-
Mining and Quarrying	-	-	-	-	-	-	-	-	-	-	-
Food and Tobacco	-	1	-	-	-	-	-	2	-	-	-
Paper, Pulp and Print	-	2	-	-	-	-	-	-	-	-	-
Wood and Wood Products	-	-	-	-	-	-	-	-	-	-	-
Construction	-	-	-	-	-	-	-	-	-	-	-
Textile and Leather	-	-	-	-	-	-	-	-	-	-	-
Non-specified	-	3	-	-	-	-	-	-	-	-	-
TRANSPORT SECTOR	**-**	**3**	**-**	**-**	**-**	**-**	**-**	**-**	**-**	**-**	**-**
Air	-	-	-	-	-	-	-	-	-	-	-
Road	-	-	-	-	-	-	-	-	-	-	-
Rail	-	3	-	-	-	-	-	-	-	-	-
Pipeline Transport	-	-	-	-	-	-	-	-	-	-	-
Internal Navigation	-	-	-	-	-	-	-	-	-	-	-
Non-specified	-	-	-	-	-	-	-	-	-	-	-
OTHER SECTORS	**-**	**283**	**-**	**-**	**19**	**-**	**23**	**3**	**-**	**-**	**-**
Agriculture	-	4	-	-	-	-	-	-	-	-	-
Comm. and Publ. Services	-	186	-	-	3	-	2	3	-	-	-
Residential	-	84	-	-	16	-	21	-	-	-	-
Non-specified	-	9	-	-	-	-	-	-	-	-	-
NON-ENERGY USE	**-**	**1**	**-**	**-**	**-**	**-**	**-**	**-**	**-**	**-**	**-**
in Industry/Trans./Energy	-	1	-	-	-	-	-	-	-	-	-
in Transport	-	-	-	-	-	-	-	-	-	-	-
in Other Sectors	-	-	-	-	-	-	-	-	-	-	-

Lithuania / Lituanie : 1996

SUPPLY AND CONSUMPTION / *APPROVISIONNEMENT ET DEMANDE*	Oil cont. / *Pétrole cont.* (1000 tonnes)										
	Refinery Gas / *Gaz de raffinerie*	LPG + Ethane / *GPL + éthane*	Motor Gasoline / *Essence moteur*	Aviation Gasoline / *Essence aviation*	Jet Fuel / *Carbu-réacteurs*	Kerosene / *Kérosène*	Gas/ Diesel / *Gazole*	Heavy Fuel Oil / *Fioul lourd*	Naphtha / *Naphta*	Petrol. Coke / *Coke de pétrole*	Other Prod. / *Autres prod.*
Production	-	234	1295	-	260	-	1242	943	-	-	137
From Other Sources	-	-	-	-	-	-	-	-	-	-	-
Imports	-	26	182	-	3	-	150	962	-	-	63
Exports	-	-183	-841	-	-229	-	-776	-244	-	-	-46
Intl. Marine Bunkers	-	-	-	-	-	-	-	-	-	-	-
Stock Changes	-	-3	29	-	-1	-	-4	-52	-	-	-54
DOMESTIC SUPPLY	**-**	**74**	**665**	**-**	**33**	**-**	**612**	**1609**	**-**	**-**	**100**
Transfers	-	-	-	-	-	-	-	-	-	-	49
Statistical Differences	-	-	-	-	-	-	-	-	-	-	-
TRANSFORMATION	**-**	**1**	**-**	**-**	**-**	**-**	**44**	**1204**	**-**	**-**	**-**
Electricity Plants	-	-	-	-	-	-	-	-	-	-	-
CHP Plants	-	-	-	-	-	-	-	614	-	-	-
Heat Plants	-	1	-	-	-	-	44	590	-	-	-
Blast Furnaces/Gas Works	-	-	-	-	-	-	-	-	-	-	-
Coke/Pat. Fuel/BKB Plants	-	-	-	-	-	-	-	-	-	-	-
Petroleum Refineries	-	-	-	-	-	-	-	-	-	-	-
Petrochemical Industry	-	-	-	-	-	-	-	-	-	-	-
Liquefaction	-	-	-	-	-	-	-	-	-	-	-
Other Transform. Sector	-	-	-	-	-	-	-	-	-	-	-
ENERGY SECTOR	**-**	**1**	**-**	**-**	**-**	**-**	**1**	**109**	**-**	**-**	**90**
Coal Mines	-	-	-	-	-	-	-	-	-	-	-
Oil and Gas Extraction	-	-	-	-	-	-	-	-	-	-	-
Petroleum Refineries	-	-	-	-	-	-	-	-	-	-	-
Electr., CHP+Heat Plants	-	-	-	-	-	-	-	-	-	-	-
Pumped Storage (Elec.)	-	-	-	-	-	-	-	-	-	-	-
Other Energy Sector	-	1	-	-	-	-	1	109	-	-	90
Distribution Losses	-	4	5	-	1	-	1	-	-	-	3
FINAL CONSUMPTION	**-**	**68**	**660**	**-**	**32**	**-**	**566**	**296**	**-**	**-**	**56**
INDUSTRY SECTOR	**-**	**1**	**5**	**-**	**-**	**-**	**43**	**133**	**-**	**-**	**-**
Iron and Steel	-	-	-	-	-	-	-	-	-	-	-
Chemical and Petrochem.	-	-	-	-	-	-	1	14	-	-	-
of which: Feedstocks	-	-	-	-	-	-	-	-	-	-	-
Non-Ferrous Metals	-	-	-	-	-	-	-	-	-	-	-
Non-Metallic Minerals	-	-	-	-	-	-	1	110	-	-	-
Transport Equipment	-	-	-	-	-	-	-	-	-	-	-
Machinery	-	-	-	-	-	-	-	-	-	-	-
Mining and Quarrying	-	-	-	-	-	-	-	-	-	-	-
Food and Tobacco	-	-	-	-	-	-	12	4	-	-	-
Paper, Pulp and Print	-	-	-	-	-	-	-	-	-	-	-
Wood and Wood Products	-	-	-	-	-	-	1	1	-	-	-
Construction	-	1	4	-	-	-	20	1	-	-	-
Textile and Leather	-	-	-	-	-	-	-	-	-	-	-
Non-specified	-	-	1	-	-	-	8	3	-	-	-
TRANSPORT SECTOR	**-**	**7**	**637**	**-**	**31**	**-**	**378**	**139**	**-**	**-**	**-**
Air	-	-	-	-	30	-	-	-	-	-	-
Road	-	7	636	-	-	-	257	-	-	-	-
Rail	-	-	1	-	-	-	81	5	-	-	-
Pipeline Transport	-	-	-	-	-	-	-	-	-	-	-
Internal Navigation	-	-	-	-	-	-	1	-	-	-	-
Non-specified	-	-	-	-	1	-	39	134	-	-	-
OTHER SECTORS	**-**	**60**	**18**	**-**	**1**	**-**	**145**	**24**	**-**	**-**	**-**
Agriculture	-	-	16	-	-	-	104	2	-	-	-
Comm. and Publ. Services	-	1	-	-	-	-	34	19	-	-	-
Residential	-	59	-	-	1	-	1	2	-	-	-
Non-specified	-	-	2	-	-	-	6	1	-	-	-
NON-ENERGY USE	**-**	**-**	**-**	**-**	**-**	**-**	**-**	**-**	**-**	**-**	**56**
in Industry/Transf./Energy	-	-	-	-	-	-	-	-	-	-	56
in Transport	-	-	-	-	-	-	-	-	-	-	-
in Other Sectors	-	-	-	-	-	-	-	-	-	-	-

Lithuania / Lituanie : 1996

SUPPLY AND CONSUMPTION / *APPROVISIONNEMENT ET DEMANDE*	Gas / *Gaz* (TJ)				Comb. Renew. & Waste / *En. Re. Comb. & Déchets* (TJ)				(GWh)	(TJ)
	Natural Gas / *Gaz naturel*	Gas Works / *Usines à gaz*	Coke Ovens / *Cokeries*	Blast Furnaces / *Hauts fourneaux*	Solid Biomass / *Biomasse solide*	Gas/Liquids from Biomass / *Gaz/Liquides tirés de biomasse*	Municipal Waste / *Déchets urbains*	Industrial Waste / *Déchets industriels*	Electricity / *Electricité*	Heat / *Chaleur*
Production	-	-	-	-	11214	-	-	98	16789	82736
From Other Sources	-	-	-	-	-	-	-	-	-	-
Imports	100749	-	-	-	30	-	-	209	4182	-
Exports	-	-	-	-	-18	-	-	-	-9341	-
Intl. Marine Bunkers	-	-	-	-	-	-	-	-	-	-
Stock Changes	135	-	-	-	-76	-	-	-25	-	-
DOMESTIC SUPPLY	**100884**	**-**	**-**	**-**	**11150**	**-**	**-**	**282**	**11630**	**82736**
Transfers	-	-	-	-	-	-	-	-	-	-
Statistical Differences	-	-	-	-	-	-	-	-	-	-
TRANSFORMATION	**55960**	**-**	**-**	**-**	**1589**	**-**	**-**	**41**	**105**	**-**
Electricity Plants	-	-	-	-	-	-	-	-	-	-
CHP Plants	31571	-	-	-	-	-	-	-	-	-
Heat Plants	24389	-	-	-	1589	-	-	41	-	-
Blast Furnaces/Gas Works	-	-	-	-	-	-	-	-	-	-
Coke/Pat. Fuel/BKB Plants	-	-	-	-	-	-	-	-	-	-
Petroleum Refineries	-	-	-	-	-	-	-	-	-	-
Petrochemical Industry	-	-	-	-	-	-	-	-	-	-
Liquefaction	-	-	-	-	-	-	-	-	-	-
Other Transform. Sector	-	-	-	-	-	-	-	-	105	-
ENERGY SECTOR	**-**	**-**	**-**	**-**	**-**	**-**	**-**	**-**	**3231**	**5994**
Coal Mines	-	-	-	-	-	-	-	-	-	-
Oil and Gas Extraction	-	-	-	-	-	-	-	-	-	-
Petroleum Refineries	-	-	-	-	-	-	-	-	411	5994
Electr., CHP+Heat Plants	-	-	-	-	-	-	-	-	1647	-
Pumped Storage (Elec.)	-	-	-	-	-	-	-	-	763	-
Other Energy Sector	-	-	-	-	-	-	-	-	410	-
Distribution Losses	2155	-	-	-	1	-	-	-	1778	18222
FINAL CONSUMPTION	**42769**	**-**	**-**	**-**	**9560**	**-**	**-**	**241**	**6516**	**58520**
INDUSTRY SECTOR	**32078**	**-**	**-**	**-**	**256**	**-**	**-**	**180**	**2520**	**18508**
Iron and Steel	14	-	-	-	-	-	-	29	46	10
Chemical and Petrochem.	27469	-	-	-	-	-	-	-	519	5584
of which: Feedstocks	*25511*	-	-	-	-	-	-	-	-	-
Non-Ferrous Metals	-	-	-	-	-	-	-	-	-	-
Non-Metallic Minerals	2216	-	-	-	73	-	-	6	232	746
Transport Equipment	17	-	-	-	-	-	-	-	50	223
Machinery	990	-	-	-	23	-	-	32	380	1025
Mining and Quarrying	206	-	-	-	-	-	-	-	21	-
Food and Tobacco	639	-	-	-	-	-	-	113	475	5824
Paper, Pulp and Print	24	-	-	-	-	-	-	-	97	1359
Wood and Wood Products	315	-	-	-	99	-	-	-	84	229
Construction	50	-	-	-	3	-	-	-	73	78
Textile and Leather	4	-	-	-	18	-	-	-	358	2240
Non-specified	134	-	-	-	40	-	-	-	185	1190
TRANSPORT SECTOR	**160**	**-**	**-**	**-**	**9**	**-**	**-**	**-**	**103**	**-**
Air	-	-	-	-	-	-	-	-	-	-
Road	160	-	-	-	-	-	-	-	-	-
Rail	-	-	-	-	-	-	-	-	28	-
Pipeline Transport	-	-	-	-	-	-	-	-	-	-
Internal Navigation	-	-	-	-	-	-	-	-	-	-
Non-specified	-	-	-	-	9	-	-	-	75	-
OTHER SECTORS	**10531**	**-**	**-**	**-**	**9295**	**-**	**-**	**61**	**3893**	**40012**
Agriculture	321	-	-	-	36	-	-	-	501	1351
Comm. and Publ. Services	1740	-	-	-	1291	-	-	59	1786	9515
Residential	8470	-	-	-	7840	-	-	-	1606	29146
Non-specified	-	-	-	-	128	-	-	2	-	-
NON-ENERGY USE	**-**	**-**	**-**	**-**	**-**	**-**	**-**	**-**	**-**	**-**
in Industry/Transf./Energy	-	-	-	-	-	-	-	-	-	-
in Transport	-	-	-	-	-	-	-	-	-	-
in Other Sectors	-	-	-	-	-	-	-	-	-	-

Lithuania / Lituanie : 1997

	Coal / *Charbon* (1000 tonnes)							Oil / *Pétrole* (1000 tonnes)			
SUPPLY AND CONSUMPTION / ***APPROVISIONNEMENT ET DEMANDE***	Coking Coal / *Charbon à coke*	Other Bit. Coal / *Autres charb. bit.*	Sub-Bit. Coal / *Charbon sous-bit.*	Lignite / *Lignite*	Peat / *Tourbe*	Oven and Gas Coke / *Coke de four/gaz*	Pat. Fuel and BKB / *Agg./briq. de lignite*	Crude Oil / *Pétrole brut*	NGL / *LGN*	Feed-stocks / *Produits d'aliment.*	Additives / *Additifs*
Production	-	-	-	-	89	-	21	212	-	-	12
From Other Sources	-	-	-	-	-	-	-	-	-	-	-
Imports	-	259	-	3	-	10	1	5074	-	701	2
Exports	-	-48	-	-	-	-	-	-193	-	-	-
Intl. Marine Bunkers	-	-	-	-	-	-	-	-	-	-	-
Stock Changes	-	45	-	3	-10	1	1	-27	-	-74	-
DOMESTIC SUPPLY	**-**	**256**	**-**	**6**	**79**	**11**	**23**	**5066**	**-**	**627**	**14**
Transfers	-	-	-	-	-	-	-	-49	-	-	-
Statistical Differences	-	-	-	-	-	-	-	-	-	-	-
TRANSFORMATION	**-**	**25**	**-**	**-**	**46**	**-**	**-**	**5012**	**-**	**627**	**14**
Electricity Plants	-	-	-	-	-	-	-	-	-	-	-
CHP Plants	-	-	-	-	-	-	-	-	-	-	-
Heat Plants	-	25	-	-	14	-	-	4	-	-	-
Blast Furnaces/Gas Works	-	-	-	-	-	-	-	-	-	-	-
Coke/Pat. Fuel/BKB Plants	-	-	-	-	32	-	-	-	-	-	-
Petroleum Refineries	-	-	-	-	-	-	-	5008	-	627	14
Petrochemical Industry	-	-	-	-	-	-	-	-	-	-	-
Liquefaction	-	-	-	-	-	-	-	-	-	-	-
Other Transform. Sector	-	-	-	-	-	-	-	-	-	-	-
ENERGY SECTOR	**-**	**-**	**-**	**-**	**-**	**-**	**-**	**-**	**-**	**-**	**-**
Coal Mines	-	-	-	-	-	-	-	-	-	-	-
Oil and Gas Extraction	-	-	-	-	-	-	-	-	-	-	-
Petroleum Refineries	-	-	-	-	-	-	-	-	-	-	-
Electr., CHP+Heat Plants	-	-	-	-	-	-	-	-	-	-	-
Pumped Storage (Elec.)	-	-	-	-	-	-	-	-	-	-	-
Other Energy Sector	-	-	-	-	-	-	-	-	-	-	-
Distribution Losses	-	1	-	-	2	-	-	1	-	-	-
FINAL CONSUMPTION	**-**	**230**	**-**	**6**	**31**	**11**	**23**	**4**	**-**	**-**	**-**
INDUSTRY SECTOR	**-**	**7**	**-**	**-**	**9**	**11**	**-**	**2**	**-**	**-**	**-**
Iron and Steel	-	-	-	-	-	2	-	-	-	-	-
Chemical and Petrochem.	-	-	-	-	-	-	-	-	-	-	-
of which: Feedstocks	-	-	-	-	-	-	-	-	-	-	-
Non-Ferrous Metals	-	-	-	-	-	-	-	-	-	-	-
Non-Metallic Minerals	-	1	-	-	9	4	-	-	-	-	-
Transport Equipment	-	-	-	-	-	-	-	-	-	-	-
Machinery	-	-	-	-	-	-	-	-	-	-	-
Mining and Quarrying	-	-	-	-	-	-	-	-	-	-	-
Food and Tobacco	-	-	-	-	-	5	-	2	-	-	-
Paper, Pulp and Print	-	2	-	-	-	-	-	-	-	-	-
Wood and Wood Products	-	-	-	-	-	-	-	-	-	-	-
Construction	-	1	-	-	-	-	-	-	-	-	-
Textile and Leather	-	-	-	-	-	-	-	-	-	-	-
Non-specified	-	3	-	-	-	-	-	-	-	-	-
TRANSPORT SECTOR	**-**	**-**	**-**	**-**	**-**	**-**	**-**	**-**	**-**	**-**	**-**
Air	-	-	-	-	-	-	-	-	-	-	-
Road	-	-	-	-	-	-	-	-	-	-	-
Rail	-	-	-	-	-	-	-	-	-	-	-
Pipeline Transport	-	-	-	-	-	-	-	-	-	-	-
Internal Navigation	-	-	-	-	-	-	-	-	-	-	-
Non-specified	-	-	-	-	-	-	-	-	-	-	-
OTHER SECTORS	**-**	**223**	**-**	**6**	**22**	**-**	**23**	**2**	**-**	**-**	**-**
Agriculture	-	2	-	-	-	-	-	-	-	-	-
Comm. and Publ. Services	-	136	-	6	3	-	2	2	-	-	-
Residential	-	85	-	-	19	-	21	-	-	-	-
Non-specified	-	-	-	-	-	-	-	-	-	-	-
NON-ENERGY USE	**-**	**-**	**-**	**-**	**-**	**-**	**-**	**-**	**-**	**-**	**-**
in Industry/Trans./Energy	-	-	-	-	-	-	-	-	-	-	-
in Transport	-	-	-	-	-	-	-	-	-	-	-
in Other Sectors	-	-	-	-	-	-	-	-	-	-	-

Lithuania / Lituanie : 1997

SUPPLY AND CONSUMPTION / *APPROVISIONNEMENT ET DEMANDE*	Oil cont. / *Pétrole cont.* (1000 tonnes)										
	Refinery Gas / *Gaz de raffinerie*	LPG + Ethane / *GPL + éthane*	Motor Gasoline / *Essence moteur*	Aviation Gasoline / *Essence aviation*	Jet Fuel / *Carbu-réacteurs*	Kerosene / *Kérosène*	Gas/ Diesel / *Gazole*	Heavy Fuel Oil / *Fioul lourd*	Naphtha / *Naphta*	Petrol. Coke / *Coke de pétrole*	Other Prod. / *Autres prod.*
Production	176	256	1664	-	225	-	1632	1374	-	84	103
From Other Sources	-	-	-	-	-	-	-	-	-	-	-
Imports	-	28	142	-	-	-	128	421	-	-	89
Exports	-	-189	-1157	-	-198	-	-1098	-443	-	-	-47
Intl. Marine Bunkers	-	-	-	-	-	-	-7	-55	-	-	-
Stock Changes	-	-1	13	-	4	-	18	13	-	-	-9
DOMESTIC SUPPLY	**176**	**94**	**662**	**-**	**31**	**-**	**673**	**1310**	**-**	**84**	**136**
Transfers	-	-	-	-	-	-	-	-	-	-	49
Statistical Differences	-	-	-	-	-	-	-	-	-	-	-
TRANSFORMATION	**-**	**1**	**-**	**-**	**-**	**-**	**29**	**1015**	**-**	**-**	**7**
Electricity Plants	-	-	-	-	-	-	-	-	-	-	-
CHP Plants	-	-	-	-	-	-	-	470	-	-	-
Heat Plants	-	1	-	-	-	-	29	545	-	-	7
Blast Furnaces/Gas Works	-	-	-	-	-	-	-	-	-	-	-
Coke/Pat. Fuel/BKB Plants	-	-	-	-	-	-	-	-	-	-	-
Petroleum Refineries	-	-	-	-	-	-	-	-	-	-	-
Petrochemical Industry	-	-	-	-	-	-	-	-	-	-	-
Liquefaction	-	-	-	-	-	-	-	-	-	-	-
Other Transform. Sector	-	-	-	-	-	-	-	-	-	-	-
ENERGY SECTOR	**174**	**-**	**-**	**-**	**-**	**-**	**2**	**134**	**-**	**84**	**-**
Coal Mines	-	-	-	-	-	-	-	-	-	-	-
Oil and Gas Extraction	-	-	-	-	-	-	-	-	-	-	-
Petroleum Refineries	-	-	-	-	-	-	-	-	-	-	-
Electr., CHP+Heat Plants	-	-	-	-	-	-	-	-	-	-	-
Pumped Storage (Elec.)	-	-	-	-	-	-	-	-	-	-	-
Other Energy Sector	174	-	-	-	-	-	2	134	-	84	-
Distribution Losses	2	3	4	-	-	-	1	-	-	-	-
FINAL CONSUMPTION	**-**	**90**	**658**	**-**	**31**	**-**	**641**	**161**	**-**	**-**	**178**
INDUSTRY SECTOR	**-**	**2**	**5**	**-**	**-**	**-**	**42**	**135**	**-**	**-**	**2**
Iron and Steel	-	-	-	-	-	-	-	-	-	-	-
Chemical and Petrochem.	-	-	-	-	-	-	1	1	-	-	-
of which: Feedstocks	-	-	-	-	-	-	-	-	-	-	-
Non-Ferrous Metals	-	-	-	-	-	-	-	-	-	-	-
Non-Metallic Minerals	-	-	-	-	-	-	1	125	-	-	-
Transport Equipment	-	-	-	-	-	-	-	-	-	-	-
Machinery	-	-	1	-	-	-	-	-	-	-	2
Mining and Quarrying	-	-	-	-	-	-	-	1	-	-	-
Food and Tobacco	-	1	-	-	-	-	11	6	-	-	-
Paper, Pulp and Print	-	-	-	-	-	-	-	-	-	-	-
Wood and Wood Products	-	-	-	-	-	-	4	-	-	-	-
Construction	-	-	3	-	-	-	20	2	-	-	-
Textile and Leather	-	-	-	-	-	-	-	-	-	-	-
Non-specified	-	1	1	-	-	-	5	-	-	-	-
TRANSPORT SECTOR	**-**	**11**	**640**	**-**	**31**	**-**	**500**	**5**	**-**	**-**	**-**
Air	-	-	-	-	31	-	-	-	-	-	-
Road	-	11	639	-	-	-	408	-	-	-	-
Rail	-	-	-	-	-	-	78	2	-	-	-
Pipeline Transport	-	-	-	-	-	-	-	-	-	-	-
Internal Navigation	-	-	-	-	-	-	1	-	-	-	-
Non-specified	-	-	1	-	-	-	13	3	-	-	-
OTHER SECTORS	**-**	**77**	**13**	**-**	**-**	**-**	**99**	**21**	**-**	**-**	**-**
Agriculture	-	-	12	-	-	-	77	2	-	-	-
Comm. and Publ. Services	-	1	1	-	-	-	22	18	-	-	-
Residential	-	76	-	-	-	-	-	1	-	-	-
Non-specified	-	-	-	-	-	-	-	-	-	-	-
NON-ENERGY USE	**-**	**-**	**-**	**-**	**-**	**-**	**-**	**-**	**-**	**-**	**176**
in Industry/Transf./Energy	-	-	-	-	-	-	-	-	-	-	146
in Transport	-	-	-	-	-	-	-	-	-	-	30
in Other Sectors	-	-	-	-	-	-	-	-	-	-	-

Lithuania / Lituanie : 1997

SUPPLY AND CONSUMPTION / *APPROVISIONNEMENT ET DEMANDE*	Gas / *Gaz* (TJ)				Comb. Renew. & Waste / *En. Re. Comb. & Déchets* (TJ)				(GWh)	(TJ)
	Natural Gas / *Gaz naturel*	Gas Works / *Usines à gaz*	Coke Ovens / *Cokeries*	Blast Furnaces / *Hauts fourneaux*	Solid Biomass / *Biomasse solide*	Gas/Liquids from Biomass / *Gaz/Liquides tirés de biomasse*	Municipal Waste / *Déchets urbains*	Industrial Waste / *Déchets industriels*	Electricity / *Electricité*	Heat / *Chaleur*
Production	-	-	-	-	21569	-	-	100	14861	76681
From Other Sources	-	-	-	-	-	-	-	-	-	-
Imports	93134	-	-	-	5	-	-	-	4525	-
Exports	-	-	-	-	-48	-	-	-	-8050	-
Intl. Marine Bunkers	-	-	-	-	-	-	-	-	-	-
Stock Changes	-	-	-	-	102	-	-	-	-	-
DOMESTIC SUPPLY	**93134**	**-**	**-**	**-**	**21628**	**-**	**-**	**100**	**11336**	**76681**
Transfers	-	-	-	-	-	-	-	-	-	-
Statistical Differences	-	-	-	-	-	-	-	-	-	-
TRANSFORMATION	**53686**	**-**	**-**	**-**	**1617**	**-**	**-**	**46**	**76**	**-**
Electricity Plants	-	-	-	-	-	-	-	-	-	-
CHP Plants	23815	-	-	-	-	-	-	-	-	-
Heat Plants	29871	-	-	-	1617	-	-	46	-	-
Blast Furnaces/Gas Works	-	-	-	-	-	-	-	-	-	-
Coke/Pat. Fuel/BKB Plants	-	-	-	-	-	-	-	-	-	-
Petroleum Refineries	-	-	-	-	-	-	-	-	-	-
Petrochemical Industry	-	-	-	-	-	-	-	-	-	-
Liquefaction	-	-	-	-	-	-	-	-	-	-
Other Transform. Sector	-	-	-	-	-	-	-	-	76	-
ENERGY SECTOR	**-**	**-**	**-**	**-**	**-**	**-**	**-**	**-**	**2939**	**6633**
Coal Mines	-	-	-	-	-	-	-	-	-	-
Oil and Gas Extraction	-	-	-	-	-	-	-	-	-	-
Petroleum Refineries	-	-	-	-	-	-	-	-	481	4877
Electr., CHP+Heat Plants	-	-	-	-	-	-	-	-	1547	-
Pumped Storage (Elec.)	-	-	-	-	-	-	-	-	663	-
Other Energy Sector	-	-	-	-	-	-	-	-	248	1756
Distribution Losses	1909	-	-	-	18	-	-	-	1585	14524
FINAL CONSUMPTION	**37539**	**-**	**-**	**-**	**19993**	**-**	**-**	**54**	**6736**	**55524**
INDUSTRY SECTOR	**28649**	**-**	**-**	**-**	**333**	**-**	**-**	**-**	**2777**	**17148**
Iron and Steel	20	-	-	-	-	-	-	-	53	10
Chemical and Petrochem.	24447	-	-	-	-	-	-	-	503	5259
of which: Feedstocks	*21921*	-	-	-	-	-	-	-	-	-
Non-Ferrous Metals	-	-	-	-	-	-	-	-	-	-
Non-Metallic Minerals	1832	-	-	-	117	-	-	-	226	600
Transport Equipment	32	-	-	-	-	-	-	-	48	110
Machinery	970	-	-	-	22	-	-	-	337	764
Mining and Quarrying	178	-	-	-	-	-	-	-	26	-
Food and Tobacco	661	-	-	-	82	-	-	-	489	5778
Paper, Pulp and Print	3	-	-	-	-	-	-	-	96	1284
Wood and Wood Products	399	-	-	-	33	-	-	-	117	286
Construction	62	-	-	-	10	-	-	-	65	51
Textile and Leather	6	-	-	-	20	-	-	-	431	2410
Non-specified	39	-	-	-	49	-	-	-	386	596
TRANSPORT SECTOR	**25**	**-**	**-**	**-**	**-**	**-**	**-**	**-**	**101**	**-**
Air	-	-	-	-	-	-	-	-	-	-
Road	25	-	-	-	-	-	-	-	-	-
Rail	-	-	-	-	-	-	-	-	23	-
Pipeline Transport	-	-	-	-	-	-	-	-	-	-
Internal Navigation	-	-	-	-	-	-	-	-	-	-
Non-specified	-	-	-	-	-	-	-	-	78	-
OTHER SECTORS	**8865**	**-**	**-**	**-**	**19660**	**-**	**-**	**54**	**3858**	**38376**
Agriculture	342	-	-	-	41	-	-	-	426	1561
Comm. and Publ. Services	1641	-	-	-	1367	-	-	54	1712	10102
Residential	6882	-	-	-	18252	-	-	-	1720	26662
Non-specified	-	-	-	-	-	-	-	-	-	51
NON-ENERGY USE	**-**	**-**	**-**	**-**	**-**	**-**	**-**	**-**	**-**	**-**
in Industry/Transf./Energy	-	-	-	-	-	-	-	-	-	-
in Transport	-	-	-	-	-	-	-	-	-	-
in Other Sectors	-	-	-	-	-	-	-	-	-	-

FYR of Macedonia / ex-République yougoslave de Macédoine

SUPPLY AND CONSUMPTION 1996	Coal (1000 tonnes)							Oil (1000 tonnes)			
	Coking Coal	Other Bit. Coal	Sub-Bit. Coal	Lignite	Peat	Oven and Gas Coke	Pat. Fuel and BKB	Crude Oil	NGL	Feed-stocks	Additives
Production	-	-	-	7145	-	-	-	-	-	-	-
Imports	49	51	-	19	-	-	-	656	-	-	-
Exports	-	-	-	-	-	-	-	-	-	-	-
Intl. Marine Bunkers	-	-	-	-	-	-	-	-	-	-	-
Stock Changes	-	-	-	24	-	-	-	-	-	-	-
DOMESTIC SUPPLY	**49**	**51**	**-**	**7188**	**-**	**-**	**-**	**656**	**-**	**-**	**-**
Transfers and Stat. Diff.	-	-	-	-	-	-	-	19	-	-	-
TRANSFORMATION	**-**	**-**	**-**	**7030**	**-**	**-**	**-**	**675**	**-**	**-**	**-**
Electricity and CHP Plants	-	-	-	7030	-	-	-	-	-	-	-
Petroleum Refineries	-	-	-	-	-	-	-	675	-	-	-
Other Transform. Sector	-	-	-	-	-	-	-	-	-	-	-
ENERGY SECTOR	**-**	**-**	**-**	**-**	**-**	**-**	**-**	**-**	**-**	**-**	**-**
DISTRIBUTION LOSSES	**-**	**-**	**-**	**-**	**-**	**-**	**-**	**-**	**-**	**-**	**-**
FINAL CONSUMPTION	**49**	**51**	**-**	**158**	**-**	**-**	**-**	**-**	**-**	**-**	**-**
INDUSTRY SECTOR	**49**	**51**	**-**	**113**	**-**	**-**	**-**	**-**	**-**	**-**	**-**
Iron and Steel	11	22	-	105	-	-	-	-	-	-	-
Chemical and Petrochem.	-	-	-	-	-	-	-	-	-	-	-
Non-Metallic Minerals	-	-	-	-	-	-	-	-	-	-	-
Non-specified	38	29	-	8	-	-	-	-	-	-	-
TRANSPORT SECTOR	**-**	**-**	**-**	**-**	**-**	**-**	**-**	**-**	**-**	**-**	**-**
Air	-	-	-	-	-	-	-	-	-	-	-
Road	-	-	-	-	-	-	-	-	-	-	-
Non-specified	-	-	-	-	-	-	-	-	-	-	-
OTHER SECTORS	**-**	**-**	**-**	**45**	**-**	**-**	**-**	**-**	**-**	**-**	**-**
Agriculture	-	-	-	-	-	-	-	-	-	-	-
Comm. and Publ. Services	-	-	-	20	-	-	-	-	-	-	-
Residential	-	-	-	25	-	-	-	-	-	-	-
Non-specified	-	-	-	-	-	-	-	-	-	-	-
NON-ENERGY USE	**-**	**-**	**-**	**-**	**-**	**-**	**-**	**-**	**-**	**-**	**-**

APPROVISIONNEMENT ET DEMANDE 1997	*Charbon* (1000 tonnes)							*Pétrole* (1000 tonnes)			
	Charbon à coke	*Autres charb. bit.*	*Charbon sous-bit.*	*Lignite*	*Tourbe*	*Coke de four/gaz*	*Agg./briq. de lignite*	*Pétrole brut*	*LGN*	*Produits d'aliment.*	*Additifs*
Production	-	-	-	6700	-	-	-	-	-	-	-
Imports	49	51	-	-	-	-	-	400	-	-	-
Exports	-	-	-	-	-	-	-	-	-	-	-
Intl. Marine Bunkers	-	-	-	-	-	-	-		-	-	-
Stock Changes	-	-	-	-	-	-	-	-	-	-	-
DOMESTIC SUPPLY	**49**	**51**	**-**	**6700**	**-**	**-**	**-**	**400**	**-**	**-**	**-**
Transfers and Stat. Diff.	-	-	-	-	-	-	-	-	-	-	-
TRANSFORMATION	**-**	**-**	**-**	**6550**	**-**	**-**	**-**	**400**	**-**	**-**	**-**
Electricity and CHP Plants	-	-	-	6550	-	-	-	-	-	-	-
Petroleum Refineries	-	-	-	-	-	-	-	400	-	-	-
Other Transform. Sector	-	-	-	-	-	-	-	-	-	-	-
ENERGY SECTOR	**-**	**-**	**-**	**-**	**-**	**-**	**-**	**-**	**-**	**-**	**-**
DISTRIBUTION LOSSES	**-**	**-**	**-**	**-**	**-**	**-**	**-**	**-**	**-**	**-**	**-**
FINAL CONSUMPTION	**49**	**51**	**-**	**150**	**-**	**-**	**-**	**-**	**-**	**-**	**-**
INDUSTRY SECTOR	**49**	**51**	**-**	**105**	**-**	**-**	**-**	**-**	**-**	**-**	**-**
Iron and Steel	11	22	-	98	-	-	-	-	-	-	-
Chemical and Petrochem.	-	-	-	-	-	-	-	-	-	-	-
Non-Metallic Minerals	-	-	-	-	-	-	-	-	-	-	-
Non-specified	38	29	-	7	-	-	-	-	-	-	-
TRANSPORT SECTOR	**-**	**-**	**-**	**-**	**-**	**-**	**-**	**-**	**-**	**-**	**-**
Air	-	-	-	-	-	-	-	-	-	-	-
Road	-	-	-	-	-	-	-	-	-	-	-
Non-specified	-	-	-	-	-	-	-	-	-	-	-
OTHER SECTORS	**-**	**-**	**-**	**45**	**-**	**-**	**-**	**-**	**-**	**-**	**-**
Agriculture	-	-	-	-	-	-	-	-	-	-	-
Comm. and Publ. Services	-	-	-	20	-	-	-	-	-	-	-
Residential	-	-	-	25	-	-	-	-	-	-	-
Non-specified	-	-	-	-	-	-	-	-	-	-	-
NON-ENERGY USE	**-**	**-**	**-**	**-**	**-**	**-**	**-**	**-**	**-**	**-**	**-**

FYR of Macedonia / ex-République yougoslave de Macédoine

SUPPLY AND CONSUMPTION 1996	Oil cont. (1000 tonnes)										
	Refinery Gas	LPG + Ethane	Motor Gasoline	Aviation Gasoline	Jet Fuel	Kerosene	Gas/ Diesel	Heavy Fuel Oil	Naphtha	Petrol. Coke	Other Prod.
Production	-	7	58	-	-	-	277	333	-	-	-
Imports	-	14	157	-	28	-	184	176	-	-	40
Exports	-	-2	-6	-	-2	-	-4	-23	-	-	-
Intl. Marine Bunkers	-	-	-	-	-	-	-	-	-	-	-
Stock Changes	-	-	-	-	-	-	-	-	-	-	-
DOMESTIC SUPPLY	**-**	**19**	**209**	**-**	**26**	**-**	**457**	**486**	**-**	**-**	**40**
Transfers and Stat. Diff.	-	-	-	-	-	-	-	-	-	-	-
TRANSFORMATION	**-**	**-**	**-**	**-**	**-**	**-**	**-**	**239**	**-**	**-**	**-**
Electricity and CHP Plants	-	-	-	-	-	-	-	30	-	-	-
Petroleum Refineries	-	-	-	-	-	-	-	-	-	-	-
Other Transform. Sector	-	-	-	-	-	-	-	209	-	-	-
ENERGY SECTOR	**-**	**-**	**-**	**-**	**-**	**-**	**-**	**-**	**-**	**-**	**-**
DISTRIBUTION LOSSES	**-**	**-**	**-**	**-**	**-**	**-**	**-**	**-**	**-**	**-**	**-**
FINAL CONSUMPTION	**-**	**19**	**209**	**-**	**26**	**-**	**457**	**247**	**-**	**-**	**40**
INDUSTRY SECTOR	**-**	**17**	**-**	**-**	**-**	**-**	**62**	**184**	**-**	**-**	**-**
Iron and Steel	-	2	-	-	-	-	2	30	-	-	-
Chemical and Petrochem.	-	-	-	-	-	-	-	59	-	-	-
Non-Metallic Minerals	-	11	-	-	-	-	4	12	-	-	-
Non-specified	-	4	-	-	-	-	56	83	-	-	-
TRANSPORT SECTOR	**-**	**-**	**207**	**-**	**26**	**-**	**330**	**-**	**-**	**-**	**-**
Air	-	-	-	-	26	-	-	-	-	-	-
Road	-	-	207	-	-	-	322	-	-	-	-
Non-specified	-	-	-	-	-	-	8	-	-	-	-
OTHER SECTORS	**-**	**2**	**2**	**-**	**-**	**-**	**65**	**63**	**-**	**-**	**-**
Agriculture	-	-	2	-	-	-	65	24	-	-	-
Comm. and Publ. Services	-	-	-	-	-	-	-	-	-	-	-
Residential	-	2	-	-	-	-	-	39	-	-	-
Non-specified	-	-	-	-	-	-	-	-	-	-	-
NON-ENERGY USE	**-**	**-**	**-**	**-**	**-**	**-**	**-**	**-**	**-**	**-**	**40**

APPROVISIONNEMENT ET DEMANDE 1997	*Pétrole cont.* (1000 tonnes)										
	Gaz de raffinerie	*GPL + éthane*	*Essence moteur*	*Essence aviation*	*Carbu- réacteurs*	*Kérosène*	*Gazole*	*Fioul lourd*	*Naphta*	*Coke de pétrole*	*Autres prod.*
Production	-	3	35	-	-	-	173	174	-	-	-
Imports	-	13	184	-	30	-	193	186	-	-	41
Exports	-	-	-5	-	-	-	-	-	-	-	-
Intl. Marine Bunkers	-	-	-	-	-	-	-	-	-	-	-
Stock Changes	-	-	-	-	-	-	-	-	-	-	-
DOMESTIC SUPPLY	**-**	**16**	**214**	**-**	**30**	**-**	**366**	**360**	**-**	**-**	**41**
Transfers and Stat. Diff.	-	-	-	-	-	-	-	-	-	-	-
TRANSFORMATION	**-**	**-**	**-**	**-**	**-**	**-**	**-**	**200**	**-**	**-**	**-**
Electricity and CHP Plants	-	-	-	-	-	-	-	20	-	-	-
Petroleum Refineries	-	-	-	-	-	-	-	-	-	-	-
Other Transform. Sector	-	-	-	-	-	-	-	180	-	-	-
ENERGY SECTOR	**-**	**-**	**-**	**-**	**-**	**-**	**-**	**-**	**-**	**-**	**-**
DISTRIBUTION LOSSES	**-**	**-**	**-**	**-**	**-**	**-**	**-**	**-**	**-**	**-**	**-**
FINAL CONSUMPTION	**-**	**16**	**214**	**-**	**30**	**-**	**366**	**160**	**-**	**-**	**41**
INDUSTRY SECTOR	**-**	**15**	**-**	**-**	**-**	**-**	**52**	**120**	**-**	**-**	**-**
Iron and Steel	-	2	-	-	-	-	2	20	-	-	-
Chemical and Petrochem.	-	-	-	-	-	-	-	39	-	-	-
Non-Metallic Minerals	-	10	-	-	-	-	3	8	-	-	-
Non-specified	-	3	-	-	-	-	47	53	-	-	-
TRANSPORT SECTOR	**-**	**-**	**211**	**-**	**30**	**-**	**264**	**-**	**-**	**-**	**-**
Air	-	-	-	-	30	-	-	-	-	-	-
Road	-	-	211	-	-	-	230	-	-	-	-
Non-specified	-	-	-	-	-	-	34	-	-	-	-
OTHER SECTORS	**-**	**1**	**3**	**-**	**-**	**-**	**50**	**40**	**-**	**-**	**-**
Agriculture	-	-	3	-	-	-	50	10	-	-	-
Comm. and Publ. Services	-	-	-	-	-	-	-	-	-	-	-
Residential	-	1	-	-	-	-	-	30	-	-	-
Non-specified	-	-	-	-	-	-	-	-	-	-	-
NON-ENERGY USE	**-**	**-**	**-**	**-**	**-**	**-**	**-**	**-**	**-**	**-**	**41**

FYR of Macedonia / ex-République yougoslave de Macédoine

SUPPLY AND CONSUMPTION 1996	Gas (TJ)				Comb. Renew. & Waste (TJ)				(GWh)	(TJ)
	Natural Gas	Gas Works	Coke Ovens	Blast Furnaces	Solid Biomass	Gas/Liquids from Biomass	Municipal Waste	Industrial Waste	Electricity	Heat
Production	-	-	-	-	7814	-	-	-	6489	6208
Imports	-	-	-	-	150	-	-	-	-	-
Exports	-	-	-	-	-	-	-	-	-	-
Intl. Marine Bunkers	-	-	-	-	-	-	-	-	-	-
Stock Changes	-	-	-	-	-	-	-	-	-	-
DOMESTIC SUPPLY	**-**	**-**	**-**	**-**	**7964**	**-**	**-**	**-**	**6489**	**6208**
Transfers and Stat. Diff.	-	-	-	-	-	-	-	-	-	-
TRANSFORMATION	**-**	**-**	**-**	**-**	**-**	**-**	**-**	**-**	**-**	**-**
Electricity and CHP Plants	-	-	-	-	-	-	-	-	-	-
Petroleum Refineries	-	-	-	-	-	-	-	-	-	-
Other Transform. Sector	-	-	-	-	-	-	-	-	-	-
ENERGY SECTOR	**-**	**-**	**-**	**-**	**-**	**-**	**-**	**-**	**583**	**123**
DISTRIBUTION LOSSES	**-**	**-**	**-**	**-**	**-**	**-**	**-**	**-**	**810**	**898**
FINAL CONSUMPTION	**-**	**-**	**-**	**-**	**7964**	**-**	**-**	**-**	**5096**	**5187**
INDUSTRY SECTOR	**-**	**-**	**-**	**-**	**-**	**-**	**-**	**-**	**2015**	**2811**
Iron and Steel	-	-	-	-	-	-	-	-	1091	132
Chemical and Petrochem.	-	-	-	-	-	-	-	-	140	624
Non-Metallic Minerals	-	-	-	-	-	-	-	-	27	16
Non-specified	-	-	-	-	-	-	-	-	757	2039
TRANSPORT SECTOR	**-**	**-**	**-**	**-**	**-**	**-**	**-**	**-**	**14**	**-**
Air	-	-	-	-	-	-	-	-	-	-
Road	-	-	-	-	-	-	-	-	-	-
Non-specified	-	-	-	-	-	-	-	-	14	-
OTHER SECTORS	**-**	**-**	**-**	**-**	**7964**	**-**	**-**	**-**	**3067**	**2376**
Agriculture	-	-	-	-	-	-	-	-	24	440
Comm. and Publ. Services	-	-	-	-	-	-	-	-	515	774
Residential	-	-	-	-	7814	-	-	-	2528	1162
Non-specified	-	-	-	-	150	-	-	-	-	-
NON-ENERGY USE	**-**	**-**	**-**	**-**	**-**	**-**	**-**	**-**	**-**	**-**

APPROVISIONNEMENT ET DEMANDE 1997	*Gaz* (TJ)				*En. Re. Comb. & Déchets* (TJ)				(GWh)	(TJ)
	Gaz naturel	*Usines à gaz*	*Cokeries*	*Hauts fourneaux*	*Biomasse solide*	*Gaz/Liquides tirés de biomasse*	*Déchets urbains*	*Déchets industriels*	*Electricité*	*Chaleur*
Production	-	-	-	-	7814	-	-	-	6719	6208
Imports	-	-	-	-	150	-	-	-	-	-
Exports	-	-	-	-	-	-	-	-	-	-
Intl. Marine Bunkers	-	-	-	-	-	-	-	-	-	-
Stock Changes	-	-	-	-	-	-	-	-	-	-
DOMESTIC SUPPLY	**-**	**-**	**-**	**-**	**7964**	**-**	**-**	**-**	**6719**	**6208**
Transfers and Stat. Diff.	-	-	-	-	-	-	-	-	-	-
TRANSFORMATION	**-**	**-**	**-**	**-**	**-**	**-**	**-**	**-**	**-**	**-**
Electricity and CHP Plants	-	-	-	-	-	-	-	-	-	-
Petroleum Refineries	-	-	-	-	-	-	-	-	-	-
Other Transform. Sector	-	-	-	-	-	-	-	-	-	-
ENERGY SECTOR	**-**	**-**	**-**	**-**	**-**	**-**	**-**	**-**	**603**	**123**
DISTRIBUTION LOSSES	**-**	**-**	**-**	**-**	**-**	**-**	**-**	**-**	**839**	**898**
FINAL CONSUMPTION	**-**	**-**	**-**	**-**	**7964**	**-**	**-**	**-**	**5277**	**5187**
INDUSTRY SECTOR	**-**	**-**	**-**	**-**	**-**	**-**	**-**	**-**	**2087**	**2811**
Iron and Steel	-	-	-	-	-	-	-	-	1130	132
Chemical and Petrochem.	-	-	-	-	-	-	-	-	145	624
Non-Metallic Minerals	-	-	-	-	-	-	-	-	28	16
Non-specified	-	-	-	-	-	-	-	-	784	2039
TRANSPORT SECTOR	**-**	**-**	**-**	**-**	**-**	**-**	**-**	**-**	**14**	**-**
Air	-	-	-	-	-	-	-	-	-	-
Road	-	-	-	-	-	-	-	-	-	-
Non-specified	-	-	-	-	-	-	-	-	14	-
OTHER SECTORS	**-**	**-**	**-**	**-**	**7964**	**-**	**-**	**-**	**3176**	**2376**
Agriculture	-	-	-	-	-	-	-	-	25	440
Comm. and Publ. Services	-	-	-	-	-	-	-	-	533	774
Residential	-	-	-	-	7814	-	-	-	2618	1162
Non-specified	-	-	-	-	150	-	-	-	-	-
NON-ENERGY USE	**-**	**-**	**-**	**-**	**-**	**-**	**-**	**-**	**-**	**-**

Malaysia / Malaisie : 1996

SUPPLY AND CONSUMPTION / *APPROVISIONNEMENT ET DEMANDE*	Coal / *Charbon* (1000 tonnes)							Oil / *Pétrole* (1000 tonnes)			
	Coking Coal / *Charbon à coke*	Other Bit. Coal / *Autres charb. bit.*	Sub-Bit. Coal / *Charbon sous-bit.*	Lignite / *Lignite*	Peat / *Tourbe*	Oven and Gas Coke / *Coke de four/gaz*	Pat. Fuel and BKB / *Agg./briq. de lignite*	Crude Oil / *Pétrole brut*	NGL / *LGN*	Feed-stocks / *Produits d'aliment.*	Additives / *Additifs*
Production	-	219	-	-	-	-	-	31648	3430	-	-
From Other Sources	-	-	-	-	-	-	-	-	-	-	-
Imports	-	2661	-	-	-	108	-	1292	-	860	-
Exports	-	-21	-	-	-	-	-	-16687	-	-	-
Intl. Marine Bunkers	-	-	-	-	-	-	-	-	-	-	-
Stock Changes	-	-33	-	-	-	-	-	-	-	-	-
DOMESTIC SUPPLY	**-**	**2826**	**-**	**-**	**-**	**108**	**-**	**16253**	**3430**	**860**	**-**
Transfers	-	-	-	-	-	-	-	-	-3430	-	-
Statistical Differences	-	-430	-	-	-	-	-	2112	-	-	-
TRANSFORMATION	**-**	**1357**	**-**	**-**	**-**	**86**	**-**	**18365**	**-**	**860**	**-**
Electricity Plants	-	1357	-	-	-	-	-	-	-	-	-
CHP Plants	-	-	-	-	-	-	-	-	-	-	-
Heat Plants	-	-	-	-	-	-	-	-	-	-	-
Blast Furnaces/Gas Works	-	-	-	-	-	86	-	-	-	-	-
Coke/Pat. Fuel/BKB Plants	-	-	-	-	-	-	-	-	-	-	-
Petroleum Refineries	-	-	-	-	-	-	-	18365	-	860	-
Petrochemical Industry	-	-	-	-	-	-	-	-	-	-	-
Liquefaction	-	-	-	-	-	-	-	-	-	-	-
Other Transform. Sector	-	-	-	-	-	-	-	-	-	-	-
ENERGY SECTOR	**-**	**-**	**-**	**-**	**-**	**-**	**-**	**-**	**-**	**-**	**-**
Coal Mines	-	-	-	-	-	-	-	-	-	-	-
Oil and Gas Extraction	-	-	-	-	-	-	-	-	-	-	-
Petroleum Refineries	-	-	-	-	-	-	-	-	-	-	-
Electr., CHP+Heat Plants	-	-	-	-	-	-	-	-	-	-	-
Pumped Storage (Elec.)	-	-	-	-	-	-	-	-	-	-	-
Other Energy Sector	-	-	-	-	-	-	-	-	-	-	-
Distribution Losses	-	-	-	-	-	-	-	-	-	-	-
FINAL CONSUMPTION	**-**	**1039**	**-**	**-**	**-**	**22**	**-**	**-**	**-**	**-**	**-**
INDUSTRY SECTOR	**-**	**1039**	**-**	**-**	**-**	**22**	**-**	**-**	**-**	**-**	**-**
Iron and Steel	-	-	-	-	-	22	-	-	-	-	-
Chemical and Petrochem.	-	-	-	-	-	-	-	-	-	-	-
of which: Feedstocks	-	-	-	-	-	-	-	-	-	-	-
Non-Ferrous Metals	-	-	-	-	-	-	-	-	-	-	-
Non-Metallic Minerals	-	1039	-	-	-	-	-	-	-	-	-
Transport Equipment	-	-	-	-	-	-	-	-	-	-	-
Machinery	-	-	-	-	-	-	-	-	-	-	-
Mining and Quarrying	-	-	-	-	-	-	-	-	-	-	-
Food and Tobacco	-	-	-	-	-	-	-	-	-	-	-
Paper, Pulp and Print	-	-	-	-	-	-	-	-	-	-	-
Wood and Wood Products	-	-	-	-	-	-	-	-	-	-	-
Construction	-	-	-	-	-	-	-	-	-	-	-
Textile and Leather	-	-	-	-	-	-	-	-	-	-	-
Non-specified	-	-	-	-	-	-	-	-	-	-	-
TRANSPORT SECTOR	**-**	**-**	**-**	**-**	**-**	**-**	**-**	**-**	**-**	**-**	**-**
Air	-	-	-	-	-	-	-	-	-	-	-
Road	-	-	-	-	-	-	-	-	-	-	-
Rail	-	-	-	-	-	-	-	-	-	-	-
Pipeline Transport	-	-	-	-	-	-	-	-	-	-	-
Internal Navigation	-	-	-	-	-	-	-	-	-	-	-
Non-specified	-	-	-	-	-	-	-	-	-	-	-
OTHER SECTORS	**-**	**-**	**-**	**-**	**-**	**-**	**-**	**-**	**-**	**-**	**-**
Agriculture	-	-	-	-	-	-	-	-	-	-	-
Comm. and Publ. Services	-	-	-	-	-	-	-	-	-	-	-
Residential	-	-	-	-	-	-	-	-	-	-	-
Non-specified	-	-	-	-	-	-	-	-	-	-	-
NON-ENERGY USE	**-**	**-**	**-**	**-**	**-**	**-**	**-**	**-**	**-**	**-**	**-**
in Industry/Trans./Energy	-	-	-	-	-	-	-	-	-	-	-
in Transport	-	-	-	-	-	-	-	-	-	-	-
in Other Sectors	-	-	-	-	-	-	-	-	-	-	-

Malaysia / Malaisie : 1996

SUPPLY AND CONSUMPTION *APPROVISIONNEMENT ET DEMANDE*	Oil cont. / *Pétrole cont.* (1000 tonnes)										
	Refinery Gas *Gaz de raffinerie*	LPG + Ethane *GPL + éthane*	Motor Gasoline *Essence moteur*	Aviation Gasoline *Essence aviation*	Jet Fuel *Carbu-réacteurs*	Kerosene *Kérosène*	Gas/ Diesel *Gazole*	Heavy Fuel Oil *Fioul lourd*	Naphtha *Naphta*	Petrol. Coke *Coke de pétrole*	Other Prod. *Autres prod.*
Production	365	396	2385	-	1574	357	6921	2533	-	-	2976
From Other Sources	-	-	-	-	-	-	-	-	-	-	-
Imports	-	643	2172	2	97	23	1335	2975	-	-	1387
Exports	-	-726	-386	-	-372	-150	-1706	-1211	-	-	-3810
Intl. Marine Bunkers	-	-	-	-	-	-	-21	-142	-	-	-5
Stock Changes	-	-	-	-	-	-	-	-	-	-	-
DOMESTIC SUPPLY	**365**	**313**	**4171**	**2**	**1299**	**230**	**6529**	**4155**	**-**	**-**	**548**
Transfers	98	2280	-	-	-	29	135	-	-	-	265
Statistical Differences	-23	-363	657	-	43	-53	-66	-126	-	-	-
TRANSFORMATION	**-**	**-**	**-**	**-**	**-**	**-**	**288**	**2308**	**-**	**-**	**-**
Electricity Plants	-	-	-	-	-	-	288	2308	-	-	-
CHP Plants	-	-	-	-	-	-	-	-	-	-	-
Heat Plants	-	-	-	-	-	-	-	-	-	-	-
Blast Furnaces/Gas Works	-	-	-	-	-	-	-	-	-	-	-
Coke/Pat. Fuel/BKB Plants	-	-	-	-	-	-	-	-	-	-	-
Petroleum Refineries	-	-	-	-	-	-	-	-	-	-	-
Petrochemical Industry	-	-	-	-	-	-	-	-	-	-	-
Liquefaction	-	-	-	-	-	-	-	-	-	-	-
Other Transform. Sector	-	-	-	-	-	-	-	-	-	-	-
ENERGY SECTOR	**440**	**-**	**-**	**-**	**-**	**-**	**-**	**45**	**-**	**-**	**-**
Coal Mines	-	-	-	-	-	-	-	-	-	-	-
Oil and Gas Extraction	-	-	-	-	-	-	-	-	-	-	-
Petroleum Refineries	440	-	-	-	-	-	-	45	-	-	-
Electr., CHP+Heat Plants	-	-	-	-	-	-	-	-	-	-	-
Pumped Storage (Elec.)	-	-	-	-	-	-	-	-	-	-	-
Other Energy Sector	-	-	-	-	-	-	-	-	-	-	-
Distribution Losses	-	-	-	-	-	-	-	-	-	-	-
FINAL CONSUMPTION	**-**	**2230**	**4828**	**2**	**1342**	**206**	**6310**	**1676**	**-**	**-**	**813**
INDUSTRY SECTOR	**-**	**239**	**72**	**-**	**-**	**13**	**3445**	**1604**	**-**	**-**	**-**
Iron and Steel	-	-	-	-	-	-	-	-	-	-	-
Chemical and Petrochem.	-	-	-	-	-	-	-	-	-	-	-
of which: Feedstocks	-	-	-	-	-	-	-	-	-	-	-
Non-Ferrous Metals	-	-	-	-	-	-	-	-	-	-	-
Non-Metallic Minerals	-	-	-	-	-	-	-	-	-	-	-
Transport Equipment	-	-	-	-	-	-	-	-	-	-	-
Machinery	-	-	-	-	-	-	-	-	-	-	-
Mining and Quarrying	-	-	-	-	-	-	-	-	-	-	-
Food and Tobacco	-	-	-	-	-	-	-	-	-	-	-
Paper, Pulp and Print	-	-	-	-	-	-	-	-	-	-	-
Wood and Wood Products	-	-	-	-	-	-	-	-	-	-	-
Construction	-	-	-	-	-	-	-	-	-	-	-
Textile and Leather	-	-	-	-	-	-	-	-	-	-	-
Non-specified	-	239	72	-	-	13	3445	1604	-	-	-
TRANSPORT SECTOR	**-**	**-**	**4752**	**2**	**1342**	**-**	**2355**	**19**	**-**	**-**	**-**
Air	-	-	-	2	1342	-	-	-	-	-	-
Road	-	-	4752	-	-	-	2355	-	-	-	-
Rail	-	-	-	-	-	-	-	-	-	-	-
Pipeline Transport	-	-	-	-	-	-	-	-	-	-	-
Internal Navigation	-	-	-	-	-	-	-	19	-	-	-
Non-specified	-	-	-	-	-	-	-	-	-	-	-
OTHER SECTORS	**-**	**742**	**4**	**-**	**-**	**193**	**510**	**53**	**-**	**-**	**-**
Agriculture	-	-	4	-	-	-	466	15	-	-	-
Comm. and Publ. Services	-	-	-	-	-	-	-	-	-	-	-
Residential	-	742	-	-	-	193	44	38	-	-	-
Non-specified	-	-	-	-	-	-	-	-	-	-	-
NON-ENERGY USE	**-**	**1249**	**-**	**-**	**-**	**-**	**-**	**-**	**-**	**-**	**813**
in Industry/Transf./Energy	-	1249	-	-	-	-	-	-	-	-	813
in Transport	-	-	-	-	-	-	-	-	-	-	-
in Other Sectors	-	-	-	-	-	-	-	-	-	-	-

Malaysia / Malaisie : 1996

SUPPLY AND CONSUMPTION / *APPROVISIONNEMENT ET DEMANDE*	Gas / *Gaz* (TJ)				Comb. Renew. & Waste / *En. Re. Comb. & Déchets* (TJ)				(GWh)	(TJ)
	Natural Gas / *Gaz naturel*	Gas Works / *Usines à gaz*	Coke Ovens / *Cokeries*	Blast Furnaces / *Hauts fourneaux*	Solid Biomass / *Biomasse solide*	Gas/Liquids from Biomass / *Gaz/Liquides tirés de biomasse*	Municipal Waste / *Déchets urbains*	Industrial Waste / *Déchets industriels*	Electricity / *Electricité*	Heat / *Chaleur*
Production	1407657	-	-	-	100794	-	-	-	51407	-
From Other Sources	-	-	-	-	-	-	-	-	-	-
Imports	-	-	-	-	189	-	-	-	128	-
Exports	-711876	-	-	-	-897	-	-	-	-140	-
Intl. Marine Bunkers	-	-	-	-	-	-	-	-	-	-
Stock Changes	-	-	-	-	-	-	-	-	-	-
DOMESTIC SUPPLY	**695781**	-	-	-	**100086**	-	-	-	**51395**	-
Transfers	-	-	-	-	-	-	-	-	-	-
Statistical Differences	5561	-	-	-	-1	-	-	-	-163	-
TRANSFORMATION	**386468**	-	-	-	**39943**	-	-	-	-	-
Electricity Plants	386468	-	-	-	-	-	-	-	-	-
CHP Plants	-	-	-	-	-	-	-	-	-	-
Heat Plants	-	-	-	-	-	-	-	-	-	-
Blast Furnaces/Gas Works	-	-	-	-	-	-	-	-	-	-
Coke/Pat. Fuel/BKB Plants	-	-	-	-	-	-	-	-	-	-
Petroleum Refineries	-	-	-	-	-	-	-	-	-	-
Petrochemical Industry	-	-	-	-	-	-	-	-	-	-
Liquefaction	-	-	-	-	-	-	-	-	-	-
Other Transform. Sector	-	-	-	-	39943	-	-	-	-	-
ENERGY SECTOR	**196667**	-	-	-	-	-	-	-	**1616**	-
Coal Mines	-	-	-	-	-	-	-	-	-	-
Oil and Gas Extraction	179917	-	-	-	-	-	-	-	-	-
Petroleum Refineries	-	-	-	-	-	-	-	-	-	-
Electr., CHP+Heat Plants	-	-	-	-	-	-	-	-	1616	-
Pumped Storage (Elec.)	-	-	-	-	-	-	-	-	-	-
Other Energy Sector	16750	-	-	-	-	-	-	-	-	-
Distribution Losses	5001	-	-	-	-	-	-	-	5709	-
FINAL CONSUMPTION	**113206**	-	-	-	**60142**	-	-	-	**43907**	-
INDUSTRY SECTOR	**95490**	-	-	-	**2772**	-	-	-	**23593**	-
Iron and Steel	6325	-	-	-	-	-	-	-	-	-
Chemical and Petrochem.	88401	-	-	-	-	-	-	-	-	-
of which: Feedstocks	*60736*	-	-	-	-	-	-	-	-	-
Non-Ferrous Metals	-	-	-	-	-	-	-	-	-	-
Non-Metallic Minerals	764	-	-	-	-	-	-	-	-	-
Transport Equipment	-	-	-	-	-	-	-	-	-	-
Machinery	-	-	-	-	-	-	-	-	-	-
Mining and Quarrying	-	-	-	-	-	-	-	-	-	-
Food and Tobacco	-	-	-	-	-	-	-	-	-	-
Paper, Pulp and Print	-	-	-	-	-	-	-	-	-	-
Wood and Wood Products	-	-	-	-	-	-	-	-	-	-
Construction	-	-	-	-	-	-	-	-	-	-
Textile and Leather	-	-	-	-	-	-	-	-	-	-
Non-specified	-	-	-	-	2772	-	-	-	23593	-
TRANSPORT SECTOR	**302**	-	-	-	-	-	-	-	-	-
Air	-	-	-	-	-	-	-	-	-	-
Road	-	-	-	-	-	-	-	-	-	-
Rail	-	-	-	-	-	-	-	-	-	-
Pipeline Transport	-	-	-	-	-	-	-	-	-	-
Internal Navigation	-	-	-	-	-	-	-	-	-	-
Non-specified	302	-	-	-	-	-	-	-	-	-
OTHER SECTORS	**17414**	-	-	-	**57370**	-	-	-	**20314**	-
Agriculture	-	-	-	-	-	-	-	-	-	-
Comm. and Publ. Services	-	-	-	-	-	-	-	-	11756	-
Residential	17414	-	-	-	57370	-	-	-	8558	-
Non-specified	-	-	-	-	-	-	-	-	-	-
NON-ENERGY USE	-	-	-	-	-	-	-	-	-	-
in Industry/Transf./Energy	-	-	-	-	-	-	-	-	-	-
in Transport	-	-	-	-	-	-	-	-	-	-
in Other Sectors	-	-	-	-	-	-	-	-	-	-

Malaysia / Malaisie : 1997

SUPPLY AND CONSUMPTION / *APPROVISIONNEMENT ET DEMANDE*	Coal / *Charbon* (1000 tonnes)							Oil / *Pétrole* (1000 tonnes)			
	Coking Coal / *Charbon à coke*	Other Bit. Coal / *Autres charb. bit.*	Sub-Bit. Coal / *Charbon sous-bit.*	Lignite / *Lignite*	Peat / *Tourbe*	Oven and Gas Coke / *Coke de four/gaz*	Pat. Fuel and BKB / *Agg./briq. de lignite*	Crude Oil / *Pétrole brut*	NGL / *LGN*	Feed-stocks / *Produits d'aliment.*	Additives / *Additifs*
Production	-	219	-	-	-	-	-	31520	5149	-	-
From Other Sources	-	-	-	-	-	-	-	-	-	-	-
Imports	-	1986	-	-	-	108	-	1257	-	1291	-
Exports	-	-13	-	-	-	-	-	-17016	-	-	-
Intl. Marine Bunkers	-	-	-	-	-	-	-	-	-	-	-
Stock Changes	-	-30	-	-	-	-	-	-	-	-	-
DOMESTIC SUPPLY	**-**	**2162**	**-**	**-**	**-**	**108**	**-**	**15761**	**5149**	**1291**	**-**
Transfers	-	-	-	-	-	-	-	-	-2907	-	-
Statistical Differences	-	156	-	-	-	-	-	1566	-2242	-	-
TRANSFORMATION	**-**	**1260**	**-**	**-**	**-**	**86**	**-**	**17327**	**-**	**1291**	**-**
Electricity Plants	-	1260	-	-	-	-	-	-	-	-	-
CHP Plants	-	-	-	-	-	-	-	-	-	-	-
Heat Plants	-	-	-	-	-	-	-	-	-	-	-
Blast Furnaces/Gas Works	-	-	-	-	-	86	-	-	-	-	-
Coke/Pat. Fuel/BKB Plants	-	-	-	-	-	-	-	-	-	-	-
Petroleum Refineries	-	-	-	-	-	-	-	17327	-	1291	-
Petrochemical Industry	-	-	-	-	-	-	-	-	-	-	-
Liquefaction	-	-	-	-	-	-	-	-	-	-	-
Other Transform. Sector	-	-	-	-	-	-	-	-	-	-	-
ENERGY SECTOR	**-**	**-**	**-**	**-**	**-**	**-**	**-**	**-**	**-**	**-**	**-**
Coal Mines	-	-	-	-	-	-	-	-	-	-	-
Oil and Gas Extraction	-	-	-	-	-	-	-	-	-	-	-
Petroleum Refineries	-	-	-	-	-	-	-	-	-	-	-
Electr., CHP+Heat Plants	-	-	-	-	-	-	-	-	-	-	-
Pumped Storage (Elec.)	-	-	-	-	-	-	-	-	-	-	-
Other Energy Sector	-	-	-	-	-	-	-	-	-	-	-
Distribution Losses	-	-	-	-	-	-	-	-	-	-	-
FINAL CONSUMPTION	**-**	**1058**	**-**	**-**	**-**	**22**	**-**	**-**	**-**	**-**	**-**
INDUSTRY SECTOR	**-**	**1058**	**-**	**-**	**-**	**22**	**-**	**-**	**-**	**-**	**-**
Iron and Steel	-	-	-	-	-	22	-	-	-	-	-
Chemical and Petrochem.	-	-	-	-	-	-	-	-	-	-	-
of which: Feedstocks	-	-	-	-	-	-	-	-	-	-	-
Non-Ferrous Metals	-	-	-	-	-	-	-	-	-	-	-
Non-Metallic Minerals	-	1058	-	-	-	-	-	-	-	-	-
Transport Equipment	-	-	-	-	-	-	-	-	-	-	-
Machinery	-	-	-	-	-	-	-	-	-	-	-
Mining and Quarrying	-	-	-	-	-	-	-	-	-	-	-
Food and Tobacco	-	-	-	-	-	-	-	-	-	-	-
Paper, Pulp and Print	-	-	-	-	-	-	-	-	-	-	-
Wood and Wood Products	-	-	-	-	-	-	-	-	-	-	-
Construction	-	-	-	-	-	-	-	-	-	-	-
Textile and Leather	-	-	-	-	-	-	-	-	-	-	-
Non-specified	-	-	-	-	-	-	-	-	-	-	-
TRANSPORT SECTOR	**-**	**-**	**-**	**-**	**-**	**-**	**-**	**-**	**-**	**-**	**-**
Air	-	-	-	-	-	-	-	-	-	-	-
Road	-	-	-	-	-	-	-	-	-	-	-
Rail	-	-	-	-	-	-	-	-	-	-	-
Pipeline Transport	-	-	-	-	-	-	-	-	-	-	-
Internal Navigation	-	-	-	-	-	-	-	-	-	-	-
Non-specified	-	-	-	-	-	-	-	-	-	-	-
OTHER SECTORS	**-**	**-**	**-**	**-**	**-**	**-**	**-**	**-**	**-**	**-**	**-**
Agriculture	-	-	-	-	-	-	-	-	-	-	-
Comm. and Publ. Services	-	-	-	-	-	-	-	-	-	-	-
Residential	-	-	-	-	-	-	-	-	-	-	-
Non-specified	-	-	-	-	-	-	-	-	-	-	-
NON-ENERGY USE	**-**	**-**	**-**	**-**	**-**	**-**	**-**	**-**	**-**	**-**	**-**
in Industry/Trans./Energy	-	-	-	-	-	-	-	-	-	-	-
in Transport	-	-	-	-	-	-	-	-	-	-	-
in Other Sectors	-	-	-	-	-	-	-	-	-	-	-

Malaysia / Malaisie : 1997

	Oil cont. / *Pétrole cont.* (1000 tonnes)										
SUPPLY AND CONSUMPTION	Refinery Gas	LPG + Ethane	Motor Gasoline	Aviation Gasoline	Jet Fuel	Kerosene	Gas/ Diesel	Heavy Fuel Oil	Naphtha	Petrol. Coke	Other Prod.
APPROVISIONNEMENT ET DEMANDE	*Gaz de raffinerie*	*GPL + éthane*	*Essence moteur*	*Essence aviation*	*Carbu-réacteurs*	*Kérosène*	*Gazole*	*Fioul lourd*	*Naphta*	*Coke de pétrole*	*Autres prod.*
Production	353	396	2250	-	1658	324	6530	2390	-	-	2078
From Other Sources	-	-	-	-	-	-	-	-	-	-	-
Imports	-	782	2652	2	62	4	1560	3212	-	-	1147
Exports	-	-795	-42	-	-335	-207	-1374	-1415	-	-	-3175
Intl. Marine Bunkers	-	-	-	-	-	-	-21	-136	-	-	-1
Stock Changes	-	-	26	-	-	-	-	-40	-	-	-9
DOMESTIC SUPPLY	**353**	**383**	**4886**	**2**	**1385**	**121**	**6695**	**4011**	**-**	**-**	**40**
Transfers	91	2359	-	-	-	30	125	-	-	-	302
Statistical Differences	-	-383	-	-	62	118	220	349	-	-	-1
TRANSFORMATION	**-**	**-**	**-**	**-**	**-**	**-**	**188**	**2433**	**-**	**-**	**-**
Electricity Plants	-	-	-	-	-	-	188	2433	-	-	-
CHP Plants	-	-	-	-	-	-	-	-	-	-	-
Heat Plants	-	-	-	-	-	-	-	-	-	-	-
Blast Furnaces/Gas Works	-	-	-	-	-	-	-	-	-	-	-
Coke/Pat. Fuel/BKB Plants	-	-	-	-	-	-	-	-	-	-	-
Petroleum Refineries	-	-	-	-	-	-	-	-	-	-	-
Petrochemical Industry	-	-	-	-	-	-	-	-	-	-	-
Liquefaction	-	-	-	-	-	-	-	-	-	-	-
Other Transform. Sector	-	-	-	-	-	-	-	-	-	-	-
ENERGY SECTOR	**444**	**-**	**-**	**-**	**-**	**-**	**-**	**39**	**-**	**-**	**-**
Coal Mines	-	-	-	-	-	-	-	-	-	-	-
Oil and Gas Extraction	-	-	-	-	-	-	-	-	-	-	-
Petroleum Refineries	444	-	-	-	-	-	-	39	-	-	-
Electr., CHP+Heat Plants	-	-	-	-	-	-	-	-	-	-	-
Pumped Storage (Elec.)	-	-	-	-	-	-	-	-	-	-	-
Other Energy Sector	-	-	-	-	-	-	-	-	-	-	-
Distribution Losses	-	-	-	-	-	-	-	-	-	-	-
FINAL CONSUMPTION	**-**	**2359**	**4886**	**2**	**1447**	**269**	**6852**	**1888**	**-**	**-**	**341**
INDUSTRY SECTOR	**-**	**254**	**73**	**-**	**-**	**17**	**3741**	**1807**	**-**	**-**	**-**
Iron and Steel	-	-	-	-	-	-	-	-	-	-	-
Chemical and Petrochem.	-	-	-	-	-	-	-	-	-	-	-
of which: Feedstocks	-	-	-	-	-	-	-	-	-	-	-
Non-Ferrous Metals	-	-	-	-	-	-	-	-	-	-	-
Non-Metallic Minerals	-	-	-	-	-	-	-	-	-	-	-
Transport Equipment	-	-	-	-	-	-	-	-	-	-	-
Machinery	-	-	-	-	-	-	-	-	-	-	-
Mining and Quarrying	-	-	-	-	-	-	-	-	-	-	-
Food and Tobacco	-	-	-	-	-	-	-	-	-	-	-
Paper, Pulp and Print	-	-	-	-	-	-	-	-	-	-	-
Wood and Wood Products	-	-	-	-	-	-	-	-	-	-	-
Construction	-	-	-	-	-	-	-	-	-	-	-
Textile and Leather	-	-	-	-	-	-	-	-	-	-	-
Non-specified	-	254	73	-	-	17	3741	1807	-	-	-
TRANSPORT SECTOR	**-**	**-**	**4809**	**2**	**1447**	**-**	**2557**	**21**	**-**	**-**	**-**
Air	-	-	-	2	1447	-	-	-	-	-	-
Road	-	-	4809	-	-	-	2557	-	-	-	-
Rail	-	-	-	-	-	-	-	-	-	-	-
Pipeline Transport	-	-	-	-	-	-	-	-	-	-	-
Internal Navigation	-	-	-	-	-	-	-	21	-	-	-
Non-specified	-	-	-	-	-	-	-	-	-	-	-
OTHER SECTORS	**-**	**788**	**4**	**-**	**-**	**252**	**554**	**60**	**-**	**-**	**-**
Agriculture	-	-	4	-	-	-	506	17	-	-	-
Comm. and Publ. Services	-	-	-	-	-	-	-	-	-	-	-
Residential	-	788	-	-	-	252	48	43	-	-	-
Non-specified	-	-	-	-	-	-	-	-	-	-	-
NON-ENERGY USE	**-**	**1317**	**-**	**-**	**-**	**-**	**-**	**-**	**-**	**-**	**341**
in Industry/Transf./Energy	-	1317	-	-	-	-	-	-	-	-	341
in Transport	-	-	-	-	-	-	-	-	-	-	-
in Other Sectors	-	-	-	-	-	-	-	-	-	-	-

Malaysia / Malaisie : 1997

SUPPLY AND CONSUMPTION / *APPROVISIONNEMENT ET DEMANDE*	Gas / *Gaz* (TJ)				Comb. Renew. & Waste / *En. Re. Comb. & Déchets* (TJ)				(GWh)	(TJ)
	Natural Gas / *Gaz naturel*	Gas Works / *Usines à gaz*	Coke Ovens / *Cokeries*	Blast Furnaces / *Hauts fourneaux*	Solid Biomass / *Biomasse solide*	Gas/Liquids from Biomass / *Gaz/Liquides tirés de biomasse*	Municipal Waste / *Déchets urbains*	Industrial Waste / *Déchets industriels*	Electricity / *Electricité*	Heat / *Chaleur*
Production	1540087	-	-	-	100794	-	-	-	57875	-
From Other Sources	-	-	-	-	-	-	-	-	-	-
Imports	-	-	-	-	189	-	-	-	83	-
Exports	-647057	-	-	-	-897	-	-	-	-75	-
Intl. Marine Bunkers	-	-	-	-	-	-	-	-	-	-
Stock Changes	-	-	-	-	-	-	-	-	-	-
DOMESTIC SUPPLY	**893030**	**-**	**-**	**-**	**100086**	**-**	**-**	**-**	**57883**	**-**
Transfers	-	-	-	-	-	-	-	-	-	-
Statistical Differences	-71479	-	-	-	-1	-	-	-	-75	-
TRANSFORMATION	**481568**	**-**	**-**	**-**	**39943**	**-**	**-**	**-**	**-**	**-**
Electricity Plants	481568	-	-	-	-	-	-	-	-	-
CHP Plants	-	-	-	-	-	-	-	-	-	-
Heat Plants	-	-	-	-	-	-	-	-	-	-
Blast Furnaces/Gas Works	-	-	-	-	-	-	-	-	-	-
Coke/Pat. Fuel/BKB Plants	-	-	-	-	-	-	-	-	-	-
Petroleum Refineries	-	-	-	-	-	-	-	-	-	-
Petrochemical Industry	-	-	-	-	-	-	-	-	-	-
Liquefaction	-	-	-	-	-	-	-	-	-	-
Other Transform. Sector	-	-	-	-	39943	-	-	-	-	-
ENERGY SECTOR	**235616**	**-**	**-**	**-**	**-**	**-**	**-**	**-**	**1509**	**-**
Coal Mines	-	-	-	-	-	-	-	-	-	-
Oil and Gas Extraction	215549	-	-	-	-	-	-	-	-	-
Petroleum Refineries	-	-	-	-	-	-	-	-	-	-
Electr., CHP+Heat Plants	-	-	-	-	-	-	-	-	1509	-
Pumped Storage (Elec.)	-	-	-	-	-	-	-	-	-	-
Other Energy Sector	20067	-	-	-	-	-	-	-	-	-
Distribution Losses	1123	-	-	-	-	-	-	-	5331	-
FINAL CONSUMPTION	**103244**	**-**	**-**	**-**	**60142**	**-**	**-**	**-**	**50968**	**-**
INDUSTRY SECTOR	**87087**	**-**	**-**	**-**	**2772**	**-**	**-**	**-**	**28163**	**-**
Iron and Steel	5768	-	-	-	-	-	-	-	-	-
Chemical and Petrochem.	80622	-	-	-	-	-	-	-	-	-
of which: Feedstocks	*55391*	-	-	-	-	-	-	-	-	-
Non-Ferrous Metals	-	-	-	-	-	-	-	-	-	-
Non-Metallic Minerals	697	-	-	-	-	-	-	-	-	-
Transport Equipment	-	-	-	-	-	-	-	-	-	-
Machinery	-	-	-	-	-	-	-	-	-	-
Mining and Quarrying	-	-	-	-	-	-	-	-	-	-
Food and Tobacco	-	-	-	-	-	-	-	-	-	-
Paper, Pulp and Print	-	-	-	-	-	-	-	-	-	-
Wood and Wood Products	-	-	-	-	-	-	-	-	-	-
Construction	-	-	-	-	-	-	-	-	-	-
Textile and Leather	-	-	-	-	-	-	-	-	-	-
Non-specified	-	-	-	-	2772	-	-	-	28163	-
TRANSPORT SECTOR	**275**	**-**	**-**	**-**	**-**	**-**	**-**	**-**	**-**	**-**
Air	-	-	-	-	-	-	-	-	-	-
Road	-	-	-	-	-	-	-	-	-	-
Rail	-	-	-	-	-	-	-	-	-	-
Pipeline Transport	-	-	-	-	-	-	-	-	-	-
Internal Navigation	-	-	-	-	-	-	-	-	-	-
Non-specified	275	-	-	-	-	-	-	-	-	-
OTHER SECTORS	**15882**	**-**	**-**	**-**	**57370**	**-**	**-**	**-**	**22804**	**-**
Agriculture	-	-	-	-	-	-	-	-	-	-
Comm. and Publ. Services	-	-	-	-	-	-	-	-	13851	-
Residential	15882	-	-	-	57370	-	-	-	8953	-
Non-specified	-	-	-	-	-	-	-	-	-	-
NON-ENERGY USE	**-**	**-**	**-**	**-**	**-**	**-**	**-**	**-**	**-**	**-**
in Industry/Transf./Energy	-	-	-	-	-	-	-	-	-	-
in Transport	-	-	-	-	-	-	-	-	-	-
in Other Sectors	-	-	-	-	-	-	-	-	-	-

Malta / Malte

SUPPLY AND CONSUMPTION 1996	Coal (1000 tonnes)							Oil (1000 tonnes)			
	Coking Coal	Other Bit. Coal	Sub-Bit. Coal	Lignite	Peat	Oven and Gas Coke	Pat. Fuel and BKB	Crude Oil	NGL	Feed-stocks	Additives
Production	-	-	-	-	-	-	-	-	-	-	-
Imports	-	-	-	-	-	-	-	-	-	-	-
Exports	-	-	-	-	-	-	-	-	-	-	-
Intl. Marine Bunkers	-	-	-	-	-	-	-	-	-	-	-
Stock Changes	-	-	-	-	-	-	-	-	-	-	-
DOMESTIC SUPPLY	-	-	-	-	-	-	-	-	-	-	-
Transfers and Stat. Diff.	-	-	-	-	-	-	-	-	-	-	-
TRANSFORMATION	-	-	-	-	-	-	-	-	-	-	-
Electricity and CHP Plants	-	-	-	-	-	-	-	-	-	-	-
Petroleum Refineries	-	-	-	-	-	-	-	-	-	-	-
Other Transform. Sector	-	-	-	-	-	-	-	-	-	-	-
ENERGY SECTOR	-	-	-	-	-	-	-	-	-	-	-
DISTRIBUTION LOSSES	-	-	-	-	-	-	-	-	-	-	-
FINAL CONSUMPTION	-	-	-	-	-	-	-	-	-	-	-
INDUSTRY SECTOR	-	-	-	-	-	-	-	-	-	-	-
Iron and Steel	-	-	-	-	-	-	-	-	-	-	-
Chemical and Petrochem.	-	-	-	-	-	-	-	-	-	-	-
Non-Metallic Minerals	-	-	-	-	-	-	-	-	-	-	-
Non-specified	-	-	-	-	-	-	-	-	-	-	-
TRANSPORT SECTOR	-	-	-	-	-	-	-	-	-	-	-
Air	-	-	-	-	-	-	-	-	-	-	-
Road	-	-	-	-	-	-	-	-	-	-	-
Non-specified	-	-	-	-	-	-	-	-	-	-	-
OTHER SECTORS	-	-	-	-	-	-	-	-	-	-	-
Agriculture	-	-	-	-	-	-	-	-	-	-	-
Comm. and Publ. Services	-	-	-	-	-	-	-	-	-	-	-
Residential	-	-	-	-	-	-	-	-	-	-	-
Non-specified	-	-	-	-	-	-	-	-	-	-	-
NON-ENERGY USE	-	-	-	-	-	-	-	-	-	-	-

APPROVISIONNEMENT ET DEMANDE 1997	*Charbon* (1000 tonnes)							*Pétrole* (1000 tonnes)			
	Charbon à coke	*Autres charb. bit.*	*Charbon sous-bit.*	*Lignite*	*Tourbe*	*Coke de four/gaz*	*Agg./briq. de lignite*	*Pétrole brut*	*LGN*	*Produits d'aliment.*	*Additifs*
Production	-	-	-	-	-	-	-	-	-	-	-
Imports	-	-	-	-	-	-	-	-	-	-	-
Exports	-	-	-	-	-	-	-	-	-	-	-
Intl. Marine Bunkers	-	-	-	-	-	-	-	-	-	-	-
Stock Changes	-	-	-	-	-	-	-	-	-	-	-
DOMESTIC SUPPLY	-	-	-	-	-	-	-	-	-	-	-
Transfers and Stat. Diff.	-	-	-	-	-	-	-	-	-	-	-
TRANSFORMATION	-	-	-	-	-	-	-	-	-	-	-
Electricity and CHP Plants	-	-	-	-	-	-	-	-	-	-	-
Petroleum Refineries	-	-	-	-	-	-	-	-	-	-	-
Other Transform. Sector	-	-	-	-	-	-	-	-	-	-	-
ENERGY SECTOR	-	-	-	-	-	-	-	-	-	-	-
DISTRIBUTION LOSSES	-	-	-	-	-	-	-	-	-	-	-
FINAL CONSUMPTION	-	-	-	-	-	-	-	-	-	-	-
INDUSTRY SECTOR	-	-	-	-	-	-	-	-	-	-	-
Iron and Steel	-	-	-	-	-	-	-	-	-	-	-
Chemical and Petrochem.	-	-	-	-	-	-	-	-	-	-	-
Non-Metallic Minerals	-	-	-	-	-	-	-	-	-	-	-
Non-specified	-	-	-	-	-	-	-	-	-	-	-
TRANSPORT SECTOR	-	-	-	-	-	-	-	-	-	-	-
Air	-	-	-	-	-	-	-	-	-	-	-
Road	-	-	-	-	-	-	-	-	-	-	-
Non-specified	-	-	-	-	-	-	-	-	-	-	-
OTHER SECTORS	-	-	-	-	-	-	-	-	-	-	-
Agriculture	-	-	-	-	-	-	-	-	-	-	-
Comm. and Publ. Services	-	-	-	-	-	-	-	-	-	-	-
Residential	-	-	-	-	-	-	-	-	-	-	-
Non-specified	-	-	-	-	-	-	-	-	-	-	-
NON-ENERGY USE	-	-	-	-	-	-	-	-	-	-	-

Malta / Malte

SUPPLY AND CONSUMPTION 1996	Oil cont. (1000 tonnes)										
	Refinery Gas	LPG + Ethane	Motor Gasoline	Aviation Gasoline	Jet Fuel	Kerosene	Gas/ Diesel	Heavy Fuel Oil	Naphtha	Petrol. Coke	Other Prod.
Production	-	-	-	-	-	-	-	-	-	-	-
Imports	-	17	77	-	107	19	190	547	-	-	14
Exports	-	-	-	-	-	-	-	-	-	-	-
Intl. Marine Bunkers	-	-	-	-	-	-	-20	-59	-	-	-
Stock Changes	-	-	-	-	-	-	-	-	-	-	-
DOMESTIC SUPPLY	**-**	**17**	**77**	**-**	**107**	**19**	**170**	**488**	**-**	**-**	**14**
Transfers and Stat. Diff.	-	-	-	-	-	-	-	-	-	-	-
TRANSFORMATION	**-**	**-**	**-**	**-**	**-**	**-**	**21**	**488**	**-**	**-**	**-**
Electricity and CHP Plants	-	-	-	-	-	-	21	488	-	-	-
Petroleum Refineries	-	-	-	-	-	-	-	-	-	-	-
Other Transform. Sector	-	-	-	-	-	-	-	-	-	-	-
ENERGY SECTOR	**-**	**-**	**-**	**-**	**-**	**-**	**-**	**-**	**-**	**-**	**-**
DISTRIBUTION LOSSES	**-**	**-**	**-**	**-**	**-**	**-**	**-**	**-**	**-**	**-**	**-**
FINAL CONSUMPTION	**-**	**17**	**77**	**-**	**107**	**19**	**149**	**-**	**-**	**-**	**14**
INDUSTRY SECTOR	**-**	**-**	**-**	**-**	**-**	**-**	**-**	**-**	**-**	**-**	**-**
Iron and Steel	-	-	-	-	-	-	-	-	-	-	-
Chemical and Petrochem.	-	-	-	-	-	-	-	-	-	-	-
Non-Metallic Minerals	-	-	-	-	-	-	-	-	-	-	-
Non-specified	-	-	-	-	-	-	-	-	-	-	-
TRANSPORT SECTOR	**-**	**-**	**77**	**-**	**107**	**-**	**149**	**-**	**-**	**-**	**-**
Air	-	-	-	-	107	-	-	-	-	-	-
Road	-	-	77	-	-	-	149	-	-	-	-
Non-specified	-	-	-	-	-	-	-	-	-	-	-
OTHER SECTORS	**-**	**17**	**-**	**-**	**-**	**19**	**-**	**-**	**-**	**-**	**14**
Agriculture	-	-	-	-	-	-	-	-	-	-	-
Comm. and Publ. Services	-	-	-	-	-	-	-	-	-	-	-
Residential	-	17	-	-	-	19	-	-	-	-	-
Non-specified	-	-	-	-	-	-	-	-	-	-	14
NON-ENERGY USE	**-**	**-**	**-**	**-**	**-**	**-**	**-**	**-**	**-**	**-**	**-**

APPROVISIONNEMENT ET DEMANDE 1997	*Pétrole cont.* (1000 tonnes)										
	Gaz de raffinerie	*GPL + éthane*	*Essence moteur*	*Essence aviation*	*Carbu- réacteurs*	*Kérosène*	*Gazole*	*Fioul lourd*	*Naphta*	*Coke de pétrole*	*Autres prod.*
Production	-	-	-	-	-	-	-	-	-	-	-
Imports	-	18	77	-	112	55	190	557	-	-	11
Exports	-	-	-	-	-	-	-	-	-	-	-
Intl. Marine Bunkers	-	-	-	-	-	-	-20	-60	-	-	-
Stock Changes	-	-	-	-	-	-	-	-	-	-	-
DOMESTIC SUPPLY	**-**	**18**	**77**	**-**	**112**	**55**	**170**	**497**	**-**	**-**	**11**
Transfers and Stat. Diff.	-	-	-	-	-	-	-	-	-	-	-
TRANSFORMATION	**-**	**-**	**-**	**-**	**-**	**-**	**21**	**497**	**-**	**-**	**-**
Electricity and CHP Plants	-	-	-	-	-	-	21	497	-	-	-
Petroleum Refineries	-	-	-	-	-	-	-	-	-	-	-
Other Transform. Sector	-	-	-	-	-	-	-	-	-	-	-
ENERGY SECTOR	**-**	**-**	**-**	**-**	**-**	**-**	**-**	**-**	**-**	**-**	**-**
DISTRIBUTION LOSSES	**-**	**-**	**-**	**-**	**-**	**-**	**-**	**-**	**-**	**-**	**-**
FINAL CONSUMPTION	**-**	**18**	**77**	**-**	**112**	**55**	**149**	**-**	**-**	**-**	**11**
INDUSTRY SECTOR	**-**	**-**	**-**	**-**	**-**	**-**	**-**	**-**	**-**	**-**	**-**
Iron and Steel	-	-	-	-	-	-	-	-	-	-	-
Chemical and Petrochem.	-	-	-	-	-	-	-	-	-	-	-
Non-Metallic Minerals	-	-	-	-	-	-	-	-	-	-	-
Non-specified	-	-	-	-	-	-	-	-	-	-	-
TRANSPORT SECTOR	**-**	**-**	**77**	**-**	**112**	**-**	**149**	**-**	**-**	**-**	**-**
Air	-	-	-	-	112	-	-	-	-	-	-
Road	-	-	77	-	-	-	149	-	-	-	-
Non-specified	-	-	-	-	-	-	-	-	-	-	-
OTHER SECTORS	**-**	**18**	**-**	**-**	**-**	**55**	**-**	**-**	**-**	**-**	**11**
Agriculture	-	-	-	-	-	-	-	-	-	-	-
Comm. and Publ. Services	-	-	-	-	-	-	-	-	-	-	-
Residential	-	18	-	-	-	55	-	-	-	-	-
Non-specified	-	-	-	-	-	-	-	-	-	-	11
NON-ENERGY USE	**-**	**-**	**-**	**-**	**-**	**-**	**-**	**-**	**-**	**-**	**-**

Malta / Malte

SUPPLY AND CONSUMPTION 1996	Gas (TJ)				Comb. Renew. & Waste (TJ)				(GWh)	(TJ)
	Natural Gas	Gas Works	Coke Ovens	Blast Furnaces	Solid Biomass	Gas/Liquids from Biomass	Municipal Waste	Industrial Waste	Electricity	Heat
Production	-	-	-	-	-	-	-	-	1658	-
Imports	-	-	-	-	-	-	-	-	-	-
Exports	-	-	-	-	-	-	-	-	-	-
Intl. Marine Bunkers	-	-	-	-	-	-	-	-	-	-
Stock Changes	-	-	-	-	-	-	-	-	-	-
DOMESTIC SUPPLY	-	-	-	-	-	-	-	-	**1658**	-
Transfers and Stat. Diff.	-	-	-	-	-	-	-	-	-	-
TRANSFORMATION	-	-	-	-	-	-	-	-	-	-
Electricity and CHP Plants	-	-	-	-	-	-	-	-	-	-
Petroleum Refineries	-	-	-	-	-	-	-	-	-	-
Other Transform. Sector	-	-	-	-	-	-	-	-	-	-
ENERGY SECTOR	-	-	-	-	-	-	-	-	**101**	-
DISTRIBUTION LOSSES	-	-	-	-	-	-	-	-	**218**	-
FINAL CONSUMPTION	-	-	-	-	-	-	-	-	**1339**	-
INDUSTRY SECTOR	-	-	-	-	-	-	-	-	**510**	-
Iron and Steel	-	-	-	-	-	-	-	-	-	-
Chemical and Petrochem.	-	-	-	-	-	-	-	-	-	-
Non-Metallic Minerals	-	-	-	-	-	-	-	-	-	-
Non-specified	-	-	-	-	-	-	-	-	510	-
TRANSPORT SECTOR	-	-	-	-	-	-	-	-	-	-
Air	-	-	-	-	-	-	-	-	-	-
Road	-	-	-	-	-	-	-	-	-	-
Non-specified	-	-	-	-	-	-	-	-	-	-
OTHER SECTORS	-	-	-	-	-	-	-	-	**829**	-
Agriculture	-	-	-	-	-	-	-	-	-	-
Comm. and Publ. Services	-	-	-	-	-	-	-	-	396	-
Residential	-	-	-	-	-	-	-	-	433	-
Non-specified	-	-	-	-	-	-	-	-	-	-
NON-ENERGY USE	-	-	-	-	-	-	-	-	-	-

APPROVISIONNEMENT ET DEMANDE 1997	*Gaz* (TJ)				*En. Re. Comb. & Déchets* (TJ)				(GWh)	(TJ)
	Gaz naturel	*Usines à gaz*	*Cokeries*	*Hauts fourneaux*	*Biomasse solide*	*Gaz/Liquides tirés de biomasse*	*Déchets urbains*	*Déchets industriels*	*Electricité*	*Chaleur*
Production	-	-	-	-	-	-	-	-	1686	-
Imports	-	-	-	-	-	-	-	-	-	-
Exports	-	-	-	-	-	-	-	-	-	-
Intl. Marine Bunkers	-	-	-	-	-	-	-	-	-	-
Stock Changes	-	-	-	-	-	-	-	-	-	-
DOMESTIC SUPPLY	-	-	-	-	-	-	-	-	**1686**	-
Transfers and Stat. Diff.	-	-	-	-	-	-	-	-	-	-
TRANSFORMATION	-	-	-	-	-	-	-	-	-	-
Electricity and CHP Plants	-	-	-	-	-	-	-	-	-	-
Petroleum Refineries	-	-	-	-	-	-	-	-	-	-
Other Transform. Sector	-	-	-	-	-	-	-	-	-	-
ENERGY SECTOR	-	-	-	-	-	-	-	-	**105**	-
DISTRIBUTION LOSSES	-	-	-	-	-	-	-	-	**221**	-
FINAL CONSUMPTION	-	-	-	-	-	-	-	-	**1360**	-
INDUSTRY SECTOR	-	-	-	-	-	-	-	-	**453**	-
Iron and Steel	-	-	-	-	-	-	-	-	-	-
Chemical and Petrochem.	-	-	-	-	-	-	-	-	-	-
Non-Metallic Minerals	-	-	-	-	-	-	-	-	-	-
Non-specified	-	-	-	-	-	-	-	-	453	-
TRANSPORT SECTOR	-	-	-	-	-	-	-	-	-	-
Air	-	-	-	-	-	-	-	-	-	-
Road	-	-	-	-	-	-	-	-	-	-
Non-specified	-	-	-	-	-	-	-	-	-	-
OTHER SECTORS	-	-	-	-	-	-	-	-	**907**	-
Agriculture	-	-	-	-	-	-	-	-	-	-
Comm. and Publ. Services	-	-	-	-	-	-	-	-	445	-
Residential	-	-	-	-	-	-	-	-	462	-
Non-specified	-	-	-	-	-	-	-	-	-	-
NON-ENERGY USE	-	-	-	-	-	-	-	-	-	-

Moldova : 1996

SUPPLY AND CONSUMPTION / *APPROVISIONNEMENT ET DEMANDE*	Coal / *Charbon* (1000 tonnes)							Oil / *Pétrole* (1000 tonnes)			
	Coking Coal / *Charbon à coke*	Other Bit. Coal / *Autres charb. bit.*	Sub-Bit. Coal / *Charbon sous-bit.*	Lignite / *Lignite*	Peat / *Tourbe*	Oven and Gas Coke / *Coke de four/gaz*	Pat. Fuel and BKB / *Agg./briq. de lignite*	Crude Oil / *Pétrole brut*	NGL / *LGN*	Feed-stocks / *Produits d'aliment.*	Additives / *Additifs*
Production	-	-	-	-	-	-	-	-	-	-	-
From Other Sources	-	-	-	-	-	-	-	-	-	-	-
Imports	-	1079	-	-	-	13	-	-	-	-	-
Exports	-	-	-	-	-	-	-	-	-	-	-
Intl. Marine Bunkers	-	-	-	-	-	-	-	-	-	-	-
Stock Changes	-	45	-	-	-	7	-	-	-	-	-
DOMESTIC SUPPLY	**-**	**1124**	**-**	**-**	**-**	**20**	**-**	**-**	**-**	**-**	**-**
Transfers	-	-	-	-	-	-	-	-	-	-	-
Statistical Differences	-	-	-	-	-	-	-	-	-	-	-
TRANSFORMATION	**-**	**775**	**-**	**-**	**-**	**5**	**-**	**-**	**-**	**-**	**-**
Electricity Plants	-	740	-	-	-	-	-	-	-	-	-
CHP Plants	-	35	-	-	-	-	-	-	-	-	-
Heat Plants	-	-	-	-	-	-	-	-	-	-	-
Blast Furnaces/Gas Works	-	-	-	-	-	-	-	-	-	-	-
Coke/Pat. Fuel/BKB Plants	-	-	-	-	-	5	-	-	-	-	-
Petroleum Refineries	-	-	-	-	-	-	-	-	-	-	-
Petrochemical Industry	-	-	-	-	-	-	-	-	-	-	-
Liquefaction	-	-	-	-	-	-	-	-	-	-	-
Other Transform. Sector	-	-	-	-	-	-	-	-	-	-	-
ENERGY SECTOR	**-**	**-**	**-**	**-**	**-**	**-**	**-**	**-**	**-**	**-**	**-**
Coal Mines	-	-	-	-	-	-	-	-	-	-	-
Oil and Gas Extraction	-	-	-	-	-	-	-	-	-	-	-
Petroleum Refineries	-	-	-	-	-	-	-	-	-	-	-
Electr., CHP+Heat Plants	-	-	-	-	-	-	-	-	-	-	-
Pumped Storage (Elec.)	-	-	-	-	-	-	-	-	-	-	-
Other Energy Sector	-	-	-	-	-	-	-	-	-	-	-
Distribution Losses	-	1	-	-	-	-	-	-	-	-	-
FINAL CONSUMPTION	**-**	**348**	**-**	**-**	**-**	**15**	**-**	**-**	**-**	**-**	**-**
INDUSTRY SECTOR	**-**	**3**	**-**	**-**	**-**	**14**	**-**	**-**	**-**	**-**	**-**
Iron and Steel	-	-	-	-	-	-	-	-	-	-	-
Chemical and Petrochem.	-	-	-	-	-	-	-	-	-	-	-
of which: Feedstocks	-	-	-	-	-	-	-	-	-	-	-
Non-Ferrous Metals	-	-	-	-	-	-	-	-	-	-	-
Non-Metallic Minerals	-	3	-	-	-	1	-	-	-	-	-
Transport Equipment	-	-	-	-	-	-	-	-	-	-	-
Machinery	-	-	-	-	-	-	-	-	-	-	-
Mining and Quarrying	-	-	-	-	-	-	-	-	-	-	-
Food and Tobacco	-	-	-	-	-	13	-	-	-	-	-
Paper, Pulp and Print	-	-	-	-	-	-	-	-	-	-	-
Wood and Wood Products	-	-	-	-	-	-	-	-	-	-	-
Construction	-	-	-	-	-	-	-	-	-	-	-
Textile and Leather	-	-	-	-	-	-	-	-	-	-	-
Non-specified	-	-	-	-	-	-	-	-	-	-	-
TRANSPORT SECTOR	**-**	**3**	**-**	**-**	**-**	**-**	**-**	**-**	**-**	**-**	**-**
Air	-	-	-	-	-	-	-	-	-	-	-
Road	-	-	-	-	-	-	-	-	-	-	-
Rail	-	3	-	-	-	-	-	-	-	-	-
Pipeline Transport	-	-	-	-	-	-	-	-	-	-	-
Internal Navigation	-	-	-	-	-	-	-	-	-	-	-
Non-specified	-	-	-	-	-	-	-	-	-	-	-
OTHER SECTORS	**-**	**342**	**-**	**-**	**-**	**1**	**-**	**-**	**-**	**-**	**-**
Agriculture	-	4	-	-	-	-	-	-	-	-	-
Comm. and Publ. Services	-	135	-	-	-	-	-	-	-	-	-
Residential	-	194	-	-	-	1	-	-	-	-	-
Non-specified	-	9	-	-	-	-	-	-	-	-	-
NON-ENERGY USE	**-**	**-**	**-**	**-**	**-**	**-**	**-**	**-**	**-**	**-**	**-**
in Industry/Trans./Energy	-	-	-	-	-	-	-	-	-	-	-
in Transport	-	-	-	-	-	-	-	-	-	-	-
in Other Sectors	-	-	-	-	-	-	-	-	-	-	-

Moldova : 1996

	Oil cont. / *Pétrole cont.* (1000 tonnes)										
SUPPLY AND CONSUMPTION ***APPROVISIONNEMENT ET DEMANDE***	Refinery Gas *Gaz de raffinerie*	LPG + Ethane *GPL + éthane*	Motor Gasoline *Essence moteur*	Aviation Gasoline *Essence aviation*	Jet Fuel *Carbu-réacteurs*	Kerosene *Kérosène*	Gas/ Diesel *Gazole*	Heavy Fuel Oil *Fioul lourd*	Naphtha *Naphta*	Petrol. Coke *Coke de pétrole*	Other Prod. *Autres prod.*
Production	-	-	-	-	-	-	-	-	-	-	-
From Other Sources	-	-	-	-	-	-	-	-	-	-	-
Imports	-	22	181	-	20	9	302	272	-	-	14
Exports	-	-	-	-	-	-	-	-	-	-	-
Intl. Marine Bunkers	-	-	-	-	-	-	-	-	-	-	-
Stock Changes	-	-	37	-	-	-1	23	71	-	-	3
DOMESTIC SUPPLY	**-**	**22**	**218**	**-**	**20**	**8**	**325**	**343**	**-**	**-**	**17**
Transfers	-	-	-	-	-	-	-	-	-	-	-
Statistical Differences	-	-	-	-	-	-	-	-	-	-	-
TRANSFORMATION	**-**	**-**	**-**	**-**	**-**	**1**	**6**	**316**	**-**	**-**	**-**
Electricity Plants	-	-	-	-	-	-	6	65	-	-	-
CHP Plants	-	-	-	-	-	-	-	251	-	-	-
Heat Plants	-	-	-	-	-	1	-	-	-	-	-
Blast Furnaces/Gas Works	-	-	-	-	-	-	-	-	-	-	-
Coke/Pat. Fuel/BKB Plants	-	-	-	-	-	-	-	-	-	-	-
Petroleum Refineries	-	-	-	-	-	-	-	-	-	-	-
Petrochemical Industry	-	-	-	-	-	-	-	-	-	-	-
Liquefaction	-	-	-	-	-	-	-	-	-	-	-
Other Transform. Sector	-	-	-	-	-	-	-	-	-	-	-
ENERGY SECTOR	**-**	**-**	**-**	**-**	**-**	**-**	**-**	**-**	**-**	**-**	**-**
Coal Mines	-	-	-	-	-	-	-	-	-	-	-
Oil and Gas Extraction	-	-	-	-	-	-	-	-	-	-	-
Petroleum Refineries	-	-	-	-	-	-	-	-	-	-	-
Electr., CHP+Heat Plants	-	-	-	-	-	-	-	-	-	-	-
Pumped Storage (Elec.)	-	-	-	-	-	-	-	-	-	-	-
Other Energy Sector	-	-	-	-	-	-	-	-	-	-	-
Distribution Losses	-	-	3	-	-	1	1	1	-	-	-
FINAL CONSUMPTION	**-**	**22**	**215**	**-**	**20**	**6**	**318**	**26**	**-**	**-**	**17**
INDUSTRY SECTOR	**-**	**-**	**1**	**-**	**-**	**2**	**15**	**10**	**-**	**-**	**-**
Iron and Steel	-	-	-	-	-	-	-	-	-	-	-
Chemical and Petrochem.	-	-	-	-	-	-	-	-	-	-	-
of which: Feedstocks	-	-	-	-	-	-	-	-	-	-	-
Non-Ferrous Metals	-	-	-	-	-	-	-	-	-	-	-
Non-Metallic Minerals	-	-	-	-	-	-	-	1	-	-	-
Transport Equipment	-	-	-	-	-	-	-	-	-	-	-
Machinery	-	-	-	-	-	-	-	-	-	-	-
Mining and Quarrying	-	-	-	-	-	-	1	-	-	-	-
Food and Tobacco	-	-	-	-	-	2	9	7	-	-	-
Paper, Pulp and Print	-	-	-	-	-	-	-	-	-	-	-
Wood and Wood Products	-	-	-	-	-	-	-	-	-	-	-
Construction	-	-	1	-	-	-	5	2	-	-	-
Textile and Leather	-	-	-	-	-	-	-	-	-	-	-
Non-specified	-	-	-	-	-	-	-	-	-	-	-
TRANSPORT SECTOR	**-**	**4**	**208**	**-**	**18**	**-**	**124**	**1**	**-**	**-**	**-**
Air	-	-	-	-	18	-	-	-	-	-	-
Road	-	4	207	-	-	-	99	-	-	-	-
Rail	-	-	-	-	-	-	25	1	-	-	-
Pipeline Transport	-	-	-	-	-	-	-	-	-	-	-
Internal Navigation	-	-	-	-	-	-	-	-	-	-	-
Non-specified	-	-	1	-	-	-	-	-	-	-	-
OTHER SECTORS	**-**	**18**	**6**	**-**	**2**	**4**	**179**	**15**	**-**	**-**	**-**
Agriculture	-	-	5	-	-	3	157	-	-	-	-
Comm. and Publ. Services	-	1	-	-	-	-	2	5	-	-	-
Residential	-	15	-	-	-	-	15	-	-	-	-
Non-specified	-	2	1	-	2	1	5	10	-	-	-
NON-ENERGY USE	**-**	**-**	**-**	**-**	**-**	**-**	**-**	**-**	**-**	**-**	**17**
in Industry/Transf./Energy	-	-	-	-	-	-	-	-	-	-	-
in Transport	-	-	-	-	-	-	-	-	-	-	8
in Other Sectors	-	-	-	-	-	-	-	-	-	-	9

Moldova : 1996

SUPPLY AND CONSUMPTION / *APPROVISIONNEMENT ET DEMANDE*	Gas / *Gaz* (TJ)				Comb. Renew. & Waste / *En. Re. Comb. & Déchets* (TJ)				(GWh)	(TJ)
	Natural Gas / *Gaz naturel*	Gas Works / *Usines à gaz*	Coke Ovens / *Cokeries*	Blast Furnaces / *Hauts fourneaux*	Solid Biomass / *Biomasse solide*	Gas/Liquids from Biomass / *Gaz/Liquides tirés de biomasse*	Municipal Waste / *Déchets urbains*	Industrial Waste / *Déchets industriels*	Electricity / *Electricité*	Heat / *Chaleur*
Production	-	-	-	-	2494	-	-	-	6122	14598
From Other Sources	-	-	-	-	-	-	-	-	-	-
Imports	136679	-	-	-	-	-	-	-	1610	-
Exports	-	-	-	-	-	-	-	-	-4	-
Intl. Marine Bunkers	-	-	-	-	-	-	-	-	-	-
Stock Changes	-476	-	-	-	88	-	-	-	-	-
DOMESTIC SUPPLY	**136203**	**-**	**-**	**-**	**2582**	**-**	**-**	**-**	**7728**	**14598**
Transfers	-	-	-	-	-	-	-	-	-	-
Statistical Differences	-	-	-	-	-	-	-	-	-	-
TRANSFORMATION	**62088**	**-**	**-**	**-**	**-**	**-**	**-**	**-**	**50**	**-**
Electricity Plants	35995	-	-	-	-	-	-	-	-	-
CHP Plants	23606	-	-	-	-	-	-	-	-	-
Heat Plants	2487	-	-	-	-	-	-	-	-	-
Blast Furnaces/Gas Works	-	-	-	-	-	-	-	-	-	-
Coke/Pat. Fuel/BKB Plants	-	-	-	-	-	-	-	-	-	-
Petroleum Refineries	-	-	-	-	-	-	-	-	-	-
Petrochemical Industry	-	-	-	-	-	-	-	-	-	-
Liquefaction	-	-	-	-	-	-	-	-	-	-
Other Transform. Sector	-	-	-	-	-	-	-	-	50	-
ENERGY SECTOR	**-**	**-**	**-**	**-**	**-**	**-**	**-**	**-**	**601**	**-**
Coal Mines	-	-	-	-	-	-	-	-	-	-
Oil and Gas Extraction	-	-	-	-	-	-	-	-	-	-
Petroleum Refineries	-	-	-	-	-	-	-	-	-	-
Electr., CHP+Heat Plants	-	-	-	-	-	-	-	-	600	-
Pumped Storage (Elec.)	-	-	-	-	-	-	-	-	1	-
Other Energy Sector	-	-	-	-	-	-	-	-	-	-
Distribution Losses	5646	-	-	-	-	-	-	-	1392	1056
FINAL CONSUMPTION	**68469**	**-**	**-**	**-**	**2582**	**-**	**-**	**-**	**5685**	**13542**
INDUSTRY SECTOR	**19803**	**-**	**-**	**-**	**-**	**-**	**-**	**-**	**1738**	**463**
Iron and Steel	-	-	-	-	-	-	-	-	565	-
Chemical and Petrochem.	-	-	-	-	-	-	-	-	2	-
of which: Feedstocks	-	-	-	-	-	-	-	-	-	-
Non-Ferrous Metals	-	-	-	-	-	-	-	-	-	-
Non-Metallic Minerals	5945	-	-	-	-	-	-	-	206	-
Transport Equipment	-	-	-	-	-	-	-	-	-	-
Machinery	793	-	-	-	-	-	-	-	41	71
Mining and Quarrying	-	-	-	-	-	-	-	-	-	-
Food and Tobacco	394	-	-	-	-	-	-	-	403	247
Paper, Pulp and Print	-	-	-	-	-	-	-	-	15	-
Wood and Wood Products	184	-	-	-	-	-	-	-	18	-
Construction	-	-	-	-	-	-	-	-	53	-
Textile and Leather	6143	-	-	-	-	-	-	-	96	96
Non-specified	6344	-	-	-	-	-	-	-	339	49
TRANSPORT SECTOR	**3718**	**-**	**-**	**-**	**-**	**-**	**-**	**-**	**125**	**-**
Air	-	-	-	-	-	-	-	-	-	-
Road	495	-	-	-	-	-	-	-	-	-
Rail	-	-	-	-	-	-	-	-	47	-
Pipeline Transport	3223	-	-	-	-	-	-	-	4	-
Internal Navigation	-	-	-	-	-	-	-	-	-	-
Non-specified	-	-	-	-	-	-	-	-	74	-
OTHER SECTORS	**44948**	**-**	**-**	**-**	**2582**	**-**	**-**	**-**	**3822**	**13079**
Agriculture	3733	-	-	-	-	-	-	-	794	21
Comm. and Publ. Services	22058	-	-	-	-	-	-	-	1080	263
Residential	15160	-	-	-	-	-	-	-	1948	12795
Non-specified	3997	-	-	-	2582	-	-	-	-	-
NON-ENERGY USE	**-**	**-**	**-**	**-**	**-**	**-**	**-**	**-**	**-**	**-**
in Industry/Transf./Energy	-	-	-	-	-	-	-	-	-	-
in Transport	-	-	-	-	-	-	-	-	-	-
in Other Sectors	-	-	-	-	-	-	-	-	-	-

Moldova : 1997

SUPPLY AND CONSUMPTION *APPROVISIONNEMENT ET DEMANDE*	Coal / *Charbon* (1000 tonnes)							Oil / *Pétrole* (1000 tonnes)			
	Coking Coal *Charbon à coke*	Other Bit. Coal *Autres charb. bit.*	Sub-Bit. Coal *Charbon sous-bit.*	Lignite *Lignite*	Peat *Tourbe*	Oven and Gas Coke *Coke de four/gaz*	Pat. Fuel and BKB *Agg./briq. de lignite*	Crude Oil *Pétrole brut*	NGL *LGN*	Feed-stocks *Produits d'aliment.*	Additives *Additifs*
Production	-	-	-	-	-	-	-	-	-	-	-
From Other Sources	-	-	-	-	-	-	-	-	-	-	-
Imports	-	372	-	-	-	8	-	-	-	-	-
Exports	-	-	-	-	-	-	-	-	-	-	-
Intl. Marine Bunkers	-	-	-	-	-	-	-	-	-	-	-
Stock Changes	-	-50	-	-	-	5	-	-	-	-	-
DOMESTIC SUPPLY	**-**	**322**	**-**	**-**	**-**	**13**	**-**	**-**	**-**	**-**	**-**
Transfers	-	-	-	-	-	-	-	-	-	-	-
Statistical Differences	-	281	-	-	-	-	-	-	-	-	-
TRANSFORMATION	**-**	**359**	**-**	**-**	**-**	**1**	**-**	**-**	**-**	**-**	**-**
Electricity Plants	-	281	-	-	-	-	-	-	-	-	-
CHP Plants	-	-	-	-	-	-	-	-	-	-	-
Heat Plants	-	-	-	-	-	-	-	-	-	-	-
Blast Furnaces/Gas Works	-	-	-	-	-	-	-	-	-	-	-
Coke/Pat. Fuel/BKB Plants	-	-	-	-	-	1	-	-	-	-	-
Petroleum Refineries	-	-	-	-	-	-	-	-	-	-	-
Petrochemical Industry	-	-	-	-	-	-	-	-	-	-	-
Liquefaction	-	-	-	-	-	-	-	-	-	-	-
Other Transform. Sector	-	78	-	-	-	-	-	-	-	-	-
ENERGY SECTOR	**-**	**-**	**-**	**-**	**-**	**-**	**-**	**-**	**-**	**-**	**-**
Coal Mines	-	-	-	-	-	-	-	-	-	-	-
Oil and Gas Extraction	-	-	-	-	-	-	-	-	-	-	-
Petroleum Refineries	-	-	-	-	-	-	-	-	-	-	-
Electr., CHP+Heat Plants	-	-	-	-	-	-	-	-	-	-	-
Pumped Storage (Elec.)	-	-	-	-	-	-	-	-	-	-	-
Other Energy Sector	-	-	-	-	-	-	-	-	-	-	-
Distribution Losses	-	-	-	-	-	-	-	-	-	-	-
FINAL CONSUMPTION	**-**	**244**	**-**	**-**	**-**	**12**	**-**	**-**	**-**	**-**	**-**
INDUSTRY SECTOR	**-**	**3**	**-**	**-**	**-**	**12**	**-**	**-**	**-**	**-**	**-**
Iron and Steel	-	-	-	-	-	-	-	-	-	-	-
Chemical and Petrochem.	-	-	-	-	-	-	-	-	-	-	-
of which: Feedstocks	-	-	-	-	-	-	-	-	-	-	-
Non-Ferrous Metals	-	-	-	-	-	-	-	-	-	-	-
Non-Metallic Minerals	-	3	-	-	-	1	-	-	-	-	-
Transport Equipment	-	-	-	-	-	-	-	-	-	-	-
Machinery	-	-	-	-	-	-	-	-	-	-	-
Mining and Quarrying	-	-	-	-	-	-	-	-	-	-	-
Food and Tobacco	-	-	-	-	-	11	-	-	-	-	-
Paper, Pulp and Print	-	-	-	-	-	-	-	-	-	-	-
Wood and Wood Products	-	-	-	-	-	-	-	-	-	-	-
Construction	-	-	-	-	-	-	-	-	-	-	-
Textile and Leather	-	-	-	-	-	-	-	-	-	-	-
Non-specified	-	-	-	-	-	-	-	-	-	-	-
TRANSPORT SECTOR	**-**	**5**	**-**	**-**	**-**	**-**	**-**	**-**	**-**	**-**	**-**
Air	-	-	-	-	-	-	-	-	-	-	-
Road	-	-	-	-	-	-	-	-	-	-	-
Rail	-	5	-	-	-	-	-	-	-	-	-
Pipeline Transport	-	-	-	-	-	-	-	-	-	-	-
Internal Navigation	-	-	-	-	-	-	-	-	-	-	-
Non-specified	-	-	-	-	-	-	-	-	-	-	-
OTHER SECTORS	**-**	**236**	**-**	**-**	**-**	**-**	**-**	**-**	**-**	**-**	**-**
Agriculture	-	2	-	-	-	-	-	-	-	-	-
Comm. and Publ. Services	-	117	-	-	-	-	-	-	-	-	-
Residential	-	111	-	-	-	-	-	-	-	-	-
Non-specified	-	6	-	-	-	-	-	-	-	-	-
NON-ENERGY USE	**-**	**-**	**-**	**-**	**-**	**-**	**-**	**-**	**-**	**-**	**-**
in Industry/Trans./Energy	-	-	-	-	-	-	-	-	-	-	-
in Transport	-	-	-	-	-	-	-	-	-	-	-
in Other Sectors	-	-	-	-	-	-	-	-	-	-	-

Moldova : 1997

SUPPLY AND CONSUMPTION / *APPROVISIONNEMENT ET DEMANDE*	Oil cont. / *Pétrole cont.* (1000 tonnes)										
	Refinery Gas / *Gaz de raffinerie*	LPG + Ethane / *GPL + éthane*	Motor Gasoline / *Essence moteur*	Aviation Gasoline / *Essence aviation*	Jet Fuel / *Carbu-réacteurs*	Kerosene / *Kérosène*	Gas/ Diesel / *Gazole*	Heavy Fuel Oil / *Fioul lourd*	Naphtha / *Naphta*	Petrol. Coke / *Coke de pétrole*	Other Prod. / *Autres prod.*
Production	-	-	-	-	-	-	-	-	-	-	-
From Other Sources	-	-	-	-	-	-	-	-	-	-	-
Imports	-	28	262	-	25	10	328	275	-	-	16
Exports	-	-	-	-	-	-	-	-	-	-	-
Intl. Marine Bunkers	-	-	-	-	-	-	-	-	-	-	-
Stock Changes	-	-1	-13	-	-2	-3	-16	-33	-	-	-
DOMESTIC SUPPLY	**-**	**27**	**249**	**-**	**23**	**7**	**312**	**242**	**-**	**-**	**16**
Transfers	-	-	-	-	-	-	-	-	-	-	-
Statistical Differences	-	-	-	-	-	-	-	-	-	-	-
TRANSFORMATION	**-**	**-**	**-**	**-**	**-**	**1**	**4**	**225**	**-**	**-**	**-**
Electricity Plants	-	-	-	-	-	-	3	47	-	-	-
CHP Plants	-	-	-	-	-	-	-	178	-	-	-
Heat Plants	-	-	-	-	-	1	1	-	-	-	-
Blast Furnaces/Gas Works	-	-	-	-	-	-	-	-	-	-	-
Coke/Pat. Fuel/BKB Plants	-	-	-	-	-	-	-	-	-	-	-
Petroleum Refineries	-	-	-	-	-	-	-	-	-	-	-
Petrochemical Industry	-	-	-	-	-	-	-	-	-	-	-
Liquefaction	-	-	-	-	-	-	-	-	-	-	-
Other Transform. Sector	-	-	-	-	-	-	-	-	-	-	-
ENERGY SECTOR	**-**	**-**	**-**	**-**	**-**	**-**	**-**	**-**	**-**	**-**	**-**
Coal Mines	-	-	-	-	-	-	-	-	-	-	-
Oil and Gas Extraction	-	-	-	-	-	-	-	-	-	-	-
Petroleum Refineries	-	-	-	-	-	-	-	-	-	-	-
Electr., CHP+Heat Plants	-	-	-	-	-	-	-	-	-	-	-
Pumped Storage (Elec.)	-	-	-	-	-	-	-	-	-	-	-
Other Energy Sector	-	-	-	-	-	-	-	-	-	-	-
Distribution Losses	-	1	5	-	1	1	7	-	-	-	-
FINAL CONSUMPTION	**-**	**26**	**244**	**-**	**22**	**5**	**301**	**17**	**-**	**-**	**16**
INDUSTRY SECTOR	**-**	**-**	**1**	**-**	**-**	**3**	**13**	**7**	**-**	**-**	**-**
Iron and Steel	-	-	-	-	-	-	-	-	-	-	-
Chemical and Petrochem.	-	-	-	-	-	-	-	-	-	-	-
of which: Feedstocks	-	-	-	-	-	-	-	-	-	-	-
Non-Ferrous Metals	-	-	-	-	-	-	-	-	-	-	-
Non-Metallic Minerals	-	-	-	-	-	-	-	1	-	-	-
Transport Equipment	-	-	-	-	-	-	-	-	-	-	-
Machinery	-	-	-	-	-	-	-	-	-	-	-
Mining and Quarrying	-	-	-	-	-	-	1	-	-	-	-
Food and Tobacco	-	-	-	-	-	3	7	4	-	-	-
Paper, Pulp and Print	-	-	-	-	-	-	-	-	-	-	-
Wood and Wood Products	-	-	-	-	-	-	-	-	-	-	-
Construction	-	-	1	-	-	-	5	2	-	-	-
Textile and Leather	-	-	-	-	-	-	-	-	-	-	-
Non-specified	-	-	-	-	-	-	-	-	-	-	-
TRANSPORT SECTOR	**-**	**5**	**239**	**-**	**21**	**-**	**110**	**1**	**-**	**-**	**-**
Air	-	-	-	-	21	-	-	-	-	-	-
Road	-	5	238	-	-	-	88	-	-	-	-
Rail	-	-	-	-	-	-	21	1	-	-	-
Pipeline Transport	-	-	-	-	-	-	-	-	-	-	-
Internal Navigation	-	-	-	-	-	-	-	-	-	-	-
Non-specified	-	-	1	-	-	-	1	-	-	-	-
OTHER SECTORS	**-**	**21**	**4**	**-**	**1**	**2**	**178**	**9**	**-**	**-**	**-**
Agriculture	-	-	3	-	-	2	146	1	-	-	-
Comm. and Publ. Services	-	-	-	-	-	-	2	1	-	-	-
Residential	-	20	-	-	-	-	28	-	-	-	-
Non-specified	-	1	1	-	1	-	2	7	-	-	-
NON-ENERGY USE	**-**	**-**	**-**	**-**	**-**	**-**	**-**	**-**	**-**	**-**	**16**
in Industry/Transf./Energy	-	-	-	-	-	-	-	-	-	-	-
in Transport	-	-	-	-	-	-	-	-	-	-	7
in Other Sectors	-	-	-	-	-	-	-	-	-	-	9

Moldova : 1997

	Gas / *Gaz* (TJ)				Comb. Renew. & Waste / *En. Re. Comb. & Déchets* (TJ)				(GWh)	(TJ)
SUPPLY AND CONSUMPTION / ***APPROVISIONNEMENT ET DEMANDE***	Natural Gas / *Gaz naturel*	Gas Works / *Usines à gaz*	Coke Ovens / *Cokeries*	Blast Furnaces / *Hauts fourneaux*	Solid Biomass / *Biomasse solide*	Gas/Liquids from Biomass / *Gaz/Liquides tirés de biomasse*	Municipal Waste / *Déchets urbains*	Industrial Waste / *Déchets industriels*	Electricity / *Electricité*	Heat / *Chaleur*
Production	-	-	-	-	2817	-	-	-	5273	13819
From Other Sources	-	-	-	-	-	-	-	-	-	-
Imports	145742	-	-	-	-	-	-	-	1979	-
Exports	-	-	-	-	-	-	-	-	-27	-
Intl. Marine Bunkers	-	-	-	-	-	-	-	-	-	-
Stock Changes	-238	-	-	-	-264	-	-	-	-	-
DOMESTIC SUPPLY	**145504**	**-**	**-**	**-**	**2553**	**-**	**-**	**-**	**7225**	**13819**
Transfers	-	-	-	-	-	-	-	-	-	-
Statistical Differences	-	-	-	-	-	-	-	-	-	-
TRANSFORMATION	**66328**	**-**	**-**	**-**	**-**	**-**	**-**	**-**	**27**	**-**
Electricity Plants	38453	-	-	-	-	-	-	-	-	-
CHP Plants	25218	-	-	-	-	-	-	-	-	-
Heat Plants	2657	-	-	-	-	-	-	-	-	-
Blast Furnaces/Gas Works	-	-	-	-	-	-	-	-	-	-
Coke/Pat. Fuel/BKB Plants	-	-	-	-	-	-	-	-	-	-
Petroleum Refineries	-	-	-	-	-	-	-	-	-	-
Petrochemical Industry	-	-	-	-	-	-	-	-	-	-
Liquefaction	-	-	-	-	-	-	-	-	-	-
Other Transform. Sector	-	-	-	-	-	-	-	-	27	-
ENERGY SECTOR	**-**	**-**	**-**	**-**	**-**	**-**	**-**	**-**	**524**	**-**
Coal Mines	-	-	-	-	-	-	-	-	-	-
Oil and Gas Extraction	-	-	-	-	-	-	-	-	-	-
Petroleum Refineries	-	-	-	-	-	-	-	-	-	-
Electr., CHP+Heat Plants	-	-	-	-	-	-	-	-	524	-
Pumped Storage (Elec.)	-	-	-	-	-	-	-	-	-	-
Other Energy Sector	-	-	-	-	-	-	-	-	-	-
Distribution Losses	6031	-	-	-	-	-	-	-	1425	1380
FINAL CONSUMPTION	**73145**	**-**	**-**	**-**	**2553**	**-**	**-**	**-**	**5249**	**12439**
INDUSTRY SECTOR	**21157**	**-**	**-**	**-**	**-**	**-**	**-**	**-**	**1802**	**-**
Iron and Steel	-	-	-	-	-	-	-	-	-	-
Chemical and Petrochem.	-	-	-	-	-	-	-	-	-	-
of which: Feedstocks	-	-	-	-	-	-	-	-	-	-
Non-Ferrous Metals	-	-	-	-	-	-	-	-	-	-
Non-Metallic Minerals	6352	-	-	-	-	-	-	-	-	-
Transport Equipment	-	-	-	-	-	-	-	-	-	-
Machinery	847	-	-	-	-	-	-	-	-	-
Mining and Quarrying	-	-	-	-	-	-	-	-	-	-
Food and Tobacco	421	-	-	-	-	-	-	-	-	-
Paper, Pulp and Print	-	-	-	-	-	-	-	-	-	-
Wood and Wood Products	196	-	-	-	-	-	-	-	-	-
Construction	-	-	-	-	-	-	-	-	-	-
Textile and Leather	6563	-	-	-	-	-	-	-	-	-
Non-specified	6778	-	-	-	-	-	-	-	1802	-
TRANSPORT SECTOR	**3971**	**-**	**-**	**-**	**-**	**-**	**-**	**-**	**118**	**-**
Air	-	-	-	-	-	-	-	-	-	-
Road	529	-	-	-	-	-	-	-	-	-
Rail	-	-	-	-	-	-	-	-	46	-
Pipeline Transport	3442	-	-	-	-	-	-	-	4	-
Internal Navigation	-	-	-	-	-	-	-	-	-	-
Non-specified	-	-	-	-	-	-	-	-	68	-
OTHER SECTORS	**48017**	**-**	**-**	**-**	**2553**	**-**	**-**	**-**	**3329**	**12439**
Agriculture	3988	-	-	-	-	-	-	-	526	-
Comm. and Publ. Services	23564	-	-	-	-	-	-	-	1036	-
Residential	16195	-	-	-	-	-	-	-	1767	12439
Non-specified	4270	-	-	-	2553	-	-	-	-	-
NON-ENERGY USE	**-**	**-**	**-**	**-**	**-**	**-**	**-**	**-**	**-**	**-**
in Industry/Transf./Energy	-	-	-	-	-	-	-	-	-	-
in Transport	-	-	-	-	-	-	-	-	-	-
in Other Sectors	-	-	-	-	-	-	-	-	-	-

Morocco / Maroc : 1996

SUPPLY AND CONSUMPTION / *APPROVISIONNEMENT ET DEMANDE*	Coal / *Charbon* (1000 tonnes)							Oil / *Pétrole* (1000 tonnes)			
	Coking Coal / *Charbon à coke*	Other Bit. Coal / *Autres charb. bit.*	Sub-Bit. Coal / *Charbon sous-bit.*	Lignite / *Lignite*	Peat / *Tourbe*	Oven and Gas Coke / *Coke de four/gaz*	Pat. Fuel and BKB / *Agg./briq. de lignite*	Crude Oil / *Pétrole brut*	NGL / *LGN*	Feed-stocks / *Produits d'aliment.*	Additives / *Additifs*
Production	-	506	-	-	-	-	-	5	-	-	-
From Other Sources	-	-	-	-	-	-	-	349	-	-	-
Imports	-	2720	-	-	-	-	-	5336	-	-	-
Exports	-	-	-	-	-	-	-	-	-	-	-
Intl. Marine Bunkers	-	-	-	-	-	-	-	-	-	-	-
Stock Changes	-	-	-	-	-	-	-	297	-	-	-
DOMESTIC SUPPLY	**-**	**3226**	**-**	**-**	**-**	**-**	**-**	**5987**	**-**	**-**	**-**
Transfers	-	-	-	-	-	-	-	-	-	-	-
Statistical Differences	-	-16	-	-	-	-	-	-	-	-	-
TRANSFORMATION	**-**	**2258**	**-**	**-**	**-**	**-**	**-**	**5987**	**-**	**-**	**-**
Electricity Plants	-	2258	-	-	-	-	-	-	-	-	-
CHP Plants	-	-	-	-	-	-	-	-	-	-	-
Heat Plants	-	-	-	-	-	-	-	-	-	-	-
Blast Furnaces/Gas Works	-	-	-	-	-	-	-	-	-	-	-
Coke/Pat. Fuel/BKB Plants	-	-	-	-	-	-	-	-	-	-	-
Petroleum Refineries	-	-	-	-	-	-	-	5987	-	-	-
Petrochemical Industry	-	-	-	-	-	-	-	-	-	-	-
Liquefaction	-	-	-	-	-	-	-	-	-	-	-
Other Transform. Sector	-	-	-	-	-	-	-	-	-	-	-
ENERGY SECTOR	**-**	**-**	**-**	**-**	**-**	**-**	**-**	**-**	**-**	**-**	**-**
Coal Mines	-	-	-	-	-	-	-	-	-	-	-
Oil and Gas Extraction	-	-	-	-	-	-	-	-	-	-	-
Petroleum Refineries	-	-	-	-	-	-	-	-	-	-	-
Electr., CHP+Heat Plants	-	-	-	-	-	-	-	-	-	-	-
Pumped Storage (Elec.)	-	-	-	-	-	-	-	-	-	-	-
Other Energy Sector	-	-	-	-	-	-	-	-	-	-	-
Distribution Losses	-	-	-	-	-	-	-	-	-	-	-
FINAL CONSUMPTION	**-**	**952**	**-**	**-**	**-**	**-**	**-**	**-**	**-**	**-**	**-**
INDUSTRY SECTOR	**-**	**952**	**-**	**-**	**-**	**-**	**-**	**-**	**-**	**-**	**-**
Iron and Steel	-	-	-	-	-	-	-	-	-	-	-
Chemical and Petrochem.	-	-	-	-	-	-	-	-	-	-	-
of which: Feedstocks	-	-	-	-	-	-	-	-	-	-	-
Non-Ferrous Metals	-	-	-	-	-	-	-	-	-	-	-
Non-Metallic Minerals	-	952	-	-	-	-	-	-	-	-	-
Transport Equipment	-	-	-	-	-	-	-	-	-	-	-
Machinery	-	-	-	-	-	-	-	-	-	-	-
Mining and Quarrying	-	-	-	-	-	-	-	-	-	-	-
Food and Tobacco	-	-	-	-	-	-	-	-	-	-	-
Paper, Pulp and Print	-	-	-	-	-	-	-	-	-	-	-
Wood and Wood Products	-	-	-	-	-	-	-	-	-	-	-
Construction	-	-	-	-	-	-	-	-	-	-	-
Textile and Leather	-	-	-	-	-	-	-	-	-	-	-
Non-specified	-	-	-	-	-	-	-	-	-	-	-
TRANSPORT SECTOR	**-**	**-**	**-**	**-**	**-**	**-**	**-**	**-**	**-**	**-**	**-**
Air	-	-	-	-	-	-	-	-	-	-	-
Road	-	-	-	-	-	-	-	-	-	-	-
Rail	-	-	-	-	-	-	-	-	-	-	-
Pipeline Transport	-	-	-	-	-	-	-	-	-	-	-
Internal Navigation	-	-	-	-	-	-	-	-	-	-	-
Non-specified	-	-	-	-	-	-	-	-	-	-	-
OTHER SECTORS	**-**	**-**	**-**	**-**	**-**	**-**	**-**	**-**	**-**	**-**	**-**
Agriculture	-	-	-	-	-	-	-	-	-	-	-
Comm. and Publ. Services	-	-	-	-	-	-	-	-	-	-	-
Residential	-	-	-	-	-	-	-	-	-	-	-
Non-specified	-	-	-	-	-	-	-	-	-	-	-
NON-ENERGY USE	**-**	**-**	**-**	**-**	**-**	**-**	**-**	**-**	**-**	**-**	**-**
in Industry/Trans./Energy	-	-	-	-	-	-	-	-	-	-	-
in Transport	-	-	-	-	-	-	-	-	-	-	-
in Other Sectors	-	-	-	-	-	-	-	-	-	-	-

Morocco / Maroc : 1996

	Oil cont. / *Pétrole cont.* (1000 tonnes)										
SUPPLY AND CONSUMPTION	Refinery Gas	LPG + Ethane	Motor Gasoline	Aviation Gasoline	Jet Fuel	Kerosene	Gas/ Diesel	Heavy Fuel Oil	Naphtha	Petrol. Coke	Other Prod.
APPROVISIONNEMENT ET DEMANDE	*Gaz de raffinerie*	*GPL + éthane*	*Essence moteur*	*Essence aviation*	*Carbu- réacteurs*	*Kérosène*	*Gazole*	*Fioul lourd*	*Naphta*	*Coke de pétrole*	*Autres prod.*
Production	94	243	394	-	231	38	2122	1755	275	-	253
From Other Sources	-	-	-	-	-	-	-	-	-	-	-
Imports	-	688	-	-	-	-	479	-	-	-	25
Exports	-	-	-	-	-	-	-	-	-306	-	-41
Intl. Marine Bunkers	-	-	-	-	-	-	-13	-	-	-	-
Stock Changes	-	4	-33	-	9	-	52	-46	31	-	-30
DOMESTIC SUPPLY	**94**	**935**	**361**	**-**	**240**	**38**	**2640**	**1709**	**-**	**-**	**207**
Transfers	-	-	-	-	-	-	-	-	-	-	-
Statistical Differences	-	16	14	-	-2	1	-	63	-	-	1
TRANSFORMATION	**-**	**-**	**-**	**-**	**-**	**-**	**145**	**1161**	**-**	**-**	**-**
Electricity Plants	-	-	-	-	-	-	145	1161	-	-	-
CHP Plants	-	-	-	-	-	-	-	-	-	-	-
Heat Plants	-	-	-	-	-	-	-	-	-	-	-
Blast Furnaces/Gas Works	-	-	-	-	-	-	-	-	-	-	-
Coke/Pat. Fuel/BKB Plants	-	-	-	-	-	-	-	-	-	-	-
Petroleum Refineries	-	-	-	-	-	-	-	-	-	-	-
Petrochemical Industry	-	-	-	-	-	-	-	-	-	-	-
Liquefaction	-	-	-	-	-	-	-	-	-	-	-
Other Transform. Sector	-	-	-	-	-	-	-	-	-	-	-
ENERGY SECTOR	**94**	**-**	**-**	**-**	**-**	**-**	**-**	**155**	**-**	**-**	**-**
Coal Mines	-	-	-	-	-	-	-	-	-	-	-
Oil and Gas Extraction	-	-	-	-	-	-	-	-	-	-	-
Petroleum Refineries	94	-	-	-	-	-	-	155	-	-	-
Electr., CHP+Heat Plants	-	-	-	-	-	-	-	-	-	-	-
Pumped Storage (Elec.)	-	-	-	-	-	-	-	-	-	-	-
Other Energy Sector	-	-	-	-	-	-	-	-	-	-	-
Distribution Losses	-	-	-	-	-	-	-	-	-	-	-
FINAL CONSUMPTION	**-**	**951**	**375**	**-**	**238**	**39**	**2495**	**456**	**-**	**-**	**208**
INDUSTRY SECTOR	**-**	**87**	**-**	**-**	**-**	**-**	**-**	**456**	**-**	**-**	**-**
Iron and Steel	-	-	-	-	-	-	-	-	-	-	-
Chemical and Petrochem.	-	-	-	-	-	-	-	-	-	-	-
of which: Feedstocks	-	-	-	-	-	-	-	-	-	-	-
Non-Ferrous Metals	-	-	-	-	-	-	-	-	-	-	-
Non-Metallic Minerals	-	-	-	-	-	-	-	-	-	-	-
Transport Equipment	-	-	-	-	-	-	-	-	-	-	-
Machinery	-	-	-	-	-	-	-	-	-	-	-
Mining and Quarrying	-	-	-	-	-	-	-	-	-	-	-
Food and Tobacco	-	-	-	-	-	-	-	-	-	-	-
Paper, Pulp and Print	-	-	-	-	-	-	-	-	-	-	-
Wood and Wood Products	-	-	-	-	-	-	-	-	-	-	-
Construction	-	-	-	-	-	-	-	-	-	-	-
Textile and Leather	-	-	-	-	-	-	-	-	-	-	-
Non-specified	-	87	-	-	-	-	-	456	-	-	-
TRANSPORT SECTOR	**-**	**-**	**375**	**-**	**238**	**-**	**150**	**-**	**-**	**-**	**-**
Air	-	-	-	-	238	-	-	-	-	-	-
Road	-	-	375	-	-	-	-	-	-	-	-
Rail	-	-	-	-	-	-	-	-	-	-	-
Pipeline Transport	-	-	-	-	-	-	-	-	-	-	-
Internal Navigation	-	-	-	-	-	-	150	-	-	-	-
Non-specified	-	-	-	-	-	-	-	-	-	-	-
OTHER SECTORS	**-**	**864**	**-**	**-**	**-**	**39**	**2345**	**-**	**-**	**-**	**-**
Agriculture	-	-	-	-	-	-	-	-	-	-	-
Comm. and Publ. Services	-	-	-	-	-	-	-	-	-	-	-
Residential	-	864	-	-	-	39	-	-	-	-	-
Non-specified	-	-	-	-	-	-	2345	-	-	-	-
NON-ENERGY USE	**-**	**-**	**-**	**-**	**-**	**-**	**-**	**-**	**-**	**-**	**208**
in Industry/Transf./Energy	-	-	-	-	-	-	-	-	-	-	208
in Transport	-	-	-	-	-	-	-	-	-	-	-
in Other Sectors	-	-	-	-	-	-	-	-	-	-	-

Morocco / Maroc : 1996

	Gas / *Gaz* (TJ)				Comb. Renew. & Waste / *En. Re. Comb. & Déchets* (TJ)				(GWh)	(TJ)
SUPPLY AND CONSUMPTION *APPROVISIONNEMENT ET DEMANDE*	Natural Gas *Gaz naturel*	Gas Works *Usines à gaz*	Coke Ovens *Cokeries*	Blast Furnaces *Hauts fourneaux*	Solid Biomass *Biomasse solide*	Gas/Liquids from Biomass *Gaz/Liquides tirés de biomasse*	Municipal Waste *Déchets urbains*	Industrial Waste *Déchets industriels*	Electricity *Electricité*	Heat *Chaleur*
Production	698	-	-	-	17032	-	-	-	12356	-
From Other Sources	-	-	-	-	-	-	-	-	-	-
Imports	-	-	-	-	-	-	-	-	129	-
Exports	-	-	-	-	-	-	-	-	-	-
Intl. Marine Bunkers	-	-	-	-	-	-	-	-	-	-
Stock Changes	-	-	-	-	-	-	-	-	-	-
DOMESTIC SUPPLY	**698**	**-**	**-**	**-**	**17032**	**-**	**-**	**-**	**12485**	**-**
Transfers	-	-	-	-	-	-	-	-	-	-
Statistical Differences	-	-	-	-	-	-	-	-	-	-
TRANSFORMATION	**-**	**-**	**-**	**-**	**-**	**-**	**-**	**-**	**-**	**-**
Electricity Plants	-	-	-	-	-	-	-	-	-	-
CHP Plants	-	-	-	-	-	-	-	-	-	-
Heat Plants	-	-	-	-	-	-	-	-	-	-
Blast Furnaces/Gas Works	-	-	-	-	-	-	-	-	-	-
Coke/Pat. Fuel/BKB Plants	-	-	-	-	-	-	-	-	-	-
Petroleum Refineries	-	-	-	-	-	-	-	-	-	-
Petrochemical Industry	-	-	-	-	-	-	-	-	-	-
Liquefaction	-	-	-	-	-	-	-	-	-	-
Other Transform. Sector	-	-	-	-	-	-	-	-	-	-
ENERGY SECTOR	**-**	**-**	**-**	**-**	**-**	**-**	**-**	**-**	**1029**	**-**
Coal Mines	-	-	-	-	-	-	-	-	75	-
Oil and Gas Extraction	-	-	-	-	-	-	-	-	23	-
Petroleum Refineries	-	-	-	-	-	-	-	-	172	-
Electr., CHP+Heat Plants	-	-	-	-	-	-	-	-	759	-
Pumped Storage (Elec.)	-	-	-	-	-	-	-	-	-	-
Other Energy Sector	-	-	-	-	-	-	-	-	-	-
Distribution Losses	-	-	-	-	-	-	-	-	490	-
FINAL CONSUMPTION	**698**	**-**	**-**	**-**	**17032**	**-**	**-**	**-**	**10966**	**-**
INDUSTRY SECTOR	**698**	**-**	**-**	**-**	**2576**	**-**	**-**	**-**	**4969**	**-**
Iron and Steel	-	-	-	-	-	-	-	-	-	-
Chemical and Petrochem.	-	-	-	-	-	-	-	-	373	-
of which: Feedstocks	-	-	-	-	-	-	-	-	-	-
Non-Ferrous Metals	-	-	-	-	-	-	-	-	-	-
Non-Metallic Minerals	-	-	-	-	-	-	-	-	648	-
Transport Equipment	-	-	-	-	-	-	-	-	-	-
Machinery	-	-	-	-	-	-	-	-	250	-
Mining and Quarrying	698	-	-	-	-	-	-	-	710	-
Food and Tobacco	-	-	-	-	-	-	-	-	501	-
Paper, Pulp and Print	-	-	-	-	-	-	-	-	376	-
Wood and Wood Products	-	-	-	-	-	-	-	-	-	-
Construction	-	-	-	-	-	-	-	-	207	-
Textile and Leather	-	-	-	-	-	-	-	-	592	-
Non-specified	-	-	-	-	2576	-	-	-	1312	-
TRANSPORT SECTOR	**-**	**-**	**-**	**-**	**-**	**-**	**-**	**-**	**204**	**-**
Air	-	-	-	-	-	-	-	-	-	-
Road	-	-	-	-	-	-	-	-	-	-
Rail	-	-	-	-	-	-	-	-	204	-
Pipeline Transport	-	-	-	-	-	-	-	-	-	-
Internal Navigation	-	-	-	-	-	-	-	-	-	-
Non-specified	-	-	-	-	-	-	-	-	-	-
OTHER SECTORS	**-**	**-**	**-**	**-**	**14456**	**-**	**-**	**-**	**5793**	**-**
Agriculture	-	-	-	-	-	-	-	-	515	-
Comm. and Publ. Services	-	-	-	-	-	-	-	-	1850	-
Residential	-	-	-	-	14456	-	-	-	3428	-
Non-specified	-	-	-	-	-	-	-	-	-	-
NON-ENERGY USE	**-**	**-**	**-**	**-**	**-**	**-**	**-**	**-**	**-**	**-**
in Industry/Transf./Energy	-	-	-	-	-	-	-	-	-	-
in Transport	-	-	-	-	-	-	-	-	-	-
in Other Sectors	-	-	-	-	-	-	-	-	-	-

Morocco / Maroc : 1997

SUPPLY AND CONSUMPTION / *APPROVISIONNEMENT ET DEMANDE*	Coal / *Charbon* (1000 tonnes)							Oil / *Pétrole* (1000 tonnes)			
	Coking Coal / *Charbon à coke*	Other Bit. Coal / *Autres charb. bit.*	Sub-Bit. Coal / *Charbon sous-bit.*	Lignite / *Lignite*	Peat / *Tourbe*	Oven and Gas Coke / *Coke de four/gaz*	Pat. Fuel and BKB / *Agg./briq. de lignite*	Crude Oil / *Pétrole brut*	NGL / *LGN*	Feed-stocks / *Produits d'aliment.*	Additives / *Additifs*
Production	-	376	-	-	-	-	-	12	-	-	-
From Other Sources	-	-	-	-	-	-	-	249	-	-	-
Imports	-	2740	-	-	-	-	-	6015	-	-	-
Exports	-	-	-	-	-	-	-	-	-	-	-
Intl. Marine Bunkers	-	-	-	-	-	-	-	-	-	-	-
Stock Changes	-	-	-	-	-	-	-	-115	-	-	-
DOMESTIC SUPPLY	**-**	**3116**	**-**	**-**	**-**	**-**	**-**	**6161**	**-**	**-**	**-**
Transfers	-	-	-	-	-	-	-	-	-	-	-
Statistical Differences	-	290	-	-	-	-	-	-	-	-	-
TRANSFORMATION	**-**	**2404**	**-**	**-**	**-**	**-**	**-**	**6161**	**-**	**-**	**-**
Electricity Plants	-	2404	-	-	-	-	-	-	-	-	-
CHP Plants	-	-	-	-	-	-	-	-	-	-	-
Heat Plants	-	-	-	-	-	-	-	-	-	-	-
Blast Furnaces/Gas Works	-	-	-	-	-	-	-	-	-	-	-
Coke/Pat. Fuel/BKB Plants	-	-	-	-	-	-	-	-	-	-	-
Petroleum Refineries	-	-	-	-	-	-	-	6161	-	-	-
Petrochemical Industry	-	-	-	-	-	-	-	-	-	-	-
Liquefaction	-	-	-	-	-	-	-	-	-	-	-
Other Transform. Sector	-	-	-	-	-	-	-	-	-	-	-
ENERGY SECTOR	**-**	**-**	**-**	**-**	**-**	**-**	**-**	**-**	**-**	**-**	**-**
Coal Mines	-	-	-	-	-	-	-	-	-	-	-
Oil and Gas Extraction	-	-	-	-	-	-	-	-	-	-	-
Petroleum Refineries	-	-	-	-	-	-	-	-	-	-	-
Electr., CHP+Heat Plants	-	-	-	-	-	-	-	-	-	-	-
Pumped Storage (Elec.)	-	-	-	-	-	-	-	-	-	-	-
Other Energy Sector	-	-	-	-	-	-	-	-	-	-	-
Distribution Losses	-	-	-	-	-	-	-	-	-	-	-
FINAL CONSUMPTION	**-**	**1002**	**-**	**-**	**-**	**-**	**-**	**-**	**-**	**-**	**-**
INDUSTRY SECTOR	**-**	**1002**	**-**	**-**	**-**	**-**	**-**	**-**	**-**	**-**	**-**
Iron and Steel	-	-	-	-	-	-	-	-	-	-	-
Chemical and Petrochem.	-	-	-	-	-	-	-	-	-	-	-
of which: Feedstocks	-	-	-	-	-	-	-	-	-	-	-
Non-Ferrous Metals	-	-	-	-	-	-	-	-	-	-	-
Non-Metallic Minerals	-	1002	-	-	-	-	-	-	-	-	-
Transport Equipment	-	-	-	-	-	-	-	-	-	-	-
Machinery	-	-	-	-	-	-	-	-	-	-	-
Mining and Quarrying	-	-	-	-	-	-	-	-	-	-	-
Food and Tobacco	-	-	-	-	-	-	-	-	-	-	-
Paper, Pulp and Print	-	-	-	-	-	-	-	-	-	-	-
Wood and Wood Products	-	-	-	-	-	-	-	-	-	-	-
Construction	-	-	-	-	-	-	-	-	-	-	-
Textile and Leather	-	-	-	-	-	-	-	-	-	-	-
Non-specified	-	-	-	-	-	-	-	-	-	-	-
TRANSPORT SECTOR	**-**	**-**	**-**	**-**	**-**	**-**	**-**	**-**	**-**	**-**	**-**
Air	-	-	-	-	-	-	-	-	-	-	-
Road	-	-	-	-	-	-	-	-	-	-	-
Rail	-	-	-	-	-	-	-	-	-	-	-
Pipeline Transport	-	-	-	-	-	-	-	-	-	-	-
Internal Navigation	-	-	-	-	-	-	-	-	-	-	-
Non-specified	-	-	-	-	-	-	-	-	-	-	-
OTHER SECTORS	**-**	**-**	**-**	**-**	**-**	**-**	**-**	**-**	**-**	**-**	**-**
Agriculture	-	-	-	-	-	-	-	-	-	-	-
Comm. and Publ. Services	-	-	-	-	-	-	-	-	-	-	-
Residential	-	-	-	-	-	-	-	-	-	-	-
Non-specified	-	-	-	-	-	-	-	-	-	-	-
NON-ENERGY USE	**-**	**-**	**-**	**-**	**-**	**-**	**-**	**-**	**-**	**-**	**-**
in Industry/Trans./Energy	-	-	-	-	-	-	-	-	-	-	-
in Transport	-	-	-	-	-	-	-	-	-	-	-
in Other Sectors	-	-	-	-	-	-	-	-	-	-	-

Morocco / Maroc : 1997

SUPPLY AND CONSUMPTION / *APPROVISIONNEMENT ET DEMANDE*	Oil cont. / *Pétrole cont.* (1000 tonnes)										
	Refinery Gas / *Gaz de raffinerie*	LPG + Ethane / *GPL + éthane*	Motor Gasoline / *Essence moteur*	Aviation Gasoline / *Essence aviation*	Jet Fuel / *Carbu- réacteurs*	Kerosene / *Kérosène*	Gas/ Diesel / *Gazole*	Heavy Fuel Oil / *Fioul lourd*	Naphtha / *Naphta*	Petrol. Coke / *Coke de pétrole*	Other Prod. / *Autres prod.*
Production	94	239	409	-	241	62	2394	1711	303	-	259
From Other Sources	-	-	-	-	-	-	-	-	-	-	-
Imports	-	745	-	-	-	-	423	-	-	-	22
Exports	-	-	-	-	-	-	-	-	-243	-	-61
Intl. Marine Bunkers	-	-	-	-	-	-	-13	-	-	-	-
Stock Changes	-	-2	-29	-	-3	-1	-94	82	-12	-	3
DOMESTIC SUPPLY	**94**	**982**	**380**	**-**	**238**	**61**	**2710**	**1793**	**48**	**-**	**223**
Transfers	-	-	-	-	-	-	-	-	-	-	-
Statistical Differences	-	25	-	-	9	-4	-89	59	-48	-	-
TRANSFORMATION	**-**	**-**	**-**	**-**	**-**	**-**	**156**	**1228**	**-**	**-**	**-**
Electricity Plants	-	-	-	-	-	-	156	1228	-	-	-
CHP Plants	-	-	-	-	-	-	-	-	-	-	-
Heat Plants	-	-	-	-	-	-	-	-	-	-	-
Blast Furnaces/Gas Works	-	-	-	-	-	-	-	-	-	-	-
Coke/Pat. Fuel/BKB Plants	-	-	-	-	-	-	-	-	-	-	-
Petroleum Refineries	-	-	-	-	-	-	-	-	-	-	-
Petrochemical Industry	-	-	-	-	-	-	-	-	-	-	-
Liquefaction	-	-	-	-	-	-	-	-	-	-	-
Other Transform. Sector	-	-	-	-	-	-	-	-	-	-	-
ENERGY SECTOR	**94**	**-**	**-**	**-**	**-**	**-**	**-**	**155**	**-**	**-**	**-**
Coal Mines	-	-	-	-	-	-	-	-	-	-	-
Oil and Gas Extraction	-	-	-	-	-	-	-	-	-	-	-
Petroleum Refineries	94	-	-	-	-	-	-	155	-	-	-
Electr., CHP+Heat Plants	-	-	-	-	-	-	-	-	-	-	-
Pumped Storage (Elec.)	-	-	-	-	-	-	-	-	-	-	-
Other Energy Sector	-	-	-	-	-	-	-	-	-	-	-
Distribution Losses	-	-	-	-	-	-	-	-	-	-	-
FINAL CONSUMPTION	**-**	**1007**	**380**	**-**	**247**	**57**	**2465**	**469**	**-**	**-**	**223**
INDUSTRY SECTOR	**-**	**92**	**-**	**-**	**-**	**-**	**-**	**469**	**-**	**-**	**-**
Iron and Steel	-	-	-	-	-	-	-	-	-	-	-
Chemical and Petrochem.	-	-	-	-	-	-	-	-	-	-	-
of which: Feedstocks	-	-	-	-	-	-	-	-	-	-	-
Non-Ferrous Metals	-	-	-	-	-	-	-	-	-	-	-
Non-Metallic Minerals	-	-	-	-	-	-	-	-	-	-	-
Transport Equipment	-	-	-	-	-	-	-	-	-	-	-
Machinery	-	-	-	-	-	-	-	-	-	-	-
Mining and Quarrying	-	-	-	-	-	-	-	-	-	-	-
Food and Tobacco	-	-	-	-	-	-	-	-	-	-	-
Paper, Pulp and Print	-	-	-	-	-	-	-	-	-	-	-
Wood and Wood Products	-	-	-	-	-	-	-	-	-	-	-
Construction	-	-	-	-	-	-	-	-	-	-	-
Textile and Leather	-	-	-	-	-	-	-	-	-	-	-
Non-specified	-	92	-	-	-	-	-	469	-	-	-
TRANSPORT SECTOR	**-**	**-**	**380**	**-**	**247**	**-**	**148**	**-**	**-**	**-**	**-**
Air	-	-	-	-	247	-	-	-	-	-	-
Road	-	-	380	-	-	-	-	-	-	-	-
Rail	-	-	-	-	-	-	-	-	-	-	-
Pipeline Transport	-	-	-	-	-	-	-	-	-	-	-
Internal Navigation	-	-	-	-	-	-	148	-	-	-	-
Non-specified	-	-	-	-	-	-	-	-	-	-	-
OTHER SECTORS	**-**	**915**	**-**	**-**	**-**	**57**	**2317**	**-**	**-**	**-**	**-**
Agriculture	-	-	-	-	-	-	-	-	-	-	-
Comm. and Publ. Services	-	-	-	-	-	-	-	-	-	-	-
Residential	-	915	-	-	-	57	-	-	-	-	-
Non-specified	-	-	-	-	-	-	2317	-	-	-	-
NON-ENERGY USE	**-**	**-**	**-**	**-**	**-**	**-**	**-**	**-**	**-**	**-**	**223**
in Industry/Transf./Energy	-	-	-	-	-	-	-	-	-	-	223
in Transport	-	-	-	-	-	-	-	-	-	-	-
in Other Sectors	-	-	-	-	-	-	-	-	-	-	-

Morocco / Maroc : 1997

SUPPLY AND CONSUMPTION / *APPROVISIONNEMENT ET DEMANDE*	Gas / *Gaz* (TJ)				Comb. Renew. & Waste / *En. Re. Comb. & Déchets* (TJ)				(GWh)	(TJ)
	Natural Gas / *Gaz naturel*	Gas Works / *Usines à gaz*	Coke Ovens / *Cokeries*	Blast Furnaces / *Hauts fourneaux*	Solid Biomass / *Biomasse solide*	Gas/Liquids from Biomass / *Gaz/Liquides tirés de biomasse*	Municipal Waste / *Déchets urbains*	Industrial Waste / *Déchets industriels*	Electricity / *Electricité*	Heat / *Chaleur*
Production	1047	-	-	-	17373	-	-	-	13128	-
From Other Sources	-	-	-	-	-	-	-	-	-	-
Imports	-	-	-	-	-	-	-	-	124	-
Exports	-	-	-	-	-	-	-	-	-	-
Intl. Marine Bunkers	-	-	-	-	-	-	-	-	-	-
Stock Changes	-	-	-	-	-	-	-	-	-	-
DOMESTIC SUPPLY	**1047**	**-**	**-**	**-**	**17373**	**-**	**-**	**-**	**13252**	**-**
Transfers	-	-	-	-	-	-	-	-	-	-
Statistical Differences	-	-	-	-	-	-	-	-	2	-
TRANSFORMATION	**-**	**-**	**-**	**-**	**-**	**-**	**-**	**-**	**-**	**-**
Electricity Plants	-	-	-	-	-	-	-	-	-	-
CHP Plants	-	-	-	-	-	-	-	-	-	-
Heat Plants	-	-	-	-	-	-	-	-	-	-
Blast Furnaces/Gas Works	-	-	-	-	-	-	-	-	-	-
Coke/Pat. Fuel/BKB Plants	-	-	-	-	-	-	-	-	-	-
Petroleum Refineries	-	-	-	-	-	-	-	-	-	-
Petrochemical Industry	-	-	-	-	-	-	-	-	-	-
Liquefaction	-	-	-	-	-	-	-	-	-	-
Other Transform. Sector	-	-	-	-	-	-	-	-	-	-
ENERGY SECTOR	**-**	**-**	**-**	**-**	**-**	**-**	**-**	**-**	**1134**	**-**
Coal Mines	-	-	-	-	-	-	-	-	58	-
Oil and Gas Extraction	-	-	-	-	-	-	-	-	26	-
Petroleum Refineries	-	-	-	-	-	-	-	-	194	-
Electr., CHP+Heat Plants	-	-	-	-	-	-	-	-	856	-
Pumped Storage (Elec.)	-	-	-	-	-	-	-	-	-	-
Other Energy Sector	-	-	-	-	-	-	-	-	-	-
Distribution Losses	-	-	-	-	-	-	-	-	560	-
FINAL CONSUMPTION	**1047**	**-**	**-**	**-**	**17373**	**-**	**-**	**-**	**11560**	**-**
INDUSTRY SECTOR	**1047**	**-**	**-**	**-**	**2628**	**-**	**-**	**-**	**5250**	**-**
Iron and Steel	-	-	-	-	-	-	-	-	-	-
Chemical and Petrochem.	-	-	-	-	-	-	-	-	393	-
of which: Feedstocks	-	-	-	-	-	-	-	-	-	-
Non-Ferrous Metals	-	-	-	-	-	-	-	-	-	-
Non-Metallic Minerals	-	-	-	-	-	-	-	-	682	-
Transport Equipment	-	-	-	-	-	-	-	-	-	-
Machinery	-	-	-	-	-	-	-	-	263	-
Mining and Quarrying	1047	-	-	-	-	-	-	-	768	-
Food and Tobacco	-	-	-	-	-	-	-	-	527	-
Paper, Pulp and Print	-	-	-	-	-	-	-	-	396	-
Wood and Wood Products	-	-	-	-	-	-	-	-	-	-
Construction	-	-	-	-	-	-	-	-	218	-
Textile and Leather	-	-	-	-	-	-	-	-	623	-
Non-specified	-	-	-	-	2628	-	-	-	1380	-
TRANSPORT SECTOR	**-**	**-**	**-**	**-**	**-**	**-**	**-**	**-**	**214**	**-**
Air	-	-	-	-	-	-	-	-	-	-
Road	-	-	-	-	-	-	-	-	-	-
Rail	-	-	-	-	-	-	-	-	214	-
Pipeline Transport	-	-	-	-	-	-	-	-	-	-
Internal Navigation	-	-	-	-	-	-	-	-	-	-
Non-specified	-	-	-	-	-	-	-	-	-	-
OTHER SECTORS	**-**	**-**	**-**	**-**	**14745**	**-**	**-**	**-**	**6096**	**-**
Agriculture	-	-	-	-	-	-	-	-	542	-
Comm. and Publ. Services	-	-	-	-	-	-	-	-	1947	-
Residential	-	-	-	-	14745	-	-	-	3607	-
Non-specified	-	-	-	-	-	-	-	-	-	-
NON-ENERGY USE	**-**	**-**	**-**	**-**	**-**	**-**	**-**	**-**	**-**	**-**
in Industry/Transf./Energy	-	-	-	-	-	-	-	-	-	-
in Transport	-	-	-	-	-	-	-	-	-	-
in Other Sectors	-	-	-	-	-	-	-	-	-	-

Mozambique

SUPPLY AND CONSUMPTION 1996	Coal (1000 tonnes)							Oil (1000 tonnes)			
	Coking Coal	Other Bit. Coal	Sub-Bit. Coal	Lignite	Peat	Oven and Gas Coke	Pat. Fuel and BKB	Crude Oil	NGL	Feed-stocks	Additives
Production	-	-	-	-	-	-	-	-	-	-	-
Imports	-	18	-	-	-	-	-	-	-	-	-
Exports	-	-	-	-	-	-	-	-	-	-	-
Intl. Marine Bunkers	-	-	-	-	-	-	-	-	-	-	-
Stock Changes	-	-	-	-	-	-	-	-	-	-	-
DOMESTIC SUPPLY	**-**	**18**	**-**	**-**	**-**	**-**	**-**	**-**	**-**	**-**	**-**
Transfers and Stat. Diff.	-	-	-	-	-	-	-	-	-	-	-
TRANSFORMATION	**-**	**-**	**-**	**-**	**-**	**-**	**-**	**-**	**-**	**-**	**-**
Electricity and CHP Plants	-	-	-	-	-	-	-	-	-	-	-
Petroleum Refineries	-	-	-	-	-	-	-	-	-	-	-
Other Transform. Sector	-	-	-	-	-	-	-	-	-	-	-
ENERGY SECTOR	**-**	**-**	**-**	**-**	**-**	**-**	**-**	**-**	**-**	**-**	**-**
DISTRIBUTION LOSSES	**-**	**-**	**-**	**-**	**-**	**-**	**-**	**-**	**-**	**-**	**-**
FINAL CONSUMPTION	**-**	**18**	**-**	**-**	**-**	**-**	**-**	**-**	**-**	**-**	**-**
INDUSTRY SECTOR	**-**	**18**	**-**	**-**	**-**	**-**	**-**	**-**	**-**	**-**	**-**
Iron and Steel	-	-	-	-	-	-	-	-	-	-	-
Chemical and Petrochem.	-	-	-	-	-	-	-	-	-	-	-
Non-Metallic Minerals	-	-	-	-	-	-	-	-	-	-	-
Non-specified	-	18	-	-	-	-	-	-	-	-	-
TRANSPORT SECTOR	**-**	**-**	**-**	**-**	**-**	**-**	**-**	**-**	**-**	**-**	**-**
Air	-	-	-	-	-	-	-	-	-	-	-
Road	-	-	-	-	-	-	-	-	-	-	-
Non-specified	-	-	-	-	-	-	-	-	-	-	-
OTHER SECTORS	**-**	**-**	**-**	**-**	**-**	**-**	**-**	**-**	**-**	**-**	**-**
Agriculture	-	-	-	-	-	-	-	-	-	-	-
Comm. and Publ. Services	-	-	-	-	-	-	-	-	-	-	-
Residential	-	-	-	-	-	-	-	-	-	-	-
Non-specified	-	-	-	-	-	-	-	-	-	-	-
NON-ENERGY USE	**-**	**-**	**-**	**-**	**-**	**-**	**-**	**-**	**-**	**-**	**-**

APPROVISIONNEMENT ET DEMANDE 1997	*Charbon* (1000 tonnes)							*Pétrole* (1000 tonnes)			
	Charbon à coke	*Autres charb. bit.*	*Charbon sous-bit.*	*Lignite*	*Tourbe*	*Coke de four/gaz*	*Agg./briq. de lignite*	*Pétrole brut*	*LGN*	*Produits d'aliment.*	*Additifs*
Production	-	-	-	-	-	-	-	-	-	-	-
Imports	-	20	-	-	-	-	-	-	-	-	-
Exports	-	-	-	-	-	-	-	-	-	-	-
Intl. Marine Bunkers	-	-	-	-	-	-	-	-	-	-	-
Stock Changes	-	-	-	-	-	-	-	-	-	-	-
DOMESTIC SUPPLY	**-**	**20**	**-**	**-**	**-**	**-**	**-**	**-**	**-**	**-**	**-**
Transfers and Stat. Diff.	-	-	-	-	-	-	-	-	-	-	-
TRANSFORMATION	**-**	**-**	**-**	**-**	**-**	**-**	**-**	**-**	**-**	**-**	**-**
Electricity and CHP Plants	-	-	-	-	-	-	-	-	-	-	-
Petroleum Refineries	-	-	-	-	-	-	-	-	-	-	-
Other Transform. Sector	-	-	-	-	-	-	-	-	-	-	-
ENERGY SECTOR	**-**	**-**	**-**	**-**	**-**	**-**	**-**	**-**	**-**	**-**	**-**
DISTRIBUTION LOSSES	**-**	**-**	**-**	**-**	**-**	**-**	**-**	**-**	**-**	**-**	**-**
FINAL CONSUMPTION	**-**	**20**	**-**	**-**	**-**	**-**	**-**	**-**	**-**	**-**	**-**
INDUSTRY SECTOR	**-**	**20**	**-**	**-**	**-**	**-**	**-**	**-**	**-**	**-**	**-**
Iron and Steel	-	-	-	-	-	-	-	-	-	-	-
Chemical and Petrochem.	-	-	-	-	-	-	-	-	-	-	-
Non-Metallic Minerals	-	-	-	-	-	-	-	-	-	-	-
Non-specified	-	20	-	-	-	-	-	-	-	-	-
TRANSPORT SECTOR	**-**	**-**	**-**	**-**	**-**	**-**	**-**	**-**	**-**	**-**	**-**
Air	-	-	-	-	-	-	-	-	-	-	-
Road	-	-	-	-	-	-	-	-	-	-	-
Non-specified	-	-	-	-	-	-	-	-	-	-	-
OTHER SECTORS	**-**	**-**	**-**	**-**	**-**	**-**	**-**	**-**	**-**	**-**	**-**
Agriculture	-	-	-	-	-	-	-	-	-	-	-
Comm. and Publ. Services	-	-	-	-	-	-	-	-	-	-	-
Residential	-	-	-	-	-	-	-	-	-	-	-
Non-specified	-	-	-	-	-	-	-	-	-	-	-
NON-ENERGY USE	**-**	**-**	**-**	**-**	**-**	**-**	**-**	**-**	**-**	**-**	**-**

Mozambique

SUPPLY AND CONSUMPTION 1996	Oil cont. (1000 tonnes)										
	Refinery Gas	LPG + Ethane	Motor Gasoline	Aviation Gasoline	Jet Fuel	Kerosene	Gas/ Diesel	Heavy Fuel Oil	Naphtha	Petrol. Coke	Other Prod.
Production	-	-	-	-	-	-	-	-	-	-	4
Imports	-	4	42	-	21	17	234	192	-	-	2
Exports	-	-	-	-	-	-2	-	-	-	-	-
Intl. Marine Bunkers	-	-	-	-	-	-	-3	-12	-	-	-
Stock Changes	-	-	-1	-	-2	2	-4	-	-	-	2
DOMESTIC SUPPLY	**-**	**4**	**41**	**-**	**19**	**17**	**227**	**180**	**-**	**-**	**8**
Transfers and Stat. Diff.	-	-1	-	-	2	-	3	2	-	-	-1
TRANSFORMATION	**-**	**-**	**-**	**-**	**-**	**-**	**8**	**102**	**-**	**-**	**-**
Electricity and CHP Plants	-	-	-	-	-	-	8	102	-	-	-
Petroleum Refineries	-	-	-	-	-	-	-	-	-	-	-
Other Transform. Sector	-	-	-	-	-	-	-	-	-	-	-
ENERGY SECTOR	**-**	**-**	**-**	**-**	**-**	**-**	**-**	**-**	**-**	**-**	**-**
DISTRIBUTION LOSSES	**-**	**-**	**-**	**-**	**-**	**-**	**-**	**-**	**-**	**-**	**-**
FINAL CONSUMPTION	**-**	**3**	**41**	**-**	**21**	**17**	**222**	**80**	**-**	**-**	**7**
INDUSTRY SECTOR	**-**	**1**	**-**	**-**	**-**	**-**	**9**	**-**	**-**	**-**	**-**
Iron and Steel	-	-	-	-	-	-	-	-	-	-	-
Chemical and Petrochem.	-	-	-	-	-	-	-	-	-	-	-
Non-Metallic Minerals	-	-	-	-	-	-	-	-	-	-	-
Non-specified	-	1	-	-	-	-	9	-	-	-	-
TRANSPORT SECTOR	**-**	**-**	**41**	**-**	**21**	**-**	**24**	**-**	**-**	**-**	**-**
Air	-	-	-	-	21	-	-	-	-	-	-
Road	-	-	41	-	-	-	24	-	-	-	-
Non-specified	-	-	-	-	-	-	-	-	-	-	-
OTHER SECTORS	**-**	**2**	**-**	**-**	**-**	**17**	**189**	**80**	**-**	**-**	**-**
Agriculture	-	-	-	-	-	-	31	-	-	-	-
Comm. and Publ. Services	-	-	-	-	-	-	158	-	-	-	-
Residential	-	2	-	-	-	17	-	-	-	-	-
Non-specified	-	-	-	-	-	-	-	80	-	-	-
NON-ENERGY USE	**-**	**-**	**-**	**-**	**-**	**-**	**-**	**-**	**-**	**-**	**7**

APPROVISIONNEMENT ET DEMANDE 1997	*Pétrole cont.* (1000 tonnes)										
	Gaz de raffinerie	*GPL + éthane*	*Essence moteur*	*Essence aviation*	*Carbu- réacteurs*	*Kérosène*	*Gazole*	*Fioul lourd*	*Naphta*	*Coke de pétrole*	*Autres prod.*
Production	-	-	-	-	-	-	-	-	-	-	5
Imports	-	5	44	-	27	26	264	332	-	-	5
Exports	-	-	-	-	-	-1	-	-	-	-	-
Intl. Marine Bunkers	-	-	-	-	-	-	-1	-63	-	-	-
Stock Changes	-	-	-2	-	-2	-	-	-	-	-	2
DOMESTIC SUPPLY	**-**	**5**	**42**	**-**	**25**	**25**	**263**	**269**	**-**	**-**	**12**
Transfers and Stat. Diff.	-	-	-	-	-	-	33	3	-	-	-5
TRANSFORMATION	**-**	**-**	**-**	**-**	**-**	**-**	**9**	**153**	**-**	**-**	**-**
Electricity and CHP Plants	-	-	-	-	-	-	9	153	-	-	-
Petroleum Refineries	-	-	-	-	-	-	-	-	-	-	-
Other Transform. Sector	-	-	-	-	-	-	-	-	-	-	-
ENERGY SECTOR	**-**	**-**	**-**	**-**	**-**	**-**	**-**	**-**	**-**	**-**	**-**
DISTRIBUTION LOSSES	**-**	**-**	**-**	**-**	**-**	**-**	**-**	**-**	**-**	**-**	**-**
FINAL CONSUMPTION	**-**	**5**	**42**	**-**	**25**	**25**	**287**	**119**	**-**	**-**	**7**
INDUSTRY SECTOR	**-**	**1**	**-**	**-**	**-**	**-**	**10**	**-**	**-**	**-**	**-**
Iron and Steel	-	-	-	-	-	-	-	-	-	-	-
Chemical and Petrochem.	-	-	-	-	-	-	-	-	-	-	-
Non-Metallic Minerals	-	-	-	-	-	-	-	-	-	-	-
Non-specified	-	1	-	-	-	-	10	-	-	-	-
TRANSPORT SECTOR	**-**	**-**	**42**	**-**	**25**	**-**	**29**	**-**	**-**	**-**	**-**
Air	-	-	-	-	25	-	-	-	-	-	-
Road	-	-	42	-	-	-	29	-	-	-	-
Non-specified	-	-	-	-	-	-	-	-	-	-	-
OTHER SECTORS	**-**	**4**	**-**	**-**	**-**	**25**	**248**	**119**	**-**	**-**	**-**
Agriculture	-	-	-	-	-	-	37	-	-	-	-
Comm. and Publ. Services	-	-	-	-	-	-	211	-	-	-	-
Residential	-	4	-	-	-	25	-	-	-	-	-
Non-specified	-	-	-	-	-	-	-	119	-	-	-
NON-ENERGY USE	**-**	**-**	**-**	**-**	**-**	**-**	**-**	**-**	**-**	**-**	**7**

Mozambique

SUPPLY AND CONSUMPTION 1996	Gas (TJ)				Comb. Renew. & Waste (TJ)				(GWh)	(TJ)
	Natural Gas	Gas Works	Coke Ovens	Blast Furnaces	Solid Biomass	Gas/Liquids from Biomass	Municipal Waste	Industrial Waste	Electricity	Heat
Production	15	-	-	-	290000	-	-	-	476	-
Imports	-	-	-	-	-	-	-	-	599	-
Exports	-	-	-	-	-	-	-	-	-	-
Intl. Marine Bunkers	-	-	-	-	-	-	-	-	-	-
Stock Changes	-	-	-	-	-	-	-	-	-	-
DOMESTIC SUPPLY	**15**	**-**	**-**	**-**	**290000**	**-**	**-**	**-**	**1075**	**-**
Transfers and Stat. Diff.	-	-	-	-	-	-	-	-	-19	-
TRANSFORMATION	**15**	**-**	**-**	**-**	**30750**	**-**	**-**	**-**	**-**	**-**
Electricity and CHP Plants	15	-	-	-	-	-	-	-	-	-
Petroleum Refineries	-	-	-	-	-	-	-	-	-	-
Other Transform. Sector	-	-	-	-	30750	-	-	-	-	-
ENERGY SECTOR	**-**	**-**	**-**	**-**	**-**	**-**	**-**	**-**	**52**	**-**
DISTRIBUTION LOSSES	**-**	**-**	**-**	**-**	**-**	**-**	**-**	**-**	**343**	**-**
FINAL CONSUMPTION	**-**	**-**	**-**	**-**	**259250**	**-**	**-**	**-**	**661**	**-**
INDUSTRY SECTOR	**-**	**-**	**-**	**-**	**20000**	**-**	**-**	**-**	**247**	**-**
Iron and Steel	-	-	-	-	-	-	-	-	-	-
Chemical and Petrochem.	-	-	-	-	-	-	-	-	-	-
Non-Metallic Minerals	-	-	-	-	-	-	-	-	-	-
Non-specified	-	-	-	-	20000	-	-	-	247	-
TRANSPORT SECTOR	**-**	**-**	**-**	**-**	**-**	**-**	**-**	**-**	**-**	**-**
Air	-	-	-	-	-	-	-	-	-	-
Road	-	-	-	-	-	-	-	-	-	-
Non-specified	-	-	-	-	-	-	-	-	-	-
OTHER SECTORS	**-**	**-**	**-**	**-**	**239250**	**-**	**-**	**-**	**414**	**-**
Agriculture	-	-	-	-	-	-	-	-	-	-
Comm. and Publ. Services	-	-	-	-	-	-	-	-	156	-
Residential	-	-	-	-	239250	-	-	-	258	-
Non-specified	-	-	-	-	-	-	-	-	-	-
NON-ENERGY USE	**-**	**-**	**-**	**-**	**-**	**-**	**-**	**-**	**-**	**-**

APPROVISIONNEMENT ET DEMANDE 1997	*Gaz* (TJ)				*En. Re. Comb. & Déchets* (TJ)				(GWh)	(TJ)
	Gaz naturel	*Usines à gaz*	*Cokeries*	*Hauts fourneaux*	*Biomasse solide*	*Gaz/Liquides tirés de biomasse*	*Déchets urbains*	*Déchets industriels*	*Electricité*	*Chaleur*
Production	21	-	-	-	290000	-	-	-	1005	-
Imports	-	-	-	-	-	-	-	-	686	-
Exports	-	-	-	-	-	-	-	-	-483	-
Intl. Marine Bunkers	-	-	-	-	-	-	-	-	-	-
Stock Changes	-	-	-	-	-	-	-	-	-	-
DOMESTIC SUPPLY	**21**	**-**	**-**	**-**	**290000**	**-**	**-**	**-**	**1208**	**-**
Transfers and Stat. Diff.	-	-	-	-	-	-	-	-	-42	-
TRANSFORMATION	**21**	**-**	**-**	**-**	**30750**	**-**	**-**	**-**	**-**	**-**
Electricity and CHP Plants	21	-	-	-	-	-	-	-	-	-
Petroleum Refineries	-	-	-	-	-	-	-	-	-	-
Other Transform. Sector	-	-	-	-	30750	-	-	-	-	-
ENERGY SECTOR	**-**	**-**	**-**	**-**	**-**	**-**	**-**	**-**	**72**	**-**
DISTRIBUTION LOSSES	**-**	**-**	**-**	**-**	**-**	**-**	**-**	**-**	**315**	**-**
FINAL CONSUMPTION	**-**	**-**	**-**	**-**	**259250**	**-**	**-**	**-**	**779**	**-**
INDUSTRY SECTOR	**-**	**-**	**-**	**-**	**20000**	**-**	**-**	**-**	**291**	**-**
Iron and Steel	-	-	-	-	-	-	-	-	-	-
Chemical and Petrochem.	-	-	-	-	-	-	-	-	-	-
Non-Metallic Minerals	-	-	-	-	-	-	-	-	-	-
Non-specified	-	-	-	-	20000	-	-	-	291	-
TRANSPORT SECTOR	**-**	**-**	**-**	**-**	**-**	**-**	**-**	**-**	**-**	**-**
Air	-	-	-	-	-	-	-	-	-	-
Road	-	-	-	-	-	-	-	-	-	-
Non-specified	-	-	-	-	-	-	-	-	-	-
OTHER SECTORS	**-**	**-**	**-**	**-**	**239250**	**-**	**-**	**-**	**488**	**-**
Agriculture	-	-	-	-	-	-	-	-	-	-
Comm. and Publ. Services	-	-	-	-	-	-	-	-	184	-
Residential	-	-	-	-	239250	-	-	-	304	-
Non-specified	-	-	-	-	-	-	-	-	-	-
NON-ENERGY USE	**-**	**-**	**-**	**-**	**-**	**-**	**-**	**-**	**-**	**-**

Myanmar

SUPPLY AND CONSUMPTION 1996	Coal (1000 tonnes)							Oil (1000 tonnes)			
	Coking Coal	Other Bit. Coal	Sub-Bit. Coal	Lignite	Peat	Oven and Gas Coke	Pat. Fuel and BKB	Crude Oil	NGL	Feed-stocks	Additives
Production	-	-	-	46	-	-	-	401	1	-	-
Imports	-	-	-	-	-	-	-	365	-	-	-
Exports	-	-	-	-	-	-	-	-	-	-	-
Intl. Marine Bunkers	-	-	-	-	-	-	-	-	-	-	-
Stock Changes	-	-	-	-	-	-	-	45	-	-	-
DOMESTIC SUPPLY	**-**	**-**	**-**	**46**	**-**	**-**	**-**	**811**	**1**	**-**	**-**
Transfers and Stat. Diff.	-	-	-	-	-	-	-	-12	-	-	-
TRANSFORMATION	**-**	**-**	**-**	**-**	**-**	**-**	**-**	**799**	**1**	**-**	**-**
Electricity and CHP Plants	-	-	-	-	-	-	-	-	-	-	-
Petroleum Refineries	-	-	-	-	-	-	-	799	1	-	-
Other Transform. Sector	-	-	-	-	-	-	-	-	-	-	-
ENERGY SECTOR	**-**	**-**	**-**	**-**	**-**	**-**	**-**	**-**	**-**	**-**	**-**
DISTRIBUTION LOSSES	**-**	**-**	**-**	**-**	**-**	**-**	**-**	**-**	**-**	**-**	**-**
FINAL CONSUMPTION	**-**	**-**	**-**	**46**	**-**	**-**	**-**	**-**	**-**	**-**	**-**
INDUSTRY SECTOR	**-**	**-**	**-**	**46**	**-**	**-**	**-**	**-**	**-**	**-**	**-**
Iron and Steel	-	-	-	-	-	-	-	-	-	-	-
Chemical and Petrochem.	-	-	-	-	-	-	-	-	-	-	-
Non-Metallic Minerals	-	-	-	-	-	-	-	-	-	-	-
Non-specified	-	-	-	46	-	-	-	-	-	-	-
TRANSPORT SECTOR	**-**	**-**	**-**	**-**	**-**	**-**	**-**	**-**	**-**	**-**	**-**
Air	-	-	-	-	-	-	-	-	-	-	-
Road	-	-	-	-	-	-	-	-	-	-	-
Non-specified	-	-	-	-	-	-	-	-	-	-	-
OTHER SECTORS	**-**	**-**	**-**	**-**	**-**	**-**	**-**	**-**	**-**	**-**	**-**
Agriculture	-	-	-	-	-	-	-	-	-	-	-
Comm. and Publ. Services	-	-	-	-	-	-	-	-	-	-	-
Residential	-	-	-	-	-	-	-	-	-	-	-
Non-specified	-	-	-	-	-	-	-	-	-	-	-
NON-ENERGY USE	**-**	**-**	**-**	**-**	**-**	**-**	**-**	**-**	**-**	**-**	**-**

APPROVISIONNEMENT ET DEMANDE 1997	*Charbon* (1000 tonnes)							*Pétrole* (1000 tonnes)			
	Charbon à coke	*Autres charb. bit.*	*Charbon sous-bit.*	*Lignite*	*Tourbe*	*Coke de four/gaz*	*Agg./briq. de lignite*	*Pétrole brut*	*LGN*	*Produits d'aliment.*	*Additifs*
Production	-	-	-	46	-	-	-	395	1	-	-
Imports	-	-	-	-	-	-	-	647	-	-	-
Exports	-	-	-	-	-	-	-	-	-	-	-
Intl. Marine Bunkers	-	-	-	-	-	-	-	-	-	-	-
Stock Changes	-	-	-	-	-	-	-	-1	-	-	-
DOMESTIC SUPPLY	**-**	**-**	**-**	**46**	**-**	**-**	**-**	**1041**	**1**	**-**	**-**
Transfers and Stat. Diff.	-	-	-	-	-	-	-	-7	-	-	-
TRANSFORMATION	**-**	**-**	**-**	**-**	**-**	**-**	**-**	**1034**	**1**	**-**	**-**
Electricity and CHP Plants	-	-	-	-	-	-	-	-	-	-	-
Petroleum Refineries	-	-	-	-	-	-	-	1034	1	-	-
Other Transform. Sector	-	-	-	-	-	-	-	-	-	-	-
ENERGY SECTOR	**-**	**-**	**-**	**-**	**-**	**-**	**-**	**-**	**-**	**-**	**-**
DISTRIBUTION LOSSES	**-**	**-**	**-**	**-**	**-**	**-**	**-**	**-**	**-**	**-**	**-**
FINAL CONSUMPTION	**-**	**-**	**-**	**46**	**-**	**-**	**-**	**-**	**-**	**-**	**-**
INDUSTRY SECTOR	**-**	**-**	**-**	**46**	**-**	**-**	**-**	**-**	**-**	**-**	**-**
Iron and Steel	-	-	-	-	-	-	-	-	-	-	-
Chemical and Petrochem.	-	-	-	-	-	-	-	-	-	-	-
Non-Metallic Minerals	-	-	-	-	-	-	-	-	-	-	-
Non-specified	-	-	-	46	-	-	-	-	-	-	-
TRANSPORT SECTOR	**-**	**-**	**-**	**-**	**-**	**-**	**-**	**-**	**-**	**-**	**-**
Air	-	-	-	-	-	-	-	-	-	-	-
Road	-	-	-	-	-	-	-	-	-	-	-
Non-specified	-	-	-	-	-	-	-	-	-	-	-
OTHER SECTORS	**-**	**-**	**-**	**-**	**-**	**-**	**-**	**-**	**-**	**-**	**-**
Agriculture	-	-	-	-	-	-	-	-	-	-	-
Comm. and Publ. Services	-	-	-	-	-	-	-	-	-	-	-
Residential	-	-	-	-	-	-	-	-	-	-	-
Non-specified	-	-	-	-	-	-	-	-	-	-	-
NON-ENERGY USE	**-**	**-**	**-**	**-**	**-**	**-**	**-**	**-**	**-**	**-**	**-**

Myanmar

SUPPLY AND CONSUMPTION 1996	Oil cont. (1000 tonnes)										
	Refinery Gas	LPG + Ethane	Motor Gasoline	Aviation Gasoline	Jet Fuel	Kerosene	Gas/ Diesel	Heavy Fuel Oil	Naphtha	Petrol. Coke	Other Prod.
Production	30	8	175	-	52	1	345	98	-	25	27
Imports	-	2	6	-	-	-	309	-	-	-	-
Exports	-	-	-	-	-	-	-	-	-	-26	-
Intl. Marine Bunkers	-	-	-	-	-	-	-5	-	-	-	-
Stock Changes	-	-6	12	-	7	1	-19	-8	-	21	1
DOMESTIC SUPPLY	**30**	**4**	**193**	**-**	**59**	**2**	**630**	**90**	**-**	**20**	**28**
Transfers and Stat. Diff.	-	7	11	-	-5	6	246	14	-	-1	1
TRANSFORMATION	**-**	**-**	**-**	**-**	**-**	**-**	**30**	**18**	**-**	**-**	**-**
Electricity and CHP Plants	-	-	-	-	-	-	30	18	-	-	-
Petroleum Refineries	-	-	-	-	-	-	-	-	-	-	-
Other Transform. Sector	-	-	-	-	-	-	-	-	-	-	-
ENERGY SECTOR	**30**	**-**	**-**	**-**	**-**	**-**	**-**	**4**	**-**	**-**	**11**
DISTRIBUTION LOSSES	**-**	**-**	**-**	**-**	**-**	**-**	**-**	**-**	**-**	**-**	**-**
FINAL CONSUMPTION	**-**	**11**	**204**	**-**	**54**	**8**	**846**	**82**	**-**	**19**	**18**
INDUSTRY SECTOR	**-**	**1**	**-**	**-**	**-**	**-**	**335**	**66**	**-**	**19**	**-**
Iron and Steel	-	-	-	-	-	-	-	-	-	-	-
Chemical and Petrochem.	-	-	-	-	-	-	-	-	-	-	-
Non-Metallic Minerals	-	-	-	-	-	-	-	-	-	-	-
Non-specified	-	1	-	-	-	-	335	66	-	19	-
TRANSPORT SECTOR	**-**	**2**	**204**	**-**	**54**	**-**	**378**	**7**	**-**	**-**	**-**
Air	-	-	-	-	54	-	-	-	-	-	-
Road	-	2	204	-	-	-	378	-	-	-	-
Non-specified	-	-	-	-	-	-	-	7	-	-	-
OTHER SECTORS	**-**	**8**	**-**	**-**	**-**	**8**	**133**	**9**	**-**	**-**	**-**
Agriculture	-	-	-	-	-	-	-	4	-	-	-
Comm. and Publ. Services	-	-	-	-	-	-	-	-	-	-	-
Residential	-	5	-	-	-	8	129	-	-	-	-
Non-specified	-	3	-	-	-	-	4	5	-	-	-
NON-ENERGY USE	**-**	**-**	**-**	**-**	**-**	**-**	**-**	**-**	**-**	**-**	**18**

APPROVISIONNEMENT ET DEMANDE 1997	*Pétrole cont.* (1000 tonnes)										
	Gaz de raffinerie	*GPL + éthane*	*Essence moteur*	*Essence aviation*	*Carbu- réacteurs*	*Kérosène*	*Gazole*	*Fioul lourd*	*Naphta*	*Coke de pétrole*	*Autres prod.*
Production	38	10	255	-	56	1	465	103	-	26	29
Imports	-	1	-	-	-	-	289	-	-	-	-
Exports	-	-	-	-	-	-	-	-	-	-9	-
Intl. Marine Bunkers	-	-	-	-	-	-	-4	-	-	-	-
Stock Changes	-	-5	-4	-	-2	1	-9	-20	-	9	-1
DOMESTIC SUPPLY	**38**	**6**	**251**	**-**	**54**	**2**	**741**	**83**	**-**	**26**	**28**
Transfers and Stat. Diff.	-	6	-3	-	-2	1	289	24	-	2	2
TRANSFORMATION	**-**	**-**	**-**	**-**	**-**	**-**	**36**	**20**	**-**	**-**	**-**
Electricity and CHP Plants	-	-	-	-	-	-	36	20	-	-	-
Petroleum Refineries	-	-	-	-	-	-	-	-	-	-	-
Other Transform. Sector	-	-	-	-	-	-	-	-	-	-	-
ENERGY SECTOR	**38**	**-**	**-**	**-**	**-**	**-**	**22**	**1**	**-**	**-**	**12**
DISTRIBUTION LOSSES	**-**	**-**	**-**	**-**	**-**	**1**	**-**	**-**	**-**	**-**	**-**
FINAL CONSUMPTION	**-**	**12**	**248**	**-**	**52**	**2**	**972**	**86**	**-**	**28**	**18**
INDUSTRY SECTOR	**-**	**1**	**-**	**-**	**-**	**-**	**385**	**66**	**-**	**28**	**-**
Iron and Steel	-	-	-	-	-	-	-	-	-	-	-
Chemical and Petrochem.	-	-	-	-	-	-	-	-	-	-	-
Non-Metallic Minerals	-	-	-	-	-	-	-	-	-	-	-
Non-specified	-	1	-	-	-	-	385	66	-	28	-
TRANSPORT SECTOR	**-**	**2**	**248**	**-**	**52**	**-**	**434**	**7**	**-**	**-**	**-**
Air	-	-	-	-	52	-	-	-	-	-	-
Road	-	2	248	-	-	-	434	-	-	-	-
Non-specified	-	-	-	-	-	-	-	7	-	-	-
OTHER SECTORS	**-**	**9**	**-**	**-**	**-**	**2**	**153**	**13**	**-**	**-**	**-**
Agriculture	-	-	-	-	-	-	-	9	-	-	-
Comm. and Publ. Services	-	-	-	-	-	-	-	-	-	-	-
Residential	-	5	-	-	-	2	148	-	-	-	-
Non-specified	-	4	-	-	-	-	5	4	-	-	-
NON-ENERGY USE	**-**	**-**	**-**	**-**	**-**	**-**	**-**	**-**	**-**	**-**	**18**

Myanmar

SUPPLY AND CONSUMPTION 1996	Gas (TJ)				Comb. Renew. & Waste (TJ)				(GWh)	(TJ)
	Natural Gas	Gas Works	Coke Ovens	Blast Furnaces	Solid Biomass	Gas/Liquids from Biomass	Municipal Waste	Industrial Waste	Electricity	Heat
Production	61495	-	-	-	423549	-	-	-	3945	-
Imports	-	-	-	-	-	-	-	-	-	-
Exports	-	-	-	-	-	-	-	-	-	-
Intl. Marine Bunkers	-	-	-	-	-	-	-	-	-	-
Stock Changes	-7322	-	-	-	-	-	-	-	-	-
DOMESTIC SUPPLY	**54173**	-	-	-	**423549**	-	-	-	**3945**	-
Transfers and Stat. Diff.	7324	-	-	-	-1	-	-	-	-	-
TRANSFORMATION	**42283**	-	-	-	**57895**	-	-	-	-	-
Electricity and CHP Plants	42283	-	-	-	-	-	-	-	-	-
Petroleum Refineries	-	-	-	-	-	-	-	-	-	-
Other Transform. Sector	-	-	-	-	57895	-	-	-	-	-
ENERGY SECTOR	**6114**	-	-	-	-	-	-	-	**190**	-
DISTRIBUTION LOSSES	-	-	-	-	-	-	-	-	**1383**	-
FINAL CONSUMPTION	**13100**	-	-	-	**365653**	-	-	-	**2372**	-
INDUSTRY SECTOR	**12899**	-	-	-	**12188**	-	-	-	**890**	-
Iron and Steel	-	-	-	-	-	-	-	-	-	-
Chemical and Petrochem.	3676	-	-	-	-	-	-	-	-	-
Non-Metallic Minerals	-	-	-	-	-	-	-	-	-	-
Non-specified	9223	-	-	-	12188	-	-	-	890	-
TRANSPORT SECTOR	**99**	-	-	-	-	-	-	-	-	-
Air	-	-	-	-	-	-	-	-	-	-
Road	-	-	-	-	-	-	-	-	-	-
Non-specified	99	-	-	-	-	-	-	-	-	-
OTHER SECTORS	**102**	-	-	-	**353465**	-	-	-	**1482**	-
Agriculture	-	-	-	-	-	-	-	-	-	-
Comm. and Publ. Services	-	-	-	-	-	-	-	-	428	-
Residential	28	-	-	-	353465	-	-	-	1054	-
Non-specified	74	-	-	-	-	-	-	-	-	-
NON-ENERGY USE	-	-	-	-	-	-	-	-	-	-

APPROVISIONNEMENT ET DEMANDE 1997	*Gaz* (TJ)				*En. Re. Comb. & Déchets* (TJ)				(GWh)	(TJ)
	Gaz naturel	*Usines à gaz*	*Cokeries*	*Hauts fourneaux*	*Biomasse solide*	*Gaz/Liquides tirés de biomasse*	*Déchets urbains*	*Déchets industriels*	*Electricité*	*Chaleur*
Production	66444	-	-	-	430070	-	-	-	4205	-
Imports	-	-	-	-	-	-	-	-	-	-
Exports	-	-	-	-	-	-	-	-	-	-
Intl. Marine Bunkers	-	-	-	-	-	-	-	-	-	-
Stock Changes	-6853	-	-	-	-	-	-	-	-	-
DOMESTIC SUPPLY	**59591**	-	-	-	**430070**	-	-	-	**4205**	-
Transfers and Stat. Diff.	6848	-	-	-	-	-	-	-	-1	-
TRANSFORMATION	**44363**	-	-	-	**58787**	-	-	-	-	-
Electricity and CHP Plants	44363	-	-	-	-	-	-	-	-	-
Petroleum Refineries	-	-	-	-	-	-	-	-	-	-
Other Transform. Sector	-	-	-	-	58787	-	-	-	-	-
ENERGY SECTOR	**6731**	-	-	-	-	-	-	-	**215**	-
DISTRIBUTION LOSSES	-	-	-	-	-	-	-	-	**1488**	-
FINAL CONSUMPTION	**15345**	-	-	-	**371283**	-	-	-	**2501**	-
INDUSTRY SECTOR	**15177**	-	-	-	**12376**	-	-	-	**919**	-
Iron and Steel	-	-	-	-	-	-	-	-	-	-
Chemical and Petrochem.	3039	-	-	-	-	-	-	-	-	-
Non-Metallic Minerals	-	-	-	-	-	-	-	-	-	-
Non-specified	12138	-	-	-	12376	-	-	-	919	-
TRANSPORT SECTOR	**78**	-	-	-	-	-	-	-	-	-
Air	-	-	-	-	-	-	-	-	-	-
Road	-	-	-	-	-	-	-	-	-	-
Non-specified	78	-	-	-	-	-	-	-	-	-
OTHER SECTORS	**90**	-	-	-	**358907**	-	-	-	**1582**	-
Agriculture	-	-	-	-	-	-	-	-	-	-
Comm. and Publ. Services	-	-	-	-	-	-	-	-	446	-
Residential	-	-	-	-	358907	-	-	-	1136	-
Non-specified	90	-	-	-	-	-	-	-	-	-
NON-ENERGY USE	-	-	-	-	-	-	-	-	-	-

Nepal / Népal

SUPPLY AND CONSUMPTION 1996	Coal (1000 tonnes)							Oil (1000 tonnes)			
	Coking Coal	Other Bit. Coal	Sub-Bit. Coal	Lignite	Peat	Oven and Gas Coke	Pat. Fuel and BKB	Crude Oil	NGL	Feed-stocks	Additives
Production	-	-	-	-	-	-	-	-	-	-	-
Imports	-	119	-	-	-	-	-	-	-	-	-
Exports	-	-	-	-	-	-	-	-	-	-	-
Intl. Marine Bunkers	-	-	-	-	-	-	-	-	-	-	-
Stock Changes	-	-	-	-	-	-	-	-	-	-	-
DOMESTIC SUPPLY	**-**	**119**	**-**	**-**	**-**	**-**	**-**	**-**	**-**	**-**	**-**
Transfers and Stat. Diff.	-	2	-	-	-	-	-	-	-	-	-
TRANSFORMATION	**-**	**-**	**-**	**-**	**-**	**-**	**-**	**-**	**-**	**-**	**-**
Electricity and CHP Plants	-	-	-	-	-	-	-	-	-	-	-
Petroleum Refineries	-	-	-	-	-	-	-	-	-	-	-
Other Transform. Sector	-	-	-	-	-	-	-	-	-	-	-
ENERGY SECTOR	**-**	**-**	**-**	**-**	**-**	**-**	**-**	**-**	**-**	**-**	**-**
DISTRIBUTION LOSSES	**-**	**-**	**-**	**-**	**-**	**-**	**-**	**-**	**-**	**-**	**-**
FINAL CONSUMPTION	**-**	**121**	**-**	**-**	**-**	**-**	**-**	**-**	**-**	**-**	**-**
INDUSTRY SECTOR	**-**	**102**	**-**	**-**	**-**	**-**	**-**	**-**	**-**	**-**	**-**
Iron and Steel	-	-	-	-	-	-	-	-	-	-	-
Chemical and Petrochem.	-	-	-	-	-	-	-	-	-	-	-
Non-Metallic Minerals	-	-	-	-	-	-	-	-	-	-	-
Non-specified	-	102	-	-	-	-	-	-	-	-	-
TRANSPORT SECTOR	**-**	**5**	**-**	**-**	**-**	**-**	**-**	**-**	**-**	**-**	**-**
Air	-	-	-	-	-	-	-	-	-	-	-
Road	-	-	-	-	-	-	-	-	-	-	-
Non-specified	-	5	-	-	-	-	-	-	-	-	-
OTHER SECTORS	**-**	**14**	**-**	**-**	**-**	**-**	**-**	**-**	**-**	**-**	**-**
Agriculture	-	-	-	-	-	-	-	-	-	-	-
Comm. and Publ. Services	-	13	-	-	-	-	-	-	-	-	-
Residential	-	1	-	-	-	-	-	-	-	-	-
Non-specified	-	-	-	-	-	-	-	-	-	-	-
NON-ENERGY USE	**-**	**-**	**-**	**-**	**-**	**-**	**-**	**-**	**-**	**-**	**-**

APPROVISIONNEMENT ET DEMANDE 1997	*Charbon* (1000 tonnes)							*Pétrole* (1000 tonnes)			
	Charbon à coke	*Autres charb. bit.*	*Charbon sous-bit.*	*Lignite*	*Tourbe*	*Coke de four/gaz*	*Agg./briq. de lignite*	*Pétrole brut*	*LGN*	*Produits d'aliment.*	*Additifs*
Production	-	-	-	-	-	-	-	-	-	-	-
Imports	-	119	-	-	-	-	-	-	-	-	-
Exports	-	-	-	-	-	-	-	-	-	-	-
Intl. Marine Bunkers	-	-	-	-	-	-	-	-	-	-	-
Stock Changes	-	-	-	-	-	-	-	-	-	-	-
DOMESTIC SUPPLY	**-**	**119**	**-**	**-**	**-**	**-**	**-**	**-**	**-**	**-**	**-**
Transfers and Stat. Diff.	-	2	-	-	-	-	-	-	-	-	-
TRANSFORMATION	**-**	**-**	**-**	**-**	**-**	**-**	**-**	**-**	**-**	**-**	**-**
Electricity and CHP Plants	-	-	-	-	-	-	-	-	-	-	-
Petroleum Refineries	-	-	-	-	-	-	-	-	-	-	-
Other Transform. Sector	-	-	-	-	-	-	-	-	-	-	-
ENERGY SECTOR	**-**	**-**	**-**	**-**	**-**	**-**	**-**	**-**	**-**	**-**	**-**
DISTRIBUTION LOSSES	**-**	**-**	**-**	**-**	**-**	**-**	**-**	**-**	**-**	**-**	**-**
FINAL CONSUMPTION	**-**	**121**	**-**	**-**	**-**	**-**	**-**	**-**	**-**	**-**	**-**
INDUSTRY SECTOR	**-**	**102**	**-**	**-**	**-**	**-**	**-**	**-**	**-**	**-**	**-**
Iron and Steel	-	-	-	-	-	-	-	-	-	-	-
Chemical and Petrochem.	-	-	-	-	-	-	-	-	-	-	-
Non-Metallic Minerals	-	-	-	-	-	-	-	-	-	-	-
Non-specified	-	102	-	-	-	-	-	-	-	-	-
TRANSPORT SECTOR	**-**	**5**	**-**	**-**	**-**	**-**	**-**	**-**	**-**	**-**	**-**
Air	-	-	-	-	-	-	-	-	-	-	-
Road	-	-	-	-	-	-	-	-	-	-	-
Non-specified	-	5	-	-	-	-	-	-	-	-	-
OTHER SECTORS	**-**	**14**	**-**	**-**	**-**	**-**	**-**	**-**	**-**	**-**	**-**
Agriculture	-	-	-	-	-	-	-	-	-	-	-
Comm. and Publ. Services	-	13	-	-	-	-	-	-	-	-	-
Residential	-	1	-	-	-	-	-	-	-	-	-
Non-specified	-	-	-	-	-	-	-	-	-	-	-
NON-ENERGY USE	**-**	**-**	**-**	**-**	**-**	**-**	**-**	**-**	**-**	**-**	**-**

Nepal / Népal

SUPPLY AND CONSUMPTION 1996	Oil cont. (1000 tonnes)										
	Refinery Gas	LPG + Ethane	Motor Gasoline	Aviation Gasoline	Jet Fuel	Kerosene	Gas/ Diesel	Heavy Fuel Oil	Naphtha	Petrol. Coke	Other Prod.
Production	-	-	-	-	-	-	-	-	-	-	-
Imports	-	19	32	-	31	170	222	31	-	-	7
Exports	-	-	-2	-	-2	-10	-	-	-	-	-
Intl. Marine Bunkers	-	-	-	-	-	-	-	-	-	-	-
Stock Changes	-	-	2	-	-	10	-	-	-	-	-
DOMESTIC SUPPLY	**-**	**19**	**32**	**-**	**29**	**170**	**222**	**31**	**-**	**-**	**7**
Transfers and Stat. Diff.	-	-	-	-	-	-	1	-	-	-	-
TRANSFORMATION	**-**	**-**	**-**	**-**	**-**	**-**	**-**	**11**	**-**	**-**	**-**
Electricity and CHP Plants	-	-	-	-	-	-	-	11	-	-	-
Petroleum Refineries	-	-	-	-	-	-	-	-	-	-	-
Other Transform. Sector	-	-	-	-	-	-	-	-	-	-	-
ENERGY SECTOR	**-**	**-**	**-**	**-**	**-**	**-**	**-**	**-**	**-**	**-**	**-**
DISTRIBUTION LOSSES	**-**	**-**	**-**	**-**	**-**	**-**	**-**	**-**	**-**	**-**	**-**
FINAL CONSUMPTION	**-**	**19**	**32**	**-**	**29**	**170**	**223**	**20**	**-**	**-**	**7**
INDUSTRY SECTOR	**-**	**-**	**-**	**-**	**-**	**9**	**78**	**18**	**-**	**-**	**-**
Iron and Steel	-	-	-	-	-	-	-	-	-	-	-
Chemical and Petrochem.	-	-	-	-	-	-	-	-	-	-	-
Non-Metallic Minerals	-	-	-	-	-	-	-	-	-	-	-
Non-specified	-	-	-	-	-	9	78	18	-	-	-
TRANSPORT SECTOR	**-**	**-**	**32**	**-**	**29**	**-**	**130**	**-**	**-**	**-**	**-**
Air	-	-	-	-	29	-	-	-	-	-	-
Road	-	-	32	-	-	-	130	-	-	-	-
Non-specified	-	-	-	-	-	-	-	-	-	-	-
OTHER SECTORS	**-**	**19**	**-**	**-**	**-**	**161**	**15**	**2**	**-**	**-**	**-**
Agriculture	-	-	-	-	-	-	13	-	-	-	-
Comm. and Publ. Services	-	2	-	-	-	25	2	2	-	-	-
Residential	-	17	-	-	-	136	-	-	-	-	-
Non-specified	-	-	-	-	-	-	-	-	-	-	-
NON-ENERGY USE	**-**	**-**	**-**	**-**	**-**	**-**	**-**	**-**	**-**	**-**	**7**

APPROVISIONNEMENT ET DEMANDE 1997	*Pétrole cont.* (1000 tonnes)										
	Gaz de raffinerie	*GPL + éthane*	*Essence moteur*	*Essence aviation*	*Carbu-réacteurs*	*Kérosène*	*Gazole*	*Fioul lourd*	*Naphta*	*Coke de pétrole*	*Autres prod.*
Production	-	-	-	-	-	-	-	-	-	-	-
Imports	-	19	32	-	31	170	222	31	-	-	7
Exports	-	-	-2	-	-2	-10	-	-	-	-	-
Intl. Marine Bunkers	-	-	-	-	-	-	-	-	-	-	-
Stock Changes	-	-	2	-	-	10	-	-	-	-	-
DOMESTIC SUPPLY	**-**	**19**	**32**	**-**	**29**	**170**	**222**	**31**	**-**	**-**	**7**
Transfers and Stat. Diff.	-	-	-	-	-	-	1	-	-	-	-
TRANSFORMATION	**-**	**-**	**-**	**-**	**-**	**-**	**-**	**11**	**-**	**-**	**-**
Electricity and CHP Plants	-	-	-	-	-	-	-	11	-	-	-
Petroleum Refineries	-	-	-	-	-	-	-	-	-	-	-
Other Transform. Sector	-	-	-	-	-	-	-	-	-	-	-
ENERGY SECTOR	**-**	**-**	**-**	**-**	**-**	**-**	**-**	**-**	**-**	**-**	**-**
DISTRIBUTION LOSSES	**-**	**-**	**-**	**-**	**-**	**-**	**-**	**-**	**-**	**-**	**-**
FINAL CONSUMPTION	**-**	**19**	**32**	**-**	**29**	**170**	**223**	**20**	**-**	**-**	**7**
INDUSTRY SECTOR	**-**	**-**	**-**	**-**	**-**	**9**	**78**	**18**	**-**	**-**	**-**
Iron and Steel	-	-	-	-	-	-	-	-	-	-	-
Chemical and Petrochem.	-	-	-	-	-	-	-	-	-	-	-
Non-Metallic Minerals	-	-	-	-	-	-	-	-	-	-	-
Non-specified	-	-	-	-	-	9	78	18	-	-	-
TRANSPORT SECTOR	**-**	**-**	**32**	**-**	**29**	**-**	**130**	**-**	**-**	**-**	**-**
Air	-	-	-	-	29	-	-	-	-	-	-
Road	-	-	32	-	-	-	130	-	-	-	-
Non-specified	-	-	-	-	-	-	-	-	-	-	-
OTHER SECTORS	**-**	**19**	**-**	**-**	**-**	**161**	**15**	**2**	**-**	**-**	**-**
Agriculture	-	-	-	-	-	-	13	-	-	-	-
Comm. and Publ. Services	-	2	-	-	-	25	2	2	-	-	-
Residential	-	17	-	-	-	136	-	-	-	-	-
Non-specified	-	-	-	-	-	-	-	-	-	-	-
NON-ENERGY USE	**-**	**-**	**-**	**-**	**-**	**-**	**-**	**-**	**-**	**-**	**7**

Nepal / Népal

SUPPLY AND CONSUMPTION 1996	Gas (TJ)				Comb. Renew. & Waste (TJ)				(GWh)	(TJ)
	Natural Gas	Gas Works	Coke Ovens	Blast Furnaces	Solid Biomass	Gas/Liquids from Biomass	Municipal Waste	Industrial Waste	Electricity	Heat
Production	-	-	-	-	270687	-	-	-	1191	-
Imports	-	-	-	-	-	-	-	-	71	-
Exports	-	-	-	-	-	-	-	-	-90	-
Intl. Marine Bunkers	-	-	-	-	-	-	-	-	-	-
Stock Changes	-	-	-	-	-	-	-	-	-	-
DOMESTIC SUPPLY	**-**	**-**	**-**	**-**	**270687**	**-**	**-**	**-**	**1172**	**-**
Transfers and Stat. Diff.	-	-	-	-	-	-	-	-	25	-
TRANSFORMATION	**-**	**-**	**-**	**-**	**-**	**-**	**-**	**-**	**-**	**-**
Electricity and CHP Plants	-	-	-	-	-	-	-	-	-	-
Petroleum Refineries	-	-	-	-	-	-	-	-	-	-
Other Transform. Sector	-	-	-	-	-	-	-	-	-	-
ENERGY SECTOR	**-**	**-**	**-**	**-**	**-**	**-**	**-**	**-**	**20**	**-**
DISTRIBUTION LOSSES	**-**	**-**	**-**	**-**	**-**	**-**	**-**	**-**	**331**	**-**
FINAL CONSUMPTION	**-**	**-**	**-**	**-**	**270687**	**-**	**-**	**-**	**846**	**-**
INDUSTRY SECTOR	**-**	**-**	**-**	**-**	**4084**	**-**	**-**	**-**	**355**	**-**
Iron and Steel	-	-	-	-	-	-	-	-	-	-
Chemical and Petrochem.	-	-	-	-	-	-	-	-	-	-
Non-Metallic Minerals	-	-	-	-	-	-	-	-	-	-
Non-specified	-	-	-	-	4084	-	-	-	355	-
TRANSPORT SECTOR	**-**	**-**	**-**	**-**	**-**	**-**	**-**	**-**	**1**	**-**
Air	-	-	-	-	-	-	-	-	-	-
Road	-	-	-	-	-	-	-	-	-	-
Non-specified	-	-	-	-	-	-	-	-	1	-
OTHER SECTORS	**-**	**-**	**-**	**-**	**266603**	**-**	**-**	**-**	**490**	**-**
Agriculture	-	-	-	-	-	-	-	-	21	-
Comm. and Publ. Services	-	-	-	-	1521	-	-	-	80	-
Residential	-	-	-	-	265082	-	-	-	333	-
Non-specified	-	-	-	-	-	-	-	-	56	-
NON-ENERGY USE	**-**	**-**	**-**	**-**	**-**	**-**	**-**	**-**	**-**	**-**

APPROVISIONNEMENT ET DEMANDE 1997	*Gaz* (TJ)				*En. Re. Comb. & Déchets* (TJ)				(GWh)	(TJ)
	Gaz naturel	*Usines à gaz*	*Cokeries*	*Hauts fourneaux*	*Biomasse solide*	*Gaz/Liquides tirés de biomasse*	*Déchets urbains*	*Déchets industriels*	*Electricité*	*Chaleur*
Production	-	-	-	-	270687	-	-	-	1226	-
Imports	-	-	-	-	-	-	-	-	71	-
Exports	-	-	-	-	-	-	-	-	-90	-
Intl. Marine Bunkers	-	-	-	-	-	-	-	-	-	-
Stock Changes	-	-	-	-	-	-	-	-	-	-
DOMESTIC SUPPLY	**-**	**-**	**-**	**-**	**270687**	**-**	**-**	**-**	**1207**	**-**
Transfers and Stat. Diff.	-	-	-	-	-	-	-	-	27	-
TRANSFORMATION	**-**	**-**	**-**	**-**	**-**	**-**	**-**	**-**	**-**	**-**
Electricity and CHP Plants	-	-	-	-	-	-	-	-	-	-
Petroleum Refineries	-	-	-	-	-	-	-	-	-	-
Other Transform. Sector	-	-	-	-	-	-	-	-	-	-
ENERGY SECTOR	**-**	**-**	**-**	**-**	**-**	**-**	**-**	**-**	**21**	**-**
DISTRIBUTION LOSSES	**-**	**-**	**-**	**-**	**-**	**-**	**-**	**-**	**341**	**-**
FINAL CONSUMPTION	**-**	**-**	**-**	**-**	**270687**	**-**	**-**	**-**	**872**	**-**
INDUSTRY SECTOR	**-**	**-**	**-**	**-**	**4084**	**-**	**-**	**-**	**366**	**-**
Iron and Steel	-	-	-	-	-	-	-	-	-	-
Chemical and Petrochem.	-	-	-	-	-	-	-	-	-	-
Non-Metallic Minerals	-	-	-	-	-	-	-	-	-	-
Non-specified	-	-	-	-	4084	-	-	-	366	-
TRANSPORT SECTOR	**-**	**-**	**-**	**-**	**-**	**-**	**-**	**-**	**1**	**-**
Air	-	-	-	-	-	-	-	-	-	-
Road	-	-	-	-	-	-	-	-	-	-
Non-specified	-	-	-	-	-	-	-	-	1	-
OTHER SECTORS	**-**	**-**	**-**	**-**	**266603**	**-**	**-**	**-**	**505**	**-**
Agriculture	-	-	-	-	-	-	-	-	22	-
Comm. and Publ. Services	-	-	-	-	1521	-	-	-	82	-
Residential	-	-	-	-	265082	-	-	-	343	-
Non-specified	-	-	-	-	-	-	-	-	58	-
NON-ENERGY USE	**-**	**-**	**-**	**-**	**-**	**-**	**-**	**-**	**-**	**-**

Netherlands Antilles / Antilles néerlandaises

SUPPLY AND CONSUMPTION 1996	Coal (1000 tonnes)							Oil (1000 tonnes)			
	Coking Coal	Other Bit. Coal	Sub-Bit. Coal	Lignite	Peat	Oven and Gas Coke	Pat. Fuel and BKB	Crude Oil	NGL	Feed-stocks	Additives
Production	-	-	-	-	-	-	-	-	-	-	-
Imports	-	-	-	-	-	-	-	14198	-	-	-
Exports	-	-	-	-	-	-	-	-520	-	-	-
Intl. Marine Bunkers	-	-	-	-	-	-	-	-	-	-	-
Stock Changes	-	-	-	-	-	-	-	-112	-	-	-
DOMESTIC SUPPLY	-	-	-	-	-	-	-	**13566**	-	-	-
Transfers and Stat. Diff.	-	-	-	-	-	-	-	-971	-	-	-
TRANSFORMATION	-	-	-	-	-	-	-	**12595**	-	-	-
Electricity and CHP Plants	-	-	-	-	-	-	-	-	-	-	-
Petroleum Refineries	-	-	-	-	-	-	-	12595	-	-	-
Other Transform. Sector	-	-	-	-	-	-	-	-	-	-	-
ENERGY SECTOR	-	-	-	-	-	-	-	-	-	-	-
DISTRIBUTION LOSSES	-	-	-	-	-	-	-	-	-	-	-
FINAL CONSUMPTION	-	-	-	-	-	-	-	-	-	-	-
INDUSTRY SECTOR	-	-	-	-	-	-	-	-	-	-	-
Iron and Steel	-	-	-	-	-	-	-	-	-	-	-
Chemical and Petrochem.	-	-	-	-	-	-	-	-	-	-	-
Non-Metallic Minerals	-	-	-	-	-	-	-	-	-	-	-
Non-specified	-	-	-	-	-	-	-	-	-	-	-
TRANSPORT SECTOR	-	-	-	-	-	-	-	-	-	-	-
Air	-	-	-	-	-	-	-	-	-	-	-
Road	-	-	-	-	-	-	-	-	-	-	-
Non-specified	-	-	-	-	-	-	-	-	-	-	-
OTHER SECTORS	-	-	-	-	-	-	-	-	-	-	-
Agriculture	-	-	-	-	-	-	-	-	-	-	-
Comm. and Publ. Services	-	-	-	-	-	-	-	-	-	-	-
Residential	-	-	-	-	-	-	-	-	-	-	-
Non-specified	-	-	-	-	-	-	-	-	-	-	-
NON-ENERGY USE	-	-	-	-	-	-	-	-	-	-	-

APPROVISIONNEMENT ET DEMANDE 1997	*Charbon* (1000 tonnes)							*Pétrole* (1000 tonnes)			
	Charbon à coke	*Autres charb. bit.*	*Charbon sous-bit.*	*Lignite*	*Tourbe*	*Coke de four/gaz*	*Agg./briq. de lignite*	*Pétrole brut*	*LGN*	*Produits d'aliment.*	*Additifs*
Production	-	-	-	-	-	-	-	-	-	-	-
Imports	-	-	-	-	-	-	-	14198	-	-	-
Exports	-	-	-	-	-	-	-	-520	-	-	-
Intl. Marine Bunkers	-	-	-	-	-	-	-	-	-	-	-
Stock Changes	-	-	-	-	-	-	-	-112	-	-	-
DOMESTIC SUPPLY	-	-	-	-	-	-	-	**13566**	-	-	-
Transfers and Stat. Diff.	-	-	-	-	-	-	-	-971	-	-	-
TRANSFORMATION	-	-	-	-	-	-	-	**12595**	-	-	-
Electricity and CHP Plants	-	-	-	-	-	-	-	-	-	-	-
Petroleum Refineries	-	-	-	-	-	-	-	12595	-	-	-
Other Transform. Sector	-	-	-	-	-	-	-	-	-	-	-
ENERGY SECTOR	-	-	-	-	-	-	-	-	-	-	-
DISTRIBUTION LOSSES	-	-	-	-	-	-	-	-	-	-	-
FINAL CONSUMPTION	-	-	-	-	-	-	-	-	-	-	-
INDUSTRY SECTOR	-	-	-	-	-	-	-	-	-	-	-
Iron and Steel	-	-	-	-	-	-	-	-	-	-	-
Chemical and Petrochem.	-	-	-	-	-	-	-	-	-	-	-
Non-Metallic Minerals	-	-	-	-	-	-	-	-	-	-	-
Non-specified	-	-	-	-	-	-	-	-	-	-	-
TRANSPORT SECTOR	-	-	-	-	-	-	-	-	-	-	-
Air	-	-	-	-	-	-	-	-	-	-	-
Road	-	-	-	-	-	-	-	-	-	-	-
Non-specified	-	-	-	-	-	-	-	-	-	-	-
OTHER SECTORS	-	-	-	-	-	-	-	-	-	-	-
Agriculture	-	-	-	-	-	-	-	-	-	-	-
Comm. and Publ. Services	-	-	-	-	-	-	-	-	-	-	-
Residential	-	-	-	-	-	-	-	-	-	-	-
Non-specified	-	-	-	-	-	-	-	-	-	-	-
NON-ENERGY USE	-	-	-	-	-	-	-	-	-	-	-

Netherlands Antilles / Antilles néerlandaises

SUPPLY AND CONSUMPTION 1996	Oil cont. (1000 tonnes)										
	Refinery Gas	LPG + Ethane	Motor Gasoline	Aviation Gasoline	Jet Fuel	Kerosene	Gas/ Diesel	Heavy Fuel Oil	Naphtha	Petrol. Coke	Other Prod.
Production	-	63	1790	20	820	111	2218	5013	-	-	2369
Imports	-	24	306	-	58	-	317	715	-	-	246
Exports	-	-32	-2022	-18	-816	-	-2030	-3823	-	-	-2226
Intl. Marine Bunkers	-	-	-	-	-	-	-212	-1512	-	-	-
Stock Changes	-	-	-	-	-	-	-	-	-	-	-
DOMESTIC SUPPLY	**-**	**55**	**74**	**2**	**62**	**111**	**293**	**393**	**-**	**-**	**389**
Transfers and Stat. Diff.	-	-	-5	-	-2	-	-	-	-	-	-324
TRANSFORMATION	**-**	**-**	**-**	**-**	**-**	**-**	**-**	**78**	**-**	**-**	**-**
Electricity and CHP Plants	-	-	-	-	-	-	-	78	-	-	-
Petroleum Refineries	-	-	-	-	-	-	-	-	-	-	-
Other Transform. Sector	-	-	-	-	-	-	-	-	-	-	-
ENERGY SECTOR	**-**	**-**	**-**	**-**	**-**	**-**	**-**	**178**	**-**	**-**	**-**
DISTRIBUTION LOSSES	**-**	**-**	**-**	**-**	**-**	**-**	**-**	**-**	**-**	**-**	**-**
FINAL CONSUMPTION	**-**	**55**	**69**	**2**	**60**	**111**	**293**	**137**	**-**	**-**	**65**
INDUSTRY SECTOR	**-**	**-**	**-**	**-**	**-**	**-**	**-**	**137**	**-**	**-**	**-**
Iron and Steel	-	-	-	-	-	-	-	-	-	-	-
Chemical and Petrochem.	-	-	-	-	-	-	-	-	-	-	-
Non-Metallic Minerals	-	-	-	-	-	-	-	-	-	-	-
Non-specified	-	-	-	-	-	-	-	137	-	-	-
TRANSPORT SECTOR	**-**	**-**	**69**	**2**	**60**	**-**	**293**	**-**	**-**	**-**	**-**
Air	-	-	-	2	60	-	-	-	-	-	-
Road	-	-	69	-	-	-	293	-	-	-	-
Non-specified	-	-	-	-	-	-	-	-	-	-	-
OTHER SECTORS	**-**	**55**	**-**	**-**	**-**	**111**	**-**	**-**	**-**	**-**	**-**
Agriculture	-	-	-	-	-	-	-	-	-	-	-
Comm. and Publ. Services	-	-	-	-	-	-	-	-	-	-	-
Residential	-	55	-	-	-	111	-	-	-	-	-
Non-specified	-	-	-	-	-	-	-	-	-	-	-
NON-ENERGY USE	**-**	**-**	**-**	**-**	**-**	**-**	**-**	**-**	**-**	**-**	**65**

APPROVISIONNEMENT ET DEMANDE 1997	*Pétrole cont.* (1000 tonnes)										
	Gaz de raffinerie	*GPL + éthane*	*Essence moteur*	*Essence aviation*	*Carbu- réacteurs*	*Kérosène*	*Gazole*	*Fioul lourd*	*Naphta*	*Coke de pétrole*	*Autres prod.*
Production	-	63	1790	20	820	111	2218	5013	-	-	2369
Imports	-	24	306	-	58	-	317	715	-	-	246
Exports	-	-32	-2022	-18	-816	-	-2030	-3823	-	-	-2226
Intl. Marine Bunkers	-	-	-	-	-	-	-212	-1512	-	-	-
Stock Changes	-	-	-	-	-	-	-	-	-	-	-
DOMESTIC SUPPLY	**-**	**55**	**74**	**2**	**62**	**111**	**293**	**393**	**-**	**-**	**389**
Transfers and Stat. Diff.	-	-	-5	-	-2	-	-	-	-	-	-324
TRANSFORMATION	**-**	**-**	**-**	**-**	**-**	**-**	**-**	**78**	**-**	**-**	**-**
Electricity and CHP Plants	-	-	-	-	-	-	-	78	-	-	-
Petroleum Refineries	-	-	-	-	-	-	-	-	-	-	-
Other Transform. Sector	-	-	-	-	-	-	-	-	-	-	-
ENERGY SECTOR	**-**	**-**	**-**	**-**	**-**	**-**	**-**	**178**	**-**	**-**	**-**
DISTRIBUTION LOSSES	**-**	**-**	**-**	**-**	**-**	**-**	**-**	**-**	**-**	**-**	**-**
FINAL CONSUMPTION	**-**	**55**	**69**	**2**	**60**	**111**	**293**	**137**	**-**	**-**	**65**
INDUSTRY SECTOR	**-**	**-**	**-**	**-**	**-**	**-**	**-**	**137**	**-**	**-**	**-**
Iron and Steel	-	-	-	-	-	-	-	-	-	-	-
Chemical and Petrochem.	-	-	-	-	-	-	-	-	-	-	-
Non-Metallic Minerals	-	-	-	-	-	-	-	-	-	-	-
Non-specified	-	-	-	-	-	-	-	137	-	-	-
TRANSPORT SECTOR	**-**	**-**	**69**	**2**	**60**	**-**	**293**	**-**	**-**	**-**	**-**
Air	-	-	-	2	60	-	-	-	-	-	-
Road	-	-	69	-	-	-	293	-	-	-	-
Non-specified	-	-	-	-	-	-	-	-	-	-	-
OTHER SECTORS	**-**	**55**	**-**	**-**	**-**	**111**	**-**	**-**	**-**	**-**	**-**
Agriculture	-	-	-	-	-	-	-	-	-	-	-
Comm. and Publ. Services	-	-	-	-	-	-	-	-	-	-	-
Residential	-	55	-	-	-	111	-	-	-	-	-
Non-specified	-	-	-	-	-	-	-	-	-	-	-
NON-ENERGY USE	**-**	**-**	**-**	**-**	**-**	**-**	**-**	**-**	**-**	**-**	**65**

Netherlands Antilles / Antilles néerlandaises

SUPPLY AND CONSUMPTION 1996	Gas (TJ)				Comb. Renew. & Waste (TJ)				(GWh)	(TJ)
	Natural Gas	Gas Works	Coke Ovens	Blast Furnaces	Solid Biomass	Gas/Liquids from Biomass	Municipal Waste	Industrial Waste	Electricity	Heat
Production	-	-	-	-	-	-	-	-	1017	-
Imports	-	-	-	-	-	-	-	-	-	-
Exports	-	-	-	-	-	-	-	-	-	-
Intl. Marine Bunkers	-	-	-	-	-	-	-	-	-	-
Stock Changes	-	-	-	-	-	-	-	-	-	-
DOMESTIC SUPPLY	-	-	-	-	-	-	-	-	**1017**	-
Transfers and Stat. Diff.	-	-	-	-	-	-	-	-	1	-
TRANSFORMATION	-	-	-	-	-	-	-	-	-	-
Electricity and CHP Plants	-	-	-	-	-	-	-	-	-	-
Petroleum Refineries	-	-	-	-	-	-	-	-	-	-
Other Transform. Sector	-	-	-	-	-	-	-	-	-	-
ENERGY SECTOR	-	-	-	-	-	-	-	-	**95**	-
DISTRIBUTION LOSSES	-	-	-	-	-	-	-	-	**126**	-
FINAL CONSUMPTION	-	-	-	-	-	-	-	-	**797**	-
INDUSTRY SECTOR	-	-	-	-	-	-	-	-	**440**	-
Iron and Steel	-	-	-	-	-	-	-	-	-	-
Chemical and Petrochem.	-	-	-	-	-	-	-	-	-	-
Non-Metallic Minerals	-	-	-	-	-	-	-	-	-	-
Non-specified	-	-	-	-	-	-	-	-	440	-
TRANSPORT SECTOR	-	-	-	-	-	-	-	-	-	-
Air	-	-	-	-	-	-	-	-	-	-
Road	-	-	-	-	-	-	-	-	-	-
Non-specified	-	-	-	-	-	-	-	-	-	-
OTHER SECTORS	-	-	-	-	-	-	-	-	**357**	-
Agriculture	-	-	-	-	-	-	-	-	-	-
Comm. and Publ. Services	-	-	-	-	-	-	-	-	-	-
Residential	-	-	-	-	-	-	-	-	-	-
Non-specified	-	-	-	-	-	-	-	-	357	-
NON-ENERGY USE	-	-	-	-	-	-	-	-	-	-

APPROVISIONNEMENT ET DEMANDE 1997	*Gaz* (TJ)				*En. Re. Comb. & Déchets* (TJ)				(GWh)	(TJ)
	Gaz naturel	*Usines à gaz*	*Cokeries*	*Hauts fourneaux*	*Biomasse solide*	*Gaz/Liquides tirés de biomasse*	*Déchets urbains*	*Déchets industriels*	*Electricité*	*Chaleur*
Production	-	-	-	-	-	-	-	-	1052	-
Imports	-	-	-	-	-	-	-	-	-	-
Exports	-	-	-	-	-	-	-	-	-	-
Intl. Marine Bunkers	-	-	-	-	-	-	-	-	-	-
Stock Changes	-	-	-	-	-	-	-	-	-	-
DOMESTIC SUPPLY	-	-	-	-	-	-	-	-	**1052**	-
Transfers and Stat. Diff.	-	-	-	-	-	-	-	-	-	-
TRANSFORMATION	-	-	-	-	-	-	-	-	-	-
Electricity and CHP Plants	-	-	-	-	-	-	-	-	-	-
Petroleum Refineries	-	-	-	-	-	-	-	-	-	-
Other Transform. Sector	-	-	-	-	-	-	-	-	-	-
ENERGY SECTOR	-	-	-	-	-	-	-	-	**98**	-
DISTRIBUTION LOSSES	-	-	-	-	-	-	-	-	**130**	-
FINAL CONSUMPTION	-	-	-	-	-	-	-	-	**824**	-
INDUSTRY SECTOR	-	-	-	-	-	-	-	-	**455**	-
Iron and Steel	-	-	-	-	-	-	-	-	-	-
Chemical and Petrochem.	-	-	-	-	-	-	-	-	-	-
Non-Metallic Minerals	-	-	-	-	-	-	-	-	-	-
Non-specified	-	-	-	-	-	-	-	-	455	-
TRANSPORT SECTOR	-	-	-	-	-	-	-	-	-	-
Air	-	-	-	-	-	-	-	-	-	-
Road	-	-	-	-	-	-	-	-	-	-
Non-specified	-	-	-	-	-	-	-	-	-	-
OTHER SECTORS	-	-	-	-	-	-	-	-	**369**	-
Agriculture	-	-	-	-	-	-	-	-	-	-
Comm. and Publ. Services	-	-	-	-	-	-	-	-	-	-
Residential	-	-	-	-	-	-	-	-	-	-
Non-specified	-	-	-	-	-	-	-	-	369	-
NON-ENERGY USE	-	-	-	-	-	-	-	-	-	-

Nicaragua

SUPPLY AND CONSUMPTION 1996	Coal (1000 tonnes)							Oil (1000 tonnes)			
	Coking Coal	Other Bit. Coal	Sub-Bit. Coal	Lignite	Peat	Oven and Gas Coke	Pat. Fuel and BKB	Crude Oil	NGL	Feed-stocks	Additives
Production	-	-	-	-	-	-	-	-	-	-	-
Imports	-	-	-	-	-	-	-	617	-	-	-
Exports	-	-	-	-	-	-	-	-	-	-	-
Intl. Marine Bunkers	-	-	-	-	-	-	-	-	-	-	-
Stock Changes	-	-	-	-	-	-	-	41	-	-	-
DOMESTIC SUPPLY	**-**	**-**	**-**	**-**	**-**	**-**	**-**	**658**	**-**	**-**	**-**
Transfers and Stat. Diff.	-	-	-	-	-	-	-	-	-	-	-
TRANSFORMATION	**-**	**-**	**-**	**-**	**-**	**-**	**-**	**651**	**-**	**-**	**-**
Electricity and CHP Plants	-	-	-	-	-	-	-	-	-	-	-
Petroleum Refineries	-	-	-	-	-	-	-	651	-	-	-
Other Transform. Sector	-	-	-	-	-	-	-	-	-	-	-
ENERGY SECTOR	**-**	**-**	**-**	**-**	**-**	**-**	**-**	**7**	**-**	**-**	**-**
DISTRIBUTION LOSSES	**-**	**-**	**-**	**-**	**-**	**-**	**-**	**-**	**-**	**-**	**-**
FINAL CONSUMPTION	**-**	**-**	**-**	**-**	**-**	**-**	**-**	**-**	**-**	**-**	**-**
INDUSTRY SECTOR	**-**	**-**	**-**	**-**	**-**	**-**	**-**	**-**	**-**	**-**	**-**
Iron and Steel	-	-	-	-	-	-	-	-	-	-	-
Chemical and Petrochem.	-	-	-	-	-	-	-	-	-	-	-
Non-Metallic Minerals	-	-	-	-	-	-	-	-	-	-	-
Non-specified	-	-	-	-	-	-	-	-	-	-	-
TRANSPORT SECTOR	**-**	**-**	**-**	**-**	**-**	**-**	**-**	**-**	**-**	**-**	**-**
Air	-	-	-	-	-	-	-	-	-	-	-
Road	-	-	-	-	-	-	-	-	-	-	-
Non-specified	-	-	-	-	-	-	-	-	-	-	-
OTHER SECTORS	**-**	**-**	**-**	**-**	**-**	**-**	**-**	**-**	**-**	**-**	**-**
Agriculture	-	-	-	-	-	-	-	-	-	-	-
Comm. and Publ. Services	-	-	-	-	-	-	-	-	-	-	-
Residential	-	-	-	-	-	-	-	-	-	-	-
Non-specified	-	-	-	-	-	-	-	-	-	-	-
NON-ENERGY USE	**-**	**-**	**-**	**-**	**-**	**-**	**-**	**-**	**-**	**-**	**-**

APPROVISIONNEMENT ET DEMANDE 1997	*Charbon* (1000 tonnes)							*Pétrole* (1000 tonnes)			
	Charbon à coke	*Autres charb. bit.*	*Charbon sous-bit.*	*Lignite*	*Tourbe*	*Coke de four/gaz*	*Agg./briq. de lignite*	*Pétrole brut*	*LGN*	*Produits d'aliment.*	*Additifs*
Production	-	-	-	-	-	-	-	-	-	-	-
Imports	-	-	-	-	-	-	-	729	-	-	-
Exports	-	-	-	-	-	-	-	-	-	-	-
Intl. Marine Bunkers	-	-	-	-	-	-	-	-	-	-	-
Stock Changes	-	-	-	-	-	-	-	101	-	-	-
DOMESTIC SUPPLY	**-**	**-**	**-**	**-**	**-**	**-**	**-**	**830**	**-**	**-**	**-**
Transfers and Stat. Diff.	-	-	-	-	-	-	-	-	-	-	-
TRANSFORMATION	**-**	**-**	**-**	**-**	**-**	**-**	**-**	**822**	**-**	**-**	**-**
Electricity and CHP Plants	-	-	-	-	-	-	-	-	-	-	-
Petroleum Refineries	-	-	-	-	-	-	-	822	-	-	-
Other Transform. Sector	-	-	-	-	-	-	-	-	-	-	-
ENERGY SECTOR	**-**	**-**	**-**	**-**	**-**	**-**	**-**	**8**	**-**	**-**	**-**
DISTRIBUTION LOSSES	**-**	**-**	**-**	**-**	**-**	**-**	**-**	**-**	**-**	**-**	**-**
FINAL CONSUMPTION	**-**	**-**	**-**	**-**	**-**	**-**	**-**	**-**	**-**	**-**	**-**
INDUSTRY SECTOR	**-**	**-**	**-**	**-**	**-**	**-**	**-**	**-**	**-**	**-**	**-**
Iron and Steel	-	-	-	-	-	-	-	-	-	-	-
Chemical and Petrochem.	-	-	-	-	-	-	-	-	-	-	-
Non-Metallic Minerals	-	-	-	-	-	-	-	-	-	-	-
Non-specified	-	-	-	-	-	-	-	-	-	-	-
TRANSPORT SECTOR	**-**	**-**	**-**	**-**	**-**	**-**	**-**	**-**	**-**	**-**	**-**
Air	-	-	-	-	-	-	-	-	-	-	-
Road	-	-	-	-	-	-	-	-	-	-	-
Non-specified	-	-	-	-	-	-	-	-	-	-	-
OTHER SECTORS	**-**	**-**	**-**	**-**	**-**	**-**	**-**	**-**	**-**	**-**	**-**
Agriculture	-	-	-	-	-	-	-	-	-	-	-
Comm. and Publ. Services	-	-	-	-	-	-	-	-	-	-	-
Residential	-	-	-	-	-	-	-	-	-	-	-
Non-specified	-	-	-	-	-	-	-	-	-	-	-
NON-ENERGY USE	**-**	**-**	**-**	**-**	**-**	**-**	**-**	**-**	**-**	**-**	**-**

Nicaragua

SUPPLY AND CONSUMPTION 1996	Oil cont. (1000 tonnes)										
	Refinery Gas	LPG + Ethane	Motor Gasoline	Aviation Gasoline	Jet Fuel	Kerosene	Gas/ Diesel	Heavy Fuel Oil	Naphtha	Petrol. Coke	Other Prod.
Production	6	15	102	-	22	13	169	249	-	-	34
Imports	-	14	19	1	2	-	123	169	-	-	6
Exports	-	-	-4	-	-	-	-	-7	-	-	-17
Intl. Marine Bunkers	-	-	-	-	-	-	-	-	-	-	-
Stock Changes	-	-	-19	-	-4	-	10	-34	-	-	2
DOMESTIC SUPPLY	**6**	**29**	**98**	**1**	**20**	**13**	**302**	**377**	**-**	**-**	**25**
Transfers and Stat. Diff.	-	-	-1	-	1	-	-	-1	-	-	-
TRANSFORMATION	**-**	**-**	**-**	**-**	**-**	**-**	**9**	**302**	**-**	**-**	**-**
Electricity and CHP Plants	-	-	-	-	-	-	9	302	-	-	-
Petroleum Refineries	-	-	-	-	-	-	-	-	-	-	-
Other Transform. Sector	-	-	-	-	-	-	-	-	-	-	-
ENERGY SECTOR	**6**	**-**	**-**	**-**	**-**	**-**	**-**	**-**	**-**	**-**	**-**
DISTRIBUTION LOSSES	**-**	**-**	**-**	**-**	**-**	**-**	**-**	**-**	**-**	**-**	**-**
FINAL CONSUMPTION	**-**	**29**	**97**	**1**	**21**	**13**	**293**	**74**	**-**	**-**	**25**
INDUSTRY SECTOR	**-**	**2**	**-**	**-**	**-**	**-**	**17**	**72**	**-**	**-**	**-**
Iron and Steel	-	-	-	-	-	-	-	-	-	-	-
Chemical and Petrochem.	-	-	-	-	-	-	-	-	-	-	-
Non-Metallic Minerals	-	-	-	-	-	-	-	-	-	-	-
Non-specified	-	2	-	-	-	-	17	72	-	-	-
TRANSPORT SECTOR	**-**	**-**	**97**	**1**	**21**	**-**	**256**	**-**	**-**	**-**	**-**
Air	-	-	-	1	21	-	-	-	-	-	-
Road	-	-	97	-	-	-	226	-	-	-	-
Non-specified	-	-	-	-	-	-	30	-	-	-	-
OTHER SECTORS	**-**	**27**	**-**	**-**	**-**	**13**	**20**	**2**	**-**	**-**	**-**
Agriculture	-	-	-	-	-	-	4	2	-	-	-
Comm. and Publ. Services	-	6	-	-	-	-	16	-	-	-	-
Residential	-	21	-	-	-	13	-	-	-	-	-
Non-specified	-	-	-	-	-	-	-	-	-	-	-
NON-ENERGY USE	**-**	**-**	**-**	**-**	**-**	**-**	**-**	**-**	**-**	**-**	**25**

APPROVISIONNEMENT ET DEMANDE 1997	*Pétrole cont.* (1000 tonnes)										
	Gaz de raffinerie	*GPL + éthane*	*Essence moteur*	*Essence aviation*	*Carbu- réacteurs*	*Kérosène*	*Gazole*	*Fioul lourd*	*Naphta*	*Coke de pétrole*	*Autres prod.*
Production	8	13	94	-	25	12	224	353	-	-	38
Imports	-	19	16	1	3	-	168	20	-	-	7
Exports	-	-	-	-	-	-	-	-	-	-	-17
Intl. Marine Bunkers	-	-	-	-	-	-	-	-	-	-	-
Stock Changes	-	3	6	-	-1	-	-42	1	-	-	2
DOMESTIC SUPPLY	**8**	**35**	**116**	**1**	**27**	**12**	**350**	**374**	**-**	**-**	**30**
Transfers and Stat. Diff.	-	-	-	-	-	-	-	-1	-	-	-
TRANSFORMATION	**-**	**-**	**-**	**-**	**-**	**-**	**30**	**289**	**-**	**-**	**-**
Electricity and CHP Plants	-	-	-	-	-	-	30	289	-	-	-
Petroleum Refineries	-	-	-	-	-	-	-	-	-	-	-
Other Transform. Sector	-	-	-	-	-	-	-	-	-	-	-
ENERGY SECTOR	**8**	**-**	**-**	**-**	**-**	**-**	**-**	**-**	**-**	**-**	**-**
DISTRIBUTION LOSSES	**-**	**-**	**-**	**-**	**-**	**-**	**-**	**-**	**-**	**-**	**-**
FINAL CONSUMPTION	**-**	**35**	**116**	**1**	**27**	**12**	**320**	**84**	**-**	**-**	**30**
INDUSTRY SECTOR	**-**	**7**	**-**	**-**	**-**	**-**	**23**	**35**	**-**	**-**	**-**
Iron and Steel	-	-	-	-	-	-	-	-	-	-	-
Chemical and Petrochem.	-	-	-	-	-	-	-	-	-	-	-
Non-Metallic Minerals	-	-	-	-	-	-	-	-	-	-	-
Non-specified	-	7	-	-	-	-	23	35	-	-	-
TRANSPORT SECTOR	**-**	**-**	**116**	**1**	**27**	**-**	**277**	**-**	**-**	**-**	**-**
Air	-	-	-	1	27	-	-	-	-	-	-
Road	-	-	116	-	-	-	248	-	-	-	-
Non-specified	-	-	-	-	-	-	29	-	-	-	-
OTHER SECTORS	**-**	**28**	**-**	**-**	**-**	**12**	**20**	**49**	**-**	**-**	**-**
Agriculture	-	-	-	-	-	-	4	2	-	-	-
Comm. and Publ. Services	-	6	-	-	-	-	16	47	-	-	-
Residential	-	22	-	-	-	12	-	-	-	-	-
Non-specified	-	-	-	-	-	-	-	-	-	-	-
NON-ENERGY USE	**-**	**-**	**-**	**-**	**-**	**-**	**-**	**-**	**-**	**-**	**30**

Nicaragua

SUPPLY AND	Gas (TJ)				Comb. Renew. & Waste (TJ)				(GWh)	(TJ)
CONSUMPTION 1996	Natural Gas	Gas Works	Coke Ovens	Blast Furnaces	Solid Biomass	Gas/Liquids from Biomass	Municipal Waste	Industrial Waste	Electricity	Heat
Production	-	-	-	-	51070	-	-	-	1920	-
Imports	-	-	-	-	-	-	-	-	14	-
Exports	-	-	-	-	-	-	-	-	-	-
Intl. Marine Bunkers	-	-	-	-	-	-	-	-	-	-
Stock Changes	-	-	-	-	-	-	-	-	-	-
DOMESTIC SUPPLY	**-**	**-**	**-**	**-**	**51070**	**-**	**-**	**-**	**1934**	**-**
Transfers and Stat. Diff.	-	-	-	-	-	-	-	-	2	-
TRANSFORMATION	**-**	**-**	**-**	**-**	**5534**	**-**	**-**	**-**	**-**	**-**
Electricity and CHP Plants	-	-	-	-	3730	-	-	-	-	-
Petroleum Refineries	-	-	-	-	-	-	-	-	-	-
Other Transform. Sector	-	-	-	-	1804	-	-	-	-	-
ENERGY SECTOR	**-**	**-**	**-**	**-**	**-**	**-**	**-**	**-**	**228**	**-**
DISTRIBUTION LOSSES	**-**	**-**	**-**	**-**	**-**	**-**	**-**	**-**	**532**	**-**
FINAL CONSUMPTION	**-**	**-**	**-**	**-**	**45536**	**-**	**-**	**-**	**1176**	**-**
INDUSTRY SECTOR	**-**	**-**	**-**	**-**	**4112**	**-**	**-**	**-**	**345**	**-**
Iron and Steel	-	-	-	-	-	-	-	-	-	-
Chemical and Petrochem.	-	-	-	-	-	-	-	-	-	-
Non-Metallic Minerals	-	-	-	-	-	-	-	-	-	-
Non-specified	-	-	-	-	4112	-	-	-	345	-
TRANSPORT SECTOR	**-**	**-**	**-**	**-**	**-**	**-**	**-**	**-**	**-**	**-**
Air	-	-	-	-	-	-	-	-	-	-
Road	-	-	-	-	-	-	-	-	-	-
Non-specified	-	-	-	-	-	-	-	-	-	-
OTHER SECTORS	**-**	**-**	**-**	**-**	**41424**	**-**	**-**	**-**	**831**	**-**
Agriculture	-	-	-	-	105	-	-	-	92	-
Comm. and Publ. Services	-	-	-	-	674	-	-	-	318	-
Residential	-	-	-	-	40645	-	-	-	421	-
Non-specified	-	-	-	-	-	-	-	-	-	-
NON-ENERGY USE	**-**	**-**	**-**	**-**	**-**	**-**	**-**	**-**	**-**	**-**

APPROVISIONNEMENT	*Gaz* (TJ)				*En. Re. Comb. & Déchets* (TJ)				(GWh)	(TJ)
ET DEMANDE 1997	*Gaz naturel*	*Usines à gaz*	*Cokeries*	*Hauts fourneaux*	*Biomasse solide*	*Gaz/Liquides tirés de biomasse*	*Déchets urbains*	*Déchets industriels*	*Electricité*	*Chaleur*
Production	-	-	-	-	55017	-	-	-	1907	-
Imports	-	-	-	-	-	-	-	-	162	-
Exports	-	-	-	-	-	-	-	-	-	-
Intl. Marine Bunkers	-	-	-	-	-	-	-	-	-	-
Stock Changes	-	-	-	-	-	-	-	-	-	-
DOMESTIC SUPPLY	**-**	**-**	**-**	**-**	**55017**	**-**	**-**	**-**	**2069**	**-**
Transfers and Stat. Diff.	-	-	-	-	38	-	-	-	-	-
TRANSFORMATION	**-**	**-**	**-**	**-**	**7505**	**-**	**-**	**-**	**-**	**-**
Electricity and CHP Plants	-	-	-	-	5584	-	-	-	-	-
Petroleum Refineries	-	-	-	-	-	-	-	-	-	-
Other Transform. Sector	-	-	-	-	1921	-	-	-	-	-
ENERGY SECTOR	**-**	**-**	**-**	**-**	**-**	**-**	**-**	**-**	**228**	**-**
DISTRIBUTION LOSSES	**-**	**-**	**-**	**-**	**-**	**-**	**-**	**-**	**505**	**-**
FINAL CONSUMPTION	**-**	**-**	**-**	**-**	**47550**	**-**	**-**	**-**	**1336**	**-**
INDUSTRY SECTOR	**-**	**-**	**-**	**-**	**4934**	**-**	**-**	**-**	**389**	**-**
Iron and Steel	-	-	-	-	-	-	-	-	-	-
Chemical and Petrochem.	-	-	-	-	-	-	-	-	-	-
Non-Metallic Minerals	-	-	-	-	-	-	-	-	-	-
Non-specified	-	-	-	-	4934	-	-	-	389	-
TRANSPORT SECTOR	**-**	**-**	**-**	**-**	**-**	**-**	**-**	**-**	**-**	**-**
Air	-	-	-	-	-	-	-	-	-	-
Road	-	-	-	-	-	-	-	-	-	-
Non-specified	-	-	-	-	-	-	-	-	-	-
OTHER SECTORS	**-**	**-**	**-**	**-**	**42616**	**-**	**-**	**-**	**947**	**-**
Agriculture	-	-	-	-	134	-	-	-	120	-
Comm. and Publ. Services	-	-	-	-	674	-	-	-	368	-
Residential	-	-	-	-	41808	-	-	-	459	-
Non-specified	-	-	-	-	-	-	-	-	-	-
NON-ENERGY USE	**-**	**-**	**-**	**-**	**-**	**-**	**-**	**-**	**-**	**-**

Nigeria / Nigéria

SUPPLY AND CONSUMPTION 1996	Coal (1000 tonnes)							Oil (1000 tonnes)			
	Coking Coal	Other Bit. Coal	Sub-Bit. Coal	Lignite	Peat	Oven and Gas Coke	Pat. Fuel and BKB	Crude Oil	NGL	Feed-stocks	Additives
Production	-	140	-	-	-	-	-	104439	3593	-	-
Imports	-	-	-	-	-	-	-	-	-	-	-
Exports	-	-	-	-	-	-	-	-94331	-	-	-
Intl. Marine Bunkers	-	-	-	-	-	-	-	-	-	-	-
Stock Changes	-	-	-	-	-	-	-	-	-	-	-
DOMESTIC SUPPLY	-	**140**	-	-	-	-	-	**10108**	**3593**	-	-
Transfers and Stat. Diff.	-	-130	-	-	-	-	-	3179	-	-	-
TRANSFORMATION	-	-	-	-	-	-	-	**13287**	**3593**	-	-
Electricity and CHP Plants	-	-	-	-	-	-	-	-	-	-	-
Petroleum Refineries	-	-	-	-	-	-	-	13287	3593	-	-
Other Transform. Sector	-	-	-	-	-	-	-	-	-	-	-
ENERGY SECTOR	-	-	-	-	-	-	-	-	-	-	-
DISTRIBUTION LOSSES	-	-	-	-	-	-	-	-	-	-	-
FINAL CONSUMPTION	-	**10**	-	-	-	-	-	-	-	-	-
INDUSTRY SECTOR	-	**10**	-	-	-	-	-	-	-	-	-
Iron and Steel	-	-	-	-	-	-	-	-	-	-	-
Chemical and Petrochem.	-	-	-	-	-	-	-	-	-	-	-
Non-Metallic Minerals	-	10	-	-	-	-	-	-	-	-	-
Non-specified	-	-	-	-	-	-	-	-	-	-	-
TRANSPORT SECTOR	-	-	-	-	-	-	-	-	-	-	-
Air	-	-	-	-	-	-	-	-	-	-	-
Road	-	-	-	-	-	-	-	-	-	-	-
Non-specified	-	-	-	-	-	-	-	-	-	-	-
OTHER SECTORS	-	-	-	-	-	-	-	-	-	-	-
Agriculture	-	-	-	-	-	-	-	-	-	-	-
Comm. and Publ. Services	-	-	-	-	-	-	-	-	-	-	-
Residential	-	-	-	-	-	-	-	-	-	-	-
Non-specified	-	-	-	-	-	-	-	-	-	-	-
NON-ENERGY USE	-	-	-	-	-	-	-	-	-	-	-

APPROVISIONNEMENT ET DEMANDE 1997	*Charbon* (1000 tonnes)							*Pétrole* (1000 tonnes)			
	Charbon à coke	*Autres charb. bit.*	*Charbon sous-bit.*	*Lignite*	*Tourbe*	*Coke de four/gaz*	*Agg./briq. de lignite*	*Pétrole brut*	*LGN*	*Produits d'aliment.*	*Additifs*
Production	-	140	-	-	-	-	-	111057	3780	-	-
Imports	-	-	-	-	-	-	-	-	-	-	-
Exports	-	-	-	-	-	-	-	-99248	-	-	-
Intl. Marine Bunkers	-	-	-	-	-	-	-	-	-	-	-
Stock Changes	-	-	-	-	-	-	-	-	-	-	-
DOMESTIC SUPPLY	-	**140**	-	-	-	-	-	**11809**	**3780**	-	-
Transfers and Stat. Diff.	-	-130	-	-	-	-	-	1400	-	-	-
TRANSFORMATION	-	-	-	-	-	-	-	**13209**	**3780**	-	-
Electricity and CHP Plants	-	-	-	-	-	-	-	-	-	-	-
Petroleum Refineries	-	-	-	-	-	-	-	13209	3780	-	-
Other Transform. Sector	-	-	-	-	-	-	-	-	-	-	-
ENERGY SECTOR	-	-	-	-	-	-	-	-	-	-	-
DISTRIBUTION LOSSES	-	-	-	-	-	-	-	-	-	-	-
FINAL CONSUMPTION	-	**10**	-	-	-	-	-	-	-	-	-
INDUSTRY SECTOR	-	**10**	-	-	-	-	-	-	-	-	-
Iron and Steel	-	-	-	-	-	-	-	-	-	-	-
Chemical and Petrochem.	-	-	-	-	-	-	-	-	-	-	-
Non-Metallic Minerals	-	10	-	-	-	-	-	-	-	-	-
Non-specified	-	-	-	-	-	-	-	-	-	-	-
TRANSPORT SECTOR	-	-	-	-	-	-	-	-	-	-	-
Air	-	-	-	-	-	-	-	-	-	-	-
Road	-	-	-	-	-	-	-	-	-	-	-
Non-specified	-	-	-	-	-	-	-	-	-	-	-
OTHER SECTORS	-	-	-	-	-	-	-	-	-	-	-
Agriculture	-	-	-	-	-	-	-	-	-	-	-
Comm. and Publ. Services	-	-	-	-	-	-	-	-	-	-	-
Residential	-	-	-	-	-	-	-	-	-	-	-
Non-specified	-	-	-	-	-	-	-	-	-	-	-
NON-ENERGY USE	-	-	-	-	-	-	-	-	-	-	-

Nigeria / Nigéria

SUPPLY AND CONSUMPTION 1996	Oil cont. (1000 tonnes)										
	Refinery Gas	LPG + Ethane	Motor Gasoline	Aviation Gasoline	Jet Fuel	Kerosene	Gas/ Diesel	Heavy Fuel Oil	Naphtha	Petrol. Coke	Other Prod.
Production	346	202	5502	-	474	2716	2961	3839	-	-	638
Imports	-	13	-	-	-	-	27	-	-	-	-
Exports	-	-191	-	-	-	-	-164	-650	-	-	-
Intl. Marine Bunkers	-	-	-	-	-	-	-80	-	-	-	-
Stock Changes	-	-	-	-	-	-	-	-	-	-	-
DOMESTIC SUPPLY	**346**	**24**	**5502**	**-**	**474**	**2716**	**2744**	**3189**	**-**	**-**	**638**
Transfers and Stat. Diff.	-	-13	-2251	-	13	-1870	-1466	-2188	-	-	1300
TRANSFORMATION	**-**	**-**	**-**	**-**	**-**	**-**	**205**	**114**	**-**	**-**	**-**
Electricity and CHP Plants	-	-	-	-	-	-	205	114	-	-	-
Petroleum Refineries	-	-	-	-	-	-	-	-	-	-	-
Other Transform. Sector	-	-	-	-	-	-	-	-	-	-	-
ENERGY SECTOR	**346**	**-**	**-**	**-**	**-**	**-**	**-**	**199**	**-**	**-**	**-**
DISTRIBUTION LOSSES	**-**	**-**	**-**	**-**	**-**	**-**	**-**	**-**	**-**	**-**	**-**
FINAL CONSUMPTION	**-**	**11**	**3251**	**-**	**487**	**846**	**1073**	**688**	**-**	**-**	**1938**
INDUSTRY SECTOR	**-**	**-**	**-**	**-**	**-**	**-**	**39**	**688**	**-**	**-**	**-**
Iron and Steel	-	-	-	-	-	-	-	-	-	-	-
Chemical and Petrochem.	-	-	-	-	-	-	-	-	-	-	-
Non-Metallic Minerals	-	-	-	-	-	-	-	-	-	-	-
Non-specified	-	-	-	-	-	-	39	688	-	-	-
TRANSPORT SECTOR	**-**	**-**	**3251**	**-**	**487**	**-**	**1034**	**-**	**-**	**-**	**-**
Air	-	-	-	-	487	-	-	-	-	-	-
Road	-	-	3251	-	-	-	994	-	-	-	-
Non-specified	-	-	-	-	-	-	40	-	-	-	-
OTHER SECTORS	**-**	**11**	**-**	**-**	**-**	**846**	**-**	**-**	**-**	**-**	**-**
Agriculture	-	-	-	-	-	-	-	-	-	-	-
Comm. and Publ. Services	-	-	-	-	-	-	-	-	-	-	-
Residential	-	11	-	-	-	846	-	-	-	-	-
Non-specified	-	-	-	-	-	-	-	-	-	-	-
NON-ENERGY USE	**-**	**-**	**-**	**-**	**-**	**-**	**-**	**-**	**-**	**-**	**1938**

APPROVISIONNEMENT ET DEMANDE 1997	*Pétrole cont.* (1000 tonnes)										
	Gaz de raffinerie	*GPL + éthane*	*Essence moteur*	*Essence aviation*	*Carbu- réacteurs*	*Kérosène*	*Gazole*	*Fioul lourd*	*Naphta*	*Coke de pétrole*	*Autres prod.*
Production	346	202	5595	-	479	2744	3002	3939	-	-	638
Imports	-	13	-	-	-	-	27	-	-	-	-
Exports	-	-191	-	-	-	-	-164	-650	-	-	-
Intl. Marine Bunkers	-	-	-	-	-	-	-80	-	-	-	-
Stock Changes	-	-	-	-	-	-	-	-	-	-	-
DOMESTIC SUPPLY	**346**	**24**	**5595**	**-**	**479**	**2744**	**2785**	**3289**	**-**	**-**	**638**
Transfers and Stat. Diff.	-	-13	-2289	-	13	-1889	-1491	-2266	-	-	1300
TRANSFORMATION	**-**	**-**	**-**	**-**	**-**	**-**	**205**	**114**	**-**	**-**	**-**
Electricity and CHP Plants	-	-	-	-	-	-	205	114	-	-	-
Petroleum Refineries	-	-	-	-	-	-	-	-	-	-	-
Other Transform. Sector	-	-	-	-	-	-	-	-	-	-	-
ENERGY SECTOR	**346**	**-**	**-**	**-**	**-**	**-**	**-**	**199**	**-**	**-**	**-**
DISTRIBUTION LOSSES	**-**	**-**	**-**	**-**	**-**	**-**	**-**	**-**	**-**	**-**	**-**
FINAL CONSUMPTION	**-**	**11**	**3306**	**-**	**492**	**855**	**1089**	**710**	**-**	**-**	**1938**
INDUSTRY SECTOR	**-**	**-**	**-**	**-**	**-**	**-**	**40**	**710**	**-**	**-**	**-**
Iron and Steel	-	-	-	-	-	-	-	-	-	-	-
Chemical and Petrochem.	-	-	-	-	-	-	-	-	-	-	-
Non-Metallic Minerals	-	-	-	-	-	-	-	-	-	-	-
Non-specified	-	-	-	-	-	-	40	710	-	-	-
TRANSPORT SECTOR	**-**	**-**	**3306**	**-**	**492**	**-**	**1049**	**-**	**-**	**-**	**-**
Air	-	-	-	-	492	-	-	-	-	-	-
Road	-	-	3306	-	-	-	1009	-	-	-	-
Non-specified	-	-	-	-	-	-	40	-	-	-	-
OTHER SECTORS	**-**	**11**	**-**	**-**	**-**	**855**	**-**	**-**	**-**	**-**	**-**
Agriculture	-	-	-	-	-	-	-	-	-	-	-
Comm. and Publ. Services	-	-	-	-	-	-	-	-	-	-	-
Residential	-	11	-	-	-	855	-	-	-	-	-
Non-specified	-	-	-	-	-	-	-	-	-	-	-
NON-ENERGY USE	**-**	**-**	**-**	**-**	**-**	**-**	**-**	**-**	**-**	**-**	**1938**

Nigeria / Nigéria

SUPPLY AND CONSUMPTION 1996	Gas (TJ)				Comb. Renew. & Waste (TJ)				(GWh)	(TJ)
	Natural Gas	Gas Works	Coke Ovens	Blast Furnaces	Solid Biomass	Gas/Liquids from Biomass	Municipal Waste	Industrial Waste	Electricity	Heat
Production	207371	-	-	-	2794010	-	-	-	14991	-
Imports	-	-	-	-	-	-	-	-	-	-
Exports	-	-	-	-	-	-	-	-	-	-
Intl. Marine Bunkers	-	-	-	-	-	-	-	-	-	-
Stock Changes	-	-	-	-	-	-	-	-	-	-
DOMESTIC SUPPLY	**207371**	**-**	**-**	**-**	**2794010**	**-**	**-**	**-**	**14991**	**-**
Transfers and Stat. Diff.	1	-	-	-	-	-	-	-	-	-
TRANSFORMATION	**74293**	**-**	**-**	**-**	**74232**	**-**	**-**	**-**	**-**	**-**
Electricity and CHP Plants	74293	-	-	-	-	-	-	-	-	-
Petroleum Refineries	-	-	-	-	-	-	-	-	-	-
Other Transform. Sector	-	-	-	-	74232	-	-	-	-	-
ENERGY SECTOR	**76701**	**-**	**-**	**-**	**-**	**-**	**-**	**-**	**455**	**-**
DISTRIBUTION LOSSES	**21480**	**-**	**-**	**-**	**-**	**-**	**-**	**-**	**4769**	**-**
FINAL CONSUMPTION	**34898**	**-**	**-**	**-**	**2719778**	**-**	**-**	**-**	**9767**	**-**
INDUSTRY SECTOR	**34898**	**-**	**-**	**-**	**275277**	**-**	**-**	**-**	**2108**	**-**
Iron and Steel	2622	-	-	-	-	-	-	-	-	-
Chemical and Petrochem.	14120	-	-	-	-	-	-	-	-	-
Non-Metallic Minerals	-	-	-	-	-	-	-	-	-	-
Non-specified	18156	-	-	-	275277	-	-	-	2108	-
TRANSPORT SECTOR	**-**	**-**	**-**	**-**	**-**	**-**	**-**	**-**	**-**	**-**
Air	-	-	-	-	-	-	-	-	-	-
Road	-	-	-	-	-	-	-	-	-	-
Non-specified	-	-	-	-	-	-	-	-	-	-
OTHER SECTORS	**-**	**-**	**-**	**-**	**2444501**	**-**	**-**	**-**	**7659**	**-**
Agriculture	-	-	-	-	-	-	-	-	-	-
Comm. and Publ. Services	-	-	-	-	-	-	-	-	2535	-
Residential	-	-	-	-	2444501	-	-	-	5124	-
Non-specified	-	-	-	-	-	-	-	-	-	-
NON-ENERGY USE	**-**	**-**	**-**	**-**	**-**	**-**	**-**	**-**	**-**	**-**

APPROVISIONNEMENT ET DEMANDE 1997	*Gaz* (TJ)				*En. Re. Comb. & Déchets* (TJ)				(GWh)	(TJ)
	Gaz naturel	*Usines à gaz*	*Cokeries*	*Hauts fourneaux*	*Biomasse solide*	*Gaz/Liquides tirés de biomasse*	*Déchets urbains*	*Déchets industriels*	*Electricité*	*Chaleur*
Production	206110	-	-	-	2880624	-	-	-	15179	-
Imports	-	-	-	-	-	-	-	-	-	-
Exports	-	-	-	-	-	-	-	-	-	-
Intl. Marine Bunkers	-	-	-	-	-	-	-	-	-	-
Stock Changes	-	-	-	-	-	-	-	-	-	-
DOMESTIC SUPPLY	**206110**	**-**	**-**	**-**	**2880624**	**-**	**-**	**-**	**15179**	**-**
Transfers and Stat. Diff.	-1	-	-	-	-	-	-	-	-	-
TRANSFORMATION	**75556**	**-**	**-**	**-**	**76533**	**-**	**-**	**-**	**-**	**-**
Electricity and CHP Plants	75556	-	-	-	-	-	-	-	-	-
Petroleum Refineries	-	-	-	-	-	-	-	-	-	-
Other Transform. Sector	-	-	-	-	76533	-	-	-	-	-
ENERGY SECTOR	**76701**	**-**	**-**	**-**	**-**	**-**	**-**	**-**	**461**	**-**
DISTRIBUTION LOSSES	**20217**	**-**	**-**	**-**	**-**	**-**	**-**	**-**	**4829**	**-**
FINAL CONSUMPTION	**33635**	**-**	**-**	**-**	**2804091**	**-**	**-**	**-**	**9889**	**-**
INDUSTRY SECTOR	**33635**	**-**	**-**	**-**	**283811**	**-**	**-**	**-**	**2134**	**-**
Iron and Steel	2527	-	-	-	-	-	-	-	-	-
Chemical and Petrochem.	13609	-	-	-	-	-	-	-	-	-
Non-Metallic Minerals	-	-	-	-	-	-	-	-	-	-
Non-specified	17499	-	-	-	283811	-	-	-	2134	-
TRANSPORT SECTOR	**-**	**-**	**-**	**-**	**-**	**-**	**-**	**-**	**-**	**-**
Air	-	-	-	-	-	-	-	-	-	-
Road	-	-	-	-	-	-	-	-	-	-
Non-specified	-	-	-	-	-	-	-	-	-	-
OTHER SECTORS	**-**	**-**	**-**	**-**	**2520280**	**-**	**-**	**-**	**7755**	**-**
Agriculture	-	-	-	-	-	-	-	-	-	-
Comm. and Publ. Services	-	-	-	-	-	-	-	-	2567	-
Residential	-	-	-	-	2520280	-	-	-	5188	-
Non-specified	-	-	-	-	-	-	-	-	-	-
NON-ENERGY USE	**-**	**-**	**-**	**-**	**-**	**-**	**-**	**-**	**-**	**-**

Oman

SUPPLY AND CONSUMPTION 1996	Coal (1000 tonnes)							Oil (1000 tonnes)			
	Coking Coal	Other Bit. Coal	Sub-Bit. Coal	Lignite	Peat	Oven and Gas Coke	Pat. Fuel and BKB	Crude Oil	NGL	Feed-stocks	Additives
Production	-	-	-	-	-	-	-	45220	85	-	-
Imports	-	-	-	-	-	-	-	-	-	-	-
Exports	-	-	-	-	-	-	-	-41459	-85	-	-
Intl. Marine Bunkers	-	-	-	-	-	-	-	-	-	-	-
Stock Changes	-	-	-	-	-	-	-	-42	-	-	-
DOMESTIC SUPPLY	-	-	-	-	-	-	-	**3719**	-	-	-
Transfers and Stat. Diff.	-	-	-	-	-	-	-	1	-	-	-
TRANSFORMATION	-	-	-	-	-	-	-	**3720**	-	-	-
Electricity and CHP Plants	-	-	-	-	-	-	-	-	-	-	-
Petroleum Refineries	-	-	-	-	-	-	-	3720	-	-	-
Other Transform. Sector	-	-	-	-	-	-	-	-	-	-	-
ENERGY SECTOR	-	-	-	-	-	-	-	-	-	-	-
DISTRIBUTION LOSSES	-	-	-	-	-	-	-	-	-	-	-
FINAL CONSUMPTION	-	-	-	-	-	-	-	-	-	-	-
INDUSTRY SECTOR	-	-	-	-	-	-	-	-	-	-	-
Iron and Steel	-	-	-	-	-	-	-	-	-	-	-
Chemical and Petrochem.	-	-	-	-	-	-	-	-	-	-	-
Non-Metallic Minerals	-	-	-	-	-	-	-	-	-	-	-
Non-specified	-	-	-	-	-	-	-	-	-	-	-
TRANSPORT SECTOR	-	-	-	-	-	-	-	-	-	-	-
Air	-	-	-	-	-	-	-	-	-	-	-
Road	-	-	-	-	-	-	-	-	-	-	-
Non-specified	-	-	-	-	-	-	-	-	-	-	-
OTHER SECTORS	-	-	-	-	-	-	-	-	-	-	-
Agriculture	-	-	-	-	-	-	-	-	-	-	-
Comm. and Publ. Services	-	-	-	-	-	-	-	-	-	-	-
Residential	-	-	-	-	-	-	-	-	-	-	-
Non-specified	-	-	-	-	-	-	-	-	-	-	-
NON-ENERGY USE	-	-	-	-	-	-	-	-	-	-	-

APPROVISIONNEMENT ET DEMANDE 1997	*Charbon* (1000 tonnes)							*Pétrole* (1000 tonnes)			
	Charbon à coke	*Autres charb. bit.*	*Charbon sous-bit.*	*Lignite*	*Tourbe*	*Coke de four/gaz*	*Agg./briq. de lignite*	*Pétrole brut*	*LGN*	*Produits d'aliment.*	*Additifs*
Production	-	-	-	-	-	-	-	45916	85	-	-
Imports	-	-	-	-	-	-	-	-	-	-	-
Exports	-	-	-	-	-	-	-	-42448	-85	-	-
Intl. Marine Bunkers	-	-	-	-	-	-	-	-	-	-	-
Stock Changes	-	-	-	-	-	-	-	-160	-	-	-
DOMESTIC SUPPLY	-	-	-	-	-	-	-	**3308**	-	-	-
Transfers and Stat. Diff.	-	-	-	-	-	-	-	1	-	-	-
TRANSFORMATION	-	-	-	-	-	-	-	**3309**	-	-	-
Electricity and CHP Plants	-	-	-	-	-	-	-	-	-	-	-
Petroleum Refineries	-	-	-	-	-	-	-	3309	-	-	-
Other Transform. Sector	-	-	-	-	-	-	-	-	-	-	-
ENERGY SECTOR	-	-	-	-	-	-	-	-	-	-	-
DISTRIBUTION LOSSES	-	-	-	-	-	-	-	-	-	-	-
FINAL CONSUMPTION	-	-	-	-	-	-	-	-	-	-	-
INDUSTRY SECTOR	-	-	-	-	-	-	-	-	-	-	-
Iron and Steel	-	-	-	-	-	-	-	-	-	-	-
Chemical and Petrochem.	-	-	-	-	-	-	-	-	-	-	-
Non-Metallic Minerals	-	-	-	-	-	-	-	-	-	-	-
Non-specified	-	-	-	-	-	-	-	-	-	-	-
TRANSPORT SECTOR	-	-	-	-	-	-	-	-	-	-	-
Air	-	-	-	-	-	-	-	-	-	-	-
Road	-	-	-	-	-	-	-	-	-	-	-
Non-specified	-	-	-	-	-	-	-	-	-	-	-
OTHER SECTORS	-	-	-	-	-	-	-	-	-	-	-
Agriculture	-	-	-	-	-	-	-	-	-	-	-
Comm. and Publ. Services	-	-	-	-	-	-	-	-	-	-	-
Residential	-	-	-	-	-	-	-	-	-	-	-
Non-specified	-	-	-	-	-	-	-	-	-	-	-
NON-ENERGY USE	-	-	-	-	-	-	-	-	-	-	-

Oman

SUPPLY AND CONSUMPTION 1996	Oil cont. (1000 tonnes)										
	Refinery Gas	LPG + Ethane	Motor Gasoline	Aviation Gasoline	Jet Fuel	Kerosene	Gas/ Diesel	Heavy Fuel Oil	Naphtha	Petrol. Coke	Other Prod.
Production	-	63	573	-	162	4	705	2115	-	-	25
Imports	-	12	39	-	-	-	25	-	-	-	16
Exports	-	-	-	-	-	-	-2	-1112	-	-	-
Intl. Marine Bunkers	-	-	-	-	-	-	-	-27	-	-	-
Stock Changes	-	-13	27	-	-10	-	-20	-	-	-	-
DOMESTIC SUPPLY	**-**	**62**	**639**	**-**	**152**	**4**	**708**	**976**	**-**	**-**	**41**
Transfers and Stat. Diff.	-	1	-	-	-	-	-52	-	-	-	-
TRANSFORMATION	**-**	**-**	**-**	**-**	**-**	**-**	**395**	**-**	**-**	**-**	**-**
Electricity and CHP Plants	-	-	-	-	-	-	395	-	-	-	-
Petroleum Refineries	-	-	-	-	-	-	-	-	-	-	-
Other Transform. Sector	-	-	-	-	-	-	-	-	-	-	-
ENERGY SECTOR	**-**	**-**	**-**	**-**	**-**	**-**	**-**	**152**	**-**	**-**	**-**
DISTRIBUTION LOSSES	**-**	**-**	**-**	**-**	**-**	**-**	**-**	**-**	**-**	**-**	**-**
FINAL CONSUMPTION	**-**	**63**	**639**	**-**	**152**	**4**	**261**	**824**	**-**	**-**	**41**
INDUSTRY SECTOR	**-**	**-**	**-**	**-**	**-**	**4**	**-**	**824**	**-**	**-**	**-**
Iron and Steel	-	-	-	-	-	-	-	-	-	-	-
Chemical and Petrochem.	-	-	-	-	-	-	-	-	-	-	-
Non-Metallic Minerals	-	-	-	-	-	-	-	-	-	-	-
Non-specified	-	-	-	-	-	4	-	824	-	-	-
TRANSPORT SECTOR	**-**	**-**	**639**	**-**	**152**	**-**	**90**	**-**	**-**	**-**	**-**
Air	-	-	-	-	152	-	-	-	-	-	-
Road	-	-	639	-	-	-	90	-	-	-	-
Non-specified	-	-	-	-	-	-	-	-	-	-	-
OTHER SECTORS	**-**	**63**	**-**	**-**	**-**	**-**	**171**	**-**	**-**	**-**	**-**
Agriculture	-	-	-	-	-	-	-	-	-	-	-
Comm. and Publ. Services	-	-	-	-	-	-	-	-	-	-	-
Residential	-	63	-	-	-	-	-	-	-	-	-
Non-specified	-	-	-	-	-	-	171	-	-	-	-
NON-ENERGY USE	**-**	**-**	**-**	**-**	**-**	**-**	**-**	**-**	**-**	**-**	**41**

APPROVISIONNEMENT ET DEMANDE 1997	*Pétrole cont.* (1000 tonnes)										
	Gaz de raffinerie	*GPL + éthane*	*Essence moteur*	*Essence aviation*	*Carbu- réacteurs*	*Kérosène*	*Gazole*	*Fioul lourd*	*Naphta*	*Coke de pétrole*	*Autres prod.*
Production	-	52	479	-	134	4	651	1902	-	-	25
Imports	-	19	170	-	39	-	77	-	-	-	16
Exports	-	-	-	-	-	-	-	-1293	-	-	-
Intl. Marine Bunkers	-	-	-	-	-	-	-	-30	-	-	-
Stock Changes	-	-19	29	-	-1	-	28	-	-	-	-
DOMESTIC SUPPLY	**-**	**52**	**678**	**-**	**172**	**4**	**756**	**579**	**-**	**-**	**41**
Transfers and Stat. Diff.	-	-	-	-	-	-	-78	-	-	-	-
TRANSFORMATION	**-**	**-**	**-**	**-**	**-**	**-**	**400**	**-**	**-**	**-**	**-**
Electricity and CHP Plants	-	-	-	-	-	-	400	-	-	-	-
Petroleum Refineries	-	-	-	-	-	-	-	-	-	-	-
Other Transform. Sector	-	-	-	-	-	-	-	-	-	-	-
ENERGY SECTOR	**-**	**-**	**-**	**-**	**-**	**-**	**-**	**137**	**-**	**-**	**-**
DISTRIBUTION LOSSES	**-**	**-**	**-**	**-**	**-**	**-**	**-**	**-**	**-**	**-**	**-**
FINAL CONSUMPTION	**-**	**52**	**678**	**-**	**172**	**4**	**278**	**442**	**-**	**-**	**41**
INDUSTRY SECTOR	**-**	**-**	**-**	**-**	**-**	**4**	**-**	**442**	**-**	**-**	**-**
Iron and Steel	-	-	-	-	-	-	-	-	-	-	-
Chemical and Petrochem.	-	-	-	-	-	-	-	-	-	-	-
Non-Metallic Minerals	-	-	-	-	-	-	-	-	-	-	-
Non-specified	-	-	-	-	-	4	-	442	-	-	-
TRANSPORT SECTOR	**-**	**-**	**678**	**-**	**172**	**-**	**96**	**-**	**-**	**-**	**-**
Air	-	-	-	-	172	-	-	-	-	-	-
Road	-	-	678	-	-	-	96	-	-	-	-
Non-specified	-	-	-	-	-	-	-	-	-	-	-
OTHER SECTORS	**-**	**52**	**-**	**-**	**-**	**-**	**182**	**-**	**-**	**-**	**-**
Agriculture	-	-	-	-	-	-	-	-	-	-	-
Comm. and Publ. Services	-	-	-	-	-	-	-	-	-	-	-
Residential	-	52	-	-	-	-	-	-	-	-	-
Non-specified	-	-	-	-	-	-	182	-	-	-	-
NON-ENERGY USE	**-**	**-**	**-**	**-**	**-**	**-**	**-**	**-**	**-**	**-**	**41**

Oman

SUPPLY AND CONSUMPTION 1996	Gas (TJ)				Comb. Renew. & Waste (TJ)				(GWh)	(TJ)
	Natural Gas	Gas Works	Coke Ovens	Blast Furnaces	Solid Biomass	Gas/Liquids from Biomass	Municipal Waste	Industrial Waste	Electricity	Heat
Production	186000	-	-	-	-	-	-	-	6802	-
Imports	-	-	-	-	-	-	-	-	-	-
Exports	-18850	-	-	-	-	-	-	-	-	-
Intl. Marine Bunkers	-	-	-	-	-	-	-	-	-	-
Stock Changes	-	-	-	-	-	-	-	-	-	-
DOMESTIC SUPPLY	**167150**	-	-	-	-	-	-	-	**6802**	-
Transfers and Stat. Diff.	-3057	-	-	-	-	-	-	-	-	-
TRANSFORMATION	**81429**	-	-	-	-	-	-	-	-	-
Electricity and CHP Plants	81429	-	-	-	-	-	-	-	-	-
Petroleum Refineries	-	-	-	-	-	-	-	-	-	-
Other Transform. Sector	-	-	-	-	-	-	-	-	-	-
ENERGY SECTOR	**53372**	-	-	-	-	-	-	-	**162**	-
DISTRIBUTION LOSSES	-	-	-	-	-	-	-	-	**1100**	-
FINAL CONSUMPTION	**29292**	-	-	-	-	-	-	-	**5540**	-
INDUSTRY SECTOR	**21515**	-	-	-	-	-	-	-	**391**	-
Iron and Steel	-	-	-	-	-	-	-	-	-	-
Chemical and Petrochem.	-	-	-	-	-	-	-	-	-	-
Non-Metallic Minerals	2399	-	-	-	-	-	-	-	-	-
Non-specified	19116	-	-	-	-	-	-	-	391	-
TRANSPORT SECTOR	-	-	-	-	-	-	-	-	-	-
Air	-	-	-	-	-	-	-	-	-	-
Road	-	-	-	-	-	-	-	-	-	-
Non-specified	-	-	-	-	-	-	-	-	-	-
OTHER SECTORS	**7777**	-	-	-	-	-	-	-	**5149**	-
Agriculture	-	-	-	-	-	-	-	-	-	-
Comm. and Publ. Services	-	-	-	-	-	-	-	-	2014	-
Residential	7777	-	-	-	-	-	-	-	3043	-
Non-specified	-	-	-	-	-	-	-	-	92	-
NON-ENERGY USE	-	-	-	-	-	-	-	-	-	-

APPROVISIONNEMENT ET DEMANDE 1997	*Gaz* (TJ)				*En. Re. Comb. & Déchets* (TJ)				(GWh)	(TJ)
	Gaz naturel	*Usines à gaz*	*Cokeries*	*Hauts fourneaux*	*Biomasse solide*	*Gaz/Liquides tirés de biomasse*	*Déchets urbains*	*Déchets industriels*	*Electricité*	*Chaleur*
Production	218657	-	-	-	-	-	-	-	7318	-
Imports	-	-	-	-	-	-	-	-	-	-
Exports	-18850	-	-	-	-	-	-	-	-	-
Intl. Marine Bunkers	-	-	-	-	-	-	-	-	-	-
Stock Changes	-	-	-	-	-	-	-	-	-	-
DOMESTIC SUPPLY	**199807**	-	-	-	-	-	-	-	**7318**	-
Transfers and Stat. Diff.	5198	-	-	-	-	-	-	-	1	-
TRANSFORMATION	**85011**	-	-	-	-	-	-	-	-	-
Electricity and CHP Plants	85011	-	-	-	-	-	-	-	-	-
Petroleum Refineries	-	-	-	-	-	-	-	-	-	-
Other Transform. Sector	-	-	-	-	-	-	-	-	-	-
ENERGY SECTOR	**62732**	-	-	-	-	-	-	-	**178**	-
DISTRIBUTION LOSSES	-	-	-	-	-	-	-	-	**1245**	-
FINAL CONSUMPTION	**57262**	-	-	-	-	-	-	-	**5896**	-
INDUSTRY SECTOR	**39322**	-	-	-	-	-	-	-	**455**	-
Iron and Steel	-	-	-	-	-	-	-	-	-	-
Chemical and Petrochem.	-	-	-	-	-	-	-	-	-	-
Non-Metallic Minerals	2202	-	-	-	-	-	-	-	-	-
Non-specified	37120	-	-	-	-	-	-	-	455	-
TRANSPORT SECTOR	-	-	-	-	-	-	-	-	-	-
Air	-	-	-	-	-	-	-	-	-	-
Road	-	-	-	-	-	-	-	-	-	-
Non-specified	-	-	-	-	-	-	-	-	-	-
OTHER SECTORS	**17940**	-	-	-	-	-	-	-	**5441**	-
Agriculture	-	-	-	-	-	-	-	-	-	-
Comm. and Publ. Services	-	-	-	-	-	-	-	-	2099	-
Residential	17940	-	-	-	-	-	-	-	3246	-
Non-specified	-	-	-	-	-	-	-	-	96	-
NON-ENERGY USE	-	-	-	-	-	-	-	-	-	-

Pakistan : 1996

SUPPLY AND CONSUMPTION *APPROVISIONNEMENT ET DEMANDE*	Coal / *Charbon* (1000 tonnes)							Oil / *Pétrole* (1000 tonnes)			
	Coking Coal *Charbon à coke*	Other Bit. Coal *Autres charb. bit.*	Sub-Bit. Coal *Charbon sous-bit.*	Lignite *Lignite*	Peat *Tourbe*	Oven and Gas Coke *Coke de four/gaz*	Pat. Fuel and BKB *Agg./briq. de lignite*	Crude Oil *Pétrole brut*	NGL *LGN*	Feed-stocks *Produits d'aliment.*	Additives *Additifs*
Production	-	3638	-	-	-	756	-	2826	121	-	62
From Other Sources	-	-	-	-	-	-	-	-	-	-	-
Imports	1080	-	-	-	-	-	-	4231	-	-	101
Exports	-	-	-	-	-	-	-	-289	-	-	-
Intl. Marine Bunkers	-	-	-	-	-	-	-	-	-	-	-
Stock Changes	-	-	-	-	-	-	-	-130	-	-	-5
DOMESTIC SUPPLY	**1080**	**3638**	**-**	**-**	**-**	**756**	**-**	**6638**	**121**	**-**	**158**
Transfers	-	-	-	-	-	-	-	-	-121	-	-
Statistical Differences	-	-	-	-	-	-	-	-	-	-	-
TRANSFORMATION	**1080**	**399**	**-**	**-**	**-**	**605**	**-**	**6638**	**-**	**-**	**158**
Electricity Plants	-	399	-	-	-	-	-	-	-	-	-
CHP Plants	-	-	-	-	-	-	-	-	-	-	-
Heat Plants	-	-	-	-	-	-	-	-	-	-	-
Blast Furnaces/Gas Works	-	-	-	-	-	605	-	-	-	-	-
Coke/Pat. Fuel/BKB Plants	1080	-	-	-	-	-	-	-	-	-	-
Petroleum Refineries	-	-	-	-	-	-	-	6638	-	-	158
Petrochemical Industry	-	-	-	-	-	-	-	-	-	-	-
Liquefaction	-	-	-	-	-	-	-	-	-	-	-
Other Transform. Sector	-	-	-	-	-	-	-	-	-	-	-
ENERGY SECTOR	**-**	**-**	**-**	**-**	**-**	**-**	**-**	**-**	**-**	**-**	**-**
Coal Mines	-	-	-	-	-	-	-	-	-	-	-
Oil and Gas Extraction	-	-	-	-	-	-	-	-	-	-	-
Petroleum Refineries	-	-	-	-	-	-	-	-	-	-	-
Electr., CHP+Heat Plants	-	-	-	-	-	-	-	-	-	-	-
Pumped Storage (Elec.)	-	-	-	-	-	-	-	-	-	-	-
Other Energy Sector	-	-	-	-	-	-	-	-	-	-	-
Distribution Losses	-	-	-	-	-	-	-	-	-	-	-
FINAL CONSUMPTION	**-**	**3239**	**-**	**-**	**-**	**151**	**-**	**-**	**-**	**-**	**-**
INDUSTRY SECTOR	**-**	**3236**	**-**	**-**	**-**	**151**	**-**	**-**	**-**	**-**	**-**
Iron and Steel	-	-	-	-	-	151	-	-	-	-	-
Chemical and Petrochem.	-	-	-	-	-	-	-	-	-	-	-
of which: Feedstocks	-	-	-	-	-	-	-	-	-	-	-
Non-Ferrous Metals	-	-	-	-	-	-	-	-	-	-	-
Non-Metallic Minerals	-	3236	-	-	-	-	-	-	-	-	-
Transport Equipment	-	-	-	-	-	-	-	-	-	-	-
Machinery	-	-	-	-	-	-	-	-	-	-	-
Mining and Quarrying	-	-	-	-	-	-	-	-	-	-	-
Food and Tobacco	-	-	-	-	-	-	-	-	-	-	-
Paper, Pulp and Print	-	-	-	-	-	-	-	-	-	-	-
Wood and Wood Products	-	-	-	-	-	-	-	-	-	-	-
Construction	-	-	-	-	-	-	-	-	-	-	-
Textile and Leather	-	-	-	-	-	-	-	-	-	-	-
Non-specified	-	-	-	-	-	-	-	-	-	-	-
TRANSPORT SECTOR	**-**	**-**	**-**	**-**	**-**	**-**	**-**	**-**	**-**	**-**	**-**
Air	-	-	-	-	-	-	-	-	-	-	-
Road	-	-	-	-	-	-	-	-	-	-	-
Rail	-	-	-	-	-	-	-	-	-	-	-
Pipeline Transport	-	-	-	-	-	-	-	-	-	-	-
Internal Navigation	-	-	-	-	-	-	-	-	-	-	-
Non-specified	-	-	-	-	-	-	-	-	-	-	-
OTHER SECTORS	**-**	**3**	**-**	**-**	**-**	**-**	**-**	**-**	**-**	**-**	**-**
Agriculture	-	-	-	-	-	-	-	-	-	-	-
Comm. and Publ. Services	-	-	-	-	-	-	-	-	-	-	-
Residential	-	3	-	-	-	-	-	-	-	-	-
Non-specified	-	-	-	-	-	-	-	-	-	-	-
NON-ENERGY USE	**-**	**-**	**-**	**-**	**-**	**-**	**-**	**-**	**-**	**-**	**-**
in Industry/Trans./Energy	-	-	-	-	-	-	-	-	-	-	-
in Transport	-	-	-	-	-	-	-	-	-	-	-
in Other Sectors	-	-	-	-	-	-	-	-	-	-	-

Pakistan : 1996

SUPPLY AND CONSUMPTION / *APPROVISIONNEMENT ET DEMANDE*	Oil cont. / *Pétrole cont.* (1000 tonnes)										
	Refinery Gas / *Gaz de raffinerie*	LPG + Ethane / *GPL + éthane*	Motor Gasoline / *Essence moteur*	Aviation Gasoline / *Essence aviation*	Jet Fuel / *Carbu-réacteurs*	Kerosene / *Kérosène*	Gas/ Diesel / *Gazole*	Heavy Fuel Oil / *Fioul lourd*	Naphtha / *Naphta*	Petrol. Coke / *Coke de pétrole*	Other Prod. / *Autres prod.*
Production	113	36	969	-	585	507	1694	1916	105	-	469
From Other Sources	-	-	-	-	-	-	-	-	-	-	-
Imports	-	48	106	-	-	88	5094	4749	-	-	-
Exports	-	-	-	-	-	-	-	-	-95	-	-38
Intl. Marine Bunkers	-	-	-	-	-	-	-3	-9	-	-	-
Stock Changes	-	-	-13	-	1	19	-185	51	-10	-	-
DOMESTIC SUPPLY	**113**	**84**	**1062**	**-**	**586**	**614**	**6600**	**6707**	**-**	**-**	**431**
Transfers	-	121	-	-	-	-	-	-	-	-	-
Statistical Differences	-	-	-	-	-	-	-	99	-	-	-82
TRANSFORMATION	**-**	**-**	**-**	**-**	**-**	**-**	**331**	**4455**	**-**	**-**	**-**
Electricity Plants	-	-	-	-	-	-	331	4455	-	-	-
CHP Plants	-	-	-	-	-	-	-	-	-	-	-
Heat Plants	-	-	-	-	-	-	-	-	-	-	-
Blast Furnaces/Gas Works	-	-	-	-	-	-	-	-	-	-	-
Coke/Pat. Fuel/BKB Plants	-	-	-	-	-	-	-	-	-	-	-
Petroleum Refineries	-	-	-	-	-	-	-	-	-	-	-
Petrochemical Industry	-	-	-	-	-	-	-	-	-	-	-
Liquefaction	-	-	-	-	-	-	-	-	-	-	-
Other Transform. Sector	-	-	-	-	-	-	-	-	-	-	-
ENERGY SECTOR	**113**	**-**	**-**	**-**	**-**	**-**	**-**	**100**	**-**	**-**	**-**
Coal Mines	-	-	-	-	-	-	-	-	-	-	-
Oil and Gas Extraction	-	-	-	-	-	-	-	-	-	-	-
Petroleum Refineries	113	-	-	-	-	-	-	100	-	-	-
Electr., CHP+Heat Plants	-	-	-	-	-	-	-	-	-	-	-
Pumped Storage (Elec.)	-	-	-	-	-	-	-	-	-	-	-
Other Energy Sector	-	-	-	-	-	-	-	-	-	-	-
Distribution Losses	-	-	-	-	-	-	-	-	-	-	-
FINAL CONSUMPTION	**-**	**205**	**1062**	**-**	**586**	**614**	**6269**	**2251**	**-**	**-**	**349**
INDUSTRY SECTOR	**-**	**-**	**-**	**-**	**-**	**-**	**237**	**2179**	**-**	**-**	**-**
Iron and Steel	-	-	-	-	-	-	-	-	-	-	-
Chemical and Petrochem.	-	-	-	-	-	-	237	-	-	-	-
of which: Feedstocks	-	-	-	-	-	-	-	-	-	-	-
Non-Ferrous Metals	-	-	-	-	-	-	-	-	-	-	-
Non-Metallic Minerals	-	-	-	-	-	-	-	1562	-	-	-
Transport Equipment	-	-	-	-	-	-	-	-	-	-	-
Machinery	-	-	-	-	-	-	-	-	-	-	-
Mining and Quarrying	-	-	-	-	-	-	-	-	-	-	-
Food and Tobacco	-	-	-	-	-	-	-	-	-	-	-
Paper, Pulp and Print	-	-	-	-	-	-	-	-	-	-	-
Wood and Wood Products	-	-	-	-	-	-	-	-	-	-	-
Construction	-	-	-	-	-	-	-	-	-	-	-
Textile and Leather	-	-	-	-	-	-	-	-	-	-	-
Non-specified	-	-	-	-	-	-	-	617	-	-	-
TRANSPORT SECTOR	**-**	**-**	**1062**	**-**	**586**	**-**	**5663**	**35**	**-**	**-**	**-**
Air	-	-	-	-	586	-	-	-	-	-	-
Road	-	-	1062	-	-	-	5463	-	-	-	-
Rail	-	-	-	-	-	-	200	35	-	-	-
Pipeline Transport	-	-	-	-	-	-	-	-	-	-	-
Internal Navigation	-	-	-	-	-	-	-	-	-	-	-
Non-specified	-	-	-	-	-	-	-	-	-	-	-
OTHER SECTORS	**-**	**205**	**-**	**-**	**-**	**614**	**369**	**37**	**-**	**-**	**-**
Agriculture	-	-	-	-	-	-	250	-	-	-	-
Comm. and Publ. Services	-	-	-	-	-	17	118	37	-	-	-
Residential	-	205	-	-	-	597	1	-	-	-	-
Non-specified	-	-	-	-	-	-	-	-	-	-	-
NON-ENERGY USE	**-**	**-**	**-**	**-**	**-**	**-**	**-**	**-**	**-**	**-**	**349**
in Industry/Transf./Energy	-	-	-	-	-	-	-	-	-	-	349
in Transport	-	-	-	-	-	-	-	-	-	-	-
in Other Sectors	-	-	-	-	-	-	-	-	-	-	-

Pakistan : 1996

SUPPLY AND CONSUMPTION / *APPROVISIONNEMENT ET DEMANDE*	Gas / *Gaz* (TJ)				Comb. Renew. & Waste / *En. Re. Comb. & Déchets* (TJ)				(GWh)	(TJ)
	Natural Gas / *Gaz naturel*	Gas Works / *Usines à gaz*	Coke Ovens / *Cokeries*	Blast Furnaces / *Hauts fourneaux*	Solid Biomass / *Biomasse solide*	Gas/Liquids from Biomass / *Gaz/Liquides tirés de biomasse*	Municipal Waste / *Déchets urbains*	Industrial Waste / *Déchets industriels*	Electricity / *Electricité*	Heat / *Chaleur*
Production	614769	-	-	-	919378	-	-	-	56956	-
From Other Sources	-	-	-	-	-	-	-	-	-	-
Imports	-	-	-	-	-	-	-	-	-	-
Exports	-	-	-	-	-	-	-	-	-	-
Intl. Marine Bunkers	-	-	-	-	-	-	-	-	-	-
Stock Changes	-	-	-	-	-	-	-	-	-	-
DOMESTIC SUPPLY	**614769**	**-**	**-**	**-**	**919378**	**-**	**-**	**-**	**56956**	**-**
Transfers	-	-	-	-	-	-	-	-	-	-
Statistical Differences	-2	-	-	-	-	-	-	-	-1	-
TRANSFORMATION	**188974**	**-**	**-**	**-**	**18368**	**-**	**-**	**-**	**-**	**-**
Electricity Plants	188974	-	-	-	-	-	-	-	-	-
CHP Plants	-	-	-	-	-	-	-	-	-	-
Heat Plants	-	-	-	-	-	-	-	-	-	-
Blast Furnaces/Gas Works	-	-	-	-	-	-	-	-	-	-
Coke/Pat. Fuel/BKB Plants	-	-	-	-	-	-	-	-	-	-
Petroleum Refineries	-	-	-	-	-	-	-	-	-	-
Petrochemical Industry	-	-	-	-	-	-	-	-	-	-
Liquefaction	-	-	-	-	-	-	-	-	-	-
Other Transform. Sector	-	-	-	-	18368	-	-	-	-	-
ENERGY SECTOR	**-**	**-**	**-**	**-**	**-**	**-**	**-**	**-**	**2039**	**-**
Coal Mines	-	-	-	-	-	-	-	-	-	-
Oil and Gas Extraction	-	-	-	-	-	-	-	-	-	-
Petroleum Refineries	-	-	-	-	-	-	-	-	-	-
Electr., CHP+Heat Plants	-	-	-	-	-	-	-	-	2039	-
Pumped Storage (Elec.)	-	-	-	-	-	-	-	-	-	-
Other Energy Sector	-	-	-	-	-	-	-	-	-	-
Distribution Losses	26401	-	-	-	-	-	-	-	13179	-
FINAL CONSUMPTION	**399392**	**-**	**-**	**-**	**901010**	**-**	**-**	**-**	**41737**	**-**
INDUSTRY SECTOR	**260883**	**-**	**-**	**-**	**99999**	**-**	**-**	**-**	**12183**	**-**
Iron and Steel	-	-	-	-	-	-	-	-	-	-
Chemical and Petrochem.	131568	-	-	-	-	-	-	-	-	-
of which: Feedstocks	*86590*	-	-	-	-	-	-	-	-	-
Non-Ferrous Metals	-	-	-	-	-	-	-	-	-	-
Non-Metallic Minerals	8241	-	-	-	-	-	-	-	-	-
Transport Equipment	-	-	-	-	-	-	-	-	-	-
Machinery	-	-	-	-	-	-	-	-	-	-
Mining and Quarrying	-	-	-	-	-	-	-	-	-	-
Food and Tobacco	-	-	-	-	-	-	-	-	-	-
Paper, Pulp and Print	-	-	-	-	-	-	-	-	-	-
Wood and Wood Products	-	-	-	-	-	-	-	-	-	-
Construction	-	-	-	-	-	-	-	-	-	-
Textile and Leather	-	-	-	-	-	-	-	-	-	-
Non-specified	121074	-	-	-	99999	-	-	-	12183	-
TRANSPORT SECTOR	**166**	**-**	**-**	**-**	**-**	**-**	**-**	**-**	**20**	**-**
Air	-	-	-	-	-	-	-	-	-	-
Road	-	-	-	-	-	-	-	-	-	-
Rail	-	-	-	-	-	-	-	-	20	-
Pipeline Transport	166	-	-	-	-	-	-	-	-	-
Internal Navigation	-	-	-	-	-	-	-	-	-	-
Non-specified	-	-	-	-	-	-	-	-	-	-
OTHER SECTORS	**138343**	**-**	**-**	**-**	**801011**	**-**	**-**	**-**	**29534**	**-**
Agriculture	-	-	-	-	-	-	-	-	6696	-
Comm. and Publ. Services	18465	-	-	-	-	-	-	-	5722	-
Residential	119878	-	-	-	801011	-	-	-	17116	-
Non-specified	-	-	-	-	-	-	-	-	-	-
NON-ENERGY USE	**-**	**-**	**-**	**-**	**-**	**-**	**-**	**-**	**-**	**-**
in Industry/Transf./Energy	-	-	-	-	-	-	-	-	-	-
in Transport	-	-	-	-	-	-	-	-	-	-
in Other Sectors	-	-	-	-	-	-	-	-	-	-

Pakistan : 1997

	Coal / *Charbon* (1000 tonnes)							Oil / *Pétrole* (1000 tonnes)			
SUPPLY AND CONSUMPTION *APPROVISIONNEMENT ET DEMANDE*	Coking Coal *Charbon à coke*	Other Bit. Coal *Autres charb. bit.*	Sub-Bit. Coal *Charbon sous-bit.*	Lignite *Lignite*	Peat *Tourbe*	Oven and Gas Coke *Coke de four/gaz*	Pat. Fuel and BKB *Agg./briq. de lignite*	Crude Oil *Pétrole brut*	NGL *LGN*	Feed-stocks *Produits d'aliment.*	Additives *Additifs*
Production	-	3159	-	-	-	672	-	2729	121	-	62
From Other Sources	-	-	-	-	-	-	-	-	-	-	-
Imports	960	-	-	-	-	-	-	4483	-	-	101
Exports	-	-	-	-	-	-	-	-139	-	-	-
Intl. Marine Bunkers	-	-	-	-	-	-	-	-	-	-	-
Stock Changes	-	-	-	-	-	-	-	-128	-	-	-
DOMESTIC SUPPLY	**960**	**3159**	**-**	**-**	**-**	**672**	**-**	**6945**	**121**	**-**	**163**
Transfers	-	-	-	-	-	-	-	-	-121	-	-
Statistical Differences	-	-	-	-	-	-	-	-	-	-	-
TRANSFORMATION	**960**	**347**	**-**	**-**	**-**	**538**	**-**	**6945**	**-**	**-**	**163**
Electricity Plants	-	347	-	-	-	-	-	-	-	-	-
CHP Plants	-	-	-	-	-	-	-	-	-	-	-
Heat Plants	-	-	-	-	-	-	-	-	-	-	-
Blast Furnaces/Gas Works	-	-	-	-	-	538	-	-	-	-	-
Coke/Pat. Fuel/BKB Plants	960	-	-	-	-	-	-	-	-	-	-
Petroleum Refineries	-	-	-	-	-	-	-	6945	-	-	163
Petrochemical Industry	-	-	-	-	-	-	-	-	-	-	-
Liquefaction	-	-	-	-	-	-	-	-	-	-	-
Other Transform. Sector	-	-	-	-	-	-	-	-	-	-	-
ENERGY SECTOR	**-**	**-**	**-**	**-**	**-**	**-**	**-**	**-**	**-**	**-**	**-**
Coal Mines	-	-	-	-	-	-	-	-	-	-	-
Oil and Gas Extraction	-	-	-	-	-	-	-	-	-	-	-
Petroleum Refineries	-	-	-	-	-	-	-	-	-	-	-
Electr., CHP+Heat Plants	-	-	-	-	-	-	-	-	-	-	-
Pumped Storage (Elec.)	-	-	-	-	-	-	-	-	-	-	-
Other Energy Sector	-	-	-	-	-	-	-	-	-	-	-
Distribution Losses	-	-	-	-	-	-	-	-	-	-	-
FINAL CONSUMPTION	**-**	**2812**	**-**	**-**	**-**	**134**	**-**	**-**	**-**	**-**	**-**
INDUSTRY SECTOR	**-**	**2809**	**-**	**-**	**-**	**134**	**-**	**-**	**-**	**-**	**-**
Iron and Steel	-	-	-	-	-	134	-	-	-	-	-
Chemical and Petrochem.	-	-	-	-	-	-	-	-	-	-	-
of which: Feedstocks	-	-	-	-	-	-	-	-	-	-	-
Non-Ferrous Metals	-	-	-	-	-	-	-	-	-	-	-
Non-Metallic Minerals	-	2809	-	-	-	-	-	-	-	-	-
Transport Equipment	-	-	-	-	-	-	-	-	-	-	-
Machinery	-	-	-	-	-	-	-	-	-	-	-
Mining and Quarrying	-	-	-	-	-	-	-	-	-	-	-
Food and Tobacco	-	-	-	-	-	-	-	-	-	-	-
Paper, Pulp and Print	-	-	-	-	-	-	-	-	-	-	-
Wood and Wood Products	-	-	-	-	-	-	-	-	-	-	-
Construction	-	-	-	-	-	-	-	-	-	-	-
Textile and Leather	-	-	-	-	-	-	-	-	-	-	-
Non-specified	-	-	-	-	-	-	-	-	-	-	-
TRANSPORT SECTOR	**-**	**-**	**-**	**-**	**-**	**-**	**-**	**-**	**-**	**-**	**-**
Air	-	-	-	-	-	-	-	-	-	-	-
Road	-	-	-	-	-	-	-	-	-	-	-
Rail	-	-	-	-	-	-	-	-	-	-	-
Pipeline Transport	-	-	-	-	-	-	-	-	-	-	-
Internal Navigation	-	-	-	-	-	-	-	-	-	-	-
Non-specified	-	-	-	-	-	-	-	-	-	-	-
OTHER SECTORS	**-**	**3**	**-**	**-**	**-**	**-**	**-**	**-**	**-**	**-**	**-**
Agriculture	-	-	-	-	-	-	-	-	-	-	-
Comm. and Publ. Services	-	-	-	-	-	-	-	-	-	-	-
Residential	-	3	-	-	-	-	-	-	-	-	-
Non-specified	-	-	-	-	-	-	-	-	-	-	-
NON-ENERGY USE	**-**	**-**	**-**	**-**	**-**	**-**	**-**	**-**	**-**	**-**	**-**
in Industry/Trans./Energy	-	-	-	-	-	-	-	-	-	-	-
in Transport	-	-	-	-	-	-	-	-	-	-	-
in Other Sectors	-	-	-	-	-	-	-	-	-	-	-

Pakistan : 1997

	Oil cont. / *Pétrole cont.* (1000 tonnes)										
SUPPLY AND CONSUMPTION	Refinery Gas	LPG + Ethane	Motor Gasoline	Aviation Gasoline	Jet Fuel	Kerosene	Gas/ Diesel	Heavy Fuel Oil	Naphtha	Petrol. Coke	Other Prod.
APPROVISIONNEMENT ET DEMANDE	*Gaz de raffinerie*	*GPL + éthane*	*Essence moteur*	*Essence aviation*	*Carbu-réacteurs*	*Kérosène*	*Gazole*	*Fioul lourd*	*Naphta*	*Coke de pétrole*	*Autres prod.*
Production	109	47	1054	-	574	710	1735	2152	92	-	431
From Other Sources	-	-	-	-	-	-	-	-	-	-	-
Imports	-	51	75	-	-	42	4924	4802	-	-	-
Exports	-	-	-	-	-	-	-	-	-92	-	-
Intl. Marine Bunkers	-	-	-	-	-	-	-3	-12	-	-	-
Stock Changes	-	-	-	-	-	-	-	48	-	-	-
DOMESTIC SUPPLY	**109**	**98**	**1129**	**-**	**574**	**752**	**6656**	**6990**	**-**	**-**	**431**
Transfers	-	121	-	-	-	-	-	-	-	-	-
Statistical Differences	-	1	-	-	-	-	-1	-1	-	-	-87
TRANSFORMATION	**-**	**-**	**-**	**-**	**-**	**-**	**260**	**5343**	**-**	**-**	**-**
Electricity Plants	-	-	-	-	-	-	260	5343	-	-	-
CHP Plants	-	-	-	-	-	-	-	-	-	-	-
Heat Plants	-	-	-	-	-	-	-	-	-	-	-
Blast Furnaces/Gas Works	-	-	-	-	-	-	-	-	-	-	-
Coke/Pat. Fuel/BKB Plants	-	-	-	-	-	-	-	-	-	-	-
Petroleum Refineries	-	-	-	-	-	-	-	-	-	-	-
Petrochemical Industry	-	-	-	-	-	-	-	-	-	-	-
Liquefaction	-	-	-	-	-	-	-	-	-	-	-
Other Transform. Sector	-	-	-	-	-	-	-	-	-	-	-
ENERGY SECTOR	**109**	**-**	**-**	**-**	**-**	**-**	**-**	**100**	**-**	**-**	**-**
Coal Mines	-	-	-	-	-	-	-	-	-	-	-
Oil and Gas Extraction	-	-	-	-	-	-	-	-	-	-	-
Petroleum Refineries	109	-	-	-	-	-	-	100	-	-	-
Electr., CHP+Heat Plants	-	-	-	-	-	-	-	-	-	-	-
Pumped Storage (Elec.)	-	-	-	-	-	-	-	-	-	-	-
Other Energy Sector	-	-	-	-	-	-	-	-	-	-	-
Distribution Losses	-	-	-	-	-	-	-	-	-	-	-
FINAL CONSUMPTION	**-**	**220**	**1129**	**-**	**574**	**752**	**6395**	**1546**	**-**	**-**	**344**
INDUSTRY SECTOR	**-**	**-**	**-**	**-**	**-**	**-**	**242**	**1497**	**-**	**-**	**-**
Iron and Steel	-	-	-	-	-	-	-	-	-	-	-
Chemical and Petrochem.	-	-	-	-	-	-	242	-	-	-	-
of which: Feedstocks	-	-	-	-	-	-	-	-	-	-	-
Non-Ferrous Metals	-	-	-	-	-	-	-	-	-	-	-
Non-Metallic Minerals	-	-	-	-	-	-	-	1073	-	-	-
Transport Equipment	-	-	-	-	-	-	-	-	-	-	-
Machinery	-	-	-	-	-	-	-	-	-	-	-
Mining and Quarrying	-	-	-	-	-	-	-	-	-	-	-
Food and Tobacco	-	-	-	-	-	-	-	-	-	-	-
Paper, Pulp and Print	-	-	-	-	-	-	-	-	-	-	-
Wood and Wood Products	-	-	-	-	-	-	-	-	-	-	-
Construction	-	-	-	-	-	-	-	-	-	-	-
Textile and Leather	-	-	-	-	-	-	-	-	-	-	-
Non-specified	-	-	-	-	-	-	-	424	-	-	-
TRANSPORT SECTOR	**-**	**-**	**1129**	**-**	**574**	**-**	**5777**	**24**	**-**	**-**	**-**
Air	-	-	-	-	574	-	-	-	-	-	-
Road	-	-	1129	-	-	-	5573	-	-	-	-
Rail	-	-	-	-	-	-	204	24	-	-	-
Pipeline Transport	-	-	-	-	-	-	-	-	-	-	-
Internal Navigation	-	-	-	-	-	-	-	-	-	-	-
Non-specified	-	-	-	-	-	-	-	-	-	-	-
OTHER SECTORS	**-**	**220**	**-**	**-**	**-**	**752**	**376**	**25**	**-**	**-**	**-**
Agriculture	-	-	-	-	-	-	255	-	-	-	-
Comm. and Publ. Services	-	-	-	-	-	21	120	25	-	-	-
Residential	-	220	-	-	-	731	1	-	-	-	-
Non-specified	-	-	-	-	-	-	-	-	-	-	-
NON-ENERGY USE	**-**	**-**	**-**	**-**	**-**	**-**	**-**	**-**	**-**	**-**	**344**
in Industry/Transf./Energy	-	-	-	-	-	-	-	-	-	-	344
in Transport	-	-	-	-	-	-	-	-	-	-	-
in Other Sectors	-	-	-	-	-	-	-	-	-	-	-

Pakistan : 1997

SUPPLY AND CONSUMPTION / *APPROVISIONNEMENT ET DEMANDE*	Gas / *Gaz* (TJ)				Comb. Renew. & Waste / *En. Re. Comb. & Déchets* (TJ)				(GWh)	(TJ)
	Natural Gas / *Gaz naturel*	Gas Works / *Usines à gaz*	Coke Ovens / *Cokeries*	Blast Furnaces / *Hauts fourneaux*	Solid Biomass / *Biomasse solide*	Gas/Liquids from Biomass / *Gaz/Liquides tirés de biomasse*	Municipal Waste / *Déchets urbains*	Industrial Waste / *Déchets industriels*	Electricity / *Electricité*	Heat / *Chaleur*
Production	625149	-	-	-	937950	-	-	-	59125	-
From Other Sources	-	-	-	-	-	-	-	-	-	-
Imports	-	-	-	-	-	-	-	-	-	-
Exports	-	-	-	-	-	-	-	-	-	-
Intl. Marine Bunkers	-	-	-	-	-	-	-	-	-	-
Stock Changes	-	-	-	-	-	-	-	-	-	-
DOMESTIC SUPPLY	**625149**	**-**	**-**	**-**	**937950**	**-**	**-**	**-**	**59125**	**-**
Transfers	-	-	-	-	-	-	-	-	-	-
Statistical Differences	-1	-	-	-	-	-	-	-	-	-
TRANSFORMATION	**182431**	**-**	**-**	**-**	**18738**	**-**	**-**	**-**	**-**	**-**
Electricity Plants	182431	-	-	-	-	-	-	-	-	-
CHP Plants	-	-	-	-	-	-	-	-	-	-
Heat Plants	-	-	-	-	-	-	-	-	-	-
Blast Furnaces/Gas Works	-	-	-	-	-	-	-	-	-	-
Coke/Pat. Fuel/BKB Plants	-	-	-	-	-	-	-	-	-	-
Petroleum Refineries	-	-	-	-	-	-	-	-	-	-
Petrochemical Industry	-	-	-	-	-	-	-	-	-	-
Liquefaction	-	-	-	-	-	-	-	-	-	-
Other Transform. Sector	-	-	-	-	18738	-	-	-	-	-
ENERGY SECTOR	**-**	**-**	**-**	**-**	**-**	**-**	**-**	**-**	**2357**	**-**
Coal Mines	-	-	-	-	-	-	-	-	-	-
Oil and Gas Extraction	-	-	-	-	-	-	-	-	-	-
Petroleum Refineries	-	-	-	-	-	-	-	-	-	-
Electr., CHP+Heat Plants	-	-	-	-	-	-	-	-	2357	-
Pumped Storage (Elec.)	-	-	-	-	-	-	-	-	-	-
Other Energy Sector	-	-	-	-	-	-	-	-	-	-
Distribution Losses	26401	-	-	-	-	-	-	-	14053	-
FINAL CONSUMPTION	**416316**	**-**	**-**	**-**	**919212**	**-**	**-**	**-**	**42715**	**-**
INDUSTRY SECTOR	**271938**	**-**	**-**	**-**	**102019**	**-**	**-**	**-**	**11982**	**-**
Iron and Steel	-	-	-	-	-	-	-	-	-	-
Chemical and Petrochem.	137143	-	-	-	-	-	-	-	-	-
of which: Feedstocks	*90259*	-	-	-	-	-	-	-	-	-
Non-Ferrous Metals	-	-	-	-	-	-	-	-	-	-
Non-Metallic Minerals	8590	-	-	-	-	-	-	-	-	-
Transport Equipment	-	-	-	-	-	-	-	-	-	-
Machinery	-	-	-	-	-	-	-	-	-	-
Mining and Quarrying	-	-	-	-	-	-	-	-	-	-
Food and Tobacco	-	-	-	-	-	-	-	-	-	-
Paper, Pulp and Print	-	-	-	-	-	-	-	-	-	-
Wood and Wood Products	-	-	-	-	-	-	-	-	-	-
Construction	-	-	-	-	-	-	-	-	-	-
Textile and Leather	-	-	-	-	-	-	-	-	-	-
Non-specified	126205	-	-	-	102019	-	-	-	11982	-
TRANSPORT SECTOR	**173**	**-**	**-**	**-**	**-**	**-**	**-**	**-**	**19**	**-**
Air	-	-	-	-	-	-	-	-	-	-
Road	-	-	-	-	-	-	-	-	-	-
Rail	-	-	-	-	-	-	-	-	19	-
Pipeline Transport	173	-	-	-	-	-	-	-	-	-
Internal Navigation	-	-	-	-	-	-	-	-	-	-
Non-specified	-	-	-	-	-	-	-	-	-	-
OTHER SECTORS	**144205**	**-**	**-**	**-**	**817193**	**-**	**-**	**-**	**30714**	**-**
Agriculture	-	-	-	-	-	-	-	-	7086	-
Comm. and Publ. Services	19247	-	-	-	-	-	-	-	5944	-
Residential	124958	-	-	-	817193	-	-	-	17684	-
Non-specified	-	-	-	-	-	-	-	-	-	-
NON-ENERGY USE	**-**	**-**	**-**	**-**	**-**	**-**	**-**	**-**	**-**	**-**
in Industry/Transf./Energy	-	-	-	-	-	-	-	-	-	-
in Transport	-	-	-	-	-	-	-	-	-	-
in Other Sectors	-	-	-	-	-	-	-	-	-	-

Panama

SUPPLY AND CONSUMPTION 1996	Coal (1000 tonnes)							Oil (1000 tonnes)			
	Coking Coal	Other Bit. Coal	Sub-Bit. Coal	Lignite	Peat	Oven and Gas Coke	Pat. Fuel and BKB	Crude Oil	NGL	Feed-stocks	Additives
Production	-	-	-	-	-	-	-	-	-	-	-
Imports	-	64	-	-	-	-	-	2078	-	-	-
Exports	-	-	-	-	-	-	-	-	-	-	-
Intl. Marine Bunkers	-	-	-	-	-	-	-	-	-	-	-
Stock Changes	-	-8	-	-	-	-	-	-9	-	-	-
DOMESTIC SUPPLY	-	**56**	-	-	-	-	-	**2069**	-	-	-
Transfers and Stat. Diff.	-	-	-	-	-	-	-	-4	-	-	-
TRANSFORMATION	-	-	-	-	-	-	-	**2065**	-	-	-
Electricity and CHP Plants	-	-	-	-	-	-	-	-	-	-	-
Petroleum Refineries	-	-	-	-	-	-	-	2065	-	-	-
Other Transform. Sector	-	-	-	-	-	-	-	-	-	-	-
ENERGY SECTOR	-	-	-	-	-	-	-	-	-	-	-
DISTRIBUTION LOSSES	-	-	-	-	-	-	-	-	-	-	-
FINAL CONSUMPTION	-	**56**	-	-	-	-	-	-	-	-	-
INDUSTRY SECTOR	-	**56**	-	-	-	-	-	-	-	-	-
Iron and Steel	-	-	-	-	-	-	-	-	-	-	-
Chemical and Petrochem.	-	-	-	-	-	-	-	-	-	-	-
Non-Metallic Minerals	-	-	-	-	-	-	-	-	-	-	-
Non-specified	-	56	-	-	-	-	-	-	-	-	-
TRANSPORT SECTOR	-	-	-	-	-	-	-	-	-	-	-
Air	-	-	-	-	-	-	-	-	-	-	-
Road	-	-	-	-	-	-	-	-	-	-	-
Non-specified	-	-	-	-	-	-	-	-	-	-	-
OTHER SECTORS	-	-	-	-	-	-	-	-	-	-	-
Agriculture	-	-	-	-	-	-	-	-	-	-	-
Comm. and Publ. Services	-	-	-	-	-	-	-	-	-	-	-
Residential	-	-	-	-	-	-	-	-	-	-	-
Non-specified	-	-	-	-	-	-	-	-	-	-	-
NON-ENERGY USE	-	-	-	-	-	-	-	-	-	-	-

APPROVISIONNEMENT ET DEMANDE 1997	*Charbon* (1000 tonnes)							*Pétrole* (1000 tonnes)			
	Charbon à coke	*Autres charb. bit.*	*Charbon sous-bit.*	*Lignite*	*Tourbe*	*Coke de four/gaz*	*Agg./briq. de lignite*	*Pétrole brut*	*LGN*	*Produits d'aliment.*	*Additifs*
Production	-	-	-	-	-	-	-	-	-	-	-
Imports	-	57	-	-	-	-	-	2017	-	-	-
Exports	-	-	-	-	-	-	-	-	-	-	-
Intl. Marine Bunkers	-	-	-	-	-	-	-	-	-	-	-
Stock Changes	-	-	-	-	-	-	-	16	-	-	-
DOMESTIC SUPPLY	-	**57**	-	-	-	-	-	**2033**	-	-	-
Transfers and Stat. Diff.	-	-	-	-	-	-	-	-	-	-	-
TRANSFORMATION	-	-	-	-	-	-	-	**2033**	-	-	-
Electricity and CHP Plants	-	-	-	-	-	-	-	-	-	-	-
Petroleum Refineries	-	-	-	-	-	-	-	2033	-	-	-
Other Transform. Sector	-	-	-	-	-	-	-	-	-	-	-
ENERGY SECTOR	-	-	-	-	-	-	-	-	-	-	-
DISTRIBUTION LOSSES	-	-	-	-	-	-	-	-	-	-	-
FINAL CONSUMPTION	-	**57**	-	-	-	-	-	-	-	-	-
INDUSTRY SECTOR	-	**57**	-	-	-	-	-	-	-	-	-
Iron and Steel	-	-	-	-	-	-	-	-	-	-	-
Chemical and Petrochem.	-	-	-	-	-	-	-	-	-	-	-
Non-Metallic Minerals	-	-	-	-	-	-	-	-	-	-	-
Non-specified	-	57	-	-	-	-	-	-	-	-	-
TRANSPORT SECTOR	-	-	-	-	-	-	-	-	-	-	-
Air	-	-	-	-	-	-	-	-	-	-	-
Road	-	-	-	-	-	-	-	-	-	-	-
Non-specified	-	-	-	-	-	-	-	-	-	-	-
OTHER SECTORS	-	-	-	-	-	-	-	-	-	-	-
Agriculture	-	-	-	-	-	-	-	-	-	-	-
Comm. and Publ. Services	-	-	-	-	-	-	-	-	-	-	-
Residential	-	-	-	-	-	-	-	-	-	-	-
Non-specified	-	-	-	-	-	-	-	-	-	-	-
NON-ENERGY USE	-	-	-	-	-	-	-	-	-	-	-

Panama

SUPPLY AND CONSUMPTION 1996	Oil cont. (1000 tonnes)										
	Refinery Gas	LPG + Ethane	Motor Gasoline	Aviation Gasoline	Jet Fuel	Kerosene	Gas/ Diesel	Heavy Fuel Oil	Naphtha	Petrol. Coke	Other Prod.
Production	59	19	255	-	2	5	555	1122	-	-	19
Imports	-	51	109	-	-	-	54	304	-	-	7
Exports	-	-	-39	-	-2	-6	-89	-95	-	-	-
Intl. Marine Bunkers	-	-	-	-	-	-	-60	-1000	-	-	-
Stock Changes	-	-	1	-	3	9	-22	22	-	-	4
DOMESTIC SUPPLY	**59**	**70**	**326**	**-**	**3**	**8**	**438**	**353**	**-**	**-**	**30**
Transfers and Stat. Diff.	-	13	-9	-	-	-	6	-2	-	-	-11
TRANSFORMATION	**-**	**-**	**-**	**-**	**-**	**-**	**67**	**247**	**-**	**-**	**-**
Electricity and CHP Plants	-	-	-	-	-	-	67	247	-	-	-
Petroleum Refineries	-	-	-	-	-	-	-	-	-	-	-
Other Transform. Sector	-	-	-	-	-	-	-	-	-	-	-
ENERGY SECTOR	**59**	**-**	**-**	**-**	**-**	**-**	**2**	**72**	**-**	**-**	**-**
DISTRIBUTION LOSSES	**-**	**-**	**-**	**-**	**-**	**-**	**-**	**-**	**-**	**-**	**-**
FINAL CONSUMPTION	**-**	**83**	**317**	**-**	**3**	**8**	**375**	**32**	**-**	**-**	**19**
INDUSTRY SECTOR	**-**	**2**	**-**	**-**	**-**	**-**	**117**	**31**	**-**	**-**	**-**
Iron and Steel	-	-	-	-	-	-	-	-	-	-	-
Chemical and Petrochem.	-	-	-	-	-	-	-	-	-	-	-
Non-Metallic Minerals	-	-	-	-	-	-	-	-	-	-	-
Non-specified	-	2	-	-	-	-	117	31	-	-	-
TRANSPORT SECTOR	**-**	**-**	**317**	**-**	**3**	**-**	**243**	**-**	**-**	**-**	**-**
Air	-	-	-	-	3	-	-	-	-	-	-
Road	-	-	317	-	-	-	243	-	-	-	-
Non-specified	-	-	-	-	-	-	-	-	-	-	-
OTHER SECTORS	**-**	**81**	**-**	**-**	**-**	**8**	**15**	**1**	**-**	**-**	**-**
Agriculture	-	-	-	-	-	-	-	-	-	-	-
Comm. and Publ. Services	-	19	-	-	-	-	15	1	-	-	-
Residential	-	62	-	-	-	8	-	-	-	-	-
Non-specified	-	-	-	-	-	-	-	-	-	-	-
NON-ENERGY USE	**-**	**-**	**-**	**-**	**-**	**-**	**-**	**-**	**-**	**-**	**19**

APPROVISIONNEMENT ET DEMANDE 1997	*Pétrole cont.* (1000 tonnes)										
	Gaz de raffinerie	*GPL + éthane*	*Essence moteur*	*Essence aviation*	*Carbu-réacteurs*	*Kérosène*	*Gazole*	*Fioul lourd*	*Naphta*	*Coke de pétrole*	*Autres prod.*
Production	42	15	261	-	-	-	569	1086	-	-	47
Imports	-	76	77	-	4	14	135	719	-	-	3
Exports	-	-	-	-	-	-	-167	-397	-	-	-
Intl. Marine Bunkers	-	-	-	-	-	-	-60	-1000	-	-	-
Stock Changes	-	-	2	-	-1	-4	96	-90	-	-	-12
DOMESTIC SUPPLY	**42**	**91**	**340**	**-**	**3**	**10**	**573**	**318**	**-**	**-**	**38**
Transfers and Stat. Diff.	-	-6	-	-	-	-2	-	-	-	-	9
TRANSFORMATION	**-**	**-**	**-**	**-**	**-**	**-**	**150**	**220**	**-**	**-**	**-**
Electricity and CHP Plants	-	-	-	-	-	-	150	220	-	-	-
Petroleum Refineries	-	-	-	-	-	-	-	-	-	-	-
Other Transform. Sector	-	-	-	-	-	-	-	-	-	-	-
ENERGY SECTOR	**42**	**-**	**-**	**-**	**-**	**-**	**2**	**67**	**-**	**-**	**-**
DISTRIBUTION LOSSES	**-**	**-**	**-**	**-**	**-**	**-**	**-**	**-**	**-**	**-**	**-**
FINAL CONSUMPTION	**-**	**85**	**340**	**-**	**3**	**8**	**421**	**31**	**-**	**-**	**47**
INDUSTRY SECTOR	**-**	**2**	**-**	**-**	**-**	**-**	**151**	**30**	**-**	**-**	**-**
Iron and Steel	-	-	-	-	-	-	-	-	-	-	-
Chemical and Petrochem.	-	-	-	-	-	-	-	-	-	-	-
Non-Metallic Minerals	-	-	-	-	-	-	-	-	-	-	-
Non-specified	-	2	-	-	-	-	151	30	-	-	-
TRANSPORT SECTOR	**-**	**-**	**340**	**-**	**3**	**-**	**251**	**-**	**-**	**-**	**-**
Air	-	-	-	-	3	-	-	-	-	-	-
Road	-	-	340	-	-	-	251	-	-	-	-
Non-specified	-	-	-	-	-	-	-	-	-	-	-
OTHER SECTORS	**-**	**83**	**-**	**-**	**-**	**8**	**19**	**1**	**-**	**-**	**-**
Agriculture	-	-	-	-	-	-	-	-	-	-	-
Comm. and Publ. Services	-	19	-	-	-	-	19	1	-	-	-
Residential	-	64	-	-	-	8	-	-	-	-	-
Non-specified	-	-	-	-	-	-	-	-	-	-	-
NON-ENERGY USE	**-**	**-**	**-**	**-**	**-**	**-**	**-**	**-**	**-**	**-**	**47**

Panama

SUPPLY AND CONSUMPTION 1996	Gas (TJ)				Comb. Renew. & Waste (TJ)				(GWh)	(TJ)
	Natural Gas	Gas Works	Coke Ovens	Blast Furnaces	Solid Biomass	Gas/Liquids from Biomass	Municipal Waste	Industrial Waste	Electricity	Heat
Production	-	-	-	-	22962	-	-	-	3915	-
Imports	-	-	-	-	-	-	-	-	95	-
Exports	-	-	-	-	-	-	-	-	-157	-
Intl. Marine Bunkers	-	-	-	-	-	-	-	-	-	-
Stock Changes	-	-	-	-	-	-	-	-	-	-
DOMESTIC SUPPLY	**-**	**-**	**-**	**-**	**22962**	**-**	**-**	**-**	**3853**	**-**
Transfers and Stat. Diff.	-	-	-	-	-	-	-	-	-	-
TRANSFORMATION	**-**	**-**	**-**	**-**	**2138**	**-**	**-**	**-**	**-**	**-**
Electricity and CHP Plants	-	-	-	-	1848	-	-	-	-	-
Petroleum Refineries	-	-	-	-	-	-	-	-	-	-
Other Transform. Sector	-	-	-	-	290	-	-	-	-	-
ENERGY SECTOR	**-**	**-**	**-**	**-**	**-**	**-**	**-**	**-**	**80**	**-**
DISTRIBUTION LOSSES	**-**	**-**	**-**	**-**	**-**	**-**	**-**	**-**	**724**	**-**
FINAL CONSUMPTION	**-**	**-**	**-**	**-**	**20824**	**-**	**-**	**-**	**3049**	**-**
INDUSTRY SECTOR	**-**	**-**	**-**	**-**	**3860**	**-**	**-**	**-**	**642**	**-**
Iron and Steel	-	-	-	-	-	-	-	-	-	-
Chemical and Petrochem.	-	-	-	-	-	-	-	-	-	-
Non-Metallic Minerals	-	-	-	-	-	-	-	-	-	-
Non-specified	-	-	-	-	3860	-	-	-	642	-
TRANSPORT SECTOR	**-**	**-**	**-**	**-**	**-**	**-**	**-**	**-**	**77**	**-**
Air	-	-	-	-	-	-	-	-	-	-
Road	-	-	-	-	-	-	-	-	-	-
Non-specified	-	-	-	-	-	-	-	-	77	-
OTHER SECTORS	**-**	**-**	**-**	**-**	**16964**	**-**	**-**	**-**	**2330**	**-**
Agriculture	-	-	-	-	-	-	-	-	-	-
Comm. and Publ. Services	-	-	-	-	28	-	-	-	-	-
Residential	-	-	-	-	16936	-	-	-	2330	-
Non-specified	-	-	-	-	-	-	-	-	-	-
NON-ENERGY USE	**-**	**-**	**-**	**-**	**-**	**-**	**-**	**-**	**-**	**-**

APPROVISIONNEMENT ET DEMANDE 1997	*Gaz (TJ)*				*En. Re. Comb. & Déchets (TJ)*				(GWh)	(TJ)
	Gaz naturel	*Usines à gaz*	*Cokeries*	*Hauts fourneaux*	*Biomasse solide*	*Gaz/Liquides tirés de biomasse*	*Déchets urbains*	*Déchets industriels*	*Electricité*	*Chaleur*
Production	-	-	-	-	23861	-	-	-	4151	-
Imports	-	-	-	-	-	-	-	-	65	-
Exports	-	-	-	-	-	-	-	-	-118	-
Intl. Marine Bunkers	-	-	-	-	-	-	-	-	-	-
Stock Changes	-	-	-	-	-	-	-	-	-	-
DOMESTIC SUPPLY	**-**	**-**	**-**	**-**	**23861**	**-**	**-**	**-**	**4098**	**-**
Transfers and Stat. Diff.	-	-	-	-	-	-	-	-	-	-
TRANSFORMATION	**-**	**-**	**-**	**-**	**2444**	**-**	**-**	**-**	**-**	**-**
Electricity and CHP Plants	-	-	-	-	2154	-	-	-	-	-
Petroleum Refineries	-	-	-	-	-	-	-	-	-	-
Other Transform. Sector	-	-	-	-	290	-	-	-	-	-
ENERGY SECTOR	**-**	**-**	**-**	**-**	**-**	**-**	**-**	**-**	**52**	**-**
DISTRIBUTION LOSSES	**-**	**-**	**-**	**-**	**-**	**-**	**-**	**-**	**914**	**-**
FINAL CONSUMPTION	**-**	**-**	**-**	**-**	**21416**	**-**	**-**	**-**	**3132**	**-**
INDUSTRY SECTOR	**-**	**-**	**-**	**-**	**4294**	**-**	**-**	**-**	**498**	**-**
Iron and Steel	-	-	-	-	-	-	-	-	-	-
Chemical and Petrochem.	-	-	-	-	-	-	-	-	-	-
Non-Metallic Minerals	-	-	-	-	-	-	-	-	-	-
Non-specified	-	-	-	-	4294	-	-	-	498	-
TRANSPORT SECTOR	**-**	**-**	**-**	**-**	**-**	**-**	**-**	**-**	**93**	**-**
Air	-	-	-	-	-	-	-	-	-	-
Road	-	-	-	-	-	-	-	-	-	-
Non-specified	-	-	-	-	-	-	-	-	93	-
OTHER SECTORS	**-**	**-**	**-**	**-**	**17122**	**-**	**-**	**-**	**2541**	**-**
Agriculture	-	-	-	-	-	-	-	-	-	-
Comm. and Publ. Services	-	-	-	-	28	-	-	-	-	-
Residential	-	-	-	-	17094	-	-	-	2541	-
Non-specified	-	-	-	-	-	-	-	-	-	-
NON-ENERGY USE	**-**	**-**	**-**	**-**	**-**	**-**	**-**	**-**	**-**	**-**

Paraguay

SUPPLY AND CONSUMPTION 1996	Coal (1000 tonnes)							Oil (1000 tonnes)			
	Coking Coal	Other Bit. Coal	Sub-Bit. Coal	Lignite	Peat	Oven and Gas Coke	Pat. Fuel and BKB	Crude Oil	NGL	Feed-stocks	Additives
Production	-	-	-	-	-	-	-	-	-	-	-
Imports	-	-	-	-	-	-	-	158	-	-	-
Exports	-	-	-	-	-	-	-	-	-	-	-
Intl. Marine Bunkers	-	-	-	-	-	-	-	-	-	-	-
Stock Changes	-	-	-	-	-	-	-	-1	-	-	-
DOMESTIC SUPPLY	-	-	-	-	-	-	-	**157**	-	-	-
Transfers and Stat. Diff.	-	-	-	-	-	-	-	-	-	-	-
TRANSFORMATION	-	-	-	-	-	-	-	**157**	-	-	-
Electricity and CHP Plants	-	-	-	-	-	-	-	-	-	-	-
Petroleum Refineries	-	-	-	-	-	-	-	157	-	-	-
Other Transform. Sector	-	-	-	-	-	-	-	-	-	-	-
ENERGY SECTOR	-	-	-	-	-	-	-	-	-	-	-
DISTRIBUTION LOSSES	-	-	-	-	-	-	-	-	-	-	-
FINAL CONSUMPTION	-	-	-	-	-	-	-	-	-	-	-
INDUSTRY SECTOR	-	-	-	-	-	-	-	-	-	-	-
Iron and Steel	-	-	-	-	-	-	-	-	-	-	-
Chemical and Petrochem.	-	-	-	-	-	-	-	-	-	-	-
Non-Metallic Minerals	-	-	-	-	-	-	-	-	-	-	-
Non-specified	-	-	-	-	-	-	-	-	-	-	-
TRANSPORT SECTOR	-	-	-	-	-	-	-	-	-	-	-
Air	-	-	-	-	-	-	-	-	-	-	-
Road	-	-	-	-	-	-	-	-	-	-	-
Non-specified	-	-	-	-	-	-	-	-	-	-	-
OTHER SECTORS	-	-	-	-	-	-	-	-	-	-	-
Agriculture	-	-	-	-	-	-	-	-	-	-	-
Comm. and Publ. Services	-	-	-	-	-	-	-	-	-	-	-
Residential	-	-	-	-	-	-	-	-	-	-	-
Non-specified	-	-	-	-	-	-	-	-	-	-	-
NON-ENERGY USE	-	-	-	-	-	-	-	-	-	-	-

APPROVISIONNEMENT ET DEMANDE 1997	*Charbon* (1000 tonnes)							*Pétrole* (1000 tonnes)			
	Charbon à coke	*Autres charb. bit.*	*Charbon sous-bit.*	*Lignite*	*Tourbe*	*Coke de four/gaz*	*Agg./briq. de lignite*	*Pétrole brut*	*LGN*	*Produits d'aliment.*	*Additifs*
Production	-	-	-	-	-	-	-	-	-	-	-
Imports	-	-	-	-	-	-	-	155	-	-	-
Exports	-	-	-	-	-	-	-	-	-	-	-
Intl. Marine Bunkers	-	-	-	-	-	-	-	-	-	-	-
Stock Changes	-	-	-	-	-	-	-	3	-	-	-
DOMESTIC SUPPLY	-	-	-	-	-	-	-	**158**	-	-	-
Transfers and Stat. Diff.	-	-	-	-	-	-	-	-	-	-	-
TRANSFORMATION	-	-	-	-	-	-	-	**158**	-	-	-
Electricity and CHP Plants	-	-	-	-	-	-	-	-	-	-	-
Petroleum Refineries	-	-	-	-	-	-	-	158	-	-	-
Other Transform. Sector	-	-	-	-	-	-	-	-	-	-	-
ENERGY SECTOR	-	-	-	-	-	-	-	-	-	-	-
DISTRIBUTION LOSSES	-	-	-	-	-	-	-	-	-	-	-
FINAL CONSUMPTION	-	-	-	-	-	-	-	-	-	-	-
INDUSTRY SECTOR	-	-	-	-	-	-	-	-	-	-	-
Iron and Steel	-	-	-	-	-	-	-	-	-	-	-
Chemical and Petrochem.	-	-	-	-	-	-	-	-	-	-	-
Non-Metallic Minerals	-	-	-	-	-	-	-	-	-	-	-
Non-specified	-	-	-	-	-	-	-	-	-	-	-
TRANSPORT SECTOR	-	-	-	-	-	-	-	-	-	-	-
Air	-	-	-	-	-	-	-	-	-	-	-
Road	-	-	-	-	-	-	-	-	-	-	-
Non-specified	-	-	-	-	-	-	-	-	-	-	-
OTHER SECTORS	-	-	-	-	-	-	-	-	-	-	-
Agriculture	-	-	-	-	-	-	-	-	-	-	-
Comm. and Publ. Services	-	-	-	-	-	-	-	-	-	-	-
Residential	-	-	-	-	-	-	-	-	-	-	-
Non-specified	-	-	-	-	-	-	-	-	-	-	-
NON-ENERGY USE	-	-	-	-	-	-	-	-	-	-	-

Paraguay

SUPPLY AND CONSUMPTION 1996	Oil cont. (1000 tonnes)										
	Refinery Gas	LPG + Ethane	Motor Gasoline	Aviation Gasoline	Jet Fuel	Kerosene	Gas/ Diesel	Heavy Fuel Oil	Naphtha	Petrol. Coke	Other Prod.
Production	-	1	26	-	-	6	71	51	-	-	-
Imports	-	66	159	3	-	2	510	-	-	-	12
Exports	-	-	-	-	-	-	-	-18	-	-	-
Intl. Marine Bunkers	-	-	-	-	-	-	-	-	-	-	-
Stock Changes	-	-1	7	-	-	-1	130	-	-	-	-
DOMESTIC SUPPLY	**-**	**66**	**192**	**3**	**-**	**7**	**711**	**33**	**-**	**-**	**12**
Transfers and Stat. Diff.	-	-	-	-	-	-	-	-	-	-	-
TRANSFORMATION	**-**	**-**	**-**	**-**	**-**	**-**	**34**	**-**	**-**	**-**	**-**
Electricity and CHP Plants	-	-	-	-	-	-	34	-	-	-	-
Petroleum Refineries	-	-	-	-	-	-	-	-	-	-	-
Other Transform. Sector	-	-	-	-	-	-	-	-	-	-	-
ENERGY SECTOR	**-**	**-**	**-**	**-**	**-**	**-**	**-**	**-**	**-**	**-**	**-**
DISTRIBUTION LOSSES	**-**	**-**	**-**	**-**	**-**	**-**	**-**	**-**	**-**	**-**	**-**
FINAL CONSUMPTION	**-**	**66**	**192**	**3**	**-**	**7**	**677**	**33**	**-**	**-**	**12**
INDUSTRY SECTOR	**-**	**-**	**-**	**-**	**-**	**-**	**1**	**33**	**-**	**-**	**-**
Iron and Steel	-	-	-	-	-	-	-	-	-	-	-
Chemical and Petrochem.	-	-	-	-	-	-	-	-	-	-	-
Non-Metallic Minerals	-	-	-	-	-	-	-	-	-	-	-
Non-specified	-	-	-	-	-	-	1	33	-	-	-
TRANSPORT SECTOR	**-**	**9**	**192**	**3**	**-**	**-**	**676**	**-**	**-**	**-**	**-**
Air	-	-	-	3	-	-	-	-	-	-	-
Road	-	-	192	-	-	-	676	-	-	-	-
Non-specified	-	9	-	-	-	-	-	-	-	-	-
OTHER SECTORS	**-**	**57**	**-**	**-**	**-**	**7**	**-**	**-**	**-**	**-**	**-**
Agriculture	-	-	-	-	-	-	-	-	-	-	-
Comm. and Publ. Services	-	-	-	-	-	-	-	-	-	-	-
Residential	-	57	-	-	-	7	-	-	-	-	-
Non-specified	-	-	-	-	-	-	-	-	-	-	-
NON-ENERGY USE	**-**	**-**	**-**	**-**	**-**	**-**	**-**	**-**	**-**	**-**	**12**

APPROVISIONNEMENT ET DEMANDE 1997	*Pétrole cont.* (1000 tonnes)										
	Gaz de raffinerie	*GPL + éthane*	*Essence moteur*	*Essence aviation*	*Carbu- réacteurs*	*Kérosène*	*Gazole*	*Fioul lourd*	*Naphta*	*Coke de pétrole*	*Autres prod.*
Production	-	1	27	-	-	5	56	56	-	-	-
Imports	-	64	144	3	-	2	693	-	-	-	12
Exports	-	-	-	-	-	-	-	-19	-	-	-
Intl. Marine Bunkers	-	-	-	-	-	-	-	-	-	-	-
Stock Changes	-	1	-1	-	-	-1	58	-	-	-	-
DOMESTIC SUPPLY	**-**	**66**	**170**	**3**	**-**	**6**	**807**	**37**	**-**	**-**	**12**
Transfers and Stat. Diff.	-	-	-1	-	-	1	-	1	-	-	-
TRANSFORMATION	**-**	**-**	**-**	**-**	**-**	**-**	**34**	**-**	**-**	**-**	**-**
Electricity and CHP Plants	-	-	-	-	-	-	34	-	-	-	-
Petroleum Refineries	-	-	-	-	-	-	-	-	-	-	-
Other Transform. Sector	-	-	-	-	-	-	-	-	-	-	-
ENERGY SECTOR	**-**	**-**	**-**	**-**	**-**	**-**	**-**	**-**	**-**	**-**	**-**
DISTRIBUTION LOSSES	**-**	**-**	**-**	**-**	**-**	**-**	**-**	**-**	**-**	**-**	**-**
FINAL CONSUMPTION	**-**	**66**	**169**	**3**	**-**	**7**	**773**	**38**	**-**	**-**	**12**
INDUSTRY SECTOR	**-**	**-**	**-**	**-**	**-**	**-**	**2**	**38**	**-**	**-**	**-**
Iron and Steel	-	-	-	-	-	-	-	-	-	-	-
Chemical and Petrochem.	-	-	-	-	-	-	-	-	-	-	-
Non-Metallic Minerals	-	-	-	-	-	-	-	-	-	-	-
Non-specified	-	-	-	-	-	-	2	38	-	-	-
TRANSPORT SECTOR	**-**	**10**	**169**	**3**	**-**	**-**	**771**	**-**	**-**	**-**	**-**
Air	-	-	-	3	-	-	-	-	-	-	-
Road	-	-	169	-	-	-	771	-	-	-	-
Non-specified	-	10	-	-	-	-	-	-	-	-	-
OTHER SECTORS	**-**	**56**	**-**	**-**	**-**	**7**	**-**	**-**	**-**	**-**	**-**
Agriculture	-	-	-	-	-	-	-	-	-	-	-
Comm. and Publ. Services	-	-	-	-	-	-	-	-	-	-	-
Residential	-	56	-	-	-	7	-	-	-	-	-
Non-specified	-	-	-	-	-	-	-	-	-	-	-
NON-ENERGY USE	**-**	**-**	**-**	**-**	**-**	**-**	**-**	**-**	**-**	**-**	**12**

Paraguay

SUPPLY AND CONSUMPTION 1996	Gas (TJ)				Comb. Renew. & Waste (TJ)				(GWh)	(TJ)
	Natural Gas	Gas Works	Coke Ovens	Blast Furnaces	Solid Biomass	Gas/Liquids from Biomass	Municipal Waste	Industrial Waste	Electricity	Heat
Production	-	-	-	-	106226	450	-	-	48200	-
Imports	-	-	-	-	-	-	-	-	-	-
Exports	-	-	-	-	-87	-	-	-	-43709	-
Intl. Marine Bunkers	-	-	-	-	-	-	-	-	-	-
Stock Changes	-	-	-	-	-	-	-	-	-	-
DOMESTIC SUPPLY	**-**	**-**	**-**	**-**	**106139**	**450**	**-**	**-**	**4491**	**-**
Transfers and Stat. Diff.	-	-	-	-	-	-	-	-	-2	-
TRANSFORMATION	**-**	**-**	**-**	**-**	**3680**	**-**	**-**	**-**	**-**	**-**
Electricity and CHP Plants	-	-	-	-	705	-	-	-	-	-
Petroleum Refineries	-	-	-	-	-	-	-	-	-	-
Other Transform. Sector	-	-	-	-	2975	-	-	-	-	-
ENERGY SECTOR	**-**	**-**	**-**	**-**	**-**	**-**	**-**	**-**	**185**	**-**
DISTRIBUTION LOSSES	**-**	**-**	**-**	**-**	**-**	**-**	**-**	**-**	**675**	**-**
FINAL CONSUMPTION	**-**	**-**	**-**	**-**	**102459**	**450**	**-**	**-**	**3629**	**-**
INDUSTRY SECTOR	**-**	**-**	**-**	**-**	**53827**	**-**	**-**	**-**	**941**	**-**
Iron and Steel	-	-	-	-	-	-	-	-	-	-
Chemical and Petrochem.	-	-	-	-	-	-	-	-	-	-
Non-Metallic Minerals	-	-	-	-	-	-	-	-	-	-
Non-specified	-	-	-	-	53827	-	-	-	941	-
TRANSPORT SECTOR	**-**	**-**	**-**	**-**	**347**	**450**	**-**	**-**	**-**	**-**
Air	-	-	-	-	-	-	-	-	-	-
Road	-	-	-	-	-	450	-	-	-	-
Non-specified	-	-	-	-	347	-	-	-	-	-
OTHER SECTORS	**-**	**-**	**-**	**-**	**48285**	**-**	**-**	**-**	**2688**	**-**
Agriculture	-	-	-	-	-	-	-	-	-	-
Comm. and Publ. Services	-	-	-	-	181	-	-	-	471	-
Residential	-	-	-	-	48104	-	-	-	2217	-
Non-specified	-	-	-	-	-	-	-	-	-	-
NON-ENERGY USE	**-**	**-**	**-**	**-**	**-**	**-**	**-**	**-**	**-**	**-**

APPROVISIONNEMENT ET DEMANDE 1997	*Gaz (TJ)*				*En. Re. Comb. & Déchets (TJ)*				(GWh)	(TJ)
	Gaz naturel	*Usines à gaz*	*Cokeries*	*Hauts fourneaux*	*Biomasse solide*	*Gaz/Liquides tirés de biomasse*	*Déchets urbains*	*Déchets industriels*	*Electricité*	*Chaleur*
Production	-	-	-	-	109266	477	-	-	50619	-
Imports	-	-	-	-	-	-	-	-	-	-
Exports	-	-	-	-	-87	-	-	-	-45673	-
Intl. Marine Bunkers	-	-	-	-	-	-	-	-	-	-
Stock Changes	-	-	-	-	-	-	-	-	-	-
DOMESTIC SUPPLY	**-**	**-**	**-**	**-**	**109179**	**477**	**-**	**-**	**4946**	**-**
Transfers and Stat. Diff.	-	-	-	-	-	-	-	-	-	-
TRANSFORMATION	**-**	**-**	**-**	**-**	**3550**	**-**	**-**	**-**	**-**	**-**
Electricity and CHP Plants	-	-	-	-	716	-	-	-	-	-
Petroleum Refineries	-	-	-	-	-	-	-	-	-	-
Other Transform. Sector	-	-	-	-	2834	-	-	-	-	-
ENERGY SECTOR	**-**	**-**	**-**	**-**	**-**	**-**	**-**	**-**	**197**	**-**
DISTRIBUTION LOSSES	**-**	**-**	**-**	**-**	**-**	**-**	**-**	**-**	**888**	**-**
FINAL CONSUMPTION	**-**	**-**	**-**	**-**	**105629**	**477**	**-**	**-**	**3861**	**-**
INDUSTRY SECTOR	**-**	**-**	**-**	**-**	**56729**	**-**	**-**	**-**	**1386**	**-**
Iron and Steel	-	-	-	-	-	-	-	-	-	-
Chemical and Petrochem.	-	-	-	-	-	-	-	-	-	-
Non-Metallic Minerals	-	-	-	-	-	-	-	-	-	-
Non-specified	-	-	-	-	56729	-	-	-	1386	-
TRANSPORT SECTOR	**-**	**-**	**-**	**-**	**347**	**477**	**-**	**-**	**-**	**-**
Air	-	-	-	-	-	-	-	-	-	-
Road	-	-	-	-	-	477	-	-	-	-
Non-specified	-	-	-	-	347	-	-	-	-	-
OTHER SECTORS	**-**	**-**	**-**	**-**	**48553**	**-**	**-**	**-**	**2475**	**-**
Agriculture	-	-	-	-	-	-	-	-	-	-
Comm. and Publ. Services	-	-	-	-	181	-	-	-	753	-
Residential	-	-	-	-	48372	-	-	-	1722	-
Non-specified	-	-	-	-	-	-	-	-	-	-
NON-ENERGY USE	**-**	**-**	**-**	**-**	**-**	**-**	**-**	**-**	**-**	**-**

Peru / Pérou

SUPPLY AND CONSUMPTION 1996	Coal (1000 tonnes)							Oil (1000 tonnes)			
	Coking Coal	Other Bit. Coal	Sub-Bit. Coal	Lignite	Peat	Oven and Gas Coke	Pat. Fuel and BKB	Crude Oil	NGL	Feed-stocks	Additives
Production	-	58	-	-	-	-	-	5934	66	-	-
Imports	51	204	-	-	-	231	-	2775	-	-	-
Exports	-	-	-	-	-	-	-	-1841	-	-	-
Intl. Marine Bunkers	-	-	-	-	-	-	-	-	-	-	-
Stock Changes	-	-11	-	-	-	-85	-	50	-	-	-
DOMESTIC SUPPLY	**51**	**251**	**-**	**-**	**-**	**146**	**-**	**6918**	**66**	**-**	**-**
Transfers and Stat. Diff.	-	-5	-	-	-	-	-	550	-50	-	-
TRANSFORMATION	**51**	**-**	**-**	**-**	**-**	**106**	**-**	**7468**	**16**	**-**	**-**
Electricity and CHP Plants	-	-	-	-	-	-	-	-	-	-	-
Petroleum Refineries	-	-	-	-	-	-	-	7468	16	-	-
Other Transform. Sector	51	-	-	-	-	106	-	-	-	-	-
ENERGY SECTOR	**-**	**-**	**-**	**-**	**-**	**-**	**-**	**-**	**-**	**-**	**-**
DISTRIBUTION LOSSES	**-**	**-**	**-**	**-**	**-**	**-**	**-**	**-**	**-**	**-**	**-**
FINAL CONSUMPTION	**-**	**246**	**-**	**-**	**-**	**40**	**-**	**-**	**-**	**-**	**-**
INDUSTRY SECTOR	**-**	**238**	**-**	**-**	**-**	**40**	**-**	**-**	**-**	**-**	**-**
Iron and Steel	-	-	-	-	-	40	-	-	-	-	-
Chemical and Petrochem.	-	-	-	-	-	-	-	-	-	-	-
Non-Metallic Minerals	-	-	-	-	-	-	-	-	-	-	-
Non-specified	-	238	-	-	-	-	-	-	-	-	-
TRANSPORT SECTOR	**-**	**-**	**-**	**-**	**-**	**-**	**-**	**-**	**-**	**-**	**-**
Air	-	-	-	-	-	-	-	-	-	-	-
Road	-	-	-	-	-	-	-	-	-	-	-
Non-specified	-	-	-	-	-	-	-	-	-	-	-
OTHER SECTORS	**-**	**8**	**-**	**-**	**-**	**-**	**-**	**-**	**-**	**-**	**-**
Agriculture	-	-	-	-	-	-	-	-	-	-	-
Comm. and Publ. Services	-	-	-	-	-	-	-	-	-	-	-
Residential	-	8	-	-	-	-	-	-	-	-	-
Non-specified	-	-	-	-	-	-	-	-	-	-	-
NON-ENERGY USE	**-**	**-**	**-**	**-**	**-**	**-**	**-**	**-**	**-**	**-**	**-**

APPROVISIONNEMENT ET DEMANDE 1997	*Charbon* (1000 tonnes)							*Pétrole* (1000 tonnes)			
	Charbon à coke	*Autres charb. bit.*	*Charbon sous-bit.*	*Lignite*	*Tourbe*	*Coke de four/gaz*	*Agg./briq. de lignite*	*Pétrole brut*	*LGN*	*Produits d'aliment.*	*Additifs*
Production	-	41	-	-	-	-	-	5824	-	-	-
Imports	51	153	-	-	-	234	-	3938	-	-	-
Exports	-	-	-	-	-	-	-	-2246	-	-	-
Intl. Marine Bunkers	-	-	-	-	-	-	-	-	-	-	-
Stock Changes	-	-	-	-	-	-28	-	602	-	-	-
DOMESTIC SUPPLY	**51**	**194**	**-**	**-**	**-**	**206**	**-**	**8118**	**-**	**-**	**-**
Transfers and Stat. Diff.	-	-	-	-	-	-4	-	-	-	-	-
TRANSFORMATION	**51**	**-**	**-**	**-**	**-**	**167**	**-**	**8118**	**-**	**-**	**-**
Electricity and CHP Plants	-	-	-	-	-	-	-	-	-	-	-
Petroleum Refineries	-	-	-	-	-	-	-	8118	-	-	-
Other Transform. Sector	51	-	-	-	-	167	-	-	-	-	-
ENERGY SECTOR	**-**	**-**	**-**	**-**	**-**	**-**	**-**	**-**	**-**	**-**	**-**
DISTRIBUTION LOSSES	**-**	**-**	**-**	**-**	**-**	**-**	**-**	**-**	**-**	**-**	**-**
FINAL CONSUMPTION	**-**	**194**	**-**	**-**	**-**	**35**	**-**	**-**	**-**	**-**	**-**
INDUSTRY SECTOR	**-**	**188**	**-**	**-**	**-**	**35**	**-**	**-**	**-**	**-**	**-**
Iron and Steel	-	-	-	-	-	35	-	-	-	-	-
Chemical and Petrochem.	-	-	-	-	-	-	-	-	-	-	-
Non-Metallic Minerals	-	-	-	-	-	-	-	-	-	-	-
Non-specified	-	188	-	-	-	-	-	-	-	-	-
TRANSPORT SECTOR	**-**	**-**	**-**	**-**	**-**	**-**	**-**	**-**	**-**	**-**	**-**
Air	-	-	-	-	-	-	-	-	-	-	-
Road	-	-	-	-	-	-	-	-	-	-	-
Non-specified	-	-	-	-	-	-	-	-	-	-	-
OTHER SECTORS	**-**	**6**	**-**	**-**	**-**	**-**	**-**	**-**	**-**	**-**	**-**
Agriculture	-	-	-	-	-	-	-	-	-	-	-
Comm. and Publ. Services	-	-	-	-	-	-	-	-	-	-	-
Residential	-	6	-	-	-	-	-	-	-	-	-
Non-specified	-	-	-	-	-	-	-	-	-	-	-
NON-ENERGY USE	**-**	**-**	**-**	**-**	**-**	**-**	**-**	**-**	**-**	**-**	**-**

Peru / Pérou

SUPPLY AND CONSUMPTION 1996	Oil cont. (1000 tonnes)										
	Refinery Gas	LPG + Ethane	Motor Gasoline	Aviation Gasoline	Jet Fuel	Kerosene	Gas/ Diesel	Heavy Fuel Oil	Naphtha	Petrol. Coke	Other Prod.
Production	55	155	1293	-	401	717	1564	2928	-	-	35
Imports	-	120	62	-	50	-	1063	-	-	-	20
Exports	-	-	-122	-	-2	-	-	-838	-	-	-
Intl. Marine Bunkers	-	-	-	-	-	-	-1	-	-	-	-
Stock Changes	-	-	-	2	-	-	-	-	-	-	-
DOMESTIC SUPPLY	**55**	**275**	**1233**	**2**	**449**	**717**	**2626**	**2090**	**-**	**-**	**55**
Transfers and Stat. Diff.	-	-7	-39	-	-	1	260	1	-	-	-
TRANSFORMATION	**-**	**-**	**-**	**-**	**-**	**-**	**553**	**474**	**-**	**-**	**-**
Electricity and CHP Plants	-	-	-	-	-	-	553	474	-	-	-
Petroleum Refineries	-	-	-	-	-	-	-	-	-	-	-
Other Transform. Sector	-	-	-	-	-	-	-	-	-	-	-
ENERGY SECTOR	**55**	**-**	**-**	**-**	**-**	**18**	**88**	**124**	**-**	**-**	**1**
DISTRIBUTION LOSSES	**-**	**-**	**-**	**-**	**-**	**-**	**-**	**-**	**-**	**-**	**-**
FINAL CONSUMPTION	**-**	**268**	**1194**	**2**	**449**	**700**	**2245**	**1493**	**-**	**-**	**54**
INDUSTRY SECTOR	**-**	**2**	**2**	**-**	**-**	**26**	**300**	**949**	**-**	**-**	**-**
Iron and Steel	-	-	2	-	-	-	156	-	-	-	-
Chemical and Petrochem.	-	-	-	-	-	-	-	-	-	-	-
Non-Metallic Minerals	-	2	-	-	-	-	-	-	-	-	-
Non-specified	-	-	-	-	-	26	144	949	-	-	-
TRANSPORT SECTOR	**-**	**-**	**1056**	**2**	**449**	**-**	**1624**	**52**	**-**	**-**	**-**
Air	-	-	-	2	449	-	-	-	-	-	-
Road	-	-	1056	-	-	-	1624	-	-	-	-
Non-specified	-	-	-	-	-	-	-	52	-	-	-
OTHER SECTORS	**-**	**266**	**136**	**-**	**-**	**674**	**321**	**492**	**-**	**-**	**-**
Agriculture	-	-	12	-	-	7	217	440	-	-	-
Comm. and Publ. Services	-	-	124	-	-	56	104	52	-	-	-
Residential	-	266	-	-	-	611	-	-	-	-	-
Non-specified	-	-	-	-	-	-	-	-	-	-	-
NON-ENERGY USE	**-**	**-**	**-**	**-**	**-**	**-**	**-**	**-**	**-**	**-**	**54**

APPROVISIONNEMENT ET DEMANDE 1997	*Pétrole cont.* (1000 tonnes)										
	Gaz de raffinerie	*GPL + éthane*	*Essence moteur*	*Essence aviation*	*Carbu- réacteurs*	*Kérosène*	*Gazole*	*Fioul lourd*	*Naphta*	*Coke de pétrole*	*Autres prod.*
Production	60	187	1348	-	309	686	2027	2849	-	-	173
Imports	-	107	-	31	125	-	1075	-	-	-	6
Exports	-	-	-157	-	-6	-	-	-1006	-	-	-20
Intl. Marine Bunkers	-	-	-	-	-	-	-21	-	-	-	-
Stock Changes	-	-	-	-20	-	-	4	79	-	-	-
DOMESTIC SUPPLY	**60**	**294**	**1191**	**11**	**428**	**686**	**3085**	**1922**	**-**	**-**	**159**
Transfers and Stat. Diff.	-	-6	-100	-	1	-	-1	-1	-	-	-7
TRANSFORMATION	**-**	**-**	**-**	**-**	**-**	**-**	**649**	**471**	**-**	**-**	**-**
Electricity and CHP Plants	-	-	-	-	-	-	649	471	-	-	-
Petroleum Refineries	-	-	-	-	-	-	-	-	-	-	-
Other Transform. Sector	-	-	-	-	-	-	-	-	-	-	-
ENERGY SECTOR	**60**	**-**	**-**	**-**	**-**	**18**	**103**	**135**	**-**	**-**	**-**
DISTRIBUTION LOSSES	**-**	**-**	**-**	**-**	**-**	**-**	**-**	**-**	**-**	**-**	**-**
FINAL CONSUMPTION	**-**	**288**	**1091**	**11**	**429**	**668**	**2332**	**1315**	**-**	**-**	**152**
INDUSTRY SECTOR	**-**	**3**	**1**	**-**	**-**	**25**	**317**	**895**	**-**	**-**	**-**
Iron and Steel	-	-	1	-	-	-	162	-	-	-	-
Chemical and Petrochem.	-	-	-	-	-	-	-	-	-	-	-
Non-Metallic Minerals	-	3	-	-	-	-	-	-	-	-	-
Non-specified	-	-	-	-	-	25	155	895	-	-	-
TRANSPORT SECTOR	**-**	**-**	**946**	**11**	**429**	**-**	**1726**	**48**	**-**	**-**	**-**
Air	-	-	-	11	429	-	-	-	-	-	-
Road	-	-	946	-	-	-	1726	-	-	-	-
Non-specified	-	-	-	-	-	-	-	48	-	-	-
OTHER SECTORS	**-**	**285**	**144**	**-**	**-**	**643**	**289**	**372**	**-**	**-**	**-**
Agriculture	-	-	10	-	-	7	201	364	-	-	-
Comm. and Publ. Services	-	-	134	-	-	60	88	8	-	-	-
Residential	-	285	-	-	-	576	-	-	-	-	-
Non-specified	-	-	-	-	-	-	-	-	-	-	-
NON-ENERGY USE	**-**	**-**	**-**	**-**	**-**	**-**	**-**	**-**	**-**	**-**	**152**

Peru / Pérou

SUPPLY AND CONSUMPTION 1996	Gas (TJ)				Comb. Renew. & Waste (TJ)				(GWh)	(TJ)
	Natural Gas	Gas Works	Coke Ovens	Blast Furnaces	Solid Biomass	Gas/Liquids from Biomass	Municipal Waste	Industrial Waste	Electricity	Heat
Production	34030	6238	-	686	176778	-	-	-	17280	-
Imports	-	-	-	-	-	-	-	-	-	-
Exports	-	-	-	-	-	-	-	-	-	-
Intl. Marine Bunkers	-	-	-	-	-	-	-	-	-	-
Stock Changes	-	-	-	-	-	-	-	-	-	-
DOMESTIC SUPPLY	**34030**	**6238**	**-**	**686**	**176778**	**-**	**-**	**-**	**17280**	**-**
Transfers and Stat. Diff.	-	-	-	-	-	-	-	-	1	-
TRANSFORMATION	**34030**	**3022**	**-**	**-**	**9998**	**-**	**-**	**-**	**-**	**-**
Electricity and CHP Plants	-	3022	-	-	1644	-	-	-	-	-
Petroleum Refineries	-	-	-	-	-	-	-	-	-	-
Other Transform. Sector	34030	-	-	-	8354	-	-	-	-	-
ENERGY SECTOR	**-**	**703**	**-**	**-**	**-**	**-**	**-**	**-**	**299**	**-**
DISTRIBUTION LOSSES	**-**	**-**	**-**	**-**	**-**	**-**	**-**	**-**	**2668**	**-**
FINAL CONSUMPTION	**-**	**2513**	**-**	**686**	**166780**	**-**	**-**	**-**	**14314**	**-**
INDUSTRY SECTOR	**-**	**1074**	**-**	**686**	**20197**	**-**	**-**	**-**	**5965**	**-**
Iron and Steel	-	-	-	686	-	-	-	-	-	-
Chemical and Petrochem.	-	-	-	-	-	-	-	-	-	-
Non-Metallic Minerals	-	-	-	-	-	-	-	-	-	-
Non-specified	-	1074	-	-	20197	-	-	-	5965	-
TRANSPORT SECTOR	**-**	**-**	**-**	**-**	**-**	**-**	**-**	**-**	**-**	**-**
Air	-	-	-	-	-	-	-	-	-	-
Road	-	-	-	-	-	-	-	-	-	-
Non-specified	-	-	-	-	-	-	-	-	-	-
OTHER SECTORS	**-**	**1439**	**-**	**-**	**146583**	**-**	**-**	**-**	**8349**	**-**
Agriculture	-	-	-	-	4445	-	-	-	1213	-
Comm. and Publ. Services	-	-	-	-	-	-	-	-	2723	-
Residential	-	1439	-	-	142138	-	-	-	4413	-
Non-specified	-	-	-	-	-	-	-	-	-	-
NON-ENERGY USE	**-**	**-**	**-**	**-**	**-**	**-**	**-**	**-**	**-**	**-**

APPROVISIONNEMENT ET DEMANDE 1997	*Gaz* (TJ)				*En. Re. Comb. & Déchets* (TJ)				(GWh)	(TJ)
	Gaz naturel	*Usines à gaz*	*Cokeries*	*Hauts fourneaux*	*Biomasse solide*	*Gaz/Liquides tirés de biomasse*	*Déchets urbains*	*Déchets industriels*	*Electricité*	*Chaleur*
Production	35759	14696	-	1081	179950	-	-	-	17954	-
Imports	-	-	-	-	-	-	-	-	2	-
Exports	-	-	-	-	-	-	-	-	-	-
Intl. Marine Bunkers	-	-	-	-	-	-	-	-	-	-
Stock Changes	-	-	-	-	-	-	-	-	-	-
DOMESTIC SUPPLY	**35759**	**14696**	**-**	**1081**	**179950**	**-**	**-**	**-**	**17956**	**-**
Transfers and Stat. Diff.	-	-	-	-	-	-	-	-	2	-
TRANSFORMATION	**35759**	**4354**	**-**	**-**	**10303**	**-**	**-**	**-**	**-**	**-**
Electricity and CHP Plants	-	4354	-	-	1853	-	-	-	-	-
Petroleum Refineries	-	-	-	-	-	-	-	-	-	-
Other Transform. Sector	35759	-	-	-	8450	-	-	-	-	-
ENERGY SECTOR	**-**	**1656**	**-**	**-**	**-**	**-**	**-**	**-**	**280**	**-**
DISTRIBUTION LOSSES	**-**	**-**	**-**	**-**	**-**	**-**	**-**	**-**	**2885**	**-**
FINAL CONSUMPTION	**-**	**8686**	**-**	**1081**	**169648**	**-**	**-**	**-**	**14793**	**-**
INDUSTRY SECTOR	**-**	**5296**	**-**	**1081**	**20468**	**-**	**-**	**-**	**8536**	**-**
Iron and Steel	-	-	-	1081	-	-	-	-	3756	-
Chemical and Petrochem.	-	-	-	-	-	-	-	-	-	-
Non-Metallic Minerals	-	-	-	-	-	-	-	-	-	-
Non-specified	-	5296	-	-	20468	-	-	-	4780	-
TRANSPORT SECTOR	**-**	**-**	**-**	**-**	**-**	**-**	**-**	**-**	**-**	**-**
Air	-	-	-	-	-	-	-	-	-	-
Road	-	-	-	-	-	-	-	-	-	-
Non-specified	-	-	-	-	-	-	-	-	-	-
OTHER SECTORS	**-**	**3390**	**-**	**-**	**149180**	**-**	**-**	**-**	**6257**	**-**
Agriculture	-	-	-	-	5243	-	-	-	582	-
Comm. and Publ. Services	-	-	-	-	-	-	-	-	717	-
Residential	-	3390	-	-	143937	-	-	-	4958	-
Non-specified	-	-	-	-	-	-	-	-	-	-
NON-ENERGY USE	**-**	**-**	**-**	**-**	**-**	**-**	**-**	**-**	**-**	**-**

Philippines : 1996

SUPPLY AND CONSUMPTION *APPROVISIONNEMENT ET DEMANDE*	Coal / *Charbon* (1000 tonnes)							Oil / *Pétrole* (1000 tonnes)			
	Coking Coal *Charbon à coke*	Other Bit. Coal *Autres charb. bit.*	Sub-Bit. Coal *Charbon sous-bit.*	Lignite *Lignite*	Peat *Tourbe*	Oven and Gas Coke *Coke de four/gaz*	Pat. Fuel and BKB *Agg./briq. de lignite*	Crude Oil *Pétrole brut*	NGL *LGN*	Feed-stocks *Produits d'aliment.*	Additives *Additifs*
Production	-	976	-	-	-	-	-	45	-	-	-
From Other Sources	-	-	-	-	-	-	-	-	-	-	-
Imports	-	2984	-	-	-	185	-	17561	-	-	-
Exports	-	-	-	-	-	-	-	-	-	-	-
Intl. Marine Bunkers	-	-	-	-	-	-	-	-	-	-	-
Stock Changes	-	-	-	-	-	-	-	22	-	-	-
DOMESTIC SUPPLY	**-**	**3960**	**-**	**-**	**-**	**185**	**-**	**17628**	**-**	**-**	**-**
Transfers	-	-	-	-	-	-	-	-	-	-	-
Statistical Differences	-	1	-	-	-	-	-	-	-	-	-
TRANSFORMATION	**-**	**1652**	**-**	**-**	**-**	**148**	**-**	**17628**	**-**	**-**	**-**
Electricity Plants	-	1652	-	-	-	-	-	-	-	-	-
CHP Plants	-	-	-	-	-	-	-	-	-	-	-
Heat Plants	-	-	-	-	-	-	-	-	-	-	-
Blast Furnaces/Gas Works	-	-	-	-	-	148	-	-	-	-	-
Coke/Pat. Fuel/BKB Plants	-	-	-	-	-	-	-	-	-	-	-
Petroleum Refineries	-	-	-	-	-	-	-	17628	-	-	-
Petrochemical Industry	-	-	-	-	-	-	-	-	-	-	-
Liquefaction	-	-	-	-	-	-	-	-	-	-	-
Other Transform. Sector	-	-	-	-	-	-	-	-	-	-	-
ENERGY SECTOR	**-**	**639**	**-**	**-**	**-**	**-**	**-**	**-**	**-**	**-**	**-**
Coal Mines	-	639	-	-	-	-	-	-	-	-	-
Oil and Gas Extraction	-	-	-	-	-	-	-	-	-	-	-
Petroleum Refineries	-	-	-	-	-	-	-	-	-	-	-
Electr., CHP+Heat Plants	-	-	-	-	-	-	-	-	-	-	-
Pumped Storage (Elec.)	-	-	-	-	-	-	-	-	-	-	-
Other Energy Sector	-	-	-	-	-	-	-	-	-	-	-
Distribution Losses	-	-	-	-	-	-	-	-	-	-	-
FINAL CONSUMPTION	**-**	**1670**	**-**	**-**	**-**	**37**	**-**	**-**	**-**	**-**	**-**
INDUSTRY SECTOR	**-**	**1670**	**-**	**-**	**-**	**37**	**-**	**-**	**-**	**-**	**-**
Iron and Steel	-	28	-	-	-	37	-	-	-	-	-
Chemical and Petrochem.	-	48	-	-	-	-	-	-	-	-	-
of which: Feedstocks	-	-	-	-	-	-	-	-	-	-	-
Non-Ferrous Metals	-	-	-	-	-	-	-	-	-	-	-
Non-Metallic Minerals	-	1500	-	-	-	-	-	-	-	-	-
Transport Equipment	-	-	-	-	-	-	-	-	-	-	-
Machinery	-	-	-	-	-	-	-	-	-	-	-
Mining and Quarrying	-	-	-	-	-	-	-	-	-	-	-
Food and Tobacco	-	94	-	-	-	-	-	-	-	-	-
Paper, Pulp and Print	-	-	-	-	-	-	-	-	-	-	-
Wood and Wood Products	-	-	-	-	-	-	-	-	-	-	-
Construction	-	-	-	-	-	-	-	-	-	-	-
Textile and Leather	-	-	-	-	-	-	-	-	-	-	-
Non-specified	-	-	-	-	-	-	-	-	-	-	-
TRANSPORT SECTOR	**-**	**-**	**-**	**-**	**-**	**-**	**-**	**-**	**-**	**-**	**-**
Air	-	-	-	-	-	-	-	-	-	-	-
Road	-	-	-	-	-	-	-	-	-	-	-
Rail	-	-	-	-	-	-	-	-	-	-	-
Pipeline Transport	-	-	-	-	-	-	-	-	-	-	-
Internal Navigation	-	-	-	-	-	-	-	-	-	-	-
Non-specified	-	-	-	-	-	-	-	-	-	-	-
OTHER SECTORS	**-**	**-**	**-**	**-**	**-**	**-**	**-**	**-**	**-**	**-**	**-**
Agriculture	-	-	-	-	-	-	-	-	-	-	-
Comm. and Publ. Services	-	-	-	-	-	-	-	-	-	-	-
Residential	-	-	-	-	-	-	-	-	-	-	-
Non-specified	-	-	-	-	-	-	-	-	-	-	-
NON-ENERGY USE	**-**	**-**	**-**	**-**	**-**	**-**	**-**	**-**	**-**	**-**	**-**
in Industry/Trans./Energy	-	-	-	-	-	-	-	-	-	-	-
in Transport	-	-	-	-	-	-	-	-	-	-	-
in Other Sectors	-	-	-	-	-	-	-	-	-	-	-

Philippines : 1996

SUPPLY AND CONSUMPTION / *APPROVISIONNEMENT ET DEMANDE*	Oil cont. / *Pétrole cont.* (1000 tonnes)										
	Refinery Gas / *Gaz de raffinerie*	LPG + Ethane / *GPL + éthane*	Motor Gasoline / *Essence moteur*	Aviation Gasoline / *Essence aviation*	Jet Fuel / *Carbu- réacteurs*	Kerosene / *Kérosène*	Gas/ Diesel / *Gazole*	Heavy Fuel Oil / *Fioul lourd*	Naphtha / *Naphta*	Petrol. Coke / *Coke de pétrole*	Other Prod. / *Autres prod.*
Production	510	390	2037	-	787	599	5136	6955	710	-	349
From Other Sources	-	-	-	-	-	-	-	-	-	-	-
Imports	-	337	27	4	20	1	444	417	-	-	12
Exports	-	-	-27	-	-45	-	-220	-432	-636	-	-
Intl. Marine Bunkers	-	-	-	-	-	-	-23	-50	-	-	-
Stock Changes	-	74	296	-	27	-10	52	-109	-27	-	-1
DOMESTIC SUPPLY	**510**	**801**	**2333**	**4**	**789**	**590**	**5389**	**6781**	**47**	**-**	**360**
Transfers	-	-	-	-	-	-	-	-	-	-	-
Statistical Differences	-	-1	1	-	-	-	542	-1021	-	-	-1
TRANSFORMATION	**-**	**-**	**-**	**-**	**-**	**-**	**2286**	**2490**	**-**	**-**	**-**
Electricity Plants	-	-	-	-	-	-	2286	2482	-	-	-
CHP Plants	-	-	-	-	-	-	-	-	-	-	-
Heat Plants	-	-	-	-	-	-	-	-	-	-	-
Blast Furnaces/Gas Works	-	-	-	-	-	-	-	8	-	-	-
Coke/Pat. Fuel/BKB Plants	-	-	-	-	-	-	-	-	-	-	-
Petroleum Refineries	-	-	-	-	-	-	-	-	-	-	-
Petrochemical Industry	-	-	-	-	-	-	-	-	-	-	-
Liquefaction	-	-	-	-	-	-	-	-	-	-	-
Other Transform. Sector	-	-	-	-	-	-	-	-	-	-	-
ENERGY SECTOR	**510**	**-**	**-**	**-**	**-**	**27**	**267**	**395**	**47**	**-**	**-**
Coal Mines	-	-	-	-	-	27	267	237	-	-	-
Oil and Gas Extraction	-	-	-	-	-	-	-	-	-	-	-
Petroleum Refineries	510	-	-	-	-	-	-	158	47	-	-
Electr., CHP+Heat Plants	-	-	-	-	-	-	-	-	-	-	-
Pumped Storage (Elec.)	-	-	-	-	-	-	-	-	-	-	-
Other Energy Sector	-	-	-	-	-	-	-	-	-	-	-
Distribution Losses	-	-	-	-	-	-	-	-	-	-	-
FINAL CONSUMPTION	**-**	**800**	**2334**	**4**	**789**	**563**	**3378**	**2875**	**-**	**-**	**359**
INDUSTRY SECTOR	**-**	**300**	**-**	**-**	**-**	**45**	**767**	**2242**	**-**	**-**	**-**
Iron and Steel	-	-	-	-	-	21	44	337	-	-	-
Chemical and Petrochem.	-	-	-	-	-	1	5	60	-	-	-
of which: Feedstocks	-	-	-	-	-	-	-	-	-	-	-
Non-Ferrous Metals	-	-	-	-	-	-	44	-	-	-	-
Non-Metallic Minerals	-	-	-	-	-	-	47	336	-	-	-
Transport Equipment	-	-	-	-	-	-	-	-	-	-	-
Machinery	-	-	-	-	-	-	-	-	-	-	-
Mining and Quarrying	-	-	-	-	-	-	-	-	-	-	-
Food and Tobacco	-	-	-	-	-	-	259	158	-	-	-
Paper, Pulp and Print	-	-	-	-	-	-	26	255	-	-	-
Wood and Wood Products	-	-	-	-	-	-	146	10	-	-	-
Construction	-	-	-	-	-	-	112	19	-	-	-
Textile and Leather	-	-	-	-	-	-	21	113	-	-	-
Non-specified	-	300	-	-	-	23	63	954	-	-	-
TRANSPORT SECTOR	**-**	**-**	**2334**	**4**	**789**	**5**	**719**	**50**	**-**	**-**	**-**
Air	-	-	-	4	789	-	-	-	-	-	-
Road	-	-	2334	-	-	-	310	-	-	-	-
Rail	-	-	-	-	-	-	-	-	-	-	-
Pipeline Transport	-	-	-	-	-	-	-	-	-	-	-
Internal Navigation	-	-	-	-	-	5	409	50	-	-	-
Non-specified	-	-	-	-	-	-	-	-	-	-	-
OTHER SECTORS	**-**	**500**	**-**	**-**	**-**	**513**	**1892**	**583**	**-**	**-**	**-**
Agriculture	-	-	-	-	-	8	1437	373	-	-	-
Comm. and Publ. Services	-	-	-	-	-	-	-	210	-	-	-
Residential	-	500	-	-	-	502	455	-	-	-	-
Non-specified	-	-	-	-	-	3	-	-	-	-	-
NON-ENERGY USE	**-**	**-**	**-**	**-**	**-**	**-**	**-**	**-**	**-**	**-**	**359**
in Industry/Transf./Energy	-	-	-	-	-	-	-	-	-	-	359
in Transport	-	-	-	-	-	-	-	-	-	-	-
in Other Sectors	-	-	-	-	-	-	-	-	-	-	-

Philippines : 1996

	Gas / *Gaz* (TJ)				Comb. Renew. & Waste / *En. Re. Comb. & Déchets* (TJ)				(GWh)	(TJ)
SUPPLY AND CONSUMPTION	Natural Gas	Gas Works	Coke Ovens	Blast Furnaces	Solid Biomass	Gas/Liquids from Biomass	Municipal Waste	Industrial Waste	Electricity	Heat
APPROVISIONNEMENT ET DEMANDE	*Gaz naturel*	*Usines à gaz*	*Cokeries*	*Hauts fourneaux*	*Biomasse solide*	*Gaz/Liquides tirés de biomasse*	*Déchets urbains*	*Déchets industriels*	*Electricité*	*Chaleur*
Production	406	202	-	-	389881	-	-	-	36663	-
From Other Sources	-	-	-	-	-	-	-	-	-	-
Imports	-	-	-	-	-	-	-	-	-	-
Exports	-	-	-	-	-	-	-	-	-	-
Intl. Marine Bunkers	-	-	-	-	-	-	-	-	-	-
Stock Changes	-	-	-	-	-	-	-	-	-	-
DOMESTIC SUPPLY	**406**	**202**	**-**	**-**	**389881**	**-**	**-**	**-**	**36663**	**-**
Transfers	-	-	-	-	-	-	-	-	-	-
Statistical Differences	-	-	-	-	-	-	-	-	-1	-
TRANSFORMATION	**406**	**-**	**-**	**-**	**38361**	**-**	**-**	**-**	**-**	**-**
Electricity Plants	406	-	-	-	-	-	-	-	-	-
CHP Plants	-	-	-	-	-	-	-	-	-	-
Heat Plants	-	-	-	-	-	-	-	-	-	-
Blast Furnaces/Gas Works	-	-	-	-	-	-	-	-	-	-
Coke/Pat. Fuel/BKB Plants	-	-	-	-	-	-	-	-	-	-
Petroleum Refineries	-	-	-	-	-	-	-	-	-	-
Petrochemical Industry	-	-	-	-	-	-	-	-	-	-
Liquefaction	-	-	-	-	-	-	-	-	-	-
Other Transform. Sector	-	-	-	-	38361	-	-	-	-	-
ENERGY SECTOR	**-**	**-**	**-**	**-**	**-**	**-**	**-**	**-**	**1340**	**-**
Coal Mines	-	-	-	-	-	-	-	-	-	-
Oil and Gas Extraction	-	-	-	-	-	-	-	-	-	-
Petroleum Refineries	-	-	-	-	-	-	-	-	-	-
Electr., CHP+Heat Plants	-	-	-	-	-	-	-	-	1340	-
Pumped Storage (Elec.)	-	-	-	-	-	-	-	-	-	-
Other Energy Sector	-	-	-	-	-	-	-	-	-	-
Distribution Losses	-	63	-	-	-	-	-	-	6128	-
FINAL CONSUMPTION	**-**	**139**	**-**	**-**	**351520**	**-**	**-**	**-**	**29194**	**-**
INDUSTRY SECTOR	**-**	**1**	**-**	**-**	**151094**	**-**	**-**	**-**	**11850**	**-**
Iron and Steel	-	-	-	-	-	-	-	-	883	-
Chemical and Petrochem.	-	-	-	-	-	-	-	-	1371	-
of which: Feedstocks	-	-	-	-	-	-	-	-	-	-
Non-Ferrous Metals	-	-	-	-	-	-	-	-	-	-
Non-Metallic Minerals	-	-	-	-	-	-	-	-	-	-
Transport Equipment	-	-	-	-	-	-	-	-	-	-
Machinery	-	-	-	-	-	-	-	-	-	-
Mining and Quarrying	-	-	-	-	-	-	-	-	-	-
Food and Tobacco	-	-	-	-	-	-	-	-	-	-
Paper, Pulp and Print	-	-	-	-	-	-	-	-	-	-
Wood and Wood Products	-	-	-	-	-	-	-	-	-	-
Construction	-	-	-	-	-	-	-	-	-	-
Textile and Leather	-	-	-	-	-	-	-	-	-	-
Non-specified	-	1	-	-	151094	-	-	-	9596	-
TRANSPORT SECTOR	**-**	**-**	**-**	**-**	**-**	**-**	**-**	**-**	**-**	**-**
Air	-	-	-	-	-	-	-	-	-	-
Road	-	-	-	-	-	-	-	-	-	-
Rail	-	-	-	-	-	-	-	-	-	-
Pipeline Transport	-	-	-	-	-	-	-	-	-	-
Internal Navigation	-	-	-	-	-	-	-	-	-	-
Non-specified	-	-	-	-	-	-	-	-	-	-
OTHER SECTORS	**-**	**138**	**-**	**-**	**200426**	**-**	**-**	**-**	**17344**	**-**
Agriculture	-	-	-	-	-	-	-	-	-	-
Comm. and Publ. Services	-	-	-	-	-	-	-	-	7072	-
Residential	-	37	-	-	186219	-	-	-	9105	-
Non-specified	-	101	-	-	14207	-	-	-	1167	-
NON-ENERGY USE	**-**	**-**	**-**	**-**	**-**	**-**	**-**	**-**	**-**	**-**
in Industry/Transf./Energy	-	-	-	-	-	-	-	-	-	-
in Transport	-	-	-	-	-	-	-	-	-	-
in Other Sectors	-	-	-	-	-	-	-	-	-	-

Philippines : 1997

SUPPLY AND CONSUMPTION / *APPROVISIONNEMENT ET DEMANDE*	Coal / *Charbon* (1000 tonnes)							Oil / *Pétrole* (1000 tonnes)			
	Coking Coal / *Charbon à coke*	Other Bit. Coal / *Autres charb. bit.*	Sub-Bit. Coal / *Charbon sous-bit.*	Lignite / *Lignite*	Peat / *Tourbe*	Oven and Gas Coke / *Coke de four/gaz*	Pat. Fuel and BKB / *Agg./briq. de lignite*	Crude Oil / *Pétrole brut*	NGL / *LGN*	Feed-stocks / *Produits d'aliment.*	Additives / *Additifs*
Production	-	930	-	-	-	-	-	16	-	-	-
From Other Sources	-	-	-	-	-	-	-	-	-	-	-
Imports	-	4248	-	-	-	185	-	17687	-	-	-
Exports	-	-	-	-	-	-	-	-1	-	-	-
Intl. Marine Bunkers	-	-	-	-	-	-	-	-	-	-	-
Stock Changes	-	-	-	-	-	-	-	254	-	-	-
DOMESTIC SUPPLY	**-**	**5178**	**-**	**-**	**-**	**185**	**-**	**17956**	**-**	**-**	**-**
Transfers	-	-	-	-	-	-	-	-	-	-	-
Statistical Differences	-	1	-	-	-	-	-	-	-	-	-
TRANSFORMATION	**-**	**2161**	**-**	**-**	**-**	**148**	**-**	**17956**	**-**	**-**	**-**
Electricity Plants	-	2161	-	-	-	-	-	-	-	-	-
CHP Plants	-	-	-	-	-	-	-	-	-	-	-
Heat Plants	-	-	-	-	-	-	-	-	-	-	-
Blast Furnaces/Gas Works	-	-	-	-	-	148	-	-	-	-	-
Coke/Pat. Fuel/BKB Plants	-	-	-	-	-	-	-	-	-	-	-
Petroleum Refineries	-	-	-	-	-	-	-	17956	-	-	-
Petrochemical Industry	-	-	-	-	-	-	-	-	-	-	-
Liquefaction	-	-	-	-	-	-	-	-	-	-	-
Other Transform. Sector	-	-	-	-	-	-	-	-	-	-	-
ENERGY SECTOR	**-**	**835**	**-**	**-**	**-**	**-**	**-**	**-**	**-**	**-**	**-**
Coal Mines	-	835	-	-	-	-	-	-	-	-	-
Oil and Gas Extraction	-	-	-	-	-	-	-	-	-	-	-
Petroleum Refineries	-	-	-	-	-	-	-	-	-	-	-
Electr., CHP+Heat Plants	-	-	-	-	-	-	-	-	-	-	-
Pumped Storage (Elec.)	-	-	-	-	-	-	-	-	-	-	-
Other Energy Sector	-	-	-	-	-	-	-	-	-	-	-
Distribution Losses	-	-	-	-	-	-	-	-	-	-	-
FINAL CONSUMPTION	**-**	**2183**	**-**	**-**	**-**	**37**	**-**	**-**	**-**	**-**	**-**
INDUSTRY SECTOR	**-**	**2183**	**-**	**-**	**-**	**37**	**-**	**-**	**-**	**-**	**-**
Iron and Steel	-	36	-	-	-	37	-	-	-	-	-
Chemical and Petrochem.	-	63	-	-	-	-	-	-	-	-	-
of which: Feedstocks	-	-	-	-	-	-	-	-	-	-	-
Non-Ferrous Metals	-	-	-	-	-	-	-	-	-	-	-
Non-Metallic Minerals	-	1961	-	-	-	-	-	-	-	-	-
Transport Equipment	-	-	-	-	-	-	-	-	-	-	-
Machinery	-	-	-	-	-	-	-	-	-	-	-
Mining and Quarrying	-	-	-	-	-	-	-	-	-	-	-
Food and Tobacco	-	123	-	-	-	-	-	-	-	-	-
Paper, Pulp and Print	-	-	-	-	-	-	-	-	-	-	-
Wood and Wood Products	-	-	-	-	-	-	-	-	-	-	-
Construction	-	-	-	-	-	-	-	-	-	-	-
Textile and Leather	-	-	-	-	-	-	-	-	-	-	-
Non-specified	-	-	-	-	-	-	-	-	-	-	-
TRANSPORT SECTOR	**-**	**-**	**-**	**-**	**-**	**-**	**-**	**-**	**-**	**-**	**-**
Air	-	-	-	-	-	-	-	-	-	-	-
Road	-	-	-	-	-	-	-	-	-	-	-
Rail	-	-	-	-	-	-	-	-	-	-	-
Pipeline Transport	-	-	-	-	-	-	-	-	-	-	-
Internal Navigation	-	-	-	-	-	-	-	-	-	-	-
Non-specified	-	-	-	-	-	-	-	-	-	-	-
OTHER SECTORS	**-**	**-**	**-**	**-**	**-**	**-**	**-**	**-**	**-**	**-**	**-**
Agriculture	-	-	-	-	-	-	-	-	-	-	-
Comm. and Publ. Services	-	-	-	-	-	-	-	-	-	-	-
Residential	-	-	-	-	-	-	-	-	-	-	-
Non-specified	-	-	-	-	-	-	-	-	-	-	-
NON-ENERGY USE	**-**	**-**	**-**	**-**	**-**	**-**	**-**	**-**	**-**	**-**	**-**
in Industry/Trans./Energy	-	-	-	-	-	-	-	-	-	-	-
in Transport	-	-	-	-	-	-	-	-	-	-	-
in Other Sectors	-	-	-	-	-	-	-	-	-	-	-

Philippines : 1997

SUPPLY AND CONSUMPTION / *APPROVISIONNEMENT ET DEMANDE*	Oil cont. / *Pétrole cont.* (1000 tonnes)										
	Refinery Gas / *Gaz de raffinerie*	LPG + Ethane / *GPL + éthane*	Motor Gasoline / *Essence moteur*	Aviation Gasoline / *Essence aviation*	Jet Fuel / *Carbu-réacteurs*	Kerosene / *Kérosène*	Gas/ Diesel / *Gazole*	Heavy Fuel Oil / *Fioul lourd*	Naphtha / *Naphta*	Petrol. Coke / *Coke de pétrole*	Other Prod. / *Autres prod.*
Production	491	456	2199	-	810	582	5385	6849	666	-	254
From Other Sources	-	-	-	-	-	-	-	-	-	-	-
Imports	-	443	337	4	80	99	537	381	-	-	108
Exports	-	-	-12	-	-	-17	-55	-610	-615	-	-9
Intl. Marine Bunkers	-	-	-	-	-	-	-43	-51	-	-	-
Stock Changes	-	-86	2	-	105	-11	95	468	-10	-	3
DOMESTIC SUPPLY	**491**	**813**	**2526**	**4**	**995**	**653**	**5919**	**7037**	**41**	**-**	**356**
Transfers	-	-	-	-	-	-	-	-	-	-	-
Statistical Differences	-	-	-	-	-1	-	598	-1058	-	-	-3
TRANSFORMATION	**-**	**-**	**-**	**-**	**-**	**-**	**2510**	**2584**	**-**	**-**	**-**
Electricity Plants	-	-	-	-	-	-	2510	2576	-	-	-
CHP Plants	-	-	-	-	-	-	-	-	-	-	-
Heat Plants	-	-	-	-	-	-	-	-	-	-	-
Blast Furnaces/Gas Works	-	-	-	-	-	-	-	8	-	-	-
Coke/Pat. Fuel/BKB Plants	-	-	-	-	-	-	-	-	-	-	-
Petroleum Refineries	-	-	-	-	-	-	-	-	-	-	-
Petrochemical Industry	-	-	-	-	-	-	-	-	-	-	-
Liquefaction	-	-	-	-	-	-	-	-	-	-	-
Other Transform. Sector	-	-	-	-	-	-	-	-	-	-	-
ENERGY SECTOR	**491**	**-**	**-**	**-**	**-**	**29**	**293**	**410**	**41**	**-**	**-**
Coal Mines	-	-	-	-	-	29	293	246	-	-	-
Oil and Gas Extraction	-	-	-	-	-	-	-	-	-	-	-
Petroleum Refineries	491	-	-	-	-	-	-	164	41	-	-
Electr., CHP+Heat Plants	-	-	-	-	-	-	-	-	-	-	-
Pumped Storage (Elec.)	-	-	-	-	-	-	-	-	-	-	-
Other Energy Sector	-	-	-	-	-	-	-	-	-	-	-
Distribution Losses	-	-	-	-	-	-	-	-	-	-	-
FINAL CONSUMPTION	**-**	**813**	**2526**	**4**	**994**	**624**	**3714**	**2985**	**-**	**-**	**353**
INDUSTRY SECTOR	**-**	**305**	**-**	**-**	**-**	**50**	**846**	**2328**	**-**	**-**	**-**
Iron and Steel	-	-	-	-	-	24	49	350	-	-	-
Chemical and Petrochem.	-	-	-	-	-	1	6	63	-	-	-
of which: Feedstocks	-	-	-	-	-	-	-	-	-	-	-
Non-Ferrous Metals	-	-	-	-	-	-	49	-	-	-	-
Non-Metallic Minerals	-	-	-	-	-	-	52	349	-	-	-
Transport Equipment	-	-	-	-	-	-	-	-	-	-	-
Machinery	-	-	-	-	-	-	-	-	-	-	-
Mining and Quarrying	-	-	-	-	-	-	-	-	-	-	-
Food and Tobacco	-	-	-	-	-	-	284	164	-	-	-
Paper, Pulp and Print	-	-	-	-	-	-	29	265	-	-	-
Wood and Wood Products	-	-	-	-	-	-	161	10	-	-	-
Construction	-	-	-	-	-	-	124	20	-	-	-
Textile and Leather	-	-	-	-	-	-	23	117	-	-	-
Non-specified	-	305	-	-	-	25	69	990	-	-	-
TRANSPORT SECTOR	**-**	**-**	**2526**	**4**	**994**	**6**	**790**	**52**	**-**	**-**	**-**
Air	-	-	-	4	994	-	-	-	-	-	-
Road	-	-	2526	-	-	-	341	-	-	-	-
Rail	-	-	-	-	-	-	-	-	-	-	-
Pipeline Transport	-	-	-	-	-	-	-	-	-	-	-
Internal Navigation	-	-	-	-	-	6	449	52	-	-	-
Non-specified	-	-	-	-	-	-	-	-	-	-	-
OTHER SECTORS	**-**	**508**	**-**	**-**	**-**	**568**	**2078**	**605**	**-**	**-**	**-**
Agriculture	-	-	-	-	-	9	1578	387	-	-	-
Comm. and Publ. Services	-	-	-	-	-	-	-	218	-	-	-
Residential	-	508	-	-	-	556	500	-	-	-	-
Non-specified	-	-	-	-	-	3	-	-	-	-	-
NON-ENERGY USE	**-**	**-**	**-**	**-**	**-**	**-**	**-**	**-**	**-**	**-**	**353**
in Industry/Transf./Energy	-	-	-	-	-	-	-	-	-	-	353
in Transport	-	-	-	-	-	-	-	-	-	-	-
in Other Sectors	-	-	-	-	-	-	-	-	-	-	-

Philippines : 1997

SUPPLY AND CONSUMPTION / *APPROVISIONNEMENT ET DEMANDE*	Gas / *Gaz* (TJ)				Comb. Renew. & Waste / *En. Re. Comb. & Déchets* (TJ)				(GWh)	(TJ)
	Natural Gas / *Gaz naturel*	Gas Works / *Usines à gaz*	Coke Ovens / *Cokeries*	Blast Furnaces / *Hauts fourneaux*	Solid Biomass / *Biomasse solide*	Gas/Liquids from Biomass / *Gaz/Liquides tirés de biomasse*	Municipal Waste / *Déchets urbains*	Industrial Waste / *Déchets industriels*	Electricity / *Electricité*	Heat / *Chaleur*
Production	217	202	-	-	393756	-	-	-	39816	-
From Other Sources	-	-	-	-	-	-	-	-	-	-
Imports	-	-	-	-	-	-	-	-	-	-
Exports	-	-	-	-	-	-	-	-	-	-
Intl. Marine Bunkers	-	-	-	-	-	-	-	-	-	-
Stock Changes	-	-	-	-	-	-	-	-	-	-
DOMESTIC SUPPLY	**217**	**202**	**-**	**-**	**393756**	**-**	**-**	**-**	**39816**	**-**
Transfers	-	-	-	-	-	-	-	-	-	-
Statistical Differences	-	-	-	-	-	-	-	-	-	-
TRANSFORMATION	**217**	**-**	**-**	**-**	**87880**	**-**	**-**	**-**	**-**	**-**
Electricity Plants	217	-	-	-	-	-	-	-	-	-
CHP Plants	-	-	-	-	-	-	-	-	-	-
Heat Plants	-	-	-	-	-	-	-	-	-	-
Blast Furnaces/Gas Works	-	-	-	-	-	-	-	-	-	-
Coke/Pat. Fuel/BKB Plants	-	-	-	-	-	-	-	-	-	-
Petroleum Refineries	-	-	-	-	-	-	-	-	-	-
Petrochemical Industry	-	-	-	-	-	-	-	-	-	-
Liquefaction	-	-	-	-	-	-	-	-	-	-
Other Transform. Sector	-	-	-	-	87880	-	-	-	-	-
ENERGY SECTOR	**-**	**-**	**-**	**-**	**-**	**-**	**-**	**-**	**1438**	**-**
Coal Mines	-	-	-	-	-	-	-	-	-	-
Oil and Gas Extraction	-	-	-	-	-	-	-	-	-	-
Petroleum Refineries	-	-	-	-	-	-	-	-	-	-
Electr., CHP+Heat Plants	-	-	-	-	-	-	-	-	1438	-
Pumped Storage (Elec.)	-	-	-	-	-	-	-	-	-	-
Other Energy Sector	-	-	-	-	-	-	-	-	-	-
Distribution Losses	-	63	-	-	-	-	-	-	6593	-
FINAL CONSUMPTION	**-**	**139**	**-**	**-**	**305876**	**-**	**-**	**-**	**31785**	**-**
INDUSTRY SECTOR	**-**	**1**	**-**	**-**	**149329**	**-**	**-**	**-**	**12568**	**-**
Iron and Steel	-	-	-	-	-	-	-	-	937	-
Chemical and Petrochem.	-	-	-	-	-	-	-	-	1454	-
of which: Feedstocks	-	-	-	-	-	-	-	-	-	-
Non-Ferrous Metals	-	-	-	-	-	-	-	-	-	-
Non-Metallic Minerals	-	-	-	-	-	-	-	-	-	-
Transport Equipment	-	-	-	-	-	-	-	-	-	-
Machinery	-	-	-	-	-	-	-	-	-	-
Mining and Quarrying	-	-	-	-	-	-	-	-	-	-
Food and Tobacco	-	-	-	-	-	-	-	-	-	-
Paper, Pulp and Print	-	-	-	-	-	-	-	-	-	-
Wood and Wood Products	-	-	-	-	-	-	-	-	-	-
Construction	-	-	-	-	-	-	-	-	-	-
Textile and Leather	-	-	-	-	-	-	-	-	-	-
Non-specified	-	1	-	-	149329	-	-	-	10177	-
TRANSPORT SECTOR	**-**	**-**	**-**	**-**	**-**	**-**	**-**	**-**	**-**	**-**
Air	-	-	-	-	-	-	-	-	-	-
Road	-	-	-	-	-	-	-	-	-	-
Rail	-	-	-	-	-	-	-	-	-	-
Pipeline Transport	-	-	-	-	-	-	-	-	-	-
Internal Navigation	-	-	-	-	-	-	-	-	-	-
Non-specified	-	-	-	-	-	-	-	-	-	-
OTHER SECTORS	**-**	**138**	**-**	**-**	**156547**	**-**	**-**	**-**	**19217**	**-**
Agriculture	-	-	-	-	-	-	-	-	-	-
Comm. and Publ. Services	-	-	-	-	-	-	-	-	7723	-
Residential	-	37	-	-	125473	-	-	-	10121	-
Non-specified	-	101	-	-	31074	-	-	-	1373	-
NON-ENERGY USE	**-**	**-**	**-**	**-**	**-**	**-**	**-**	**-**	**-**	**-**
in Industry/Transf./Energy	-	-	-	-	-	-	-	-	-	-
in Transport	-	-	-	-	-	-	-	-	-	-
in Other Sectors	-	-	-	-	-	-	-	-	-	-

Qatar

SUPPLY AND CONSUMPTION 1996	Coal (1000 tonnes)							Oil (1000 tonnes)			
	Coking Coal	Other Bit. Coal	Sub-Bit. Coal	Lignite	Peat	Oven and Gas Coke	Pat. Fuel and BKB	Crude Oil	NGL	Feed-stocks	Additives
Production	-	-	-	-	-	-	-	18606	2730	-	-
Imports	-	-	-	-	-	-	-	-	-	-	-
Exports	-	-	-	-	-	-	-	-17371	-1500	-	-
Intl. Marine Bunkers	-	-	-	-	-	-	-	-	-	-	-
Stock Changes	-	-	-	-	-	-	-	1828	-	-	-
DOMESTIC SUPPLY	**-**	**-**	**-**	**-**	**-**	**-**	**-**	**3063**	**1230**	**-**	**-**
Transfers and Stat. Diff.	-	-	-	-	-	-	-	-	-1230	-	-
TRANSFORMATION	**-**	**-**	**-**	**-**	**-**	**-**	**-**	**3063**	**-**	**-**	**-**
Electricity and CHP Plants	-	-	-	-	-	-	-	-	-	-	-
Petroleum Refineries	-	-	-	-	-	-	-	3063	-	-	-
Other Transform. Sector	-	-	-	-	-	-	-	-	-	-	-
ENERGY SECTOR	**-**	**-**	**-**	**-**	**-**	**-**	**-**	**-**	**-**	**-**	**-**
DISTRIBUTION LOSSES	**-**	**-**	**-**	**-**	**-**	**-**	**-**	**-**	**-**	**-**	**-**
FINAL CONSUMPTION	**-**	**-**	**-**	**-**	**-**	**-**	**-**	**-**	**-**	**-**	**-**
INDUSTRY SECTOR	**-**	**-**	**-**	**-**	**-**	**-**	**-**	**-**	**-**	**-**	**-**
Iron and Steel	-	-	-	-	-	-	-	-	-	-	-
Chemical and Petrochem.	-	-	-	-	-	-	-	-	-	-	-
Non-Metallic Minerals	-	-	-	-	-	-	-	-	-	-	-
Non-specified	-	-	-	-	-	-	-	-	-	-	-
TRANSPORT SECTOR	**-**	**-**	**-**	**-**	**-**	**-**	**-**	**-**	**-**	**-**	**-**
Air	-	-	-	-	-	-	-	-	-	-	-
Road	-	-	-	-	-	-	-	-	-	-	-
Non-specified	-	-	-	-	-	-	-	-	-	-	-
OTHER SECTORS	**-**	**-**	**-**	**-**	**-**	**-**	**-**	**-**	**-**	**-**	**-**
Agriculture	-	-	-	-	-	-	-	-	-	-	-
Comm. and Publ. Services	-	-	-	-	-	-	-	-	-	-	-
Residential	-	-	-	-	-	-	-	-	-	-	-
Non-specified	-	-	-	-	-	-	-	-	-	-	-
NON-ENERGY USE	**-**	**-**	**-**	**-**	**-**	**-**	**-**	**-**	**-**	**-**	**-**

APPROVISIONNEMENT ET DEMANDE 1997	*Charbon* (1000 tonnes)							*Pétrole* (1000 tonnes)			
	Charbon à coke	*Autres charb. bit.*	*Charbon sous-bit.*	*Lignite*	*Tourbe*	*Coke de four/gaz*	*Agg./briq. de lignite*	*Pétrole brut*	*LGN*	*Produits d'aliment.*	*Additifs*
Production	-	-	-	-	-	-	-	26168	2599	-	-
Imports	-	-	-	-	-	-	-	-	-	-	-
Exports	-	-	-	-	-	-	-	-22010	-1372	-	-
Intl. Marine Bunkers	-	-	-	-	-	-	-	-	-	-	-
Stock Changes	-	-	-	-	-	-	-	-1095	-	-	-
DOMESTIC SUPPLY	**-**	**-**	**-**	**-**	**-**	**-**	**-**	**3063**	**1227**	**-**	**-**
Transfers and Stat. Diff.	-	-	-	-	-	-	-	-	-1227	-	-
TRANSFORMATION	**-**	**-**	**-**	**-**	**-**	**-**	**-**	**3063**	**-**	**-**	**-**
Electricity and CHP Plants	-	-	-	-	-	-	-	-	-	-	-
Petroleum Refineries	-	-	-	-	-	-	-	3063	-	-	-
Other Transform. Sector	-	-	-	-	-	-	-	-	-	-	-
ENERGY SECTOR	**-**	**-**	**-**	**-**	**-**	**-**	**-**	**-**	**-**	**-**	**-**
DISTRIBUTION LOSSES	**-**	**-**	**-**	**-**	**-**	**-**	**-**	**-**	**-**	**-**	**-**
FINAL CONSUMPTION	**-**	**-**	**-**	**-**	**-**	**-**	**-**	**-**	**-**	**-**	**-**
INDUSTRY SECTOR	**-**	**-**	**-**	**-**	**-**	**-**	**-**	**-**	**-**	**-**	**-**
Iron and Steel	-	-	-	-	-	-	-	-	-	-	-
Chemical and Petrochem.	-	-	-	-	-	-	-	-	-	-	-
Non-Metallic Minerals	-	-	-	-	-	-	-	-	-	-	-
Non-specified	-	-	-	-	-	-	-	-	-	-	-
TRANSPORT SECTOR	**-**	**-**	**-**	**-**	**-**	**-**	**-**	**-**	**-**	**-**	**-**
Air	-	-	-	-	-	-	-	-	-	-	-
Road	-	-	-	-	-	-	-	-	-	-	-
Non-specified	-	-	-	-	-	-	-	-	-	-	-
OTHER SECTORS	**-**	**-**	**-**	**-**	**-**	**-**	**-**	**-**	**-**	**-**	**-**
Agriculture	-	-	-	-	-	-	-	-	-	-	-
Comm. and Publ. Services	-	-	-	-	-	-	-	-	-	-	-
Residential	-	-	-	-	-	-	-	-	-	-	-
Non-specified	-	-	-	-	-	-	-	-	-	-	-
NON-ENERGY USE	**-**	**-**	**-**	**-**	**-**	**-**	**-**	**-**	**-**	**-**	**-**

Qatar

SUPPLY AND CONSUMPTION 1996	Oil cont. (1000 tonnes)										
	Refinery Gas	LPG + Ethane	Motor Gasoline	Aviation Gasoline	Jet Fuel	Kerosene	Gas/ Diesel	Heavy Fuel Oil	Naphtha	Petrol. Coke	Other Prod.
Production	52	89	608	-	456	21	684	1009	-	-	-
Imports	-	-	-	-	-	-	-	-	-	-	-
Exports	-	-1293	-174	-	-289	-	-407	-1008	-	-	-
Intl. Marine Bunkers	-	-	-	-	-	-	-	-	-	-	-
Stock Changes	-	-	-44	-	8	-	-	-	-	-	-
DOMESTIC SUPPLY	**52**	**-1204**	**390**	**-**	**175**	**21**	**277**	**1**	**-**	**-**	**-**
Transfers and Stat. Diff.	-	1230	-	-	-1	-	-	-	-	-	-
TRANSFORMATION	**-**	**-**	**-**	**-**	**-**	**-**	**-**	**-**	**-**	**-**	**-**
Electricity and CHP Plants	-	-	-	-	-	-	-	-	-	-	-
Petroleum Refineries	-	-	-	-	-	-	-	-	-	-	-
Other Transform. Sector	-	-	-	-	-	-	-	-	-	-	-
ENERGY SECTOR	**52**	**-**	**-**	**-**	**-**	**-**	**-**	**1**	**-**	**-**	**-**
DISTRIBUTION LOSSES	**-**	**-**	**-**	**-**	**-**	**-**	**-**	**-**	**-**	**-**	**-**
FINAL CONSUMPTION	**-**	**26**	**390**	**-**	**174**	**21**	**277**	**-**	**-**	**-**	**-**
INDUSTRY SECTOR	**-**	**26**	**-**	**-**	**-**	**-**	**-**	**-**	**-**	**-**	**-**
Iron and Steel	-	-	-	-	-	-	-	-	-	-	-
Chemical and Petrochem.	-	26	-	-	-	-	-	-	-	-	-
Non-Metallic Minerals	-	-	-	-	-	-	-	-	-	-	-
Non-specified	-	-	-	-	-	-	-	-	-	-	-
TRANSPORT SECTOR	**-**	**-**	**390**	**-**	**174**	**-**	**277**	**-**	**-**	**-**	**-**
Air	-	-	-	-	174	-	-	-	-	-	-
Road	-	-	390	-	-	-	277	-	-	-	-
Non-specified	-	-	-	-	-	-	-	-	-	-	-
OTHER SECTORS	**-**	**-**	**-**	**-**	**-**	**21**	**-**	**-**	**-**	**-**	**-**
Agriculture	-	-	-	-	-	-	-	-	-	-	-
Comm. and Publ. Services	-	-	-	-	-	-	-	-	-	-	-
Residential	-	-	-	-	-	21	-	-	-	-	-
Non-specified	-	-	-	-	-	-	-	-	-	-	-
NON-ENERGY USE	**-**	**-**	**-**	**-**	**-**	**-**	**-**	**-**	**-**	**-**	**-**

APPROVISIONNEMENT ET DEMANDE 1997	*Pétrole cont.* (1000 tonnes)										
	Gaz de raffinerie	*GPL + éthane*	*Essence moteur*	*Essence aviation*	*Carbu-réacteurs*	*Kérosène*	*Gazole*	*Fioul lourd*	*Naphta*	*Coke de pétrole*	*Autres prod.*
Production	52	69	525	-	413	21	684	899	-	-	-
Imports	-	-	-	-	-	-	-	-	-	-	-
Exports	-	-1267	-111	-	-216	-	-353	-898	-	-	-
Intl. Marine Bunkers	-	-	-	-	-	-	-	-	-	-	-
Stock Changes	-	-	-	-	-	-	-	-	-	-	-
DOMESTIC SUPPLY	**52**	**-1198**	**414**	**-**	**197**	**21**	**331**	**1**	**-**	**-**	**-**
Transfers and Stat. Diff.	-	1227	-	-	-	-	-	-	-	-	-
TRANSFORMATION	**-**	**-**	**-**	**-**	**-**	**-**	**-**	**-**	**-**	**-**	**-**
Electricity and CHP Plants	-	-	-	-	-	-	-	-	-	-	-
Petroleum Refineries	-	-	-	-	-	-	-	-	-	-	-
Other Transform. Sector	-	-	-	-	-	-	-	-	-	-	-
ENERGY SECTOR	**52**	**-**	**-**	**-**	**-**	**-**	**-**	**1**	**-**	**-**	**-**
DISTRIBUTION LOSSES	**-**	**-**	**-**	**-**	**-**	**-**	**-**	**-**	**-**	**-**	**-**
FINAL CONSUMPTION	**-**	**29**	**414**	**-**	**197**	**21**	**331**	**-**	**-**	**-**	**-**
INDUSTRY SECTOR	**-**	**29**	**-**	**-**	**-**	**-**	**-**	**-**	**-**	**-**	**-**
Iron and Steel	-	-	-	-	-	-	-	-	-	-	-
Chemical and Petrochem.	-	29	-	-	-	-	-	-	-	-	-
Non-Metallic Minerals	-	-	-	-	-	-	-	-	-	-	-
Non-specified	-	-	-	-	-	-	-	-	-	-	-
TRANSPORT SECTOR	**-**	**-**	**414**	**-**	**197**	**-**	**331**	**-**	**-**	**-**	**-**
Air	-	-	-	-	197	-	-	-	-	-	-
Road	-	-	414	-	-	-	331	-	-	-	-
Non-specified	-	-	-	-	-	-	-	-	-	-	-
OTHER SECTORS	**-**	**-**	**-**	**-**	**-**	**21**	**-**	**-**	**-**	**-**	**-**
Agriculture	-	-	-	-	-	-	-	-	-	-	-
Comm. and Publ. Services	-	-	-	-	-	-	-	-	-	-	-
Residential	-	-	-	-	-	21	-	-	-	-	-
Non-specified	-	-	-	-	-	-	-	-	-	-	-
NON-ENERGY USE	**-**	**-**	**-**	**-**	**-**	**-**	**-**	**-**	**-**	**-**	**-**

Qatar

SUPPLY AND CONSUMPTION 1996	Gas (TJ)				Comb. Renew. & Waste (TJ)				(GWh)	(TJ)
	Natural Gas	Gas Works	Coke Ovens	Blast Furnaces	Solid Biomass	Gas/Liquids from Biomass	Municipal Waste	Industrial Waste	Electricity	Heat
Production	516490	-	-	-	-	-	-	-	6575	-
Imports	-	-	-	-	63	-	-	-	-	-
Exports	-	-	-	-	-	-	-	-	-	-
Intl. Marine Bunkers	-	-	-	-	-	-	-	-	-	-
Stock Changes	-	-	-	-	-	-	-	-	-	-
DOMESTIC SUPPLY	**516490**	**-**	**-**	**-**	**63**	**-**	**-**	**-**	**6575**	**-**
Transfers and Stat. Diff.	-	-	-	-	-	-	-	-	-1	-
TRANSFORMATION	**78896**	**-**	**-**	**-**	**-**	**-**	**-**	**-**	**-**	**-**
Electricity and CHP Plants	78896	-	-	-	-	-	-	-	-	-
Petroleum Refineries	-	-	-	-	-	-	-	-	-	-
Other Transform. Sector	-	-	-	-	-	-	-	-	-	-
ENERGY SECTOR	**216284**	**-**	**-**	**-**	**-**	**-**	**-**	**-**	**-**	**-**
DISTRIBUTION LOSSES	**-**	**-**	**-**	**-**	**-**	**-**	**-**	**-**	**391**	**-**
FINAL CONSUMPTION	**221310**	**-**	**-**	**-**	**63**	**-**	**-**	**-**	**6183**	**-**
INDUSTRY SECTOR	**221310**	**-**	**-**	**-**	**-**	**-**	**-**	**-**	**1422**	**-**
Iron and Steel	16557	-	-	-	-	-	-	-	1422	-
Chemical and Petrochem.	185952	-	-	-	-	-	-	-	-	-
Non-Metallic Minerals	5841	-	-	-	-	-	-	-	-	-
Non-specified	12960	-	-	-	-	-	-	-	-	-
TRANSPORT SECTOR	**-**	**-**	**-**	**-**	**-**	**-**	**-**	**-**	**-**	**-**
Air	-	-	-	-	-	-	-	-	-	-
Road	-	-	-	-	-	-	-	-	-	-
Non-specified	-	-	-	-	-	-	-	-	-	-
OTHER SECTORS	**-**	**-**	**-**	**-**	**63**	**-**	**-**	**-**	**4761**	**-**
Agriculture	-	-	-	-	-	-	-	-	-	-
Comm. and Publ. Services	-	-	-	-	-	-	-	-	618	-
Residential	-	-	-	-	63	-	-	-	-	-
Non-specified	-	-	-	-	-	-	-	-	4143	-
NON-ENERGY USE	**-**	**-**	**-**	**-**	**-**	**-**	**-**	**-**	**-**	**-**

APPROVISIONNEMENT ET DEMANDE 1997	*Gaz* (TJ)				*En. Re. Comb. & Déchets* (TJ)				(GWh)	(TJ)
	Gaz naturel	*Usines à gaz*	*Cokeries*	*Hauts fourneaux*	*Biomasse solide*	*Gaz/Liquides tirés de biomasse*	*Déchets urbains*	*Déchets industriels*	*Electricité*	*Chaleur*
Production	674830	-	-	-	-	-	-	-	6868	-
Imports	-	-	-	-	63	-	-	-	-	-
Exports	-107822	-	-	-	-	-	-	-	-	-
Intl. Marine Bunkers	-	-	-	-	-	-	-	-	-	-
Stock Changes	-	-	-	-	-	-	-	-	-	-
DOMESTIC SUPPLY	**567008**	**-**	**-**	**-**	**63**	**-**	**-**	**-**	**6868**	**-**
Transfers and Stat. Diff.	-1	-	-	-	-	-	-	-	1	-
TRANSFORMATION	**82416**	**-**	**-**	**-**	**-**	**-**	**-**	**-**	**-**	**-**
Electricity and CHP Plants	82416	-	-	-	-	-	-	-	-	-
Petroleum Refineries	-	-	-	-	-	-	-	-	-	-
Other Transform. Sector	-	-	-	-	-	-	-	-	-	-
ENERGY SECTOR	**239513**	**-**	**-**	**-**	**-**	**-**	**-**	**-**	**-**	**-**
DISTRIBUTION LOSSES	**-**	**-**	**-**	**-**	**-**	**-**	**-**	**-**	**412**	**-**
FINAL CONSUMPTION	**245078**	**-**	**-**	**-**	**63**	**-**	**-**	**-**	**6457**	**-**
INDUSTRY SECTOR	**245078**	**-**	**-**	**-**	**-**	**-**	**-**	**-**	**1485**	**-**
Iron and Steel	18335	-	-	-	-	-	-	-	1485	-
Chemical and Petrochem.	205923	-	-	-	-	-	-	-	-	-
Non-Metallic Minerals	6468	-	-	-	-	-	-	-	-	-
Non-specified	14352	-	-	-	-	-	-	-	-	-
TRANSPORT SECTOR	**-**	**-**	**-**	**-**	**-**	**-**	**-**	**-**	**-**	**-**
Air	-	-	-	-	-	-	-	-	-	-
Road	-	-	-	-	-	-	-	-	-	-
Non-specified	-	-	-	-	-	-	-	-	-	-
OTHER SECTORS	**-**	**-**	**-**	**-**	**63**	**-**	**-**	**-**	**4972**	**-**
Agriculture	-	-	-	-	-	-	-	-	-	-
Comm. and Publ. Services	-	-	-	-	-	-	-	-	646	-
Residential	-	-	-	-	63	-	-	-	-	-
Non-specified	-	-	-	-	-	-	-	-	4326	-
NON-ENERGY USE	**-**	**-**	**-**	**-**	**-**	**-**	**-**	**-**	**-**	**-**

Romania / Roumanie : 1996

SUPPLY AND CONSUMPTION / *APPROVISIONNEMENT ET DEMANDE*	Coal / *Charbon* (1000 tonnes)							Oil / *Pétrole* (1000 tonnes)			
	Coking Coal / *Charbon à coke*	Other Bit. Coal / *Autres charb. bit.*	Sub-Bit. Coal / *Charbon sous-bit.*	Lignite / *Lignite*	Peat / *Tourbe*	Oven and Gas Coke / *Coke de four/gaz*	Pat. Fuel and BKB / *Agg./briq. de lignite*	Crude Oil / *Pétrole brut*	NGL / *LGN*	Feed-stocks / *Produits d'aliment.*	Additives / *Additifs*
Production	312	1011	-	40546	2	3153	-	6626	226	-	113
From Other Sources	-	-	-	-	-	-	-	-	-	-	-
Imports	3996	9	-	729	-	83	-	7156	-	-	1
Exports	-	-	-	-	-	-219	-	-	-	-	-
Intl. Marine Bunkers	-	-	-	-	-	-	-	-	-	-	-
Stock Changes	356	-70	-	-1381	2	99	-	-79	-	-	3
DOMESTIC SUPPLY	**4664**	**950**	**-**	**39894**	**4**	**3116**	**-**	**13703**	**226**	**-**	**117**
Transfers	-	-	-	-	-	-	-	-	-	-	-
Statistical Differences	-45	-4	-	-35	-1	-35	5	-5	-65	-	-
TRANSFORMATION	**4573**	**765**	**-**	**38674**	**-**	**1815**	**1**	**13426**	**132**	**-**	**117**
Electricity Plants	-	-	-	12785	-	-	-	-	-	-	-
CHP Plants	-	694	-	25510	-	-	-	-	-	-	-
Heat Plants	5	71	-	379	-	-	1	-	-	-	-
Blast Furnaces/Gas Works	-	-	-	-	-	1815	-	-	-	-	-
Coke/Pat. Fuel/BKB Plants	4568	-	-	-	-	-	-	-	-	-	-
Petroleum Refineries	-	-	-	-	-	-	-	13426	132	-	117
Petrochemical Industry	-	-	-	-	-	-	-	-	-	-	-
Liquefaction	-	-	-	-	-	-	-	-	-	-	-
Other Transform. Sector	-	-	-	-	-	-	-	-	-	-	-
ENERGY SECTOR	**16**	**68**	**-**	**155**	**-**	**-**	**1**	**89**	**12**	**-**	**-**
Coal Mines	16	68	-	143	-	-	-	-	-	-	-
Oil and Gas Extraction	-	-	-	-	-	-	-	89	-	-	-
Petroleum Refineries	-	-	-	-	-	-	-	-	-	-	-
Electr., CHP+Heat Plants	-	-	-	12	-	-	1	-	-	-	-
Pumped Storage (Elec.)	-	-	-	-	-	-	-	-	-	-	-
Other Energy Sector	-	-	-	-	-	-	-	-	12	-	-
Distribution Losses	-	3	-	295	-	-	-	183	17	-	-
FINAL CONSUMPTION	**30**	**110**	**-**	**735**	**3**	**1266**	**3**	**-**	**-**	**-**	**-**
INDUSTRY SECTOR	**29**	**14**	**-**	**178**	**-**	**698**	**1**	**-**	**-**	**-**	**-**
Iron and Steel	4	-	-	-	-	454	-	-	-	-	-
Chemical and Petrochem.	24	-	-	-	-	174	-	-	-	-	-
of which: Feedstocks	-	-	-	-	-	-	-	-	-	-	-
Non-Ferrous Metals	-	-	-	-	-	-	-	-	-	-	-
Non-Metallic Minerals	1	-	-	52	-	28	-	-	-	-	-
Transport Equipment	-	-	-	-	-	-	-	-	-	-	-
Machinery	-	1	-	6	-	20	-	-	-	-	-
Mining and Quarrying	-	8	-	16	-	-	-	-	-	-	-
Food and Tobacco	-	2	-	47	-	22	-	-	-	-	-
Paper, Pulp and Print	-	1	-	-	-	-	-	-	-	-	-
Wood and Wood Products	-	1	-	-	-	-	-	-	-	-	-
Construction	-	-	-	55	-	-	-	-	-	-	-
Textile and Leather	-	-	-	2	-	-	-	-	-	-	-
Non-specified	-	1	-	-	-	-	1	-	-	-	-
TRANSPORT SECTOR	**1**	**1**	**-**	**2**	**-**	**-**	**-**	**-**	**-**	**-**	**-**
Air	-	-	-	-	-	-	-	-	-	-	-
Road	-	-	-	-	-	-	-	-	-	-	-
Rail	1	1	-	2	-	-	-	-	-	-	-
Pipeline Transport	-	-	-	-	-	-	-	-	-	-	-
Internal Navigation	-	-	-	-	-	-	-	-	-	-	-
Non-specified	-	-	-	-	-	-	-	-	-	-	-
OTHER SECTORS	**-**	**80**	**-**	**555**	**3**	**-**	**2**	**-**	**-**	**-**	**-**
Agriculture	-	-	-	4	3	-	-	-	-	-	-
Comm. and Publ. Services	-	-	-	-	-	-	-	-	-	-	-
Residential	-	78	-	527	-	-	-	-	-	-	-
Non-specified	-	2	-	24	-	-	2	-	-	-	-
NON-ENERGY USE	**-**	**15**	**-**	**-**	**-**	**568**	**-**	**-**	**-**	**-**	**-**
in Industry/Trans./Energy	-	15	-	-	-	568	-	-	-	-	-
in Transport	-	-	-	-	-	-	-	-	-	-	-
in Other Sectors	-	-	-	-	-	-	-	-	-	-	-

Romania / Roumanie : 1996

SUPPLY AND CONSUMPTION *APPROVISIONNEMENT ET DEMANDE*	Oil cont. / *Pétrole cont.* (1000 tonnes)										
	Refinery Gas *Gaz de raffinerie*	LPG + Ethane *GPL + éthane*	Motor Gasoline *Essence moteur*	Aviation Gasoline *Essence aviation*	Jet Fuel *Carbu- réacteurs*	Kerosene *Kérosène*	Gas/ Diesel *Gazole*	Heavy Fuel Oil *Fioul lourd*	Naphtha *Naphta*	Petrol. Coke *Coke de pétrole*	Other Prod. *Autres prod.*
Production	800	264	3125	5	92	67	4201	2402	541	551	1198
From Other Sources	-	-	-	-	-	-	-	-	-	-	-
Imports	-	9	140	-	-	-	14	3003	-	-	59
Exports	-	-	-1930	-	-58	-9	-1252	-182	-14	-223	-62
Intl. Marine Bunkers	-	-	-	-	-	-	-	-	-	-	-
Stock Changes	2	6	147	-	5	2	81	-30	-2	22	26
DOMESTIC SUPPLY	**802**	**279**	**1482**	**5**	**39**	**60**	**3044**	**5193**	**525**	**350**	**1221**
Transfers	-	-	-	-	-	-4	-42	-60	-	-	110
Statistical Differences	-61	-121	-38	-3	45	-25	504	94	-348	-42	-258
TRANSFORMATION	**59**	**-**	**-**	**-**	**-**	**-**	**9**	**3677**	**-**	**-**	**163**
Electricity Plants	2	-	-	-	-	-	4	282	-	-	-
CHP Plants	44	-	-	-	-	-	-	3074	-	-	38
Heat Plants	13	-	-	-	-	-	5	321	-	-	125
Blast Furnaces/Gas Works	-	-	-	-	-	-	-	-	-	-	-
Coke/Pat. Fuel/BKB Plants	-	-	-	-	-	-	-	-	-	-	-
Petroleum Refineries	-	-	-	-	-	-	-	-	-	-	-
Petrochemical Industry	-	-	-	-	-	-	-	-	-	-	-
Liquefaction	-	-	-	-	-	-	-	-	-	-	-
Other Transform. Sector	-	-	-	-	-	-	-	-	-	-	-
ENERGY SECTOR	**563**	**-**	**4**	**-**	**-**	**2**	**98**	**534**	**128**	**16**	**246**
Coal Mines	-	-	-	-	-	-	32	2	-	-	4
Oil and Gas Extraction	-	-	1	-	-	1	44	-	-	-	11
Petroleum Refineries	504	-	-	-	-	-	-	-	-	-	-
Electr., CHP+Heat Plants	-	-	3	-	-	-	22	22	-	-	12
Pumped Storage (Elec.)	-	-	-	-	-	-	-	-	-	-	-
Other Energy Sector	59	-	-	-	-	1	-	510	128	16	219
Distribution Losses	-	-	31	-	-	-	20	-	-	-	123
FINAL CONSUMPTION	**119**	**158**	**1409**	**2**	**84**	**29**	**3379**	**1016**	**49**	**292**	**541**
INDUSTRY SECTOR	**119**	**1**	**52**	**-**	**-**	**5**	**601**	**920**	**48**	**292**	**128**
Iron and Steel	-	-	-	-	-	-	7	153	-	100	1
Chemical and Petrochem.	119	-	9	-	-	-	2	140	42	-	14
of which: Feedstocks	-	-	-	-	-	-	-	-	-	-	-
Non-Ferrous Metals	-	-	-	-	-	-	-	-	-	-	-
Non-Metallic Minerals	-	-	-	-	-	-	4	270	-	192	-
Transport Equipment	-	-	-	-	-	-	-	-	-	-	-
Machinery	-	1	3	-	-	3	328	90	1	-	2
Mining and Quarrying	-	-	-	-	-	-	5	15	-	-	-
Food and Tobacco	-	-	1	-	-	-	16	121	4	-	8
Paper, Pulp and Print	-	-	-	-	-	-	-	65	-	-	-
Wood and Wood Products	-	-	-	-	-	-	3	12	-	-	-
Construction	-	-	30	-	-	1	229	1	-	-	2
Textile and Leather	-	-	4	-	-	1	1	31	1	-	-
Non-specified	-	-	5	-	-	-	6	22	-	-	101
TRANSPORT SECTOR	**-**	**-**	**1320**	**2**	**84**	**6**	**2253**	**92**	**1**	**-**	**3**
Air	-	-	-	2	84	3	-	-	-	-	-
Road	-	-	1320	-	-	2	1903	-	1	-	3
Rail	-	-	-	-	-	1	286	-	-	-	-
Pipeline Transport	-	-	-	-	-	-	5	-	-	-	-
Internal Navigation	-	-	-	-	-	-	59	92	-	-	-
Non-specified	-	-	-	-	-	-	-	-	-	-	-
OTHER SECTORS	**-**	**157**	**37**	**-**	**-**	**18**	**525**	**4**	**-**	**-**	**-**
Agriculture	-	-	15	-	-	-	445	1	-	-	-
Comm. and Publ. Services	-	-	-	-	-	-	34	-	-	-	-
Residential	-	156	-	-	-	18	-	1	-	-	-
Non-specified	-	1	22	-	-	-	46	2	-	-	-
NON-ENERGY USE	**-**	**-**	**-**	**-**	**-**	**-**	**-**	**-**	**-**	**-**	**410**
in Industry/Transf./Energy	-	-	-	-	-	-	-	-	-	-	322
in Transport	-	-	-	-	-	-	-	-	-	-	50
in Other Sectors	-	-	-	-	-	-	-	-	-	-	38

Romania / Roumanie : 1996

SUPPLY AND CONSUMPTION / *APPROVISIONNEMENT ET DEMANDE*	Gas / *Gaz* (TJ)				Comb. Renew. & Waste / *En. Re. Comb. & Déchets* (TJ)				(GWh)	(TJ)
	Natural Gas / *Gaz naturel*	Gas Works / *Usines à gaz*	Coke Ovens / *Cokeries*	Blast Furnaces / *Hauts fourneaux*	Solid Biomass / *Biomasse solide*	Gas/Liquids from Biomass / *Gaz/Liquides tirés de biomasse*	Municipal Waste / *Déchets urbains*	Industrial Waste / *Déchets industriels*	Electricity / *Electricité*	Heat / *Chaleur*
Production	640302	-	21161	26402	204367	-	-	1735	61350	296264
From Other Sources	-	-	-	-	-	-	-	-	-	-
Imports	263046	-	-	-	-	-	-	-	2242	-
Exports	-	-	-	-	-347	-	-	-	-1435	-
Intl. Marine Bunkers	-	-	-	-	-	-	-	-	-	-
Stock Changes	-	-	-	-	-146	-	-	-	-	-
DOMESTIC SUPPLY	**903348**	**-**	**21161**	**26402**	**203874**	**-**	**-**	**1735**	**62157**	**296264**
Transfers	-	-	-	-	-	-	-	-	-	-
Statistical Differences	-3549	-	-228	358	2666	-	-	-	-	21354
TRANSFORMATION	**364394**	**-**	**1432**	**7683**	**787**	**-**	**-**	**1735**	**-**	**-**
Electricity Plants	46102	-	-	1107	-	-	-	-	-	-
CHP Plants	241229	-	1432	6283	754	-	-	513	-	-
Heat Plants	77063	-	-	293	1729	-	-	1222	-	-
Blast Furnaces/Gas Works	-	-	-	-	-	-	-	-	-	-
Coke/Pat. Fuel/BKB Plants	-	-	-	-	66	-	-	-	-	-
Petroleum Refineries	-	-	-	-	-	-	-	-	-	-
Petrochemical Industry	-	-	-	-	-	-	-	-	-	-
Liquefaction	-	-	-	-	-	-	-	-	-	-
Other Transform. Sector	-	-	-	-	-1762	-	-	-	-	-
ENERGY SECTOR	**49061**	**-**	**2509**	**-**	**293**	**-**	**-**	**-**	**15247**	**39567**
Coal Mines	456	-	-	-	76	-	-	-	1723	850
Oil and Gas Extraction	32066	-	-	-	-	-	-	-	2107	389
Petroleum Refineries	10906	-	-	-	-	-	-	-	1351	13084
Electr., CHP+Heat Plants	5177	-	-	-	217	-	-	-	8827	23389
Pumped Storage (Elec.)	-	-	-	-	-	-	-	-	-	-
Other Energy Sector	456	-	2509	-	-	-	-	-	1239	1855
Distribution Losses	22203	-	63	717	59	-	-	-	7183	43000
FINAL CONSUMPTION	**464141**	**-**	**16929**	**18360**	**205401**	**-**	**-**	**-**	**39727**	**235051**
INDUSTRY SECTOR	**364926**	**-**	**16929**	**18360**	**9551**	**-**	**-**	**-**	**24512**	**53485**
Iron and Steel	62940	-	15789	17774	205	-	-	-	7352	4966
Chemical and Petrochem.	167282	-	1140	586	557	-	-	-	5158	23175
of which: Feedstocks	*45137*	-	-	-	-	-	-	-	-	-
Non-Ferrous Metals	-	-	-	-	-	-	-	-	-	-
Non-Metallic Minerals	34027	-	-	-	58	-	-	-	1749	1130
Transport Equipment	-	-	-	-	-	-	-	-	-	-
Machinery	31139	-	-	-	59	-	-	-	3842	7319
Mining and Quarrying	2051	-	-	-	-	-	-	-	999	88
Food and Tobacco	39816	-	-	-	791	-	-	-	1304	6829
Paper, Pulp and Print	7032	-	-	-	2344	-	-	-	752	1461
Wood and Wood Products	2083	-	-	-	3428	-	-	-	283	1206
Construction	2181	-	-	-	176	-	-	-	640	1574
Textile and Leather	9799	-	-	-	87	-	-	-	752	3421
Non-specified	6576	-	-	-	1846	-	-	-	1681	2316
TRANSPORT SECTOR	**4102**	**-**	**-**	**-**	**235**	**-**	**-**	**-**	**2326**	**-**
Air	-	-	-	-	-	-	-	-	-	-
Road	3939	-	-	-	-	-	-	-	-	-
Rail	-	-	-	-	211	-	-	-	1612	-
Pipeline Transport	-	-	-	-	-	-	-	-	63	-
Internal Navigation	-	-	-	-	-	-	-	-	-	-
Non-specified	163	-	-	-	24	-	-	-	651	-
OTHER SECTORS	**95113**	**-**	**-**	**-**	**195615**	**-**	**-**	**-**	**12889**	**181566**
Agriculture	2441	-	-	-	468	-	-	-	1332	8148
Comm. and Publ. Services	-	-	-	-	-	-	-	-	-	-
Residential	79631	-	-	-	190567	-	-	-	8122	134072
Non-specified	13041	-	-	-	4580	-	-	-	3435	39346
NON-ENERGY USE	**-**	**-**	**-**	**-**	**-**	**-**	**-**	**-**	**-**	**-**
in Industry/Transf./Energy	-	-	-	-	-	-	-	-	-	-
in Transport	-	-	-	-	-	-	-	-	-	-
in Other Sectors	-	-	-	-	-	-	-	-	-	-

Romania / Roumanie : 1997

SUPPLY AND CONSUMPTION	Coal / *Charbon* (1000 tonnes)							Oil / *Pétrole* (1000 tonnes)			
APPROVISIONNEMENT ET DEMANDE	Coking Coal *Charbon à coke*	Other Bit. Coal *Autres charb. bit.*	Sub-Bit. Coal *Charbon sous-bit.*	Lignite *Lignite*	Peat *Tourbe*	Oven and Gas Coke *Coke de four/gaz*	Pat. Fuel and BKB *Agg./briq. de lignite*	Crude Oil *Pétrole brut*	NGL *LGN*	Feed-stocks *Produits d'aliment.*	Additives *Additifs*
Production	324	1428	-	32055	-	3316	-	6517	234	-	104
From Other Sources	-	-	-	-	-	-	-	-	-	-	-
Imports	5014	235	-	-	-	189	-	6245	-	-	1
Exports	-	-	-	-	-	-146	-	-	-	-	-
Intl. Marine Bunkers	-	-	-	-	-	-	-	-	-	-	-
Stock Changes	-364	15	-	-551	-	-38	-	19	-5	-	-1
DOMESTIC SUPPLY	**4974**	**1678**	**-**	**31504**	**-**	**3321**	**-**	**12781**	**229**	**-**	**104**
Transfers	-	-	-	-	-	-	-	-	-	-	-
Statistical Differences	-97	-280	-	-61	-	63	-	-91	-65	-	-
TRANSFORMATION	**4833**	**1252**	**-**	**30468**	**-**	**2071**	**-**	**12431**	**127**	**-**	**104**
Electricity Plants	-	697	-	8251	-	-	-	-	-	-	-
CHP Plants	-	476	-	21984	-	-	-	-	-	-	-
Heat Plants	-	79	-	233	-	-	-	-	-	-	-
Blast Furnaces/Gas Works	-	-	-	-	-	2071	-	-	-	-	-
Coke/Pat. Fuel/BKB Plants	4833	-	-	-	-	-	-	-	-	-	-
Petroleum Refineries	-	-	-	-	-	-	-	12431	127	-	104
Petrochemical Industry	-	-	-	-	-	-	-	-	-	-	-
Liquefaction	-	-	-	-	-	-	-	-	-	-	-
Other Transform. Sector	-	-	-	-	-	-	-	-	-	-	-
ENERGY SECTOR	**42**	**22**	**-**	**119**	**-**	**-**	**-**	**148**	**37**	**-**	**-**
Coal Mines	13	21	-	108	-	-	-	-	-	-	-
Oil and Gas Extraction	-	-	-	-	-	-	-	148	-	-	-
Petroleum Refineries	-	-	-	-	-	-	-	-	-	-	-
Electr., CHP+Heat Plants	-	1	-	11	-	-	-	-	-	-	-
Pumped Storage (Elec.)	-	-	-	-	-	-	-	-	-	-	-
Other Energy Sector	29	-	-	-	-	-	-	-	37	-	-
Distribution Losses	-	4	-	293	-	-	-	109	-	-	-
FINAL CONSUMPTION	**2**	**120**	**-**	**563**	**-**	**1313**	**-**	**2**	**-**	**-**	**-**
INDUSTRY SECTOR	**1**	**16**	**-**	**116**	**-**	**615**	**-**	**2**	**-**	**-**	**-**
Iron and Steel	-	-	-	-	-	518	-	-	-	-	-
Chemical and Petrochem.	-	6	-	15	-	62	-	-	-	-	-
of which: Feedstocks	-	-	-	-	-	-	-	-	-	-	-
Non-Ferrous Metals	-	-	-	-	-	-	-	-	-	-	-
Non-Metallic Minerals	1	-	-	54	-	3	-	-	-	-	-
Transport Equipment	-	-	-	-	-	-	-	-	-	-	-
Machinery	-	1	-	11	-	17	-	-	-	-	-
Mining and Quarrying	-	5	-	1	-	-	-	-	-	-	-
Food and Tobacco	-	1	-	27	-	15	-	-	-	-	-
Paper, Pulp and Print	-	-	-	2	-	-	-	-	-	-	-
Wood and Wood Products	-	1	-	-	-	-	-	-	-	-	-
Construction	-	-	-	2	-	-	-	2	-	-	-
Textile and Leather	-	1	-	3	-	-	-	-	-	-	-
Non-specified	-	1	-	1	-	-	-	-	-	-	-
TRANSPORT SECTOR	**1**	**3**	**-**	**4**	**-**	**-**	**-**	**-**	**-**	**-**	**-**
Air	-	-	-	-	-	-	-	-	-	-	-
Road	-	-	-	-	-	-	-	-	-	-	-
Rail	1	3	-	4	-	-	-	-	-	-	-
Pipeline Transport	-	-	-	-	-	-	-	-	-	-	-
Internal Navigation	-	-	-	-	-	-	-	-	-	-	-
Non-specified	-	-	-	-	-	-	-	-	-	-	-
OTHER SECTORS	**-**	**87**	**-**	**443**	**-**	**-**	**-**	**-**	**-**	**-**	**-**
Agriculture	-	-	-	3	-	-	-	-	-	-	-
Comm. and Publ. Services	-	-	-	-	-	-	-	-	-	-	-
Residential	-	84	-	438	-	-	-	-	-	-	-
Non-specified	-	3	-	2	-	-	-	-	-	-	-
NON-ENERGY USE	**-**	**14**	**-**	**-**	**-**	**698**	**-**	**-**	**-**	**-**	**-**
in Industry/Trans./Energy	-	14	-	-	-	698	-	-	-	-	-
in Transport	-	-	-	-	-	-	-	-	-	-	-
in Other Sectors	-	-	-	-	-	-	-	-	-	-	-

Romania / Roumanie : 1997

SUPPLY AND CONSUMPTION / *APPROVISIONNEMENT ET DEMANDE*	Oil cont. / *Pétrole cont.* (1000 tonnes)										
	Refinery Gas / *Gaz de raffinerie*	LPG + Ethane / *GPL + éthane*	Motor Gasoline / *Essence moteur*	Aviation Gasoline / *Essence aviation*	Jet Fuel / *Carbu-réacteurs*	Kerosene / *Kérosène*	Gas/ Diesel / *Gazole*	Heavy Fuel Oil / *Fioul lourd*	Naphtha / *Naphta*	Petrol. Coke / *Coke de pétrole*	Other Prod. / *Autres prod.*
Production	884	242	3205	3	88	66	3953	2079	433	442	907
From Other Sources	-	-	-	-	-	-	-	-	-	-	-
Imports	-	6	191	-	79	-	110	3567	2	-	47
Exports	-	-	-1575	-	-	-4	-1031	-1	-	-217	-46
Intl. Marine Bunkers	-	-	-	-	-	-	-	-	-	-	-
Stock Changes	3	1	-294	-	45	-5	-81	-472	-10	-47	3
DOMESTIC SUPPLY	**887**	**249**	**1527**	**3**	**212**	**57**	**2951**	**5173**	**425**	**178**	**911**
Transfers	-	-	-	-	-	-	-17	-47	-179	-	250
Statistical Differences	-57	-	34	-	-92	9	208	127	-204	102	425
TRANSFORMATION	**59**	**-**	**-**	**-**	**-**	**-**	**7**	**3719**	**-**	**-**	**184**
Electricity Plants	1	-	-	-	-	-	3	127	-	-	-
CHP Plants	58	-	-	-	-	-	-	3300	-	-	-
Heat Plants	-	-	-	-	-	-	4	292	-	-	184
Blast Furnaces/Gas Works	-	-	-	-	-	-	-	-	-	-	-
Coke/Pat. Fuel/BKB Plants	-	-	-	-	-	-	-	-	-	-	-
Petroleum Refineries	-	-	-	-	-	-	-	-	-	-	-
Petrochemical Industry	-	-	-	-	-	-	-	-	-	-	-
Liquefaction	-	-	-	-	-	-	-	-	-	-	-
Other Transform. Sector	-	-	-	-	-	-	-	-	-	-	-
ENERGY SECTOR	**535**	**-**	**6**	**-**	**-**	**24**	**70**	**385**	**-**	**21**	**396**
Coal Mines	-	-	-	-	-	-	25	3	-	-	-
Oil and Gas Extraction	-	-	2	-	-	-	36	-	-	-	-
Petroleum Refineries	535	-	-	-	-	-	-	-	-	-	-
Electr., CHP+Heat Plants	-	-	3	-	-	-	9	19	-	-	-
Pumped Storage (Elec.)	-	-	-	-	-	-	-	-	-	-	-
Other Energy Sector	-	-	1	-	-	24	-	363	-	21	396
Distribution Losses	-	-	-	-	-	4	-	-	-	-	140
FINAL CONSUMPTION	**236**	**249**	**1555**	**3**	**120**	**38**	**3065**	**1149**	**42**	**259**	**866**
INDUSTRY SECTOR	**236**	**3**	**36**	**-**	**-**	**8**	**299**	**892**	**42**	**259**	**297**
Iron and Steel	-	-	1	-	-	-	8	148	-	121	5
Chemical and Petrochem.	235	-	-	-	-	2	1	239	34	-	53
of which: Feedstocks	-	-	-	-	-	-	-	-	*2*	-	-
Non-Ferrous Metals	-	-	-	-	-	-	-	-	-	-	-
Non-Metallic Minerals	-	-	-	-	-	-	8	201	-	137	7
Transport Equipment	-	-	-	-	-	-	-	-	-	-	-
Machinery	-	2	4	-	-	5	19	57	1	1	31
Mining and Quarrying	-	-	-	-	-	-	12	14	-	-	6
Food and Tobacco	1	1	20	-	-	-	36	129	6	-	102
Paper, Pulp and Print	-	-	-	-	-	-	1	46	-	-	2
Wood and Wood Products	-	-	1	-	-	-	6	21	-	-	5
Construction	-	-	1	-	-	1	176	1	-	-	34
Textile and Leather	-	-	9	-	-	-	27	20	1	-	39
Non-specified	-	-	-	-	-	-	5	16	-	-	13
TRANSPORT SECTOR	**-**	**-**	**1429**	**3**	**120**	**8**	**2056**	**239**	**-**	**-**	**13**
Air	-	-	2	3	120	3	-	-	-	-	-
Road	-	-	1425	-	-	4	1664	-	-	-	13
Rail	-	-	1	-	-	1	285	-	-	-	-
Pipeline Transport	-	-	-	-	-	-	1	-	-	-	-
Internal Navigation	-	-	1	-	-	-	106	239	-	-	-
Non-specified	-	-	-	-	-	-	-	-	-	-	-
OTHER SECTORS	**-**	**246**	**90**	**-**	**-**	**22**	**710**	**18**	**-**	**-**	**180**
Agriculture	-	-	7	-	-	-	497	1	-	-	15
Comm. and Publ. Services	-	-	-	-	-	-	-	-	-	-	-
Residential	-	246	-	-	-	19	-	13	-	-	116
Non-specified	-	-	83	-	-	3	213	4	-	-	49
NON-ENERGY USE	**-**	**-**	**-**	**-**	**-**	**-**	**-**	**-**	**-**	**-**	**376**
in Industry/Transf./Energy	-	-	-	-	-	-	-	-	-	-	249
in Transport	-	-	-	-	-	-	-	-	-	-	71
in Other Sectors	-	-	-	-	-	-	-	-	-	-	56

Romania / Roumanie : 1997

	Gas / *Gaz* (TJ)				Comb. Renew. & Waste / *En. Re. Comb. & Déchets* (TJ)				(GWh)	(TJ)
SUPPLY AND CONSUMPTION / ***APPROVISIONNEMENT ET DEMANDE***	Natural Gas / *Gaz naturel*	Gas Works / *Usines à gaz*	Coke Ovens / *Cokeries*	Blast Furnaces / *Hauts fourneaux*	Solid Biomass / *Biomasse solide*	Gas/Liquids from Biomass / *Gaz/Liquides tirés de biomasse*	Municipal Waste / *Déchets urbains*	Industrial Waste / *Déchets industriels*	Electricity / *Electricité*	Heat / *Chaleur*
Production	553958	-	22092	30182	140672	-	-	1188	57148	284612
From Other Sources	-	-	-	-	-	-	-	-	-	-
Imports	187503	-	-	-	-	-	-	-	1038	-
Exports	-	-	-	-	-347	-	-	-	-817	-
Intl. Marine Bunkers	-	-	-	-	-	-	-	-	-	-
Stock Changes	-	-	-	-	-221	-	-	-	-	-
DOMESTIC SUPPLY	**741461**	**-**	**22092**	**30182**	**140104**	**-**	**-**	**1188**	**57369**	**284612**
Transfers	-	-	-	-	-	-	-	-	-	-
Statistical Differences	33275	-	670	-210	-1130	-	-	-	-285	-1693
TRANSFORMATION	**270522**	**-**	**1803**	**5543**	**-870**	**-**	**-**	**1188**	**-**	**-**
Electricity Plants	26489	-	-	2030	125	-	-	-	-	-
CHP Plants	170471	-	1803	3513	157	-	-	727	-	-
Heat Plants	73562	-	-	-	572	-	-	461	-	-
Blast Furnaces/Gas Works	-	-	-	-	-	-	-	-	-	-
Coke/Pat. Fuel/BKB Plants	-	-	-	-	38	-	-	-	-	-
Petroleum Refineries	-	-	-	-	-	-	-	-	-	-
Petrochemical Industry	-	-	-	-	-	-	-	-	-	-
Liquefaction	-	-	-	-	-	-	-	-	-	-
Other Transform. Sector	-	-	-	-	-1762	-	-	-	-	-
ENERGY SECTOR	**84171**	**-**	**5708**	**3591**	**190**	**-**	**-**	**-**	**12074**	**39754**
Coal Mines	102	-	-	-	164	-	-	-	1589	615
Oil and Gas Extraction	70996	-	-	-	-	-	-	-	1975	-
Petroleum Refineries	10250	-	-	-	-	-	-	-	1361	12673
Electr., CHP+Heat Plants	2309	-	-	-	26	-	-	-	3764	25372
Pumped Storage (Elec.)	-	-	-	-	-	-	-	-	-	-
Other Energy Sector	514	-	5708	3591	-	-	-	-	3385	1094
Distribution Losses	6141	-	102	1588	-	-	-	-	6581	38627
FINAL CONSUMPTION	**413902**	**-**	**15149**	**19250**	**139654**	**-**	**-**	**-**	**38429**	**204538**
INDUSTRY SECTOR	**302124**	**-**	**15149**	**19250**	**12489**	**-**	**-**	**-**	**25125**	**41459**
Iron and Steel	52941	-	15149	13515	88	-	-	-	7912	3305
Chemical and Petrochem.	121495	-	-	-	813	-	-	-	4815	14572
of which: Feedstocks	*55458*	-	-	-	-	-	-	-	-	-
Non-Ferrous Metals	-	-	-	-	-	-	-	-	-	-
Non-Metallic Minerals	35719	-	-	3543	45	-	-	-	1719	887
Transport Equipment	-	-	-	-	-	-	-	-	-	-
Machinery	38533	-	-	2192	70	-	-	-	3688	8024
Mining and Quarrying	2809	-	-	-	27	-	-	-	1071	20
Food and Tobacco	26368	-	-	-	914	-	-	-	1449	5445
Paper, Pulp and Print	6519	-	-	-	2534	-	-	-	776	1619
Wood and Wood Products	1404	-	-	-	4815	-	-	-	344	1572
Construction	1541	-	-	-	240	-	-	-	674	1073
Textile and Leather	11012	-	-	-	144	-	-	-	905	3198
Non-specified	3783	-	-	-	2799	-	-	-	1772	1744
TRANSPORT SECTOR	**1400**	**-**	**-**	**-**	**412**	**-**	**-**	**-**	**2230**	**-**
Air	-	-	-	-	-	-	-	-	-	-
Road	1298	-	-	-	-	-	-	-	-	-
Rail	-	-	-	-	189	-	-	-	1517	-
Pipeline Transport	-	-	-	-	-	-	-	-	59	-
Internal Navigation	-	-	-	-	-	-	-	-	-	-
Non-specified	102	-	-	-	223	-	-	-	654	-
OTHER SECTORS	**110378**	**-**	**-**	**-**	**126753**	**-**	**-**	**-**	**11074**	**163079**
Agriculture	3325	-	-	-	1291	-	-	-	1791	5270
Comm. and Publ. Services	-	-	-	-	-	-	-	-	-	-
Residential	95720	-	-	-	121839	-	-	-	7946	144803
Non-specified	11333	-	-	-	3623	-	-	-	1337	13006
NON-ENERGY USE	**-**	**-**	**-**	**-**	**-**	**-**	**-**	**-**	**-**	**-**
in Industry/Transf./Energy	-	-	-	-	-	-	-	-	-	-
in Transport	-	-	-	-	-	-	-	-	-	-
in Other Sectors	-	-	-	-	-	-	-	-	-	-

Russia / Russie : 1996

SUPPLY AND CONSUMPTION / *APPROVISIONNEMENT ET DEMANDE*	Coal / *Charbon* (1000 tonnes)							Oil / *Pétrole* (1000 tonnes)			
	Coking Coal / *Charbon à coke*	Other Bit. Coal / *Autres charb. bit.*	Sub-Bit. Coal / *Charbon sous-bit.*	Lignite / *Lignite*	Peat / *Tourbe*	Oven and Gas Coke / *Coke de four/gaz*	Pat. Fuel and BKB / *Agg./briq. de lignite*	Crude Oil / *Pétrole brut*	NGL / *LGN*	Feed-stocks / *Produits d'aliment.*	Additives / *Additifs*
Production	55276	111240	-	90179	4103	26930	603	299498	-	-	-
From Other Sources	-	-	-	-	-	-	-	-	-	-	-
Imports	1708	18373	-	-	-	108	16	6737	-	-	-
Exports	-6538	-18803	-	-2059	-38	-495	-	-126451	-	-	-
Intl. Marine Bunkers	-	-	-	-	-	-	-	-	-	-	-
Stock Changes	987	13331	-	1837	997	110	34	1327	-	-	-
DOMESTIC SUPPLY	**51433**	**124141**	**-**	**89957**	**5062**	**26653**	**653**	**181111**	**-**	**-**	**-**
Transfers	-	-	-	-	-	-	-	-	-	-	-
Statistical Differences	-12935	-721	-	-2705	-287	-2231	-	-1537	-	-	-
TRANSFORMATION	**38498**	**82047**	**-**	**74564**	**3288**	**15021**	**55**	**176471**	**-**	**-**	**-**
Electricity Plants	-	-	-	-	-	3	-	-	-	-	-
CHP Plants	-	62191	-	50822	2399	-	20	596	-	-	-
Heat Plants	-	19856	-	22425	564	355	35	-	-	-	-
Blast Furnaces/Gas Works	-	-	-	-	-	14633	-	-	-	-	-
Coke/Pat. Fuel/BKB Plants	38498	-	-	1317	325	30	-	-	-	-	-
Petroleum Refineries	-	-	-	-	-	-	-	175875	-	-	-
Petrochemical Industry	-	-	-	-	-	-	-	-	-	-	-
Liquefaction	-	-	-	-	-	-	-	-	-	-	-
Other Transform. Sector	-	-	-	-	-	-	-	-	-	-	-
ENERGY SECTOR	**-**	**548**	**-**	**192**	**109**	**128**	**62**	**540**	**-**	**-**	**-**
Coal Mines	-	431	-	167	109	128	35	-	-	-	-
Oil and Gas Extraction	-	-	-	-	-	-	-	540	-	-	-
Petroleum Refineries	-	-	-	-	-	-	-	-	-	-	-
Electr., CHP+Heat Plants	-	117	-	25	-	-	27	-	-	-	-
Pumped Storage (Elec.)	-	-	-	-	-	-	-	-	-	-	-
Other Energy Sector	-	-	-	-	-	-	-	-	-	-	-
Distribution Losses	-	8000	-	4042	69	18	-	-	-	-	-
FINAL CONSUMPTION	**-**	**32825**	**-**	**8454**	**1309**	**9255**	**536**	**2563**	**-**	**-**	**-**
INDUSTRY SECTOR	**-**	**3077**	**-**	**2714**	**22**	**6533**	**32**	**94**	**-**	**-**	**-**
Iron and Steel	-	586	-	433	-	3658	4	-	-	-	-
Chemical and Petrochem.	-	79	-	10	-	389	1	-	-	-	-
of which: Feedstocks	-	-	-	-	-	-	-	-	-	-	-
Non-Ferrous Metals	-	273	-	547	1	879	1	5	-	-	-
Non-Metallic Minerals	-	720	-	722	-	271	-	-	-	-	-
Transport Equipment	-	-	-	-	-	-	-	-	-	-	-
Machinery	-	465	-	160	8	1088	1	5	-	-	-
Mining and Quarrying	-	-	-	-	-	-	-	-	-	-	-
Food and Tobacco	-	170	-	75	-	143	4	5	-	-	-
Paper, Pulp and Print	-	-	-	-	-	-	-	1	-	-	-
Wood and Wood Products	-	38	-	139	3	9	1	-	-	-	-
Construction	-	644	-	527	3	6	2	56	-	-	-
Textile and Leather	-	55	-	92	5	1	-	-	-	-	-
Non-specified	-	47	-	9	2	89	18	22	-	-	-
TRANSPORT SECTOR	**-**	**650**	**-**	**205**	**1**	**3**	**28**	**9**	**-**	**-**	**-**
Air	-	-	-	-	-	-	-	-	-	-	-
Road	-	-	-	-	-	-	-	-	-	-	-
Rail	-	557	-	146	1	-	27	-	-	-	-
Pipeline Transport	-	-	-	-	-	-	-	8	-	-	-
Internal Navigation	-	8	-	11	-	3	1	-	-	-	-
Non-specified	-	85	-	48	-	-	-	1	-	-	-
OTHER SECTORS	**-**	**28405**	**-**	**5388**	**1286**	**198**	**473**	**928**	**-**	**-**	**-**
Agriculture	-	3946	-	1534	8	4	7	68	-	-	-
Comm. and Publ. Services	-	-	-	-	-	-	-	-	-	-	-
Residential	-	24459	-	3854	1140	60	466	118	-	-	-
Non-specified	-	-	-	-	138	134	-	742	-	-	-
NON-ENERGY USE	**-**	**693**	**-**	**147**	**-**	**2521**	**3**	**1532**	**-**	**-**	**-**
in Industry/Trans./Energy	-	693	-	147	-	2521	3	1532	-	-	-
in Transport	-	-	-	-	-	-	-	-	-	-	-
in Other Sectors	-	-	-	-	-	-	-	-	-	-	-

Russia / Russie : 1996

SUPPLY AND CONSUMPTION / *APPROVISIONNEMENT ET DEMANDE*	Oil cont. / *Pétrole cont.* (1000 tonnes)										
	Refinery Gas / *Gaz de raffinerie*	LPG + Ethane / *GPL + éthane*	Motor Gasoline / *Essence moteur*	Aviation Gasoline / *Essence aviation*	Jet Fuel / *Carbu-réacteurs*	Kerosene / *Kérosène*	Gas/ Diesel / *Gazole*	Heavy Fuel Oil / *Fioul lourd*	Naphtha / *Naphta*	Petrol. Coke / *Coke de pétrole*	Other Prod. / *Autres prod.*
Production	4109	4883	26780	38	8616	125	46682	65594	-	625	15680
From Other Sources	-	-	-	-	-	-	-	-	-	-	-
Imports	-	-	1304	3	47	-	663	490	-	117	-
Exports	-	-2	-3699	-2	-1319	-	-23957	-25127	-	-8	-2493
Intl. Marine Bunkers	-	-	-	-	-	-	-	-	-	-	-
Stock Changes	-	19	316	11	79	6	1227	374	-	-	-3
DOMESTIC SUPPLY	**4109**	**4900**	**24701**	**50**	**7423**	**131**	**24615**	**41331**	**-**	**734**	**13184**
Transfers	-	-	-	-	-	-	-	-	-	-	-
Statistical Differences	-	-	-	-	-	-	-	-	-	-	-
TRANSFORMATION	**1528**	**552**	**-**	**-**	**-**	**9**	**674**	**35053**	**-**	**-**	**1704**
Electricity Plants	-	-	-	-	-	-	-	-	-	-	-
CHP Plants	1528	552	-	-	-	9	-	-	-	-	1704
Heat Plants	-	-	-	-	-	-	674	35053	-	-	-
Blast Furnaces/Gas Works	-	-	-	-	-	-	-	-	-	-	-
Coke/Pat. Fuel/BKB Plants	-	-	-	-	-	-	-	-	-	-	-
Petroleum Refineries	-	-	-	-	-	-	-	-	-	-	-
Petrochemical Industry	-	-	-	-	-	-	-	-	-	-	-
Liquefaction	-	-	-	-	-	-	-	-	-	-	-
Other Transform. Sector	-	-	-	-	-	-	-	-	-	-	-
ENERGY SECTOR	**1876**	**18**	**1079**	**-**	**-**	**3**	**76**	**662**	**-**	**-**	**94**
Coal Mines	-	-	-	-	-	-	-	-	-	-	-
Oil and Gas Extraction	-	-	1079	-	-	-	-	-	-	-	-
Petroleum Refineries	1876	-	-	-	-	-	-	-	-	-	-
Electr., CHP+Heat Plants	-	-	-	-	-	-	-	-	-	-	-
Pumped Storage (Elec.)	-	-	-	-	-	-	-	-	-	-	-
Other Energy Sector	-	18	-	-	-	3	76	662	-	-	94
Distribution Losses	-	-	-	-	-	-	-	-	-	-	-
FINAL CONSUMPTION	**705**	**4330**	**23622**	**50**	**7423**	**119**	**23865**	**5616**	**-**	**734**	**11386**
INDUSTRY SECTOR	**678**	**1812**	**3433**	**6**	**218**	**37**	**2365**	**1333**	**-**	**-**	**3210**
Iron and Steel	1	4	127	-	-	3	52	102	-	-	292
Chemical and Petrochem.	643	1559	193	1	218	22	39	38	-	-	1982
of which: Feedstocks	*44*	*1244*	*65*	*1*	-	*13*	*10*	*4*	-	-	*1763*
Non-Ferrous Metals	2	157	137	1	-	2	123	68	-	-	219
Non-Metallic Minerals	-	-	-	-	-	-	-	-	-	-	-
Transport Equipment	-	-	-	-	-	-	-	-	-	-	-
Machinery	3	29	528	3	-	2	62	74	-	-	101
Mining and Quarrying	-	-	-	-	-	-	-	-	-	-	-
Food and Tobacco	1	9	200	-	-	-	277	272	-	-	7
Paper, Pulp and Print	-	-	-	-	-	-	-	-	-	-	-
Wood and Wood Products	-	4	259	-	-	-	220	108	-	-	-
Construction	9	43	1448	-	-	7	1571	543	-	-	517
Textile and Leather	19	-	74	-	-	-	2	36	-	-	92
Non-specified	-	7	467	1	-	1	19	92	-	-	-
TRANSPORT SECTOR	**15**	**140**	**11998**	**41**	**7204**	**34**	**9760**	**2056**	**-**	**-**	**572**
Air	-	-	-	41	7204	-	-	-	-	-	-
Road	2	100	11510	-	-	2	6076	76	-	-	147
Rail	-	9	348	-	-	-	2079	992	-	-	239
Pipeline Transport	11	7	114	-	-	-	4	14	-	-	22
Internal Navigation	-	12	26	-	-	-	1038	744	-	-	112
Non-specified	2	12	-	-	-	32	563	230	-	-	52
OTHER SECTORS	**12**	**1881**	**8150**	**-**	**-**	**47**	**11734**	**2215**	**-**	**-**	**677**
Agriculture	7	269	8150	-	-	16	6145	670	-	-	26
Comm. and Publ. Services	-	-	-	-	-	-	-	-	-	-	-
Residential	5	1532	-	-	-	25	1608	1545	-	-	606
Non-specified	-	80	-	-	-	6	3981	-	-	-	45
NON-ENERGY USE	**-**	**497**	**41**	**3**	**1**	**1**	**6**	**12**	**-**	**734**	**6927**
in Industry/Transf./Energy	-	497	41	3	1	1	6	12	-	734	6927
in Transport	-	-	-	-	-	-	-	-	-	-	-
in Other Sectors	-	-	-	-	-	-	-	-	-	-	-

Russia / Russie : 1996

SUPPLY AND CONSUMPTION / *APPROVISIONNEMENT ET DEMANDE*	Gas / *Gaz* (TJ)				Comb. Renew. & Waste / *En. Re. Comb. & Déchets* (TJ)				(GWh)	(TJ)
	Natural Gas / *Gaz naturel*	Gas Works / *Usines à gaz*	Coke Ovens / *Cokeries*	Blast Furnaces / *Hauts fourneaux*	Solid Biomass / *Biomasse solide*	Gas/Liquids from Biomass / *Gaz/Liquides tirés de biomasse*	Municipal Waste / *Déchets urbains*	Industrial Waste / *Déchets industriels*	Electricity / *Electricité*	Heat / *Chaleur*
Production	22339444	-	159070	231610	649342	-	-	71041	847211	6709300
From Other Sources	-	-	-	-	3341	-	-	-	-	-
Imports	130000	-	-	-	29	-	-	-	12356	-
Exports	-7382444	-	-	-	-5217	-	-	-6799	-31846	-
Intl. Marine Bunkers	-	-	-	-	-	-	-	-	-	-
Stock Changes	-525778	-	-	-	1114	-	-	-	-	-
DOMESTIC SUPPLY	**14561222**	**-**	**159070**	**231610**	**648609**	**-**	**-**	**64242**	**827721**	**6709300**
Transfers	-	-	-	-	-	-	-	-	-	-
Statistical Differences	-117111	-	-	-	-147	-	-	-3605	-	-
TRANSFORMATION	**8774667**	**-**	**43873**	**53093**	**256296**	**-**	**-**	**16207**	**-**	**-**
Electricity Plants	-	-	-	-	-	-	-	-	-	-
CHP Plants	6840667	-	-	36899	91117	-	-	9701	-	-
Heat Plants	1934000	-	42082	16194	164504	-	-	6506	-	-
Blast Furnaces/Gas Works	-	-	1791	-	-	-	-	-	-	-
Coke/Pat. Fuel/BKB Plants	-	-	-	-	-	-	-	-	-	-
Petroleum Refineries	-	-	-	-	-	-	-	-	-	-
Petrochemical Industry	-	-	-	-	-	-	-	-	-	-
Liquefaction	-	-	-	-	-	-	-	-	-	-
Other Transform. Sector	-	-	-	-	675	-	-	-	-	-
ENERGY SECTOR	**1123778**	**-**	**4490**	**5230**	**-**	**-**	**-**	**1788**	**142284**	**521400**
Coal Mines	1123333	-	-	-	-	-	-	-	9607	66200
Oil and Gas Extraction	-	-	-	-	-	-	-	-	41285	157300
Petroleum Refineries	-	-	-	-	-	-	-	-	11762	213900
Electr., CHP+Heat Plants	445	-	-	5230	-	-	-	-	61024	-
Pumped Storage (Elec.)	-	-	-	-	-	-	-	-	-	-
Other Energy Sector	-	-	4490	-	-	-	-	1788	18606	84000
Distribution Losses	310888	-	-	-	5950	-	-	29	84457	214500
FINAL CONSUMPTION	**4234778**	**-**	**110707**	**173287**	**386216**	**-**	**-**	**42613**	**600980**	**5973400**
INDUSTRY SECTOR	**1604445**	**-**	**107391**	**170789**	**26406**	**-**	**-**	**36810**	**266667**	**2349700**
Iron and Steel	508778	-	107339	159808	235	-	-	234	50414	262800
Chemical and Petrochem.	621778	-	-	-	117	-	-	36254	37218	559200
of which: Feedstocks	*509667*	-	-	-	-	-	-	-	-	-
Non-Ferrous Metals	74778	-	-	10813	-	-	-	-	83250	188000
Non-Metallic Minerals	-	-	-	-	-	-	-	-	-	-
Transport Equipment	-	-	-	-	-	-	-	-	-	-
Machinery	91000	-	52	168	88	-	-	-	40300	526800
Mining and Quarrying	-	-	-	-	1260	-	-	-	-	-
Food and Tobacco	19889	-	-	-	3635	-	-	-	10334	225400
Paper, Pulp and Print	-	-	-	-	15913	-	-	-	12704	224300
Wood and Wood Products	50667	-	-	-	-	-	-	88	2930	-
Construction	230778	-	-	-	1817	-	-	234	21756	221600
Textile and Leather	1333	-	-	-	-	-	-	-	3874	67700
Non-specified	5444	-	-	-	3341	-	-	-	3887	73900
TRANSPORT SECTOR	**669333**	**-**	**-**	**-**	**323**	**-**	**-**	**29**	**64933**	**-**
Air	-	-	-	-	-	-	-	-	-	-
Road	3333	-	-	-	-	-	-	-	-	-
Rail	-	-	-	-	-	-	-	-	30525	-
Pipeline Transport	661333	-	-	-	-	-	-	-	18890	-
Internal Navigation	-	-	-	-	-	-	-	-	-	-
Non-specified	4667	-	-	-	323	-	-	29	15518	-
OTHER SECTORS	**1961000**	**-**	**33**	**1931**	**359487**	**-**	**-**	**5774**	**269380**	**3623700**
Agriculture	173889	-	-	-	28135	-	-	29	48706	253300
Comm. and Publ. Services	-	-	-	-	175377	-	-	5745	-	154700
Residential	1719667	-	33	-	138713	-	-	-	144825	2766800
Non-specified	67444	-	-	1931	17262	-	-	-	75849	448900
NON-ENERGY USE	**-**	**-**	**3283**	**567**	**-**	**-**	**-**	**-**	**-**	**-**
in Industry/Transf./Energy	-	-	3283	567	-	-	-	-	-	-
in Transport	-	-	-	-	-	-	-	-	-	-
in Other Sectors	-	-	-	-	-	-	-	-	-	-

Russia / Russie : 1997

SUPPLY AND CONSUMPTION / *APPROVISIONNEMENT ET DEMANDE*	Coal / *Charbon* (1000 tonnes)							Oil / *Pétrole* (1000 tonnes)			
	Coking Coal / *Charbon à coke*	Other Bit. Coal / *Autres charb. bit.*	Sub-Bit. Coal / *Charbon sous-bit.*	Lignite / *Lignite*	Peat / *Tourbe*	Oven and Gas Coke / *Coke de four/gaz*	Pat. Fuel and BKB / *Agg./briq. de lignite*	Crude Oil / *Pétrole brut*	NGL / *LGN*	Feed-stocks / *Produits d'aliment.*	Additives / *Additifs*
Production	52400	106800	-	85200	3224	25374	577	304000	-	-	-
From Other Sources	-	-	-	-	-	-	-	-	-	-	-
Imports	1355	13725	-	-	-	314	16	7500	-	-	-
Exports	-5295	-15885	-	-1720	-	-1314	-	-126000	-	-	-
Intl. Marine Bunkers	-	-	-	-	-	-	-	-	-	-	-
Stock Changes	-	-	-	-	-	-	-	-2540	-	-	-
DOMESTIC SUPPLY	**48460**	**104640**	**-**	**83480**	**3224**	**24374**	**593**	**182960**	**-**	**-**	**-**
Transfers	-	-	-	-	-	-	-	-	-	-	-
Statistical Differences	-12187	-	-	-	-	-	-	-2500	-	-	-
TRANSFORMATION	**36273**	**69562**	**-**	**71340**	**2276**	**14806**	**50**	**177348**	**-**	**-**	**-**
Electricity Plants	-	-	-	-	-	3	-	-	-	-	-
CHP Plants	-	52728	-	48625	1268	-	18	598	-	-	-
Heat Plants	-	16834	-	21455	629	369	32	-	-	-	-
Blast Furnaces/Gas Works	-	-	-	-	-	14403	-	-	-	-	-
Coke/Pat. Fuel/BKB Plants	36273	-	-	1260	379	31	-	-	-	-	-
Petroleum Refineries	-	-	-	-	-	-	-	176750	-	-	-
Petrochemical Industry	-	-	-	-	-	-	-	-	-	-	-
Liquefaction	-	-	-	-	-	-	-	-	-	-	-
Other Transform. Sector	-	-	-	-	-	-	-	-	-	-	-
ENERGY SECTOR	**-**	**465**	**-**	**184**	**45**	**133**	**56**	**541**	**-**	**-**	**-**
Coal Mines	-	366	-	160	45	133	32	-	-	-	-
Oil and Gas Extraction	-	-	-	-	-	-	-	541	-	-	-
Petroleum Refineries	-	-	-	-	-	-	-	-	-	-	-
Electr., CHP+Heat Plants	-	99	-	24	-	-	24	-	-	-	-
Pumped Storage (Elec.)	-	-	-	-	-	-	-	-	-	-	-
Other Energy Sector	-	-	-	-	-	-	-	-	-	-	-
Distribution Losses	-	6783	-	3867	53	19	-	-	-	-	-
FINAL CONSUMPTION	**-**	**27830**	**-**	**8089**	**850**	**9416**	**487**	**2571**	**-**	**-**	**-**
INDUSTRY SECTOR	**-**	**2609**	**-**	**2597**	**11**	**6588**	**29**	**94**	**-**	**-**	**-**
Iron and Steel	-	497	-	414	-	3601	4	-	-	-	-
Chemical and Petrochem.	-	67	-	10	-	404	1	-	-	-	-
of which: Feedstocks	-	-	-	-	-	-	-	-	-	-	-
Non-Ferrous Metals	-	232	-	523	-	913	1	5	-	-	-
Non-Metallic Minerals	-	610	-	691	-	282	-	-	-	-	-
Transport Equipment	-	-	-	-	-	-	-	-	-	-	-
Machinery	-	394	-	153	5	1130	1	5	-	-	-
Mining and Quarrying	-	-	-	-	-	-	-	-	-	-	-
Food and Tobacco	-	144	-	72	-	149	3	5	-	-	-
Paper, Pulp and Print	-	-	-	-	-	-	-	1	-	-	-
Wood and Wood Products	-	32	-	133	2	9	1	-	-	-	-
Construction	-	546	-	504	2	6	2	56	-	-	-
Textile and Leather	-	47	-	88	-	1	-	-	-	-	-
Non-specified	-	40	-	9	2	93	16	22	-	-	-
TRANSPORT SECTOR	**-**	**551**	**-**	**196**	**1**	**3**	**25**	**9**	**-**	**-**	**-**
Air	-	-	-	-	-	-	-	-	-	-	-
Road	-	-	-	-	-	-	-	-	-	-	-
Rail	-	472	-	140	1	-	24	-	-	-	-
Pipeline Transport	-	-	-	-	-	-	-	8	-	-	-
Internal Navigation	-	7	-	10	-	3	1	-	-	-	-
Non-specified	-	72	-	46	-	-	-	1	-	-	-
OTHER SECTORS	**-**	**24083**	**-**	**5155**	**838**	**206**	**430**	**931**	**-**	**-**	**-**
Agriculture	-	3346	-	1468	7	4	7	68	-	-	-
Comm. and Publ. Services	-	-	-	-	-	-	-	-	-	-	-
Residential	-	20737	-	3687	778	63	423	119	-	-	-
Non-specified	-	-	-	-	53	139	-	744	-	-	-
NON-ENERGY USE	**-**	**587**	**-**	**141**	**-**	**2619**	**3**	**1537**	**-**	**-**	**-**
in Industry/Trans./Energy	-	587	-	141	-	2619	3	1537	-	-	-
in Transport	-	-	-	-	-	-	-	-	-	-	-
in Other Sectors	-	-	-	-	-	-	-	-	-	-	-

Russia / Russie : 1997

SUPPLY AND CONSUMPTION / *APPROVISIONNEMENT ET DEMANDE*	Oil cont. / *Pétrole cont.* (1000 tonnes)										
	Refinery Gas / *Gaz de raffinerie*	LPG + Ethane / *GPL + éthane*	Motor Gasoline / *Essence moteur*	Aviation Gasoline / *Essence aviation*	Jet Fuel / *Carbu-réacteurs*	Kerosene / *Kérosène*	Gas/ Diesel / *Gazole*	Heavy Fuel Oil / *Fioul lourd*	Naphtha / *Naphta*	Petrol. Coke / *Coke de pétrole*	Other Prod. / *Autres prod.*
Production	4400	5250	27200	40	7900	130	47200	64800	-	680	16400
From Other Sources	-	-	-	-	-	-	-	-	-	-	-
Imports	-	-	1100	5	350	-	500	500	-	500	345
Exports	-	-1000	-5600	-5	-1250	-	-23000	-26000	-	-10	-3135
Intl. Marine Bunkers	-	-	-	-	-	-	-	-	-	-	-
Stock Changes	-	-	580	-	-	-	380	-500	-	-	-
DOMESTIC SUPPLY	**4400**	**4250**	**23280**	**40**	**7000**	**130**	**25080**	**38800**	**-**	**1170**	**13610**
Transfers	-	-	-	-	-	-	-	-	-	-	-
Statistical Differences	-	-	-	-	-	-	-	-	-	-	-
TRANSFORMATION	**1642**	**594**	**-**	**-**	**-**	**9**	**687**	**32906**	**-**	**-**	**1810**
Electricity Plants	-	-	-	-	-	-	-	-	-	-	-
CHP Plants	1642	594	-	-	-	9	-	-	-	-	1810
Heat Plants	-	-	-	-	-	-	687	32906	-	-	-
Blast Furnaces/Gas Works	-	-	-	-	-	-	-	-	-	-	-
Coke/Pat. Fuel/BKB Plants	-	-	-	-	-	-	-	-	-	-	-
Petroleum Refineries	-	-	-	-	-	-	-	-	-	-	-
Petrochemical Industry	-	-	-	-	-	-	-	-	-	-	-
Liquefaction	-	-	-	-	-	-	-	-	-	-	-
Other Transform. Sector	-	-	-	-	-	-	-	-	-	-	-
ENERGY SECTOR	**2000**	**21**	**1017**	**-**	**-**	**3**	**77**	**622**	**-**	**-**	**100**
Coal Mines	-	-	-	-	-	-	-	-	-	-	-
Oil and Gas Extraction	-	-	1017	-	-	-	-	-	-	-	-
Petroleum Refineries	2000	-	-	-	-	-	-	-	-	-	-
Electr., CHP+Heat Plants	-	-	-	-	-	-	-	-	-	-	-
Pumped Storage (Elec.)	-	-	-	-	-	-	-	-	-	-	-
Other Energy Sector	-	21	-	-	-	3	77	622	-	-	100
Distribution Losses	-	-	-	-	-	-	-	-	-	-	-
FINAL CONSUMPTION	**758**	**3635**	**22263**	**40**	**7000**	**118**	**24316**	**5272**	**-**	**1170**	**11700**
INDUSTRY SECTOR	**729**	**1676**	**3235**	**5**	**205**	**37**	**2410**	**1252**	**-**	**-**	**3410**
Iron and Steel	1	-	120	-	-	3	53	96	-	-	310
Chemical and Petrochem.	691	1568	182	1	205	22	40	36	-	-	2106
of which: Feedstocks	*47*	*1246*	*61*	*1*	*-*	*13*	*10*	*4*	*-*	*-*	*1790*
Non-Ferrous Metals	2	13	129	1	-	2	125	64	-	-	233
Non-Metallic Minerals	-	-	-	-	-	-	-	-	-	-	-
Transport Equipment	-	-	-	-	-	-	-	-	-	-	-
Machinery	3	31	498	2	-	2	63	69	-	-	107
Mining and Quarrying	-	-	-	-	-	-	-	-	-	-	-
Food and Tobacco	1	9	188	-	-	-	282	255	-	-	7
Paper, Pulp and Print	-	-	-	-	-	-	-	-	-	-	-
Wood and Wood Products	-	3	244	-	-	-	224	101	-	-	-
Construction	10	46	1364	-	-	7	1601	510	-	-	549
Textile and Leather	21	1	70	-	-	-	2	34	-	-	98
Non-specified	-	5	440	1	-	1	20	87	-	-	-
TRANSPORT SECTOR	**16**	**134**	**11308**	**33**	**6794**	**34**	**9944**	**1930**	**-**	**-**	**608**
Air	-	-	-	33	6794	-	-	-	-	-	-
Road	2	107	10848	-	-	2	6191	71	-	-	156
Rail	-	6	328	-	-	-	2118	931	-	-	254
Pipeline Transport	12	5	107	-	-	-	4	13	-	-	24
Internal Navigation	-	11	25	-	-	-	1057	699	-	-	119
Non-specified	2	5	-	-	-	32	574	216	-	-	55
OTHER SECTORS	**13**	**1822**	**7681**	**-**	**-**	**46**	**11956**	**2079**	**-**	**-**	**719**
Agriculture	8	336	7681	-	-	16	6261	629	-	-	27
Comm. and Publ. Services	-	-	-	-	-	-	-	-	-	-	-
Residential	5	1486	-	-	-	24	1639	1450	-	-	644
Non-specified	-	-	-	-	-	6	4056	-	-	-	48
NON-ENERGY USE	**-**	**3**	**39**	**2**	**1**	**1**	**6**	**11**	**-**	**1170**	**6963**
in Industry/Transf./Energy	-	3	39	2	1	1	6	11	-	1170	6963
in Transport	-	-	-	-	-	-	-	-	-	-	-
in Other Sectors	-	-	-	-	-	-	-	-	-	-	-

Russia / Russie : 1997

SUPPLY AND CONSUMPTION / *APPROVISIONNEMENT ET DEMANDE*	Gas / *Gaz* (TJ)				Comb. Renew. & Waste / *En. Re. Comb. & Déchets* (TJ)				(GWh)	(TJ)
	Natural Gas / *Gaz naturel*	Gas Works / *Usines à gaz*	Coke Ovens / *Cokeries*	Blast Furnaces / *Hauts fourneaux*	Solid Biomass / *Biomasse solide*	Gas/Liquids from Biomass / *Gaz/Liquides tirés de biomasse*	Municipal Waste / *Déchets urbains*	Industrial Waste / *Déchets industriels*	Electricity / *Electricité*	Heat / *Chaleur*
Production	21457609	-	149879	227970	649342	-	-	71041	834100	6038400
From Other Sources	-	-	-	-	3341	-	-	-	-	-
Imports	87108	-	-	-	29	-	-	-	7100	-
Exports	-7132494	-	-	-	-5217	-	-	-6799	-26800	-
Intl. Marine Bunkers	-	-	-	-	-	-	-	-	-	-
Stock Changes	-	-	-	-	1114	-	-	-	-	-
DOMESTIC SUPPLY	**14412223**	**-**	**149879**	**227970**	**648609**	**-**	**-**	**64242**	**814400**	**6038400**
Transfers	-	-	-	-	-	-	-	-	-	-
Statistical Differences	-	-	-	-	-147	-	-	-3605	-	-
TRANSFORMATION	**8755295**	**-**	**45012**	**52186**	**256296**	**-**	**-**	**16207**	**-**	**-**
Electricity Plants	-	-	-	-	-	-	-	-	-	-
CHP Plants	6825565	-	-	36269	91117	-	-	9701	-	-
Heat Plants	1929730	-	43175	15917	164504	-	-	6506	-	-
Blast Furnaces/Gas Works	-	-	1837	-	-	-	-	-	-	-
Coke/Pat. Fuel/BKB Plants	-	-	-	-	-	-	-	-	-	-
Petroleum Refineries	-	-	-	-	-	-	-	-	-	-
Petrochemical Industry	-	-	-	-	-	-	-	-	-	-
Liquefaction	-	-	-	-	-	-	-	-	-	-
Other Transform. Sector	-	-	-	-	675	-	-	-	-	-
ENERGY SECTOR	**1121297**	**-**	**4607**	**5141**	**-**	**-**	**-**	**1788**	**146592**	**464960**
Coal Mines	1120853	-	-	-	-	-	-	-	9898	59030
Oil and Gas Extraction	-	-	-	-	-	-	-	-	42535	140270
Petroleum Refineries	-	-	-	-	-	-	-	-	12118	190750
Electr., CHP+Heat Plants	444	-	-	5141	-	-	-	-	62872	-
Pumped Storage (Elec.)	-	-	-	-	-	-	-	-	-	-
Other Energy Sector	-	-	4607	-	-	-	-	1788	19169	74910
Distribution Losses	310202	-	-	-	5950	-	-	29	81440	181150
FINAL CONSUMPTION	**4225429**	**-**	**100260**	**170643**	**386216**	**-**	**-**	**42613**	**586368**	**5392290**
INDUSTRY SECTOR	**1600903**	**-**	**96858**	**168188**	**26406**	**-**	**-**	**36810**	**264470**	**2156920**
Iron and Steel	507655	-	96805	157395	235	-	-	234	49999	241240
Chemical and Petrochem.	620405	-	-	-	117	-	-	36254	36911	513320
of which: Feedstocks	*508542*	-	-	-	-	-	-	-	-	-
Non-Ferrous Metals	74613	-	-	10628	-	-	-	-	82564	172570
Non-Metallic Minerals	-	-	-	-	-	-	-	-	-	-
Transport Equipment	-	-	-	-	-	-	-	-	-	-
Machinery	90799	-	53	165	88	-	-	-	39968	483580
Mining and Quarrying	-	-	-	-	1260	-	-	-	-	-
Food and Tobacco	19845	-	-	-	3635	-	-	-	10249	206910
Paper, Pulp and Print	-	-	-	-	15913	-	-	-	12599	205900
Wood and Wood Products	50555	-	-	-	-	-	-	88	2906	-
Construction	230269	-	-	-	1817	-	-	234	21577	203420
Textile and Leather	1330	-	-	-	-	-	-	-	3842	62140
Non-specified	5432	-	-	-	3341	-	-	-	3855	67840
TRANSPORT SECTOR	**667855**	**-**	**-**	**-**	**323**	**-**	**-**	**29**	**63500**	**-**
Air	-	-	-	-	-	-	-	-	-	-
Road	3325	-	-	-	-	-	-	-	-	-
Rail	-	-	-	-	-	-	-	-	29851	-
Pipeline Transport	659873	-	-	-	-	-	-	-	18473	-
Internal Navigation	-	-	-	-	-	-	-	-	-	-
Non-specified	4657	-	-	-	323	-	-	29	15176	-
OTHER SECTORS	**1956671**	**-**	**34**	**1898**	**359487**	**-**	**-**	**5774**	**258398**	**3235370**
Agriculture	173505	-	-	-	28135	-	-	29	44283	226160
Comm. and Publ. Services	-	-	-	-	175377	-	-	5745	-	138120
Residential	1715871	-	34	-	138713	-	-	-	141316	2470300
Non-specified	67295	-	-	1898	17262	-	-	-	72799	400790
NON-ENERGY USE	**-**	**-**	**3368**	**557**	**-**	**-**	**-**	**-**	**-**	**-**
in Industry/Transf./Energy	-	-	3368	557	-	-	-	-	-	-
in Transport	-	-	-	-	-	-	-	-	-	-
in Other Sectors	-	-	-	-	-	-	-	-	-	-

Saudi Arabia / Arabie saoudite

SUPPLY AND CONSUMPTION 1996	Coal (1000 tonnes)							Oil (1000 tonnes)			
	Coking Coal	Other Bit. Coal	Sub-Bit. Coal	Lignite	Peat	Oven and Gas Coke	Pat. Fuel and BKB	Crude Oil	NGL	Feed-stocks	Additives
Production	-	-	-	-	-	-	-	407280	24951	-	-
Imports	-	-	-	-	-	-	-	-	-	-	-
Exports	-	-	-	-	-	-	-	-313898	-13151	-	-
Intl. Marine Bunkers	-	-	-	-	-	-	-	-	-	-	-
Stock Changes	-	-	-	-	-	-	-	-	-	-	-
DOMESTIC SUPPLY	**-**	**-**	**-**	**-**	**-**	**-**	**-**	**93382**	**11800**	**-**	**-**
Transfers and Stat. Diff.	-	-	-	-	-	-	-	816	-7056	-	-
TRANSFORMATION	**-**	**-**	**-**	**-**	**-**	**-**	**-**	**94198**	**-**	**-**	**-**
Electricity and CHP Plants	-	-	-	-	-	-	-	8716	-	-	-
Petroleum Refineries	-	-	-	-	-	-	-	85482	-	-	-
Other Transform. Sector	-	-	-	-	-	-	-	-	-	-	-
ENERGY SECTOR	**-**	**-**	**-**	**-**	**-**	**-**	**-**	**-**	**-**	**-**	**-**
DISTRIBUTION LOSSES	**-**	**-**	**-**	**-**	**-**	**-**	**-**	**-**	**-**	**-**	**-**
FINAL CONSUMPTION	**-**	**-**	**-**	**-**	**-**	**-**	**-**	**-**	**4744**	**-**	**-**
INDUSTRY SECTOR	**-**	**-**	**-**	**-**	**-**	**-**	**-**	**-**	**4744**	**-**	**-**
Iron and Steel	-	-	-	-	-	-	-	-	-	-	-
Chemical and Petrochem.	-	-	-	-	-	-	-	-	4744	-	-
Non-Metallic Minerals	-	-	-	-	-	-	-	-	-	-	-
Non-specified	-	-	-	-	-	-	-	-	-	-	-
TRANSPORT SECTOR	**-**	**-**	**-**	**-**	**-**	**-**	**-**	**-**	**-**	**-**	**-**
Air	-	-	-	-	-	-	-	-	-	-	-
Road	-	-	-	-	-	-	-	-	-	-	-
Non-specified	-	-	-	-	-	-	-	-	-	-	-
OTHER SECTORS	**-**	**-**	**-**	**-**	**-**	**-**	**-**	**-**	**-**	**-**	**-**
Agriculture	-	-	-	-	-	-	-	-	-	-	-
Comm. and Publ. Services	-	-	-	-	-	-	-	-	-	-	-
Residential	-	-	-	-	-	-	-	-	-	-	-
Non-specified	-	-	-	-	-	-	-	-	-	-	-
NON-ENERGY USE	**-**	**-**	**-**	**-**	**-**	**-**	**-**	**-**	**-**	**-**	**-**

APPROVISIONNEMENT ET DEMANDE 1997	*Charbon* (1000 tonnes)							*Pétrole* (1000 tonnes)			
	Charbon à coke	*Autres charb. bit.*	*Charbon sous-bit.*	*Lignite*	*Tourbe*	*Coke de four/gaz*	*Agg./briq. de lignite*	*Pétrole brut*	*LGN*	*Produits d'aliment.*	*Additifs*
Production	-	-	-	-	-	-	-	417326	25057	-	-
Imports	-	-	-	-	-	-	-	-	-	-	-
Exports	-	-	-	-	-	-	-	-325499	-13313	-	-
Intl. Marine Bunkers	-	-	-	-	-	-	-	-	-	-	-
Stock Changes	-	-	-	-	-	-	-	-	-	-	-
DOMESTIC SUPPLY	**-**	**-**	**-**	**-**	**-**	**-**	**-**	**91827**	**11744**	**-**	**-**
Transfers and Stat. Diff.	-	-	-	-	-	-	-	-	-5794	-	-
TRANSFORMATION	**-**	**-**	**-**	**-**	**-**	**-**	**-**	**91827**	**-**	**-**	**-**
Electricity and CHP Plants	-	-	-	-	-	-	-	9293	-	-	-
Petroleum Refineries	-	-	-	-	-	-	-	82534	-	-	-
Other Transform. Sector	-	-	-	-	-	-	-	-	-	-	-
ENERGY SECTOR	**-**	**-**	**-**	**-**	**-**	**-**	**-**	**-**	**-**	**-**	**-**
DISTRIBUTION LOSSES	**-**	**-**	**-**	**-**	**-**	**-**	**-**	**-**	**-**	**-**	**-**
FINAL CONSUMPTION	**-**	**-**	**-**	**-**	**-**	**-**	**-**	**-**	**5950**	**-**	**-**
INDUSTRY SECTOR	**-**	**-**	**-**	**-**	**-**	**-**	**-**	**-**	**5950**	**-**	**-**
Iron and Steel	-	-	-	-	-	-	-	-	-	-	-
Chemical and Petrochem.	-	-	-	-	-	-	-	-	5950	-	-
Non-Metallic Minerals	-	-	-	-	-	-	-	-	-	-	-
Non-specified	-	-	-	-	-	-	-	-	-	-	-
TRANSPORT SECTOR	**-**	**-**	**-**	**-**	**-**	**-**	**-**	**-**	**-**	**-**	**-**
Air	-	-	-	-	-	-	-	-	-	-	-
Road	-	-	-	-	-	-	-	-	-	-	-
Non-specified	-	-	-	-	-	-	-	-	-	-	-
OTHER SECTORS	**-**	**-**	**-**	**-**	**-**	**-**	**-**	**-**	**-**	**-**	**-**
Agriculture	-	-	-	-	-	-	-	-	-	-	-
Comm. and Publ. Services	-	-	-	-	-	-	-	-	-	-	-
Residential	-	-	-	-	-	-	-	-	-	-	-
Non-specified	-	-	-	-	-	-	-	-	-	-	-
NON-ENERGY USE	**-**	**-**	**-**	**-**	**-**	**-**	**-**	**-**	**-**	**-**	**-**

Saudi Arabia / Arabie saoudite

SUPPLY AND CONSUMPTION 1996	Oil cont. (1000 tonnes)										
	Refinery Gas	LPG + Ethane	Motor Gasoline	Aviation Gasoline	Jet Fuel	Kerosene	Gas/ Diesel	Heavy Fuel Oil	Naphtha	Petrol. Coke	Other Prod.
Production	1851	1240	11844	-	4296	3849	25735	27114	5182	-	712
Imports	-	-	-	-	-	-	-	-	-	-	-
Exports	-	-375	-9341	-	-1607	-3641	-8513	-19051	-5169	-	-
Intl. Marine Bunkers	-	-	-	-	-	-	-	-1936	-	-	-
Stock Changes	-	-	-	-	-	-	-	-	-	-	-
DOMESTIC SUPPLY	**1851**	**865**	**2503**	**-**	**2689**	**208**	**17222**	**6127**	**13**	**-**	**712**
Transfers and Stat. Diff.	-	-45	7056	-	-233	8	-1313	-81	-13	-	272
TRANSFORMATION	**-**	**-**	**-**	**-**	**-**	**-**	**5448**	**90**	**-**	**-**	**-**
Electricity and CHP Plants	-	-	-	-	-	-	5448	90	-	-	-
Petroleum Refineries	-	-	-	-	-	-	-	-	-	-	-
Other Transform. Sector	-	-	-	-	-	-	-	-	-	-	-
ENERGY SECTOR	**1851**	**-**	**-**	**-**	**-**	**-**	**358**	**1845**	**-**	**-**	**-**
DISTRIBUTION LOSSES	**-**	**-**	**-**	**-**	**-**	**-**	**-**	**-**	**-**	**-**	**-**
FINAL CONSUMPTION	**-**	**820**	**9559**	**-**	**2456**	**216**	**10103**	**4111**	**-**	**-**	**984**
INDUSTRY SECTOR	**-**	**-**	**-**	**-**	**-**	**-**	**-**	**4111**	**-**	**-**	**-**
Iron and Steel	-	-	-	-	-	-	-	-	-	-	-
Chemical and Petrochem.	-	-	-	-	-	-	-	-	-	-	-
Non-Metallic Minerals	-	-	-	-	-	-	-	-	-	-	-
Non-specified	-	-	-	-	-	-	-	4111	-	-	-
TRANSPORT SECTOR	**-**	**-**	**9559**	**-**	**2456**	**-**	**-**	**-**	**-**	**-**	**-**
Air	-	-	-	-	2456	-	-	-	-	-	-
Road	-	-	9559	-	-	-	-	-	-	-	-
Non-specified	-	-	-	-	-	-	-	-	-	-	-
OTHER SECTORS	**-**	**820**	**-**	**-**	**-**	**216**	**10103**	**-**	**-**	**-**	**-**
Agriculture	-	-	-	-	-	-	-	-	-	-	-
Comm. and Publ. Services	-	-	-	-	-	-	-	-	-	-	-
Residential	-	820	-	-	-	216	-	-	-	-	-
Non-specified	-	-	-	-	-	-	10103	-	-	-	-
NON-ENERGY USE	**-**	**-**	**-**	**-**	**-**	**-**	**-**	**-**	**-**	**-**	**984**

APPROVISIONNEMENT ET DEMANDE 1997	*Pétrole cont.* (1000 tonnes)										
	Gaz de raffinerie	*GPL + éthane*	*Essence moteur*	*Essence aviation*	*Carbu- réacteurs*	*Kérosène*	*Gazole*	*Fioul lourd*	*Naphta*	*Coke de pétrole*	*Autres prod.*
Production	1787	1240	11309	-	4092	3666	24514	24834	4948	-	731
Imports	-	-	-	-	-	-	-	-	-	-	-
Exports	-	-299	-7460	-	-1283	-3451	-6798	-17588	-4948	-	-
Intl. Marine Bunkers	-	-	-	-	-	-	-	-1936	-	-	-
Stock Changes	-	-	-	-	-	-	-	-	-	-	-
DOMESTIC SUPPLY	**1787**	**941**	**3849**	**-**	**2809**	**215**	**17716**	**5310**	**-**	**-**	**731**
Transfers and Stat. Diff.	-	-83	5794	-	-361	-	-1170	-70	-	-	299
TRANSFORMATION	**-**	**-**	**-**	**-**	**-**	**-**	**5604**	**78**	**-**	**-**	**-**
Electricity and CHP Plants	-	-	-	-	-	-	5604	78	-	-	-
Petroleum Refineries	-	-	-	-	-	-	-	-	-	-	-
Other Transform. Sector	-	-	-	-	-	-	-	-	-	-	-
ENERGY SECTOR	**1787**	**-**	**-**	**-**	**-**	**-**	**368**	**1599**	**-**	**-**	**-**
DISTRIBUTION LOSSES	**-**	**-**	**-**	**-**	**-**	**-**	**-**	**-**	**-**	**-**	**-**
FINAL CONSUMPTION	**-**	**858**	**9643**	**-**	**2448**	**215**	**10574**	**3563**	**-**	**-**	**1030**
INDUSTRY SECTOR	**-**	**-**	**-**	**-**	**-**	**-**	**-**	**3563**	**-**	**-**	**-**
Iron and Steel	-	-	-	-	-	-	-	-	-	-	-
Chemical and Petrochem.	-	-	-	-	-	-	-	-	-	-	-
Non-Metallic Minerals	-	-	-	-	-	-	-	-	-	-	-
Non-specified	-	-	-	-	-	-	-	3563	-	-	-
TRANSPORT SECTOR	**-**	**-**	**9643**	**-**	**2448**	**-**	**-**	**-**	**-**	**-**	**-**
Air	-	-	-	-	2448	-	-	-	-	-	-
Road	-	-	9643	-	-	-	-	-	-	-	-
Non-specified	-	-	-	-	-	-	-	-	-	-	-
OTHER SECTORS	**-**	**858**	**-**	**-**	**-**	**215**	**10574**	**-**	**-**	**-**	**-**
Agriculture	-	-	-	-	-	-	-	-	-	-	-
Comm. and Publ. Services	-	-	-	-	-	-	-	-	-	-	-
Residential	-	858	-	-	-	215	-	-	-	-	-
Non-specified	-	-	-	-	-	-	10574	-	-	-	-
NON-ENERGY USE	**-**	**-**	**-**	**-**	**-**	**-**	**-**	**-**	**-**	**-**	**1030**

Saudi Arabia / Arabie saoudite

SUPPLY AND CONSUMPTION 1996	Gas (TJ)				Comb. Renew. & Waste (TJ)				(GWh)	(TJ)
	Natural Gas	Gas Works	Coke Ovens	Blast Furnaces	Solid Biomass	Gas/Liquids from Biomass	Municipal Waste	Industrial Waste	Electricity	Heat
Production	1665902	-	-	-	-	-	-	-	97819	-
Imports	-	-	-	-	187	-	-	-	-	-
Exports	-	-	-	-	-	-	-	-	-	-
Intl. Marine Bunkers	-	-	-	-	-	-	-	-	-	-
Stock Changes	-	-	-	-	-	-	-	-	-	-
DOMESTIC SUPPLY	**1665902**	-	-	-	**187**	-	-	-	**97819**	-
Transfers and Stat. Diff.	-28687	-	-	-	-	-	-	-	-	-
TRANSFORMATION	**300694**	-	-	-	-	-	-	-	-	-
Electricity and CHP Plants	300694	-	-	-	-	-	-	-	-	-
Petroleum Refineries	-	-	-	-	-	-	-	-	-	-
Other Transform. Sector	-	-	-	-	-	-	-	-	-	-
ENERGY SECTOR	**797675**	-	-	-	-	-	-	-	**12380**	-
DISTRIBUTION LOSSES	-	-	-	-	-	-	-	-	**8200**	-
FINAL CONSUMPTION	**538846**	-	-	-	**187**	-	-	-	**77239**	-
INDUSTRY SECTOR	-	-	-	-	-	-	-	-	**10129**	-
Iron and Steel	-	-	-	-	-	-	-	-	-	-
Chemical and Petrochem.	-	-	-	-	-	-	-	-	-	-
Non-Metallic Minerals	-	-	-	-	-	-	-	-	-	-
Non-specified	-	-	-	-	-	-	-	-	10129	-
TRANSPORT SECTOR	-	-	-	-	-	-	-	-	-	-
Air	-	-	-	-	-	-	-	-	-	-
Road	-	-	-	-	-	-	-	-	-	-
Non-specified	-	-	-	-	-	-	-	-	-	-
OTHER SECTORS	**538846**	-	-	-	**187**	-	-	-	**67110**	-
Agriculture	-	-	-	-	-	-	-	-	1603	-
Comm. and Publ. Services	-	-	-	-	-	-	-	-	23365	-
Residential	-	-	-	-	187	-	-	-	42142	-
Non-specified	538846	-	-	-	-	-	-	-	-	-
NON-ENERGY USE	-	-	-	-	-	-	-	-	-	-

APPROVISIONNEMENT ET DEMANDE 1997	*Gaz* (TJ)				*En. Re. Comb. & Déchets* (TJ)				(GWh)	(TJ)
	Gaz naturel	*Usines à gaz*	*Cokeries*	*Hauts fourneaux*	*Biomasse solide*	*Gaz/Liquides tirés de biomasse*	*Déchets urbains*	*Déchets industriels*	*Electricité*	*Chaleur*
Production	1748915	-	-	-	-	-	-	-	103801	-
Imports	-	-	-	-	173	-	-	-	-	-
Exports	-	-	-	-	-	-	-	-	-	-
Intl. Marine Bunkers	-	-	-	-	-	-	-	-	-	-
Stock Changes	-	-	-	-	-	-	-	-	-	-
DOMESTIC SUPPLY	**1748915**	-	-	-	**173**	-	-	-	**103801**	-
Transfers and Stat. Diff.	-30116	-	-	-	-	-	-	-	-1	-
TRANSFORMATION	**315678**	-	-	-	-	-	-	-	-	-
Electricity and CHP Plants	315678	-	-	-	-	-	-	-	-	-
Petroleum Refineries	-	-	-	-	-	-	-	-	-	-
Other Transform. Sector	-	-	-	-	-	-	-	-	-	-
ENERGY SECTOR	**837424**	-	-	-	-	-	-	-	**13137**	-
DISTRIBUTION LOSSES	-	-	-	-	-	-	-	-	**8701**	-
FINAL CONSUMPTION	**565697**	-	-	-	**173**	-	-	-	**81962**	-
INDUSTRY SECTOR	-	-	-	-	-	-	-	-	**10748**	-
Iron and Steel	-	-	-	-	-	-	-	-	-	-
Chemical and Petrochem.	-	-	-	-	-	-	-	-	-	-
Non-Metallic Minerals	-	-	-	-	-	-	-	-	-	-
Non-specified	-	-	-	-	-	-	-	-	10748	-
TRANSPORT SECTOR	-	-	-	-	-	-	-	-	-	-
Air	-	-	-	-	-	-	-	-	-	-
Road	-	-	-	-	-	-	-	-	-	-
Non-specified	-	-	-	-	-	-	-	-	-	-
OTHER SECTORS	**565697**	-	-	-	**173**	-	-	-	**71214**	-
Agriculture	-	-	-	-	-	-	-	-	1701	-
Comm. and Publ. Services	-	-	-	-	-	-	-	-	24794	-
Residential	-	-	-	-	173	-	-	-	44719	-
Non-specified	565697	-	-	-	-	-	-	-	-	-
NON-ENERGY USE	-	-	-	-	-	-	-	-	-	-

Senegal / Sénégal

SUPPLY AND CONSUMPTION 1996	Coal (1000 tonnes)							Oil (1000 tonnes)			
	Coking Coal	Other Bit. Coal	Sub-Bit. Coal	Lignite	Peat	Oven and Gas Coke	Pat. Fuel and BKB	Crude Oil	NGL	Feed-stocks	Additives
Production	-	-	-	-	-	-	-	1	-	-	-
Imports	-	-	-	-	-	-	-	643	-	-	-
Exports	-	-	-	-	-	-	-	-	-	-	-
Intl. Marine Bunkers	-	-	-	-	-	-	-	-	-	-	-
Stock Changes	-	-	-	-	-	-	-	32	-	-	-
DOMESTIC SUPPLY	-	-	-	-	-	-	-	**676**	-	-	-
Transfers and Stat. Diff.	-	-	-	-	-	-	-	-	-	-	-
TRANSFORMATION	-	-	-	-	-	-	-	**676**	-	-	-
Electricity and CHP Plants	-	-	-	-	-	-	-	-	-	-	-
Petroleum Refineries	-	-	-	-	-	-	-	676	-	-	-
Other Transform. Sector	-	-	-	-	-	-	-	-	-	-	-
ENERGY SECTOR	-	-	-	-	-	-	-	-	-	-	-
DISTRIBUTION LOSSES	-	-	-	-	-	-	-	-	-	-	-
FINAL CONSUMPTION	-	-	-	-	-	-	-	-	-	-	-
INDUSTRY SECTOR	-	-	-	-	-	-	-	-	-	-	-
Iron and Steel	-	-	-	-	-	-	-	-	-	-	-
Chemical and Petrochem.	-	-	-	-	-	-	-	-	-	-	-
Non-Metallic Minerals	-	-	-	-	-	-	-	-	-	-	-
Non-specified	-	-	-	-	-	-	-	-	-	-	-
TRANSPORT SECTOR	-	-	-	-	-	-	-	-	-	-	-
Air	-	-	-	-	-	-	-	-	-	-	-
Road	-	-	-	-	-	-	-	-	-	-	-
Non-specified	-	-	-	-	-	-	-	-	-	-	-
OTHER SECTORS	-	-	-	-	-	-	-	-	-	-	-
Agriculture	-	-	-	-	-	-	-	-	-	-	-
Comm. and Publ. Services	-	-	-	-	-	-	-	-	-	-	-
Residential	-	-	-	-	-	-	-	-	-	-	-
Non-specified	-	-	-	-	-	-	-	-	-	-	-
NON-ENERGY USE	-	-	-	-	-	-	-	-	-	-	-

APPROVISIONNEMENT ET DEMANDE 1997	*Charbon* (1000 tonnes)							*Pétrole* (1000 tonnes)			
	Charbon à coke	*Autres charb. bit.*	*Charbon sous-bit.*	*Lignite*	*Tourbe*	*Coke de four/gaz*	*Agg./briq. de lignite*	*Pétrole brut*	*LGN*	*Produits d'aliment.*	*Additifs*
Production	-	-	-	-	-	-	-	-	-	-	-
Imports	-	-	-	-	-	-	-	771	-	-	-
Exports	-	-	-	-	-	-	-	-	-	-	-
Intl. Marine Bunkers	-	-	-	-	-	-	-	-	-	-	-
Stock Changes	-	-	-	-	-	-	-	-12	-	-	-
DOMESTIC SUPPLY	-	-	-	-	-	-	-	**759**	-	-	-
Transfers and Stat. Diff.	-	-	-	-	-	-	-	-	-	-	-
TRANSFORMATION	-	-	-	-	-	-	-	**759**	-	-	-
Electricity and CHP Plants	-	-	-	-	-	-	-	-	-	-	-
Petroleum Refineries	-	-	-	-	-	-	-	759	-	-	-
Other Transform. Sector	-	-	-	-	-	-	-	-	-	-	-
ENERGY SECTOR	-	-	-	-	-	-	-	-	-	-	-
DISTRIBUTION LOSSES	-	-	-	-	-	-	-	-	-	-	-
FINAL CONSUMPTION	-	-	-	-	-	-	-	-	-	-	-
INDUSTRY SECTOR	-	-	-	-	-	-	-	-	-	-	-
Iron and Steel	-	-	-	-	-	-	-	-	-	-	-
Chemical and Petrochem.	-	-	-	-	-	-	-	-	-	-	-
Non-Metallic Minerals	-	-	-	-	-	-	-	-	-	-	-
Non-specified	-	-	-	-	-	-	-	-	-	-	-
TRANSPORT SECTOR	-	-	-	-	-	-	-	-	-	-	-
Air	-	-	-	-	-	-	-	-	-	-	-
Road	-	-	-	-	-	-	-	-	-	-	-
Non-specified	-	-	-	-	-	-	-	-	-	-	-
OTHER SECTORS	-	-	-	-	-	-	-	-	-	-	-
Agriculture	-	-	-	-	-	-	-	-	-	-	-
Comm. and Publ. Services	-	-	-	-	-	-	-	-	-	-	-
Residential	-	-	-	-	-	-	-	-	-	-	-
Non-specified	-	-	-	-	-	-	-	-	-	-	-
NON-ENERGY USE	-	-	-	-	-	-	-	-	-	-	-

Senegal / Sénégal

SUPPLY AND CONSUMPTION 1996	Oil cont. (1000 tonnes)										
	Refinery Gas	LPG + Ethane	Motor Gasoline	Aviation Gasoline	Jet Fuel	Kerosene	Gas/ Diesel	Heavy Fuel Oil	Naphtha	Petrol. Coke	Other Prod.
Production	6	5	90	-	98	13	213	232	-	-	5
Imports	-	64	16	-	84	-	90	149	-	-	13
Exports	-	-2	-28	-	-10	-4	-9	-	-	-	-1
Intl. Marine Bunkers	-	-	-	-	-	-	-58	-	-	-	-
Stock Changes	-	-	-	-	28	1	-	-	-	-	2
DOMESTIC SUPPLY	**6**	**67**	**78**	**-**	**200**	**10**	**236**	**381**	**-**	**-**	**19**
Transfers and Stat. Diff.	-	1	-4	-	-	-	22	-	-	-	-
TRANSFORMATION	**-**	**-**	**-**	**-**	**-**	**-**	**17**	**285**	**-**	**-**	**-**
Electricity and CHP Plants	-	-	-	-	-	-	17	285	-	-	-
Petroleum Refineries	-	-	-	-	-	-	-	-	-	-	-
Other Transform. Sector	-	-	-	-	-	-	-	-	-	-	-
ENERGY SECTOR	**6**	**-**	**-**	**-**	**-**	**-**	**-**	**-**	**-**	**-**	**-**
DISTRIBUTION LOSSES	**-**	**-**	**-**	**-**	**1**	**-**	**1**	**-**	**-**	**-**	**-**
FINAL CONSUMPTION	**-**	**68**	**74**	**-**	**199**	**10**	**240**	**96**	**-**	**-**	**19**
INDUSTRY SECTOR	**-**	**-**	**-**	**-**	**-**	**1**	**33**	**75**	**-**	**-**	**-**
Iron and Steel	-	-	-	-	-	-	-	-	-	-	-
Chemical and Petrochem.	-	-	-	-	-	-	-	-	-	-	-
Non-Metallic Minerals	-	-	-	-	-	-	-	-	-	-	-
Non-specified	-	-	-	-	-	1	33	75	-	-	-
TRANSPORT SECTOR	**-**	**-**	**74**	**-**	**199**	**-**	**193**	**-**	**-**	**-**	**-**
Air	-	-	-	-	199	-	-	-	-	-	-
Road	-	-	50	-	-	-	193	-	-	-	-
Non-specified	-	-	24	-	-	-	-	-	-	-	-
OTHER SECTORS	**-**	**68**	**-**	**-**	**-**	**9**	**14**	**21**	**-**	**-**	**-**
Agriculture	-	-	-	-	-	-	14	21	-	-	-
Comm. and Publ. Services	-	-	-	-	-	-	-	-	-	-	-
Residential	-	68	-	-	-	9	-	-	-	-	-
Non-specified	-	-	-	-	-	-	-	-	-	-	-
NON-ENERGY USE	**-**	**-**	**-**	**-**	**-**	**-**	**-**	**-**	**-**	**-**	**19**

APPROVISIONNEMENT ET DEMANDE 1997	*Pétrole cont.* (1000 tonnes)										
	Gaz de raffinerie	*GPL + éthane*	*Essence moteur*	*Essence aviation*	*Carbu- réacteurs*	*Kérosène*	*Gazole*	*Fioul lourd*	*Naphta*	*Coke de pétrole*	*Autres prod.*
Production	6	10	113	-	115	13	242	251	-	-	5
Imports	-	68	-	-	45	-	104	169	-	-	13
Exports	-	-2	-28	-	-10	-4	-9	-8	-	-	-1
Intl. Marine Bunkers	-	-	-	-	-	-	-58	-	-	-	-
Stock Changes	-	-	-	-	53	3	-	-	-	-	2
DOMESTIC SUPPLY	**6**	**76**	**85**	**-**	**203**	**12**	**279**	**412**	**-**	**-**	**19**
Transfers and Stat. Diff.	-	-	-7	-	-	-	43	6	-	-	-
TRANSFORMATION	**-**	**-**	**-**	**-**	**-**	**-**	**40**	**308**	**-**	**-**	**-**
Electricity and CHP Plants	-	-	-	-	-	-	40	308	-	-	-
Petroleum Refineries	-	-	-	-	-	-	-	-	-	-	-
Other Transform. Sector	-	-	-	-	-	-	-	-	-	-	-
ENERGY SECTOR	**6**	**-**	**-**	**-**	**-**	**-**	**-**	**-**	**-**	**-**	**-**
DISTRIBUTION LOSSES	**-**	**-**	**-**	**-**	**-**	**-**	**-**	**1**	**-**	**-**	**-**
FINAL CONSUMPTION	**-**	**76**	**78**	**-**	**203**	**12**	**282**	**109**	**-**	**-**	**19**
INDUSTRY SECTOR	**-**	**-**	**-**	**-**	**-**	**1**	**49**	**85**	**-**	**-**	**-**
Iron and Steel	-	-	-	-	-	-	-	-	-	-	-
Chemical and Petrochem.	-	-	-	-	-	-	-	-	-	-	-
Non-Metallic Minerals	-	-	-	-	-	-	-	-	-	-	-
Non-specified	-	-	-	-	-	1	49	85	-	-	-
TRANSPORT SECTOR	**-**	**-**	**78**	**-**	**203**	**-**	**212**	**-**	**-**	**-**	**-**
Air	-	-	-	-	203	-	-	-	-	-	-
Road	-	-	50	-	-	-	212	-	-	-	-
Non-specified	-	-	28	-	-	-	-	-	-	-	-
OTHER SECTORS	**-**	**76**	**-**	**-**	**-**	**11**	**21**	**24**	**-**	**-**	**-**
Agriculture	-	-	-	-	-	-	21	24	-	-	-
Comm. and Publ. Services	-	-	-	-	-	-	-	-	-	-	-
Residential	-	76	-	-	-	11	-	-	-	-	-
Non-specified	-	-	-	-	-	-	-	-	-	-	-
NON-ENERGY USE	**-**	**-**	**-**	**-**	**-**	**-**	**-**	**-**	**-**	**-**	**19**

Senegal / Sénégal

SUPPLY AND CONSUMPTION 1996	Gas (TJ)				Comb. Renew. & Waste (TJ)				(GWh)	(TJ)
	Natural Gas	Gas Works	Coke Ovens	Blast Furnaces	Solid Biomass	Gas/Liquids from Biomass	Municipal Waste	Industrial Waste	Electricity	Heat
Production	1685	-	-	-	66176	-	-	-	1175	-
Imports	-	-	-	-	-	-	-	-	-	-
Exports	-	-	-	-	-	-	-	-	-	-
Intl. Marine Bunkers	-	-	-	-	-	-	-	-	-	-
Stock Changes	-	-	-	-	-	-	-	-	-	-
DOMESTIC SUPPLY	**1685**	**-**	**-**	**-**	**66176**	**-**	**-**	**-**	**1175**	**-**
Transfers and Stat. Diff.	-	-	-	-	-	-	-	-	-38	-
TRANSFORMATION	**-**	**-**	**-**	**-**	**15717**	**-**	**-**	**-**	**-**	**-**
Electricity and CHP Plants	-	-	-	-	-	-	-	-	-	-
Petroleum Refineries	-	-	-	-	-	-	-	-	-	-
Other Transform. Sector	-	-	-	-	15717	-	-	-	-	-
ENERGY SECTOR	**-**	**-**	**-**	**-**	**-**	**-**	**-**	**-**	**66**	**-**
DISTRIBUTION LOSSES	**-**	**-**	**-**	**-**	**-**	**-**	**-**	**-**	**187**	**-**
FINAL CONSUMPTION	**1685**	**-**	**-**	**-**	**50459**	**-**	**-**	**-**	**884**	**-**
INDUSTRY SECTOR	**1685**	**-**	**-**	**-**	**6721**	**-**	**-**	**-**	**508**	**-**
Iron and Steel	-	-	-	-	-	-	-	-	-	-
Chemical and Petrochem.	-	-	-	-	-	-	-	-	-	-
Non-Metallic Minerals	-	-	-	-	-	-	-	-	-	-
Non-specified	1685	-	-	-	6721	-	-	-	508	-
TRANSPORT SECTOR	**-**	**-**	**-**	**-**	**-**	**-**	**-**	**-**	**-**	**-**
Air	-	-	-	-	-	-	-	-	-	-
Road	-	-	-	-	-	-	-	-	-	-
Non-specified	-	-	-	-	-	-	-	-	-	-
OTHER SECTORS	**-**	**-**	**-**	**-**	**43738**	**-**	**-**	**-**	**376**	**-**
Agriculture	-	-	-	-	-	-	-	-	45	-
Comm. and Publ. Services	-	-	-	-	-	-	-	-	135	-
Residential	-	-	-	-	43738	-	-	-	196	-
Non-specified	-	-	-	-	-	-	-	-	-	-
NON-ENERGY USE	**-**	**-**	**-**	**-**	**-**	**-**	**-**	**-**	**-**	**-**

APPROVISIONNEMENT ET DEMANDE 1997	*Gaz* (TJ)				*En. Re. Comb. & Déchets* (TJ)				(GWh)	(TJ)
	Gaz naturel	*Usines à gaz*	*Cokeries*	*Hauts fourneaux*	*Biomasse solide*	*Gaz/Liquides tirés de biomasse*	*Déchets urbains*	*Déchets industriels*	*Electricité*	*Chaleur*
Production	928	-	-	-	68426	-	-	-	1261	-
Imports	-	-	-	-	-	-	-	-	-	-
Exports	-	-	-	-	-	-	-	-	-	-
Intl. Marine Bunkers	-	-	-	-	-	-	-	-	-	-
Stock Changes	-	-	-	-	-	-	-	-	-	-
DOMESTIC SUPPLY	**928**	**-**	**-**	**-**	**68426**	**-**	**-**	**-**	**1261**	**-**
Transfers and Stat. Diff.	-	-	-	-	1	-	-	-	-40	-
TRANSFORMATION	**-**	**-**	**-**	**-**	**16251**	**-**	**-**	**-**	**-**	**-**
Electricity and CHP Plants	-	-	-	-	-	-	-	-	-	-
Petroleum Refineries	-	-	-	-	-	-	-	-	-	-
Other Transform. Sector	-	-	-	-	16251	-	-	-	-	-
ENERGY SECTOR	**-**	**-**	**-**	**-**	**-**	**-**	**-**	**-**	**67**	**-**
DISTRIBUTION LOSSES	**-**	**-**	**-**	**-**	**-**	**-**	**-**	**-**	**211**	**-**
FINAL CONSUMPTION	**928**	**-**	**-**	**-**	**52176**	**-**	**-**	**-**	**943**	**-**
INDUSTRY SECTOR	**928**	**-**	**-**	**-**	**6950**	**-**	**-**	**-**	**545**	**-**
Iron and Steel	-	-	-	-	-	-	-	-	-	-
Chemical and Petrochem.	-	-	-	-	-	-	-	-	-	-
Non-Metallic Minerals	-	-	-	-	-	-	-	-	-	-
Non-specified	928	-	-	-	6950	-	-	-	545	-
TRANSPORT SECTOR	**-**	**-**	**-**	**-**	**-**	**-**	**-**	**-**	**-**	**-**
Air	-	-	-	-	-	-	-	-	-	-
Road	-	-	-	-	-	-	-	-	-	-
Non-specified	-	-	-	-	-	-	-	-	-	-
OTHER SECTORS	**-**	**-**	**-**	**-**	**45226**	**-**	**-**	**-**	**398**	**-**
Agriculture	-	-	-	-	-	-	-	-	48	-
Comm. and Publ. Services	-	-	-	-	-	-	-	-	143	-
Residential	-	-	-	-	45226	-	-	-	207	-
Non-specified	-	-	-	-	-	-	-	-	-	-
NON-ENERGY USE	**-**	**-**	**-**	**-**	**-**	**-**	**-**	**-**	**-**	**-**

Singapore / Singapour

SUPPLY AND CONSUMPTION 1996	Coal (1000 tonnes)							Oil (1000 tonnes)			
	Coking Coal	Other Bit. Coal	Sub-Bit. Coal	Lignite	Peat	Oven and Gas Coke	Pat. Fuel and BKB	Crude Oil	NGL	Feed-stocks	Additives
Production	-	-	-	-	-	-	-	-	-	-	-
Imports	-	-	-	1	-	12	-	55192	-	-	-
Exports	-	-	-	-	-	-14	-	-	-	-	-
Intl. Marine Bunkers	-	-	-	-	-	-	-	-	-	-	-
Stock Changes	-	-	-	-	-	-	-	-	-	-	-
DOMESTIC SUPPLY	**-**	**-**	**-**	**1**	**-**	**-2**	**-**	**55192**	**-**	**-**	**-**
Transfers and Stat. Diff.	-	-	-	-1	-	2	-	790	-	-	-
TRANSFORMATION	**-**	**-**	**-**	**-**	**-**	**-**	**-**	**55982**	**-**	**-**	**-**
Electricity and CHP Plants	-	-	-	-	-	-	-	-	-	-	-
Petroleum Refineries	-	-	-	-	-	-	-	55982	-	-	-
Other Transform. Sector	-	-	-	-	-	-	-	-	-	-	-
ENERGY SECTOR	**-**	**-**	**-**	**-**	**-**	**-**	**-**	**-**	**-**	**-**	**-**
DISTRIBUTION LOSSES	**-**	**-**	**-**	**-**	**-**	**-**	**-**	**-**	**-**	**-**	**-**
FINAL CONSUMPTION	**-**	**-**	**-**	**-**	**-**	**-**	**-**	**-**	**-**	**-**	**-**
INDUSTRY SECTOR	**-**	**-**	**-**	**-**	**-**	**-**	**-**	**-**	**-**	**-**	**-**
Iron and Steel	-	-	-	-	-	-	-	-	-	-	-
Chemical and Petrochem.	-	-	-	-	-	-	-	-	-	-	-
Non-Metallic Minerals	-	-	-	-	-	-	-	-	-	-	-
Non-specified	-	-	-	-	-	-	-	-	-	-	-
TRANSPORT SECTOR	**-**	**-**	**-**	**-**	**-**	**-**	**-**	**-**	**-**	**-**	**-**
Air	-	-	-	-	-	-	-	-	-	-	-
Road	-	-	-	-	-	-	-	-	-	-	-
Non-specified	-	-	-	-	-	-	-	-	-	-	-
OTHER SECTORS	**-**	**-**	**-**	**-**	**-**	**-**	**-**	**-**	**-**	**-**	**-**
Agriculture	-	-	-	-	-	-	-	-	-	-	-
Comm. and Publ. Services	-	-	-	-	-	-	-	-	-	-	-
Residential	-	-	-	-	-	-	-	-	-	-	-
Non-specified	-	-	-	-	-	-	-	-	-	-	-
NON-ENERGY USE	**-**	**-**	**-**	**-**	**-**	**-**	**-**	**-**	**-**	**-**	**-**

APPROVISIONNEMENT ET DEMANDE 1997	*Charbon* (1000 tonnes)							*Pétrole* (1000 tonnes)			
	Charbon à coke	*Autres charb. bit.*	*Charbon sous-bit.*	*Lignite*	*Tourbe*	*Coke de four/gaz*	*Agg./briq. de lignite*	*Pétrole brut*	*LGN*	*Produits d'aliment.*	*Additifs*
Production	-	-	-	-	-	-	-	-	-	-	-
Imports	-	-	-	-	-	9	-	55507	-	-	-
Exports	-	-	-	-	-	-9	-	-14	-	-	-
Intl. Marine Bunkers	-	-	-	-	-	-	-	-	-	-	-
Stock Changes	-	-	-	-	-	-	-	-	-	-	-
DOMESTIC SUPPLY	**-**	**-**	**-**	**-**	**-**	**-**	**-**	**55493**	**-**	**-**	**-**
Transfers and Stat. Diff.	-	-	-	-	-	-	-	-	-	-	-
TRANSFORMATION	**-**	**-**	**-**	**-**	**-**	**-**	**-**	**55493**	**-**	**-**	**-**
Electricity and CHP Plants	-	-	-	-	-	-	-	-	-	-	-
Petroleum Refineries	-	-	-	-	-	-	-	55493	-	-	-
Other Transform. Sector	-	-	-	-	-	-	-	-	-	-	-
ENERGY SECTOR	**-**	**-**	**-**	**-**	**-**	**-**	**-**	**-**	**-**	**-**	**-**
DISTRIBUTION LOSSES	**-**	**-**	**-**	**-**	**-**	**-**	**-**	**-**	**-**	**-**	**-**
FINAL CONSUMPTION	**-**	**-**	**-**	**-**	**-**	**-**	**-**	**-**	**-**	**-**	**-**
INDUSTRY SECTOR	**-**	**-**	**-**	**-**	**-**	**-**	**-**	**-**	**-**	**-**	**-**
Iron and Steel	-	-	-	-	-	-	-	-	-	-	-
Chemical and Petrochem.	-	-	-	-	-	-	-	-	-	-	-
Non-Metallic Minerals	-	-	-	-	-	-	-	-	-	-	-
Non-specified	-	-	-	-	-	-	-	-	-	-	-
TRANSPORT SECTOR	**-**	**-**	**-**	**-**	**-**	**-**	**-**	**-**	**-**	**-**	**-**
Air	-	-	-	-	-	-	-	-	-	-	-
Road	-	-	-	-	-	-	-	-	-	-	-
Non-specified	-	-	-	-	-	-	-	-	-	-	-
OTHER SECTORS	**-**	**-**	**-**	**-**	**-**	**-**	**-**	**-**	**-**	**-**	**-**
Agriculture	-	-	-	-	-	-	-	-	-	-	-
Comm. and Publ. Services	-	-	-	-	-	-	-	-	-	-	-
Residential	-	-	-	-	-	-	-	-	-	-	-
Non-specified	-	-	-	-	-	-	-	-	-	-	-
NON-ENERGY USE	**-**	**-**	**-**	**-**	**-**	**-**	**-**	**-**	**-**	**-**	**-**

Singapore / Singapour

SUPPLY AND CONSUMPTION 1996	Oil cont. (1000 tonnes)										
	Refinery Gas	LPG + Ethane	Motor Gasoline	Aviation Gasoline	Jet Fuel	Kerosene	Gas/ Diesel	Heavy Fuel Oil	Naphtha	Petrol. Coke	Other Prod.
Production	582	875	5231	-	9029	278	15615	11802	4923	-	2008
Imports	-	5	1582	1	824	710	4159	21241	1072	10	386
Exports	-	-730	-6772	-	-6685	-1105	-16618	-10274	-4005	-	-1840
Intl. Marine Bunkers	-	-	-	-	-	-	-2000	-12280	-	-	-
Stock Changes	-	-	-	-	-	-	-	-	-	-	-
DOMESTIC SUPPLY	**582**	**150**	**41**	**1**	**3168**	**-117**	**1156**	**10489**	**1990**	**10**	**554**
Transfers and Stat. Diff.	-	87	625	-	-608	160	-	-3943	-2	5	4
TRANSFORMATION	**-**	**94**	**-**	**-**	**-**	**-**	**45**	**3840**	**-**	**-**	**-**
Electricity and CHP Plants	-	-	-	-	-	-	45	3840	-	-	-
Petroleum Refineries	-	-	-	-	-	-	-	-	-	-	-
Other Transform. Sector	-	94	-	-	-	-	-	-	-	-	-
ENERGY SECTOR	**582**	**-**	**-**	**-**	**-**	**-**	**-**	**2706**	**-**	**15**	**-**
DISTRIBUTION LOSSES	**-**	**-**	**-**	**-**	**-**	**-**	**-**	**-**	**-**	**-**	**-**
FINAL CONSUMPTION	**-**	**143**	**666**	**1**	**2560**	**43**	**1111**	**-**	**1988**	**-**	**558**
INDUSTRY SECTOR	**-**	**143**	**-**	**-**	**-**	**43**	**-**	**-**	**1988**	**-**	**-**
Iron and Steel	-	-	-	-	-	-	-	-	-	-	-
Chemical and Petrochem.	-	143	-	-	-	-	-	-	1988	-	-
Non-Metallic Minerals	-	-	-	-	-	-	-	-	-	-	-
Non-specified	-	-	-	-	-	43	-	-	-	-	-
TRANSPORT SECTOR	**-**	**-**	**666**	**1**	**2560**	**-**	**1111**	**-**	**-**	**-**	**-**
Air	-	-	-	1	2560	-	-	-	-	-	-
Road	-	-	666	-	-	-	1111	-	-	-	-
Non-specified	-	-	-	-	-	-	-	-	-	-	-
OTHER SECTORS	**-**	**-**	**-**	**-**	**-**	**-**	**-**	**-**	**-**	**-**	**-**
Agriculture	-	-	-	-	-	-	-	-	-	-	-
Comm. and Publ. Services	-	-	-	-	-	-	-	-	-	-	-
Residential	-	-	-	-	-	-	-	-	-	-	-
Non-specified	-	-	-	-	-	-	-	-	-	-	-
NON-ENERGY USE	**-**	**-**	**-**	**-**	**-**	**-**	**-**	**-**	**-**	**-**	**558**

APPROVISIONNEMENT ET DEMANDE 1997	*Pétrole cont.* (1000 tonnes)										
	Gaz de raffinerie	*GPL + éthane*	*Essence moteur*	*Essence aviation*	*Carbu- réacteurs*	*Kérosène*	*Gazole*	*Fioul lourd*	*Naphta*	*Coke de pétrole*	*Autres prod.*
Production	582	965	5032	-	7995	278	15238	11699	3765	-	2557
Imports	-	9	1528	1	1394	513	5124	21591	2140	10	414
Exports	-	-760	-5870	-	-6585	-471	-16262	-10563	-3487	-	-2437
Intl. Marine Bunkers	-	-	-	-	-	-	-2887	-13304	-	-	-
Stock Changes	-	-	-	-	-	-	-	-	-	-	-
DOMESTIC SUPPLY	**582**	**214**	**690**	**1**	**2804**	**320**	**1213**	**9423**	**2418**	**10**	**534**
Transfers and Stat. Diff.	-	-	-	-	1	-277	-	-2483	-	5	24
TRANSFORMATION	**-**	**-**	**-**	**-**	**-**	**-**	**18**	**4234**	**-**	**-**	**-**
Electricity and CHP Plants	-	-	-	-	-	-	18	4234	-	-	-
Petroleum Refineries	-	-	-	-	-	-	-	-	-	-	-
Other Transform. Sector	-	-	-	-	-	-	-	-	-	-	-
ENERGY SECTOR	**582**	**-**	**-**	**-**	**-**	**-**	**-**	**2706**	**-**	**15**	**-**
DISTRIBUTION LOSSES	**-**	**-**	**-**	**-**	**-**	**-**	**-**	**-**	**-**	**-**	**-**
FINAL CONSUMPTION	**-**	**214**	**690**	**1**	**2805**	**43**	**1195**	**-**	**2418**	**-**	**558**
INDUSTRY SECTOR	**-**	**214**	**-**	**-**	**-**	**43**	**-**	**-**	**2418**	**-**	**-**
Iron and Steel	-	-	-	-	-	-	-	-	-	-	-
Chemical and Petrochem.	-	214	-	-	-	-	-	-	2418	-	-
Non-Metallic Minerals	-	-	-	-	-	-	-	-	-	-	-
Non-specified	-	-	-	-	-	43	-	-	-	-	-
TRANSPORT SECTOR	**-**	**-**	**690**	**1**	**2805**	**-**	**1195**	**-**	**-**	**-**	**-**
Air	-	-	-	1	2805	-	-	-	-	-	-
Road	-	-	690	-	-	-	1195	-	-	-	-
Non-specified	-	-	-	-	-	-	-	-	-	-	-
OTHER SECTORS	**-**	**-**	**-**	**-**	**-**	**-**	**-**	**-**	**-**	**-**	**-**
Agriculture	-	-	-	-	-	-	-	-	-	-	-
Comm. and Publ. Services	-	-	-	-	-	-	-	-	-	-	-
Residential	-	-	-	-	-	-	-	-	-	-	-
Non-specified	-	-	-	-	-	-	-	-	-	-	-
NON-ENERGY USE	**-**	**-**	**-**	**-**	**-**	**-**	**-**	**-**	**-**	**-**	**558**

Singapore / Singapour

SUPPLY AND CONSUMPTION 1996	Gas (TJ)				Comb. Renew. & Waste (TJ)				(GWh)	(TJ)
	Natural Gas	Gas Works	Coke Ovens	Blast Furnaces	Solid Biomass	Gas/Liquids from Biomass	Municipal Waste	Industrial Waste	Electricity	Heat
Production	-	4254	-	-	-	-	-	-	24100	-
Imports	60211	-	-	-	-	-	-	-	-	-
Exports	-	-	-	-	-	-	-	-	-	-
Intl. Marine Bunkers	-	-	-	-	-	-	-	-	-	-
Stock Changes	-	-	-	-	-	-	-	-	-	-
DOMESTIC SUPPLY	**60211**	**4254**	**-**	**-**	**-**	**-**	**-**	**-**	**24100**	**-**
Transfers and Stat. Diff.	-	12	-	-	-	-	-	-	2299	-
TRANSFORMATION	**60211**	**-**	**-**	**-**	**-**	**-**	**-**	**-**	**-**	**-**
Electricity and CHP Plants	60211	-	-	-	-	-	-	-	-	-
Petroleum Refineries	-	-	-	-	-	-	-	-	-	-
Other Transform. Sector	-	-	-	-	-	-	-	-	-	-
ENERGY SECTOR	**-**	**-**	**-**	**-**	**-**	**-**	**-**	**-**	**3407**	**-**
DISTRIBUTION LOSSES	**-**	**170**	**-**	**-**	**-**	**-**	**-**	**-**	**1084**	**-**
FINAL CONSUMPTION	**-**	**4096**	**-**	**-**	**-**	**-**	**-**	**-**	**21908**	**-**
INDUSTRY SECTOR	**-**	**2097**	**-**	**-**	**-**	**-**	**-**	**-**	**9787**	**-**
Iron and Steel	-	-	-	-	-	-	-	-	859	-
Chemical and Petrochem.	-	-	-	-	-	-	-	-	515	-
Non-Metallic Minerals	-	-	-	-	-	-	-	-		-
Non-specified	-	2097	-	-	-	-	-	-	8413	-
TRANSPORT SECTOR	**-**	**-**	**-**	**-**	**-**	**-**	**-**	**-**	**236**	**-**
Air	-	-	-	-	-	-	-	-	-	-
Road	-	-	-	-	-	-	-	-	-	-
Non-specified	-	-	-	-	-	-	-	-	236	-
OTHER SECTORS	**-**	**1999**	**-**	**-**	**-**	**-**	**-**	**-**	**11885**	**-**
Agriculture	-	-	-	-	-	-	-	-	25	-
Comm. and Publ. Services	-	-	-	-	-	-	-	-	7224	-
Residential	-	1999	-	-	-	-	-	-	4636	-
Non-specified	-	-	-	-	-	-	-	-	-	-
NON-ENERGY USE	**-**	**-**	**-**	**-**	**-**	**-**	**-**	**-**	**-**	**-**

APPROVISIONNEMENT ET DEMANDE 1997	*Gaz (TJ)*				*En. Re. Comb. & Déchets (TJ)*				(GWh)	(TJ)
	Gaz naturel	*Usines à gaz*	*Cokeries*	*Hauts fourneaux*	*Biomasse solide*	*Gaz/Liquides tirés de biomasse*	*Déchets urbains*	*Déchets industriels*	*Electricité*	*Chaleur*
Production	-	4394	-	-	-	-	-	-	26898	-
Imports	59350	-	-	-	-	-	-	-	-	-
Exports	-	-	-	-	-	-	-	-	-	-
Intl. Marine Bunkers	-	-	-	-	-	-	-	-	-	-
Stock Changes	-	-	-	-	-	-	-	-	-	-
DOMESTIC SUPPLY	**59350**	**4394**	**-**	**-**	**-**	**-**	**-**	**-**	**26898**	**-**
Transfers and Stat. Diff.	-	189	-	-	-	-	-	-	2711	-
TRANSFORMATION	**59350**	**-**	**-**	**-**	**-**	**-**	**-**	**-**	**-**	**-**
Electricity and CHP Plants	59350	-	-	-	-	-	-	-	-	-
Petroleum Refineries	-	-	-	-	-	-	-	-	-	-
Other Transform. Sector	-	-	-	-	-	-	-	-	-	-
ENERGY SECTOR	**-**	**-**	**-**	**-**	**-**	**-**	**-**	**-**	**3836**	**-**
DISTRIBUTION LOSSES	**-**	**425**	**-**	**-**	**-**	**-**	**-**	**-**	**1118**	**-**
FINAL CONSUMPTION	**-**	**4158**	**-**	**-**	**-**	**-**	**-**	**-**	**24655**	**-**
INDUSTRY SECTOR	**-**	**2097**	**-**	**-**	**-**	**-**	**-**	**-**	**11077**	**-**
Iron and Steel	-	-	-	-	-	-	-	-	972	-
Chemical and Petrochem.	-	-	-	-	-	-	-	-	583	-
Non-Metallic Minerals	-	-	-	-	-	-	-	-	-	-
Non-specified	-	2097	-	-	-	-	-	-	9522	-
TRANSPORT SECTOR	**-**	**-**	**-**	**-**	**-**	**-**	**-**	**-**	**248**	**-**
Air	-	-	-	-	-	-	-	-	-	-
Road	-	-	-	-	-	-	-	-	-	-
Non-specified	-	-	-	-	-	-	-	-	248	-
OTHER SECTORS	**-**	**2061**	**-**	**-**	**-**	**-**	**-**	**-**	**13330**	**-**
Agriculture	-	-	-	-	-	-	-	-	25	-
Comm. and Publ. Services	-	-	-	-	-	-	-	-	8102	-
Residential	-	2061	-	-	-	-	-	-	5203	-
Non-specified	-	-	-	-	-	-	-	-	-	-
NON-ENERGY USE	**-**	**-**	**-**	**-**	**-**	**-**	**-**	**-**	**-**	**-**

Slovak Republic / République slovaque : 1996

SUPPLY AND CONSUMPTION / *APPROVISIONNEMENT ET DEMANDE*	Coal / *Charbon* (1000 tonnes)							Oil / *Pétrole* (1000 tonnes)			
	Coking Coal / *Charbon à coke*	Other Bit. Coal / *Autres charb. bit.*	Sub-Bit. Coal / *Charbon sous-bit.*	Lignite / *Lignite*	Peat / *Tourbe*	Oven and Gas Coke / *Coke de four/gaz*	Pat. Fuel and BKB / *Agg./briq. de lignite*	Crude Oil / *Pétrole brut*	NGL / *LGN*	Feed-stocks / *Produits d'aliment.*	Additives / *Additifs*
Production	-	-	-	3829	-	1708	-	71	1	-	-
From Other Sources	-	-	-	-	-	-	-	-	-	-	-
Imports	2850	2660	-	3392	-	153	5	5341	-	-	-
Exports	-	-10	-	-27	-	-114	-	-	-	-	-
Intl. Marine Bunkers	-	-	-	-	-	-	-	-	-	-	-
Stock Changes	5	-448	-	-52	-	65	-	-160	-	-	-
DOMESTIC SUPPLY	**2855**	**2202**	**-**	**7142**	**-**	**1812**	**5**	**5252**	**1**	**-**	**-**
Transfers	-	-	-	-	-	-	-	-	-1	-	-
Statistical Differences	-	-	-	-	-	-	-	-38	-	-	-
TRANSFORMATION	**2227**	**1517**	**-**	**3782**	**-**	**1271**	**-**	**5214**	**-**	**-**	**-**
Electricity Plants	-	1517	-	3715	-	-	-	-	-	-	-
CHP Plants	-	-	-	-	-	-	-	-	-	-	-
Heat Plants	-	-	-	67	-	7	-	-	-	-	-
Blast Furnaces/Gas Works	-	-	-	-	-	1264	-	-	-	-	-
Coke/Pat. Fuel/BKB Plants	2227	-	-	-	-	-	-	-	-	-	-
Petroleum Refineries	-	-	-	-	-	-	-	5214	-	-	-
Petrochemical Industry	-	-	-	-	-	-	-	-	-	-	-
Liquefaction	-	-	-	-	-	-	-	-	-	-	-
Other Transform. Sector	-	-	-	-	-	-	-	-	-	-	-
ENERGY SECTOR	**-**	**-**	**-**	**-**	**-**	**-**	**-**	**-**	**-**	**-**	**-**
Coal Mines	-	-	-	-	-	-	-	-	-	-	-
Oil and Gas Extraction	-	-	-	-	-	-	-	-	-	-	-
Petroleum Refineries	-	-	-	-	-	-	-	-	-	-	-
Electr., CHP+Heat Plants	-	-	-	-	-	-	-	-	-	-	-
Pumped Storage (Elec.)	-	-	-	-	-	-	-	-	-	-	-
Other Energy Sector	-	-	-	-	-	-	-	-	-	-	-
Distribution Losses	-	1	-	6	-	-	-	-	-	-	-
FINAL CONSUMPTION	**628**	**684**	**-**	**3354**	**-**	**541**	**5**	**-**	**-**	**-**	**-**
INDUSTRY SECTOR	**394**	**623**	**-**	**1614**	**-**	**465**	**2**	**-**	**-**	**-**	**-**
Iron and Steel	389	399	-	99	-	316	-	-	-	-	-
Chemical and Petrochem.	-	11	-	487	-	50	-	-	-	-	-
of which: Feedstocks	-	-	-	-	-	-	-	-	-	-	-
Non-Ferrous Metals	-	-	-	-	-	-	-	-	-	-	-
Non-Metallic Minerals	4	152	-	36	-	82	-	-	-	-	-
Transport Equipment	-	-	-	134	-	-	-	-	-	-	-
Machinery	-	18	-	200	-	3	2	-	-	-	-
Mining and Quarrying	-	-	-	10	-	4	-	-	-	-	-
Food and Tobacco	-	20	-	114	-	8	-	-	-	-	-
Paper, Pulp and Print	-	21	-	279	-	-	-	-	-	-	-
Wood and Wood Products	-	-	-	46	-	-	-	-	-	-	-
Construction	1	1	-	26	-	-	-	-	-	-	-
Textile and Leather	-	-	-	152	-	1	-	-	-	-	-
Non-specified	-	1	-	31	-	1	-	-	-	-	-
TRANSPORT SECTOR	**-**	**-**	**-**	**-**	**-**	**-**	**-**	**-**	**-**	**-**	**-**
Air	-	-	-	-	-	-	-	-	-	-	-
Road	-	-	-	-	-	-	-	-	-	-	-
Rail	-	-	-	-	-	-	-	-	-	-	-
Pipeline Transport	-	-	-	-	-	-	-	-	-	-	-
Internal Navigation	-	-	-	-	-	-	-	-	-	-	-
Non-specified	-	-	-	-	-	-	-	-	-	-	-
OTHER SECTORS	**234**	**61**	**-**	**1740**	**-**	**76**	**3**	**-**	**-**	**-**	**-**
Agriculture	1	3	-	72	-	5	-	-	-	-	-
Comm. and Publ. Services	232	55	-	754	-	50	-	-	-	-	-
Residential	1	3	-	914	-	21	3	-	-	-	-
Non-specified	-	-	-	-	-	-	-	-	-	-	-
NON-ENERGY USE	**-**	**-**	**-**	**-**	**-**	**-**	**-**	**-**	**-**	**-**	**-**
in Industry/Trans./Energy	-	-	-	-	-	-	-	-	-	-	-
in Transport	-	-	-	-	-	-	-	-	-	-	-
in Other Sectors	-	-	-	-	-	-	-	-	-	-	-

Slovak Republic / République slovaque : 1996

SUPPLY AND CONSUMPTION *APPROVISIONNEMENT ET DEMANDE*	Oil cont. / *Pétrole cont.* (1000 tonnes)										
	Refinery Gas *Gaz de raffinerie*	LPG + Ethane *GPL + éthane*	Motor Gasoline *Essence moteur*	Aviation Gasoline *Essence aviation*	Jet Fuel *Carbu-réacteurs*	Kerosene *Kérosène*	Gas/ Diesel *Gazole*	Heavy Fuel Oil *Fioul lourd*	Naphtha *Naphta*	Petrol. Coke *Coke de pétrole*	Other Prod. *Autres prod.*
Production	75	53	800	-	70	34	1724	1117	670	-	671
From Other Sources	-	-	-	-	-	-	-	-	-	-	-
Imports	-	13	8	-	-	-	-	13	-	-	-
Exports	-	-35	-352	-	-32	-29	-909	-491	-	-	-20
Intl. Marine Bunkers	-	-	-	-	-	-	-	-	-	-	-
Stock Changes	-	1	3	-	-	-	-9	23	-	-	-
DOMESTIC SUPPLY	**75**	**32**	**459**	**-**	**38**	**5**	**806**	**662**	**670**	**-**	**651**
Transfers	-	1	-	-	-	-	-	-	-	-	-
Statistical Differences	-	-	-	-	-	-	-	-	-	-	-
TRANSFORMATION	**-**	**-**	**-**	**-**	**-**	**-**	**-**	**286**	**-**	**-**	**-**
Electricity Plants	-	-	-	-	-	-	-	-	-	-	-
CHP Plants	-	-	-	-	-	-	-	243	-	-	-
Heat Plants	-	-	-	-	-	-	-	43	-	-	-
Blast Furnaces/Gas Works	-	-	-	-	-	-	-	-	-	-	-
Coke/Pat. Fuel/BKB Plants	-	-	-	-	-	-	-	-	-	-	-
Petroleum Refineries	-	-	-	-	-	-	-	-	-	-	-
Petrochemical Industry	-	-	-	-	-	-	-	-	-	-	-
Liquefaction	-	-	-	-	-	-	-	-	-	-	-
Other Transform. Sector	-	-	-	-	-	-	-	-	-	-	-
ENERGY SECTOR	**-**	**-**	**-**	**-**	**-**	**-**	**-**	**-**	**-**	**-**	**-**
Coal Mines	-	-	-	-	-	-	-	-	-	-	-
Oil and Gas Extraction	-	-	-	-	-	-	-	-	-	-	-
Petroleum Refineries	-	-	-	-	-	-	-	-	-	-	-
Electr., CHP+Heat Plants	-	-	-	-	-	-	-	-	-	-	-
Pumped Storage (Elec.)	-	-	-	-	-	-	-	-	-	-	-
Other Energy Sector	-	-	-	-	-	-	-	-	-	-	-
Distribution Losses	-	-	-	-	-	-	-	-	-	-	-
FINAL CONSUMPTION	**75**	**33**	**459**	**-**	**38**	**5**	**806**	**376**	**670**	**-**	**651**
INDUSTRY SECTOR	**75**	**-**	**-**	**-**	**-**	**1**	**-**	**345**	**670**	**-**	**-**
Iron and Steel	-	-	-	-	-	-	-	45	-	-	-
Chemical and Petrochem.	-	-	-	-	-	-	-	19	670	-	-
of which: Feedstocks	-	-	-	-	-	-	-	-	*670*	-	-
Non-Ferrous Metals	-	-	-	-	-	-	-	-	-	-	-
Non-Metallic Minerals	-	-	-	-	-	-	-	27	-	-	-
Transport Equipment	-	-	-	-	-	1	-	7	-	-	-
Machinery	-	-	-	-	-	-	-	13	-	-	-
Mining and Quarrying	-	-	-	-	-	-	-	1	-	-	-
Food and Tobacco	-	-	-	-	-	-	-	38	-	-	-
Paper, Pulp and Print	-	-	-	-	-	-	-	89	-	-	-
Wood and Wood Products	-	-	-	-	-	-	-	4	-	-	-
Construction	-	-	-	-	-	-	-	6	-	-	-
Textile and Leather	-	-	-	-	-	-	-	69	-	-	-
Non-specified	75	-	-	-	-	-	-	27	-	-	-
TRANSPORT SECTOR	**-**	**-**	**450**	**-**	**38**	**-**	**627**	**-**	**-**	**-**	**-**
Air	-	-	-	-	38	-	-	-	-	-	-
Road	-	-	450	-	-	-	627	-	-	-	-
Rail	-	-	-	-	-	-	-	-	-	-	-
Pipeline Transport	-	-	-	-	-	-	-	-	-	-	-
Internal Navigation	-	-	-	-	-	-	-	-	-	-	-
Non-specified	-	-	-	-	-	-	-	-	-	-	-
OTHER SECTORS	**-**	**33**	**9**	**-**	**-**	**4**	**179**	**31**	**-**	**-**	**-**
Agriculture	-	-	9	-	-	4	179	7	-	-	-
Comm. and Publ. Services	-	8	-	-	-	-	-	24	-	-	-
Residential	-	25	-	-	-	-	-	-	-	-	-
Non-specified	-	-	-	-	-	-	-	-	-	-	-
NON-ENERGY USE	**-**	**-**	**-**	**-**	**-**	**-**	**-**	**-**	**-**	**-**	**651**
in Industry/Transf./Energy	-	-	-	-	-	-	-	-	-	-	651
in Transport	-	-	-	-	-	-	-	-	-	-	-
in Other Sectors	-	-	-	-	-	-	-	-	-	-	-

Slovak Republic / République slovaque : 1996

SUPPLY AND CONSUMPTION / *APPROVISIONNEMENT ET DEMANDE*	Gas / *Gaz* (TJ)				Comb. Renew. & Waste / *En. Re. Comb. & Déchets* (TJ)				(GWh)	(TJ)
	Natural Gas / *Gaz naturel*	Gas Works / *Usines à gaz*	Coke Ovens / *Cokeries*	Blast Furnaces / *Hauts fourneaux*	Solid Biomass / *Biomasse solide*	Gas/Liquids from Biomass / *Gaz/Liquides tirés de biomasse*	Municipal Waste / *Déchets urbains*	Industrial Waste / *Déchets industriels*	Electricity / *Electricité*	Heat / *Chaleur*
Production	11360	-	11788	15856	3184	-	-	-	25290	41198
From Other Sources	-	-	-	-	-	-	-	-	-	-
Imports	237880	-	-	-	-	-	-	-	5945	-
Exports	-	-	-	-	-	-	-	-	-2353	-
Intl. Marine Bunkers	-	-	-	-	-	-	-	-	-	-
Stock Changes	9129	-	-	-	-35	-	-	-	-	-
DOMESTIC SUPPLY	**258369**	**-**	**11788**	**15856**	**3149**	**-**	**-**	**-**	**28882**	**41198**
Transfers	-	-	-	-	-	-	-	-	-	-
Statistical Differences	-150	-	-	-	-	-	-	-	-	-
TRANSFORMATION	**40648**	**-**	**1049**	**1355**	**2916**	**-**	**-**	**-**	**1000**	**-**
Electricity Plants	-	-	-	-	-	-	-	-	-	-
CHP Plants	31217	-	1049	1355	-	-	-	-	-	-
Heat Plants	9431	-	-	-	-	-	-	-	-	-
Blast Furnaces/Gas Works	-	-	-	-	-	-	-	-	-	-
Coke/Pat. Fuel/BKB Plants	-	-	-	-	-	-	-	-	-	-
Petroleum Refineries	-	-	-	-	-	-	-	-	-	-
Petrochemical Industry	-	-	-	-	-	-	-	-	-	-
Liquefaction	-	-	-	-	-	-	-	-	-	-
Other Transform. Sector	-	-	-	-	2916	-	-	-	1000	-
ENERGY SECTOR	**-**	**-**	**1049**	**3825**	**-**	**-**	**-**	**-**	**2378**	**5550**
Coal Mines	-	-	-	-	-	-	-	-	-	-
Oil and Gas Extraction	-	-	-	-	-	-	-	-	-	-
Petroleum Refineries	-	-	-	-	-	-	-	-	-	-
Electr., CHP+Heat Plants	-	-	-	-	-	-	-	-	2084	-
Pumped Storage (Elec.)	-	-	-	-	-	-	-	-	294	-
Other Energy Sector	-	-	1049	3825	-	-	-	-	-	5550
Distribution Losses	6428	-	250	822	-	-	-	-	2025	2091
FINAL CONSUMPTION	**211143**	**-**	**9440**	**9854**	**233**	**-**	**-**	**-**	**23479**	**33557**
INDUSTRY SECTOR	**115405**	**-**	**9440**	**9854**	**2**	**-**	**-**	**-**	**10501**	**1010**
Iron and Steel	11698	-	9395	-	-	-	-	-	3502	-
Chemical and Petrochem.	55535	-	-	-	-	-	-	-	1556	-
of which: Feedstocks	*27778*	-	-	-	-	-	-	-	-	-
Non-Ferrous Metals	-	-	-	-	-	-	-	-	-	-
Non-Metallic Minerals	12691	-	-	-	-	-	-	-	1094	-
Transport Equipment	2374	-	-	-	-	-	-	-	268	246
Machinery	6863	-	45	-	-	-	-	-	903	230
Mining and Quarrying	3281	-	-	-	-	-	-	-	142	-
Food and Tobacco	10403	-	-	-	-	-	-	-	1225	82
Paper, Pulp and Print	2849	-	-	-	-	-	-	-	736	-
Wood and Wood Products	1079	-	-	-	-	-	-	-	222	164
Construction	42	-	-	-	2	-	-	-	167	41
Textile and Leather	3151	-	-	-	-	-	-	-	352	82
Non-specified	5439	-	-	9854	-	-	-	-	334	165
TRANSPORT SECTOR	**-**	**-**	**-**	**-**	**7**	**-**	**-**	**-**	**984**	**-**
Air	-	-	-	-	-	-	-	-	-	-
Road	-	-	-	-	-	-	-	-	-	-
Rail	-	-	-	-	7	-	-	-	984	-
Pipeline Transport	-	-	-	-	-	-	-	-	-	-
Internal Navigation	-	-	-	-	-	-	-	-	-	-
Non-specified	-	-	-	-	-	-	-	-	-	-
OTHER SECTORS	**95738**	**-**	**-**	**-**	**224**	**-**	**-**	**-**	**11994**	**32547**
Agriculture	2297	-	-	-	147	-	-	-	854	905
Comm. and Publ. Services	35863	-	-	-	44	-	-	-	5689	16053
Residential	57578	-	-	-	33	-	-	-	5451	15589
Non-specified	-	-	-	-	-	-	-	-	-	-
NON-ENERGY USE	**-**	**-**	**-**	**-**	**-**	**-**	**-**	**-**	**-**	**-**
in Industry/Transf./Energy	-	-	-	-	-	-	-	-	-	-
in Transport	-	-	-	-	-	-	-	-	-	-
in Other Sectors	-	-	-	-	-	-	-	-	-	-

Slovak Republic / République slovaque : 1997

SUPPLY AND CONSUMPTION / *APPROVISIONNEMENT ET DEMANDE*	Coal / *Charbon* (1000 tonnes)							Oil / *Pétrole* (1000 tonnes)			
	Coking Coal / *Charbon à coke*	Other Bit. Coal / *Autres charb. bit.*	Sub-Bit. Coal / *Charbon sous-bit.*	Lignite / *Lignite*	Peat / *Tourbe*	Oven and Gas Coke / *Coke de four/gaz*	Pat. Fuel and BKB / *Agg./briq. de lignite*	Crude Oil / *Pétrole brut*	NGL / *LGN*	Feed-stocks / *Produits d'aliment.*	Additives / *Additifs*
Production	-	-	-	3915	-	1730	-	64	1	-	-
From Other Sources	-	-	-	-	-	-	-	-	-	-	-
Imports	2568	2668	-	2196	-	146	5	5270	-	-	-
Exports	-	-	-	-8	-	-21	-	-	-	-	-
Intl. Marine Bunkers	-	-	-	-	-	-	-	-	-	-	-
Stock Changes	-2	-359	-	93	-	38	3	-8	-	-	-
DOMESTIC SUPPLY	**2566**	**2309**	**-**	**6196**	**-**	**1893**	**8**	**5326**	**1**	**-**	**-**
Transfers	-	-	-	-	-	-	-	-	-1	-	-
Statistical Differences	-	-1	-	-1	-	-	-	73	-	-	-
TRANSFORMATION	**2271**	**1467**	**-**	**3815**	**-**	**1313**	**-**	**5399**	**-**	**-**	**-**
Electricity Plants	-	1467	-	3755	-	-	-	-	-	-	-
CHP Plants	-	-	-	-	-	-	-	-	-	-	-
Heat Plants	-	-	-	60	-	6	-	-	-	-	-
Blast Furnaces/Gas Works	-	-	-	-	-	1307	-	-	-	-	-
Coke/Pat. Fuel/BKB Plants	2271	-	-	-	-	-	-	-	-	-	-
Petroleum Refineries	-	-	-	-	-	-	-	5399	-	-	-
Petrochemical Industry	-	-	-	-	-	-	-	-	-	-	-
Liquefaction	-	-	-	-	-	-	-	-	-	-	-
Other Transform. Sector	-	-	-	-	-	-	-	-	-	-	-
ENERGY SECTOR	**-**	**-**	**-**	**-**	**-**	**-**	**-**	**-**	**-**	**-**	**-**
Coal Mines	-	-	-	-	-	-	-	-	-	-	-
Oil and Gas Extraction	-	-	-	-	-	-	-	-	-	-	-
Petroleum Refineries	-	-	-	-	-	-	-	-	-	-	-
Electr., CHP+Heat Plants	-	-	-	-	-	-	-	-	-	-	-
Pumped Storage (Elec.)	-	-	-	-	-	-	-	-	-	-	-
Other Energy Sector	-	-	-	-	-	-	-	-	-	-	-
Distribution Losses	-	1	-	7	-	-	-	-	-	-	-
FINAL CONSUMPTION	**295**	**840**	**-**	**2373**	**-**	**580**	**8**	**-**	**-**	**-**	**-**
INDUSTRY SECTOR	**293**	**706**	**-**	**1318**	**-**	**480**	**1**	**-**	**-**	**-**	**-**
Iron and Steel	289	452	-	81	-	327	-	-	-	-	-
Chemical and Petrochem.	-	13	-	398	-	51	-	-	-	-	-
of which: Feedstocks	-	-	-	-	-	-	-	-	-	-	-
Non-Ferrous Metals	-	-	-	-	-	-	-	-	-	-	-
Non-Metallic Minerals	3	172	-	30	-	84	-	-	-	-	-
Transport Equipment	-	-	-	110	-	-	-	-	-	-	-
Machinery	-	20	-	163	-	3	1	-	-	-	-
Mining and Quarrying	-	-	-	8	-	4	-	-	-	-	-
Food and Tobacco	-	23	-	93	-	9	-	-	-	-	-
Paper, Pulp and Print	-	24	-	228	-	-	-	-	-	-	-
Wood and Wood Products	-	-	-	37	-	-	-	-	-	-	-
Construction	1	1	-	21	-	-	-	-	-	-	-
Textile and Leather	-	-	-	124	-	1	-	-	-	-	-
Non-specified	-	1	-	25	-	1	-	-	-	-	-
TRANSPORT SECTOR	**-**	**-**	**-**	**-**	**-**	**-**	**-**	**-**	**-**	**-**	**-**
Air	-	-	-	-	-	-	-	-	-	-	-
Road	-	-	-	-	-	-	-	-	-	-	-
Rail	-	-	-	-	-	-	-	-	-	-	-
Pipeline Transport	-	-	-	-	-	-	-	-	-	-	-
Internal Navigation	-	-	-	-	-	-	-	-	-	-	-
Non-specified	-	-	-	-	-	-	-	-	-	-	-
OTHER SECTORS	**2**	**134**	**-**	**1055**	**-**	**100**	**7**	**-**	**-**	**-**	**-**
Agriculture	1	3	-	56	-	8	3	-	-	-	-
Comm. and Publ. Services	-	127	-	550	-	74	-	-	-	-	-
Residential	1	4	-	449	-	18	4	-	-	-	-
Non-specified	-	-	-	-	-	-	-	-	-	-	-
NON-ENERGY USE	**-**	**-**	**-**	**-**	**-**	**-**	**-**	**-**	**-**	**-**	**-**
in Industry/Trans./Energy	-	-	-	-	-	-	-	-	-	-	-
in Transport	-	-	-	-	-	-	-	-	-	-	-
in Other Sectors	-	-	-	-	-	-	-	-	-	-	-

Slovak Republic / République slovaque : 1997

SUPPLY AND CONSUMPTION / *APPROVISIONNEMENT ET DEMANDE*	Oil cont. / *Pétrole cont.* (1000 tonnes)										
	Refinery Gas / *Gaz de raffinerie*	LPG + Ethane / *GPL + éthane*	Motor Gasoline / *Essence moteur*	Aviation Gasoline / *Essence aviation*	Jet Fuel / *Carbu-réacteurs*	Kerosene / *Kérosène*	Gas/ Diesel / *Gazole*	Heavy Fuel Oil / *Fioul lourd*	Naphtha / *Naphta*	Petrol. Coke / *Coke de pétrole*	Other Prod. / *Autres prod.*
Production	75	53	832	-	62	26	1782	1146	670	-	741
From Other Sources	-	-	-	-	-	-	-	-	-	-	-
Imports	-	15	92	-	-	-	57	41	-	-	-
Exports	-	-34	-397	-	-27	-18	-1052	-610	-	-	-20
Intl. Marine Bunkers	-	-	-	-	-	-	-	-	-	-	-
Stock Changes	-	-	21	-	-	-7	-44	-12	-	-	-
DOMESTIC SUPPLY	**75**	**34**	**548**	**-**	**35**	**1**	**743**	**565**	**670**	**-**	**721**
Transfers	-	1	-	-	-	-	-	-	-	-	-
Statistical Differences	-	-	-	-	-	-	-	-	-	-	-
TRANSFORMATION	**-**	**-**	**-**	**-**	**-**	**-**	**-**	**172**	**-**	**-**	**-**
Electricity Plants	-	-	-	-	-	-	-	-	-	-	-
CHP Plants	-	-	-	-	-	-	-	152	-	-	-
Heat Plants	-	-	-	-	-	-	-	20	-	-	-
Blast Furnaces/Gas Works	-	-	-	-	-	-	-	-	-	-	-
Coke/Pat. Fuel/BKB Plants	-	-	-	-	-	-	-	-	-	-	-
Petroleum Refineries	-	-	-	-	-	-	-	-	-	-	-
Petrochemical Industry	-	-	-	-	-	-	-	-	-	-	-
Liquefaction	-	-	-	-	-	-	-	-	-	-	-
Other Transform. Sector	-	-	-	-	-	-	-	-	-	-	-
ENERGY SECTOR	**-**	**-**	**15**	**-**	**-**	**-**	**1**	**27**	**-**	**-**	**-**
Coal Mines	-	-	-	-	-	-	-	-	-	-	-
Oil and Gas Extraction	-	-	-	-	-	-	-	-	-	-	-
Petroleum Refineries	-	-	-	-	-	-	-	-	-	-	-
Electr., CHP+Heat Plants	-	-	-	-	-	-	-	-	-	-	-
Pumped Storage (Elec.)	-	-	-	-	-	-	-	-	-	-	-
Other Energy Sector	-	-	15	-	-	-	1	27	-	-	-
Distribution Losses	-	-	-	-	-	-	-	-	-	-	-
FINAL CONSUMPTION	**75**	**35**	**533**	**-**	**35**	**1**	**742**	**366**	**670**	**-**	**721**
INDUSTRY SECTOR	**75**	**5**	**-**	**-**	**-**	**1**	**-**	**296**	**670**	**-**	**-**
Iron and Steel	-	-	-	-	-	-	-	38	-	-	-
Chemical and Petrochem.	-	-	-	-	-	-	-	17	670	-	-
of which: Feedstocks	-	-	-	-	-	-	-	-	*670*	-	-
Non-Ferrous Metals	-	-	-	-	-	-	-	-	-	-	-
Non-Metallic Minerals	-	-	-	-	-	-	-	22	-	-	-
Transport Equipment	-	-	-	-	-	1	-	6	-	-	-
Machinery	-	-	-	-	-	-	-	11	-	-	-
Mining and Quarrying	-	-	-	-	-	-	-	1	-	-	-
Food and Tobacco	-	-	-	-	-	-	-	33	-	-	-
Paper, Pulp and Print	-	-	-	-	-	-	-	77	-	-	-
Wood and Wood Products	-	-	-	-	-	-	-	3	-	-	-
Construction	-	-	-	-	-	-	-	5	-	-	-
Textile and Leather	-	-	-	-	-	-	-	60	-	-	-
Non-specified	75	5	-	-	-	-	-	23	-	-	-
TRANSPORT SECTOR	**-**	**-**	**523**	**-**	**35**	**-**	**566**	**-**	**-**	**-**	**-**
Air	-	-	-	-	35	-	-	-	-	-	-
Road	-	-	523	-	-	-	566	-	-	-	-
Rail	-	-	-	-	-	-	-	-	-	-	-
Pipeline Transport	-	-	-	-	-	-	-	-	-	-	-
Internal Navigation	-	-	-	-	-	-	-	-	-	-	-
Non-specified	-	-	-	-	-	-	-	-	-	-	-
OTHER SECTORS	**-**	**30**	**10**	**-**	**-**	**-**	**176**	**70**	**-**	**-**	**-**
Agriculture	-	1	10	-	-	-	176	8	-	-	-
Comm. and Publ. Services	-	6	-	-	-	-	-	62	-	-	-
Residential	-	23	-	-	-	-	-	-	-	-	-
Non-specified	-	-	-	-	-	-	-	-	-	-	-
NON-ENERGY USE	**-**	**-**	**-**	**-**	**-**	**-**	**-**	**-**	**-**	**-**	**721**
in Industry/Transf./Energy	-	-	-	-	-	-	-	-	-	-	721
in Transport	-	-	-	-	-	-	-	-	-	-	-
in Other Sectors	-	-	-	-	-	-	-	-	-	-	-

Slovak Republic / République slovaque : 1997

	Gas / *Gaz* (TJ)				Comb. Renew. & Waste / *En. Re. Comb. & Déchets* (TJ)				(GWh)	(TJ)
SUPPLY AND CONSUMPTION	Natural Gas	Gas Works	Coke Ovens	Blast Furnaces	Solid Biomass	Gas/Liquids from Biomass	Municipal Waste	Industrial Waste	Electricity	Heat
APPROVISIONNEMENT ET DEMANDE	*Gaz naturel*	*Usines à gaz*	*Cokeries*	*Hauts fourneaux*	*Biomasse solide*	*Gaz/Liquides tirés de biomasse*	*Déchets urbains*	*Déchets industriels*	*Electricité*	*Chaleur*
Production	10451	-	12919	16256	3454	-	-	-	24547	36339
From Other Sources	-	-	-	-	-	-	-	-	-	-
Imports	242593	-	-	-	-	-	-	-	6825	-
Exports	-	-	-	-	-	-	-	-	-2743	-
Intl. Marine Bunkers	-	-	-	-	-	-	-	-	-	-
Stock Changes	9139	-	-	-	24	-	-	-	-	-
DOMESTIC SUPPLY	**262183**	**-**	**12919**	**16256**	**3478**	**-**	**-**	**-**	**28629**	**36339**
Transfers	-	-	-	-	-	-	-	-	-	-
Statistical Differences	-162	-	-	-	-	-	-	-	-	-
TRANSFORMATION	**45411**	**-**	**1167**	**1458**	**3283**	**-**	**-**	**-**	**1688**	**-**
Electricity Plants	-	-	-	-	-	-	-	-	-	-
CHP Plants	33636	-	1167	1458	-	-	-	-	-	-
Heat Plants	11775	-	-	-	-	-	-	-	-	-
Blast Furnaces/Gas Works	-	-	-	-	-	-	-	-	-	-
Coke/Pat. Fuel/BKB Plants	-	-	-	-	-	-	-	-	-	-
Petroleum Refineries	-	-	-	-	-	-	-	-	-	-
Petrochemical Industry	-	-	-	-	-	-	-	-	-	-
Liquefaction	-	-	-	-	-	-	-	-	-	-
Other Transform. Sector	-	-	-	-	3283	-	-	-	1688	-
ENERGY SECTOR	**-**	**-**	**1170**	**4006**	**-**	**-**	**-**	**-**	**2016**	**5000**
Coal Mines	-	-	-	-	-	-	-	-	-	-
Oil and Gas Extraction	-	-	-	-	-	-	-	-	-	-
Petroleum Refineries	-	-	-	-	-	-	-	-	-	-
Electr., CHP+Heat Plants	-	-	-	-	-	-	-	-	1721	-
Pumped Storage (Elec.)	-	-	-	-	-	-	-	-	295	-
Other Energy Sector	-	-	1170	4006	-	-	-	-	-	5000
Distribution Losses	4144	-	230	184	-	-	-	-	2085	2000
FINAL CONSUMPTION	**212466**	**-**	**10352**	**10608**	**195**	**-**	**-**	**-**	**22840**	**29339**
INDUSTRY SECTOR	**113621**	**-**	**10352**	**10608**	**3**	**-**	**-**	**-**	**10053**	**1251**
Iron and Steel	11535	-	10302	-	-	-	-	-	3353	-
Chemical and Petrochem.	54590	-	-	-	-	-	-	-	1490	-
of which: Feedstocks	*27222*	-	-	-	-	-	-	-	-	-
Non-Ferrous Metals	-	-	-	-	-	-	-	-	-	-
Non-Metallic Minerals	12513	-	-	-	-	-	-	-	-	-
Transport Equipment	2341	-	-	-	-	-	-	-	1047	-
Machinery	6767	-	50	-	-	-	-	-	256	305
Mining and Quarrying	3234	-	-	-	-	-	-	-	864	285
Food and Tobacco	10257	-	-	-	-	-	-	-	136	-
Paper, Pulp and Print	2809	-	-	-	-	-	-	-	1173	102
Wood and Wood Products	1064	-	-	-	-	-	-	-	704	-
Construction	42	-	-	-	3	-	-	-	213	203
Textile and Leather	3107	-	-	-	-	-	-	-	183	51
Non-specified	5362	-	-	10608	-	-	-	-	337	102
									297	203
TRANSPORT SECTOR	**-**	**-**	**-**	**-**	**-**	**-**	**-**	**-**	**1010**	**-**
Air	-	-	-	-	-	-	-	-	-	-
Road	-	-	-	-	-	-	-	-	-	-
Rail	-	-	-	-	-	-	-	-	1010	-
Pipeline Transport	-	-	-	-	-	-	-	-	-	-
Internal Navigation	-	-	-	-	-	-	-	-	-	-
Non-specified	-	-	-	-	-	-	-	-	-	-
OTHER SECTORS	**98845**	**-**	**-**	**-**	**192**	**-**	**-**	**-**	**11777**	**28088**
Agriculture	2363	-	-	-	129	-	-	-	1136	934
Comm. and Publ. Services	32586	-	-	-	49	-	-	-	5134	11334
Residential	63896	-	-	-	14	-	-	-	5507	15820
Non-specified	-	-	-	-	-	-	-	-	-	-
NON-ENERGY USE	**-**	**-**	**-**	**-**	**-**	**-**	**-**	**-**	**-**	**-**
in Industry/Transf./Energy	-	-	-	-	-	-	-	-	-	-
in Transport	-	-	-	-	-	-	-	-	-	-
in Other Sectors	-	-	-	-	-	-	-	-	-	-

Slovenia / Slovénie : 1996

	Coal / *Charbon* (1000 tonnes)							Oil / *Pétrole* (1000 tonnes)			
SUPPLY AND CONSUMPTION	Coking Coal	Other Bit. Coal	Sub-Bit. Coal	Lignite	Peat	Oven and Gas Coke	Pat. Fuel and BKB	Crude Oil	NGL	Feed-stocks	Additives
APPROVISIONNEMENT ET DEMANDE	*Charbon à coke*	*Autres charb. bit.*	*Charbon sous-bit.*	*Lignite*	*Tourbe*	*Coke de four/gaz*	*Agg./briq. de lignite*	*Pétrole brut*	*LGN*	*Produits d'aliment.*	*Additifs*
Production	-	-	830	3937	-	-	-	1	-	-	-
From Other Sources	-	-	-	-	-	-	-	-	-	-	-
Imports	-	22	385	21	-	62	-	455	-	69	-
Exports	-	-	-	-2	-	-	-	-	-	-	-
Intl. Marine Bunkers	-	-	-	-	-	-	-	-	-	-	-
Stock Changes	-	4	-87	-149	-	2	-	-1	-	1	-
DOMESTIC SUPPLY	**-**	**26**	**1128**	**3807**	**-**	**64**	**-**	**455**	**-**	**70**	**-**
Transfers	-	-	-	-	-	-	-	-	-	-	-
Statistical Differences	-	-	-	-	-	-	-	27	-	-	-
TRANSFORMATION	**-**	**-**	**1075**	**3641**	**-**	**25**	**-**	**482**	**-**	**70**	**-**
Electricity Plants	-	-	529	-	-	-	-	-	-	-	-
CHP Plants	-	-	545	3641	-	-	-	-	-	-	-
Heat Plants	-	-	1	-	-	-	-	-	-	-	-
Blast Furnaces/Gas Works	-	-	-	-	-	25	-	-	-	-	-
Coke/Pat. Fuel/BKB Plants	-	-	-	-	-	-	-	-	-	-	-
Petroleum Refineries	-	-	-	-	-	-	-	482	-	70	-
Petrochemical Industry	-	-	-	-	-	-	-	-	-	-	-
Liquefaction	-	-	-	-	-	-	-	-	-	-	-
Other Transform. Sector	-	-	-	-	-	-	-	-	-	-	-
ENERGY SECTOR	**-**	**-**	**5**	**-**	**-**	**-**	**-**	**-**	**-**	**-**	**-**
Coal Mines	-	-	5	-	-	-	-	-	-	-	-
Oil and Gas Extraction	-	-	-	-	-	-	-	-	-	-	-
Petroleum Refineries	-	-	-	-	-	-	-	-	-	-	-
Electr., CHP+Heat Plants	-	-	-	-	-	-	-	-	-	-	-
Pumped Storage (Elec.)	-	-	-	-	-	-	-	-	-	-	-
Other Energy Sector	-	-	-	-	-	-	-	-	-	-	-
Distribution Losses	-	-	-	-	-	-	-	-	-	-	-
FINAL CONSUMPTION	**-**	**26**	**48**	**166**	**-**	**39**	**-**	**-**	**-**	**-**	**-**
INDUSTRY SECTOR	**-**	**26**	**10**	**51**	**-**	**39**	**-**	**-**	**-**	**-**	**-**
Iron and Steel	-	-	-	-	-	6	-	-	-	-	-
Chemical and Petrochem.	-	-	-	-	-	-	-	-	-	-	-
of which: Feedstocks	-	-	-	-	-	-	-	-	-	-	-
Non-Ferrous Metals	-	2	-	-	-	-	-	-	-	-	-
Non-Metallic Minerals	-	23	-	-	-	32	-	-	-	-	-
Transport Equipment	-	-	1	-	-	-	-	-	-	-	-
Machinery	-	-	-	-	-	-	-	-	-	-	-
Mining and Quarrying	-	-	-	-	-	-	-	-	-	-	-
Food and Tobacco	-	-	1	-	-	1	-	-	-	-	-
Paper, Pulp and Print	-	-	-	51	-	-	-	-	-	-	-
Wood and Wood Products	-	-	-	-	-	-	-	-	-	-	-
Construction	-	-	-	-	-	-	-	-	-	-	-
Textile and Leather	-	-	6	-	-	-	-	-	-	-	-
Non-specified	-	1	2	-	-	-	-	-	-	-	-
TRANSPORT SECTOR	**-**	**-**	**1**	**-**	**-**	**-**	**-**	**-**	**-**	**-**	**-**
Air	-	-	-	-	-	-	-	-	-	-	-
Road	-	-	-	-	-	-	-	-	-	-	-
Rail	-	-	1	-	-	-	-	-	-	-	-
Pipeline Transport	-	-	-	-	-	-	-	-	-	-	-
Internal Navigation	-	-	-	-	-	-	-	-	-	-	-
Non-specified	-	-	-	-	-	-	-	-	-	-	-
OTHER SECTORS	**-**	**-**	**37**	**115**	**-**	**-**	**-**	**-**	**-**	**-**	**-**
Agriculture	-	-	-	-	-	-	-	-	-	-	-
Comm. and Publ. Services	-	-	15	40	-	-	-	-	-	-	-
Residential	-	-	22	75	-	-	-	-	-	-	-
Non-specified	-	-	-	-	-	-	-	-	-	-	-
NON-ENERGY USE	**-**	**-**	**-**	**-**	**-**	**-**	**-**	**-**	**-**	**-**	**-**
in Industry/Trans./Energy	-	-	-	-	-	-	-	-	-	-	-
in Transport	-	-	-	-	-	-	-	-	-	-	-
in Other Sectors	-	-	-	-	-	-	-	-	-	-	-

Slovenia / Slovénie : 1996

SUPPLY AND CONSUMPTION / *APPROVISIONNEMENT ET DEMANDE*	Oil cont. / *Pétrole cont.* (1000 tonnes)										
	Refinery Gas / *Gaz de raffinerie*	LPG + Ethane / *GPL + éthane*	Motor Gasoline / *Essence moteur*	Aviation Gasoline / *Essence aviation*	Jet Fuel / *Carbu-réacteurs*	Kerosene / *Kérosène*	Gas/ Diesel / *Gazole*	Heavy Fuel Oil / *Fioul lourd*	Naphtha / *Naphta*	Petrol. Coke / *Coke de pétrole*	Other Prod. / *Autres prod.*
Production	-	-	79	-	-	-	189	121	125	-	7
From Other Sources	-	-	-	-	-	-	-	-	-	-	-
Imports	-	43	834	1	20	-	1085	152	-	-	-
Exports	-	-	-1	-	-3	-	-2	-	-47	-	-
Intl. Marine Bunkers	-	-	-	-	-	-	-	-	-	-	-
Stock Changes	-	-4	12	-	-	-	11	-3	-2	-	-
DOMESTIC SUPPLY	**-**	**39**	**924**	**1**	**17**	**-**	**1283**	**270**	**76**	**-**	**7**
Transfers	-	-	-	-	-	-	-	-	-	-	-
Statistical Differences	-	-	-	-	-	-	-	-	-	-	-
TRANSFORMATION	**-**	**-**	**-**	**-**	**-**	**-**	**7**	**140**	**-**	**-**	**-**
Electricity Plants	-	-	-	-	-	-	3	-	-	-	-
CHP Plants	-	-	-	-	-	-	1	128	-	-	-
Heat Plants	-	-	-	-	-	-	3	12	-	-	-
Blast Furnaces/Gas Works	-	-	-	-	-	-	-	-	-	-	-
Coke/Pat. Fuel/BKB Plants	-	-	-	-	-	-	-	-	-	-	-
Petroleum Refineries	-	-	-	-	-	-	-	-	-	-	-
Petrochemical Industry	-	-	-	-	-	-	-	-	-	-	-
Liquefaction	-	-	-	-	-	-	-	-	-	-	-
Other Transform. Sector	-	-	-	-	-	-	-	-	-	-	-
ENERGY SECTOR	**-**	**-**	**1**	**-**	**-**	**-**	**-**	**19**	**-**	**-**	**-**
Coal Mines	-	-	-	-	-	-	-	-	-	-	-
Oil and Gas Extraction	-	-	-	-	-	-	-	6	-	-	-
Petroleum Refineries	-	-	1	-	-	-	-	13	-	-	-
Electr., CHP+Heat Plants	-	-	-	-	-	-	-	-	-	-	-
Pumped Storage (Elec.)	-	-	-	-	-	-	-	-	-	-	-
Other Energy Sector	-	-	-	-	-	-	-	-	-	-	-
Distribution Losses	-	-	-	-	-	-	-	-	-	-	-
FINAL CONSUMPTION	**-**	**39**	**923**	**1**	**17**	**-**	**1276**	**111**	**76**	**-**	**7**
INDUSTRY SECTOR	**-**	**16**	**-**	**-**	**-**	**-**	**34**	**84**	**76**	**-**	**3**
Iron and Steel	-	2	-	-	-	-	1	-	-	-	-
Chemical and Petrochem.	-	-	-	-	-	-	2	8	76	-	3
of which: Feedstocks	-	-	-	-	-	-	-	-	*76*	-	*3*
Non-Ferrous Metals	-	1	-	-	-	-	-	2	-	-	-
Non-Metallic Minerals	-	4	-	-	-	-	4	33	-	-	-
Transport Equipment	-	1	-	-	-	-	3	3	-	-	-
Machinery	-	2	-	-	-	-	6	6	-	-	-
Mining and Quarrying	-	1	-	-	-	-	2	-	-	-	-
Food and Tobacco	-	1	-	-	-	-	10	11	-	-	-
Paper, Pulp and Print	-	3	-	-	-	-	1	-	-	-	-
Wood and Wood Products	-	-	-	-	-	-	-	5	-	-	-
Construction	-	-	-	-	-	-	-	-	-	-	-
Textile and Leather	-	-	-	-	-	-	3	13	-	-	-
Non-specified	-	1	-	-	-	-	2	3	-	-	-
TRANSPORT SECTOR	**-**	**-**	**923**	**1**	**17**	**-**	**487**	**-**	**-**	**-**	**-**
Air	-	-	-	1	17	-	-	-	-	-	-
Road	-	-	923	-	-	-	475	-	-	-	-
Rail	-	-	-	-	-	-	12	-	-	-	-
Pipeline Transport	-	-	-	-	-	-	-	-	-	-	-
Internal Navigation	-	-	-	-	-	-	-	-	-	-	-
Non-specified	-	-	-	-	-	-	-	-	-	-	-
OTHER SECTORS	**-**	**23**	**-**	**-**	**-**	**-**	**755**	**27**	**-**	**-**	**-**
Agriculture	-	-	-	-	-	-	-	-	-	-	-
Comm. and Publ. Services	-	5	-	-	-	-	304	-	-	-	-
Residential	-	18	-	-	-	-	451	27	-	-	-
Non-specified	-	-	-	-	-	-	-	-	-	-	-
NON-ENERGY USE	**-**	**-**	**-**	**-**	**-**	**-**	**-**	**-**	**-**	**-**	**4**
in Industry/Transf./Energy	-	-	-	-	-	-	-	-	-	-	4
in Transport	-	-	-	-	-	-	-	-	-	-	-
in Other Sectors	-	-	-	-	-	-	-	-	-	-	-

Slovenia / Slovénie : 1996

SUPPLY AND CONSUMPTION / *APPROVISIONNEMENT ET DEMANDE*	Gas / *Gaz* (TJ)				Comb. Renew. & Waste / *En. Re. Comb. & Déchets* (TJ)				(GWh)	(TJ)
	Natural Gas / *Gaz naturel*	Gas Works / *Usines à gaz*	Coke Ovens / *Cokeries*	Blast Furnaces / *Hauts fourneaux*	Solid Biomass / *Biomasse solide*	Gas/Liquids from Biomass / *Gaz/Liquides tirés de biomasse*	Municipal Waste / *Déchets urbains*	Industrial Waste / *Déchets industriels*	Electricity / *Electricité*	Heat / *Chaleur*
Production	466	-	-	-	9792	-	-	-	12770	9702
From Other Sources	-	-	-	-	-	-	-	-	-	-
Imports	30039	-	-	-	1236	-	-	-	855	-
Exports	-	-	-	-	-	-	-	-	-2516	-
Intl. Marine Bunkers	-	-	-	-	-	-	-	-	-	-
Stock Changes	-	-	-	-	-31	-	-	-	-	-
DOMESTIC SUPPLY	**30505**	**-**	**-**	**-**	**10997**	**-**	**-**	**-**	**11109**	**9702**
Transfers	-	-	-	-	-	-	-	-	-	-
Statistical Differences	939	-	-	-	-	-	-	-	-	-
TRANSFORMATION	**6000**	**-**	**-**	**-**	**96**	**-**	**-**	**-**	**-**	**-**
Electricity Plants	80	-	-	-	-	-	-	-	-	-
CHP Plants	2930	-	-	-	-	-	-	-	-	-
Heat Plants	2990	-	-	-	96	-	-	-	-	-
Blast Furnaces/Gas Works	-	-	-	-	-	-	-	-	-	-
Coke/Pat. Fuel/BKB Plants	-	-	-	-	-	-	-	-	-	-
Petroleum Refineries	-	-	-	-	-	-	-	-	-	-
Petrochemical Industry	-	-	-	-	-	-	-	-	-	-
Liquefaction	-	-	-	-	-	-	-	-	-	-
Other Transform. Sector	-	-	-	-	-	-	-	-	-	-
ENERGY SECTOR	**47**	**-**	**-**	**-**	**-**	**-**	**-**	**-**	**894**	**1123**
Coal Mines	-	-	-	-	-	-	-	-	113	-
Oil and Gas Extraction	-	-	-	-	-	-	-	-	-	-
Petroleum Refineries	47	-	-	-	-	-	-	-	26	-
Electr., CHP+Heat Plants	-	-	-	-	-	-	-	-	755	1123
Pumped Storage (Elec.)	-	-	-	-	-	-	-	-	-	-
Other Energy Sector	-	-	-	-	-	-	-	-	-	-
Distribution Losses	-	-	-	-	-	-	-	-	726	-
FINAL CONSUMPTION	**25397**	**-**	**-**	**-**	**10901**	**-**	**-**	**-**	**9489**	**8579**
INDUSTRY SECTOR	**22677**	**-**	**-**	**-**	**2556**	**-**	**-**	**-**	**4778**	**1285**
Iron and Steel	3608	-	-	-	-	-	-	-	644	110
Chemical and Petrochem.	5141	-	-	-	-	-	-	-	409	145
of which: Feedstocks	*2764*	-	-	-	-	-	-	-	-	-
Non-Ferrous Metals	135	-	-	-	-	-	-	-	1187	-
Non-Metallic Minerals	4350	-	-	-	-	-	-	-	232	30
Transport Equipment	519	-	-	-	-	-	-	-	2	-
Machinery	943	-	-	-	-	-	-	-	689	-
Mining and Quarrying	-	-	-	-	-	-	-	-	-	-
Food and Tobacco	1366	-	-	-	-	-	-	-	226	-
Paper, Pulp and Print	1406	-	-	-	-	-	-	-	561	-
Wood and Wood Products	390	-	-	-	-	-	-	-	244	-
Construction	-	-	-	-	-	-	-	-	198	-
Textile and Leather	1660	-	-	-	-	-	-	-	289	236
Non-specified	3159	-	-	-	2556	-	-	-	97	764
TRANSPORT SECTOR	**-**	**-**	**-**	**-**	**-**	**-**	**-**	**-**	**160**	**-**
Air	-	-	-	-	-	-	-	-	-	-
Road	-	-	-	-	-	-	-	-	-	-
Rail	-	-	-	-	-	-	-	-	160	-
Pipeline Transport	-	-	-	-	-	-	-	-	-	-
Internal Navigation	-	-	-	-	-	-	-	-	-	-
Non-specified	-	-	-	-	-	-	-	-	-	-
OTHER SECTORS	**2720**	**-**	**-**	**-**	**8345**	**-**	**-**	**-**	**4551**	**7294**
Agriculture	-	-	-	-	-	-	-	-	-	-
Comm. and Publ. Services	815	-	-	-	2141	-	-	-	1922	2218
Residential	1905	-	-	-	6204	-	-	-	2629	5076
Non-specified	-	-	-	-	-	-	-	-	-	-
NON-ENERGY USE	**-**	**-**	**-**	**-**	**-**	**-**	**-**	**-**	**-**	**-**
in Industry/Transf./Energy	-	-	-	-	-	-	-	-	-	-
in Transport	-	-	-	-	-	-	-	-	-	-
in Other Sectors	-	-	-	-	-	-	-	-	-	-

Slovenia / Slovénie : 1997

SUPPLY AND CONSUMPTION *APPROVISIONNEMENT ET DEMANDE*	Coal / *Charbon* (1000 tonnes)							Oil / *Pétrole* (1000 tonnes)			
	Coking Coal *Charbon à coke*	Other Bit. Coal *Autres charb. bit.*	Sub-Bit. Coal *Charbon sous-bit.*	Lignite *Lignite*	Peat *Tourbe*	Oven and Gas Coke *Coke de four/gaz*	Pat. Fuel and BKB *Agg./briq. de lignite*	Crude Oil *Pétrole brut*	NGL *LGN*	Feed-stocks *Produits d'aliment.*	Additives *Additifs*
Production	-	-	811	4142	-	-	-	1	-	-	-
From Other Sources	-	-	-	-	-	-	-	-	-	-	-
Imports	-	19	300	7	-	72	-	545	-	77	-
Exports	-	-	-1	-1	-	-	-	-	-	-	-
Intl. Marine Bunkers	-	-	-	-	-	-	-	-	-	-	-
Stock Changes	-	-	58	51	-	-1	-	-12	-	-4	-
DOMESTIC SUPPLY	**-**	**19**	**1168**	**4199**	**-**	**71**	**-**	**534**	**-**	**73**	**-**
Transfers	-	-	-	-	-	-	-	-	-	-	-
Statistical Differences	-	-	-	-	-	-	-	-	-	-	-
TRANSFORMATION	**-**	**-**	**1137**	**4086**	**-**	**20**	**-**	**534**	**-**	**73**	**-**
Electricity Plants	-	-	625	-	-	-	-	-	-	-	-
CHP Plants	-	-	512	4086	-	-	-	-	-	-	-
Heat Plants	-	-	-	-	-	-	-	-	-	-	-
Blast Furnaces/Gas Works	-	-	-	-	-	20	-	-	-	-	-
Coke/Pat. Fuel/BKB Plants	-	-	-	-	-	-	-	-	-	-	-
Petroleum Refineries	-	-	-	-	-	-	-	534	-	73	-
Petrochemical Industry	-	-	-	-	-	-	-	-	-	-	-
Liquefaction	-	-	-	-	-	-	-	-	-	-	-
Other Transform. Sector	-	-	-	-	-	-	-	-	-	-	-
ENERGY SECTOR	**-**	**-**	**3**	**-**	**-**	**-**	**-**	**-**	**-**	**-**	**-**
Coal Mines	-	-	3	-	-	-	-	-	-	-	-
Oil and Gas Extraction	-	-	-	-	-	-	-	-	-	-	-
Petroleum Refineries	-	-	-	-	-	-	-	-	-	-	-
Electr., CHP+Heat Plants	-	-	-	-	-	-	-	-	-	-	-
Pumped Storage (Elec.)	-	-	-	-	-	-	-	-	-	-	-
Other Energy Sector	-	-	-	-	-	-	-	-	-	-	-
Distribution Losses	-	-	-	-	-	-	-	-	-	-	-
FINAL CONSUMPTION	**-**	**19**	**28**	**113**	**-**	**51**	**-**	**-**	**-**	**-**	**-**
INDUSTRY SECTOR	**-**	**19**	**7**	**46**	**-**	**51**	**-**	**-**	**-**	**-**	**-**
Iron and Steel	-	-	-	-	-	5	-	-	-	-	-
Chemical and Petrochem.	-	-	-	-	-	-	-	-	-	-	-
of which: Feedstocks	-	-	-	-	-	-	-	-	-	-	-
Non-Ferrous Metals	-	2	-	-	-	-	-	-	-	-	-
Non-Metallic Minerals	-	16	-	-	-	44	-	-	-	-	-
Transport Equipment	-	-	1	-	-	1	-	-	-	-	-
Machinery	-	-	-	-	-	-	-	-	-	-	-
Mining and Quarrying	-	-	-	-	-	-	-	-	-	-	-
Food and Tobacco	-	-	1	-	-	1	-	-	-	-	-
Paper, Pulp and Print	-	-	-	46	-	-	-	-	-	-	-
Wood and Wood Products	-	-	-	-	-	-	-	-	-	-	-
Construction	-	-	-	-	-	-	-	-	-	-	-
Textile and Leather	-	-	3	-	-	-	-	-	-	-	-
Non-specified	-	1	2	-	-	-	-	-	-	-	-
TRANSPORT SECTOR	**-**	**-**	**1**	**-**	**-**	**-**	**-**	**-**	**-**	**-**	**-**
Air	-	-	-	-	-	-	-	-	-	-	-
Road	-	-	-	-	-	-	-	-	-	-	-
Rail	-	-	1	-	-	-	-	-	-	-	-
Pipeline Transport	-	-	-	-	-	-	-	-	-	-	-
Internal Navigation	-	-	-	-	-	-	-	-	-	-	-
Non-specified	-	-	-	-	-	-	-	-	-	-	-
OTHER SECTORS	**-**	**-**	**20**	**67**	**-**	**-**	**-**	**-**	**-**	**-**	**-**
Agriculture	-	-	-	-	-	-	-	-	-	-	-
Comm. and Publ. Services	-	-	8	23	-	-	-	-	-	-	-
Residential	-	-	12	44	-	-	-	-	-	-	-
Non-specified	-	-	-	-	-	-	-	-	-	-	-
NON-ENERGY USE	**-**	**-**	**-**	**-**	**-**	**-**	**-**	**-**	**-**	**-**	**-**
in Industry/Trans./Energy	-	-	-	-	-	-	-	-	-	-	-
in Transport	-	-	-	-	-	-	-	-	-	-	-
in Other Sectors	-	-	-	-	-	-	-	-	-	-	-

Slovenia / Slovénie : 1997

SUPPLY AND CONSUMPTION / *APPROVISIONNEMENT ET DEMANDE*	Oil cont. / *Pétrole cont.* (1000 tonnes)										
	Refinery Gas / *Gaz de raffinerie*	LPG + Ethane / *GPL + éthane*	Motor Gasoline / *Essence moteur*	Aviation Gasoline / *Essence aviation*	Jet Fuel / *Carbu-réacteurs*	Kerosene / *Kérosène*	Gas/ Diesel / *Gazole*	Heavy Fuel Oil / *Fioul lourd*	Naphtha / *Naphta*	Petrol. Coke / *Coke de pétrole*	Other Prod. / *Autres prod.*
Production	-	-	89	-	-	-	200	173	74	-	5
From Other Sources	-	-	-	-	-	-	-	-	-	-	-
Imports	-	54	853	1	21	-	1134	56	-	-	-
Exports	-	-	-	-	-3	-	-	-3	-63	-	-1
Intl. Marine Bunkers	-	-	-	-	-	-	-	-	-	-	-
Stock Changes	-	-	-28	-	-	-	-31	-2	-	-	-
DOMESTIC SUPPLY	**-**	**54**	**914**	**1**	**18**	**-**	**1303**	**224**	**11**	**-**	**4**
Transfers	-	-	-	-	-	-	-	-	-	-	-
Statistical Differences	-	-	-	-	-	-	-	1	-	-	-
TRANSFORMATION	**-**	**-**	**-**	**-**	**-**	**-**	**7**	**56**	**-**	**-**	**-**
Electricity Plants	-	-	-	-	-	-	3	-	-	-	-
CHP Plants	-	-	-	-	-	-	1	51	-	-	-
Heat Plants	-	-	-	-	-	-	3	5	-	-	-
Blast Furnaces/Gas Works	-	-	-	-	-	-	-	-	-	-	-
Coke/Pat. Fuel/BKB Plants	-	-	-	-	-	-	-	-	-	-	-
Petroleum Refineries	-	-	-	-	-	-	-	-	-	-	-
Petrochemical Industry	-	-	-	-	-	-	-	-	-	-	-
Liquefaction	-	-	-	-	-	-	-	-	-	-	-
Other Transform. Sector	-	-	-	-	-	-	-	-	-	-	-
ENERGY SECTOR	**-**	**-**	**1**	**-**	**-**	**-**	**-**	**22**	**-**	**-**	**-**
Coal Mines	-	-	-	-	-	-	-	-	-	-	-
Oil and Gas Extraction	-	-	-	-	-	-	-	7	-	-	-
Petroleum Refineries	-	-	1	-	-	-	-	15	-	-	-
Electr., CHP+Heat Plants	-	-	-	-	-	-	-	-	-	-	-
Pumped Storage (Elec.)	-	-	-	-	-	-	-	-	-	-	-
Other Energy Sector	-	-	-	-	-	-	-	-	-	-	-
Distribution Losses	-	-	-	-	-	-	-	-	-	-	-
FINAL CONSUMPTION	**-**	**54**	**913**	**1**	**18**	**-**	**1296**	**147**	**11**	**-**	**4**
INDUSTRY SECTOR	**-**	**11**	**-**	**-**	**-**	**-**	**36**	**103**	**11**	**-**	**3**
Iron and Steel	-	2	-	-	-	-	-	-	-	-	-
Chemical and Petrochem.	-	-	-	-	-	-	2	12	11	-	3
of which: Feedstocks	-	-	-	-	-	-	-	-	*11*	-	*3*
Non-Ferrous Metals	-	1	-	-	-	-	1	3	-	-	-
Non-Metallic Minerals	-	5	-	-	-	-	5	28	-	-	-
Transport Equipment	-	-	-	-	-	-	2	2	-	-	-
Machinery	-	2	-	-	-	-	7	7	-	-	-
Mining and Quarrying	-	-	-	-	-	-	1	1	-	-	-
Food and Tobacco	-	1	-	-	-	-	10	24	-	-	-
Paper, Pulp and Print	-	-	-	-	-	-	1	-	-	-	-
Wood and Wood Products	-	-	-	-	-	-	-	8	-	-	-
Construction	-	-	-	-	-	-	-	-	-	-	-
Textile and Leather	-	-	-	-	-	-	4	12	-	-	-
Non-specified	-	-	-	-	-	-	3	6	-	-	-
TRANSPORT SECTOR	**-**	**-**	**913**	**1**	**18**	**-**	**518**	**-**	**-**	**-**	**-**
Air	-	-	-	1	18	-	-	-	-	-	-
Road	-	-	913	-	-	-	506	-	-	-	-
Rail	-	-	-	-	-	-	12	-	-	-	-
Pipeline Transport	-	-	-	-	-	-	-	-	-	-	-
Internal Navigation	-	-	-	-	-	-	-	-	-	-	-
Non-specified	-	-	-	-	-	-	-	-	-	-	-
OTHER SECTORS	**-**	**43**	**-**	**-**	**-**	**-**	**742**	**44**	**-**	**-**	**-**
Agriculture	-	-	-	-	-	-	-	-	-	-	-
Comm. and Publ. Services	-	8	-	-	-	-	297	44	-	-	-
Residential	-	35	-	-	-	-	445	-	-	-	-
Non-specified	-	-	-	-	-	-	-	-	-	-	-
NON-ENERGY USE	**-**	**-**	**-**	**-**	**-**	**-**	**-**	**-**	**-**	**-**	**1**
in Industry/Transf./Energy	-	-	-	-	-	-	-	-	-	-	1
in Transport	-	-	-	-	-	-	-	-	-	-	-
in Other Sectors	-	-	-	-	-	-	-	-	-	-	-

Slovenia / Slovénie : 1997

SUPPLY AND CONSUMPTION / *APPROVISIONNEMENT ET DEMANDE*	Gas / *Gaz* (TJ)				Comb. Renew. & Waste / *En. Re. Comb. & Déchets* (TJ)				(GWh)	(TJ)
	Natural Gas / *Gaz naturel*	Gas Works / *Usines à gaz*	Coke Ovens / *Cokeries*	Blast Furnaces / *Hauts fourneaux*	Solid Biomass / *Biomasse solide*	Gas/Liquids from Biomass / *Gaz/Liquides tirés de biomasse*	Municipal Waste / *Déchets urbains*	Industrial Waste / *Déchets industriels*	Electricity / *Electricité*	Heat / *Chaleur*
Production	432	-	-	-	9792	-	-	-	13166	9066
From Other Sources	-	-	-	-	-	-	-	-	-	-
Imports	32847	-	-	-	1236	-	-	-	824	-
Exports	-	-	-	-	-	-	-	-	-2520	-
Intl. Marine Bunkers	-	-	-	-	-	-	-	-	-	-
Stock Changes	-	-	-	-	-33	-	-	-	-	-
DOMESTIC SUPPLY	**33279**	**-**	**-**	**-**	**10995**	**-**	**-**	**-**	**11470**	**9066**
Transfers	-	-	-	-	-	-	-	-	-	-
Statistical Differences	147	-	-	-	-	-	-	-	-	-
TRANSFORMATION	**3630**	**-**	**-**	**-**	**94**	**-**	**-**	**-**	**-**	**-**
Electricity Plants	98	-	-	-	-	-	-	-	-	-
CHP Plants	612	-	-	-	-	-	-	-	-	-
Heat Plants	2920	-	-	-	94	-	-	-	-	-
Blast Furnaces/Gas Works	-	-	-	-	-	-	-	-	-	-
Coke/Pat. Fuel/BKB Plants	-	-	-	-	-	-	-	-	-	-
Petroleum Refineries	-	-	-	-	-	-	-	-	-	-
Petrochemical Industry	-	-	-	-	-	-	-	-	-	-
Liquefaction	-	-	-	-	-	-	-	-	-	-
Other Transform. Sector	-	-	-	-	-	-	-	-	-	-
ENERGY SECTOR	**40**	**-**	**-**	**-**	**-**	**-**	**-**	**-**	**942**	**938**
Coal Mines	-	-	-	-	-	-	-	-	94	-
Oil and Gas Extraction	-	-	-	-	-	-	-	-	-	-
Petroleum Refineries	40	-	-	-	-	-	-	-	30	-
Electr., CHP+Heat Plants	-	-	-	-	-	-	-	-	818	938
Pumped Storage (Elec.)	-	-	-	-	-	-	-	-	-	-
Other Energy Sector	-	-	-	-	-	-	-	-	-	-
Distribution Losses	-	-	-	-	-	-	-	-	687	-
FINAL CONSUMPTION	**29756**	**-**	**-**	**-**	**10901**	**-**	**-**	**-**	**9841**	**8128**
INDUSTRY SECTOR	**27065**	**-**	**-**	**-**	**2556**	**-**	**-**	**-**	**4726**	**1113**
Iron and Steel	5170	-	-	-	-	-	-	-	748	106
Chemical and Petrochem.	8385	-	-	-	-	-	-	-	406	155
of which: Feedstocks	*4449*	-	-	-	-	-	-	-	-	-
Non-Ferrous Metals	749	-	-	-	-	-	-	-	1210	-
Non-Metallic Minerals	4052	-	-	-	-	-	-	-	431	31
Transport Equipment	830	-	-	-	-	-	-	-	98	-
Machinery	695	-	-	-	-	-	-	-	504	-
Mining and Quarrying	28	-	-	-	-	-	-	-	26	-
Food and Tobacco	1378	-	-	-	-	-	-	-	220	-
Paper, Pulp and Print	2873	-	-	-	-	-	-	-	376	-
Wood and Wood Products	344	-	-	-	-	-	-	-	160	-
Construction	-	-	-	-	-	-	-	-	-	-
Textile and Leather	1420	-	-	-	-	-	-	-	286	216
Non-specified	1141	-	-	-	2556	-	-	-	261	605
TRANSPORT SECTOR	**-**	**-**	**-**	**-**	**-**	**-**	**-**	**-**	**164**	**-**
Air	-	-	-	-	-	-	-	-	-	-
Road	-	-	-	-	-	-	-	-	-	-
Rail	-	-	-	-	-	-	-	-	164	-
Pipeline Transport	-	-	-	-	-	-	-	-	-	-
Internal Navigation	-	-	-	-	-	-	-	-	-	-
Non-specified	-	-	-	-	-	-	-	-	-	-
OTHER SECTORS	**2691**	**-**	**-**	**-**	**8345**	**-**	**-**	**-**	**4951**	**7015**
Agriculture	-	-	-	-	-	-	-	-	-	-
Comm. and Publ. Services	807	-	-	-	2141	-	-	-	2267	2124
Residential	1884	-	-	-	6204	-	-	-	2684	4891
Non-specified	-	-	-	-	-	-	-	-	-	-
NON-ENERGY USE	**-**	**-**	**-**	**-**	**-**	**-**	**-**	**-**	**-**	**-**
in Industry/Transf./Energy	-	-	-	-	-	-	-	-	-	-
in Transport	-	-	-	-	-	-	-	-	-	-
in Other Sectors	-	-	-	-	-	-	-	-	-	-

South Africa / Afrique du Sud : 1996

SUPPLY AND CONSUMPTION / *APPROVISIONNEMENT ET DEMANDE*	Coal / *Charbon* (1000 tonnes)							Oil / *Pétrole* (1000 tonnes)			
	Coking Coal / *Charbon à coke*	Other Bit. Coal / *Autres charb. bit.*	Sub-Bit. Coal / *Charbon sous-bit.*	Lignite / *Lignite*	Peat / *Tourbe*	Oven and Gas Coke / *Coke de four/gaz*	Pat. Fuel and BKB / *Agg./briq. de lignite*	Crude Oil / *Pétrole brut*	NGL / *LGN*	Feed-stocks / *Produits d'aliment.*	Additives / *Additifs*
Production	3533	202829	-	-	-	2839	-	-	393	-	-
From Other Sources	-	-	-	-	-	-	-	-	-	-	-
Imports	425	-	-	-	-	-	-	13972	-	-	-
Exports	-	-60224	-	-	-	-	-	-	-	-	-
Intl. Marine Bunkers	-	-	-	-	-	-	-	-	-	-	-
Stock Changes	-	2837	-	-	-	-	-	-	-	-	-
DOMESTIC SUPPLY	**3958**	**145442**	-	-	-	**2839**	-	**13972**	**393**	-	-
Transfers	-	-	-	-	-	-	-	-	-	-	-
Statistical Differences	-	766	-	-	-	-	-	1634	-	-	-
TRANSFORMATION	**3958**	**124401**	-	-	-	**2271**	-	**13972**	**393**	-	-
Electricity Plants	-	90893	-	-	-	-	-	-	-	-	-
CHP Plants	-	-	-	-	-	-	-	-	-	-	-
Heat Plants	-	-	-	-	-	-	-	-	-	-	-
Blast Furnaces/Gas Works	-	4982	-	-	-	2271	-	-	-	-	-
Coke/Pat. Fuel/BKB Plants	3958	-	-	-	-	-	-	-	-	-	-
Petroleum Refineries	-	-	-	-	-	-	-	21223	393	-	-
Petrochemical Industry	-	-	-	-	-	-	-	-	-	-	-
Liquefaction	-	28526	-	-	-	-	-	-7251	-	-	-
Other Transform. Sector	-	-	-	-	-	-	-	-	-	-	-
ENERGY SECTOR	-	-	-	-	-	-	-	**1634**	-	-	-
Coal Mines	-	-	-	-	-	-	-	-	-	-	-
Oil and Gas Extraction	-	-	-	-	-	-	-	-	-	-	-
Petroleum Refineries	-	-	-	-	-	-	-	1634	-	-	-
Electr., CHP+Heat Plants	-	-	-	-	-	-	-	-	-	-	-
Pumped Storage (Elec.)	-	-	-	-	-	-	-	-	-	-	-
Other Energy Sector	-	-	-	-	-	-	-	-	-	-	-
Distribution Losses	-	-	-	-	-	-	-	-	-	-	-
FINAL CONSUMPTION	-	**21807**	-	-	-	**568**	-	-	-	-	-
INDUSTRY SECTOR	-	**10415**	-	-	-	**568**	-	-	-	-	-
Iron and Steel	-	2709	-	-	-	568	-	-	-	-	-
Chemical and Petrochem.	-	-	-	-	-	-	-	-	-	-	-
of which: Feedstocks	-	-	-	-	-	-	-	-	-	-	-
Non-Ferrous Metals	-	1665	-	-	-	-	-	-	-	-	-
Non-Metallic Minerals	-	1206	-	-	-	-	-	-	-	-	-
Transport Equipment	-	-	-	-	-	-	-	-	-	-	-
Machinery	-	-	-	-	-	-	-	-	-	-	-
Mining and Quarrying	-	555	-	-	-	-	-	-	-	-	-
Food and Tobacco	-	-	-	-	-	-	-	-	-	-	-
Paper, Pulp and Print	-	-	-	-	-	-	-	-	-	-	-
Wood and Wood Products	-	-	-	-	-	-	-	-	-	-	-
Construction	-	-	-	-	-	-	-	-	-	-	-
Textile and Leather	-	-	-	-	-	-	-	-	-	-	-
Non-specified	-	4280	-	-	-	-	-	-	-	-	-
TRANSPORT SECTOR	-	**23**	-	-	-	-	-	-	-	-	-
Air	-	-	-	-	-	-	-	-	-	-	-
Road	-	-	-	-	-	-	-	-	-	-	-
Rail	-	23	-	-	-	-	-	-	-	-	-
Pipeline Transport	-	-	-	-	-	-	-	-	-	-	-
Internal Navigation	-	-	-	-	-	-	-	-	-	-	-
Non-specified	-	-	-	-	-	-	-	-	-	-	-
OTHER SECTORS	-	**3744**	-	-	-	-	-	-	-	-	-
Agriculture	-	242	-	-	-	-	-	-	-	-	-
Comm. and Publ. Services	-	1302	-	-	-	-	-	-	-	-	-
Residential	-	2200	-	-	-	-	-	-	-	-	-
Non-specified	-	-	-	-	-	-	-	-	-	-	-
NON-ENERGY USE	-	**7625**	-	-	-	-	-	-	-	-	-
in Industry/Trans./Energy	-	7625	-	-	-	-	-	-	-	-	-
in Transport	-	-	-	-	-	-	-	-	-	-	-
in Other Sectors	-	-	-	-	-	-	-	-	-	-	-

South Africa / Afrique du Sud : 1996

SUPPLY AND CONSUMPTION / *APPROVISIONNEMENT ET DEMANDE*	Oil cont. / *Pétrole cont.* (1000 tonnes)										
	Refinery Gas / *Gaz de raffinerie*	LPG + Ethane / *GPL + éthane*	Motor Gasoline / *Essence moteur*	Aviation Gasoline / *Essence aviation*	Jet Fuel / *Carbu-réacteurs*	Kerosene / *Kérosène*	Gas/ Diesel / *Gazole*	Heavy Fuel Oil / *Fioul lourd*	Naphtha / *Naphta*	Petrol. Coke / *Coke de pétrole*	Other Prod. / *Autres prod.*
Production	71	256	7995	104	1432	902	6469	3128	-	-	839
From Other Sources	-	-	-	-	-	-	-	-	-	-	-
Imports	-	-	465	13	30	30	217	78	-	-	137
Exports	-	-	-669	-99	-192	-192	-1631	-195	-	-	-192
Intl. Marine Bunkers	-	-	-	-	-	-	-335	-2404	-	-	-20
Stock Changes	-	-	-	-	-	-	-	-	-	-	-
DOMESTIC SUPPLY	**71**	**256**	**7791**	**18**	**1270**	**740**	**4720**	**607**	**-**	**-**	**764**
Transfers	-	-	-	-	-	-	-	-	-	-	-
Statistical Differences	-	-	-	-	-	-1	1	-3	-	-	-16
TRANSFORMATION	**-**	**-**	**-**	**-**	**-**	**-**	**-**	**100**	**-**	**-**	**-**
Electricity Plants	-	-	-	-	-	-	-	100	-	-	-
CHP Plants	-	-	-	-	-	-	-	-	-	-	-
Heat Plants	-	-	-	-	-	-	-	-	-	-	-
Blast Furnaces/Gas Works	-	-	-	-	-	-	-	-	-	-	-
Coke/Pat. Fuel/BKB Plants	-	-	-	-	-	-	-	-	-	-	-
Petroleum Refineries	-	-	-	-	-	-	-	-	-	-	-
Petrochemical Industry	-	-	-	-	-	-	-	-	-	-	-
Liquefaction	-	-	-	-	-	-	-	-	-	-	-
Other Transform. Sector	-	-	-	-	-	-	-	-	-	-	-
ENERGY SECTOR	**-**	**-**	**-**	**-**	**-**	**-**	**-**	**7**	**-**	**-**	**-**
Coal Mines	-	-	-	-	-	-	-	-	-	-	-
Oil and Gas Extraction	-	-	-	-	-	-	-	-	-	-	-
Petroleum Refineries	-	-	-	-	-	-	-	7	-	-	-
Electr., CHP+Heat Plants	-	-	-	-	-	-	-	-	-	-	-
Pumped Storage (Elec.)	-	-	-	-	-	-	-	-	-	-	-
Other Energy Sector	-	-	-	-	-	-	-	-	-	-	-
Distribution Losses	-	-	-	-	-	-	-	-	-	-	-
FINAL CONSUMPTION	**71**	**256**	**7791**	**18**	**1270**	**739**	**4721**	**497**	**-**	**-**	**748**
INDUSTRY SECTOR	**71**	**116**	**7**	**-**	**-**	**128**	**661**	**397**	**-**	**-**	**-**
Iron and Steel	-	-	-	-	-	-	-	167	-	-	-
Chemical and Petrochem.	71	-	-	-	-	-	-	14	-	-	-
of which: Feedstocks	*49*	-	-	-	-	-	-	-	-	-	-
Non-Ferrous Metals	-	-	-	-	-	-	-	-	-	-	-
Non-Metallic Minerals	-	-	-	-	-	-	-	73	-	-	-
Transport Equipment	-	-	-	-	-	-	-	-	-	-	-
Machinery	-	-	-	-	-	-	-	11	-	-	-
Mining and Quarrying	-	2	-	-	-	7	389	15	-	-	-
Food and Tobacco	-	-	-	-	-	-	-	27	-	-	-
Paper, Pulp and Print	-	-	-	-	-	-	-	-	-	-	-
Wood and Wood Products	-	-	-	-	-	-	-	3	-	-	-
Construction	-	1	7	-	-	11	272	1	-	-	-
Textile and Leather	-	-	-	-	-	-	-	7	-	-	-
Non-specified	-	113	-	-	-	110	-	79	-	-	-
TRANSPORT SECTOR	**-**	**-**	**7720**	**18**	**1270**	**13**	**2856**	**-**	**-**	**-**	**-**
Air	-	-	-	18	1270	-	-	-	-	-	-
Road	-	-	7718	-	-	-	2694	-	-	-	-
Rail	-	-	-	-	-	-	160	-	-	-	-
Pipeline Transport	-	-	-	-	-	-	-	-	-	-	-
Internal Navigation	-	-	-	-	-	-	-	-	-	-	-
Non-specified	-	-	2	-	-	13	2	-	-	-	-
OTHER SECTORS	**-**	**140**	**64**	**-**	**-**	**598**	**1204**	**100**	**-**	**-**	**-**
Agriculture	-	15	64	-	-	55	1204	2	-	-	-
Comm. and Publ. Services	-	47	-	-	-	2	-	-	-	-	-
Residential	-	78	-	-	-	541	-	-	-	-	-
Non-specified	-	-	-	-	-	-	-	98	-	-	-
NON-ENERGY USE	**-**	**-**	**-**	**-**	**-**	**-**	**-**	**-**	**-**	**-**	**748**
in Industry/Transf./Energy	-	-	-	-	-	-	-	-	-	-	748
in Transport	-	-	-	-	-	-	-	-	-	-	-
in Other Sectors	-	-	-	-	-	-	-	-	-	-	-

South Africa / Afrique du Sud : 1996

SUPPLY AND CONSUMPTION / *APPROVISIONNEMENT ET DEMANDE*	Gas / *Gaz* (TJ)				Comb. Renew. & Waste / *En. Re. Comb. & Déchets* (TJ)				(GWh)	(TJ)
	Natural Gas / *Gaz naturel*	Gas Works / *Usines à gaz*	Coke Ovens / *Cokeries*	Blast Furnaces / *Hauts fourneaux*	Solid Biomass / *Biomasse solide*	Gas/Liquids from Biomass / *Gaz/Liquides tirés de biomasse*	Municipal Waste / *Déchets urbains*	Industrial Waste / *Déchets industriels*	Electricity / *Electricité*	Heat / *Chaleur*
Production	71814	29965	22937	32208	496296	-	-	-	200566	-
From Other Sources	-	-	-	-	-	-	-	-	-	-
Imports	-	-	-	-	-	-	-	-	29	-
Exports	-	-	-	-	-9153	-	-	-	-5579	-
Intl. Marine Bunkers	-	-	-	-	-	-	-	-	-	-
Stock Changes	-	-	-	-	-	-	-	-	-	-
DOMESTIC SUPPLY	**71814**	**29965**	**22937**	**32208**	**487143**	-	-	-	**195016**	-
Transfers	-	-	-	-	-	-	-	-	-	-
Statistical Differences	-	-1	-4	-7361	-	-	-	-	639	-
TRANSFORMATION	**71814**	-	-	-	**127125**	-	-	-	-	-
Electricity Plants	-	-	-	-	-	-	-	-	-	-
CHP Plants	-	-	-	-	-	-	-	-	-	-
Heat Plants	-	-	-	-	-	-	-	-	-	-
Blast Furnaces/Gas Works	-	-	-	-	-	-	-	-	-	-
Coke/Pat. Fuel/BKB Plants	-	-	-	-	-	-	-	-	-	-
Petroleum Refineries	-	-	-	-	-	-	-	-	-	-
Petrochemical Industry	-	-	-	-	-	-	-	-	-	-
Liquefaction	71814	-	-	-	-	-	-	-	-	-
Other Transform. Sector	-	-	-	-	127125	-	-	-	-	-
ENERGY SECTOR	-	**53**	-	-	-	-	-	-	**31759**	-
Coal Mines	-	-	-	-	-	-	-	-	2732	-
Oil and Gas Extraction	-	-	-	-	-	-	-	-	-	-
Petroleum Refineries	-	53	-	-	-	-	-	-	13423	-
Electr., CHP+Heat Plants	-	-	-	-	-	-	-	-	12563	-
Pumped Storage (Elec.)	-	-	-	-	-	-	-	-	3041	-
Other Energy Sector	-	-	-	-	-	-	-	-	-	-
Distribution Losses	-	-	-	-	-	-	-	-	15294	-
FINAL CONSUMPTION	-	**29911**	**22933**	**24847**	**360018**	-	-	-	**148602**	-
INDUSTRY SECTOR	-	**28588**	**22933**	**24847**	**65088**	-	-	-	**89904**	-
Iron and Steel	-	10748	22933	24847	-	-	-	-	15630	-
Chemical and Petrochem.	-	2866	-	-	-	-	-	-	2524	-
of which: Feedstocks	-	-	-	-	-	-	-	-	-	-
Non-Ferrous Metals	-	972	-	-	-	-	-	-	13046	-
Non-Metallic Minerals	-	5740	-	-	-	-	-	-	1143	-
Transport Equipment	-	201	-	-	-	-	-	-	9	-
Machinery	-	4734	-	-	-	-	-	-	115	-
Mining and Quarrying	-	325	-	-	-	-	-	-	34831	-
Food and Tobacco	-	949	-	-	-	-	-	-	503	-
Paper, Pulp and Print	-	388	-	-	-	-	-	-	969	-
Wood and Wood Products	-	271	-	-	-	-	-	-	590	-
Construction	-	-	-	-	-	-	-	-	16	-
Textile and Leather	-	101	-	-	-	-	-	-	491	-
Non-specified	-	1293	-	-	65088	-	-	-	20037	-
TRANSPORT SECTOR	-	**14**	-	-	-	-	-	-	**4275**	-
Air	-	14	-	-	-	-	-	-	13	-
Road	-	-	-	-	-	-	-	-	8	-
Rail	-	-	-	-	-	-	-	-	3446	-
Pipeline Transport	-	-	-	-	-	-	-	-	59	-
Internal Navigation	-	-	-	-	-	-	-	-	-	-
Non-specified	-	-	-	-	-	-	-	-	749	-
OTHER SECTORS	-	**1309**	-	-	**294930**	-	-	-	**54423**	-
Agriculture	-	-	-	-	-	-	-	-	5103	-
Comm. and Publ. Services	-	839	-	-	-	-	-	-	19768	-
Residential	-	470	-	-	294930	-	-	-	29552	-
Non-specified	-	-	-	-	-	-	-	-	-	-
NON-ENERGY USE	-	-	-	-	-	-	-	-	-	-
in Industry/Transf./Energy	-	-	-	-	-	-	-	-	-	-
in Transport	-	-	-	-	-	-	-	-	-	-
in Other Sectors	-	-	-	-	-	-	-	-	-	-

South Africa / Afrique du Sud : 1997

SUPPLY AND CONSUMPTION *APPROVISIONNEMENT ET DEMANDE*	Coal / *Charbon* (1000 tonnes)							Oil / *Pétrole* (1000 tonnes)			
	Coking Coal *Charbon à coke*	Other Bit. Coal *Autres charb. bit.*	Sub-Bit. Coal *Charbon sous-bit.*	Lignite *Lignite*	Peat *Tourbe*	Oven and Gas Coke *Coke de four/gaz*	Pat. Fuel and BKB *Agg./briq. de lignite*	Crude Oil *Pétrole brut*	NGL *LGN*	Feed-stocks *Produits d'aliment.*	Additives *Additifs*
Production	3647	216426	-	-	-	2839	-	-	393	-	-
From Other Sources	-	-	-	-	-	-	-	-	-	-	-
Imports	425	-	-	-	-	-	-	12616	-	-	-
Exports	-	-64200	-	-	-	-	-	-	-	-	-
Intl. Marine Bunkers	-	-	-	-	-	-	-	-	-	-	-
Stock Changes	-	-2445	-	-	-	-	-	-	-	-	-
DOMESTIC SUPPLY	**4072**	**149781**	-	-	-	**2839**	-	**12616**	**393**	-	-
Transfers	-	-	-	-	-	-	-	-	-	-	-
Statistical Differences	-	2327	-	-	-	-	-	1424	-	-	-
TRANSFORMATION	**4072**	**128780**	-	-	-	**2271**	-	**12616**	**393**	-	-
Electricity Plants	-	95651	-	-	-	-	-	-	-	-	-
CHP Plants	-	-	-	-	-	-	-	-	-	-	-
Heat Plants	-	-	-	-	-	-	-	-	-	-	-
Blast Furnaces/Gas Works	-	5363	-	-	-	2271	-	-	-	-	-
Coke/Pat. Fuel/BKB Plants	4072	-	-	-	-	-	-	-	-	-	-
Petroleum Refineries	-	-	-	-	-	-	-	19867	393	-	-
Petrochemical Industry	-	-	-	-	-	-	-	-	-	-	-
Liquefaction	-	27766	-	-	-	-	-	-7251	-	-	-
Other Transform. Sector	-	-	-	-	-	-	-	-	-	-	-
ENERGY SECTOR	-	-	-	-	-	-	-	**1424**	-	-	-
Coal Mines	-	-	-	-	-	-	-	-	-	-	-
Oil and Gas Extraction	-	-	-	-	-	-	-	-	-	-	-
Petroleum Refineries	-	-	-	-	-	-	-	1424	-	-	-
Electr., CHP+Heat Plants	-	-	-	-	-	-	-	-	-	-	-
Pumped Storage (Elec.)	-	-	-	-	-	-	-	-	-	-	-
Other Energy Sector	-	-	-	-	-	-	-	-	-	-	-
Distribution Losses	-	-	-	-	-	-	-	-	-	-	-
FINAL CONSUMPTION	-	**23328**	-	-	-	**568**	-	-	-	-	-
INDUSTRY SECTOR	-	**11862**	-	-	-	**568**	-	-	-	-	-
Iron and Steel	-	1758	-	-	-	568	-	-	-	-	-
Chemical and Petrochem.	-	-	-	-	-	-	-	-	-	-	-
of which: Feedstocks	-	-	-	-	-	-	-	-	-	-	-
Non-Ferrous Metals	-	1356	-	-	-	-	-	-	-	-	-
Non-Metallic Minerals	-	1072	-	-	-	-	-	-	-	-	-
Transport Equipment	-	-	-	-	-	-	-	-	-	-	-
Machinery	-	-	-	-	-	-	-	-	-	-	-
Mining and Quarrying	-	1248	-	-	-	-	-	-	-	-	-
Food and Tobacco	-	-	-	-	-	-	-	-	-	-	-
Paper, Pulp and Print	-	-	-	-	-	-	-	-	-	-	-
Wood and Wood Products	-	-	-	-	-	-	-	-	-	-	-
Construction	-	-	-	-	-	-	-	-	-	-	-
Textile and Leather	-	-	-	-	-	-	-	-	-	-	-
Non-specified	-	6428	-	-	-	-	-	-	-	-	-
TRANSPORT SECTOR	-	**2**	-	-	-	-	-	-	-	-	-
Air	-	-	-	-	-	-	-	-	-	-	-
Road	-	-	-	-	-	-	-	-	-	-	-
Rail	-	2	-	-	-	-	-	-	-	-	-
Pipeline Transport	-	-	-	-	-	-	-	-	-	-	-
Internal Navigation	-	-	-	-	-	-	-	-	-	-	-
Non-specified	-	-	-	-	-	-	-	-	-	-	-
OTHER SECTORS	-	**3847**	-	-	-	-	-	-	-	-	-
Agriculture	-	241	-	-	-	-	-	-	-	-	-
Comm. and Publ. Services	-	1356	-	-	-	-	-	-	-	-	-
Residential	-	2250	-	-	-	-	-	-	-	-	-
Non-specified	-	-	-	-	-	-	-	-	-	-	-
NON-ENERGY USE	-	**7617**	-	-	-	-	-	-	-	-	-
in Industry/Trans./Energy	-	7617	-	-	-	-	-	-	-	-	-
in Transport	-	-	-	-	-	-	-	-	-	-	-
in Other Sectors	-	-	-	-	-	-	-	-	-	-	-

South Africa / Afrique du Sud : 1997

SUPPLY AND CONSUMPTION / *APPROVISIONNEMENT ET DEMANDE*	Oil cont. / *Pétrole cont.* (1000 tonnes)										
	Refinery Gas / *Gaz de raffinerie*	LPG + Ethane / *GPL + éthane*	Motor Gasoline / *Essence moteur*	Aviation Gasoline / *Essence aviation*	Jet Fuel / *Carbu- réacteurs*	Kerosene / *Kérosène*	Gas/ Diesel / *Gazole*	Heavy Fuel Oil / *Fioul lourd*	Naphtha / *Naphta*	Petrol. Coke / *Coke de pétrole*	Other Prod. / *Autres prod.*
Production	64	286	7898	20	1409	780	5147	2337	-	-	716
From Other Sources	-	-	-	-	-	-	-	-	-	-	-
Imports	-	-	-	-	-	-	-	-	-	-	-
Exports	-	-	-	-	-	-	-	-	-	-	-
Intl. Marine Bunkers	-	-	-	-	-	-	-343	-1802	-	-	-
Stock Changes	-	-	-	-	-	-	-	-	-	-	-
DOMESTIC SUPPLY	**64**	**286**	**7898**	**20**	**1409**	**780**	**4804**	**535**	**-**	**-**	**716**
Transfers	-	-	-	-	-	-	-	-	-	-	-
Statistical Differences	-	-1	-	-	-	-1	-	-63	-	-	-
TRANSFORMATION	**-**	**-**	**-**	**-**	**-**	**-**	**-**	**88**	**-**	**-**	**-**
Electricity Plants	-	-	-	-	-	-	-	88	-	-	-
CHP Plants	-	-	-	-	-	-	-	-	-	-	-
Heat Plants	-	-	-	-	-	-	-	-	-	-	-
Blast Furnaces/Gas Works	-	-	-	-	-	-	-	-	-	-	-
Coke/Pat. Fuel/BKB Plants	-	-	-	-	-	-	-	-	-	-	-
Petroleum Refineries	-	-	-	-	-	-	-	-	-	-	-
Petrochemical Industry	-	-	-	-	-	-	-	-	-	-	-
Liquefaction	-	-	-	-	-	-	-	-	-	-	-
Other Transform. Sector	-	-	-	-	-	-	-	-	-	-	-
ENERGY SECTOR	**-**	**-**	**-**	**-**	**-**	**-**	**-**	**-**	**-**	**-**	**-**
Coal Mines	-	-	-	-	-	-	-	-	-	-	-
Oil and Gas Extraction	-	-	-	-	-	-	-	-	-	-	-
Petroleum Refineries	-	-	-	-	-	-	-	-	-	-	-
Electr., CHP+Heat Plants	-	-	-	-	-	-	-	-	-	-	-
Pumped Storage (Elec.)	-	-	-	-	-	-	-	-	-	-	-
Other Energy Sector	-	-	-	-	-	-	-	-	-	-	-
Distribution Losses	-	-	-	-	-	-	-	-	-	-	-
FINAL CONSUMPTION	**64**	**285**	**7898**	**20**	**1409**	**779**	**4804**	**384**	**-**	**-**	**716**
INDUSTRY SECTOR	**64**	**128**	**7**	**-**	**-**	**163**	**696**	**350**	**-**	**-**	**-**
Iron and Steel	-	-	-	-	-	-	-	147	-	-	-
Chemical and Petrochem.	64	-	-	-	-	-	-	12	-	-	-
of which: Feedstocks	*44*	-	-	-	-	-	-	-	-	-	-
Non-Ferrous Metals	-	-	-	-	-	-	-	-	-	-	-
Non-Metallic Minerals	-	-	-	-	-	-	-	64	-	-	-
Transport Equipment	-	-	-	-	-	-	-	-	-	-	-
Machinery	-	-	-	-	-	-	-	10	-	-	-
Mining and Quarrying	-	2	-	-	-	10	411	13	-	-	-
Food and Tobacco	-	-	-	-	-	-	-	24	-	-	-
Paper, Pulp and Print	-	-	-	-	-	-	-	-	-	-	-
Wood and Wood Products	-	-	-	-	-	-	-	3	-	-	-
Construction	-	-	5	-	-	7	283	1	-	-	-
Textile and Leather	-	-	-	-	-	-	-	6	-	-	-
Non-specified	-	126	2	-	-	146	2	70	-	-	-
TRANSPORT SECTOR	**-**	**-**	**7830**	**20**	**1409**	**11**	**3015**	**3**	**-**	**-**	**-**
Air	-	-	-	20	1409	-	-	-	-	-	-
Road	-	-	7447	-	-	-	2855	-	-	-	-
Rail	-	-	-	-	-	-	160	-	-	-	-
Pipeline Transport	-	-	-	-	-	-	-	-	-	-	-
Internal Navigation	-	-	-	-	-	-	-	-	-	-	-
Non-specified	-	-	383	-	-	11	-	3	-	-	-
OTHER SECTORS	**-**	**157**	**61**	**-**	**-**	**605**	**1093**	**31**	**-**	**-**	**-**
Agriculture	-	6	61	-	-	60	1093	9	-	-	-
Comm. and Publ. Services	-	57	-	-	-	2	-	22	-	-	-
Residential	-	94	-	-	-	542	-	-	-	-	-
Non-specified	-	-	-	-	-	1	-	-	-	-	-
NON-ENERGY USE	**-**	**-**	**-**	**-**	**-**	**-**	**-**	**-**	**-**	**-**	**716**
in Industry/Transf./Energy	-	-	-	-	-	-	-	-	-	-	716
in Transport	-	-	-	-	-	-	-	-	-	-	-
in Other Sectors	-	-	-	-	-	-	-	-	-	-	-

South Africa / Afrique du Sud : 1997

SUPPLY AND CONSUMPTION / *APPROVISIONNEMENT ET DEMANDE*	Gas / *Gaz* (TJ)				Comb. Renew. & Waste / *En. Re. Comb. & Déchets* (TJ)				(GWh)	(TJ)
	Natural Gas / *Gaz naturel*	Gas Works / *Usines à gaz*	Coke Ovens / *Cokeries*	Blast Furnaces / *Hauts fourneaux*	Solid Biomass / *Biomasse solide*	Gas/Liquids from Biomass / *Gaz/Liquides tirés de biomasse*	Municipal Waste / *Déchets urbains*	Industrial Waste / *Déchets industriels*	Electricity / *Electricité*	Heat / *Chaleur*
Production	71814	29902	22576	28329	504237	-	-	-	210352	-
From Other Sources	-	-	-	-	-	-	-	-	-	-
Imports	-	-	-	-	-	-	-	-	5	-
Exports	-	-	-	-	-9299	-	-	-	-6617	-
Intl. Marine Bunkers	-	-	-	-	-	-	-	-	-	-
Stock Changes	-	-	-	-	-	-	-	-	-	-
DOMESTIC SUPPLY	**71814**	**29902**	**22576**	**28329**	**494938**	**-**	**-**	**-**	**203740**	**-**
Transfers	-	-	-	-	-	-	-	-	-	-
Statistical Differences	-	-	-4	-7361	-1	-	-	-	-1072	-
TRANSFORMATION	**71814**	**-**	**-**	**-**	**129159**	**-**	**-**	**-**	**-**	**-**
Electricity Plants	-	-	-	-	-	-	-	-	-	-
CHP Plants	-	-	-	-	-	-	-	-	-	-
Heat Plants	-	-	-	-	-	-	-	-	-	-
Blast Furnaces/Gas Works	-	-	-	-	-	-	-	-	-	-
Coke/Pat. Fuel/BKB Plants	-	-	-	-	-	-	-	-	-	-
Petroleum Refineries	-	-	-	-	-	-	-	-	-	-
Petrochemical Industry	-	-	-	-	-	-	-	-	-	-
Liquefaction	71814	-	-	-	-	-	-	-	-	-
Other Transform. Sector	-	-	-	-	129159	-	-	-	-	-
ENERGY SECTOR	**-**	**7**	**-**	**-**	**-**	**-**	**-**	**-**	**32020**	**-**
Coal Mines	-	-	-	-	-	-	-	-	2848	-
Oil and Gas Extraction	-	-	-	-	-	-	-	-	-	-
Petroleum Refineries	-	7	-	-	-	-	-	-	12908	-
Electr., CHP+Heat Plants	-	-	-	-	-	-	-	-	12691	-
Pumped Storage (Elec.)	-	-	-	-	-	-	-	-	3573	-
Other Energy Sector	-	-	-	-	-	-	-	-	-	-
Distribution Losses	-	-	-	-	-	-	-	-	16094	-
FINAL CONSUMPTION	**-**	**29895**	**22572**	**20968**	**365778**	**-**	**-**	**-**	**154554**	**-**
INDUSTRY SECTOR	**-**	**28469**	**22572**	**20968**	**66129**	**-**	**-**	**-**	**91460**	**-**
Iron and Steel	-	8372	22572	20968	-	-	-	-	17875	-
Chemical and Petrochem.	-	2948	-	-	-	-	-	-	2432	-
of which: Feedstocks	-	-	-	-	-	-	-	-	-	-
Non-Ferrous Metals	-	1230	-	-	-	-	-	-	14584	-
Non-Metallic Minerals	-	5918	-	-	-	-	-	-	1188	-
Transport Equipment	-	174	-	-	-	-	-	-	11	-
Machinery	-	5414	-	-	-	-	-	-	127	-
Mining and Quarrying	-	549	-	-	-	-	-	-	30390	-
Food and Tobacco	-	1181	-	-	-	-	-	-	539	-
Paper, Pulp and Print	-	393	-	-	-	-	-	-	1029	-
Wood and Wood Products	-	879	-	-	-	-	-	-	596	-
Construction	-	-	-	-	-	-	-	-	17	-
Textile and Leather	-	20	-	-	-	-	-	-	514	-
Non-specified	-	1391	-	-	66129	-	-	-	22158	-
TRANSPORT SECTOR	**-**	**12**	**-**	**-**	**-**	**-**	**-**	**-**	**4562**	**-**
Air	-	12	-	-	-	-	-	-	14	-
Road	-	-	-	-	-	-	-	-	8	-
Rail	-	-	-	-	-	-	-	-	3385	-
Pipeline Transport	-	-	-	-	-	-	-	-	66	-
Internal Navigation	-	-	-	-	-	-	-	-	-	-
Non-specified	-	-	-	-	-	-	-	-	1089	-
OTHER SECTORS	**-**	**1414**	**-**	**-**	**299649**	**-**	**-**	**-**	**58532**	**-**
Agriculture	-	-	-	-	-	-	-	-	5640	-
Comm. and Publ. Services	-	902	-	-	-	-	-	-	22170	-
Residential	-	512	-	-	299649	-	-	-	30722	-
Non-specified	-	-	-	-	-	-	-	-	-	-
NON-ENERGY USE	**-**	**-**	**-**	**-**	**-**	**-**	**-**	**-**	**-**	**-**
in Industry/Transf./Energy	-	-	-	-	-	-	-	-	-	-
in Transport	-	-	-	-	-	-	-	-	-	-
in Other Sectors	-	-	-	-	-	-	-	-	-	-

Sri Lanka

SUPPLY AND CONSUMPTION 1996	Coal (1000 tonnes)							Oil (1000 tonnes)			
	Coking Coal	Other Bit. Coal	Sub-Bit. Coal	Lignite	Peat	Oven and Gas Coke	Pat. Fuel and BKB	Crude Oil	NGL	Feed-stocks	Additives
Production	-	-	-	-	-	-	-	-	-	-	-
Imports	-	1	-	-	-	-	-	2040	-	-	-
Exports	-	-	-	-	-	-	-	-	-	-	-
Intl. Marine Bunkers	-	-	-	-	-	-	-	-	-	-	-
Stock Changes	-	-	-	-	-	-	-	4	-	-	-
DOMESTIC SUPPLY	**-**	**1**	**-**	**-**	**-**	**-**	**-**	**2044**	**-**	**-**	**-**
Transfers and Stat. Diff.	-	-	-	-	-	-	-	-	-	-	-
TRANSFORMATION	**-**	**-**	**-**	**-**	**-**	**-**	**-**	**2044**	**-**	**-**	**-**
Electricity and CHP Plants	-	-	-	-	-	-	-	-	-	-	-
Petroleum Refineries	-	-	-	-	-	-	-	2044	-	-	-
Other Transform. Sector	-	-	-	-	-	-	-	-	-	-	-
ENERGY SECTOR	**-**	**-**	**-**	**-**	**-**	**-**	**-**	**-**	**-**	**-**	**-**
DISTRIBUTION LOSSES	**-**	**-**	**-**	**-**	**-**	**-**	**-**	**-**	**-**	**-**	**-**
FINAL CONSUMPTION	**-**	**1**	**-**	**-**	**-**	**-**	**-**	**-**	**-**	**-**	**-**
INDUSTRY SECTOR	**-**	**-**	**-**	**-**	**-**	**-**	**-**	**-**	**-**	**-**	**-**
Iron and Steel	-	-	-	-	-	-	-	-	-	-	-
Chemical and Petrochem.	-	-	-	-	-	-	-	-	-	-	-
Non-Metallic Minerals	-	-	-	-	-	-	-	-	-	-	-
Non-specified	-	-	-	-	-	-	-	-	-	-	-
TRANSPORT SECTOR	**-**	**1**	**-**	**-**	**-**	**-**	**-**	**-**	**-**	**-**	**-**
Air	-	-	-	-	-	-	-	-	-	-	-
Road	-	-	-	-	-	-	-	-	-	-	-
Non-specified	-	1	-	-	-	-	-	-	-	-	-
OTHER SECTORS	**-**	**-**	**-**	**-**	**-**	**-**	**-**	**-**	**-**	**-**	**-**
Agriculture	-	-	-	-	-	-	-	-	-	-	-
Comm. and Publ. Services	-	-	-	-	-	-	-	-	-	-	-
Residential	-	-	-	-	-	-	-	-	-	-	-
Non-specified	-	-	-	-	-	-	-	-	-	-	-
NON-ENERGY USE	**-**	**-**	**-**	**-**	**-**	**-**	**-**	**-**	**-**	**-**	**-**

APPROVISIONNEMENT ET DEMANDE 1997	*Charbon* (1000 tonnes)							*Pétrole* (1000 tonnes)			
	Charbon à coke	*Autres charb. bit.*	*Charbon sous-bit.*	*Lignite*	*Tourbe*	*Coke de four/gaz*	*Agg./briq. de lignite*	*Pétrole brut*	*LGN*	*Produits d'aliment.*	*Additifs*
Production	-	-	-	-	-	-	-	-	-	-	-
Imports	-	1	-	-	-	-	-	2082	-	-	-
Exports	-	-	-	-	-	-	-	-	-	-	-
Intl. Marine Bunkers	-	-	-	-	-	-	-	-	-	-	-
Stock Changes	-	-	-	-	-	-	-	-	-	-	-
DOMESTIC SUPPLY	**-**	**1**	**-**	**-**	**-**	**-**	**-**	**2082**	**-**	**-**	**-**
Transfers and Stat. Diff.	-	-	-	-	-	-	-	-	-	-	-
TRANSFORMATION	**-**	**-**	**-**	**-**	**-**	**-**	**-**	**2082**	**-**	**-**	**-**
Electricity and CHP Plants	-	-	-	-	-	-	-	-	-	-	-
Petroleum Refineries	-	-	-	-	-	-	-	2082	-	-	-
Other Transform. Sector	-	-	-	-	-	-	-	-	-	-	-
ENERGY SECTOR	**-**	**-**	**-**	**-**	**-**	**-**	**-**	**-**	**-**	**-**	**-**
DISTRIBUTION LOSSES	**-**	**-**	**-**	**-**	**-**	**-**	**-**	**-**	**-**	**-**	**-**
FINAL CONSUMPTION	**-**	**1**	**-**	**-**	**-**	**-**	**-**	**-**	**-**	**-**	**-**
INDUSTRY SECTOR	**-**	**-**	**-**	**-**	**-**	**-**	**-**	**-**	**-**	**-**	**-**
Iron and Steel	-	-	-	-	-	-	-	-	-	-	-
Chemical and Petrochem.	-	-	-	-	-	-	-	-	-	-	-
Non-Metallic Minerals	-	-	-	-	-	-	-	-	-	-	-
Non-specified	-	-	-	-	-	-	-	-	-	-	-
TRANSPORT SECTOR	**-**	**1**	**-**	**-**	**-**	**-**	**-**	**-**	**-**	**-**	**-**
Air	-	-	-	-	-	-	-	-	-	-	-
Road	-	-	-	-	-	-	-	-	-	-	-
Non-specified	-	1	-	-	-	-	-	-	-	-	-
OTHER SECTORS	**-**	**-**	**-**	**-**	**-**	**-**	**-**	**-**	**-**	**-**	**-**
Agriculture	-	-	-	-	-	-	-	-	-	-	-
Comm. and Publ. Services	-	-	-	-	-	-	-	-	-	-	-
Residential	-	-	-	-	-	-	-	-	-	-	-
Non-specified	-	-	-	-	-	-	-	-	-	-	-
NON-ENERGY USE	**-**	**-**	**-**	**-**	**-**	**-**	**-**	**-**	**-**	**-**	**-**

Sri Lanka

SUPPLY AND CONSUMPTION 1996	Oil cont. (1000 tonnes)										
	Refinery Gas	LPG + Ethane	Motor Gasoline	Aviation Gasoline	Jet Fuel	Kerosene	Gas/ Diesel	Heavy Fuel Oil	Naphtha	Petrol. Coke	Other Prod.
Production	37	61	189	-	69	201	586	782	-	-	40
Imports	-	68	-	-	167	-	594	-	-	-	12
Exports	-	-	-	-	-	-	-	-79	-	-	-
Intl. Marine Bunkers	-	-	-	-	-	-	-49	-336	-	-	-
Stock Changes	-	-	10	-	-40	28	68	28	-	-	3
DOMESTIC SUPPLY	**37**	**129**	**199**	**-**	**196**	**229**	**1199**	**395**	**-**	**-**	**55**
Transfers and Stat. Diff.	-	-	-	-	-	-	-	1	-	-	-
TRANSFORMATION	**-**	**-**	**-**	**-**	**-**	**-**	**177**	**119**	**-**	**-**	**-**
Electricity and CHP Plants	-	-	-	-	-	-	177	119	-	-	-
Petroleum Refineries	-	-	-	-	-	-	-	-	-	-	-
Other Transform. Sector	-	-	-	-	-	-	-	-	-	-	-
ENERGY SECTOR	**37**	**-**	**-**	**-**	**-**	**-**	**-**	**33**	**-**	**-**	**-**
DISTRIBUTION LOSSES	**-**	**42**	**1**	**-**	**1**	**-**	**1**	**1**	**-**	**-**	**-**
FINAL CONSUMPTION	**-**	**87**	**198**	**-**	**195**	**229**	**1021**	**243**	**-**	**-**	**55**
INDUSTRY SECTOR	**-**	**10**	**-**	**-**	**-**	**4**	**198**	**223**	**-**	**-**	**-**
Iron and Steel	-	-	-	-	-	-	-	-	-	-	-
Chemical and Petrochem.	-	-	-	-	-	-	-	-	-	-	-
Non-Metallic Minerals	-	-	-	-	-	-	-	-	-	-	-
Non-specified	-	10	-	-	-	4	198	223	-	-	-
TRANSPORT SECTOR	**-**	**-**	**198**	**-**	**195**	**-**	**823**	**20**	**-**	**-**	**-**
Air	-	-	-	-	195	-	-	-	-	-	-
Road	-	-	198	-	-	-	797	-	-	-	-
Non-specified	-	-	-	-	-	-	26	20	-	-	-
OTHER SECTORS	**-**	**77**	**-**	**-**	**-**	**225**	**-**	**-**	**-**	**-**	**-**
Agriculture	-	-	-	-	-	13	-	-	-	-	-
Comm. and Publ. Services	-	13	-	-	-	-	-	-	-	-	-
Residential	-	64	-	-	-	212	-	-	-	-	-
Non-specified	-	-	-	-	-	-	-	-	-	-	-
NON-ENERGY USE	**-**	**-**	**-**	**-**	**-**	**-**	**-**	**-**	**-**	**-**	**55**

APPROVISIONNEMENT ET DEMANDE 1997	*Pétrole cont.* (1000 tonnes)										
	Gaz de raffinerie	*GPL + éthane*	*Essence moteur*	*Essence aviation*	*Carbu- réacteurs*	*Kérosène*	*Gazole*	*Fioul lourd*	*Naphta*	*Coke de pétrole*	*Autres prod.*
Production	37	61	161	-	73	147	532	610	117	-	40
Imports	-	51	23	-	207	10	848	-	-	-	-
Exports	-	-	-	-	-	-	-	-80	-117	-	-
Intl. Marine Bunkers	-	-	-	-	-	-	-37	-221	-	-	-
Stock Changes	-	-	-	-	-5	3	-19	-	-	-	-
DOMESTIC SUPPLY	**37**	**112**	**184**	**-**	**275**	**160**	**1324**	**309**	**-**	**-**	**40**
Transfers and Stat. Diff.	-	-	10	-	-	-	-	-	-	-	-
TRANSFORMATION	**-**	**-**	**-**	**-**	**-**	**-**	**361**	**66**	**-**	**-**	**-**
Electricity and CHP Plants	-	-	-	-	-	-	361	66	-	-	-
Petroleum Refineries	-	-	-	-	-	-	-	-	-	-	-
Other Transform. Sector	-	-	-	-	-	-	-	-	-	-	-
ENERGY SECTOR	**37**	**-**	**-**	**-**	**-**	**-**	**-**	**19**	**-**	**-**	**-**
DISTRIBUTION LOSSES	**-**	**-**	**-**	**-**	**-**	**-**	**-**	**-**	**-**	**-**	**-**
FINAL CONSUMPTION	**-**	**112**	**194**	**-**	**275**	**160**	**963**	**224**	**-**	**-**	**40**
INDUSTRY SECTOR	**-**	**13**	**-**	**-**	**-**	**4**	**179**	**207**	**-**	**-**	**-**
Iron and Steel	-	-	-	-	-	-	-	-	-	-	-
Chemical and Petrochem.	-	-	-	-	-	-	-	-	-	-	-
Non-Metallic Minerals	-	-	-	-	-	-	-	-	-	-	-
Non-specified	-	13	-	-	-	4	179	207	-	-	-
TRANSPORT SECTOR	**-**	**-**	**194**	**-**	**275**	**-**	**784**	**17**	**-**	**-**	**-**
Air	-	-	-	-	275	-	-	-	-	-	-
Road	-	-	194	-	-	-	760	-	-	-	-
Non-specified	-	-	-	-	-	-	24	17	-	-	-
OTHER SECTORS	**-**	**99**	**-**	**-**	**-**	**156**	**-**	**-**	**-**	**-**	**-**
Agriculture	-	-	-	-	-	13	-	-	-	-	-
Comm. and Publ. Services	-	17	-	-	-	-	-	-	-	-	-
Residential	-	82	-	-	-	143	-	-	-	-	-
Non-specified	-	-	-	-	-	-	-	-	-	-	-
NON-ENERGY USE	**-**	**-**	**-**	**-**	**-**	**-**	**-**	**-**	**-**	**-**	**40**

Sri Lanka

SUPPLY AND CONSUMPTION 1996	Gas (TJ)				Comb. Renew. & Waste (TJ)				(GWh)	(TJ)
	Natural Gas	Gas Works	Coke Ovens	Blast Furnaces	Solid Biomass	Gas/Liquids from Biomass	Municipal Waste	Industrial Waste	Electricity	Heat
Production	-	-	-	-	164334	-	-	-	4530	-
Imports	-	-	-	-	-	-	-	-	-	-
Exports	-	-	-	-	-	-	-	-	-	-
Intl. Marine Bunkers	-	-	-	-	-	-	-	-	-	-
Stock Changes	-	-	-	-	-1	-	-	-	-	-
DOMESTIC SUPPLY	**-**	**-**	**-**	**-**	**164333**	**-**	**-**	**-**	**4530**	**-**
Transfers and Stat. Diff.	-	-	-	-	-	-	-	-	-1	-
TRANSFORMATION	**-**	**-**	**-**	**-**	**1**	**-**	**-**	**-**	**-**	**-**
Electricity and CHP Plants	-	-	-	-	-	-	-	-	-	-
Petroleum Refineries	-	-	-	-	-	-	-	-	-	-
Other Transform. Sector	-	-	-	-	1	-	-	-	-	-
ENERGY SECTOR	**-**	**-**	**-**	**-**	**-**	**-**	**-**	**-**	**39**	**-**
DISTRIBUTION LOSSES	**-**	**-**	**-**	**-**	**-**	**-**	**-**	**-**	**774**	**-**
FINAL CONSUMPTION	**-**	**-**	**-**	**-**	**164332**	**-**	**-**	**-**	**3716**	**-**
INDUSTRY SECTOR	**-**	**-**	**-**	**-**	**26827**	**-**	**-**	**-**	**1650**	**-**
Iron and Steel	-	-	-	-	-	-	-	-	-	-
Chemical and Petrochem.	-	-	-	-	-	-	-	-	-	-
Non-Metallic Minerals	-	-	-	-	-	-	-	-	-	-
Non-specified	-	-	-	-	26827	-	-	-	1650	-
TRANSPORT SECTOR	**-**	**-**	**-**	**-**	**-**	**-**	**-**	**-**	**-**	**-**
Air	-	-	-	-	-	-	-	-	-	-
Road	-	-	-	-	-	-	-	-	-	-
Non-specified	-	-	-	-	-	-	-	-	-	-
OTHER SECTORS	**-**	**-**	**-**	**-**	**137505**	**-**	**-**	**-**	**2066**	**-**
Agriculture	-	-	-	-	-	-	-	-	-	-
Comm. and Publ. Services	-	-	-	-	5400	-	-	-	781	-
Residential	-	-	-	-	132105	-	-	-	1285	-
Non-specified	-	-	-	-	-	-	-	-	-	-
NON-ENERGY USE	**-**	**-**	**-**	**-**	**-**	**-**	**-**	**-**	**-**	**-**

APPROVISIONNEMENT ET DEMANDE 1997	*Gaz* (TJ)				*En. Re. Comb. & Déchets* (TJ)				(GWh)	(TJ)
	Gaz naturel	*Usines à gaz*	*Cokeries*	*Hauts fourneaux*	*Biomasse solide*	*Gaz/Liquides tirés de biomasse*	*Déchets urbains*	*Déchets industriels*	*Electricité*	*Chaleur*
Production	-	-	-	-	169554	-	-	-	5145	-
Imports	-	-	-	-	-	-	-	-	-	-
Exports	-	-	-	-	-	-	-	-	-	-
Intl. Marine Bunkers	-	-	-	-	-	-	-	-	-	-
Stock Changes	-	-	-	-	1	-	-	-	-	-
DOMESTIC SUPPLY	**-**	**-**	**-**	**-**	**169555**	**-**	**-**	**-**	**5145**	**-**
Transfers and Stat. Diff.	-	-	-	-	-	-	-	-	1	-
TRANSFORMATION	**-**	**-**	**-**	**-**	**8**	**-**	**-**	**-**	**-**	**-**
Electricity and CHP Plants	-	-	-	-	-	-	-	-	-	-
Petroleum Refineries	-	-	-	-	-	-	-	-	-	-
Other Transform. Sector	-	-	-	-	8	-	-	-	-	-
ENERGY SECTOR	**-**	**-**	**-**	**-**	**-**	**-**	**-**	**-**	**39**	**-**
DISTRIBUTION LOSSES	**-**	**-**	**-**	**-**	**-**	**-**	**-**	**-**	**900**	**-**
FINAL CONSUMPTION	**-**	**-**	**-**	**-**	**169547**	**-**	**-**	**-**	**4207**	**-**
INDUSTRY SECTOR	**-**	**-**	**-**	**-**	**30677**	**-**	**-**	**-**	**1819**	**-**
Iron and Steel	-	-	-	-	-	-	-	-	-	-
Chemical and Petrochem.	-	-	-	-	-	-	-	-	-	-
Non-Metallic Minerals	-	-	-	-	-	-	-	-	-	-
Non-specified	-	-	-	-	30677	-	-	-	1819	-
TRANSPORT SECTOR	**-**	**-**	**-**	**-**	**-**	**-**	**-**	**-**	**-**	**-**
Air	-	-	-	-	-	-	-	-	-	-
Road	-	-	-	-	-	-	-	-	-	-
Non-specified	-	-	-	-	-	-	-	-	-	-
OTHER SECTORS	**-**	**-**	**-**	**-**	**138870**	**-**	**-**	**-**	**2388**	**-**
Agriculture	-	-	-	-	-	-	-	-	-	-
Comm. and Publ. Services	-	-	-	-	5454	-	-	-	771	-
Residential	-	-	-	-	133416	-	-	-	1617	-
Non-specified	-	-	-	-	-	-	-	-	-	-
NON-ENERGY USE	**-**	**-**	**-**	**-**	**-**	**-**	**-**	**-**	**-**	**-**

Sudan / Soudan

SUPPLY AND CONSUMPTION 1996	Coal (1000 tonnes)							Oil (1000 tonnes)			
	Coking Coal	Other Bit. Coal	Sub-Bit. Coal	Lignite	Peat	Oven and Gas Coke	Pat. Fuel and BKB	Crude Oil	NGL	Feed-stocks	Additives
Production	-	-	-	-	-	-	-	102	-	-	-
Imports	-	-	-	-	-	-	-	712	-	-	-
Exports	-	-	-	-	-	-	-	-	-	-	-
Intl. Marine Bunkers	-	-	-	-	-	-	-	-	-	-	-
Stock Changes	-	-	-	-	-	-	-	-88	-	-	-
DOMESTIC SUPPLY	-	-	-	-	-	-	-	**726**	-	-	-
Transfers and Stat. Diff.	-	-	-	-	-	-	-	-	-	-	-
TRANSFORMATION	-	-	-	-	-	-	-	**726**	-	-	-
Electricity and CHP Plants	-	-	-	-	-	-	-	-	-	-	-
Petroleum Refineries	-	-	-	-	-	-	-	726	-	-	-
Other Transform. Sector	-	-	-	-	-	-	-	-	-	-	-
ENERGY SECTOR	-	-	-	-	-	-	-	-	-	-	-
DISTRIBUTION LOSSES	-	-	-	-	-	-	-	-	-	-	-
FINAL CONSUMPTION	-	-	-	-	-	-	-	-	-	-	-
INDUSTRY SECTOR	-	-	-	-	-	-	-	-	-	-	-
Iron and Steel	-	-	-	-	-	-	-	-	-	-	-
Chemical and Petrochem.	-	-	-	-	-	-	-	-	-	-	-
Non-Metallic Minerals	-	-	-	-	-	-	-	-	-	-	-
Non-specified	-	-	-	-	-	-	-	-	-	-	-
TRANSPORT SECTOR	-	-	-	-	-	-	-	-	-	-	-
Air	-	-	-	-	-	-	-	-	-	-	-
Road	-	-	-	-	-	-	-	-	-	-	-
Non-specified	-	-	-	-	-	-	-	-	-	-	-
OTHER SECTORS	-	-	-	-	-	-	-	-	-	-	-
Agriculture	-	-	-	-	-	-	-	-	-	-	-
Comm. and Publ. Services	-	-	-	-	-	-	-	-	-	-	-
Residential	-	-	-	-	-	-	-	-	-	-	-
Non-specified	-	-	-	-	-	-	-	-	-	-	-
NON-ENERGY USE	-	-	-	-	-	-	-	-	-	-	-

APPROVISIONNEMENT ET DEMANDE 1997	*Charbon* (1000 tonnes)							*Pétrole* (1000 tonnes)			
	Charbon à coke	*Autres charb. bit.*	*Charbon sous-bit.*	*Lignite*	*Tourbe*	*Coke de four/gaz*	*Agg./briq. de lignite*	*Pétrole brut*	*LGN*	*Produits d'aliment.*	*Additifs*
Production	-	-	-	-	-	-	-	254	-	-	-
Imports	-	-	-	-	-	-	-	688	-	-	-
Exports	-	-	-	-	-	-	-	-	-	-	-
Intl. Marine Bunkers	-	-	-	-	-	-	-	-	-	-	-
Stock Changes	-	-	-	-	-	-	-	2	-	-	-
DOMESTIC SUPPLY	-	-	-	-	-	-	-	**944**	-	-	-
Transfers and Stat. Diff.	-	-	-	-	-	-	-	-	-	-	-
TRANSFORMATION	-	-	-	-	-	-	-	**944**	-	-	-
Electricity and CHP Plants	-	-	-	-	-	-	-	-	-	-	-
Petroleum Refineries	-	-	-	-	-	-	-	944	-	-	-
Other Transform. Sector	-	-	-	-	-	-	-	-	-	-	-
ENERGY SECTOR	-	-	-	-	-	-	-	-	-	-	-
DISTRIBUTION LOSSES	-	-	-	-	-	-	-	-	-	-	-
FINAL CONSUMPTION	-	-	-	-	-	-	-	-	-	-	-
INDUSTRY SECTOR	-	-	-	-	-	-	-	-	-	-	-
Iron and Steel	-	-	-	-	-	-	-	-	-	-	-
Chemical and Petrochem.	-	-	-	-	-	-	-	-	-	-	-
Non-Metallic Minerals	-	-	-	-	-	-	-	-	-	-	-
Non-specified	-	-	-	-	-	-	-	-	-	-	-
TRANSPORT SECTOR	-	-	-	-	-	-	-	-	-	-	-
Air	-	-	-	-	-	-	-	-	-	-	-
Road	-	-	-	-	-	-	-	-	-	-	-
Non-specified	-	-	-	-	-	-	-	-	-	-	-
OTHER SECTORS	-	-	-	-	-	-	-	-	-	-	-
Agriculture	-	-	-	-	-	-	-	-	-	-	-
Comm. and Publ. Services	-	-	-	-	-	-	-	-	-	-	-
Residential	-	-	-	-	-	-	-	-	-	-	-
Non-specified	-	-	-	-	-	-	-	-	-	-	-
NON-ENERGY USE	-	-	-	-	-	-	-	-	-	-	-

Sudan / Soudan

SUPPLY AND CONSUMPTION 1996	Oil cont. (1000 tonnes)										
	Refinery Gas	LPG + Ethane	Motor Gasoline	Aviation Gasoline	Jet Fuel	Kerosene	Gas/ Diesel	Heavy Fuel Oil	Naphtha	Petrol. Coke	Other Prod.
Production	4	9	137	-	14	27	210	288	-	-	-
Imports	-	17	76	20	9	16	480	27	24	-	106
Exports	-	-	-	-	-	-	-	-	-	-	-
Intl. Marine Bunkers	-	-	-	-	-	-	-8	-	-	-	-
Stock Changes	-	-	-	-	10	18	-	-	-	-	-
DOMESTIC SUPPLY	**4**	**26**	**213**	**20**	**33**	**61**	**682**	**315**	**24**	**-**	**106**
Transfers and Stat. Diff.	-	-	-1	-	-1	1	-	-	-	-	-2
TRANSFORMATION	**-**	**-**	**-**	**-**	**-**	**-**	**58**	**264**	**-**	**-**	**-**
Electricity and CHP Plants	-	-	-	-	-	-	58	264	-	-	-
Petroleum Refineries	-	-	-	-	-	-	-	-	-	-	-
Other Transform. Sector	-	-	-	-	-	-	-	-	-	-	-
ENERGY SECTOR	**4**	**-**	**-**	**-**	**-**	**-**	**-**	**-**	**-**	**-**	**-**
DISTRIBUTION LOSSES	**-**	**-**	**-**	**-**	**-**	**-**	**-**	**-**	**-**	**-**	**-**
FINAL CONSUMPTION	**-**	**26**	**212**	**20**	**32**	**62**	**624**	**51**	**24**	**-**	**104**
INDUSTRY SECTOR	**-**	**-**	**2**	**-**	**-**	**-**	**-**	**51**	**24**	**-**	**-**
Iron and Steel	-	-	-	-	-	-	-	-	-	-	-
Chemical and Petrochem.	-	-	-	-	-	-	-	-	24	-	-
Non-Metallic Minerals	-	-	-	-	-	-	-	-	-	-	-
Non-specified	-	-	2	-	-	-	-	51	-	-	-
TRANSPORT SECTOR	**-**	**-**	**208**	**20**	**32**	**-**	**624**	**-**	**-**	**-**	**-**
Air	-	-	-	20	32	-	-	-	-	-	-
Road	-	-	208	-	-	-	624	-	-	-	-
Non-specified	-	-	-	-	-	-	-	-	-	-	-
OTHER SECTORS	**-**	**26**	**2**	**-**	**-**	**62**	**-**	**-**	**-**	**-**	**-**
Agriculture	-	-	2	-	-	-	-	-	-	-	-
Comm. and Publ. Services	-	-	-	-	-	-	-	-	-	-	-
Residential	-	26	-	-	-	62	-	-	-	-	-
Non-specified	-	-	-	-	-	-	-	-	-	-	-
NON-ENERGY USE	**-**	**-**	**-**	**-**	**-**	**-**	**-**	**-**	**-**	**-**	**104**

APPROVISIONNEMENT ET DEMANDE 1997	*Pétrole cont.* (1000 tonnes)										
	Gaz de raffinerie	*GPL + éthane*	*Essence moteur*	*Essence aviation*	*Carbu- réacteurs*	*Kérosène*	*Gazole*	*Fioul lourd*	*Naphta*	*Coke de pétrole*	*Autres prod.*
Production	5	5	121	-	21	40	242	462	-	-	-
Imports	-	20	116	20	10	19	530	27	24	-	106
Exports	-	-	-	-	-	-	-	-	-	-	-
Intl. Marine Bunkers	-	-	-	-	-	-	-8	-	-	-	-
Stock Changes	-	-	-	-	1	3	-	-	-	-	-
DOMESTIC SUPPLY	**5**	**25**	**237**	**20**	**32**	**62**	**764**	**489**	**24**	**-**	**106**
Transfers and Stat. Diff.	-	-	-	-	-	-	-	-	-	-	-2
TRANSFORMATION	**-**	**-**	**-**	**-**	**-**	**-**	**65**	**290**	**-**	**-**	**-**
Electricity and CHP Plants	-	-	-	-	-	-	65	290	-	-	-
Petroleum Refineries	-	-	-	-	-	-	-	-	-	-	-
Other Transform. Sector	-	-	-	-	-	-	-	-	-	-	-
ENERGY SECTOR	**5**	**-**	**-**	**-**	**-**	**-**	**-**	**-**	**-**	**-**	**-**
DISTRIBUTION LOSSES	**-**	**-**	**-**	**-**	**-**	**-**	**-**	**-**	**-**	**-**	**-**
FINAL CONSUMPTION	**-**	**25**	**237**	**20**	**32**	**62**	**699**	**199**	**24**	**-**	**104**
INDUSTRY SECTOR	**-**	**-**	**3**	**-**	**-**	**-**	**-**	**199**	**24**	**-**	**-**
Iron and Steel	-	-	-	-	-	-	-	-	-	-	-
Chemical and Petrochem.	-	-	-	-	-	-	-	-	24	-	-
Non-Metallic Minerals	-	-	-	-	-	-	-	-	-	-	-
Non-specified	-	-	3	-	-	-	-	199	-	-	-
TRANSPORT SECTOR	**-**	**-**	**231**	**20**	**32**	**-**	**699**	**-**	**-**	**-**	**-**
Air	-	-	-	20	32	-	-	-	-	-	-
Road	-	-	231	-	-	-	699	-	-	-	-
Non-specified	-	-	-	-	-	-	-	-	-	-	-
OTHER SECTORS	**-**	**25**	**3**	**-**	**-**	**62**	**-**	**-**	**-**	**-**	**-**
Agriculture	-	-	3	-	-	-	-	-	-	-	-
Comm. and Publ. Services	-	-	-	-	-	-	-	-	-	-	-
Residential	-	25	-	-	-	62	-	-	-	-	-
Non-specified	-	-	-	-	-	-	-	-	-	-	-
NON-ENERGY USE	**-**	**-**	**-**	**-**	**-**	**-**	**-**	**-**	**-**	**-**	**104**

Sudan / Soudan

SUPPLY AND CONSUMPTION 1996	Gas (TJ)				Comb. Renew. & Waste (TJ)				(GWh)	(TJ)
	Natural Gas	Gas Works	Coke Ovens	Blast Furnaces	Solid Biomass	Gas/Liquids from Biomass	Municipal Waste	Industrial Waste	Electricity	Heat
Production	-	-	-	-	394057	-	-	-	2063	-
Imports	-	-	-	-	-	-	-	-	-	-
Exports	-	-	-	-	-	-	-	-	-	-
Intl. Marine Bunkers	-	-	-	-	-	-	-	-	-	-
Stock Changes	-	-	-	-	-	-	-	-	-	-
DOMESTIC SUPPLY	**-**	**-**	**-**	**-**	**394057**	**-**	**-**	**-**	**2063**	**-**
Transfers and Stat. Diff.	-	-	-	-	-	-	-	-	-	-
TRANSFORMATION	**-**	**-**	**-**	**-**	**220327**	**-**	**-**	**-**	**-**	**-**
Electricity and CHP Plants	-	-	-	-	-	-	-	-	-	-
Petroleum Refineries	-	-	-	-	-	-	-	-	-	-
Other Transform. Sector	-	-	-	-	220327	-	-	-	-	-
ENERGY SECTOR	**-**	**-**	**-**	**-**	**-**	**-**	**-**	**-**	**17**	**-**
DISTRIBUTION LOSSES	**-**	**-**	**-**	**-**	**-**	**-**	**-**	**-**	**650**	**-**
FINAL CONSUMPTION	**-**	**-**	**-**	**-**	**173730**	**-**	**-**	**-**	**1396**	**-**
INDUSTRY SECTOR	**-**	**-**	**-**	**-**	**7091**	**-**	**-**	**-**	**388**	**-**
Iron and Steel	-	-	-	-	-	-	-	-	-	-
Chemical and Petrochem.	-	-	-	-	-	-	-	-	-	-
Non-Metallic Minerals	-	-	-	-	-	-	-	-	-	-
Non-specified	-	-	-	-	7091	-	-	-	388	-
TRANSPORT SECTOR	**-**	**-**	**-**	**-**	**-**	**-**	**-**	**-**	**-**	**-**
Air	-	-	-	-	-	-	-	-	-	-
Road	-	-	-	-	-	-	-	-	-	-
Non-specified	-	-	-	-	-	-	-	-	-	-
OTHER SECTORS	**-**	**-**	**-**	**-**	**166639**	**-**	**-**	**-**	**1008**	**-**
Agriculture	-	-	-	-	-	-	-	-	32	-
Comm. and Publ. Services	-	-	-	-	4052	-	-	-	202	-
Residential	-	-	-	-	162587	-	-	-	774	-
Non-specified	-	-	-	-	-	-	-	-	-	-
NON-ENERGY USE	**-**	**-**	**-**	**-**	**-**	**-**	**-**	**-**	**-**	**-**

APPROVISIONNEMENT ET DEMANDE 1997	*Gaz (TJ)*				*En. Re. Comb. & Déchets (TJ)*				(GWh)	(TJ)
	Gaz naturel	*Usines à gaz*	*Cokeries*	*Hauts fourneaux*	*Biomasse solide*	*Gaz/Liquides tirés de biomasse*	*Déchets urbains*	*Déchets industriels*	*Electricité*	*Chaleur*
Production	-	-	-	-	399180	-	-	-	1966	-
Imports	-	-	-	-	-	-	-	-	-	-
Exports	-	-	-	-	-	-	-	-	-	-
Intl. Marine Bunkers	-	-	-	-	-	-	-	-	-	-
Stock Changes	-	-	-	-	-	-	-	-	-	-
DOMESTIC SUPPLY	**-**	**-**	**-**	**-**	**399180**	**-**	**-**	**-**	**1966**	**-**
Transfers and Stat. Diff.	-	-	-	-	-	-	-	-	-	-
TRANSFORMATION	**-**	**-**	**-**	**-**	**223191**	**-**	**-**	**-**	**-**	**-**
Electricity and CHP Plants	-	-	-	-	-	-	-	-	-	-
Petroleum Refineries	-	-	-	-	-	-	-	-	-	-
Other Transform. Sector	-	-	-	-	223191	-	-	-	-	-
ENERGY SECTOR	**-**	**-**	**-**	**-**	**-**	**-**	**-**	**-**	**17**	**-**
DISTRIBUTION LOSSES	**-**	**-**	**-**	**-**	**-**	**-**	**-**	**-**	**605**	**-**
FINAL CONSUMPTION	**-**	**-**	**-**	**-**	**175989**	**-**	**-**	**-**	**1344**	**-**
INDUSTRY SECTOR	**-**	**-**	**-**	**-**	**7184**	**-**	**-**	**-**	**433**	**-**
Iron and Steel	-	-	-	-	-	-	-	-	-	-
Chemical and Petrochem.	-	-	-	-	-	-	-	-	-	-
Non-Metallic Minerals	-	-	-	-	-	-	-	-	-	-
Non-specified	-	-	-	-	7184	-	-	-	433	-
TRANSPORT SECTOR	**-**	**-**	**-**	**-**	**-**	**-**	**-**	**-**	**-**	**-**
Air	-	-	-	-	-	-	-	-	-	-
Road	-	-	-	-	-	-	-	-	-	-
Non-specified	-	-	-	-	-	-	-	-	-	-
OTHER SECTORS	**-**	**-**	**-**	**-**	**168805**	**-**	**-**	**-**	**911**	**-**
Agriculture	-	-	-	-	-	-	-	-	36	-
Comm. and Publ. Services	-	-	-	-	4104	-	-	-	211	-
Residential	-	-	-	-	164701	-	-	-	664	-
Non-specified	-	-	-	-	-	-	-	-	-	-
NON-ENERGY USE	**-**	**-**	**-**	**-**	**-**	**-**	**-**	**-**	**-**	**-**

Syria / Syrie

SUPPLY AND CONSUMPTION 1996	Coal (1000 tonnes)							Oil (1000 tonnes)			
	Coking Coal	Other Bit. Coal	Sub-Bit. Coal	Lignite	Peat	Oven and Gas Coke	Pat. Fuel and BKB	Crude Oil	NGL	Feed-stocks	Additives
Production	-	-	-	-	-	-	-	30745	-	-	-
Imports	-	-	-	-	-	2	-	-	-	-	-
Exports	-	-	-	-	-	-	-	-17651	-	-	-
Intl. Marine Bunkers	-	-	-	-	-	-	-	-	-	-	-
Stock Changes	-	-	-	-	-	-	-	-375	-	-	-
DOMESTIC SUPPLY	**-**	**-**	**-**	**-**	**-**	**2**	**-**	**12719**	**-**	**-**	**-**
Transfers and Stat. Diff.	-	-	-	-	-	-	-	-	-	-	-
TRANSFORMATION	**-**	**-**	**-**	**-**	**-**	**2**	**-**	**12719**	**-**	**-**	**-**
Electricity and CHP Plants	-	-	-	-	-	-	-	-	-	-	-
Petroleum Refineries	-	-	-	-	-	-	-	12719	-	-	-
Other Transform. Sector	-	-	-	-	-	2	-	-	-	-	-
ENERGY SECTOR	**-**	**-**	**-**	**-**	**-**	**-**	**-**	**-**	**-**	**-**	**-**
DISTRIBUTION LOSSES	**-**	**-**	**-**	**-**	**-**	**-**	**-**	**-**	**-**	**-**	**-**
FINAL CONSUMPTION	**-**	**-**	**-**	**-**	**-**	**-**	**-**	**-**	**-**	**-**	**-**
INDUSTRY SECTOR	**-**	**-**	**-**	**-**	**-**	**-**	**-**	**-**	**-**	**-**	**-**
Iron and Steel	-	-	-	-	-	-	-	-	-	-	-
Chemical and Petrochem.	-	-	-	-	-	-	-	-	-	-	-
Non-Metallic Minerals	-	-	-	-	-	-	-	-	-	-	-
Non-specified	-	-	-	-	-	-	-	-	-	-	-
TRANSPORT SECTOR	**-**	**-**	**-**	**-**	**-**	**-**	**-**	**-**	**-**	**-**	**-**
Air	-	-	-	-	-	-	-	-	-	-	-
Road	-	-	-	-	-	-	-	-	-	-	-
Non-specified	-	-	-	-	-	-	-	-	-	-	-
OTHER SECTORS	**-**	**-**	**-**	**-**	**-**	**-**	**-**	**-**	**-**	**-**	**-**
Agriculture	-	-	-	-	-	-	-	-	-	-	-
Comm. and Publ. Services	-	-	-	-	-	-	-	-	-	-	-
Residential	-	-	-	-	-	-	-	-	-	-	-
Non-specified	-	-	-	-	-	-	-	-	-	-	-
NON-ENERGY USE	**-**	**-**	**-**	**-**	**-**	**-**	**-**	**-**	**-**	**-**	**-**

APPROVISIONNEMENT ET DEMANDE 1997	*Charbon* (1000 tonnes)							*Pétrole* (1000 tonnes)			
	Charbon à coke	*Autres charb. bit.*	*Charbon sous-bit.*	*Lignite*	*Tourbe*	*Coke de four/gaz*	*Agg./briq. de lignite*	*Pétrole brut*	*LGN*	*Produits d'aliment.*	*Additifs*
Production	-	-	-	-	-	-	-	29741	-	-	-
Imports	-	-	-	-	-	2	-	-	-	-	-
Exports	-	-	-	-	-	-	-	-17022	-	-	-
Intl. Marine Bunkers	-	-	-	-	-	-	-	-	-	-	-
Stock Changes	-	-	-	-	-	-	-	-	-	-	-
DOMESTIC SUPPLY	**-**	**-**	**-**	**-**	**-**	**2**	**-**	**12719**	**-**	**-**	**-**
Transfers and Stat. Diff.	-	-	-	-	-	-	-	-	-	-	-
TRANSFORMATION	**-**	**-**	**-**	**-**	**-**	**2**	**-**	**12719**	**-**	**-**	**-**
Electricity and CHP Plants	-	-	-	-	-	-	-	-	-	-	-
Petroleum Refineries	-	-	-	-	-	-	-	12719	-	-	-
Other Transform. Sector	-	-	-	-	-	2	-	-	-	-	-
ENERGY SECTOR	**-**	**-**	**-**	**-**	**-**	**-**	**-**	**-**	**-**	**-**	**-**
DISTRIBUTION LOSSES	**-**	**-**	**-**	**-**	**-**	**-**	**-**	**-**	**-**	**-**	**-**
FINAL CONSUMPTION	**-**	**-**	**-**	**-**	**-**	**-**	**-**	**-**	**-**	**-**	**-**
INDUSTRY SECTOR	**-**	**-**	**-**	**-**	**-**	**-**	**-**	**-**	**-**	**-**	**-**
Iron and Steel	-	-	-	-	-	-	-	-	-	-	-
Chemical and Petrochem.	-	-	-	-	-	-	-	-	-	-	-
Non-Metallic Minerals	-	-	-	-	-	-	-	-	-	-	-
Non-specified	-	-	-	-	-	-	-	-	-	-	-
TRANSPORT SECTOR	**-**	**-**	**-**	**-**	**-**	**-**	**-**	**-**	**-**	**-**	**-**
Air	-	-	-	-	-	-	-	-	-	-	-
Road	-	-	-	-	-	-	-	-	-	-	-
Non-specified	-	-	-	-	-	-	-	-	-	-	-
OTHER SECTORS	**-**	**-**	**-**	**-**	**-**	**-**	**-**	**-**	**-**	**-**	**-**
Agriculture	-	-	-	-	-	-	-	-	-	-	-
Comm. and Publ. Services	-	-	-	-	-	-	-	-	-	-	-
Residential	-	-	-	-	-	-	-	-	-	-	-
Non-specified	-	-	-	-	-	-	-	-	-	-	-
NON-ENERGY USE	**-**	**-**	**-**	**-**	**-**	**-**	**-**	**-**	**-**	**-**	**-**

Syria / Syrie

SUPPLY AND CONSUMPTION 1996	Oil cont. (1000 tonnes)										
	Refinery Gas	LPG + Ethane	Motor Gasoline	Aviation Gasoline	Jet Fuel	Kerosene	Gas/ Diesel	Heavy Fuel Oil	Naphtha	Petrol. Coke	Other Prod.
Production	-	286	1591	-	131	169	4088	5216	244	-	389
Imports	-	82	-	-	31	78	345	-	-	-	-
Exports	-	-	-490	-	-	-	-	-998	-139	-	-
Intl. Marine Bunkers	-	-	-	-	-	-	-	-	-	-	-
Stock Changes	-	-	-	-	-	-	-	-	-	-	-
DOMESTIC SUPPLY	**-**	**368**	**1101**	**-**	**162**	**247**	**4433**	**4218**	**105**	**-**	**389**
Transfers and Stat. Diff.	-	-	-	-	-	-	-	1031	-	-	-
TRANSFORMATION	**-**	**-**	**-**	**-**	**-**	**-**	**368**	**2255**	**-**	**-**	**-**
Electricity and CHP Plants	-	-	-	-	-	-	368	2255	-	-	-
Petroleum Refineries	-	-	-	-	-	-	-	-	-	-	-
Other Transform. Sector	-	-	-	-	-	-	-	-	-	-	-
ENERGY SECTOR	**-**	**-**	**-**	**-**	**-**	**-**	**-**	**463**	**-**	**-**	**-**
DISTRIBUTION LOSSES	**-**	**-**	**-**	**-**	**-**	**-**	**-**	**-**	**-**	**-**	**-**
FINAL CONSUMPTION	**-**	**368**	**1101**	**-**	**162**	**247**	**4065**	**2531**	**105**	**-**	**389**
INDUSTRY SECTOR	**-**	**-**	**-**	**-**	**-**	**-**	**-**	**979**	**105**	**-**	**-**
Iron and Steel	-	-	-	-	-	-	-	-	-	-	-
Chemical and Petrochem.	-	-	-	-	-	-	-	-	105	-	-
Non-Metallic Minerals	-	-	-	-	-	-	-	-	-	-	-
Non-specified	-	-	-	-	-	-	-	979	-	-	-
TRANSPORT SECTOR	**-**	**-**	**1101**	**-**	**162**	**-**	**-**	**-**	**-**	**-**	**-**
Air	-	-	-	-	162	-	-	-	-	-	-
Road	-	-	1101	-	-	-	-	-	-	-	-
Non-specified	-	-	-	-	-	-	-	-	-	-	-
OTHER SECTORS	**-**	**368**	**-**	**-**	**-**	**247**	**4065**	**1552**	**-**	**-**	**-**
Agriculture	-	-	-	-	-	-	-	-	-	-	-
Comm. and Publ. Services	-	-	-	-	-	-	-	-	-	-	-
Residential	-	368	-	-	-	247	-	-	-	-	-
Non-specified	-	-	-	-	-	-	4065	1552	-	-	-
NON-ENERGY USE	**-**	**-**	**-**	**-**	**-**	**-**	**-**	**-**	**-**	**-**	**389**

APPROVISIONNEMENT ET DEMANDE 1997	*Pétrole cont.* (1000 tonnes)										
	Gaz de raffinerie	*GPL + éthane*	*Essence moteur*	*Essence aviation*	*Carbu- réacteurs*	*Kérosène*	*Gazole*	*Fioul lourd*	*Naphta*	*Coke de pétrole*	*Autres prod.*
Production	-	286	1591	-	192	169	4088	5216	244	-	389
Imports	-	111	-	-	32	78	478	-	-	-	-
Exports	-	-	-455	-	-	-	-	-1215	-136	-	-
Intl. Marine Bunkers	-	-	-	-	-	-	-	-	-	-	-
Stock Changes	-	-	-	-	-	-	-	-	-	-	-
DOMESTIC SUPPLY	**-**	**397**	**1136**	**-**	**224**	**247**	**4566**	**4001**	**108**	**-**	**389**
Transfers and Stat. Diff.	-	-	-	-	-	-	348	978	-	-	-
TRANSFORMATION	**-**	**-**	**-**	**-**	**-**	**-**	**379**	**2139**	**-**	**-**	**-**
Electricity and CHP Plants	-	-	-	-	-	-	379	2139	-	-	-
Petroleum Refineries	-	-	-	-	-	-	-	-	-	-	-
Other Transform. Sector	-	-	-	-	-	-	-	-	-	-	-
ENERGY SECTOR	**-**	**-**	**-**	**-**	**-**	**-**	**-**	**439**	**-**	**-**	**-**
DISTRIBUTION LOSSES	**-**	**-**	**-**	**-**	**-**	**-**	**-**	**-**	**-**	**-**	**-**
FINAL CONSUMPTION	**-**	**397**	**1136**	**-**	**224**	**247**	**4535**	**2401**	**108**	**-**	**389**
INDUSTRY SECTOR	**-**	**-**	**-**	**-**	**-**	**-**	**-**	**929**	**108**	**-**	**-**
Iron and Steel	-	-	-	-	-	-	-	-	-	-	-
Chemical and Petrochem.	-	-	-	-	-	-	-	-	108	-	-
Non-Metallic Minerals	-	-	-	-	-	-	-	-	-	-	-
Non-specified	-	-	-	-	-	-	-	929	-	-	-
TRANSPORT SECTOR	**-**	**-**	**1136**	**-**	**224**	**-**	**-**	**-**	**-**	**-**	**-**
Air	-	-	-	-	224	-	-	-	-	-	-
Road	-	-	1136	-	-	-	-	-	-	-	-
Non-specified	-	-	-	-	-	-	-	-	-	-	-
OTHER SECTORS	**-**	**397**	**-**	**-**	**-**	**247**	**4535**	**1472**	**-**	**-**	**-**
Agriculture	-	-	-	-	-	-	-	-	-	-	-
Comm. and Publ. Services	-	-	-	-	-	-	-	-	-	-	-
Residential	-	397	-	-	-	247	-	-	-	-	-
Non-specified	-	-	-	-	-	-	4535	1472	-	-	-
NON-ENERGY USE	**-**	**-**	**-**	**-**	**-**	**-**	**-**	**-**	**-**	**-**	**389**

Syria / Syrie

SUPPLY AND CONSUMPTION 1996	Gas (TJ)				Comb. Renew. & Waste (TJ)				(GWh)	(TJ)
	Natural Gas	Gas Works	Coke Ovens	Blast Furnaces	Solid Biomass	Gas/Liquids from Biomass	Municipal Waste	Industrial Waste	Electricity	Heat
Production	104022	-	-	-	200	-	-	-	16885	-
Imports	-	-	-	-	-	-	-	-	-	-
Exports	-	-	-	-	-2	-	-	-	-	-
Intl. Marine Bunkers	-	-	-	-	-	-	-	-	-	-
Stock Changes	-	-	-	-	-	-	-	-	-	-
DOMESTIC SUPPLY	**104022**	**-**	**-**	**-**	**198**	**-**	**-**	**-**	**16885**	**-**
Transfers and Stat. Diff.	2	-	-	-	-	-	-	-	-	-
TRANSFORMATION	**57001**	**-**	**-**	**-**	**20**	**-**	**-**	**-**	**-**	**-**
Electricity and CHP Plants	57001	-	-	-	-	-	-	-	-	-
Petroleum Refineries	-	-	-	-	-	-	-	-	-	-
Other Transform. Sector	-	-	-	-	20	-	-	-	-	-
ENERGY SECTOR	**3965**	**-**	**-**	**-**	**-**	**-**	**-**	**-**	**6007**	**-**
DISTRIBUTION LOSSES	**-**	**-**	**-**	**-**	**-**	**-**	**-**	**-**	**-**	**-**
FINAL CONSUMPTION	**43058**	**-**	**-**	**-**	**178**	**-**	**-**	**-**	**10878**	**-**
INDUSTRY SECTOR	**38274**	**-**	**-**	**-**	**-**	**-**	**-**	**-**	**5037**	**-**
Iron and Steel	-	-	-	-	-	-	-	-	-	-
Chemical and Petrochem.	-	-	-	-	-	-	-	-	-	-
Non-Metallic Minerals	-	-	-	-	-	-	-	-	-	-
Non-specified	38274	-	-	-	-	-	-	-	5037	-
TRANSPORT SECTOR	**-**	**-**	**-**	**-**	**-**	**-**	**-**	**-**	**-**	**-**
Air	-	-	-	-	-	-	-	-	-	-
Road	-	-	-	-	-	-	-	-	-	-
Non-specified	-	-	-	-	-	-	-	-	-	-
OTHER SECTORS	**4784**	**-**	**-**	**-**	**178**	**-**	**-**	**-**	**5841**	**-**
Agriculture	-	-	-	-	-	-	-	-	-	-
Comm. and Publ. Services	-	-	-	-	-	-	-	-	-	-
Residential	-	-	-	-	-	-	-	-	5841	-
Non-specified	4784	-	-	-	178	-	-	-	-	-
NON-ENERGY USE	**-**	**-**	**-**	**-**	**-**	**-**	**-**	**-**	**-**	**-**

APPROVISIONNEMENT ET DEMANDE 1997	*Gaz* (TJ)				*En. Re. Comb. & Déchets* (TJ)				(GWh)	(TJ)
	Gaz naturel	*Usines à gaz*	*Cokeries*	*Hauts fourneaux*	*Biomasse solide*	*Gaz/Liquides tirés de biomasse*	*Déchets urbains*	*Déchets industriels*	*Electricité*	*Chaleur*
Production	96142	-	-	-	200	-	-	-	17950	-
Imports	-	-	-	-	-	-	-	-	-	-
Exports	-	-	-	-	-2	-	-	-	-	-
Intl. Marine Bunkers	-	-	-	-	-	-	-	-	-	-
Stock Changes	-	-	-	-	-	-	-	-	-	-
DOMESTIC SUPPLY	**96142**	**-**	**-**	**-**	**198**	**-**	**-**	**-**	**17950**	**-**
Transfers and Stat. Diff.	2	-	-	-	-	-	-	-	1	-
TRANSFORMATION	**52683**	**-**	**-**	**-**	**20**	**-**	**-**	**-**	**-**	**-**
Electricity and CHP Plants	52683	-	-	-	-	-	-	-	-	-
Petroleum Refineries	-	-	-	-	-	-	-	-	-	-
Other Transform. Sector	-	-	-	-	20	-	-	-	-	-
ENERGY SECTOR	**3665**	**-**	**-**	**-**	**-**	**-**	**-**	**-**	**6386**	**-**
DISTRIBUTION LOSSES	**-**	**-**	**-**	**-**	**-**	**-**	**-**	**-**	**-**	**-**
FINAL CONSUMPTION	**39796**	**-**	**-**	**-**	**178**	**-**	**-**	**-**	**11565**	**-**
INDUSTRY SECTOR	**35374**	**-**	**-**	**-**	**-**	**-**	**-**	**-**	**5355**	**-**
Iron and Steel	-	-	-	-	-	-	-	-	-	-
Chemical and Petrochem.	-	-	-	-	-	-	-	-	-	-
Non-Metallic Minerals	-	-	-	-	-	-	-	-	-	-
Non-specified	35374	-	-	-	-	-	-	-	5355	-
TRANSPORT SECTOR	**-**	**-**	**-**	**-**	**-**	**-**	**-**	**-**	**-**	**-**
Air	-	-	-	-	-	-	-	-	-	-
Road	-	-	-	-	-	-	-	-	-	-
Non-specified	-	-	-	-	-	-	-	-	-	-
OTHER SECTORS	**4422**	**-**	**-**	**-**	**178**	**-**	**-**	**-**	**6210**	**-**
Agriculture	-	-	-	-	-	-	-	-	-	-
Comm. and Publ. Services	-	-	-	-	-	-	-	-	-	-
Residential	-	-	-	-	-	-	-	-	6210	-
Non-specified	4422	-	-	-	178	-	-	-	-	-
NON-ENERGY USE	**-**	**-**	**-**	**-**	**-**	**-**	**-**	**-**	**-**	**-**

Tajikistan / Tadjikistan

SUPPLY AND CONSUMPTION 1996	Coal (1000 tonnes)							Oil (1000 tonnes)			
	Coking Coal	Other Bit. Coal	Sub-Bit. Coal	Lignite	Peat	Oven and Gas Coke	Pat. Fuel and BKB	Crude Oil	NGL	Feed-stocks	Additives
Production	-	-	20	-	-	-	-	20	1	-	-
Imports	-	100	-	-	-	-	-	-	-	-	-
Exports	-	-	-	-	-	-	-	-	-	-	-
Intl. Marine Bunkers	-	-	-	-	-	-	-	-	-	-	-
Stock Changes	-	-	-	-	-	-	-	-	-	-	-
DOMESTIC SUPPLY	-	100	20	-	-	-	-	20	1	-	-
Transfers and Stat. Diff.	-	-	-	-	-	-	-	-	-	-	-
TRANSFORMATION	-	-	-	-	-	-	-	20	1	-	-
Electricity and CHP Plants	-	-	-	-	-	-	-	-	-	-	-
Petroleum Refineries	-	-	-	-	-	-	-	20	1	-	-
Other Transform. Sector	-	-	-	-	-	-	-	-	-	-	-
ENERGY SECTOR	-	-	-	-	-	-	-	-	-	-	-
DISTRIBUTION LOSSES	-	-	-	-	-	-	-	-	-	-	-
FINAL CONSUMPTION	-	100	20	-	-	-	-	-	-	-	-
INDUSTRY SECTOR	-	-	-	-	-	-	-	-	-	-	-
Iron and Steel	-	-	-	-	-	-	-	-	-	-	-
Chemical and Petrochem.	-	-	-	-	-	-	-	-	-	-	-
Non-Metallic Minerals	-	-	-	-	-	-	-	-	-	-	-
Non-specified	-	-	-	-	-	-	-	-	-	-	-
TRANSPORT SECTOR	-	-	-	-	-	-	-	-	-	-	-
Air	-	-	-	-	-	-	-	-	-	-	-
Road	-	-	-	-	-	-	-	-	-	-	-
Non-specified	-	-	-	-	-	-	-	-	-	-	-
OTHER SECTORS	-	100	20	-	-	-	-	-	-	-	-
Agriculture	-	-	-	-	-	-	-	-	-	-	-
Comm. and Publ. Services	-	-	-	-	-	-	-	-	-	-	-
Residential	-	-	-	-	-	-	-	-	-	-	-
Non-specified	-	100	20	-	-	-	-	-	-	-	-
NON-ENERGY USE	-	-	-	-	-	-	-	-	-	-	-

APPROVISIONNEMENT ET DEMANDE 1997	*Charbon* (1000 tonnes)							*Pétrole* (1000 tonnes)			
	Charbon à coke	*Autres charb. bit.*	*Charbon sous-bit.*	*Lignite*	*Tourbe*	*Coke de four/gaz*	*Agg./briq. de lignite*	*Pétrole brut*	*LGN*	*Produits d'aliment.*	*Additifs*
Production	-	-	10	-	-	-	-	25	1	-	-
Imports	-	100	-	-	-	-	-	-	-	-	-
Exports	-	-	-	-	-	-	-	-	-	-	-
Intl. Marine Bunkers	-	-	-	-	-	-	-	-	-	-	-
Stock Changes	-	-	-	-	-	-	-	-	-	-	-
DOMESTIC SUPPLY	-	100	10	-	-	-	-	25	1	-	-
Transfers and Stat. Diff.	-	-	-	-	-	-	-	-	-	-	-
TRANSFORMATION	-	-	-	-	-	-	-	25	1	-	-
Electricity and CHP Plants	-	-	-	-	-	-	-	-	-	-	-
Petroleum Refineries	-	-	-	-	-	-	-	25	1	-	-
Other Transform. Sector	-	-	-	-	-	-	-	-	-	-	-
ENERGY SECTOR	-	-	-	-	-	-	-	-	-	-	-
DISTRIBUTION LOSSES	-	-	-	-	-	-	-	-	-	-	-
FINAL CONSUMPTION	-	100	10	-	-	-	-	-	-	-	-
INDUSTRY SECTOR	-	-	-	-	-	-	-	-	-	-	-
Iron and Steel	-	-	-	-	-	-	-	-	-	-	-
Chemical and Petrochem.	-	-	-	-	-	-	-	-	-	-	-
Non-Metallic Minerals	-	-	-	-	-	-	-	-	-	-	-
Non-specified	-	-	-	-	-	-	-	-	-	-	-
TRANSPORT SECTOR	-	-	-	-	-	-	-	-	-	-	-
Air	-	-	-	-	-	-	-	-	-	-	-
Road	-	-	-	-	-	-	-	-	-	-	-
Non-specified	-	-	-	-	-	-	-	-	-	-	-
OTHER SECTORS	-	100	10	-	-	-	-	-	-	-	-
Agriculture	-	-	-	-	-	-	-	-	-	-	-
Comm. and Publ. Services	-	-	-	-	-	-	-	-	-	-	-
Residential	-	-	-	-	-	-	-	-	-	-	-
Non-specified	-	100	10	-	-	-	-	-	-	-	-
NON-ENERGY USE	-	-	-	-	-	-	-	-	-	-	-

Tajikistan / Tadjikistan

SUPPLY AND CONSUMPTION 1996	Oil cont. (1000 tonnes)										
	Refinery Gas	LPG + Ethane	Motor Gasoline	Aviation Gasoline	Jet Fuel	Kerosene	Gas/ Diesel	Heavy Fuel Oil	Naphtha	Petrol. Coke	Other Prod.
Production	-	-	-	-	-	-	-	-	21	-	-
Imports	-	7	996	-	5	-	75	33	-	-	43
Exports	-	-	-	-	-	-	-	-	-21	-	-
Intl. Marine Bunkers	-	-	-	-	-	-	-	-	-	-	-
Stock Changes	-	-	-	-	-	-	-	-	-	-	-
DOMESTIC SUPPLY	**-**	**7**	**996**	**-**	**5**	**-**	**75**	**33**	**-**	**-**	**43**
Transfers and Stat. Diff.	-	-	-	-	-	-	-	-	-	-	-
TRANSFORMATION	**-**	**-**	**-**	**-**	**-**	**-**	**-**	**-**	**-**	**-**	**-**
Electricity and CHP Plants	-	-	-	-	-	-	-	-	-	-	-
Petroleum Refineries	-	-	-	-	-	-	-	-	-	-	-
Other Transform. Sector	-	-	-	-	-	-	-	-	-	-	-
ENERGY SECTOR	**-**	**-**	**-**	**-**	**-**	**-**	**-**	**-**	**-**	**-**	**-**
DISTRIBUTION LOSSES	**-**	**-**	**-**	**-**	**-**	**-**	**-**	**-**	**-**	**-**	**-**
FINAL CONSUMPTION	**-**	**7**	**996**	**-**	**5**	**-**	**75**	**33**	**-**	**-**	**43**
INDUSTRY SECTOR	**-**	**-**	**-**	**-**	**-**	**-**	**-**	**-**	**-**	**-**	**-**
Iron and Steel	-	-	-	-	-	-	-	-	-	-	-
Chemical and Petrochem.	-	-	-	-	-	-	-	-	-	-	-
Non-Metallic Minerals	-	-	-	-	-	-	-	-	-	-	-
Non-specified	-	-	-	-	-	-	-	-	-	-	-
TRANSPORT SECTOR	**-**	**-**	**996**	**-**	**5**	**-**	**-**	**-**	**-**	**-**	**-**
Air	-	-	-	-	5	-	-	-	-	-	-
Road	-	-	996	-	-	-	-	-	-	-	-
Non-specified	-	-	-	-	-	-	-	-	-	-	-
OTHER SECTORS	**-**	**7**	**-**	**-**	**-**	**-**	**75**	**33**	**-**	**-**	**43**
Agriculture	-	-	-	-	-	-	-	-	-	-	-
Comm. and Publ. Services	-	-	-	-	-	-	-	-	-	-	-
Residential	-	-	-	-	-	-	-	-	-	-	-
Non-specified	-	7	-	-	-	-	75	33	-	-	43
NON-ENERGY USE	**-**	**-**	**-**	**-**	**-**	**-**	**-**	**-**	**-**	**-**	**-**

APPROVISIONNEMENT ET DEMANDE 1997	*Pétrole cont.* (1000 tonnes)										
	Gaz de raffinerie	*GPL + éthane*	*Essence moteur*	*Essence aviation*	*Carbu- réacteurs*	*Kérosène*	*Gazole*	*Fioul lourd*	*Naphta*	*Coke de pétrole*	*Autres prod.*
Production	-	-	-	-	-	-	-	-	26	-	-
Imports	-	7	996	-	5	-	75	33	-	-	43
Exports	-	-	-	-	-	-	-	-	-26	-	-
Intl. Marine Bunkers	-	-	-	-	-	-	-	-	-	-	-
Stock Changes	-	-	-	-	-	-	-	-	-	-	-
DOMESTIC SUPPLY	**-**	**7**	**996**	**-**	**5**	**-**	**75**	**33**	**-**	**-**	**43**
Transfers and Stat. Diff.	-	-	-	-	-	-	-	-	-	-	-
TRANSFORMATION	**-**	**-**	**-**	**-**	**-**	**-**	**-**	**-**	**-**	**-**	**-**
Electricity and CHP Plants	-	-	-	-	-	-	-	-	-	-	-
Petroleum Refineries	-	-	-	-	-	-	-	-	-	-	-
Other Transform. Sector	-	-	-	-	-	-	-	-	-	-	-
ENERGY SECTOR	**-**	**-**	**-**	**-**	**-**	**-**	**-**	**-**	**-**	**-**	**-**
DISTRIBUTION LOSSES	**-**	**-**	**-**	**-**	**-**	**-**	**-**	**-**	**-**	**-**	**-**
FINAL CONSUMPTION	**-**	**7**	**996**	**-**	**5**	**-**	**75**	**33**	**-**	**-**	**43**
INDUSTRY SECTOR	**-**	**-**	**-**	**-**	**-**	**-**	**-**	**-**	**-**	**-**	**-**
Iron and Steel	-	-	-	-	-	-	-	-	-	-	-
Chemical and Petrochem.	-	-	-	-	-	-	-	-	-	-	-
Non-Metallic Minerals	-	-	-	-	-	-	-	-	-	-	-
Non-specified	-	-	-	-	-	-	-	-	-	-	-
TRANSPORT SECTOR	**-**	**-**	**996**	**-**	**5**	**-**	**-**	**-**	**-**	**-**	**-**
Air	-	-	-	-	5	-	-	-	-	-	-
Road	-	-	996	-	-	-	-	-	-	-	-
Non-specified	-	-	-	-	-	-	-	-	-	-	-
OTHER SECTORS	**-**	**7**	**-**	**-**	**-**	**-**	**75**	**33**	**-**	**-**	**43**
Agriculture	-	-	-	-	-	-	-	-	-	-	-
Comm. and Publ. Services	-	-	-	-	-	-	-	-	-	-	-
Residential	-	-	-	-	-	-	-	-	-	-	-
Non-specified	-	7	-	-	-	-	75	33	-	-	43
NON-ENERGY USE	**-**	**-**	**-**	**-**	**-**	**-**	**-**	**-**	**-**	**-**	**-**

Tajikistan / Tadjikistan

SUPPLY AND CONSUMPTION 1996	Gas (TJ)				Comb. Renew. & Waste (TJ)				(GWh)	(TJ)
	Natural Gas	Gas Works	Coke Ovens	Blast Furnaces	Solid Biomass	Gas/Liquids from Biomass	Municipal Waste	Industrial Waste	Electricity	Heat
Production	1862	-	-	-	-	-	-	-	15000	-
Imports	41800	-	-	-	-	-	-	-	4930	-
Exports	-	-	-	-	-	-	-	-	-4610	-
Intl. Marine Bunkers	-	-	-	-	-	-	-	-	-	-
Stock Changes	-	-	-	-	-	-	-	-	-	-
DOMESTIC SUPPLY	**43662**	-	-	-	-	-	-	-	**15320**	-
Transfers and Stat. Diff.	-	-	-	-	-	-	-	-	-	-
TRANSFORMATION	**20553**	-	-	-	-	-	-	-	-	-
Electricity and CHP Plants	20553	-	-	-	-	-	-	-	-	-
Petroleum Refineries	-	-	-	-	-	-	-	-	-	-
Other Transform. Sector	-	-	-	-	-	-	-	-	-	-
ENERGY SECTOR	-	-	-	-	-	-	-	-	-	-
DISTRIBUTION LOSSES	-	-	-	-	-	-	-	-	**1761**	-
FINAL CONSUMPTION	**23109**	-	-	-	-	-	-	-	**13559**	-
INDUSTRY SECTOR	-	-	-	-	-	-	-	-	**6750**	-
Iron and Steel	-	-	-	-	-	-	-	-	-	-
Chemical and Petrochem.	-	-	-	-	-	-	-	-	223	-
Non-Metallic Minerals	-	-	-	-	-	-	-	-	-	-
Non-specified	-	-	-	-	-	-	-	-	6527	-
TRANSPORT SECTOR	-	-	-	-	-	-	-	-	**79**	-
Air	-	-	-	-	-	-	-	-	-	-
Road	-	-	-	-	-	-	-	-	-	-
Non-specified	-	-	-	-	-	-	-	-	79	-
OTHER SECTORS	**23109**	-	-	-	-	-	-	-	**6730**	-
Agriculture	-	-	-	-	-	-	-	-	4482	-
Comm. and Publ. Services	-	-	-	-	-	-	-	-	-	-
Residential	-	-	-	-	-	-	-	-	1980	-
Non-specified	23109	-	-	-	-	-	-	-	268	-
NON-ENERGY USE	-	-	-	-	-	-	-	-	-	-

APPROVISIONNEMENT ET DEMANDE 1997	*Gaz* (TJ)				*En. Re. Comb. & Déchets* (TJ)				(GWh)	(TJ)
	Gaz naturel	*Usines à gaz*	*Cokeries*	*Hauts fourneaux*	*Biomasse solide*	*Gaz/Liquides tirés de biomasse*	*Déchets urbains*	*Déchets industriels*	*Electricité*	*Chaleur*
Production	1596	-	-	-	-	-	-	-	14000	-
Imports	38000	-	-	-	-	-	-	-	4900	-
Exports	-	-	-	-	-	-	-	-	-4100	-
Intl. Marine Bunkers	-	-	-	-	-	-	-	-	-	-
Stock Changes	-	-	-	-	-	-	-	-	-	-
DOMESTIC SUPPLY	**39596**	-	-	-	-	-	-	-	**14800**	-
Transfers and Stat. Diff.	-	-	-	-	-	-	-	-	-	-
TRANSFORMATION	**18620**	-	-	-	-	-	-	-	-	-
Electricity and CHP Plants	18620	-	-	-	-	-	-	-	-	-
Petroleum Refineries	-	-	-	-	-	-	-	-	-	-
Other Transform. Sector	-	-	-	-	-	-	-	-	-	-
ENERGY SECTOR	-	-	-	-	-	-	-	-	-	-
DISTRIBUTION LOSSES	-	-	-	-	-	-	-	-	**1700**	-
FINAL CONSUMPTION	**20976**	-	-	-	-	-	-	-	**13100**	-
INDUSTRY SECTOR	-	-	-	-	-	-	-	-	**6520**	-
Iron and Steel	-	-	-	-	-	-	-	-	-	-
Chemical and Petrochem.	-	-	-	-	-	-	-	-	215	-
Non-Metallic Minerals	-	-	-	-	-	-	-	-	-	-
Non-specified	-	-	-	-	-	-	-	-	6305	-
TRANSPORT SECTOR	-	-	-	-	-	-	-	-	**80**	-
Air	-	-	-	-	-	-	-	-	-	-
Road	-	-	-	-	-	-	-	-	-	-
Non-specified	-	-	-	-	-	-	-	-	80	-
OTHER SECTORS	**20976**	-	-	-	-	-	-	-	**6500**	-
Agriculture	-	-	-	-	-	-	-	-	4330	-
Comm. and Publ. Services	-	-	-	-	-	-	-	-	-	-
Residential	-	-	-	-	-	-	-	-	1910	-
Non-specified	20976	-	-	-	-	-	-	-	260	-
NON-ENERGY USE	-	-	-	-	-	-	-	-	-	-

Tanzania / Tanzanie

SUPPLY AND CONSUMPTION 1996	Coal (1000 tonnes)							Oil (1000 tonnes)			
	Coking Coal	Other Bit. Coal	Sub-Bit. Coal	Lignite	Peat	Oven and Gas Coke	Pat. Fuel and BKB	Crude Oil	NGL	Feed-stocks	Additives
Production	1	4	-	-	-	-	-	-	-	-	-
Imports	-	-	-	-	-	-	-	589	-	-	-
Exports	-	-	-	-	-	-	-	-	-	-	-
Intl. Marine Bunkers	-	-	-	-	-	-	-	-	-	-	-
Stock Changes	-	-	-	-	-	-	-	-	-	-	-
DOMESTIC SUPPLY	**1**	**4**	**-**	**-**	**-**	**-**	**-**	**589**	**-**	**-**	**-**
Transfers and Stat. Diff.	-	-	-	-	-	-	-	-1	-	-	-
TRANSFORMATION	**1**	**-**	**-**	**-**	**-**	**-**	**-**	**588**	**-**	**-**	**-**
Electricity and CHP Plants	-	-	-	-	-	-	-	-	-	-	-
Petroleum Refineries	-	-	-	-	-	-	-	588	-	-	-
Other Transform. Sector	1	-	-	-	-	-	-	-	-	-	-
ENERGY SECTOR	**-**	**-**	**-**	**-**	**-**	**-**	**-**	**-**	**-**	**-**	**-**
DISTRIBUTION LOSSES	**-**	**-**	**-**	**-**	**-**	**-**	**-**	**-**	**-**	**-**	**-**
FINAL CONSUMPTION	**-**	**4**	**-**	**-**	**-**	**-**	**-**	**-**	**-**	**-**	**-**
INDUSTRY SECTOR	**-**	**4**	**-**	**-**	**-**	**-**	**-**	**-**	**-**	**-**	**-**
Iron and Steel	-	-	-	-	-	-	-	-	-	-	-
Chemical and Petrochem.	-	-	-	-	-	-	-	-	-	-	-
Non-Metallic Minerals	-	1	-	-	-	-	-	-	-	-	-
Non-specified	-	3	-	-	-	-	-	-	-	-	-
TRANSPORT SECTOR	**-**	**-**	**-**	**-**	**-**	**-**	**-**	**-**	**-**	**-**	**-**
Air	-	-	-	-	-	-	-	-	-	-	-
Road	-	-	-	-	-	-	-	-	-	-	-
Non-specified	-	-	-	-	-	-	-	-	-	-	-
OTHER SECTORS	**-**	**-**	**-**	**-**	**-**	**-**	**-**	**-**	**-**	**-**	**-**
Agriculture	-	-	-	-	-	-	-	-	-	-	-
Comm. and Publ. Services	-	-	-	-	-	-	-	-	-	-	-
Residential	-	-	-	-	-	-	-	-	-	-	-
Non-specified	-	-	-	-	-	-	-	-	-	-	-
NON-ENERGY USE	**-**	**-**	**-**	**-**	**-**	**-**	**-**	**-**	**-**	**-**	**-**

APPROVISIONNEMENT ET DEMANDE 1997	*Charbon* (1000 tonnes)							*Pétrole* (1000 tonnes)			
	Charbon à coke	*Autres charb. bit.*	*Charbon sous-bit.*	*Lignite*	*Tourbe*	*Coke de four/gaz*	*Agg./briq. de lignite*	*Pétrole brut*	*LGN*	*Produits d'aliment.*	*Additifs*
Production	1	4	-	-	-	-	-	-	-	-	-
Imports	-	-	-	-	-	-	-	589	-	-	-
Exports	-	-	-	-	-	-	-	-	-	-	-
Intl. Marine Bunkers	-	-	-	-	-	-	-	-	-	-	-
Stock Changes	-	-	-	-	-	-	-	-	-	-	-
DOMESTIC SUPPLY	**1**	**4**	**-**	**-**	**-**	**-**	**-**	**589**	**-**	**-**	**-**
Transfers and Stat. Diff.	-	-	-	-	-	-	-	-1	-	-	-
TRANSFORMATION	**1**	**-**	**-**	**-**	**-**	**-**	**-**	**588**	**-**	**-**	**-**
Electricity and CHP Plants	-	-	-	-	-	-	-	-	-	-	-
Petroleum Refineries	-	-	-	-	-	-	-	588	-	-	-
Other Transform. Sector	1	-	-	-	-	-	-	-	-	-	-
ENERGY SECTOR	**-**	**-**	**-**	**-**	**-**	**-**	**-**	**-**	**-**	**-**	**-**
DISTRIBUTION LOSSES	**-**	**-**	**-**	**-**	**-**	**-**	**-**	**-**	**-**	**-**	**-**
FINAL CONSUMPTION	**-**	**4**	**-**	**-**	**-**	**-**	**-**	**-**	**-**	**-**	**-**
INDUSTRY SECTOR	**-**	**4**	**-**	**-**	**-**	**-**	**-**	**-**	**-**	**-**	**-**
Iron and Steel	-	-	-	-	-	-	-	-	-	-	-
Chemical and Petrochem.	-	-	-	-	-	-	-	-	-	-	-
Non-Metallic Minerals	-	1	-	-	-	-	-	-	-	-	-
Non-specified	-	3	-	-	-	-	-	-	-	-	-
TRANSPORT SECTOR	**-**	**-**	**-**	**-**	**-**	**-**	**-**	**-**	**-**	**-**	**-**
Air	-	-	-	-	-	-	-	-	-	-	-
Road	-	-	-	-	-	-	-	-	-	-	-
Non-specified	-	-	-	-	-	-	-	-	-	-	-
OTHER SECTORS	**-**	**-**	**-**	**-**	**-**	**-**	**-**	**-**	**-**	**-**	**-**
Agriculture	-	-	-	-	-	-	-	-	-	-	-
Comm. and Publ. Services	-	-	-	-	-	-	-	-	-	-	-
Residential	-	-	-	-	-	-	-	-	-	-	-
Non-specified	-	-	-	-	-	-	-	-	-	-	-
NON-ENERGY USE	**-**	**-**	**-**	**-**	**-**	**-**	**-**	**-**	**-**	**-**	**-**

Tanzania / Tanzanie

SUPPLY AND CONSUMPTION 1996	Oil cont. (1000 tonnes)										
	Refinery Gas	LPG + Ethane	Motor Gasoline	Aviation Gasoline	Jet Fuel	Kerosene	Gas/ Diesel	Heavy Fuel Oil	Naphtha	Petrol. Coke	Other Prod.
Production	14	6	105	-	33	45	148	246	-	-	2
Imports	-	-	11	8	9	27	102	-	-	-	14
Exports	-	-	-4	-	-	-	-12	-13	-	-	-
Intl. Marine Bunkers	-	-	-	-	-	-	-1	-22	-	-	-
Stock Changes	-	-	-	-	-	-	-	-	-	-	-
DOMESTIC SUPPLY	**14**	**6**	**112**	**8**	**42**	**72**	**237**	**211**	**-**	**-**	**16**
Transfers and Stat. Diff.	-	-	-14	-	-14	-	-36	-106	-	-	-
TRANSFORMATION	**-**	**-**	**-**	**-**	**-**	**-**	**80**	**-**	**-**	**-**	**-**
Electricity and CHP Plants	-	-	-	-	-	-	80	-	-	-	-
Petroleum Refineries	-	-	-	-	-	-	-	-	-	-	-
Other Transform. Sector	-	-	-	-	-	-	-	-	-	-	-
ENERGY SECTOR	**14**	**-**	**-**	**-**	**-**	**-**	**-**	**-**	**-**	**-**	**-**
DISTRIBUTION LOSSES	**-**	**-**	**-**	**-**	**-**	**-**	**-**	**-**	**-**	**-**	**-**
FINAL CONSUMPTION	**-**	**6**	**98**	**8**	**28**	**72**	**121**	**105**	**-**	**-**	**16**
INDUSTRY SECTOR	**-**	**-**	**-**	**-**	**-**	**-**	**-**	**105**	**-**	**-**	**-**
Iron and Steel	-	-	-	-	-	-	-	-	-	-	-
Chemical and Petrochem.	-	-	-	-	-	-	-	-	-	-	-
Non-Metallic Minerals	-	-	-	-	-	-	-	-	-	-	-
Non-specified	-	-	-	-	-	-	-	105	-	-	-
TRANSPORT SECTOR	**-**	**-**	**98**	**8**	**28**	**-**	**121**	**-**	**-**	**-**	**-**
Air	-	-	-	8	28	-	-	-	-	-	-
Road	-	-	98	-	-	-	121	-	-	-	-
Non-specified	-	-	-	-	-	-	-	-	-	-	-
OTHER SECTORS	**-**	**6**	**-**	**-**	**-**	**72**	**-**	**-**	**-**	**-**	**-**
Agriculture	-	-	-	-	-	-	-	-	-	-	-
Comm. and Publ. Services	-	-	-	-	-	-	-	-	-	-	-
Residential	-	6	-	-	-	72	-	-	-	-	-
Non-specified	-	-	-	-	-	-	-	-	-	-	-
NON-ENERGY USE	**-**	**-**	**-**	**-**	**-**	**-**	**-**	**-**	**-**	**-**	**16**

APPROVISIONNEMENT ET DEMANDE 1997	*Pétrole cont.* (1000 tonnes)										
	Gaz de raffinerie	*GPL + éthane*	*Essence moteur*	*Essence aviation*	*Carbu- réacteurs*	*Kérosène*	*Gazole*	*Fioul lourd*	*Naphta*	*Coke de pétrole*	*Autres prod.*
Production	14	6	105	-	33	45	148	246	-	-	2
Imports	-	-	11	8	9	27	102	-	-	-	14
Exports	-	-	-4	-	-	-	-12	-13	-	-	-
Intl. Marine Bunkers	-	-	-	-	-	-	-1	-22	-	-	-
Stock Changes	-	-	-	-	-	-	-	-	-	-	-
DOMESTIC SUPPLY	**14**	**6**	**112**	**8**	**42**	**72**	**237**	**211**	**-**	**-**	**16**
Transfers and Stat. Diff.	-	-	-14	-	-14	-	-	-106	-	-	-
TRANSFORMATION	**-**	**-**	**-**	**-**	**-**	**-**	**160**	**-**	**-**	**-**	**-**
Electricity and CHP Plants	-	-	-	-	-	-	160	-	-	-	-
Petroleum Refineries	-	-	-	-	-	-	-	-	-	-	-
Other Transform. Sector	-	-	-	-	-	-	-	-	-	-	-
ENERGY SECTOR	**14**	**-**	**-**	**-**	**-**	**-**	**-**	**-**	**-**	**-**	**-**
DISTRIBUTION LOSSES	**-**	**-**	**-**	**-**	**-**	**-**	**-**	**-**	**-**	**-**	**-**
FINAL CONSUMPTION	**-**	**6**	**98**	**8**	**28**	**72**	**77**	**105**	**-**	**-**	**16**
INDUSTRY SECTOR	**-**	**-**	**-**	**-**	**-**	**-**	**-**	**105**	**-**	**-**	**-**
Iron and Steel	-	-	-	-	-	-	-	-	-	-	-
Chemical and Petrochem.	-	-	-	-	-	-	-	-	-	-	-
Non-Metallic Minerals	-	-	-	-	-	-	-	-	-	-	-
Non-specified	-	-	-	-	-	-	-	105	-	-	-
TRANSPORT SECTOR	**-**	**-**	**98**	**8**	**28**	**-**	**77**	**-**	**-**	**-**	**-**
Air	-	-	-	8	28	-	-	-	-	-	-
Road	-	-	98	-	-	-	77	-	-	-	-
Non-specified	-	-	-	-	-	-	-	-	-	-	-
OTHER SECTORS	**-**	**6**	**-**	**-**	**-**	**72**	**-**	**-**	**-**	**-**	**-**
Agriculture	-	-	-	-	-	-	-	-	-	-	-
Comm. and Publ. Services	-	-	-	-	-	-	-	-	-	-	-
Residential	-	6	-	-	-	72	-	-	-	-	-
Non-specified	-	-	-	-	-	-	-	-	-	-	-
NON-ENERGY USE	**-**	**-**	**-**	**-**	**-**	**-**	**-**	**-**	**-**	**-**	**16**

Tanzania / Tanzanie

SUPPLY AND CONSUMPTION 1996	Gas (TJ)				Comb. Renew. & Waste (TJ)				(GWh)	(TJ)
	Natural Gas	Gas Works	Coke Ovens	Blast Furnaces	Solid Biomass	Gas/Liquids from Biomass	Municipal Waste	Industrial Waste	Electricity	Heat
Production	-	-	-	-	550738	-	-	-	1991	-
Imports	-	-	-	-	-	-	-	-	34	-
Exports	-	-	-	-	-	-	-	-	-	-
Intl. Marine Bunkers	-	-	-	-	-	-	-	-	-	-
Stock Changes	-	-	-	-	-	-	-	-	-	-
DOMESTIC SUPPLY	**-**	**-**	**-**	**-**	**550738**	**-**	**-**	**-**	**2025**	**-**
Transfers and Stat. Diff.	-	-	-	-	-	-	-	-	-	-
TRANSFORMATION	**-**	**-**	**-**	**-**	**44792**	**-**	**-**	**-**	**-**	**-**
Electricity and CHP Plants	-	-	-	-	-	-	-	-	-	-
Petroleum Refineries	-	-	-	-	-	-	-	-	-	-
Other Transform. Sector	-	-	-	-	44792	-	-	-	-	-
ENERGY SECTOR	**-**	**-**	**-**	**-**	**-**	**-**	**-**	**-**	**-**	**-**
DISTRIBUTION LOSSES	**-**	**-**	**-**	**-**	**-**	**-**	**-**	**-**	**232**	**-**
FINAL CONSUMPTION	**-**	**-**	**-**	**-**	**505946**	**-**	**-**	**-**	**1793**	**-**
INDUSTRY SECTOR	**-**	**-**	**-**	**-**	**55990**	**-**	**-**	**-**	**485**	**-**
Iron and Steel	-	-	-	-	-	-	-	-	-	-
Chemical and Petrochem.	-	-	-	-	-	-	-	-	-	-
Non-Metallic Minerals	-	-	-	-	-	-	-	-	-	-
Non-specified	-	-	-	-	55990	-	-	-	485	-
TRANSPORT SECTOR	**-**	**-**	**-**	**-**	**-**	**-**	**-**	**-**	**-**	**-**
Air	-	-	-	-	-	-	-	-	-	-
Road	-	-	-	-	-	-	-	-	-	-
Non-specified	-	-	-	-	-	-	-	-	-	-
OTHER SECTORS	**-**	**-**	**-**	**-**	**449956**	**-**	**-**	**-**	**1308**	**-**
Agriculture	-	-	-	-	16797	-	-	-	81	-
Comm. and Publ. Services	-	-	-	-	-	-	-	-	468	-
Residential	-	-	-	-	412799	-	-	-	687	-
Non-specified	-	-	-	-	20360	-	-	-	72	-
NON-ENERGY USE	**-**	**-**	**-**	**-**	**-**	**-**	**-**	**-**	**-**	**-**

APPROVISIONNEMENT ET DEMANDE 1997	*Gaz* (TJ)				*En. Re. Comb. & Déchets* (TJ)				(GWh)	(TJ)
	Gaz naturel	*Usines à gaz*	*Cokeries*	*Hauts fourneaux*	*Biomasse solide*	*Gaz/Liquides tirés de biomasse*	*Déchets urbains*	*Déchets industriels*	*Electricité*	*Chaleur*
Production	-	-	-	-	561202	-	-	-	1934	-
Imports	-	-	-	-	-	-	-	-	44	-
Exports	-	-	-	-	-	-	-	-	-	-
Intl. Marine Bunkers	-	-	-	-	-	-	-	-	-	-
Stock Changes	-	-	-	-	-	-	-	-	-	-
DOMESTIC SUPPLY	**-**	**-**	**-**	**-**	**561202**	**-**	**-**	**-**	**1978**	**-**
Transfers and Stat. Diff.	-	-	-	-	-	-	-	-	1	-
TRANSFORMATION	**-**	**-**	**-**	**-**	**45644**	**-**	**-**	**-**	**-**	**-**
Electricity and CHP Plants	-	-	-	-	-	-	-	-	-	-
Petroleum Refineries	-	-	-	-	-	-	-	-	-	-
Other Transform. Sector	-	-	-	-	45644	-	-	-	-	-
ENERGY SECTOR	**-**	**-**	**-**	**-**	**-**	**-**	**-**	**-**	**-**	**-**
DISTRIBUTION LOSSES	**-**	**-**	**-**	**-**	**-**	**-**	**-**	**-**	**278**	**-**
FINAL CONSUMPTION	**-**	**-**	**-**	**-**	**515559**	**-**	**-**	**-**	**1701**	**-**
INDUSTRY SECTOR	**-**	**-**	**-**	**-**	**57054**	**-**	**-**	**-**	**460**	**-**
Iron and Steel	-	-	-	-	-	-	-	-	-	-
Chemical and Petrochem.	-	-	-	-	-	-	-	-	-	-
Non-Metallic Minerals	-	-	-	-	-	-	-	-	-	-
Non-specified	-	-	-	-	57054	-	-	-	460	-
TRANSPORT SECTOR	**-**	**-**	**-**	**-**	**-**	**-**	**-**	**-**	**-**	**-**
Air	-	-	-	-	-	-	-	-	-	-
Road	-	-	-	-	-	-	-	-	-	-
Non-specified	-	-	-	-	-	-	-	-	-	-
OTHER SECTORS	**-**	**-**	**-**	**-**	**458505**	**-**	**-**	**-**	**1241**	**-**
Agriculture	-	-	-	-	17116	-	-	-	77	-
Comm. and Publ. Services	-	-	-	-	-	-	-	-	444	-
Residential	-	-	-	-	420642	-	-	-	651	-
Non-specified	-	-	-	-	20747	-	-	-	69	-
NON-ENERGY USE	**-**	**-**	**-**	**-**	**-**	**-**	**-**	**-**	**-**	**-**

Thailand / Thaïlande : 1996

SUPPLY AND CONSUMPTION / *APPROVISIONNEMENT ET DEMANDE*	Coal / *Charbon* (1000 tonnes)							Oil / *Pétrole* (1000 tonnes)			
	Coking Coal / *Charbon à coke*	Other Bit. Coal / *Autres charb. bit.*	Sub-Bit. Coal / *Charbon sous-bit.*	Lignite / *Lignite*	Peat / *Tourbe*	Oven and Gas Coke / *Coke de four/gaz*	Pat. Fuel and BKB / *Agg./briq. de lignite*	Crude Oil / *Pétrole brut*	NGL / *LGN*	Feed-stocks / *Produits d'aliment.*	Additives / *Additifs*
Production	-	3	-	21474	-	-	-	2856	2103	-	-
From Other Sources	-	-	-	-	-	-	-	-	-	-	-
Imports	-	3750	-	-	-	125	80	30390	-	-	-
Exports	-	-	-	-	-	-	-	-1073	-22	-	-
Intl. Marine Bunkers	-	-	-	-	-	-	-	-	-	-	-
Stock Changes	-	2	-	-491	-	-	-	-175	-	-	-
DOMESTIC SUPPLY	**-**	**3755**	**-**	**20983**	**-**	**125**	**80**	**31998**	**2081**	**-**	**-**
Transfers	-	-	-	-	-	-	-	-	-1954	-	-
Statistical Differences	-	-	-	-	-	-	-	-	-	-	-
TRANSFORMATION	**-**	**32**	**-**	**16432**	**-**	**100**	**-**	**31665**	**93**	**-**	**-**
Electricity Plants	-	32	-	16432	-	-	-	-	-	-	-
CHP Plants	-	-	-	-	-	-	-	-	-	-	-
Heat Plants	-	-	-	-	-	-	-	-	-	-	-
Blast Furnaces/Gas Works	-	-	-	-	-	100	-	-	-	-	-
Coke/Pat. Fuel/BKB Plants	-	-	-	-	-	-	-	-	-	-	-
Petroleum Refineries	-	-	-	-	-	-	-	31665	93	-	-
Petrochemical Industry	-	-	-	-	-	-	-	-	-	-	-
Liquefaction	-	-	-	-	-	-	-	-	-	-	-
Other Transform. Sector	-	-	-	-	-	-	-	-	-	-	-
ENERGY SECTOR	**-**	**-**	**-**	**-**	**-**	**-**	**-**	**-**	**-**	**-**	**-**
Coal Mines	-	-	-	-	-	-	-	-	-	-	-
Oil and Gas Extraction	-	-	-	-	-	-	-	-	-	-	-
Petroleum Refineries	-	-	-	-	-	-	-	-	-	-	-
Electr., CHP+Heat Plants	-	-	-	-	-	-	-	-	-	-	-
Pumped Storage (Elec.)	-	-	-	-	-	-	-	-	-	-	-
Other Energy Sector	-	-	-	-	-	-	-	-	-	-	-
Distribution Losses	-	-	-	-	-	-	-	100	2	-	-
FINAL CONSUMPTION	**-**	**3723**	**-**	**4551**	**-**	**25**	**80**	**233**	**32**	**-**	**-**
INDUSTRY SECTOR	**-**	**3723**	**-**	**4544**	**-**	**25**	**80**	**233**	**32**	**-**	**-**
Iron and Steel	-	-	-	-	-	25	-	-	-	-	-
Chemical and Petrochem.	-	-	-	-	-	-	-	233	32	-	-
of which: Feedstocks	-	-	-	-	-	-	-	*233*	*32*	-	-
Non-Ferrous Metals	-	-	-	-	-	-	-	-	-	-	-
Non-Metallic Minerals	-	3150	-	3048	-	-	-	-	-	-	-
Transport Equipment	-	-	-	-	-	-	-	-	-	-	-
Machinery	-	-	-	-	-	-	-	-	-	-	-
Mining and Quarrying	-	-	-	-	-	-	-	-	-	-	-
Food and Tobacco	-	197	-	209	-	-	-	-	-	-	-
Paper, Pulp and Print	-	-	-	-	-	-	-	-	-	-	-
Wood and Wood Products	-	-	-	-	-	-	-	-	-	-	-
Construction	-	-	-	-	-	-	-	-	-	-	-
Textile and Leather	-	-	-	-	-	-	-	-	-	-	-
Non-specified	-	376	-	1287	-	-	80	-	-	-	-
TRANSPORT SECTOR	**-**	**-**	**-**	**-**	**-**	**-**	**-**	**-**	**-**	**-**	**-**
Air	-	-	-	-	-	-	-	-	-	-	-
Road	-	-	-	-	-	-	-	-	-	-	-
Rail	-	-	-	-	-	-	-	-	-	-	-
Pipeline Transport	-	-	-	-	-	-	-	-	-	-	-
Internal Navigation	-	-	-	-	-	-	-	-	-	-	-
Non-specified	-	-	-	-	-	-	-	-	-	-	-
OTHER SECTORS	**-**	**-**	**-**	**-**	**-**	**-**	**-**	**-**	**-**	**-**	**-**
Agriculture	-	-	-	-	-	-	-	-	-	-	-
Comm. and Publ. Services	-	-	-	-	-	-	-	-	-	-	-
Residential	-	-	-	-	-	-	-	-	-	-	-
Non-specified	-	-	-	-	-	-	-	-	-	-	-
NON-ENERGY USE	**-**	**-**	**-**	**7**	**-**	**-**	**-**	**-**	**-**	**-**	**-**
in Industry/Trans./Energy	-	-	-	7	-	-	-	-	-	-	-
in Transport	-	-	-	-	-	-	-	-	-	-	-
in Other Sectors	-	-	-	-	-	-	-	-	-	-	-

Thailand / Thaïlande : 1996

SUPPLY AND CONSUMPTION / *APPROVISIONNEMENT ET DEMANDE*	Oil cont. / *Pétrole cont.* (1000 tonnes)										
	Refinery Gas / *Gaz de raffinerie*	LPG + Ethane / *GPL + éthane*	Motor Gasoline / *Essence moteur*	Aviation Gasoline / *Essence aviation*	Jet Fuel / *Carbu-réacteurs*	Kerosene / *Kérosène*	Gas/ Diesel / *Gazole*	Heavy Fuel Oil / *Fioul lourd*	Naphtha / *Naphta*	Petrol. Coke / *Coke de pétrole*	Other Prod. / *Autres prod.*
Production	318	888	5897	-	2668	156	11896	7899	-	-	356
From Other Sources	-	-	-	-	-	-	-	-	-	-	-
Imports	-	14	257	10	54	-	4415	2577	-	-	215
Exports	-	-197	-894	-	-147	-31	-921	-954	-	-	-
Intl. Marine Bunkers	-	-	-	-	-	-	-119	-640	-	-	-
Stock Changes	-	-8	-122	-	-93	-47	-81	-426	-	-	-
DOMESTIC SUPPLY	**318**	**697**	**5138**	**10**	**2482**	**78**	**15190**	**8456**	**-**	**-**	**571**
Transfers	-	1954	-	-	-	-	-	-	-	-	-
Statistical Differences	-	-1	-79	-	-	-	2	-1	-	-	-
TRANSFORMATION	**-**	**-**	**-**	**-**	**-**	**-**	**1140**	**4769**	**-**	**-**	**-**
Electricity Plants	-	-	-	-	-	-	1140	4769	-	-	-
CHP Plants	-	-	-	-	-	-	-	-	-	-	-
Heat Plants	-	-	-	-	-	-	-	-	-	-	-
Blast Furnaces/Gas Works	-	-	-	-	-	-	-	-	-	-	-
Coke/Pat. Fuel/BKB Plants	-	-	-	-	-	-	-	-	-	-	-
Petroleum Refineries	-	-	-	-	-	-	-	-	-	-	-
Petrochemical Industry	-	-	-	-	-	-	-	-	-	-	-
Liquefaction	-	-	-	-	-	-	-	-	-	-	-
Other Transform. Sector	-	-	-	-	-	-	-	-	-	-	-
ENERGY SECTOR	**318**	**-**	**-**	**-**	**-**	**-**	**-**	**-**	**-**	**-**	**-**
Coal Mines	-	-	-	-	-	-	-	-	-	-	-
Oil and Gas Extraction	-	-	-	-	-	-	-	-	-	-	-
Petroleum Refineries	318	-	-	-	-	-	-	-	-	-	-
Electr., CHP+Heat Plants	-	-	-	-	-	-	-	-	-	-	-
Pumped Storage (Elec.)	-	-	-	-	-	-	-	-	-	-	-
Other Energy Sector	-	-	-	-	-	-	-	-	-	-	-
Distribution Losses	-	13	-	-	-	-	-	-	-	-	-
FINAL CONSUMPTION	**-**	**2637**	**5059**	**10**	**2482**	**78**	**14052**	**3686**	**-**	**-**	**571**
INDUSTRY SECTOR	**-**	**1331**	**129**	**-**	**-**	**45**	**901**	**3575**	**-**	**-**	**-**
Iron and Steel	-	63	-	-	-	6	44	237	-	-	-
Chemical and Petrochem.	-	1081	14	-	-	15	52	353	-	-	-
of which: Feedstocks	-	*260*	-	-	-	-	-	-	-	-	-
Non-Ferrous Metals	-	-	-	-	-	-	-	-	-	-	-
Non-Metallic Minerals	-	62	34	-	-	5	64	675	-	-	-
Transport Equipment	-	-	-	-	-	-	-	-	-	-	-
Machinery	-	61	12	-	-	5	35	60	-	-	-
Mining and Quarrying	-	-	-	-	-	-	32	12	-	-	-
Food and Tobacco	-	21	26	-	-	-	201	510	-	-	-
Paper, Pulp and Print	-	3	5	-	-	1	23	174	-	-	-
Wood and Wood Products	-	-	3	-	-	-	17	14	-	-	-
Construction	-	-	26	-	-	-	202	85	-	-	-
Textile and Leather	-	14	2	-	-	2	9	570	-	-	-
Non-specified	-	26	7	-	-	11	222	885	-	-	-
TRANSPORT SECTOR	**-**	**134**	**4882**	**10**	**2482**	**-**	**11461**	**-**	**-**	**-**	**-**
Air	-	-	-	10	2482	-	-	-	-	-	-
Road	-	134	4863	-	-	-	11225	-	-	-	-
Rail	-	-	-	-	-	-	132	-	-	-	-
Pipeline Transport	-	-	-	-	-	-	-	-	-	-	-
Internal Navigation	-	-	19	-	-	-	104	-	-	-	-
Non-specified	-	-	-	-	-	-	-	-	-	-	-
OTHER SECTORS	**-**	**1172**	**48**	**-**	**-**	**33**	**1690**	**111**	**-**	**-**	**-**
Agriculture	-	2	48	-	-	1	1685	33	-	-	-
Comm. and Publ. Services	-	-	-	-	-	-	-	-	-	-	-
Residential	-	1170	-	-	-	32	5	78	-	-	-
Non-specified	-	-	-	-	-	-	-	-	-	-	-
NON-ENERGY USE	**-**	**-**	**-**	**-**	**-**	**-**	**-**	**-**	**-**	**-**	**571**
in Industry/Transf./Energy	-	-	-	-	-	-	-	-	-	-	571
in Transport	-	-	-	-	-	-	-	-	-	-	-
in Other Sectors	-	-	-	-	-	-	-	-	-	-	-

Thailand / Thaïlande : 1996

SUPPLY AND CONSUMPTION / *APPROVISIONNEMENT ET DEMANDE*	Gas / *Gaz* (TJ)				Comb. Renew. & Waste / *En. Re. Comb. & Déchets* (TJ)				(GWh)	(TJ)
	Natural Gas / *Gaz naturel*	Gas Works / *Usines à gaz*	Coke Ovens / *Cokeries*	Blast Furnaces / *Hauts fourneaux*	Solid Biomass / *Biomasse solide*	Gas/Liquids from Biomass / *Gaz/Liquides tirés de biomasse*	Municipal Waste / *Déchets urbains*	Industrial Waste / *Déchets industriels*	Electricity / *Electricité*	Heat / *Chaleur*
Production	473969	-	-	-	894944	-	-	-	87467	-
From Other Sources	-	-	-	-	-	-	-	-	-	-
Imports	-	-	-	-	716	-	-	-	806	-
Exports	-	-	-	-	-258	-	-	-	-89	-
Intl. Marine Bunkers	-	-	-	-	-	-	-	-	-	-
Stock Changes	-	-	-	-	-	-	-	-	-	-
DOMESTIC SUPPLY	**473969**	-	-	-	**895402**	-	-	-	**88184**	-
Transfers	-	-	-	-	-	-	-	-	-	-
Statistical Differences	1	-	-	-	-	-	-	-	-2	-
TRANSFORMATION	**351438**	-	-	-	**373808**	-	-	-	-	-
Electricity Plants	351438	-	-	-	8921	-	-	-	-	-
CHP Plants	-	-	-	-	-	-	-	-	-	-
Heat Plants	-	-	-	-	-	-	-	-	-	-
Blast Furnaces/Gas Works	-	-	-	-	-	-	-	-	-	-
Coke/Pat. Fuel/BKB Plants	-	-	-	-	-	-	-	-	-	-
Petroleum Refineries	-	-	-	-	-	-	-	-	-	-
Petrochemical Industry	-	-	-	-	-	-	-	-	-	-
Liquefaction	-	-	-	-	-	-	-	-	-	-
Other Transform. Sector	-	-	-	-	364887	-	-	-	-	-
ENERGY SECTOR	**83222**	-	-	-	-	-	-	-	**3226**	-
Coal Mines	-	-	-	-	-	-	-	-	-	-
Oil and Gas Extraction	-	-	-	-	-	-	-	-	-	-
Petroleum Refineries	-	-	-	-	-	-	-	-	-	-
Electr., CHP+Heat Plants	-	-	-	-	-	-	-	-	3226	-
Pumped Storage (Elec.)	-	-	-	-	-	-	-	-	-	-
Other Energy Sector	83222	-	-	-	-	-	-	-	-	-
Distribution Losses	-	-	-	-	-	-	-	-	7603	-
FINAL CONSUMPTION	**39310**	-	-	-	**521594**	-	-	-	**77353**	-
INDUSTRY SECTOR	**39095**	-	-	-	**185673**	-	-	-	**34645**	-
Iron and Steel	-	-	-	-	-	-	-	-	3340	-
Chemical and Petrochem.	6743	-	-	-	7090	-	-	-	5657	-
of which: Feedstocks	*6743*	-	-	-	-	-	-	-	-	-
Non-Ferrous Metals	-	-	-	-	-	-	-	-	-	-
Non-Metallic Minerals	530	-	-	-	7918	-	-	-	5766	-
Transport Equipment	-	-	-	-	-	-	-	-	-	-
Machinery	-	-	-	-	-	-	-	-	5824	-
Mining and Quarrying	-	-	-	-	-	-	-	-	-	-
Food and Tobacco	-	-	-	-	170250	-	-	-	5231	-
Paper, Pulp and Print	-	-	-	-	-	-	-	-	1210	-
Wood and Wood Products	-	-	-	-	415	-	-	-	660	-
Construction	-	-	-	-	-	-	-	-	-	-
Textile and Leather	-	-	-	-	-	-	-	-	5780	-
Non-specified	31822	-	-	-	-	-	-	-	1177	-
TRANSPORT SECTOR	**215**	-	-	-	-	-	-	-	-	-
Air	-	-	-	-	-	-	-	-	-	-
Road	215	-	-	-	-	-	-	-	-	-
Rail	-	-	-	-	-	-	-	-	-	-
Pipeline Transport	-	-	-	-	-	-	-	-	-	-
Internal Navigation	-	-	-	-	-	-	-	-	-	-
Non-specified	-	-	-	-	-	-	-	-	-	-
OTHER SECTORS	-	-	-	-	**335921**	-	-	-	**42708**	-
Agriculture	-	-	-	-	-	-	-	-	124	-
Comm. and Publ. Services	-	-	-	-	-	-	-	-	25782	-
Residential	-	-	-	-	335921	-	-	-	16047	-
Non-specified	-	-	-	-	-	-	-	-	755	-
NON-ENERGY USE	-	-	-	-	-	-	-	-	-	-
in Industry/Transf./Energy	-	-	-	-	-	-	-	-	-	-
in Transport	-	-	-	-	-	-	-	-	-	-
in Other Sectors	-	-	-	-	-	-	-	-	-	-

Thailand / Thaïlande : 1997

	Coal / *Charbon* (1000 tonnes)							Oil / *Pétrole* (1000 tonnes)			
SUPPLY AND CONSUMPTION ***APPROVISIONNEMENT ET DEMANDE***	Coking Coal *Charbon à coke*	Other Bit. Coal *Autres charb. bit.*	Sub-Bit. Coal *Charbon sous-bit.*	Lignite *Lignite*	Peat *Tourbe*	Oven and Gas Coke *Coke de four/gaz*	Pat. Fuel and BKB *Agg./briq. de lignite*	Crude Oil *Pétrole brut*	NGL *LGN*	Feed-stocks *Produits d'aliment.*	Additives *Additifs*
Production	-	1	-	23393	-	-	-	3304	2094	-	-
From Other Sources	-	-	-	-	-	-	-	-	-	-	-
Imports	-	3074	-	-	-	83	126	34931	-	-	-
Exports	-	-	-	-	-	-	-	-1042	-85	-	-
Intl. Marine Bunkers	-	-	-	-	-	-	-	-	-	-	-
Stock Changes	-	4	-	-779	-	-	-	-142	-	-	-
DOMESTIC SUPPLY	**-**	**3079**	**-**	**22614**	**-**	**83**	**126**	**37051**	**2009**	**-**	**-**
Transfers	-	-	-	-	-	-	-	-	-1895	-	-
Statistical Differences	-	-	-	-1	-	-	-	-	-	-	-
TRANSFORMATION	**-**	**42**	**-**	**18144**	**-**	**66**	**13**	**36707**	**13**	**-**	**-**
Electricity Plants	-	42	-	18144	-	-	13	-	-	-	-
CHP Plants	-	-	-	-	-	-	-	-	-	-	-
Heat Plants	-	-	-	-	-	-	-	-	-	-	-
Blast Furnaces/Gas Works	-	-	-	-	-	66	-	-	-	-	-
Coke/Pat. Fuel/BKB Plants	-	-	-	-	-	-	-	-	-	-	-
Petroleum Refineries	-	-	-	-	-	-	-	36707	13	-	-
Petrochemical Industry	-	-	-	-	-	-	-	-	-	-	-
Liquefaction	-	-	-	-	-	-	-	-	-	-	-
Other Transform. Sector	-	-	-	-	-	-	-	-	-	-	-
ENERGY SECTOR	**-**	**-**	**-**	**-**	**-**	**-**	**-**	**-**	**-**	**-**	**-**
Coal Mines	-	-	-	-	-	-	-	-	-	-	-
Oil and Gas Extraction	-	-	-	-	-	-	-	-	-	-	-
Petroleum Refineries	-	-	-	-	-	-	-	-	-	-	-
Electr., CHP+Heat Plants	-	-	-	-	-	-	-	-	-	-	-
Pumped Storage (Elec.)	-	-	-	-	-	-	-	-	-	-	-
Other Energy Sector	-	-	-	-	-	-	-	-	-	-	-
Distribution Losses	-	-	-	-	-	-	-	67	14	-	-
FINAL CONSUMPTION	**-**	**3037**	**-**	**4469**	**-**	**17**	**113**	**277**	**87**	**-**	**-**
INDUSTRY SECTOR	**-**	**3037**	**-**	**4467**	**-**	**17**	**113**	**277**	**87**	**-**	**-**
Iron and Steel	-	-	-	-	-	17	-	-	-	-	-
Chemical and Petrochem.	-	-	-	-	-	-	-	277	87	-	-
of which: Feedstocks	-	-	-	-	-	-	-	*277*	*87*	-	-
Non-Ferrous Metals	-	-	-	-	-	-	-	-	-	-	-
Non-Metallic Minerals	-	2569	-	2997	-	-	-	-	-	-	-
Transport Equipment	-	-	-	-	-	-	-	-	-	-	-
Machinery	-	-	-	-	-	-	-	-	-	-	-
Mining and Quarrying	-	-	-	-	-	-	-	-	-	-	-
Food and Tobacco	-	161	-	205	-	-	-	-	-	-	-
Paper, Pulp and Print	-	-	-	-	-	-	-	-	-	-	-
Wood and Wood Products	-	-	-	-	-	-	-	-	-	-	-
Construction	-	-	-	-	-	-	-	-	-	-	-
Textile and Leather	-	-	-	-	-	-	-	-	-	-	-
Non-specified	-	307	-	1265	-	-	113	-	-	-	-
TRANSPORT SECTOR	**-**	**-**	**-**	**-**	**-**	**-**	**-**	**-**	**-**	**-**	**-**
Air	-	-	-	-	-	-	-	-	-	-	-
Road	-	-	-	-	-	-	-	-	-	-	-
Rail	-	-	-	-	-	-	-	-	-	-	-
Pipeline Transport	-	-	-	-	-	-	-	-	-	-	-
Internal Navigation	-	-	-	-	-	-	-	-	-	-	-
Non-specified	-	-	-	-	-	-	-	-	-	-	-
OTHER SECTORS	**-**	**-**	**-**	**-**	**-**	**-**	**-**	**-**	**-**	**-**	**-**
Agriculture	-	-	-	-	-	-	-	-	-	-	-
Comm. and Publ. Services	-	-	-	-	-	-	-	-	-	-	-
Residential	-	-	-	-	-	-	-	-	-	-	-
Non-specified	-	-	-	-	-	-	-	-	-	-	-
NON-ENERGY USE	**-**	**-**	**-**	**2**	**-**	**-**	**-**	**-**	**-**	**-**	**-**
in Industry/Trans./Energy	-	-	-	2	-	-	-	-	-	-	-
in Transport	-	-	-	-	-	-	-	-	-	-	-
in Other Sectors	-	-	-	-	-	-	-	-	-	-	-

Thailand / Thaïlande : 1997

SUPPLY AND CONSUMPTION / *APPROVISIONNEMENT ET DEMANDE*	Oil cont. / *Pétrole cont.* (1000 tonnes)										
	Refinery Gas / *Gaz de raffinerie*	LPG + Ethane / *GPL + éthane*	Motor Gasoline / *Essence moteur*	Aviation Gasoline / *Essence aviation*	Jet Fuel / *Carbu- réacteurs*	Kerosene / *Kérosène*	Gas/ Diesel / *Gazole*	Heavy Fuel Oil / *Fioul lourd*	Naphtha / *Naphta*	Petrol. Coke / *Coke de pétrole*	Other Prod. / *Autres prod.*
Production	367	1002	6866	-	2851	103	14794	8217	-	-	616
From Other Sources	-	-	-	-	-	-	-	-	-	-	-
Imports	-	2	32	10	17	29	1725	889	-	-	112
Exports	-	-471	-1414	-	-208	-11	-1680	-672	-	-	-
Intl. Marine Bunkers	-	-	-	-	-	-	-102	-707	-	-	-
Stock Changes	-	-61	-14	-	-4	-53	255	127	-	-	-9
DOMESTIC SUPPLY	**367**	**472**	**5470**	**10**	**2656**	**68**	**14992**	**7854**	**-**	**-**	**719**
Transfers	-	1895	-	-	-	-	-	-	-	-	-
Statistical Differences	-	-1	-86	-	1	2	1	-1	-	-	-
TRANSFORMATION	**-**	**-**	**-**	**-**	**-**	**-**	**632**	**4413**	**-**	**-**	**-**
Electricity Plants	-	-	-	-	-	-	632	4413	-	-	-
CHP Plants	-	-	-	-	-	-	-	-	-	-	-
Heat Plants	-	-	-	-	-	-	-	-	-	-	-
Blast Furnaces/Gas Works	-	-	-	-	-	-	-	-	-	-	-
Coke/Pat. Fuel/BKB Plants	-	-	-	-	-	-	-	-	-	-	-
Petroleum Refineries	-	-	-	-	-	-	-	-	-	-	-
Petrochemical Industry	-	-	-	-	-	-	-	-	-	-	-
Liquefaction	-	-	-	-	-	-	-	-	-	-	-
Other Transform. Sector	-	-	-	-	-	-	-	-	-	-	-
ENERGY SECTOR	**367**	**-**	**-**	**-**	**-**	**-**	**-**	**-**	**-**	**-**	**-**
Coal Mines	-	-	-	-	-	-	-	-	-	-	-
Oil and Gas Extraction	-	-	-	-	-	-	-	-	-	-	-
Petroleum Refineries	367	-	-	-	-	-	-	-	-	-	-
Electr., CHP+Heat Plants	-	-	-	-	-	-	-	-	-	-	-
Pumped Storage (Elec.)	-	-	-	-	-	-	-	-	-	-	-
Other Energy Sector	-	-	-	-	-	-	-	-	-	-	-
Distribution Losses	-	85	-	-	-	-	-	-	-	-	-
FINAL CONSUMPTION	**-**	**2281**	**5384**	**10**	**2657**	**70**	**14361**	**3440**	**-**	**-**	**719**
INDUSTRY SECTOR	**-**	**943**	**30**	**-**	**-**	**42**	**815**	**3411**	**-**	**-**	**-**
Iron and Steel	-	68	-	-	-	4	36	228	-	-	-
Chemical and Petrochem.	-	680	3	-	-	24	128	349	-	-	-
of which: Feedstocks	-	*142*	-	-	-	-	-	-	-	-	-
Non-Ferrous Metals	-	-	-	-	-	-	-	-	-	-	-
Non-Metallic Minerals	-	82	6	-	-	3	60	571	-	-	-
Transport Equipment	-	-	-	-	-	-	-	-	-	-	-
Machinery	-	52	5	-	-	4	28	46	-	-	-
Mining and Quarrying	-	-	-	-	-	-	27	18	-	-	-
Food and Tobacco	-	17	5	-	-	-	163	553	-	-	-
Paper, Pulp and Print	-	2	1	-	-	1	16	185	-	-	-
Wood and Wood Products	-	-	1	-	-	-	10	18	-	-	-
Construction	-	-	-	-	-	-	266	101	-	-	-
Textile and Leather	-	11	-	-	-	2	8	527	-	-	-
Non-specified	-	31	9	-	-	4	73	815	-	-	-
TRANSPORT SECTOR	**-**	**112**	**5307**	**10**	**2657**	**-**	**12124**	**-**	**-**	**-**	**-**
Air	-	-	-	10	2657	-	-	-	-	-	-
Road	-	112	5303	-	-	-	11913	-	-	-	-
Rail	-	-	-	-	-	-	114	-	-	-	-
Pipeline Transport	-	-	-	-	-	-	-	-	-	-	-
Internal Navigation	-	-	4	-	-	-	97	-	-	-	-
Non-specified	-	-	-	-	-	-	-	-	-	-	-
OTHER SECTORS	**-**	**1226**	**47**	**-**	**-**	**28**	**1422**	**29**	**-**	**-**	**-**
Agriculture	-	2	47	-	-	1	1416	8	-	-	-
Comm. and Publ. Services	-	-	-	-	-	-	-	-	-	-	-
Residential	-	1224	-	-	-	27	6	21	-	-	-
Non-specified	-	-	-	-	-	-	-	-	-	-	-
NON-ENERGY USE	**-**	**-**	**-**	**-**	**-**	**-**	**-**	**-**	**-**	**-**	**719**
in Industry/Transf./Energy	-	-	-	-	-	-	-	-	-	-	719
in Transport	-	-	-	-	-	-	-	-	-	-	-
in Other Sectors	-	-	-	-	-	-	-	-	-	-	-

Thailand / Thaïlande : 1997

	Gas / *Gaz* (TJ)				Comb. Renew. & Waste / *En. Re. Comb. & Déchets* (TJ)				(GWh)	(TJ)
SUPPLY AND CONSUMPTION *APPROVISIONNEMENT ET DEMANDE*	Natural Gas *Gaz naturel*	Gas Works *Usines à gaz*	Coke Ovens *Cokeries*	Blast Furnaces *Hauts fourneaux*	Solid Biomass *Biomasse solide*	Gas/Liquids from Biomass *Gaz/Liquides tirés de biomasse*	Municipal Waste *Déchets urbains*	Industrial Waste *Déchets industriels*	Electricity *Electricité*	Heat *Chaleur*
Production	577301	-	-	-	864706	-	-	-	93253	-
From Other Sources	-	-	-	-	-	-	-	-	-	-
Imports	-	-	-	-	429	-	-	-	746	-
Exports	-	-	-	-	-143	-	-	-	-104	-
Intl. Marine Bunkers	-	-	-	-	-	-	-	-	-	-
Stock Changes	-	-	-	-	-	-	-	-	-	-
DOMESTIC SUPPLY	**577301**	**-**	**-**	**-**	**864992**	**-**	**-**	**-**	**93895**	**-**
Transfers	-	-	-	-	-	-	-	-	-	-
Statistical Differences	1	-	-	-	-	-	-	-	-	-
TRANSFORMATION	**453495**	**-**	**-**	**-**	**357716**	**-**	**-**	**-**	**-**	**-**
Electricity Plants	453495	-	-	-	16039	-	-	-	-	-
CHP Plants	-	-	-	-	-	-	-	-	-	-
Heat Plants	-	-	-	-	-	-	-	-	-	-
Blast Furnaces/Gas Works	-	-	-	-	-	-	-	-	-	-
Coke/Pat. Fuel/BKB Plants	-	-	-	-	-	-	-	-	-	-
Petroleum Refineries	-	-	-	-	-	-	-	-	-	-
Petrochemical Industry	-	-	-	-	-	-	-	-	-	-
Liquefaction	-	-	-	-	-	-	-	-	-	-
Other Transform. Sector	-	-	-	-	341677	-	-	-	-	-
ENERGY SECTOR	**84023**	**-**	**-**	**-**	**-**	**-**	**-**	**-**	**3378**	**-**
Coal Mines	-	-	-	-	-	-	-	-	-	-
Oil and Gas Extraction	-	-	-	-	-	-	-	-	-	-
Petroleum Refineries	-	-	-	-	-	-	-	-	-	-
Electr., CHP+Heat Plants	-	-	-	-	-	-	-	-	3378	-
Pumped Storage (Elec.)	-	-	-	-	-	-	-	-	-	-
Other Energy Sector	84023	-	-	-	-	-	-	-	-	-
Distribution Losses	-	-	-	-	-	-	-	-	8088	-
FINAL CONSUMPTION	**39784**	**-**	**-**	**-**	**508646**	**-**	**-**	**-**	**82429**	**-**
INDUSTRY SECTOR	**39552**	**-**	**-**	**-**	**195280**	**-**	**-**	**-**	**34541**	**-**
Iron and Steel	-	-	-	-	-	-	-	-	3404	-
Chemical and Petrochem.	6008	-	-	-	5631	-	-	-	5433	-
of which: Feedstocks	*6008*	-	-	-	-	-	-	-	-	-
Non-Ferrous Metals	-	-	-	-	-	-	-	-	-	-
Non-Metallic Minerals	309	-	-	-	9668	-	-	-	5689	-
Transport Equipment	-	-	-	-	-	-	-	-	-	-
Machinery	-	-	-	-	-	-	-	-	5736	-
Mining and Quarrying	-	-	-	-	-	-	-	-	-	-
Food and Tobacco	-	-	-	-	179449	-	-	-	5436	-
Paper, Pulp and Print	-	-	-	-	-	-	-	-	1211	-
Wood and Wood Products	-	-	-	-	532	-	-	-	630	-
Construction	-	-	-	-	-	-	-	-	-	-
Textile and Leather	-	-	-	-	-	-	-	-	5811	-
Non-specified	33235	-	-	-	-	-	-	-	1191	-
TRANSPORT SECTOR	**232**	**-**	**-**	**-**	**-**	**-**	**-**	**-**	**-**	**-**
Air	-	-	-	-	-	-	-	-	-	-
Road	232	-	-	-	-	-	-	-	-	-
Rail	-	-	-	-	-	-	-	-	-	-
Pipeline Transport	-	-	-	-	-	-	-	-	-	-
Internal Navigation	-	-	-	-	-	-	-	-	-	-
Non-specified	-	-	-	-	-	-	-	-	-	-
OTHER SECTORS	**-**	**-**	**-**	**-**	**313366**	**-**	**-**	**-**	**47888**	**-**
Agriculture	-	-	-	-	-	-	-	-	165	-
Comm. and Publ. Services	-	-	-	-	-	-	-	-	29204	-
Residential	-	-	-	-	313366	-	-	-	17667	-
Non-specified	-	-	-	-	-	-	-	-	852	-
NON-ENERGY USE	**-**	**-**	**-**	**-**	**-**	**-**	**-**	**-**	**-**	**-**
in Industry/Transf./Energy	-	-	-	-	-	-	-	-	-	-
in Transport	-	-	-	-	-	-	-	-	-	-
in Other Sectors	-	-	-	-	-	-	-	-	-	-

Trinidad-&-Tobago / Trinité-et-Tobago

SUPPLY AND CONSUMPTION 1996	Coal (1000 tonnes)							Oil (1000 tonnes)			
	Coking Coal	Other Bit. Coal	Sub-Bit. Coal	Lignite	Peat	Oven and Gas Coke	Pat. Fuel and BKB	Crude Oil	NGL	Feed-stocks	Additives
Production	-	-	-	-	-	-	-	6659	414	-	-
Imports	-	-	-	-	-	-	-	2153	-	-	-
Exports	-	-	-	-	-	-	-	-3039	-414	-	-
Intl. Marine Bunkers	-	-	-	-	-	-	-	-	-	-	-
Stock Changes	-	-	-	-	-	-	-	-108	-	-	-
DOMESTIC SUPPLY	**-**	**-**	**-**	**-**	**-**	**-**	**-**	**5665**	**-**	**-**	**-**
Transfers and Stat. Diff.	-	-	-	-	-	-	-	-1	-	-	-
TRANSFORMATION	**-**	**-**	**-**	**-**	**-**	**-**	**-**	**5664**	**-**	**-**	**-**
Electricity and CHP Plants	-	-	-	-	-	-	-	-	-	-	-
Petroleum Refineries	-	-	-	-	-	-	-	5664	-	-	-
Other Transform. Sector	-	-	-	-	-	-	-	-	-	-	-
ENERGY SECTOR	**-**	**-**	**-**	**-**	**-**	**-**	**-**	**-**	**-**	**-**	**-**
DISTRIBUTION LOSSES	**-**	**-**	**-**	**-**	**-**	**-**	**-**	**-**	**-**	**-**	**-**
FINAL CONSUMPTION	**-**	**-**	**-**	**-**	**-**	**-**	**-**	**-**	**-**	**-**	**-**
INDUSTRY SECTOR	**-**	**-**	**-**	**-**	**-**	**-**	**-**	**-**	**-**	**-**	**-**
Iron and Steel	-	-	-	-	-	-	-	-	-	-	-
Chemical and Petrochem.	-	-	-	-	-	-	-	-	-	-	-
Non-Metallic Minerals	-	-	-	-	-	-	-	-	-	-	-
Non-specified	-	-	-	-	-	-	-	-	-	-	-
TRANSPORT SECTOR	**-**	**-**	**-**	**-**	**-**	**-**	**-**	**-**	**-**	**-**	**-**
Air	-	-	-	-	-	-	-	-	-	-	-
Road	-	-	-	-	-	-	-	-	-	-	-
Non-specified	-	-	-	-	-	-	-	-	-	-	-
OTHER SECTORS	**-**	**-**	**-**	**-**	**-**	**-**	**-**	**-**	**-**	**-**	**-**
Agriculture	-	-	-	-	-	-	-	-	-	-	-
Comm. and Publ. Services	-	-	-	-	-	-	-	-	-	-	-
Residential	-	-	-	-	-	-	-	-	-	-	-
Non-specified	-	-	-	-	-	-	-	-	-	-	-
NON-ENERGY USE	**-**	**-**	**-**	**-**	**-**	**-**	**-**	**-**	**-**	**-**	**-**

APPROVISIONNEMENT ET DEMANDE 1997	*Charbon* (1000 tonnes)							*Pétrole* (1000 tonnes)			
	Charbon à coke	*Autres charb. bit.*	*Charbon sous-bit.*	*Lignite*	*Tourbe*	*Coke de four/gaz*	*Agg./briq. de lignite*	*Pétrole brut*	*LGN*	*Produits d'aliment.*	*Additifs*
Production	-	-	-	-	-	-	-	6378	284	-	-
Imports	-	-	-	-	-	-	-	2039	-	-	-
Exports	-	-	-	-	-	-	-	-2867	-284	-	-
Intl. Marine Bunkers	-	-	-	-	-	-	-	-	-	-	-
Stock Changes	-	-	-	-	-	-	-	-	-	-	-
DOMESTIC SUPPLY	**-**	**-**	**-**	**-**	**-**	**-**	**-**	**5550**	**-**	**-**	**-**
Transfers and Stat. Diff.	-	-	-	-	-	-	-	-	-	-	-
TRANSFORMATION	**-**	**-**	**-**	**-**	**-**	**-**	**-**	**5550**	**-**	**-**	**-**
Electricity and CHP Plants	-	-	-	-	-	-	-	-	-	-	-
Petroleum Refineries	-	-	-	-	-	-	-	5550	-	-	-
Other Transform. Sector	-	-	-	-	-	-	-	-	-	-	-
ENERGY SECTOR	**-**	**-**	**-**	**-**	**-**	**-**	**-**	**-**	**-**	**-**	**-**
DISTRIBUTION LOSSES	**-**	**-**	**-**	**-**	**-**	**-**	**-**	**-**	**-**	**-**	**-**
FINAL CONSUMPTION	**-**	**-**	**-**	**-**	**-**	**-**	**-**	**-**	**-**	**-**	**-**
INDUSTRY SECTOR	**-**	**-**	**-**	**-**	**-**	**-**	**-**	**-**	**-**	**-**	**-**
Iron and Steel	-	-	-	-	-	-	-	-	-	-	-
Chemical and Petrochem.	-	-	-	-	-	-	-	-	-	-	-
Non-Metallic Minerals	-	-	-	-	-	-	-	-	-	-	-
Non-specified	-	-	-	-	-	-	-	-	-	-	-
TRANSPORT SECTOR	**-**	**-**	**-**	**-**	**-**	**-**	**-**	**-**	**-**	**-**	**-**
Air	-	-	-	-	-	-	-	-	-	-	-
Road	-	-	-	-	-	-	-	-	-	-	-
Non-specified	-	-	-	-	-	-	-	-	-	-	-
OTHER SECTORS	**-**	**-**	**-**	**-**	**-**	**-**	**-**	**-**	**-**	**-**	**-**
Agriculture	-	-	-	-	-	-	-	-	-	-	-
Comm. and Publ. Services	-	-	-	-	-	-	-	-	-	-	-
Residential	-	-	-	-	-	-	-	-	-	-	-
Non-specified	-	-	-	-	-	-	-	-	-	-	-
NON-ENERGY USE	**-**	**-**	**-**	**-**	**-**	**-**	**-**	**-**	**-**	**-**	**-**

Trinidad-&-Tobago / Trinité-et-Tobago

SUPPLY AND CONSUMPTION 1996	Oil cont. (1000 tonnes)										
	Refinery Gas	LPG + Ethane	Motor Gasoline	Aviation Gasoline	Jet Fuel	Kerosene	Gas/ Diesel	Heavy Fuel Oil	Naphtha	Petrol. Coke	Other Prod.
Production	143	84	741	-	502	13	1044	2555	-	-	56
Imports	-	-	157	-	22	-	55	-	-	-	2
Exports	-	-4	-462	-	-244	-	-714	-2441	-	-	-41
Intl. Marine Bunkers	-	-	-	-	-	-	-12	-40	-	-	-
Stock Changes	-	-	-164	-	-167	-	-158	-68	-	-	-8
DOMESTIC SUPPLY	**143**	**80**	**272**	**-**	**113**	**13**	**215**	**6**	**-**	**-**	**9**
Transfers and Stat. Diff.	-	-26	15	-	-52	4	2	-	-	-	12
TRANSFORMATION	**-**	**-**	**-**	**-**	**-**	**-**	**-**	**-**	**-**	**-**	**-**
Electricity and CHP Plants	-	-	-	-	-	-	-	-	-	-	-
Petroleum Refineries	-	-	-	-	-	-	-	-	-	-	-
Other Transform. Sector	-	-	-	-	-	-	-	-	-	-	-
ENERGY SECTOR	**143**	**-**	**1**	**-**	**-**	**-**	**-**	**-**	**-**	**-**	**-**
DISTRIBUTION LOSSES	**-**	**-**	**-**	**-**	**-**	**-**	**-**	**-**	**-**	**-**	**-**
FINAL CONSUMPTION	**-**	**54**	**286**	**-**	**61**	**17**	**217**	**6**	**-**	**-**	**21**
INDUSTRY SECTOR	**-**	**10**	**-**	**-**	**-**	**12**	**61**	**2**	**-**	**-**	**-**
Iron and Steel	-	-	-	-	-	-	-	-	-	-	-
Chemical and Petrochem.	-	-	-	-	-	-	-	-	-	-	-
Non-Metallic Minerals	-	-	-	-	-	-	-	-	-	-	-
Non-specified	-	10	-	-	-	12	61	2	-	-	-
TRANSPORT SECTOR	**-**	**-**	**286**	**-**	**61**	**-**	**155**	**1**	**-**	**-**	**-**
Air	-	-	-	-	61	-	-	-	-	-	-
Road	-	-	286	-	-	-	155	-	-	-	-
Non-specified	-	-	-	-	-	-	-	1	-	-	-
OTHER SECTORS	**-**	**44**	**-**	**-**	**-**	**5**	**1**	**3**	**-**	**-**	**-**
Agriculture	-	-	-	-	-	-	-	-	-	-	-
Comm. and Publ. Services	-	25	-	-	-	-	1	-	-	-	-
Residential	-	19	-	-	-	5	-	3	-	-	-
Non-specified	-	-	-	-	-	-	-	-	-	-	-
NON-ENERGY USE	**-**	**-**	**-**	**-**	**-**	**-**	**-**	**-**	**-**	**-**	**21**

APPROVISIONNEMENT ET DEMANDE 1997	*Pétrole cont.* (1000 tonnes)										
	Gaz de raffinerie	*GPL + éthane*	*Essence moteur*	*Essence aviation*	*Carbu- réacteurs*	*Kérosène*	*Gazole*	*Fioul lourd*	*Naphta*	*Coke de pétrole*	*Autres prod.*
Production	155	78	872	-	508	13	1123	2517	-	-	25
Imports	-	-	-	-	22	-	55	-	-	-	2
Exports	-	-4	-473	-	-247	-	-730	-2367	-	-	-26
Intl. Marine Bunkers	-	-	-	-	-	-	-12	-40	-	-	-
Stock Changes	-	-	-90	-	-158	-	-158	-42	-	-	-
DOMESTIC SUPPLY	**155**	**74**	**309**	**-**	**125**	**13**	**278**	**68**	**-**	**-**	**1**
Transfers and Stat. Diff.	-	-21	-	-	-58	6	-	-60	-	-	16
TRANSFORMATION	**-**	**-**	**-**	**-**	**-**	**-**	**-**	**-**	**-**	**-**	**-**
Electricity and CHP Plants	-	-	-	-	-	-	-	-	-	-	-
Petroleum Refineries	-	-	-	-	-	-	-	-	-	-	-
Other Transform. Sector	-	-	-	-	-	-	-	-	-	-	-
ENERGY SECTOR	**155**	**-**	**1**	**-**	**-**	**-**	**4**	**-**	**-**	**-**	**-**
DISTRIBUTION LOSSES	**-**	**-**	**-**	**-**	**-**	**-**	**-**	**-**	**-**	**-**	**-**
FINAL CONSUMPTION	**-**	**53**	**308**	**-**	**67**	**19**	**274**	**8**	**-**	**-**	**17**
INDUSTRY SECTOR	**-**	**10**	**-**	**-**	**-**	**13**	**77**	**3**	**-**	**-**	**-**
Iron and Steel	-	-	-	-	-	-	-	-	-	-	-
Chemical and Petrochem.	-	-	-	-	-	-	-	-	-	-	-
Non-Metallic Minerals	-	-	-	-	-	-	-	-	-	-	-
Non-specified	-	10	-	-	-	13	77	3	-	-	-
TRANSPORT SECTOR	**-**	**-**	**308**	**-**	**67**	**-**	**196**	**1**	**-**	**-**	**-**
Air	-	-	-	-	67	-	-	-	-	-	-
Road	-	-	308	-	-	-	196	-	-	-	-
Non-specified	-	-	-	-	-	-	-	1	-	-	-
OTHER SECTORS	**-**	**43**	**-**	**-**	**-**	**6**	**1**	**4**	**-**	**-**	**-**
Agriculture	-	-	-	-	-	-	-	-	-	-	-
Comm. and Publ. Services	-	24	-	-	-	-	1	-	-	-	-
Residential	-	19	-	-	-	6	-	4	-	-	-
Non-specified	-	-	-	-	-	-	-	-	-	-	-
NON-ENERGY USE	**-**	**-**	**-**	**-**	**-**	**-**	**-**	**-**	**-**	**-**	**17**

Trinidad-&-Tobago / Trinité-et-Tobago

SUPPLY AND CONSUMPTION 1996	Gas (TJ)				Comb. Renew. & Waste (TJ)				(GWh)	(TJ)
	Natural Gas	Gas Works	Coke Ovens	Blast Furnaces	Solid Biomass	Gas/Liquids from Biomass	Municipal Waste	Industrial Waste	Electricity	Heat
Production	298397	-	-	-	1232	-	-	-	4541	-
Imports	-	-	-	-	-	-	-	-	-	-
Exports	-	-	-	-	-	-	-	-	-	-
Intl. Marine Bunkers	-	-	-	-	-	-	-	-	-	-
Stock Changes	-	-	-	-	-	-	-	-	-	-
DOMESTIC SUPPLY	**298397**	-	-	-	**1232**	-	-	-	**4541**	-
Transfers and Stat. Diff.	-681	-	-	-	-	-	-	-	39	-
TRANSFORMATION	**68261**	-	-	-	**1232**	-	-	-	-	-
Electricity and CHP Plants	68261	-	-	-	1232	-	-	-	-	-
Petroleum Refineries	-	-	-	-	-	-	-	-	-	-
Other Transform. Sector	-	-	-	-	-	-	-	-	-	-
ENERGY SECTOR	**38411**	-	-	-	-	-	-	-	**186**	-
DISTRIBUTION LOSSES	-	-	-	-	-	-	-	-	**450**	-
FINAL CONSUMPTION	**191044**	-	-	-	-	-	-	-	**3944**	-
INDUSTRY SECTOR	**191044**	-	-	-	-	-	-	-	**2566**	-
Iron and Steel	17883	-	-	-	-	-	-	-	-	-
Chemical and Petrochem.	135432	-	-	-	-	-	-	-	-	-
Non-Metallic Minerals	5888	-	-	-	-	-	-	-	-	-
Non-specified	31841	-	-	-	-	-	-	-	2566	-
TRANSPORT SECTOR	-	-	-	-	-	-	-	-	-	-
Air	-	-	-	-	-	-	-	-	-	-
Road	-	-	-	-	-	-	-	-	-	-
Non-specified	-	-	-	-	-	-	-	-	-	-
OTHER SECTORS	-	-	-	-	-	-	-	-	**1378**	-
Agriculture	-	-	-	-	-	-	-	-	-	-
Comm. and Publ. Services	-	-	-	-	-	-	-	-	360	-
Residential	-	-	-	-	-	-	-	-	1018	-
Non-specified	-	-	-	-	-	-	-	-	-	-
NON-ENERGY USE	-	-	-	-	-	-	-	-	-	-

APPROVISIONNEMENT ET DEMANDE 1997	*Gaz (TJ)*				*En. Re. Comb. & Déchets (TJ)*				*(GWh)*	*(TJ)*
	Gaz naturel	*Usines à gaz*	*Cokeries*	*Hauts fourneaux*	*Biomasse solide*	*Gaz/Liquides tirés de biomasse*	*Déchets urbains*	*Déchets industriels*	*Electricité*	*Chaleur*
Production	317727	-	-	-	1332	-	-	-	4988	-
Imports	-	-	-	-	-	-	-	-	-	-
Exports	-	-	-	-	-	-	-	-	-	-
Intl. Marine Bunkers	-	-	-	-	-	-	-	-	-	-
Stock Changes	-	-	-	-	-	-	-	-	-	-
DOMESTIC SUPPLY	**317727**	-	-	-	**1332**	-	-	-	**4988**	-
Transfers and Stat. Diff.	1461	-	-	-	-	-	-	-	-147	-
TRANSFORMATION	**75011**	-	-	-	**1332**	-	-	-	-	-
Electricity and CHP Plants	75011	-	-	-	1332	-	-	-	-	-
Petroleum Refineries	-	-	-	-	-	-	-	-	-	-
Other Transform. Sector	-	-	-	-	-	-	-	-	-	-
ENERGY SECTOR	**38894**	-	-	-	-	-	-	-	**137**	-
DISTRIBUTION LOSSES	-	-	-	-	-	-	-	-	**401**	-
FINAL CONSUMPTION	**205283**	-	-	-	-	-	-	-	**4303**	-
INDUSTRY SECTOR	**205283**	-	-	-	-	-	-	-	**2882**	-
Iron and Steel	19216	-	-	-	-	-	-	-	-	-
Chemical and Petrochem.	145526	-	-	-	-	-	-	-	-	-
Non-Metallic Minerals	6327	-	-	-	-	-	-	-	-	-
Non-specified	34214	-	-	-	-	-	-	-	2882	-
TRANSPORT SECTOR	-	-	-	-	-	-	-	-	-	-
Air	-	-	-	-	-	-	-	-	-	-
Road	-	-	-	-	-	-	-	-	-	-
Non-specified	-	-	-	-	-	-	-	-	-	-
OTHER SECTORS	-	-	-	-	-	-	-	-	**1421**	-
Agriculture	-	-	-	-	-	-	-	-	-	-
Comm. and Publ. Services	-	-	-	-	-	-	-	-	413	-
Residential	-	-	-	-	-	-	-	-	1008	-
Non-specified	-	-	-	-	-	-	-	-	-	-
NON-ENERGY USE	-	-	-	-	-	-	-	-	-	-

Tunisia / Tunisie : 1996

SUPPLY AND CONSUMPTION	Coal / *Charbon* (1000 tonnes)							Oil / *Pétrole* (1000 tonnes)			
	Coking Coal	Other Bit. Coal	Sub-Bit. Coal	Lignite	Peat	Oven and Gas Coke	Pat. Fuel and BKB	Crude Oil	NGL	Feed-stocks	Additives
APPROVISIONNEMENT ET DEMANDE	*Charbon à coke*	*Autres charb. bit.*	*Charbon sous-bit.*	*Lignite*	*Tourbe*	*Coke de four/gaz*	*Agg./briq. de lignite*	*Pétrole brut*	*LGN*	*Produits d'aliment.*	*Additifs*
Production	-	-	-	-	-	-	-	4170	121	-	-
From Other Sources	-	-	-	-	-	-	-	-	-	-	-
Imports	-	-	-	-	-	106	-	922	-	-	-
Exports	-	-	-	-	-	-	-	-3298	-	-	-
Intl. Marine Bunkers	-	-	-	-	-	-	-	-	-	-	-
Stock Changes	-	-	-	-	-	-	-	-14	-	-	-
DOMESTIC SUPPLY	**-**	**-**	**-**	**-**	**-**	**106**	**-**	**1780**	**121**	**-**	**-**
Transfers	-	-	-	-	-	-	-	-	-121	-	-
Statistical Differences	-	-	-	-	-	-	-	35	-	-	-
TRANSFORMATION	**-**	**-**	**-**	**-**	**-**	**85**	**-**	**1815**	**-**	**-**	**-**
Electricity Plants	-	-	-	-	-	-	-	-	-	-	-
CHP Plants	-	-	-	-	-	-	-	-	-	-	-
Heat Plants	-	-	-	-	-	-	-	-	-	-	-
Blast Furnaces/Gas Works	-	-	-	-	-	85	-	-	-	-	-
Coke/Pat. Fuel/BKB Plants	-	-	-	-	-	-	-	-	-	-	-
Petroleum Refineries	-	-	-	-	-	-	-	1815	-	-	-
Petrochemical Industry	-	-	-	-	-	-	-	-	-	-	-
Liquefaction	-	-	-	-	-	-	-	-	-	-	-
Other Transform. Sector	-	-	-	-	-	-	-	-	-	-	-
ENERGY SECTOR	**-**	**-**	**-**	**-**	**-**	**-**	**-**	**-**	**-**	**-**	**-**
Coal Mines	-	-	-	-	-	-	-	-	-	-	-
Oil and Gas Extraction	-	-	-	-	-	-	-	-	-	-	-
Petroleum Refineries	-	-	-	-	-	-	-	-	-	-	-
Electr., CHP+Heat Plants	-	-	-	-	-	-	-	-	-	-	-
Pumped Storage (Elec.)	-	-	-	-	-	-	-	-	-	-	-
Other Energy Sector	-	-	-	-	-	-	-	-	-	-	-
Distribution Losses	-	-	-	-	-	-	-	-	-	-	-
FINAL CONSUMPTION	**-**	**-**	**-**	**-**	**-**	**21**	**-**	**-**	**-**	**-**	**-**
INDUSTRY SECTOR	**-**	**-**	**-**	**-**	**-**	**21**	**-**	**-**	**-**	**-**	**-**
Iron and Steel	-	-	-	-	-	21	-	-	-	-	-
Chemical and Petrochem.	-	-	-	-	-	-	-	-	-	-	-
of which: Feedstocks	-	-	-	-	-	-	-	-	-	-	-
Non-Ferrous Metals	-	-	-	-	-	-	-	-	-	-	-
Non-Metallic Minerals	-	-	-	-	-	-	-	-	-	-	-
Transport Equipment	-	-	-	-	-	-	-	-	-	-	-
Machinery	-	-	-	-	-	-	-	-	-	-	-
Mining and Quarrying	-	-	-	-	-	-	-	-	-	-	-
Food and Tobacco	-	-	-	-	-	-	-	-	-	-	-
Paper, Pulp and Print	-	-	-	-	-	-	-	-	-	-	-
Wood and Wood Products	-	-	-	-	-	-	-	-	-	-	-
Construction	-	-	-	-	-	-	-	-	-	-	-
Textile and Leather	-	-	-	-	-	-	-	-	-	-	-
Non-specified	-	-	-	-	-	-	-	-	-	-	-
TRANSPORT SECTOR	**-**	**-**	**-**	**-**	**-**	**-**	**-**	**-**	**-**	**-**	**-**
Air	-	-	-	-	-	-	-	-	-	-	-
Road	-	-	-	-	-	-	-	-	-	-	-
Rail	-	-	-	-	-	-	-	-	-	-	-
Pipeline Transport	-	-	-	-	-	-	-	-	-	-	-
Internal Navigation	-	-	-	-	-	-	-	-	-	-	-
Non-specified	-	-	-	-	-	-	-	-	-	-	-
OTHER SECTORS	**-**	**-**	**-**	**-**	**-**	**-**	**-**	**-**	**-**	**-**	**-**
Agriculture	-	-	-	-	-	-	-	-	-	-	-
Comm. and Publ. Services	-	-	-	-	-	-	-	-	-	-	-
Residential	-	-	-	-	-	-	-	-	-	-	-
Non-specified	-	-	-	-	-	-	-	-	-	-	-
NON-ENERGY USE	**-**	**-**	**-**	**-**	**-**	**-**	**-**	**-**	**-**	**-**	**-**
in Industry/Trans./Energy	-	-	-	-	-	-	-	-	-	-	-
in Transport	-	-	-	-	-	-	-	-	-	-	-
in Other Sectors	-	-	-	-	-	-	-	-	-	-	-

Tunisia / Tunisie : 1996

SUPPLY AND CONSUMPTION / *APPROVISIONNEMENT ET DEMANDE*	Oil cont. / *Pétrole cont.* (1000 tonnes)										
	Refinery Gas / *Gaz de raffinerie*	LPG + Ethane / *GPL + éthane*	Motor Gasoline / *Essence moteur*	Aviation Gasoline / *Essence aviation*	Jet Fuel / *Carbu- réacteurs*	Kerosene / *Kérosène*	Gas/ Diesel / *Gazole*	Heavy Fuel Oil / *Fioul lourd*	Naphtha / *Naphta*	Petrol. Coke / *Coke de pétrole*	Other Prod. / *Autres prod.*
Production	34	15	325	-	-	131	571	643	59	-	37
From Other Sources	-	-	-	-	-	-	-	-	-	-	-
Imports	-	197	50	-	245	37	822	838	-	-	118
Exports	-	-6	-	-	-	-	-28	-590	-51	-	-50
Intl. Marine Bunkers	-	-	-	-	-	-	-	-19	-	-	-
Stock Changes	-	-5	-51	-	4	-	-25	-17	-8	-	42
DOMESTIC SUPPLY	**34**	**201**	**324**	**-**	**249**	**168**	**1340**	**855**	**-**	**-**	**147**
Transfers	-	121	-	-	-	-	-	-	-	-	-
Statistical Differences	-	1	-	-	-	-	-1	-12	-	-	4
TRANSFORMATION	**-**	**-**	**-**	**-**	**-**	**-**	**7**	**309**	**-**	**-**	**-**
Electricity Plants	-	-	-	-	-	-	7	309	-	-	-
CHP Plants	-	-	-	-	-	-	-	-	-	-	-
Heat Plants	-	-	-	-	-	-	-	-	-	-	-
Blast Furnaces/Gas Works	-	-	-	-	-	-	-	-	-	-	-
Coke/Pat. Fuel/BKB Plants	-	-	-	-	-	-	-	-	-	-	-
Petroleum Refineries	-	-	-	-	-	-	-	-	-	-	-
Petrochemical Industry	-	-	-	-	-	-	-	-	-	-	-
Liquefaction	-	-	-	-	-	-	-	-	-	-	-
Other Transform. Sector	-	-	-	-	-	-	-	-	-	-	-
ENERGY SECTOR	**34**	**-**	**-**	**-**	**-**	**-**	**-**	**34**	**-**	**-**	**-**
Coal Mines	-	-	-	-	-	-	-	-	-	-	-
Oil and Gas Extraction	-	-	-	-	-	-	-	-	-	-	-
Petroleum Refineries	34	-	-	-	-	-	-	34	-	-	-
Electr., CHP+Heat Plants	-	-	-	-	-	-	-	-	-	-	-
Pumped Storage (Elec.)	-	-	-	-	-	-	-	-	-	-	-
Other Energy Sector	-	-	-	-	-	-	-	-	-	-	-
Distribution Losses	-	-	-	-	-	-	-	-	-	-	-
FINAL CONSUMPTION	**-**	**323**	**324**	**-**	**249**	**168**	**1332**	**500**	**-**	**-**	**151**
INDUSTRY SECTOR	**-**	**10**	**-**	**-**	**-**	**6**	**165**	**431**	**-**	**-**	**-**
Iron and Steel	-	-	-	-	-	-	-	-	-	-	-
Chemical and Petrochem.	-	-	-	-	-	-	-	-	-	-	-
of which: Feedstocks	-	-	-	-	-	-	-	-	-	-	-
Non-Ferrous Metals	-	-	-	-	-	-	-	-	-	-	-
Non-Metallic Minerals	-	-	-	-	-	-	-	-	-	-	-
Transport Equipment	-	-	-	-	-	-	-	-	-	-	-
Machinery	-	-	-	-	-	-	-	-	-	-	-
Mining and Quarrying	-	-	-	-	-	-	-	-	-	-	-
Food and Tobacco	-	-	-	-	-	-	-	-	-	-	-
Paper, Pulp and Print	-	-	-	-	-	-	-	-	-	-	-
Wood and Wood Products	-	-	-	-	-	-	-	-	-	-	-
Construction	-	-	-	-	-	-	-	-	-	-	-
Textile and Leather	-	-	-	-	-	-	-	-	-	-	-
Non-specified	-	10	-	-	-	6	165	431	-	-	-
TRANSPORT SECTOR	**-**	**2**	**324**	**-**	**249**	**-**	**714**	**18**	**-**	**-**	**-**
Air	-	-	-	-	249	-	-	-	-	-	-
Road	-	2	324	-	-	-	714	-	-	-	-
Rail	-	-	-	-	-	-	-	-	-	-	-
Pipeline Transport	-	-	-	-	-	-	-	-	-	-	-
Internal Navigation	-	-	-	-	-	-	-	-	-	-	-
Non-specified	-	-	-	-	-	-	-	18	-	-	-
OTHER SECTORS	**-**	**311**	**-**	**-**	**-**	**162**	**453**	**51**	**-**	**-**	**-**
Agriculture	-	-	-	-	-	-	247	50	-	-	-
Comm. and Publ. Services	-	49	-	-	-	8	183	1	-	-	-
Residential	-	262	-	-	-	154	23	-	-	-	-
Non-specified	-	-	-	-	-	-	-	-	-	-	-
NON-ENERGY USE	**-**	**-**	**-**	**-**	**-**	**-**	**-**	**-**	**-**	**-**	**151**
in Industry/Transf./Energy	-	-	-	-	-	-	-	-	-	-	151
in Transport	-	-	-	-	-	-	-	-	-	-	-
in Other Sectors	-	-	-	-	-	-	-	-	-	-	-

Tunisia / Tunisie : 1996

SUPPLY AND CONSUMPTION / *APPROVISIONNEMENT ET DEMANDE*	Gas / *Gaz* (TJ)				Comb. Renew. & Waste / *En. Re. Comb. & Déchets* (TJ)				(GWh)	(TJ)
	Natural Gas / *Gaz naturel*	Gas Works / *Usines à gaz*	Coke Ovens / *Cokeries*	Blast Furnaces / *Hauts fourneaux*	Solid Biomass / *Biomasse solide*	Gas/Liquids from Biomass / *Gaz/Liquides tirés de biomasse*	Municipal Waste / *Déchets urbains*	Industrial Waste / *Déchets industriels*	Electricity / *Electricité*	Heat / *Chaleur*
Production	33780	-	-	-	48360	-	-	-	7533	-
From Other Sources	-	-	-	-	-	-	-	-	-	-
Imports	56858	-	-	-	-	-	-	-	35	-
Exports	-	-	-	-	-	-	-	-	-	-
Intl. Marine Bunkers	-	-	-	-	-	-	-	-	-	-
Stock Changes	-	-	-	-	-	-	-	-	-	-
DOMESTIC SUPPLY	**90638**	-	-	-	**48360**	-	-	-	**7568**	-
Transfers	-	-	-	-	-	-	-	-	-	-
Statistical Differences	2652	-	-	-	-1	-	-	-	141	-
TRANSFORMATION	**71375**	-	-	-	**7122**	-	-	-	-	-
Electricity Plants	71375	-	-	-	-	-	-	-	-	-
CHP Plants	-	-	-	-	-	-	-	-	-	-
Heat Plants	-	-	-	-	-	-	-	-	-	-
Blast Furnaces/Gas Works	-	-	-	-	-	-	-	-	-	-
Coke/Pat. Fuel/BKB Plants	-	-	-	-	-	-	-	-	-	-
Petroleum Refineries	-	-	-	-	-	-	-	-	-	-
Petrochemical Industry	-	-	-	-	-	-	-	-	-	-
Liquefaction	-	-	-	-	-	-	-	-	-	-
Other Transform. Sector	-	-	-	-	7122	-	-	-	-	-
ENERGY SECTOR	-	-	-	-	-	-	-	-	**786**	-
Coal Mines	-	-	-	-	-	-	-	-	-	-
Oil and Gas Extraction	-	-	-	-	-	-	-	-	-	-
Petroleum Refineries	-	-	-	-	-	-	-	-	-	-
Electr., CHP+Heat Plants	-	-	-	-	-	-	-	-	786	-
Pumped Storage (Elec.)	-	-	-	-	-	-	-	-	-	-
Other Energy Sector	-	-	-	-	-	-	-	-	-	-
Distribution Losses	-	-	-	-	-	-	-	-	801	-
FINAL CONSUMPTION	**21915**	-	-	-	**41237**	-	-	-	**6122**	-
INDUSTRY SECTOR	**16797**	-	-	-	-	-	-	-	**2805**	-
Iron and Steel	-	-	-	-	-	-	-	-	176	-
Chemical and Petrochem.	-	-	-	-	-	-	-	-	139	-
of which: Feedstocks	-	-	-	-	-	-	-	-	-	-
Non-Ferrous Metals	-	-	-	-	-	-	-	-	-	-
Non-Metallic Minerals	-	-	-	-	-	-	-	-	886	-
Transport Equipment	-	-	-	-	-	-	-	-	-	-
Machinery	-	-	-	-	-	-	-	-	-	-
Mining and Quarrying	-	-	-	-	-	-	-	-	251	-
Food and Tobacco	-	-	-	-	-	-	-	-	303	-
Paper, Pulp and Print	-	-	-	-	-	-	-	-	102	-
Wood and Wood Products	-	-	-	-	-	-	-	-	-	-
Construction	-	-	-	-	-	-	-	-	-	-
Textile and Leather	-	-	-	-	-	-	-	-	328	-
Non-specified	16797	-	-	-	-	-	-	-	620	-
TRANSPORT SECTOR	-	-	-	-	-	-	-	-	**148**	-
Air	-	-	-	-	-	-	-	-	-	-
Road	-	-	-	-	-	-	-	-	-	-
Rail	-	-	-	-	-	-	-	-	-	-
Pipeline Transport	-	-	-	-	-	-	-	-	-	-
Internal Navigation	-	-	-	-	-	-	-	-	-	-
Non-specified	-	-	-	-	-	-	-	-	148	-
OTHER SECTORS	**5118**	-	-	-	**41237**	-	-	-	**3169**	-
Agriculture	-	-	-	-	-	-	-	-	374	-
Comm. and Publ. Services	1489	-	-	-	-	-	-	-	1125	-
Residential	3629	-	-	-	41237	-	-	-	1670	-
Non-specified	-	-	-	-	-	-	-	-	-	-
NON-ENERGY USE	-	-	-	-	-	-	-	-	-	-
in Industry/Transf./Energy	-	-	-	-	-	-	-	-	-	-
in Transport	-	-	-	-	-	-	-	-	-	-
in Other Sectors	-	-	-	-	-	-	-	-	-	-

Tunisia / Tunisie : 1997

SUPPLY AND CONSUMPTION / *APPROVISIONNEMENT ET DEMANDE*	Coal / *Charbon* (1000 tonnes)							Oil / *Pétrole* (1000 tonnes)			
	Coking Coal / *Charbon à coke*	Other Bit. Coal / *Autres charb. bit.*	Sub-Bit. Coal / *Charbon sous-bit.*	Lignite / *Lignite*	Peat / *Tourbe*	Oven and Gas Coke / *Coke de four/gaz*	Pat. Fuel and BKB / *Agg./briq. de lignite*	Crude Oil / *Pétrole brut*	NGL / *LGN*	Feed-stocks / *Produits d'aliment.*	Additives / *Additifs*
Production	-	-	-	-	-	-	-	3789	107	-	-
From Other Sources	-	-	-	-	-	-	-	-	-	-	-
Imports	-	-	-	-	-	106	-	915	-	-	-
Exports	-	-	-	-	-	-	-	-2780	-	-	-
Intl. Marine Bunkers	-	-	-	-	-	-	-	-	-	-	-
Stock Changes	-	-	-	-	-	-	-	-116	-	-	-
DOMESTIC SUPPLY	**-**	**-**	**-**	**-**	**-**	**106**	**-**	**1808**	**107**	**-**	**-**
Transfers	-	-	-	-	-	-	-	-	-107	-	-
Statistical Differences	-	-	-	-	-	-	-	201	-	-	-
TRANSFORMATION	**-**	**-**	**-**	**-**	**-**	**85**	**-**	**2009**	**-**	**-**	**-**
Electricity Plants	-	-	-	-	-	-	-	-	-	-	-
CHP Plants	-	-	-	-	-	-	-	-	-	-	-
Heat Plants	-	-	-	-	-	-	-	-	-	-	-
Blast Furnaces/Gas Works	-	-	-	-	-	85	-	-	-	-	-
Coke/Pat. Fuel/BKB Plants	-	-	-	-	-	-	-	-	-	-	-
Petroleum Refineries	-	-	-	-	-	-	-	2009	-	-	-
Petrochemical Industry	-	-	-	-	-	-	-	-	-	-	-
Liquefaction	-	-	-	-	-	-	-	-	-	-	-
Other Transform. Sector	-	-	-	-	-	-	-	-	-	-	-
ENERGY SECTOR	**-**	**-**	**-**	**-**	**-**	**-**	**-**	**-**	**-**	**-**	**-**
Coal Mines	-	-	-	-	-	-	-	-	-	-	-
Oil and Gas Extraction	-	-	-	-	-	-	-	-	-	-	-
Petroleum Refineries	-	-	-	-	-	-	-	-	-	-	-
Electr., CHP+Heat Plants	-	-	-	-	-	-	-	-	-	-	-
Pumped Storage (Elec.)	-	-	-	-	-	-	-	-	-	-	-
Other Energy Sector	-	-	-	-	-	-	-	-	-	-	-
Distribution Losses	-	-	-	-	-	-	-	-	-	-	-
FINAL CONSUMPTION	**-**	**-**	**-**	**-**	**-**	**21**	**-**	**-**	**-**	**-**	**-**
INDUSTRY SECTOR	**-**	**-**	**-**	**-**	**-**	**21**	**-**	**-**	**-**	**-**	**-**
Iron and Steel	-	-	-	-	-	21	-	-	-	-	-
Chemical and Petrochem.	-	-	-	-	-	-	-	-	-	-	-
of which: Feedstocks	-	-	-	-	-	-	-	-	-	-	-
Non-Ferrous Metals	-	-	-	-	-	-	-	-	-	-	-
Non-Metallic Minerals	-	-	-	-	-	-	-	-	-	-	-
Transport Equipment	-	-	-	-	-	-	-	-	-	-	-
Machinery	-	-	-	-	-	-	-	-	-	-	-
Mining and Quarrying	-	-	-	-	-	-	-	-	-	-	-
Food and Tobacco	-	-	-	-	-	-	-	-	-	-	-
Paper, Pulp and Print	-	-	-	-	-	-	-	-	-	-	-
Wood and Wood Products	-	-	-	-	-	-	-	-	-	-	-
Construction	-	-	-	-	-	-	-	-	-	-	-
Textile and Leather	-	-	-	-	-	-	-	-	-	-	-
Non-specified	-	-	-	-	-	-	-	-	-	-	-
TRANSPORT SECTOR	**-**	**-**	**-**	**-**	**-**	**-**	**-**	**-**	**-**	**-**	**-**
Air	-	-	-	-	-	-	-	-	-	-	-
Road	-	-	-	-	-	-	-	-	-	-	-
Rail	-	-	-	-	-	-	-	-	-	-	-
Pipeline Transport	-	-	-	-	-	-	-	-	-	-	-
Internal Navigation	-	-	-	-	-	-	-	-	-	-	-
Non-specified	-	-	-	-	-	-	-	-	-	-	-
OTHER SECTORS	**-**	**-**	**-**	**-**	**-**	**-**	**-**	**-**	**-**	**-**	**-**
Agriculture	-	-	-	-	-	-	-	-	-	-	-
Comm. and Publ. Services	-	-	-	-	-	-	-	-	-	-	-
Residential	-	-	-	-	-	-	-	-	-	-	-
Non-specified	-	-	-	-	-	-	-	-	-	-	-
NON-ENERGY USE	**-**	**-**	**-**	**-**	**-**	**-**	**-**	**-**	**-**	**-**	**-**
in Industry/Trans./Energy	-	-	-	-	-	-	-	-	-	-	-
in Transport	-	-	-	-	-	-	-	-	-	-	-
in Other Sectors	-	-	-	-	-	-	-	-	-	-	-

Tunisia / Tunisie : 1997

SUPPLY AND CONSUMPTION / *APPROVISIONNEMENT ET DEMANDE*	Oil cont. / *Pétrole cont.* (1000 tonnes)										
	Refinery Gas / *Gaz de raffinerie*	LPG + Ethane / *GPL + éthane*	Motor Gasoline / *Essence moteur*	Aviation Gasoline / *Essence aviation*	Jet Fuel / *Carbu- réacteurs*	Kerosene / *Kérosène*	Gas/ Diesel / *Gazole*	Heavy Fuel Oil / *Fioul lourd*	Naphtha / *Naphta*	Petrol. Coke / *Coke de pétrole*	Other Prod. / *Autres prod.*
Production	34	13	339	-	-	107	595	646	234	-	41
From Other Sources	-	-	-	-	-	-	-	-	-	-	-
Imports	-	226	201	-	292	64	939	684	-	-	129
Exports	-	-4	-	-	-	-	-	-565	-224	-	-35
Intl. Marine Bunkers	-	-	-	-	-	-	-	-18	-	-	-
Stock Changes	-	-	-200	-	-9	8	-100	61	-11	-	8
DOMESTIC SUPPLY	**34**	**235**	**340**	**-**	**283**	**179**	**1434**	**808**	**-1**	**-**	**143**
Transfers	-	107	-	-	-	-	-	-	-	-	-
Statistical Differences	-	-	-	-	-	-1	-	-1	1	-	-
TRANSFORMATION	**-**	**-**	**-**	**-**	**-**	**-**	**11**	**360**	**-**	**-**	**-**
Electricity Plants	-	-	-	-	-	-	11	360	-	-	-
CHP Plants	-	-	-	-	-	-	-	-	-	-	-
Heat Plants	-	-	-	-	-	-	-	-	-	-	-
Blast Furnaces/Gas Works	-	-	-	-	-	-	-	-	-	-	-
Coke/Pat. Fuel/BKB Plants	-	-	-	-	-	-	-	-	-	-	-
Petroleum Refineries	-	-	-	-	-	-	-	-	-	-	-
Petrochemical Industry	-	-	-	-	-	-	-	-	-	-	-
Liquefaction	-	-	-	-	-	-	-	-	-	-	-
Other Transform. Sector	-	-	-	-	-	-	-	-	-	-	-
ENERGY SECTOR	**34**	**-**	**-**	**-**	**-**	**-**	**-**	**34**	**-**	**-**	**-**
Coal Mines	-	-	-	-	-	-	-	-	-	-	-
Oil and Gas Extraction	-	-	-	-	-	-	-	-	-	-	-
Petroleum Refineries	34	-	-	-	-	-	-	34	-	-	-
Electr., CHP+Heat Plants	-	-	-	-	-	-	-	-	-	-	-
Pumped Storage (Elec.)	-	-	-	-	-	-	-	-	-	-	-
Other Energy Sector	-	-	-	-	-	-	-	-	-	-	-
Distribution Losses	-	-	-	-	-	-	-	-	-	-	-
FINAL CONSUMPTION	**-**	**342**	**340**	**-**	**283**	**178**	**1423**	**413**	**-**	**-**	**143**
INDUSTRY SECTOR	**-**	**11**	**-**	**-**	**-**	**6**	**181**	**383**	**-**	**-**	**-**
Iron and Steel	-	-	-	-	-	-	-	-	-	-	-
Chemical and Petrochem.	-	-	-	-	-	-	-	-	-	-	-
of which: Feedstocks	-	-	-	-	-	-	-	-	-	-	-
Non-Ferrous Metals	-	-	-	-	-	-	-	-	-	-	-
Non-Metallic Minerals	-	-	-	-	-	-	-	-	-	-	-
Transport Equipment	-	-	-	-	-	-	-	-	-	-	-
Machinery	-	-	-	-	-	-	-	-	-	-	-
Mining and Quarrying	-	-	-	-	-	-	-	-	-	-	-
Food and Tobacco	-	-	-	-	-	-	-	-	-	-	-
Paper, Pulp and Print	-	-	-	-	-	-	-	-	-	-	-
Wood and Wood Products	-	-	-	-	-	-	-	-	-	-	-
Construction	-	-	-	-	-	-	-	-	-	-	-
Textile and Leather	-	-	-	-	-	-	-	-	-	-	-
Non-specified	-	11	-	-	-	6	181	383	-	-	-
TRANSPORT SECTOR	**-**	**6**	**340**	**-**	**283**	**-**	**766**	**6**	**-**	**-**	**-**
Air	-	-	-	-	283	-	-	-	-	-	-
Road	-	6	340	-	-	-	766	-	-	-	-
Rail	-	-	-	-	-	-	-	-	-	-	-
Pipeline Transport	-	-	-	-	-	-	-	-	-	-	-
Internal Navigation	-	-	-	-	-	-	-	-	-	-	-
Non-specified	-	-	-	-	-	-	-	6	-	-	-
OTHER SECTORS	**-**	**325**	**-**	**-**	**-**	**172**	**476**	**24**	**-**	**-**	**-**
Agriculture	-	-	-	-	-	-	256	23	-	-	-
Comm. and Publ. Services	-	49	-	-	-	9	195	1	-	-	-
Residential	-	276	-	-	-	163	25	-	-	-	-
Non-specified	-	-	-	-	-	-	-	-	-	-	-
NON-ENERGY USE	**-**	**-**	**-**	**-**	**-**	**-**	**-**	**-**	**-**	**-**	**143**
in Industry/Transf./Energy	-	-	-	-	-	-	-	-	-	-	143
in Transport	-	-	-	-	-	-	-	-	-	-	-
in Other Sectors	-	-	-	-	-	-	-	-	-	-	-

Tunisia / Tunisie : 1997

SUPPLY AND CONSUMPTION / *APPROVISIONNEMENT ET DEMANDE*	Gas / *Gaz* (TJ)				Comb. Renew. & Waste / *En. Re. Comb. & Déchets* (TJ)				(GWh)	(TJ)
	Natural Gas / *Gaz naturel*	Gas Works / *Usines à gaz*	Coke Ovens / *Cokeries*	Blast Furnaces / *Hauts fourneaux*	Solid Biomass / *Biomasse solide*	Gas/Liquids from Biomass / *Gaz/Liquides tirés de biomasse*	Municipal Waste / *Déchets urbains*	Industrial Waste / *Déchets industriels*	Electricity / *Electricité*	Heat / *Chaleur*
Production	68211	-	-	-	49085	-	-	-	7977	-
From Other Sources	-	-	-	-	-	-	-	-	-	-
Imports	28894	-	-	-	-	-	-	-	35	-
Exports	-	-	-	-	-	-	-	-	-	-
Intl. Marine Bunkers	-	-	-	-	-	-	-	-	-	-
Stock Changes	-	-	-	-	-	-	-	-	-	-
DOMESTIC SUPPLY	**97105**	**-**	**-**	**-**	**49085**	**-**	**-**	**-**	**8012**	**-**
Transfers	-	-	-	-	-	-	-	-	-	-
Statistical Differences	-139	-	-	-	-1	-	-	-	308	-
TRANSFORMATION	**74493**	**-**	**-**	**-**	**7229**	**-**	**-**	**-**	**-**	**-**
Electricity Plants	74493	-	-	-	-	-	-	-	-	-
CHP Plants	-	-	-	-	-	-	-	-	-	-
Heat Plants	-	-	-	-	-	-	-	-	-	-
Blast Furnaces/Gas Works	-	-	-	-	-	-	-	-	-	-
Coke/Pat. Fuel/BKB Plants	-	-	-	-	-	-	-	-	-	-
Petroleum Refineries	-	-	-	-	-	-	-	-	-	-
Petrochemical Industry	-	-	-	-	-	-	-	-	-	-
Liquefaction	-	-	-	-	-	-	-	-	-	-
Other Transform. Sector	-	-	-	-	7229	-	-	-	-	-
ENERGY SECTOR	**-**	**-**	**-**	**-**	**-**	**-**	**-**	**-**	**903**	**-**
Coal Mines	-	-	-	-	-	-	-	-	-	-
Oil and Gas Extraction	-	-	-	-	-	-	-	-	-	-
Petroleum Refineries	-	-	-	-	-	-	-	-	-	-
Electr., CHP+Heat Plants	-	-	-	-	-	-	-	-	903	-
Pumped Storage (Elec.)	-	-	-	-	-	-	-	-	-	-
Other Energy Sector	-	-	-	-	-	-	-	-	-	-
Distribution Losses	-	-	-	-	-	-	-	-	882	-
FINAL CONSUMPTION	**22473**	**-**	**-**	**-**	**41855**	**-**	**-**	**-**	**6535**	**-**
INDUSTRY SECTOR	**17122**	**-**	**-**	**-**	**-**	**-**	**-**	**-**	**2906**	**-**
Iron and Steel	-	-	-	-	-	-	-	-	182	-
Chemical and Petrochem.	-	-	-	-	-	-	-	-	144	-
of which: Feedstocks	-	-	-	-	-	-	-	-	-	-
Non-Ferrous Metals	-	-	-	-	-	-	-	-	-	-
Non-Metallic Minerals	-	-	-	-	-	-	-	-	918	-
Transport Equipment	-	-	-	-	-	-	-	-	-	-
Machinery	-	-	-	-	-	-	-	-	-	-
Mining and Quarrying	-	-	-	-	-	-	-	-	260	-
Food and Tobacco	-	-	-	-	-	-	-	-	314	-
Paper, Pulp and Print	-	-	-	-	-	-	-	-	106	-
Wood and Wood Products	-	-	-	-	-	-	-	-	-	-
Construction	-	-	-	-	-	-	-	-	-	-
Textile and Leather	-	-	-	-	-	-	-	-	340	-
Non-specified	17122	-	-	-	-	-	-	-	642	-
TRANSPORT SECTOR	**-**	**-**	**-**	**-**	**-**	**-**	**-**	**-**	**148**	**-**
Air	-	-	-	-	-	-	-	-	-	-
Road	-	-	-	-	-	-	-	-	-	-
Rail	-	-	-	-	-	-	-	-	-	-
Pipeline Transport	-	-	-	-	-	-	-	-	-	-
Internal Navigation	-	-	-	-	-	-	-	-	-	-
Non-specified	-	-	-	-	-	-	-	-	148	-
OTHER SECTORS	**5351**	**-**	**-**	**-**	**41855**	**-**	**-**	**-**	**3481**	**-**
Agriculture	-	-	-	-	-	-	-	-	472	-
Comm. and Publ. Services	1582	-	-	-	-	-	-	-	1213	-
Residential	3769	-	-	-	41855	-	-	-	1796	-
Non-specified	-	-	-	-	-	-	-	-	-	-
NON-ENERGY USE	**-**	**-**	**-**	**-**	**-**	**-**	**-**	**-**	**-**	**-**
in Industry/Transf./Energy	-	-	-	-	-	-	-	-	-	-
in Transport	-	-	-	-	-	-	-	-	-	-
in Other Sectors	-	-	-	-	-	-	-	-	-	-

Turkmenistan / Turkménistan

SUPPLY AND CONSUMPTION 1996	Coal (1000 tonnes)							Oil (1000 tonnes)			
	Coking Coal	Other Bit. Coal	Sub-Bit. Coal	Lignite	Peat	Oven and Gas Coke	Pat. Fuel and BKB	Crude Oil	NGL	Feed-stocks	Additives
Production	-	-	-	-	-	-	-	4025	307	-	-
Imports	-	100	-	-	-	-	-	100	-	-	-
Exports	-	-	-	-	-	-	-	-200	-	-	-
Intl. Marine Bunkers	-	-	-	-	-	-	-	-	-	-	-
Stock Changes	-	-	-	-	-	-	-	-	-	-	-
DOMESTIC SUPPLY	**-**	**100**	**-**	**-**	**-**	**-**	**-**	**3925**	**307**	**-**	**-**
Transfers and Stat. Diff.	-	-	-	-	-	-	-	-	-	-	-
TRANSFORMATION	**-**	**-**	**-**	**-**	**-**	**-**	**-**	**3925**	**307**	**-**	**-**
Electricity and CHP Plants	-	-	-	-	-	-	-	-	-	-	-
Petroleum Refineries	-	-	-	-	-	-	-	3925	-	-	-
Other Transform. Sector	-	-	-	-	-	-	-	-	307	-	-
ENERGY SECTOR	**-**	**-**	**-**	**-**	**-**	**-**	**-**	**-**	**-**	**-**	**-**
DISTRIBUTION LOSSES	**-**	**-**	**-**	**-**	**-**	**-**	**-**	**-**	**-**	**-**	**-**
FINAL CONSUMPTION	**-**	**100**	**-**	**-**	**-**	**-**	**-**	**-**	**-**	**-**	**-**
INDUSTRY SECTOR	**-**	**-**	**-**	**-**	**-**	**-**	**-**	**-**	**-**	**-**	**-**
Iron and Steel	-	-	-	-	-	-	-	-	-	-	-
Chemical and Petrochem.	-	-	-	-	-	-	-	-	-	-	-
Non-Metallic Minerals	-	-	-	-	-	-	-	-	-	-	-
Non-specified	-	-	-	-	-	-	-	-	-	-	-
TRANSPORT SECTOR	**-**	**-**	**-**	**-**	**-**	**-**	**-**	**-**	**-**	**-**	**-**
Air	-	-	-	-	-	-	-	-	-	-	-
Road	-	-	-	-	-	-	-	-	-	-	-
Non-specified	-	-	-	-	-	-	-	-	-	-	-
OTHER SECTORS	**-**	**100**	**-**	**-**	**-**	**-**	**-**	**-**	**-**	**-**	**-**
Agriculture	-	-	-	-	-	-	-	-	-	-	-
Comm. and Publ. Services	-	-	-	-	-	-	-	-	-	-	-
Residential	-	-	-	-	-	-	-	-	-	-	-
Non-specified	-	100	-	-	-	-	-	-	-	-	-
NON-ENERGY USE	**-**	**-**	**-**	**-**	**-**	**-**	**-**	**-**	**-**	**-**	**-**

APPROVISIONNEMENT ET DEMANDE 1997	*Charbon* (1000 tonnes)							*Pétrole* (1000 tonnes)			
	Charbon à coke	*Autres charb. bit.*	*Charbon sous-bit.*	*Lignite*	*Tourbe*	*Coke de four/gaz*	*Agg./briq. de lignite*	*Pétrole brut*	*LGN*	*Produits d'aliment.*	*Additifs*
Production	-	-	-	-	-	-	-	4481	219	-	-
Imports	-	-	-	-	-	-	-	700	-	-	-
Exports	-	-	-	-	-	-	-	-200	-	-	-
Intl. Marine Bunkers	-	-	-	-	-	-	-	-	-	-	-
Stock Changes	-	-	-	-	-	-	-	-	-	-	-
DOMESTIC SUPPLY	**-**	**-**	**-**	**-**	**-**	**-**	**-**	**4981**	**219**	**-**	**-**
Transfers and Stat. Diff.	-	-	-	-	-	-	-	-	-	-	-
TRANSFORMATION	**-**	**-**	**-**	**-**	**-**	**-**	**-**	**4981**	**219**	**-**	**-**
Electricity and CHP Plants	-	-	-	-	-	-	-	-	-	-	-
Petroleum Refineries	-	-	-	-	-	-	-	4981	-	-	-
Other Transform. Sector	-	-	-	-	-	-	-	-	219	-	-
ENERGY SECTOR	**-**	**-**	**-**	**-**	**-**	**-**	**-**	**-**	**-**	**-**	**-**
DISTRIBUTION LOSSES	**-**	**-**	**-**	**-**	**-**	**-**	**-**	**-**	**-**	**-**	**-**
FINAL CONSUMPTION	**-**	**-**	**-**	**-**	**-**	**-**	**-**	**-**	**-**	**-**	**-**
INDUSTRY SECTOR	**-**	**-**	**-**	**-**	**-**	**-**	**-**	**-**	**-**	**-**	**-**
Iron and Steel	-	-	-	-	-	-	-	-	-	-	-
Chemical and Petrochem.	-	-	-	-	-	-	-	-	-	-	-
Non-Metallic Minerals	-	-	-	-	-	-	-	-	-	-	-
Non-specified	-	-	-	-	-	-	-	-	-	-	-
TRANSPORT SECTOR	**-**	**-**	**-**	**-**	**-**	**-**	**-**	**-**	**-**	**-**	**-**
Air	-	-	-	-	-	-	-	-	-	-	-
Road	-	-	-	-	-	-	-	-	-	-	-
Non-specified	-	-	-	-	-	-	-	-	-	-	-
OTHER SECTORS	**-**	**-**	**-**	**-**	**-**	**-**	**-**	**-**	**-**	**-**	**-**
Agriculture	-	-	-	-	-	-	-	-	-	-	-
Comm. and Publ. Services	-	-	-	-	-	-	-	-	-	-	-
Residential	-	-	-	-	-	-	-	-	-	-	-
Non-specified	-	-	-	-	-	-	-	-	-	-	-
NON-ENERGY USE	**-**	**-**	**-**	**-**	**-**	**-**	**-**	**-**	**-**	**-**	**-**

Turkmenistan / Turkménistan

SUPPLY AND CONSUMPTION 1996	Oil cont. (1000 tonnes)										
	Refinery Gas	LPG + Ethane	Motor Gasoline	Aviation Gasoline	Jet Fuel	Kerosene	Gas/ Diesel	Heavy Fuel Oil	Naphtha	Petrol. Coke	Other Prod.
Production	193	-	677	-	-	-	1253	1301	-	-	-
Imports	-	72	-	-	-	-	-	-	-	-	-
Exports	-	-	-194	-	-	-	-329	-485	-	-	-
Intl. Marine Bunkers	-	-	-	-	-	-	-	-	-	-	-
Stock Changes	-	-	-	-	-	-	-	-	-	-	-
DOMESTIC SUPPLY	**193**	**72**	**483**	-	-	-	**924**	**816**	-	-	-
Transfers and Stat. Diff.	-	-	-	-	-	-	-	-	-	-	-
TRANSFORMATION	-	-	-	-	-	-	-	-	-	-	-
Electricity and CHP Plants	-	-	-	-	-	-	-	-	-	-	-
Petroleum Refineries	-	-	-	-	-	-	-	-	-	-	-
Other Transform. Sector	-	-	-	-	-	-	-	-	-	-	-
ENERGY SECTOR	**193**	-	-	-	-	-	-	-	-	-	-
DISTRIBUTION LOSSES	-	-	-	-	-	-	-	-	-	-	-
FINAL CONSUMPTION	-	**72**	**483**	-	-	-	**924**	**816**	-	-	-
INDUSTRY SECTOR	-	-	-	-	-	-	-	-	-	-	-
Iron and Steel	-	-	-	-	-	-	-	-	-	-	-
Chemical and Petrochem.	-	-	-	-	-	-	-	-	-	-	-
Non-Metallic Minerals	-	-	-	-	-	-	-	-	-	-	-
Non-specified	-	-	-	-	-	-	-	-	-	-	-
TRANSPORT SECTOR	-	-	**483**	-	-	-	-	-	-	-	-
Air	-	-	-	-	-	-	-	-	-	-	-
Road	-	-	483	-	-	-	-	-	-	-	-
Non-specified	-	-	-	-	-	-	-	-	-	-	-
OTHER SECTORS	-	**72**	-	-	-	-	**924**	**816**	-	-	-
Agriculture	-	-	-	-	-	-	-	-	-	-	-
Comm. and Publ. Services	-	-	-	-	-	-	-	-	-	-	-
Residential	-	-	-	-	-	-	-	-	-	-	-
Non-specified	-	72	-	-	-	-	924	816	-	-	-
NON-ENERGY USE	-	-	-	-	-	-	-	-	-	-	-

APPROVISIONNEMENT ET DEMANDE 1997	*Pétrole cont.* (1000 tonnes)										
	Gaz de raffinerie	*GPL + éthane*	*Essence moteur*	*Essence aviation*	*Carbu-réacteurs*	*Kérosène*	*Gazole*	*Fioul lourd*	*Naphta*	*Coke de pétrole*	*Autres prod.*
Production	250	-	766	-	-	-	1520	1600	-	-	-
Imports	-	53	-	-	-	-	-	-	-	-	-
Exports	-	-	-323	-	-	-	-642	-642	-	-	-
Intl. Marine Bunkers	-	-	-	-	-	-	-	-	-	-	-
Stock Changes	-	-	-	-	-	-	-	-	-	-	-
DOMESTIC SUPPLY	**250**	**53**	**443**	-	-	-	**878**	**958**	-	-	-
Transfers and Stat. Diff.	-	-	-	-	-	-	-	-	-	-	-
TRANSFORMATION	-	-	-	-	-	-	-	-	-	-	-
Electricity and CHP Plants	-	-	-	-	-	-	-	-	-	-	-
Petroleum Refineries	-	-	-	-	-	-	-	-	-	-	-
Other Transform. Sector	-	-	-	-	-	-	-	-	-	-	-
ENERGY SECTOR	**250**	-	-	-	-	-	-	-	-	-	-
DISTRIBUTION LOSSES	-	-	-	-	-	-	-	-	-	-	-
FINAL CONSUMPTION	-	**53**	**443**	-	-	-	**878**	**958**	-	-	-
INDUSTRY SECTOR	-	-	-	-	-	-	-	-	-	-	-
Iron and Steel	-	-	-	-	-	-	-	-	-	-	-
Chemical and Petrochem.	-	-	-	-	-	-	-	-	-	-	-
Non-Metallic Minerals	-	-	-	-	-	-	-	-	-	-	-
Non-specified	-	-	-	-	-	-	-	-	-	-	-
TRANSPORT SECTOR	-	-	**443**	-	-	-	-	-	-	-	-
Air	-	-	-	-	-	-	-	-	-	-	-
Road	-	-	443	-	-	-	-	-	-	-	-
Non-specified	-	-	-	-	-	-	-	-	-	-	-
OTHER SECTORS	-	**53**	-	-	-	-	**878**	**958**	-	-	-
Agriculture	-	-	-	-	-	-	-	-	-	-	-
Comm. and Publ. Services	-	-	-	-	-	-	-	-	-	-	-
Residential	-	-	-	-	-	-	-	-	-	-	-
Non-specified	-	53	-	-	-	-	878	958	-	-	-
NON-ENERGY USE	-	-	-	-	-	-	-	-	-	-	-

Turkmenistan / Turkménistan

SUPPLY AND CONSUMPTION 1996	Gas (TJ)				Comb. Renew. & Waste (TJ)				(GWh)	(TJ)
	Natural Gas	Gas Works	Coke Ovens	Blast Furnaces	Solid Biomass	Gas/Liquids from Biomass	Municipal Waste	Industrial Waste	Electricity	Heat
Production	1326361	11580	-	-	-	-	-	-	10100	-
Imports	-	-	-	-	-	-	-	-	950	-
Exports	-904800	-	-	-	-	-	-	-	-3750	-
Intl. Marine Bunkers	-	-	-	-	-	-	-	-	-	-
Stock Changes	-	-	-	-	-	-	-	-	-	-
DOMESTIC SUPPLY	**421561**	**11580**	**-**	**-**	**-**	**-**	**-**	**-**	**7300**	**-**
Transfers and Stat. Diff.	-	-	-	-	-	-	-	-	-	-
TRANSFORMATION	**146878**	**-**	**-**	**-**	**-**	**-**	**-**	**-**	**-**	**-**
Electricity and CHP Plants	146878	-	-	-	-	-	-	-	-	-
Petroleum Refineries	-	-	-	-	-	-	-	-	-	-
Other Transform. Sector	-	-	-	-	-	-	-	-	-	-
ENERGY SECTOR	**40623**	**-**	**-**	**-**	**-**	**-**	**-**	**-**	**1478**	**-**
DISTRIBUTION LOSSES	**-**	**-**	**-**	**-**	**-**	**-**	**-**	**-**	**1131**	**-**
FINAL CONSUMPTION	**234060**	**11580**	**-**	**-**	**-**	**-**	**-**	**-**	**4691**	**-**
INDUSTRY SECTOR	**-**	**-**	**-**	**-**	**-**	**-**	**-**	**-**	**1695**	**-**
Iron and Steel	-	-	-	-	-	-	-	-	1	-
Chemical and Petrochem.	-	-	-	-	-	-	-	-	549	-
Non-Metallic Minerals	-	-	-	-	-	-	-	-	-	-
Non-specified	-	-	-	-	-	-	-	-	1145	-
TRANSPORT SECTOR	**-**	**-**	**-**	**-**	**-**	**-**	**-**	**-**	**121**	**-**
Air	-	-	-	-	-	-	-	-	-	-
Road	-	-	-	-	-	-	-	-	-	-
Non-specified	-	-	-	-	-	-	-	-	121	-
OTHER SECTORS	**234060**	**11580**	**-**	**-**	**-**	**-**	**-**	**-**	**2875**	**-**
Agriculture	-	-	-	-	-	-	-	-	1489	-
Comm. and Publ. Services	-	-	-	-	-	-	-	-	-	-
Residential	-	-	-	-	-	-	-	-	986	-
Non-specified	234060	11580	-	-	-	-	-	-	400	-
NON-ENERGY USE	**-**	**-**	**-**	**-**	**-**	**-**	**-**	**-**	**-**	**-**

APPROVISIONNEMENT ET DEMANDE 1997	*Gaz* (TJ)				*En. Re. Comb. & Déchets* (TJ)				(GWh)	(TJ)
	Gaz naturel	*Usines à gaz*	*Cokeries*	*Hauts fourneaux*	*Biomasse solide*	*Gaz/Liquides tirés de biomasse*	*Déchets urbains*	*Déchets industriels*	*Electricité*	*Chaleur*
Production	652210	8260	-	-	-	-	-	-	9400	-
Imports	-	-	-	-	-	-	-	-	950	-
Exports	-245050	-	-	-	-	-	-	-	-3600	-
Intl. Marine Bunkers	-	-	-	-	-	-	-	-	-	-
Stock Changes	-	-	-	-	-	-	-	-	-	-
DOMESTIC SUPPLY	**407160**	**8260**	**-**	**-**	**-**	**-**	**-**	**-**	**6750**	**-**
Transfers and Stat. Diff.	-	-	-	-	-	-	-	-	-	-
TRANSFORMATION	**141752**	**-**	**-**	**-**	**-**	**-**	**-**	**-**	**-**	**-**
Electricity and CHP Plants	141752	-	-	-	-	-	-	-	-	-
Petroleum Refineries	-	-	-	-	-	-	-	-	-	-
Other Transform. Sector	-	-	-	-	-	-	-	-	-	-
ENERGY SECTOR	**39208**	**-**	**-**	**-**	**-**	**-**	**-**	**-**	**1350**	**-**
DISTRIBUTION LOSSES	**-**	**-**	**-**	**-**	**-**	**-**	**-**	**-**	**1050**	**-**
FINAL CONSUMPTION	**226200**	**8260**	**-**	**-**	**-**	**-**	**-**	**-**	**4350**	**-**
INDUSTRY SECTOR	**-**	**-**	**-**	**-**	**-**	**-**	**-**	**-**	**1570**	**-**
Iron and Steel	-	-	-	-	-	-	-	-	1	-
Chemical and Petrochem.	-	-	-	-	-	-	-	-	509	-
Non-Metallic Minerals	-	-	-	-	-	-	-	-	-	-
Non-specified	-	-	-	-	-	-	-	-	1060	-
TRANSPORT SECTOR	**-**	**-**	**-**	**-**	**-**	**-**	**-**	**-**	**110**	**-**
Air	-	-	-	-	-	-	-	-	-	-
Road	-	-	-	-	-	-	-	-	-	-
Non-specified	-	-	-	-	-	-	-	-	110	-
OTHER SECTORS	**226200**	**8260**	**-**	**-**	**-**	**-**	**-**	**-**	**2670**	**-**
Agriculture	-	-	-	-	-	-	-	-	1385	-
Comm. and Publ. Services	-	-	-	-	-	-	-	-	-	-
Residential	-	-	-	-	-	-	-	-	915	-
Non-specified	226200	8260	-	-	-	-	-	-	370	-
NON-ENERGY USE	**-**	**-**	**-**	**-**	**-**	**-**	**-**	**-**	**-**	**-**

Ukraine : 1996

SUPPLY AND CONSUMPTION / *APPROVISIONNEMENT ET DEMANDE*	Coal / *Charbon* (1000 tonnes)							Oil / *Pétrole* (1000 tonnes)			
	Coking Coal / *Charbon à coke*	Other Bit. Coal / *Autres charb. bit.*	Sub-Bit. Coal / *Charbon sous-bit.*	Lignite / *Lignite*	Peat / *Tourbe*	Oven and Gas Coke / *Coke de four/gaz*	Pat. Fuel and BKB / *Agg./briq. de lignite*	Crude Oil / *Pétrole brut*	NGL / *LGN*	Feed-stocks / *Produits d'aliment.*	Additives / *Additifs*
Production	28951	45181	-	1587	1026	14800	2952	4098	384	-	-
From Other Sources	-	-	-	-	-	-	-	-	-	-	-
Imports	5585	5217	-	1352	-	788	164	9237	78	-	-
Exports	-31	-2258	-	-1	-3	-384	-	-	-	-	-
Intl. Marine Bunkers	-	-	-	-	-	-	-	-	-	-	-
Stock Changes	-	-	-	-	-	-	-	-	-	-	-
DOMESTIC SUPPLY	**34505**	**48140**	**-**	**2938**	**1023**	**15204**	**3116**	**13335**	**462**	**-**	**-**
Transfers	-	-	-	-	-	-	-	-	-	-	-
Statistical Differences	-	-	-	-	-	-	-	-	-	-	-
TRANSFORMATION	**34505**	**34392**	**-**	**938**	**-**	**12163**	**-**	**13335**	**462**	**-**	**-**
Electricity Plants	-	-	-	938	-	-	-	-	-	-	-
CHP Plants	-	27160	-	-	-	-	-	-	-	-	-
Heat Plants	-	3082	-	-	-	-	-	-	-	-	-
Blast Furnaces/Gas Works	-	-	-	-	-	12163	-	-	-	-	-
Coke/Pat. Fuel/BKB Plants	34505	4150	-	-	-	-	-	-	-	-	-
Petroleum Refineries	-	-	-	-	-	-	-	13335	462	-	-
Petrochemical Industry	-	-	-	-	-	-	-	-	-	-	-
Liquefaction	-	-	-	-	-	-	-	-	-	-	-
Other Transform. Sector	-	-	-	-	-	-	-	-	-	-	-
ENERGY SECTOR	**-**	**207**	**-**	**-**	**-**	**-**	**-**	**-**	**-**	**-**	**-**
Coal Mines	-	207	-	-	-	-	-	-	-	-	-
Oil and Gas Extraction	-	-	-	-	-	-	-	-	-	-	-
Petroleum Refineries	-	-	-	-	-	-	-	-	-	-	-
Electr., CHP+Heat Plants	-	-	-	-	-	-	-	-	-	-	-
Pumped Storage (Elec.)	-	-	-	-	-	-	-	-	-	-	-
Other Energy Sector	-	-	-	-	-	-	-	-	-	-	-
Distribution Losses	-	-	-	-	-	-	-	-	-	-	-
FINAL CONSUMPTION	**-**	**13541**	**-**	**2000**	**1023**	**3041**	**3116**	**-**	**-**	**-**	**-**
INDUSTRY SECTOR	**-**	**9006**	**-**	**-**	**-**	**3041**	**-**	**-**	**-**	**-**	**-**
Iron and Steel	-	-	-	-	-	3041	-	-	-	-	-
Chemical and Petrochem.	-	-	-	-	-	-	-	-	-	-	-
of which: Feedstocks	-	-	-	-	-	-	-	-	-	-	-
Non-Ferrous Metals	-	-	-	-	-	-	-	-	-	-	-
Non-Metallic Minerals	-	-	-	-	-	-	-	-	-	-	-
Transport Equipment	-	-	-	-	-	-	-	-	-	-	-
Machinery	-	-	-	-	-	-	-	-	-	-	-
Mining and Quarrying	-	-	-	-	-	-	-	-	-	-	-
Food and Tobacco	-	-	-	-	-	-	-	-	-	-	-
Paper, Pulp and Print	-	-	-	-	-	-	-	-	-	-	-
Wood and Wood Products	-	-	-	-	-	-	-	-	-	-	-
Construction	-	-	-	-	-	-	-	-	-	-	-
Textile and Leather	-	-	-	-	-	-	-	-	-	-	-
Non-specified	-	9006	-	-	-	-	-	-	-	-	-
TRANSPORT SECTOR	**-**	**-**	**-**	**-**	**-**	**-**	**-**	**-**	**-**	**-**	**-**
Air	-	-	-	-	-	-	-	-	-	-	-
Road	-	-	-	-	-	-	-	-	-	-	-
Rail	-	-	-	-	-	-	-	-	-	-	-
Pipeline Transport	-	-	-	-	-	-	-	-	-	-	-
Internal Navigation	-	-	-	-	-	-	-	-	-	-	-
Non-specified	-	-	-	-	-	-	-	-	-	-	-
OTHER SECTORS	**-**	**4535**	**-**	**2000**	**1023**	**-**	**3116**	**-**	**-**	**-**	**-**
Agriculture	-	-	-	-	-	-	-	-	-	-	-
Comm. and Publ. Services	-	-	-	-	-	-	-	-	-	-	-
Residential	-	4535	-	2000	1023	-	3116	-	-	-	-
Non-specified	-	-	-	-	-	-	-	-	-	-	-
NON-ENERGY USE	**-**	**-**	**-**	**-**	**-**	**-**	**-**	**-**	**-**	**-**	**-**
in Industry/Trans./Energy	-	-	-	-	-	-	-	-	-	-	-
in Transport	-	-	-	-	-	-	-	-	-	-	-
in Other Sectors	-	-	-	-	-	-	-	-	-	-	-

Ukraine : 1996

SUPPLY AND CONSUMPTION / *APPROVISIONNEMENT ET DEMANDE*	Oil cont. / *Pétrole cont.* (1000 tonnes)										
	Refinery Gas / *Gaz de raffinerie*	LPG + Ethane / *GPL + éthane*	Motor Gasoline / *Essence moteur*	Aviation Gasoline / *Essence aviation*	Jet Fuel / *Carbu- réacteurs*	Kerosene / *Kérosène*	Gas/ Diesel / *Gazole*	Heavy Fuel Oil / *Fioul lourd*	Naphtha / *Naphta*	Petrol. Coke / *Coke de pétrole*	Other Prod. / *Autres prod.*
Production	185	158	2138	-	665	11	3793	5141	-	-	1052
From Other Sources	-	-	-	-	-	-	-	-	-	-	-
Imports	-	1544	1686	37	82	-	1814	2864	-	-	619
Exports	-	-	-239	-	-	-	-552	-1753	-	-	-23
Intl. Marine Bunkers	-	-	-	-	-	-	-	-	-	-	-
Stock Changes	-	-	-	-	-	-	-	-	-	-	-
DOMESTIC SUPPLY	**185**	**1702**	**3585**	**37**	**747**	**11**	**5055**	**6252**	**-**	**-**	**1648**
Transfers	-	-	-	-	-	-	-	-	-	-	-
Statistical Differences	-	-	-	-	-	-	-	-	-	-	-
TRANSFORMATION	**-**	**-**	**-**	**-**	**-**	**-**	**75**	**2768**	**-**	**-**	**260**
Electricity Plants	-	-	-	-	-	-	75	1865	-	-	260
CHP Plants	-	-	-	-	-	-	-	-	-	-	-
Heat Plants	-	-	-	-	-	-	-	903	-	-	-
Blast Furnaces/Gas Works	-	-	-	-	-	-	-	-	-	-	-
Coke/Pat. Fuel/BKB Plants	-	-	-	-	-	-	-	-	-	-	-
Petroleum Refineries	-	-	-	-	-	-	-	-	-	-	-
Petrochemical Industry	-	-	-	-	-	-	-	-	-	-	-
Liquefaction	-	-	-	-	-	-	-	-	-	-	-
Other Transform. Sector	-	-	-	-	-	-	-	-	-	-	-
ENERGY SECTOR	**185**	**-**	**-**	**-**	**-**	**-**	**-**	**224**	**-**	**-**	**-**
Coal Mines	-	-	-	-	-	-	-	-	-	-	-
Oil and Gas Extraction	-	-	-	-	-	-	-	-	-	-	-
Petroleum Refineries	185	-	-	-	-	-	-	224	-	-	-
Electr., CHP+Heat Plants	-	-	-	-	-	-	-	-	-	-	-
Pumped Storage (Elec.)	-	-	-	-	-	-	-	-	-	-	-
Other Energy Sector	-	-	-	-	-	-	-	-	-	-	-
Distribution Losses	-	-	-	-	-	-	-	-	-	-	-
FINAL CONSUMPTION	**-**	**1702**	**3585**	**37**	**747**	**11**	**4980**	**3260**	**-**	**-**	**1388**
INDUSTRY SECTOR	**-**	**-**	**-**	**-**	**-**	**-**	**945**	**3260**	**-**	**-**	**-**
Iron and Steel	-	-	-	-	-	-	-	-	-	-	-
Chemical and Petrochem.	-	-	-	-	-	-	-	-	-	-	-
of which: Feedstocks	-	-	-	-	-	-	-	-	-	-	-
Non-Ferrous Metals	-	-	-	-	-	-	-	-	-	-	-
Non-Metallic Minerals	-	-	-	-	-	-	-	-	-	-	-
Transport Equipment	-	-	-	-	-	-	-	-	-	-	-
Machinery	-	-	-	-	-	-	-	-	-	-	-
Mining and Quarrying	-	-	-	-	-	-	-	-	-	-	-
Food and Tobacco	-	-	-	-	-	-	-	-	-	-	-
Paper, Pulp and Print	-	-	-	-	-	-	-	-	-	-	-
Wood and Wood Products	-	-	-	-	-	-	-	-	-	-	-
Construction	-	-	-	-	-	-	-	-	-	-	-
Textile and Leather	-	-	-	-	-	-	-	-	-	-	-
Non-specified	-	-	-	-	-	-	945	3260	-	-	-
TRANSPORT SECTOR	**-**	**-**	**3585**	**37**	**747**	**-**	**1791**	**-**	**-**	**-**	**-**
Air	-	-	-	37	747	-	-	-	-	-	-
Road	-	-	3585	-	-	-	-	-	-	-	-
Rail	-	-	-	-	-	-	1791	-	-	-	-
Pipeline Transport	-	-	-	-	-	-	-	-	-	-	-
Internal Navigation	-	-	-	-	-	-	-	-	-	-	-
Non-specified	-	-	-	-	-	-	-	-	-	-	-
OTHER SECTORS	**-**	**1702**	**-**	**-**	**-**	**-**	**2244**	**-**	**-**	**-**	**-**
Agriculture	-	-	-	-	-	-	2244	-	-	-	-
Comm. and Publ. Services	-	-	-	-	-	-	-	-	-	-	-
Residential	-	1702	-	-	-	-	-	-	-	-	-
Non-specified	-	-	-	-	-	-	-	-	-	-	-
NON-ENERGY USE	**-**	**-**	**-**	**-**	**-**	**11**	**-**	**-**	**-**	**-**	**1388**
in Industry/Transf./Energy	-	-	-	-	-	11	-	-	-	-	1388
in Transport	-	-	-	-	-	-	-	-	-	-	-
in Other Sectors	-	-	-	-	-	-	-	-	-	-	-

Ukraine : 1996

SUPPLY AND CONSUMPTION *APPROVISIONNEMENT ET DEMANDE*	Gas / *Gaz* (TJ)				Comb. Renew. & Waste / *En. Re. Comb. & Déchets* (TJ)				(GWh)	(TJ)
	Natural Gas *Gaz naturel*	Gas Works *Usines à gaz*	Coke Ovens *Cokeries*	Blast Furnaces *Hauts fourneaux*	Solid Biomass *Biomasse solide*	Gas/Liquids from Biomass *Gaz/Liquides tirés de biomasse*	Municipal Waste *Déchets urbains*	Industrial Waste *Déchets industriels*	Electricity *Electricité*	Heat *Chaleur*
Production	718280	-	-	-	10564	-	-	-	182986	626043
From Other Sources	-	-	-	-	-	-	-	-	-	-
Imports	2778224	-	-	-	-	-	-	-	4166	-
Exports	-31099	-	-	-	-	-	-	-	-6183	-
Intl. Marine Bunkers	-	-	-	-	-	-	-	-	-	-
Stock Changes	-	-	-	-	-	-	-	-	-	-
DOMESTIC SUPPLY	**3465405**	-	-	-	**10564**	-	-	-	**180969**	**626043**
Transfers	-	-	-	-	-	-	-	-	-	-
Statistical Differences	-	-	-	-	-	-	-	-	-	-
TRANSFORMATION	**1181252**	-	-	-	-	-	-	-	-	-
Electricity Plants	438077	-	-	-	-	-	-	-	-	-
CHP Plants	-	-	-	-	-	-	-	-	-	-
Heat Plants	739975	-	-	-	-	-	-	-	-	-
Blast Furnaces/Gas Works	-	-	-	-	-	-	-	-	-	-
Coke/Pat. Fuel/BKB Plants	-	-	-	-	-	-	-	-	-	-
Petroleum Refineries	3200	-	-	-	-	-	-	-	-	-
Petrochemical Industry	-	-	-	-	-	-	-	-	-	-
Liquefaction	-	-	-	-	-	-	-	-	-	-
Other Transform. Sector	-	-	-	-	-	-	-	-	-	-
ENERGY SECTOR	**63993**	-	-	-	-	-	-	-	**26778**	**5000**
Coal Mines	-	-	-	-	-	-	-	-	9655	-
Oil and Gas Extraction	63993	-	-	-	-	-	-	-	368	-
Petroleum Refineries	-	-	-	-	-	-	-	-	967	-
Electr., CHP+Heat Plants	-	-	-	-	-	-	-	-	14036	5000
Pumped Storage (Elec.)	-	-	-	-	-	-	-	-	-	-
Other Energy Sector	-	-	-	-	-	-	-	-	1752	-
Distribution Losses	63993	-	-	-	-	-	-	-	24996	187528
FINAL CONSUMPTION	**2156167**	-	-	-	**10564**	-	-	-	**129195**	**433515**
INDUSTRY SECTOR	**812630**	-	-	-	-	-	-	-	**65888**	**279946**
Iron and Steel	-	-	-	-	-	-	-	-	20176	-
Chemical and Petrochem.	-	-	-	-	-	-	-	-	8413	-
of which: Feedstocks	-	-	-	-	-	-	-	-	-	-
Non-Ferrous Metals	-	-	-	-	-	-	-	-	3407	-
Non-Metallic Minerals	-	-	-	-	-	-	-	-	2884	-
Transport Equipment	-	-	-	-	-	-	-	-	965	-
Machinery	-	-	-	-	-	-	-	-	6616	-
Mining and Quarrying	-	-	-	-	-	-	-	-	8004	-
Food and Tobacco	-	-	-	-	-	-	-	-	4581	-
Paper, Pulp and Print	-	-	-	-	-	-	-	-	574	-
Wood and Wood Products	-	-	-	-	-	-	-	-	328	-
Construction	-	-	-	-	-	-	-	-	1498	-
Textile and Leather	-	-	-	-	-	-	-	-	639	-
Non-specified	812630	-	-	-	-	-	-	-	7803	279946
TRANSPORT SECTOR	-	-	-	-	-	-	-	-	**9754**	-
Air	-	-	-	-	-	-	-	-	-	-
Road	-	-	-	-	-	-	-	-	-	-
Rail	-	-	-	-	-	-	-	-	4696	-
Pipeline Transport	-	-	-	-	-	-	-	-	1664	-
Internal Navigation	-	-	-	-	-	-	-	-	-	-
Non-specified	-	-	-	-	-	-	-	-	3394	-
OTHER SECTORS	**1343537**	-	-	-	**10564**	-	-	-	**53553**	**153569**
Agriculture	16818	-	-	-	-	-	-	-	10994	17614
Comm. and Publ. Services	442526	-	-	-	-	-	-	-	9661	-
Residential	884193	-	-	-	-	-	-	-	-	135955
Non-specified	-	-	-	-	10564	-	-	-	32898	-
NON-ENERGY USE	-	-	-	-	-	-	-	-	-	-
in Industry/Transf./Energy	-	-	-	-	-	-	-	-	-	-
in Transport	-	-	-	-	-	-	-	-	-	-
in Other Sectors	-	-	-	-	-	-	-	-	-	-

Ukraine : 1997

SUPPLY AND CONSUMPTION / *APPROVISIONNEMENT ET DEMANDE*	Coal / *Charbon* (1000 tonnes)							Oil / *Pétrole* (1000 tonnes)			
	Coking Coal / *Charbon à coke*	Other Bit. Coal / *Autres charb. bit.*	Sub-Bit. Coal / *Charbon sous-bit.*	Lignite / *Lignite*	Peat / *Tourbe*	Oven and Gas Coke / *Coke de four/gaz*	Pat. Fuel and BKB / *Agg./briq. de lignite*	Crude Oil / *Pétrole brut*	NGL / *LGN*	Feed-stocks / *Produits d'aliment.*	Additives / *Additifs*
Production	31838	43676	-	1433	765	15000	2881	4128	439	-	-
From Other Sources	-	-	-	-	-	-	-	-	-	-	-
Imports	3562	4807	-	482	1	181	-	8957	5	-	-
Exports	-44	-2330	-	-	-2	-359	-	-5	-	-	-
Intl. Marine Bunkers	-	-	-	-	-	-	-	-	-	-	-
Stock Changes	-	-	-	-	-	-	-	-	-	-	-
DOMESTIC SUPPLY	**35356**	**46153**	**-**	**1915**	**764**	**14822**	**2881**	**13080**	**444**	**-**	**-**
Transfers	-	-	-	-	-	-	-	-	-	-	-
Statistical Differences	-	-	-	-	-	-	-	-	-	-	-
TRANSFORMATION	**35356**	**32973**	**-**	**915**	**-**	**11858**	**-**	**13080**	**444**	**-**	**-**
Electricity Plants	-	-	-	915	-	-	-	-	-	-	-
CHP Plants	-	26039	-	-	-	-	-	-	-	-	-
Heat Plants	-	2955	-	-	-	-	-	-	-	-	-
Blast Furnaces/Gas Works	-	-	-	-	-	11858	-	-	-	-	-
Coke/Pat. Fuel/BKB Plants	35356	3979	-	-	-	-	-	-	-	-	-
Petroleum Refineries	-	-	-	-	-	-	-	13080	444	-	-
Petrochemical Industry	-	-	-	-	-	-	-	-	-	-	-
Liquefaction	-	-	-	-	-	-	-	-	-	-	-
Other Transform. Sector	-	-	-	-	-	-	-	-	-	-	-
ENERGY SECTOR	**-**	**198**	**-**	**-**	**-**	**-**	**-**	**-**	**-**	**-**	**-**
Coal Mines	-	198	-	-	-	-	-	-	-	-	-
Oil and Gas Extraction	-	-	-	-	-	-	-	-	-	-	-
Petroleum Refineries	-	-	-	-	-	-	-	-	-	-	-
Electr., CHP+Heat Plants	-	-	-	-	-	-	-	-	-	-	-
Pumped Storage (Elec.)	-	-	-	-	-	-	-	-	-	-	-
Other Energy Sector	-	-	-	-	-	-	-	-	-	-	-
Distribution Losses	-	-	-	-	-	-	-	-	-	-	-
FINAL CONSUMPTION	**-**	**12982**	**-**	**1000**	**764**	**2964**	**2881**	**-**	**-**	**-**	**-**
INDUSTRY SECTOR	**-**	**8634**	**-**	**-**	**-**	**2964**	**-**	**-**	**-**	**-**	**-**
Iron and Steel	-	-	-	-	-	2964	-	-	-	-	-
Chemical and Petrochem.	-	-	-	-	-	-	-	-	-	-	-
of which: Feedstocks	-	-	-	-	-	-	-	-	-	-	-
Non-Ferrous Metals	-	-	-	-	-	-	-	-	-	-	-
Non-Metallic Minerals	-	-	-	-	-	-	-	-	-	-	-
Transport Equipment	-	-	-	-	-	-	-	-	-	-	-
Machinery	-	-	-	-	-	-	-	-	-	-	-
Mining and Quarrying	-	-	-	-	-	-	-	-	-	-	-
Food and Tobacco	-	-	-	-	-	-	-	-	-	-	-
Paper, Pulp and Print	-	-	-	-	-	-	-	-	-	-	-
Wood and Wood Products	-	-	-	-	-	-	-	-	-	-	-
Construction	-	-	-	-	-	-	-	-	-	-	-
Textile and Leather	-	-	-	-	-	-	-	-	-	-	-
Non-specified	-	8634	-	-	-	-	-	-	-	-	-
TRANSPORT SECTOR	**-**	**-**	**-**	**-**	**-**	**-**	**-**	**-**	**-**	**-**	**-**
Air	-	-	-	-	-	-	-	-	-	-	-
Road	-	-	-	-	-	-	-	-	-	-	-
Rail	-	-	-	-	-	-	-	-	-	-	-
Pipeline Transport	-	-	-	-	-	-	-	-	-	-	-
Internal Navigation	-	-	-	-	-	-	-	-	-	-	-
Non-specified	-	-	-	-	-	-	-	-	-	-	-
OTHER SECTORS	**-**	**4348**	**-**	**1000**	**764**	**-**	**2881**	**-**	**-**	**-**	**-**
Agriculture	-	-	-	-	-	-	-	-	-	-	-
Comm. and Publ. Services	-	-	-	-	-	-	-	-	-	-	-
Residential	-	4348	-	1000	764	-	2881	-	-	-	-
Non-specified	-	-	-	-	-	-	-	-	-	-	-
NON-ENERGY USE	**-**	**-**	**-**	**-**	**-**	**-**	**-**	**-**	**-**	**-**	**-**
in Industry/Trans./Energy	-	-	-	-	-	-	-	-	-	-	-
in Transport	-	-	-	-	-	-	-	-	-	-	-
in Other Sectors	-	-	-	-	-	-	-	-	-	-	-

Ukraine : 1997

SUPPLY AND CONSUMPTION / *APPROVISIONNEMENT ET DEMANDE*	Oil cont. / *Pétrole cont.* (1000 tonnes)										
	Refinery Gas / *Gaz de raffinerie*	LPG + Ethane / *GPL + éthane*	Motor Gasoline / *Essence moteur*	Aviation Gasoline / *Essence aviation*	Jet Fuel / *Carbu-réacteurs*	Kerosene / *Kérosène*	Gas/ Diesel / *Gazole*	Heavy Fuel Oil / *Fioul lourd*	Naphtha / *Naphta*	Petrol. Coke / *Coke de pétrole*	Other Prod. / *Autres prod.*
Production	181	147	2803	-	620	11	3779	4326	-	-	981
From Other Sources	-	-	-	-	-	-	-	-	-	-	-
Imports	-	1437	437	36	23	4	1146	2501	-	-	461
Exports	-	-	-40	-	-5	-5	-130	-1165	-	-	-42
Intl. Marine Bunkers	-	-	-	-	-	-	-	-	-	-	-
Stock Changes	-	-	-	-	-	-	-	-	-	-	-
DOMESTIC SUPPLY	**181**	**1584**	**3200**	**36**	**638**	**10**	**4795**	**5662**	**-**	**-**	**1400**
Transfers	-	-	-	-	-	-	-	-	-	-	-
Statistical Differences	-	-	-	-	-	-	-	-	-	-	-
TRANSFORMATION	**-**	**-**	**-**	**-**	**-**	**-**	**71**	**2507**	**-**	**-**	**222**
Electricity Plants	-	-	-	-	-	-	71	1689	-	-	222
CHP Plants	-	-	-	-	-	-	-	-	-	-	-
Heat Plants	-	-	-	-	-	-	-	818	-	-	-
Blast Furnaces/Gas Works	-	-	-	-	-	-	-	-	-	-	-
Coke/Pat. Fuel/BKB Plants	-	-	-	-	-	-	-	-	-	-	-
Petroleum Refineries	-	-	-	-	-	-	-	-	-	-	-
Petrochemical Industry	-	-	-	-	-	-	-	-	-	-	-
Liquefaction	-	-	-	-	-	-	-	-	-	-	-
Other Transform. Sector	-	-	-	-	-	-	-	-	-	-	-
ENERGY SECTOR	**181**	**-**	**-**	**-**	**-**	**-**	**-**	**203**	**-**	**-**	**-**
Coal Mines	-	-	-	-	-	-	-	-	-	-	-
Oil and Gas Extraction	-	-	-	-	-	-	-	-	-	-	-
Petroleum Refineries	181	-	-	-	-	-	-	203	-	-	-
Electr., CHP+Heat Plants	-	-	-	-	-	-	-	-	-	-	-
Pumped Storage (Elec.)	-	-	-	-	-	-	-	-	-	-	-
Other Energy Sector	-	-	-	-	-	-	-	-	-	-	-
Distribution Losses	-	-	-	-	-	-	-	-	-	-	-
FINAL CONSUMPTION	**-**	**1584**	**3200**	**36**	**638**	**10**	**4724**	**2952**	**-**	**-**	**1178**
INDUSTRY SECTOR	**-**	**-**	**-**	**-**	**-**	**-**	**896**	**2952**	**-**	**-**	**-**
Iron and Steel	-	-	-	-	-	-	-	-	-	-	-
Chemical and Petrochem.	-	-	-	-	-	-	-	-	-	-	-
of which: Feedstocks	-	-	-	-	-	-	-	-	-	-	-
Non-Ferrous Metals	-	-	-	-	-	-	-	-	-	-	-
Non-Metallic Minerals	-	-	-	-	-	-	-	-	-	-	-
Transport Equipment	-	-	-	-	-	-	-	-	-	-	-
Machinery	-	-	-	-	-	-	-	-	-	-	-
Mining and Quarrying	-	-	-	-	-	-	-	-	-	-	-
Food and Tobacco	-	-	-	-	-	-	-	-	-	-	-
Paper, Pulp and Print	-	-	-	-	-	-	-	-	-	-	-
Wood and Wood Products	-	-	-	-	-	-	-	-	-	-	-
Construction	-	-	-	-	-	-	-	-	-	-	-
Textile and Leather	-	-	-	-	-	-	-	-	-	-	-
Non-specified	-	-	-	-	-	-	896	2952	-	-	-
TRANSPORT SECTOR	**-**	**-**	**3200**	**36**	**638**	**-**	**1699**	**-**	**-**	**-**	**-**
Air	-	-	-	36	638	-	-	-	-	-	-
Road	-	-	3200	-	-	-	-	-	-	-	-
Rail	-	-	-	-	-	-	1699	-	-	-	-
Pipeline Transport	-	-	-	-	-	-	-	-	-	-	-
Internal Navigation	-	-	-	-	-	-	-	-	-	-	-
Non-specified	-	-	-	-	-	-	-	-	-	-	-
OTHER SECTORS	**-**	**1584**	**-**	**-**	**-**	**-**	**2129**	**-**	**-**	**-**	**-**
Agriculture	-	-	-	-	-	-	2129	-	-	-	-
Comm. and Publ. Services	-	-	-	-	-	-	-	-	-	-	-
Residential	-	1584	-	-	-	-	-	-	-	-	-
Non-specified	-	-	-	-	-	-	-	-	-	-	-
NON-ENERGY USE	**-**	**-**	**-**	**-**	**-**	**10**	**-**	**-**	**-**	**-**	**1178**
in Industry/Transf./Energy	-	-	-	-	-	10	-	-	-	-	1178
in Transport	-	-	-	-	-	-	-	-	-	-	-
in Other Sectors	-	-	-	-	-	-	-	-	-	-	-

Ukraine : 1997

	Gas / *Gaz* (TJ)				Comb. Renew. & Waste / *En. Re. Comb. & Déchets* (TJ)				(GWh)	(TJ)
SUPPLY AND CONSUMPTION	Natural Gas	Gas Works	Coke Ovens	Blast Furnaces	Solid Biomass	Gas/Liquids from Biomass	Municipal Waste	Industrial Waste	Electricity	Heat
APPROVISIONNEMENT ET DEMANDE	*Gaz naturel*	*Usines à gaz*	*Cokeries*	*Hauts fourneaux*	*Biomasse solide*	*Gaz/Liquides tirés de biomasse*	*Déchets urbains*	*Déchets industriels*	*Electricité*	*Chaleur*
Production	707472	-	-	-	10289	-	-	-	178002	609015
From Other Sources	-	-	-	-	-	-	-	-	-	-
Imports	2433131	-	-	-	-	-	-	-	9719	-
Exports	-54784	-	-	-	-	-	-	-	-9873	-
Intl. Marine Bunkers	-	-	-	-	-	-	-	-	-	-
Stock Changes	-	-	-	-	-	-	-	-	-	-
DOMESTIC SUPPLY	**3085819**	**-**	**-**	**-**	**10289**	**-**	**-**	**-**	**177848**	**609015**
Transfers	-	-	-	-	-	-	-	-	-	-
Statistical Differences	-	-	-	-	-	-	-	-	-	-
TRANSFORMATION	**990913**	**-**	**-**	**-**	**-**	**-**	**-**	**-**	**-**	**-**
Electricity Plants	367217	-	-	-	-	-	-	-	-	-
CHP Plants	-	-	-	-	-	-	-	-	-	-
Heat Plants	620262	-	-	-	-	-	-	-	-	-
Blast Furnaces/Gas Works	-	-	-	-	-	-	-	-	-	-
Coke/Pat. Fuel/BKB Plants	-	-	-	-	-	-	-	-	-	-
Petroleum Refineries	3434	-	-	-	-	-	-	-	-	-
Petrochemical Industry	-	-	-	-	-	-	-	-	-	-
Liquefaction	-	-	-	-	-	-	-	-	-	-
Other Transform. Sector	-	-	-	-	-	-	-	-	-	-
ENERGY SECTOR	**67934**	**-**	**-**	**-**	**-**	**-**	**-**	**-**	**25290**	**5000**
Coal Mines	-	-	-	-	-	-	-	-	9036	-
Oil and Gas Extraction	67934	-	-	-	-	-	-	-	381	-
Petroleum Refineries	-	-	-	-	-	-	-	-	1032	-
Electr., CHP+Heat Plants	-	-	-	-	-	-	-	-	13166	5000
Pumped Storage (Elec.)	-	-	-	-	-	-	-	-	-	-
Other Energy Sector	-	-	-	-	-	-	-	-	1675	-
Distribution Losses	67973	-	-	-	-	-	-	-	28407	182292
FINAL CONSUMPTION	**1958999**	**-**	**-**	**-**	**10289**	**-**	**-**	**-**	**124151**	**421723**
INDUSTRY SECTOR	**722026**	**-**	**-**	**-**	**-**	**-**	**-**	**-**	**64822**	**272331**
Iron and Steel	-	-	-	-	-	-	-	-	22347	-
Chemical and Petrochem.	-	-	-	-	-	-	-	-	7569	-
of which: Feedstocks	-	-	-	-	-	-	-	-	-	-
Non-Ferrous Metals	-	-	-	-	-	-	-	-	3390	-
Non-Metallic Minerals	-	-	-	-	-	-	-	-	2717	-
Transport Equipment	-	-	-	-	-	-	-	-	887	-
Machinery	-	-	-	-	-	-	-	-	6049	-
Mining and Quarrying	-	-	-	-	-	-	-	-	8724	-
Food and Tobacco	-	-	-	-	-	-	-	-	3896	-
Paper, Pulp and Print	-	-	-	-	-	-	-	-	536	-
Wood and Wood Products	-	-	-	-	-	-	-	-	309	-
Construction	-	-	-	-	-	-	-	-	1450	-
Textile and Leather	-	-	-	-	-	-	-	-	573	-
Non-specified	722026	-	-	-	-	-	-	-	6375	272331
TRANSPORT SECTOR	**-**	**-**	**-**	**-**	**-**	**-**	**-**	**-**	**9545**	**-**
Air	-	-	-	-	-	-	-	-	-	-
Road	-	-	-	-	-	-	-	-	-	-
Rail	-	-	-	-	-	-	-	-	4499	-
Pipeline Transport	-	-	-	-	-	-	-	-	1745	-
Internal Navigation	-	-	-	-	-	-	-	-	-	-
Non-specified	-	-	-	-	-	-	-	-	3301	-
OTHER SECTORS	**1236973**	**-**	**-**	**-**	**10289**	**-**	**-**	**-**	**49784**	**149392**
Agriculture	14983	-	-	-	-	-	-	-	9184	17135
Comm. and Publ. Services	407330	-	-	-	-	-	-	-	9277	-
Residential	814660	-	-	-	-	-	-	-	-	132257
Non-specified	-	-	-	-	10289	-	-	-	31323	-
NON-ENERGY USE	**-**	**-**	**-**	**-**	**-**	**-**	**-**	**-**	**-**	**-**
in Industry/Transf./Energy	-	-	-	-	-	-	-	-	-	-
in Transport	-	-	-	-	-	-	-	-	-	-
in Other Sectors	-	-	-	-	-	-	-	-	-	-

United Arab Emirates / Emirats arabes unis

SUPPLY AND CONSUMPTION 1996	Coal (1000 tonnes)							Oil (1000 tonnes)			
	Coking Coal	Other Bit. Coal	Sub-Bit. Coal	Lignite	Peat	Oven and Gas Coke	Pat. Fuel and BKB	Crude Oil	NGL	Feed-stocks	Additives
Production	-	-	-	-	-	-	-	107352	11417	-	-
Imports	-	-	-	-	-	-	-	-	-	-	-
Exports	-	-	-	-	-	-	-	-93374	-6882	-	-
Intl. Marine Bunkers	-	-	-	-	-	-	-	-	-	-	-
Stock Changes	-	-	-	-	-	-	-	-	-	-	-
DOMESTIC SUPPLY	-	-	-	-	-	-	-	**13978**	**4535**	-	-
Transfers and Stat. Diff.	-	-	-	-	-	-	-	-2048	-4535	-	-
TRANSFORMATION	-	-	-	-	-	-	-	**11930**	-	-	-
Electricity and CHP Plants	-	-	-	-	-	-	-	-	-	-	-
Petroleum Refineries	-	-	-	-	-	-	-	11930	-	-	-
Other Transform. Sector	-	-	-	-	-	-	-	-	-	-	-
ENERGY SECTOR	-	-	-	-	-	-	-	-	-	-	-
DISTRIBUTION LOSSES	-	-	-	-	-	-	-	-	-	-	-
FINAL CONSUMPTION	-	-	-	-	-	-	-	-	-	-	-
INDUSTRY SECTOR	-	-	-	-	-	-	-	-	-	-	-
Iron and Steel	-	-	-	-	-	-	-	-	-	-	-
Chemical and Petrochem.	-	-	-	-	-	-	-	-	-	-	-
Non-Metallic Minerals	-	-	-	-	-	-	-	-	-	-	-
Non-specified	-	-	-	-	-	-	-	-	-	-	-
TRANSPORT SECTOR	-	-	-	-	-	-	-	-	-	-	-
Air	-	-	-	-	-	-	-	-	-	-	-
Road	-	-	-	-	-	-	-	-	-	-	-
Non-specified	-	-	-	-	-	-	-	-	-	-	-
OTHER SECTORS	-	-	-	-	-	-	-	-	-	-	-
Agriculture	-	-	-	-	-	-	-	-	-	-	-
Comm. and Publ. Services	-	-	-	-	-	-	-	-	-	-	-
Residential	-	-	-	-	-	-	-	-	-	-	-
Non-specified	-	-	-	-	-	-	-	-	-	-	-
NON-ENERGY USE	-	-	-	-	-	-	-	-	-	-	-

APPROVISIONNEMENT ET DEMANDE 1997	*Charbon* (1000 tonnes)							*Pétrole* (1000 tonnes)			
	Charbon à coke	*Autres charb. bit.*	*Charbon sous-bit.*	*Lignite*	*Tourbe*	*Coke de four/gaz*	*Agg./briq. de lignite*	*Pétrole brut*	*LGN*	*Produits d'aliment.*	*Additifs*
Production	-	-	-	-	-	-	-	108605	12223	-	-
Imports	-	-	-	-	-	-	-	-	-	-	-
Exports	-	-	-	-	-	-	-	-95365	-7171	-	-
Intl. Marine Bunkers	-	-	-	-	-	-	-	-	-	-	-
Stock Changes	-	-	-	-	-	-	-	-	-	-	-
DOMESTIC SUPPLY	-	-	-	-	-	-	-	**13240**	**5052**	-	-
Transfers and Stat. Diff.	-	-	-	-	-	-	-	-	-5052	-	-
TRANSFORMATION	-	-	-	-	-	-	-	**13240**	-	-	-
Electricity and CHP Plants	-	-	-	-	-	-	-	-	-	-	-
Petroleum Refineries	-	-	-	-	-	-	-	13240	-	-	-
Other Transform. Sector	-	-	-	-	-	-	-	-	-	-	-
ENERGY SECTOR	-	-	-	-	-	-	-	-	-	-	-
DISTRIBUTION LOSSES	-	-	-	-	-	-	-	-	-	-	-
FINAL CONSUMPTION	-	-	-	-	-	-	-	-	-	-	-
INDUSTRY SECTOR	-	-	-	-	-	-	-	-	-	-	-
Iron and Steel	-	-	-	-	-	-	-	-	-	-	-
Chemical and Petrochem.	-	-	-	-	-	-	-	-	-	-	-
Non-Metallic Minerals	-	-	-	-	-	-	-	-	-	-	-
Non-specified	-	-	-	-	-	-	-	-	-	-	-
TRANSPORT SECTOR	-	-	-	-	-	-	-	-	-	-	-
Air	-	-	-	-	-	-	-	-	-	-	-
Road	-	-	-	-	-	-	-	-	-	-	-
Non-specified	-	-	-	-	-	-	-	-	-	-	-
OTHER SECTORS	-	-	-	-	-	-	-	-	-	-	-
Agriculture	-	-	-	-	-	-	-	-	-	-	-
Comm. and Publ. Services	-	-	-	-	-	-	-	-	-	-	-
Residential	-	-	-	-	-	-	-	-	-	-	-
Non-specified	-	-	-	-	-	-	-	-	-	-	-
NON-ENERGY USE	-	-	-	-	-	-	-	-	-	-	-

United Arab Emirates / Emirats arabes unis

SUPPLY AND CONSUMPTION 1996	Oil cont. (1000 tonnes)										
	Refinery Gas	LPG + Ethane	Motor Gasoline	Aviation Gasoline	Jet Fuel	Kerosene	Gas/ Diesel	Heavy Fuel Oil	Naphtha	Petrol. Coke	Other Prod.
Production	366	404	1464	-	2474	80	2775	1860	983	-	47
Imports	-	-	-	-	-	-	100	12765	-	-	-
Exports	-	-4500	-343	-	-2358	-51	-438	-1858	-687	-	-46
Intl. Marine Bunkers	-	-	-	-	-	-	-90	-10677	-	-	-
Stock Changes	-	-	-	-	-	-	-	-	-	-	-
DOMESTIC SUPPLY	**366**	**-4096**	**1121**	**-**	**116**	**29**	**2347**	**2090**	**296**	**-**	**1**
Transfers and Stat. Diff.	-	4464	-48	-	93	3	18	-	-296	-	-1
TRANSFORMATION	**-**	**-**	**-**	**-**	**-**	**-**	**120**	**1500**	**-**	**-**	**-**
Electricity and CHP Plants	-	-	-	-	-	-	120	1500	-	-	-
Petroleum Refineries	-	-	-	-	-	-	-	-	-	-	-
Other Transform. Sector	-	-	-	-	-	-	-	-	-	-	-
ENERGY SECTOR	**366**	**-**	**-**	**-**	**-**	**-**	**-**	**169**	**-**	**-**	**-**
DISTRIBUTION LOSSES	**-**	**-**	**-**	**-**	**-**	**-**	**-**	**-**	**-**	**-**	**-**
FINAL CONSUMPTION	**-**	**368**	**1073**	**-**	**209**	**32**	**2245**	**421**	**-**	**-**	**-**
INDUSTRY SECTOR	**-**	**-**	**-**	**-**	**-**	**-**	**-**	**421**	**-**	**-**	**-**
Iron and Steel	-	-	-	-	-	-	-	-	-	-	-
Chemical and Petrochem.	-	-	-	-	-	-	-	-	-	-	-
Non-Metallic Minerals	-	-	-	-	-	-	-	-	-	-	-
Non-specified	-	-	-	-	-	-	-	421	-	-	-
TRANSPORT SECTOR	**-**	**-**	**1073**	**-**	**209**	**-**	**2245**	**-**	**-**	**-**	**-**
Air	-	-	-	-	209	-	-	-	-	-	-
Road	-	-	1073	-	-	-	2245	-	-	-	-
Non-specified	-	-	-	-	-	-	-	-	-	-	-
OTHER SECTORS	**-**	**368**	**-**	**-**	**-**	**32**	**-**	**-**	**-**	**-**	**-**
Agriculture	-	-	-	-	-	-	-	-	-	-	-
Comm. and Publ. Services	-	-	-	-	-	-	-	-	-	-	-
Residential	-	368	-	-	-	32	-	-	-	-	-
Non-specified	-	-	-	-	-	-	-	-	-	-	-
NON-ENERGY USE	**-**	**-**	**-**	**-**	**-**	**-**	**-**	**-**	**-**	**-**	**-**

APPROVISIONNEMENT ET DEMANDE 1997	*Pétrole cont.* (1000 tonnes)										
	Gaz de raffinerie	*GPL + éthane*	*Essence moteur*	*Essence aviation*	*Carbu- réacteurs*	*Kérosène*	*Gazole*	*Fioul lourd*	*Naphta*	*Coke de pétrole*	*Autres prod.*
Production	406	462	1504	-	2971	96	3170	1982	1072	-	101
Imports	-	-	-	-	-	-	100	12765	-	-	-
Exports	-	-5129	-333	-	-2779	-67	-1397	-1965	-816	-	-100
Intl. Marine Bunkers	-	-	-	-	-	-	-90	-10677	-	-	-
Stock Changes	-	-	-	-	-	-	-	-	-	-	-
DOMESTIC SUPPLY	**406**	**-4667**	**1171**	**-**	**192**	**29**	**1783**	**2105**	**256**	**-**	**1**
Transfers and Stat. Diff.	-	5052	-	-	-	-	14	-	-256	-	-
TRANSFORMATION	**-**	**-**	**-**	**-**	**-**	**-**	**91**	**1511**	**-**	**-**	**-**
Electricity and CHP Plants	-	-	-	-	-	-	91	1511	-	-	-
Petroleum Refineries	-	-	-	-	-	-	-	-	-	-	-
Other Transform. Sector	-	-	-	-	-	-	-	-	-	-	-
ENERGY SECTOR	**406**	**-**	**-**	**-**	**-**	**-**	**-**	**170**	**-**	**-**	**-**
DISTRIBUTION LOSSES	**-**	**-**	**-**	**-**	**-**	**-**	**-**	**-**	**-**	**-**	**-**
FINAL CONSUMPTION	**-**	**385**	**1171**	**-**	**192**	**29**	**1706**	**424**	**-**	**-**	**1**
INDUSTRY SECTOR	**-**	**-**	**-**	**-**	**-**	**-**	**-**	**424**	**-**	**-**	**-**
Iron and Steel	-	-	-	-	-	-	-	-	-	-	-
Chemical and Petrochem.	-	-	-	-	-	-	-	-	-	-	-
Non-Metallic Minerals	-	-	-	-	-	-	-	-	-	-	-
Non-specified	-	-	-	-	-	-	-	424	-	-	-
TRANSPORT SECTOR	**-**	**-**	**1171**	**-**	**192**	**-**	**1706**	**-**	**-**	**-**	**-**
Air	-	-	-	-	192	-	-	-	-	-	-
Road	-	-	1171	-	-	-	1706	-	-	-	-
Non-specified	-	-	-	-	-	-	-	-	-	-	-
OTHER SECTORS	**-**	**385**	**-**	**-**	**-**	**29**	**-**	**-**	**-**	**-**	**-**
Agriculture	-	-	-	-	-	-	-	-	-	-	-
Comm. and Publ. Services	-	-	-	-	-	-	-	-	-	-	-
Residential	-	385	-	-	-	29	-	-	-	-	-
Non-specified	-	-	-	-	-	-	-	-	-	-	-
NON-ENERGY USE	**-**	**-**	**-**	**-**	**-**	**-**	**-**	**-**	**-**	**-**	**1**

United Arab Emirates / Emirats arabes unis

SUPPLY AND CONSUMPTION 1996	Gas (TJ)				Comb. Renew. & Waste (TJ)				(GWh)	(TJ)
	Natural Gas	Gas Works	Coke Ovens	Blast Furnaces	Solid Biomass	Gas/Liquids from Biomass	Municipal Waste	Industrial Waste	Electricity	Heat
Production	1298788	-	-	-	-	-	-	-	19737	-
Imports	9310	-	-	-	730	-	-	-	-	-
Exports	-261399	-	-	-	-	-	-	-	-	-
Intl. Marine Bunkers	-	-	-	-	-	-	-	-	-	-
Stock Changes	-	-	-	-	-	-	-	-	-	-
DOMESTIC SUPPLY	**1046699**	-	-	-	**730**	-	-	-	**19737**	-
Transfers and Stat. Diff.	-6837	-	-	-	-	-	-	-	-	-
TRANSFORMATION	**208028**	-	-	-	-	-	-	-	-	-
Electricity and CHP Plants	208028	-	-	-	-	-	-	-	-	-
Petroleum Refineries	-	-	-	-	-	-	-	-	-	-
Other Transform. Sector	-	-	-	-	-	-	-	-	-	-
ENERGY SECTOR	**750396**	-	-	-	-	-	-	-	-	-
DISTRIBUTION LOSSES	-	-	-	-	-	-	-	-	-	-
FINAL CONSUMPTION	**81438**	-	-	-	**730**	-	-	-	**19737**	-
INDUSTRY SECTOR	**81438**	-	-	-	-	-	-	-	-	-
Iron and Steel	-	-	-	-	-	-	-	-	-	-
Chemical and Petrochem.	81438	-	-	-	-	-	-	-	-	-
Non-Metallic Minerals	-	-	-	-	-	-	-	-	-	-
Non-specified	-	-	-	-	-	-	-	-	-	-
TRANSPORT SECTOR	-	-	-	-	-	-	-	-	-	-
Air	-	-	-	-	-	-	-	-	-	-
Road	-	-	-	-	-	-	-	-	-	-
Non-specified	-	-	-	-	-	-	-	-	-	-
OTHER SECTORS	-	-	-	-	**730**	-	-	-	**19737**	-
Agriculture	-	-	-	-	-	-	-	-	-	-
Comm. and Publ. Services	-	-	-	-	-	-	-	-	-	-
Residential	-	-	-	-	-	-	-	-	-	-
Non-specified	-	-	-	-	730	-	-	-	19737	-
NON-ENERGY USE	-	-	-	-	-	-	-	-	-	-

APPROVISIONNEMENT ET DEMANDE 1997	*Gaz (TJ)*				*En. Re. Comb. & Déchets (TJ)*				(GWh)	(TJ)
	Gaz naturel	*Usines à gaz*	*Cokeries*	*Hauts fourneaux*	*Biomasse solide*	*Gaz/Liquides tirés de biomasse*	*Déchets urbains*	*Déchets industriels*	*Electricité*	*Chaleur*
Production	1421684	-	-	-	-	-	-	-	20571	-
Imports	9310	-	-	-	730	-	-	-	-	-
Exports	-327786	-	-	-	-	-	-	-	-	-
Intl. Marine Bunkers	-	-	-	-	-	-	-	-	-	-
Stock Changes	-	-	-	-	-	-	-	-	-	-
DOMESTIC SUPPLY	**1103208**	-	-	-	**730**	-	-	-	**20571**	-
Transfers and Stat. Diff.	-7943	-	-	-	-	-	-	-	-	-
TRANSFORMATION	**219111**	-	-	-	-	-	-	-	-	-
Electricity and CHP Plants	219111	-	-	-	-	-	-	-	-	-
Petroleum Refineries	-	-	-	-	-	-	-	-	-	-
Other Transform. Sector	-	-	-	-	-	-	-	-	-	-
ENERGY SECTOR	**790377**	-	-	-	-	-	-	-	-	-
DISTRIBUTION LOSSES	-	-	-	-	-	-	-	-	-	-
FINAL CONSUMPTION	**85777**	-	-	-	**730**	-	-	-	**20571**	-
INDUSTRY SECTOR	**85777**	-	-	-	-	-	-	-	-	-
Iron and Steel	-	-	-	-	-	-	-	-	-	-
Chemical and Petrochem.	85777	-	-	-	-	-	-	-	-	-
Non-Metallic Minerals	-	-	-	-	-	-	-	-	-	-
Non-specified	-	-	-	-	-	-	-	-	-	-
TRANSPORT SECTOR	-	-	-	-	-	-	-	-	-	-
Air	-	-	-	-	-	-	-	-	-	-
Road	-	-	-	-	-	-	-	-	-	-
Non-specified	-	-	-	-	-	-	-	-	-	-
OTHER SECTORS	-	-	-	-	**730**	-	-	-	**20571**	-
Agriculture	-	-	-	-	-	-	-	-	-	-
Comm. and Publ. Services	-	-	-	-	-	-	-	-	-	-
Residential	-	-	-	-	-	-	-	-	-	-
Non-specified	-	-	-	-	730	-	-	-	20571	-
NON-ENERGY USE	-	-	-	-	-	-	-	-	-	-

Uruguay

SUPPLY AND CONSUMPTION 1996	Coal (1000 tonnes)							Oil (1000 tonnes)			
	Coking Coal	Other Bit. Coal	Sub-Bit. Coal	Lignite	Peat	Oven and Gas Coke	Pat. Fuel and BKB	Crude Oil	NGL	Feed-stocks	Additives
Production	-	-	-	-	-	-	-	-	-	-	-
Imports	-	-	-	-	-	-	-	1497	-	-	-
Exports	-	-	-	-	-	-	-	-	-	-	-
Intl. Marine Bunkers	-	-	-	-	-	-	-	-	-	-	-
Stock Changes	-	-	-	-	-	-	-	135	-	-	-
DOMESTIC SUPPLY	**-**	**-**	**-**	**-**	**-**	**-**	**-**	**1632**	**-**	**-**	**-**
Transfers and Stat. Diff.	-	-	-	-	-	-	-	-	-	-	-
TRANSFORMATION	**-**	**-**	**-**	**-**	**-**	**-**	**-**	**1631**	**-**	**-**	**-**
Electricity and CHP Plants	-	-	-	-	-	-	-	-	-	-	-
Petroleum Refineries	-	-	-	-	-	-	-	1631	-	-	-
Other Transform. Sector	-	-	-	-	-	-	-	-	-	-	-
ENERGY SECTOR	**-**	**-**	**-**	**-**	**-**	**-**	**-**	**-**	**-**	**-**	**-**
DISTRIBUTION LOSSES	**-**	**-**	**-**	**-**	**-**	**-**	**-**	**1**	**-**	**-**	**-**
FINAL CONSUMPTION	**-**	**-**	**-**	**-**	**-**	**-**	**-**	**-**	**-**	**-**	**-**
INDUSTRY SECTOR	**-**	**-**	**-**	**-**	**-**	**-**	**-**	**-**	**-**	**-**	**-**
Iron and Steel	-	-	-	-	-	-	-	-	-	-	-
Chemical and Petrochem.	-	-	-	-	-	-	-	-	-	-	-
Non-Metallic Minerals	-	-	-	-	-	-	-	-	-	-	-
Non-specified	-	-	-	-	-	-	-	-	-	-	-
TRANSPORT SECTOR	**-**	**-**	**-**	**-**	**-**	**-**	**-**	**-**	**-**	**-**	**-**
Air	-	-	-	-	-	-	-	-	-	-	-
Road	-	-	-	-	-	-	-	-	-	-	-
Non-specified	-	-	-	-	-	-	-	-	-	-	-
OTHER SECTORS	**-**	**-**	**-**	**-**	**-**	**-**	**-**	**-**	**-**	**-**	**-**
Agriculture	-	-	-	-	-	-	-	-	-	-	-
Comm. and Publ. Services	-	-	-	-	-	-	-	-	-	-	-
Residential	-	-	-	-	-	-	-	-	-	-	-
Non-specified	-	-	-	-	-	-	-	-	-	-	-
NON-ENERGY USE	**-**	**-**	**-**	**-**	**-**	**-**	**-**	**-**	**-**	**-**	**-**

APPROVISIONNEMENT ET DEMANDE 1997	*Charbon* (1000 tonnes)							*Pétrole* (1000 tonnes)			
	Charbon à coke	*Autres charb. bit.*	*Charbon sous-bit.*	*Lignite*	*Tourbe*	*Coke de four/gaz*	*Agg./briq. de lignite*	*Pétrole brut*	*LGN*	*Produits d'aliment.*	*Additifs*
Production	-	-	-	-	-	-	-	-	-	-	-
Imports	-	-	-	-	-	1	-	1465	-	-	-
Exports	-	-	-	-	-	-	-	-	-	-	-
Intl. Marine Bunkers	-	-	-	-	-	-	-	-	-	-	-
Stock Changes	-	-	-	-	-	-	-	-71	-	-	-
DOMESTIC SUPPLY	**-**	**-**	**-**	**-**	**-**	**1**	**-**	**1394**	**-**	**-**	**-**
Transfers and Stat. Diff.	-	-	-	-	-	-1	-	-	-	-	-
TRANSFORMATION	**-**	**-**	**-**	**-**	**-**	**-**	**-**	**1394**	**-**	**-**	**-**
Electricity and CHP Plants	-	-	-	-	-	-	-	-	-	-	-
Petroleum Refineries	-	-	-	-	-	-	-	1394	-	-	-
Other Transform. Sector	-	-	-	-	-	-	-	-	-	-	-
ENERGY SECTOR	**-**	**-**	**-**	**-**	**-**	**-**	**-**	**-**	**-**	**-**	**-**
DISTRIBUTION LOSSES	**-**	**-**	**-**	**-**	**-**	**-**	**-**	**-**	**-**	**-**	**-**
FINAL CONSUMPTION	**-**	**-**	**-**	**-**	**-**	**-**	**-**	**-**	**-**	**-**	**-**
INDUSTRY SECTOR	**-**	**-**	**-**	**-**	**-**	**-**	**-**	**-**	**-**	**-**	**-**
Iron and Steel	-	-	-	-	-	-	-	-	-	-	-
Chemical and Petrochem.	-	-	-	-	-	-	-	-	-	-	-
Non-Metallic Minerals	-	-	-	-	-	-	-	-	-	-	-
Non-specified	-	-	-	-	-	-	-	-	-	-	-
TRANSPORT SECTOR	**-**	**-**	**-**	**-**	**-**	**-**	**-**	**-**	**-**	**-**	**-**
Air	-	-	-	-	-	-	-	-	-	-	-
Road	-	-	-	-	-	-	-	-	-	-	-
Non-specified	-	-	-	-	-	-	-	-	-	-	-
OTHER SECTORS	**-**	**-**	**-**	**-**	**-**	**-**	**-**	**-**	**-**	**-**	**-**
Agriculture	-	-	-	-	-	-	-	-	-	-	-
Comm. and Publ. Services	-	-	-	-	-	-	-	-	-	-	-
Residential	-	-	-	-	-	-	-	-	-	-	-
Non-specified	-	-	-	-	-	-	-	-	-	-	-
NON-ENERGY USE	**-**	**-**	**-**	**-**	**-**	**-**	**-**	**-**	**-**	**-**	**-**

Uruguay

SUPPLY AND CONSUMPTION 1996	Oil cont. (1000 tonnes)										
	Refinery Gas	LPG + Ethane	Motor Gasoline	Aviation Gasoline	Jet Fuel	Kerosene	Gas/ Diesel	Heavy Fuel Oil	Naphtha	Petrol. Coke	Other Prod.
Production	33	37	236	-	41	36	608	509	16	-	64
Imports	-	60	114	3	-	-	357	103	-	-	17
Exports	-	-	-27	-	-34	-	-	-	-	-	-
Intl. Marine Bunkers	-	-	-	-	-	-	-184	-195	-	-	-
Stock Changes	-	-1	-7	-	-2	-3	-74	3	-	-	-
DOMESTIC SUPPLY	**33**	**96**	**316**	**3**	**5**	**33**	**707**	**420**	**16**	**-**	**81**
Transfers and Stat. Diff.	-	-	-1	-	-	-1	-2	-2	-	-	-1
TRANSFORMATION	**-**	**-**	**-**	**-**	**-**	**-**	**51**	**159**	**16**	**-**	**-**
Electricity and CHP Plants	-	-	-	-	-	-	51	159	-	-	-
Petroleum Refineries	-	-	-	-	-	-	-	-	-	-	-
Other Transform. Sector	-	-	-	-	-	-	-	-	16	-	-
ENERGY SECTOR	**33**	**6**	**-**	**-**	**-**	**-**	**-**	**71**	**-**	**-**	**-**
DISTRIBUTION LOSSES	**-**	**-**	**3**	**-**	**-**	**1**	**-**	**1**	**-**	**-**	**-**
FINAL CONSUMPTION	**-**	**90**	**312**	**3**	**5**	**31**	**654**	**187**	**-**	**-**	**80**
INDUSTRY SECTOR	**-**	**2**	**-**	**-**	**-**	**1**	**10**	**158**	**-**	**-**	**-**
Iron and Steel	-	-	-	-	-	-	-	-	-	-	-
Chemical and Petrochem.	-	-	-	-	-	-	-	-	-	-	-
Non-Metallic Minerals	-	-	-	-	-	-	-	-	-	-	-
Non-specified	-	2	-	-	-	1	10	158	-	-	-
TRANSPORT SECTOR	**-**	**-**	**300**	**3**	**5**	**-**	**421**	**-**	**-**	**-**	**-**
Air	-	-	-	3	5	-	-	-	-	-	-
Road	-	-	300	-	-	-	421	-	-	-	-
Non-specified	-	-	-	-	-	-	-	-	-	-	-
OTHER SECTORS	**-**	**88**	**12**	**-**	**-**	**30**	**223**	**29**	**-**	**-**	**-**
Agriculture	-	-	10	-	-	-	179	-	-	-	-
Comm. and Publ. Services	-	-	-	-	-	-	29	6	-	-	-
Residential	-	88	-	-	-	30	10	23	-	-	-
Non-specified	-	-	2	-	-	-	5	-	-	-	-
NON-ENERGY USE	**-**	**-**	**-**	**-**	**-**	**-**	**-**	**-**	**-**	**-**	**80**

APPROVISIONNEMENT ET DEMANDE 1997	*Pétrole cont.* (1000 tonnes)										
	Gaz de raffinerie	*GPL + éthane*	*Essence moteur*	*Essence aviation*	*Carbu-réacteurs*	*Kérosène*	*Gazole*	*Fioul lourd*	*Naphta*	*Coke de pétrole*	*Autres prod.*
Production	14	62	268	-	37	21	431	429	11	-	89
Imports	-	40	83	3	4	-	384	119	-	-	18
Exports	-	-	-12	-	-39	-	-	-	-2	-	-1
Intl. Marine Bunkers	-	-	-	-	-	-	-128	-179	-	-	-
Stock Changes	-	-2	-15	-	3	6	46	44	1	-	-4
DOMESTIC SUPPLY	**14**	**100**	**324**	**3**	**5**	**27**	**733**	**413**	**10**	**-**	**102**
Transfers and Stat. Diff.	-	-	-2	-	-1	-2	-1	1	1	-	-1
TRANSFORMATION	**-**	**4**	**-**	**-**	**-**	**-**	**31**	**124**	**11**	**-**	**-**
Electricity and CHP Plants	-	-	-	-	-	-	31	124	-	-	-
Petroleum Refineries	-	-	-	-	-	-	-	-	-	-	-
Other Transform. Sector	-	4	-	-	-	-	-	-	11	-	-
ENERGY SECTOR	**14**	**3**	**-**	**-**	**-**	**-**	**-**	**64**	**-**	**-**	**17**
DISTRIBUTION LOSSES	**-**	**1**	**2**	**-**	**-**	**-**	**1**	**1**	**-**	**-**	**-**
FINAL CONSUMPTION	**-**	**92**	**320**	**3**	**4**	**25**	**700**	**225**	**-**	**-**	**84**
INDUSTRY SECTOR	**-**	**3**	**-**	**-**	**-**	**1**	**10**	**195**	**-**	**-**	**-**
Iron and Steel	-	-	-	-	-	-	-	-	-	-	-
Chemical and Petrochem.	-	-	-	-	-	-	-	-	-	-	-
Non-Metallic Minerals	-	-	-	-	-	-	-	-	-	-	-
Non-specified	-	3	-	-	-	1	10	195	-	-	-
TRANSPORT SECTOR	**-**	**-**	**308**	**3**	**4**	**-**	**477**	**-**	**-**	**-**	**-**
Air	-	-	-	3	4	-	-	-	-	-	-
Road	-	-	308	-	-	-	477	-	-	-	-
Non-specified	-	-	-	-	-	-	-	-	-	-	-
OTHER SECTORS	**-**	**89**	**12**	**-**	**-**	**24**	**213**	**30**	**-**	**-**	**-**
Agriculture	-	-	10	-	-	-	168	-	-	-	-
Comm. and Publ. Services	-	1	-	-	-	-	36	6	-	-	-
Residential	-	88	-	-	-	24	9	24	-	-	-
Non-specified	-	-	2	-	-	-	-	-	-	-	-
NON-ENERGY USE	**-**	**-**	**-**	**-**	**-**	**-**	**-**	**-**	**-**	**-**	**84**

Uruguay

SUPPLY AND CONSUMPTION 1996	Gas (TJ)				Comb. Renew. & Waste (TJ)				(GWh)	(TJ)
	Natural Gas	Gas Works	Coke Ovens	Blast Furnaces	Solid Biomass	Gas/Liquids from Biomass	Municipal Waste	Industrial Waste	Electricity	Heat
Production	-	-	-	-	21977	-	-	-	6670	-
Imports	-	-	-	-	31	-	-	-	308	-
Exports	-	-	-	-	-	-	-	-	-343	-
Intl. Marine Bunkers	-	-	-	-		-	-	-		
Stock Changes	-	-	-	-	-	-	-	-	-	-
DOMESTIC SUPPLY	**-**	**-**	**-**	**-**	**22008**	**-**	**-**	**-**	**6635**	**-**
Transfers and Stat. Diff.	-	-	-	-	-	-	-	-	-1	-
TRANSFORMATION	**-**	**-**	**-**	**-**	**576**	**-**	**-**	**-**	**-**	**-**
Electricity and CHP Plants	-	-	-	-	515	-	-	-	-	-
Petroleum Refineries	-	-	-	-	-	-	-	-	-	-
Other Transform. Sector	-	-	-	-	61	-	-	-	-	-
ENERGY SECTOR	**-**	**-**	**-**	**-**	**-**	**-**	**-**	**-**	**108**	**-**
DISTRIBUTION LOSSES	**-**	**-**	**-**	**-**	**-**	**-**	**-**	**-**	**1322**	**-**
FINAL CONSUMPTION	**-**	**-**	**-**	**-**	**21432**	**-**	**-**	**-**	**5204**	**-**
INDUSTRY SECTOR	**-**	**-**	**-**	**-**	**8742**	**-**	**-**	**-**	**1285**	**-**
Iron and Steel	-	-	-	-	-	-	-	-	-	-
Chemical and Petrochem.	-	-	-	-	-	-	-	-	-	-
Non-Metallic Minerals	-	-	-	-	-	-	-	-	-	-
Non-specified	-	-	-	-	8742	-	-	-	1285	-
TRANSPORT SECTOR	**-**	**-**	**-**	**-**	**-**	**-**	**-**	**-**	**-**	**-**
Air	-	-	-	-	-	-	-	-	-	-
Road	-	-	-	-	-	-	-	-	-	-
Non-specified	-	-	-	-	-	-	-	-	-	-
OTHER SECTORS	**-**	**-**	**-**	**-**	**12690**	**-**	**-**	**-**	**3919**	**-**
Agriculture	-	-	-	-	-	-	-	-	66	-
Comm. and Publ. Services	-	-	-	-	113	-	-	-	1456	-
Residential	-	-	-	-	12577	-	-	-	2397	-
Non-specified	-	-	-	-	-	-	-	-	-	-
NON-ENERGY USE	**-**	**-**	**-**	**-**	**-**	**-**	**-**	**-**	**-**	**-**

APPROVISIONNEMENT ET DEMANDE 1997	*Gaz* (TJ)				*En. Re. Comb. & Déchets* (TJ)				(GWh)	(TJ)
	Gaz naturel	*Usines à gaz*	*Cokeries*	*Hauts fourneaux*	*Biomasse solide*	*Gaz/Liquides tirés de biomasse*	*Déchets urbains*	*Déchets industriels*	*Electricité*	*Chaleur*
Production	-	-	-	-	22105	-	-	-	7149	-
Imports	-	1	-	-	31	-	-	-	271	-
Exports	-	-	-	-	-	-	-	-	-415	-
Intl. Marine Bunkers	-	-	-	-	-	-	-	-	-	-
Stock Changes	-	-	-	-	-	-	-	-	-	-
DOMESTIC SUPPLY	**-**	**1**	**-**	**-**	**22136**	**-**	**-**	**-**	**7005**	**-**
Transfers and Stat. Diff.	-	-	-	-	-	-	-	-	25	-
TRANSFORMATION	**-**	**-**	**-**	**-**	**628**	**-**	**-**	**-**	**-**	**-**
Electricity and CHP Plants	-	-	-	-	535	-	-	-	-	-
Petroleum Refineries	-	-	-	-	-	-	-	-	-	-
Other Transform. Sector	-	-	-	-	93	-	-	-	-	-
ENERGY SECTOR	**-**	**-**	**-**	**-**	**-**	**-**	**-**	**-**	**107**	**-**
DISTRIBUTION LOSSES	**-**	**-**	**-**	**-**	**-**	**-**	**-**	**-**	**1340**	**-**
FINAL CONSUMPTION	**-**	**1**	**-**	**-**	**21511**	**-**	**-**	**-**	**5583**	**-**
INDUSTRY SECTOR	**-**	**1**	**-**	**-**	**8821**	**-**	**-**	**-**	**1300**	**-**
Iron and Steel	-	1	-	-	-	-	-	-	-	-
Chemical and Petrochem.	-	-	-	-	-	-	-	-	-	-
Non-Metallic Minerals	-	-	-	-	-	-	-	-	-	-
Non-specified	-	-	-	-	8821	-	-	-	1300	-
TRANSPORT SECTOR	**-**	**-**	**-**	**-**	**-**	**-**	**-**	**-**	**-**	**-**
Air	-	-	-	-	-	-	-	-	-	-
Road	-	-	-	-	-	-	-	-	-	-
Non-specified	-	-	-	-	-	-	-	-	-	-
OTHER SECTORS	**-**	**-**	**-**	**-**	**12690**	**-**	**-**	**-**	**4283**	**-**
Agriculture	-	-	-	-	-	-	-	-	190	-
Comm. and Publ. Services	-	-	-	-	113	-	-	-	1632	-
Residential	-	-	-	-	12577	-	-	-	2461	-
Non-specified	-	-	-	-	-	-	-	-	-	-
NON-ENERGY USE	**-**	**-**	**-**	**-**	**-**	**-**	**-**	**-**	**-**	**-**

Uzbekistan / Ouzbékistan : 1996

SUPPLY AND CONSUMPTION *APPROVISIONNEMENT ET DEMANDE*	Coal / *Charbon* (1000 tonnes)							Oil / *Pétrole* (1000 tonnes)			
	Coking Coal *Charbon à coke*	Other Bit. Coal *Autres charb. bit.*	Sub-Bit. Coal *Charbon sous-bit.*	Lignite *Lignite*	Peat *Tourbe*	Oven and Gas Coke *Coke de four/gaz*	Pat. Fuel and BKB *Agg./briq. de lignite*	Crude Oil *Pétrole brut*	NGL *LGN*	Feed-stocks *Produits d'aliment.*	Additives *Additifs*
Production	-	74	-	2763	-	-	42	5198	2519	-	-
From Other Sources	-	-	-	-	-	-	-	-	-	-	-
Imports	-	5	-	46	-	-	-	4	-	-	8
Exports	-	-	-	-8	-	-	-	-289	-494	-	-
Intl. Marine Bunkers	-	-	-	-	-	-	-	-	-	-	-
Stock Changes	-	5	-	512	-	-	-	68	-6	-	-
DOMESTIC SUPPLY	**-**	**84**	**-**	**3313**	**-**	**-**	**42**	**4981**	**2019**	**-**	**8**
Transfers	-	-	-	-	-	-	-	-27	-	-	-
Statistical Differences	-	-	-	-	-	-	-	-28	-4	-	-
TRANSFORMATION	**-**	**39**	**-**	**2300**	**-**	**-**	**-**	**4703**	**1996**	**-**	**8**
Electricity Plants	-	-	-	1228	-	-	-	-	-	-	-
CHP Plants	-	-	-	1065	-	-	-	-	-	-	-
Heat Plants	-	-	-	7	-	-	-	-	-	-	-
Blast Furnaces/Gas Works	-	-	-	-	-	-	-	-	-	-	-
Coke/Pat. Fuel/BKB Plants	-	39	-	-	-	-	-	-	-	-	-
Petroleum Refineries	-	-	-	-	-	-	-	4703	1996	-	8
Petrochemical Industry	-	-	-	-	-	-	-	-	-	-	-
Liquefaction	-	-	-	-	-	-	-	-	-	-	-
Other Transform. Sector	-	-	-	-	-	-	-	-	-	-	-
ENERGY SECTOR	**-**	**-**	**-**	**37**	**-**	**-**	**-**	**14**	**2**	**-**	**-**
Coal Mines	-	-	-	15	-	-	-	-	-	-	-
Oil and Gas Extraction	-	-	-	-	-	-	-	14	2	-	-
Petroleum Refineries	-	-	-	-	-	-	-	-	-	-	-
Electr., CHP+Heat Plants	-	-	-	22	-	-	-	-	-	-	-
Pumped Storage (Elec.)	-	-	-	-	-	-	-	-	-	-	-
Other Energy Sector	-	-	-	-	-	-	-	-	-	-	-
Distribution Losses	-	9	-	135	-	-	-	65	17	-	-
FINAL CONSUMPTION	**-**	**36**	**-**	**841**	**-**	**-**	**42**	**144**	**-**	**-**	**-**
INDUSTRY SECTOR	**-**	**24**	**-**	**177**	**-**	**-**	**-**	**144**	**-**	**-**	**-**
Iron and Steel	-	-	-	-	-	-	-	-	-	-	-
Chemical and Petrochem.	-	-	-	-	-	-	-	144	-	-	-
of which: Feedstocks	-	-	-	-	-	-	-	*144*	-	-	-
Non-Ferrous Metals	-	-	-	-	-	-	-	-	-	-	-
Non-Metallic Minerals	-	-	-	-	-	-	-	-	-	-	-
Transport Equipment	-	-	-	-	-	-	-	-	-	-	-
Machinery	-	-	-	-	-	-	-	-	-	-	-
Mining and Quarrying	-	-	-	-	-	-	-	-	-	-	-
Food and Tobacco	-	-	-	-	-	-	-	-	-	-	-
Paper, Pulp and Print	-	-	-	-	-	-	-	-	-	-	-
Wood and Wood Products	-	-	-	-	-	-	-	-	-	-	-
Construction	-	-	-	-	-	-	-	-	-	-	-
Textile and Leather	-	-	-	-	-	-	-	-	-	-	-
Non-specified	-	24	-	177	-	-	-	-	-	-	-
TRANSPORT SECTOR	**-**	**-**	**-**	**-**	**-**	**-**	**-**	**-**	**-**	**-**	**-**
Air	-	-	-	-	-	-	-	-	-	-	-
Road	-	-	-	-	-	-	-	-	-	-	-
Rail	-	-	-	-	-	-	-	-	-	-	-
Pipeline Transport	-	-	-	-	-	-	-	-	-	-	-
Internal Navigation	-	-	-	-	-	-	-	-	-	-	-
Non-specified	-	-	-	-	-	-	-	-	-	-	-
OTHER SECTORS	**-**	**12**	**-**	**664**	**-**	**-**	**42**	**-**	**-**	**-**	**-**
Agriculture	-	-	-	20	-	-	-	-	-	-	-
Comm. and Publ. Services	-	-	-	-	-	-	-	-	-	-	-
Residential	-	-	-	56	-	-	-	-	-	-	-
Non-specified	-	12	-	588	-	-	42	-	-	-	-
NON-ENERGY USE	**-**	**-**	**-**	**-**	**-**	**-**	**-**	**-**	**-**	**-**	**-**
in Industry/Trans./Energy	-	-	-	-	-	-	-	-	-	-	-
in Transport	-	-	-	-	-	-	-	-	-	-	-
in Other Sectors	-	-	-	-	-	-	-	-	-	-	-

Uzbekistan / Ouzbékistan : 1996

SUPPLY AND CONSUMPTION / *APPROVISIONNEMENT ET DEMANDE*	Oil cont. / *Pétrole cont.* (1000 tonnes)										
	Refinery Gas / *Gaz de raffinerie*	LPG + Ethane / *GPL + éthane*	Motor Gasoline / *Essence moteur*	Aviation Gasoline / *Essence aviation*	Jet Fuel / *Carbu-réacteurs*	Kerosene / *Kérosène*	Gas/ Diesel / *Gazole*	Heavy Fuel Oil / *Fioul lourd*	Naphtha / *Naphta*	Petrol. Coke / *Coke de pétrole*	Other Prod. / *Autres prod.*
Production	198	14	1202	5	300	55	1861	2181	-	25	730
From Other Sources	-	-	-	-	-	-	-	-	-	-	-
Imports	-	1	6	-	19	-	14	-	-	-	2
Exports	-	-	-42	-	-	-	-62	-63	-	-25	-175
Intl. Marine Bunkers	-	-	-	-	-	-	-	-	-	-	-
Stock Changes	-	-1	12	-	-12	-9	133	-2	-	-	-16
DOMESTIC SUPPLY	**198**	**14**	**1178**	**5**	**307**	**46**	**1946**	**2116**	**-**	**-**	**541**
Transfers	-	27	-	-	-	-	-	-	-	-	-
Statistical Differences	-	-	-	-	-	-	-	-	-	-	-
TRANSFORMATION	**-**	**-**	**-**	**-**	**-**	**3**	**5**	**1977**	**-**	**-**	**8**
Electricity Plants	-	-	-	-	-	-	5	630	-	-	-
CHP Plants	-	-	-	-	-	-	-	977	-	-	-
Heat Plants	-	-	-	-	-	3	-	370	-	-	8
Blast Furnaces/Gas Works	-	-	-	-	-	-	-	-	-	-	-
Coke/Pat. Fuel/BKB Plants	-	-	-	-	-	-	-	-	-	-	-
Petroleum Refineries	-	-	-	-	-	-	-	-	-	-	-
Petrochemical Industry	-	-	-	-	-	-	-	-	-	-	-
Liquefaction	-	-	-	-	-	-	-	-	-	-	-
Other Transform. Sector	-	-	-	-	-	-	-	-	-	-	-
ENERGY SECTOR	**198**	**-**	**-**	**-**	**-**	**-**	**-**	**30**	**-**	**-**	**-**
Coal Mines	-	-	-	-	-	-	-	-	-	-	-
Oil and Gas Extraction	-	-	-	-	-	-	-	-	-	-	-
Petroleum Refineries	198	-	-	-	-	-	-	-	-	-	-
Electr., CHP+Heat Plants	-	-	-	-	-	-	-	30	-	-	-
Pumped Storage (Elec.)	-	-	-	-	-	-	-	-	-	-	-
Other Energy Sector	-	-	-	-	-	-	-	-	-	-	-
Distribution Losses	-	-	-	-	-	-	-	-	-	-	-
FINAL CONSUMPTION	**-**	**41**	**1178**	**5**	**307**	**43**	**1941**	**109**	**-**	**-**	**533**
INDUSTRY SECTOR	**-**	**1**	**64**	**-**	**-**	**14**	**181**	**6**	**-**	**-**	**3**
Iron and Steel	-	-	-	-	-	-	-	-	-	-	-
Chemical and Petrochem.	-	-	-	-	-	-	-	-	-	-	-
of which: Feedstocks	-	-	-	-	-	-	-	-	-	-	-
Non-Ferrous Metals	-	-	-	-	-	-	-	-	-	-	-
Non-Metallic Minerals	-	-	-	-	-	-	-	6	-	-	-
Transport Equipment	-	-	-	-	-	-	-	-	-	-	-
Machinery	-	-	2	-	-	-	-	-	-	-	-
Mining and Quarrying	-	-	-	-	-	-	-	-	-	-	-
Food and Tobacco	-	-	-	-	-	-	-	-	-	-	-
Paper, Pulp and Print	-	-	-	-	-	-	-	-	-	-	-
Wood and Wood Products	-	-	-	-	-	-	-	-	-	-	-
Construction	-	-	62	-	-	-	181	-	-	-	-
Textile and Leather	-	-	-	-	-	14	-	-	-	-	3
Non-specified	-	1	-	-	-	-	-	-	-	-	-
TRANSPORT SECTOR	**-**	**15**	**1063**	**5**	**307**	**-**	**625**	**-**	**-**	**-**	**-**
Air	-	-	-	5	307	-	-	-	-	-	-
Road	-	15	1063	-	-	-	494	-	-	-	-
Rail	-	-	-	-	-	-	129	-	-	-	-
Pipeline Transport	-	-	-	-	-	-	-	-	-	-	-
Internal Navigation	-	-	-	-	-	-	-	-	-	-	-
Non-specified	-	-	-	-	-	-	2	-	-	-	-
OTHER SECTORS	**-**	**25**	**51**	**-**	**-**	**29**	**1135**	**103**	**-**	**-**	**12**
Agriculture	-	-	46	-	-	11	1004	-	-	-	1
Comm. and Publ. Services	-	-	-	-	-	-	-	-	-	-	-
Residential	-	12	-	-	-	1	-	-	-	-	1
Non-specified	-	13	5	-	-	17	131	103	-	-	10
NON-ENERGY USE	**-**	**-**	**-**	**-**	**-**	**-**	**-**	**-**	**-**	**-**	**518**
in Industry/Transf./Energy	-	-	-	-	-	-	-	-	-	-	365
in Transport	-	-	-	-	-	-	-	-	-	-	-
in Other Sectors	-	-	-	-	-	-	-	-	-	-	153

Uzbekistan / Ouzbékistan : 1996

SUPPLY AND CONSUMPTION / *APPROVISIONNEMENT ET DEMANDE*	Gas / *Gaz* (TJ) Natural Gas / *Gaz naturel*	Gas Works / *Usines à gaz*	Coke Ovens / *Cokeries*	Blast Furnaces / *Hauts fourneaux*	Comb. Renew. & Waste / *En. Re. Comb. & Déchets* (TJ) Solid Biomass / *Biomasse solide*	Gas/Liquids from Biomass / *Gaz/Liquides tirés de biomasse*	Municipal Waste / *Déchets urbains*	Industrial Waste / *Déchets industriels*	(GWh) Electricity / *Electricité*	(TJ) Heat / *Chaleur*
Production	1783203	-	-	-	10	-	-	-	45419	107874
From Other Sources	-	-	-	-	-	-	-	-	-	-
Imports	159133	-	-	-	-	-	-	-	13918	-
Exports	-343880	-	-	-	-	-	-	-	-12826	-
Intl. Marine Bunkers	-	-	-	-	-	-	-	-	-	-
Stock Changes	-19550	-	-	-	-	-	-	-	-	-
DOMESTIC SUPPLY	**1578906**	-	-	-	**10**	-	-	-	**46511**	**107874**
Transfers	-	-	-	-	-	-	-	-	-	-
Statistical Differences	-	-	-	-	-	-	-	-	-	-
TRANSFORMATION	**470780**	-	-	-	-	-	-	-	-	-
Electricity Plants	193665	-	-	-	-	-	-	-	-	-
CHP Plants	213691	-	-	-	-	-	-	-	-	-
Heat Plants	63424	-	-	-	-	-	-	-	-	-
Blast Furnaces/Gas Works	-	-	-	-	-	-	-	-	-	-
Coke/Pat. Fuel/BKB Plants	-	-	-	-	-	-	-	-	-	-
Petroleum Refineries	-	-	-	-	-	-	-	-	-	-
Petrochemical Industry	-	-	-	-	-	-	-	-	-	-
Liquefaction	-	-	-	-	-	-	-	-	-	-
Other Transform. Sector	-	-	-	-	-	-	-	-	-	-
ENERGY SECTOR	**77520**	-	-	-	-	-	-	-	**3961**	-
Coal Mines	-	-	-	-	-	-	-	-	-	-
Oil and Gas Extraction	15231	-	-	-	-	-	-	-	-	-
Petroleum Refineries	-	-	-	-	-	-	-	-	-	-
Electr., CHP+Heat Plants	6896	-	-	-	-	-	-	-	2662	-
Pumped Storage (Elec.)	-	-	-	-	-	-	-	-	-	-
Other Energy Sector	55393	-	-	-	-	-	-	-	1299	-
Distribution Losses	81196	-	-	-	-	-	-	-	4061	-
FINAL CONSUMPTION	**949410**	-	-	-	**10**	-	-	-	**38489**	**107874**
INDUSTRY SECTOR	**243919**	-	-	-	-	-	-	-	**14070**	-
Iron and Steel	-	-	-	-	-	-	-	-	-	-
Chemical and Petrochem.	59940	-	-	-	-	-	-	-	-	-
of which: Feedstocks	*59940*	-	-	-	-	-	-	-	-	-
Non-Ferrous Metals	-	-	-	-	-	-	-	-	-	-
Non-Metallic Minerals	-	-	-	-	-	-	-	-	-	-
Transport Equipment	-	-	-	-	-	-	-	-	-	-
Machinery	-	-	-	-	-	-	-	-	-	-
Mining and Quarrying	-	-	-	-	-	-	-	-	-	-
Food and Tobacco	-	-	-	-	-	-	-	-	-	-
Paper, Pulp and Print	-	-	-	-	-	-	-	-	-	-
Wood and Wood Products	-	-	-	-	-	-	-	-	-	-
Construction	-	-	-	-	-	-	-	-	-	-
Textile and Leather	-	-	-	-	-	-	-	-	-	-
Non-specified	183979	-	-	-	-	-	-	-	14070	-
TRANSPORT SECTOR	**62365**	-	-	-	-	-	-	-	**1398**	-
Air	-	-	-	-	-	-	-	-	-	-
Road	2842	-	-	-	-	-	-	-	-	-
Rail	-	-	-	-	-	-	-	-	135	-
Pipeline Transport	59523	-	-	-	-	-	-	-	890	-
Internal Navigation	-	-	-	-	-	-	-	-	-	-
Non-specified	-	-	-	-	-	-	-	-	373	-
OTHER SECTORS	**643126**	-	-	-	**10**	-	-	-	**23021**	**107874**
Agriculture	8487	-	-	-	-	-	-	-	12740	-
Comm. and Publ. Services	58652	-	-	-	-	-	-	-	3157	-
Residential	575987	-	-	-	-	-	-	-	7124	-
Non-specified	-	-	-	-	10	-	-	-	-	107874
NON-ENERGY USE	-	-	-	-	-	-	-	-	-	-
in Industry/Transf./Energy	-	-	-	-	-	-	-	-	-	-
in Transport	-	-	-	-	-	-	-	-	-	-
in Other Sectors	-	-	-	-	-	-	-	-	-	-

Uzbekistan / Ouzbékistan : 1997

SUPPLY AND CONSUMPTION / *APPROVISIONNEMENT ET DEMANDE*	Coal / *Charbon* (1000 tonnes)							Oil / *Pétrole* (1000 tonnes)			
	Coking Coal / *Charbon à coke*	Other Bit. Coal / *Autres charb. bit.*	Sub-Bit. Coal / *Charbon sous-bit.*	Lignite / *Lignite*	Peat / *Tourbe*	Oven and Gas Coke / *Coke de four/gaz*	Pat. Fuel and BKB / *Agg./briq. de lignite*	Crude Oil / *Pétrole brut*	NGL / *LGN*	Feed-stocks / *Produits d'aliment.*	Additives / *Additifs*
Production	-	59	-	2888	-	-	45	5456	2668	-	-
From Other Sources	-	-	-	-	-	-	-	-	-	-	-
Imports	-	-	-	27	-	-	-	-	-	-	7
Exports	-	-	-	-30	-	-	-	-324	-589	-	-
Intl. Marine Bunkers	-	-	-	-	-	-	-	-	-	-	-
Stock Changes	-	16	-	-168	-	-	-	17	24	-	-
DOMESTIC SUPPLY	**-**	**75**	**-**	**2717**	**-**	**-**	**45**	**5149**	**2103**	**-**	**7**
Transfers	-	-	-	-	-	-	-	-32	-	-	-
Statistical Differences	-	-	-	-	-	-	-	-13	-10	-	-
TRANSFORMATION	**-**	**42**	**-**	**2223**	**-**	**-**	**-**	**4897**	**2078**	**-**	**7**
Electricity Plants	-	-	-	1191	-	-	-	-	-	-	-
CHP Plants	-	-	-	1028	-	-	-	-	-	-	-
Heat Plants	-	-	-	4	-	-	-	-	-	-	-
Blast Furnaces/Gas Works	-	-	-	-	-	-	-	-	-	-	-
Coke/Pat. Fuel/BKB Plants	-	42	-	-	-	-	-	-	-	-	-
Petroleum Refineries	-	-	-	-	-	-	-	4897	2078	-	7
Petrochemical Industry	-	-	-	-	-	-	-	-	-	-	-
Liquefaction	-	-	-	-	-	-	-	-	-	-	-
Other Transform. Sector	-	-	-	-	-	-	-	-	-	-	-
ENERGY SECTOR	**-**	**-**	**-**	**6**	**-**	**-**	**-**	**14**	**1**	**-**	**-**
Coal Mines	-	-	-	-	-	-	-	-	-	-	-
Oil and Gas Extraction	-	-	-	-	-	-	-	14	1	-	-
Petroleum Refineries	-	-	-	-	-	-	-	-	-	-	-
Electr., CHP+Heat Plants	-	-	-	6	-	-	-	-	-	-	-
Pumped Storage (Elec.)	-	-	-	-	-	-	-	-	-	-	-
Other Energy Sector	-	-	-	-	-	-	-	-	-	-	-
Distribution Losses	-	8	-	28	-	-	-	74	14	-	-
FINAL CONSUMPTION	**-**	**25**	**-**	**460**	**-**	**-**	**45**	**119**	**-**	**-**	**-**
INDUSTRY SECTOR	**-**	**17**	**-**	**131**	**-**	**-**	**-**	**119**	**-**	**-**	**-**
Iron and Steel	-	-	-	-	-	-	-	-	-	-	-
Chemical and Petrochem.	-	-	-	-	-	-	-	119	-	-	-
of which: Feedstocks	-	-	-	-	-	-	-	*119*	-	-	-
Non-Ferrous Metals	-	-	-	-	-	-	-	-	-	-	-
Non-Metallic Minerals	-	-	-	-	-	-	-	-	-	-	-
Transport Equipment	-	-	-	-	-	-	-	-	-	-	-
Machinery	-	-	-	-	-	-	-	-	-	-	-
Mining and Quarrying	-	-	-	-	-	-	-	-	-	-	-
Food and Tobacco	-	-	-	-	-	-	-	-	-	-	-
Paper, Pulp and Print	-	-	-	-	-	-	-	-	-	-	-
Wood and Wood Products	-	-	-	-	-	-	-	-	-	-	-
Construction	-	-	-	-	-	-	-	-	-	-	-
Textile and Leather	-	-	-	-	-	-	-	-	-	-	-
Non-specified	-	17	-	131	-	-	-	-	-	-	-
TRANSPORT SECTOR	**-**	**-**	**-**	**-**	**-**	**-**	**-**	**-**	**-**	**-**	**-**
Air	-	-	-	-	-	-	-	-	-	-	-
Road	-	-	-	-	-	-	-	-	-	-	-
Rail	-	-	-	-	-	-	-	-	-	-	-
Pipeline Transport	-	-	-	-	-	-	-	-	-	-	-
Internal Navigation	-	-	-	-	-	-	-	-	-	-	-
Non-specified	-	-	-	-	-	-	-	-	-	-	-
OTHER SECTORS	**-**	**8**	**-**	**329**	**-**	**-**	**45**	**-**	**-**	**-**	**-**
Agriculture	-	-	-	9	-	-	-	-	-	-	-
Comm. and Publ. Services	-	-	-	-	-	-	-	-	-	-	-
Residential	-	-	-	34	-	-	-	-	-	-	-
Non-specified	-	8	-	286	-	-	45	-	-	-	-
NON-ENERGY USE	**-**	**-**	**-**	**-**	**-**	**-**	**-**	**-**	**-**	**-**	**-**
in Industry/Trans./Energy	-	-	-	-	-	-	-	-	-	-	-
in Transport	-	-	-	-	-	-	-	-	-	-	-
in Other Sectors	-	-	-	-	-	-	-	-	-	-	-

Uzbekistan / Ouzbékistan : 1997

SUPPLY AND CONSUMPTION / *APPROVISIONNEMENT ET DEMANDE*	Oil cont. / *Pétrole cont.* (1000 tonnes)										
	Refinery Gas / *Gaz de raffinerie*	LPG + Ethane / *GPL + éthane*	Motor Gasoline / *Essence moteur*	Aviation Gasoline / *Essence aviation*	Jet Fuel / *Carbu-réacteurs*	Kerosene / *Kérosène*	Gas/ Diesel / *Gazole*	Heavy Fuel Oil / *Fioul lourd*	Naphtha / *Naphta*	Petrol. Coke / *Coke de pétrole*	Other Prod. / *Autres prod.*
Production	221	10	1344	3	272	62	2035	1979	-	46	795
From Other Sources	-	-	-	-	-	-	-	-	-	-	-
Imports	-	-	-	-	33	-	-	-	-	-	1
Exports	-	-	-29	-	-5	-	-245	-4	-	-	-83
Intl. Marine Bunkers	-	-	-	-	-	-	-	-	-	-	-
Stock Changes	-	-	92	-	-26	7	221	73	-	-46	24
DOMESTIC SUPPLY	**221**	**10**	**1407**	**3**	**274**	**69**	**2011**	**2048**	**-**	**-**	**737**
Transfers	-	32	-	-	-	-	-	-	-	-	-
Statistical Differences	-	-	-	-	-	-	-	-	-	-	-
TRANSFORMATION	**-**	**-**	**-**	**-**	**-**	**6**	**6**	**2013**	**-**	**-**	**8**
Electricity Plants	-	-	-	-	-	-	6	737	-	-	-
CHP Plants	-	-	-	-	-	-	-	954	-	-	-
Heat Plants	-	-	-	-	-	6	-	322	-	-	8
Blast Furnaces/Gas Works	-	-	-	-	-	-	-	-	-	-	-
Coke/Pat. Fuel/BKB Plants	-	-	-	-	-	-	-	-	-	-	-
Petroleum Refineries	-	-	-	-	-	-	-	-	-	-	-
Petrochemical Industry	-	-	-	-	-	-	-	-	-	-	-
Liquefaction	-	-	-	-	-	-	-	-	-	-	-
Other Transform. Sector	-	-	-	-	-	-	-	-	-	-	-
ENERGY SECTOR	**221**	**-**	**-**	**-**	**-**	**-**	**-**	**27**	**-**	**-**	**-**
Coal Mines	-	-	-	-	-	-	-	-	-	-	-
Oil and Gas Extraction	-	-	-	-	-	-	-	-	-	-	-
Petroleum Refineries	221	-	-	-	-	-	-	-	-	-	-
Electr., CHP+Heat Plants	-	-	-	-	-	-	-	27	-	-	-
Pumped Storage (Elec.)	-	-	-	-	-	-	-	-	-	-	-
Other Energy Sector	-	-	-	-	-	-	-	-	-	-	-
Distribution Losses	-	-	-	-	-	-	-	-	-	-	-
FINAL CONSUMPTION	**-**	**42**	**1407**	**3**	**274**	**63**	**2005**	**8**	**-**	**-**	**729**
INDUSTRY SECTOR	**-**	**-**	**63**	**-**	**-**	**12**	**196**	**5**	**-**	**-**	**3**
Iron and Steel	-	-	-	-	-	-	-	-	-	-	-
Chemical and Petrochem.	-	-	-	-	-	-	-	-	-	-	-
of which: Feedstocks	-	-	-	-	-	-	-	-	-	-	-
Non-Ferrous Metals	-	-	-	-	-	-	-	-	-	-	-
Non-Metallic Minerals	-	-	-	-	-	-	-	5	-	-	-
Transport Equipment	-	-	-	-	-	-	-	-	-	-	-
Machinery	-	-	4	-	-	-	-	-	-	-	-
Mining and Quarrying	-	-	-	-	-	-	-	-	-	-	-
Food and Tobacco	-	-	-	-	-	-	-	-	-	-	-
Paper, Pulp and Print	-	-	-	-	-	-	-	-	-	-	-
Wood and Wood Products	-	-	-	-	-	-	-	-	-	-	-
Construction	-	-	58	-	-	-	196	-	-	-	-
Textile and Leather	-	-	-	-	-	12	-	-	-	-	3
Non-specified	-	-	1	-	-	-	-	-	-	-	-
TRANSPORT SECTOR	**-**	**14**	**1309**	**3**	**274**	**-**	**539**	**-**	**-**	**-**	**-**
Air	-	-	-	3	274	-	-	-	-	-	-
Road	-	14	1309	-	-	-	418	-	-	-	-
Rail	-	-	-	-	-	-	121	-	-	-	-
Pipeline Transport	-	-	-	-	-	-	-	-	-	-	-
Internal Navigation	-	-	-	-	-	-	-	-	-	-	-
Non-specified	-	-	-	-	-	-	-	-	-	-	-
OTHER SECTORS	**-**	**28**	**35**	**-**	**-**	**51**	**1270**	**3**	**-**	**-**	**14**
Agriculture	-	-	30	-	-	11	1061	1	-	-	1
Comm. and Publ. Services	-	-	-	-	-	-	-	-	-	-	-
Residential	-	14	1	-	-	1	6	-	-	-	1
Non-specified	-	14	4	-	-	39	203	2	-	-	12
NON-ENERGY USE	**-**	**-**	**-**	**-**	**-**	**-**	**-**	**-**	**-**	**-**	**712**
in Industry/Transf./Energy	-	-	-	-	-	-	-	-	-	-	537
in Transport	-	-	-	-	-	-	-	-	-	-	-
in Other Sectors	-	-	-	-	-	-	-	-	-	-	175

Uzbekistan / Ouzbékistan : 1997

	Gas / *Gaz* (TJ)				Comb. Renew. & Waste / *En. Re. Comb. & Déchets* (TJ)				(GWh)	(TJ)
SUPPLY AND CONSUMPTION *APPROVISIONNEMENT ET DEMANDE*	Natural Gas *Gaz naturel*	Gas Works *Usines à gaz*	Coke Ovens *Cokeries*	Blast Furnaces *Hauts fourneaux*	Solid Biomass *Biomasse solide*	Gas/Liquids from Biomass *Gaz/Liquides tirés de biomasse*	Municipal Waste *Déchets urbains*	Industrial Waste *Déchets industriels*	Electricity *Electricité*	Heat *Chaleur*
Production	1848372	-	-	-	10	-	-	-	46054	100100
From Other Sources	-	-	-	-	-	-	-	-	-	-
Imports	104573	-	-	-	-	-	-	-	12418	-
Exports	-374986	-	-	-	-	-	-	-	-11489	-
Intl. Marine Bunkers	-	-	-	-	-	-	-	-	-	-
Stock Changes	5987	-	-	-	-	-	-	-	-	-
DOMESTIC SUPPLY	**1583946**	**-**	**-**	**-**	**10**	**-**	**-**	**-**	**46983**	**100100**
Transfers	-	-	-	-	-	-	-	-	-	-
Statistical Differences	-	-	-	-	-	-	-	-	-	-
TRANSFORMATION	**478523**	**-**	**-**	**-**	**-**	**-**	**-**	**-**	**-**	**-**
Electricity Plants	207343	-	-	-	-	-	-	-	-	-
CHP Plants	211978	-	-	-	-	-	-	-	-	-
Heat Plants	59202	-	-	-	-	-	-	-	-	-
Blast Furnaces/Gas Works	-	-	-	-	-	-	-	-	-	-
Coke/Pat. Fuel/BKB Plants	-	-	-	-	-	-	-	-	-	-
Petroleum Refineries	-	-	-	-	-	-	-	-	-	-
Petrochemical Industry	-	-	-	-	-	-	-	-	-	-
Liquefaction	-	-	-	-	-	-	-	-	-	-
Other Transform. Sector	-	-	-	-	-	-	-	-	-	-
ENERGY SECTOR	**75285**	**-**	**-**	**-**	**-**	**-**	**-**	**-**	**4035**	**-**
Coal Mines	-	-	-	-	-	-	-	-	-	-
Oil and Gas Extraction	15724	-	-	-	-	-	-	-	-	-
Petroleum Refineries	-	-	-	-	-	-	-	-	-	-
Electr., CHP+Heat Plants	8676	-	-	-	-	-	-	-	2682	-
Pumped Storage (Elec.)	-	-	-	-	-	-	-	-	-	-
Other Energy Sector	50885	-	-	-	-	-	-	-	1353	-
Distribution Losses	65851	-	-	-	-	-	-	-	4008	-
FINAL CONSUMPTION	**964287**	**-**	**-**	**-**	**10**	**-**	**-**	**-**	**38940**	**100100**
INDUSTRY SECTOR	**190102**	**-**	**-**	**-**	**-**	**-**	**-**	**-**	**14921**	**-**
Iron and Steel	-	-	-	-	-	-	-	-	-	-
Chemical and Petrochem.	59963	-	-	-	-	-	-	-	-	-
of which: Feedstocks	*59963*	-	-	-	-	-	-	-	-	-
Non-Ferrous Metals	-	-	-	-	-	-	-	-	-	-
Non-Metallic Minerals	-	-	-	-	-	-	-	-	-	-
Transport Equipment	-	-	-	-	-	-	-	-	-	-
Machinery	-	-	-	-	-	-	-	-	-	-
Mining and Quarrying	-	-	-	-	-	-	-	-	-	-
Food and Tobacco	-	-	-	-	-	-	-	-	-	-
Paper, Pulp and Print	-	-	-	-	-	-	-	-	-	-
Wood and Wood Products	-	-	-	-	-	-	-	-	-	-
Construction	-	-	-	-	-	-	-	-	-	-
Textile and Leather	-	-	-	-	-	-	-	-	-	-
Non-specified	130139	-	-	-	-	-	-	-	14921	-
TRANSPORT SECTOR	**57478**	**-**	**-**	**-**	**-**	**-**	**-**	**-**	**1282**	**-**
Air	-	-	-	-	-	-	-	-	-	-
Road	2539	-	-	-	-	-	-	-	-	-
Rail	-	-	-	-	-	-	-	-	156	-
Pipeline Transport	54939	-	-	-	-	-	-	-	820	-
Internal Navigation	-	-	-	-	-	-	-	-	-	-
Non-specified	-	-	-	-	-	-	-	-	306	-
OTHER SECTORS	**716707**	**-**	**-**	**-**	**10**	**-**	**-**	**-**	**22737**	**100100**
Agriculture	6328	-	-	-	-	-	-	-	12647	-
Comm. and Publ. Services	118403	-	-	-	-	-	-	-	3019	-
Residential	591976	-	-	-	-	-	-	-	7071	-
Non-specified	-	-	-	-	10	-	-	-	-	100100
NON-ENERGY USE	**-**	**-**	**-**	**-**	**-**	**-**	**-**	**-**	**-**	**-**
in Industry/Transf./Energy	-	-	-	-	-	-	-	-	-	-
in Transport	-	-	-	-	-	-	-	-	-	-
in Other Sectors	-	-	-	-	-	-	-	-	-	-

Venezuela / Vénézuela : 1996

SUPPLY AND CONSUMPTION / *APPROVISIONNEMENT ET DEMANDE*	Coal / *Charbon* (1000 tonnes)							Oil / *Pétrole* (1000 tonnes)			
	Coking Coal / *Charbon à coke*	Other Bit. Coal / *Autres charb. bit.*	Sub-Bit. Coal / *Charbon sous-bit.*	Lignite / *Lignite*	Peat / *Tourbe*	Oven and Gas Coke / *Coke de four/gaz*	Pat. Fuel and BKB / *Agg./briq. de lignite*	Crude Oil / *Pétrole brut*	NGL / *LGN*	Feed-stocks / *Produits d'aliment.*	Additives / *Additifs*
Production	-	3486	-	-	-	-	-	156274	4861	-	-
From Other Sources	-	-	-	-	-	-	-	-	-	-	-
Imports	-	294	-	-	-	-	-	-	-	-	-
Exports	-	-3452	-	-	-	-	-	-99640	-	-	-
Intl. Marine Bunkers	-	-	-	-	-	-	-	-	-	-	-
Stock Changes	-	-	-	-	-	-	-	-962	-	-	-
DOMESTIC SUPPLY	**-**	**328**	**-**	**-**	**-**	**-**	**-**	**55672**	**4861**	**-**	**-**
Transfers	-	-	-	-	-	-	-	-	-4861	-	-
Statistical Differences	-	-	-	-	-	-	-	-791	-	-	-
TRANSFORMATION	**-**	**-**	**-**	**-**	**-**	**-**	**-**	**51344**	**-**	**-**	**-**
Electricity Plants	-	-	-	-	-	-	-	-	-	-	-
CHP Plants	-	-	-	-	-	-	-	-	-	-	-
Heat Plants	-	-	-	-	-	-	-	-	-	-	-
Blast Furnaces/Gas Works	-	-	-	-	-	-	-	-	-	-	-
Coke/Pat. Fuel/BKB Plants	-	-	-	-	-	-	-	-	-	-	-
Petroleum Refineries	-	-	-	-	-	-	-	51344	-	-	-
Petrochemical Industry	-	-	-	-	-	-	-	-	-	-	-
Liquefaction	-	-	-	-	-	-	-	-	-	-	-
Other Transform. Sector	-	-	-	-	-	-	-	-	-	-	-
ENERGY SECTOR	**-**	**-**	**-**	**-**	**-**	**-**	**-**	**-**	**-**	**-**	**-**
Coal Mines	-	-	-	-	-	-	-	-	-	-	-
Oil and Gas Extraction	-	-	-	-	-	-	-	-	-	-	-
Petroleum Refineries	-	-	-	-	-	-	-	-	-	-	-
Electr., CHP+Heat Plants	-	-	-	-	-	-	-	-	-	-	-
Pumped Storage (Elec.)	-	-	-	-	-	-	-	-	-	-	-
Other Energy Sector	-	-	-	-	-	-	-	-	-	-	-
Distribution Losses	-	-	-	-	-	-	-	3537	-	-	-
FINAL CONSUMPTION	**-**	**328**	**-**	**-**	**-**	**-**	**-**	**-**	**-**	**-**	**-**
INDUSTRY SECTOR	**-**	**328**	**-**	**-**	**-**	**-**	**-**	**-**	**-**	**-**	**-**
Iron and Steel	-	-	-	-	-	-	-	-	-	-	-
Chemical and Petrochem.	-	-	-	-	-	-	-	-	-	-	-
of which: Feedstocks	-	-	-	-	-	-	-	-	-	-	-
Non-Ferrous Metals	-	-	-	-	-	-	-	-	-	-	-
Non-Metallic Minerals	-	328	-	-	-	-	-	-	-	-	-
Transport Equipment	-	-	-	-	-	-	-	-	-	-	-
Machinery	-	-	-	-	-	-	-	-	-	-	-
Mining and Quarrying	-	-	-	-	-	-	-	-	-	-	-
Food and Tobacco	-	-	-	-	-	-	-	-	-	-	-
Paper, Pulp and Print	-	-	-	-	-	-	-	-	-	-	-
Wood and Wood Products	-	-	-	-	-	-	-	-	-	-	-
Construction	-	-	-	-	-	-	-	-	-	-	-
Textile and Leather	-	-	-	-	-	-	-	-	-	-	-
Non-specified	-	-	-	-	-	-	-	-	-	-	-
TRANSPORT SECTOR	**-**	**-**	**-**	**-**	**-**	**-**	**-**	**-**	**-**	**-**	**-**
Air	-	-	-	-	-	-	-	-	-	-	-
Road	-	-	-	-	-	-	-	-	-	-	-
Rail	-	-	-	-	-	-	-	-	-	-	-
Pipeline Transport	-	-	-	-	-	-	-	-	-	-	-
Internal Navigation	-	-	-	-	-	-	-	-	-	-	-
Non-specified	-	-	-	-	-	-	-	-	-	-	-
OTHER SECTORS	**-**	**-**	**-**	**-**	**-**	**-**	**-**	**-**	**-**	**-**	**-**
Agriculture	-	-	-	-	-	-	-	-	-	-	-
Comm. and Publ. Services	-	-	-	-	-	-	-	-	-	-	-
Residential	-	-	-	-	-	-	-	-	-	-	-
Non-specified	-	-	-	-	-	-	-	-	-	-	-
NON-ENERGY USE	**-**	**-**	**-**	**-**	**-**	**-**	**-**	**-**	**-**	**-**	**-**
in Industry/Trans./Energy	-	-	-	-	-	-	-	-	-	-	-
in Transport	-	-	-	-	-	-	-	-	-	-	-
in Other Sectors	-	-	-	-	-	-	-	-	-	-	-

Venezuela / Vénézuela : 1996

SUPPLY AND CONSUMPTION / *APPROVISIONNEMENT ET DEMANDE*	Oil cont. / *Pétrole cont.* (1000 tonnes)										
	Refinery Gas / *Gaz de raffinerie*	LPG + Ethane / *GPL + éthane*	Motor Gasoline / *Essence moteur*	Aviation Gasoline / *Essence aviation*	Jet Fuel / *Carbu-réacteurs*	Kerosene / *Kérosène*	Gas/ Diesel / *Gazole*	Heavy Fuel Oil / *Fioul lourd*	Naphtha / *Naphta*	Petrol. Coke / *Coke de pétrole*	Other Prod. / *Autres prod.*
Production	337	373	13299	19	4992	334	13873	13332	-	-	2700
From Other Sources	-	-	-	-	-	-	-	-	-	-	-
Imports	-	-	-	-	-	-	-	-	-	-	-
Exports	-	-1915	-6028	-	-4026	-	-9730	-11357	-	-	-2985
Intl. Marine Bunkers	-	-	-	-	-	-	-71	-594	-	-	-
Stock Changes	-	-	-	-	-36	-30	-	-	-	-	-
DOMESTIC SUPPLY	**337**	**-1542**	**7271**	**19**	**930**	**304**	**4072**	**1381**	**-**	**-**	**-285**
Transfers	-	3196	847	-	-	-	-	818	-	-	-
Statistical Differences	-	308	-	-	-188	-29	135	-1050	-	-	1781
TRANSFORMATION	**-**	**-**	**-**	**-**	**-**	**-**	**409**	**378**	**-**	**-**	**-**
Electricity Plants	-	-	-	-	-	-	409	378	-	-	-
CHP Plants	-	-	-	-	-	-	-	-	-	-	-
Heat Plants	-	-	-	-	-	-	-	-	-	-	-
Blast Furnaces/Gas Works	-	-	-	-	-	-	-	-	-	-	-
Coke/Pat. Fuel/BKB Plants	-	-	-	-	-	-	-	-	-	-	-
Petroleum Refineries	-	-	-	-	-	-	-	-	-	-	-
Petrochemical Industry	-	-	-	-	-	-	-	-	-	-	-
Liquefaction	-	-	-	-	-	-	-	-	-	-	-
Other Transform. Sector	-	-	-	-	-	-	-	-	-	-	-
ENERGY SECTOR	**337**	**13**	**-**	**-**	**-**	**2**	**408**	**68**	**-**	**-**	**-**
Coal Mines	-	-	-	-	-	-	-	-	-	-	-
Oil and Gas Extraction	-	-	-	-	-	-	408	23	-	-	-
Petroleum Refineries	337	13	-	-	-	2	-	45	-	-	-
Electr., CHP+Heat Plants	-	-	-	-	-	-	-	-	-	-	-
Pumped Storage (Elec.)	-	-	-	-	-	-	-	-	-	-	-
Other Energy Sector	-	-	-	-	-	-	-	-	-	-	-
Distribution Losses	-	-	-	-	-	-	-	-	-	-	-
FINAL CONSUMPTION	**-**	**1949**	**8118**	**19**	**742**	**273**	**3390**	**703**	**-**	**-**	**1496**
INDUSTRY SECTOR	**-**	**1175**	**149**	**-**	**-**	**68**	**750**	**211**	**-**	**-**	**-**
Iron and Steel	-	-	-	-	-	-	65	-	-	-	-
Chemical and Petrochem.	-	1002	98	-	-	-	-	-	-	-	-
of which: Feedstocks	-	-	-	-	-	-	-	-	-	-	-
Non-Ferrous Metals	-	-	4	-	-	-	-	-	-	-	-
Non-Metallic Minerals	-	-	-	-	-	-	-	-	-	-	-
Transport Equipment	-	-	-	-	-	-	-	-	-	-	-
Machinery	-	-	-	-	-	-	-	-	-	-	-
Mining and Quarrying	-	-	-	-	-	-	-	-	-	-	-
Food and Tobacco	-	-	-	-	-	-	-	-	-	-	-
Paper, Pulp and Print	-	-	-	-	-	-	-	-	-	-	-
Wood and Wood Products	-	-	-	-	-	-	-	-	-	-	-
Construction	-	-	-	-	-	-	-	-	-	-	-
Textile and Leather	-	-	-	-	-	-	-	-	-	-	-
Non-specified	-	173	47	-	-	68	685	211	-	-	-
TRANSPORT SECTOR	**-**	**41**	**7969**	**19**	**742**	**-**	**1791**	**6**	**-**	**-**	**-**
Air	-	-	-	19	742	-	-	-	-	-	-
Road	-	41	7969	-	-	-	1791	-	-	-	-
Rail	-	-	-	-	-	-	-	6	-	-	-
Pipeline Transport	-	-	-	-	-	-	-	-	-	-	-
Internal Navigation	-	-	-	-	-	-	-	-	-	-	-
Non-specified	-	-	-	-	-	-	-	-	-	-	-
OTHER SECTORS	**-**	**733**	**-**	**-**	**-**	**205**	**849**	**486**	**-**	**-**	**-**
Agriculture	-	-	-	-	-	-	447	101	-	-	-
Comm. and Publ. Services	-	-	-	-	-	13	402	317	-	-	-
Residential	-	733	-	-	-	192	-	-	-	-	-
Non-specified	-	-	-	-	-	-	-	68	-	-	-
NON-ENERGY USE	**-**	**-**	**-**	**-**	**-**	**-**	**-**	**-**	**-**	**-**	**1496**
in Industry/Transf./Energy	-	-	-	-	-	-	-	-	-	-	1496
in Transport	-	-	-	-	-	-	-	-	-	-	-
in Other Sectors	-	-	-	-	-	-	-	-	-	-	-

Venezuela / Vénézuela : 1996

SUPPLY AND CONSUMPTION / *APPROVISIONNEMENT ET DEMANDE*	Gas / *Gaz* (TJ)				Comb. Renew. & Waste / *En. Re. Comb. & Déchets* (TJ)				(GWh)	(TJ)
	Natural Gas / *Gaz naturel*	Gas Works / *Usines à gaz*	Coke Ovens / *Cokeries*	Blast Furnaces / *Hauts fourneaux*	Solid Biomass / *Biomasse solide*	Gas/Liquids from Biomass / *Gaz/Liquides tirés de biomasse*	Municipal Waste / *Déchets urbains*	Industrial Waste / *Déchets industriels*	Electricity / *Electricité*	Heat / *Chaleur*
Production	1222415	-	-	-	22647	-	-	-	75372	-
From Other Sources	-	-	-	-	-	-	-	-	-	-
Imports	-	-	-	-	-	-	-	-	2	-
Exports	-	-	-	-	-	-	-	-	-151	-
Intl. Marine Bunkers	-	-	-	-	-	-	-	-	-	-
Stock Changes	-	-	-	-	-	-	-	-	-	-
DOMESTIC SUPPLY	**1222415**	-	-	-	**22647**	-	-	-	**75223**	-
Transfers	-	-	-	-	-	-	-	-	-	-
Statistical Differences	-32174	-	-	-	-	-	-	-	-912	-
TRANSFORMATION	**264626**	-	-	-	**600**	-	-	-	-	-
Electricity Plants	264626	-	-	-	-	-	-	-	-	-
CHP Plants	-	-	-	-	-	-	-	-	-	-
Heat Plants	-	-	-	-	-	-	-	-	-	-
Blast Furnaces/Gas Works	-	-	-	-	-	-	-	-	-	-
Coke/Pat. Fuel/BKB Plants	-	-	-	-	-	-	-	-	-	-
Petroleum Refineries	-	-	-	-	-	-	-	-	-	-
Petrochemical Industry	-	-	-	-	-	-	-	-	-	-
Liquefaction	-	-	-	-	-	-	-	-	-	-
Other Transform. Sector	-	-	-	-	600	-	-	-	-	-
ENERGY SECTOR	**444602**	-	-	-	-	-	-	-	**3231**	-
Coal Mines	-	-	-	-	-	-	-	-	-	-
Oil and Gas Extraction	391427	-	-	-	-	-	-	-	1387	-
Petroleum Refineries	34768	-	-	-	-	-	-	-	653	-
Electr., CHP+Heat Plants	-	-	-	-	-	-	-	-	878	-
Pumped Storage (Elec.)	-	-	-	-	-	-	-	-	-	-
Other Energy Sector	18407	-	-	-	-	-	-	-	313	-
Distribution Losses	-	-	-	-	-	-	-	-	15345	-
FINAL CONSUMPTION	**481013**	-	-	-	**22047**	-	-	-	**55735**	-
INDUSTRY SECTOR	**445115**	-	-	-	**13846**	-	-	-	**27598**	-
Iron and Steel	128104	-	-	-	-	-	-	-	6373	-
Chemical and Petrochem.	143435	-	-	-	-	-	-	-	1121	-
of which: Feedstocks	-	-	-	-	-	-	-	-	-	-
Non-Ferrous Metals	23905	-	-	-	-	-	-	-	11025	-
Non-Metallic Minerals	41575	-	-	-	-	-	-	-	352	-
Transport Equipment	-	-	-	-	-	-	-	-	-	-
Machinery	-	-	-	-	-	-	-	-	-	-
Mining and Quarrying	-	-	-	-	-	-	-	-	-	-
Food and Tobacco	-	-	-	-	13846	-	-	-	-	-
Paper, Pulp and Print	-	-	-	-	-	-	-	-	-	-
Wood and Wood Products	-	-	-	-	-	-	-	-	-	-
Construction	-	-	-	-	-	-	-	-	-	-
Textile and Leather	-	-	-	-	-	-	-	-	-	-
Non-specified	108096	-	-	-	-	-	-	-	8727	-
TRANSPORT SECTOR	-	-	-	-	-	-	-	-	**80**	-
Air	-	-	-	-	-	-	-	-	-	-
Road	-	-	-	-	-	-	-	-	-	-
Rail	-	-	-	-	-	-	-	-	80	-
Pipeline Transport	-	-	-	-	-	-	-	-	-	-
Internal Navigation	-	-	-	-	-	-	-	-	-	-
Non-specified	-	-	-	-	-	-	-	-	-	-
OTHER SECTORS	**35898**	-	-	-	**8201**	-	-	-	**28057**	-
Agriculture	-	-	-	-	-	-	-	-	-	-
Comm. and Publ. Services	25249	-	-	-	-	-	-	-	15267	-
Residential	10649	-	-	-	8201	-	-	-	12790	-
Non-specified	-	-	-	-	-	-	-	-	-	-
NON-ENERGY USE	-	-	-	-	-	-	-	-	-	-
in Industry/Transf./Energy	-	-	-	-	-	-	-	-	-	-
in Transport	-	-	-	-	-	-	-	-	-	-
in Other Sectors	-	-	-	-	-	-	-	-	-	-

Venezuela / Vénézuela : 1997

SUPPLY AND CONSUMPTION / *APPROVISIONNEMENT ET DEMANDE*	Coal / *Charbon* (1000 tonnes)							Oil / *Pétrole* (1000 tonnes)			
	Coking Coal / *Charbon à coke*	Other Bit. Coal / *Autres charb. bit.*	Sub-Bit. Coal / *Charbon sous-bit.*	Lignite / *Lignite*	Peat / *Tourbe*	Oven and Gas Coke / *Coke de four/gaz*	Pat. Fuel and BKB / *Agg./briq. de lignite*	Crude Oil / *Pétrole brut*	NGL / *LGN*	Feed-stocks / *Produits d'aliment.*	Additives / *Additifs*
Production	-	5552	-	-	-	-	-	161876	5028	-	-
From Other Sources	-	-	-	-	-	-	-	-	-	-	-
Imports	-	318	-	-	-	-	-	-	-	-	-
Exports	-	-5451	-	-	-	-	-	-107057	-	-	-
Intl. Marine Bunkers	-	-	-	-	-	-	-	-	-	-	-
Stock Changes	-	-	-	-	-	-	-	2182	-	-	-
DOMESTIC SUPPLY	**-**	**419**	**-**	**-**	**-**	**-**	**-**	**57001**	**5028**	**-**	**-**
Transfers	-	-	-	-	-	-	-	-	-5028	-	-
Statistical Differences	-	-	-	-	-	-	-	-1207	-	-	-
TRANSFORMATION	**-**	**-**	**-**	**-**	**-**	**-**	**-**	**55794**	**-**	**-**	**-**
Electricity Plants	-	-	-	-	-	-	-	-	-	-	-
CHP Plants	-	-	-	-	-	-	-	-	-	-	-
Heat Plants	-	-	-	-	-	-	-	-	-	-	-
Blast Furnaces/Gas Works	-	-	-	-	-	-	-	-	-	-	-
Coke/Pat. Fuel/BKB Plants	-	-	-	-	-	-	-	-	-	-	-
Petroleum Refineries	-	-	-	-	-	-	-	55794	-	-	-
Petrochemical Industry	-	-	-	-	-	-	-	-	-	-	-
Liquefaction	-	-	-	-	-	-	-	-	-	-	-
Other Transform. Sector	-	-	-	-	-	-	-	-	-	-	-
ENERGY SECTOR	**-**	**-**	**-**	**-**	**-**	**-**	**-**	**-**	**-**	**-**	**-**
Coal Mines	-	-	-	-	-	-	-	-	-	-	-
Oil and Gas Extraction	-	-	-	-	-	-	-	-	-	-	-
Petroleum Refineries	-	-	-	-	-	-	-	-	-	-	-
Electr., CHP+Heat Plants	-	-	-	-	-	-	-	-	-	-	-
Pumped Storage (Elec.)	-	-	-	-	-	-	-	-	-	-	-
Other Energy Sector	-	-	-	-	-	-	-	-	-	-	-
Distribution Losses	-	-	-	-	-	-	-	-	-	-	-
FINAL CONSUMPTION	**-**	**419**	**-**	**-**	**-**	**-**	**-**	**-**	**-**	**-**	**-**
INDUSTRY SECTOR	**-**	**419**	**-**	**-**	**-**	**-**	**-**	**-**	**-**	**-**	**-**
Iron and Steel	-	-	-	-	-	-	-	-	-	-	-
Chemical and Petrochem.	-	-	-	-	-	-	-	-	-	-	-
of which: Feedstocks	-	-	-	-	-	-	-	-	-	-	-
Non-Ferrous Metals	-	-	-	-	-	-	-	-	-	-	-
Non-Metallic Minerals	-	419	-	-	-	-	-	-	-	-	-
Transport Equipment	-	-	-	-	-	-	-	-	-	-	-
Machinery	-	-	-	-	-	-	-	-	-	-	-
Mining and Quarrying	-	-	-	-	-	-	-	-	-	-	-
Food and Tobacco	-	-	-	-	-	-	-	-	-	-	-
Paper, Pulp and Print	-	-	-	-	-	-	-	-	-	-	-
Wood and Wood Products	-	-	-	-	-	-	-	-	-	-	-
Construction	-	-	-	-	-	-	-	-	-	-	-
Textile and Leather	-	-	-	-	-	-	-	-	-	-	-
Non-specified	-	-	-	-	-	-	-	-	-	-	-
TRANSPORT SECTOR	**-**	**-**	**-**	**-**	**-**	**-**	**-**	**-**	**-**	**-**	**-**
Air	-	-	-	-	-	-	-	-	-	-	-
Road	-	-	-	-	-	-	-	-	-	-	-
Rail	-	-	-	-	-	-	-	-	-	-	-
Pipeline Transport	-	-	-	-	-	-	-	-	-	-	-
Internal Navigation	-	-	-	-	-	-	-	-	-	-	-
Non-specified	-	-	-	-	-	-	-	-	-	-	-
OTHER SECTORS	**-**	**-**	**-**	**-**	**-**	**-**	**-**	**-**	**-**	**-**	**-**
Agriculture	-	-	-	-	-	-	-	-	-	-	-
Comm. and Publ. Services	-	-	-	-	-	-	-	-	-	-	-
Residential	-	-	-	-	-	-	-	-	-	-	-
Non-specified	-	-	-	-	-	-	-	-	-	-	-
NON-ENERGY USE	**-**	**-**	**-**	**-**	**-**	**-**	**-**	**-**	**-**	**-**	**-**
in Industry/Trans./Energy	-	-	-	-	-	-	-	-	-	-	-
in Transport	-	-	-	-	-	-	-	-	-	-	-
in Other Sectors	-	-	-	-	-	-	-	-	-	-	-

Venezuela / Vénézuela : 1997

SUPPLY AND CONSUMPTION / *APPROVISIONNEMENT ET DEMANDE*	Oil cont. / *Pétrole cont.* (1000 tonnes)										
	Refinery Gas / *Gaz de raffinerie*	LPG + Ethane / *GPL + éthane*	Motor Gasoline / *Essence moteur*	Aviation Gasoline / *Essence aviation*	Jet Fuel / *Carbu- réacteurs*	Kerosene / *Kérosène*	Gas/ Diesel / *Gazole*	Heavy Fuel Oil / *Fioul lourd*	Naphtha / *Naphta*	Petrol. Coke / *Coke de pétrole*	Other Prod. / *Autres prod.*
Production	358	396	15072	18	5128	342	15035	14448	-	-	2731
From Other Sources	-	-	-	-	-	-	-	-	-	-	-
Imports	-	-	-	-	-	-	-	-	-	-	-
Exports	-	-1286	-6832	-	-4099	-	-10545	-12309	-	-	-2985
Intl. Marine Bunkers	-	-	-	-	-	-	-71	-594	-	-	-
Stock Changes	-	-	-	-	-	-43	-16	-16	-	-	-
DOMESTIC SUPPLY	**358**	**-890**	**8240**	**18**	**1029**	**299**	**4403**	**1529**	**-**	**-**	**-254**
Transfers	-	3342	960	-	-	-	-	726	-	-	-
Statistical Differences	-	-	1	-	-269	-35	155	-1010	-	-	1750
TRANSFORMATION	**-**	**-**	**-**	**-**	**-**	**-**	**443**	**409**	**-**	**-**	**-**
Electricity Plants	-	-	-	-	-	-	443	409	-	-	-
CHP Plants	-	-	-	-	-	-	-	-	-	-	-
Heat Plants	-	-	-	-	-	-	-	-	-	-	-
Blast Furnaces/Gas Works	-	-	-	-	-	-	-	-	-	-	-
Coke/Pat. Fuel/BKB Plants	-	-	-	-	-	-	-	-	-	-	-
Petroleum Refineries	-	-	-	-	-	-	-	-	-	-	-
Petrochemical Industry	-	-	-	-	-	-	-	-	-	-	-
Liquefaction	-	-	-	-	-	-	-	-	-	-	-
Other Transform. Sector	-	-	-	-	-	-	-	-	-	-	-
ENERGY SECTOR	**358**	**15**	**-**	**-**	**-**	**-**	**442**	**74**	**-**	**-**	**-**
Coal Mines	-	-	-	-	-	-	-	-	-	-	-
Oil and Gas Extraction	-	-	-	-	-	-	442	25	-	-	-
Petroleum Refineries	358	15	-	-	-	-	-	49	-	-	-
Electr., CHP+Heat Plants	-	-	-	-	-	-	-	-	-	-	-
Pumped Storage (Elec.)	-	-	-	-	-	-	-	-	-	-	-
Other Energy Sector	-	-	-	-	-	-	-	-	-	-	-
Distribution Losses	-	-	-	-	-	-	-	-	-	-	-
FINAL CONSUMPTION	**-**	**2437**	**9201**	**18**	**760**	**264**	**3673**	**762**	**-**	**-**	**1496**
INDUSTRY SECTOR	**-**	**1469**	**169**	**-**	**-**	**66**	**812**	**229**	**-**	**-**	**-**
Iron and Steel	-	-	-	-	-	-	70	-	-	-	-
Chemical and Petrochem.	-	1253	111	-	-	-	-	-	-	-	-
of which: Feedstocks	-	-	-	-	-	-	-	-	-	-	-
Non-Ferrous Metals	-	-	5	-	-	-	-	-	-	-	-
Non-Metallic Minerals	-	-	-	-	-	-	-	-	-	-	-
Transport Equipment	-	-	-	-	-	-	-	-	-	-	-
Machinery	-	-	-	-	-	-	-	-	-	-	-
Mining and Quarrying	-	-	-	-	-	-	-	-	-	-	-
Food and Tobacco	-	-	-	-	-	-	-	-	-	-	-
Paper, Pulp and Print	-	-	-	-	-	-	-	-	-	-	-
Wood and Wood Products	-	-	-	-	-	-	-	-	-	-	-
Construction	-	-	-	-	-	-	-	-	-	-	-
Textile and Leather	-	-	-	-	-	-	-	-	-	-	-
Non-specified	-	216	53	-	-	66	742	229	-	-	-
TRANSPORT SECTOR	**-**	**51**	**9032**	**18**	**760**	**-**	**1941**	**7**	**-**	**-**	**-**
Air	-	-	-	18	760	-	-	-	-	-	-
Road	-	51	9032	-	-	-	1941	-	-	-	-
Rail	-	-	-	-	-	-	-	7	-	-	-
Pipeline Transport	-	-	-	-	-	-	-	-	-	-	-
Internal Navigation	-	-	-	-	-	-	-	-	-	-	-
Non-specified	-	-	-	-	-	-	-	-	-	-	-
OTHER SECTORS	**-**	**917**	**-**	**-**	**-**	**198**	**920**	**526**	**-**	**-**	**-**
Agriculture	-	-	-	-	-	-	484	109	-	-	-
Comm. and Publ. Services	-	-	-	-	-	13	436	343	-	-	-
Residential	-	917	-	-	-	185	-	-	-	-	-
Non-specified	-	-	-	-	-	-	-	74	-	-	-
NON-ENERGY USE	**-**	**-**	**-**	**-**	**-**	**-**	**-**	**-**	**-**	**-**	**1496**
in Industry/Transf./Energy	-	-	-	-	-	-	-	-	-	-	1496
in Transport	-	-	-	-	-	-	-	-	-	-	-
in Other Sectors	-	-	-	-	-	-	-	-	-	-	-

Venezuela / Vénézuela : 1997

SUPPLY AND CONSUMPTION *APPROVISIONNEMENT ET DEMANDE*	Gas / *Gaz* (TJ)				Comb. Renew. & Waste / *En. Re. Comb. & Déchets* (TJ)				(GWh)	(TJ)
	Natural Gas *Gaz naturel*	Gas Works *Usines à gaz*	Coke Ovens *Cokeries*	Blast Furnaces *Hauts fourneaux*	Solid Biomass *Biomasse solide*	Gas/Liquids from Biomass *Gaz/Liquides tirés de biomasse*	Municipal Waste *Déchets urbains*	Industrial Waste *Déchets industriels*	Electricity *Electricité*	Heat *Chaleur*
Production	1347691	-	-	-	22647	-	-	-	74865	-
From Other Sources	-	-	-	-	-	-	-	-	-	-
Imports	-	-	-	-	-	-	-	-	-	-
Exports	-	-	-	-	-	-	-	-	-	-
Intl. Marine Bunkers	-	-	-	-	-	-	-	-	-	-
Stock Changes	-	-	-	-	-	-	-	-	-	-
DOMESTIC SUPPLY	**1347691**	**-**	**-**	**-**	**22647**	**-**	**-**	**-**	**74865**	**-**
Transfers	-	-	-	-	-	-	-	-	-	-
Statistical Differences	-29961	-	-	-	-	-	-	-	1065	-
TRANSFORMATION	**211256**	**-**	**-**	**-**	**600**	**-**	**-**	**-**	**-**	**-**
Electricity Plants	211256	-	-	-	-	-	-	-	-	-
CHP Plants	-	-	-	-	-	-	-	-	-	-
Heat Plants	-	-	-	-	-	-	-	-	-	-
Blast Furnaces/Gas Works	-	-	-	-	-	-	-	-	-	-
Coke/Pat. Fuel/BKB Plants	-	-	-	-	-	-	-	-	-	-
Petroleum Refineries	-	-	-	-	-	-	-	-	-	-
Petrochemical Industry	-	-	-	-	-	-	-	-	-	-
Liquefaction	-	-	-	-	-	-	-	-	-	-
Other Transform. Sector	-	-	-	-	600	-	-	-	-	-
ENERGY SECTOR	**589688**	**-**	**-**	**-**	**-**	**-**	**-**	**-**	**3352**	**-**
Coal Mines	-	-	-	-	-	-	-	-	-	-
Oil and Gas Extraction	519160	-	-	-	-	-	-	-	1439	-
Petroleum Refineries	46114	-	-	-	-	-	-	-	677	-
Electr., CHP+Heat Plants	-	-	-	-	-	-	-	-	911	-
Pumped Storage (Elec.)	-	-	-	-	-	-	-	-	-	-
Other Energy Sector	24414	-	-	-	-	-	-	-	325	-
Distribution Losses	-	-	-	-	-	-	-	-	15916	-
FINAL CONSUMPTION	**516786**	**-**	**-**	**-**	**22047**	**-**	**-**	**-**	**56662**	**-**
INDUSTRY SECTOR	**478218**	**-**	**-**	**-**	**13846**	**-**	**-**	**-**	**26412**	**-**
Iron and Steel	137631	-	-	-	-	-	-	-	6099	-
Chemical and Petrochem.	154102	-	-	-	-	-	-	-	1073	-
of which: Feedstocks	-	-	-	-	-	-	-	-	-	-
Non-Ferrous Metals	25683	-	-	-	-	-	-	-	10551	-
Non-Metallic Minerals	44667	-	-	-	-	-	-	-	337	-
Transport Equipment	-	-	-	-	-	-	-	-	-	-
Machinery	-	-	-	-	-	-	-	-	-	-
Mining and Quarrying	-	-	-	-	-	-	-	-	-	-
Food and Tobacco	-	-	-	-	13846	-	-	-	-	-
Paper, Pulp and Print	-	-	-	-	-	-	-	-	-	-
Wood and Wood Products	-	-	-	-	-	-	-	-	-	-
Construction	-	-	-	-	-	-	-	-	-	-
Textile and Leather	-	-	-	-	-	-	-	-	-	-
Non-specified	116135	-	-	-	-	-	-	-	8352	-
TRANSPORT SECTOR	**-**	**-**	**-**	**-**	**-**	**-**	**-**	**-**	**832**	**-**
Air	-	-	-	-	-	-	-	-	-	-
Road	-	-	-	-	-	-	-	-	752	-
Rail	-	-	-	-	-	-	-	-	80	-
Pipeline Transport	-	-	-	-	-	-	-	-	-	-
Internal Navigation	-	-	-	-	-	-	-	-	-	-
Non-specified	-	-	-	-	-	-	-	-	-	-
OTHER SECTORS	**38568**	**-**	**-**	**-**	**8201**	**-**	**-**	**-**	**29418**	**-**
Agriculture	-	-	-	-	-	-	-	-	-	-
Comm. and Publ. Services	27127	-	-	-	-	-	-	-	16008	-
Residential	11441	-	-	-	8201	-	-	-	13410	-
Non-specified	-	-	-	-	-	-	-	-	-	-
NON-ENERGY USE	**-**	**-**	**-**	**-**	**-**	**-**	**-**	**-**	**-**	**-**
in Industry/Transf./Energy	-	-	-	-	-	-	-	-	-	-
in Transport	-	-	-	-	-	-	-	-	-	-
in Other Sectors	-	-	-	-	-	-	-	-	-	-

Vietnam / Viêt-Nam

SUPPLY AND CONSUMPTION 1996	Coal (1000 tonnes)							Oil (1000 tonnes)			
	Coking Coal	Other Bit. Coal	Sub-Bit. Coal	Lignite	Peat	Oven and Gas Coke	Pat. Fuel and BKB	Crude Oil	NGL	Feed-stocks	Additives
Production	-	11209	-	-	-	-	-	8601	-	-	-
Imports	-	-	-	-	-	-	-	-	-	-	-
Exports	-	-4394	-	-	-	-	-	-8561	-	-	-
Intl. Marine Bunkers	-	-	-	-	-	-	-	-	-	-	-
Stock Changes	-	-	-	-	-	-	-	-	-	-	-
DOMESTIC SUPPLY	-	**6815**	-	-	-	-	-	**40**	-	-	-
Transfers and Stat. Diff.	-	-	-	-	-	-	-	-	-	-	-
TRANSFORMATION	-	**1654**	-	-	-	-	-	**40**	-	-	-
Electricity and CHP Plants	-	1654	-	-	-	-	-	-	-	-	-
Petroleum Refineries	-	-	-	-	-	-	-	40	-	-	-
Other Transform. Sector	-	-	-	-	-	-	-	-	-	-	-
ENERGY SECTOR	-	-	-	-	-	-	-	-	-	-	-
DISTRIBUTION LOSSES	-	-	-	-	-	-	-	-	-	-	-
FINAL CONSUMPTION	-	**5161**	-	-	-	-	-	-	-	-	-
INDUSTRY SECTOR	-	**3826**	-	-	-	-	-	-	-	-	-
Iron and Steel	-	-	-	-	-	-	-	-	-	-	-
Chemical and Petrochem.	-	-	-	-	-	-	-	-	-	-	-
Non-Metallic Minerals	-	-	-	-	-	-	-	-	-	-	-
Non-specified	-	3826	-	-	-	-	-	-	-	-	-
TRANSPORT SECTOR	-	**10**	-	-	-	-	-	-	-	-	-
Air	-	-	-	-	-	-	-	-	-	-	-
Road	-	-	-	-	-	-	-	-	-	-	-
Non-specified	-	10	-	-	-	-	-	-	-	-	-
OTHER SECTORS	-	**1325**	-	-	-	-	-	-	-	-	-
Agriculture	-	71	-	-	-	-	-	-	-	-	-
Comm. and Publ. Services	-	-	-	-	-	-	-	-	-	-	-
Residential	-	1009	-	-	-	-	-	-	-	-	-
Non-specified	-	245	-	-	-	-	-	-	-	-	-
NON-ENERGY USE	-	-	-	-	-	-	-	-	-	-	-

APPROVISIONNEMENT ET DEMANDE 1997	*Charbon* (1000 tonnes)							*Pétrole* (1000 tonnes)			
	Charbon à coke	*Autres charb. bit.*	*Charbon sous-bit.*	*Lignite*	*Tourbe*	*Coke de four/gaz*	*Agg./briq. de lignite*	*Pétrole brut*	*LGN*	*Produits d'aliment.*	*Additifs*
Production	-	13116	-	-	-	-	-	9702	-	-	-
Imports	-	-	-	-	-	-	-	-	-	-	-
Exports	-	-4235	-	-	-	-	-	-9662	-	-	-
Intl. Marine Bunkers	-	-	-	-	-	-	-	-	-	-	-
Stock Changes	-	-	-	-	-	-	-	-	-	-	-
DOMESTIC SUPPLY	-	**8881**	-	-	-	-	-	**40**	-	-	-
Transfers and Stat. Diff.	-	1	-	-	-	-	-	-	-	-	-
TRANSFORMATION	-	**2777**	-	-	-	-	-	**40**	-	-	-
Electricity and CHP Plants	-	2777	-	-	-	-	-	-	-	-	-
Petroleum Refineries	-	-	-	-	-	-	-	40	-	-	-
Other Transform. Sector	-	-	-	-	-	-	-	-	-	-	-
ENERGY SECTOR	-	-	-	-	-	-	-	-	-	-	-
DISTRIBUTION LOSSES	-	-	-	-	-	-	-	-	-	-	-
FINAL CONSUMPTION	-	**6105**	-	-	-	-	-	-	-	-	-
INDUSTRY SECTOR	-	**4663**	-	-	-	-	-	-	-	-	-
Iron and Steel	-	-	-	-	-	-	-	-	-	-	-
Chemical and Petrochem.	-	-	-	-	-	-	-	-	-	-	-
Non-Metallic Minerals	-	-	-	-	-	-	-	-	-	-	-
Non-specified	-	4663	-	-	-	-	-	-	-	-	-
TRANSPORT SECTOR	-	**9**	-	-	-	-	-	-	-	-	-
Air	-	-	-	-	-	-	-	-	-	-	-
Road	-	-	-	-	-	-	-	-	-	-	-
Non-specified	-	9	-	-	-	-	-	-	-	-	-
OTHER SECTORS	-	**1433**	-	-	-	-	-	-	-	-	-
Agriculture	-	90	-	-	-	-	-	-	-	-	-
Comm. and Publ. Services	-	-	-	-	-	-	-	-	-	-	-
Residential	-	1072	-	-	-	-	-	-	-	-	-
Non-specified	-	271	-	-	-	-	-	-	-	-	-
NON-ENERGY USE	-	-	-	-	-	-	-	-	-	-	-

Vietnam / Viêt-Nam

SUPPLY AND CONSUMPTION 1996	Oil cont. (1000 tonnes)										
	Refinery Gas	LPG + Ethane	Motor Gasoline	Aviation Gasoline	Jet Fuel	Kerosene	Gas/ Diesel	Heavy Fuel Oil	Naphtha	Petrol. Coke	Other Prod.
Production	2	-	7	-	2	3	17	10	-	-	-
Imports	-	53	1244	-	252	328	2766	1211	-	-	201
Exports	-	-	-	-	-	-	-	-	-	-	-
Intl. Marine Bunkers	-	-	-	-	-	-	-	-	-	-	-
Stock Changes	-	-	-	-	-	-	-	-	-	-	-
DOMESTIC SUPPLY	**2**	**53**	**1251**	**-**	**254**	**331**	**2783**	**1221**	**-**	**-**	**201**
Transfers and Stat. Diff.	-	-	-1	-	-	-	-	1	-	-	-
TRANSFORMATION	**-**	**-**	**-**	**-**	**-**	**-**	**173**	**246**	**-**	**-**	**-**
Electricity and CHP Plants	-	-	-	-	-	-	173	246	-	-	-
Petroleum Refineries	-	-	-	-	-	-	-	-	-	-	-
Other Transform. Sector	-	-	-	-	-	-	-	-	-	-	-
ENERGY SECTOR	**2**	**-**	**-**	**-**	**-**	**-**	**-**	**-**	**-**	**-**	**-**
DISTRIBUTION LOSSES	**-**	**-**	**-**	**-**	**-**	**-**	**-**	**-**	**-**	**-**	**-**
FINAL CONSUMPTION	**-**	**53**	**1250**	**-**	**254**	**331**	**2610**	**976**	**-**	**-**	**201**
INDUSTRY SECTOR	**-**	**-**	**-**	**-**	**-**	**-**	**42**	**728**	**-**	**-**	**-**
Iron and Steel	-	-	-	-	-	-	-	-	-	-	-
Chemical and Petrochem.	-	-	-	-	-	-	-	-	-	-	-
Non-Metallic Minerals	-	-	-	-	-	-	-	-	-	-	-
Non-specified	-	-	-	-	-	-	42	728	-	-	-
TRANSPORT SECTOR	**-**	**-**	**1250**	**-**	**254**	**-**	**2098**	**117**	**-**	**-**	**-**
Air	-	-	-	-	254	-	-	-	-	-	-
Road	-	-	1250	-	-	-	2098	-	-	-	-
Non-specified	-	-	-	-	-	-	-	117	-	-	-
OTHER SECTORS	**-**	**53**	**-**	**-**	**-**	**331**	**470**	**131**	**-**	**-**	**-**
Agriculture	-	-	-	-	-	-	235	9	-	-	-
Comm. and Publ. Services	-	-	-	-	-	-	-	-	-	-	-
Residential	-	53	-	-	-	331	28	15	-	-	-
Non-specified	-	-	-	-	-	-	207	107	-	-	-
NON-ENERGY USE	**-**	**-**	**-**	**-**	**-**	**-**	**-**	**-**	**-**	**-**	**201**

APPROVISIONNEMENT ET DEMANDE 1997	*Pétrole cont.* (1000 tonnes)										
	Gaz de raffinerie	*GPL + éthane*	*Essence moteur*	*Essence aviation*	*Carbu- réacteurs*	*Kérosène*	*Gazole*	*Fioul lourd*	*Naphta*	*Coke de pétrole*	*Autres prod.*
Production	-	-	7	-	2	3	17	10	-	-	1
Imports	-	120	1599	-	277	291	3599	1507	-	-	117
Exports	-	-	-	-	-	-	-	-	-	-	-
Intl. Marine Bunkers	-	-	-	-	-	-	-	-	-	-	-
Stock Changes	-	-	-	-	-	-	-	-	-	-	-
DOMESTIC SUPPLY	**-**	**120**	**1606**	**-**	**279**	**294**	**3616**	**1517**	**-**	**-**	**118**
Transfers and Stat. Diff.	-	-	-1	-	-	-	-	1	-	-	-
TRANSFORMATION	**-**	**-**	**-**	**-**	**-**	**-**	**300**	**288**	**-**	**-**	**-**
Electricity and CHP Plants	-	-	-	-	-	-	300	288	-	-	-
Petroleum Refineries	-	-	-	-	-	-	-	-	-	-	-
Other Transform. Sector	-	-	-	-	-	-	-	-	-	-	-
ENERGY SECTOR	**-**	**-**	**-**	**-**	**-**	**-**	**-**	**-**	**-**	**-**	**-**
DISTRIBUTION LOSSES	**-**	**-**	**-**	**-**	**-**	**-**	**-**	**-**	**-**	**-**	**-**
FINAL CONSUMPTION	**-**	**120**	**1605**	**-**	**279**	**294**	**3316**	**1230**	**-**	**-**	**118**
INDUSTRY SECTOR	**-**	**-**	**-**	**-**	**-**	**-**	**53**	**918**	**-**	**-**	**-**
Iron and Steel	-	-	-	-	-	-	-	-	-	-	-
Chemical and Petrochem.	-	-	-	-	-	-	-	-	-	-	-
Non-Metallic Minerals	-	-	-	-	-	-	-	-	-	-	-
Non-specified	-	-	-	-	-	-	53	918	-	-	-
TRANSPORT SECTOR	**-**	**-**	**1605**	**-**	**279**	**-**	**2666**	**147**	**-**	**-**	**-**
Air	-	-	-	-	279	-	-	-	-	-	-
Road	-	-	1605	-	-	-	2666	-	-	-	-
Non-specified	-	-	-	-	-	-	-	147	-	-	-
OTHER SECTORS	**-**	**120**	**-**	**-**	**-**	**294**	**597**	**165**	**-**	**-**	**-**
Agriculture	-	-	-	-	-	-	298	12	-	-	-
Comm. and Publ. Services	-	-	-	-	-	-	-	-	-	-	-
Residential	-	120	-	-	-	294	263	18	-	-	-
Non-specified	-	-	-	-	-	-	36	135	-	-	-
NON-ENERGY USE	**-**	**-**	**-**	**-**	**-**	**-**	**-**	**-**	**-**	**-**	**118**

Vietnam / Viêt-Nam

SUPPLY AND CONSUMPTION 1996	Gas (TJ)				Comb. Renew. & Waste (TJ)				(GWh)	(TJ)
	Natural Gas	Gas Works	Coke Ovens	Blast Furnaces	Solid Biomass	Gas/Liquids from Biomass	Municipal Waste	Industrial Waste	Electricity	Heat
Production	145106	-	-	-	919378	-	-	-	16945	-
Imports	-	-	-	-	-	-	-	-	-	-
Exports	-	-	-	-	-	-	-	-	-	-
Intl. Marine Bunkers	-	-	-	-	-	-	-	-	-	-
Stock Changes	-	-	-	-	-	-	-	-	-	-
DOMESTIC SUPPLY	**145106**	-	-	-	**919378**	-	-	-	**16945**	-
Transfers and Stat. Diff.	-	-	-	-	-	-	-	-	-	-
TRANSFORMATION	**9579**	-	-	-	**29320**	-	-	-	-	-
Electricity and CHP Plants	9579	-	-	-	-	-	-	-	-	-
Petroleum Refineries	-	-	-	-	-	-	-	-	-	-
Other Transform. Sector	-	-	-	-	29320	-	-	-	-	-
ENERGY SECTOR	-	-	-	-	-	-	-	-	**367**	-
DISTRIBUTION LOSSES	-	-	-	-	-	-	-	-	**3203**	-
FINAL CONSUMPTION	**135527**	-	-	-	**890058**	-	-	-	**13375**	-
INDUSTRY SECTOR	**135527**	-	-	-	-	-	-	-	**5504**	-
Iron and Steel	-	-	-	-	-	-	-	-	-	-
Chemical and Petrochem.	-	-	-	-	-	-	-	-	-	-
Non-Metallic Minerals	-	-	-	-	-	-	-	-	-	-
Non-specified	135527	-	-	-	-	-	-	-	5504	-
TRANSPORT SECTOR	-	-	-	-	-	-	-	-	**114**	-
Air	-	-	-	-	-	-	-	-	-	-
Road	-	-	-	-	-	-	-	-	-	-
Non-specified	-	-	-	-	-	-	-	-	114	-
OTHER SECTORS	-	-	-	-	**890058**	-	-	-	**7757**	-
Agriculture	-	-	-	-	-	-	-	-	722	-
Comm. and Publ. Services	-	-	-	-	-	-	-	-	-	-
Residential	-	-	-	-	890058	-	-	-	6055	-
Non-specified	-	-	-	-	-	-	-	-	980	-
NON-ENERGY USE	-	-	-	-	-	-	-	-	-	-

APPROVISIONNEMENT ET DEMANDE 1997	*Gaz* (TJ)				*En. Re. Comb. & Déchets* (TJ)				(GWh)	(TJ)
	Gaz naturel	*Usines à gaz*	*Cokeries*	*Hauts fourneaux*	*Biomasse solide*	*Gaz/Liquides tirés de biomasse*	*Déchets urbains*	*Déchets industriels*	*Electricité*	*Chaleur*
Production	154053	-	-	-	937950	-	-	-	19151	-
Imports	-	-	-	-	-	-	-	-	-	-
Exports	-	-	-	-	-	-	-	-	-	-
Intl. Marine Bunkers	-	-	-	-	-	-	-	-	-	-
Stock Changes	-	-	-	-	-	-	-	-	-	-
DOMESTIC SUPPLY	**154053**	-	-	-	**937950**	-	-	-	**19151**	-
Transfers and Stat. Diff.	-	-	-	-	1	-	-	-	1	-
TRANSFORMATION	**10169**	-	-	-	**29912**	-	-	-	-	-
Electricity and CHP Plants	10169	-	-	-	-	-	-	-	-	-
Petroleum Refineries	-	-	-	-	-	-	-	-	-	-
Other Transform. Sector	-	-	-	-	29912	-	-	-	-	-
ENERGY SECTOR	-	-	-	-	-	-	-	-	**465**	-
DISTRIBUTION LOSSES	-	-	-	-	-	-	-	-	**3384**	-
FINAL CONSUMPTION	**143884**	-	-	-	**908039**	-	-	-	**15303**	-
INDUSTRY SECTOR	**143884**	-	-	-	-	-	-	-	**6163**	-
Iron and Steel	-	-	-	-	-	-	-	-	-	-
Chemical and Petrochem.	-	-	-	-	-	-	-	-	-	-
Non-Metallic Minerals	-	-	-	-	-	-	-	-	-	-
Non-specified	143884	-	-	-	-	-	-	-	6163	-
TRANSPORT SECTOR	-	-	-	-	-	-	-	-	**128**	-
Air	-	-	-	-	-	-	-	-	-	-
Road	-	-	-	-	-	-	-	-	-	-
Non-specified	-	-	-	-	-	-	-	-	128	-
OTHER SECTORS	-	-	-	-	**908039**	-	-	-	**9012**	-
Agriculture	-	-	-	-	-	-	-	-	691	-
Comm. and Publ. Services	-	-	-	-	-	-	-	-	-	-
Residential	-	-	-	-	908039	-	-	-	7221	-
Non-specified	-	-	-	-	-	-	-	-	1100	-
NON-ENERGY USE	-	-	-	-	-	-	-	-	-	-

Yemen / Yémen

SUPPLY AND CONSUMPTION 1996	Coal (1000 tonnes)							Oil (1000 tonnes)			
	Coking Coal	Other Bit. Coal	Sub-Bit. Coal	Lignite	Peat	Oven and Gas Coke	Pat. Fuel and BKB	Crude Oil	NGL	Feed-stocks	Additives
Production	-	-	-	-	-	-	-	17333	368	-	-
Imports	-	-	-	-	-	-	-	-	-	-	-
Exports	-	-	-	-	-	-	-	-14223	-	-	-
Intl. Marine Bunkers	-	-	-	-	-	-	-	-	-	-	-
Stock Changes	-	-	-	-	-	-	-	1440	-	-	-
DOMESTIC SUPPLY	-	-	-	-	-	-	-	**4550**	**368**	-	-
Transfers and Stat. Diff.	-	-	-	-	-	-	-	-	-368	-	-
TRANSFORMATION	-	-	-	-	-	-	-	**4550**	-	-	-
Electricity and CHP Plants	-	-	-	-	-	-	-	-	-	-	-
Petroleum Refineries	-	-	-	-	-	-	-	4550	-	-	-
Other Transform. Sector	-	-	-	-	-	-	-	-	-	-	-
ENERGY SECTOR	-	-	-	-	-	-	-	-	-	-	-
DISTRIBUTION LOSSES	-	-	-	-	-	-	-	-	-	-	-
FINAL CONSUMPTION	-	-	-	-	-	-	-	-	-	-	-
INDUSTRY SECTOR	-	-	-	-	-	-	-	-	-	-	-
Iron and Steel	-	-	-	-	-	-	-	-	-	-	-
Chemical and Petrochem.	-	-	-	-	-	-	-	-	-	-	-
Non-Metallic Minerals	-	-	-	-	-	-	-	-	-	-	-
Non-specified	-	-	-	-	-	-	-	-	-	-	-
TRANSPORT SECTOR	-	-	-	-	-	-	-	-	-	-	-
Air	-	-	-	-	-	-	-	-	-	-	-
Road	-	-	-	-	-	-	-	-	-	-	-
Non-specified	-	-	-	-	-	-	-	-	-	-	-
OTHER SECTORS	-	-	-	-	-	-	-	-	-	-	-
Agriculture	-	-	-	-	-	-	-	-	-	-	-
Comm. and Publ. Services	-	-	-	-	-	-	-	-	-	-	-
Residential	-	-	-	-	-	-	-	-	-	-	-
Non-specified	-	-	-	-	-	-	-	-	-	-	-
NON-ENERGY USE	-	-	-	-	-	-	-	-	-	-	-

APPROVISIONNEMENT ET DEMANDE 1997	*Charbon* (1000 tonnes)							*Pétrole* (1000 tonnes)			
	Charbon à coke	*Autres charb. bit.*	*Charbon sous-bit.*	*Lignite*	*Tourbe*	*Coke de four/gaz*	*Agg./briq. de lignite*	*Pétrole brut*	*LGN*	*Produits d'aliment.*	*Additifs*
Production	-	-	-	-	-	-	-	18124	408	-	-
Imports	-	-	-	-	-	-	-	-	-	-	-
Exports	-	-	-	-	-	-	-	-11778	-	-	-
Intl. Marine Bunkers	-	-	-	-	-	-	-	-	-	-	-
Stock Changes	-	-	-	-	-	-	-	-1818	-	-	-
DOMESTIC SUPPLY	-	-	-	-	-	-	-	**4528**	**408**	-	-
Transfers and Stat. Diff.	-	-	-	-	-	-	-	-	-408	-	-
TRANSFORMATION	-	-	-	-	-	-	-	**4528**	-	-	-
Electricity and CHP Plants	-	-	-	-	-	-	-	-	-	-	-
Petroleum Refineries	-	-	-	-	-	-	-	4528	-	-	-
Other Transform. Sector	-	-	-	-	-	-	-	-	-	-	-
ENERGY SECTOR	-	-	-	-	-	-	-	-	-	-	-
DISTRIBUTION LOSSES	-	-	-	-	-	-	-	-	-	-	-
FINAL CONSUMPTION	-	-	-	-	-	-	-	-	-	-	-
INDUSTRY SECTOR	-	-	-	-	-	-	-	-	-	-	-
Iron and Steel	-	-	-	-	-	-	-	-	-	-	-
Chemical and Petrochem.	-	-	-	-	-	-	-	-	-	-	-
Non-Metallic Minerals	-	-	-	-	-	-	-	-	-	-	-
Non-specified	-	-	-	-	-	-	-	-	-	-	-
TRANSPORT SECTOR	-	-	-	-	-	-	-	-	-	-	-
Air	-	-	-	-	-	-	-	-	-	-	-
Road	-	-	-	-	-	-	-	-	-	-	-
Non-specified	-	-	-	-	-	-	-	-	-	-	-
OTHER SECTORS	-	-	-	-	-	-	-	-	-	-	-
Agriculture	-	-	-	-	-	-	-	-	-	-	-
Comm. and Publ. Services	-	-	-	-	-	-	-	-	-	-	-
Residential	-	-	-	-	-	-	-	-	-	-	-
Non-specified	-	-	-	-	-	-	-	-	-	-	-
NON-ENERGY USE	-	-	-	-	-	-	-	-	-	-	-

Yemen / Yémen

SUPPLY AND CONSUMPTION 1996	Oil cont. (1000 tonnes)										
	Refinery Gas	LPG + Ethane	Motor Gasoline	Aviation Gasoline	Jet Fuel	Kerosene	Gas/ Diesel	Heavy Fuel Oil	Naphtha	Petrol. Coke	Other Prod.
Production	-	19	1045	-	381	114	944	1547	224	-	60
Imports	-	-	-	1	-	-	-	-	-	-	-
Exports	-	-	-	-	-294	-	-224	-1015	-224	-	-
Intl. Marine Bunkers	-	-	-	-	-	-	-40	-60	-	-	-
Stock Changes	-	-	-	-	-	-	-	-	-	-	-
DOMESTIC SUPPLY	**-**	**19**	**1045**	**1**	**87**	**114**	**680**	**472**	**-**	**-**	**60**
Transfers and Stat. Diff.	-	368	-	-	-	-	-	-	-	-	-
TRANSFORMATION	**-**	**-**	**-**	**-**	**-**	**-**	**253**	**180**	**-**	**-**	**-**
Electricity and CHP Plants	-	-	-	-	-	-	253	180	-	-	-
Petroleum Refineries	-	-	-	-	-	-	-	-	-	-	-
Other Transform. Sector	-	-	-	-	-	-	-	-	-	-	-
ENERGY SECTOR	**-**	**-**	**-**	**-**	**-**	**-**	**-**	**129**	**-**	**-**	**-**
DISTRIBUTION LOSSES	**-**	**-**	**-**	**-**	**-**	**-**	**-**	**-**	**-**	**-**	**-**
FINAL CONSUMPTION	**-**	**387**	**1045**	**1**	**87**	**114**	**427**	**163**	**-**	**-**	**60**
INDUSTRY SECTOR	**-**	**-**	**-**	**-**	**-**	**-**	**-**	**163**	**-**	**-**	**-**
Iron and Steel	-	-	-	-	-	-	-	-	-	-	-
Chemical and Petrochem.	-	-	-	-	-	-	-	-	-	-	-
Non-Metallic Minerals	-	-	-	-	-	-	-	-	-	-	-
Non-specified	-	-	-	-	-	-	-	163	-	-	-
TRANSPORT SECTOR	**-**	**-**	**1045**	**1**	**87**	**-**	**427**	**-**	**-**	**-**	**-**
Air	-	-	-	1	87	-	-	-	-	-	-
Road	-	-	1045	-	-	-	427	-	-	-	-
Non-specified	-	-	-	-	-	-	-	-	-	-	-
OTHER SECTORS	**-**	**387**	**-**	**-**	**-**	**114**	**-**	**-**	**-**	**-**	**-**
Agriculture	-	-	-	-	-	-	-	-	-	-	-
Comm. and Publ. Services	-	-	-	-	-	-	-	-	-	-	-
Residential	-	387	-	-	-	114	-	-	-	-	-
Non-specified	-	-	-	-	-	-	-	-	-	-	-
NON-ENERGY USE	**-**	**-**	**-**	**-**	**-**	**-**	**-**	**-**	**-**	**-**	**60**

APPROVISIONNEMENT ET DEMANDE 1997	*Pétrole cont.* (1000 tonnes)										
	Gaz de raffinerie	*GPL + éthane*	*Essence moteur*	*Essence aviation*	*Carbu-réacteurs*	*Kérosène*	*Gazole*	*Fioul lourd*	*Naphta*	*Coke de pétrole*	*Autres prod.*
Production	-	20	1055	-	381	114	944	1531	224	-	44
Imports	-	-	-	1	-	-	-	-	-	-	-
Exports	-	-	-	-	-288	-	-194	-980	-224	-	-
Intl. Marine Bunkers	-	-	-	-	-	-	-40	-60	-	-	-
Stock Changes	-	-	-	-	-	-	-	-	-	-	-
DOMESTIC SUPPLY	**-**	**20**	**1055**	**1**	**93**	**114**	**710**	**491**	**-**	**-**	**44**
Transfers and Stat. Diff.	-	408	-	-	-1	-	-1	-1	-	-	-
TRANSFORMATION	**-**	**-**	**-**	**-**	**-**	**-**	**263**	**187**	**-**	**-**	**-**
Electricity and CHP Plants	-	-	-	-	-	-	263	187	-	-	-
Petroleum Refineries	-	-	-	-	-	-	-	-	-	-	-
Other Transform. Sector	-	-	-	-	-	-	-	-	-	-	-
ENERGY SECTOR	**-**	**-**	**-**	**-**	**-**	**-**	**-**	**134**	**-**	**-**	**-**
DISTRIBUTION LOSSES	**-**	**-**	**-**	**-**	**-**	**-**	**-**	**-**	**-**	**-**	**-**
FINAL CONSUMPTION	**-**	**428**	**1055**	**1**	**92**	**114**	**446**	**169**	**-**	**-**	**44**
INDUSTRY SECTOR	**-**	**-**	**-**	**-**	**-**	**-**	**-**	**169**	**-**	**-**	**-**
Iron and Steel	-	-	-	-	-	-	-	-	-	-	-
Chemical and Petrochem.	-	-	-	-	-	-	-	-	-	-	-
Non-Metallic Minerals	-	-	-	-	-	-	-	-	-	-	-
Non-specified	-	-	-	-	-	-	-	169	-	-	-
TRANSPORT SECTOR	**-**	**-**	**1055**	**1**	**92**	**-**	**446**	**-**	**-**	**-**	**-**
Air	-	-	-	1	92	-	-	-	-	-	-
Road	-	-	1055	-	-	-	446	-	-	-	-
Non-specified	-	-	-	-	-	-	-	-	-	-	-
OTHER SECTORS	**-**	**428**	**-**	**-**	**-**	**114**	**-**	**-**	**-**	**-**	**-**
Agriculture	-	-	-	-	-	-	-	-	-	-	-
Comm. and Publ. Services	-	-	-	-	-	-	-	-	-	-	-
Residential	-	428	-	-	-	114	-	-	-	-	-
Non-specified	-	-	-	-	-	-	-	-	-	-	-
NON-ENERGY USE	**-**	**-**	**-**	**-**	**-**	**-**	**-**	**-**	**-**	**-**	**44**

Yemen / Yémen

SUPPLY AND CONSUMPTION 1996	Gas (TJ)				Comb. Renew. & Waste (TJ)				(GWh)	(TJ)
	Natural Gas	Gas Works	Coke Ovens	Blast Furnaces	Solid Biomass	Gas/Liquids from Biomass	Municipal Waste	Industrial Waste	Electricity	Heat
Production	-	-	-	-	3240	-	-	-	2334	-
Imports	-	-	-	-	-	-	-	-	-	-
Exports	-	-	-	-	-	-	-	-	-	-
Intl. Marine Bunkers	-	-	-	-	-	-	-	-	-	-
Stock Changes	-	-	-	-	-	-	-	-	-	-
DOMESTIC SUPPLY	**-**	**-**	**-**	**-**	**3240**	**-**	**-**	**-**	**2334**	**-**
Transfers and Stat. Diff.	-	-	-	-	-	-	-	-	-	-
TRANSFORMATION	**-**	**-**	**-**	**-**	**1680**	**-**	**-**	**-**	**-**	**-**
Electricity and CHP Plants	-	-	-	-	-	-	-	-	-	-
Petroleum Refineries	-	-	-	-	-	-	-	-	-	-
Other Transform. Sector	-	-	-	-	1680	-	-	-	-	-
ENERGY SECTOR	**-**	**-**	**-**	**-**	**-**	**-**	**-**	**-**	**251**	**-**
DISTRIBUTION LOSSES	**-**	**-**	**-**	**-**	**-**	**-**	**-**	**-**	**600**	**-**
FINAL CONSUMPTION	**-**	**-**	**-**	**-**	**1560**	**-**	**-**	**-**	**1483**	**-**
INDUSTRY SECTOR	**-**	**-**	**-**	**-**	**-**	**-**	**-**	**-**	**-**	**-**
Iron and Steel	-	-	-	-	-	-	-	-	-	-
Chemical and Petrochem.	-	-	-	-	-	-	-	-	-	-
Non-Metallic Minerals	-	-	-	-	-	-	-	-	-	-
Non-specified	-	-	-	-	-	-	-	-	-	-
TRANSPORT SECTOR	**-**	**-**	**-**	**-**	**-**	**-**	**-**	**-**	**-**	**-**
Air	-	-	-	-	-	-	-	-	-	-
Road	-	-	-	-	-	-	-	-	-	-
Non-specified	-	-	-	-	-	-	-	-	-	-
OTHER SECTORS	**-**	**-**	**-**	**-**	**1560**	**-**	**-**	**-**	**1483**	**-**
Agriculture	-	-	-	-	-	-	-	-	-	-
Comm. and Publ. Services	-	-	-	-	-	-	-	-	-	-
Residential	-	-	-	-	-	-	-	-	-	-
Non-specified	-	-	-	-	1560	-	-	-	1483	-
NON-ENERGY USE	**-**	**-**	**-**	**-**	**-**	**-**	**-**	**-**	**-**	**-**

APPROVISIONNEMENT ET DEMANDE 1997	*Gaz* (TJ)				*En. Re. Comb. & Déchets* (TJ)				(GWh)	(TJ)
	Gaz naturel	*Usines à gaz*	*Cokeries*	*Hauts fourneaux*	*Biomasse solide*	*Gaz/Liquides tirés de biomasse*	*Déchets urbains*	*Déchets industriels*	*Electricité*	*Chaleur*
Production	-	-	-	-	3240	-	-	-	2358	-
Imports	-	-	-	-	-	-	-	-	-	-
Exports	-	-	-	-	-	-	-	-	-	-
Intl. Marine Bunkers	-	-	-	-	-	-	-	-	-	-
Stock Changes	-	-	-	-	-	-	-	-	-	-
DOMESTIC SUPPLY	**-**	**-**	**-**	**-**	**3240**	**-**	**-**	**-**	**2358**	**-**
Transfers and Stat. Diff.	-	-	-	-	-	-	-	-	-	-
TRANSFORMATION	**-**	**-**	**-**	**-**	**1680**	**-**	**-**	**-**	**-**	**-**
Electricity and CHP Plants	-	-	-	-	-	-	-	-	-	-
Petroleum Refineries	-	-	-	-	-	-	-	-	-	-
Other Transform. Sector	-	-	-	-	1680	-	-	-	-	-
ENERGY SECTOR	**-**	**-**	**-**	**-**	**-**	**-**	**-**	**-**	**254**	**-**
DISTRIBUTION LOSSES	**-**	**-**	**-**	**-**	**-**	**-**	**-**	**-**	**606**	**-**
FINAL CONSUMPTION	**-**	**-**	**-**	**-**	**1560**	**-**	**-**	**-**	**1498**	**-**
INDUSTRY SECTOR	**-**	**-**	**-**	**-**	**-**	**-**	**-**	**-**	**-**	**-**
Iron and Steel	-	-	-	-	-	-	-	-	-	-
Chemical and Petrochem.	-	-	-	-	-	-	-	-	-	-
Non-Metallic Minerals	-	-	-	-	-	-	-	-	-	-
Non-specified	-	-	-	-	-	-	-	-	-	-
TRANSPORT SECTOR	**-**	**-**	**-**	**-**	**-**	**-**	**-**	**-**	**-**	**-**
Air	-	-	-	-	-	-	-	-	-	-
Road	-	-	-	-	-	-	-	-	-	-
Non-specified	-	-	-	-	-	-	-	-	-	-
OTHER SECTORS	**-**	**-**	**-**	**-**	**1560**	**-**	**-**	**-**	**1498**	**-**
Agriculture	-	-	-	-	-	-	-	-	-	-
Comm. and Publ. Services	-	-	-	-	-	-	-	-	-	-
Residential	-	-	-	-	-	-	-	-	-	-
Non-specified	-	-	-	-	1560	-	-	-	1498	-
NON-ENERGY USE	**-**	**-**	**-**	**-**	**-**	**-**	**-**	**-**	**-**	**-**

Federal Republic of Yugoslavia / République fédérative de Yougoslavie

SUPPLY AND CONSUMPTION 1996	Coal (1000 tonnes)							Oil (1000 tonnes)			
	Coking Coal	Other Bit. Coal	Sub-Bit. Coal	Lignite	Peat	Oven and Gas Coke	Pat. Fuel and BKB	Crude Oil	NGL	Feed-stocks	Additives
Production	-	78	-	38367	-	-	-	1030	-	-	-
Imports	-	60	-	-	-	-	-	1321	-	-	-
Exports	-	-	-	-	-	-	-	-	-	-	-
Intl. Marine Bunkers	-	-	-	-	-	-	-	-	-	-	-
Stock Changes	-	-	-	-	-	-	-	-	-	-	-
DOMESTIC SUPPLY	**-**	**138**	**-**	**38367**	**-**	**-**	**-**	**2351**	**-**	**-**	**-**
Transfers and Stat. Diff.	-	-	-	-	-	-	-	-	-	-	-
TRANSFORMATION	**-**	**12**	**-**	**35000**	**-**	**-**	**-**	**2351**	**-**	**-**	**-**
Electricity and CHP Plants	-	12	-	35000	-	-	-	-	-	-	-
Petroleum Refineries	-	-	-	-	-	-	-	2351	-	-	-
Other Transform. Sector	-	-	-	-	-	-	-	-	-	-	-
ENERGY SECTOR	**-**	**-**	**-**	**-**	**-**	**-**	**-**	**-**	**-**	**-**	**-**
DISTRIBUTION LOSSES	**-**	**-**	**-**	**-**	**-**	**-**	**-**	**-**	**-**	**-**	**-**
FINAL CONSUMPTION	**-**	**126**	**-**	**3367**	**-**	**-**	**-**	**-**	**-**	**-**	**-**
INDUSTRY SECTOR	**-**	**-**	**-**	**2300**	**-**	**-**	**-**	**-**	**-**	**-**	**-**
Iron and Steel	-	-	-	-	-	-	-	-	-	-	-
Chemical and Petrochem.	-	-	-	-	-	-	-	-	-	-	-
Non-Metallic Minerals	-	-	-	-	-	-	-	-	-	-	-
Non-specified	-	-	-	2300	-	-	-	-	-	-	-
TRANSPORT SECTOR	**-**	**-**	**-**	**-**	**-**	**-**	**-**	**-**	**-**	**-**	**-**
Air	-	-	-	-	-	-	-	-	-	-	-
Road	-	-	-	-	-	-	-	-	-	-	-
Non-specified	-	-	-	-	-	-	-	-	-	-	-
OTHER SECTORS	**-**	**126**	**-**	**1067**	**-**	**-**	**-**	**-**	**-**	**-**	**-**
Agriculture	-	-	-	-	-	-	-	-	-	-	-
Comm. and Publ. Services	-	-	-	-	-	-	-	-	-	-	-
Residential	-	-	-	-	-	-	-	-	-	-	-
Non-specified	-	126	-	1067	-	-	-	-	-	-	-
NON-ENERGY USE	**-**	**-**	**-**	**-**	**-**	**-**	**-**	**-**	**-**	**-**	**-**

APPROVISIONNEMENT ET DEMANDE 1997	*Charbon* (1000 tonnes)							*Pétrole* (1000 tonnes)			
	Charbon à coke	*Autres charb. bit.*	*Charbon sous-bit.*	*Lignite*	*Tourbe*	*Coke de four/gaz*	*Agg./briq. de lignite*	*Pétrole brut*	*LGN*	*Produits d'aliment.*	*Additifs*
Production	-	93	-	40563	-	-	-	979	-	-	-
Imports	-	52	-	-	-	-	-	2292	-	-	-
Exports	-	-	-	-	-	-	-	-	-	-	-
Intl. Marine Bunkers	-	-	-	-	-	-	-	-	-	-	-
Stock Changes	-	-	-	-	-	-	-	-	-	-	-
DOMESTIC SUPPLY	**-**	**145**	**-**	**40563**	**-**	**-**	**-**	**3271**	**-**	**-**	**-**
Transfers and Stat. Diff.	-	-	-	-	-	-	-	-	-	-	-
TRANSFORMATION	**-**	**15**	**-**	**37000**	**-**	**-**	**-**	**3271**	**-**	**-**	**-**
Electricity and CHP Plants	-	15	-	37000	-	-	-	-	-	-	-
Petroleum Refineries	-	-	-	-	-	-	-	3271	-	-	-
Other Transform. Sector	-	-	-	-	-	-	-	-	-	-	-
ENERGY SECTOR	**-**	**-**	**-**	**-**	**-**	**-**	**-**	**-**	**-**	**-**	**-**
DISTRIBUTION LOSSES	**-**	**-**	**-**	**-**	**-**	**-**	**-**	**-**	**-**	**-**	**-**
FINAL CONSUMPTION	**-**	**130**	**-**	**3563**	**-**	**-**	**-**	**-**	**-**	**-**	**-**
INDUSTRY SECTOR	**-**	**-**	**-**	**2400**	**-**	**-**	**-**	**-**	**-**	**-**	**-**
Iron and Steel	-	-	-	-	-	-	-	-	-	-	-
Chemical and Petrochem.	-	-	-	-	-	-	-	-	-	-	-
Non-Metallic Minerals	-	-	-	-	-	-	-	-	-	-	-
Non-specified	-	-	-	2400	-	-	-	-	-	-	-
TRANSPORT SECTOR	**-**	**-**	**-**	**-**	**-**	**-**	**-**	**-**	**-**	**-**	**-**
Air	-	-	-	-	-	-	-	-	-	-	-
Road	-	-	-	-	-	-	-	-	-	-	-
Non-specified	-	-	-	-	-	-	-	-	-	-	-
OTHER SECTORS	**-**	**130**	**-**	**1163**	**-**	**-**	**-**	**-**	**-**	**-**	**-**
Agriculture	-	-	-	-	-	-	-	-	-	-	-
Comm. and Publ. Services	-	-	-	-	-	-	-	-	-	-	-
Residential	-	-	-	-	-	-	-	-	-	-	-
Non-specified	-	130	-	1163	-	-	-	-	-	-	-
NON-ENERGY USE	**-**	**-**	**-**	**-**	**-**	**-**	**-**	**-**	**-**	**-**	**-**

Federal Republic of Yugoslavia / République fédérative de Yougoslavie

SUPPLY AND CONSUMPTION 1996	Oil cont. (1000 tonnes)										
	Refinery Gas	LPG + Ethane	Motor Gasoline	Aviation Gasoline	Jet Fuel	Kerosene	Gas/ Diesel	Heavy Fuel Oil	Naphtha	Petrol. Coke	Other Prod.
Production	-	27	334	-	55	37	307	552	226	-	319
Imports	-	-	100	-	-	-	33	220	-	-	-
Exports	-	-	-	-	-	-	-	-	-	-	-
Intl. Marine Bunkers	-	-	-	-	-	-	-	-	-	-	-
Stock Changes	-	-	-	-	-	-	-	-	-	-	-
DOMESTIC SUPPLY	**-**	**27**	**434**	**-**	**55**	**37**	**340**	**772**	**226**	**-**	**319**
Transfers and Stat. Diff.	-	-	-	-	-	-	-	-	-	-	-
TRANSFORMATION	**-**	**-**	**-**	**-**	**-**	**-**	**-**	**301**	**-**	**-**	**-**
Electricity and CHP Plants	-	-	-	-	-	-	-	301	-	-	-
Petroleum Refineries	-	-	-	-	-	-	-	-	-	-	-
Other Transform. Sector	-	-	-	-	-	-	-	-	-	-	-
ENERGY SECTOR	**-**	**-**	**-**	**-**	**-**	**-**	**-**	**-**	**-**	**-**	**-**
DISTRIBUTION LOSSES	**-**	**-**	**-**	**-**	**-**	**-**	**-**	**-**	**-**	**-**	**-**
FINAL CONSUMPTION	**-**	**27**	**434**	**-**	**55**	**37**	**340**	**471**	**226**	**-**	**319**
INDUSTRY SECTOR	**-**	**27**	**-**	**-**	**-**	**-**	**107**	**353**	**226**	**-**	**-**
Iron and Steel	-	-	-	-	-	-	-	-	-	-	-
Chemical and Petrochem.	-	-	-	-	-	-	-	-	226	-	-
Non-Metallic Minerals	-	-	-	-	-	-	-	-	-	-	-
Non-specified	-	27	-	-	-	-	107	353	-	-	-
TRANSPORT SECTOR	**-**	**-**	**434**	**-**	**55**	**-**	**233**	**-**	**-**	**-**	**-**
Air	-	-	-	-	55	-	-	-	-	-	-
Road	-	-	434	-	-	-	233	-	-	-	-
Non-specified	-	-	-	-	-	-	-	-	-	-	-
OTHER SECTORS	**-**	**-**	**-**	**-**	**-**	**37**	**-**	**118**	**-**	**-**	**69**
Agriculture	-	-	-	-	-	-	-	-	-	-	-
Comm. and Publ. Services	-	-	-	-	-	-	-	-	-	-	-
Residential	-	-	-	-	-	37	-	-	-	-	-
Non-specified	-	-	-	-	-	-	-	118	-	-	69
NON-ENERGY USE	**-**	**-**	**-**	**-**	**-**	**-**	**-**	**-**	**-**	**-**	**250**

APPROVISIONNEMENT ET DEMANDE 1997	*Pétrole cont.* (1000 tonnes)										
	Gaz de raffinerie	*GPL + éthane*	*Essence moteur*	*Essence aviation*	*Carbu- réacteurs*	*Kérosène*	*Gazole*	*Fioul lourd*	*Naphta*	*Coke de pétrole*	*Autres prod.*
Production	-	47	470	-	95	54	520	809	385	-	292
Imports	-	-	200	-	-	-	-	100	-	-	-
Exports	-	-	-	-	-	-	-	-	-	-	-
Intl. Marine Bunkers	-	-	-	-	-	-	-	-	-	-	-
Stock Changes	-	-	-	-	-	-	-	-	-	-	-
DOMESTIC SUPPLY	**-**	**47**	**670**	**-**	**95**	**54**	**520**	**909**	**385**	**-**	**292**
Transfers and Stat. Diff.	-	-	-	-	-	-	-	-	-	-	-
TRANSFORMATION	**-**	**-**	**-**	**-**	**-**	**-**	**-**	**350**	**-**	**-**	**-**
Electricity and CHP Plants	-	-	-	-	-	-	-	350	-	-	-
Petroleum Refineries	-	-	-	-	-	-	-	-	-	-	-
Other Transform. Sector	-	-	-	-	-	-	-	-	-	-	-
ENERGY SECTOR	**-**	**-**	**-**	**-**	**-**	**-**	**-**	**-**	**-**	**-**	**-**
DISTRIBUTION LOSSES	**-**	**-**	**-**	**-**	**-**	**-**	**-**	**-**	**-**	**-**	**-**
FINAL CONSUMPTION	**-**	**47**	**670**	**-**	**95**	**54**	**520**	**559**	**385**	**-**	**292**
INDUSTRY SECTOR	**-**	**47**	**-**	**-**	**-**	**-**	**200**	**400**	**385**	**-**	**-**
Iron and Steel	-	-	-	-	-	-	-	-	-	-	-
Chemical and Petrochem.	-	-	-	-	-	-	-	-	385	-	-
Non-Metallic Minerals	-	-	-	-	-	-	-	-	-	-	-
Non-specified	-	47	-	-	-	-	200	400	-	-	-
TRANSPORT SECTOR	**-**	**-**	**670**	**-**	**95**	**-**	**320**	**-**	**-**	**-**	**-**
Air	-	-	-	-	95	-	-	-	-	-	-
Road	-	-	670	-	-	-	320	-	-	-	-
Non-specified	-	-	-	-	-	-	-	-	-	-	-
OTHER SECTORS	**-**	**-**	**-**	**-**	**-**	**54**	**-**	**159**	**-**	**-**	**119**
Agriculture	-	-	-	-	-	-	-	-	-	-	-
Comm. and Publ. Services	-	-	-	-	-	-	-	-	-	-	-
Residential	-	-	-	-	-	54	-	-	-	-	-
Non-specified	-	-	-	-	-	-	-	159	-	-	119
NON-ENERGY USE	**-**	**-**	**-**	**-**	**-**	**-**	**-**	**-**	**-**	**-**	**173**

Federal Republic of Yugoslavia / République fédérative de Yougoslavie

SUPPLY AND CONSUMPTION 1996	Gas (TJ)				Comb. Renew. & Waste (TJ)				(GWh)	(TJ)
	Natural Gas	Gas Works	Coke Ovens	Blast Furnaces	Solid Biomass	Gas/Liquids from Biomass	Municipal Waste	Industrial Waste	Electricity	Heat
Production	25297	-	-	-	8791	-	-	-	38093	19000
Imports	79170	-	-	-	-	-	-	-	-	-
Exports	-	-	-	-	-	-	-	-	-156	-
Intl. Marine Bunkers	-	-	-	-	-	-	-	-	-	-
Stock Changes	-	-	-	-	-	-	-	-	-	-
DOMESTIC SUPPLY	**104467**	**-**	**-**	**-**	**8791**	**-**	**-**	**-**	**37937**	**19000**
Transfers and Stat. Diff.	-	-	-	-	-	-	-	-	-	-
TRANSFORMATION	**12070**	**-**	**-**	**-**	**-**	**-**	**-**	**-**	**-**	**-**
Electricity and CHP Plants	12070	-	-	-	-	-	-	-	-	-
Petroleum Refineries	-	-	-	-	-	-	-	-	-	-
Other Transform. Sector	-	-	-	-	-	-	-	-	-	-
ENERGY SECTOR	**1038**	**-**	**-**	**-**	**-**	**-**	**-**	**-**	**2983**	**-**
DISTRIBUTION LOSSES	**-**	**-**	**-**	**-**	**-**	**-**	**-**	**-**	**3525**	**-**
FINAL CONSUMPTION	**91359**	**-**	**-**	**-**	**8791**	**-**	**-**	**-**	**31429**	**19000**
INDUSTRY SECTOR	**30100**	**-**	**-**	**-**	**-**	**-**	**-**	**-**	**6227**	**-**
Iron and Steel	-	-	-	-	-	-	-	-	339	-
Chemical and Petrochem.	-	-	-	-	-	-	-	-	404	-
Non-Metallic Minerals	-	-	-	-	-	-	-	-	-	-
Non-specified	30100	-	-	-	-	-	-	-	5484	-
TRANSPORT SECTOR	**-**	**-**	**-**	**-**	**-**	**-**	**-**	**-**	**270**	**-**
Air	-	-	-	-	-	-	-	-	-	-
Road	-	-	-	-	-	-	-	-	54	-
Non-specified	-	-	-	-	-	-	-	-	216	-
OTHER SECTORS	**61259**	**-**	**-**	**-**	**8791**	**-**	**-**	**-**	**24932**	**19000**
Agriculture	-	-	-	-	-	-	-	-	182	-
Comm. and Publ. Services	-	-	-	-	-	-	-	-	292	-
Residential	-	-	-	-	-	-	-	-	16444	-
Non-specified	61259	-	-	-	8791	-	-	-	8014	19000
NON-ENERGY USE	**-**	**-**	**-**	**-**	**-**	**-**	**-**	**-**	**-**	**-**

APPROVISIONNEMENT ET DEMANDE 1997	*Gaz* (TJ)				*En. Re. Comb. & Déchets* (TJ)				(GWh)	(TJ)
	Gaz naturel	*Usines à gaz*	*Cokeries*	*Hauts fourneaux*	*Biomasse solide*	*Gaz/Liquides tirés de biomasse*	*Déchets urbains*	*Déchets industriels*	*Electricité*	*Chaleur*
Production	25937	-	-	-	8791	-	-	-	40312	19000
Imports	77813	-	-	-	-	-	-	-	-	-
Exports	-	-	-	-	-	-	-	-	-	-
Intl. Marine Bunkers	-	-	-	-	-	-	-	-	-	-
Stock Changes	-	-	-	-	-	-	-	-	-	-
DOMESTIC SUPPLY	**103750**	**-**	**-**	**-**	**8791**	**-**	**-**	**-**	**40312**	**19000**
Transfers and Stat. Diff.	-	-	-	-	-	-	-	-	-	-
TRANSFORMATION	**11988**	**-**	**-**	**-**	**-**	**-**	**-**	**-**	**-**	**-**
Electricity and CHP Plants	11988	-	-	-	-	-	-	-	-	-
Petroleum Refineries	-	-	-	-	-	-	-	-	-	-
Other Transform. Sector	-	-	-	-	-	-	-	-	-	-
ENERGY SECTOR	**1056**	**-**	**-**	**-**	**-**	**-**	**-**	**-**	**3170**	**-**
DISTRIBUTION LOSSES	**-**	**-**	**-**	**-**	**-**	**-**	**-**	**-**	**3746**	**-**
FINAL CONSUMPTION	**90706**	**-**	**-**	**-**	**8791**	**-**	**-**	**-**	**33396**	**19000**
INDUSTRY SECTOR	**29858**	**-**	**-**	**-**	**-**	**-**	**-**	**-**	**6617**	**-**
Iron and Steel	-	-	-	-	-	-	-	-	360	-
Chemical and Petrochem.	-	-	-	-	-	-	-	-	429	-
Non-Metallic Minerals	-	-	-	-	-	-	-	-	-	-
Non-specified	29858	-	-	-	-	-	-	-	5828	-
TRANSPORT SECTOR	**-**	**-**	**-**	**-**	**-**	**-**	**-**	**-**	**287**	**-**
Air	-	-	-	-	-	-	-	-	-	-
Road	-	-	-	-	-	-	-	-	57	-
Non-specified	-	-	-	-	-	-	-	-	230	-
OTHER SECTORS	**60848**	**-**	**-**	**-**	**8791**	**-**	**-**	**-**	**26492**	**19000**
Agriculture	-	-	-	-	-	-	-	-	193	-
Comm. and Publ. Services	-	-	-	-	-	-	-	-	310	-
Residential	-	-	-	-	-	-	-	-	17473	-
Non-specified	60848	-	-	-	8791	-	-	-	8516	19000
NON-ENERGY USE	**-**	**-**	**-**	**-**	**-**	**-**	**-**	**-**	**-**	**-**

Former Yugoslavia / ex-Yougoslavie

SUPPLY AND CONSUMPTION 1996	Coal (1000 tonnes)							Oil (1000 tonnes)			
	Coking Coal	Other Bit. Coal	Sub-Bit. Coal	Lignite	Peat	Oven and Gas Coke	Pat. Fuel and BKB	Crude Oil	NGL	Feed-stocks	Additives
Production	-	142	830	51091	-	-	-	2500	227	-	-
Imports	49	184	385	187	-	83	-	6025	-	69	-
Exports	-	-	-	-2	-	-	-	-	-	-	-
Intl. Marine Bunkers	-	-	-	-	-	-	-	-	-	-	-
Stock Changes	-	6	-87	-125	-	7	-	-134	-	1	-
DOMESTIC SUPPLY	**49**	**332**	**1128**	**51151**	**-**	**90**	**-**	**8391**	**227**	**70**	**-**
Transfers and Stat. Diff.	-	-	-	-	-	-	-	46	-227	-	-
TRANSFORMATION	**-**	**67**	**1075**	**47027**	**-**	**25**	**-**	**8350**	**-**	**70**	**-**
Electricity and CHP Plants	-	67	1074	47027	-	-	-	-	-	-	-
Petroleum Refineries	-	-	-	-	-	-	-	8350	-	70	-
Other Transform. Sector	-	-	1	-	-	25	-	-	-	-	-
ENERGY SECTOR	**-**	**1**	**5**	**-**	**-**	**-**	**-**	**87**	**-**	**-**	**-**
DISTRIBUTION LOSSES	**-**	**-**	**-**	**-**	**-**	**-**	**-**	**-**	**-**	**-**	**-**
FINAL CONSUMPTION	**49**	**264**	**48**	**4124**	**-**	**65**	**-**	**-**	**-**	**-**	**-**
INDUSTRY SECTOR	**49**	**138**	**10**	**2570**	**-**	**65**	**-**	**-**	**-**	**-**	**-**
Iron and Steel	11	22	-	105	-	15	-	-	-	-	-
Chemical and Petrochem.	-	-	-	13	-	-	-	-	-	-	-
Non-Metallic Minerals	-	82	-	5	-	45	-	-	-	-	-
Non-specified	38	34	10	2447	-	5	-	-	-	-	-
TRANSPORT SECTOR	**-**	**-**	**1**	**290**	**-**	**-**	**-**	**-**	**-**	**-**	**-**
Air	-	-	-	-	-	-	-	-	-	-	-
Road	-	-	-	-	-	-	-	-	-	-	-
Non-specified	-	-	1	290	-	-	-	-	-	-	-
OTHER SECTORS	**-**	**126**	**37**	**1264**	**-**	**-**	**-**	**-**	**-**	**-**	**-**
Agriculture	-	-	-	-	-	-	-	-	-	-	-
Comm. and Publ. Services	-	-	15	74	-	-	-	-	-	-	-
Residential	-	-	22	123	-	-	-	-	-	-	-
Non-specified	-	126	-	1067	-	-	-	-	-	-	-
NON-ENERGY USE	**-**	**-**	**-**	**-**	**-**	**-**	**-**	**-**	**-**	**-**	**-**

APPROVISIONNEMENT ET DEMANDE 1997	*Charbon* (1000 tonnes)							*Pétrole* (1000 tonnes)			
	Charbon à coke	*Autres charb. bit.*	*Charbon sous-bit.*	*Lignite*	*Tourbe*	*Coke de four/gaz*	*Agg./briq. de lignite*	*Pétrole brut*	*LGN*	*Produits d'aliment.*	*Additifs*
Production	-	142	811	53045	-	-	-	2476	252	-	-
Imports	49	247	300	140	-	105	-	6913	-	77	-
Exports	-	-3	-1	-1	-	-	-	-17	-	-	-
Intl. Marine Bunkers	-	-	-	-	-	-	-	-	-	-	-
Stock Changes	-	115	58	51	-	-4	-	-55	-	-4	-
DOMESTIC SUPPLY	**49**	**501**	**1168**	**53235**	**-**	**101**	**-**	**9317**	**252**	**73**	**-**
Transfers and Stat. Diff.	-	-	-	-	-	-	-	-	-252	-	-
TRANSFORMATION	**-**	**245**	**1137**	**49003**	**-**	**20**	**-**	**9317**	**-**	**73**	**-**
Electricity and CHP Plants	-	245	1137	49003	-	-	-	-	-	-	-
Petroleum Refineries	-	-	-	-	-	-	-	9317	-	73	-
Other Transform. Sector	-	-	-	-	-	20	-	-	-	-	-
ENERGY SECTOR	**-**	**-**	**3**	**-**	**-**	**-**	**-**	**-**	**-**	**-**	**-**
DISTRIBUTION LOSSES	**-**	**-**	**-**	**-**	**-**	**-**	**-**	**-**	**-**	**-**	**-**
FINAL CONSUMPTION	**49**	**256**	**28**	**4232**	**-**	**81**	**-**	**-**	**-**	**-**	**-**
INDUSTRY SECTOR	**49**	**126**	**7**	**2635**	**-**	**81**	**-**	**-**	**-**	**-**	**-**
Iron and Steel	11	22	-	98	-	22	-	-	-	-	-
Chemical and Petrochem.	-	-	-	6	-	-	-	-	-	-	-
Non-Metallic Minerals	-	69	-	4	-	50	-	-	-	-	-
Non-specified	38	35	7	2527	-	9	-	-	-	-	-
TRANSPORT SECTOR	**-**	**-**	**1**	**290**	**-**	**-**	**-**	**-**	**-**	**-**	**-**
Air	-	-	-	-	-	-	-	-	-	-	-
Road	-	-	-	-	-	-	-	-	-	-	-
Non-specified	-	-	1	290	-	-	-	-	-	-	-
OTHER SECTORS	**-**	**130**	**20**	**1307**	**-**	**-**	**-**	**-**	**-**	**-**	**-**
Agriculture	-	-	-	-	-	-	-	-	-	-	-
Comm. and Publ. Services	-	-	8	55	-	-	-	-	-	-	-
Residential	-	-	12	89	-	-	-	-	-	-	-
Non-specified	-	130	-	1163	-	-	-	-	-	-	-
NON-ENERGY USE	**-**	**-**	**-**	**-**	**-**	**-**	**-**	**-**	**-**	**-**	**-**

Former Yugoslavia / ex-Yougoslavie

SUPPLY AND CONSUMPTION 1996	Oil cont. (1000 tonnes)										
	Refinery Gas	LPG + Ethane	Motor Gasoline	Aviation Gasoline	Jet Fuel	Kerosene	Gas/ Diesel	Heavy Fuel Oil	Naphtha	Petrol. Coke	Other Prod.
Production	195	200	1495	-	138	46	2207	2346	608	46	578
Imports	-	4	1008	1	84	-	1157	737	298	-	43
Exports	-	-93	-333	-	-14	-2	-266	-161	-221	-8	-102
Intl. Marine Bunkers	-	-	-	-	-	-	-12	-17	-	-	-
Stock Changes	-	6	112	-	4	-	286	-55	-2	-1	-19
DOMESTIC SUPPLY	**195**	**117**	**2282**	**1**	**212**	**44**	**3372**	**2850**	**683**	**37**	**500**
Transfers and Stat. Diff.	-	177	-	-	-	-	-	-	50	-	-
TRANSFORMATION	**5**	**11**	**-**	**-**	**-**	**-**	**46**	**1515**	**61**	**-**	**-**
Electricity and CHP Plants	5	-	-	-	-	-	38	1203	20	-	-
Petroleum Refineries	-	-	-	-	-	-	-	-	-	-	-
Other Transform. Sector	-	11	-	-	-	-	8	312	41	-	-
ENERGY SECTOR	**190**	**30**	**1**	**-**	**-**	**-**	**1**	**247**	**-**	**37**	**-**
DISTRIBUTION LOSSES	**-**	**-**	**87**	**-**	**-**	**-**	**62**	**36**	**223**	**-**	**-**
FINAL CONSUMPTION	**-**	**253**	**2194**	**1**	**212**	**44**	**3263**	**1052**	**449**	**-**	**500**
INDUSTRY SECTOR	**-**	**162**	**8**	**-**	**-**	**-**	**292**	**824**	**449**	**-**	**48**
Iron and Steel	-	5	-	-	-	-	6	35	7	-	16
Chemical and Petrochem.	-	94	-	-	-	-	4	119	435	-	3
Non-Metallic Minerals	-	18	-	-	-	-	11	111	-	-	-
Non-specified	-	45	8	-	-	-	271	559	7	-	29
TRANSPORT SECTOR	**-**	**12**	**2177**	**1**	**212**	**-**	**1751**	**8**	**-**	**-**	**-**
Air	-	-	-	1	212	-	-	-	-	-	-
Road	-	12	2176	-	-	-	1654	2	-	-	-
Non-specified	-	-	1	-	-	-	97	6	-	-	-
OTHER SECTORS	**-**	**79**	**9**	**-**	**-**	**44**	**1220**	**220**	**-**	**-**	**69**
Agriculture	-	3	9	-	-	-	203	29	-	-	-
Comm. and Publ. Services	-	7	-	-	-	-	393	1	-	-	-
Residential	-	69	-	-	-	44	624	72	-	-	-
Non-specified	-	-	-	-	-	-	-	118	-	-	69
NON-ENERGY USE	**-**	**-**	**-**	**-**	**-**	**-**	**-**	**-**	**-**	**-**	**383**

APPROVISIONNEMENT ET DEMANDE 1997	*Pétrole cont.* (1000 tonnes)										
	Gaz de raffinerie	*GPL + éthane*	*Essence moteur*	*Essence aviation*	*Carbu- réacteurs*	*Kérosène*	*Gazole*	*Fioul lourd*	*Naphta*	*Coke de pétrole*	*Autres prod.*
Production	202	208	1683	-	190	59	2385	2640	641	28	620
Imports	-	9	1023	1	81	-	1303	389	287	-	52
Exports	-	-86	-64	-	-6	-	-	-27	-290	-13	-94
Intl. Marine Bunkers	-	-	-	-	-	-	-7	-17	-	-	-
Stock Changes	-	1	-51	-	-4	1	-33	-76	2	-	5
DOMESTIC SUPPLY	**202**	**132**	**2591**	**1**	**261**	**60**	**3648**	**2909**	**640**	**15**	**583**
Transfers and Stat. Diff.	-	198	-	-	-	-	-	1	54	-	-
TRANSFORMATION	**4**	**14**	**-**	**-**	**-**	**-**	**29**	**1415**	**61**	**-**	**-**
Electricity and CHP Plants	4	-	-	-	-	-	20	1133	20	-	-
Petroleum Refineries	-	-	-	-	-	-	-	-	-	-	-
Other Transform. Sector	-	14	-	-	-	-	9	282	41	-	-
ENERGY SECTOR	**198**	**-**	**1**	**-**	**-**	**-**	**4**	**303**	**-**	**-**	**-**
DISTRIBUTION LOSSES	**-**	**-**	**87**	**-**	**-**	**-**	**62**	**36**	**223**	**-**	**-**
FINAL CONSUMPTION	**-**	**316**	**2503**	**1**	**261**	**60**	**3553**	**1156**	**410**	**15**	**583**
INDUSTRY SECTOR	**-**	**194**	**11**	**-**	**-**	**-**	**410**	**874**	**410**	**15**	**48**
Iron and Steel	-	5	-	-	-	-	5	25	7	-	16
Chemical and Petrochem.	-	102	-	-	-	-	4	123	396	-	3
Non-Metallic Minerals	-	18	-	-	-	-	14	123	-	15	-
Non-specified	-	69	11	-	-	-	387	603	7	-	29
TRANSPORT SECTOR	**-**	**12**	**2483**	**1**	**261**	**-**	**1867**	**11**	**-**	**-**	**-**
Air	-	-	-	1	261	-	-	-	-	-	-
Road	-	12	2483	-	-	-	1734	-	-	-	-
Non-specified	-	-	-	-	-	-	133	11	-	-	-
OTHER SECTORS	**-**	**110**	**9**	**-**	**-**	**60**	**1276**	**271**	**-**	**-**	**119**
Agriculture	-	2	9	-	-	-	216	12	-	-	-
Comm. and Publ. Services	-	11	-	-	-	1	406	51	-	-	-
Residential	-	97	-	-	-	59	654	49	-	-	-
Non-specified	-	-	-	-	-	-	-	159	-	-	119
NON-ENERGY USE	**-**	**-**	**-**	**-**	**-**	**-**	**-**	**-**	**-**	**-**	**416**

Former Yugoslavia / ex-Yougoslavie

SUPPLY AND CONSUMPTION 1996	Gas (TJ)				Comb. Renew. & Waste (TJ)				(GWh)	(TJ)
	Natural Gas	Gas Works	Coke Ovens	Blast Furnaces	Solid Biomass	Gas/Liquids from Biomass	Municipal Waste	Industrial Waste	Electricity	Heat
Production	93616	436	-	-	46577	-	-	100	70103	49613
Imports	152445	-	-	-	-	-	-	-	718	-
Exports	-	-	-	-	-	-	-	-	-	-
Intl. Marine Bunkers	-	-	-	-	-	-	-	-	-	-
Stock Changes	-426	-	-	-	1355	-	-	-	-	-
DOMESTIC SUPPLY	**245635**	**436**	**-**	**-**	**47932**	**-**	**-**	**100**	**70821**	**49613**
Transfers and Stat. Diff.	7523	-	-	-	-	-	-	-	-	-
TRANSFORMATION	**53957**	**-**	**-**	**-**	**96**	**-**	**-**	**100**	**-**	**-**
Electricity and CHP Plants	33495	-	-	-	-	-	-	100	-	-
Petroleum Refineries	-	-	-	-	-	-	-	-	-	-
Other Transform. Sector	20462	-	-	-	96	-	-	-	-	-
ENERGY SECTOR	**11923**	**-**	**-**	**-**	**-**	**-**	**-**	**-**	**5303**	**2152**
DISTRIBUTION LOSSES	**4330**	**11**	**-**	**-**	**-**	**-**	**-**	**-**	**7481**	**2816**
FINAL CONSUMPTION	**182948**	**425**	**-**	**-**	**47836**	**-**	**-**	**-**	**58037**	**44645**
INDUSTRY SECTOR	**95648**	**148**	**-**	**-**	**2556**	**-**	**-**	**-**	**15671**	**7799**
Iron and Steel	5479	26	-	-	-	-	-	-	2310	242
Chemical and Petrochem.	30231	-	-	-	-	-	-	-	1495	1490
Non-Metallic Minerals	11641	44	-	-	-	-	-	-	685	156
Non-specified	48297	78	-	-	2556	-	-	-	11181	5911
TRANSPORT SECTOR	**-**	**-**	**-**	**-**	**-**	**-**	**-**	**-**	**686**	**-**
Air	-	-	-	-	-	-	-	-	-	-
Road	-	-	-	-	-	-	-	-	54	-
Non-specified	-	-	-	-	-	-	-	-	632	-
OTHER SECTORS	**87300**	**277**	**-**	**-**	**45280**	**-**	**-**	**-**	**41680**	**36846**
Agriculture	958	-	-	-	-	-	-	-	275	467
Comm. and Publ. Services	5470	70	-	-	2141	-	-	-	5153	3817
Residential	19613	207	-	-	34198	-	-	-	26499	12596
Non-specified	61259	-	-	-	8941	-	-	-	9753	19966
NON-ENERGY USE	**-**	**-**	**-**	**-**	**-**	**-**	**-**	**-**	**-**	**-**

APPROVISIONNEMENT ET DEMANDE 1997	*Gaz* (TJ)				*En. Re. Comb. & Déchets* (TJ)				(GWh)	(TJ)
	Gaz naturel	*Usines à gaz*	*Cokeries*	*Hauts fourneaux*	*Biomasse solide*	*Gaz/Liquides tirés de biomasse*	*Déchets urbains*	*Déchets industriels*	*Electricité*	*Chaleur*
Production	91623	464	-	-	46442	-	-	10	72085	48567
Imports	160223	-	-	-	-	-	-	-	2457	-
Exports	-	-	-	-	-	-	-	-	-	-
Intl. Marine Bunkers	-	-	-	-	-	-	-	-	-	-
Stock Changes	-464	-	-	-	1353	-	-	-	-	-
DOMESTIC SUPPLY	**251382**	**464**	**-**	**-**	**47795**	**-**	**-**	**10**	**74542**	**48567**
Transfers and Stat. Diff.	6731	-	-	-	-	-	-	-	-	-
TRANSFORMATION	**52303**	**-**	**-**	**-**	**94**	**-**	**-**	**10**	**-**	**-**
Electricity and CHP Plants	31987	-	-	-	-	-	-	10	-	-
Petroleum Refineries	-	-	-	-	-	-	-	-	-	-
Other Transform. Sector	20316	-	-	-	94	-	-	-	-	-
ENERGY SECTOR	**11527**	**-**	**-**	**-**	**-**	**-**	**-**	**-**	**5663**	**2320**
DISTRIBUTION LOSSES	**3225**	**49**	**-**	**-**	**-**	**-**	**-**	**-**	**7606**	**2736**
FINAL CONSUMPTION	**191058**	**415**	**-**	**-**	**47701**	**-**	**-**	**-**	**61273**	**43511**
INDUSTRY SECTOR	**103853**	**175**	**-**	**-**	**2556**	**-**	**-**	**-**	**16442**	**7112**
Iron and Steel	6979	19	-	-	-	-	-	-	2514	238
Chemical and Petrochem.	35798	-	-	-	-	-	-	-	1516	1737
Non-Metallic Minerals	11667	32	-	-	-	-	-	-	907	172
Non-specified	49409	124	-	-	2556	-	-	-	11505	4965
TRANSPORT SECTOR	**-**	**-**	**-**	**-**	**-**	**-**	**-**	**-**	**713**	**-**
Air	-	-	-	-	-	-	-	-	-	-
Road	-	-	-	-	-	-	-	-	57	-
Non-specified	-	-	-	-	-	-	-	-	656	-
OTHER SECTORS	**87205**	**240**	**-**	**-**	**45145**	**-**	**-**	**-**	**44118**	**36399**
Agriculture	855	-	-	-	-	-	-	-	278	440
Comm. and Publ. Services	5120	30	-	-	2141	-	-	-	5620	3796
Residential	20382	210	-	-	34063	-	-	-	27965	12197
Non-specified	60848	-	-	-	8941	-	-	-	10255	19966
NON-ENERGY USE	**-**	**-**	**-**	**-**	**-**	**-**	**-**	**-**	**-**	**-**

Zambia / Zambie

SUPPLY AND CONSUMPTION 1996	Coal (1000 tonnes)							Oil (1000 tonnes)			
	Coking Coal	Other Bit. Coal	Sub-Bit. Coal	Lignite	Peat	Oven and Gas Coke	Pat. Fuel and BKB	Crude Oil	NGL	Feed-stocks	Additives
Production	24	155	-	-	-	32	-	-	-	-	-
Imports	-	-	-	-	-	-	-	588	-	-	-
Exports	-	-7	-	-	-	-	-	-	-	-	-
Intl. Marine Bunkers	-	-	-	-	-	-	-	-	-	-	-
Stock Changes	-	-	-	-	-	-	-	-	-	-	-
DOMESTIC SUPPLY	**24**	**148**	**-**	**-**	**-**	**32**	**-**	**588**	**-**	**-**	**-**
Transfers and Stat. Diff.	-	-	-	-	-	-	-	-	-	-	-
TRANSFORMATION	**24**	**30**	**-**	**-**	**-**	**26**	**-**	**588**	**-**	**-**	**-**
Electricity and CHP Plants	-	6	-	-	-	-	-	-	-	-	-
Petroleum Refineries	-	-	-	-	-	-	-	588	-	-	-
Other Transform. Sector	24	24	-	-	-	26	-	-	-	-	-
ENERGY SECTOR	**-**	**-**	**-**	**-**	**-**	**-**	**-**	**-**	**-**	**-**	**-**
DISTRIBUTION LOSSES	**-**	**-**	**-**	**-**	**-**	**-**	**-**	**-**	**-**	**-**	**-**
FINAL CONSUMPTION	**-**	**118**	**-**	**-**	**-**	**6**	**-**	**-**	**-**	**-**	**-**
INDUSTRY SECTOR	**-**	**118**	**-**	**-**	**-**	**6**	**-**	**-**	**-**	**-**	**-**
Iron and Steel	-	-	-	-	-	6	-	-	-	-	-
Chemical and Petrochem.	-	-	-	-	-	-	-	-	-	-	-
Non-Metallic Minerals	-	-	-	-	-	-	-	-	-	-	-
Non-specified	-	118	-	-	-	-	-	-	-	-	-
TRANSPORT SECTOR	**-**	**-**	**-**	**-**	**-**	**-**	**-**	**-**	**-**	**-**	**-**
Air	-	-	-	-	-	-	-	-	-	-	-
Road	-	-	-	-	-	-	-	-	-	-	-
Non-specified	-	-	-	-	-	-	-	-	-	-	-
OTHER SECTORS	**-**	**-**	**-**	**-**	**-**	**-**	**-**	**-**	**-**	**-**	**-**
Agriculture	-	-	-	-	-	-	-	-	-	-	-
Comm. and Publ. Services	-	-	-	-	-	-	-	-	-	-	-
Residential	-	-	-	-	-	-	-	-	-	-	-
Non-specified	-	-	-	-	-	-	-	-	-	-	-
NON-ENERGY USE	**-**	**-**	**-**	**-**	**-**	**-**	**-**	**-**	**-**	**-**	**-**

APPROVISIONNEMENT ET DEMANDE 1997	*Charbon* (1000 tonnes)							*Pétrole* (1000 tonnes)			
	Charbon à coke	*Autres charb. bit.*	*Charbon sous-bit.*	*Lignite*	*Tourbe*	*Coke de four/gaz*	*Agg./briq. de lignite*	*Pétrole brut*	*LGN*	*Produits d'aliment.*	*Additifs*
Production	24	155	-	-	-	32	-	-	-	-	-
Imports	-	-	-	-	-	-	-	588	-	-	-
Exports	-	-7	-	-	-	-	-	-	-	-	-
Intl. Marine Bunkers	-	-	-	-	-	-	-	-	-	-	-
Stock Changes	-	-	-	-	-	-	-	-	-	-	-
DOMESTIC SUPPLY	**24**	**148**	**-**	**-**	**-**	**32**	**-**	**588**	**-**	**-**	**-**
Transfers and Stat. Diff.	-	-	-	-	-	-	-	-	-	-	-
TRANSFORMATION	**24**	**30**	**-**	**-**	**-**	**26**	**-**	**588**	**-**	**-**	**-**
Electricity and CHP Plants	-	6	-	-	-	-	-	-	-	-	-
Petroleum Refineries	-	-	-	-	-	-	-	588	-	-	-
Other Transform. Sector	24	24	-	-	-	26	-	-	-	-	-
ENERGY SECTOR	**-**	**-**	**-**	**-**	**-**	**-**	**-**	**-**	**-**	**-**	**-**
DISTRIBUTION LOSSES	**-**	**-**	**-**	**-**	**-**	**-**	**-**	**-**	**-**	**-**	**-**
FINAL CONSUMPTION	**-**	**118**	**-**	**-**	**-**	**6**	**-**	**-**	**-**	**-**	**-**
INDUSTRY SECTOR	**-**	**118**	**-**	**-**	**-**	**6**	**-**	**-**	**-**	**-**	**-**
Iron and Steel	-	-	-	-	-	6	-	-	-	-	-
Chemical and Petrochem.	-	-	-	-	-	-	-	-	-	-	-
Non-Metallic Minerals	-	-	-	-	-	-	-	-	-	-	-
Non-specified	-	118	-	-	-	-	-	-	-	-	-
TRANSPORT SECTOR	**-**	**-**	**-**	**-**	**-**	**-**	**-**	**-**	**-**	**-**	**-**
Air	-	-	-	-	-	-	-	-	-	-	-
Road	-	-	-	-	-	-	-	-	-	-	-
Non-specified	-	-	-	-	-	-	-	-	-	-	-
OTHER SECTORS	**-**	**-**	**-**	**-**	**-**	**-**	**-**	**-**	**-**	**-**	**-**
Agriculture	-	-	-	-	-	-	-	-	-	-	-
Comm. and Publ. Services	-	-	-	-	-	-	-	-	-	-	-
Residential	-	-	-	-	-	-	-	-	-	-	-
Non-specified	-	-	-	-	-	-	-	-	-	-	-
NON-ENERGY USE	**-**	**-**	**-**	**-**	**-**	**-**	**-**	**-**	**-**	**-**	**-**

Zambia / Zambie

SUPPLY AND CONSUMPTION 1996	Oil cont. (1000 tonnes)										
	Refinery Gas	LPG + Ethane	Motor Gasoline	Aviation Gasoline	Jet Fuel	Kerosene	Gas/ Diesel	Heavy Fuel Oil	Naphtha	Petrol. Coke	Other Prod.
Production	15	10	120	-	40	31	219	95	-	-	32
Imports	-	-	-	3	-	-	10	-	-	-	-
Exports	-	-	-10	-	-	-	-30	-	-	-	-
Intl. Marine Bunkers	-	-	-	-	-	-	-	-	-	-	-
Stock Changes	-	-	-	-	-	-	-	-	-	-	-
DOMESTIC SUPPLY	**15**	**10**	**110**	**3**	**40**	**31**	**199**	**95**	**-**	**-**	**32**
Transfers and Stat. Diff.	-	-	-	-	-5	-	-2	-	-	-	-
TRANSFORMATION	**-**	**-**	**-**	**-**	**-**	**-**	**1**	**-**	**-**	**-**	**-**
Electricity and CHP Plants	-	-	-	-	-	-	1	-	-	-	-
Petroleum Refineries	-	-	-	-	-	-	-	-	-	-	-
Other Transform. Sector	-	-	-	-	-	-	-	-	-	-	-
ENERGY SECTOR	**15**	**-**	**-**	**-**	**-**	**-**	**-**	**21**	**-**	**-**	**-**
DISTRIBUTION LOSSES	**-**	**-**	**-**	**-**	**-**	**-**	**-**	**-**	**-**	**-**	**-**
FINAL CONSUMPTION	**-**	**10**	**110**	**3**	**35**	**31**	**196**	**74**	**-**	**-**	**32**
INDUSTRY SECTOR	**-**	**10**	**-**	**-**	**-**	**-**	**36**	**74**	**-**	**-**	**-**
Iron and Steel	-	-	-	-	-	-	-	-	-	-	-
Chemical and Petrochem.	-	-	-	-	-	-	-	-	-	-	-
Non-Metallic Minerals	-	-	-	-	-	-	-	-	-	-	-
Non-specified	-	10	-	-	-	-	36	74	-	-	-
TRANSPORT SECTOR	**-**	**-**	**110**	**3**	**35**	**-**	**70**	**-**	**-**	**-**	**-**
Air	-	-	-	3	35	-	-	-	-	-	-
Road	-	-	110	-	-	-	70	-	-	-	-
Non-specified	-	-	-	-	-	-	-	-	-	-	-
OTHER SECTORS	**-**	**-**	**-**	**-**	**-**	**31**	**90**	**-**	**-**	**-**	**-**
Agriculture	-	-	-	-	-	-	15	-	-	-	-
Comm. and Publ. Services	-	-	-	-	-	-	56	-	-	-	-
Residential	-	-	-	-	-	-	-	-	-	-	-
Non-specified	-	-	-	-	-	31	19	-	-	-	-
NON-ENERGY USE	**-**	**-**	**-**	**-**	**-**	**-**	**-**	**-**	**-**	**-**	**32**

APPROVISIONNEMENT ET DEMANDE 1997	*Pétrole cont.* (1000 tonnes)										
	Gaz de raffinerie	*GPL + éthane*	*Essence moteur*	*Essence aviation*	*Carbu-réacteurs*	*Kérosène*	*Gazole*	*Fioul lourd*	*Naphta*	*Coke de pétrole*	*Autres prod.*
Production	15	10	120	-	40	31	219	95	-	-	32
Imports	-	-	-	3	-	-	10	-	-	-	-
Exports	-	-	-10	-	-	-	-30	-	-	-	-
Intl. Marine Bunkers	-	-	-	-	-	-	-	-	-	-	-
Stock Changes	-	-	-	-	-	-	-	-	-	-	-
DOMESTIC SUPPLY	**15**	**10**	**110**	**3**	**40**	**31**	**199**	**95**	**-**	**-**	**32**
Transfers and Stat. Diff.	-	-	-	-	-5	-	-2	-	-	-	-
TRANSFORMATION	**-**	**-**	**-**	**-**	**-**	**-**	**1**	**-**	**-**	**-**	**-**
Electricity and CHP Plants	-	-	-	-	-	-	1	-	-	-	-
Petroleum Refineries	-	-	-	-	-	-	-	-	-	-	-
Other Transform. Sector	-	-	-	-	-	-	-	-	-	-	-
ENERGY SECTOR	**15**	**-**	**-**	**-**	**-**	**-**	**-**	**21**	**-**	**-**	**-**
DISTRIBUTION LOSSES	**-**	**-**	**-**	**-**	**-**	**-**	**-**	**-**	**-**	**-**	**-**
FINAL CONSUMPTION	**-**	**10**	**110**	**3**	**35**	**31**	**196**	**74**	**-**	**-**	**32**
INDUSTRY SECTOR	**-**	**10**	**-**	**-**	**-**	**-**	**36**	**74**	**-**	**-**	**-**
Iron and Steel	-	-	-	-	-	-	-	-	-	-	-
Chemical and Petrochem.	-	-	-	-	-	-	-	-	-	-	-
Non-Metallic Minerals	-	-	-	-	-	-	-	-	-	-	-
Non-specified	-	10	-	-	-	-	36	74	-	-	-
TRANSPORT SECTOR	**-**	**-**	**110**	**3**	**35**	**-**	**70**	**-**	**-**	**-**	**-**
Air	-	-	-	3	35	-	-	-	-	-	-
Road	-	-	110	-	-	-	70	-	-	-	-
Non-specified	-	-	-	-	-	-	-	-	-	-	-
OTHER SECTORS	**-**	**-**	**-**	**-**	**-**	**31**	**90**	**-**	**-**	**-**	**-**
Agriculture	-	-	-	-	-	-	15	-	-	-	-
Comm. and Publ. Services	-	-	-	-	-	-	56	-	-	-	-
Residential	-	-	-	-	-	-	-	-	-	-	-
Non-specified	-	-	-	-	-	31	19	-	-	-	-
NON-ENERGY USE	**-**	**-**	**-**	**-**	**-**	**-**	**-**	**-**	**-**	**-**	**32**

Zambia / Zambie

SUPPLY AND CONSUMPTION 1996	Gas (TJ)				Comb. Renew. & Waste (TJ)				(GWh)	(TJ)
	Natural Gas	Gas Works	Coke Ovens	Blast Furnaces	Solid Biomass	Gas/Liquids from Biomass	Municipal Waste	Industrial Waste	Electricity	Heat
Production	-	-	-	-	195264	-	-	-	7795	-
Imports	-	-	-	-	-	-	-	-	20	-
Exports	-	-	-	-	-	-	-	-	-1500	-
Intl. Marine Bunkers	-	-	-	-	-	-	-	-	-	-
Stock Changes	-	-	-	-	-	-	-	-	-	-
DOMESTIC SUPPLY	-	-	-	-	**195264**	-	-	-	**6315**	-
Transfers and Stat. Diff.	-	-	-	-	-	-	-	-	-1	-
TRANSFORMATION	-	-	-	-	**52630**	-	-	-	-	-
Electricity and CHP Plants	-	-	-	-	-	-	-	-	-	-
Petroleum Refineries	-	-	-	-	-	-	-	-	-	-
Other Transform. Sector	-	-	-	-	52630	-	-	-	-	-
ENERGY SECTOR	-	-	-	-	-	-	-	-	**270**	-
DISTRIBUTION LOSSES	-	-	-	-	-	-	-	-	**884**	-
FINAL CONSUMPTION	-	-	-	-	**142634**	-	-	-	**5160**	-
INDUSTRY SECTOR	-	-	-	-	**21357**	-	-	-	**3998**	-
Iron and Steel	-	-	-	-	-	-	-	-	-	-
Chemical and Petrochem.	-	-	-	-	-	-	-	-	-	-
Non-Metallic Minerals	-	-	-	-	-	-	-	-	-	-
Non-specified	-	-	-	-	21357	-	-	-	3998	-
TRANSPORT SECTOR	-	-	-	-	-	-	-	-	-	-
Air	-	-	-	-	-	-	-	-	-	-
Road	-	-	-	-	-	-	-	-	-	-
Non-specified	-	-	-	-	-	-	-	-	-	-
OTHER SECTORS	-	-	-	-	**121277**	-	-	-	**1162**	-
Agriculture	-	-	-	-	-	-	-	-	200	-
Comm. and Publ. Services	-	-	-	-	-	-	-	-	390	-
Residential	-	-	-	-	121277	-	-	-	572	-
Non-specified	-	-	-	-	-	-	-	-	-	-
NON-ENERGY USE	-	-	-	-	-	-	-	-	-	-

APPROVISIONNEMENT ET DEMANDE 1997	*Gaz* (TJ)				*En. Re. Comb. & Déchets* (TJ)				(GWh)	(TJ)
	Gaz naturel	*Usines à gaz*	*Cokeries*	*Hauts fourneaux*	*Biomasse solide*	*Gaz/Liquides tirés de biomasse*	*Déchets urbains*	*Déchets industriels*	*Electricité*	*Chaleur*
Production	-	-	-	-	199560	-	-	-	8006	-
Imports	-	-	-	-	-	-	-	-	-	-
Exports	-	-	-	-	-	-	-	-	-1500	-
Intl. Marine Bunkers	-	-	-	-	-	-	-	-	-	-
Stock Changes	-	-	-	-	-	-	-	-	-	-
DOMESTIC SUPPLY	-	-	-	-	**199560**	-	-	-	**6506**	-
Transfers and Stat. Diff.	-	-	-	-	-	-	-	-	-2	-
TRANSFORMATION	-	-	-	-	**53788**	-	-	-	-	-
Electricity and CHP Plants	-	-	-	-	-	-	-	-	-	-
Petroleum Refineries	-	-	-	-	-	-	-	-	-	-
Other Transform. Sector	-	-	-	-	53788	-	-	-	-	-
ENERGY SECTOR	-	-	-	-	-	-	-	-	**278**	-
DISTRIBUTION LOSSES	-	-	-	-	-	-	-	-	**910**	-
FINAL CONSUMPTION	-	-	-	-	**145772**	-	-	-	**5316**	-
INDUSTRY SECTOR	-	-	-	-	**21827**	-	-	-	**4118**	-
Iron and Steel	-	-	-	-	-	-	-	-	-	-
Chemical and Petrochem.	-	-	-	-	-	-	-	-	-	-
Non-Metallic Minerals	-	-	-	-	-	-	-	-	-	-
Non-specified	-	-	-	-	21827	-	-	-	4118	-
TRANSPORT SECTOR	-	-	-	-	-	-	-	-	-	-
Air	-	-	-	-	-	-	-	-	-	-
Road	-	-	-	-	-	-	-	-	-	-
Non-specified	-	-	-	-	-	-	-	-	-	-
OTHER SECTORS	-	-	-	-	**123945**	-	-	-	**1198**	-
Agriculture	-	-	-	-	-	-	-	-	206	-
Comm. and Publ. Services	-	-	-	-	-	-	-	-	402	-
Residential	-	-	-	-	123945	-	-	-	590	-
Non-specified	-	-	-	-	-	-	-	-	-	-
NON-ENERGY USE	-	-	-	-	-	-	-	-	-	-

Zimbabwe

SUPPLY AND CONSUMPTION 1996	Coal (1000 tonnes)							Oil (1000 tonnes)			
	Coking Coal	Other Bit. Coal	Sub-Bit. Coal	Lignite	Peat	Oven and Gas Coke	Pat. Fuel and BKB	Crude Oil	NGL	Feed-stocks	Additives
Production	571	4103	-	-	-	540	-	-	-	-	-
Imports	-	-	-	-	-	20	-	-	-	-	-
Exports	-	-151	-	-	-	-30	-	-	-	-	-
Intl. Marine Bunkers	-	-	-	-	-	-	-	-	-	-	-
Stock Changes	-	-	-	-	-	-	-	-	-	-	-
DOMESTIC SUPPLY	**571**	**3952**	**-**	**-**	**-**	**530**	**-**	**-**	**-**	**-**	**-**
Transfers and Stat. Diff.	-	-263	-	-	-	-	-	-	-	-	-
TRANSFORMATION	**571**	**2547**	**-**	**-**	**-**	**348**	**-**	**-**	**-**	**-**	**-**
Electricity and CHP Plants	-	2547	-	-	-	-	-	-	-	-	-
Petroleum Refineries	-	-	-	-	-	-	-	-	-	-	-
Other Transform. Sector	571	-	-	-	-	348	-	-	-	-	-
ENERGY SECTOR	**-**	**-**	**-**	**-**	**-**	**41**	**-**	**-**	**-**	**-**	**-**
DISTRIBUTION LOSSES	**-**	**-**	**-**	**-**	**-**	**-**	**-**	**-**	**-**	**-**	**-**
FINAL CONSUMPTION	**-**	**1142**	**-**	**-**	**-**	**141**	**-**	**-**	**-**	**-**	**-**
INDUSTRY SECTOR	**-**	**322**	**-**	**-**	**-**	**140**	**-**	**-**	**-**	**-**	**-**
Iron and Steel	-	132	-	-	-	87	-	-	-	-	-
Chemical and Petrochem.	-	-	-	-	-	-	-	-	-	-	-
Non-Metallic Minerals	-	-	-	-	-	-	-	-	-	-	-
Non-specified	-	190	-	-	-	53	-	-	-	-	-
TRANSPORT SECTOR	**-**	**30**	**-**	**-**	**-**	**1**	**-**	**-**	**-**	**-**	**-**
Air	-	-	-	-	-	-	-	-	-	-	-
Road	-	-	-	-	-	-	-	-	-	-	-
Non-specified	-	30	-	-	-	1	-	-	-	-	-
OTHER SECTORS	**-**	**790**	**-**	**-**	**-**	**-**	**-**	**-**	**-**	**-**	**-**
Agriculture	-	548	-	-	-	-	-	-	-	-	-
Comm. and Publ. Services	-	231	-	-	-	-	-	-	-	-	-
Residential	-	11	-	-	-	-	-	-	-	-	-
Non-specified	-	-	-	-	-	-	-	-	-	-	-
NON-ENERGY USE	**-**	**-**	**-**	**-**	**-**	**-**	**-**	**-**	**-**	**-**	**-**

APPROVISIONNEMENT ET DEMANDE 1997	*Charbon* (1000 tonnes)							*Pétrole* (1000 tonnes)			
	Charbon à coke	*Autres charb. bit.*	*Charbon sous-bit.*	*Lignite*	*Tourbe*	*Coke de four/gaz*	*Agg./briq. de lignite*	*Pétrole brut*	*LGN*	*Produits d'aliment.*	*Additifs*
Production	571	3838	-	-	-	540	-	-	-	-	-
Imports	-	-	-	-	-	20	-	-	-	-	-
Exports	-	-112	-	-	-	-30	-	-	-	-	-
Intl. Marine Bunkers	-	-	-	-	-	-	-	-	-	-	-
Stock Changes	-	-	-	-	-	-	-	-	-	-	-
DOMESTIC SUPPLY	**571**	**3726**	**-**	**-**	**-**	**530**	**-**	**-**	**-**	**-**	**-**
Transfers and Stat. Diff.	-	-404	-	-	-	-	-	-	-	-	-
TRANSFORMATION	**571**	**2270**	**-**	**-**	**-**	**348**	**-**	**-**	**-**	**-**	**-**
Electricity and CHP Plants	-	2270	-	-	-	-	-	-	-	-	-
Petroleum Refineries	-	-	-	-	-	-	-	-	-	-	-
Other Transform. Sector	571	-	-	-	-	348	-	-	-	-	-
ENERGY SECTOR	**-**	**-**	**-**	**-**	**-**	**41**	**-**	**-**	**-**	**-**	**-**
DISTRIBUTION LOSSES	**-**	**-**	**-**	**-**	**-**	**-**	**-**	**-**	**-**	**-**	**-**
FINAL CONSUMPTION	**-**	**1052**	**-**	**-**	**-**	**141**	**-**	**-**	**-**	**-**	**-**
INDUSTRY SECTOR	**-**	**285**	**-**	**-**	**-**	**140**	**-**	**-**	**-**	**-**	**-**
Iron and Steel	-	91	-	-	-	87	-	-	-	-	-
Chemical and Petrochem.	-	-	-	-	-	-	-	-	-	-	-
Non-Metallic Minerals	-	-	-	-	-	-	-	-	-	-	-
Non-specified	-	194	-	-	-	53	-	-	-	-	-
TRANSPORT SECTOR	**-**	**31**	**-**	**-**	**-**	**1**	**-**	**-**	**-**	**-**	**-**
Air	-	-	-	-	-	-	-	-	-	-	-
Road	-	-	-	-	-	-	-	-	-	-	-
Non-specified	-	31	-	-	-	1	-	-	-	-	-
OTHER SECTORS	**-**	**736**	**-**	**-**	**-**	**-**	**-**	**-**	**-**	**-**	**-**
Agriculture	-	489	-	-	-	-	-	-	-	-	-
Comm. and Publ. Services	-	236	-	-	-	-	-	-	-	-	-
Residential	-	11	-	-	-	-	-	-	-	-	-
Non-specified	-	-	-	-	-	-	-	-	-	-	-
NON-ENERGY USE	**-**	**-**	**-**	**-**	**-**	**-**	**-**	**-**	**-**	**-**	**-**

Zimbabwe

SUPPLY AND CONSUMPTION 1996	Oil cont. (1000 tonnes)										
	Refinery Gas	LPG + Ethane	Motor Gasoline	Aviation Gasoline	Jet Fuel	Kerosene	Gas/ Diesel	Heavy Fuel Oil	Naphtha	Petrol. Coke	Other Prod.
Production	-	-	-	-	-	-	-	-	-	-	-
Imports	-	10	450	5	180	42	750	-	-	-	26
Exports	-	-	-	-	-	-	-	-	-	-	-
Intl. Marine Bunkers	-	-	-	-	-	-	-	-	-	-	-
Stock Changes	-	-	-30	-	-	-	-	-	-	-	1
DOMESTIC SUPPLY	**-**	**10**	**420**	**5**	**180**	**42**	**750**	**-**	**-**	**-**	**27**
Transfers and Stat. Diff.	-	2	15	-	-67	-	-116	-	-	-	-
TRANSFORMATION	**-**	**-**	**-**	**-**	**-**	**-**	**-**	**-**	**-**	**-**	**-**
Electricity and CHP Plants	-	-	-	-	-	-	-	-	-	-	-
Petroleum Refineries	-	-	-	-	-	-	-	-	-	-	-
Other Transform. Sector	-	-	-	-	-	-	-	-	-	-	-
ENERGY SECTOR	**-**	**-**	**-**	**-**	**-**	**-**	**20**	**-**	**-**	**-**	**-**
DISTRIBUTION LOSSES	**-**	**-**	**-**	**-**	**-**	**-**	**-**	**-**	**-**	**-**	**-**
FINAL CONSUMPTION	**-**	**12**	**435**	**5**	**113**	**42**	**614**	**-**	**-**	**-**	**27**
INDUSTRY SECTOR	**-**	**8**	**15**	**-**	**-**	**4**	**73**	**-**	**-**	**-**	**-**
Iron and Steel	-	-	-	-	-	-	-	-	-	-	-
Chemical and Petrochem.	-	-	-	-	-	-	-	-	-	-	-
Non-Metallic Minerals	-	-	-	-	-	-	-	-	-	-	-
Non-specified	-	8	15	-	-	4	73	-	-	-	-
TRANSPORT SECTOR	**-**	**-**	**350**	**5**	**113**	**-**	**302**	**-**	**-**	**-**	**-**
Air	-	-	-	5	113	-	-	-	-	-	-
Road	-	-	350	-	-	-	302	-	-	-	-
Non-specified	-	-	-	-	-	-	-	-	-	-	-
OTHER SECTORS	**-**	**4**	**70**	**-**	**-**	**38**	**239**	**-**	**-**	**-**	**-**
Agriculture	-	-	10	-	-	2	85	-	-	-	-
Comm. and Publ. Services	-	-	-	-	-	-	-	-	-	-	-
Residential	-	2	-	-	-	25	-	-	-	-	-
Non-specified	-	2	60	-	-	11	154	-	-	-	-
NON-ENERGY USE	**-**	**-**	**-**	**-**	**-**	**-**	**-**	**-**	**-**	**-**	**27**

APPROVISIONNEMENT ET DEMANDE 1997	*Pétrole cont.* (1000 tonnes)										
	Gaz de raffinerie	*GPL + éthane*	*Essence moteur*	*Essence aviation*	*Carbu- réacteurs*	*Kérosène*	*Gazole*	*Fioul lourd*	*Naphta*	*Coke de pétrole*	*Autres prod.*
Production	-	-	-	-	-	-	-	-	-	-	-
Imports	-	10	450	5	180	42	750	-	-	-	26
Exports	-	-	-	-	-	-	-	-	-	-	-
Intl. Marine Bunkers	-	-	-	-	-	-	-	-	-	-	-
Stock Changes	-	-	-30	-	-	-	-	-	-	-	1
DOMESTIC SUPPLY	**-**	**10**	**420**	**5**	**180**	**42**	**750**	**-**	**-**	**-**	**27**
Transfers and Stat. Diff.	-	2	15	-	-67	-	-116	-	-	-	-
TRANSFORMATION	**-**	**-**	**-**	**-**	**-**	**-**	**-**	**-**	**-**	**-**	**-**
Electricity and CHP Plants	-	-	-	-	-	-	-	-	-	-	-
Petroleum Refineries	-	-	-	-	-	-	-	-	-	-	-
Other Transform. Sector	-	-	-	-	-	-	-	-	-	-	-
ENERGY SECTOR	**-**	**-**	**-**	**-**	**-**	**-**	**20**	**-**	**-**	**-**	**-**
DISTRIBUTION LOSSES	**-**	**-**	**-**	**-**	**-**	**-**	**-**	**-**	**-**	**-**	**-**
FINAL CONSUMPTION	**-**	**12**	**435**	**5**	**113**	**42**	**614**	**-**	**-**	**-**	**27**
INDUSTRY SECTOR	**-**	**8**	**15**	**-**	**-**	**4**	**73**	**-**	**-**	**-**	**-**
Iron and Steel	-	-	-	-	-	-	-	-	-	-	-
Chemical and Petrochem.	-	-	-	-	-	-	-	-	-	-	-
Non-Metallic Minerals	-	-	-	-	-	-	-	-	-	-	-
Non-specified	-	8	15	-	-	4	73	-	-	-	-
TRANSPORT SECTOR	**-**	**-**	**350**	**5**	**113**	**-**	**302**	**-**	**-**	**-**	**-**
Air	-	-	-	5	113	-	-	-	-	-	-
Road	-	-	350	-	-	-	302	-	-	-	-
Non-specified	-	-	-	-	-	-	-	-	-	-	-
OTHER SECTORS	**-**	**4**	**70**	**-**	**-**	**38**	**239**	**-**	**-**	**-**	**-**
Agriculture	-	-	10	-	-	2	85	-	-	-	-
Comm. and Publ. Services	-	-	-	-	-	-	-	-	-	-	-
Residential	-	2	-	-	-	25	-	-	-	-	-
Non-specified	-	2	60	-	-	11	154	-	-	-	-
NON-ENERGY USE	**-**	**-**	**-**	**-**	**-**	**-**	**-**	**-**	**-**	**-**	**27**

Zimbabwe

SUPPLY AND CONSUMPTION 1996	Gas (TJ)				Comb. Renew. & Waste (TJ)				(GWh)	(TJ)
	Natural Gas	Gas Works	Coke Ovens	Blast Furnaces	Solid Biomass	Gas/Liquids from Biomass	Municipal Waste	Industrial Waste	Electricity	Heat
Production	-	-	2500	-	218592	-	-	-	7323	-
Imports	-	-	-	-	-	-	-	-	3172	-
Exports	-	-	-	-	-	-	-	-	-	-
Intl. Marine Bunkers	-	-	-	-	-	-	-	-	-	-
Stock Changes	-	-	-	-	-	-	-	-	-	-
DOMESTIC SUPPLY	**-**	**-**	**2500**	**-**	**218592**	**-**	**-**	**-**	**10495**	**-**
Transfers and Stat. Diff.	-	-	-	-	-	-	-	-	494	-
TRANSFORMATION	**-**	**-**	**-**	**-**	**-**	**-**	**-**	**-**	**-**	**-**
Electricity and CHP Plants	-	-	-	-	-	-	-	-	-	-
Petroleum Refineries	-	-	-	-	-	-	-	-	-	-
Other Transform. Sector	-	-	-	-	-	-	-	-	-	-
ENERGY SECTOR	**-**	**-**	**-**	**-**	**-**	**-**	**-**	**-**	**190**	**-**
DISTRIBUTION LOSSES	**-**	**-**	**-**	**-**	**-**	**-**	**-**	**-**	**752**	**-**
FINAL CONSUMPTION	**-**	**-**	**2500**	**-**	**218592**	**-**	**-**	**-**	**10047**	**-**
INDUSTRY SECTOR	**-**	**-**	**2500**	**-**	**4048**	**-**	**-**	**-**	**5699**	**-**
Iron and Steel	-	-	2500	-	-	-	-	-	-	-
Chemical and Petrochem.	-	-	-	-	-	-	-	-	-	-
Non-Metallic Minerals	-	-	-	-	-	-	-	-	-	-
Non-specified	-	-	-	-	4048	-	-	-	5699	-
TRANSPORT SECTOR	**-**	**-**	**-**	**-**	**-**	**-**	**-**	**-**	**-**	**-**
Air	-	-	-	-	-	-	-	-	-	-
Road	-	-	-	-	-	-	-	-	-	-
Non-specified	-	-	-	-	-	-	-	-	-	-
OTHER SECTORS	**-**	**-**	**-**	**-**	**214544**	**-**	**-**	**-**	**4348**	**-**
Agriculture	-	-	-	-	11132	-	-	-	988	-
Comm. and Publ. Services	-	-	-	-	-	-	-	-	1528	-
Residential	-	-	-	-	203412	-	-	-	1832	-
Non-specified	-	-	-	-	-	-	-	-	-	-
NON-ENERGY USE	**-**	**-**	**-**	**-**	**-**	**-**	**-**	**-**	**-**	**-**

APPROVISIONNEMENT ET DEMANDE 1997	*Gaz* (TJ)				*En. Re. Comb. & Déchets* (TJ)				(GWh)	(TJ)
	Gaz naturel	*Usines à gaz*	*Cokeries*	*Hauts fourneaux*	*Biomasse solide*	*Gaz/Liquides tirés de biomasse*	*Déchets urbains*	*Déchets industriels*	*Electricité*	*Chaleur*
Production	-	-	2500	-	221434	-	-	-	7298	-
Imports	-	-	-	-	-	-	-	-	4013	-
Exports	-	-	-	-	-	-	-	-	-	-
Intl. Marine Bunkers	-	-	-	-	-	-	-	-	-	-
Stock Changes	-	-	-	-	-	-	-	-	-	-
DOMESTIC SUPPLY	**-**	**-**	**2500**	**-**	**221434**	**-**	**-**	**-**	**11311**	**-**
Transfers and Stat. Diff.	-	-	-	-	-	-	-	-	414	-
TRANSFORMATION	**-**	**-**	**-**	**-**	**-**	**-**	**-**	**-**	**-**	**-**
Electricity and CHP Plants	-	-	-	-	-	-	-	-	-	-
Petroleum Refineries	-	-	-	-	-	-	-	-	-	-
Other Transform. Sector	-	-	-	-	-	-	-	-	-	-
ENERGY SECTOR	**-**	**-**	**-**	**-**	**-**	**-**	**-**	**-**	**210**	**-**
DISTRIBUTION LOSSES	**-**	**-**	**-**	**-**	**-**	**-**	**-**	**-**	**978**	**-**
FINAL CONSUMPTION	**-**	**-**	**2500**	**-**	**221434**	**-**	**-**	**-**	**10537**	**-**
INDUSTRY SECTOR	**-**	**-**	**2500**	**-**	**4101**	**-**	**-**	**-**	**5670**	**-**
Iron and Steel	-	-	2500	-	-	-	-	-	-	-
Chemical and Petrochem.	-	-	-	-	-	-	-	-	-	-
Non-Metallic Minerals	-	-	-	-	-	-	-	-	-	-
Non-specified	-	-	-	-	4101	-	-	-	5670	-
TRANSPORT SECTOR	**-**	**-**	**-**	**-**	**-**	**-**	**-**	**-**	**-**	**-**
Air	-	-	-	-	-	-	-	-	-	-
Road	-	-	-	-	-	-	-	-	-	-
Non-specified	-	-	-	-	-	-	-	-	-	-
OTHER SECTORS	**-**	**-**	**-**	**-**	**217333**	**-**	**-**	**-**	**4867**	**-**
Agriculture	-	-	-	-	11277	-	-	-	1055	-
Comm. and Publ. Services	-	-	-	-	-	-	-	-	1821	-
Residential	-	-	-	-	206056	-	-	-	1991	-
Non-specified	-	-	-	-	-	-	-	-	-	-
NON-ENERGY USE	**-**	**-**	**-**	**-**	**-**	**-**	**-**	**-**	**-**	**-**

SUMMARY TABLES

TABLEAUX RECAPITULATIFS

Production of Coking Coal (1000 tonnes)
Production de charbon à coke (1000 tonnes)
Erzeugung von Kokskohle (1000 Tonnen)

Produzione di carbone siderurgico (1000 tonnellate)

原料炭の生産量（千トン）

Producción de carbón coquizable (1000 toneladas)

Производство коксующихся углей (тыс. т)

	1971	1973	1978	1983	1984	1985	1986	1987	1988
Algérie	-	-	-	23	23	23	11	15	8
Afrique du Sud	-	-	9 718	9 206	10 803	11 142	12 115	10 830	11 187
Tanzanie	-	-	-	-	-	-	-	-	-
Zambie	-	-	115	62	56	50	50	50	65
Zimbabwe	-	-	946	216	245	270	300	373	237
Afrique	-	-	10 779	9 507	11 127	11 485	12 476	11 268	11 497
Brésil	-	-	1 316	1 170	1 303	1 408	1 330	992	1 230
Colombie	-	-	682	706	785	684	891	963	1 085
Amérique latine	-	-	1 998	1 876	2 088	2 092	2 221	1 955	2 315
Inde	-	-	15 335	16 876	16 531	15 425	12 328	10 134	9 735
RPD de Corée	-	-	4 958	4 800	4 600	4 500	4 500	4 500	4 500
Asie	-	-	20 293	21 676	21 131	19 925	16 828	14 634	14 235
Rép. populaire de Chine	-	-	52 604	66 235	72 034	75 338	82 769	91 060	92 943
Chine	-	-	52 604	66 235	72 034	75 338	82 769	91 060	92 943
Bulgarie	-	-	-	-	-	-	-	-	-
Roumanie	-	-	2 134	3 318	3 984	3 825	4 110	3 300	3 500
Europe non-OCDE	-	-	2 134	3 318	3 984	3 825	4 110	3 300	3 500
Kazakhstan	-	-	-	-	-	-	-	-	-
Russie	-	-	-	-	-	-	-	-	-
Tadjikistan	-	-	-	-	-	-	-	-	-
Ukraine	-	-	-	-	-	-	-	-	-
Ex-URSS	-	-	139 250	133 872	131 985	134 985	137 551	142 985	144 985
Iran	-	-	377	259	320	324	280	230	230
Moyen-Orient	-	-	377	259	320	324	280	230	230
Total non-OCDE	-	-	227 435	236 743	242 669	247 974	256 235	265 432	269 705
OCDE Amérique du N.	-	-	109 066	97 977	118 641	117 547	107 828	107 094	124 185
OCDE Pacifique	-	-	48 943	49 834	54 670	60 173	58 549	61 203	62 170
OCDE Europe	-	-	100 223	88 455	79 988	81 266	81 857	75 122	72 136
Total OCDE	-	-	258 232	236 266	253 299	258 986	248 234	243 419	258 491
Monde	-	-	485 667	473 009	495 968	506 960	504 469	508 851	528 196

Production of Coking Coal (1000 tonnes)

Production de charbon à coke (1000 tonnes)

Erzeugung von Kokskohle (1000 Tonnen)

Produzione di carbone siderurgico (1000 tonnellate)

原料炭の生産量（千トン）

Producción de carbón coquizable (1000 toneladas)

Производство коксующихся углей (тыс. т)

1989	1990	1991	1992	1993	1994	1995	1996	1997	
7	7	47	52	-	-	-	-	-	Algeria
9 799	9 308	9 187	9 857	4 702	4 283	3 860	3 533	3 647	South Africa
1	1	1	1	1	1	1	1	1	Tanzania
65	60	47	54	40	24	25	24	24	Zambia
200	224	200	220	552	571	571	571	571	Zimbabwe
10 072	9 600	9 482	10 184	5 295	4 879	4 457	4 129	4 243	**Africa**
1 052	498	230	125	58	119	106	133	90	Brazil
905	657	835	1 036	1 703	1 701	1 743	1 776	1 671	Colombia
1 957	1 155	1 065	1 161	1 761	1 820	1 849	1 909	1 761	**Latin America**
12 722	9 390	9 980	9 490	8 650	7 780	6 930	6 436	5 837	India
4 500	4 250	4 000	3 750	3 500	3 250	3 000	2 769	2 686	DPR of Korea
17 222	13 640	13 980	13 240	12 150	11 030	9 930	9 205	8 523	**Asia**
99 511	110 076	112 208	114 162	120 929	139 479	183 964	184 558	185 189	People's Rep. of China
99 511	110 076	112 208	114 162	120 929	139 479	183 964	184 558	185 189	**China**
-	-	61	11	-	-	-	-	-	Bulgaria
3 685	1 482	523	1 027	539	444	349	312	324	Romania
3 685	1 482	584	1 038	539	444	349	312	324	**Non-OECD Europe**
-	-	-	20 855	19 222	15 491	11 107	10 986	10 526	Kazakhstan
-	-	-	70 900	62 419	56 582	60 616	55 276	52 400	Russia
-	-	-	-	-	3	7	-	-	Tajikistan
-	-	-	53 726	42 546	36 390	29 595	28 951	31 838	Ukraine
143 985	141 062	100 000	145 481	124 187	108 466	101 325	95 213	94 764	**Former USSR**
150	50	50	-	850	485	448	402	402	Iran
150	50	50	-	850	485	448	402	402	**Middle East**
276 582	277 065	237 369	285 266	265 711	266 603	302 322	295 728	295 206	**Non-OECD Total**
127 413	123 881	120 458	106 508	100 782	101 338	107 410	108 249	106 941	OECD North America
64 582	65 959	66 146	71 594	79 021	78 192	81 058	83 439	85 126	OECD Pacific
69 680	64 653	55 816	55 109	75 676	70 213	73 163	66 663	65 418	OECD Europe
261 675	254 493	242 420	233 211	255 479	249 743	261 631	258 351	257 485	**OECD Total**
538 257	531 558	479 789	518 477	521 190	516 346	563 953	554 079	552 691	**World**

Production of Other Bituminous Coal (1000 tonnes)
Production d'autres charbons bitumineux (1000 tonnes)

Erzeugung von sonstiger bituminöser Kohle (1000 Tonnen)

Produzione di altri carboni bituminosi (1000 tonnellate)

その他歴青炭の生産量（千トン）

Producción de otro carbón bituminoso (1000 toneladas)

Производство других битуминозных углей (тыс. т)

	1971	1973	1978	1983	1984	1985	1986	1987	1988
RD du Congo	-	-	107	113	121	121	119	122	123
Maroc	-	-	720	751	838	775	775	634	636
Mozambique	-	-	149	58	40	35	4	43	45
Nigéria	-	-	219	53	76	140	144	145	82
Afrique du Sud	-	-	80 640	136 394	152 097	162 358	164 585	165 770	170 213
Tanzanie	-	-	1	10	10	15	4	3	3
Zambie	-	-	500	391	455	461	507	413	559
Zimbabwe	-	-	1 567	2 295	2 473	2 835	3 247	4 490	4 328
Autre Afrique	-	-	480	627	668	754	795	908	929
Afrique	-	-	84 383	140 692	156 778	167 494	170 180	172 528	176 918
Argentine	-	-	434	486	509	400	365	373	511
Brésil	-	-	3 266	5 567	6 216	6 304	6 061	5 892	6 101
Chili	-	-	1 129	990	1 184	1 291	1 633	1 562	1 926
Colombie	-	-	3 552	4 486	5 852	8 290	9 846	12 491	14 192
Pérou	-	-	26	90	103	127	147	107	130
Vénézuela	-	-	83	37	51	42	57	62	1 091
Amérique latine	-	-	8 490	11 656	13 915	16 454	18 109	20 487	23 951
Inde	-	-	86 615	121 344	130 879	138 875	153 362	169 716	184 475
Indonésie	-	-	264	486	1 085	2 000	2 559	3 027	4 613
RPD de Corée	-	-	30 042	33 200	33 400	34 500	35 000	35 000	35 500
Malaisie	-	-	-	-	-	-	-	-	21
Myanmar	-	-	28	36	44	43	51	39	30
Pakistan	-	-	1 251	1 609	1 869	2 238	2 202	2 261	2 750
Philippines	-	-	255	1 020	1 216	1 261	1 239	1 169	1 336
Taipei chinois	-	-	2 884	2 236	2 011	1 858	1 725	1 499	1 225
Thailande	-	-	-	-	-	-	-	-	-
Viêt-Nam	-	-	6 000	6 300	5 000	5 594	6 122	6 332	6 059
Autre Asie	-	-	218	145	148	151	160	167	138
Asie	-	-	127 557	166 376	175 652	186 520	202 420	219 210	236 147
Rép. populaire de Chine	-	-	565 396	648 295	717 196	796 946	811 270	836 905	886 933
Chine	-	-	565 396	648 295	717 196	796 946	811 270	836 905	886 933
Bulgarie	-	-	273	243	223	223	207	198	196
Roumanie	-	-	5 284	4 475	4 474	4 832	4 586	5 799	5 642
Croatie	-	-	-	-	-	-	-	-	-
RF de Yougoslavie	-	-	-	-	-	-	-	-	-
Ex-Yougoslavie	-	-	471	392	388	400	407	379	363
Europe non-OCDE	-	-	6 028	5 110	5 085	5 455	5 200	6 376	6 201
Géorgie	-	-	-	-	-	-	-	-	-
Kazakhstan	-	-	-	-	-	-	-	-	-
Kirghizistan	-	-	-	-	-	-	-	-	-
Russie	-	-	-	-	-	-	-	-	-
Ukraine	-	-	-	-	-	-	-	-	-
Ouzbékistan	-	-	-	-	-	-	-	-	-
Ex-URSS	-	-	417 850	424 128	424 015	434 015	450 449	452 015	454 015

Production of Other Bituminous Coal (1000 tonnes)

Production d'autres charbons bitumineux (1000 tonnes)

Erzeugung von sonstiger bituminöser Kohle (1000 Tonnen)

Produzione di altri carboni bituminosi (1000 tonnellate)

その他歴青炭の生産量 (千トン)

Producción de otro carbón bituminoso (1000 toneladas)

Производство других битуминозных углей (тыс. т)

1989	1990	1991	1992	1993	1994	1995	1996	1997	
125	126	128	128	92	93	93	93	93	DR of Congo
504	526	551	576	604	650	650	506	376	Morocco
50	40	42	40	40	40	38	-	-	Mozambique
84	90	138	100	120	130	140	140	140	Nigeria
166 501	165 492	169 013	164 543	183 512	191 522	202 351	202 829	216 426	South Africa
2	3	3	3	3	3	4	4	4	Tanzania
332	377	312	349	268	161	159	155	155	Zambia
4 391	4 733	5 400	5 328	4 733	4 351	4 119	4 103	3 838	Zimbabwe
969	1 126	1 095	1 173	1 112	1 254	1 160	1 067	1 067	Other Africa
172 958	172 513	176 682	172 240	190 484	198 204	208 714	208 897	222 099	**Africa**
514	270	292	202	167	348	304	310	251	Argentina
5 619	4 097	4 958	4 606	4 537	5 015	5 093	4 672	5 557	Brazil
1 949	2 184	2 208	1 626	1 355	1 182	1 038	1 004	1 044	Chile
17 997	19 811	22 765	20 864	19 470	20 964	23 997	28 289	28 996	Colombia
100	102	60	90	98	74	143	58	41	Peru
2 125	2 189	2 521	3 072	3 891	4 741	4 640	3 486	5 552	Venezuela
28 304	28 653	32 804	30 460	29 518	32 324	35 215	37 819	41 441	**Latin America**
188 188	202 220	219 370	229 000	240 030	249 990	266 490	279 150	291 333	India
8 812	10 486	13 715	22 357	29 328	32 275	41 145	50 157	55 102	Indonesia
36 000	33 750	31 500	29 250	27 000	24 750	23 000	21 367	20 726	DPR of Korea
123	114	180	190	260	145	121	219	219	Malaysia
29	40	44	30	32	36	-	-	-	Myanmar
2 536	2 745	3 054	3 099	3 267	3 534	3 043	3 638	3 159	Pakistan
1 328	1 243	1 262	1 661	1 675	1 231	1 120	976	930	Philippines
784	472	403	335	328	285	235	147	99	Chinese Taipei
-	-	-	22	16	12	5	3	1	Thailand
5 136	5 130	5 204	5 232	5 575	6 157	6 550	11 209	13 116	Vietnam
129	107	96	10	9	8	7	6	6	Other Asia
243 065	256 307	274 828	291 186	307 520	318 423	341 716	366 872	384 691	**Asia**
954 632	969 807	975 198	1 002 218	1 029 742	1 100 423	1 176 767	1 212 141	1 187 631	People's Rep. of China
954 632	969 807	975 198	1 002 218	1 029 742	1 100 423	1 176 767	1 212 141	1 187 631	**China**
193	143	67	237	263	173	194	138	102	Bulgaria
4 615	2 964	3 313	3 071	685	921	799	1 011	1 428	Romania
-	-	-	120	105	96	75	64	49	*Croatia*
-	-	-	102	73	82	71	78	93	*FR of Yugoslavia*
293	292	260	222	178	178	146	142	142	Former Yugoslavia
5 101	3 399	3 640	3 530	1 126	1 272	1 139	1 291	1 672	**Non-OECD Europe**
-	-	-	200	82	44	43	23	5	Georgia
-	-	-	101 536	87 982	84 320	68 508	62 254	59 648	Kazakhstan
-	-	-	1 040	736	746	463	370	500	Kyrgyzstan
-	-	-	139 532	130 717	120 172	116 302	111 240	106 800	Russia
-	-	-	74 148	69 054	54 910	53 933	45 181	43 676	Ukraine
-	-	-	200	152	200	74	74	59	Uzbekistan
433 015	401 938	397 968	316 656	288 723	260 392	239 323	219 142	210 688	**Former USSR**

Production of Other Bituminous Coal (1000 tonnes)

Production d'autres charbons bitumineux (1000 tonnes)

Erzeugung von sonstiger bituminöser Kohle (1000 Tonnen)

Produzione di altri carboni bituminosi (1000 tonnellate)

その他歴青炭の生産量（千トン）

Producción de otro carbón bituminoso (1000 toneladas)

Производство других битуминозных углей (тыс. т)

	1971	1973	1978	1983	1984	1985	1986	1987	1988
Iran	-	-	523	541	921	928	982	1 010	1 030
Moyen-Orient	-	-	523	541	921	928	982	1 010	1 030
Total non-OCDE	-	-	1 210 227	1 396 798	1 493 562	1 607 812	1 658 610	1 708 531	1 785 195
OCDE Amérique du N.	-	-	400 185	447 838	510 223	484 544	492 266	509 679	498 964
OCDE Pacifique	-	-	52 198	76 895	78 922	87 342	103 503	109 538	94 534
OCDE Europe	-	-	375 676	379 112	315 468	358 467	368 086	364 202	360 078
Total OCDE	-	-	828 059	903 845	904 613	930 353	963 855	983 419	953 576
Monde	-	-	2 038 286	2 300 643	2 398 175	2 538 165	2 622 465	2 691 950	2 738 771

Production of Other Bituminous Coal (1000 tonnes)

Production d'autres charbons bitumineux (1000 tonnes)

Erzeugung von sonstiger bituminöser Kohle (1000 Tonnen)

Produzione di altri carboni bituminosi (1000 tonnellate)

その他歴青炭の生産量（千トン）

Producción de otro carbón bituminoso (1000 toneladas)

Производство других битуминозных углей (тыс. т)

1989	1990	1991	1992	1993	1994	1995	1996	1997	
1 050	1 250	1 350	1 500	673	415	594	520	520	Iran
1 050	1 250	1 350	1 500	673	415	594	520	520	**Middle East**
1 838 125	1 833 867	1 862 470	1 817 790	1 847 786	1 911 453	2 003 468	2 046 682	2 048 742	**Non-OECD Total**
515 764	548 800	515 205	521 913	463 289	522 459	493 888	510 017	533 607	OECD North America
100 710	102 238	105 026	106 044	97 775	93 721	103 564	102 140	110 720	OECD Pacific
340 276	302 374	294 830	278 660	229 996	211 618	215 236	212 350	208 494	OECD Europe
956 750	953 412	915 061	906 617	791 060	827 798	812 688	824 507	852 821	**OECD Total**
2 794 875	2 787 279	2 777 531	2 724 407	2 638 846	2 739 251	2 816 156	2 871 189	2 901 563	**World**

Production of Sub-Bituminous Coal (1000 tonnes)

Production de charbons sous-bitumineux (1000 tonnes)

Erzeugung von subbituminöser Kohle (1000 Tonnen)

Produzione di carbone sub-bituminoso (1000 tonnellate)

亜歴青炭の生産量 (千トン)

Producción de carbón sub-bituminoso (1000 toneladas)

Производство полубитуминозных углей (тыс. т)

	1971	1973	1978	1983	1984	1985	1986	1987	1988
RPD de Corée	-	-	9 000	11 000	11 000	12 000	12 500	12 500	12 500
Asie	-	-	9 000	11 000	11 000	12 000	12 500	12 500	12 500
Bulgarie	-	-	-	-	-	-	-	-	-
Slovénie	-	-	-	-	-	-	-	-	-
Ex-Yougoslavie	-	-	-	-	-	-	-	-	-
Europe non-OCDE	-	-	-	-	-	-	-	-	-
Tadjikistan	-	-	-	-	-	-	-	-	-
Ex-URSS	-	-	-	-	-	-	-	-	-
Total non-OCDE	-	-	9 000	11 000	11 000	12 000	12 500	12 500	12 500
OCDE Amérique du N.	-	-	96 054	152 446	179 232	193 127	192 711	202 971	225 760
OCDE Pacifique	-	-	7 387	10 714	11 316	11 192	13 773	16 630	15 851
OCDE Europe	-	-	92 684	103 117	104 496	101 667	102 033	100 572	97 604
Total OCDE	-	-	196 125	266 277	295 044	305 986	308 517	320 173	339 215
Monde	-	-	205 125	277 277	306 044	317 986	321 017	332 673	351 715

Production of Sub-Bituminous Coal (1000 tonnes)

Production de charbons sous-bitumineux (1000 tonnes)

Erzeugung von subbituminöser Kohle (1000 Tonnen)

Produzione di carbone sub-bituminoso (1000 tonnellate)

亜歴青炭の生産量（千トン）

Producción de carbón sub-bituminoso (1000 toneladas)

Производство полубитуминозных углей (тыс. т)

1989	1990	1991	1992	1993	1994	1995	1996	1997	
13 000	12 250	11 500	10 750	10 000	9 300	8 500	7 769	7 536	DPR of Korea
13 000	12 250	11 500	10 750	10 000	9 300	8 500	7 769	7 536	**Asia**
-	-	3 092	3 352	3 419	3 155	3 187	3 060	2 677	Bulgaria
-	-	-	1 323	1 201	1 079	967	830	811	*Slovenia*
-	-	-	1 323	1 201	1 079	967	830	811	Former Yugoslavia
-	-	3 092	4 675	4 620	4 234	4 154	3 890	3 488	**Non-OECD Europe**
-	-	-	-	-	103	27	20	10	Tajikistan
-	-	-	-	-	103	27	20	10	**Former USSR**
13 000	12 250	14 592	15 425	14 620	13 637	12 681	11 679	11 034	**Non-OECD Total**
233 793	245 931	257 043	256 273	277 952	304 914	330 826	342 286	347 265	OECD North America
16 044	18 544	19 086	19 828	19 745	21 854	21 668	22 523	22 895	OECD Pacific
92 382	84 242	64 944	72 212	70 966	63 972	61 339	63 781	61 706	OECD Europe
342 219	348 717	341 073	348 313	368 663	390 740	413 833	428 590	431 866	**OECD Total**
355 219	360 967	355 665	363 738	383 283	404 377	426 514	440 269	442 900	**World**

Production of Lignite (1000 tonnes)

Production de lignite (1000 tonnes)

Erzeugung von Braunkohle (1000 Tonnen)

Produzione di lignite (1000 tonnellate)

亜炭の生産量（千トン）

Producción de lignito (1000 toneladas)

Производство лигнита (тыс. т)

	1971	1973	1978	1983	1984	1985	1986	1987	1988
Chili	-	-	30	40	40	35	35	35	35
Amérique latine	-	-	30	40	40	35	35	35	35
Inde	-	-	3 270	6 640	7 110	7 776	9 600	11 270	12 590
Myanmar	-	-	3	35	43	43	47	35	35
Philippines	-	-	-	4	3	4	4	3	3
Thailande	-	-	684	2 156	2 493	5 188	5 476	6 901	7 259
Asie	-	-	3 957	8 835	9 649	13 011	15 127	18 209	19 887
Albanie	-	-	1 200	1 850	2 010	2 150	2 167	2 134	2 184
Bulgarie	-	-	25 531	32 147	33 433	30 657	35 016	36 621	33 951
Roumanie	-	-	21 845	36 730	35 822	37 924	38 824	42 425	49 612
République slovaque	-	-	5 804	5 819	5 785	5 751	5 374	5 589	5 639
Bosnie-Herzegovine	-	-	-	-	-	-	-	-	-
Croatie	-	-	-	-	-	-	-	-	-
Ex-RYM	-	-	-	-	-	-	-	-	-
Slovénie	-	-	-	-	-	-	-	-	-
RF de Yougoslavie	-	-	-	-	-	-	-	-	-
Ex-Yougoslavie	-	-	39 021	59 000	64 684	68 072	68 381	70 754	70 498
Europe non-OCDE	-	-	93 401	135 546	141 734	144 554	149 762	157 523	161 884
Estonie	-	-	-	-	-	-	-	-	-
Kazakhstan	-	-	-	-	-	-	-	-	-
Kirghizistan	-	-	-	-	-	-	-	-	-
Russie	-	-	-	-	-	-	-	-	-
Tadjikistan	-	-	-	-	-	-	-	-	-
Ukraine	-	-	-	-	-	-	-	-	-
Ouzbékistan	-	-	-	-	-	-	-	-	-
Ex-URSS	-	-	166 500	158 000	156 000	157 000	163 000	165 000	172 000
Israël	-	-	-	-	-	-	-	-	-
Moyen-Orient	-	-	-	-	-	-	-	-	-
Total non-OCDE	-	-	263 888	302 421	307 423	314 600	327 924	340 767	353 806
OCDE Amérique du N.	-	-	36 227	60 693	67 134	72 451	77 548	81 166	89 350
OCDE Pacifique	-	-	30 661	34 937	33 425	38 627	36 270	41 890	43 572
OCDE Europe	-	-	492 686	546 677	582 376	611 153	620 046	623 456	615 766
Total OCDE	-	-	559 574	642 307	682 935	722 231	733 864	746 512	748 688
Monde	-	-	823 462	944 728	990 358	1 036 831	1 061 788	1 087 279	1 102 494

Production of Lignite (1000 tonnes)

Production de lignite (1000 tonnes)

Erzeugung von Braunkohle (1000 Tonnen)

Produzione di lignite (1000 tonnellate)

亜炭の生産量（千トン）

Producción de lignito (1000 toneladas)

Производство лигнита (тыс. т)

1989	1990	1991	1992	1993	1994	1995	1996	1997	
38	40	14	-	-	-	-	-	-	Chile
38	40	14	-	-	-	-	-	-	**Latin America**
12 360	14 070	15 990	16 620	18 100	19 310	22 150	22 640	23 050	India
35	38	39	39	39	40	41	46	46	Myanmar
3	3	3	3	3	3	3	-	-	Philippines
8 901	12 421	14 689	15 335	15 530	17 083	18 416	21 474	23 393	Thailand
21 299	26 532	30 721	31 997	33 672	36 436	40 610	44 160	46 489	**Asia**
2 193	2 071	1 087	800	600	169	163	101	70	Albania
34 105	31 544	25 231	26 735	25 350	25 429	27 449	28 104	26 929	Bulgaria
53 043	33 737	28 578	34 272	38 527	39 182	39 973	40 546	32 055	Romania
5 269	4 766	4 119	3 497	3 547	3 634	3 759	3 829	3 915	Slovak Republic
-	-	-	15 000	13 000	1 400	1 640	1 640	1 640	*Bosnia-Herzegovina*
-	-	-	-	11	7	7	2	-	*Croatia*
-	-	-	6 978	6 917	6 860	7 249	7 145	6 700	*FYROM*
-	-	-	4 251	3 920	3 775	3 917	3 937	4 142	*Slovenia*
-	-	-	40 003	37 360	38 269	39 939	38 367	40 563	*FR of Yugoslavia*
70 861	75 556	71 066	66 232	61 208	50 311	52 752	51 091	53 045	Former Yugoslavia
165 471	147 674	130 081	131 536	129 232	118 725	124 096	123 671	116 014	**Non-OECD Europe**
-	-	-	16 976	12 498	12 192	11 011	12 297	11 895	Estonia
-	-	-	4 152	4 669	4 814	3 740	3 591	2 473	Kazakhstan
-	-	-	1 111	985	550	280	280	280	Kyrgyzstan
-	-	-	126 832	112 778	95 292	85 984	90 179	85 200	Russia
-	-	-	200	200	-	-	-	-	Tajikistan
-	-	-	5 779	4 100	3 100	2 296	1 587	1 433	Ukraine
-	-	-	4 481	3 655	3 600	2 980	2 763	2 888	Uzbekistan
164 000	160 000	149 500	159 531	138 885	119 548	106 291	110 697	104 169	**Former USSR**
-	303	336	355	399	381	470	421	421	Israel
-	303	336	355	399	381	470	421	421	**Middle East**
350 808	334 549	310 652	323 419	302 188	275 090	271 467	278 949	267 093	**Non-OECD Total**
89 222	89 321	87 466	91 730	91 282	90 591	89 210	90 738	89 980	OECD North America
48 448	46 151	49 559	50 902	47 824	49 004	50 945	53 926	58 388	OECD Pacific
626 827	561 121	482 190	446 901	422 666	410 565	394 885	391 610	382 766	OECD Europe
764 497	696 593	619 215	589 533	561 772	550 160	535 040	536 274	531 134	**OECD Total**
1 115 305	1 031 142	929 867	912 952	863 960	825 250	806 507	815 223	798 227	**World**

Production of Peat (1000 tonnes)

Production de tourbe (1000 tonnes)

Erzeugung von Torf (1000 Tonnen)

Produzione di torba (1000 tonnellate)

泥炭の生産量（千トン）

Producción de turba (1000 toneladas)

Производство торфа (тыс. т)

	1971	1973	1978	1983	1984	1985	1986	1987	1988
Autre Afrique	1	1	2	13	14	10	12	18	12
Afrique	1	1	2	13	14	10	12	18	12
Roumanie	-	-	-	-	-	-	-	-	-
Europe non-OCDE	-	-	-	-	-	-	-	-	-
Bélarus	-	-	-	-	-	-	-	-	-
Estonie	-	-	-	-	-	-	-	-	-
Lettonie	-	-	-	-	-	-	-	-	-
Lituanie	-	-	-	-	-	-	-	-	-
Russie	-	-	-	-	-	-	-	-	-
Ukraine	-	-	-	-	-	-	-	-	-
Ex-URSS	61 808	66 549	31 184	29 249	22 193	22 762	27 314	17 754	25 038
Total non-OCDE	61 809	66 550	31 186	29 262	22 207	22 772	27 326	17 772	25 050
OCDE Amérique du N.	-	-	-	-	-	-	-	-	-
OCDE Europe	7 635	6 337	8 050	11 142	12 103	8 276	15 558	12 334	12 150
Total OCDE	7 635	6 337	8 050	11 142	12 103	8 276	15 558	12 334	12 150
Monde	69 444	72 887	39 236	40 404	34 310	31 048	42 884	30 106	37 200

Production of Peat (1000 tonnes)

Production de tourbe (1000 tonnes)

Erzeugung von Torf (1000 Tonnen)

Produzione di torba (1000 tonnellate)

泥炭の生産量（千トン）

Producción de turba (1000 toneladas)

Производство торфа (тыс. т)

1989	1990	1991	1992	1993	1994	1995	1996	1997	
14	11	10	12	12	12	12	12	12	Other Africa
14	11	10	12	12	12	12	12	12	**Africa**
-	-	-	-	26	19	6	2	-	Romania
-	-	-	-	26	19	6	2	-	**Non-OECD Europe**
-	-	-	6 262	2 909	3 477	3 133	2 846	2 763	Belarus
-	-	-	797	465	645	583	700	571	Estonia
-	-	-	1 995	334	637	346	389	391	Latvia
-	-	-	109	52	91	62	77	89	Lithuania
-	-	-	5 533	7 651	2 928	4 401	4 103	3 224	Russia
-	-	-	2 990	2 990	2 096	1 577	1 026	765	Ukraine
24 128	17 641	16 719	17 686	14 401	9 874	10 102	9 141	7 803	**Former USSR**
24 142	17 652	16 729	17 698	14 439	9 905	10 120	9 155	7 815	**Non-OECD Total**
-	-	-	11	24	-	-	-	-	OECD North America
18 061	15 402	10 712	12 607	10 611	14 771	17 151	15 470	14 584	OECD Europe
18 061	15 402	10 712	12 618	10 635	14 771	17 151	15 470	14 584	**OECD Total**
42 203	33 054	27 441	30 316	25 074	24 676	27 271	24 625	22 399	**World**

Production of Crude Oil, NGL and Additives (1000 tonnes)

Production de pétrole brut, LGN et additifs (1000 tonnes)

Erzeugung von Rohöl, Kondensaten und Additiven (1000 Tonnen)

Produzione petrolio grezzo, LNG e additivi (1000 tonnellate)

原油、NGL等随伴物の生産量（千トン）

Producción de petróleo crudo, liquidos de gas natural y aditivos (1000 toneladas)

Производство сырой нефти, газовых конденсатов и присадок (тыс. т)

	1971	1973	1978	1983	1984	1985	1986	1987	1988
Algérie	38 000	51 118	58 021	46 251	49 050	50 421	51 851	52 432	53 043
Angola	5 721	8 154	6 469	8 790	10 074	11 452	13 926	17 315	22 276
Bénin	-	-	-	23	5	325	365	312	238
Cameroun	-	-	505	5 531	7 021	8 426	8 273	8 244	8 183
Congo	14	2 091	2 570	5 365	6 025	5 937	5 951	6 317	7 038
RD du Congo	-	-	1 145	1 271	1 375	1 672	1 621	1 561	1 495
Egypte	14 962	8 504	25 193	36 223	41 611	44 825	40 915	46 002	44 000
Gabon	5 785	7 598	10 355	7 711	7 810	8 520	8 172	7 666	7 790
Ghana	-	-	-	62	33	14	-	-	-
Côte d'Ivoire	-	-	-	1 021	1 003	1 008	965	798	525
Libye	132 535	106 172	97 061	53 535	53 499	51 461	50 827	49 629	52 659
Maroc	23	42	24	17	16	22	24	18	20
Nigéria	75 524	101 412	93 550	61 025	68 741	74 031	72 435	66 288	73 237
Sénégal	-	-	-	-	-	-	-	-	3
Afrique du Sud	-	-	-	-	-	-	-	-	-
Soudan	-	-	-	-	-	-	-	-	-
Tunisie	4 109	3 878	4 944	5 532	5 480	5 408	5 249	5 038	4 997
Autre Afrique	-	-	-	-	-	-	-	-	-
Afrique	276 673	288 969	299 837	232 357	251 743	263 522	260 574	261 620	275 504
Argentine	21 918	21 938	23 549	26 290	25 850	24 805	23 487	23 194	23 419
Bolivie	1 728	2 216	1 555	1 365	1 249	1 188	1 162	1 186	1 142
Brésil	8 719	8 715	8 294	16 937	23 725	28 088	30 437	28 797	28 251
Chili	1 756	1 801	1 059	2 123	2 089	1 931	1 820	1 756	1 387
Colombie	11 376	9 758	6 867	8 032	8 794	9 295	15 585	20 097	19 585
Cuba	137	157	330	848	770	876	947	903	724
Equateur	196	10 631	10 487	12 269	13 358	14 401	14 820	8 901	15 310
Guatemala	-	-	3	349	235	146	247	183	184
Pérou	3 140	3 567	7 640	8 602	9 432	9 547	8 997	8 156	7 071
Trinité-et-Tobago	6 468	8 291	11 446	7 969	8 749	9 086	8 713	7 960	7 798
Vénézuela	187 237	178 678	115 474	97 007	97 373	90 518	97 281	93 022	102 610
Autre Amérique latine	-	2	67	123	218	250	276	284	344
Amérique latine	242 675	245 754	186 771	181 914	191 842	190 131	203 772	194 439	207 825
Bangladesh	-	5	6	-	-	80	92	110	115
Brunei	6 403	11 369	11 796	9 869	9 513	8 620	8 453	7 474	7 224
Inde	7 185	7 198	11 271	25 356	28 194	30 302	31 795	30 920	32 672
Indonésie	44 071	66 112	79 272	65 353	73 650	65 565	70 531	70 086	67 695
Malaisie	3 276	4 340	10 722	18 702	21 836	21 954	24 829	24 623	26 324
Myanmar	872	981	1 353	1 439	1 586	1 452	1 431	980	733
Pakistan	415	419	479	641	660	1 284	1 931	2 018	2 152
Philippines	-	-	-	671	540	368	433	267	290
Taipei chinois	111	148	218	234	231	217	191	217	247
Thailande	13	6	11	631	1 111	2 171	2 221	2 072	2 348
Viêt-Nam	-	-	-	-	-	-	-	-	681
Autre Asie	1	2	1	8	7	8	7	6	3
Asie	62 347	90 580	115 129	122 904	137 328	132 021	141 914	138 773	140 484
Rép. populaire de Chine	39 410	53 610	104 050	106 068	114 613	124 895	130 688	134 140	137 046
Chine	39 410	53 610	104 050	106 068	114 613	124 895	130 688	134 140	137 046

Production of Crude Oil, NGL and Additives (1000 tonnes)

Production de pétrole brut, LGN et additifs (1000 tonnes)

Erzeugung von Rohöl, Kondensaten und Additiven (1000 Tonnen)

Produzione petrolio grezzo, LNG e additivi (1000 tonnellate)

原油、NGL 等随伴物の生産量 (千トン)

Producción de petróleo crudo, liquidos de gas natural y aditivos (1000 toneladas)

Производство сырой нефти, газовых конденсатов и присадок (тыс. т)

1989	1990	1991	1992	1993	1994	1995	1996	1997	
54 101	56 830	56 128	56 062	56 110	55 369	55 719	58 722	59 642	Algeria
22 642	23 553	24 731	23 715	24 189	24 237	25 437	34 689	34 966	Angola
191	206	197	136	154	129	95	72	66	Benin
8 262	8 429	7 637	7 045	6 855	6 488	5 964	5 984	6 269	Cameroon
7 969	8 029	8 054	8 654	9 408	9 562	9 275	10 370	12 495	Congo
1 497	1 410	1 372	1 238	1 280	1 345	1 371	1 371	1 306	DR of Congo
44 287	45 499	45 418	45 869	47 527	46 497	46 742	45 080	43 937	Egypt
10 137	13 352	14 648	14 489	15 481	14 732	18 131	18 277	18 462	Gabon
-	-	-	-	12	304	304	355	355	Ghana
106	98	60	293	324	335	345	884	789	Ivory Coast
53 980	66 198	74 207	70 427	66 976	67 780	70 376	70 105	71 146	Libya
13	15	12	11	10	8	5	5	12	Morocco
86 658	88 322	92 144	95 575	96 509	96 217	97 541	108 032	114 837	Nigeria
2	1	1	1	1	1	1	1	-	Senegal
-	-	-	-	393	393	393	393	393	South Africa
-	-	-	-	-	-	-	102	254	Sudan
5 020	4 571	5 269	5 279	5 020	4 450	4 352	4 291	3 896	Tunisia
-	-	-	-	221	242	339	857	2 167	Other Africa
294 865	316 513	329 878	328 794	330 470	328 089	336 390	359 590	370 992	**Africa**
24 339	25 570	25 827	29 870	31 999	35 573	38 616	41 951	44 350	Argentina
1 203	1 311	1 375	1 263	1 312	1 510	1 693	1 736	1 813	Bolivia
30 205	32 258	31 962	32 182	33 484	34 786	35 727	40 378	43 232	Brazil
1 264	1 114	1 020	884	866	860	732	575	491	Chile
20 585	22 753	22 090	22 887	23 764	23 731	30 433	32 641	33 861	Colombia
726	671	527	882	1 108	1 299	1 471	1 476	1 714	Cuba
14 392	14 895	15 674	16 789	18 104	19 974	20 160	20 096	21 068	Ecuador
182	197	185	281	337	366	468	795	1 065	Guatemala
6 486	6 393	5 724	5 772	6 295	6 345	6 072	6 000	5 824	Peru
7 699	7 797	7 385	7 397	6 720	7 098	7 136	7 073	6 662	Trinidad and Tobago
99 468	105 378	119 888	124 029	115 327	131 053	154 891	161 135	166 904	Venezuela
368	375	425	431	432	432	436	393	393	Other Latin America
206 917	218 712	232 082	242 667	239 748	263 027	297 835	314 249	327 377	**Latin America**
119	109	116	120	124	105	10	10	7	Bangladesh
7 603	7 712	8 269	9 258	8 965	9 242	9 076	8 588	8 540	Brunei
34 967	34 792	33 120	30 195	28 951	33 213	37 180	34 142	36 763	India
72 135	73 387	80 318	76 163	76 129	76 268	75 517	77 043	75 842	Indonesia
28 671	30 412	31 502	31 931	33 543	35 469	36 637	35 078	36 669	Malaysia
732	722	760	723	675	681	474	402	396	Myanmar
2 251	2 721	3 359	3 219	3 084	2 883	2 813	3 009	2 912	Pakistan
262	237	160	404	471	262	146	45	16	Philippines
132	171	103	66	61	62	57	55	48	Chinese Taipei
2 169	2 618	3 028	3 401	3 496	4 051	3 912	4 959	5 398	Thailand
1 490	2 701	3 921	5 501	6 312	6 901	8 502	8 601	9 702	Vietnam
4	4 504	5 003	5 300	5 400	5 500	5 000	4 000	4 000	Other Asia
150 535	160 086	169 659	166 281	167 211	174 637	179 324	175 932	180 293	**Asia**
137 654	138 306	140 992	141 747	145 174	146 082	150 044	157 334	160 741	People's Rep. of China
137 654	138 306	140 992	141 747	145 174	146 082	150 044	157 334	160 741	**China**

Production of Crude Oil, NGL and Additives (1000 tonnes)

Production de pétrole brut, LGN et additifs (1000 tonnes)

Erzeugung von Rohöl, Kondensaten und Additiven (1000 Tonnen)

Produzione petrolio grezzo, LNG e additivi (1000 tonnellate)

原油、NGL等随伴物の生産量（千トン）

Producción de petróleo crudo, liquidos de gas natural y aditivos (1000 toneladas)

Производство сырой нефти, газовых конденсатов и присадок (тыс. т)

	1971	1973	1978	1983	1984	1985	1986	1987	1988
Albanie	1 657	2 107	2 200	1 200	1 388	1 188	1 205	1 182	1 167
Bulgarie	305	190	246	150	150	200	100	90	80
Roumanie	13 793	14 287	13 724	11 593	11 453	10 718	10 125	9 504	9 389
République slovaque	163	130	64	37	35	66	87	96	98
Croatie	-	-	-	-	-	-	-	-	-
Slovénie	-	-	-	-	-	-	-	-	-
RF de Yougoslavie	-	-	-	-	-	-	-	-	-
Ex-Yougoslavie	2 961	3 332	4 076	4 193	4 059	4 152	4 152	3 864	3 684
Europe non-OCDE	18 879	20 046	20 310	17 173	17 085	16 324	15 669	14 736	14 418
Azerbaïdjan	-	-	-	-	-	-	-	-	-
Bélarus	-	-	-	-	-	-	-	-	-
Géorgie	-	-	-	-	-	-	-	-	-
Kazakhstan	-	-	-	-	-	-	-	-	-
Kirghizistan	-	-	-	-	-	-	-	-	-
Lituanie	-	-	-	-	-	-	-	-	-
Russie	-	-	-	-	-	-	-	-	-
Tadjikistan	-	-	-	-	-	-	-	-	-
Turkménistan	-	-	-	-	-	-	-	-	-
Ukraine	-	-	-	-	-	-	-	-	-
Ouzbékistan	-	-	-	-	-	-	-	-	-
Ex-URSS	377 100	429 100	567 500	616 300	612 700	595 291	614 753	624 177	624 324
Bahrein	3 739	3 411	2 758	2 372	2 375	2 355	2 395	2 383	2 432
Iran	226 946	293 159	263 734	123 050	109 100	111 750	91 500	116 110	112 244
Irak	83 266	99 542	127 080	48 059	59 246	70 656	83 976	103 694	129 356
Israël	5 738	6 005	24	12	9	9	12	13	18
Jordanie	-	-	-	-	-	-	15	21	16
Koweit	163 393	152 506	109 659	54 907	59 143	54 491	74 752	70 573	75 334
Oman	14 598	14 598	16 021	19 782	21 175	25 333	28 580	29 732	31 698
Qatar	20 645	27 502	23 382	15 008	20 526	15 623	16 399	14 919	16 331
Arabie saoudite	238 703	380 179	424 553	265 500	244 700	182 100	260 600	222 400	268 500
Syrie	5 289	5 543	9 924	9 359	9 417	9 236	9 267	12 364	14 164
Emirats arabes unis	51 000	76 650	91 915	59 202	62 280	61 730	70 460	76 250	82 050
Yémen	-	-	-	-	-	-	-	948	8 094
Moyen-Orient	813 317	1 059 095	1 069 050	597 251	587 971	533 283	637 956	649 407	740 237
Total non-OCDE	1 830 401	2 187 154	2 362 647	1 873 967	1 913 282	1 855 467	2 005 326	2 017 292	2 139 838
OCDE Amérique du N.	623 660	634 606	627 383	711 823	730 920	732 110	708 021	705 132	699 651
OCDE Pacifique	15 103	20 060	23 031	20 284	24 376	28 547	29 464	29 183	29 368
OCDE Europe	22 205	22 848	90 330	172 128	188 100	195 017	200 191	205 359	204 233
Total OCDE	660 968	677 514	740 744	904 235	943 396	955 674	937 676	939 674	933 252
Monde	2 491 369	2 864 668	3 103 391	2 778 202	2 856 678	2 811 141	2 943 002	2 956 966	3 073 090
Pour mémoire: OPEP	1 261 320	1 533 030	1 483 701	888 897	897 308	828 346	940 612	935 403	1 033 059

Production of Crude Oil, NGL and Additives (1000 tonnes)

Production de pétrole brut, LGN et additifs (1000 tonnes)

Erzeugung von Rohöl, Kondensaten und Additiven (1000 Tonnen)

Produzione petrolio grezzo, LNG e additivi (1000 tonnellate)

原油、NGL等随伴物の生産量（千トン）

Producción de petróleo crudo, liquidos de gas natural y aditivos (1000 toneladas)

Производство сырой нефти, газовых конденсатов и присадок (тыс. т)

1989	1990	1991	1992	1993	1994	1995	1996	1997	
1 226	1 162	884	592	582	545	522	489	361	Albania
80	60	58	53	43	36	43	32	28	Bulgaria
9 173	7 929	6 791	6 827	6 931	6 974	7 040	6 965	6 855	Romania
99	75	120	92	69	70	76	72	65	Slovak Republic
-	-	-	2 115	2 139	1 823	1 761	1 696	1 748	*Croatia*
-	-	-	2	2	2	2	1	1	*Slovenia*
-	-	-	1 165	1 148	1 077	1 066	1 030	979	*FR of Yugoslavia*
3 396	3 145	3 069	3 282	3 289	2 902	2 829	2 727	2 728	Former Yugoslavia
13 974	12 371	10 922	10 846	10 914	10 527	10 510	10 285	10 037	**Non-OECD Europe**
-	-	-	11 566	10 619	9 563	9 162	9 100	9 022	Azerbaijan
-	-	-	2 000	2 005	2 000	1 932	1 860	1 822	Belarus
-	-	-	100	100	74	47	128	134	Georgia
-	-	-	27 773	23 000	20 279	20 450	22 975	25 449	Kazakhstan
-	-	-	113	88	88	89	100	85	Kyrgyzstan
-	-	-	64	73	93	128	155	224	Lithuania
-	-	-	399 337	351 496	315 767	305 107	299 498	304 000	Russia
-	-	-	60	51	33	25	21	26	Tajikistan
-	-	-	4 950	4 154	4 060	4 427	4 332	4 700	Turkmenistan
-	-	-	4 474	4 248	4 200	4 058	4 482	4 567	Ukraine
-	-	-	3 293	3 944	5 500	7 530	7 717	8 124	Uzbekistan
607 254	570 753	512 332	453 730	399 778	361 657	352 955	350 368	358 153	**Former USSR**
2 226	2 398	2 421	2 493	2 440	2 421	2 375	2 329	2 328	Bahrain
143 110	155 900	167 619	173 936	182 870	180 949	183 018	183 812	181 116	Iran
139 399	99 156	14 590	21 266	23 648	26 006	27 185	28 540	57 000	Iraq
16	13	11	9	28	25	28	25	26	Israel
-	-	-	3	1	2	2	2	-	Jordan
91 939	61 575	9 632	54 786	96 539	105 162	106 556	106 758	109 867	Kuwait
32 696	34 926	36 080	37 776	39 699	41 398	43 689	45 305	46 001	Oman
19 322	20 514	20 103	22 189	21 398	20 744	20 856	21 336	28 767	Qatar
267 400	337 950	430 750	439 967	433 794	428 721	430 844	432 231	442 383	Saudi Arabia
17 500	20 630	24 600	25 750	26 767	29 000	31 329	30 745	29 741	Syria
96 340	105 670	120 010	114 215	108 490	116 520	115 481	118 769	120 828	United Arab Emirates
9 136	9 459	9 921	8 996	11 048	17 138	17 475	17 701	18 532	Yemen
819 084	848 191	835 737	901 386	946 722	968 086	978 838	987 553	1 036 589	**Middle East**
2 230 283	2 264 932	2 231 602	2 245 451	2 240 017	2 252 105	2 305 896	2 355 311	2 444 182	**Non-OECD Total**
668 717	656 545	662 865	661 145	654 944	650 496	646 934	654 318	666 546	OECD North America
26 927	29 891	30 217	29 642	29 424	27 732	29 242	29 227	30 220	OECD Pacific
199 556	206 541	218 485	234 685	247 489	289 938	302 476	320 253	318 938	OECD Europe
895 200	892 977	911 567	925 472	931 857	968 166	978 652	1 003 798	1 015 704	**OECD Total**
3 125 483	3 157 909	3 143 169	3 170 923	3 171 874	3 220 271	3 284 548	3 359 109	3 459 886	**World**
1 123 852	1 170 880	1 185 389	1 248 615	1 277 790	1 304 789	1 337 984	1 366 483	1 428 332	***Memo: OPEC***

Production of Natural Gas (TJ)
Production de gaz naturel (TJ)
Erzeugung von Erdgas (TJ)
Produzione di gas naturale (TJ)
天然ガスの生産量 (TJ)
Producción de gas natural (TJ)
Производство природного газа (ТДж)

	1971	1973	1978	1983	1984	1985	1986	1987	1988
Algérie	111 757	188 055	500 460	1 324 095	1 317 721	1 435 166	1 527 577	1 724 031	1 802 324
Angola	1 678	2 536	2 758	3 997	4 499	4 499	4 997	5 998	6 098
Congo	2 842	607	88	80	80	126	80	84	88
Egypte	3 269	3 390	26 611	108 450	139 030	173 602	200 251	222 526	240 758
Gabon	3 618	18 465	1 438	3 626	1 536	1 972	3 512	4 486	6 824
Libye	60 648	159 045	199 724	144 722	179 441	202 846	184 903	189 176	213 461
Maroc	2 009	2 721	3 177	3 248	3 244	3 666	3 549	2 884	2 947
Mozambique	-	-	-	-	-	-	-	-	-
Nigéria	8 011	16 491	36 138	120 819	116 842	137 921	124 129	139 373	138 122
Sénégal	-	-	-	-	-	-	-	11	314
Afrique du Sud	-	-	-	-	-	-	-	-	-
Tunisie	42	5 240	13 172	19 977	19 630	18 789	17 466	14 930	13 812
Autre Afrique	37	37	37	42	16	15	10	4	7
Afrique	193 911	396 587	783 603	1 729 056	1 782 039	1 978 602	2 066 474	2 303 503	2 424 755
Argentine	262 985	267 382	296 230	493 976	535 920	556 005	592 737	610 961	731 504
Bolivie	2 744	78 727	85 105	118 536	115 753	118 381	117 183	116 565	121 628
Brésil	5 781	5 981	36 158	70 698	86 519	104 172	125 256	138 889	143 123
Chili	1 289	18 202	48 792	30 096	30 780	30 742	28 120	27 284	38 342
Colombie	52 018	65 840	101 370	158 869	161 515	161 307	160 562	165 219	163 159
Cuba	67	565	427	306	126	266	228	912	836
Pérou	20 664	19 433	27 822	21 972	29 630	46 263	48 999	48 784	45 995
Trinité-et-Tobago	74 540	73 991	97 041	157 188	151 630	149 777	156 618	163 743	193 474
Vénézuela	459 600	555 731	612 996	756 079	858 099	851 663	871 199	867 255	790 281
Autre Amérique latine	117	117	374	428	760	993	1 028	865	1 025
Amérique latine	879 805	1 085 969	1 306 315	1 808 148	1 970 732	2 019 569	2 101 930	2 140 477	2 229 367
Bangladesh	17 596	24 322	34 041	78 963	91 565	101 823	116 016	137 327	153 195
Brunei	1 934	61 811	281 935	362 712	340 611	334 509	322 410	340 348	334 908
Inde	28 168	29 197	65 810	121 202	146 626	178 137	247 677	285 320	339 597
Indonésie	11 261	15 456	414 000	786 158	1 134 139	1 350 057	1 441 376	1 537 346	1 634 248
Malaisie	3 315	4 642	38 749	174 874	296 091	447 940	602 434	651 373	672 440
Myanmar	2 578	3 976	10 644	20 095	26 971	36 708	41 044	42 233	42 649
Pakistan	109 874	133 215	189 134	298 575	282 724	300 616	310 648	401 750	413 627
Philippines	-	-	-	-	-	-	-	-	-
Taipei chinois	42 244	56 701	73 882	55 421	55 767	49 996	45 554	46 074	52 553
Thailande	-	-	-	57 360	86 453	133 740	180 636	180 636	213 985
Viêt-Nam	-	-	-	2 726	2 367	1 506	1 530	1 527	1 147
Autre Asie	96 601	102 111	90 223	99 482	105 940	111 240	111 444	70 950	58 938
Asie	313 571	431 431	1 198 418	2 057 568	2 569 254	3 046 272	3 420 769	3 694 884	3 917 287
Rép. populaire de Chine	145 756	233 032	534 158	476 197	492 186	504 000	536 353	541 420	555 842
Chine	145 756	233 032	534 158	476 197	492 186	504 000	536 353	541 420	555 842
Albanie	4 914	7 413	13 804	14 997	14 997	14 997	14 997	14 997	15 997
Bulgarie	11 506	7 810	1 134	2 213	1 836	800	670	521	391
Roumanie	1 044 291	1 130 968	1 474 669	1 536 547	1 525 552	1 458 152	1 475 663	1 402 488	1 379 495
République slovaque	23 047	18 006	22 285	9 500	14 955	14 473	15 827	17 889	24 044
Croatie	-	-	-	-	-	-	-	-	-
Slovénie	-	-	-	-	-	-	-	-	-
RF de Yougoslavie	-	-	-	-	-	-	-	-	-
Ex-Yougoslavie	46 534	61 724	82 019	77 783	72 870	93 374	93 542	99 401	114 868
Europe non-OCDE	1 130 292	1 225 921	1 593 911	1 641 040	1 630 210	1 581 796	1 600 699	1 535 296	1 534 795

Production of Natural Gas (TJ)

Production de gaz naturel (TJ)

Erzeugung von Erdgas (TJ)

Produzione di gas naturale (TJ)

天然ガスの生産量 (TJ)

Producción de gas natural (TJ)

Производство природного газа (ТДж)

1989	1990	1991	1992	1993	1994	1995	1996	1997	
1 943 326	2 061 394	2 225 098	2 303 000	2 302 857	2 117 113	2 396 659	2 558 440	2 949 330	Algeria
6 500	20 520	22 040	21 660	21 280	19 760	21 280	21 280	21 660	Angola
80	90	83	115	113	200	130	131	131	Congo
301 860	313 188	345 040	371 600	438 683	461 889	479 075	503 495	515 137	Egypt
3 300	3 200	3 200	3 100	3 200	2 853	3 109	3 317	3 162	Gabon
265 000	240 000	263 000	276 000	259 000	260 295	258 213	261 471	267 580	Libya
2 233	2 014	1 377	847	828	880	884	698	1 047	Morocco
-	-	-	6	6	6	13	15	21	Mozambique
161 500	152 000	185 350	195 000	213 000	208 740	204 611	207 371	206 110	Nigeria
299	219	170	105	498	761	1 843	1 685	928	Senegal
-	70 000	71 000	71 500	79 793	79 793	79 793	71 814	71 814	South Africa
13 830	15 402	12 099	10 979	8 234	8 096	6 188	33 780	68 211	Tunisia
5	6	5	6	7	7	7	7	7	Other Africa
2 697 933	2 878 033	3 128 462	3 253 918	3 327 499	3 160 393	3 451 805	3 663 504	4 105 138	**Africa**
781 630	740 462	796 735	902 243	945 840	994 230	1 079 797	1 195 328	1 305 680	Argentina
121 666	124 178	124 217	127 425	129 082	136 568	123 389	116 023	139 720	Bolivia
152 941	155 267	151 916	168 016	178 262	184 906	198 399	222 269	246 045	Brazil
59 812	64 182	55 594	65 567	63 216	65 976	64 255	64 720	75 334	Chile
164 899	170 806	172 718	169 563	173 216	178 686	187 329	202 081	248 458	Colombia
1 292	1 281	1 125	794	874	752	657	733	850	Cuba
38 975	37 295	35 028	31 592	33 223	34 686	33 620	34 030	35 759	Peru
193 568	201 953	213 891	219 750	217 586	265 957	269 604	298 397	317 727	Trinidad and Tobago
842 681	938 482	945 566	900 319	981 051	1 047 288	1 167 160	1 222 415	1 347 691	Venezuela
1 104	1 132	880	849	1 022	861	1 023	1 136	1 136	Other Latin America
2 358 568	2 435 038	2 497 670	2 586 118	2 723 372	2 909 910	3 125 233	3 357 132	3 718 400	**Latin America**
171 366	172 789	177 604	205 991	220 770	237 517	264 857	274 361	269 618	Bangladesh
340 416	352 914	361 890	367 021	368 914	379 424	415 463	412 233	412 149	Brunei
413 553	471 217	526 309	608 797	602 923	647 768	725 500	805 113	828 972	India
1 769 481	2 045 547	2 176 978	2 332 980	2 397 568	2 721 696	2 761 181	2 931 762	2 933 615	Indonesia
727 805	720 455	855 746	914 098	956 263	974 130	1 108 319	1 407 657	1 540 087	Malaysia
41 263	35 359	33 355	32 820	40 550	50 632	56 717	61 495	66 444	Myanmar
470 264	469 925	488 907	513 126	523 818	558 852	573 792	614 769	625 149	Pakistan
-	-	-	-	-	249	272	406	217	Philippines
52 943	49 148	36 762	31 870	30 843	33 818	35 088	33 574	32 036	Chinese Taipei
213 740	232 811	288 503	308 689	345 550	382 973	404 652	473 969	577 301	Thailand
307	121	100	-	-	-	19 678	145 106	154 053	Vietnam
12 754	11 167	10 920	10 373	10 230	9 930	9 750	9 500	9 500	Other Asia
4 213 892	4 561 453	4 957 074	5 325 765	5 497 429	5 996 989	6 375 269	7 169 945	7 449 141	**Asia**
586 471	595 644	625 621	634 012	726 278	760 630	777 436	871 307	983 449	People's Rep. of China
586 471	595 644	625 621	634 012	726 278	760 630	777 436	871 307	983 449	**China**
12 997	9 482	5 593	4 051	3 500	1 976	1 089	894	700	Albania
312	453	377	1 405	2 537	2 113	1 841	1 534	1 307	Bulgaria
1 239 397	1 065 811	933 074	819 293	779 681	689 594	672 012	640 302	553 958	Romania
21 789	16 031	11 322	15 000	9 186	10 588	12 412	11 360	10 451	Slovak Republic
-	-	-	68 514	77 862	68 096	74 723	67 853	65 254	*Croatia*
-	-	-	600	479	451	652	466	432	*Slovenia*
-	-	-	32 148	32 708	31 312	33 440	25 297	25 937	*FR of Yugoslavia*
110 267	100 976	99 035	101 262	111 049	99 859	108 815	93 616	91 623	Former Yugoslavia
1 384 762	1 192 753	1 049 401	941 011	905 953	804 130	796 169	747 706	658 039	**Non-OECD Europe**

Production of Natural Gas (TJ)

Production de gaz naturel (TJ)

Erzeugung von Erdgas (TJ)

Produzione di gas naturale (TJ)

天然ガスの生産量 (TJ)

Producción de gas natural (TJ)

Производство природного газа (ТДж)

	1971	1973	1978	1983	1984	1985	1986	1987	1988
Azerbaïdjan	-	-	-	-	-	-	-	-	-
Bélarus	-	-	-	-	-	-	-	-	-
Géorgie	-	-	-	-	-	-	-	-	-
Kazakhstan	-	-	-	-	-	-	-	-	-
Kirghizistan	-	-	-	-	-	-	-	-	-
Russie	-	-	-	-	-	-	-	-	-
Tadjikistan	-	-	-	-	-	-	-	-	-
Turkménistan	-	-	-	-	-	-	-	-	-
Ukraine	-	-	-	-	-	-	-	-	-
Ouzbékistan	-	-	-	-	-	-	-	-	-
Ex-URSS	8 175 896	9 092 601	14 322 941	20 296 676	22 148 172	24 199 704	25 818 600	27 358 412	28 977 308
Bahrein	35 326	62 494	101 084	137 209	143 475	175 791	210 250	200 046	215 859
Iran	577 600	725 800	708 700	418 000	513 000	554 800	577 600	608 000	788 500
Irak	35 340	45 980	64 600	103 163	114 300	121 250	172 996	291 021	375 958
Israël	4 880	2 097	2 210	2 478	2 051	1 837	1 444	1 624	1 461
Jordanie	-	-	-	-	-	-	-	-	-
Koweit	130 629	146 057	215 620	145 831	160 254	163 770	223 426	206 672	183 223
Oman	-	-	5 990	26 328	37 240	42 375	52 141	57 433	60 156
Qatar	38 380	60 040	56 240	180 500	225 340	209 000	219 640	220 020	237 120
Arabie saoudite	66 441	90 302	235 912	495 994	733 074	733 074	982 630	1 045 437	1 134 271
Syrie	-	-	1 348	2 990	5 050	6 000	15 171	14 937	35 334
Emirats arabes unis	40 750	48 782	216 600	319 200	418 000	501 600	577 600	642 200	661 200
Moyen-Orient	929 346	1 181 552	1 608 304	1 831 693	2 351 784	2 509 497	3 032 898	3 287 390	3 693 082
Total non-OCDE	11 768 577	13 647 093	21 347 650	29 840 378	32 944 377	35 839 440	38 577 723	40 861 382	43 332 436
OCDE Amérique du N.	26 301 914	26 734 297	24 071 675	21 465 161	23 140 565	22 334 868	21 529 971	22 382 518	23 434 304
OCDE Pacifique	188 142	276 933	452 396	640 381	687 248	754 228	825 982	840 559	872 960
OCDE Europe	4 079 611	5 807 089	7 604 918	7 327 059	7 481 883	7 745 982	7 594 023	7 866 825	7 421 843
Total OCDE	30 569 667	32 818 319	32 128 989	29 432 601	31 309 696	30 835 078	29 949 976	31 089 902	31 729 107
Monde	42 338 244	46 465 412	53 476 639	59 272 979	64 254 073	66 674 518	68 527 699	71 951 284	75 061 543

Production of Natural Gas (TJ)

Production de gaz naturel (TJ)

Erzeugung von Erdgas (TJ)

Produzione di gas naturale (TJ)

天然ガスの生産量 (TJ)

Producción de gas natural (TJ)

Производство природного газа (ТДж)

1989	1990	1991	1992	1993	1994	1995	1996	1997	
-	-	-	296 774	256 549	240 488	250 479	237 698	224 692	Azerbaijan
-	-	-	11 272	11 239	11 354	10 273	9 617	9 501	Belarus
-	-	-	1 430	1 885	377	377	113	113	Georgia
-	-	-	304 241	252 590	169 198	223 033	245 955	305 898	Kazakhstan
-	-	-	2 714	1 583	1 470	1 405	1 015	1 561	Kyrgyzstan
-	-	-	24 128 000	23 158 851	22 375 556	22 153 111	22 339 444	21 457 609	Russia
-	-	-	2 444	1 664	1 254	1 482	1 862	1 596	Tajikistan
-	-	-	2 040 403	2 477 600	1 360 514	1 352 800	1 326 361	652 210	Turkmenistan
-	-	-	787 251	723 840	689 910	708 642	718 280	707 472	Ukraine
-	-	-	1 626 514	1 711 330	1 745 504	1 774 223	1 783 203	1 848 372	Uzbekistan
29 958 880	30 540 161	30 681 816	29 201 043	28 597 131	26 595 625	26 475 825	26 663 548	25 209 024	**Former USSR**
214 100	221 000	211 140	212 164	226 167	221 644	235 018	245 157	237 915	Bahrain
843 600	952 415	1 267 263	1 381 396	1 452 236	1 589 982	1 519 142	1 590 376	1 786 762	Iran
434 481	234 135	65 021	108 680	122 048	151 706	150 796	154 126	172 203	Iraq
1 464	1 316	967	953	1 005	903	903	572	605	Israel
2 378	4 724	4 571	4 975	5 853	8 194	8 776	8 171	8 776	Jordan
218 098	200 000	75 000	140 000	174 420	192 036	192 036	206 194	207 299	Kuwait
61 999	113 322	116 437	142 267	172 852	180 063	178 646	186 000	218 657	Oman
224 960	237 510	287 651	475 774	508 950	508 950	508 950	516 490	674 830	Qatar
1 160 000	1 180 000	1 248 000	1 292 000	1 364 200	1 432 410	1 532 920	1 665 902	1 748 915	Saudi Arabia
57 837	63 687	74 334	76 888	76 100	82 492	104 022	104 022	96 142	Syria
775 200	763 800	904 400	843 600	870 200	961 571	1 122 153	1 298 788	1 421 684	United Arab Emirates
3 994 117	3 971 909	4 254 784	4 678 697	4 974 031	5 329 951	5 553 362	5 975 798	6 573 788	**Middle East**
45 194 623	46 174 991	47 194 828	46 620 564	46 751 693	45 557 628	46 555 099	48 448 940	48 696 979	**Non-OECD Total**
23 893 591	24 666 914	24 749 623	25 306 430	26 084 066	27 444 015	27 453 832	28 040 844	28 281 412	OECD North America
891 305	1 061 382	1 112 672	1 198 996	1 263 356	1 327 863	1 431 581	1 489 547	1 501 322	OECD Pacific
7 656 878	7 604 411	8 207 734	8 361 878	8 880 692	9 045 100	9 422 765	10 854 773	10 713 318	OECD Europe
32 441 774	33 332 707	34 070 029	34 867 304	36 228 114	37 816 978	38 308 178	40 385 164	40 496 052	**OECD Total**
77 636 397	79 507 698	81 264 857	81 487 868	82 979 807	83 374 606	84 863 277	88 834 104	89 193 031	**World**

Production of Combustible Renewables and Waste (TJ)

Production d'énergies renouvelables combustibles et déchets (TJ)

Erzeugung von erneuerbaren Brennstoffen und Abfällen (TJ)

Produzione di energia da combustibili rinnovabili e da rifiuti (TJ)

可燃性再生可能エネルギー及び廃棄物の生産量 (TJ)

Producción de combustibles renovables y desechos (TJ)

Производство возобновляемых видов топлива и отходов (ТДж)

	1971	1973	1978	1983	1984	1985	1986	1987	1988
Algérie	10 940	11 575	13 109	15 326	15 814	16 147	16 684	17 162	17 631
Angola	133 542	135 265	145 360	158 557	161 231	164 009	166 807	169 829	173 147
Bénin	41 970	43 662	48 406	54 582	56 015	57 506	59 022	60 594	62 196
Cameroun	98 931	103 013	117 596	134 131	137 397	140 711	144 174	147 825	151 676
Congo	20 235	21 205	23 482	25 920	26 484	27 100	27 769	28 459	29 183
RD du Congo	233 072	246 090	284 325	332 231	343 279	354 740	366 661	379 096	391 957
Egypte	27 257	28 468	30 494	34 836	36 368	39 517	40 924	42 427	43 522
Ethiopie	353 660	370 917	419 286	476 698	489 328	502 344	516 793	532 586	549 822
Gabon	20 368	21 263	23 755	26 416	27 006	27 624	28 147	28 666	29 214
Ghana	87 400	96 000	113 302	128 616	133 100	138 000	143 100	148 500	154 000
Côte d'Ivoire	68 142	73 019	86 900	103 683	107 423	111 312	115 304	119 388	123 586
Kenya	265 434	276 972	310 692	352 852	361 879	370 642	378 821	386 787	394 545
Libye	4 063	4 415	5 236	5 236	5 236	5 236	5 236	5 236	5 236
Maroc	5 275	5 558	8 664	12 093	12 298	12 503	12 708	12 884	13 089
Mozambique	304 855	301 187	301 349	303 132	302 418	300 765	297 413	293 845	290 308
Nigéria	1 421 913	1 493 666	1 701 819	1 961 087	2 017 829	2 074 773	2 132 032	2 189 941	2 248 623
Sénégal	34 549	36 513	41 659	47 151	48 349	49 619	50 944	52 323	53 751
Afrique du Sud	196 800	215 000	245 000	310 000	344 000	349 000	355 000	382 000	404 000
Soudan	236 454	246 015	279 234	324 912	326 372	333 729	340 306	346 576	352 587
Tanzanie	374 052	376 166	387 435	420 859	430 029	439 450	448 535	457 920	467 584
Tunisie	28 245	29 090	32 449	36 945	37 667	38 808	40 023	41 019	41 912
Zambie	103 368	107 507	120 140	143 331	148 438	152 919	155 817	158 770	161 806
Zimbabwe	127 788	132 901	146 442	166 066	170 710	175 426	180 008	184 591	189 147
Autre Afrique	1 015 946	1 052 485	1 171 128	1 315 383	1 346 381	1 378 660	1 412 222	1 447 085	1 483 319
Afrique	5 214 259	5 427 952	6 057 262	6 890 043	7 085 051	7 260 540	7 434 450	7 633 509	7 831 841
Argentine	94 738	87 793	87 690	86 349	87 151	85 134	74 185	79 546	76 369
Bolivie	8 150	9 348	12 064	37 904	37 850	38 529	39 134	27 458	30 580
Brésil	1 487 643	1 510 589	1 488 925	1 694 764	1 850 840	1 872 749	1 839 833	1 923 166	1 876 355
Chili	58 422	55 478	69 501	86 047	91 699	93 302	96 997	101 337	105 131
Colombie	185 183	142 485	196 856	206 474	206 756	216 492	222 059	220 127	227 234
Costa Rica	24 066	24 618	24 723	24 794	30 785	32 085	30 933	31 550	31 273
Cuba	137 486	127 082	161 934	162 539	167 193	155 590	164 106	166 692	180 417
République dominicaine	48 856	48 691	54 141	70 253	76 362	52 513	52 873	53 595	43 332
Equateur	46 295	43 985	41 157	44 170	45 971	44 464	42 329	43 124	46 558
El Salvador	50 670	53 985	63 683	63 044	64 289	65 200	52 016	56 239	50 172
Guatemala	80 073	84 716	110 377	101 421	103 622	106 050	108 625	112 321	114 652
Haïti	57 608	61 138	71 971	66 718	67 016	67 580	50 053	50 912	51 877
Honduras	41 515	43 054	48 361	55 179	55 674	56 770	56 744	59 093	60 116
Jamaïque	11 147	10 152	10 182	9 306	10 870	10 023	10 155	10 273	10 531
Nicaragua	30 108	31 988	36 108	37 541	40 369	42 593	40 821	41 802	44 497
Panama	14 118	14 119	17 492	19 266	19 454	18 257	17 734	17 281	17 597
Paraguay	48 486	52 339	58 541	70 092	69 208	70 689	73 784	86 868	90 563
Pérou	149 413	152 110	150 394	154 422	156 050	159 596	160 507	161 264	157 576
Trinité-et-Tobago	866	790	744	535	535	1 069	1 778	2 109	1 720
Uruguay	16 443	16 860	18 834	21 764	22 713	24 360	26 604	26 569	24 359
Vénézuela	17 147	16 589	15 431	15 624	16 106	18 918	21 416	22 723	20 063
Autre Amérique latine	19 827	20 123	24 418	29 070	29 016	28 476	28 853	25 989	23 648
Amérique latine	2 628 260	2 608 032	2 763 527	3 057 276	3 249 529	3 260 439	3 211 539	3 320 038	3 284 620

Production of Combustible Renewables and Waste (TJ)

Production d'énergies renouvelables combustibles et déchets (TJ)

Erzeugung von erneuerbaren Brennstoffen und Abfällen (TJ)

Produzione di energia da combustibili rinnovabili e da rifiuti (TJ)

可燃性再生可能エネルギー及び廃棄物の生産量 (TJ)

Producción de combustibles renovables y desechos (TJ)

Производство возобновляемых видов топлива и отходов (ТДж)

1989	1990	1991	1992	1993	1994	1995	1996	1997	
18 100	18 549	18 989	19 438	19 878	20 327	20 786	21 264	21 732	Algeria
176 833	180 978	185 199	189 515	193 925	198 000	203 000	210 714	216 825	Angola
63 839	65 497	67 134	68 758	70 395	72 000	73 500	75 044	76 620	Benin
155 726	159 987	164 362	168 863	173 490	178 000	183 000	188 307	193 768	Cameroon
29 928	30 693	31 476	32 252	33 009	34 000	34 500	35 294	36 106	Congo
405 176	418 675	432 476	446 590	461 018	476 000	491 000	506 712	522 927	DR of Congo
42 418	44 272	46 938	47 599	48 770	51 805	50 585	51 597	52 577	Egypt
568 611	589 075	608 403	628 391	610 212	627 000	644 000	660 100	675 282	Ethiopia
29 792	31 097	32 244	33 216	33 984	34 000	35 000	35 525	36 022	Gabon
159 800	163 300	172 200	178 700	185 600	192 800	198 000	202 752	207 415	Ghana
128 229	133 005	137 693	142 291	146 768	151 100	155 000	159 960	164 599	Ivory Coast
402 102	409 498	416 700	423 923	431 140	438 000	445 000	452 565	461 164	Kenya
5 236	5 236	5 236	5 236	5 236	5 236	5 236	5 236	5 236	Libya
13 294	13 255	15 982	15 997	16 152	16 251	16 698	17 032	17 373	Morocco
287 107	284 652	284 166	285 636	289 037	294 000	290 000	290 000	290 000	Mozambique
2 308 276	2 369 215	2 432 000	2 497 578	2 565 617	2 636 000	2 710 000	2 794 010	2 880 624	Nigeria
55 240	56 783	58 249	59 684	61 081	62 439	64 000	66 176	68 426	Senegal
434 000	443 000	452 000	460 000	469 000	478 000	488 000	496 296	504 237	South Africa
358 378	364 000	369 067	374 164	379 305	384 000	389 000	394 057	399 180	Sudan
477 564	487 819	497 955	508 421	519 145	530 000	541 000	550 738	561 202	Tanzania
42 420	43 433	44 412	45 313	46 157	47 000	47 645	48 360	49 085	Tunisia
164 942	168 224	172 653	177 244	182 016	187 000	192 000	195 264	199 560	Zambia
193 620	197 970	201 950	205 720	209 264	213 000	216 000	218 592	221 434	Zimbabwe
1 520 876	1 559 879	1 589 966	1 631 664	1 655 361	1 700 000	1 745 000	1 790 370	1 836 920	Other Africa
8 041 507	8 238 092	8 437 450	8 646 193	8 805 560	9 025 958	9 237 950	9 465 965	9 698 314	**Africa**
65 443	72 093	75 408	77 713	82 438	96 920	111 196	112 664	110 851	Argentina
29 104	31 580	33 016	32 165	32 631	33 283	34 132	32 413	36 672	Bolivia
1 867 001	1 690 000	1 659 045	1 631 754	1 616 737	1 698 892	1 666 389	1 645 754	1 757 636	Brazil
105 848	112 117	125 348	139 060	129 033	135 410	145 034	155 261	154 569	Chile
231 446	243 676	238 368	246 934	252 833	262 598	292 309	288 427	220 566	Colombia
29 648	31 030	31 051	28 739	14 795	15 018	15 351	15 660	11 274	Costa Rica
203 730	227 371	226 487	226 878	227 510	228 106	228 471	229 138	232 144	Cuba
41 484	42 592	42 547	54 348	54 292	54 235	54 360	54 544	54 785	Dominican Republic
45 134	44 840	46 180	47 060	49 823	49 530	51 213	50 913	47 608	Ecuador
48 948	51 110	54 363	55 768	57 906	58 952	76 832	77 965	80 660	El Salvador
119 561	122 305	124 880	126 639	123 148	120 931	122 876	124 866	126 879	Guatemala
52 209	50 805	51 898	55 254	55 088	56 716	57 379	65 949	53 647	Haiti
61 125	62 732	62 939	63 083	63 233	63 383	63 532	63 744	72 204	Honduras
12 509	18 840	15 179	15 486	17 115	22 188	20 512	22 452	24 501	Jamaica
44 452	47 239	46 836	45 970	46 823	47 900	49 965	51 070	55 017	Nicaragua
17 065	16 954	17 130	17 252	19 494	20 590	21 739	22 962	23 861	Panama
92 864	93 925	97 874	91 650	88 854	94 301	100 775	106 676	109 743	Paraguay
159 102	165 238	166 922	169 592	169 812	172 662	174 715	176 778	179 950	Peru
1 586	2 034	1 737	1 743	1 615	1 860	1 070	1 232	1 332	Trinidad and Tobago
24 549	23 778	24 550	25 293	24 747	23 615	22 367	21 977	22 105	Uruguay
21 418	22 378	23 196	23 501	23 050	23 603	22 647	22 647	22 647	Venezuela
23 479	24 213	24 336	24 093	24 108	24 640	26 121	26 361	25 696	Other Latin America
3 297 705	3 196 850	3 189 290	3 199 975	3 175 085	3 305 333	3 358 985	3 369 453	3 424 347	**Latin America**

Production of Combustible Renewables and Waste (TJ)

Production d'énergies renouvelables combustibles et déchets (TJ)

Erzeugung von erneuerbaren Brennstoffen und Abfällen (TJ)

Produzione di energia da combustibili rinnovabili e da rifiuti (TJ)

可燃性再生可能エネルギー及び廃棄物の生産量 *(TJ)*

Producción de combustibles renovables y desechos (TJ)

Производство возобновляемых видов топлива и отходов (ТДж)

	1971	1973	1978	1983	1984	1985	1986	1987	1988
Bangladesh	399 809	429 525	484 023	536 663	548 939	561 612	576 947	590 541	603 775
Brunei	664	694	781	772	772	772	772	772	772
Inde	5 056 264	5 290 474	5 928 712	6 553 324	6 682 171	6 797 788	6 928 534	7 059 289	7 153 440
Indonésie	1 151 476	1 209 156	1 352 941	1 502 709	1 533 381	1 562 905	1 593 576	1 623 026	1 651 256
RPD de Corée	28 523	30 281	34 676	37 490	37 998	38 515	38 359	38 740	39 140
Malaisie	53 079	56 334	64 335	71 983	74 537	75 334	77 043	81 013	83 142
Myanmar	266 124	279 829	306 851	334 406	340 498	347 350	355 583	364 908	376 938
Népal	104 712	109 295	123 631	201 240	205 137	208 257	212 961	217 923	223 066
Pakistan	445 097	474 496	553 261	639 682	658 930	678 064	698 542	719 397	740 471
Philippines	256 417	265 836	318 839	362 558	372 252	387 363	392 348	394 390	396 297
Sri Lanka	114 668	116 707	126 421	138 779	148 489	149 461	154 428	158 961	162 406
Taipei chinois	3 390	4 122	918	772	664	606	586	567	547
Thailande	318 287	331 320	436 307	365 424	398 205	450 081	485 208	511 962	535 074
Viêt-Nam	522 498	547 592	614 642	677 007	689 399	703 452	719 954	737 890	755 970
Autre Asie	79 307	83 399	96 489	100 428	99 300	98 876	101 228	102 093	103 802
Asie	8 800 315	9 229 060	10 442 827	11 523 237	11 790 672	12 060 436	12 336 069	12 601 472	12 826 096
Rép. populaire de Chine	6 458 822	6 772 393	7 342 365	7 787 382	7 836 222	7 894 774	7 983 778	8 072 335	8 163 445
Hong-Kong, Chine	1 299	1 368	1 495	1 690	1 690	1 758	1 758	1 758	1 817
Chine	6 460 121	6 773 761	7 343 860	7 789 072	7 837 912	7 896 532	7 985 536	8 074 093	8 165 262
Albanie	15 707	15 707	15 707	15 707	15 707	15 707	15 707	15 707	15 707
Bulgarie	11 145	10 012	8 996	16 928	16 869	17 280	17 113	8 850	7 970
Chypre	382	382	342	242	277	282	282	272	252
Roumanie	58 061	57 319	50 247	44 591	40 557	49 974	45 265	39 913	37 529
République slovaque	8 498	7 619	7 756	5 187	5 099	5 402	5 402	5 763	5 988
Bosnie-Herzegovine	-	-	-	-	-	-	-	-	-
Croatie	-	-	-	-	-	-	-	-	-
Ex-RYM	-	-	-	-	-	-	-	-	-
Slovénie	-	-	-	-	-	-	-	-	-
RF de Yougoslavie	-	-	-	-	-	-	-	-	-
Ex-Yougoslavie	69 949	37 372	32 547	39 111	41 260	40 840	42 315	37 021	37 714
Europe non-OCDE	163 742	128 411	115 595	121 766	119 769	129 485	126 084	107 526	105 160
Arménie	-	-	-	-	-	-	-	-	-
Azerbaïdjan	-	-	-	-	-	-	-	-	-
Bélarus	-	-	-	-	-	-	-	-	-
Estonie	-	-	-	-	-	-	-	-	-
Géorgie	-	-	-	-	-	-	-	-	-
Kazakhstan	-	-	-	-	-	-	-	-	-
Kirghizistan	-	-	-	-	-	-	-	-	-
Lettonie	-	-	-	-	-	-	-	-	-
Lituanie	-	-	-	-	-	-	-	-	-
Moldova	-	-	-	-	-	-	-	-	-
Russie	-	-	-	-	-	-	-	-	-
Ukraine	-	-	-	-	-	-	-	-	-
Ouzbékistan	-	-	-	-	-	-	-	-	-
Ex-URSS	843 869	815 545	759 873	784 290	832 148	847 502	788 207	842 960	791 889

Production of Combustible Renewables and Waste (TJ)

Production d'énergies renouvelables combustibles et déchets (TJ)

Erzeugung von erneuerbaren Brennstoffen und Abfällen (TJ)

Produzione di energia da combustibili rinnovabili e da rifiuti (TJ)

可燃性再生可能エネルギー及び廃棄物の生産量 (TJ)

Producción de combustibles renovables y desechos (TJ)

Производство возобновляемых видов топлива и отходов (ТДж)

1989	1990	1991	1992	1993	1994	1995	1996	1997	
616 258	623 015	632 831	640 429	646 957	654 000	660 000	665 910	671 327	Bangladesh
772	772	772	772	772	772	772	772	772	Brunei
7 253 657	7 362 496	7 526 756	7 624 816	7 737 372	7 820 000	7 900 000	7 976 282	8 073 745	India
1 679 485	1 707 715	1 736 018	1 763 026	1 790 035	1 817 044	1 845 273	1 867 465	1 867 465	Indonesia
39 551	39 971	40 410	40 860	41 309	41 768	42 227	42 676	42 676	DPR of Korea
85 535	87 939	89 957	92 155	94 294	96 564	98 684	100 794	100 794	Malaysia
383 419	390 002	397 953	401 861	407 239	412 000	417 000	423 549	430 070	Myanmar
227 105	233 456	240 421	246 178	256 028	262 283	263 048	270 687	270 687	Nepal
762 712	785 890	808 477	829 040	854 286	878 000	901 000	919 378	937 950	Pakistan
402 263	412 161	415 820	424 461	422 257	431 016	430 100	389 881	393 756	Philippines
165 071	164 184	164 639	170 780	164 700	152 654	152 385	164 334	169 554	Sri Lanka
547	547	537	537	537	537	537	537	537	Chinese Taipei
600 597	618 977	639 994	681 737	742 320	811 026	847 073	894 944	864 706	Thailand
774 026	791 461	808 883	826 270	843 597	861 000	878 000	919 378	937 950	Vietnam
101 943	103 353	105 515	107 991	111 097	114 668	118 466	122 334	122 334	Other Asia
13 092 941	13 321 939	13 608 983	13 850 913	14 112 800	14 353 332	14 554 565	14 758 921	14 884 323	**Asia**
8 279 229	8 392 235	8 473 543	8 521 378	8 565 497	8 605 000	8 630 000	8 674 758	8 721 205	People's Rep. of China
1 817	1 817	1 817	1 827	1 885	1 885	1 954	2 012	2 012	Hong Kong, China
8 281 046	8 394 052	8 475 360	8 523 205	8 567 382	8 606 885	8 631 954	8 676 770	8 723 217	**China**
15 707	15 199	15 199	15 199	5 255	2 500	2 500	2 500	2 500	Albania
15 131	15 000	5 900	6 945	6 639	7 175	9 160	10 471	10 492	Bulgaria
252	257	232	686	661	653	635	596	602	Cyprus
27 273	25 212	29 949	44 553	58 344	49 772	72 285	206 102	141 860	Romania
5 372	6 965	5 939	4 933	4 884	7 157	3 196	3 184	3 454	Slovak Republic
-	-	-	6 838	6 838	6 838	6 500	6 500	6 500	*Bosnia-Herzegovina*
-	-	-	10 770	10 190	10 860	11 170	13 780	13 555	*Croatia*
-	-	-	7 844	8 088	7 814	7 814	7 814	7 814	*FYROM*
-	-	-	9 912	9 792	9 792	9 792	9 792	9 792	*Slovenia*
-	-	-	8 996	8 996	8 791	8 791	8 791	8 791	*FR of Yugoslavia*
39 512	32 322	32 020	44 360	43 904	44 095	44 067	46 677	46 452	Former Yugoslavia
103 247	94 955	89 239	116 676	119 687	111 352	131 843	269 530	205 360	**Non-OECD Europe**
-	-	-	39	39	39	39	39	39	Armenia
-	-	-	68	68	68	68	68	68	Azerbaijan
-	-	-	25 000	24 000	23 000	22 000	21 596	28 687	Belarus
-	-	-	7 709	7 539	12 989	14 521	24 745	26 176	Estonia
-	-	-	1 954	1 954	1 465	1 465	1 465	1 465	Georgia
-	-	-	4 800	3 300	3 200	3 300	3 100	3 100	Kazakhstan
-	-	-	186	137	166	156	150	150	Kyrgyzstan
-	-	-	20 093	20 108	24 520	16 844	31 814	54 585	Latvia
-	-	-	7 987	10 455	9 795	9 953	11 312	21 669	Lithuania
-	-	-	1 500	1 300	1 200	1 100	2 494	2 817	Moldova
-	-	-	720 383	720 383	720 383	720 383	720 383	720 383	Russia
-	-	-	12 454	11 398	11 120	10 842	10 564	10 289	Ukraine
-	-	-	10	10	10	10	10	10	Uzbekistan
728 110	800 000	800 000	802 183	800 691	807 955	800 681	827 740	869 438	**Former USSR**

Production of Combustible Renewables and Waste (TJ)

Production d'énergies renouvelables combustibles et déchets (TJ)

Erzeugung von erneuerbaren Brennstoffen und Abfällen (TJ)

Produzione di energia da combustibili rinnovabili e da rifiuti (TJ)

可燃性再生可能エネルギー及び廃棄物の生産量 (TJ)

Producción de combustibles renovables y desechos (TJ)

Производство возобновляемых видов топлива и отходов (ТДж)

	1971	1973	1978	1983	1984	1985	1986	1987	1988
Iran	13 873	14 645	24 257	26 291	27 092	28 527	28 703	27 206	29 629
Irak	932	868	1 029	791	791	850	850	908	908
Israël	156	127	107	107	107	107	107	107	107
Jordanie	39	39	29	49	49	49	49	49	59
Koweit	46	40	18	21	21	21	21	21	21
Liban	4 034	4 239	4 513	4 357	4 366	4 366	4 376	4 386	4 396
Syrie	30	142	75	215	147	147	147	147	137
Yémen	2 040	2 100	2 400	2 700	2 760	2 880	2 940	3 060	3 120
Moyen-Orient	21 150	22 200	32 428	34 531	35 333	36 947	37 193	35 884	38 377
Total non-OCDE	24 131 716	25 004 961	27 515 372	30 200 215	30 950 414	31 491 881	31 919 078	32 615 482	33 043 245
OCDE Amérique du N.	2 129 642	2 240 726	2 739 805	3 114 413	3 420 154	3 389 486	3 477 112	3 761 835	3 721 931
OCDE Pacifique	148 975	147 710	174 926	387 594	405 758	419 551	408 688	422 026	445 049
OCDE Europe	890 894	938 116	1 093 714	1 374 108	1 434 433	1 458 123	1 462 811	1 468 940	1 555 676
Total OCDE	3 169 511	3 326 552	4 008 445	4 876 115	5 260 345	5 267 160	5 348 611	5 652 801	5 722 656
Monde	27 301 227	28 331 513	31 523 817	35 076 330	36 210 759	36 759 041	37 267 689	38 268 283	38 765 901

Avant 1978 les données pour les pays de l'OCDE sont incomplètes.

Production of Combustible Renewables and Waste (TJ)

Production d'énergies renouvelables combustibles et déchets (TJ)

Erzeugung von erneuerbaren Brennstoffen und Abfällen (TJ)

Produzione di energia da combustibili rinnovabili e da rifiuti (TJ)

可燃性再生可能エネルギー及び廃棄物の生産量 *(TJ)*

Producción de combustibles renovables y desechos (TJ)

Производство возобновляемых видов топлива и отходов (ТДж)

1989	1990	1991	1992	1993	1994	1995	1996	1997	
27 658	28 591	29 021	30 143	30 391	30 700	30 100	32 914	32 914	Iran
908	967	967	1 026	1 026	1 050	1 100	1 100	1 100	Iraq
68	127	127	127	127	127	127	127	127	Israel
59	59	68	68	68	78	78	78	91	Jordan
21	21	21	-	-	-	-	-	-	Kuwait
4 484	4 181	4 308	4 405	4 601	4 800	4 960	5 060	5 152	Lebanon
156	127	225	244	200	200	200	200	200	Syria
3 240	3 240	3 240	3 240	3 240	3 240	3 240	3 240	3 240	Yemen
36 594	37 313	37 977	39 253	39 653	40 195	39 805	42 719	42 824	**Middle East**
33 581 150	34 083 201	34 638 299	35 178 398	35 620 858	36 251 010	36 755 783	37 411 098	37 847 823	**Non-OECD Total**
3 564 489	3 273 965	3 366 261	3 756 091	3 493 893	3 579 173	3 718 135	3 750 244	3 615 551	OECD North America
466 195	474 598	479 433	460 630	485 562	513 790	542 451	580 986	630 000	OECD Pacific
1 945 606	1 906 926	1 937 906	1 968 698	2 117 536	2 171 576	2 230 784	2 302 089	2 398 436	OECD Europe
5 976 290	5 655 489	5 783 600	6 185 419	6 096 991	6 264 539	6 491 370	6 633 319	6 643 987	**OECD Total**
39 557 440	39 738 690	40 421 899	41 363 817	41 717 849	42 515 549	43 247 153	44 044 417	44 491 810	**World**

Prior to 1978 data for OECD countries are incomplete.

Production of Charcoal (TJ)

Production de charbon de bois (TJ)

Erzeugung von Holzkohle (TJ)

Produzione di carbone di legna (TJ)

木炭の生産量 (TJ)

Producción de carbón vegetal (TJ)

Производство древесного угля (ТДж)

	1991	1992	1993	1994	1995	1996	1997	
Angola	-	-	-	22 000	22 500	23 355	24 032	Angola
Bénin	-	-	-	500	500	510	521	Benin
Cameroun	-	-	-	2 700	2 700	2 778	2 859	Cameroon
Congo	-	-	-	450	450	460	471	Congo
RD du Congo	-	-	-	7 250	7 500	7 740	7 988	DR of Congo
Ethiopie	-	-	-	7 200	7 400	7 585	7 759	Ethiopia
Ghana	-	-	-	15 700	16 100	16 486	16 865	Ghana
Côte d'Ivoire	-	-	-	18 800	19 250	19 866	20 442	Ivory Coast
Kenya	-	-	-	41 500	42 000	42 714	43 526	Kenya
Mozambique	-	-	-	10 500	10 250	10 250	10 250	Mozambique
Nigéria	-	-	-	23 300	24 000	24 744	25 511	Nigeria
Sénégal	-	-	-	3 732	3 800	3 929	4 063	Senegal
Afrique du Sud	-	-	-	41 000	41 000	41 697	42 364	South Africa
Soudan	-	-	-	71 500	72 500	73 443	74 398	Sudan
Tanzanie	-	-	-	14 500	14 500	14 761	15 041	Tanzania
Tunisie	-	-	-	4 600	4 677	4 747	4 818	Tunisia
Zambie	-	-	-	17 000	17 250	17 543	17 929	Zambia
Autre Afrique	-	-	-	34 500	35 000	35 910	36 844	Other Africa
Afrique	-	-	-	336 732	341 377	348 518	355 681	**Africa**
Argentine	-	-	-	9 253	10 178	10 368	8 194	Argentina
Bolivie	-	-	-	61	61	91	91	Bolivia
Brésil	-	-	-	226 658	218 191	192 379	195 896	Brazil
Colombie	-	-	-	2 923	2 964	3 007	5 987	Colombia
Costa Rica	-	-	-	136	134	136	136	Costa Rica
Cuba	-	-	-	2 168	2 197	2 226	2 381	Cuba
République dominicaine	-	-	-	8 843	9 448	10 081	10 629	Dominican Republic
El Salvador	-	-	-	490	490	517	517	El Salvador
Guatemala	-	-	-	918	975	946	975	Guatemala
Haiti	-	-	-	9 764	9 764	10 082	10 429	Haiti
Honduras	-	-	-	356	377	419	419	Honduras
Jamaïque	-	-	-	2 051	2 138	2 224	2 282	Jamaica
Nicaragua	-	-	-	880	880	909	938	Nicaragua
Panama	-	-	-	85	85	85	85	Panama
Paraguay	-	-	-	6 875	6 500	7 078	7 280	Paraguay
Pérou	-	-	-	5 434	3 478	5 543	5 597	Peru
Uruguay	-	-	-	94	63	63	63	Uruguay
Vénézuela	-	-	-	200	200	200	200	Venezuela
Autre Amérique latine	-	-	-	87	87	87	87	Other Latin America
Amérique latine	-	-	-	277 276	268 210	246 441	252 186	**Latin America**
Indonésie	-	-	-	3 929	3 987	3 885	3 885	Indonesia
Malaisie	-	-	-	12 740	13 028	13 315	13 315	Malaysia
Myanmar	-	-	-	9 900	10 000	10 157	10 313	Myanmar
Pakistan	-	-	-	5 500	6 000	6 122	6 246	Pakistan
Philippines	-	-	-	-	12 700	12 720	29 140	Philippines
Sri Lanka	-	-	-	368	4	3	-	Sri Lanka
Thailande	-	-	-	186 355	191 797	205 797	189 940	Thailand
Viêt-Nam	-	-	-	16 300	17 000	17 801	18 161	Vietnam
Asie	-	-	-	235 092	254 516	269 800	271 000	**Asia**

Production of Charcoal (TJ)

Production de charbon de bois (TJ)

Erzeugung von Holzkohle (TJ)

Produzione di carbone di legna (TJ)

木炭の生産量 （TJ）

Producción de carbón vegetal (TJ)

Производство древесного угля (ТДж)

	1991	1992	1993	1994	1995	1996	1997	
Chypre	-	-	-	142	136	115	101	Cyprus
Roumanie	-	-	-	1 618	1 733	1 762	1 762	Romania
Europe non-OCDE	-	-	-	1 760	1 869	1 877	1 863	**Non-OECD Europe**
Russie	-	-	-	175	175	175	175	Russia
Ex-URSS	-	-	-	175	175	175	175	**Former USSR**
Iran	-	-	-	2 687	2 744	2 773	2 666	Iran
Irak	-	-	-	433	462	462	462	Iraq
Liban	-	-	-	607	636	636	665	Lebanon
Syrie	-	-	-	20	20	20	20	Syria
Yémen	-	-	-	1 560	1 560	1 560	1 560	Yemen
Moyen-Orient	-	-	-	5 307	5 422	5 451	5 373	**Middle East**
Total non-OCDE	-	-	-	856 342	871 569	872 262	886 278	**Non-OECD Total**
Monde	-	-	-	856 342	871 569	872 262	886 278	**World**

Production of Electricity from Fossil Fuels (GWh)

Production d'électricité à partir de combustibles fossiles (GWh)

Elektrizitätserzeugung aus fossilen Energieträgern (GWh)

Produzione di energia elettrica da combustibili fossili (GWh)

化石燃料発電量 *(GWh)*

Producción de electricidad a partir de combustibles fósiles (GWh)

Производство электроэнергии из ископаемых видов топлива (ГВт.ч)

	1971	1973	1978	1983	1984	1985	1986	1987	1988
Algérie	1 899	2 054	5 230	9 983	10 730	11 628	12 731	12 223	13 783
Angola	137	170	57	205	205	105	105	109	114
Bénin	-	9	5	21	61	24	26	26	20
Cameroun	29	51	78	75	37	53	47	57	51
Congo	33	46	6	4	4	4	4	6	6
RD du Congo	108	80	80	139	140	144	147	149	144
Egypte	2 957	2 950	5 830	16 062	19 416	22 795	24 183	29 187	30 580
Ethiopie	289	258	161	175	165	156	139	110	102
Gabon	114	160	140	182	185	194	195	197	201
Ghana	35	38	52	20	32	30	30	190	34
Côte d'Ivoire	449	628	1 251	631	1 316	484	668	1 197	1 069
Kenya	288	385	309	164	225	139	259	384	126
Libye	508	1 127	3 363	7 150	9 200	11 844	13 280	15 600	16 551
Maroc	770	1 683	2 972	6 205	6 555	6 859	7 121	7 176	8 048
Mozambique	450	450	72	181	154	175	137	156	171
Nigéria	242	749	2 324	6 724	6 367	7 340	7 135	7 981	7 646
Sénégal	347	407	586	714	774	772	769	831	857
Afrique du Sud	54 535	63 405	82 589	120 527	130 775	135 445	136 030	142 815	143 083
Soudan	250	183	320	280	298	492	466	480	486
Tanzanie	182	286	175	110	140	130	112	121	126
Tunisie	874	1 106	2 053	3 522	3 814	3 911	4 227	4 436	4 906
Zambie	308	271	139	85	89	77	87	73	76
Zimbabwe	1 177	1 685	707	692	1 080	1 928	2 833	5 237	5 366
Autre Afrique	1 773	1 998	2 755	3 220	3 505	3 445	3 755	4 161	4 425
Afrique	67 754	80 179	111 254	177 071	195 267	208 174	214 486	232 902	237 971
Argentine	21 935	23 525	22 631	20 783	20 206	18 555	17 490	20 723	28 680
Bolivie	111	158	384	488	481	518	567	611	659
Brésil	6 734	4 964	8 377	7 005	6 843	7 877	15 270	11 801	10 426
Chili	3 935	3 327	3 368	3 374	3 832	3 157	2 960	2 777	4 917
Colombie	2 714	4 503	5 051	8 813	9 127	8 210	7 833	7 875	8 525
Costa Rica	120	216	447	84	63	60	58	133	147
Cuba	4 123	4 947	7 479	10 427	11 125	11 041	11 953	12 380	13 188
République dominicaine	478	1 585	1 974	2 279	2 364	2 865	2 750	2 921	2 175
Equateur	610	821	1 814	2 582	1 303	1 279	1 000	814	815
El Salvador	205	398	192	70	69	120	73	331	253
Guatemala	393	594	1 188	200	271	332	58	145	129
Haiti	43	26	43	202	210	210	105	158	127
Honduras	149	127	112	168	161	96	49	60	69
Jamaïque	1 447	1 883	1 343	1 458	1 439	1 436	1 522	1 659	1 641
Antilles néerlandaises	710	775	825	883	818	760	637	692	700
Nicaragua	435	330	805	574	338	349	469	537	492
Panama	862	1 195	874	1 476	822	572	539	692	407
Paraguay	37	55	284	147	53	18	22	18	27
Pérou	1 372	1 588	2 289	2 375	2 881	2 500	2 691	2 951	2 957
Trinité-et-Tobago	956	1 076	1 528	2 889	2 993	3 006	3 265	3 450	3 458
Uruguay	901	963	1 404	154	109	102	80	260	1 527
Vénézuela	7 900	9 852	13 653	24 701	25 028	26 332	25 934	23 289	23 570
Autre Amérique latine	1 996	2 614	3 226	3 411	3 427	3 595	3 949	4 307	4 607
Amérique latine	58 166	65 522	79 291	94 543	93 963	92 990	99 274	98 584	109 496

Production of Electricity from Fossil Fuels (GWh)

Production d'électricité à partir de combustibles fossiles (GWh)

Elektrizitätserzeugung aus fossilen Energieträgern (GWh)

Produzione di energia elettrica da combustibili fossili (GWh)

化石燃料発電量 (GWh)

Producción de electricidad a partir de combustibles fósiles (GWh)

Производство электроэнергии из ископаемых видов топлива (ГВт.ч)

1989	1990	1991	1992	1993	1994	1995	1996	1997	
15 098	15 970	17 052	18 087	19 062	19 717	19 521	20 515	21 610	Algeria
109	72	162	107	60	60	60	103	103	Angola
20	21	23	25	26	52	33	47	50	Benin
42	41	39	35	30	32	32	34	36	Cameroon
6	6	6	6	6	6	6	6	6	Congo
32	25	22	19	105	118	235	130	130	DR of Congo
31 675	33 546	36 186	37 278	39 755	41 947	44 023	42 889	45 669	Egypt
94	118	121	126	111	123	141	148	66	Ethiopia
201	210	207	207	212	215	215	236	267	Gabon
29	10	7	1	22	28	2	7	7	Ghana
607	557	606	668	1 102	1 314	1 143	1 143	1 191	Ivory Coast
109	231	160	147	131	210	334	355	355	Kenya
16 700	16 800	16 800	16 950	17 000	17 800	18 000	18 180	18 180	Libya
7 867	8 408	7 979	8 755	9 467	10 126	11 302	10 418	11 066	Morocco
188	170	148	91	48	29	29	30	214	Mozambique
8 672	8 177	8 236	8 775	8 933	9 969	8 983	9 491	9 586	Nigeria
888	913	937	1 021	1 008	1 041	1 103	1 174	1 261	Senegal
148 462	155 926	157 192	157 776	165 835	171 182	174 987	185 252	193 005	South Africa
470	557	623	545	595	737	892	989	924	Sudan
90	79	96	166	177	211	302	243	485	Tanzania
5 122	5 490	5 638	6 115	6 249	6 674	7 267	7 466	7 933	Tunisia
95	39	41	40	42	42	42	42	42	Zambia
5 240	5 571	5 361	5 221	5 497	5 524	5 463	5 236	5 520	Zimbabwe
4 533	4 698	4 988	5 127	5 452	5 507	5 548	5 488	5 521	Other Africa
246 349	257 635	262 630	267 288	280 925	292 664	299 663	309 622	323 227	**Africa**
28 854	22 245	26 619	28 074	27 293	26 211	30 588	39 026	35 683	Argentina
796	904	965	1 177	1 152	1 177	1 293	1 156	1 083	Bolivia
10 764	8 626	9 439	10 531	10 245	10 620	12 387	15 065	16 314	Brazil
7 644	8 771	6 284	5 507	6 690	8 178	9 473	13 712	14 844	Chile
7 765	8 648	9 463	11 225	10 422	9 325	11 420	9 027	13 016	Colombia
82	86	208	589	474	829	843	446	171	Costa Rica
13 912	13 297	11 730	9 930	9 853	10 873	11 492	11 987	12 508	Cuba
2 701	3 277	3 311	3 620	4 294	4 503	4 722	5 552	5 967	Dominican Republic
815	1 360	1 896	2 210	1 594	1 605	3 297	3 011	2 820	Ecuador
166	154	610	575	896	1 311	1 445	1 085	1 706	El Salvador
119	92	107	813	697	755	845	878	917	Guatemala
153	123	119	85	90	48	242	333	405	Haiti
62	16	9	125	227	321	246	9	48	Honduras
1 583	1 786	1 854	2 011	3 505	4 430	5 430	5 631	5 814	Jamaica
737	790	801	853	909	971	1 009	1 017	1 052	Netherlands Antilles
365	567	629	822	760	916	1 035	1 169	1 248	Nicaragua
473	572	909	1 022	1 108	1 272	1 156	1 069	1 355	Panama
4	7	5	8	8	4	58	61	61	Paraguay
3 048	2 857	2 537	3 210	2 893	1 862	3 828	3 434	4 055	Peru
3 402	3 546	3 695	3 946	3 789	4 037	4 274	4 508	4 952	Trinidad and Tobago
1 797	395	856	938	650	111	411	863	620	Uruguay
22 954	22 333	18 792	20 165	21 899	19 933	21 992	21 528	17 713	Venezuela
4 474	4 816	5 048	5 240	5 422	5 487	5 667	5 706	5 952	Other Latin America
112 670	105 268	105 886	112 676	114 870	114 779	133 153	146 273	148 304	**Latin America**

Production of Electricity from Fossil Fuels (GWh)

Production d'électricité à partir de combustibles fossiles (GWh)

Elektrizitätserzeugung aus fossilen Energieträgern (GWh)

Produzione di energia elettrica da combustibili fossili (GWh)

化石燃料発電量 (GWh)

Producción de electricidad a partir de combustibles fósiles (GWh)

Производство электроэнергии из ископаемых видов топлива (ГВт.ч)

	1971	1973	1978	1983	1984	1985	1986	1987	1988
Bangladesh	855	1 073	1 713	2 771	3 069	3 789	4 350	5 070	5 866
Brunei	150	200	361	682	750	766	776	1 043	1 098
Inde	37 161	41 418	60 188	97 476	111 164	127 369	142 398	166 486	177 607
Indonésie	1 620	2 021	4 235	11 367	12 442	13 685	14 288	16 383	18 702
RPD de Corée	5 310	7 500	12 500	15 000	18 000	20 000	21 000	21 000	24 000
Malaisie	2 749	3 665	7 345	11 000	10 302	10 821	11 601	12 068	13 061
Myanmar	214	245	507	682	879	1 116	1 203	1 296	1 291
Népal	19	23	21	30	30	28	25	7	6
Pakistan	3 789	3 718	4 702	8 104	8 723	10 416	11 355	12 951	16 147
Philippines	7 212	10 812	12 728	13 409	11 370	12 284	11 204	12 863	13 431
Singapour	2 585	3 719	5 898	8 704	9 490	9 960	10 640	11 909	13 113
Sri Lanka	66	323	19	897	170	69	7	530	202
Taipei chinois	12 720	17 336	28 210	24 402	23 195	19 901	27 969	28 930	38 925
Thailande	3 035	5 091	10 527	15 197	16 943	19 383	19 163	24 577	28 686
Viêt-Nam	1 686	1 930	2 800	2 763	3 179	3 592	4 120	4 667	4 994
Autre Asie	1 345	2 635	2 497	2 330	2 252	2 434	2 739	2 959	2 683
Asie	80 516	101 709	154 251	214 814	231 958	255 613	282 838	322 739	359 812
Rép. populaire de Chine	108 400	128 800	211 952	265 080	290 610	318 320	355 000	397 260	436 060
Hong-Kong, Chine	5 574	6 799	10 370	16 467	17 918	19 230	21 406	23 746	25 501
Chine	113 974	135 599	222 322	281 547	308 528	337 550	376 406	421 006	461 561
Albanie	525	575	700	580	442	203	1 706	1 045	394
Bulgarie	18 846	19 386	22 673	26 979	28 677	26 265	27 424	28 499	26 395
Chypre	665	830	919	1 221	1 250	1 319	1 423	1 501	1 647
Gibraltar	47	49	52	61	61	63	65	70	74
Malte	305	365	459	675	700	784	850	944	1 030
Roumanie	34 959	39 232	53 641	60 222	60 341	58 218	63 590	61 655	60 546
République slovaque	9 389	10 745	13 154	11 250	11 153	10 446	10 338	9 677	9 263
Bosnie-Herzegovine	-	-	-	-	-	-	-	-	-
Croatie	-	-	-	-	-	-	-	-	-
Ex-RYM	-	-	-	-	-	-	-	-	-
Slovénie	-	-	-	-	-	-	-	-	-
RF de Yougoslavie	-	-	-	-	-	-	-	-	-
Ex-Yougoslavie	13 865	18 668	26 051	40 805	42 673	46 479	46 393	50 044	53 645
Europe non-OCDE	78 601	89 850	117 649	141 793	145 297	143 777	151 789	153 435	152 994
Arménie	-	-	-	-	-	-	-	-	-
Azerbaïdjan	-	-	-	-	-	-	-	-	-
Bélarus	-	-	-	-	-	-	-	-	-
Estonie	-	-	-	-	-	-	-	-	-
Géorgie	-	-	-	-	-	-	-	-	-
Kazakhstan	-	-	-	-	-	-	-	-	-
Kirghizistan	-	-	-	-	-	-	-	-	-
Lettonie	-	-	-	-	-	-	-	-	-
Lituanie	-	-	-	-	-	-	-	-	-
Moldova	-	-	-	-	-	-	-	-	-
Russie	-	-	-	-	-	-	-	-	-
Tadjikistan	-	-	-	-	-	-	-	-	-
Turkménistan	-	-	-	-	-	-	-	-	-
Ukraine	-	-	-	-	-	-	-	-	-
Ouzbékistan	-	-	-	-	-	-	-	-	-
Ex-URSS	645 600	757 500	956 200	1 103 900	1 122 900	1 139 497	1 200 462	1 234 875	1 236 320

Production of Electricity from Fossil Fuels (GWh)

Production d'électricité à partir de combustibles fossiles (GWh)

Elektrizitätserzeugung aus fossilen Energieträgern (GWh)

Produzione di energia elettrica da combustibili fossili (GWh)

化石燃料発電量 (GWh)

Producción de electricidad a partir de combustibles fósiles (GWh)

Производство электроэнергии из ископаемых видов топлива (ГВт.ч)

1989	1990	1991	1992	1993	1994	1995	1996	1997	
6 195	6 848	7 432	8 098	8 598	8 937	10 434	10 735	11 139	Bangladesh
1 131	1 172	1 269	1 408	1 547	1 663	1 966	2 123	2 407	Brunei
201 959	211 610	237 293	256 050	280 402	297 125	337 462	357 546	378 470	India
21 097	28 503	30 053	31 525	36 704	43 488	44 973	55 809	66 264	Indonesia
24 500	24 500	21 750	14 000	14 000	13 500	13 000	12 519	12 143	DPR of Korea
15 219	19 024	22 128	24 954	29 861	32 570	39 232	46 221	54 581	Malaysia
1 254	1 285	1 437	1 478	1 680	1 980	2 431	2 294	2 550	Myanmar
22	1	3	57	77	92	155	117	117	Nepal
17 601	20 455	22 388	26 400	27 116	30 713	30 186	33 267	37 921	Pakistan
13 772	13 717	14 752	15 917	16 115	18 222	21 198	23 055	26 505	Philippines
14 136	15 714	16 921	17 674	18 962	20 849	22 244	23 458	26 188	Singapore
56	5	261	640	183	298	350	1 278	1 698	Sri Lanka
49 097	49 425	58 889	64 361	75 398	82 174	91 082	88 445	107 458	Chinese Taipei
31 836	39 199	45 599	52 859	59 703	66 662	73 083	79 860	83 894	Thailand
3 946	3 337	2 973	1 546	1 892	2 011	1 468	2 325	2 968	Vietnam
2 755	2 803	2 921	2 824	2 841	2 951	3 244	3 108	3 112	Other Asia
404 576	437 598	486 069	519 791	575 079	623 235	692 508	742 160	817 415	**Asia**
466 410	494 480	552 460	622 965	683 877	745 927	804 316	877 712	924 070	People's Rep. of China
27 361	28 938	31 807	34 907	35 950	26 743	27 916	28 442	28 945	Hong Kong, China
493 771	523 418	584 267	657 872	719 827	772 670	832 232	906 154	953 015	**China**
533	341	168	131	170	133	210	242	209	Albania
27 071	25 598	23 292	21 995	22 082	21 330	22 214	21 715	22 116	Bulgaria
1 831	1 974	2 077	2 404	2 581	2 681	2 473	2 592	2 711	Cyprus
81	82	82	88	88	88	88	93	93	Gibraltar
1 100	1 100	1 419	1 490	1 500	1 541	1 632	1 658	1 686	Malta
62 173	52 898	42 215	42 438	42 639	42 090	42 570	44 209	34 228	Romania
9 281	9 516	9 163	8 965	8 530	8 051	9 643	9 496	9 392	Slovak Republic
-	-	-	9 200	9 700	671	783	783	783	*Bosnia-Herzegovina*
-	-	-	4 547	4 995	3 339	3 593	3 310	4 384	*Croatia*
-	-	-	5 217	4 658	4 816	5 313	5 639	5 819	*FYROM*
-	-	-	4 704	4 714	4 622	4 629	4 535	5 055	*Slovenia*
-	-	-	25 145	23 442	23 401	25 956	26 596	28 145	*FR of Yugoslavia*
55 289	58 484	51 304	48 813	47 509	36 849	40 274	40 863	44 186	Former Yugoslavia
157 359	149 993	129 720	126 324	125 099	112 763	119 104	120 868	114 621	**Non-OECD Europe**
-	-	-	5 960	2 002	2 144	3 338	2 318	3 032	Armenia
-	-	-	17 926	16 700	15 742	15 488	15 550	15 280	Azerbaijan
-	-	-	37 578	33 350	31 378	24 898	23 712	26 036	Belarus
-	-	-	11 830	9 117	9 148	8 685	9 096	9 207	Estonia
-	-	-	5 005	3 116	2 090	1 590	1 185	1 128	Georgia
-	-	-	75 835	69 815	57 218	58 328	51 326	45 501	Kazakhstan
-	-	-	2 692	2 188	1 208	1 231	1 503	1 376	Kyrgyzstan
-	-	-	1 314	1 049	1 135	1 042	1 263	1 548	Latvia
-	-	-	3 758	1 469	1 597	1 325	1 973	2 068	Lithuania
-	-	-	10 989	9 890	7 950	5 744	5 756	4 891	Moldova
-	-	-	711 201	632 112	574 853	580 805	580 303	560 642	Russia
-	-	-	894	623	291	172	175	175	Tajikistan
-	-	-	13 179	12 632	10 492	9 796	10 095	9 395	Turkmenistan
-	-	-	170 723	143 426	121 747	113 345	94 576	88 537	Ukraine
-	-	-	44 630	41 791	40 644	41 265	38 894	40 277	Uzbekistan
1 264 605	1 261 500	1 212 758	1 113 514	979 280	877 637	867 052	837 725	809 093	**Former USSR**

Production of Electricity from Fossil Fuels (GWh)

Production d'électricité à partir de combustibles fossiles (GWh)

Elektrizitätserzeugung aus fossilen Energieträgern (GWh)

Produzione di energia elettrica da combustibili fossili (GWh)

化石燃料発電量 (GWh)

Producción de electricidad a partir de combustibles fósiles (GWh)

Производство электроэнергии из ископаемых видов топлива (ГВт.ч)

	1971	1973	1978	1983	1984	1985	1986	1987	1988
Bahrein	430	500	1 212	2 216	2 417	2 937	2 970	3 316	3 320
Iran	5 221	8 551	13 197	26 806	30 844	33 670	34 054	37 807	40 289
Irak	2 600	3 229	7 127	15 585	18 849	20 363	21 687	19 910	20 850
Israël	7 639	8 720	11 874	14 578	14 909	15 701	16 277	17 491	19 215
Jordanie	230	315	704	1 918	2 265	2 495	2 952	3 467	3 236
Koweit	3 087	4 138	7 446	12 830	14 196	15 689	17 216	18 092	19 598
Liban	536	1 313	1 500	3 040	3 215	3 276	3 610	3 990	3 400
Oman	13	47	468	1 615	2 016	2 498	3 180	3 392	3 773
Qatar	316	419	1 349	3 235	3 563	3 949	4 303	4 371	4 501
Arabie saoudite	2 082	2 949	10 551	35 190	40 069	44 311	47 646	50 649	57 229
Syrie	1 294	1 406	776	4 314	4 433	5 038	6 309	5 866	4 807
Emirats arabes unis	203	720	3 764	10 141	11 000	12 000	12 814	13 657	14 840
Yémen	209	206	360	752	856	895	1 117	1 165	1 318
Moyen-Orient	23 860	32 513	60 328	132 220	148 632	162 822	174 135	183 173	196 376
Total non-OCDE	1 068 471	1 262 872	1 701 295	2 145 888	2 246 545	2 340 423	2 499 390	2 646 714	2 754 530
OCDE Amérique du N.	1 470 253	1 689 133	1 852 193	1 946 837	2 041 151	2 078 047	2 053 320	2 166 448	2 275 747
OCDE Pacifique	342 226	457 808	531 183	539 268	569 187	560 870	555 519	587 223	632 548
OCDE Europe	1 009 653	1 181 175	1 319 836	1 272 808	1 255 999	1 259 730	1 276 549	1 297 650	1 275 506
Total OCDE	2 822 132	3 328 116	3 703 212	3 758 913	3 866 337	3 898 647	3 885 388	4 051 321	4 183 801
Monde	3 890 603	4 590 988	5 404 507	5 904 801	6 112 882	6 239 070	6 384 778	6 698 035	6 938 331

Production of Electricity from Fossil Fuels (GWh)

Production d'électricité à partir de combustibles fossiles (GWh)

Elektrizitätserzeugung aus fossilen Energieträgern (GWh)

Produzione di energia elettrica da combustibili fossili (GWh)

化石燃料発電量 (GWh)

Producción de electricidad a partir de combustibles fósiles (GWh)

Производство электроэнергии из ископаемых видов топлива (ГВт.ч)

1989	1990	1991	1992	1993	1994	1995	1996	1997	
3 400	3 490	3 495	3 896	4 245	4 550	4 750	4 771	4 924	Bahrain
45 190	53 019	57 070	59 089	66 191	74 574	77 694	83 475	88 418	Iran
21 260	21 400	19 910	24 600	25 700	27 440	28 430	28 430	28 980	Iraq
20 297	20 722	21 450	24 565	25 890	28 215	30 277	32 435	35 033	Israel
3 416	3 626	3 716	4 406	4 738	5 061	5 597	6 035	6 250	Jordan
21 179	20 610	10 780	16 786	20 178	22 798	23 726	25 475	27 091	Kuwait
2 000	1 000	2 440	3 051	4 000	4 367	4 564	6 167	7 614	Lebanon
3 927	4 501	4 628	5 117	5 832	6 197	6 460	6 802	7 318	Oman
4 624	4 818	4 653	5 153	5 525	5 814	5 976	6 575	6 868	Qatar
61 568	64 899	69 212	74 009	82 183	91 019	93 898	97 819	103 801	Saudi Arabia
5 728	5 963	5 930	5 176	5 929	8 382	8 356	9 941	7 912	Syria
15 612	17 081	17 222	17 460	17 578	18 870	19 070	19 737	20 571	United Arab Emirates
1 345	1 663	1 802	1 953	2 051	2 159	2 369	2 334	2 358	Yemen
209 546	222 792	222 308	245 261	270 040	299 446	311 167	329 996	347 138	**Middle East**
2 888 876	2 958 204	3 003 638	3 042 726	3 065 120	3 093 194	3 254 879	3 392 798	3 512 813	**Non-OECD Total**
2 452 509	2 412 781	2 448 430	2 488 867	2 582 779	2 658 261	2 684 864	2 742 459	2 860 953	OECD North America
682 392	734 751	754 423	782 120	764 273	873 769	875 243	907 336	932 263	OECD Pacific
1 358 716	1 373 392	1 398 044	1 369 651	1 330 060	1 370 290	1 421 224	1 471 428	1 471 067	OECD Europe
4 493 617	4 520 924	4 600 897	4 640 638	4 677 112	4 902 320	4 981 331	5 121 223	5 264 283	**OECD Total**
7 382 493	7 479 128	7 604 535	7 683 364	7 742 232	7 995 514	8 236 210	8 514 021	8 777 096	**World**

Production of Nuclear Electricity (GWh)
Production d'électricité d'origine nucléaire (GWh)
Elektrizitätserzeugung in Kernkraftwerken (GWh)
Produzione di energia nucleotermoelettrica (GWh)
原子力発電量(GWh)
Producción de electricidad nuclear (GWh)
Производство атомной электроэнергии (ГВт.ч)

	1971	1973	1978	1983	1984	1985	1986	1987	1988
Afrique du Sud	-	-	-	-	3 925	5 315	8 803	6 167	10 493
Afrique	-	-	-	-	3 925	5 315	8 803	6 167	10 493
Argentine	-	-	2 896	3 405	4 641	5 766	5 711	6 465	5 798
Brésil	-	-	-	-	1 643	3 381	144	973	608
Amérique latine	-	-	2 896	3 405	6 284	9 147	5 855	7 438	6 406
Inde	1 189	2 396	2 770	3 546	4 075	4 982	5 022	5 035	5 817
Pakistan	104	304	231	228	324	346	430	502	254
Taipei chinois	-	-	2 670	18 904	24 589	28 727	26 941	33 128	30 651
Asie	1 293	2 700	5 671	22 678	28 988	34 055	32 393	38 665	36 722
Rép. populaire de Chine	-	-	-	-	-	-	-	-	-
Chine	-	-	-	-	-	-	-	-	-
Bulgarie	-	-	5 910	12 318	12 735	13 131	12 070	12 436	16 030
Roumanie	-	-	-	-	-	-	-	-	-
République slovaque	-	232	19	6 150	7 239	9 382	11 716	11 513	11 474
Slovénie	-	-	-	-	-	-	-	-	-
Ex-Yougoslavie	-	-	-	3 916	4 420	4 053	4 019	4 495	4 135
Europe non-OCDE	-	232	5 929	22 384	24 394	26 566	27 805	28 444	31 639
Arménie	-	-	-	-	-	-	-	-	-
Lituanie	-	-	-	-	-	-	-	-	-
Russie	-	-	-	-	-	-	-	-	-
Ukraine	-	-	-	-	-	-	-	-	-
Ex-URSS	6 100	12 000	50 000	110 000	142 000	167 000	161 000	189 000	215 700
Total non-OCDE	7 393	14 932	64 496	158 467	205 591	242 083	235 856	269 714	300 960
OCDE Amérique du N.	44 819	104 421	324 189	359 908	399 502	467 233	510 147	559 847	641 458
OCDE Pacifique	8 000	9 707	61 637	123 256	146 056	176 323	196 615	227 072	218 760
OCDE Europe	50 874	74 136	175 670	393 498	504 195	606 459	658 713	681 172	730 071
Total OCDE	103 693	188 264	561 496	876 662	1 049 753	1 250 015	1 365 475	1 468 091	1 590 289
Monde	111 086	203 196	625 992	1 035 129	1 255 344	1 492 098	1 601 331	1 737 805	1 891 249

Le Kazakhstan a produit une quantité inconnue d' électricité d' origine nucléaire de 1992 à 1994.

Production of Nuclear Electricity (GWh)

Production d'électricité d'origine nucléaire (GWh)

Elektrizitätserzeugung in Kernkraftwerken (GWh)

Produzione di energia nucleotermoelettrica (GWh)

原子力発電量(GWh)

Producción de electricidad nuclear (GWh)

Производство атомной электроэнергии (ГВт.ч)

1989	1990	1991	1992	1993	1994	1995	1996	1997	
11 099	8 449	9 144	9 288	7 255	9 697	11 301	11 775	12 647	South Africa
11 099	8 449	9 144	9 288	7 255	9 697	11 301	11 775	12 647	**Africa**
5 039	7 281	7 756	7 081	7 750	8 235	7 066	7 459	7 892	Argentina
1 830	2 237	1 442	1 759	442	55	2 519	2 429	3 169	Brazil
6 869	9 518	9 198	8 840	8 192	8 290	9 585	9 888	11 061	**Latin America**
4 625	6 141	5 525	6 726	5 398	5 648	7 600	8 400	10 100	India
30	293	385	418	582	497	511	483	346	Pakistan
28 276	32 866	35 290	33 845	34 354	34 871	35 316	37 788	36 269	Chinese Taipei
32 931	39 300	41 200	40 989	40 334	41 016	43 427	46 671	46 715	**Asia**
-	-	-	-	1 604	13 906	12 833	14 339	14 418	People's Rep. of China
-	-	-	-	1 604	13 906	12 833	14 339	14 418	**China**
14 566	14 665	13 184	11 552	13 973	15 335	17 261	18 082	17 751	Bulgaria
-	-	-	-	-	-	-	1 386	5 400	Romania
12 157	12 036	11 689	11 050	11 022	12 135	11 437	11 261	10 797	Slovak Republic
-	-	-	3 971	3 957	4 609	4 779	4 562	5 019	*Slovenia*
4 688	4 622	4 952	3 971	3 957	4 609	4 779	4 562	5 019	Former Yugoslavia
31 411	31 323	29 825	26 573	28 952	32 079	33 477	35 291	38 967	**Non-OECD Europe**
-	-	-	-	-	-	304	2 324	1 600	Armenia
-	-	-	14 638	12 260	7 706	11 822	13 942	12 024	Lithuania
-	-	-	119 626	119 186	97 820	99 532	109 026	109 000	Russia
-	-	-	73 732	75 243	68 848	70 523	79 577	79 433	Ukraine
212 600	211 500	212 655	207 996	206 689	174 374	182 181	204 869	202 057	**Former USSR**
294 910	300 090	302 022	293 686	293 026	279 362	292 804	322 833	325 865	**Non-OECD Total**
641 409	687 412	738 570	740 469	746 741	790 993	820 093	815 857	759 347	OECD North America
230 234	255 159	269 771	279 789	307 394	327 777	358 283	376 124	396 142	OECD Pacific
772 536	770 141	795 963	809 560	843 858	842 842	860 417	901 775	911 765	OECD Europe
1 644 179	1 712 712	1 804 304	1 829 818	1 897 993	1 961 612	2 038 793	2 093 756	2 067 254	**OECD Total**
1 939 089	2 012 802	2 106 326	2 123 504	2 191 019	2 240 974	2 331 597	2 416 589	2 393 119	**World**

Kazakhstan produced an unkown quantity of nuclear electricity in 1992 - 1994.

Production of Hydro Electricity (GWh)

Production d'électricité d'origine hydraulique (GWh)

Elektrizitätserzeugung in Wasserkraftwerken (GWh)

Produzione di energia idroelettrica (GWh)

水力発電量(GWh)

Producción de electricidad hidráulica (GWh)

Производство гидроэлектроэнергии (ГВт.ч)

	1971	1973	1978	1983	1984	1985	1986	1987	1988
Algérie	330	752	250	235	452	646	250	499	183
Angola	605	814	528	578	601	701	701	701	701
Cameroun	1 100	1 068	1 259	2 089	2 178	2 388	2 431	2 451	2 572
Congo	55	50	49	229	231	310	266	275	286
RD du Congo	3 437	3 768	4 000	4 479	4 695	5 027	5 259	5 242	5 248
Egypte	5 041	5 156	9 320	9 817	9 633	8 663	9 281	8 658	9 000
Ethiopie	304	333	406	630	690	675	717	721	712
Gabon	-	5	296	547	610	667	693	698	709
Ghana	2 909	3 872	3 721	2 548	1 799	2 987	4 372	4 676	4 808
Côte d'Ivoire	139	168	167	977	374	1 358	1 418	877	1 240
Kenya	339	408	1 073	1 740	1 724	2 016	1 680	1 911	2 323
Maroc	1 520	1 192	1 416	481	366	486	643	825	936
Mozambique	224	191	804	236	244	375	173	180	166
Nigéria	1 574	1 858	2 324	1 989	2 668	3 091	3 630	3 284	4 008
Afrique du Sud	112	985	1 907	2 552	2 554	2 731	3 408	3 391	4 565
Soudan	165	347	450	653	695	737	872	899	892
Tanzanie	309	296	520	739	782	885	1 034	1 151	1 252
Tunisie	52	73	29	29	66	108	51	113	47
Zambie	908	3 097	7 953	10 251	9 972	10 247	10 014	8 612	8 409
Zimbabwe	2 425	3 487	5 545	3 472	3 040	2 857	2 371	1 108	2 755
Autre Afrique	817	1 103	1 340	2 646	2 927	3 150	3 257	3 433	3 251
Afrique	22 365	29 023	43 357	46 917	46 301	50 105	52 521	49 705	54 063
Argentine	1 544	2 994	7 752	18 419	19 880	20 646	21 015	21 905	15 791
Bolivie	898	972	1 259	1 143	1 164	1 142	1 114	1 061	1 163
Brésil	43 199	57 890	102 746	151 472	166 595	178 374	182 418	185 603	199 090
Chili	4 397	5 319	6 822	8 933	9 332	10 358	11 273	12 341	11 470
Colombie	6 663	7 939	12 035	15 269	16 977	18 290	21 608	23 279	24 425
Costa Rica	1 028	1 131	1 478	2 786	3 004	2 756	2 891	3 000	3 046
Cuba	110	62	83	62	70	54	59	44	73
République dominicaine	590	583	818	1 034	1 268	1 258	1 525	1 689	1 436
Equateur	440	435	791	1 721	3 233	3 279	4 012	4 570	4 826
El Salvador	522	492	666	803	869	941	1 028	917	1 089
Guatemala	260	233	283	1 275	1 276	1 279	1 722	1 766	1 950
Haiti	27	84	190	197	200	216	318	320	392
Honduras	242	359	647	982	1 049	1 315	1 438	1 756	1 907
Jamaïque	126	99	115	150	150	150	150	125	101
Nicaragua	203	371	325	326	320	368	406	488	468
Panama	83	102	719	866	1 492	1 929	2 096	2 032	2 199
Paraguay	154	280	249	655	991	4 029	11 842	18 524	19 934
Pérou	4 283	4 769	6 199	8 111	8 704	9 396	10 058	10 881	10 425
Uruguay	1 469	1 556	1 631	7 192	7 108	6 452	7 294	7 263	5 421
Vénézuela	5 390	6 225	12 232	18 063	20 243	22 648	25 159	30 843	34 203
Autre Amérique latine	1 093	1 017	893	759	898	1 017	810	376	546
Amérique latine	72 721	92 912	157 933	240 218	264 823	285 897	308 236	328 783	339 955

Production of Hydro Electricity (GWh)

Production d'électricité d'origine hydraulique (GWh)

Elektrizitätserzeugung in Wasserkraftwerken (GWh)

Produzione di energia idroelettrica (GWh)

水力発電量(GWh)

Producción de electricidad hidráulica (GWh)

Производство гидроэлектроэнергии (ГВт.ч)

1989	1990	1991	1992	1993	1994	1995	1996	1997	
226	135	293	199	353	166	193	135	75	Algeria
710	725	772	840	890	895	900	925	1 006	Angola
2 664	2 656	2 671	2 702	2 802	2 688	2 753	2 868	3 092	Cameroon
391	502	476	422	425	429	432	429	425	Congo
6 905	5 625	5 259	6 054	5 780	5 427	5 945	6 131	5 880	DR of Congo
9 974	9 932	9 900	9 700	9 680	10 000	10 810	11 555	11 987	Egypt
721	1 062	1 082	1 078	1 213	1 117	1 130	1 119	1 170	Ethiopia
675	705	707	712	710	718	725	734	740	Gabon
5 231	5 721	6 109	6 602	6 313	6 077	6 107	5 980	6 148	Ghana
1 593	1 464	1 288	116	1 048	971	1 727	1 872	2 022	Ivory Coast
2 469	2 477	2 770	2 796	2 993	3 068	3 123	3 295	3 490	Kenya
1 157	1 220	1 226	964	443	840	605	1 938	2 062	Morocco
284	284	323	325	344	366	379	446	791	Mozambique
4 140	4 387	5 931	6 059	5 572	5 562	5 500	5 500	5 593	Nigeria
3 798	2 851	3 784	2 085	1 491	2 591	1 803	3 539	4 700	South Africa
895	958	1 038	1 089	1 091	1 121	972	1 074	1 042	Sudan
1 419	1 549	1 726	1 650	1 698	1 492	1 539	1 748	1 449	Tanzania
34	44	105	65	64	40	39	67	44	Tunisia
6 687	6 291	7 793	7 740	7 743	7 743	7 748	7 754	7 964	Zambia
3 329	3 790	3 563	3 016	1 970	2 011	2 343	2 087	1 777	Zimbabwe
3 457	3 606	3 552	3 685	3 726	3 751	3 906	3 942	3 979	Other Africa
56 759	55 984	60 368	57 899	56 349	57 073	58 679	63 138	65 436	**Africa**
16 677	17 619	15 994	19 146	23 575	26 903	26 658	22 985	28 157	Argentina
1 167	1 180	1 256	1 177	1 455	1 610	1 688	2 037	2 306	Bolivia
204 691	206 708	217 781	223 342	235 064	242 705	253 904	265 769	279 064	Brazil
9 603	8 928	13 128	16 735	17 194	16 978	18 400	16 878	18 944	Chile
26 582	27 366	27 594	22 042	27 625	31 830	31 931	35 281	32 799	Colombia
3 328	3 382	3 656	3 584	3 957	3 943	3 554	3 812	4 808	Costa Rica
82	91	105	81	82	49	74	95	93	Cuba
521	395	558	1 934	1 552	1 650	1 754	1 265	1 338	Dominican Republic
4 954	5 013	5 117	4 989	5 838	6 652	5 229	6 249	6 775	Ecuador
1 195	1 440	845	983	1 168	1 201	1 240	1 633	1 194	El Salvador
2 100	2 150	2 288	1 980	2 184	2 191	2 480	3 153	3 760	Guatemala
406	457	332	321	274	233	253	272	200	Haiti
2 000	2 287	2 317	2 195	2 309	1 986	2 452	3 050	3 246	Honduras
127	110	120	109	90	111	120	128	121	Jamaica
572	403	337	257	483	383	407	431	407	Nicaragua
2 181	2 213	1 906	2 314	2 274	2 072	2 334	2 816	2 766	Panama
24 307	27 158	29 305	27 116	31 408	36 394	42 092	48 035	50 452	Paraguay
10 518	10 170	11 231	9 690	11 676	12 816	12 938	13 324	13 215	Peru
3 903	7 009	6 113	7 922	7 299	7 468	5 856	5 768	6 486	Uruguay
34 666	36 976	44 533	47 262	47 469	51 274	51 440	53 844	57 152	Venezuela
1 413	1 228	1 284	1 307	1 313	1 317	1 324	1 390	1 449	Other Latin America
350 993	362 283	385 800	394 486	424 289	449 766	466 128	488 215	514 732	**Latin America**

Production of Hydro Electricity (GWh)

Production d'électricité d'origine hydraulique (GWh)

Elektrizitätserzeugung in Wasserkraftwerken (GWh)

Produzione di energia idroelettrica (GWh)

水力発電量(GWh)

Producción de electricidad hidráulica (GWh)

Производство гидроэлектроэнергии (ГВт.ч)

	1971	1973	1978	1983	1984	1985	1986	1987	1988
Bangladesh	175	331	506	662	897	739	450	517	675
Inde	28 034	28 982	47 172	49 972	53 966	51 039	53 859	47 462	57 884
Indonésie	1 425	1 603	1 299	1 816	2 118	2 990	4 935	5 183	5 892
RPD de Corée	11 600	12 500	19 500	26 000	27 000	28 000	29 000	29 200	29 000
Malaisie	1 046	1 118	896	1 733	3 419	3 733	4 081	4 918	5 674
Myanmar	477	576	736	993	1 011	1 003	1 042	1 024	935
Népal	67	81	153	286	306	334	440	550	563
Pakistan	3 679	4 355	7 442	11 365	12 826	12 241	13 804	15 250	16 690
Philippines	1 933	2 374	2 811	3 968	5 278	5 553	6 017	5 247	6 264
Sri Lanka	834	708	1 366	1 217	2 091	2 395	2 645	2 177	2 597
Taipei chinois	3 091	3 399	4 966	4 988	4 429	6 926	7 419	7 118	6 682
Thailande	2 048	1 880	2 110	3 660	4 081	3 691	5 554	4 075	3 778
Viêt-Nam	614	420	800	1 223	1 599	1 477	1 407	1 384	1 791
Autre Asie	911	865	1 254	1 506	1 911	2 032	2 305	3 019	3 692
Asie	55 934	59 192	91 011	109 389	120 932	122 153	132 958	127 124	142 117
Rép. populaire de Chine	30 000	38 000	44 600	86 360	86 780	92 370	94 530	100 010	109 150
Chine	30 000	38 000	44 600	86 360	86 780	92 370	94 530	100 010	109 150
Albanie	700	1 127	2 400	3 200	3 275	2 944	3 400	3 350	3 590
Bulgarie	2 170	2 570	2 909	3 351	3 260	2 236	2 326	2 538	2 596
Roumanie	4 495	7 547	10 614	10 038	11 326	12 713	10 810	11 209	13 622
République slovaque	1 476	1 322	2 259	2 223	1 970	2 678	2 115	2 439	2 303
Bosnie-Herzegovine	-	-	-	-	-	-	-	-	-
Croatie	-	-	-	-	-	-	-	-	-
Ex-RYM	-	-	-	-	-	-	-	-	-
Slovénie	-	-	-	-	-	-	-	-	-
RF de Yougoslavie	-	-	-	-	-	-	-	-	-
Ex-Yougoslavie	15 644	16 394	25 199	21 850	25 915	24 270	27 504	26 253	25 871
Europe non-OCDE	24 485	28 960	43 381	40 662	45 746	44 841	46 155	45 789	47 982
Arménie	-	-	-	-	-	-	-	-	-
Azerbaïdjan	-	-	-	-	-	-	-	-	-
Bélarus	-	-	-	-	-	-	-	-	-
Estonie	-	-	-	-	-	-	-	-	-
Géorgie	-	-	-	-	-	-	-	-	-
Kazakhstan	-	-	-	-	-	-	-	-	-
Kirghizistan	-	-	-	-	-	-	-	-	-
Lettonie	-	-	-	-	-	-	-	-	-
Lituanie	-	-	-	-	-	-	-	-	-
Moldova	-	-	-	-	-	-	-	-	-
Russie	-	-	-	-	-	-	-	-	-
Tadjikistan	-	-	-	-	-	-	-	-	-
Turkménistan	-	-	-	-	-	-	-	-	-
Ukraine	-	-	-	-	-	-	-	-	-
Ouzbékistan	-	-	-	-	-	-	-	-	-
Ex-URSS	126 000	122 300	169 700	180 000	203 000	214 403	215 738	219 825	231 880

Production of Hydro Electricity (GWh)

Production d'électricité d'origine hydraulique (GWh)

Elektrizitätserzeugung in Wasserkraftwerken (GWh)

Produzione di energia idroelettrica (GWh)

水力発電量(GWh)

Producción de electricidad hidráulica (GWh)

Производство гидроэлектроэнергии (ГВт.ч)

1989	1990	1991	1992	1993	1994	1995	1996	1997	
920	884	838	796	608	847	372	739	719	Bangladesh
62 074	71 656	72 774	69 885	70 478	82 727	72 717	69 072	74 775	India
7 450	6 489	7 396	9 836	8 845	6 846	8 674	8 941	5 990	Indonesia
29 000	29 000	31 750	24 000	24 000	23 500	23 000	22 517	21 842	DPR of Korea
5 580	3 988	4 407	4 360	4 872	6 523	6 221	5 186	3 294	Malaysia
1 240	1 193	1 240	1 518	1 705	1 614	1 624	1 651	1 655	Myanmar
542	712	870	870	804	835	849	1 074	1 109	Nepal
16 970	16 925	18 303	18 647	21 112	19 436	22 858	23 206	20 858	Pakistan
6 485	6 062	5 145	4 252	4 994	5 946	6 199	7 074	6 074	Philippines
2 802	3 145	3 116	2 900	3 796	4 089	4 452	3 252	3 447	Sri Lanka
6 682	8 188	5 508	8 351	6 719	8 887	8 879	9 044	9 567	Chinese Taipei
5 570	4 975	4 586	4 238	3 702	4 514	6 712	7 340	7 200	Thailand
3 838	5 385	6 327	8 145	8 767	10 259	12 209	14 620	16 183	Vietnam
3 722	3 717	3 546	3 409	3 410	3 519	3 625	3 681	3 846	Other Asia
152 875	162 319	165 806	161 207	163 812	179 542	178 391	177 397	176 559	**Asia**
118 400	126 720	125 090	131 230	152 775	168 250	190 577	187 966	195 983	People's Rep. of China
118 400	126 720	125 090	131 230	152 775	168 250	190 577	187 966	195 983	**China**
3 590	2 848	3 518	3 226	3 314	3 771	4 204	5 684	5 391	Albania
2 691	1 878	2 441	2 063	1 942	1 468	2 314	2 919	2 936	Bulgaria
12 629	10 982	14 234	11 700	12 768	13 046	16 693	15 755	17 509	Romania
2 631	2 515	1 894	2 332	3 865	4 554	5 226	4 533	4 358	Slovak Republic
-	-	-	3 000	2 000	1 250	1 420	1 420	1 420	*Bosnia-Herzegovina*
-	-	-	4 341	4 345	4 930	5 265	7 228	5 299	*Croatia*
-	-	-	848	522	695	801	850	900	*FYROM*
-	-	-	3 411	3 021	3 399	3 241	3 673	3 092	*Slovenia*
-	-	-	11 343	10 015	11 127	11 220	11 497	12 167	*FR of Yugoslavia*
23 491	19 799	25 629	22 943	19 903	21 401	21 947	24 668	22 878	Former Yugoslavia
45 032	38 022	47 716	42 264	41 792	44 240	50 384	53 559	53 072	**Non-OECD Europe**
-	-	-	3 044	4 293	3 514	1 919	1 572	1 389	Armenia
-	-	-	1 747	2 400	1 829	1 556	1 538	1 520	Azerbaijan
-	-	-	17	19	19	20	16	21	Belarus
-	-	-	1	1	3	2	2	3	Estonia
-	-	-	6 515	7 034	4 713	5 310	6 041	6 044	Georgia
-	-	-	6 866	7 629	9 179	8 331	7 331	6 499	Kazakhstan
-	-	-	9 200	9 085	11 724	11 118	12 255	11 224	Kyrgyzstan
-	-	-	2 520	2 875	3 305	2 937	1 860	2 952	Latvia
-	-	-	311	393	718	751	874	769	Lithuania
-	-	-	259	375	278	324	366	382	Moldova
-	-	-	172 594	174 627	176 959	177 256	155 326	158 000	Russia
-	-	-	15 928	17 118	16 691	14 596	14 825	13 825	Tajikistan
-	-	-	4	5	4	4	5	5	Turkmenistan
-	-	-	8 069	11 237	12 327	10 150	8 833	10 032	Ukraine
-	-	-	6 281	7 358	7 156	6 188	6 525	5 777	Uzbekistan
223 900	233 000	234 670	233 356	244 449	248 419	240 462	217 369	218 442	**Former USSR**

Production of Hydro Electricity (GWh)

Production d'électricité d'origine hydraulique (GWh)

Elektrizitätserzeugung in Wasserkraftwerken (GWh)

Produzione di energia idroelettrica (GWh)

水力発電量(GWh)

Producción de electricidad hidráulica (GWh)

Производство гидроэлектроэнергии (ГВт.ч)

	1971	1973	1978	1983	1984	1985	1986	1987	1988
Iran	2 679	2 989	6 249	6 203	5 750	5 550	7 517	8 390	7 311
Irak	200	290	708	600	610	610	600	2 600	2 600
Israël	-	-	-	-	-	-	-	-	-
Jordanie	-	-	-	-	-	-	3	19	27
Liban	839	478	800	560	585	585	560	610	600
Syrie	51	17	2 095	2 802	2 877	2 860	1 633	2 123	4 807
Moyen-Orient	3 769	3 774	9 852	10 165	9 822	9 605	10 313	13 742	15 345
Total non-OCDE	335 274	374 161	559 834	713 711	777 404	819 374	860 451	884 978	940 492
OCDE Amérique du N.	440 469	476 344	534 849	622 210	634 151	613 917	624 465	586 819	553 570
OCDE Pacifique	113 175	99 078	107 123	123 829	110 325	126 216	127 393	123 027	137 575
OCDE Europe	329 708	348 813	420 904	455 126	458 946	457 170	436 950	478 732	511 967
Total OCDE	883 352	924 235	1 062 876	1 201 165	1 203 422	1 197 303	1 188 808	1 188 578	1 203 112
Monde	1 218 626	1 298 396	1 622 710	1 914 876	1 980 826	2 016 677	2 049 259	2 073 556	2 143 604

Production of Hydro Electricity (GWh)

Production d'électricité d'origine hydraulique (GWh)

Elektrizitätserzeugung in Wasserkraftwerken (GWh)

Produzione di energia idroelettrica (GWh)

水力発電量(GWh)

Producción de electricidad hidráulica (GWh)

Производство гидроэлектроэнергии (ГВт.ч)

1989	1990	1991	1992	1993	1994	1995	1996	1997	
7 522	6 083	7 056	9 330	9 823	7 445	7 275	7 376	7 376	Iran
2 600	2 600	900	700	600	560	570	570	581	Iraq
-	-	20	76	67	58	65	62	65	Israel
17	11	7	15	22	14	18	22	22	Jordan
500	500	560	720	720	817	717	798	901	Lebanon
4 727	5 648	6 249	7 386	6 709	6 800	6 944	6 944	10 038	Syria
15 366	14 842	14 792	18 227	17 941	15 694	15 589	15 772	18 983	**Middle East**
963 325	993 170	1 034 242	1 038 669	1 101 407	1 162 984	1 200 210	1 203 416	1 243 207	**Non-OECD Total**
589 417	609 357	639 488	617 537	652 985	634 278	700 880	764 229	736 165	OECD North America
139 737	140 416	149 872	130 877	151 908	122 238	140 192	136 704	146 292	OECD Pacific
444 900	460 743	460 702	491 360	506 124	506 562	514 743	496 760	512 704	OECD Europe
1 174 054	1 210 516	1 250 062	1 239 774	1 311 017	1 263 078	1 355 815	1 397 693	1 395 161	**OECD Total**
2 137 379	2 203 686	2 284 304	2 278 443	2 412 424	2 426 062	2 556 025	2 601 109	2 638 368	**World**

Production of Geothermal Electricity (GWh)

Production d'électricité d'origine géothermique (GWh)

Geothermische Elektrizitätserzeugung (GWh)

Produzione di energia geotermoelettrica (GWh)

地熱発電量(GWh)

Producción de electricidad geotérmica (GWh)

Производство геотермальной электроэнергии (ГВт.ч)

	1971	1973	1978	1983	1984	1985	1986	1987	1988
Ethiopie	-	-	-	55	55	57	59	61	63
Kenya	-	-	-	-	-	-	368	359	323
Afrique	-	-	-	55	55	57	427	420	386
Costa Rica	-	-	-	-	-	-	-	-	-
El Salvador	-	-	600	671	680	652	576	652	643
Nicaragua	-	-	-	1	272	301	300	234	190
Amérique latine	-	-	600	672	952	953	876	886	833
Indonésie	-	-	-	209	217	224	232	719	1 012
Philippines	-	-	3	4 077	4 532	4 929	4 576	4 532	4 844
Asie	-	-	3	4 286	4 749	5 153	4 808	5 251	5 856
Russie	-	-	-	-	-	-	-	-	-
Ex-URSS	-	-	-	-	-	-	-	-	-
Total non-OCDE	-	-	603	5 013	5 756	6 163	6 111	6 557	7 075
OCDE Amérique du N.	586	2 612	3 740	7 494	9 266	11 526	14 320	15 840	15 551
OCDE Pacifique	1 256	1 512	1 816	2 621	2 594	2 633	2 558	2 593	2 595
OCDE Europe	2 677	2 506	2 514	2 907	3 059	2 883	3 031	3 293	3 417
Total OCDE	4 519	6 630	8 070	13 022	14 919	17 042	19 909	21 726	21 563
Monde	4 519	6 630	8 673	18 035	20 675	23 205	26 020	28 283	28 638

Production of Geothermal Electricity (GWh)

Production d'électricité d'origine géothermique (GWh)

Geothermische Elektrizitätserzeugung (GWh)

Produzione di energia geotermoelettrica (GWh)

地熱発電量(GWh)

Producción de electricidad geotérmica (GWh)

Производство геотермальной электроэнергии (ГВт.ч)

1989	1990	1991	1992	1993	1994	1995	1996	1997	
64	66	67	63	75	62	57	49	104	Ethiopia
323	326	297	272	272	261	290	390	393	Kenya
387	392	364	335	347	323	347	439	497	**Africa**
-	-	-	-	-	-	468	510	544	Costa Rica
671	626	848	824	730	653	674	680	721	El Salvador
381	386	458	468	406	360	310	277	209	Nicaragua
1 052	1 012	1 306	1 292	1 136	1 013	1 452	1 467	1 474	**Latin America**
1 007	1 125	1 049	1 084	1 090	1 935	2 175	2 312	2 578	Indonesia
5 316	5 466	5 757	5 700	5 668	6 297	6 134	6 534	7 237	Philippines
6 323	6 591	6 806	6 784	6 758	8 232	8 309	8 846	9 815	**Asia**
-	-	-	29	28	28	30	28	28	Russia
-	-	-	29	28	28	30	28	28	**Former USSR**
7 762	7 995	8 476	8 440	8 269	9 596	10 138	10 780	11 814	**Non-OECD Total**
19 623	21 136	21 702	22 972	23 651	23 077	20 610	21 475	20 373	OECD North America
3 293	3 951	3 979	4 059	3 936	4 275	5 222	5 814	5 766	OECD Pacific
3 510	3 626	3 571	3 770	4 005	3 789	3 854	4 241	4 414	OECD Europe
26 426	28 713	29 252	30 801	31 592	31 141	29 686	31 530	30 553	**OECD Total**
34 188	36 708	37 728	39 241	39 861	40 737	39 824	42 310	42 367	**World**

Production of Solar, Wind, Tide and Wave Electricity (GWh)

Production d'électricité d'origine solaire, éolienne, marémotrice et houlomotrice (GWh)

Elektrizitätserzeugung aus Sonnenenergie, Wind, Gezeiten und Wellen (GWh)

Produzione di energia elettrica solare, eolica e energia da moto ondoso (GWh)

太陽光、風力、潮力、波力発電量 (GWh)

Producción de electricidad solar, eólica, maremotriz y de oleaje (GWh)

Производство электроэнергии солнца, ветра, приливов и волн (ГВт.ч)

	1971	1973	1978	1983	1984	1985	1986	1987	1988
Costa Rica	-	-	-	-	-	-	-	-	-
Pérou	-	-	-	-	-	-	-	-	-
Amérique latine	-	-	-	-	-	-	-	-	-
Inde	-	-	-	-	-	-	2	3	6
Asie	-	-	-	-	-	-	2	3	6
Lettonie	-	-	-	-	-	-	-	-	-
Ex-URSS	-	-	-	-	-	-	-	-	-
Jordanie	-	-	-	-	-	-	-	-	-
Moyen-Orient	-	-	-	-	-	-	-	-	-
Total non-OCDE	-	-	-	-	-	-	2	3	6
OCDE Amérique du N.	-	-	-	3	20	42	53	50	33
OCDE Pacifique	-	-	-	-	-	-	1	2	1
OCDE Europe	501	559	473	633	642	661	721	761	897
Total OCDE	501	559	473	636	662	703	775	813	931
Monde	501	559	473	636	662	703	777	816	937

Production of Solar, Wind, Tide and Wave Electricity (GWh)

Production d'électricité d'origine solaire, éolienne, marémotrice et houlomotrice (GWh)

Elektrizitätserzeugung aus Sonnenenergie, Wind, Gezeiten und Wellen (GWh)

Produzione di energia elettrica solare, eolica e energia da moto ondoso (GWh)

太陽光、風力、潮力、波力発電量 (GWh)

Producción de electricidad solar, eólica, maremotriz y de oleaje (GWh)

Производство электроэнергии солнца, ветра, приливов и волн (ГВт.ч)

1989	1990	1991	1992	1993	1994	1995	1996	1997	
-	-	-	-	-	-	-	23	76	Costa Rica
-	-	-	-	-	-	-	408	556	Peru
-	-	-	-	-	-	-	431	632	**Latin America**
6	32	39	52	57	57	57	57	57	India
6	32	39	52	57	57	57	57	57	**Asia**
-	-	-	-	-	-	-	1	1	Latvia
-	-	-	-	-	-	-	1	1	**Former USSR**
-	1	1	1	1	1	1	1	1	Jordan
-	1	1	1	1	1	1	1	1	**Middle East**
6	33	40	53	58	58	58	490	691	**Non-OECD Total**
2 720	2 991	3 468	3 704	4 021	4 408	4 125	4 418	4 389	OECD North America
1	1	1	1	1	35	59	74	86	OECD Pacific
1 081	1 327	1 624	2 079	2 870	3 584	4 448	5 415	7 965	OECD Europe
3 802	4 319	5 093	5 784	6 892	8 027	8 632	9 907	12 440	**OECD Total**
3 808	4 352	5 133	5 837	6 950	8 085	8 690	10 397	13 131	**World**

Production of Electricity from Combustible Renewables and Waste (GWh)
Production d'électricité à partir d'énergies renouv. combustibles et de déchets (GWh)

Elektrizitätserzeugung aus erneuerbaren Brennstoffen und Abfällen (GWh)

Produzione di energia elettrica da combustibili rinnovabili e da rifiuti (GWh)

可燃性再生可能エネルギー及び廃棄物からの発電量(GWh)

Producción de Electricidad a partir de combustibles renovables y desechos (GWh)

Производство электроэнергии из возобновляемых видов топлива и отходов (ГВт.ч)

	1971	1973	1978	1983	1984	1985	1986	1987	1988
Argentine	145	142	155	396	239	298	307	307	307
Bolivie	15	21	32	43	41	37	39	49	50
Brésil	945	1 000	1 360	4 013	4 309	4 049	4 297	4 955	4 824
Chili	192	120	170	317	333	525	511	519	528
Colombie	123	154	123	161	257	133	219	155	251
Cuba	788	694	919	1 062	1 097	1 104	1 155	1 170	1 281
République dominicaine	-	78	82	96	137	52	52	53	29
El Salvador	16	22	31	27	25	25	27	18	22
Guatemala	36	47	125	58	58	58	61	67	74
Haiti	12	12	12	12	12	15	20	16	16
Jamaïque	103	205	59	59	56	50	50	71	89
Nicaragua	19	16	33	39	44	42	36	41	37
Panama	5	5	29	148	152	102	111	101	135
Paraguay	45	44	36	42	46	42	48	56	27
Pérou	294	298	277	188	185	219	200	178	178
Trinité-et-Tobago	35	29	29	16	13	17	26	31	27
Uruguay	11	11	12	20	27	46	56	56	52
Autre Amérique latine	108	110	140	127	123	121	131	110	86
Amérique latine	2 892	3 008	3 624	6 824	7 154	6 935	7 346	7 953	8 013
Thailande	-	-	-	-	-	-	-	-	-
Asie	-	-	-	-	-	-	-	-	-
Roumanie	-	-	-	-	-	-	-	-	-
Croatie	-	-	-	-	-	-	-	-	-
Ex-Yougoslavie	-	-	-	-	-	-	-	-	-
Europe non-OCDE	-	-	-	-	-	-	-	-	-
Estonie	-	-	-	-	-	-	-	-	-
Russie	-	-	-	-	-	-	-	-	-
Ex-URSS	22 700	22 800	26 000	24 100	24 100	23 100	21 800	21 200	21 100
Total non-OCDE	25 592	25 808	29 624	30 924	31 254	30 035	29 146	29 153	29 113
OCDE Amérique du N.	258	297	1 488	2 367	2 696	3 138	3 095	3 862	4 258
OCDE Pacifique	263	338	752	10 830	12 202	12 774	14 101	15 151	16 760
OCDE Europe	5 870	6 517	8 912	10 011	10 865	11 492	12 470	14 375	14 504
Total OCDE	6 391	7 152	11 152	23 208	25 763	27 404	29 666	33 388	35 522
Monde	31 983	32 960	40 776	54 132	57 017	57 439	58 812	62 541	64 635

Production of Electricity from Combustible Renewables and Waste (GWh)

Production d'électricité à partir d'énergies renouv. combustibles et de déchets (GWh)

Elektrizitätserzeugung aus erneuerbaren Brennstoffen und Abfällen (GWh)

Produzione di energia elettrica da combustibili rinnovabili e da rifiuti (GWh)

可燃性再生可能エネルギー及び廃棄物からの発電量(GWh)

Producción de Electricidad a partir de combustibles renovables y desechos (GWh)

Производство электроэнергии из возобновляемых видов топлива и отходов (ГВт.ч)

1989	1990	1991	1992	1993	1994	1995	1996	1997	
267	267	250	220	240	240	289	289	215	Argentina
47	49	54	58	39	37	39	39	44	Bolivia
4 452	5 250	5 704	6 099	6 220	6 696	6 839	7 941	8 755	Brazil
564	673	549	120	120	120	154	200	206	Chile
255	143	134	178	172	204	265	297	300	Colombia
1 246	1 637	1 412	1 527	1 069	1 042	893	1 154	1 486	Cuba
33	26	26	27	28	29	30	30	30	Dominican Republic
24	22	17	30	16	5	5	20	22	El Salvador
94	88	98	9	213	215	216	216	220	Guatemala
16	17	17	17	18	21	20	20	20	Haiti
93	120	80	79	196	234	279	279	320	Jamaica
34	43	47	36	32	27	43	43	43	Nicaragua
72	86	82	96	62	17	29	30	30	Panama
22	20	18	17	33	17	86	104	106	Paraguay
171	136	133	144	109	108	114	114	128	Peru
26	31	25	30	28	32	33	33	36	Trinidad and Tobago
50	40	50	39	29	40	40	39	42	Uruguay
83	87	88	89	89	90	90	90	90	Other Latin America
7 549	8 735	8 784	8 815	8 713	9 174	9 464	10 938	12 093	**Latin America**
-	-	-	-	-	-	265	265	2 158	Thailand
-	-	-	-	-	-	265	265	2 158	**Asia**
-	-	-	57	69	-	3	-	11	Romania
-	-	-	6	19	6	5	10	2	*Croatia*
-	-	-	6	19	6	5	10	2	Former Yugoslavia
-	-	-	63	88	6	8	10	13	**Non-OECD Europe**
-	-	-	-	-	-	6	5	8	Estonia
-	-	-	5 000	30 634	26 254	2 404	2 528	6 430	Russia
20 895	21 000	21 000	5 000	30 634	26 254	2 410	2 533	6 438	**Former USSR**
28 444	29 735	29 784	13 878	39 435	35 434	12 147	13 746	20 702	**Non-OECD Total**
56 958	68 430	56 462	69 577	65 002	66 856	66 530	68 252	67 914	OECD North America
17 930	18 021	18 897	20 986	21 230	22 365	23 317	24 159	25 992	OECD Pacific
15 157	15 357	15 977	23 579	28 049	31 397	34 424	35 411	40 559	OECD Europe
90 045	101 808	91 336	114 142	114 281	120 618	124 271	127 822	134 465	**OECD Total**
118 489	131 543	121 120	128 020	153 716	156 052	136 418	141 568	155 167	**World**

Total Production of Electricity (GWh)

Production totale d'électricité (GWh)

Gesamterzeugung von Elektrizität (GWh)

Produzione totale di energia elettrica (GWh)

総発電量(GWh)

Producción total de electricidad (GWh)

Общее производство электроэнергии (ГВт.ч)

	1971	1973	1978	1983	1984	1985	1986	1987	1988
Algérie	2 229	2 806	5 480	10 218	11 182	12 274	12 981	12 722	13 966
Angola	742	984	585	783	806	806	806	810	815
Bénin	-	9	5	21	61	24	26	26	20
Cameroun	1 075	1 119	1 337	2 164	2 215	2 441	2 478	2 508	2 623
Congo	88	96	55	233	235	314	270	281	292
RD du Congo	3 545	3 848	4 080	4 618	4 835	5 171	5 406	5 391	5 392
Egypte	7 998	8 106	15 150	25 879	29 049	31 458	33 464	37 845	39 580
Ethiopie	593	591	567	860	910	888	915	892	877
Gabon	114	165	436	729	795	861	888	895	910
Ghana	2 944	3 910	3 773	2 568	1 831	3 017	4 402	4 866	4 842
Côte d'Ivoire	588	796	1 418	1 608	1 690	1 842	2 086	2 074	2 309
Kenya	627	793	1 382	1 904	1 949	2 155	2 307	2 654	2 772
Libye	508	1 127	3 363	7 150	9 200	11 844	13 280	15 600	16 551
Maroc	2 290	2 875	4 388	6 686	6 921	7 345	7 764	8 001	8 984
Mozambique	674	641	876	417	398	550	310	336	337
Nigéria	1 816	2 607	4 648	8 713	9 035	10 431	10 765	11 265	11 654
Sénégal	347	407	586	714	773	772	768	830	857
Afrique du Sud	54 647	64 390	84 496	123 079	137 254	143 491	148 241	152 373	158 141
Soudan	495	610	865	933	993	1 229	1 338	1 379	1 378
Tanzanie	491	582	695	849	922	1 015	1 146	1 272	1 378
Tunisie	926	1 179	2 082	3 551	3 880	4 019	4 278	4 549	4 953
Zambie	1 216	3 368	8 092	10 336	10 061	10 324	10 101	8 685	8 485
Zimbabwe	3 602	5 172	6 253	4 164	4 120	4 784	5 204	6 345	8 121
Autre Afrique	2 590	3 101	4 095	5 866	6 432	6 595	7 012	7 594	7 676
Afrique	90 145	109 282	154 707	224 043	245 547	263 650	276 236	289 193	302 913
Argentine	23 624	26 661	33 434	43 003	44 966	45 265	44 523	49 400	50 576
Bolivie	1 024	1 151	1 675	1 674	1 686	1 697	1 720	1 721	1 872
Brésil	50 878	63 854	112 483	162 490	179 390	193 681	202 129	203 332	214 948
Chili	8 524	8 766	10 360	12 624	13 497	14 040	14 744	15 637	16 915
Colombie	9 500	12 596	17 209	24 243	26 361	26 633	29 660	31 309	33 201
Costa Rica	1 148	1 347	1 925	2 870	3 067	2 816	2 949	3 133	3 193
Cuba	5 021	5 703	8 481	11 551	12 292	12 199	13 167	13 594	14 542
République dominicaine	1 068	2 246	2 874	3 409	3 769	4 175	4 327	4 663	3 640
Equateur	1 050	1 256	2 605	4 303	4 536	4 558	5 012	5 384	5 641
El Salvador	743	912	1 489	1 571	1 643	1 738	1 704	1 918	2 007
Guatemala	689	874	1 596	1 533	1 605	1 669	1 841	1 978	2 153
Haiti	82	106	245	411	422	441	443	494	535
Honduras	391	486	759	1 150	1 210	1 411	1 487	1 816	1 976
Jamaïque	1 676	2 187	1 517	1 667	1 645	1 636	1 722	1 855	1 831
Antilles néerlandaises	710	775	825	883	818	760	637	692	700
Nicaragua	657	717	1 163	940	974	1 060	1 211	1 300	1 187
Panama	950	1 302	1 622	2 490	2 466	2 603	2 746	2 825	2 741
Paraguay	236	379	569	844	1 090	4 089	11 912	18 598	19 988
Pérou	5 949	6 655	8 765	10 674	11 770	12 115	12 949	14 010	13 560
Trinité-et-Tobago	991	1 105	1 557	2 905	3 006	3 023	3 291	3 481	3 485
Uruguay	2 381	2 530	3 047	7 366	7 244	6 604	7 430	7 579	6 999
Vénézuela	13 290	16 077	25 885	42 764	45 271	48 980	51 093	54 132	57 773
Autre Amérique latine	3 197	3 741	4 259	4 297	4 448	4 733	4 890	4 793	5 239
Amérique latine	133 779	161 426	244 344	345 662	373 176	395 926	421 587	443 644	464 702

La production totale d'électricité peut être supérieure à la somme des productions par source d'énergie.

Total Production of Electricity (GWh)

Production totale d'électricité (GWh)

Gesamterzeugung von Elektrizität (GWh)

Produzione totale di energia elettrica (GWh)

総発電量(GWh)

Producción total de electricidad (GWh)

Общее производство электроэнергии (ГВт.ч)

1989	1990	1991	1992	1993	1994	1995	1996	1997	
15 324	16 105	17 345	18 286	19 415	19 883	19 714	20 650	21 685	Algeria
819	797	934	947	950	955	960	1 027	1 109	Angola
20	21	23	25	26	52	33	47	50	Benin
2 706	2 697	2 710	2 737	2 832	2 720	2 785	2 902	3 128	Cameroon
397	508	482	428	431	435	438	435	431	Congo
6 937	5 650	5 281	6 073	5 885	5 545	6 180	6 261	6 010	DR of Congo
41 649	43 478	46 086	46 978	49 435	51 947	54 833	54 444	57 656	Egypt
879	1 246	1 270	1 267	1 399	1 302	1 328	1 316	1 340	Ethiopia
876	915	914	919	922	933	940	970	1 007	Gabon
5 260	5 731	6 116	6 603	6 335	6 105	6 109	5 987	6 155	Ghana
2 200	2 021	1 894	784	2 150	2 285	2 870	3 015	3 213	Ivory Coast
2 901	3 034	3 227	3 215	3 396	3 539	3 747	4 040	4 238	Kenya
16 700	16 800	16 800	16 950	17 000	17 800	18 000	18 180	18 180	Libya
9 024	9 628	9 205	9 719	9 910	10 966	11 907	12 356	13 128	Morocco
472	454	471	416	392	395	408	476	1 005	Mozambique
12 812	12 564	14 167	14 834	14 505	15 531	14 483	14 991	15 179	Nigeria
888	913	937	1 021	1 007	1 041	1 102	1 175	1 261	Senegal
163 359	167 226	170 120	169 149	174 581	183 470	188 091	200 566	210 352	South Africa
1 365	1 515	1 661	1 634	1 686	1 858	1 864	2 063	1 966	Sudan
1 509	1 628	1 822	1 816	1 875	1 703	1 841	1 991	1 934	Tanzania
5 156	5 534	5 743	6 180	6 313	6 714	7 306	7 533	7 977	Tunisia
6 782	6 330	7 834	7 780	7 785	7 785	7 790	7 795	8 006	Zambia
8 569	9 361	8 924	8 237	7 467	7 535	7 806	7 323	7 298	Zimbabwe
7 990	8 304	8 540	8 812	9 178	9 258	9 454	9 430	9 500	Other Africa
314 594	322 460	332 506	334 810	344 875	359 757	369 989	384 973	401 808	**Africa**
50 837	47 412	50 619	54 521	58 858	61 589	64 601	69 759	71 947	Argentina
2 010	2 133	2 275	2 412	2 646	2 824	3 020	3 232	3 433	Bolivia
221 737	222 821	234 366	241 731	251 971	260 076	275 649	291 204	307 302	Brazil
17 811	18 374	19 961	22 364	23 626	25 276	28 027	30 790	33 994	Chile
34 602	36 157	37 191	33 445	38 219	41 359	43 616	44 605	46 115	Colombia
3 410	3 468	3 864	4 173	4 431	4 772	4 865	4 791	5 599	Costa Rica
15 240	15 025	13 247	11 538	11 004	11 964	12 459	13 236	14 087	Cuba
3 255	3 698	3 895	5 581	5 874	6 182	6 506	6 847	7 335	Dominican Republic
5 768	6 373	7 013	7 199	7 432	8 257	8 526	9 260	9 595	Ecuador
2 056	2 242	2 320	2 412	2 810	3 170	3 364	3 418	3 643	El Salvador
2 313	2 330	2 493	2 802	3 094	3 161	3 541	4 247	4 897	Guatemala
575	597	468	423	382	302	515	625	625	Haiti
2 062	2 303	2 326	2 320	2 536	2 307	2 698	3 059	3 294	Honduras
1 803	2 016	2 054	2 199	3 791	4 775	5 829	6 038	6 255	Jamaica
737	790	801	853	909	971	1 009	1 017	1 052	Netherlands Antilles
1 352	1 399	1 471	1 583	1 681	1 686	1 795	1 920	1 907	Nicaragua
2 726	2 871	2 897	3 432	3 444	3 361	3 519	3 915	4 151	Panama
24 333	27 185	29 328	27 141	31 449	36 415	42 236	48 200	50 619	Paraguay
13 737	13 162	13 901	13 044	14 679	14 785	16 880	17 280	17 954	Peru
3 428	3 577	3 720	3 976	3 817	4 069	4 307	4 541	4 988	Trinidad and Tobago
5 750	7 444	7 019	8 899	7 978	7 619	6 307	6 670	7 149	Uruguay
57 620	59 309	63 325	67 427	69 368	71 207	73 432	75 372	74 865	Venezuela
5 970	6 131	6 420	6 636	6 824	6 894	7 081	7 186	7 491	Other Latin America
479 132	486 817	510 974	526 111	556 823	583 021	619 782	657 212	688 297	**Latin America**

Total electricity production may be greater than the sum of production by energy source.

Total Production of Electricity (GWh)
Production totale d'électricité (GWh)

Gesamterzeugung von Elektrizität (GWh)

Produzione totale di energia elettrica (GWh)

総発電量(GWh)

Producción total de electricidad (GWh)

Общее производство электроэнергии (ГВт.ч)

	1971	1973	1978	1983	1984	1985	1986	1987	1988
Bangladesh	1 030	1 404	2 219	3 433	3 966	4 528	4 800	5 587	6 541
Brunei	150	200	361	682	750	766	776	1 043	1 098
Inde	66 384	72 796	110 130	150 994	169 205	183 390	201 281	218 986	241 314
Indonésie	3 045	3 624	5 534	13 392	14 777	16 899	19 455	22 285	25 606
RPD de Corée	16 910	20 000	32 000	41 000	45 000	48 000	50 000	50 200	53 000
Malaisie	3 795	4 783	8 241	12 733	13 721	14 554	15 682	16 986	18 735
Myanmar	691	821	1 243	1 675	1 890	2 119	2 245	2 320	2 226
Népal	86	104	174	316	336	362	465	557	569
Pakistan	7 572	8 377	12 375	19 697	21 873	23 003	25 589	28 703	33 091
Philippines	9 145	13 186	15 542	21 454	21 180	22 766	21 797	22 642	24 539
Singapour	2 585	3 719	5 898	8 704	9 490	9 960	10 640	11 909	13 113
Sri Lanka	900	1 031	1 385	2 114	2 261	2 464	2 652	2 707	2 799
Taipei chinois	15 811	20 735	35 846	48 294	52 213	55 554	62 329	69 176	76 258
Thailande	5 083	6 971	12 637	18 857	21 024	23 074	24 717	28 652	32 464
Viêt-Nam	2 300	2 350	3 600	3 986	4 778	5 069	5 527	6 051	6 785
Autre Asie	2 256	3 500	3 751	3 836	4 163	4 466	5 044	5 978	6 375
Asie	137 743	163 601	250 936	351 167	386 627	416 974	452 999	493 782	544 513
Rép. populaire de Chine	138 400	166 800	256 552	351 440	377 390	410 690	449 530	497 270	545 210
Hong-Kong, Chine	5 574	6 799	10 370	16 467	17 918	19 230	21 406	23 746	25 501
Chine	143 974	173 599	266 922	367 907	395 308	429 920	470 936	521 016	570 711
Albanie	1 225	1 702	3 100	3 780	3 717	3 147	5 106	4 395	3 984
Bulgarie	21 016	21 956	31 492	42 648	44 672	41 632	41 820	43 473	45 021
Chypre	665	830	919	1 221	1 250	1 319	1 423	1 501	1 647
Gibraltar	47	49	52	61	61	63	65	70	74
Malte	305	365	459	675	700	784	850	944	1 030
Roumanie	39 454	46 779	64 255	70 260	71 667	71 818	75 478	74 077	75 322
République slovaque	10 865	12 299	15 432	19 623	20 362	22 506	24 169	23 629	23 040
Bosnie-Herzegovine	-	-	-	-	-	-	-	-	-
Croatie	-	-	-	-	-	-	-	-	-
Ex-RYM	-	-	-	-	-	-	-	-	-
Slovénie	-	-	-	-	-	-	-	-	-
RF de Yougoslavie	-	-	-	-	-	-	-	-	-
Ex-Yougoslavie	29 509	35 062	51 250	66 571	73 008	74 802	77 916	80 792	83 651
Europe non-OCDE	103 086	119 042	166 959	204 839	215 437	216 071	226 827	228 881	233 769
Arménie	-	-	-	-	-	-	-	-	-
Azerbaïdjan	-	-	-	-	-	-	-	-	-
Bélarus	-	-	-	-	-	-	-	-	-
Estonie	-	-	-	-	-	-	-	-	-
Géorgie	-	-	-	-	-	-	-	-	-
Kazakhstan	-	-	-	-	-	-	-	-	-
Kirghizistan	-	-	-	-	-	-	-	-	-
Lettonie	-	-	-	-	-	-	-	-	-
Lituanie	-	-	-	-	-	-	-	-	-
Moldova	-	-	-	-	-	-	-	-	-
Russie	-	-	-	-	-	-	-	-	-
Tadjikistan	-	-	-	-	-	-	-	-	-
Turkménistan	-	-	-	-	-	-	-	-	-
Ukraine	-	-	-	-	-	-	-	-	-
Ouzbékistan	-	-	-	-	-	-	-	-	-
Ex-URSS	800 400	914 600	1 201 900	1 418 000	1 492 000	1 544 000	1 599 000	1 664 900	1 705 000

La production totale d'électricité peut être supérieure à la somme des productions par source d'énergie.

Total Production of Electricity (GWh)

Production totale d'électricité (GWh)

Gesamterzeugung von Elektrizität (GWh)

Produzione totale di energia elettrica (GWh)

総発電量(GWh)

Producción total de electricidad (GWh)

Общее производство электроэнергии (ГВт.ч)

1989	1990	1991	1992	1993	1994	1995	1996	1997	
7 115	7 732	8 270	8 894	9 206	9 784	10 806	11 474	11 858	Bangladesh
1 131	1 172	1 269	1 408	1 547	1 663	1 966	2 123	2 407	Brunei
268 664	289 439	315 631	332 713	356 335	385 557	417 836	435 075	463 402	India
29 554	36 117	38 498	42 445	46 639	52 269	55 822	67 062	74 832	Indonesia
53 500	53 500	53 500	38 000	38 000	37 000	36 000	35 036	33 985	DPR of Korea
20 799	23 012	26 535	29 314	34 733	39 093	45 453	51 407	57 875	Malaysia
2 494	2 478	2 677	2 996	3 385	3 594	4 055	3 945	4 205	Myanmar
564	713	873	927	881	927	1 004	1 191	1 226	Nepal
34 601	37 673	41 076	45 465	48 810	50 646	53 555	56 956	59 125	Pakistan
25 573	25 245	25 654	25 869	26 777	30 465	33 531	36 663	39 816	Philippines
14 136	15 714	16 921	17 674	18 962	20 849	22 244	24 100	26 898	Singapore
2 858	3 150	3 377	3 540	3 979	4 387	4 802	4 530	5 145	Sri Lanka
84 055	90 479	99 687	106 558	116 471	125 930	135 277	135 277	153 294	Chinese Taipei
37 406	44 175	50 186	57 098	63 406	71 177	80 061	87 467	93 253	Thailand
7 784	8 722	9 300	9 691	10 659	12 270	13 677	16 945	19 151	Vietnam
6 477	6 520	6 467	6 233	6 251	6 470	6 869	6 789	6 958	Other Asia
596 711	645 841	699 921	728 825	786 041	852 081	922 958	976 040	1 053 430	**Asia**
584 810	621 200	677 550	754 195	838 256	928 083	1 007 726	1 080 017	1 134 471	People's Rep. of China
27 361	28 938	31 807	34 907	35 950	26 743	27 916	28 442	28 945	Hong Kong, China
612 171	650 138	709 357	789 102	874 206	954 826	1 035 642	1 108 459	1 163 416	**China**
4 123	3 189	3 686	3 357	3 484	3 904	4 414	5 926	5 600	Albania
44 328	42 141	38 917	35 610	37 997	38 133	41 789	42 716	42 803	Bulgaria
1 831	1 974	2 077	2 404	2 581	2 681	2 473	2 592	2 711	Cyprus
81	82	82	88	88	88	88	93	93	Gibraltar
1 100	1 100	1 419	1 490	1 500	1 541	1 632	1 658	1 686	Malta
75 851	64 309	56 803	54 195	55 476	55 136	59 266	61 350	57 148	Romania
24 069	24 067	22 746	22 347	23 417	24 740	26 306	25 290	24 547	Slovak Republic
-	-	-	12 200	11 700	1 921	2 203	2 203	2 203	*Bosnia-Herzegovina*
-	-	-	8 894	9 359	8 275	8 863	10 548	9 685	*Croatia*
-	-	-	6 065	5 180	5 511	6 114	6 489	6 719	*FYROM*
-	-	-	12 086	11 692	12 630	12 649	12 770	13 166	*Slovenia*
-	-	-	36 488	33 457	34 528	37 176	38 093	40 312	*FR of Yugoslavia*
83 468	82 905	81 885	75 733	71 388	62 865	67 005	70 103	72 085	Former Yugoslavia
234 851	219 767	207 615	195 224	195 931	189 088	202 973	209 728	206 673	**Non-OECD Europe**
-	-	-	9 004	6 295	5 658	5 561	6 214	6 021	Armenia
-	-	-	19 673	19 100	17 571	17 044	17 088	16 800	Azerbaijan
-	-	-	37 595	33 369	31 397	24 918	23 728	26 057	Belarus
-	-	-	11 831	9 118	9 151	8 693	9 103	9 218	Estonia
-	-	-	11 520	10 150	6 803	6 900	7 226	7 172	Georgia
-	-	-	82 701	77 444	66 397	66 659	58 657	52 000	Kazakhstan
-	-	-	11 892	11 273	12 932	12 349	13 758	12 600	Kyrgyzstan
-	-	-	3 834	3 924	4 440	3 979	3 124	4 501	Latvia
-	-	-	18 707	14 122	10 021	13 898	16 789	14 861	Lithuania
-	-	-	11 248	10 265	8 228	6 068	6 122	5 273	Moldova
-	-	-	1 008 450	956 587	875 914	860 027	847 211	834 100	Russia
-	-	-	16 822	17 741	16 982	14 768	15 000	14 000	Tajikistan
-	-	-	13 183	12 637	10 496	9 800	10 100	9 400	Turkmenistan
-	-	-	252 524	229 906	202 922	194 018	182 986	178 002	Ukraine
-	-	-	50 911	49 149	47 800	47 453	45 419	46 054	Uzbekistan
1 722 000	1 727 000	1 681 083	1 559 895	1 461 080	1 326 712	1 292 135	1 262 525	1 236 059	**Former USSR**

Total electricity production may be greater than the sum of production by energy source.

Total Production of Electricity (GWh)

Production totale d'électricité (GWh)

Gesamterzeugung von Elektrizität (GWh)

Produzione totale di energia elettrica (GWh)

総発電量(GWh)

Producción total de electricidad (GWh)

Общее производство электроэнергии (ГВт.ч)

	1971	1973	1978	1983	1984	1985	1986	1987	1988
Bahrein	430	500	1 212	2 216	2 417	2 937	2 970	3 316	3 320
Iran	7 900	11 540	19 446	33 009	36 594	39 220	41 571	46 197	47 600
Irak	2 800	3 519	7 835	16 185	19 459	20 973	22 287	22 510	23 450
Israël	7 639	8 720	11 874	14 578	14 909	15 701	16 277	17 491	19 215
Jordanie	230	315	704	1 918	2 265	2 495	2 955	3 486	3 263
Koweit	3 087	4 138	7 446	12 830	14 196	15 689	17 216	18 092	19 598
Liban	1 375	1 791	2 300	3 600	3 800	3 861	4 170	4 600	4 000
Oman	13	47	468	1 615	2 016	2 498	3 180	3 392	3 773
Qatar	316	419	1 349	3 235	3 563	3 949	4 303	4 371	4 501
Arabie saoudite	2 082	2 949	10 551	35 190	40 069	44 311	47 646	50 649	57 229
Syrie	1 345	1 423	2 871	7 116	7 310	7 898	7 942	7 989	9 614
Emirats arabes unis	203	720	3 764	10 141	11 000	12 000	12 814	13 657	14 840
Yémen	209	206	360	752	856	895	1 117	1 165	1 318
Moyen-Orient	27 629	36 287	70 180	142 385	158 454	172 427	184 448	196 915	211 721
Total non-OCDE	1 436 756	1 677 837	2 355 948	3 054 003	3 266 549	3 438 968	3 632 033	3 838 331	4 033 329
OCDE Amérique du N.	1 956 385	2 272 807	2 716 459	2 938 819	3 086 786	3 173 903	3 205 400	3 332 866	3 490 617
OCDE Pacifique	464 920	568 443	702 511	799 804	840 364	878 816	896 187	955 068	1 008 239
OCDE Europe	1 399 283	1 613 706	1 928 309	2 134 983	2 233 706	2 338 395	2 388 434	2 475 983	2 536 362
Total OCDE	3 820 588	4 454 956	5 347 279	5 873 606	6 160 856	6 391 114	6 490 021	6 763 917	7 035 218
Monde	5 257 344	6 132 793	7 703 227	8 927 609	9 427 405	9 830 082	10 122 054	10 602 248	11 068 547

La production totale d'électricité peut être supérieure à la somme des productions par source d'énergie.

Total Production of Electricity (GWh)

Production totale d'électricité (GWh)

Gesamterzeugung von Elektrizität (GWh)

Produzione totale di energia elettrica (GWh)

総発電量(GWh)

Producción total de electricidad (GWh)

Общее производство электроэнергии (ГВт.ч)

1989	1990	1991	1992	1993	1994	1995	1996	1997	
3 400	3 490	3 495	3 896	4 245	4 550	4 750	4 771	4 924	Bahrain
52 712	59 102	64 126	68 419	76 014	82 019	84 969	90 851	95 794	Iran
23 860	24 000	20 810	25 300	26 300	28 000	29 000	29 000	29 561	Iraq
20 297	20 722	21 470	24 641	25 957	28 273	30 342	32 497	35 098	Israel
3 433	3 638	3 724	4 422	4 761	5 076	5 616	6 058	6 273	Jordan
21 179	20 610	10 780	16 786	20 178	22 798	23 726	25 475	27 091	Kuwait
2 500	1 500	3 000	3 771	4 720	5 184	5 281	6 965	8 515	Lebanon
3 927	4 501	4 628	5 117	5 832	6 197	6 460	6 802	7 318	Oman
4 624	4 818	4 653	5 153	5 525	5 814	5 976	6 575	6 868	Qatar
61 568	64 899	69 212	74 009	82 183	91 019	93 898	97 819	103 801	Saudi Arabia
10 455	11 611	12 179	12 562	12 638	15 182	15 300	16 885	17 950	Syria
15 612	17 081	17 222	17 460	17 578	18 870	19 070	19 737	20 571	United Arab Emirates
1 345	1 663	1 802	1 953	2 051	2 159	2 369	2 334	2 358	Yemen
224 912	237 635	237 101	263 489	287 982	315 141	326 757	345 769	366 122	**Middle East**
4 184 371	4 289 658	4 378 557	4 397 456	4 506 938	4 580 626	4 770 236	4 944 706	5 115 805	**Non-OECD Total**
3 762 636	3 802 107	3 908 120	3 943 126	4 075 179	4 177 873	4 297 102	4 416 690	4 449 141	OECD North America
1 073 587	1 152 299	1 196 945	1 217 849	1 248 762	1 350 473	1 402 345	1 450 240	1 506 555	OECD Pacific
2 595 900	2 624 586	2 675 881	2 700 312	2 715 322	2 758 771	2 839 720	2 915 913	2 949 839	OECD Europe
7 432 123	7 578 992	7 780 946	7 861 287	8 039 263	8 287 117	8 539 167	8 782 843	8 905 535	**OECD Total**
11 616 494	11 868 650	12 159 503	12 258 743	12 546 201	12 867 743	13 309 403	13 727 549	14 021 340	**World**

Total electricity production may be greater than the sum of production by energy source.

Production of Heat (TJ)
Production de chaleur (TJ)
Erzeugung von Wärme (TJ)
Produzione di calore (TJ)
総発熱量 (TJ)
Producción de calor (TJ)
Производство теплоэнергии (ТДж)

	1971	1973	1978	1983	1984	1985	1986	1987	1988
Rép. populaire de Chine	-	-	-	334 170	354 478	382 277	413 204	463 720	477 142
Chine	-	-	-	334 170	354 478	382 277	413 204	463 720	477 142
Albanie	-	-	-	-	-	-	-	-	-
Bulgarie	28 043	33 065	93 415	77 835	73 399	64 500	69 100	91 100	63 300
Roumanie	150 000	150 000	170 000	200 000	200 000	213 916	228 201	238 066	263 002
République slovaque	30 623	31 899	35 300	26 847	28 151	30 069	30 088	31 343	31 757
Bosnie-Herzegovine	-	-	-	-	-	-	-	-	-
Croatie	-	-	-	-	-	-	-	-	-
Ex-RYM	-	-	-	-	-	-	-	-	-
Slovénie	-	-	-	-	-	-	-	-	-
RF de Yougoslavie	-	-	-	-	-	-	-	-	-
Ex-Yougoslavie	-	-	-	112 858	101 490	88 502	93 777	88 460	78 312
Europe non-OCDE	208 666	214 964	298 715	417 540	403 040	396 987	421 166	448 969	436 371
Arménie	-	-	-	-	-	-	-	-	-
Azerbaïdjan	-	-	-	-	-	-	-	-	-
Bélarus	-	-	-	-	-	-	-	-	-
Estonie	-	-	-	-	-	-	-	-	-
Géorgie	-	-	-	-	-	-	-	-	-
Kazakhstan	-	-	-	-	-	-	-	-	-
Kirghizistan	-	-	-	-	-	-	-	-	-
Lettonie	-	-	-	-	-	-	-	-	-
Lituanie	-	-	-	-	-	-	-	-	-
Moldova	-	-	-	-	-	-	-	-	-
Russie	-	-	-	-	-	-	-	-	-
Ukraine	-	-	-	-	-	-	-	-	-
Ouzbékistan	-	-	-	-	-	-	-	-	-
Ex-URSS	2 310 396	2 595 010	3 335 844	4 316 591	4 546 865	4 848 314	4 978 105	5 112 083	5 250 247
Total non-OCDE	2 519 062	2 809 974	3 634 559	5 068 301	5 304 383	5 627 578	5 812 475	6 024 772	6 163 760
OCDE Amérique du N.	-	4 003	17 011	133 956	132 339	117 844	114 003	116 937	95 866
OCDE Pacifique	-	1 160	8 264	8 192	8 217	9 222	9 275	9 641	9 629
OCDE Europe	862 026	969 847	1 396 139	1 594 722	1 653 620	1 791 130	1 772 577	1 828 199	1 732 312
Total OCDE	862 026	975 010	1 421 414	1 736 870	1 794 176	1 918 196	1 895 855	1 954 777	1 837 807
Monde	3 381 088	3 784 984	5 055 973	6 805 171	7 098 559	7 545 774	7 708 330	7 979 549	8 001 567

Production of Heat (TJ)

Production de chaleur (TJ)

Erzeugung von Wärme (TJ)

Produzione di calore (TJ)

総発熱量 (TJ)

Producción de calor (TJ)

Производство теплоэнергии (ТДж)

1989	1990	1991	1992	1993	1994	1995	1996	1997	
529 298	626 364	690 157	724 508	945 442	1 119 551	1 072 002	1 148 301	1 175 695	People's Rep. of China
529 298	626 364	690 157	724 508	945 442	1 119 551	1 072 002	1 148 301	1 175 695	**China**
-	4 605	4 600	4 600	4 600	1 443	1 189	1 189	1 189	Albania
200 000	210 056	183 725	147 733	136 317	126 671	133 463	140 120	123 844	Bulgaria
269 994	258 112	202 056	501 307	509 537	279 432	286 999	296 264	284 612	Romania
32 023	35 880	38 111	36 997	32 142	31 500	35 676	41 198	36 339	Slovak Republic
-	-	-	10 000	10 000	428	966	966	966	*Bosnia-Herzegovina*
-	-	-	13 371	12 691	12 067	13 219	13 737	13 327	*Croatia*
-	-	-	4 293	8 636	6 623	6 208	6 208	6 208	*FYROM*
-	-	-	8 274	8 455	8 118	8 917	9 702	9 066	*Slovenia*
-	-	-	20 000	20 000	19 000	19 000	19 000	19 000	*FR of Yugoslavia*
73 743	75 000	65 000	55 938	59 782	46 236	48 310	49 613	48 567	Former Yugoslavia
575 760	583 653	493 492	746 575	742 378	485 282	505 637	528 384	494 551	**Non-OECD Europe**
-	-	-	11 179	3 926	2 883	3 285	3 762	3 762	Armenia
-	-	-	-	15 200	50 125	40 866	31 067	31 067	Azerbaijan
-	-	-	400 000	393 559	326 570	305 636	326 570	334 944	Belarus
-	-	-	63 256	48 352	46 758	30 625	33 132	32 593	Estonia
-	-	-	42 705	29 740	19 950	19 950	19 950	19 950	Georgia
-	-	-	300	300	284	256	256	256	Kazakhstan
-	-	-	20 000	6 378	4 454	18 556	19 162	19 162	Kyrgyzstan
-	-	-	60 074	54 891	45 871	42 616	54 925	46 540	Latvia
-	-	-	128 251	102 149	85 970	81 476	82 736	76 681	Lithuania
-	-	-	25 653	20 360	15 792	14 881	14 598	13 819	Moldova
-	-	-	8 700 000	9 466 604	8 631 400	8 052 800	6 709 300	6 038 400	Russia
-	-	-	1 014 527	858 882	795 495	663 814	626 043	609 015	Ukraine
-	-	-	115 974	111 655	112 033	108 692	107 874	100 100	Uzbekistan
5 283 742	5 191 632	4 867 551	10 581 919	11 111 996	10 137 585	9 383 453	8 029 375	7 326 289	**Former USSR**
6 388 800	6 401 649	6 051 200	12 053 002	12 799 816	11 742 418	10 961 092	9 706 060	8 996 535	**Non-OECD Total**
118 824	102 315	321 007	393 512	419 385	406 888	424 893	436 926	447 940	OECD North America
10 759	10 727	11 808	13 680	14 498	50 714	61 789	73 163	81 996	OECD Pacific
1 727 432	1 740 723	1 752 234	1 718 018	1 761 963	1 739 167	1 681 217	1 801 096	1 757 211	OECD Europe
1 857 015	1 853 765	2 085 049	2 125 210	2 195 846	2 196 769	2 167 899	2 311 185	2 287 147	**OECD Total**
8 245 815	8 255 414	8 136 249	14 178 212	14 995 662	13 939 187	13 128 991	12 017 245	11 283 682	**World**

Refinery Output of Petroleum Products (1000 tonnes)

Production de produits pétroliers en raffineries (1000 tonnes)

Raffinerieausstoß von Ölprodukten (1000 Tonnen)

Produzione di prodotti petroliferi nelle raffinerie (1000 tonnellate)

石油製品の生産量（千トン）

Producción de productos petrolíferos en refinerías (1000 toneladas)

Производство нефтепродуктов нефтеперерабатующими заводами (тыс. т)

	1971	1973	1978	1983	1984	1985	1986	1987	1988
Algérie	2 416	6 127	5 441	19 337	20 976	20 316	21 750	20 845	22 040
Angola	661	725	960	1 105	1 147	1 435	1 332	1 325	1 346
Cameroun	-	-	-	1 224	1 257	1 139	1 138	1 150	1 203
Congo	-	-	-	624	513	467	513	509	501
RD du Congo	669	719	200	460	416	320	189	305	351
Egypte	5 003	6 851	11 753	18 163	19 597	20 119	20 921	22 050	22 238
Ethiopie	664	616	586	738	715	711	737	759	658
Gabon	1 002	1 058	1 688	988	924	741	733	584	628
Ghana	870	992	1 153	619	764	918	891	829	909
Côte d'Ivoire	808	1 154	1 618	1 849	1 864	1 941	1 866	1 836	1 870
Kenya	2 546	2 658	2 594	1 998	2 063	1 988	1 931	2 118	2 031
Libye	435	1 608	5 217	5 602	6 558	10 379	10 948	12 341	13 723
Maroc	1 485	2 224	2 969	4 484	4 625	4 752	4 487	4 515	5 051
Mozambique	785	739	599	286	120	-	-	-	-
Nigéria	1 949	2 762	2 633	7 680	8 401	8 140	7 234	9 106	10 612
Sénégal	553	670	762	285	349	303	477	524	711
Afrique du Sud	12 668	13 106	13 147	14 266	14 912	14 469	14 160	14 786	16 052
Soudan	682	1 145	980	709	587	604	778	462	772
Tanzanie	755	785	547	530	534	569	565	606	606
Tunisie	1 117	1 036	1 181	1 496	1 588	1 563	1 577	1 657	1 691
Zambie	-	403	777	693	624	580	548	573	579
Autre Afrique	880	854	611	738	494	719	645	1 752	1 764
Afrique	35 948	46 232	55 416	83 874	89 028	92 173	93 420	98 632	105 336
Argentine	23 096	23 491	24 215	23 817	23 275	22 652	21 183	20 855	21 011
Bolivie	647	729	1 121	1 010	988	1 006	937	988	1 034
Brésil	24 879	35 490	49 614	47 260	50 095	50 478	54 357	55 443	52 953
Chili	4 655	4 628	4 669	3 801	3 768	3 739	4 025	4 220	4 775
Colombie	7 679	8 121	7 986	9 351	9 214	9 239	10 781	10 899	11 872
Costa Rica	400	382	410	314	400	398	614	606	596
Cuba	4 402	5 436	6 565	7 015	6 639	6 829	6 765	7 067	7 862
République dominicaine	-	1 162	1 483	1 660	1 703	1 435	1 710	1 573	1 559
Equateur	1 350	1 570	4 094	3 741	4 407	4 326	4 870	4 159	5 856
El Salvador	442	584	695	568	596	634	632	691	670
Guatemala	788	919	812	582	679	667	485	621	620
Honduras	496	532	326	280	330	299	160	251	339
Jamaïque	1 617	1 761	983	985	566	879	706	596	713
Antilles néerlandaises	40 927	43 644	26 931	22 114	19 470	8 420	8 065	9 566	9 872
Nicaragua	475	627	606	519	408	481	496	519	514
Panama	3 701	3 313	2 314	1 608	1 435	1 216	918	1 345	945
Paraguay	176	226	293	206	147	193	207	232	274
Pérou	4 188	4 656	5 873	7 244	8 290	8 259	8 156	8 308	8 122
Trinité-et-Tobago	20 224	19 583	11 765	4 235	4 048	4 256	4 162	4 916	4 376
Uruguay	1 675	1 654	1 821	1 208	1 173	1 013	990	1 206	1 129
Vénézuela	64 366	67 167	51 313	44 361	43 504	45 137	44 411	41 705	47 109
Autre Amérique latine	9 908	12 948	8 556	8 189	6 520	1 562	747	777	775
Amérique latine	216 091	238 623	212 445	190 068	187 655	173 118	175 377	176 543	182 976

Refinery Output of Petroleum Products (1000 tonnes)

Production de produits pétroliers en raffineries (1000 tonnes)

Raffinerieausstoß von Ölprodukten (1000 Tonnen)

Produzione di prodotti petroliferi nelle raffinerie (1000 tonnellate)

石油製品の生産量（千トン）

Producción de productos petrolíferos en refinerías (1000 toneladas)

Производство нефтепродуктов нефтеперерабатующими заводами (тыс. т)

1989	1990	1991	1992	1993	1994	1995	1996	1997	
21 187	21 374	21 436	21 501	21 617	19 673	20 295	19 251	20 582	Algeria
1 367	1 387	1 383	1 374	1 371	1 363	1 351	1 524	1 880	Angola
1 213	1 199	885	821	812	853	851	848	873	Cameroon
615	628	645	605	650	634	614	625	625	Congo
334	349	360	333	364	372	342	342	342	DR of Congo
22 482	23 648	23 997	23 600	24 901	25 919	26 667	27 615	27 921	Egypt
662	655	680	757	750	756	758	765	426	Ethiopia
799	419	363	625	655	666	681	721	717	Gabon
982	817	1 041	1 089	872	1 145	1 154	1 161	1 161	Ghana
1 813	1 860	1 838	1 642	1 910	2 001	2 017	2 135	2 312	Ivory Coast
2 148	2 198	2 053	2 217	2 083	2 036	1 809	1 733	1 623	Kenya
13 258	13 249	13 396	13 463	12 996	14 230	14 522	14 730	14 861	Libya
5 430	5 559	5 417	5 976	6 031	6 410	6 028	5 405	5 712	Morocco
-	6	6	4	4	5	4	4	5	Mozambique
11 471	12 849	13 902	13 521	13 283	12 997	13 933	16 678	16 945	Nigeria
586	675	529	601	548	171	679	662	755	Senegal
16 264	17 551	17 691	17 838	18 857	20 076	22 059	21 196	18 657	South Africa
539	809	957	724	312	617	710	689	896	Sudan
570	584	590	589	589	592	598	599	599	Tanzania
1 699	1 788	1 759	1 670	1 695	1 669	1 800	1 815	2 009	Tunisia
568	533	546	510	547	543	554	562	562	Zambia
1 510	1 152	1 207	1 297	1 313	1 322	1 333	1 344	1 373	Other Africa
105 497	109 289	110 681	110 757	112 160	114 050	118 759	120 404	120 836	**Africa**
21 755	22 362	22 926	24 320	22 943	22 416	21 275	23 274	24 192	Argentina
1 098	1 159	1 251	1 245	1 252	1 363	1 474	1 551	1 608	Bolivia
53 409	53 122	51 134	52 989	53 291	56 672	55 611	59 366	63 986	Brazil
5 761	5 978	6 003	6 249	6 744	7 078	7 677	7 987	8 464	Chile
11 131	11 053	11 640	11 666	12 842	11 876	12 978	13 999	13 541	Colombia
641	442	351	530	538	568	738	624	613	Costa Rica
8 300	6 846	4 749	1 715	1 961	1 705	1 755	2 122	2 202	Cuba
1 781	1 509	1 800	1 938	1 884	1 971	2 038	2 106	2 201	Dominican Republic
5 646	6 044	6 327	5 995	6 006	6 437	6 493	7 076	7 196	Ecuador
623	652	745	803	901	832	745	751	890	El Salvador
607	450	566	774	760	812	828	785	822	Guatemala
371	331	321	285	-	-	-	-	-	Honduras
849	1 005	861	1 186	702	737	999	1 043	1 093	Jamaica
9 649	10 004	9 782	11 592	12 269	12 354	12 380	12 404	12 404	Netherlands Antilles
578	622	622	671	650	674	585	610	767	Nicaragua
920	1 139	1 090	1 701	1 720	1 070	1 219	2 036	2 020	Panama
268	300	268	299	248	254	200	155	145	Paraguay
7 355	7 302	7 471	7 510	7 547	7 174	6 751	7 148	7 639	Peru
3 987	4 264	5 207	5 149	4 781	5 110	4 753	5 138	5 291	Trinidad and Tobago
1 109	1 142	1 195	1 112	368	-	1 275	1 580	1 362	Uruguay
46 168	48 715	50 348	48 050	51 638	52 813	49 502	49 259	53 528	Venezuela
851	957	965	947	985	975	1 002	998	1 022	Other Latin America
182 857	185 398	185 622	186 726	190 030	192 891	190 278	200 012	210 986	**Latin America**

Refinery Output of Petroleum Products (1000 tonnes)

Production de produits pétroliers en raffineries (1000 tonnes)

Raffinerieausstoß von Ölprodukten (1000 Tonnen)

Produzione di prodotti petroliferi nelle raffinerie (1000 tonnellate)

石油製品の生産量（千トン）

Producción de productos petrolíferos en refinerías (1000 toneladas)

Производство нефтепродуктов нефтеперерабатующими заводами (тыс. т)

	1971	1973	1978	1983	1984	1985	1986	1987	1988
Bangladesh	732	609	1 037	941	1 130	1 026	984	1 050	1 063
Brunei	-	-	-	41	252	241	240	243	258
Inde	19 290	20 215	25 683	34 177	34 968	40 558	44 506	47 033	46 799
Indonésie	8 989	10 130	13 923	12 324	21 461	22 956	25 888	30 686	32 328
RPD de Corée	-	-	1 220	2 223	2 387	2 552	2 715	2 715	2 715
Malaisie	3 746	3 858	5 740	6 523	7 278	7 039	7 286	7 363	7 907
Myanmar	1 262	974	1 202	1 162	1 214	1 173	1 200	796	717
Pakistan	3 291	3 360	3 965	4 721	4 824	5 201	5 452	5 646	5 925
Philippines	8 406	8 645	9 443	8 837	7 861	6 449	7 186	8 249	9 284
Singapour	16 489	22 799	27 101	37 316	34 213	30 241	32 847	33 824	32 014
Sri Lanka	1 523	1 733	1 456	1 512	1 761	1 603	1 646	1 682	1 743
Taipei chinois	5 477	9 004	17 103	18 681	19 784	19 935	19 894	21 623	22 718
Thailande	4 889	6 956	7 906	7 807	7 620	7 711	8 265	8 331	8 741
Viêt-Nam	-	-	-	-	-	-	-	1	42
Autre Asie	-	-	-	-	-	-	-	-	-
Asie	74 094	88 283	115 779	136 265	144 753	146 685	158 109	169 242	172 254
Rép. populaire de Chine	32 300	41 600	67 161	80 218	82 315	85 685	92 316	97 495	101 645
Chine	32 300	41 600	67 161	80 218	82 315	85 685	92 316	97 495	101 645
Albanie	1 477	1 646	2 145	1 180	1 260	1 111	1 134	1 130	1 074
Bulgarie	7 742	9 359	11 851	11 572	11 159	11 464	11 303	11 492	11 950
Chypre	-	663	431	551	552	442	539	612	717
Roumanie	16 178	17 986	25 927	22 743	23 546	23 724	25 524	28 639	28 810
République slovaque	5 361	6 043	7 411	6 769	6 637	6 614	6 661	6 825	6 585
Croatie	-	-	-	-	-	-	-	-	-
Ex-RYM	-	-	-	-	-	-	-	-	-
Slovénie	-	-	-	-	-	-	-	-	-
RF de Yougoslavie	-	-	-	-	-	-	-	-	-
Ex-Yougoslavie	8 007	9 029	14 301	13 668	12 927	12 714	14 666	15 266	16 890
Europe non-OCDE	38 765	44 726	62 066	56 483	56 081	56 069	59 827	63 964	66 026
Azerbaïdjan	-	-	-	-	-	-	-	-	-
Bélarus	-	-	-	-	-	-	-	-	-
Estonie	-	-	-	-	-	-	-	-	-
Géorgie	-	-	-	-	-	-	-	-	-
Kazakhstan	-	-	-	-	-	-	-	-	-
Kirghizistan	-	-	-	-	-	-	-	-	-
Lituanie	-	-	-	-	-	-	-	-	-
Russie	-	-	-	-	-	-	-	-	-
Tadjikistan	-	-	-	-	-	-	-	-	-
Turkménistan	-	-	-	-	-	-	-	-	-
Ukraine	-	-	-	-	-	-	-	-	-
Ouzbékistan	-	-	-	-	-	-	-	-	-
Ex-URSS	289 600	332 100	420 690	456 700	451 300	449 000	446 000	461 300	461 500

Refinery Output of Petroleum Products (1000 tonnes)

Production de produits pétroliers en raffineries (1000 tonnes)

Raffinerieausstoß von Ölprodukten (1000 Tonnen)

Produzione di prodotti petroliferi nelle raffinerie (1000 tonnellate)

石油製品の生産量（千トン）

Producción de productos petrolíferos en refinerías (1000 toneladas)

Производство нефтепродуктов нефтеперерабатующими заводами (тыс. т)

1989	1990	1991	1992	1993	1994	1995	1996	1997	
1 135	1 096	1 135	1 049	1 378	1 288	1 169	1 005	1 015	Bangladesh
273	285	315	378	389	403	460	450	476	Brunei
50 858	50 301	49 594	53 007	52 604	55 356	57 969	62 644	65 810	India
33 085	36 603	37 496	39 575	39 778	39 225	39 756	43 192	43 966	Indonesia
2 875	2 875	2 380	1 884	1 388	1 101	1 101	1 112	1 078	DPR of Korea
8 643	9 743	10 060	9 824	10 705	13 681	16 503	17 507	15 979	Malaysia
707	713	671	745	774	846	929	761	983	Myanmar
5 686	5 711	6 416	6 330	6 219	6 309	5 931	6 394	6 904	Pakistan
9 557	10 853	10 410	11 670	11 440	11 809	15 455	17 473	17 692	Philippines
33 781	40 264	42 538	44 063	47 737	47 612	46 629	50 343	48 111	Singapore
1 158	1 714	1 560	1 546	1 659	1 923	1 875	1 965	1 778	Sri Lanka
24 676	23 576	23 859	23 907	26 392	27 325	31 359	33 053	34 167	Chinese Taipei
10 705	11 259	11 674	14 431	16 638	19 291	22 064	30 078	34 816	Thailand
42	42	42	42	1	1	-	41	40	Vietnam
-	-	50	50	50	50	50	50	50	Other Asia
183 181	195 035	198 200	208 501	217 152	226 220	241 250	266 068	272 865	**Asia**
105 193	107 204	113 530	119 693	127 546	128 381	139 334	148 628	160 585	People's Rep. of China
105 193	107 204	113 530	119 693	127 546	128 381	139 334	148 628	160 585	**China**
1 067	1 055	1 007	753	631	554	456	434	350	Albania
12 001	8 003	4 263	2 420	5 388	6 595	7 255	6 722	5 830	Bulgaria
654	626	747	724	778	902	826	758	1 038	Cyprus
28 423	22 618	14 736	12 955	13 217	14 752	14 559	13 246	12 302	Romania
8 138	6 434	5 184	4 557	4 366	4 950	4 954	5 214	5 387	Slovak Republic
-	-	-	3 960	4 975	4 925	5 294	4 806	5 058	*Croatia*
-	-	-	884	902	164	111	675	385	*FYROM*
-	-	-	593	558	373	593	521	541	*Slovenia*
-	-	-	1 509	1 265	1 197	1 335	1 857	2 672	*FR of Yugoslavia*
16 047	14 859	12 259	6 946	7 700	6 659	7 333	7 859	8 656	Former Yugoslavia
66 330	53 595	38 196	28 355	32 080	34 412	35 383	34 233	33 563	**Non-OECD Europe**
-	-	-	11 279	10 409	9 161	8 888	7 956	7 522	Azerbaijan
-	-	-	20 120	13 735	11 847	12 262	11 200	11 072	Belarus
-	-	-	273	178	301	313	351	380	Estonia
-	-	-	714	300	201	41	13	24	Georgia
-	-	-	17 412	14 891	11 821	10 784	10 946	9 012	Kazakhstan
-	-	-	-	-	-	70	91	110	Kyrgyzstan
-	-	-	4 096	5 185	4 104	3 471	4 111	5 514	Lithuania
-	-	-	248 531	223 624	187 129	179 459	173 132	174 000	Russia
-	-	-	56	39	32	25	21	26	Tajikistan
-	-	-	5 650	3 813	3 499	3 200	3 424	4 136	Turkmenistan
-	-	-	36 667	22 282	19 362	16 550	13 143	12 848	Ukraine
-	-	-	6 454	6 211	5 813	6 607	6 571	6 767	Uzbekistan
456 000	430 230	418 260	351 252	300 667	253 270	241 670	230 959	231 411	**Former USSR**

Refinery Output of Petroleum Products (1000 tonnes)

Production de produits pétroliers en raffineries (1000 tonnes)

Raffinerieausstoß von Ölprodukten (1000 Tonnen)

Produzione di prodotti petroliferi nelle raffinerie (1000 tonnellate)

石油製品の生産量（千トン）

Producción de productos petrolíferos en refinerías (1000 toneladas)

Производство нефтепродуктов нефтеперерабатующими заводами (тыс. т)

	1971	1973	1978	1983	1984	1985	1986	1987	1988
Bahrein	12 640	12 042	12 540	8 930	10 098	9 141	12 233	12 061	12 222
Iran	27 437	29 111	36 567	27 594	33 198	35 368	35 714	34 908	35 903
Irak	3 578	4 035	9 015	15 418	17 516	18 634	19 473	20 169	21 491
Israël	5 207	6 113	7 392	7 297	6 746	6 590	6 844	7 458	7 046
Jordanie	554	671	1 370	2 416	2 456	2 410	2 236	2 386	2 321
Koweit	21 437	19 878	20 400	25 077	24 912	28 221	30 702	34 249	39 593
Liban	2 034	2 420	1 829	490	735	615	1 030	1 150	540
Oman	-	-	-	2 236	2 437	2 844	3 307	2 747	2 835
Qatar	21	20	304	432	732	1 249	1 580	1 744	2 084
Arabie saoudite	27 514	28 244	30 804	42 808	42 475	50 400	55 121	66 121	70 154
Syrie	2 112	1 963	4 090	9 183	10 159	10 041	10 090	11 132	11 511
Emirats arabes unis	-	-	570	5 785	6 930	6 880	7 206	7 291	7 666
Yémen	3 505	2 852	1 845	4 119	4 362	3 716	3 296	4 996	5 015
Moyen-Orient	106 039	107 349	126 726	151 785	162 756	176 109	188 832	206 412	218 381
Total non-OCDE	792 837	898 913	1 060 283	1 155 393	1 173 888	1 178 839	1 213 881	1 273 588	1 308 118
OCDE Amérique du N.	689 448	773 666	916 952	760 083	788 677	784 786	817 234	832 181	854 246
OCDE Pacifique	215 293	265 838	271 128	220 574	227 127	213 149	209 480	206 069	220 541
OCDE Europe	667 702	776 413	723 477	599 704	601 223	586 786	625 195	622 060	646 211
Total OCDE	1 572 443	1 815 917	1 911 557	1 580 361	1 617 027	1 584 721	1 651 909	1 660 310	1 720 998
Monde	2 365 280	2 714 830	2 971 840	2 735 754	2 790 915	2 763 560	2 865 790	2 933 898	3 029 116

Refinery Output of Petroleum Products (1000 tonnes)

Production de produits pétroliers en raffineries (1000 tonnes)

Raffinerieausstoß von Ölprodukten (1000 Tonnen)

Produzione di prodotti petroliferi nelle raffinerie (1000 tonnellate)

石油製品の生産量（千トン）

Producción de productos petrolíferos en refinerías (1000 toneladas)

Производство нефтепродуктов нефтеперерабатующими заводами (тыс. т)

1989	1990	1991	1992	1993	1994	1995	1996	1997	
12 274	12 498	12 606	12 981	12 331	12 128	12 378	11 139	11 660	Bahrain
39 804	43 168	44 822	46 143	52 595	54 583	56 010	56 188	55 526	Iran
22 406	19 865	13 365	17 520	20 101	22 120	21 673	21 216	21 732	Iraq
7 333	8 164	8 404	10 117	11 381	11 535	10 910	9 633	11 021	Israel
2 330	2 654	2 370	2 910	2 883	3 006	3 162	3 252	3 465	Jordan
46 288	24 185	2 837	16 791	20 395	33 689	37 553	35 755	40 261	Kuwait
-	92	530	429	-	-	-	-	-	Lebanon
2 946	3 360	3 189	2 923	2 405	3 632	3 638	3 647	3 247	Oman
2 437	2 892	2 574	2 690	2 424	2 819	2 646	2 919	2 663	Qatar
69 052	78 477	69 512	74 178	74 192	75 143	72 080	81 823	77 121	Saudi Arabia
11 179	11 317	11 676	11 587	11 416	11 630	11 704	12 114	12 175	Syria
7 904	8 222	8 484	8 276	8 348	9 053	9 919	10 453	11 764	United Arab Emirates
3 624	4 655	4 510	5 180	3 497	3 504	3 478	4 334	4 313	Yemen
227 577	219 549	184 879	211 725	221 968	242 842	245 151	252 473	254 948	**Middle East**
1 326 635	1 300 300	1 249 368	1 217 009	1 201 603	1 192 066	1 211 825	1 252 777	1 285 194	**Non-OECD Total**
867 803	879 961	871 359	879 565	900 679	909 305	910 450	927 960	949 641	OECD North America
234 273	251 622	278 350	302 564	314 545	328 470	339 795	350 341	377 879	OECD Pacific
648 789	655 382	664 024	682 868	694 998	702 866	700 577	720 507	734 279	OECD Europe
1 750 865	1 786 965	1 813 733	1 864 997	1 910 222	1 940 641	1 950 822	1 998 808	2 061 799	**OECD Total**
3 077 500	3 087 265	3 063 101	3 082 006	3 111 825	3 132 707	3 162 647	3 251 585	3 346 993	**World**

Net Imports of Coking Coal (1000 tonnes)

Importations nettes de charbon à coke (1000 tonnes)

Nettoimporte von Kokskohle (1000 Tonnen)

Importazioni nette di carbone siderurgico (1000 tonnellate)

原料炭の純輸入量（千トン）

Importaciones netas de carbón coquizable (1000 toneladas)

Чистый импорт коксующихся углей (тыс. т)

	1971	1973	1978	1983	1984	1985	1986	1987	1988
Algérie	-	-	7	1 189	1 320	1 379	1 427	1 705	1 596
Egypte	-	-	857	980	1 100	1 200	1 200	1 250	1 400
Afrique du Sud	-	-	- 2 700	- 3 306	- 4 803	- 5 142	- 5 867	- 4 755	- 5 016
Afrique	-	-	- 1 836	- 1 137	- 2 383	- 2 563	- 3 240	- 1 800	- 2 020
Argentine	-	-	814	479	539	764	1 157	1 254	1 383
Brésil	-	-	3 547	5 459	7 576	8 049	8 442	9 660	9 282
Chili	-	-	185	307	342	372	424	430	359
Colombie	-	-	- 100	- 200	- 200	- 200	- 300	- 400	- 500
Pérou	-	-	41	46	47	13	27	36	20
Amérique latine	-	-	4 487	6 091	8 304	8 998	9 750	10 980	10 544
Inde	-	-	220	539	350	2 030	2 100	2 670	3 500
RPD de Corée	-	-	442	2 000	2 200	2 500	2 500	2 500	2 500
Pakistan	-	-	16	520	491	716	852	918	853
Taipei chinois	-	-	1 386	2 663	2 460	2 562	2 771	2 984	4 328
Asie	-	-	2 064	5 722	5 501	7 808	8 223	9 072	11 181
Rép. populaire de Chine	-	-	- 300	- 2 300	- 2 400	- 2 300	- 2 200	- 3 400	- 4 150
Chine	-	-	- 300	- 2 300	- 2 400	- 2 300	- 2 200	- 3 400	- 4 150
Albanie	-	-	25	30	32	33	35	35	37
Bulgarie	-	-	1 921	1 814	1 694	1 553	1 651	1 651	1 400
Roumanie	-	-	3 600	3 800	4 000	4 000	3 800	4 700	4 500
République slovaque	-	-	2 180	2 063	2 512	2 530	2 548	2 621	2 793
Croatie	-	-	-	-	-	-	-	-	-
Ex-RYM	-	-	-	-	-	-	-	-	-
Ex-Yougoslavie	-	-	1 447	4 546	4 854	4 689	4 837	3 980	4 269
Europe non-OCDE	-	-	9 173	12 253	13 092	12 805	12 871	12 987	12 999
Ex-URSS	-	-	- 10 000	- 8 000	- 8 000	- 11 000	- 13 000	- 19 000	- 21 000
Iran	-	-	100	60	50	100	150	200	200
Moyen-Orient	-	-	100	60	50	100	150	200	200
Total non-OCDE	-	-	3 688	12 689	14 164	13 848	12 554	9 039	7 754
OCDE Amérique du N.	-	-	- 34 426	- 53 442	- 65 918	- 70 420	- 65 427	- 63 770	- 77 560
OCDE Pacifique	-	-	19 386	27 384	30 684	25 847	24 856	23 044	25 769
OCDE Europe	-	-	17 359	22 184	31 242	37 151	35 038	37 631	41 328
Total OCDE	-	-	2 319	- 3 874	- 3 992	- 7 422	- 5 533	- 3 095	- 10 463
Monde	-	-	6 007	8 815	10 172	6 426	7 021	5 944	- 2 709

Un chiffre négatif correspond à des exportations nettes.

La ligne Monde montre la divergence entre le total des exportations et des importations mondiales.

Les divergences dans les échanges de charbon s'expliquent principalement par des différences de classification.

Net Imports of Coking Coal (1000 tonnes)

Importations nettes de charbon à coke (1000 tonnes)

Nettoimporte von Kokskohle (1000 Tonnen)

Importazioni nette di carbone siderurgico (1000 tonnellate)

原料炭の純輸入量 (千トン)

Importaciones netas de carbón coquizable (1000 toneladas)

Чистый импорт коксующихся углей (тыс. т)

1989	1990	1991	1992	1993	1994	1995	1996	1997	
1 350	913	1 059	1 262	805	761	. 761	761	552	Algeria
1 448	1 339	1 206	1 154	1 500	1 852	1 540	1 540	1 540	Egypt
- 4 161	- 3 633	- 3 523	- 5 088	-	-	360	425	425	South Africa
- 1 363	- 1 381	- 1 258	- 2 672	2 305	2 613	2 661	2 726	2 517	**Africa**
1 121	1 081	883	992	966	1 238	753	1 049	881	Argentina
9 552	10 146	10 758	10 399	10 975	11 212	11 790	12 847	12 256	Brazil
434	477	602	594	665	659	704	713	685	Chile
- 300	- 100	- 100	- 300	- 800	- 800	- 800	- 800	- 800	Colombia
60	45	51	52	53	47	51	51	51	Peru
10 867	11 649	12 194	11 737	11 859	12 356	12 498	13 860	13 073	**Latin America**
4 210	5 610	5 820	6 060	6 800	7 870	8 470	9 050	9 010	India
2 500	2 500	2 500	2 500	2 500	2 500	2 500	2 525	2 449	DPR of Korea
895	900	971	985	994	1 094	1 096	1 080	960	Pakistan
4 921	4 237	4 350	3 998	4 125	4 161	4 580	4 112	6 300	Chinese Taipei
12 526	13 247	13 641	13 543	14 419	15 625	16 646	16 767	18 719	**Asia**
- 3 200	- 3 100	- 3 600	- 3 300	-	-	-	-	-	People's Rep. of China
- 3 200	- 3 100	- 3 600	- 3 300	-	-	-	-	-	**China**
37	62	60	-	-	-	-	-	-	Albania
1 208	1 100	933	1 143	1 258	1 527	1 734	1 438	1 683	Bulgaria
4 400	3 600	2 100	4 563	2 615	4 005	4 675	3 996	5 014	Romania
3 101	3 132	2 720	2 517	2 612	2 727	2 746	2 850	2 568	Slovak Republic
-	-	-	613	490	317	-	-	-	*Croatia*
-	-	-	73	44	47	54	49	49	*FYROM*
3 501	3 297	2 200	686	534	364	54	49	49	Former Yugoslavia
12 247	11 191	8 013	8 909	7 019	8 623	9 209	8 333	9 314	**Non-OECD Europe**
- 20 000	- 17 000	-	-	231	- 2 880	- 2 415	- 2 225	- 2 400	**Former USSR**
300	400	400	500	-	400	552	598	598	Iran
300	400	400	500	-	400	552	598	598	**Middle East**
11 377	15 006	29 390	28 717	35 833	36 737	39 151	40 059	41 821	**Non-OECD Total**
- 81 857	- 79 210	- 82 501	- 71 426	- 64 312	- 65 816	- 71 690	- 71 925	- 72 947	OECD North America
23 345	21 940	23 515	13 349	9 091	7 659	7 437	5 906	3 418	OECD Pacific
40 605	43 366	44 424	43 488	30 201	31 722	30 672	36 224	40 051	OECD Europe
- 17 907	- 13 904	- 14 562	- 14 589	- 25 020	- 26 435	- 33 581	- 29 795	- 29 478	**OECD Total**
- 6 530	1 102	14 828	14 128	10 813	10 302	5 570	10 264	12 343	**World**

A negative number shows net exports.
The row World shows the discrepancy between total world exports and imports.
Discrepancies in coal trade are mainly due to inconsistent classifications.

Net Imports of Other Bituminous Coal (1000 tonnes)

Importations nettes d'autres charbons bitumineux (1000 tonnes)

Nettoimporte von sonstiger bituminöser Kohle (1000 Tonnen)

Importazioni nette di altri carboni bituminosi (1000 tonnellate)

その他歴青炭の純輸入量 （千トン）

Importaciones netas de otro carbón bituminoso (1000 toneladas)

Чистый импорт других битуминозных углей (тыс. т)

	1971	1973	1978	1983	1984	1985	1986	1987	1988
RD du Congo	-	-	60	34	36	35	41	43	43
Egypte	-	-	-	-	-	-	-	-	-
Ghana	-	-	2	2	2	2	2	3	3
Kenya	-	-	50	32	83	90	85	92	113
Maroc	-	-	- 41	244	151	385	767	1 015	1 068
Mozambique	-	-	98	123	85	71	62	20	20
Nigéria	-	-	- 4	-	- 7	- 7	- 40	- 45	- 40
Afrique du Sud	-	-	- 12 689	- 27 914	- 34 054	- 42 488	- 40 318	- 36 945	- 39 154
Tunisie	-	-	27	28	24	21	12	16	29
Zambie	-	-	-	-	-	-	- 10	-	-
Zimbabwe	-	-	- 197	- 98	- 138	- 78	- 13	- 25	- 34
Autre Afrique	-	-	62	77	80	71	100	76	74
Afrique	-	-	- 12 632	- 27 472	- 33 738	- 41 898	- 39 312	- 35 750	- 37 878
Argentine	-	-	-	- 33	-	-	-	-	-
Brésil	-	-	-	-	-	-	- 107	-	-
Chili	-	-	-	-	-	-	35	3	-
Colombie	-	-	- 117	- 301	- 757	- 3 336	- 5 457	- 9 267	- 10 327
Costa Rica	-	-	1	1	1	1	1	1	1
Cuba	-	-	91	90	92	126	119	85	95
République dominicaine	-	-	-	-	-	224	28	232	97
Haiti	-	-	-	57	60	61	18	18	36
Jamaïque	-	-	-	-	-	-	-	-	8
Panama	-	-	-	3	9	60	39	39	16
Pérou	-	-	-	-	-	-	- 21	-	- 30
Uruguay	-	-	2	-	-	-	-	1	-
Vénézuela	-	-	- 21	-	-	-	-	-	- 1 007
Amérique latine	-	-	- 44	- 183	- 595	- 2 864	- 5 345	- 8 888	- 11 111
Bangladesh	-	-	319	163	62	98	148	233	240
Inde	-	-	- 270	- 80	- 120	- 210	- 160	130	-
Indonésie	-	-	- 29	- 423	- 882	- 1 066	330	708	324
RPD de Corée	-	-	- 27	- 50	- 50	- 50	- 50	- 50	- 50
Malaisie	-	-	14	345	365	489	353	434	688
Myanmar	-	-	210	180	175	180	180	40	40
Népal	-	-	16	42	100	80	53	85	89
Philippines	-	-	20	202	513	1 240	956	615	1 320
Sri Lanka	-	-	-	-	-	1	-	-	-
Taipei chinois	-	-	-	3 742	5 053	7 530	7 914	11 038	13 148
Thailande	-	-	52	172	227	212	183	223	304
Viêt-Nam	-	-	- 1 430	- 402	- 470	- 604	- 620	- 200	- 314
Autre Asie	-	-	68	116	125	194	191	176	195
Asie	-	-	- 1 057	4 007	5 098	8 094	9 478	13 432	15 984
Rép. populaire de Chine	-	-	- 380	- 2 120	- 2 065	- 3 163	- 2 687	- 4 834	- 9 803
Hong-Kong, Chine	-	-	8	3 418	4 462	5 523	6 392	8 009	9 267
Chine	-	-	- 372	1 298	2 397	2 360	3 705	3 175	- 536

Un chiffre négatif correspond à des exportations nettes.

Net Imports of Other Bituminous Coal (1000 tonnes)

Importations nettes d'autres charbons bitumineux (1000 tonnes)

Nettoimporte von sonstiger bituminöser Kohle (1000 Tonnen)

Importazioni nette di altri carboni bituminosi (1000 tonnellate)

その他歴青炭の純輸入量（千トン）

Importaciones netas de otro carbón bituminoso (1000 toneladas)

Чистый импорт других битуминозных углей (тыс. т)

1989	1990	1991	1992	1993	1994	1995	1996	1997	
45	45	46	42	42	43	43	43	43	DR of Congo
2	1	1	1	-	-	-	-	-	Egypt
3	3	3	3	3	3	3	3	3	Ghana
131	151	151	158	131	109	94	89	92	Kenya
1 304	1 226	1 398	1 189	1 276	1 473	1 931	2 720	2 740	Morocco
22	18	20	20	20	20	18	18	20	Mozambique
- 40	- 35	- 35	- 35	- 20	-	-	-	-	Nigeria
- 42 770	- 46 267	- 43 834	- 46 971	- 51 711	- 54 838	- 59 676	- 60 224	- 64 200	South Africa
30	15	12	15	14	-	-	-	-	Tunisia
-	- 5	- 18	- 15	- 19	- 15	- 7	- 7	- 7	Zambia
- 30	10	- 25	- 11	-	- 225	- 199	- 151	- 112	Zimbabwe
141	138	113	112	108	88	113	114	119	Other Africa
- 41 162	- 44 700	- 42 168	- 45 492	- 50 156	- 53 342	- 57 680	- 57 395	- 61 302	**Africa**
-	- 1	- 15	- 7	-	-	-	-	-	Argentina
-	-	-	-	-	107	-	-	-	Brazil
1 049	1 183	888	285	423	1 271	1 504	2 882	3 940	Chile
- 13 162	- 14 492	- 16 197	- 15 762	- 16 862	- 17 637	- 17 474	- 23 981	- 25 696	Colombia
1	1	-	-	-	-	-	-	-	Costa Rica
214	153	106	47	70	87	77	1	42	Cuba
10	9	57	139	120	104	111	128	139	Dominican Republic
6	12	31	26	-	-	-	-	-	Haiti
56	62	40	40	49	53	55	64	67	Jamaica
18	32	46	50	75	61	62	64	57	Panama
-	-	148	138	311	321	256	204	153	Peru
-	-	1	1	-	-	-	-	-	Uruguay
- 1 689	- 1 834	- 2 119	- 2 622	- 3 600	- 4 387	- 4 294	- 3 158	- 5 133	Venezuela
- 13 497	- 14 875	- 17 014	- 17 665	- 19 414	- 20 020	- 19 703	- 23 796	- 26 431	**Latin America**
250	563	180	169	63	-	-	-	-	Bangladesh
40	-	- 10	70	200	280	310	270	380	India
- 1 461	- 4 110	- 6 933	- 15 604	- 17 932	- 20 997	- 30 684	- 36 032	- 41 089	Indonesia
- 50	- 50	- 475	- 450	- 500	- 500	- 500	- 505	- 490	DPR of Korea
1 759	2 270	1 912	1 985	1 620	2 132	2 100	2 640	1 973	Malaysia
40	40	40	-	-	-	-	-	-	Myanmar
81	12	67	92	100	111	121	119	119	Nepal
987	1 364	1 356	676	725	1 049	1 032	2 984	4 248	Philippines
2	8	2	2	2	2	1	1	1	Sri Lanka
11 859	14 231	14 032	18 094	21 169	22 507	24 101	26 976	29 952	Chinese Taipei
392	250	436	471	968	1 417	2 308	3 750	3 074	Thailand
- 527	- 745	- 945	- 1 306	- 1 736	- 2 060	- 1 802	- 4 394	- 4 235	Vietnam
219	200	211	207	208	204	206	212	222	Other Asia
13 591	14 033	9 873	4 406	4 887	4 145	- 2 807	- 3 979	- 5 845	**Asia**
- 9 848	- 12 177	- 15 033	- 15 133	- 18 387	- 22 985	- 26 982	- 33 267	- 28 717	People's Rep. of China
9 927	8 928	9 635	10 229	11 830	8 451	9 109	6 769	5 711	Hong Kong, China
79	- 3 249	- 5 398	- 4 904	- 6 557	- 14 534	- 17 873	- 26 498	- 23 006	**China**

A negative number shows net exports.

Net Imports of Other Bituminous Coal (1000 tonnes)

Importations nettes d'autres charbons bitumineux (1000 tonnes)

Nettoimporte von sonstiger bituminöser Kohle (1000 Tonnen)

Importazioni nette di altri carboni bituminosi (1000 tonnellate)

その他歴青炭の純輸入量（千トン）

Importaciones netas de otro carbón bituminoso (1000 toneladas)

Чистый импорт других битуминозных углей (тыс. т)

	1971	1973	1978	1983	1984	1985	1986	1987	1988
Albanie	-	-	143	170	178	187	195	195	203
Bulgarie	-	-	4 280	5 050	5 219	6 501	5 317	5 536	5 074
Chypre	-	-	-	-	52	74	55	151	91
Malte	-	-	-	52	94	212	146	182	252
Roumanie	-	-	1 069	1 984	1 300	2 048	2 524	2 382	3 585
République slovaque	-	-	3 971	3 503	3 423	3 507	3 827	3 660	3 643
Croatie	-	-	-	-	-	-	-	-	-
Ex-RYM	-	-	-	-	-	-	-	-	-
Slovénie	-	-	-	-	-	-	-	-	-
RF de Yougoslavie	-	-	-	-	-	-	-	-	-
Ex-Yougoslavie	-	-	-	-	-	-	-	-	-
Europe non-OCDE	-	-	9 463	10 759	10 266	12 529	12 064	12 106	12 848
Ex-URSS	-	-	- 8 100	- 5 400	- 6 300	- 8 800	- 9 900	- 7 800	- 11 700
Iran	-	-	-	-	-	-	-	-	-
Israël	-	-	-	2 136	2 669	3 063	3 084	3 518	3 536
Liban	-	-	1	-	-	-	-	-	-
Syrie	-	-	1	3	2	-	-	-	-
Moyen-Orient	-	-	2	2 139	2 671	3 063	3 084	3 518	3 536
Total non-OCDE	-	-	- 12 740	- 14 852	- 20 201	- 27 516	- 26 226	- 20 207	- 28 857
OCDE Amérique du N.	-	-	667	- 17 458	- 13 354	- 24 066	- 22 459	- 19 267	- 20 968
OCDE Pacifique	-	-	- 1 677	2 939	2 648	2 872	- 4 278	- 7 525	- 3 509
OCDE Europe	-	-	- 176	22 106	26 430	42 877	44 553	45 603	40 696
Total OCDE	-	-	- 1 186	7 587	15 724	21 683	17 816	18 811	16 219
Monde	-	-	- 13 926	- 7 265	- 4 477	- 5 833	- 8 410	- 1 396	- 12 638

Un chiffre négatif correspond à des exportations nettes.

La ligne Monde montre la divergence entre le total des exportations et des importations mondiales.

Les divergences dans les échanges de charbon s'expliquent principalement par des différences de classification.

Net Imports of Other Bituminous Coal (1000 tonnes)

Importations nettes d'autres charbons bitumineux (1000 tonnes)

Nettoimporte von sonstiger bituminöser Kohle (1000 Tonnen)

Importazioni nette di altri carboni bituminosi (1000 tonnellate)

その他歴青炭の純輸入量（千トン）

Importaciones netas de otro carbón bituminoso (1000 toneladas)

Чистый импорт других битуминозных углей (тыс. т)

1989	1990	1991	1992	1993	1994	1995	1996	1997	
203	240	140	50	-	-	-	-	-	Albania
5 032	4 690	3 595	2 531	2 977	1 834	1 719	2 275	2 029	Bulgaria
102	97	97	26	33	30	26	17	26	Cyprus
303	300	243	205	300	210	37	-	-	Malta
1 408	1 381	871	1 223	51	30	30	9	235	Romania
3 629	2 734	2 440	2 428	2 512	2 441	2 182	2 650	2 668	Slovak Republic
-	-	-	146	143	60	106	51	122	*Croatia*
-	-	-	12	65	30	44	51	51	*FYROM*
-	-	-	14	20	23	20	22	19	*Slovenia*
-	-	-	1 000	60	50	60	60	52	*FR of Yugoslavia*
-	-	-	1 172	288	163	230	184	244	Former Yugoslavia
10 677	9 442	7 386	7 635	6 161	4 708	4 224	5 135	5 202	**Non-OECD Europe**
- 7 700	- 9 400	- 22 088	- 20 400	- 23 761	- 16 020	- 18 348	- 15 330	- 14 500	**Former USSR**
-	-	-	-	-	-	- 50	- 13	- 13	Iran
3 695	3 998	3 921	5 432	5 381	6 121	7 090	7 156	8 085	Israel
-	-	-	-	111	112	180	200	200	Lebanon
-	-	-	-	-	-	-	-	-	Syria
3 695	3 998	3 921	5 432	5 492	6 233	7 220	7 343	8 272	**Middle East**
- 34 317	- 44 751	- 65 488	- 70 988	- 83 348	- 88 830	- 104 967	- 114 520	- 117 610	**Non-OECD Total**
- 25 322	- 30 920	- 34 914	- 32 013	- 14 475	- 12 038	- 22 460	- 21 893	- 13 643	OECD North America
2 804	1 102	1 113	1 921	9 104	18 253	21 429	26 988	28 285	OECD Pacific
50 180	65 345	83 600	86 076	76 938	71 771	78 919	82 611	85 015	OECD Europe
27 662	35 527	49 799	55 984	71 567	77 986	77 888	87 706	99 657	**OECD Total**
- 6 655	- 9 224	- 15 689	- 15 004	- 11 781	- 10 844	- 27 079	- 26 814	- 17 953	**World**

A negative number shows net exports.

The row World shows the discrepancy between total world exports and imports.

Discrepancies in coal trade are mainly due to inconsistent classifications.

Net Imports of Sub-Bituminous Coal (1000 tonnes)

Importations nettes de charbons sous-bitumineux (1000 tonnes)

Nettoimporte von subbituminöser Kohle (1000 Tonnen)

Importazioni nette di carbone sub-bituminoso (1000 tonnellate)

亜歴青炭の純輸入量（千トン）

Importaciones netas de otro carbón sub-bituminoso (1000 toneladas)

Чистый импорт полубитуминозных углей (тыс. т)

	1971	1973	1978	1983	1984	1985	1986	1987	1988
Slovénie	-	-	-	-	-	-	-	-	-
Ex-Yougoslavie	-	-	-	-	-	-	-	-	-
Europe non-OCDE	-	-	-	-	-	-	-	-	-
Total non-OCDE	-	-	-	-	-	-	-	-	-
OCDE Amérique du N.	-	-	-	-	-	-	-	-	-
OCDE Pacifique	-	-	-	-	-	-	-	-	-
OCDE Europe	-	-	- 6 778	- 9 331	- 9 414	- 9 345	- 10 318	- 10 410	- 9 705
Total OCDE	-	-	- 6 778	- 9 331	- 9 414	- 9 345	- 10 318	- 10 410	- 9 705
Monde	-	-	- 6 778	- 9 331	- 9 414	- 9 345	- 10 318	- 10 410	- 9 705

Un chiffre négatif correspond à des exportations nettes.

La ligne Monde montre la divergence entre le total des exportations et des importations mondiales.

Les divergences dans les échanges de charbon s'expliquent principalement par des différences de classification.

Net Imports of Sub-Bituminous Coal (1000 tonnes)
Importations nettes de charbons sous-bitumineux (1000 tonnes)
Nettoimporte von subbituminöser Kohle (1000 Tonnen)

Importazioni nette di carbone sub-bituminoso (1000 tonnellate)

亜歴青炭の純輸入量 (千トン)

Importaciones netas de otro carbón sub-bituminoso (1000 toneladas)

Чистый импорт полубитуминозных углей (тыс. т)

1989	1990	1991	1992	1993	1994	1995	1996	1997	
-	-	-	118	297	228	329	385	299	*Slovenia*
-	-	-	118	297	228	329	385	299	Former Yugoslavia
-	-	-	118	297	228	329	385	299	**Non-OECD Europe**
-	-	-	118	297	228	329	385	299	**Non-OECD Total**
-	-	-	-	- 1 931	- 2 170	- 3 086	- 3 323	- 3 681	OECD North America
-	-	-	-	- 7	-	-	-	-	OECD Pacific
- 9 051	- 8 405	- 4 850	- 7 923	- 8 061	- 5 772	- 5 565	- 5 040	- 4 347	OECD Europe
- 9 051	- 8 405	- 4 850	- 7 923	- 9 999	- 7 942	- 8 651	- 8 363	- 8 028	**OECD Total**
- 9 051	- 8 405	- 4 850	- 7 805	- 9 702	- 7 714	- 8 322	- 7 978	- 7 729	**World**

A negative number shows net exports.

The row World shows the discrepancy between total world exports and imports.

Discrepancies in coal trade are mainly due to inconsistent classifications.

Net Imports of Lignite (1000 tonnes)
Importations nettes de lignite (1000 tonnes)
Nettoimporte von Braunkohle (1000 Tonnen)
Importazioni nette di lignite (1000 tonnellate)
亜炭の純輸入量（千トン）
Importaciones netas de lignito (1000 toneladas)
Чистый импорт лигнита (тыс. т)

	1971	1973	1978	1983	1984	1985	1986	1987	1988
Singapour	-	-	2	2	1	2	1	1	1
Asie	-	-	2	2	1	2	1	1	1
Albanie	-	-	-	-	-	-	-	-	-
Bulgarie	-	-	-	-	-	-	-	21	-
Roumanie	-	-	430	-	210	480	1 002	1 586	1 953
République slovaque	-	-	7 560	7 338	7 455	8 072	8 831	9 209	9 037
Croatie	-	-	-	-	-	-	-	-	-
Ex-RYM	-	-	-	-	-	-	-	-	-
Slovénie	-	-	-	-	-	-	-	-	-
Ex-Yougoslavie	-	-	- 376	- 1 033	- 123	- 390	- 351	- 33	- 47
Europe non-OCDE	-	-	7 614	6 305	7 542	8 162	9 482	10 783	10 943
Ex-URSS	-	-	-	15	15	15	15	15	-
Total non-OCDE	-	-	7 616	6 322	7 558	8 179	9 498	10 799	10 944
OCDE Amérique du N.	-	-	- 83	- 114	- 8	- 30	- 37	- 32	- 7
OCDE Europe	-	-	1 559	3 187	3 053	3 445	3 424	2 741	2 258
Total OCDE	-	-	1 476	3 073	3 045	3 415	3 387	2 709	2 251
Monde	-	-	9 092	9 395	10 603	11 594	12 885	13 508	13 195

Un chiffre négatif correspond à des exportations nettes.

La ligne Monde montre la divergence entre le total des exportations et des importations mondiales.

Les divergences dans les échanges de charbon s'expliquent principalement par des différences de classification.

Net Imports of Lignite (1000 tonnes)

Importations nettes de lignite (1000 tonnes)

Nettoimporte von Braunkohle (1000 Tonnen)

Importazioni nette di lignite (1000 tonnellate)

亜炭の純輸入量（千トン）

Importaciones netas de lignito (1000 toneladas)

Чистый импорт лигнита (тыс. т)

1989	1990	1991	1992	1993	1994	1995	1996	1997	
1	2	2	1	1	1	1	1	-	Singapore
1	2	2	1	1	1	1	1	-	**Asia**
-	- 228	- 100	- 100	- 8	-	-	-	-	Albania
-	-	-	-	-	-	-	-	-	Bulgaria
2 539	3 451	2 181	1 295	541	197	29	729	-	Romania
8 223	6 753	6 767	5 791	5 558	3 656	3 413	3 365	2 188	Slovak Republic
-	-	-	204	208	173	181	147	133	*Croatia*
-	-	-	130	17	-	19	19	-	*FYROM*
-	-	-	- 36	- 8	- 6	9	19	6	*Slovenia*
93	- 37	-	298	217	167	209	185	139	Former Yugoslavia
10 855	9 939	8 848	7 284	6 308	4 020	3 651	4 279	2 327	**Non-OECD Europe**
-	-	-	-	-	-	-	-	-	**Former USSR**
10 856	9 941	8 850	7 285	6 309	4 021	3 652	4 280	2 327	**Non-OECD Total**
- 10	- 79	- 55	- 54	- 27	- 50	2	- 9	2	OECD North America
2 416	2 606	3 926	3 656	3 269	2 831	2 524	2 854	3 048	OECD Europe
2 406	2 527	3 871	3 602	3 242	2 781	2 526	2 845	3 050	**OECD Total**
13 262	12 468	12 721	10 887	9 551	6 802	6 178	7 125	5 377	**World**

A negative number shows net exports.

The row World shows the discrepancy between total world exports and imports.

Discrepancies in coal trade are mainly due to inconsistent classifications.

Net Imports of Crude Oil, NGL, Refinery Feedstocks and Additives (1000 tonnes)
Import. nettes de pétr. brut, LGN, produits d'aliment. des raffin. et additifs (1000 tonnes)

Nettoimporte von Rohöl, Kondensaten, Raffinerie-Feedstocks und Additiven (1000 Tonnen)

Importazioni nette di petrolio grezzo, GNL, prodotti intermedi e additivi (1000 tonnellate)

原油、NGL等随伴物の純輸入量（千トン）

Import. netas de petróleo crudo, líquidos de gas natural, prod. de alim. de refinerias y aditivos (1000 toneladas)

Чистый имп. сырой нефти, газ. конденсатов, нефтезаводского сырья и присадок (тыс. т)

	1971	1973	1978	1983	1984	1985	1986	1987	1988
Algérie	- 34 300	- 44 856	- 49 088	- 26 749	- 26 610	- 27 769	- 26 641	- 27 405	- 26 742
Angola	- 4 747	- 7 323	- 5 500	- 7 422	- 8 657	- 9 931	- 12 440	- 15 600	- 20 741
Bénin	-	-	-	- 23	- 5	- 325	- 365	- 312	- 238
Cameroun	-	-	- 505	- 4 287	- 5 744	- 7 278	- 7 127	- 7 084	- 6 920
Congo	- 20	- 1 461	- 2 000	- 4 796	- 5 407	- 5 470	- 5 513	- 5 416	- 5 749
RD du Congo	679	732	- 930	- 807	- 955	- 1 352	- 1 430	- 1 268	- 1 115
Egypte	- 9 915	- 1 610	- 13 251	- 17 999	- 22 413	- 24 734	- 20 828	- 24 046	- 20 632
Ethiopie	665	624	612	733	731	670	815	812	732
Gabon	- 4 842	- 6 500	- 8 609	- 6 768	- 6 907	- 8 113	- 7 448	- 6 867	- 7 225
Ghana	900	1 069	1 146	406	809	888	841	883	835
Côte d'Ivoire	751	1 255	1 551	423	455	767	901	922	1 583
Kenya	2 541	2 696	2 369	1 985	1 874	1 981	2 006	2 131	2 042
Libye	- 132 369	- 106 278	- 91 617	- 46 392	- 45 458	- 39 331	- 38 834	- 35 276	- 37 946
Maroc	1 651	2 094	2 815	4 245	4 644	4 805	4 507	4 833	5 050
Mozambique	836	792	549	300	130	-	-	-	-
Nigéria	- 73 317	- 97 975	- 90 183	- 53 080	- 57 185	- 69 616	- 64 847	- 56 419	- 60 978
Sénégal	563	666	758	308	384	133	648	529	757
Afrique du Sud	13 220	13 665	14 965	12 486	13 423	13 750	13 250	12 756	11 223
Soudan	970	1 100	1 058	740	617	639	811	480	798
Tanzanie	944	879	566	560	565	610	570	550	553
Tunisie	- 2 622	- 2 644	- 3 567	- 3 557	- 3 821	- 3 746	- 3 575	- 3 641	- 3 270
Zambie	-	658	780	774	615	594	588	594	619
Autre Afrique	778	648	983	1 110	706	791	1 145	1 715	1 868
Afrique	- 237 634	- 241 769	- 237 098	- 147 810	- 158 209	- 172 037	- 162 966	- 157 129	- 165 496
Argentine	2 221	2 978	2 192	-	-	- 461	- 101	- 102	- 410
Bolivie	- 1 062	- 1 489	- 364	- 131	- 39	-	- 40	-	-
Brésil	19 804	34 587	45 078	36 064	32 174	26 990	29 758	30 619	31 420
Chili	3 308	3 141	3 686	1 881	1 968	2 057	2 596	2 675	4 074
Colombie	- 3 512	- 1 340	1 227	1 922	1 362	938	- 4 351	- 7 336	- 7 035
Costa Rica	417	420	463	345	387	417	635	629	629
Cuba	5 438	5 993	7 269	7 843	5 937	6 217	5 975	6 741	7 505
République dominicaine	-	1 227	1 571	1 707	1 778	1 710	1 804	1 896	1 803
Equateur	1 163	- 8 932	- 6 316	- 8 363	- 8 520	- 9 576	- 9 916	- 4 776	- 9 583
El Salvador	480	677	793	644	641	679	708	748	693
Guatemala	837	946	794	297	557	609	269	435	473
Honduras	557	607	423	321	449	326	241	334	440
Jamaïque	1 528	1 830	965	971	632	855	784	625	705
Antilles néerlandaises	39 868	43 441	27 525	23 347	19 800	8 350	8 050	9 315	9 858
Nicaragua	480	594	622	525	433	485	532	505	517
Panama	4 010	3 717	2 382	1 716	1 500	1 257	975	1 360	958
Paraguay	180	240	320	179	179	157	205	275	268
Pérou	1 381	1 729	- 1 388	- 1 132	- 1 000	- 1 264	- 650	121	936
Trinité-et-Tobago	13 623	10 913	383	- 3 806	- 4 255	- 4 217	- 3 958	- 3 192	- 3 439
Uruguay	1 773	1 747	2 132	1 207	1 224	1 065	1 337	1 126	1 118
Vénézuela	- 121 000	- 111 364	- 65 822	- 52 969	- 54 568	- 44 727	- 50 622	- 47 695	- 52 436
Autre Amérique latine	12 494	15 530	9 854	7 790	6 209	1 616	634	810	740
Amérique latine	- 16 012	7 192	33 789	20 358	6 848	- 6 517	- 15 135	- 4 887	- 10 766

Un chiffre négatif correspond à des exportations nettes.

Net Imports of Crude Oil, NGL, Refinery Feedstocks and Additives (1000 tonnes)

Import. nettes de pétr. brut, LGN, produits d'aliment. des raffin. et additifs (1000 tonnes)

Nettoimporte von Rohöl, Kondensaten, Raffinerie-Feedstocks und Additiven (1000 Tonnen)

Importazioni nette di petrolio grezzo, GNL, prodotti intermedi e additivi (1000 tonnellate)

原油、NGL 等随伴物の純輸入量 (千トン)

Import. netas de petróleo crudo, líquidos de gas natural, prod. de alim. de refinerias y aditivos (1000 toneladas)

Чистый имп. сырой нефти, газ. конденсатов, нефтезаводского сырья и присадок (тыс. т)

1989	1990	1991	1992	1993	1994	1995	1996	1997	
- 28 959	- 31 120	- 30 474	- 29 697	- 30 736	- 30 751	- 30 910	- 34 293	- 33 130	Algeria
- 21 077	- 22 000	- 23 250	- 22 140	- 22 583	- 22 488	- 23 822	- 33 024	- 32 983	Angola
- 191	- 206	- 197	- 136	- 154	- 129	- 95	- 75	- 64	Benin
- 6 990	- 7 171	- 6 706	- 6 184	- 6 003	- 5 591	- 5 072	- 5 104	- 5 361	Cameroon
- 6 701	- 7 425	- 7 315	- 7 938	- 9 002	- 8 621	- 7 806	- 8 996	- 11 720	Congo
- 1 136	- 1 035	- 997	- 900	- 902	- 965	- 960	- 960	- 895	DR of Congo
- 19 942	- 19 615	- 18 170	- 20 091	- 19 610	- 8 346	- 8 369	- 6 728	- 6 143	Egypt
735	700	740	775	745	748	755	758	463	Ethiopia
- 8 584	- 11 909	- 13 819	- 13 546	- 14 757	- 15 432	- 16 770	- 17 363	- 17 539	Gabon
914	817	988	942	690	1 089	1 095	1 098	1 098	Ghana
2 002	1 952	1 885	1 949	1 813	1 842	1 805	2 041	1 719	Ivory Coast
2 101	2 178	2 059	2 235	2 274	2 173	1 680	1 413	1 834	Kenya
- 46 000	- 54 300	- 58 815	- 56 926	- 53 450	- 54 198	- 53 138	- 53 909	- 54 691	Libya
5 472	5 681	5 139	6 455	6 352	6 625	6 377	5 336	6 015	Morocco
-	-	-	-	-	-	-	-	-	Mozambique
- 74 387	- 75 556	- 78 519	- 81 299	- 90 298	- 78 609	- 83 250	- 94 331	- 99 248	Nigeria
583	642	506	643	546	303	659	643	771	Senegal
10 000	12 378	10 828	11 813	11 705	13 820	15 604	13 972	12 616	South Africa
557	837	1 013	750	327	641	761	712	688	Sudan
553	557	574	502	578	580	588	589	589	Tanzania
- 3 200	- 2 839	- 3 275	- 3 648	- 3 016	- 2 512	- 2 607	- 2 376	- 1 865	Tunisia
609	557	557	563	568	568	578	588	588	Zambia
1 477	1 190	1 270	1 325	1 140	1 130	1 060	533	- 782	Other Africa
- 192 164	- 205 687	- 215 978	- 214 553	- 223 773	- 198 123	- 201 837	- 229 476	- 238 040	**Africa**
- 593	- 892	- 1 019	- 3 029	- 4 308	- 9 975	- 13 310	- 15 940	- 16 169	Argentina
-	-	-	- 17	- 62	- 67	-	-	-	Bolivia
29 200	28 427	26 186	26 390	25 713	27 958	24 563	28 490	29 401	Brazil
4 508	5 260	5 408	5 867	6 133	6 831	7 368	7 597	8 224	Chile
- 8 323	- 9 508	- 8 752	- 9 222	- 10 285	- 9 718	- 16 113	- 16 607	- 17 867	Colombia
660	445	349	552	566	548	720	622	626	Costa Rica
7 505	6 308	4 420	1 411	1 630	1 386	1 199	1 636	1 940	Cuba
1 888	1 585	1 873	1 913	1 954	1 995	2 080	2 110	2 214	Dominican Republic
- 9 105	- 8 851	- 8 894	- 10 626	- 11 386	- 12 401	- 12 667	- 12 054	- 13 376	Ecuador
663	697	811	865	991	817	801	785	804	El Salvador
462	385	515	543	399	601	360	76	- 135	Guatemala
452	425	408	419	-	-	-	-	-	Honduras
686	1 095	908	1 191	740	800	945	1 030	1 125	Jamaica
9 424	10 495	11 100	12 450	13 715	13 693	13 680	13 678	13 678	Netherlands Antilles
569	609	616	679	652	558	574	617	729	Nicaragua
962	1 219	1 168	1 769	1 749	1 170	1 203	2 078	2 017	Panama
317	307	288	285	213	236	195	158	155	Paraguay
880	723	1 624	1 389	1 148	1 068	923	934	1 692	Peru
- 3 580	- 2 975	- 1 679	- 1 379	- 1 184	- 1 947	- 2 009	- 1 300	- 1 112	Trinidad and Tobago
1 077	1 122	1 349	1 322	144	181	1 429	1 497	1 465	Uruguay
- 47 940	- 54 589	- 60 672	- 60 569	- 66 574	- 87 607	- 99 313	- 99 640	- 107 057	Venezuela
743	932	885	871	933	858	905	906	906	Other Latin America
- 9 545	- 16 781	- 23 108	- 26 926	- 37 119	- 63 015	- 86 467	- 83 327	- 90 740	**Latin America**

A negative number shows net exports.

Net Imports of Crude Oil, NGL, Refinery Feedstocks and Additives (1000 tonnes)

Import. nettes de pétr. brut, LGN, produits d'aliment. des raffin. et additifs (1000 tonnes)

Nettoimporte von Rohöl, Kondensaten, Raffinerie-Feedstocks und Additiven (1000 Tonnen)

Importazioni nette di petrolio grezzo, GNL, prodotti intermedi e additivi (1000 tonnellate)

原油、NGL等随伴物の純輸入量（千トン）

Import. netas de petróleo crudo, líquidos de gas natural, prod. de alim. de refinerias y aditivos (1000 toneladas)

Чистый имп. сырой нефти, газ. конденсатов, нефтезаводского сырья и присадок (тыс. т)

	1971	1973	1978	1983	1984	1985	1986	1987	1988
Bangladesh	850	719	1 064	1 030	1 104	992	983	1 000	1 212
Brunei	- 5 889	- 11 597	- 11 967	- 9 810	- 9 251	- 8 410	- 8 087	- 7 274	- 7 081
Inde	12 688	13 425	14 892	10 833	7 860	12 767	14 481	17 959	17 690
Indonésie	- 32 416	- 49 237	- 67 346	- 53 628	- 52 117	- 41 248	- 40 720	- 35 557	- 33 348
RPD de Corée	-	-	1 240	2 260	2 420	2 590	2 760	2 800	2 800
Malaisie	824	- 333	- 5 076	- 11 795	- 14 123	- 14 765	- 17 389	- 16 490	- 17 927
Myanmar	266	-	-	-	-	-	-	-	-
Pakistan	3 057	3 044	3 461	4 186	4 294	4 017	3 797	3 713	3 801
Philippines	9 027	9 266	10 022	8 221	9 821	6 725	6 922	8 301	9 769
Singapour	16 309	22 320	27 900	39 941	35 866	29 248	31 916	34 500	32 839
Sri Lanka	1 536	1 747	1 449	1 491	1 772	1 657	1 638	1 704	1 848
Taipei chinois	5 805	9 075	17 417	17 188	17 468	16 114	16 116	17 308	18 250
Thailande	5 413	7 643	8 203	7 728	6 685	6 348	6 685	7 423	6 833
Viêt-Nam	-	-	-	-	-	-	-	-	- 646
Autre Asie	-	-	-	-	-	-	-	-	-
Asie	17 470	6 072	1 259	17 645	11 799	16 035	19 102	35 387	36 040
Rép. populaire de Chine	- 213	- 1 834	- 10 623	- 14 824	- 22 047	- 29 780	- 28 042	- 26 725	- 25 190
Chine	- 213	- 1 834	- 10 623	- 14 824	- 22 047	- 29 780	- 28 042	- 26 725	- 25 190
Albanie	- 143	- 412	-	-	-	-	-	-	-
Bulgarie	7 547	9 652	12 644	12 650	12 650	12 400	13 300	12 700	13 100
Chypre	-	656	451	475	594	466	559	616	707
Roumanie	2 858	4 143	12 937	12 395	13 534	14 626	17 047	21 366	20 957
République slovaque	5 470	6 684	8 843	8 026	7 924	7 942	8 075	8 159	7 923
Croatie	-	-	-	-	-	-	-	-	-
Ex-RYM	-	-	-	-	-	-	-	-	-
Slovénie	-	-	-	-	-	-	-	-	-
RF de Yougoslavie	-	-	-	-	-	-	-	-	-
Ex-Yougoslavie	4 747	8 226	10 380	8 968	9 693	8 661	10 436	11 870	13 230
Europe non-OCDE	20 479	28 949	45 255	42 514	44 395	44 095	49 417	54 711	55 917
Azerbaïdjan	-	-	-	-	-	-	-	-	-
Bélarus	-	-	-	-	-	-	-	-	-
Géorgie	-	-	-	-	-	-	-	-	-
Kazakhstan	-	-	-	-	-	-	-	-	-
Kirghizistan	-	-	-	-	-	-	-	-	-
Lituanie	-	-	-	-	-	-	-	-	-
Russie	-	-	-	-	-	-	-	-	-
Tadjikistan	-	-	-	-	-	-	-	-	-
Turkménistan	-	-	-	-	-	-	-	-	-
Ukraine	-	-	-	-	-	-	-	-	-
Ouzbékistan	-	-	-	-	-	-	-	-	-
Ex-URSS	- 69 700	- 72 100	- 112 920	- 119 700	- 120 500	- 104 600	- 118 100	- 122 600	- 124 500

Un chiffre négatif correspond à des exportations nettes.

Ex-URSS: la somme des républiques peut être différente du total.

Net Imports of Crude Oil, NGL, Refinery Feedstocks and Additives (1000 tonnes)

Import. nettes de pétr. brut, LGN, produits d'aliment. des raffin. et additifs (1000 tonnes)

Nettoimporte von Rohöl, Kondensaten, Raffinerie-Feedstocks und Additiven (1000 Tonnen)

Importazioni nette di petrolio grezzo, GNL, prodotti intermedi e additivi (1000 tonnellate)

原油、NGL等随伴物の純輸入量（千トン）

Import. netas de petróleo crudo, líquidos de gas natural, prod. de alim. de refinerias y aditivos (1000 toneladas)

Чистый имп. сырой нефти, газ. конденсатов, нефтезаводского сырья и присадок (тыс. т)

1989	1990	1991	1992	1993	1994	1995	1996	1997	
1 253	1 008	1 078	1 018	1 129	1 236	909	752	763	Bangladesh
- 7 402	- 7 578	- 8 221	- 8 954	- 8 739	- 9 010	- 8 728	- 8 374	- 7 999	Brunei
18 609	20 154	21 141	29 115	29 445	27 590	26 831	33 499	31 773	India
- 35 763	- 33 748	- 36 121	- 32 109	- 30 030	- 34 258	- 31 448	- 29 107	- 30 105	Indonesia
2 900	2 900	2 400	1 900	1 400	1 110	1 110	1 121	1 087	DPR of Korea
- 19 440	- 20 276	- 20 962	- 21 134	- 18 927	- 17 939	- 17 100	- 14 535	- 14 468	Malaysia
-	-	72	108	215	329	479	365	647	Myanmar
3 467	3 412	3 752	3 788	3 642	4 013	3 645	4 043	4 445	Pakistan
9 777	10 981	10 466	11 827	11 074	11 211	15 371	17 561	17 686	Philippines
36 591	42 661	45 025	46 314	52 627	55 776	51 442	55 192	55 493	Singapore
1 284	1 758	1 636	1 296	1 799	1 913	1 915	2 040	2 082	Sri Lanka
21 493	21 987	21 674	23 619	25 510	24 632	29 878	32 522	32 366	Chinese Taipei
9 013	9 641	9 647	12 416	14 847	16 979	21 239	29 295	33 804	Thailand
- 1 450	- 2 661	- 3 881	- 5 461	- 6 312	- 6 901	- 8 502	- 8 561	- 9 662	Vietnam
-	- 4 500	- 4 900	- 5 220	- 5 320	- 5 420	- 4 945	- 3 945	- 3 945	Other Asia
40 332	45 739	42 806	58 523	72 360	71 261	82 096	111 868	113 967	**Asia**
- 21 128	- 21 067	- 16 625	- 10 147	- 3 763	- 6 145	- 1 137	2 215	15 641	People's Rep. of China
- 21 128	- 21 067	- 16 625	- 10 147	- 3 763	- 6 145	- 1 137	2 215	15 641	**China**
-	-	-	-	-	-	-	5	6	Albania
13 600	8 166	4 430	2 215	5 744	6 944	7 973	6 996	5 888	Bulgaria
680	624	763	749	789	907	797	804	1 039	Cyprus
21 809	16 058	8 398	6 572	7 581	8 122	8 657	7 157	6 246	Romania
7 926	6 169	4 924	4 299	4 495	4 762	5 390	5 341	5 270	Slovak Republic
-	-	-	2 159	3 247	3 580	3 959	3 513	3 659	*Croatia*
-	-	-	1 000	933	112	159	656	400	*FYROM*
-	-	-	587	570	395	595	524	622	*Slovenia*
-	-	-	400	200	200	300	1 321	2 292	*FR of Yugoslavia*
13 100	11 995	9 506	4 085	4 941	4 232	4 860	6 094	6 973	Former Yugoslavia
57 115	43 012	28 021	17 981	23 559	25 022	27 830	26 317	25 422	**Non-OECD Europe**
-	-	-	- 1 945	- 812	853	62	-	- 40	Azerbaijan
-	-	-	18 689	12 094	11 050	11 355	10 345	10 061	Belarus
-	-	-	672	207	200	-	- 113	- 104	Georgia
-	-	-	- 7 800	- 8 477	- 7 600	- 9 472	- 11 236	- 15 782	Kazakhstan
-	-	-	- 112	- 88	- 88	- 22	- 9	25	Kyrgyzstan
-	-	-	4 070	5 161	3 952	3 554	4 048	5 584	Lithuania
-	-	-	- 117 387	- 117 256	- 121 897	- 113 832	- 119 714	- 118 500	Russia
-	-	-	-	-	- 10	-	-	-	Tajikistan
-	-	-	1 119	- 69	419	-	- 100	500	Turkmenistan
-	-	-	34 360	19 679	15 800	13 300	9 315	8 957	Ukraine
-	-	-	3 106	3 999	1 503	- 368	- 771	- 906	Uzbekistan
- 113 700	- 108 000	- 60 375	- 62 958	- 83 535	- 96 148	- 97 291	- 111 478	- 115 130	**Former USSR**

A negative number shows net exports.

Former USSR: data for individual republics may not add to the total.

Net Imports of Crude Oil, NGL, Refinery Feedstocks and Additives (1000 tonnes)

Import. nettes de pétr. brut, LGN, produits d'aliment. des raffin. et additifs (1000 tonnes)

Nettoimporte von Rohöl, Kondensaten, Raffinerie-Feedstocks und Additiven (1000 Tonnen)

Importazioni nette di petrolio grezzo, GNL, prodotti intermedi e additivi (1000 tonnellate)

原油、NGL等随伴物の純輸入量（千トン）

Import. netas de petróleo crudo, líquidos de gas natural, prod. de alim. de refinerias y aditivos (1000 toneladas)

Чистый имп. сырой нефти, газ. конденсатов, нефтезаводского сырья и присадок (тыс. т)

	1971	1973	1978	1983	1984	1985	1986	1987	1988
Bahrein	9 022	8 812	9 513	6 418	7 922	7 067	9 804	9 595	9 806
Iran	- 197 994	- 262 650	- 186 456	- 94 547	- 74 823	- 75 582	- 55 625	- 80 681	- 75 944
Irak	- 79 572	- 94 991	- 116 949	- 32 182	- 40 706	- 50 691	- 63 551	- 82 189	- 106 375
Israël	500	2 700	7 975	7 326	6 993	6 292	7 100	6 900	8 400
Jordanie	592	706	1 332	2 540	2 616	2 486	2 245	2 495	2 412
Koweit	- 140 627	- 133 365	- 88 383	- 26 194	- 30 220	- 21 378	- 39 435	- 31 991	- 30 948
Liban	2 060	2 413	1 829	500	760	625	1 120	1 250	600
Oman	- 14 461	- 14 598	- 16 021	- 17 971	- 18 807	- 22 954	- 26 234	- 27 579	- 29 608
Qatar	- 20 545	- 27 485	- 23 395	- 13 133	- 17 493	- 13 686	- 15 333	- 12 533	- 14 262
Arabie saoudite	- 210 396	- 351 025	- 392 501	- 215 084	- 193 897	- 122 046	- 195 688	- 143 333	- 181 976
Syrie	- 2 750	- 3 288	- 5 869	211	1 096	622	1 833	- 1 493	- 2 246
Emirats arabes unis	- 51 000	- 76 650	- 90 890	- 47 079	- 50 100	- 49 500	- 58 130	- 63 850	- 68 940
Yémen	3 748	3 124	1 930	4 130	4 410	3 750	3 353	4 052	- 3 094
Moyen-Orient	- 701 423	- 946 297	- 897 885	- 425 065	- 402 249	- 334 995	- 428 541	- 419 357	- 492 175
Total non-OCDE	- 987 033	-1 219 787	-1 178 223	- 626 882	- 639 963	- 587 799	- 684 265	- 640 600	- 726 170
OCDE Amérique du N.	87 221	161 233	344 553	105 279	112 039	103 253	162 225	183 293	201 326
OCDE Pacifique	213 202	270 931	265 812	214 564	217 926	197 970	198 690	191 201	205 615
OCDE Europe	655 931	756 882	624 348	385 150	376 235	361 347	391 658	383 378	407 358
Total OCDE	956 354	1 189 046	1 234 713	704 993	706 200	662 570	752 573	757 872	814 299
Monde	- 30 679	- 30 741	56 490	78 111	66 237	74 771	68 308	117 272	88 129
Pour mémoire: OPEP	-1 093 536	-1 355 876	-1 262 630	- 661 037	- 643 177	- 555 574	- 649 426	- 616 929	- 689 895

Un chiffre négatif correspond à des exportations nettes.

La ligne Monde montre la divergence entre le total des exportations et des importations mondiales.

Net Imports of Crude Oil, NGL, Refinery Feedstocks and Additives (1000 tonnes)

Import. nettes de pétr. brut, LGN, produits d'aliment. des raffin. et additifs (1000 tonnes)

Nettoimporte von Rohöl, Kondensaten, Raffinerie-Feedstocks und Additiven (1000 Tonnen)

Importazioni nette di petrolio grezzo, GNL, prodotti intermedi e additivi (1000 tonnellate)

原油、NGL等随伴物の純輸入量（千トン）

Import. netas de petróleo crudo, líquidos de gas natural, prod. de alim. de refinerias y aditivos (1000 toneladas)

Чистый имп. сырой нефти, газ. конденсатов, нефтезаводского сырья и присадок (тыс. т)

1989	1990	1991	1992	1993	1994	1995	1996	1997	
10 073	9 994	10 224	10 687	10 079	9 983	10 206	10 944	11 994	Bahrain
- 99 600	- 110 000	- 120 177	- 126 095	- 129 116	- 120 376	- 121 021	- 119 631	- 119 829	Iran
- 115 700	- 78 821	- 2 127	- 2 973	- 2 899	- 2 938	- 3 119	- 4 327	- 31 593	Iraq
8 400	8 300	8 355	10 982	12 329	11 199	11 123	10 090	11 543	Israel
2 452	2 704	2 350	2 978	2 954	2 978	3 161	3 272	3 489	Jordan
- 37 662	- 34 181	- 6 669	- 36 427	- 72 826	- 63 900	- 60 000	- 63 641	- 60 896	Kuwait
-	100	540	436	-	-	-	-	-	Lebanon
- 30 214	- 32 037	- 32 767	- 35 231	- 37 530	- 37 860	- 39 997	- 41 544	- 42 533	Oman
- 16 952	- 16 440	- 16 355	- 18 896	- 17 803	- 16 999	- 17 034	- 18 871	- 23 382	Qatar
- 184 870	- 249 675	- 341 490	- 343 026	- 336 481	- 335 073	- 336 766	- 327 049	- 338 812	Saudi Arabia
- 6 060	- 9 240	- 12 780	- 13 053	- 14 275	- 17 012	- 18 109	- 17 651	- 17 022	Syria
- 83 190	- 92 900	- 106 808	- 100 973	- 95 222	- 100 086	- 98 274	- 100 256	- 102 536	United Arab Emirates
- 5 136	- 4 459	- 5 139	- 3 458	- 7 144	- 12 605	- 14 026	- 14 223	- 11 778	Yemen
- 558 459	- 606 655	- 622 843	- 655 049	- 687 934	- 682 689	- 683 856	- 682 887	- 721 355	**Middle East**
- 797 549	- 869 439	- 868 102	- 893 129	- 940 205	- 949 837	- 960 662	- 966 768	-1 010 235	**Non-OECD Total**
246 495	260 763	242 466	257 448	298 198	313 645	308 056	319 273	343 877	OECD North America
227 724	243 104	266 264	290 198	303 400	321 822	325 340	337 538	363 882	OECD Pacific
414 300	417 428	413 226	424 073	411 390	381 424	365 930	371 542	379 773	OECD Europe
888 519	921 295	921 956	971 719	1 012 988	1 016 891	999 326	1 028 353	1 087 532	**OECD Total**
90 970	51 856	53 854	78 590	72 783	67 054	38 664	61 585	77 297	**World**
- 771 023	- 831 330	- 858 227	- 888 990	- 925 435	- 924 795	- 934 273	- 945 055	-1 001 279	***Memo: OPEC***

A negative number shows net exports.

The row World shows the discrepancy between total world exports and imports.

Net Imports of Petroleum Products (1000 tonnes)
Importations nettes de produits pétroliers (1000 tonnes)

Nettoimporte von Ölprodukten (1000 Tonnen)

Importazioni nette di prodotti petroliferi (1000 tonnellate)

石油製品の純輸入量（千トン）

Importaciones netas de productos petrolíferos (1000 toneladas)

Чистый импорт нефтепродуктов (тыс. т)

	1971	1973	1978	1983	1984	1985	1986	1987	1988
Algérie	- 118	- 2 933	- 1 120	- 12 388	- 14 460	- 14 079	- 16 858	- 16 557	- 18 001
Angola	172	183	62	- 90	- 110	- 115	- 122	- 127	- 117
Bénin	101	134	132	134	138	169	149	152	132
Cameroun	286	314	531	- 355	- 321	- 200	- 237	- 262	- 312
Congo	158	164	215	- 327	- 238	- 207	- 230	- 248	- 266
RD du Congo	127	139	624	645	647	672	692	709	725
Egypte	1 197	- 237	- 691	260	394	- 359	- 1 047	- 939	- 1 560
Ethiopie	- 96	- 75	- 119	- 134	- 130	- 117	- 39	57	177
Gabon	- 535	- 512	- 811	- 17	- 22	- 11	- 4	58	53
Ghana	- 171	- 197	- 250	- 42	- 91	- 188	- 135	- 27	2
Côte d'Ivoire	58	- 191	- 249	- 499	- 443	- 590	- 507	- 411	- 475
Kenya	- 884	- 964	- 471	- 267	- 277	- 265	- 231	- 161	78
Libye	189	- 262	- 1 849	- 399	- 1 316	- 4 734	- 5 585	- 6 340	- 7 930
Maroc	485	267	799	31	- 119	- 68	11	- 103	- 119
Mozambique	- 1	- 11	- 46	256	385	458	482	490	491
Nigéria	- 134	- 304	3 937	1 495	236	418	1 248	- 570	- 1 641
Sénégal	886	891	499	599	606	648	429	382	118
Afrique du Sud	93	523	525	-	-	-	-	-	-
Soudan	471	464	58	577	638	846	633	641	809
Tanzanie	- 98	91	203	139	128	130	156	156	166
Tunisie	81	400	1 157	1 414	1 160	1 049	1 218	803	1 177
Zambie	482	278	28	21	15	17	- 24	- 25	- 17
Zimbabwe	539	679	666	741	725	752	779	731	898
Autre Afrique	2 430	2 559	3 205	3 817	4 067	3 911	4 121	4 287	4 387
Afrique	5 718	1 400	7 035	- 4 389	- 8 388	- 11 863	- 15 101	- 17 304	- 21 225
Argentine	918	788	- 198	- 1 980	- 1 399	- 3 366	- 1 138	1 074	307
Bolivie	- 23	3	- 12	- 46	- 22	- 15	- 13	- 10	- 6
Brésil	1 027	- 1 004	- 445	- 1 530	- 4 739	- 3 369	- 1 613	- 2 138	- 1 240
Chili	447	275	22	508	546	382	625	244	203
Colombie	- 1 460	- 1 544	- 727	- 1 551	- 1 906	- 1 820	- 2 104	- 3 053	- 3 496
Costa Rica	38	160	401	269	228	237	81	127	185
Cuba	2 038	2 837	3 162	3 246	3 739	3 968	3 854	3 697	3 508
République dominicaine	1 139	425	434	585	637	475	632	862	1 073
Equateur	- 29	- 95	- 998	157	- 508	- 202	- 671	312	- 1 220
El Salvador	19	10	-	- 40	- 74	- 55	- 42	8	32
Guatemala	76	86	575	462	464	522	474	457	514
Haiti	130	128	225	203	216	220	250	271	287
Honduras	- 194	- 211	95	249	195	225	314	306	267
Jamaïque	388	1 126	1 364	963	1 239	771	860	1 078	1 051
Antilles néerlandaises	- 33 699	- 36 246	- 21 627	- 17 760	- 15 246	- 4 919	- 4 801	- 6 620	- 6 804
Nicaragua	42	- 25	190	164	165	179	180	263	206
Panama	3 165	3 276	935	166	168	332	629	474	537
Paraguay	28	31	152	226	347	290	336	357	341
Pérou	498	122	- 198	- 1 526	- 2 289	- 2 678	- 2 349	- 1 995	- 1 401
Trinité-et-Tobago	- 17 408	- 16 153	- 9 865	- 2 966	- 2 910	- 3 163	- 3 310	- 3 831	- 3 542
Uruguay	81	92	95	34	36	35	121	111	450
Vénézuela	- 52 362	- 56 702	- 36 525	- 24 114	- 25 434	- 26 486	- 28 340	- 22 744	- 30 191
Autre Amérique latine	- 5 412	- 7 629	- 3 568	- 4 444	- 2 753	1 781	2 655	3 125	3 129
Amérique latine	- 100 553	- 110 250	- 66 513	- 48 725	- 49 300	- 36 656	- 33 370	- 27 625	- 35 810

Un chiffre négatif correspond à des exportations nettes.

Net Imports of Petroleum Products (1000 tonnes)

Importations nettes de produits pétroliers (1000 tonnes)

Nettoimporte von Ölprodukten (1000 Tonnen)

Importazioni nette di prodotti petroliferi (1000 tonnellate)

石油製品の純輸入量（千トン）

Importaciones netas de productos petrolíferos (1000 toneladas)

Чистый импорт нефтепродуктов (тыс. т)

1989	1990	1991	1992	1993	1994	1995	1996	1997	
- 16 459	- 17 200	- 16 597	- 17 159	- 16 553	- 16 280	- 16 663	- 16 442	- 17 044	Algeria
- 126	- 123	- 121	- 117	- 118	- 114	- 116	- 102	- 247	Angola
109	97	78	80	85	88	89	296	299	Benin
- 242	- 255	-	-	- 6	- 8	- 3	4	- 3	Cameroon
- 293	- 307	- 316	- 277	- 318	- 302	- 293	- 295	- 295	Congo
727	797	813	754	849	861	909	909	909	DR of Congo
- 1 729	- 1 557	- 2 978	- 3 081	- 4 191	- 6 098	- 5 155	- 4 156	- 3 480	Egypt
183	268	262	320	377	427	520	612	392	Ethiopia
- 237	- 77	27	- 26	- 15	- 92	- 143	- 133	- 53	Gabon
- 16	105	- 149	- 118	159	- 20	- 11	- 21	- 21	Ghana
- 396	- 656	- 553	- 514	- 599	- 610	- 599	- 672	- 880	Ivory Coast
67	- 56	- 44	- 193	- 19	210	388	729	751	Kenya
- 7 456	- 7 547	- 7 699	- 7 771	- 6 844	- 6 860	- 6 394	- 5 919	- 6 129	Libya
- 35	- 70	317	15	335	527	411	845	886	Morocco
473	458	412	433	462	463	470	510	702	Mozambique
- 1 820	- 2 919	- 2 860	- 2 382	- 541	- 794	- 859	- 965	- 965	Nigeria
286	187	297	321	322	733	254	362	337	Senegal
-	-	-	-	- 2 315	- 2 732	- 2 815	- 2 200	-	South Africa
893	1 046	705	867	833	1 044	836	775	872	Sudan
172	195	174	168	137	139	139	142	142	Tanzania
1 491	1 498	1 851	1 921	2 053	1 642	1 312	1 582	1 707	Tunisia
- 27	- 26	- 27	- 24	- 36	- 27	- 29	- 27	- 27	Zambia
833	772	935	1 026	1 101	1 287	1 504	1 463	1 463	Zimbabwe
4 654	4 627	4 723	4 816	4 969	5 080	5 192	5 240	5 450	Other Africa
- 18 948	- 20 743	- 20 750	- 20 941	- 19 873	- 21 436	- 21 056	- 17 463	- 15 234	**Africa**
- 2 450	- 3 935	- 3 428	- 3 666	- 2 503	- 1 654	- 850	- 1 796	- 3 498	Argentina
- 1	- 10	- 15	47	124	95	192	133	235	Bolivia
- 965	- 153	1 647	2 188	7 960	5 805	8 153	9 081	9 481	Brazil
356	305	553	835	874	1 380	1 538	2 324	2 123	Chile
- 2 594	- 2 114	- 2 739	- 1 395	- 1 393	- 1 032	- 893	- 1 887	- 1 508	Colombia
169	469	614	796	819	927	776	781	826	Costa Rica
3 444	3 883	3 731	4 599	3 944	4 367	5 003	4 702	5 092	Cuba
1 007	1 308	989	1 328	1 283	1 387	1 440	1 494	1 630	Dominican Republic
- 1 107	- 1 154	- 1 256	- 834	- 1 284	- 1 248	- 917	- 721	- 720	Ecuador
92	79	193	237	308	549	809	638	539	El Salvador
530	715	777	841	985	1 023	1 297	1 318	1 386	Guatemala
316	309	284	289	216	60	311	353	468	Haiti
319	298	360	463	789	913	1 098	1 162	1 116	Honduras
1 296	1 452	1 557	1 415	1 912	1 871	1 979	2 031	2 122	Jamaica
- 6 535	- 6 793	- 7 043	- 8 680	- 9 345	- 9 251	- 9 265	- 9 301	- 9 301	Netherlands Antilles
41	24	4	63	64	86	287	306	217	Nicaragua
637	754	954	623	431	1 214	1 189	294	464	Panama
404	354	351	456	568	796	875	734	899	Paraguay
- 1 672	- 1 513	- 1 933	- 1 984	- 1 909	- 1 356	253	353	155	Peru
- 3 150	- 3 395	- 4 217	- 4 266	- 4 142	- 4 371	- 4 362	- 3 670	- 3 768	Trinidad and Tobago
462	210	294	574	1 357	1 418	522	593	597	Uruguay
- 31 115	- 30 935	- 34 365	- 33 193	- 31 207	- 33 654	- 35 384	- 36 041	- 38 056	Venezuela
3 175	3 247	3 216	3 031	3 039	3 081	3 076	3 101	3 191	Other Latin America
- 37 341	- 36 595	- 39 472	- 36 233	- 27 110	- 27 594	- 22 873	- 24 018	- 26 310	**Latin America**

A negative number shows net exports.

Net Imports of Petroleum Products (1000 tonnes)

Importations nettes de produits pétroliers (1000 tonnes)

Nettoimporte von Ölprodukten (1000 Tonnen)

Importazioni nette di prodotti petroliferi (1000 tonnellate)

石油製品の純輸入量 (千トン)

Importaciones netas de productos petrolíferos (1000 toneladas)

Чистый импорт нефтепродуктов (тыс. т)

	1971	1973	1978	1983	1984	1985	1986	1987	1988
Bangladesh	97	169	288	368	365	530	709	631	685
Brunei	62	67	98	108	10	14	12	6	8
Inde	1 797	3 579	3 868	2 815	5 043	2 561	847	184	3 256
Indonésie	- 1 063	- 834	1 642	10 136	- 160	- 1 092	- 3 916	- 7 600	- 7 558
RPD de Corée	683	765	580	490	490	490	490	520	520
Malaisie	586	581	1 315	2 960	2 024	2 379	2 205	2 218	2 095
Myanmar	11	48	- 12	- 18	- 17	- 18	- 13	- 6	- 6
Népal	54	70	88	125	144	151	175	178	209
Pakistan	- 238	39	434	1 554	2 033	2 182	2 300	2 991	3 579
Philippines	- 465	- 174	999	1 882	343	414	475	1 056	268
Singapour	- 9 877	- 10 265	- 16 356	- 26 902	- 24 201	- 18 792	- 17 040	- 13 777	- 10 423
Sri Lanka	- 228	- 108	- 8	139	- 91	- 86	- 74	134	- 87
Taipei chinois	1 443	1 201	1 008	450	439	- 49	2 581	1 708	4 584
Thailande	840	506	2 287	2 855	3 620	2 323	2 061	3 151	4 256
Viêt-Nam	5 660	5 654	933	1 874	1 839	1 872	2 099	2 432	2 473
Autre Asie	1 609	2 053	1 923	2 131	2 048	2 281	2 177	2 530	2 413
Asie	971	3 351	- 913	967	- 6 071	- 4 840	- 4 912	- 3 644	6 272
Rép. populaire de Chine	- 40	16	- 1 505	- 4 573	- 5 314	- 5 729	- 2 894	- 2 228	- 1 435
Hong-Kong, Chine	4 044	4 861	6 187	5 787	4 919	5 054	5 462	4 861	5 524
Chine	4 004	4 877	4 682	1 214	- 395	- 675	2 568	2 633	4 089
Albanie	- 666	- 946	- 1 054	- 66	- 166	- 208	- 205	- 176	- 117
Bulgarie	2 147	1 490	1 464	- 538	- 852	- 970	- 728	- 997	- 1 169
Chypre	629	192	401	482	388	449	465	630	561
Gibraltar	217	233	173	162	152	328	411	448	580
Malte	330	343	385	343	461	298	597	600	616
Roumanie	- 4 946	- 4 593	- 7 217	- 8 844	- 10 240	- 9 418	- 10 164	- 11 600	- 12 916
République slovaque	- 1 260	- 1 386	- 1 798	- 1 852	- 1 988	- 2 017	- 2 195	- 2 399	- 2 302
Bosnie-Herzegovine	-	-	-	-	-	-	-	-	-
Croatie	-	-	-	-	-	-	-	-	-
Ex-RYM	-	-	-	-	-	-	-	-	-
Slovénie	-	-	-	-	-	-	-	-	-
RF de Yougoslavie	-	-	-	-	-	-	-	-	-
Ex-Yougoslavie	507	882	898	461	327	842	323	70	45
Europe non-OCDE	- 3 042	- 3 785	- 6 748	- 9 852	- 11 918	- 10 696	- 11 496	- 13 424	- 14 702
Arménie	-	-	-	-	-	-	-	-	-
Azerbaïdjan	-	-	-	-	-	-	-	-	-
Bélarus	-	-	-	-	-	-	-	-	-
Estonie	-	-	-	-	-	-	-	-	-
Géorgie	-	-	-	-	-	-	-	-	-
Kazakhstan	-	-	-	-	-	-	-	-	-
Kirghizistan	-	-	-	-	-	-	-	-	-
Lettonie	-	-	-	-	-	-	-	-	-
Lituanie	-	-	-	-	-	-	-	-	-
Moldova	-	-	-	-	-	-	-	-	-
Russie	-	-	-	-	-	-	-	-	-
Tadjikistan	-	-	-	-	-	-	-	-	-
Turkménistan	-	-	-	-	-	-	-	-	-
Ukraine	-	-	-	-	-	-	-	-	-
Ouzbékistan	-	-	-	-	-	-	-	-	-
Ex-URSS	- 28 700	- 31 500	- 43 400	- 53 200	- 49 800	- 47 350	- 54 350	- 57 575	- 59 425

Un chiffre négatif correspond à des exportations nettes.

Ex-URSS: la somme des républiques peut être différente du total.

Net Imports of Petroleum Products (1000 tonnes)

Importations nettes de produits pétroliers (1000 tonnes)

Nettoimporte von Ölprodukten (1000 Tonnen)

Importazioni nette di prodotti petroliferi (1000 tonnellate)

石油製品の純輸入量（千トン）

Importaciones netas de productos petrolíferos (1000 toneladas)

Чистый импорт нефтепродуктов (тыс. т)

1989	1990	1991	1992	1993	1994	1995	1996	1997	
774	879	527	729	750	936	1 327	1 630	1 623	Bangladesh
13	10	7	16	20	22	39	88	102	Brunei
4 261	5 359	7 799	7 140	8 776	10 941	16 816	17 729	17 714	India
- 7 255	- 7 008	- 7 051	- 7 125	- 3 622	- 4 955	- 3 273	- 3 737	3 959	Indonesia
620	577	596	688	711	711	505	372	361	DPR of Korea
2 257	3 028	3 571	3 757	4 477	3 246	- 705	273	2 078	Malaysia
- 6	- 29	- 23	57	53	55	162	291	281	Myanmar
196	190	237	294	289	422	500	498	498	Nepal
4 185	5 056	4 119	5 125	6 412	7 732	8 428	9 952	9 802	Pakistan
1 114	601	1 052	1 609	2 134	3 094	714	- 98	671	Philippines
- 15 194	- 17 808	- 17 950	- 16 935	- 19 587	- 20 923	- 19 935	- 18 039	- 13 711	Singapore
228	- 109	- 10	386	270	140	539	762	942	Sri Lanka
4 002	6 341	5 637	5 983	5 551	8 682	6 853	4 744	4 829	Chinese Taipei
5 148	7 543	7 882	7 891	8 678	8 261	8 894	4 398	- 1 640	Thailand
2 277	2 764	2 483	3 178	3 900	4 309	4 059	6 055	7 510	Vietnam
2 489	2 591	2 497	2 277	2 251	2 153	2 180	2 186	2 262	Other Asia
5 109	9 985	11 373	15 070	21 063	24 826	27 103	27 104	37 281	**Asia**
37	- 2 204	- 1 041	2 829	14 856	11 377	11 723	13 926	22 476	People's Rep. of China
6 359	6 194	6 300	8 097	8 570	8 865	9 245	9 463	9 452	Hong Kong, China
6 396	3 990	5 259	10 926	23 426	20 242	20 968	23 389	31 928	**China**
- 126	14	- 72	- 55	14	61	90	99	105	Albania
- 1 360	315	1 619	3 077	938	- 987	- 1 461	- 1 143	- 884	Bulgaria
844	947	944	1 154	1 258	1 247	1 258	1 403	1 117	Cyprus
496	515	934	944	980	985	987	987	987	Gibraltar
616	617	707	762	769	748	890	971	1 020	Malta
- 13 291	- 4 779	- 726	- 577	- 939	- 3 277	- 1 749	- 505	1 128	Romania
- 2 456	- 1 424	- 1 138	- 809	- 1 286	- 1 663	- 1 691	- 1 834	- 1 953	Slovak Republic
-	-	-	536	716	859	859	859	859	*Bosnia-Herzegovina*
-	-	-	- 812	- 1 830	- 1 722	- 1 742	- 1 301	- 1 287	*Croatia*
-	-	-	79	220	706	808	562	642	*FYROM*
-	-	-	979	1 324	1 679	1 606	2 082	2 049	*Slovenia*
-	-	-	455	160	160	160	353	300	*FR of Yugoslavia*
- 143	560	937	1 284	577	1 623	1 541	2 132	2 565	Former Yugoslavia
- 15 420	- 3 235	3 205	5 733	2 324	- 1 204	15	2 533	4 083	**Non-OECD Europe**
-	-	-	2 427	1 211	391	278	154	154	Armenia
-	-	-	- 2 952	- 2 129	- 1 686	- 2 198	- 2 159	- 2 004	Azerbaijan
-	-	-	404	133	- 1 404	- 2 240	- 3 030	- 2 335	Belarus
-	-	-	1 323	1 530	1 325	1 025	1 037	907	Estonia
-	-	-	603	444	96	100	878	941	Georgia
-	-	-	3 915	1 482	346	479	- 1 288	- 603	Kazakhstan
-	-	-	1 909	1 165	417	529	576	378	Kyrgyzstan
-	-	-	2 717	2 491	2 454	2 059	2 309	1 556	Latvia
-	-	-	- 99	- 1 362	- 691	134	- 933	- 2 324	Lithuania
-	-	-	2 872	1 866	1 132	1 076	820	944	Moldova
-	-	-	- 41 509	- 40 523	- 43 288	- 43 423	- 53 983	- 56 700	Russia
-	-	-	5 533	3 446	1 127	1 134	1 138	1 133	Tajikistan
-	-	-	- 1 002	- 1 253	- 925	- 901	- 936	- 1 554	Turkmenistan
-	-	-	1 751	4 986	3 442	7 421	6 079	4 658	Ukraine
-	-	-	1 851	1 369	997	- 130	- 325	- 332	Uzbekistan
- 56 508	- 49 494	- 43 387	- 16 560	- 21 437	- 38 411	- 41 825	- 55 645	- 59 510	**Former USSR**

A negative number shows net exports.

Former USSR: data for individual republics may not add to the total.

Net Imports of Petroleum Products (1000 tonnes)

Importations nettes de produits pétroliers (1000 tonnes)

Nettoimporte von Ölprodukten (1000 Tonnen)

Importazioni nette di prodotti petroliferi (1000 tonnellate)

石油製品の純輸入量 (千トン)

Importaciones netas de productos petrolíferos (1000 toneladas)

Чистый импорт нефтепродуктов (тыс. т)

	1971	1973	1978	1983	1984	1985	1986	1987	1988
Bahrein	- 11 044	- 10 431	- 11 850	- 8 223	- 9 138	- 8 030	- 11 306	- 11 076	- 11 211
Iran	- 10 911	- 7 750	- 7 329	5 708	5 806	4 708	6 878	8 823	8 144
Irak	- 63	54	- 1 929	- 2 754	- 4 616	- 4 866	- 5 709	- 5 697	- 6 504
Israël	- 101	- 284	- 535	- 505	- 559	- 107	85	89	1 696
Jordanie	16	- 38	- 13	- 38	144	404	665	733	628
Koweit	- 15 209	- 13 816	- 14 580	- 20 470	- 21 088	- 22 816	- 27 747	- 31 890	- 37 666
Liban	- 42	- 39	226	1 790	1 434	1 937	1 593	1 515	1 495
Oman	1 335	1 326	1 151	- 1 251	- 1 478	- 1 480	- 1 436	- 1 028	- 1 398
Qatar	58	114	66	- 412	- 935	- 1 206	- 1 523	- 1 710	- 2 128
Arabie saoudite	- 11 152	- 9 749	- 9 647	- 8 514	- 12 882	- 15 571	- 21 945	- 34 435	- 39 336
Syrie	211	180	394	- 1 121	- 1 283	- 2 468	- 2 054	- 1 675	- 2 643
Emirats arabes unis	217	435	1 633	- 2 513	- 3 606	- 1 538	149	82	842
Yémen	- 2 730	- 1 905	- 235	- 2 104	- 2 219	- 1 157	- 778	- 2 197	- 2 120
Moyen-Orient	- 49 415	- 41 903	- 42 648	- 40 407	- 50 420	- 52 190	- 63 128	- 78 466	- 90 201
Total non-OCDE	- 171 017	- 177 810	- 148 505	- 154 392	- 176 292	- 164 270	- 179 789	- 195 405	- 211 002
OCDE Amérique du N.	103 321	129 112	64 484	21 837	34 647	22 735	29 100	30 037	38 452
OCDE Pacifique	28 506	24 047	29 188	35 628	36 377	39 683	46 316	57 911	61 622
OCDE Europe	21 062	12 637	41 850	47 800	53 093	53 293	50 474	51 803	33 696
Total OCDE	152 889	165 796	135 522	105 265	124 117	115 711	125 890	139 751	133 770
Monde	- 18 128	- 12 014	- 12 983	- 49 127	- 52 175	- 48 559	- 53 899	- 55 654	- 77 232
Pour mémoire: OPEP	- 90 548	- 91 747	- 65 701	- 54 225	- 78 455	- 87 262	- 103 348	- 118 638	- 141 969

Un chiffre négatif correspond à des exportations nettes.

La ligne Monde montre la divergence entre le total des exportations et des importations mondiales.

Net Imports of Petroleum Products (1000 tonnes)

Importations nettes de produits pétroliers (1000 tonnes)

Nettoimporte von Ölprodukten (1000 Tonnen)

Importazioni nette di prodotti petroliferi (1000 tonnellate)

石油製品の純輸入量（千トン）

Importaciones netas de productos petrolíferos (1000 toneladas)

Чистый импорт нефтепродуктов (тыс. т)

1989	1990	1991	1992	1993	1994	1995	1996	1997	
- 11 472	- 11 592	- 12 014	- 12 338	- 11 838	- 11 740	- 12 155	- 11 507	- 10 997	Bahrain
6 579	6 415	7 603	6 482	2 458	1 943	2 144	4 170	4 757	Iran
- 6 440	- 3 540	- 510	- 2 465	- 2 515	- 3 039	- 2 840	- 3 121	- 2 670	Iraq
1 199	592	306	- 602	- 1 645	- 1 773	362	769	214	Israel
625	781	776	842	902	952	906	988	1 058	Jordan
- 45 094	- 19 877	- 1 061	- 12 745	- 17 790	- 30 660	- 34 260	- 32 628	- 36 530	Kuwait
1 675	2 023	2 210	2 248	3 291	3 640	4 176	4 276	4 759	Lebanon
- 1 374	- 1 317	- 87	- 41	568	- 1 321	- 861	- 1 022	- 972	Oman
- 1 664	- 2 899	- 2 465	- 3 064	- 2 757	- 3 050	- 2 929	- 3 171	- 2 845	Qatar
- 42 595	- 49 427	- 40 508	- 44 438	- 42 520	- 42 272	- 42 246	- 47 697	- 41 827	Saudi Arabia
- 2 863	- 1 936	- 1 479	- 1 688	- 1 403	- 1 313	- 1 064	- 1 091	- 1 107	Syria
1 675	1 556	5 006	4 885	4 981	4 870	3 419	2 584	279	United Arab Emirates
- 836	- 2 161	- 1 633	- 1 906	- 1 195	- 1 093	- 846	- 1 756	- 1 685	Yemen
- 100 585	- 81 382	- 43 856	- 64 830	- 69 463	- 84 856	- 86 194	- 89 206	- 87 566	**Middle East**
- 217 297	- 177 474	- 127 628	- 106 835	- 91 070	- 128 433	- 123 862	- 133 306	- 115 328	**Non-OECD Total**
34 130	17 461	3 799	6 288	- 2 685	6 771	- 6 695	5 426	7 433	OECD North America
65 508	67 355	50 257	46 897	44 169	51 240	50 704	54 018	33 078	OECD Pacific
37 989	33 010	33 804	21 022	14 746	8 968	18 234	18 549	14 573	OECD Europe
137 627	117 826	87 860	74 207	56 230	66 979	62 243	77 993	55 084	**OECD Total**
- 79 670	- 59 648	- 39 768	- 32 628	- 34 840	- 61 454	- 61 619	- 55 313	- 60 244	**World**
- 151 644	- 133 381	- 100 507	- 118 975	- 116 910	- 134 751	- 139 285	- 142 967	- 137 071	***Memo: OPEC***

A negative number shows net exports.

The row World shows the discrepancy between total world exports and imports.

Net Imports of Natural Gas (TJ)
Importations nettes de gaz naturel (TJ)

Nettoimporte von Erdgas (TJ)

Importazioni nette di gas naturale (TJ)

天然ガスの純輸入量 *(TJ)*

Importaciones netas de gas natural (TJ)

Чистый импорт природного газа (ТДж)

	1971	1973	1978	1983	1984	1985	1986	1987	1988
Algérie	- 58 484	- 107 986	- 288 381	- 791 980	- 832 077	- 948 434	- 933 052	-1 140 856	-1 164 675
Libye	- 18 793	- 119 789	- 153 407	- 30 604	- 44 329	- 44 245	- 38 988	- 33 710	- 41 328
Tunisie	-	-	-	-	13 963	24 226	13 276	36 238	28 043
Afrique	- 77 277	- 227 775	- 441 788	- 822 584	- 862 443	- 968 453	- 958 764	-1 138 328	-1 177 960
Argentine	-	67 511	93 814	93 000	92 613	91 805	90 720	88 525	92 806
Bolivie	-	- 74 747	- 75 520	- 94 728	- 93 994	- 94 226	- 94 148	- 90 168	- 94 767
Chili	-	-	- 26 220	-	-	-	-	-	-
Amérique latine	-	- 7 236	- 7 926	- 1 728	- 1 381	- 2 421	- 3 428	- 1 643	- 1 961
Brunei	-	- 59 070	- 277 900	- 284 877	- 290 153	- 279 602	- 278 546	- 284 877	- 292 263
Indonésie	-	-	- 199 347	- 526 318	- 778 813	- 818 705	- 822 846	- 909 047	-1 003 855
Malaisie	-	-	- 594	- 110 254	- 209 886	- 236 210	- 315 777	- 365 414	- 380 719
Singapour	-	-	-	-	-	-	-	-	-
Taipei chinois	-	-	-	-	-	-	-	-	-
Autre Asie	- 87 634	- 95 365	- 84 129	- 87 000	- 87 000	- 87 000	- 87 000	- 63 000	- 50 000
Asie	- 87 634	- 154 435	- 561 970	-1 008 449	-1 365 852	-1 421 517	-1 504 169	-1 622 338	-1 726 837
Rép. populaire de Chine	-	-	-	-	-	-	-	-	-
Hong-Kong, Chine	-	-	-	-	-	-	-	-	-
Chine	-	-	-	-	-	-	-	-	-
Bulgarie	-	-	106 467	184 945	194 944	213 078	221 840	237 171	236 162
Roumanie	- 7 555	- 7 555	30 416	66 482	71 480	70 811	110 445	119 404	146 128
République slovaque	37 848	54 618	120 235	190 908	184 524	182 259	182 779	183 076	180 032
Bosnie-Herzegovine	-	-	-	-	-	-	-	-	-
Croatie	-	-	-	-	-	-	-	-	-
Ex-RYM	-	-	-	-	-	-	-	-	-
Slovénie	-	-	-	-	-	-	-	-	-
RF de Yougoslavie	-	-	-	-	-	-	-	-	-
Ex-Yougoslavie	-	-	-	104 625	130 596	137 075	146 032	164 247	153 849
Europe non-OCDE	30 293	47 063	257 118	546 960	581 544	603 223	661 096	703 898	716 171
Arménie	-	-	-	-	-	-	-	-	-
Azerbaïdjan	-	-	-	-	-	-	-	-	-
Bélarus	-	-	-	-	-	-	-	-	-
Estonie	-	-	-	-	-	-	-	-	-
Géorgie	-	-	-	-	-	-	-	-	-
Kazakhstan	-	-	-	-	-	-	-	-	-
Kirghizistan	-	-	-	-	-	-	-	-	-
Lettonie	-	-	-	-	-	-	-	-	-
Lituanie	-	-	-	-	-	-	-	-	-
Moldova	-	-	-	-	-	-	-	-	-
Russie	-	-	-	-	-	-	-	-	-
Tadjikistan	-	-	-	-	-	-	-	-	-
Turkménistan	-	-	-	-	-	-	-	-	-
Ukraine	-	-	-	-	-	-	-	-	-
Ouzbékistan	-	-	-	-	-	-	-	-	-
Ex-URSS	134 947	176 827	-1 023 732	-2 215 010	-2 480 251	-2 550 050	-2 964 193	-3 182 894	-3 270 222

Un chiffre négatif correspond à des exportations nettes.

Ex-URSS: la somme des républiques peut être différente du total.

Net Imports of Natural Gas (TJ)

Importations nettes de gaz naturel (TJ)

Nettoimporte von Erdgas (TJ)

Importazioni nette di gas naturale (TJ)

天然ガスの純輸入量 (TJ)

Importaciones netas de gas natural (TJ)

Чистый импорт природного газа (ТДж)

1989	1990	1991	1992	1993	1994	1995	1996	1997	
-1 314 423	-1 379 178	-1 503 852	-1 560 651	-1 567 048	-1 412 898	-1 665 922	-1 825 284	-2 162 014	Algeria
- 48 000	- 42 000	- 74 000	- 86 000	- 75 000	- 69 375	- 69 861	- 56 264	- 56 264	Libya
30 000	42 020	27 734	45 775	46 566	69 374	83 333	56 858	28 894	Tunisia
-1 332 423	-1 379 158	-1 550 118	-1 600 876	-1 595 482	-1 412 899	-1 652 450	-1 824 690	-2 189 384	**Africa**
92 503	84 695	84 120	82 124	80 774	87 149	82 077	81 891	37 124	Argentina
- 95 733	- 95 462	- 94 380	- 92 100	- 90 617	- 93 925	- 88 947	- 82 293	- 89 240	Bolivia
-	-	-	-	-	-	-	-	25 346	Chile
- 3 230	- 10 767	- 10 260	- 9 976	- 9 843	- 6 776	- 6 870	- 402	- 26 770	**Latin America**
- 291 208	- 291 208	- 290 153	- 293 318	- 301 759	- 314 420	- 340 797	- 338 687	- 336 577	Brunei
-1 013 113	-1 121 462	-1 225 212	-1 296 124	-1 324 766	-1 434 447	-1 357 846	-1 450 383	-1 467 535	Indonesia
- 393 745	- 404 537	- 385 092	- 384 301	- 461 195	- 489 810	- 570 631	- 711 876	- 647 057	Malaysia
-	-	-	20 097	54 382	64 802	67 500	60 211	59 350	Singapore
-	35 465	85 819	89 637	96 291	120 545	137 393	142 400	174 343	Chinese Taipei
- 3 000	-	-	-	-	-	-	-	-	Other Asia
-1 701 066	-1 781 742	-1 814 638	-1 864 009	-1 937 047	-2 053 330	-2 064 381	-2 298 335	-2 217 476	**Asia**
-	-	-	-	-	-	- 1 086	- 68 823	- 107 462	People's Rep. of China
-	-	-	-	-	-	1 086	68 823	107 462	Hong Kong, China
-	-	-	-	-	-	-	-	-	**China**
228 802	227 353	209 228	188 931	175 690	173 913	212 258	220 040	179 192	Bulgaria
277 701	275 754	173 989	166 688	168 206	173 770	223 038	263 046	187 503	Romania
194 889	260 714	223 987	221 560	199 640	201 238	212 780	237 880	242 593	Slovak Republic
-	-	-	16 000	5 473	9 834	9 834	9 834	9 834	*Bosnia-Herzegovina*
-	-	-	27 398	30 020	28 234	10 408	33 402	39 729	*Croatia*
-	-	-	10 000	10 500	-	-	-	-	*FYROM*
-	-	-	25 164	24 943	26 491	31 412	30 039	32 847	*Slovenia*
-	-	-	44 840	36 546	-	7 560	79 170	77 813	*FR of Yugoslavia*
157 992	180 151	184 421	123 402	107 482	64 559	59 214	152 445	160 223	Former Yugoslavia
859 384	943 972	791 625	700 581	651 018	613 480	707 290	873 411	769 511	**Non-OECD Europe**
-	-	-	70 680	30 628	32 870	53 200	41 470	51 649	Armenia
-	-	-	162 637	94 778	98 435	20 019	905	-	Azerbaijan
-	-	-	677 771	629 735	552 182	522 594	554 032	627 260	Belarus
-	-	-	33 348	16 546	23 763	27 096	29 585	28 747	Estonia
-	-	-	182 717	138 284	92 591	33 930	29 368	35 589	Georgia
-	-	-	388 762	238 339	208 745	247 312	118 868	- 17 342	Kazakhstan
-	-	-	68 426	51 385	32 271	33 050	40 074	22 241	Kyrgyzstan
-	-	-	99 298	35 309	37 798	46 491	40 692	49 249	Latvia
-	-	-	134 073	72 511	80 518	94 502	100 749	93 134	Lithuania
-	-	-	136 163	126 333	118 325	119 118	136 679	145 742	Moldova
-	-	-	-6 940 834	-6 279 921	-6 924 444	-7 062 222	-7 252 444	-7 045 386	Russia
-	-	-	63 032	52 394	28 956	32 528	41 800	38 000	Tajikistan
-	-	-	-1 783 948	-2 113 256	- 865 640	- 831 440	- 904 800	- 245 050	Turkmenistan
-	-	-	3 364 763	2 880 770	2 494 496	2 477 224	2 747 125	2 378 347	Ukraine
-	-	-	- 37 555	- 53 580	- 156 788	- 159 436	- 184 747	- 270 413	Uzbekistan
-3 798 277	-4 042 188	-3 826 289	-3 605 840	-3 674 084	-3 867 348	-4 415 218	-4 655 950	-4 399 590	**Former USSR**

A negative number shows net exports.

Former USSR: data for individual republics may not add to the total.

Net Imports of Natural Gas (TJ)

Importations nettes de gaz naturel (TJ)

Nettoimporte von Erdgas (TJ)

Importazioni nette di gas naturale (TJ)

天然ガスの純輸入量 (TJ)

Importaciones netas de gas natural (TJ)

Чистый импорт природного газа (ТДж)

	1971	1973	1978	1983	1984	1985	1986	1987	1988
Iran	- 196 049	- 317 010	- 275 500	-	-	-	-	-	-
Irak	-	-	-	-	-	-	- 24 700	- 104 500	- 117 800
Koweit	-	-	-	-	-	-	49 987	107 220	109 970
Oman	-	-	-	-	-	-	-	-	-
Qatar	-	-	-	-	-	-	-	-	-
Emirats arabes unis	-	-	- 64 600	- 91 200	- 106 400	- 117 800	- 110 200	- 110 200	- 121 600
Moyen-Orient	- 196 049	- 317 010	- 340 100	- 91 200	- 106 400	- 117 800	- 84 913	- 107 480	- 129 430
Total non-OCDE	- 195 720	- 482 566	-2 118 398	-3 592 011	-4 234 783	-4 457 018	-4 854 371	-5 348 785	-5 590 239
OCDE Amérique du N.	- 31 451	- 33 113	50 896	80 583	- 35 624	- 46 389	- 61 733	- 66 063	- 55 432
OCDE Pacifique	54 387	129 581	625 219	1 044 004	1 433 158	1 534 416	1 580 057	1 679 138	1 786 552
OCDE Europe	210 713	415 668	1 419 499	2 143 651	2 477 628	2 699 438	3 106 513	3 437 646	3 557 228
Total OCDE	233 649	512 136	2 095 614	3 268 238	3 875 162	4 187 465	4 624 837	5 050 721	5 288 348
Monde	37 929	29 570	- 22 784	- 323 773	- 359 621	- 269 553	- 229 534	- 298 064	- 301 891

Un chiffre négatif correspond à des exportations nettes.

La ligne Monde montre la divergence entre le total des exportations et des importations mondiales.

Net Imports of Natural Gas (TJ)

Importations nettes de gaz naturel (TJ)

Nettoimporte von Erdgas (TJ)

Importazioni nette di gas naturale (TJ)

天然ガスの純輸入量 (TJ)

Importaciones netas de gas natural (TJ)

Чистый импорт природного газа (ТДж)

1989	1990	1991	1992	1993	1994	1995	1996	1997	
-	- 57 000	- 114 380	-	- 19 000	- 30 400	- 3 800	- 3 800	- 3 800	Iran
- 136 800	- 76 000	-	-	-	-	-	-	-	Iraq
140 000	76 000	-	-	-	-	-	-	-	Kuwait
-	-	-	-	-	-	- 18 850	- 18 850	- 18 850	Oman
-	-	-	-	-	-	-	-	- 107 822	Qatar
- 117 800	- 121 600	- 133 000	- 129 200	- 125 400	- 155 409	- 244 507	- 252 089	- 318 476	United Arab Emirates
- 114 600	- 178 600	- 247 380	- 129 200	- 144 400	- 185 809	- 267 157	- 274 739	- 448 948	**Middle East**
-6 090 212	-6 448 483	-6 657 060	-6 509 320	-6 709 838	-6 912 682	-7 698 786	-8 180 705	-8 512 657	**Non-OECD Total**
- 36 662	48 372	30 891	- 3 775	51 635	- 15 168	- 31 147	18 930	23 129	OECD North America
1 916 515	1 954 208	2 037 803	2 075 322	2 100 560	2 296 558	2 327 094	2 636 172	2 698 174	OECD Pacific
3 944 464	4 246 533	4 344 086	4 360 112	4 404 927	4 480 890	5 118 561	5 346 231	5 502 163	OECD Europe
5 824 317	6 249 113	6 412 780	6 431 659	6 557 122	6 762 280	7 414 508	8 001 333	8 223 466	**OECD Total**
- 265 895	- 199 370	- 244 280	- 77 661	- 152 716	- 150 402	- 284 278	- 179 372	- 289 191	**World**

A negative number shows net exports.

The row World shows the discrepancy between total world exports and imports.

Net Imports of Electricity (GWh)
Importations nettes d'électricité (GWh)
Nettoimporte von Elektrizität (GWh)

Importazioni nette di energia elettrica (GWh)

電力の純輸入量(GWh)

Importaciones netas de electricidad (GWh)

Чистый импорт электроэнергии (ГВт.ч)

	1971	1973	1978	1983	1984	1985	1986	1987	1988
Algérie	1	- 1	1	- 18	71	36	84	- 50	- 52
Bénin	33	50	164	159	102	149	141	153	193
Congo	-	-	71	57	53	55	129	148	149
RD du Congo	- 26	- 26	- 55	- 101	- 128	- 147	- 184	- 187	- 109
Ghana	-	- 100	- 217	- 491	- 622	- 611	- 664	- 494	- 211
Côte d'Ivoire	-	-	-	178	383	217	-	-	-
Kenya	293	303	217	179	217	215	220	176	110
Maroc	-	-	-	-	-	-	-	-	-
Mozambique	2	153	77	106	37	- 12	222	273	183
Nigéria	-	-	- 45	- 128	- 135	- 140	-	-	-
Afrique du Sud	- 14	- 194	6 609	3 746	- 405	- 314	- 1 223	- 1 187	- 997
Tanzanie	-	-	-	-	-	-	-	-	-
Tunisie	-	-	-	17	- 73	29	8	- 4	- 25
Zambie	3 430	2 029	- 1 731	- 3 021	- 2 901	- 2 903	- 3 080	- 1 500	- 1 500
Zimbabwe	176	94	53	3 548	3 433	3 996	3 308	2 536	855
Autre Afrique	21	101	226	192	105	187	200	287	361
Afrique	3 916	2 409	5 370	4 423	137	757	- 839	151	- 1 043
Argentine	- 10	50	74	- 6	- 5	- 6	3 151	3 058	2 186
Bolivie	-	-	-	-	-	-	-	-	-
Brésil	- 18	- 17	- 127	- 243	- 86	1 913	10 292	16 804	17 943
Chili	2	3	1	-	-	-	-	-	-
Colombie	-	-	26	39	30	5	5	-	-
Costa Rica	-	-	-	- 488	- 430	- 59	78	170	190
El Salvador	-	-	-	-	-	-	88	9	35
Guatemala	-	-	-	-	-	-	- 88	- 11	-
Nicaragua	-	-	- 1	330	265	187	69	80	76
Panama	- 46	- 50	- 13	20	58	22	- 18	119	19
Paraguay	-	- 77	63	245	54	- 2 821	- 10 462	- 16 813	- 17 960
Pérou	-	-	-	-	-	-	-	-	-
Uruguay	35	27	29	- 3 597	- 3 332	- 2 678	- 3 152	- 2 955	- 2 091
Vénézuela	-	-	- 19	-	-	-	-	-	-
Autre Amérique latine	-	-	-	-	-	-	-	-	-
Amérique latine	- 37	- 64	33	- 3 700	- 3 446	- 3 437	- 37	461	398
Inde	-	- 11	- 44	- 85	- 105	- 91	103	855	1 200
Malaisie	-	-	12	35	81	58	-	- 23	- 70
Népal	1	-	27	57	61	65	32	10	52
Singapour	-	-	- 48	- 55	- 73	- 50	-	-	-
Thailande	- 41	156	216	676	688	703	741	398	410
Autre Asie	-	-	4	4	9	31	- 292	- 1 092	- 1 392
Asie	- 40	145	167	632	661	716	584	148	200
Rép. populaire de Chine	-	-	-	430	770	1 070	1 168	1 250	1 470
Hong-Kong, Chine	-	-	-	- 367	- 740	- 1 050	- 1 208	- 1 354	- 1 434
Chine	-	-	-	63	30	20	- 40	- 104	36

Un chiffre négatif correspond à des exportations nettes.

Net Imports of Electricity (GWh)

Importations nettes d'électricité (GWh)

Nettoimporte von Elektrizität (GWh)

Importazioni nette di energia elettrica (GWh)

電力の純輸入量 (GWh)

Importaciones netas de electricidad (GWh)

Чистый импорт электроэнергии (ГВт.ч)

1989	1990	1991	1992	1993	1994	1995	1996	1997	
- 15	- 60	- 661	- 928	- 1 241	- 1 124	- 273	- 142	- 1	Algeria
183	196	215	214	233	211	256	264	238	Benin
61	63	65	108	110	112	115	115	114	Congo
- 533	- 5	- 42	- 143	- 142	- 140	- 138	- 140	- 140	DR of Congo
- 521	- 761	- 805	- 893	- 765	- 387	- 393	- 397	- 397	Ghana
-	294	367	484	91	34	34	34	34	Ivory Coast
112	181	134	240	273	264	172	149	144	Kenya
-	103	641	932	1 027	793	243	129	124	Morocco
116	166	301	391	422	465	501	599	203	Mozambique
-	-	- 92	- 92	- 104	- 138	-	-	-	Nigeria
- 1 120	- 1 263	- 1 627	- 1 466	- 2 489	- 2 625	- 2 851	- 5 550	- 6 612	South Africa
-	-	-	-	-	15	35	34	44	Tanzania
28	- 7	21	- 4	216	329	81	35	35	Tunisia
- 1 500	- 1 480	- 1 480	- 1 480	- 1 471	- 1 480	- 1 480	- 1 480	- 1 500	Zambia
835	332	1 144	2 027	1 214	2 009	2 312	3 172	4 013	Zimbabwe
445	472	568	562	600	719	769	1 127	1 186	Other Africa
- 1 909	- 1 769	- 1 251	- 48	- 2 026	- 943	- 617	- 2 051	- 2 515	**Africa**
1 295	2 605	1 872	3 407	2 256	1 756	2 116	3 361	5 186	Argentina
8	9	9	12	13	16	10	- 2	4	Bolivia
22 106	26 538	27 080	24 014	27 550	31 767	35 352	36 558	40 470	Brazil
-	-	-	-	-	-	-	-	-	Chile
226	200	213	348	303	280	370	161	199	Colombia
143	163	20	- 64	- 3	- 6	29	124	- 134	Costa Rica
4	1	5	53	79	11	35	21	88	El Salvador
-	-	-	-	-	-	-	-	-	Guatemala
12	66	91	31	- 47	- 21	-	14	162	Nicaragua
72	114	136	143	86	83	87	- 62	- 53	Panama
- 20 525	- 24 970	- 26 975	- 24 618	- 28 120	- 32 600	- 37 827	- 43 709	- 45 673	Paraguay
-	-	-	-	-	3	-	-	2	Peru
- 1 154	- 2 538	- 1 792	- 3 385	- 2 241	- 1 661	- 45	- 35	- 144	Uruguay
-	-	- 200	- 360	-	- 131	- 195	- 149	-	Venezuela
-	-	3	17	12	12	16	16	17	Other Latin America
2 187	2 188	462	- 402	- 112	- 491	- 52	- 3 702	124	**Latin America**
1 297	1 376	1 453	1 206	1 240	1 423	1 545	1 545	1 545	India
- 164	- 58	- 23	- 23	23	47	- 23	- 12	8	Malaysia
100	37	- 52	- 5	36	51	74	- 19	- 19	Nepal
-	-	-	-	- 52	- 91	-	-	-	Singapore
620	621	555	440	596	826	620	717	642	Thailand
- 1 392	- 1 392	- 1 397	- 1 311	- 1 312	- 1 324	- 1 351	- 1 366	- 1 433	Other Asia
461	584	536	307	531	932	865	865	743	**Asia**
1 720	1 840	2 850	2 850	4 389	- 2 046	- 5 386	- 3 588	- 7 115	People's Rep. of China
- 1 771	- 1 799	- 3 061	- 4 962	- 4 054	6 495	6 063	7 247	7 317	Hong Kong, China
- 51	41	- 211	- 2 112	335	4 449	677	3 659	202	**China**

A negative number shows net exports.

Net Imports of Electricity (GWh)

Importations nettes d'électricité (GWh)

Nettoimporte von Elektrizität (GWh)

Importazioni nette di energia elettrica (GWh)

電力の純輸入量(GWh)

Importaciones netas de electricidad (GWh)

Чистый импорт электроэнергии (ГВт.ч)

	1971	1973	1978	1983	1984	1985	1986	1987	1988
Albanie	-	- 267	- 133	- 575	- 600	- 625	- 650	- 650	- 650
Bulgarie	218	3 256	3 813	2 361	2 392	4 304	3 972	4 349	4 146
Gibraltar	-	-	-	-	-	-	-	-	-
Roumanie	- 3 155	- 3 548	- 1 812	669	1 956	3 259	4 430	5 150	7 199
République slovaque	2 381	2 840	2 849	5 071	5 054	4 185	3 256	4 733	5 833
Bosnie-Herzegovine	-	-	-	-	-	-	-	-	-
Croatie	-	-	-	-	-	-	-	-	-
Ex-RYM	-	-	-	-	-	-	-	-	-
Slovénie	-	-	-	-	-	-	-	-	-
RF de Yougoslavie	-	-	-	-	-	-	-	-	-
Ex-Yougoslavie	- 162	- 49	- 652	1 055	- 1 130	627	466	375	- 1 412
Europe non-OCDE	- 718	2 232	4 065	8 581	7 672	11 750	11 474	13 957	15 116
Arménie	-	-	-	-	-	-	-	-	-
Azerbaïdjan	-	-	-	-	-	-	-	-	-
Bélarus	-	-	-	-	-	-	-	-	-
Estonie	-	-	-	-	-	-	-	-	-
Géorgie	-	-	-	-	-	-	-	-	-
Kazakhstan	-	-	-	-	-	-	-	-	-
Kirghizistan	-	-	-	-	-	-	-	-	-
Lettonie	-	-	-	-	-	-	-	-	-
Lituanie	-	-	-	-	-	-	-	-	-
Moldova	-	-	-	-	-	-	-	-	-
Russie	-	-	-	-	-	-	-	-	-
Tadjikistan	-	-	-	-	-	-	-	-	-
Turkménistan	-	-	-	-	-	-	-	-	-
Ukraine	-	-	-	-	-	-	-	-	-
Ouzbékistan	-	-	-	-	-	-	-	-	-
Ex-URSS	- 6 700	- 9 700	- 12 200	- 23 900	- 24 700	- 28 900	- 29 000	- 34 600	- 38 900
Israël	- 33	- 57	- 163	- 242	- 254	- 300	- 369	- 359	- 535
Jordanie	33	57	-	-	-	- 21	- 215	- 334	-
Liban	-	-	50	40	30	40	30	40	-
Syrie	- 71	-	-	- 137	- 135	- 140	- 130	- 130	-
Moyen-Orient	- 71	-	- 113	- 339	- 359	- 421	- 684	- 783	- 535
Total non-OCDE	- 3 650	- 4 978	- 2 678	- 14 240	- 20 005	- 19 515	- 18 542	- 20 770	- 24 728
OCDE Amérique du N.	110	558	71	- 833	484	505	554	458	2 096
OCDE Europe	5 003	2 905	2 384	13 560	15 566	16 551	17 404	20 101	22 662
Total OCDE	5 113	3 463	2 455	12 727	16 050	17 056	17 958	20 559	24 758
Monde	1 463	- 1 515	- 223	- 1 513	- 3 955	- 2 459	- 584	- 211	30

Un chiffre négatif correspond à des exportations nettes.

La ligne Monde montre la divergence entre le total des exportations et des importations mondiales.

Ex-URSS: la somme des républiques peut être différente du total.

Net Imports of Electricity (GWh)

Importations nettes d'électricité (GWh)

Nettoimporte von Elektrizität (GWh)

Importazioni nette di energia elettrica (GWh)

電力の純輸入量(GWh)

Importaciones netas de electricidad (GWh)

Чистый импорт электроэнергии (ГВт.ч)

1989	1990	1991	1992	1993	1994	1995	1996	1997	
- 607	206	- 1 173	- 510	- 141	- 185	- 74	200	200	Albania
4 389	3 790	2 124	2 705	110	- 72	- 160	- 449	- 3 550	Bulgaria
- 1	- 1	- 1	-	- 5	-	-	-	-	Gibraltar
7 811	9 476	7 047	4 203	1 873	725	299	807	221	Romania
5 586	5 196	4 338	3 468	1 113	438	1 383	3 592	4 082	Slovak Republic
-	-	-	-	-	160	205	205	205	*Bosnia-Herzegovina*
-	-	-	2 787	2 326	3 565	3 496	2 330	3 948	*Croatia*
-	-	-	274	607	167	-	-	-	*FYROM*
-	-	-	- 1 813	- 1 418	- 1 934	- 1 652	- 1 661	- 1 696	*Slovenia*
-	-	-	- 400	-	-	-	- 156	-	*FR of Yugoslavia*
- 408	- 359	- 300	848	1 515	1 958	2 049	718	2 457	Former Yugoslavia
16 770	18 308	12 035	10 714	4 465	2 864	3 497	4 868	3 410	**Non-OECD Europe**
-	-	-	283	114	16	13	-	-	Armenia
-	-	-	- 513	100	260	399	442	800	Azerbaijan
-	-	-	6 504	6 005	3 820	7 159	8 543	7 620	Belarus
-	-	-	- 3 238	- 1 596	- 1 191	- 760	- 860	- 974	Estonia
-	-	-	1 016	711	800	670	86	191	Georgia
-	-	-	14 173	6 000	13 031	7 393	6 854	6 700	Kazakhstan
-	-	-	- 2 088	- 1 023	- 2 505	- 1 368	- 2 080	- 1 700	Kyrgyzstan
-	-	-	4 077	2 502	1 818	2 256	3 227	1 823	Latvia
-	-	-	- 5 303	- 2 731	1 144	- 2 678	- 5 159	- 3 525	Lithuania
-	-	-	- 768	83	615	1 870	1 606	1 952	Moldova
-	-	-	- 16 242	- 18 732	- 20 496	- 19 605	- 19 490	- 19 700	Russia
-	-	-	832	- 1 172	- 501	662	320	800	Tajikistan
-	-	-	- 4 295	- 3 186	- 2 650	- 2 020	- 2 800	- 2 650	Turkmenistan
-	-	-	- 5 085	- 1 543	- 1 042	- 2 952	- 2 017	- 154	Ukraine
-	-	-	- 490	- 407	- 400	- 1 290	1 092	929	Uzbekistan
- 39 300	- 35 000	- 21 629	- 11 137	- 14 875	- 7 281	- 10 251	- 10 236	- 7 888	**Former USSR**
- 413	- 456	- 744	- 615	- 630	- 735	- 914	- 979	- 1 061	Israel
-	-	-	- 67	- 46	-	-	- 2	-	Jordan
-	-	-	-	-	-	292	683	608	Lebanon
-	-	-	-	-	-	-	-	-	Syria
- 413	- 456	- 744	- 682	- 676	- 735	- 622	- 298	- 453	**Middle East**
- 22 255	- 16 104	- 10 802	- 3 360	- 12 358	- 1 205	- 6 503	- 6 895	- 6 377	**Non-OECD Total**
295	156	2 443	2 247	- 95	- 106	812	451	4 225	OECD North America
20 284	17 751	11 205	2 042	4 473	3 313	2 792	5 444	5 839	OECD Europe
20 579	17 907	13 648	4 289	4 378	3 207	3 604	5 895	10 064	**OECD Total**
- 1 676	1 803	2 846	929	- 7 980	2 002	- 2 899	- 1 000	3 687	**World**

A negative number shows net exports.

The row World shows the discrepancy between total world exports and imports.

Former USSR: data for individual republics may not add to the total.

Final Consumption of Coking Coal (1000 tonnes)
Consommation finale de charbon à coke (1000 tonnes)

Endverbrauch von Kokskohle (1000 Tonnen)

Consumo finale di carbone siderurgico (1000 tonnellate)

原料炭の最終消費量（千トン）

Consumo final de carbón coquizable (1000 toneladas)

Конечное потребление коксующихся углей (тыс. т)

	1971	1973	1978	1983	1984	1985	1986	1987	1988
Brésil	-	-	-	-	-	-	-	-	-
Amérique latine	-	-	-	-	-	-	-	-	-
Roumanie	-	-	-	-	-	-	-	-	-
République slovaque	-	-	27	25	31	32	31	32	34
Ex-RYM	-	-	-	-	-	-	-	-	-
Ex-Yougoslavie	-	-	-	-	-	-	-	-	-
Europe non-OCDE	-	-	27	25	31	32	31	32	34
Kazakhstan	-	-	-	-	-	-	-	-	-
Tadjikistan	-	-	-	-	-	-	-	-	-
Ex-URSS	-	-	-	-	-	-	-	-	-
Total non-OCDE	-	-	27	25	31	32	31	32	34
OCDE Pacifique	-	-	121	10	38	264	749	742	580
OCDE Europe	-	-	801	815	808	2 748	1 680	1 502	1 756
Total OCDE	-	-	922	825	846	3 012	2 429	2 244	2 336
Monde	-	-	949	850	877	3 044	2 460	2 276	2 370

Ex-URSS: les séries antérieures à 1990-1992 ne sont pas comparables aux années récentes; 1991 a été estimé.

Final Consumption of Coking Coal (1000 tonnes)

Consommation finale de charbon à coke (1000 tonnes)

Endverbrauch von Kokskohle (1000 Tonnen)

Consumo finale di carbone siderurgico (1000 tonnellate)

原料炭の最終消費量（千トン）

Consumo final de carbón coquizable (1000 toneladas)

Конечное потребление коксующихся углей (тыс. т)

1989	1990	1991	1992	1993	1994	1995	1996	1997	
-	-	-	-	235	354	834	1 596	1 928	Brazil
-	-	-	-	235	354	834	1 596	1 928	**Latin America**
-	-	-	915	37	6	121	30	2	Romania
38	6	28	-	166	323	314	628	295	Slovak Republic
-	-	-	73	74	66	70	49	49	*FYROM*
-	-	77	73	74	66	70	49	49	Former Yugoslavia
38	6	105	988	277	395	505	707	346	**Non-OECD Europe**
-	-	-	15 985	11 381	12 116	9 204	8 076	7 106	Kazakhstan
-	-	-	-	-	3	7	-	-	Tajikistan
-	-	-	15 985	11 381	12 119	9 211	8 076	7 106	**Former USSR**
38	6	105	16 973	11 893	12 868	10 550	10 379	9 380	**Non-OECD Total**
914	1 298	1 699	1 338	1 584	1 860	1 922	1 913	2 190	OECD Pacific
1 463	2 259	3 710	2 978	1 939	1 707	1 606	2 018	3 704	OECD Europe
2 377	3 557	5 409	4 316	3 523	3 567	3 528	3 931	5 894	**OECD Total**
2 415	3 563	5 514	21 289	15 416	16 435	14 078	14 310	15 274	**World**

Former USSR: series up to 1990-1992 are not comparable with recent years; 1991 is estimated.

Final Consumption of Other Bituminous Coal (1000 tonnes)

Consommation finale d'autres charbons bitumineux (1000 tonnes)

Endverbrauch von sonstiger bituminöser Kohle (1000 Tonnen)

Consumo finale di altri carboni bituminosi (1000 tonnellate)

その他歴青炭の最終消費量（千トン）

Consumo final de otro carbón bituminoso (1000 toneladas)

Конечное потребление других битуминозных углей (тыс. т)

	1971	1973	1978	1983	1984	1985	1986	1987	1988
RD du Congo	-	-	167	147	157	156	160	165	166
Ghana	-	-	2	2	2	2	2	3	3
Kenya	-	-	50	32	83	90	85	92	113
Maroc	-	-	28	248	282	523	495	439	447
Mozambique	-	-	247	144	94	65	39	32	28
Nigéria	-	-	159	44	64	91	100	96	34
Afrique du Sud	-	-	24 428	20 013	20 836	21 288	20 111	20 631	21 596
Tanzanie	-	-	1	7	10	15	4	3	3
Tunisie	-	-	27	28	24	21	12	16	29
Zambie	-	-	450	415	380	401	444	500	541
Zimbabwe	-	-	1 110	1 893	1 638	1 706	1 765	1 786	1 733
Autre Afrique	-	-	261	337	298	338	426	431	435
Afrique	-	-	26 930	23 310	23 868	24 696	23 643	24 194	25 128
Argentine	-	-	34	29	131	14	78	196	268
Brésil	-	-	379	3 236	3 163	3 427	3 726	3 911	3 380
Chili	-	-	527	432	559	632	724	667	732
Colombie	-	-	2 410	2 564	2 589	2 799	2 900	2 973	2 971
Costa Rica	-	-	1	1	1	1	1	1	1
Cuba	-	-	91	90	81	126	119	85	95
Haiti	-	-	-	51	54	61	18	18	31
Jamaïque	-	-	-	-	-	-	-	-	8
Panama	-	-	-	-	-	38	38	38	18
Pérou	-	-	26	75	89	114	107	100	80
Uruguay	-	-	2	-	-	-	-	1	-
Vénézuela	-	-	63	37	51	42	57	62	71
Amérique latine	-	-	3 533	6 515	6 718	7 254	7 768	8 052	7 655
Bangladesh	-	-	250	163	62	98	148	233	240
Inde	-	-	46 664	56 511	58 083	73 631	69 782	70 526	77 275
Indonésie	-	-	137	233	302	331	290	300	760
RPD de Corée	-	-	23 616	25 471	24 136	24 212	24 200	24 200	24 650
Malaisie	-	-	14	345	365	489	353	434	608
Myanmar	-	-	218	198	204	203	206	59	52
Népal	-	-	16	42	100	80	53	85	89
Pakistan	-	-	1 241	1 576	1 845	2 206	2 176	2 242	2 730
Philippines	-	-	171	667	813	954	743	958	1 155
Sri Lanka	-	-	-	-	-	-	-	-	-
Taipei chinois	-	-	1 956	2 947	3 122	3 220	3 441	3 643	4 385
Thailande	-	-	52	172	227	212	183	223	304
Viêt-Nam	-	-	3 172	3 426	3 244	3 208	3 518	3 882	2 884
Autre Asie	-	-	286	261	273	345	351	343	334
Asie	-	-	77 793	92 012	92 776	109 189	105 444	107 128	115 466
Rép. populaire de Chine	-	-	424 824	458 393	504 779	537 425	566 657	591 973	613 526
Hong-Kong, Chine	-	-	6	-	-	-	-	-	-
Chine	-	-	424 830	458 393	504 779	537 425	566 657	591 973	613 526

Final Consumption of Other Bituminous Coal (1000 tonnes)

Consommation finale d'autres charbons bitumineux (1000 tonnes)

Endverbrauch von sonstiger bituminöser Kohle (1000 Tonnen)

Consumo finale di altri carboni bituminosi (1000 tonnellate)

その他歴青炭の最終消費量（千トン)

Consumo final de otro carbón bituminoso (1000 toneladas)

Конечное потребление других битуминозных углей (тыс. т)

1989	1990	1991	1992	1993	1994	1995	1996	1997	
170	169	174	170	134	136	136	136	136	DR of Congo
3	3	3	3	3	3	3	3	3	Ghana
131	151	151	158	131	109	97	89	92	Kenya
772	604	786	731	739	760	758	952	1 002	Morocco
34	34	32	41	54	60	56	18	20	Mozambique
36	46	46	55	60	70	10	10	10	Nigeria
22 385	21 690	20 333	19 038	18 526	18 701	23 628	21 807	23 328	South Africa
2	3	3	3	3	3	4	4	4	Tanzania
30	15	12	15	14	-	-	-	-	Tunisia
312	297	234	267	200	119	124	118	118	Zambia
1 641	2 056	2 040	1 988	1 918	943	1 033	1 142	1 052	Zimbabwe
436	400	223	364	360	149	133	134	135	Other Africa
25 952	25 468	24 037	22 833	22 142	21 053	25 982	24 413	25 900	**Africa**
84	25	25	74	92	32	51	-	-	Argentina
2 705	2 189	2 845	2 159	1 802	1 893	1 463	1 277	1 009	Brazil
802	733	788	900	790	628	584	763	1 483	Chile
3 140	3 502	3 984	3 475	3 405	3 144	2 785	2 945	3 020	Colombia
1	1	-	-	-	-	-	-	-	Costa Rica
167	153	117	107	86	70	63	12	42	Cuba
12	12	25	26	-	-	-	-	-	Haiti
56	62	9	65	71	53	55	64	67	Jamaica
19	32	46	58	63	52	54	56	57	Panama
126	107	229	236	401	421	376	246	194	Peru
-	-	1	1	-	-	-	-	-	Uruguay
436	355	402	450	291	354	346	328	419	Venezuela
7 548	7 171	8 471	7 551	7 001	6 647	5 777	5 691	6 291	**Latin America**
250	563	180	169	63	-	-	-	-	Bangladesh
74 108	80 570	82 360	83 650	81 750	88 980	84 060	80 990	83 643	India
1 517	1 804	1 404	1 555	1 737	2 209	1 868	5 122	4 453	Indonesia
24 900	22 700	20 025	20 800	18 500	16 250	16 250	16 413	15 921	DPR of Korea
903	834	718	775	780	972	920	1 039	1 058	Malaysia
51	60	61	31	32	35	-	-	-	Myanmar
81	12	67	92	100	111	121	121	121	Nepal
2 602	3 103	3 030	3 059	3 220	3 490	3 002	3 239	2 812	Pakistan
724	795	1 243	1 004	1 093	961	907	1 670	2 183	Philippines
1	8	1	1	1	1	5	1	1	Sri Lanka
4 170	4 308	4 601	5 325	5 768	5 981	5 795	6 145	5 896	Chinese Taipei
392	250	436	489	986	1 424	2 305	3 723	3 037	Thailand
2 642	2 834	3 295	3 227	3 236	3 196	3 364	5 161	6 105	Vietnam
348	307	307	217	217	212	213	218	228	Other Asia
112 689	118 148	117 728	120 394	117 483	123 822	118 810	123 842	125 458	**Asia**
616 531	597 762	588 675	567 033	582 441	588 573	618 949	613 457	548 067	People's Rep. of China
-	-	-	-	-	-	-	-	-	Hong Kong, China
616 531	597 762	588 675	567 033	582 441	588 573	618 949	613 457	548 067	**China**

Final Consumption of Other Bituminous Coal (1000 tonnes)

Consommation finale d'autres charbons bitumineux (1000 tonnes)

Endverbrauch von sonstiger bituminöser Kohle (1000 Tonnen)

Consumo finale di altri carboni bituminosi (1000 tonnellate)

その他歴青炭の最終消費量（千トン）

Consumo final de otro carbón bituminoso (1000 toneladas)

Конечное потребление других битуминозных углей (тыс. т)

	1971	1973	1978	1983	1984	1985	1986	1987	1988
Albanie	-	-	143	170	178	187	195	195	203
Bulgarie	-	-	2 253	2 573	2 722	4 004	2 804	3 014	70
Chypre	-	-	-	-	52	74	55	151	91
Roumanie	-	-	1 572	1 195	282	2 759	3 134	4 213	5 174
République slovaque	-	-	1 312	1 157	1 189	1 181	1 209	1 212	1 194
Croatie	-	-	-	-	-	-	-	-	-
Ex-RYM	-	-	-	-	-	-	-	-	-
Slovénie	-	-	-	-	-	-	-	-	-
RF de Yougoslavie	-	-	-	-	-	-	-	-	-
Ex-Yougoslavie	-	-	572	137	288	89	259	236	75
Europe non-OCDE	-	-	5 852	5 232	4 711	8 294	7 656	9 021	6 807
Arménie	-	-	-	-	-	-	-	-	-
Azerbaïdjan	-	-	-	-	-	-	-	-	-
Bélarus	-	-	-	-	-	-	-	-	-
Estonie	-	-	-	-	-	-	-	-	-
Géorgie	-	-	-	-	-	-	-	-	-
Kazakhstan	-	-	-	-	-	-	-	-	-
Kirghizistan	-	-	-	-	-	-	-	-	-
Lettonie	-	-	-	-	-	-	-	-	-
Lituanie	-	-	-	-	-	-	-	-	-
Moldova	-	-	-	-	-	-	-	-	-
Russie	-	-	-	-	-	-	-	-	-
Tadjikistan	-	-	-	-	-	-	-	-	-
Turkménistan	-	-	-	-	-	-	-	-	-
Ukraine	-	-	-	-	-	-	-	-	-
Ouzbékistan	-	-	-	-	-	-	-	-	-
Ex-URSS	-	-	243 465	226 795	216 687	206 551	196 377	206 945	212 228
Iran	-	-	523	541	921	928	982	1 010	1 030
Israël	-	-	-	-	-	-	-	162	137
Liban	-	-	1	-	-	-	-	-	-
Syrie	-	-	1	3	2	-	-	-	-
Moyen-Orient	-	-	525	544	923	928	982	1 172	1 167
Total non-OCDE	-	-	782 928	812 801	850 462	894 337	908 527	948 485	981 977
OCDE Amérique du N.	-	-	62 375	61 571	67 152	64 955	64 207	62 667	63 732
OCDE Pacifique	-	-	23 219	36 805	40 677	44 024	45 145	43 080	44 573
OCDE Europe	-	-	90 885	93 432	94 980	103 441	101 766	104 163	100 326
Total OCDE	-	-	176 479	191 808	202 809	212 420	211 118	209 910	208 631
Monde	-	-	959 407	1 004 609	1 053 271	1 106 757	1 119 645	1 158 395	1 190 608

Ex-URSS: les séries antérieures à 1990-1992 ne sont pas comparables aux années récentes; 1991 a été estimé.

Final Consumption of Other Bituminous Coal (1000 tonnes)

Consommation finale d'autres charbons bitumineux (1000 tonnes)

Endverbrauch von sonstiger bituminöser Kohle (1000 Tonnen)

Consumo finale di altri carboni bituminosi (1000 tonnellate)

その他歴青炭の最終消費量（千トン）

Consumo final de otro carbón bituminoso (1000 toneladas)

Конечное потребление других битуминозных углей (тыс. т)

1989	1990	1991	1992	1993	1994	1995	1996	1997	
203	240	160	50	-	-	-	-	-	Albania
31	60	123	272	178	209	135	125	91	Bulgaria
102	97	97	26	31	27	20	17	19	Cyprus
700	648	1 173	414	108	105	104	110	120	Romania
1 260	1 077	1 196	1 106	1 046	988	1 033	684	840	Slovak Republic
-	-	-	83	68	60	46	61	56	*Croatia*
-	-	-	12	68	61	72	51	51	*FYROM*
-	-	-	13	17	22	20	26	19	*Slovenia*
-	-	-	1 070	121	120	119	126	130	*FR of Yugoslavia*
247	212	110	1 178	274	263	257	264	256	Former Yugoslavia
2 543	2 334	2 859	3 046	1 637	1 592	1 549	1 200	1 326	**Non-OECD Europe**
-	-	-	141	3	36	3	5	5	Armenia
-	-	-	27	7	8	6	6	6	Azerbaijan
-	-	-	1 112	1 182	894	808	778	557	Belarus
-	-	-	113	33	9	54	88	71	Estonia
-	-	-	406	288	152	71	72	5	Georgia
-	-	-	7 827	9 450	7 921	7 762	5 084	4 463	Kazakhstan
-	-	-	990	893	1 520	395	581	1 390	Kyrgyzstan
-	-	-	233	294	312	169	163	156	Latvia
-	-	-	596	586	414	318	303	230	Lithuania
-	-	-	640	419	435	244	348	244	Moldova
-	-	-	54 573	60 494	46 277	39 340	32 825	27 830	Russia
-	-	-	413	176	-	-	100	100	Tajikistan
-	-	-	600	135	-	-	100	-	Turkmenistan
-	-	-	28 877	33 879	22 955	18 274	13 541	12 982	Ukraine
-	-	-	1 381	688	788	42	36	25	Uzbekistan
209 517	194 312	149 208	97 929	108 527	81 721	67 486	54 030	48 064	**Former USSR**
1 050	1 250	1 350	1 500	607	632	750	750	750	Iran
160	18	11	11	10	8	12	26	28	Israel
-	-	-	-	111	112	180	200	200	Lebanon
-	-	-	-	-	-	-	-	-	Syria
1 210	1 268	1 361	1 511	728	752	942	976	978	**Middle East**
975 990	946 463	892 339	820 297	839 959	824 160	839 495	823 609	756 084	**Non-OECD Total**
61 402	62 125	60 156	29 167	31 603	28 760	27 004	26 898	29 564	OECD North America
41 814	40 563	37 686	34 090	30 959	28 712	25 601	27 750	26 944	OECD Pacific
92 111	74 378	76 474	72 296	74 387	67 336	70 642	74 253	67 596	OECD Europe
195 327	177 066	174 316	135 553	136 949	124 808	123 247	128 901	124 104	**OECD Total**
1 171 317	1 123 529	1 066 655	955 850	976 908	948 968	962 742	952 510	880 188	**World**

Former USSR: series up to 1990-1992 are not comparable with recent years; 1991 is estimated.

Final Consumption of Sub-Bituminous Coal (1000 tonnes)

Consommation finale de charbons sous-bitumineux (1000 tonnes)

Endverbrauch von subbituminöser Kohle (1000 Tonnen)

Consumo finale di carbone sub-bituminoso (1000 tonnellate)

亜歴青炭の最終消費量 （千トン）

Consumo final de carbón sub-bituminoso (1000 toneladas)

Конечное потребление полубитуминозных углей (тыс. т)

	1971	1973	1978	1983	1984	1985	1986	1987	1988
RPD de Corée	-	-	9 000	11 000	11 000	12 000	12 500	12 500	12 500
Asie	-	-	9 000	11 000	11 000	12 000	12 500	12 500	12 500
Bulgarie	-	-	-	-	-	-	-	-	-
Slovénie	-	-	-	-	-	-	-	-	-
Ex-Yougoslavie	-	-	-	-	-	-	-	-	-
Europe non-OCDE	-	-	-	-	-	-	-	-	-
Tadjikistan	-	-	-	-	-	-	-	-	-
Ex-URSS	-	-	-	-	-	-	-	-	-
Total non-OCDE	-	-	9 000	11 000	11 000	12 000	12 500	12 500	12 500
OCDE Amérique du N.	-	-	2 810	4 492	5 137	5 240	5 218	5 313	5 663
OCDE Pacifique	-	-	2 182	2 305	2 861	3 111	3 223	3 265	3 208
OCDE Europe	-	-	44 454	40 070	42 236	41 096	42 220	45 287	42 604
Total OCDE	-	-	49 446	46 867	50 234	49 447	50 661	53 865	51 475
Monde	-	-	58 446	57 867	61 234	61 447	63 161	66 365	63 975

Ex-URSS: les séries antérieures à 1990-1992 ne sont pas comparables aux années récentes; 1991 a été estimé.

Final Consumption of Sub-Bituminous Coal (1000 tonnes)

Consommation finale de charbons sous-bitumineux (1000 tonnes)

Endverbrauch von subbituminöser Kohle (1000 Tonnen)

Consumo finale di carbone sub-bituminoso (1000 tonnellate)

亜歴青炭の最終消費量 （千トン）

Consumo final de carbón sub-bituminoso (1000 toneladas)

Конечное потребление полубитуминозных углей (тыс. т)

1989	1990	1991	1992	1993	1994	1995	1996	1997	
13 000	12 250	11 500	10 750	10 000	9 300	8 500	7 769	7 536	DPR of Korea
13 000	12 250	11 500	10 750	10 000	9 300	8 500	7 769	7 536	**Asia**
-	-	510	548	527	420	388	588	384	Bulgaria
-	-	-	242	199	128	69	48	28	*Slovenia*
-	-	-	242	199	128	69	48	28	Former Yugoslavia
-	-	510	790	726	548	457	636	412	**Non-OECD Europe**
-	-	-	-	-	103	27	20	10	Tajikistan
-	-	-	-	-	103	27	20	10	**Former USSR**
13 000	12 250	12 010	11 540	10 726	9 951	8 984	8 425	7 958	**Non-OECD Total**
6 253	6 713	7 102	3 988	4 122	3 493	3 712	3 727	2 531	OECD North America
3 593	3 972	4 065	4 350	4 454	4 286	4 235	4 242	4 045	OECD Pacific
35 531	32 315	13 255	20 135	10 738	10 027	7 716	3 365	4 075	OECD Europe
45 377	43 000	24 422	28 473	19 314	17 806	15 663	11 334	10 651	**OECD Total**
58 377	55 250	36 432	40 013	30 040	27 757	24 647	19 759	18 609	**World**

Former USSR: series up to 1990-1992 are not comparable with recent years; 1991 is estimated.

Final Consumption of Lignite (1000 tonnes)

Consommation finale de lignite (1000 tonnes)

Endverbrauch von Braunkohle (1000 Tonnen)

Consumo finale di lignite (1000 tonnellate)

亜炭の最終消費量 (千トン)

Consumo final de lignito (1000 toneladas)

Конечное потребление лигнита (тыс. т)

	1971	1973	1978	1983	1984	1985	1986	1987	1988
Inde	-	-	601	344	504	1 024	639	1 754	1 853
Myanmar	-	-	3	35	43	43	47	35	35
Philippines	-	-	-	4	3	4	4	3	3
Thailande	-	-	132	347	361	535	741	1 096	1 303
Asie	-	-	736	730	911	1 606	1 431	2 888	3 194
Albanie	-	-	1 014	1 573	1 719	1 859	1 811	1 911	1 910
Bulgarie	-	-	3 655	4 500	4 499	4 292	4 902	5 000	2 800
Roumanie	-	-	4 015	10 801	7 990	5 426	2 883	5 696	2 632
République slovaque	-	-	6 672	6 381	6 547	6 926	7 232	7 256	7 076
Bosnie-Herzegovine	-	-	-	-	-	-	-	-	-
Croatie	-	-	-	-	-	-	-	-	-
Ex-RYM	-	-	-	-	-	-	-	-	-
Slovénie	-	-	-	-	-	-	-	-	-
RF de Yougoslavie	-	-	-	-	-	-	-	-	-
Ex-Yougoslavie	-	-	9 084	9 908	8 297	8 292	6 635	6 865	3 514
Europe non-OCDE	-	-	24 440	33 163	29 052	26 795	23 463	26 728	17 932
Estonie	-	-	-	-	-	-	-	-	-
Kazakhstan	-	-	-	-	-	-	-	-	-
Kirghizistan	-	-	-	-	-	-	-	-	-
Lituanie	-	-	-	-	-	-	-	-	-
Russie	-	-	-	-	-	-	-	-	-
Tadjikistan	-	-	-	-	-	-	-	-	-
Ukraine	-	-	-	-	-	-	-	-	-
Ouzbékistan	-	-	-	-	-	-	-	-	-
Ex-URSS	-	-	-	-	-	-	-	-	-
Total non-OCDE	-	-	25 176	33 893	29 963	28 401	24 894	29 616	21 126
OCDE Amérique du N.	-	-	2 665	3 017	4 419	7 067	7 972	8 388	8 263
OCDE Pacifique	-	-	477	526	509	532	473	340	520
OCDE Europe	-	-	40 229	49 748	60 273	67 184	64 516	68 360	64 780
Total OCDE	-	-	43 371	53 291	65 201	74 783	72 961	77 088	73 563
Monde	-	-	68 547	87 184	95 164	103 184	97 855	106 704	94 689

Ex-URSS: les séries antérieures à 1990-1992 ne sont pas comparables aux années récentes; 1991 a été estimé.

Final Consumption of Lignite (1000 tonnes)

Consommation finale de lignite (1000 tonnes)

Endverbrauch von Braunkohle (1000 Tonnen)

Consumo finale di lignite (1000 tonnellate)

亜炭の最終消費量 (千トン)

Consumo final de lignito (1000 toneladas)

Конечное потребление лигнита (тыс. т)

1989	1990	1991	1992	1993	1994	1995	1996	1997	
3 287	2 802	3 900	4 140	4 509	4 310	4 950	5 060	5 152	India
35	38	39	39	39	40	41	46	46	Myanmar
3	3	3	3	3	3	3	-	-	Philippines
1 794	2 582	2 818	3 147	4 263	4 916	4 915	4 551	4 469	Thailand
5 119	5 425	6 760	7 329	8 814	9 269	9 909	9 657	9 667	**Asia**
1 910	1 395	631	350	242	169	163	101	70	Albania
2 505	963	200	218	215	141	123	184	128	Bulgaria
5 079	2 267	1 531	1 398	1 034	477	297	735	563	Romania
8 506	7 983	7 451	6 239	5 364	3 855	3 352	3 354	2 373	Slovak Republic
-	-	-	5 000	5 000	337	290	290	290	*Bosnia-Herzegovina*
-	-	-	195	211	172	177	143	116	*Croatia*
-	-	-	162	201	213	160	158	150	*FYROM*
-	-	-	366	245	180	170	166	113	*Slovenia*
-	-	-	5 829	3 314	3 389	3 539	3 367	3 563	*FR of Yugoslavia*
4 907	5 163	4 626	11 552	8 971	4 291	4 336	4 124	4 232	Former Yugoslavia
22 907	17 771	14 439	19 757	15 826	8 933	8 271	8 498	7 366	**Non-OECD Europe**
-	-	-	337	264	439	593	596	434	Estonia
-	-	-	4 152	4 669	4 559	3 695	3 446	2 443	Kazakhstan
-	-	-	1 111	985	550	287	332	330	Kyrgyzstan
-	-	-	-	-	-	-	-	6	Lithuania
-	-	-	15 139	13 162	9 327	7 933	8 454	8 089	Russia
-	-	-	200	200	-	-	-	-	Tajikistan
-	-	-	4 779	3 121	2 269	2 000	2 000	1 000	Ukraine
-	-	-	575	795	704	846	841	460	Uzbekistan
-	-	-	26 293	23 196	17 848	15 354	15 669	12 762	**Former USSR**
28 026	23 196	21 199	53 379	47 836	36 050	33 534	33 824	29 795	**Non-OECD Total**
8 716	8 163	7 975	6 478	6 078	6 047	6 307	6 351	5 788	OECD North America
327	248	263	271	281	302	304	305	329	OECD Pacific
61 927	49 761	36 757	28 980	23 811	18 361	18 347	17 476	17 677	OECD Europe
70 970	58 172	44 995	35 729	30 170	24 710	24 958	24 132	23 794	**OECD Total**
98 996	81 368	66 194	89 108	78 006	60 760	58 492	57 956	53 589	**World**

Former USSR: series up to 1990-1992 are not comparable with recent years; 1991 is estimated.

Final Consumption of Oil (1000 tonnes)

Consommation finale de pétrole (1000 tonnes)

Endverbrauch von Öl (1000 Tonnen)

Consumo finale di petrolio (1000 tonnellate)

石油の最終消費量（千トン）

Consumo final de petróleo (1000 toneladas)

Конечное потребление нефти и нефтепродуктов (тыс. т)

	1971	1973	1978	1983	1984	1985	1986	1987	1988
Algérie	1 607	2 283	3 701	5 994	6 698	6 975	7 170	7 293	7 216
Angola	523	683	654	459	370	535	426	396	412
Bénin	101	124	126	116	131	161	141	126	122
Cameroun	276	301	511	799	860	856	832	825	866
Congo	158	164	212	261	237	246	243	221	200
RD du Congo	628	700	749	1 014	972	907	794	925	987
Egypte	4 939	5 176	7 562	11 839	13 082	13 832	13 433	14 235	14 491
Ethiopie	361	379	390	461	462	474	576	698	720
Gabon	135	170	566	589	546	561	506	434	478
Ghana	622	649	813	645	629	675	686	717	872
Côte d'Ivoire	659	694	937	947	837	819	897	940	942
Kenya	1 037	1 192	1 498	1 380	1 582	1 643	1 743	1 805	1 787
Libye	438	891	1 897	3 475	3 407	3 334	3 154	3 279	3 154
Maroc	1 706	2 077	2 972	3 020	2 977	2 989	2 958	2 929	3 084
Mozambique	349	359	331	307	279	310	331	334	337
Nigéria	1 496	2 353	6 243	8 600	7 674	7 655	7 102	7 239	7 624
Sénégal	369	437	484	542	567	563	539	512	490
Afrique du Sud	8 653	10 213	11 691	12 245	12 992	12 589	12 380	13 006	14 099
Soudan	1 070	1 529	959	1 202	1 143	1 282	1 254	970	1 373
Tanzanie	433	507	496	412	408	429	450	450	488
Tunisie	892	1 062	1 624	2 085	2 149	2 196	2 105	1 958	2 201
Zambie	462	637	687	630	571	540	477	491	501
Zimbabwe	539	680	658	750	707	758	797	807	839
Autre Afrique	2 263	2 535	2 634	2 676	2 641	2 750	2 816	4 113	4 181
Afrique	29 716	35 795	48 395	60 448	61 921	63 079	61 810	64 703	67 464
Argentine	16 212	16 756	17 523	17 608	17 566	16 201	16 229	17 575	16 047
Bolivie	551	653	976	920	854	831	848	912	923
Brésil	23 139	31 266	44 893	39 116	38 745	41 354	44 622	45 930	44 433
Chili	3 795	3 820	3 816	3 814	3 855	3 765	4 010	4 176	4 678
Colombie	5 409	5 836	6 707	7 498	7 410	7 554	7 619	8 345	8 399
Costa Rica	354	435	660	526	565	611	629	665	692
Cuba	4 235	5 605	6 168	6 375	6 435	6 929	6 766	6 974	7 202
République dominicaine	750	877	1 080	1 128	1 177	982	1 121	1 328	1 396
Equateur	1 000	1 133	2 372	2 951	3 266	3 614	3 783	3 687	4 060
El Salvador	348	434	614	472	482	506	506	545	616
Guatemala	643	711	891	721	770	790	797	878	940
Haiti	107	121	204	161	166	169	185	193	205
Honduras	295	352	459	505	531	523	538	596	666
Jamaïque	1 408	2 068	1 952	1 441	1 353	1 202	1 143	1 171	1 188
Antilles néerlandaises	2 159	2 248	1 234	1 013	933	860	773	752	693
Nicaragua	396	498	548	477	480	481	524	559	494
Panama	478	523	618	510	568	582	618	613	533
Paraguay	170	218	366	419	464	490	508	535	554
Pérou	4 073	4 280	4 855	4 608	4 685	4 656	5 032	5 565	5 461
Trinité-et-Tobago	407	515	758	827	868	820	755	699	681
Uruguay	1 325	1 330	1 354	1 082	976	908	922	992	1 010
Vénézuela	5 838	6 918	10 865	12 435	12 054	12 798	13 343	14 411	14 248
Autre Amérique latine	1 685	2 113	1 904	1 638	1 822	1 718	1 862	2 224	2 150
Amérique latine	74 777	88 710	110 817	106 245	106 025	108 344	113 133	119 325	117 269

Final Consumption of Oil (1000 tonnes)

Consommation finale de pétrole (1000 tonnes)

Endverbrauch von Öl (1000 Tonnen)

Consumo finale di petrolio (1000 tonnellate)

石油の最終消費量（千トン）

Consumo final de petróleo (1000 toneladas)

Конечное потребление нефти и нефтепродуктов (тыс. т)

1989	1990	1991	1992	1993	1994	1995	1996	1997	
7 482	7 308	7 889	7 770	8 007	7 386	7 179	7 011	8 409	Algeria
391	401	396	401	402	399	403	488	732	Angola
102	89	71	72	76	81	83	316	308	Benin
948	919	851	793	782	815	810	799	807	Cameroon
264	254	264	249	257	251	274	278	278	Congo
975	1 057	1 083	1 000	1 122	1 140	1 154	1 154	1 154	DR of Congo
14 630	15 026	14 177	14 546	12 729	12 411	14 154	15 052	16 069	Egypt
763	796	817	826	859	906	975	1 114	681	Ethiopia
361	290	334	346	384	375	411	444	523	Gabon
932	885	838	899	961	1 023	1 063	1 059	1 059	Ghana
923	770	764	726	758	776	813	862	888	Ivory Coast
1 870	1 822	1 728	1 798	1 825	1 903	1 858	2 022	1 934	Kenya
3 593	3 592	3 757	3 742	4 212	4 883	5 359	5 421	5 627	Libya
3 421	3 567	3 705	4 213	4 176	4 565	4 706	4 762	4 848	Morocco
322	366	322	390	435	396	378	391	510	Mozambique
7 878	6 534	7 965	10 511	8 344	7 631	7 521	8 294	8 401	Nigeria
536	558	548	577	513	564	626	706	779	Senegal
14 461	14 874	14 823	14 670	14 098	14 668	16 017	16 111	16 359	South Africa
1 283	1 659	1 470	1 393	864	1 417	1 256	1 155	1 402	Sudan
496	521	514	508	467	471	464	454	410	Tanzania
2 455	2 497	2 412	2 560	2 834	2 834	2 886	3 047	3 122	Tunisia
493	473	485	452	466	481	485	491	491	Zambia
785	917	987	962	1 004	1 112	1 254	1 248	1 248	Zimbabwe
4 409	4 235	4 434	4 589	4 761	4 755	4 837	4 891	5 073	Other Africa
69 773	69 410	70 634	73 993	70 336	71 243	74 966	77 570	81 112	**Africa**
15 200	15 073	15 450	16 049	17 571	17 970	18 240	19 269	19 309	Argentina
979	1 012	1 038	1 046	1 079	1 134	1 255	1 356	1 398	Bolivia
45 831	46 684	47 408	48 719	50 514	53 277	57 200	61 269	65 753	Brazil
5 105	5 304	5 599	6 144	6 722	7 164	7 818	8 373	9 172	Chile
8 726	8 869	8 840	9 545	10 467	11 108	11 978	12 335	12 570	Colombia
821	816	838	1 111	1 163	1 275	1 281	1 278	1 349	Costa Rica
7 578	6 546	4 742	3 677	3 554	3 816	3 881	4 262	4 326	Cuba
1 599	1 579	1 532	1 909	1 769	1 885	1 958	2 032	2 182	Dominican Republic
3 974	3 991	4 029	3 827	3 789	4 040	4 275	4 651	4 987	Ecuador
629	673	716	875	933	994	1 155	1 114	1 166	El Salvador
981	1 051	1 098	1 180	1 316	1 388	1 669	1 667	1 762	Guatemala
229	239	222	217	200	52	260	279	321	Haiti
742	691	703	805	785	924	1 119	1 135	1 152	Honduras
1 565	1 778	1 841	1 857	955	1 323	1 260	1 345	1 434	Jamaica
624	596	639	674	712	800	802	792	792	Netherlands Antilles
474	458	436	467	471	527	524	553	625	Nicaragua
554	582	620	670	681	757	817	837	935	Panama
624	625	614	750	839	997	1 094	990	1 068	Paraguay
4 874	4 787	4 751	4 632	4 957	4 966	5 971	6 405	6 286	Peru
616	654	725	703	611	603	623	662	746	Trinidad and Tobago
1 026	1 017	1 073	1 131	1 189	1 248	1 254	1 362	1 453	Uruguay
13 412	13 744	14 406	14 993	16 842	17 870	17 547	16 690	18 611	Venezuela
2 466	2 449	2 384	2 182	2 212	2 295	2 323	2 385	2 462	Other Latin America
118 629	119 218	119 704	123 163	129 331	136 413	144 304	151 041	159 859	**Latin America**

Final Consumption of Oil (1000 tonnes)
Consommation finale de pétrole (1000 tonnes)
Endverbrauch von Öl (1000 Tonnen)

Consumo finale di petrolio (1000 tonnellate)

石油の最終消費量（千トン）

Consumo final de petróleo (1000 toneladas)

Конечное потребление нефти и нефтепродуктов (тыс. т)

	1971	1973	1978	1983	1984	1985	1986	1987	1988
Bangladesh	569	604	993	1 121	1 140	1 243	1 239	1 319	1 342
Brunei	61	67	97	216	205	220	225	237	261
Inde	17 865	20 350	24 700	32 324	34 498	37 310	39 684	42 418	45 740
Indonésie	6 142	7 413	13 373	18 063	17 266	17 870	17 919	18 571	20 337
RPD de Corée	683	765	1 732	2 594	2 748	2 902	3 059	3 090	3 090
Malaisie	3 497	3 374	4 582	6 549	6 571	6 625	6 758	7 162	7 633
Myanmar	1 177	947	1 064	999	1 043	1 000	1 029	639	616
Népal	45	61	78	112	131	136	166	191	217
Pakistan	2 960	3 109	4 031	5 197	5 825	6 201	6 626	7 114	7 553
Philippines	5 227	5 870	6 927	6 036	5 086	4 576	4 850	5 548	6 023
Singapour	1 023	1 381	2 294	2 877	3 131	3 777	3 834	4 239	4 263
Sri Lanka	797	1 013	1 034	1 084	1 141	1 079	1 085	1 085	1 126
Taipei chinois	3 714	5 751	10 508	12 417	13 304	14 095	15 374	16 488	17 505
Thailande	4 873	6 100	7 467	8 436	9 164	9 204	9 748	10 863	12 213
Viêt-Nam	5 660	5 654	933	1 570	1 540	1 524	1 719	2 005	2 059
Autre Asie	1 391	1 622	1 285	1 126	1 198	1 376	1 185	1 411	1 351
Asie	55 684	64 081	81 098	100 721	103 991	109 138	114 500	122 380	131 329
Rép. populaire de Chine	34 402	41 215	71 580	58 856	62 260	65 710	71 174	77 040	81 175
Hong-Kong, Chine	1 991	1 938	3 127	3 040	3 202	2 953	3 455	3 539	4 208
Chine	36 393	43 153	74 707	61 896	65 462	68 663	74 629	80 579	85 383
Albanie	583	468	866	862	834	635	656	678	712
Bulgarie	8 758	9 317	7 971	6 821	6 471	5 922	5 964	5 752	7 565
Chypre	423	549	520	629	647	629	719	769	842
Gibraltar	27	27	20	23	21	26	32	38	48
Malte	156	182	162	173	213	119	237	254	251
Roumanie	9 218	10 472	12 235	8 527	8 087	10 001	9 928	11 519	11 031
République slovaque	2 870	3 255	3 318	4 249	4 113	4 048	3 924	3 887	3 786
Bosnie-Herzegovine	-	-	-	-	-	-	-	-	-
Croatie	-	-	-	-	-	-	-	-	-
Ex-RYM	-	-	-	-	-	-	-	-	-
Slovénie	-	-	-	-	-	-	-	-	-
RF de Yougoslavie	-	-	-	-	-	-	-	-	-
Ex-Yougoslavie	7 718	8 782	12 050	10 510	9 190	9 700	10 815	10 796	11 364
Europe non-OCDE	29 753	33 052	37 142	31 794	29 576	31 080	32 275	33 693	35 599
Arménie	-	-	-	-	-	-	-	-	-
Azerbaïdjan	-	-	-	-	-	-	-	-	-
Bélarus	-	-	-	-	-	-	-	-	-
Estonie	-	-	-	-	-	-	-	-	-
Géorgie	-	-	-	-	-	-	-	-	-
Kazakhstan	-	-	-	-	-	-	-	-	-
Kirghizistan	-	-	-	-	-	-	-	-	-
Lettonie	-	-	-	-	-	-	-	-	-
Lituanie	-	-	-	-	-	-	-	-	-
Moldova	-	-	-	-	-	-	-	-	-
Russie	-	-	-	-	-	-	-	-	-
Tadjikistan	-	-	-	-	-	-	-	-	-
Turkménistan	-	-	-	-	-	-	-	-	-
Ukraine	-	-	-	-	-	-	-	-	-
Ouzbékistan	-	-	-	-	-	-	-	-	-
Ex-URSS	198 700	225 100	279 500	299 300	299 200	296 416	300 912	304 617	312 548

Ex-URSS: les séries antérieures à 1990-1992 ne sont pas comparables aux années récentes; 1991 a été estimé.

Final Consumption of Oil (1000 tonnes)

Consommation finale de pétrole (1000 tonnes)

Endverbrauch von Öl (1000 Tonnen)

Consumo finale di petrolio (1000 tonnellate)

石油の最終消費量（千トン）

Consumo final de petróleo (1000 toneladas)

Конечное потребление нефти и нефтепродуктов (тыс. т)

1989	1990	1991	1992	1993	1994	1995	1996	1997	
1 531	1 584	1 535	1 656	1 819	1 972	2 349	2 440	2 463	Bangladesh
272	281	298	346	370	392	426	464	499	Brunei
49 652	50 309	53 722	56 408	57 913	61 875	69 192	74 867	77 310	India
23 009	24 345	25 479	26 814	29 212	31 675	33 957	36 914	41 188	Indonesia
3 330	3 287	2 992	2 772	2 483	1 707	1 481	1 287	1 248	DPR of Korea
8 465	9 795	10 647	11 320	13 363	14 496	15 713	17 407	18 044	Malaysia
606	591	552	629	753	873	1 138	1 242	1 418	Myanmar
188	179	236	292	302	398	500	500	500	Nepal
7 854	7 937	7 903	8 230	9 325	9 724	10 216	11 336	10 960	Pakistan
6 843	7 181	6 992	7 884	8 943	9 519	10 438	11 102	12 013	Philippines
4 583	5 472	5 298	5 068	5 578	6 194	6 643	7 070	7 924	Singapore
1 106	1 139	1 145	1 344	1 434	1 588	1 756	2 028	1 968	Sri Lanka
18 500	19 475	19 822	21 097	22 049	24 035	24 702	25 884	26 517	Chinese Taipei
14 329	16 107	16 660	18 406	20 312	22 825	25 829	28 840	29 286	Thailand
1 978	2 433	2 118	2 765	3 311	3 716	3 884	5 675	6 962	Vietnam
1 529	1 650	1 644	1 423	1 405	1 346	1 445	1 426	1 464	Other Asia
143 775	151 765	157 043	166 454	178 572	192 335	209 669	228 482	239 764	**Asia**
84 805	84 459	91 281	97 808	105 219	109 799	122 698	137 614	153 485	People's Rep. of China
4 371	4 554	4 357	5 635	5 890	6 597	6 537	6 394	6 714	Hong Kong, China
89 176	89 013	95 638	103 443	111 109	116 396	129 235	144 008	160 199	**China**
669	834	682	523	468	590	453	447	386	Albania
7 591	6 142	3 244	3 304	3 730	3 564	3 472	3 437	2 980	Bulgaria
873	878	1 067	1 147	1 141	1 176	1 237	1 291	1 288	Cyprus
44	46	62	72	72	85	83	83	83	Gibraltar
251	252	281	288	294	294	317	383	422	Malta
9 898	8 936	8 374	7 601	5 105	5 927	5 964	7 078	7 584	Romania
5 171	4 523	3 758	3 326	2 585	3 084	3 027	3 113	3 178	Slovak Republic
-	-	-	436	418	318	318	318	318	*Bosnia-Herzegovina*
-	-	-	2 083	2 154	2 363	2 401	2 293	2 647	*Croatia*
-	-	-	765	759	687	643	998	827	*FYROM*
-	-	-	1 424	1 734	1 869	2 078	2 450	2 444	*Slovenia*
-	-	-	1 599	1 204	1 147	1 265	1 909	2 622	*FR of Yugoslavia*
10 652	10 457	9 634	6 307	6 269	6 384	6 705	7 968	8 858	Former Yugoslavia
35 149	32 068	27 102	22 568	19 664	21 104	21 258	23 800	24 779	**Non-OECD Europe**
-	-	-	1 227	851	228	163	91	91	Armenia
-	-	-	4 935	3 453	3 448	2 737	2 000	1 735	Azerbaijan
-	-	-	11 808	8 107	6 051	5 548	5 318	5 465	Belarus
-	-	-	726	732	815	876	957	918	Estonia
-	-	-	936	539	205	95	739	807	Georgia
-	-	-	18 147	13 433	9 440	9 224	7 599	6 788	Kazakhstan
-	-	-	1 909	1 165	417	599	667	488	Kyrgyzstan
-	-	-	1 814	1 640	1 518	1 272	1 057	1 179	Latvia
-	-	-	2 772	1 763	1 529	1 638	1 684	1 763	Lithuania
-	-	-	1 200	865	678	685	624	631	Moldova
-	-	-	136 762	117 654	77 447	86 635	80 413	78 843	Russia
-	-	-	5 595	3 483	1 159	1 159	1 159	1 159	Tajikistan
-	-	-	4 418	2 360	2 377	2 119	2 295	2 332	Turkmenistan
-	-	-	30 125	20 754	18 115	19 323	15 710	14 322	Ukraine
-	-	-	6 090	5 531	5 195	4 674	4 301	4 650	Uzbekistan
313 118	303 755	300 115	228 464	182 330	128 622	136 747	124 614	121 171	**Former USSR**

Former USSR: series up to 1990-1992 are not comparable with recent years; 1991 is estimated.

Final Consumption of Oil (1000 tonnes)

Consommation finale de pétrole (1000 tonnes)

Endverbrauch von Öl (1000 Tonnen)

Consumo finale di petrolio (1000 tonnellate)

石油の最終消費量（千トン）

Consumo final de petróleo (1000 toneladas)

Конечное потребление нефти и нефтепродуктов (тыс. т)

	1971	1973	1978	1983	1984	1985	1986	1987	1988
Bahrein	208	289	639	703	724	725	724	721	744
Iran	7 599	10 823	21 030	24 777	29 587	29 376	31 166	34 528	34 451
Irak	1 939	2 253	5 056	8 555	9 214	9 714	10 295	10 298	11 183
Israël	3 154	3 640	3 966	4 306	4 203	4 165	4 633	4 958	5 064
Jordanie	394	465	1 001	1 770	1 735	1 799	1 906	1 902	1 909
Koweit	633	717	1 695	2 758	2 132	2 701	2 152	2 115	2 267
Liban	1 477	1 772	1 444	1 514	1 427	1 618	1 547	1 632	1 106
Oman	81	82	446	713	911	980	934	837	882
Qatar	78	133	354	523	498	537	501	532	520
Arabie saoudite	2 203	3 287	11 404	31 135	23 480	24 914	25 826	24 705	26 289
Syrie	1 861	1 752	4 186	6 099	6 668	6 312	6 417	6 999	6 834
Emirats arabes unis	201	402	2 260	3 830	3 485	4 036	4 040	4 106	3 774
Yémen	292	484	790	1 419	1 426	1 691	1 762	1 883	1 977
Moyen-Orient	20 120	26 099	54 271	88 102	85 490	88 568	91 903	95 216	97 000
Total non-OCDE	445 143	515 990	685 930	748 506	751 665	765 288	789 162	820 513	846 592
OCDE Amérique du N.	697 407	766 782	842 083	711 357	740 378	742 672	756 886	778 994	801 046
OCDE Pacifique	174 580	205 201	212 564	192 681	202 971	201 582	208 727	217 236	232 146
OCDE Europe	516 092	585 572	581 511	491 931	493 045	494 434	512 939	517 584	527 170
Total OCDE	1 388 079	1 557 555	1 636 158	1 395 969	1 436 394	1 438 688	1 478 552	1 513 814	1 560 362
Monde	1 833 222	2 073 545	2 322 088	2 144 475	2 188 059	2 203 976	2 267 714	2 334 327	2 406 954

Final Consumption of Oil (1000 tonnes)

Consommation finale de pétrole (1000 tonnes)

Endverbrauch von Öl (1000 Tonnen)

Consumo finale di petrolio (1000 tonnellate)

石油の最終消費量（千トン）

Consumo final de petróleo (1000 toneladas)

Конечное потребление нефти и нефтепродуктов (тыс. т)

1989	1990	1991	1992	1993	1994	1995	1996	1997	
767	843	723	735	734	798	796	788	950	Bahrain
35 824	38 109	40 938	42 433	44 155	45 524	46 798	48 783	49 160	Iran
12 344	12 436	10 035	10 791	13 032	14 192	13 965	13 638	14 005	Iraq
5 065	5 349	5 458	5 906	6 469	6 820	7 690	8 054	8 307	Israel
1 941	2 194	2 023	2 355	2 438	2 529	2 624	2 736	2 929	Jordan
2 408	2 108	564	4 004	3 298	3 445	3 861	3 809	4 334	Kuwait
1 175	1 216	1 630	1 807	2 266	2 461	2 920	2 739	3 028	Lebanon
941	1 533	2 721	2 452	2 080	1 647	2 079	1 984	1 667	Oman
1 377	609	747	635	667	742	804	888	992	Qatar
25 795	28 325	29 805	30 539	32 753	32 647	30 873	32 993	34 281	Saudi Arabia
6 819	7 351	7 925	8 029	8 094	8 495	8 663	8 968	9 437	Syria
3 944	3 813	3 869	4 241	4 672	4 910	4 701	4 348	3 908	United Arab Emirates
1 938	1 664	2 211	2 490	1 914	2 041	2 289	2 284	2 349	Yemen
100 338	105 550	108 649	116 417	122 572	126 251	128 063	132 012	135 347	**Middle East**
869 958	870 779	878 885	834 502	813 914	792 364	844 242	881 527	922 231	**Non-OECD Total**
800 880	790 587	768 705	791 662	801 141	827 458	829 799	854 752	871 909	OECD North America
242 972	256 879	267 391	281 862	289 122	301 874	315 920	325 282	333 227	OECD Pacific
521 197	525 219	539 021	545 145	544 941	548 671	558 668	571 698	576 356	OECD Europe
1 565 049	1 572 685	1 575 117	1 618 669	1 635 204	1 678 003	1 704 387	1 751 732	1 781 492	**OECD Total**
2 435 007	2 443 464	2 454 002	2 453 171	2 449 118	2 470 367	2 548 629	2 633 259	2 703 723	**World**

Final Consumption of Natural Gas (TJ)

Consommation finale de gaz naturel (TJ)

Endverbrauch von Erdgas (TJ)

Consumo finale di gas naturale (TJ)

天然ガスの最終消費量 (TJ)

Consumo final de gas natural (TJ)

Конечное потребление природного газа (ТДж)

	1971	1973	1978	1983	1984	1985	1986	1987	1988
Algérie	11 455	13 900	42 645	64 787	71 764	88 397	109 100	118 500	119 800
Angola	1 678	2 536	2 758	3 997	4 499	4 499	4 997	5 998	6 098
Congo	2 256	-	-	-	-	-	-	-	-
Egypte	-	-	15 101	53 755	66 801	68 070	71 149	81 935	81 120
Gabon	-	-	-	1 450	1 546	1 984	3 512	4 486	6 824
Libye	-	-	30 408	55 000	60 000	67 000	69 000	70 000	51 000
Maroc	2 009	2 721	3 177	3 248	3 244	3 666	3 549	2 884	2 947
Nigéria	1 411	1 352	1 398	5 717	14 343	17 188	18 149	21 296	26 957
Sénégal	-	-	-	-	-	-	-	11	314
Tunisie	38	331	2 888	4 817	5 085	5 902	7 969	13 268	7 124
Autre Afrique	37	37	37	42	16	15	10	4	7
Afrique	18 884	20 877	98 412	192 813	227 298	256 721	287 435	318 382	302 191
Argentine	134 479	171 102	221 648	324 083	360 389	355 784	377 651	421 248	454 345
Bolivie	-	155	2 203	1 739	2 087	2 783	3 208	3 672	3 710
Brésil	1 521	4 613	29 029	57 811	64 563	75 338	76 074	89 335	96 360
Chili	155	1 786	4 560	5 510	5 890	5 852	6 384	6 118	16 872
Colombie	4 905	9 832	25 485	32 711	36 216	37 682	41 199	40 998	44 002
Cuba	67	565	427	306	126	266	228	874	722
Trinité-et-Tobago	21 363	24 192	39 896	77 526	76 243	76 099	82 371	83 510	111 442
Vénézuela	108 198	202 311	279 811	342 688	399 112	421 136	377 354	407 927	328 007
Autre Amérique latine	117	117	374	428	760	993	675	592	479
Amérique latine	270 805	414 673	603 433	842 802	945 386	975 933	965 144	1 054 274	1 055 939
Bangladesh	9 049	15 403	21 075	46 851	51 917	55 365	64 185	71 460	79 972
Inde	9 218	8 617	34 299	59 508	73 570	112 884	149 155	156 615	238 889
Indonésie	4 637	5 630	64 584	155 884	185 804	251 208	256 651	265 396	262 699
Malaisie	293	419	1 415	2 055	5 659	23 489	41 490	51 616	57 588
Myanmar	762	1 175	3 145	5 938	7 969	10 845	12 126	12 478	12 602
Pakistan	69 684	83 917	119 992	204 917	201 729	208 309	212 680	246 773	256 774
Taipei chinois	30 295	43 741	67 810	45 322	45 889	41 088	37 485	38 929	41 913
Thailande	-	-	-	1 329	8 116	7 452	3 638	1 653	2 516
Viêt-Nam	-	-	-	-	-	-	-	-	-
Autre Asie	8 967	6 746	6 094	12 482	18 940	24 240	24 444	7 950	8 938
Asie	132 905	165 648	318 414	534 286	599 593	734 880	801 854	852 870	961 891
Rép. populaire de Chine	65 591	104 864	293 789	271 446	290 160	334 020	327 419	347 295	357 709
Hong-Kong, Chine	-	-	-	-	-	-	-	-	-
Chine	65 591	104 864	293 789	271 446	290 160	334 020	327 419	347 295	357 709
Albanie	4 914	7 413	13 804	14 997	14 997	14 997	14 997	14 997	15 997
Bulgarie	11 506	7 810	107 601	187 158	196 780	171 632	177 181	187 692	99 898
Roumanie	605 483	646 572	1 170 245	1 283 973	1 290 428	1 015 452	1 035 853	1 029 288	1 046 281
République slovaque	54 763	65 312	128 169	180 227	179 391	176 922	178 607	180 728	183 525
Bosnie-Herzegovine	-	-	-	-	-	-	-	-	-
Croatie	-	-	-	-	-	-	-	-	-
Ex-RYM	-	-	-	-	-	-	-	-	-
Slovénie	-	-	-	-	-	-	-	-	-
RF de Yougoslavie	-	-	-	-	-	-	-	-	-
Ex-Yougoslavie	37 152	45 773	68 010	92 866	126 176	137 380	144 085	192 123	147 526
Europe non-OCDE	713 818	772 880	1 487 829	1 759 221	1 807 772	1 516 383	1 550 723	1 604 828	1 493 227

Final Consumption of Natural Gas (TJ)

Consommation finale de gaz naturel (TJ)

Endverbrauch von Erdgas (TJ)

Consumo finale di gas naturale (TJ)

天然ガスの最終消費量 (TJ)

Consumo final de gas natural (TJ)

Конечное потребление природного газа (ТДж)

	1989	1990	1991	1992	1993	1994	1995	1996	1997
Algeria	109 500	113 500	134 600	141 508	229 946	220 456	223 245	225 899	234 180
Angola	6 500	20 520	22 040	21 660	21 280	19 760	21 280	21 280	21 660
Congo	-	-	-	-	-	-	-	-	-
Egypt	120 396	112 408	114 200	119 776	144 086	132 958	143 811	146 452	154 128
Gabon	30	30	30	30	30	25	27	29	28
Libya	52 000	60 000	65 000	65 000	67 000	69 144	68 314	73 526	78 696
Morocco	2 177	2 014	1 289	847	828	880	884	698	1 047
Nigeria	31 825	33 326	30 326	32 500	34 600	37 714	34 433	34 898	33 635
Senegal	299	219	170	105	498	761	1 843	1 685	928
Tunisia	9 830	14 611	15 775	17 537	18 654	19 588	21 775	21 915	22 473
Other Africa	5	6	5	6	7	7	7	7	7
Africa	332 562	356 634	383 435	398 969	516 929	501 293	515 619	526 389	546 782
Argentina	393 440	415 804	452 381	475 340	537 363	568 399	593 663	597 106	642 315
Bolivia	6 996	7 382	8 194	9 740	11 001	12 307	13 495	11 898	16 084
Brazil	101 247	103 389	108 784	118 181	121 305	129 117	137 446	155 267	173 738
Chile	37 278	40 736	31 616	40 789	40 013	42 882	42 006	43 665	75 347
Colombia	45 383	48 703	51 084	52 588	54 869	58 898	62 223	66 067	68 170
Cuba	988	1 250	1 033	566	810	639	543	608	734
Trinidad and Tobago	113 555	116 965	124 213	127 403	124 498	167 374	169 198	191 044	205 283
Venezuela	335 513	388 024	444 103	419 622	434 318	436 823	477 219	481 013	516 786
Other Latin America	636	742	490	459	670	548	671	857	857
Latin America	1 035 036	1 122 995	1 221 898	1 244 688	1 324 847	1 416 987	1 496 464	1 547 525	1 699 314
Bangladesh	85 747	83 384	81 241	99 385	106 328	121 497	139 161	143 095	138 129
India	292 008	279 961	288 792	333 680	320 862	340 969	381 885	423 792	438 558
Indonesia	260 932	324 005	338 804	377 878	333 368	338 995	373 495	411 888	437 360
Malaysia	49 792	52 256	52 334	63 638	79 844	86 667	87 419	113 206	103 244
Myanmar	12 190	10 447	8 451	8 371	10 709	13 347	13 341	13 100	15 345
Pakistan	278 550	278 603	294 339	299 379	330 428	353 905	368 335	399 392	416 316
Chinese Taipei	42 170	43 767	52 061	58 505	59 733	63 434	71 352	74 823	75 648
Thailand	4 781	11 051	15 053	18 455	20 607	24 556	32 901	39 310	39 784
Vietnam	-	-	-	-	-	-	9 881	135 527	143 884
Other Asia	9 754	11 167	10 920	10 373	10 230	9 930	9 750	9 500	9 500
Asia	1 035 924	1 094 641	1 141 995	1 269 664	1 272 109	1 353 300	1 487 520	1 763 633	1 817 768
People's Rep. of China	384 348	404 103	434 761	447 356	447 782	455 840	466 712	577 738	505 331
Hong Kong, China	-	-	-	-	-	-	1 086	68 823	107 462
China	384 348	404 103	434 761	447 356	447 782	455 840	467 798	646 561	612 793
Albania	12 597	9 059	5 126	3 584	3 080	1 976	1 089	894	700
Bulgaria	103 693	106 508	90 855	75 096	73 858	83 265	97 520	98 539	81 778
Romania	1 050 489	923 608	696 575	342 355	332 571	430 410	471 352	464 141	413 902
Slovak Republic	193 498	214 116	190 334	193 586	187 090	174 262	189 428	211 143	212 466
Bosnia-Herzegovina	-	-	-	16 000	-	-	-	-	-
Croatia	-	-	-	58 938	59 063	60 397	62 822	66 192	70 596
FYROM	-	-	-	10 000	10 500	-	-	-	-
Slovenia	-	-	-	18 546	18 369	21 825	23 187	25 397	29 756
FR of Yugoslavia	-	-	-	61 400	62 979	28 472	37 280	91 359	90 706
Former Yugoslavia	134 074	273 295	215 337	164 884	150 911	110 694	123 289	182 948	191 058
Non-OECD Europe	1 494 351	1 526 586	1 198 227	779 505	747 510	800 607	882 678	957 665	899 904

Final Consumption of Natural Gas (TJ)

Consommation finale de gaz naturel (TJ)

Endverbrauch von Erdgas (TJ)

Consumo finale di gas naturale (TJ)

天然ガスの最終消費量 (TJ)

Consumo final de gas natural (TJ)

Конечное потребление природного газа (ТДж)

	1971	1973	1978	1983	1984	1985	1986	1987	1988
Arménie	-	-	-	-	-	-	-	-	-
Azerbaïdjan	-	-	-	-	-	-	-	-	-
Bélarus	-	-	-	-	-	-	-	-	-
Estonie	-	-	-	-	-	-	-	-	-
Géorgie	-	-	-	-	-	-	-	-	-
Kazakhstan	-	-	-	-	-	-	-	-	-
Kirghizistan	-	-	-	-	-	-	-	-	-
Lettonie	-	-	-	-	-	-	-	-	-
Lituanie	-	-	-	-	-	-	-	-	-
Moldova	-	-	-	-	-	-	-	-	-
Russie	-	-	-	-	-	-	-	-	-
Tadjikistan	-	-	-	-	-	-	-	-	-
Turkménistan	-	-	-	-	-	-	-	-	-
Ukraine	-	-	-	-	-	-	-	-	-
Ouzbékistan	-	-	-	-	-	-	-	-	-
Ex-URSS	4 695 208	5 407 165	7 226 618	9 612 150	10 077 430	10 603 194	11 063 827	11 589 600	12 096 708
Bahrein	8 371	27 369	43 408	51 189	54 089	77 481	101 754	90 641	104 516
Iran	376 386	383 782	389 252	330 624	403 403	464 824	477 629	395 460	503 020
Irak	35 340	45 980	64 600	103 163	114 300	121 250	148 296	186 521	258 158
Israël	4 232	2 097	2 210	2 478	2 051	1 838	1 444	1 623	1 461
Koweit	89 804	92 772	115 474	104 746	86 447	113 374	158 698	171 698	128 251
Oman	-	-	-	796	11 950	17 111	6 986	9 090	13 194
Qatar	17 674	28 065	21 185	71 909	92 680	81 922	85 198	84 976	92 824
Arabie saoudite	14 783	14 783	14 783	92 312	75 946	110 510	314 093	430 534	430 039
Syrie	-	-	-	-	-	-	8 463	7 985	23 336
Emirats arabes unis	-	-	-	-	24 983	23 372	25 967	29 299	33 484
Moyen-Orient	546 590	594 848	650 912	757 217	865 849	1 011 682	1 328 528	1 407 827	1 588 283
Total non-OCDE	6 443 801	7 480 955	10 679 407	13 969 935	14 813 488	15 432 813	16 324 930	17 175 076	17 855 948
OCDE Amérique du N.	17 908 538	18 518 244	16 970 340	15 892 753	16 917 149	16 427 363	15 597 639	16 129 494	17 471 245
OCDE Pacifique	128 971	171 325	275 813	367 708	401 797	447 889	462 161	485 502	496 746
OCDE Europe	2 829 555	4 359 958	6 855 427	7 379 668	7 774 213	8 181 707	8 321 593	8 822 853	8 722 432
Total OCDE	20 867 064	23 049 527	24 101 580	23 640 129	25 093 159	25 056 959	24 381 393	25 437 849	26 690 423
Monde	27 310 865	30 530 482	34 780 987	37 610 064	39 906 647	40 489 772	40 706 323	42 612 925	44 546 371

Ex-URSS: les séries antérieures à 1990-1992 ne sont pas comparables aux années récentes; 1991 a été estimé.

Final Consumption of Natural Gas (TJ)

Consommation finale de gaz naturel (TJ)

Endverbrauch von Erdgas (TJ)

Consumo finale di gas naturale (TJ)

天然ガスの最終消費量 (TJ)

Consumo final de gas natural (TJ)

Конечное потребление природного газа (ТДж)

1989	1990	1991	1992	1993	1994	1995	1996	1997	
-	-	-	50 353	18 212	20 670	33 437	9 594	10 292	Armenia
-	-	-	287 310	232 497	231 214	182 695	167 011	161 733	Azerbaijan
-	-	-	174 542	165 998	122 934	145 489	142 283	141 163	Belarus
-	-	-	13 076	6 884	11 301	14 098	14 337	14 067	Estonia
-	-	-	128 636	103 072	60 320	22 469	16 173	21 602	Georgia
-	-	-	411 396	282 448	217 378	270 498	209 800	165 955	Kazakhstan
-	-	-	35 740	23 034	14 363	14 673	17 481	9 365	Kyrgyzstan
-	-	-	28 183	17 165	14 722	13 603	21 635	15 801	Latvia
-	-	-	51 534	29 840	32 783	39 390	42 769	37 539	Lithuania
-	-	-	39 442	60 055	57 834	57 834	68 469	73 145	Moldova
-	-	-	7 132 727	5 969 417	5 501 111	5 830 889	4 234 778	4 225 429	Russia
-	-	-	34 511	28 590	15 980	18 010	23 109	20 976	Tajikistan
-	-	-	142 395	202 294	274 764	289 470	234 060	226 200	Turkmenistan
-	-	-	2 258 268	2 226 260	1 880 740	1 956 892	2 156 167	1 958 999	Ukraine
-	-	-	884 902	950 018	961 540	948 953	949 410	964 287	Uzbekistan
12 374 594	12 890 966	12 237 812	11 673 015	10 315 784	9 417 654	9 838 400	8 307 076	8 046 553	**Former USSR**
102 100	107 500	102 060	103 349	122 930	113 816	120 684	128 096	122 801	Bahrain
507 828	377 568	653 667	811 188	847 564	870 995	1 021 584	1 187 379	1 335 803	Iran
297 681	158 135	65 021	108 680	122 048	151 706	150 796	154 126	172 203	Iraq
1 464	1 316	967	953	1 005	899	754	377	395	Israel
178 098	136 000	-	-	-	-	-	-	-	Kuwait
13 846	30 495	32 690	54 418	77 230	79 665	56 431	29 292	57 262	Oman
85 927	90 879	117 239	209 345	223 867	222 111	221 129	221 310	245 078	Qatar
397 475	417 321	408 636	412 606	437 712	445 800	495 832	538 846	565 697	Saudi Arabia
35 270	34 908	27 202	29 004	31 500	34 146	43 058	43 058	39 796	Syria
36 300	37 000	50 780	56 520	58 330	62 821	68 412	81 438	85 777	United Arab Emirates
1 655 989	1 391 122	1 458 262	1 786 063	1 922 186	1 981 959	2 178 680	2 383 922	2 624 812	**Middle East**
18 312 804	18 787 047	18 076 390	17 599 260	16 547 147	15 927 640	16 867 159	16 132 771	16 247 926	**Non-OECD Total**
17 257 914	16 766 350	16 710 698	16 935 194	17 710 826	17 892 053	18 340 177	19 082 523	18 781 521	OECD North America
521 604	554 138	587 055	606 359	653 638	713 064	792 928	889 140	967 477	OECD Pacific
8 950 793	9 125 555	9 839 040	9 805 529	10 161 401	10 178 236	10 807 256	11 776 977	11 427 469	OECD Europe
26 730 311	26 446 043	27 136 793	27 347 082	28 525 865	28 783 353	29 940 361	31 748 640	31 176 467	**OECD Total**
45 043 115	45 233 090	45 213 183	44 946 342	45 073 012	44 710 993	46 807 520	47 881 411	47 424 393	**World**

Former USSR: series up to 1990-1992 are not comparable with recent years; 1991 is estimated.

Consumption of Electricity (GWh)
Consommation d'électricité (GWh)
Verbrauch von Elektrizität (GWh)
Consumo di energia elettrica (GWh)
電力消費量(GWh)
Consumo de electricidad (GWh)
Потребление электроэнергии (ГВт.ч)

	1971	1973	1978	1983	1984	1985	1986	1987	1988
Algérie	1 991	2 494	5 050	8 810	9 604	10 464	11 211	10 827	11 779
Angola	556	738	439	587	604	604	604	607	611
Bénin	32	58	168	149	135	143	137	140	167
Cameroun	1 019	1 080	1 265	2 020	1 996	2 253	2 167	2 228	2 323
Congo	87	95	125	289	287	368	398	428	439
RD du Congo	3 337	3 630	3 703	3 921	4 116	4 427	4 623	4 598	4 733
Egypte	7 216	7 282	13 309	23 526	25 024	24 487	29 867	32 255	33 914
Ethiopie	552	551	535	780	842	809	835	881	864
Gabon	112	162	432	723	787	852	878	883	897
Ghana	2 809	3 728	3 542	2 060	1 198	2 382	3 727	4 340	4 601
Côte d'Ivoire	517	697	1 226	1 666	2 029	1 853	1 752	1 742	1 940
Kenya	808	959	1 396	1 789	1 888	2 060	2 192	2 463	2 407
Libye	508	1 127	3 363	7 150	9 200	11 844	13 280	15 600	16 551
Maroc	2 047	2 594	3 928	6 029	6 226	6 627	7 060	7 297	8 149
Mozambique	476	559	671	471	372	401	473	540	441
Nigéria	1 534	2 184	4 086	6 607	5 967	6 471	7 793	7 862	8 264
Sénégal	324	374	517	642	690	673	678	711	740
Afrique du Sud	50 768	59 870	84 337	117 452	128 480	134 565	140 874	143 197	149 360
Soudan	373	491	744	697	961	1 055	1 198	1 242	1 119
Tanzanie	424	503	601	692	727	787	909	946	1 061
Tunisie	803	1 036	1 833	3 111	3 290	3 501	3 756	4 027	4 364
Zambie	4 411	5 174	5 838	6 708	6 660	6 565	6 183	6 335	6 135
Zimbabwe	3 560	4 962	5 942	7 282	7 108	8 308	8 019	8 373	8 452
Autre Afrique	2 591	3 173	4 247	5 839	6 234	6 417	6 871	7 452	7 428
Afrique	86 855	103 521	147 297	209 000	224 425	237 916	255 485	264 974	276 739
Argentine	21 214	24 111	29 589	36 846	38 682	38 945	39 755	44 307	45 274
Bolivie	919	1 035	1 512	1 465	1 477	1 488	1 499	1 488	1 628
Brésil	44 132	55 844	98 793	143 911	160 001	173 563	187 070	192 757	203 899
Chili	7 569	7 780	9 249	10 794	11 678	12 112	12 796	13 331	14 409
Colombie	8 386	10 932	14 493	20 446	22 226	22 430	24 978	26 395	25 587
Costa Rica	1 148	1 347	1 925	2 382	2 637	2 757	3 027	3 303	3 383
Cuba	4 519	5 133	7 554	10 133	10 794	10 621	11 433	11 749	12 436
République dominicaine	1 068	1 955	2 269	2 270	2 815	3 082	3 199	3 222	2 617
Equateur	903	1 080	2 221	3 524	3 641	3 744	3 884	4 303	4 408
El Salvador	644	802	1 315	1 357	1 423	1 495	1 539	1 643	1 713
Guatemala	666	851	1 446	1 349	1 400	1 456	1 511	1 679	1 753
Haiti	59	83	187	300	309	323	351	350	367
Honduras	354	426	628	957	1 015	1 235	1 257	1 516	1 621
Jamaïque	1 552	1 969	1 317	1 382	1 369	1 356	1 432	1 527	1 467
Antilles néerlandaises	603	658	701	750	695	646	525	580	588
Nicaragua	603	624	1 023	1 097	1 111	1 098	1 095	1 177	1 071
Panama	819	1 096	1 407	2 168	2 114	2 156	2 197	2 384	2 230
Paraguay	209	279	567	924	1 046	1 081	1 270	1 498	1 728
Pérou	5 399	6 018	7 768	9 664	10 593	10 702	11 396	12 154	11 636
Trinité-et-Tobago	991	1 105	1 557	2 905	3 006	3 023	3 291	3 481	3 157
Uruguay	2 033	1 969	2 604	3 107	3 132	3 224	3 367	3 561	3 874
Vénézuela	12 172	14 251	22 496	36 457	38 566	40 995	42 096	44 670	48 400
Autre Amérique latine	3 134	3 646	4 100	4 157	4 234	4 537	4 712	4 565	5 000
Amérique latine	119 096	142 994	214 721	298 345	323 964	342 069	363 680	381 640	398 246

Consumption of Electricity (GWh)

Consommation d'électricité (GWh)

Verbrauch von Elektrizität (GWh)

Consumo di energia elettrica (GWh)

電力消費量(GWh)

Consumo de electricidad (GWh)

Потребление электроэнергии (ГВт.ч)

1989	1990	1991	1992	1993	1994	1995	1996	1997	
13 109	13 845	14 484	14 688	15 145	15 545	16 102	16 694	18 506	Algeria
614	598	667	677	680	684	687	735	794	Angola
162	172	189	210	216	225	246	270	250	Benin
2 402	2 345	2 334	2 360	2 457	2 213	2 177	2 315	2 517	Cameroon
458	570	546	535	540	546	552	546	541	Congo
5 672	4 922	4 944	5 700	5 204	5 205	5 840	5 916	5 673	DR of Congo
36 015	38 459	40 766	41 555	43 728	45 950	48 503	48 159	51 000	Egypt
866	1 231	1 255	1 252	1 384	1 287	1 313	1 301	1 325	Ethiopia
771	810	814	814	816	847	846	873	906	Gabon
4 706	4 927	5 263	5 657	5 521	5 705	5 703	5 577	5 745	Ghana
1 848	1 945	1 899	1 065	1 882	1 948	2 439	2 561	2 727	Ivory Coast
2 558	2 729	2 854	2 945	3 103	3 234	3 321	3 550	3 675	Kenya
16 700	16 800	16 800	16 950	17 000	17 800	18 000	18 180	18 180	Libya
8 174	8 910	9 385	10 339	10 517	11 302	11 678	11 995	12 692	Morocco
507	548	742	746	728	662	714	732	893	Mozambique
9 304	7 833	9 063	9 693	8 695	11 065	9 876	10 222	10 350	Nigeria
735	781	796	873	859	931	949	988	1 050	Senegal
153 469	155 988	157 518	156 741	159 333	166 814	173 708	179 722	187 646	South Africa
1 075	1 282	1 264	1 295	1 271	1 345	1 362	1 413	1 361	Sudan
1 151	1 303	1 428	1 441	1 433	1 456	1 623	1 793	1 700	Tanzania
4 611	4 926	5 140	5 527	5 838	6 329	6 630	6 767	7 130	Tunisia
4 632	4 164	5 484	5 410	5 423	5 423	5 427	5 431	5 596	Zambia
8 898	9 133	9 436	9 814	8 242	8 998	9 524	9 743	10 333	Zimbabwe
7 879	8 225	8 554	8 813	9 268	9 436	9 710	10 002	10 102	Other Africa
286 316	292 446	301 625	305 100	309 283	324 950	336 930	345 485	360 692	**Africa**
43 992	41 784	43 908	46 533	48 625	51 753	55 075	60 361	64 568	Argentina
1 757	1 829	1 963	2 082	2 353	2 484	2 673	2 851	3 067	Bolivia
212 380	217 658	225 372	230 472	241 165	249 828	264 853	277 645	295 524	Brazil
15 785	16 428	17 732	19 990	21 124	22 506	25 100	28 102	30 772	Chile
27 304	28 652	29 612	27 306	30 288	32 501	34 425	35 023	36 131	Colombia
3 272	3 344	3 592	3 791	4 084	4 389	4 526	4 529	5 045	Costa Rica
12 955	12 848	11 195	9 711	9 067	9 737	10 010	10 660	11 689	Cuba
2 348	2 791	2 918	4 035	4 380	4 610	4 852	5 106	5 299	Dominican Republic
4 454	4 896	5 362	5 617	5 908	6 629	6 844	7 345	7 432	Ecuador
1 740	1 880	2 004	2 116	2 430	2 679	2 969	2 977	3 251	El Salvador
1 903	1 958	2 120	2 407	2 694	2 752	3 083	3 698	4 263	Guatemala
391	414	323	295	231	163	254	287	358	Haiti
1 637	1 834	1 801	1 728	1 855	1 695	2 026	2 386	2 494	Honduras
1 393	1 658	1 671	1 760	3 791	4 257	5 197	5 385	5 579	Jamaica
625	678	689	743	799	851	884	891	922	Netherlands Antilles
1 104	1 220	1 243	1 267	1 247	1 211	1 298	1 402	1 564	Nicaragua
2 146	2 282	2 324	2 443	2 649	2 749	2 940	3 129	3 184	Panama
1 837	2 131	2 200	2 392	2 854	3 269	3 743	3 816	4 058	Paraguay
11 838	10 818	11 426	10 721	12 065	12 155	13 874	14 612	15 071	Peru
3 056	3 278	3 257	3 522	3 502	3 611	3 892	4 091	4 587	Trinidad and Tobago
3 753	3 870	4 212	4 309	4 623	4 759	5 089	5 313	5 665	Uruguay
47 856	48 633	51 027	54 916	56 685	57 104	57 946	59 878	58 949	Venezuela
5 695	5 849	6 094	6 311	6 484	6 541	6 684	6 792	7 081	Other Latin America
409 221	416 733	432 045	444 467	468 903	488 233	518 237	546 279	576 553	**Latin America**

Consumption of Electricity (GWh)
Consommation d'électricité (GWh)
Verbrauch von Elektrizität (GWh)

Consumo di energia elettrica (GWh)

電力消費量(GWh)

Consumo de electricidad (GWh)

Потребление электроэнергии (ГВт.ч)

	1971	1973	1978	1983	1984	1985	1986	1987	1988
Bangladesh	710	1 042	1 651	2 537	2 866	3 040	3 534	3 771	4 173
Brunei	150	200	343	651	716	731	740	995	1 048
Inde	55 522	59 854	90 727	123 220	137 886	149 105	163 600	177 610	196 482
Indonésie	2 530	3 010	4 404	10 681	11 735	13 561	15 614	18 107	21 280
RPD de Corée	15 320	18 120	28 992	37 100	40 750	43 470	45 000	45 200	48 000
Malaisie	3 464	4 347	7 437	11 461	12 551	13 103	14 056	15 096	16 643
Myanmar	553	666	1 020	1 192	1 313	1 595	1 626	1 665	1 503
Népal	74	74	151	263	282	315	346	408	423
Pakistan	5 584	6 389	8 913	14 757	16 384	18 336	20 398	22 454	25 919
Philippines	8 688	12 562	14 740	18 650	18 047	19 020	18 199	18 304	20 191
Singapour	2 440	3 506	5 574	8 166	8 896	9 422	10 104	11 440	12 616
Sri Lanka	735	881	1 178	1 811	1 886	2 073	2 243	2 285	2 383
Taipei chinois	14 520	18 913	33 345	45 196	49 019	52 681	59 392	65 474	72 373
Thailande	4 568	6 485	11 886	17 514	19 456	21 117	23 047	26 085	29 560
Viêt-Nam	1 794	1 833	2 808	3 260	3 866	4 141	4 423	4 949	5 515
Autre Asie	2 170	3 386	3 559	3 516	3 829	4 145	4 441	4 562	4 670
Asie	118 822	141 268	216 728	299 975	329 482	355 855	386 763	418 405	462 779
Rép. populaire de Chine	138 400	166 800	256 552	324 410	348 800	381 330	417 488	462 400	508 730
Hong-Kong, Chine	5 250	6 009	9 107	14 031	15 041	15 923	17 660	21 430	22 918
Chine	143 650	172 809	265 659	338 441	363 841	397 253	435 148	483 830	531 648
Albanie	1 164	1 363	2 819	3 045	2 957	2 361	4 294	3 583	3 161
Bulgarie	19 571	23 131	32 157	40 963	42 763	41 920	41 622	43 409	44 373
Chypre	615	774	865	1 149	1 175	1 243	1 334	1 411	1 540
Gibraltar	45	47	50	59	59	61	63	62	66
Malte	305	365	459	615	635	710	770	857	935
Roumanie	33 395	39 732	58 495	67 115	69 732	71 222	75 795	75 175	78 226
République slovaque	12 290	14 046	16 962	23 020	23 708	24 990	25 666	26 559	27 043
Bosnie-Herzegovine	-	-	-	-	-	-	-	-	-
Croatie	-	-	-	-	-	-	-	-	-
Ex-RYM	-	-	-	-	-	-	-	-	-
Slovénie	-	-	-	-	-	-	-	-	-
RF de Yougoslavie	-	-	-	-	-	-	-	-	-
Ex-Yougoslavie	26 127	31 260	45 608	61 815	65 990	67 899	70 645	73 736	75 956
Europe non-OCDE	93 512	110 718	157 415	197 781	207 019	210 406	220 189	224 792	231 300
Arménie	-	-	-	-	-	-	-	-	-
Azerbaïdjan	-	-	-	-	-	-	-	-	-
Bélarus	-	-	-	-	-	-	-	-	-
Estonie	-	-	-	-	-	-	-	-	-
Géorgie	-	-	-	-	-	-	-	-	-
Kazakhstan	-	-	-	-	-	-	-	-	-
Kirghizistan	-	-	-	-	-	-	-	-	-
Lettonie	-	-	-	-	-	-	-	-	-
Lituanie	-	-	-	-	-	-	-	-	-
Moldova	-	-	-	-	-	-	-	-	-
Russie	-	-	-	-	-	-	-	-	-
Tadjikistan	-	-	-	-	-	-	-	-	-
Turkménistan	-	-	-	-	-	-	-	-	-
Ukraine	-	-	-	-	-	-	-	-	-
Ouzbékistan	-	-	-	-	-	-	-	-	-
Ex-URSS	730 500	832 300	1 091 900	1 278 800	1 341 200	1 381 400	1 432 700	1 487 700	1 526 200

Consumption of Electricity (GWh)
Consommation d'électricité (GWh)

Verbrauch von Elektrizität (GWh)

Consumo di energia elettrica (GWh)

電力消費量(GWh)

Consumo de electricidad (GWh)

Потребление электроэнергии (ГВт.ч)

1989	1990	1991	1992	1993	1994	1995	1996	1997	
5 093	5 135	5 324	6 522	7 411	8 010	9 011	9 637	10 062	Bangladesh
1 079	1 118	1 192	1 320	1 483	1 615	1 897	2 080	2 378	Brunei
216 701	238 238	260 917	277 706	297 795	317 411	344 723	358 893	382 177	India
24 878	30 663	32 809	37 402	40 542	45 625	48 215	58 782	66 216	Indonesia
48 500	48 500	28 500	6 080	6 080	5 920	5 760	5 612	5 444	DPR of Korea
18 378	20 745	22 779	26 500	28 919	36 059	41 058	45 686	52 552	Malaysia
1 638	1 823	1 712	1 875	2 173	2 342	2 510	2 562	2 717	Myanmar
484	543	588	702	715	713	774	841	866	Nepal
27 652	29 865	32 922	35 381	37 979	39 099	41 192	43 777	45 072	Pakistan
21 776	22 020	22 451	21 778	22 862	25 725	28 109	30 535	33 223	Philippines
13 596	15 184	16 136	16 849	18 035	19 838	21 258	23 016	25 780	Singapore
2 367	2 624	2 765	2 946	3 272	3 613	3 936	3 756	4 245	Sri Lanka
79 243	85 124	93 931	100 639	110 556	119 495	128 198	127 621	146 398	Chinese Taipei
34 297	40 130	45 334	51 647	58 834	65 136	74 178	80 581	85 807	Thailand
6 099	6 610	6 970	7 358	8 159	9 586	10 956	13 742	15 767	Vietnam
4 764	4 820	4 735	4 592	4 609	4 809	5 261	5 161	5 251	Other Asia
506 545	553 142	579 065	599 297	649 424	704 996	767 036	812 282	883 955	**Asia**
545 200	579 580	631 650	701 582	780 735	866 430	927 888	1 002 172	1 037 699	People's Rep. of China
22 385	23 833	25 316	26 148	27 727	29 184	29 856	31 635	32 245	Hong Kong, China
567 585	603 413	656 966	727 730	808 462	895 614	957 744	1 033 807	1 069 944	**China**
3 297	2 828	2 063	2 397	2 866	1 876	2 094	3 021	2 860	Albania
44 058	41 488	35 702	33 367	33 333	33 316	36 203	36 690	33 243	Bulgaria
1 724	1 863	1 943	2 246	2 451	2 556	2 358	2 443	2 538	Cyprus
76	76	76	85	83	88	88	93	93	Gibraltar
1 000	1 000	1 309	1 370	1 380	1 400	1 480	1 440	1 465	Malta
79 027	67 856	57 986	52 873	51 721	50 796	52 827	54 974	50 788	Romania
27 784	27 436	25 261	23 752	22 663	23 198	25 974	26 857	26 544	Slovak Republic
-	-	-	12 200	11 700	1 675	1 896	1 896	1 896	*Bosnia-Herzegovina*
-	-	-	10 194	10 205	10 256	10 699	10 970	11 811	*Croatia*
-	-	-	5 900	5 186	4 999	5 351	5 679	5 880	*FYROM*
-	-	-	9 611	9 607	10 160	10 308	10 383	10 783	*Slovenia*
-	-	-	32 685	30 150	31 108	33 722	34 412	36 566	*FR of Yugoslavia*
75 711	74 794	75 542	70 590	66 848	58 198	61 976	63 340	66 936	Former Yugoslavia
232 677	217 341	199 882	186 680	181 345	171 428	183 000	188 858	184 467	**Non-OECD Europe**
-	-	-	6 763	4 019	3 433	3 379	3 865	4 770	Armenia
-	-	-	16 374	15 400	14 354	13 485	13 745	13 800	Azerbaijan
-	-	-	39 974	35 327	31 380	28 441	28 514	29 876	Belarus
-	-	-	7 564	6 052	6 433	6 160	6 533	6 734	Estonia
-	-	-	9 736	8 331	5 903	5 870	6 238	6 200	Georgia
-	-	-	88 023	74 170	69 865	63 906	56 535	50 657	Kazakhstan
-	-	-	8 616	8 466	8 190	7 524	7 127	6 623	Kyrgyzstan
-	-	-	6 898	5 250	4 974	4 963	4 895	5 000	Latvia
-	-	-	11 702	9 185	9 184	9 212	9 852	9 751	Lithuania
-	-	-	9 281	8 901	7 316	6 510	6 336	5 800	Moldova
-	-	-	908 115	850 142	769 972	756 947	743 264	732 960	Russia
-	-	-	16 454	14 373	14 326	13 636	13 559	13 100	Tajikistan
-	-	-	7 793	8 274	6 868	6 574	6 169	5 700	Turkmenistan
-	-	-	224 662	206 002	180 141	172 235	155 973	149 441	Ukraine
-	-	-	45 651	44 284	43 064	42 020	42 450	42 975	Uzbekistan
1 541 100	1 550 000	1 518 528	1 407 606	1 298 176	1 175 403	1 140 862	1 105 055	1 083 387	**Former USSR**

Consumption of Electricity (GWh)
Consommation d'électricité (GWh)

Verbrauch von Elektrizität (GWh)

Consumo di energia elettrica (GWh)

電力消費量(GWh)

Consumo de electricidad (GWh)

Потребление электроэнергии (ГВт.ч)

	1971	1973	1978	1983	1984	1985	1986	1987	1988
Bahrein	430	500	1 212	2 216	2 417	2 937	2 970	3 316	3 320
Iran	7 199	10 533	17 404	28 850	31 984	34 279	36 334	40 397	41 603
Irak	2 660	3 343	7 443	15 376	18 486	19 924	21 173	21 332	22 250
Israël	7 024	8 187	11 114	13 601	13 972	14 753	15 223	16 411	17 890
Jordanie	237	338	574	1 664	1 961	2 160	2 343	2 673	2 780
Koweit	2 810	3 807	6 740	11 654	12 930	14 364	15 763	16 592	18 098
Liban	1 237	1 612	2 115	3 276	3 447	3 511	3 780	4 176	3 600
Oman	11	40	399	1 378	1 720	2 132	2 714	2 895	3 220
Qatar	316	419	1 213	2 945	3 263	3 734	4 088	4 151	4 271
Arabie saoudite	1 927	2 716	9 381	31 177	36 986	41 904	45 866	48 906	51 531
Syrie	1 158	1 283	2 406	5 968	6 068	6 583	6 674	6 709	7 360
Emirats arabes unis	189	670	3 501	9 431	10 230	11 160	11 904	12 657	13 840
Yémen	209	206	360	693	776	823	972	1 041	1 185
Moyen-Orient	25 407	33 654	63 862	128 229	144 240	158 264	169 804	181 256	190 948
Total non-OCDE	1 317 842	1 537 264	2 157 582	2 750 571	2 934 171	3 083 163	3 263 769	3 442 597	3 617 860
OCDE Amérique du N.	1 789 307	2 079 898	2 473 291	2 691 663	2 869 343	2 943 602	2 989 260	3 145 986	3 296 548
OCDE Pacifique	432 675	528 815	662 578	758 730	801 277	833 158	851 119	906 767	960 575
OCDE Europe	1 302 702	1 501 891	1 796 939	1 994 650	2 091 138	2 186 838	2 241 103	2 324 636	2 386 674
Total OCDE	3 524 684	4 110 604	4 932 808	5 445 043	5 761 758	5 963 598	6 081 482	6 377 389	6 643 797
Monde	4 842 526	5 647 868	7 090 390	8 195 614	8 695 929	9 046 761	9 345 251	9 819 986	10 261 657

Consumption of Electricity (GWh)

Consommation d'électricité (GWh)

Verbrauch von Elektrizität (GWh)

Consumo di energia elettrica (GWh)

電力消費量(GWh)

Consumo de electricidad (GWh)

Потребление электроэнергии (ГВт.ч)

1989	1990	1991	1992	1993	1994	1995	1996	1997	
3 400	3 490	3 495	3 700	4 044	4 386	4 579	4 599	4 747	Bahrain
42 468	47 838	56 559	60 151	61 521	66 256	68 661	72 472	75 145	Iran
22 660	22 800	19 610	25 300	26 300	28 000	29 000	29 000	29 561	Iraq
18 948	19 285	19 659	22 901	24 242	26 352	28 167	30 213	31 047	Israel
2 930	3 330	3 387	3 969	4 291	4 640	5 088	5 462	5 658	Jordan
19 679	18 768	9 280	15 286	20 178	22 798	23 726	25 475	27 091	Kuwait
2 250	1 400	2 800	3 371	3 964	4 355	4 681	6 730	8 000	Lebanon
3 351	3 841	3 950	4 316	4 909	5 209	5 495	5 702	6 073	Oman
4 394	4 568	4 381	4 873	5 199	5 455	5 314	6 184	6 456	Qatar
55 201	58 973	63 632	67 492	74 171	82 182	85 890	89 619	95 100	Saudi Arabia
8 179	8 573	8 978	9 262	12 638	15 182	15 300	16 885	17 950	Syria
14 612	16 081	16 222	16 460	17 578	18 870	19 070	19 737	20 571	United Arab Emirates
1 212	1 471	1 586	1 738	1 826	1 664	1 760	1 734	1 752	Yemen
199 284	210 418	213 539	238 819	260 861	285 349	296 731	313 812	329 151	**Middle East**
3 742 728	3 843 493	3 901 650	3 909 699	3 976 454	4 045 973	4 200 540	4 345 578	4 488 149	**Non-OECD Total**
3 396 906	3 474 521	3 630 870	3 653 676	3 771 761	3 876 360	3 985 474	4 125 131	4 169 194	OECD North America
1 021 573	1 099 032	1 140 825	1 161 241	1 190 802	1 290 636	1 339 905	1 388 824	1 441 300	OECD Pacific
2 444 406	2 470 522	2 514 907	2 534 951	2 538 629	2 569 411	2 638 688	2 710 967	2 751 817	OECD Europe
6 862 885	7 044 075	7 286 602	7 349 868	7 501 192	7 736 407	7 964 067	8 224 922	8 362 311	**OECD Total**
10 605 613	10 887 568	11 188 252	11 259 567	11 477 646	11 782 380	12 164 607	12 570 500	12 850 460	**World**

Industry Consumption of Coking Coal (1000 tonnes)

Consommation industrielle de charbon à coke (1000 tonnes)

Industrieverbrauch von Kokskohle (1000 Tonnen)

Consumo di carbone siderurgico nell'industria (1000 tonnellate)

原料炭の産業用消費量（千トン）

Consumo industrial de carbón coquizable (1000 toneladas)

Потребление коксующихся углей промышленным сектором (тыс. т)

	1971	1973	1978	1983	1984	1985	1986	1987	1988
Brésil	-	-	-	-	-	-	-	-	-
Amérique latine	-	-	-	-	-	-	-	-	-
Roumanie	-	-	-	-	-	-	-	-	-
République slovaque	-	-	27	25	31	32	31	32	34
Ex-RYM	-	-	-	-	-	-	-	-	-
Ex-Yougoslavie	-	-	-	-	-	-	-	-	-
Europe non-OCDE	-	-	27	25	31	32	31	32	34
Kazakhstan	-	-	-	-	-	-	-	-	-
Ex-URSS	-	-	-	-	-	-	-	-	-
Total non-OCDE	-	-	27	25	31	32	31	32	34
OCDE Pacifique	-	-	116	10	38	264	749	738	558
OCDE Europe	-	-	116	212	412	2 194	1 433	1 319	1 403
Total OCDE	-	-	232	222	450	2 458	2 182	2 057	1 961
Monde	-	-	259	247	481	2 490	2 213	2 089	1 995

Ex-URSS: les séries antérieures à 1990-1992 ne sont pas comparables aux années récentes; 1991 a été estimé.

Industry Consumption of Coking Coal (1000 tonnes)

Consommation industrielle de charbon à coke (1000 tonnes)

Industrieverbrauch von Kokskohle (1000 Tonnen)

Consumo di carbone siderurgico nell'industria (1000 tonnellate)

原料炭の産業用消費量（千トン）

Consumo industrial de carbón coquizable (1000 toneladas)

Потребление коксующихся углей промышленным сектором (тыс. т)

1989	1990	1991	1992	1993	1994	1995	1996	1997	
-	-	-	-	235	354	834	1 596	1 928	Brazil
-	-	-	-	235	354	834	1 596	1 928	**Latin America**
-	-	-	5	12	6	94	29	1	Romania
38	6	28	-	158	296	314	394	293	Slovak Republic
-	-	-	71	74	66	70	49	49	*FYROM*
-	-	75	71	74	66	70	49	49	Former Yugoslavia
38	6	103	76	244	368	478	472	343	**Non-OECD Europe**
-	-	-	15 985	11 381	12 116	9 204	8 076	7 106	Kazakhstan
-	-	-	15 985	11 381	12 116	9 204	8 076	7 106	**Former USSR**
38	6	103	16 061	11 860	12 838	10 516	10 144	9 377	**Non-OECD Total**
865	1 238	1 699	1 338	1 584	1 860	1 922	1 913	2 190	OECD Pacific
1 292	2 110	3 416	2 232	1 418	1 307	1 199	1 578	3 293	OECD Europe
2 157	3 348	5 115	3 570	3 002	3 167	3 121	3 491	5 483	**OECD Total**
2 195	3 354	5 218	19 631	14 862	16 005	13 637	13 635	14 860	**World**

Former USSR: series up to 1990-1992 are not comparable with recent years; 1991 is estimated.

Industry Consumption of Other Bituminous Coal (1000 tonnes)

Consommation industrielle d'autres charbons bitumineux (1000 tonnes)

Industrieverbrauch von sonstiger bituminöser Kohle (1000 Tonnen)

Consumo di altri carboni bituminosi nell'industria (1000 tonnellate)

その他歴青炭の産業用消費量（千トン）

Consumo industrial de otro carbón bituminoso (1000 toneladas)

Потребление других битуминозных углей промышленным сектором (тыс. т)

	1971	1973	1978	1983	1984	1985	1986	1987	1988
RD du Congo	-	-	167	147	157	156	160	165	166
Kenya	-	-	10	32	83	90	80	89	113
Maroc	-	-	28	248	282	523	495	439	447
Mozambique	-	-	247	144	94	65	39	32	28
Nigéria	-	-	124	33	53	86	75	80	31
Afrique du Sud	-	-	15 797	13 832	13 350	14 104	13 558	14 322	14 863
Tanzanie	-	-	1	7	10	15	4	3	3
Tunisie	-	-	27	28	24	21	12	16	29
Zambie	-	-	435	400	365	386	425	500	541
Zimbabwe	-	-	514	1 234	1 008	981	931	966	933
Autre Afrique	-	-	225	231	265	291	247	251	280
Afrique	-	-	17 575	16 336	15 691	16 718	16 026	16 863	17 434
Argentine	-	-	31	29	131	14	78	196	268
Brésil	-	-	340	3 186	3 115	3 401	3 712	3 898	3 365
Chili	-	-	378	413	540	614	714	643	702
Colombie	-	-	2 183	2 267	2 289	2 514	2 621	2 696	2 691
Costa Rica	-	-	1	1	1	1	1	1	1
Cuba	-	-	91	90	79	126	119	85	95
Haiti	-	-	-	51	54	61	18	18	31
Jamaïque	-	-	-	-	-	-	-	-	8
Panama	-	-	-	-	-	38	38	38	18
Pérou	-	-	26	75	89	114	107	100	80
Uruguay	-	-	2	-	-	-	-	1	-
Vénézuela	-	-	63	37	51	42	57	62	71
Amérique latine	-	-	3 115	6 149	6 349	6 925	7 465	7 738	7 330
Bangladesh	-	-	250	163	62	98	148	233	240
Inde	-	-	31 554	44 141	46 313	61 861	60 202	61 086	68 763
Indonésie	-	-	100	223	291	283	290	300	760
RPD de Corée	-	-	23 616	25 471	24 136	24 212	24 200	24 200	24 650
Malaisie	-	-	14	345	365	489	353	434	608
Myanmar	-	-	218	198	204	203	206	59	52
Népal	-	-	14	40	94	76	50	55	59
Pakistan	-	-	1 187	1 546	1 811	2 174	2 148	2 221	2 705
Philippines	-	-	165	666	813	954	743	958	1 155
Taipei chinois	-	-	1 658	2 790	2 968	3 118	3 366	3 580	4 353
Thailande	-	-	52	172	227	212	183	223	304
Viêt-Nam	-	-	-	2 066	2 246	2 576	2 850	3 146	2 321
Autre Asie	-	-	-	261	273	345	351	343	333
Asie	-	-	58 828	78 082	79 803	96 601	95 090	96 838	106 303
Rép. populaire de Chine	-	-	-	263 821	295 575	312 848	338 003	356 266	363 511
Hong-Kong, Chine	-	-	6	-	-	-	-	-	-
Chine	-	-	6	263 821	295 575	312 848	338 003	356 266	363 511

Rép. populaire de Chine: jusqu'en 1978 la ventilation de la consommation finale par secteur est incomplète.

Industry Consumption of Other Bituminous Coal (1000 tonnes)

Consommation industrielle d'autres charbons bitumineux (1000 tonnes)

Industrieverbrauch von sonstiger bituminöser Kohle (1000 Tonnen)

Consumo di altri carboni bituminosi nell'industria (1000 tonnellate)

その他歴青炭の産業用消費量（千トン）

Consumo industrial de otro carbón bituminoso (1000 toneladas)

Потребление других битуминозных углей промышленным сектором (тыс. т)

1989	1990	1991	1992	1993	1994	1995	1996	1997	
170	169	174	170	134	136	136	136	136	DR of Congo
131	151	151	158	131	109	97	89	92	Kenya
772	604	786	731	739	760	758	952	1 002	Morocco
34	34	32	41	54	60	56	18	20	Mozambique
33	43	43	53	60	70	10	10	10	Nigeria
14 166	13 547	11 501	10 756	8 005	8 474	9 005	10 415	11 862	South Africa
2	3	3	3	3	3	4	4	4	Tanzania
30	15	12	15	14	-	-	-	-	Tunisia
312	297	234	267	200	119	124	118	118	Zambia
845	1 190	1 150	1 150	1 096	275	333	322	285	Zimbabwe
264	226	48	209	205	14	9	10	11	Other Africa
16 759	16 279	14 134	13 553	10 641	10 020	10 532	12 074	13 540	**Africa**
84	25	25	74	92	32	51	-	-	Argentina
2 691	2 178	2 839	2 159	1 802	1 893	1 463	1 277	1 009	Brazil
771	647	640	755	692	536	526	635	1 326	Chile
2 845	3 219	3 701	3 189	3 122	2 900	2 579	2 728	2 809	Colombia
1	1	-	-	-	-	-	-	-	Costa Rica
167	153	117	107	86	70	63	12	42	Cuba
12	12	25	26	-	-	-	-	-	Haiti
56	62	9	65	71	53	55	64	67	Jamaica
19	32	46	58	63	52	54	56	57	Panama
91	70	206	207	388	408	364	238	188	Peru
-	-	1	1	-	-	-	-	-	Uruguay
436	355	402	450	291	354	346	328	419	Venezuela
7 173	6 754	8 011	7 091	6 607	6 298	5 501	5 338	5 917	**Latin America**
250	563	180	169	63	-	-	-	-	Bangladesh
66 923	74 130	77 300	79 310	79 750	88 310	83 790	80 850	83 583	India
1 407	1 804	1 404	1 555	1 737	2 209	1 868	5 122	4 453	Indonesia
24 900	22 700	20 025	20 800	18 500	16 250	16 250	16 413	15 921	DPR of Korea
903	834	718	775	780	972	920	1 039	1 058	Malaysia
51	60	61	31	32	35	-	-	-	Myanmar
50	12	59	91	94	98	102	102	102	Nepal
2 586	3 096	3 026	3 052	3 217	3 487	2 999	3 236	2 809	Pakistan
724	795	1 243	1 004	1 093	961	907	1 670	2 183	Philippines
4 170	4 308	4 601	5 325	5 768	5 981	5 795	6 145	5 896	Chinese Taipei
392	250	436	489	986	1 424	2 305	3 723	3 037	Thailand
2 163	2 357	2 847	2 793	2 503	2 429	2 623	3 826	4 663	Vietnam
348	307	307	217	216	211	213	218	228	Other Asia
104 867	111 216	112 207	115 611	114 739	122 367	117 772	122 344	123 933	**Asia**
372 430	357 817	352 265	353 341	372 151	394 340	404 639	402 992	362 229	People's Rep. of China
-	-	-	-	-	-	-	-	-	Hong Kong, China
372 430	357 817	352 265	353 341	372 151	394 340	404 639	402 992	362 229	**China**

People's Rep. of China: up to 1978 the breakdown of final consumption by sector is incomplete.

Industry Consumption of Other Bituminous Coal (1000 tonnes)

Consommation industrielle d'autres charbons bitumineux (1000 tonnes)

Industrieverbrauch von sonstiger bituminöser Kohle (1000 Tonnen)

Consumo di altri carboni bituminosi nell'industria (1000 tonnellate)

その他歴青炭の産業用消費量 (千トン)

Consumo industrial de otro carbón bituminoso (1000 toneladas)

Потребление других битуминозных углей промышленным сектором (тыс. т)

	1971	1973	1978	1983	1984	1985	1986	1987	1988
Albanie	-	-	143	170	178	187	195	195	203
Bulgarie	-	-	1 910	2 323	2 472	3 754	2 554	2 733	70
Chypre	-	-	-	-	52	74	55	151	91
Roumanie	-	-	-	-	-	1 460	1 705	1 812	1 747
République slovaque	-	-	1 237	1 091	1 121	1 114	1 140	1 143	1 126
Croatie	-	-	-	-	-	-	-	-	-
Ex-RYM	-	-	-	-	-	-	-	-	-
Slovénie	-	-	-	-	-	-	-	-	-
Ex-Yougoslavie	-	-	172	137	288	89	259	236	75
Europe non-OCDE	-	-	3 462	3 721	4 111	6 678	5 908	6 270	3 312
Azerbaïdjan	-	-	-	-	-	-	-	-	-
Bélarus	-	-	-	-	-	-	-	-	-
Estonie	-	-	-	-	-	-	-	-	-
Kazakhstan	-	-	-	-	-	-	-	-	-
Kirghizistan	-	-	-	-	-	-	-	-	-
Lettonie	-	-	-	-	-	-	-	-	-
Lituanie	-	-	-	-	-	-	-	-	-
Moldova	-	-	-	-	-	-	-	-	-
Russie	-	-	-	-	-	-	-	-	-
Ukraine	-	-	-	-	-	-	-	-	-
Ouzbékistan	-	-	-	-	-	-	-	-	-
Ex-URSS	-	-	92 144	75 473	65 964	59 422	52 884	65 742	70 640
Iran	-	-	523	541	921	928	982	1 010	1 030
Israël	-	-	-	-	-	-	-	162	137
Liban	-	-	1	-	-	-	-	-	-
Syrie	-	-	1	3	2	-	-	-	-
Moyen-Orient	-	-	525	544	923	928	982	1 172	1 167
Total non-OCDE	-	-	175 655	444 126	468 416	500 120	516 358	550 889	569 697
OCDE Amérique du N.	-	-	43 377	45 027	51 127	51 829	51 446	50 635	51 067
OCDE Pacifique	-	-	5 575	16 973	18 545	20 256	20 303	18 974	21 198
OCDE Europe	-	-	35 314	42 627	45 166	48 941	46 861	48 470	48 154
Total OCDE	-	-	84 266	104 627	114 838	121 026	118 610	118 079	120 419
Monde	-	-	259 921	548 753	583 254	621 146	634 968	668 968	690 116

Ex-URSS: les séries antérieures à 1990-1992 ne sont pas comparables aux années récentes; 1991 a été estimé.

Industry Consumption of Other Bituminous Coal (1000 tonnes)
Consommation industrielle d'autres charbons bitumineux (1000 tonnes)

Industrieverbrauch von sonstiger bituminöser Kohle (1000 Tonnen)

Consumo di altri carboni bituminosi nell'industria (1000 tonnellate)

その他歴青炭の産業用消費量（千トン）

Consumo industrial de otro carbón bituminoso (1000 toneladas)

Потребление других битуминозных углей промышленным сектором (тыс. т)

1989	1990	1991	1992	1993	1994	1995	1996	1997	
203	240	160	50	-	-	-	-	-	Albania
31	60	71	57	79	132	131	120	86	Bulgaria
102	97	97	26	31	27	20	17	19	Cyprus
232	136	37	30	8	10	7	14	16	Romania
1 058	860	1 083	985	789	699	781	623	706	Slovak Republic
-	-	-	83	68	60	46	61	56	*Croatia*
-	-	-	12	68	61	72	51	51	*FYROM*
-	-	-	13	17	22	20	26	19	*Slovenia*
247	212	110	108	153	143	138	138	126	Former Yugoslavia
1 873	1 605	1 558	1 256	1 060	1 011	1 077	912	953	**Non-OECD Europe**
-	-	-	-	1	-	-	-	-	Azerbaijan
-	-	-	64	43	37	29	21	18	Belarus
-	-	-	56	2	1	44	37	10	Estonia
-	-	-	7 827	9 450	7 921	7 762	5 084	4 463	Kazakhstan
-	-	-	990	893	1 520	395	581	1 390	Kyrgyzstan
-	-	-	25	21	30	13	11	12	Latvia
-	-	-	44	68	54	21	16	7	Lithuania
-	-	-	57	15	10	6	3	3	Moldova
-	-	-	5 879	12 243	8 062	5 181	3 077	2 609	Russia
-	-	-	14 983	21 373	13 015	12 154	9 006	8 634	Ukraine
-	-	-	-	-	-	30	24	17	Uzbekistan
68 872	66 484	51 051	29 925	44 109	30 650	25 635	17 860	17 163	**Former USSR**
1 050	1 250	1 350	1 500	607	632	750	750	750	Iran
160	18	11	11	10	8	12	26	28	Israel
-	-	-	-	111	112	180	200	200	Lebanon
-	-	-	-	-	-	-	-	-	Syria
1 210	1 268	1 361	1 511	728	752	942	976	978	**Middle East**
573 184	561 423	540 587	522 288	550 035	565 438	566 098	562 496	524 713	**Non-OECD Total**
49 256	48 063	46 378	25 466	28 520	25 607	24 148	24 048	26 270	OECD North America
21 363	21 480	22 378	22 841	22 968	23 873	22 451	25 671	25 448	OECD Pacific
46 026	44 070	40 821	39 843	40 389	38 714	43 282	48 000	43 416	OECD Europe
116 645	113 613	109 577	88 150	91 877	88 194	89 881	97 719	95 134	**OECD Total**
689 829	675 036	650 164	610 438	641 912	653 632	655 979	660 215	619 847	**World**

Former USSR: series up to 1990-1992 are not comparable with recent years; 1991 is estimated.

Industry Consumption of Sub-Bituminous Coal (1000 tonnes)

Consommation industrielle de charbons sous-bitumineux (1000 tonnes)

Industrieverbrauch von subbituminöser Kohle (1000 Tonnen)

Consumo di carbone sub-bituminoso nell'industria (1000 tonnellate)

亜瀝青炭の産業用消費量 (千トン)

Consumo industrial de carbón sub-bituminoso (1000 toneladas)

Потребление полубитуминозных углей промышленным сектором (тыс. т)

	1971	1973	1978	1983	1984	1985	1986	1987	1988
RPD de Corée	-	-	9 000	11 000	11 000	12 000	12 500	12 500	12 500
Asie	-	-	9 000	11 000	11 000	12 000	12 500	12 500	12 500
Bulgarie	-	-	-	-	-	-	-	-	-
Slovénie	-	-	-	-	-	-	-	-	-
Ex-Yougoslavie	-	-	-	-	-	-	-	-	-
Europe non-OCDE	-	-	-	-	-	-	-	-	-
Total non-OCDE	-	-	9 000	11 000	11 000	12 000	12 500	12 500	12 500
OCDE Amérique du N.	-	-	2 674	4 416	5 062	5 155	5 150	5 265	5 614
OCDE Pacifique	-	-	1 791	1 922	2 378	2 715	2 873	2 897	2 961
OCDE Europe	-	-	21 027	18 378	19 440	18 407	18 882	20 383	19 256
Total OCDE	-	-	25 492	24 716	26 880	26 277	26 905	28 545	27 831
Monde	-	-	34 492	35 716	37 880	38 277	39 405	41 045	40 331

Industry Consumption of Sub-Bituminous Coal (1000 tonnes)

Consommation industrielle de charbons sous-bitumineux (1000 tonnes)

Industrieverbrauch von subbituminöser Kohle (1000 Tonnen)

Consumo di carbone sub-bituminoso nell'industria (1000 tonnellate)

亜歴青炭の産業用消費量（千トン）

Consumo industrial de carbón sub-bituminoso (1000 toneladas)

Потребление полубитуминозных углей промышленным сектором (тыс. т)

1989	1990	1991	1992	1993	1994	1995	1996	1997	
13 000	12 250	11 500	10 750	10 000	9 300	8 500	7 769	7 536	DPR of Korea
13 000	12 250	11 500	10 750	10 000	9 300	8 500	7 769	7 536	**Asia**
-	-	128	108	73	77	80	72	47	Bulgaria
-	-	-	64	56	25	14	10	7	*Slovenia*
-	-	-	64	56	25	14	10	7	Former Yugoslavia
-	-	128	172	129	102	94	82	54	**Non-OECD Europe**
13 000	12 250	11 628	10 922	10 129	9 402	8 594	7 851	7 590	**Non-OECD Total**
6 203	6 663	7 053	3 950	4 079	3 453	3 685	3 685	2 498	OECD North America
3 282	3 663	3 723	4 051	3 984	3 851	3 774	3 804	3 652	OECD Pacific
16 268	14 867	6 052	9 106	4 924	4 630	3 560	1 557	1 828	OECD Europe
25 753	25 193	16 828	17 107	12 987	11 934	11 019	9 046	7 978	**OECD Total**
38 753	37 443	28 456	28 029	23 116	21 336	19 613	16 897	15 568	**World**

Industry Consumption of Lignite (1000 tonnes)

Consommation industrielle de lignite (1000 tonnes)

Industrieverbrauch von Braunkohle (1000 Tonnen)

Consumo di lignite nell'industria (1000 tonnellate)

亜炭の産業用消費量（千トン）

Consumo industrial de lignito (1000 toneladas)

Потребление лигнита промышленным сектором (тыс. т)

	1971	1973	1978	1983	1984	1985	1986	1987	1988
Inde	-	-	601	344	504	1 024	639	1 754	1 853
Myanmar	-	-	3	35	43	43	47	35	35
Philippines	-	-	-	4	3	4	4	3	3
Thailande	-	-	132	347	361	535	741	1 096	1 303
Asie	-	-	736	730	911	1 606	1 431	2 888	3 194
Albanie	-	-	600	925	1 005	1 075	950	1 050	1 006
Bulgarie	-	-	2 753	3 536	3 535	3 372	3 852	4 000	2 000
Roumanie	-	-	-	-	-	-	-	-	-
République slovaque	-	-	3 820	3 653	3 748	3 965	4 140	4 154	4 051
Croatie	-	-	-	-	-	-	-	-	-
Ex-RYM	-	-	-	-	-	-	-	-	-
Slovénie	-	-	-	-	-	-	-	-	-
RF de Yougoslavie	-	-	-	-	-	-	-	-	-
Ex-Yougoslavie	-	-	3 300	1 439	1 424	1 242	1 361	1 336	1 300
Europe non-OCDE	-	-	10 473	9 553	9 712	9 654	10 303	10 540	8 357
Estonie	-	-	-	-	-	-	-	-	-
Kazakhstan	-	-	-	-	-	-	-	-	-
Kirghizistan	-	-	-	-	-	-	-	-	-
Russie	-	-	-	-	-	-	-	-	-
Ouzbékistan	-	-	-	-	-	-	-	-	-
Ex-URSS	-	-	-	-	-	-	-	-	-
Total non-OCDE	-	-	11 209	10 283	10 623	11 260	11 734	13 428	11 551
OCDE Amérique du N.	-	-	2 464	2 719	4 127	6 808	7 720	8 209	8 066
OCDE Pacifique	-	-	424	478	471	438	390	257	427
OCDE Europe	-	-	27 504	29 685	35 809	38 868	36 239	37 303	37 618
Total OCDE	-	-	30 392	32 882	40 407	46 114	44 349	45 769	46 111
Monde	-	-	41 601	43 165	51 030	57 374	56 083	59 197	57 662

Ex-URSS: les séries antérieures à 1990-1992 ne sont pas comparables aux années récentes; 1991 a été estimé.

Industry Consumption of Lignite (1000 tonnes)

Consommation industrielle de lignite (1000 tonnes)

Industrieverbrauch von Braunkohle (1000 Tonnen)

Consumo di lignite nell'industria (1000 tonnellate)

亜炭の産業用消費量（千トン）

Consumo industrial de lignito (1000 toneladas)

Потребление лигнита промышленным сектором (тыс. т)

1989	1990	1991	1992	1993	1994	1995	1996	1997	
3 287	2 802	3 900	4 140	4 509	4 310	4 950	5 060	5 152	India
35	38	39	39	39	40	41	46	46	Myanmar
3	3	3	3	3	3	3	-	-	Philippines
1 794	2 582	2 818	3 114	4 248	4 916	4 915	4 544	4 467	Thailand
5 119	5 425	6 760	7 296	8 799	9 269	9 909	9 650	9 665	**Asia**
1 006	843	321	-	-	-	-	-	-	Albania
1 805	271	19	8	8	7	9	5	3	Bulgaria
2 887	184	185	119	123	261	166	178	116	Romania
3 641	3 347	3 284	3 107	2 166	1 879	1 693	1 614	1 318	Slovak Republic
-	-	-	137	129	128	141	106	84	*Croatia*
-	-	-	122	169	166	115	113	105	*FYROM*
-	-	-	94	66	42	58	51	46	*Slovenia*
-	-	-	386	2 255	2 300	2 400	2 300	2 400	*FR of Yugoslavia*
1 300	1 666	764	739	2 619	2 636	2 714	2 570	2 635	Former Yugoslavia
10 639	6 311	4 573	3 973	4 916	4 783	4 582	4 367	4 072	**Non-OECD Europe**
-	-	-	336	260	368	500	498	342	Estonia
-	-	-	4 152	4 669	4 559	3 695	3 446	2 443	Kazakhstan
-	-	-	1 111	985	550	287	332	330	Kyrgyzstan
-	-	-	2 025	6 148	2 739	2 193	2 714	2 597	Russia
-	-	-	-	-	-	196	177	131	Uzbekistan
-	-	-	7 624	12 062	8 216	6 871	7 167	5 843	**Former USSR**
15 758	11 736	11 333	18 893	25 777	22 268	21 362	21 184	19 580	**Non-OECD Total**
8 548	8 046	7 966	6 465	6 072	6 041	6 127	6 137	5 614	OECD North America
263	174	181	164	170	192	206	214	232	OECD Pacific
37 447	30 715	22 041	16 683	12 701	9 288	9 399	8 922	8 989	OECD Europe
46 258	38 935	30 188	23 312	18 943	15 521	15 732	15 273	14 835	**OECD Total**
62 016	50 671	41 521	42 205	44 720	37 789	37 094	36 457	34 415	**World**

Former USSR: series up to 1990-1992 are not comparable with recent years; 1991 is estimated.

Industry Consumption of Oil (1000 tonnes)

Consommation industrielle de pétrole (1000 tonnes)

Industrieverbrauch von Öl (1000 Tonnen)

Consumo di petrolio nell'industria (1000 tonnellate)

石油の産業用消費量（千トン）

Consumo industrial de petróleo (1000 toneladas)

Потребление нефти и нефтепродуктов промышленным сектором (тыс. т)

	1971	1973	1978	1983	1984	1985	1986	1987	1988
Algérie	102	316	318	501	474	507	412	417	392
Angola	56	74	101	81	75	61	30	6	7
Bénin	4	1	4	11	10	10	10	8	8
Cameroun	26	23	52	57	60	68	68	56	54
Congo	32	27	8	19	18	18	18	-	-
RD du Congo	1	2	1	1	1	1	46	48	49
Egypte	2 135	2 147	3 237	4 849	5 249	5 342	4 785	6 349	6 616
Ethiopie	87	91	72	66	67	74	103	102	131
Gabon	127	135	344	334	317	299	198	184	200
Ghana	110	112	127	76	88	67	87	80	79
Côte d'Ivoire	225	235	204	244	64	126	153	152	165
Kenya	189	195	286	276	306	304	311	377	344
Libye	-	63	168	584	619	685	648	560	480
Maroc	584	755	1 146	1 053	1 024	969	870	740	464
Mozambique	-	-	-	-	-	-	-	-	-
Nigéria	212	375	833	1 346	948	1 029	733	539	794
Sénégal	103	111	116	119	139	138	129	139	123
Afrique du Sud	1 900	2 252	2 320	3 140	3 227	3 106	2 973	2 966	3 186
Soudan	188	525	190	252	210	215	244	142	179
Tanzanie	76	106	90	84	86	100	99	96	97
Tunisie	227	259	352	576	579	608	602	395	585
Zambie	149	259	313	259	196	152	136	135	120
Zimbabwe	49	60	49	58	85	89	103	89	97
Autre Afrique	16	18	104	170	207	201	212	217	265
Afrique	6 598	8 141	10 435	14 156	14 049	14 169	12 970	13 797	14 435
Argentine	3 749	3 088	3 109	2 643	3 005	2 660	1 830	2 362	2 200
Bolivie	134	144	137	108	75	54	31	21	43
Brésil	6 879	10 024	15 642	10 389	10 189	10 610	11 101	12 278	11 472
Chili	1 171	1 221	1 295	1 058	1 096	1 060	1 104	1 150	1 290
Colombie	1 432	1 346	1 136	1 016	951	861	904	1 072	1 146
Costa Rica	106	131	173	137	150	177	161	165	135
Cuba	1 810	2 211	2 602	2 545	2 505	2 296	2 171	2 313	2 376
République dominicaine	211	285	391	388	355	151	232	250	397
Equateur	151	133	346	643	651	783	770	789	699
El Salvador	109	139	178	117	123	112	119	133	161
Guatemala	276	221	271	143	141	142	165	184	204
Haiti	48	51	87	39	40	40	44	46	45
Honduras	93	114	159	166	198	168	157	187	221
Jamaïque	625	1 197	1 136	843	812	624	527	530	495
Antilles néerlandaises	1 491	1 637	531	387	359	344	327	313	257
Nicaragua	78	109	95	109	116	113	121	118	112
Panama	26	36	71	107	123	181	199	199	131
Paraguay	11	39	6	23	11	19	20	27	40
Pérou	687	950	1 268	1 073	1 078	1 162	1 201	990	844
Trinité-et-Tobago	28	25	37	48	21	6	1	11	81
Uruguay	422	417	425	228	200	160	178	183	182
Vénézuela	711	741	558	1 337	1 204	1 207	1 571	2 347	1 751
Autre Amérique latine	16	15	22	30	30	34	54	174	170
Amérique latine	20 264	24 274	29 675	23 577	23 433	22 964	22 988	25 842	24 452

Industry Consumption of Oil (1000 tonnes)

Consommation industrielle de pétrole (1000 tonnes)

Industrieverbrauch von Öl (1000 Tonnen)

Consumo di petrolio nell'industria (1000 tonnellate)

石油の産業用消費量（千トン）

Consumo industrial de petróleo (1000 toneladas)

Потребление нефти и нефтепродуктов промышленным сектором (тыс. т)

1989	1990	1991	1992	1993	1994	1995	1996	1997	
315	4	-	-	51	52	41	42	42	Algeria
7	7	7	7	7	7	6	7	15	Angola
10	10	7	9	10	11	11	19	19	Benin
58	54	51	45	46	43	48	51	49	Cameroon
1	2	2	5	4	4	4	4	4	Congo
56	60	61	56	62	62	58	58	58	DR of Congo
6 803	7 070	6 418	6 844	5 245	4 780	6 028	6 424	7 004	Egypt
154	174	178	184	154	174	199	206	103	Ethiopia
90	47	49	89	99	92	98	114	128	Gabon
110	86	87	91	92	99	107	103	103	Ghana
170	176	136	141	149	152	159	136	141	Ivory Coast
319	335	332	349	398	388	355	411	377	Kenya
782	850	970	880	960	1 062	1 238	1 243	1 404	Libya
679	620	584	706	387	520	538	543	561	Morocco
-	8	8	7	9	9	10	10	11	Mozambique
800	695	647	588	603	649	583	727	750	Nigeria
66	76	79	88	71	98	113	109	135	Senegal
3 194	3 370	3 319	3 050	1 283	1 366	1 266	1 380	1 408	South Africa
200	238	295	226	24	270	164	77	226	Sudan
102	105	109	109	105	105	105	105	105	Tanzania
745	782	694	760	641	586	559	612	581	Tunisia
120	115	120	109	115	116	118	120	120	Zambia
103	95	213	68	73	72	99	100	100	Zimbabwe
271	279	269	260	254	264	237	240	246	Other Africa
15 155	15 258	14 635	14 671	10 842	10 981	12 144	12 841	13 690	**Africa**
1 906	2 009	1 895	1 803	1 660	1 657	1 343	1 522	1 385	Argentina
51	70	73	70	66	69	78	82	77	Bolivia
11 915	11 739	11 595	12 413	12 888	13 765	14 394	15 310	17 061	Brazil
1 454	1 427	1 426	1 588	1 779	1 791	2 074	2 242	2 692	Chile
1 153	1 323	1 070	1 346	1 145	1 346	1 521	1 566	1 672	Colombia
201	217	208	185	186	209	215	198	224	Costa Rica
2 610	2 854	1 944	1 620	1 734	2 020	2 024	2 225	2 223	Cuba
290	251	238	267	308	326	338	349	373	Dominican Republic
718	883	878	653	586	619	683	667	758	Ecuador
159	162	177	234	234	249	305	278	288	El Salvador
217	229	243	269	331	347	409	389	402	Guatemala
54	49	49	51	41	6	44	49	77	Haiti
260	243	249	284	279	291	364	484	491	Honduras
898	146	161	159	135	218	186	190	195	Jamaica
209	174	174	174	137	137	136	137	137	Netherlands Antilles
95	103	78	81	76	98	89	91	65	Nicaragua
141	125	167	176	164	197	217	150	183	Panama
40	51	46	78	64	104	82	34	40	Paraguay
688	687	634	516	537	600	945	1 279	1 241	Peru
86	93	101	87	66	70	77	85	103	Trinidad and Tobago
173	167	171	173	158	145	148	171	209	Uruguay
1 670	1 650	1 781	1 513	2 309	2 727	2 513	2 353	2 745	Venezuela
171	149	148	134	134	138	84	85	89	Other Latin America
25 159	24 801	23 506	23 874	25 017	27 129	28 269	29 936	32 730	**Latin America**

Industry Consumption of Oil (1000 tonnes)

Consommation industrielle de pétrole (1000 tonnes)

Industrieverbrauch von Öl (1000 Tonnen)

Consumo di petrolio nell'industria (1000 tonnellate)

石油の産業用消費量（千トン）

Consumo industrial de petróleo (1000 toneladas)

Потребление нефти и нефтепродуктов промышленным сектором (тыс. т)

	1971	1973	1978	1983	1984	1985	1986	1987	1988
Bangladesh	179	237	272	240	229	151	98	60	39
Brunei	22	22	17	42	32	36	32	33	38
Inde	4 544	5 806	6 819	7 967	8 685	9 193	9 411	9 386	10 072
Indonésie	1 398	1 639	3 604	5 292	4 645	5 098	4 621	4 527	4 917
RPD de Corée	90	95	326	484	528	572	619	620	620
Malaisie	1 917	1 496	1 914	2 569	2 430	2 344	2 200	2 384	2 495
Myanmar	363	256	384	370	355	349	310	200	174
Népal	-	1	1	5	5	7	18	13	10
Pakistan	153	247	223	391	689	815	946	1 227	1 225
Philippines	1 574	2 012	2 483	2 121	1 607	1 408	1 455	1 734	1 898
Singapour	75	101	235	248	606	1 175	1 111	1 383	1 376
Sri Lanka	95	175	179	209	197	100	92	87	109
Taipei chinois	1 167	2 004	5 440	5 976	6 528	6 993	7 819	8 270	8 552
Thailande	1 324	1 570	1 858	1 628	1 622	1 621	1 637	1 865	1 918
Viêt-Nam	-	-	-	558	642	383	402	413	372
Autre Asie	40	49	57	123	119	120	109	106	100
Asie	12 941	15 710	23 812	28 223	28 919	30 365	30 880	32 308	33 915
Rép. populaire de Chine	6 697	8 776	22 719	20 710	21 086	24 018	24 452	27 361	28 973
Hong-Kong, Chine	772	654	1 461	718	872	751	774	875	1 023
Chine	7 469	9 430	24 180	21 428	21 958	24 769	25 226	28 236	29 996
Albanie	213	50	-	212	189	117	157	187	112
Bulgarie	-	-	-	-	-	-	-	-	2 349
Chypre	125	164	210	206	179	149	212	209	267
Malte	-	-	-	-	-	-	-	-	-
Roumanie	724	754	1 061	-	-	1 171	1 328	2 206	2 302
République slovaque	804	882	995	2 360	2 234	2 226	2 108	2 044	1 914
Bosnie-Herzegovine	-	-	-	-	-	-	-	-	-
Croatie	-	-	-	-	-	-	-	-	-
Ex-RYM	-	-	-	-	-	-	-	-	-
Slovénie	-	-	-	-	-	-	-	-	-
RF de Yougoslavie	-	-	-	-	-	-	-	-	-
Ex-Yougoslavie	2 320	3 205	4 175	3 232	2 936	2 808	3 055	3 014	3 251
Europe non-OCDE	4 186	5 055	6 441	6 010	5 538	6 471	6 860	7 660	10 195
Arménie	-	-	-	-	-	-	-	-	-
Azerbaïdjan	-	-	-	-	-	-	-	-	-
Bélarus	-	-	-	-	-	-	-	-	-
Estonie	-	-	-	-	-	-	-	-	-
Géorgie	-	-	-	-	-	-	-	-	-
Lettonie	-	-	-	-	-	-	-	-	-
Lituanie	-	-	-	-	-	-	-	-	-
Moldova	-	-	-	-	-	-	-	-	-
Russie	-	-	-	-	-	-	-	-	-
Ukraine	-	-	-	-	-	-	-	-	-
Ouzbékistan	-	-	-	-	-	-	-	-	-
Ex-URSS	57 000	67 100	82 900	81 700	81 000	76 816	76 812	76 117	76 648

Ex-URSS: les séries antérieures à 1990-1992 ne sont pas comparables aux années récentes; 1991 a été estimé.

Rép. populaire de Chine: jusqu'en 1978 la ventilation de la consommation finale par secteur est incomplète.

Industry Consumption of Oil (1000 tonnes)

Consommation industrielle de pétrole (1000 tonnes)

Industrieverbrauch von Öl (1000 Tonnen)

Consumo di petrolio nell'industria (1000 tonnellate)

石油の産業用消費量（千トン）

Consumo industrial de petróleo (1000 toneladas)

Потребление нефти и нефтепродуктов промышленным сектором (тыс. т)

1989	1990	1991	1992	1993	1994	1995	1996	1997	
30	25	57	72	130	197	205	214	206	Bangladesh
42	44	45	56	58	62	67	76	88	Brunei
11 603	12 173	12 297	12 608	12 477	13 376	14 516	16 253	17 699	India
4 835	5 243	5 295	5 586	6 332	6 824	7 245	7 650	10 643	Indonesia
700	700	608	496	364	299	243	198	192	DPR of Korea
2 878	3 403	3 672	3 769	4 924	4 708	4 888	5 373	5 892	Malaysia
167	130	131	215	254	273	378	421	480	Myanmar
9	3	5	8	9	29	105	105	105	Nepal
1 291	1 297	1 147	1 163	1 479	1 654	1 869	2 416	1 739	Pakistan
2 109	2 034	1 966	2 091	2 332	2 640	3 323	3 354	3 529	Philippines
1 460	1 702	1 570	1 383	1 521	1 519	1 881	2 174	2 675	Singapore
132	123	117	191	207	198	172	435	403	Sri Lanka
8 346	8 642	8 580	8 786	9 056	10 240	10 523	10 826	11 451	Chinese Taipei
2 316	2 730	2 943	3 560	3 887	4 790	5 224	6 246	5 605	Thailand
374	569	478	1 192	1 171	1 315	1 374	770	971	Vietnam
100	103	103	103	96	98	20	20	20	Other Asia
36 392	38 921	39 014	41 279	44 297	48 222	52 033	56 531	61 698	**Asia**
30 803	29 706	31 244	33 350	33 012	36 585	37 350	40 169	40 324	People's Rep. of China
1 018	641	260	676	328	333	254	201	163	Hong Kong, China
31 821	30 347	31 504	34 026	33 340	36 918	37 604	40 370	40 487	**China**
152	303	195	111	138	209	69	61	29	Albania
2 287	2 067	757	1 180	1 182	1 077	1 194	1 368	1 341	Bulgaria
233	135	326	335	351	363	375	419	391	Cyprus
-	-	-	-	-	-	23	-	-	Malta
2 865	2 082	2 486	1 487	964	1 914	1 950	2 166	2 074	Romania
2 937	2 106	1 829	1 805	1 183	1 382	1 146	1 091	1 047	Slovak Republic
-	-	-	80	179	59	59	59	59	*Bosnia-Herzegovina*
-	-	-	562	580	628	614	535	520	*Croatia*
-	-	-	339	208	141	134	263	187	*FYROM*
-	-	-	184	177	205	214	213	164	*Slovenia*
-	-	-	665	518	489	537	713	1 032	*FR of Yugoslavia*
3 022	2 879	2 437	1 830	1 662	1 522	1 558	1 783	1 962	Former Yugoslavia
11 496	9 572	8 030	6 748	5 480	6 467	6 315	6 888	6 844	**Non-OECD Europe**
-	-	-	200	240	-	-	-	-	Armenia
-	-	-	-	302	196	163	157	150	Azerbaijan
-	-	-	1 507	2 024	1 707	1 585	1 444	1 540	Belarus
-	-	-	170	116	156	267	255	203	Estonia
-	-	-	-	-	-	-	59	62	Georgia
-	-	-	181	195	117	141	64	283	Latvia
-	-	-	976	217	208	195	185	188	Lithuania
-	-	-	40	89	33	31	28	24	Moldova
-	-	-	25 624	28 436	17 507	16 846	13 186	13 053	Russia
-	-	-	9 164	6 244	5 172	5 637	4 205	3 848	Ukraine
-	-	-	273	-	-	148	413	398	Uzbekistan
74 451	72 542	74 258	38 135	37 863	25 096	25 013	19 996	19 749	**Former USSR**

Former USSR: series up to 1990-1992 are not comparable with recent years; 1991 is estimated.

People's Rep. of China: up to 1978 the breakdown of final consumption by sector is incomplete.

Industry Consumption of Oil (1000 tonnes)

Consommation industrielle de pétrole (1000 tonnes)

Industrieverbrauch von Öl (1000 Tonnen)

Consumo di petrolio nell'industria (1000 tonnellate)

石油の産業用消費量（千トン）

Consumo industrial de petróleo (1000 toneladas)

Потребление нефти и нефтепродуктов промышленным сектором (тыс. т)

	1971	1973	1978	1983	1984	1985	1986	1987	1988
Iran	1 286	1 748	3 495	2 069	5 175	5 040	7 009	8 065	8 035
Irak	65	32	90	34	310	244	379	-	100
Israël	833	944	1 215	1 276	1 319	1 318	1 502	1 511	1 560
Jordanie	47	74	116	361	327	386	431	420	382
Koweit	-	-	143	621	21	541	6	6	191
Liban	464	494	431	44	36	51	50	97	67
Oman	6	9	13	11	11	9	10	10	5
Qatar	-	-	6	5	-	49	-	24	-
Arabie saoudite	632	1 179	3 618	15 159	5 844	7 012	7 612	7 433	6 632
Syrie	511	249	1 572	1 931	2 015	2 084	2 020	2 045	1 758
Emirats arabes unis	-	-	-	455	92	478	344	350	300
Yémen	-	4	107	220	224	379	362	317	304
Moyen-Orient	3 844	4 733	10 806	22 186	15 374	17 591	19 725	20 278	19 334
Total non-OCDE	112 302	134 443	188 249	197 280	190 271	193 145	195 461	204 238	208 975
OCDE Amérique du N.	112 634	131 228	142 872	114 043	122 669	117 317	119 779	121 296	119 994
OCDE Pacifique	85 855	98 157	86 803	64 258	67 833	66 080	66 692	69 588	73 148
OCDE Europe	173 691	191 955	170 667	125 857	118 442	114 012	115 562	115 473	115 293
Total OCDE	372 180	421 340	400 342	304 158	308 944	297 409	302 033	306 357	308 435
Monde	484 482	555 783	588 591	501 438	499 215	490 554	497 494	510 595	517 410

Industry Consumption of Oil (1000 tonnes)

Consommation industrielle de pétrole (1000 tonnes)

Industrieverbrauch von Öl (1000 Tonnen)

Consumo di petrolio nell'industria (1000 tonnellate)

石油の産業用消費量 (千トン)

Consumo industrial de petróleo (1000 toneladas)

Потребление нефти и нефтепродуктов промышленным сектором (тыс. т)

1989	1990	1991	1992	1993	1994	1995	1996	1997	
7 533	130	140	140	6 707	6 803	7 158	7 733	7 523	Iran
800	900	2 090	1 310	1 900	2 409	2 310	2 217	2 279	Iraq
1 447	1 431	1 480	1 337	1 407	1 532	1 821	1 923	2 348	Israel
383	416	367	415	463	477	500	505	536	Jordan
200	200	50	2 154	1 361	1 324	1 624	1 498	1 797	Kuwait
70	105	100	70	411	547	726	627	827	Lebanon
10	511	1 666	1 432	970	525	958	828	446	Oman
862	21	184	22	23	24	24	26	29	Qatar
7 093	7 450	7 560	7 677	8 024	7 648	8 028	8 855	9 513	Saudi Arabia
670	650	750	810	832	948	1 047	1 084	1 037	Syria
350	350	350	500	643	629	444	421	424	United Arab Emirates
310	108	328	403	136	145	145	163	169	Yemen
19 728	12 272	15 065	16 270	22 877	23 011	24 785	25 880	26 928	**Middle East**
214 202	203 713	206 012	175 003	179 716	177 824	186 163	192 442	202 126	**Non-OECD Total**
117 829	114 571	105 604	111 154	109 857	114 248	110 757	115 600	116 828	OECD North America
76 658	80 596	83 605	88 453	91 319	94 572	97 930	99 552	104 493	OECD Pacific
111 079	105 509	106 348	104 388	100 530	102 999	107 713	105 758	107 782	OECD Europe
305 566	300 676	295 557	303 995	301 706	311 819	316 400	320 910	329 103	**OECD Total**
519 768	504 389	501 569	478 998	481 422	489 643	502 563	513 352	531 229	**World**

Industry Consumption of Natural Gas (TJ)

Consommation industrielle de gaz naturel (TJ)

Industrieverbrauch von Erdgas (TJ)

Consumo di gas naturale nell'industria (TJ)

天然ガスの産業用消費量 (TJ)

Consumo industrial de gas natural (TJ)

Потребление природного газа промышленным сектором (ТДж)

	1971	1973	1978	1983	1984	1985	1986	1987	1988
Algérie	8 379	9 531	29 055	38 883	42 353	54 951	64 800	70 200	68 700
Angola	1 678	2 536	2 758	3 997	4 499	4 499	4 997	5 998	6 098
Congo	2 256	-	-	-	-	-	-	-	-
Egypte	-	-	15 101	53 131	65 763	66 722	69 420	79 935	78 620
Gabon	-	-	-	1 450	1 546	1 972	2 263	2 224	3 125
Libye	-	-	30 408	55 000	60 000	67 000	69 000	70 000	51 000
Maroc	2 009	2 721	3 177	3 248	3 244	3 666	3 549	2 884	2 947
Nigéria	1 411	1 352	1 398	5 717	14 343	17 188	18 149	21 296	26 957
Sénégal	-	-	-	-	-	-	-	11	314
Tunisie	38	331	2 888	4 817	5 085	5 902	7 425	12 364	6 203
Afrique	15 771	16 471	84 785	166 243	196 833	221 900	239 603	264 912	243 964
Argentine	79 484	107 298	130 036	170 017	174 499	174 996	185 697	206 278	207 589
Bolivie	-	155	2 203	1 739	2 087	2 783	3 208	3 556	3 401
Brésil	1 521	4 613	29 029	57 811	64 563	75 338	76 074	89 241	96 220
Chili	38	76	380	266	304	304	304	380	10 640
Colombie	4 905	9 832	25 410	32 026	34 898	36 247	39 063	38 113	40 106
Cuba	67	565	427	306	126	266	228	874	722
Trinité-et-Tobago	21 363	24 192	39 896	77 526	76 243	76 099	82 371	83 510	111 442
Vénézuela	86 780	176 413	248 004	296 898	355 882	379 137	353 917	378 089	301 641
Autre Amérique latine	-	-	135	130	200	393	39	39	39
Amérique latine	194 158	323 144	475 520	636 719	708 802	745 563	740 901	800 080	771 800
Bangladesh	8 978	15 248	19 325	39 273	43 585	46 238	54 764	61 244	68 375
Inde	8 403	7 674	31 898	56 421	70 140	108 640	144 096	150 913	233 358
Indonésie	4 637	4 968	52 330	133 804	182 602	95 087	97 134	102 809	95 180
Malaisie	-	-	46	46	3 696	21 342	39 029	48 242	54 796
Myanmar	755	1 164	3 116	5 883	7 895	10 745	12 014	12 363	12 485
Pakistan	65 540	78 450	104 363	168 404	161 193	162 048	161 299	183 924	188 370
Taipei chinois	28 881	40 670	57 355	29 300	27 180	21 195	16 650	17 052	18 420
Thailande	-	-	-	1 329	8 116	7 452	3 638	1 653	2 515
Viêt-Nam	-	-	-	-	-	-	-	-	-
Asie	117 194	148 174	268 433	434 460	504 407	472 747	528 624	578 200	673 499
Rép. populaire de Chine	65 591	104 864	293 789	264 034	269 104	314 921	296 237	311 053	291 834
Hong-Kong, Chine	-	-	-	-	-	-	-	-	-
Chine	65 591	104 864	293 789	264 034	269 104	314 921	296 237	311 053	291 834
Albanie	-	-	-	-	-	-	-	-	-
Bulgarie	-	-	-	-	-	171 632	177 181	187 692	99 898
Roumanie	605 483	532 592	1 170 245	1 283 973	1 290 428	873 008	897 612	887 519	919 990
République slovaque	32 162	38 358	75 274	105 848	105 357	103 907	104 896	106 142	107 785
Croatie	-	-	-	-	-	-	-	-	-
Ex-RYM	-	-	-	-	-	-	-	-	-
Slovénie	-	-	-	-	-	-	-	-	-
RF de Yougoslavie	-	-	-	-	-	-	-	-	-
Ex-Yougoslavie	32 786	40 013	56 964	85 709	111 824	119 692	126 657	140 524	104 914
Europe non-OCDE	670 431	610 963	1 302 483	1 475 530	1 507 609	1 268 239	1 306 346	1 321 877	1 232 587

Rép. populaire de Chine: jusqu'en 1978 la ventilation de la consommation finale par secteur est incomplète.

Industry Consumption of Natural Gas (TJ)

Consommation industrielle de gaz naturel (TJ)

Industrieverbrauch von Erdgas (TJ)

Consumo di gas naturale nell'industria (TJ)

天然ガスの産業用消費量 (TJ)

Consumo industrial de gas natural (TJ)

Потребление природного газа промышленным сектором (ТДж)

1989	1990	1991	1992	1993	1994	1995	1996	1997	
55 200	58 500	64 300	67 194	127 861	113 950	111 902	106 971	116 415	Algeria
6 500	20 520	22 040	21 660	21 280	19 760	21 280	21 280	21 660	Angola
-	-	-	-	-	-	-	-	-	Congo
117 396	109 408	110 750	116 060	140 086	101 251	105 948	106 058	111 492	Egypt
30	30	30	30	30	25	27	29	28	Gabon
52 000	60 000	65 000	65 000	67 000	69 144	68 314	73 526	78 696	Libya
2 177	2 014	1 289	847	828	880	884	698	1 047	Morocco
31 825	33 326	30 326	32 500	34 600	37 714	34 433	34 898	33 635	Nigeria
299	219	170	105	498	761	1 843	1 685	928	Senegal
8 830	11 866	12 517	13 816	14 514	15 308	16 843	16 797	17 122	Tunisia
274 257	295 883	306 422	317 212	406 697	358 793	361 474	361 942	381 023	**Africa**
175 151	181 757	204 967	215 847	218 686	249 442	273 404	274 428	290 166	Argentina
6 648	7 382	8 155	9 701	10 962	12 229	13 417	11 820	16 006	Bolivia
101 014	102 969	108 272	117 809	118 979	125 348	131 956	148 986	164 991	Brazil
30 818	33 675	24 261	32 958	31 962	34 718	33 523	34 973	66 028	Chile
40 783	42 502	42 640	41 770	41 370	45 171	47 062	48 677	46 999	Colombia
988	1 250	1 033	566	810	616	448	547	673	Cuba
113 555	116 965	124 213	127 403	124 498	167 374	169 198	191 044	205 283	Trinidad and Tobago
312 240	359 852	412 241	361 722	401 422	402 407	441 590	445 115	478 218	Venezuela
39	39	39	39	40	40	39	50	50	Other Latin America
781 236	846 391	925 821	907 815	948 729	1 037 345	1 110 637	1 155 640	1 268 414	**Latin America**
72 642	69 819	67 173	83 936	89 371	100 521	116 301	117 765	111 167	Bangladesh
286 906	274 001	281 075	321 161	307 571	327 078	366 327	406 527	419 730	India
70 758	309 258	316 986	352 711	306 247	303 252	323 327	352 049	366 163	Indonesia
47 698	50 767	50 753	62 150	78 774	76 616	73 739	95 490	87 087	Malaysia
12 078	10 351	8 317	8 200	10 544	13 202	13 188	12 899	15 177	Myanmar
200 933	200 971	208 185	208 115	232 285	247 485	245 134	260 883	271 938	Pakistan
16 805	18 057	24 772	29 179	30 533	32 127	38 071	39 959	40 423	Chinese Taipei
4 778	11 049	15 051	18 454	20 580	24 354	32 762	39 095	39 552	Thailand
-	-	-	-	-	-	9 881	135 527	143 884	Vietnam
712 598	944 273	972 312	1 083 906	1 075 905	1 124 635	1 218 730	1 460 194	1 495 121	**Asia**
312 161	319 623	353 003	356 154	327 313	356 164	372 104	476 373	401 483	People's Rep. of China
-	-	-	-	-	-	1 086	68 823	107 462	Hong Kong, China
312 161	319 623	353 003	356 154	327 313	356 164	373 190	545 196	508 945	**China**
10 000	7 639	4 349	2 807	2 380	1 623	895	735	583	Albania
103 693	105 743	90 623	74 774	73 547	82 923	96 585	96 901	81 202	Bulgaria
909 942	779 979	532 075	252 697	242 600	336 689	366 965	364 926	302 124	Romania
113 642	133 541	105 800	104 791	99 238	76 064	100 891	115 405	113 621	Slovak Republic
-	-	-	45 562	42 081	44 004	42 705	42 871	46 930	*Croatia*
-	-	-	6 000	6 250	-	-	-	-	*FYROM*
-	-	-	16 908	16 663	20 291	20 900	22 677	27 065	*Slovenia*
-	-	-	24 082	21 602	9 770	12 800	30 100	29 858	*FR of Yugoslavia*
105 480	30 795	92 916	92 552	86 596	74 065	76 405	95 648	103 853	Former Yugoslavia
1 242 757	1 057 697	825 763	527 621	504 361	571 364	641 741	673 615	601 383	**Non-OECD Europe**

People's Rep. of China: up to 1978 the breakdown of final consumption by sector is incomplete.

Industry Consumption of Natural Gas (TJ)

Consommation industrielle de gaz naturel (TJ)

Industrieverbrauch von Erdgas (TJ)

Consumo di gas naturale nell'industria (TJ)

天然ガスの産業用消費量 (TJ)

Consumo industrial de gas natural (TJ)

Потребление природного газа промышленным сектором (ТДж)

	1971	1973	1978	1983	1984	1985	1986	1987	1988
Arménie	-	-	-	-	-	-	-	-	-
Azerbaïdjan	-	-	-	-	-	-	-	-	-
Bélarus	-	-	-	-	-	-	-	-	-
Estonie	-	-	-	-	-	-	-	-	-
Géorgie	-	-	-	-	-	-	-	-	-
Lettonie	-	-	-	-	-	-	-	-	-
Lituanie	-	-	-	-	-	-	-	-	-
Moldova	-	-	-	-	-	-	-	-	-
Russie	-	-	-	-	-	-	-	-	-
Ukraine	-	-	-	-	-	-	-	-	-
Ouzbékistan	-	-	-	-	-	-	-	-	-
Ex-URSS	3 592 369	4 132 154	5 444 394	6 504 524	6 765 103	7 039 643	7 295 568	7 565 454	7 839 943
Bahrein	8 371	27 369	43 408	51 189	54 089	77 481	101 754	90 641	104 516
Iran	376 386	383 782	389 252	330 624	403 403	464 824	477 629	395 460	286 612
Irak	35 340	45 980	64 600	103 163	114 300	121 250	148 296	186 521	258 158
Israël	4 232	2 097	2 185	2 461	2 034	1 821	1 427	1 615	1 453
Koweit	89 804	92 772	115 474	104 746	86 447	113 374	158 698	171 698	128 251
Oman	-	-	-	403	9 909	14 792	6 982	8 898	9 846
Qatar	17 674	28 065	21 185	71 909	92 680	81 922	85 198	84 976	92 824
Arabie saoudite	14 783	14 783	14 783	34 334	75 946	110 510	130 031	-	-
Syrie	-	-	-	-	-	-	-	-	-
Emirats arabes unis	-	-	-	-	24 983	23 372	25 967	29 299	33 484
Moyen-Orient	546 590	594 848	650 887	698 829	863 791	1 009 346	1 135 982	969 108	915 144
Total non-OCDE	5 202 104	5 930 618	8 520 291	10 180 339	10 815 649	11 072 359	11 543 261	11 810 684	11 968 771
OCDE Amérique du N.	8 463 968	9 118 124	7 571 515	7 025 124	7 658 377	7 338 862	6 847 240	7 297 490	7 823 356
OCDE Pacifique	111 156	138 617	221 213	279 954	307 780	345 717	353 285	365 517	369 359
OCDE Europe	1 791 914	2 554 827	3 512 033	3 440 224	3 636 740	3 656 289	3 602 643	3 872 042	3 940 244
Total OCDE	10 367 038	11 811 568	11 304 761	10 745 302	11 602 897	11 340 868	10 803 168	11 535 049	12 132 959
Monde	15 569 142	17 742 186	19 825 052	20 925 641	22 418 546	22 413 227	22 346 429	23 345 733	24 101 730

Ex-URSS: les séries antérieures à 1990-1992 ne sont pas comparables aux années récentes; 1991 a été estimé.

Industry Consumption of Natural Gas (TJ)
Consommation industrielle de gaz naturel (TJ)
Industrieverbrauch von Erdgas (TJ)

Consumo di gas naturale nell'industria (TJ)

天然ガスの産業用消費量 (TJ)

Consumo industrial de gas natural (TJ)

Потребление природного газа промышленным сектором (ТДж)

1989	1990	1991	1992	1993	1994	1995	1996	1997	
-	-	-	18 413	6 941	6 672	10 771	4 955	5 316	Armenia
-	-	-	-	162 563	158 755	125 390	69 952	70 250	Azerbaijan
-	-	-	98 472	93 659	54 495	78 943	76 008	73 150	Belarus
-	-	-	8 731	3 914	8 110	11 865	12 167	11 942	Estonia
-	-	-	-	74 872	29 406	18 209	11 574	16 324	Georgia
-	-	-	15 231	7 215	11 053	7 696	10 089	9 962	Latvia
-	-	-	28 735	14 713	21 071	28 741	32 078	28 649	Lithuania
-	-	-	19 344	17 759	16 733	16 728	19 803	21 157	Moldova
-	-	-	3 861 800	2 243 294	1 825 556	1 896 555	1 604 445	1 600 903	Russia
-	-	-	1 281 046	1 216 956	761 766	762 802	812 630	722 026	Ukraine
-	-	-	-	-	-	312 722	243 919	190 102	Uzbekistan
7 978 454	8 332 006	7 909 844	5 331 772	3 841 886	2 893 617	3 270 422	2 897 620	2 749 781	**Former USSR**
102 100	107 500	102 060	103 349	122 930	113 816	120 684	128 096	122 801	Bahrain
289 351	264 924	402 602	541 188	558 893	536 458	596 024	706 609	794 936	Iran
297 681	158 135	65 021	108 680	122 048	151 706	150 796	154 126	172 203	Iraq
1 456	1 308	959	949	1 001	894	749	372	390	Israel
178 098	136 000	-	-	-	-	-	-	-	Kuwait
10 628	27 258	29 451	50 852	73 215	75 146	52 046	21 515	39 322	Oman
85 927	90 879	117 239	209 345	223 867	222 111	221 129	221 310	245 078	Qatar
-	-	-	-	-	-	-	-	-	Saudi Arabia
-	-	17 000	27 000	28 000	30 352	38 274	38 274	35 374	Syria
36 300	37 000	50 780	56 520	58 330	62 821	68 412	81 438	85 777	United Arab Emirates
1 001 541	823 004	785 112	1 097 883	1 188 284	1 193 304	1 248 114	1 351 740	1 495 881	**Middle East**
12 303 004	12 618 877	12 078 277	9 622 363	8 293 175	7 535 222	8 224 308	8 445 947	8 500 548	**Non-OECD Total**
7 292 871	7 313 947	6 982 921	6 913 476	7 244 936	7 414 779	7 667 263	7 743 304	7 674 474	OECD North America
381 167	389 066	399 964	391 182	408 108	438 181	461 779	496 569	528 044	OECD Pacific
4 132 884	4 128 023	4 046 037	4 088 033	4 127 606	4 207 779	4 431 141	4 594 458	4 693 429	OECD Europe
11 806 922	11 831 036	11 428 922	11 392 691	11 780 650	12 060 739	12 560 183	12 834 331	12 895 947	**OECD Total**
24 109 926	24 449 913	23 507 199	21 015 054	20 073 825	19 595 961	20 784 491	21 280 278	21 396 495	**World**

Former USSR: series up to 1990-1992 are not comparable with recent years; 1991 is estimated.

Industry Consumption of Electricity (GWh)
Consommation industrielle d'électricité (GWh)
Industrieverbrauch von Elektrizität (GWh)
Consumo di energia elettrica nell'industria (GWh)
電力の産業用消費量 (GWh)
Consumo industrial de electricidad (GWh)
Потребление электроэнергии промышленным сектором (ГВт.ч)

	1971	1973	1978	1983	1984	1985	1986	1987	1988
Algérie	793	944	1 975	3 810	3 939	4 291	5 117	4 330	4 650
Angola	129	172	102	137	141	141	141	141	142
Bénin	16	21	30	75	68	72	64	59	83
Cameroun	884	837	825	1 246	1 183	1 289	1 228	1 200	1 296
Congo	42	46	60	142	141	142	182	208	214
RD du Congo	-	-	2 450	2 656	2 690	2 567	2 516	2 520	2 525
Egypte	4 113	4 203	8 141	11 422	11 708	12 708	13 798	14 710	15 390
Ethiopie	365	359	316	435	468	400	424	518	463
Gabon	-	32	179	247	349	377	350	383	385
Ghana	2 439	3 238	2 826	1 292	548	1 448	2 745	3 148	3 459
Côte d'Ivoire	132	183	386	329	900	585	852	838	942
Kenya	360	434	711	1 144	1 206	1 354	1 476	1 522	1 573
Maroc	975	1 218	1 849	2 710	2 844	2 919	3 201	3 163	3 785
Mozambique	201	236	283	197	154	167	186	252	164
Nigéria	590	850	1 312	2 042	1 728	1 902	2 457	2 576	2 377
Sénégal	181	226	316	349	391	435	409	412	427
Afrique du Sud	32 529	35 557	51 297	61 372	68 123	71 359	74 205	76 512	79 499
Soudan	206	234	360	167	182	186	170	230	239
Tanzanie	151	179	214	247	259	281	324	338	345
Tunisie	459	579	963	1 429	1 528	1 619	1 706	1 858	1 961
Zambie	3 619	4 202	4 661	5 349	5 316	5 225	4 911	5 123	4 773
Zimbabwe	2 206	3 407	3 983	4 566	4 736	5 101	5 318	5 382	5 586
Autre Afrique	251	317	1 010	1 362	1 470	1 938	1 829	2 000	2 150
Afrique	50 641	57 474	84 249	102 725	110 072	116 506	123 609	127 423	132 428
Argentine	10 370	12 313	14 465	19 376	19 969	19 255	20 383	21 395	23 302
Bolivie	152	174	337	872	837	802	709	593	639
Brésil	22 302	29 514	54 473	75 304	87 189	96 233	104 361	104 911	111 453
Chili	4 908	4 757	5 759	6 588	7 259	7 486	7 841	8 137	9 030
Colombie	2 994	3 449	4 726	5 871	6 675	6 161	6 677	7 455	8 414
Costa Rica	309	365	568	612	739	765	821	850	847
Cuba	2 008	2 281	3 372	4 863	5 190	5 023	5 491	5 679	5 975
République dominicaine	298	675	826	884	1 302	1 383	1 256	1 526	744
Equateur	290	320	628	1 326	1 360	1 395	1 407	1 360	1 384
El Salvador	274	352	589	465	469	480	472	515	527
Guatemala	349	360	583	438	442	483	507	499	639
Haiti	11	23	81	161	166	173	174	157	165
Honduras	184	229	307	440	467	584	569	739	776
Jamaïque	1 205	1 507	1 513	670	657	771	727	872	855
Antilles néerlandaises	319	337	360	377	337	313	247	277	282
Nicaragua	272	304	398	349	347	319	306	340	299
Panama	234	255	273	479	468	331	332	359	276
Paraguay	100	131	281	352	371	225	244	450	568
Pérou	3 668	4 151	5 253	5 601	6 494	6 391	6 808	7 124	6 777
Trinité-et-Tobago	340	514	696	1 635	1 526	1 525	1 742	1 887	1 782
Uruguay	663	649	978	1 073	1 103	1 140	1 235	1 273	1 362
Vénézuela	4 122	4 539	7 605	14 873	16 618	18 554	18 932	21 813	23 713
Autre Amérique latine	113	144	205	271	291	307	326	351	371
Amérique latine	55 485	67 343	104 276	142 880	160 276	170 099	181 567	188 562	200 180

Industry Consumption of Electricity (GWh)
Consommation industrielle d'électricité (GWh)
Industrieverbrauch von Elektrizität (GWh)
Consumo di energia elettrica nell'industria (GWh)
電力の産業用消費量(GWh)
Consumo industrial de electricidad (GWh)
Потребление электроэнергии промышленным сектором (ГВт.ч)

1989	1990	1991	1992	1993	1994	1995	1996	1997	
5 265	5 455	5 625	5 584	6 220	6 089	6 216	6 259	7 685	Algeria
143	123	191	200	200	201	202	216	233	Angola
76	87	102	121	111	109	121	130	121	Benin
1 364	1 359	1 342	1 304	1 376	1 295	1 279	1 315	1 428	Cameroon
238	293	288	282	285	288	291	288	285	Congo
3 020	3 031	3 300	3 400	4 234	4 321	3 757	3 806	3 650	DR of Congo
16 282	17 232	18 266	18 619	19 593	20 589	21 732	21 578	22 851	Egypt
470	677	690	688	600	812	814	815	830	Ethiopia
380	390	395	395	398	420	421	434	451	Gabon
3 527	3 558	3 714	3 865	3 969	3 591	3 401	3 384	3 486	Ghana
919	914	892	501	885	916	1 147	1 204	1 281	Ivory Coast
1 626	1 754	1 832	1 835	1 915	1 955	1 999	2 137	2 263	Kenya
3 629	4 016	4 180	4 513	4 411	4 736	4 815	4 969	5 250	Morocco
185	177	311	294	266	244	260	247	291	Mozambique
2 676	1 905	2 235	2 183	2 067	2 042	2 037	2 108	2 134	Nigeria
459	437	450	461	507	464	505	508	545	Senegal
81 352	82 341	82 430	79 454	75 706	75 681	80 657	89 904	91 460	South Africa
190	204	221	231	231	416	378	388	433	Sudan
384	399	465	391	389	395	439	485	460	Tanzania
2 079	2 231	2 315	2 393	2 510	2 651	2 762	2 805	2 906	Tunisia
3 465	2 793	4 018	3 990	3 992	3 992	3 995	3 998	4 118	Zambia
5 765	5 660	4 955	5 329	4 517	4 958	5 375	5 699	5 670	Zimbabwe
2 104	2 188	2 037	1 979	1 570	1 608	1 252	923	923	Other Africa
135 598	137 224	140 254	138 012	135 952	137 773	143 855	153 600	158 754	**Africa**
22 043	17 511	17 596	18 187	19 580	20 139	21 514	22 276	24 105	Argentina
701	728	820	850	1 115	1 118	1 181	1 143	1 235	Bolivia
114 543	112 339	115 041	116 586	122 462	126 177	127 171	129 755	135 702	Brazil
9 868	10 094	11 333	13 192	13 747	14 438	16 485	18 349	20 276	Chile
8 523	7 959	8 411	8 450	9 736	10 692	11 054	11 332	11 283	Colombia
711	740	759	816	998	1 110	1 005	1 021	1 096	Costa Rica
6 079	5 674	4 521	3 916	3 331	3 533	3 541	3 976	5 258	Cuba
721	616	674	826	1 066	1 122	1 181	1 243	1 392	Dominican Republic
1 407	1 535	1 686	1 640	1 814	2 059	2 083	2 262	2 361	Ecuador
509	569	588	628	731	774	830	842	905	El Salvador
598	788	789	823	908	928	1 040	1 247	1 438	Guatemala
169	179	133	104	84	62	100	113	125	Haiti
745	852	767	596	637	570	581	778	732	Honduras
748	208	289	316	2 841	2 710	3 308	3 425	3 548	Jamaica
302	331	337	365	397	420	436	440	455	Netherlands Antilles
325	355	333	288	249	255	299	345	389	Nicaragua
321	374	419	451	529	562	603	642	498	Panama
559	624	518	637	725	774	928	941	1 386	Paraguay
7 038	4 535	4 789	4 494	4 511	4 988	5 466	5 965	8 536	Peru
1 734	1 916	1 932	2 089	2 062	2 209	2 263	2 566	2 882	Trinidad and Tobago
1 430	1 480	1 582	1 533	1 546	1 547	1 310	1 285	1 300	Uruguay
23 445	24 607	25 566	26 891	26 109	25 878	27 143	27 598	26 412	Venezuela
360	382	399	402	414	422	425	433	451	Other Latin America
202 879	194 396	199 282	204 080	215 592	222 487	229 947	237 977	251 765	**Latin America**

Industry Consumption of Electricity (GWh)

Consommation industrielle d'électricité (GWh)

Industrieverbrauch von Elektrizität (GWh)

Consumo di energia elettrica nell'industria (GWh)

電力の産業用消費量(GWh)

Consumo industrial de electricidad (GWh)

Потребление электроэнергии промышленным сектором (ГВт.ч)

	1971	1973	1978	1983	1984	1985	1986	1987	1988
Bangladesh	475	781	1 227	1 569	1 682	1 678	1 843	1 907	2 040
Brunei	71	108	148	138	152	155	157	211	223
Inde	34 713	37 210	50 929	61 766	67 643	72 277	76 329	75 718	83 289
Indonésie	511	638	1 155	3 436	4 011	4 874	6 183	7 402	9 087
RPD de Corée	7 155	8 460	13 536	17 550	19 125	20 385	21 100	21 245	22 500
Malaisie	1 814	2 378	3 749	4 910	5 282	5 159	5 494	6 099	7 010
Myanmar	302	359	546	585	645	942	919	904	737
Népal	12	7	34	74	149	107	143	170	166
Pakistan	2 857	3 662	4 589	5 572	5 862	6 249	7 288	8 012	8 973
Philippines	3 397	4 743	6 918	8 723	8 555	9 007	5 843	7 751	8 566
Singapour	920	1 349	2 607	2 835	3 270	3 398	3 513	4 021	4 592
Sri Lanka	379	497	535	753	857	872	926	867	905
Taipei chinois	9 029	11 688	20 586	25 442	27 622	28 227	32 236	34 807	37 946
Thailande	2 998	3 939	5 091	7 787	8 537	9 110	9 956	11 099	12 952
Viêt-Nam	-	-	-	1 738	2 027	2 113	2 196	2 384	2 590
Autre Asie	-	-	1 898	1 910	2 152	2 366	2 322	2 300	2 425
Asie	64 633	75 819	113 548	144 788	157 571	166 919	176 448	184 897	204 001
Rép. populaire de Chine	-	-	-	224 700	240 860	271 770	300 120	314 320	346 530
Hong-Kong, Chine	2 030	2 419	3 549	4 671	5 149	5 226	5 941	6 399	6 647
Chine	2 030	2 419	3 549	229 371	246 009	276 996	306 061	320 719	353 177
Albanie	-	-	-	-	-	-	1 226	1 055	884
Bulgarie	12 399	14 036	17 431	20 913	21 647	18 894	21 372	20 485	21 047
Chypre	177	231	237	272	296	206	244	273	294
Malte	-	-	-	-	-	-	-	-	-
Roumanie	22 440	26 710	40 318	51 548	54 731	44 356	47 480	47 012	48 766
République slovaque	7 281	8 321	10 049	13 286	13 745	14 488	14 849	15 288	15 604
Croatie	-	-	-	-	-	-	-	-	-
Ex-RYM	-	-	-	-	-	-	-	-	-
Slovénie	-	-	-	-	-	-	-	-	-
RF de Yougoslavie	-	-	-	-	-	-	-	-	-
Ex-Yougoslavie	13 535	15 749	23 986	30 223	35 348	36 714	37 941	37 116	34 734
Europe non-OCDE	55 832	65 047	92 021	116 242	125 767	114 658	123 112	121 229	121 329
Arménie	-	-	-	-	-	-	-	-	-
Azerbaïdjan	-	-	-	-	-	-	-	-	-
Bélarus	-	-	-	-	-	-	-	-	-
Estonie	-	-	-	-	-	-	-	-	-
Géorgie	-	-	-	-	-	-	-	-	-
Kazakhstan	-	-	-	-	-	-	-	-	-
Kirghizistan	-	-	-	-	-	-	-	-	-
Lettonie	-	-	-	-	-	-	-	-	-
Lituanie	-	-	-	-	-	-	-	-	-
Moldova	-	-	-	-	-	-	-	-	-
Russie	-	-	-	-	-	-	-	-	-
Tadjikistan	-	-	-	-	-	-	-	-	-
Turkménistan	-	-	-	-	-	-	-	-	-
Ukraine	-	-	-	-	-	-	-	-	-
Ouzbékistan	-	-	-	-	-	-	-	-	-
Ex-URSS	451 400	508 900	639 100	697 300	730 000	742 100	756 600	783 300	805 100

Ex-URSS: les séries antérieures à 1990-1992 ne sont pas comparables aux années récentes; 1991 a été estimé.

Rép. populaire de Chine: jusqu'en 1978 la ventilation de la consommation finale par secteur est incomplète.

Industry Consumption of Electricity (GWh)

Consommation industrielle d'électricité (GWh)

Industrieverbrauch von Elektrizität (GWh)

Consumo di energia elettrica nell'industria (GWh)

電力の産業用消費量(GWh)

Consumo industrial de electricidad (GWh)

Потребление электроэнергии промышленным сектором (ГВт.ч)

1989	1990	1991	1992	1993	1994	1995	1996	1997	
2 564	2 716	2 616	4 201	5 140	5 623	6 440	6 982	7 486	Bangladesh
229	238	248	257	280	283	291	281	263	Brunei
100 373	109 319	115 891	121 520	126 144	133 930	143 701	149 608	159 313	India
11 418	14 166	16 026	17 755	21 881	22 465	25 125	28 410	31 270	Indonesia
23 000	23 000	-	-	-	-	-	-	-	DPR of Korea
8 024	9 651	10 698	13 221	15 151	18 221	21 233	23 593	28 163	Malaysia
805	862	713	766	829	843	918	890	919	Myanmar
182	178	207	246	274	304	328	355	366	Nepal
9 455	10 337	11 263	12 598	13 044	12 635	12 528	12 183	11 982	Pakistan
9 764	8 982	9 339	8 646	9 427	10 683	11 228	11 850	12 568	Philippines
5 099	5 638	7 234	7 446	7 930	8 725	9 349	9 787	11 077	Singapore
909	910	1 049	1 057	1 223	1 560	1 674	1 650	1 819	Sri Lanka
41 063	42 783	46 270	49 155	51 805	55 160	59 139	62 002	67 914	Chinese Taipei
15 432	17 929	19 814	20 407	22 374	28 921	32 860	34 645	34 541	Thailand
2 623	2 850	3 107	3 192	3 476	3 840	4 512	5 504	6 163	Vietnam
2 523	2 487	2 380	2 375	2 374	2 404	1 440	1 360	1 360	Other Asia
233 463	252 046	246 855	262 842	281 352	305 597	330 766	349 100	375 204	**Asia**
365 610	381 860	416 490	459 311	449 968	495 608	534 609	584 381	559 020	People's Rep. of China
6 994	6 926	6 958	6 721	6 454	5 955	5 618	5 538	5 268	Hong Kong, China
372 604	388 786	423 448	466 032	456 422	501 563	540 227	589 919	564 288	**China**
1 500	1 374	1 056	1 242	1 504	527	506	797	755	Albania
19 507	18 552	13 998	11 952	10 914	11 461	12 167	12 258	11 737	Bulgaria
318	334	334	363	381	394	397	403	395	Cyprus
-	-	236	257	260	488	489	510	453	Malta
49 044	38 553	30 881	25 489	23 559	21 878	23 343	24 512	25 125	Romania
16 005	15 008	11 082	11 235	9 953	10 128	9 146	10 501	10 053	Slovak Republic
-	-	-	3 420	3 064	3 054	2 747	2 651	3 012	*Croatia*
-	-	-	2 540	2 263	2 130	1 899	2 015	2 087	*FYROM*
-	-	-	4 662	4 553	4 950	4 931	4 778	4 726	*Slovenia*
-	-	-	10 930	5 841	6 060	6 102	6 227	6 617	*FR of Yugoslavia*
37 052	38 650	22 075	21 552	15 721	16 194	15 679	15 671	16 442	Former Yugoslavia
123 426	112 471	79 662	72 090	62 292	61 070	61 727	64 652	64 960	**Non-OECD Europe**
-	-	-	1 807	828	668	981	815	724	Armenia
-	-	-	4 461	4 400	5 698	4 849	4 796	4 453	Azerbaijan
-	-	-	19 952	14 976	11 925	10 617	10 785	12 569	Belarus
-	-	-	2 229	1 384	1 710	1 751	1 907	2 115	Estonia
-	-	-	5 441	4 126	2 770	2 944	876	828	Georgia
-	-	-	57 560	44 927	24 158	23 336	20 854	18 686	Kazakhstan
-	-	-	3 460	2 506	1 958	1 479	1 589	1 493	Kyrgyzstan
-	-	-	2 613	1 439	1 492	1 427	1 697	1 484	Latvia
-	-	-	4 525	3 116	2 795	2 806	2 520	2 777	Lithuania
-	-	-	2 971	3 000	2 579	1 847	1 738	1 802	Moldova
-	-	-	469 801	383 544	325 763	293 252	266 667	264 470	Russia
-	-	-	9 400	7 068	6 802	6 876	6 750	6 520	Tajikistan
-	-	-	1 929	2 191	1 821	1 806	1 695	1 570	Turkmenistan
-	-	-	108 637	95 077	76 656	71 208	65 888	64 822	Ukraine
-	-	-	19 122	18 163	17 664	14 400	14 070	14 921	Uzbekistan
799 500	792 700	793 254	713 908	586 745	484 459	439 579	402 647	399 234	**Former USSR**

Former USSR: series up to 1990-1992 are not comparable with recent years; 1991 is estimated.

People's Rep. of China: up to 1978 the breakdown of final consumption by sector is incomplete.

Industry Consumption of Electricity (GWh)

Consommation industrielle d'électricité (GWh)

Industrieverbrauch von Elektrizität (GWh)

Consumo di energia elettrica nell'industria (GWh)

電力の産業用消費量(GWh)

Consumo industrial de electricidad (GWh)

Потребление электроэнергии промышленным сектором (ГВт.ч)

	1971	1973	1978	1983	1984	1985	1986	1987	1988
Bahrein	172	170	220	230	230	300	320	320	350
Iran	1 903	3 766	5 881	8 413	11 548	13 398	16 052	11 500	10 614
Irak	904	1 137	3 275	5 458	6 563	7 073	7 516	7 942	9 050
Israël	1 983	2 308	3 237	3 978	4 130	4 203	4 390	4 840	4 834
Jordanie	80	110	210	715	851	903	906	1 061	1 040
Koweit	300	360	595	673	700	712	740	765	1 050
Liban	-	-	-	-	-	-	-	-	-
Oman	-	3	33	113	141	175	222	238	264
Qatar	-	-	-	452	473	499	471	461	485
Arabie saoudite	1 281	1 745	1 481	1 795	2 922	6 200	5 269	5 585	5 935
Syrie	774	876	1 442	3 404	3 213	3 532	3 710	3 728	3 785
Yémen	24	32	118	33	25	31	37	66	70
Moyen-Orient	7 421	10 507	16 492	25 264	30 796	37 026	39 633	36 506	37 477
Total non-OCDE	687 442	787 509	1 053 235	1 458 570	1 560 491	1 624 304	1 707 030	1 762 636	1 853 692
OCDE Amérique du N.	684 057	769 752	897 398	872 187	962 463	970 104	945 573	986 316	1 042 324
OCDE Pacifique	272 660	329 041	376 315	359 095	382 322	393 787	396 058	419 061	448 397
OCDE Europe	633 725	730 893	823 458	833 843	877 706	894 710	906 973	935 009	970 137
Total OCDE	1 590 442	1 829 686	2 097 171	2 065 125	2 222 491	2 258 601	2 248 604	2 340 386	2 460 858
Monde	2 277 884	2 617 195	3 150 406	3 523 695	3 782 982	3 882 905	3 955 634	4 103 022	4 314 550

Industry Consumption of Electricity (GWh)

Consommation industrielle d'électricité (GWh)

Industrieverbrauch von Elektrizität (GWh)

Consumo di energia elettrica nell'industria (GWh)

電力の産業用消費量(GWh)

Consumo industrial de electricidad (GWh)

Потребление электроэнергии промышленным сектором (ГВт.ч)

1989	1990	1991	1992	1993	1994	1995	1996	1997	
400	420	400	410	613	668	698	701	723	Bahrain
8 466	10 220	15 053	17 899	15 572	19 497	20 450	21 721	22 481	Iran
9 000	-	-	-	-	-	-	-	-	Iraq
5 068	5 289	5 348	5 687	5 990	6 601	7 141	7 826	8 004	Israel
1 097	1 189	1 181	1 342	1 449	1 519	1 677	1 773	1 837	Jordan
1 130	1 049	-	-	-	-	-	-	-	Kuwait
-	-	-	-	1 189	1 350	1 224	1 759	2 100	Lebanon
274	315	339	360	410	402	372	391	455	Oman
502	600	696	799	755	800	751	1 422	1 485	Qatar
7 185	7 964	8 177	8 394	8 745	9 581	9 625	10 129	10 748	Saudi Arabia
3 894	4 201	4 236	3 840	3 828	4 529	4 564	5 037	5 355	Syria
-	-	-	-	-	-	-	-	-	Yemen
37 016	31 247	35 430	38 731	38 551	44 947	46 502	50 759	53 188	**Middle East**
1 904 486	1 908 870	1 918 185	1 895 695	1 776 906	1 757 896	1 792 603	1 848 654	1 867 393	**Non-OECD Total**
1 065 163	1 086 858	1 179 075	1 205 129	1 220 403	1 271 719	1 293 167	1 338 601	1 372 378	OECD North America
475 358	505 297	518 340	518 299	522 036	567 814	585 700	607 453	629 479	OECD Pacific
1 000 097	990 914	975 896	970 685	961 378	971 954	1 001 481	1 009 190	1 044 178	OECD Europe
2 540 618	2 583 069	2 673 311	2 694 113	2 703 817	2 811 487	2 880 348	2 955 244	3 046 035	**OECD Total**
4 445 104	4 491 939	4 591 496	4 589 808	4 480 723	4 569 383	4 672 951	4 803 898	4 913 428	**World**

Consumption of Oil in Transport (1000 tonnes)

Consommation de pétrole dans les transports (1000 tonnes)

Ölverbrauch im Verkehrssektor (1000 Tonnen)

Consumo di petrolio nel settore dei trasporti (1000 tonnellate)

運輸部門における石油の消費量（千トン）

Consumo de petróleo en el transporte (1000 toneladas)

Потребление нефти и нефтепродуктов в транспорте (тыс. т)

	1971	1973	1978	1983	1984	1985	1986	1987	1988
Algérie	853	1 074	1 892	3 167	3 593	3 956	4 236	4 293	4 280
Angola	394	529	503	313	213	391	318	305	313
Bénin	84	104	95	92	108	133	122	107	102
Cameroun	232	264	385	525	549	599	571	577	640
Congo	116	125	173	219	196	193	193	209	189
RD du Congo	519	550	606	687	674	628	529	611	625
Egypte	1 312	1 468	2 102	3 447	3 855	4 171	4 110	3 753	3 807
Ethiopie	218	224	238	281	298	313	355	429	447
Gabon	8	22	195	160	102	129	194	146	162
Ghana	373	377	467	327	365	432	414	397	537
Côte d'Ivoire	321	340	526	485	463	468	466	476	513
Kenya	617	741	869	827	947	1 000	1 084	1 057	1 045
Libye	297	632	1 360	2 244	2 265	2 198	2 116	2 239	2 190
Maroc	663	790	1 002	1 044	1 020	1 044	1 078	1 121	586
Mozambique	137	135	108	98	89	70	73	76	78
Nigéria	918	1 328	3 975	5 409	5 005	4 763	4 460	4 368	4 764
Sénégal	233	275	319	377	375	371	359	317	303
Afrique du Sud	5 164	6 367	7 647	7 574	8 149	7 898	7 835	8 417	9 096
Soudan	751	892	704	877	783	931	876	694	1 061
Tanzanie	252	283	263	238	240	243	239	244	272
Tunisie	400	474	727	776	794	806	783	833	882
Zambie	204	251	253	228	232	236	226	228	245
Zimbabwe	343	443	447	554	520	562	578	531	611
Autre Afrique	628	769	1 131	1 476	1 484	1 521	1 672	2 178	2 244
Afrique	15 037	18 457	25 987	31 425	32 319	33 056	32 887	33 606	34 992
Argentine	7 644	8 340	8 855	10 014	9 525	8 589	9 469	9 903	8 708
Bolivie	294	356	624	540	526	527	587	651	654
Brésil	12 504	16 708	21 692	19 985	19 763	20 911	22 800	22 472	22 007
Chili	1 862	1 717	1 795	2 144	2 145	2 114	2 209	2 364	2 668
Colombie	2 411	2 757	4 008	4 776	4 854	4 997	5 043	5 222	5 674
Costa Rica	221	275	433	357	379	390	434	463	526
Cuba	1 460	1 832	2 096	2 014	2 054	2 745	2 722	2 813	2 929
République dominicaine	477	527	608	628	710	683	756	934	842
Equateur	677	786	1 443	1 661	1 862	1 979	2 053	1 988	2 378
El Salvador	191	228	359	289	305	336	324	353	391
Guatemala	265	365	453	396	429	452	434	508	546
Haiti	52	64	104	105	108	110	120	128	141
Honduras	140	166	198	245	247	266	287	308	332
Jamaïque	647	672	387	398	360	385	373	429	467
Antilles néerlandaises	520	484	574	473	446	391	323	338	350
Nicaragua	241	278	348	281	282	275	299	325	282
Panama	344	374	456	324	362	334	349	341	333
Paraguay	132	147	327	352	405	409	429	449	465
Pérou	1 828	2 166	2 136	2 299	2 312	2 168	2 301	2 601	2 613
Trinité-et-Tobago	298	395	441	582	607	605	634	598	477
Uruguay	548	573	558	475	430	418	422	442	453
Vénézuela	4 110	5 004	8 216	8 975	8 751	8 951	9 018	9 299	9 618
Autre Amérique latine	408	296	486	610	642	624	583	750	804
Amérique latine	37 274	44 510	56 597	57 923	57 504	58 659	61 969	63 679	63 658

Consumption of Oil in Transport (1000 tonnes)
Consommation de pétrole dans les transports (1000 tonnes)
Ölverbrauch im Verkehrssektor (1000 Tonnen)

Consumo di petrolio nel settore dei trasporti (1000 tonnellate)

運輸部門における石油の消費量（千トン）

Consumo de petróleo en el transporte (1000 toneladas)

Потребление нефти и нефтепродуктов в транспорте (тыс. т)

1989	1990	1991	1992	1993	1994	1995	1996	1997	
4 405	4 465	4 574	2 575	2 676	2 546	2 407	2 395	2 148	Algeria
290	292	288	298	297	296	297	366	594	Angola
81	68	56	51	56	58	59	257	254	Benin
630	603	543	543	573	615	593	599	606	Cameroon
214	203	208	185	193	186	198	203	203	Congo
636	650	659	603	677	690	705	705	705	DR of Congo
3 620	3 752	3 726	3 716	3 756	3 970	4 289	4 697	4 949	Egypt
462	460	469	471	521	544	579	689	440	Ethiopia
217	179	178	173	179	181	195	211	275	Gabon
545	538	480	534	598	633	662	661	661	Ghana
491	435	451	430	454	464	487	542	551	Ivory Coast
1 114	1 115	1 101	1 116	1 096	1 174	1 157	1 251	1 223	Kenya
2 253	2 162	2 187	2 230	2 592	3 087	3 306	3 337	3 351	Libya
594	630	568	613	771	758	758	763	775	Morocco
78	105	99	86	95	102	78	86	96	Mozambique
4 797	4 050	5 243	6 324	5 486	4 662	4 528	4 772	4 847	Nigeria
365	374	363	376	340	360	391	466	493	Senegal
9 422	9 702	9 744	9 965	10 211	10 606	11 924	11 877	12 288	South Africa
948	1 277	1 023	1 022	689	985	927	884	982	Sudan
272	291	277	283	271	274	265	255	211	Tanzania
919	963	939	1 021	1 146	1 212	1 261	1 307	1 401	Tunisia
240	219	223	203	205	214	214	218	218	Zambia
597	575	479	625	562	625	772	770	770	Zimbabwe
2 362	2 365	2 251	2 338	2 562	2 501	2 320	2 330	2 427	Other Africa
35 552	35 473	36 129	35 781	36 006	36 743	38 372	39 641	40 468	**Africa**
8 256	8 202	9 154	9 583	10 561	11 179	11 455	12 029	12 293	Argentina
691	719	737	729	759	791	876	954	962	Bolivia
22 886	23 589	24 758	24 950	25 917	27 089	29 934	32 627	35 061	Brazil
2 864	2 975	3 186	3 412	3 736	4 122	4 497	4 912	5 229	Chile
5 768	5 845	6 075	6 438	6 514	6 205	7 318	7 470	7 615	Colombia
564	543	562	834	868	930	935	934	964	Costa Rica
3 050	1 741	1 279	900	736	688	740	880	938	Cuba
1 022	889	881	1 042	936	974	1 003	1 035	1 081	Dominican Republic
2 298	2 186	2 160	2 161	2 158	2 246	2 349	2 689	2 843	Ecuador
402	434	461	547	595	632	716	698	746	El Salvador
553	598	622	659	723	767	952	957	1 019	Guatemala
151	158	146	141	131	36	181	197	202	Haiti
363	336	339	379	367	502	602	509	513	Honduras
461	471	464	451	509	606	691	751	816	Jamaica
323	319	362	387	324	416	417	424	424	Netherlands Antilles
252	251	273	287	291	330	358	375	421	Nicaragua
342	377	371	400	415	452	486	563	594	Panama
523	514	506	606	706	822	938	880	953	Paraguay
2 309	2 491	2 539	2 621	2 807	2 619	3 093	3 183	3 160	Peru
457	493	516	510	442	457	479	503	572	Trinidad and Tobago
474	476	506	544	614	692	686	729	792	Uruguay
9 123	9 406	9 796	9 834	10 251	10 637	11 204	10 568	11 809	Venezuela
830	827	825	797	791	862	875	870	881	Other Latin America
63 962	63 840	66 518	68 212	71 151	74 054	80 785	84 737	89 888	**Latin America**

Consumption of Oil in Transport (1000 tonnes)

Consommation de pétrole dans les transports (1000 tonnes)

Ölverbrauch im Verkehrssektor (1000 Tonnen)

Consumo di petrolio nel settore dei trasporti (1000 tonnellate)

運輸部門における石油の消費量（千トン）

Consumo de petróleo en el transporte (1000 toneladas)

Потребление нефти и нефтепродуктов в транспорте (тыс. т)

	1971	1973	1978	1983	1984	1985	1986	1987	1988
Bangladesh	98	92	243	399	402	492	491	516	552
Brunei	27	33	71	149	152	160	172	183	198
Inde	6 554	7 297	10 284	14 312	15 149	16 583	17 991	18 810	19 819
Indonésie	2 579	2 974	4 868	6 559	6 684	6 865	7 283	8 012	8 685
RPD de Corée	563	635	1 305	1 900	2 000	2 100	2 200	2 220	2 220
Malaisie	1 425	1 651	1 865	3 110	3 214	3 383	3 607	3 826	4 109
Myanmar	495	421	535	589	647	607	671	397	402
Népal	23	25	41	61	70	72	79	101	122
Pakistan	1 359	1 366	2 058	3 101	3 364	3 522	3 723	4 198	4 481
Philippines	2 264	2 349	2 254	1 795	1 713	1 641	1 735	1 963	2 144
Singapour	704	880	1 595	2 074	2 141	2 086	2 232	2 348	2 401
Sri Lanka	426	515	577	654	732	764	768	787	795
Taipei chinois	1 137	1 938	2 815	3 816	4 052	4 267	4 611	5 110	5 706
Thailande	2 264	3 065	3 685	4 697	5 394	5 584	5 956	6 846	7 955
Viêt-Nam	3 169	2 920	185	514	497	870	990	1 138	1 246
Autre Asie	555	544	687	854	942	955	864	984	931
Asie	23 642	26 705	33 068	44 584	47 153	49 951	53 373	57 439	61 766
Rép. populaire de Chine	4 465	6 110	9 562	18 586	20 340	22 991	25 802	27 955	30 325
Hong-Kong, Chine	852	1 006	1 235	2 069	2 070	1 962	2 423	2 401	2 956
Chine	5 317	7 116	10 797	20 655	22 410	24 953	28 225	30 356	33 281
Albanie	233	252	360	300	425	346	324	335	370
Bulgarie	1 552	1 525	1 620	1 600	1 594	1 579	1 572	1 588	2 634
Chypre	243	313	246	335	381	400	420	464	479
Gibraltar	16	15	11	14	13	16	23	29	36
Malte	123	142	121	140	177	93	212	217	215
Roumanie	2 030	2 291	2 596	1 729	1 656	1 405	1 308	1 821	1 700
République slovaque	1 367	1 576	1 490	868	875	867	890	918	965
Bosnie-Herzegovine	-	-	-	-	-	-	-	-	-
Croatie	-	-	-	-	-	-	-	-	-
Ex-RYM	-	-	-	-	-	-	-	-	-
Slovénie	-	-	-	-	-	-	-	-	-
RF de Yougoslavie	-	-	-	-	-	-	-	-	-
Ex-Yougoslavie	3 563	3 634	4 798	4 161	3 984	4 504	5 549	4 962	5 467
Europe non-OCDE	9 127	9 748	11 242	9 147	9 105	9 210	10 298	10 334	11 866
Arménie	-	-	-	-	-	-	-	-	-
Azerbaïdjan	-	-	-	-	-	-	-	-	-
Bélarus	-	-	-	-	-	-	-	-	-
Estonie	-	-	-	-	-	-	-	-	-
Géorgie	-	-	-	-	-	-	-	-	-
Kazakhstan	-	-	-	-	-	-	-	-	-
Kirghizistan	-	-	-	-	-	-	-	-	-
Lettonie	-	-	-	-	-	-	-	-	-
Lituanie	-	-	-	-	-	-	-	-	-
Moldova	-	-	-	-	-	-	-	-	-
Russie	-	-	-	-	-	-	-	-	-
Tadjikistan	-	-	-	-	-	-	-	-	-
Turkménistan	-	-	-	-	-	-	-	-	-
Ukraine	-	-	-	-	-	-	-	-	-
Ouzbékistan	-	-	-	-	-	-	-	-	-
Ex-URSS	73 100	81 200	100 000	111 100	112 600	113 600	117 800	120 100	121 600

Ex-URSS: les séries antérieures à 1990-1992 ne sont pas comparables aux années récentes; 1991 a été estimé.

Rép. populaire de Chine: jusqu'en 1978 la ventilation de la consommation finale par secteur est incomplète.

Consumption of Oil in Transport (1000 tonnes)

Consommation de pétrole dans les transports (1000 tonnes)

Ölverbrauch im Verkehrssektor (1000 Tonnen)

Consumo di petrolio nel settore dei trasporti (1000 tonnellate)

運輸部門における石油の消費量（千トン）

Consumo de petróleo en el transporte (1000 toneladas)

Потребление нефти и нефтепродуктов в транспорте (тыс. т)

1989	1990	1991	1992	1993	1994	1995	1996	1997	
593	609	653	736	789	845	1 028	1 073	1 084	Bangladesh
201	212	230	255	270	289	313	341	365	Brunei
21 949	22 641	25 756	27 263	28 907	31 078	35 233	37 884	39 323	India
9 448	10 812	11 609	12 306	13 140	14 735	16 137	17 953	18 922	Indonesia
2 370	2 327	2 139	2 067	1 947	1 257	1 105	972	943	DPR of Korea
4 416	5 117	5 614	5 939	6 949	7 683	7 541	8 470	8 836	Malaysia
392	418	382	333	398	478	596	645	743	Myanmar
99	101	137	165	170	170	191	191	191	Nepal
4 637	4 719	4 924	5 518	6 255	6 563	6 800	7 346	7 504	Pakistan
2 422	2 503	2 304	2 656	2 984	3 148	3 451	3 901	4 372	Philippines
2 614	3 066	3 049	3 111	3 561	4 152	4 255	4 338	4 691	Singapore
755	784	789	858	917	1 080	1 146	1 236	1 270	Sri Lanka
6 509	7 099	7 511	8 724	9 341	10 079	10 668	11 067	11 323	Chinese Taipei
9 443	10 581	10 844	11 790	13 522	15 078	17 470	18 969	20 210	Thailand
1 169	1 384	1 171	1 134	1 552	1 669	1 748	3 719	4 697	Vietnam
905	935	897	838	826	742	646	659	663	Other Asia
67 922	73 308	78 009	83 693	91 528	99 046	108 328	118 764	125 137	**Asia**
31 517	32 235	36 737	39 325	46 833	42 724	49 845	53 425	60 509	People's Rep. of China
3 098	3 668	3 894	4 733	5 346	6 010	6 045	5 952	6 184	Hong Kong, China
34 615	35 903	40 631	44 058	52 179	48 734	55 890	59 377	66 693	**China**
306	265	226	171	204	240	270	281	286	Albania
2 796	2 377	1 354	1 375	1 604	1 478	1 523	1 418	1 028	Bulgaria
541	609	652	690	655	678	728	733	750	Cyprus
32	34	51	57	57	70	68	68	68	Gibraltar
215	215	243	248	270	270	283	333	338	Malta
3 644	3 921	3 443	3 319	2 881	3 004	2 798	3 761	3 868	Romania
954	933	784	822	848	1 045	1 193	1 115	1 124	Slovak Republic
-	-	-	356	239	259	259	259	259	*Bosnia-Herzegovina*
-	-	-	915	998	1 079	1 133	1 189	1 336	*Croatia*
-	-	-	257	474	322	316	563	505	*FYROM*
-	-	-	840	1 017	1 135	1 264	1 428	1 450	*Slovenia*
-	-	-	580	366	352	386	722	1 085	*FR of Yugoslavia*
5 146	5 163	4 618	2 948	3 094	3 147	3 358	4 161	4 635	Former Yugoslavia
13 634	13 517	11 371	9 630	9 613	9 932	10 221	11 870	12 097	**Non-OECD Europe**
-	-	-	750	455	100	71	39	39	Armenia
-	-	-	1 470	1 551	1 699	1 464	1 274	953	Azerbaijan
-	-	-	4 562	3 421	2 431	1 890	1 741	1 796	Belarus
-	-	-	353	417	493	464	510	531	Estonia
-	-	-	401	244	71	46	505	476	Georgia
-	-	-	4 815	3 736	2 889	2 648	2 593	2 203	Kazakhstan
-	-	-	472	289	115	293	319	178	Kyrgyzstan
-	-	-	959	858	772	791	786	760	Latvia
-	-	-	1 054	1 068	966	1 124	1 192	1 187	Lithuania
-	-	-	738	461	369	374	355	376	Moldova
-	-	-	69 040	47 450	30 200	33 499	31 829	30 810	Russia
-	-	-	4 778	3 005	1 001	1 001	1 001	1 001	Tajikistan
-	-	-	713	527	493	451	483	443	Turkmenistan
-	-	-	10 204	8 389	7 279	7 275	6 160	5 573	Ukraine
-	-	-	2 185	1 995	1 978	1 542	2 015	2 139	Uzbekistan
123 900	116 085	140 512	102 494	73 866	50 856	52 933	50 802	48 465	**Former USSR**

Former USSR: series up to 1990-1992 are not comparable with recent years; 1991 is estimated.
People's Rep. of China: up to 1978 the breakdown of final consumption by sector is incomplete.

Consumption of Oil in Transport (1000 tonnes)

Consommation de pétrole dans les transports (1000 tonnes)

Ölverbrauch im Verkehrssektor (1000 Tonnen)

Consumo di petrolio nel settore dei trasporti (1000 tonnellate)

運輸部門における石油の消費量（千トン）

Consumo de petróleo en el transporte (1000 toneladas)

Потребление нефти и нефтепродуктов в транспорте (тыс. т)

	1971	1973	1978	1983	1984	1985	1986	1987	1988
Bahrein	193	277	597	645	663	661	647	657	688
Iran	1 610	2 694	4 958	4 757	5 313	5 761	10 469	11 319	11 098
Irak	1 037	1 282	3 237	6 061	6 459	6 899	7 244	7 460	7 675
Israël	1 593	1 896	2 030	2 221	2 125	2 237	2 194	2 402	2 414
Jordanie	197	260	610	957	936	950	1 060	1 075	1 093
Koweit	575	633	1 348	1 914	1 861	1 883	1 901	1 847	1 862
Liban	921	1 182	924	1 350	1 276	1 407	1 326	1 380	914
Oman	34	41	269	437	532	579	610	529	544
Qatar	74	127	343	512	488	473	486	493	505
Arabie saoudite	723	1 016	3 613	10 984	11 114	11 855	11 936	11 719	8 105
Syrie	775	973	1 921	2 373	2 580	2 597	2 642	2 874	2 860
Emirats arabes unis	193	392	2 239	3 319	3 333	3 494	3 631	3 681	3 359
Yémen	237	426	580	1 000	989	1 053	1 142	1 225	1 301
Moyen-Orient	8 162	11 199	22 669	36 530	37 669	39 849	45 288	46 661	42 418
Total non-OCDE	171 659	198 935	260 360	311 364	318 760	329 278	349 840	362 175	369 581
OCDE Amérique du N.	378 385	424 261	479 623	453 029	466 997	471 295	482 528	498 487	519 743
OCDE Pacifique	48 489	56 129	71 103	75 431	78 485	80 442	84 010	88 269	93 968
OCDE Europe	148 814	170 696	201 871	210 827	217 424	221 312	232 700	242 513	256 089
Total OCDE	575 688	651 086	752 597	739 287	762 906	773 049	799 238	829 269	869 800
Monde	747 347	850 021	1 012 957	1 050 651	1 081 666	1 102 327	1 149 078	1 191 444	1 239 381

Consumption of Oil in Transport (1000 tonnes)

Consommation de pétrole dans les transports (1000 tonnes)

Ölverbrauch im Verkehrssektor (1000 Tonnen)

Consumo di petrolio nel settore dei trasporti (1000 tonnellate)

運輸部門における石油の消費量（千トン）

Consumo de petróleo en el transporte (1000 toneladas)

Потребление нефти и нефтепродуктов в транспорте (тыс. т)

1989	1990	1991	1992	1993	1994	1995	1996	1997	
709	772	656	662	666	725	723	712	874	Bahrain
12 000	6 637	7 390	7 655	17 229	17 842	18 341	18 491	19 155	Iran
8 066	8 416	6 550	7 618	8 375	8 797	8 737	8 592	8 826	Iraq
2 482	2 609	2 666	2 739	2 953	3 131	3 414	3 580	3 472	Israel
1 146	1 096	972	1 064	1 103	1 132	1 192	1 282	1 364	Jordan
2 000	1 701	432	1 744	1 815	1 920	2 028	2 115	2 316	Kuwait
980	660	825	1 148	1 333	1 285	1 553	1 501	1 440	Lebanon
578	836	835	790	860	842	851	881	946	Oman
500	578	559	603	631	698	760	841	942	Qatar
7 455	9 245	10 280	9 666	10 612	11 503	11 313	12 015	12 091	Saudi Arabia
1 436	1 610	1 582	1 390	1 324	1 327	1 219	1 263	1 360	Syria
3 459	3 318	3 388	3 446	3 642	3 854	3 827	3 527	3 069	United Arab Emirates
1 261	1 262	1 545	1 675	1 319	1 392	1 614	1 560	1 594	Yemen
42 072	38 740	37 680	40 200	51 862	54 448	55 572	56 360	57 449	**Middle East**
381 657	376 866	410 850	384 068	386 205	373 813	402 101	421 551	440 197	**Non-OECD Total**
526 820	525 723	518 898	530 939	539 821	558 724	569 485	582 695	594 758	OECD North America
100 936	106 588	110 654	116 164	119 523	126 538	133 357	139 765	143 156	OECD Pacific
266 034	273 805	277 183	285 924	291 714	294 938	299 554	308 817	313 433	OECD Europe
893 790	906 116	906 735	933 027	951 058	980 200	1 002 396	1 031 277	1 051 347	**OECD Total**
1 275 447	1 282 982	1 317 585	1 317 095	1 337 263	1 354 013	1 404 497	1 452 828	1 491 544	**World**

Consumption of Electricity in Transport (GWh)

Consommation d'électricité dans les transports (GWh)

Elektrizitätsverbrauch im Verkehrssektor (GWh)

Consumo di energia elettrica nel settore dei trasporti (GWh)

運輸部門における電力の消費量(GWh)

Consumo de electricidad en el transporte (GWh)

Потребление электроэнергии в транспорте (ГВт.ч)

	1971	1973	1978	1983	1984	1985	1986	1987	1988
Algérie	19	28	32	32	35	39	180	198	163
Maroc	76	85	103	114	115	150	153	165	183
Afrique du Sud	3 278	2 896	3 517	4 339	4 595	4 587	4 501	4 049	4 120
Tunisie	-	-	-	69	74	79	83	87	87
Afrique	3 373	3 009	3 652	4 554	4 819	4 855	4 917	4 499	4 553
Argentine	290	292	257	279	269	267	267	337	337
Brésil	619	601	675	1 059	1 112	1 146	1 158	1 181	1 200
Chili	216	193	211	221	225	226	223	233	219
Colombie	-	-	-	3	3	3	3	3	3
Costa Rica	12	11	9	12	10	8	8	7	7
Cuba	-	-	-	-	-	6	5	6	5
Panama	-	-	-	84	88	75	-	-	-
Paraguay	-	-	-	-	-	1	1	1	1
Vénézuela	-	-	-	-	-	-	-	175	186
Amérique latine	1 137	1 097	1 152	1 658	1 707	1 732	1 665	1 943	1 958
Inde	1 643	1 531	2 307	2 878	3 064	3 315	3 543	3 985	4 067
Népal	-	-	-	-	-	-	-	-	-
Pakistan	28	28	31	44	38	37	36	38	40
Philippines	-	-	-	-	-	-	-	26	27
Singapour	-	-	-	-	-	-	-	29	107
Taipei chinois	4	6	55	253	257	255	280	263	255
Viêt-Nam	-	-	-	-	-	-	-	-	38
Asie	1 675	1 565	2 393	3 175	3 359	3 607	3 859	4 341	4 534
Rép. populaire de Chine	-	-	-	3 580	4 140	6 340	6 690	7 670	8 950
Hong-Kong, Chine	-	-	-	-	-	-	-	603	593
Chine	-	-	-	3 580	4 140	6 340	6 690	8 273	9 543
Bulgarie	-	-	-	-	1 347	1 319	1 340	1 441	1 488
Chypre	34	38	4	7	8	28	30	36	43
Roumanie	-	-	-	-	-	2 430	2 580	2 583	2 824
République slovaque	554	633	764	1 063	1 095	1 080	1 092	1 103	1 163
Croatie	-	-	-	-	-	-	-	-	-
Ex-RYM	-	-	-	-	-	-	-	-	-
Slovénie	-	-	-	-	-	-	-	-	-
RF de Yougoslavie	-	-	-	-	-	-	-	-	-
Ex-Yougoslavie	533	717	830	1 196	1 224	1 237	1 493	1 231	1 298
Europe non-OCDE	1 121	1 388	1 598	2 266	3 674	6 094	6 535	6 394	6 816

Rép. populaire de Chine: jusqu'en 1978 la ventilation de la consommation finale par secteur est incomplète.

Consumption of Electricity in Transport (GWh)
Consommation d'électricité dans les transports (GWh)
Elektrizitätsverbrauch im Verkehrssektor (GWh)

Consumo di energia elettrica nel settore dei trasporti (GWh)

運輸部門における電力の消費量(GWh)

Consumo de electricidad en el transporte (GWh)

Потребление электроэнергии в транспорте (ГВт.ч)

1989	1990	1991	1992	1993	1994	1995	1996	1997	
213	185	200	192	314	298	323	338	354	Algeria
178	201	216	269	184	189	199	204	214	Morocco
4 229	3 958	3 685	3 568	4 017	4 388	4 290	4 275	4 562	South Africa
92	103	107	122	127	132	138	148	148	Tunisia
4 712	4 447	4 208	4 151	4 642	5 007	4 950	4 965	5 278	**Africa**
226	315	247	278	275	299	307	420	478	Argentina
1 293	1 194	1 081	1 192	1 200	1 176	1 211	1 150	1 160	Brazil
200	212	223	226	250	218	200	201	211	Chile
3	4	-	-	-	-	-	-	43	Colombia
15	15	14	5	5	6	-	-	-	Costa Rica
5	89	77	55	53	52	64	69	91	Cuba
60	63	56	22	11	67	72	77	93	Panama
1	1	1	1	1	1	1	-	-	Paraguay
219	277	196	194	145	124	100	80	832	Venezuela
2 022	2 170	1 895	1 973	1 940	1 943	1 955	1 997	2 908	**Latin America**
4 097	4 112	4 519	5 068	5 620	5 886	6 315	6 575	7 002	India
-	2	2	2	1	1	1	1	1	Nepal
35	38	33	29	27	27	22	20	19	Pakistan
29	29	21	25	22	-	-	-	-	Philippines
126	186	186	186	186	198	198	236	248	Singapore
380	434	429	437	458	463	466	513	589	Chinese Taipei
44	51	38	40	45	87	93	114	128	Vietnam
4 711	4 852	5 228	5 787	6 359	6 662	7 095	7 459	7 987	**Asia**
9 870	10 590	9 290	11 847	14 623	16 404	10 913	11 846	15 318	People's Rep. of China
-	-	-	-	-	-	-	-	-	Hong Kong, China
9 870	10 590	9 290	11 847	14 623	16 404	10 913	11 846	15 318	**China**
1 317	1 305	1 206	1 040	726	636	803	811	657	Bulgaria
47	51	53	25	27	31	33	37	21	Cyprus
2 921	2 614	1 786	2 827	2 207	1 880	2 173	2 326	2 230	Romania
1 182	1 164	1 439	950	1 125	1 467	1 379	984	1 010	Slovak Republic
-	-	-	219	229	238	240	242	248	*Croatia*
-	-	-	29	21	-	13	14	14	*FYROM*
-	-	-	147	146	147	170	160	164	*Slovenia*
-	-	-	342	255	264	265	270	287	*FR of Yugoslavia*
1 412	1 059	740	737	651	649	688	686	713	Former Yugoslavia
6 879	6 193	5 224	5 579	4 736	4 663	5 076	4 844	4 631	**Non-OECD Europe**

People's Rep. of China: up to 1978 the breakdown of final consumption by sector is incomplete.

Consumption of Electricity in Transport (GWh)

Consommation d'électricité dans les transports (GWh)

Elektrizitätsverbrauch im Verkehrssektor (GWh)

Consumo di energia elettrica nel settore dei trasporti (GWh)

運輸部門における電力の消費量(GWh)

Consumo de electricidad en el transporte (GWh)

Потребление электроэнергии в транспорте (ГВт.ч)

	1971	1973	1978	1983	1984	1985	1986	1987	1988
Arménie	-	-	-	-	-	-	-	-	-
Azerbaïdjan	-	-	-	-	-	-	-	-	-
Bélarus	-	-	-	-	-	-	-	-	-
Estonie	-	-	-	-	-	-	-	-	-
Géorgie	-	-	-	-	-	-	-	-	-
Kazakhstan	-	-	-	-	-	-	-	-	-
Kirghizistan	-	-	-	-	-	-	-	-	-
Lettonie	-	-	-	-	-	-	-	-	-
Lituanie	-	-	-	-	-	-	-	-	-
Moldova	-	-	-	-	-	-	-	-	-
Russie	-	-	-	-	-	-	-	-	-
Tadjikistan	-	-	-	-	-	-	-	-	-
Turkménistan	-	-	-	-	-	-	-	-	-
Ukraine	-	-	-	-	-	-	-	-	-
Ouzbékistan	-	-	-	-	-	-	-	-	-
Ex-URSS	48 800	53 800	70 100	79 600	80 800	82 000	83 200	84 400	85 600
Israël	-	-	-	-	-	-	-	-	481
Moyen-Orient	-	-	-	-	-	-	-	-	481
Total non-OCDE	56 106	60 859	78 895	94 833	98 499	104 628	106 866	109 850	113 485
OCDE Amérique du N.	6 715	7 878	4 979	5 489	6 430	7 344	7 514	7 718	8 156
OCDE Pacifique	12 297	14 058	16 166	17 597	17 582	18 233	18 762	19 304	20 543
OCDE Europe	37 231	38 994	44 139	48 974	50 853	53 019	54 711	55 850	56 815
Total OCDE	56 243	60 930	65 284	72 060	74 865	78 596	80 987	82 872	85 514
Monde	112 349	121 789	144 179	166 893	173 364	183 224	187 853	192 722	198 999

Ex-URSS: les séries antérieures à 1990-1992 ne sont pas comparables aux années récentes; 1991 a été estimé.

Consumption of Electricity in Transport (GWh)

Consommation d'électricité dans les transports (GWh)

Elektrizitätsverbrauch im Verkehrssektor (GWh)

Consumo di energia elettrica nel settore dei trasporti (GWh)

運輸部門における電力の消費量(GWh)

Consumo de electricidad en el transporte (GWh)

Потребление электроэнергии в транспорте (ГВт.ч)

1989	1990	1991	1992	1993	1994	1995	1996	1997	
-	-	-	312	187	178	182	173	152	Armenia
-	-	-	854	700	551	481	455	422	Azerbaijan
-	-	-	965	2 029	2 173	1 832	1 824	1 715	Belarus
-	-	-	330	147	120	116	105	108	Estonia
-	-	-	619	608	405	430	250	231	Georgia
-	-	-	6 020	4 969	4 375	3 841	3 432	3 075	Kazakhstan
-	-	-	90	142	128	133	127	119	Kyrgyzstan
-	-	-	292	221	230	203	178	176	Latvia
-	-	-	141	93	102	96	103	101	Lithuania
-	-	-	211	92	83	135	125	118	Moldova
-	-	-	86 774	76 721	68 396	65 160	64 933	63 500	Russia
-	-	-	-	-	98	80	79	80	Tajikistan
-	-	-	155	154	130	129	121	110	Turkmenistan
-	-	-	12 698	11 941	10 873	10 777	9 754	9 545	Ukraine
-	-	-	1 256	1 090	1 111	1 366	1 398	1 282	Uzbekistan
86 500	87 000	87 061	110 717	99 094	88 953	84 961	83 057	80 734	**Former USSR**
518	532	558	608	676	738	798	873	893	Israel
518	532	558	608	676	738	798	873	893	**Middle East**
115 212	115 784	113 464	140 662	132 070	124 370	115 748	115 041	117 749	**Non-OECD Total**
8 128	8 205	7 938	8 285	8 383	8 588	8 707	8 759	9 190	OECD North America
21 701	22 815	23 543	23 888	24 155	24 761	25 177	25 227	25 662	OECD Pacific
57 813	60 944	61 768	63 969	65 534	66 805	67 641	69 719	69 697	OECD Europe
87 642	91 964	93 249	96 142	98 072	100 154	101 525	103 705	104 549	**OECD Total**
202 854	207 748	206 713	236 804	230 142	224 524	217 273	218 746	222 298	**World**

Former USSR: series up to 1990-1992 are not comparable with recent years; 1991 is estimated.

PETROLEUM PRODUCTS

PRODUITS PETROLIERS

Oil Products Consumption
Consommation de produits pétroliers

World

	1971	1973	1978	1983	1984	1985	1986	1987	1988
Thousand tonnes									
NGL/LPG/Ethane	83 393	100 064	111 182	133 178	145 854	148 586	152 285	163 633	167 018
Naphtha	78 708	101 250	102 298	91 395	93 945	93 646	98 671	103 056	109 833
Motor Gasoline	493 267	557 492	658 187	648 031	657 437	662 576	683 868	706 029	728 863
Aviation Fuels	104 864	115 076	123 513	130 320	139 114	143 571	151 093	157 689	163 947
Kerosene	70 993	75 220	78 393	73 364	74 703	73 496	72 814	74 463	77 141
Gas Diesel	499 711	592 243	689 786	653 843	672 779	684 677	705 122	724 388	749 622
Heavy Fuel Oil	648 649	746 138	812 532	619 390	599 693	556 669	557 127	543 079	542 267
Other Products	155 982	192 393	240 816	215 607	218 094	218 664	229 439	234 986	248 859
Refinery Fuel	120 926	148 502	155 176	141 429	142 498	144 542	153 250	157 556	163 027
Sub-Total	2 256 493	2 628 378	2 971 883	2 706 557	2 744 117	2 726 427	2 803 669	2 864 879	2 950 577
Bunkers	122 258	134 268	115 871	89 157	91 047	97 845	103 644	101 853	107 658
Total	2 378 751	2 762 646	3 087 754	2 795 714	2 835 164	2 824 272	2 907 313	2 966 732	3 058 235
Thousand barrels/day									
NGL/LPG/Ethane	2 648.1	3 173.2	3 504.7	4 215.5	4 612.4	4 728.6	4 841.2	5 200.1	5 294.1
Naphtha	1 868.3	2 404.0	2 425.6	2 175.6	2 227.0	2 224.7	2 345.7	2 448.1	2 600.6
Motor Gasoline	11 507.3	13 005.2	15 352.6	15 108.3	15 283.3	15 443.1	15 935.3	16 452.0	16 937.0
Aviation Fuels	2 288.3	2 508.6	2 698.1	2 845.4	3 028.5	3 133.6	3 296.4	3 438.2	3 563.8
Kerosene	1 509.0	1 599.1	1 665.3	1 557.7	1 581.6	1 560.4	1 545.9	1 581.0	1 633.2
Gas Diesel	10 212.7	12 103.8	14 096.8	13 362.1	13 711.5	13 992.2	14 410.2	14 804.0	15 277.9
Heavy Fuel Oil	11 705.5	13 469.2	14 693.2	11 217.2	10 827.2	10 080.8	10 091.7	9 837.0	9 798.1
Other Products	2 879.7	3 574.6	4 576.2	4 032.6	4 062.0	4 074.6	4 262.0	4 357.7	4 616.0
Refinery Fuel	2 350.3	2 876.6	3 022.1	2 765.2	2 768.3	2 829.4	2 999.6	3 083.6	3 185.0
Sub-Total	46 969.3	54 714.2	62 034.7	57 279.5	58 101.9	58 067.3	59 728.1	61 201.6	62 905.7
Bunkers	2 240.9	2 461.1	2 134.6	1 644.6	1 676.1	1 804.2	1 911.0	1 878.9	1 979.6
Total	49 210.2	57 175.3	64 169.2	58 924.1	59 778.0	59 871.6	61 639.1	63 080.5	64 885.2

	1989	1990	1991	1992	1993	1994	1995	1996	1997
Thousand tonnes									
NGL/LPG/Ethane	171 153	166 910	176 920	169 915	172 283	182 241	189 765	198 365	201 446
Naphtha	115 029	114 303	114 362	127 198	129 424	134 694	141 734	146 963	161 920
Motor Gasoline	743 408	748 081	749 875	750 902	761 690	768 450	785 495	799 699	810 407
Aviation Fuels	170 664	171 622	167 391	165 869	167 954	175 341	180 036	186 677	192 364
Kerosene	75 687	74 493	75 976	74 826	76 940	77 400	82 107	87 498	86 519
Gas Diesel	759 146	768 336	783 006	806 686	822 570	822 783	851 657	885 479	907 125
Heavy Fuel Oil	532 784	515 376	513 481	512 882	500 938	479 903	459 417	455 565	454 121
Other Products	247 193	255 759	245 499	242 436	236 207	241 607	243 561	251 437	262 983
Refinery Fuel	169 707	169 524	166 563	170 014	164 413	163 583	165 743	171 535	173 523
Sub-Total	2 984 771	2 984 404	2 993 073	3 020 728	3 032 419	3 046 002	3 099 515	3 183 218	3 250 408
Bunkers	107 274	121 430	127 531	128 865	127 886	129 061	130 857	132 222	135 350
Total	3 092 045	3 105 834	3 120 604	3 149 593	3 160 305	3 175 063	3 230 372	3 315 440	3 385 758
Thousand barrels/day									
NGL/LPG/Ethane	5 437.5	5 425.3	5 767.3	5 466.9	5 590.0	5 881.3	6 125.9	6 398.9	6 530.0
Naphtha	2 728.5	2 710.5	2 710.4	2 994.7	3 055.9	3 183.7	3 350.0	3 464.6	3 822.1
Motor Gasoline	17 320.9	17 430.9	17 471.3	17 407.1	17 771.8	17 907.2	18 279.0	18 587.8	18 865.3
Aviation Fuels	3 719.2	3 740.2	3 646.1	3 606.0	3 652.9	3 816.5	3 923.1	4 049.9	4 192.8
Kerosene	1 606.7	1 581.2	1 612.8	1 591.0	1 653.6	1 661.8	1 759.0	1 863.0	1 842.4
Gas Diesel	15 514.3	15 702.1	16 001.8	16 445.9	16 800.5	16 839.5	17 436.2	18 101.2	18 604.5
Heavy Fuel Oil	9 650.3	9 335.5	9 300.4	9 271.3	8 984.5	8 660.1	8 362.4	8 254.1	8 146.2
Other Products	4 587.6	4 733.1	4 550.3	4 575.4	4 438.3	4 537.6	4 574.5	4 783.4	4 970.1
Refinery Fuel	3 336.6	3 330.2	3 271.0	3 369.6	3 197.3	3 145.0	3 185.7	3 242.3	3 364.3
Sub-Total	63 901.7	63 989.1	64 331.4	64 727.8	65 144.8	65 632.9	66 995.8	68 745.1	70 337.7
Bunkers	1 976.9	2 246.4	2 357.3	2 340.2	2 437.6	2 416.7	2 388.4	2 441.4	2 594.4
Total	65 878.6	66 235.5	66 688.7	67 068.0	67 582.5	68 049.5	69 384.3	71 186.5	72 932.1

Oil Products Consumption
Consommation de produits pétroliers

OECD Total

	1971	1973	1978	1983	1984	1985	1986	1987	1988
Thousand tonnes									
NGL/LPG/Ethane	70 506	82 979	86 561	98 069	108 696	109 439	107 373	115 378	116 164
Naphtha	74 636	92 922	89 162	75 876	75 268	71 632	77 356	79 394	83 199
Motor Gasoline	393 521	443 111	504 374	477 056	481 500	482 631	496 567	511 799	528 068
Aviation Fuels	70 449	76 947	82 048	83 662	91 551	95 383	101 284	105 762	111 441
Kerosene	37 979	41 660	40 379	32 107	33 584	32 280	32 534	32 474	34 639
Gas Diesel	377 570	442 643	485 769	403 817	418 082	425 862	436 932	444 678	462 004
Heavy Fuel Oil	464 517	521 280	494 120	281 937	278 694	243 583	247 041	234 477	238 969
Other Products	105 996	132 476	146 308	119 889	125 388	124 802	131 116	133 325	141 672
Refinery Fuel	96 373	120 791	112 363	95 722	94 796	94 953	100 464	103 013	106 914
Sub-Total	1 691 547	1 954 809	2 041 084	1 668 135	1 707 559	1 680 565	1 730 667	1 760 300	1 823 070
Bunkers	64 832	73 473	74 249	55 810	54 846	56 353	59 714	58 631	61 221
Total	1 756 379	2 028 282	2 115 333	1 723 945	1 762 405	1 736 918	1 790 381	1 818 931	1 884 291
Thousand barrels/day									
NGL/LPG/Ethane	2 259.5	2 660.3	2 763.2	3 154.0	3 490.7	3 539.6	3 486.4	3 746.9	3 767.2
Naphtha	1 773.5	2 210.0	2 119.7	1 814.2	1 793.2	1 712.0	1 849.3	1 897.1	1 982.1
Motor Gasoline	9 176.9	10 332.8	11 760.7	11 121.1	11 193.8	11 251.0	11 575.5	11 930.6	12 276.0
Aviation Fuels	1 536.5	1 676.9	1 795.5	1 830.1	1 996.3	2 085.0	2 212.6	2 308.5	2 424.6
Kerosene	809.0	887.4	859.2	682.8	712.1	686.4	691.8	690.6	734.4
Gas Diesel	7 716.4	9 046.2	9 927.0	8 252.0	8 520.1	8 702.5	8 928.8	9 087.2	9 415.5
Heavy Fuel Oil	8 345.7	9 366.3	8 883.3	5 059.8	4 986.1	4 368.0	4 433.7	4 206.1	4 279.0
Other Products	1 921.5	2 423.6	2 747.1	2 195.2	2 292.6	2 279.5	2 383.8	2 414.8	2 574.5
Refinery Fuel	1 872.0	2 336.1	2 171.3	1 849.6	1 815.9	1 833.4	1 944.8	1 991.4	2 064.0
Sub-Total	35 410.9	40 939.7	43 027.0	35 958.8	36 800.9	36 457.5	37 506.7	38 273.2	39 517.2
Bunkers	1 184.2	1 342.8	1 363.5	1 026.7	1 007.6	1 037.4	1 099.3	1 078.7	1 122.4
Total	36 595.1	42 282.5	44 390.5	36 985.5	37 808.5	37 494.9	38 606.0	39 351.9	40 639.6

	1989	1990	1991	1992	1993	1994	1995	1996	1997
Thousand tonnes									
NGL/LPG/Ethane	118 621	112 693	121 322	126 577	126 680	131 965	133 750	138 664	139 459
Naphtha	84 831	84 960	84 818	93 683	92 380	99 706	107 847	110 802	121 779
Motor Gasoline	536 353	538 752	539 054	553 529	562 150	571 974	580 237	587 724	595 207
Aviation Fuels	116 429	120 174	117 499	118 607	121 385	127 550	130 670	136 692	140 177
Kerosene	32 906	31 587	32 383	33 727	35 943	36 528	39 483	43 149	41 759
Gas Diesel	461 539	467 154	477 635	488 170	498 829	506 476	519 569	543 013	544 162
Heavy Fuel Oil	244 849	227 768	221 620	223 916	216 266	221 401	203 268	197 890	195 032
Other Products	139 048	146 747	145 471	144 470	139 366	149 792	144 386	147 429	154 722
Refinery Fuel	110 244	112 029	112 111	115 721	116 045	118 159	118 248	122 173	121 251
Sub-Total	1 844 820	1 841 864	1 851 913	1 898 400	1 909 044	1 963 551	1 977 458	2 027 536	2 053 548
Bunkers	62 900	74 721	77 288	79 702	78 269	77 041	79 756	79 108	80 337
Total	1 907 720	1 916 585	1 929 201	1 978 102	1 987 313	2 040 592	2 057 214	2 106 644	2 133 885
Thousand barrels/day									
NGL/LPG/Ethane	3 851.7	3 789.0	4 088.7	4 107.2	4 152.4	4 294.9	4 357.3	4 510.7	4 571.7
Naphtha	2 025.3	2 027.2	2 022.4	2 216.3	2 193.2	2 369.0	2 560.9	2 624.8	2 887.4
Motor Gasoline	12 502.2	12 557.1	12 563.5	12 825.6	13 127.6	13 335.8	13 503.1	13 669.2	13 857.1
Aviation Fuels	2 539.0	2 619.3	2 560.0	2 579.4	2 639.0	2 776.2	2 848.2	2 964.4	3 056.4
Kerosene	699.5	671.4	688.4	721.8	784.2	795.1	855.1	925.0	893.1
Gas Diesel	9 431.7	9 546.4	9 760.5	9 953.7	10 183.1	10 374.1	10 648.2	11 120.1	11 185.4
Heavy Fuel Oil	4 396.5	4 087.6	3 974.9	4 013.0	3 790.8	3 943.8	3 689.2	3 565.9	3 419.4
Other Products	2 521.6	2 660.7	2 649.8	2 744.4	2 597.6	2 806.3	2 716.7	2 816.4	2 914.9
Refinery Fuel	2 146.3	2 181.3	2 183.1	2 250.4	2 219.4	2 229.0	2 225.9	2 242.3	2 299.9
Sub-Total	40 113.8	40 140.0	40 491.3	41 411.8	41 687.4	42 924.1	43 404.6	44 438.8	45 085.2
Bunkers	1 157.2	1 383.7	1 430.4	1 434.6	1 517.3	1 451.3	1 441.0	1 457.5	1 571.4
Total	41 271.0	41 523.7	41 921.6	42 846.4	43 204.7	44 375.4	44 845.6	45 896.3	46 656.6

Oil Products Consumption
Consommation de produits pétroliers

OECD Europe

	1971	1973	1978	1983	1984	1985	1986	1987	1988
Thousand tonnes									
NGL/LPG/Ethane	12 079	14 708	16 821	19 838	21 266	21 771	22 177	24 215	23 383
Naphtha	31 961	41 675	39 067	41 756	39 778	38 436	41 960	42 163	43 834
Motor Gasoline	83 035	95 282	110 021	112 134	114 266	113 528	118 789	122 051	126 784
Aviation Fuels	16 368	18 928	21 028	21 245	22 135	23 053	24 399	25 812	27 802
Kerosene	7 920	8 959	5 990	3 648	3 600	3 908	3 945	4 175	3 898
Gas Diesel	194 557	226 380	236 987	199 195	202 517	210 055	218 700	218 851	219 900
Heavy Fuel Oil	220 115	245 040	215 779	127 604	127 830	110 414	105 384	99 832	95 137
Other Products	37 318	41 949	34 892	29 825	30 522	30 304	30 857	33 093	35 210
Refinery Fuel	31 002	35 123	37 488	31 127	29 669	29 485	31 124	31 295	33 142
Sub-Total	634 355	728 044	718 073	586 372	591 583	580 954	597 335	601 487	609 090
Bunkers	39 498	43 771	36 385	27 605	26 580	28 780	33 843	32 915	34 211
Total	673 853	771 815	754 458	613 977	618 163	609 734	631 178	634 402	643 301
Thousand barrels/day									
NGL/LPG/Ethane	390.8	476.6	543.7	643.6	689.1	705.7	721.1	787.2	758.4
Naphtha	779.3	1 016.2	952.6	1 018.2	967.3	937.2	1 023.1	1 028.1	1 065.9
Motor Gasoline	1 922.3	2 205.8	2 547.1	2 596.0	2 638.1	2 628.3	2 750.0	2 825.6	2 927.1
Aviation Fuels	355.5	410.5	456.3	461.0	479.0	500.1	529.1	559.4	600.0
Kerosene	171.0	193.4	129.3	78.8	77.5	84.4	85.2	90.1	83.9
Gas Diesel	3 976.4	4 626.8	4 843.6	4 071.2	4 127.8	4 293.2	4 469.9	4 473.0	4 482.1
Heavy Fuel Oil	3 889.7	4 330.2	3 813.1	2 254.9	2 252.7	1 951.2	1 862.3	1 764.2	1 676.6
Other Products	689.2	772.7	625.1	492.7	513.7	515.1	509.1	556.0	600.4
Refinery Fuel	608.2	684.1	731.2	609.8	578.8	579.6	611.8	614.9	652.8
Sub-Total	12 782.4	14 716.4	14 641.9	12 226.2	12 324.2	12 194.6	12 561.6	12 698.4	12 847.2
Bunkers	716.9	794.5	664.8	506.3	488.3	529.2	622.1	604.4	625.2
Total	13 499.3	15 510.9	15 306.7	12 732.4	12 812.4	12 723.8	13 183.7	13 302.8	13 472.4

	1989	1990	1991	1992	1993	1994	1995	1996	1997
Thousand tonnes									
NGL/LPG/Ethane	23 191	23 235	25 234	24 978	25 173	26 434	26 535	27 372	27 638
Naphtha	43 362	42 785	41 510	42 450	40 501	42 507	45 732	45 284	47 575
Motor Gasoline	129 495	132 925	134 641	138 324	137 929	136 741	136 203	137 500	137 154
Aviation Fuels	29 383	30 237	29 536	31 164	32 595	34 168	35 992	37 742	39 757
Kerosene	3 743	3 631	4 106	4 095	4 216	4 338	4 155	4 978	5 027
Gas Diesel	214 743	219 011	230 056	231 719	235 413	232 729	238 248	250 512	248 292
Heavy Fuel Oil	96 061	91 385	92 936	93 836	87 857	86 786	87 546	82 753	76 537
Other Products	35 520	35 160	36 958	37 800	36 354	39 666	41 443	41 376	46 239
Refinery Fuel	34 239	34 104	33 797	36 065	37 293	38 045	39 187	40 396	39 528
Sub-Total	609 737	612 473	628 774	640 431	637 331	641 414	655 041	667 913	667 747
Bunkers	34 379	36 154	35 737	36 204	37 055	35 713	36 997	39 240	42 800
Total	644 116	648 627	664 511	676 635	674 386	677 127	692 038	707 153	710 547
Thousand barrels/day									
NGL/LPG/Ethane	753.1	750.1	817.1	809.6	821.1	854.7	861.8	881.4	888.0
Naphtha	1 057.3	1 043.3	1 012.2	1 032.5	988.0	1 036.6	1 115.2	1 101.9	1 160.7
Motor Gasoline	2 997.9	3 077.3	3 117.0	3 193.4	3 193.2	3 165.7	3 153.2	3 174.5	3 175.7
Aviation Fuels	635.6	654.1	638.6	671.8	704.5	738.2	778.0	813.1	858.9
Kerosene	80.8	78.4	88.6	88.2	91.1	93.7	89.2	107.2	108.6
Gas Diesel	4 389.0	4 476.2	4 702.0	4 724.5	4 813.5	4 758.3	4 875.7	5 111.6	5 081.9
Heavy Fuel Oil	1 697.5	1 614.9	1 642.3	1 637.9	1 532.8	1 516.0	1 543.6	1 462.0	1 335.9
Other Products	602.2	592.0	631.8	612.1	601.1	668.6	701.4	708.3	801.0
Refinery Fuel	680.0	680.2	673.3	749.5	772.4	790.5	802.1	801.1	809.4
Sub-Total	12 893.4	12 966.5	13 322.9	13 519.7	13 517.6	13 622.2	13 920.2	14 161.2	14 220.1
Bunkers	630.4	662.3	654.7	657.5	673.5	653.0	667.0	707.4	771.8
Total	13 523.7	13 628.8	13 977.6	14 177.1	14 191.1	14 275.2	14 587.2	14 868.6	14 991.9

Oil Products Consumption
Consommation de produits pétroliers

OECD North America

	1971	1973	1978	1983	1984	1985	1986	1987	1988
Thousand tonnes									
NGL/LPG/Ethane	48 958	56 188	52 398	57 559	64 706	65 612	62 785	67 989	70 001
Naphtha	20 255	23 435	19 670	14 309	14 471	12 016	12 710	13 128	13 202
Motor Gasoline	284 388	317 088	356 735	325 340	326 940	328 222	335 772	346 855	356 721
Aviation Fuels	50 515	53 158	54 989	55 253	61 775	64 381	68 187	70 543	73 419
Kerosene	15 673	13 951	11 776	8 222	7 524	6 994	6 271	6 149	6 287
Gas Diesel	154 100	177 892	200 980	155 138	162 720	162 106	161 859	165 780	175 662
Heavy Fuel Oil	140 576	162 544	173 834	86 142	84 562	77 301	89 974	85 072	89 915
Other Products	46 935	55 470	81 097	63 854	67 648	70 110	73 400	73 968	75 107
Refinery Fuel	56 701	74 076	60 466	53 077	54 157	54 483	58 697	60 530	61 989
Sub-Total	818 101	933 802	1 011 945	818 894	844 503	841 225	869 655	890 014	922 303
Bunkers	7 620	9 629	23 343	19 739	19 503	18 737	18 301	18 337	19 781
Total	825 721	943 431	1 035 288	838 633	864 006	859 962	887 956	908 351	942 084
Thousand barrels/day									
NGL/LPG/Ethane	1 567.8	1 800.0	1 685.1	1 874.3	2 101.8	2 146.9	2 064.0	2 234.0	2 302.8
Naphtha	472.1	546.2	458.6	334.6	337.8	281.6	297.9	307.7	308.5
Motor Gasoline	6 644.6	7 408.5	8 334.6	7 600.1	7 616.6	7 667.4	7 843.7	8 102.6	8 310.3
Aviation Fuels	1 103.2	1 160.2	1 207.5	1 212.7	1 351.1	1 411.6	1 493.8	1 543.9	1 602.4
Kerosene	332.9	296.4	250.4	174.9	159.6	148.7	133.4	130.8	133.3
Gas Diesel	3 149.0	3 635.1	4 106.4	3 169.4	3 315.2	3 311.8	3 306.8	3 386.9	3 579.1
Heavy Fuel Oil	2 561.6	2 961.6	3 163.3	1 560.7	1 526.9	1 397.5	1 628.4	1 537.4	1 621.3
Other Products	817.0	971.7	1 543.1	1 215.7	1 273.9	1 316.4	1 380.5	1 376.8	1 394.4
Refinery Fuel	1 101.1	1 434.0	1 151.0	1 012.3	1 024.5	1 040.1	1 124.8	1 155.5	1 178.4
Sub-Total	17 749.3	20 213.7	21 900.0	18 154.8	18 707.3	18 721.9	19 273.2	19 775.7	20 430.6
Bunkers	141.0	177.8	430.8	364.0	358.2	345.1	337.3	337.9	363.8
Total	17 890.3	20 391.5	22 330.7	18 518.8	19 065.6	19 067.1	19 610.6	20 113.6	20 794.4

	1989	1990	1991	1992	1993	1994	1995	1996	1997
Thousand tonnes									
NGL/LPG/Ethane	70 445	61 993	66 839	71 319	71 082	75 641	76 880	80 098	80 931
Naphtha	14 292	13 314	11 298	13 102	12 212	13 716	13 645	15 877	16 603
Motor Gasoline	359 336	355 442	353 043	361 942	369 468	377 347	384 384	388 267	394 553
Aviation Fuels	75 621	77 735	75 279	74 373	75 157	78 624	78 011	81 695	82 453
Kerosene	5 359	3 085	2 813	2 597	3 127	2 989	3 298	3 591	3 808
Gas Diesel	175 259	171 162	166 085	171 699	176 314	182 318	185 773	193 802	198 547
Heavy Fuel Oil	92 412	75 460	68 598	66 834	68 632	69 103	53 292	54 707	58 906
Other Products	71 878	78 404	76 220	74 718	74 257	77 836	74 571	77 883	83 600
Refinery Fuel	63 668	65 054	64 253	64 987	63 547	64 274	62 799	64 960	63 828
Sub-Total	928 270	901 649	884 428	901 571	913 796	941 848	932 653	960 880	983 229
Bunkers	20 732	30 647	32 361	33 575	29 585	28 977	30 541	28 917	25 184
Total	949 002	932 296	916 789	935 146	943 381	970 825	963 194	989 797	1 008 413
Thousand barrels/day									
NGL/LPG/Ethane	2 324.8	2 195.5	2 371.4	2 361.2	2 381.6	2 504.6	2 544.8	2 654.8	2 708.1
Naphtha	335.1	311.8	264.8	297.4	282.7	318.4	316.6	373.4	385.3
Motor Gasoline	8 393.7	8 302.3	8 245.9	8 389.8	8 652.1	8 814.4	8 955.8	9 049.2	9 196.9
Aviation Fuels	1 654.2	1 699.0	1 644.9	1 622.2	1 637.1	1 713.0	1 703.9	1 774.8	1 806.2
Kerosene	113.9	65.6	59.8	61.5	86.8	82.5	86.8	86.5	86.3
Gas Diesel	3 580.6	3 496.8	3 392.9	3 500.4	3 585.6	3 742.3	3 814.5	3 992.6	4 109.4
Heavy Fuel Oil	1 670.3	1 361.1	1 236.3	1 223.5	1 166.1	1 229.9	1 006.3	1 003.6	996.7
Other Products	1 335.3	1 454.8	1 425.7	1 534.6	1 462.4	1 537.6	1 487.6	1 588.4	1 697.1
Refinery Fuel	1 221.7	1 246.8	1 233.0	1 239.9	1 171.4	1 154.6	1 133.4	1 146.2	1 139.8
Sub-Total	20 629.6	20 133.7	19 874.7	20 230.5	20 425.9	21 097.3	21 049.6	21 669.4	22 125.7
Bunkers	382.8	574.8	605.8	595.7	630.8	570.0	547.9	549.2	574.1
Total	21 012.4	20 708.6	20 480.5	20 826.2	21 056.6	21 667.3	21 597.5	22 218.7	22 699.8

Oil Products Consumption
Consommation de produits pétroliers

OECD Pacific

	1971	1973	1978	1983	1984	1985	1986	1987	1988
Thousand tonnes									
NGL/LPG/Ethane	9 469	12 083	17 342	20 672	22 724	22 056	22 411	23 174	22 780
Naphtha	22 420	27 812	30 425	19 811	21 019	21 180	22 686	24 103	26 163
Motor Gasoline	26 098	30 741	37 618	39 582	40 294	40 881	42 006	42 893	44 563
Aviation Fuels	3 566	4 861	6 031	7 164	7 641	7 949	8 698	9 407	10 220
Kerosene	14 386	18 750	22 613	20 237	22 460	21 378	22 318	22 150	24 454
Gas Diesel	28 913	38 371	47 802	49 484	52 845	53 701	56 373	60 047	66 442
Heavy Fuel Oil	103 826	113 696	104 507	68 191	66 302	55 868	51 683	49 573	53 917
Other Products	21 743	35 057	30 319	26 210	27 218	24 388	26 859	26 264	31 355
Refinery Fuel	8 670	11 592	14 409	11 518	10 970	10 985	10 643	11 188	11 783
Sub-Total	239 091	292 963	311 066	262 869	271 473	258 386	263 677	268 799	291 677
Bunkers	17 714	20 073	14 521	8 466	8 763	8 836	7 570	7 379	7 229
Total	256 805	313 036	325 587	271 335	280 236	267 222	271 247	276 178	298 906
Thousand barrels/day									
NGL/LPG/Ethane	300.9	383.7	534.4	636.1	699.8	687.0	701.3	725.7	706.0
Naphtha	522.1	647.7	708.5	461.4	488.1	493.2	528.3	561.3	607.6
Motor Gasoline	609.9	718.4	879.1	925.0	939.1	955.4	981.7	1 002.4	1 038.6
Aviation Fuels	77.8	106.2	131.7	156.3	166.3	173.3	189.7	205.2	222.2
Kerosene	305.1	397.6	479.5	429.1	475.0	453.3	473.3	469.7	517.1
Gas Diesel	590.9	784.2	977.0	1 011.4	1 077.1	1 097.6	1 152.2	1 227.3	1 354.3
Heavy Fuel Oil	1 894.5	2 074.6	1 906.9	1 244.3	1 206.5	1 019.4	943.0	904.5	981.1
Other Products	415.2	679.1	578.9	486.8	505.0	448.0	494.2	482.0	579.7
Refinery Fuel	162.7	217.9	289.1	227.4	212.5	213.6	208.2	221.0	232.8
Sub-Total	4 879.2	6 009.5	6 485.1	5 577.8	5 769.4	5 540.9	5 671.9	5 799.0	6 239.5
Bunkers	326.3	370.6	267.9	156.5	161.2	163.1	139.8	136.4	133.4
Total	5 205.5	6 380.1	6 753.1	5 734.2	5 930.6	5 704.1	5 811.7	5 935.4	6 372.9

	1989	1990	1991	1992	1993	1994	1995	1996	1997
Thousand tonnes									
NGL/LPG/Ethane	24 985	27 465	29 249	30 280	30 425	29 890	30 335	31 194	30 890
Naphtha	27 177	28 861	32 010	38 131	39 667	43 483	48 470	49 641	57 601
Motor Gasoline	47 522	50 385	51 370	53 263	54 753	57 886	59 650	61 957	63 500
Aviation Fuels	11 425	12 202	12 684	13 070	13 633	14 758	16 667	17 255	17 967
Kerosene	23 804	24 871	25 464	27 035	28 600	29 201	32 030	34 580	32 924
Gas Diesel	71 537	76 981	81 494	84 752	87 102	91 429	95 548	98 699	97 323
Heavy Fuel Oil	56 376	60 923	60 086	63 246	59 777	65 512	62 430	60 430	59 589
Other Products	31 650	33 183	32 293	31 952	28 755	32 290	28 372	28 170	24 883
Refinery Fuel	12 337	12 871	14 061	14 669	15 205	15 840	16 262	16 817	17 895
Sub-Total	306 813	327 742	338 711	356 398	357 917	380 289	389 764	398 743	402 572
Bunkers	7 789	7 920	9 190	9 923	11 629	12 351	12 218	10 951	12 353
Total	314 602	335 662	347 901	366 321	369 546	392 640	401 982	409 694	414 925
Thousand barrels/day									
NGL/LPG/Ethane	773.9	843.4	900.2	936.4	949.8	935.6	950.8	974.5	975.5
Naphtha	632.9	672.1	745.4	886.5	922.5	1 013.9	1 129.0	1 149.5	1 341.4
Motor Gasoline	1 110.6	1 177.5	1 200.5	1 242.4	1 282.4	1 355.7	1 394.1	1 445.5	1 484.5
Aviation Fuels	249.1	266.1	276.5	285.4	297.4	325.0	366.3	376.5	391.3
Kerosene	504.8	527.4	540.0	572.1	606.2	618.8	679.1	731.3	698.3
Gas Diesel	1 462.1	1 573.4	1 665.6	1 728.9	1 784.0	1 873.5	1 958.0	2 015.9	1 994.1
Heavy Fuel Oil	1 028.7	1 111.6	1 096.4	1 151.6	1 091.9	1 198.0	1 139.3	1 100.3	1 086.9
Other Products	584.2	613.9	592.3	597.6	534.1	600.2	527.7	519.6	416.8
Refinery Fuel	244.6	254.3	276.8	261.0	275.6	283.8	290.4	295.0	350.6
Sub-Total	6 590.8	7 039.8	7 293.7	7 661.7	7 743.9	8 204.6	8 434.8	8 608.2	8 739.3
Bunkers	144.0	146.6	169.9	181.4	213.0	228.3	226.0	200.9	225.5
Total	6 734.8	7 186.3	7 463.6	7 843.1	7 957.0	8 432.9	8 660.8	8 809.1	8 964.9

Oil Products Consumption
Consommation de produits pétroliers

Non-OECD Total

	1971	1973	1978	1983	1984	1985	1986	1987	1988
Thousand tonnes									
NGL/LPG/Ethane	12 887	17 085	24 621	35 109	37 158	39 147	44 912	48 255	50 854
Naphtha	4 072	8 328	13 136	15 519	18 677	22 014	21 315	23 662	26 634
Motor Gasoline	99 746	114 381	153 813	170 975	175 937	179 945	187 301	194 230	200 795
Aviation Fuels	34 415	38 129	41 465	46 658	47 563	48 188	49 809	51 927	52 506
Kerosene	33 014	33 560	38 014	41 257	41 119	41 216	40 280	41 989	42 502
Gas Diesel	122 141	149 600	204 017	250 026	254 697	258 815	268 190	279 710	287 618
Heavy Fuel Oil	184 132	224 858	318 412	337 453	320 999	313 086	310 086	308 602	303 298
Other Products	49 986	59 917	94 508	95 718	92 706	93 862	98 323	101 661	107 187
Refinery Fuel	24 553	27 711	42 813	45 707	47 702	49 589	52 786	54 543	56 113
Sub-Total	564 946	673 569	930 799	1 038 422	1 036 558	1 045 862	1 073 002	1 104 579	1 127 507
Bunkers	57 426	60 795	41 622	33 347	36 201	41 492	43 930	43 222	46 437
Total	622 372	734 364	972 421	1 071 769	1 072 759	1 087 354	1 116 932	1 147 801	1 173 944
Thousand barrels/day									
NGL/LPG/Ethane	388.7	512.9	741.5	1 061.5	1 121.7	1 189.0	1 354.9	1 453.1	1 526.9
Naphtha	94.8	193.9	305.9	361.4	433.8	512.7	496.4	551.0	618.5
Motor Gasoline	2 330.4	2 672.4	3 591.9	3 987.2	4 089.6	4 192.1	4 359.8	4 521.4	4 661.0
Aviation Fuels	751.8	831.6	902.6	1 015.3	1 032.1	1 048.6	1 083.8	1 129.7	1 139.2
Kerosene	700.1	711.7	806.1	874.9	869.6	874.0	854.2	890.4	898.8
Gas Diesel	2 496.4	3 057.6	4 169.8	5 110.1	5 191.4	5 289.8	5 481.4	5 716.8	5 862.4
Heavy Fuel Oil	3 359.8	4 102.9	5 809.9	6 157.4	5 841.1	5 712.7	5 658.0	5 630.9	5 519.0
Other Products	958.2	1 151.0	1 829.2	1 837.5	1 769.4	1 795.1	1 878.2	1 942.8	2 041.4
Refinery Fuel	478.3	540.6	850.8	915.7	952.4	996.0	1 054.8	1 092.2	1 121.1
Sub-Total	11 558.4	13 774.6	19 007.6	21 320.8	21 301.0	21 609.9	22 221.4	22 928.4	23 388.4
Bunkers	1 056.6	1 118.3	771.1	617.8	668.5	766.8	811.7	800.2	857.2
Total	12 615.0	14 892.9	19 778.7	21 938.6	21 969.4	22 376.7	23 033.1	23 728.6	24 245.6

	1989	1990	1991	1992	1993	1994	1995	1996	1997
Thousand tonnes									
NGL/LPG/Ethane	52 532	54 217	55 598	43 338	45 603	50 276	56 015	59 701	61 987
Naphtha	30 198	29 343	29 544	33 515	37 044	34 988	33 887	36 161	40 141
Motor Gasoline	207 055	209 329	210 821	197 373	199 540	196 476	205 258	211 975	215 200
Aviation Fuels	54 235	51 448	49 892	47 262	46 569	47 791	49 366	49 985	52 187
Kerosene	42 781	42 906	43 593	41 099	40 997	40 872	42 624	44 349	44 760
Gas Diesel	297 607	301 182	305 371	318 516	323 741	316 307	332 088	342 466	362 963
Heavy Fuel Oil	287 935	287 608	291 861	288 966	284 672	258 502	256 149	257 675	259 089
Other Products	108 145	109 012	100 028	97 966	96 841	91 815	99 175	104 008	108 261
Refinery Fuel	59 463	57 495	54 452	54 293	48 368	45 424	47 495	49 362	52 272
Sub-Total	1 139 951	1 142 540	1 141 160	1 122 328	1 123 375	1 082 451	1 122 057	1 155 682	1 196 860
Bunkers	44 374	46 709	50 243	49 163	49 617	52 020	51 101	53 114	55 013
Total	1 184 325	1 189 249	1 191 403	1 171 491	1 172 992	1 134 471	1 173 158	1 208 796	1 251 873
Thousand barrels/day									
NGL/LPG/Ethane	1 585.8	1 636.3	1 678.6	1 359.8	1 437.6	1 586.4	1 768.6	1 888.2	1 958.3
Naphtha	703.2	683.3	688.0	778.4	862.7	814.8	789.2	839.8	934.8
Motor Gasoline	4 818.7	4 873.8	4 907.8	4 581.5	4 644.2	4 571.4	4 775.9	4 918.6	5 008.2
Aviation Fuels	1 180.3	1 121.0	1 086.1	1 026.6	1 013.9	1 040.4	1 074.8	1 085.5	1 136.4
Kerosene	907.2	909.8	924.4	869.1	869.4	866.8	903.9	938.0	949.3
Gas Diesel	6 082.6	6 155.7	6 241.3	6 492.2	6 617.3	6 465.5	6 788.0	6 981.0	7 419.1
Heavy Fuel Oil	5 253.8	5 247.9	5 325.5	5 258.2	5 193.7	4 716.3	4 673.2	4 688.2	4 726.8
Other Products	2 066.0	2 072.4	1 900.5	1 831.0	1 840.8	1 731.3	1 857.8	1 967.0	2 055.3
Refinery Fuel	1 190.4	1 148.9	1 087.9	1 119.2	977.9	916.1	959.7	1 000.0	1 064.4
Sub-Total	23 787.9	23 849.1	23 840.1	23 316.0	23 457.4	22 708.8	23 591.2	24 306.3	25 252.6
Bunkers	819.7	862.7	926.9	905.6	920.3	965.4	947.4	983.9	1 023.0
Total	24 607.6	24 711.8	24 767.1	24 221.6	24 377.8	23 674.2	24 538.7	25 290.2	26 275.6

Oil Products Consumption
Consommation de produits pétroliers

Africa

	1971	1973	1978	1983	1984	1985	1986	1987	1988
Thousand tonnes									
NGL/LPG/Ethane	623	847	1 528	2 492	2 795	2 827	2 997	3 390	3 457
Naphtha	90	75	87	452	488	622	647	575	542
Motor Gasoline	7 721	9 108	12 509	15 052	15 695	15 949	16 310	17 431	18 446
Aviation Fuels	1 970	2 440	3 449	4 114	4 147	4 292	3 990	4 309	4 329
Kerosene	2 753	2 993	3 802	4 930	5 091	5 275	5 633	6 001	5 991
Gas Diesel	9 806	12 223	17 757	23 104	23 245	23 292	22 554	23 245	24 236
Heavy Fuel Oil	7 884	9 232	12 356	17 100	17 451	17 392	16 913	17 380	18 541
Other Products	1 920	2 226	3 518	5 331	5 458	5 109	5 224	5 439	5 498
Refinery Fuel	1 356	1 558	2 177	3 080	3 174	3 247	3 499	3 643	3 691
Sub-Total	34 123	40 702	57 183	75 655	77 544	78 005	77 767	81 413	84 731
Bunkers	7 028	6 114	4 095	5 137	5 279	5 015	5 147	5 051	5 507
Total	41 151	46 816	61 278	80 792	82 823	83 020	82 914	86 464	90 238
Thousand barrels/day									
NGL/LPG/Ethane	19.8	26.9	48.6	79.2	88.6	89.8	95.2	107.7	109.6
Naphtha	2.1	1.7	2.0	10.5	11.3	14.5	15.1	13.4	12.6
Motor Gasoline	180.4	212.9	292.3	351.8	365.8	372.7	381.2	407.4	429.9
Aviation Fuels	43.2	53.4	75.3	89.7	90.2	93.6	87.1	94.0	94.2
Kerosene	58.4	63.5	80.6	104.5	107.7	111.9	119.5	127.3	126.7
Gas Diesel	200.4	249.8	362.9	472.2	473.8	476.1	461.0	475.1	494.0
Heavy Fuel Oil	143.9	168.5	225.5	312.0	317.6	317.3	308.6	317.1	337.4
Other Products	36.2	41.9	67.0	98.9	101.3	94.7	96.0	100.2	101.2
Refinery Fuel	26.0	30.1	42.2	60.7	62.5	64.3	69.3	72.3	73.5
Sub-Total	710.4	848.6	1 196.5	1 579.6	1 618.7	1 634.9	1 632.9	1 714.4	1 779.0
Bunkers	130.9	114.3	77.2	96.3	99.1	94.5	97.0	95.1	103.4
Total	841.3	962.9	1 273.7	1 675.9	1 717.8	1 729.4	1 729.9	1 809.5	1 882.4

	1989	1990	1991	1992	1993	1994	1995	1996	1997
Thousand tonnes									
NGL/LPG/Ethane	3 835	3 754	4 263	4 490	4 915	5 325	5 607	5 636	5 994
Naphtha	736	804	932	846	857	868	973	982	1 016
Motor Gasoline	19 041	18 423	19 226	19 892	19 783	19 981	21 200	21 548	21 697
Aviation Fuels	4 691	4 613	4 427	4 699	5 010	5 266	5 622	6 282	6 673
Kerosene	6 149	5 965	5 767	5 737	5 267	4 720	4 665	4 824	4 859
Gas Diesel	24 105	24 418	24 503	25 627	26 421	27 128	27 900	28 544	30 887
Heavy Fuel Oil	18 374	18 420	18 614	18 496	16 747	16 894	17 321	17 689	18 874
Other Products	5 749	5 618	5 909	7 275	5 357	6 164	6 389	6 537	6 468
Refinery Fuel	3 722	3 652	3 740	3 821	4 905	5 120	5 311	5 378	5 124
Sub-Total	86 402	85 667	87 381	90 883	89 262	91 466	94 988	97 420	101 592
Bunkers	5 697	6 215	6 489	7 072	6 698	7 882	8 453	8 554	7 767
Total	92 099	91 882	93 870	97 955	95 960	99 348	103 441	105 974	109 359
Thousand barrels/day									
NGL/LPG/Ethane	121.9	119.3	135.5	142.3	159.3	172.0	180.9	181.3	193.2
Naphtha	17.1	18.7	21.7	19.6	20.0	20.2	22.7	22.8	23.7
Motor Gasoline	445.0	430.5	449.3	463.6	462.3	467.0	495.4	502.2	507.1
Aviation Fuels	102.6	101.1	97.0	102.7	109.7	115.3	123.0	137.0	145.9
Kerosene	130.4	126.5	122.3	121.3	111.7	100.1	98.9	102.0	103.0
Gas Diesel	492.7	499.1	500.8	522.3	540.0	554.5	570.2	581.8	631.3
Heavy Fuel Oil	335.3	336.1	339.6	336.6	305.6	308.3	316.1	321.9	344.4
Other Products	106.2	103.5	109.3	135.1	99.7	116.1	120.0	122.3	121.1
Refinery Fuel	74.3	72.8	74.7	76.3	105.3	110.3	114.6	115.5	109.7
Sub-Total	1 825.4	1 807.7	1 850.3	1 919.8	1 913.5	1 963.5	2 041.8	2 086.8	2 179.4
Bunkers	107.1	116.1	121.1	131.4	125.2	147.2	157.9	159.1	145.1
Total	1 932.5	1 923.8	1 971.3	2 051.3	2 038.7	2 110.8	2 199.8	2 245.9	2 324.5

Oil Products Consumption
Consommation de produits pétroliers

Latin America

	1971	1973	1978	1983	1984	1985	1986	1987	1988
Thousand tonnes									
NGL/LPG/Ethane	3 549	4 079	5 318	7 323	7 507	8 166	8 467	8 855	9 198
Naphtha	550	1 546	2 542	4 199	4 459	5 095	5 047	5 493	5 240
Motor Gasoline	22 471	26 490	31 144	33 089	33 464	34 039	37 215	38 276	38 780
Aviation Fuels	2 964	3 371	4 904	5 128	5 000	4 905	5 481	5 848	5 439
Kerosene	4 413	4 674	4 787	4 196	4 013	3 807	4 000	4 004	4 093
Gas Diesel	20 159	24 656	33 330	37 186	37 173	37 582	40 540	43 163	42 459
Heavy Fuel Oil	32 003	36 034	41 196	30 060	28 206	26 626	27 962	29 389	28 897
Other Products	4 223	4 693	7 699	7 365	8 372	9 688	10 129	10 423	10 526
Refinery Fuel	7 105	8 180	7 860	6 922	7 227	6 160	7 258	7 158	8 184
Sub-Total	97 437	113 723	138 780	135 468	135 421	136 068	146 099	152 609	152 816
Bunkers	14 675	14 161	9 941	8 575	8 414	8 255	7 019	7 167	7 119
Total	112 112	127 884	148 721	144 043	143 835	144 323	153 118	159 776	159 935
Thousand barrels/day									
NGL/LPG/Ethane	112.8	129.6	169.0	232.3	237.5	259.1	268.6	281.0	291.1
Naphtha	12.8	36.0	59.2	97.8	103.6	118.6	117.5	127.9	121.7
Motor Gasoline	524.5	618.4	725.1	764.8	769.1	782.3	852.3	876.8	885.1
Aviation Fuels	65.4	74.2	107.3	112.0	108.9	107.1	119.7	127.6	118.4
Kerosene	93.6	99.1	101.5	89.0	84.9	80.7	84.8	84.9	86.6
Gas Diesel	412.0	503.9	681.2	760.0	757.7	768.1	828.6	882.2	865.4
Heavy Fuel Oil	583.9	657.5	751.7	548.5	513.3	485.8	510.2	536.2	525.8
Other Products	76.8	85.1	143.4	137.3	152.2	177.5	184.4	190.0	191.4
Refinery Fuel	137.7	158.4	152.1	136.7	142.1	122.9	142.7	140.7	160.0
Sub-Total	2 019.5	2 362.3	2 890.5	2 878.4	2 869.1	2 902.2	3 108.9	3 247.3	3 245.5
Bunkers	271.1	260.9	184.8	159.0	155.6	153.1	130.4	133.3	132.0
Total	2 290.6	2 623.1	3 075.3	3 037.4	3 024.7	3 055.3	3 239.4	3 380.6	3 377.5

	1989	1990	1991	1992	1993	1994	1995	1996	1997
Thousand tonnes									
NGL/LPG/Ethane	9 457	9 694	9 959	11 382	11 332	12 555	12 991	13 311	14 177
Naphtha	5 620	5 779	5 282	5 500	5 057	5 776	5 430	5 297	6 406
Motor Gasoline	39 937	39 828	41 126	41 967	43 255	45 149	48 166	49 827	51 482
Aviation Fuels	5 440	5 635	5 594	5 600	5 856	6 179	6 836	7 418	7 998
Kerosene	4 055	3 809	3 573	3 391	3 246	2 993	2 989	3 088	2 930
Gas Diesel	43 190	43 252	44 310	45 659	48 084	53 356	56 091	59 039	62 368
Heavy Fuel Oil	28 107	26 260	25 045	24 028	25 262	24 363	25 443	27 963	28 619
Other Products	9 390	9 538	12 730	10 817	12 195	13 152	16 274	17 100	14 239
Refinery Fuel	8 132	7 905	8 293	8 670	10 198	7 658	6 784	7 307	7 897
Sub-Total	153 328	151 700	155 912	157 014	164 485	171 181	181 004	190 350	196 116
Bunkers	6 875	7 072	7 379	7 763	8 134	8 208	7 807	7 033	7 411
Total	160 203	158 772	163 291	164 777	172 619	179 389	188 811	197 383	203 527
Thousand barrels/day									
NGL/LPG/Ethane	300.1	307.7	316.1	360.2	359.6	398.5	412.3	421.3	449.9
Naphtha	130.9	134.6	123.0	127.7	117.8	134.5	126.5	123.0	149.2
Motor Gasoline	913.1	912.6	942.1	959.6	991.6	1 034.6	1 104.5	1 139.3	1 181.9
Aviation Fuels	118.8	123.0	122.1	121.9	127.7	134.8	149.0	161.2	174.3
Kerosene	86.0	80.8	75.8	71.7	68.8	63.5	63.4	65.3	62.1
Gas Diesel	882.7	884.0	905.6	930.6	982.8	1 090.5	1 146.4	1 203.4	1 274.7
Heavy Fuel Oil	512.9	479.2	457.0	437.2	460.9	444.5	464.2	508.8	522.2
Other Products	171.8	175.7	236.5	198.5	222.9	243.2	301.4	316.1	262.8
Refinery Fuel	160.1	156.4	165.0	171.5	204.0	152.6	134.4	144.5	155.7
Sub-Total	3 276.3	3 253.8	3 343.1	3 379.0	3 536.0	3 696.7	3 902.1	4 083.0	4 232.9
Bunkers	128.1	131.8	137.6	144.6	152.6	153.8	145.6	131.1	138.6
Total	3 404.4	3 385.6	3 480.6	3 523.6	3 688.6	3 850.5	4 047.7	4 214.0	4 371.5

Oil Products Consumption
Consommation de produits pétroliers

Asia (excluding China)

	1971	1973	1978	1983	1984	1985	1986	1987	1988
Thousand tonnes									
NGL/LPG/Ethane	631	930	1 632	2 607	2 952	3 891	4 394	4 792	5 431
Naphtha	1 146	1 586	3 678	4 428	5 676	6 562	6 962	6 879	7 337
Motor Gasoline	9 070	10 083	11 966	13 792	14 602	15 169	16 242	17 861	19 461
Aviation Fuels	5 334	6 063	5 125	5 960	6 182	6 396	6 890	7 420	7 861
Kerosene	8 200	8 632	12 117	14 388	14 348	14 433	14 885	15 506	16 267
Gas Diesel	18 386	21 136	28 898	41 672	42 745	44 406	45 719	49 811	53 908
Heavy Fuel Oil	19 519	25 266	35 090	35 464	33 500	31 482	31 971	33 541	37 067
Other Products	4 092	5 586	5 265	5 588	5 230	5 973	6 322	6 786	7 303
Refinery Fuel	2 733	3 244	4 659	6 684	7 234	7 854	8 461	8 842	9 163
Sub-Total	69 111	82 526	108 430	130 583	132 469	136 166	141 846	151 438	163 798
Bunkers	4 648	6 628	6 849	6 783	7 041	6 513	10 406	10 709	12 534
Total	73 759	89 154	115 279	137 366	139 510	142 679	152 252	162 147	176 332
Thousand barrels/day									
NGL/LPG/Ethane	20.1	29.6	51.9	82.1	92.8	122.6	138.5	151.0	170.8
Naphtha	26.7	36.9	85.7	103.1	131.8	152.8	162.1	160.2	170.4
Motor Gasoline	212.0	235.6	279.6	322.3	340.3	354.5	379.6	417.4	453.6
Aviation Fuels	118.3	133.5	111.7	129.8	134.2	139.2	149.9	161.4	170.5
Kerosene	173.9	183.0	256.9	305.1	303.4	306.1	315.6	328.8	344.0
Gas Diesel	375.8	432.0	590.6	851.7	871.3	907.6	934.4	1 018.1	1 098.8
Heavy Fuel Oil	356.2	461.0	640.3	647.1	609.6	574.4	583.4	612.0	674.5
Other Products	75.0	104.0	96.4	101.4	94.8	108.9	115.1	123.8	132.3
Refinery Fuel	52.6	62.1	90.8	132.5	142.9	155.6	166.8	174.7	181.1
Sub-Total	1 410.4	1 677.7	2 203.8	2 675.1	2 721.0	2 821.7	2 945.5	3 147.4	3 396.0
Bunkers	86.3	123.4	128.5	126.7	130.3	120.8	192.3	199.0	232.0
Total	1 496.7	1 801.0	2 332.3	2 801.8	2 851.3	2 942.5	3 137.8	3 346.4	3 627.9

	1989	1990	1991	1992	1993	1994	1995	1996	1997
Thousand tonnes									
NGL/LPG/Ethane	6 152	6 724	7 125	7 612	8 323	9 795	11 509	13 525	13 317
Naphtha	7 849	7 739	7 118	7 004	7 277	8 477	9 045	9 555	11 291
Motor Gasoline	21 363	23 088	23 653	25 221	27 597	30 134	32 499	35 302	36 517
Aviation Fuels	8 431	9 349	8 938	10 002	11 045	12 518	13 148	14 341	15 323
Kerosene	17 184	18 097	17 838	17 910	18 294	18 781	19 589	20 475	21 060
Gas Diesel	59 740	63 543	68 905	74 745	83 686	86 588	93 976	102 285	107 740
Heavy Fuel Oil	39 981	45 785	46 186	48 937	51 890	53 077	56 904	58 094	63 634
Other Products	9 054	8 826	8 785	8 982	9 187	9 488	10 970	12 128	10 888
Refinery Fuel	8 652	10 219	10 069	10 521	10 738	11 399	11 691	12 349	12 575
Sub-Total	178 406	193 370	198 617	210 934	228 037	240 257	259 331	278 054	292 345
Bunkers	12 814	14 630	13 632	16 766	15 377	15 961	16 112	19 014	21 304
Total	191 220	208 000	212 249	227 700	243 414	256 218	275 443	297 068	313 649
Thousand barrels/day									
NGL/LPG/Ethane	194.0	212.0	224.7	239.4	263.2	314.6	369.8	438.4	428.8
Naphtha	182.8	180.2	165.8	162.7	169.5	197.4	210.6	221.9	262.9
Motor Gasoline	499.2	539.6	552.8	587.8	645.2	704.5	759.7	823.0	853.7
Aviation Fuels	183.4	203.3	194.4	216.9	240.1	272.1	285.8	310.8	333.0
Kerosene	364.4	383.8	378.3	378.8	388.0	398.3	415.5	433.1	446.7
Gas Diesel	1 221.0	1 298.7	1 408.3	1 523.5	1 711.0	1 770.4	1 921.4	2 085.5	2 202.8
Heavy Fuel Oil	729.5	835.4	842.7	890.5	946.2	967.9	1 037.7	1 056.5	1 160.4
Other Products	165.5	160.0	158.4	161.4	165.9	171.3	198.5	218.7	196.5
Refinery Fuel	171.8	201.4	197.7	205.6	209.9	223.2	230.0	243.1	248.7
Sub-Total	3 711.5	4 014.4	4 123.0	4 366.4	4 739.0	5 019.7	5 429.0	5 831.0	6 133.5
Bunkers	235.8	269.8	251.5	308.0	283.7	295.4	298.1	352.1	396.5
Total	3 947.3	4 284.2	4 374.5	4 674.4	5 022.7	5 315.0	5 727.1	6 183.1	6 530.0

Oil Products Consumption
Consommation de produits pétroliers

China

	1971	1973	1978	1983	1984	1985	1986	1987	1988
Thousand tonnes									
NGL/LPG/Ethane	46	72	897	1 636	1 676	1 748	2 194	2 327	2 501
Naphtha	26	32	1 103	3 259	3 429	5 124	4 084	5 726	7 327
Motor Gasoline	4 579	6 242	9 727	11 280	12 391	14 187	15 191	16 437	18 194
Aviation Fuels	447	523	633	1 240	1 396	1 441	1 530	1 729	1 977
Kerosene	3 210	3 161	3 509	3 394	3 441	3 395	3 447	3 265	3 005
Gas Diesel	10 111	13 385	18 415	17 575	18 585	20 288	22 427	24 200	27 071
Heavy Fuel Oil	12 915	16 987	27 783	28 661	28 561	27 486	28 548	29 170	30 168
Other Products	9 461	12 866	30 972	18 595	18 693	17 155	18 865	18 785	18 229
Refinery Fuel	1 600	1 800	4 500	3 254	3 251	4 707	5 295	5 541	5 787
Sub-Total	42 395	55 068	97 539	88 894	91 423	95 531	101 581	107 180	114 259
Bunkers	642	1 179	551	995	317	1 003	838	911	1 075
Total	43 037	56 247	98 090	89 889	91 740	96 534	102 419	108 091	115 334
Thousand barrels/day									
NGL/LPG/Ethane	1.5	2.3	28.5	52.0	53.1	55.6	69.7	74.0	79.3
Naphtha	0.6	0.7	25.7	75.9	79.6	119.3	95.1	133.3	170.2
Motor Gasoline	107.0	145.9	227.3	263.6	288.8	331.5	355.0	384.1	424.0
Aviation Fuels	9.7	11.4	13.8	27.1	30.4	31.5	33.4	37.8	43.0
Kerosene	68.1	67.0	74.4	72.0	72.8	72.0	73.1	69.2	63.5
Gas Diesel	206.7	273.6	376.4	359.2	378.8	414.7	458.4	494.6	551.8
Heavy Fuel Oil	235.7	310.0	506.9	523.0	519.7	501.5	520.9	532.3	549.0
Other Products	182.9	250.5	611.3	358.4	357.7	328.9	361.5	359.7	348.3
Refinery Fuel	31.4	35.8	85.8	63.0	62.9	93.9	105.5	110.4	114.8
Sub-Total	843.4	1 097.1	1 950.0	1 794.3	1 843.8	1 948.8	2 072.6	2 195.3	2 343.8
Bunkers	11.7	21.5	10.1	18.5	6.1	18.6	15.6	17.1	20.0
Total	855.1	1 118.7	1 960.1	1 812.7	1 849.8	1 967.5	2 088.3	2 212.4	2 363.8

	1989	1990	1991	1992	1993	1994	1995	1996	1997
Thousand tonnes									
NGL/LPG/Ethane	2 704	2 745	3 211	3 671	4 429	4 995	6 959	8 478	8 948
Naphtha	8 150	8 631	9 909	10 319	11 950	13 195	14 975	16 409	17 526
Motor Gasoline	18 813	19 078	22 365	24 752	30 983	27 086	29 273	31 750	33 274
Aviation Fuels	2 215	2 792	2 983	3 581	4 134	4 730	5 670	5 879	6 637
Kerosene	2 982	2 661	2 587	2 589	2 548	2 582	2 755	3 124	3 080
Gas Diesel	28 776	28 446	31 859	34 251	40 693	39 517	45 675	49 198	56 594
Heavy Fuel Oil	30 303	29 870	29 746	31 861	34 513	31 714	33 346	35 614	36 104
Other Products	17 800	16 019	14 830	15 388	13 387	20 256	17 451	22 953	29 023
Refinery Fuel	7 166	7 152	7 134	7 880	7 045	6 322	7 763	8 143	10 524
Sub-Total	118 909	117 394	124 624	134 292	149 682	150 397	163 867	181 548	201 710
Bunkers	1 367	1 457	1 244	1 527	4 206	4 337	3 175	3 349	3 101
Total	120 276	118 851	125 868	135 819	153 888	154 734	167 042	184 897	204 811
Thousand barrels/day									
NGL/LPG/Ethane	85.9	87.2	102.0	116.3	140.8	158.7	221.2	268.7	284.4
Naphtha	189.8	201.0	230.8	239.6	278.3	307.3	348.7	381.1	408.1
Motor Gasoline	439.7	445.9	522.7	576.9	724.1	633.0	684.1	740.0	777.6
Aviation Fuels	48.3	60.9	65.0	77.8	89.8	102.8	123.4	127.7	144.5
Kerosene	63.2	56.4	54.9	54.8	54.0	54.8	58.4	66.1	65.3
Gas Diesel	588.1	581.4	651.1	698.1	831.7	807.7	933.5	1 002.8	1 156.7
Heavy Fuel Oil	552.9	545.0	542.8	579.8	629.7	578.7	608.5	648.1	658.8
Other Products	340.8	305.3	277.8	287.0	265.6	383.8	324.9	449.3	568.3
Refinery Fuel	140.5	140.8	140.4	155.0	150.5	136.0	162.2	174.1	226.5
Sub-Total	2 449.3	2 424.0	2 587.5	2 785.4	3 164.6	3 162.7	3 465.0	3 857.7	4 290.3
Bunkers	25.7	27.3	23.4	28.4	80.0	82.5	60.6	64.4	59.9
Total	2 475.0	2 451.2	2 610.8	2 813.8	3 244.6	3 245.2	3 525.5	3 922.2	4 350.1

Oil Products Consumption
Consommation de produits pétroliers

Non-OECD Europe

	1971	1973	1978	1983	1984	1985	1986	1987	1988
Thousand tonnes									
NGL/LPG/Ethane	319	519	683	830	848	768	812	888	891
Naphtha	420	880	1 317	717	668	474	647	828	1 049
Motor Gasoline	5 402	5 866	6 989	5 421	5 349	4 996	5 433	5 763	6 221
Aviation Fuels	543	600	709	703	888	886	936	985	1 055
Kerosene	1 188	1 262	1 299	908	907	791	877	883	943
Gas Diesel	9 547	10 850	13 537	12 840	11 882	12 138	12 612	12 830	12 797
Heavy Fuel Oil	12 308	14 487	20 587	17 239	16 028	17 177	18 905	19 327	19 789
Other Products	3 035	3 089	4 230	4 515	3 681	3 878	3 968	4 209	4 084
Refinery Fuel	1 777	2 081	3 115	3 254	3 545	3 444	3 688	3 814	3 838
Sub-Total	34 539	39 634	52 466	46 427	43 796	44 552	47 878	49 527	50 667
Bunkers	242	228	185	171	432	568	716	752	859
Total	34 781	39 862	52 651	46 598	44 228	45 120	48 594	50 279	51 526
Thousand barrels/day									
NGL/LPG/Ethane	10.1	16.5	21.7	26.4	26.9	24.4	25.8	28.2	28.2
Naphtha	9.8	20.5	30.7	16.7	15.5	11.0	15.1	19.3	24.4
Motor Gasoline	126.2	137.1	163.3	126.7	124.7	116.8	127.0	134.7	145.0
Aviation Fuels	11.9	13.1	15.5	15.3	19.3	19.3	20.4	21.4	22.9
Kerosene	25.2	26.8	27.5	19.3	19.2	16.8	18.6	18.7	19.9
Gas Diesel	195.1	221.8	276.7	262.4	242.2	248.1	257.8	262.2	260.8
Heavy Fuel Oil	224.6	264.3	375.6	314.6	291.7	313.4	345.0	352.7	360.1
Other Products	56.9	57.4	78.4	83.7	67.5	71.5	72.3	77.5	74.5
Refinery Fuel	34.4	40.5	61.4	64.6	69.8	68.0	72.9	75.6	75.9
Sub-Total	694.3	798.0	1 050.8	929.6	876.6	889.2	954.7	990.2	1 011.7
Bunkers	4.6	4.4	3.5	3.2	8.4	10.9	13.6	14.3	16.2
Total	698.9	802.3	1 054.4	932.8	884.9	900.1	968.3	1 004.5	1 027.9

	1989	1990	1991	1992	1993	1994	1995	1996	1997
Thousand tonnes									
NGL/LPG/Ethane	848	888	1 016	641	644	608	680	619	808
Naphtha	2 078	1 307	1 285	1 542	2 521	1 847	1 973	2 258	2 079
Motor Gasoline	6 353	6 523	5 779	5 370	4 531	4 954	4 959	5 568	5 771
Aviation Fuels	1 371	1 209	970	1 024	1 128	1 041	1 133	917	1 008
Kerosene	673	595	468	517	200	130	171	186	259
Gas Diesel	11 968	11 200	8 930	8 979	8 321	7 666	7 943	9 644	9 363
Heavy Fuel Oil	19 282	19 818	15 986	12 017	12 014	10 713	10 576	11 510	11 338
Other Products	5 001	5 110	4 122	2 931	2 627	3 413	3 753	3 925	4 391
Refinery Fuel	3 983	3 689	2 770	1 418	1 503	1 795	1 484	1 321	1 274
Sub-Total	51 557	50 339	41 326	34 439	33 489	32 167	32 672	35 948	36 291
Bunkers	861	857	1 253	1 236	1 201	1 252	1 276	1 292	1 397
Total	52 418	51 196	42 579	35 675	34 690	33 419	33 948	37 240	37 688
Thousand barrels/day									
NGL/LPG/Ethane	27.0	28.2	31.4	21.1	21.6	20.6	22.9	20.8	27.0
Naphtha	48.4	30.4	29.9	35.8	58.7	43.0	45.9	52.4	48.4
Motor Gasoline	148.5	152.4	135.1	125.2	105.9	115.8	115.9	129.8	134.9
Aviation Fuels	29.8	26.3	21.1	22.2	24.5	22.6	24.6	19.9	21.9
Kerosene	14.3	12.6	9.9	10.9	4.2	2.8	3.6	3.9	5.5
Gas Diesel	244.6	228.9	182.5	183.0	170.1	156.7	162.3	196.6	191.4
Heavy Fuel Oil	351.8	361.6	291.7	218.7	219.2	195.5	193.0	209.4	206.9
Other Products	92.8	96.2	76.5	54.9	49.7	64.6	70.7	72.6	81.7
Refinery Fuel	79.3	74.8	55.6	28.8	30.8	36.8	30.9	27.4	26.5
Sub-Total	1 036.4	1 011.6	833.7	700.5	684.8	658.2	670.0	732.9	744.1
Bunkers	16.3	16.3	23.4	23.0	22.3	23.3	23.8	24.0	26.0
Total	1 052.8	1 027.8	857.1	723.4	707.1	681.5	693.8	756.9	770.1

Oil Products Consumption
Consommation de produits pétroliers

Former USSR

	1971	1973	1978	1983	1984	1985	1986	1987	1988
Thousand tonnes									
NGL/LPG/Ethane	7 000	9 800	13 200	18 000	18 500	16 916	20 012	21 817	22 748
Naphtha	1 400	3 800	3 800	2 000	3 500	3 700	3 400	3 600	4 700
Motor Gasoline	46 300	51 200	70 500	75 000	76 300	75 900	77 400	78 500	79 500
Aviation Fuels	21 900	22 900	22 200	24 600	24 900	25 200	25 700	26 300	27 000
Kerosene	9 700	8 500	5 400	4 700	4 600	4 500	4 400	4 300	4 200
Gas Diesel	47 500	58 800	70 800	80 300	81 400	82 200	84 800	85 600	85 400
Heavy Fuel Oil	92 000	113 298	165 800	178 900	171 400	165 400	156 300	149 900	138 600
Other Products	25 300	29 000	38 900	43 054	40 754	41 255	41 554	43 553	49 553
Refinery Fuel	5 600	6 300	15 340	15 700	15 800	15 600	14 900	15 100	14 900
Sub-Total	256 700	303 598	405 940	442 254	437 154	430 671	428 466	428 670	426 601
Bunkers	4 300	4 700	4 900	4 600	4 700	4 500	4 700	4 700	4 800
Total	261 000	308 298	410 840	446 854	441 854	435 171	433 166	433 370	431 401
Thousand barrels/day									
NGL/LPG/Ethane	201.6	281.4	378.6	518.9	531.6	490.1	575.0	624.5	648.2
Naphtha	32.6	88.5	88.5	46.6	81.3	86.2	79.2	83.8	109.2
Motor Gasoline	1 082.0	1 196.5	1 647.6	1 752.7	1 778.2	1 773.8	1 808.8	1 834.5	1 852.8
Aviation Fuels	475.8	497.5	482.3	534.5	539.5	547.5	558.4	571.4	585.0
Kerosene	205.7	180.2	114.5	99.7	97.3	95.4	93.3	91.2	88.8
Gas Diesel	970.8	1 201.8	1 447.0	1 641.2	1 659.1	1 680.0	1 733.2	1 749.5	1 740.7
Heavy Fuel Oil	1 678.7	2 067.3	3 025.3	3 264.3	3 118.9	3 018.0	2 851.9	2 735.2	2 522.1
Other Products	492.9	565.0	757.8	838.4	792.0	803.5	809.6	848.0	960.5
Refinery Fuel	112.8	127.4	320.8	328.3	330.3	327.6	313.0	317.7	313.2
Sub-Total	5 252.9	6 205.7	8 262.4	9 024.6	8 928.3	8 822.0	8 822.4	8 855.9	8 820.5
Bunkers	78.5	85.8	89.4	83.9	85.5	82.1	85.8	85.8	87.3
Total	5 331.4	6 291.4	8 351.8	9 108.5	9 013.8	8 904.1	8 908.1	8 941.6	8 907.8

	1989	1990	1991	1992	1993	1994	1995	1996	1997
Thousand tonnes									
NGL/LPG/Ethane	22 351	22 942	23 080	7 560	6 507	6 218	7 205	7 313	6 577
Naphtha	4 600	3 900	3 900	7 032	8 240	3 577	-	-	-
Motor Gasoline	80 500	79 500	76 000	54 668	45 544	39 926	38 966	37 485	35 160
Aviation Fuels	27 300	21 585	21 200	17 664	14 691	12 589	11 628	9 738	8 996
Kerosene	4 200	4 200	5 000	2 526	1 598	1 208	1 142	774	797
Gas Diesel	86 400	85 000	80 199	79 429	64 049	48 284	47 622	39 082	39 357
Heavy Fuel Oil	121 600	116 500	123 459	118 703	109 086	84 829	73 501	66 074	60 537
Other Products	49 722	51 428	40 777	38 365	37 454	22 979	29 109	24 969	26 191
Refinery Fuel	17 200	14 500	13 700	12 389	3 644	2 185	3 712	3 755	3 832
Sub-Total	413 873	399 555	387 315	338 336	290 813	221 795	212 885	189 190	181 447
Bunkers	4 700	4 600	4 600	157	158	101	90	93	164
Total	418 573	404 155	391 915	338 493	290 971	221 896	212 975	189 283	181 611
Thousand barrels/day									
NGL/LPG/Ethane	639.1	655.3	659.1	239.6	206.8	197.6	229.0	231.8	209.0
Naphtha	107.1	90.8	90.8	163.3	191.9	83.3	-	-	-
Motor Gasoline	1 881.3	1 857.9	1 776.1	1 274.1	1 064.4	933.1	910.6	873.6	821.7
Aviation Fuels	593.1	470.0	460.6	383.2	319.6	273.8	252.9	211.3	195.7
Kerosene	89.1	89.1	106.0	53.4	33.9	25.6	24.2	16.4	16.9
Gas Diesel	1 765.9	1 737.3	1 639.1	1 619.0	1 309.1	986.8	973.3	796.6	804.4
Heavy Fuel Oil	2 218.8	2 125.7	2 252.7	2 160.0	1 990.4	1 547.8	1 341.1	1 202.3	1 104.6
Other Products	966.0	988.5	789.7	718.1	711.2	431.7	549.7	473.1	495.7
Refinery Fuel	361.2	303.5	286.0	298.2	77.1	45.2	79.3	80.0	82.3
Sub-Total	8 621.5	8 318.1	8 060.1	6 908.9	5 904.4	4 525.0	4 360.2	3 885.1	3 730.3
Bunkers	85.8	83.9	83.9	3.0	3.1	1.9	1.7	1.8	3.1
Total	8 707.3	8 402.0	8 144.0	6 911.9	5 907.4	4 526.9	4 361.9	3 886.9	3 733.3

Oil Products Consumption
Consommation de produits pétroliers

Middle East

	1971	1973	1978	1983	1984	1985	1986	1987	1988
Thousand tonnes									
NGL/LPG/Ethane	719	838	1 363	2 221	2 880	4 831	6 036	6 186	6 628
Naphtha	440	409	609	464	457	437	528	561	439
Motor Gasoline	4 203	5 392	10 978	17 341	18 136	19 705	19 510	19 962	20 193
Aviation Fuels	1 257	2 232	4 445	4 913	5 050	5 068	5 282	5 336	4 845
Kerosene	3 550	4 338	7 100	8 741	8 719	9 015	7 038	8 030	8 003
Gas Diesel	6 632	8 550	21 280	37 349	39 667	38 909	39 538	40 861	41 747
Heavy Fuel Oil	7 503	9 554	15 600	30 029	25 853	27 523	29 487	29 895	30 236
Other Products	1 955	2 457	3 924	11 270	10 518	10 804	12 261	12 466	11 994
Refinery Fuel	4 382	4 548	5 162	6 813	7 471	8 577	9 685	10 445	10 550
Sub-Total	30 641	38 318	70 461	119 141	118 751	124 869	129 365	133 742	134 635
Bunkers	25 891	27 785	15 101	7 086	10 018	15 638	15 104	13 932	14 543
Total	56 532	66 103	85 562	126 227	128 769	140 507	144 469	147 674	149 178
Thousand barrels/day									
NGL/LPG/Ethane	22.9	26.6	43.3	70.6	91.3	147.4	182.0	186.7	199.7
Naphtha	10.2	9.5	14.2	10.8	10.6	10.2	12.3	13.1	10.2
Motor Gasoline	98.2	126.0	256.6	405.3	422.7	460.5	455.9	466.5	470.6
Aviation Fuels	27.4	48.6	96.8	106.8	109.6	110.3	115.0	116.2	105.2
Kerosene	75.3	92.0	150.6	185.4	184.4	191.2	149.2	170.3	169.2
Gas Diesel	135.5	174.7	434.9	763.4	808.5	795.2	808.1	835.1	850.9
Heavy Fuel Oil	136.9	174.3	284.6	547.9	470.4	502.2	538.0	545.5	550.2
Other Products	37.6	47.1	75.0	219.3	204.1	210.1	239.3	243.7	233.2
Refinery Fuel	83.4	86.3	97.7	129.7	142.0	163.8	184.4	200.9	202.6
Sub-Total	627.5	785.3	1 453.6	2 439.2	2 443.6	2 590.9	2 684.4	2 777.9	2 791.9
Bunkers	473.5	508.1	277.6	130.3	183.5	286.8	276.9	255.7	266.2
Total	1 101.0	1 293.4	1 731.2	2 569.4	2 627.1	2 877.7	2 961.3	3 033.6	3 058.2

	1989	1990	1991	1992	1993	1994	1995	1996	1997
Thousand tonnes									
NGL/LPG/Ethane	7 185	7 470	6 944	7 684	9 199	10 080	10 126	10 493	11 932
Naphtha	1 165	1 183	1 118	1 272	1 142	1 248	1 491	1 660	1 823
Motor Gasoline	21 048	22 889	22 672	25 503	27 847	29 246	30 195	30 495	31 299
Aviation Fuels	4 787	6 265	5 780	4 692	4 705	5 468	5 329	5 410	5 552
Kerosene	7 538	7 579	8 360	8 429	9 844	10 458	11 313	11 878	11 775
Gas Diesel	43 428	45 323	46 665	49 826	52 487	53 768	52 881	54 674	56 654
Heavy Fuel Oil	30 288	30 955	32 825	34 924	35 160	36 912	39 058	40 731	39 983
Other Products	11 429	12 473	12 875	14 506	16 888	17 063	16 167	16 722	17 295
Refinery Fuel	10 608	10 378	8 746	9 594	10 335	10 945	10 750	11 109	11 046
Sub-Total	137 476	144 515	145 985	156 430	167 607	175 188	177 310	183 172	187 359
Bunkers	12 060	11 878	15 646	14 642	13 843	14 279	14 188	13 779	13 869
Total	149 536	156 393	161 631	171 072	181 450	189 467	191 498	196 951	201 228
Thousand barrels/day									
NGL/LPG/Ethane	217.8	226.6	209.8	232.7	279.4	305.3	306.7	317.0	359.6
Naphtha	27.1	27.5	26.0	29.5	26.6	29.1	34.7	38.6	42.5
Motor Gasoline	491.9	534.9	529.8	594.4	650.8	683.5	705.7	710.7	731.5
Aviation Fuels	104.2	136.4	125.9	102.0	102.5	119.1	116.1	117.6	121.0
Kerosene	159.8	160.7	177.3	178.3	208.7	221.8	239.9	251.2	249.7
Gas Diesel	887.6	926.3	953.8	1 015.6	1 072.7	1 098.9	1 080.8	1 114.4	1 157.9
Heavy Fuel Oil	552.7	564.8	598.9	635.5	641.5	673.5	712.7	741.2	729.6
Other Products	223.0	243.1	252.4	283.8	331.8	335.0	311.7	321.6	334.0
Refinery Fuel	203.2	199.2	168.6	183.8	200.2	212.1	208.4	215.4	215.0
Sub-Total	2 867.4	3 019.6	3 042.5	3 255.5	3 514.3	3 678.2	3 716.6	3 827.5	3 940.7
Bunkers	220.9	217.5	286.1	267.3	253.4	261.3	259.7	251.5	253.9
Total	3 088.3	3 237.1	3 328.7	3 522.8	3 767.7	3 939.4	3 976.4	4 079.0	4 194.6

Oil Products Consumption
Consommation de produits pétroliers

Albania

	1971	1973	1978	1983	1984	1985	1986	1987	1988
Thousand tonnes									
NGL/LPG/Ethane	-	-	-	-	-	-	-	-	-
Naphtha	-	-	-	-	-	-	-	-	-
Motor Gasoline	88	78	140	125	200	121	99	110	145
Aviation Fuels	-	-	-	-	-	-	-	-	-
Kerosene	15	35	60	95	80	55	55	56	60
Gas Diesel	145	174	220	175	225	225	225	225	225
Heavy Fuel Oil	290	315	400	380	365	300	345	375	300
Other Products	239	64	215	255	140	117	120	100	170
Refinery Fuel	34	34	56	84	84	85	85	84	73
Sub-Total	811	700	1 091	1 114	1 094	903	929	950	973
Bunkers	-	-	-	-	-	-	-	-	-
Total	811	700	1 091	1 114	1 094	903	929	950	973
Thousand barrels/day									
NGL/LPG/Ethane	-	-	-	-	-	-	-	-	-
Naphtha	-	-	-	-	-	-	-	-	-
Motor Gasoline	2.1	1.8	3.3	2.9	4.7	2.8	2.3	2.6	3.4
Aviation Fuels	-	-	-	-	-	-	-	-	-
Kerosene	0.3	0.7	1.3	2.0	1.7	1.2	1.2	1.2	1.3
Gas Diesel	3.0	3.6	4.5	3.6	4.6	4.6	4.6	4.6	4.6
Heavy Fuel Oil	5.3	5.7	7.3	6.9	6.6	5.5	6.3	6.8	5.5
Other Products	4.6	1.2	4.1	4.9	2.7	2.2	2.3	1.9	3.3
Refinery Fuel	0.7	0.7	1.1	1.7	1.7	1.8	1.8	1.7	1.5
Sub-Total	15.9	13.7	21.5	22.1	22.0	18.1	18.4	18.9	19.5
Bunkers	-	-	-	-	-	-	-	-	-
Total	15.9	13.7	21.5	22.1	22.0	18.1	18.4	18.9	19.5

	1989	1990	1991	1992	1993	1994	1995	1996	1997
Thousand tonnes									
NGL/LPG/Ethane	-	-	-	-	-	-	-	-	-
Naphtha	-	-	-	-	-	-	-	-	-
Motor Gasoline	81	100	101	90	120	155	164	173	179
Aviation Fuels	-	-	-	-	-	-	-	-	-
Kerosene	61	56	61	41	29	29	74	69	57
Gas Diesel	225	238	180	116	120	123	106	108	107
Heavy Fuel Oil	340	412	320	176	202	206	43	46	39
Other Products	150	210	200	200	97	112	109	97	43
Refinery Fuel	84	73	73	75	75	75	50	40	30
Sub-Total	941	1 089	935	698	643	700	546	533	455
Bunkers	-	-	-	-	-	-	-	-	-
Total	941	1 089	935	698	643	700	546	533	455
Thousand barrels/day									
NGL/LPG/Ethane	-	-	-	-	-	-	-	-	-
Naphtha	-	-	-	-	-	-	-	-	-
Motor Gasoline	1.9	2.3	2.4	2.1	2.8	3.6	3.8	4.0	4.2
Aviation Fuels	-	-	-	-	-	-	-	-	-
Kerosene	1.3	1.2	1.3	0.9	0.6	0.6	1.6	1.5	1.2
Gas Diesel	4.6	4.9	3.7	2.4	2.5	2.5	2.2	2.2	2.2
Heavy Fuel Oil	6.2	7.5	5.8	3.2	3.7	3.8	0.8	0.8	0.7
Other Products	2.9	4.0	3.8	3.8	1.8	2.1	1.7	1.5	0.7
Refinery Fuel	1.7	1.5	1.5	1.6	1.6	1.6	1.1	0.9	0.7
Sub-Total	18.6	21.5	18.5	13.9	12.9	14.1	11.2	10.9	9.6
Bunkers	-	-	-	-	-	-	-	-	-
Total	18.6	21.5	18.5	13.9	12.9	14.1	11.2	10.9	9.6

Oil Products Consumption
Consommation de produits pétroliers

Algeria

	1971	1973	1978	1983	1984	1985	1986	1987	1988
Thousand tonnes									
NGL/LPG/Ethane	173	300	551	991	1 161	1 087	1 165	1 239	1 199
Naphtha	26	8	-	-	-	-	-	-	-
Motor Gasoline	449	538	958	1 529	1 658	1 785	1 897	1 963	1 987
Aviation Fuels	92	128	243	370	394	415	383	355	347
Kerosene	84	83	58	32	33	33	30	29	27
Gas Diesel	771	1 000	1 739	2 731	3 025	3 267	3 320	3 332	3 303
Heavy Fuel Oil	221	225	199	238	191	184	190	155	137
Other Products	169	199	435	471	597	556	553	558	556
Refinery Fuel	35	66	63	243	273	267	292	279	302
Sub-Total	2 020	2 547	4 246	6 605	7 332	7 594	7 830	7 910	7 858
Bunkers	196	158	275	406	428	372	248	248	310
Total	2 216	2 705	4 521	7 011	7 760	7 966	8 078	8 158	8 168
Thousand barrels/day									
NGL/LPG/Ethane	5.5	9.5	17.5	31.5	36.8	34.5	37.0	39.4	38.0
Naphtha	0.6	0.2	-	-	-	-	-	-	-
Motor Gasoline	10.5	12.6	22.4	35.7	38.6	41.7	44.3	45.9	46.3
Aviation Fuels	2.0	2.8	5.3	8.0	8.5	9.0	8.3	7.7	7.5
Kerosene	1.8	1.8	1.2	0.7	0.7	0.7	0.6	0.6	0.6
Gas Diesel	15.8	20.4	35.5	55.8	61.7	66.8	67.9	68.1	67.3
Heavy Fuel Oil	4.0	4.1	3.6	4.3	3.5	3.4	3.5	2.8	2.5
Other Products	3.3	3.9	8.7	9.0	11.4	10.7	9.6	9.7	9.6
Refinery Fuel	0.8	1.4	1.4	5.3	6.0	5.9	6.4	6.1	6.6
Sub-Total	44.3	56.7	95.7	150.5	167.2	172.6	177.7	180.3	178.5
Bunkers	3.7	3.0	5.1	7.6	8.0	7.0	4.7	4.7	5.9
Total	48.0	59.8	100.8	158.1	175.2	179.7	182.4	185.0	184.3

	1989	1990	1991	1992	1993	1994	1995	1996	1997
Thousand tonnes									
NGL/LPG/Ethane	1 363	1 277	1 545	1 563	1 836	1 940	1 885	1 642	1 601
Naphtha	-	-	-	-	-	-	-	-	-
Motor Gasoline	2 109	2 182	2 268	2 269	2 328	2 210	2 060	2 023	1 747
Aviation Fuels	354	345	303	306	309	302	304	302	330
Kerosene	15	12	10	10	11	12	16	10	10
Gas Diesel	3 288	3 256	3 536	3 438	3 365	3 039	3 032	2 937	4 600
Heavy Fuel Oil	135	124	101	84	78	36	29	23	33
Other Products	558	473	437	419	891	998	965	790	669
Refinery Fuel	287	197	286	292	675	624	592	658	573
Sub-Total	8 109	7 866	8 486	8 381	9 493	9 161	8 883	8 385	9 563
Bunkers	444	438	381	375	334	358	375	334	230
Total	8 553	8 304	8 867	8 756	9 827	9 519	9 258	8 719	9 793
Thousand barrels/day									
NGL/LPG/Ethane	43.3	40.6	49.1	49.5	61.4	64.4	62.6	54.8	53.6
Naphtha	-	-	-	-	-	-	-	-	-
Motor Gasoline	49.3	51.0	53.0	52.9	54.4	51.6	48.1	47.1	40.8
Aviation Fuels	7.7	7.5	6.6	6.6	6.7	6.6	6.6	6.5	7.2
Kerosene	0.3	0.3	0.2	0.2	0.2	0.3	0.3	0.2	0.2
Gas Diesel	67.2	66.5	72.3	70.1	68.8	62.1	62.0	59.9	94.0
Heavy Fuel Oil	2.5	2.3	1.8	1.5	1.4	0.7	0.5	0.4	0.6
Other Products	9.7	8.3	7.6	7.3	17.7	20.5	19.7	16.0	13.5
Refinery Fuel	6.3	4.3	6.3	6.4	14.9	13.8	13.1	14.5	12.7
Sub-Total	186.3	180.7	196.9	194.5	225.6	219.9	213.0	199.4	222.6
Bunkers	8.4	8.2	7.2	7.0	6.3	6.8	7.1	6.3	4.4
Total	194.6	189.0	204.1	201.5	231.8	226.7	220.1	205.7	227.0

Oil Products Consumption
Consommation de produits pétroliers

Angola

	1971	1973	1978	1983	1984	1985	1986	1987	1988
Thousand tonnes									
NGL/LPG/Ethane	16	21	13	20	18	21	18	18	18
Naphtha	-	-	-	6	5	7	5	6	7
Motor Gasoline	67	119	100	85	75	84	75	75	70
Aviation Fuels	72	74	93	70	65	173	150	145	150
Kerosene	23	25	21	25	40	46	40	45	48
Gas Diesel	262	344	322	180	95	156	115	107	115
Heavy Fuel Oil	93	120	200	200	195	179	150	120	120
Other Products	34	34	16	20	24	16	20	22	26
Refinery Fuel	44	51	75	94	100	108	97	97	105
Sub-Total	611	788	840	700	617	790	670	635	659
Bunkers	249	151	232	380	490	600	600	653	655
Total	860	939	1 072	1 080	1 107	1 390	1 270	1 288	1 314
Thousand barrels/day									
NGL/LPG/Ethane	0.5	0.7	0.4	0.6	0.6	0.7	0.6	0.6	0.6
Naphtha	-	-	-	0.1	0.1	0.2	0.1	0.1	0.2
Motor Gasoline	1.6	2.8	2.3	2.0	1.7	2.0	1.8	1.8	1.6
Aviation Fuels	1.6	1.6	2.0	1.5	1.4	3.8	3.3	3.2	3.3
Kerosene	0.5	0.5	0.4	0.5	0.8	1.0	0.8	1.0	1.0
Gas Diesel	5.4	7.0	6.6	3.7	1.9	3.2	2.4	2.2	2.3
Heavy Fuel Oil	1.7	2.2	3.6	3.6	3.5	3.3	2.7	2.2	2.2
Other Products	0.6	0.6	0.3	0.4	0.4	0.3	0.4	0.4	0.5
Refinery Fuel	0.9	1.1	1.6	2.0	2.1	2.3	2.0	2.0	2.2
Sub-Total	12.7	16.5	17.3	14.5	12.7	16.5	14.0	13.4	13.8
Bunkers	4.6	2.8	4.4	7.1	9.4	11.4	11.4	12.4	12.4
Total	17.3	19.3	21.7	21.6	22.0	27.9	25.4	25.7	26.2

	1989	1990	1991	1992	1993	1994	1995	1996	1997
Thousand tonnes									
NGL/LPG/Ethane	20	22	21	19	19	19	22	27	35
Naphtha	7	7	7	7	7	7	6	7	15
Motor Gasoline	70	71	70	68	69	70	73	89	99
Aviation Fuels	155	157	155	160	162	160	158	178	291
Kerosene	50	52	52	50	52	50	50	56	56
Gas Diesel	85	86	85	90	85	85	85	127	262
Heavy Fuel Oil	105	118	115	110	115	115	115	203	217
Other Products	24	28	28	27	27	27	28	32	32
Refinery Fuel	105	96	99	90	92	94	95	107	110
Sub-Total	621	637	632	621	628	627	632	826	1 117
Bunkers	685	682	690	690	680	675	660	660	660
Total	1 306	1 319	1 322	1 311	1 308	1 302	1 292	1 486	1 777
Thousand barrels/day									
NGL/LPG/Ethane	0.6	0.7	0.7	0.6	0.6	0.6	0.7	0.9	1.1
Naphtha	0.2	0.2	0.2	0.2	0.2	0.2	0.1	0.2	0.3
Motor Gasoline	1.6	1.7	1.6	1.6	1.6	1.6	1.7	2.1	2.3
Aviation Fuels	3.4	3.4	3.4	3.5	3.5	3.5	3.4	3.9	6.3
Kerosene	1.1	1.1	1.1	1.1	1.1	1.1	1.1	1.2	1.2
Gas Diesel	1.7	1.8	1.7	1.8	1.7	1.7	1.7	2.6	5.4
Heavy Fuel Oil	1.9	2.2	2.1	2.0	2.1	2.1	2.1	3.7	4.0
Other Products	0.4	0.5	0.5	0.5	0.5	0.5	0.5	0.6	0.6
Refinery Fuel	2.2	2.0	2.1	1.9	1.9	2.0	2.0	2.2	2.3
Sub-Total	13.1	13.5	13.3	13.1	13.2	13.2	13.4	17.2	23.5
Bunkers	13.0	13.0	13.1	13.1	12.9	12.8	12.5	12.5	12.5
Total	26.2	26.4	26.5	26.2	26.2	26.1	25.9	29.7	36.0

Oil Products Consumption
Consommation de produits pétroliers

Argentina

	1971	1973	1978	1983	1984	1985	1986	1987	1988
Thousand tonnes									
NGL/LPG/Ethane	1 012	969	1 107	1 320	1 352	1 352	1 364	1 420	1 443
Naphtha	148	210	495	644	707	634	588	557	526
Motor Gasoline	4 022	4 524	4 339	5 043	4 784	4 312	4 453	4 743	4 260
Aviation Fuels	371	385	634	643	736	645	729	719	660
Kerosene	795	737	668	564	609	443	466	494	452
Gas Diesel	5 849	5 894	6 759	7 284	6 827	6 820	7 142	7 515	6 857
Heavy Fuel Oil	8 814	8 860	8 090	4 958	4 582	3 078	2 840	3 602	3 911
Other Products	1 637	1 686	1 415	1 147	1 605	1 548	1 693	1 736	1 702
Refinery Fuel	689	819	1 002	850	763	909	1 640	1 601	1 515
Sub-Total	23 337	24 084	24 509	22 453	21 965	19 741	20 915	22 387	21 326
Bunkers	208	140	355	714	699	645	452	637	562
Total	23 545	24 224	24 864	23 167	22 664	20 386	21 367	23 024	21 888
Thousand barrels/day									
NGL/LPG/Ethane	32.2	30.8	35.2	41.5	42.4	42.5	42.9	44.7	45.3
Naphtha	3.4	4.9	11.5	15.0	16.4	14.8	13.7	13.0	12.2
Motor Gasoline	94.0	105.7	101.4	117.9	111.5	100.8	104.1	110.8	99.3
Aviation Fuels	8.2	8.5	13.8	14.0	16.0	14.1	15.9	15.6	14.4
Kerosene	16.9	15.6	14.2	12.0	12.9	9.4	9.9	10.5	9.6
Gas Diesel	119.5	120.5	138.1	148.9	139.2	139.4	146.0	153.6	139.8
Heavy Fuel Oil	160.8	161.7	147.6	90.5	83.4	56.2	51.8	65.7	71.2
Other Products	27.6	28.1	24.5	19.9	26.6	25.5	27.6	28.6	28.2
Refinery Fuel	14.2	16.8	20.1	16.8	15.2	18.0	31.3	30.6	28.9
Sub-Total	476.8	492.5	506.4	476.4	463.5	420.6	443.1	473.2	448.8
Bunkers	4.1	2.6	6.7	13.4	13.1	12.2	8.7	12.2	10.9
Total	480.9	495.1	513.1	489.8	476.6	432.8	451.8	485.4	459.6

	1989	1990	1991	1992	1993	1994	1995	1996	1997
Thousand tonnes									
NGL/LPG/Ethane	1 405	1 283	1 287	1 530	1 542	1 531	1 638	1 668	1 693
Naphtha	589	569	552	550	188	330	121	136	107
Motor Gasoline	4 058	4 206	4 367	4 734	4 767	4 882	4 731	4 630	4 305
Aviation Fuels	666	653	609	584	716	901	946	1 078	1 217
Kerosene	436	503	402	372	369	341	327	339	260
Gas Diesel	6 530	6 218	6 606	6 821	7 456	8 108	8 571	9 448	9 596
Heavy Fuel Oil	2 925	2 166	2 617	2 300	2 259	1 660	996	1 386	997
Other Products	1 479	1 253	1 338	1 341	2 164	1 591	1 789	1 840	1 811
Refinery Fuel	1 525	1 424	1 799	1 905	1 831	1 154	974	1 006	1 094
Sub-Total	19 613	18 275	19 577	20 137	21 292	20 498	20 093	21 531	21 080
Bunkers	850	708	479	485	382	435	563	578	578
Total	20 463	18 983	20 056	20 622	21 674	20 933	20 656	22 109	21 658
Thousand barrels/day									
NGL/LPG/Ethane	44.2	40.4	40.5	48.0	48.5	48.2	51.5	52.3	53.2
Naphtha	13.7	13.3	12.9	12.8	4.4	7.7	2.8	3.2	2.5
Motor Gasoline	94.8	98.3	102.1	110.3	111.4	114.1	110.6	107.9	100.6
Aviation Fuels	14.5	14.2	13.3	12.7	15.6	19.6	20.6	23.4	26.5
Kerosene	9.2	10.7	8.5	7.9	7.8	7.2	6.9	7.2	5.5
Gas Diesel	133.5	127.1	135.0	139.0	152.4	165.7	175.2	192.6	196.1
Heavy Fuel Oil	53.4	39.5	47.8	41.9	41.2	30.3	18.2	25.2	18.2
Other Products	24.6	20.9	22.2	21.9	35.2	26.1	28.9	29.6	29.2
Refinery Fuel	29.3	27.7	34.7	36.8	35.6	23.0	19.5	20.1	21.3
Sub-Total	417.2	392.0	416.9	431.2	452.1	441.9	434.2	461.5	453.1
Bunkers	16.3	13.5	9.3	9.4	7.3	8.3	10.7	11.0	11.0
Total	433.5	405.5	426.2	440.6	459.4	450.3	445.0	472.4	464.1

Oil Products Consumption
Consommation de produits pétroliers

Armenia

	1971	1973	1978	1983	1984	1985	1986	1987	1988
Thousand tonnes									
NGL/LPG/Ethane	-	-	-	-	-	-	-	-	-
Naphtha	-	-	-	-	-	-	-	-	-
Motor Gasoline	-	-	-	-	-	-	-	-	-
Aviation Fuels	-	-	-	-	-	-	-	-	-
Kerosene	-	-	-	-	-	-	-	-	-
Gas Diesel	-	-	-	-	-	-	-	-	-
Heavy Fuel Oil	-	-	-	-	-	-	-	-	-
Other Products	-	-	-	-	-	-	-	-	-
Refinery Fuel	-	-	-	-	-	-	-	-	-
Sub-Total	-	-	-	-	-	-	-	-	-
Bunkers	-	-	-	-	-	-	-	-	-
Total	-	-	-	-	-	-	-	-	-
Thousand barrels/day									
NGL/LPG/Ethane	-	-	-	-	-	-	-	-	-
Naphtha	-	-	-	-	-	-	-	-	-
Motor Gasoline	-	-	-	-	-	-	-	-	-
Aviation Fuels	-	-	-	-	-	-	-	-	-
Kerosene	-	-	-	-	-	-	-	-	-
Gas Diesel	-	-	-	-	-	-	-	-	-
Heavy Fuel Oil	-	-	-	-	-	-	-	-	-
Other Products	-	-	-	-	-	-	-	-	-
Refinery Fuel	-	-	-	-	-	-	-	-	-
Sub-Total	-	-	-	-	-	-	-	-	-
Bunkers	-	-	-	-	-	-	-	-	-
Total	-	-	-	-	-	-	-	-	-

	1989	1990	1991	1992	1993	1994	1995	1996	1997
Thousand tonnes									
NGL/LPG/Ethane	-	-	-	10	4	1	1	1	1
Naphtha	-	-	-	-	-	-	-	-	-
Motor Gasoline	-	-	-	600	390	52	37	20	20
Aviation Fuels	-	-	-	150	65	48	34	19	19
Kerosene	-	-	-	7	5	5	4	2	2
Gas Diesel	-	-	-	213	100	86	61	34	34
Heavy Fuel Oil	-	-	-	1 400	600	163	115	63	63
Other Products	-	-	-	47	47	36	26	15	15
Refinery Fuel	-	-	-	-	-	-	-	-	-
Sub-Total	-	-	-	2 427	1 211	391	278	154	154
Bunkers	-	-	-	-	-	-	-	-	-
Total	-	-	-	2 427	1 211	391	278	154	154
Thousand barrels/day									
NGL/LPG/Ethane	-	-	-	0.3	0.1	0.0	0.0	0.0	0.0
Naphtha	-	-	-	-	-	-	-	-	-
Motor Gasoline	-	-	-	14.0	9.1	1.2	0.9	0.5	0.5
Aviation Fuels	-	-	-	3.3	1.4	1.0	0.7	0.4	0.4
Kerosene	-	-	-	0.1	0.1	0.1	0.1	0.0	0.0
Gas Diesel	-	-	-	4.3	2.0	1.8	1.2	0.7	0.7
Heavy Fuel Oil	-	-	-	25.5	10.9	3.0	2.1	1.1	1.2
Other Products	-	-	-	0.9	0.9	0.7	0.5	0.3	0.3
Refinery Fuel	-	-	-	-	-	-	-	-	-
Sub-Total	-	-	-	48.4	24.7	7.8	5.6	3.1	3.1
Bunkers	-	-	-	-	-	-	-	-	-
Total	-	-	-	48.4	24.7	7.8	5.6	3.1	3.1

Oil Products Consumption
Consommation de produits pétroliers

Azerbaijan

	1971	1973	1978	1983	1984	1985	1986	1987	1988
Thousand tonnes									
NGL/LPG/Ethane	-	-	-	-	-	-	-	-	-
Naphtha	-	-	-	-	-	-	-	-	-
Motor Gasoline	-	-	-	-	-	-	-	-	-
Aviation Fuels	-	-	-	-	-	-	-	-	-
Kerosene	-	-	-	-	-	-	-	-	-
Gas Diesel	-	-	-	-	-	-	-	-	-
Heavy Fuel Oil	-	-	-	-	-	-	-	-	-
Other Products	-	-	-	-	-	-	-	-	-
Refinery Fuel	-	-	-	-	-	-	-	-	-
Sub-Total	-	-	-	-	-	-	-	-	-
Bunkers	-	-	-	-	-	-	-	-	-
Total	-	-	-	-	-	-	-	-	-
Thousand barrels/day									
NGL/LPG/Ethane	-	-	-	-	-	-	-	-	-
Naphtha	-	-	-	-	-	-	-	-	-
Motor Gasoline	-	-	-	-	-	-	-	-	-
Aviation Fuels	-	-	-	-	-	-	-	-	-
Kerosene	-	-	-	-	-	-	-	-	-
Gas Diesel	-	-	-	-	-	-	-	-	-
Heavy Fuel Oil	-	-	-	-	-	-	-	-	-
Other Products	-	-	-	-	-	-	-	-	-
Refinery Fuel	-	-	-	-	-	-	-	-	-
Sub-Total	-	-	-	-	-	-	-	-	-
Bunkers	-	-	-	-	-	-	-	-	-
Total	-	-	-	-	-	-	-	-	-

	1989	1990	1991	1992	1993	1994	1995	1996	1997
Thousand tonnes									
NGL/LPG/Ethane	-	-	-	-	2	5	1	4	3
Naphtha	-	-	-	-	-	-	-	-	-
Motor Gasoline	-	-	-	1 024	1 066	1 269	950	737	492
Aviation Fuels	-	-	-	446	470	419	505	528	451
Kerosene	-	-	-	346	359	434	405	116	150
Gas Diesel	-	-	-	1 854	952	824	588	302	339
Heavy Fuel Oil	-	-	-	3 049	4 996	4 081	4 093	3 927	3 900
Other Products	-	-	-	1 265	165	223	48	83	83
Refinery Fuel	-	-	-	343	270	220	100	100	100
Sub-Total	-	-	-	8 327	8 280	7 475	6 690	5 797	5 518
Bunkers	-	-	-	-	-	-	-	-	-
Total	-	-	-	8 327	8 280	7 475	6 690	5 797	5 518
Thousand barrels/day									
NGL/LPG/Ethane	-	-	-	-	0.1	0.2	0.0	0.1	0.1
Naphtha	-	-	-	-	-	-	-	-	-
Motor Gasoline	-	-	-	23.9	24.9	29.7	22.2	17.2	11.5
Aviation Fuels	-	-	-	9.7	10.2	9.1	11.0	11.4	9.8
Kerosene	-	-	-	7.3	7.6	9.2	8.6	2.5	3.2
Gas Diesel	-	-	-	37.8	19.5	16.8	12.0	6.2	6.9
Heavy Fuel Oil	-	-	-	55.5	91.2	74.5	74.7	71.5	71.2
Other Products	-	-	-	24.1	2.9	4.0	0.8	1.3	1.3
Refinery Fuel	-	-	-	7.5	5.9	4.8	2.2	2.2	2.2
Sub-Total	-	-	-	165.8	162.2	148.3	131.5	112.3	106.2
Bunkers	-	-	-	-	-	-	-	-	-
Total	-	-	-	165.8	162.2	148.3	131.5	112.3	106.2

Oil Products Consumption
Consommation de produits pétroliers

Bahrain

	1971	1973	1978	1983	1984	1985	1986	1987	1988
Thousand tonnes									
NGL/LPG/Ethane	6	4	10	15	17	18	18	18	20
Naphtha	-	-	-	-	-	-	-	-	-
Motor Gasoline	27	37	87	150	161	173	179	186	198
Aviation Fuels	136	204	386	374	392	383	380	390	405
Kerosene	9	8	8	9	10	10	12	16	16
Gas Diesel	30	36	124	121	110	105	88	81	85
Heavy Fuel Oil	-	-	-	-	-	-	-	-	-
Other Products	-	-	24	34	34	36	47	30	20
Refinery Fuel	319	305	307	221	256	232	307	308	258
Sub-Total	527	594	946	924	980	957	1 031	1 029	1 002
Bunkers	1 060	1 016	406	175	191	206	206	239	220
Total	1 587	1 610	1 352	1 099	1 171	1 163	1 237	1 268	1 222
Thousand barrels/day									
NGL/LPG/Ethane	0.2	0.1	0.3	0.5	0.5	0.6	0.6	0.6	0.6
Naphtha	-	-	-	-	-	-	-	-	-
Motor Gasoline	0.6	0.9	2.0	3.5	3.8	4.0	4.2	4.3	4.6
Aviation Fuels	3.0	4.4	8.4	8.1	8.5	8.3	8.3	8.5	8.8
Kerosene	0.2	0.2	0.2	0.2	0.2	0.2	0.3	0.3	0.3
Gas Diesel	0.6	0.7	2.5	2.5	2.2	2.1	1.8	1.7	1.7
Heavy Fuel Oil	-	-	-	-	-	-	-	-	-
Other Products	-	-	0.4	0.7	0.7	0.7	0.9	0.6	0.4
Refinery Fuel	7.0	6.7	6.7	4.8	5.6	5.1	6.7	6.8	5.6
Sub-Total	11.6	13.0	20.6	20.3	21.5	21.1	22.7	22.7	22.1
Bunkers	19.6	18.7	7.6	3.4	3.7	4.0	4.0	4.6	4.2
Total	31.1	31.7	28.1	23.6	25.2	25.0	26.7	27.3	26.4

	1989	1990	1991	1992	1993	1994	1995	1996	1997
Thousand tonnes									
NGL/LPG/Ethane	21	20	22	23	23	24	24	24	24
Naphtha	-	-	-	-	-	-	-	-	-
Motor Gasoline	211	224	230	249	264	274	329	344	521
Aviation Fuels	411	453	333	304	302	346	293	271	252
Kerosene	17	21	23	25	25	27	20	23	23
Gas Diesel	87	95	93	109	100	105	101	97	101
Heavy Fuel Oil	-	-	-	-	-	-	-	-	-
Other Products	20	30	22	25	20	22	29	29	29
Refinery Fuel	100	100	75	50	50	50	50	49	53
Sub-Total	867	943	798	785	784	848	846	837	1 003
Bunkers	210	178	140	109	93	110	110	110	110
Total	1 077	1 121	938	894	877	958	956	947	1 113
Thousand barrels/day									
NGL/LPG/Ethane	0.7	0.6	0.7	0.7	0.7	0.8	0.8	0.8	0.8
Naphtha	-	-	-	-	-	-	-	-	-
Motor Gasoline	4.9	5.2	5.4	5.8	6.2	6.4	7.7	8.0	12.2
Aviation Fuels	8.9	9.8	7.2	6.6	6.6	7.5	6.4	5.9	5.5
Kerosene	0.4	0.4	0.5	0.5	0.5	0.6	0.4	0.5	0.5
Gas Diesel	1.8	1.9	1.9	2.2	2.0	2.1	2.1	2.0	2.1
Heavy Fuel Oil	-	-	-	-	-	-	-	-	-
Other Products	0.4	0.6	0.4	0.5	0.4	0.4	0.6	0.6	0.6
Refinery Fuel	2.2	2.2	1.6	1.1	1.1	1.1	1.1	1.1	1.2
Sub-Total	19.2	20.9	17.8	17.4	17.5	18.9	19.0	18.7	22.7
Bunkers	4.1	3.4	2.7	2.1	1.8	2.1	2.1	2.1	2.1
Total	23.3	24.3	20.4	19.5	19.3	21.0	21.0	20.8	24.8

Oil Products Consumption
Consommation de produits pétroliers

Bangladesh

	1971	1973	1978	1983	1984	1985	1986	1987	1988
Thousand tonnes									
NGL/LPG/Ethane	-	-	-	4	6	7	9	10	9
Naphtha	36	46	15	-	-	-	-	-	-
Motor Gasoline	43	31	53	35	48	49	62	62	72
Aviation Fuels	19	18	33	54	65	70	77	72	81
Kerosene	265	219	382	330	322	351	385	369	384
Gas Diesel	134	197	316	485	506	582	659	685	696
Heavy Fuel Oil	179	249	359	336	395	324	346	296	284
Other Products	17	44	38	31	78	92	79	121	156
Refinery Fuel	43	41	58	54	61	56	56	55	55
Sub-Total	736	845	1 254	1 329	1 481	1 531	1 673	1 670	1 737
Bunkers	21	31	32	25	34	22	12	10	10
Total	757	876	1 286	1 354	1 515	1 553	1 685	1 680	1 747
Thousand barrels/day									
NGL/LPG/Ethane	-	-	-	0.1	0.2	0.2	0.3	0.3	0.3
Naphtha	0.8	1.1	0.3	-	-	-	-	-	-
Motor Gasoline	1.0	0.7	1.2	0.8	1.1	1.1	1.4	1.4	1.7
Aviation Fuels	0.4	0.4	0.7	1.2	1.4	1.5	1.7	1.6	1.8
Kerosene	5.6	4.6	8.1	7.0	6.8	7.4	8.2	7.8	8.1
Gas Diesel	2.7	4.0	6.5	9.9	10.3	11.9	13.5	14.0	14.2
Heavy Fuel Oil	3.3	4.5	6.6	6.1	7.2	5.9	6.3	5.4	5.2
Other Products	0.3	0.9	0.7	0.6	1.5	1.8	1.5	2.3	3.0
Refinery Fuel	0.9	0.8	1.2	1.1	1.2	1.1	1.1	1.1	1.1
Sub-Total	15.1	17.1	25.3	26.8	29.7	31.0	34.0	34.0	35.3
Bunkers	0.4	0.6	0.6	0.5	0.6	0.4	0.2	0.2	0.2
Total	15.5	17.7	25.9	27.3	30.4	31.4	34.2	34.2	35.5

	1989	1990	1991	1992	1993	1994	1995	1996	1997
Thousand tonnes									
NGL/LPG/Ethane	9	9	8	8	9	13	15	15	18
Naphtha	-	-	6	16	28	73	69	72	73
Motor Gasoline	93	98	107	120	138	153	187	195	198
Aviation Fuels	78	86	78	74	77	90	95	99	102
Kerosene	420	475	388	358	375	331	434	453	468
Gas Diesel	856	833	807	920	971	1 016	1 247	1 302	1 309
Heavy Fuel Oil	207	177	129	157	218	242	253	264	240
Other Products	153	223	171	183	200	254	256	256	256
Refinery Fuel	55	50	50	50	62	69	73	62	59
Sub-Total	1 871	1 951	1 744	1 886	2 078	2 241	2 629	2 718	2 723
Bunkers	10	15	27	8	6	12	7	7	7
Total	1 881	1 966	1 771	1 894	2 084	2 253	2 636	2 725	2 730
Thousand barrels/day									
NGL/LPG/Ethane	0.3	0.3	0.3	0.3	0.3	0.4	0.5	0.5	0.6
Naphtha	-	-	0.1	0.4	0.7	1.7	1.6	1.7	1.7
Motor Gasoline	2.2	2.3	2.5	2.8	3.2	3.6	4.4	4.5	4.6
Aviation Fuels	1.7	1.9	1.7	1.6	1.7	2.0	2.1	2.1	2.2
Kerosene	8.9	10.1	8.2	7.6	8.0	7.0	9.2	9.6	9.9
Gas Diesel	17.5	17.0	16.5	18.8	19.8	20.8	25.5	26.5	26.8
Heavy Fuel Oil	3.8	3.2	2.4	2.9	4.0	4.4	4.6	4.8	4.4
Other Products	2.9	4.3	3.3	3.5	3.8	4.9	4.9	4.9	4.9
Refinery Fuel	1.1	1.0	1.0	1.0	1.2	1.4	1.5	1.2	1.2
Sub-Total	38.4	40.1	36.0	38.7	42.7	46.1	54.2	55.9	56.3
Bunkers	0.2	0.3	0.5	0.1	0.1	0.2	0.1	0.1	0.1
Total	38.6	40.4	36.5	38.9	42.8	46.3	54.3	56.0	56.4

Oil Products Consumption
Consommation de produits pétroliers

Belarus

	1971	1973	1978	1983	1984	1985	1986	1987	1988
Thousand tonnes									
NGL/LPG/Ethane	-	-	-	-	-	-	-	-	-
Naphtha	-	-	-	-	-	-	-	-	-
Motor Gasoline	-	-	-	-	-	-	-	-	-
Aviation Fuels	-	-	-	-	-	-	-	-	-
Kerosene	-	-	-	-	-	-	-	-	-
Gas Diesel	-	-	-	-	-	-	-	-	-
Heavy Fuel Oil	-	-	-	-	-	-	-	-	-
Other Products	-	-	-	-	-	-	-	-	-
Refinery Fuel	-	-	-	-	-	-	-	-	-
Sub-Total	-	-	-	-	-	-	-	-	-
Bunkers	-	-	-	-	-	-	-	-	-
Total	-	-	-	-	-	-	-	-	-
Thousand barrels/day									
NGL/LPG/Ethane	-	-	-	-	-	-	-	-	-
Naphtha	-	-	-	-	-	-	-	-	-
Motor Gasoline	-	-	-	-	-	-	-	-	-
Aviation Fuels	-	-	-	-	-	-	-	-	-
Kerosene	-	-	-	-	-	-	-	-	-
Gas Diesel	-	-	-	-	-	-	-	-	-
Heavy Fuel Oil	-	-	-	-	-	-	-	-	-
Other Products	-	-	-	-	-	-	-	-	-
Refinery Fuel	-	-	-	-	-	-	-	-	-
Sub-Total	-	-	-	-	-	-	-	-	-
Bunkers	-	-	-	-	-	-	-	-	-
Total	-	-	-	-	-	-	-	-	-

	1989	1990	1991	1992	1993	1994	1995	1996	1997
Thousand tonnes									
NGL/LPG/Ethane	-	-	-	338	336	359	343	316	315
Naphtha	-	-	-	380	-	-	-	-	-
Motor Gasoline	-	-	-	2 235	1 558	1 094	1 269	1 305	1 247
Aviation Fuels	-	-	-	900	444	444	-	-	-
Kerosene	-	-	-	46	21	30	15	16	24
Gas Diesel	-	-	-	2 893	3 039	1 966	1 805	1 900	1 916
Heavy Fuel Oil	-	-	-	9 027	7 063	6 135	5 178	4 711	3 705
Other Products	-	-	-	4 264	1 519	1 288	1 365	1 189	1 223
Refinery Fuel	-	-	-	800	543	491	499	500	500
Sub-Total	-	-	-	20 883	14 523	11 807	10 474	9 937	8 930
Bunkers	-	-	-	-	-	-	-	-	-
Total	-	-	-	20 883	14 523	11 807	10 474	9 937	8 930
Thousand barrels/day									
NGL/LPG/Ethane	-	-	-	10.7	10.7	11.4	10.9	10.0	10.0
Naphtha	-	-	-	8.8	-	-	-	-	-
Motor Gasoline	-	-	-	52.1	36.4	25.6	29.7	30.4	29.1
Aviation Fuels	-	-	-	19.5	9.6	9.6	-	-	-
Kerosene	-	-	-	1.0	0.4	0.6	0.3	0.3	0.5
Gas Diesel	-	-	-	59.0	62.1	40.2	36.9	38.7	39.2
Heavy Fuel Oil	-	-	-	164.3	128.9	111.9	94.5	85.7	67.6
Other Products	-	-	-	79.5	28.5	24.3	25.7	22.3	22.9
Refinery Fuel	-	-	-	17.5	11.9	10.8	10.9	10.9	11.0
Sub-Total	-	-	-	412.3	288.6	234.4	208.8	198.5	180.2
Bunkers	-	-	-	-	-	-	-	-	-
Total	-	-	-	412.3	288.6	234.4	208.8	198.5	180.2

Oil Products Consumption
Consommation de produits pétroliers

Benin

	1971	1973	1978	1983	1984	1985	1986	1987	1988
Thousand tonnes									
NGL/LPG/Ethane	-	-	-	-	-	-	-	-	-
Naphtha	-	-	-	-	-	-	-	-	-
Motor Gasoline	21	27	37	23	41	60	58	41	45
Aviation Fuels	5	4	10	19	18	19	19	19	16
Kerosene	13	19	27	13	13	18	9	11	12
Gas Diesel	58	79	51	66	54	60	51	52	46
Heavy Fuel Oil	4	4	6	11	10	10	10	8	8
Other Products	-	-	-	-	-	-	-	-	-
Refinery Fuel	-	-	-	-	-	-	-	-	-
Sub-Total	101	133	131	132	136	167	147	131	127
Bunkers	-	-	-	-	-	-	-	-	-
Total	101	133	131	132	136	167	147	131	127
Thousand barrels/day									
NGL/LPG/Ethane	-	-	-	-	-	-	-	-	-
Naphtha	-	-	-	-	-	-	-	-	-
Motor Gasoline	0.5	0.6	0.9	0.5	1.0	1.4	1.4	1.0	1.0
Aviation Fuels	0.1	0.1	0.2	0.4	0.4	0.4	0.4	0.4	0.3
Kerosene	0.3	0.4	0.6	0.3	0.3	0.4	0.2	0.2	0.3
Gas Diesel	1.2	1.6	1.0	1.3	1.1	1.2	1.0	1.1	0.9
Heavy Fuel Oil	0.1	0.1	0.1	0.2	0.2	0.2	0.2	0.1	0.1
Other Products	-	-	-	-	-	-	-	-	-
Refinery Fuel	-	-	-	-	-	-	-	-	-
Sub-Total	2.1	2.8	2.8	2.8	2.9	3.6	3.2	2.8	2.7
Bunkers	-	-	-	-	-	-	-	-	-
Total	2.1	2.8	2.8	2.8	2.9	3.6	3.2	2.8	2.7

	1989	1990	1991	1992	1993	1994	1995	1996	1997
Thousand tonnes									
NGL/LPG/Ethane	-	-	-	-	-	-	-	1	1
Naphtha	-	-	-	-	-	-	-	-	-
Motor Gasoline	23	16	10	10	10	10	10	105	126
Aviation Fuels	20	16	15	14	19	20	21	56	49
Kerosene	11	11	8	12	10	12	13	39	34
Gas Diesel	43	42	37	33	34	33	34	102	85
Heavy Fuel Oil	10	10	7	9	10	11	11	19	19
Other Products	-	-	-	-	-	-	-	-	-
Refinery Fuel	-	-	-	-	-	-	-	-	-
Sub-Total	107	95	77	78	83	86	89	322	314
Bunkers	-	-	-	-	-	-	-	5	5
Total	107	95	77	78	83	86	89	327	319
Thousand barrels/day									
NGL/LPG/Ethane	-	-	-	-	-	-	-	0.0	0.0
Naphtha	-	-	-	-	-	-	-	-	-
Motor Gasoline	0.5	0.4	0.2	0.2	0.2	0.2	0.2	2.4	2.9
Aviation Fuels	0.4	0.3	0.3	0.3	0.4	0.4	0.5	1.2	1.1
Kerosene	0.2	0.2	0.2	0.3	0.2	0.3	0.3	0.8	0.7
Gas Diesel	0.9	0.9	0.8	0.7	0.7	0.7	0.7	2.1	1.7
Heavy Fuel Oil	0.2	0.2	0.1	0.2	0.2	0.2	0.2	0.3	0.3
Other Products	-	-	-	-	-	-	-	-	-
Refinery Fuel	-	-	-	-	-	-	-	-	-
Sub-Total	2.3	2.0	1.6	1.6	1.7	1.8	1.9	6.9	6.8
Bunkers	-	-	-	-	-	-	-	0.1	0.1
Total	2.3	2.0	1.6	1.6	1.7	1.8	1.9	7.0	6.9

Oil Products Consumption
Consommation de produits pétroliers

Bolivia

	1971	1973	1978	1983	1984	1985	1986	1987	1988
Thousand tonnes									
NGL/LPG/Ethane	8	14	60	151	147	154	169	181	172
Naphtha	-	-	-	-	-	-	-	-	-
Motor Gasoline	225	261	319	335	321	326	340	380	379
Aviation Fuels	27	32	95	81	85	82	81	79	78
Kerosene	109	127	141	90	84	73	43	40	33
Gas Diesel	92	118	267	204	220	219	272	263	255
Heavy Fuel Oil	147	157	151	113	75	54	20	9	31
Other Products	6	12	14	16	7	8	8	9	11
Refinery Fuel	10	11	23	67	76	76	80	85	77
Sub-Total	624	732	1 070	1 057	1 015	992	1 013	1 046	1 036
Bunkers	-	-	-	-	-	-	-	-	-
Total	624	732	1 070	1 057	1 015	992	1 013	1 046	1 036
Thousand barrels/day									
NGL/LPG/Ethane	0.3	0.4	1.9	4.8	4.7	4.9	5.4	5.8	5.5
Naphtha	-	-	-	-	-	-	-	-	-
Motor Gasoline	5.3	6.1	7.5	7.8	7.5	7.6	7.9	8.9	8.8
Aviation Fuels	0.6	0.7	2.1	1.8	1.9	1.8	1.8	1.7	1.7
Kerosene	2.3	2.7	3.0	1.9	1.8	1.5	0.9	0.8	0.7
Gas Diesel	1.9	2.4	5.5	4.2	4.5	4.5	5.6	5.4	5.2
Heavy Fuel Oil	2.7	2.9	2.8	2.1	1.4	1.0	0.4	0.2	0.6
Other Products	0.1	0.2	0.3	0.3	0.1	0.1	0.2	0.2	0.2
Refinery Fuel	0.2	0.2	0.5	1.4	1.6	1.6	1.6	1.7	1.7
Sub-Total	13.3	15.7	23.4	24.2	23.3	23.0	23.7	24.7	24.3
Bunkers	-	-	-	-	-	-	-	-	-
Total	13.3	15.7	23.4	24.2	23.3	23.0	23.7	24.7	24.3

	1989	1990	1991	1992	1993	1994	1995	1996	1997
Thousand tonnes									
NGL/LPG/Ethane	167	175	180	190	200	217	235	249	291
Naphtha	-	-	-	-	-	-	-	-	-
Motor Gasoline	392	390	370	360	361	376	401	428	456
Aviation Fuels	86	89	85	87	93	95	108	128	136
Kerosene	37	30	29	23	21	21	24	25	23
Gas Diesel	288	326	376	393	438	462	530	570	534
Heavy Fuel Oil	27	25	23	15	4	3	2	1	1
Other Products	19	14	11	14	7	9	11	11	13
Refinery Fuel	76	82	81	81	87	96	109	105	118
Sub-Total	1 092	1 131	1 155	1 163	1 211	1 279	1 420	1 517	1 572
Bunkers	-	-	-	-	-	-	-	-	-
Total	1 092	1 131	1 155	1 163	1 211	1 279	1 420	1 517	1 572
Thousand barrels/day									
NGL/LPG/Ethane	5.3	5.6	5.7	6.0	6.4	6.9	7.5	7.9	9.2
Naphtha	-	-	-	-	-	-	-	-	-
Motor Gasoline	9.2	9.1	8.6	8.4	8.4	8.8	9.4	10.0	10.7
Aviation Fuels	1.9	2.0	1.9	1.9	2.0	2.1	2.4	2.8	3.0
Kerosene	0.8	0.6	0.6	0.5	0.4	0.4	0.5	0.5	0.5
Gas Diesel	5.9	6.7	7.7	8.0	9.0	9.4	10.8	11.6	10.9
Heavy Fuel Oil	0.5	0.5	0.4	0.3	0.1	0.1	0.0	0.0	0.0
Other Products	0.4	0.3	0.2	0.3	0.1	0.2	0.2	0.2	0.2
Refinery Fuel	1.6	1.8	1.8	1.8	1.9	2.1	2.4	2.3	2.6
Sub-Total	25.5	26.4	26.9	27.1	28.3	30.0	33.1	35.3	37.1
Bunkers	-	-	-	-	-	-	-	-	-
Total	25.5	26.4	26.9	27.1	28.3	30.0	33.1	35.3	37.1

Oil Products Consumption
Consommation de produits pétroliers

Bosnia-Herzegovina

	1971	1973	1978	1983	1984	1985	1986	1987	1988
Thousand tonnes									
NGL/LPG/Ethane	-	-	-	-	-	-	-	-	-
Naphtha	-	-	-	-	-	-	-	-	-
Motor Gasoline	-	-	-	-	-	-	-	-	-
Aviation Fuels	-	-	-	-	-	-	-	-	-
Kerosene	-	-	-	-	-	-	-	-	-
Gas Diesel	-	-	-	-	-	-	-	-	-
Heavy Fuel Oil	-	-	-	-	-	-	-	-	-
Other Products	-	-	-	-	-	-	-	-	-
Refinery Fuel	-	-	-	-	-	-	-	-	-
Sub-Total	-	-	-	-	-	-	-	-	-
Bunkers	-	-	-	-	-	-	-	-	-
Total	-	-	-	-	-	-	-	-	-
Thousand barrels/day									
NGL/LPG/Ethane	-	-	-	-	-	-	-	-	-
Naphtha	-	-	-	-	-	-	-	-	-
Motor Gasoline	-	-	-	-	-	-	-	-	-
Aviation Fuels	-	-	-	-	-	-	-	-	-
Kerosene	-	-	-	-	-	-	-	-	-
Gas Diesel	-	-	-	-	-	-	-	-	-
Heavy Fuel Oil	-	-	-	-	-	-	-	-	-
Other Products	-	-	-	-	-	-	-	-	-
Refinery Fuel	-	-	-	-	-	-	-	-	-
Sub-Total	-	-	-	-	-	-	-	-	-
Bunkers	-	-	-	-	-	-	-	-	-
Total	-	-	-	-	-	-	-	-	-

	1989	1990	1991	1992	1993	1994	1995	1996	1997
Thousand tonnes									
NGL/LPG/Ethane	-	-	-	-	-	-	-	-	-
Naphtha	-	-	-	-	964	298	298	298	298
Motor Gasoline	-	-	-	100	96	115	115	115	115
Aviation Fuels	-	-	-	16	30	36	36	36	36
Kerosene	-	-	-	-	-	-	-	-	-
Gas Diesel	-	-	-	320	224	269	269	269	269
Heavy Fuel Oil	-	-	-	100	80	96	96	96	96
Other Products	-	-	-	-	38	45	45	45	45
Refinery Fuel	-	-	-	-	-	-	-	-	-
Sub-Total	-	-	-	536	1 432	859	859	859	859
Bunkers	-	-	-	-	-	-	-	-	-
Total	-	-	-	536	1 432	859	859	859	859
Thousand barrels/day									
NGL/LPG/Ethane	-	-	-	-	-	-	-	-	-
Naphtha	-	-	-	-	22.4	6.9	6.9	6.9	6.9
Motor Gasoline	-	-	-	2.3	2.2	2.7	2.7	2.7	2.7
Aviation Fuels	-	-	-	0.3	0.7	0.8	0.8	0.8	0.8
Kerosene	-	-	-	-	-	-	-	-	-
Gas Diesel	-	-	-	6.5	4.6	5.5	5.5	5.5	5.5
Heavy Fuel Oil	-	-	-	1.8	1.5	1.8	1.8	1.7	1.8
Other Products	-	-	-	-	0.7	0.9	0.9	0.9	0.9
Refinery Fuel	-	-	-	-	-	-	-	-	-
Sub-Total	-	-	-	11.0	32.1	18.5	18.5	18.5	18.5
Bunkers	-	-	-	-	-	-	-	-	-
Total	-	-	-	11.0	32.1	18.5	18.5	18.5	18.5

Oil Products Consumption
Consommation de produits pétroliers

Brazil

	1971	1973	1978	1983	1984	1985	1986	1987	1988
Thousand tonnes									
NGL/LPG/Ethane	1 310	1 608	2 283	3 307	3 280	3 531	3 902	4 208	4 296
Naphtha	103	1 022	1 724	3 009	3 080	3 803	3 798	4 319	4 087
Motor Gasoline	6 724	8 817	9 717	9 573	10 171	11 296	13 971	13 413	13 895
Aviation Fuels	720	1 035	1 463	1 752	1 648	1 729	2 149	2 261	1 724
Kerosene	525	549	591	514	343	313	330	232	308
Gas Diesel	5 983	8 078	13 384	15 128	15 547	16 578	18 411	19 387	18 897
Heavy Fuel Oil	8 382	10 559	14 942	7 621	7 084	7 191	8 683	8 614	8 512
Other Products	1 254	1 667	3 655	3 668	3 981	4 924	4 805	4 523	4 091
Refinery Fuel	1 010	1 479	1 885	1 695	1 779	1 753	1 837	1 966	3 268
Sub-Total	26 011	34 814	49 644	46 267	46 913	51 118	57 886	58 923	59 078
Bunkers	-	-	1 892	2 935	3 206	3 003	2 617	2 561	2 541
Total	26 011	34 814	51 536	49 202	50 119	54 121	60 503	61 484	61 619
Thousand barrels/day									
NGL/LPG/Ethane	41.6	51.1	72.6	105.1	104.0	112.2	124.0	133.7	136.2
Naphtha	2.4	23.8	40.1	70.1	71.5	88.6	88.4	100.6	94.9
Motor Gasoline	156.7	205.6	224.6	215.6	226.6	251.1	309.5	296.2	305.6
Aviation Fuels	15.8	22.7	32.0	38.2	35.8	37.7	46.8	49.3	37.5
Kerosene	11.1	11.6	12.5	10.9	7.3	6.6	7.0	4.9	6.5
Gas Diesel	122.3	165.1	273.5	309.2	316.9	338.8	376.3	396.2	385.2
Heavy Fuel Oil	152.9	192.7	272.6	139.1	128.9	131.2	158.4	157.2	154.9
Other Products	24.0	32.0	70.1	70.3	76.1	94.4	92.2	86.7	78.2
Refinery Fuel	18.5	27.2	34.7	31.9	33.1	32.9	34.4	36.9	61.8
Sub-Total	545.5	731.8	1 032.8	990.3	1 000.2	1 093.6	1 237.1	1 261.7	1 260.7
Bunkers	-	-	34.9	54.1	59.0	55.4	48.2	47.2	46.6
Total	545.5	731.8	1 067.7	1 044.4	1 059.2	1 149.0	1 285.3	1 308.9	1 307.4

	1989	1990	1991	1992	1993	1994	1995	1996	1997
Thousand tonnes									
NGL/LPG/Ethane	4 496	4 703	4 669	4 986	4 962	5 095	5 398	5 697	5 910
Naphtha	4 355	4 363	4 242	4 581	4 733	5 325	5 173	4 998	6 141
Motor Gasoline	15 207	14 952	15 837	15 585	16 307	17 609	19 435	21 401	21 935
Aviation Fuels	1 819	1 722	1 803	1 696	1 791	1 837	2 135	2 279	2 565
Kerosene	294	261	230	188	166	141	129	113	69
Gas Diesel	19 363	19 011	19 774	20 269	20 998	22 012	23 579	24 421	25 920
Heavy Fuel Oil	8 005	7 819	7 683	8 275	8 629	8 688	9 216	10 552	10 880
Other Products	3 966	4 384	3 849	4 027	4 012	4 578	4 738	5 077	5 474
Refinery Fuel	3 301	3 220	2 939	3 045	3 218	3 431	3 276	3 532	4 016
Sub-Total	60 806	60 435	61 026	62 652	64 816	68 716	73 079	78 070	82 910
Bunkers	1 800	1 600	1 770	1 784	2 394	2 471	1 850	1 352	1 708
Total	62 606	62 035	62 796	64 436	67 210	71 187	74 929	79 422	84 618
Thousand barrels/day									
NGL/LPG/Ethane	142.9	149.5	148.4	158.0	157.7	161.9	171.6	180.6	187.8
Naphtha	101.4	101.6	98.8	106.4	110.2	124.0	120.5	116.1	143.0
Motor Gasoline	335.6	331.6	351.5	345.1	362.1	391.3	433.3	477.1	491.7
Aviation Fuels	39.7	37.5	39.3	36.8	39.0	40.0	46.5	49.5	55.9
Kerosene	6.2	5.5	4.9	4.0	3.5	3.0	2.7	2.4	1.5
Gas Diesel	395.7	388.6	404.1	413.1	429.2	449.9	481.9	497.8	529.8
Heavy Fuel Oil	146.1	142.7	140.2	150.6	157.5	158.5	168.2	192.0	198.5
Other Products	76.1	84.1	73.8	77.0	76.9	87.8	90.9	97.1	105.0
Refinery Fuel	62.8	61.1	56.2	58.1	61.2	64.9	61.7	66.6	75.9
Sub-Total	1 306.4	1 302.1	1 317.2	1 349.2	1 397.4	1 481.4	1 577.2	1 679.1	1 789.0
Bunkers	33.2	29.6	32.6	33.0	45.0	46.4	34.3	25.2	32.1
Total	1 339.7	1 331.7	1 349.9	1 382.1	1 442.3	1 527.8	1 611.5	1 704.3	1 821.1

Oil Products Consumption
Consommation de produits pétroliers

Brunei

	1971	1973	1978	1983	1984	1985	1986	1987	1988
Thousand tonnes									
NGL/LPG/Ethane	5	5	5	7	8	9	9	9	10
Naphtha	1	5	4	4	4	5	4	4	4
Motor Gasoline	26	27	54	92	102	113	120	126	131
Aviation Fuels	1	6	17	15	18	14	23	25	30
Kerosene	1	1	1	5	5	3	3	3	3
Gas Diesel	22	17	14	80	60	65	59	61	72
Heavy Fuel Oil	-	-	-	1	-	1	1	1	1
Other Products	6	6	3	13	9	11	9	9	11
Refinery Fuel	-	-	-	-	1	1	1	1	1
Sub-Total	62	67	98	217	207	222	229	239	263
Bunkers	-	-	-	-	-	-	-	-	-
Total	62	67	98	217	207	222	229	239	263
Thousand barrels/day									
NGL/LPG/Ethane	0.2	0.2	0.2	0.2	0.3	0.3	0.3	0.3	0.3
Naphtha	0.0	0.1	0.1	0.1	0.1	0.1	0.1	0.1	0.1
Motor Gasoline	0.6	0.6	1.3	2.2	2.4	2.6	2.8	2.9	3.1
Aviation Fuels	0.0	0.1	0.4	0.3	0.4	0.3	0.5	0.5	0.7
Kerosene	0.0	0.0	0.0	0.1	0.1	0.1	0.1	0.1	0.1
Gas Diesel	0.5	0.3	0.3	1.6	1.2	1.3	1.2	1.2	1.5
Heavy Fuel Oil	-	-	-	0.0	-	0.0	0.0	0.0	0.0
Other Products	0.1	0.1	0.1	0.2	0.2	0.2	0.2	0.2	0.2
Refinery Fuel	-	-	-	-	0.0	0.0	0.0	0.0	0.0
Sub-Total	1.4	1.5	2.2	4.8	4.6	5.0	5.2	5.4	5.9
Bunkers	-	-	-	-	-	-	-	-	-
Total	1.4	1.5	2.2	4.8	4.6	5.0	5.2	5.4	5.9

	1989	1990	1991	1992	1993	1994	1995	1996	1997
Thousand tonnes									
NGL/LPG/Ethane	10	10	11	12	13	14	16	17	18
Naphtha	5	6	4	4	5	6	6	6	6
Motor Gasoline	134	140	151	157	164	176	187	195	200
Aviation Fuels	29	33	37	45	50	56	63	72	85
Kerosene	4	3	3	3	5	5	5	5	3
Gas Diesel	78	80	87	110	114	120	130	150	173
Heavy Fuel Oil	1	1	1	1	1	1	1	1	1
Other Products	14	11	8	18	22	20	27	27	24
Refinery Fuel	1	1	8	25	24	27	48	55	59
Sub-Total	276	285	310	375	398	425	483	528	569
Bunkers	-	-	-	-	-	-	-	-	-
Total	276	285	310	375	398	425	483	528	569
Thousand barrels/day									
NGL/LPG/Ethane	0.3	0.3	0.4	0.4	0.4	0.4	0.5	0.5	0.6
Naphtha	0.1	0.1	0.1	0.1	0.1	0.1	0.1	0.1	0.1
Motor Gasoline	3.1	3.3	3.5	3.7	3.8	4.1	4.4	4.5	4.7
Aviation Fuels	0.6	0.7	0.8	1.0	1.1	1.2	1.4	1.6	1.8
Kerosene	0.1	0.1	0.1	0.1	0.1	0.1	0.1	0.1	0.1
Gas Diesel	1.6	1.6	1.8	2.2	2.3	2.5	2.7	3.1	3.5
Heavy Fuel Oil	0.0	0.0	0.0	0.0	0.0	0.0	0.0	0.0	0.0
Other Products	0.2	0.2	0.1	0.3	0.4	0.3	0.5	0.5	0.4
Refinery Fuel	0.0	0.0	0.2	0.5	0.5	0.6	1.1	1.2	1.3
Sub-Total	6.2	6.4	6.9	8.3	8.8	9.4	10.7	11.6	12.5
Bunkers	-	-	-	-	-	-	-	-	-
Total	6.2	6.4	6.9	8.3	8.8	9.4	10.7	11.6	12.5

Oil Products Consumption
Consommation de produits pétroliers

Bulgaria

	1971	1973	1978	1983	1984	1985	1986	1987	1988
Thousand tonnes									
NGL/LPG/Ethane	7	23	29	155	156	150	155	173	168
Naphtha	-	-	-	-	-	-	-	-	-
Motor Gasoline	1 352	1 325	1 320	1 300	1 257	1 213	1 210	1 218	1 383
Aviation Fuels	200	200	300	300	337	366	362	370	416
Kerosene	143	155	200	220	220	220	220	220	220
Gas Diesel	2 527	3 000	4 000	3 000	2 751	2 572	2 881	2 643	2 801
Heavy Fuel Oil	4 983	5 290	6 175	4 710	4 187	4 245	4 383	4 176	4 465
Other Products	282	366	666	709	763	732	789	757	774
Refinery Fuel	395	490	625	640	635	635	635	635	600
Sub-Total	9 889	10 849	13 315	11 034	10 306	10 133	10 635	10 192	10 827
Bunkers	-	-	-	-	261	229	286	299	303
Total	9 889	10 849	13 315	11 034	10 567	10 362	10 921	10 491	11 130
Thousand barrels/day									
NGL/LPG/Ethane	0.2	0.7	0.9	4.9	4.9	4.8	4.9	5.5	5.3
Naphtha	-	-	-	-	-	-	-	-	-
Motor Gasoline	31.6	31.0	30.8	30.4	29.3	28.3	28.3	28.5	32.2
Aviation Fuels	4.3	4.3	6.5	6.5	7.3	8.0	7.9	8.0	9.0
Kerosene	3.0	3.3	4.2	4.7	4.7	4.7	4.7	4.7	4.7
Gas Diesel	51.6	61.3	81.8	61.3	56.1	52.6	58.9	54.0	57.1
Heavy Fuel Oil	90.9	96.5	112.7	85.9	76.2	77.5	80.0	76.2	81.2
Other Products	5.1	6.4	11.4	12.3	13.3	12.8	13.8	13.2	13.5
Refinery Fuel	7.2	8.9	11.4	11.7	11.6	11.6	11.6	11.6	10.9
Sub-Total	194.0	212.5	259.8	217.7	203.3	200.1	210.0	201.7	214.0
Bunkers	-	-	-	-	5.2	4.6	5.6	5.9	5.9
Total	194.0	212.5	259.8	217.7	208.4	204.7	215.6	207.5	219.9

	1989	1990	1991	1992	1993	1994	1995	1996	1997
Thousand tonnes									
NGL/LPG/Ethane	170	174	148	32	51	68	70	67	87
Naphtha	-	-	-	520	584	484	595	678	673
Motor Gasoline	1 533	1 332	673	830	978	984	1 083	932	609
Aviation Fuels	444	293	187	308	398	329	324	220	227
Kerosene	165	145	75	2	2	2	1	-	-
Gas Diesel	2 697	2 063	1 281	1 394	1 432	1 180	1 087	1 169	1 016
Heavy Fuel Oil	4 178	3 347	2 576	2 319	2 099	2 120	1 762	1 614	1 507
Other Products	748	678	311	131	156	129	112	374	277
Refinery Fuel	600	468	333	156	216	219	272	250	184
Sub-Total	10 535	8 500	5 584	5 692	5 916	5 515	5 306	5 304	4 580
Bunkers	310	326	302	272	263	265	275	239	340
Total	10 845	8 826	5 886	5 964	6 179	5 780	5 581	5 543	4 920
Thousand barrels/day									
NGL/LPG/Ethane	5.4	5.5	4.7	1.0	1.6	2.2	2.2	2.1	2.8
Naphtha	-	-	-	12.1	13.6	11.3	13.9	15.7	15.7
Motor Gasoline	35.8	31.1	15.7	19.3	22.9	23.0	25.3	21.7	14.2
Aviation Fuels	9.6	6.4	4.1	6.7	8.7	7.2	7.0	4.8	4.9
Kerosene	3.5	3.1	1.6	0.0	0.0	0.0	0.0	-	-
Gas Diesel	55.1	42.2	26.2	28.4	29.3	24.1	22.2	23.8	20.8
Heavy Fuel Oil	76.2	61.1	47.0	42.2	38.3	38.7	32.2	29.4	27.5
Other Products	13.0	12.6	5.8	2.7	3.2	2.6	2.3	7.3	5.4
Refinery Fuel	10.9	9.2	6.6	3.1	4.4	4.5	5.5	4.9	3.7
Sub-Total	209.7	171.0	111.7	115.5	121.9	113.5	110.6	109.7	95.0
Bunkers	6.0	6.3	5.7	5.1	4.9	5.0	5.2	4.5	6.4
Total	215.7	177.4	117.4	120.6	126.8	118.5	115.8	114.2	101.4

Oil Products Consumption
Consommation de produits pétroliers

Cameroon

	1971	1973	1978	1983	1984	1985	1986	1987	1988
Thousand tonnes									
NGL/LPG/Ethane	2	2	4	11	13	16	21	21	22
Naphtha	-	-	-	-	-	-	-	-	-
Motor Gasoline	93	101	142	212	230	267	270	299	272
Aviation Fuels	53	61	59	48	48	48	48	48	50
Kerosene	16	12	51	80	87	106	105	104	111
Gas Diesel	96	115	204	292	299	312	270	244	332
Heavy Fuel Oil	26	23	52	57	60	68	68	56	54
Other Products	-	-	19	126	151	67	67	67	39
Refinery Fuel	-	-	-	39	41	46	47	46	-
Sub-Total	286	314	531	865	929	930	896	885	880
Bunkers	-	-	-	4	7	9	5	3	11
Total	286	314	531	869	936	939	901	888	891
Thousand barrels/day									
NGL/LPG/Ethane	0.1	0.1	0.1	0.4	0.4	0.5	0.7	0.7	0.7
Naphtha	-	-	-	-	-	-	-	-	-
Motor Gasoline	2.2	2.4	3.3	5.0	5.4	6.2	6.3	7.0	6.3
Aviation Fuels	1.2	1.3	1.3	1.0	1.0	1.0	1.0	1.0	1.1
Kerosene	0.3	0.3	1.1	1.7	1.8	2.2	2.2	2.2	2.3
Gas Diesel	2.0	2.4	4.2	6.0	6.1	6.4	5.5	5.0	6.8
Heavy Fuel Oil	0.5	0.4	0.9	1.0	1.1	1.2	1.2	1.0	1.0
Other Products	-	-	0.4	2.4	2.9	1.3	1.3	1.3	0.7
Refinery Fuel	-	-	-	0.7	0.7	0.8	0.9	0.8	-
Sub-Total	6.2	6.8	11.3	18.2	19.5	19.8	19.1	19.0	18.9
Bunkers	-	-	-	0.1	0.1	0.2	0.1	0.1	0.2
Total	6.2	6.8	11.3	18.3	19.6	20.0	19.3	19.1	19.2

	1989	1990	1991	1992	1993	1994	1995	1996	1997
Thousand tonnes									
NGL/LPG/Ethane	23	23	19	18	13	11	10	18	24
Naphtha	-	-	-	-	-	-	-	-	-
Motor Gasoline	242	218	189	206	241	249	231	234	233
Aviation Fuels	51	49	48	47	39	58	53	56	51
Kerosene	192	194	196	147	105	106	114	100	97
Gas Diesel	349	347	317	300	301	317	318	318	331
Heavy Fuel Oil	58	54	51	45	46	43	48	51	49
Other Products	45	45	42	40	45	40	45	31	31
Refinery Fuel	-	-	-	-	-	-	-	-	-
Sub-Total	960	930	862	803	790	824	819	808	816
Bunkers	10	13	23	19	15	22	28	44	53
Total	970	943	885	822	805	846	847	852	869
Thousand barrels/day									
NGL/LPG/Ethane	0.7	0.7	0.6	0.6	0.4	0.4	0.3	0.6	0.8
Naphtha	-	-	-	-	-	-	-	-	-
Motor Gasoline	5.7	5.1	4.4	4.8	5.6	5.8	5.4	5.5	5.4
Aviation Fuels	1.1	1.1	1.0	1.0	0.8	1.3	1.2	1.2	1.1
Kerosene	4.1	4.1	4.2	3.1	2.2	2.2	2.4	2.1	2.1
Gas Diesel	7.1	7.1	6.5	6.1	6.2	6.5	6.5	6.5	6.8
Heavy Fuel Oil	1.1	1.0	0.9	0.8	0.8	0.8	0.9	0.9	0.9
Other Products	0.8	0.8	0.8	0.7	0.8	0.8	0.8	0.6	0.6
Refinery Fuel	-	-	-	-	-	-	-	-	-
Sub-Total	20.6	19.9	18.4	17.2	16.9	17.7	17.5	17.3	17.6
Bunkers	0.2	0.3	0.5	0.4	0.3	0.4	0.6	0.9	1.1
Total	20.8	20.2	18.9	17.6	17.3	18.1	18.1	18.2	18.7

Oil Products Consumption
Consommation de produits pétroliers

Chile

	1971	1973	1978	1983	1984	1985	1986	1987	1988
Thousand tonnes									
NGL/LPG/Ethane	347	402	461	430	436	426	434	470	510
Naphtha	45	60	51	34	43	33	32	35	36
Motor Gasoline	1 348	1 197	1 009	1 087	1 058	998	1 038	1 094	1 220
Aviation Fuels	138	123	165	155	161	156	196	214	217
Kerosene	417	512	320	157	149	116	147	171	188
Gas Diesel	752	785	1 044	1 337	1 383	1 416	1 534	1 600	1 815
Heavy Fuel Oil	1 539	1 434	1 411	1 078	1 011	943	935	880	976
Other Products	26	10	18	11	11	8	17	20	20
Refinery Fuel	286	320	374	292	285	290	288	324	349
Sub-Total	4 898	4 843	4 853	4 581	4 537	4 386	4 621	4 808	5 331
Bunkers	195	189	86	32	23	29	46	43	51
Total	5 093	5 032	4 939	4 613	4 560	4 415	4 667	4 851	5 382
Thousand barrels/day									
NGL/LPG/Ethane	11.0	12.8	14.7	13.7	13.8	13.5	13.8	14.9	16.2
Naphtha	1.0	1.4	1.2	0.8	1.0	0.8	0.7	0.8	0.8
Motor Gasoline	31.5	28.0	23.6	25.4	24.7	23.3	24.3	25.6	28.4
Aviation Fuels	3.1	2.7	3.6	3.4	3.5	3.4	4.3	4.7	4.7
Kerosene	8.8	10.9	6.8	3.3	3.2	2.5	3.1	3.6	4.0
Gas Diesel	15.4	16.0	21.3	27.3	28.2	28.9	31.4	32.7	37.0
Heavy Fuel Oil	28.1	26.2	25.7	19.7	18.4	17.2	17.1	16.1	17.8
Other Products	0.5	0.2	0.3	0.2	0.2	0.2	0.3	0.4	0.4
Refinery Fuel	5.7	6.4	7.7	6.0	5.8	5.9	5.9	6.6	7.1
Sub-Total	105.1	104.6	104.9	99.7	98.7	95.6	100.8	105.3	116.3
Bunkers	3.6	3.4	1.6	0.6	0.4	0.5	0.8	0.8	0.9
Total	108.7	108.0	106.5	100.3	99.1	96.2	101.6	106.1	117.3

	1989	1990	1991	1992	1993	1994	1995	1996	1997
Thousand tonnes									
NGL/LPG/Ethane	545	567	606	673	737	754	828	898	937
Naphtha	38	41	43	48	53	40	47	62	54
Motor Gasoline	1 329	1 374	1 422	1 567	1 653	1 857	2 009	2 151	2 237
Aviation Fuels	236	283	273	318	363	348	401	441	557
Kerosene	231	177	221	263	278	273	271	320	319
Gas Diesel	2 075	2 250	2 301	2 357	2 665	2 911	3 215	3 494	3 968
Heavy Fuel Oil	1 112	1 108	1 122	1 238	1 352	1 441	1 608	1 694	1 752
Other Products	30	30	30	30	30	-	-	-	-
Refinery Fuel	379	401	421	463	453	482	524	672	646
Sub-Total	5 975	6 231	6 439	6 957	7 584	8 106	8 903	9 732	10 470
Bunkers	117	183	185	227	297	316	383	254	308
Total	6 092	6 414	6 624	7 184	7 881	8 422	9 286	9 986	10 778
Thousand barrels/day									
NGL/LPG/Ethane	17.3	18.0	19.3	21.3	23.4	24.0	26.3	28.5	29.8
Naphtha	0.9	1.0	1.0	1.1	1.2	0.9	1.1	1.4	1.3
Motor Gasoline	31.1	32.1	33.2	36.5	38.6	43.4	47.0	50.1	52.3
Aviation Fuels	5.1	6.2	5.9	6.9	7.9	7.6	8.7	9.6	12.1
Kerosene	4.9	3.8	4.7	5.6	5.9	5.8	5.7	6.8	6.8
Gas Diesel	42.4	46.0	47.0	48.0	54.5	59.5	65.7	71.2	81.1
Heavy Fuel Oil	20.3	20.2	20.5	22.5	24.7	26.3	29.3	30.8	32.0
Other Products	0.6	0.6	0.6	0.6	0.6	-	-	-	-
Refinery Fuel	7.7	8.2	8.6	9.5	9.4	10.0	11.0	14.1	13.9
Sub-Total	130.3	136.0	140.8	152.0	166.2	177.5	194.9	212.5	229.1
Bunkers	2.1	3.3	3.4	4.1	5.4	5.8	7.0	4.6	5.6
Total	132.4	139.3	144.2	156.2	171.6	183.2	201.9	217.2	234.7

Oil Products Consumption
Consommation de produits pétroliers

People's Republic of China

	1971	1973	1978	1983	1984	1985	1986	1987	1988
Thousand tonnes									
NGL/LPG/Ethane	-	-	800	1 499	1 530	1 597	2 020	2 153	2 318
Naphtha	-	-	1 040	3 110	3 240	4 920	3 830	5 560	7 020
Motor Gasoline	4 465	6 110	9 562	11 070	12 193	13 996	15 002	16 241	17 965
Aviation Fuels	-	-	-	480	578	632	600	747	756
Kerosene	3 000	3 000	3 335	3 261	3 314	3 282	3 324	3 139	2 895
Gas Diesel	9 445	12 580	17 335	16 472	17 527	19 323	21 120	22 974	25 562
Heavy Fuel Oil	11 055	15 120	24 418	26 324	26 569	25 848	27 219	28 195	29 181
Other Products	9 392	12 782	30 895	18 498	18 585	17 045	18 748	18 665	18 126
Refinery Fuel	1 600	1 800	4 500	3 254	3 251	4 707	5 295	5 541	5 787
Sub-Total	38 957	51 392	91 885	83 968	86 787	91 350	97 158	103 215	109 610
Bunkers	-	-	-	-	-	-	-	-	-
Total	38 957	51 392	91 885	83 968	86 787	91 350	97 158	103 215	109 610
Thousand barrels/day									
NGL/LPG/Ethane	-	-	25.4	47.6	48.5	50.8	64.2	68.4	73.5
Naphtha	-	-	24.2	72.4	75.2	114.6	89.2	129.5	163.0
Motor Gasoline	104.3	142.8	223.5	258.7	284.2	327.1	350.6	379.6	418.7
Aviation Fuels	-	-	-	10.6	12.7	13.9	13.2	16.4	16.6
Kerosene	63.6	63.6	70.7	69.2	70.1	69.6	70.5	66.6	61.2
Gas Diesel	193.0	257.1	354.3	336.7	357.2	394.9	431.7	469.6	521.0
Heavy Fuel Oil	201.7	275.9	445.5	480.3	483.5	471.6	496.7	514.5	531.0
Other Products	181.6	249.0	609.8	356.7	355.7	326.9	359.3	357.5	346.4
Refinery Fuel	31.4	35.8	85.8	63.0	62.9	93.9	105.5	110.4	114.8
Sub-Total	775.7	1 024.2	1 839.3	1 695.2	1 749.9	1 863.2	1 980.8	2 112.3	2 246.2
Bunkers	-	-	-	-	-	-	-	-	-
Total	775.7	1 024.2	1 839.3	1 695.2	1 749.9	1 863.2	1 980.8	2 112.3	2 246.2

	1989	1990	1991	1992	1993	1994	1995	1996	1997
Thousand tonnes									
NGL/LPG/Ethane	2 515	2 542	3 009	3 456	4 234	4 807	6 773	8 317	8 801
Naphtha	7 915	8 432	9 631	9 874	11 487	12 674	14 447	15 892	17 073
Motor Gasoline	18 568	18 841	22 095	24 457	30 661	26 746	28 939	31 415	32 930
Aviation Fuels	797	1 011	1 402	1 489	1 761	2 000	2 751	2 856	3 435
Kerosene	2 885	2 574	2 509	2 507	2 479	2 518	2 693	3 069	3 027
Gas Diesel	27 330	26 785	29 806	31 898	38 034	36 566	42 867	46 594	53 940
Heavy Fuel Oil	29 310	29 336	29 305	30 956	33 965	31 518	33 114	35 514	36 019
Other Products	17 644	15 864	14 706	15 245	13 241	20 067	17 276	22 768	28 710
Refinery Fuel	7 166	7 152	7 134	7 880	7 045	6 322	7 763	8 143	10 524
Sub-Total	114 130	112 537	119 597	127 762	142 907	143 218	156 623	174 568	194 459
Bunkers	-	-	-	-	2 631	2 631	872	972	952
Total	114 130	112 537	119 597	127 762	145 538	145 849	157 495	175 540	195 411
Thousand barrels/day									
NGL/LPG/Ethane	79.9	80.8	95.6	109.5	134.6	152.8	215.3	263.6	279.7
Naphtha	184.3	196.4	224.3	229.3	267.5	295.1	336.4	369.1	397.6
Motor Gasoline	433.9	440.3	516.4	570.0	716.5	625.1	676.3	732.2	769.6
Aviation Fuels	17.5	22.2	30.7	32.5	38.3	43.5	60.0	62.2	75.0
Kerosene	61.2	54.6	53.2	53.0	52.6	53.4	57.1	64.9	64.2
Gas Diesel	558.6	547.4	609.2	650.2	777.4	747.3	876.1	949.7	1 102.4
Heavy Fuel Oil	534.8	535.3	534.7	563.3	619.7	575.1	604.2	646.2	657.2
Other Products	338.0	302.5	275.5	284.4	263.0	380.3	321.7	445.9	562.5
Refinery Fuel	140.5	140.8	140.4	155.0	150.5	136.0	162.2	174.1	226.5
Sub-Total	2 348.7	2 320.2	2 480.0	2 647.3	3 020.0	3 008.6	3 309.4	3 707.9	4 134.7
Bunkers	-	-	-	-	50.6	50.6	17.1	19.9	19.5
Total	2 348.7	2 320.2	2 480.0	2 647.3	3 070.6	3 059.2	3 326.6	3 727.8	4 154.2

Oil Products Consumption
Consommation de produits pétroliers

Chinese Taipei

	1971	1973	1978	1983	1984	1985	1986	1987	1988
Thousand tonnes									
NGL/LPG/Ethane	167	336	593	788	848	909	1 043	1 115	1 217
Naphtha	-	-	1 057	1 601	2 079	2 412	2 731	2 654	2 724
Motor Gasoline	488	687	1 221	1 832	1 999	2 084	2 269	2 631	3 116
Aviation Fuels	469	879	644	616	500	553	575	661	654
Kerosene	25	28	32	51	63	80	106	88	91
Gas Diesel	627	974	2 013	2 485	2 555	2 595	2 786	3 007	3 226
Heavy Fuel Oil	4 141	6 138	10 712	8 237	7 118	5 853	6 682	6 592	8 857
Other Products	415	470	494	619	672	998	1 027	1 233	1 317
Refinery Fuel	327	420	960	1 571	1 706	1 773	1 877	1 872	1 745
Sub-Total	6 659	9 932	17 726	17 800	17 540	17 257	19 096	19 853	22 947
Bunkers	123	401	554	279	583	548	893	983	1 259
Total	6 782	10 333	18 280	18 079	18 123	17 805	19 989	20 836	24 206
Thousand barrels/day									
NGL/LPG/Ethane	5.3	10.7	18.8	24.3	26.1	28.0	32.1	34.4	37.4
Naphtha	-	-	24.6	37.3	48.3	56.2	63.6	61.8	63.3
Motor Gasoline	11.4	16.1	28.5	42.8	46.6	48.7	53.0	61.5	72.6
Aviation Fuels	10.3	19.2	14.1	13.5	10.9	12.1	12.6	14.4	14.2
Kerosene	0.5	0.6	0.7	1.1	1.3	1.7	2.2	1.9	1.9
Gas Diesel	12.8	19.9	41.1	50.8	52.1	53.0	56.9	61.5	65.8
Heavy Fuel Oil	75.6	112.0	195.5	150.3	129.5	106.8	121.9	120.3	161.2
Other Products	8.0	9.0	8.6	11.1	12.1	18.2	18.8	22.8	24.2
Refinery Fuel	6.1	7.8	17.7	31.4	34.3	35.9	38.0	37.8	35.1
Sub-Total	129.9	195.2	349.6	362.6	361.1	360.6	399.2	416.3	475.6
Bunkers	2.5	8.2	11.3	5.7	10.8	10.1	16.4	18.1	23.0
Total	132.4	203.4	360.9	368.3	371.9	370.7	415.6	434.4	498.7

	1989	1990	1991	1992	1993	1994	1995	1996	1997
Thousand tonnes									
NGL/LPG/Ethane	1 295	1 331	1 312	1 388	1 401	1 443	1 481	1 509	1 526
Naphtha	2 688	2 690	2 465	2 426	2 581	3 443	3 429	3 568	3 851
Motor Gasoline	3 631	4 051	4 354	4 819	5 191	5 556	5 889	6 197	6 314
Aviation Fuels	782	900	913	1 145	1 151	1 462	1 703	1 838	1 873
Kerosene	70	39	32	29	39	43	31	40	29
Gas Diesel	3 549	3 752	3 950	4 018	4 342	4 750	4 764	4 617	4 243
Heavy Fuel Oil	10 078	10 276	11 150	10 648	11 403	11 338	12 577	12 063	12 848
Other Products	1 585	1 661	1 704	1 715	1 843	1 953	1 999	2 222	2 292
Refinery Fuel	1 666	1 513	1 099	1 074	1 201	1 403	1 490	1 541	1 586
Sub-Total	25 344	26 213	26 979	27 262	29 152	31 391	33 363	33 595	34 562
Bunkers	1 660	1 566	1 481	2 036	1 940	2 136	2 438	2 364	2 816
Total	27 004	27 779	28 460	29 298	31 092	33 527	35 801	35 959	37 378
Thousand barrels/day									
NGL/LPG/Ethane	39.9	41.0	40.4	42.7	43.2	44.5	46.5	47.2	47.9
Naphtha	62.6	62.6	57.4	56.3	60.1	80.2	79.9	82.9	89.7
Motor Gasoline	84.9	94.7	101.8	112.3	121.3	129.8	137.6	144.4	147.6
Aviation Fuels	17.0	19.6	19.9	24.8	25.0	31.8	37.0	39.8	40.7
Kerosene	1.5	0.8	0.7	0.6	0.8	0.9	0.7	0.8	0.6
Gas Diesel	72.5	76.7	80.7	81.9	88.7	97.1	97.4	94.1	86.7
Heavy Fuel Oil	183.9	187.5	203.4	193.8	208.1	206.9	229.5	219.5	234.4
Other Products	29.2	30.7	31.4	31.3	33.5	35.7	36.6	40.6	42.0
Refinery Fuel	33.7	31.3	22.9	22.3	25.1	29.2	30.8	32.0	33.0
Sub-Total	525.2	544.9	558.6	566.0	605.9	656.0	695.9	701.4	722.6
Bunkers	30.6	28.8	27.2	37.3	35.7	39.2	44.8	43.6	51.8
Total	555.8	573.7	585.8	603.3	641.6	695.3	740.6	745.0	774.4

Oil Products Consumption
Consommation de produits pétroliers

Colombia

	1971	1973	1978	1983	1984	1985	1986	1987	1988
Thousand tonnes									
NGL/LPG/Ethane	270	338	224	345	389	393	411	365	379
Naphtha	246	222	185	223	215	210	210	210	190
Motor Gasoline	1 987	2 254	3 065	3 594	3 732	3 861	3 925	4 159	4 448
Aviation Fuels	247	284	432	519	452	445	405	433	497
Kerosene	433	410	375	238	220	225	206	285	231
Gas Diesel	888	905	1 032	1 373	1 363	1 374	1 353	1 528	1 612
Heavy Fuel Oil	1 448	1 549	1 421	917	748	668	630	673	202
Other Products	263	240	227	441	477	497	551	765	1 272
Refinery Fuel	302	250	336	531	413	422	294	251	314
Sub-Total	6 084	6 452	7 297	8 181	8 009	8 095	7 985	8 669	9 145
Bunkers	300	250	73	71	55	62	60	94	57
Total	6 384	6 702	7 370	8 252	8 064	8 157	8 045	8 763	9 202
Thousand barrels/day									
NGL/LPG/Ethane	8.6	10.7	7.1	11.0	12.3	12.5	13.1	11.6	12.0
Naphtha	5.7	5.2	4.3	5.2	5.0	4.9	4.9	4.9	4.4
Motor Gasoline	46.4	52.7	71.6	84.0	87.0	90.2	91.7	97.2	103.7
Aviation Fuels	5.5	6.3	9.5	11.4	9.9	9.7	8.9	9.5	10.9
Kerosene	9.2	8.7	8.0	5.0	4.7	4.8	4.4	6.0	4.9
Gas Diesel	18.1	18.5	21.1	28.1	27.8	28.1	27.7	31.2	32.9
Heavy Fuel Oil	26.4	28.3	25.9	16.7	13.6	12.2	11.5	12.3	3.7
Other Products	5.1	4.6	4.4	8.5	9.2	9.6	10.6	14.8	24.5
Refinery Fuel	6.2	5.1	7.0	11.3	8.7	9.3	6.2	5.2	6.4
Sub-Total	131.3	140.1	158.9	181.2	178.1	181.3	179.0	192.7	203.3
Bunkers	5.7	4.7	1.4	1.4	1.1	1.2	1.2	1.8	1.1
Total	137.0	144.8	160.3	182.6	179.2	182.6	180.1	194.5	204.5

	1989	1990	1991	1992	1993	1994	1995	1996	1997
Thousand tonnes									
NGL/LPG/Ethane	416	419	438	526	556	544	545	642	492
Naphtha	230	250	-	200	-	-	-	-	-
Motor Gasoline	4 734	4 682	4 809	5 176	5 131	5 334	5 546	5 562	5 732
Aviation Fuels	488	521	517	578	567	676	708	694	627
Kerosene	195	269	242	273	284	155	163	158	133
Gas Diesel	1 644	1 699	1 821	1 687	1 847	2 105	2 259	2 542	2 593
Heavy Fuel Oil	128	113	121	171	102	85	178	108	188
Other Products	967	1 000	1 031	1 142	2 100	2 323	2 649	2 700	2 877
Refinery Fuel	327	354	521	739	1 042	575	663	699	704
Sub-Total	9 129	9 307	9 500	10 492	11 629	11 797	12 711	13 105	13 346
Bunkers	46	105	115	125	119	148	183	189	170
Total	9 175	9 412	9 615	10 617	11 748	11 945	12 894	13 294	13 516
Thousand barrels/day									
NGL/LPG/Ethane	13.2	13.3	13.9	16.7	17.7	17.3	17.3	20.3	15.6
Naphtha	5.4	5.8	-	4.6	-	-	-	-	-
Motor Gasoline	110.6	109.4	112.4	120.6	119.9	124.7	129.6	129.6	134.0
Aviation Fuels	10.7	11.4	11.3	12.6	12.4	14.7	15.4	15.1	13.7
Kerosene	4.1	5.7	5.1	5.8	6.0	3.3	3.5	3.3	2.8
Gas Diesel	33.6	34.7	37.2	34.4	37.8	43.0	46.2	51.8	53.0
Heavy Fuel Oil	2.3	2.1	2.2	3.1	1.9	1.6	3.2	2.0	3.4
Other Products	18.7	19.3	19.9	22.0	40.4	44.7	51.0	51.8	55.4
Refinery Fuel	6.8	7.2	10.3	14.3	21.0	12.3	14.3	15.0	15.1
Sub-Total	205.4	209.0	212.4	234.1	257.0	261.6	280.5	289.0	293.0
Bunkers	0.9	2.1	2.3	2.5	2.4	3.0	3.7	3.8	3.5
Total	206.3	211.1	214.7	236.6	259.4	264.6	284.2	292.8	296.5

Oil Products Consumption

Consommation de produits pétroliers

Congo

	1971	1973	1978	1983	1984	1985	1986	1987	1988
Thousand tonnes									
NGL/LPG/Ethane	2	2	1	8	8	5	8	5	4
Naphtha	-	-	-	-	-	-	-	-	-
Motor Gasoline	26	28	35	57	52	53	58	61	51
Aviation Fuels	20	22	27	42	37	32	24	13	13
Kerosene	8	10	18	2	2	2	2	2	2
Gas Diesel	70	75	128	142	128	132	133	135	125
Heavy Fuel Oil	32	27	3	10	10	10	10	-	-
Other Products	-	-	-	-	-	12	8	5	5
Refinery Fuel	-	-	-	30	30	5	30	40	35
Sub-Total	158	164	212	291	267	251	273	261	235
Bunkers	-	-	3	8	8	9	10	-	-
Total	158	164	215	299	275	260	283	261	235
Thousand barrels/day									
NGL/LPG/Ethane	0.1	0.1	0.0	0.3	0.3	0.2	0.3	0.2	0.1
Naphtha	-	-	-	-	-	-	-	-	-
Motor Gasoline	0.6	0.7	0.8	1.3	1.2	1.2	1.4	1.4	1.2
Aviation Fuels	0.4	0.5	0.6	0.9	0.8	0.7	0.5	0.3	0.3
Kerosene	0.2	0.2	0.4	0.0	0.0	0.0	0.0	0.0	0.0
Gas Diesel	1.4	1.5	2.6	2.9	2.6	2.7	2.7	2.8	2.5
Heavy Fuel Oil	0.6	0.5	0.1	0.2	0.2	0.2	0.2	-	-
Other Products	-	-	-	-	-	0.2	0.2	0.1	0.1
Refinery Fuel	-	-	-	0.5	0.5	0.1	0.5	0.7	0.6
Sub-Total	3.3	3.4	4.5	6.2	5.7	5.3	5.8	5.5	4.9
Bunkers	-	-	0.1	0.1	0.1	0.2	0.2	-	-
Total	3.3	3.4	4.5	6.3	5.8	5.5	6.0	5.5	4.9

	1989	1990	1991	1992	1993	1994	1995	1996	1997
Thousand tonnes									
NGL/LPG/Ethane	4	4	5	4	4	4	3	3	3
Naphtha	-	-	-	-	-	-	-	-	-
Motor Gasoline	55	40	41	42	45	37	52	53	53
Aviation Fuels	61	66	65	64	65	66	67	69	69
Kerosene	34	34	34	35	34	34	49	48	48
Gas Diesel	99	98	103	80	84	84	80	82	82
Heavy Fuel Oil	1	2	2	5	4	4	4	4	4
Other Products	11	11	15	20	22	23	20	20	20
Refinery Fuel	42	41	41	42	42	40	38	38	38
Sub-Total	307	296	306	292	300	292	313	317	317
Bunkers	-	5	6	13	8	8	8	10	10
Total	307	301	312	305	308	300	321	327	327
Thousand barrels/day									
NGL/LPG/Ethane	0.1	0.1	0.2	0.1	0.1	0.1	0.1	0.1	0.1
Naphtha	-	-	-	-	-	-	-	-	-
Motor Gasoline	1.3	0.9	1.0	1.0	1.1	0.9	1.2	1.2	1.2
Aviation Fuels	1.5	1.6	1.6	1.5	1.6	1.6	1.6	1.6	1.7
Kerosene	0.7	0.7	0.7	0.7	0.7	0.7	1.0	1.0	1.0
Gas Diesel	2.0	2.0	2.1	1.6	1.7	1.7	1.6	1.7	1.7
Heavy Fuel Oil	0.0	0.0	0.0	0.1	0.1	0.1	0.1	0.1	0.1
Other Products	0.2	0.2	0.3	0.4	0.4	0.4	0.4	0.4	0.4
Refinery Fuel	0.8	0.7	0.7	0.8	0.8	0.7	0.7	0.7	0.7
Sub-Total	6.6	6.4	6.6	6.2	6.4	6.2	6.7	6.8	6.8
Bunkers	-	0.1	0.1	0.2	0.1	0.1	0.1	0.2	0.2
Total	6.6	6.4	6.7	6.5	6.6	6.4	6.9	7.0	7.0

Oil Products Consumption
Consommation de produits pétroliers

Democratic Republic of Congo

	1971	1973	1978	1983	1984	1985	1986	1987	1988
Thousand tonnes									
NGL/LPG/Ethane	1	9	8	9	10	10	10	11	11
Naphtha	-	-	-	-	-	-	-	-	-
Motor Gasoline	134	159	168	190	192	175	176	182	185
Aviation Fuels	88	91	126	127	120	126	120	124	130
Kerosene	69	72	29	81	85	75	62	88	84
Gas Diesel	297	300	339	402	395	360	267	338	340
Heavy Fuel Oil	48	77	116	230	195	180	180	196	245
Other Products	-	-	-	17	18	24	23	31	34
Refinery Fuel	28	30	8	18	16	12	8	12	16
Sub-Total	665	738	794	1 074	1 031	962	846	982	1 045
Bunkers	131	120	30	31	32	30	35	32	31
Total	796	858	824	1 105	1 063	992	881	1 014	1 076
Thousand barrels/day									
NGL/LPG/Ethane	0.0	0.3	0.3	0.3	0.3	0.3	0.3	0.4	0.3
Naphtha	-	-	-	-	-	-	-	-	-
Motor Gasoline	3.1	3.7	3.9	4.4	4.5	4.1	4.1	4.3	4.3
Aviation Fuels	1.9	2.0	2.8	2.8	2.6	2.8	2.6	2.7	2.8
Kerosene	1.5	1.5	0.6	1.7	1.8	1.6	1.3	1.9	1.8
Gas Diesel	6.1	6.1	6.9	8.2	8.1	7.4	5.5	6.9	6.9
Heavy Fuel Oil	0.9	1.4	2.1	4.2	3.5	3.3	3.3	3.6	4.5
Other Products	-	-	-	0.3	0.3	0.5	0.4	0.6	0.7
Refinery Fuel	0.6	0.6	0.2	0.4	0.3	0.2	0.2	0.2	0.3
Sub-Total	14.1	15.7	16.8	22.3	21.5	20.1	17.7	20.5	21.6
Bunkers	2.4	2.2	0.5	0.6	0.6	0.5	0.6	0.6	0.6
Total	16.5	17.9	17.3	22.9	22.1	20.7	18.4	21.1	22.2

	1989	1990	1991	1992	1993	1994	1995	1996	1997
Thousand tonnes									
NGL/LPG/Ethane	11	12	13	12	13	13	13	13	13
Naphtha	-	-	-	-	-	-	-	-	-
Motor Gasoline	185	207	208	193	220	226	235	235	235
Aviation Fuels	133	142	147	134	150	153	159	159	159
Kerosene	85	77	90	83	96	97	113	113	113
Gas Diesel	345	328	332	301	335	339	341	341	341
Heavy Fuel Oil	208	283	292	274	294	295	280	280	280
Other Products	49	49	44	42	57	60	59	59	59
Refinery Fuel	14	15	17	16	16	16	16	16	16
Sub-Total	1 030	1 113	1 143	1 055	1 181	1 199	1 216	1 216	1 216
Bunkers	31	31	35	29	32	34	35	35	35
Total	1 061	1 144	1 178	1 084	1 213	1 233	1 251	1 251	1 251
Thousand barrels/day									
NGL/LPG/Ethane	0.4	0.4	0.4	0.4	0.4	0.4	0.4	0.4	0.4
Naphtha	-	-	-	-	-	-	-	-	-
Motor Gasoline	4.3	4.8	4.9	4.5	5.1	5.3	5.5	5.5	5.5
Aviation Fuels	2.9	3.1	3.2	2.9	3.3	3.4	3.5	3.5	3.5
Kerosene	1.8	1.6	1.9	1.8	2.0	2.1	2.4	2.4	2.4
Gas Diesel	7.1	6.7	6.8	6.1	6.8	6.9	7.0	7.0	7.0
Heavy Fuel Oil	3.8	5.2	5.3	5.0	5.4	5.4	5.1	5.1	5.1
Other Products	0.9	0.9	0.8	0.8	1.1	1.2	1.1	1.1	1.1
Refinery Fuel	0.3	0.3	0.3	0.3	0.3	0.3	0.3	0.3	0.3
Sub-Total	21.5	23.1	23.7	21.8	24.5	24.9	25.3	25.3	25.3
Bunkers	0.6	0.6	0.6	0.5	0.6	0.6	0.6	0.6	0.6
Total	22.0	23.7	24.4	22.3	25.1	25.5	26.0	25.9	26.0

Oil Products Consumption
Consommation de produits pétroliers

Costa Rica

	1971	1973	1978	1983	1984	1985	1986	1987	1988
Thousand tonnes									
NGL/LPG/Ethane	7	10	17	13	15	15	16	17	20
Naphtha	-	-	-	-	-	-	-	-	-
Motor Gasoline	97	126	155	114	126	128	140	168	181
Aviation Fuels	7	9	19	14	12	13	21	26	33
Kerosene	21	21	25	13	15	14	13	13	10
Gas Diesel	179	232	398	279	285	294	319	335	366
Heavy Fuel Oil	70	97	162	99	111	134	118	123	111
Other Products	-	-	19	13	14	25	13	18	12
Refinery Fuel	29	30	16	28	31	30	44	39	41
Sub-Total	410	525	811	573	609	653	684	739	774
Bunkers	-	-	-	-	-	-	-	-	-
Total	410	525	811	573	609	653	684	739	774
Thousand barrels/day									
NGL/LPG/Ethane	0.2	0.3	0.5	0.4	0.5	0.5	0.5	0.5	0.6
Naphtha	-	-	-	-	-	-	-	-	-
Motor Gasoline	2.3	2.9	3.6	2.7	2.9	3.0	3.3	3.9	4.2
Aviation Fuels	0.2	0.2	0.4	0.3	0.3	0.3	0.5	0.6	0.7
Kerosene	0.4	0.4	0.5	0.3	0.3	0.3	0.3	0.3	0.2
Gas Diesel	3.7	4.7	8.1	5.7	5.8	6.0	6.5	6.8	7.5
Heavy Fuel Oil	1.3	1.8	3.0	1.8	2.0	2.4	2.2	2.2	2.0
Other Products	-	-	0.4	0.2	0.3	0.5	0.2	0.3	0.2
Refinery Fuel	0.5	0.5	0.3	0.5	0.6	0.6	0.8	0.7	0.8
Sub-Total	8.6	11.0	16.9	11.9	12.7	13.5	14.3	15.5	16.3
Bunkers	-	-	-	-	-	-	-	-	-
Total	8.6	11.0	16.9	11.9	12.7	13.5	14.3	15.5	16.3

	1989	1990	1991	1992	1993	1994	1995	1996	1997
Thousand tonnes									
NGL/LPG/Ethane	23	23	26	28	30	35	40	46	50
Naphtha	-	-	-	12	6	5	4	2	2
Motor Gasoline	206	208	226	441	428	430	423	437	447
Aviation Fuels	36	42	41	63	74	96	101	94	104
Kerosene	9	8	9	7	3	8	8	7	8
Gas Diesel	386	386	429	533	548	724	714	627	583
Heavy Fuel Oil	167	163	166	198	194	199	206	179	192
Other Products	14	8	2	14	18	39	33	26	31
Refinery Fuel	42	38	20	25	20	25	31	30	30
Sub-Total	883	876	919	1 321	1 321	1 561	1 560	1 448	1 447
Bunkers	-	-	-	-	-	-	-	-	-
Total	883	876	919	1 321	1 321	1 561	1 560	1 448	1 447
Thousand barrels/day									
NGL/LPG/Ethane	0.7	0.7	0.8	0.9	1.0	1.1	1.3	1.5	1.6
Naphtha	-	-	-	0.3	0.1	0.1	0.1	0.0	0.0
Motor Gasoline	4.8	4.9	5.3	10.3	10.0	10.0	9.9	10.2	10.4
Aviation Fuels	0.8	0.9	0.9	1.4	1.6	2.1	2.2	2.0	2.3
Kerosene	0.2	0.2	0.2	0.1	0.1	0.2	0.2	0.1	0.2
Gas Diesel	7.9	7.9	8.8	10.9	11.2	14.8	14.6	12.8	11.9
Heavy Fuel Oil	3.0	3.0	3.0	3.6	3.5	3.6	3.8	3.3	3.5
Other Products	0.3	0.2	0.0	0.2	0.3	0.7	0.6	0.5	0.5
Refinery Fuel	0.8	0.7	0.4	0.5	0.4	0.5	0.6	0.5	0.5
Sub-Total	18.5	18.4	19.4	28.1	28.2	33.1	33.1	30.9	31.0
Bunkers	-	-	-	-	-	-	-	-	-
Total	18.5	18.4	19.4	28.1	28.2	33.1	33.1	30.9	31.0

Oil Products Consumption
Consommation de produits pétroliers

Croatia

	1971	1973	1978	1983	1984	1985	1986	1987	1988
Thousand tonnes									
NGL/LPG/Ethane	-	-	-	-	-	-	-	-	-
Naphtha	-	-	-	-	-	-	-	-	-
Motor Gasoline	-	-	-	-	-	-	-	-	-
Aviation Fuels	-	-	-	-	-	-	-	-	-
Kerosene	-	-	-	-	-	-	-	-	-
Gas Diesel	-	-	-	-	-	-	-	-	-
Heavy Fuel Oil	-	-	-	-	-	-	-	-	-
Other Products	-	-	-	-	-	-	-	-	-
Refinery Fuel	-	-	-	-	-	-	-	-	-
Sub-Total	-	-	-	-	-	-	-	-	-
Bunkers	-	-	-	-	-	-	-	-	-
Total	-	-	-	-	-	-	-	-	-
Thousand barrels/day									
NGL/LPG/Ethane	-	-	-	-	-	-	-	-	-
Naphtha	-	-	-	-	-	-	-	-	-
Motor Gasoline	-	-	-	-	-	-	-	-	-
Aviation Fuels	-	-	-	-	-	-	-	-	-
Kerosene	-	-	-	-	-	-	-	-	-
Gas Diesel	-	-	-	-	-	-	-	-	-
Heavy Fuel Oil	-	-	-	-	-	-	-	-	-
Other Products	-	-	-	-	-	-	-	-	-
Refinery Fuel	-	-	-	-	-	-	-	-	-
Sub-Total	-	-	-	-	-	-	-	-	-
Bunkers	-	-	-	-	-	-	-	-	-
Total	-	-	-	-	-	-	-	-	-

	1989	1990	1991	1992	1993	1994	1995	1996	1997
Thousand tonnes									
NGL/LPG/Ethane	-	-	184	164	190	194	206	179	213
Naphtha	-	-	115	129	142	162	153	133	-
Motor Gasoline	-	-	591	511	497	545	575	600	678
Aviation Fuels	-	-	32	25	63	90	85	78	82
Kerosene	-	-	4	2	10	1	8	7	6
Gas Diesel	-	-	953	909	928	986	1 061	1 023	1 190
Heavy Fuel Oil	-	-	858	954	963	804	1 032	998	1 039
Other Products	-	-	215	153	158	214	208	181	220
Refinery Fuel	-	-	498	395	476	490	558	485	479
Sub-Total	-	-	3 450	3 242	3 427	3 486	3 886	3 684	3 907
Bunkers	-	-	-	-	-	45	33	29	24
Total	-	-	3 450	3 242	3 427	3 531	3 919	3 713	3 931
Thousand barrels/day									
NGL/LPG/Ethane	-	-	6.9	6.0	7.3	7.5	8.0	6.9	8.1
Naphtha	-	-	2.7	3.0	3.3	3.8	3.6	3.1	-
Motor Gasoline	-	-	13.8	11.9	11.6	12.7	13.4	14.0	15.8
Aviation Fuels	-	-	0.7	0.5	1.4	2.0	1.8	1.7	1.8
Kerosene	-	-	0.1	0.0	0.2	0.0	0.2	0.1	0.1
Gas Diesel	-	-	19.5	18.5	19.0	20.2	21.7	20.9	24.3
Heavy Fuel Oil	-	-	15.7	17.4	17.6	14.7	18.8	18.2	19.0
Other Products	-	-	4.0	2.8	3.0	4.1	4.0	3.5	3.9
Refinery Fuel	-	-	9.8	7.6	9.3	9.9	11.3	9.8	9.5
Sub-Total	-	-	73.0	67.7	72.6	74.8	82.8	78.1	82.5
Bunkers	-	-	-	-	-	0.9	0.6	0.6	0.5
Total	-	-	73.0	67.7	72.6	75.7	83.5	78.7	83.0

Oil Products Consumption
Consommation de produits pétroliers

Cuba

	1971	1973	1978	1983	1984	1985	1986	1987	1988
Thousand tonnes									
NGL/LPG/Ethane	58	67	102	116	121	124	127	126	133
Naphtha	8	32	76	115	205	202	207	186	184
Motor Gasoline	951	1 093	1 333	1 369	1 371	1 241	1 263	1 250	1 319
Aviation Fuels	88	111	306	184	204	214	182	220	231
Kerosene	399	461	447	626	657	662	669	632	653
Gas Diesel	1 145	2 167	1 714	1 959	2 239	2 191	2 184	2 533	2 364
Heavy Fuel Oil	3 427	3 581	4 893	5 334	5 079	5 313	5 372	5 278	5 747
Other Products	100	161	240	270	276	476	603	722	712
Refinery Fuel	157	169	200	256	329	243	267	258	273
Sub-Total	6 333	7 842	9 311	10 229	10 481	10 666	10 874	11 205	11 616
Bunkers	-	-	-	-	190	219	228	96	122
Total	6 333	7 842	9 311	10 229	10 671	10 885	11 102	11 301	11 738
Thousand barrels/day									
NGL/LPG/Ethane	1.8	2.1	3.2	3.7	3.8	3.9	4.0	4.0	4.2
Naphtha	0.2	0.7	1.8	2.7	4.8	4.7	4.8	4.3	4.3
Motor Gasoline	22.0	25.3	30.9	31.7	31.6	28.7	29.2	28.8	30.4
Aviation Fuels	1.9	2.4	6.7	4.0	4.5	4.7	4.0	4.8	5.0
Kerosene	8.5	9.8	9.5	13.3	13.9	14.0	14.2	13.4	13.8
Gas Diesel	23.4	44.3	35.0	40.0	45.6	44.8	44.6	51.8	48.2
Heavy Fuel Oil	62.5	65.3	89.3	97.3	92.4	96.9	98.0	96.3	104.6
Other Products	1.7	2.7	4.0	4.5	4.6	8.3	10.6	12.7	12.5
Refinery Fuel	3.0	3.2	3.8	4.8	6.2	4.7	5.1	5.0	5.2
Sub-Total	125.1	156.0	184.2	202.1	207.4	210.8	214.5	221.1	228.2
Bunkers	-	-	-	-	3.5	4.1	4.2	1.8	2.3
Total	125.1	156.0	184.2	202.1	210.9	214.9	218.8	222.9	230.4

	1989	1990	1991	1992	1993	1994	1995	1996	1997
Thousand tonnes									
NGL/LPG/Ethane	138	130	114	94	64	75	89	87	88
Naphtha	194	341	220	93	62	61	70	83	91
Motor Gasoline	1 393	1 169	766	547	446	451	462	503	531
Aviation Fuels	204	326	293	216	196	168	176	261	279
Kerosene	652	640	620	438	310	260	237	223	216
Gas Diesel	2 485	2 354	1 818	1 523	1 299	1 436	1 511	1 640	1 676
Heavy Fuel Oil	6 009	5 510	4 433	3 224	3 038	3 267	3 513	3 710	4 055
Other Products	683	572	363	766	1 302	1 547	1 487	1 584	1 560
Refinery Fuel	272	91	75	39	27	23	38	40	36
Sub-Total	12 030	11 133	8 702	6 940	6 744	7 288	7 583	8 131	8 532
Bunkers	108	245	231	168	101	105	85	105	113
Total	12 138	11 378	8 933	7 108	6 845	7 393	7 668	8 236	8 645
Thousand barrels/day									
NGL/LPG/Ethane	4.4	4.1	3.6	3.0	2.0	2.4	2.8	2.8	2.8
Naphtha	4.5	7.9	5.1	2.2	1.4	1.4	1.6	1.9	2.1
Motor Gasoline	32.2	27.0	17.6	12.5	10.2	10.3	10.6	11.5	12.2
Aviation Fuels	4.5	7.1	6.4	4.7	4.3	3.7	3.8	5.7	6.1
Kerosene	13.8	13.6	13.1	9.3	6.6	5.5	5.0	4.7	4.6
Gas Diesel	50.8	48.1	37.2	31.0	26.5	29.3	30.9	33.4	34.3
Heavy Fuel Oil	109.6	100.5	80.9	58.7	55.4	59.6	64.1	67.5	74.0
Other Products	12.0	10.1	6.4	13.5	23.3	27.7	26.6	28.2	27.9
Refinery Fuel	5.2	2.0	1.6	0.8	0.6	0.5	0.8	0.9	0.8
Sub-Total	237.0	220.5	172.0	135.7	130.4	140.4	146.2	156.6	164.7
Bunkers	2.0	4.5	4.2	3.1	1.9	2.0	1.6	1.9	2.1
Total	239.0	225.0	176.3	138.8	132.2	142.3	147.8	158.5	166.8

Oil Products Consumption
Consommation de produits pétroliers

Cyprus

	1971	1973	1978	1983	1984	1985	1986	1987	1988
Thousand tonnes									
NGL/LPG/Ethane	24	29	31	40	41	42	43	46	46
Naphtha	-	-	-	-	-	-	-	-	-
Motor Gasoline	108	137	100	111	118	124	130	140	147
Aviation Fuels	49	75	61	106	135	143	148	169	170
Kerosene	14	13	7	8	8	7	7	9	10
Gas Diesel	133	156	131	182	197	204	218	239	250
Heavy Fuel Oil	285	365	437	483	460	449	523	583	626
Other Products	17	30	26	40	38	31	37	41	40
Refinery Fuel	-	37	17	24	21	10	17	27	30
Sub-Total	630	842	810	994	1 018	1 010	1 123	1 254	1 319
Bunkers	2	10	13	34	37	34	36	68	68
Total	632	852	823	1 028	1 055	1 044	1 159	1 322	1 387
Thousand barrels/day									
NGL/LPG/Ethane	0.8	0.9	1.0	1.3	1.3	1.3	1.4	1.5	1.5
Naphtha	-	-	-	-	-	-	-	-	-
Motor Gasoline	2.5	3.2	2.3	2.6	2.8	2.9	3.0	3.3	3.4
Aviation Fuels	1.1	1.6	1.3	2.3	2.9	3.1	3.2	3.7	3.7
Kerosene	0.3	0.3	0.1	0.2	0.2	0.1	0.1	0.2	0.2
Gas Diesel	2.7	3.2	2.7	3.7	4.0	4.2	4.5	4.9	5.1
Heavy Fuel Oil	5.2	6.7	8.0	8.8	8.4	8.2	9.5	10.6	11.4
Other Products	0.3	0.5	0.5	0.7	0.7	0.6	0.7	0.7	0.7
Refinery Fuel	-	0.7	0.3	0.5	0.4	0.2	0.4	0.6	0.6
Sub-Total	12.9	17.2	16.3	20.1	20.6	20.6	22.8	25.4	26.6
Bunkers	0.0	0.2	0.3	0.7	0.7	0.7	0.7	1.3	1.3
Total	12.9	17.3	16.5	20.7	21.3	21.3	23.5	26.7	27.8

	1989	1990	1991	1992	1993	1994	1995	1996	1997
Thousand tonnes									
NGL/LPG/Ethane	45	49	49	55	51	50	51	51	52
Naphtha	-	-	-	-	-	-	-	-	-
Motor Gasoline	154	163	170	172	169	180	183	186	191
Aviation Fuels	209	236	280	272	231	237	260	249	245
Kerosene	10	12	12	17	16	17	17	18	20
Gas Diesel	274	308	311	378	392	404	446	465	489
Heavy Fuel Oil	640	577	685	763	797	837	759	814	813
Other Products	44	73	121	135	182	180	191	217	227
Refinery Fuel	30	29	29	30	26	38	30	28	30
Sub-Total	1 406	1 447	1 657	1 822	1 864	1 943	1 937	2 028	2 067
Bunkers	86	58	56	59	50	62	69	91	99
Total	1 492	1 505	1 713	1 881	1 914	2 005	2 006	2 119	2 166
Thousand barrels/day									
NGL/LPG/Ethane	1.4	1.6	1.6	1.7	1.6	1.6	1.6	1.6	1.7
Naphtha	-	-	-	-	-	-	-	-	-
Motor Gasoline	3.6	3.8	4.0	4.0	4.0	4.2	4.3	4.3	4.5
Aviation Fuels	4.5	5.1	6.1	5.9	5.0	5.1	5.6	5.4	5.3
Kerosene	0.2	0.3	0.3	0.4	0.3	0.4	0.4	0.4	0.4
Gas Diesel	5.6	6.3	6.4	7.7	8.0	8.3	9.1	9.5	10.0
Heavy Fuel Oil	11.7	10.5	12.5	13.9	14.5	15.3	13.8	14.8	14.8
Other Products	0.7	1.3	1.9	2.1	2.9	2.9	3.0	3.4	3.6
Refinery Fuel	0.6	0.6	0.6	0.6	0.5	0.8	0.6	0.6	0.6
Sub-Total	28.4	29.5	33.2	36.3	36.9	38.5	38.5	40.0	40.9
Bunkers	1.6	1.1	1.1	1.1	0.9	1.2	1.3	1.7	1.9
Total	30.1	30.6	34.3	37.4	37.8	39.6	39.8	41.7	42.7

Oil Products Consumption
Consommation de produits pétroliers

Dominican Republic

	1971	1973	1978	1983	1984	1985	1986	1987	1988
Thousand tonnes									
NGL/LPG/Ethane	35	46	60	99	100	141	105	114	126
Naphtha	-	-	-	-	-	-	-	-	-
Motor Gasoline	315	385	401	362	416	479	487	625	500
Aviation Fuels	25	26	38	71	82	52	84	92	107
Kerosene	19	14	14	9	9	15	25	28	25
Gas Diesel	229	173	383	396	488	346	465	605	677
Heavy Fuel Oil	516	859	995	1 281	1 229	835	1 148	1 017	1 184
Other Products	-	-	-	-	-	-	-	-	-
Refinery Fuel	-	-	-	-	-	-	-	-	-
Sub-Total	1 139	1 503	1 891	2 218	2 324	1 868	2 314	2 481	2 619
Bunkers	-	-	-	-	-	-	-	-	-
Total	1 139	1 503	1 891	2 218	2 324	1 868	2 314	2 481	2 619
Thousand barrels/day									
NGL/LPG/Ethane	1.1	1.5	1.9	3.1	3.2	4.5	3.3	3.6	4.0
Naphtha	-	-	-	-	-	-	-	-	-
Motor Gasoline	7.4	9.0	9.4	8.5	9.7	11.2	11.4	14.6	11.7
Aviation Fuels	0.5	0.6	0.8	1.5	1.8	1.1	1.8	2.0	2.3
Kerosene	0.4	0.3	0.3	0.2	0.2	0.3	0.5	0.6	0.5
Gas Diesel	4.7	3.5	7.8	8.1	9.9	7.1	9.5	12.4	13.8
Heavy Fuel Oil	9.4	15.7	18.2	23.4	22.4	15.2	20.9	18.6	21.5
Other Products	-	-	-	-	-	-	-	-	-
Refinery Fuel	-	-	-	-	-	-	-	-	-
Sub-Total	23.5	30.5	38.4	44.8	47.1	39.4	47.5	51.7	53.8
Bunkers	-	-	-	-	-	-	-	-	-
Total	23.5	30.5	38.4	44.8	47.1	39.4	47.5	51.7	53.8

	1989	1990	1991	1992	1993	1994	1995	1996	1997
Thousand tonnes									
NGL/LPG/Ethane	128	109	178	222	259	302	321	340	425
Naphtha	-	-	-	-	-	-	-	-	-
Motor Gasoline	726	692	638	781	574	586	597	609	621
Aviation Fuels	32	36	34	48	46	52	55	59	62
Kerosene	147	156	144	202	192	218	233	247	263
Gas Diesel	660	747	687	882	867	898	929	960	995
Heavy Fuel Oil	1 085	1 112	1 126	1 173	1 238	1 312	1 352	1 392	1 477
Other Products	-	-	-	-	-	-	-	-	-
Refinery Fuel	-	-	-	-	-	-	-	-	-
Sub-Total	2 778	2 852	2 807	3 308	3 176	3 368	3 487	3 607	3 843
Bunkers	-	-	-	-	-	-	-	-	-
Total	2 778	2 852	2 807	3 308	3 176	3 368	3 487	3 607	3 843
Thousand barrels/day									
NGL/LPG/Ethane	4.1	3.5	5.7	7.0	8.2	9.6	10.2	10.8	13.5
Naphtha	-	-	-	-	-	-	-	-	-
Motor Gasoline	17.0	16.2	14.9	18.2	13.4	13.7	14.0	14.2	14.5
Aviation Fuels	0.7	0.8	0.7	1.0	1.0	1.1	1.2	1.3	1.3
Kerosene	3.1	3.3	3.1	4.3	4.1	4.6	4.9	5.2	5.6
Gas Diesel	13.5	15.3	14.0	18.0	17.7	18.4	19.0	19.6	20.3
Heavy Fuel Oil	19.8	20.3	20.5	21.3	22.6	23.9	24.7	25.3	27.0
Other Products	-	-	-	-	-	-	-	-	-
Refinery Fuel	-	-	-	-	-	-	-	-	-
Sub-Total	58.1	59.3	58.9	69.9	67.0	71.3	73.9	76.4	82.2
Bunkers	-	-	-	-	-	-	-	-	-
Total	58.1	59.3	58.9	69.9	67.0	71.3	73.9	76.4	82.2

Oil Products Consumption
Consommation de produits pétroliers

Ecuador

	1971	1973	1978	1983	1984	1985	1986	1987	1988
Thousand tonnes									
NGL/LPG/Ethane	7	14	63	139	161	185	214	250	292
Naphtha	-	-	-	163	198	201	199	173	203
Motor Gasoline	427	495	974	936	957	975	1 025	981	1 266
Aviation Fuels	86	125	104	118	134	145	165	162	171
Kerosene	50	51	360	268	289	291	310	179	173
Gas Diesel	263	333	668	773	805	928	968	1 042	1 048
Heavy Fuel Oil	319	317	667	1 104	1 029	1 208	1 012	1 033	946
Other Products	32	45	126	111	224	251	277	276	313
Refinery Fuel	47	58	168	278	281	80	231	160	140
Sub-Total	1 231	1 438	3 130	3 890	4 078	4 264	4 401	4 256	4 552
Bunkers	90	100	7	18	44	37	43	109	100
Total	1 321	1 538	3 137	3 908	4 122	4 301	4 444	4 365	4 652
Thousand barrels/day									
NGL/LPG/Ethane	0.2	0.4	2.0	4.4	5.1	5.9	6.8	7.9	9.3
Naphtha	-	-	-	3.8	4.6	4.7	4.6	4.0	4.7
Motor Gasoline	10.0	11.6	22.8	21.9	22.3	22.8	24.0	22.9	29.5
Aviation Fuels	1.9	2.7	2.3	2.6	2.9	3.2	3.6	3.5	3.7
Kerosene	1.1	1.1	7.6	5.7	6.1	6.2	6.6	3.8	3.7
Gas Diesel	5.4	6.8	13.7	15.8	16.4	19.0	19.8	21.3	21.4
Heavy Fuel Oil	5.8	5.8	12.2	20.1	18.7	22.0	18.5	18.8	17.2
Other Products	0.6	0.9	2.3	2.1	4.2	4.7	5.1	5.1	5.7
Refinery Fuel	0.9	1.1	3.2	5.3	5.4	1.6	4.3	3.1	2.6
Sub-Total	25.9	30.4	65.9	81.7	85.7	90.0	93.3	90.6	97.7
Bunkers	1.6	1.9	0.1	0.3	0.8	0.7	0.8	2.0	1.8
Total	27.5	32.2	66.1	82.0	86.5	90.7	94.1	92.6	99.6

	1989	1990	1991	1992	1993	1994	1995	1996	1997
Thousand tonnes									
NGL/LPG/Ethane	322	338	385	417	449	483	553	576	631
Naphtha	200	200	210	-	-	-	-	-	-
Motor Gasoline	1 285	1 236	1 353	1 380	1 300	1 258	1 222	1 330	1 303
Aviation Fuels	166	185	180	177	196	177	182	215	219
Kerosene	203	181	129	108	52	46	30	30	30
Gas Diesel	1 018	1 187	1 159	1 318	1 285	1 602	1 786	2 106	2 209
Heavy Fuel Oil	886	960	1 048	1 014	886	883	1 123	1 043	1 158
Other Products	262	241	207	165	189	278	230	254	271
Refinery Fuel	84	100	104	78	56	55	46	31	31
Sub-Total	4 426	4 628	4 775	4 657	4 413	4 782	5 172	5 585	5 852
Bunkers	139	185	240	466	454	515	366	287	316
Total	4 565	4 813	5 015	5 123	4 867	5 297	5 538	5 872	6 168
Thousand barrels/day									
NGL/LPG/Ethane	10.2	10.7	12.2	13.2	14.3	15.4	17.6	18.3	20.1
Naphtha	4.7	4.7	4.9	-	-	-	-	-	-
Motor Gasoline	30.0	28.9	31.6	32.2	30.4	29.4	28.6	31.0	30.5
Aviation Fuels	3.6	4.0	3.9	3.9	4.3	3.9	4.0	4.7	4.8
Kerosene	4.3	3.8	2.7	2.3	1.1	1.0	0.6	0.6	0.6
Gas Diesel	20.8	24.3	23.7	26.9	26.3	32.7	36.5	42.9	45.1
Heavy Fuel Oil	16.2	17.5	19.1	18.5	16.2	16.1	20.5	19.0	21.1
Other Products	4.7	4.4	3.8	3.2	3.6	5.3	4.4	4.9	5.2
Refinery Fuel	1.7	2.0	2.0	1.5	1.1	1.1	1.0	0.6	0.6
Sub-Total	96.2	100.3	104.0	101.5	97.2	104.8	113.1	121.9	128.0
Bunkers	2.5	3.4	4.5	8.7	8.5	9.6	6.7	5.2	5.8
Total	98.7	103.7	108.6	110.2	105.7	114.5	119.9	127.1	133.7

Oil Products Consumption
Consommation de produits pétroliers

Egypt

	1971	1973	1978	1983	1984	1985	1986	1987	1988
Thousand tonnes									
NGL/LPG/Ethane	106	118	297	520	580	615	666	739	722
Naphtha	-	-	23	-	-	-	-	8	53
Motor Gasoline	570	674	959	1 610	1 809	1 958	2 025	2 140	2 175
Aviation Fuels	103	167	193	389	406	447	367	553	442
Kerosene	977	1 112	1 389	1 994	2 118	2 220	2 328	2 450	2 412
Gas Diesel	1 222	1 200	1 859	3 720	4 178	4 058	3 875	4 058	4 173
Heavy Fuel Oil	2 754	2 732	4 420	7 258	7 722	7 860	7 656	8 437	8 580
Other Products	282	218	344	677	871	1 024	1 178	1 114	1 104
Refinery Fuel	167	228	396	609	658	683	724	755	749
Sub-Total	6 181	6 449	9 880	16 777	18 342	18 865	18 819	20 254	20 410
Bunkers	19	165	705	1 735	1 776	1 522	1 629	1 700	1 670
Total	6 200	6 614	10 585	18 512	20 118	20 387	20 448	21 954	22 080
Thousand barrels/day									
NGL/LPG/Ethane	3.4	3.8	9.4	16.5	18.4	19.5	21.2	23.5	22.9
Naphtha	-	-	0.5	-	-	-	-	0.2	1.2
Motor Gasoline	13.3	15.8	22.4	37.6	42.2	45.8	47.3	50.0	50.7
Aviation Fuels	2.3	3.7	4.2	8.5	8.8	9.7	8.0	12.0	9.6
Kerosene	20.7	23.6	29.5	42.3	44.8	47.1	49.4	52.0	51.0
Gas Diesel	25.0	24.5	38.0	76.0	85.2	82.9	79.2	82.9	85.1
Heavy Fuel Oil	50.3	49.9	80.7	132.4	140.5	143.4	139.7	153.9	156.1
Other Products	5.4	4.2	6.6	11.6	15.0	17.6	20.4	19.2	19.2
Refinery Fuel	3.3	4.6	7.9	12.2	13.1	13.7	14.4	15.0	14.9
Sub-Total	123.7	129.9	199.2	337.2	368.0	379.8	379.6	408.8	410.7
Bunkers	0.3	3.0	13.0	32.2	33.0	28.3	30.2	31.6	30.9
Total	124.0	132.9	212.2	369.4	400.9	408.1	409.7	440.3	441.6

	1989	1990	1991	1992	1993	1994	1995	1996	1997
Thousand tonnes									
NGL/LPG/Ethane	865	845	874	976	1 050	1 240	1 454	1 610	1 865
Naphtha	-	-	-	-	-	-	-	-	-
Motor Gasoline	2 141	2 172	2 082	1 957	1 877	1 878	1 951	2 013	2 080
Aviation Fuels	469	480	444	459	479	567	602	813	864
Kerosene	2 386	2 333	2 165	2 004	1 629	1 426	1 334	1 266	1 197
Gas Diesel	4 316	4 486	4 042	4 605	4 713	5 052	5 625	6 049	6 483
Heavy Fuel Oil	8 343	8 562	8 512	8 389	6 779	6 058	6 938	7 330	8 169
Other Products	1 131	1 202	1 183	1 186	1 232	1 242	1 349	1 340	1 402
Refinery Fuel	742	770	800	810	820	830	855	874	908
Sub-Total	20 393	20 850	20 102	20 386	18 579	18 293	20 108	21 295	22 968
Bunkers	1 700	1 700	1 710	1 410	1 410	2 444	2 501	3 076	3 029
Total	22 093	22 550	21 812	21 796	19 989	20 737	22 609	24 371	25 997
Thousand barrels/day									
NGL/LPG/Ethane	27.5	26.9	27.8	30.9	33.4	39.4	46.2	51.0	59.3
Naphtha	-	-	-	-	-	-	-	-	-
Motor Gasoline	50.0	50.8	48.7	45.6	43.9	43.9	45.6	46.9	48.6
Aviation Fuels	10.2	10.4	9.6	9.9	10.4	12.3	13.1	17.6	18.8
Kerosene	50.6	49.5	45.9	42.4	34.5	30.2	28.3	26.8	25.4
Gas Diesel	88.2	91.7	82.6	93.9	96.3	103.3	115.0	123.3	132.5
Heavy Fuel Oil	152.2	156.2	155.3	152.7	123.7	110.5	126.6	133.4	149.1
Other Products	19.9	21.0	20.7	20.7	21.6	21.7	23.5	23.4	24.6
Refinery Fuel	14.8	15.4	16.0	16.2	16.5	16.7	17.2	17.5	18.1
Sub-Total	413.4	421.8	406.7	412.3	380.3	378.0	415.4	439.9	476.3
Bunkers	31.5	31.5	31.6	26.3	26.4	45.3	46.3	56.6	55.9
Total	444.9	453.3	438.3	438.6	406.7	423.3	461.8	496.5	532.2

Oil Products Consumption
Consommation de produits pétroliers

El Salvador

	1971	1973	1978	1983	1984	1985	1986	1987	1988
Thousand tonnes									
NGL/LPG/Ethane	10	14	27	27	29	30	32	35	46
Naphtha	-	-	-	-	-	-	-	-	-
Motor Gasoline	98	114	159	131	134	141	139	153	163
Aviation Fuels	11	15	21	21	27	33	25	20	33
Kerosene	36	35	30	19	18	22	22	19	18
Gas Diesel	103	130	219	164	174	191	189	219	235
Heavy Fuel Oil	151	227	192	118	112	121	117	193	185
Other Products	-	18	24	17	13	11	11	9	16
Refinery Fuel	24	26	32	21	22	25	24	26	24
Sub-Total	433	579	704	518	529	574	559	674	720
Bunkers	-	-	-	-	-	-	-	-	-
Total	433	579	704	518	529	574	559	674	720
Thousand barrels/day									
NGL/LPG/Ethane	0.3	0.4	0.9	0.9	0.9	1.0	1.0	1.1	1.5
Naphtha	-	-	-	-	-	-	-	-	-
Motor Gasoline	2.3	2.7	3.7	3.1	3.1	3.3	3.2	3.6	3.8
Aviation Fuels	0.2	0.3	0.5	0.5	0.6	0.7	0.5	0.4	0.7
Kerosene	0.8	0.7	0.6	0.4	0.4	0.5	0.5	0.4	0.4
Gas Diesel	2.1	2.7	4.5	3.4	3.5	3.9	3.9	4.5	4.8
Heavy Fuel Oil	2.8	4.1	3.5	2.2	2.0	2.2	2.1	3.5	3.4
Other Products	-	0.3	0.5	0.3	0.2	0.2	0.2	0.2	0.3
Refinery Fuel	0.5	0.5	0.6	0.4	0.4	0.5	0.5	0.5	0.5
Sub-Total	9.0	11.9	14.7	11.0	11.3	12.3	12.0	14.2	15.3
Bunkers	-	-	-	-	-	-	-	-	-
Total	9.0	11.9	14.7	11.0	11.3	12.3	12.0	14.2	15.3

	1989	1990	1991	1992	1993	1994	1995	1996	1997
Thousand tonnes									
NGL/LPG/Ethane	50	51	56	63	76	86	99	96	105
Naphtha	-	-	-	-	-	-	-	-	-
Motor Gasoline	168	166	173	213	236	257	285	284	308
Aviation Fuels	27	34	25	41	43	44	47	48	48
Kerosene	19	17	17	18	17	17	17	18	18
Gas Diesel	233	261	374	413	513	644	730	513	549
Heavy Fuel Oil	173	173	252	295	286	300	328	382	387
Other Products	16	22	20	27	26	29	34	45	32
Refinery Fuel	10	7	16	10	12	16	17	18	19
Sub-Total	696	731	933	1 080	1 209	1 393	1 557	1 404	1 466
Bunkers	-	-	-	-	-	-	-	-	-
Total	696	731	933	1 080	1 209	1 393	1 557	1 404	1 466
Thousand barrels/day									
NGL/LPG/Ethane	1.6	1.6	1.8	2.0	2.4	2.7	3.1	3.0	3.3
Naphtha	-	-	-	-	-	-	-	-	-
Motor Gasoline	3.9	3.9	4.0	5.0	5.5	6.0	6.7	6.6	7.2
Aviation Fuels	0.6	0.7	0.5	0.9	0.9	1.0	1.0	1.0	1.0
Kerosene	0.4	0.4	0.4	0.4	0.4	0.4	0.4	0.4	0.4
Gas Diesel	4.8	5.3	7.6	8.4	10.5	13.2	14.9	10.5	11.2
Heavy Fuel Oil	3.2	3.2	4.6	5.4	5.2	5.5	6.0	7.0	7.1
Other Products	0.3	0.4	0.4	0.5	0.5	0.6	0.7	0.9	0.6
Refinery Fuel	0.2	0.2	0.3	0.2	0.3	0.4	0.4	0.4	0.4
Sub-Total	15.0	15.7	19.7	22.8	25.7	29.6	33.1	29.7	31.3
Bunkers	-	-	-	-	-	-	-	-	-
Total	15.0	15.7	19.7	22.8	25.7	29.6	33.1	29.7	31.3

Oil Products Consumption
Consommation de produits pétroliers

Estonia

	1971	1973	1978	1983	1984	1985	1986	1987	1988
Thousand tonnes									
NGL/LPG/Ethane	-	-	-	-	-	-	-	-	-
Naphtha	-	-	-	-	-	-	-	-	-
Motor Gasoline	-	-	-	-	-	-	-	-	-
Aviation Fuels	-	-	-	-	-	-	-	-	-
Kerosene	-	-	-	-	-	-	-	-	-
Gas Diesel	-	-	-	-	-	-	-	-	-
Heavy Fuel Oil	-	-	-	-	-	-	-	-	-
Other Products	-	-	-	-	-	-	-	-	-
Refinery Fuel	-	-	-	-	-	-	-	-	-
Sub-Total	-	-	-	-	-	-	-	-	-
Bunkers	-	-	-	-	-	-	-	-	-
Total	-	-	-	-	-	-	-	-	-
Thousand barrels/day									
NGL/LPG/Ethane	-	-	-	-	-	-	-	-	-
Naphtha	-	-	-	-	-	-	-	-	-
Motor Gasoline	-	-	-	-	-	-	-	-	-
Aviation Fuels	-	-	-	-	-	-	-	-	-
Kerosene	-	-	-	-	-	-	-	-	-
Gas Diesel	-	-	-	-	-	-	-	-	-
Heavy Fuel Oil	-	-	-	-	-	-	-	-	-
Other Products	-	-	-	-	-	-	-	-	-
Refinery Fuel	-	-	-	-	-	-	-	-	-
Sub-Total	-	-	-	-	-	-	-	-	-
Bunkers	-	-	-	-	-	-	-	-	-
Total	-	-	-	-	-	-	-	-	-

	1989	1990	1991	1992	1993	1994	1995	1996	1997
Thousand tonnes									
NGL/LPG/Ethane	-	-	-	13	8	11	7	7	9
Naphtha	-	-	-	-	-	-	-	-	-
Motor Gasoline	-	-	-	228	235	287	248	281	305
Aviation Fuels	-	-	-	12	45	15	18	16	22
Kerosene	-	-	-	-	-	49	23	-	-
Gas Diesel	-	-	-	332	337	342	313	374	378
Heavy Fuel Oil	-	-	-	879	898	766	550	549	484
Other Products	-	-	-	41	44	51	43	46	42
Refinery Fuel	-	-	-	-	-	-	-	-	-
Sub-Total	-	-	-	1 505	1 567	1 521	1 202	1 273	1 240
Bunkers	-	-	-	157	158	101	90	93	102
Total	-	-	-	1 662	1 725	1 622	1 292	1 366	1 342
Thousand barrels/day									
NGL/LPG/Ethane	-	-	-	0.4	0.3	0.4	0.2	0.2	0.3
Naphtha	-	-	-	-	-	-	-	-	-
Motor Gasoline	-	-	-	5.3	5.5	6.7	5.8	6.5	7.1
Aviation Fuels	-	-	-	0.3	1.0	0.3	0.4	0.3	0.5
Kerosene	-	-	-	-	-	1.0	0.5	-	-
Gas Diesel	-	-	-	6.8	6.9	7.0	6.4	7.6	7.7
Heavy Fuel Oil	-	-	-	16.0	16.4	14.0	10.0	10.0	8.8
Other Products	-	-	-	0.7	0.8	0.9	0.8	0.8	0.7
Refinery Fuel	-	-	-	-	-	-	-	-	-
Sub-Total	-	-	-	29.5	30.8	30.3	24.1	25.5	25.2
Bunkers	-	-	-	3.0	3.1	1.9	1.7	1.8	1.9
Total	-	-	-	32.5	33.9	32.2	25.8	27.3	27.1

Oil Products Consumption
Consommation de produits pétroliers

Ethiopia

	1971	1973	1978	1983	1984	1985	1986	1987	1988
Thousand tonnes									
NGL/LPG/Ethane	3	4	3	5	5	5	6	5	5
Naphtha	-	-	-	-	-	-	-	-	-
Motor Gasoline	82	84	85	123	109	116	123	124	119
Aviation Fuels	45	42	31	77	106	108	115	156	163
Kerosene	4	8	10	12	10	7	8	13	12
Gas Diesel	190	188	219	204	186	182	242	324	349
Heavy Fuel Oil	103	107	55	75	83	95	98	84	90
Other Products	29	31	38	34	34	33	33	43	33
Refinery Fuel	35	22	26	44	36	37	44	39	35
Sub-Total	491	486	467	574	569	583	669	788	806
Bunkers	22	5	1	10	10	10	13	4	7
Total	513	491	468	584	579	593	682	792	813
Thousand barrels/day									
NGL/LPG/Ethane	0.1	0.1	0.1	0.2	0.2	0.2	0.2	0.2	0.2
Naphtha	-	-	-	-	-	-	-	-	-
Motor Gasoline	1.9	2.0	2.0	2.9	2.5	2.7	2.9	2.9	2.8
Aviation Fuels	1.0	0.9	0.7	1.7	2.3	2.4	2.5	3.4	3.5
Kerosene	0.1	0.2	0.2	0.3	0.2	0.1	0.2	0.3	0.3
Gas Diesel	3.9	3.8	4.5	4.2	3.8	3.7	4.9	6.6	7.1
Heavy Fuel Oil	1.9	2.0	1.0	1.4	1.5	1.7	1.8	1.5	1.6
Other Products	0.5	0.5	0.7	0.6	0.6	0.6	0.6	0.8	0.6
Refinery Fuel	0.7	0.4	0.5	0.9	0.7	0.7	0.9	0.7	0.7
Sub-Total	10.1	10.0	9.7	12.0	11.9	12.2	14.0	16.4	16.8
Bunkers	0.4	0.1	0.0	0.2	0.2	0.2	0.2	0.1	0.1
Total	10.5	10.1	9.7	12.2	12.0	12.4	14.2	16.5	16.9

	1989	1990	1991	1992	1993	1994	1995	1996	1997
Thousand tonnes									
NGL/LPG/Ethane	5	5	5	6	7	7	7	7	3
Naphtha	-	-	-	-	-	-	-	-	-
Motor Gasoline	119	134	145	135	157	147	151	180	147
Aviation Fuels	169	168	175	167	188	200	205	266	272
Kerosene	12	12	13	13	12	13	13	12	8
Gas Diesel	380	381	366	397	421	452	499	555	136
Heavy Fuel Oil	91	111	136	135	86	105	125	118	101
Other Products	38	37	43	41	45	47	48	52	52
Refinery Fuel	35	44	36	36	24	29	33	32	26
Sub-Total	849	892	919	930	940	1 000	1 081	1 222	745
Bunkers	7	14	15	147	187	186	169	153	70
Total	856	906	934	1 077	1 127	1 186	1 250	1 375	815
Thousand barrels/day									
NGL/LPG/Ethane	0.2	0.2	0.2	0.2	0.2	0.2	0.2	0.2	0.1
Naphtha	-	-	-	-	-	-	-	-	-
Motor Gasoline	2.8	3.1	3.4	3.1	3.7	3.4	3.5	4.2	3.4
Aviation Fuels	3.7	3.7	3.8	3.6	4.1	4.4	4.5	5.8	5.9
Kerosene	0.3	0.3	0.3	0.3	0.3	0.3	0.3	0.3	0.2
Gas Diesel	7.8	7.8	7.5	8.1	8.6	9.2	10.2	11.3	2.8
Heavy Fuel Oil	1.7	2.0	2.5	2.5	1.6	1.9	2.3	2.1	1.8
Other Products	0.7	0.7	0.8	0.8	0.8	0.9	0.9	1.0	1.0
Refinery Fuel	0.7	0.8	0.7	0.7	0.5	0.5	0.6	0.6	0.5
Sub-Total	17.7	18.5	19.1	19.2	19.7	20.9	22.5	25.5	15.7
Bunkers	0.1	0.3	0.3	2.7	3.4	3.4	3.1	2.8	1.3
Total	17.8	18.8	19.4	21.9	23.1	24.3	25.6	28.3	17.0

Oil Products Consumption
Consommation de produits pétroliers

Gabon

	1971	1973	1978	1983	1984	1985	1986	1987	1988
Thousand tonnes									
NGL/LPG/Ethane	-	-	-	1	5	5	5	4	4
Naphtha	-	-	-	71	78	77	78	66	54
Motor Gasoline	-	12	13	19	33	38	40	29	36
Aviation Fuels	8	10	22	26	26	27	46	47	54
Kerosene	-	13	19	81	95	98	81	69	79
Gas Diesel	-	-	288	226	143	155	303	261	264
Heavy Fuel Oil	127	135	231	227	178	207	102	106	124
Other Products	-	-	8	13	27	30	28	31	43
Refinery Fuel	-	-	-	-	-	-	-	-	-
Sub-Total	135	170	581	664	585	637	683	613	658
Bunkers	65	65	60	71	70	72	80	68	71
Total	200	235	641	735	655	709	763	681	729
Thousand barrels/day									
NGL/LPG/Ethane	-	-	-	0.0	0.2	0.2	0.2	0.1	0.1
Naphtha	-	-	-	1.7	1.8	1.8	1.8	1.5	1.3
Motor Gasoline	-	0.3	0.3	0.4	0.8	0.9	0.9	0.7	0.8
Aviation Fuels	0.2	0.2	0.5	0.6	0.6	0.6	1.0	1.0	1.2
Kerosene	-	0.3	0.4	1.7	2.0	2.1	1.7	1.5	1.7
Gas Diesel	-	-	5.9	4.6	2.9	3.2	6.2	5.3	5.4
Heavy Fuel Oil	2.3	2.5	4.2	4.1	3.2	3.8	1.9	1.9	2.3
Other Products	-	-	0.1	0.2	0.5	0.6	0.5	0.6	0.8
Refinery Fuel	-	-	-	-	-	-	-	-	-
Sub-Total	2.5	3.2	11.4	13.4	12.0	13.0	14.2	12.7	13.5
Bunkers	1.2	1.2	1.1	1.3	1.3	1.4	1.5	1.3	1.3
Total	3.7	4.5	12.6	14.8	13.3	14.4	15.8	14.0	14.9

	1989	1990	1991	1992	1993	1994	1995	1996	1997
Thousand tonnes									
NGL/LPG/Ethane	9	10	12	12	12	13	13	15	17
Naphtha	-	-	-	10	10	10	10	10	10
Motor Gasoline	50	40	39	34	33	31	35	34	41
Aviation Fuels	68	62	57	56	51	67	62	67	93
Kerosene	25	28	31	30	30	27	30	29	30
Gas Diesel	209	163	172	176	200	174	206	232	298
Heavy Fuel Oil	30	-	-	28	32	32	29	37	32
Other Products	10	18	56	34	54	54	65	64	59
Refinery Fuel	40	20	22	27	28	29	30	30	29
Sub-Total	441	341	389	407	450	437	480	518	609
Bunkers	134	-	-	192	188	136	64	70	54
Total	575	341	389	599	638	573	544	588	663
Thousand barrels/day									
NGL/LPG/Ethane	0.3	0.3	0.4	0.4	0.4	0.4	0.4	0.5	0.5
Naphtha	-	-	-	0.2	0.2	0.2	0.2	0.2	0.2
Motor Gasoline	1.2	0.9	0.9	0.8	0.8	0.7	0.8	0.8	1.0
Aviation Fuels	1.5	1.4	1.3	1.2	1.1	1.5	1.4	1.5	2.1
Kerosene	0.5	0.6	0.7	0.6	0.6	0.6	0.6	0.6	0.6
Gas Diesel	4.3	3.3	3.5	3.6	4.1	3.6	4.2	4.7	6.1
Heavy Fuel Oil	0.5	-	-	0.5	0.6	0.6	0.5	0.7	0.6
Other Products	0.2	0.3	1.1	0.6	1.0	1.0	1.2	1.2	1.1
Refinery Fuel	0.8	0.4	0.5	0.6	0.6	0.6	0.7	0.7	0.6
Sub-Total	9.3	7.3	8.3	8.6	9.5	9.2	10.1	10.8	12.8
Bunkers	2.5	-	-	3.5	3.4	2.5	1.2	1.3	1.0
Total	11.8	7.3	8.3	12.1	12.9	11.7	11.3	12.1	13.8

Oil Products Consumption
Consommation de produits pétroliers

Georgia

	1971	1973	1978	1983	1984	1985	1986	1987	1988
Thousand tonnes									
NGL/LPG/Ethane	-	-	-	-	-	-	-	-	-
Naphtha	-	-	-	-	-	-	-	-	-
Motor Gasoline	-	-	-	-	-	-	-	-	-
Aviation Fuels	-	-	-	-	-	-	-	-	-
Kerosene	-	-	-	-	-	-	-	-	-
Gas Diesel	-	-	-	-	-	-	-	-	-
Heavy Fuel Oil	-	-	-	-	-	-	-	-	-
Other Products	-	-	-	-	-	-	-	-	-
Refinery Fuel	-	-	-	-	-	-	-	-	-
Sub-Total	-	-	-	-	-	-	-	-	-
Bunkers	-	-	-	-	-	-	-	-	-
Total	-	-	-	-	-	-	-	-	-
Thousand barrels/day									
NGL/LPG/Ethane	-	-	-	-	-	-	-	-	-
Naphtha	-	-	-	-	-	-	-	-	-
Motor Gasoline	-	-	-	-	-	-	-	-	-
Aviation Fuels	-	-	-	-	-	-	-	-	-
Kerosene	-	-	-	-	-	-	-	-	-
Gas Diesel	-	-	-	-	-	-	-	-	-
Heavy Fuel Oil	-	-	-	-	-	-	-	-	-
Other Products	-	-	-	-	-	-	-	-	-
Refinery Fuel	-	-	-	-	-	-	-	-	-
Sub-Total	-	-	-	-	-	-	-	-	-
Bunkers	-	-	-	-	-	-	-	-	-
Total	-	-	-	-	-	-	-	-	-

	1989	1990	1991	1992	1993	1994	1995	1996	1997
Thousand tonnes									
NGL/LPG/Ethane	-	-	-	-	1	1	1	15	39
Naphtha	-	-	-	-	-	-	-	-	-
Motor Gasoline	-	-	-	392	234	69	44	438	472
Aviation Fuels	-	-	-	9	10	2	2	46	48
Kerosene	-	-	-	85	42	19	7	30	31
Gas Diesel	-	-	-	255	138	64	23	184	166
Heavy Fuel Oil	-	-	-	516	268	115	44	97	85
Other Products	-	-	-	60	44	20	6	18	54
Refinery Fuel	-	-	-	16	16	16	14	14	12
Sub-Total	-	-	-	1 333	753	306	141	842	907
Bunkers	-	-	-	-	-	-	-	-	-
Total	-	-	-	1 333	753	306	141	842	907
Thousand barrels/day									
NGL/LPG/Ethane	-	-	-	-	0.0	0.0	0.0	0.5	1.2
Naphtha	-	-	-	-	-	-	-	-	-
Motor Gasoline	-	-	-	9.1	5.5	1.6	1.0	10.2	11.0
Aviation Fuels	-	-	-	0.2	0.2	0.0	0.0	1.0	1.0
Kerosene	-	-	-	1.8	0.9	0.4	0.1	0.6	0.7
Gas Diesel	-	-	-	5.2	2.8	1.3	0.5	3.8	3.4
Heavy Fuel Oil	-	-	-	9.4	4.9	2.1	0.8	1.8	1.6
Other Products	-	-	-	1.1	0.8	0.4	0.1	0.3	1.0
Refinery Fuel	-	-	-	0.4	0.4	0.4	0.3	0.3	0.3
Sub-Total	-	-	-	27.2	15.5	6.2	2.9	18.5	20.2
Bunkers	-	-	-	-	-	-	-	-	-
Total	-	-	-	27.2	15.5	6.2	2.9	18.5	20.2

Oil Products Consumption

Consommation de produits pétroliers

Ghana

	1971	1973	1978	1983	1984	1985	1986	1987	1988
Thousand tonnes									
NGL/LPG/Ethane	3	4	7	3	5	5	7	5	5
Naphtha	-	-	-	-	-	-	-	-	-
Motor Gasoline	189	199	257	191	177	215	221	215	278
Aviation Fuels	42	26	33	26	30	32	33	50	51
Kerosene	80	95	135	151	73	78	92	112	108
Gas Diesel	222	236	277	183	251	286	275	238	338
Heavy Fuel Oil	79	80	86	37	48	27	46	40	39
Other Products	18	21	34	62	51	38	37	75	74
Refinery Fuel	37	49	36	15	30	32	27	-	-
Sub-Total	670	710	865	668	665	713	738	735	893
Bunkers	51	59	30	15	20	25	28	27	29
Total	721	769	895	683	685	738	766	762	922
Thousand barrels/day									
NGL/LPG/Ethane	0.1	0.1	0.2	0.1	0.2	0.2	0.2	0.2	0.2
Naphtha	-	-	-	-	-	-	-	-	-
Motor Gasoline	4.4	4.7	6.0	4.5	4.1	5.0	5.2	5.0	6.5
Aviation Fuels	0.9	0.6	0.7	0.6	0.7	0.7	0.7	1.1	1.1
Kerosene	1.7	2.0	2.9	3.2	1.5	1.7	2.0	2.4	2.3
Gas Diesel	4.5	4.8	5.7	3.7	5.1	5.8	5.6	4.9	6.9
Heavy Fuel Oil	1.4	1.5	1.6	0.7	0.9	0.5	0.8	0.7	0.7
Other Products	0.4	0.4	0.7	1.2	1.0	0.7	0.7	1.4	1.4
Refinery Fuel	0.7	1.0	0.8	0.3	0.7	0.7	0.6	-	-
Sub-Total	14.2	15.0	18.5	14.3	14.1	15.3	15.8	15.7	19.0
Bunkers	1.0	1.2	0.6	0.3	0.4	0.5	0.6	0.6	0.6
Total	15.2	16.2	19.1	14.6	14.5	15.8	16.4	16.2	19.6

	1989	1990	1991	1992	1993	1994	1995	1996	1997
Thousand tonnes									
NGL/LPG/Ethane	7	6	10	15	22	30	30	30	30
Naphtha	-	-	-	-	-	-	-	-	-
Motor Gasoline	326	325	243	306	341	324	357	357	357
Aviation Fuels	51	52	51	53	56	56	57	57	57
Kerosene	136	122	120	127	130	130	132	132	132
Gas Diesel	289	275	306	294	328	394	390	390	390
Heavy Fuel Oil	69	46	45	47	45	49	57	53	53
Other Products	75	79	81	76	58	59	60	60	60
Refinery Fuel	-	-	-	61	55	56	55	55	55
Sub-Total	953	905	856	979	1 035	1 098	1 138	1 134	1 134
Bunkers	28	21	22	22	23	24	24	25	25
Total	981	926	878	1 001	1 058	1 122	1 162	1 159	1 159
Thousand barrels/day									
NGL/LPG/Ethane	0.2	0.2	0.3	0.5	0.7	1.0	1.0	1.0	1.0
Naphtha	-	-	-	-	-	-	-	-	-
Motor Gasoline	7.6	7.6	5.7	7.1	8.0	7.6	8.3	8.3	8.3
Aviation Fuels	1.1	1.1	1.1	1.2	1.2	1.2	1.3	1.2	1.3
Kerosene	2.9	2.6	2.5	2.7	2.8	2.8	2.8	2.8	2.8
Gas Diesel	5.9	5.6	6.3	6.0	6.7	8.1	8.0	7.9	8.0
Heavy Fuel Oil	1.3	0.8	0.8	0.9	0.8	0.9	1.0	1.0	1.0
Other Products	1.4	1.5	1.5	1.4	1.1	1.1	1.1	1.1	1.1
Refinery Fuel	-	-	-	1.3	1.2	1.2	1.2	1.2	1.2
Sub-Total	20.4	19.5	18.3	21.1	22.5	23.8	24.7	24.6	24.6
Bunkers	0.6	0.4	0.5	0.4	0.5	0.5	0.5	0.5	0.5
Total	21.0	19.9	18.7	21.5	23.0	24.3	25.2	25.1	25.1

Oil Products Consumption
Consommation de produits pétroliers

Gibraltar

	1971	1973	1978	1983	1984	1985	1986	1987	1988
Thousand tonnes									
NGL/LPG/Ethane	-	-	-	-	-	-	-	-	-
Naphtha	-	-	-	-	-	-	-	-	-
Motor Gasoline	4	4	3	4	4	5	4	8	12
Aviation Fuels	5	7	5	3	3	4	3	5	8
Kerosene	-	-	-	-	-	-	-	-	-
Gas Diesel	10	6	4	10	8	9	18	16	16
Heavy Fuel Oil	8	9	10	14	15	15	15	15	15
Other Products	11	12	9	9	8	10	9	9	12
Refinery Fuel	-	-	-	-	-	-	-	-	-
Sub-Total	38	38	31	40	38	43	49	53	63
Bunkers	179	195	142	122	114	285	362	355	458
Total	217	233	173	162	152	328	411	408	521
Thousand barrels/day									
NGL/LPG/Ethane	-	-	-	-	-	-	-	-	-
Naphtha	-	-	-	-	-	-	-	-	-
Motor Gasoline	0.1	0.1	0.1	0.1	0.1	0.1	0.1	0.2	0.3
Aviation Fuels	0.1	0.2	0.1	0.1	0.1	0.1	0.1	0.1	0.2
Kerosene	-	-	-	-	-	-	-	-	-
Gas Diesel	0.2	0.1	0.1	0.2	0.2	0.2	0.4	0.3	0.3
Heavy Fuel Oil	0.1	0.2	0.2	0.3	0.3	0.3	0.3	0.3	0.3
Other Products	0.2	0.2	0.2	0.2	0.2	0.2	0.2	0.2	0.2
Refinery Fuel	-	-	-	-	-	-	-	-	-
Sub-Total	0.8	0.8	0.6	0.8	0.7	0.9	1.0	1.1	1.3
Bunkers	3.4	3.7	2.7	2.3	2.1	5.3	6.7	6.6	8.5
Total	4.2	4.5	3.3	3.1	2.9	6.1	7.7	7.6	9.7

	1989	1990	1991	1992	1993	1994	1995	1996	1997
Thousand tonnes									
NGL/LPG/Ethane	-	-	-	-	-	-	-	-	-
Naphtha	-	-	-	-	-	-	-	-	-
Motor Gasoline	12	12	12	16	16	16	16	16	16
Aviation Fuels	8	7	8	4	4	4	4	4	4
Kerosene	-	-	-	-	-	-	-	-	-
Gas Diesel	12	15	31	37	37	50	48	48	48
Heavy Fuel Oil	17	25	17	17	30	50	50	50	50
Other Products	12	12	11	15	15	15	15	15	15
Refinery Fuel	-	-	-	-	-	-	-	-	-
Sub-Total	61	71	79	89	102	135	133	133	133
Bunkers	435	443	855	855	858	850	854	854	854
Total	496	514	934	944	960	985	987	987	987
Thousand barrels/day									
NGL/LPG/Ethane	-	-	-	-	-	-	-	-	-
Naphtha	-	-	-	-	-	-	-	-	-
Motor Gasoline	0.3	0.3	0.3	0.4	0.4	0.4	0.4	0.4	0.4
Aviation Fuels	0.2	0.2	0.2	0.1	0.1	0.1	0.1	0.1	0.1
Kerosene	-	-	-	-	-	-	-	-	-
Gas Diesel	0.2	0.3	0.6	0.8	0.8	1.0	1.0	1.0	1.0
Heavy Fuel Oil	0.3	0.5	0.3	0.3	0.5	0.9	0.9	0.9	0.9
Other Products	0.2	0.2	0.2	0.3	0.3	0.3	0.3	0.3	0.3
Refinery Fuel	-	-	-	-	-	-	-	-	-
Sub-Total	1.2	1.4	1.6	1.8	2.1	2.7	2.6	2.6	2.6
Bunkers	8.1	8.2	15.8	15.8	15.9	15.7	15.8	15.8	15.8
Total	9.3	9.6	17.4	17.6	17.9	18.4	18.5	18.4	18.5

Oil Products Consumption
Consommation de produits pétroliers

Guatemala

	1971	1973	1978	1983	1984	1985	1986	1987	1988
Thousand tonnes									
NGL/LPG/Ethane	20	27	39	55	58	69	82	83	89
Naphtha	-	-	-	-	-	-	-	-	-
Motor Gasoline	166	210	298	235	247	250	235	271	271
Aviation Fuels	48	62	48	37	39	38	34	43	40
Kerosene	55	64	50	55	59	55	45	34	34
Gas Diesel	161	214	365	250	293	324	260	322	351
Heavy Fuel Oil	284	283	376	211	235	233	111	139	145
Other Products	32	36	79	92	123	122	55	64	83
Refinery Fuel	37	45	39	28	32	32	25	26	28
Sub-Total	803	941	1 294	963	1 086	1 123	847	982	1 041
Bunkers	57	65	118	125	120	120	115	120	125
Total	860	1 006	1 412	1 088	1 206	1 243	962	1 102	1 166
Thousand barrels/day									
NGL/LPG/Ethane	0.6	0.9	1.2	1.7	1.8	2.2	2.6	2.6	2.8
Naphtha	-	-	-	-	-	-	-	-	-
Motor Gasoline	3.9	4.9	7.0	5.5	5.8	5.8	5.5	6.3	6.3
Aviation Fuels	1.1	1.4	1.1	0.8	0.9	0.8	0.8	0.9	0.9
Kerosene	1.2	1.4	1.1	1.2	1.2	1.2	1.0	0.7	0.7
Gas Diesel	3.3	4.4	7.5	5.1	6.0	6.6	5.3	6.6	7.2
Heavy Fuel Oil	5.2	5.2	6.9	3.9	4.3	4.3	2.0	2.5	2.6
Other Products	0.6	0.7	1.4	1.7	2.3	2.3	1.0	1.2	1.5
Refinery Fuel	0.7	0.9	0.8	0.6	0.6	0.6	0.5	0.5	0.5
Sub-Total	16.5	19.6	26.9	20.5	22.9	23.8	18.7	21.4	22.6
Bunkers	1.1	1.3	2.4	2.6	2.4	2.5	2.4	2.5	2.5
Total	17.7	20.9	29.3	23.0	25.3	26.3	21.0	23.9	25.1

	1989	1990	1991	1992	1993	1994	1995	1996	1997
Thousand tonnes									
NGL/LPG/Ethane	94	100	101	110	120	123	140	153	159
Naphtha	-	-	-	-	-	-	-	-	-
Motor Gasoline	292	314	357	360	366	390	494	526	550
Aviation Fuels	39	41	37	40	40	38	43	47	50
Kerosene	36	38	35	36	39	38	43	46	49
Gas Diesel	364	383	470	651	522	573	666	630	684
Heavy Fuel Oil	142	152	162	196	393	400	494	503	515
Other Products	71	77	84	90	110	126	125	68	73
Refinery Fuel	27	26	26	26	17	18	19	20	21
Sub-Total	1 065	1 131	1 272	1 509	1 607	1 706	2 024	1 993	2 101
Bunkers	120	120	120	120	120	120	120	120	120
Total	1 185	1 251	1 392	1 629	1 727	1 826	2 144	2 113	2 221
Thousand barrels/day									
NGL/LPG/Ethane	3.0	3.2	3.2	3.5	3.8	3.9	4.4	4.8	5.1
Naphtha	-	-	-	-	-	-	-	-	-
Motor Gasoline	6.8	7.3	8.2	8.3	8.6	9.1	11.5	12.3	12.9
Aviation Fuels	0.9	0.9	0.8	0.9	0.9	0.8	0.9	1.0	1.1
Kerosene	0.8	0.8	0.7	0.8	0.8	0.8	0.9	1.0	1.0
Gas Diesel	7.4	7.8	9.6	13.3	10.7	11.7	13.6	12.8	14.0
Heavy Fuel Oil	2.6	2.8	3.0	3.6	7.2	7.3	9.0	9.2	9.4
Other Products	1.3	1.4	1.6	1.7	2.1	2.4	2.3	1.3	1.4
Refinery Fuel	0.5	0.5	0.5	0.5	0.4	0.4	0.4	0.4	0.5
Sub-Total	23.3	24.7	27.6	32.4	34.3	36.4	43.2	42.8	45.3
Bunkers	2.5	2.5	2.5	2.4	2.5	2.5	2.5	2.4	2.5
Total	25.7	27.2	30.0	34.9	36.8	38.9	45.7	45.3	47.7

Oil Products Consumption

Consommation de produits pétroliers

Haiti

	1971	1973	1978	1983	1984	1985	1986	1987	1988
Thousand tonnes									
NGL/LPG/Ethane	1	1	2	4	4	5	4	4	4
Naphtha	-	-	-	-	-	-	-	-	-
Motor Gasoline	25	28	48	43	44	43	45	50	53
Aviation Fuels	6	8	12	11	12	14	19	20	23
Kerosene	4	3	8	12	13	14	21	19	21
Gas Diesel	54	48	76	97	101	101	93	102	115
Heavy Fuel Oil	35	36	70	42	43	42	58	70	64
Other Products	-	-	-	-	-	-	-	-	-
Refinery Fuel	1	-	1	-	-	-	-	-	-
Sub-Total	126	124	217	209	217	219	240	265	280
Bunkers	-	-	-	-	-	-	-	-	-
Total	126	124	217	209	217	219	240	265	280
Thousand barrels/day									
NGL/LPG/Ethane	0.0	0.0	0.1	0.1	0.1	0.2	0.1	0.1	0.1
Naphtha	-	-	-	-	-	-	-	-	-
Motor Gasoline	0.6	0.7	1.1	1.0	1.0	1.0	1.1	1.2	1.2
Aviation Fuels	0.1	0.2	0.3	0.2	0.3	0.3	0.4	0.4	0.5
Kerosene	0.1	0.1	0.2	0.3	0.3	0.3	0.4	0.4	0.4
Gas Diesel	1.1	1.0	1.6	2.0	2.1	2.1	1.9	2.1	2.3
Heavy Fuel Oil	0.6	0.7	1.3	0.8	0.8	0.8	1.1	1.3	1.2
Other Products	-	-	-	-	-	-	-	-	-
Refinery Fuel	0.0	-	0.0	-	-	-	-	-	-
Sub-Total	2.6	2.6	4.5	4.4	4.5	4.6	5.0	5.5	5.8
Bunkers	-	-	-	-	-	-	-	-	-
Total	2.6	2.6	4.5	4.4	4.5	4.6	5.0	5.5	5.8

	1989	1990	1991	1992	1993	1994	1995	1996	1997
Thousand tonnes									
NGL/LPG/Ethane	7	8	7	6	6	6	9	6	6
Naphtha	-	-	-	-	-	-	-	-	-
Motor Gasoline	59	60	56	54	49	16	67	78	85
Aviation Fuels	22	23	20	12	11	-	22	19	19
Kerosene	24	23	18	14	16	5	25	28	35
Gas Diesel	133	128	123	119	114	31	167	198	209
Heavy Fuel Oil	64	68	60	45	21	2	17	24	68
Other Products	-	7	6	9	8	-	4	2	9
Refinery Fuel	-	-	-	-	-	-	-	-	-
Sub-Total	309	317	290	259	225	60	311	355	431
Bunkers	-	-	-	-	-	-	-	-	-
Total	309	317	290	259	225	60	311	355	431
Thousand barrels/day									
NGL/LPG/Ethane	0.2	0.3	0.2	0.2	0.2	0.2	0.3	0.2	0.2
Naphtha	-	-	-	-	-	-	-	-	-
Motor Gasoline	1.4	1.4	1.3	1.3	1.1	0.4	1.6	1.8	2.0
Aviation Fuels	0.5	0.5	0.4	0.3	0.2	-	0.5	0.4	0.4
Kerosene	0.5	0.5	0.4	0.3	0.3	0.1	0.5	0.6	0.7
Gas Diesel	2.7	2.6	2.5	2.4	2.3	0.6	3.4	4.0	4.3
Heavy Fuel Oil	1.2	1.2	1.1	0.8	0.4	0.0	0.3	0.4	1.2
Other Products	-	0.1	0.1	0.2	0.2	-	0.1	0.0	0.2
Refinery Fuel	-	-	-	-	-	-	-	-	-
Sub-Total	6.5	6.6	6.1	5.4	4.8	1.3	6.7	7.5	9.0
Bunkers	-	-	-	-	-	-	-	-	-
Total	6.5	6.6	6.1	5.4	4.8	1.3	6.7	7.5	9.0

Oil Products Consumption
Consommation de produits pétroliers

Honduras

	1971	1973	1978	1983	1984	1985	1986	1987	1988
Thousand tonnes									
NGL/LPG/Ethane	5	7	9	10	12	12	13	13	16
Naphtha	-	-	-	-	-	-	-	-	-
Motor Gasoline	87	96	103	100	103	104	113	123	128
Aviation Fuels	7	9	17	28	21	38	25	27	30
Kerosene	32	34	42	41	35	32	57	60	67
Gas Diesel	149	175	234	287	302	261	257	293	321
Heavy Fuel Oil	57	65	86	92	103	76	80	87	112
Other Products	-	-	-	-	-	-	-	-	-
Refinery Fuel	25	23	27	17	26	24	12	17	15
Sub-Total	362	409	518	575	602	547	557	620	689
Bunkers	-	-	-	-	-	-	-	-	-
Total	362	409	518	575	602	547	557	620	689
Thousand barrels/day									
NGL/LPG/Ethane	0.2	0.2	0.3	0.3	0.4	0.4	0.4	0.4	0.5
Naphtha	-	-	-	-	-	-	-	-	-
Motor Gasoline	2.0	2.2	2.4	2.3	2.4	2.4	2.6	2.9	3.0
Aviation Fuels	0.2	0.2	0.4	0.6	0.5	0.8	0.5	0.6	0.7
Kerosene	0.7	0.7	0.9	0.9	0.7	0.7	1.2	1.3	1.4
Gas Diesel	3.0	3.6	4.8	5.9	6.2	5.3	5.3	6.0	6.5
Heavy Fuel Oil	1.0	1.2	1.6	1.7	1.9	1.4	1.5	1.6	2.0
Other Products	-	-	-	-	-	-	-	-	-
Refinery Fuel	0.5	0.4	0.5	0.3	0.5	0.5	0.2	0.4	0.3
Sub-Total	7.6	8.6	10.8	12.0	12.5	11.5	11.8	13.1	14.5
Bunkers	-	-	-	-	-	-	-	-	-
Total	7.6	8.6	10.8	12.0	12.5	11.5	11.8	13.1	14.5

	1989	1990	1991	1992	1993	1994	1995	1996	1997
Thousand tonnes									
NGL/LPG/Ethane	13	12	13	13	13	23	36	38	43
Naphtha	-	-	-	-	-	-	-	-	-
Motor Gasoline	141	134	138	159	145	214	229	212	215
Aviation Fuels	31	30	30	40	38	24	24	24	24
Kerosene	69	67	67	90	89	57	55	45	46
Gas Diesel	354	315	315	332	331	476	619	491	494
Heavy Fuel Oil	143	137	144	176	172	133	160	329	334
Other Products	-	-	-	-	-	-	-	-	-
Refinery Fuel	11	7	6	7	-	-	-	-	-
Sub-Total	762	702	713	817	788	927	1 123	1 139	1 156
Bunkers	-	-	-	-	-	-	-	-	-
Total	762	702	713	817	788	927	1 123	1 139	1 156
Thousand barrels/day									
NGL/LPG/Ethane	0.4	0.4	0.4	0.4	0.4	0.7	1.1	1.2	1.4
Naphtha	-	-	-	-	-	-	-	-	-
Motor Gasoline	3.3	3.1	3.2	3.7	3.4	5.0	5.4	4.9	5.0
Aviation Fuels	0.7	0.7	0.7	0.9	0.8	0.5	0.5	0.5	0.5
Kerosene	1.5	1.4	1.4	1.9	1.9	1.2	1.2	1.0	1.0
Gas Diesel	7.2	6.4	6.4	6.8	6.8	9.7	12.7	10.0	10.1
Heavy Fuel Oil	2.6	2.5	2.6	3.2	3.1	2.4	2.9	6.0	6.1
Other Products	-	-	-	-	-	-	-	-	-
Refinery Fuel	0.2	0.1	0.1	0.1	-	-	-	-	-
Sub-Total	15.9	14.7	14.9	17.0	16.4	19.6	23.8	23.6	24.1
Bunkers	-	-	-	-	-	-	-	-	-
Total	15.9	14.7	14.9	17.0	16.4	19.6	23.8	23.6	24.1

Oil Products Consumption
Consommation de produits pétroliers

Hong Kong, China

	1971	1973	1978	1983	1984	1985	1986	1987	1988
Thousand tonnes									
NGL/LPG/Ethane	46	72	97	137	146	151	174	174	183
Naphtha	26	32	63	149	189	204	254	166	307
Motor Gasoline	114	132	165	210	198	191	189	196	229
Aviation Fuels	447	523	633	760	818	809	930	982	1 221
Kerosene	210	161	174	133	127	113	123	126	110
Gas Diesel	666	805	1 080	1 103	1 058	965	1 307	1 226	1 509
Heavy Fuel Oil	1 860	1 867	3 365	2 337	1 992	1 638	1 329	975	987
Other Products	69	84	77	97	108	110	117	120	103
Refinery Fuel	-	-	-	-	-	-	-	-	-
Sub-Total	3 438	3 676	5 654	4 926	4 636	4 181	4 423	3 965	4 649
Bunkers	642	1 179	551	995	317	1 003	838	911	1 075
Total	4 080	4 855	6 205	5 921	4 953	5 184	5 261	4 876	5 724
Thousand barrels/day									
NGL/LPG/Ethane	1.5	2.3	3.1	4.4	4.6	4.8	5.5	5.5	5.8
Naphtha	0.6	0.7	1.5	3.5	4.4	4.8	5.9	3.9	7.1
Motor Gasoline	2.7	3.1	3.9	4.9	4.6	4.5	4.4	4.6	5.3
Aviation Fuels	9.7	11.4	13.8	16.5	17.7	17.6	20.2	21.3	26.5
Kerosene	4.5	3.4	3.7	2.8	2.7	2.4	2.6	2.7	2.3
Gas Diesel	13.6	16.5	22.1	22.5	21.6	19.7	26.7	25.1	30.8
Heavy Fuel Oil	33.9	34.1	61.4	42.6	36.2	29.9	24.3	17.8	18.0
Other Products	1.3	1.6	1.4	1.8	2.0	2.0	2.2	2.2	1.9
Refinery Fuel	-	-	-	-	-	-	-	-	-
Sub-Total	67.7	73.0	110.7	99.0	93.8	85.6	91.8	83.0	97.6
Bunkers	11.7	21.5	10.1	18.5	6.1	18.6	15.6	17.1	20.0
Total	79.5	94.5	120.8	117.5	99.9	104.2	107.4	100.1	117.6

	1989	1990	1991	1992	1993	1994	1995	1996	1997
Thousand tonnes									
NGL/LPG/Ethane	189	203	202	215	195	188	186	161	147
Naphtha	235	199	278	445	463	521	528	517	453
Motor Gasoline	245	237	270	295	322	340	334	335	344
Aviation Fuels	1 418	1 781	1 581	2 092	2 373	2 730	2 919	3 023	3 202
Kerosene	97	87	78	82	69	64	62	55	53
Gas Diesel	1 446	1 661	2 053	2 353	2 659	2 951	2 808	2 604	2 654
Heavy Fuel Oil	993	534	441	905	548	196	232	100	85
Other Products	156	155	124	143	146	189	175	185	313
Refinery Fuel	-	-	-	-	-	-	-	-	-
Sub-Total	4 779	4 857	5 027	6 530	6 775	7 179	7 244	6 980	7 251
Bunkers	1 367	1 457	1 244	1 527	1 575	1 706	2 303	2 377	2 149
Total	6 146	6 314	6 271	8 057	8 350	8 885	9 547	9 357	9 400
Thousand barrels/day									
NGL/LPG/Ethane	6.0	6.5	6.4	6.8	6.2	6.0	5.9	5.1	4.7
Naphtha	5.5	4.6	6.5	10.3	10.8	12.1	12.3	12.0	10.5
Motor Gasoline	5.7	5.5	6.3	6.9	7.5	7.9	7.8	7.8	8.0
Aviation Fuels	30.8	38.7	34.3	45.3	51.6	59.3	63.4	65.5	69.6
Kerosene	2.1	1.8	1.7	1.7	1.5	1.4	1.3	1.2	1.1
Gas Diesel	29.6	33.9	42.0	48.0	54.3	60.3	57.4	53.1	54.2
Heavy Fuel Oil	18.1	9.7	8.0	16.5	10.0	3.6	4.2	1.8	1.6
Other Products	2.9	2.9	2.3	2.6	2.7	3.5	3.2	3.4	5.9
Refinery Fuel	-	-	-	-	-	-	-	-	-
Sub-Total	100.6	103.7	107.5	138.1	144.5	154.1	155.6	149.8	155.6
Bunkers	25.7	27.3	23.4	28.4	29.4	31.9	43.4	44.5	40.3
Total	126.3	131.0	130.8	166.5	173.9	186.0	199.0	194.4	195.9

Oil Products Consumption
Consommation de produits pétroliers

India

	1971	1973	1978	1983	1984	1985	1986	1987	1988
Thousand tonnes									
NGL/LPG/Ethane	202	263	408	702	886	1 275	1 529	1 740	1 979
Naphtha	1 107	1 454	2 377	2 696	3 153	3 107	3 194	2 875	3 258
Motor Gasoline	1 553	1 630	1 499	1 834	2 026	2 247	2 428	2 735	2 974
Aviation Fuels	824	872	1 174	1 207	1 323	1 441	1 588	1 627	1 691
Kerosene	3 595	3 461	3 952	5 392	5 838	6 168	6 522	7 088	7 595
Gas Diesel	5 385	6 510	9 464	13 554	14 485	15 858	16 852	18 385	19 704
Heavy Fuel Oil	4 770	5 729	6 260	7 527	7 786	7 955	8 016	8 150	8 360
Other Products	2 090	2 433	2 245	2 504	2 335	2 642	2 846	2 946	3 293
Refinery Fuel	1 077	1 128	1 409	1 879	2 005	2 314	2 606	2 504	2 454
Sub-Total	20 603	23 480	28 788	37 295	39 837	43 007	45 581	48 050	51 308
Bunkers	230	234	183	200	204	109	122	228	278
Total	20 833	23 714	28 971	37 495	40 041	43 116	45 703	48 278	51 586
Thousand barrels/day									
NGL/LPG/Ethane	6.4	8.4	13.0	22.3	28.1	40.4	48.4	55.1	62.6
Naphtha	25.8	33.9	55.4	62.8	73.2	72.4	74.4	67.0	75.7
Motor Gasoline	36.3	38.1	35.0	42.9	47.2	52.5	56.7	63.9	69.3
Aviation Fuels	18.0	19.0	25.6	26.3	28.7	31.4	34.6	35.4	36.7
Kerosene	76.2	73.4	83.8	114.3	123.5	130.8	138.3	150.3	160.6
Gas Diesel	110.1	133.1	193.4	277.0	295.2	324.1	344.4	375.8	401.6
Heavy Fuel Oil	87.0	104.5	114.2	137.3	141.7	145.2	146.3	148.7	152.1
Other Products	37.1	43.3	40.1	44.6	41.8	47.3	50.9	52.4	58.2
Refinery Fuel	20.9	21.9	27.3	36.5	38.7	44.8	50.4	48.7	47.7
Sub-Total	417.8	475.5	587.8	764.0	818.1	888.8	944.4	997.2	1 064.5
Bunkers	4.3	4.3	3.5	3.8	3.8	2.1	2.3	4.3	5.2
Total	422.0	479.8	591.3	767.7	821.9	890.9	946.7	1 001.5	1 069.7

	1989	1990	1991	1992	1993	1994	1995	1996	1997
Thousand tonnes									
NGL/LPG/Ethane	2 362	2 643	2 862	3 159	3 272	3 823	4 168	4 861	4 818
Naphtha	3 421	3 397	3 418	3 380	3 310	3 258	3 701	3 623	4 754
Motor Gasoline	3 370	3 583	3 587	3 592	3 761	4 037	4 561	4 965	4 967
Aviation Fuels	1 767	1 711	1 563	1 582	1 705	1 894	2 072	2 191	2 228
Kerosene	8 089	8 423	8 412	8 573	8 618	8 931	9 334	9 577	9 869
Gas Diesel	21 820	21 346	24 702	26 277	27 830	30 269	34 354	36 850	38 485
Heavy Fuel Oil	8 700	8 855	8 765	9 207	8 914	9 287	9 590	10 954	11 238
Other Products	3 608	3 673	3 644	3 873	3 556	3 617	4 684	5 181	4 286
Refinery Fuel	2 678	2 706	2 621	2 793	2 830	3 093	3 214	3 456	3 614
Sub-Total	55 815	56 337	59 574	62 436	63 796	68 209	75 678	81 658	84 259
Bunkers	235	414	393	454	513	539	539	539	526
Total	56 050	56 751	59 967	62 890	64 309	68 748	76 217	82 197	84 785
Thousand barrels/day									
NGL/LPG/Ethane	74.8	83.6	90.5	99.6	103.6	120.7	131.8	152.8	152.0
Naphtha	79.7	79.1	79.6	78.5	77.1	75.9	86.2	84.1	110.7
Motor Gasoline	78.8	83.7	83.8	83.7	87.9	94.3	106.6	115.7	116.1
Aviation Fuels	38.4	37.2	34.0	34.3	37.1	41.2	45.1	47.5	48.5
Kerosene	171.5	178.6	178.4	181.3	182.7	189.4	197.9	202.5	209.3
Gas Diesel	446.0	436.3	504.9	535.6	568.8	618.6	702.1	751.1	786.6
Heavy Fuel Oil	158.7	161.6	159.9	167.5	162.7	169.5	175.0	199.3	205.1
Other Products	63.9	65.3	64.8	68.6	63.4	64.2	83.8	92.1	76.0
Refinery Fuel	52.1	52.6	51.0	54.2	55.0	60.2	62.5	67.1	70.7
Sub-Total	1 163.9	1 178.1	1 246.9	1 303.4	1 338.2	1 434.0	1 591.0	1 712.4	1 774.9
Bunkers	4.4	8.0	7.6	8.7	9.8	10.3	10.3	10.3	10.0
Total	1 168.3	1 186.1	1 254.5	1 312.1	1 348.0	1 444.3	1 601.3	1 722.6	1 784.9

Oil Products Consumption
Consommation de produits pétroliers

Indonesia

	1971	1973	1978	1983	1984	1985	1986	1987	1988
Thousand tonnes									
NGL/LPG/Ethane	3	13	46	86	110	145	198	217	234
Naphtha	-	-	-	-	-	334	340	350	350
Motor Gasoline	1 236	1 425	2 393	2 889	2 967	3 029	3 295	3 578	3 833
Aviation Fuels	143	170	551	483	496	498	488	645	708
Kerosene	2 430	2 978	5 367	6 199	5 837	5 674	5 635	5 602	5 803
Gas Diesel	2 433	2 972	5 306	8 200	8 292	8 169	7 783	8 363	9 307
Heavy Fuel Oil	992	1 109	2 079	3 052	2 736	3 186	2 710	2 926	2 715
Other Products	439	1 233	483	360	258	382	415	420	633
Refinery Fuel	181	201	318	711	1 035	1 279	1 609	1 670	1 784
Sub-Total	7 857	10 101	16 543	21 980	21 731	22 696	22 473	23 771	25 367
Bunkers	226	428	274	148	268	220	160	175	237
Total	8 083	10 529	16 817	22 128	21 999	22 916	22 633	23 946	25 604
Thousand barrels/day									
NGL/LPG/Ethane	0.1	0.4	1.5	2.7	3.5	4.6	6.3	6.9	7.4
Naphtha	-	-	-	-	-	7.8	7.9	8.2	8.1
Motor Gasoline	28.9	33.3	55.9	67.5	69.1	70.8	77.0	83.6	89.3
Aviation Fuels	3.2	3.7	12.0	10.5	10.8	10.8	10.6	14.0	15.4
Kerosene	51.5	63.2	113.8	131.5	123.4	120.3	119.5	118.8	122.7
Gas Diesel	49.7	60.7	108.4	167.6	169.0	167.0	159.1	170.9	189.7
Heavy Fuel Oil	18.1	20.2	37.9	55.7	49.8	58.1	49.4	53.4	49.4
Other Products	8.8	24.9	9.7	7.1	5.2	7.6	8.2	8.2	12.1
Refinery Fuel	3.5	3.9	6.1	13.2	19.3	23.9	29.9	31.1	33.1
Sub-Total	163.8	210.4	345.4	455.9	450.1	470.9	468.0	495.1	527.3
Bunkers	4.2	7.9	5.1	2.8	4.9	4.0	3.1	3.2	4.4
Total	168.1	218.3	350.5	458.7	455.1	474.9	471.1	498.3	531.7

	1989	1990	1991	1992	1993	1994	1995	1996	1997
Thousand tonnes									
NGL/LPG/Ethane	265	318	359	349	469	553	639	721	565
Naphtha	667	409	76	104	151	409	144	298	189
Motor Gasoline	4 184	4 687	4 982	5 309	5 495	6 150	6 776	7 396	7 435
Aviation Fuels	776	904	893	1 023	1 157	1 292	1 389	1 604	1 678
Kerosene	6 022	6 380	6 561	6 811	7 072	7 250	7 518	7 948	8 093
Gas Diesel	9 956	11 419	12 538	13 688	15 866	15 294	15 986	17 353	19 466
Heavy Fuel Oil	2 775	3 740	4 332	4 271	4 656	3 553	3 465	3 723	7 444
Other Products	1 541	620	768	585	869	937	997	1 130	1 130
Refinery Fuel	1 508	1 993	1 984	1 988	2 035	1 952	2 053	1 958	1 948
Sub-Total	27 694	30 470	32 493	34 128	37 770	37 390	38 967	42 131	47 948
Bunkers	251	541	248	254	243	203	89	344	344
Total	27 945	31 011	32 741	34 382	38 013	37 593	39 056	42 475	48 292
Thousand barrels/day									
NGL/LPG/Ethane	8.4	10.1	11.4	11.1	14.9	17.6	20.3	22.9	18.0
Naphtha	15.5	9.5	1.8	2.4	3.5	9.5	3.4	6.9	4.4
Motor Gasoline	97.8	109.5	116.4	123.7	128.4	143.7	158.4	172.4	173.8
Aviation Fuels	16.9	19.7	19.4	22.2	25.2	28.1	30.2	34.8	36.5
Kerosene	127.7	135.3	139.1	144.0	150.0	153.7	159.4	168.1	171.6
Gas Diesel	203.5	233.4	256.3	279.0	324.3	312.6	326.7	353.7	397.9
Heavy Fuel Oil	50.6	68.2	79.0	77.7	85.0	64.8	63.2	67.7	135.8
Other Products	29.6	10.7	13.4	10.1	15.3	16.7	17.8	20.3	20.4
Refinery Fuel	28.2	38.6	38.1	38.4	38.9	37.3	39.3	37.9	37.8
Sub-Total	578.2	635.0	675.0	708.6	785.4	784.1	818.8	884.7	996.1
Bunkers	4.7	10.3	4.6	4.7	4.5	3.7	1.7	6.3	6.3
Total	582.9	645.3	679.6	713.3	789.9	787.9	820.4	891.0	1 002.4

Oil Products Consumption
Consommation de produits pétroliers

Iran

	1971	1973	1978	1983	1984	1985	1986	1987	1988
Thousand tonnes									
NGL/LPG/Ethane	402	451	494	530	869	850	750	741	830
Naphtha	311	346	-	-	-	-	-	-	10
Motor Gasoline	1 300	1 854	3 771	4 377	4 752	5 166	4 444	4 736	4 713
Aviation Fuels	310	840	1 187	380	561	595	489	524	420
Kerosene	1 713	2 321	4 740	6 291	6 450	6 752	4 780	5 736	5 671
Gas Diesel	2 884	3 947	7 713	11 095	12 411	11 556	13 315	14 261	14 137
Heavy Fuel Oil	1 588	2 217	4 481	5 379	9 029	10 086	12 297	11 876	12 193
Other Products	640	767	1 307	2 004	1 401	1 479	2 080	2 285	2 471
Refinery Fuel	1 559	1 608	2 042	1 526	1 852	1 942	2 837	2 222	2 102
Sub-Total	10 707	14 351	25 735	31 582	37 325	38 426	40 992	42 381	42 547
Bunkers	5 819	7 010	3 443	1 720	1 700	1 650	1 600	1 350	1 500
Total	16 526	21 361	29 178	33 302	39 025	40 076	42 592	43 731	44 047
Thousand barrels/day									
NGL/LPG/Ethane	12.8	14.3	15.7	16.8	27.5	27.0	23.8	23.6	26.3
Naphtha	7.2	8.1	-	-	-	-	-	-	0.2
Motor Gasoline	30.4	43.3	88.1	102.3	110.8	120.7	103.9	110.7	109.8
Aviation Fuels	6.8	18.3	25.9	8.3	12.3	13.1	10.8	11.6	9.3
Kerosene	36.3	49.2	100.5	133.4	136.4	143.2	101.4	121.6	119.9
Gas Diesel	58.9	80.7	157.6	226.8	253.0	236.2	272.1	291.5	288.1
Heavy Fuel Oil	29.0	40.5	81.8	98.1	164.3	184.0	224.4	216.7	221.9
Other Products	12.3	14.7	25.1	38.4	26.8	28.4	39.9	43.8	47.3
Refinery Fuel	28.4	29.3	37.3	27.8	33.7	35.4	51.8	40.5	38.3
Sub-Total	222.1	298.4	532.0	652.1	764.8	788.0	828.0	860.0	861.1
Bunkers	106.5	128.1	62.9	31.6	31.4	30.7	29.6	25.2	28.0
Total	328.6	426.6	594.9	683.7	796.1	818.7	857.6	885.1	889.1

	1989	1990	1991	1992	1993	1994	1995	1996	1997
Thousand tonnes									
NGL/LPG/Ethane	1 290	1 370	1 600	1 693	1 667	1 694	1 689	1 816	1 866
Naphtha	10	130	140	140	140	140	253	348	351
Motor Gasoline	5 451	6 076	6 800	7 050	7 912	8 254	8 438	8 010	8 297
Aviation Fuels	410	561	590	605	592	610	613	685	687
Kerosene	5 140	5 300	5 975	6 058	7 203	7 653	8 519	9 058	8 798
Gas Diesel	14 716	16 146	17 220	17 549	18 217	18 745	19 398	20 291	21 070
Heavy Fuel Oil	12 401	11 900	11 900	12 463	12 723	12 816	13 242	14 248	13 401
Other Products	2 371	3 000	3 300	3 350	3 655	3 665	2 969	3 060	3 067
Refinery Fuel	2 387	2 400	2 200	2 200	2 091	2 114	2 182	2 288	2 193
Sub-Total	44 176	46 883	49 725	51 108	54 200	55 691	57 303	59 804	59 730
Bunkers	2 207	2 700	2 700	1 517	851	831	851	555	555
Total	46 383	49 583	52 425	52 625	55 051	56 522	58 154	60 359	60 285
Thousand barrels/day									
NGL/LPG/Ethane	41.0	43.5	50.8	53.7	53.0	53.8	53.7	57.6	59.3
Naphtha	0.2	3.0	3.3	3.3	3.3	3.3	5.9	8.1	8.2
Motor Gasoline	127.4	142.0	158.9	164.3	184.9	192.9	197.2	186.7	193.9
Aviation Fuels	9.1	12.5	13.1	13.4	13.1	13.5	13.6	15.2	15.2
Kerosene	109.0	112.4	126.7	128.1	152.7	162.3	180.6	191.6	186.6
Gas Diesel	300.8	330.0	351.9	357.7	372.3	383.1	396.5	413.6	430.6
Heavy Fuel Oil	226.3	217.1	217.1	226.8	232.2	233.8	241.6	259.3	244.5
Other Products	45.5	57.5	63.3	64.1	70.1	70.3	50.7	52.1	52.4
Refinery Fuel	43.6	43.8	40.1	40.0	39.5	39.9	41.2	43.3	41.8
Sub-Total	902.8	961.9	1 025.3	1 051.3	1 121.0	1 152.9	1 181.0	1 227.3	1 232.4
Bunkers	40.3	49.3	49.3	27.8	15.7	15.3	15.7	10.3	10.3
Total	943.1	1 011.1	1 074.6	1 079.1	1 136.7	1 168.3	1 196.7	1 237.6	1 242.8

Oil Products Consumption
Consommation de produits pétroliers

Iraq

	1971	1973	1978	1983	1984	1985	1986	1987	1988
Thousand tonnes									
NGL/LPG/Ethane	42	57	325	459	528	629	740	861	1 046
Naphtha	65	32	90	-	-	-	-	-	-
Motor Gasoline	419	465	963	2 072	2 173	2 345	2 500	2 520	2 650
Aviation Fuels	75	117	284	294	336	354	629	745	675
Kerosene	649	635	889	900	800	760	750	790	800
Gas Diesel	543	700	1 990	3 695	3 950	4 200	4 115	4 195	4 350
Heavy Fuel Oil	1 343	1 580	1 719	3 125	3 013	3 425	3 000	3 400	3 400
Other Products	299	422	797	1 101	1 117	1 182	1 877	1 649	1 963
Refinery Fuel	149	176	191	1 089	1 387	1 509	1 434	1 436	1 434
Sub-Total	3 584	4 184	7 248	12 735	13 304	14 404	15 045	15 596	16 318
Bunkers	84	80	120	140	150	150	100	200	130
Total	3 668	4 264	7 368	12 875	13 454	14 554	15 145	15 796	16 448
Thousand barrels/day									
NGL/LPG/Ethane	1.3	1.8	10.3	14.6	16.7	20.0	23.5	27.4	33.2
Naphtha	1.5	0.7	2.1	-	-	-	-	-	-
Motor Gasoline	9.8	10.9	22.5	48.4	50.6	54.8	58.4	58.9	61.8
Aviation Fuels	1.6	2.6	6.2	6.4	7.3	7.7	13.7	16.2	14.6
Kerosene	13.8	13.5	18.9	19.1	16.9	16.1	15.9	16.8	16.9
Gas Diesel	11.1	14.3	40.7	75.5	80.5	85.8	84.1	85.7	88.7
Heavy Fuel Oil	24.5	28.8	31.4	57.0	54.8	62.5	54.7	62.0	61.9
Other Products	5.7	7.8	14.9	20.4	20.5	21.7	35.8	31.2	37.0
Refinery Fuel	2.7	3.2	3.5	21.0	26.7	29.1	27.8	27.8	27.7
Sub-Total	72.0	83.6	150.4	262.5	274.2	297.8	314.0	326.0	341.7
Bunkers	1.5	1.5	2.2	2.6	2.7	2.7	1.8	3.6	2.4
Total	73.6	85.1	152.5	265.0	276.9	300.5	315.8	329.6	344.1

	1989	1990	1991	1992	1993	1994	1995	1996	1997
Thousand tonnes									
NGL/LPG/Ethane	1 178	1 020	300	518	1 182	1 186	1 166	1 131	1 159
Naphtha	700	600	450	600	500	500	492	477	489
Motor Gasoline	2 750	2 800	2 000	2 460	2 740	2 970	2 990	2 858	2 936
Aviation Fuels	716	916	250	358	385	440	425	412	423
Kerosene	800	600	560	675	885	1 012	978	948	973
Gas Diesel	4 600	4 700	4 300	4 800	5 250	5 387	5 322	5 322	5 467
Heavy Fuel Oil	3 400	3 600	3 840	4 010	4 700	5 309	5 218	5 140	5 287
Other Products	1 582	1 778	1 335	2 120	2 490	2 588	2 574	2 550	2 612
Refinery Fuel	1 420	1 269	870	1 130	1 420	1 670	1 634	1 599	1 643
Sub-Total	17 146	17 283	13 905	16 671	19 552	21 062	20 799	20 437	20 989
Bunkers	130	130	50	-	-	-	-	-	-
Total	17 276	17 413	13 955	16 671	19 552	21 062	20 799	20 437	20 989
Thousand barrels/day									
NGL/LPG/Ethane	37.4	32.4	9.5	16.4	37.6	37.7	37.1	35.8	36.8
Naphtha	16.3	14.0	10.5	13.9	11.6	11.6	11.5	11.1	11.4
Motor Gasoline	64.3	65.4	46.7	57.3	64.0	69.4	69.9	66.6	68.6
Aviation Fuels	15.6	19.9	5.4	7.8	8.4	9.6	9.2	8.9	9.2
Kerosene	17.0	12.7	11.9	14.3	18.8	21.5	20.7	20.0	20.6
Gas Diesel	94.0	96.1	87.9	97.8	107.3	110.1	108.8	108.5	111.7
Heavy Fuel Oil	62.0	65.7	70.1	73.0	85.8	96.9	95.2	93.5	96.5
Other Products	29.3	33.3	25.7	41.3	48.9	50.6	50.4	49.8	51.2
Refinery Fuel	27.5	24.5	16.9	21.8	27.5	32.2	31.5	30.7	31.7
Sub-Total	363.4	364.1	284.6	343.6	409.8	439.6	434.2	425.1	437.7
Bunkers	2.4	2.4	0.9	-	-	-	-	-	-
Total	365.7	366.4	285.5	343.6	409.8	439.6	434.2	425.1	437.7

Oil Products Consumption
Consommation de produits pétroliers

Israel

	1971	1973	1978	1983	1984	1985	1986	1987	1988
Thousand tonnes									
NGL/LPG/Ethane	111	123	126	140	135	127	201	223	225
Naphtha	64	31	206	274	273	226	317	383	412
Motor Gasoline	613	675	748	932	932	1 010	1 115	1 246	1 321
Aviation Fuels	-	-	-	-	-	-	-	-	-
Kerosene	575	789	730	704	662	640	630	705	683
Gas Diesel	770	866	951	989	914	843	989	1 078	1 091
Heavy Fuel Oil	2 606	2 957	3 700	3 354	2 927	2 851	2 859	3 118	3 562
Other Products	329	416	352	214	171	125	146	166	198
Refinery Fuel	162	179	233	200	187	560	578	632	634
Sub-Total	5 230	6 036	7 046	6 807	6 201	6 382	6 835	7 551	8 126
Bunkers	-	-	-	-	-	113	118	136	116
Total	5 230	6 036	7 046	6 807	6 201	6 495	6 953	7 687	8 242
Thousand barrels/day									
NGL/LPG/Ethane	3.5	3.9	4.0	4.4	4.3	4.0	6.4	7.1	7.1
Naphtha	1.5	0.7	4.8	6.4	6.3	5.3	7.4	8.9	9.6
Motor Gasoline	14.3	15.8	17.5	21.8	21.7	23.6	26.1	29.1	30.8
Aviation Fuels	-	-	-	-	-	-	-	-	-
Kerosene	12.2	16.7	15.5	14.9	14.0	13.6	13.4	15.0	14.4
Gas Diesel	15.7	17.7	19.4	20.2	18.6	17.2	20.2	22.0	22.2
Heavy Fuel Oil	47.6	54.0	67.5	61.2	53.3	52.0	52.2	56.9	64.8
Other Products	6.4	8.1	6.9	4.1	3.3	2.4	2.8	3.2	3.8
Refinery Fuel	3.0	3.3	4.3	3.6	3.4	10.2	10.5	11.5	11.5
Sub-Total	104.2	120.2	139.8	136.7	124.9	128.3	138.9	153.7	164.3
Bunkers	-	-	-	-	-	2.1	2.2	2.6	2.2
Total	104.2	120.2	139.8	136.7	124.9	130.5	141.2	156.3	166.5

	1989	1990	1991	1992	1993	1994	1995	1996	1997
Thousand tonnes									
NGL/LPG/Ethane	209	228	215	215	240	264	296	315	301
Naphtha	385	403	478	482	450	550	645	730	875
Motor Gasoline	1 400	1 484	1 546	1 650	1 720	1 834	1 970	2 029	1 903
Aviation Fuels	-	-	-	-	-	-	-	-	-
Kerosene	720	735	710	725	838	852	936	993	1 104
Gas Diesel	1 164	1 291	1 351	1 527	1 514	1 739	2 049	1 996	1 788
Heavy Fuel Oil	3 524	3 443	3 269	3 468	3 463	3 726	3 944	3 742	4 182
Other Products	184	291	283	349	534	488	540	527	424
Refinery Fuel	722	712	721	745	856	794	637	500	500
Sub-Total	8 308	8 587	8 573	9 161	9 615	10 247	11 017	10 832	11 077
Bunkers	130	122	153	191	191	203	207	94	181
Total	8 438	8 709	8 726	9 352	9 806	10 450	11 224	10 926	11 258
Thousand barrels/day									
NGL/LPG/Ethane	6.6	7.2	6.8	6.8	7.6	8.4	9.4	9.9	9.5
Naphtha	9.0	9.4	11.1	11.2	10.5	12.8	15.0	17.0	20.4
Motor Gasoline	32.7	34.7	36.1	38.5	40.2	42.9	46.0	47.3	44.5
Aviation Fuels	-	-	-	-	-	-	-	-	-
Kerosene	15.3	15.6	15.1	15.3	17.8	18.1	19.8	21.0	23.4
Gas Diesel	23.8	26.4	27.6	31.1	30.9	35.5	41.9	40.7	36.5
Heavy Fuel Oil	64.3	62.8	59.6	63.1	63.2	68.0	72.0	68.1	76.3
Other Products	3.5	5.6	5.4	6.7	10.2	9.4	10.4	10.1	8.1
Refinery Fuel	13.2	13.0	13.2	13.6	15.6	14.5	11.6	9.1	9.1
Sub-Total	168.4	174.7	175.0	186.3	196.0	209.5	226.1	223.1	227.9
Bunkers	2.4	2.3	2.9	3.6	3.6	3.9	4.0	1.8	3.5
Total	170.8	177.0	177.9	189.9	199.7	213.3	230.1	224.9	231.4

Oil Products Consumption
Consommation de produits pétroliers

Ivory Coast

	1971	1973	1978	1983	1984	1985	1986	1987	1988
Thousand tonnes									
NGL/LPG/Ethane	6	7	11	17	23	25	19	18	22
Naphtha	-	-	-	-	-	-	-	-	-
Motor Gasoline	160	177	246	219	200	201	198	190	194
Aviation Fuels	42	50	78	81	87	91	87	87	109
Kerosene	39	39	49	66	70	72	66	62	62
Gas Diesel	257	268	410	470	376	372	377	391	365
Heavy Fuel Oil	282	331	483	338	318	199	147	150	200
Other Products	24	32	63	75	168	80	139	180	130
Refinery Fuel	21	43	72	81	78	104	98	100	105
Sub-Total	831	947	1 412	1 347	1 320	1 144	1 131	1 178	1 187
Bunkers	20	15	2	199	196	233	168	170	220
Total	851	962	1 414	1 546	1 516	1 377	1 299	1 348	1 407
Thousand barrels/day									
NGL/LPG/Ethane	0.2	0.2	0.4	0.5	0.7	0.8	0.6	0.6	0.7
Naphtha	-	-	-	-	-	-	-	-	-
Motor Gasoline	3.7	4.1	5.7	5.1	4.7	4.7	4.6	4.4	4.5
Aviation Fuels	0.9	1.1	1.7	1.8	1.9	2.0	1.9	1.9	2.4
Kerosene	0.8	0.8	1.0	1.4	1.5	1.5	1.4	1.3	1.3
Gas Diesel	5.3	5.5	8.4	9.6	7.7	7.6	7.7	8.0	7.4
Heavy Fuel Oil	5.1	6.0	8.8	6.2	5.8	3.6	2.7	2.7	3.6
Other Products	0.5	0.6	1.2	1.4	3.2	1.5	2.7	3.5	2.5
Refinery Fuel	0.4	0.9	1.4	1.6	1.5	2.1	1.9	2.0	2.1
Sub-Total	16.9	19.3	28.7	27.6	27.0	23.8	23.5	24.4	24.5
Bunkers	0.4	0.3	0.0	3.8	3.7	4.5	3.3	3.3	4.3
Total	17.3	19.5	28.7	31.4	30.7	28.3	26.8	27.7	28.8

	1989	1990	1991	1992	1993	1994	1995	1996	1997
Thousand tonnes									
NGL/LPG/Ethane	22	22	25	25	21	23	23	35	37
Naphtha	-	-	-	-	-	-	-	-	-
Motor Gasoline	180	166	169	155	101	103	108	156	150
Aviation Fuels	102	84	85	85	81	83	87	100	104
Kerosene	60	59	64	56	52	53	56	65	65
Gas Diesel	351	339	327	338	522	533	560	552	573
Heavy Fuel Oil	218	205	211	180	199	203	213	148	154
Other Products	130	51	57	47	52	53	55	55	64
Refinery Fuel	105	105	105	95	116	118	124	106	112
Sub-Total	1 168	1 031	1 043	981	1 144	1 169	1 226	1 217	1 259
Bunkers	209	164	204	193	174	177	174	172	172
Total	1 377	1 195	1 247	1 174	1 318	1 346	1 400	1 389	1 431
Thousand barrels/day									
NGL/LPG/Ethane	0.7	0.7	0.8	0.8	0.7	0.7	0.7	1.1	1.2
Naphtha	-	-	-	-	-	-	-	-	-
Motor Gasoline	4.2	3.9	4.0	3.6	2.4	2.4	2.5	3.6	3.5
Aviation Fuels	2.2	1.8	1.8	1.8	1.8	1.8	1.9	2.2	2.3
Kerosene	1.3	1.3	1.4	1.2	1.1	1.1	1.2	1.4	1.4
Gas Diesel	7.2	6.9	6.7	6.9	10.7	10.9	11.4	11.3	11.7
Heavy Fuel Oil	4.0	3.7	3.9	3.3	3.6	3.7	3.9	2.7	2.8
Other Products	2.5	1.0	1.1	0.9	1.0	1.0	1.1	1.1	1.2
Refinery Fuel	2.1	2.1	2.1	1.9	2.3	2.3	2.4	2.1	2.3
Sub-Total	24.1	21.4	21.6	20.4	23.5	24.0	25.2	25.4	26.3
Bunkers	4.1	3.0	3.7	3.5	3.2	3.2	3.2	3.2	3.2
Total	28.2	24.4	25.4	23.9	26.7	27.3	28.4	28.6	29.5

Oil Products Consumption
Consommation de produits pétroliers

Jamaica

	1971	1973	1978	1983	1984	1985	1986	1987	1988
Thousand tonnes									
NGL/LPG/Ethane	24	38	39	34	33	32	34	36	37
Naphtha	-	-	-	-	-	-	-	-	-
Motor Gasoline	330	236	206	200	181	176	184	197	207
Aviation Fuels	137	150	87	113	99	126	118	134	135
Kerosene	22	22	56	34	38	46	52	55	63
Gas Diesel	254	359	185	191	175	169	173	171	215
Heavy Fuel Oil	1 066	1 847	1 745	1 302	1 232	1 060	961	1 010	998
Other Products	40	39	26	29	21	16	12	17	23
Refinery Fuel	74	80	14	13	13	13	12	33	35
Sub-Total	1 947	2 771	2 358	1 916	1 792	1 638	1 546	1 653	1 713
Bunkers	51	117	37	32	13	12	20	21	13
Total	1 998	2 888	2 395	1 948	1 805	1 650	1 566	1 674	1 726
Thousand barrels/day									
NGL/LPG/Ethane	0.8	1.2	1.2	1.1	1.0	1.0	1.1	1.1	1.2
Naphtha	-	-	-	-	-	-	-	-	-
Motor Gasoline	7.7	5.5	4.8	4.7	4.2	4.1	4.3	4.6	4.8
Aviation Fuels	3.0	3.3	1.9	2.5	2.2	2.7	2.6	2.9	2.9
Kerosene	0.5	0.5	1.2	0.7	0.8	1.0	1.1	1.2	1.3
Gas Diesel	5.2	7.3	3.8	3.9	3.6	3.5	3.5	3.5	4.4
Heavy Fuel Oil	19.5	33.7	31.8	23.8	22.4	19.3	17.5	18.4	18.2
Other Products	0.7	0.7	0.5	0.5	0.4	0.3	0.2	0.3	0.4
Refinery Fuel	1.4	1.6	0.3	0.3	0.3	0.3	0.3	0.7	0.7
Sub-Total	38.7	53.7	45.5	37.4	34.9	32.2	30.6	32.7	33.9
Bunkers	0.9	2.1	0.7	0.6	0.3	0.2	0.4	0.4	0.3
Total	39.7	55.9	46.2	38.0	35.1	32.5	31.0	33.1	34.2

	1989	1990	1991	1992	1993	1994	1995	1996	1997
Thousand tonnes									
NGL/LPG/Ethane	40	43	44	48	48	55	58	65	72
Naphtha	-	-	-	-	-	-	-	-	-
Motor Gasoline	229	234	247	233	275	288	347	375	407
Aviation Fuels	144	148	142	150	155	155	185	210	238
Kerosene	55	34	30	42	43	51	61	70	79
Gas Diesel	291	315	272	251	309	473	485	499	512
Heavy Fuel Oil	1 343	1 607	1 715	1 762	1 727	1 584	1 839	1 889	1 935
Other Products	18	69	53	18	14	12	12	12	12
Refinery Fuel	15	60	-	6	3	2	2	2	2
Sub-Total	2 135	2 510	2 503	2 510	2 574	2 620	2 989	3 122	3 257
Bunkers	30	36	42	44	37	37	37	37	37
Total	2 165	2 546	2 545	2 554	2 611	2 657	3 026	3 159	3 294
Thousand barrels/day									
NGL/LPG/Ethane	1.3	1.4	1.4	1.5	1.5	1.7	1.8	2.1	2.3
Naphtha	-	-	-	-	-	-	-	-	-
Motor Gasoline	5.4	5.5	5.8	5.4	6.4	6.7	8.1	8.7	9.5
Aviation Fuels	3.1	3.2	3.1	3.3	3.4	3.4	4.0	4.6	5.2
Kerosene	1.2	0.7	0.6	0.9	0.9	1.1	1.3	1.5	1.7
Gas Diesel	5.9	6.4	5.6	5.1	6.3	9.7	9.9	10.2	10.5
Heavy Fuel Oil	24.5	29.3	31.3	32.1	31.5	28.9	33.6	34.4	35.3
Other Products	0.3	1.3	1.0	0.3	0.2	0.2	0.2	0.2	0.2
Refinery Fuel	0.3	1.2	-	0.2	0.1	0.0	0.0	0.0	0.0
Sub-Total	42.0	49.0	48.8	48.8	50.4	51.7	59.0	61.6	64.7
Bunkers	0.6	0.7	0.9	0.9	0.8	0.8	0.8	0.8	0.8
Total	42.6	49.8	49.6	49.7	51.2	52.5	59.7	62.4	65.4

Oil Products Consumption
Consommation de produits pétroliers

Jordan

	1971	1973	1978	1983	1984	1985	1986	1987	1988
Thousand tonnes									
NGL/LPG/Ethane	14	17	42	75	83	77	85	90	99
Naphtha	-	-	-	-	-	-	-	-	-
Motor Gasoline	83	107	229	296	309	309	311	314	319
Aviation Fuels	45	50	131	230	232	215	175	174	174
Kerosene	102	92	137	158	141	127	136	136	152
Gas Diesel	122	165	407	696	683	737	783	784	805
Heavy Fuel Oil	69	116	179	727	919	962	1 177	1 273	1 127
Other Products	34	22	96	126	132	133	138	153	147
Refinery Fuel	32	42	58	109	102	144	111	162	172
Sub-Total	501	611	1 279	2 417	2 601	2 704	2 916	3 086	2 995
Bunkers	-	-	-	-	-	-	9	32	-
Total	501	611	1 279	2 417	2 601	2 704	2 925	3 118	2 995
Thousand barrels/day									
NGL/LPG/Ethane	0.4	0.5	1.3	2.4	2.6	2.4	2.7	2.9	3.1
Naphtha	-	-	-	-	-	-	-	-	-
Motor Gasoline	1.9	2.5	5.4	6.9	7.2	7.2	7.3	7.3	7.4
Aviation Fuels	1.0	1.1	2.9	5.0	5.1	4.7	3.9	3.8	3.8
Kerosene	2.2	2.0	2.9	3.4	3.0	2.7	2.9	2.9	3.2
Gas Diesel	2.5	3.4	8.3	14.2	13.9	15.1	16.0	16.0	16.4
Heavy Fuel Oil	1.3	2.1	3.3	13.3	16.7	17.6	21.5	23.2	20.5
Other Products	0.6	0.4	1.6	2.1	2.2	2.2	2.3	2.6	2.5
Refinery Fuel	0.6	0.8	1.1	2.0	1.9	2.7	2.1	3.1	3.3
Sub-Total	10.4	12.7	26.7	49.3	52.6	54.7	58.6	61.8	60.3
Bunkers	-	-	-	-	-	-	0.2	0.6	-
Total	10.4	12.7	26.7	49.3	52.6	54.7	58.8	62.4	60.3

	1989	1990	1991	1992	1993	1994	1995	1996	1997
Thousand tonnes									
NGL/LPG/Ethane	100	121	131	155	167	186	205	221	243
Naphtha	-	-	-	-	-	-	-	-	-
Motor Gasoline	322	400	384	421	435	455	484	510	535
Aviation Fuels	224	229	163	210	221	236	252	305	330
Kerosene	141	159	158	284	236	230	207	193	216
Gas Diesel	809	847	822	852	892	981	1 031	1 104	1 180
Heavy Fuel Oil	1 114	1 185	1 212	1 479	1 542	1 584	1 688	1 751	1 859
Other Products	111	132	123	149	169	161	166	172	182
Refinery Fuel	179	190	218	183	246	259	196	203	216
Sub-Total	3 000	3 263	3 211	3 733	3 908	4 092	4 229	4 459	4 761
Bunkers	-	-	-	-	-	-	-	-	-
Total	3 000	3 263	3 211	3 733	3 908	4 092	4 229	4 459	4 761
Thousand barrels/day									
NGL/LPG/Ethane	3.2	3.8	4.2	4.9	5.3	5.9	6.5	7.0	7.7
Naphtha	-	-	-	-	-	-	-	-	-
Motor Gasoline	7.5	9.3	9.0	9.8	10.2	10.6	11.3	11.9	12.5
Aviation Fuels	4.9	5.0	3.6	4.6	4.8	5.2	5.5	6.6	7.2
Kerosene	3.0	3.4	3.4	6.0	5.0	4.9	4.4	4.1	4.6
Gas Diesel	16.5	17.3	16.8	17.4	18.2	20.1	21.1	22.5	24.1
Heavy Fuel Oil	20.3	21.6	22.1	26.9	28.1	28.9	30.8	31.9	33.9
Other Products	1.9	2.2	2.1	2.5	2.9	2.7	2.8	2.9	3.1
Refinery Fuel	3.4	3.6	4.2	3.5	4.7	4.9	3.8	3.9	4.2
Sub-Total	60.8	66.4	65.2	75.7	79.2	83.2	86.2	90.8	97.3
Bunkers	-	-	-	-	-	-	-	-	-
Total	60.8	66.4	65.2	75.7	79.2	83.2	86.2	90.8	97.3

Oil Products Consumption
Consommation de produits pétroliers

Kazakhstan

	1971	1973	1978	1983	1984	1985	1986	1987	1988
Thousand tonnes									
NGL/LPG/Ethane	-	-	-	-	-	-	-	-	-
Naphtha	-	-	-	-	-	-	-	-	-
Motor Gasoline	-	-	-	-	-	-	-	-	-
Aviation Fuels	-	-	-	-	-	-	-	-	-
Kerosene	-	-	-	-	-	-	-	-	-
Gas Diesel	-	-	-	-	-	-	-	-	-
Heavy Fuel Oil	-	-	-	-	-	-	-	-	-
Other Products	-	-	-	-	-	-	-	-	-
Refinery Fuel	-	-	-	-	-	-	-	-	-
Sub-Total	-	-	-	-	-	-	-	-	-
Bunkers	-	-	-	-	-	-	-	-	-
Total	-	-	-	-	-	-	-	-	-
Thousand barrels/day									
NGL/LPG/Ethane	-	-	-	-	-	-	-	-	-
Naphtha	-	-	-	-	-	-	-	-	-
Motor Gasoline	-	-	-	-	-	-	-	-	-
Aviation Fuels	-	-	-	-	-	-	-	-	-
Kerosene	-	-	-	-	-	-	-	-	-
Gas Diesel	-	-	-	-	-	-	-	-	-
Heavy Fuel Oil	-	-	-	-	-	-	-	-	-
Other Products	-	-	-	-	-	-	-	-	-
Refinery Fuel	-	-	-	-	-	-	-	-	-
Sub-Total	-	-	-	-	-	-	-	-	-
Bunkers	-	-	-	-	-	-	-	-	-
Total	-	-	-	-	-	-	-	-	-

	1989	1990	1991	1992	1993	1994	1995	1996	1997
Thousand tonnes									
NGL/LPG/Ethane	-	-	-	700	216	157	130	117	115
Naphtha	-	-	-	-	-	-	-	-	-
Motor Gasoline	-	-	-	3 833	3 405	2 404	2 228	2 214	1 898
Aviation Fuels	-	-	-	982	331	485	420	379	305
Kerosene	-	-	-	1 001	511	362	455	410	370
Gas Diesel	-	-	-	5 786	4 408	3 172	3 943	2 161	2 196
Heavy Fuel Oil	-	-	-	5 557	4 762	4 362	3 368	3 210	2 530
Other Products	-	-	-	3 088	2 160	660	352	702	630
Refinery Fuel	-	-	-	730	580	565	367	465	365
Sub-Total	-	-	-	21 677	16 373	12 167	11 263	9 658	8 409
Bunkers	-	-	-	-	-	-	-	-	-
Total	-	-	-	21 677	16 373	12 167	11 263	9 658	8 409
Thousand barrels/day									
NGL/LPG/Ethane	-	-	-	22.2	6.9	5.0	4.1	3.7	3.7
Naphtha	-	-	-	-	-	-	-	-	-
Motor Gasoline	-	-	-	89.3	79.6	56.2	52.1	51.6	44.4
Aviation Fuels	-	-	-	21.3	7.2	10.5	9.1	8.2	6.6
Kerosene	-	-	-	21.2	10.8	7.7	9.6	8.7	7.8
Gas Diesel	-	-	-	117.9	90.1	64.8	80.6	44.0	44.9
Heavy Fuel Oil	-	-	-	101.1	86.9	79.6	61.5	58.4	46.2
Other Products	-	-	-	59.3	41.4	12.6	6.7	13.4	12.1
Refinery Fuel	-	-	-	13.9	11.1	10.8	7.1	8.9	7.0
Sub-Total	-	-	-	446.3	334.0	247.2	230.8	196.9	172.6
Bunkers	-	-	-	-	-	-	-	-	-
Total	-	-	-	446.3	334.0	247.2	230.8	196.9	172.6

Oil Products Consumption
Consommation de produits pétroliers

Kenya

	1971	1973	1978	1983	1984	1985	1986	1987	1988
Thousand tonnes									
NGL/LPG/Ethane	8	12	17	20	22	22	19	26	27
Naphtha	-	-	-	-	-	-	-	-	-
Motor Gasoline	197	233	291	256	318	348	395	421	425
Aviation Fuels	186	261	335	256	265	267	269	256	261
Kerosene	46	53	78	82	111	121	138	167	185
Gas Diesel	249	301	353	412	445	473	510	580	547
Heavy Fuel Oil	399	411	464	346	406	377	458	373	358
Other Products	34	31	67	52	70	71	69	87	70
Refinery Fuel	88	78	106	61	61	68	71	83	78
Sub-Total	1 207	1 380	1 711	1 485	1 698	1 747	1 929	1 993	1 951
Bunkers	476	458	219	120	160	145	158	169	183
Total	1 683	1 838	1 930	1 605	1 858	1 892	2 087	2 162	2 134
Thousand barrels/day									
NGL/LPG/Ethane	0.3	0.4	0.5	0.6	0.7	0.7	0.6	0.8	0.9
Naphtha	-	-	-	-	-	-	-	-	-
Motor Gasoline	4.6	5.4	6.8	6.0	7.4	8.1	9.2	9.8	9.9
Aviation Fuels	4.1	5.7	7.3	5.6	5.8	5.8	5.9	5.6	5.7
Kerosene	1.0	1.1	1.7	1.7	2.3	2.6	2.9	3.5	3.9
Gas Diesel	5.1	6.2	7.2	8.4	9.1	9.7	10.4	11.9	11.1
Heavy Fuel Oil	7.3	7.5	8.5	6.3	7.4	6.9	8.4	6.8	6.5
Other Products	0.6	0.5	1.2	0.9	1.3	1.3	1.3	1.6	1.3
Refinery Fuel	1.7	1.5	2.1	1.2	1.2	1.4	1.4	1.7	1.6
Sub-Total	24.6	28.4	35.3	30.8	35.1	36.4	40.1	41.7	40.9
Bunkers	8.9	8.6	4.1	2.2	3.0	2.7	3.0	3.2	3.4
Total	33.4	36.9	39.5	33.0	38.1	39.1	43.1	44.9	44.3

	1989	1990	1991	1992	1993	1994	1995	1996	1997
Thousand tonnes									
NGL/LPG/Ethane	26	27	25	27	25	28	31	31	31
Naphtha	-	-	-	-	-	-	-	-	-
Motor Gasoline	477	398	363	347	352	352	339	399	391
Aviation Fuels	280	272	239	316	357	481	438	447	435
Kerosene	211	184	175	175	165	173	158	162	149
Gas Diesel	569	592	590	599	577	564	626	674	662
Heavy Fuel Oil	336	343	336	341	369	389	324	393	356
Other Products	56	59	60	63	59	42	55	49	43
Refinery Fuel	86	89	75	80	181	187	184	173	184
Sub-Total	2 041	1 964	1 863	1 948	2 085	2 216	2 155	2 328	2 251
Bunkers	158	178	149	140	166	157	161	220	220
Total	2 199	2 142	2 012	2 088	2 251	2 373	2 316	2 548	2 471
Thousand barrels/day									
NGL/LPG/Ethane	0.8	0.9	0.8	0.9	0.8	0.9	1.0	1.0	1.0
Naphtha	-	-	-	-	-	-	-	-	-
Motor Gasoline	11.1	9.3	8.5	8.1	8.2	8.2	7.9	9.3	9.1
Aviation Fuels	6.1	5.9	5.2	6.9	7.8	10.5	9.5	9.7	9.5
Kerosene	4.5	3.9	3.7	3.7	3.5	3.7	3.4	3.4	3.2
Gas Diesel	11.6	12.1	12.1	12.2	11.8	11.5	12.8	13.7	13.5
Heavy Fuel Oil	6.1	6.3	6.1	6.2	6.7	7.1	5.9	7.2	6.5
Other Products	1.0	1.0	1.1	1.1	1.1	0.8	1.0	0.9	0.8
Refinery Fuel	1.8	1.8	1.5	1.6	3.7	3.8	3.8	3.5	3.7
Sub-Total	43.1	41.2	39.0	40.7	43.6	46.4	45.2	48.7	47.3
Bunkers	2.9	3.3	2.8	2.6	3.1	2.9	3.0	4.1	4.1
Total	46.0	44.5	41.8	43.3	46.7	49.3	48.2	52.8	51.4

Oil Products Consumption
Consommation de produits pétroliers

Dem. People's Rep. of Korea

	1971	1973	1978	1983	1984	1985	1986	1987	1988
Thousand tonnes									
NGL/LPG/Ethane	-	-	-	-	-	-	-	-	-
Naphtha	-	-	-	-	-	-	-	-	-
Motor Gasoline	300	350	700	850	900	950	1 000	1 020	1 020
Aviation Fuels	-	-	-	-	-	-	-	-	-
Kerosene	30	35	101	210	220	230	240	250	250
Gas Diesel	263	285	605	1 050	1 100	1 150	1 200	1 200	1 200
Heavy Fuel Oil	90	95	326	484	528	572	619	620	620
Other Products	-	-	-	-	-	-	-	-	-
Refinery Fuel	-	-	68	119	129	140	146	145	145
Sub-Total	683	765	1 800	2 713	2 877	3 042	3 205	3 235	3 235
Bunkers	-	-	-	-	-	-	-	-	-
Total	683	765	1 800	2 713	2 877	3 042	3 205	3 235	3 235
Thousand barrels/day									
NGL/LPG/Ethane	-	-	-	-	-	-	-	-	-
Naphtha	-	-	-	-	-	-	-	-	-
Motor Gasoline	7.0	8.2	16.4	19.9	21.0	22.2	23.4	23.8	23.8
Aviation Fuels	-	-	-	-	-	-	-	-	-
Kerosene	0.6	0.7	2.1	4.5	4.7	4.9	5.1	5.3	5.3
Gas Diesel	5.4	5.8	12.4	21.5	22.4	23.5	24.5	24.5	24.5
Heavy Fuel Oil	1.6	1.7	5.9	8.8	9.6	10.4	11.3	11.3	11.3
Other Products	-	-	-	-	-	-	-	-	-
Refinery Fuel	-	-	1.4	2.4	2.6	2.8	2.9	2.9	2.9
Sub-Total	14.7	16.5	38.2	57.0	60.2	63.8	67.2	67.9	67.7
Bunkers	-	-	-	-	-	-	-	-	-
Total	14.7	16.5	38.2	57.0	60.2	63.8	67.2	67.9	67.7

	1989	1990	1991	1992	1993	1994	1995	1996	1997
Thousand tonnes									
NGL/LPG/Ethane	-	-	-	-	-	-	-	-	-
Naphtha	-	-	-	-	-	-	-	-	-
Motor Gasoline	1 070	1 027	867	746	626	531	450	381	370
Aviation Fuels	-	-	-	-	-	-	-	-	-
Kerosene	260	260	245	209	172	151	133	117	113
Gas Diesel	1 300	1 300	1 272	1 321	1 321	726	655	591	573
Heavy Fuel Oil	700	700	608	496	364	299	243	198	192
Other Products	-	-	-	-	-	-	-	-	-
Refinery Fuel	165	165	154	143	131	125	125	126	122
Sub-Total	3 495	3 452	3 146	2 915	2 614	1 832	1 606	1 413	1 370
Bunkers	-	-	-	-	-	-	-	-	-
Total	3 495	3 452	3 146	2 915	2 614	1 832	1 606	1 413	1 370
Thousand barrels/day									
NGL/LPG/Ethane	-	-	-	-	-	-	-	-	-
Naphtha	-	-	-	-	-	-	-	-	-
Motor Gasoline	25.0	24.0	20.3	17.4	14.6	12.4	10.5	8.9	8.6
Aviation Fuels	-	-	-	-	-	-	-	-	-
Kerosene	5.5	5.5	5.2	4.4	3.6	3.2	2.8	2.5	2.4
Gas Diesel	26.6	26.6	26.0	26.9	27.0	14.8	13.4	12.0	11.7
Heavy Fuel Oil	12.8	12.8	11.1	9.0	6.6	5.5	4.4	3.6	3.5
Other Products	-	-	-	-	-	-	-	-	-
Refinery Fuel	3.3	3.3	3.0	2.8	2.5	2.4	2.4	2.4	2.3
Sub-Total	73.1	72.1	65.6	60.5	54.4	38.3	33.5	29.4	28.6
Bunkers	-	-	-	-	-	-	-	-	-
Total	73.1	72.1	65.6	60.5	54.4	38.3	33.5	29.4	28.6

Oil Products Consumption
Consommation de produits pétroliers

Kuwait

	1971	1973	1978	1983	1984	1985	1986	1987	1988
Thousand tonnes									
NGL/LPG/Ethane	-	-	50	197	439	296	283	210	136
Naphtha	-	-	-	-	-	-	-	-	-
Motor Gasoline	338	382	760	1 063	1 102	1 127	1 147	1 169	1 233
Aviation Fuels	102	101	308	341	339	314	304	288	309
Kerosene	30	30	35	25	24	26	32	39	30
Gas Diesel	138	160	550	2 001	901	1 321	586	476	711
Heavy Fuel Oil	-	-	-	851	1 546	2 117	1 345	1 167	574
Other Products	28	54	119	1 229	1 269	1 078	1 202	798	109
Refinery Fuel	1 083	943	966	1 243	1 230	1 361	1 456	1 517	1 617
Sub-Total	1 719	1 670	2 788	6 950	6 850	7 640	6 355	5 664	4 719
Bunkers	4 075	4 295	2 417	791	739	766	766	766	766
Total	5 794	5 965	5 205	7 741	7 589	8 406	7 121	6 430	5 485
Thousand barrels/day									
NGL/LPG/Ethane	-	-	1.6	6.3	13.9	9.4	9.0	6.7	4.3
Naphtha	-	-	-	-	-	-	-	-	-
Motor Gasoline	7.9	8.9	17.8	24.8	25.7	26.3	26.8	27.3	28.7
Aviation Fuels	2.2	2.2	6.7	7.4	7.3	6.8	6.6	6.3	6.7
Kerosene	0.6	0.6	0.7	0.5	0.5	0.6	0.7	0.8	0.6
Gas Diesel	2.8	3.3	11.2	40.9	18.4	27.0	12.0	9.7	14.5
Heavy Fuel Oil	-	-	-	15.5	28.1	38.6	24.5	21.3	10.4
Other Products	0.5	0.9	2.0	24.0	24.7	20.9	23.4	15.4	1.9
Refinery Fuel	20.6	17.9	18.3	23.6	23.3	25.5	27.4	28.7	30.5
Sub-Total	34.6	33.9	58.4	143.1	141.9	155.2	130.4	116.2	97.7
Bunkers	74.5	78.6	44.3	14.5	13.5	14.0	14.0	14.0	14.0
Total	109.1	112.4	102.7	157.6	155.4	169.2	144.5	130.3	111.7

	1989	1990	1991	1992	1993	1994	1995	1996	1997
Thousand tonnes									
NGL/LPG/Ethane	150	150	82	110	126	90	93	86	99
Naphtha	-	-	-	-	-	-	-	-	-
Motor Gasoline	1 250	1 000	279	1 161	1 307	1 367	1 475	1 558	1 700
Aviation Fuels	450	401	49	301	213	331	371	389	425
Kerosene	30	50	30	26	22	22	24	25	27
Gas Diesel	700	1 300	554	804	923	795	648	614	697
Heavy Fuel Oil	500	800	720	2 332	1 533	1 751	2 258	2 152	2 609
Other Products	88	57	45	110	142	452	92	85	95
Refinery Fuel	1 615	1 245	124	729	1 141	1 446	1 503	1 442	1 716
Sub-Total	4 783	5 003	1 883	5 573	5 407	6 254	6 464	6 351	7 368
Bunkers	600	300	20	70	130	170	190	190	190
Total	5 383	5 303	1 903	5 643	5 537	6 424	6 654	6 541	7 558
Thousand barrels/day									
NGL/LPG/Ethane	4.8	4.8	2.6	3.5	4.0	2.9	3.0	2.7	3.1
Naphtha	-	-	-	-	-	-	-	-	-
Motor Gasoline	29.2	23.4	6.5	27.1	30.5	31.9	34.5	36.3	39.7
Aviation Fuels	9.8	8.7	1.1	6.5	4.6	7.2	8.1	8.4	9.2
Kerosene	0.6	1.1	0.6	0.6	0.5	0.5	0.5	0.5	0.6
Gas Diesel	14.3	26.6	11.3	16.4	18.9	16.2	13.2	12.5	14.2
Heavy Fuel Oil	9.1	14.6	13.1	42.4	28.0	32.0	41.2	39.2	47.6
Other Products	1.5	1.0	0.9	2.2	2.8	8.9	1.8	1.6	1.8
Refinery Fuel	30.5	23.5	2.4	13.4	21.7	27.7	28.7	27.5	32.7
Sub-Total	99.8	103.6	38.5	112.0	111.0	127.2	130.9	128.8	149.0
Bunkers	10.9	5.5	0.4	1.3	2.4	3.2	3.6	3.5	3.6
Total	110.8	109.0	38.9	113.3	113.4	130.4	134.5	132.3	152.6

Oil Products Consumption
Consommation de produits pétroliers

Kyrgyzstan

	1971	1973	1978	1983	1984	1985	1986	1987	1988
Thousand tonnes									
NGL/LPG/Ethane	-	-	-	-	-	-	-	-	-
Naphtha	-	-	-	-	-	-	-	-	-
Motor Gasoline	-	-	-	-	-	-	-	-	-
Aviation Fuels	-	-	-	-	-	-	-	-	-
Kerosene	-	-	-	-	-	-	-	-	-
Gas Diesel	-	-	-	-	-	-	-	-	-
Heavy Fuel Oil	-	-	-	-	-	-	-	-	-
Other Products	-	-	-	-	-	-	-	-	-
Refinery Fuel	-	-	-	-	-	-	-	-	-
Sub-Total	-	-	-	-	-	-	-	-	-
Bunkers	-	-	-	-	-	-	-	-	-
Total	-	-	-	-	-	-	-	-	-
Thousand barrels/day									
NGL/LPG/Ethane	-	-	-	-	-	-	-	-	-
Naphtha	-	-	-	-	-	-	-	-	-
Motor Gasoline	-	-	-	-	-	-	-	-	-
Aviation Fuels	-	-	-	-	-	-	-	-	-
Kerosene	-	-	-	-	-	-	-	-	-
Gas Diesel	-	-	-	-	-	-	-	-	-
Heavy Fuel Oil	-	-	-	-	-	-	-	-	-
Other Products	-	-	-	-	-	-	-	-	-
Refinery Fuel	-	-	-	-	-	-	-	-	-
Sub-Total	-	-	-	-	-	-	-	-	-
Bunkers	-	-	-	-	-	-	-	-	-
Total	-	-	-	-	-	-	-	-	-

	1989	1990	1991	1992	1993	1994	1995	1996	1997
Thousand tonnes									
NGL/LPG/Ethane	-	-	-	76	35	13	-	-	-
Naphtha	-	-	-	-	-	-	-	-	-
Motor Gasoline	-	-	-	412	273	102	212	230	115
Aviation Fuels	-	-	-	60	16	13	81	89	63
Kerosene	-	-	-	-	-	-	-	-	-
Gas Diesel	-	-	-	460	236	59	132	144	100
Heavy Fuel Oil	-	-	-	498	255	95	92	100	88
Other Products	-	-	-	404	350	135	82	104	122
Refinery Fuel	-	-	-	-	-	-	-	-	-
Sub-Total	-	-	-	1 910	1 165	417	599	667	488
Bunkers	-	-	-	-	-	-	-	-	-
Total	-	-	-	1 910	1 165	417	599	667	488
Thousand barrels/day									
NGL/LPG/Ethane	-	-	-	2.4	1.1	0.4	-	-	-
Naphtha	-	-	-	-	-	-	-	-	-
Motor Gasoline	-	-	-	9.6	6.4	2.4	5.0	5.4	2.7
Aviation Fuels	-	-	-	1.3	0.3	0.3	1.8	2.0	1.4
Kerosene	-	-	-	-	-	-	-	-	-
Gas Diesel	-	-	-	9.4	4.8	1.2	2.7	2.9	2.0
Heavy Fuel Oil	-	-	-	9.1	4.7	1.7	1.7	1.8	1.6
Other Products	-	-	-	7.7	6.7	2.6	1.6	2.0	2.3
Refinery Fuel	-	-	-	-	-	-	-	-	-
Sub-Total	-	-	-	39.5	24.0	8.6	12.7	14.1	10.1
Bunkers	-	-	-	-	-	-	-	-	-
Total	-	-	-	39.5	24.0	8.6	12.7	14.1	10.1

Oil Products Consumption
Consommation de produits pétroliers

Latvia

	1971	1973	1978	1983	1984	1985	1986	1987	1988
Thousand tonnes									
NGL/LPG/Ethane	-	-	-	-	-	-	-	-	-
Naphtha	-	-	-	-	-	-	-	-	-
Motor Gasoline	-	-	-	-	-	-	-	-	-
Aviation Fuels	-	-	-	-	-	-	-	-	-
Kerosene	-	-	-	-	-	-	-	-	-
Gas Diesel	-	-	-	-	-	-	-	-	-
Heavy Fuel Oil	-	-	-	-	-	-	-	-	-
Other Products	-	-	-	-	-	-	-	-	-
Refinery Fuel	-	-	-	-	-	-	-	-	-
Sub-Total	-	-	-	-	-	-	-	-	-
Bunkers	-	-	-	-	-	-	-	-	-
Total	-	-	-	-	-	-	-	-	-
Thousand barrels/day									
NGL/LPG/Ethane	-	-	-	-	-	-	-	-	-
Naphtha	-	-	-	-	-	-	-	-	-
Motor Gasoline	-	-	-	-	-	-	-	-	-
Aviation Fuels	-	-	-	-	-	-	-	-	-
Kerosene	-	-	-	-	-	-	-	-	-
Gas Diesel	-	-	-	-	-	-	-	-	-
Heavy Fuel Oil	-	-	-	-	-	-	-	-	-
Other Products	-	-	-	-	-	-	-	-	-
Refinery Fuel	-	-	-	-	-	-	-	-	-
Sub-Total	-	-	-	-	-	-	-	-	-
Bunkers	-	-	-	-	-	-	-	-	-
Total	-	-	-	-	-	-	-	-	-

	1989	1990	1991	1992	1993	1994	1995	1996	1997
Thousand tonnes									
NGL/LPG/Ethane	-	-	-	58	61	52	34	35	38
Naphtha	-	-	-	-	-	-	-	-	-
Motor Gasoline	-	-	-	493	478	457	412	434	374
Aviation Fuels	-	-	-	90	26	21	29	34	38
Kerosene	-	-	-	84	8	7	10	4	4
Gas Diesel	-	-	-	790	722	564	502	431	404
Heavy Fuel Oil	-	-	-	1 089	1 116	1 415	1 017	1 141	832
Other Products	-	-	-	191	28	-	15	3	41
Refinery Fuel	-	-	-	-	-	-	-	-	-
Sub-Total	-	-	-	2 795	2 439	2 516	2 019	2 082	1 731
Bunkers	-	-	-	-	-	-	-	-	-
Total	-	-	-	2 795	2 439	2 516	2 019	2 082	1 731
Thousand barrels/day									
NGL/LPG/Ethane	-	-	-	1.8	1.9	1.7	1.1	1.1	1.2
Naphtha	-	-	-	-	-	-	-	-	-
Motor Gasoline	-	-	-	11.5	11.2	10.7	9.6	10.1	8.7
Aviation Fuels	-	-	-	2.0	0.6	0.5	0.6	0.7	0.8
Kerosene	-	-	-	1.8	0.2	0.1	0.2	0.1	0.1
Gas Diesel	-	-	-	16.1	14.8	11.5	10.3	8.8	8.3
Heavy Fuel Oil	-	-	-	19.8	20.4	25.8	18.6	20.8	15.2
Other Products	-	-	-	3.7	0.5	-	0.3	0.1	0.8
Refinery Fuel	-	-	-	-	-	-	-	-	-
Sub-Total	-	-	-	56.6	49.5	50.3	40.7	41.7	35.1
Bunkers	-	-	-	-	-	-	-	-	-
Total	-	-	-	56.6	49.5	50.3	40.7	41.7	35.1

Oil Products Consumption
Consommation de produits pétroliers

Lebanon

	1971	1973	1978	1983	1984	1985	1986	1987	1988
Thousand tonnes									
NGL/LPG/Ethane	68	76	70	80	75	90	91	95	75
Naphtha	-	-	-	-	-	-	-	-	-
Motor Gasoline	403	517	473	616	640	739	666	680	479
Aviation Fuels	264	347	201	191	101	121	151	180	85
Kerosene	24	20	19	40	40	70	80	60	50
Gas Diesel	254	318	250	543	535	547	509	520	350
Heavy Fuel Oil	632	907	902	805	760	974	1 090	1 097	967
Other Products	-	-	-	-	-	-	-	-	-
Refinery Fuel	105	206	130	25	38	31	56	63	30
Sub-Total	1 750	2 391	2 045	2 300	2 189	2 572	2 643	2 695	2 036
Bunkers	230	21	10	-	-	-	-	-	-
Total	1 980	2 412	2 055	2 300	2 189	2 572	2 643	2 695	2 036
Thousand barrels/day									
NGL/LPG/Ethane	2.2	2.4	2.2	2.5	2.4	2.9	2.9	3.0	2.4
Naphtha	-	-	-	-	-	-	-	-	-
Motor Gasoline	9.4	12.1	11.1	14.4	14.9	17.3	15.6	15.9	11.2
Aviation Fuels	5.8	7.6	4.4	4.2	2.2	2.6	3.3	3.9	1.8
Kerosene	0.5	0.4	0.4	0.8	0.8	1.5	1.7	1.3	1.1
Gas Diesel	5.2	6.5	5.1	11.1	10.9	11.2	10.4	10.6	7.1
Heavy Fuel Oil	11.5	16.6	16.5	14.7	13.8	17.8	19.9	20.0	17.6
Other Products	-	-	-	-	-	-	-	-	-
Refinery Fuel	1.9	3.8	2.4	0.5	0.7	0.6	1.0	1.2	0.5
Sub-Total	36.5	49.3	42.0	48.2	45.8	53.8	54.7	55.9	41.7
Bunkers	4.2	0.4	0.2	-	-	-	-	-	-
Total	40.7	49.7	42.2	48.2	45.8	53.8	54.7	55.9	41.7

	1989	1990	1991	1992	1993	1994	1995	1996	1997
Thousand tonnes									
NGL/LPG/Ethane	75	91	95	128	130	153	116	124	141
Naphtha	-	-	-	-	-	-	-	-	-
Motor Gasoline	580	600	750	1 041	1 200	1 123	1 327	1 379	1 310
Aviation Fuels	50	60	75	107	120	146	208	107	109
Kerosene	50	60	50	9	4	4	4	4	4
Gas Diesel	350	300	560	430	662	795	1 022	930	1 376
Heavy Fuel Oil	570	999	1 100	870	1 125	1 352	1 432	1 623	1 731
Other Products	-	-	-	22	50	67	67	109	88
Refinery Fuel	-	5	20	20	-	-	-	-	-
Sub-Total	1 675	2 115	2 650	2 627	3 291	3 640	4 176	4 276	4 759
Bunkers	-	-	10	10	-	-	-	-	-
Total	1 675	2 115	2 660	2 637	3 291	3 640	4 176	4 276	4 759
Thousand barrels/day									
NGL/LPG/Ethane	2.4	2.9	3.0	4.1	4.1	4.9	3.7	3.9	4.5
Naphtha	-	-	-	-	-	-	-	-	-
Motor Gasoline	13.6	14.0	17.5	24.3	28.0	26.2	31.0	32.1	30.6
Aviation Fuels	1.1	1.3	1.6	2.3	2.6	3.2	4.5	2.3	2.4
Kerosene	1.1	1.3	1.1	0.2	0.1	0.1	0.1	0.1	0.1
Gas Diesel	7.2	6.1	11.4	8.8	13.5	16.2	20.9	19.0	28.1
Heavy Fuel Oil	10.4	18.2	20.1	15.8	20.5	24.7	26.1	29.5	31.6
Other Products	-	-	-	0.4	1.0	1.3	1.3	2.1	1.7
Refinery Fuel	-	0.1	0.4	0.4	-	-	-	-	-
Sub-Total	35.6	43.9	55.1	56.2	69.9	76.6	87.6	89.0	98.9
Bunkers	-	-	0.2	0.2	-	-	-	-	-
Total	35.6	43.9	55.3	56.4	69.9	76.6	87.6	89.0	98.9

Oil Products Consumption
Consommation de produits pétroliers

Libya

	1971	1973	1978	1983	1984	1985	1986	1987	1988
Thousand tonnes									
NGL/LPG/Ethane	37	20	63	80	100	110	100	240	240
Naphtha	-	-	-	290	320	450	480	410	330
Motor Gasoline	210	302	670	961	948	968	1 001	979	1 000
Aviation Fuels	87	112	177	409	362	335	305	366	290
Kerosene	65	74	110	72	73	94	85	64	62
Gas Diesel	16	578	1 304	2 040	1 864	1 567	1 416	1 494	1 600
Heavy Fuel Oil	176	217	722	1 696	1 955	1 734	1 715	1 980	2 181
Other Products	23	75	130	495	350	247	205	176	182
Refinery Fuel	12	14	180	279	286	335	462	590	630
Sub-Total	626	1 392	3 356	6 322	6 258	5 840	5 769	6 299	6 515
Bunkers	3	2	12	17	8	12	15	20	50
Total	629	1 394	3 368	6 339	6 266	5 852	5 784	6 319	6 565
Thousand barrels/day									
NGL/LPG/Ethane	1.2	0.6	2.0	2.5	3.2	3.5	3.2	7.6	7.6
Naphtha	-	-	-	6.8	7.4	10.5	11.2	9.5	7.7
Motor Gasoline	4.9	7.1	15.7	22.5	22.1	22.6	23.4	22.9	23.3
Aviation Fuels	1.9	2.4	3.9	8.9	7.8	7.3	6.6	8.0	6.3
Kerosene	1.4	1.6	2.3	1.5	1.5	2.0	1.8	1.4	1.3
Gas Diesel	0.3	11.8	26.7	41.7	38.0	32.0	28.9	30.5	32.6
Heavy Fuel Oil	3.2	4.0	13.2	30.9	35.6	31.6	31.3	36.1	39.7
Other Products	0.4	1.3	2.4	8.4	6.1	4.4	3.8	3.2	3.3
Refinery Fuel	0.2	0.3	3.6	5.8	5.9	6.9	9.6	12.4	13.1
Sub-Total	13.6	29.1	69.7	129.0	127.7	120.8	119.8	131.6	135.0
Bunkers	0.1	0.0	0.2	0.3	0.1	0.2	0.3	0.4	0.9
Total	13.6	29.1	69.9	129.3	127.8	121.1	120.1	132.0	135.9

	1989	1990	1991	1992	1993	1994	1995	1996	1997
Thousand tonnes									
NGL/LPG/Ethane	208	250	290	300	340	360	408	411	422
Naphtha	632	700	820	710	810	820	926	933	959
Motor Gasoline	1 031	1 040	1 050	1 063	1 207	1 477	1 675	1 687	1 775
Aviation Fuels	302	202	212	227	237	288	291	318	299
Kerosene	150	180	180	192	190	232	234	256	271
Gas Diesel	1 620	1 620	1 625	1 640	1 898	2 322	2 340	2 332	2 277
Heavy Fuel Oil	2 029	2 000	2 000	2 020	2 050	2 542	2 975	2 973	3 108
Other Products	200	150	130	140	130	142	173	174	179
Refinery Fuel	620	620	620	620	620	709	716	726	726
Sub-Total	6 792	6 762	6 927	6 912	7 482	8 892	9 738	9 810	10 016
Bunkers	70	80	80	70	90	80	90	90	90
Total	6 862	6 842	7 007	6 982	7 572	8 972	9 828	9 900	10 106
Thousand barrels/day									
NGL/LPG/Ethane	6.6	7.9	9.2	9.5	10.8	11.4	13.0	13.0	13.4
Naphtha	14.7	16.3	19.1	16.5	18.9	19.1	21.6	21.7	22.3
Motor Gasoline	24.1	24.3	24.5	24.8	28.2	34.5	39.1	39.3	41.5
Aviation Fuels	6.6	4.4	4.6	4.9	5.2	6.3	6.3	6.9	6.5
Kerosene	3.2	3.8	3.8	4.1	4.0	4.9	5.0	5.4	5.7
Gas Diesel	33.1	33.1	33.2	33.4	38.8	47.5	47.8	47.5	46.5
Heavy Fuel Oil	37.0	36.5	36.5	36.8	37.4	46.4	54.3	54.1	56.7
Other Products	3.6	2.7	2.3	2.4	2.3	2.5	3.0	3.0	3.1
Refinery Fuel	13.0	13.0	13.0	13.0	13.0	14.9	15.1	15.2	15.3
Sub-Total	141.9	142.0	146.3	145.3	158.5	187.5	205.1	206.2	211.1
Bunkers	1.3	1.5	1.5	1.3	1.6	1.5	1.6	1.6	1.6
Total	143.2	143.5	147.7	146.6	160.2	188.9	206.8	207.8	212.8

Oil Products Consumption
Consommation de produits pétroliers

Lithuania

	1971	1973	1978	1983	1984	1985	1986	1987	1988
Thousand tonnes									
NGL/LPG/Ethane	-	-	-	-	-	-	-	-	-
Naphtha	-	-	-	-	-	-	-	-	-
Motor Gasoline	-	-	-	-	-	-	-	-	-
Aviation Fuels	-	-	-	-	-	-	-	-	-
Kerosene	-	-	-	-	-	-	-	-	-
Gas Diesel	-	-	-	-	-	-	-	-	-
Heavy Fuel Oil	-	-	-	-	-	-	-	-	-
Other Products	-	-	-	-	-	-	-	-	-
Refinery Fuel	-	-	-	-	-	-	-	-	-
Sub-Total	-	-	-	-	-	-	-	-	-
Bunkers	-	-	-	-	-	-	-	-	-
Total	-	-	-	-	-	-	-	-	-
Thousand barrels/day									
NGL/LPG/Ethane	-	-	-	-	-	-	-	-	-
Naphtha	-	-	-	-	-	-	-	-	-
Motor Gasoline	-	-	-	-	-	-	-	-	-
Aviation Fuels	-	-	-	-	-	-	-	-	-
Kerosene	-	-	-	-	-	-	-	-	-
Gas Diesel	-	-	-	-	-	-	-	-	-
Heavy Fuel Oil	-	-	-	-	-	-	-	-	-
Other Products	-	-	-	-	-	-	-	-	-
Refinery Fuel	-	-	-	-	-	-	-	-	-
Sub-Total	-	-	-	-	-	-	-	-	-
Bunkers	-	-	-	-	-	-	-	-	-
Total	-	-	-	-	-	-	-	-	-

	1989	1990	1991	1992	1993	1994	1995	1996	1997
Thousand tonnes									
NGL/LPG/Ethane	-	-	-	98	96	93	93	74	94
Naphtha	-	-	-	-	-	-	-	-	-
Motor Gasoline	-	-	-	675	545	442	607	665	662
Aviation Fuels	-	-	-	66	37	39	40	33	31
Kerosene	-	-	-	-	-	-	-	-	-
Gas Diesel	-	-	-	994	773	638	621	612	673
Heavy Fuel Oil	-	-	-	2 247	2 239	2 027	1 544	1 609	1 310
Other Products	-	-	-	296	214	280	507	161	454
Refinery Fuel	-	-	-	-	-	-	-	-	-
Sub-Total	-	-	-	4 376	3 904	3 519	3 412	3 154	3 224
Bunkers	-	-	-	-	-	-	-	-	62
Total	-	-	-	4 376	3 904	3 519	3 412	3 154	3 286
Thousand barrels/day									
NGL/LPG/Ethane	-	-	-	3.1	3.1	3.0	3.0	2.3	3.0
Naphtha	-	-	-	-	-	-	-	-	-
Motor Gasoline	-	-	-	15.7	12.7	10.3	14.2	15.5	15.5
Aviation Fuels	-	-	-	1.4	0.8	0.8	0.9	0.7	0.7
Kerosene	-	-	-	-	-	-	-	-	-
Gas Diesel	-	-	-	20.3	15.8	13.0	12.7	12.5	13.8
Heavy Fuel Oil	-	-	-	40.9	40.9	37.0	28.2	29.3	23.9
Other Products	-	-	-	5.6	4.0	5.3	9.7	2.9	8.6
Refinery Fuel	-	-	-	-	-	-	-	-	-
Sub-Total	-	-	-	87.0	77.3	69.4	68.6	63.3	65.4
Bunkers	-	-	-	-	-	-	-	-	1.1
Total	-	-	-	87.0	77.3	69.4	68.6	63.3	66.5

Oil Products Consumption
Consommation de produits pétroliers

Former Yugoslav Republic of Macedonia

	1971	1973	1978	1983	1984	1985	1986	1987	1988
Thousand tonnes									
NGL/LPG/Ethane	-	-	-	-	-	-	-	-	-
Naphtha	-	-	-	-	-	-	-	-	-
Motor Gasoline	-	-	-	-	-	-	-	-	-
Aviation Fuels	-	-	-	-	-	-	-	-	-
Kerosene	-	-	-	-	-	-	-	-	-
Gas Diesel	-	-	-	-	-	-	-	-	-
Heavy Fuel Oil	-	-	-	-	-	-	-	-	-
Other Products	-	-	-	-	-	-	-	-	-
Refinery Fuel	-	-	-	-	-	-	-	-	-
Sub-Total	-	-	-	-	-	-	-	-	-
Bunkers	-	-	-	-	-	-	-	-	-
Total	-	-	-	-	-	-	-	-	-
Thousand barrels/day									
NGL/LPG/Ethane	-	-	-	-	-	-	-	-	-
Naphtha	-	-	-	-	-	-	-	-	-
Motor Gasoline	-	-	-	-	-	-	-	-	-
Aviation Fuels	-	-	-	-	-	-	-	-	-
Kerosene	-	-	-	-	-	-	-	-	-
Gas Diesel	-	-	-	-	-	-	-	-	-
Heavy Fuel Oil	-	-	-	-	-	-	-	-	-
Other Products	-	-	-	-	-	-	-	-	-
Refinery Fuel	-	-	-	-	-	-	-	-	-
Sub-Total	-	-	-	-	-	-	-	-	-
Bunkers	-	-	-	-	-	-	-	-	-
Total	-	-	-	-	-	-	-	-	-

	1989	1990	1991	1992	1993	1994	1995	1996	1997
Thousand tonnes									
NGL/LPG/Ethane	-	41	42	35	18	15	12	19	16
Naphtha	-	-	-	-	-	-	-	-	-
Motor Gasoline	-	161	163	163	182	241	207	209	214
Aviation Fuels	-	5	7	7	18	19	31	26	30
Kerosene	-	-	-	-	-	-	-	-	-
Gas Diesel	-	258	250	250	395	88	220	457	366
Heavy Fuel Oil	-	452	425	425	370	377	297	486	360
Other Products	-	84	45	45	37	41	48	40	41
Refinery Fuel	-	52	35	35	-	-	-	-	-
Sub-Total	-	1 053	967	960	1 020	781	815	1 237	1 027
Bunkers	-	-	-	-	-	-	-	-	-
Total	-	1 053	967	960	1 020	781	815	1 237	1 027
Thousand barrels/day									
NGL/LPG/Ethane	-	1.3	1.3	1.1	0.6	0.5	0.4	0.6	0.5
Naphtha	-	-	-	-	-	-	-	-	-
Motor Gasoline	-	3.8	3.8	3.8	4.3	5.6	4.8	4.9	5.0
Aviation Fuels	-	0.1	0.2	0.2	0.4	0.4	0.7	0.6	0.7
Kerosene	-	-	-	-	-	-	-	-	-
Gas Diesel	-	5.3	5.1	5.1	8.1	1.8	4.5	9.3	7.5
Heavy Fuel Oil	-	8.2	7.8	7.7	6.8	6.9	5.4	8.8	6.6
Other Products	-	1.5	0.8	0.8	0.6	0.7	0.8	0.7	0.7
Refinery Fuel	-	0.9	0.6	0.6	-	-	-	-	-
Sub-Total	-	21.2	19.6	19.3	20.7	15.9	16.6	24.9	20.9
Bunkers	-	-	-	-	-	-	-	-	-
Total	-	21.2	19.6	19.3	20.7	15.9	16.6	24.9	20.9

Oil Products Consumption
Consommation de produits pétroliers

Malaysia

	1971	1973	1978	1983	1984	1985	1986	1987	1988
Thousand tonnes									
NGL/LPG/Ethane	56	32	111	159	172	211	249	304	348
Naphtha	-	-	-	-	-	-	-	-	-
Motor Gasoline	663	817	962	1 673	1 833	1 989	2 073	2 187	2 333
Aviation Fuels	132	167	210	348	360	377	416	422	416
Kerosene	72	139	356	341	345	300	291	261	248
Gas Diesel	1 600	1 695	1 990	3 459	3 173	3 072	2 996	3 161	3 455
Heavy Fuel Oil	1 553	1 338	2 571	2 998	2 902	2 749	2 724	2 636	2 670
Other Products	27	56	380	414	472	458	475	474	460
Refinery Fuel	185	160	155	154	197	252	121	117	155
Sub-Total	4 288	4 404	6 735	9 546	9 454	9 408	9 345	9 562	10 085
Bunkers	34	23	47	60	60	99	74	94	63
Total	4 322	4 427	6 782	9 606	9 514	9 507	9 419	9 656	10 148
Thousand barrels/day									
NGL/LPG/Ethane	1.8	1.0	3.5	5.1	5.5	6.7	7.9	9.7	11.0
Naphtha	-	-	-	-	-	-	-	-	-
Motor Gasoline	15.5	19.1	22.5	39.1	42.7	46.5	48.4	51.1	54.4
Aviation Fuels	2.9	3.6	4.6	7.6	7.8	8.2	9.0	9.2	9.0
Kerosene	1.5	2.9	7.5	7.2	7.3	6.4	6.2	5.5	5.2
Gas Diesel	32.7	34.6	40.7	70.7	64.7	62.8	61.2	64.6	70.4
Heavy Fuel Oil	28.3	24.4	46.9	54.7	52.8	50.2	49.7	48.1	48.6
Other Products	0.5	1.0	7.1	7.6	8.6	8.3	8.6	8.7	8.5
Refinery Fuel	3.6	3.1	3.0	3.1	4.0	5.2	2.5	2.4	3.2
Sub-Total	86.8	89.9	135.8	195.1	193.4	194.2	193.6	199.3	210.4
Bunkers	0.6	0.4	0.9	1.2	1.2	2.0	1.4	1.7	1.2
Total	87.5	90.3	136.8	196.2	194.6	196.1	195.0	201.0	211.5

	1989	1990	1991	1992	1993	1994	1995	1996	1997
Thousand tonnes									
NGL/LPG/Ethane	365	493	560	693	982	1 046	2 086	2 230	2 359
Naphtha	-	-	-	-	-	-	-	-	-
Motor Gasoline	2 429	2 711	2 990	3 154	3 923	4 429	4 332	4 828	4 886
Aviation Fuels	441	599	668	680	932	1 041	1 124	1 344	1 449
Kerosene	201	195	168	143	143	159	172	206	269
Gas Diesel	3 956	4 381	4 788	5 106	5 616	6 125	5 985	6 598	7 040
Heavy Fuel Oil	2 783	3 914	3 439	3 214	3 507	3 215	3 611	3 984	4 321
Other Products	563	607	668	710	616	644	756	813	341
Refinery Fuel	150	161	194	211	211	254	445	485	483
Sub-Total	10 888	13 061	13 475	13 911	15 930	16 913	18 511	20 488	21 148
Bunkers	67	94	112	115	64	238	168	168	158
Total	10 955	13 155	13 587	14 026	15 994	17 151	18 679	20 656	21 306
Thousand barrels/day									
NGL/LPG/Ethane	11.6	15.7	17.8	22.0	31.6	33.7	67.9	72.5	76.8
Naphtha	-	-	-	-	-	-	-	-	-
Motor Gasoline	56.8	63.4	69.9	73.5	91.9	103.7	101.5	112.8	114.5
Aviation Fuels	9.6	13.0	14.5	14.7	20.2	22.6	24.4	29.1	31.4
Kerosene	4.3	4.1	3.6	3.0	3.1	3.4	3.7	4.4	5.8
Gas Diesel	80.9	89.5	97.9	104.1	115.4	125.9	123.0	135.2	144.7
Heavy Fuel Oil	50.8	71.4	62.8	58.5	63.4	58.1	65.3	71.8	78.1
Other Products	10.3	11.0	12.0	12.8	11.0	11.5	13.5	14.4	6.1
Refinery Fuel	3.1	3.4	4.1	4.5	4.5	5.4	9.6	10.4	10.4
Sub-Total	227.3	271.6	282.5	293.0	341.1	364.3	408.8	450.7	467.8
Bunkers	1.3	1.8	2.2	2.2	1.2	4.4	3.1	3.1	2.9
Total	228.5	273.4	284.7	295.2	342.3	368.7	411.9	453.8	470.7

Oil Products Consumption
Consommation de produits pétroliers

Malta

	1971	1973	1978	1983	1984	1985	1986	1987	1988
Thousand tonnes									
NGL/LPG/Ethane	5	5	25	13	14	8	13	15	15
Naphtha	-	-	-	-	-	-	-	-	-
Motor Gasoline	39	46	41	45	41	20	56	60	65
Aviation Fuels	57	63	57	45	82	47	67	72	70
Kerosene	28	35	16	10	18	10	12	15	15
Gas Diesel	27	33	23	50	54	26	104	100	105
Heavy Fuel Oil	113	138	193	155	228	159	313	301	310
Other Products	-	-	-	10	4	8	-	7	6
Refinery Fuel	-	-	-	-	-	-	-	-	-
Sub-Total	269	320	355	328	441	278	565	570	586
Bunkers	61	23	30	15	20	20	32	30	30
Total	330	343	385	343	461	298	597	600	616
Thousand barrels/day									
NGL/LPG/Ethane	0.2	0.2	0.8	0.4	0.4	0.3	0.4	0.5	0.5
Naphtha	-	-	-	-	-	-	-	-	-
Motor Gasoline	0.9	1.1	1.0	1.1	1.0	0.5	1.3	1.4	1.5
Aviation Fuels	1.2	1.4	1.2	1.0	1.8	1.0	1.5	1.6	1.5
Kerosene	0.6	0.7	0.3	0.2	0.4	0.2	0.3	0.3	0.3
Gas Diesel	0.6	0.7	0.5	1.0	1.1	0.5	2.1	2.0	2.1
Heavy Fuel Oil	2.1	2.5	3.5	2.8	4.1	2.9	5.7	5.5	5.6
Other Products	-	-	-	0.2	0.1	0.1	-	0.1	0.1
Refinery Fuel	-	-	-	-	-	-	-	-	-
Sub-Total	5.5	6.5	7.3	6.7	8.9	5.5	11.3	11.4	11.7
Bunkers	1.2	0.4	0.6	0.3	0.4	0.4	0.6	0.6	0.6
Total	6.7	7.0	7.9	7.0	9.3	5.9	11.9	12.0	12.3

	1989	1990	1991	1992	1993	1994	1995	1996	1997
Thousand tonnes									
NGL/LPG/Ethane	15	15	16	18	18	18	16	17	18
Naphtha	-	-	-	-	-	-	-	-	-
Motor Gasoline	65	65	67	70	85	85	73	77	77
Aviation Fuels	70	70	80	79	79	79	111	107	112
Kerosene	15	16	16	16	-	-	18	19	55
Gas Diesel	105	105	127	112	120	120	120	170	170
Heavy Fuel Oil	310	310	296	323	431	410	468	488	497
Other Products	6	6	6	6	6	6	-	14	11
Refinery Fuel	-	-	-	-	-	-	-	-	-
Sub-Total	586	587	608	624	739	718	806	892	940
Bunkers	30	30	40	50	30	30	45	79	80
Total	616	617	648	674	769	748	851	971	1 020
Thousand barrels/day									
NGL/LPG/Ethane	0.5	0.5	0.5	0.6	0.6	0.6	0.5	0.5	0.6
Naphtha	-	-	-	-	-	-	-	-	-
Motor Gasoline	1.5	1.5	1.6	1.6	2.0	2.0	1.7	1.8	1.8
Aviation Fuels	1.5	1.5	1.7	1.7	1.7	1.7	2.4	2.3	2.4
Kerosene	0.3	0.3	0.3	0.3	-	-	0.4	0.4	1.2
Gas Diesel	2.1	2.1	2.6	2.3	2.5	2.5	2.5	3.5	3.5
Heavy Fuel Oil	5.7	5.7	5.4	5.9	7.9	7.5	8.5	8.9	9.1
Other Products	0.1	0.1	0.1	0.1	0.1	0.1	-	0.3	0.2
Refinery Fuel	-	-	-	-	-	-	-	-	-
Sub-Total	11.7	11.8	12.3	12.5	14.7	14.3	16.0	17.7	18.7
Bunkers	0.6	0.6	0.8	1.0	0.6	0.6	0.9	1.5	1.5
Total	12.3	12.4	13.0	13.5	15.3	14.9	16.9	19.1	20.2

Oil Products Consumption
Consommation de produits pétroliers

Moldova

	1971	1973	1978	1983	1984	1985	1986	1987	1988
Thousand tonnes									
NGL/LPG/Ethane	-	-	-	-	-	-	-	-	-
Naphtha	-	-	-	-	-	-	-	-	-
Motor Gasoline	-	-	-	-	-	-	-	-	-
Aviation Fuels	-	-	-	-	-	-	-	-	-
Kerosene	-	-	-	-	-	-	-	-	-
Gas Diesel	-	-	-	-	-	-	-	-	-
Heavy Fuel Oil	-	-	-	-	-	-	-	-	-
Other Products	-	-	-	-	-	-	-	-	-
Refinery Fuel	-	-	-	-	-	-	-	-	-
Sub-Total	-	-	-	-	-	-	-	-	-
Bunkers	-	-	-	-	-	-	-	-	-
Total	-	-	-	-	-	-	-	-	-
Thousand barrels/day									
NGL/LPG/Ethane	-	-	-	-	-	-	-	-	-
Naphtha	-	-	-	-	-	-	-	-	-
Motor Gasoline	-	-	-	-	-	-	-	-	-
Aviation Fuels	-	-	-	-	-	-	-	-	-
Kerosene	-	-	-	-	-	-	-	-	-
Gas Diesel	-	-	-	-	-	-	-	-	-
Heavy Fuel Oil	-	-	-	-	-	-	-	-	-
Other Products	-	-	-	-	-	-	-	-	-
Refinery Fuel	-	-	-	-	-	-	-	-	-
Sub-Total	-	-	-	-	-	-	-	-	-
Bunkers	-	-	-	-	-	-	-	-	-
Total	-	-	-	-	-	-	-	-	-

	1989	1990	1991	1992	1993	1994	1995	1996	1997
Thousand tonnes									
NGL/LPG/Ethane	-	-	-	100	41	20	20	22	27
Naphtha	-	-	-	-	-	-	-	-	-
Motor Gasoline	-	-	-	405	221	213	225	218	249
Aviation Fuels	-	-	-	33	21	13	15	20	23
Kerosene	-	-	-	-	20	5	2	8	7
Gas Diesel	-	-	-	598	469	391	383	325	312
Heavy Fuel Oil	-	-	-	1 745	1 298	465	365	343	242
Other Products	-	-	-	22	14	17	18	17	16
Refinery Fuel	-	-	-	-	-	-	-	-	-
Sub-Total	-	-	-	2 903	2 084	1 124	1 028	953	876
Bunkers	-	-	-	-	-	-	-	-	-
Total	-	-	-	2 903	2 084	1 124	1 028	953	876
Thousand barrels/day									
NGL/LPG/Ethane	-	-	-	3.2	1.3	0.6	0.6	0.7	0.9
Naphtha	-	-	-	-	-	-	-	-	-
Motor Gasoline	-	-	-	9.4	5.2	5.0	5.3	5.1	5.8
Aviation Fuels	-	-	-	0.7	0.5	0.3	0.3	0.4	0.5
Kerosene	-	-	-	-	0.4	0.1	0.0	0.2	0.1
Gas Diesel	-	-	-	12.2	9.6	8.0	7.8	6.6	6.4
Heavy Fuel Oil	-	-	-	31.8	23.7	8.5	6.7	6.2	4.4
Other Products	-	-	-	0.4	0.3	0.3	0.4	0.3	0.3
Refinery Fuel	-	-	-	-	-	-	-	-	-
Sub-Total	-	-	-	57.7	40.9	22.8	21.1	19.6	18.4
Bunkers	-	-	-	-	-	-	-	-	-
Total	-	-	-	57.7	40.9	22.8	21.1	19.6	18.4

Oil Products Consumption
Consommation de produits pétroliers

Morocco

	1971	1973	1978	1983	1984	1985	1986	1987	1988
Thousand tonnes									
NGL/LPG/Ethane	92	126	244	338	353	383	415	454	496
Naphtha	-	-	-	-	-	-	-	-	-
Motor Gasoline	330	373	393	350	338	329	336	348	364
Aviation Fuels	110	154	181	199	207	221	207	207	212
Kerosene	77	74	78	62	59	54	45	53	47
Gas Diesel	521	639	1 035	1 171	1 165	1 224	1 306	1 371	1 483
Heavy Fuel Oil	607	839	1 426	1 964	2 088	2 065	1 866	1 702	1 820
Other Products	115	125	147	148	131	126	142	153	151
Refinery Fuel	62	99	150	215	228	228	226	226	236
Sub-Total	1 914	2 429	3 654	4 447	4 569	4 630	4 543	4 514	4 809
Bunkers	77	103	52	19	14	12	12	13	12
Total	1 991	2 532	3 706	4 466	4 583	4 642	4 555	4 527	4 821
Thousand barrels/day									
NGL/LPG/Ethane	2.9	4.0	7.8	10.7	11.2	12.2	13.2	14.4	15.7
Naphtha	-	-	-	-	-	-	-	-	-
Motor Gasoline	7.7	8.7	9.2	8.2	7.9	7.7	7.9	8.1	8.5
Aviation Fuels	2.4	3.4	3.9	4.3	4.5	4.8	4.5	4.5	4.6
Kerosene	1.6	1.6	1.7	1.3	1.2	1.1	1.0	1.1	1.0
Gas Diesel	10.6	13.1	21.2	23.9	23.7	25.0	26.7	28.0	30.2
Heavy Fuel Oil	11.1	15.3	26.0	35.8	38.0	37.7	34.0	31.1	33.1
Other Products	2.2	2.4	2.8	2.8	2.5	2.4	2.7	2.9	2.9
Refinery Fuel	1.2	2.0	3.0	4.3	4.5	4.5	4.5	4.5	4.6
Sub-Total	39.8	50.4	75.5	91.4	93.6	95.4	94.4	94.7	100.7
Bunkers	1.5	2.0	1.0	0.4	0.3	0.2	0.2	0.3	0.2
Total	41.3	52.3	76.5	91.8	93.8	95.7	94.7	94.9	100.9

	1989	1990	1991	1992	1993	1994	1995	1996	1997
Thousand tonnes									
NGL/LPG/Ethane	548	512	634	719	786	850	897	951	1 007
Naphtha	-	-	-	27	-	-	-	-	-
Motor Gasoline	370	381	385	400	402	402	378	375	380
Aviation Fuels	214	249	183	213	220	207	231	238	247
Kerosene	47	48	45	46	46	42	48	39	57
Gas Diesel	1 529	1 706	1 804	2 118	2 352	2 578	2 615	2 640	2 621
Heavy Fuel Oil	2 087	2 011	1 984	2 267	2 136	2 414	1 951	1 617	1 697
Other Products	177	213	241	238	197	232	230	208	223
Refinery Fuel	250	250	240	250	260	265	259	249	249
Sub-Total	5 222	5 370	5 516	6 278	6 399	6 990	6 609	6 317	6 481
Bunkers	20	20	13	14	14	13	13	13	13
Total	5 242	5 390	5 529	6 292	6 413	7 003	6 622	6 330	6 494
Thousand barrels/day									
NGL/LPG/Ethane	17.4	16.3	20.1	22.8	25.0	27.0	28.5	30.1	32.0
Naphtha	-	-	-	0.6	-	-	-	-	-
Motor Gasoline	8.6	8.9	9.0	9.3	9.4	9.4	8.8	8.7	8.9
Aviation Fuels	4.6	5.4	4.0	4.6	4.8	4.5	5.0	5.2	5.4
Kerosene	1.0	1.0	1.0	1.0	1.0	0.9	1.0	0.8	1.2
Gas Diesel	31.3	34.9	36.9	43.2	48.1	52.7	53.4	53.8	53.6
Heavy Fuel Oil	38.1	36.7	36.2	41.3	39.0	44.0	35.6	29.4	31.0
Other Products	3.2	3.8	4.4	4.3	3.5	4.1	4.1	3.7	4.0
Refinery Fuel	4.9	4.9	4.7	4.9	5.1	5.2	5.1	4.9	4.9
Sub-Total	109.2	111.9	116.2	131.9	135.8	147.9	141.6	136.7	140.9
Bunkers	0.4	0.4	0.3	0.3	0.3	0.3	0.3	0.3	0.3
Total	109.6	112.3	116.5	132.2	136.1	148.2	141.9	136.9	141.1

Oil Products Consumption
Consommation de produits pétroliers

Mozambique

	1971	1973	1978	1983	1984	1985	1986	1987	1988
Thousand tonnes									
NGL/LPG/Ethane	9	10	7	10	10	10	10	10	4
Naphtha	-	-	-	-	-	-	-	-	-
Motor Gasoline	100	98	78	75	66	40	33	35	36
Aviation Fuels	37	37	30	23	23	30	40	41	42
Kerosene	41	41	55	29	12	5	4	4	5
Gas Diesel	102	94	98	125	112	170	195	195	200
Heavy Fuel Oil	179	167	186	161	169	173	175	175	176
Other Products	34	40	6	2	5	-	-	-	-
Refinery Fuel	33	32	25	12	5	-	-	-	-
Sub-Total	535	519	485	437	402	428	457	460	463
Bunkers	244	208	68	105	103	30	25	30	28
Total	779	727	553	542	505	458	482	490	491
Thousand barrels/day									
NGL/LPG/Ethane	0.3	0.3	0.2	0.3	0.3	0.3	0.3	0.3	0.1
Naphtha	-	-	-	-	-	-	-	-	-
Motor Gasoline	2.3	2.3	1.8	1.8	1.5	0.9	0.8	0.8	0.8
Aviation Fuels	0.8	0.8	0.7	0.5	0.5	0.7	0.9	0.9	0.9
Kerosene	0.9	0.9	1.2	0.6	0.3	0.1	0.1	0.1	0.1
Gas Diesel	2.1	1.9	2.0	2.6	2.3	3.5	4.0	4.0	4.1
Heavy Fuel Oil	3.3	3.0	3.4	2.9	3.1	3.2	3.2	3.2	3.2
Other Products	0.6	0.7	0.1	0.0	0.1	-	-	-	-
Refinery Fuel	0.6	0.6	0.5	0.2	0.1	-	-	-	-
Sub-Total	10.8	10.5	9.8	8.9	8.1	8.6	9.2	9.3	9.3
Bunkers	4.6	3.9	1.3	2.0	1.9	0.6	0.5	0.6	0.6
Total	15.4	14.4	11.1	10.9	10.1	9.3	9.7	9.9	9.8

	1989	1990	1991	1992	1993	1994	1995	1996	1997
Thousand tonnes									
NGL/LPG/Ethane	4	4	5	4	4	3	3	3	5
Naphtha	-	-	-	-	-	-	-	-	-
Motor Gasoline	38	40	41	37	38	37	38	41	42
Aviation Fuels	40	41	36	27	35	43	18	21	25
Kerosene	5	23	5	31	12	19	14	17	25
Gas Diesel	205	210	174	234	290	230	239	230	296
Heavy Fuel Oil	156	146	144	130	137	151	155	182	272
Other Products	-	7	8	8	6	6	6	7	7
Refinery Fuel	-	-	-	-	-	-	-	-	-
Sub-Total	448	471	413	471	522	489	473	501	672
Bunkers	25	28	9	7	12	14	15	15	64
Total	473	499	422	478	534	503	488	516	736
Thousand barrels/day									
NGL/LPG/Ethane	0.1	0.1	0.2	0.1	0.1	0.1	0.1	0.1	0.2
Naphtha	-	-	-	-	-	-	-	-	-
Motor Gasoline	0.9	0.9	1.0	0.9	0.9	0.9	0.9	1.0	1.0
Aviation Fuels	0.9	0.9	0.8	0.6	0.8	0.9	0.4	0.5	0.5
Kerosene	0.1	0.5	0.1	0.7	0.3	0.4	0.3	0.4	0.5
Gas Diesel	4.2	4.3	3.6	4.8	5.9	4.7	4.9	4.7	6.1
Heavy Fuel Oil	2.8	2.7	2.6	2.4	2.5	2.8	2.8	3.3	5.0
Other Products	-	0.1	0.2	0.2	0.1	0.1	0.1	0.1	0.1
Refinery Fuel	-	-	-	-	-	-	-	-	-
Sub-Total	9.0	9.5	8.3	9.5	10.6	9.9	9.5	10.0	13.4
Bunkers	0.5	0.6	0.2	0.1	0.2	0.3	0.3	0.3	1.2
Total	9.5	10.1	8.5	9.7	10.8	10.1	9.8	10.3	14.5

Oil Products Consumption
Consommation de produits pétroliers

Myanmar

	1971	1973	1978	1983	1984	1985	1986	1987	1988
Thousand tonnes									
NGL/LPG/Ethane	5	7	4	3	3	4	5	4	4
Naphtha	-	-	-	-	-	-	-	-	-
Motor Gasoline	162	174	220	238	260	235	261	177	153
Aviation Fuels	30	34	30	35	37	40	40	20	20
Kerosene	287	241	113	19	18	7	8	5	5
Gas Diesel	324	231	326	371	410	397	430	269	271
Heavy Fuel Oil	183	77	203	210	213	194	215	120	115
Other Products	228	221	245	218	205	228	175	163	114
Refinery Fuel	52	40	49	50	51	50	53	32	29
Sub-Total	1 271	1 025	1 190	1 144	1 197	1 155	1 187	790	711
Bunkers	2	1	-	-	-	-	-	-	-
Total	1 273	1 026	1 190	1 144	1 197	1 155	1 187	790	711
Thousand barrels/day									
NGL/LPG/Ethane	0.2	0.2	0.1	0.1	0.1	0.1	0.2	0.1	0.1
Naphtha	-	-	-	-	-	-	-	-	-
Motor Gasoline	3.8	4.1	5.1	5.6	6.1	5.5	6.1	4.1	3.6
Aviation Fuels	0.7	0.7	0.7	0.8	0.8	0.9	0.9	0.4	0.4
Kerosene	6.1	5.1	2.4	0.4	0.4	0.1	0.2	0.1	0.1
Gas Diesel	6.6	4.7	6.7	7.6	8.4	8.1	8.8	5.5	5.5
Heavy Fuel Oil	3.3	1.4	3.7	3.8	3.9	3.5	3.9	2.2	2.1
Other Products	4.4	4.2	4.7	4.1	3.9	4.4	3.3	3.1	2.2
Refinery Fuel	1.1	0.9	1.1	1.1	1.1	1.1	1.2	0.7	0.6
Sub-Total	26.2	21.4	24.5	23.5	24.6	23.7	24.5	16.3	14.6
Bunkers	0.0	0.0	-	-	-	-	-	-	-
Total	26.2	21.4	24.5	23.5	24.6	23.7	24.5	16.3	14.6

	1989	1990	1991	1992	1993	1994	1995	1996	1997
Thousand tonnes									
NGL/LPG/Ethane	9	9	9	10	12	13	9	11	12
Naphtha	-	-	-	-	-	-	-	-	-
Motor Gasoline	152	125	123	153	169	194	209	204	248
Aviation Fuels	20	28	25	28	33	44	46	54	52
Kerosene	5	2	1	9	11	12	13	8	3
Gas Diesel	262	305	275	362	450	541	787	876	1 008
Heavy Fuel Oil	115	118	115	142	137	94	108	100	106
Other Products	109	68	69	19	95	137	30	37	46
Refinery Fuel	29	29	31	184	59	67	49	45	73
Sub-Total	701	684	648	907	966	1 102	1 251	1 335	1 548
Bunkers	-	-	-	27	1	5	3	5	4
Total	701	684	648	934	967	1 107	1 254	1 340	1 552
Thousand barrels/day									
NGL/LPG/Ethane	0.3	0.3	0.3	0.3	0.4	0.4	0.3	0.3	0.4
Naphtha	-	-	-	-	-	-	-	-	-
Motor Gasoline	3.6	2.9	2.9	3.6	4.0	4.5	4.9	4.8	5.8
Aviation Fuels	0.4	0.6	0.5	0.6	0.7	1.0	1.0	1.2	1.1
Kerosene	0.1	0.0	0.0	0.2	0.2	0.3	0.3	0.2	0.1
Gas Diesel	5.4	6.2	5.6	7.4	9.2	11.1	16.1	17.9	20.6
Heavy Fuel Oil	2.1	2.2	2.1	2.6	2.5	1.7	2.0	1.8	1.9
Other Products	2.1	1.3	1.3	0.4	1.8	2.6	0.5	0.6	0.8
Refinery Fuel	0.6	0.6	0.7	3.7	1.2	1.4	1.0	0.9	1.5
Sub-Total	14.5	14.2	13.4	18.7	20.1	23.0	26.1	27.7	32.2
Bunkers	-	-	-	0.5	0.0	0.1	0.1	0.1	0.1
Total	14.5	14.2	13.4	19.2	20.1	23.1	26.1	27.8	32.3

Oil Products Consumption
Consommation de produits pétroliers

Nepal

	1971	1973	1978	1983	1984	1985	1986	1987	1988
Thousand tonnes									
NGL/LPG/Ethane	-	-	1	1	1	3	2	5	5
Naphtha	-	-	-	-	-	-	-	-	-
Motor Gasoline	10	10	8	11	12	13	14	15	16
Aviation Fuels	3	4	11	15	19	18	23	21	24
Kerosene	16	28	26	28	36	38	43	55	64
Gas Diesel	25	27	41	57	62	66	72	78	98
Heavy Fuel Oil	-	1	1	6	6	9	15	15	9
Other Products	-	-	-	7	8	4	4	4	3
Refinery Fuel	-	-	-	-	-	-	-	-	-
Sub-Total	54	70	88	125	144	151	173	193	219
Bunkers	-	-	-	-	-	-	-	-	-
Total	54	70	88	125	144	151	173	193	219
Thousand barrels/day									
NGL/LPG/Ethane	-	-	0.0	0.0	0.0	0.1	0.1	0.2	0.2
Naphtha	-	-	-	-	-	-	-	-	-
Motor Gasoline	0.2	0.2	0.2	0.3	0.3	0.3	0.3	0.4	0.4
Aviation Fuels	0.1	0.1	0.2	0.3	0.4	0.4	0.5	0.5	0.5
Kerosene	0.3	0.6	0.6	0.6	0.8	0.8	0.9	1.2	1.4
Gas Diesel	0.5	0.6	0.8	1.2	1.3	1.3	1.5	1.6	2.0
Heavy Fuel Oil	-	0.0	0.0	0.1	0.1	0.2	0.3	0.3	0.2
Other Products	-	-	-	0.1	0.1	0.1	0.1	0.1	0.1
Refinery Fuel	-	-	-	-	-	-	-	-	-
Sub-Total	1.1	1.5	1.9	2.6	3.0	3.2	3.6	4.1	4.6
Bunkers	-	-	-	-	-	-	-	-	-
Total	1.1	1.5	1.9	2.6	3.0	3.2	3.6	4.1	4.6

	1989	1990	1991	1992	1993	1994	1995	1996	1997
Thousand tonnes									
NGL/LPG/Ethane	3	1	4	6	7	13	19	19	19
Naphtha	-	-	-	-	-	-	-	-	-
Motor Gasoline	14	13	17	19	19	25	32	32	32
Aviation Fuels	17	13	15	19	21	27	29	29	29
Kerosene	64	63	76	96	98	134	170	170	170
Gas Diesel	86	83	117	146	155	187	223	223	223
Heavy Fuel Oil	7	4	6	10	12	31	31	31	31
Other Products	3	3	3	5	5	6	7	7	7
Refinery Fuel	-	-	-	-	-	-	-	-	-
Sub-Total	194	180	238	301	317	423	511	511	511
Bunkers	-	-	-	-	-	-	-	-	-
Total	194	180	238	301	317	423	511	511	511
Thousand barrels/day									
NGL/LPG/Ethane	0.1	0.0	0.1	0.2	0.2	0.4	0.6	0.6	0.6
Naphtha	-	-	-	-	-	-	-	-	-
Motor Gasoline	0.3	0.3	0.4	0.4	0.4	0.6	0.7	0.7	0.7
Aviation Fuels	0.4	0.3	0.3	0.4	0.5	0.6	0.6	0.6	0.6
Kerosene	1.4	1.3	1.6	2.0	2.1	2.8	3.6	3.6	3.6
Gas Diesel	1.8	1.7	2.4	3.0	3.2	3.8	4.6	4.5	4.6
Heavy Fuel Oil	0.1	0.1	0.1	0.2	0.2	0.6	0.6	0.6	0.6
Other Products	0.1	0.1	0.1	0.1	0.1	0.1	0.1	0.1	0.1
Refinery Fuel	-	-	-	-	-	-	-	-	-
Sub-Total	4.1	3.8	5.0	6.3	6.7	8.9	10.8	10.8	10.8
Bunkers	-	-	-	-	-	-	-	-	-
Total	4.1	3.8	5.0	6.3	6.7	8.9	10.8	10.8	10.8

Oil Products Consumption
Consommation de produits pétroliers

Netherlands Antilles

	1971	1973	1978	1983	1984	1985	1986	1987	1988
Thousand tonnes									
NGL/LPG/Ethane	79	57	82	72	65	60	60	38	35
Naphtha	-	-	-	-	-	-	-	-	-
Motor Gasoline	67	78	179	77	100	100	100	99	98
Aviation Fuels	49	55	46	46	46	41	36	37	37
Kerosene	45	46	20	34	28	30	30	30	18
Gas Diesel	404	351	349	350	300	250	187	202	215
Heavy Fuel Oil	1 670	1 835	742	613	570	540	492	492	404
Other Products	24	24	27	47	35	35	33	33	33
Refinery Fuel	2 390	2 640	1 600	1 040	1 080	460	508	408	335
Sub-Total	4 728	5 086	3 045	2 279	2 224	1 516	1 446	1 339	1 175
Bunkers	2 500	2 312	2 259	2 100	1 950	1 985	1 675	1 600	1 600
Total	7 228	7 398	5 304	4 379	4 174	3 501	3 121	2 939	2 775
Thousand barrels/day									
NGL/LPG/Ethane	2.5	1.8	2.6	2.3	2.1	1.9	1.9	1.2	1.1
Naphtha	-	-	-	-	-	-	-	-	-
Motor Gasoline	1.6	1.8	4.2	1.8	2.3	2.3	2.3	2.3	2.3
Aviation Fuels	1.1	1.2	1.0	1.0	1.0	0.9	0.8	0.8	0.8
Kerosene	1.0	1.0	0.4	0.7	0.6	0.6	0.6	0.6	0.4
Gas Diesel	8.3	7.2	7.1	7.2	6.1	5.1	3.8	4.1	4.4
Heavy Fuel Oil	30.5	33.5	13.5	11.2	10.4	9.9	9.0	9.0	7.4
Other Products	0.5	0.5	0.5	0.9	0.7	0.7	0.6	0.6	0.6
Refinery Fuel	43.6	48.2	29.2	19.0	19.7	8.4	9.3	7.4	6.1
Sub-Total	88.9	95.1	58.6	44.0	42.8	29.8	28.4	26.1	23.0
Bunkers	45.8	42.5	41.6	38.9	35.9	36.6	30.9	29.5	29.4
Total	134.8	137.6	100.2	82.9	78.7	66.4	59.3	55.7	52.5

	1989	1990	1991	1992	1993	1994	1995	1996	1997
Thousand tonnes									
NGL/LPG/Ethane	39	45	45	45	52	51	52	55	55
Naphtha	-	-	-	-	-	-	-	-	-
Motor Gasoline	85	85	85	85	43	59	61	69	69
Aviation Fuels	37	37	49	49	59	60	61	62	62
Kerosene	20	25	25	30	101	108	108	111	111
Gas Diesel	201	197	228	253	222	297	295	293	293
Heavy Fuel Oil	329	274	274	274	215	215	214	215	215
Other Products	33	33	33	38	98	88	89	65	65
Refinery Fuel	272	227	227	227	178	179	178	178	178
Sub-Total	1 016	923	966	1 001	968	1 057	1 058	1 048	1 048
Bunkers	1 650	1 675	1 725	1 725	1 697	1 714	1 720	1 724	1 724
Total	2 666	2 598	2 691	2 726	2 665	2 771	2 778	2 772	2 772
Thousand barrels/day									
NGL/LPG/Ethane	1.2	1.4	1.4	1.4	1.7	1.6	1.7	1.7	1.7
Naphtha	-	-	-	-	-	-	-	-	-
Motor Gasoline	2.0	2.0	2.0	2.0	1.0	1.4	1.4	1.6	1.6
Aviation Fuels	0.8	0.8	1.1	1.1	1.3	1.3	1.3	1.3	1.4
Kerosene	0.4	0.5	0.5	0.6	2.1	2.3	2.3	2.3	2.4
Gas Diesel	4.1	4.0	4.7	5.2	4.5	6.1	6.0	6.0	6.0
Heavy Fuel Oil	6.0	5.0	5.0	5.0	3.9	3.9	3.9	3.9	3.9
Other Products	0.6	0.6	0.6	0.7	1.8	1.6	1.6	1.2	1.3
Refinery Fuel	5.0	4.1	4.1	4.1	3.2	3.3	3.2	3.2	3.2
Sub-Total	20.2	18.6	19.4	20.1	19.6	21.5	21.5	21.4	21.5
Bunkers	30.5	31.0	32.0	31.9	31.4	31.7	31.8	31.8	31.9
Total	50.7	49.6	51.5	52.0	51.0	53.2	53.4	53.3	53.4

Oil Products Consumption
Consommation de produits pétroliers

Nicaragua

	1971	1973	1978	1983	1984	1985	1986	1987	1988
Thousand tonnes									
NGL/LPG/Ethane	13	16	16	19	16	17	19	19	16
Naphtha	-	-	-	-	-	-	-	-	-
Motor Gasoline	112	140	167	111	110	107	118	120	107
Aviation Fuels	24	23	25	22	21	21	19	36	27
Kerosene	17	18	16	16	12	15	22	15	13
Gas Diesel	163	175	226	211	216	213	231	245	227
Heavy Fuel Oil	168	180	319	242	211	218	275	269	250
Other Products	18	43	32	12	11	19	19	36	29
Refinery Fuel	10	12	12	10	8	10	10	10	10
Sub-Total	525	607	813	643	605	620	713	750	679
Bunkers	-	-	-	-	-	-	-	-	-
Total	525	607	813	643	605	620	713	750	679
Thousand barrels/day									
NGL/LPG/Ethane	0.4	0.5	0.5	0.6	0.5	0.5	0.6	0.6	0.5
Naphtha	-	-	-	-	-	-	-	-	-
Motor Gasoline	2.6	3.3	3.9	2.6	2.6	2.5	2.8	2.8	2.5
Aviation Fuels	0.5	0.5	0.6	0.5	0.5	0.5	0.4	0.8	0.6
Kerosene	0.4	0.4	0.3	0.3	0.3	0.3	0.5	0.3	0.3
Gas Diesel	3.3	3.6	4.6	4.3	4.4	4.4	4.7	5.0	4.6
Heavy Fuel Oil	3.1	3.3	5.8	4.4	3.8	4.0	5.0	4.9	4.5
Other Products	0.3	0.8	0.6	0.2	0.2	0.4	0.4	0.7	0.5
Refinery Fuel	0.2	0.3	0.3	0.2	0.2	0.2	0.2	0.2	0.2
Sub-Total	10.9	12.6	16.6	13.2	12.4	12.7	14.6	15.3	13.8
Bunkers	-	-	-	-	-	-	-	-	-
Total	10.9	12.6	16.6	13.2	12.4	12.7	14.6	15.3	13.8

	1989	1990	1991	1992	1993	1994	1995	1996	1997
Thousand tonnes									
NGL/LPG/Ethane	17	17	20	21	22	26	28	29	35
Naphtha	-	-	-	-	-	-	-	-	-
Motor Gasoline	82	82	104	102	91	99	97	97	116
Aviation Fuels	28	25	16	14	18	21	23	22	28
Kerosene	9	9	12	12	13	14	13	13	12
Gas Diesel	203	210	197	214	226	281	306	302	350
Heavy Fuel Oil	186	241	233	288	270	285	272	376	373
Other Products	64	38	29	44	48	21	21	25	30
Refinery Fuel	12	12	13	14	13	14	12	13	16
Sub-Total	601	634	624	709	701	761	772	877	960
Bunkers	-	-	-	-	-	-	-	-	-
Total	601	634	624	709	701	761	772	877	960
Thousand barrels/day									
NGL/LPG/Ethane	0.5	0.5	0.6	0.7	0.7	0.8	0.9	0.9	1.1
Naphtha	-	-	-	-	-	-	-	-	-
Motor Gasoline	1.9	1.9	2.4	2.4	2.1	2.3	2.3	2.3	2.7
Aviation Fuels	0.6	0.5	0.4	0.3	0.4	0.5	0.5	0.5	0.6
Kerosene	0.2	0.2	0.3	0.3	0.3	0.3	0.3	0.3	0.3
Gas Diesel	4.1	4.3	4.0	4.4	4.6	5.7	6.3	6.2	7.2
Heavy Fuel Oil	3.4	4.4	4.3	5.2	4.9	5.2	5.0	6.8	6.8
Other Products	1.2	0.7	0.5	0.8	0.9	0.4	0.4	0.5	0.6
Refinery Fuel	0.3	0.3	0.3	0.3	0.3	0.3	0.3	0.3	0.3
Sub-Total	12.2	12.8	12.8	14.3	14.2	15.5	15.8	17.7	19.5
Bunkers	-	-	-	-	-	-	-	-	-
Total	12.2	12.8	12.8	14.3	14.2	15.5	15.8	17.7	19.5

Oil Products Consumption

Consommation de produits pétroliers

Nigeria

	1971	1973	1978	1983	1984	1985	1986	1987	1988
Thousand tonnes									
NGL/LPG/Ethane	12	17	32	60	67	86	88	87	84
Naphtha	-	-	-	-	-	-	-	-	-
Motor Gasoline	509	700	2 362	2 900	2 900	2 900	2 885	3 057	3 312
Aviation Fuels	77	121	315	420	411	422	300	280	280
Kerosene	216	257	671	1 266	1 304	1 307	1 528	1 630	1 551
Gas Diesel	450	704	1 851	2 770	2 275	2 103	1 823	1 670	1 754
Heavy Fuel Oil	274	438	753	1 118	804	1 015	762	689	938
Other Products	88	290	486	366	269	280	307	376	255
Refinery Fuel	62	69	92	421	426	433	481	483	533
Sub-Total	1 688	2 596	6 562	9 321	8 456	8 546	8 174	8 272	8 707
Bunkers	5	13	45	95	100	110	372	204	307
Total	1 693	2 609	6 607	9 416	8 556	8 656	8 546	8 476	9 014
Thousand barrels/day									
NGL/LPG/Ethane	0.4	0.5	1.0	1.9	2.1	2.7	2.8	2.8	2.7
Naphtha	-	-	-	-	-	-	-	-	-
Motor Gasoline	11.9	16.4	55.2	67.8	67.6	67.8	67.4	71.4	77.2
Aviation Fuels	1.7	2.6	6.9	9.1	8.9	9.2	6.5	6.1	6.1
Kerosene	4.6	5.5	14.2	26.8	27.6	27.7	32.4	34.6	32.8
Gas Diesel	9.2	14.4	37.8	56.6	46.4	43.0	37.3	34.1	35.8
Heavy Fuel Oil	5.0	8.0	13.7	20.4	14.6	18.5	13.9	12.6	17.1
Other Products	1.7	5.6	9.3	7.0	5.1	5.4	5.9	7.2	4.9
Refinery Fuel	1.2	1.4	1.8	8.3	8.4	8.6	9.4	9.5	10.7
Sub-Total	35.7	54.3	140.0	198.0	180.7	182.8	175.6	178.3	187.2
Bunkers	0.1	0.2	0.9	1.8	1.9	2.1	7.1	3.9	5.8
Total	35.7	54.5	140.9	199.9	182.7	185.0	182.8	182.1	192.9

	1989	1990	1991	1992	1993	1994	1995	1996	1997
Thousand tonnes									
NGL/LPG/Ethane	69	40	56	49	36	21	10	11	11
Naphtha	-	-	-	-	-	-	-	-	-
Motor Gasoline	3 576	2 625	3 380	3 969	3 336	3 016	2 966	3 251	3 306
Aviation Fuels	278	302	354	363	369	306	432	487	492
Kerosene	1 503	1 298	1 253	1 442	1 194	764	750	846	855
Gas Diesel	1 599	1 396	1 787	2 280	2 054	1 596	1 378	1 278	1 294
Heavy Fuel Oil	950	877	774	744	662	757	619	802	824
Other Products	453	451	766	2 108	1 025	1 535	1 650	1 938	1 938
Refinery Fuel	533	530	535	540	431	392	455	545	545
Sub-Total	8 961	7 519	8 905	11 495	9 107	8 387	8 260	9 158	9 265
Bunkers	340	186	354	331	269	236	226	80	80
Total	9 301	7 705	9 259	11 826	9 376	8 623	8 486	9 238	9 345
Thousand barrels/day									
NGL/LPG/Ethane	2.2	1.3	1.8	1.6	1.1	0.7	0.3	0.3	0.4
Naphtha	-	-	-	-	-	-	-	-	-
Motor Gasoline	83.6	61.3	79.0	92.5	78.0	70.5	69.3	75.8	77.3
Aviation Fuels	6.0	6.6	7.7	7.9	8.0	6.6	9.4	10.6	10.7
Kerosene	31.9	27.5	26.6	30.5	25.3	16.2	15.9	17.9	18.1
Gas Diesel	32.7	28.5	36.5	46.5	42.0	32.6	28.2	26.0	26.4
Heavy Fuel Oil	17.3	16.0	14.1	13.5	12.1	13.8	11.3	14.6	15.0
Other Products	8.7	8.6	14.5	40.2	19.4	29.4	31.6	37.0	37.1
Refinery Fuel	10.8	10.7	10.8	10.9	8.7	8.0	9.4	11.2	11.2
Sub-Total	193.1	160.6	191.0	243.5	194.7	177.8	175.3	193.4	196.2
Bunkers	6.4	3.6	6.7	6.2	5.1	4.5	4.3	1.6	1.6
Total	199.5	164.2	197.7	249.7	199.8	182.3	179.6	195.0	197.9

Oil Products Consumption
Consommation de produits pétroliers

Oman

	1971	1973	1978	1983	1984	1985	1986	1987	1988
Thousand tonnes									
NGL/LPG/Ethane	-	-	-	13	18	23	29	25	27
Naphtha	-	-	-	-	-	-	-	-	-
Motor Gasoline	30	37	165	296	352	397	421	402	415
Aviation Fuels	4	4	104	141	180	182	189	127	129
Kerosene	6	9	13	11	11	9	10	10	5
Gas Diesel	45	46	215	372	452	525	520	508	540
Heavy Fuel Oil	-	-	-	-	-	-	-	-	-
Other Products	-	-	72	166	123	106	49	54	54
Refinery Fuel	-	-	-	74	88	92	100	79	96
Sub-Total	85	96	569	1 073	1 224	1 334	1 318	1 205	1 266
Bunkers	1 250	1 230	637	112	96	114	120	150	180
Total	1 335	1 326	1 206	1 185	1 320	1 448	1 438	1 355	1 446
Thousand barrels/day									
NGL/LPG/Ethane	-	-	-	0.4	0.6	0.7	0.9	0.8	0.9
Naphtha	-	-	-	-	-	-	-	-	-
Motor Gasoline	0.7	0.9	3.9	6.9	8.2	9.3	9.8	9.4	9.7
Aviation Fuels	0.1	0.1	2.3	3.1	3.9	4.0	4.1	2.8	2.8
Kerosene	0.1	0.2	0.3	0.2	0.2	0.2	0.2	0.2	0.1
Gas Diesel	0.9	0.9	4.4	7.6	9.2	10.7	10.6	10.4	11.0
Heavy Fuel Oil	-	-	-	-	-	-	-	-	-
Other Products	-	-	1.4	3.2	2.4	2.0	1.0	1.1	1.1
Refinery Fuel	-	-	-	1.4	1.6	1.7	1.8	1.4	1.7
Sub-Total	1.8	2.1	12.2	22.8	26.1	28.6	28.5	26.1	27.3
Bunkers	22.8	22.4	11.6	2.0	1.7	2.1	2.2	2.7	3.3
Total	24.6	24.5	23.8	24.9	27.8	30.7	30.7	28.8	30.5

	1989	1990	1991	1992	1993	1994	1995	1996	1997
Thousand tonnes									
NGL/LPG/Ethane	30	66	82	73	57	78	63	63	52
Naphtha	-	-	-	-	-	-	-	-	-
Motor Gasoline	438	478	521	560	589	593	617	639	678
Aviation Fuels	140	296	256	163	192	164	147	152	172
Kerosene	10	4	5	5	5	4	4	4	4
Gas Diesel	536	433	459	513	595	641	675	656	678
Heavy Fuel Oil	-	507	1 661	1 427	965	521	954	824	442
Other Products	70	24	27	30	43	41	41	41	41
Refinery Fuel	100	122	130	112	94	147	150	152	137
Sub-Total	1 324	1 930	3 141	2 883	2 540	2 189	2 651	2 531	2 204
Bunkers	200	20	21	22	247	25	27	27	30
Total	1 524	1 950	3 162	2 905	2 787	2 214	2 678	2 558	2 234
Thousand barrels/day									
NGL/LPG/Ethane	1.0	2.1	2.6	2.3	1.8	2.5	2.0	2.0	1.7
Naphtha	-	-	-	-	-	-	-	-	-
Motor Gasoline	10.2	11.2	12.2	13.1	13.8	13.9	14.4	14.9	15.8
Aviation Fuels	3.0	6.4	5.6	3.5	4.2	3.6	3.2	3.3	3.7
Kerosene	0.2	0.1	0.1	0.1	0.1	0.1	0.1	0.1	0.1
Gas Diesel	11.0	8.9	9.4	10.5	12.2	13.1	13.8	13.4	13.9
Heavy Fuel Oil	-	9.3	30.3	26.0	17.6	9.5	17.4	15.0	8.1
Other Products	1.4	0.4	0.5	0.5	0.8	0.8	0.8	0.8	0.8
Refinery Fuel	1.8	2.2	2.4	2.0	1.7	2.7	2.7	2.8	2.5
Sub-Total	28.6	40.5	63.0	58.0	52.1	46.0	54.4	52.1	46.5
Bunkers	3.6	0.4	0.4	0.4	4.5	0.5	0.5	0.5	0.5
Total	32.2	40.9	63.4	58.4	56.6	46.5	54.9	52.6	47.0

Oil Products Consumption
Consommation de produits pétroliers

Pakistan

	1971	1973	1978	1983	1984	1985	1986	1987	1988
Thousand tonnes									
NGL/LPG/Ethane	4	5	11	67	70	70	75	100	130
Naphtha	-	-	-	-	-	-	-	-	-
Motor Gasoline	261	315	445	612	755	784	823	872	940
Aviation Fuels	368	339	440	474	465	459	499	464	475
Kerosene	422	436	644	604	690	760	811	877	876
Gas Diesel	1 122	1 248	1 636	2 694	2 653	2 784	2 980	3 499	3 975
Heavy Fuel Oil	735	711	584	1 195	1 662	1 943	2 070	2 123	2 374
Other Products	147	133	307	306	297	346	373	360	383
Refinery Fuel	97	97	114	135	148	155	166	170	168
Sub-Total	3 156	3 284	4 181	6 087	6 740	7 301	7 797	8 465	9 321
Bunkers	92	158	81	48	46	25	13	10	15
Total	3 248	3 442	4 262	6 135	6 786	7 326	7 810	8 475	9 336
Thousand barrels/day									
NGL/LPG/Ethane	0.1	0.2	0.4	2.1	2.2	2.2	2.4	3.2	4.1
Naphtha	-	-	-	-	-	-	-	-	-
Motor Gasoline	6.1	7.4	10.4	14.3	17.6	18.3	19.2	20.4	21.9
Aviation Fuels	8.0	7.4	9.6	10.3	10.1	10.0	10.8	10.1	10.3
Kerosene	8.9	9.2	13.7	12.8	14.6	16.1	17.2	18.6	18.5
Gas Diesel	22.9	25.5	33.4	55.1	54.1	56.9	60.9	71.5	81.0
Heavy Fuel Oil	13.4	13.0	10.7	21.8	30.2	35.5	37.8	38.7	43.2
Other Products	2.8	2.6	5.9	5.6	5.4	6.3	6.8	6.5	6.9
Refinery Fuel	1.9	1.9	2.3	2.7	2.9	3.1	3.3	3.4	3.3
Sub-Total	64.3	67.1	86.2	124.7	137.1	148.4	158.4	172.4	189.3
Bunkers	1.7	2.9	1.5	0.9	0.8	0.5	0.2	0.2	0.3
Total	66.0	70.0	87.7	125.6	138.0	148.8	158.7	172.6	189.6

	1989	1990	1991	1992	1993	1994	1995	1996	1997
Thousand tonnes									
NGL/LPG/Ethane	131	127	147	131	135	122	171	205	220
Naphtha	-	-	-	-	-	-	-	-	-
Motor Gasoline	989	821	829	899	989	992	983	1 062	1 129
Aviation Fuels	466	455	504	540	568	553	556	586	574
Kerosene	989	1 133	962	632	640	608	602	614	752
Gas Diesel	4 286	4 277	4 305	5 036	5 501	5 912	6 149	6 600	6 655
Heavy Fuel Oil	2 438	3 101	3 271	3 875	4 274	5 102	5 629	6 706	6 889
Other Products	377	212	319	334	376	338	361	349	344
Refinery Fuel	180	180	185	190	210	212	205	213	209
Sub-Total	9 856	10 306	10 522	11 637	12 693	13 839	14 656	16 335	16 772
Bunkers	22	34	18	13	15	17	15	12	15
Total	9 878	10 340	10 540	11 650	12 708	13 856	14 671	16 347	16 787
Thousand barrels/day									
NGL/LPG/Ethane	4.2	4.0	4.7	4.2	4.3	3.9	5.4	6.5	7.0
Naphtha	-	-	-	-	-	-	-	-	-
Motor Gasoline	23.1	19.2	19.4	21.0	23.1	23.2	23.0	24.8	26.4
Aviation Fuels	10.1	9.9	11.0	11.7	12.3	12.0	12.1	12.7	12.5
Kerosene	21.0	24.0	20.4	13.4	13.6	12.9	12.8	13.0	15.9
Gas Diesel	87.6	87.4	88.0	102.6	112.4	120.8	125.7	134.5	136.0
Heavy Fuel Oil	44.5	56.6	59.7	70.5	78.0	93.1	102.7	122.0	125.7
Other Products	6.9	3.7	5.7	5.9	6.6	6.0	6.4	6.1	6.1
Refinery Fuel	3.6	3.6	3.7	3.8	4.2	4.3	4.1	4.3	4.2
Sub-Total	200.9	208.4	212.4	233.0	254.6	276.2	292.1	323.9	333.8
Bunkers	0.4	0.6	0.3	0.2	0.3	0.3	0.3	0.2	0.3
Total	201.3	209.0	212.8	233.3	254.9	276.5	292.4	324.1	334.1

Oil Products Consumption
Consommation de produits pétroliers

Panama

	1971	1973	1978	1983	1984	1985	1986	1987	1988
Thousand tonnes									
NGL/LPG/Ethane	23	29	40	48	50	51	55	58	56
Naphtha	-	-	-	-	-	-	-	-	-
Motor Gasoline	234	256	259	182	205	203	210	198	191
Aviation Fuels	6	6	2	2	2	2	2	1	1
Kerosene	8	5	4	3	4	3	3	4	3
Gas Diesel	169	209	264	343	296	310	313	332	267
Heavy Fuel Oil	330	368	305	343	228	173	183	227	132
Other Products	13	15	9	30	32	12	12	6	4
Refinery Fuel	141	160	123	64	65	91	76	97	75
Sub-Total	924	1 048	1 006	1 015	882	845	854	923	729
Bunkers	5 808	5 264	2 220	690	740	745	725	845	710
Total	6 732	6 312	3 226	1 705	1 622	1 590	1 579	1 768	1 439
Thousand barrels/day									
NGL/LPG/Ethane	0.7	0.9	1.3	1.5	1.6	1.6	1.7	1.8	1.8
Naphtha	-	-	-	-	-	-	-	-	-
Motor Gasoline	5.5	6.0	6.1	4.3	4.8	4.7	4.9	4.6	4.5
Aviation Fuels	0.1	0.1	0.0	0.0	0.0	0.0	0.0	0.0	0.0
Kerosene	0.2	0.1	0.1	0.1	0.1	0.1	0.1	0.1	0.1
Gas Diesel	3.5	4.3	5.4	7.0	6.0	6.3	6.4	6.8	5.4
Heavy Fuel Oil	6.0	6.7	5.6	6.3	4.1	3.2	3.3	4.1	2.4
Other Products	0.2	0.3	0.2	0.6	0.6	0.2	0.2	0.1	0.1
Refinery Fuel	2.9	3.2	2.4	1.3	1.3	1.8	1.5	1.9	1.5
Sub-Total	19.1	21.6	21.0	21.0	18.6	18.0	18.2	19.5	15.7
Bunkers	107.1	96.8	41.3	12.7	13.6	13.7	13.3	15.5	13.0
Total	126.2	118.4	62.3	33.7	32.2	31.7	31.5	35.1	28.7

	1989	1990	1991	1992	1993	1994	1995	1996	1997
Thousand tonnes									
NGL/LPG/Ethane	55	62	65	68	68	73	79	83	85
Naphtha	-	-	-	-	-	-	-	-	-
Motor Gasoline	192	214	223	240	259	287	301	317	340
Aviation Fuels	1	2	2	2	2	3	2	3	3
Kerosene	4	5	5	6	8	8	8	8	8
Gas Diesel	283	296	380	404	402	438	450	442	571
Heavy Fuel Oil	146	158	180	209	222	273	270	279	251
Other Products	6	7	4	10	13	12	13	19	47
Refinery Fuel	77	77	79	97	101	104	106	133	111
Sub-Total	764	821	938	1 036	1 075	1 198	1 229	1 284	1 416
Bunkers	860	1 040	1 110	1 160	1 060	1 060	1 060	1 060	1 060
Total	1 624	1 861	2 048	2 196	2 135	2 258	2 289	2 344	2 476
Thousand barrels/day									
NGL/LPG/Ethane	1.7	2.0	2.1	2.2	2.2	2.3	2.5	2.6	2.7
Naphtha	-	-	-	-	-	-	-	-	-
Motor Gasoline	4.5	5.0	5.2	5.6	6.1	6.7	7.0	7.4	7.9
Aviation Fuels	0.0	0.0	0.0	0.0	0.0	0.1	0.0	0.1	0.1
Kerosene	0.1	0.1	0.1	0.1	0.2	0.2	0.2	0.2	0.2
Gas Diesel	5.8	6.1	7.8	8.2	8.2	9.0	9.2	9.0	11.7
Heavy Fuel Oil	2.7	2.9	3.3	3.8	4.1	5.0	4.9	5.1	4.6
Other Products	0.1	0.1	0.1	0.2	0.2	0.2	0.2	0.4	0.9
Refinery Fuel	1.5	1.5	1.5	1.9	2.0	2.0	2.1	2.6	2.2
Sub-Total	16.4	17.7	20.1	22.0	22.9	25.5	26.2	27.3	30.2
Bunkers	15.8	19.1	20.4	21.2	19.5	19.5	19.5	19.4	19.5
Total	32.2	36.8	40.4	43.3	42.4	44.9	45.7	46.8	49.7

Oil Products Consumption
Consommation de produits pétroliers

Paraguay

	1971	1973	1978	1983	1984	1985	1986	1987	1988
Thousand tonnes									
NGL/LPG/Ethane	-	8	5	28	29	32	30	33	28
Naphtha	-	-	-	-	-	-	-	-	-
Motor Gasoline	80	83	123	77	101	109	122	134	147
Aviation Fuels	6	8	19	13	22	23	35	32	33
Kerosene	18	13	19	11	11	18	15	13	13
Gas Diesel	47	59	195	279	301	297	301	316	322
Heavy Fuel Oil	33	61	56	30	21	20	24	34	45
Other Products	9	11	9	5	8	12	14	13	12
Refinery Fuel	11	14	19	13	9	12	13	10	10
Sub-Total	204	257	445	456	502	523	554	585	610
Bunkers	-	-	-	-	-	-	-	-	-
Total	204	257	445	456	502	523	554	585	610
Thousand barrels/day									
NGL/LPG/Ethane	-	0.3	0.2	0.9	0.9	1.0	1.0	1.0	0.9
Naphtha	-	-	-	-	-	-	-	-	-
Motor Gasoline	1.9	1.9	2.9	1.8	2.3	2.5	2.8	3.1	3.4
Aviation Fuels	0.1	0.2	0.4	0.3	0.5	0.5	0.8	0.7	0.7
Kerosene	0.4	0.3	0.4	0.2	0.2	0.4	0.3	0.3	0.3
Gas Diesel	1.0	1.2	4.0	5.7	6.1	6.1	6.2	6.5	6.6
Heavy Fuel Oil	0.6	1.1	1.0	0.5	0.4	0.4	0.4	0.6	0.8
Other Products	0.2	0.2	0.2	0.1	0.1	0.2	0.2	0.2	0.2
Refinery Fuel	0.2	0.3	0.3	0.2	0.2	0.2	0.2	0.2	0.2
Sub-Total	4.3	5.4	9.4	9.8	10.8	11.3	11.9	12.6	13.0
Bunkers	-	-	-	-	-	-	-	-	-
Total	4.3	5.4	9.4	9.8	10.8	11.3	11.9	12.6	13.0

	1989	1990	1991	1992	1993	1994	1995	1996	1997
Thousand tonnes									
NGL/LPG/Ethane	41	50	54	59	63	65	65	66	66
Naphtha	-	-	-	-	-	-	-	-	-
Motor Gasoline	137	146	155	165	189	216	246	209	187
Aviation Fuels	35	3	3	3	3	3	3	3	3
Kerosene	14	8	8	8	7	8	8	7	7
Gas Diesel	374	383	365	477	553	632	728	711	807
Heavy Fuel Oil	40	49	45	61	46	86	81	33	38
Other Products	12	12	12	10	10	9	12	12	12
Refinery Fuel	10	10	-	8	7	-	-	-	-
Sub-Total	663	661	642	791	878	1 019	1 143	1 041	1 120
Bunkers	-	-	-	-	-	-	-	-	-
Total	663	661	642	791	878	1 019	1 143	1 041	1 120
Thousand barrels/day									
NGL/LPG/Ethane	1.3	1.6	1.7	1.9	2.0	2.1	2.1	2.1	2.1
Naphtha	-	-	-	-	-	-	-	-	-
Motor Gasoline	3.1	3.4	3.6	3.8	4.4	5.0	5.7	4.8	4.3
Aviation Fuels	0.8	0.1	0.1	0.1	0.1	0.1	0.1	0.1	0.1
Kerosene	0.3	0.2	0.2	0.2	0.1	0.2	0.2	0.1	0.1
Gas Diesel	7.6	7.8	7.5	9.7	11.3	12.9	14.9	14.5	16.5
Heavy Fuel Oil	0.7	0.9	0.8	1.1	0.8	1.6	1.5	0.6	0.7
Other Products	0.2	0.2	0.2	0.2	0.2	0.2	0.2	0.2	0.2
Refinery Fuel	0.2	0.2	-	0.1	0.1	-	-	-	-
Sub-Total	14.3	14.3	14.0	17.1	19.0	22.0	24.6	22.5	24.0
Bunkers	-	-	-	-	-	-	-	-	-
Total	14.3	14.3	14.0	17.1	19.0	22.0	24.6	22.5	24.0

Oil Products Consumption
Consommation de produits pétroliers

Peru

	1971	1973	1978	1983	1984	1985	1986	1987	1988
Thousand tonnes									
NGL/LPG/Ethane	62	81	116	112	121	118	137	144	161
Naphtha	-	-	-	-	-	-	-	-	-
Motor Gasoline	1 224	1 419	1 243	1 231	1 222	1 118	1 194	1 313	1 338
Aviation Fuels	235	255	345	380	336	233	255	274	243
Kerosene	527	593	722	807	785	806	933	1 045	1 111
Gas Diesel	837	875	1 188	1 357	1 489	1 441	1 482	1 604	1 590
Heavy Fuel Oil	1 525	1 482	1 784	1 352	1 423	1 483	1 563	1 754	1 687
Other Products	107	106	142	89	116	140	172	161	103
Refinery Fuel	138	142	211	220	246	139	205	201	209
Sub-Total	4 655	4 953	5 751	5 548	5 738	5 478	5 941	6 496	6 442
Bunkers	29	36	67	86	82	77	67	63	60
Total	4 684	4 989	5 818	5 634	5 820	5 555	6 008	6 559	6 502
Thousand barrels/day									
NGL/LPG/Ethane	2.0	2.6	3.7	3.6	3.8	3.8	4.4	4.6	5.1
Naphtha	-	-	-	-	-	-	-	-	-
Motor Gasoline	28.6	33.2	29.0	28.8	28.5	26.1	27.9	30.7	31.2
Aviation Fuels	5.2	5.6	7.5	8.3	7.3	5.1	5.5	6.0	5.3
Kerosene	11.2	12.6	15.3	17.1	16.6	17.1	19.8	22.2	23.5
Gas Diesel	17.1	17.9	24.3	27.7	30.4	29.5	30.3	32.8	32.4
Heavy Fuel Oil	27.8	27.0	32.6	24.7	25.9	27.1	28.5	32.0	30.7
Other Products	2.1	2.0	2.7	1.7	2.2	2.7	3.3	3.1	2.0
Refinery Fuel	2.8	2.8	4.2	4.4	5.0	2.9	4.1	4.0	4.1
Sub-Total	96.7	103.7	119.3	116.2	119.6	114.1	123.8	135.2	134.3
Bunkers	0.6	0.7	1.3	1.7	1.6	1.5	1.3	1.2	1.2
Total	97.2	104.4	120.6	117.9	121.2	115.6	125.1	136.4	135.4

	1989	1990	1991	1992	1993	1994	1995	1996	1997
Thousand tonnes									
NGL/LPG/Ethane	159	175	189	201	211	239	262	268	288
Naphtha	-	-	-	-	-	-	-	-	-
Motor Gasoline	1 097	1 232	1 188	1 092	1 160	1 075	1 095	1 194	1 091
Aviation Fuels	204	266	267	269	285	308	400	451	440
Kerosene	957	743	711	723	706	683	683	718	686
Gas Diesel	1 601	1 580	1 545	1 967	1 946	2 105	2 770	2 886	3 084
Heavy Fuel Oil	1 571	1 562	1 392	1 015	1 189	1 224	1 679	1 967	1 786
Other Products	72	66	83	67	89	91	78	54	152
Refinery Fuel	174	123	119	117	124	122	143	180	195
Sub-Total	5 835	5 747	5 494	5 451	5 710	5 847	7 110	7 718	7 722
Bunkers	60	11	25	69	66	64	131	1	21
Total	5 895	5 758	5 519	5 520	5 776	5 911	7 241	7 719	7 743
Thousand barrels/day									
NGL/LPG/Ethane	5.1	5.6	6.0	6.4	6.7	7.6	8.3	8.5	9.2
Naphtha	-	-	-	-	-	-	-	-	-
Motor Gasoline	25.6	28.8	27.8	25.5	27.1	25.1	25.6	27.8	25.5
Aviation Fuels	4.4	5.8	5.8	5.8	6.2	6.7	8.7	9.8	9.6
Kerosene	20.3	15.8	15.1	15.3	15.0	14.5	14.5	15.2	14.5
Gas Diesel	32.7	32.3	31.6	40.1	39.8	43.0	56.6	58.8	63.0
Heavy Fuel Oil	28.7	28.5	25.4	18.5	21.7	22.3	30.6	35.8	32.6
Other Products	1.4	1.3	1.6	1.3	1.7	1.7	1.5	1.0	2.9
Refinery Fuel	3.4	2.4	2.3	2.4	2.5	2.4	2.8	3.5	3.8
Sub-Total	121.6	120.4	115.6	115.2	120.7	123.4	148.7	160.4	161.1
Bunkers	1.2	0.2	0.5	1.4	1.3	1.3	2.7	0.0	0.4
Total	122.8	120.6	116.1	116.6	122.0	124.8	151.3	160.4	161.5

Oil Products Consumption
Consommation de produits pétroliers

Philippines

	1971	1973	1978	1983	1984	1985	1986	1987	1988
Thousand tonnes									
NGL/LPG/Ethane	119	158	223	205	186	181	211	205	302
Naphtha	2	33	170	59	42	92	51	43	15
Motor Gasoline	1 864	1 937	1 791	1 076	1 017	990	1 072	1 189	1 289
Aviation Fuels	250	280	350	341	360	346	346	421	478
Kerosene	419	428	469	331	312	268	292	324	356
Gas Diesel	1 520	1 739	2 155	2 504	2 254	2 072	2 137	2 355	2 529
Heavy Fuel Oil	3 188	4 047	5 384	5 022	3 847	3 099	2 817	4 147	4 363
Other Products	240	311	302	306	193	118	156	179	199
Refinery Fuel	125	206	271	520	429	392	402	906	909
Sub-Total	7 727	9 139	11 115	10 364	8 640	7 558	7 484	9 769	10 440
Bunkers	412	261	187	343	80	-	-	-	-
Total	8 139	9 400	11 302	10 707	8 720	7 558	7 484	9 769	10 440
Thousand barrels/day									
NGL/LPG/Ethane	3.8	5.0	7.1	6.5	5.9	5.8	6.7	6.5	9.6
Naphtha	0.0	0.8	4.0	1.4	1.0	2.1	1.2	1.0	0.3
Motor Gasoline	43.6	45.3	41.9	25.1	23.7	23.1	25.1	27.8	30.0
Aviation Fuels	5.5	6.1	7.6	7.4	7.8	7.5	7.5	9.2	10.4
Kerosene	8.9	9.1	9.9	7.0	6.6	5.7	6.2	6.9	7.5
Gas Diesel	31.1	35.5	44.0	51.2	45.9	42.3	43.7	48.1	51.5
Heavy Fuel Oil	58.2	73.8	98.2	91.6	70.0	56.5	51.4	75.7	79.4
Other Products	4.6	5.8	5.7	5.6	3.6	2.2	2.9	3.3	3.7
Refinery Fuel	2.4	4.0	5.4	11.0	9.1	8.4	8.6	18.7	18.6
Sub-Total	157.9	185.4	223.8	206.9	173.6	153.7	153.3	197.2	211.2
Bunkers	7.6	4.8	3.6	6.6	1.5	-	-	-	-
Total	165.5	190.2	227.4	213.6	175.1	153.7	153.3	197.2	211.2

	1989	1990	1991	1992	1993	1994	1995	1996	1997
Thousand tonnes									
NGL/LPG/Ethane	361	396	409	469	533	615	719	800	813
Naphtha	28	-	-	-	-	-	-	-	-
Motor Gasoline	1 442	1 507	1 299	1 453	1 623	1 767	2 080	2 334	2 526
Aviation Fuels	544	506	494	564	591	633	624	793	998
Kerosene	433	462	463	506	547	552	564	590	653
Gas Diesel	3 025	3 616	3 784	4 986	6 813	5 592	5 697	5 931	6 517
Heavy Fuel Oil	4 678	4 614	3 304	4 256	4 527	5 151	5 672	5 602	5 815
Other Products	240	262	299	286	305	310	319	359	353
Refinery Fuel	488	491	473	512	519	538	648	715	696
Sub-Total	11 239	11 854	10 525	13 032	15 458	15 158	16 323	17 124	18 371
Bunkers	-	-	-	-	-	-	73	73	94
Total	11 239	11 854	10 525	13 032	15 458	15 158	16 396	17 197	18 465
Thousand barrels/day									
NGL/LPG/Ethane	11.5	12.6	13.0	14.9	16.9	19.5	22.9	25.4	25.8
Naphtha	0.7	-	-	-	-	-	-	-	-
Motor Gasoline	33.7	35.2	30.4	33.9	37.9	41.3	48.6	54.4	59.0
Aviation Fuels	11.8	11.0	10.7	12.2	12.9	13.8	13.6	17.2	21.7
Kerosene	9.2	9.8	9.8	10.7	11.6	11.7	12.0	12.5	13.8
Gas Diesel	61.8	73.9	77.3	101.6	139.2	114.3	116.4	120.9	133.2
Heavy Fuel Oil	85.4	84.2	60.3	77.4	82.6	94.0	103.5	101.9	106.1
Other Products	4.5	4.9	5.6	5.4	5.8	5.9	6.0	6.8	6.7
Refinery Fuel	10.3	10.4	10.0	10.8	11.0	11.3	13.7	15.1	14.7
Sub-Total	228.8	242.0	217.2	267.0	317.9	311.8	336.6	354.2	381.1
Bunkers	-	-	-	-	-	-	1.4	1.4	1.8
Total	228.8	242.0	217.2	267.0	317.9	311.8	338.0	355.5	383.0

Oil Products Consumption
Consommation de produits pétroliers

Qatar

	1971	1973	1978	1983	1984	1985	1986	1987	1988
Thousand tonnes									
NGL/LPG/Ethane	-	-	6	5	-	49	-	24	-
Naphtha	-	-	-	-	-	-	-	-	-
Motor Gasoline	39	62	126	234	234	235	247	255	271
Aviation Fuels	-	21	77	74	83	75	93	101	85
Kerosene	4	6	5	6	10	15	15	15	15
Gas Diesel	35	44	142	205	185	177	163	156	169
Heavy Fuel Oil	-	-	-	-	-	-	-	-	-
Other Products	17	22	84	22	18	4	4	2	-
Refinery Fuel	1	1	14	16	5	33	45	52	63
Sub-Total	96	156	454	562	535	588	567	605	603
Bunkers	-	-	-	-	-	-	-	-	-
Total	96	156	454	562	535	588	567	605	603
Thousand barrels/day									
NGL/LPG/Ethane	-	-	0.2	0.2	-	1.6	-	0.8	-
Naphtha	-	-	-	-	-	-	-	-	-
Motor Gasoline	0.9	1.4	2.9	5.5	5.5	5.5	5.8	6.0	6.3
Aviation Fuels	-	0.5	1.7	1.6	1.8	1.6	2.0	2.2	1.8
Kerosene	0.1	0.1	0.1	0.1	0.2	0.3	0.3	0.3	0.3
Gas Diesel	0.7	0.9	2.9	4.2	3.8	3.6	3.3	3.2	3.4
Heavy Fuel Oil	-	-	-	-	-	-	-	-	-
Other Products	0.4	0.5	1.7	0.5	0.4	0.1	0.1	0.0	-
Refinery Fuel	0.0	0.0	0.3	0.3	0.1	0.7	0.9	1.1	1.3
Sub-Total	2.1	3.4	9.8	12.3	11.7	13.4	12.5	13.5	13.2
Bunkers	-	-	-	-	-	-	-	-	-
Total	2.1	3.4	9.8	12.3	11.7	13.4	12.5	13.5	13.2

	1989	1990	1991	1992	1993	1994	1995	1996	1997
Thousand tonnes									
NGL/LPG/Ethane	59	21	21	22	23	24	24	26	29
Naphtha	-	-	-	-	-	-	-	-	-
Motor Gasoline	260	303	307	323	339	347	371	390	414
Aviation Fuels	90	109	86	99	98	131	135	174	197
Kerosene	15	10	4	10	13	20	20	21	21
Gas Diesel	170	166	166	181	194	220	254	277	331
Heavy Fuel Oil	803	-	163	-	-	-	-	-	-
Other Products	-	-	-	-	-	-	-	-	-
Refinery Fuel	40	40	40	45	52	51	47	53	53
Sub-Total	1 437	649	787	680	719	793	851	941	1 045
Bunkers	-	-	-	-	-	-	-	-	-
Total	1 437	649	787	680	719	793	851	941	1 045
Thousand barrels/day									
NGL/LPG/Ethane	1.9	0.7	0.7	0.7	0.7	0.8	0.8	0.8	0.9
Naphtha	-	-	-	-	-	-	-	-	-
Motor Gasoline	6.1	7.1	7.2	7.5	7.9	8.1	8.7	9.1	9.7
Aviation Fuels	2.0	2.4	1.9	2.1	2.1	2.8	2.9	3.8	4.3
Kerosene	0.3	0.2	0.1	0.2	0.3	0.4	0.4	0.4	0.4
Gas Diesel	3.5	3.4	3.4	3.7	4.0	4.5	5.2	5.6	6.8
Heavy Fuel Oil	14.7	-	3.0	-	-	-	-	-	-
Other Products	-	-	-	-	-	-	-	-	-
Refinery Fuel	0.9	0.9	0.9	1.0	1.1	1.1	1.0	1.2	1.2
Sub-Total	29.2	14.6	17.0	15.3	16.1	17.8	19.0	20.9	23.2
Bunkers	-	-	-	-	-	-	-	-	-
Total	29.2	14.6	17.0	15.3	16.1	17.8	19.0	20.9	23.2

Oil Products Consumption
Consommation de produits pétroliers

Romania

	1971	1973	1978	1983	1984	1985	1986	1987	1988
Thousand tonnes									
NGL/LPG/Ethane	203	245	242	218	210	185	207	216	211
Naphtha	420	481	743	-	-	-	-	-	-
Motor Gasoline	2 009	2 270	2 581	1 729	1 656	1 405	1 308	1 821	1 700
Aviation Fuels	21	21	15	-	-	-	-	-	-
Kerosene	981	1 016	1 005	564	570	487	571	563	625
Gas Diesel	2 657	3 402	4 599	4 147	4 094	4 071	3 692	4 295	3 935
Heavy Fuel Oil	2 580	3 440	6 461	4 561	4 076	5 487	6 690	7 294	6 737
Other Products	1 600	1 679	1 850	1 315	1 281	1 258	1 479	1 319	1 207
Refinery Fuel	761	841	1 214	1 365	1 384	1 388	1 462	1 512	1 479
Sub-Total	11 232	13 395	18 710	13 899	13 271	14 281	15 409	17 020	15 894
Bunkers	-	-	-	-	-	-	-	-	-
Total	11 232	13 395	18 710	13 899	13 271	14 281	15 409	17 020	15 894
Thousand barrels/day									
NGL/LPG/Ethane	6.5	7.8	7.7	6.9	6.7	5.9	6.6	6.9	6.7
Naphtha	9.8	11.2	17.3	-	-	-	-	-	-
Motor Gasoline	47.0	53.1	60.3	40.4	38.6	32.8	30.6	42.6	39.6
Aviation Fuels	0.5	0.5	0.4	-	-	-	-	-	-
Kerosene	20.8	21.5	21.3	12.0	12.1	10.3	12.1	11.9	13.2
Gas Diesel	54.3	69.5	94.0	84.8	83.4	83.2	75.5	87.8	80.2
Heavy Fuel Oil	47.1	62.8	117.9	83.2	74.2	100.1	122.1	133.1	122.6
Other Products	29.7	31.0	34.1	23.3	22.5	22.2	25.5	23.0	20.4
Refinery Fuel	15.8	17.4	25.1	28.6	28.9	29.0	30.6	31.7	31.0
Sub-Total	231.3	274.8	378.1	279.2	266.3	283.6	302.9	336.9	313.8
Bunkers	-	-	-	-	-	-	-	-	-
Total	231.3	274.8	378.1	279.2	266.3	283.6	302.9	336.9	313.8

	1989	1990	1991	1992	1993	1994	1995	1996	1997
Thousand tonnes									
NGL/LPG/Ethane	204	221	222	224	230	177	245	187	286
Naphtha	-	-	-	-	107	122	56	177	42
Motor Gasoline	1 689	2 083	1 744	2 150	1 125	1 232	1 052	1 444	1 561
Aviation Fuels	245	227	161	242	248	165	181	86	123
Kerosene	411	357	284	354	114	35	23	31	66
Gas Diesel	4 009	4 028	3 436	3 630	2 856	2 562	2 414	3 506	3 142
Heavy Fuel Oil	6 674	8 162	6 895	4 759	5 209	4 073	4 592	5 227	5 253
Other Products	1 537	1 284	939	1 490	1 435	1 977	2 298	1 890	2 420
Refinery Fuel	1 668	1 616	1 153	591	689	951	558	504	535
Sub-Total	16 437	17 978	14 834	13 440	12 013	11 294	11 419	13 052	13 428
Bunkers	-	-	-	-	-	-	-	-	-
Total	16 437	17 978	14 834	13 440	12 013	11 294	11 419	13 052	13 428
Thousand barrels/day									
NGL/LPG/Ethane	6.5	7.0	7.1	7.1	7.2	5.5	7.7	5.9	9.0
Naphtha	-	-	-	-	2.5	2.8	1.3	4.1	1.0
Motor Gasoline	39.5	48.7	40.8	50.1	26.3	28.8	24.6	33.7	36.5
Aviation Fuels	5.3	4.9	3.5	5.2	5.4	3.6	3.9	1.9	2.7
Kerosene	8.7	7.6	6.0	7.5	2.4	0.7	0.5	0.7	1.4
Gas Diesel	81.9	82.3	70.2	74.0	58.4	52.4	49.3	71.5	64.2
Heavy Fuel Oil	121.8	148.9	125.8	86.6	95.0	74.3	83.8	95.1	95.8
Other Products	27.3	23.0	16.5	27.7	27.7	37.8	44.4	35.4	46.0
Refinery Fuel	35.2	34.4	24.2	12.4	14.5	19.6	12.1	11.0	11.7
Sub-Total	326.2	356.9	294.0	270.7	239.4	225.6	227.7	259.2	268.4
Bunkers	-	-	-	-	-	-	-	-	-
Total	326.2	356.9	294.0	270.7	239.4	225.6	227.7	259.2	268.4

Oil Products Consumption
Consommation de produits pétroliers

Russia

	1971	1973	1978	1983	1984	1985	1986	1987	1988
Thousand tonnes									
NGL/LPG/Ethane	-	-	-	-	-	-	-	-	-
Naphtha	-	-	-	-	-	-	-	-	-
Motor Gasoline	-	-	-	-	-	-	-	-	-
Aviation Fuels	-	-	-	-	-	-	-	-	-
Kerosene	-	-	-	-	-	-	-	-	-
Gas Diesel	-	-	-	-	-	-	-	-	-
Heavy Fuel Oil	-	-	-	-	-	-	-	-	-
Other Products	-	-	-	-	-	-	-	-	-
Refinery Fuel	-	-	-	-	-	-	-	-	-
Sub-Total	-	-	-	-	-	-	-	-	-
Bunkers	-	-	-	-	-	-	-	-	-
Total	-	-	-	-	-	-	-	-	-
Thousand barrels/day									
NGL/LPG/Ethane	-	-	-	-	-	-	-	-	-
Naphtha	-	-	-	-	-	-	-	-	-
Motor Gasoline	-	-	-	-	-	-	-	-	-
Aviation Fuels	-	-	-	-	-	-	-	-	-
Kerosene	-	-	-	-	-	-	-	-	-
Gas Diesel	-	-	-	-	-	-	-	-	-
Heavy Fuel Oil	-	-	-	-	-	-	-	-	-
Other Products	-	-	-	-	-	-	-	-	-
Refinery Fuel	-	-	-	-	-	-	-	-	-
Sub-Total	-	-	-	-	-	-	-	-	-
Bunkers	-	-	-	-	-	-	-	-	-
Total	-	-	-	-	-	-	-	-	-

	1989	1990	1991	1992	1993	1994	1995	1996	1997
Thousand tonnes									
NGL/LPG/Ethane	-	-	-	3 487	4 410	3 955	4 204	4 900	4 250
Naphtha	-	-	-	6 379	8 240	3 577	-	-	-
Motor Gasoline	-	-	-	30 996	27 543	26 323	26 078	24 701	23 280
Aviation Fuels	-	-	-	14 100	11 951	9 953	9 271	7 473	7 040
Kerosene	-	-	-	882	616	283	143	131	130
Gas Diesel	-	-	-	52 374	41 573	30 456	29 476	24 615	25 080
Heavy Fuel Oil	-	-	-	70 274	71 063	54 931	46 278	41 331	38 800
Other Products	-	-	-	23 416	29 640	16 925	23 448	19 850	20 890
Refinery Fuel	-	-	-	9 095	1 284	-	1 809	1 876	2 000
Sub-Total	-	-	-	211 003	196 320	146 403	140 707	124 877	121 470
Bunkers	-	-	-	-	-	-	-	-	-
Total	-	-	-	211 003	196 320	146 403	140 707	124 877	121 470
Thousand barrels/day									
NGL/LPG/Ethane	-	-	-	110.5	140.2	125.7	133.6	155.3	135.1
Naphtha	-	-	-	148.1	191.9	83.3	-	-	-
Motor Gasoline	-	-	-	722.4	643.7	615.2	609.4	575.7	544.1
Aviation Fuels	-	-	-	305.9	259.9	216.5	201.6	162.0	153.1
Kerosene	-	-	-	18.7	13.1	6.0	3.0	2.8	2.8
Gas Diesel	-	-	-	1 067.5	849.7	622.5	602.4	501.7	512.6
Heavy Fuel Oil	-	-	-	1 278.8	1 296.7	1 002.3	844.4	752.1	708.0
Other Products	-	-	-	432.8	560.3	315.4	440.9	376.0	394.8
Refinery Fuel	-	-	-	230.8	28.1	-	39.6	41.0	43.8
Sub-Total	-	-	-	4 315.5	3 983.5	2 986.8	2 875.1	2 566.7	2 494.2
Bunkers	-	-	-	-	-	-	-	-	-
Total	-	-	-	4 315.5	3 983.5	2 986.8	2 875.1	2 566.7	2 494.2

Oil Products Consumption
Consommation de produits pétroliers

Saudi Arabia

	1971	1973	1978	1983	1984	1985	1986	1987	1988
Thousand tonnes									
NGL/LPG/Ethane	51	77	195	435	399	2 338	3 476	3 496	3 704
Naphtha	-	-	-	-	-	-	-	-	-
Motor Gasoline	524	700	2 480	5 650	5 830	6 327	6 354	6 199	6 213
Aviation Fuels	199	316	1 133	1 983	1 884	1 928	1 886	1 820	1 892
Kerosene	133	155	132	177	181	208	206	143	195
Gas Diesel	809	1 081	5 079	11 951	13 541	12 819	12 477	12 428	13 394
Heavy Fuel Oil	632	1 179	3 618	11 567	2 858	2 521	2 913	2 685	2 913
Other Products	385	559	853	5 911	5 775	6 109	6 296	6 883	6 566
Refinery Fuel	825	925	937	1 425	1 313	1 720	1 834	2 933	3 233
Sub-Total	3 558	4 992	14 427	39 099	31 781	33 970	35 442	36 587	38 110
Bunkers	13 004	13 803	7 205	2 302	5 221	9 098	7 464	6 083	5 221
Total	16 562	18 795	21 632	41 401	37 002	43 068	42 906	42 670	43 331
Thousand barrels/day									
NGL/LPG/Ethane	1.6	2.4	6.2	13.8	12.6	68.2	100.7	101.2	107.0
Naphtha	-	-	-	-	-	-	-	-	-
Motor Gasoline	12.2	16.4	58.0	132.0	135.9	147.9	148.5	144.9	144.8
Aviation Fuels	4.3	6.9	24.6	43.1	40.8	41.9	41.0	39.5	41.0
Kerosene	2.8	3.3	2.8	3.8	3.8	4.4	4.4	3.0	4.1
Gas Diesel	16.5	22.1	103.8	244.3	276.0	262.0	255.0	254.0	273.0
Heavy Fuel Oil	11.5	21.5	66.0	211.1	52.0	46.0	53.2	49.0	53.0
Other Products	7.6	11.0	16.7	117.0	114.0	121.1	124.9	137.3	130.5
Refinery Fuel	16.5	18.3	18.6	28.2	26.1	34.7	36.8	59.0	64.7
Sub-Total	73.2	101.9	296.7	793.2	661.3	726.2	764.4	788.0	818.1
Bunkers	237.5	252.2	132.7	42.0	95.0	166.0	136.2	111.0	95.0
Total	310.7	354.1	429.5	835.2	756.3	892.2	900.5	899.0	913.1

	1989	1990	1991	1992	1993	1994	1995	1996	1997
Thousand tonnes									
NGL/LPG/Ethane	3 597	3 857	3 851	3 964	4 654	5 342	5 342	5 564	6 808
Naphtha	-	-	-	-	-	-	-	-	-
Motor Gasoline	5 766	6 650	6 902	7 754	8 537	9 004	8 900	9 559	9 643
Aviation Fuels	1 689	2 595	3 378	1 912	2 075	2 499	2 413	2 456	2 448
Kerosene	171	279	434	230	236	236	212	216	215
Gas Diesel	13 847	13 749	14 238	15 602	16 636	16 782	14 819	15 909	16 546
Heavy Fuel Oil	3 343	3 343	2 960	3 077	2 795	3 069	3 449	4 201	3 641
Other Products	6 535	6 693	7 397	7 953	9 379	9 135	9 255	9 700	10 323
Refinery Fuel	3 135	3 335	3 388	3 405	3 405	3 405	3 265	3 696	3 386
Sub-Total	38 083	40 501	42 548	43 897	47 717	49 472	47 655	51 301	53 010
Bunkers	1 918	1 863	2 521	2 638	2 521	2 083	1 936	1 936	1 936
Total	40 001	42 364	45 069	46 535	50 238	51 555	49 591	53 237	54 946
Thousand barrels/day									
NGL/LPG/Ethane	103.8	111.7	111.5	114.8	135.0	154.7	154.7	160.8	196.8
Naphtha	-	-	-	-	-	-	-	-	-
Motor Gasoline	134.8	155.4	161.3	180.7	199.5	210.4	208.0	222.8	225.4
Aviation Fuels	36.7	56.4	73.4	41.4	45.1	54.3	52.4	53.2	53.2
Kerosene	3.6	5.9	9.2	4.9	5.0	5.0	4.5	4.6	4.6
Gas Diesel	283.0	281.0	291.0	318.0	340.0	343.0	302.9	324.3	338.2
Heavy Fuel Oil	61.0	61.0	54.0	56.0	51.0	56.0	62.9	76.4	66.4
Other Products	130.5	133.5	147.6	158.2	187.1	182.2	184.9	193.2	206.2
Refinery Fuel	62.7	66.7	68.0	68.2	68.4	68.4	65.6	74.0	68.3
Sub-Total	816.1	871.6	916.1	942.2	1 031.1	1 074.0	1 035.9	1 109.3	1 159.1
Bunkers	35.0	34.0	46.0	48.0	46.0	38.0	35.3	35.2	35.3
Total	851.1	905.6	962.1	990.2	1 077.1	1 112.1	1 071.2	1 144.5	1 194.4

Oil Products Consumption
Consommation de produits pétroliers

Senegal

	1971	1973	1978	1983	1984	1985	1986	1987	1988
Thousand tonnes									
NGL/LPG/Ethane	3	3	4	12	13	16	18	22	26
Naphtha	-	-	-	-	-	-	-	-	-
Motor Gasoline	83	92	123	110	102	92	89	86	84
Aviation Fuels	95	113	118	143	132	135	134	112	131
Kerosene	11	11	13	11	11	11	11	11	11
Gas Diesel	89	106	131	191	218	250	220	229	196
Heavy Fuel Oil	184	210	255	280	294	314	318	317	299
Other Products	11	28	20	10	15	12	9	12	12
Refinery Fuel	19	19	35	8	12	10	15	15	7
Sub-Total	495	582	699	765	797	840	814	804	766
Bunkers	962	983	523	121	124	105	88	57	33
Total	1 457	1 565	1 222	886	921	945	902	861	799
Thousand barrels/day									
NGL/LPG/Ethane	0.1	0.1	0.1	0.4	0.4	0.5	0.6	0.7	0.8
Naphtha	-	-	-	-	-	-	-	-	-
Motor Gasoline	1.9	2.2	2.9	2.6	2.4	2.2	2.1	2.0	2.0
Aviation Fuels	2.1	2.5	2.6	3.1	2.9	2.9	2.9	2.4	2.8
Kerosene	0.2	0.2	0.3	0.2	0.2	0.2	0.2	0.2	0.2
Gas Diesel	1.8	2.2	2.7	3.9	4.4	5.1	4.5	4.7	4.0
Heavy Fuel Oil	3.4	3.8	4.7	5.1	5.4	5.7	5.8	5.8	5.4
Other Products	0.2	0.5	0.4	0.2	0.3	0.2	0.2	0.2	0.2
Refinery Fuel	0.4	0.4	0.7	0.2	0.2	0.2	0.3	0.3	0.2
Sub-Total	10.1	11.8	14.2	15.7	16.2	17.1	16.6	16.4	15.7
Bunkers	18.1	18.4	10.1	2.4	2.4	2.1	1.8	1.1	0.7
Total	28.2	30.2	24.4	18.1	18.6	19.2	18.3	17.5	16.3

	1989	1990	1991	1992	1993	1994	1995	1996	1997
Thousand tonnes									
NGL/LPG/Ethane	29	33	38	42	45	50	57	68	76
Naphtha	-	-	-	-	-	-	-	-	-
Motor Gasoline	80	79	76	75	68	67	72	74	78
Aviation Fuels	149	145	128	129	103	128	145	200	203
Kerosene	10	10	10	10	10	10	10	10	12
Gas Diesel	199	213	220	250	251	245	247	258	322
Heavy Fuel Oil	336	336	328	357	319	360	367	381	418
Other Products	12	12	12	11	12	18	19	19	19
Refinery Fuel	4	7	4	6	6	2	6	6	6
Sub-Total	819	835	816	880	814	880	923	1 016	1 134
Bunkers	45	36	1	17	5	22	29	58	58
Total	864	871	817	897	819	902	952	1 074	1 192
Thousand barrels/day									
NGL/LPG/Ethane	0.9	1.0	1.2	1.3	1.4	1.6	1.8	2.2	2.4
Naphtha	-	-	-	-	-	-	-	-	-
Motor Gasoline	1.9	1.8	1.8	1.7	1.6	1.6	1.7	1.7	1.8
Aviation Fuels	3.2	3.2	2.8	2.8	2.2	2.8	3.2	4.3	4.4
Kerosene	0.2	0.2	0.2	0.2	0.2	0.2	0.2	0.2	0.3
Gas Diesel	4.1	4.4	4.5	5.1	5.1	5.0	5.0	5.3	6.6
Heavy Fuel Oil	6.1	6.1	6.0	6.5	5.8	6.6	6.7	6.9	7.6
Other Products	0.2	0.2	0.2	0.2	0.2	0.3	0.4	0.4	0.4
Refinery Fuel	0.1	0.2	0.1	0.1	0.1	0.0	0.1	0.1	0.1
Sub-Total	16.8	17.1	16.8	18.0	16.8	18.1	19.1	21.1	23.6
Bunkers	0.9	0.7	0.0	0.3	0.1	0.4	0.6	1.2	1.2
Total	17.6	17.8	16.8	18.3	16.9	18.6	19.7	22.3	24.8

Oil Products Consumption
Consommation de produits pétroliers

Singapore

	1971	1973	1978	1983	1984	1985	1986	1987	1988
Thousand tonnes									
NGL/LPG/Ethane	4	6	45	71	72	432	387	390	390
Naphtha	-	48	55	65	397	611	641	953	986
Motor Gasoline	213	276	300	378	403	370	410	436	463
Aviation Fuels	221	314	777	1 001	992	1 012	1 136	1 218	1 296
Kerosene	105	79	55	56	60	60	60	60	60
Gas Diesel	353	334	552	717	752	717	715	724	682
Heavy Fuel Oil	771	1 091	1 672	2 226	2 374	2 466	2 528	2 674	3 121
Other Products	155	335	407	493	321	450	412	438	416
Refinery Fuel	500	790	1 107	1 356	1 328	1 243	1 220	1 160	1 560
Sub-Total	2 322	3 273	4 970	6 363	6 699	7 361	7 509	8 053	8 974
Bunkers	2 877	4 383	4 882	5 098	5 057	4 897	8 360	8 346	9 884
Total	5 199	7 656	9 852	11 461	11 756	12 258	15 869	16 399	18 858
Thousand barrels/day									
NGL/LPG/Ethane	0.1	0.2	1.4	2.3	2.3	13.7	12.3	12.4	12.4
Naphtha	-	1.1	1.3	1.5	9.2	14.2	14.9	22.2	22.9
Motor Gasoline	5.0	6.5	7.0	8.8	9.4	8.6	9.6	10.2	10.8
Aviation Fuels	4.8	6.8	16.9	21.8	21.5	22.0	24.7	26.5	28.1
Kerosene	2.2	1.7	1.2	1.2	1.3	1.3	1.3	1.3	1.3
Gas Diesel	7.2	6.8	11.3	14.7	15.3	14.7	14.6	14.8	13.9
Heavy Fuel Oil	14.1	19.9	30.5	40.6	43.2	45.0	46.1	48.8	56.8
Other Products	2.9	6.3	7.6	9.1	5.9	8.4	7.6	8.1	7.6
Refinery Fuel	9.1	14.4	22.2	27.2	26.5	25.0	24.5	23.3	32.0
Sub-Total	45.4	63.7	99.4	127.1	134.5	152.9	155.6	167.5	185.7
Bunkers	53.1	80.9	90.4	94.3	93.4	90.7	154.3	155.2	183.0
Total	98.6	144.6	189.8	221.4	227.9	243.5	309.9	322.7	368.7

	1989	1990	1991	1992	1993	1994	1995	1996	1997
Thousand tonnes									
NGL/LPG/Ethane	430	393	365	189	189	189	143	237	214
Naphtha	1 040	1 237	1 149	1 074	1 202	1 288	1 696	1 988	2 418
Motor Gasoline	483	447	476	513	535	556	656	666	690
Aviation Fuels	1 329	1 784	1 497	1 611	2 072	2 533	2 474	2 561	2 806
Kerosene	60	60	35	35	40	42	42	43	43
Gas Diesel	842	972	1 081	992	959	1 130	1 206	1 156	1 213
Heavy Fuel Oil	3 244	4 500	4 474	4 605	5 365	5 467	5 575	3 840	4 234
Other Products	429	666	619	574	496	523	507	558	558
Refinery Fuel	1 560	2 740	3 085	3 150	3 225	3 389	3 036	3 303	3 303
Sub-Total	9 417	12 799	12 781	12 743	14 083	15 117	15 335	14 352	15 479
Bunkers	9 700	10 981	10 389	12 810	11 438	11 477	11 419	14 280	16 191
Total	19 117	23 780	23 170	25 553	25 521	26 594	26 754	28 632	31 670
Thousand barrels/day									
NGL/LPG/Ethane	13.7	12.5	11.6	6.0	6.0	6.0	4.5	7.5	6.8
Naphtha	24.2	28.8	26.8	24.9	28.0	30.0	39.5	46.2	56.3
Motor Gasoline	11.3	10.4	11.1	12.0	12.5	13.0	15.3	15.5	16.1
Aviation Fuels	28.9	38.8	32.5	34.9	45.0	55.0	53.8	55.5	61.0
Kerosene	1.3	1.3	0.7	0.7	0.8	0.9	0.9	0.9	0.9
Gas Diesel	17.2	19.9	22.1	20.2	19.6	23.1	24.6	23.6	24.8
Heavy Fuel Oil	59.2	82.1	81.6	83.8	97.9	99.8	101.7	69.9	77.3
Other Products	7.9	12.3	11.5	10.6	9.3	9.8	9.6	10.5	10.5
Refinery Fuel	32.1	52.6	59.0	59.3	60.8	63.9	57.3	62.2	62.4
Sub-Total	195.7	258.6	257.0	252.5	280.0	301.5	307.3	291.7	316.1
Bunkers	178.1	201.7	191.1	234.7	210.6	212.2	211.0	264.2	301.8
Total	373.8	460.3	448.2	487.2	490.6	513.7	518.3	556.0	617.8

Oil Products Consumption
Consommation de produits pétroliers

Slovak Republic

	1971	1973	1978	1983	1984	1985	1986	1987	1988
Thousand tonnes									
NGL/LPG/Ethane	6	14	17	18	19	18	17	21	17
Naphtha	-	-	-	-	-	-	-	-	-
Motor Gasoline	459	532	550	386	421	408	436	452	493
Aviation Fuels	-	-	-	-	-	-	-	-	-
Kerosene	7	8	11	11	11	12	12	20	13
Gas Diesel	1 049	1 205	1 168	1 220	1 164	1 171	1 169	1 199	1 225
Heavy Fuel Oil	1 106	1 201	1 278	2 677	2 495	2 487	2 316	2 195	2 011
Other Products	243	295	294	460	490	438	425	428	419
Refinery Fuel	281	339	446	123	123	123	118	118	110
Sub-Total	3 151	3 594	3 764	4 895	4 723	4 657	4 493	4 433	4 288
Bunkers	-	-	-	-	-	-	-	-	-
Total	3 151	3 594	3 764	4 895	4 723	4 657	4 493	4 433	4 288
Thousand barrels/day									
NGL/LPG/Ethane	0.2	0.4	0.5	0.6	0.6	0.6	0.5	0.7	0.5
Naphtha	-	-	-	-	-	-	-	-	-
Motor Gasoline	10.7	12.4	12.9	9.0	9.8	9.5	10.2	10.6	11.5
Aviation Fuels	-	-	-	-	-	-	-	-	-
Kerosene	0.1	0.2	0.2	0.2	0.2	0.3	0.3	0.4	0.3
Gas Diesel	21.4	24.6	23.9	24.9	23.7	23.9	23.9	24.5	25.0
Heavy Fuel Oil	20.2	21.9	23.3	48.8	45.4	45.4	42.3	40.1	36.6
Other Products	4.7	5.7	5.6	9.2	9.8	8.8	8.6	8.6	8.4
Refinery Fuel	5.2	6.6	8.6	2.5	2.5	2.5	2.4	2.4	2.2
Sub-Total	62.6	71.8	75.0	95.3	92.0	91.0	88.1	87.2	84.5
Bunkers	-	-	-	-	-	-	-	-	-
Total	62.6	71.8	75.0	95.3	92.0	91.0	88.1	87.2	84.5

	1989	1990	1991	1992	1993	1994	1995	1996	1997
Thousand tonnes									
NGL/LPG/Ethane	16	15	77	45	37	38	34	33	35
Naphtha	1 038	517	505	684	560	626	670	670	670
Motor Gasoline	489	469	396	388	414	482	504	459	548
Aviation Fuels	-	-	-	-	-	26	39	38	35
Kerosene	11	9	20	60	2	21	2	5	1
Gas Diesel	1 208	1 124	878	842	701	735	852	806	743
Heavy Fuel Oil	1 824	1 535	1 422	1 212	1 020	920	642	662	565
Other Products	944	1 171	926	496	283	487	498	726	796
Refinery Fuel	138	109	122	118	13	13	-	-	-
Sub-Total	5 668	4 949	4 346	3 845	3 030	3 348	3 241	3 399	3 393
Bunkers	-	-	-	-	-	-	-	-	-
Total	5 668	4 949	4 346	3 845	3 030	3 348	3 241	3 399	3 393
Thousand barrels/day									
NGL/LPG/Ethane	0.5	0.5	2.4	1.4	1.2	1.2	1.1	1.0	1.1
Naphtha	24.2	12.0	11.8	15.9	13.0	14.6	15.6	15.6	15.6
Motor Gasoline	11.4	11.0	9.3	9.0	9.7	11.3	11.8	10.7	12.8
Aviation Fuels	-	-	-	-	-	0.6	0.8	0.8	0.8
Kerosene	0.2	0.2	0.4	1.3	0.0	0.4	0.0	0.1	0.0
Gas Diesel	24.7	23.0	17.9	17.2	14.3	15.0	17.4	16.4	15.2
Heavy Fuel Oil	33.3	28.0	25.9	22.1	18.6	16.8	11.7	12.0	10.3
Other Products	18.6	22.9	18.0	9.7	5.3	9.2	9.0	13.3	14.7
Refinery Fuel	2.8	2.2	2.5	2.5	0.3	0.3	-	-	-
Sub-Total	115.7	99.8	88.3	79.0	62.4	69.4	67.4	70.0	70.5
Bunkers	-	-	-	-	-	-	-	-	-
Total	115.7	99.8	88.3	79.0	62.4	69.4	67.4	70.0	70.5

Oil Products Consumption
Consommation de produits pétroliers

Slovenia

	1971	1973	1978	1983	1984	1985	1986	1987	1988
Thousand tonnes									
NGL/LPG/Ethane	-	-	-	53	57	44	43	42	43
Naphtha	-	-	-	17	2	17	11	23	11
Motor Gasoline	-	-	-	296	325	363	434	482	520
Aviation Fuels	-	-	-	19	23	25	32	29	28
Kerosene	-	-	-	-	-	-	-	-	-
Gas Diesel	-	-	-	745	728	693	760	742	759
Heavy Fuel Oil	-	-	-	236	236	239	255	272	259
Other Products	-	-	-	-	-	-	-	1	5
Refinery Fuel	-	-	-	6	5	6	6	11	13
Sub-Total	-	-	-	1 372	1 376	1 387	1 541	1 602	1 638
Bunkers	-	-	-	-	-	-	-	-	-
Total	-	-	-	1 372	1 376	1 387	1 541	1 602	1 638
Thousand barrels/day									
NGL/LPG/Ethane	-	-	-	1.7	1.8	1.4	1.4	1.3	1.4
Naphtha	-	-	-	0.4	0.0	0.4	0.3	0.5	0.3
Motor Gasoline	-	-	-	6.9	7.6	8.5	10.1	11.3	12.1
Aviation Fuels	-	-	-	0.4	0.5	0.5	0.7	0.6	0.6
Kerosene	-	-	-	-	-	-	-	-	-
Gas Diesel	-	-	-	15.2	14.8	14.2	15.5	15.2	15.5
Heavy Fuel Oil	-	-	-	4.3	4.3	4.4	4.7	5.0	4.7
Other Products	-	-	-	-	-	-	-	0.0	0.1
Refinery Fuel	-	-	-	0.1	0.1	0.1	0.1	0.2	0.2
Sub-Total	-	-	-	29.1	29.1	29.5	32.8	34.1	34.9
Bunkers	-	-	-	-	-	-	-	-	-
Total	-	-	-	29.1	29.1	29.5	32.8	34.1	34.9

	1989	1990	1991	1992	1993	1994	1995	1996	1997
Thousand tonnes									
NGL/LPG/Ethane	39	34	40	43	39	38	35	39	54
Naphtha	19	6	21	9	-	-	28	76	11
Motor Gasoline	530	565	526	580	689	765	821	923	913
Aviation Fuels	27	26	10	11	17	18	20	18	19
Kerosene	-	-	-	-	-	-	-	-	-
Gas Diesel	788	777	831	681	870	915	1 060	1 283	1 303
Heavy Fuel Oil	272	262	201	219	248	280	244	257	210
Other Products	3	3	2	11	7	6	5	7	4
Refinery Fuel	14	14	15	18	8	9	16	14	16
Sub-Total	1 692	1 687	1 646	1 572	1 878	2 031	2 229	2 617	2 530
Bunkers	-	-	-	-	-	-	-	-	-
Total	1 692	1 687	1 646	1 572	1 878	2 031	2 229	2 617	2 530
Thousand barrels/day									
NGL/LPG/Ethane	1.2	1.1	1.3	1.4	1.2	1.2	1.1	1.2	1.7
Naphtha	0.4	0.1	0.5	0.2	-	-	0.7	1.8	0.3
Motor Gasoline	12.4	13.2	12.3	13.5	16.1	17.9	19.2	21.5	21.3
Aviation Fuels	0.6	0.6	0.2	0.2	0.4	0.4	0.4	0.4	0.4
Kerosene	-	-	-	-	-	-	-	-	-
Gas Diesel	16.1	15.9	17.0	13.9	17.8	18.7	21.7	26.2	26.6
Heavy Fuel Oil	5.0	4.8	3.7	4.0	4.5	5.1	4.5	4.7	3.8
Other Products	0.1	0.1	0.0	0.2	0.1	0.1	0.1	0.1	0.1
Refinery Fuel	0.3	0.3	0.3	0.3	0.1	0.2	0.3	0.3	0.3
Sub-Total	36.0	36.0	35.2	33.7	40.3	43.6	47.9	56.1	54.6
Bunkers	-	-	-	-	-	-	-	-	-
Total	36.0	36.0	35.2	33.7	40.3	43.6	47.9	56.1	54.6

Oil Products Consumption
Consommation de produits pétroliers

South Africa

	1971	1973	1978	1983	1984	1985	1986	1987	1988
Thousand tonnes									
NGL/LPG/Ethane	106	133	159	183	193	176	177	197	221
Naphtha	50	53	50	50	50	50	50	50	50
Motor Gasoline	3 459	3 975	4 159	4 519	4 867	4 678	4 764	5 251	5 687
Aviation Fuels	329	360	564	590	616	606	565	577	640
Kerosene	535	535	535	307	340	341	365	401	452
Gas Diesel	2 566	3 222	4 127	3 953	4 288	4 185	3 992	4 125	4 431
Heavy Fuel Oil	862	1 198	820	483	580	633	547	522	470
Other Products	746	737	1 280	2 170	2 070	1 930	1 930	1 893	2 160
Refinery Fuel	604	645	778	811	808	770	770	770	750
Sub-Total	9 257	10 858	12 472	13 066	13 812	13 369	13 160	13 786	14 861
Bunkers	3 504	2 771	1 200	1 200	1 100	1 100	1 000	1 000	1 191
Total	12 761	13 629	13 672	14 266	14 912	14 469	14 160	14 786	16 052
Thousand barrels/day									
NGL/LPG/Ethane	3.4	4.2	5.1	5.8	6.1	5.6	5.6	6.3	7.0
Naphtha	1.2	1.2	1.2	1.2	1.2	1.2	1.2	1.2	1.2
Motor Gasoline	80.8	92.9	97.2	105.6	113.4	109.3	111.3	122.7	132.5
Aviation Fuels	7.2	7.9	12.4	12.9	13.4	13.2	12.3	12.6	13.9
Kerosene	11.3	11.3	11.3	6.5	7.2	7.2	7.7	8.5	9.6
Gas Diesel	52.4	65.9	84.3	80.8	87.4	85.5	81.6	84.3	90.3
Heavy Fuel Oil	15.7	21.9	15.0	8.8	10.6	11.6	10.0	9.5	8.6
Other Products	13.8	13.7	24.0	40.9	38.9	36.4	36.4	35.8	40.7
Refinery Fuel	11.0	11.8	14.2	14.8	14.7	14.1	14.1	14.1	13.6
Sub-Total	196.9	230.8	264.7	277.3	292.9	284.1	280.2	294.9	317.4
Bunkers	64.4	51.0	22.3	22.3	20.5	20.5	18.7	18.7	22.1
Total	261.3	281.8	287.0	299.6	313.4	304.6	298.9	313.6	339.5

	1989	1990	1991	1992	1993	1994	1995	1996	1997
Thousand tonnes									
NGL/LPG/Ethane	249	240	255	253	253	268	267	256	285
Naphtha	50	50	50	50	-	-	-	-	-
Motor Gasoline	5 921	6 251	6 442	6 469	6 784	7 099	8 068	7 791	7 898
Aviation Fuels	690	706	706	819	979	965	1 122	1 288	1 429
Kerosene	478	559	556	512	694	733	702	739	779
Gas Diesel	4 443	4 387	4 261	4 026	4 195	4 333	4 622	4 721	4 804
Heavy Fuel Oil	481	548	476	397	559	601	573	597	472
Other Products	2 149	2 133	2 077	2 144	634	774	759	819	780
Refinery Fuel	750	750	750	750	1 427	1 613	1 731	1 641	1 424
Sub-Total	15 211	15 624	15 573	15 420	15 525	16 386	17 844	17 852	17 871
Bunkers	1 053	1 927	2 118	2 742	2 442	2 564	3 123	2 759	2 145
Total	16 264	17 551	17 691	18 162	17 967	18 950	20 967	20 611	20 016
Thousand barrels/day									
NGL/LPG/Ethane	7.9	7.6	8.1	8.0	8.0	8.5	8.5	8.1	9.1
Naphtha	1.2	1.2	1.2	1.2	-	-	-	-	-
Motor Gasoline	138.4	146.1	150.5	150.8	158.5	165.9	188.5	181.6	184.6
Aviation Fuels	15.0	15.4	15.4	17.8	21.3	21.0	24.4	28.0	31.1
Kerosene	10.1	11.9	11.8	10.8	14.7	15.5	14.9	15.6	16.5
Gas Diesel	90.8	89.7	87.1	82.1	85.7	88.6	94.5	96.2	98.2
Heavy Fuel Oil	8.8	10.0	8.7	7.2	10.2	11.0	10.5	10.9	8.6
Other Products	40.6	40.3	39.4	40.5	11.5	14.5	14.2	15.3	14.6
Refinery Fuel	13.7	13.7	13.7	13.6	33.3	37.7	40.4	38.2	33.3
Sub-Total	326.5	335.8	335.8	332.0	343.4	362.6	395.9	393.8	395.9
Bunkers	19.7	35.6	39.1	50.3	45.2	47.7	58.0	51.0	39.9
Total	346.1	371.4	374.9	382.4	388.7	410.3	453.9	444.8	435.8

Oil Products Consumption
Consommation de produits pétroliers

Sri Lanka

	1971	1973	1978	1983	1984	1985	1986	1987	1988
Thousand tonnes									
NGL/LPG/Ethane	-	-	5	8	11	13	16	19	20
Naphtha	-	-	-	-	-	-	-	-	-
Motor Gasoline	114	123	121	117	119	122	131	140	158
Aviation Fuels	18	51	92	75	116	119	113	115	100
Kerosene	270	275	238	186	157	165	157	153	163
Gas Diesel	297	355	368	724	560	525	524	649	569
Heavy Fuel Oil	112	257	184	252	200	110	94	129	135
Other Products	6	48	35	27	44	37	52	39	39
Refinery Fuel	93	84	67	54	65	60	65	61	64
Sub-Total	910	1 193	1 110	1 443	1 272	1 151	1 152	1 305	1 248
Bunkers	385	433	338	211	335	328	394	412	362
Total	1 295	1 626	1 448	1 654	1 607	1 479	1 546	1 717	1 610
Thousand barrels/day									
NGL/LPG/Ethane	-	-	0.2	0.3	0.3	0.4	0.5	0.6	0.6
Naphtha	-	-	-	-	-	-	-	-	-
Motor Gasoline	2.7	2.9	2.8	2.7	2.8	2.9	3.1	3.3	3.7
Aviation Fuels	0.4	1.1	2.0	1.6	2.5	2.6	2.5	2.5	2.2
Kerosene	5.7	5.8	5.0	3.9	3.3	3.5	3.3	3.2	3.4
Gas Diesel	6.1	7.3	7.5	14.8	11.4	10.7	10.7	13.3	11.6
Heavy Fuel Oil	2.0	4.7	3.4	4.6	3.6	2.0	1.7	2.4	2.5
Other Products	0.1	0.9	0.7	0.5	0.8	0.7	1.0	0.7	0.7
Refinery Fuel	1.9	1.7	1.3	1.1	1.3	1.2	1.3	1.2	1.3
Sub-Total	18.9	24.4	22.9	29.6	26.2	24.0	24.1	27.2	26.0
Bunkers	7.1	8.0	6.3	3.9	6.2	6.1	7.3	7.6	6.6
Total	26.0	32.4	29.3	33.4	32.3	30.1	31.3	34.8	32.7

	1989	1990	1991	1992	1993	1994	1995	1996	1997
Thousand tonnes									
NGL/LPG/Ethane	29	34	38	44	52	64	106	129	112
Naphtha	-	-	-	-	-	-	-	-	-
Motor Gasoline	177	179	159	165	173	184	188	199	194
Aviation Fuels	49	57	70	81	72	146	172	196	275
Kerosene	162	167	161	189	191	208	224	229	160
Gas Diesel	531	548	590	786	744	804	879	1 199	1 324
Heavy Fuel Oil	144	125	146	252	252	236	155	363	290
Other Products	28	31	40	42	45	38	214	55	40
Refinery Fuel	63	76	64	52	59	70	66	70	56
Sub-Total	1 183	1 217	1 268	1 611	1 588	1 750	2 004	2 440	2 451
Bunkers	312	391	312	304	352	345	344	385	258
Total	1 495	1 608	1 580	1 915	1 940	2 095	2 348	2 825	2 709
Thousand barrels/day									
NGL/LPG/Ethane	0.9	1.1	1.2	1.4	1.7	2.0	3.4	4.1	3.6
Naphtha	-	-	-	-	-	-	-	-	-
Motor Gasoline	4.1	4.2	3.7	3.8	4.0	4.3	4.4	4.6	4.5
Aviation Fuels	1.1	1.2	1.5	1.8	1.6	3.2	3.7	4.2	6.0
Kerosene	3.4	3.5	3.4	4.0	4.1	4.4	4.8	4.8	3.4
Gas Diesel	10.9	11.2	12.1	16.0	15.2	16.4	18.0	24.4	27.1
Heavy Fuel Oil	2.6	2.3	2.7	4.6	4.6	4.3	2.8	6.6	5.3
Other Products	0.5	0.6	0.8	0.8	0.9	0.7	4.1	1.1	0.8
Refinery Fuel	1.3	1.5	1.3	1.0	1.2	1.4	1.4	1.4	1.2
Sub-Total	24.8	25.7	26.6	33.5	33.2	36.8	42.5	51.3	51.7
Bunkers	5.8	7.2	5.8	5.6	6.5	6.4	6.4	7.1	4.8
Total	30.6	32.9	32.4	39.1	39.7	43.2	48.9	58.4	56.5

Oil Products Consumption
Consommation de produits pétroliers

Sudan

	1971	1973	1978	1983	1984	1985	1986	1987	1988
Thousand tonnes									
NGL/LPG/Ethane	4	5	4	8	7	8	10	9	11
Naphtha	-	-	-	17	16	18	17	19	19
Motor Gasoline	97	103	158	201	179	190	202	143	243
Aviation Fuels	107	119	54	65	65	68	64	61	60
Kerosene	107	81	24	39	37	31	36	31	31
Gas Diesel	561	682	514	630	559	704	640	521	804
Heavy Fuel Oil	242	564	270	305	269	322	346	246	291
Other Products	31	33	36	25	105	96	87	92	88
Refinery Fuel	4	5	5	4	3	3	4	2	4
Sub-Total	1 153	1 592	1 065	1 294	1 240	1 440	1 406	1 124	1 551
Bunkers	-	2	4	9	5	5	4	5	5
Total	1 153	1 594	1 069	1 303	1 245	1 445	1 410	1 129	1 556
Thousand barrels/day									
NGL/LPG/Ethane	0.1	0.2	0.1	0.3	0.2	0.3	0.3	0.3	0.3
Naphtha	-	-	-	0.4	0.4	0.4	0.4	0.4	0.4
Motor Gasoline	2.3	2.4	3.7	4.7	4.2	4.4	4.7	3.3	5.7
Aviation Fuels	2.4	2.7	1.2	1.4	1.4	1.5	1.4	1.3	1.3
Kerosene	2.3	1.7	0.5	0.8	0.8	0.7	0.8	0.7	0.7
Gas Diesel	11.5	13.9	10.5	12.9	11.4	14.4	13.1	10.6	16.4
Heavy Fuel Oil	4.4	10.3	4.9	5.6	4.9	5.9	6.3	4.5	5.3
Other Products	0.6	0.6	0.7	0.5	2.0	1.8	1.7	1.8	1.7
Refinery Fuel	0.1	0.1	0.1	0.1	0.1	0.1	0.1	0.0	0.1
Sub-Total	23.6	31.9	21.7	26.6	25.3	29.4	28.8	23.0	31.9
Bunkers	-	0.0	0.1	0.2	0.1	0.1	0.1	0.1	0.1
Total	23.6	32.0	21.8	26.8	25.4	29.5	28.8	23.1	32.0

	1989	1990	1991	1992	1993	1994	1995	1996	1997
Thousand tonnes									
NGL/LPG/Ethane	10	14	16	14	11	17	20	26	25
Naphtha	20	20	22	20	22	23	23	24	24
Motor Gasoline	219	218	190	230	161	206	199	212	237
Aviation Fuels	63	36	41	42	45	50	50	52	52
Kerosene	31	32	37	36	37	40	41	62	62
Gas Diesel	715	1 060	849	802	575	768	735	682	764
Heavy Fuel Oil	272	343	399	315	192	422	366	315	489
Other Products	91	95	97	92	101	103	102	104	104
Refinery Fuel	3	4	5	4	2	3	4	4	5
Sub-Total	1 424	1 822	1 656	1 555	1 146	1 632	1 540	1 481	1 762
Bunkers	5	7	7	7	7	7	8	8	8
Total	1 429	1 829	1 663	1 562	1 153	1 639	1 548	1 489	1 770
Thousand barrels/day									
NGL/LPG/Ethane	0.3	0.4	0.5	0.4	0.4	0.5	0.6	0.8	0.8
Naphtha	0.5	0.5	0.5	0.5	0.5	0.5	0.5	0.6	0.6
Motor Gasoline	5.1	5.1	4.4	5.4	3.8	4.8	4.7	4.9	5.5
Aviation Fuels	1.4	0.8	0.9	0.9	1.0	1.1	1.1	1.2	1.2
Kerosene	0.7	0.7	0.8	0.8	0.8	0.8	0.9	1.3	1.3
Gas Diesel	14.6	21.7	17.4	16.3	11.8	15.7	15.0	13.9	15.6
Heavy Fuel Oil	5.0	6.3	7.3	5.7	3.5	7.7	6.7	5.7	8.9
Other Products	1.8	1.8	1.9	1.8	1.9	2.0	2.0	2.0	2.0
Refinery Fuel	0.1	0.1	0.1	0.1	0.0	0.1	0.1	0.1	0.1
Sub-Total	29.3	37.3	33.8	31.9	23.7	33.3	31.6	30.5	36.0
Bunkers	0.1	0.1	0.1	0.1	0.1	0.1	0.2	0.2	0.2
Total	29.4	37.5	33.9	32.0	23.8	33.5	31.7	30.7	36.2

Oil Products Consumption
Consommation de produits pétroliers

Syria

	1971	1973	1978	1983	1984	1985	1986	1987	1988
Thousand tonnes									
NGL/LPG/Ethane	23	27	31	197	231	248	265	289	293
Naphtha	-	-	313	190	184	211	211	178	17
Motor Gasoline	212	281	533	691	736	788	828	943	960
Aviation Fuels	75	159	238	257	284	215	295	266	300
Kerosene	244	215	282	240	203	161	212	178	172
Gas Diesel	820	689	2 000	2 732	2 990	3 055	2 920	3 192	2 927
Heavy Fuel Oil	577	542	768	2 553	2 992	2 175	2 520	3 046	3 276
Other Products	223	195	220	463	478	552	372	346	366
Refinery Fuel	89	79	150	475	564	489	450	529	399
Sub-Total	2 263	2 187	4 535	7 798	8 662	7 894	8 073	8 967	8 710
Bunkers	-	-	-	-	-	-	-	-	-
Total	2 263	2 187	4 535	7 798	8 662	7 894	8 073	8 967	8 710
Thousand barrels/day									
NGL/LPG/Ethane	0.7	0.9	1.0	6.3	7.3	7.9	8.4	9.2	9.3
Naphtha	-	-	7.3	4.4	4.3	4.9	4.9	4.1	0.4
Motor Gasoline	5.0	6.6	12.5	16.1	17.2	18.4	19.4	22.0	22.4
Aviation Fuels	1.6	3.5	5.2	5.6	6.2	4.7	6.4	5.8	6.5
Kerosene	5.2	4.6	6.0	5.1	4.3	3.4	4.5	3.8	3.6
Gas Diesel	16.8	14.1	40.9	55.8	60.9	62.4	59.7	65.2	59.7
Heavy Fuel Oil	10.5	9.9	14.0	46.6	54.4	39.7	46.0	55.6	59.6
Other Products	4.3	3.7	4.2	8.9	9.1	10.6	7.1	6.6	7.0
Refinery Fuel	1.6	1.4	2.7	8.7	10.3	8.9	8.2	9.7	7.3
Sub-Total	45.7	44.6	93.7	157.5	174.0	160.9	164.6	182.0	175.7
Bunkers	-	-	-	-	-	-	-	-	-
Total	45.7	44.6	93.7	157.5	174.0	160.9	164.6	182.0	175.7

	1989	1990	1991	1992	1993	1994	1995	1996	1997
Thousand tonnes									
NGL/LPG/Ethane	256	264	270	280	288	319	356	368	397
Naphtha	70	50	50	50	52	58	101	105	108
Motor Gasoline	1 100	1 220	1 258	1 060	1 110	1 090	1 063	1 101	1 136
Aviation Fuels	336	390	324	330	214	237	156	162	224
Kerosene	230	252	260	226	231	245	239	247	247
Gas Diesel	3 200	3 305	3 620	4 095	4 160	4 210	4 283	4 433	4 914
Heavy Fuel Oil	2 353	3 100	3 702	3 270	3 885	4 350	4 624	4 786	4 540
Other Products	400	400	300	350	348	386	376	389	389
Refinery Fuel	400	400	400	410	420	420	447	463	439
Sub-Total	8 345	9 381	10 184	10 071	10 708	11 315	11 645	12 054	12 394
Bunkers	-	-	-	-	-	-	-	-	-
Total	8 345	9 381	10 184	10 071	10 708	11 315	11 645	12 054	12 394
Thousand barrels/day									
NGL/LPG/Ethane	8.1	8.4	8.6	8.9	9.2	10.1	11.3	11.7	12.6
Naphtha	1.6	1.2	1.2	1.2	1.2	1.4	2.4	2.4	2.5
Motor Gasoline	25.7	28.5	29.4	24.7	25.9	25.5	24.8	25.7	26.5
Aviation Fuels	7.3	8.5	7.0	7.2	4.6	5.1	3.4	3.5	4.9
Kerosene	4.9	5.3	5.5	4.8	4.9	5.2	5.1	5.2	5.2
Gas Diesel	65.4	67.5	74.0	83.5	85.0	86.0	87.5	90.4	100.4
Heavy Fuel Oil	42.9	56.6	67.5	59.5	70.9	79.4	84.4	87.1	82.8
Other Products	7.7	7.7	5.8	6.7	6.7	7.4	7.2	7.4	7.5
Refinery Fuel	7.3	7.3	7.3	7.5	7.7	7.7	8.2	8.4	8.0
Sub-Total	171.0	191.0	206.3	203.8	216.1	227.8	234.2	241.8	250.5
Bunkers	-	-	-	-	-	-	-	-	-
Total	171.0	191.0	206.3	203.8	216.1	227.8	234.2	241.8	250.5

Oil Products Consumption
Consommation de produits pétroliers

Tajikistan

	1971	1973	1978	1983	1984	1985	1986	1987	1988
Thousand tonnes									
NGL/LPG/Ethane	-	-	-	-	-	-	-	-	-
Naphtha	-	-	-	-	-	-	-	-	-
Motor Gasoline	-	-	-	-	-	-	-	-	-
Aviation Fuels	-	-	-	-	-	-	-	-	-
Kerosene	-	-	-	-	-	-	-	-	-
Gas Diesel	-	-	-	-	-	-	-	-	-
Heavy Fuel Oil	-	-	-	-	-	-	-	-	-
Other Products	-	-	-	-	-	-	-	-	-
Refinery Fuel	-	-	-	-	-	-	-	-	-
Sub-Total	-	-	-	-	-	-	-	-	-
Bunkers	-	-	-	-	-	-	-	-	-
Total	-	-	-	-	-	-	-	-	-
Thousand barrels/day									
NGL/LPG/Ethane	-	-	-	-	-	-	-	-	-
Naphtha	-	-	-	-	-	-	-	-	-
Motor Gasoline	-	-	-	-	-	-	-	-	-
Aviation Fuels	-	-	-	-	-	-	-	-	-
Kerosene	-	-	-	-	-	-	-	-	-
Gas Diesel	-	-	-	-	-	-	-	-	-
Heavy Fuel Oil	-	-	-	-	-	-	-	-	-
Other Products	-	-	-	-	-	-	-	-	-
Refinery Fuel	-	-	-	-	-	-	-	-	-
Sub-Total	-	-	-	-	-	-	-	-	-
Bunkers	-	-	-	-	-	-	-	-	-
Total	-	-	-	-	-	-	-	-	-

	1989	1990	1991	1992	1993	1994	1995	1996	1997
Thousand tonnes									
NGL/LPG/Ethane	-	-	-	-	22	7	7	7	7
Naphtha	-	-	-	-	-	-	-	-	-
Motor Gasoline	-	-	-	4 763	2 988	996	996	996	996
Aviation Fuels	-	-	-	15	17	5	5	5	5
Kerosene	-	-	-	-	-	-	-	-	-
Gas Diesel	-	-	-	282	226	75	75	75	75
Heavy Fuel Oil	-	-	-	309	100	33	33	33	33
Other Products	-	-	-	226	130	43	43	43	43
Refinery Fuel	-	-	-	-	-	-	-	-	-
Sub-Total	-	-	-	5 595	3 483	1 159	1 159	1 159	1 159
Bunkers	-	-	-	-	-	-	-	-	-
Total	-	-	-	5 595	3 483	1 159	1 159	1 159	1 159
Thousand barrels/day									
NGL/LPG/Ethane	-	-	-	-	0.7	0.2	0.2	0.2	0.2
Naphtha	-	-	-	-	-	-	-	-	-
Motor Gasoline	-	-	-	111.0	69.8	23.3	23.3	23.2	23.3
Aviation Fuels	-	-	-	0.3	0.4	0.1	0.1	0.1	0.1
Kerosene	-	-	-	-	-	-	-	-	-
Gas Diesel	-	-	-	5.7	4.6	1.5	1.5	1.5	1.5
Heavy Fuel Oil	-	-	-	5.6	1.8	0.6	0.6	0.6	0.6
Other Products	-	-	-	4.3	2.5	0.8	0.8	0.8	0.8
Refinery Fuel	-	-	-	-	-	-	-	-	-
Sub-Total	-	-	-	127.0	79.8	26.6	26.6	26.5	26.6
Bunkers	-	-	-	-	-	-	-	-	-
Total	-	-	-	127.0	79.8	26.6	26.6	26.5	26.6

Oil Products Consumption
Consommation de produits pétroliers

Tanzania

	1971	1973	1978	1983	1984	1985	1986	1987	1988
Thousand tonnes									
NGL/LPG/Ethane	5	5	6	5	7	7	7	6	6
Naphtha	-	-	-	-	-	-	-	-	-
Motor Gasoline	109	120	100	82	85	87	85	90	92
Aviation Fuels	27	27	52	38	40	40	37	35	61
Kerosene	58	75	94	55	57	58	93	85	97
Gas Diesel	184	216	177	191	189	191	192	194	194
Heavy Fuel Oil	72	101	84	80	81	95	95	96	97
Other Products	44	41	47	33	22	25	16	19	16
Refinery Fuel	19	15	11	17	15	15	15	15	15
Sub-Total	518	600	571	501	496	518	540	540	578
Bunkers	17	20	30	25	26	26	25	23	24
Total	535	620	601	526	522	544	565	563	602
Thousand barrels/day									
NGL/LPG/Ethane	0.2	0.2	0.2	0.2	0.2	0.2	0.2	0.2	0.2
Naphtha	-	-	-	-	-	-	-	-	-
Motor Gasoline	2.5	2.8	2.3	1.9	2.0	2.0	2.0	2.1	2.1
Aviation Fuels	0.6	0.6	1.1	0.8	0.9	0.9	0.8	0.8	1.3
Kerosene	1.2	1.6	2.0	1.2	1.2	1.2	2.0	1.8	2.1
Gas Diesel	3.8	4.4	3.6	3.9	3.9	3.9	3.9	4.0	4.0
Heavy Fuel Oil	1.3	1.8	1.5	1.5	1.5	1.7	1.7	1.8	1.8
Other Products	0.8	0.8	0.9	0.6	0.4	0.5	0.3	0.4	0.3
Refinery Fuel	0.4	0.3	0.2	0.4	0.3	0.3	0.3	0.3	0.3
Sub-Total	10.9	12.5	12.0	10.5	10.4	10.8	11.3	11.3	12.1
Bunkers	0.3	0.4	0.5	0.5	0.5	0.5	0.5	0.4	0.4
Total	11.2	12.9	12.5	10.9	10.8	11.3	11.8	11.7	12.5

	1989	1990	1991	1992	1993	1994	1995	1996	1997
Thousand tonnes									
NGL/LPG/Ethane	6	7	7	5	6	6	6	6	6
Naphtha	-	-	-	-	-	-	-	-	-
Motor Gasoline	85	92	91	99	108	108	98	98	98
Aviation Fuels	64	70	70	67	46	49	49	36	36
Kerosene	100	102	104	96	70	70	72	72	72
Gas Diesel	199	207	194	195	195	195	197	201	237
Heavy Fuel Oil	102	105	109	109	105	105	105	105	105
Other Products	16	16	17	15	15	16	16	16	16
Refinery Fuel	15	15	15	14	14	14	14	14	14
Sub-Total	587	614	607	600	559	563	557	548	584
Bunkers	24	26	22	22	22	22	23	23	23
Total	611	640	629	622	581	585	580	571	607
Thousand barrels/day									
NGL/LPG/Ethane	0.2	0.2	0.2	0.2	0.2	0.2	0.2	0.2	0.2
Naphtha	-	-	-	-	-	-	-	-	-
Motor Gasoline	2.0	2.2	2.1	2.3	2.5	2.5	2.3	2.3	2.3
Aviation Fuels	1.4	1.5	1.5	1.5	1.0	1.1	1.1	0.8	0.8
Kerosene	2.1	2.2	2.2	2.0	1.5	1.5	1.5	1.5	1.5
Gas Diesel	4.1	4.2	4.0	4.0	4.0	4.0	4.0	4.1	4.8
Heavy Fuel Oil	1.9	1.9	2.0	2.0	1.9	1.9	1.9	1.9	1.9
Other Products	0.3	0.3	0.3	0.3	0.3	0.3	0.3	0.3	0.3
Refinery Fuel	0.3	0.3	0.3	0.3	0.3	0.3	0.3	0.3	0.3
Sub-Total	12.3	12.9	12.7	12.5	11.7	11.8	11.7	11.4	12.2
Bunkers	0.4	0.5	0.4	0.4	0.4	0.4	0.4	0.4	0.4
Total	12.7	13.3	13.1	12.9	12.1	12.2	12.1	11.8	12.6

Oil Products Consumption

Consommation de produits pétroliers

Thailand

	1971	1973	1978	1983	1984	1985	1986	1987	1988
Thousand tonnes									
NGL/LPG/Ethane	50	77	162	474	548	604	635	646	758
Naphtha	-	-	-	-	-	-	-	-	-
Motor Gasoline	939	1 104	1 656	1 488	1 525	1 527	1 654	1 930	2 114
Aviation Fuels	400	724	628	914	965	990	1 096	1 202	1 387
Kerosene	153	169	220	426	232	121	114	103	100
Gas Diesel	2 181	2 628	3 251	3 700	4 423	4 655	4 835	5 335	6 134
Heavy Fuel Oil	1 630	2 378	3 849	3 032	2 797	2 041	2 088	1 968	2 342
Other Products	280	255	230	214	260	170	200	292	166
Refinery Fuel	53	77	83	81	79	139	139	149	92
Sub-Total	5 686	7 412	10 079	10 329	10 829	10 247	10 761	11 625	13 093
Bunkers	65	86	123	275	284	202	295	375	364
Total	5 751	7 498	10 202	10 604	11 113	10 449	11 056	12 000	13 457
Thousand barrels/day									
NGL/LPG/Ethane	1.6	2.4	5.1	15.1	17.4	19.2	20.2	20.5	24.0
Naphtha	-	-	-	-	-	-	-	-	-
Motor Gasoline	21.9	25.8	38.7	34.8	35.5	35.7	38.7	45.1	49.3
Aviation Fuels	8.7	15.8	13.7	19.9	20.9	21.5	23.8	26.1	30.1
Kerosene	3.2	3.6	4.7	9.0	4.9	2.6	2.4	2.2	2.1
Gas Diesel	44.6	53.7	66.4	75.6	90.2	95.1	98.8	109.0	125.0
Heavy Fuel Oil	29.7	43.4	70.2	55.3	50.9	37.2	38.1	35.9	42.6
Other Products	4.7	4.2	3.8	3.6	4.3	2.8	3.3	5.3	2.8
Refinery Fuel	1.2	1.7	1.8	1.8	1.7	3.1	3.1	3.4	2.0
Sub-Total	115.6	150.6	204.5	215.1	225.9	217.3	228.5	247.5	277.9
Bunkers	1.2	1.6	2.2	5.1	5.3	3.8	5.5	7.0	6.8
Total	116.8	152.2	206.8	220.2	231.1	221.1	233.9	254.5	284.7

	1989	1990	1991	1992	1993	1994	1995	1996	1997
Thousand tonnes									
NGL/LPG/Ethane	855	928	1 008	1 124	1 219	1 845	1 892	2 684	2 467
Naphtha	-	-	-	-	-	-	-	-	-
Motor Gasoline	2 409	2 670	2 824	3 147	3 575	4 071	4 590	5 059	5 384
Aviation Fuels	1 638	1 782	1 719	2 121	2 142	2 341	2 389	2 492	2 667
Kerosene	94	97	88	90	86	92	80	78	70
Gas Diesel	7 245	8 361	8 483	8 818	10 228	11 279	13 274	15 192	14 993
Heavy Fuel Oil	3 057	4 544	5 261	6 276	6 928	7 692	8 739	8 455	7 853
Other Products	304	722	412	613	727	521	610	904	1 063
Refinery Fuel	107	112	119	147	172	200	239	318	367
Sub-Total	15 709	19 216	19 914	22 336	25 077	28 041	31 813	35 182	34 864
Bunkers	491	528	580	673	733	914	939	759	809
Total	16 200	19 744	20 494	23 009	25 810	28 955	32 752	35 941	35 673
Thousand barrels/day									
NGL/LPG/Ethane	27.2	29.5	32.0	35.6	38.7	63.6	63.9	94.8	83.8
Naphtha	-	-	-	-	-	-	-	-	-
Motor Gasoline	56.3	62.4	66.0	73.3	83.5	95.1	107.3	117.9	125.8
Aviation Fuels	35.6	38.7	37.4	46.0	46.6	50.9	51.9	54.0	58.0
Kerosene	2.0	2.1	1.9	1.9	1.8	2.0	1.7	1.7	1.5
Gas Diesel	148.1	170.9	173.4	179.7	209.0	230.5	271.3	309.7	306.4
Heavy Fuel Oil	55.8	82.9	96.0	114.2	126.4	140.4	159.5	153.9	143.3
Other Products	5.5	13.7	7.3	11.2	13.4	9.2	10.7	16.1	18.9
Refinery Fuel	2.3	2.5	2.6	3.2	3.8	4.4	5.2	7.0	8.0
Sub-Total	332.8	402.6	416.6	465.1	523.2	596.0	671.5	755.0	745.7
Bunkers	9.1	9.8	10.8	12.5	13.6	17.0	17.5	14.1	15.0
Total	341.9	412.4	427.3	477.6	536.9	613.1	689.0	769.0	760.7

Oil Products Consumption
Consommation de produits pétroliers

Trinidad-and-Tobago

	1971	1973	1978	1983	1984	1985	1986	1987	1988
Thousand tonnes									
NGL/LPG/Ethane	15	30	29	46	56	47	47	44	77
Naphtha	-	-	-	-	-	-	-	-	-
Motor Gasoline	172	227	270	375	388	400	407	402	375
Aviation Fuels	67	38	40	64	64	69	84	50	50
Kerosene	30	30	30	15	13	11	15	8	8
Gas Diesel	62	134	132	175	162	136	143	126	121
Heavy Fuel Oil	38	29	45	20	37	10	10	26	7
Other Products	221	35	273	146	185	171	71	47	34
Refinery Fuel	544	541	417	70	75	102	105	123	111
Sub-Total	1 149	1 064	1 236	911	980	946	882	826	783
Bunkers	1 656	1 658	1 001	461	168	99	27	87	55
Total	2 805	2 722	2 237	1 372	1 148	1 045	909	913	838
Thousand barrels/day									
NGL/LPG/Ethane	0.5	1.0	0.9	1.5	1.8	1.5	1.5	1.4	2.4
Naphtha	-	-	-	-	-	-	-	-	-
Motor Gasoline	4.0	5.3	6.3	8.8	9.0	9.3	9.5	9.4	8.7
Aviation Fuels	1.5	0.8	0.9	1.4	1.4	1.5	1.8	1.1	1.1
Kerosene	0.6	0.6	0.6	0.3	0.3	0.2	0.3	0.2	0.2
Gas Diesel	1.3	2.7	2.7	3.6	3.3	2.8	2.9	2.6	2.5
Heavy Fuel Oil	0.7	0.5	0.8	0.4	0.7	0.2	0.2	0.5	0.1
Other Products	4.2	0.7	5.2	2.8	3.5	3.3	1.3	0.9	0.6
Refinery Fuel	11.3	11.2	8.4	1.5	1.6	2.2	2.3	2.7	2.4
Sub-Total	24.1	22.8	25.9	20.2	21.6	21.0	19.9	18.7	18.1
Bunkers	30.8	30.7	18.9	8.5	3.1	1.9	0.5	1.7	1.0
Total	54.8	53.5	44.8	28.6	24.7	22.9	20.4	20.4	19.1

	1989	1990	1991	1992	1993	1994	1995	1996	1997
Thousand tonnes									
NGL/LPG/Ethane	47	47	49	53	49	51	58	54	53
Naphtha	-	-	-	-	-	-	-	-	-
Motor Gasoline	360	347	344	331	308	281	282	286	308
Aviation Fuels	40	62	62	78	54	48	55	61	67
Kerosene	7	5	7	7	7	14	16	17	19
Gas Diesel	128	143	177	176	159	186	197	217	274
Heavy Fuel Oil	11	31	28	30	8	18	6	6	8
Other Products	26	21	59	29	27	17	17	21	17
Refinery Fuel	114	102	120	190	142	167	143	144	160
Sub-Total	733	758	846	894	754	782	774	806	906
Bunkers	40	35	30	37	53	52	52	52	52
Total	773	793	876	931	807	834	826	858	958
Thousand barrels/day									
NGL/LPG/Ethane	1.5	1.5	1.6	1.7	1.6	1.6	1.8	1.7	1.7
Naphtha	-	-	-	-	-	-	-	-	-
Motor Gasoline	8.4	8.1	8.0	7.7	7.2	6.6	6.6	6.7	7.2
Aviation Fuels	0.9	1.3	1.3	1.7	1.2	1.0	1.2	1.3	1.5
Kerosene	0.1	0.1	0.1	0.1	0.1	0.3	0.3	0.4	0.4
Gas Diesel	2.6	2.9	3.6	3.6	3.3	3.8	4.0	4.4	5.6
Heavy Fuel Oil	0.2	0.6	0.5	0.5	0.1	0.3	0.1	0.1	0.1
Other Products	0.5	0.4	1.1	0.5	0.5	0.3	0.3	0.4	0.3
Refinery Fuel	2.5	2.2	2.6	4.2	3.1	3.7	3.1	3.1	3.5
Sub-Total	16.7	17.2	19.0	20.0	17.1	17.6	17.6	18.1	20.3
Bunkers	0.8	0.7	0.6	0.7	1.0	1.0	1.0	1.0	1.0
Total	17.5	17.8	19.5	20.7	18.0	18.6	18.5	19.1	21.3

Oil Products Consumption
Consommation de produits pétroliers

Tunisia

	1971	1973	1978	1983	1984	1985	1986	1987	1988
Thousand tonnes									
NGL/LPG/Ethane	20	29	72	130	149	168	181	191	233
Naphtha	9	9	12	15	15	15	12	10	22
Motor Gasoline	99	118	145	189	203	220	220	216	238
Aviation Fuels	122	141	244	137	121	99	92	145	164
Kerosene	72	84	106	128	133	136	139	143	142
Gas Diesel	330	411	765	998	1 007	1 020	983	981	934
Heavy Fuel Oil	481	469	624	1 175	989	840	1 129	741	1 108
Other Products	42	46	71	79	81	63	4	4	4
Refinery Fuel	53	38	43	44	33	56	56	56	56
Sub-Total	1 228	1 345	2 082	2 895	2 731	2 617	2 816	2 487	2 901
Bunkers	19	17	12	7	5	4	2	-	-
Total	1 247	1 362	2 094	2 902	2 736	2 621	2 818	2 487	2 901
Thousand barrels/day									
NGL/LPG/Ethane	0.6	0.9	2.3	4.1	4.7	5.3	5.8	6.1	7.4
Naphtha	0.2	0.2	0.3	0.3	0.3	0.3	0.3	0.2	0.5
Motor Gasoline	2.3	2.8	3.4	4.4	4.7	5.1	5.1	5.0	5.5
Aviation Fuels	2.7	3.1	5.3	3.0	2.6	2.2	2.0	3.2	3.6
Kerosene	1.5	1.8	2.2	2.7	2.8	2.9	2.9	3.0	3.0
Gas Diesel	6.7	8.4	15.6	20.4	20.5	20.8	20.1	20.1	19.0
Heavy Fuel Oil	8.8	8.6	11.4	21.4	18.0	15.3	20.6	13.5	20.2
Other Products	0.8	0.9	1.4	1.5	1.5	1.2	0.1	0.1	0.1
Refinery Fuel	1.0	0.8	0.9	0.9	0.7	1.1	1.1	1.1	1.1
Sub-Total	24.7	27.3	42.7	58.8	56.0	54.4	58.0	52.3	60.4
Bunkers	0.4	0.3	0.2	0.1	0.1	0.1	0.0	-	-
Total	25.1	27.7	43.0	58.9	56.1	54.5	58.1	52.3	60.4

	1989	1990	1991	1992	1993	1994	1995	1996	1997
Thousand tonnes									
NGL/LPG/Ethane	258	287	287	297	282	286	302	323	342
Naphtha	20	20	25	14	-	-	-	-	-
Motor Gasoline	246	263	272	289	299	301	312	324	340
Aviation Fuels	172	179	132	187	212	244	242	249	283
Kerosene	152	142	144	153	164	152	165	168	178
Gas Diesel	1 006	1 031	1 053	1 065	1 207	1 230	1 262	1 339	1 434
Heavy Fuel Oil	1 247	1 147	1 493	1 376	1 368	989	793	809	773
Other Products	4	5	6	6	164	145	141	151	143
Refinery Fuel	56	65	56	54	62	64	68	68	68
Sub-Total	3 161	3 139	3 468	3 441	3 758	3 411	3 285	3 431	3 561
Bunkers	-	22	29	25	19	19	19	19	18
Total	3 161	3 161	3 497	3 466	3 777	3 430	3 304	3 450	3 579
Thousand barrels/day									
NGL/LPG/Ethane	8.2	9.1	9.1	9.4	9.0	9.1	9.6	10.2	10.9
Naphtha	0.5	0.5	0.6	0.3	-	-	-	-	-
Motor Gasoline	5.7	6.1	6.4	6.7	7.0	7.0	7.3	7.6	7.9
Aviation Fuels	3.7	3.9	2.9	4.1	4.6	5.3	5.3	5.4	6.1
Kerosene	3.2	3.0	3.1	3.2	3.5	3.2	3.5	3.6	3.8
Gas Diesel	20.6	21.1	21.5	21.7	24.7	25.1	25.8	27.3	29.3
Heavy Fuel Oil	22.8	20.9	27.2	25.0	25.0	18.0	14.5	14.7	14.1
Other Products	0.1	0.1	0.1	0.1	3.1	2.8	2.7	2.9	2.7
Refinery Fuel	1.1	1.3	1.1	1.1	1.2	1.3	1.4	1.4	1.4
Sub-Total	65.9	66.1	72.0	71.7	78.1	71.9	70.0	73.0	76.3
Bunkers	-	0.4	0.5	0.5	0.3	0.3	0.3	0.3	0.3
Total	65.9	66.5	72.5	72.2	78.4	72.2	70.3	73.3	76.6

Oil Products Consumption
Consommation de produits pétroliers

Turkmenistan

	1971	1973	1978	1983	1984	1985	1986	1987	1988
Thousand tonnes									
NGL/LPG/Ethane	-	-	-	-	-	-	-	-	-
Naphtha	-	-	-	-	-	-	-	-	-
Motor Gasoline	-	-	-	-	-	-	-	-	-
Aviation Fuels	-	-	-	-	-	-	-	-	-
Kerosene	-	-	-	-	-	-	-	-	-
Gas Diesel	-	-	-	-	-	-	-	-	-
Heavy Fuel Oil	-	-	-	-	-	-	-	-	-
Other Products	-	-	-	-	-	-	-	-	-
Refinery Fuel	-	-	-	-	-	-	-	-	-
Sub-Total	-	-	-	-	-	-	-	-	-
Bunkers	-	-	-	-	-	-	-	-	-
Total	-	-	-	-	-	-	-	-	-
Thousand barrels/day									
NGL/LPG/Ethane	-	-	-	-	-	-	-	-	-
Naphtha	-	-	-	-	-	-	-	-	-
Motor Gasoline	-	-	-	-	-	-	-	-	-
Aviation Fuels	-	-	-	-	-	-	-	-	-
Kerosene	-	-	-	-	-	-	-	-	-
Gas Diesel	-	-	-	-	-	-	-	-	-
Heavy Fuel Oil	-	-	-	-	-	-	-	-	-
Other Products	-	-	-	-	-	-	-	-	-
Refinery Fuel	-	-	-	-	-	-	-	-	-
Sub-Total	-	-	-	-	-	-	-	-	-
Bunkers	-	-	-	-	-	-	-	-	-
Total	-	-	-	-	-	-	-	-	-

	1989	1990	1991	1992	1993	1994	1995	1996	1997
Thousand tonnes									
NGL/LPG/Ethane	-	-	-	297	254	805	968	379	272
Naphtha	-	-	-	-	-	-	-	-	-
Motor Gasoline	-	-	-	713	527	493	451	483	443
Aviation Fuels	-	-	-	-	-	-	-	-	-
Kerosene	-	-	-	-	-	-	-	-	-
Gas Diesel	-	-	-	1 164	999	945	864	924	878
Heavy Fuel Oil	-	-	-	2 541	834	834	763	816	958
Other Products	-	-	-	-	-	-	-	-	-
Refinery Fuel	-	-	-	230	200	197	180	193	250
Sub-Total	-	-	-	4 945	2 814	3 274	3 226	2 795	2 801
Bunkers	-	-	-	-	-	-	-	-	-
Total	-	-	-	4 945	2 814	3 274	3 226	2 795	2 801
Thousand barrels/day									
NGL/LPG/Ethane	-	-	-	8.1	7.0	22.5	26.7	10.7	7.7
Naphtha	-	-	-	-	-	-	-	-	-
Motor Gasoline	-	-	-	16.6	12.3	11.5	10.5	11.3	10.4
Aviation Fuels	-	-	-	-	-	-	-	-	-
Kerosene	-	-	-	-	-	-	-	-	-
Gas Diesel	-	-	-	23.7	20.4	19.3	17.7	18.8	17.9
Heavy Fuel Oil	-	-	-	46.2	15.2	15.2	13.9	14.8	17.5
Other Products	-	-	-	-	-	-	-	-	-
Refinery Fuel	-	-	-	5.0	4.4	4.3	3.9	4.2	5.5
Sub-Total	-	-	-	99.7	59.3	72.9	72.8	59.8	58.9
Bunkers	-	-	-	-	-	-	-	-	-
Total	-	-	-	99.7	59.3	72.9	72.8	59.8	58.9

Oil Products Consumption
Consommation de produits pétroliers

Ukraine

	1971	1973	1978	1983	1984	1985	1986	1987	1988
Thousand tonnes									
NGL/LPG/Ethane	-	-	-	-	-	-	-	-	-
Naphtha	-	-	-	-	-	-	-	-	-
Motor Gasoline	-	-	-	-	-	-	-	-	-
Aviation Fuels	-	-	-	-	-	-	-	-	-
Kerosene	-	-	-	-	-	-	-	-	-
Gas Diesel	-	-	-	-	-	-	-	-	-
Heavy Fuel Oil	-	-	-	-	-	-	-	-	-
Other Products	-	-	-	-	-	-	-	-	-
Refinery Fuel	-	-	-	-	-	-	-	-	-
Sub-Total	-	-	-	-	-	-	-	-	-
Bunkers	-	-	-	-	-	-	-	-	-
Total	-	-	-	-	-	-	-	-	-
Thousand barrels/day									
NGL/LPG/Ethane	-	-	-	-	-	-	-	-	-
Naphtha	-	-	-	-	-	-	-	-	-
Motor Gasoline	-	-	-	-	-	-	-	-	-
Aviation Fuels	-	-	-	-	-	-	-	-	-
Kerosene	-	-	-	-	-	-	-	-	-
Gas Diesel	-	-	-	-	-	-	-	-	-
Heavy Fuel Oil	-	-	-	-	-	-	-	-	-
Other Products	-	-	-	-	-	-	-	-	-
Refinery Fuel	-	-	-	-	-	-	-	-	-
Sub-Total	-	-	-	-	-	-	-	-	-
Bunkers	-	-	-	-	-	-	-	-	-
Total	-	-	-	-	-	-	-	-	-

	1989	1990	1991	1992	1993	1994	1995	1996	1997
Thousand tonnes									
NGL/LPG/Ethane	-	-	-	2 681	1 275	1 439	2 281	1 702	1 584
Naphtha	-	-	-	-	-	-	-	-	-
Motor Gasoline	-	-	-	5 797	4 142	3 836	3 984	3 585	3 200
Aviation Fuels	-	-	-	718	1 202	1 043	891	784	674
Kerosene	-	-	-	75	16	14	12	11	10
Gas Diesel	-	-	-	8 826	7 344	6 308	6 775	5 055	4 795
Heavy Fuel Oil	-	-	-	17 402	11 675	7 882	8 080	6 028	5 459
Other Products	-	-	-	3 747	1 920	1 703	1 410	1 648	1 400
Refinery Fuel	-	-	-	929	621	579	538	409	384
Sub-Total	-	-	-	40 175	28 195	22 804	23 971	19 222	17 506
Bunkers	-	-	-	-	-	-	-	-	-
Total	-	-	-	40 175	28 195	22 804	23 971	19 222	17 506
Thousand barrels/day									
NGL/LPG/Ethane	-	-	-	85.0	40.5	45.7	72.5	53.9	50.3
Naphtha	-	-	-	-	-	-	-	-	-
Motor Gasoline	-	-	-	135.1	96.8	89.6	93.1	83.6	74.8
Aviation Fuels	-	-	-	15.6	26.1	22.7	19.4	17.1	14.7
Kerosene	-	-	-	1.6	0.3	0.3	0.3	0.2	0.2
Gas Diesel	-	-	-	179.9	150.1	128.9	138.5	103.0	98.0
Heavy Fuel Oil	-	-	-	316.7	213.0	143.8	147.4	109.7	99.6
Other Products	-	-	-	70.9	37.6	32.7	27.0	31.3	26.8
Refinery Fuel	-	-	-	18.5	12.5	11.6	10.7	8.1	7.7
Sub-Total	-	-	-	823.2	577.0	475.3	508.8	407.0	372.2
Bunkers	-	-	-	-	-	-	-	-	-
Total	-	-	-	823.2	577.0	475.3	508.8	407.0	372.2

Oil Products Consumption
Consommation de produits pétroliers

United Arab Emirates

	1971	1973	1978	1983	1984	1985	1986	1987	1988
Thousand tonnes									
NGL/LPG/Ethane	-	-	6	21	25	29	30	40	80
Naphtha	-	-	-	-	-	-	-	-	-
Motor Gasoline	70	110	437	684	620	770	890	840	929
Aviation Fuels	20	41	325	518	530	540	540	560	200
Kerosene	8	10	15	35	35	35	35	35	35
Gas Diesel	110	260	1 522	2 256	2 300	2 300	2 300	2 400	2 350
Heavy Fuel Oil	2	2	68	1 338	1 473	1 920	1 810	1 816	1 800
Other Products	-	-	-	-	-	-	-	-	-
Refinery Fuel	-	-	66	290	330	340	350	400	400
Sub-Total	210	423	2 439	5 142	5 313	5 934	5 955	6 091	5 794
Bunkers	5	10	203	1 546	1 531	3 141	4 341	4 576	6 000
Total	215	433	2 642	6 688	6 844	9 075	10 296	10 667	11 794
Thousand barrels/day									
NGL/LPG/Ethane	-	-	0.2	0.7	0.8	0.9	1.0	1.3	2.5
Naphtha	-	-	-	-	-	-	-	-	-
Motor Gasoline	1.6	2.6	10.2	16.0	14.5	18.0	20.8	19.6	21.7
Aviation Fuels	0.4	0.9	7.1	11.3	11.5	11.7	11.7	12.2	4.3
Kerosene	0.2	0.2	0.3	0.7	0.7	0.7	0.7	0.7	0.7
Gas Diesel	2.2	5.3	31.1	46.1	46.9	47.0	47.0	49.1	47.9
Heavy Fuel Oil	0.0	0.0	1.2	24.4	26.8	35.0	33.0	33.1	32.8
Other Products	-	-	-	-	-	-	-	-	-
Refinery Fuel	-	-	1.4	5.6	6.6	6.8	7.0	8.0	8.2
Sub-Total	4.5	9.0	51.5	104.8	107.7	120.3	121.3	124.0	118.1
Bunkers	0.1	0.2	3.9	28.4	28.2	57.6	79.5	83.8	109.5
Total	4.6	9.2	55.4	133.2	135.9	177.9	200.8	207.9	227.6

	1989	1990	1991	1992	1993	1994	1995	1996	1997
Thousand tonnes									
NGL/LPG/Ethane	100	110	96	260	350	390	396	368	385
Naphtha	-	-	-	-	-	-	-	-	-
Motor Gasoline	1 030	850	838	850	880	1 078	1 158	1 073	1 171
Aviation Fuels	200	200	220	220	210	240	223	209	192
Kerosene	35	35	35	35	37	37	34	32	29
Gas Diesel	2 349	2 388	2 450	2 496	2 672	2 656	2 566	2 365	1 797
Heavy Fuel Oil	1 850	1 850	1 850	2 000	2 143	2 129	1 944	1 921	1 935
Other Products	-	-	-	-	-	-	-	-	1
Refinery Fuel	400	450	450	450	450	474	524	535	576
Sub-Total	5 964	5 883	5 939	6 311	6 742	7 004	6 845	6 503	6 086
Bunkers	6 245	6 165	9 711	9 710	9 700	10 757	10 767	10 767	10 767
Total	12 209	12 048	15 650	16 021	16 442	17 761	17 612	17 270	16 853
Thousand barrels/day									
NGL/LPG/Ethane	3.2	3.5	3.1	8.2	11.1	12.4	12.6	11.7	12.2
Naphtha	-	-	-	-	-	-	-	-	-
Motor Gasoline	24.1	19.9	19.6	19.8	20.6	25.2	27.1	25.0	27.4
Aviation Fuels	4.3	4.3	4.8	4.8	4.6	5.2	4.8	4.5	4.2
Kerosene	0.7	0.7	0.7	0.7	0.8	0.8	0.7	0.7	0.6
Gas Diesel	48.0	48.8	50.1	50.9	54.6	54.3	52.4	48.2	36.7
Heavy Fuel Oil	33.8	33.8	33.8	36.4	39.1	38.8	35.5	35.0	35.3
Other Products	-	-	-	-	-	-	-	-	0.0
Refinery Fuel	8.2	9.3	9.3	9.3	9.3	9.8	10.9	11.1	12.0
Sub-Total	122.3	120.3	121.3	130.1	140.1	146.6	144.0	136.1	128.4
Bunkers	114.2	112.8	177.5	177.0	177.2	196.5	196.7	196.1	196.7
Total	236.5	233.1	298.8	307.1	317.3	343.0	340.7	332.2	325.1

Oil Products Consumption
Consommation de produits pétroliers

Uruguay

	1971	1973	1978	1983	1984	1985	1986	1987	1988
Thousand tonnes									
NGL/LPG/Ethane	37	35	38	46	44	45	47	53	64
Naphtha	-	-	11	11	11	12	13	13	14
Motor Gasoline	257	225	203	180	175	174	177	187	200
Aviation Fuels	23	22	30	12	10	10	12	15	12
Kerosene	200	193	144	82	68	55	51	51	52
Gas Diesel	345	364	455	449	409	386	378	391	412
Heavy Fuel Oil	710	772	822	310	264	246	238	292	558
Other Products	89	86	112	62	48	43	49	72	68
Refinery Fuel	57	57	65	58	50	45	39	70	69
Sub-Total	1 718	1 754	1 880	1 210	1 079	1 016	1 004	1 144	1 449
Bunkers	88	105	82	86	110	105	78	96	95
Total	1 806	1 859	1 962	1 296	1 189	1 121	1 082	1 240	1 544
Thousand barrels/day									
NGL/LPG/Ethane	1.2	1.1	1.2	1.5	1.4	1.4	1.5	1.7	2.0
Naphtha	-	-	0.3	0.3	0.3	0.3	0.3	0.3	0.3
Motor Gasoline	6.0	5.3	4.7	4.2	4.1	4.1	4.1	4.4	4.7
Aviation Fuels	0.5	0.5	0.7	0.3	0.2	0.2	0.3	0.3	0.3
Kerosene	4.2	4.1	3.1	1.7	1.4	1.2	1.1	1.1	1.1
Gas Diesel	7.1	7.4	9.3	9.2	8.3	7.9	7.7	8.0	8.4
Heavy Fuel Oil	13.0	14.1	15.0	5.7	4.8	4.5	4.3	5.3	10.2
Other Products	1.6	1.5	2.0	1.2	0.9	0.8	0.9	1.4	1.3
Refinery Fuel	1.1	1.1	1.3	1.1	1.0	0.9	0.8	1.4	1.3
Sub-Total	34.6	35.1	37.5	25.1	22.4	21.3	21.1	23.8	29.6
Bunkers	1.7	2.0	1.6	1.6	2.1	2.0	1.5	1.9	1.9
Total	36.3	37.1	39.1	26.7	24.5	23.3	22.6	25.7	31.5

	1989	1990	1991	1992	1993	1994	1995	1996	1997
Thousand tonnes									
NGL/LPG/Ethane	67	70	75	78	80	85	83	96	100
Naphtha	14	15	15	16	15	15	15	16	11
Motor Gasoline	214	215	228	243	266	291	304	315	322
Aviation Fuels	12	11	15	16	17	18	15	8	7
Kerosene	50	49	53	46	43	35	33	32	25
Gas Diesel	449	429	452	649	652	612	616	705	732
Heavy Fuel Oil	616	287	386	286	237	179	255	347	350
Other Products	66	56	69	65	62	61	56	81	84
Refinery Fuel	70	71	81	98	31	11	71	104	95
Sub-Total	1 558	1 203	1 374	1 497	1 403	1 307	1 448	1 704	1 726
Bunkers	89	115	139	199	306	183	368	379	307
Total	1 647	1 318	1 513	1 696	1 709	1 490	1 816	2 083	2 033
Thousand barrels/day									
NGL/LPG/Ethane	2.1	2.2	2.4	2.5	2.5	2.7	2.6	3.0	3.2
Naphtha	0.3	0.3	0.3	0.4	0.3	0.3	0.3	0.4	0.3
Motor Gasoline	5.0	5.0	5.3	5.7	6.2	6.8	7.1	7.3	7.5
Aviation Fuels	0.3	0.2	0.3	0.4	0.4	0.4	0.3	0.2	0.2
Kerosene	1.1	1.0	1.1	1.0	0.9	0.7	0.7	0.7	0.5
Gas Diesel	9.2	8.8	9.2	13.2	13.3	12.5	12.6	14.4	15.0
Heavy Fuel Oil	11.2	5.2	7.0	5.2	4.3	3.3	4.7	6.3	6.4
Other Products	1.3	1.1	1.3	1.2	1.2	1.2	1.1	1.5	1.6
Refinery Fuel	1.4	1.4	1.6	1.9	0.6	0.2	1.4	2.0	1.8
Sub-Total	31.8	25.4	28.7	31.4	29.8	28.1	30.8	35.9	36.4
Bunkers	1.8	2.3	2.7	3.8	6.1	3.6	6.9	7.3	5.9
Total	33.6	27.6	31.4	35.2	35.9	31.7	37.8	43.2	42.3

Oil Products Consumption
Consommation de produits pétroliers

Uzbekistan

	1971	1973	1978	1983	1984	1985	1986	1987	1988
Thousand tonnes									
NGL/LPG/Ethane	-	-	-	-	-	-	-	-	-
Naphtha	-	-	-	-	-	-	-	-	-
Motor Gasoline	-	-	-	-	-	-	-	-	-
Aviation Fuels	-	-	-	-	-	-	-	-	-
Kerosene	-	-	-	-	-	-	-	-	-
Gas Diesel	-	-	-	-	-	-	-	-	-
Heavy Fuel Oil	-	-	-	-	-	-	-	-	-
Other Products	-	-	-	-	-	-	-	-	-
Refinery Fuel	-	-	-	-	-	-	-	-	-
Sub-Total	-	-	-	-	-	-	-	-	-
Bunkers	-	-	-	-	-	-	-	-	-
Total	-	-	-	-	-	-	-	-	-
Thousand barrels/day									
NGL/LPG/Ethane	-	-	-	-	-	-	-	-	-
Naphtha	-	-	-	-	-	-	-	-	-
Motor Gasoline	-	-	-	-	-	-	-	-	-
Aviation Fuels	-	-	-	-	-	-	-	-	-
Kerosene	-	-	-	-	-	-	-	-	-
Gas Diesel	-	-	-	-	-	-	-	-	-
Heavy Fuel Oil	-	-	-	-	-	-	-	-	-
Other Products	-	-	-	-	-	-	-	-	-
Refinery Fuel	-	-	-	-	-	-	-	-	-
Sub-Total	-	-	-	-	-	-	-	-	-
Bunkers	-	-	-	-	-	-	-	-	-
Total	-	-	-	-	-	-	-	-	-

	1989	1990	1991	1992	1993	1994	1995	1996	1997
Thousand tonnes									
NGL/LPG/Ethane	-	-	-	-	-	-	53	60	57
Naphtha	-	-	-	273	-	-	-	-	-
Motor Gasoline	-	-	-	2 102	1 939	1 889	1 225	1 178	1 407
Aviation Fuels	-	-	-	83	56	89	317	312	277
Kerosene	-	-	-	-	-	-	66	46	69
Gas Diesel	-	-	-	2 608	2 733	2 394	2 061	1 946	2 011
Heavy Fuel Oil	-	-	-	2 170	1 919	1 525	1 981	2 116	2 048
Other Products	-	-	-	1 000	925	898	808	764	944
Refinery Fuel	-	-	-	246	130	117	205	198	221
Sub-Total	-	-	-	8 482	7 702	6 912	6 716	6 620	7 034
Bunkers	-	-	-	-	-	-	-	-	-
Total	-	-	-	8 482	7 702	6 912	6 716	6 620	7 034
Thousand barrels/day									
NGL/LPG/Ethane	-	-	-	-	-	-	1.6	1.8	1.7
Naphtha	-	-	-	6.3	-	-	-	-	-
Motor Gasoline	-	-	-	49.0	45.3	44.1	28.6	27.5	32.9
Aviation Fuels	-	-	-	1.8	1.2	1.9	6.9	6.8	6.0
Kerosene	-	-	-	-	-	-	1.4	1.0	1.5
Gas Diesel	-	-	-	53.2	55.9	48.9	42.1	39.7	41.1
Heavy Fuel Oil	-	-	-	39.5	35.0	27.8	36.1	38.5	37.4
Other Products	-	-	-	19.1	17.8	17.3	15.3	14.5	18.0
Refinery Fuel	-	-	-	4.6	2.8	2.6	4.5	4.3	4.8
Sub-Total	-	-	-	173.5	158.1	142.7	136.6	134.0	143.4
Bunkers	-	-	-	-	-	-	-	-	-
Total	-	-	-	173.5	158.1	142.7	136.6	134.0	143.4

Oil Products Consumption
Consommation de produits pétroliers

Venezuela

	1971	1973	1978	1983	1984	1985	1986	1987	1988
Thousand tonnes									
NGL/LPG/Ethane	158	220	431	810	899	1 230	1 064	1 068	1 120
Naphtha	-	-	-	-	-	-	-	-	-
Motor Gasoline	3 176	3 715	6 097	7 227	7 052	7 053	7 058	7 675	7 467
Aviation Fuels	257	365	544	518	446	440	476	522	644
Kerosene	471	468	463	385	388	375	374	389	396
Gas Diesel	1 165	1 633	2 730	3 485	2 845	2 493	2 842	2 879	3 081
Heavy Fuel Oil	1 230	1 388	1 400	2 771	2 587	2 780	2 835	3 063	2 178
Other Products	344	432	1 179	1 051	1 019	1 197	1 533	1 704	1 807
Refinery Fuel	1 123	1 304	1 296	1 371	1 644	1 404	1 548	1 453	1 286
Sub-Total	7 924	9 525	14 140	17 618	16 880	16 972	17 730	18 753	17 979
Bunkers	2 699	2 814	872	521	550	539	465	473	783
Total	10 623	12 339	15 012	18 139	17 430	17 511	18 195	19 226	18 762
Thousand barrels/day									
NGL/LPG/Ethane	5.0	7.0	13.7	25.7	28.5	39.1	33.8	33.9	35.5
Naphtha	-	-	-	-	-	-	-	-	-
Motor Gasoline	74.2	86.8	142.5	168.9	164.4	164.8	164.9	179.4	174.0
Aviation Fuels	5.7	8.0	11.9	11.3	9.7	9.6	10.4	11.4	14.0
Kerosene	10.0	9.9	9.8	8.2	8.2	8.0	7.9	8.2	8.4
Gas Diesel	23.8	33.4	55.8	71.2	58.0	51.0	58.1	58.8	62.8
Heavy Fuel Oil	22.4	25.3	25.5	50.6	47.1	50.7	51.7	55.9	39.6
Other Products	6.6	8.3	22.6	20.2	18.3	21.6	27.3	30.6	32.2
Refinery Fuel	23.2	27.3	26.5	29.4	34.8	29.9	33.3	31.2	27.7
Sub-Total	170.9	206.0	308.4	385.4	369.0	374.8	387.6	409.5	394.2
Bunkers	49.5	51.6	16.1	9.7	10.2	10.0	8.7	8.8	14.5
Total	220.5	257.7	324.5	395.1	379.2	384.8	396.2	418.2	408.7

	1989	1990	1991	1992	1993	1994	1995	1996	1997
Thousand tonnes									
NGL/LPG/Ethane	1 110	1 162	1 234	1 822	1 596	2 498	2 227	1 949	2 437
Naphtha	-	-	-	-	-	-	-	-	-
Motor Gasoline	6 929	7 044	7 411	7 488	8 265	8 242	8 852	8 118	9 201
Aviation Fuels	634	664	660	720	716	680	709	761	778
Kerosene	325	315	302	270	262	260	262	273	264
Gas Diesel	2 891	3 186	3 296	2 931	3 651	5 282	3 854	4 207	4 558
Heavy Fuel Oil	2 504	2 058	1 327	1 284	2 265	1 611	1 202	1 104	1 196
Other Products	1 425	1 471	5 282	2 754	1 700	2 150	4 706	5 033	1 496
Refinery Fuel	1 331	1 470	1 643	1 492	2 833	1 181	431	397	422
Sub-Total	17 149	17 370	21 155	18 761	21 288	21 904	22 243	21 842	20 352
Bunkers	758	791	943	933	823	763	665	665	665
Total	17 907	18 161	22 098	19 694	22 111	22 667	22 908	22 507	21 017
Thousand barrels/day									
NGL/LPG/Ethane	35.3	36.9	39.2	57.7	50.7	79.4	70.8	61.8	77.5
Naphtha	-	-	-	-	-	-	-	-	-
Motor Gasoline	161.9	164.6	173.2	174.5	193.2	192.6	206.9	189.2	215.0
Aviation Fuels	13.8	14.5	14.4	15.7	15.6	14.8	15.5	16.5	17.0
Kerosene	6.9	6.7	6.4	5.7	5.6	5.5	5.6	5.8	5.6
Gas Diesel	59.1	65.1	67.4	59.7	74.6	108.0	78.8	85.7	93.2
Heavy Fuel Oil	45.7	37.6	24.2	23.4	41.3	29.4	21.9	20.1	21.8
Other Products	26.1	27.0	99.7	51.1	31.5	40.6	88.8	94.8	27.7
Refinery Fuel	28.8	31.8	35.9	32.4	60.2	25.5	9.3	8.6	9.2
Sub-Total	377.6	384.2	460.3	420.2	472.7	495.8	497.5	482.5	467.0
Bunkers	14.1	14.7	17.5	17.3	15.3	14.2	12.3	12.3	12.3
Total	391.6	398.8	477.8	437.5	488.1	510.0	509.8	494.8	479.2

Oil Products Consumption
Consommation de produits pétroliers

Vietnam

	1971	1973	1978	1983	1984	1985	1986	1987	1988
Thousand tonnes									
NGL/LPG/Ethane	5	14	3	5	5	-	-	-	-
Naphtha	-	-	-	-	-	-	-	-	-
Motor Gasoline	922	920	185	312	303	330	343	401	505
Aviation Fuels	2 247	2 000	-	99	86	78	115	120	134
Kerosene	50	50	80	165	167	153	150	182	185
Gas Diesel	1 664	1 460	295	811	776	843	874	1 043	1 029
Heavy Fuel Oil	747	1 200	325	442	462	470	556	613	584
Other Products	25	10	45	40	40	-	61	73	74
Refinery Fuel	-	-	-	-	-	-	-	-	2
Sub-Total	5 660	5 654	933	1 874	1 839	1 874	2 099	2 432	2 513
Bunkers	-	-	-	-	-	-	-	-	-
Total	5 660	5 654	933	1 874	1 839	1 874	2 099	2 432	2 513
Thousand barrels/day									
NGL/LPG/Ethane	0.2	0.4	0.1	0.2	0.2	-	-	-	-
Naphtha	-	-	-	-	-	-	-	-	-
Motor Gasoline	21.5	21.5	4.3	7.3	7.1	7.7	8.0	9.4	11.8
Aviation Fuels	50.8	44.8	-	2.2	1.9	1.7	2.5	2.6	2.9
Kerosene	1.1	1.1	1.7	3.5	3.5	3.2	3.2	3.9	3.9
Gas Diesel	34.0	29.8	6.0	16.6	15.8	17.2	17.9	21.3	21.0
Heavy Fuel Oil	13.6	21.9	5.9	8.1	8.4	8.6	10.1	11.2	10.6
Other Products	0.5	0.2	0.9	0.8	0.8	-	1.2	1.4	1.4
Refinery Fuel	-	-	-	-	-	-	-	-	0.0
Sub-Total	121.7	119.7	18.9	38.5	37.6	38.5	42.9	49.8	51.7
Bunkers	-	-	-	-	-	-	-	-	-
Total	121.7	119.7	18.9	38.5	37.6	38.5	42.9	49.8	51.7

	1989	1990	1991	1992	1993	1994	1995	1996	1997
Thousand tonnes									
NGL/LPG/Ethane	-	-	-	-	-	11	12	53	120
Naphtha	-	-	-	-	-	-	-	-	-
Motor Gasoline	431	653	521	630	874	972	1 040	1 250	1 605
Aviation Fuels	111	102	104	173	166	177	178	254	279
Kerosene	215	214	174	159	191	201	203	331	294
Gas Diesel	974	1 242	1 094	1 311	1 923	2 005	1 782	2 783	3 616
Heavy Fuel Oil	518	564	601	945	747	784	671	1 222	1 518
Other Products	65	29	28	-	-	160	173	201	118
Refinery Fuel	2	2	2	2	-	-	-	2	-
Sub-Total	2 316	2 806	2 524	3 220	3 901	4 310	4 059	6 096	7 550
Bunkers	-	-	-	-	-	-	-	-	-
Total	2 316	2 806	2 524	3 220	3 901	4 310	4 059	6 096	7 550
Thousand barrels/day									
NGL/LPG/Ethane	-	-	-	-	-	0.4	0.4	1.7	3.8
Naphtha	-	-	-	-	-	-	-	-	-
Motor Gasoline	10.1	15.3	12.2	14.7	20.4	22.7	24.3	29.1	37.5
Aviation Fuels	2.4	2.2	2.3	3.7	3.6	3.8	3.9	5.5	6.1
Kerosene	4.6	4.5	3.7	3.4	4.1	4.3	4.3	7.0	6.2
Gas Diesel	19.9	25.4	22.4	26.7	39.3	41.0	36.4	56.7	73.9
Heavy Fuel Oil	9.5	10.3	11.0	17.2	13.6	14.3	12.2	22.2	27.7
Other Products	1.3	0.6	0.5	-	-	3.1	3.4	3.9	2.3
Refinery Fuel	0.0	0.0	0.0	0.0	-	-	-	0.0	-
Sub-Total	47.7	58.3	52.0	65.8	81.0	89.6	84.9	126.2	157.5
Bunkers	-	-	-	-	-	-	-	-	-
Total	47.7	58.3	52.0	65.8	81.0	89.6	84.9	126.2	157.5

Oil Products Consumption
Consommation de produits pétroliers

Yemen

	1971	1973	1978	1983	1984	1985	1986	1987	1988
Thousand tonnes									
NGL/LPG/Ethane	2	6	8	54	61	57	68	74	93
Naphtha	-	-	-	-	-	-	-	-	-
Motor Gasoline	145	165	206	280	295	319	408	472	492
Aviation Fuels	27	32	71	130	128	146	151	161	171
Kerosene	53	48	95	145	152	202	140	167	179
Gas Diesel	72	238	337	693	695	724	773	782	838
Heavy Fuel Oil	54	54	165	330	336	492	476	417	424
Other Products	-	-	-	-	-	-	50	100	100
Refinery Fuel	58	84	68	120	119	124	127	112	112
Sub-Total	411	627	950	1 752	1 786	2 064	2 193	2 285	2 409
Bunkers	364	320	660	300	390	400	380	400	410
Total	775	947	1 610	2 052	2 176	2 464	2 573	2 685	2 819
Thousand barrels/day									
NGL/LPG/Ethane	0.1	0.2	0.3	1.7	1.9	1.8	2.2	2.4	2.9
Naphtha	-	-	-	-	-	-	-	-	-
Motor Gasoline	3.4	3.9	4.8	6.5	6.9	7.5	9.5	11.0	11.5
Aviation Fuels	0.6	0.7	1.5	2.8	2.8	3.2	3.3	3.5	3.7
Kerosene	1.1	1.0	2.0	3.1	3.2	4.3	3.0	3.5	3.8
Gas Diesel	1.5	4.9	6.9	14.2	14.2	14.8	15.8	16.0	17.1
Heavy Fuel Oil	1.0	1.0	3.0	6.0	6.1	9.0	8.7	7.6	7.7
Other Products	-	-	-	-	-	-	1.0	1.9	1.9
Refinery Fuel	1.1	1.5	1.2	2.2	2.2	2.3	2.3	2.0	2.0
Sub-Total	8.7	13.1	19.8	36.5	37.2	42.8	45.7	48.0	50.7
Bunkers	6.8	6.0	12.2	5.7	7.3	7.5	7.1	7.5	7.7
Total	15.5	19.1	32.0	42.2	44.6	50.3	52.8	55.5	58.3

	1989	1990	1991	1992	1993	1994	1995	1996	1997
Thousand tonnes									
NGL/LPG/Ethane	120	152	179	243	292	330	356	387	428
Naphtha	-	-	-	-	-	-	-	-	-
Motor Gasoline	490	804	857	924	814	857	1 073	1 045	1 055
Aviation Fuels	71	55	56	83	83	88	93	88	93
Kerosene	179	74	116	121	109	116	116	114	114
Gas Diesel	900	603	832	868	672	712	713	680	709
Heavy Fuel Oil	430	228	448	528	286	305	305	343	356
Other Products	68	68	43	48	58	58	58	60	44
Refinery Fuel	110	110	110	115	110	115	115	129	134
Sub-Total	2 368	2 094	2 641	2 930	2 424	2 581	2 829	2 846	2 933
Bunkers	420	400	320	375	110	100	100	100	100
Total	2 788	2 494	2 961	3 305	2 534	2 681	2 929	2 946	3 033
Thousand barrels/day									
NGL/LPG/Ethane	3.8	4.8	5.7	7.7	9.3	10.5	11.3	12.3	13.6
Naphtha	-	-	-	-	-	-	-	-	-
Motor Gasoline	11.5	18.8	20.0	21.5	19.0	20.0	25.1	24.4	24.7
Aviation Fuels	1.5	1.2	1.2	1.8	1.8	1.9	2.0	1.9	2.0
Kerosene	3.8	1.6	2.5	2.6	2.3	2.5	2.5	2.4	2.4
Gas Diesel	18.4	12.3	17.0	17.7	13.7	14.6	14.6	13.9	14.5
Heavy Fuel Oil	7.8	4.2	8.2	9.6	5.2	5.6	5.6	6.2	6.5
Other Products	1.3	1.3	0.7	0.8	1.0	1.0	1.0	1.0	0.7
Refinery Fuel	2.0	2.0	2.0	2.1	2.0	2.1	2.1	2.3	2.4
Sub-Total	50.2	46.2	57.3	63.8	54.3	58.1	64.1	64.4	66.9
Bunkers	7.9	7.5	6.0	7.0	2.1	1.9	1.9	1.9	1.9
Total	58.1	53.7	63.3	70.8	56.5	60.0	66.0	66.3	68.8

Oil Products Consumption
Consommation de produits pétroliers

Federal Republic of Yugoslavia

	1971	1973	1978	1983	1984	1985	1986	1987	1988
Thousand tonnes									
NGL/LPG/Ethane	-	-	-	-	-	-	-	-	-
Naphtha	-	-	-	-	-	-	-	-	-
Motor Gasoline	-	-	-	-	-	-	-	-	-
Aviation Fuels	-	-	-	-	-	-	-	-	-
Kerosene	-	-	-	-	-	-	-	-	-
Gas Diesel	-	-	-	-	-	-	-	-	-
Heavy Fuel Oil	-	-	-	-	-	-	-	-	-
Other Products	-	-	-	-	-	-	-	-	-
Refinery Fuel	-	-	-	-	-	-	-	-	-
Sub-Total	-	-	-	-	-	-	-	-	-
Bunkers	-	-	-	-	-	-	-	-	-
Total	-	-	-	-	-	-	-	-	-
Thousand barrels/day									
NGL/LPG/Ethane	-	-	-	-	-	-	-	-	-
Naphtha	-	-	-	-	-	-	-	-	-
Motor Gasoline	-	-	-	-	-	-	-	-	-
Aviation Fuels	-	-	-	-	-	-	-	-	-
Kerosene	-	-	-	-	-	-	-	-	-
Gas Diesel	-	-	-	-	-	-	-	-	-
Heavy Fuel Oil	-	-	-	-	-	-	-	-	-
Other Products	-	-	-	-	-	-	-	-	-
Refinery Fuel	-	-	-	-	-	-	-	-	-
Sub-Total	-	-	-	-	-	-	-	-	-
Bunkers	-	-	-	-	-	-	-	-	-
Total	-	-	-	-	-	-	-	-	-

	1989	1990	1991	1992	1993	1994	1995	1996	1997
Thousand tonnes									
NGL/LPG/Ethane	-	-	55	25	10	10	11	27	47
Naphtha	-	-	599	200	164	155	173	226	385
Motor Gasoline	-	-	462	300	160	154	166	434	670
Aviation Fuels	-	-	100	60	40	38	42	55	95
Kerosene	-	-	36	25	27	25	28	37	54
Gas Diesel	-	-	473	310	246	234	260	340	520
Heavy Fuel Oil	-	-	1 162	750	565	540	591	772	909
Other Products	-	-	309	249	213	201	224	319	292
Refinery Fuel	-	-	-	-	-	-	-	-	-
Sub-Total	-	-	3 196	1 919	1 425	1 357	1 495	2 210	2 972
Bunkers	-	-	-	-	-	-	-	-	-
Total	-	-	3 196	1 919	1 425	1 357	1 495	2 210	2 972
Thousand barrels/day									
NGL/LPG/Ethane	-	-	1.7	0.8	0.3	0.3	0.4	0.9	1.5
Naphtha	-	-	13.9	4.6	3.8	3.6	4.0	5.2	9.0
Motor Gasoline	-	-	10.8	7.0	3.7	3.6	3.9	10.1	15.7
Aviation Fuels	-	-	2.2	1.3	0.9	0.8	0.9	1.2	2.1
Kerosene	-	-	0.8	0.5	0.6	0.5	0.6	0.8	1.1
Gas Diesel	-	-	9.7	6.3	5.0	4.8	5.3	6.9	10.6
Heavy Fuel Oil	-	-	21.2	13.6	10.3	9.9	10.8	14.0	16.6
Other Products	-	-	5.8	4.7	4.0	3.8	4.3	6.0	5.3
Refinery Fuel	-	-	-	-	-	-	-	-	-
Sub-Total	-	-	66.1	38.9	28.7	27.3	30.1	45.2	61.9
Bunkers	-	-	-	-	-	-	-	-	-
Total	-	-	66.1	38.9	28.7	27.3	30.1	45.2	61.9

Oil Products Consumption
Consommation de produits pétroliers

Former Yugoslavia

	1971	1973	1978	1983	1984	1985	1986	1987	1988
Thousand tonnes									
NGL/LPG/Ethane	74	203	339	386	408	365	377	417	434
Naphtha	-	399	574	717	668	474	647	828	1 049
Motor Gasoline	1 343	1 474	2 254	1 721	1 652	1 700	2 190	1 954	2 276
Aviation Fuels	211	234	271	249	331	326	356	369	391
Kerosene	-	-	-	-	-	-	-	-	-
Gas Diesel	2 999	2 874	3 392	4 056	3 389	3 860	4 305	4 113	4 240
Heavy Fuel Oil	2 943	3 729	5 633	4 259	4 202	4 035	4 320	4 388	5 325
Other Products	643	643	1 170	1 717	957	1 284	1 109	1 548	1 456
Refinery Fuel	306	340	757	1 018	1 298	1 203	1 371	1 438	1 546
Sub-Total	8 519	9 896	14 390	14 123	12 905	13 247	14 675	15 055	16 717
Bunkers	-	-	-	-	-	-	-	-	-
Total	8 519	9 896	14 390	14 123	12 905	13 247	14 675	15 055	16 717
Thousand barrels/day									
NGL/LPG/Ethane	2.4	6.5	10.8	12.3	12.9	11.6	12.0	13.3	13.8
Naphtha	-	9.3	13.4	16.7	15.5	11.0	15.1	19.3	24.4
Motor Gasoline	31.4	34.4	52.7	40.2	38.5	39.7	51.2	45.7	53.0
Aviation Fuels	4.6	5.1	5.9	5.4	7.2	7.1	7.8	8.0	8.5
Kerosene	-	-	-	-	-	-	-	-	-
Gas Diesel	61.3	58.7	69.3	82.9	69.1	78.9	88.0	84.1	86.4
Heavy Fuel Oil	53.7	68.0	102.8	77.7	76.5	73.6	78.8	80.1	96.9
Other Products	12.3	12.3	22.4	32.9	18.3	24.6	21.3	29.7	27.8
Refinery Fuel	5.6	6.2	14.9	19.7	24.7	22.9	26.2	27.6	29.6
Sub-Total	171.3	200.6	292.2	287.8	262.7	269.5	300.3	307.6	340.4
Bunkers	-	-	-	-	-	-	-	-	-
Total	171.3	200.6	292.2	287.8	262.7	269.5	300.3	307.6	340.4

	1989	1990	1991	1992	1993	1994	1995	1996	1997
Thousand tonnes									
NGL/LPG/Ethane	398	414	504	267	257	257	264	264	330
Naphtha	1 040	790	780	338	1 270	615	652	733	694
Motor Gasoline	2 330	2 299	2 616	1 654	1 624	1 820	1 884	2 281	2 590
Aviation Fuels	395	376	254	119	168	201	214	213	262
Kerosene	-	-	-	27	37	26	36	44	60
Gas Diesel	3 438	3 319	2 686	2 470	2 663	2 492	2 870	3 372	3 648
Heavy Fuel Oil	5 299	5 450	3 775	2 448	2 226	2 097	2 260	2 609	2 614
Other Products	1 560	1 676	1 608	458	453	507	530	592	602
Refinery Fuel	1 463	1 394	1 060	448	484	499	574	499	495
Sub-Total	15 923	15 718	13 283	8 229	9 182	8 514	9 284	10 607	11 295
Bunkers	-	-	-	-	-	45	33	29	24
Total	15 923	15 718	13 283	8 229	9 182	8 559	9 317	10 636	11 319
Thousand barrels/day									
NGL/LPG/Ethane	12.6	13.2	15.1	9.2	9.4	9.5	9.8	9.6	11.8
Naphtha	24.2	18.4	18.2	7.9	29.6	14.3	15.2	17.0	16.2
Motor Gasoline	54.5	53.7	61.1	38.5	38.0	42.5	44.0	53.2	60.5
Aviation Fuels	8.6	8.2	5.5	2.6	3.7	4.4	4.7	4.6	5.7
Kerosene	-	-	-	0.6	0.8	0.6	0.8	0.9	1.3
Gas Diesel	70.3	67.8	54.9	50.3	54.4	50.9	58.7	68.7	74.6
Heavy Fuel Oil	96.7	99.4	68.9	44.5	40.6	38.3	41.2	47.5	47.7
Other Products	29.9	32.1	30.2	8.5	8.6	9.6	10.1	11.2	10.9
Refinery Fuel	28.0	26.8	20.1	8.6	9.5	10.1	11.6	10.1	9.8
Sub-Total	324.8	319.7	274.0	170.7	194.4	180.2	196.0	222.8	238.4
Bunkers	-	-	-	-	-	0.9	0.6	0.6	0.5
Total	324.8	319.7	274.0	170.7	194.4	181.0	196.7	223.3	238.9

Oil Products Consumption
Consommation de produits pétroliers

Zambia

	1971	1973	1978	1983	1984	1985	1986	1987	1988
Thousand tonnes									
NGL/LPG/Ethane	1	3	3	7	4	4	3	3	4
Naphtha	5	5	-	-	-	-	-	-	-
Motor Gasoline	146	171	144	111	117	117	105	107	118
Aviation Fuels	14	21	55	54	52	52	46	51	54
Kerosene	4	8	27	35	34	35	25	31	38
Gas Diesel	271	320	269	275	225	191	188	185	185
Heavy Fuel Oil	1	85	167	110	102	95	90	90	77
Other Products	40	39	37	53	52	61	35	39	40
Refinery Fuel	-	21	41	35	35	35	32	35	35
Sub-Total	482	673	743	680	621	590	524	541	551
Bunkers	-	-	-	-	-	-	-	-	-
Total	482	673	743	680	621	590	524	541	551
Thousand barrels/day									
NGL/LPG/Ethane	0.0	0.1	0.1	0.2	0.1	0.1	0.1	0.1	0.1
Naphtha	0.1	0.1	-	-	-	-	-	-	-
Motor Gasoline	3.4	4.0	3.4	2.6	2.7	2.7	2.5	2.5	2.8
Aviation Fuels	0.3	0.5	1.2	1.2	1.1	1.1	1.0	1.1	1.2
Kerosene	0.1	0.2	0.6	0.7	0.7	0.7	0.5	0.7	0.8
Gas Diesel	5.5	6.5	5.5	5.6	4.6	3.9	3.8	3.8	3.8
Heavy Fuel Oil	0.0	1.6	3.0	2.0	1.9	1.7	1.6	1.6	1.4
Other Products	0.8	0.7	0.7	1.0	1.0	1.1	0.6	0.7	0.7
Refinery Fuel	-	0.4	0.8	0.7	0.7	0.7	0.6	0.7	0.7
Sub-Total	10.3	14.1	15.3	14.1	12.8	12.2	10.9	11.2	11.5
Bunkers	-	-	-	-	-	-	-	-	-
Total	10.3	14.1	15.3	14.1	12.8	12.2	10.9	11.2	11.5

	1989	1990	1991	1992	1993	1994	1995	1996	1997
Thousand tonnes									
NGL/LPG/Ethane	4	10	10	9	9	10	10	10	10
Naphtha	-	-	-	-	-	-	-	-	-
Motor Gasoline	110	106	109	99	101	105	107	110	110
Aviation Fuels	53	40	41	41	40	40	38	38	38
Kerosene	38	29	30	27	32	30	32	31	31
Gas Diesel	192	185	190	173	181	195	195	197	197
Heavy Fuel Oil	75	70	72	72	73	70	72	74	74
Other Products	36	33	33	31	31	32	32	32	32
Refinery Fuel	35	34	34	34	34	35	36	36	36
Sub-Total	543	507	519	486	501	517	522	528	528
Bunkers	-	-	-	-	-	-	-	-	-
Total	543	507	519	486	501	517	522	528	528
Thousand barrels/day									
NGL/LPG/Ethane	0.1	0.3	0.3	0.3	0.3	0.3	0.3	0.3	0.3
Naphtha	-	-	-	-	-	-	-	-	-
Motor Gasoline	2.6	2.5	2.5	2.3	2.4	2.5	2.5	2.6	2.6
Aviation Fuels	1.2	0.9	0.9	0.9	0.9	0.9	0.8	0.8	0.8
Kerosene	0.8	0.6	0.6	0.6	0.7	0.6	0.7	0.7	0.7
Gas Diesel	3.9	3.8	3.9	3.5	3.7	4.0	4.0	4.0	4.0
Heavy Fuel Oil	1.4	1.3	1.3	1.3	1.3	1.3	1.3	1.3	1.4
Other Products	0.7	0.6	0.6	0.6	0.6	0.6	0.6	0.6	0.6
Refinery Fuel	0.7	0.7	0.7	0.7	0.7	0.7	0.7	0.7	0.7
Sub-Total	11.3	10.6	10.9	10.1	10.5	10.8	10.9	11.0	11.1
Bunkers	-	-	-	-	-	-	-	-	-
Total	11.3	10.6	10.9	10.1	10.5	10.8	10.9	11.0	11.1

Oil Products Consumption
Consommation de produits pétroliers

Zimbabwe

	1971	1973	1978	1983	1984	1985	1986	1987	1988
Thousand tonnes									
NGL/LPG/Ethane	4	5	6	3	3	5	5	4	6
Naphtha	-	-	-	-	-	-	-	-	-
Motor Gasoline	184	234	189	210	171	185	198	164	188
Aviation Fuels	27	44	79	92	91	105	123	100	116
Kerosene	42	46	30	31	30	35	51	35	36
Gas Diesel	240	301	315	375	395	411	402	487	476
Heavy Fuel Oil	-	-	-	1	1	1	-	-	-
Other Products	42	50	39	38	16	16	18	17	17
Refinery Fuel	-	-	-	-	-	-	-	-	-
Sub-Total	539	680	658	750	707	758	797	807	839
Bunkers	-	-	-	-	-	-	-	-	-
Total	539	680	658	750	707	758	797	807	839
Thousand barrels/day									
NGL/LPG/Ethane	0.1	0.2	0.2	0.1	0.1	0.2	0.2	0.1	0.2
Naphtha	-	-	-	-	-	-	-	-	-
Motor Gasoline	4.3	5.5	4.4	4.9	4.0	4.3	4.6	3.8	4.4
Aviation Fuels	0.6	1.0	1.7	2.0	2.0	2.3	2.7	2.2	2.5
Kerosene	0.9	1.0	0.6	0.7	0.6	0.7	1.1	0.7	0.8
Gas Diesel	4.9	6.2	6.4	7.7	8.1	8.4	8.2	10.0	9.7
Heavy Fuel Oil	-	-	-	0.0	0.0	0.0	-	-	-
Other Products	0.8	1.0	0.7	0.7	0.3	0.3	0.3	0.3	0.3
Refinery Fuel	-	-	-	-	-	-	-	-	-
Sub-Total	11.6	14.7	14.2	16.1	15.1	16.2	17.1	17.2	17.9
Bunkers	-	-	-	-	-	-	-	-	-
Total	11.6	14.7	14.2	16.1	15.1	16.2	17.1	17.2	17.9

	1989	1990	1991	1992	1993	1994	1995	1996	1997
Thousand tonnes									
NGL/LPG/Ethane	4	6	6	6	6	7	12	12	12
Naphtha	-	-	-	-	-	-	-	-	-
Motor Gasoline	191	238	206	261	285	290	429	435	435
Aviation Fuels	104	79	139	79	72	98	110	118	118
Kerosene	43	51	43	38	51	54	54	42	42
Gas Diesel	426	516	592	584	598	731	645	634	634
Heavy Fuel Oil	-	-	-	-	-	-	-	-	-
Other Products	17	27	27	25	25	26	27	27	27
Refinery Fuel	-	-	-	-	-	-	-	-	-
Sub-Total	785	917	1 013	993	1 037	1 206	1 277	1 268	1 268
Bunkers	-	-	-	-	-	-	-	-	-
Total	785	917	1 013	993	1 037	1 206	1 277	1 268	1 268
Thousand barrels/day									
NGL/LPG/Ethane	0.1	0.2	0.2	0.2	0.2	0.2	0.4	0.4	0.4
Naphtha	-	-	-	-	-	-	-	-	-
Motor Gasoline	4.5	5.6	4.8	6.1	6.7	6.8	10.0	10.1	10.2
Aviation Fuels	2.3	1.7	3.0	1.7	1.6	2.1	2.4	2.6	2.6
Kerosene	0.9	1.1	0.9	0.8	1.1	1.1	1.1	0.9	0.9
Gas Diesel	8.7	10.5	12.1	11.9	12.2	14.9	13.2	12.9	13.0
Heavy Fuel Oil	-	-	-	-	-	-	-	-	-
Other Products	0.3	0.5	0.5	0.5	0.5	0.5	0.5	0.5	0.5
Refinery Fuel	-	-	-	-	-	-	-	-	-
Sub-Total	16.8	19.6	21.6	21.2	22.2	25.7	27.7	27.4	27.5
Bunkers	-	-	-	-	-	-	-	-	-
Total	16.8	19.6	21.6	21.2	22.2	25.7	27.7	27.4	27.5

Oil Products Consumption
Consommation de produits pétroliers

Other Africa

	1971	1973	1978	1983	1984	1985	1986	1987	1988
Thousand tonnes									
NGL/LPG/Ethane	10	12	16	51	39	38	39	76	87
Naphtha	-	-	2	3	4	5	5	6	7
Motor Gasoline	407	471	697	830	825	843	856	1 215	1 247
Aviation Fuels	182	255	330	413	425	394	416	481	493
Kerosene	166	166	175	276	264	292	290	361	377
Gas Diesel	782	844	982	1 357	1 373	1 463	1 459	1 733	1 682
Heavy Fuel Oil	638	672	734	700	703	709	755	1 097	1 129
Other Products	114	156	195	365	331	302	316	445	459
Refinery Fuel	33	34	35	-	-	-	-	-	-
Sub-Total	2 332	2 610	3 166	3 995	3 964	4 046	4 136	5 414	5 481
Bunkers	968	799	592	560	597	584	630	625	670
Total	3 300	3 409	3 758	4 555	4 561	4 630	4 766	6 039	6 151
Thousand barrels/day									
NGL/LPG/Ethane	0.3	0.4	0.5	1.6	1.2	1.2	1.2	2.4	2.8
Naphtha	-	-	0.0	0.1	0.1	0.1	0.1	0.1	0.2
Motor Gasoline	9.5	11.0	16.3	19.4	19.2	19.7	20.0	28.4	29.1
Aviation Fuels	4.0	5.6	7.3	9.1	9.4	8.7	9.2	10.6	10.8
Kerosene	3.5	3.5	3.7	5.9	5.6	6.2	6.2	7.7	8.0
Gas Diesel	16.0	17.3	20.1	27.7	28.0	29.9	29.8	35.4	34.3
Heavy Fuel Oil	11.6	12.3	13.4	12.8	12.8	12.9	13.8	20.0	20.5
Other Products	2.2	3.0	3.7	6.9	6.3	5.7	6.0	8.4	8.7
Refinery Fuel	0.6	0.6	0.6	-	-	-	-	-	-
Sub-Total	47.8	53.6	65.6	83.5	82.5	84.5	86.2	113.1	114.3
Bunkers	18.7	15.6	11.4	10.6	11.4	11.1	12.1	11.9	12.8
Total	66.5	69.2	77.0	94.1	93.9	95.6	98.3	125.0	127.1

	1989	1990	1991	1992	1993	1994	1995	1996	1997
Thousand tonnes									
NGL/LPG/Ethane	91	98	105	115	115	119	124	127	133
Naphtha	7	7	8	8	8	8	8	8	8
Motor Gasoline	1 197	1 121	1 157	1 179	1 220	1 236	1 256	1 272	1 339
Aviation Fuels	649	671	601	644	696	635	681	667	677
Kerosene	375	373	402	412	441	441	465	510	536
Gas Diesel	1 649	1 494	1 541	1 609	1 660	1 639	1 629	1 673	1 764
Heavy Fuel Oil	1 035	979	1 027	1 062	1 089	1 143	1 172	1 175	1 175
Other Products	467	424	449	462	475	490	485	490	509
Refinery Fuel	-	-	-	-	-	-	-	-	-
Sub-Total	5 470	5 167	5 290	5 491	5 704	5 711	5 820	5 922	6 141
Bunkers	709	637	621	607	601	684	708	685	705
Total	6 179	5 804	5 911	6 098	6 305	6 395	6 528	6 607	6 846
Thousand barrels/day									
NGL/LPG/Ethane	2.9	3.1	3.3	3.6	3.7	3.8	3.9	4.0	4.2
Naphtha	0.2	0.2	0.2	0.2	0.2	0.2	0.2	0.2	0.2
Motor Gasoline	28.0	26.2	27.0	27.5	28.5	28.9	29.4	29.6	31.3
Aviation Fuels	14.5	15.1	13.5	14.4	15.6	14.3	15.3	14.9	15.2
Kerosene	8.0	7.9	8.5	8.7	9.4	9.4	9.9	10.8	11.4
Gas Diesel	33.7	30.5	31.5	32.8	33.9	33.5	33.3	34.1	36.1
Heavy Fuel Oil	18.9	17.9	18.7	19.3	19.9	20.9	21.4	21.4	21.4
Other Products	8.9	8.0	8.5	8.7	9.0	9.2	9.2	9.2	9.6
Refinery Fuel	-	-	-	-	-	-	-	-	-
Sub-Total	114.9	108.9	111.3	115.2	120.1	120.1	122.5	124.3	129.4
Bunkers	13.6	12.2	11.9	11.6	11.5	13.1	13.7	13.1	13.5
Total	128.5	121.1	123.2	126.8	131.5	133.2	136.1	137.4	142.9

Oil Products Consumption
Consommation de produits pétroliers

Other Asia

	1971	1973	1978	1983	1984	1985	1986	1987	1988
Thousand tonnes									
NGL/LPG/Ethane	11	14	15	27	26	28	26	28	25
Naphtha	-	-	-	3	1	1	1	-	-
Motor Gasoline	276	257	358	355	333	337	287	362	344
Aviation Fuels	209	205	168	283	380	381	355	387	367
Kerosene	60	65	81	45	46	55	68	86	84
Gas Diesel	436	464	566	781	684	856	817	997	961
Heavy Fuel Oil	428	846	581	444	474	510	490	531	517
Other Products	17	31	51	36	38	37	38	35	39
Refinery Fuel	-	-	-	-	-	-	-	-	-
Sub-Total	1 437	1 882	1 820	1 974	1 982	2 205	2 082	2 426	2 337
Bunkers	181	189	148	96	90	63	83	76	62
Total	1 618	2 071	1 968	2 070	2 072	2 268	2 165	2 502	2 399
Thousand barrels/day									
NGL/LPG/Ethane	0.4	0.4	0.5	0.9	0.8	0.9	0.8	0.9	0.8
Naphtha	-	-	-	0.1	0.0	0.0	0.0	-	-
Motor Gasoline	6.5	6.0	8.4	8.3	7.8	7.9	6.7	8.5	8.0
Aviation Fuels	4.7	4.6	3.7	6.2	8.3	8.3	7.8	8.4	8.0
Kerosene	1.3	1.4	1.7	1.0	1.0	1.2	1.4	1.8	1.8
Gas Diesel	8.9	9.5	11.6	16.0	13.9	17.5	16.7	20.4	19.6
Heavy Fuel Oil	7.8	15.4	10.6	8.1	8.6	9.3	8.9	9.7	9.4
Other Products	0.3	0.6	1.0	0.7	0.7	0.7	0.7	0.7	0.7
Refinery Fuel	-	-	-	-	-	-	-	-	-
Sub-Total	29.8	37.9	37.4	41.1	41.1	45.8	43.1	50.3	48.3
Bunkers	3.5	3.6	2.9	1.9	1.8	1.3	1.7	1.5	1.2
Total	33.3	41.5	40.3	43.0	42.9	47.0	44.8	51.9	49.6

	1989	1990	1991	1992	1993	1994	1995	1996	1997
Thousand tonnes									
NGL/LPG/Ethane	28	32	33	30	30	31	33	34	36
Naphtha	-	-	-	-	-	-	-	-	-
Motor Gasoline	355	376	367	345	342	341	339	339	339
Aviation Fuels	384	389	358	316	308	229	234	228	228
Kerosene	96	124	69	68	66	62	64	66	71
Gas Diesel	974	1 028	1 032	868	853	838	858	864	902
Heavy Fuel Oil	536	552	584	582	585	585	584	588	614
Other Products	35	38	33	25	32	30	30	29	30
Refinery Fuel	-	-	-	-	-	-	-	-	-
Sub-Total	2 408	2 539	2 476	2 234	2 216	2 116	2 142	2 148	2 220
Bunkers	66	66	72	72	72	75	78	78	82
Total	2 474	2 605	2 548	2 306	2 288	2 191	2 220	2 226	2 302
Thousand barrels/day									
NGL/LPG/Ethane	0.9	1.0	1.0	1.0	1.0	1.0	1.0	1.1	1.1
Naphtha	-	-	-	-	-	-	-	-	-
Motor Gasoline	8.3	8.8	8.6	8.0	8.0	8.0	7.9	7.9	7.9
Aviation Fuels	8.4	8.5	7.8	6.9	6.7	5.0	5.1	5.0	5.0
Kerosene	2.0	2.6	1.5	1.4	1.4	1.3	1.4	1.4	1.5
Gas Diesel	19.9	21.0	21.1	17.7	17.4	17.1	17.5	17.6	18.4
Heavy Fuel Oil	9.8	10.1	10.7	10.6	10.7	10.7	10.7	10.7	11.2
Other Products	0.7	0.7	0.6	0.5	0.6	0.6	0.6	0.5	0.6
Refinery Fuel	-	-	-	-	-	-	-	-	-
Sub-Total	49.9	52.7	51.3	46.0	45.8	43.6	44.2	44.2	45.8
Bunkers	1.3	1.3	1.4	1.4	1.4	1.5	1.6	1.6	1.6
Total	51.3	54.0	52.7	47.5	47.2	45.1	45.8	45.7	47.4

Oil Products Consumption
Consommation de produits pétroliers

Other Latin America

	1971	1973	1978	1983	1984	1985	1986	1987	1988
Thousand tonnes									
NGL/LPG/Ethane	48	48	68	92	90	97	101	76	78
Naphtha	-	-	-	-	-	-	-	-	-
Motor Gasoline	347	511	477	507	466	445	471	541	567
Aviation Fuels	379	225	412	324	341	336	329	431	413
Kerosene	180	268	242	203	166	173	151	188	203
Gas Diesel	866	1 245	1 063	815	953	844	1 043	1 153	1 096
Heavy Fuel Oil	44	48	522	109	192	200	257	504	512
Other Products	8	27	73	108	166	173	181	192	181
Refinery Fuel	-	-	-	-	-	-	-	-	-
Sub-Total	1 872	2 372	2 857	2 158	2 374	2 268	2 533	3 085	3 050
Bunkers	994	1 111	872	704	464	578	401	322	245
Total	2 866	3 483	3 729	2 862	2 838	2 846	2 934	3 407	3 295
Thousand barrels/day									
NGL/LPG/Ethane	1.5	1.5	2.2	2.9	2.9	3.1	3.2	2.4	2.5
Naphtha	-	-	-	-	-	-	-	-	-
Motor Gasoline	8.1	11.9	11.1	11.8	10.9	10.4	11.0	12.6	13.2
Aviation Fuels	8.4	5.0	9.0	7.1	7.5	7.4	7.2	9.5	9.0
Kerosene	3.8	5.7	5.1	4.3	3.5	3.7	3.2	4.0	4.3
Gas Diesel	17.7	25.4	21.7	16.7	19.4	17.3	21.3	23.6	22.3
Heavy Fuel Oil	0.8	0.9	9.5	2.0	3.5	3.6	4.7	9.2	9.3
Other Products	0.1	0.5	0.8	1.1	1.4	1.5	1.6	1.8	1.6
Refinery Fuel	-	-	-	-	-	-	-	-	-
Sub-Total	40.5	51.0	59.5	45.9	49.0	46.9	52.3	63.1	62.3
Bunkers	18.6	20.6	16.2	13.0	8.6	10.6	7.4	5.9	4.5
Total	59.1	71.5	75.7	58.9	57.6	57.6	59.7	69.0	66.8

	1989	1990	1991	1992	1993	1994	1995	1996	1997
Thousand tonnes									
NGL/LPG/Ethane	78	105	124	129	129	138	148	150	156
Naphtha	-	-	-	-	-	-	-	-	-
Motor Gasoline	622	646	629	631	636	651	680	696	716
Aviation Fuels	453	432	431	399	373	427	435	450	465
Kerosene	262	246	257	215	220	232	235	240	250
Gas Diesel	1 236	1 248	1 145	1 039	1 081	1 068	1 114	1 137	1 177
Heavy Fuel Oil	495	487	508	499	509	515	432	444	463
Other Products	161	157	165	157	168	171	170	171	173
Refinery Fuel	3	3	3	3	3	3	1	3	3
Sub-Total	3 310	3 324	3 262	3 072	3 119	3 205	3 215	3 291	3 403
Bunkers	208	223	225	221	225	225	224	230	232
Total	3 518	3 547	3 487	3 293	3 344	3 430	3 439	3 521	3 635
Thousand barrels/day									
NGL/LPG/Ethane	2.5	3.3	3.9	4.1	4.1	4.4	4.7	4.8	5.0
Naphtha	-	-	-	-	-	-	-	-	-
Motor Gasoline	14.5	15.1	14.7	14.7	14.9	15.2	15.9	16.2	16.7
Aviation Fuels	10.0	9.5	9.5	8.8	8.3	9.5	9.6	9.9	10.3
Kerosene	5.6	5.2	5.5	4.5	4.7	4.9	5.0	5.1	5.3
Gas Diesel	25.3	25.5	23.4	21.2	22.1	21.8	22.8	23.2	24.1
Heavy Fuel Oil	9.0	8.9	9.3	9.1	9.3	9.4	7.9	8.1	8.4
Other Products	1.2	1.1	1.3	1.1	1.3	1.4	1.3	1.4	1.4
Refinery Fuel	0.1	0.1	0.1	0.1	0.1	0.1	0.0	0.1	0.1
Sub-Total	68.1	68.8	67.6	63.6	64.6	66.6	67.2	68.6	71.2
Bunkers	3.9	4.2	4.2	4.1	4.2	4.2	4.2	4.3	4.4
Total	72.0	73.0	71.8	67.7	68.9	70.8	71.4	73.0	75.6

INTERNATIONAL ENERGY AGENCY
ENERGY STATISTICS DIVISION
POSSIBLE STAFF VACANCIES

The Division is responsible for statistical support and advice to the policy and operational Divisions of the International Energy Agency. It also produces a wide range of annual and quarterly publications complemented by a data service on microcomputer diskettes. For these purposes, the Division maintains, on a central computer and an expanding network of microcomputers, extensive international databases covering most aspects of energy supply and use.

- Vacancies for statistical assistants occur from time to time. Typically their work includes:
- Gathering and vetting data from questionnaires and publications, discussions on data issues with respondents to questionnaires in national administrations and fuel companies.
- Managing energy databases on a mainframe computer and microcomputers in order to maintain accuracy and timeliness of output.
- Preparing computer procedures for the production of tables, reports and analyses. Seasonal adjustment of data and analysis of trends and market movements.
- Preparing studies on an ad-hoc basis as required by other Divisions of the International Energy Agency.

Nationals of any OECD Member Country are eligible for appointment. Basic salaries range from 15 400 to 20 400 French francs per month, depending on qualifications. The possibilities for advancement are good for candidates with appropriate qualifications and experience. Tentative enquiries about future vacancies are welcomed from men and women with relevant qualifications and experience. Applications in French or English, specifying the reference "ENERSTAT" and enclosing a curriculum vitae, should be sent to:

Human Resources Management Division
OECD, 2, rue André-Pascal,
75775 Paris CEDEX 16,
France

AGENCE INTERNATIONALE DE L'ENERGIE
DIVISION DES STATISTIQUES DE L'ENERGIE
VACANCES D'EMPLOI EVENTUELLES

La Division est chargée de fournir une aide et des conseils dans le domaine statistique aux Divisions administratives et opérationnelles de l'Agence internationale de l'énergie. En outre, elle diffuse une large gamme de publications annuelles et trimestrielles complétées par un service de données sur disquettes pour micro-ordinateur. A cet effet, la Division tient à jour, sur un ordinateur central et un réseau de plus en plus étendu de micro-ordinateurs, de vastes bases de données internationales portant sur la plupart des aspects de l'offre et de la consommation d'énergie.

Des postes d'assistant statisticien sont susceptibles de se libérer de temps à autre. Les fonctions dévolues aux titulaires de ces postes sont notamment les suivantes :

- Rassembler et valider les données tirées de questionnaires et de publications, ainsi que d'échanges de vues sur les données avec les personnes des Administrations nationales ou des entreprises du secteur de l'énergie qui répondent aux questionnaires.
- Gérer des bases de données relatives à l'énergie sur un ordinateur central et des micro-ordinateurs en vue de s'assurer de l'exactitude et de l'actualisation des données de sortie.
- Mettre au point des procédures informatiques pour la réalisation de tableaux, rapports et analyses. Procéder à l'ajustement saisonnier des données et analyses relatives aux tendances et aux fluctuations du marché.
- Effectuer des études en fonction des besoins des autres Divisions de l'Agence internationale de l'énergie.

Ces postes sont ouverts aux ressortissants des pays Membres de l'OCDE. Les traitements de base sont compris entre 15 400 et 20 400 francs français par mois, suivant les qualifications. Les candidats possédant les qualifications et l'expérience appropriées se verront offrir des perspectives de promotion. Les demandes de renseignements sur les postes susceptibles de se libérer qui émanent de personnes dotées des qualifications et de l'expérience voulues seront les bienvenues. Les candidatures, rédigées en français ou en anglais et accompagnées d'un curriculum vitae, doivent être envoyées, sous la référence "ENERSTAT", à l'adresse suivante :

Division de la Gestion des Ressources Humaines
OCDE, 2 rue André-Pascal,
75775 Paris CEDEX 16,
France

MULTILINGUAL PULLOUT

French
German

1.	Production	Produktion
2.	Autres sources	Andere Quellen
3.	Importations	Importe
4.	Exportations	Exporte
5.	Soutages maritimes internationaux	Bunker
6.	Variations des stocks	Bestandsveränderungen
7.	**APPROVISIONNEMENT INTERIEUR**	**INLANDSVERSORGUNG**
8.	Transferts	Transfer
9.	Ecarts statistiques	Stat. Differenzen
10.	**SECTEUR TRANSFORMATION**	**UMWANDLUNGSBEREICH**
11.	Centrales électriques	Elektrizitätstwerke
12.	Centrales de cogénération	Elektrizitäts- und Heizkraftwerke
13.	Centrales calogènes	Heizkraftwerke
14.	Hauts-fourneaux et usines à gaz	Hochofen und Gaswerke
15.	Cokeries/fabriques d'agglomérés/fabriques de briquettes de lignite	Koks- und Brickettfabriken
16.	Raffineries de pétrole	Raffinerien
17.	Industrie pétrochimique	Petrochemische Industrie
18.	Liquéfaction	Verflüssigung
19.	Secteur transformation - autres	Sonstige im Umwandlungsbereich
20.	**SECTEUR ENERGIE**	**ENERGIESEKTOR**
21.	Mines de charbon	Kohlenbergwerke
22.	Extraction de pétrole et de gaz	Erdöl- und Erdgasgewinnung
23.	Raffineries de pétrole	Raffinerien
24.	Centrales électriques, centrales de cogénération et centrales calogènes	Wärmekraftwerke
25.	Accumulation par pompage (Electricité)	Pumpspeicherwerke
26.	Secteurs énergie - autres	Sonstige Energiesektoren
27.	Pertes de distribution	Verteilungsverluste
28.	**CONSOMMATION FINALE**	**ENDENERGIEVERBRAUCH**
29.	**SECTEUR INDUSTRIE**	**INDUSTRIE**
30.	Sidérurgie	Eisen- und Stahlindustrie
31.	Industrie chimique et pétrochimique	Chemische Industrie
32.	*Dont : produits d'alimentation*	*Davon: Feedstocks*
33.	Métaux non ferreux	Ne-Metallerzeugung
34.	Produits mineraux non métalliques	Glas- und Keramikindustrie
35.	Matériel de transport	Fahrzeugbau
36.	Construction mécanique	Maschinenbau
37.	Industrie extractives	Bergbau und Steinbrüche
38.	Industrie alimentaire et tabacs	Nahrungs- und Genußmittel
39.	Papier, pâte à papier et imprimerie	Zellstoff, Papier, Pappeerzeugung
40.	Bois et produits dérivés	Holz und Holzprodukte
41.	Construction	Baugewerbe
42.	Textiles et cuir	Textil- und Lederindustrie
43.	Non spécifiés	Sonstige
44.	**SECTEUR TRANSPORTS**	**VERKEHRSSEKTOR**
45.	Transport aérien	Luftverkehr
46.	Transport routier	Straßenverkehr
47.	Transport ferroviaire	Schienenverkehr
48.	Transport par conduites	Rohrleitungen
49.	Navigation intérieure	Binnenschiffahrt
50.	Non spécifiés	Sonstige
51.	**AUTRES SECTEURS**	**ANDERE SEKTOREN**
52.	Agriculture	Landwirtschaft
53.	Commerce et services publics	Handel- und öffentliche Einrichtungen
54.	Résidentiel	Wohungssektor
55.	Non spécifiés	Sonstige
56.	**UTILISATIONS NON ENERGETIQUES**	**NICHTENERGETISCHER VERBRAUCH**
57.	Dans l'industrie/transformation/énergie	In Industrie/Umwandlung/Energiesektor
58.	Dans les transports	In Verkehr
59.	Dans les autres secteurs	In anderen Sektoren

MULTILINGUAL PULLOUT

Italian
Japanese

MULTILINGUAL PULLOUT

Spanish
Russian

MULTILINGUAL PULLOUT

Chinese

INTERNATIONAL ENERGY AGENCY

NINE ANNUAL PUBLICATIONS

Coal Information 1998

Issued annually since 1983, this publication provides comprehensive information on current world coal market trends and long-term prospects. Compiled in cooperation with the Coal Industry Advisory Board, it contains thorough analysis and current country-specific statistics for OECD Member countries and selected non-OECD countries on coal prices, demand, trade, production, productive capacity, emissions standards for coal-fired boilers, coal ports, coal-fired power stations and coal data for non-OECD countries. This publication is a key reference tool for all sectors of the coal industry as well as for OECD Member country governments. *Published July 1999.*

Electricity Information 1998

This publication brings together in one volume the IEA's data on electricity and heat supply and demand in the OECD. The report presents a comprehensive picture of electricity capacity and production, consumption, trade and prices for the OECD regions and individual countries in over 20 separate tables for each OECD country. Detailed data on the fuels used for electricity and heat production are also presented.
Published July 1999.

Natural Gas Information 1998

A detailed reference work on gas supply and demand, covering not only the OECD countries but also the rest of the world. Contains essential information on LNG and pipeline trade, gas reserves, storage capacity and prices. The main part of the book, however, concentrates on OECD countries, showing a detailed gas supply and demand balance for each individual country and for the three OECD regions: North America, Europe and Asia-Pacific, as well as a breakdown of gas consumption by end-user. Import and export data are reported by source and destination. *Published July 1999.*

Oil Information 1998

A comprehensive reference book on current developments in oil supply and demand. The first part of this publication contains key data on world production, trade, prices and consumption of major oil product groups, with time series back to the early 1970s. The second part gives a more detailed and comprehensive picture of oil supply, demand, trade, production and consumption by end-user for each OECD country individually and for the OECD regions. Trade data are reported extensively by origin and destination. *Published July 1999.*

Energy Statistics of OECD Countries 1996-1997

No other publication offers such in-depth statistical coverage. It is intended for anyone involved in analytical or policy work related to energy issues. It contains data on energy supply and consumption in original units for coal, oil, natural gas, combustible renewables/wastes and products derived from these primary fuels, as well as for electricity and heat. Data are presented for the two most recent years available in detailed supply and consumption tables. Historical tables summarise data on production, trade and final consumption. Each issue includes definitions of products and flows and explanatory notes on the individual country data. *Published June 1999.*

Energy Balances of OECD Countries 1996-1997

A companion volume to *Energy Statistics of OECD Countries*, this publication presents standardised energy balances expressed in million tonnes of oil equivalent. Energy supply and consumption data are divided by main fuel: coal, oil, gas, nuclear, hydro, geothermal/solar, combustible renewables/wastes, electricity and heat. This allows for easy comparison of the contributions each fuel makes to the economy and their interrelationships through the conversion of one fuel to another. All of this is essential for estimating total energy supply, forecasting, energy conservation, and analysing the potential for interfuel substitution. Complete energy balances are presented for the two most recent years available. Historical tables summarise key energy and economic indicators as well as data on production, trade and final consumption. Each issue includes definitions of products and flows and explanatory notes on the individual country data as well as conversion factors from original units to tonnes of oil equivalent. *Published June 1999.*

Energy Statistics of non-OECD Countries 1996-1997

This new publication offers the same in-depth statistical coverage as the homonymous publication covering OECD countries. It includes data in original units for 103 individual countries and nine main regions. The consistency of OECD and non-OECD countries' detailed statistics provides an accurate picture of the global energy situation. For a description of the content, please see *Energy Statistics of OECD Countries* above. *Published August 1999.*

Energy Balances of non-OECD Countries 1996-1997

A companion volume to the new publication *Energy Statistics of Non-OECD Countries*, this publication presents energy balances in million tonnes of oil equivalent and key economic and energy indicators for 103 individual countries and nine main regions. It offers the same statistical coverage as the homonymous publication covering OECD Countries, and thus provides an accurate picture of the global energy situation. It includes most of the historical series as well as the energy indicators and energy balances previously published in *Energy Statistics and Balances of Non-OECD Countries*. For a description of the content, please see *Energy Balances of OECD Countries* above. *Published August 1999.*

CO_2 Emissions from Fuel Combustion - 1999 Edition

In order for nations to tackle the problem of climate change, they need accurate greenhouse gas emissions data. This publication provides a new basis for comparative analysis of CO_2 emissions from fossil fuel combustion, a major source of anthropogenic emissions. The data in this book are designed to assist in understanding the evolution of these emissions from 1971 to 1997 on a country, regional and worldwide basis. They should help in the preparation and the follow-up to the Fifth Conference of the Parties (COP-5) meeting under the U.N. Climate Convention in Bonn in October/November 1999. Emissions were calculated using IEA energy databases and the default methods and emissions factors from the *Revised 1996 IPCC Guidelines for National Greenhouse Gas Inventories. Published October 1999.*

TWO QUARTERLIES

Oil, Gas, Coal and Electricity, Quarterly Statistics

Oil statistics cover OECD production, trade (by origin and destination), refinery intake and output, stock changes and consumption for crude oil, NGL and nine selected oil product groups. Statistics for natural gas show OECD supply, consumption and trade (by origin and destination). Coal data cover the main OECD and world-wide producers of hard and brown coal and major exporters and importers of steam and coking coal. Trade data for the main OECD countries are reported by origin and destination. Electricity statistics cover production (by major fuel category), consumption and trade for 28 OECD countries. Quarterly data on world oil and coal production are included, as well as world steam and coking coal trade.

Energy Prices and Taxes

This publication responds to the needs of the energy industry and OECD governments for up-to-date information on prices and taxes in national and international energy markets. It contains for OECD countries and certain non-OECD countries prices at all market levels: import prices, industry prices and consumer prices. The statistics cover the main petroleum products, gas, coal and electricity, giving for imported products an average price both for importing country and country of origin. Every issue includes full notes on sources and methods and a description of price mechanisms in each country.

For more information on the IEA statistics publications, please feel free to contact Ms. Sharon Michel in the Energy Statistics Division, Tel: (33 1) 40 57 66 25; Fax: (33 1) 40 57 66 49.

STATISTICS ELECTRONIC EDITIONS

To complement its publications, the Energy Statistics Division produces diskettes containing the complete databases which are used for preparing the statistics publications. State-of-the-art software allows you to access and manipulate all these data in a very user-friendly manner and includes graphic and mapping facilities.

The diskette service includes:

Annual Diskettes

- Energy Statistics of OECD Countries, 1960-1997
- Energy Balances of OECD Countries, 1960-1997
- Energy Statistics of non-OECD Countries, 1971-1997
- Energy Balances of non-OECD Countries, 1971-1997
- CO_2 Emissions from Fuel Combustion
- Natural Gas Information
- Oil Information
- Coal Information
- Electricity Information
- IEA Energy Technology R&D Statistics 74-95

Quarterly Diskettes

- Energy Prices and Taxes.

Monthly Diskettes

- The IEA Monthly Oil Data Service (see box below).

The IEA Monthly Oil Data Service

Diskettes

The IEA Monthly Oil Data Service provides the detailed database of historical and projected information which is used in preparing the IEA's monthly Oil Market Report (OMR) and includes all the information previously contained in the Monthly Oil Statistics (MOS) diskettes. The IEA Monthly Oil Data Service comprises three packages:

- Supply, Demand, Balances and Stocks;
- Trade;
- Field-by-Field Supply;

available separately or combined, either as a subscriber service on the Internet or on diskettes.

Internet

The Internet version is available two days after the official release of the Oil Market Report. Diskettes are mailed four days afterwards.

CD-ROM and INTERNET (http://www.iea.org)

Annual Oil and Gas Statistics, Energy Statistics and Balances of OECD and Non-OECD Countries are also available on **CD Rom** in the OECD Statistical Compendium (for additional information on the CD-ROM contact the OECD Publications Service, Tel: (33 1) 45 24 82 00; Fax: (33 1) 49 10 42 76. Moreover, the IEA site on **Internet** contains key energy indicators by country, graphs on the world and OECD's energy situation evolution from 1971 to the most recent year available, as well as selected databases for demonstration. The IEA site can be accessed at: **http://www.iea.org**. For more information, please feel free to contact the Energy Statistics Division of the IEA by Fax: (33 1) 40 57 66 49 or Phone: (33 1) 40 57 66 25.

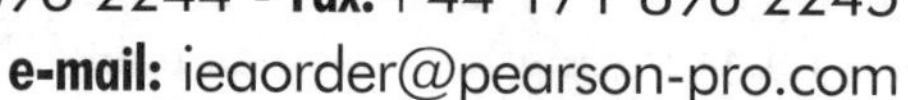

INTERNATIONAL ENERGY AGENCY

IEA Publications, PO Box 2722, London, W1A 5BL, UK
Telephone:+44 171 896 2244 - **Fax:**+44 171 896 2245
e-mail: ieaorder@pearson-pro.com

▶ *Special Discount: Order three or more products to claim a 20% discount.*

I would like to order the following publications (please tick)

ANNUAL PUBLICATIONS - 1999 Edition		QTY	PRICE	TOTAL
☐ Energy Statistics of OECD Countries 1996-1997			$110.00	
☐ Energy Balances of OECD Countries 1996-1997			$110.00	
☐ Energy Statistics of Non-OECD Countries 1996-1997	Formerly, these 2 volumes were contained in *Energy Statistics and Balances of Non-OECD Countries.*		$110.00	
☐ Energy Balances of Non-OECD Countries 1996-1997			$110.00	
☐ Coal Information 1998			$200.00	
☐ Electricity Information 1998			$130.00	
☐ Natural Gas Information 1998			$150.00	
☐ Oil Information 1998			$150.00	
☐ CO_2 Emissions from Fuel Combustion 1971-1997			$150.00	

Please enter my subscription as indicated below (please tick)

QUARTERLY PUBLICATIONS	QTY	SINGLE COPY	ANNUAL	TOTAL
☐ Energy Prices and Taxes		$100.00	$300.00	
☐ Oil, Gas, Coal and Electricity Statistics		$100.00	$300.00	

Sub Total	
Postage and packing ($15.00)	
Less discount	
Total	

MONEY-BACK GUARANTEE:
Refunds are given on one-off publications returned in resaleable condition by registered post within 7 days of receipt.

Delivery details (If your billing address differs from your delivery address please advise us)

Name

Position | Organisation

Address

Country | Postcode

Telephone | Fax

As all products are delivered by express courier service, please enter your telephone number above.

Payment details

☐ I enclose a cheque payable to IEA Publications for the sum of $________

☐ Please debit my credit card (tick choice).

☐ Access/Mastercard ☐ Diners ☐ VISA ☐ AMEX

Card no:

Signature: Expiry date:

EU registered companies	Non-EU registered companies
☐ Please send me an invoice along with the publications. I enclose a copy of my company headed stationery with this order form. To avoid extra charges EU companies (except UK) must supply VAT/TVA/MOMS/MST/IVA/FPA numbers:	☐ Please send me a pro-forma invoice. The publications will be forwarded to me on receipt of payment.

Data Protection Act: The information you provide will be held on our database and may be used to keep you informed of our and our associated companies' products and for selected third party mailings.

FOR FAST PROCESSING OF YOUR ORDER, FAX THIS FORM TO CUSTOMER SERVICES ON +44 171 896 2245

INTERNATIONAL ENERGY AGENCY

IEA Publications, PO Box 2722, London, W1A 5BL, UK
Telephone:+44 171 896 2244 - **Fax:**+44 171 896 2245
e-mail: ieaorder@pearson-pro.com

▶ *Special Discount: Order three or more products to claim a 20% discount.*

I would like to order the following diskettes (please tick)

ANNUAL DISKETTES* - 1999 Edition	QTY	PRICE	TOTAL
☐ Energy Statistics of OECD Countries (1960-1997)		$500.00	
☐ Energy Balances of OECD Countries (1960-1997)		$500.00	
☐ Energy Statistics of Non-OECD Countries (1971-1997)		$500.00	
☐ Energy Balances of Non-OECD Countries (1971-1997)		$500.00	
☐ *Combined subscription of the above four series*		*$1 200.00*	
☐ Coal Information 1998		$500.00	
☐ Electricity Information 1998		$500.00	
☐ Natural Gas Information 1998		$500.00	
☐ Oil Information 1998		$500.00	
☐ CO_2 Emissions from Fuel Combustion (1960/71-1997)		$500.00	
☐ IEA Energy Technology R&D Statistics (1974-1995)		$500.00	

Please enter my subscription as indicated below (please tick)

QUARTERLY DISKETTES*	QTY	PRICE	TOTAL
☐ Energy Prices and Taxes (four quarters)		$800.00	

*Prices are for single user licence. Please contact us for pricing information on multi-user licences.

Sub Total	
Postage and packing ($15.00)	
Less discount	
Total	

MONEY-BACK GUARANTEE:
Refunds are given on one-off publications returned in resaleable condition by registered post within 7 days of receipt.

Delivery details (If your billing address differs from your delivery address please advise us)

Name
Position Organisation
Address

Country Postcode
Telephone Fax

As all products are delivered by express courier service, please enter your telephone number above.

Payment details

☐ I enclose a cheque payable to IEA Publications for the sum of $________
☐ Please debit my credit card (tick choice).

☐ Access/Mastercard ☐ Diners ☐ VISA ☐ AMEX

Card no:
Signature: Expiry date:

EU registered companies	Non-EU registered companies
☐ Please send me an invoice along with the publications. I enclose a copy of my company headed stationery with this order form. To avoid extra charges EU companies (except UK) must supply VAT/TVA/MOMS/MST/IVA/FPA numbers:	☐ Please send me a pro-forma invoice. The publications will be forwarded to me on receipt of payment.

Data Protection Act: The information you provide will be held on our database and may be used to keep you informed of our and our associated companies' products and for selected third party mailings.

FOR FAST PROCESSING OF YOUR ORDER, FAX THIS FORM TO CUSTOMER SERVICES ON +44 171 896 2245

International Energy Agency, 9 rue de la Fédération, 75739 Paris CEDEX 15
PRINTED IN FRANCE BY CHIRAT
(61 99 15 3 P) ISBN 92-64-05864-8 - 1999

Photo de couverture : COMSTOCK